ALGEBRA'S COMMON GRAPHS

Identity Function

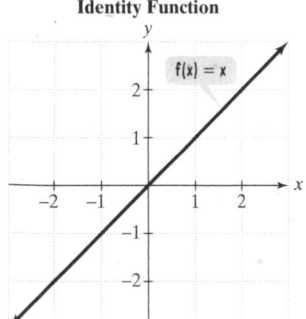

Standard Quadratic Function

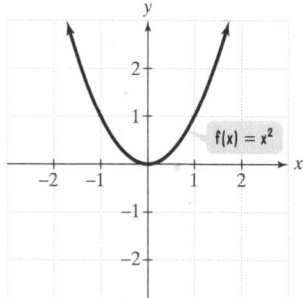

Standard Cubic Function

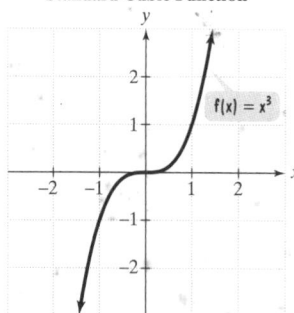

Absolute Value Function

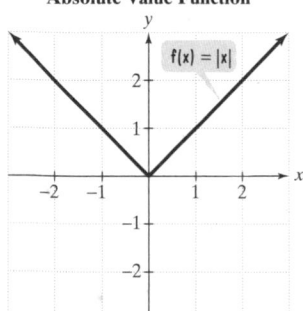

Square Root Function

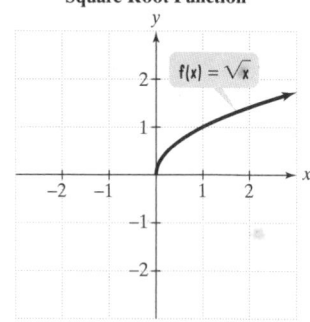

Greatest Integer Function

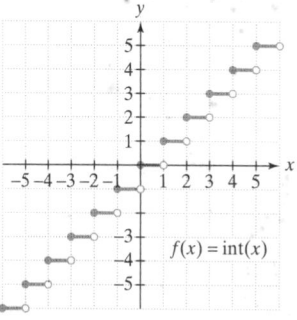

TRANSFORMATIONS

In each case, c represents a positive real number.

Function		Draw the graph of f and:
Vertical translations	$\begin{cases} y = f(x) + c \\ y = f(x) - c \end{cases}$	Shift f upward c units. Shift f downward c units.
Horizontal translations	$\begin{cases} y = f(x - c) \\ y = f(x + c) \end{cases}$	Shift f to the right c units. Shift f to the left c units.
Reflections	$\begin{cases} y = -f(x) \\ y = f(-x) \end{cases}$	Reflect f about the x-axis. Reflect f about the y-axis.
Stretching or Shrinking	$\begin{cases} y = cf(x); c > 1 \\ y = cf(x); 0 < x < 1 \end{cases}$	Stretch f, multiplying each of its y-values by c. Shrink f, multiplying each of its y-values by c.

DISTANCE AND MIDPOINT FORMULAS

1. The distance from (x_1, y_1) to (x_2, y_2) is
$$\sqrt{(x_2 - x_1)^2 + (y_2 - y_1)^2}.$$

2. The midpoint of the line segment with endpoints (x_1, y_1) and (x_2, y_2) is
$$\left(\frac{x_1 + x_2}{2}, \frac{y_1 + y_2}{2} \right).$$

QUADRATIC FORMULA

The solutions to $ax^2 + bx + c = 0$ with $a \neq 0$ are
$$x = \frac{-b \pm \sqrt{b^2 - 4ac}}{2a}.$$

FUNCTIONS

1. Linear Function: $f(x) = mx + b$
Graph is a line with slope m and y-intercept b.

2. Quadratic Function: $f(x) = ax^2 + bx + c, a \neq 0$

Graph is a parabola with vertex at $x = -\dfrac{b}{2a}$.

Quadratic Function: $f(x) = a(x - h)^2 + k$
In this form, the parabola's vertex is (h, k).

3. nth-Degree Polynomial Function: $f(x) =$
$a_n x^n + a_{n-1} x^{n-1} + a_{n-2} x^{n-2} + \cdots + a_1 x + a_0, a_n \neq 0$
For n odd and $a_n > 0$, graph falls to the left and rises to the right.
For n odd and $a_n < 0$, graph rises to the left and falls to the right.
For n even and $a_n > 0$, graph rises to the left and to the right.
For n even and $a_n < 0$, graph falls to the left and to the right.

4. Rational Function: $f(x) = \dfrac{p(x)}{q(x)}$, $p(x)$ and $q(x)$ are polynomials, $q(x) \neq 0$

5. Exponential Function: $f(x) = b^x, b > 0, b \neq 1$
Graphs:

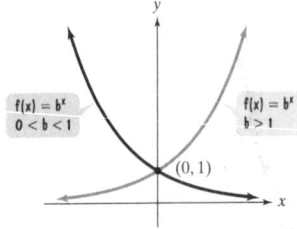

6. Logarithmic Function: $f(x) = \log_b x, b > 0, b \neq 1$
$y = \log_b x$ is equivalent to $x = b^y$.
Graph:

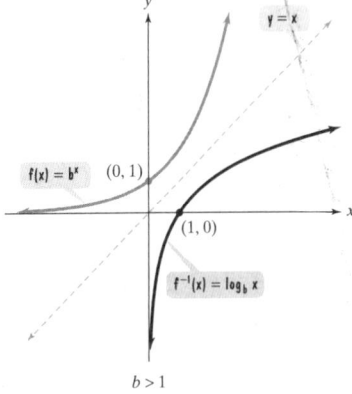

PROPERTIES OF LOGARITHMS

1. $\log_b(MN) = \log_b M + \log_b N$

2. $\log_b\left(\dfrac{M}{N}\right) = \log_b M - \log_b N$

3. $\log_b M^p = p \log_b M$

4. $\log_b M = \dfrac{\log_a M}{\log_a b} = \dfrac{\ln M}{\ln b} = \dfrac{\log M}{\log b}$

5. $\log_b b^x = x; \quad \ln e^x = x$

6. $b^{\log_b x} = x; \quad e^{\ln x} = x$

INVERSE OF A 2 × 2 MATRIX

If $A = \begin{bmatrix} a & b \\ c & d \end{bmatrix}$, then $A^{-1} = \dfrac{1}{ad - bc} \begin{bmatrix} d & -b \\ -c & a \end{bmatrix}$, where $ad - bc \neq 0$.

CRAMER'S RULE

If

$$a_{11}x_1 + a_{12}x_2 + a_{13}x_3 + \cdots + a_{1n}x_n = b_1$$
$$a_{21}x_1 + a_{22}x_2 + a_{23}x_3 + \cdots + a_{2n}x_n = b_2$$
$$a_{31}x_1 + a_{32}x_2 + a_{33}x_3 + \cdots + a_{3n}x_n = b_3$$
$$\vdots$$
$$a_{n1}x_1 + a_{n2}x_2 + a_{n3}x_3 + \cdots + a_{nn}x_n = b_n$$

then $x_i = \dfrac{D_i}{D}, D \neq 0$.

D: determinant of the system's coefficients

D_i: determinant in which coefficients of x_i are replaced by $b_1, b_2, b_3, \ldots, b_n$.

CONIC SECTIONS

Circle

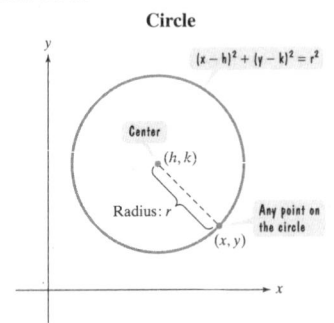

Ellipse

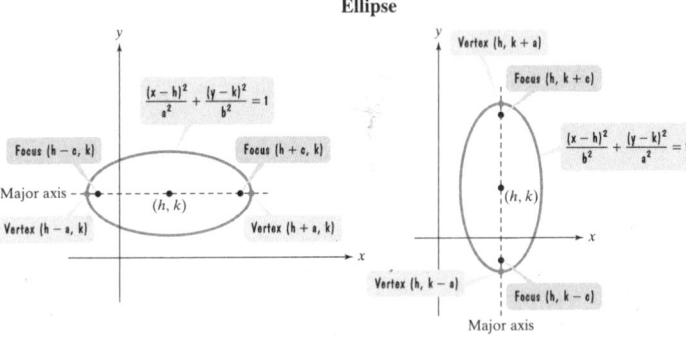

Hyperbola

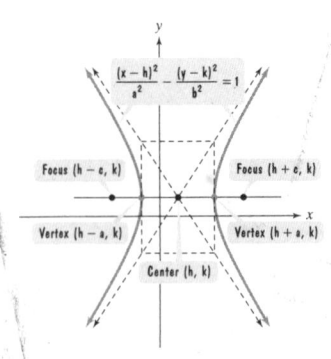

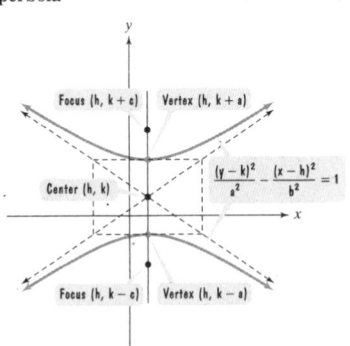

(continued on inside back cover)

Algebra & Trigonometry

Robert Blitzer
Miami-Dade Community College

Prentice
Hall

PRENTICE HALL
Upper Saddle River, NJ 07458

Library of Congress Cataloging-in-Publication Data

Blitzer, Robert.
   Algebra & trigonometry / Robert Blitzer.
     p. cm.
   Includes index.
   ISBN 0-13-089332-3
    1. Algebra.   2. Trigonometry I. Title: Algebra and trigonometry. II. Title.
   QA152.2.B583 2001
   512'.13--dc21   00-051657

Editor-in-Chief and Acquisitions Editor: *Sally Yagan*
Senior Development Editor: *Shana Ederer*
Editor-in-Chief, Development: *Carol Trueheart*
Editorial/Production Supervision: *Bayani Mendoza de Leon*
Vice President/Director of Production and Manufacturing: *David W. Riccardi*
Senior Managing Editor: *Linda Mihatov Behrens*
Executive Managing Editor: *Kathleen Schiaparelli*
Manufacturing Buyer: *Alan Fischer*
Manufacturing Manager: *Trudy Pisciotti*
Director of Marketing: *John Tweeddale*
Senior Marketing Manager: *Patrice Lumumba Jones*
Marketing Assistant: *Vince Jansen*
Associate Editor, Mathematics/Statistics Media: *Audra J. Walsh*
Director of Creative Services: *Paul Belfanti*
Art Director: *Maureen Eide*
Assistant to the Art Director: *John Christiana*
Art Editor: *Grace Hazeldine*
Photo Researcher: *Melinda Alexander*
Photo Editor: *Beth Boyd*
Cover Designer: *Maureen Eide*
Interior Design and Layout: *Lorraine Castellano*
Editorial Assistant: *Meisha Welch*
Cover image: *Telegraph Colour Library/FPG International*

©2001 by Prentice-Hall, Inc.
Upper Saddle River, New Jersey 07458

Printed in the United States of America

10 9 8 7 6 5 4 3 2 1

ISBN 0-13-089332-3

Prentice-Hall International (UK) Limited, *London*
Prentice-Hall of Australia Pty. Limited, *Sydney*
Prentice-Hall Canada Inc., *Toronto*
Prentice-Hall Hispanoamericana, S.A., *Mexico*
Prentice-Hall of India Private Limited, *New Delhi*
Prentice-Hall of Japan, Inc., *Tokyo*
Pearson Education Asia Pte. Ltd.
Editora Prentice-Hall do Brasil, Ltda., *Rio de Janeiro*

Contents

Chapter 2

Chapter 3

Chapter 4

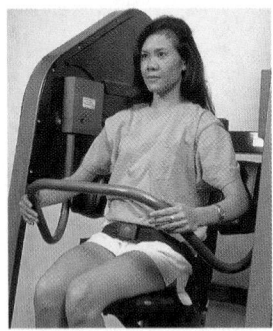

Chapter 5

Chapter 6

Chapter 7

Chapter 8

Chapter 9

Matrices and Determinants　737

Chapter 10

Conic Sections and Analytic Geometry　808

Chapter 11

Sequences, Induction, and Probability　891

Appendix

Where Did That Come From? Selected Proofs　A1

Preface

Today's college algebra and trigonometry students are a diverse group. Some are going on into calculus or other math sequences, whereas others will complete their math requirements with this course. This text is designed and written to help both sets of students succeed. The book has three fundamental goals: First, to help students acquire a solid foundation in algebra and trigonometry, preparing them for other college courses such as calculus, business calculus, finite mathematics, and computer science; second, to show students how algebra and trigonometry can model and solve authentic real-world problems; and third, to enable students to develop problem-solving skills, fostering critical thinking within a varied and interesting setting.

A source of frustration for me and my colleagues is that very few students read their textbook. When I ask students why they do not take full advantage of the text, their responses generally fall into two categories:

- "I cannot follow the explanations."
- "The applications are not interesting."

I thought about both of these objections in writing every page of this book.

"I can't follow the explanations." For many of my students, textbook explanations are too compressed. The chapters in *Algebra & Trigonometry* have been written to make them extremely accessible. Every section contains a range of simple, intermediate, and challenging examples. Voice balloons allow for specific annotations in examples, further clarifying procedures and concepts.

"The applications are not interesting." One of the things I enjoy most about teaching in a large urban community college is the diversity of who my students are and what interests them. Real-world data that celebrate this variety are used to bring relevance to examples, discussions, and applications. I selected all updated real-world data to be interesting and intriguing to students. By connecting algebra and trigonometry to the whole spectrum of their interests, it is my intent to show students that their world is profoundly mathematical and, indeed, π is in the sky.

Student Supplements

Student Solutions Manual (0-13-089785-X); (8978E-0) Includes fully worked out solutions to most of the odd-numbered exercises in the text as well as all exercises in chapter tests and all review exercises.

MathPak Integrated Learning Environment (0-13-089384-6); (8938D-2) Contains the College Algebra MathPro 4.0 along with a passcode-protected Website specifically designed to accompany this text. This product combines the series' key supplements into a comprehensive, easy-to-navigate package. Materials on the Website include but are not limited to: Section-by-section reading quizzes, Section-by-Section Powerpoint downloads, additional chapter projects, Chapter Quizzes and Tests, Student Solutions Manuals presented by chapter (exactly what is in the print version), Chapter Destinations and to interesting math Websites, and Graphing Calculator Manuals for the full line of TI's, Sharp, HP, and Casio Calculators.

Review Videos (0-13-089387-0); (8938G-5) Section-by-section videos written by and highlighting Jacquelyn White of St. Leo College. Each segment covers approximately 20 minutes of the key concepts and examples for each section. Each set of videos comes with a permissions letter allowing the school to duplicate for specific campus needs.

Precalculus Investigations/Simundza, et al. (013-010954-1); (1095D-6) A three year NSF-funded project integrates an applied approach to the topics in the Precalculus curriculum via applied projects. The investigations reflect the AMATYC and NCTM Standards in both curriculum content and pedagogy.

Companion Website www.prenhall.com/blitzer This CW address will lead to the bridge page for all of the Blitzer titles. On the CW sites (which are different than the MathPak sites) are the following: Chapter Quizzes, Chapter Tests, Projects, Graphing Calculator Manual, Destinations, and PowerPoints.

WebCT/Blackboard Contains all the materials from the MathPak website (i.e., no MathPro) plus testing materials. Can be made available in Blackboard on adoption.

Instructor Supplements

Instructor's Resource Manual (0-13-089784-1); (8978D-3) IRM contains the full solutions to the even-numbered exercises in the text.

TestGen-EQ WIN/MAC CD (0-13-089386-2); (8938F-7) New to Prentice Hall Mathematics is the use of TestGen EQ for our mathematics testing. TestGen-EQ is a fully algorithmic, easy-to-use software program written and based on the section objectives in the text.

Test Item File (0-13-089783-3); (8978C-5) A hard-copy version of materials derived from the TestGen-EQ program.

Acknowledgments

I wish to express my appreciation to all the reviewers for their helpful criticisms and suggestions, frequently transmitted with wit, humor, and intelligence. In particular, I would like to thank the following for reviewing *College Algebra* 2e and *Algebra & Trigonometry* 1e:

Celeste Hernandez	*Richland College*
Christopher N. Hay-Jahans	*University of South Dakota*
Cynthia Glickman	*Community College of Southern Nevada*
Dan Van Peursem	*University of South Dakota*
David White	*The Victoria College*
David L. Gross	*University of Connecticut*
Debra A. Pharo	*Northwestern Michigan College*
Diana Colt	*University of Minnesota-Duluth*
Donald Gordon	*Manatee Community College*
Joel K. Haack	*University of Northern Iowa*
Kayoko Yates Barnhill	*Clark College*
Lloyd Best	*Pacific Union College*
Nancy Raye Johnson	*Manatee Community College*
Richard E. Van Lommel	*California State University-Sacramento*
Sudhir Kumar Goel	*Valdosta State University*
Winfield A. Ihlow	*SUNY College at Oswego*
Yvelyne Germain-McCarthy	*University of New Orleans*

Additional acknowledgments are extended to Jacquelyn White for creating the dynamic video tape series covering every section of the book; the team at Laurel Technical Services for the Herculean task of solving all the book's exercises, preparing the answer section and solutions manuals, as well as serving as accuracy checker; Melinda Alexander, photo researcher, for obtaining the book's photographs; the team of graphic artists and mathematicians at Scientific Illustrators, whose superb illustrations and graphs provide visual support to the verbal portions of the text; Prepare Inc., the book's compositor, for inputting hundreds of pages with hardly an error; and Bayani Mendoza de Leon, whose talents as production editor contributed to the book's wonderful look.

Most of all, I wish to thank Sally Yagan and Shana Ederer. Shana, my development editor, contributed invaluable edits and suggestions that resulted in a finished product that is both accessible and up-to-date. Her influence on this book is extraordinary, guiding and coordinating every detail of this project. Sally, editor-in-chief of mathematics at Prentice Hall, is the key person in making this book a reality, and I am grateful to have had an editor with her experience, insight, and professionalism.

Sally Yagan and Shana Ederer are members of the terrific team at Prentice Hall who made this book possible, including Prentice Hall Co-President Tim Bozik and ESM President Paul Corey. Thank you Patrice Lumumba Jones, Senior Marketing Manager, for your innovative marketing efforts as well as the Prentice Hall sales force for your confidence in and enthusiasm for the book.

To the Student

I've written this book so that you can learn about the power of algebra and trigonometry and how it relates directly to your life outside the classroom. All concepts are carefully explained, important definitions and procedures are set off in boxes, and worked-out examples that present solutions in a step-by-step manner appear in every section. Each example is followed by a similar matched problem, called a Check Point, for you to try so that you can actively participate in the learning process as you read the book. (Answers to all Check Points appear in the back of the book.) Study Tips offer hints and suggestions and often point out common errors to avoid. A great deal of attention has been given to applying algebra and trigonometry to your life to make your learning experience both interesting and relevant.

As you begin your studies, I would like to offer some specific suggestions for using this book and for being successful in this course:

1. **Attend all lectures.** No book is intended to be a substitute for valuable insights and interactions that occur in the classroom. In addition to arriving for lecture on time and being prepared, you will find it useful to read the section before it is covered in lecture. This will give you a clear idea of the new material that will be discussed.

2. **Read the book.** Read each section with pen (or pencil) in hand. Move through the illustrative examples with great care. These worked-out examples provide a model for doing exercises in the exercise sets. As you proceed through the reading, do not give up if you do not understand every single word. Things will become clearer as you read on and see how various procedures are applied to specific worked-out examples.

3. **Work problems every day and check your answers.** The way to learn mathematics is by doing mathematics, which means working the Check Points and assigned exercises in the exercise sets. The more exercises you work, the better you will understand the material.

4. **Prepare for chapter exams.** After completing a chapter, study the summary, work the exercises in the Chapter Review, and work the exercises in the Chapter Test. Answers to all these exercises are given in the back of the book.

5. **Use the supplements available with this book.** A solutions manual containing worked-out solutions to the book's odd-numbered exercises and all review exercises, a dynamic web page, and video tapes created for every section of the book are among the supplements created to help you tap into the power of mathematics. Ask you instructor or bookstore what supplements are available and where you can find them.

It is my hope that that you will enjoy the pages of this book as you empower yourself with the algebra and trigonometry needed to succeed in college, your career, and in your life.

Regards,

Bob

Robert Blitzer

A Guide to Using This Text

Relevant Chapter Openers

Every chapter highlights a scenario from everyday life and how the algebra relates to it. These scenarios are revisited later in the chapter.

Functions and Graphs

Chapter 2

The cost of mailing a package depends on its weight. The probability that you and another person in a room share the same birthday depends on the number of people in the room. In both these situations, the relationship between variables can be illustrated with the notion of a *function*. Understanding this concept will give you a new perspective on many ordinary situations.

'Tis the season and you've waited until the last minute to mail your holiday gifts. Your only option is overnight express mail. You realize that the cost of mailing a gift depends on its weight, but the mailing costs seem somewhat odd. Your packages that weigh 1.1 pounds, 1.5 pounds, and 2 pounds cost $15.75 each to send overnight. Packages that weigh 2.01 pounds and 3 pounds cost you $18.50 each. Finally, your heaviest gift is barely over 3 pounds and its mailing cost is $21.25. What sort of system is this in which costs increase by $2.75, stepping from $15.75 to $18.50 and from $18.50 to $21.25?

167

Page 167

Section Objectives

The learning objectives focus the students' study habits and also are the foundation for the algorithms found in MathPro (tutorial software) and in the Test-Gen-EQ (test generator software). Objectives reappear in the margin at their point of use.

Section Openers

Each and every section opens with a unique application of algebra in students' lives outside the classroom. These scenarios are revisited later in the section.

SECTION 5.7   Inverse Trigonometric Functions

Objectives

1. Understand and use the inverse sine function.
2. Understand and use the inverse cosine function.
3. Understand and use the inverse tangent function.
4. Use a calculator to evaluate inverse trigonometric functions.
5. Find exact values of composite functions with inverse trigonometric functions.

You watched *The Matrix* on video and were impressed by the elaborate computer-generated effects. The movie is being shown again at a local theater, where you can experience its stunning visual force on a large screen. Where in the theater should you sit to maximize the film's visual impact? In this section you will see how an inverse trigonometric function can enhance your movie-going experiences.

Study Tip

Here are some helpful things to remember from our discussion of inverse functions in Section 2.6.

- If no horizontal line intersects the graph of a function more than once, the function is one-to-one and has an inverse function.
- If the point (a, b) is on the graph of f, then the point (b, a) is on the graph of the inverse function, denoted f^{-1}. The graph of f^{-1} is a reflection of the graph of f about the line $y = x$.

Page 498

Current Real-World Data

Relevant current data is used to illustrate the power of the algebra to real issues and contemporary information. It is used throughout the examples, exercises, and discussions. The data was selected to be interesting and intriguing to students.

Discovery

These exercises found in the side columns of the text encourage students to explore problems in order to better understand them and their solutions. These are a great way to stimulate class-time exploration of concepts.

Discovery

The study of how changing a function's equation can affect its graph can be explored with a graphing utility. Use your graphing utility to verify the hand-drawn graphs as you read this section.

Page 221

xiii

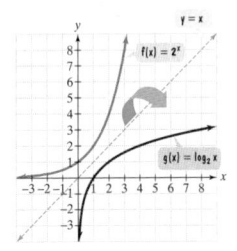

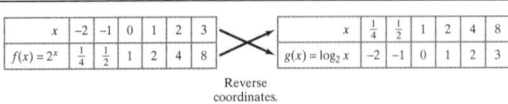

Reverse
coordinates.

Figure 4.6 The graphs of $f(x) = 2^x$
and its inverse function

We now plot the ordered pairs in both tables, connecting them with smooth curves. Figure 4.6 shows the graphs of $f(x) = 2^x$ and its inverse function $g(x) = \log_2 x$. The graph of the inverse can also be drawn by reflecting the graph of $f(x) = 2^x$ about the line $y = x$.

Check Point 6 Graph $f(x) = 3^x$ and $g(x) = \log_3 x$ in the same rectangular coordinate system.

Figure 4.7 illustrates the relationship between the graph of the exponential function, shown in blue and its inverse, the logarithmic function, shown in red, for bases greater than 1 and for bases between 0 and 1.

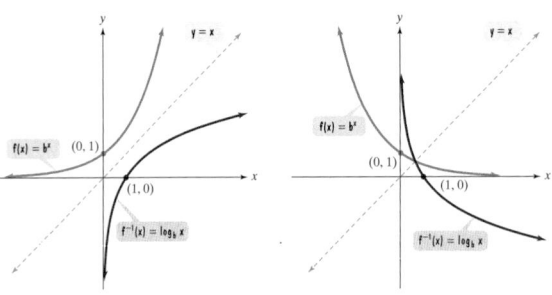

Figure 4.7 Gra

Carbon Dating and Artistic Development

The artistic community was electrified by the discovery in 1995 of spectacular cave paintings in a limestone cavern in France. Carbon dating of the charcoal from the site showed that the images, created by artists of remarkable talent, were 30,000 years old, making them the oldest cave paintings ever found. The artists seemed to have used the cavern's natural contours to heighten a sense of perspective. The quality of the painting suggests that the art of early humans did not mature steadily from primitive to sophisticated in any simple linear fashion.

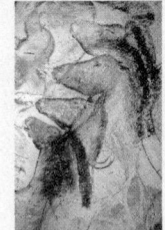

Our next example involves exponential decay and its use in determining the age of fossils and artifacts. The method is based on considering the percentage of carbon-14 remaining in the fossil or artifact. Carbon-14 decays exponentially with a *half-life* of approximately 5715 years. The **half-life** of a substance is the time required for half of a given sample to disintegrate. Thus, after 5715 years a given amount of carbon-14 will have decayed to half the original amount. Carbon dating is useful for artifacts or fossils up to 80,000 years old. Older objects do not have enough carbon-14 left to date age accurately.

EXAMPLE 2 Carbon-14 Dating: The Dead Sea Scrolls

a. Use the fact that after 5715 years a given amount of carbon-14 will have decayed to half the original amount to find the exponential decay model for carbon-14.

b. In 1947, earthenware jars containing what are known as the Dead Sea Scrolls were found by an Arab Bedouin herdsman. Analysis indicated that the scroll wrappings contained 76% of their original carbon-14. Estimate the age of the Dead Sea Scrolls.

Solution We begin with the exponential decay model $A = A_0 e^{kt}$. We know that $k < 0$ because the problem involves the decay of carbon-14. After 5715 years ($t = 5715$), the amount of carbon-14 present, A, is half the original amount A_0. Thus we can substitute $\frac{A_0}{2}$ for A in the exponential decay model. This will enable us to find k, the decay rate.

a. $\dfrac{A_0}{2} = A_0 e^{k5715}$ After 5715 years ($t = 5715$), $A = \frac{A_0}{2}$ (because the amount present, A, is half the original amount, A_0).

$\dfrac{1}{2} = e^{5715k}$ Divide both sides of the equation by A_0.

$\ln \dfrac{1}{2} = \ln e^{5715k}$ Take the natural logarithm of both sides.

$\ln \dfrac{1}{2} = 5715k$ $\ln e^x = x$

$k = \dfrac{\ln \dfrac{1}{2}}{5715} \approx -0.000121$ Solve for k.

Substituting for k in the decay model, the model for carbon-14 is $A = A_0 e^{-0.000121t}$.

Check
Point
1

Find the domain and the range of the relation
$\{(20, 157.4), (30, 231.8), (100, 752.6), (200, 1496.6)\}.$

As you worked Checkpoint 1, did you wonder if the numbers in each ordered pair represented anything? Think snakes! The first number in each ordered pair is a snake's tail length, in millimeters, and the second number is its body length, also in millimeters. Consider, for example, the ordered pair (30, 231.8).

$(30, \ 231.8)$

| A snake whose tail length is 30 millimeters | has a body length of 231.8 millimeters. |

The relation in the snake example can be pictured as follows:

20	→ 157.4
30	→ 231.8
100	→ 752.6
200	→ 1496.6

Domain   Range

Figure 2.19 The graph of a relation showing a correspondence between a snake's tail length and its body length

A scatter plot, like the one shown in Figure 2.19, is a way to represent the relation.

Body Length (millimeters)

(200, 1496.6)

(100, 752.6)

(30, 231.8)

(20, 157.4)

Tail Length (millimeters)

Check Points

Check Points offer students the opportunity to test their understanding of the example by working a similar exercise while they are reading the material. The answers to all the Check Points are given in the answer section

Study Tips

Study Tip boxes offer suggestions for problem solving, point out common student errors, and provide informal tips and suggestions. These invaluable hints appear in abundance throughout the book.

Study Tip

The word *range* can mean many things, from a chain of mountains to a cooking stove. For functions, it means the set of images of the domain. For graphing utilities, it means the setting used for the viewing rectangle. Try not to confuse these meanings.

Study Tip

The notation $f(x)$ does *not* mean "f times x." The notation describes the value of the function at x.

Thorough, yet Optional Technology

Although the use of graphing utilities is optional, they are utilized in technology boxes to enable students to visualize, discover, and explore algebraic concepts. The use of graphing utilities is also reinforced in the technology exercises appearing in the exercise sets for those who want this option. With the book's early introduction to graphing, students can look at the calculator screens in the technology boxes and gain an increased understanding of an example's solution even if they are not using a graphing utility in the course.

Step 3 **Find the values of y for the five key points.** Take a few minutes and use your calculator to evaluate the function at each value of x from step 2. Show that the key points are

$$\left(-\frac{\pi}{4}, \frac{1}{2}\right), \quad \left(-\frac{\pi}{8}, 0\right), \quad \left(0, -\frac{1}{2}\right), \quad \left(\frac{\pi}{8}, 0\right), \quad \text{and} \quad \left(\frac{\pi}{4}, \frac{1}{2}\right).$$

maximum point   x-intercept   minimum point   y-intercept   maximum point

Step 4 **Connect the five key points with a smooth curve and graph one complete cycle of the given function.** The key points and the graph of $y = \frac{1}{2}\cos(4x + \pi)$ are shown in Figure 5.53.

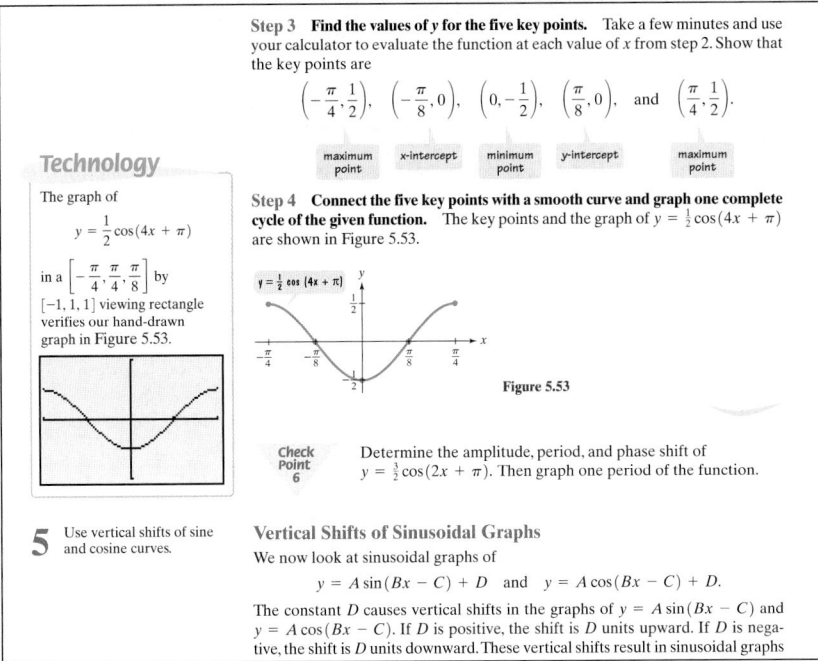

$y = \frac{1}{2}\cos(4x + \pi)$

Figure 5.53

Technology

The graph of

$$y = \frac{1}{2}\cos(4x + \pi)$$

in a $\left[-\frac{\pi}{4}, \frac{\pi}{4}, \frac{\pi}{8}\right]$ by $[-1, 1, 1]$ viewing rectangle verifies our hand-drawn graph in Figure 5.53.

Check
Point
6

Determine the amplitude, period, and phase shift of $y = \frac{3}{2}\cos(2x + \pi)$. Then graph one period of the function.

5 Use vertical shifts of sine and cosine curves.

Vertical Shifts of Sinusoidal Graphs

We now look at sinusoidal graphs of

$$y = A\sin(Bx - C) + D \quad \text{and} \quad y = A\cos(Bx - C) + D.$$

The constant D causes vertical shifts in the graphs of $y = A\sin(Bx - C)$ and $y = A\cos(Bx - C)$. If D is positive, the shift is D units upward. If D is negative, the shift is D units downward. These vertical shifts result in sinusoidal graphs

Exercise Sets

An extensive collection of exercises is included in all end-of-section and end-of-chapter materials. Within each category type, the exercises are organized by level. The category types found are: Practice Exercises, Application Exercises, Writing in Mathematics, Technology Exercises, Critical Thinking Exercises, and Group Exercises.

EXERCISE SET 5.6

Practice Exercises

In Exercises 1–4, the graph of a tangent function is given. Select the equation for each graph from the following options.

$$y = \tan\left(x + \frac{\pi}{2}\right), \quad y = \tan(x + \pi), \quad y = -\tan x, \quad y = -\tan\left(x - \frac{\pi}{2}\right)$$

1. 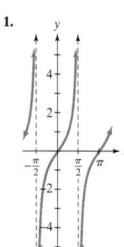   **2.** 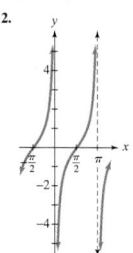   **3.** 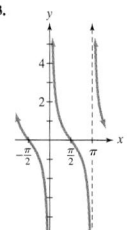   **4.**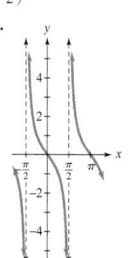

Pages 495-498

Application Exercises

45. An ambulance with a rotating beacon of light is parked 12 feet from a building. The function

$$d = 12 \tan 2\pi t$$

describes the distance, d, in feet, of the rotating beacon from point C after t seconds.

a. Graph the function on the interval $[0, 2]$.

b. For what values of t in $[0, 2]$ is the function undefined? What does this mean in terms of the rotating beacon in the figure shown?

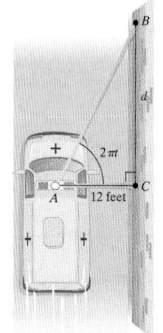

Technology Exercises

In working Exercises 59–62, describe what happens at the asymptotes on the graphing utility. Compare the graphs in the connected and dot modes.

59. Use a graphing utility to verify any two of the tangent curves that you drew by hand in Exercises 5–12.

60. Use a graphing utility to verify any two of the cotangent curves that you drew by hand in Exercises 17–24.

61. Use a graphing utility to verify any two of the cosecant curves that you drew by hand in Exercises 29–44.

62. Use a graphing utility to verify any two of the secant curves that you drew by hand in Exercises 29–44.

In Exercises 63–68, use a graphing utility to graph each function. Use a range setting so that the graph is shown for at least two periods.

63. $y = \tan \frac{x}{4}$ **64.** $y = \tan 4x$

65. $y = \cot 2x$ **66.** $y = \cot \frac{x}{2}$

67. $y = \frac{1}{2} \tan \pi x$ **68.** $y = \frac{1}{2} \tan(\pi x + 1)$

In Exercises 69–72, use a graphing utility to graph each pair of functions in the same viewing rectangle. Use a range setting so that the graphs are shown for at least two periods.

69. $y = 0.8 \sin \frac{x}{2}$ and $y = 0.8 \csc \frac{x}{2}$

70. $y = 0.8 \sin \frac{x}{2}$ and $y = 0.8 \csc \frac{x}{2}$

71. $y = 4 \cos\left(2x - \frac{\pi}{6}\right)$ and $y = 4 \sec\left(2x - \frac{\pi}{6}\right)$

72. $y = -3.5 \cos\left(\pi x - \frac{\pi}{6}\right)$ and $y = -3.5 \sec\left(\pi x - \frac{\pi}{6}\right)$

73. Carbon dioxide particles in our atmosphere trap heat and raise the planet's temperature. The resultant gradually increasing temperature is called the greenhouse effect.

Writing in Mathematics

51. Without drawing a graph, describe the behavior of the basic tangent curve.

52. If you are given the equation of a tangent function, how do you find consecutive asymptotes?

53. If you are given the equation of a tangent function, how do you identify an x-intercept?

54. Without drawing a graph, describe the behavior of the basic cotangent curve.

55. If you are given the equation of a cotangent function, how do you find consecutive asymptotes?

56. Explain how to determine the range of $y = \csc x$ from the graph. What is the range?

57. Explain how to use a sine curve to obtain a cosecant curve. Why can the same procedure be used to obtain a secant curve from a cosine curve?

58. Scientists record brain activity by attaching electrodes to the scalp and then connecting these electrodes to a machine. The record of brain activity recorded with this machine is shown in the three graphs at the top of the next column. Which trigonometric functions would be most appropriate for describing the oscillations in brain activity? Describe similarities and differences among these functions when modeling brain activity when awake, during dreaming sleep, and during non-dreaming sleep.

Critical Thinking Exercises

In Exercises 75–76, write an equation for each blue graph.

75.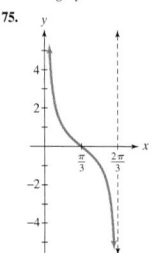

Group Exercise

62. Music and mathematics have been linked over the centuries. Group members should research and present a seminar to the class on music and mathematics. Be sure to include the role of trigonometric functions in the music-mathematics link.

Page 525

xvi

End-of-Chapter Materials
Chapter Summaries and Review Exercises

Each section has its own focused summary and review exercises. These provide students with a good review for a chapter test.

CHAPTER SUMMARY, REVIEW, AND TEST

Summary

4.1 Exponential Functions

a. The exponential function with base b is defined by $f(x) = b^x$, where $b > 0$ and $b \neq 1$.

b. Characteristics of exponential functions and graphs for $0 < b < 1$ and $b > 1$ are shown in the box on page 354.

c. Transformations involving exponential functions are summarized in Table 4.1 on page 354.

d. The natural exponential function: $f(x) = e^x$. The irrational number e is called the natural base, where $e \approx 2.7183$.

e. Formulas for compound interest: After t years, the balance A in an account with principal P and annual interest rate r (in decimal form) is given by one of the following formulas:

  1. For n compoundings per year: $A = P\left(1 + \dfrac{r}{n}\right)^{nt}$

  2. For continuous compounding: $A = Pe^{rt}$

4.2 Logarithmic Functions

a. Definition of the logarithmic function: For $x > 0$ and $b > 0$, $b \neq 1$, $y = \log_b x$ is equivalent to $b^y = x$. The function $f(x) = \log_b x$ is the logarithmic function with base b. This function is the inverse function of the exponential function with base b.

b. Graphs of logarithmic functions for $b > 1$ and $0 < b < 1$ are shown in Figure 4.7 on page 366. Characteristics of the graphs are summarized in the box that follows the figure.

c. Transformations involving logarithmic functions are summarized in Table 4.3 on page 367.

d. The domain of a logarithmic function is the set of all positive real numbers. The domain of $f(x) = \log_b(x + c)$ consists of all x for which $x + c > 0$.

e. Common and natural logarithms: $f(x) = \log x$ means $f(x) = \log_{10} x$ and is the common logarithmic function. $f(x) = \ln x$ means $f(x) = \log_e x$ and is the natural logarithmic function.

f. Basic Logarithmic Properties

Base b $(b > 0, b \neq 1)$	Base 10 (Common Logarithms)	Base e (Natural Logarithms)
$\log_b 1 = 0$	$\log 1 = 0$	$\ln 1 = 0$
$\log_b b = 1$	$\log 10 = 1$	$\ln e = 1$
$\log_b b^x = x$	$\log 10^x = x$	$\ln e^x = x$
$b^{\log_b x} = x$	$10^{\log x} = x$	$e^{\ln x} = x$

Pages 408–414

Chapter 4 Test

1. Graph $f(x) = 2^x$ and $g(x) = 2^{x+1}$ in the same rectangular coordinate system.

2. Graph $f(x) = \log_2 x$ and $g(x) = \log_2(x - 1)$ in the same rectangular coordinate system.

3. Write in exponential form: $\log_5 125 = 3$.

4. Write in logarithmic form: $\sqrt{36} = 6$.

5. Find the domain of $f(x) = \ln(3 - x)$.

In Exercises 6–7, use properties of logarithms to expand each logarithmic expression as much as possible. Where possible, evaluate logarithmic expressions without using a calculator.

6. $\log_4(64x^5)$

7. $\log_3 \dfrac{\sqrt[3]{x}}{81}$

In Exercises 8–9, write each expression as a single logarithm.

8. $6 \log x + 2 \log y$

9. $\ln 7 - 3 \ln x$

10. Use a calculator to evaluate $\log_{15} 71$ to four decimal places.

In Exercises 11–16, solve each equation.

11. $5^x = 1.4$

12. $400e^{0.005x} = 1600$

13. $e^{2x} - 6e^x + 5 = 0$

14. $\log_6(4x - 1) = 3$

15. $\log x + \log(x + 15) = 2$

16. $2 \ln 3x = 8$

17. Suppose you have $3000 to invest. Which investment yields the greater return over 10 years: 6.5% compounded semiannually or 6% compounded continuously? How much more (to the nearest dollar) is yielded by the better investment?

18. On the decibel scale, the loudness of a sound, in decibels, is given by $D = 10 \log \dfrac{I}{I_0}$, where I is the intensity of the sound, in watts per meter^2, and I_0 is the intensity of a sound barely audible to the human ear. If the intensity of a sound is $10^{12}I_0$, what is its loudness in decibels? (Such a sound is potentially damaging to the ear.)

19. The percentage of married men in the United States who are employed is modeled by $P = 89.18e^{-0.004t}$. The model indicates that P% of married men were employed t years after 1959.

  a. What percentage of married men were employed in 1959?

  b. Is the percentage of married men who are employed increasing or decreasing? Explain.

  c. In what year were 77% of U.S. married men employed?

Cumulative Review Exercises (Chapters 1–4)

Solve each equation in Exercises 1–5.

1. $|3x - 4| = 2$

2. $\sqrt{2x - 5} - \sqrt{x - 3} = 1$

3. $x^4 + x^3 - 3x^2 - x + 2 = 0$

4. $e^{5x} - 32 = 96$

5. $\log_2(x + 5) + \log_2(x - 1) = 4$

Solve each inequality in Exercises 6–7. Express the answer in interval notation.

6. $14 - 5x \geq -6$

7. $|2x - 4| \leq 2$

8. Write the point-slope form and the slope-intercept form of the line passing through $(1, 3)$ and $(3, -3)$.

9. If $f(x) = x^2$ and $g(x) = x + 2$, find $(f \circ g)(x)$ and $(g \circ f)(x)$.

10. If $f(x) = 2x - 7$, find $f^{-1}(x)$.

11. Divide $x^3 + 5x^2 + 3x - 10$ by $x + 2$.

12. Use the Rational Zero Theorem to list all possible rational zeros for $f(x) = 4x^3 - 7x - 3$.

13. The value of y varies directly as the square of x. If $x = 3$ when $y = 12$, find y when $x = 15$.

14. Solve $x^3 - 4x^2 + 6x - 4 = 0$ given that $1 + i$ is a root.

In Exercises 15–18, graph each equation.

15. $(x - 3)^2 + (y + 2)^2 = 4$

16. $f(x) = (x - 2)^2 - 1$

17. $f(x) = \dfrac{x^2 - 1}{x^2 - 4}$

18. $f(x) = (x - 2)^2(x + 1)$

19. You are paid time-and-a-half for each hour worked over 40 hours a week. Last week you worked 50 hours and earned $660. What is your normal hourly salary?

20. The formula $F = 1 - k \ln(t + 1)$ models the fraction of people, F, who remember all the words in a list of nonsense words t hours after memorizing the list. After 3 hours only half the people could remember all the words. Determine the value of k and then predict the fraction of people in the group who will remember all the words after 6 hours.

Applications Index

Prerequisites: Fundamental Concepts of Algebra

This chapter reviews fundamental concepts of algebra that are prerequisites for the study of college algebra. Algebra, like all of mathematics, provides the tools to help you recognize, classify, and explore the hidden patterns of your world, revealing its underling structure. Throughout the new millennium, literacy in algebra will be a prerequisite for functioning in a meaningful way personally, professionally, and as a citizen.

Listening to the radio on the way to work, you hear candidates in the upcoming election discussing the problem of the country's 5.5 trillion dollar deficit. It seems like this is a real problem, but then you realize that you don't really know what that number means. How can you look at this deficit in the proper perspective? If the national debt were evenly divided among all citizens of the country, how much would each citizen have to pay? Does the deficit seem like such a significant problem now?

SECTION P.1 *Real Numbers and Algebraic Expressions*

Objectives

1. Recognize subsets of the real numbers.
2. Use inequality symbols.
3. Evaluate absolute value.
4. Use absolute value to express distance.
5. Evaluate algebraic expressions.
6. Identity properties of the real numbers.
7. Simplify algebraic expressions.

The U.N. building is designed with three golden rectangles.

The United Nations Building in New York was designed to represent its mission of promoting world harmony. Viewed from the front, the building looks like three rectangles stacked upon each other. In each rectangle, the ratio of the width to height is $\sqrt{5} + 1$ to 2, approximately 1.618 to 1. The ancient Greeks believed that such a rectangle, called a **golden rectangle**, was the most visually pleasing of all rectangles.

The ratio 1.618 to 1 is approximate because $\sqrt{5}$ is an irrational number, a special kind of real number. Irrational? Real? Let's make sense of all this by describing the kinds of numbers you will encounter in this course.

1 Recognize subsets of the real numbers.

The Set of Real Numbers

Before we describe the set of real numbers, let's be sure you are familiar with some basic ideas about sets. A **set** is a collection of objects whose contents can be clearly determined. The objects in a set are called the **elements** of the set. For example, the set of numbers used for counting can be represented by

$$\{1, 2, 3, 4, 5, \ldots\}.$$

The braces, { }, indicate that we are representing a set. This form of representing a set uses commas to separate the elements of the set. The set of numbers used for counting is called the set of **natural numbers**. The three dots after the 5 indicate that there is no final element and that the listing goes on forever.

The sets that make up the real numbers are summarized in Table P.1. We refer to these sets as **subsets** of the real numbers, meaning that all elements in each subset are also elements in the set of real numbers.

Notice the use of the symbol $\approx$ in the examples of irrational numbers. The symbol means "is approximately equal to." Thus,

$$\sqrt{2} \approx 1.414214.$$

We can verify that this is only an approximation by multiplying 1.414214 by itself. The product is very close to but not exactly 2:

$$1.414214 \times 1.414214 = 2.0000012378.$$

Technology

A calculator with a square root key gives a decimal approximation for $\sqrt{2}$, not the exact value.

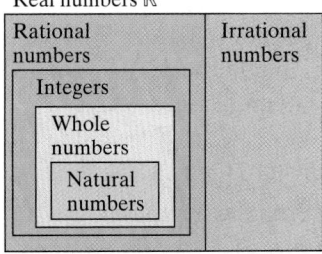

Real numbers ℝ

This diagram shows that every real number is rational or irrational.

Table P.1 Important Subsets of the Real Numbers

Name	Description	Examples
Natural numbers ℕ	$\{1, 2, 3, 4, 5, \ldots\}$ These numbers are used for counting.	$2, 3, 5, 17$
Whole numbers 𝕎	$\{0, 1, 2, 3, 4, 5, \ldots\}$ The whole numbers add 0 to the set of natural numbers.	$0, 2, 3, 5, 17$
Integers ℤ	$\{\ldots, -5, -4, -3, -2, -1, 0, 1, 2, 3, 4, 5, \ldots\}$ The integers add the negatives of the natural numbers to the set of whole numbers.	$-17, -5, -3, -2, 0,$ $2, 3, 5, 17$
Rational numbers ℚ	These numbers can he expressed as an integer divided by a nonzero integer: $\frac{a}{b}$: a and b are integers: $b \neq 0$. Rational numbers can be expressed as terminating or repeating decimals.	$-17 = \frac{-17}{1}, -5 = \frac{-5}{1}, -3, -2,$ $0, 2, 3, 5, 17,$ $\frac{2}{5} = 0.4, \frac{-2}{3} = -0.6666\cdots = -0.\overline{6}$
Irrational numbers 𝕀	This is the set of numbers whose decimal representations are neither terminating nor repeating. Irrational numbers cannot be expressed as a quotient of integers.	$\sqrt{2} \approx 1.414214$ $-\sqrt{3} \approx -1.73205$ $\pi \approx 3.142$ $-\frac{\pi}{2} \approx -1.571$

The set of **real numbers** is formed by combining the rational numbers and the irrational numbers. Thus, every real number is either rational or irrational.

The Real Number Line

The **real number line** is a graph used to represent the set of real numbers. An arbitrary point, called the **origin**, is labeled 0; units to the right of the origin are **positive** and units to the left of the origin are **negative**. The real number line is shown in Figure P.1.

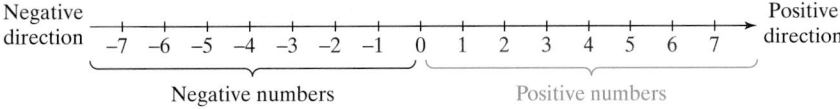

Figure P.1 The real number line

Real numbers are **graphed** on a number line by placing a dot at the correct location for each number. The integers are easiest to locate. In Figure P.2 we've graphed the integers $-3, 0,$ and 4.

Figure P.2 Graphing $-3, 0,$ and 4 on number line

Every real number corresponds to a point on the number line and every point on the number line corresponds to a real number. We say there is a **one-to-one correspondence** between all the real numbers and all points on a real number line. If you draw a point on the real number line corresponding to a real number, you are **plotting** the real number. In Figure P.2, we are plotting the real numbers $-3, 0,$ and 4.

2 Use inequality symbols.

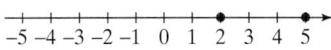

Figure P.3

Study Tip

The symbols $<$ and $>$ always point to the lesser of the two real numbers when the inequality is true.

$2 < 5$ The symbol points to 2, the lesser number.

$5 > 2$ The symbol points to 2, the lesser number.

Ordering the Real Numbers

On the real number line, the real numbers increase from left to right. The lesser of two real numbers is the one farther to the left on a number line. The greater of two real numbers is the one farther to the right on a number line.

Look at the number line in Figure P.3. The integers 2 and 5 are plotted. Observe that 2 is to the left of 5 on the number line. This means that 2 is less than 5:

$2 < 5$:   2 is less than 5 because 2 is to the *left* of 5 on the number line.

In Figure P.3, we can also observe that 5 is to the right of 2 on the number line. This means that 5 is greater than 2.

$5 > 2$:   5 is greater than 2 because 5 is to the right of 2 on the number line.

The symbols $<$ and $>$ are called **inequality symbols**. They may be combined with an equal sign, as shown in the following table.

Symbols	Meaning	Example	Explanation
$a \leq b$	a is less than or equal to b.	$3 \leq 7$	Because $3 < 7$
		$7 \leq 7$	Because $7 = 7$
$b \geq a$	b is greater than or equal to a.	$7 \geq 3$	Because $7 > 3$
		$-5 \geq -5$	Because $-5 = -5$

3 Evaluate absolute value.

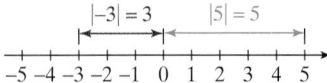

Figure P.4 Absolute value as the distance from 0

Absolute Value

Absolute value describes the distance from 0 on a real number line. If a represents a real number, the symbol $|a|$ represents its absolute value, read "the absolute value of a." For example, the real number line in Figure P.4 shows that

$$|-3| = 3 \quad \text{and} \quad |5| = 5.$$

The absolute value of -3 is 3 because -3 is 3 units from 0 on the number line. The absolute value of 5 is 5 because 5 is 5 units from 0 on the number line. The absolute value of a positive real number or 0 is the number itself. The absolute value of a negative real number, such as -3, is positive.

We can define the absolute value of the real number x without referring to a number line. The algebraic definition of the absolute value of x is given as follows:

Definition of Absolute Value

$$|x| = \begin{cases} x & \text{if } x \geq 0 \\ -x & \text{if } x < 0 \end{cases}$$

If x is nonnegative (that is $x \geq 0$), the absolute value of x is the number itself. For example:

$$|5| = 5 \qquad |\pi| = \pi \qquad \left|\frac{1}{3}\right| = \frac{1}{3} \qquad |0| = 0$$

   Zero is the only number whose absolute value is 0.

If x is a negative number (that is, $x < 0$), the absolute value of x is the opposite of x. This makes the absolute value positive. For example,

$$|-3| = -(-3) = 3 \qquad |-\pi| = -(-\pi) = \pi \qquad \left|-\frac{1}{3}\right| = -\left(-\frac{1}{3}\right) = \frac{1}{3}.$$

This middle step is usually omitted.

EXAMPLE 1 Evaluating Absolute Value

Rewrite each expression without absolute value bars.

a. $\left|\sqrt{3} - 1\right|$ **b.** $|2 - \pi|$ **c.** $\dfrac{|x|}{x}$ if $x < 0$

Solution

a. Because $\sqrt{3} \approx 1.7$, the expression inside the absolute value bars is positive. The absolute value of a positive number is the number itself. Thus,
$$\left|\sqrt{3} - 1\right| = \sqrt{3} - 1.$$

b. Because $\pi \approx 3.14$, the number inside the absolute value bars is negative. The absolute value of x when $x < 0$ is $-x$. Thus,
$$|2 - \pi| = -(2 - \pi) = \pi - 2.$$

c. If $x < 0$, then $|x| = -x$. Thus,
$$\frac{|x|}{x} = \frac{-x}{x} = -1.$$

> **Check Point 1** Rewrite each expression without absolute value bars.
>
> **a.** $\left|1 - \sqrt{2}\right|$ **b.** $|\pi - 3|$ **c.** $\dfrac{|x|}{x}$ if $x > 0$

Next, we list several basic properties of absolute value. Each of these properties can be derived from the definition of absolute value.

Properties of Absolute Value

For all real numbers a and b,

1. $|a| \geq 0$ **2.** $|-a| = |a|$ **3.** $a \leq |a|$

4. $|ab| = |a||b|$ **5.** $\left|\dfrac{a}{b}\right| = \dfrac{|a|}{|b|}$, $b \neq 0$

6. $|a + b| \leq |a| + |b|$ (called the triangle inequality)

4 Use absolute value to express distance.

Distance Between Points on a Real Number Line

Absolute value is used to find the distance between two points on a real number line. If a and b are any real numbers, the **distance between a and b** is the absolute value of their difference. For example, the distance between 4 and 10 is 6. Using absolute value, we find this distance in one of two ways:

$$|10 - 4| = |6| = 6 \quad \text{or} \quad |4 - 10| = |-6| = 6.$$

The distance between 4 and 10 on the real number line is 6.

Notice that we obtain the same distance regardless of the order in which we subtract.

> ### Distance Between Two Points on the Real Number Line
>
> If a and b are any two points on a real number line, then the distance between a and b is given by
>
> $$|a - b| \quad \text{or} \quad |b - a|.$$

EXAMPLE 2   Distance Between Two Points on a Number Line

Find the distance between -5 and 3 on the real number line.

Solution   Because the distance between a and b is given by $|a - b|$, the distance between -5 and 3 is

$$|-5 - 3| = |-8| = 8.$$

$a = -5$ $b = 3$

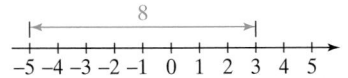

Figure P.5 The distance between -5 and 3 is 8.

Figure P.5 verifies that there are 8 units between -5 and 3 on the real number line. We obtain the same distance if we reverse the order of the subtraction:

$$|3 - (-5)| = |8| = 8.$$

Check Point 2   Find the distance between -4 and 5 on the real number line.

Algebraic Expressions

Algebra uses letters, such as x and y, to represent real numbers. Such letters are called **variables**. For example, imagine that you are basking in the sun on the beach. We can let x represent the number of minutes that you can stay in the sun without burning with no sunscreen. With a number 6 sunscreen, exposure time without burning is six times as long, or 6 times x. This can be written $6 \cdot x$, but it is usually expressed as $6x$. Placing a number and a letter next to one another indicates multiplication.

Notice that $6x$ combines the number 6 and the variable x using the operation of multiplication. A combination of variables and numbers using the operations of addition, subtraction, multiplication, or division, as well as powers or roots, is called an **algebraic expression**. Here are some examples of algebraic expressions:

$$x + 6, \quad x - 6, \quad 6x, \quad \frac{x}{6}, \quad 3x + 5, \quad \sqrt{x} + 7.$$

5 Evaluate algebraic expressions.

Evaluating Algebraic Expressions

Evaluating an algebraic expression means to find the value of the expression for a given value of the variable. For example, we can evaluate $6x$ (from the sunscreen example) when $x = 15$. We substitute 15 for x. We obtain $6 \cdot 15$, or 90. This means if you can stay in the sun for 15 minutes without burning when you don't put on any lotion, then with a number 6 lotion, you can "cook" for 90 minutes without burning.

Many algebraic expressions involve more than one operation. Evaluating an algebraic expression without a calculator involves carefully applying the following order of operations agreement.

The Order of Operations Agreement

1. Perform operations within the innermost parentheses and work outward. If the algebraic expression involves division, treat the numerator and the denominator as if they were each enclosed in parentheses.

2. Perform multiplication or division as they occur, working from left to right.

3. Perform addition or subtraction as they occur, working from left to right.

EXAMPLE 3   Evaluating an Algebraic Expression

The algebraic expression $2.35x + 179.5$ describes the population of the United States, in millions, x years after 1980. Evaluate the expression when $x = 20$. Describe what the answer means in practical terms.

Solution   We begin by substituting 20 for x. Because $x = 20$, we will be finding the U.S. population 20 years after 1980, in the year 2000.

$$2.35x + 179.5$$

Replace x by 20.

$$= 2.35(20) + 179.5$$

$$= 47 + 179.5 \qquad \text{Perform the multiplication: } 2.35(20) = 47.$$

$$= 226.5 \qquad \text{Perform the addition.}$$

Thus, in 2000 the population of the United States was 226.5 million.

Check Point 3   Evaluate: $2.35x + 179.5$ when $x = 100$.
Describe what your answer means in practical terms.

6 Identify properties of the real numbers.

Properties of Real Numbers and Algebraic Expressions

When you use your calculator to add two real numbers, you can enter them in any order. The fact that two real numbers can be added in any order is called the **commutative property of addition**. You probably use this property, as well as other properties of real numbers listed in Table P.2 on the next page, without giving it much thought. The properties of the real numbers are especially useful when working with algebraic expressions. For each property listed in Table P.2, a, b, and c represent real numbers, variables, or algebraic expressions.

Table P.2 Properties of the Real Numbers

Name	Meaning	Examples
Commutative Property of Addition	Two real numbers can be added in any order. $a + b = b + a$	• $13 + 7 = 7 + 13$ • $13x + 7 = 7 + 13x$
Commutative Property of Multiplication	Two real numbers can be multiplied in any order. $ab = ba$	• $\sqrt{2} \cdot \sqrt{5} = \sqrt{5} \cdot \sqrt{2}$ • $x \cdot 6 = 6x$
Associative Property of Addition	If three real numbers are added, it makes no difference which two are added first. $(a + b) + c = a + (b + c)$	• $3 + (8 + x) = (3 + 8) + x$ $= 11 + x$
Associative Property of Multiplication	If three real numbers are multiplied, it makes no difference which two are multiplied first. $(a \cdot b) \cdot c = a \cdot (b \cdot c)$	• $-2(3x) = (-2 \cdot 3)x = -6x$
Distributive Property of Multiplication over Addition	Multiplication distributes over addition. $a \cdot (b + c) = a \cdot b + a \cdot c$	• $7(4 + \sqrt{3}) = 7 \cdot 4 + 7 \cdot \sqrt{3}$ $= 28 + 7\sqrt{3}$ • $5(3x + 7) = 5 \cdot 3x + 5 \cdot 7$ $= 15x + 35$
Identity Property of Addition	Zero can be deleted from a sum. $a + 0 = a$ $0 + a = a$	• $\sqrt{3} + 0 = \sqrt{3}$ • $0 + 6x = 6x$
Identity Property of Multiplication	One can be deleted from a product. $a \cdot 1 = a$ $1 \cdot a = a$	• $1 \cdot \pi = \pi$ • $13x \cdot 1 = 13x$
Inverse Property of Addition	The sum of a real number and its additive inverse gives 0, the additive identity. $a + (-a) = 0$ $(-a) + a = 0$	• $\sqrt{5} + (-\sqrt{5}) = 0$ • $6x + (-6x) = 0$ • $(-4y) + 4y = 0$
Inverse Property of Multiplication	The product of a nonzero real number and its multiplicative inverse gives 1, the multiplicative identity. $a \cdot \dfrac{1}{a} = 1, \quad a \neq 0$ $\dfrac{1}{a} \cdot a = 1, \quad a \neq 0$	• $7 \cdot \frac{1}{7} = 1$ • $\left(\dfrac{1}{x - 3}\right)(x - 3) = 1, \ x \neq 3$

Commutative Words and Sentences

The commutative property states that a change in order produces no change in the answer. The words and sentences listed here are commutative; they read the same from left to right and from right to left!

dad	Draw, o coward!	Revolting is error. Resign it, lover.
repaper	Dennis sinned.	Naomi, did I moan?
never odd or even	Ma is a nun, as I am.	Al lets Della call Ed Stella.

The properties in Table P.2 apply to the operations of addition and multiplication. Subtraction and division are defined in terms of addition and multiplication.

Definitions of Subtraction and Division

Let a and b represent real numbers.

Subtraction: $a - b = a + (-b)$

We call $-b$ the **additive inverse** or **opposite** of b.

Division: $a \div b = a \cdot \frac{1}{b}$, where $b \neq 0$

We call $\frac{1}{b}$ the **multiplicative inverse** or **reciprocal** of b. The quotient of a and b, $a \div b$, can be written in the form $\frac{a}{b}$, where a is the **numerator** and b the **denominator** of the fraction.

Because subtraction is defined in terms of adding an inverse, the distributive property can be applied to subtraction:

$$a(b - c) = ab - ac$$

$$(b - c)a = ba - ca.$$

For example,

$$4(2x - 5) = 4 \cdot 2x - 4 \cdot 5 = 8x - 20.$$

7 Simplify algebraic expressions.

Simplifying Algebraic Expressions

The **terms** of an algebraic expression are those parts that are separated by addition. For example, consider the algebraic expression

$$7x - 9y - 3,$$

which can be expressed as

$$7x + (-9y) + (-3).$$

This expression contains three terms, namely $7x$, $-9y$, and -3.

The numerical part of a term is called its **numerical coefficient**. In the term $7x$, the 7 is the numerical coefficient. In the term $-9y$, the -9 is the numerical coefficient.

A term that consists of just a number is called a **constant term**. The constant term of $7x - 9y - 3$ is -3.

A term indicates a product. The expressions that are multiplied to form the term are called its **factors**. **Like terms** have the same variable factors with the same exponents on the variables. For example, $7x$ and $3x$ are like terms because they have the same variable factor, x. The distributive property (in reverse) can be used to add these terms:

$$7x + 3x = (7 + 3)x = 10x.$$

Study Tip

To add like terms, add their numerical coefficients. Use this result as the numerical coefficient of the terms' common variable(s).

An algebraic expression is **simplified** when parentheses have been removed and like terms have been combined.

EXAMPLE 4 Simplifying an Algebraic Expression

Simplify: $6(2x - 4y) + 10(4x + 3y)$.

Solution

$$6(2x - 4y) + 10(4x + 3y)$$

$$= 6 \cdot 2x - 6 \cdot 4y + 10 \cdot 4x + 10 \cdot 3y \quad \text{Use the distributive property to remove the parentheses.}$$

$$= 12x - 24y + 40x + 30y \quad \text{Multiply.}$$

$$= (12x + 40x) + (30y - 24y) \quad \text{Group like terms.}$$

$$= 52x + 6y \quad \text{Combine like terms.}$$

Check Point 4   Simplify: $7(4x - 3y) + 2(5x + y)$.

Properties of Negatives

The distributive property can be extended to cover more than two terms within parentheses. For example,

> This sign represents subtraction.

> This sign tells us that −3 is negative.

$$-3(4x - 2y + 6) = -3 \cdot 4x - (-3) \cdot 2y - 3 \cdot 6$$

$$= -12x - (-6y) - 18$$

$$= -12x + 6y - 18$$

The voice balloons illustrate that negative signs can appear side by side. They can represent the operation of subtraction or the fact that a real number is negative. Here is a list of properties of negatives and how they are applied to algebraic expressions.

Properties of Negatives

Let a and b represent real numbers, variables, or algebraic expressions.

Property	Examples
1. $(-1)a = -a$	$(-1)4xy = -4xy$
2. $-(-a) = a$	$-(-6y) = 6y$
3. $(-a)b = -ab$	$(-7)4xy = -7 \cdot 4xy = -28xy$
4. $a(-b) = -ab$	$5x(-3y) = -5x \cdot 3y = -15xy$
5. $-(a + b) = -a - b$	$-(7x + 6y) = -7x - 6y$
6. $-(a - b) = -a + b$	$-(3x - 7y) = -3x + 7y$
$\qquad = b - a$	$\qquad = 7y - 3x$

Do you notice that properties 5 and 6 in the box are related? In general, expressions within parentheses that are preceded by a negative can be simplified by dropping the parentheses and changing the sign of every term inside the parentheses.
For example,

$$-(3x - 2y + 5z - 6) = -3x + 2y - 5z + 6.$$

EXERCISE SET P.1

✓ Practice Exercises

In Exercises 1–4, list all numbers from the given set that are a. natural numbers, b. whole numbers, c. integers, d. rational numbers, e. irrational numbers.

1. $\{-9, -\frac{4}{5}, 0, 0.25, \sqrt{3}, 9.2, \sqrt{100}\}$

2. $\{-7, -0.\overline{6}, 0, \sqrt{49}, \sqrt{50}\}$

3. $\{-11, -\frac{5}{6}, 0, 0.75, \sqrt{5}, \pi, \sqrt{64}\}$

4. $\{-5, -0.\overline{3}, 0, \sqrt{2}, \sqrt{4}\}$

5. Give an example of a whole number that is not a natural number.

6. Give an example of a rational number that is not an integer.

7. Give an example of a number that is an integer, a whole number, and a natural number.

8. Give an example of a number that is a rational number, an integer, and a real number.

Determine whether each statement in Exercises 9–14 is true or false.

9. $-13 \le -2$ **10.** $-6 > 2$

11. $4 \ge -7$ **12.** $-13 < -5$

13. $-\pi \ge -\pi$ **14.** $-3 > -13$

In Exercises 15–22, rewrite each expression without absolute value bars.

15. $|300|$ **16.** $|-203|$

17. $|12 - \pi|$ **18.** $|7 - \pi|$

19. $|\sqrt{2} - 5|$ **20.** $|\sqrt{5} - 13|$

21. $\dfrac{-3}{|-3|}$ **22.** $\dfrac{-7}{|-7|}$

In Exercises 23–30, express the distance between the given numbers using absolute value. Then find the distance by evaluating the absolute value expression.

23. 2 and 17   **24.** 4 and 15

25. -2 and 5   **26.** -6 and 8

27. -19 and -4 **28.** -26 and -3

29. -3.6 and -1.4 **30.** -5.4 and -1.2

In Exercises 31–38, evaluate each algebraic expression for the given value of the variable.

31. $5x + 7$; $x = 4$ **32.** $9x + 6$; $x = 5$

33. $4(x + 3) - 11$; $x = -5$

34. $6(x + 5) - 13$; $x = -7$

35. $\dfrac{5}{9}(F - 32)$; $F = 77$ **36.** $\dfrac{5}{9}(F - 32)$; $F = 50$

37. $\dfrac{5(x + 2)}{2x - 14}$; $x = 10$ **38.** $\dfrac{7(x - 3)}{2x - 16}$; $x = 9$

In Exercises 39–46, state the name of the property illustrated.

39. $6 + (-4) = (-4) + 6$

40. $11 \cdot (7 + 4) = 11 \cdot 7 + 11 \cdot 4$

41. $6 + (2 + 7) = (6 + 2) + 7$

42. $6 \cdot (2 \cdot 3) = 6 \cdot (3 \cdot 2)$

43. $(2 + 3) + (4 + 5) = (4 + 5) + (2 + 3)$

44. $7 \cdot (11 \cdot 8) = (11 \cdot 8) \cdot 7$

45. $2(-8 + 6) = -16 + 12$

46. $-8(3 + 11) = -24 + (-88)$

In Exercises 47–52, simplify each algebraic expression.

47. $5(3x + 4) - 4$ **48.** $2(5x + 4) - 3$

49. $5(3x - 2) + 12x$ **50.** $2(5x - 1) + 14x$

51. $7(3y - 5) + 2(4y + 3)$ **52.** $4(2y - 6) + 3(5y + 10)$

In Exercises 53–58, write each algebraic expression without parentheses.

53. $-(-14x)$ **54.** $-(-17y)$

55. $-(2x - 3y - 6)$ **56.** $-(5x - 13y - 1)$

57. $\frac{1}{3}(3x) + [(4y) + (-4y)]$ **58.** $\frac{1}{2}(2y) + [(-7x) + 7x]$

☆ Application Exercises

59. Are first putting on your left shoe and then putting on your right shoe commutative?

60. Are first getting undressed and then taking a shower commutative?

61. Give an example of two things that you do that are not commutative.

62. Give an example of two things that you do that are commutative.

63. The algebraic expression $962x + 18{,}667$ describes average yearly earnings in United States x years after 1990. Evaluate the algebraic expression when $x = 7$. Describe what the answer means in practical terms.

64. The algebraic expression $1527x + 31{,}290$ describes average yearly earnings for elementary and secondary teachers in the United States x years after 1990. Evaluate the algebraic expression when $x = 10$. Describe what the answer means in practical terms.

65. The optimum heart rate is the rate that a person should achieve during exercise for the exercise to be most beneficial. The algebraic expression

$$0.6(220 - a)$$

describes a person's optimum heart rate in beats per minute, where a represents the age of the person.
   a. Use the distributive property to rewrite the algebraic expression without parentheses.
   b. Use each form of the algebraic expression to determine the optimum heart rate for a 20-year-old runner.

Writing in Mathematics

Writing about mathematics will help you to learn mathematics. For all writing exercises in this book, use complete sentences to respond to the question. Some writing exercises can be answered in a sentence; others require a paragraph or two. You can decide how much you need to write as long as your writing clearly and directly answers the question in the exercise. Standard references such as a dictionary and a thesaurus should be helpful.

66. How do the whole numbers differ from the natural numbers?

67. Can a real number be both rational and irrational? Explain your answer.

68. If you are given two real numbers, explain how to determine which one is the lesser.

69. How can $\dfrac{|x|}{x}$ be equal to 1 or -1?

70. What is an algebraic expression? Give an example with your explanation.

71. Why is $3(x + 7) - 4x$ not simplified? What must be done to simplify the expression?

72. You can transpose the letters in the word "conversation" to form the phrase "voices rant on." From "total abstainers" we can form "sit not at ale bars." What two algebraic properties do each of these transpositions (called anagrams) remind you of? Explain your answer.

Critical Thinking Exercises

73. Which one of the following statements is true?
   a. Every rational number is an integer.
   b. Some whole numbers are not integers.
   c. Some rational numbers are not positive.
   d. Irrational numbers cannot be negative.

74. Which of the following is true?
   a. The term x has no numerical coefficient.
   b. $5 + 3(x - 4) = 8(x - 4) = 8x - 32$
   c. $-x - x = -x + (-x) = 0$
   d. $x - 0.02(x + 200) = 0.98x - 4$

In Exercises 75–77, insert either $<$ or $>$ in the box between the numbers to make the statement true.

75. $\sqrt{2} \ \Box \ 1.5$

76. $-\pi \ \Box \ -3.5$

77. $-\dfrac{3.14}{2} \ \Box \ -\dfrac{\pi}{2}$

78. A business that manufactures small alarm clocks has weekly fixed costs of $5000. The average cost per clock for the business to manufacture x clocks is described by

$$\frac{0.5x + 5000}{x}.$$

   a. Find the average cost when $x = 100, 1000,$ and $10{,}000$.
   b. Like all other businesses, the alarm clock manufacturer must make a profit. To do this, each clock must be sold for at least 50¢ more than what it costs to manufacture. Due to competition from a larger company, the clocks can be sold for $1.50 each and no more. Our small manufacturer can only produce 2000 clocks weekly. Does this business have much of a future? Explain.

SECTION P.2 *Exponents and Scientific Notation*

Objectives

1. Understand and use integer exponents.
2. Use properties of exponents.
3. Simplify exponential expressions.
4. Use scientific notation.

1 Understand and use integer exponents.

Although people do a great deal of talking, the total output since the beginning of gabble to the present day, including all baby talk, love songs, and congressional debates, only amounts to about 10 million billion words. This can be expressed as 16 factors of 10, or 10^{16} words.

Exponents such as 2, 3, 4, and so on are used to indicate repeated multiplication. For example,

$$10^2 = 10 \cdot 10 = 100,$$
$$10^3 = 10 \cdot 10 \cdot 10 = 1000, \quad 10^4 = 10 \cdot 10 \cdot 10 \cdot 10 = 10,000.$$

The 10 that is repeated when multiplying is called the **base**. The small numbers above and to the right of the base are called **exponents**. The exponent tells the number of times the base is to be used when multiplying. In 10^3, the base is 10 and the exponent is 3.

Any number with an exponent of 1 is the number itself. Thus, $10^1 = 10$.

Multiplications that are expressed in exponential notation are read as follows:

10^1: "ten to the first power"

10^2: "ten to the second power" or "ten squared"

10^3: "ten to the third power" or "ten cubed"

10^4: "ten to the fourth power"

10^5: "ten to the fifth power"

Any real number can be used as the base. Thus,

$$7^2 = 7 \cdot 7 = 49 \text{ and } (-3)^4 = (-3)(-3)(-3)(-3) = 81.$$

The bases are 7 and -3, respectively. Do not confuse $(-3)^4$ and -3^4.

$$-3^4 = -3 \cdot 3 \cdot 3 \cdot 3 = -81$$

The negative is not taken to the power because it is not inside parentheses.

Powers Of Ten

$10 = 10^1$
$100 = 10^2$
$1000 = 10^3$
$10,000 = 10^4$
$100,000 = 10^5$
$1,000,000 = 10^6$ **million**
$10,000,000 = 10^7$
$100,000,000 = 10^8$
$1,000,000,000 = 10^9$ **billion**

Technology

You can use a calculator to evaluate exponential expressions. For example, to evaluate 5^3, press the following keys:

Scientific Calculator
5 $\boxed{x^y}$ 3 $\boxed{=}$

Graphing Calculator
5 $\boxed{\wedge}$ 3 $\boxed{\text{ENTER}}$

Although calculators have special keys to evaluate powers of ten and squaring bases, you can always use one of the sequences shown here.

than 10 multiplied by some power of 10. It is customary to use the multiplication symbol, ×, rather than a dot in scientific notation.

Here are two examples of numbers in scientific notation:

Each day, 2.6×10^7 pounds of dust from the atmosphere settle on Earth.

The diameter of a hydrogen atom is 1.016×10^{-8} centimeter.

We can use the exponent on the 10 to change a number in scientific notation to decimal notation. If the exponent is *positive*, move the decimal point in the number to the *right* the same number of places as the exponent. If the exponent is *negative*, move the decimal point in the number to the *left* the same number of places as the exponent.

EXAMPLE 9 Converting from Scientific to Decimal Notation

Write each number in decimal notation.

    a. 2.6×10^7 **b.** 1.016×10^{-8}

Solution

    a. We express 2.6×10^7 in decimal notation by moving the decimal point in 2.6 seven places to the right. We need to add six zeros.

$$2.6 \times 10^7 = 26,000,000$$

    b. We express 1.016×10^{-8} in decimal notation by moving the decimal point in 1.016 eight places to the left. We need to add seven zeros to the right of the decimal point.

$$1.016 \times 10^{-8} = 0.00000001016$$

> **Check Point 9**
>
> Write each number in decimal notation.
>
>   **a.** 7.4×10^9 **b.** 3.017×10^{-6}

To convert from decimal notation to scientific notation, we reverse the procedure of Example 9.

- Move the decimal point in the given number to obtain a number greater than or equal to 1 and less than 10.

- The number of places the decimal point moves gives the exponent on 10; the exponent is positive if the given number is greater than 10 and negative if the given number is between 0 and 1.

EXAMPLE 10 Converting from Decimal Notation to Scientific Notation

Write each number in scientific notation.

    a. 4,600,000   **b.** 0.00023

Solution

    a. $4,600,000 = 4.6 \times 10^?$ *Decimal point moves 6 places.* → 4.6×10^6

    b. $0.00023 = 2.3 \times 10^{-?}$ *Decimal point moves 4 places.* → 2.3×10^{-4}

Check Point 10

Write each number in scientific notation.

a. 7,410,000,000   **b.** 0.000000092

Computations with Scientific Notation

The product and quotient rules for exponents can be used to multiply or divide numbers that are expressed in scientific notation. For example, here's how to find the product of 3.4×10^9 and 2×10^{-5}.

$$(3.4 \times 10^9)(2 \times 10^{-5}) = (3.4 \times 2) \times (10^9 \times 10^{-5})$$
$$= 6.8 \times 10^{9+(-5)}$$
$$= 6.8 \times 10^4 \quad \text{or} \quad 68,000$$

In our next example, we use the quotient of two numbers in scientific notation to help put a number into perspective. The number is our national debt. The United States began accumulating large deficits in the 1980s. To finance the deficit, the government had borrowed $5.5 trillion as of the end of 1998. The graph in Figure P.6 shows the national debt increasing over time.

Technology

On a graphing calculator, you can use the EE (enter exponent) key to find the product of 3.4×10^9 and 2×10^{-5}:

3.4 [EE] 9 [×] 2 [EE] [(−)] 5 [ENTER]

If your calculator is in the scientific notation mode, it will display 6.8 E4; in the normal mode it will display 68000.

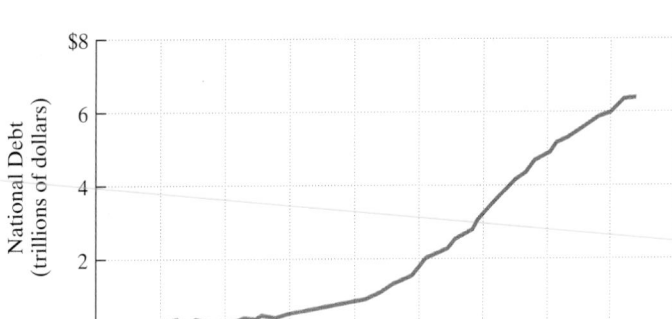

Figure P.6 The national debt

Source: Office of Management and Budget

EXAMPLE 11   The National Debt

As of the end of 1998, the national debt was $5.5 trillion, or 5.5×10^{12} dollars. At that time, the U.S. population was approximately 270,000,000 (270 million), or 2.7×10^8. If the national debt were evenly divided among every individual in the United States, how much would each citizen have to pay?

Technology

Here is the keystroke sequence for solving Example 11 using a graphing calculator:

5.5 [EE] 12 [÷] 2.7 [EE] 8 [ENTER]

Solution   The amount each citizen must pay is the total debt, 5.5×10^{12} dollars, divided by the number of citizens, 2.7×10^8.

$$\frac{5.5 \times 10^{12}}{2.7 \times 10^8} = \left(\frac{5.5}{2.7}\right) \times \left(\frac{10^{12}}{10^8}\right)$$
$$\approx 2.04 \times 10^{12-8}$$
$$= 2.04 \times 10^4$$
$$= 20,400$$

Every U.S. citizen would have to pay about $20,400 to the federal government to pay off the national debt. A family of three would owe $61,200!

Check Point 11

Approximately 2×10^4 people run in the New York City Marathon each year. Each runner runs a distance of 26 miles. Write the total distance covered by all the runners (assuming that each person completes the marathon) in scientific notation.

EXERCISE SET P.2

Practice Exercises

Evaluate each exponential expression in Exercises 1–22.

1. $5^2 \cdot 2$ **2.** $6^2 \cdot 2$

3. $(-2)^6$ **4.** $(-2)^4$

5. -2^6 **6.** -2^4

7. $(-3)^0$ **8.** $(-9)^0$

9. -3^0 **10.** -9^0

11. 4^{-3} **12.** 2^{-6}

13. $2^2 \cdot 2^3$ **14.** $3^3 \cdot 3^2$

15. $(2^2)^3$ **16.** $(3^3)^2$

17. $\dfrac{2^8}{2^4}$ **18.** $\dfrac{3^8}{3^4}$

19. $3^{-3} \cdot 3$ **20.** $2^{-3} \cdot 2$

21. $\dfrac{2^3}{2^7}$ **22.** $\dfrac{3^4}{3^7}$

Simplify each exponential expression in Exercises 23–60.

23. $x^{-2}y$ **24.** xy^{-3}

25. $x^0 y^5$ **26.** $x^7 y^0$

27. $x^3 \cdot x^7$ **28.** $x^{11} \cdot x^5$

29. $x^{-5} \cdot x^{10}$ **30.** $x^{-6} \cdot x^{12}$

31. $(x^3)^7$ **32.** $(x^{11})^5$

33. $(x^{-5})^3$ **34.** $(x^{-6})^4$

35. $\dfrac{x^{14}}{x^7}$ **36.** $\dfrac{x^{30}}{x^{10}}$

37. $\dfrac{x^{14}}{x^{-7}}$ **38.** $\dfrac{x^{30}}{x^{-10}}$

39. $(8x^3)^2$ **40.** $(6x^4)^2$

41. $\left(-\dfrac{4}{x}\right)^3$ **42.** $\left(-\dfrac{6}{y}\right)^3$

43. $(-3x^2y^5)^2$ **44.** $(-3x^4y^6)^3$

45. $(3x^4)(2x^7)$ **46.** $(11x^5)(9x^{12})$

47. $(-9x^3y)(-2x^6y^4)$ **48.** $(-5x^4y)(-6x^7y^{11})$

49. $\dfrac{8x^{20}}{2x^4}$ **50.** $\dfrac{20x^{24}}{10x^6}$

51. $\dfrac{25a^{13}b^4}{-5a^2b^3}$ **52.** $\dfrac{35a^{14}b^6}{-7a^7b^3}$

53. $\dfrac{14b^7}{7b^{14}}$ **54.** $\dfrac{20b^{10}}{10b^{20}}$

55. $(4x^3)^{-2}$ **56.** $(10x^2)^{-3}$

57. $\dfrac{24x^3y^5}{32x^7y^{-9}}$ **58.** $\dfrac{10x^4y^9}{30x^{12}y^{-3}}$

59. $\left(\dfrac{5x^3}{y}\right)^{-2}$ **60.** $\left(\dfrac{3x^4}{y}\right)^{-3}$

In Exercises 61–68, write each number in decimal notation.

61. 4.7×10^3 **62.** 9.12×10^5

63. 4×10^6 **64.** 7×10^6

65. 7.86×10^{-4} **66.** 4.63×10^{-5}

67. 3.18×10^{-6} **68.** 5.84×10^{-7}

In Exercises 69–76, write each number in scientific notation.

69. 3600   **70.** 2700

71. 220,000,000   **72.** 370,000,000,000

73. 0.027   **74.** 0.014

75. 0.000763   **76.** 0.000972

In Exercises 77–84, perform the indicated operation and express the answer in decimal notation.

77. $(2 \times 10^3)(3 \times 10^2)$ **78.** $(5 \times 10^2)(4 \times 10^4)$

79. $(4.1 \times 10^2)(3 \times 10^{-4})$ **80.** $(1.2 \times 10^3)(2 \times 10^{-5})$

81. $\dfrac{12 \times 10^6}{4 \times 10^2}$ **82.** $\dfrac{20 \times 10^{26}}{10 \times 10^{15}}$

83. $\dfrac{6.3 \times 10^3}{3 \times 10^5}$ **84.** $\dfrac{9.6 \times 10^2}{3 \times 10^{-3}}$

Application Exercises

In Exercises 85–88, use 10^6 for one million and 10^9 for one billion to rewrite the number in each statement in scientific notation.

85. In 1999, the U.S. government collected $1,694,300 million.

86. In 1999, the U.S. government spent $1,751,800 million.

87. The federal government is expected to provide nearly $60 billion in student aid in 2002.

88. In 1998, U.S. consumers spent $5,493.7 billion.

89. If the population of the United States is 2.7×10^8 and each person spends about $120 per year on ice cream, express the total annual spending on ice cream in scientific notation.

Writing in Mathematics

90. Describe what it means to raise a number to a power. In your description, include a discussion of the difference between -5^2 and $(-5)^2$.

91. Explain the product rule for exponents. Use $2^3 \cdot 2^5$ in your explanation.

92. Explain the power rule for exponents. Use $(3^2)^4$ in your explanation.

93. Explain the quotient rule for exponents. Use $\dfrac{5^8}{5^2}$ in your explanation.

94. Why is $(-3x^2)(2x^{-5})$ not simplified? What must be done to simplify the expression?

95. How do you know if a number is written in scientific notation?

96. Explain how to convert from scientific to decimal notation and give an example.

97. Explain how to convert from decimal to scientific notation and give an example.

Critical Thinking Exercises

98. Which one of the following is true?
    a. $4^{-2} < 4^{-3}$
    b. $5^{-2} > 2^{-5}$
    c. $(-2)^4 = 2^{-4}$
    d. $5^2 \cdot 5^{-2} > 2^5 \cdot 2^{-5}$

99. The mad Dr. Frankenstein has gathered enough bits and pieces (so to speak) for $2^{-1} + 2^{-2}$ of his creature-to-be. Write a fraction that represents the amount of his creature that must still be obtained.

100. If $b^A = MN$, $b^C = M$, and $b^D = N$, what is the relationship among A, C, and D?

Group Exercise

101. **Putting Numbers into Perspective**. A large number can be put into perspective by comparing it with another number. For example, we put the $5.5 trillion national debt into perspective by comparing it to the number of U.S. citizens. The total distance covered by all the runners in the New York City Marathon (checkpoint Example 11 on page 22) can be put into perspective by comparing this distance with, say, the distance from New York to San Francisco.

For this project, each group member should consult an almanac, a newspaper, or the World Wide Web to find a number greater than one million. Explain to other members of the group the context in which the large number is used. Express the number in scientific notation. Then put the number into perspective by comparing it with another number.

SECTION P.3   *Radicals and Rational Exponents*

Objectives

1. Evaluate square roots.
2. Use the product rule to simplify square roots.
3. Use the quotient rule to simplify square roots.
4. Add and subtract square roots.
5. Rationalize denominators.
6. Evaluate and perform operations with higher roots.
7. Understand and use rational exponents.

What is the maximum speed at which a racing cyclist can turn a corner without tipping over? The answer, in miles per hour, is given by the algebraic expression $4\sqrt{x}$, where x is the radius of the corner, in feet. Algebraic expressions containing roots describe phenomena as diverse as a wild animal's territorial area, evaporation on a lake's surface, and Albert Einstein's bizarre concept of how an astronaut moving close to the speed of light would barely age relative to friends watching from Earth. No description of your world can be complete without roots and radicals. In this section, we review the basics of radical expressions and the use of rational exponents to indicate radicals.

1 Evaluate square roots.

Square Roots

The **principal square root** of a nonnegative real number b, written $\sqrt{b}$, is that number whose square equals b. For example,

$$\sqrt{100} = 10 \text{ because } 10^2 = 100 \quad \text{and} \quad \sqrt{0} = 0 \text{ because } 0^2 = 0.$$

Observe that the principal square root of a positive number is positive and the principal square root of 0 is 0.

The symbol $\sqrt{}$ that we use to denote the principal square root is called a **radical sign**. The number under the radical sign is called the **radicand**. Together we refer to the radical sign and its radicand as a **radical**.

The following definition summarizes our discussion.

> ### Definition of the Principal Square Root
> If a is a nonnegative real number, the nonnegative number b such that $b^2 = a$, denoted by $b = \sqrt{a}$, is the **principal square root** of a.

In the real number system, negative numbers do not have square roots. For example, $\sqrt{-9}$ is not a real number because there is no real number whose square is -9.

If a number is nonnegative ($a \geq 0$), then $\left(\sqrt{a}\right)^2 = a$. For example,

$$\left(\sqrt{2}\right)^2 = 2, \quad \left(\sqrt{3}\right)^2 = 3, \quad \left(\sqrt{4}\right)^2 = 4, \quad \text{and} \left(\sqrt{5}\right)^2 = 5.$$

A number that is the square of a rational number is called a **perfect square**. For example,

64 is a perfect square because $64 = 8^2$.

$\dfrac{1}{9}$ is a perfect square because $\dfrac{1}{9} = \left(\dfrac{1}{3}\right)^2$.

The following rule can be used to find square roots of perfect squares.

> ### Square Roots of Perfect Squares
> $$\sqrt{a^2} = |a|$$

For example, $\sqrt{6^2} = 6$ and $\sqrt{(-6)^2} = |-6| = 6$.

2 Use the product rule to simplify square roots.

The Product Rule for Square Roots

A square root is **simplified** when its radicand has no factors other than 1 that are perfect squares. For example, $\sqrt{500}$ is not simplified because it can be expressed as $\sqrt{100 \cdot 5}$ and $\sqrt{100}$ is a perfect square. The **product rule for square roots** can be used to simplify $\sqrt{500}$.

> ### The Product Rule for Square Roots
> If a and b represent nonnegative real numbers, then
> $$\sqrt{ab} = \sqrt{a}\sqrt{b} \text{ and } \sqrt{a}\sqrt{b} = \sqrt{ab}.$$
> The square root of a product is the product of the square roots.

Example 1 shows how the product rule is used to remove from the square root any perfect squares that occur as factors.

EXAMPLE 1 **Using the Product Rule to Simplify Square Roots**

Simplify: **a.** $\sqrt{500}$ **b.** $\sqrt{6x} \cdot \sqrt{3x}$

Solution

a. $\sqrt{500} = \sqrt{100 \cdot 5}$ 100 is the largest perfect square factor of 500.

$\qquad\quad = \sqrt{100}\sqrt{5}$ $\sqrt{ab} = \sqrt{a}\sqrt{b}$

$\qquad\quad = 10\sqrt{5}$ $\sqrt{100} = 10$

b. $\sqrt{6x} \cdot \sqrt{3x} = \sqrt{6x \cdot 3x}$ $\sqrt{a}\sqrt{b} = \sqrt{ab}$

$\qquad\qquad\quad = \sqrt{18x^2}$ Multiply.

$\qquad\qquad\quad = \sqrt{9x^2 \cdot 2}$ 9 is the largest perfect square factor of 18.

$\qquad\qquad\quad = \sqrt{9x^2}\sqrt{2}$ $\sqrt{ab} = \sqrt{a}\sqrt{b}$

$\qquad\qquad\quad = \sqrt{9}\sqrt{x^2}\sqrt{2}$ Split $\sqrt{9x^2}$ into two square roots.

$\qquad\qquad\quad = 3|x|\sqrt{2}$ $\sqrt{9} = 3$ (because $3^2 = 9$) and $\sqrt{x^2} = |x|$.

Check Point 1   Simplify: **a.** $\sqrt{3^2}$ **b.** $\sqrt{5x} \cdot \sqrt{10x}$

③ Use the quotient rule to simplify square roots.

The Quotient Rule for Square Roots

Another property for square roots involves division.

> **The Quotient Rule for Square Roots**
>
> If a and b represent nonnegative real numbers and $b \neq 0$, then
>
> $$\frac{\sqrt{a}}{\sqrt{b}} = \sqrt{\frac{a}{b}} \text{ and } \sqrt{\frac{a}{b}} = \frac{\sqrt{a}}{\sqrt{b}}.$$
>
> The square root of a quotient is the quotient of the square roots.

EXAMPLE 2 **Using the Quotient Rule to Simplify Square Roots**

Simplify: **a.** $\sqrt{\dfrac{100}{9}}$ **b.** $\dfrac{\sqrt{48x^3}}{\sqrt{6x}}$

Solution

a. $\sqrt{\dfrac{100}{9}} = \dfrac{\sqrt{100}}{\sqrt{9}} = \dfrac{10}{3}$

b. $\dfrac{\sqrt{48x^3}}{\sqrt{6x}} = \sqrt{\dfrac{48x^3}{6x}} = \sqrt{8x^2} = \sqrt{4x^2}\sqrt{2} = \sqrt{4}\sqrt{x^2}\sqrt{2} = 2|x|\sqrt{2}$

Check Point 2 Simplify: **a.** $\sqrt{\dfrac{25}{16}}$ **b.** $\dfrac{\sqrt{150x^3}}{\sqrt{2x}}$

4 Add and subtract square roots.

A Radical Idea: Time Is Relative

What does travel in space have to do with radicals? Imagine that in the future we will be able to travel at velocities approaching the speed of light (approximately 186,000 miles per second). According to Einstein's theory of relativity, time would pass more quickly on Earth than it would in the moving spaceship. The expression

$$R_f \sqrt{1 - \left(\dfrac{v}{c}\right)^2}$$

gives the aging rate of an astronaut relative to the aging rate of a friend on Earth, R_f. In the expression, v is the astronaut's speed and c is the speed of light. As the astronaut's speed approaches the speed of light, we can substitute c for v:

$$R_f \sqrt{1 - \left(\dfrac{v}{c}\right)^2} \quad \text{Let } v = c.$$
$$= R_f \sqrt{1 - 1^2}$$
$$= R_f \sqrt{0} = 0$$

Close to the speed of light, the astronaut's aging rate relative to a friend on earth is nearly 0. What does this mean? As we age here on Earth, the space traveler would barely get older. The space traveler would return to a futuristic world in which friends and loved ones would be long dead.

Adding and Subtracting Square Roots

Two or more square roots can be combined provided that they have the same radicand. Such radicals are called **like radicals**. For example,

$$7\sqrt{11} + 6\sqrt{11} = (7 + 6)\sqrt{11} = 13\sqrt{11}.$$

EXAMPLE 3   Adding and Subtracting Like Radicals

Add or subtract as indicated:

a. $7\sqrt{2} + 5\sqrt{2}$ **b.** $\sqrt{5x} - 7\sqrt{5x}$

Solution

a. $7\sqrt{2} + 5\sqrt{2} = (7 + 5)\sqrt{2}$ Apply the distributive property.
$$= 12\sqrt{2} \qquad \text{Simplify.}$$

b. $\sqrt{5x} - 7\sqrt{5x} = 1\sqrt{5x} - 7\sqrt{5x}$ Write $\sqrt{5x}$ as $1\sqrt{5x}$.
$$= (1 - 7)\sqrt{5x} \qquad \text{Apply the distributive property.}$$
$$= -6\sqrt{5x} \qquad \text{Simplify.}$$

Check Point 3 Add or subtract as indicated:

a. $8\sqrt{13} + 9\sqrt{13}$ **b.** $\sqrt{17x} - 20\sqrt{17x}$

In some cases, radicals can be combined once they have been simplified. For example, to add $\sqrt{2}$ and $\sqrt{8}$, we can write $\sqrt{8}$ as $\sqrt{4 \cdot 2}$ because 4 is a perfect square factor of 8.

$$\sqrt{2} + \sqrt{8} = \sqrt{2} + \sqrt{4 \cdot 2} = 1\sqrt{2} + 2\sqrt{2} = (1 + 2)\sqrt{2} = 3\sqrt{2}$$

EXAMPLE 4   Combining Radicals That First Require Simplification

Add or subtract as indicated:

a. $7\sqrt{3} + \sqrt{12}$ **b.** $4\sqrt{50x} - 6\sqrt{32x}$

Solution

a. $7\sqrt{3} + \sqrt{12}$
$$= 7\sqrt{3} + \sqrt{4 \cdot 3} \quad \text{Split 12 into two factors such that one is a perfect square.}$$
$$= 7\sqrt{3} + 2\sqrt{3} \quad \sqrt{4 \cdot 3} = \sqrt{4}\sqrt{3} = 2\sqrt{3}$$
$$= (7 + 2)\sqrt{3} \quad \text{Apply the distributive property. You will find that this step is usually done mentally.}$$
$$= 9\sqrt{3} \quad \text{Simplify.}$$

b. $4\sqrt{50x} - 6\sqrt{32x}$

$= 4\sqrt{25 \cdot 2x} - 6\sqrt{16 \cdot 2x}$ 25 is the largest perfect square factor of 50 and 16 is the largest perfect square factor of 32.

$= 4 \cdot 5\sqrt{2x} - 6 \cdot 4\sqrt{2x}$ $\sqrt{25 \cdot 2} = \sqrt{25}\sqrt{2} = 5\sqrt{2}$ and $\sqrt{16 \cdot 2} = \sqrt{16}\sqrt{2} = 4\sqrt{2}$

$= 20\sqrt{2x} - 24\sqrt{2x}$ Multiply.

$= (20 - 24)\sqrt{2x}$ Apply the distributive property.

$= -4\sqrt{2x}$ Simplify.

Check Point 4

Add or subtract as indicated:

a. $5\sqrt{27} + \sqrt{12}$ **b.** $6\sqrt{18x} - 4\sqrt{8x}$

5   Rationalize denominators.

Rationalizing Denominators

You can use a calculator to compare the approximate values for $\dfrac{1}{\sqrt{3}}$ and $\dfrac{\sqrt{3}}{3}$.

The two approximations are the same. This is not a coincidence:

$$\frac{1}{\sqrt{3}} = \frac{1}{\sqrt{3}} \cdot \boxed{\frac{\sqrt{3}}{\sqrt{3}}} = \frac{\sqrt{3}}{\sqrt{9}} = \frac{\sqrt{3}}{3}.$$

Any number divided by itself is 1. Multiplication by 1 does not change the value of $\dfrac{1}{\sqrt{3}}$.

This process involves rewriting a radical to remove the square root from the denominator without changing the value of the radical. The process is called **rationalizing the denominator**. If the denominator contains the square root of a natural number that is not a perfect square, multiply the numerator and denominator by the smallest number that produces the square root of a perfect square in the denominator.

EXAMPLE 5   Rationalizing Denominators

Rationalize the denominator: **a.** $\dfrac{15}{\sqrt{6}}$ **b.** $\dfrac{12}{\sqrt{8}}$

Solution

a. If we multiply numerator and denominator by $\sqrt{6}$, the denominator becomes $\sqrt{6} \cdot \sqrt{6} = \sqrt{36} = 6$. Therefore, we multiply by 1, choosing $\dfrac{\sqrt{6}}{\sqrt{6}}$ for 1:

$$\frac{15}{\sqrt{6}} = \frac{15}{\sqrt{6}} \cdot \frac{\sqrt{6}}{\sqrt{6}} = \frac{15\sqrt{6}}{\sqrt{36}} = \frac{15\sqrt{6}}{6} = \frac{5\sqrt{6}}{2}$$

Multiply by 1.   Simplify: $\dfrac{15}{6} = \dfrac{15 \div 3}{6 \div 3} = \dfrac{5}{2}$.

b. The *smallest* number that will produce a perfect square in the denominator of $\dfrac{12}{\sqrt{8}}$ is $\sqrt{2}$, because $\sqrt{8} \cdot \sqrt{2} = \sqrt{16} = 4$. We multiply by 1, choosing $\dfrac{\sqrt{2}}{\sqrt{2}}$ for 1.

$$\frac{12}{\sqrt{8}} = \frac{12}{\sqrt{8}} \cdot \frac{\sqrt{2}}{\sqrt{2}} = \frac{12\sqrt{2}}{\sqrt{16}} = \frac{12\sqrt{2}}{4} = 3\sqrt{2}$$

Check Point 5   Rationalize the denominator: **a.** $\dfrac{5}{\sqrt{3}}$ **b.** $\dfrac{6}{\sqrt{12}}$

How can we rationalize a denominator if the denominator contains two terms? In general,

$$(\sqrt{a} + \sqrt{b})(\sqrt{a} - \sqrt{b}) = (\sqrt{a})^2 - (\sqrt{b})^2 = a - b.$$

Notice that the product does not contain a radical. Here are some specific examples.

The Denominator Contains:	**Multiply by:**	**The New Denominator Contains:**
$7 + \sqrt{5}$	$7 - \sqrt{5}$	$7^2 - (\sqrt{5})^2 = 49 - 5 = 44$
$\sqrt{3} - 6$	$\sqrt{3} + 6$	$(\sqrt{3})^2 - 6^2 = 3 - 36 = -33$
$\sqrt{7} + \sqrt{3}$	$\sqrt{7} - \sqrt{3}$	$(\sqrt{7})^2 - (\sqrt{3})^2 = 7 - 3 = 4$

EXAMPLE 6   Rationalizing a Denominator Containing Two Terms

Rationalize the denominator: $\dfrac{7}{5 + \sqrt{3}}$.

Solution   If we multiply the numerator and denominator by $5 - \sqrt{3}$, the denominator will not contain a radical. Therefore, we multiply by 1, choosing $\dfrac{5 - \sqrt{3}}{5 - \sqrt{3}}$ for 1:

$$\frac{7}{5 + \sqrt{3}} = \underbrace{\frac{7}{5 + \sqrt{3}} \cdot \frac{5 - \sqrt{3}}{5 - \sqrt{3}}}_{\text{Multiply by 1.}} = \frac{7(5 - \sqrt{3})}{5^2 - (\sqrt{3})^2} = \frac{7(5 - \sqrt{3})}{25 - 3}$$

$$= \underbrace{\frac{7(5 - \sqrt{3})}{22} \quad \text{or} \quad \frac{35 - 7\sqrt{3}}{22}}_{\substack{\text{In either form of the} \\ \text{answer, there is no} \\ \text{radical in the denominator.}}}.$$

<table>
<tr><td>

Check Point 6

</td><td>

Rationalize the denominator: $\dfrac{8}{4 + \sqrt{5}}$.

</td></tr>
</table>

6 Evaluate and perform operations with higher roots.

Other Kinds of Roots

We define the **principal nth root** of a real number a, symbolized by $\sqrt[n]{a}$, as follows:

Definition of the Principal nth Root of a Real Number

$$\sqrt[n]{a} = b \text{ means that } b^n = a.$$

If n, the **index**, is even, then a is nonnegative ($a \geq 0$) and b is also nonnegative ($b \geq 0$). If n is odd, a and b can be any real numbers.

For example,

$$\sqrt[3]{64} = 4 \text{ because } 4^3 = 64 \quad \text{and} \quad \sqrt[5]{-32} = -2 \text{ because } (-2)^5 = -32.$$

The same vocabulary that we learned for square roots applies to nth roots. The symbol $\sqrt[n]{a}$ is called a **radical** and a is called the **radicand**.

A number that is the nth power of a rational number is called a **perfect nth power**. For example, 8 is a perfect third power, or perfect cube, because $8 = 2^3$. In general, one of the following rules can be used to find nth roots of perfect nth powers.

Finding nth Roots of Perfect nth Powers

If n is odd, $\sqrt[n]{a^n} = a$.

If n is even, $\sqrt[n]{a^n} = |a|$.

For example,

$$\sqrt[3]{(-2)^3} = -2 \quad \text{and} \quad \sqrt[4]{(-2)^4} = |-2| = 2.$$

> Absolute value is not needed with odd roots, but is necessary with even roots.

The Product and Quotient Rules for Other Roots

The product and quotient rules apply to cube roots, fourth roots, and all higher roots.

The Product and Quotient Rules for nth Roots

For all real numbers, where the indicated roots represent real numbers,

$$\sqrt[n]{a} \cdot \sqrt[n]{b} = \sqrt[n]{ab} \quad \text{and} \quad \frac{\sqrt[n]{a}}{\sqrt[n]{b}} = \sqrt[n]{\frac{a}{b}}, \quad b \neq 0.$$

EXAMPLE 7 **Simplifying, Multiplying, and Dividing Higher Roots**

Simplify: **a.** $\sqrt[3]{24}$ **b.** $\sqrt[4]{8} \cdot \sqrt[4]{4}$ **c.** $\sqrt[4]{\dfrac{81}{16}}$

Solution

a. $\sqrt[3]{24} = \sqrt[3]{8 \cdot 3}$ Find the largest *perfect cube* that is a factor of 24. $\sqrt[3]{8} = 2$, so 8 is a perfect cube and is the largest perfect cube factor of 24.

$\qquad\quad = \sqrt[3]{8} \cdot \sqrt[3]{3}$ $\sqrt[n]{ab} = \sqrt[n]{a}\,\sqrt[n]{b}$

$\qquad\quad = 2\sqrt[3]{3}$

b. $\sqrt[4]{8} \cdot \sqrt[4]{4} = \sqrt[4]{8 \cdot 4}$ $\sqrt[n]{a} \cdot \sqrt[n]{b} = \sqrt[n]{ab}$

$\qquad\qquad\quad = \sqrt[4]{32}$ Find the largest *perfect fourth power* that is a factor of 32.

$\qquad\qquad\quad = \sqrt[4]{16 \cdot 2}$ $\sqrt[4]{16} = 2$, so 16 is a perfect fourth power and is the largest perfect fourth power that is a factor of 32.

$\qquad\qquad\quad = \sqrt[4]{16} \cdot \sqrt[4]{2}$ $\sqrt[n]{ab} = \sqrt[n]{a} \cdot \sqrt[n]{b}$

$\qquad\qquad\quad = 2\sqrt[4]{2}$

c. $\sqrt[4]{\dfrac{81}{16}} = \dfrac{\sqrt[4]{81}}{\sqrt[4]{16}}$ $\sqrt[n]{\dfrac{a}{b}} = \dfrac{\sqrt[n]{a}}{\sqrt[n]{b}}$

$\qquad\quad = \dfrac{3}{2}$ $\sqrt[4]{81} = 3$ because $3^4 = 81$ and $\sqrt[4]{16} = 2$ because $2^4 = 16$.

> **Check Point 7**   Simplify: **a.** $\sqrt[3]{40}$ **b.** $\sqrt[5]{8} \cdot \sqrt[5]{8}$ **c.** $\sqrt[3]{\dfrac{125}{27}}$

We have seen that adding and subtracting square roots often involves simplifying terms. The same idea applies to adding and subtracting *n*th roots.

EXAMPLE 8 **Combining Cube Roots**

Subtract: $5\sqrt[3]{16} - 11\sqrt[3]{2}$.

Solution

$5\sqrt[3]{16} - 11\sqrt[3]{2}$

$= 5\sqrt[3]{8 \cdot 2} - 11\sqrt[3]{2}$ Because $\sqrt[3]{8} = 2$, 8 is the largest perfect cube that is a factor of 16.

$= 5 \cdot 2\sqrt[3]{2} - 11\sqrt[3]{2}$ $\sqrt[3]{8 \cdot 2} = \sqrt[3]{8}\,\sqrt[3]{2} = 2\sqrt[3]{2}$

$= 10\sqrt[3]{2} - 11\sqrt[3]{2}$ Multiply.

$= (10 - 11)\sqrt[3]{2}$ Apply the distributive property.

$= -1\sqrt[3]{2}$ or $-\sqrt[3]{2}$ Simplify.

> **Check Point 8**   Subtract: $3\sqrt[3]{81} - 4\sqrt[3]{3}$.

7 Understand and use rational exponents.

Rational Exponents

Animals in the wild have regions to which they confine their movement, called their territorial area. Territorial area, in square miles, is related to an animal's body weight. If an animal weighs W pounds, its territorial area is

$$W^{141/100}$$

square miles.

W to the *what* power?! How can we interpret the information given by this algebraic expression?

In the last part of this section, we turn our attention to rational exponents such as $\frac{141}{100}$ and their relationship to roots of real numbers.

Definition of Rational Exponents

If $\sqrt[n]{a}$ represents a real number and $n \geq 2$ is an integer, then

$$a^{1/n} = \sqrt[n]{a}.$$

Furthermore,

$$a^{-1/n} = \frac{1}{a^{1/n}} = \frac{1}{\sqrt[n]{a}}, a \neq 0.$$

EXAMPLE 9   Using the Definition of $a^{1/n}$

Simplify: **a.** $64^{1/2}$ **b.** $8^{1/3}$ **c.** $64^{-1/3}$

Solution

a. $64^{1/2} = \sqrt{64} = 8$ **b.** $8^{1/3} = \sqrt[3]{8} = 2$

c. $64^{-1/3} = \dfrac{1}{64^{1/3}} = \dfrac{1}{\sqrt[3]{64}} = \dfrac{1}{4}$

Check Point 9   Simplify: **a.** $81^{1/2}$ **b.** $27^{1/3}$ **c.** $32^{-1/5}$

Note that every rational exponent in Example 9 has a numerator of 1 or −1. We now define rational exponents with any integer in the numerator.

Definition of Rational Exponents

If $\sqrt[n]{a}$ represents a real number, $\dfrac{m}{n}$ is a rational number reduced to lowest terms, and $n \geq 2$ is an integer, then

$$a^{m/n} = \left(\sqrt[n]{a}\right)^m = \sqrt[n]{a^m}.$$

The exponent m/n consists of two parts: the denominator n is the root and the numerator m is the exponent. Furthermore,

$$a^{-m/n} = \frac{1}{a^{m/n}}.$$

EXAMPLE 10 Using the Definition of $a^{m/n}$

Simplify: **a.** $27^{2/3}$ **b.** $9^{3/2}$ **c.** $16^{-3/4}$

Solution

a. $27^{2/3} = \left(\sqrt[3]{27}\right)^2 = 3^2 = 9$

> The denominator of $\frac{2}{3}$ is the root and the numerator is the exponent.

b. $9^{3/2} = \left(\sqrt{9}\right)^3 = 3^3 = 27$

c. $16^{-3/4} = \dfrac{1}{16^{3/4}} = \dfrac{1}{\left(\sqrt[4]{16}\right)^3} = \dfrac{1}{2^3} = \dfrac{1}{8}$

> **Check Point 10**   Simplify: **a.** $4^{3/2}$ **b.** $32^{-2/5}$

Properties of exponents can be applied to expressions containing rational exponents.

EXAMPLE 11 Simplifying Expressions with Rational Exponents

Simplify using properties of exponents:

a. $\left(5x^{1/2}\right)\left(7x^{3/4}\right)$ **b.** $\dfrac{32x^{5/3}}{16x^{3/4}}$

Solution

a. $\left(5x^{1/2}\right)\left(7x^{3/4}\right) = 5 \cdot 7x^{1/2} \cdot x^{3/4}$ Group factors with the same base.

$= 35x^{(1/2)+(3/4)}$ When multiplying expressions with the same base, add the exponents.

$= 35x^{5/4}$ $\frac{1}{2} + \frac{3}{4} = \frac{2}{4} + \frac{3}{4} = \frac{5}{4}$

b. $\dfrac{32x^{5/3}}{16x^{3/4}} = \left(\dfrac{32}{16}\right)\left(\dfrac{x^{5/3}}{x^{3/4}}\right)$ Group factors with the same base.

$= 2x^{(5/3)-(3/4)}$ When dividing expressions with the same base, subtract the exponents.

$= 2x^{11/12}$ $\frac{5}{3} - \frac{3}{4} = \frac{20}{12} - \frac{9}{12} = \frac{11}{12}$

> **Check Point 11**   Simplify: **a.** $\left(2x^{4/3}\right)\left(5x^{8/3}\right)$ **b.** $\dfrac{20x^4}{5x^{3/2}}$

Rational exponents are sometimes useful for simplifying radicals by reducing their index.

EXAMPLE 12   Reducing the Index of a Radical

Simplify: $\sqrt[9]{x^3}$.

Solution $\sqrt[9]{x^3} = x^{3/9} = x^{1/3} = \sqrt[3]{x}$

Check Point 12   Simplify: $\sqrt[6]{x^3}$.

EXERCISE SET P.3

Practice Exercises

Evaluate each expression in Exercises 1–7 or indicate that the root is not a real number.

1. $\sqrt{36}$ **2.** $\sqrt{25}$

3. $\sqrt{-36}$ **4.** $\sqrt{-25}$

5. $\sqrt{(-13)^2}$ **6.** $\sqrt{(-17)^2}$

Use the product rule to simplify the expressions in Exercises 7–16.

7. $\sqrt{50}$ **8.** $\sqrt{27}$

9. $\sqrt{45x^2}$ **10.** $\sqrt{125x^2}$

11. $\sqrt{2x} \cdot \sqrt{6x}$ **12.** $\sqrt{10x} \cdot \sqrt{8x}$

13. $\sqrt{x^3}$ **14.** $\sqrt{y^3}$

15. $\sqrt{2x^2} \cdot \sqrt{6x}$ **16.** $\sqrt{6x} \cdot \sqrt{3x^2}$

Use the quotient rule to simplify the expressions in Exercises 17–24.

17. $\sqrt{\dfrac{1}{81}}$ **18.** $\sqrt{\dfrac{1}{49}}$

19. $\sqrt{\dfrac{49}{16}}$ **20.** $\sqrt{\dfrac{121}{9}}$

21. $\dfrac{\sqrt{48x^3}}{\sqrt{3x}}$ **22.** $\dfrac{\sqrt{72x^3}}{\sqrt{8x}}$

23. $\dfrac{\sqrt{150x^4}}{\sqrt{3x}}$ **24.** $\dfrac{\sqrt{24x^4}}{\sqrt{3x}}$

In Exercises 25–34, add or subtract terms whenever possible.

25. $7\sqrt{3} + 6\sqrt{3}$ **26.** $8\sqrt{5} + 11\sqrt{5}$

27. $6\sqrt{17x} - 8\sqrt{17x}$ **28.** $4\sqrt{13x} - 6\sqrt{13x}$

29. $\sqrt{8} + 3\sqrt{2}$ **30.** $\sqrt{20} + 6\sqrt{5}$

31. $\sqrt{50x} - \sqrt{8x}$ **32.** $\sqrt{63x} - \sqrt{28x}$

33. $3\sqrt{18} + 5\sqrt{50}$ **34.** $4\sqrt{12} - 2\sqrt{75}$

In Exercises 35–44, rationalize the denominator.

35. $\dfrac{1}{\sqrt{7}}$ **36.** $\dfrac{2}{\sqrt{10}}$

37. $\dfrac{\sqrt{2}}{\sqrt{5}}$ **38.** $\dfrac{\sqrt{7}}{\sqrt{3}}$

39. $\dfrac{13}{3 + \sqrt{11}}$ **40.** $\dfrac{3}{3 + \sqrt{7}}$

41. $\dfrac{7}{\sqrt{5} - 2}$ **42.** $\dfrac{5}{\sqrt{3} - 1}$

43. $\dfrac{6}{\sqrt{5} + \sqrt{3}}$ **44.** $\dfrac{11}{\sqrt{7} - \sqrt{3}}$

Evaluate each expression in Exercises 45–54 or indicate that the root is not a real number.

45. $\sqrt[3]{125}$ **46.** $\sqrt[3]{8}$

47. $\sqrt[3]{-8}$ **48.** $\sqrt[3]{-125}$

49. $\sqrt[4]{-16}$ **50.** $\sqrt[4]{-81}$

51. $\sqrt[4]{(-3)^4}$ **52.** $\sqrt[4]{(-2)^4}$

53. $\sqrt[5]{(-3)^5}$ **54.** $\sqrt[5]{(-2)^5}$

Simplify the radical expressions in Exercises 55–62.

55. $\sqrt[3]{32}$ **56.** $\sqrt[3]{150}$

57. $\sqrt[3]{x^4}$ **58.** $\sqrt[3]{x^5}$

59. $\sqrt[3]{9} \cdot \sqrt[3]{6}$ **60.** $\sqrt[3]{12} \cdot \sqrt[3]{4}$

61. $\dfrac{\sqrt[5]{64x^6}}{\sqrt[5]{2x}}$ **62.** $\dfrac{\sqrt[4]{162x^5}}{\sqrt[4]{2x}}$

In Exercises 63–70, evaluate each expression without using a calculator.

63. $36^{1/2}$ **64.** $121^{1/2}$

65. $8^{1/3}$ **66.** $27^{1/3}$

67. $125^{2/3}$ **68.** $8^{2/3}$

69. $32^{-4/5}$ **70.** $16^{-5/2}$

In Exercises 71–78, simplify using properties of exponents.

71. $\left(7x^{1/3}\right)\left(2x^{1/4}\right)$ **72.** $\left(3x^{2/3}\right)\left(4x^{3/4}\right)$

73. $\dfrac{20x^{1/2}}{5x^{1/4}}$ **74.** $\dfrac{72x^{3/4}}{9x^{1/3}}$

75. $\left(x^{2/3}\right)^3$ **76.** $\left(x^{4/5}\right)^5$

77. $\left(25x^4y^6\right)^{1/2}$ **78.** $\left(125x^9y^6\right)^{1/3}$

In Exercises 79–84, simplify by reducing the index of the radical.

79. $\sqrt[4]{5^2}$ **80.** $\sqrt[4]{7^2}$

81. $\sqrt[3]{x^6}$ **82.** $\sqrt[4]{x^{12}}$

83. $\sqrt[6]{x^4}$ **84.** $\sqrt[9]{x^6}$

 Application Exercises

85. The algebraic expression $2\sqrt{5L}$ is used to estimate the speed of a car prior to an accident, in miles per hour, based on the length of its skid marks L, in feet. Find the speed of a car that left skid marks 40 feet long, and write the answer in simplified radical form.

86. The time, in seconds, that it takes an object to fall a distance d, in feet, is given by the algebraic expression $\sqrt{\dfrac{d}{16}}$. Find how long it will take a ball dropped from the top of a building 320 feet tall to hit the ground. Write the answer in simplified radical form.

87. The early Greeks believed that the most pleasing of all rectangles were golden rectangles whose ratio of width to height is

$$\frac{w}{h} = \frac{2}{\sqrt{5} - 1}.$$

Rationalize the denominator for this ratio and then use a calculator to approximate the answer correct to the nearest hundredth.

88. The amount of evaporation, in inches per day, of a large body of water can be described by the algebraic expression

$$\frac{w}{20\sqrt{a}}$$

where

     a = surface area of the water in square miles

     w = average wind speed of the air over the water, in miles per hour.

Determine the evaporation on a lake whose surface area is 9 square miles on a day when the wind speed over the water is 10 miles per hour.

89. In the Peanuts cartoon shown above, Woodstock appears to be working steps mentally. Fill in the missing steps that show how to go from $\dfrac{7\sqrt{2 \cdot 2 \cdot 3}}{6}$ to $\dfrac{7}{3}\sqrt{3}$.

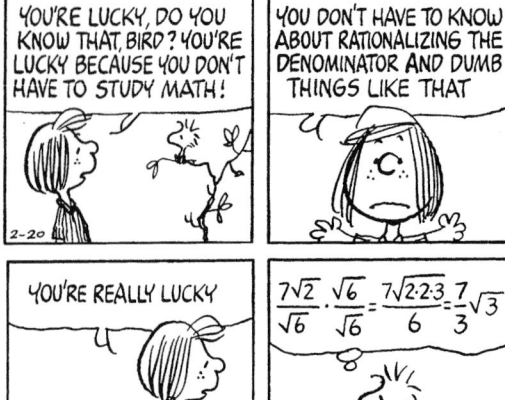

PEANUTS reprinted by permission of United Feature Syndicate, Inc.

90. The algebraic expression $63.25x^{1/4}$ describes the average sale price, in thousands of dollars, of single-family homes in the U.S. Midwest x years after 1981. Evaluate the algebraic expression when $x = 16$. Describe what the answer means in practical terms.

91. The algebraic expression $0.07d^{3/2}$ describes the duration of a storm, in hours, whose diameter is d miles. Evaluate the algebraic expression when $d = 9$. Describe what the answer means in practical terms.

 **Writing in Mathematics**

92. Explain how to simplify $\sqrt{10} \cdot \sqrt{5}$.

93. Explain how to add $\sqrt{3} + \sqrt{12}$.

94. Describe what it means to rationalize a denominator. Use both $\dfrac{1}{\sqrt{5}}$ and $\dfrac{1}{5 + \sqrt{5}}$ in your explanation.

95. What difference is there in simplifying $\sqrt[3]{(-5)^3}$ and $\sqrt[4]{(-5)^4}$?

96. What does $a^{m/n}$ mean?

97. Describe the kinds of numbers that have rational fifth roots.

98. Why must a and b represent nonnegative numbers when we write $\sqrt{a} \cdot \sqrt{b} = \sqrt{ab}$? Is it necessary to use this restriction in the case of $\sqrt[3]{a} \cdot \sqrt[3]{b} = \sqrt[3]{ab}$? Explain.

 Technology Exercises

99. The algebraic expression $60.19x^{0.025}$ describes the expected lifespan of African American men x years after 1969. Use a calculator to find the expected lifespan from 1970 through 2000. During what year did the expected lifespan of African American men first exceed 65 years?

100. The territorial area of an animal in the wild is defined to be the area of the region to which the animal confines its movements. The algebraic expression $W^{1.41}$ describes the territorial area, in square miles, of an animal that weighs W pounds. Use a calculator to find the territorial area of animals weighing 25, 50, 150, 200, 250, and 300 pounds. What do the values that you obtain with your calculator indicate about the relationship between body weight and territorial area?

Critical Thinking Exercises

101. Which one of the following is true?
  a. Neither $(-8)^{1/2}$ nor $(-8)^{1/3}$ represent real numbers.
  b. $\sqrt{x^2 + y^2} = x + y$
  c. $8^{-1/3} = -2$ **d.** $2^{1/2} \cdot 2^{1/2} = 2$

In Exercises 102–103, fill in each box to make the statement true.

102. $(5 + \sqrt{\square})(5 - \sqrt{\square}) = 22$

103. $\sqrt{\square x^{\square}} = 5x^7$

104. Find exact value of $\sqrt{13 + \sqrt{2} + \dfrac{7}{3 + \sqrt{2}}}$ without the use of a calculator.

105. Place the correct symbol, $>$ or $<$, in the box between each of the given numbers. *Do not use a calculator.* Then check your result with a calculator.

  a. $3^{1/2} \:\square\: 3^{1/3}$ **b.** $\sqrt{7} + \sqrt{18} \:\square\: \sqrt{7 + 18}$

SECTION P.4 *Polynomials*

Objectives

1. Understand the vocabulary of polynomials.
2. Add and subtract polynomials.
3. Multiply polynomials.
4. Use FOIL in polynomial multiplication.
5. Use special products in polynomial multiplication.
6. Perform operations with polynomials in several variables.

Runny nose? Sneezing? You are probably familiar with the unpleasant onset of a cold. We "catch cold" when the cold virus enters our bodies, where it multiplies. Fortunately, at a certain point the virus begins to die. The algebraic expression $-0.75x^4 + 3x^3 + 5$ describes the billions of viral particles in our bodies after x days of invasion. The expression enables mathematicians to determine the day on which there is a maximum number of viral particles and, consequently, the day we feel sickest.

The algebraic expression $-0.75x^4 + 3x^3 + 5$ is an example of a polynomial. A **polynomial** is a single term or the sum of two or more terms containing variables with whole number exponents. This particular polynomial contains three terms. Equations containing polynomials are used in such diverse areas as science, business, medicine, psychology, and sociology. In this section, we review basic ideas about polynomials and their operations.

1 Understand the vocabulary of polynomials.

The Vocabulary of Polynomials

Consider the polynomial

$$7x^3 - 9x^2 + 13x - 6.$$

We can express this polynomial as

$$7x^3 + (-9x^2) + 13x + (-6).$$

The polynomial contains four terms. It is customary to write the terms in the order of descending powers of the variables. This is the **standard form** of a polynomial.

We begin this section by limiting our discussion to polynomials containing only one variable. Each term of a polynomial in x is of the form ax^n. The **degree** of ax^n is n. For example, the degree of the term $7x^3$ is 3.

Study Tip

We can express 0 in many ways, including $0x$, $0x^2$, and $0x^3$. It is impossible to assign a single exponent on the variable. This is why 0 has no defined degree.

The Degree of ax^n

If $a \neq 0$, the degree of ax^n is n. The degree of a nonzero constant is 0. The constant 0 has no defined degree.

Here is an example of a polynomial and the degree of each of its four terms.

$$6x^4 - 3x^3 + 2x - 5$$

| degree 4 | degree 3 | degree 1 | degree of nonzero constant: 0 |

Notice that the exponent on x for the term $2x$ is understood to be 1: $2x^1$. For this reason, the degree of $2x$ is 1. You can think of -5 as $-5x^0$; thus, its degree is 0.

A polynomial with exactly one term is called a **monomial**. A **binomial** is a polynomial that has two terms, each with a different exponent. A **trinomial** is a polynomial with three terms, each with a different exponent. Polynomials with four or more terms have no special names.

The **degree of a polynomial** is the highest degree of all the terms of the polynomial. For example, $4x^2 + 3x$ is a binomial of degree 2 because the degree of the first term is 2, and the degree of the other term is less than 2. Also, $7x^5 - 2x^2 + 4$ is a trinomial of degree 5 because the degree of the first term is 5, and the degrees of the other terms are less than 5.

Up to now, we have used x to represent the variable in a polynomial. However, any letter can be used. For example,

$7x^5 - 3x^3 + 8$ is a polynomial (in x) of degree 5.
$6y^3 + 4y^2 - y + 3$ is a polynomial (in y) of degree 3.
$z^7 + \sqrt{2}$ is a polynomial (in z) of degree 7.

Not every algebraic expression is a polynomial. Algebraic expressions whose variables do not contain whole number exponents such as

$$3x^{-2} + 7 \quad \text{and} \quad 5x^{3/2} + 9x^{1/2} + 2$$

are not polynomials. Furthermore, a quotient of polynomials such as

$$\frac{x^2 + 2x + 5}{x^3 - 7x^2 + 9x - 3}$$

is not a polynomial because the form of a polynomial involves only addition and subtraction of terms, not division.

We can tie together the threads of our discussion with the formal definition of a polynomial in one variable. In this definition, the coefficients of the terms are represented by a_n (read "a sub n"), a_{n-1} (read "a sub n minus 1"), a_{n-2}, and so on. The small letters to the lower right of each a are called **subscripts** and are *not exponents*. Subscripts are used to distinguish one constant from another when a large and undetermined number of such constants are needed.

Definition of a Polynomial in x

A **polynomial in x** is an algebraic expression of the form

$$a_n x^n + a_{n-1} x^{n-1} + a_{n-2} x^{n-2} + \cdots + a_1 x + a_0,$$

where $a_n, a_{n-1}, a_{n-2}, \ldots, a_1$ and a_0 are real numbers, $a_n \neq 0$, and n is a nonnegative integer. The polynomial is of **degree n**, a_n is the **leading coefficient**, and a_0 is the **constant term**.

---

2 Add and subtract polynomials.

Adding and Subtracting Polynomials

Polynomials are added and subtracted by combining like terms. For example, we can combine the monomials $-9x^3$ and $13x^3$ using addition as follows:

$$-9x^3 + 13x^3 = (-9 + 13)x^3 = 4x^3.$$

EXAMPLE 1   Adding and Subtracting Polynomials

Perform the indicated operations and simplify:

a. $\left(-9x^3 + 7x^2 - 5x + 3\right) + \left(13x^3 + 2x^2 - 8x - 6\right)$
b. $\left(7x^3 - 8x^2 + 9x - 6\right) - \left(2x^3 - 6x^2 - 3x + 9\right)$

Solution

a. $\left(-9x^3 + 7x^2 - 5x + 3\right) + \left(13x^3 + 2x^2 - 8x - 6\right)$
$= \left(-9x^3 + 13x^3\right) + \left(7x^2 + 2x^2\right)$ Group like terms.
$\quad + (-5x - 8x) + (3 - 6)$
$= 4x^3 + 9x^2 + (-13x) + (-3)$ Combine like terms.
$= 4x^3 + 9x^2 - 13x - 3$

b. $\left(7x^3 - 8x^2 + 9x - 6\right) - \left(2x^3 - 6x^2 - 3x + 9\right)$
$= \left(7x^3 - 8x^2 + 9x - 6\right) + \left(-2x^3 + 6x^2 + 3x - 9\right)$ Rewrite subtraction as addition of the additive inverse. Be sure to change the sign of each term inside parentheses preceded by the negative sign.

$= \left(7x^3 - 2x^3\right) + \left(-8x^2 + 6x^2\right)$ Group like terms.
$\quad + (9x + 3x) + (-6 - 9)$
$= 5x^3 + \left(-2x^2\right) + 12x + (-15)$ Combine like terms.
$= 5x^3 - 2x^2 + 12x - 15$

Study Tip

You can also arrange like terms in columns and combine vertically:

$$\begin{array}{r} 7x^3 - 8x^2 + 9x - 6 \\ -2x^3 + 6x^2 + 3x - 9 \\ \hline 5x^3 - 2x^2 + 12x - 15 \end{array}$$

The like terms can be combined by adding their coefficients.

Check Point 1

Perform the indicated operations and simplify:

a. $(-17x^3 + 4x^2 - 11x - 5) + (16x^3 - 3x^2 + 3x - 15)$
b. $(13x^3 - 9x^2 - 7x + 1) - (-7x^3 + 2x^2 - 5x + 9)$

3 Multiply polynomials.

Multiplying Polynomials

The product of two monomials is obtained by using properties of exponents. For example,

$$(-8x^6)(5x^3) = -8 \cdot 5x^{6+3} = -40x^9$$

Multiply coefficients and add exponents.

Furthermore, we can use the distributive property to multiply a monomial and a polynomial that is not a monomial. For example,

$$3x^4(2x^3 - 7x + 3) = 3x^4 \cdot 2x^3 - 3x^4 \cdot 7x + 3x^4 \cdot 3 = 6x^7 - 21x^5 + 9x^4.$$

monomial   trinomial

How do we multiply two polynomials if neither is a monomial? For example, consider

$$(2x + 3)(x^2 + 4x + 5).$$

binomial   trinomial

One way to perform this multiplication is to distribute $2x$ throughout the trinomial

$$2x(x^2 + 4x + 5)$$

and 3 throughout the trinomial

$$3(x^2 + 4x + 5).$$

Then combine the like terms that result. In general, the product of two polynomials is the polynomial obtained by multiplying each term of one polynomial by each term of the other polynomial and then combining like terms.

EXAMPLE 2   Multiplying a Binomial and a Trinomial

Multiply: $(2x + 3)(x^2 + 4x + 5).$

Solution

$(2x + 3)(x^2 + 4x + 5)$
$= 2x(x^2 + 4x + 5) + 3(x^2 + 4x + 5)$ Use the distributive property to multiply the trinomial by each term of the binomial.

$= 2x \cdot x^2 + 2x \cdot 4x + 2x \cdot 5 + 3x^2 + 3 \cdot 4x + 3 \cdot 5$ Use the distributive property.

$= 2x^3 + 8x^2 + 10x + 3x^2 + 12x + 15$ Multiply the monomials.

$= 2x^3 + 11x^2 + 22x + 15$ Combine like terms.

Another method for solving Example 2 is to use a vertical format similar to that used for multiplying whole numbers.

$$x^2 + 4x + 5$$
$$\underline{2x + 3}$$

Write like terms in the same column.

$$\underline{3x^2 + 12x + 15} \quad 3(x^2 + 4x + 5)$$
$$\underline{2x^3 + 8x^2 + 10x} \quad 2x(x^2 + 4x + 5)$$
$$2x^3 + 11x^2 + 22x + 15 \quad \text{Combine like terms.}$$

Check Point 2

Multiply: $(5x - 2)(3x^2 - 5x + 4)$.

---

4 Use FOIL in polynomial multiplication.

The Product of Two Binomials: FOIL

Frequently we need to find the product of two binomials. We can use a method called FOIL, which is based on the distributive property, to do so. For example, we can find the product of the binomials $3x + 2$ and $4x + 5$ as follows:

$$(3x + 2)(4x + 5) = 3x(4x + 5) + 2(4x + 5) \quad \text{First, distribute } 3x \text{ over}$$
$$\text{4x + 5. Then distribute 2.}$$
$$= 3x(4x) + 3x(5) + 2(4x) + 2(5)$$
$$= 12x^2 + 15x + 8x + 10.$$

Two binomials can be quickly multiplied by using the FOIL method, in which F represents the product of the **first** terms in each binomial, O represents the product of the **outside** terms, I represents the product of the two **inside** terms, and L represents the product of the **last**, or second, terms in each binomial.

$$(3x + 2)(4x + 5) = 12x^2 + 15x + 8x + 10$$
$$= 12x^2 + 23x + 10 \quad \text{Combine like terms.}$$

In general, here's how to use the FOIL method to find the product of $ax + b$ and $cx + d$:

Using the FOIL Method to Multiply Binomials

$$(ax + b)(cx + d) = ax \cdot cx + ax \cdot d + b \cdot cx + b \cdot d$$

| | Product of First terms | Product of Outside terms | Product of Inside terms | Product of Last terms |

EXAMPLE 3 Using the FOIL Method

Multiply: $(3x + 4)(5x - 3)$.

Solution

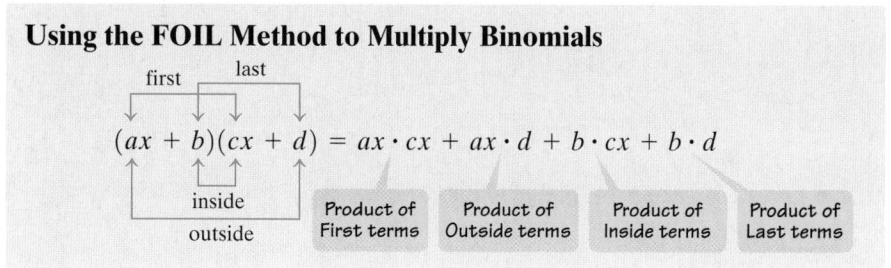

$$(3x + 4)(5x - 3) = 3x \cdot 5x + 3x(-3) + 4 \cdot 5x + 4(-3)$$
$$= 15x^2 - 9x + 20x - 12$$
$$= 15x^2 + 11x - 12 \quad \text{Combine like terms.}$$

> **Check Point 3**   Multiply: $(7x - 5)(4x - 3)$.

5   Use special products in polynomial multiplication.

Multiplying the Sum and Difference of Two Terms

We can use the FOIL method to multiply $A + B$ and $A - B$ as follows:

$$\begin{array}{cccc} \text{F} & \text{O} & \text{I} & \text{L} \end{array}$$

$$(A + B)(A - B) = A^2 - AB + AB - B^2 = A^2 - B^2.$$

Notice that the outside and inside products have a sum of 0 and the terms cancel. The FOIL multiplication provides us with a quick rule for multiplying the sum and difference of two terms, referred to as a special-product formula.

The Product of the Sum and Difference of Two Terms

$$(A + B)(A - B) = A^2 - B^2$$

The product of the sum and the difference of the same two terms   is   The square of the first term minus the square of the second term.

EXAMPLE 4   Finding the Product of the Sum and Difference of Two Terms

Find each product by using the preceding rule.

   a. $(4y + 3)(4y - 3)$ **b.** $(5a^4 + 6)(5a^4 - 6)$

Solution   Use the special-product formula shown.

$$(A + B)(A - B) \quad = \quad A^2 \quad - \quad B^2$$

First term squared   −   Second term squared   =   Product

a. $(4y + 3)(4y - 3) \quad = \quad (4y)^2 \quad - \quad 3^2 \quad = 16y^2 - 9$

b. $(5a^4 + 6)(5a^4 - 6) \quad = \quad (5a^4)^2 \quad - \quad 6^2 \quad = 25a^8 - 36$

> **Check Point 4**   Find each product:
>
>   **a.** $(7x + 8)(7x - 8)$ **b.** $(2y^3 - 5)(2y^3 + 5)$

The Square of a Binomial

Let us find $(A + B)^2$, the square of a binomial sum. To do so, we begin with the FOIL method and look for a general rule.

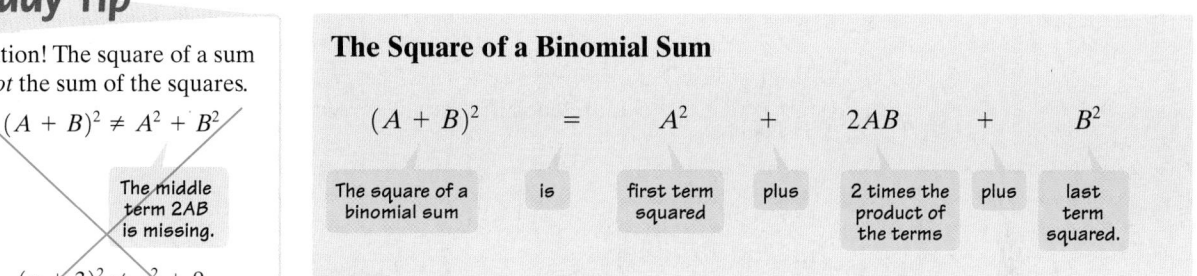

$$(A + B)^2 = (A + B)(A + B) = A \cdot A + A \cdot B + A \cdot B + B \cdot B$$
$$= A^2 + 2AB + B^2$$

This result implies the following rule, which is another example of a special-product formula.

Study Tip

Caution! The square of a sum is *not* the sum of the squares.

$(A + B)^2 \neq A^2 + B^2$

The middle term $2AB$ is missing.

$(x + 3)^2 \neq x^2 + 9$

Incorrect!

Show that $(x + 3)^2$ and $x^2 + 9$ are not equal by substituting 5 for x in each expression and simplifying.

The Square of a Binomial Sum

$(A + B)^2$	$=$	A^2	$+$	$2AB$	$+$	B^2
The square of a binomial sum	is	first term squared	plus	2 times the product of the terms	plus	last term squared.

EXAMPLE 5 Finding the Square of a Binomial Sum

Square each binomial using the preceding rule.

   a. $(x + 3)^2$ **b.** $(3x + 7)^2$

Solution   Use the special-product formula shown.

$(A + B)^2 = \quad A^2 \quad + \quad 2AB \quad + \quad B^2$

	(First Term)^2	+	2 · Product of the Terms	+	(Last Term)^2	= Product
a. $(x + 3)^2 =$	x^2	+	$2 \cdot x \cdot 3$	+	3^2	$= x^2 + 6x + 9$
b. $(3x + 7)^2 =$	$(3x)^2$	+	$2(3x)(7)$	+	7^2	$= 9x^2 + 42x + 49$

Check Point 5   Square each binomial:

   a. $(x + 10)^2$ **b.** $(5x + 4)^2$

Using the FOIL method on $(A - B)^2$, the square of a binomial difference, we obtain the following rule.

The Square of a Binomial Difference

$(A - B)^2$	$=$	A^2	$-$	$2AB$	$+$	B^2
The square of a binomial difference	is	first term squared	minus	2 times the product of the terms	plus	last term squared.

EXAMPLE 6   Finding the Square of a Binomial Difference

Square each binomial using the preceding rule.

a. $(x - 4)^2$ **b.** $(5y - 6)^2$

Solution   Use the special-product formula shown.

$$(A - B)^2 = A^2 - 2AB + B^2$$

	(First Term)^2	$-$	2 · Product of the Terms	$+$	(Last Term)^2	= Product
a. $(x - 4)^2 =$	x^2	$-$	$2 \cdot x \cdot 4$	$+$	4^2	$= x^2 - 8x + 16$
b. $(5y - 6)^2 =$	$(5y)^2$	$-$	$2(5y)(6)$	$+$	6^2	$= 25y^2 - 60y + 36$

Check Point 6

Square each binomial:

a. $(x - 9)^2$ **b.** $(7x - 3)^2$

Special Products

There are several products that occur so frequently that it's convenient to memorize the form or pattern of these formulas.

> **Special Products**
>
> Let A and B represent real numbers, variables, or algebraic expressions.
>
> | **Special Product** | **Example** |
> |---|---|
> | *Sum and Difference of Two Terms* | |
> | $(A + B)(A - B) = A^2 - B^2$ | $(2x + 3)(2x - 3) = (2x)^2 - 3^2$ |
> | | $= 4x^2 - 9$ |
> | *Squaring a Binomial* | |
> | $(A + B)^2 = A^2 + 2AB + B^2$ | $(y + 5)^2 = y^2 + 2 \cdot y \cdot 5 + 5^2$ |
> | | $= y^2 + 10y + 25$ |
> | $(A - B)^2 = A^2 - 2AB + B^2$ | $(3x - 4)^2$ |
> | | $= (3x)^2 - 2 \cdot 3x \cdot 4 + 4^2$ |
> | | $= 9x^2 - 24x + 16$ |
> | *Cubing a Binomial* | |
> | $(A + B)^3 = A^3 + 3A^2B + 3AB^2 + B^3$ | $(x + 4)^3$ |
> | | $= x^3 + 3x^2(4) + 3x(4)^2 + 4^3$ |
> | | $= x^3 + 12x^2 + 48x + 64$ |
> | $(A - B)^3 = A^3 - 3A^2B + 3AB^2 - B^3$ | $(x - 2)^3$ |
> | | $= x^3 - 3x^2(2) + 3x(2)^2 - 2^3$ |
> | | $= x^3 - 6x^2 + 12x - 8$ |

Study Tip

Although it's convenient to memorize these forms, the FOIL method can be used on all five examples in the box. To cube $x + 4$, you can first square $x + 4$ using FOIL and then multiply this result by $x + 4$. In short, you do not necessarily have to utilize these special formulas. What is the advantage of knowing and using these forms?

6 Perform operations with polynomials in several variables.

Polynomials in Several Variables

The next time you visit the lumber yard and go rummaging through piles of wood, think *polynomials*, although polynomials a bit different from those we have encountered so far. The construction industry uses a polynomial in two variables to determine the number of board feet that can be manufactured from a tree with a diameter of x inches and a length of y feet. This polynomial is

$$\tfrac{1}{4}x^2y - 2xy + 4y.$$

In general, a **polynomial in two variables**, x and y, contains the sum of one or more monomials in the form ax^ny^m. The constant a is the **coefficient**. The exponents n and m represent whole numbers. The **degree** of the monomial ax^ny^m is $n + m$. We'll use the polynomial from the construction industry to illustrate these ideas.

The coefficients are $\tfrac{1}{4}$, -2, and 4.

$$\tfrac{1}{4}x^2y \qquad -2xy \qquad +4y$$

Degree of monomial: $2 + 1 = 3$	Degree of monomial: $1 + 1 = 2$	Degree of monomial: $0 + 1 = 1$

The degree of a polynomial in two variables is the highest degree of all its terms. For the preceding polynomial, the degree is 3.

Polynomials containing two or more variables can be added, subtracted, and multiplied just like polynomials that contain only one variable.

EXAMPLE 7   Subtracting Polynomials in Two Variables

Subtract as indicated:

$$\left(5x^3 - 9x^2y + 3xy^2 - 4\right) - \left(3x^3 - 6x^2y - 2xy^2 + 3\right)$$

Solution

$\left(5x^3 - 9x^2y + 3xy^2 - 4\right) - \left(3x^3 - 6x^2y - 2xy^2 + 3\right)$

$= \left(5x^3 - 9x^2y + 3xy^2 - 4\right) + \left(-3x^3 + 6x^2y + 2xy^2 - 3\right)$

> Change the sign of each term in the second polynomial and add the two polynomials.

$= \left(5x^3 - 3x^3\right) + \left(-9x^2y + 6x^2y\right) + \left(3xy^2 + 2xy^2\right) + \left(-4 - 3\right)$

> Group like terms.

$= 2x^3 - 3x^2y + 5xy^2 - 7$ Combine like terms by combining coefficients and keeping the same variable factors.

Check Point 7   Subtract: $\left(x^3 - 4x^2y + 5xy^2 - y^3\right) - \left(x^3 - 6x^2y + y^3\right)$.

EXAMPLE 8   Multiplying Polynomials in Two Variables

Multiply: **a.** $(x + 4y)(3x - 5y)$ **b.** $(5x + 3y)^2$

Solution   We will perform the multiplication in part (a) using the FOIL method. We will multiply in part (b) using the formula for the square of a binomial, $(A + B)^2$.

a. $(x + 4y)(3x - 5y)$ Multiply these binomials using the FOIL method.

F   O   I   L

$$= (x)(3x) + (x)(-5y) + (4y)(3x) + (4y)(-5y)$$
$$= 3x^2 - 5xy + 12xy - 20y^2$$
$$= 3x^2 + 7xy - 20y^2 \quad \text{Combine like terms.}$$

$$(A + B)^2 = A^2 + 2 \cdot A \cdot B + B^2$$

b. $(5x + 3y)^2 = (5x)^2 + 2(5x)(3y) + (3y)^2$
$$= 25x^2 + 30xy + 9y^2$$

Check Point 8   Multiply:

a. $(7x - 6y)(3x - y)$ **b.** $(x^2 + 5y)^2$

EXERCISE SET P.4

Practice Exercises

In Exercises 1–4, is the algebraic expression a polynomial? If it is, write the polynomial in standard form.

1. $2x + 3x^2 - 5$
2. $2x + 3x^{-1} - 5$
3. $\dfrac{2x + 3}{x}$
4. $x^2 - x^3 + x^4 - 5$

In Exercises 5–8, find the degree of the polynomial.

5. $3x^2 - 5x + 4$
6. $-4x^3 + 7x^2 - 11$
7. $x^2 - 4x^3 + 9x - 12x^4 + 63$
8. $x^2 - 8x^3 + 15x^4 + 91$

In Exercises 9–14, perform the indicated operations. Write the resulting polynomial in standard form and indicate its degree.

9. $(-6x^3 + 5x^2 - 8x + 9) + (17x^3 + 2x^2 - 4x - 13)$
10. $(-7x^3 + 6x^2 - 11x + 13) + (19x^3 - 11x^2 + 7x - 17)$
11. $(17x^3 - 5x^2 + 4x - 3) - (5x^3 - 9x^2 - 8x + 11)$
12. $(18x^4 - 2x^3 - 7x + 8) - (9x^4 - 6x^3 - 5x + 7)$
13. $(5x^2 - 7x - 8) + (2x^2 - 3x + 7) - (x^2 - 4x - 3)$
14. $(8x^2 + 7x - 5) - (3x^2 - 4x) - (-6x^3 - 5x^2 + 3)$

In Exercises 15–54, find each product.

15. $(x + 1)(x^2 - x + 1)$
16. $(x + 5)(x^2 - 5x + 25)$
17. $(2x - 3)(x^2 - 3x + 5)$
18. $(2x - 1)(x^2 - 4x + 3)$
19. $(x + 7)(x + 3)$
20. $(x + 8)(x + 5)$
21. $(x - 5)(x + 3)$
22. $(x - 1)(x + 2)$
23. $(3x + 5)(2x + 1)$
24. $(7x + 4)(3x + 1)$
25. $(2x - 3)(5x + 3)$
26. $(2x - 5)(7x + 2)$

27. $(5x^2 - 4)(3x^2 - 7)$
28. $(7x^2 - 2)(3x^2 - 5)$
29. $(x + 3)(x - 3)$
30. $(x + 5)(x - 5)$
31. $(3x + 2)(3x - 2)$
32. $(2x + 5)(2x - 5)$
33. $(5 - 7x)(5 + 7x)$
34. $(4 - 3x)(4 + 3x)$
35. $(4x^2 + 5x)(4x^2 - 5x)$
36. $(3x^2 + 4x)(3x^2 - 4x)$
37. $(x + 2)^2$
38. $(x + 5)^2$
39. $(2x + 3)^2$
40. $(3x + 2)^2$
41. $(x - 3)^2$
42. $(x - 4)^2$
43. $(4x^2 - 1)^2$
44. $(5x^2 - 3)^2$
45. $(7 - 2x)^2$
46. $(9 - 5x)^2$
47. $(x + 1)^3$
48. $(x + 2)^3$
49. $(2x + 3)^3$
50. $(3x + 4)^3$
51. $(x - 3)^3$
52. $(x - 1)^3$
53. $(3x - 4)^3$
54. $(2x - 3)^3$

In Exercises 55–62, perform the indicated operations. Indicate the degree of the resulting polynomial.

55. $(5x^2y - 3xy) + (2x^2y - xy)$
56. $(-2x^2y + xy) + (4x^2y + 7xy)$
57. $(4x^2y + 8xy + 11) + (-2x^2y + 5xy + 2)$
58. $(7x^4y^2 - 5x^2y^2 + 3xy) + (-18x^4y^2 - 6x^2y^2 - xy)$
59. $(x^3 + 7xy - 5y^2) - (6x^3 - xy + 4y^2)$
60. $(x^4 - 7xy - 5y^3) - (6x^4 - 3xy + 4y^3)$
61. $(3x^4y^2 + 5x^3y - 3y) - (2x^4y^2 - 3x^3y - 4y + 6x)$
62. $(5x^4y^2 + 6x^3y - 7y) - (3x^4y^2 - 5x^3y - 6y + 8x)$

In Exercises 63–76, find each product.

63. $(x + 5y)(7x + 3y)$
64. $(x + 9y)(6x + 7y)$
65. $(x - 3y)(2x + 7y)$
66. $(3x - y)(2x + 5y)$

67. $(3xy - 1)(5xy + 2)$ **68.** $(7x^2y + 1)(2x^2y - 3)$
69. $(7x + 5y)^2$ **70.** $(9x + 7y)^2$
71. $(x^2y^2 - 3)^2$ **72.** $(x^2y^2 - 5)^2$
73. $(x - y)(x^2 + xy + y^2)$ **74.** $(x + y)(x^2 - xy + y^2)$
75. $(3x + 5y)(3x - 5y)$ **76.** $(7x + 3y)(7x - 3y)$

 Application Exercises

77. The polynomial $0.018x^2 - 0.757x + 9.047$ describes the amount, in thousands of dollars, that a person earning x thousand dollars a year feels underpaid. Evaluate the polynomial when $x = 40$. Describe what the answer means in practical terms.

78. The polynomial $104.5x^2 - 1501.5x + 6016$ describes the death rate per year per 100,000 men for men averaging x hours of sleep each night. Evaluate the polynomial when $x = 10$. Describe what the answer means in practical terms.

79. The polynomial $-0.02A^2 + 2A + 22$ is used by coaches to get athletes fired up so that they can perform well. The polynomial represents the performance level related to various levels of enthusiasm, from $A = 1$ (almost no enthusiasm) to $A = 100$ (maximum level of enthusiasm). Evaluate the polynomial when $A = 20$, $A = 50$, and $A = 80$. Describe what happens to performance as we get more and more fired up.

80. The polynomial

$$0.0001x^3 - 0.0043x^2 + 0.089x + 2.66$$

describes the number of pounds of waste produced each day by every American x years after 1960. (The bar graph illustrates daily waste production for eight years.) Evaluate the polynomial when $x = 10$. Describe what the answer means in practical terms.

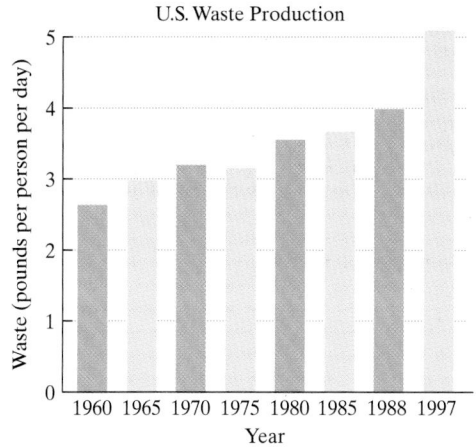

U.S. Waste Production

Source: U.S. Environmental Protection Agency.

81. The number of people who catch a cold t weeks after January 1 is $5t - 3t^2 + t^3$. The number of people who recover t weeks after January 1 is $t - t^2 + \frac{1}{3}t^3$. Write a polynomial in standard form for the number of people who are still ill with a cold t weeks after January 1.

82. The weekly cost, in thousands of dollars, of producing x stereo headphones is $30x + 50$. The weekly revenue, in thousands of dollars, of selling x stereo headphones is $90x^2 - x$. Write a polynomial in standard form for the weekly profit, in thousands of dollars, for producing and selling x stereo headphones.

In Exercises 83–84, write a polynomial in standard form that represents the area of the shaded region of each figure.

83.
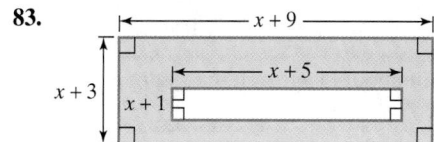

84.
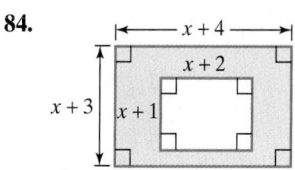

 **Writing in Mathematics**

85. What is a polynomial in x?

86. Explain how to subtract polynomials.

87. Explain how to multiply two binomials using the FOIL method. Give an example with your explanation.

88. Explain how to find the product of the sum and difference of two terms. Give an example with your explanation.

89. Explain how to square a binomial difference. Give an example with your explanation.

90. Explain how to find the degree of a polynomial in two variables.

91. For Exercise 79, explain why performance levels do what they do as we get more and more fired up. If possible, describe an example of a time when you were too enthused and thus did poorly at something you were hoping to do well.

Technology Exercises

92. The common cold is caused by a rhinovirus. The polynomial

$$-0.75x^4 + 3x^3 + 5$$

describes the billions of viral particles in our bodies after x days of invasion. Use a calculator to find the number of viral particles after 0 days (the time of the cold's onset), 1 day, 2 days, 3 days, and 4 days. After how many days is the number of viral particles at a maximum and consequently the day we feel the sickest? By when should we feel completely better?

93. The polynomial $-3.08x^2 + 40.35x + 305.89$ describes the annual number of aggravated assaults in the United States per 100,000 people x years after 1986. Use a calculator to find the number of aggravated assaults per 100,000 people from 1986 to 2000. For this time period, during what year was the number of aggravated assaults per 100,000 people the greatest?

Critical Thinking Exercises

In Exercises 94–97, perform the indicated operations.

94. $(x - y)^2 - (x + y)^2$

95. $[(7x + 5) + 4y][(7x + 5) - 4y]$

96. $[(3x + y) + 1]^2$

97. $(x + y)(x - y)(x^2 + y^2)$

98. Express the area of the plane figure shown as a polynomial in standard form.

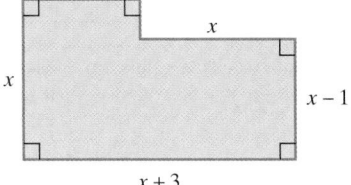

SECTION P.5 *Factoring Polynomials*

Objectives

1. Factor out the greatest common factor of a polynomial.
2. Factor by grouping.
3. Factor trinomials.
4. Factor the difference of squares.
5. Factor perfect square trinomials.
6. Factor the sum and difference of cubes.
7. Factor completely.

A two-year-old boy is asked, "Do you have a brother?" He answers, "Yes." "What is your brother's name?" "Tom." Asked if Tom has a brother, the two-year-old replies, "No." The child can go in the direction from self to brother, but he cannot reverse this direction and move from brother back to self.

As our intellects develop, we learn to reverse the direction of our thinking. Reversibility of thought is found throughout algebra. For example, we can multiply polynomials and show that

$$(2x + 1)(3x - 2) = 6x^2 - x - 2.$$

We can also reverse this process and express the resulting polynomial as

$$6x^2 - x - 2 = (2x + 1)(3x - 2).$$

Factoring is the process of writing a polynomial as the product of two or more polynomials. The factors of $6x^2 - x - 2$ are $2x + 1$ and $3x - 2$.

In this section, we will be **factoring over the set of integers**, meaning that the coefficients in the factors are integers. Polynomials that cannot be factored using integer coefficients are called **irreducible over the integers**, or **prime**.

The goal in factoring a polynomial is to use one or more factoring techniques until each of the polynomial's factors is prime or irreducible. In this situation, the polynomial is said to be **factored completely**.

We will now discuss basic techniques for factoring polynomials.

1 Factor out the greatest common factor of a polynomial.

Common Factors

In any factoring problem, the first step is to look for the **greatest common factor**. The greatest common factor is an expression of the highest degree that divides each term of the polynomial. The distributive property in the reverse direction

$$ab + ac = a(b + c)$$

can be used to factor out the greatest common factor.

EXAMPLE 1 Factoring out the Greatest Common Factor

Factor: **a.** $18x^3 + 27x^2$ **b.** $x^2(x + 3) + 5(x + 3)$

Solution

a. We begin by determining the greatest common factor. 9 is the greatest integer that divides 18 and 27. Furthermore, x^2 is the greatest expression that divides x^3 and x^2. Thus, the greatest common factor of the two terms in the polynomial is $9x^2$.

$18x^3 + 27x^2$

$= 9x^2(2x) + 9x^2(3)$ Express each term with the greatest common factor as a factor.

$= 9x^2(2x + 3)$ Factor out the greatest common factor.

b. In this situation, the greatest common factor is the common binomial factor $(x + 3)$. We factor out this common factor as follows.

$x^2(x + 3) + 5(x + 3) = (x + 3)(x^2 + 5)$ Factor out the common binomial factor.

Check Point 1   Factor:

a. $10x^3 - 4x^2$ **b.** $2x(x - 7) + 3(x - 7)$

2 Factor by grouping.

Factoring by Grouping

Some polynomials have only a greatest common factor of 1. However, by a suitable rearrangement of the terms, it still may be possible to factor. This process, called **factoring by grouping**, is illustrated in Example 2.

EXAMPLE 2 Factoring by Grouping

Factor: $x^3 + 4x^2 + 3x + 12$.

Solution   Group terms that have a common factor:

$$\boxed{x^3 + 4x^2} \; + \; \boxed{3x + 12}.$$

Common factor is x^2.   Common factor is 3.

Discovery

In Example 2, group the terms as follows:

$$(x^3 + 3x) + (4x^2 + 12).$$

Factor out the greatest common factor from each group and complete the factoring process. Describe what happens. What can you conclude?

We now factor the given polynomial as follows.

$$x^3 + 4x^2 + 3x + 12$$
$$= (x^3 + 4x^2) + (3x + 12) \quad \text{Group terms with common factors.}$$
$$= x^2(x + 4) + 3(x + 4) \quad \text{Factor out the greatest common factor from the grouped terms. The remaining two terms have } x + 4 \text{ as a common binomial factor.}$$
$$= (x + 4)(x^2 + 3) \quad \text{Factor } (x + 4) \text{ out of both terms.}$$

Thus, $x^3 + 4x^2 + 3x + 12 = (x + 4)(x^2 + 3)$. Check the factorization by multiplying the right side of the equation using the FOIL method. If the factorization is correct, you will obtain the original polynomial.

Check Point 2   Factor: $x^3 + 5x^2 - 2x - 10.$

3   Factor trinomials.

Factoring Trinomials

To factor a trinomial of the form $ax^2 + bx + c$, a little trial and error may be necessary.

A Strategy for Factoring $ax^2 + bx + c$

(Assume, for the moment, that there is no greatest common factor.)

1. Find two **First** terms whose product is ax^2:

$$(\Box x + \quad)(\Box x + \quad) = ax^2 + bx + c$$

2. Find two **Last** terms whose product is c:

$$(x + \Box)(x + \Box) = ax^2 + bx + c$$

3. By trial and error, perform steps 1 and 2 until the sum of the **Outside** product and **Inside** product is bx:

$$(\Box x + \Box)(\Box x + \Box) = ax^2 + bx + c$$

(sum of O + I)

If no such combinations exist, the polynomial is prime.

EXAMPLE 3   Factoring Trinomials Whose Leading Coefficients Are 1

Factor: **a.** $x^2 + 6x + 8$ **b.** $x^2 + 3x - 18$

Solution

Factors of 8	8, 1	4, 2	−8, −1	−4, −2
Sum of Factors	9	6	−9	−6

This is the desired sum.

a. The factors of the first term are x and x:

$$(x \quad)(x \quad)$$

To find the second term of each factor, we must find two numbers whose product is 8 and whose sum is 6.

From the table in the margin, we see that 4 and 2 are the required integers. Thus,

$$x^2 + 6x + 8 = (x + 4)(x + 2) \quad \text{or} \quad (x + 2)(x + 4).$$

b. We begin with

$$x^2 + 3x - 18 = (x \quad)(x \quad).$$

To find the second term of each factor, we must find two numbers whose product is −18 and whose sum is 3. From the table in the margin, we see that 6 and −3 are the required integers. Thus,

$$x^2 + 3x - 18 = (x + 6)(x - 3)$$
$$\text{or} \quad (x - 3)(x + 6).$$

Factors of −18	18, −1	−18, 1	9, −2	−9, 2	6, −3	−6, 3
Sum of factors	17	−17	7	−7	3	−3

This is the desired sum.

Check Point 3

Factor:

a. $x^2 + 13x + 40$ **b.** $x^2 - 5x - 14$

EXAMPLE 4   Factoring a Trinomial Whose Leading Coefficient Is Not 1

Factor: $8x^2 - 10x - 3$.

Solution

Step 1   Find two First terms whose product is $8x^2$.

$$8x^2 - 10x - 3 \stackrel{?}{=} (8x \quad)(x \quad)$$
$$8x^2 - 10x - 3 \stackrel{?}{=} (4x \quad)(2x \quad)$$

Step 2   Find two Last terms whose product is − 3.   The possible factors are 1(−3) and −1(3).

Step 3   Try various combinations of these factors.   The correct factorization of $8x^2 - 10x - 3$ is the one in which the sum of the Outside and Inside products is equal to −10x. Here is a list of the possible factors.

Possible Factors of $8x^2 - 10x - 3$	Sum of Outside and Inside Products (Should Equal − 10x)
$(8x + 1)(x - 3)$	$-24x + x = -23x$
$(8x - 3)(x + 1)$	$8x - 3x = 5x$
$(8x - 1)(x + 3)$	$24x - x = 23x$
$(8x + 3)(x - 1)$	$-8x + 3x = -5x$
$(4x + 1)(2x - 3)$	$-12x + 2x = -10x$
$(4x - 3)(2x + 1)$	$4x - 6x = -2x$
$(4x - 1)(2x + 3)$	$12x - 2x = 10x$
$(4x + 3)(2x - 1)$	$-4x + 6x = 2x$

This is the required middle term.

Thus,

$$8x^2 - 10x - 3 = (4x + 1)(2x - 3) \quad \text{or} \quad (2x - 3)(4x + 1).$$

Show that this factorization is correct by multiplying the factors using the FOIL method. You should obtain the original trinomial.

Check Point 4 Factor: $6x^2 + 19x - 7$.

4 Factor the difference of squares.

Factoring the Difference of Two Squares

A method for factoring the difference of two squares is obtained by reversing the special product for the sum and difference of two terms.

> ### The Difference of Two Squares
> If A and B are real numbers, variables, or algebraic expressions, then
> $$A^2 - B^2 = (A + B)(A - B).$$
> In words: The difference of the squares of two terms factors as the product of a sum and the difference of those terms.

EXAMPLE 5   Factoring the Difference of Two Squares

Factor: **a.** $x^2 - 4$ **b.** $81x^2 - 49$

Solution   We must express each term as the square of some monomial. Then we use the formula for factoring $A^2 - B^2$.

a. $x^2 - 4^2 = x^2 - 2^2 = (x + 2)\ (x - 2)$

$$A^2 - B^2 = (A + B)(A - B)$$

b. $81x^2 - 49 = (9x)^2 - 7^2 = (9x + 7)(9x - 7)$

Check Point 5 Factor:

a. $x^2 - 81$ **b.** $36x^2 - 25$

We have seen that a polynomial is factored completely when it is written as the product of prime polynomials. To be sure that you have factored completely, check to see whether the factors can be factored.

EXAMPLE 6   A Repeated Factorization

Factor completely: $x^4 - 81$.

Study Tip

Factoring $x^4 - 81$ as
$$(x^2 + 9)(x^2 - 9)$$
is not a complete factorization. The second factor $x^2 - 9$ is itself a difference of two squares and can be factored.

Solution

$x^4 - 81 = (x^2)^2 - 9^2$	Express as the difference of two squares.
$= (x^2 + 9)(x^2 - 9)$	The factors are the sum and difference of the squared terms.
$= (x^2 + 9)(x^2 - 3^2)$	The factor $x^2 - 9$ is the difference of two squares and can be factored.
$= (x^2 + 9)(x + 3)(x - 3)$	The factors of $x^2 - 9$ are the sum and difference of the squared terms.

| Check Point 6 | Factor completely: $81x^4 - 16$. |

5 Factor perfect square trinomials.

Factoring Perfect Square Trinomials

Our next factoring technique is obtained by reversing the special products for squaring binomials. The trinomials that are factored using this technique are called **perfect square trinomials**.

Factoring Perfect Square Trinomials

Let A and B be real numbers, variables, or algebraic expressions.

1. $A^2 + 2AB + B^2 = (A + B)^2$

Same sign

2. $A^2 - 2AB + B^2 = (A - B)^2$

Same sign

The two items in the box show that perfect square trinomials come in two forms: one in which the middle term is positive and one in which the middle term is negative. Here's how to recognize a perfect square trinomial:

1. The first and last terms are positive perfect squares.
2. The middle term is twice the product of the square roots of the first and last terms.

EXAMPLE 7   Factoring Perfect Square Trinomials

Factor: **a.** $x^2 + 6x + 9$ **b.** $25x^2 - 60x + 36$

Solution

a. $x^2 + 6x + 9 = x^2 + 2 \cdot x \cdot 3 + 3^2 = (x + 3)^2$ The middle term has a positive sign.

$$A^2 + 2AB + B^2 = (A + B)^2$$

b. We suspect that $25x^2 - 60x + 36$ is a perfect square trinomial because $25x^2 = (5x)^2$ and $36 = 6^2$. The middle term can be expressed as twice the product of $5x$ and 6.

$$25x^2 - 60x + 36 = (5x)^2 - 2 \cdot 5x \cdot 6 + 6^2 = (5x - 6)^2$$

$$A^2 - 2AB + B^2 = (A - B)^2$$

| Check Point 7 | Factor: |

a. $x^2 + 14x + 49$ **b.** $16x^2 - 56x + 49$

6 Factor the sum and difference of cubes.

Factoring the Sum and Difference of Two Cubes

We can use the following formulas to factor the sum or the difference of two cubes.

Factoring the Sum and Difference of Two Cubes

1. Factoring the Sum of Two Cubes

$$A^3 + B^3 = (A + B)(A^2 - AB + B^2)$$

2. Factoring the Difference of Two Cubes

$$A^3 - B^3 = (A - B)(A^2 + AB + B^2)$$

EXAMPLE 8   Factoring Sums and Differences of Two Cubes

Factor: **a.** $x^3 + 8$ **b.** $64x^3 - 125$

Solution

a. $x^3 + 8 = x^3 + 2^3 = (x + 2)(x^2 - x \cdot 2 + 2^2) = (x + 2)(x^2 - 2x + 4)$

$$A^3 + B^3 = (A + B)(A^2 - AB + B^2)$$

b. $64x^3 - 125 = (4x)^3 - 5^3 = (4x - 5)[(4x)^2 + (4x)(5) + 5^2]$

$$A^3 - B^3 = (A - B)(A^2 + AB + B^2)$$

$$= (4x - 5)(16x^2 + 20x + 25)$$

Check Point 8   Factor:

a. $x^3 + 1$ **b.** $125x^3 - 8$

7 Factor completely.

Factoring Completely

Some polynomials can be factored using more than one technique. Always begin by trying to factor out the greatest common factor.

EXAMPLE 9   Factoring Completely

Factor: **a.** $2a^3 + 8a^2 + 8a$ **b.** $x^3 - 5x^2 - 4x + 20$

Solution

a. $2a^3 + 8a^2 + 8a$

$= 2a(a^2 + 4a + 4)$ Factor out the greatest common factor.

$= 2a(a + 2)^2$ Factor the perfect square trinomial.

b. $x^3 - 5x^2 - 4x + 20$
$= (x^3 - 5x^2) + (-4x + 20)$ Group the terms with common factors.
$= x^2(x - 5) - 4(x - 5)$ Factor from each group.
$= (x - 5)(x^2 - 4)$ Factor out the common binomial factor, $(x - 5)$.
$= (x - 5)(x + 2)(x - 2)$ Factor completely by factoring $x^2 - 4$ as the difference of two squares.

Check Point 9 Factor:

a. $2x^3 - 24x^2 + 72x$ **b.** $x^3 - 4x^2 - 9x + 36$

EXERCISE SET P.5

Practice Exercises

In Exercises 1–10, factor out the greatest common factor.

1. $18x + 27$ **2.** $16x - 24$
3. $3x^2 + 6x$ **4.** $4x^2 - 8x$
5. $9x^4 - 18x^3 + 27x^2$ **6.** $6x^4 - 18x^3 + 12x^2$
7. $x(x + 5) + 3(x + 5)$ **8.** $x(2x + 1) + 4(2x + 1)$
9. $x^2(x - 3) + 12(x - 3)$ **10.** $x^2(2x + 5) + 17(2x + 5)$

In Exercises 11–16, factor by grouping.

11. $x^3 - 2x^2 + 5x - 10$ **12.** $x^3 - 3x^2 + 4x - 12$
13. $x^3 - x^2 + 2x - 2$ **14.** $x^3 + 6x^2 - 2x - 12$
15. $3x^3 - 2x^2 - 6x + 4$ **16.** $x^3 - x^2 - 5x + 5$

In Exercises 17–30, factor each trinomial, or state that the trinomial is prime.

17. $x^2 + 5x + 6$ **18.** $x^2 + 8x + 15$
19. $x^2 - 2x - 15$ **20.** $x^2 - 4x - 5$
21. $x^2 - 8x + 15$ **22.** $x^2 - 14x + 45$
23. $3x^2 - x - 2$ **24.** $2x^2 + 5x - 3$
25. $3x^2 - 25x - 28$ **26.** $3x^2 - 2x - 5$
27. $6x^2 - 11x + 4$ **28.** $6x^2 - 17x + 12$
29. $4x^2 + 16x + 15$ **30.** $8x^2 + 33x + 4$

In Exercises 31–40, factor the difference of two squares.

31. $x^2 - 100$ **32.** $x^2 - 144$
33. $36x^2 - 49$ **34.** $64x^2 - 81$
35. $9x^2 - 25y^2$ **36.** $36x^2 - 49y^2$
37. $x^4 - 16$ **38.** $x^4 - 1$
39. $16x^4 - 81$ **40.** $81x^4 - 1$

In Exercises 41–48, factor any perfect square trinomials, or state that the polynomial is prime.

41. $x^2 + 2x + 1$ **42.** $x^2 + 4x + 4$
43. $x^2 - 14x + 49$ **44.** $x^2 - 10x + 25$
45. $4x^2 + 4x + 1$ **46.** $25x^2 + 10x + 1$
47. $9x^2 - 6x + 1$ **48.** $64x^2 - 16x + 1$

In Exercises 49–56, factor using the formula for the sum or difference of two cubes.

49. $x^3 + 27$ **50.** $x^3 + 64$
51. $x^3 - 64$ **52.** $x^3 - 27$
53. $8x^3 - 1$ **54.** $27x^3 - 1$
55. $64x^3 + 27$ **56.** $8x^3 + 125$

In Exercises 57–76, factor completely, or state that the polynomial is prime.

57. $3x^3 - 3x$ **58.** $5x^3 - 45x$
59. $4x^2 - 4x - 24$ **60.** $6x^2 - 18x - 60$
61. $2x^4 - 162$ **62.** $7x^4 - 7$
63. $x^3 + 2x^2 - 9x - 18$ **64.** $x^3 + 3x^2 - 25x - 75$
65. $2x^2 - 2x - 112$ **66.** $6x^2 - 6x - 12$
67. $x^3 - 4x$ **68.** $9x^3 - 9x$
69. $x^2 + 64$ **70.** $x^2 + 36$
71. $x^3 + 2x^2 - 4x - 8$ **72.** $x^3 + 2x^2 - x - 2$
73. $y^5 - 81y$ **74.** $y^5 - 16y$
75. $20y^4 - 45y^2$ **76.** $48y^4 - 3y^2$

Application Exercises

77. You dive directly upward from a board that is 32 feet high. After t seconds, your height above the water is described by the polynomial $-16t^2 + 16t + 32$. Factor the polynomial completely.

78. If x represents a positive integer, factor $x^3 + 3x^2 + 2x$ to show that the trinomial represents the product of three consecutive integers.

In Exercises 79–80, find the formula for the area of the shaded region and express it in factored form.

79.

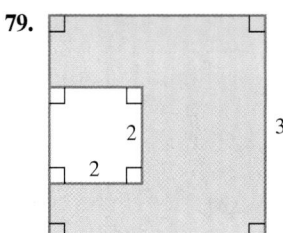

80.

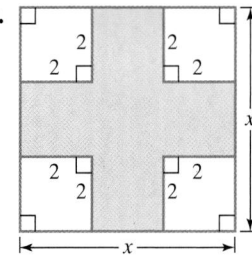

Writing in Mathematics

81. Use an example and explain how to factor out the greatest common factor of a polynomial.

82. Suppose that a polynomial contains four terms. Explain how to use factoring by grouping to factor the polynomial.

83. Explain how to factor $3x^2 + 10x + 8$.

84. Explain how to factor the difference of two squares. Provide an example with your explanation.

85. What is a perfect square trinomial and how is it factored?

86. Explain how to factor $x^3 + 1$.

87. What does it mean to factor completely?

88. For Exercise 77, explain how to use your factored polynomial to figure out how many seconds it will take for you to hit the water.

Critical Thinking Exercises

89. Which one of the following is true?
   a. Because $x^2 + 1$ is irreducible over the integers, it follows that $x^3 + 1$ is also irreducible.
   b. One correct factored form for
   $x^2 - 4x + 3$ is $x(x - 4) + 3$.
   c. $x^3 - 64 = (x - 4)^3$
   d. None of the above is true.

In Exercises 90–93, factor completely.

90. $x^{2n} + 6x^n + 8$ **91.** $-x^2 - 4x + 5$

92. $x^2 + 8x + 16 - 25y^2$ **93.** $x^4 - y^4 - 2x^3y + 2xy^3$

In Exercises 94–95, find all integers b so that the trinomial can be factored.

94. $x^2 + bx + 15$ **95.** $x^2 + 4x + b$

Group Exercise

96. Without looking at any factoring problems in the book, create five factoring problems. Make sure that some of your problems require at least two factoring strategies. Next, exchange problems with another person in your group. Work to factor your partner's problems. Evaluate the problems as you work: Are they too easy? Too difficult? Can the polynomials really be factored? Share your response with the person who wrote the problems. Finally, grade each other's work in factoring the polynomials. Each factoring problem is worth 20 points. You may award partial credit. If you take off points, explain why points are deducted and how you decided to take off a particular number of points for the error(s) that you found.

SECTION P.6 Rational Expressions

Objectives

1. Specify numbers that must be excluded from the domain of rational expressions.

2. Simplify rational expressions.

3. Multiply rational expressions.

4. Divide rational expressions.

5. Add and subtract rational expressions.

6. Simplify complex rational expressions.

How do we describe the costs of reducing environmental pollution? We often use algebraic expressions involving quotients of polynomials. For example, the algebraic expression

$$\frac{250x}{100 - x}$$

describes the cost, in millions of dollars, to remove x percent of the pollutants that are discharged into a river. Removing a modest percentage of pollutants, say

Discovery

What happens if you try substituting 100 for x in

$$\frac{250x}{100 - x} \; ?$$

What does this tell you about the cost of cleaning up all of the river's pollutants?

40%, is far less costly than removing a substantially greater percentage, such as 95%. We see this by evaluating the algebraic expression for $x = 40$ and $x = 95$.

Evaluating $\dfrac{250x}{100 - x}$ for

$x = 40$:	$x = 95$:
Cost is $\dfrac{250(40)}{100 - 40} \approx 167$	Cost is $\dfrac{250(95)}{100 - 95} = 4750$

The cost increases from approximately \$167 million to a possibly prohibitive \$4750 million, or \$4.75 billion. Costs spiral upward as the percentage of removed pollutants increases.

Many algebraic expressions that describe costs of environmental projects are examples of rational expressions. First we will define rational expressions. Then we will review how to perform operations with such expressions.

1 Specify numbers that must be excluded from the domain of rational expressions.

Rational Expressions

A **rational expression** is the quotient of two polynomials. Some examples are

$$\frac{x - 2}{4}, \quad \frac{4}{x - 2}, \quad \frac{x}{x^2 - 1}, \quad \text{and} \quad \frac{x^2 + 1}{x^2 + 2x - 3}.$$

The set of real numbers for which an algebraic expression is defined is the **domain** of the expression. Because rational expressions indicate division and division by zero is undefined, we must exclude numbers from a rational expression's domain that make the denominator zero.

EXAMPLE 1 Excluding Numbers from the Domain

Find all the numbers that must be excluded from the domain of each rational expression.

a. $\dfrac{4}{x - 2}$ **b.** $\dfrac{x}{x^2 - 1}$

Solution To determine the numbers that must be excluded from each domain, examine the denominators.

a. $\dfrac{4}{x - 2}$ **b.** $\dfrac{x}{x^2 - 1} = \dfrac{x}{(x + 1)(x - 1)}$

This denominator would equal zero if $x = 2$.

This factor would equal zero if $x = -1$.

This factor would equal zero if $x = 1$.

For the rational expression in part (a), we must exclude 2 from the domain. For the rational expression in part (b), we must exclude both -1 and 1 from the domain. These excluded numbers are often written to the right of a rational expression.

$$\frac{4}{x - 2}, x \neq 2 \qquad \frac{x}{x^2 - 1}, x \neq -1, x \neq 1$$

Check Point 1

Find all the numbers that must be excluded from each rational expression's domain.

a. $\dfrac{7}{x+5}$ **b.** $\dfrac{x}{x^2-36}$

2 Simplify rational expressions.

Simplifying Rational Expressions

A rational expression is **simplified** if its numerator and denominator have no common factors other than 1 or −1. The following procedure can be used to simplify rational expressions.

> **Simplifying Rational Expressions**
>
> 1. Factor the numerator and denominator completely.
> 2. Divide both the numerator and denominator by the common factors.

EXAMPLE 2   Simplifying Rational Expressions

Simplify: **a.** $\dfrac{x^3+x^2}{x+1}$ **b.** $\dfrac{x^2+6x+5}{x^2-25}$

Solution

a. $\dfrac{x^3+x^2}{x+1} = \dfrac{x^2(x+1)}{x+1}$ Factor the numerator. Because the denominator is x + 1, x ≠ −1.

$$= \dfrac{x^2\overset{1}{\cancel{(x+1)}}}{\underset{1}{\cancel{x+1}}}$$ Divide out the common factor of x + 1.

$$= x^2,\ x \ne -1$$ Denominators of 1 need not be written because $\frac{a}{1} = a$.

b. $\dfrac{x^2+6x+5}{x^2-25} = \dfrac{(x+5)(x+1)}{(x+5)(x-5)}$ Factor the numerator and denominator. Because the denominator is (x + 5)(x − 5), x ≠ −5 and x ≠ 5.

$$= \dfrac{\overset{1}{\cancel{(x+5)}}(x+1)}{\underset{1}{\cancel{(x+5)}}(x-5)}$$ Divide out the common factor of x + 5.

$$= \dfrac{x+1}{x-5},\quad x \ne -5 \text{ and } x \ne 5$$

Check Point 2

Simplify:

a. $\dfrac{x^3+3x^2}{x+3}$ **b.** $\dfrac{x^2-1}{x^2+2x+1}$

3 Multiply rational expressions.

Multiplying Rational Expressions

The product of two rational expressions is the product of their numerators over the product of their denominators. Here is a step-by-step procedure for multiplying rational expressions.

Multiplying Rational Expressions

1. Factor all numerators and denominators completely.

2. Divide both the numerator and denominator by common factors.

3. Multiply the remaining factors in the numerator and multiply the remaining factors in the denominator.

EXAMPLE 3   Multiplying Rational Expressions

Multiply and simplify:

$$\frac{x - 7}{x - 1} \cdot \frac{x^2 - 1}{3x - 21}.$$

Solution

$$\frac{x - 7}{x - 1} \cdot \frac{x^2 - 1}{3x - 21}$$

$$= \frac{x - 7}{x - 1} \cdot \frac{(x + 1)(x - 1)}{3(x - 7)}$$

Factor all numerators and denominators. Because the denominator has factors of $x - 1$ and $x - 7$, $x \neq 1$ and $x \neq 7$.

$$= \frac{\overset{1}{\cancel{x - 7}}}{\underset{1}{\cancel{x - 1}}} \cdot \frac{(x + 1)\overset{1}{\cancel{(x - 1)}}}{3\underset{1}{\cancel{(x - 7)}}}$$

Divide both the numerator and the denominator by common factors.

$$= \frac{x + 1}{3}, x \neq 1, x \neq 7$$

Multiply the remaining factors in the numerator and denominator.

These excluded numbers from the domain must also be excluded from the simplified expression's domain.

Check Point 3   Multiply and simplify:

$$\frac{x + 3}{x^2 - 4} \cdot \frac{x^2 - x - 6}{x^2 + 6x + 9}.$$

4 Divide rational expressions.

Dividing Rational Expressions

We find the quotient of two rational expressions by inverting the divisor and multiplying.

EXAMPLE 4   Dividing Rational Expressions

Divide and simplify:

$$\frac{x^2 - 2x - 8}{x^2 - 9} \div \frac{x - 4}{x + 3}.$$

Solution

$$\frac{x^2 - 2x - 8}{x^2 - 9} \div \frac{x - 4}{x + 3}$$

$$= \frac{x^2 - 2x - 8}{x^2 - 9} \cdot \frac{x + 3}{x - 4}$$ Invert the divisor and multiply.

$$= \frac{(x - 4)(x + 2)}{(x + 3)(x - 3)} \cdot \frac{x + 3}{x - 4}$$ Factor throughout. For nonzero denominators, $x \neq -3$, $x \neq 3$, and $x \neq 4$.

$$= \frac{\overset{1}{\cancel{(x - 4)}}(x + 2)}{\cancel{(x + 3)}(x - 3)} \cdot \frac{\overset{1}{\cancel{(x + 3)}}}{\underset{1}{\cancel{(x - 4)}}}$$ Divide both the numerator and denominator by common factors.

$$= \frac{x + 2}{x - 3}, x \neq -3, x \neq 3, x \neq 4$$ Multiply the remaining factors in the numerator and the denominator.

Check Point 4 Divide and simplify:

$$\frac{x^2 - 2x + 1}{x^3 + x} \div \frac{x^2 + x - 2}{3x^2 + 3}.$$

5 Add and subtract rational expressions.

Adding and Subtracting Rational Expressions with the Same Denominator

We add or subtract rational expressions with the same denominator by (1) adding or subtracting the numerators, (2) placing this result over the common denominator, and (3) simplifying, if possible.

EXAMPLE 5 Subtracting Rational Expressions with the Same Denominator

Subtract: $\dfrac{5x + 1}{x^2 - 9} - \dfrac{4x - 2}{x^2 - 9}.$

Study Tip

Example 5 shows that when a numerator is being subtracted, we must subtract every term in that expression.

Solution

$$\frac{5x + 1}{x^2 - 9} - \frac{4x - 2}{x^2 - 9} = \frac{5x + 1 - (4x - 2)}{x^2 - 9}$$ Subtract numerators and include parentheses to indicate that both terms are subtracted. Place this difference over the common denominator.

$$= \frac{5x + 1 - 4x + 2}{x^2 - 9}$$ Remove parentheses and then change the sign of each term.

$$= \frac{x + 3}{x^2 - 9}$$ Combine like terms.

$$= \frac{\overset{1}{\cancel{x + 3}}}{\cancel{(x + 3)}(x - 3)}$$ Factor and simplify ($x \neq -3$ and $x \neq 3$).

$$= \frac{1}{x - 3}, x \neq -3, x \neq 3$$

Check Point 5 Subtract: $\dfrac{x}{x + 1} - \dfrac{3x + 2}{x + 1}.$

Adding and Subtracting Rational Expressions with Different Denominators

Rational expressions that have no common factors in their denominators can be added or subtracted using one of the following properties:

$$\frac{a}{b} + \frac{c}{d} = \frac{ad + bc}{bd} \qquad \frac{a}{b} - \frac{c}{d} = \frac{ad - bc}{bd}, b \neq 0, d \neq 0.$$

The least common denominator, bd, is the product of the distinct factors in the two denominators.

EXAMPLE 6   Subtracting Rational Expressions Having No Common Factors in Their Denominators

Subtract: $\dfrac{x + 2}{2x - 3} - \dfrac{4}{x + 3}$.

Solution   We need to find the least common denominator. This is the product of the distinct factors in each denominator, namely $(2x - 3)(x + 3)$. We can therefore use the subtraction property given above as follows:

$$\frac{a}{b} - \frac{c}{d} = \frac{ad - bc}{bd}$$

$$\frac{x + 2}{2x - 3} - \frac{4}{x + 3} = \frac{(x + 2)(x + 3) - (2x - 3)4}{(2x - 3)(x + 3)} \qquad \begin{array}{l}\textit{Observe that}\\ a = x + 2, b = 2x - 3,\\ c = 4, \text{ and } d = x + 3.\end{array}$$

$$= \frac{x^2 + 5x + 6 - (8x - 12)}{(2x - 3)(x + 3)} \qquad \textit{Multiply.}$$

$$= \frac{x^2 + 5x + 6 - 8x + 12}{(2x - 3)(x + 3)} \qquad \begin{array}{l}\textit{Remove parentheses and}\\ \textit{then change the sign of}\\ \textit{each term.}\end{array}$$

$$= \frac{x^2 - 3x + 18}{(2x - 3)(x + 3)}, x \neq \frac{3}{2}, x \neq -3 \qquad \begin{array}{l}\textit{Combine like terms in the}\\ \textit{numerator.}\end{array}$$

Check Point 6   Add: $\dfrac{3}{x + 1} + \dfrac{5}{x - 1}$.

When adding and subtracting rational expressions that have different denominators with one or more common factors in the denominators, it is efficient to find the least common denominator first.

Finding the Least Common Denominator

1. Factor each denominator completely.
2. List the factors of the first denominator.
3. Add to the list in step 2 any factors of the second denominator that do not appear in the list.
4. Form the product of each different factor from the list in step 3. This product is the least common denominator.

EXAMPLE 7 Finding the Least Common Denominator

Find the least common denominator that is needed to add or subtract the rational expressions

$$\frac{7}{5x^2 + 15x} \quad \text{and} \quad \frac{9}{x^2 + 6x + 9}.$$

Solution

Step 1 Factor each denominator completely.

$$5x^2 + 15x = 5x(x + 3)$$

$$x^2 + 6x + 9 = (x + 3)^2$$

Step 2 List the factors of the first denominator.

$$5, x, (x + 3)$$

Step 3 Add any unlisted factors from the second denominator. The second denominator is $(x + 3)^2$ or $(x + 3)(x + 3)$. One factor of $x + 3$ is already in our list, but the other factor is not. We add $x + 3$ to the list. We have

$$5, x, (x + 3), (x + 3).$$

Step 4 The least common denominator is the product of all factors in the final list. Thus,

$$5x(x + 3)(x + 3)$$

is the least common denominator.

> **Check Point 7** What is the least common denominator for denominators of $x^2 - 6x + 9$ and $x^2 - 9$?

Finding the least common denominator for two (or more) rational expressions is the first step needed to add or subtract the expressions.

Adding and Subtracting Rational Expressions That Have Different Denominators With Shared Factors

1. Find the least common denominator.

2. Write all rational expressions in terms of the least common denominator. To do so, multiply both the numerator and the denominator of each rational expression by any factor(s) needed to convert the denominator into the least common denominator.

3. Add or subtract the numerators, placing the resulting expression over the least common denominator.

4. If necessary, simplify the resulting rational expression.

EXAMPLE 8   Adding Rational Expressions with Different Denominators

Add: $\dfrac{x + 3}{x^2 + x - 2} + \dfrac{2}{x^2 - 1}$.

Solution

Step 1   Find the least common denominator.   Start by factoring the denominators.

$$x^2 + x - 2 = (x + 2)(x - 1)$$
$$x^2 - 1 = (x + 1)(x - 1)$$

The factors of the first denominator are $x + 2$ and $x - 1$. The only factor from the second denominator that is unlisted is $x + 1$. Thus, the least common denominator is

$$(x + 2)(x - 1)(x + 1).$$

Step 2   Write all rational expressions in terms of the least common denominator. We do so by multiplying both the numerator and the denominator by any factor(s) needed to convert the denominator into the least common denominator.

$\dfrac{x + 3}{x^2 + x - 2} + \dfrac{2}{x^2 - 1}$

$= \dfrac{x + 3}{(x + 2)(x - 1)} + \dfrac{2}{(x + 1)(x - 1)}$

The least common denominator is $(x + 2)(x - 1)(x + 1)$.

$= \dfrac{(x + 3)(x + 1)}{(x + 2)(x - 1)(x + 1)} + \dfrac{2(x + 2)}{(x + 2)(x - 1)(x + 1)}$

Rewrite each rational expression with the least common denominator. Multiply the numerator and the denominator by whatever extra factors are required to form $(x + 2)(x - 1)(x + 1)$.

Step 3   Add numerators, putting this sum over the least common denominator.

$= \dfrac{(x + 3)(x + 1) + 2(x + 2)}{(x + 2)(x - 1)(x + 1)}$

$= \dfrac{x^2 + 4x + 3 + 2x + 4}{(x + 2)(x - 1)(x + 1)}$

Multiply in the numerator.

$= \dfrac{x^2 + 6x + 7}{(x + 2)(x - 1)(x + 1)}, \; x \ne -2, x \ne 1, x \ne -1$

Combine like terms in the numerator.

Step 4   If necessary, simplify.   Because the numerator is prime, no further simplification is possible.

Check Point 8   Subtract: $\dfrac{x}{x^2 - 10x + 25} - \dfrac{x - 4}{2x - 10}$.

6 Simplify complex rational expressions.

Complex Rational Expressions

Complex rational expressions have numerators or denominators containing one or more rational expressions. Here are two examples of such expressions:

$$\frac{1 + \dfrac{1}{x}}{1 - \dfrac{1}{x}}$$

Separate rational expressions occur in the numerator and denominator.

$$\frac{\dfrac{1}{x + h} - \dfrac{1}{x}}{h}$$

Separate rational expressions occur in the numerator.

One method for simplifying a complex rational expression is to combine its numerator into a single expression and combine its denominator into a single expression. Then perform the division by inverting the denominator and multiplying.

EXAMPLE 9   Simplifying a Complex Rational Expression

Simplify: $\dfrac{1 + \dfrac{1}{x}}{1 - \dfrac{1}{x}}$.

Solution

$$\frac{1 + \dfrac{1}{x}}{1 - \dfrac{1}{x}} = \frac{\dfrac{x}{x} + \dfrac{1}{x}}{\dfrac{x}{x} - \dfrac{1}{x}}, x \neq 0$$

The terms in the numerator and in the denominator are each combined by performing the addition and subtraction. The least common denominator is x.

$$= \frac{\dfrac{x + 1}{x}}{\dfrac{x - 1}{x}}$$

Perform the addition in the numerator and the subtraction in the denominator.

$$= \frac{x + 1}{x} \div \frac{x - 1}{x}$$

Rewrite the main fraction bar as ÷.

$$= \frac{x + 1}{x} \cdot \frac{x}{x - 1}$$

Invert the divisor and multiply ($x \neq 0$ and $x \neq 1$).

$$= \frac{x + 1}{\overset{1}{\cancel{x}}} \cdot \frac{\overset{1}{\cancel{x}}}{x - 1}$$

Divide both the numerator and denominator by the common factor, x.

$$= \frac{x + 1}{x - 1}, x \neq 0, x \neq 1$$

Multiply the remaining factors in the numerator and in the denominator.

Check Point 9   Simplify: $\dfrac{\dfrac{1}{x} - \dfrac{3}{2}}{\dfrac{1}{x} + \dfrac{3}{4}}$.

A second method for simplifying a complex rational expression is to find the least common denominator of all the rational expressions in its numerator and denominator. Then multiply each term in its numerator and denominator by this least common denominator. Here we use this method to simplify the complex rational expression in Example 9.

$$\frac{1 + \dfrac{1}{x}}{1 - \dfrac{1}{x}} = \frac{\left(1 + \dfrac{1}{x}\right)}{\left(1 - \dfrac{1}{x}\right)} \cdot \frac{x}{x}$$

The least common denominator of all the rational expressions is x. Multiply the numerator and denominator by x. Because $\dfrac{x}{x} = 1$, we are not changing the complex fraction $(x \neq 0)$.

$$= \frac{1 \cdot x + \dfrac{1}{x} \cdot x}{1 \cdot x - \dfrac{1}{x} \cdot x}$$

Use the distributive property. Be sure to distribute x to every term.

$$= \frac{x + 1}{x - 1}, x \neq 0, x \neq 1$$

Multiply. The complex rational expression is now simplified.

EXERCISE SET P.6

Practice Exercises

In Exercises 1–6, find all numbers that must be excluded from the domain of each rational expression.

1. $\dfrac{7}{x - 3}$

2. $\dfrac{13}{x + 9}$

3. $\dfrac{x + 5}{x^2 - 25}$

4. $\dfrac{x + 7}{x^2 - 49}$

5. $\dfrac{x - 1}{x^2 + 11x + 10}$

6. $\dfrac{x - 3}{x^2 + 4x - 45}$

In Exercises 7–14, simplify each rational expression. Find all numbers that must be excluded from the domain of the simplified rational expression.

7. $\dfrac{3x - 9}{x^2 - 6x + 9}$

8. $\dfrac{4x - 8}{x^2 - 4x + 4}$

9. $\dfrac{x^2 - 12x + 36}{4x - 24}$

10. $\dfrac{x^2 - 8x + 16}{3x - 12}$

11. $\dfrac{y^2 + 7y - 18}{y^2 - 3y + 2}$

12. $\dfrac{y^2 - 4y - 5}{y^2 + 5y + 4}$

13. $\dfrac{x^2 + 12x + 36}{x^2 - 36}$

14. $\dfrac{x^2 - 14x + 49}{x^2 - 49}$

In Exercises 15–30, multiply or divide as indicated.

15. $\dfrac{x - 2}{3x + 9} \cdot \dfrac{2x + 6}{2x - 4}$

16. $\dfrac{6x + 9}{3x - 15} \cdot \dfrac{x - 5}{4x + 6}$

17. $\dfrac{x^2 - 9}{x^2} \cdot \dfrac{x^2 - 3x}{x^2 + x - 12}$

18. $\dfrac{x^2 - 4}{x^2 - 4x + 4} \cdot \dfrac{2x - 4}{x + 2}$

19. $\dfrac{x^2 - 5x + 6}{x^2 - 2x - 3} \cdot \dfrac{x^2 - 1}{x^2 - 4}$

20. $\dfrac{x^2 + 5x + 6}{x^2 + x - 6} \cdot \dfrac{x^2 - 9}{x^2 - x - 6}$

21. $\dfrac{x^3 - 8}{x^2 - 4} \cdot \dfrac{x + 2}{3x}$

22. $\dfrac{x^2 + 6x + 9}{x^3 + 27} \cdot \dfrac{1}{x + 3}$

23. $\dfrac{x + 1}{3} \div \dfrac{3x + 3}{7}$

24. $\dfrac{x + 5}{7} \div \dfrac{4x + 20}{9}$

25. $\dfrac{x^2 - 4}{x} \div \dfrac{x + 2}{x - 2}$

26. $\dfrac{x^2 - 4}{x - 2} \div \dfrac{x + 2}{4x - 8}$

27. $\dfrac{4x^2 + 10}{x - 3} \div \dfrac{6x^2 + 15}{x^2 - 9}$

28. $\dfrac{x^2 + x}{x^2 - 4} \div \dfrac{x^2 - 1}{x^2 + 5x + 6}$

29. $\dfrac{x^2 - 25}{2x - 2} \div \dfrac{x^2 + 10x + 25}{x^2 + 4x - 5}$

30. $\dfrac{x^2 - 4}{x^2 + 3x - 10} \div \dfrac{x^2 + 5x + 6}{x^2 + 8x + 15}$

In Exercises 31–50, add or subtract as indicated.

31. $\dfrac{4x + 1}{6x + 5} + \dfrac{8x + 9}{6x + 5}$

32. $\dfrac{3x + 2}{3x + 4} + \dfrac{3x + 6}{3x + 4}$

33. $\dfrac{x^2 - 2x}{x^2 + 3x} + \dfrac{x^2 + x}{x^2 + 3x}$

34. $\dfrac{x^2 - 4x}{x^2 - x - 6} + \dfrac{4x - 4}{x^2 - x - 6}$

35. $\dfrac{4x - 10}{x - 2} - \dfrac{x - 4}{x - 2}$

36. $\dfrac{2x + 3}{3x - 6} - \dfrac{3 - x}{3x - 6}$

37. $\dfrac{x^2 + 3x}{x^2 + x - 12} - \dfrac{x^2 - 12}{x^2 + x - 12}$

38. $\dfrac{x^2 - 4x}{x^2 - x - 6} - \dfrac{x - 6}{x^2 - x - 6}$

39. $\dfrac{3}{x + 4} + \dfrac{6}{x + 5}$

40. $\dfrac{8}{x - 2} + \dfrac{2}{x - 3}$

41. $\dfrac{3}{x + 1} - \dfrac{3}{x}$

42. $\dfrac{4}{x} - \dfrac{3}{x + 3}$

43. $\dfrac{2x}{x+2} + \dfrac{x+2}{x-2}$

44. $\dfrac{3x}{x-3} - \dfrac{x+4}{x+2}$

45. $\dfrac{x+5}{x-5} + \dfrac{x-5}{x+5}$

46. $\dfrac{x+3}{x-3} + \dfrac{x-3}{x+3}$

47. $\dfrac{4}{x^2+6x+9} + \dfrac{4}{x+3}$

48. $\dfrac{3}{5x+2} + \dfrac{5x}{25x^2-4}$

49. $\dfrac{3x}{x^2+3x-10} - \dfrac{2x}{x^2+x-6}$

50. $\dfrac{x}{x^2-2x-24} - \dfrac{x}{x^2-7x+6}$

In Exercise 51–60, simplify each complex rational expression.

51. $\dfrac{\dfrac{x}{3}-1}{x-3}$

52. $\dfrac{\dfrac{x}{4}-1}{x-4}$

53. $\dfrac{1+\dfrac{1}{x}}{3-\dfrac{1}{x}}$

54. $\dfrac{8+\dfrac{1}{x}}{4-\dfrac{1}{x}}$

55. $\dfrac{\dfrac{1}{x}+\dfrac{1}{y}}{x+y}$

56. $\dfrac{1-\dfrac{1}{x}}{xy}$

57. $\dfrac{x-\dfrac{x}{x+3}}{x+2}$

58. $\dfrac{x-3}{x-\dfrac{3}{x-2}}$

59. $\dfrac{\dfrac{3}{x-2}-\dfrac{4}{x+2}}{\dfrac{7}{x^2-4}}$

60. $\dfrac{\dfrac{x}{x-2}+1}{\dfrac{3}{x^2-4}+1}$

 Application Exercises

61. The polynomial $-0.14t^2 + 0.51t + 31.6$ describes the U.S. population (in millions) age 65 and older t years after 1990. The polynomial $0.54t^2 + 12.64t + 107.1$ describes the total yearly cost of Medicare (in billions of dollars) t years after 1990. Write a rational expression that describes the average cost of Medicare per person age 65 or older t years after 1990.

62. The polynomial

$$6t^4 - 207t^3 + 2128t^2 - 6622t + 15,220$$

describes the annual number of drug convictions in the United States t years after 1984. The polynomial

$$28t^4 - 711t^3 + 5963t^2 - 1695t + 27,424$$

describes the annual number of drug arrests in the United States t years after 1984. Write a rational expression that describes the conviction rate for drug arrests in the United States t years after 1984.

63. The rational expression

$$\dfrac{130x}{100-x}$$

describes the cost, in millions of dollars, to inoculate x percent of the population against a particular strain of flu.

a. Evaluate the expression for $x = 40$, $x = 80$, and $x = 90$. Describe the meaning of each evaluation in terms of percentage inoculated and cost.

b. For what value of x is the expression undefined?

c. What happens to the cost as x approaches 100%? How can you interpret this observation?

64. Doctors use the rational expression

$$\dfrac{DA}{A+12}$$

to determine the dosage of a drug prescribed for children. In this expression, A = child's age, and D = adult dosage. What is the difference in the child's dosage for a 7-year-old child and a 3-year-old child? Express the answer as a single rational expression in terms of D. Then describe what your answer means in terms of the variables in the rational expression.

65. The average speed on a round-trip commute having a one-way distance d is given by the complex rational expression

$$\dfrac{2d}{\dfrac{d}{r_1}+\dfrac{d}{r_2}}$$

in which r_1 and r_2 are the speeds on the outgoing and return trips, respectively. Simplify the expression. Then find the average speed for a person who drives from home to work at 30 miles per hour and returns on the same route averaging 20 miles per hour. Explain why the answer is not 25 miles per hour.

 **Writing in Mathematics**

66. What is a rational expression?

67. Explain how to determine what numbers must be excluded from the domain of a rational expression.

68. Explain how to simplify a rational expression.

69. Explain how to multiply rational expressions.

70. Explain how to divide rational expressions.

71. Explain how to add or subtract rational expressions with the same denominators.

72. Explain how to add rational expressions having no common factors in their denominators. Use $\dfrac{3}{x+5} + \dfrac{7}{x+2}$ in your explanation.

73. Explain how to find the least common denominator for denominators of $x^2 - 100$ and $x^2 - 20x + 100$.

74. Describe two ways to simplify $\dfrac{\dfrac{3}{x} + \dfrac{2}{x^2}}{\dfrac{1}{x^2} + \dfrac{2}{x}}$.

Explain the error in Exercises 75–77. Then rewrite the right side of the equation to correct the error that now exists.

75. $\dfrac{1}{a} + \dfrac{1}{b} = \dfrac{1}{a + b}$

76. $\dfrac{1}{x} + 7 = \dfrac{1}{x + 7}$

77. $\dfrac{a}{x} + \dfrac{a}{b} = \dfrac{a}{x + b}$

78. A politician claims that each year the conviction rate for drug arrests in the United States is increasing. Explain how to use the polynomials in Exercise 62 to verify this claim.

 Technology Exercise

79. The polynomial

$$413.48t^2 + 185.72t + 24{,}031.95$$

describes the amount of Medicaid payments, in millions of dollars, t years after 1980. The polynomial

$$0.004t^2 + 0.02t^3 + 0.01t^2 - 0.24t + 21.66$$

describes the annual number of Medicaid recipients in the United States t years after 1980. Use a calculator to find the amount paid per recipient of Medicaid each year from 1988 to 2000. In what year did the amount paid per recipient fall below $900?

 **Critical Thinking Exercises**

80. Which one of the following is true?

a. $\dfrac{x^2 - 25}{x - 5} = x - 5$

b. $\dfrac{x}{y} \div \dfrac{y}{x} = 1$, if $x \ne 0$ and $y \ne 0$.

c. The least common denominator needed to find $\dfrac{1}{x} + \dfrac{1}{x + 3}$ is $x + 3$.

d. The rational expression

$$\frac{x^2 - 16}{x - 4}$$

is not defined for $x = 4$. However, as x gets closer and closer to 4, the value of the expression approaches 8.

In Exercises 81–82, find the missing expression.

81. $\dfrac{3x}{x - 5} + \dfrac{\boxed{}}{5 - x} = \dfrac{7x + 1}{x - 5}$

82. $\dfrac{4}{x - 2} - \boxed{} = \dfrac{2x + 8}{(x - 2)(x + 1)}$

83. In one short sentence, five words or less, explain what

$$\frac{\dfrac{1}{x} + \dfrac{1}{x^2} + \dfrac{1}{x^3}}{\dfrac{1}{x^4} + \dfrac{1}{x^5} + \dfrac{1}{x^6}}$$

does to each number x.

SECTION P.7   *Complex Numbers*

Objectives

1. Add and subtract complex numbers.

2. Multiply complex numbers.

3. Divide complex numbers.

4. Perform operations with square roots of negative numbers.

Who is this kid warning us about our eyeballs turning black if we attempt to find the square root of −9? Don't believe what you hear on the street. Although square roots of negative numbers are not real numbers, they do play a significant role in algebra. In this section, we move beyond the real numbers and discuss square roots with negative radicands.

The Imaginary Unit i

In Chapter 1, we'll be studying equations whose solutions involve the square roots of negative numbers. Because the square of a real number is never negative, there is no real number x such that $x^2 = -1$. To provide a setting in which such equations have solutions, mathematicians invented an expanded system of numbers, the complex numbers. The imaginary number i, defined to be a solution to the equation $x^2 = -1$, is the basis of this new set.

The Imaginary Unit i

The imaginary unit i is defined as

$$i = \sqrt{-1}, \quad \text{where} \quad i^2 = -1.$$

Using the imaginary unit i, we can express the square root of any negative number as a real multiple of i. For example,

$$\sqrt{-25} = i\sqrt{25} = 5i.$$

We can check this result by squaring $5i$ and obtaining −25.

$$(5i)^2 = 5^2 i^2 = 25(-1) = -25$$

A new system of numbers, called **complex numbers**, is based on adding multiples of i, such as $5i$, to the real numbers.

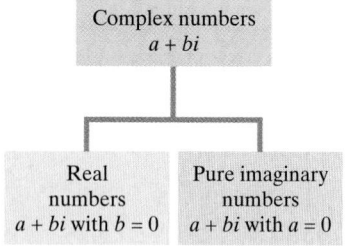

Figure P.7 The complex number system

Complex Numbers

The set of all numbers in the form

$$a + bi$$

with real numbers a and b, and i, the imaginary unit, is called the set of **complex numbers**. The real number a is called the **real part**, and the real number b is called the **imaginary part**, of the complex number $a + bi$. If $a = 0$ and $b \neq 0$, then the complex number bi is called a **pure imaginary number** (Figure P.7).

A complex number is said to be **simplified** if it is expressed in the **standard form** $a + bi$. If b is a radical, we usually write i before b. For example, we write $7 + i\sqrt{5}$ rather than $7 + \sqrt{5}i$, which could easily be confused with $7 + \sqrt{5i}$.

Expressed in standard form, two complex numbers are equal if and only if their real parts are equal and their imaginary parts are equal.

Equality of Complex Numbers

$a + bi = c + di$ if and only if $a = c$ and $b = d$.

1 Add and subtract complex numbers.

Operations with Complex Numbers

The form of a complex number $a + bi$ is like the binomial $a + bx$. Consequently, we can add, subtract, and multiply complex numbers using the same methods we used for binomials, remembering that $i^2 = -1$.

Adding and Subtracting Complex Numbers

1. $(a + bi) + (c + di) = (a + c) + (b + d)i$
   In words, this says that you add complex numbers by adding their real parts, adding their imaginary parts, and expressing the sum as a complex number.
2. $(a + bi) - (c + di) = (a - c) + (b - d)i$
   In words, this says that you subtract complex numbers by subtracting their real parts, subtracting their imaginary parts, and expressing the difference as a complex number.

EXAMPLE 1   Adding and Subtracting Complex Numbers

Perform the indicated operations, writing the result in standard form.

a. $(5 - 11i) + (7 + 4i)$ **b.** $(-5 + 7i) - (-11 - 6i)$

Solution

a. $(5 - 11i) + (7 + 4i)$
$= 5 - 11i + 7 + 4i$ Remove the parentheses.
$= 5 + 7 - 11i + 4i$ Group real and imaginary terms.
$= (5 + 7) + (-11 + 4)i$
$= 12 - 7i$ Add real parts and add imaginary parts.

b. $(-5 + 7i) - (-11 - 6i)$
$= -5 + 7i + 11 + 6i$ Remove the parentheses.
$= -5 + 11 + 7i + 6i$ Group real and imaginary terms.
$= (-5 + 11) + (7 + 6)i$
$= 6 + 13i$

> **Study Tip**
>
> The following examples, using the same integers as in Example 1, show how operations with complex numbers are just like operations with polynomials.
>
> **a.** $(5 - 11x) + (7 + 4x)$
> $= 12 - 7x$
> **b.** $(-5 + 7x) - (-11 - 6x)$
> $= -5 + 7x + 11 + 6x$
> $= 6 + 13x$

Check Point 1   Add or subtract as indicated.

a. $(5 - 2i) + (3 + 3i)$ **b.** $(2 + 6i) - (12 - 4i)$

2 Multiply complex numbers.

Multiplication of complex numbers is performed the same way as multiplication of polynomials, using the distributive property and the FOIL method. After completing the multiplication, we replace i^2 with -1. This idea is illustrated in the next example.

EXAMPLE 2   Multiplying Complex Numbers

Find the products: **a.** $4i(3 - 5i)$ **b.** $(7 - 3i)(-2 - 5i)$

Solution

a. $4i(3 - 5i) = 4i(3) - 4i(5i)$ Distribute 4i throughout the parentheses.

$= 12i - 20i^2$ Multiply.

$= 12i - 20(-1)$ Replace i^2 with −1.

$= 20 + 12i$ Simplify to 12i + 20 and write in standard form.

b. $(7 - 3i)(-2 - 5i)$

F   O   I   L

$= -14 - 35i + 6i + 15i^2$ Use the FOIL method.

$= -14 - 35i + 6i + 15(-1)$ $i^2 = -1$

$= -14 - 15 - 35i + 6i$ Group real and imaginary terms.

$= -29 - 29i$ Combine real and imaginary terms.

Check Point 2

Find the products:

a. $7i(2 - 9i)$ **b.** $(5 + 4i)(6 - 7i)$

---

3 Divide complex numbers.

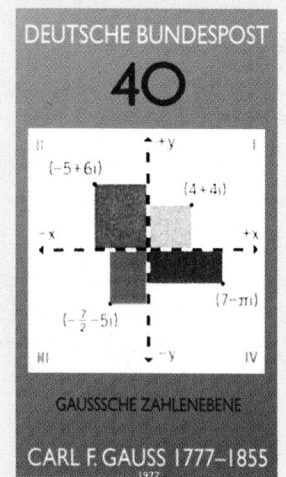

Complex Numbers on a Postage Stamp

This stamp honors the work done by the German mathematician Carl Friedrich Gauss (1777–1855) with complex numbers. Gauss represented complex numbers as points in the plane.

Complex Conjugates and Division

It is possible to multiply complex numbers and obtain a real number. This occurs when we multiply $a + bi$ and $a - bi$.

F   O   I   L

$(a + bi)(a - bi) = a^2 - abi + abi - b^2i^2$ Use the FOIL method.

$= a^2 - b^2(-1)$ $i^2 = -1$

$= a^2 + b^2$ Notice that this product eliminates i.

For the complex number $a + bi$, we define its **complex conjugate** to be $a - bi$. The multiplication of complex conjugates results in a real number.

> ### Conjugate of a Complex Number
> The **complex conjugate** of the number $a + bi$ is $a - bi$, and the complex conjugate of $a - bi$ is $a + bi$. The multiplication of complex conjugates gives a real number.
> $$(a + bi)(a - bi) = a^2 + b^2$$
> $$(a - bi)(a + bi) = a^2 + b^2$$

Complex conjugates are used to divide complex numbers. By multiplying the numerator and the denominator of the division by the complex conjugate of the denominator, you will obtain a real number in the denominator.

EXAMPLE 3 Using Complex Conjugates to Divide Complex Numbers

Divide: $7 + 4i$ by $2 - 5i$.

Solution We first write the problem as $\dfrac{7 + 4i}{2 - 5i}$. The complex conjugate of the denominator, $2 - 5i$, is $2 + 5i$, so we multiply the numerator and the denominator by $2 + 5i$.

$$\frac{7 + 4i}{2 - 5i} = \frac{(7 + 4i)}{(2 - 5i)} \cdot \frac{(2 + 5i)}{(2 + 5i)}$$ Multiply the numerator and the denominator by the complex conjugate of the denominator.

F   O   I   L

$$= \frac{14 + 35i + 8i + 20i^2}{2^2 + 5^2}$$ Use the FOIL method in the numerator and $(a - bi)(a + bi) = a^2 + b^2$ in the denominator.

$$= \frac{14 + 43i + 20(-1)}{29}$$ Replace i^2 by -1.

$$= \frac{-6 + 43i}{29}$$ Combine real terms in the numerator.

$$= -\frac{6}{29} + \frac{43}{29} i$$

Observe that the quotient is expressed in the standard form $a + bi$, with $a = -\frac{6}{29}$ and $b = \frac{43}{29}$.

Check Point 3   Divide: $\dfrac{5 + 4i}{4 - 2i}$.

4 Perform operations with square roots of negative numbers.

Roots of Negative Numbers

The square of $4i$ and the square of $-4i$ both result in -16.

$$(4i)^2 = 16i^2 = 16(-1) = -16 \qquad (-4i)^2 = 16i^2 = -16$$

Consequently, in the complex number system -16 has two square roots, namely, $4i$ and $-4i$. We call $4i$ the **principal square root** of -16.

Principal Square Root of a Negative Number

For any positive number real number b, the **principal square root** of the negative number $-b$ is defined by

$$\sqrt{-b} = i\sqrt{b}.$$

EXAMPLE 4 Operations Involving Square Roots of Negative Numbers

Perform the indicated operations and write the result in standard form:

a. $\sqrt{-18} - \sqrt{-8}$ **b.** $\left(-1 + \sqrt{-5}\right)^2$ **c.** $\dfrac{-25 + \sqrt{-50}}{15}$

Study Tip

Do not apply the properties

$$\sqrt{b}\,\sqrt{c} = \sqrt{bc}$$

and

$$\frac{\sqrt{b}}{\sqrt{c}} = \sqrt{\frac{b}{c}}$$

to the pure imaginary numbers because these properties can only be used when b and c are positive.

Correct:

$$\sqrt{-25}\,\sqrt{-4} = i\sqrt{25}\,i\sqrt{4}$$
$$= (5i)(2i)$$
$$= 10i^2$$
$$= 10(-1)$$
$$= -10$$

Incorrect:

$$\cancel{\sqrt{-25}\,\sqrt{-4} = \sqrt{(-25)(-4)}}$$
$$\cancel{= \sqrt{100}}$$
$$\cancel{= 10}$$

One way to avoid confusion is to represent square roots of negative numbers in terms of i before performing any operations.

Solution Begin by expressing all square roots of negative numbers in terms of i.

a. $\sqrt{-18} - \sqrt{-8} = i\sqrt{18} - i\sqrt{8} = i\sqrt{9\cdot 2} - i\sqrt{4\cdot 2}$
$$= 3i\sqrt{2} - 2i\sqrt{2} = i\sqrt{2}$$

$$(A+B)^2 = A^2 + 2\,AB + B^2$$

b. $\left(-1 + \sqrt{-5}\right)^2 = \left(-1 + i\sqrt{5}\right)^2 = (-1)^2 + 2(-1)(i\sqrt{5}) + (i\sqrt{5})^2$
$$= 1 - 2i\sqrt{5} + 5i^2$$
$$= 1 - 2i\sqrt{5} + 5(-1)$$
$$= -4 - 2i\sqrt{5}$$

c. $\dfrac{-25 + \sqrt{-50}}{15}$

$$= \frac{-25 + i\sqrt{50}}{15} \qquad \sqrt{-b} = i\sqrt{b}$$

$$= \frac{-25 + 5i\sqrt{2}}{15} \qquad \sqrt{50} = \sqrt{25\cdot 2} = 5\sqrt{2}$$

$$= \frac{-25}{15} + \frac{5i\sqrt{2}}{15} \qquad \text{Write the complex number in standard form.}$$

$$= -\frac{5}{3} + i\frac{\sqrt{2}}{3} \qquad \text{Simplify.}$$

Check Point 4 Perform the indicated operations and write the result in standard form.

a. $\sqrt{-27} + \sqrt{-48}$ **b.** $\left(-2 + \sqrt{-3}\right)^2$ **c.** $\dfrac{-14 + \sqrt{-12}}{2}$

EXERCISE SET P.7

Practice Exercises

In Exercises 1–8, add or subtract as indicated and write the result in standard form.

1. $(7 + 2i) + (1 - 4i)$ **2.** $(-2 + 6i) + (4 - i)$
3. $(3 + 2i) - (5 - 7i)$ **4.** $(-7 + 5i) - (-9 - 11i)$
5. $6 - (-5 + 4i) - (-13 - 11i)$
6. $7 - (-9 + 2i) - (-17 - 6i)$
7. $8i - (14 - 9i)$ **8.** $15i - (12 - 11i)$

In Exercises 9–20, find each product and write the result in standard form.

9. $-3i(7i - 5)$ **10.** $-8i(2i - 7)$
11. $(-5 + 4i)(3 + 7i)$ **12.** $(-4 - 8i)(3 + 9i)$
13. $(7 - 5i)(-2 - 3i)$ **14.** $(8 - 4i)(-3 + 9i)$
15. $(3 + 5i)(3 - 5i)$ **16.** $(2 + 7i)(2 - 7i)$
17. $(-5 + 3i)(-5 - 3i)$ **18.** $(-7 - 4i)(-7 + 4i)$
19. $(2 + 3i)^2$ **20.** $(5 - 2i)^2$

In Exercises 21–28, divide and express the result in standard form.

21. $\dfrac{2}{3 - i}$ **22.** $\dfrac{3}{4 + i}$
23. $\dfrac{2i}{1 + i}$ **24.** $\dfrac{5i}{2 - i}$
25. $\dfrac{8i}{4 - 3i}$ **26.** $\dfrac{-6i}{3 + 2i}$
27. $\dfrac{2 + 3i}{2 + i}$ **28.** $\dfrac{3 - 4i}{4 + 3i}$

In Exercises 29–44, perform the indicated operations and write the result in standard form.

29. $\sqrt{-64} - \sqrt{-25}$ **30.** $\sqrt{-81} - \sqrt{-144}$
31. $5\sqrt{-16} + 3\sqrt{-81}$ **32.** $5\sqrt{-8} + 3\sqrt{-18}$
33. $\left(-2 + \sqrt{-4}\right)^2$ **34.** $\left(-5 - \sqrt{-9}\right)^2$
35. $\left(-3 - \sqrt{-7}\right)^2$ **36.** $\left(-2 + \sqrt{-11}\right)^2$

37. $\dfrac{-8 + \sqrt{-32}}{24}$

38. $\dfrac{-12 + \sqrt{-28}}{32}$

39. $\dfrac{-6 - \sqrt{-12}}{48}$

40. $\dfrac{-15 - \sqrt{-18}}{33}$

41. $\sqrt{-8}\left(\sqrt{-3} - \sqrt{5}\right)$

42. $\sqrt{-12}\left(\sqrt{-4} - \sqrt{2}\right)$

43. $\left(3\sqrt{-5}\right)\left(-4\sqrt{-12}\right)$

44. $\left(3\sqrt{-7}\right)\left(2\sqrt{-8}\right)$

Writing in Mathematics

45. What is i?

46. Explain how to add complex numbers. Provide an example with your explanation.

47. Explain how to multiply complex numbers and give an example.

48. What is the complex conjugate of $2 + 3i$? What happens when you multiply this complex number by its complex conjugate?

49. Explain how to divide complex numbers. Provide an example with your explanation.

50. A stand-up comedian uses algebra in some jokes, including one about a telephone recording that announces "You have just reached an imaginary number. Please multiply by i and dial again." Explain the joke.

Explain the error in Exercises 51–52.

51. $\sqrt{-9} + \sqrt{-16} = \sqrt{-25} = i\sqrt{25} = 5i$

52. $\left(\sqrt{-9}\right)^2 = \sqrt{-9} \cdot \sqrt{-9} = \sqrt{81} = 9$

 **Critical Thinking Exercises**

53. Which one of the following is true?

   a. Some irrational numbers are not complex numbers.

   b. $(3 + 7i)(3 - 7i)$ is an imaginary number.

   c. $\dfrac{7 + 3i}{5 + 3i} = \dfrac{7}{5}$

   d. In the complex number system, $x^2 + y^2$ (the sum of two squares) can be factored as $(x + yi)(x - yi)$.

In Exercises 54–56, perform the indicated operations and write the result in standard form.

54. $(8 + 9i)(2 - i) - (1 - i)(1 + i)$

55. $\dfrac{4}{(2 + i)(3 - i)}$

56. $\dfrac{1 + i}{1 + 2i} + \dfrac{1 - i}{1 - 2i}$

57. Evaluate $x^2 - 2x + 2$ for $x = 1 + i$.

SECTION P.8 *Graphs and Graphing Utilities*

Objectives

1. Plot points in the rectangular coordinate system.
2. Graph equations in the rectangular coordinate system.
3. Interpret information about a graphing utility's viewing rectangle.
4. Use a graph to determine intercepts.
5. Find the distance between two points.
6. Find the midpoint of a line segment.
7. Interpret information given by graphs.

The beginning of the seventeenth century was a time of innovative ideas and enormous intellectual progress in Europe. English theatergoers enjoyed a succession of exciting new plays by Shakespeare. William Harvey proposed the radical notion that the heart was a pump for blood rather than the center of emotion. Galileo, with his new-fangled invention called the telescope, supported the theory of Polish astronomer Copernicus that the sun, not the Earth, was the center of the solar system. Monteverdi was writing the world's first grand operas. French mathematicians Pascal and Fermat invented a new field of mathematics called probability theory.

Into this arena of intellectual electricity stepped French aristocrat René Descartes (1596–1650). Descartes, propelled by the creativity surrounding him, developed a new branch of mathematics that brought together algebra and geometry in a unified way—a way that visualized numbers as points on a graph, equations as geometric figures, and geometric figures as equations. This new branch of mathematics, called *analytic geometry*, established Descartes as one of the founders of modern thought and among the most original mathematicians and philosophers of any age. We begin this section by looking at Descartes's deceptively simple idea, called the **rectangular coordinate system** or (in his honor) the **Cartesian coordinate system**.

1 Plot points in the rectangular coordinate system.

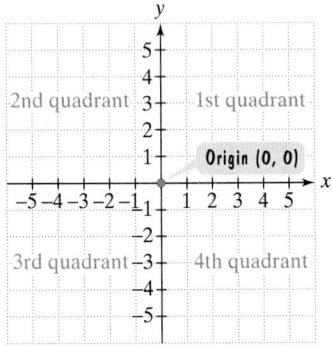

Figure P.8 The rectangular coordinate system

Points and Ordered Pairs

Descartes used two number lines that intersect at right angles at their zero points, as shown in Figure P.8. The horizontal number line is the **x-axis**. The vertical number line is the **y-axis**. The point of intersection of these axes is the **origin**. Positive numbers are shown to the right and above the origin. Negative numbers are shown to the left and below the origin. The axes divide the plane into four quarters, called **quadrants**. The points located on the axes are not in any quadrant.

Each point in the rectangular coordinate system corresponds to an **ordered pair** of real numbers, (x, y). Examples of such pairs are $(4, 2)$ and $(-5, -3)$. The first number in each pair, called the **x-coordinate**, denotes the distance and direction from the origin along the x-axis. The second number, called the **y-coordinate**, denotes vertical distance and direction along a line parallel to the y-axis or along the y-axis itself.

Figure P.9 shows how we **plot**, or locate, the points corresponding to the ordered pairs $(4, 2)$ and $(-5, -3)$. We plot $(4, 2)$ by going 4 units from 0 to the right along the x-axis. Then we go 2 units up parallel to the y-axis. We plot $(-5, -3)$ by going 5 units from 0 to the left along the x-axis and 3 units down parallel to the y-axis. The phrase "the point corresponding to the ordered pair $(-5, -3)$" is often abbreviated as "the point $(-5, -3)$."

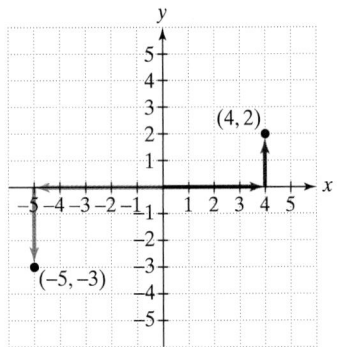

Figure P.9 Plotting $(4, 2)$ and $(-5, -3)$

EXAMPLE 1 Plotting Points in the Rectangular Coordinate System

Plot the points $A(-3, 5)$, $B(2, -4)$, $C(5, 0)$, $D(-5, -3)$, $E(0, 4)$, and $F(0, 0)$.

Solution See Figure P.10. We plot the points in the following way:

$A(-3, 5)$:	3 units left, 5 units up
$B(2, -4)$:	2 units right, 4 units down
$C(5, 0)$:	5 units right, 0 units up or down
$D(-5, -3)$:	5 units left, 3 units down
$E(0, 4)$:	0 units right or left, 4 units up
$F(0, 0)$:	0 units right or left, 0 units up or down

The origin is represented by $(0, 0)$.

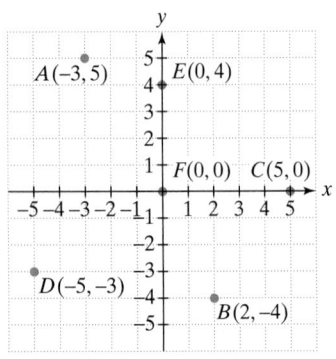

Figure P.10 Plotting points

Check Point 1   Plot the points

$A(-2, 4)$, $B(4, -2)$, $C(-3, 0)$, and $D(0, -3)$.

2 Graph equations in the rectangular coordinate system.

Graphs of Equations

A relationship between two quantities can be expressed as an **equation in two variables**, such as

$$y = x^2 - 4.$$

A **solution** to this equation is an ordered pair of real numbers with the following property: When the x-coordinate is substituted for x and the y-coordinate is substituted for y in the equation, we obtain a true statement. For example, if we let $x = 3$, then $y = 3^2 - 4 = 9 - 4 = 5$. The ordered pair $(3, 5)$ is a solution to the equation $y = x^2 - 4$. We also say that $(3, 5)$ **satisfies** the equation.

We can generate as many ordered-pair solutions as desired to $y = x^2 - 4$ by substituting numbers for x and then finding the values for y. The **graph of the equation** is the set of all points whose coordinates satisfy the equation.

One method for graphing an equation such as $y = x^2 - 4$ is the **point-plotting method**. First, we find several ordered pairs that are solutions to the equation. Next, we plot these ordered pairs as points in the rectangular coordinate system. Finally, we connect the points with a smooth curve or line. This often gives us a picture of all ordered pairs that satisfy the equation.

EXAMPLE 2 Graphing an Equation Using the Point-Plotting Method

Graph $y = x^2 - 4$. Select integers for x, starting with -3 and ending with 3.

Solution For each value of x we find the corresponding value for y.

x	$y = x^2 - 4$	Ordered Pair (x, y)
-3	$y = (-3)^2 - 4 = 9 - 4 = 5$	$(-3, 5)$
-2	$y = (-2)^2 - 4 = 4 - 4 = 0$	$(-2, 0)$
-1	$y = (-1)^2 - 4 = 1 - 4 = -3$	$(-1, -3)$
0	$y = 0^2 - 4 = 0 - 4 = -4$	$(0, -4)$
1	$y = 1^2 - 4 = 1 - 4 = -3$	$(1, -3)$
2	$y = 2^2 - 4 = 4 - 4 = 0$	$(2, 0)$
3	$y = 3^2 - 4 = 9 - 4 = 5$	$(3, 5)$

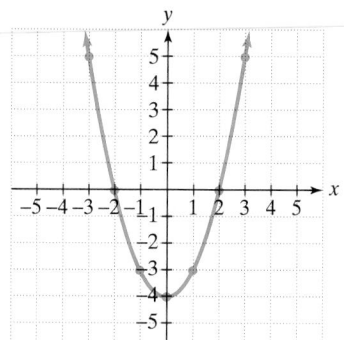

Figure P.11 The graph of $y = x^2 - 4$

Now we plot the seven points and join them with a smooth curve, as shown in Figure P.11. The graph of $y = x^2 - 4$ is a curve where the part of the graph to the right of the y-axis is a reflection of the part to the left of it and vice versa. The arrows on the left and the right of the curve indicate that it extends indefinitely in both directions.

Check Point 2 Graph $y = 2x - 4$. Select integers for x, starting with -1 and ending with 3.

3 Interpret information about a graphing utility's viewing rectangle.

Graphing calculators or graphing software packages for computers are referred to as **graphing utilities** or graphers. A graphing utility is a powerful tool that quickly generates the graph of an equation in two variables. Figure P.12 on page 74 shows two such graphs for the equations in Example 2 and the checkpoint example.

What differences do you notice between these graphs and the graphs that we (and you) drew by hand? They do seem a bit "jittery." Arrows do not appear

125.

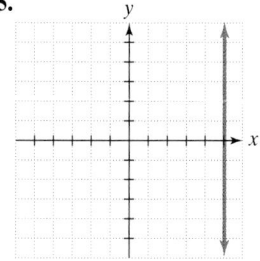

In Exercises 126–127, a. Plot the points. b. Find the distance between the points. c. Find the midpoint of the line segment joining the points.

126. $(-1, 0), (2, 4)$ **127.** $(2, -3), (4, 2)$

128. The line graphs show the number of applicants to U.S. law schools and medical schools.

    a. For the period shown, when did the number of law school applicants reach a maximum? What is a reasonable estimate for the number of applicants during that year?

b. Find an estimate for the number of medical school applicants in 1990.

c. Estimate the coordinates of point A. Then interpret the coordinates in terms of the information given by the graph.

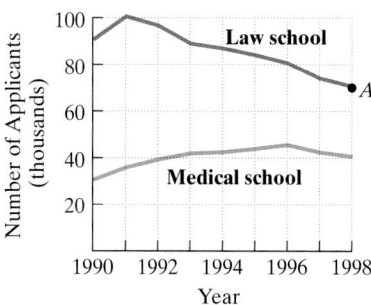

Number of Applicants to U.S. Law and Medical Schools

Source: U.S. Education Department

Chapter P Test

1. List all the rational numbers in this set.
$$\left\{ -7, -\tfrac{4}{5}, 0, 0.25, \sqrt{3}, \sqrt{4}, \tfrac{22}{7}, \pi \right\}$$

In Exercises 2–3, state the name of the property illustrated.

2. $3(2 + 5) = 3(5 + 2)$ **3.** $6(7 + 4) = 6 \cdot 7 + 6 \cdot 4$

4. Express in scientific notation: 0.00076.

Simplify each expression in Exercises 5–11.

5. $9(10x - 2y) - 5(x - 4y + 3)$

6. $\dfrac{30x^3 y^4}{6x^9 y^{-4}}$

7. $\sqrt{6r}\,\sqrt{3r}$

8. $4\sqrt{50} - 3\sqrt{18}$

9. $\dfrac{3}{5 + \sqrt{2}}$

10. $\sqrt[3]{16x^4}$

11. $\dfrac{x^2 + 2x - 3}{x^2 - 3x + 2}$

12. Evaluate: $27^{-5/3}$

In Exercises 13–14, find each product.

13. $(2x - 5)(x^2 - 4x + 3)$ **14.** $(5x + 3y)^2$

In Exercises 15–19, factor completely, or state that the polynomial is prime.

15. $x^2 - 9x + 18$ **16.** $x^3 + 2x^2 + 3x + 6$

17. $25x^2 - 9$ **18.** $36x^2 - 84x + 49$

19. $y^3 - 125$

In Exercises 20–23, perform the operations and simplify, if possible.

20. $\dfrac{2x + 8}{x - 3} \div \dfrac{x^2 + 5x + 4}{x^2 - 9}$ **21.** $\dfrac{x}{x + 3} + \dfrac{5}{x - 3}$

22. $\dfrac{2x + 3}{x^2 - 7x + 12} - \dfrac{2}{x - 3}$ **23.** $\dfrac{\dfrac{1}{x} - \dfrac{1}{3}}{\dfrac{1}{x}}$

In Exercises 24–26, perform the indicated operations and write the result in standard form.

24. $(6 - 7i)(2 + 5i)$ **25.** $\dfrac{5}{2 - i}$

26. $2\sqrt{-49} + 3\sqrt{-64}$

27. Graph $y = x^2 - 4$ by letting x equal integers from -3 through 3.

28. Find the distance between $(2, 9)$ and $(6, 3)$ in simplified radical form.

Equations, Inequalities, and Mathematical Models

Formulas like those that describe the height a child will attain as an adult are frequently obtained from actual data. Formulas can be used to explain what is happening in the present and to make predictions about what might occur in the future. Knowing how to create and use formulas will help you recognize patterns, logic, and order in a world that can appear chaotic to the untrained eye. In many ways, algebra will provide you with a new way of looking at your world.

Sitting in the biology department office, you overhear two of the professors discussing the possible adult heights of their respective children. Looking at the blackboard that they've been writing on, you see that there are formulas that can estimate the height a child will attain as an adult. If the child is x years old and h inches tall, that child's adult height, H, in inches, is approximated by one of the following formulas.

Girls: $H = \dfrac{h}{0.00028x^3 - 0.0071x^2 + 0.0926x + 0.3524}$

Boys: $H = \dfrac{h}{0.00011x^3 - 0.0032x^2 + 0.0604x + 0.3796}$

SECTION 1.1 *Linear Equations*

Objectives

1. Solve linear equations in one variable.
2. Solve equations with constants in denominators.
3. Solve equations with variables in denominators.
4. Recognize identities, conditional equations, and inconsistent equations.

Unfortunately, many of us have been fined for driving over the speed limit. The amount of the fine depends on how fast we are speeding. Suppose that a highway has a speed limit of 60 miles per hour. The amount that speeders are fined, F, is described by the statement of equality

$$F = 10x - 600,$$

where x is the speed in miles per hour. We can use this statement to determine the fine, F, for a speeder traveling at, say, 70 miles per hour. We substitute 70 for x in the given statement and then find the corresponding value for F.

$$F = 10(70) - 600 = 700 - 600 = 100$$

Thus, a person caught driving 70 miles per hour gets a $100 fine.

A friend, whom we shall call Leadfoot, borrows your car and returns a few hours later with a $400 speeding fine. Leadfoot is furious, protesting that the car was barely driven over the speed limit. Should you believe Leadfoot?

In order to decide if Leadfoot is telling the truth, use $F = 10x - 600$. Leadfoot was fined $400, so substitute 400 for F:

$$400 = 10x - 600.$$

In Example 1, we will find the value for x. This variable represents Leadfoot's speed, which resulted in the $400 fine.

An **equation** consists of two algebraic expressions joined by an equal sign. Thus, $400 = 10x - 600$ is an example of an equation. The equal sign divides the equation into two parts, the left side and the right side:

$$\boxed{400} \quad = \quad \boxed{10x - 600}$$

Left side   Right side

The two sides of an equation can be reversed. So, we can also express this equation as

$$10x - 600 = 400.$$

The form of this equation is $ax + b = c$, with $a = 10$, $b = -600$, and $c = 400$. Any equation in this form is called a **linear equation in one variable**. The exponent on the variable in such an equation is 1. In this section, we will study how to solve linear equations.

1 Solve linear equations in one variable.

Solving Linear Equations in One Variable

We begin by restating the definition of a linear equation in one variable.

Definition of a Linear Equation

A **linear equation in one variable** x is an equation that can be written in the form

$$ax + b = c$$

where a and b are real numbers and $a \neq 0$.

An example of a linear equation in one variable is $4x + 12 = 0$. **Solving an equation** in x involves determining all values of x that result in a true statement when substituted into the equation. Such values are **solutions** or **roots** of the equation. For example, substitute -3 into $4x + 12 = 0$. We obtain $4(-3) + 12 = 0$, or $-12 + 12 = 0$. This simplifies to the true statement $0 = 0$. Thus, -3 is a solution of the equation $4x + 12 = 0$. We also say that -3 **satisfies** the equation $4x + 12 = 0$, because when we substitute -3 for x, a true statement results. The set of all such solutions is called the equation's **solution set**. For example, the solution set of the equation $4x + 12 = 0$ is $\{-3\}$.

Equations that have the same solution set are called **equivalent equations**. For example, the equations $4x + 12 = 0$, $4x = -12$, and $x = -3$ are equivalent equations because the solution set for each is $\{-3\}$. To solve a linear equation in x, we transform the equation into an equivalent equation one or more times. Our final equivalent equation should be in the form $x = d$, where d is a real number. By inspection, we can see that the solution set for this equation is $\{d\}$.

To generate equivalent equations, we will use the following principles.

Study Tip

We can solve equations such as $3(x - 6) = 5x$ for a variable. However, we cannot solve for a variable in an algebraic expression such as $3(x - 6)$. We *simplify* algebraic expressions.

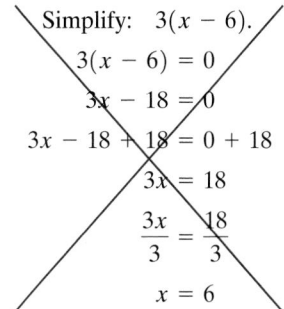

Correct

Simplify: $3(x - 6)$.

$3(x - 6) = 3x - 18$

Incorrect

Simplify: $3(x - 6)$.

$3(x - 6) = 0$

$3x - 18 = 0$

$3x - 18 + 18 = 0 + 18$

$3x = 18$

$\dfrac{3x}{3} = \dfrac{18}{3}$

$x = 6$

Generating Equivalent Equations

An equation can be transformed into an **equivalent equation** by one or more of the following operations.

Example

1. Simplify an expression by removing grouping symbols and combining like terms.

$$3(x - 6) = 6x - x$$
$$3x - 18 = 5x$$

> Subtract $3x$ from both sides of the equation.

2. Add (or subtract) the same real number or variable expression on *both* sides of the equation.

$$3x - 18 = 5x$$
$$3x - 18 - 3x = 5x - 3x$$
$$-18 = 2x$$

3. Multiply (or divide) on *both* sides of the equation by the same *nonzero* quantity.

$$-18 = 2x$$
$$\dfrac{-18}{2} = \dfrac{2x}{2}$$
$$-9 = x$$

> Divide both sides of the equation by 2.

4. Interchange the two sides of the equation.

$$-9 = x$$
$$x = -9$$

If you look closely at the equations in the box, you will notice that we have solved the equation $3(x - 6) = 6x - x$. The final equation, with x isolated by itself on the left side, shows that $\{-9\}$ is the solution set. The idea in solving a linear equation is to get the variable by itself on one side of the equal sign and a number by itself on the other side.

**EXAMPLE 1   Solving a Linear Equation
(Is Leadfoot Telling the Truth?)**

Solve the equation: $10x - 600 = 400$.

Solution   Remember that x represents Leadfoot's speed that resulted in the $400 fine. Our goal is to get x by itself on the left side. We do this by adding 600 to both sides to get $10x$ by itself. Then we isolate x from $10x$ by dividing both sides of the equation by 10.

$$10x - 600 = 400 \qquad \text{This is the given equation.}$$
$$10x - 600 + 600 = 400 + 600 \qquad \text{Add 600 to both sides.}$$
$$10x = 1000 \qquad \text{Combine like terms.}$$
$$\frac{10x}{10} = \frac{1000}{10} \qquad \text{Divide both sides by 10.}$$
$$x = 100$$

Can this possibly be correct? Was Leadfoot doing 100 miles per hour in the car he borrowed from you? To find out, check the proposed solution, 100, in the original equation. In other words, evaluate when $x = 100$.

Check

$$10x - 600 = 400 \qquad \text{This is the original equation.}$$
$$10(100) - 600 \overset{?}{=} 400 \qquad \text{Substitute 100 for x.}$$
$$1000 - 600 \overset{?}{=} 400 \qquad \text{Multiply:}\quad 10(100) = 1000.$$
$$400 = 400 \qquad \text{Subtract:}\quad 1000 - 600 = 400.$$

The true statement $400 = 400$ indicates that 100 is the solution. This verifies that the solution set is $\{100\}$. Leadfoot was doing an outrageous 100 miles per hour, and lied by claiming that your car was barely driven over the speed limit.

Check Point 1   Solve and check: $5x - 8 = 72$.

We now present a step-by-step procedure for solving a linear equation in one variable. Not all of these steps are necessary to solve every equation.

Solving a Linear Equation

1. Simplify the algebraic expression on each side.
2. Collect all the variable terms on one side and all the constant terms on the other side.
3. Isolate the variable and solve.
4. Check the proposed solution in the original equation.

Study Tip

If your proposed solution is incorrect, you will get a false statement when you check your answer. For example, 65 is not a solution of $10x - 600 = 400$. Look what happens when we substitute 65 for x:

$$10x - 600 = 400$$
$$10(65) - 600 \overset{?}{=} 400$$
$$650 - 600 \overset{?}{=} 400$$
$$50 = 400 \qquad \text{False.}$$

The compact, symbolic notation of algebra enables us to use a clear step-by-step method for solving equations, designed to avoid the confusion shown in the painting.

EXAMPLE 2 Solving a Linear Equation

Solve the equation: $2(x - 3) - 17 = 13 - 3(x + 2)$.

Solution

Step 1 **Simplify the algebraic expression on each side.**

$$2(x - 3) - 17 = 13 - 3(x + 2) \quad \text{This is the given equation.}$$
$$2x - 6 - 17 = 13 - 3x - 6 \quad \text{Use the distributive property.}$$
$$2x - 23 = -3x + 7 \quad \text{Combine like terms.}$$

Discovery

Solve the equation in Example 2 by collecting terms with the variable on the right and numerical terms on the left. What do you observe?

Step 2 **Collect variable terms on one side and constant terms on the other side.** We will collect variable terms on the left by adding $3x$ to both sides. We will collect the numbers on the right by adding 23 to both sides.

$$2x - 23 + 3x = -3x + 7 + 3x \quad \text{Add } 3x \text{ to both sides.}$$
$$5x - 23 = 7 \quad \text{Simplify.}$$
$$5x - 23 + 23 = 7 + 23 \quad \text{Add 23 to both sides.}$$
$$5x = 30 \quad \text{Simplify.}$$

Step 3 **Isolate the variable and solve.** We isolate the variable x by dividing both sides by 5.

$$\frac{5x}{5} = \frac{30}{5} \quad \text{Divide both sides by 5.}$$
$$x = 6 \quad \text{Simplify.}$$

Step 4 **Check the proposed solution in the original equation.** Substitute 6 for x in the original equation.

$$2(x - 3) - 17 = 13 - 3(x + 2) \quad \text{This is the original equation.}$$
$$2(6 - 3) - 17 \stackrel{?}{=} 13 - 3(6 + 2) \quad \text{Substitute 6 for x.}$$
$$2(3) - 17 \stackrel{?}{=} 13 - 3(8) \quad \text{Simplify inside parentheses.}$$
$$6 - 17 \stackrel{?}{=} 13 - 24 \quad \text{Multiply.}$$
$$-11 = -11 \checkmark \quad \text{This true statement indicates that 6 is the solution.}$$

The solution set is $\{6\}$.

Check Point 2 Solve and check: $4(2x + 1) - 29 = 3(2x - 5)$.

2 Solve equations with constants in denominators.

Linear Equations with Fractions

Equations are easier to solve when they do not contain fractions. How do we solve equations involving fractions? We begin by multiplying both sides of the equation by the least common denominator. The least common denominator is the smallest number that all the denominators will divide into. Example 3 shows how this is done.

Technology

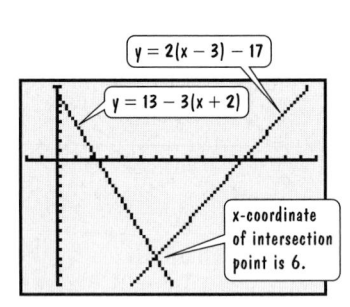

$y = 2(x - 3) - 17$

$y = 13 - 3(x + 2)$

x-coordinate
of intersection
point is 6.

You can use a graphing utility to check the solution to a linear equation in one variable. **Graph the left side and graph the right side. The solution is the x-coordinate of the point where the graphs intersect.** For example, to verify that 6 is the solution of

$$2(x - 3) - 17 = 13 - 3(x + 2),$$

graph these two equations in the same viewing rectangle:

$$y = 2(x - 3) - 17$$

$$\text{and} \quad y = 13 - 3(x + 2)$$

Choose a large enough range setting so that you can see where the graphs intersect. The viewing rectangle on the left shows that the x-coordinate of the intersection point is 6, verifying that {6} is the solution set for the given equation.

EXAMPLE 3   Solving a Linear Equation Involving Fractions

Solve the equation: $\dfrac{3x}{2} = \dfrac{x}{5} - \dfrac{39}{5}.$

Solution   The denominators are 2, 5, and 5. The smallest number that is divisible by 2, 5, and 5 is 10. We begin by multiplying both sides of the equation by 10, the least common denominator.

$$\frac{3x}{2} = \frac{x}{5} - \frac{39}{5} \qquad \text{\textit{This is the given equation.}}$$

$$10 \cdot \frac{3x}{2} = 10\left(\frac{x}{5} - \frac{39}{5}\right) \qquad \text{\textit{Multiply both sides by 10.}}$$

$$10 \cdot \frac{3x}{2} = 10 \cdot \frac{x}{5} - 10 \cdot \frac{39}{5} \qquad \text{\textit{Use the distributive property. Be sure to multiply all terms by 10.}}$$

$$\overset{5}{\cancel{10}} \cdot \frac{3x}{\underset{1}{\cancel{2}}} = \overset{2}{\cancel{10}} \cdot \frac{x}{\underset{1}{\cancel{5}}} - \overset{2}{\cancel{10}} \cdot \frac{39}{\underset{1}{\cancel{5}}} \qquad \text{\textit{Divide out common factors in the multiplication.}}$$

$$15x = 2x - 78 \qquad \text{\textit{Complete the multiplication. The fractions are now cleared.}}$$

At this point, we have an equation similar to those we previously solved. Collect the variable terms on one side and the constant terms on the other side.

$$15x - 2x = 2x - 2x - 78 \qquad \text{\textit{Subtract 2x to get the x-terms on the left.}}$$

$$13x = -78 \qquad \text{\textit{Simplify.}}$$

Isolate x by dividing both sides by 13.

$$\frac{13x}{13} = \frac{-78}{13} \qquad \text{\textit{Divide both sides by 13.}}$$

$$x = -6 \qquad \text{\textit{Simplify.}}$$

Check the proposed solution in the original equation. Substitute −6 for x in the original equation. You should obtain −9 = −9. This true statement verifies the solution set is {−6}.

Check Point 3 Solve and check: $\dfrac{x}{4} = \dfrac{2x}{3} + \dfrac{5}{6}$.

3 Solve equations with variables in denominators.

Equations Involving Rational Expressions

In Example 3 we solved a linear equation with constants in denominators. Now, let's consider an equation such as

$$\frac{1}{x} = \frac{1}{5} + \frac{3}{2x}.$$

Can you see how this equation differs from the fractional equation that we solved earlier? The variable, x, appears in two of the denominators. The procedure for solving this equation still involves multiplying each side by the least common denominator. However, we must avoid any values of the variable that make a denominator zero. For example, examine the denominators in the equation

$$\frac{1}{x} = \frac{1}{5} + \frac{3}{2x}.$$

This denominator would equal zero if x = 0. This denominator would equal zero if x = 0.

We see that x cannot equal zero. With this in mind, let's solve the equation.

EXAMPLE 4 Solving an Equation Involving Rational Expressions

Solve: $\dfrac{1}{x} = \dfrac{1}{5} + \dfrac{3}{2x}$.

Solution   The denominators are $x, 5$, and $2x$. The least common denominator is $10x$. We begin by multiplying both sides of the equation by $10x$. We will also write the restriction that x cannot equal zero to the right of the equation.

$$\frac{1}{x} = \frac{1}{5} + \frac{3}{2x}, \quad x \neq 0 \qquad \text{This is the given equation.}$$

$$10x \cdot \frac{1}{x} = 10x\left(\frac{1}{5} + \frac{3}{2x}\right) \qquad \text{Multiply both sides by 10x.}$$

$$10x \cdot \frac{1}{x} = 10x \cdot \frac{1}{5} + 10x \cdot \frac{3}{2x} \qquad \text{Use the distributive property. Be sure to multiply all terms by 10x.}$$

$$10x \cdot \frac{1}{x} = \overset{2}{\cancel{10}}x \cdot \frac{1}{\underset{1}{\cancel{5}}} + \overset{5}{\cancel{10}}x \cdot \frac{3}{\underset{1}{\cancel{2}}x} \qquad \text{Divide out common factors in the multiplication.}$$

$$10 = 2x + 15 \qquad \text{Simplify.}$$

Observe that the resulting equation,

$$10 = 2x + 15,$$

is now cleared of fractions. With the variable term, $2x$, already on the right, we will collect constant terms on the left by subtracting 15 from both sides.

$$10 - 15 = 2x + 15 - 15 \qquad \text{Subtract 15 from both sides.}$$

$$-5 = 2x \qquad \text{Simplify.}$$

Finally, we isolate the variable, x, by dividing both sides by 2.

$$\frac{-5}{2} = \frac{2x}{2} \qquad \text{Divide both sides by 2.}$$

$$-\frac{5}{2} = x \qquad \text{Simplify.}$$

We check our solution by substituting $-\frac{5}{2}$ into the original equation or by using a calculator. With a calculator, evaluate each side of the equation for $x = -\frac{5}{2}$, or for $x = -2.5$. Note that the original restriction that $x \neq 0$ is met. The solution set is $\left\{ -\frac{5}{2} \right\}$.

| Check Point 4 | Solve: $\dfrac{5}{2x} = \dfrac{17}{18} - \dfrac{1}{3x}$. |

EXAMPLE 5 Solving an Equation Involving Rational Expressions

Solve: $\dfrac{x}{x - 3} = \dfrac{3}{x - 3} + 9$.

Solution We must avoid any values of the variable x that make a denominator zero.

$$\frac{x}{x - 3} = \frac{3}{x - 3} + 9$$

> These denominators are zero if $x - 3 = 0$, or in other words, if $x = 3$.

We see that x cannot equal 3. With denominators of $x - 3$, $x - 3$, and 1, the least common denominator is $x - 3$. We multiply both sides of the equation by $x - 3$. We also write the restriction that x cannot equal 3 to the right of the equation.

$$\frac{x}{x - 3} = \frac{3}{x - 3} + 9, \quad x \neq 3 \qquad \text{This is the given equation.}$$

$$(x - 3) \cdot \frac{x}{x - 3} = (x - 3)\left[\frac{3}{x - 3} + 9 \right] \qquad \text{Multiply both sides by } x - 3.$$

$$(x - 3) \cdot \frac{x}{x - 3} = (x - 3) \cdot \frac{3}{x - 3} + (x - 3) \cdot 9 \qquad \text{Use the distributive property.}$$

$$\cancel{(x - 3)} \cdot \frac{x}{\cancel{x - 3}} = \cancel{(x - 3)} \cdot \frac{3}{\cancel{x - 3}} + (x - 3) \cdot 9 \qquad \text{Divide out common factors in the multiplications.}$$

$$x = 3 + (x - 3) \cdot 9 \qquad \text{Simplify.}$$

The resulting equation, which can be expressed as

$$x = 3 + 9(x - 3),$$

is cleared of fractions. We now solve for x.

$$x = 3 + 9x - 27 \qquad \text{Use the distributive property.}$$

$$x = 9x - 24 \qquad \text{Combine numerical terms.}$$

$$x - 9x = 9x - 24 - 9x \qquad \text{Subtract 9x from both sides.}$$

$$-8x = -24 \qquad \text{Simplify.}$$

$$\frac{-8x}{-8} = \frac{-24}{-8} \qquad \text{Solve for x, dividing both sides by } -8.$$

$$x = 3 \qquad \text{Simplify.}$$

The proposed solution, 3, is *not* a solution because of the restriction that $x \neq 3$. There is *no solution to this equation.* The solution set for this equation contains no elements and is called the empty set, written $\varnothing$.

> **Check Point 5**   Solve: $\dfrac{x}{x-2} = \dfrac{2}{x-2} - \dfrac{2}{3}$.

Types of Equations

4 Recognize identities, conditional equations, and inconsistent equations.

We tend to place things in categories, allowing us to order and structure the world. For example, you can categorize yourself by your age group, your ethnicity, your academic major, or your gender. Equations can be placed into categories that depend on their solution sets.

An equation that is true for all real numbers for which both sides are defined is called an **identity**. An example of an identity is

$$x + 3 = x + 2 + 1.$$

Every number plus 3 is equal to that number plus 2 plus 1. Therefore, the solution set to this equation is the set of all real numbers. Another example of an identity is

$$\frac{2x}{x} = 2.$$

Because division by 0 is undefined, this equation is true for all real number values of x except 0. The solution set is the set of nonzero real numbers.

An equation that is not an identity but that is true for at least one real number is called a **conditional equation**. The equation $10x - 600 = 400$ is an example of a conditional equation. The equation is not an identity and is true only if $x = 100$.

An **inconsistent equation** is an equation that is not true for even one real number. An example of an inconsistent equation is

$$x = x + 7.$$

There is no number that is equal to itself plus 7. Some inconsistent equations are less obvious than this. Consider the equation in Example 5,

$$\frac{x}{x-3} = \frac{2}{x-3} + 9.$$

This equation is not true for any real number and has no solution. Thus, it is inconsistent.

EXAMPLE 6 Categorizing an Equation

Determine whether the equation

$$2(x + 1) = 2x + 3$$

is an identity, a conditional equation, or an inconsistent equation.

Solution   Let's see what happens if we try solving the equation. Applying the distributive property on the left side, we obtain

$$2x + 2 = 2x + 3.$$

Does something look strange? Can doubling a number and increasing the product by 2 give the same result as doubling the same number and increasing the product by 3? No. Let's continue solving the equation by subtracting $2x$ from both sides.

$$2x + 2 - 2x = 2x + 3 - 2x$$
$$2 = 3$$

The false statement $2 = 3$ verifies that the given equation is inconsistent.

Check Point 6   Determine whether the equation

$$2(x + 1) = 2x + 2$$

is an identity, a conditional equation, or an inconsistent equation.

EXERCISE SET 1.1

 Practice Exercises

In Exercises 1–16, solve and check each linear equation.

1. $5x - 8 = 72$

2. $6x - 3 = 63$

3. $11x - (6x - 5) = 40$

4. $5x - (2x - 10) = 35$

5. $2x - 7 = 6 + x$

6. $3x + 5 = 2x + 13$

7. $7x + 4 = x + 16$

8. $13x + 14 = 12x - 5$

9. $3(x - 2) + 7 = 2(x + 5)$

10. $2(x - 1) + 3 = x - 3(x + 1)$

11. $3(x - 4) - 4(x - 3) = x + 3 - (x - 2)$

12. $2 - (7x + 5) = 13 - 3x$

13. $16 = 3(x - 1) - (x - 7)$

14. $5x - (2x + 2) = x + (3x - 5)$

15. $25 - [2 + 5y - 3(y + 2)] =$
$$-3(2y - 5) - [5(y - 1) - 3y + 3]$$

16. $45 - [4 - 2y - 4(y + 7)] =$
$$-4(1 + 3y) - [4 - 3(y + 2) - 2(2y - 5)]$$

Exercises 17–30 contain equations with constants in denominators. Solve each equation by multiplying both sides by the least common denominator, thereby clearing fractions.

17. $\dfrac{x}{3} = \dfrac{x}{2} - 2$

18. $\dfrac{x}{5} = \dfrac{x}{6} + 1$

19. $20 - \dfrac{x}{3} = \dfrac{x}{2}$

20. $\dfrac{x}{5} - \dfrac{1}{2} = \dfrac{x}{6}$

21. $\dfrac{3x}{5} = \dfrac{2x}{3} + 1$

22. $\dfrac{x}{2} = \dfrac{3x}{4} + 5$

23. $\dfrac{3x}{5} - x = \dfrac{x}{10} - \dfrac{5}{2}$

24. $2x - \dfrac{2x}{7} = \dfrac{x}{2} + \dfrac{17}{2}$

25. $\dfrac{x + 3}{6} = \dfrac{3}{8} + \dfrac{x - 5}{4}$

26. $\dfrac{x + 1}{4} = \dfrac{1}{6} + \dfrac{2 - x}{3}$

27. $\dfrac{x}{4} = 2 + \dfrac{x - 3}{3}$

28. $5 + \dfrac{x - 2}{3} = \dfrac{x + 3}{8}$

29. $\dfrac{x + 1}{3} = 5 - \dfrac{x + 2}{7}$

30. $\dfrac{3x}{5} - \dfrac{x - 3}{2} = \dfrac{x + 2}{3}$

*Exercises 31–50 contain equations with variables in denominators. For each equation, **a.** Write the value or values of the variable that make a denominator zero. These are the restrictions on the variable. **b.** Keeping the restrictions in mind, solve the equation by multiplying both sides by the least common denominator.*

31. $\dfrac{4}{x} = \dfrac{5}{2x} + 3$

32. $\dfrac{5}{x} = \dfrac{10}{3x} + 4$

33. $\dfrac{2}{x} + 3 = \dfrac{5}{2x} + \dfrac{13}{4}$

34. $\dfrac{7}{2x} - \dfrac{5}{3x} = \dfrac{22}{3}$

SECTION 1.2 Formulas and Ap

Objectives

1. Solve problems using formulas.
2. Use linear equations to solve problems.
3. Solve for a variable in a formula.

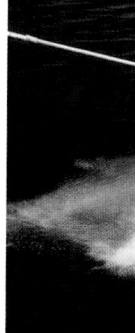

Could you li
proach 850,0
could live to
be used to m
equations.

1 Solve problems using formulas.

Formulas a

The graph in
birth. For exa
in 1980. Find
senting fema
cy for women

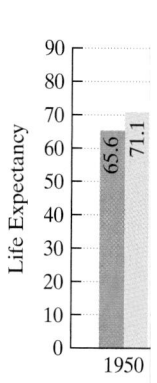

90
80
70
60
50
40
30
20
10
0

Life Expectancy

65.6   71.1

1950

Source: U.S. Bur

The dat

where the va
1950. This eq

---

35. $\dfrac{2}{3x} + \dfrac{1}{4} = \dfrac{11}{6x} - \dfrac{1}{3}$

36. $\dfrac{5}{2x} - \dfrac{8}{9} = \dfrac{1}{18} - \dfrac{1}{3x}$

37. $\dfrac{x-2}{2x} + 1 = \dfrac{x+1}{x}$

38. $\dfrac{4}{x} = \dfrac{9}{5} - \dfrac{7x-4}{5x}$

39. $\dfrac{1}{x-1} + 5 = \dfrac{11}{x-1}$

40. $\dfrac{3}{x+4} - 7 = \dfrac{-4}{x+4}$

41. $\dfrac{8x}{x+1} = 4 - \dfrac{8}{x+1}$

42. $\dfrac{2}{x-2} = \dfrac{x}{x-2} - 2$

43. $\dfrac{3}{2x-2} + \dfrac{1}{2} = \dfrac{2}{x-1}$

44. $\dfrac{3}{x+3} = \dfrac{5}{2x+6} + \dfrac{1}{x-2}$

45. $\dfrac{3}{x+2} + \dfrac{2}{x-2} = \dfrac{8}{(x+2)(x-2)}$

46. $\dfrac{5}{x+2} + \dfrac{3}{x-2} = \dfrac{12}{(x+2)(x-2)}$

47. $\dfrac{2}{x+1} - \dfrac{1}{x-1} = \dfrac{2x}{x^2-1}$

48. $\dfrac{4}{x+5} + \dfrac{2}{x-5} = \dfrac{32}{x^2-25}$

49. $\dfrac{1}{x-4} - \dfrac{5}{x+2} = \dfrac{6}{x^2-2x-8}$

50. $\dfrac{6}{x+3} - \dfrac{5}{x-2} = \dfrac{-20}{x^2+x-6}$

In Exercises 51–58, determine whether each equation is an identity, a conditional equation, or an inconsistent equation.

51. $4(x-7) = 4x - 28$

52. $4(x-7) = 4x + 28$

53. $2x + 3 = 2x - 3$

54. $\dfrac{7x}{x} = 7$

55. $4x + 5x = 8x$

56. $8x + 2x = 9x$

57. $\dfrac{2x}{x-3} = \dfrac{6}{x-3} + 4$

58. $\dfrac{3}{x-3} = \dfrac{x}{x-3} + 3$

The equations in Exercises 59–68 combine the types of equations we have discussed in this section. Solve each equation or state that it is true for all real numbers or no real numbers.

59. $\dfrac{x+5}{2} - 4 = \dfrac{2x-1}{3}$

60. $\dfrac{x+2}{7} = 5 - \dfrac{x+1}{3}$

61. $\dfrac{2}{x-2} = 3 + \dfrac{x}{x-2}$

62. $\dfrac{6}{x+3} + 2 = \dfrac{-2x}{x+3}$

63. $8x - (3x+2) + 10 = 3x$

64. $2(x+2) + 2x = 4(x+1)$

65. $\dfrac{2}{x} + \dfrac{1}{2} = \dfrac{3}{4}$

66. $\dfrac{3}{x} - \dfrac{1}{6} = \dfrac{1}{3}$

67. $\dfrac{4}{x-2} + \dfrac{3}{x+5} = \dfrac{7}{(x+5)(x-2)}$

68. $\dfrac{1}{x-1} = \dfrac{1}{(2x+3)(x-1)} + \dfrac{4}{2x+3}$

 Application Exercises

69. The equation $d = 5000c - 525{,}000$ describes the relationship between the annual number of deaths (d) in the United States from heart disease and the average cholesterol level (c) of blood. (Cholesterol level, c, is expressed in milligrams per deciliter of blood.)
  a. In 1990, 500,000 Americans died from heart disease. Substitute 500,000 for d in the given equation and then solve for c to determine the average cholesterol level in 1990.
  b. Suppose that the average cholesterol level for people in the United States could be reduced to 180. Substitute 180 for c in the given formula and then compute the value for d to determine the number of annual deaths from heart disease with this reduced cholesterol level. Compared to the number of deaths in 1990, how many lives would be saved by this cholesterol reduction?

70. There is a relationship between the vocabulary of a child and its age. The equation $60A - V = 900$ describes this relationship, where A is the age of the child in months and V is the number of words that the child uses. Suppose that a child uses 1500 words. Substitute 1500 for V in the equation to determine the child's age in months.

71. The equation

$$p = 15 + \frac{15d}{33}$$

describes the pressure of sea water (p, in pounds per square foot) at a depth of d feet below the surface. The record depth for breath-held diving, by Francisco Ferreras (Cuba) off Grand-Bahama Island, on November 14,1993, involved pressure of 201 pounds per square foot. To what depth did Ferreras descend on this ill-advised venture? (He was underwater for 2 minutes and 9 seconds!)

72. The equation $P = -0.5d + 100$ describes the percentage (P) of lost hikers found in search and rescue missions when members of the search team walk parallel to one another separated by a distance of d yards. If a search and rescue team finds 70% of lost hikers, substitute 70 for P in the equation and find the parallel distance of separation between members of the search party.

 **Writing in Mathematics**

73. What is a linear equation in one variable? Give an example of this type of equation.

74. What does it mean to solve an equation?

75. What is the solution set of an equation?

76. What are equivalent equations? Give an example.

77. What is the difference between solving an equation such as $2(x-4) + 5x = 34$ and simplifying an algebraic expression such as $2(x-4) + 5x$? If there is a difference, which topic should be taught first? Why?

78. Suppose that you solve $\frac{x}{5} - \frac{x}{2} = 1$ by multipl
sides by 20, rather than the least common denor
5 and 2 (namely, 10). Describe what happens.
the correct solution, why do you think we clear
tion of fractions by multiplying by the *least*
denominator?

79. Suppose you are an algebra teacher grading the
solution on an examination:

$$-3(x - 6) = 2 - x$$
$$-3x - 18 = 2 - x$$
$$-2x - 18 = 2$$
$$-2x = -16$$
$$x = 8$$

You should note that 8 checks, and the solution
The student who worked the problem therefo
full credit. Can you find any errors in the solutic
credit is 10 points, how many points should you
student? Justify your position.

80. Explain how to determine the restrictions on the
for the equation

$$\frac{3}{x + 5} + \frac{4}{x - 2} = \frac{7}{(x + 5)(x - 2)}.$$

81. What is an identity? Give an example.

82. What is a conditional equation? Give an examp

83. What is an inconsistent equation? Give an exar

Technology Exercises

*For Exercises 84–87, use your graphing utilit
graph each side of the equations in the same
rectangle. Based on the resulting graph, label each eq
as conditional, inconsistent, or an identity. If the equa
conditional, use the x-coordinate of the intersection p
find the solution set. Verify this value by direct substi
into the equation.*

84. $2(x - 6) + 3x = x + 6$

85. $9x + 3 - 3x = 2(3x + 1)$

over 5 feet. What is the recommended weight for a man
who is 6 feet, 3 inches tall?

*In Exercises 27–56, use the five-step strategy given in the box
on page 99 to solve each problem.*

27. During the 1998 baseball season, Mark McGwire hit four
more home runs than Sammy Sosa. Combined, the two
athletes hit 136 home runs. Determine the number of
home runs hit by McGwire and Sosa.

28. In 1999, the most populous countries in the world were
China and India. In that year, China's population exceeded India's by 269 million. Combined, the two countries had a population of 2265 million. Determine the 1999
population for China and India.

29. The first Super Bowl was played between the Green Bay
Packers and the Kansas City Chiefs in 1967. Only once, in
1991, were the winning and losing scores in the Super
Bowl consecutive integers. If the sum of the scores was
39, what were the scores?

30. The longest-lived U.S. presidents are John Adams (age
90), Herbert Hoover (also 90), and Harry Truman (88).
Behind them are James Madison, Thomas Jefferson, and
Richard Nixon. The latter three men lived a total of 249
years, and their ages at the time of death form consecutive odd integers. For how long did Nixon, Jefferson, and
Madison live?

31. The graph reflects fear of crime in ten selected countries.
The percentage of people feeling unsafe in their neighborhoods after dark in the United States exceeds twice
that of Sweden by 14%. In the two countries combined,
54.5% of the public feel unsafe walking in their neighborhood after dark. Find the percentage of people feeling
unsafe after dark for Sweden and the United States.

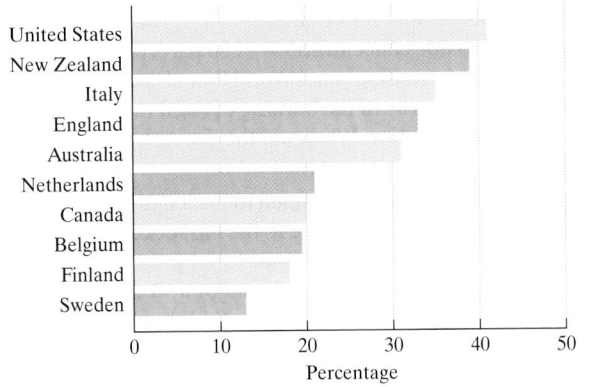

**Percentage of the Public Feeling Unsafe When Walking in
Their Own Area After Dark**

Source: Ministry of Justice, The Netherlands

32. In 1999, Americans in 68 urban areas wasted almost 7 billion gallons of fuel sitting in traffic. The graph at the top
of the next column shows the number of hours in traffic
per year for the average motorist in ten cities. The average motorist in Los Angeles spends 32 hours less than

twice that of the average motorist in Miami stuck in traffic each year. In the two cities combined, 139 hours are
spent by the average motorist per year in traffic. How
many hours are wasted in traffic by the average motorist
in Los Angeles and Miami?

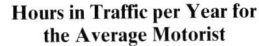
**Hours in Traffic per Year for
the Average Motorist**

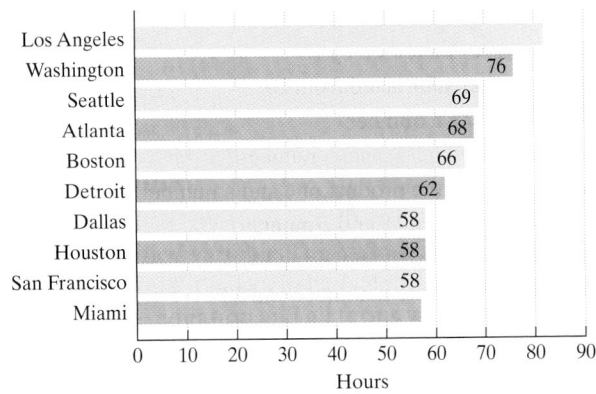

Source: Texas Transportation Institute

33. A car rental agency charges $200 per week plus $0.15 per
mile to rent a car. How many miles can you travel in one
week for $320?

34. A car rental agency charges $180 per week plus $0.25 per
mile to rent a car. How many miles can you travel in one
week for $395?

35. The average weight for female infants at birth is 7 pounds,
with a monthly weight gain of 1.5 pounds. After how many
months does a baby girl weigh 16 pounds?

36. In 1995, the average yearly salary for teachers in the United States was $38,556. If the salary increases by $1496 per
year, in which year will the salary reach $56,508?

37. Answer the question in the following *Peanuts* cartoon
strip. (*Note:* You may not use the answer given in the
cartoon!)

PEANUTS reprinted by permission of United
Features Syndicate, Inc.

38. Every year, approximately 1760 Americans suffer spinal cord injuries due to falls. This represents 22% of the total number of Americans who suffer spinal cord injuries yearly. Determine the number of Americans who suffer spinal cord injuries each year.

Causes of U.S. Spinal Cord Injuries

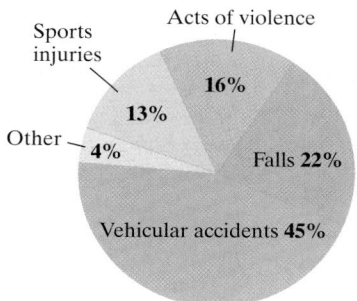

Sports injuries

Acts of violence

16%

13%

Other 4%

Falls 22%

Vehicular accidents 45%

Source: U.S. News and World Report

39. The bus fare in a city is $1.25. People who use the bus have the option of purchasing a monthly coupon book for $21.00. With the coupon book, the fare is reduced to $0.50.
  a. Let x represent the number of times in a month the bus is used. Write algebraic expressions for the total monthly costs of using the bus x times both with and without the coupon book.
  b. Determine the number of times in a month the bus must be used so that the total monthly cost without the coupon book is the same as the total monthly cost with the coupon book.

40. A coupon book for a bridge costs $21 per month. The toll for the bridge is normally $2.50, but it is reduced to $1 for people who have purchased the coupon book.
  a. Let x represent the number of times in a month the bridge is used. Write algebraic expressions for the total monthly costs of using the bridge x times both with and without the coupon book.
  b. Determine the number of times in a month the bridge must be crossed so that the total monthly cost without the coupon book is the same as the total monthly cost with the coupon book.

41. You inherit $25,000 with the stipulation that for the first year the money must be invested in two stocks paying 9% and 12% annual interest, respectively. How much should be invested at each rate if the total interest earned for the year is to be $2250?

42. You inherit $18,750 with the stipulation that for the first year the money must be invested in two stocks paying 10% and 12% annual interest, respectively. How much should be invested at each rate if the total interest earned for the year is to be $2117?

43. The length of the rectangular tennis court at Wimbledon is 6 feet longer than twice the width. If the court's perimeter is 228 feet, what are the court's dimensions?

44. The length of a rectangular basketball court is 6 feet less than twice the width. If the court's perimeter is 288 feet, what are the court's dimensions?

45. A bookcase is to be constructed as shown in the figure. The length is to be 3 times the height. If 60 feet of lumber is available for the entire unit, find the length and height of the bookcase.

x

$3x$

46. The height of the bookcase in the figure is 3 feet longer than the length of a shelf. If 18 feet of lumber is available for the entire unit, find the length and height of the unit.

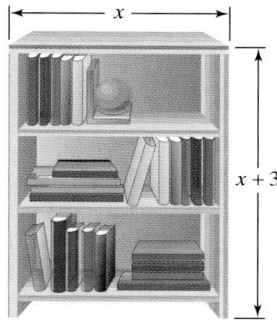

x

$x + 3$

47. An automobile repair shop charged a customer $448, listing $63 for parts and the remainder for labor. If the cost of labor is $35 per hour, how many hours of labor did it take to repair the car?

48. A repair bill on a yacht came to $1603, including $532 for parts and the remainder for labor. If the cost of labor is $63 per hour, how many hours of labor did it take to repair the yacht?

49. After a 35% price reduction, a graphing calculator sold for $81.90. What was the calculator's price before the reduction?

50. After a 12% price reduction, a car sold for $17,600. What was the car's price before the reduction?

51. Inclusive of a 6.5% sales tax, a television sold for $788.10. Find the price of the television before the tax was added.

52. Inclusive of a 6.5% sales tax, a car sold for $17,466. Find the price of the car before the tax was added.

53. Markup is the amount added to the dealer's cost of an item to arrive at the selling price of that item. The selling price of a refrigerator is $584. If the markup is 25% of the dealer's cost, what is the dealer's cost of the refrigerator?

54. A calculator costs a dealer $80. Determine the selling price if the markup is 20% of the selling price.

55. An HMO pamphlet contains the following recommended weight for women: "Give yourself 100 pounds for the first 5 feet plus 5 pounds for every inch over 5 feet tall."

Using this description, what height corresponds to an ideal weight of 135 pounds?

56. A job pays an annual salary of $33,150, which includes a holiday bonus of $750. If paychecks are issued twice a month, what is the gross amount for each paycheck?

In Exercises 57–72, solve each formula for the specified variable. Do you recognize the formula? If so, what does it describe?

57. $A = lw$ for w

58. $D = RT$ for R

59. $A = \frac{1}{2}bh$ for b

60. $V = \frac{1}{3}Bh$ for B

61. $I = Prt$ for P

62. $C = 2\pi r$ for r

63. $E = mc^2$ for m

64. $V = \pi r^2 h$ for h

65. $T = D + pm$ for p

66. $P = C + MC$ for M

67. $A = \frac{1}{2}h(a + b)$ for a

68. $A = \frac{1}{2}h(a + b)$ for b

69. $S = P + Prt$ for r

70. $S = P + Prt$ for t

71. $B = \dfrac{F}{S - V}$ for S

72 $S = \dfrac{C}{1 - r}$ for r

Writing in Mathematics

73. What is a formula?

74. We discussed formulas in this section after we considered procedures for solving linear equations. Doesn't working with a formula simply mean substituting given numbers into the formula and using the order of operations? Is it necessary to know how to solve equations to work with formulas? Explain.

75. In your own words, describe a step-by-step approach for solving algebraic word problems.

76. Did you have some difficulties solving some of the problems that were assigned in this exercise set? Discuss what you did if this happened to you. Did your course of action enhance your ability to solve algebraic word problems?

Technology Exercises

77. The average hourly rate (y) for public school cafeteria workers in the United States in 1980 was $3.82 per hour. This rate has increased steadily by $0.30 per hour each year since 1980.
   a. Write a mathematical model for the hourly rate x years after 1980.
   b. Use a graphing utility to graph the model in a $[0, 12, 1]$ by $[0, 8, 1]$ viewing rectangle.
   c. Use the trace feature to trace along the curve to determine in what year the hourly wage was $6.22.

d. Verify your observation in part (c) algebraically by setting the model equal to 6.22 and solving for x.

78. A tennis club offers two payment options. Members can pay a monthly fee of $30 plus $5 per hour for court rental time. The second option has no monthly fee, but court time costs $7.50 per hour.
   a. Write a mathematical model representing total monthly costs for each option for x hours of court rental time.
   b. Use a graphing utility to graph the two models in a $[0, 15, 1]$ by $[0, 120, 6]$ viewing rectangle.
   c. Use your utility's trace or intersection feature to determine where the two graphs intersect. Describe what the coordinates of this intersection point represent in practical terms.
   d. Verify part (c) using an algebraic approach by setting the two models equal to one another and determining how many hours one has to rent the court so that the two plans result in identical monthly costs.

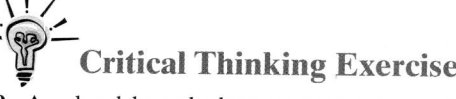

Critical Thinking Exercises

79. A school board plans to merge two schools into one school of 1000 students in which 42% of the students will be African American. One of the schools has a 10% African American student body and the other has a 90% African American student body. What is the student population in each of the two schools?

80. The price of a dress is reduced by 40%. When the dress still does not sell, it is reduced by 40% of the reduced price. If the price of the dress after both reductions is $72, what was the original price?

81. In a film, the actor Charles Coburn plays an elderly "uncle" character criticized for marrying a woman when he is 3 times her age. He wittily replies, "Ah, but in 20 years time I shall only be twice her age." How old is the "uncle" and the woman?

82. Suppose that we agree to pay you 8¢ for every problem in this chapter that you solve correctly and fine you 5¢ for every problem done incorrectly. If at the end of 26 problems we do not owe each other any money, how many problems did you solve correctly?

83. It was wartime when the Ricardos found out Mrs. Ricardo was pregnant. Ricky Ricardo was drafted and made out a will, deciding that $14,000 in a savings account was to be divided between his wife and his child-to-be. Rather strangely, and certainly with gender bias, Ricky stipulated that if the child were a boy, he would get twice the amount of the mother's portion. If it were a girl, the mother would get twice the amount the girl was to receive. We'll never know what Ricky was thinking of, for (as fate would have it) he did not return from war. Mrs. Ricardo gave birth to twins—a boy and a girl. How was the money divided?

84. Solve for P: $A = P + Prt$.

Group Exercise

85. One of the best ways to learn how to *solve* a word problem in algebra is to *design* word problems of your own. Creating a word problem makes you very aware of precisely how much information is needed to solve the problem. You must also focus on the best way to present information to a reader and on how much information to give. As you write your problem, you gain skills that will help you solve problems created by others.

The group should design five different word problems that can be solved using an algebraic equation. All of the problems should be on different topics. For example, the group should not have more than one problem on simple interest. The group should turn in both the problems and their algebraic solutions.

SECTION 1.3   *Quadratic Equations*

Objectives

1. Solve quadratic equations by factoring.
2. Solve quadratic equations by the square root method.
3. Solve quadratic equations by completing the square.
4. Solve quadratic equations using the quadratic formula.
5. Use the discriminant to determine the kinds of solutions.
6. Determine the most efficient method to use when solving a quadratic equation.
7. Solve problems modeled by quadratic equations.

The crocodile, an endangered species, was the subject of a protection program at Florida's Everglades National Park. Park rangers used the formula

$$P = -10x^2 + 475x + 3500$$

to estimate the crocodile population, P, after x years of the protection program. Their goal was to bring the population up to 7250. To find out how long the program had to be continued for this to occur, we need to substitute 7250 for P in the formula and solve for x:

$$7250 = -10x^2 + 475x + 3500.$$

Do you see how this equation differs from a linear equation? The exponent on x is 2. Solving such an equation involves finding the set of numbers that will make the equation a true statement. In this section, we study a number of methods for solving equations in the form $ax^2 + bx + c = 0$. We also look at applications of these equations.

The Standard Form of a Quadratic Equation

We begin by defining a quadratic equation.

Definition of a Quadratic Equation

A **quadratic equation** in x is an equation that can be written in the **standard form**

$$ax^2 + bx + c = 0$$

where a, b, and c are real numbers with $a \neq 0$. A quadratic equation in x is also called a **second-degree polynomial equation** in x.

An example of a quadratic equation in standard form is $x^2 - 7x + 10 = 0$. The coefficient of x^2 is $1(a = 1)$, the coefficient of x is $-7(b = -7)$, and the constant term is $10(c = 10)$.

1 Solve quadratic equations by factoring.

Solving Quadratic Equations by Factoring

We can factor the left side of the quadratic equation $x^2 - 7x + 10 = 0$. We obtain $(x - 5)(x - 2) = 0$. If a quadratic equation has zero on one side and a factored expression on the other side, it can be solved using the **zero-product principle**.

The Zero-Product Principle

If the product of two algebraic expressions is zero, then at least one of the factors is equal to zero.

$$\text{If } AB = 0, \quad \text{then } A = 0 \text{ or } B = 0.$$

For example, consider the equation $(x - 5)(x - 2) = 0$. According to the zero-product principle, this product can be zero only if at least one of the factors is zero. We set each individual factor equal to zero and solve each resulting equation for x.

$$(x - 5)(x - 2) = 0$$

$$x - 5 = 0 \quad \text{or} \quad x - 2 = 0$$

$$x = 5 \qquad\qquad x = 2$$

We can check each of these proposed solutions in the original quadratic equation, $x^2 - 7x + 10 = 0$.

Check 5:	**Check 2:**
$5^2 - 7 \cdot 5 + 10 \stackrel{?}{=} 0$	$2^2 - 7 \cdot 2 + 10 \stackrel{?}{=} 0$
$25 - 35 + 10 \stackrel{?}{=} 0$	$4 - 14 + 10 \stackrel{?}{=} 0$
$0 = 0 \checkmark$	$0 = 0 \checkmark$

The resulting true statements indicate that the solutions are 5 and 2. The solution set is $\{5, 2\}$. Note that with a quadratic equation, we can have two solutions, compared to the linear equation that had one.

Solving a Quadratic Equation by Factoring

1. If necessary, rewrite the equation in the form $ax^2 + bx + c = 0$, moving all terms to one side, thereby obtaining zero on the other side.
2. Factor.

3. Apply the zero-product principle, setting each factor equal to zero.

4. Solve the equations in step 3.

5. Check the solutions in the original equation.

EXAMPLE 1 Solving Quadratic Equations by Factoring

Solve by factoring and then using the zero-product principle.

a. $4x^2 - 2x = 0$ **b.** $2x^2 + 7x = 4$

Solution

a. We begin with $4x^2 - 2x = 0$.

Step 1 Move all terms to one side and obtain zero on the other side. All terms are already on the left and zero is on the other side, so we can skip this step.

Step 2 Factor. We factor out $2x$ from the two terms on the left side.

$$4x^2 - 2x = 0 \quad \text{This is the given equation.}$$

$$2x(2x - 1) = 0 \quad \text{Factor.}$$

Steps 3 and 4 Set each factor equal to zero and solve the resulting equations.

$$2x = 0 \quad \text{or} \quad 2x - 1 = 0$$
$$x = 0 \qquad\qquad 2x = 1$$
$$x = \tfrac{1}{2}$$

Step 5 Check the solutions in the original equation.

Check 0:	**Check $\tfrac{1}{2}$:**
$4x^2 - 2x = 0$	$4x^2 - 2x = 0$
$4 \cdot 0^2 - 2 \cdot 0 \overset{?}{=} 0$	$4\left(\tfrac{1}{2}\right)^2 - 2\left(\tfrac{1}{2}\right) \overset{?}{=} 0$
$0 - 0 \overset{?}{=} 0$	$4\left(\tfrac{1}{4}\right) - 2\left(\tfrac{1}{2}\right) \overset{?}{=} 0$
$0 = 0 \checkmark$	$1 - 1 \overset{?}{=} 0$
	$0 = 0 \checkmark$

The solution set is $\left\{0, \tfrac{1}{2}\right\}$.

b. Next, we solve $2x^2 + 7x = 4$.

Step 1 Move all terms to one side and obtain zero on the other side. Subtract 4 from both sides and write the equation in standard form.

$$2x^2 + 7x - 4 = 4 - 4$$
$$2x^2 + 7x - 4 = 0$$

Step 2 Factor.

$$2x^2 + 7x - 4 = 0$$
$$(2x - 1)(x + 4) = 0$$

Steps 3 and 4 **Set each factor equal to zero and solve each resulting equation.**

$$2x - 1 = 0 \quad \text{or} \quad x + 4 = 0$$
$$2x = 1 \qquad\qquad x = -4$$
$$x = \tfrac{1}{2}$$

Step 5 **Check the solutions in the original equation.**

Check $\tfrac{1}{2}$:

$$2x^2 + 7x = 4$$
$$2\left(\tfrac{1}{2}\right)^2 + 7\left(\tfrac{1}{2}\right) \overset{?}{=} 4$$
$$\tfrac{1}{2} + \tfrac{7}{2} \overset{?}{=} 4$$
$$4 = 4 \checkmark$$

Check -4:

$$2x^2 + 7x = 4$$
$$2(-4)^2 + 7(-4) \overset{?}{=} 4$$
$$32 + (-28) \overset{?}{=} 4$$
$$4 = 4 \checkmark$$

The solution set is $\left\{-4, \tfrac{1}{2}\right\}$.

Check Point 1

Solve by factoring and then using the zero-product principle.

a. $3x^2 - 9x = 0$ **b.** $2x^2 + x = 1$

Technology

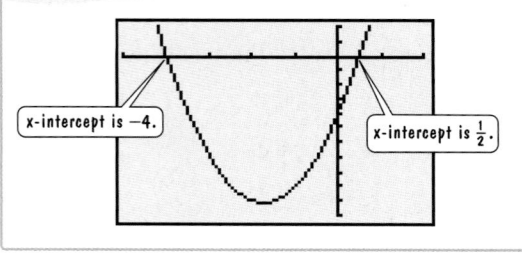

x-intercept is −4.

x-intercept is $\tfrac{1}{2}$.

You can use a graphing utility to check the real solutions to a quadratic equation. **The solutions to $ax^2 + bx + c = 0$ correspond to the x-intercepts for the graph of $y = ax^2 + bx + c$.** For example, to check the solutions of $2x^2 + 7x = 4$, or $2x^2 + 7x - 4 = 0$, graph $y = 2x^2 + 7x - 4 = 0$, as shown on the left. Note that it is important to have all nonzero terms on one side of the quadratic equation before entering it into the graphing utility. The x-intercepts are -4 and $\tfrac{1}{2}$, verifying $\left\{-4, \tfrac{1}{2}\right\}$ as the solution set.

2 Solve quadratic equations by the square root method.

Solving Quadratic Equations by the Square Root Method

Quadratic equations of the form $u^2 = d$, where $d > 0$ and u is an algebraic expression, can be solved by the **square root method**. First, isolate the squared expression u^2 on one side of the equation and the number d on the other side. Then take the square root of both sides. Remember, there are two numbers whose square is d. One number is positive and one is negative.

We can use factoring to verify that $u^2 = d$ has two solutions.

$$u^2 = d \qquad \text{\small This is the given equation.}$$

$$u^2 - d = 0 \qquad \text{\small Move all terms to one side and obtain}$$
$$\text{\small zero on the other side.}$$

$$(u + \sqrt{d})(u - \sqrt{d}) = 0 \qquad \text{\small Factor.}$$

$$u + \sqrt{d} = 0 \quad \text{or} \quad u - \sqrt{d} = 0 \qquad \text{\small Set each factor equal to zero.}$$
$$u = -\sqrt{d} \qquad\qquad u = \sqrt{d} \qquad \text{\small Solve the resulting equations.}$$

Because the solutions differ only in sign, we can write them in abbreviated notation as $u = \pm\sqrt{d}$. We read this as "u equals positive or negative the square root of d."

Now that we have verified these solutions, we can solve $u^2 = d$ directly by taking square roots. This process is called **the square root method**.

The Square Root Method

If u is an algebraic expression and d is a positive real number, then $u^2 = d$ has exactly two solutions:

$$\text{If } u^2 = d, \quad \text{then } u = \sqrt{d} \text{ or } u = -\sqrt{d}.$$

Equivalently,

$$\text{If } u^2 = d, \quad \text{then } u = \pm\sqrt{d}.$$

EXAMPLE 2 **Solving Quadratic Equations by the Square Root Method**

Solve by the square root method:

a. $4x^2 = 20$ **b.** $(x - 2)^2 = 6$

Solution

a. In order to apply the square root method, we need a squared expression by itself on one side of the equation.

$$4x^2 = 20$$

We want x^2 by itself.

We can get x^2 by itself if we divide both sides by 4.

$$\frac{4x^2}{4} = \frac{20}{4}$$

$$x^2 = 5$$

Now, we can apply the square root method.

$$x = \pm\sqrt{5}$$

By checking both values in the original equation, we can confirm that the solution set is $\{-\sqrt{5}, \sqrt{5}\}$.

b. $(x - 2)^2 = 6$

The squared expression is by itself.

With the squared expression by itself, we can apply the square root method.

$$x - 2 = \pm\sqrt{6}$$

We solve for x by adding 2 to both sides.

$$x = 2 \pm \sqrt{6}$$

By checking both values in the original equation, we can confirm that the solution set is $\{2 + \sqrt{6}, 2 - \sqrt{6}\}$.

Check Point 2

Solve by the square root method:

a. $3x^2 = 21$ **b.** $(x + 5)^2 = 11$

3 Solve quadratic equations by completing the square.

Completing the Square

How do we solve an equation in the form $ax^2 + bx + c = 0$ if the equation cannot be factored? We cannot use the zero-product principle in such a case. However, we can convert the equation into an equivalent equation that can be solved using the square root method. This is accomplished by **completing the square**.

Completing the Square

If $x^2 + bx$ is a binomial, then by adding $\left(\dfrac{b}{2}\right)^2$, which is the square of half the coefficient of x, a perfect square trinomial will result. That is,

$$x^2 + bx + \left(\frac{b}{2}\right)^2 = \left(x + \frac{b}{2}\right)^2.$$

EXAMPLE 3   Completing the Square

What term should be added to the binomial $x^2 + 8x$ so that it becomes a perfect square trinomial?

Solution   The term that should be added is the square of half the coefficient of x. The coefficient of x is 8. Thus, we will add $\left(\frac{8}{2}\right)^2 = 4^2$. A perfect square trinomial is the result.

$$x^2 + 8x + 4^2 = x^2 + 8x + 16 = (x + 4)^2$$

(half)^2

Check Point 3

What term should be added to the binomial $x^2 - 14x$ so that it becomes a perfect square trinomial? Factor the trinomial.

We can solve any quadratic equation by completing the square. If the coefficient of the x^2 term is one, we add the square of half the coefficient of x to both sides of the equation. **When you add a constant term to one side of the equation to complete the square, be certain to add the same constant to the other side of the equation.** These ideas are illustrated in Example 4.

EXAMPLE 4 Solving a Quadratic Equation by Completing the Square

Solve by completing the square: $x^2 - 6x + 2 = 0$.

Solution We begin the procedure by isolating the binomial, $x^2 - 6x$, so that we can complete the square. Thus, we subtract 2 from both sides of the equation.

$$x^2 - 6x + 2 = 0$$

$$x^2 - 6x + 2 - 2 = 0 - 2$$

$$x^2 - 6x = -2$$

> We need to add a constant to this binomial that will make it a perfect square trinomial.

What constant should we add? Add the square of half the coefficient of x.

$$x^2 - 6x = -2$$

> -6 is the coefficient of x.
>
> $\left(\dfrac{-6}{2}\right)^2 = (-3)^2 = 9$

Thus, we need to add 9 to $x^2 - 6x$. In order to keep the equation balanced, we must add 9 to both sides.

$$x^2 - 6x = -2 \qquad \text{This is the quadratic equation with the binomial isolated.}$$

$$x^2 - 6x + 9 = -2 + 9 \qquad \text{Complete the square, adding 9 to both sides.}$$

> In this step we have converted our equation into one that can be solved by the square root method.

$$(x - 3)^2 = 7 \qquad \text{Factor the perfect square trinomial.}$$

$$x - 3 = \pm\sqrt{7} \qquad \text{Apply the square root method.}$$

$$x = 3 \pm \sqrt{7} \qquad \text{Add 3 to both sides.}$$

The solution set is $\{3 + \sqrt{7}, 3 - \sqrt{7}\}$.

Check Point 4 Solve by completing the square: $x^2 - 2x - 2 = 0$.

If the coefficient of the x^2 term in a quadratic equation is not one, you must divide each side of the equation by this coefficient before completing the square. For example, to solve $3x^2 - 2x - 4 = 0$ by completing the square, first divide every term by 3:

$$\frac{3x^2}{3} - \frac{2x}{3} - \frac{4}{3} = \frac{0}{3}$$

$$x^2 - \frac{2}{3}x - \frac{4}{3} = 0.$$

Now that the coefficient of x^2 is one, we can solve by completing the square using the method of Example 4.

4 Solve quadratic equations using the quadratic formula.

Solving Quadratic Equations Using the Quadratic Formula

We can use the method of completing the square to derive a formula that can be used to solve all quadratic equations. The derivation given here also shows a particular quadratic equation, $3x^2 - 2x - 4 = 0$, to specifically illustrate each of the steps.

Deriving the Quadratic Formula

Standard Form of a Quadratic Equation	Comment	A Specific Example
$ax^2 + bx + c = 0, \quad a \neq 0$	This is the given equation.	$3x^2 - 2x - 4 = 0$
$x^2 + \frac{b}{a}x + \frac{c}{a} = 0$	Divide both sides by the coefficient of x^2.	$x^2 - \frac{2}{3}x - \frac{4}{3} = 0$
$x^2 + \frac{b}{a}x = -\frac{c}{a}$	Isolate the binomial by adding $-\frac{c}{a}$ on both sides.	$x^2 - \frac{2}{3}x = \frac{4}{3}$
$x^2 + \frac{b}{a}x + \left(\frac{b}{2a}\right)^2 = -\frac{c}{a} + \left(\frac{b}{2a}\right)^2$ $\underbrace{\qquad}_{(\text{half})^2}$	Complete the square. Add the square of half the coefficient of x to both sides.	$x^2 - \frac{2}{3}x + \left(\frac{1}{3}\right)^2 = \frac{4}{3} + \left(\frac{1}{3}\right)^2$ $\underbrace{\qquad}_{(\text{half})^2}$
$x^2 + \frac{b}{a}x + \frac{b^2}{4a^2} = -\frac{c}{a} + \frac{b^2}{4a^2}$		$x^2 - \frac{2}{3}x + \frac{1}{9} = \frac{4}{3} + \frac{1}{9}$
$\left(x + \frac{b}{2a}\right)^2 = -\frac{c}{a}\cdot\frac{4a}{4a} + \frac{b^2}{4a^2}$	Factor on the left and obtain a common denominator on the right.	$\left(x - \frac{1}{3}\right)^2 = \frac{4}{3}\cdot\frac{3}{3} + \frac{1}{9}$
$\left(x + \frac{b}{2a}\right)^2 = \frac{-4ac + b^2}{4a^2}$	Add fractions on the right.	$\left(x - \frac{1}{3}\right)^2 = \frac{12 + 1}{9}$
$\left(x + \frac{b}{2a}\right)^2 = \frac{b^2 - 4ac}{4a^2}$		$\left(x - \frac{1}{3}\right)^2 = \frac{13}{9}$
$x + \frac{b}{2a} = \pm\sqrt{\frac{b^2 - 4ac}{4a^2}}$	Apply the square root method.	$x - \frac{1}{3} = \pm\sqrt{\frac{13}{9}}$
$x + \frac{b}{2a} = \pm\frac{\sqrt{b^2 - 4ac}}{2|a|}$	Take the square root of the quotient, simplifying the denominator.	$x - \frac{1}{3} = \pm\frac{\sqrt{13}}{3}$
$x = \frac{-b}{2a} \pm \frac{\sqrt{b^2 - 4ac}}{2|a|}$	Solve for x by subtracting $\frac{b}{2a}$ from both sides.	$x = \frac{1}{3} \pm \frac{\sqrt{13}}{3}$
$x = \frac{-b \pm \sqrt{b^2 - 4ac}}{2|a|}$	Combine fractions on the right.	$x = \frac{1 \pm \sqrt{13}}{3}$

Because the same real numbers are represented by $\pm 2|a|$ and $\pm 2a$, we can omit the absolute value sign in the last step. The resulting formula is called the **quadratic formula**.

To Die at Twenty

Can the equations

$$7x^5 + 12x^3 - 9x + 4 = 0$$

and

$$8x^6 - 7x^5 + 4x^3 - 19 = 0$$

be solved using a formula similar to the quadratic formula? The first equation has five solutions and the second has six solutions, but they cannot be found using a formula. How do we know? In 1832, a 20-year-old Frenchman, Evariste Galois, wrote down a proof showing that there is no general formula to solve equations when the exponent on the variable is 5 or greater. Galois was jailed as a political activist several times while still a teenager. The day after his brilliant proof he fought a duel over a woman. The duel was a political setup. As he lay dying, Galois told his brother, Alfred, of the manuscript that contained his proof: "Mathematical manuscripts are in my room. On the table. Take care of my work. Make it known. Important. Don't cry, Alfred. I need all my courage—to die at twenty." (Our source is Leopold Infeld's biography of Galois, *Whom the Gods Love.* Some historians, however, dispute the story of Galois's ironic death the very day after his algebraic proof. Mathematical truths seem more reliable than historical ones!)

The Quadratic Formula

The solutions of a quadratic equation in standard form $ax^2 + bx + c = 0$, with $a \neq 0$, are given by the **quadratic formula**

$$x = \frac{-b \pm \sqrt{b^2 - 4ac}}{2a}.$$

x equals negative b, plus or minus the square root of $b^2 - 4ac$, all divided by 2a.

To use the quadratic formula, write the quadratic equation in standard form if necessary. Then determine the numerical values for a (the coefficient of the squared term), b (the coefficient of the x term), and c (the constant term). Substitute the values of a, b, and c in the quadratic formula and evaluate the expression. The $\pm$ sign indicates that there are two solutions of the equation.

EXAMPLE 5 Solving a Quadratic Equation Using the Quadratic Formula

Solve using the quadratic formula: $2x^2 - 6x + 1 = 0$.

Solution The given equation is in standard form. Begin by identifying the values for a, b, and c.

$$2x^2 - 6x + 1 = 0$$

$a = 2$ $b = -6$ $c = 1$

$$x = \frac{-b \pm \sqrt{b^2 - 4ac}}{2a}$$

Use the quadratic formula: $a = 2$, $b = -6$, and $c = 1$.

$$= \frac{-(-6) \pm \sqrt{(-6)^2 - 4(2)(1)}}{2 \cdot 2}$$

Substitute the values for a, b, and c.

$$= \frac{6 \pm \sqrt{36 - 8}}{4}$$

$-(-6) = 6$ and $(-6)^2 = (-6)(-6) = 36$.

$$= \frac{6 \pm \sqrt{28}}{2}$$

Complete the subtraction under the radical.

$$= \frac{6 \pm 2\sqrt{7}}{4}$$

$\sqrt{28} = \sqrt{4 \cdot 7} = \sqrt{4}\sqrt{7} = 2\sqrt{7}$.

$$= \frac{2(3 \pm \sqrt{7})}{4}$$

Factor out 2 from the numerator.

$$= \frac{3 \pm \sqrt{7}}{2}$$

Divide the numerator and denominator by 2.

The solution set is $\left\{ \dfrac{3 + \sqrt{7}}{2}, \dfrac{3 - \sqrt{7}}{2} \right\}$.

Check Point 5 Solve using the quadratic formula:

$$2x^2 + 2x - 1 = 0.$$

We have seen that a graphing utility can be used to check the solutions to the quadratic equation $ax^2 + bx + c = 0$. The x-intercepts of the graph of

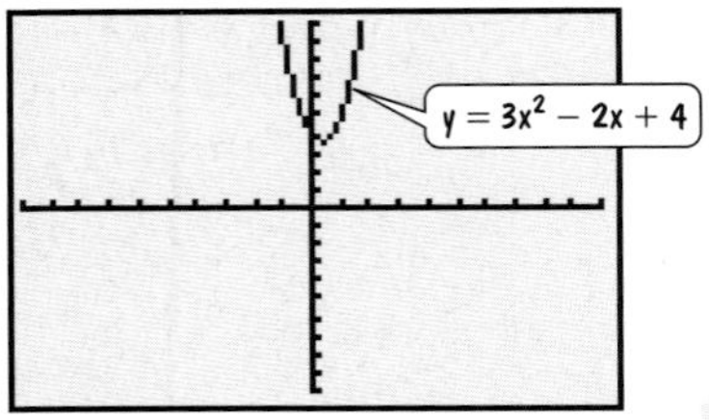

Figure 1.3 This graph has no x-intercepts.

Study Tip

See Section P.7, pages 65–70, to review complex numbers.

$y = ax^2 + bx + c$ are the solutions. However, take a look at the graph of $y = 3x^2 - 2x + 4$, shown in Figure 1.3. Notice that the graph has no x-intercepts. Can you guess what this means about the solutions of the quadratic equation $3x^2 - 2x + 4 = 0$? If you're not sure, we'll answer this question in the next example.

EXAMPLE 6   Solving a Quadratic Equation Using the Quadratic Formula

Solve using the quadratic-formula: $3x^2 - 2x + 4 = 0$.

Solution   The given equation is in standard form. Begin by identifying the values for a, b, and c.

$$3x^2 - 2x + 4 = 0$$

$a = 3$ $b = -2$ $c = 4$

$$x = \frac{-b \pm \sqrt{b^2 - 4ac}}{2a}$$

Use the quadratic formula $a = 3$, $b = -2$, and $c = 4$.

$$= \frac{-(-2) \pm \sqrt{(-2)^2 - 4(3)(4)}}{2(3)}$$

Substitute the values for a, b, and c.

$$= \frac{2 \pm \sqrt{4 - 48}}{6}$$

$-(-2) = 2$ and $(-2)^2 = (-2)(-2) = 4$.

$$= \frac{2 \pm \sqrt{-44}}{6}$$

Because the number under the radical sign is negative, the solutions will not be real numbers.

$$= \frac{2 \pm 2i\sqrt{11}}{6}$$

$\sqrt{-44} = \sqrt{4(11)(-1)}$
$= 2i\sqrt{11}$

$$= \frac{2(1 \pm i\sqrt{11})}{6}$$

Factor 2 from the numerator.

$$= \frac{1 \pm i\sqrt{11}}{3}$$

Divide numerator and denominator by 2.

You can check that these solutions are correct using operations with complex numbers. The solutions are complex conjugates and the solution set is

$$\left\{\frac{1 + i\sqrt{11}}{3}, \frac{1 - i\sqrt{11}}{3}\right\}.$$

Hence, **complex imaginary solutions mean that the graph will not have any x-intercepts.**

Check Point 6   Solve using the quadratic formula:

$$x^2 - 2x + 2 = 0.$$

5   Use the discriminant to determine the kinds of solutions.

The Discriminant

The quantity $b^2 - 4ac$, which appears under the radical sign in the quadratic formula, is called the **discriminant**. In Example 5 the discriminant was 28, a positive number that is not a perfect square. The equation had two solutions that were irrational numbers. In Example 6 the discriminant was -44, a negative number. The equation had solutions involving the imaginary number i. In this case our graph had no x-intercepts.

These observations are generalized in Table 1.3.

Table 1.3 The Discriminant and the Nature of the Solutions to $ax^2 + bx + c = 0$

Discriminant $b^2 - 4ac$	Kinds of Solutions to $ax^2 + bx + c = 0$	Graph of $y = ax^2 + bx + c$
$b^2 - 4ac > 0$	two unequal real solutions	Two x-intercepts
$b^2 - 4ac = 0$	one real solution (a repeated solution)	One x-intercept
$b^2 - 4ac < 0$	No real solution; two complex imaginary solutions	No x-intercepts

EXAMPLE 7 Using the Discriminant

Compute the discriminant of $4x^2 - 8x + 1 = 0$. What does the discriminant indicate about the kinds of solutions?

Solution   Begin by identifying the values for $a, b,$ and c.

$$4x^2 - 8x + 1 = 0$$

$$a = 4 \qquad b = -8 \qquad c = 1$$

Now, compute $b^2 - 4ac$, the discriminant.

$$b^2 - 4ac = (-8)^2 - 4 \cdot 4 \cdot 1 = 64 - 16 = 48$$

The discriminant is 48. Because the discriminant is positive, the equation $4x^2 - 8x + 1 = 0$ has two unequal real solutions.

Check Point 7
Compute the discriminant of $3x^2 - 2x + 5 = 0$. What does the discriminant indicate about the kinds of solutions?

6 Determine the most efficient method to use when solving a quadratic equation.

Determining Which Method to Use

All quadratic equations can be solved by the quadratic formula. However, if an equation is in the form $u^2 = d$, such as $x^2 = 5$ or $(2x + 3)^2 = 8$, it is faster to use the square root method, taking the square root of both sides. If the equation is not in the form $u^2 = d$, write the quadratic equation in standard form $(ax^2 + bx + c = 0)$. Try to solve the equation by the factoring method. If $ax^2 + bx + c$ cannot be factored, then solve the quadratic equation by the quadratic formula.

Because we used the method of completing the square to derive the quadratic formula, we no longer need it for solving quadratic equations. However, we will use completing the square later in the book to help graph certain kinds of equations.

Table 1.4 summarizes our observations about which technique to use when solving a quadratic equation.

Table 1.4   Determining the Most Efficient Technique to Use When Solving a Quadratic Equation

Description and Form of the Quadratic Equation	Most Efficient Solution Method	Example
$ax^2 + bx + c = 0$ and $ax^2 + bx + c$ can be factored easily.	Factor and use the zero-product principle.	$3x^2 + 5x - 2 = 0$ $(3x - 1)(x + 2) = 0$ $3x - 1 = 0$ or $x + 2 = 0$ $x = \frac{1}{3}$ $x = -2$
$ax^2 + c = 0$ The quadratic equation has no linear (x) term. $(b = 0)$	Solve for x^2 and apply the square root method.	$4x^2 - 7 = 0$ $4x^2 = 7$ $x^2 = \frac{7}{4}$ $x = \pm\frac{\sqrt{7}}{2}$
$u^2 = d$; u is a first-degree polynomial.	Use the square root method.	$(x + 4)^2 = 5$ $x + 4 = \pm\sqrt{5}$ $x = -4 \pm \sqrt{5}$
$ax^2 + bx + c = 0$ and $ax^2 + bx + c$ cannot be factored or the factoring is too difficult.	Use the quadratic formula: $x = \frac{-b \pm \sqrt{b^2 - 4ac}}{2a}$	$x^2 - 2x - 6 = 0$ $x = \frac{2 \pm \sqrt{4 - 4(1)(-6)}}{2(1)}$ $= \frac{2 \pm \sqrt{28}}{2} = \frac{2 \pm \sqrt{4}\sqrt{7}}{2}$ $= \frac{2 \pm 2\sqrt{7}}{2} = \frac{2(1 \pm \sqrt{7})}{2}$ $= 1 \pm \sqrt{7}$

7 Solve problems modeled by quadratic equations.

Applications

It's been one of those days! Traffic is really backed up on the highway. Finally, you see the source of the traffic jam—a minor fender-bender. Still stuck in traffic, you notice that the driver appears to be quite young. This might seem like a strange observation. After all, what does a driver's age have to do with his or her chance of getting into an accident?

Oddly enough a driver's age does have something to do with his or her chance of getting into a car accident. The formula $N = 0.4x^2 - 36x + 1000$ approximates the number of accidents, N, per 50 million miles driven, for a driver who is x years old. The formula models data for drivers ages 16 to 74 years, inclusively. Notice that this formula contains an expression in the form $ax^2 + bx + c$ on the right. If a formula contains such an expression, we can write and solve a quadratic equation to answer questions about the variable x. Our next example shows how this is done.

EXAMPLE 8 Using a Quadratic Equation to Answer a Question About the Variable in a Formula

Use the formula $N = 0.4x^2 - 36x + 1000$ to answer this question: What is the age of a driver predicted to have 312 accidents per 50 million miles driven?

Solution We must find x, a driver's age, with $N = 312$ accidents per 50 million miles driven. Use the formula and substitute 312 for N.

$$N = 0.4x^2 - 36x + 1000 \qquad 312 = 0.4x^2 - 36x + 1000$$

Substitute 312 for N.

Let's write the quadratic equation on the right in standard form. Subtract 312 from both sides.

$$312 - 312 = 0.4x^2 - 36x + 1000 - 312$$
$$0 = 0.4x^2 - 36x + 688$$

Equivalently, $0.4x^2 - 36x + 1000 = 0$. The most efficient technique for solving this equation is the quadratic formula. Identify the values for a, b, and c.

$$0.4x^2 - 36x + 688 = 0$$

$a = 0.4 \quad b = -36 \quad c = 688$

Now, substitute these values into the quadratic formula.

$$x = \frac{-b \pm \sqrt{b^2 - 4ac}}{2a} = \frac{-(-36) \pm \sqrt{(-36)^2 - 4(0.4)(688)}}{2(0.4)}$$

$$= \frac{36 \pm \sqrt{195.2}}{0.8}$$

Thus,

$$x = \frac{36 + \sqrt{195.2}}{0.8} \quad \text{or} \quad x = \frac{36 - \sqrt{195.2}}{0.8}$$

$$\approx 62 \qquad\qquad\qquad \approx 28 \quad \text{*Use a calculator to obtain an approximation to the nearest whole number.*}$$

Drivers who are about 28 and 62 years old are predicted to have 312 accidents per 50 million miles driven.

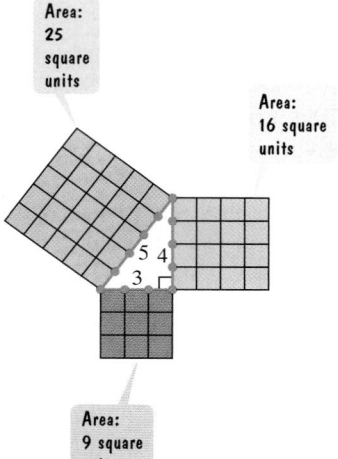

Area: 25 square units

Area: 16 square units

5 4 3

Area: 9 square units

Figure 1.4 The area of the large square equals the sum of the areas of the smaller squares.

Check Point 8 As we mentioned in the introduction to this section, rangers at a national park used the formula

$$P = -10x^2 + 475x + 3500$$

to estimate the crocodile population, P, after x years of a protection program. How many years will it take to bring the population up to 7250?

In our next example, we will be using the *Pythagorean Theorem* to obtain a verbal model. The ancient Greek philosopher and mathematician Pythagoras (approximately 582–500 B.C.) founded a school whose motto was "All is number." Pythagoras is best remembered for his work with the **right triangle**, a triangle with one angle measuring 90°. The side opposite the 90° angle is called the **hypotenuse**. The other sides are called **legs**. Pythagoras found that if he constructed squares on each of the legs, as well as a larger square on the hypotenuse, the sum of the areas of the smaller squares is equal to the area of the larger square. This is illustrated in Figure 1.4.

This relationship is usually stated in terms of the lengths of the three sides of a right triangle and is called the **Pythagorean Theorem**.

The Pythagorean Theorem

The sum of the squares of the lengths of the legs of a right triangle equals the square of the length of the hypotenuse.

If the legs have lengths a and b, and the hypotenuse has length c, then

$$a^2 + b^2 = c^2.$$

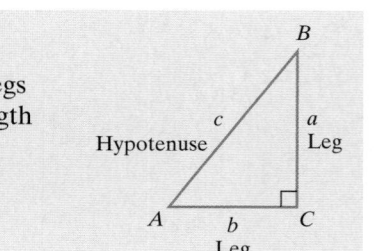

EXAMPLE 9   Using the Pythagorean Theorem

In a 25-inch television set, the length of the screen's diagonal is 25 inches. If the screen's height is 15 inches, what is its width?

Solution   Figure 1.5 shows a right triangle that is formed by the height, width, and diagonal. We can find w, the screen's width, using the Pythagorean Theorem.

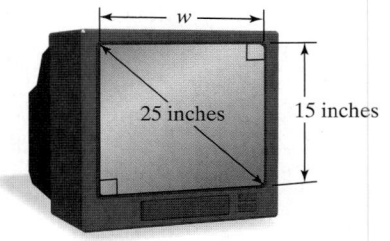

Figure 1.5 A right triangle is formed by the television's height, width, and diagonal.

$(\text{Leg})^2$	plus	$(\text{Leg})^2$	equals	$(\text{Hypotenuse})^2$	
w^2	$+$	15^2	$=$	25^2	This is the equation resulting from the Pythagorean Theorem.

The equation $w^2 + 15^2 = 25^2$ can be solved most efficiently by the square root method.

$$w^2 + 15^2 = 25^2 \qquad \text{This is the equation that models the verbal conditions.}$$
$$w^2 + 225 = 625 \qquad \text{Square 15 and 25.}$$
$$w^2 + 225 - 225 = 625 - 225 \qquad \text{Isolate } w^2 \text{ by subtracting 225 from both sides.}$$
$$w^2 = 400 \qquad \text{Simplify.}$$
$$w = \pm\sqrt{400} \qquad \text{Apply the square root method.}$$
$$w = \pm 20$$

Because w represents the width of the television's screen, this dimension must be positive. We reject -20. Thus, the width of the television is 20 inches.

Check Point 9 What is the width in a 15-inch television set whose height is 9 inches?

EXERCISE SET 1.3

 Practice Exercises

Solve each equation in Exercises 1–14 by factoring and then using the zero-product principle.

1. $x^2 - 3x - 10 = 0$ **2.** $x^2 - 13x + 36 = 0$
3. $x^2 = 8x - 15$ **4.** $x^2 = -11x - 10$
5. $6x^2 + 11x - 10 = 0$ **6.** $9x^2 + 9x + 2 = 0$
7. $3x^2 - 2x = 8$ **8.** $4x^2 - 13x = -3$
9. $3x^2 + 12x = 0$ **10.** $5x^2 - 20x = 0$
11. $2x(x - 3) = 5x^2 - 7x$ **12.** $16x(x - 2) = 8x - 25$
13. $7 - 7x = (3x + 2)(x - 1)$
14. $10x - 1 = (2x + 1)^2$

Solve each equation in Exercises 15–26 by the square root method.

15. $3x^2 = 27$ **16.** $5x^2 = 45$

17. $5x^2 + 1 = 51$ **18.** $3x^2 - 1 = 47$
19. $(x + 2)^2 = 25$ **20.** $(x - 3)^2 = 36$
21. $(3x + 2)^2 = 9$ **22.** $(4x - 1)^2 = 16$
23. $(5x - 1)^2 = 7$ **24.** $(8x - 3)^2 = 5$
25. $(3x - 4)^2 = 8$ **26.** $(2x + 8)^2 = 27$

In Exercises 27–38, determine the constant that should be added to the binomial so that it becomes a perfect square trinomial. Then factor the trinomial.

27. $x^2 + 12x$ **28.** $x^2 + 16x$
29. $x^2 - 10x$ **30.** $x^2 - 14x$
31. $x^2 + 3x$ **32.** $x^2 + 5x$
33. $x^2 - 7x$ **34.** $x^2 - 9x$
35. $x^2 - \dfrac{2}{3}x$ **36.** $x^2 + \dfrac{4}{5}x$

37. $x^2 - \dfrac{1}{3}x$ **38.** $x^2 - \dfrac{1}{4}x$

Solve each equation in Exercises 39–54 by completing the square.

39. $x^2 + 6x = 7$ **40.** $x^2 + 6x = -8$

41. $x^2 - 2x = 2$ **42.** $x^2 + 4x = 12$

43. $x^2 - 6x - 11 = 0$ **44.** $x^2 - 2x - 5 = 0$

45. $x^2 + 4x + 1 = 0$ **46.** $x^2 + 6x - 5 = 0$

47. $x^2 + 3x - 1 = 0$ **48.** $x^2 - 3x - 5 = 0$

49. $2x^2 - 7x + 3 = 0$ **50.** $2x^2 + 5x - 3 = 0$

51. $4x^2 - 4x - 1 = 0$ **52.** $2x^2 - 4x - 1 = 0$

53. $3x^2 - 2x - 2 = 0$ **54.** $3x^2 - 5x - 10 = 0$

Solve each equation in Exercises 55–64 using the quadratic formula.

55. $x^2 + 8x + 15 = 0$ **56.** $x^2 + 8x + 12 = 0$

57. $x^2 + 5x + 3 = 0$ **58.** $x^2 + 5x + 2 = 0$

59. $3x^2 - 3x - 4 = 0$ **60.** $5x^2 + x - 2 = 0$

61. $4x^2 = 2x + 7$ **62.** $3x^2 = 6x - 1$

63. $x^2 - 6x + 10 = 0$ **64.** $x^2 - 2x + 17 = 0$

Compute the discriminant of each equation in Exercises 65–72. What does the discriminant indicate about the kinds of solutions?

65. $x^2 - 4x - 5 = 0$ **66.** $4x^2 - 2x + 3 = 0$

67. $2x^2 - 11x + 3 = 0$ **68.** $2x^2 + 11x - 6 = 0$

69. $x^2 - 2x + 1 = 0$ **70.** $3x^2 = 2x - 1$

71. $x^2 - 3x - 7 = 0$ **72.** $3x^2 + 4x - 2 = 0$

Solve each equation in Exercises 73–94 by the method of your choice.

73. $2x^2 - x = 1$ **74.** $3x^2 - 4x = 4$

75. $5x^2 + 2 = 11x$ **76.** $5x^2 = 6 - 13x$

77. $3x^2 = 60$ **78.** $2x^2 = 250$

79. $x^2 - 2x = 1$ **80.** $2x^2 + 3x = 1$

81. $(2x + 3)(x + 4) = 1$ **82.** $(2x - 5)(x + 1) = 2$

83. $(3x - 4)^2 = 16$ **84.** $(2x + 7)^2 = 25$

85. $3x^2 - 12x + 12 = 0$ **86.** $9 - 6x + x^2 = 0$

87. $4x^2 - 16 = 0$ **88.** $3x^2 - 27 = 0$

89. $x^2 - 6x + 13 = 0$ **90.** $x^2 - 4x + 29 = 0$

91. $x^2 = 4x - 7$ **92.** $5x^2 = 2x - 3$

93. $2x^2 - 7x = 0$ **94.** $2x^2 + 5x = 3$

⭐ Application Exercises

95. The formula $M = 0.0075x^2 - 0.2676x + 14.8$ models the fuel efficiency of passenger cars, M, in miles per gallon, x years after 1940. Environmentalists pressured automo-bile manufacturers for a fuel efficiency of 45 miles per gallon by the year 2000. In which year will fuel efficiency reach 45 miles per gallon according to the formula?

96. The formula $N = 0.036x^2 - 2.8x + 58.14$ models the number of deaths per year, N, per thousand people, for people who are x years old, where $40 \leq x \leq 60$. Find, to the nearest whole number, the age at which 12 people per 1000 die annually.

The Internet is the world's largest communications network. Although millions of computer owners access the Internet using phone lines, the process of downloading files can be slow and tedious. Cable-TV modems dramatically speed up this process. By contrast to phone modems, which transmit 56,000 bits per second, cable modems are capable of trans-mitting 10 million bits per second. The graph shows the mil-lions of Internet users in the United States with this new technology. The data can be modeled by the formula $N = 0.4x^2 + 0.5$, where N represents the millions of people in the United States using cable modems x years after 1996. Use this formula to solve Exercises 97–98.

Number of People in the United States Using Cable TV Modems

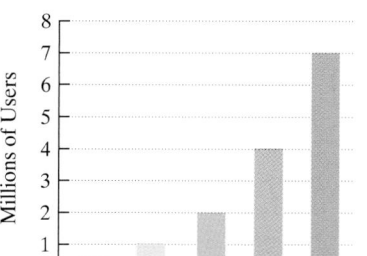

Source: The New York Times

97. According to the formula, in which year will 4.1 million Americans use cable TV modems? How well does the formula describe the actual number of users for that year?

98. According to the formula, in which year will 20.1 mil-lion Americans use cable TV modems?

99. The formula $N = 29{,}035t^2 + 429{,}200$ describes the lead-ing golf winnings in the United States t years after 1983. The leading golf winner for one of the years modeled by the formula was Greg Norman, who won \$690,515. In what year did this occur?

100. The weight of a human fetus is given by the formula $W = 3t^2$, where W is the weight in grams and t is the time in weeks, $t \geq 0$ and $t \leq 39$. After how many weeks does the fetus weigh 300 grams?

The data and the accompanying graph show the number of inmates in U.S. state and federal prisons from 1980 through 1998. The data can be modeled by the formula
$N = 2x^2 + 22x + 320$, *in which N represents the number of inmates, in thousands, in U.S. state and federal prisons x years after 1980. Use this formula to solve Exercises 101–102.*

Number of Inmates in U.S. State and Federal Prisons

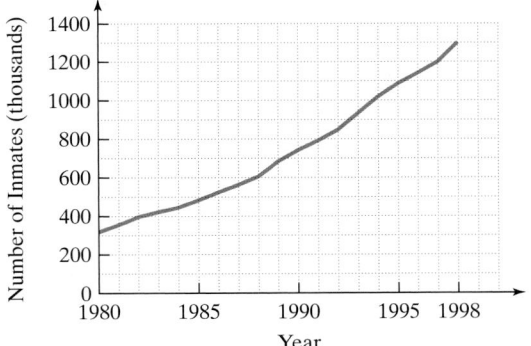

Source: U.S. Justice Department

Year	Number of Inmates
1980	315,974
1981	353,167
1982	394,374
1983	419,820
1984	443,398
1985	480,568
1986	522,084
1987	560,812
1988	603,732
1989	680,907
1990	739,980
1991	789,610
1992	846,277
1993	932,074
1994	1,016,691
1995	1,085,022
1996	1,138,984
1997	1,197,590
1998	1,302,019

101. According to the formula, in which year were there 740 thousand inmates in U.S. state and federal prisons? What was the actual number for that year? How well does the formula describe the actual number of inmates for that year?

102. According to the formula, in which year were there 1100 thousand inmates in U.S. state and federal prisons? What was the actual number for that year? How well does the formula describe the actual number of inmates for that year?

The data and the accompanying graph show the cumulative number of deaths from AIDS in the United States from 1990 through 1998. The data can be modeled by the formula
$N = -1.65x^2 + 51.8x + 111.44$, *in which N represents the cumulative number of U.S. AIDS deaths, in thousands, x years after 1990. Use this formula to solve Exercises 103–104.*

Cumulative Number of Deaths from AIDS in the United States

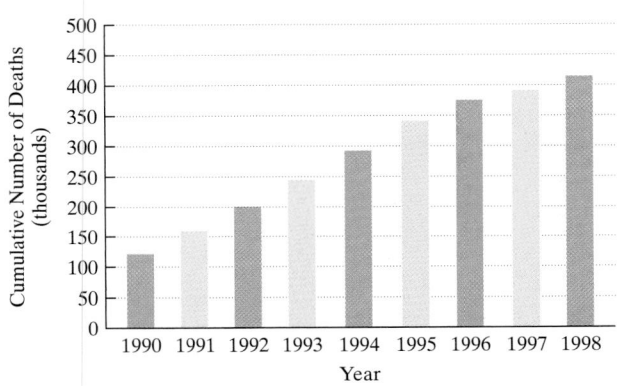

Source: Centers for Disease Control

Year	Cumulative Number of Deaths from AIDS, in Thousands
1990	121
1991	158
1992	199
1993	243
1994	292
1995	340
1996	375
1997	390
1998	414

103. According to the formula, in which year did the cumulative number of deaths from AIDS in the United States reach 330 thousand? What was the actual number for that year? How well does the formula describe the situation for that year?

104. According to the formula, in which year will the total number of U.S. AIDS deaths reach 500 thousand?

105. A baseball diamond is actually a square with 90-foot sides. What is the distance from home plate to second base?

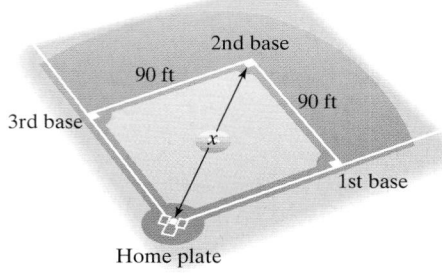

2nd base

90 ft

90 ft

3rd base

x

1st base

Home plate

106. A 20-foot ladder is 15 feet from the house. How far up the house does the ladder reach?

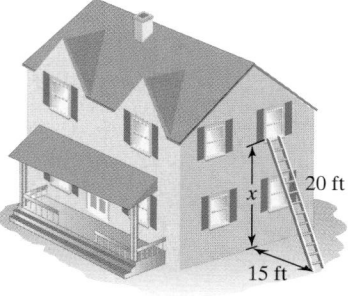

20 ft

x

15 ft

107. An 8-foot tree is supported by two wires that extend from the top of the tree to a point on the ground located 15 feet from the base of the tree. Find the total length of the two support wires.

108. A vertical pole is supported by three wires. Each wire is 13 yards long and is anchored 5 yards from the base of the pole. How far up the pole will the wires be attached?

109. The length of a rectangular garden is 5 feet greater than the width. The area of the garden is 300 square feet. Find the length and the width.

110. A rectangular parking lot has a length that is 3 yards greater than the width. The area of the rectangular lot is 180 square yards. Find the length and the width.

111. A machine produces open boxes using square sheets of metal. The figure illustrates that the machine cuts equal-sized squares measuring 2 inches on a side from the corners and then shapes the metal into an open box by turning up the sides. If each box must have a volume of 200 cubic inches, find the size of the length and width of the open box.

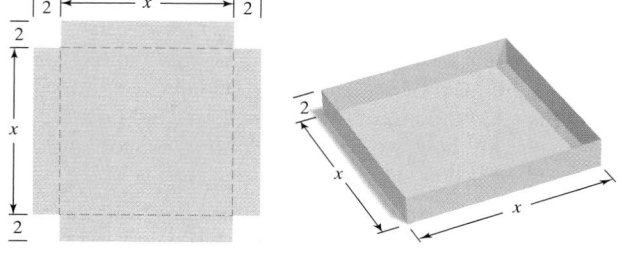

| 2 |←—— x ——→| 2 |

2

x

2

2

x

x

112. A machine produces open boxes using square sheets of metal. The machine cuts equal-sized squares measuring 3 inches on a side from the corners and then shapes the metal into an open box by turning up the sides. If each box must have a volume of 75 cubic inches, find the size of the length and width of the open box.

Writing in Mathematics

113. What is a quadratic equation?

114. Explain how to solve $x^2 + 6x + 8 = 0$ using factoring and the zero-product principle.

115. Explain how to solve $x^2 + 6x + 8 = 0$ by completing the square.

116. Explain how to solve $x^2 + 6x + 8 = 0$ using the quadratic formula.

117. How is the quadratic formula derived?

118. What is the discriminant and what information does it provide about a quadratic equation?

119. If you are given a quadratic equation, how do you determine which method to use to solve it?

120. If $(x + 2)(x - 4) = 0$ indicates that $x + 2 = 0$ or $x - 4 = 0$, explain why $(x + 2)(x - 4) = 6$ does not mean $x + 2 = 6$ or $x - 4 = 6$. Could we solve the equation using $x + 2 = 3$ and $x - 4 = 2$ because $3 \cdot 2 = 6$?

Technology Exercises

121. If you have access to a calculator that solves quadratic equations, consult the owner's manual to determine how to use this feature. Then use your calculator to solve any five of the equations in Exercises 55–64.

122. Graph the formula in Exercise 95,

$$y = 0.0075x^2 - 0.2676x + 14.8$$

using a $[0, 80, 20]$ by $[0, 50, 10]$ viewing rectangle. Move the cursor along the graph, using the $\boxed{\text{TRACE}}$ and $\boxed{\text{ZOOM}}$ features to estimate the year in which automobile fuel efficiency was the poorest. What was the gas mileage in that year? Use the Internet or your library to find pictures of the most popular cars for that year and write a sentence relating the most popular cars of the time with the fuel efficiency indicated by the formula's graph.

Critical Thinking Exercises

123. Which one of the following is true?
   a. The equation $(2x - 3)^2 = 25$ is equivalent to $2x - 3 = 5$.
   b. Every quadratic equation has two distinct numbers in its solution set.

c. A quadratic equation whose coefficients are real numbers can never have a solution set containing one real number and one complex nonreal number.

d. The equation $ax^2 + c = 0$ cannot be solved by the quadratic formula.

124. Solve the equation: $x^2 + 2\sqrt{3}x - 9 = 0$.

125. Write a quadratic equation in standard form whose solution set is $\{-3, 5\}$.

126. A person throws a rock upward from the edge of an 80-foot cliff. The height, h, in feet, of the rock above the water at the bottom of the cliff after t seconds is described by the formula

$$h = -16t^2 + 64t + 80.$$

How long will it take for the rock to reach the water?

127. The personnel manager of a roller skate company knows that the company's weekly revenue is a function of the price of each pair of skates, modeled by $R = -2x^2 + 36x$,

where x represents the dollar price of a pair of skates and R represents weekly revenue in tens of thousands of dollars. A job applicant promises the personnel manager an advertising campaign guaranteed to generate $190,000 in weekly revenue. Substitute 19 for R in the given model, compute the discriminant, and then explain why the applicant will or will not be hired in the advertising department.

Group Exercise

128. Each group member should find an algebraic formula that contains an expression in the form $ax^2 + bx + c$ on one side that he or she finds intriguing. Consult college algebra books or liberal arts mathematics books to do so. Group members should select four of the formulas. For each formula selected, write and solve a problem similar to Exercises 95 and 96 in this exercise set.

SECTION 1.4 *Other Types of Equations*

Objectives

1. Solve polynomial equations by factoring.
2. Solve radical equations.
3. Solve equations with rational exponents.
4. Solve equations that are quadratic in form.
5. Solve equations involving absolute value.

The Galápagos Islands are a volcanic chain of islands lying 600 miles west of Ecuador. They are famed for their extraordinary wildlife, which includes a rare flightless cormorant, marine iguanas, and giant tortoises weighing more than 600 pounds. It was here that naturalist Charles Darwin began to formulate his theory of evolution. Darwin made an enormous collection of the islands' plant species. The formula

$$S = 28.5\sqrt[3]{x}$$

describes the number of plant species, S, on the various islands of the Galápagos chain in terms of the area, x, in square miles, of a particular island.

How can we find the area of a Galápagos island with 57 species of plants? Substitute 57 for S in the formula and solve for x:

$$57 = 28.5\sqrt[3]{x}.$$

The resulting equation contains a variable in the radicand and is called a *radical equation*. In this section, in addition to radical equations, we will show you how to solve certain kinds of polynomial equations, equations involving rational exponents, and equations involving absolute value.

1 Solve polynomial equations by factoring.

Polynomial Equations

The linear and quadratic equations that we studied in the first three sections of this chapter can be thought of as polynomial equations of degrees 1 and 2, respectively. By contrast, consider the following polynomial equations of degree greater than 2.

$$3x^4 = 27x^2 \qquad\qquad x^3 + x^2 = 4x + 4$$

This equation is of degree 4 because 4 is the largest exponent.

This equation is of degree 3 because 3 is the largest exponent.

We can solve these equations by moving all terms to one side, thereby obtaining zero on the other side. We then use factoring and the zero-product principle.

EXAMPLE 1   Solving a Polynomial Equation by Factoring

Solve by factoring: $3x^4 = 27x^2$.

Solution

Step 1   Move all terms to one side and obtain zero on the other side.   Subtract $27x^2$ from both sides.

$$3x^4 - 27x^2 = 27x^2 - 27x^2$$
$$3x^4 - 27x^2 = 0$$

Step 2   Factor.   We can factor $3x^2$ from each term.

$$3x^4 - 27x^2 = 0$$
$$3x^2(x^2 - 9) = 0$$

Steps 3 and 4   Set each factor equal to zero and solve the resulting equations.

$$3x^2 = 0 \qquad \text{or} \qquad x^2 - 9 = 0$$
$$x^2 = 0 \qquad\qquad x^2 = 9$$
$$x = \pm\sqrt{0} \qquad\qquad x = \pm\sqrt{9}$$
$$x = 0 \qquad\qquad x = \pm 3$$

Step 5   Check the solutions in the original equation.   Check the three solutions, 0, -3, and 3, by substituting them into the original equation. Can you verify that the solution set is $\{-3, 0, 3\}$?

Study Tip

In solving $3x^4 = 27x^2$, be careful not to divide both sides by x^2. If you do, you'll lose 0 as a solution. In general, do not divide both sides of an equation by a variable because that variable might take on the value 0 and you cannot divide by 0.

Check Point 1   Solve by factoring: $4x^4 = 12x^2$.

EXAMPLE 2   Solving a Polynomial Equation by Factoring

Solve by factoring: $x^3 + x^2 = 4x + 4$.

Solution

Step 1   Move all terms to one side and obtain zero on the other side. Subtract $4x + 4$ from both sides.

$$x^3 + x^2 - 4x - 4 = 4x + 4 - 4x - 4$$
$$x^3 + x^2 - 4x - 4 = 0$$

Step 2   Factor. Use factoring by grouping. Group terms that have a common factor.

$$\boxed{x^3 + x^2} + \boxed{-4x - 4} = 0$$

Common factor is x^2.   Common factor is -4.

$$x^2(x + 1) - 4(x + 1) = 0 \qquad \text{Factor } x^2 \text{ from the first two terms and } -4 \text{ from the last two terms.}$$

$$(x + 1)(x^2 - 4) = 0 \qquad \text{Factor out the common binomial, } x + 1, \text{ from each term.}$$

Steps 3 and 4   Set each factor equal to zero and solve the resulting equations.

$$x + 1 = 0 \qquad \text{or} \qquad x^2 - 4 = 0$$
$$x = -1 \qquad\qquad x^2 = 4$$
$$\qquad\qquad x = \pm\sqrt{4} = \pm2$$

Step 5   Check the solutions in the original equation. Check the three solutions, $-1, -2,$ and $2,$ by substituting them into the original equation. Can you verify that the solution set is $\{-2, -1, 2\}$?

Technology

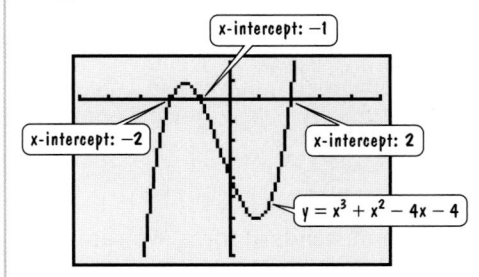

You can use a graphing utility to check the solutions to $x^3 + x^2 - 4x - 4 = 0$. Graph $y = x^3 + x^2 - 4x - 4$, as shown on the left. The x-intercepts are $-2, -1,$ and $2,$ corresponding to the equation's solutions.

Check Point 2   Solve by factoring: $2x^3 + 3x^2 = 8x + 12$.

② Solve radical equations.

Equations Involving Radicals

A **radical equation** is an equation in which the variable occurs in a square root, cube root, or any higher root. An example of a radical equation is

$$28.5\sqrt[3]{x} = 57.$$

The variable occurs in a cube root.

This equation can be used to find the area, x, of a Galápagos island with 57 species of plants. First, we isolate the radical by dividing both sides of the equation by 28.5.

$$\frac{28.5\sqrt[3]{x}}{28.5} = \frac{57}{28.5}$$

$$\sqrt[3]{x} = 2$$

Next we eliminate the radical by raising each side of the equation to a power equal to the index of the radical. Because the index is 3, we cube both sides of the equation.

$$\left(\sqrt[3]{x}\right)^3 = 2^3$$

$$x = 8$$

Thus, a Galápagos island with 57 species of plants has an area of 8 square miles.

The Galápagos equation shows that solving equations involving radicals involves raising both sides of the equation to a power equal to the radical's index. All solutions of the original equation are also solutions of the resulting equation. However, the resulting equation may have some extra solutions that do not satisfy the original equation. Because the resulting equation may not be equivalent to the original equation, we must check each proposed solution by substituting it into the original equation. Let's see exactly how this works.

EXAMPLE 3 Solving an Equation Involving a Radical

Solve: $x + \sqrt{26 - 11x} = 4$.

Solution   To solve this equation, we isolate the radical expression $\sqrt{26 - 11x}$ on one side of the equation. By squaring both sides of the equation, we can then eliminate the square root.

$$x + \sqrt{26 - 11x} = 4 \qquad \text{This is the given equation.}$$

$$x + \sqrt{26 - 11x} - x = 4 - x \qquad \text{Isolate the radical by subtracting } x \text{ from both sides.}$$

$$\sqrt{26 - 11x} = 4 - x \qquad \text{Simplify.}$$

$$\left(\sqrt{26 - 11x}\right)^2 = (4 - x)^2 \qquad \text{Square both sides.}$$

$$26 - 11x = 16 - 8x + x^2 \qquad \text{Use } (A - B)^2 = A^2 - 2AB + B^2 \text{ to square } 4 - x.$$

Next, we need to write this quadratic equation in standard form. We can obtain zero on the left side by subtracting 26 and adding $11x$ on both sides.

$$26 - 26 - 11x + 11x = 16 - 26 - 8x + 11x + x^2$$

$$0 = x^2 + 3x - 10 \qquad \text{Simplify.}$$

$$0 = (x + 5)(x - 2) \qquad \text{Factor.}$$

$$x + 5 = 0 \quad \text{or} \quad x - 2 = 0 \qquad \text{Set each factor equal to zero.}$$

$$x = -5 \qquad\qquad x = 2 \qquad \text{Solve for x.}$$

We have not completed the solution process. Although -5 and 2 satisfy the squared equation, there is no guarantee that they satisfy the original equation. Thus, we must check the proposed solutions. We can do this using a graphing utility (see the technology box in the margin) or by substituting both proposed solutions into the given equation

$$x + \sqrt{26 - 11x} = 4.$$

Study Tip

Be sure to square *both sides* of an equation. Do *not* square each term.

Correct:

$$\left(\sqrt{26 - 11}\right)^2 = (4 - x)^2$$

Incorrect:

$$\left(\sqrt{26 - 11}\right)^2 = 4^2 - x^2$$

Technology

The graph of

$$y = x + \sqrt{26 - 11x} - 4$$

is shown in a $[-10, 3, 1]$ by $[-4, 3, 1]$ viewing rectangle. The x-intercepts are -5 and 2, verifying $\{-5, 2\}$ as the solution set.

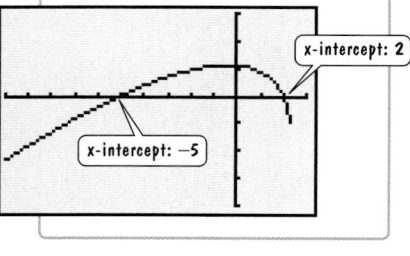

Check −5:

$$-5 + \sqrt{26 - 11(-5)} \stackrel{?}{=} 4$$

$$-5 + \sqrt{81} \stackrel{?}{=} 4$$

$$-5 + 9 \stackrel{?}{=} 4$$

$$4 = 4 \checkmark$$

Check 2:

$$2 + \sqrt{26 - 11(2)} \stackrel{?}{=} 4 \qquad \text{Substitute } -5 \text{ and } 2, \text{ respectively.}$$

$$2 + \sqrt{4} \stackrel{?}{=} 4 \qquad \text{Simplify.}$$

$$2 + 2 \stackrel{?}{=} 4$$

$$4 = 4 \checkmark \qquad \text{Both } -5 \text{ and } 2 \text{ are solutions.}$$

The solution set is $\{-5, 2\}$.

Check Point 3   Solve and check: $\sqrt{6x + 7} - x = 2.$

Study Tip

Don't forget to check for extraneous solutions when solving equations by raising both sides to an even power. Here's a simple example:

$$x = 4$$

$$x^2 = 16 \qquad \text{Square both sides.}$$

$$x = \pm\sqrt{16} \qquad \text{Use the square root method.}$$

$$x = \pm 4$$

However, −4 does not check in $x = 4$. Thus, −4 is an extraneous solution.

When solving a radical equation, extra solutions may be introduced when you raise both sides of the equation to an even power. Such solutions are called **extraneous solutions**.

The solution of radical equations with two or more square root expressions involves isolating a radical, squaring both sides, and then repeating this process. Let's consider an equation containing two square root expressions.

EXAMPLE 4   Solving an Equation Involving Two Radicals

Solve: $\sqrt{3x + 1} - \sqrt{x + 4} = 1.$

Solution

$$\sqrt{3x + 1} - \sqrt{x + 4} = 1 \qquad \text{This is the given equation.}$$

$$\sqrt{3x + 1} = \sqrt{x + 4} + 1 \qquad \text{Isolate one of the radicals by adding } \sqrt{x + 4} \text{ to both sides.}$$

$$\left(\sqrt{3x + 1}\right)^2 = \left(\sqrt{x + 4} + 1\right)^2 \qquad \text{Square both sides.}$$

Squaring the expression on the right side of the equation can be a bit tricky. We need to use the formula

$$(A + B)^2 = A^2 + 2AB + B^2.$$

Focusing on just the right side, here is how the squaring is done.

$$(A + B)^2 = A^2 + 2 \quad A \quad B + B^2$$

$$\left(\sqrt{x + 4} + 1\right)^2 = \left(\sqrt{x + 4}\right)^2 + 2 \cdot \sqrt{x + 4} \cdot 1 + 1^2$$

This simplifies to $x + 4 + 2\sqrt{x + 4} + 1$. Thus, our equation can be written as follows.

$$3x + 1 = x + 4 + 2\sqrt{x + 4} + 1$$

$$3x + 1 = x + 5 + 2\sqrt{x + 4} \qquad \text{Combine numerical terms on the right.}$$

$$2x - 4 = 2\sqrt{x + 4} \qquad \text{Isolate } 2\sqrt{x + 4}, \text{ the radical term, by subtracting } x + 5 \text{ from both sides.}$$

$$x - 2 = \sqrt{x + 4} \qquad \text{Divide both sides by 2.}$$

$$(x - 2)^2 = \left(\sqrt{x + 4}\right)^2 \qquad \text{Square both sides.}$$

$$x^2 - 4x + 4 = x + 4$$ Multiply.

$$x^2 - 5x = 0$$ Write the quadratic equation in standard form by subtracting x + 4 from both sides.

$$x(x - 5) = 0$$ Factor.

$$x = 0 \quad \text{or} \quad x - 5 = 0$$ Set each factor equal to zero.

$$x = 0 \qquad\qquad x = 5$$ Solve for x.

Complete the solution process by checking both proposed solutions. We can do this using a graphing utility (see the technology box in the margin) or by substituting both proposed solutions in the given equation.

Technology

The graph of

$$y = \sqrt{3x + 1} - \sqrt{x + 4} - 1$$

has only one x-intercept at 5. This verifies that the solution set for the given equation is $\{5\}$.

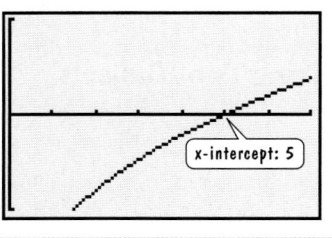

x-intercept: 5

Check 0:

$$\sqrt{3x + 1} - \sqrt{x + 4} = 1$$
$$\sqrt{3 \cdot 0 + 1} - \sqrt{0 + 4} \overset{?}{=} 1$$
$$\sqrt{1} - \sqrt{4} \overset{?}{=} 1$$
$$1 - 2 \overset{?}{=} 1$$
$$-1 = 1 \quad \text{False}$$

Check 5:

$$\sqrt{3x + 1} - \sqrt{x + 4} = 1$$
$$\sqrt{3 \cdot 5 + 1} - \sqrt{5 + 4} \overset{?}{=} 1$$
$$\sqrt{16} - \sqrt{9} \overset{?}{=} 1$$
$$4 - 3 \overset{?}{=} 1$$
$$1 = 1 \quad \checkmark$$

The check indicates that 0 is not a solution. It is an extraneous solution brought about by squaring each side of the equation. The only solution is 5, and the solution set is $\{5\}$.

Check Point 4 Solve and check: $\sqrt{x + 5} - \sqrt{x - 3} = 2$.

Radicals and Windchill

The way that we perceive the temperature on a cold day depends on both air temperature and wind speed. The windchill temperature is what the air temperature would have to be with no wind to achieve the same chilling effect on the skin. The formula that describes windchill temperature, W, in terms of the velocity of the wind, v, in miles per hour, and the actual air temperature, t, in degrees Fahrenheit, is

$$W = 91.4 - \frac{(10.5 + 6.7\sqrt{v} - 0.45v)(457 - 5t)}{110}.$$

Use your calculator to describe how cold the air temperature feels (that is, the windchill temperature) when the temperature is 15° Fahrenheit and the wind is 5 miles per hour. Contrast this with a temperature of 40° Fahrenheit and a wind blowing at 50 miles per hour.

3 Solve equations with rational exponents.

Because $\sqrt[n]{b}$ can be expressed as $b^{1/n}$, radical equations can be written using rational exponents. For example, the Galápagos equation

$$28.5\sqrt[3]{x} = 57$$

can be written

$$28.5x^{1/3} = 57.$$

We solve this equation exactly as we did when it was expressed in radical form. First, isolate $x^{1/3}$.

$$\frac{28.5x^{1/3}}{28.5} = \frac{57}{28.5}$$

$$x^{1/3} = 2$$

Complete the solution process by raising both sides to the third power.

$$\left(x^{1/3}\right)^3 = 2^3$$

$$x = 8$$

In general, a radical equation with rational exponents can be solved by: (1) isolating the expression with the rational exponent, and (2) raising both sides of the equation to a power that is the reciprocal of the rational exponent. Be sure to *complete the solution process* by *checking all proposed solutions in the original equation* to find out if they are actual solutions or extraneous solutions.

EXAMPLE 5 **Solving an Equation Involving a Rational Exponent**

Solve: $3x^{3/4} - 6 = 0$.

Solution Our goal is to isolate $x^{3/4}$. Then we can raise both sides of the equation to the $\frac{4}{3}$ power because $\frac{4}{3}$ is the reciprocal of $\frac{3}{4}$.

$$3x^{3/4} - 6 = 0 \qquad \text{This is the given equation; we will isolate } x^{3/4}.$$

$$3x^{3/4} = 6 \qquad \text{Add 6 to both sides.}$$

$$\frac{3x^{3/4}}{3} = \frac{6}{3} \qquad \text{Divide both sides by 3.}$$

$$x^{3/4} = 2 \qquad \text{Simplify.}$$

$$\left(x^{3/4}\right)^{4/3} = 2^{4/3} \qquad \text{Raise both sides to the } \tfrac{4}{3} \text{ power.}$$

$$x = 2^{4/3} \qquad \text{Simplify the left side: } \left(x^{3/4}\right)^{4/3} = x^{\frac{3 \cdot 4}{4 \cdot 3}} = x^{\frac{12}{12}} = x^1 = x.$$

The proposed solution is $2^{4/3}$. Complete the solution process by checking this value in the given equation.

$$3x^{3/4} - 6 = 0 \qquad \text{This is the original equation.}$$

$$3\left(2^{4/3}\right)^{3/4} - 6 \stackrel{?}{=} 0 \qquad \text{Substitute the proposed solution.}$$

$$3 \cdot 2 - 6 \stackrel{?}{=} 0 \qquad \left(2^{4/3}\right)^{3/4} = 2^{\frac{4 \cdot 3}{3 \cdot 4}} = 2^{\frac{12}{12}} = 2^1 = 2.$$

$$0 = 0 \checkmark \qquad \text{Thus, } 2^{4/3} \text{ is a solution.}$$

The solution is $2^{4/3} = \sqrt[3]{2^4} \approx 2.52$. The solution set is $\left\{2^{4/3}\right\}$.

Check Point 5   Solve and check: $5x^{3/2} - 25 = 0$.

4 Solve equations that are quadratic in form.

Equations That Are Quadratic in Form

Some equations that are not quadratic can be written as quadratic equations using an appropriate substitution. Here are some examples.

Given Equation	Substitution	New Equation
$x^4 - 8x^2 - 9 = 0$ or $(x^2)^2 - 8x^2 - 9 = 0$	$t = x^2$	$t^2 - 8t - 9 = 0$
$5x^{2/3} + 11x^{1/3} + 2 = 0$ or $5(x^{1/3})^2 + 11x^{1/3} + 2 = 0$	$t = x^{1/3}$	$5t^2 + 11t + 2 = 0$

An equation that is **quadratic in form** is one that can be expressed as a quadratic equation using an appropriate substitution. Both of the preceding given equations are quadratic in form.

For equations that are quadratic in form, the exponent in one of the terms is half that of the other term. By letting t equal the variable to the half power, a quadratic equation in t will result. Now it's easy. Solve this quadratic equation for t. Finally, use your substitution to find the values for the variable in the given equation. Example 6 shows how this is done.

EXAMPLE 6   Solving an Equation That Is Quadratic in Form

Solve: $x^4 - 8x^2 - 9 = 0$.

Solution   Notice that the exponent on x^2 is half that of the exponent on x^4. We let t equal the variable to the power that is half of 4. Thus,

$$\text{let } t = x^2.$$

Now we write the given equation as a quadratic equation in t and solve for t.

$x^4 - 8x^2 - 9 = 0$ — This is the given equation.

$(x^2)^2 - 8x^2 - 9 = 0$ — The given equation contains x^2 and x^2 squared.

$t^2 - 8t - 9 = 0$ — Replace x^2 by t.

$(t - 9)(t + 1) = 0$ — Factor.

$t - 9 = 0$ or $t + 1 = 0$ — Apply the zero-product principle.

$t = 9$ $t = -1$ — Solve for t.

We're not done! Why not? We were asked to solve for x and we have values for t. We use the original substitution, $t = x^2$, to solve for x. Replace t by x^2 in each equation shown.

$$x^2 = 9 \qquad x^2 = -1$$
$$x = \pm\sqrt{9} \qquad x = \pm\sqrt{-1}$$
$$x = \pm 3 \qquad x = \pm i$$

The solution set is $\{-3, 3, -i, i\}$.

> **Check Point 6** Solve: $x^4 - 5x^2 + 6 = 0$.

EXAMPLE 7   Solving an Equation That Is Quadratic in Form

Solve: $5x^{2/3} + 11x^{1/3} + 2 = 0$.

Solution   Notice that the exponent on $x^{1/3}$ is half that of the exponent on $x^{2/3}$. We let t equal the variable to the power that is half of 4. Thus,

$$\text{let } t = x^{1/3}.$$

Now we write the given equation as a quadratic equation in t and solve for t.

$$5x^{2/3} + 11x^{1/3} + 2 = 0 \qquad \text{This is the given equation.}$$

$$5(x^{1/3})^2 + 11(x^{1/3}) + 2 = 0 \qquad \text{The given equation contains } x^{1/3} \text{ and } x^{1/3} \text{ squared.}$$

$$5t^2 + 11t + 2 = 0 \qquad \text{Replace } x^{1/3} \text{ by } t.$$

$$(5t + 1)(t + 2) = 0 \qquad \text{Factor.}$$

$$5t + 1 = 0 \quad \text{ or } \quad t + 2 = 0 \qquad \text{Set each factor equal to 0.}$$

$$5t = -1 \qquad\qquad t = -2 \quad \text{Solve for } t.$$

$$t = -\tfrac{1}{5}$$

Use the original substitution, $t = x^{1/3}$, to solve for x. Replace t by $x^{1/3}$ in each of the preceding equations.

$$x^{1/3} = -\frac{1}{5} \qquad\qquad x^{1/3} = -2 \qquad \text{Replace } t \text{ with } x^{1/3}.$$

$$(x^{1/3})^3 = \left(-\frac{1}{5}\right)^3 \qquad (x^{1/3})^3 = (-2)^3 \qquad \text{Solve for } x \text{ by cubing both sides of each equation.}$$

$$x = -\frac{1}{125} \qquad\qquad x = -8$$

Check these values to verify that the solution set is $\left\{-\frac{1}{125}, -8\right\}$.

> **Check Point 7** Solve: $3x^{2/3} - 11x^{1/3} - 4 = 0$.

5 Solve equations involving absolute value.

Equations Involving Absolute Value

We solve equations containing absolute value using the fact that the expression inside the absolute value bars can be either positive or negative. For example, the equation

$$|2x - 3| = 11$$

is satisfied if $2x - 3$ is either 11 or -11, resulting in the two equations

$$2x - 3 = 11 \quad \text{or} \quad 2x - 3 = -11.$$

Rewriting an Absolute Value Equation without Absolute Value Bars

If c is a positive real number and X represents any algebraic expression, then $|X| = c$ is equivalent to $X = c$ or $X = -c$.

EXAMPLE 8 Solving an Equation Involving Absolute Value

Solve: $|2x - 3| = 11$.

Solution

$\|2x - 3\| = 11$		This is the given equation.
$2x - 3 = 11$ or $2x - 3 = -11$		Rewrite the equation without absolute value bars.
$2x = 14$ $2x = -8$		Add 3 to both sides of each equation.
$x = 7$ $x = -4$		Divide both sides of each equation by 2.

Discovery

Graph $y = |2x - 3|$ and $y = 11$ in a $[-10, 10, 1]$ by $[-1, 15, 1]$ viewing rectangle. How is the solution set of $|2x - 3| = 11$, namely $\{-4, 7\}$, shown by the graphs?

Check

$\|2x - 3\| = 11$		This is the original equation.
$\|2(7) - 3\| \stackrel{?}{=} 11$ $\|2(-4) - 3\| \stackrel{?}{=} 11$		Substitute the proposed solutions.
$\|14 - 3\| \stackrel{?}{=} 11$ $\|-8 - 3\| \stackrel{?}{=} 11$		Perform operations inside the absolute value bars.
$\|11\| \stackrel{?}{=} 11$ $\|-11\| \stackrel{?}{=} 11$		
$11 = 11$ ✓ $11 = 11$ ✓		These true statements indicate that 7 and -4 are solutions.

The solution set is $\{-4, 7\}$.

Check Point 8 Solve: $|2x - 1| = 5$.

EXERCISE SET 1.4

✓ Practice Exercises

Solve each polynomial equation in Exercises 1–10 by factoring and then using the zero-product principle.

1. $3x^4 - 48x^2 = 0$
2. $5x^4 - 20x^2 = 0$
3. $2x^4 = 16x$
4. $3x^4 = 81x$
5. $3x^3 + 2x^2 = 12x + 8$
6. $4x^3 - 12x^2 = 9x - 27$
7. $2x - 3 = 8x^3 - 12x^2$
8. $x + 1 = 9x^3 + 9x^2$
9. $4y^3 - 2 = y - 8y^2$
10. $9y^3 + 8 = 4y + 18y^2$

Solve each radical equation in Exercises 11–28. Check all proposed solutions.

11. $\sqrt{3x + 18} = x$
12. $\sqrt{20 - 8x} = x$
13. $\sqrt{x + 3} = x - 3$
14. $\sqrt{x + 10} = x - 2$
15. $\sqrt{2x + 13} = x + 7$
16. $\sqrt{6x + 1} = x - 1$
17. $x - \sqrt{2x + 5} = 5$
18. $x - \sqrt{x + 11} = 1$
19. $\sqrt{3x + 10} = x + 4$
20. $\sqrt{x} - 3 = x - 9$
21. $\sqrt{x + 8} - \sqrt{x - 4} = 2$
22. $\sqrt{x + 5} - \sqrt{x - 3} = 2$
23. $\sqrt{x - 5} - \sqrt{x - 8} = 3$
24. $\sqrt{2x - 3} - \sqrt{x - 2} = 1$
25. $\sqrt{2x + 3} + \sqrt{x - 2} = 2$
26. $\sqrt{x + 2} + \sqrt{3x + 7} = 1$
27. $\sqrt{3\sqrt{x + 1}} = \sqrt{3x - 5}$
28. $\sqrt{1 + 4\sqrt{x}} = 1 + \sqrt{x}$

Solve and check each equation with rational exponents in Exercises 29–36.

29. $x^{3/2} = 8$
30. $x^{3/2} = 27$
31. $(x - 4)^{3/2} = 27$
32. $(x + 5)^{3/2} = 8$
33. $6x^{5/2} - 12 = 0$
34. $8x^{5/3} - 24 = 0$
35. $(x^2 - x - 4)^{3/4} - 2 = 6$
36. $(x^2 - 3x + 3)^{3/2} - 1 = 0$

Solve each equation in Exercises 37–56 by making an appropriate substitution.

37. $x^4 - 5x^2 + 4 = 0$
38. $x^4 - 13x^2 + 36 = 0$
39. $9x^4 = 25x^2 - 16$
40. $4x^4 = 13x^2 - 9$
41. $x^6 + 8x^3 + 15 = 0$
42. $x^6 + 5x^3 + 6 = 0$
43. $5x^6 + x^3 = 18$
44. $3x^6 - 4x^3 = 15$

45. $x^{2/3} - x^{1/3} - 6 = 0$ **46.** $2x^{2/3} + 7x^{1/3} - 15 = 0$

47. $x^{3/2} - 2x^{3/4} + 1 = 0$ **48.** $x^{2/5} + x^{1/5} - 6 = 0$

49. $2x - 3x^{1/2} + 1 = 0$ **50.** $x + 3x^{1/2} - 4 = 0$

51. $(x - 5)^2 - 4(x - 5) - 21 = 0$

52. $(x + 3)^2 + 7(x + 3) - 18 = 0$

53. $(x^2 - x)^2 - 14(x^2 - x) + 24 = 0$

54. $(x^2 - 2x)^2 - 11(x^2 - 2x) + 24 = 0$

55. $\left(y - \dfrac{8}{y}\right)^2 + 5\left(y - \dfrac{8}{y}\right) - 14 = 0$

56. $\left(y - \dfrac{10}{y}\right)^2 + 6\left(y - \dfrac{10}{y}\right) - 27 = 0$

Solve each equation in Exercises 57–62 by first rewriting the equation as two equations without absolute value bars.

57. $|x| = 8$ **58.** $|x| = 6$

59. $|x - 2| = 7$ **60.** $|x + 1| = 5$

61. $|2x - 1| = 5$ **62.** $|2x - 3| = 11$

Solve each equation in Exercises 63–72 by the method of your choice.

63. $x + 2\sqrt{x} - 3 = 0$ **64.** $x^3 + 3x^2 - 4x - 12 = 0$

65. $(x + 4)^{3/2} = 8$ **66.** $(x^2 - 1)^2 - 2(x^2 - 1) = 3$

67. $\sqrt{4x + 15} - 2x = 0$ **68.** $x^{2/5} - 1 = 0$

69. $|x^2 + 2x - 36| = 12$

70. $\sqrt{3x + 1} - \sqrt{x - 1} = 2$

71. $x^3 - 2x^2 = x - 2$ **72.** $|x^2 + 6x + 1| = 8$

 Application Exercises

73. For a group of 50,000 births, the number of people, N, surviving to age x is modeled by the formula $N = 5000\sqrt{100 - x}$. To what age will 40,000 people in the group survive?

74. Psychologists use the formula $N = 2\sqrt{Q} - 9$ to determine the number of nonsense syllables, N, that a subject with an IQ of Q can repeat. If a subject can repeat 14 nonsense syllables, what is that person's IQ? Round to the nearest whole number.

75. The formula $N = 1220\sqrt[3]{x + 42} + 4900$ models the number of congressional aides in the House of Representatives x years after 1930. In what year were approximately 9780 aides assigned to the House of Representatives?

76. Police use the model $v = \sqrt{24L}$ to estimate the speed of a car (v, in miles per hour) prior to an accident based on the length of its skid marks (L, in feet) on dry pavement. If a car is traveling at 60 miles per hour, what is the length of its skid marks?

77. In Albert Einstein's special theory of relativity, time slows down from the point of view of an observer watching an object moving at a velocity close to the speed of light. The formula

$$T = \frac{T_0}{\sqrt{1 - \dfrac{v^2}{c^2}}}$$

relates the passage of time T on a futuristic starship moving at v miles per second to the passage of time for a stationary observer on Earth, T_0. (c, the speed of light, is approximately 186,000 miles per second.) Suppose that 4 hours pass on the starship, but for you, on Earth, 2 hours have passed. How fast is the starship traveling?

78. The graph shows the remaining life expectancies for males and females in the United States in 1998. The formula $E = \sqrt{0.66A^2 - 110.55A + 4680.24}$ is an approximate model for some of the data in the graph, where A represents current age and E stands for remaining life expectancy (in years).

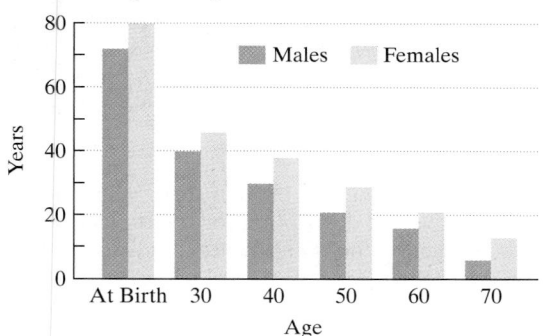

Remaining Life Expectancies in the United States in 1998

Source: Department of Health and Human Services

a. Is the formula a better model for males or for females?

b. If $E = 60$, find A and describe what your result means in terms of the variables modeled by the formula.

79. Laser Records marketing research department determines that weekly demand for a boxed set of CDs by the Jumping Artichokes depends on the price per set, modeled by the formula $p = 30 - \sqrt{0.01x + 1}$, where p represents the price of the set and x represents the number of sets sold each week at price p.

a. At what price will there be no demand for the CD sets?

b. Approximately how many CD sets will sell weekly at a price of $27.76?

Use the Pythagorean Theorem to solve Exercises 80–81.

80. Two vertical poles of lengths 6 feet and 8 feet stand 10 feet apart (see the figure on page 138). A cable reaches from the top of one pole to some point on the ground between the poles and then to the top of the other pole. Where should this point be located to use 18 feet of cable?

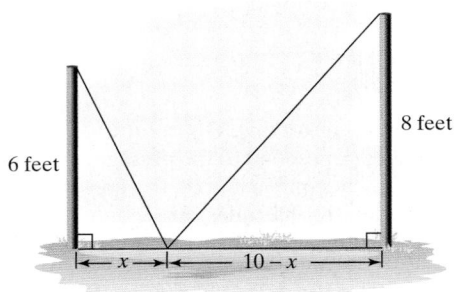

6 feet

8 feet

x $10 - x$

81. Twelve miles separate towns A and B, located 6 miles and 3 miles, respectively, from a major expressway. Two new roads are to be built from A to the expressway and then to B. (See the figure.)

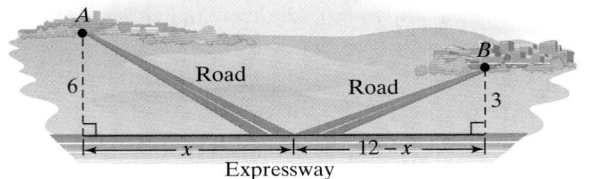

A

6

Road

Road

B

3

x $12 - x$

Expressway

a. Find x if the length of the new roads is 15 miles.
b. Write a verbal description for the road crew telling them where to position the new roads based on your answer to part (a).

Writing in Mathematics

82. Without actually solving the equation, give a general description of how to solve $x^3 - 5x^2 - x + 5 = 0$
83. In solving $\sqrt{3x + 4} - \sqrt{2x + 4} = 2$, why is it a good idea to isolate a radical term? What if we don't do this and simply square each side? Describe what happens.
84. What is an extraneous solution to a radical equation?
85. Explain how to recognize an equation that is quadratic in form. Provide two original examples with your explanation.
86. Describe two methods for solving this equation: $x - 5\sqrt{x} + 4 = 0$.
87. Explain how to solve an equation involving absolute value.
88. Explain why the procedure that you explained in Exercise 87 does not apply to the equation $|x - 2| = -3$. What is the solution set for this equation?

89. Reread Exercise 79. Suppose you are writing a report on the relationship between the price of the CDs and the units that will sell. Assume that you are the only one in the company who can understand the formula $p = 30 - \sqrt{0.01x + 1}$. Consequently, you must describe the relationship between p and x strictly using words. Write a description, minimizing the use of mathematical terminology.

Technology Exercises

Use a graphing utility to solve the equations in Exercises 90–94. Check by direct substitution.

90. $x^3 + 3x^2 - x - 3 = 0$ **91.** $-x^4 + 4x^3 - 4x^2 = 0$
92. $x^4 - 2x^2 + 1 = 0$ **93.** $\sqrt{2x + 13} - x - 5 = 0$
94. $\sqrt{4 - x} - \sqrt{x + 6} - 2 = 0$

95. Use a graphing utility to graph the formula in Exercise 73. In particular, graph $y = 5000\sqrt{100 - x}$ in a $[0, 100, 10]$ by $[0, 50,000, 5000]$ viewing rectangle. Then use the $\boxed{\text{TRACE}}$ feature to trace along the curve until you reach the point that visually shows the solution to Exercise 73.

96. Use a graphing utility to graph the formula in Exercise 74. In particular, graph $y = 2\sqrt{x} - 9$ in a $[0, 180, 10]$ by $[0, 20, 1]$ viewing rectangle. Then use the $\boxed{\text{TRACE}}$ feature to trace along the curve until you reach the point that visually shows the solution to Exercise 74.

Critical Thinking Exercises

97. Which one of the following is true?
a. Squaring both sides of $\sqrt{y + 4} + \sqrt{y - 1} = 5$ leads to $y + 4 + y - 1 = 25$, an equation with no radicals.
b. The equation $(x^2 - 2x)^9 - 5(x^2 - 2x)^3 + 6 = 0$ is quadratic in form and should be solved by letting
$$t = (x^2 - 2x)^3.$$
c. If a radical equation has two proposed solutions and one of these values is not a solution, the other value is also not a solution.
d. None of these statements is true.

98. Solve: $\sqrt{6x - 2} = \sqrt{2x + 3} - \sqrt{4x - 1}$.

99. Solve *without* squaring both sides:
$$5 - \frac{2}{x} = \sqrt{5 - \frac{2}{x}}.$$

100. Solve for x: $\sqrt[3]{x\sqrt{x}} = 9$.

101. Solve for x: $x^{5/6} + x^{2/3} - 2x^{1/2} = 0$.

SECTION 1.5 *Linear Inequalities*

Objectives

1. Graph an inequality's solution set.
2. Use set-builder and interval notations.
3. Use properties of inequalities to solve inequalities.
4. Solve compound inequalities.
5. Solve inequalities involving absolute value.

Rent-a-Heap, a car rental company, charges $125 per week plus $0.20 per mile to rent one of their cars. Suppose you are limited by how much money you can spend for the week: You can spend at most $335. If we let x represent the number of miles you drive the heap in a week, we can write an inequality that models the given conditions.

The weekly charge of $125	plus	the charge of $0.20 per mile for x miles	must be less than or equal to	$335.
125	+	0.20x	≤	335

Using the commutative property of addition, we can express this inequality as $0.20x + 125 \le 335$. The form of this inequality is $ax + b \le c$, with $a = 0.20$, $b = 125$, and $c = 335$. Any inequality in this form is called a **linear inequality in one variable**. The greatest exponent on the variable in such an equality is 1. The symbol between $ax + b$ and c can be $\le$ (is less than or equal to), $<$ (is less than), $\ge$ (is greater than or equal to), or $>$ (is greater than).

In this section, we will study how to solve linear inequalities such as $0.20x + 125 \le 335$. **Solving an inequality** is the process of finding the set of numbers that make the inequality a true statement. These numbers are called the **solutions** of the inequality, and we say that they **satisfy** the inequality. The set of all solutions is called the **solution set** of the inequality. We begin by discussing how to graph and how to represent these solution sets.

1 Graph an inequality's solution set.

Graphs of Inequalities; Interval Notation

There are infinitely many solutions to the inequality $x > -4$, namely all real numbers that are greater than -4. Although we cannot list all the solutions, we can make a drawing on a number line that represents these solutions. Such a drawing is called the **graph of the inequality**.

Graphs of solutions to linear inequalities are shown on a number line by shading all points representing numbers that are solutions. Parentheses indicate endpoints that are not solutions. Square brackets indicate endpoints that are solutions.

EXAMPLE 1 Graphing Inequalities

Graph the solutions of:

    a. $x < 3$ **b.** $x \geq -1$ **c.** $-1 < x \leq 3$.

Study Tip

Because an inequality symbol points to the smaller number, $x < 3$ (x is less than 3) may be expressed as $3 > x$ (3 is greater than x).

Solution

a. The solutions of $x < 3$ are all real numbers that are less than 3. They are graphed on a number line by shading all points to the left of 3. The parenthesis at 3 indicates that 3 is not a solution, but numbers such as 2.9999 and 2.6 are. The arrow shows that the graph extends indefinitely to the left.

b. The solutions of $x \geq -1$ are all real numbers that are greater than or equal to -1. We shade all points to the right of -1 and the point for -1 itself. The bracket at -1 shows that -1 is a solution for the given inequality. The arrow shows that the graph extends indefinitely to the right.

c. The inequality $-1 < x \leq 3$ is read "-1 is less than x *and* x is less than or equal to 3," or "x is greater than -1 *and* less than or equal to 3." The solutions of $-1 < x \leq 3$ are all real numbers between -1 and 3, not including -1 but including 3. The parenthesis at -1 indicates that -1 is not a solution. By contrast, the bracket at 3 shows that 3 is a solution. Shading indicates the other solutions.

Check Point 1

Graph the solutions of:

    a. $x \leq 2$ **b.** $x > -4$ **c.** $2 \leq x < 6$

2 Use set-builder and interval notations.

Now that we know how to graph the solution set for an inequality such as $x > -4$, let's see how to represent the solution set. One method is with **set-builder notation**. Using this method, the solution set for $x > -4$ can be expressed as

$$\{x \mid x > -4\}.$$

The set of all x   such that

We read this as "the set of all real numbers x such that x is greater than -4."

Another method used to represent solution sets for inequalities is **interval notation**. Using this notation, the solution set for $x > -4$ is expressed as $(-4, \infty)$. The parenthesis at -4 indicates that -4 is not included in the interval. The infinity symbol, ∞, does not represent a real number. It indicates that the interval extends indefinitely to the right.

Table 1.5 lists nine possible types of intervals used to describe subsets of real numbers.

Table 1.5   Intervals on the Real Number Line

Let a and b be real numbers such that $a < b$.		
Interval Notation	Set-Builder Notation	Graph
(a, b)	$\{x \mid a < x < b\}$	
$[a, b]$	$\{x \mid a \leq x \leq b\}$	
$[a, b)$	$\{x \mid a \leq x < b\}$	
$(a, b]$	$\{x \mid a < x \leq b\}$	
(a, ∞)	$\{x \mid x > a\}$	
$[a, \infty)$	$\{x \mid x \geq a\}$	
$(-\infty, b)$	$\{x \mid x < b\}$	
$(-\infty, b]$	$\{x \mid x \leq b\}$	
$(-\infty, \infty)$	$\mathbb{R}$ (set of all real numbers)	

EXAMPLE 2   Intervals and Inequalities

Express the intervals in terms of inequalities and graph:

   a. $(-1, 4]$ **b.** $[2.5, 4]$ **c.** $(-4, \infty)$

Solution

   a. $(-1, 4] = \{x \mid -1 < x \leq 4\}$

   b. $[2.5, 4] = \{x \mid 2.5 \leq x \leq 4\}$

   c. $(-4, \infty) = \{x \mid x > -4\}$

Check Point 2   Express the intervals in terms of inequalities and graph:

   a. $[-2, 5)$ **b.** $[1, 3.5]$ **c.** $(-\infty, -1)$

3 Use properties of inequalities to solve inequalities.

Solving Linear Inequalities

Back to our question: How many miles can you drive on your Rent-a-Heap car if you can spend at most $335 per week? We answer the question by solving

$$0.20x + 125 \leq 335$$

for x. The solution procedure is nearly identical to that for solving

$$0.20x + 125 = 335.$$

Our goal is to get x by itself on the left side. We do this by isolating $0.20x$, subtracting 125 from both sides:

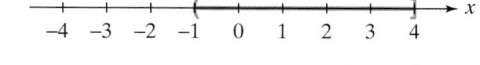

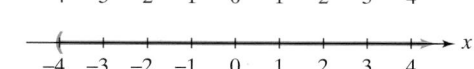

$$0.20x + 125 \leq 335$$
$$0.20x + 125 - 125 \leq 335 - 125$$
$$0.20x \leq 210$$

Finally, we isolate x from $0.20x$ by dividing both sides of the inequality by 0.20:

$$\frac{0.20x}{0.20} \leq \frac{210}{0.20}$$

$$x \leq 1050$$

With at most \$335 per week to spend, you can travel at most 1050 miles.

We started with the inequality $0.20x + 125 \leq 335$ and obtained the inequality $x \leq 1050$ in the final step. Both of these inequalities have the same solution set, namely $\{x \mid x \leq 1050\}$. Inequalities such as these, with the same solution set, are said to be **equivalent**.

We isolated x from $0.20x$ by dividing both sides of $0.20x \leq 210$ by 0.20, a positive number. Let's see what happens if we divide both sides of an inequality by a negative number. Consider the inequality $10 < 14$. Divide 10 and 14 by -2:

$$\frac{10}{-2} = -5 \quad \text{and} \quad \frac{14}{-2} = -7.$$

Because -5 lies to the right of -7 on the number line, -5 is greater than -7:

$$-5 > -7.$$

Notice that the direction of the inequality symbol is reversed:

$$10 < 14$$

$$-5 > -7$$

In general, **when we multiply or divide both sides of an inequality by a negative number, the direction of the inequality symbol is reversed**. When we reverse the direction of the inequality symbol, we say that we change the *sense* of the inequality.

We can isolate a variable in a linear inequality the same way we can isolate a variable in a linear equation. The following properties are used to create equivalent inequalities.

Properties of Inequalities

Property	The Property in Words	Example
Addition and Subtraction Properties If $a < b$, then $a + c < b + c$. If $a < b$, then $a - c < b - c$.	If the same quantity is added to or subtracted from both sides of an inequality, the resulting inequality is equivalent to the original one.	$2x + 3 < 7$ Subtract 3: $2x + 3 - 3 < 7 - 3$ Simplify: $2x < 4$
Positive Multiplication and Division Properties If $a < b$ and c is positive, then $ac < bc$. If $a < b$ and c is positive, then $\dfrac{a}{c} < \dfrac{b}{c}$.	If we multiply or divide both sides of an inequality by the same positive quantity, the resulting inequality is equivalent to the original one.	$2x < 4$ Divide by 2: $\dfrac{2x}{2} < \dfrac{4}{2}$ Simplify: $x < 2$
Negative Multiplication and Division Properties If $a < b$ and c is negative, then $ac > bc$. If $a < b$ and c is negative, then $\dfrac{a}{c} > \dfrac{b}{c}$.	If we multiply or divide both sides of an inequality by the same negative quantity and reverse the direction of the inequality symbol, the result is an equivalent inequality.	$-4x < 20$ Divide by -4 and reverse the sense of the inequality: $\dfrac{-4x}{-4} > \dfrac{20}{-4}$ Simplify: $x > -5$

EXAMPLE 3   Solving a Linear Inequality

Solve and graph the solution set on a number line:

$$3 - 2x < 11.$$

Solution

$3 - 2x \ < \ 11$	This is the given inequality.
$3 - 2x - 3 \ < \ 11 - 3$	Subtract 3 from both sides.
$-2x \ < \ 8$	Simplify.
$\dfrac{-2x}{-2} \ > \ \dfrac{8}{-2}$	Divide both sides by -2 and reverse the sense of the inequality.
$x \ > \ -4$	Simplify.

The solution set consists of all real numbers that are greater than -4, expressed as $\{x \mid x > -4\}$ in set-builder notation. The interval notation for this solution set is $(-4, \infty)$. The graph of the solution set is shown as follows:

```
       (——+——+——+——+——+——+——+——>  x
      -4  -3  -2  -1   0   1   2   3   4
```

> **Check Point 3** Solve and graph the solution set on a number line:
>
> $$2 - 3x \le 5.$$

EXAMPLE 4   Solving a Linear Inequality

Solve and graph the solution set: $7x + 15 \ge 13x + 51.$

Solution   We will collect variable terms on the left and constant terms on the right.

$7x + 15 \ \ge \ 13x + 51$	This is the given inequality.
$7x + 15 - 13x \ \ge \ 13x + 51 - 13x$	Subtract 13x from both sides.
$-6x + 15 \ \ge \ 51$	Simplify.
$-6x + 15 - 15 \ \ge \ 51 - 15$	Subtract 15 from both sides.
$-6x \ \ge \ 36$	Simplify.
$\dfrac{-6x}{-6} \ \le \ \dfrac{36}{-6}$	Divide both sides by -6 and reverse the sense of the inequality.
$x \ \le \ -6$	Simplify.

The solution set consists of all real numbers that are less than or equal to -6, expressed as $\{x \mid x \le -6\}$. The interval notation for this solution set is $(-\infty, -6]$. The graph of the solution set is shown as follows:

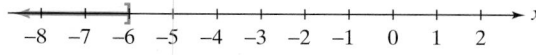

```
   <——+——+——]——+——+——+——+——+——+——+——>  x
     -8  -7  -6  -5  -4  -3  -2  -1   0   1   2
```

> **Check Point 4** Solve and graph the solution set: $6 - 3x \le 5x - 2.$

Study Tip

You can solve

$$7x + 15 \ge 13x + 51$$

by isolating x on the right side. Subtract $7x$ from both sides:

$$7x + 15 - 7x$$
$$\ge 13x + 51 - 7x$$
$$15 \ge 6x + 51$$

Now subtract 51 from both sides.

$$15 - 51 \ge 6x + 51 - 51$$
$$-36 \ge 6x$$

Finally, divide both sides by 6.

$$\frac{-36}{6} \ge \frac{6x}{6}$$

$$-6 \ge x$$

This last inequality means the same thing as

$$x \le -6.$$

Technology

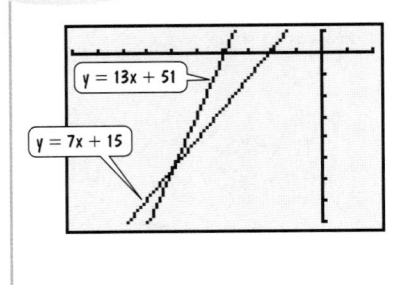

You can use a graphing utility to verify that $(-\infty, -6]$ is the solution set for

$$7x + 15 \geq 13x + 51.$$

For what values of x does the graph of y = 7x + 15 | lie above or on | the graph of y = 13x + 51?

The graphs are shown on the left in a $[-10, 2, 1]$ by $[-40, 5, 5]$ viewing rectangle. Notice that the graph of $y = 7x + 15$ lies above or on the graph of $y = 13x + 51$ when $x \leq -6$.

---

4 Solve compound inequalities.

Solving Compound Inequalities

We now consider two inequalities such as

$$-3 < 2x + 1 \text{ and } 2x + 1 \leq 3$$

expressed as a **compound inequality**

$$-3 < 2x + 1 \leq 3.$$

This double inequality form enables us to solve both inequalities at once. With three parts to a compound inequality, our goal is to **isolate x in the middle**.

EXAMPLE 5   Solving a Compound Inequality

Solve and graph the solution set:

$$-3 < 2x + 1 \leq 3.$$

Solution   We would like to isolate x in the middle. We can do this by first subtracting 1 from all three parts of the compound inequality. Then we isolate x from $2x$ by dividing all three parts of the inequality by 2.

$-3 < 2x + 1 \leq 3$	This is the given inequality.
$-3 - 1 < 2x + 1 - 1 \leq 3 - 1$	Subtract 1 from all three parts.
$-4 < 2x \leq 2$	Simplify.
$\frac{-4}{2} < \frac{2x}{2} \leq \frac{2}{2}$	Divide each part by 2.
$-2 < x \leq 1$	Simplify.

The solution set consists of all real numbers greater than -2 and less than or equal to 1, represented by $\{x \mid -2 < x \leq 1\}$ in set-builder notation and $(-2, 1]$ in interval notation. The graph is shown as follows:

$$\xleftarrow{\hspace{1cm}} \underset{\substack{-5 \;\; -4 \;\; -3 \;\; -2 \;\; -1 \;\;\; 0 \;\;\; 1 \;\;\; 2 \;\;\; 3 \;\;\; 4 \;\;\; 5}}{\rule{4cm}{0pt}} \xrightarrow{\hspace{1cm}} x$$

Check Point 5   Solve and graph the solution set: $1 \leq 2x + 3 < 11$.

5 Solve inequalities involving absolute value.

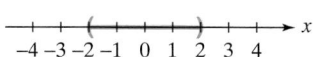

Figure 1.6 $|x| < 2$, so $-2 < x < 2$.

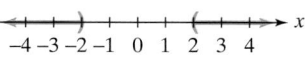

Figure 1.7 $|x| > 2$, so $x < -2$ or $x > 2$.

Study Tip

In the $|X| < c$ case, we have one compound inequality to solve. In the $|X| > c$ case, we have two separate inequalities to solve.

Solving Inequalities with Absolute Value

We have seen that $|x|$ describes the distance of x from zero on a real number line. We can use this geometric interpretation to solve an inequality such as

$$|x| < 2.$$

This means that the distance of x from 0 is *less than* 2, as shown in Figure 1.6. The interval shows values of x that lie less than 2 units from 0. Thus, x can lie between -2 and 2. That is, x is greater than -2 and less than 2. We write $(-2, 2)$ or $\{x|-2 < x < 2\}$.

   Some absolute value inequalities use the "greater than" symbol. For example, $|x| > 2$ means that the distance of x from 0 is *greater than* 2, as shown in Figure 1.7. Thus, x can be less than -2 *or* greater than 2. We write $x < -2$ or $x > 2$.

   These observations suggest the following principles for solving inequalities with absolute value.

Solving an Absolute Value Inequality

If X is an algebraic expression and c is a positive number,

   1. The solutions of $|X| < c$ are the numbers that satisfy $-c < X < c$.
   2. The solutions of $|X| > c$ are the numbers that satisfy $X < -c$ or $X > c$.

These rules are valid if $<$ is replaced by $\leq$ and $>$ is replaced by $\geq$.

EXAMPLE 6 **Solving an Absolute Value Inequality with $<$**

Solve and graph: $|x - 4| < 3$.

Solution

$$|X| < c \quad \text{means} \quad -c < X < c.$$

$$|x - 4| < 3 \quad \text{means} \quad -3 < x - 4 < 3$$

We solve the compound inequality by adding 4 to all three parts.

$$-3 < x - 4 < 3$$
$$-3 + 4 < x - 4 + 4 < 3 + 4$$
$$1 < x < 7$$

The solution set is all real numbers greater than 1 and less than 7, denoted by $\{x|1 < x < 7\}$ or $(1, 7)$. The graph of the solution set is shown as follows:

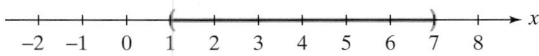

Check Point 6   Solve and graph: $|x - 2| < 5$.

EXAMPLE 7   **Solving an Absolute Value Inequality with ≥**

Solve and graph: $|2x + 3| \geq 5$.

Solution

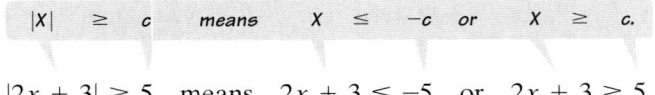

$$|X| \geq c \quad \text{means} \quad X \leq -c \quad \text{or} \quad X \geq c.$$

$$|2x + 3| \geq 5 \quad \text{means} \quad 2x + 3 \leq -5 \quad \text{or} \quad 2x + 3 \geq 5$$

We solve each of these inequalities separately.

$2x + 3 \leq -5$	or	$2x + 3 \geq 5$	These are the inequalities without absolute value bars.
$2x + 3 - 3 \leq -5 - 3$	or	$2x + 3 - 3 \geq 5 - 3$	Subtract 3 from both sides.
$2x \leq -8$	or	$2x \geq 2$	Simplify.
$\dfrac{2x}{2} \leq \dfrac{-8}{2}$	or	$\dfrac{2x}{2} \geq \dfrac{2}{2}$	Divide both sides by 2.
$x \leq -4$	or	$x \geq 1$	Simplify.

Study Tip

The graph of the solution set for $|X| > c$ will be divided into two intervals. The graph of the solution set for $|X| < c$ will be a single interval.

The solution set is $\{x \mid x \leq -4 \text{ or } x \geq 1\}$, that is, all x in $(-\infty, -4]$ or $[1, \infty)$. The graph of the solution set is shown as follows:

$$\xleftarrow{\hspace{1em}} \!\! \underset{-7 \; -6 \; -5 \; -4 \; -3 \; -2 \; -1 \;\; 0 \;\; 1 \;\; 2 \;\; 3}{\rule{0pt}{0pt}} \!\! \xrightarrow{\hspace{1em}} x$$

Check Point 7

Solve and graph: $|2x - 5| \geq 3$.

Applications

Our next example shows how to use an inequality to select the better deal between two pricing options. We will use our five-step strategy for solving problems using mathematical models.

EXAMPLE 8   **Creating and Comparing Mathematical Models**

Acme Car rental agency charges $4 a day plus $0.15 a mile, whereas Interstate rental agency charges $20 a day and $0.05 a mile. How many miles must be driven to make the daily cost of an Acme rental a better deal than an Interstate rental?

Solution

Step 1   Let x represent one of the quantities.   We are looking for the number of miles that must be driven in a day to make Acme the better deal. Thus,

let x = the number of miles driven in a day.

Step 2   Represent other quantities in terms of x.   We are not asked to find another quantity, so we can skip this step.

Step 3 Write an inequality in x that describes the conditions.

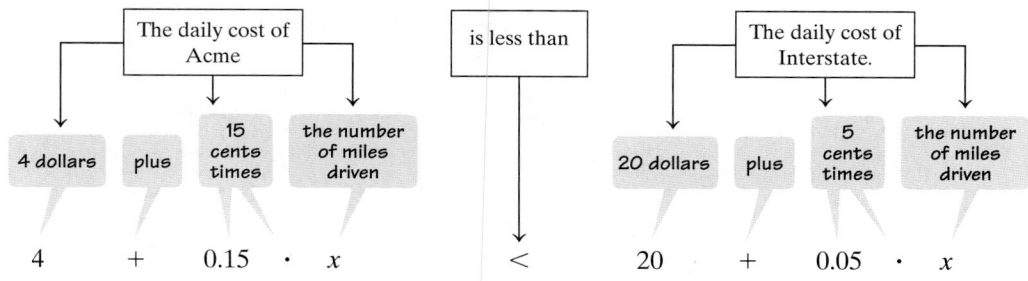

The daily cost of Acme			is less than	The daily cost of Interstate.		
4 dollars	plus	15 cents times	the number of miles driven			
				20 dollars	plus	5 cents times
						the number of miles driven

$$4 \quad + \quad 0.15 \cdot x \qquad\qquad < \qquad\qquad 20 \quad + \quad 0.05 \cdot x$$

Step 4 Solve the inequality and answer the question.

$$4 + 0.15x < 20 + 0.05x \qquad \text{This is the inequality that models the verbal conditions.}$$

$$4 + 0.15x - 0.05x < 20 + 0.05x - 0.05x \qquad \text{Subtract 0.05x from both sides.}$$

$$4 + 0.1x < 20 \qquad \text{Simplify.}$$

$$4 + 0.1x - 4 < 20 - 4 \qquad \text{Subtract 4 from both sides.}$$

$$0.1x < 16 \qquad \text{Simplify.}$$

$$\frac{0.1x}{0.1} < \frac{16}{0.1} \qquad \text{Divide both sides by 0.1.}$$

$$x < 160 \qquad \text{Simplify.}$$

Thus, driving fewer than 160 miles per day makes Acme the better deal.

Step 5 Check the proposed solution in the original wording of the problem.
One way to do this is to take a mileage less than 160 miles per day to see if Acme is the better deal. Suppose that 150 miles are driven in a day.

$$\text{Cost for Acme} = 4 + 0.15(150) = 26.50$$

$$\text{Cost for Interstate} = 20 + 0.05(150) = 27.50$$

and Acme has a lower daily cost, making Acme the better deal.

> **Check Point 8**
>
> A car can be rented from Basic Rental for $260 per week with no extra charge for mileage. Continental charges $80 per week plus 25 cents for each mile driven to rent the same car. How many miles should be driven in a week to make the rental cost for Basic Rental a better deal than Continental's?

EXERCISE SET 1.5

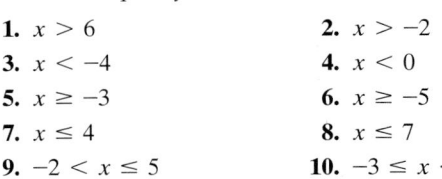

Practice Exercises

In Exercises 1–12, graph the solutions of each inequality on a number line.

1. $x > 6$
2. $x > -2$
3. $x < -4$
4. $x < 0$
5. $x \geq -3$
6. $x \geq -5$
7. $x \leq 4$
8. $x \leq 7$
9. $-2 < x \leq 5$
10. $-3 \leq x < 7$
11. $-1 < x < 4$
12. $-7 \leq x \leq 0$

In Exercises 13–26, express each interval in terms of an inequality and graph the interval on a number line.

13. $(1, 6]$
14. $(-2, 4]$
15. $[-5, 2)$
16. $[-4, 3)$
17. $[-3, 1]$
18. $[-2, 5]$
19. $(2, \infty)$
20. $(3, \infty)$
21. $[-3, \infty)$
22. $[-5, \infty)$
23. $(-\infty, 3)$
24. $(-\infty, 2)$
25. $(-\infty, 5.5)$
26. $(-\infty, 3.5]$

Solve each linear inequality in Exercises 27–48 and graph the solution set on a number line.

27. $5x + 11 < 26$
28. $2x + 5 < 17$
29. $3x - 7 \geq 13$
30. $8x - 2 \geq 14$
31. $-9x \geq 36$
32. $-5x \leq 30$
33. $8x - 11 \leq 3x - 13$
34. $18x + 45 \leq 12x - 8$
35. $4(x + 1) + 2 \geq 3x + 6$
36. $8x + 3 > 3(2x + 1) + x + 5$
37. $2x - 11 < -3(x + 2)$
38. $-4(x + 2) > 3x + 20$
39. $1 - (x + 3) \geq 4 - 2x$
40. $5(3 - x) \leq 3x - 1$
41. $\dfrac{x}{4} - \dfrac{3}{5} \leq \dfrac{x}{2} + 1$
42. $\dfrac{3x}{10} + 1 \geq \dfrac{1}{5} - \dfrac{x}{10}$
43. $1 - \dfrac{x}{2} > 4$
44. $7 - \dfrac{4}{5}x < \dfrac{3}{5}$
45. $\dfrac{x - 4}{6} \geq \dfrac{x - 2}{9} + \dfrac{5}{18}$
46. $\dfrac{4x - 3}{6} + 2 \geq \dfrac{2x - 1}{12}$
47. $4(3x - 2) - 3x < 3(1 + 3x) - 7$
48. $3(x - 8) - 2(10 - x) > 5(x - 1)$

Solve each inequality in Exercises 49–56 by isolating the variable by itself in the middle. Graph the solution set on a number line.

49. $6 < x + 3 < 8$
50. $7 < x + 5 < 11$

51. $-3 \leq x - 2 < 1$
52. $-6 < x - 4 \leq 1$
53. $-11 < 2x - 1 \leq -5$
54. $3 \leq 4x - 3 < 19$
55. $-3 \leq \dfrac{2}{3}x - 5 < -1$
56. $-6 \leq \dfrac{1}{2}x - 4 < -3$

Solve each inequality in Exercises 57–84 by first rewriting each one as an equivalent inequality without absolute value bars. Graph the solution set on a number line.

57. $|x| < 3$
58. $|x| < 5$
59. $|x - 1| \leq 2$
60. $|x + 3| \leq 4$
61. $|2x - 6| < 8$
62. $|3x + 5| < 17$
63. $|2(x - 1) + 4| \leq 8$
64. $|3(x - 1) + 2| \leq 20$
65. $\left|\dfrac{2y + 6}{3}\right| < 2$
66. $\left|\dfrac{3(x - 1)}{4}\right| < 6$
67. $|x| > 3$
68. $|x| > 5$
69. $|x - 1| \geq 2$
70. $|x + 3| \geq 4$
71. $|3x - 8| > 7$
72. $|5x - 2| > 13$
73. $\left|\dfrac{2x + 2}{4}\right| \geq 2$
74. $\left|\dfrac{3x - 3}{9}\right| \geq 1$
75. $\left|3 - \dfrac{2}{3}x\right| > 5$
76. $\left|3 - \dfrac{3}{4}x\right| > 9$
77. $3|x - 1| + 2 \geq 8$
78. $-2|4 - x| \geq -4$
79. $3 < |2x - 1|$
80. $5 \geq |4 - x|$
81. $12 < \left|-2x + \dfrac{6}{7}\right| + \dfrac{3}{7}$
82. $1 < \left|x - \dfrac{11}{3}\right| + \dfrac{7}{3}$
83. $4 + \left|3 - \dfrac{x}{3}\right| \geq 9$
84. $\left|2 - \dfrac{x}{2}\right| - 1 \leq 1$

Application Exercises

The list on the next page shown ranks the ten best-educated cities in the United States measured by the percentage of the population with 16 or more years of education. Let x represent the percentage of the population with 16 or more years of education. In Exercises 85–90 write the name or names of the city or cities described by the given inequality or interval.

85. $x \geq 34.4\%$
86. $x > 35.0\%$
87. $x < 30.0\%$
88. $x \leq 30.3\%$

89. [30.0%, 34.4%]   **90.** (30.6%, 35.0%]

Most Educated

City	% with 16 + Years of Education
1. Raleigh, NC	40.6%
2. Seattle, WA	37.9
3. San Francisco, CA	35.0
4. Austin, TX	34.4
5. Washington, DC	33.3
6. Lexington-Fayette, KY	30.6
7. Minneapolis, MN	30.3
8. Boston, MA	30.0
9. Arlington, TX	30.0
10. San Diego, CA	29.8

Source: U.S. Census Bureau

91. The bar graph shows the number of people in the United States with various disorders of mental illness. If x represents millions of people, which disorders are described by $11 \leq 3x - 4 \leq 56$ cases?

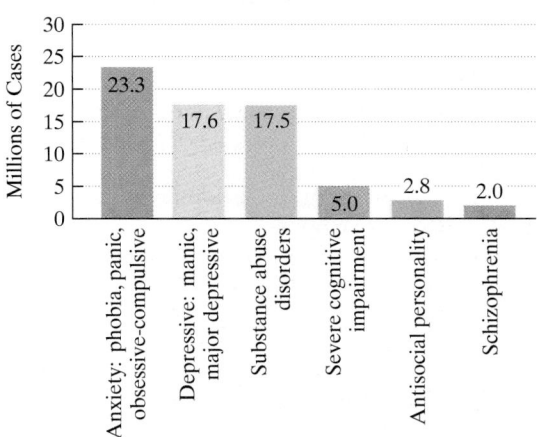

Mental Illness in America, by Disorder

Source: National Institute of Mental Health.

92. The bar graph at the top of the next column shows revenues, in billions of dollars, for the pet pharmaceutical industry for three selected years. Let x represent revenues in billions of dollars per year. Which year or years shown in the graph is/are described by the inequality

$$5 < 4x - 1 < 11?$$

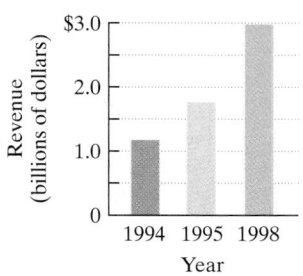

Pet Pharmaceutical Industry Revenues

Source: American Veterinary Medical Association

93. Using data from 1996–1998, the number of liposuctions in the United States can be modeled by the formula $y = 30x + 113$, where y is the number of liposuctions, in thousands, x years after 1996. According to this formula, when will the number of liposuctions exceed 623 thousand?

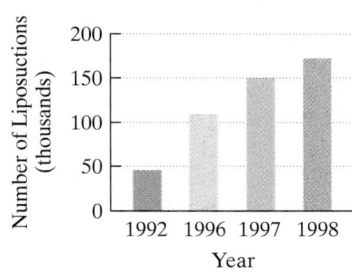

Liposuctions: U.S. Men and Women

Source: U.S. Department of Health and Human Services

94. Lower interest rates have fueled larger mortgage loans. The formula $y = 3.5x + 58$ models the data shown in the graph, where y is the size of the loan, in thousands of dollars, x years after 1980. According to this formula, when will the average mortgage loan exceed $142 thousand?

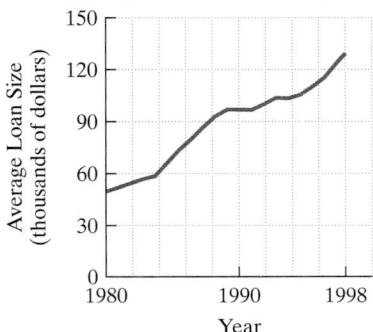

Average Mortgage Loan

Source: Mortgage Bankers Association of America

95. Using current trends, future costs of Medicare can be modeled by $C = 18x + 250$, where x represents the number of years after 2000 and C represents the cost of Medicare, in billions of dollars. Use a compound inequality to determine the years when Medicare costs will range from 322 to 412 billion dollars.

96. The formula for converting Celsius temperature (C) to Fahrenheit (F) is $F = \frac{9}{5}C + 32$. If Fahrenheit temperature exceeds 77°, what does this mean in terms of Celsius temperature?

97. A local bank charges $8 per month plus 5¢ per check. The credit union charges $2 per month plus 8¢ per check. How many checks should be written each month to make the credit union a better deal?

98. A city commission has proposed two tax bills. The first bill requires that a homeowner pay $1800 plus 3% of the assessed home value in taxes, The second bill requires taxes of $200 plus 8% of the assessed home value. What price range of home assessment would make the first bill a better deal?

99. On two examinations, you have grades of 86 and 88. There is an optional final examination, which counts as one grade. You decide to take the final in order to get a course grade of A, meaning a final average of at least 90.
   a. What must you get on the final to earn an A in the course?
   b. By taking the final, if you do poorly, you might risk the B that you have in the course based on the first two exam grades. If your final average is less than 80, you will lose your B in the course. Describe the grades on the final that will cause this to happen.

100. A company that manufactures small clocks has fixed costs of $75,000 per month. It costs the company $3 to manufacture each clock. If the clocks sell for $18 each, how many should be manufactured and sold monthly to make a profit?

101. A company that manufactures running shoes has fixed costs of $65,000 per month. It costs the company $20 to manufacture each pair of running shoes. If the shoes sell for $85 a pair, how many should be manufactured and sold monthly to make a profit?

102. If a coin is tossed 100 times, we would expect approximately 50 of the outcomes to be heads. It can be demonstrated that a coin is unfair if h, the number of outcomes that result in heads, satisfies $\left|\dfrac{h - 50}{5}\right| \geq 1.645$. Describe the number of outcomes that determine an unfair coin that is tossed 100 times.

Writing in Mathematics

103. When graphing the solutions of an inequality, what does a parenthesis signify? What does a bracket signify?

104. When solving an inequality, when is it necessary to change the sense of the inequality? Give an example.

105. Describe ways in which solving a linear inequality is similar to solving a linear equation.

106. Describe ways in which solving a linear inequality is different than solving a linear equation.

107. What is a compound inequality and how is it solved?

108. Describe how to solve an absolute value inequality involving the symbol $<$. Give an example.

109. Describe how to solve an absolute value inequality involving the symbol $>$. Give an example.

110. Explain why $|x| < -4$ has no solution.

111. Describe the solution set of $|x| > -4$.

Technology Exercises

In Exercises 112–113, solve each inequality using a graphing utility. Graph each side separately. Then determine the values of x for which the graph on the left side lies above the graph on the right side.

112. $-3(x - 6) > 2x - 2$ **113.** $-2(x + 4) > 6x + 16$

Use the same technique employed in Exercises 112–113 to solve each inequality in Exercises 114–115. In each case, what conclusion can you draw? What happens if you try solving the inequalities algebraically?

114. $12x - 10 > 2(x - 4) + 10x$

115. $2x + 3 > 3(2x - 4) - 4x$

116. A bank offers two checking account plans. Plan A has a base service charge of $4.00 per month plus 10¢ per check. Plan B charges a base service charge of $2.00 per month plus 15¢ per check.
   a. Write models for the total monthly costs for each plan if x checks are written.
   b. Use a graphing utility to graph the models in the same viewing rectangle. Use a $[0, 50, 1]$ by $[0, 10, 1]$ viewing rectangle.
   c. Use the graphs (and the TRACE or intersection feature) to determine for what number of checks per month plan A will be better than plan B.
   d. Verify the result of part (c) algebraically by solving an inequality.

Critical Thinking Exercises

117. Which one of the following is true?
   a. The first step in solving $|2x - 3| > -7$ is to rewrite the inequality as $2x - 3 > -7$ or $2x - 3 < 7$.
   b. The smallest real number in the solution set of $2x > 6$ is 4.
   c. All irrational numbers satisfy $|x - 4| > 0$.
   d. None of these statements is true.

118. What's wrong with this argument? Suppose x and y represent two real numbers, where $x > y$:

$2 > 1$	This is a true statement.
$2(y - x) > 1(y - x)$	Multiply both sides by $y - x$.
$2y - 2x > y - x$	Use the distributive property.
$y > x$	Subtract y from both sides.
	Add $2x$ to both sides.

The final inequality, $y > x$, is impossible because we were initially given $x > y$.

119. The graphs of $y = 6$, $y = 3(-x - 5) - 9$, and $y = 0$ are shown in the figure. The graph was obtained using a graphing utility with x ranging from -12 to 1 and y ranging from -2 to 8. Use the graph to write the solution set for the compound inequality

$$0 < 3(-x - 5) - 9 < 6.$$

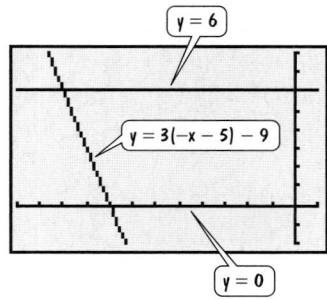

120. The percentage, p, of defective products manufactured by a company is given as $|p - 0.3\%| \leq 0.2\%$. If 100,000 products are manufactured and the company offers a \$5 refund for each defective product, describe the company's cost for refunds.

Group Exercise

121. Each group member should research one situation that provides two different pricing options. These can involve areas such as public transportation options (with or without coupon books) or long-distance telephone plans or anything of interest. Be sure to bring in all the details for each option. At a second group meeting, select the two pricing situations that are most interesting and relevant. Using each situation, write a word problem about selecting the better of the two options. The word problem should be one that can be solved using a linear inequality. The group should turn in the two problems and their solutions.

SECTION 1.6 Quadratic and Rational Inequalities

Objectives

1. Solve quadratic inequalities.
2. Solve rational inequalities.
3. Solve problems modeled by nonlinear inequalities.

Not afraid of heights and cutting-edge excitement? How about sky diving? Behind your exhilarating experience is the world of algebra. After you jump from the airplane, your height above the ground at every instant of your fall can be described by a formula involving a variable that is squared. At some point, you'll need to open your parachute. How can you determine when you must do so? Let x represent the number of seconds you are falling. You can compute when to open the parachute by solving an inequality that takes on the form $ax^2 + bx + c < 0$. Such an inequality is called a **quadratic inequality**.

Definition of a Quadratic Inequality

A **quadratic inequality** is any inequality that can be put in one of the forms

$$ax^2 + bx + c < 0 \qquad ax^2 + bx + c > 0$$
$$ax^2 + bx + c \leq 0 \qquad ax^2 + bx + c \geq 0$$

where a, b, and c are real numbers and $a \neq 0$.

In this section we establish the basic techniques for solving quadratic inequalities. We will use these techniques to solve inequalities containing quotients, called **rational inequalities**. Finally, we will consider a formula that models the position of any free-falling object. As a sky diver, you could be that free-falling object!

1 Solve quadratic inequalities.

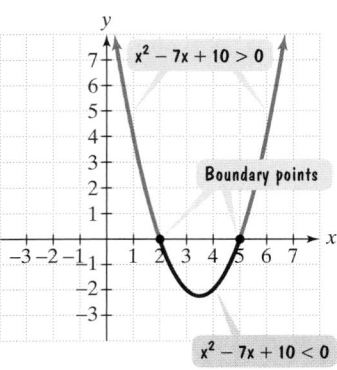

Figure 1.8 The graph of $y = x^2 - 7x + 10$. The blue parts of the graph lie above the x-axis: $x^2 - 7x + 10 > 0$. By contrast, the red part lies below the x-axis: $x^2 - 7x + 10 < 0$.

Solving Quadratic Inequalities

Graphs can help us to estimate the solutions of quadratic inequalities. The graph of $y = x^2 - 7x + 10$ is shown in Figure 1.8. The x-intercepts, 2 and 5, are **boundary points** between where the graph lies above the x-axis, shown in blue, and where the graph lies below the x-axis, shown in red. These boundary points play a critical role in solving quadratic inequalities.

Procedure for Solving Quadratic Inequalities

1. Express the inequality in the standard form
$$ax^2 + bx + c > 0 \quad \text{or} \quad ax^2 + bx + c < 0.$$
2. Solve the equation $ax^2 + bx + c = 0$. The real solutions are the **boundary points**.
3. Locate these boundary points on a number line, thereby dividing the number line into **test intervals**.
4. Choose one representative number within each test interval. If substituting that value into the original inequality produces a true statement, then all real numbers in the test interval belong to the solution set. If substituting that value into the original inequality produces a false statement, then no real numbers in the test interval belong to the solution set.
5. Write the solution set, selecting the interval(s) that produced a true statement. The graph of the solution set on a number line usually appears as

This procedure is valid if $<$ is replaced by $\leq$ and $>$ is replaced by $\geq$.

EXAMPLE 1   Solving a Quadratic Inequality

Solve and graph the solution set on a real number line: $x^2 - 7x + 10 < 0$.

Solution

Step 1   Write the inequality in standard form.   The inequality is given in this form, so this step has been done for us.

Step 2   Solve the related quadratic equation.   This equation is obtained by replacing the inequality sign by an equal sign. Thus, we will solve $x^2 - 7x + 10 = 0$.

$$x^2 - 7x + 10 = 0 \quad \text{This is the related quadratic equation.}$$
$$(x - 2)(x - 5) = 0 \quad \text{Factor.}$$
$$x - 2 = 0 \quad \text{or} \quad x - 5 = 0 \quad \text{Set each factor equal to 0.}$$
$$x = 2 \quad \text{or} \quad x = 5 \quad \text{Solve for x.}$$

The boundary points are 2 and 5.

Step 3   Locate the boundary points on a number line.   The number line with the boundary points is shown as follows:

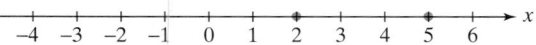

The boundary points divide the number line into three test intervals, namely $(-\infty, 2)$, $(2, 5)$, and $(5, \infty)$.

Step 4   Take one representative number within each test interval and substitute that number into the original inequality.

Test Interval	Representative Number	Substitute into $x^2 - 7x + 10 < 0$	Conclusion
$(-\infty, 2)$	0	$0^2 - 7 \cdot 0 + 10 \overset{?}{<} 0$ $10 < 0,$ False	$(-\infty, 2)$ does not belong to the solution set.
$(2, 5)$	3	$3^2 - 7 \cdot 3 + 10 \overset{?}{<} 0$ $9 - 21 + 10 \overset{?}{<} 0$ $-2 < 0,$ True	$(2, 5)$ belongs to the solution set.
$(5, \infty)$	6	$6^2 - 7 \cdot 6 + 10 \overset{?}{<} 0$ $36 - 42 + 10 \overset{?}{<} 0$ $4 < 0,$ False	$(5, \infty)$ does not belong to the solution set.

Step 5   The solution set is the interval that produced a true statement.   Our analysis shows that the solution set is the interval $(2, 5)$. The graph in Figure 1.9 confirms that $x^2 - 7x + 10 < 0$ (lies below the x-axis) in this interval. The graph of the solution set on a number line is shown as follows:

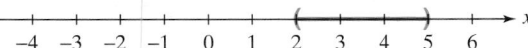

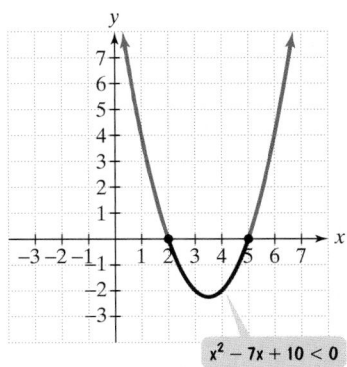

$$x^2 - 7x + 10 < 0$$

Figure 1.9 The graph lies below the x-axis between the boundary points 2 and 5, in the interval $(2, 5)$

Check Point 1   Solve and graph the solution set:

$$x^2 + 2x - 3 < 0.$$

EXAMPLE 2   Solving a Quadratic Inequality

Solve and graph the solution set on a real number line: $2x^2 + x \geq 15$.

Solution

Step 1   Write the inequality in standard form.   We can write $2x^2 + x \geq 15$ by subtracting 15 from both sides. This will give us zero on the right.

$$2x^2 + x - 15 \geq 15 - 15$$

$$2x^2 + x - 15 \geq 0$$

Step 2   Solve the related quadratic equation.   This equation is obtained by replacing the inequality sign by an equal sign. Thus, we will solve $2x^2 + x - 15 = 0$.

$$2x^2 + x - 15 = 0 \quad \text{This is the related quadratic equation.}$$

$$(2x - 5)(x + 3) = 0 \quad \text{Factor.}$$

$$2x - 5 = 0 \quad \text{or} \quad x + 3 = 0 \quad \text{Set each factor equal to 0.}$$

$$x = \tfrac{5}{2} \quad \text{or} \quad x = -3 \quad \text{Solve for x.}$$

The boundary points are -3 and $\tfrac{5}{2}$.

Step 3   Locate the boundary points on a number line.   The number line with the boundary points is shown as follows:

The boundary points divide the number line into three test intervals. Including the boundary points (because of the given greater than or equal to sign), the intervals are $(-\infty, -3]$, $[-3, \tfrac{5}{2}]$, and $[\tfrac{5}{2}, \infty)$.

Step 4   Take one representative number within each test interval and substitute that number into the original inequality.

Test Interval	Representative Number	Substitute into $2x^2 + x \geq 15$	Conclusion
$(-\infty, -3]$	-4	$2(-4)^2 + (-4) \overset{?}{\geq} 15$ $28 \geq 15$, True	$(-\infty, -3]$ belongs to the solution set.
$\left[-3, \dfrac{5}{2}\right]$	0	$2 \cdot 0^2 + 0 \overset{?}{\geq} 15$ $0 \geq 15$, False	$\left[-3, \dfrac{5}{2}\right]$ does not belong to the solution set.
$\left[\dfrac{5}{2}, \infty\right)$	3	$2 \cdot 3^2 + 3 \overset{?}{\geq} 15$ $21 \geq 15$, True	$\left[\dfrac{5}{2}, \infty\right)$ belongs to the solution set.

Step 5   The solution set are the intervals that produced a true statement.   Our analysis shows that the solution set is

$$(-\infty, -3] \text{ or } \left[\frac{5}{2}, \infty\right).$$

Technology

The solution set for
$$2x^2 + x \geq 15$$
or, equivalently,
$$2x^2 + x - 15 \geq 0$$
can be verified with a graphing utility. The graph of $y = 2x^2 + x - 15$ was obtained using a $[-10, 10, 1]$ by $[-16, 6, 1]$ viewing rectangle. The graph lies above or on the x-axis, representing $\geq$, for all x in $(-\infty, -3]$ or $[\tfrac{5}{2}, \infty)$.

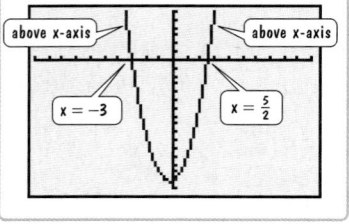

The graph of the solution set on a number line is shown as follows:

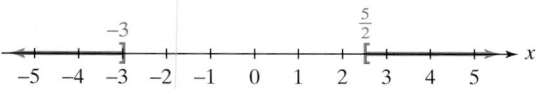

> **Check Point 2**
>
> Solve and graph the solution set: $x^2 - x \geq 20$.

2 Solve rational inequalities.

Solving Rational Inequalities

Inequalities that involve quotients can be solved in the same manner as quadratic inequalities. For example, the inequalities

$$(x + 3)(x - 7) > 0 \quad \text{and} \quad \frac{x + 3}{x - 7} > 0$$

are similar in that both are positive under the same conditions. To be positive, each of these inequalities must have two positive linear expressions

$$x + 3 > 0 \quad \text{and} \quad x - 7 > 0$$

or two negative linear expressions

$$x + 3 < 0 \quad \text{and} \quad x - 7 < 0.$$

Consequently, we use boundary points to divide the number line into test intervals. Then we select one representative number in each interval to determine whether that interval belongs to the solution set. Example 3 illustrates how this is done.

EXAMPLE 3   Using Test Numbers to Solve a Rational Inequality

Solve and graph the solution set: $\dfrac{x + 3}{x - 7} > 0$.

Solution   We begin by finding values of x that make the numerator and denominator 0.

$$x + 3 = 0 \qquad x - 7 = 0 \qquad \text{Set the numerator and denominator equal to 0.}$$

$$x = -3 \qquad\quad x = 7 \qquad \text{Solve.}$$

The boundary points are -3 and 7. We locate these numbers on a number line as follows:

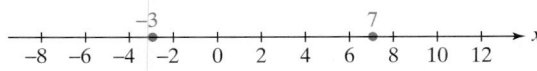

These boundary points divide the number line into three test intervals, namely $(-\infty, -3)$, $(-3, 7)$, and $(7, \infty)$. Now, we take one representative number from each test interval and substitute that number into the original inequality.

Test Interval	Representative Number	Substitute into $\dfrac{x+3}{x-7} > 0$	Conclusion
$(-\infty, -3)$	-4	$\dfrac{-4+3}{-4-7} \overset{?}{>} 0$ $\dfrac{-1}{-11} \overset{?}{>} 0$ $\dfrac{1}{11} > 0$, True	$(-\infty, -3)$ belongs to the solution set.
$(-3, 7)$	0	$\dfrac{0+3}{0-7} \overset{?}{>} 0$ $-\dfrac{3}{7} > 0$, False	$(-3, 7)$ does not belong to the solution set.
$(7, \infty)$	8	$\dfrac{8+3}{8-7} \overset{?}{>} 0$ $11 > 0$, True	$(7, \infty)$ belongs to the solution set.

Our analysis shows that the solution set is

$$(-\infty, -3) \quad \text{or} \quad (7, \infty).$$

The graph of the solution set on a number line is shown as follows:

> **Check Point 3** Solve and graph the solution set: $\dfrac{x-5}{x+2} > 0.$

The first step in solving a rational inequality is to bring all terms to one side, obtaining zero on the other side. Then express the nonzero side as a single quotient. At this point, we follow the same procedure as in Example 3.

EXAMPLE 4   Solving a Rational Inequality

Solve and graph the solution set: $\dfrac{x+1}{x+3} \leq 2.$

Solution

Step 1   Express the inequality so that one side is zero and the other side is a single quotient.   We subtract 2 from both sides to obtain zero on the right.

$$\frac{x+1}{x+3} \leq 2 \qquad \text{This is the given inequality.}$$

$$\frac{x+1}{x+3} - 2 \leq 0 \qquad \text{Subtract 2 from both sides, obtaining 0 on the right.}$$

Study Tip

Do not begin solving

$$\frac{x+1}{x+3} \leq 2$$

by multiplying both sides by $x + 3$. We do not know if $x + 3$ is positive or negative. Thus, we do not know whether or not to reverse the sense of the inequality.

$$\frac{x+1}{x+3} - \frac{2(x+3)}{x+3} \leq 0$$ The least common denominator is x + 3. Express 2 in terms of this denominator.

$$\frac{x+1-2(x+3)}{x+3} \leq 0$$ Subtract rational expressions.

$$\frac{x+1-2x-6}{x+3} \leq 0$$ Apply the distributive property.

$$\frac{-x-5}{x+3} \leq 0$$ Simplify.

Step 2   Find boundary points by setting the numerator and the denominator equal to zero.

$$-x - 5 = 0 \qquad x + 3 = 0$$ Set the numerator and denominator equal to 0. These are the values that make the previous quotient zero or undefined.

$$x = -5 \qquad x = -3$$ Solve for x.

The boundary points are -5 and -3. Because equality is included in the given less-than-or-equal-to symbol, we include the value of x that causes the quotient $\frac{-x-5}{x+3}$ to be zero. Thus, -5 is included in the solution set. By contrast, we do not include -3 in the solution set because -3 makes the denominator zero.

Step 3   Locate boundary points on a number line.   The number line, with the boundary points, is shown as follows:

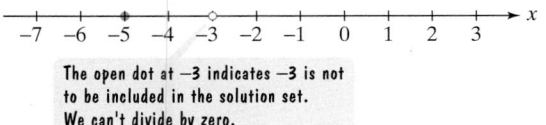

The open dot at −3 indicates −3 is not to be included in the solution set. We can't divide by zero.

The boundary points divide the number line into three test intervals, namely $(-\infty, -5]$, $[-5, -3)$, and $(-3, \infty)$.

Step 4   Take one representative number within each test interval and substitute that number into the original inequality.

Test Interval	Representative Number	Substitute into $\frac{x+1}{x+3} \leq 2$	Conclusion
$(-\infty, -5]$	-6	$\frac{-6+1}{-6+3} \overset{?}{\leq} 2$ $\frac{5}{3} \leq 2$, True	$(-\infty, -5]$ belongs to the solution set.
$[-5, -3)$	-4	$\frac{-4+1}{-4+3} \overset{?}{\leq} 2$ $3 \leq 2$, False	$[-5, -3)$ does not belong to the solution set.
$(-3, \infty)$	0	$\frac{0+1}{0+3} \overset{?}{\leq} 2$ $\frac{1}{3} \leq 2$, True	$(-3, \infty)$ belongs to the solution set.

Discovery

Because $(x + 3)^2$ is positive, it is possible so solve

$$\frac{x + 1}{x + 3} \leq 2$$

by first multiplying both sides by $(x + 3)^2$ (where $x \neq -3$). This will not reverse the sense of the inequality and will clear the fraction. Try using this solution method and compare it to the one on pages 156–158.

3 Solve problems modeled by nonlinear inequalities.

Step 5   **The solution set consists of the intervals that produced a true statement.**   Our analysis shows that the solution set is

$$(-\infty, -5] \quad \text{or} \quad (-3, \infty).$$

The graph of the solution set on a number line is shown as follows:

$$\xleftarrow{\hspace{2cm}} \begin{array}{ccccccccccc} & & & & & & & & & & \\ -7 & -6 & -5 & -4 & -3 & -2 & -1 & 0 & 1 & 2 & 3 \end{array} \xrightarrow{\hspace{1cm}} x$$

Check Point 4   Solve and graph the solution set: $\dfrac{2x}{x + 1} \leq 1.$

Applications

Intriguing signs point out that the world is profoundly mathematical. For example, did you know that every time you throw an object vertically upward, its changing height above the ground can be described by a mathematical formula? The same formula can be used to describe objects that are falling, such as the sky divers shown in the opening to this section.

The Position Formula for a Free-Falling Object Near Earth's Surface

An object that is falling or vertically projected into the air has its height in feet above the ground given by

$$s = -16t^2 + v_0 t + s_0$$

where s is the height in feet, v_0 is the original velocity (initial velocity) of the object in feet per second, t is the time that the object is in motion in seconds, and s_0 is the original height (initial height) of the object in feet.

In Example 5, we solve a quadratic inequality in a problem about the position of a free-falling object.

EXAMPLE 5   Using the Position Model

A ball is thrown vertically upward from the top of the Leaning Tower of Pisa (176 feet high) with an initial velocity of 96 feet per second (Figure 1.10). During which time period will the ball's height exceed that of the tower?

$t = 0$
$s_0 = 176$
$v_0 = 96$

176 feet

Figure 1.10 Throwing a ball from 176 feet with a velocity of 96 feet per second

Solution

$s = -16t^2 + v_0 t + s_0$ This is the position formula for a free-falling object.

$s = -16t^2 + 96t + 176$ Because v_0 (initial velocity) = 96 and s_0 (initial position) =176, substitute these values into the formula.

| When will the ball's height | exceed that | of the tower? |

$$-16t^2 + 96t + 176 \qquad > \qquad 176$$

$-16t^2 + 96t + 176 > 176$ This is the inequality implied by the problem's question. We must find t.

$-16t^2 + 96t > 0$ Subtract 176 from both sides.

$$-16t^2 + 96t = 0 \qquad \text{Solve the related quadratic equation.}$$
$$-16t(t - 6) = 0 \qquad \text{Factor.}$$
$$-16t = 0 \quad \text{or} \quad t - 6 = 0 \qquad \text{Set each factor equal to 0.}$$
$$t = 0 \qquad\qquad t = 6 \qquad \text{Solve for } t. \text{ The boundary points are 0 and 6.}$$

Locate these values on a number line, with $t \geq 0$.

The intervals are $(0, 6)$ and $(6, \infty)$, although the time interval should not extend to infinity but rather to the value of t when the ball hits the ground. (By setting $-16t^2 + 96t + 176$ equal to zero, we find $t \approx 7.47$; the ball hits the ground after approximately 7.47 seconds.)

We use $(0, 6)$ and $(6, 7.47)$ for our test intervals.

Test Interval	Representative Number	Substitute into $-16t^2 + 96t > 0$	Conclusion
$(0, 6)$	1	$-16 \cdot 1^2 + 96 \cdot 1 \overset{?}{>} 0$ $80 > 0, \text{True}$	$(0, 6)$ belongs to the solution set.
$(6, 7.47)$	7	$-16 \cdot 7^2 + 96 \cdot 7 \overset{?}{>} 0$ $-112 > 0, \text{False}$	$(6, 7.47)$ does not belong to the solution set.

The ball's height exceeds that of the tower between 0 and 6 seconds, excluding $t = 0$ and $t = 6$.

Check Point 5

An object is propelled straight up from ground level with an initial velocity of 80 feet per second. Its height at time t is described by

$$s = -16t^2 + 80t,$$

where the height, s, is measured in feet and the time, t, is measured in seconds. In which time interval will the object be more than 64 feet above the ground?

EXERCISE SET 1.6

Practice Exercises

Solve each quadratic inequality in Exercises 1–26, and graph the solution set on a real number line. Express each solution set in interval notation.

1. $(x - 4)(x + 2) > 0$
2. $(x + 3)(x - 5) > 0$
3. $(x - 7)(x + 3) \leq 0$
4. $(x + 1)(x - 7) \leq 0$
5. $x^2 - 5x + 4 > 0$
6. $x^2 - 4x + 3 < 0$
7. $x^2 + 5x + 4 > 0$
8. $x^2 + x - 6 > 0$
9. $x^2 - 6x + 9 < 0$
10. $x^2 - 2x + 1 > 0$
11. $x^2 - 6x + 8 \leq 0$
12. $x^2 - 2x - 3 \geq 0$
13. $3x^2 + 10x - 8 \leq 0$
14. $9x^2 + 3x - 2 \geq 0$

15. $2x^2 + x < 15$
16. $6x^2 + x > 1$
17. $4x^2 + 7x < -3$
18. $3x^2 + 16x < -5$
19. $5x \leq 2 - 3x^2$
20. $4x^2 + 1 \geq 4x$
21. $x^2 - 4x \geq 0$
22. $x^2 + 2x < 0$
23. $2x^2 + 3x > 0$
24. $3x^2 - 5x \leq 0$
25. $-x^2 + x \geq 0$
26. $-x^2 + 2x \geq 0$

Solve each rational inequality in Exercises 27–42, and graph the solution set on a real number line. Express each solution set in interval notation.

27. $\dfrac{x - 4}{x + 3} > 0$
28. $\dfrac{x + 5}{x - 2} > 0$

29. $\dfrac{x + 3}{x + 4} < 0$

30. $\dfrac{x + 5}{x + 2} < 0$

31. $\dfrac{-x + 2}{x - 4} \geq 0$

32. $\dfrac{-x - 3}{x + 2} \leq 0$

33. $\dfrac{4 - 2x}{3x + 4} \leq 0$

34. $\dfrac{3x + 5}{6 - 2x} \geq 0$

35. $\dfrac{x}{x - 3} > 0$

36. $\dfrac{x + 4}{x} > 0$

37. $\dfrac{x + 1}{x + 3} < 2$

38. $\dfrac{x}{x - 1} > 2$

39. $\dfrac{x + 4}{2x - 1} \leq 3$

40. $\dfrac{1}{x - 3} < 1$

41. $\dfrac{x - 2}{x + 2} \leq 2$

42. $\dfrac{x}{x + 2} \geq 2$

 Application Exercises

Use the position formula

$$s = -16t^2 + v_0 t + s_0$$

$\left(v_0 = \text{initial velocity}, s_0 = \text{initial position}, t = \text{time}\right)$

to answer Exercises 43–46.

43. A projectile is fired straight upward from ground level with an initial velocity of 80 feet per second. During which interval of time will the projectile's height exceed 96 feet?

44. A projectile is fired straight upward from ground level with an initial velocity of 128 feet per second. During which interval of time will the projectile's height exceed 128 feet?

45. A ball is thrown upward with a velocity of 64 feet per second from the top edge of a building 80 feet high. For how long is the ball higher than 96 feet?

46. A diver leaps into the air at 20 feet per second from a diving board that is 10 feet above the water. For how many seconds is the diver at least 12 feet above the water?

47. The formula

$$H = \frac{15}{8}x^2 - 30x + 200$$

describes heart rate, H, in beats per minute, x minutes after a strenuous workout.
a. What is the heart rate immediately following the workout?
b. Describe the interval of time after a strenous workout in which heart rate exceeds 110 beats per minute.

48. The data in the bar graph at the top of the next column can be modeled by the formula

$$y = 0.6x^2 - 2.9x + 3.2$$

where y represents the number of U.S. cellular telephone subscribers, in hundreds of thousands, x years after 1985.

When will the number of cellular subscribers exceed 185,200,000 or 185.2 hundred thousand?

U.S. Cellular Telephone Subscribers

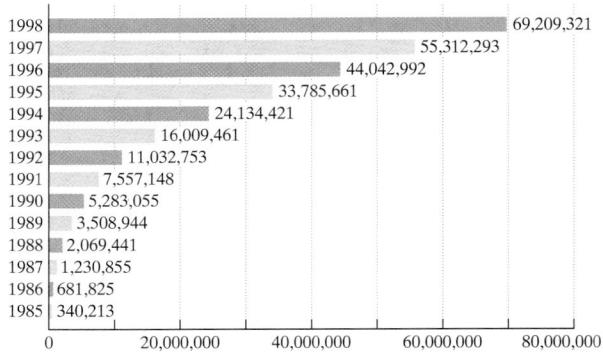

Source: Cellular Telephone Industry of America

49. The data in the bar graph shown below can be modeled by the formula

$$y = -0.22x^2 + 4.32x + 26$$

where y represents the number of international visitors, in millions, x years after 1986. According to this model, when will the number of international visitors to the United States be less than 41.3 million?

International Visitors to the United States

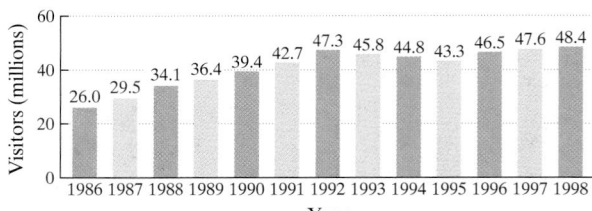

Source: U.S. Department of Commerce

50. The average cost per unit, $(\bar{C})$, of producing x units of a product is modeled by

$$\bar{C} = \frac{150{,}000 + 0.25x}{x}.$$

How many units must be produced so that the average cost of producing each unit does not exceed $1.75?

51. The cost of removing $p\%$ of the bacteria from a river is given by the formula

$$C = \frac{4p}{100 - p}$$

where C is measured in hundreds of thousands of dollars. If less than $600,000 is spent, what percentage of the bacteria can be removed?

 **Writing in Mathematics**

52. What is a quadratic inequality?
53. What is a rational inequality?

54. Describe similarities and differences between the solutions of

$$(x - 2)(x + 5) \geq 0 \quad \text{and} \quad \frac{x - 2}{x + 5} \geq 0.$$

 Technology Exercises

Solve each inequality in Exercises 55–60 using a graphing utility.

55. $x^2 + 3x - 10 > 0$ **56.** $2x^2 + 5x - 3 \leq 0$

57. $\dfrac{x - 4}{x - 1} \leq 0$ **58.** $\dfrac{x + 2}{x - 3} \leq 2$

59. $\dfrac{1}{x + 1} \leq \dfrac{2}{x + 4}$ **60.** $x^3 + 2x^2 - 5x - 6 > 0$

61. In a study of the winter moth in Nova Scotia, the number of eggs, N, in a female moth depended on her abdominal width, W, in millimeters, approximated by $N = 14W^3 - 17W^2 - 6W + 34$, where $1.5 \leq W \leq 3.5$. Graph the model on your graphing utility and use the $\boxed{\text{TRACE}}$ and $\boxed{\text{ZOOM}}$ features to describe the abdominal width of a moth with more than 46 eggs.

 **Critical Thinking Exercises**

62. Which one of the following is true?
   a. The solution set to $x^2 > 25$ is $(5, \infty)$.
   b. The inequality $\dfrac{x - 2}{x + 3} < 2$ can be solved by multiplying both sides by $x + 3$, resulting in the equivalent inequality $x - 2 < 2(x + 3)$.
   c. $(x + 3)(x - 1) \geq 0$ and $\dfrac{x + 3}{x - 1} \geq 0$ have the same solution set.
   d. None of these statements is true.

63. Write a quadratic inequality whose solution set is $[-3, 5]$.

64. Write a rational inequality whose solution set is $(-\infty, -4)$ or $[3, \infty)$.

In Exercises 65–68, use inspection to describe each inequality's solution set. Do not solve any of the inequalities.

65. $(x - 2)^2 > 0$ **66.** $(x - 2)^2 \leq 0$

67. $(x - 2)^2 < -1$ **68.** $\dfrac{1}{(x - 2)^2} > 0$

69. The graphing calculator screen at the top of the next column shows the graph of $y = 4x^2 - 8x + 7$.

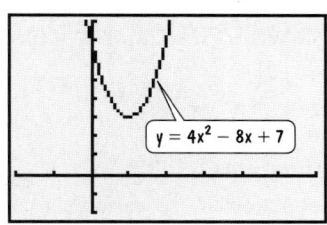

a. Use the graph to describe the solution set for $4x^2 - 8x + 7 > 0$.
b. Use the graph to describe the solution set for $4x^2 - 8x + 7 < 0$.
c. Use an algebraic approach to verify each of your descriptions in parts (a) and (b).

70. The graphing calculator screen shows the graph of $y = \sqrt{27 - 3x^2}$. Write and solve a quadratic inequality that explains why the graph only appears for $-3 \leq x \leq 3$.

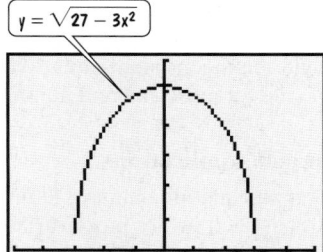

 Group Exercise

71. This exercise is intended as a group learning experience and is appropriate for groups of three to five people. Before working on the various parts of the problem, reread the description of the position formula on page 158.
   a. Drop a ball from a height of 3 feet, 6 feet, and 12 feet. Record the number of seconds it takes for the ball to hit the ground.
   b. For each of the three initial positions, use the position formula to determine the time required for the ball to hit the ground.
   c. What factors might result in differences between the times that you recorded and the times indicated by the formula?
   d. What appears to be happening to the time required for a free-falling object to hit the ground as its initial height is doubled? Verify this observation algebraically and with a graphing utility.
   e. Repeat part (a) using a sheet of paper rather than a ball. What differences do you observe? What factor seems to be ignored in the position formula?
   f. What is meant by the acceleration of gravity and how does this number appear in the position formula for a free-falling object?

CHAPTER SUMMARY, REVIEW, AND TEST

Summary

1.1 Linear Equations

a. A linear equation in one variable x can be written in the form $ax + b = c, a \neq 0$.

b. The procedure for solving a linear equation is given in the box on page 88.

c. If an equation contains fractions, begin by multiplying both sides by the least common denominator, thereby clearing fractions.

d. If an equation contains rational expressions with variable denominators, avoid in the solution set any values of the variable that make a denominator zero.

e. An identity is an equation that is true for all real numbers for which both sides are defined. A conditional equation is not an identity and is true for at least one real number. An inconsistent equation is an equation that is not true for even one real number.

1.2 Formulas and Applications

a. A formula is an equation that uses letters to express relationships between two or more variables.

b. A mathematical model is a formula or algebraic equation that can be formed from a verbal model or from actual data.

c. A five-step procedure for solving problems using mathematical models is given in the box on page 99.

1.3 Quadratic Equations

a. A quadratic equation in x can be written in the standard form $ax^2 + bx + c = 0, a \neq 0$.

b. The procedure for solving a quadratic equation by factoring and the zero-product principle is given in the box on pages 110–111.

c. The procedure for solving a quadratic equation by the square root method is given in the box on page 113.

d. All quadratic equations can be solved by completing the square. Isolate the binomial with the two variable terms on one side of the equation. If the coefficient of the x^2 term is not one, divide each side of the equation by this coefficient. Then add the square of half the coefficient of x to both sides.

e. All quadratic equations can be solved by the quadratic formula

$$x = \frac{-b \pm \sqrt{b^2 - 4ac}}{2a}.$$

The formula is derived by completing the square of the equation $ax^2 + bx + c = 0$.

f. The discriminant, $b^2 - 4ac$, indicates the kinds of solutions to the quadratic equation $ax^2 + bx + c = 0$, shown in Table 1.3 on page 119.

g. Table 1.4 on page 120 shows the most efficient technique to use when solving a quadratic equation.

1.4 Other Types of Equations

a. Some polynomial equations of degree 3 or greater can be solved by moving all terms to one side, obtaining zero on the other side, factoring, and using the zero-product principle. Factoring by grouping is often used.

b. A radical equation is an equation in which the variable occurs in a square root, cube root, and so on. A radical equation can be solved by isolating the radical and raising both sides of the equation to a power equal to the radical's index. When raising both sides to an even power, check all proposed solutions in the original equation. Eliminate extraneous solutions from the solution set.

c. A radical equation with rational exponents can be solved by isolating the expression with the rational exponent and raising both sides of the equation to a power that is the reciprocal of the rational exponent. Check for possible extraneous solutions.

d. An equation is quadratic in form if it can be written in the form $at^2 + bt + c = 0$, where t is an algebraic expression and $a \neq 0$. Solve for t and use the substitution that resulted in this equation to find the values for the variable in the given equation.

e. Absolute value equations in the form $|X| = c, c > 0$, can be solved by rewriting the equation without absolute value bars: $X = c$ or $X = -c$.

1.5 Linear Inequalities

a. A linear inequality in one variable x can be expressed as $ax + b \leq c$, $ax + b < c$, $ax + b \geq c$, or $ax + b > c$, $a \neq 0$.

b. Graphs of solutions to inequalities are shown on a number line by shading all points representing numbers that are solutions. Parentheses exclude endpoints and square brackets include endpoints.

c. Solution sets to inequalities can be expressed in set-builder or interval notation. Table 1.5 on page 141 compares the notations.

d. A linear inequality is solved using a procedure similar to solving a linear equation. However, when multiplying or dividing by a negative number, reverse the sense of the inequality.

e. A compound inequality with three parts can be solved by isolating x in the middle.

f. Inequalities involving absolute value can be solved by rewriting the inequalities without absolute value bars. The ways to do this are shown in the box on page 145.

1.6 Quadratic and Rational Inequalities

a. A quadratic inequality can be expressed as
$ax^2 + bx + c < 0$, $\quad ax^2 + bx + c > 0$,
$ax^2 + bx + c \le 0$, $\quad$ or $\quad ax^2 + bx + c \ge 0$, $\quad a \ne 0$.

b. A procedure for solving quadratic inequalities is given in the box on page 152.

c. Inequalities involving quotients are called rational inequalities. The procedure for solving such inequalities begins with expressing them so that one side is zero and the other side is a single quotient. Find boundary points by setting the numerator and denominator equal to zero. Then follow a procedure similar to that for solving quadratic inequalities.

Review Exercises

1.1

In Exercises 1–6, solve and check each linear equation.

1. $2x - 5 = 7$
2. $5x + 20 = 3x$
3. $7(x - 4) = x + 2$
4. $1 - 2(6 - x) = 3x + 2$
5. $2(x - 4) + 3(x + 5) = 2x - 2$
6. $2x - 4(5x + 1) = 3x + 17$

Exercises 7–11 contain equations with constants in denominators. Solve each equation and check by the method of your choice.

7. $\dfrac{2x}{3} = \dfrac{x}{6} + 1$

8. $\dfrac{x}{2} - \dfrac{1}{10} = \dfrac{x}{5} + \dfrac{1}{2}$

9. $\dfrac{2x}{3} = 6 - \dfrac{x}{4}$

10. $\dfrac{x}{4} = 2 + \dfrac{x - 3}{3}$

11. $\dfrac{3x + 1}{3} - \dfrac{13}{2} = \dfrac{1 - x}{4}$

Exercises 12–15 contain equations with variables in denominators. a. List the value or values representing restriction(s) on the variable. b. Solve the equation.

12. $\dfrac{9}{4} - \dfrac{1}{2x} = \dfrac{4}{x}$

13. $\dfrac{7}{x - 5} + 2 = \dfrac{x + 2}{x - 5}$

14. $\dfrac{1}{x - 1} - \dfrac{1}{x + 1} = \dfrac{2}{x^2 - 1}$

15. $\dfrac{4}{x + 2} + \dfrac{2}{x - 4} = \dfrac{30}{x^2 - 2x - 8}$

In Exercises 16–18, determine whether each equation is an identity, a conditional equation, or an inconsistent equation.

16. $\dfrac{1}{x + 5} = 0$

17. $7x + 13 = 4x - 10 + 3x + 23$

18. $7x + 13 = 3x - 10 + 2x + 23$

19. The formula $M = 420x + 720$ models the data for the amount of money lost to credit card fraud worldwide, M, expressed in millions of dollars, x years after 1989. In which year did losses amount to 4080 million dollars?

1.2

20. Suppose you were to list in order, from least to most, the family income for every U.S. family. The median income is the income in the middle of this list of ranked data. This income can be modeled by the formula
$$I = 1321.7(x - 1980) + 21,153.$$
In this formula, I represents median family income in the United States and x is the actual year, beginning in 1980. When was the median income $47,587?

In Exercises 21–27, use the five-step strategy given in the box on page 99 to solve each problem.

21. The bus fare in a city is $1.50. People who use the bus have the option of purchasing a monthly coupon book for $25.00. With the coupon book, the fare is reduced to $0.25. Determine the number of times in a month the bus must be used so that the total monthly cost without the coupon book is the same as the total monthly cost with the coupon book.

22. Los Angeles has more unhealthy air days per year than any other U.S. city. On average, the number of unhealthy air days per year in Los Angeles exceeds five times that of New York City by 29 days. If Los Angeles and New York combined have 185 unhealthy air days per year, determine the number of unhealthy days for the two cities.

23. You inherit $10,000 with the stipulation that for the first year the money must be invested in two stocks paying 8% and 12% annual interest, respectively. How much should be invested at each rate if the total interest earned for the year is to be $950?

24. The length of a rectangular football field is 14 meters more than twice the width. If the perimeter is 346 meters, find the field's dimensions.

25. A salesperson earns $300 per week plus 5% commission of sales. How much must be sold to earn $800 in a week?

26. After a 45% price reduction, a VCR sold for $247.50. What was the price before the reduction?

27. A study entitled *Performing Arts—The Economic Dilemma* documents the relationship between the number of

concerts given by a major orchestra and the attendance per concert. For each additional concert given per year, attendance per concert drops by approximately eight people. If 50 concerts are given, attendance per concert is 2987 people. How many concerts should be given to ensure an audience of 2627 people at each concert?

In Exercises 28–29, solve each formula for the specified variable.

28. $V = \dfrac{1}{3} Bh$ for h

29. $F = f(1 - M)$ for M

1.3

Solve each equation in Exercises 30–31 by factoring and then using the zero-product principle.

30. $2x^2 + 15x = 8$

31. $5x^2 + 20x = 0$

Solve each equation in Exercises 32–33 by the square root method.

32. $2x^2 - 3 = 125$

33. $(3x - 4)^2 = 18$

In Exercises 34–35, determine the constant that should be added to the binomial so that it becomes a perfect square trinomial. Then factor the trinomial.

34. $x^2 + 20x$

35. $x^2 - 3x$

Solve each equation in Exercises 36–37 by completing the square.

36. $x^2 - 12x + 27 = 0$

37. $3x^2 - 12x + 11 = 0$

Solve each equation in Exercises 38–40 using the quadratic formula.

38. $x^2 = 2x + 4$

39. $x^2 - 2x + 19 = 0$

40. $2x^2 = 3 - 4x$

Compute the discriminant of each equation in Exercises 41–42. What does the discriminant indicate about the kinds of solutions?

41. $x^2 - 4x + 13 = 0$

42. $9x^2 = 2 - 3x$

Solve each equation in Exercises 43–48 by the method of your choice.

43. $2x^2 - 11x + 5 = 0$

44. $(3x + 5)(x - 3) = 5$

45. $3x^2 - 7x + 1 = 0$

46. $x^2 - 9 = 0$

47. $(x - 3)^2 - 25 = 0$

48. $3x^2 - x + 2 = 0$

49. The weight of a human fetus is modeled by the formula $W = 3t^2$, where W is the weight in grams and t is the time in weeks, $0 \le t \le 39$. After how many weeks does the fetus weigh 1200 grams?

50. The formula $N = 0.337x^2 - 2.265x + 3.962$ models the number of mountain bike owners N, in millions, in the United States x years after 1980, where $3 \le x \le 20$. In which year (to the nearest year) did 10.9 million Americans own mountain bikes?

51. A billboard is 15 feet longer than it is high and has space for 324 square feet of advertising. What are the billboard's dimensions?

52. A building casts a shadow that is double the length of its height. If the distance from the end of the shadow to the top of the building is 300 meters, how high is the building?

1.4

Solve each polynomial equation in Exercises 53–54.

53. $2x^4 = 50x^2$

54. $2x^3 - x^2 - 18x + 9 = 0$

Solve each radical equation in Exercises 55–56.

55. $\sqrt{2x - 3} + x = 3$

56. $\sqrt{x - 4} + \sqrt{x + 1} = 5$

Solve the equations with rational exponents in Exercises 57–58.

57. $3x^{3/4} - 24 = 0$

58. $(x - 7)^{3/2} = 125$

Solve each equation in Exercises 59–60 by making an appropriate substitution.

59. $x^4 - 5x^2 + 4 = 0$

60. $x^{1/2} + 3x^{1/4} - 10 = 0$

Solve the equations containing absolute value in Exercises 61–62.

61. $|2x + 1| = 7$

62. $2|x - 3| - 7 = 10$

Solve each equation in Exercises 63–66 by the method of your choice.

63. $3x^{4/3} - 5x^{2/3} + 2 = 0$

64. $2\sqrt{x - 1} = x$

65. $|2x - 5| - 3 = 0$

66. $x^3 + 2x^2 = 9x + 18$

67. The distance to the horizon that you can see, measured in miles, on the top of a mountain H feet high is modeled by the formula $D = \sqrt{2H}$. You've hiked to the top of a mountain with views extending 50 miles to the horizon. How high is the mountain?

1.5

In Exercises 68–70, graph the solutions of each inequality on a number line.

68. $x > 5$

69. $x \le 1$

70. $-3 \le x < 0$

In Exercises 71–73, express each interval in terms of an inequality, and graph the interval on a number line.

71. $(-2, 3]$

72. $[-1.5, 2]$

73. $(-1, \infty)$

Solve each linear inequality in Exercises 74–79 and graph the solution set on a number line.

74. $-6x + 3 \le 15$

75. $6x - 9 \ge -4x - 3$

76. $\dfrac{x}{3} - \dfrac{3}{4} - 1 > \dfrac{x}{2}$

77. $6x + 5 > -2(x - 3) - 25$

78. $3(2x - 1) - 2(x - 4) \ge 7 + 2(3 + 4x)$

79. $7 < 2x + 3 \le 9$

Solve each inequality in Exercises 80–82 by first rewriting each one as an equivalent inequality without absolute value bars. Graph the solution set on a number line.

80. $|2x + 3| \le 15$

81. $\left| \dfrac{2x + 6}{3} \right| > 2$

82. $|2x + 5| - 7 \ge -6$

The graph indicates that the United States has the world's highest incarceration rate. If x represents the incarceration rate per 100,000 population, list the country or countries that satisfy each inequality in Exercises 83–84.

Countries with the Highest Incarceration Rate

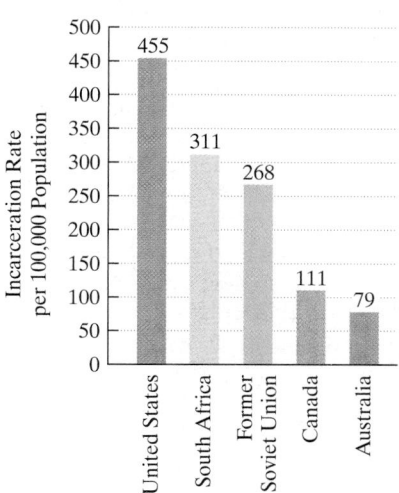

Source: FBI

83. $437 \leq 4x - 7 \leq 1229$ **84.** $|x - 320| > 80$

85. Approximately 90% of the population sleeps h hours daily, where h is modeled by the inequality $|h - 6.5| \leq 1$. Write a sentence describing the range for the number of hours that most people sleep. Do *not* use the phrase "absolute value" in your description.

86. The formula for converting Fahrenheit temperature (F) to Celsius temperature (C) is $C = \frac{5}{9}(F - 32)$. If Celsius temperature ranges from 15° to 35°, inclusively, what is the range for the Fahrenheit temperature?

87. A person can choose between two charges on a checking account. The first method involves a fixed cost of $11 per month plus 6¢ for each check written. The second method involves a fixed cost of $4 per month plus 20¢ for each check written. How many checks should be written to make the first method a better deal?

88. In 1984, approximately 1644 thousand turntables were sold in the United States. This number has decreased by around 82 thousand each year since then. By contrast, annual sales of compact disc players has increased by 496 thousand each year, with approximately 284 thousand units sold in 1984. What was the first year in which the sales of compact disc players exceeded those of turntables?

1.6

Solve each quadratic inequality in Exercises 89–90, and graph the solution set on a real number line. Express each solution set in interval notation.

89. $2x^2 + 7x \leq 4$ **90.** $2x^2 > 6x - 3$

Solve each rational inequality in Exercises 91–92, and graph the solution set on a real number line. Express each solution set in interval notation.

91. $\dfrac{x - 6}{x + 2} > 0$ **92.** $\dfrac{x + 3}{x - 4} \leq 5$

93. Use the position formula

$$s = -16t^2 + v_0 t + s_0$$

initial velocity   initial height

to solve this problem. A projectile is fired vertically upward from ground level with an initial velocity of 48 feet per second. During which time period will the projectile's height exceed 32 feet?

Chapter 1 Test

Find the solution set for each equation in Exercises 1–13.

1. $7(x - 2) = 4(x + 1) - 21$

2. $\dfrac{2x - 3}{4} = \dfrac{x - 4}{2} - \dfrac{x + 1}{4}$

3. $\dfrac{2}{x - 3} - \dfrac{4}{x + 3} = \dfrac{8}{x^2 - 9}$

4. $2x^2 - 3x - 2 = 0$ **5.** $(3x - 1)^2 = 75$

6. $x(x - 2) = 4$ **7.** $4x^2 = 8x - 5$

8. $x^3 - 4x^2 - x + 4 = 0$ **9.** $\sqrt{x - 3} + 5 = x$

10. $\sqrt{x + 4} + \sqrt{x - 1} = 5$ **11.** $5x^{3/2} - 10 = 0$

12. $x^{2/3} - 9x^{1/3} + 8 = 0$ **13.** $\left|\dfrac{2}{3}x - 6\right| = 2$

Solve each inequality in Exercises 14–19. Express the answer in interval notation and graph the solution set on a number line.

14. $3(x + 4) \geq 5x - 12$ **15.** $\dfrac{x}{6} + \dfrac{1}{8} \leq \dfrac{x}{2} - \dfrac{3}{4}$

16. $-3 \leq \dfrac{2x + 5}{3} < 6$ **17.** $|3x + 2| \geq 3$

18. $x^2 < x + 12$ **19.** $\dfrac{2x + 1}{x - 3} > 3$

20. The monthly benefit (B) of a retirement plan is given by

$$B = \frac{2}{5}w + \frac{1}{125}n$$

where

    w = an employee's average monthly salary

    n = the number of years an employee worked for the company

a. Solve the formula for n.

b. Use your answer from part (a) to find the number of years an employee whose average monthly salary is $800 must work to receive a monthly benefit of $512.

21. Approximate population and growth figures for two states are given as follows.

	1980 Population (in Thousands)	Yearly Growth (in Thousands)
Arizona	2795	89
South Carolina	3071	43

When did Arizona have the same population as South Carolina?

22. The formula $y = 420x + 720$ describes the amount of money, y, in millions of dollars, lost to credit card fraud worldwide x years after 1989. In which year did losses amount to 4080 million dollars?

23. With a 9% raise, a physical therapist will earn $45,780 annually. What is the therapist's salary prior to this raise?

24. You invest $6000, part at 9% and the remainder at 6%. If the total yearly interest from the two investments is $480, find the amount invested at each rate.

25. A vertical pole is to be supported by a wire that is 26 feet long and anchored 24 feet from the base of the pole. How far up the pole should the wire be attached?

26. A student has grades on three examinations of 76, 80, and 72. What must the student earn on a fourth examination in order to have an average of at least 80?

27. A computer online service charges a flat monthly rate of $20 or a monthly rate of $5 plus 15 cents for every hour spent online. How many hours online each month will make the second option a better deal?

Functions and Graphs

The cost of mailing a package depends on its weight. The probability that you and another person in a room share the same birthday depends on the number of people in the room. In both these situations, the relationship between variables can be illustrated with the notion of a *function*. Understanding this concept will give you a new perspective on many ordinary situations.

'Tis the season and you've waited until the last minute to mail your holiday gifts. Your only option is overnight express mail. You realize that the cost of mailing a gift depends on its weight, but the mailing costs seem somewhat odd. Your packages that weigh 1.1 pounds, 1.5 pounds, and 2 pounds cost $15.75 each to send overnight. Packages that weigh 2.01 pounds and 3 pounds cost you $18.50 each. Finally, your heaviest gift is barely over 3 pounds and its mailing cost is $21.25. What sort of system is this in which costs increase by $2.75, stepping from $15.75 to $18.50 and from $18.50 to $21.25?

SECTION 2.1 Lines and Slope

Objectives

1. Compute a line's slope.
2. Write the point-slope equation of a line.
3. Write and graph the slope-intercept equation of a line.
4. Recognize equations of horizontal and vertical lines.
5. Recognize and use the general form of a line's equation.
6. Model data with linear equations.

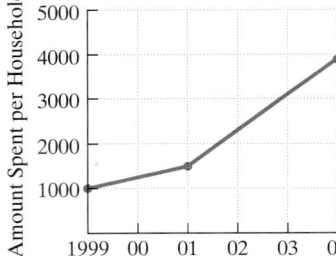

Online Spending: Yearly Spending per Online Household

Source: Forrester Research

Figure 2.1

Good news: Projections indicate that in the next decades we'll live longer and move somewhere warmer where we'll shop online and chat on our tiny video cell phones. Figure 2.1 shows projected online shopping per U.S. online household through 2004. The graph is composed of two line segments. The segment on the right is steeper than the one on the left. This shows that online shopping is expected to increase more per year in 2001–2004 than in 1999–2001.

Data often fall on or near a line. In this section we will use equations to model such data and make predictions. We begin with a discussion of a line's steepness.

The Slope of a Line

Mathematicians have developed a useful measure of the steepness of a line, called the **slope** of the line. Slope compares the vertical change (the **rise**) to the horizontal change (the **run**) when moving from one fixed point to another along the line. To calculate the slope of a line, mathematicians use a ratio comparing the change in y (the rise) to the change in x (the run).

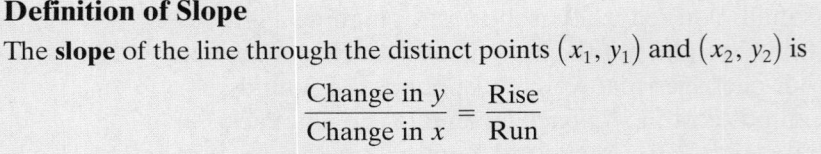

Definition of Slope

The **slope** of the line through the distinct points (x_1, y_1) and (x_2, y_2) is

$$\frac{\text{Change in } y}{\text{Change in } x} = \frac{\text{Rise}}{\text{Run}}$$

$$= \frac{y_2 - y_1}{x_2 - x_1}$$

where $x_2 - x_1 \neq 0$.

It is common notation to let the letter m represent the slope of a line. The letter m is used because it is the first letter of the French verb *monter*, meaning to rise, or to ascend.

1 Compute a line's slope.

EXAMPLE 1 Using the Definition of Slope

Find the slope of the line passing through each pair of points.

a. $(-3, -1)$ and $(-2, 4)$ **b.** $(-3, 4)$ and $(2, -2)$

Solution

a. Let $(x_1, y_1) = (-3, -1)$ and $(x_2, y_2) = (-2, 4)$. We obtain a slope of

$$m = \frac{\text{Change in } y}{\text{Change in } x} = \frac{y_2 - y_1}{x_2 - x_1} = \frac{4 - (-1)}{-2 - (-3)} = \frac{5}{1} = 5.$$

The situation is illustrated in Figure 2.2(a). The slope of the line is 5, indicating that there is a vertical change, a rise, of 5 units for each horizontal change, a run, of 1 unit. The slope is positive, and the line rises from left to right.

Study Tip

When computing slope, it makes no difference which point you call (x_1, y_1) and which point you call (x_2, y_2). If we let $(x_1, y_1) = (-2, 4)$ and $(x_2, y_2) = (-3, -1)$, the slope is still 5:

$$m = \frac{y_2 - y_1}{x_2 - x_1} = \frac{-1 - 4}{-3 - (-2)} = \frac{-5}{-1} = 5.$$

However, you should not subtract in one order in the numerator $(y_2 - y_1)$ and then in a different order in the denominator $(x_1 - x_2)$. The slope is *not*

$$\frac{-1 - 4}{-2 - (-3)} = \frac{-5}{1} = -5. \quad \text{Incorrect.}$$

b. We can let $(x_1, y_1) = (-3, 4)$ and $(x_2, y_2) = (2, -2)$. The slope of the line shown in Figure 2.2(b) is computed as follows:

$$m = \frac{-2 - 4}{2 - (-3)} = \frac{-6}{5} = -\frac{6}{5}.$$

The slope of the line is $-\frac{6}{5}$. For every vertical change of -6 units (6 units down), there is a corresponding horizontal change of 5 units. The slope is negative and the line falls from left to right.

Slope and the Streets of San Francisco

San Francisco's Filbert Street has a slope of 0.613, meaning that for every horizontal distance of 100 feet, the street ascends 61.3 feet vertically. With its 31.5° angle of inclination, the street is too steep to pave and is only accessible by wooden stairs.

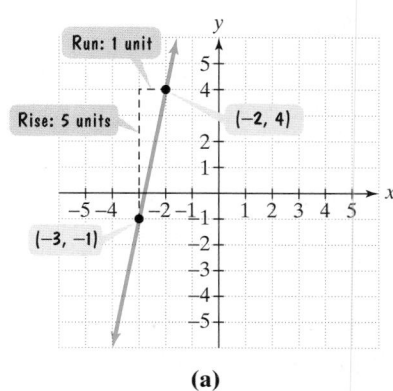

(a)

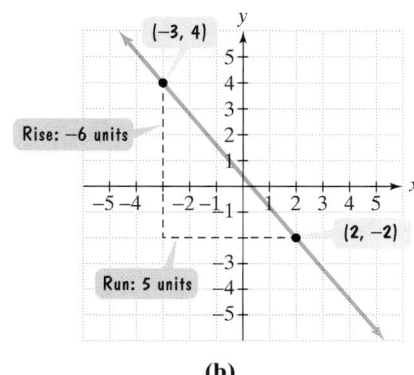

(b)

Figure 2.2 Visualizing Slope

Check Point 1

Find the slope of the line passing through each pair of points.

a. $(-3, 4)$ and $(-4, -2)$ **b.** $(4, -2)$ and $(-1, 5)$

Example 1 illustrates that a line with a positive slope is rising from left to right and a line with a negative slope is falling from left to right. By contrast, a horizontal line neither rises nor falls and has a slope of zero. A vertical line has no horizontal change, so $x_2 - x_1 = 0$ in the formula for slope. Because we cannot divide by zero, the slope of a vertical line is undefined. This discussion is summarized in Table 2.1.

Table 2.1 Possibilities for a Line's Slope

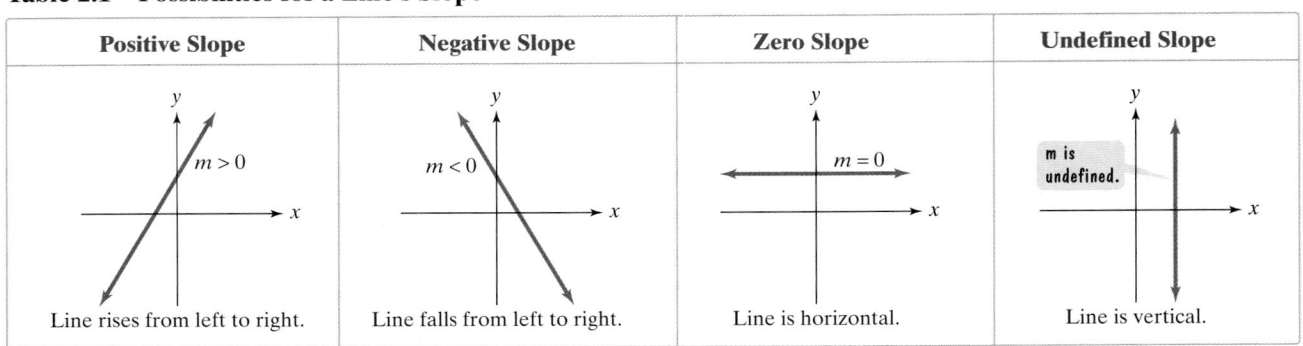

Positive Slope	Negative Slope	Zero Slope	Undefined Slope
$m > 0$	$m < 0$	$m = 0$	m is undefined.
Line rises from left to right.	Line falls from left to right.	Line is horizontal.	Line is vertical.

2 Write the point-slope equation of a line.

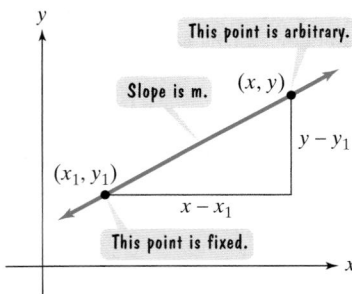

Figure 2.3 A line passing through (x_1, y_1) with slope m

The Point-Slope Form of the Equation of a Line

We can use the slope of a line to obtain various forms of the line's equation. For example, consider a nonvertical line that has a slope of m and contains the point (x_1, y_1). Now, let (x, y) represent any other point on the line, shown in Figure 2.3. Keep in mind that the point (x, y) is arbitrary and is not in one fixed position. By contrast, the point (x_1, y_1) is fixed. Regardless of where the point (x, y) is located, the shape of the triangle in Figure 2.3 remains the same. Thus, the ratio for slope stays a constant m. This means that for all points along the line,

$$m = \frac{y - y_1}{x - x_1}, \quad x \neq x_1.$$

We can clear the fraction by multiplying both sides by $x - x_1$, the least common denominator.

$$m(x - x_1) = \frac{y - y_1}{x - x_1} \cdot x - x_1$$

$$m(x - x_1) = y - y_1 \qquad \text{Simplify.}$$

Now, if we reverse the two sides, we obtain the **point-slope form** of the equation of a line.

> **Point-Slope Form of the Equation of a Line**
>
> The **point-slope equation** of a nonvertical line of slope m that passes through the point (x_1, y_1) is
>
> $$y - y_1 = m(x - x_1).$$

For example, an equation of the line passing through $(1, 1)$ with a slope of $2 \, (m = 2)$ is

$$y - 1 = 2(x - 1).$$

After we obtain the point-slope form of a line, it is customary to express the equation with y isolated on one side of the equal sign. Example 2 illustrates how this is done.

EXAMPLE 2   Writing the Point-Slope Equation of a Line

Write the point-slope form of the equation of the line passing through $(-1, 3)$ with a slope of 4. Then solve the equation for y.

Solution   We use the point-slope equation of a line with $m = 4$, $x_1 = -1$, and $y_1 = 3$.

$$y - y_1 = m(x - x_1) \qquad \text{This is the point-slope form of the equation.}$$
$$y - 3 = 4[x - (-1)] \qquad \text{Substitute the given values.}$$
$$y - 3 = 4(x + 1) \qquad \text{We now have the point-slope form of the equation for the given line.}$$

We can solve this equation for y by applying the distributive property on the right side.

$$y - 3 = 4x + 4$$

Finally, we add 3 to both sides.

$$y = 4x + 7$$

Check Point 2   Write the point-slope form of the equation of the line passing through $(2, -5)$ with a slope of 6. Then solve the equation for y.

EXAMPLE 3   Writing the Point-Slope Equation of a Line

Write the point-slope form of the equation of the line passing through the points $(4, -3)$ and $(-2, 6)$. (See Figure 2.4.) Then solve the equation for y.

Solution   To use the point-slope form, we need to find the slope. The slope is the change in the y-coordinates divided by the corresponding change in the x-coordinates.

$$m = \frac{6 - (-3)}{-2 - 4} = \frac{9}{-6} = -\frac{3}{2} \qquad \text{This is the definition of slope using } (4, -3) \text{ and } (-2, 6).$$

We can take either point on the line to be (x_1, y_1). Let's use $(x_1, y_1) = (4, -3)$. Now, we are ready to write the point-slope equation.

$$y - y_1 = m(x - x_1) \qquad \text{This is the point-slope form of the equation.}$$
$$y - (-3) = -\tfrac{3}{2}(x - 4) \qquad \text{Substitute: } (x_1, y_1) = (4, -3) \text{ and } m = -\tfrac{3}{2}.$$
$$y + 3 = -\tfrac{3}{2}(x - 4) \qquad \text{Simplify.}$$

We now have the point-slope form of the equation of the line shown in Figure 2.4. Now, we solve this equation for y.

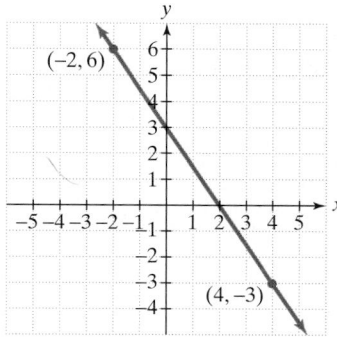

Figure 2.4 Write the point-slope equation of this line.

Discovery

You can use either point for (x_1, y_1) when you write a line's point-slope equation. Rework Example 3 using $(-2, 6)$ for (x_1, y_1). Once you solve for y, you should obtain the same equation as the one shown in the last line of the solution on page 172.

$$y + 3 = -\tfrac{3}{2}(x - 4) \quad \text{This is the point-slope form of the equation.}$$
$$y + 3 = -\tfrac{3}{2}x + 6 \quad \text{Use the distributive property.}$$
$$y = -\tfrac{3}{2}x + 3 \quad \text{Subtract 3 from both sides.}$$

Check Point 3 Write the point-slope form of the equation of the line passing through the points $(-2, -1)$ and $(-1, -6)$. Then solve the equation for y.

3 Write and graph the slope-intercept equation of a line.

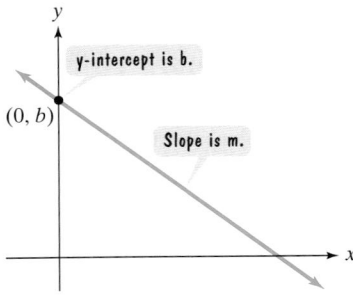

Figure 2.5 A line with slope m and y-intercept b

The Slope-Intercept Form of the Equation of a Line

Let's write the point-slope form of the equation of a line whose y-intercept is b with slope m. The line is shown in Figure 2.5. Because the y-intercept is b, the line intersects the y-axis at $(0, b)$. We use the point-slope form with $x_1 = 0$ and $y_1 = b$.

$$y - y_1 = m(x - x_1)$$

Let $y_1 = b$.   Let $x_1 = 0$.

We obtain

$$y - b = m(x - 0).$$

Simplifying on the right side gives us

$$y - b = mx.$$

Finally, we solve for y by adding b to both sides.

$$y = mx + b$$

Thus, if a line's equation is written with y isolated on one side, the x-coefficient is the line's slope and the constant term is the y-intercept. This form of a line's equation is called the **slope-intercept form** of a line.

Slope-Intercept Form of the Equation of a Line

The **slope-intercept equation** of a nonvertical line with slope m and y-intercept b is

$$y = mx + b.$$

EXAMPLE 4 Graphing by Using the Slope and y-Intercept

Graph the line whose equation is $y = \tfrac{2}{3}x + 2$.

Solution The equation of the line is in the form $y = mx + b$. We can find the slope, m, by identifying the coefficient of x. We can find the y-intercept, b, by identifying the constant term.

$$y = \frac{2}{3}x + 2$$

The slope is $\tfrac{2}{3}$.   The y-intercept is 2.

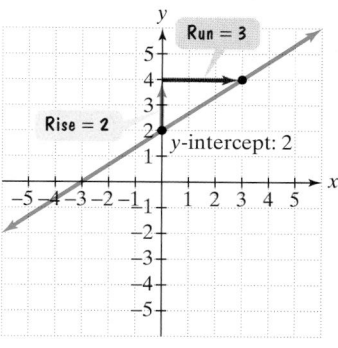

Figure 2.6 The graph of $y = \frac{2}{3}x + 2$

We need two points in order to graph the line. We can use the y-intercept, 2, to obtain the first point $(0, 2)$. Plot this point on the y-axis, shown in Figure 2.6.

We know the slope and one point on the line. We can use the slope, $\frac{2}{3}$, to determine a second point on the line. By definition,

$$m = \frac{2}{3} = \frac{\text{Rise}}{\text{Run}}.$$

We plot the second point on the line by starting at $(0, 2)$, the first point. Based on the slope, we move 2 units *up* (the rise) and 3 units to the *right* (the run). This puts us at a second point on the line, shown in Figure 2.6.

We use a straightedge to draw a line through the two points. The graph of $y = \frac{2}{3}x + 2$ is shown in Figure 2.6.

Graphing $y = mx + b$ by Using the Slope and y-Intercept

1. Plot the y-intercept on the y-axis. This is the point $(0, b)$.
2. Obtain a second point using the slope, m. Write m as a fraction, and use rise over run starting at the y-intercept to plot this point.
3. Use a straightedge to draw a line through the two points. Draw arrowheads at the ends of the line to show that the line continues indefinitely in both directions.

Check Point 4 Graph the line whose equation is $y = \dfrac{3}{5}x + 1$.

---

④ Recognize equations of horizontal and vertical lines.

Equations of Horizontal and Vertical Lines

Some things change very little. For example, from 1985 to the present, the number of Americans participating in downhill skiing has remained relatively constant, indicated by the graph shown in Figure 2.7. Shown in the figure is a horizontal line that passes through or near most of the data points.

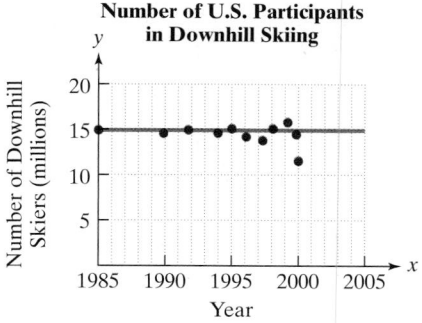

Source: National Ski Areas Association   **Figure 2.7**

We can use $y = mx + b$, the slope-intercept form of a line's equation, to write the equation of the horizontal line in Figure 2.7. We need the line's slope,

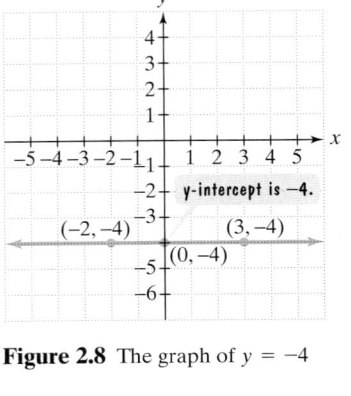

Figure 2.8 The graph of $y = -4$

m, and its y-intercept, b. Because the line is horizontal, $m = 0$. The line intersects the y-axis at 15, so $b = 15$. Thus, an equation that models the number of participants in downhill skiing for the period shown is

$$y = 0x + 15, \quad \text{or} \quad y = 15.$$

The popularity of downhill skiing remained relatively constant in the United States from 1985 to 2000 at approximately 15 million participants each year.

In general, if a line is horizontal, its slope is zero: $m = 0$. Thus, the equation $y = mx + b$ becomes $y = b$, where b is the y-intercept. For example, the graph of $y = -4$ is a horizontal line with a y-intercept of -4. The graph is shown in Figure 2.8. Three of the points along the line are shown and labeled. No matter what the x-coordinate is, the corresponding y-coordinate for every point on line is -4.

Equation of a Horizontal Line

A horizontal line is given by an equation of the form

$$y = b$$

where b is the y-intercept.

Next, let's see what we can discover about a vertical line by looking at an example.

EXAMPLE 5   Graphing a Vertical Line

Graph $x = 5$ in the rectangular coordinate system.

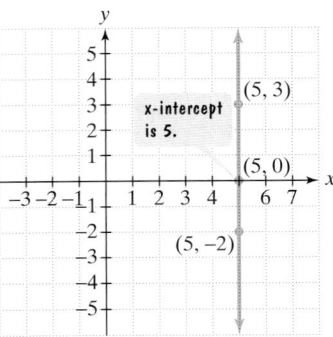

Figure 2.9 The graph $x = 5$

Solution   All points on the graph of $x = 5$ have a value of x that is always 5. No matter what the y-coordinate is, the corresponding x-coordinate for every point on the line is 5. Let us select three of the possible values of y: $-2, 0$, and 3. So, three of the points on the graph of $x = 5$ are $(5, -2), (5, 0)$, and $(5, 3)$. Plot each of these three points. Drawing a line that passes through the three points gives the vertical line shown in Figure 2.9.

Equation of a Vertical Line

A vertical line is given by an equation of the form

$$x = a$$

where a is the x-intercept.

Check Point 5   Graph $x = -1$ in the rectangular coordinate system.

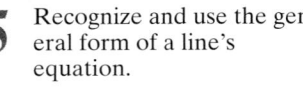

5 Recognize and use the general form of a line's equation.

The General Form of the Equation of a Line

The vertical line whose equation is $x = 5$ cannot be written in slope-intercept form, $y = mx + b$, because its slope is undefined. However, every line has an equation that can be expressed in the form $Ax + By + C = 0$. For example,

$x = 5$ can be expressed as $1x + 0y - 5 = 0$, or $x - 5 = 0$. The equation $Ax + By + C = 0$ is called the **general form** of the equation of a line.

> **General Form of the Equation of a Line**
>
> Every line has an equation that can be written in the **general form**
> $$Ax + By + C = 0$$
> where A, B, and C are three real numbers, and A and B are not both zero.

If the equation of a line is given in general form, it is possible to find the slope, m, and the y-intercept, b, for the line. We solve the equation for y, transforming it into the slope-intercept form $y = mx + b$. In this form, the coefficient of x is the slope of the line, and the constant term is its y-intercept.

EXAMPLE 6 Finding the Slope and the *y*-Intercept

Find the slope and the y-intercept of the line whose equation is $2x - 3y + 6 = 0$.

Solution The equation is given in general form. We begin by rewriting it in the form $y = mx + b$. We need to solve for y.

$$2x - 3y + 6 = 0 \qquad \text{This is the given equation.}$$
$$2x + 6 = 3y \qquad \text{To isolate the y-term, add 3y on both sides.}$$
$$3y = 2x + 6 \qquad \text{Reverse the two sides. (This step is optional.)}$$
$$y = \frac{2}{3}x + 2 \qquad \text{Divide both sides by 3.}$$

The coefficient of x, $\frac{2}{3}$, is the slope and the constant term, 2, is the y-intercept. This is the form of the equation that we graphed in Example 4 on page 173.

> **Check Point 6** Find the slope and the y-intercept of the line whose equation is $3x + 6y - 12 = 0$. Then use the y-intercept and the slope to graph the equation.

We've covered a lot of territory. Let's take a moment to summarize the various forms for equations of lines.

> **Equations of Lines**
>
> 1. Point-slope form: $\qquad\qquad\qquad y - y_1 = m(x - x_1)$
> 2. Slope-intercept form: $\qquad\qquad\quad y = mx + b$
> 3. Horizontal line: $\qquad\qquad\qquad\quad\; y = b$
> 4. Vertical line: $\qquad\qquad\qquad\qquad\; x = a$
> 5. General form: $\qquad\qquad\quad\; Ax + By + C = 0$

6 Model data with linear equations.

Applications

Linear equations are useful for modeling data that fall on or near a line. For example, Table 2.2 on page 176 gives the population of the United States, in millions, in the indicated year. The data are displayed as a set of five points in Figure 2.10.

Table 2.2

Year	x (Year after 1960)	y (U.S. Population) (in millions)
1960	0	179.3
1970	10	203.3
1980	20	226.5
1990	30	250.0
1998	38	268.9

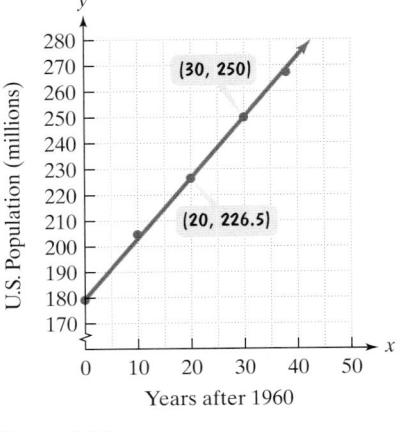

Figure 2.10

Data presented in a visual form as a set of points is called a **scatter plot**. Also shown in Figure 2.10 is a line that passes through or near the five points. By writing the equation of this line, we can obtain a model of the data and make predictions about the population of the United States in the future.

Technology

You can use a graphing utility to obtain a model for a scatter plot in which the data points fall on or near a straight line. The line that best fits the data is called the **regression line**. After entering the data in Table 2.2, a graphing utility displays a scatter plot of the data and the regression line.

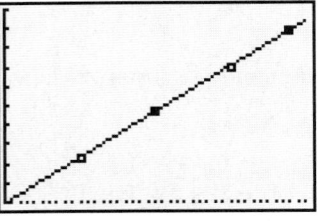

[0, 40, 1] by [180, 280, 10]

Also displayed is the regression line's equation.

EXAMPLE 7 Modeling U.S. Population

Write the equation of the line shown in Figure 2.10. Use the equation to predict U.S. population in 2010.

Solution The line in Figure 2.10 passes through (20, 226.5) and (30, 250). We start by finding the slope.

$$m = \frac{\text{change in } y}{\text{change in } x} = \frac{250 - 226.5}{30 - 20} = \frac{23.5}{10} = 2.35$$

Now, we write the line's equation.

$$y - y_1 = m(x - x_1) \qquad \text{Begin with the point-slope form.}$$
$$y - 250 = 2.35(x - 30) \qquad \text{Either ordered pair can be } (x_1, y_1).$$
$$\text{Let } (x_1, y_1) = (30, 250). \text{ From above, } m = 2.35.$$
$$y - 250 = 2.35x - 70.5 \qquad \text{Apply the distributive property on the right.}$$
$$y = 2.35x + 179.5 \qquad \text{Add 250 to both sides and solve for } y.$$

The linear equation that models U.S. population, y, in millions, x years after 1960 is

$$y = 2.35x + 179.5.$$

Now, let's use this equation to predict U.S. population in 2010. Because 2010 is 50 years after 1960, substitute 50 for x and compute y.

$$y = 2.35(50) + 179.5 = 297$$

Our equation predicts that the population of the United States in the year 2010 will be 297 million. (The projected figure from the U.S. Census Bureau is 297.716 million.)

If an equation in slope-intercept form describes some real-world situation, the slope and the y-intercept can be interpreted in terms of that situation. For example, $y = 2.35x + 179.5$ models U.S. population, y, in millions, x years after 1960. The slope, 2.35, indicates that U.S. population is increasing by 2.35 million people per year. 179.5 is the y-intercept. At the beginning, in 1960, U.S. population was 179.5 million.

Check Point 7

Use the data points $(10, 203.3)$ and $(20, 226.5)$ from Table 2.2 to write an equation that models U.S. population x years after 1960. Use the equation to predict U.S. population in 2020.

EXERCISE SET 2.1

✓ Practice Exercises

In Exercises 1–10, find the slope of the line passing through each pair of points or state that the slope is undefined. Then indicate whether the line through the points rises, falls, is horizontal, or is vertical.

1. $(4, 7)$ and $(8, 10)$
2. $(2, 1)$ and $(3, 4)$
3. $(-2, 1)$ and $(2, 2)$
4. $(-1, 3)$ and $(2, 4)$
5. $(4, -2)$ and $(3, -2)$
6. $(4, -1)$ and $(3, -1)$
7. $(-2, 4)$ and $(-1, -1)$
8. $(6, -4)$ and $(4, -2)$
9. $(5, 3)$ and $(5, -2)$
10. $(3, -4)$ and $(3, 5)$

In Exercises 11–38, use the given conditions to write an equation for each line in point-slope form and slope-intercept form.

11. Slope $= 2$, passing through $(3, 5)$
12. Slope $= 4$, passing through $(1, 3)$
13. Slope $= 6$, passing through $(-2, 5)$
14. Slope $= 8$, passing through $(4, -1)$
15. Slope $= -3$, passing through $(-2, -3)$
16. Slope $= -5$, passing through $(-4, -2)$
17. Slope $= -4$, passing through $(-4, 0)$
18. Slope $= -2$, passing through $(0, -3)$
19. Slope $= -1$, passing through $\left(-\frac{1}{2}, -2\right)$
20. Slope $= -1$, passing through $\left(-4, -\frac{1}{4}\right)$
21. Slope $= \frac{1}{2}$, passing through the origin
22. Slope $= \frac{1}{3}$, passing through the origin
23. Slope $= -\frac{2}{3}$, passing through $(6, -2)$
24. Slope $= -\frac{3}{5}$, passing through $(10, -4)$
25. Passing through $(1, 2)$ and $(5, 10)$
26. Passing through $(3, 5)$ and $(8, 15)$
27. Passing through $(-3, 0)$ and $(0, 3)$
28. Passing through $(-2, 0)$ and $(0, 2)$
29. Passing through $(-3, -1)$ and $(2, 4)$
30. Passing through $(-2, -4)$ and $(1, -1)$
31. Passing through $(-3, -2)$ and $(3, 6)$
32. Passing through $(-3, 6)$ and $(3, -2)$

33. Passing through $(-3, -1)$ and $(4, -1)$
34. Passing through $(-2, -5)$ and $(6, -5)$
35. Passing through $(2, 4)$ with x-intercept $= -2$
36. Passing through $(1, -3)$ with x-intereept $= -1$
37. x-intercept $= -\frac{1}{2}$ and y-intercept $= 4$
38. x-intercept $= 4$ and y-intercept $= -2$

In Exercises 39–46, give the slope and y-intercept of each line whose equation is given. Then graph the line.

39. $y = 2x + 1$
40. $y = 3x + 2$
41. $y = -2x + 1$
42. $y = -3x + 2$
43. $y = \dfrac{3}{4}x - 2$
44. $y = \dfrac{3}{4}x - 3$
45. $y = -\dfrac{3}{5}x + 7$
46. $y = -\dfrac{2}{5}x + 6$

In Exercises 47–52, graph each equation in the rectangular coordinate system.

47. $y = -2$
48. $y = 4$
49. $x = -3$
50. $x = 5$
51. $y = 0$
52. $x = 0$

In Exercises 53–60,
 a. *Rewrite the given equation in slope-intercept form.*
 b. *Give the slope and y-intercept.*
 c. *Graph the equation.*

53. $3x + y - 5 = 0$
54. $4x + y - 6 = 0$
55. $2x + 3y - 18 = 0$
56. $4x + 6y + 12 = 0$
57. $8x - 4y - 12 = 0$
58. $6x - 5y - 20 = 0$
59. $3x - 9 = 0$
60. $4y + 28 = 0$

Application Exercises

61. As shown in the graph on page 178, the percentage of people in the United States satisfied with their lives remains relatively constant for all age groups. If x represents a person's age and y represents the percentage of people satisfied with their lives at that age, write an equation that reasonably models the data.

Percentage of People in the U.S. Satisfied with Their Lives

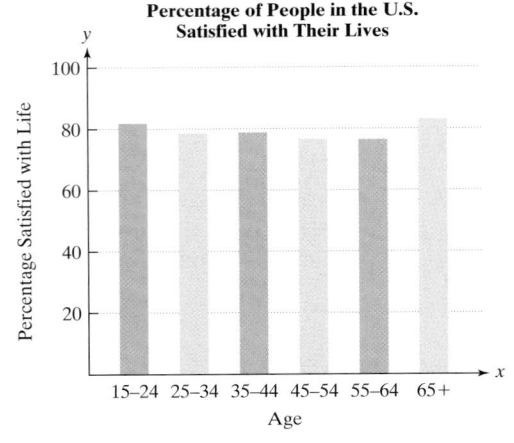

Source: *Culture Shift in Advanced Industrial Society*, Princeton University Press

62. The graph shows the life expectancy in years for U.S. women whose year of birth is indicated on the *x*-axis. Find the slope of the line passing through the points whose coordinates are shown on the graph. Describe what the slope represents.

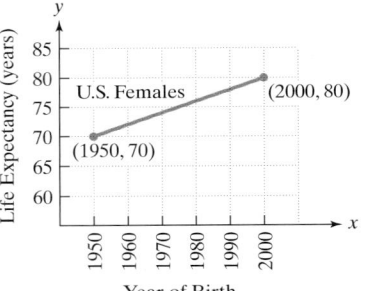

63. The graph shows U.S. population projections from 2000 through 2050. Use the equation $y = 2.35x + 179.5$, in which *x* is the number of years after 1960 and *y* is the U.S. population, in millions, to determine how well the equation models the projections for 2030, 2040, and 2050.

U.S. Population Projections: 2000–2050

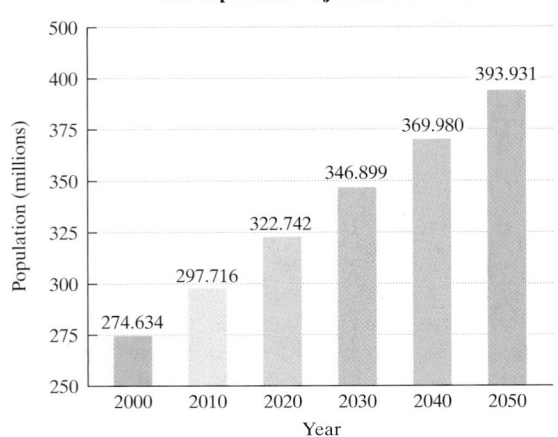

Source: U.S. Census Bureau

64. The figure shows projected online shopping per U.S. online household through 2004.

Online Spending: Yearly Spending per Online Household

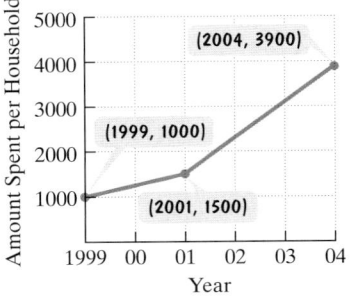

Source: Forrester Research

a. Find the slope of the line passing through (1999, 1000) and (2001, 1500). What does this represent in terms of the increase in online shopping per year?

b. Repeat part (a) for the data points (2001, 1500) and (2004, 3900).

c. Write the point-slope form of the equation of the line passing through (2001, 1500) and (2004, 3900).

d. Use the equation in part (c) to write the slope-intercept form of the equation.

e. Use the equation from part (c) to project the amount that will be spent shopping online per online household in 2010.

In Exercises 65–66, two measurements are given for variables having a linear relationship. For each exercise, write the point-slope form of the equation of the line on which these measurements fall. Then use the point-slope form of the equation to write the slope-intercept form of the equation. Finally, use this equation to answer the question.

65.

x (Number of Years after 1990)	y (Total Consumer Spending in the United States, in Billions of Dollars)
3	4459.2
7	5493.7

Source: U.S. Commerce Department

How much will consumers in the United States spend in the year 2020?

66.

x (Number of Years after 1985)	y (Total of All Health-Care Expenditures in the United States, in Billions of Dollars)
3	546
5	666

Source: U.S. Health Care Financing Administration

What will health-care expenditures in the United States be in the year 2010?

67. A business discovers a linear relationship between the number of shirts it can sell and the price per shirt. In particular, 20,000 shirts can be sold at $19 each, and 2000 of the same shirts can be sold at $55 each. Write the slope-intercept equation of the *demand line* through the ordered pairs (20,000 shirts, $19) and (2000 shirts, $55). Then determine the number of shirts that can be sold at $50 each.

68. In 1965, radioactive wastes seeping into the Columbia River exposed citizens of eight Oregon counties and the city of Portland to radioactive contamination. In an article in the *Journal of Environmenial Health* (May–June, 1965), the authors formulated an index that measured the proximity of the residents to the contamination. The ordered pair for Columbia County (6.4, 178) indicates that its index is 6.4 and there are 178 cancer deaths per 100,000 residents. The corresponding ordered pair for Clatsop County is (8.3, 210). What is the predicted number of cancer deaths for Portland, with an index of 11.6?

69. Is there a relationship between education and prejudice? With increased education, does a person's level of prejudice tend to decrease? The scatter plot shows ten data points, each representing the number of years of school completed and the score on a test measuring prejudice for each subject. Higher scores on this 1-to-10 test indicate greater prejudice. Also shown is the regression line, the line that best fits the data. Use two points on this line to write both its point-slope and slope-intercept equations. Then use the slope-intercept equation to predict the score on the prejudice test for a person with seven years of education.

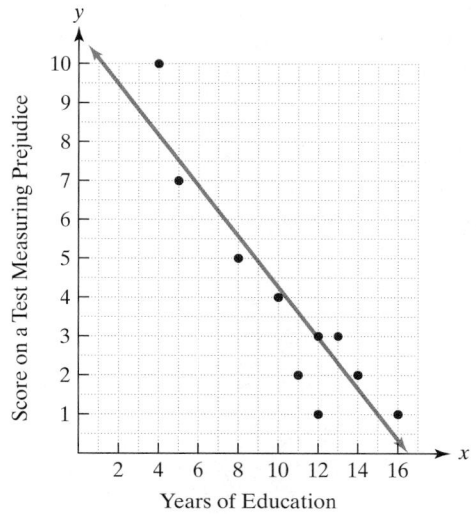

70. The scatter plot at the top of the next column shows the relationship between the percentage of married women of child-bearing age using contraceptives and the births per woman in selected countries. Also shown is the regression line. Use two points on this line to write both its point-slope and slope-intercept equations. Then find the number of births per woman if 90% of married women of child-bearing age use contraceptives.

Contraceptive Prevalence and Births per Woman, Selected Countries

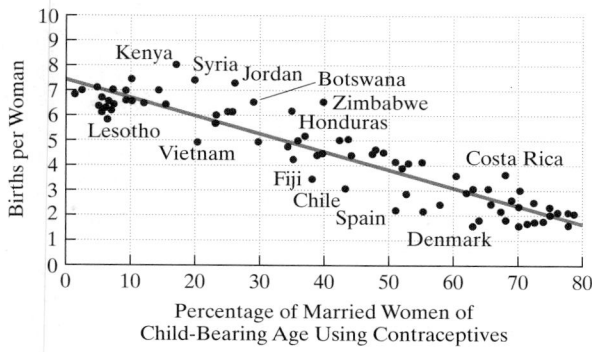

Source: Population Reference Bureau

Writing in Mathematics

71. What is the slope of a line and how is it found?

72. Describe how to write the equation of a line if two points along the line are known.

73. Explain how to derive the slope-intercept form of a line's equation, $y = mx + b$, from the point-slope form

$$y - y_1 = m(x - x_1).$$

74. Explain how to graph the equation $x = 2$. Can this equation be expressed in slope-intercept form? Explain.

75. Explain how to use the general form of a line's equation to find the line's slope and y-intercept.

76. Look back at Figure 2.1 on page 168. Do you think that the line through the points corresponding to 2001 and 2004 will describe online spending per online household in 2040? Explain your answer.

77. Take a second look at the scatter plot in Exercise 69. Although there is a relationship between education and prejudice, we cannot necessarily conclude that increased education causes a person's level of prejudice to decrease. Offer two or more possible explanations for the data in the scatter plot.

Technology Exercises

Use a graphing utility to graph each equation in Exercises 78–81. Then use the TRACE *feature to trace along the line and find the coordinates of two points. Use these points to compute the line's slope. Check your result by using the coefficient of x in the line's equation.*

78. $y = 2x + 4$

79. $y = -3x + 6$

80. $y = -\frac{1}{2}x - 5$

81. $y = \frac{3}{4}x - 2$

82. a. Use the statistical menu of your graphing utility to enter the ten data points shown in the scatter plot in Exercise 69.

b. Use the $\boxed{\text{DRAW}}$ menu and the scatter plot capability to draw a scatter plot of the data points like the one shown in Exercise 69.

c. Select the linear regression option. Your utility should give you values for a and b for the equation of the regression line, $y = ax + b$. You may also be given a *correlation coefficient*, r. Values of r close to 1 indicate that the points can be described by a linear relationship and the regression line has a positive slope. Values of r close to -1 indicate that the points can be described by a linear relationship and the regression line has a negative slope. Values of r close to 0 indicate no linear relationship between the variables.

d. Use the appropriate sequence (consult your manual) to graph the regression equation on top of the points in the scatter plot.

Critical Thinking Exercises

83. Which one of the following is true?

a. A linear equation with nonnegative slope has a graph that rises from left to right.

b. The equations $y = 4x$ and $y = -4x$ have graphs that are perpendicular lines.

c. The line whose equation is $5x + 6y - 30 = 0$ passes through the point $(6, 0)$ and has slope $-\frac{5}{6}$.

d. The graph of $y = 7$ in the rectangular coordinate system is the single point $(7, 0)$.

84. Prove that the equation of a line passing through $(a, 0)$ and $(0, b)$ $(a \neq 0, b \neq 0)$ can be written in the form $\frac{x}{a} + \frac{y}{b} = 1$. Why is this called the *intercept form* of a line?

85. Use the figure at the top of the next column to make the following lists.

a. List the slopes m_1, m_2, m_3, and m_4 in order of decreasing size.

b. List the y-intercepts b_1, b_2, b_3, and b_4 in order of decreasing size.

86. Excited about the success of celebrity stamps, post office officials were rumored to have put forth a plan to institute two new types of thermometers. On these new scales, $°E$ represents degrees Elvis and $°M$ represents degrees Madonna. If it is known that $40°E = 25°M$, $280°E = 125°M$, and degrees Elvis is linearly related to degrees Madonna, write an equation expressing E in terms of M.

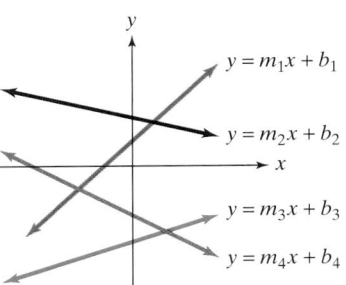

Group Activity Exercise

87. Group members should consult an almanac, newspaper, magazine, or the Internet to find data that lie approximately on or near a straight line. Working by hand or using a graphing utility, construct a scatter plot for the data. If working by hand, draw a line that approximately fits the data and then write its equation. If using a graphing utility, obtain the equation of the regression line. Then use the equation of the line to make a prediction about what might happen in the future. Are there circumstances that might affect the accuracy of this prediction? List some of these circumstances.

SECTION 2.2 *Parallel and Perpendicular Lines and Circles*

Objectives

1. Find slopes and equations of parallel and perpendicular lines.

2. Write the standard form of a circle's equation.

3. Give the center and radius of a circle whose equation is in standard form.

4. Convert the general form of a circle's equation to standard form.

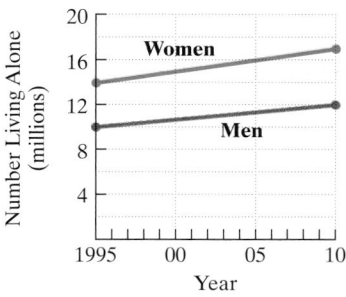

Source: Forrester Research

Figure 2.11

1 Find slopes and equations of parallel and perpendicular lines.

A best guess at the look of our nation in the next decades indicates that the number of men and women living alone will increase each year. Figure 2.11 shows that by 2010, approximately 12 million men and 17 million women will be living alone. Can you tell by the line graphs in the figure if the yearly increase for women is the same as the yearly increase for men? We begin this section by showing how we can use the slope of each line to answer this question.

Parallel and Perpendicular Lines

Two nonintersecting lines that lie in the same plane are parallel. If two lines do not intersect, the ratio of the vertical change to the horizontal change is the same for each line. Because two parallel lines have the same "steepness," they must have the same slope.

> **Slope and Parallel Lines**
>
> 1. If two nonvertical lines are parallel, then they have the same slope.
> 2. If two distinct nonvertical lines have the same slope, then they are parallel.
> 3. Two distinct vertical lines, both with undefined slopes, are parallel.

EXAMPLE 1 Writing Equations of a Line Parallel to a Given Line

Write an equation of the line passing through $(-3, 2)$ and parallel to the line whose equation is $y = 2x + 1$. Express the equation in point-slope form and slope-intercept form.

Solution The situation is illustrated in Figure 2.12. We are looking for the equation of the line shown on the left. How do we obtain this equation? Notice that the line passes through the point $(-3, 2)$. Using the point-slope form of the line's equation, we have $x_1 = -3$ and $y_1 = 2$.

$$y - y_1 = m(x - x_1)$$

$$y_1 = 2 \qquad x_1 = -3$$

The equation of this line is given: $y = 2x + 1$.

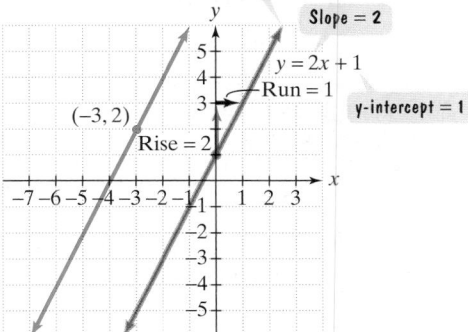

We must write the equation of this line.

Figure 2.12 Writing equations of a line parallel to a given line

The equation of this line is given: y = 2x + 1.

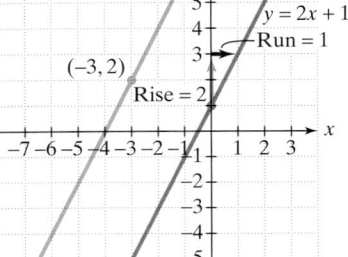

We must write the equation of this line.

Figure 2.12 Shown again so that you do not have to turn back a page

Now, the only thing missing from the equation is m, the slope of the line on the left. Do we know anything about the slope of either line in Figure 2.12? The answer is yes; we know the slope of the line on the right, whose equation is given.

$$y = 2x + 1$$

The slope of the line on the right in Figure 2.12 is 2.

Parallel lines have the same slope. Because the slope of the line with the given equation is 2, $m = 2$ for the line whose equation we must write.

$$y - y_1 = m(x - x_1)$$

$$y_1 = 2 \qquad m = 2 \qquad x_1 = -3$$

The point-slope form of the line's equation is

$$y - 2 = 2[x - (-3)] \text{ or}$$
$$y - 2 = 2(x + 3).$$

Solving for y, we obtain the slope-intercept form of the equation.

$$y - 2 = 2x + 6 \qquad \text{Apply the distributive property.}$$
$$y = 2x + 8 \qquad \text{Add 2 to both sides. This is the slope-intercept form,}$$
$$y = mx + b, \text{ of the equation.}$$

Check Point 1 Write an equation of the line passing through $(-2, 5)$ and parallel to the line whose equation is $y = 3x + 1$. Express the equation in point-slope form and slope-intercept form.

Two lines that intersect at a right angle (90°) are said to be **perpendicular**, shown in Figure 2.13. There is a relationship between the slopes of perpendicular lines.

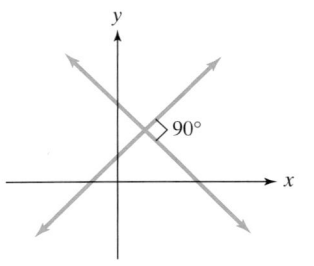

Figure 2.13 Perpendicular lines

Slope and Perpendicular Lines

1. If two nonvertical lines are perpendicular, then the product of their slopes is -1.
2. If the product of the slopes of two lines is -1, then the lines are perpendicular.
3. A horizontal line having zero slope is perpendicular to a vertical line having undefined slope.

An equivalent way of stating this relationship is to say that one line is perpendicular to another line if its slope is the *negative reciprocal* of the slope of the other. For example, if a line has slope 5, any line having slope $-\frac{1}{5}$ is perpendicular to it. Similarly, if a line has slope $-\frac{3}{4}$, any line having slope $\frac{4}{3}$ is perpendicular to it.

EXAMPLE 2 Finding the Slope of a Line Perpendicular to a Given Line

Find the slope of any line that is perpendicular to the line whose equation is $x + 4y - 8 = 0$.

Solution   We begin by writing the equation of the given line in slope-intercept form. Solve for y.

$$x + 4y - 8 = 0 \qquad \text{This is the given equation.}$$
$$4y = -x + 8 \qquad \text{To isolate the y-term, subtract x and add 8 on both sides.}$$
$$y = -\tfrac{1}{4}x + 2 \qquad \text{Divide both sides by 4.}$$

Slope is $-\dfrac{1}{4}$.

The given line has slope $-\tfrac{1}{4}$. Any line perpendicular to this line has a slope that is the negative reciprocal of $-\tfrac{1}{4}$. Thus, the slope of any perpendicular line is 4.

Check Point 2   Find the slope of any line that is perpendicular to the line whose equation is $x + 3y - 12 = 0$.

Circles

It's a good idea to know your way around a circle. Clocks, angles, maps, and compasses are based on circles. Circles occur everywhere in nature: in ripples on water, patterns on a butterfly's wings, and cross sections of trees. Some consider the circle to be the most pleasing of all shapes.

The rectangular coordinate system gives us a unique way of knowing a circle. It enables us to translate a circle's geometric definition into an algebraic equation. We begin with this geometric definition.

> **Definition of a Circle**
>
> A **circle** is the set of all points in a plane that are equidistant from a fixed point called the **center**. The fixed distance from the circle's center to any point on the circle is called the **radius**.

Figure 2.14 on page 184 is our starting point for obtaining a circle's equation. We've placed the circle into a rectangular coordinate system. The circle's center is (h, k) and its radius is r. We let (x, y) represent the coordinates of any point on the circle.

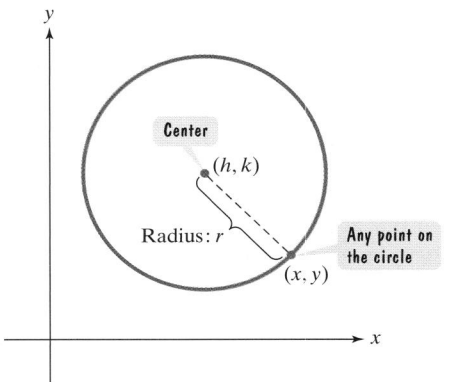

Figure 2.14 A circle centered at (h, k) with radius r

What does the geometric definition of a circle tell us about point (x, y) in Figure 2.14? The point is on the circle if and only if its distance from the center is r. We can use the distance formula to express this idea algebraically:

The distance between (x, y) and (h, k) — is always — r.

$$\sqrt{(x - h)^2 + (y - k)^2} = r$$

Squaring both sides of this equation yields the **standard form of the equation of a circle**.

The Standard Form of the Equation of a Circle

The **standard form of the equation of a circle** with center (h, k) and radius r is

$$(x - h)^2 + (y - k)^2 = r^2.$$

---

2 Write the standard form of a circle's equation.

EXAMPLE 3 Finding the Standard Form of a Circle's Equation

Write the standard form of the equation of the circle with center $(0, 0)$ and radius 2. Graph the circle.

Solution   The center is $(0, 0)$. Because the center is represented as (h, k) in the standard form of the equation, $h = 0$ and $k = 0$. The radius is 2, so we will let $r = 2$ in the equation.

$(x - h)^2 + (y - k)^2 = r^2$ This is the standard form of a circle's equation.

$(x - 0)^2 + (y - 0)^2 = 2^2$ Substitute 0 for h, 0 for k, and 2 for r.

$x^2 + y^2 = 4$ Simplify.

The standard form of the equation of the circle is $x^2 + y^2 = 4$. Figure 2.15 shows the graph.

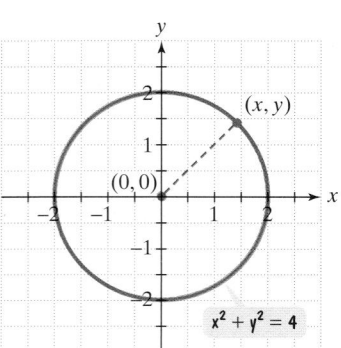

Figure 2.15 The graph of $x^2 + y^2 = 4$

Check Point 3   Write the standard form of the equation of the circle with center $(0, 0)$ and radius 4.

Technology

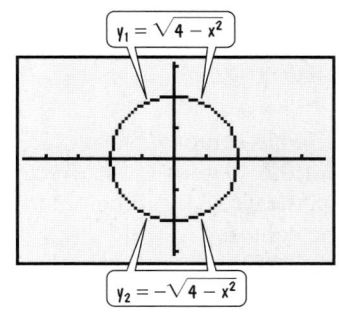

To graph a circle with a graphing utility, first solve the equation for y.

$$x^2 + y^2 = 4$$
$$y^2 = 4 - x^2$$
$$y = \pm\sqrt{4 - x^2}$$

Graph the two equations

$$y_1 = \sqrt{4 - x^2} \quad \text{and} \quad y_2 = -\sqrt{4 - x^2}$$

in the same viewing rectangle. The graph of $y_1 = \sqrt{4 - x^2}$ is the top semicircle be-cause y is always positive. The graph of $y_2 = -\sqrt{4 - x^2}$ is the bottom semicircle be-cause y is always negative. Use a ZOOM SQUARE setting so that the circle looks like a circle. (Many graphing utilities have problems connecting the two semicircles because the segments directly across horizontally from the center become nearly vertical.)

Example 3 and Checkpoint 3 involved circles centered at the origin. The standard form of the equation of all such circles is $x^2 + y^2 = r^2$, where r is the circle's radius. Now, let's consider a circle whose center is not at the origin.

EXAMPLE 4   Finding the Standard Form of a Circle's Equation

Write the standard form of the equation of the circle with center $(-2, 3)$ and radius 4.

Solution   The center is $(-2, 3)$. Because the center is represented as (h, k) in the standard form of the equation, $h = -2$ and $k = 3$. The radius is 4, so we will let $r = 4$ in the equation.

$$(x - h)^2 + (y - k)^2 = r^2 \qquad \text{This is the standard form of a circle's equation.}$$
$$[x - (-2)]^2 + (y - 3)^2 = 4^2 \qquad \text{Substitute } -2 \text{ for } h, 3 \text{ for } k, \text{ and } 4 \text{ for } r.$$
$$(x + 2)^2 + (y - 3)^2 = 16 \qquad \text{Simplify.}$$

The standard form of the equation of the circle is $(x + 2)^2 + (y - 3)^2 = 16$.

Check Point 4   Write the standard form of the equation of the circle with cen-ter $(5, -6)$ and radius 10.

3 Give the center and radius of a circle whose equation is in standard form.

EXAMPLE 5   Using the Standard Form of a Circle's Equation to Graph the Circle

Find the center and radius of the circle whose equation is

$$(x - 2)^2 + (y + 4)^2 = 9$$

and graph the equation.

Solution   In order to graph the circle, we need to know its center, (h, k), and its radius r. We can find the values for h, k, and r by comparing the given equa-tion to the standard form of the equation of a circle.

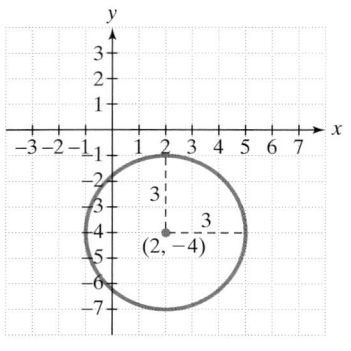

Figure 2.16 The graph of $(x - 2)^2 + (y + 4)^2 = 9$

$$(x - 2)^2 + (y + 4)^2 = 9$$
$$(x - 2)^2 + (y - (-4))^2 = 3^2$$

This is $(x - h)^2$, with $h = 2$.	This is $(y - k)^2$, with $k = -4$.	This is r^2, with $r = 3$.

We see that $h = 2$, $k = -4$, and $r = 3$. Thus, the circle has center $(h, k) = (2, -4)$ and a radius of 3 units. To graph this circle, first plot the center $(2, -4)$. Because the radius is 3, you can locate at least four points on the circle by going out three units to the right, to the left, up, and down from the center.

Two such points, to the right and to the left of $(2, -4)$, are $(5, -4)$ and $(-1, -4)$, respectively.

Using these points, we obtain the graph in Figure 2.16.

Check Point 5 Find the center and radius of the circle whose equation is
$$(x + 3)^2 + (y - 1)^2 = 4$$
and graph the equation.

If we square $x - 2$ and $y + 4$ in the standard form of the equation of Example 5, we obtain another form for the circle's equation.

$$(x - 2)^2 + (y + 4)^2 = 9 \quad \text{This is the standard form of the equation from Example 3.}$$

$$x^2 - 4x + 4 + y^2 + 8y + 16 = 9 \quad \text{Square } x - 2 \text{ and } y + 4.$$

$$x^2 + y^2 - 4x + 8y + 20 = 9 \quad \text{Combine numerical terms and rearrange terms.}$$

$$x^2 + y^2 - 4x + 8y + 11 = 0 \quad \text{Subtract 9 from both sides.}$$

This result suggests that an equation in the form $x^2 + y^2 + Dx + Ey + F = 0$ can represent a circle. This is called the **general form of the equation of a circle**.

The General Form of the Equation of a Circle

The **general form of the equation of a circle is**
$$x^2 + y^2 + Dx + Ey + F = 0.$$

---

4 Convert the general form of a circle's equation to standard form.

We can convert the general form of the equation of a circle to the standard form $(x - h)^2 + (y - k)^2 = r^2$. We do so by completing the square on x and y. Let's see how this is done.

EXAMPLE 6 **Converting the General Form of a Circle's Equation to Standard Form and Graphing the Circle**

Write in standard form and graph: $x^2 + y^2 + 4x - 6y - 23 = 0$.

Solution Because we plan to complete the square on both x and y, let's rearrange terms so that x-terms are arranged in descending order, y-terms are arranged in descending order, and the constant term appears on the right.

$$x^2 + y^2 + 4x - 6y - 23 = 0$$

This is the given equation.

$$(x^2 + 4x \quad) + (y^2 - 6y \quad) = 23$$

Rewrite in anticipation of completing the square.

$$(x^2 + 4x + 4) + (y^2 - 6y + 9) = 23 + 4 + 9$$

Complete the square on x: $\frac{1}{2} \cdot 4 = 2$ and $2^2 = 4$, so add 4 to both sides. Complete the square on y: $\frac{1}{2}(-6) = -3$ and $(-3)^2 = 9$, so add 9 to both sides.

> Remember that numbers added on the left side must also be added on the right side.

$$(x + 2)^2 + (y - 3)^2 = 36$$

Factor on the left and add on the right.

This last equation is in standard form. We can identify the circle's center and radius by comparing this equation to the standard form of the equation of a circle, $(x - h)^2 + (y - k)^2 = r^2$.

$$(x + 2)^2 + (y - 3)^2 = 36$$

$$\left(x - (-2)\right)^2 + (y - 3)^2 = 6^2$$

This is $(x - h)^2$, with $h = -2$.　　This is $(y - k)^2$, with $k = 3$.　　This is r^2, with $r = 6$.

We use the center, $(h, k) = (-2, 3)$, and the radius, $r = 6$, to graph the circle. The graph is shown in Figure 2.17.

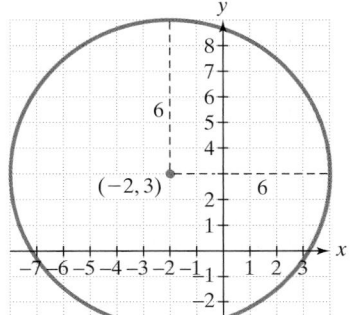

Figure 2.17 The graph of $(x + 2)^2 + (y - 3)^2 = 36$

Technology

To graph $x^2 + y^2 + 4x - 6y - 23 = 0$, rewrite the equation as a quadratic equation in y.

$$y^2 - 6y + (x^2 + 4x - 23) = 0$$

Now solve for y using the quadratic formula, with $a = 1$, $b = -6$, and $c = x^2 + 4x - 23$.

$$y = \frac{-b \pm \sqrt{b^2 - 4ac}}{2a} = \frac{-(-6) \pm \sqrt{(-6)^2 - 4 \cdot 1(x^2 + 4x - 23)}}{2 \cdot 1} = \frac{6 \pm \sqrt{36 - 4(x^2 + 4x - 23)}}{2}$$

Because we will enter these equations, there is no need to simplify. Enter

$$y_1 = \frac{6 + \sqrt{36 - 4(x^2 + 4x - 23)}}{2}$$

and

$$y_2 = \frac{6 - \sqrt{36 - 4(x^2 + 4x - 23)}}{2}.$$

Use a ZOOM SQUARE setting. The graph is shown on the right.

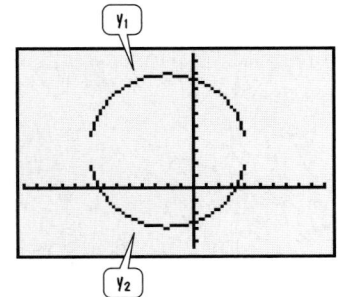

Check Point 6

Write in standard form and graph:

$$x^2 + y^2 + 4x - 4y - 1 = 0.$$

EXERCISE SET 2.2

✔ Practice Exercises

In Exercises 1–16, the equation of a line is given. Find the slope of a line that is (a) parallel to the line with the given equation; and (b) perpendicular to the line with the given equation.

1. $y = 5x$

2. $y = 3x$

3. $y = -7x$

4. $y = -9x$

5. $y = \frac{1}{2}x + 3$

6. $y = \frac{1}{4}x - 5$

7. $y = -\frac{2}{5}x - 1$

8. $y = -\frac{3}{7}x - 2$

9. $4x + y = 7$

10. $8x + y = 11$

11. $2x + 4y - 8 = 0$

12. $3x + 2y - 6 = 0$

13. $2x - 3y - 5 = 0$

14. $3x - 4y + 7 = 0$

15. $x = 6$

16. $y = 9$

In Exercises 17–20, write an equation for line L in point-slope form and slope-intercept form.

17.

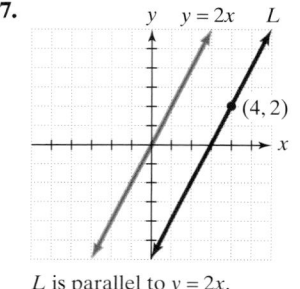

L is parallel to $y = 2x$.

18.

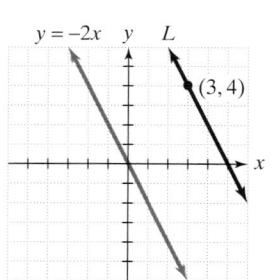

L is parallel to $y = -2x$.

19.

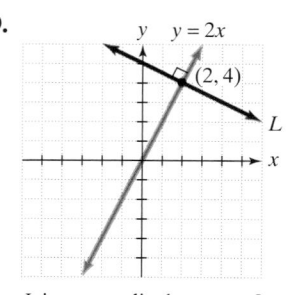

L is perpendicular to $y = 2x$.

20.

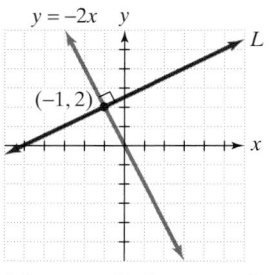

L is perpendicular to $y = -2x$.

In Exercises 21–28, use the given conditions to write an equation for each line in point-slope form and slope-intercept form.

21. Passing through $(-8, -10)$ and parallel to the line whose equation is $y = -4x + 3$

22. Passing through $(-2, -7)$ and parallel to the line whose equation is $y = -5x + 4$

23. Passing through $(2, -3)$ and perpendicular to the line whose equation is $y = \frac{1}{5}x + 6$

24. Passing through $(-4, 2)$ and perpendicular to the line whose equation is $y = \frac{1}{3}x + 7$

25. Passing through $(-2, 2)$ and parallel to the line whose equation is $2x - 3y - 7 = 0$

26. Passing through $(-1, 3)$ and parallel to the line whose equation is $3x - 2y - 5 = 0$

27. Passing through $(4, -7)$ and perpendicular to the line whose equation is $x - 2y - 3 = 0$

28. Passing through $(5, -9)$ and perpendicular to the line whose equation is $x + 7y - 12 = 0$

In Exercises 29–38, write the standard form of the equation of the circle with the given center and radius.

29. Center $(0, 0)$, $r = 7$ **30.** Center $(0, 0)$, $r = 8$

31. Center $(3, 2)$, $r = 5$ **32.** Center $(2, -1)$, $r = 4$

33. Center $(-1, 4)$, $r = 2$ **34.** Center $(-3, 5)$, $r = 3$

35. Center $(-3, -1)$, $r = \sqrt{3}$ **36.** Center $(-5, -3)$, $r = \sqrt{5}$

37. Center $(-4, 0)$, $r = 10$ **38.** Center $(-2, 0)$, $r = 6$

In Exercises 39–46, give the center and radius of the circle described by the equation and graph each equation.

39. $x^2 + y^2 = 16$ **40.** $x^2 + y^2 = 49$

41. $(x - 3)^2 + (y - 1)^2 = 36$

42. $(x - 2)^2 + (y - 3)^2 = 16$

43. $(x + 3)^2 + (y - 2)^2 = 4$

44. $(x + 1)^2 + (y - 4)^2 = 25$

45. $(x + 2)^2 + (y + 2)^2 = 4$

46. $(x + 4)^2 + (y + 5)^2 = 36$

In Exercises 47–54, complete the square and write the equation in standard form. Then give the center and radius of each circle and graph the equation.

47. $x^2 + y^2 + 6x + 2y + 6 = 0$
48. $x^2 + y^2 + 8x + 4y + 16 = 0$
49. $x^2 + y^2 - 10x - 6y - 30 = 0$
50. $x^2 + y^2 - 4x - 12y - 9 = 0$
51. $x^2 + y^2 + 8x - 2y - 8 = 0$
52. $x^2 + y^2 + 12x - 6y - 4 = 0$
53. $x^2 - 2x + y^2 - 15 = 0$
54. $x^2 + y^2 - 6y - 7 = 0$

 Application Exercises

55. The line graph shows the number of people in the United States projected to live alone through 2010.

**Number of People in the U.S.
Projected to Live Alone**

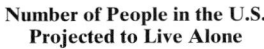
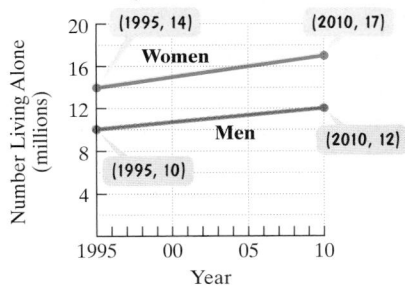

Source: Forrester Research

   a. Find the slope of the line for U.S. women.
   b. Find the slope of the line for U.S. men.
   c. Are the lines parallel? What does this mean in terms of the yearly increase for women and the yearly increase for men?

56. The amount spent annually in college bookstores in the United States can be modeled by $y = 0.19x + 1.67$, where x represents the number of years since 1982 and y represents the amount spent in billions of dollars. If the graph of this equation is parallel to a line representing the amount spent annually in bookstores in the United States since 1982, what does this mean in terms of the yearly increase for spending on books?

57. We refer to the driveway in the figure shown at the top of the next column as being *circular*, meaning that it is bounded by two circles. The figure indicates that the radius of the larger circle is 52 feet and the radius of the smaller circle is 38 feet. All points on the circular driveway satisfy the following compound inequality:

all points on the smaller circle	$\leq$	all points (x, y) on the driveway	$\leq$	all points on the larger circle.

   a. Rewrite the left portion of this inequality by writing the equation of the smaller circle.

   b. Rewrite the right portion of this inequality by writing the equation of the larger circle.

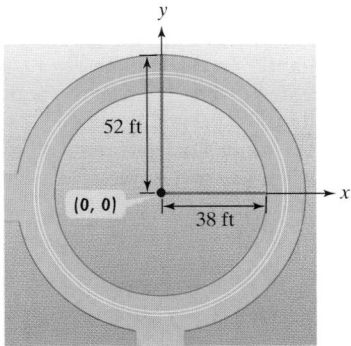

58. The circle formed by the middle lane of a circular running track can be described algebraically by $x^2 + y^2 = 4$, where all measurements are in miles. If you run around the track twice, approximately how many miles have you covered?

 **Writing in Mathematics**

59. If two lines are parallel, describe the relationship between their slopes.

60. If two lines are perpendicular, describe the relationship between their slopes.

61. The number of multiple births in the United States (twins, triplets, etc.) per 1000 live births can be modeled by $y = 0.463x + 18.888$, where x represents the number of years since 1980 and y represents multiple births per 1000 live births. Explain why the equation of this line cannot be parallel to the line representing the number of births in the United States since 1980.

62. If you know a point on a line and you know the equation of a line perpendicular to this line, explain how to write the line's equation.

63. What is a circle? Without using variables, describe how your definition of a circle can be used to obtain a form of its equation.

64. Give an example of a circle's equation in standard form. Describe how to find the center and radius for this circle.

65. How is the standard form of a circle's equation obtained from its general form?

66. Does $(x - 3)^2 + (y - 5)^2 = 0$ represent the equation of a circle? If not, describe the graph of this equation.

67. Does $(x - 3)^2 + (y - 5)^2 = -25$ represent the equation of a circle? What sort of set is the graph of this equation?

Technology Exercises

68. The lines whose equations are $y = \frac{1}{3}x + 1$ and $y = -3x - 2$ are perpendicular because the product of their slopes, $\frac{1}{3}$ and -3, respectively, is -1.

   a. Use a graphing utility to graph the equations. Do the lines appear to be perpendicular?

   b. Now use the zoom square feature of your utility. Describe what happens to the graphs. Explain why this is so.

In Exercises 69–71, use a graphing utility to graph each circle whose equation is given.

69. $x^2 + y^2 = 25$

70. $(y + 1)^2 = 36 - (x - 3)^2$

71. $x^2 + 10x + y^2 - 4y - 20 = 0$

Critical Thinking Exercises

72. Which one of the following is true?

   a. The equation of the circle whose center is at the origin with radius 16 is $x^2 + y^2 = 16$.

   b. The graph of $(x - 3)^2 + (y + 5)^2 = 36$ is a circle with a radius 6 centered at $(-3, 5)$.

   c. The graph of $(x - 4) + (y + 6) = 25$ is a circle with a radius 5 centered at $(4, -6)$.

   d. None of the above is true.

In Exercises 73–74, write the point-slope form and the slope-intercept form of the equation for each line described.

73. Having an x-intercept of -3 and perpendicular to the line passing through $(0, 0)$ and $(6, -2)$

74. Perpendicular to $3x - 2y = 4$ with the same y-intercept

In Exercises 75–76, write the standard form and the general form of the equation of each circle.

75. Center at $(3, -5)$ and passing through the point $(-2, 1)$

76. Passing through $(-7, 2)$ and $(1, 2)$; these points lie on the line that passes through the circle's center.

77. Find the area of the region bounded by the graphs of $(x - 2)^2 + (y + 3)^2 = 25$ and $(x - 2)^2 + (y + 3)^2 = 36$.

78. A **tangent line** to a circle is a line that intersects the circle at exactly one point. The tangent line is perpendicular to the radius of the circle at this point of contact. Write the point-slope equation of a line tangent to the circle whose equation is $x^2 + y^2 = 25$ at the point $(3, -4)$.

SECTION 2.3   *Introduction to Functions*

Objectives

1. Find the domain and range of a relation.

2. Determine whether a relation is a function.

3. Determine whether an equation represents a function.

4. Evaluate a function.

5. Understand and use piecewise functions.

6. Find the domain of a function.

7. Interpret function values for functions that model data.

Enjoy talking on the phone? In 1999, nearly 80 million Americans were chatting up a storm on their mobile phones, an increase of 300% from 1994, when only 20 million Americans were using mobile phones. And who can blame them? The graph in Figure 2.18 shows the decrease in the monthly average U.S. mobile-phone bills from 1994 through 1998. With video mobile phones by 2020, there seems to be no limit to the ways in which we keep in touch.

U.S. Mobile-Phone Bills

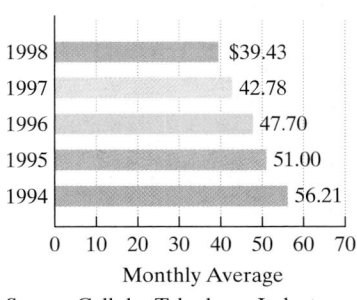

Monthly Average

Source: Cellular Telephone Industry of America

Figure 2.18

1 Find the domain and range of a relation.

If we let *x* represent a year and *y* the monthly average mobile-phone bill, the graph in Figure 2.18 shows a correspondence between the two variables *x* and *y*. We can write this correspondence using a set of ordered pairs:

$$\{(1994, 56.21), (1995, 51.00), (1996, 47.70), (1997, 42.78), (1998, 39.43)\}$$

The mathematical term for a set of ordered pairs is a **relation**.

Definition of a Relation

A **relation** is any set of ordered pairs. The set of all first components of the ordered pairs is called the **domain** of the relation, and the set of all second components is called the **range** of the relation.

EXAMPLE 1   Analyzing U.S. Mobile-Phone Bills as a Relation

Find the domain and range of the relation

$$\{(1994, 56.21), (1995, 51.00), (1996, 47.70), (1997, 42.78), (1998, 39.43)\}.$$

Solution   The domain is the set of all first components. Thus, the domain is

$$\{1994, 1995, 1996, 1997, 1998\}.$$

The range is the set of all second components. Thus, the range is

$$\{56.21, 51.00, 47.70, 42.78, 39.43\}.$$

> **Check Point 1**   Find the domain and the range of the relation
> $$\{(20, 157.4), (30, 231.8), (100, 752.6), (200, 1496.6)\}.$$

As you worked Checkpoint 1, did you wonder if the numbers in each ordered pair represented anything? Think snakes! The first number in each ordered pair is a snake's tail length, in millimeters, and the second number is its body length, also in millimeters. Consider, for example, the ordered pair (30, 231.8).

$$(30, \quad 231.8)$$

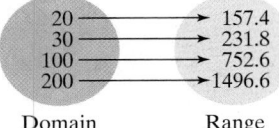

A snake whose tail length is 30 millimeters   has a body length of 231.8 millimeters.

The relation in the snake example can be pictured as follows:

```
        20 ─────────→ 157.4
        30 ─────────→ 231.8
       100 ─────────→ 752.6
       200 ─────────→1496.6

     Domain           Range
```

A scatter plot, like the one shown in Figure 2.19, is a way to represent the relation.

Figure 2.19 The graph of a relation showing a correspondence between a snake's tail length and its body length

y-axis: Body Length (millimeters) — 1000, 2000
x-axis: Tail Length (millimeters) — 100, 200

(200, 1496.6)
(100, 752.6)
(30, 231.8)
(20, 157.4)

Functions

The SAT is the test that everyone loves to hate. The scatter plot in Figure 2.20 on page 192 shows a relation indicating a correspondence between SAT scores and grade point averages for the first year in college for a group of randomly selected college students. The domain is the set of SAT scores for the students. The range is the set of their grade point averages. Is it possible for two students with the same SAT score to have different grade point averages? Look for two or more data points that are aligned vertically. We see that there are two students who have the same SAT score, 700, but their grade point averages are different. One student has a grade point average of approximately 2.4 and the other a grade point average of approximately 3.7. These students are represented by the following ordered pairs:

$$(700, \ 2.4) \qquad (700, \ 3.7).$$

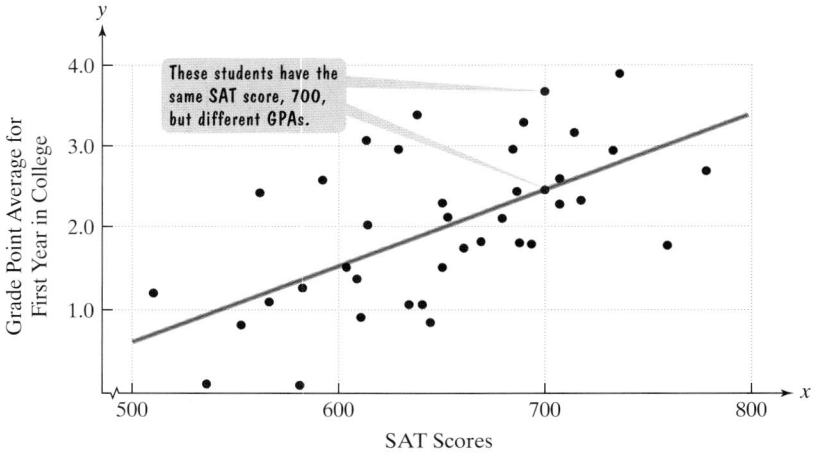

Figure 2.20

A relation in which each member of the domain corresponds to exactly one member of the range is a **function**. The relation in Figure 2.20, the SAT–grade point average scatter plot, is not a function because at least one member of the domain corresponds to two members of the range.

$$(700, \ 2.4) \qquad (700, \ 3.7)$$

The member of the domain, 700, corresponds to two members of the range, 2.4 and 3.7. Because a function is a relation in which **no two ordered pairs have the same first component and different second components**, the ordered pairs (700, 2.4) and (700, 3.7) are not ordered pairs of a function.

Same first components

$$(700, \ 2.4) \qquad (700, \ 3.7)$$

Different second components

Definition of a Function

A **function** is a correspondence between two sets X and Y that assigns to each element x of set X exactly one element y of set Y. For each element x in X, the corresponding element y in Y is called the **value** of the function at x. The set X is called the **domain** of the function, and the set of all function values, Y, is called the **range** of the function.

2 Determine whether a relation is a function.

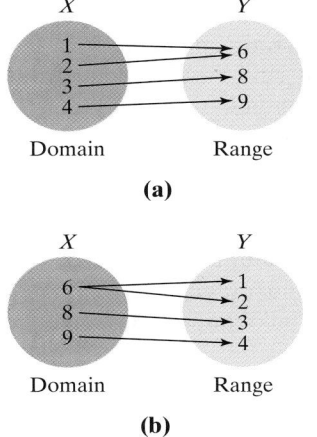

X   Y

1 ⟶ 6
2 ⟶ 8
3 ⟶ 9
4

Domain   Range

(a)

X   Y

6 ⟶ 1
8 ⟶ 2
9 ⟶ 3
    4

Domain   Range

(b)

Figure 2.21

Study Tip

The word *range* can mean many things, from a chain of mountains to a cooking stove. For functions, it means the set of all function values. For graphing utilities, it means the setting used for the viewing rectangle. Try not to confuse these meanings.

EXAMPLE 2 Determining Whether a Relation is a Function

Determine whether each relation is a function.

a. $\{(1, 6), (2, 6), (3, 8), (4, 9)\}$ **b.** $\{(6, 1), (6, 2), (8, 3), (9, 4)\}$

Solution We begin by making a figure for each relation that shows set X, the domain, and set Y, the range, shown in Figure 2.21.

a. Figure 2.21(a) shows that every element in the domain corresponds to exactly one element in the range. The element 1 in the domain corresponds to the element 6 in the range. Furthermore, 2 corresponds to 6, 3 corresponds to 8, and 4 corresponds to 9. No two ordered pairs in the given relation have the same first component and different second components. Thus, the relation is a function.

b. Figure 2.21(b) shows that 6 corresponds to both 1 and 2. If any element in the domain corresponds to more than one element in the range, the relation is not a function. This relation is not a function; two ordered pairs have the same first component and different second components.

Same first components

$(6, 1)$ $(6, 2)$

Different second components

Look at Figure 2.21 again. The fact that 1 and 2 in the domain have the same image, 6, in the range does not violate the definition of a function. A function can have two different first components with the same second component. By contrast, a relation is not a function when two different ordered pairs have the same first component and different second components. Thus, the relation in Example 2(b) is not a function.

Check Point 2 Determine whether each relation is a function.
a. $\{(1, 2), (3, 4), (5, 6), (5, 8)\}$
b. $\{(1, 2), (3, 4), (6, 5), (8, 5)\}$

Functions as Equations

Functions are usually given in terms of equations rather than as sets of ordered pairs. Earlier we noted that, for a particular snake, its total body length is a function of its tail length. The function is modeled by the equation

$$y = 7.44x + 8.6.$$

The variable x represents the snake's tail length, in millimeters. The variable y represents the snake's total body length, in millimeters. The variable y is a function of the variable x. For each value of x, there is one and only one value of y. The variable x is called the **independent variable** because it can be assigned any value from the domain. Thus, x can be assigned any positive number representing the snake's tail length. The variable y is called the **dependent variable** because its value depends on x. A snake's total body length depends on its tail length. The value of the dependent variable, y, is calculated after selecting a value for the independent variable, x.

3 Determine whether an equation represents a function.

We have seen that not every set of ordered pairs defines a function. Similarly, not all equations with the variables x and y define a function. If an equation is solved for y and more than one value of y can be obtained for a given x, then the equation does not define y as a function of x.

EXAMPLE 3   Determining Whether an Equation Represents a Function

Determine whether each equation defines y as a function of x.

a. $x^2 + y = 4$ **b.** $x^2 + y^2 = 4$

Solution   Solve each equation for y in terms of x. If two or more values of y can be obtained for a given x, the equation is not a function.

a.
$$x^2 + y = 4 \qquad \text{This is the given equation.}$$
$$x^2 + y - x^2 = 4 - x^2 \qquad \text{Solve for y by subtracting } x^2 \text{ from both sides.}$$
$$y = 4 - x^2 \qquad \text{Simplify.}$$

From this last equation we can see that for each value of x, there is one and only one value of y. For example, if $x = 1$, then $y = 4 - 1^2 = 3$. The equation defines y as a function of x.

b.
$$x^2 + y^2 = 4 \qquad \text{This given equation describes a circle.}$$
$$x^2 + y^2 - x^2 = 4 - x^2 \qquad \text{Isolate } y^2 \text{ by subtracting } x^2 \text{ from both sides.}$$
$$y^2 = 4 - x^2 \qquad \text{Simplify.}$$
$$y = \pm\sqrt{4 - x^2} \qquad \text{Apply the square root method.}$$

The $\pm$ in this last equation shows that for certain values of x (all values between -2 and 2), there are two values of y. For example, if $x = 1$, then $y = \pm\sqrt{4 - 1^2} = \pm\sqrt{3}$. For this reason, the equation does not define y as a function of x.

Check Point 3   Solve each equation for y and then determine whether the equation defines y as a function of x.
a. $2x + y = 6$ **b.** $x^2 + y^2 = 1$

4 Evaluate a function.

Function Notation

When an equation represents a function, the function is often named by a letter such as $f, g, h, F, G,$ or H. Any letter can be used to name a function. Suppose that f names a function. Think of the domain as the set of the function's inputs and the range as the set of the function's outputs. As shown in Figure 2.22, the input is represented by x and the output by $f(x)$. The special notation $f(x)$, read "f of x" or "f at x," represents the **value of the function at the number x.**

Input x

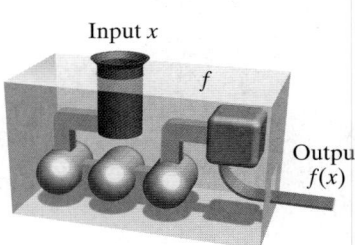

Output
$f(x)$

Figure 2.22 A function as a machine with inputs and outputs

Let's make this clearer by considering a specific example. We know that the equation $y = 4 - x^2$ defines y as a function of x. We'll name the function f. Now, we can apply our new function notation.

Input	Output	Equation	We read this equation as "f of x equals $4 - x^2$."
x	$f(x)$	$f(x) = 4 - x^2$	

Suppose that we are interested in finding $f(3)$, the function's output when the input is 3. To find the value of the function at 3, we substitute 3 for x. We are **evaluating the function** at 3.

$$f(x) = 4 - x^2 \quad \text{This is the given function.}$$
$$f(3) = 4 - 3^2 \quad \text{The input is 3.}$$
$$= 4 - 9$$
$$= -5$$

Study Tip

The notation $f(x)$ does *not* mean "f times x." The notation describes the value of the function at x.

The statement $f(3) = -5$ tells us that the value of the function at 3 is -5. When the function's input is 3, its output is -5. To find other function values, such as $f(-2), f(5),$ or $f(7)$, substitute the specified input values for x into the function's equation.

If a function is named f and x represents the independent variable, the notation $f(x)$ corresponds to the y-value for a given x. Thus,

$$f(x) = 4 - x^2 \quad \text{and} \quad y = 4 - x^2$$

define the same function. This function may be written as

$$y = f(x) = 4 - x^2.$$

EXAMPLE 4 Evaluating a Function

If $f(x) = x^2 + 3x + 5$, evaluate:

    a. $f(2)$ **b.** $f(x + 3)$ **c.** $f(-x)$

Solution We substitute 2, $x + 3$, and $-x$ for x in the definition of f. When replacing x with a variable or an algebraic expression, you might find it helpful to think of the function's equation as

$$f(\boxed{x}) = \boxed{x}^2 + 3\boxed{x} + 5.$$

a. We find $f(2)$ by substituting 2 for x in the equation.

$$f(\boxed{2}) = \boxed{2}^2 + 3 \cdot \boxed{2} + 5 = 4 + 6 + 5 = 15$$

Thus, $f(2) = 15$.

b. We find $f(x + 3)$ by substituting $x + 3$ for x in the equation.

$$f(\boxed{x + 3}) = \boxed{(x + 3)}^2 + 3\boxed{(x + 3)} + 5$$

Equivalently,

$$f(x + 3) = (x + 3)^2 + 3(x + 3) + 5$$
$$= x^2 + 6x + 9 + 3x + 9 + 5 \qquad \text{Square } x + 3 \text{ using}$$
$$(A + B)^2 = A^2 + 2AB + B^2.$$
$$\text{Distribute 3 throughout the parentheses.}$$
$$= x^2 + 9x + 23. \qquad \text{Combine like terms.}$$

c. We find $f(-x)$ by substituting $-x$ for x in the equation.

$$f(\boxed{-x}) = \boxed{(-x)}^2 + 3\boxed{(-x)} + 5$$

Equivalently,

$$f(-x) = (-x)^2 + 3(-x) + 5$$
$$= x^2 - 3x + 5.$$

Discovery

Using $f(x) = x^2 + 3x + 5$ and the answers in parts (b) and (c):

1. Is $f(x + 3)$ equal to $f(x) + f(3)$?
2. Is $f(-x)$ equal to $-f(x)$?

Check Point 4 If $f(x) = x^2 - 2x + 7$, evaluate:

    a. $f(-5)$ **b.** $f(x + 4)$ **c.** $f(-x)$

5   Understand and use piecewise functions.

Piecewise Functions

The early part of the twentieth century was the golden age of immigration in America. More than 13 million people migrated to the United States between 1900 and 1914. By 1910, foreign-born residents accounted for 15% of the total U.S. population. The graph in Figure 2.23 shows the percentage of Americans who were foreign born throughout the twentieth century.

We can model the data from 1910 through 2000 with two functions, one from 1910 through 1970, years in which the percentage was decreasing, and one from 1970 through 2000, years in which the percentage was increasing. These two trends can be approximated by the function

$$P(t) = \begin{cases} -\dfrac{11}{60}t + 15 & \text{if } 0 \le t < 60 \\[2ex] \dfrac{1}{5}t - 8 & \text{if } 60 \le t \le 90 \end{cases}$$

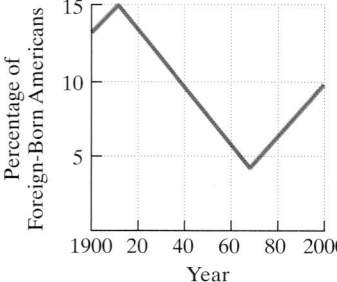

Percentage of Americans Who Are Foreign Born

Source: U.S. Census Bureau

Figure 2.23

in which t represents the number of years after 1910 and $P(t)$ is the percentage of foreign-born Americans. A function such as this that is defined by two (or more) equations over a specified domain is called a **piecewise function**.

EXAMPLE 5 Evaluating a Piecewise Function

Use the function $P(t)$, described previously, to find and interpret:

    a. $P(30)$ **b.** $P(80)$

Solution

    a. To find $P(30)$, we let $t = 30$. Because 30 is less than 60, we use the first line of the piecewise function.

$$P(t) = -\tfrac{11}{60} t + 15 \qquad \text{This is the function's equation for } 0 \le t < 60.$$
$$P(30) = -\tfrac{11}{60} \cdot 30 + 15 \qquad \text{Replace } t \text{ with } 60.$$
$$= 9.5$$

This means that 30 years after 1910, in 1940, 9.5% of Americans were foreign born.

    b. To find $P(80)$, we let $t = 80$. Because 80 is between 60 and 90, we use the second line of the piecewise function.

$$P(t) = \frac{1}{5} t - 8 \qquad \text{This is the function's equation for } 60 \le t \le 90.$$
$$P(80) = \tfrac{1}{5} \cdot 80 - 8 \qquad \text{Replace } t \text{ with } 80.$$
$$= 8$$

This means that 80 years after 1910, in 1990, 8% of Americans were foreign born.

Check Point 5   If $f(x) = \begin{cases} x^2 + 3 & \text{if } x < 0 \\ 5x + 3 & \text{if } x \ge 0 \end{cases}$, find:

    a. $f(-5)$ **b.** $f(6)$

6 Find the domain of a function.

The Domain of a Function

Let's reconsider the function that models the percentage of foreign-born Americans t years after 1910, up through and including 2000. The domain of this function is

$$\{0, \quad 1, \quad 2, \quad 3, \quad \dots, \quad 90\}.$$

O years after 1910 is 1910.   3 years after 1910 is 1913.   90 years after 1910 brings the domain up to the year 2000.

Functions that model data often have their domains explicitly given along with the function's equation. However, for most functions, only an equation is given, and the domain is not specified. In cases like this, the domain of f is the largest set of real numbers for which the value of $f(x)$ is a real number. For example, consider the function

$$f(x) = \frac{1}{x - 3}.$$

Because division by 0 is undefined (and not a real number), the denominator $x - 3$ cannot be 0. Thus, x cannot equal 3. The domain of the function consists of all real numbers other than 3, represented by $\{x \mid x \neq 3\}$.

Just as the domain of a function must exclude real numbers that cause division by zero, it must also exclude real numbers that result in an even root of a negative number. For example, consider the function

$$g(x) = \sqrt{x}.$$

The equation tells us to take the square root of x. Because only nonnegative numbers have real square roots, the expression under the radical sign, x, must be greater than or equal to 0. The domain of g is $\{x \mid x \geq 0\}$ or the interval $[0, \infty)$.

Finding a Function's Domain

If a function f does not model data or verbal conditions, its domain is the largest set of real numbers for which the value of $f(x)$ is a real number. Exclude from a function's domain real numbers that cause division by zero and real numbers that result in an even root of a negative number.

EXAMPLE 6 Finding the Domain of a Function

Find the domain of each function:

a. $f(x) = x^2 - 7x$ **b.** $g(x) = \dfrac{6x}{x^2 - 9}$ **c.** $h(x) = \sqrt{3x + 12}$

Solution

a. The function $f(x) = x^2 - 7x$ contains neither division nor an even root. The domain of f is the set of all real numbers.

b. The function $g(x) = \dfrac{6x}{x^2 - 9}$ contains division. Because division by 0 is undefined, we must exclude from the domain values of x that cause $x^2 - 9$ to be 0. Thus, x cannot equal -3 or 3. The domain of function g is $\{x \mid x \neq -3, x \neq 3\}$.

c. The function $h(x) = \sqrt{3x + 12}$ contains an even root. Because only nonnegative numbers have real square roots, the quantity under the radical sign, $3x + 12$, must be greater than or equal to 0.

$$3x + 12 \geq 0$$
$$3x \geq -12$$
$$x \geq -4$$

The domain of h is $\{x \mid x \geq -4\}$ or the interval $[-4, \infty)$.

Technology

You can graph a function and visually determine its domain. For example, $h(x) = \sqrt{3x + 12}$, or $y = \sqrt{3x + 12}$, appears only for $x \geq -4$, verifying $[-4, \infty)$ as the domain.

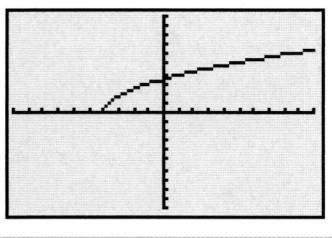

Check Point 6   Find the domain of each function:

a. $f(x) = x^2 + 3x - 17$ **b.** $g(x) = \dfrac{5x}{x^2 - 49}$

c. $h(x) = \sqrt{9x - 27}$

7 Interpret function values for functions that model data.

Applications

Like equations, functions can be obtained from verbal models or from actual data. We'll have lots of practice doing this throughout the book. For now, let's make sure that we can find and interpret function values for functions that were obtained from modeling data.

EXAMPLE 7 Evaluating a Function and Interpreting the Result

The function

$$f(x) = 0.0075x^2 - 0.2672x + 14.8$$

models the average number of miles per gallon of U.S. automobiles, y, x years after 1940. Find and interpret $f(18)$.

Solution

$$f(x) = 0.0075x^2 - 0.2672x + 14.8 \qquad \text{This is the given function.}$$
$$f(18) = 0.0075(18)^2 - 0.2672(18) + 14.8 \qquad \text{Replace each occurrence of } x \text{ by 18.}$$
$$= 12.4204$$

We see that $f(18) = 12.4204$. Because 18 represents the number of years after 1940, this means that in 1958, U.S. automobiles averaged approximately 12.4 miles per gallon.

Check Point 7 Use the function in Example 7 to find and interpret $f(50)$.

EXERCISE SET 2.3

Practice Exercises

In Exercises 1–8, determine whether each relation is a function. Give the domain and range for each relation.

1. $\{(1, 2), (3, 4), (5, 5)\}$ **2.** $\{(4, 5), (6, 7), (8, 8)\}$
3. $\{(3, 4), (3, 5), (4, 4), (4, 5)\}$
4. $\{(5, 6), (5, 7) (6, 6), (6, 7)\}$
5. $\{(-3, -3), (-2, -2), (-1, -1), (0, 0)\}$
6. $\{(-7, -7), (-5, -5), (-3, -3), (0, 0)\}$
7. $\{(1, 4), (1, 5), (1, 6)\}$ **8.** $\{(4, 1), (5, 1), (6, 1)\}$

In Exercises 9–20, determine whether each equation defines y as a function of x.

9. $x + y = 16$ **10.** $x + y = 25$
11. $x^2 + y = 16$ **12.** $x^2 + y = 25$
13. $x^2 + y^2 = 16$ **14.** $x^2 + y^2 = 25$
15. $x = y^2$ **16.** $4x = y^2$
17. $y = \sqrt{x + 4}$ **18.** $y = -\sqrt{x + 4}$
19. $x + y^3 = 8$ **20.** $x + y^3 = 27$

In Exercises 21–32, evaluate each function at the given values of the independent variable and simplify.

21. $f(x) = 4x + 5$
  a. $f(6)$ **b.** $f(x + 1)$ **c.** $f(-x)$
22. $f(x) = 3x + 7$
  a. $f(4)$ **b.** $f(x + 1)$ **c.** $f(-x)$
23. $g(x) = x^2 + 2x + 3$
  a. $g(-1)$ **b.** $g(x + 5)$ **c.** $g(-x)$
24. $g(x) = x^2 - 10x - 3$
  a. $g(-1)$ **b.** $g(x + 2)$ **c.** $g(-x)$
25. $h(x) = x^4 - x^2 + 1$
  a. $h(2)$ **b.** $h(-1)$
  c. $h(-x)$ **d.** $h(3a)$
26. $h(x) = x^3 - x + 1$
  a. $h(3)$ **b.** $h(-2)$
  c. $h(-x)$ **d.** $h(3a)$
27. $f(r) = \sqrt{r + 6} + 3$
  a. $f(-6)$ **b.** $f(10)$ **c.** $f(x - 6)$
28. $f(r) = \sqrt{25 - r} - 6$
  a. $f(16)$ **b.** $f(-24)$ **c.** $f(25 - 2x)$

29. $f(x) = \dfrac{4x^2 - 1}{x^2}$

    a. $f(2)$ **b.** $f(-2)$ **c.** $f(-x)$

30. $f(x) = \dfrac{4x^3 + 1}{x^3}$

    a. $f(2)$ **b.** $f(-2)$ **c.** $f(-x)$

31. $f(x) = \dfrac{x}{|x|}$

    a. $f(6)$ **b.** $f(-6)$ **c.** $f(r^2)$

32. $f(x) = \dfrac{|x + 3|}{x + 3}$

    a. $f(5)$ **b.** $f(-5)$ **c.** $f(-9 - x)$

In Exercises 33–44, find and simplify:

    a. $f(a)$

    b. $f(a + h)$

    c. $\dfrac{f(a + h) - f(a)}{h}, h \neq 0$

    d. $f(a) + f(h)$

33. $f(x) = 4x$ **34.** $f(x) = 7x$

35. $f(x) = 3x + 7$ **36.** $f(x) = 6x + 1$

37. $f(x) = -5x - 3$ **38.** $f(x) = -8x - 9$

39. $f(x) = x^2$ **40.** $f(x) = 2x^2$

41. $f(x) = 6$ **42.** $f(x) = 7$

43. $f(x) = \dfrac{1}{x}$ **44.** $f(x) = \dfrac{1}{2x}$

In Exercises 45–50, evaluate each piecewise function at the given values of the independent variable.

45. $f(x) = \begin{cases} 3x + 5 & \text{if } x < 0 \\ 4x + 7 & \text{if } x \geq 0 \end{cases}$

    a. $f(-2)$ **b.** $f(0)$ **c.** $f(3)$

46. $f(x) = \begin{cases} 6x - 1 & \text{if } x < 0 \\ 7x + 3 & \text{if } x \geq 0 \end{cases}$

    a. $f(-3)$ **b.** $f(0)$ **c.** $f(4)$

47. $g(x) = \begin{cases} x + 3 & \text{if } x \geq -3 \\ -(x + 3) & \text{if } x < -3 \end{cases}$

    a. $g(0)$ **b.** $g(-6)$ **c.** $g(-3)$

48. $g(x) = \begin{cases} x + 5 & \text{if } x \geq -5 \\ -(x + 5) & \text{if } x < -5 \end{cases}$

    a. $g(0)$ **b.** $g(-6)$ **c.** $g(-5)$

49. $h(x) = \begin{cases} \dfrac{x^2 - 9}{x - 3} & \text{if } x \neq 3 \\ 6 & \text{if } x = 3 \end{cases}$

    a. $h(5)$ **b.** $h(0)$ **c.** $h(3)$

50. $h(x) = \begin{cases} \dfrac{x^2 - 25}{x - 5} & \text{if } x \neq 5 \\ 10 & \text{if } x = 5 \end{cases}$

    a. $h(7)$ **b.** $h(0)$ **c.** $h(5)$

In Exercises 51–72, find the domain of each function.

51. $f(x) = 4x^2 - 3x + 1$ **52.** $f(x) = 8x^2 - 5x + 2$

53. $g(x) = \dfrac{3}{x - 4}$ **54.** $g(x) = \dfrac{2}{x + 5}$

55. $h(x) = \dfrac{7x}{x^2 - 16}$ **56.** $h(x) = \dfrac{12x}{x^2 - 36}$

57. $f(x) = \dfrac{2}{(x + 3)(x - 7)}$ **58.** $f(x) = \dfrac{15}{(x + 8)(x - 3)}$

59. $H(r) = \dfrac{4}{r^2 + 11r + 24}$ **60.** $H(r) = \dfrac{5}{6r^2 + r - 2}$

61. $f(t) = \dfrac{3}{t^2 + 4}$ **62.** $f(t) = \dfrac{5}{t^2 + 9}$

63. $f(x) = \sqrt{x - 3}$ **64.** $f(x) = \sqrt{x + 2}$

65. $f(x) = \dfrac{1}{\sqrt{x - 3}}$ **66.** $f(x) = \dfrac{1}{\sqrt{x + 2}}$

67. $g(x) = \sqrt{5x + 35}$ **68.** $g(x) = \sqrt{7x - 70}$

69. $f(x) = \sqrt{24 - 2x}$ **70.** $f(x) = \sqrt{84 - 6x}$

71. $f(x) = \sqrt{x^2 - 5x - 14}$ **72.** $f(x) = \sqrt{x^2 - 5x - 24}$

Application Exercises

It seems that Phideau's medical bills are costing us an arm and a paw. The graph shows veterinary costs, in billions of dollars, for dogs and cats in five selected years. Use the graph to solve Exercises 73–74.

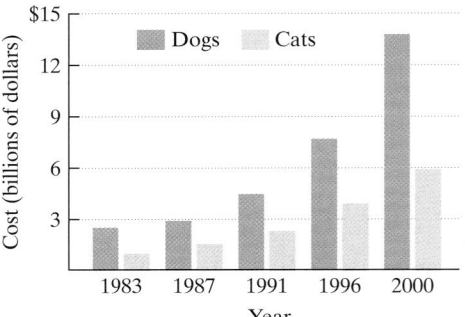

Veterinary Costs in the U.S.

Source: American Veterinary Medical Association

73. Write five ordered pairs that approximate veterinary costs for dogs for the years shown. Find the domain and the range of the relation. Is this relation a function? Explain your answer.

74. Write five ordered pairs that approximate veterinary costs for cats for the years shown. Find the domain and the range of the relation. Is this relation a function? Explain your answer.

The number of women enrolled in U.S. colleges can be modeled by the function $f(x) = 0.07x + 4.1$, where x represents the number of years since 1984 and $f(x)$ represents enrollment in millions. The number of men enrolled in U.S. colleges can be modeled by the function $g(x) = 0.01x + 3.9$, where x represents the number of years since 1984 and $g(x)$ represents enrollment, in millions. In Exercises 75–78, use these functions to find and interpret:

75. $f(16)$ **76.** $g(16)$

77. $f(20) - g(20)$ **78.** $f(25) - g(25)$

Enrollment in U.S. Colleges

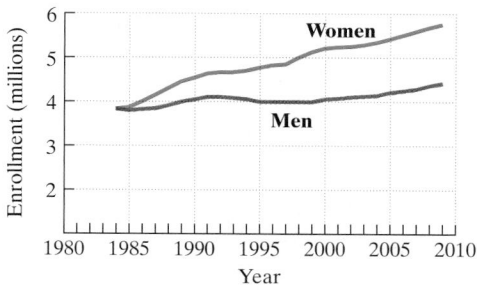

Source: Department of Education

The number of lawyers in the United States can be modeled by the function

$$f(x) = \begin{cases} 6.5x + 200 & \text{if } 0 \le x < 23 \\ 26.2x - 252 & \text{if } x \ge 23 \end{cases}$$

where x represents the number of years since 1951 and $f(x)$ represents the number of lawyers, in thousands. In Exercises 79–82, use this function to find and interpret:

79. $f(0)$ **80.** $f(10)$

81. $f(50)$ **82.** $f(60)$

Number of U.S. Lawyers

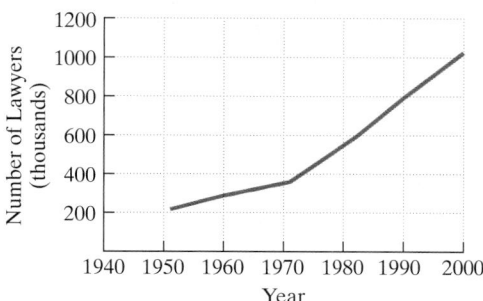

Source: Hudson Institute

83. On the average, infant girls weigh 7 pounds at birth and gain 1.5 pounds each month for the first six months. The function $f(x) = 1.5x + 7$ models this, where x represents the infant's age, in months, $x \le 6$, and $f(x)$ describes the baby's weight, in pounds. Use the function to find $f(0)$, $f(2)$, $f(4)$, and $f(6)$. Describe what these results mean. Identify each of your computations as an appropriate point on the graph at the top of the next column.

Average Weight for Infant Girls

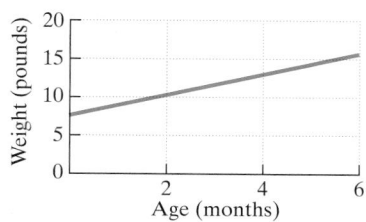

We're eating more and getting heavier. The number of calories consumed each day per person in the United States can be modeled by

$$f(x) = 0.89x^2 - 1.93x + 3306.27$$

where x is the number of years since 1974 and $f(x)$ represents the number of calories consumed each day per person. In Exercises 84–87, use this function to find and interpret:

84. $f(0)$ **85.** $f(10)$

86. $f(15) - f(0)$ **87.** $f(15) - f(10)$

Getting Heavier: Percentage of U.S. Men and Women Overweight

Men	1999	2025
Overweight	62%	73%
Obese	19%	37%
Women		
Overweight	47%	75%
Obese	19%	35%

Source: Beth Israel Medical Center

88. During a particular year, the taxes owed by a married person filing separately with an adjusted gross income of x dollars is given by the piecewise function

$$T(x) = \begin{cases} 0.15x & \text{if } 0 \le x < 17{,}900 \\ 0.28(x - 17{,}900) + 2685 & \text{if } 17{,}900 \le x < 43{,}250 \\ 0.31(x - 43{,}250) + 9783 & \text{if } x \ge 43{,}250 \end{cases}$$

Find and interpret $T(70{,}000) - T(40{,}000)$.

89. The function

$$f(x) = \begin{cases} 0.0005x^2 + 0.025x + 8.8 & \text{if } 0 \le x < 30 \\ 0.0202x^2 - 1.58x + 39.2 & \text{if } x \ge 30 \end{cases}$$

models the average number of miles (in thousands) driven per car in the United States, per year, x years after 1940. Find and interpret

a. $f(15)$ **b.** $f(50)$

90. A car was purchased for $22,500. The value of the car decreases by $3200 per year for the first seven years. Write a function V that describes the value of the car after x years, where $0 \le x \le 7$. Then find and interpret $V(3)$.

91. A car was purchased for $17,900. The value of the car decreases by $2100 per year for the first six years. Write a function V that describes the value of the car after x years, where $0 \leq x \leq 6$. Then find and interpret $V(4)$.

Writing in Mathematics

92. If a relation is represented by a set of ordered pairs, explain how to determine whether the relation is a function.

93. How do you determine if an equation in x and y defines y as a function of x?

94. A student in introductory algebra hears that functions are studied in subsequent algebra courses. The student asks you what a function is. Provide the student with a clear, relatively concise response.

95. Describe one advantage of using $f(x)$ rather than y in a function's equation.

96. What is a piecewise function?

97. How is the domain of a function determined?

98. For people filing a single return, federal income tax is a function of adjusted gross income because for each value of adjusted gross income there is a specific tax to be paid. On the other hand, the price of a house is not a function of the lot size on which the house sits because houses on same-sized lots can sell for many different prices.
  a. Describe an everyday situation between variables that is a function.
  b. Describe an everyday situation between variables that is not a function.

Technology Exercises

Use a graphing utility to find the domain of each function in Exercises 99–101. Then verify your observation algebraically.

99. $f(x) = \sqrt{x - 1}$ **100.** $g(x) = \sqrt{2x + 6}$

101. $h(x) = \sqrt{15 - 3x}$

102. Graph $y = 0.89x^2 - 1.93x + 3306.27$, the model for caloric consumption used in Exercises 84–87, in a $[0, 15, 1]$ by $[3200, 3500, 15]$ viewing rectangle. Describe one bit of information revealed by the graph of the function that is not obvious by looking at its equation.

Critical Thinking Exercises

103. Write a function defined by an equation in x whose domain is $\{x \mid x \neq -4, x \neq 11\}$.

104. Write a function defined by an equation in x whose domain is $[-6, \infty)$.

105. Give an example of an equation that does not define y as a function of x but that does define x as a function of y.

106. If $f(x) = ax^2 + bx + c$ and $r_1 = \dfrac{-b + \sqrt{b^2 - 4ac}}{2a}$, find $f(r_1)$ without doing any algebra and explain how you arrived at your result.

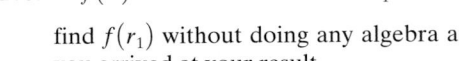

Group Exercise

107. Almanacs, newspapers, magazines, and the Internet contain bar graphs and line graphs that describe how things are changing over time. For example, the graph in Figure 2.18 on page 191 shows how mobile-phone bills are changing over time. Find a bar or line graph showing yearly changes that you find intriguing. Describe to the group what interests you about this data. The group should select their two favorite graphs. For each graph selected:
  a. Rewrite the data so that they are presented as a relation in the form of a set of ordered pairs.
  b. Determine whether the relation in part (a) is a function. Explain why the relation is a function, or why it is not.

SECTION 2.4 *Graphs of Functions*

Objectives

1. Graph functions by plotting points.
2. Obtain information about a function from its graph.
3. Use the vertical line test to identify functions.
4. Identify intervals on which a function increases, decreases, or is constant.
5. Identify even or odd functions and recognize their symmetries.
6. Recognize graphs of common functions.
7. Graph step functions.
8. Obtain information from graphs of functions in applied situations.

Have you ever seen a gas-guzzling car from the 1950s, with its huge fins and over-stated design? The worst year for automobile fuel efficiency was 1958, when cars averaged a dismal 12.4 miles per gallon. We ended the last section with a function that modeled automobile fuel efficiency over time. If we graph the function's equation, we will get a much better idea of the relationship between time and fuel efficiency. In this section we will learn how to use the graph of a function to obtain useful information about the function.

Graphs of Functions

A graph enables us to visualize a function's behavior. The graph shows the relationship between the function's two variables more clearly than the function's equation does. The **graph of a function** is the graph of its ordered pairs. For example, the graph of $f(x) = \sqrt{x}$ is the set of points (x, y) in the rectangular coordinate system satisfying the equation $y = \sqrt{x}$. Thus, one way to graph a function is by plotting several of its ordered pairs and drawing a line or smooth curve through them. With the function's graph, we can picture its domain on the x-axis and its range on the y-axis. Our first example illustrates how this is done.

1 Graph functions by plotting points.

EXAMPLE 1 Graphing a Function by Plotting Points

Graph $f(x) = x^2 + 1$. To do so, use integer values of x from the set $\{-3, -2, -1, 0, 1, 2, 3\}$ to obtain seven ordered pairs. Plot each ordered pair and draw a smooth curve through the points. Use the graph to specify the function's domain and range.

Solution The graph of $f(x) = x^2 + 1$ is, by definition, the graph of $y = x^2 + 1$. We begin by setting up a partial table of coordinates.

x	$f(x) = x^2 + 1$	(x, y) or $(x, f(x))$
-3	$f(-3) = (-3)^2 + 1 = 10$	$(-3, 10)$
-2	$f(-2) = (-2)^2 + 1 = 5$	$(-2, 5)$
-1	$f(-1) = (-1)^2 + 1 = 2$	$(-1, 2)$
0	$f(0) = 0^2 + 1 = 1$	$(0, 1)$
1	$f(1) = 1^2 + 1 = 2$	$(1, 2)$
2	$f(2) = 2^2 + 1 = 5$	$(2, 5)$
3	$f(3) = 3^2 + 1 = 10$	$(3, 10)$

Range $[1, \infty)$

Domain $(-\infty, \infty)$

Figure 2.24 The graph of $f(x) = x^2 + 1$

Now, we plot the seven points and draw a smooth curve through them, as shown in Figure 2.24. The graph of f has a cuplike shape. The points on the graph of f have x-coordinates that extend indefinitely far to the left and to the right. Thus, the domain consists of all real numbers, represented by $(-\infty, \infty)$. By contrast, the points on the graph have y-coordinates that start at 1 and extend indefinitely upward. Thus, the range consists of all real numbers greater than or equal to 1, represented by $[1, \infty)$.

Check Point 1

Graph $f(x) = x^2 - 2$, using integers from -3 to 3 for x in the partial table of coordinates. Use the graph to specify the function's domain and range.

Technology

Does your graphing utility have a $\boxed{\text{TABLE}}$ feature? If so, you can use it to create tables of coordinates for a function. You will need to enter the equation of the function and specify the starting value for x $\boxed{\text{TblStart}}$ and the increment between successive x-values $\boxed{\Delta\text{Tbl}}$. For the table of coordinates in Example 1, we start the table at $x = -3$ and increment by 1. Using the up- or down-arrow keys, you can scroll through the table and determine as many ordered pairs of the graph as desired.

2 Obtain information about a function from its graph.

Obtaining Information from Graphs

You can obtain information about a function from its graph. At the right or left of a graph, you will find closed dots, open dots, or arrows.

- A closed dot indicates that the graph does not extend beyond this point and the point belongs to the graph.

- An open dot indicates that the graph does not extend beyond this point and the point does not belong to the graph.

- An arrow indicates that the graph extends indefinitely in the direction in which the arrow points.

EXAMPLE 2   Obtaining Information from a Function's Graph

Use the graph of the function f, shown in Figure 2.25, to answer the following questions.

a. What are the function values $f(-1)$ and $f(1)$?

b. What is the domain of f?

c. What is the range of f?

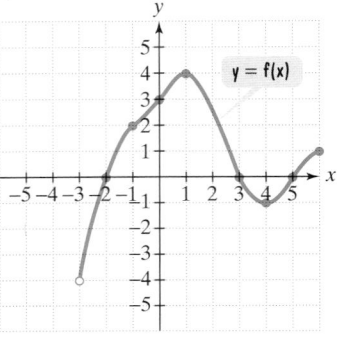

Figure 2.25

Solution

a. Because $(-1, 2)$ is a point on the graph of f, the y-coordinate, 2, is the value of the function at the x-coordinate, -1. Thus, $f(-1) = 2$. Similarly, because $(1, 4)$ is also a point on the graph of f, this indicates that $f(1) = 4$.

b. The open dot on the left shows that $x = -3$ is not in the domain of f. By contrast, the closed dot on the right shows that $x = 6$ is in the domain of f. We determine the domain of f by noticing that the points on the graph of f have x-coordinates between -3, excluding -3, and 6, including 6. For each number x between -3 and 6, there is a point $(x, f(x))$ on the graph. Thus, the domain of f is $\{x | -3 < x \le 6\}$ or the interval $(-3, 6]$.

c. The points on the graph all have y-coordinates between -4, not including -4, and 4, including 4. The graph does not extend below $y = -4$ or above $y = 4$. Thus, the range of f is $\{y | -4 < y \le 4\}$ or the interval $(-4, 4]$.

Check Point 2 Use the graph of function f, shown below, to find $f(4)$, the domain, and the range.

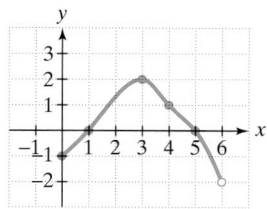

Figure 2.26 illustrates how we can identify a graph's intercepts. To find the x-intercepts, look for the points at which the graph crosses the x-axis. There are three such points: $(-2, 0)$, $(3, 0)$, and $(5, 0)$. Thus, the x-intercepts are -2, 3, and 5. We express this in function notation by writing $f(-2) = 0$, $f(3) = 0$, and $f(5) = 0$.

To find the y-intercept, look for the point at which the graph crosses the y-axis. This occurs at $(0, 3)$. Thus, the y-intercept is 3. We express this in function notation by writing $f(0) = 3$.

By the definition of a function, for each value of x we can have at most one value for y. What does this mean in terms of intercepts? A function can have more than one x-intercept but at most one y-intercept.

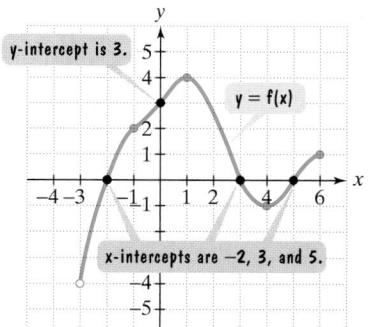

Figure 2.26 Identifying intercepts

3 Use the vertical line test to identify functions.

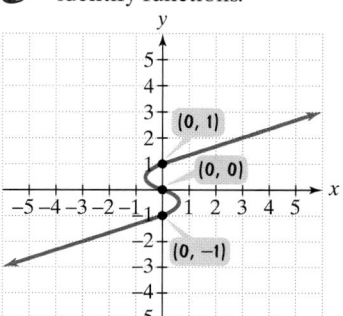

Figure 2.27 y is not a function of x because 0 is paired with three values of y, namely, 1, 0, and -1.

The Vertical Line Test

Not every graph in the rectangular coordinate system is the graph of a function. The definition of a function specifies that no value of x can be paired with two or more different values of y. Consequently, if a graph contains two or more different points with the same first coordinate, the graph cannot represent a function. This is illustrated in Figure 2.27. Observe that points sharing a common first coordinate are vertically above or below each other.

This observation is the basis of a useful test for determining whether a graph defines y as a function of x. The test is called the **vertical line test**.

The Vertical Line Test for Functions

If any vertical line intersects a graph in more than one point, the graph does not define y as a function of x.

EXAMPLE 3 Using the Vertical Line Test

Use the vertical line test to identify graphs in which y is a function of x.

a.
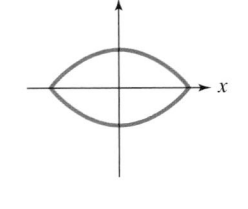

b.
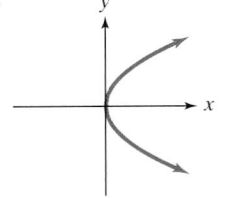

c.

d.

Solution y is a function of x for the graphs in (b) and (c).

a.
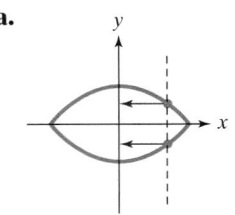

b.
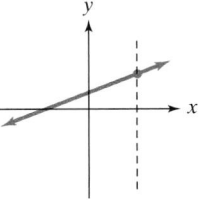

c.

d.
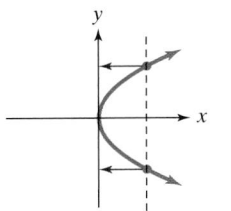

y **is not a function** of x.
Two values of y
correspond to an x-value.

y **is a function** of x.

y **is a function** of x.

y **is not a function** of x.
Two values of y
correspond to an x-value.

Check Point 3 Use the vertical line test to identify graphs in which y is a function of x.

a.

b.

c.

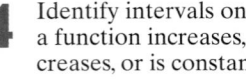

---

4 Identify intervals on which a function increases, decreases, or is constant.

Increasing and Decreasing Functions

Too late for that flu shot now! It's only 8 A.M. and you're feeling lousy. Your temperature is 101°F. Fascinated by the way that algebra models the world (your author is projecting a bit here), you decide to construct graphs showing your body temperature as a function of the time of day. You decide to let x represent the number of hours after 8 A.M. and $f(x)$ your temperature at time x.

At 8 A.M. your temperature is 101°F and you are not feeling well. However, your temperature starts to decrease. It reaches normal (98.6°F) by 11 A.M. Feeling energized, you construct the graph shown on the right, indicating decreasing temperature for $\{x \mid 0 < x < 3\}$ or on the interval $(0, 3)$.

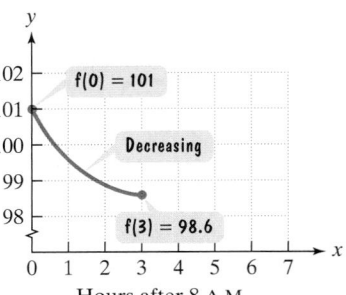

Temperature decreases on $(0, 3)$,
reaching 98.6° by 11 A.M.

Did creating that first graph drain you of your energy? Your temperature starts to rise after 11 A.M. By 1 P.M., 5 hours after 8 A.M., your temperature reaches 100°F. However, you keep plotting points on your graph. At right, we can see that your temperature increases for $\{x \mid 3 < x < 5\}$ or on the interval $(3, 5)$.

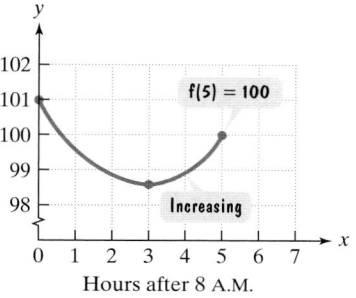

Temperature increases in $(3, 5)$.

By 3 P.M., your temperature is no worse than it was at 1 P.M.: It is still 100°F. (Of course, it's no better, either.) Your temperature remained the same, or constant, for $\{x \mid 5 < x < 7\}$ or on the interval $(5, 7)$.

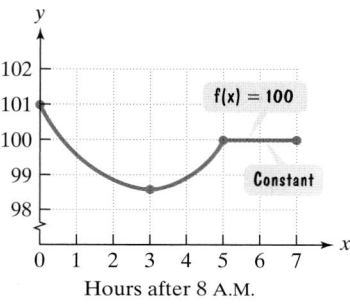

Temperature remains constant at 100° in $(5, 7)$.

The time-temperature flu scenario illustrates that a function f is increasing when its graph rises, decreasing when its graph falls, and remains constant when it neither rises nor falls. Let's now provide a more algebraic description for these intuitive concepts.

Increasing, Decreasing, and Constant Functions

1. A function is **increasing** on an interval if for any x_1 and x_2 in the interval, where $x_1 < x_2$, then $f(x_1) < f(x_2)$.
2. A function is **decreasing** on an interval if for any x_1 and x_2 in the interval, where $x_1 < x_2$, then $f(x_1) > f(x_2)$.
3. A function is **constant** on an interval if for any x_1 and x_2 in the interval, where $x_1 < x_2$, then $f(x_1) = f(x_2)$.

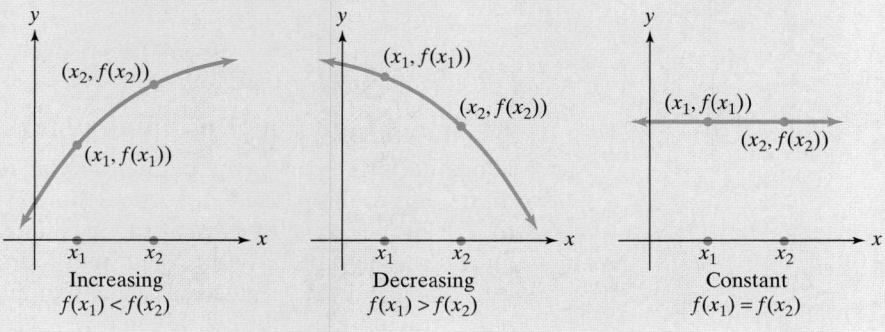

EXAMPLE 4   Intervals on Which a Function Increases, Decreases, or Is Constant

Describe the increasing, decreasing, or constant behavior of each function whose graph is shown.

a.

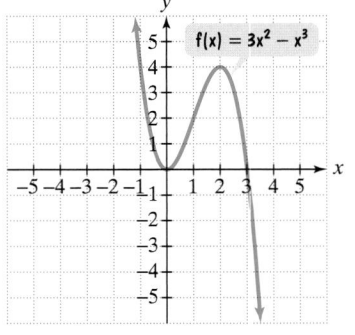

$f(x) = 3x^2 - x^3$

b.

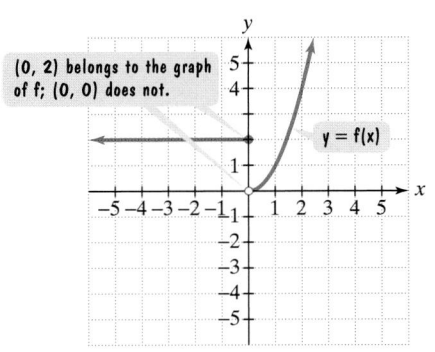

(0, 2) belongs to the graph of f; (0, 0) does not.

$y = f(x)$

Solution

a. The function is decreasing on the interval $(-\infty, 0)$, increasing on the interval $(0, 2)$, and decreasing on the interval $(2, \infty)$.

b. Although the function's equations are not given, the graph indicates that the function is defined in two pieces. The part of the graph to the left of the y-axis shows that the function is constant on the interval $(-\infty, 0)$. The part to the right of the y-axis shows that the function is increasing on the interval $(0, \infty)$.

Check Point 4

Describe the increasing, decreasing, or constant behavior of the function whose graph is shown.

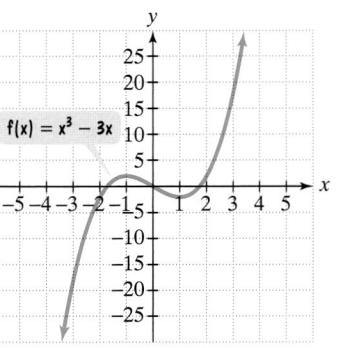

$f(x) = x^3 - 3x$

5 Identify even or odd functions and recognize their symmetries.

Even and Odd Functions and Symmetry

Is beauty in the eye of the beholder? Or are there certain objects (or people) that are so well balanced and proportioned that they are universally pleasing to the eye? What constitutes an attractive human face? In Figure 2.28, we've drawn lines between paired features and marked the midpoints. Notice how the features line up almost perfectly. Each half of the face is a mirror image of the other half through the white vertical line.

Did you know that graphs of some equations exhibit exactly the kind of symmetry shown by the attractive face in Figure 2.28? The word *symmetry* comes from the Greek *symmetria*, meaning "the same measure." We can identify graphs with symmetry by looking at a function's equation and determining if the function is *even* or *odd*.

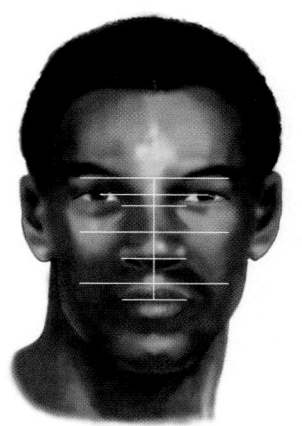

Figure 2.28 To most people, an attractive face is one in which each half is an almost perfect mirror image of the other half.

Definition of Even and Odd Functions

The function f is an **even function** if

$$f(-x) = f(x) \quad \text{for all } x \text{ in the domain of } f.$$

The right side of the equation of an even function does not change if x is replaced with $-x$.

The function f is an **odd function** if

$$f(-x) = -f(x) \quad \text{for all } x \text{ in the domain of } f.$$

Every term in the right side of the equation of an odd function changes sign if x is replaced by $-x$.

EXAMPLE 5 Identifying Even or Odd Functions

Identify each of the following functions as even, odd, or neither.

a. $f(x) = x^3$ **b.** $g(x) = x^4 - 2x^2$ **c.** $h(x) = x^2 + 2x + 1$

Solution   In each case, replace x with $-x$ and simplify. If the right side of the equation stays the same, the function is even. If every term on the right changes sign, the function is odd.

a. We use the given function's equation, $f(x) = x^3$, to find $f(-x)$.

Use $f(x) = x^3$.

Replace x with $-x$.   Replace x with $-x$.

$$f(-x) = (-x)^3 = (-x)(-x)(-x) = -x^3$$

There is only one term in the equation $f(x) = x^3$, and the term changed signs when we replaced x with $-x$. Because $f(-x) = -f(x)$, f is an odd function.

b. We use the given function's equation, $g(x) = x^4 - 2x^2$, to find $g(-x)$.

Use $g(x) = x^4 - 2x^2$.

Replace x with $-x$.

$$g(-x) = (-x)^4 - 2(-x)^2 = (-x)(-x)(-x)(-x) - 2(-x)(-x)$$
$$= x^4 - 2x^2.$$

The right side of the equation of the given function, $g(x) = x^4 - 2x^2$, did not change when we replaced x with $-x$. Because $g(-x) = g(x)$, g is an even function.

c. We use the given function's equation, $h(x) = x^2 + 2x + 1$, to find $h(-x)$.

Use $h(x) = x^2 + 2x + 1$.

Replace x with −x.

$$h(-x) = (-x)^2 + 2(-x) + 1 = x^2 - 2x + 1$$

The right side of the equation of the given function, $h(x) = x^2 + 2x + 1$, changed when we replaced x with $-x$. Thus, $h(-x) \neq h(x)$, so h is not an even function. The sign of *each* of the three terms in the equation for $h(x)$ did not change when we replaced x with $-x$. Only the second term changed signs. Thus, $h(-x) \neq -h(x)$, so h is not an odd function. We conclude that h is neither an even nor an odd function.

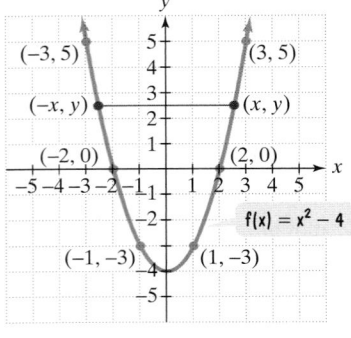

Figure 2.29 y-axis symmetry with $f(-x) = f(x)$

Check Point 5 Determine whether each of the following functions is even, odd, or neither.

a. $f(x) = x^2 + 6$ **b.** $g(x) = 7x^3 - x$ **c.** $h(x) = x^5 + 1$

Now, let's see what even and odd functions tell us about a function's graph. Begin with the even function $f(x) = x^2 - 4$, shown in Figure 2.29. The function is even because

$$f(-x) = (-x)^2 - 4 = x^2 - 4 = f(x).$$

Examine the pairs of points shown, such as $(3, 5)$ and $(-3, 5)$. Notice that we obtain the same y-coordinate whenever we evaluate the function at a value of x and the value of x equal to its opposite. Like the attractive face, each half of the graph is a mirror image of the other half through the y-axis. If we were to fold the paper along the y-axis, the two halves of the graph would coincide. This causes the graph to be *symmetric with respect to the y-axis*. A graph is **symmetric with respect to the y-axis** if, for every point (x, y) on the graph, the point $(-x, y)$ is also on the graph. All even functions have graphs with this kind of symmetry.

Even Functions and y-Axis Symmetry

The graph of an even function in which $f(-x) = f(x)$ is symmetric with respect to the y-axis.

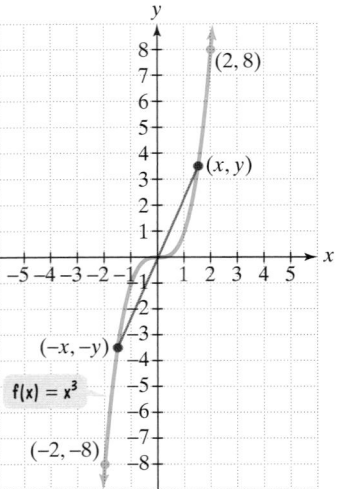

Figure 2.30 Origin symmetry with $f(-x) = -f(x)$

Now, consider the graph of the function $f(x) = x^3$. In Example 5, we saw that $f(-x) = -f(x)$, so this is an odd function. Although the graph in Figure 2.30 is not symmetric with respect to the y-axis, it is symmetric in another way. Look at the pairs of points, such as $(2, 8)$ and $(-2, -8)$. For each point (x, y) on the graph, the point $(-x, -y)$ is also on the graph. The points $(2, 8)$ and $(-2, -8)$ are reflections of one another in the origin. This means that

- the points are the same distance from the origin, and
- the points lie on a line through the origin.

A graph is **symmetric with respect to the origin** if, for every point (x, y) on the graph, the point $(-x, -y)$ is also on the graph. Observe that the first- and third-quadrant portions of $f(x) = x^3$ are reflections of one another in the origin. Notice that $f(x)$ and $f(-x)$ have opposite signs, so that $f(-x) = -f(x)$. All odd functions have graphs with origin symmetry.

Odd Functions and Origin Symmetry

The graph of an odd function in which $f(-x) = -f(x)$ is symmetric with respect to the origin.

Symmetry and Your World

Origin symmetry and y-axis symmetry are just two examples of a powerful concept whose workings appear in art, nature, the sciences, poetry, and architecture. The two halves of a bridge span, the wings of a bird or an aircraft, the blades of a propeller—all have symmetry. Many flowers have rotational symmetry. The flower on the far left has fivefold rotational symmetry; after five equal turns, the flower is restored to its original position. By contrast, the flower on the immediate left has sixfold rotational symmetry.

Sixfold rotational symmetry

Fivefold rotational symmetry

Flowers with rotational symmetries

---

6 Recognize graphs of common functions.

Graphs of Common Functions

Table 2.3 on page 212 gives names to six frequently encountered functions in algebra. The table shows each function's graph and lists characteristics of the function. Study the shape of each graph and take a few minutes to verify the function's characteristics from its graph. Knowing these graphs is essential for understanding later graphing techniques.

Discovery

Use a graphing utility to verify the six graphs shown in Table 2.3.

Table 2.3 Algebra's Common Graphs

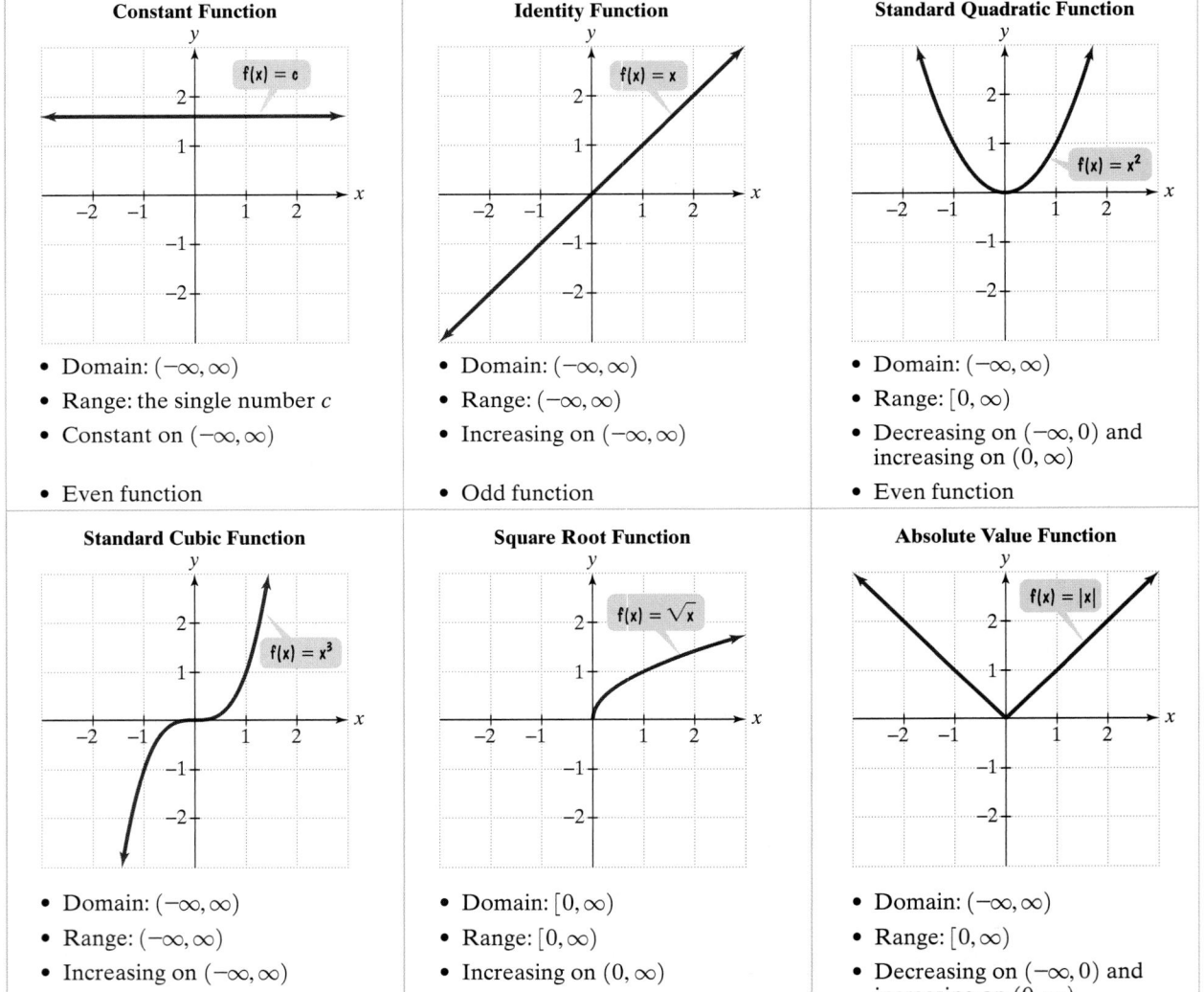

Constant Function

$f(x) = c$

- Domain: $(-\infty, \infty)$
- Range: the single number c
- Constant on $(-\infty, \infty)$

- Even function

Identity Function

$f(x) = x$

- Domain: $(-\infty, \infty)$
- Range: $(-\infty, \infty)$
- Increasing on $(-\infty, \infty)$

- Odd function

Standard Quadratic Function

$f(x) = x^2$

- Domain: $(-\infty, \infty)$
- Range: $[0, \infty)$
- Decreasing on $(-\infty, 0)$ and increasing on $(0, \infty)$
- Even function

Standard Cubic Function

$f(x) = x^3$

- Domain: $(-\infty, \infty)$
- Range: $(-\infty, \infty)$
- Increasing on $(-\infty, \infty)$

- Odd function

Square Root Function

$f(x) = \sqrt{x}$

- Domain: $[0, \infty)$
- Range: $[0, \infty)$
- Increasing on $(0, \infty)$

- Neither even nor odd

Absolute Value Function

$f(x) = |x|$

- Domain: $(-\infty, \infty)$
- Range: $[0, \infty)$
- Decreasing on $(-\infty, 0)$ and increasing on $(0, \infty)$
- Even function

7 Graph step functions.

Table 2.4 Cost of First-Class Mail (Effective January 10, 1999)

Weight Not Over	Cost
1 ounce	$0.33
2 ounces	0.55
3 ounces	0.77
4 ounces	0.99
5 ounces	1.21

Source: U.S. Postal Service

Step Functions

Have you ever mailed a letter that seemed heavier than usual? Perhaps you worried that the letter would not have enough postage. Costs for mailing a letter weighing up to 5 ounces are given in Table 2.4. If your letter weighs an ounce or less, the cost is $0.33. If your letter weighs 1.05 ounces, 1.50 ounces, 1.90 ounces, or 2.00 ounces, the cost "steps" to $0.55. The cost does not take on any value between $0.33 and $0.55. If your letter weighs 2.05 ounces, 2.50 ounces, 2.90 ounces, or 3 ounces, the cost "steps" to $0.77. Cost increases are $0.22 per step.

Now, let's see what the graph of the function that models this situation looks like. Let

x = the weight of the letter in ounces, and

$y = f(x)$ = the cost of mailing a letter weighing x ounces.

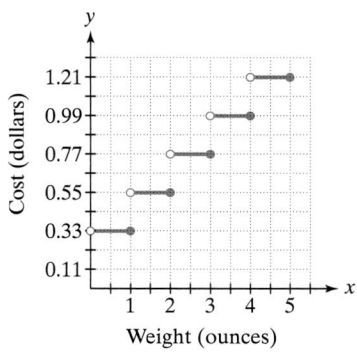

Figure 2.31

The graph is shown in Figure 2.31. Notice how it consists of a series of steps that jump vertically 22 units at each integer. The graph is constant between each pair of consecutive integers.

Mathematicians have a function that describes situations where function values graphically form discontinuous steps. The function is called the **greatest integer function**, symbolized by $\text{int}(x)$ or $[\![x]\!]$. And what is $\text{int}(x)$?

$$\text{int}(x) = \text{the greatest integer that is less than or equal to } x.$$

For example,

$$\text{int}(1) = 1, \quad \text{int}(1.3) = 1, \quad \text{int}(1.5) = 1, \quad \text{int}(1.9) = 1.$$

1 is the greatest integer that is less than or equal to 1, 1.3, 1.5, and 1.9.

Here are some additional examples:

$$\text{int}(2) = 2, \quad \text{int}(2.3) = 2, \quad \text{int}(2.5) = 2, \quad \text{int}(2.9) = 2.$$

2 is the greatest integer that is less than or equal to 2, 2.3, 2.5, and 2.9.

Notice how we jumped from 1 to 2 in the function values for $\text{int}(x)$. In particular,

$$\text{If } 1 \leq x < 2, \quad \text{then} \quad \text{int}(x) = 1.$$
$$\text{If } 2 \leq x < 3, \quad \text{then} \quad \text{int}(x) = 2.$$

The graph of $f(x) = \text{int}(x)$ is shown in Figure 2.32. The graph of the greatest integer function jumps vertically one unit at each integer. However, the graph is constant between each pair of consecutive integers. The rightmost of the horizontal steps shown in the graph illustrates that

$$\text{If } 5 \leq x < 6, \quad \text{then} \quad \text{int}(x) = 5.$$

In general,

$$\text{If } n \leq x < n + 1, \text{ where } n \text{ is an integer,} \quad \text{then} \quad \text{int}(x) = n.$$

By contrast to the graph for the cost of first-class mail, the graph of the greatest integer function includes the point on the left of each horizontal step, but does not include the point on the right. The domain of $f(x) = \text{int}(x)$ is the set of all real numbers, $(-\infty, \infty)$. The range is the set of all integers.

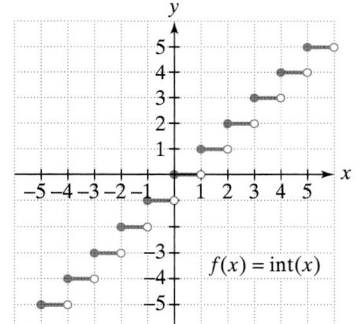

Figure 2.32 The graph of the greatest integer function

Technology

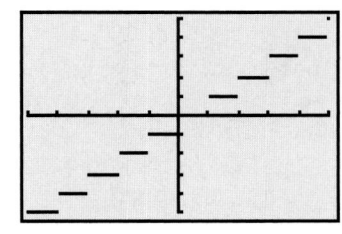

The graph of $f(x) = \text{int}(x)$, shown on the left, was obtained with a graphing utility. By graphing in "dot" mode, we can see the discontinuities at the integers. By looking at the graph, it is impossible to tell that, for each step, the point on the left is included and the point on the right is not. We must trace along the graph to obtain such information.

8 Obtain information from graphs of functions in applied situations.

Applications

We return to the function that models the fuel efficiency of U.S. automobiles over time. In the next example, we'll see what we can learn about this function from its graph.

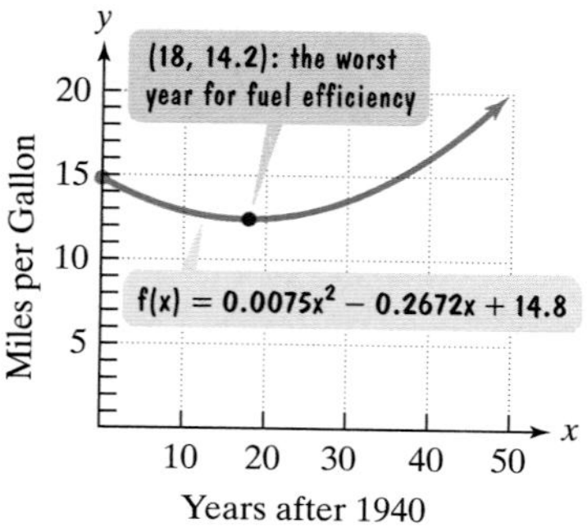

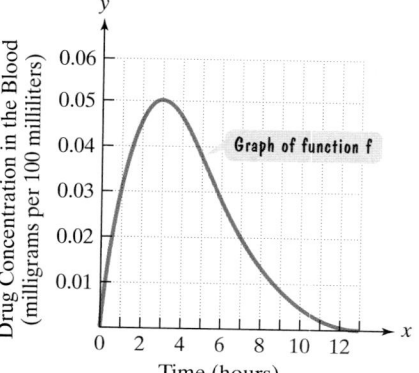

Figure 2.33 Fuel efficiency of U.S. automobiles over time

EXAMPLE 6   A Graph for the Fuel Efficiency Function

We have seen the function

$$f(x) = 0.0075x^2 - 0.2672x + 14.8$$

that models the average number of miles per gallon of U.S. automobiles, $f(x)$, x years after 1940. The graph of this function is shown as a continuous curve in Figure 2.33. (It can also be shown as a series of points, each point representing a year and miles per gallon for that year.)

  a. On which interval is f decreasing, and what does this mean?
  b. On which interval is f increasing, and what does this mean?

Solution   Note the voice balloon pointing to $(18, 14.2)$. It tells us that 18 years after 1940, in 1958, fuel efficiency was at its lowest point ever—a dismal 14.2 miles per gallon. This information, and the shape of the graph, enables us to find where f is decreasing and where it is increasing.

  a. Function f is decreasing on the interval $(0, 18)$.
      This means that fuel efficiency was decreasing from 1940 through 1958.
  b. Function f is increasing on the interval $(18, 50)$.
      This means that fuel efficiency was increasing from 1958 through 1990.

Check Point 6

When a person receives a drug injected into a muscle, the concentration of the drug in the body, measured in milligrams per 100 milliliters, is a function of the time elapsed since the injection, measured in hours. Figure 2.34 shows the graph of such a function, where x = hours since the injection and $f(x)$ = drug concentration at time x.

  a. On which interval is f increasing, and what does this mean?
  b. On which interval is f decreasing, and what does this mean?
  c. What is the drug's maximum concentration and when does this occur?
  d. What happens by the end of 13 hours?

Figure 2.34 Concentration of a drug as a function of time

EXERCISE SET 2.4

Practice Exercises

Graph the function in Exercises 1–14. Use the integer values of x given to the right of the function to obtain ordered pairs. Use the graph to specify the function's domain and range.

1. $f(x) = x^2 + 2$ $\quad x = -3, -2, -1, 0, 1, 2, 3$
2. $f(x) = x^2 - 1$ $\quad x = -3, -2, -1, 0, 1, 2, 3$
3. $g(x) = \sqrt{x} - 1$ $\quad x = 0, 1, 4, 9$
4. $g(x) = \sqrt{x} + 2$ $\quad x = 0, 1, 4, 9$
5. $h(x) = \sqrt{x - 1}$ $\quad x = 1, 2, 5, 10$
6. $h(x) = \sqrt{x + 2}$ $\quad x = -2, -1, 2, 7$
7. $f(x) = |x| - 1$ $\quad x = -3, -2, -1, 0, 1, 2, 3$
8. $f(x) = |x| + 1$ $\quad x = -3, -2, -1, 0, 1, 2, 3$
9. $g(x) = |x - 1|$ $\quad x = -3, -2, -1, 0, 1, 2, 3$
10. $g(x) = |x + 1|$ $\quad x = -3, -2, -1, 0, 1, 2, 3$
11. $f(x) = 5$ $\quad x = -3, -2, -1, 0, 1, 2, 3$
12. $f(x) = 3$ $\quad x = -3, -2, -1, 0, 1, 2, 3$
13. $f(x) = x^3 - 2$ $\quad x = -2, -1, 0, 1, 2$
14. $f(x) = x^3 + 2$ $\quad x = -2, -1, 0, 1, 2$

In Exercises 15–30, use the graph to determine **a.** *the function's domain;* **b.** *the function's range;* **c.** *the x-intercepts, if any;* **d.** *the y-intercept, if any; and* **e.** *the function values indicated below some of the graphs.*

15.
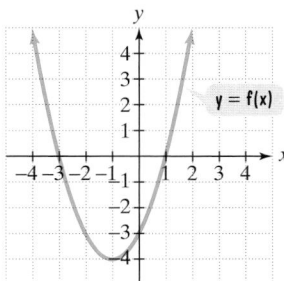

16.
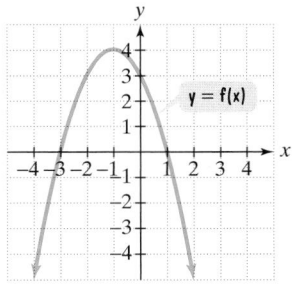

17.
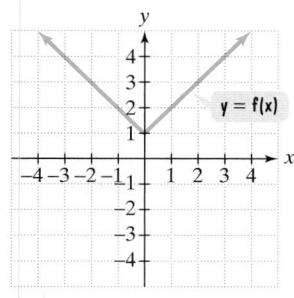
$f(-1) = ?$ $f(3) = ?$

18.
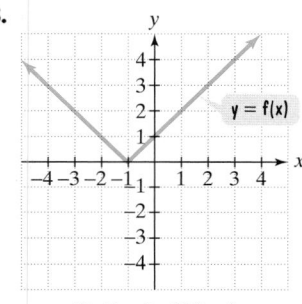
$f(-4) = ?$ $f(3) = ?$

19.
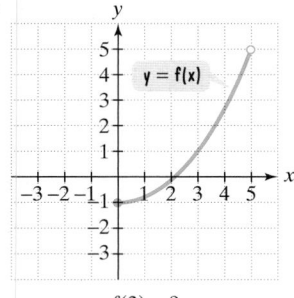
$f(3) = ?$

20.
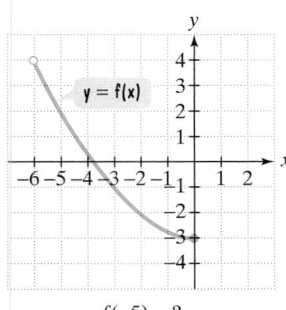
$f(-5) = ?$

21.

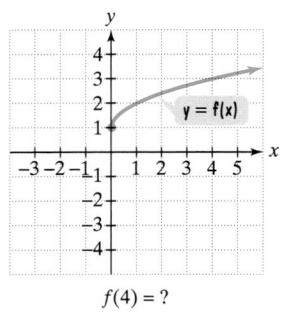

$f(4) = ?$

22.

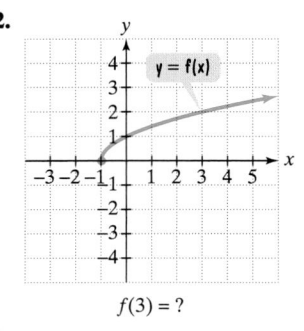

$f(3) = ?$

23.

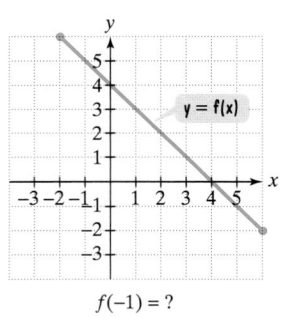

$f(-1) = ?$

24.

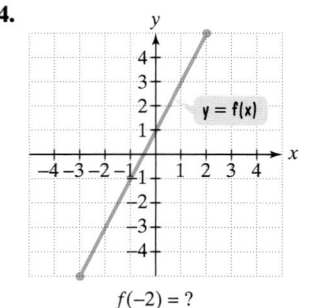

$f(-2) = ?$

25.

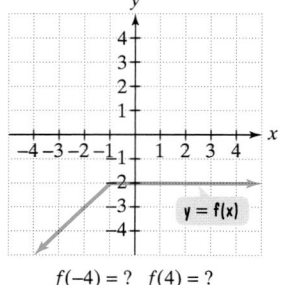

$f(-4) = ?$ $f(4) = ?$

26.

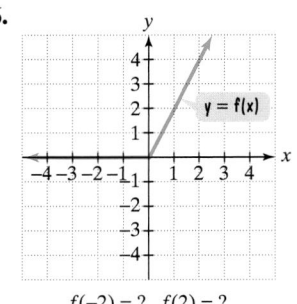

$f(-2) = ?$ $f(2) = ?$

27.

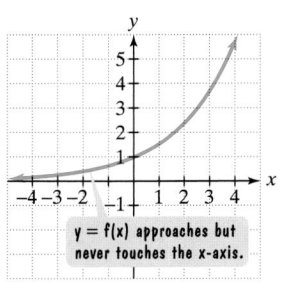

$y = f(x)$ approaches but never touches the x-axis.

28.

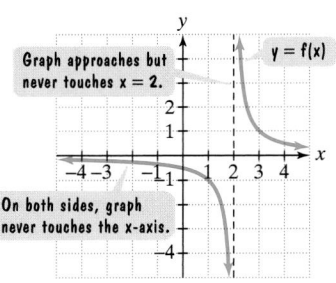

Graph approaches but never touches x = 2.

On both sides, graph never touches the x-axis.

29.

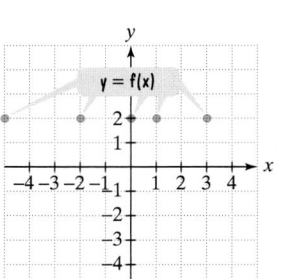

30.

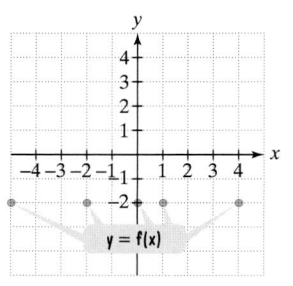

In Exercises 31–38, use the vertical line test to identify graphs in which y is a function of x.

31.

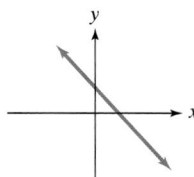

32.

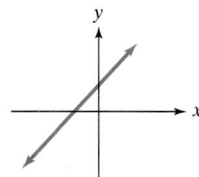

33.

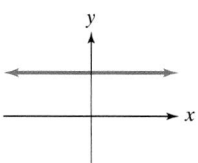

34.

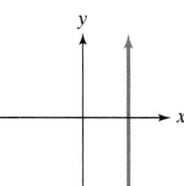

35.

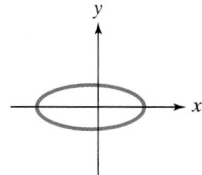

36.

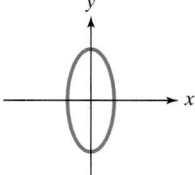

37.

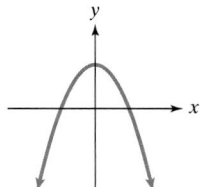

38.

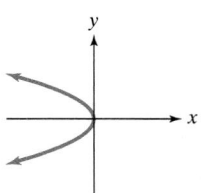

In Exercises 39–50, use the graph to determine:
   a. *intervals on which the function is increasing, if any.*
   b. *intervals on which the function is decreasing, if any.*
   c. *intervals on which the function is constant, if any.*

39. Use the graph in Exercise 15.
40. Use the graph in Exercise 16.
41. Use the graph in Exercise 21.
42. Use the graph in Exercise 22.
43. Use the graph in Exercise 23.
44. Use the graph in Exercise 24.
45. Use the graph in Exercise 25.
46. Use the graph in Exercise 26.

47.

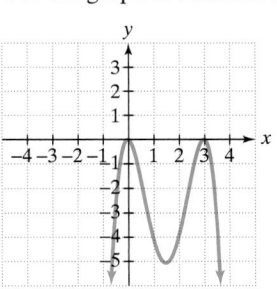

48.

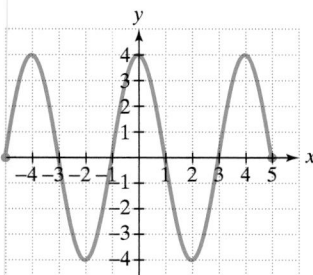

49.

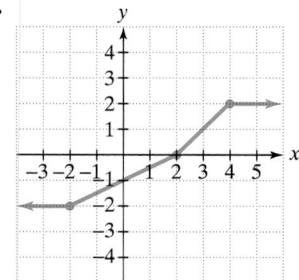

50.

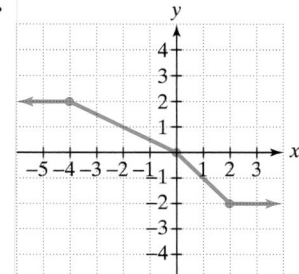

In Exercises 51–62, determine whether each function is even, odd, or neither.

51. $f(x) = x^3 + x$
52. $f(x) = x^3 - x$
53. $g(x) = x^2 + x$
54. $g(x) = x^2 - x$
55. $h(x) = x^2 - x^4$
56. $h(x) = 2x^2 + x^4$
57. $f(x) = x^2 - x^4 + 1$
58. $f(x) = 2x^2 + x^4 + 1$
59. $f(x) = \frac{1}{5}x^6 - 3x^2$
60. $f(x) = 2x^3 - 6x^5$
61. $f(x) = x\sqrt{1 - x^2}$
62. $f(x) = x^2\sqrt{1 - x^2}$

In Exercises 63–66, use possible symmetry to determine whether each graph is the graph of an even function, an odd function, or a function that is neither even nor odd.

63.

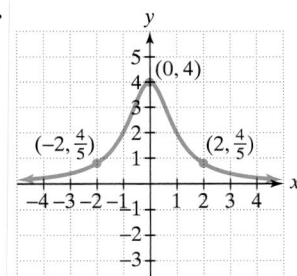

64.

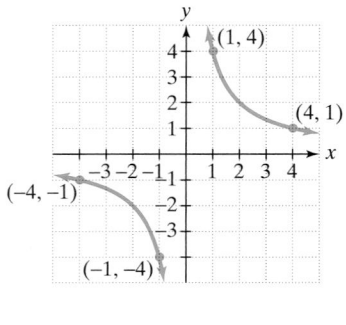

65.

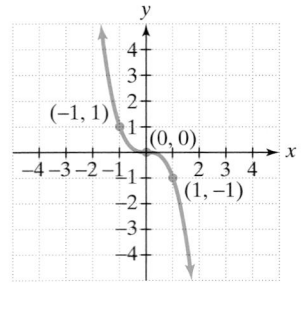

66.

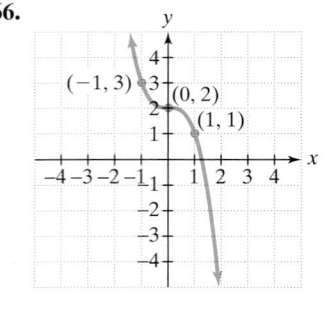

In Exercises 67–72, use the shape of each graph to name the function. Select from the following names: constant function, identity function, standard quadratic function, standard cubic function, square root function, absolute value function, greatest integer function.

67.

68.

69.

70.

71.

72.

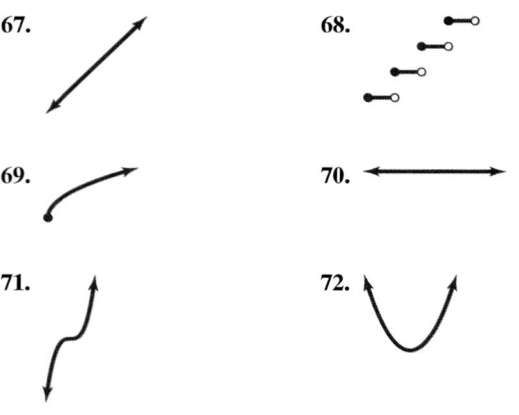

In Exercises 73–78, if $f(x) = int(x)$, find each function value.

73. $f(1.06)$

74. $f(2.99)$

75. $f\left(\frac{1}{3}\right)$

76. $f(-1.5)$

77. $f(-2.3)$

78. $f(-99.001)$

Application Exercises

The graph shows the function $y = f(x)$, where $f(x)$ is defense spending in year x for $1988 \leq x \leq 1998$. Use the graph to solve Exercises 79–82.

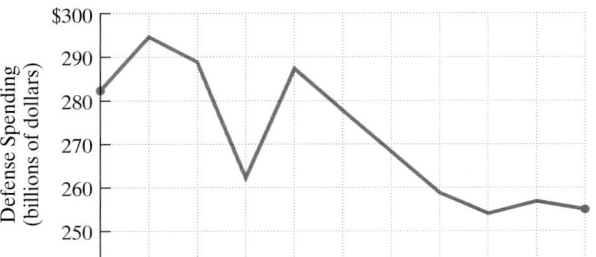

Source: Office of Management and Budget

79. Estimate the function value $f(1989)$. What is significant about this function value?

80. Estimate the function value $f(1996)$. What is significant about this function value?

81. In which time intervals is defense spending increasing?

82. In which time intervals is defense spending decreasing?

83. The function $f(x) = 0.4x^2 - 36x + 1000$ models the number of accidents per 50 million miles driven as a function of age, x, in years, where $16 \leq x \leq 74$. The graph of f is shown.

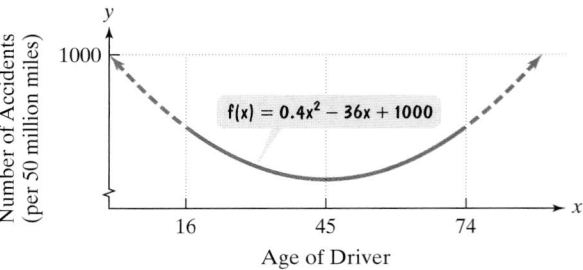

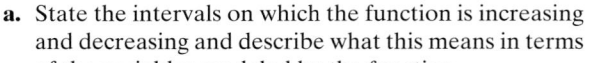

a. State the intervals on which the function is increasing and decreasing and describe what this means in terms of the variables modeled by the function.

b. For what value of x does the graph reach its lowest point? What is the minimum value of y? Describe the practical significance of this minimum value.

c. The domain of f is $[16, 74]$. Use the function's equation to determine the range of f. What is the practical significance of this range in terms of the meaning of $f(x)$ in the given function?

84. Based on a study by Vance Tucker (*Scientific American*, May 1969) the power expenditure of migratory birds in flight is a function of their flying speed, x, in miles per hour, modeled by $f(x) = 0.67x^2 - 27.74x + 387$. Power expenditure, $f(x)$, is measured in calories, and migratory birds generally fly between 12 and 30 miles per hour. The graph of f is shown in the figure, with a domain of $[12, 30]$.

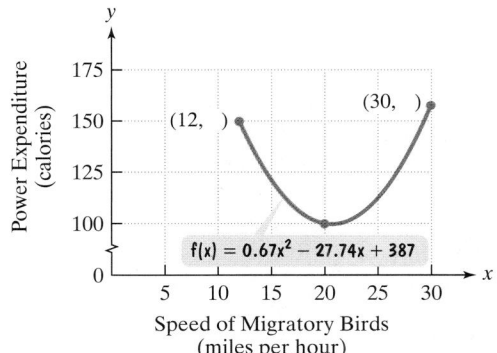

Speed of Migratory Birds
(miles per hour)

a. State the intervals on which the function is increasing and decreasing and describe what this means in terms of the variables modeled by the function.

b. For what approximate value of x does the graph reach its lowest point? What is the minimum value of y? Describe the practical significance of this minimum value.

c. The domain of f is $[12, 30]$. Use the function's equation to find the range of f. What is the practical significance of this range in terms of the meaning of $f(x)$ in the given function?

85. The cost of a telephone call between two cities is $0.10 for the first minute and $0.05 for each additional minute or portion of a minute. Draw a graph of the cost, C, in dollars, of the phone call as a function of time, t, in minutes, on the interval $(0, 5]$.

86. A cargo service charges a flat fee of $4 plus $1 for each pound or fraction of a pound to mail a package. Let $C(x)$ represent the cost to mail a package that weighs x pounds. Graph the cost function on the interval $(0, 5]$.

87. Researchers at Yale University have suggested that levels of passion and commitment in human relations are functions of time. Based on the shapes of the graphs shown, which do you think depicts passion and which represents commitment? Explain how you arrived at your answer.

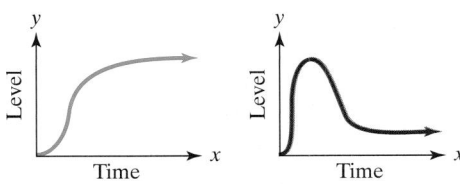

88. Although the level of air pollution varies from day to day and from hour to hour, during the summer the level of air pollution is a function of the time of day. The function

$$f(x) = 0.1x^2 - 0.4x + 0.6$$

describes the level of air pollution (in parts per million [ppm]), where x corresponds to the number of hours after 9 A.M.

a. Construct a table of values using integers from 0 to 5 for x, and graph the function from 0 to 5.

b. Researchers have determined that a level of 0.3 ppm of pollutants in the air can be hazardous to your health. Based on the graph, at what time of day between 9 A.M. and 2 P.M. should runners exercise to avoid unsafe air?

Writing in Mathematics

89. Discuss one disadvantage to using point plotting as a method for graphing functions.

90. Explain how to use a function's graph to find the function's domain and range.

91. Explain how the vertical line test is used to determine whether a graph is a function.

92. What does it mean if function f is increasing on an interval?

93. If you are given a function's equation, how do you determine if the function is even, odd, or neither?

94. If you are given a function's graph, how do you determine if the function is even, odd, or neither?

95. What is a step function? Give an example of an everyday situation that can be modeled using such a function. Do not use the cost-of-mail example.

96. Explain how to find int(-3.000004).

Technology Exercises

97. The function $f(x) = -0.00002x^3 + 0.008x^2 - 0.3x + 6.95$ models the number of annual physician visits by a person of age x.

a. Use a graphing utility to graph the function in a $[0, 100, 5]$ by $[0, 60, 3]$ viewing rectangle.

b. What does the shape of the graph indicate about the relationship between one's age and the number of annual physician visits?

c. Use the ⎡TRACE⎤ or minimum function capability to find the coordinates of the lowest point on the graph of the function. What does this mean?

98. The function

$$C(x) = \begin{cases} -0.35x + 220 & \text{for } 0 \le x \le 20 \\ -0.80x + 229 & \text{for } x > 20 \end{cases}$$

describes the number of milligrams of cholesterol per deciliter of blood for American adults x years after 1960.

a. Graph the function using a graphing utility in a [0, 100, 20] by [0, 250, 50] viewing rectangle.

b. What does the graph indicate about cholesterol level from 1960 to the present?

c. Was the goal of lowering cholesterol to a level under 200 reached before the year 2000?

In Exercises 99–104, use a graphing utility to graph each function. Use a [−5, 5, 1] by [−5, 5, 1] viewing rectangle. Then find the intervals on which the function is increasing, decreasing, or constant.

99. $f(x) = x^3 - 6x^2 + 9x + 1$ **100.** $g(x) = |4 - x^2|$

101. $h(x) = |x - 2| + |x + 2|$ **102.** $f(x) = x^{1/3}(x - 4)$

103. $g(x) = x^{2/3}$ **104.** $h(x) = 2 - x^{2/5}$

105. a. Graph the functions $f(x) = x^n$ for $n = 2, 4$, and 6 in a [−2, 2, 1] by [−1, 3, 1] viewing rectangle.

b. Graph the functions $f(x) = x^n$ for $n = 1, 3$, and 5 in a [−2, 2, 1] by [−2, 2, 1] viewing rectangle.

c. If n is even, where is the graph of $f(x) = x^n$ increasing and where is it decreasing?

d. If n is odd, what can you conclude about the graph of $f(x) = x^n$ in terms of increasing or decreasing behavior?

e. Graph all six functions in a [−1, 3, 1] by [−1, 3, 1] viewing rectangle. What do you observe about the graphs in terms of how flat or how steep they are?

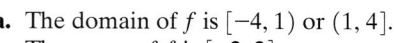

Critical Thinking Exercises

106. Which one of the following is true based on the graph of f in the figure?

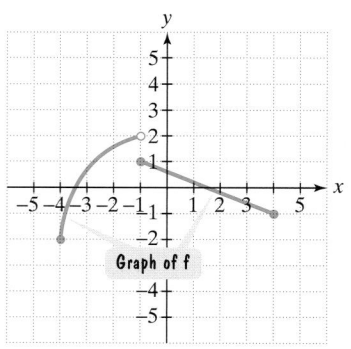

Graph of f

a. The domain of f is [−4, 1) or (1, 4].

b. The range of f is [−2, 2].

c. $f(-1) - f(4) = 2$

d. $f(0) = 2.1$

107. Describe a situation that can be modeled by the graph shown.

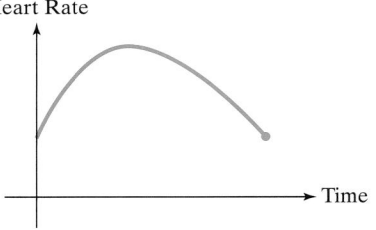

108. Sketch the graph of f using the following properties. (More than one correct graph is possible.) f is a piece-wise function that is decreasing on $(-\infty, 2)$, $f(2) = 0$, f is increasing on $(2, \infty)$, and the range of f is $[0, \infty)$.

109. Define a piecewise function on the intervals $(-\infty, 2]$, $(2, 5)$, and $[5, \infty)$ that does not "jump" at 2 or 5 such that one piece is a constant function, another piece is an increasing function, and the third piece is a decreasing function.

110. Suppose that $h(x) = \dfrac{f(x)}{g(x)}$. The function f can be even, odd, or neither. The same is true for the function g.

a. Under what conditions is h definitely an even function?

b. Under what conditions is h definitely an odd function?

111. Take another look at the cost of first-class mail and its graph (Table 2.4 and Figure 2.31 on pages 212–213). Change the description of the heading in the left column of Table 2.4 so that the graph includes the point on the left of each horizontal step, but does not include the point on the right.

Group Exercise

112. In Exercise 87, passion and commitment are graphed over time. For this activity, you will be creating a graph of a particular experience that involved your feelings of love, anger, sadness. or any other emotion you choose. The horizontal axis should be labeled time and the vertical axis the emotion you are graphing. You will not be using your algebra skills to create your graph; however, you should try to make the graph as precise as possible. You may use negative numbers on the vertical axis, if appropriate. After each group member has created a graph, pool together all of the graphs and study them to see if there are any similarities in the graphs for a particular emotion or for all emotions.

SECTION 2.5 *Transformations and Combinations of Functions*

Objectives

1. Use vertical shifts to graph functions.
2. Use horizontal shifts to graph functions.
3. Use reflections to graph functions.
4. Use vertical stretching and shrinking to graph functions.
5. Graph functions involving a sequence of transformations.
6. Combine functions arithmetically, specifying domains.

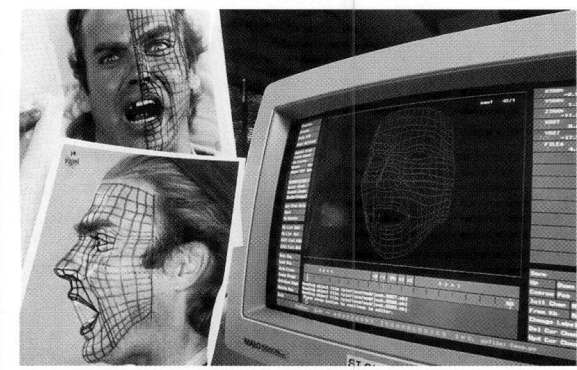

Have you seen *Terminator 2*, *The Mask*, or *The Matrix*? These were among the first films to use spectacular effects in which a character or object having one shape was transformed in a fluid fashion into a quite different shape. The name for such a transformation is **morphing**. The effect allows a real actor to be seamlessly transformed into a computer-generated animation. The animation can be made to perform impossible feats before it is morphed back to the conventionally filmed image.

Like transformed movie images, the graph of one function can be turned into the graph of a different function. To do this, we need to rely on a function's equation. Knowing that a graph is a transformation of a familiar graph makes graphing easier.

Discovery

The study of how changing a function's equation can affect its graph can be explored with a graphing utility. Use your graphing utility to verify the hand-drawn graphs as you read this section.

1 Use vertical shifts to graph functions.

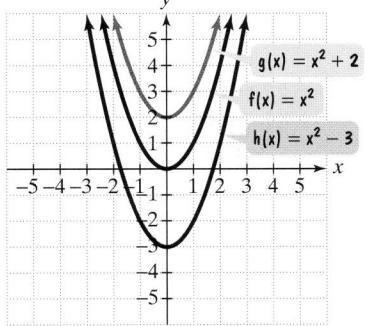

Figure 2.35 Vertical shifts

The graph of $f(x) = x^2$ can be gradually morphed into the graph of $g(x) = x^2 + 2$ by using animation to graph $f(x) = x^2 + c$ for $0 \le c \le 2$. By selecting many values for c, we can create an animated sequence in which change appears to occur continuously.

Vertical Shifts

Let's begin by looking at three graphs whose shapes are the same. Figure 2.35 shows the graphs. The black graph in the middle is the standard quadratic function $f(x) = x^2$. Now, look at the blue graph on the top. The equation of this graph, $g(x) = x^2 + 2$, adds 2 to the right side of $f(x) = x^2$. What effect does this have on the graph of f? It shifts the graph vertically up by 2 units.

$$g(x) = x^2 + 2 = f(x) + 2$$

The graph of g shifts the graph of f up 2 units.

Finally, look at the red graph on the bottom of Figure 2.35. The equation of this graph, $h(x) = x^2 - 3$, subtracts 3 from the right side of $f(x) = x^2$. What effect does this have on the graph of f? It shifts the graph vertically down by 3 units.

$$h(x) = x^2 - 3 = f(x) - 3$$

The graph of h shifts the graph of f down 3 units.

In general, if c is positive, $y = f(x) + c$ shifts the graph of f upward c units and $y = f(x) - c$ shifts the graph of f downward c units. These are called **vertical shifts** of the graph of f.

Vertical Shifts

Let f be a function and c a positive real number.

- The graph of $y = f(x) + c$ is the graph of $y = f(x)$ shifted c units vertically upward.
- The graph of $y = f(x) - c$ is the graph of $y = f(x)$ shifted c units vertically downward.

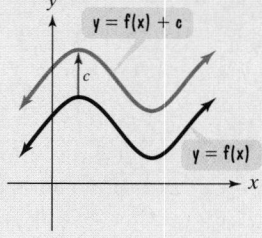
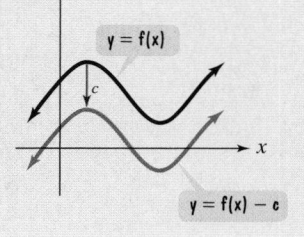

EXAMPLE 1   Vertical Shift Down

Use the graph of $f(x) = |x|$ to obtain the graph of $g(x) = |x| - 4$.

Solution   The graph of $g(x) = |x| - 4$ has the same shape as the graph of $f(x) = |x|$. However, it is shifted down vertically 4 units. We have constructed a table showing some of the coordinates for f and g. The graphs of f and g are shown in Figure 2.36.

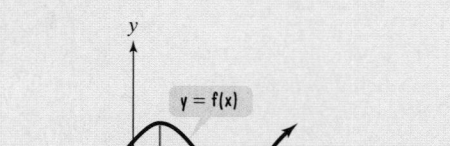

x	$y = f(x) = |x|$	$y = g(x)$ $= |x| - 4 = f(x) - 4$
-2	$|-2| = 2$	$|-2| - 4 = -2$
-1	$|-1| = 1$	$|-1| - 4 = -3$
0	$|0| = 0$	$|0| - 4 = -4$
1	$|1| = 1$	$|1| - 4 = -3$
2	$|2| = 2$	$|2| - 4 = -2$

Figure 2.36

Check Point 1   Use the graph of $f(x) = |x|$ to obtain the graph of $g(x) = |x| + 3$.

2 Use horizontal shifts to graph functions.

Horizontal Shifts

We return to the graph of $f(x) = x^2$, the standard quadratic function. In Figure 2.37, the graph of function f is in the middle of the three graphs. Note that there are graphs to the right and left of f. By contrast to the vertical shift situation,

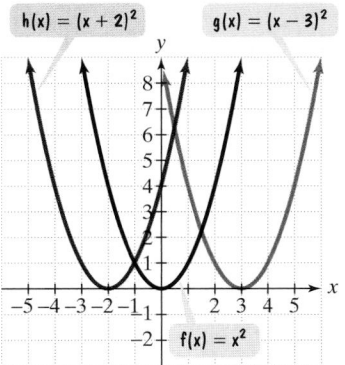

Figure 2.37 Horizontal shifts

this time there are graphs to the left and to the right of the graph of f. Look at the blue graph on the right. The equation of this graph, $g(x) = (x - 3)^2$, subtracts 3 from each value of x in the domain of $f(x) = x^2$. What effect does this have on the graph of f? It shifts the graph horizontally to the right by 3 units.

$$g(x) = (x - 3)^2 = f(x - 3)$$

The graph of g shifts the graph of f 3 units to the right.

Now, look at the red graph on the left in Figure 2.37. The equation of this graph, $h(x) = (x + 2)^2$, adds 2 to each value of x in the domain of $f(x) = x^2$. What effect does this have on the graph of f? It shifts the graph horizontally to the left by 2 units.

$$h(x) = (x + 2)^2 = f(x + 2)$$

The graph of h shifts the graph of f 2 units to the left.

In general, if c is positive, $y = f(x + c)$ shifts the graph of f to the left c units and $y = f(x - c)$ shifts the graph of f to the right c units. These are called **horizontal shifts** of the graph of f.

Study Tip

We know that positive numbers are to the right of zero on a number line and negative numbers are to the left of zero. This positive-negative orientation does not apply to horizontal shifts. A *positive* number causes a shift to the *left* and a *negative* number causes a shift to the *right*.

Horizontal Shifts

Let f be a function and c a positive real number.
- The graph of $y = f(x + c)$ is the graph of $y = f(x)$ shifted to the left c units.
- The graph of $y = f(x - c)$ is the graph of $y = f(x)$ shifted to the right c units.

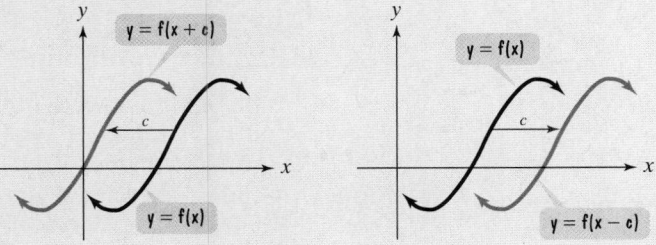

EXAMPLE 2   Horizontal Shift to the Left

Use the graph of $f(x) = \sqrt{x}$ to obtain the graph of $g(x) = \sqrt{x + 5}$.

Solution   Compare the equations for $f(x) = \sqrt{x}$ and $g(x) = \sqrt{x + 5}$. The equation for g adds 5 to each value of x in the domain of f.

$$y = g(x) = \sqrt{x + 5} = f(x + 5)$$

The graph of g shifts the graph of f 5 units to the left.

The graph of $g(x) = \sqrt{x + 5}$ has the same shape as the graph of $f(x) = \sqrt{x}$. However, it is shifted horizontally to the left 5 units. We have created tables

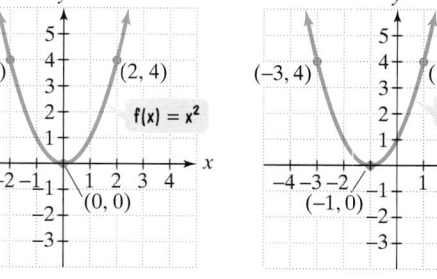

Figure 2.38 Shifting $f(x) = \sqrt{x}$ five units left

showing some of the coordinates for f and g. As shown in Figure 2.38, every point in the graph of g is exactly 5 units to the left of a corresponding point on the graph of f.

x	$y = f(x) = \sqrt{x}$
0	$\sqrt{0} = 0$
1	$\sqrt{1} = 1$
4	$\sqrt{4} = 2$

x	$y = g(x) = \sqrt{x + 5}$
-5	$\sqrt{-5 + 5} = \sqrt{0} = 0$
-4	$\sqrt{-4 + 5} = \sqrt{1} = 1$
-1	$\sqrt{-1 + 5} = \sqrt{4} = 2$

Check Point 2 Use the graph of $f(x) = \sqrt{x}$ to obtain the graph of $g(x) = \sqrt{x - 4}$.

Some functions can be graphed by combining horizontal and vertical shifts. The function should be a variation of a function whose equation you know how to graph, such as the standard quadratic function, the standard cubic function, the square root function, or the absolute value function.

In our next example, we will use the graph of the standard quadratic function $f(x) = x^2$ to obtain the graph of $h(x) = (x + 1)^2 - 3$. We will graph three functions:

$$f(x) = x^2 \qquad g(x) = (x + 1)^2 \qquad h(x) = (x + 1)^2 - 3$$

Start by graphing the standard quadratic function.

Shift the graph of f horizontally one unit to the left.

Shift the graph of g vertically down 3 units.

EXAMPLE 3   Combining Horizontal and Vertical Shifts

Use the graph of $f(x) = x^2$ to obtain the graph of $h(x) = (x + 1)^2 - 3$.

Solution

Step 1   Graph $f(x) = x^2$. The graph of the standard quadratic function is shown in Figure 2.39(a). We've identified three points on the graph.

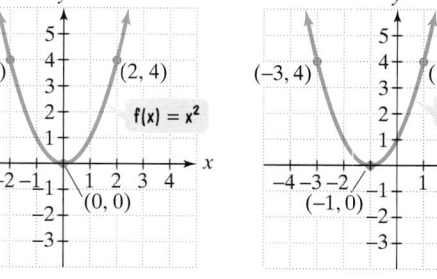

(a) The graph of $f(x) = x^2$

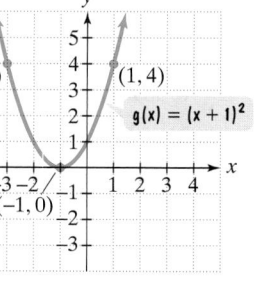

(b) The graph of $g(x) = (x + 1)^2$

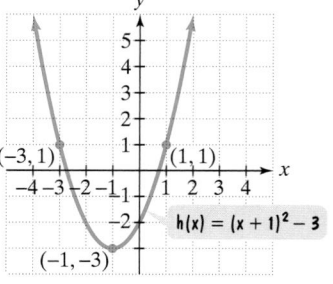

(c) The graph of $h(x) = (x + 1)^2 - 3$

Figure 2.39

Discovery

Work Example 3 by first shifting the graph of $f(x) = x^2$ three units down, graphing $g(x) = x^2 - 3$. Now, shift this graph one unit left to graph $h(x) = (x + 1)^2 - 3$. Did you obtain the graph in Figure 2.39(c)? What can you conclude?

Step 2 **Graph $g(x) = (x + 1)^2$.** Because we add 1 to each value of x in the domain of the standard quadratic function $f(x) = x^2$, we shift the graph of f horizontally one unit to the left. This is shown in Figure 2.39(b). Notice that every point in the graph in Figure 2.39(b) has an x-coordinate that is one less than the x-coordinate for the corresponding point in the graph in Figure 2.39(a).

Step 3 **Graph $h(x) = (x + 1)^2 - 3$.** Because we subtract 3, we shift the graph in Figure 2.39(b) vertically down 3 units. The graph is shown in Figure 2.39(c). Notice that every point in the graph in Figure 2.39(c) has a y-coordinate that is three less than the y-coordinate of the corresponding point in the graph in Figure 2.39(b).

Check Point 3 Use the graph of $f(x) = \sqrt{x}$ to obtain the graph of $h(x) = \sqrt{x - 1} - 2$.

3 Use reflections to graph functions.

Reflections of Graphs

This photograph shows a reflection of an old bridge in a Maryland river. This perfect reflection occurs because the surface of the water is absolutely still. A mild breeze rippling the water's surface would distort the reflection.

Is it possible for graphs to have mirror-like qualities? Yes. Figure 2.40 shows the graphs of $f(x) = x^2$ and $g(x) = -x^2$. The graph of g is a **reflection about the x-axis** of the graph of f. In general, the graph of $y = -f(x)$ reflects the graph of f about the x-axis. Thus, the graph of g is a reflection of the graph of f about the x-axis because

$$g(x) = -x^2 = -f(x).$$

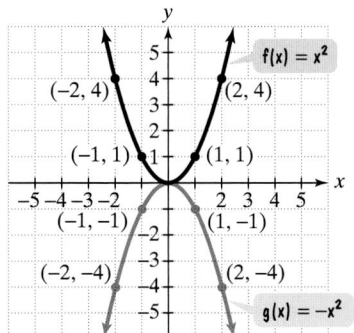

Figure 2.40 Reflections about the x-axis

Reflection About the x-Axis

The graph of $y = -f(x)$ is the graph of $y = f(x)$ reflected about the x-axis.

EXAMPLE 4   Reflection about the x-Axis

Use the graph of $f(x) = \sqrt{x}$ to obtain the graph of $g(x) = -\sqrt{x}$.

Solution   Compare the equations for $f(x) = \sqrt{x}$ and $g(x) = -\sqrt{x}$. The graph of g is a reflection about the x-axis of the graph of f because

$$g(x) = -\sqrt{x} = -f(x).$$

We have created a table showing some of the coordinates for f and g. The graphs of f and g are shown in Figure 2.41.

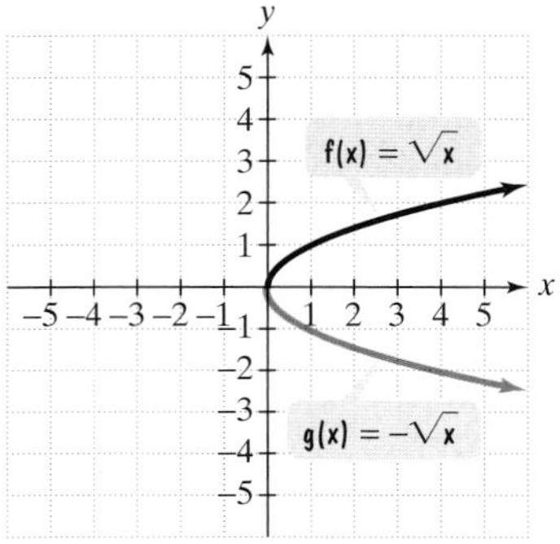

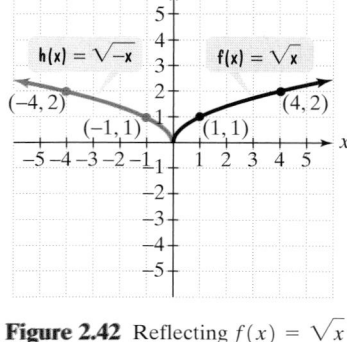

Figure 2.41 Reflecting $f(x) = \sqrt{x}$ about the x-axis

x	$f(x) = \sqrt{x}$	$g(x) = -\sqrt{x}$
0	$\sqrt{0} = 0$	$-\sqrt{0} = 0$
1	$\sqrt{1} = 1$	$-\sqrt{1} = -1$
4	$\sqrt{4} = 2$	$-\sqrt{4} = -2$

> **Check Point 4**   Use the graph of $f(x) = |x|$ to obtain the graph of $g(x) = -|x|$.

It is also possible to reflect graphs about the y-axis.

Reflection about the y-Axis

The graph of $y = f(-x)$ is the graph of $y = f(x)$ reflected about the y-axis.

EXAMPLE 5   Reflection about the y-Axis

Use the graph of $f(x) = \sqrt{x}$ to obtain the graph of $h(x) = \sqrt{-x}$.

Solution   Compare the equations for $f(x) = \sqrt{x}$ and $h(x) = \sqrt{-x}$. The graph of h is a reflection about the y-axis of the graph of f because

$$h(x) = \sqrt{-x} = f(-x).$$

We have created tables showing some of the coordinates for f and h. The graphs of f and h are shown in Figure 2.42.

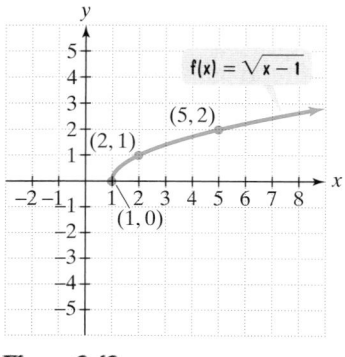

Figure 2.42 Reflecting $f(x) = \sqrt{x}$ about the y-axis

x	$f(x) = \sqrt{x}$
0	$\sqrt{0} = 0$
1	$\sqrt{1} = 1$
4	$\sqrt{4} = 2$

x	$h(x) = \sqrt{-x}$
0	$\sqrt{-0} = \sqrt{0} = 0$
-1	$\sqrt{-(-1)} = \sqrt{1} = 1$
-4	$\sqrt{-(-4)} = \sqrt{4} = 2$

Figure 2.43

> **Check Point 5**   Use the graph of $f(x) = \sqrt{x-1}$ in Figure 2.43 to obtain the graph of $h(x) = \sqrt{-x-1}$.

4 Use vertical stretching and shrinking to graph functions.

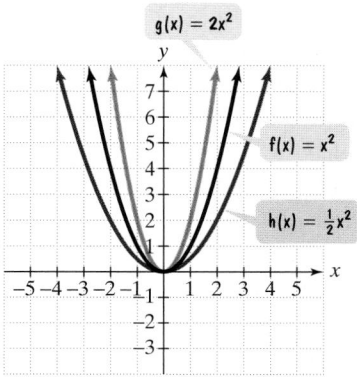

Figure 2.44 Stretching and shrinking $f(x) = x^2$

Vertical Stretching and Shrinking

Morphing does much more than move an image horizontally, vertically, or about an axis. An object having one shape is transformed into a different shape. Horizontal shifts, vertical shifts, and reflections do not change the basic shape of a graph. How can we shrink and stretch graphs, thereby altering their basic shapes?

Look at the three graphs in Figure 2.44. The black graph in the middle is the graph of the standard quadratic function, $f(x) = x^2$. Now, look at the blue graph on the top. The equation of this graph is $g(x) = 2x^2$. Thus, for each x, the y-coordinate of g is 2 times as large as the corresponding y-coordinate on the graph of f. The result is a narrower graph. We say that the graph of g is obtained by vertically *stretching* the graph of f. Now, look at the red graph on the bottom. The equation of this graph is $h(x) = \frac{1}{2}x^2$, or $h(x) = \frac{1}{2}f(x)$. Thus, for each x, the y-coordinate of h is one-half as large as the corresponding y-coordinate on the graph of f. The result is a wider graph. We say that the graph of h is obtained by vertically *shrinking* the graph of f.

These observations can be summarized as follows.

> **Stretching and Shrinking Graphs**
>
> Let f be a function and c a positive real number.
> - If $c > 1$, the graph of $y = cf(x)$ is the graph of $y = f(x)$ vertically stretched by multiplying each of its y-coordinates by c.
> - If $0 < c < 1$, the graph of $y = cf(x)$ is the graph of $y = f(x)$ vertically shrunk by multiplying each of its y-coordinates by c.

EXAMPLE 6   Vertically Stretching a Graph

Use the graph of $f(x) = |x|$ to obtain the graph of $g(x) = 2|x|$.

Solution   The graph of $g(x) = 2|x|$ is obtained by vertically stretching the graph of $f(x) = |x|$. We have constructed a table showing some of the coordinates for f and g. Observe that the y-coordinate on the graph of g is twice as large as the corresponding y-coordinate on the graph of f. The graphs of f and g are shown in Figure 2.45.

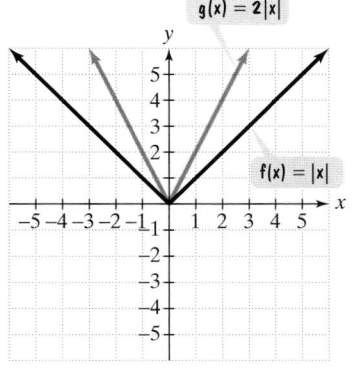

Figure 2.45 Stretching $f(x) = |x|$

x	$f(x) = |x|$	$g(x) = 2|x| = 2f(x)$
-2	$|-2| = 2$	$2|-2| = 4$
-1	$|-1| = 1$	$2|-1| = 2$
0	$|0| = 0$	$2|0| = 0$
1	$|1| = 1$	$2|1| = 2$
2	$|2| = 2$	$2|2| = 4$

Check Point 6   Use the graph of $f(x) = |x|$ to obtain the graph of $g(x) = 3|x|$.

EXAMPLE 7 Vertically Shrinking a Graph

Use the graph of $f(x) = |x|$ to obtain the graph of $h(x) = \frac{1}{2}|x|$.

Solution   The graph of $h(x) = \frac{1}{2}|x|$ is obtained by vertically shrinking the graph of $f(x) = |x|$. We have constructed a table showing some of the coordinates for f and h. Observe that the y-coordinate on the graph of h is one-half the corresponding y-coordinate on the graph of f. The graphs of f and h are shown in Figure 2.46.

Figure 2.46 Shrinking $f(x) = |x|$

x	$f(x) = |x|$	$h(x) = \frac{1}{2}|x| = \frac{1}{2}f(x)$
-2	$|-2| = 2$	$\frac{1}{2}|-2| = 1$
-1	$|-1| = 1$	$\frac{1}{2}|-1| = \frac{1}{2}$
0	$|0| = 0$	$\frac{1}{2}|0| = 0$
1	$|1| = 1$	$\frac{1}{2}|1| = \frac{1}{2}$
2	$|2| = 2$	$\frac{1}{2}|2| = 1$

> **Check Point 7**   Use the graph of $f(x) = |x|$ to obtain the graph of $h(x) = \frac{1}{4}|x|$.

5 Graph functions involving a sequence of transformations.

Sequences of Transformations

Table 2.5 summarizes the procedures for transforming the graph of $y = f(x)$.

Table 2.5   Summary of Transformations
In each case, c represents a positive real number.

To Graph:	Draw the Graph of f and:	Changes in the Equation of $y = f(x)$
Vertical shifts $y = f(x) + c$ $y = f(x) - c$	Raise the graph of f by c units. Lower the graph of f by c units.	c is added to $f(x)$. c is subtracted from $f(x)$.
Horizontal shifts $y = f(x + c)$ $y = f(x - c)$	Shift the graph of f to the left c units. Shift the graph of f to the right c units.	x is replaced by $x + c$. x is replaced by $x - c$.
Reflection about the x-axis $y = -f(x)$	Reflect the graph of f about the x-axis.	$f(x)$ is multiplied by -1.
Reflection about the y-axis $y = f(-x)$	Reflect the graph of f about the y-axis.	x is replaced by $-x$.
Vertical stretching or shrinking $y = cf(x), c > 1$	Multiply each y-coordinate of $y = f(x)$ by c, vertically stretching the graph of f.	$f(x)$ is multiplied by $c, c > 1$.
$y = cf(x), 0 < c < 1$	Multiply each y-coordinate of $y = f(x)$ by c, vertically shrinking the graph of f.	$f(x)$ is multiplied by $c, 0 < c < 1$.

A function involving more than one transformation can be graphed by performing transformations in the following order.

1. Horizontal shifting
2. Vertical stretching or shrinking
3. Reflecting
4. Vertical shifting

EXAMPLE 8 Graphing Using a Sequence of Transformations

Use the graph of $f(x) = \sqrt{x}$ to graph $g(x) = \sqrt{1 - x} + 3$.

Solution The following sequence of steps is illustrated in Figure 2.47. We begin with the graph of $f(x) = \sqrt{x}$.

Step 1 Horizontal Shifting Graph $y = \sqrt{x + 1}$. Because x is replaced by $x + 1$, the graph of $f(x) = \sqrt{x}$ is shifted 1 unit to the left.

Step 2 Vertical Stretching or Shrinking Because the equation $y = \sqrt{x + 1}$ is not multiplied by a constant in $g(x) = \sqrt{1 - x} + 3$, no stretching or shrinking is involved.

Step 3 Reflecting We are interested in graphing $y = \sqrt{1 - x} + 3$, or $y = \sqrt{-x + 1} + 3$. We have now graphed $y = \sqrt{x + 1}$. We can graph $y = \sqrt{-x + 1}$ by noting that x is replaced by $-x$. Thus, we graph $y = \sqrt{-x + 1}$ by reflecting the graph of $y = \sqrt{x + 1}$ about the y-axis.

Step 4 Vertical Shifting We can use the graph of $y = \sqrt{1 - x}$ to get the graph of $g(x) = \sqrt{1 - x} + 3$. Because 3 is added, shift the graph of $y = \sqrt{1 - x}$ up by 3 units.

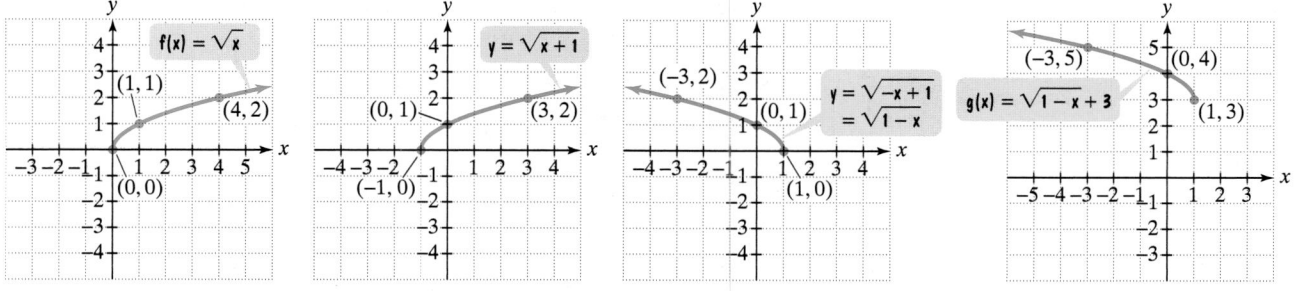

Figure 2.47 Using $f(x) = \sqrt{x}$ to graph $g(x) = \sqrt{1 - x} + 3$.

Check Point 8 Use the graph of $f(x) = x^2$ to graph $g(x) = -(x - 2)^2 + 3$.

6 Combine functions arithmetically, specifying domains.

Combinations of Functions

The graph in Figure 2.48 on page 230 shows the number of wars by region from 1990 through 1997. Have you seen data displayed in this style? The numbers on the bars do not represent the total number of wars for each year. Rather, they

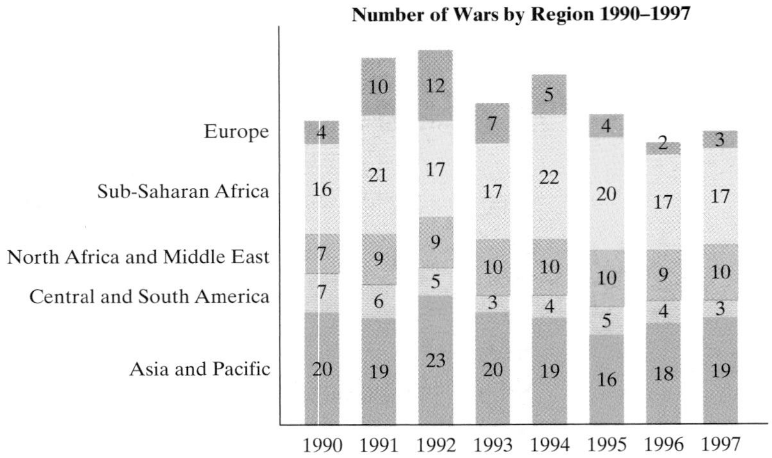

Number of Wars by Region 1990–1997

Figure 2.48
Source: International Peace Research Institute, Oslo

represent the number of wars by region for the year. Each bar's height represents the *total number* of wars per year. For example, in 1990 there were

$$20 + 7 + 7 + 16 + 4$$

or 54 wars. We can think of this addition as the addition of function values. We do this by introducing the following functions.

Let $f(x)$ = the number of wars in Asia and the Pacific in year x.

Let $g(x)$ = the number of wars in Central and South America in year x.

Using Figure 2.48, we see that

> Look at the green portion of each bar.

$$f(1990) = 20 \quad f(1991) = 19 \quad f(1992) = 23 \quad f(1993) = 20, \text{ and so on}$$

and

> Look at the light blue portion of each bar.

$$g(1990) = 7 \quad g(1991) = 6 \quad g(1992) = 5 \quad g(1993) = 3, \text{ and so on.}$$

We can add these function values by introducing a new function, $f + g$, defined by the addition of $f(x)$ and $g(x)$. Thus,

$$(f + g)(x) = f(x) + g(x) = \text{the number of wars in Asia and the Pacific in year } x \text{ plus the number of wars in Central and South America in year } x.$$

For example,

$$(f + g)(1990) = f(1990) + g(1990) = 20 + 7 = 27.$$

> In 1990, there were a total of 27 wars in Asia, the Pacific, Central America, and South America combined.

We can also subtract these function values by introducing a new function, $f - g$, defined by the subtraction of $f(x)$ and $g(x)$. Thus,

$$(f - g)(x) = f(x) - g(x) = \text{the number of wars in Asia and the Pacific in year } x \text{ minus the number of wars in Central and South America in year } x.$$

For example,

$$(f - g)(1990) = f(1990) - g(1990) = 20 - 7 = 13.$$

> In 1990, there were 13 more wars in Asia and the Pacific than in Central and South America.

Suppose that we have data for Asia and the Pacific for the years 1990–1993 only. The domain of f under these conditions is $\{1990, 1991, 1992, 1993\}$. Further suppose that we have data for Central and South America for the years 1992–1995 only. The domain of g under these conditions is $\{1992, 1993, 1994, 1995\}$. We can add or subtract f and g only for the values of x that are in the domain of f *and* in the domain of g. These common values of x are 1992 and 1993. Functions can be combined only if there are numbers that are common to the domains of both functions.

We see that functions, like numbers, can be added and subtracted. Functions can also be multiplied or divided. Because functions are usually given as equations, we perform these operations by performing operations with the algebraic expressions that appear on the right side of the equations. For example, we can combine the following two functions using addition:

$$f(x) = 2x + 1 \quad \text{and} \quad g(x) = x^2 - 4.$$

To do so, we add the terms to the right of the equal sign for $f(x)$ to the terms to the right of the equal sign for $g(x)$. Here is how it's done:

$$
\begin{aligned}
(f + g)(x) &= f(x) + g(x) \\
&= (2x + 1) + (x^2 - 4) \quad &&\text{Add terms for f(x) and g(x).} \\
&= 2x - 3 + x^2 \quad &&\text{Combine like terms.} \\
&= x^2 + 2x - 3 \quad &&\text{Arrange terms in descending powers of x.}
\end{aligned}
$$

The name of this new function is $f + g$. Thus, the sum $f + g$ is the function defined by $(f + g)(x) = x^2 + 2x - 3$. The domain of $f + g$ consists of the numbers x that are in the domain of f and in the domain of g. Because neither f nor g contains division or even roots, the domain of each function is the set of all real numbers. Thus, the domain of $f + g$ is also the set of all real numbers.

EXAMPLE 9 Finding the Sum of Two Functions

Let $f(x) = x^2 - 3$ and $g(x) = 4x + 5$. Find

  a. $(f + g)(x)$ **b.** $(f + g)(3)$

Solution

  a. $(f + g)(x) = f(x) + g(x) = (x^2 - 3) + (4x + 5) = x^2 + 4x + 2$. Thus, $(f + g)(x) = x^2 + 4x + 2$.

  b. We find $(f + g)(3)$ by substituting 3 for x in the equation for $f + g$.

$$(f + g)(x) = x^2 + 4x + 2 \quad \text{This is the equation for } f + g.$$

Substitute 3 for x.

$$(f + g)(3) = 3^2 + 4 \cdot 3 + 2 = 9 + 12 + 2 = 23$$

Check Point 9

Let $f(x) = 3x^2 + 4x - 1$ and $g(x) = 2x + 7$. Find

  a. $(f + g)(x)$ **b.** $(f + g)(4)$

Here is a general definition for function addition.

The Sum of Functions

Let f and g be two functions. The **sum $f + g$** is the function defined by

$$(f + g)(x) = f(x) + g(x).$$

The domain of $f + g$ is the set of all real numbers that are common to the domain of f and the domain of g.

EXAMPLE 10 Adding Functions and Determining the Domain

Let $f(x) = \sqrt{x + 3}$ and $g(x) = \sqrt{x - 2}$. Find:

a. $(f + g)(x)$ **b.** the domain of $f + g$

Solution

a. $(f + g)(x) = f(x) + g(x) = \sqrt{x + 3} + \sqrt{x - 2}$.

b. The domain of $f + g$ is the set of all real numbers that are common to the domain of f and the domain of g. Thus, we must find the domains of f and g. We will do so for f first.

Note that $f(x) = \sqrt{x + 3}$ is a function involving the square root of $x + 3$. Because the square root of a negative quantity is not a real number, the value of $x + 3$ must be nonnegative. Thus, the domain of f is all x such that $x + 3 \geq 0$. Equivalently, the the domain is $\{x \mid x \geq -3\}$, or $[-3, \infty)$.

Likewise, $g(x) = \sqrt{x - 2}$ is also a square root function. Because the square root of a negative quantity is not a real number, the value of $x - 2$ must be nonnegative. Thus, the domain of g is all x such that $x - 2 \geq 0$. Equivalently, the domain is $\{x \mid x \geq 2\}$, or $[2, \infty)$.

Now, we can use a number line to determine the the domain of $f + g$. Figure 2.49 shows the domain of f in blue and the domain of g in red. Can you see that all real numbers greater than or equal to 2 are common to both domains? This is shown in purple on the number line. Thus, the domain of $f + g$ is $[2, \infty)$.

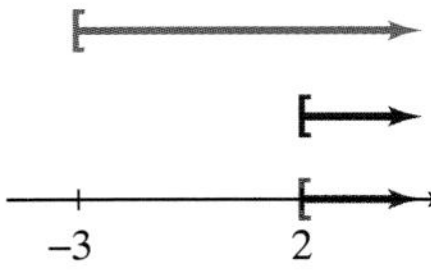

Domain of f
Domain of g
Domain of $f + g$

−3   2

Figure 2.49 Finding the domain of the sum $f + g$

Technology

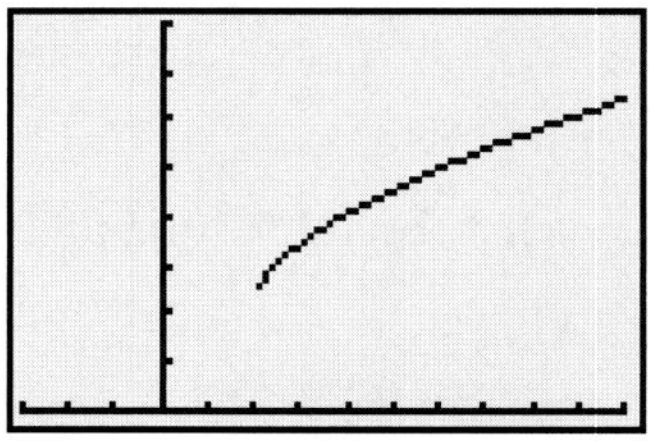

The graph on the left is the graph of

$$y = \sqrt{x + 3} + \sqrt{x - 2}$$

in a $[-3, 10, 1]$ by $[0, 8, 1]$ viewing rectangle. The graph reveals what we discovered algebraically in Example 10(b). The domain of this function is $[2, \infty)$.

Check Point 10

Let $f(x) = \sqrt{x - 3}$ and $g(x) = \sqrt{x + 1}$. Find:

a. $(f + g)(x)$ **b.** the domain of $f + g$

We can also combine functions using subtraction, multiplication, and division by performing operations with the algebraic expressions that appear on the right side of the equations. For example, the functions $f(x) = x + 3$ and

$g(x) = x - 1$ can be combined to form the difference, product, and quotient of f and g. Here's how it's done.

Difference: f − g
$$(f - g)(x) = f(x) - g(x)$$
$$= x + 3 - (x - 1) = x + 3 - x + 1 = 4$$

Product: fg
$$(fg)(x) = f(x) \cdot g(x)$$
$$= (x + 3)(x - 1) = x^2 + 2x - 3$$

Quotient: $\frac{f}{g}$
$$\left(\frac{f}{g}\right)(x) = \frac{f(x)}{g(x)} = \frac{x + 3}{x - 1}, \quad x \neq 1$$

Just like the domain for $f + g$, the domain for each of these functions consists of all real numbers that are common to the domains of f and g. In the case of the quotient function $\dfrac{f(x)}{g(x)}$, we must remember not to divide by 0, so we add the further restriction that $g(x) \neq 0$.

The following definitions summarize our discussion.

Definitions: Sum, Difference, Product, and Quotient of Functions

Let f and g be two functions. The **sum** $f + g$, the **difference** $f - g$, the **product** fg, and the **quotient** $\frac{f}{g}$ are functions whose domains are the set of all real numbers common to the domains of f and g, defined as follows:

1. Sum: $\qquad (f + g)(x) = f(x) + g(x)$

2. Difference: $\quad (f - g)(x) = f(x) - g(x)$

3. Product: $\qquad (fg)(x) = f(x) \cdot g(x)$

4. Quotient: $\qquad \left(\dfrac{f}{g}\right)(x) = \dfrac{f(x)}{g(x)}$, provided $g(x) \neq 0$

EXAMPLE 11   Combining Functions

If $f(x) = 2x - 1$ and $g(x) = x^2 + x - 2$, find:

a. $(f - g)(x)$ **b.** $(fg)(x)$ **c.** $\left(\frac{f}{g}\right)(x)$

Determine the domain for each function.

Solution

a. $(f - g)(x) = f(x) - g(x)$ This is the definition of the difference $f - g$.

$\qquad = (2x - 1) - (x^2 + x - 2)$ Subtract $g(x)$ from $f(x)$.

$\qquad = 2x - 1 - x^2 - x + 2$ Perform the subtraction.

$\qquad = -x^2 + x + 1$ Combine like terms and arrange terms in descending powers of x.

b. $(fg)(x) = (2x - 1)(x^2 + x - 2)$ *This is the definition of the product fg.*

$\qquad\qquad = 2x(x^2 + x - 2) - 1(x^2 + x - 2)$ *Multiply each term in the second factor by 2x and −1, respectively.*

$\qquad\qquad = 2x^3 + 2x^2 - 4x - x^2 - x + 2$ *Use the distributive property.*

$\qquad\qquad = 2x^3 + (2x^2 - x^2) + (-4x - x) + 2$ *Rearrange terms so that like terms are adjacent.*

$\qquad\qquad = 2x^3 + x^2 - 5x + 2$ *Combine like terms.*

c. $\left(\dfrac{f}{g}\right)(x) = \dfrac{f(x)}{g(x)}$ *This is the definition of the quotient $\dfrac{f}{g}$.*

$\qquad\qquad = \dfrac{2x - 1}{x^2 + x - 2}$ *Divide the algebraic expressions for $f(x)$ and $g(x)$.*

Because the equations for f and g do not involve division or contain even roots, the domain of both f and g is the set of all real numbers. Thus, the domain of $f - g$ and fg is the set of all real numbers. However, for $\frac{f}{g}$, the denominator cannot equal zero. We can factor the denominator as follows:

$$\left(\frac{f}{g}\right)(x) = \frac{2x - 1}{x^2 + x - 2} = \frac{2x - 1}{(x + 2)(x - 1)}$$

Because $x + 2 \neq 0$, $x \neq -2$.   Because $x - 1 \neq 0$, $x \neq 1$.

We see that the domain for $\dfrac{f}{g}$ is the set of all real numbers except -2 and 1: $\{x \mid x \neq -2 \text{ and } x \neq 1\}$.

Check Point 11

If $f(x) = x - 5$ and $g(x) = x^2 - 1$, find:

a. $(f - g)(x)$ **b.** $(fg)(x)$ **c.** $\left(\dfrac{f}{g}\right)(x)$

Determine the domain for each function.

EXERCISE SET 2.5

Practice Exercises

In Exercises 1–10, begin by graphing the standard quadratic function, $f(x) = x^2$. Then use transformations of this graph to graph the given function.

1. $g(x) = x^2 - 2$

2. $g(x) = x^2 - 1$

3. $g(x) = (x - 2)^2$

4. $g(x) = (x - 1)^2$

5. $h(x) = -(x - 2)^2$

6. $h(x) = -(x - 1)^2$

7. $h(x) = (x - 2)^2 + 1$

8. $h(x) = (x - 1)^2 + 2$

9. $g(x) = 2(x - 2)^2$

10. $g(x) = \frac{1}{2}(x - 1)^2$

In Exercises 11–22, begin by graphing the square root function, $f(x) = \sqrt{x}$. Then use transformations of this graph to graph the given function.

11. $g(x) = \sqrt{x} + 2$

12. $g(x) = \sqrt{x} + 1$

13. $g(x) = \sqrt{x + 2}$

14. $g(x) = \sqrt{x + 1}$

15. $h(x) = -\sqrt{x + 2}$

16. $h(x) = -\sqrt{x + 1}$

17. $h(x) = \sqrt{-x + 2}$

18. $h(x) = \sqrt{-x + 1}$

19. $g(x) = \frac{1}{2}\sqrt{x + 2}$

20. $g(x) = 2\sqrt{x + 1}$

21. $h(x) = \sqrt{x + 2} - 2$

22. $h(x) = \sqrt{x + 1} - 1$

In Exercises 23–34, begin by graphing the absolute value function, $f(x) = |x|$. Then use transformations of this graph to graph the given function.

23. $g(x) = |x| + 4$

24. $g(x) = |x| + 3$

25. $g(x) = |x + 4|$

26. $g(x) = |x + 3|$

27. $h(x) = |x + 4| - 2$

28. $h(x) = |x + 3| - 2$

29. $h(x) = -|x + 4|$

30. $h(x) = -|x + 3|$

31. $g(x) = -|x + 4| + 1$

32. $g(x) = -|x + 4| + 2$

33. $h(x) = 2|x + 4|$

34. $h(x) = 2|x + 3|$

In Exercises 35–44, begin by graphing the standard cubic function, $f(x) = x^3$. Then use transformations of this graph to graph the given function.

35. $g(x) = x^3 - 3$

36. $g(x) = x^3 - 2$

37. $g(x) = (x - 3)^3$

38. $g(x) = (x - 2)^3$

39. $h(x) = -x^3$

40. $h(x) = -(x - 2)^3$

41. $h(x) = \frac{1}{2}x^3$

42. $h(x) = \frac{1}{4}x^3$

43. $r(x) = (x - 3)^3 + 2$

44. $r(x) = (x - 2)^3 + 1$

In Exercises 45–52, use the graph of the function f to sketch the graph of the given function g.

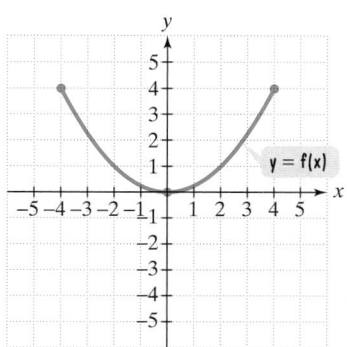

45. $g(x) = f(x) + 1$

46. $g(x) = f(x) + 2$

47. $g(x) = f(x + 1)$

48. $g(x) = f(x + 2)$

49. $g(x) = -f(x)$

50. $g(x) = \frac{1}{2}f(x)$

51. $g(x) = \frac{1}{2}f(x + 1)$

52. $g(x) = -f(x + 2)$

53. If $f(x) = 2x^2 - 5$ and $g(x) = 3x + 7$, find:
  a. $(f + g)(x)$ **b.** $(f + g)(4)$.

54. If $f(x) = 3x^2 - 2x + 1$ and $g(x) = 4x - 1$, find:
  a. $(f + g)(x)$ **b.** $(f + g)(5)$.

55. Let $f(x) = \sqrt{x - 6}$ and $g(x) = \sqrt{x + 2}$. Find:
  a. $(f + g)(x)$ **b.** the domain of $f + g$.

56. Let $f(x) = \sqrt{x - 8}$ and $g(x) = \sqrt{x + 5}$. Find:
  a. $(f + g)(x)$ **b.** the domain of $f + g$.

In Exercises 57–68, find $f + g, f - g, fg,$ and $\frac{f}{g}$. Determine the domain for each function.

57. $f(x) = 2x + 3, \quad g(x) = x - 1$

58. $f(x) = 3x - 4, \quad g(x) = x + 2$

59. $f(x) = x - 5, \quad g(x) = 3x^2$

60. $f(x) = x - 6, \quad g(x) = 5x^2$

61. $f(x) = 2x^2 - x - 3, \quad g(x) = x + 1$

62. $f(x) = 6x^2 - x - 1, \quad g(x) = x - 1$

63. $f(x) = \sqrt{x}, \quad g(x) = x - 4$

64. $f(x) = \dfrac{1}{x}, \quad g(x) = x - 5$

65. $f(x) = 2 + \dfrac{1}{x}, \quad g(x) = \dfrac{1}{x}$

66. $f(x) = 6 - \dfrac{1}{x}, \quad g(x) = \dfrac{1}{x}$

67. $f(x) = \sqrt{x + 4}, \quad g(x) = \sqrt{x - 1}$

68. $f(x) = \sqrt{x + 6}, \quad g(x) = \sqrt{x - 3}$

⭐ Application Exercises

Consider the following functions:

$f(x) =$ population of the world's more-developed regions in year x

$g(x) =$ the population of the world's less-developed regions in year x

$h(x) =$ total world population in year x.

Use these functions and the graph shown to answer Exercises 69–72.

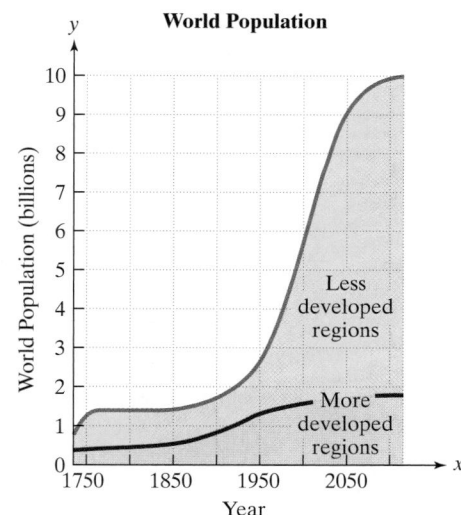

World Population

Source: Population Reference Bureau

69. What does the function $f + g$ represent?

70. What does the function $h - g$ represent?

71. Use the graph to estimate $(f + g)(2000)$.

72. Use the graph to estimate $(h - g)(2000)$.

73. A company that sells radios has yearly fixed costs of $600,000. It costs the company $45 to produce each radio. Each radio will sell for $65. The company's costs and revenue are modeled by the functions

$$C(x) = 600,000 + 45x \qquad \text{This function models the company's costs.}$$

$$R(x) = 65x \qquad \text{This function models the company's revenue.}$$

Find and interpret $(R - C)(20,000)$, $(R - C)(30,000)$ and $(R - C)(40,000)$.

74. The function $f(t) = -0.14t^2 + 0.51t + 31.6$ models the U.S. population, $f(t)$, in millions, ages 65 and older t years after 1990. The function $g(t) = 0.54t^2 + 12.64t + 107.1$ models the total yearly cost of Medicare, $g(t)$, in billions of dollars, t years after 1990.

a. What does the function $\dfrac{g}{f}$ represent?

b. Find and interpret $\dfrac{g}{f}(10)$.

75. Consider two functions M and F that represent the number of male and female members of the House of Representatives for the years 1977, 1981, 1991, 1994, and 1999. Sketch the graphs of M and F in the same rectangular coordinate system, using the data in the bar graph. Each graph should consist of five points whose first coordinates are the years and second coordinates are the numbers of representatives, male or female. Now, add to the graphs in your coordinate system the graph of $M + F$. What constant function do you obtain? What is the significance of this constant?

Gender Breakdown of the House of Representatives for Five Selected Years

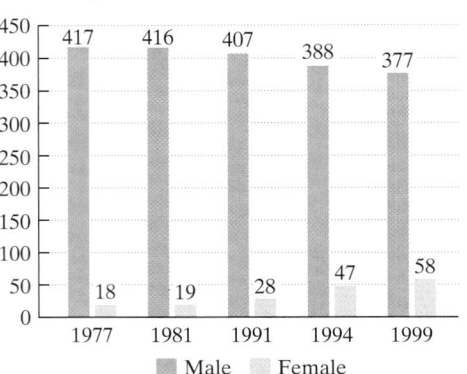

76. A department store has two locations in a city. From 1998 through 2002, the profits for each of the store's two branches are modeled by the functions $f(x) = -0.44x + 13.62$ and $g(x) = 0.51x + 11.14$. In each model, x represents the number of years after 1998 and f and g represent the profit in millions of dollars.

a. What is the slope for f? Describe what this means.

b. What is the slope for g? Describe what this means.

c. Find $f + g$. What is the slope for this function? What does this mean?

Writing in Mathematics

77. What must be done to a function's equation so that its graph is shifted vertically upward?

78. What must be done to a function's equation so that its graph is shifted horizontally to the right?

79. What must be done to a function's equation so that its graph is reflected about the x-axis?

80. What must be done to a function's equation so that its graph is reflected about the y-axis?

81. What must be done to a function's equation so that its graph is stretched?

82. If the equations of two functions are given, explain how to obtain the quotient function and its domain.

83. A company's profit is given by the function $y = P(x)$, where x represents the amount spent on advertising and P represents weekly profits, both expressed in hundreds of dollars.

a. Describe a situation that might occur in the company that would result in the graph of its profit function undergoing a vertical shift.

b. Now, consider the function $y = D(x)$, where x represents the amount spent on advertising and D represents weekly profits, both expressed in dollars rather than hundreds of dollars. If D and P are both graphed on the same axes, describe the relationship between the two graphs.

Technology Exercises

84. a. Use a graphing utility to graph $f(x) = x^2 + 1$.

b. Graph $f(x) = x^2 + 1$, $g(x) = f(2x)$, $h(x) = f(3x)$, and $k(x) = f(4x)$ on the same viewing rectangle.

c. Describe the relationship among the graphs of f, g, h, and k with emphasis on different values of x for points on all four graphs that give the same y-coordinate.

d. Generalize by describing the relationship between the graph of f and the graph of g, where $g(x) = f(cx)$ for $c > 1$.

e. Try out your generalization by sketching the graphs of $f(cx)$ for $c = 1$, $c = 2$, $c = 3$, and $c = 4$ for a function of your choice.

85. a. Use a graphing utility to graph $f(x) = x^2 + 1$.

b. Graph $f(x) = x^2 + 1$, and $g(x) = f(\frac{1}{2}x)$, and $h(x) = f(\frac{1}{4}x)$ on the same viewing rectangle.

c. Describe the relationship among the graphs of f, g, and h with emphasis on different values of x for points on all three graphs that give the same y-coordinate.

d. Generalize by describing the relationship between the graph of f and the graph of g, where $g(x) = f(cx)$ for $0 < c < 1$.

e. Try out your generalization by sketching the graphs of $f(cx)$ for $c = 1$, and $c = \frac{1}{2}$, and $c = \frac{1}{4}$ for a function of your choice.

Critical Thinking Exercises

86. Which one of the following is true?

a. If $f(x) = |x|$ and $g(x) = |x + 3| + 3$, then the graph of g is a translation of three units to the right and three units upward of the graph of f.

b. If $f(x) = -\sqrt{x}$ and $g(x) = \sqrt{-x}$, then f and g have identical graphs.

c. If $f(x) = x^2$ and $g(x) = 5(x^2 - 2)$, then the graph of g can be obtained from the graph of f by stretching f five units followed by a downward shift of two units.

d. If $f(x) = x^3$ and $g(x) = -(x - 3)^3 - 4$, then the graph of g can be obtained from the graph of f by moving f three units to the right, reflecting in the x-axis, and then moving the resulting graph down four units.

In Exercises 87–90, functions f and g are graphed in the same rectangular coordinate system. If g is obtained from f through a sequence of transformations, find an equation for g.

87.

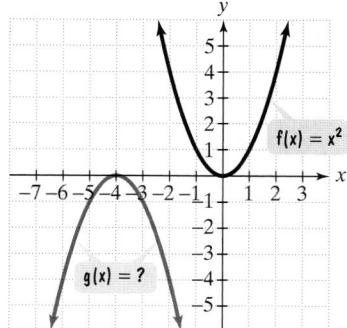

88.

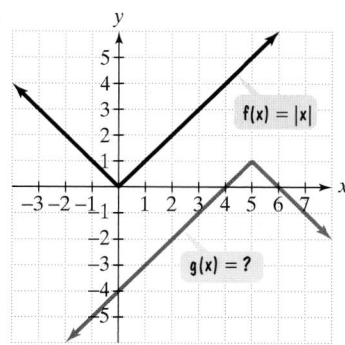

89.

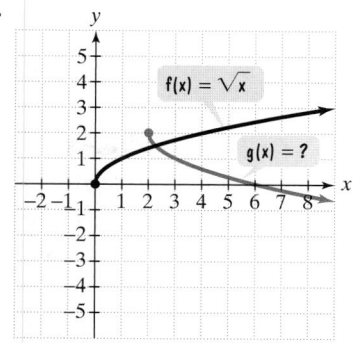

90.

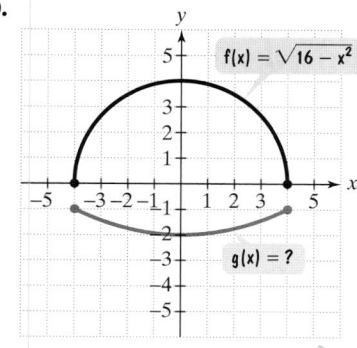

For Exercises 91–94, assume that (a, b) is a point on the graph of f. What is the corresponding point on the graph of each of the following functions?

91. $y = f(-x)$

92. $y = 2f(x)$

93. $y = f(x - 3)$

94. $y = f(x) - 3$

 Group Exercises

95. Consult an almanac, newspaper, magazine, or the Internet to find data displayed in a bar graph in the style of Figure 2.48 on page 230. Using the two bar graphs that group members find most interesting, introduce two or more functions that are related to the graphs. Then write and solve a problem involving function addition and function subtraction for each selected graph. If you are not sure where to begin, reread page 230 or look at Exercises 69–72 in this exercise set.

96. This activity is a group research project on morphing and should result in a presentation made by group members to the entire class. Be sure to include morphing images that will intrigue class members. You should have no problem finding an array of fascinating images online. Also include a discussion of films using spectacular morphing effects. Rent videos of these films and show appropriate excerpts.

SECTION 2.6 Composite and Inverse Functions

Objectives

1. Form composite functions.
2. Verify inverse functions.
3. Find the inverse of a function.
4. Use the horizontal line test to determine if a function has an inverse function.
5. Use the graph of a one-to-one function to graph its inverse function.

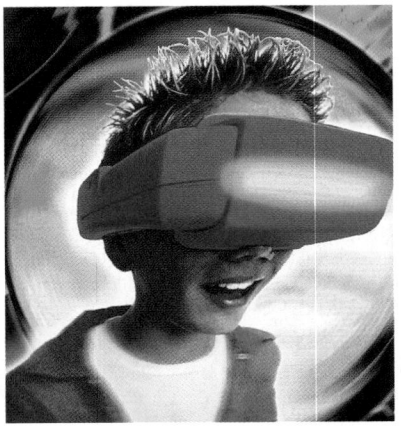

The time: the not-too-distant future. Your computer and its two-dimensional monitor have been replaced by a virtual reality system. You put on a headset and a virtual reality sensory suit. Suddenly you are skiing down a mountain. You feel the wind in your hair, the frost on your eyebrows, and the gentle heat of the sun on your face. Your heart races as you leap off cliffs, feeling the shudder through your ski boots as you race down the mountain.

Would you like to purchase a computer capable of helping you engage in virtual skiing? Luckily, your local computer store is having a sale right now. The models that are on sale cost either $300 less than the regular price or 85% of the regular price. If x represents the computer's regular price, both discounts can be described with the following functions.

$$f(x) = x - 300 \qquad\qquad g(x) = 0.85x$$

The computer is on sale for $300 less than its regular price.

The computer is on sale for 85% of its regular price.

At the store, you bargain with the salesperson. Eventually, she makes an offer you can't refuse: The sale price is 85% of the regular price followed by a $300 reduction:

$$0.85x - 300$$

85% of the regular price

followed by a $300 reduction

In terms of functions f and g, this offer can be obtained by taking the output of $g(x) = 0.85x$, namely $0.85x$, and using it as the input of f:

$$f(x) = x - 300$$

Replace x by 0.85 x, the output of g(x) = 0.85x.

$$f(0.85x) = 0.85x - 300.$$

Because $0.85x$ is $g(x)$, we can write this last equation as

$$f(g(x)) = 0.85x - 300.$$

We read this equation as "f of g of x is equal to $0.85x - 300$." We call $f(g(x))$ the **composition of the function f with g**, or a **composite function**. This composite function is written $f \circ g$. Thus,

$$(f \circ g)(x) = f(g(x)) = 0.85x - 300.$$

Like all functions, we can evaluate $f \circ g$ for a specified value of x in the function's domain. For example, here's how to find the value of this function at 1400:

$$(f \circ g)(x) = 0.85x - 300 \qquad \text{This composite function describes the offer you cannot refuse.}$$

Replace x by 1400.

$$(f \circ g)(1400) = 0.85(1400) - 300 = 1190 - 300 = 890.$$

This means that a computer that regularly sells for $1400 is on sale for $890 subject to both discounts.

In this section, we will focus on the composition of two functions. We will also study functions whose composition have a special relationship.

Before you run out to buy a new computer, let's generalize our discussion of the computer's double discount and define the composition of any two functions.

1 Form composite functions.

The Composition of Functions

The **composition of the function f with g** is denoted by $f \circ g$ and is defined by the equation

$$(f \circ g)(x) = f(g(x)).$$

The domain of the **composite function $f \circ g$** is the set of all x such that

1. x is in the domain of g and
2. $g(x)$ is in the domain of f.

The composition of f with g, $f \circ g$, is pictured as a machine with inputs and outputs in Figure 2.50. The diagram indicates that the output of g, or $g(x)$, becomes the input for "machine" f. If $g(x)$ is not in the domain of f, it cannot be input into machine f, and so $g(x)$ must be discarded.

Inputs, x

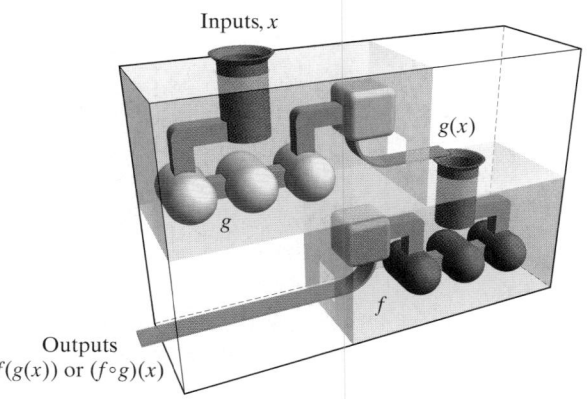

$g(x)$

g

f

Outputs
$f(g(x))$ or $(f \circ g)(x)$

Figure 2.50 Inputting one function into a second function

240 • Chapter 2 • Functions and Graphs

EXAMPLE 1   Forming Composite Functions

Given $f(x) = 3x - 4$ and $g(x) = x^2 + 6$, find:

a. $(f \circ g)(x)$ **b.** $(g \circ f)(x)$.

Solution

a. We begin with $(f \circ g)(x)$, the composition of f with g. Because $(f \circ g)(x)$ means $f(g(x))$, we must replace each occurrence of x in the equation for f by $g(x)$.

$$f(x) = 3x - 4 \qquad \text{This is the given equation for } f.$$

Replace x by $g(x)$.

$$(f \circ g)(x) = f(g(x)) = 3g(x) - 4$$
$$= 3(x^2 + 6) - 4 \qquad \text{Because } g(x) = x^2 + 6, \text{ replace } g(x) \text{ with } x^2 + 6$$
$$= 3x^2 + 18 - 4 \qquad \text{Use the distributive property.}$$
$$= 3x^2 + 14 \qquad \text{Simplify.}$$

Thus, $(f \circ g)(x) = 3x^2 + 14$.

b. Next, we find $(g \circ f)(x)$, the composition of g with f. Because $(g \circ f)(x)$ means $g(f(x))$, we must replace each occurrence of x in the equation for g by $f(x)$.

$$g(x) = x^2 + 6 \qquad \text{This is the given equation for } g.$$

Replace x by $f(x)$.

$$(g \circ f)(x) = g(f(x)) = (f(x))^2 + 6$$
$$= (3x - 4)^2 + 6 \qquad \text{Because } f(x) = 3x - 4, \text{ replace } f(x) \text{ with } 3x - 4.$$
$$= 9x^2 - 24x + 16 + 6 \qquad \text{Use } (A - B)^2 = A^2 - 2AB + B^2 \text{ to square } 3x - 4.$$
$$= 9x^2 - 24x + 22 \qquad \text{Simplify.}$$

Thus, $(g \circ f)(x) = 9x^2 - 24x + 22$. Notice that $(f \circ g)(x)$ is not the same function as $(g \circ f)(x)$.

Check Point 1   Given $f(x) = 5x + 6$ and $g(x) = x^2 - 1$, find:

a. $(f \circ g)(x)$ **b.** $(g \circ f)(x)$.

Inverse Functions

Here are two functions that describe situations related to the price of a computer, x:

$$f(x) = x - 300 \qquad g(x) = x + 300.$$

Function f subtracts $300 from the computer's price and function g adds $300 to the computer's price. Let's see what $f(g(x))$ does. Put $g(x)$ into f:

$$f(x) = x - 300 \qquad \text{This is the given equation for } f.$$

Replace x by g(x).

$$f(g(x)) = g(x) - 300$$
$$= x + 300 - 300 \qquad \text{Because } g(x) = x + 300, \text{ replace } g(x) \text{ by } x + 300.$$
$$= x \qquad \text{This is the computer's original price.}$$

Using $f(g(x))$, the computer's price, x, went through two changes: the first, an increase; the second, a decrease:

$$x + 300 - 300.$$

The final price of the computer, x, is identical to its starting price, x.

In general, if the changes made to x by function g are undone by the changes made by function f, then

$$f(g(x)) = x.$$

Assume, also, that this "undoing" takes place in the other direction:

$$g(f(x)) = x.$$

Under these conditions, we say that each function is the **inverse function** of the other. The fact that g is the inverse of f is expressed by renaming g as f^{-1}, read "f-inverse." For example, the inverse functions

$$f(x) = x - 300 \qquad g(x) = x + 300$$

are usually named as follows:

$$f(x) = x - 300 \qquad f^{-1}(x) = x + 300.$$

With these ideas in mind, we present the formal definition of the inverse of a function.

Study Tip

The notation f^{-1} represents the inverse function of f. The -1 is *not* an exponent. The notation f^{-1} does *not* mean $\frac{1}{f}$:

$$f^{-1} \neq \frac{1}{f}$$

Definition of the Inverse of a Function

Let f and g be two functions such that

$$f(g(x)) = x \qquad \text{for every } x \text{ in the domain of } g$$

and

$$g(f(x)) = x \qquad \text{for every } x \text{ in the domain of } f.$$

The function g is the **inverse** of the function f, and is denoted by f^{-1} (read "f-inverse"). Thus, $f(f^{-1}(x)) = x$ and $f^{-1}(f(x)) = x$. The domain of f is equal to the range of f^{-1}, and vice versa.

2 Verify inverse functions.

EXAMPLE 2 Verifying Inverse Functions

Show that each function is an inverse of the other:

$$f(x) = 5x \qquad \text{and} \qquad g(x) = \frac{x}{5}.$$

Solution To show that f and g are inverses of each other, we must show that $f(g(x)) = x$ and $g(f(x)) = x$. We begin with $f(g(x))$.

$$f(x) = 5x \qquad \text{This is the given equation for f.}$$

Replace x by g(x).

$$f(g(x)) = 5g(x) = 5\left(\frac{x}{5}\right) = x$$

Next, we find $g(f(x))$.

$$g(x) = \frac{x}{5} \qquad \text{This is the given equation for g.}$$

Replace x by f(x).

$$g(f(x)) = \frac{f(x)}{5} = \frac{5x}{5} = x$$

Because g is the inverse of f (and vice versa), we can use inverse notation and write

$$f(x) = 5x \qquad \text{and} \qquad f^{-1}(x) = \frac{x}{5}.$$

Notice how f^{-1} undoes the change produced by f: f changes x by multiplying by 5 and f^{-1} undoes this by dividing by 5.

Check Point 2 Show that each function is an inverse of the other:

$$f(x) = 7x \qquad \text{and} \qquad g(x) = \frac{x}{7}$$

EXAMPLE 3 Verifying Inverse Functions

Show that each function is an inverse of the other:

$$f(x) = 3x + 2 \qquad \text{and} \qquad g(x) = \frac{x - 2}{3}.$$

Solution To show that f and g are inverses of each other, we must show that $f(g(x)) = x$ and $g(f(x)) = x$. We begin with $f(g(x))$.

$$f(x) = 3x + 2 \qquad \text{This is the equation for f.}$$

Replace x by g(x).

$$f(g(x)) = 3g(x) + 2 = 3\left(\frac{x - 2}{3}\right) + 2 = x - 2 + 2 = x$$

Next, we find $g(f(x))$.

$$g(x) = \frac{x - 2}{3} \qquad \text{This is the equation for g.}$$

Replace x by f(x).

$$g(f(x)) = \frac{f(x) - 2}{3} = \frac{(3x + 2) - 2}{3} = \frac{3x}{3} = x$$

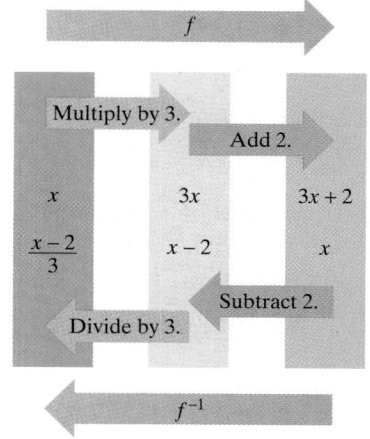

<figure>**Figure 2.51** f^{-1} undoes the changes produced by f.</figure>

3 Find the inverse of a function.

Study Tip

The procedure for finding a function's inverse uses a *switch-and-solve* strategy. Switch x and y, then solve for y.

Because g is the inverse of f (and vice versa), we can use inverse notation and write

$$f(x) = 3x + 2 \quad \text{and} \quad f^{-1}(x) = \frac{x - 2}{3}.$$

Notice how f^{-1} undoes the changes produced by f: f changes x by *multiplying* by 3 and *adding* 2, and f^{-1} undoes this by *subtracting* 2 and *dividing* by 3. This "undoing" process is illustrated in Figure 2.51.

> **Check Point 3** Show that each function is an inverse of the other:
> $$f(x) = 4x - 7 \quad \text{and} \quad g(x) = \frac{x + 7}{4}.$$

Finding the Inverse of a Function

The definition of the inverse of a function tells us that the domain of f is equal to the range of f^{-1}, and vice versa. This means that if the function f is the set of ordered pairs (x, y), then the inverse of f is the set of ordered pairs (y, x). If a function is defined by an equation, we can obtain the equation for f^{-1}, the inverse of f, by interchanging the role of x and y in the equation for function f.

Finding the Inverse of a Function

The equation for the inverse of a function f can be found as follows.

1. Replace $f(x)$ by y in the equation for $f(x)$.
2. Interchange x and y.
3. Solve for y. If this equation does not define y as a function of x, the function f does not have an inverse function and this procedure ends. If this equation does define y as a function of x, the function f has an inverse function.
4. If f has an inverse function, replace y in step 3 by $f^{-1}(x)$. We can verify our result by showing that $f(f^{-1}(x)) = x$ and $f^{-1}(f(x)) = x$.

EXAMPLE 4 Finding the Inverse of a Function

Find the inverse of $f(x) = 7x - 5$.

Solution

Step 1 Replace $f(x)$ by y:
$$y = 7x - 5$$

Step 2 Interchange x and y:
$$x = 7y - 5 \quad \text{This is the inverse function.}$$

Step 3 Solve for y:
$$x + 5 = 7y \quad \text{Add 5 to both sides.}$$
$$\frac{x + 5}{7} = y \quad \text{Divide both sides by 7.}$$

Step 4 Replace y by f⁻¹(x):

$$f^{-1}(x) = \frac{x+5}{7}$$ *The equation is written with f⁻¹ on the left.*

Discovery

In Example 4, we found that if $f(x) = 7x - 5$, then

$$f^{-1}(x) = \frac{x+5}{7}.$$

Verify this result by showing that

$$f(f^{-1}(x)) = x$$

and

$$f^{-1}(f(x)) = x.$$

Thus, $f(x) = 7x - 5$ and $f^{-1}(x) = \dfrac{x+5}{7}$.

The inverse function, f^{-1}, undoes the changes produced by f. f changes x by multiplying by 7 and subtracting 5. f^{-1} undoes this by adding 5 and dividing by 7.

Check Point 4 Find the inverse of $f(x) = 2x + 7$.

EXAMPLE 5 Finding the Equation of the Inverse

Find the inverse of $f(x) = x^3 + 1$.

Solution

Step 1 Replace f(x) with y: $y = x^3 + 1$

Step 2 Interchange x and y: $x = y^3 + 1$

Step 3 Solve for y:
$$x - 1 = y^3$$
$$\sqrt[3]{x-1} = \sqrt[3]{y^3}$$
$$\sqrt[3]{x-1} = y$$

Step 4 Replace y with f⁻¹(x): $f^{-1}(x) = \sqrt[3]{x-1}$.

Thus, the inverse of $f(x) = x^3 + 1$ is $f^{-1}(x) = \sqrt[3]{x-1}$.

Check Point 5 Find the inverse of $f(x) = 4x^3 - 1$.

4 Use the horizontal line test to determine if a function has an inverse function.

The Horizontal Line Test and One-to-One Functions

Let's see what happens if we try to find the inverse of the standard quadratic function $f(x) = x^2$.

Step 1 Replace f(x) with y: $y = x^2$.

Step 2 Interchange x and y: $x = y^2$.

Step 3 Solve for y: We apply the square root method to solve $y^2 = x$ for y. We obtain

$$y = \pm\sqrt{x}.$$

The $\pm$ in this last equation shows that for certain values of x (all positive real numbers), there are two values of y. Because this equation does not represent y as a function of x, the standard quadratic function does not have an inverse function.

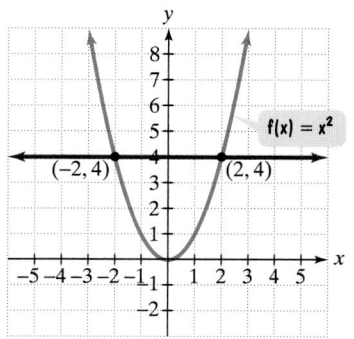

Figure 2.52 The horizontal line intersects the graph twice.

Discovery

How might you restrict the domain of $f(x) = x^2$, graphed in Figure 2.52, so that the remaining portion of the graph passes the horizontal line test?

Can we look at the graph of a function and tell if it represents a function with an inverse? Yes. The graph of the standard quadratic function is shown in Figure 2.52. Four units above the x-axis, a horizontal line is drawn. This line intersects the graph at two of its points, $(-2, 4)$ and $(2, 4)$. Because inverse functions have ordered pairs with the coordinates reversed, let's see what happens if we reverse these coordinates. We obtain $(4, -2)$ and $(4, 2)$. A function provides exactly one output for each input. However, the input 4 is associated with two outputs, -2 and 2. The points $(4, -2)$ and $(4, 2)$ do not define a function.

If any horizontal line, such as the one in Figure 2.52, intersects a graph at two or more points, these points will not define a function when their coordinates are reversed. This suggests the **horizontal line test** for inverse functions.

The Horizontal Line Test For Inverse Functions

A function f has an inverse that is a function, f^{-1}, if there is no horizontal line that intersects the graph of the function f at more than one point.

EXAMPLE 6   Applying the Horizontal Line Test

Which of the following graphs represent functions that have inverse functions?

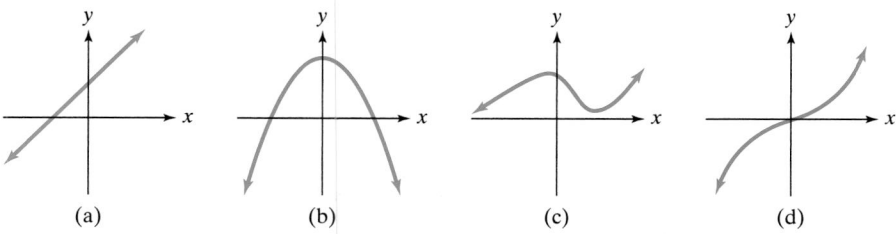

Solution   Notice that horizontal lines can be drawn in parts (b) and (c) that intersect the graphs more than once. These graphs do not pass the horizontal line test. These are not the graphs of functions with inverse functions. By contrast, no horizontal line can be drawn in parts (a) and (d) that intersect the graphs more than once. These graphs pass the horizontal line test. Thus, the graphs in parts (a) and (d) represent functions that have inverse functions.

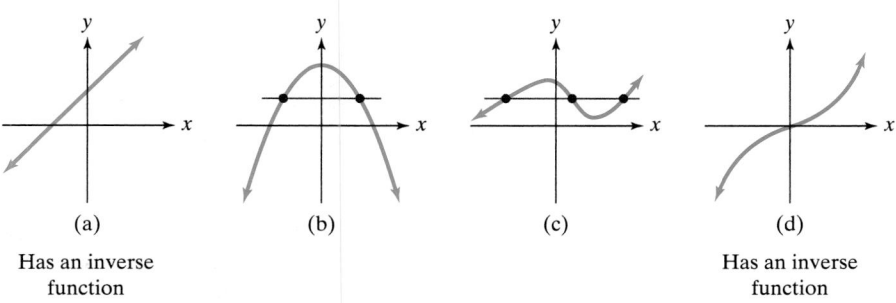

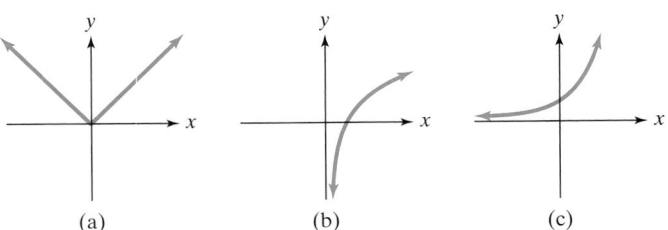

Check Point 6

Which of the following graphs represent functions that have inverse functions?

(a)   (b)   (c)

A function passes the horizontal line test when no two different ordered pairs have the same second component. This means that if $x_1 \neq x_2$, then $f(x_1) \neq f(x_2)$. Such a function is called a **one-to-one function**. Thus, a one-to-one function is a function in which no two different ordered pairs have the same second component. Only one-to-one functions have inverse functions. Any function that passes the horizontal line test is a one-to-one function. Any one-to-one function has a graph that passes the horizontal line test.

5 Use the graph of a one-to-one function to graph its inverse function.

Graphs of f and f^{-1}

There is a relationship between the graph of a one-to-one function f and its inverse f^{-1}. Because inverse functions have ordered pairs with the coordinates reversed, if the point (a, b) is on the graph of f, then the point (b, a) is on the graph of f^{-1}. The points (a, b) and (b, a) are symmetric with respect to the line $y = x$. Thus, **the graph of f^{-1} is a reflection of the graph of f about the line $y = x$.** This is illustrated in Figure 2.53.

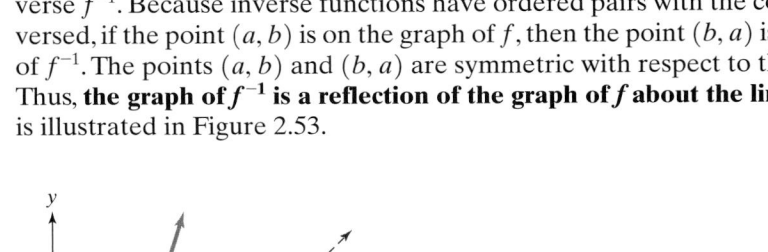
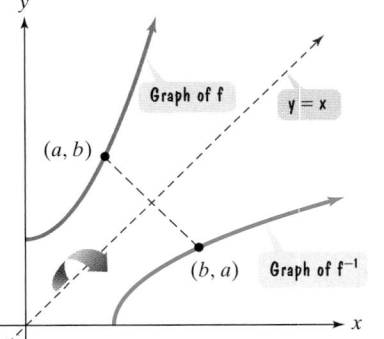

Figure 2.53 The graph of f^{-1} as a reflection of f about $y = x$

EXAMPLE 7   Graphing the Inverse Function

Use the graph of f in Figure 2.54 to draw the graph of its inverse function.

Solution   We begin by noting that no horizontal line intersects the graph of f at more than one point, so f does have an inverse function. Because the points $(-3, -2), (-1, 0)$, and $(4, 2)$ are on the graph of f, the graph of the inverse function, f^{-1}, has points with these ordered pairs reversed. Thus, $(-2, -3), (0, -1)$, and $(2, 4)$ are on the graph of f^{-1}. We can use these points to graph f^{-1}. The graph of f^{-1} is shown in Figure 2.55. Note that the graph of f^{-1} is the reflection of the graph of f about the line $y = x$.

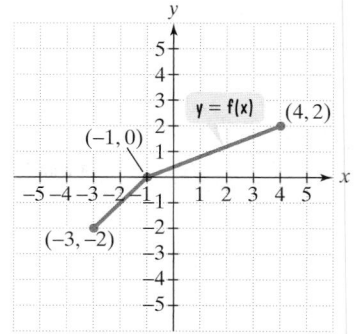

Figure 2.54

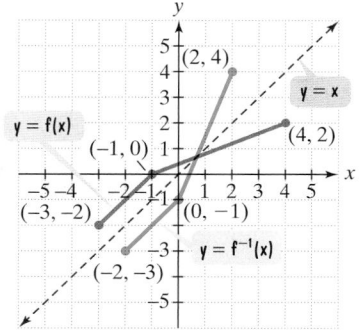

Figure 2.55 The graph of f and f^{-1}

**Check
Point
7**

Use the graph of f in the figure below to draw the graph of its inverse function.

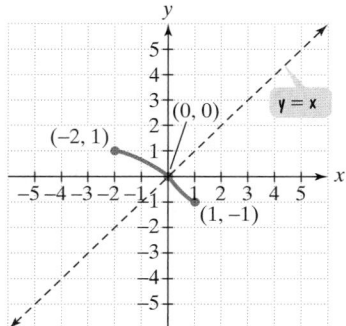

EXERCISE SET 2.6

✓ Practice Exercises

In Exercises 1–14, find
a. $(f \circ g)(x)$;
b. $(g \circ f)(x)$;
c. $(f \circ g)(2)$.

1. $f(x) = 2x$, $g(x) = x + 7$
2. $f(x) = 3x$, $g(x) = x - 5$
3. $f(x) = x + 4$, $g(x) = 2x + 1$
4. $f(x) = 5x + 2$, $g(x) = 3x - 4$
5. $f(x) = 4x - 3$, $g(x) = 5x^2 - 2$
6. $f(x) = 7x + 1$, $g(x) = 2x^2 - 9$
7. $f(x) = x^2 + 2$, $g(x) = x^2 - 2$
8. $f(x) = x^2 + 1$, $g(x) = x^2 - 3$
9. $f(x) = \sqrt{x}$, $g(x) = x - 1$
10. $f(x) = \sqrt{x}$, $g(x) = x + 2$
11. $f(x) = 2x - 3$, $g(x) = \dfrac{x + 3}{2}$

12. $f(x) = 6x - 3$, $g(x) = \dfrac{x + 3}{6}$
13. $f(x) = \dfrac{1}{x}$, $g(x) = \dfrac{1}{x}$
14. $f(x) = \dfrac{1}{x}$, $g(x) = \dfrac{2}{x}$

In Exercises 15–24, find $f(g(x))$ and $g(f(x))$ and determine whether each pair of functions f and g are inverses of each other.

15. $f(x) = 4x$ and $g(x) = \dfrac{x}{4}$

16. $f(x) = 6x$ and $g(x) = \dfrac{x}{6}$

17. $f(x) = 3x + 8$ and $g(x) = \dfrac{x - 8}{3}$

18. $f(x) = 4x + 9$ and $g(x) = \dfrac{x - 9}{4}$

19. $f(x) = 5x - 9$ and $g(x) = \dfrac{x + 5}{9}$

20. $f(x) = 3x - 7$ and $g(x) = \dfrac{x + 3}{7}$

21. $f(x) = \dfrac{3}{x - 4}$ and $g(x) = \dfrac{3}{x} + 4$

22. $f(x) = \dfrac{2}{x - 5}$ and $g(x) = \dfrac{2}{x} + 5$

23. $f(x) = -x$ and $g(x) = -x$

24. $f(x) = \sqrt[3]{x - 4}$ and $g(x) = x^3 + 4$

The functions in Exercises 25–44 are all one-to-one. For each function:

  a. *Find an equation for $f^{-1}(x)$, the inverse function.*
  b. *Verify that your equation is correct by showing that $f(f^{-1}(x)) = x$ and $f^{-1}(f(x)) = x$.*

25. $f(x) = x + 3$ **26.** $f(x) = x + 5$

27. $f(x) = 2x$ **28.** $f(x) = 4x$

29. $f(x) = 2x + 3$ **30.** $f(x) = 3x - 1$

31. $f(x) = x^3 + 2$ **32.** $f(x) = x^3 - 1$

33. $f(x) = (x + 2)^3$ **34.** $f(x) = (x - 1)^3$

35. $f(x) = \dfrac{1}{x}$ **36.** $f(x) = \dfrac{2}{x}$

37. $f(x) = \sqrt{x}$ **38.** $f(x) = \sqrt[3]{x}$

39. $f(x) = x^2 + 1$, for $x \geq 0$

40. $f(x) = x^2 - 1$, for $x \geq 0$

41. $f(x) = \dfrac{2x + 1}{x - 3}$ **42.** $f(x) = \dfrac{2x - 3}{x + 1}$

43. $f(x) = \sqrt[3]{x - 4} + 3$ **44.** $f(x) = x^{3/5}$

Which graphs in Exercises 45–50 represent functions that have inverse functions?

45.
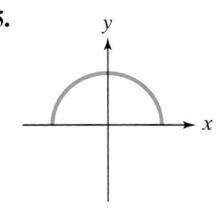

46.
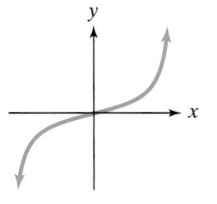

47.
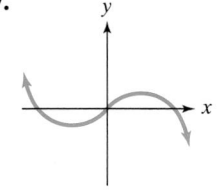

48.
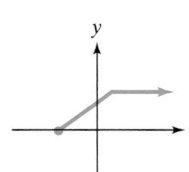

49.
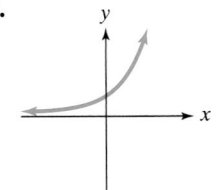

50.
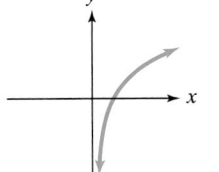

In Exercises 51–54, use the graph of f to draw the graph of its inverse function.

51.
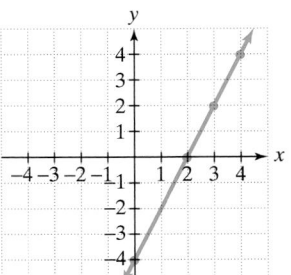

52.
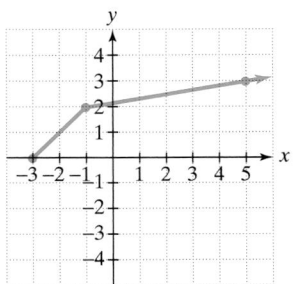

53.
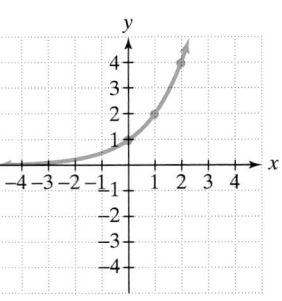

54.
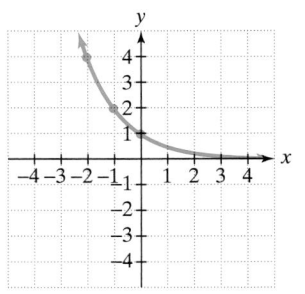

Application Exercises

55. The regular price of a computer is x dollars. Let $f(x) = x - 400$ and $g(x) = 0.75x$.
    a. Describe what the functions f and g model in terms of the price of the computer.
    b. Find $(f \circ g)(x)$ and describe what this models in terms of the price of the computer.
    c. Repeat part (b) for $(g \circ f)(x)$.
    d. Which composite function models the greater discount on the computer, $f \circ g$ or $g \circ f$? Explain.
    e. Find f^{-1} and describe what this models in terms of the price of the computer.

56. The regular price of a pair of jeans is x dollars. Let $f(x) = x - 5$ and $g(x) = 0.6x$.
    a. Describe what functions f and g model in terms of the price of the jeans.
    b. Find $(f \circ g)(x)$ and describe what this models in terms of the price of the jeans.
    c. Repeat part (b) for $(g \circ f)(x)$.
    d. Which composite function models the greater discount on the jeans, $f \circ g$ or $g \circ f$? Explain.
    e. Find f^{-1} and describe what this models in terms of the price of the jeans.

57. The graph represents the probability of two people in the same room sharing a birthday as a function of the number of people in the room. Call the function f.

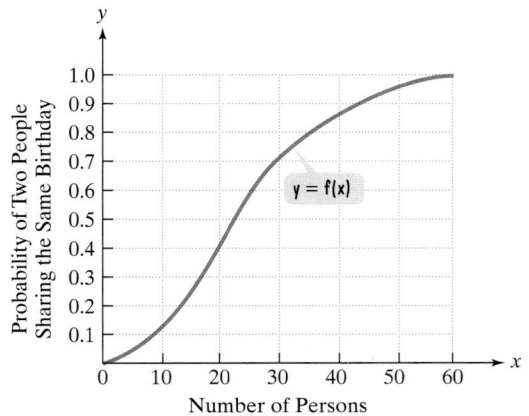

    a. Explain why f has an inverse that is a function.
    b. Describe in practical terms the meaning of $f^{-1}(0.25)$, $f^{-1}(0.5)$, and $f^{-1}(0.7)$.

58. The line graph shown at the top of the next column is based on data from the World Health Organization.
    a. Explain why f has an inverse that is a function.
    b. Describe in practical terms the meaning of $f^{-1}(20)$.

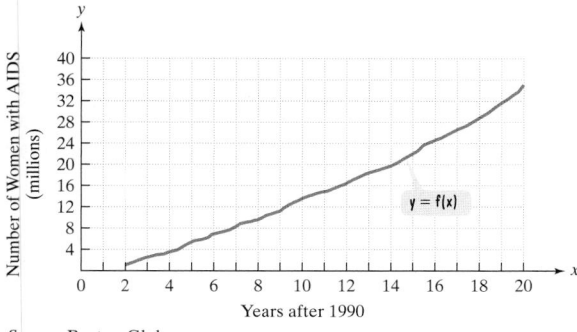

Source: Boston Globe

The graph shows the average age at which women in the United States marry for the first time over a 110-year period. Use the graph to solve Exercises 59–60.

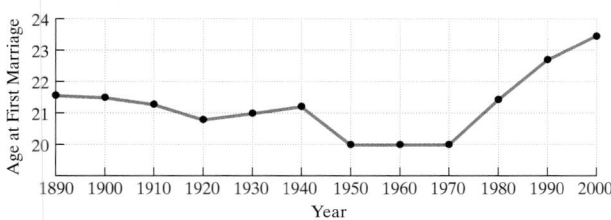

Source: U.S. Census Bureau

59. Does this graph have an inverse that is a function? What does this mean about the average age at which U.S. women marry during the period shown?

60. Identify two or more years in which U.S. women married for the first time at the same average age. What is this average age?

Writing in Mathematics

61. Describe a procedure for finding $(f \circ g)(x)$.

62. Explain how to determine if two functions are inverses of each other.

63. Describe how to find the inverse of a one-to-one function.

64. What is the horizontal line test and what does it indicate?

65. Describe how to use the graph of a one-to-one function to draw the graph of its inverse function.

66. How can a graphing utility be used to visually determine if two functions are inverses of each other?

67. Consider the following function:

(The Beatles, 20),   (Elvis Presley, 18),

(Michael Jackson, 13)

(Mariah Carey, 13),   (The Supremes, 12).

The domain is the set of the five recording artists with the most number 1 singles in the United States. The range is the set of the number of number 1 singles for each artist. Reverse each of the five ordered pairs. Is the resulting relation a function? Describe what this means in terms of whether or not the given function is one-to-one. (*Source: The Popular Music Database*)

Technology Exercises

In Exercises 68–76, use a graphing utility to graph the function. Use the graph to determine whether the function has an inverse that is a function (that is, whether the function is one-to-one).

68. $f(x) = x^2 - 1$

69. $f(x) = \sqrt[3]{2 - x}$

70. $f(x) = \dfrac{x^3}{2}$

71. $f(x) = \dfrac{x^4}{4}$

72. $f(x) = \text{int}(x - 2)$

73. $f(x) = |x - 2|$

74. $f(x) = (x - 1)^3$

75. $f(x) = -\sqrt{16 - x^2}$

76. $f(x) = x^3 + x + 1$

In Exercises 77–79, use a graphing utility to graph f and g in the same viewing rectangle. In addition, graph the line y = x and visually determine if f and g are inverses.

77. $f(x) = 4x + 4, \quad g(x) = 0.25x - 1$

78. $f(x) = \dfrac{1}{x} + 2, \quad g(x) = \dfrac{1}{x - 2}$

79. $f(x) = \sqrt[3]{x} - 2, \quad g(x) = (x + 2)^3$

Critical Thinking Exercises

80. Which one of the following is true?
   a. The inverse of $\{(1, 4), (2, 7)\}$ is $\{(2, 7), (1, 4)\}$.

b. The function $f(x) = 5$ is one-to-one.

c. If $f(x) = 3x$, then $f^{-1}(x) = \dfrac{1}{3x}$.

d. The domain of f is the same as the range of f^{-1}.

81. If $h(x) = \sqrt{3x^2 + 5}$, find functions f and g so that $h(x) = (f \circ g)(x)$.

82. If $f(x) = 3x$ and $g(x) = x + 5$, find $(f \circ g)^{-1}(x)$ and $(g^{-1} \circ f^{-1})(x)$.

83. Show that

$$f(x) = \frac{3x - 2}{5x - 3}$$

is its own inverse.

84. Consider the two functions defined by $f(x) = m_1 x + b_1$ and $g(x) = m_2 x + b_2$. Prove that the slope of the composite function of f with g is equal to the product of the slopes of the two functions.

Group Exercise

85. In Tom Stoppard's play *Arcadia*, the characters dream and talk about mathematics, including ideas involving graphing, composite functions, symmetry, and lack of symmetry in things that are tangled, mysterious, and unpredictable. Group members should rent and view the movie. Present a report on the ideas discussed by the characters that are related to concepts that we studied in this chapter. Bring in a copy of the video and show appropriate excerpts.

CHAPTER SUMMARY, REVIEW, AND TEST

Summary

2.1 Lines and Slope

a. The slope, m, of the line through (x_1, y_1) and (x_2, y_2) is
$$m = \frac{y_2 - y_1}{x_2 - x_1}.$$

b. Equations of lines include point-slope form, $y - y_1 = m(x - x_1)$, slope-intercept form, $y = mx + b$, and general form, $Ax + By + C = 0$. The equation of a horizontal line is $y = b$; a vertical line is $x = a$.

2.2 Parallel and Perpendicular Lines; Circles

a. Parallel lines have equal slopes. Perpendicular lines have slopes that are negative reciprocals.

b. The standard form of the equation of a circle with center (h, k) and radius r is $(x - h)^2 + (y - k)^2 = r^2$.

c. The general form of the equation of a circle is $x^2 + y^2 + Dx + Ey + F = 0$.

d. To convert from the general form to the standard form of a circle's equation, complete the square on x and y.

2.3 Introduction to Functions

a. A relation is any set of ordered pairs. The set of first components is the domain and the set of second components is the range.

b. A function is a correspondence between two sets X (the domain) and Y (the range) that assigns to each element x in the domain exactly one element y in the range. If any element in a relation's domain corresponds to more than one element in the range, the relation is not a function.

c. Functions are usually given in terms of equations involving x and y, in which x is the independent variable and y is the dependent variable. If an equation is solved for y and more than one value of y can be obtained for a given x, then the equation does not define y as a function of x. If an equation defines a function, $f(x)$, the value of the function at x, often replaces y.

d. If a function f does not model data or verbal conditions, its domain is the largest set of real numbers for which the value of $f(x)$ is a real number. Exclude from the function's domain real numbers that cause division by zero and real numbers that result in an even root of a negative number.

2.4 Graphs of Functions

a. The graph of a function is the graph of its ordered pairs.

b. The vertical line test for functions: If any vertical line intersects a graph in more than one point, the graph does not define y as a function of x.

c. A function is increasing on intervals where its graph rises, decreasing on intervals where it falls, and constant on intervals where it neither rises nor falls. Precise definitions are given in the box on page 207.

d. The graph of an even function in which $f(-x) = f(x)$ is symmetric with respect to the y-axis. The graph of an odd function in which $f(-x) = -f(x)$ is symmetric with respect to the origin.

e. Table 2.3 on page 212 shows the graphs of the constant function, $f(x) = c$, the identity function, $f(x) = x$, the standard quadratic function, $f(x) = x^2$, the standard cubic function, $f(x) = x^3$, the square root function, $f(x) = \sqrt{x}$, and the absolute value function, $f(x) = |x|$. The table also lists characteristics of each function.

f. The graph of $f(x) = \text{int}(x)$, where $\text{int}(x)$ is the greatest integer that is less than or equal to x, has function values that form discontinuous steps, shown in Figure 2.32 on page 213. If $n \le x < n + 1$, where n is an integer, then $\text{int}(x) = n$.

2.5 Transformations and Combinations of Functions

a. Table 2.5 on page 228 summarizes how to graph a function using vertical shifts, $y = f(x) \pm c$, horizontal shifts, $y = f(x \pm c)$, reflections about the x-axis, $y = -f(x)$, reflections about the y-axis, $y = f(-x)$, vertical stretching, $y = cf(x)$, $c > 1$, and vertical shrinking, $y = cf(x)$, $0 < c < 1$.

b. A function involving more than one transformation can be graphed in the following order: (1) horizontal shifting; (2) vertical stretching or shrinking; (3) reflecting; (4) vertical shifting.

c. When functions are given as equations, they can be added, subtracted, multiplied, or divided by performing operations with the algebraic expressions that appear on the right side of the equations. Definitions for the sum $f + g$, the difference $f - g$, the product fg, and the quotient $\frac{f}{g}$ functions are given in the box on page 233.

2.6 Composite and Inverse Functions

a. The composition of functions f and g, $f \circ g$, is defined by $(f \circ g)(x) = f(g(x))$. The domain of the composite function $f \circ g$ is given in the box on page 239. This composite function is obtained by replacing each occurrence of x in the equation for f by $g(x)$.

b. If $f(g(x)) = x$ and $g(f(x)) = x$, function g is the inverse of function f, denoted f^{-1} and read "f inverse." Thus, to show that f and g are inverses of each other, one must show $f(g(x)) = x$ and $g(f(x)) = x$.

c. The procedure for finding a function's inverse uses a switch-and-solve strategy. Switch x and y, then solve for y. The procedure is given in the box on page 243.

d. The horizontal line test for inverse functions: A function f has an inverse that is a function, f^{-1}, if there is no horizontal line that intersects the graph of the function f at more than one point.

e. A one-to-one function is one in which no two different ordered pairs have the same second component. Only one-to-one functions have inverse functions.

f. If the point (a, b) is on the graph of f, then the point (b, a) is on the graph of f^{-1}. The graph of f^{-1} is a reflection of the graph of f about the line $y = x$.

Review Exercises

2.1

In Exercises 1–4, find the slope of the line passing through each pair of points or state that the slope is undefined. Then indicate whether the line through the points rises, falls, is horizontal, or is vertical.

1. $(3, 2)$ and $(5, 1)$ **2.** $(-1, -2)$ and $(-3, -4)$

3. $\left(-3, \frac{1}{4}\right)$ and $\left(6, \frac{1}{4}\right)$ **4.** $(-2, 5)$ and $(-2, 10)$

In Exercises 5–6, use the given conditions to write an equation for each line in point-slope form and slope-intercept form.

5. Passing through $(-3, 2)$ with a slope of -6

6. Passing through $(1, 6)$ and $(-1, 2)$

In Exercises 7–10, give the slope and y-intercept of each line whose equation is given. Then graph the line.

7. $y = \frac{2}{5}x - 1$ **8.** $y = -4x + 5$

9. $2x + 3y + 6 = 0$ **10.** $2y - 8 = 0$

11. In 1900, the typical surfboard was 16 feet long. Since then, they have become shorter and shorter. Here are two data measurements for a typical surfboard's length. (A scatter plot of all such data measurements through 1980 would show all data points on or near a straight line.)

x (Years since 1900)	y (Average Surfboard Length, in Feet)
0	16
30	12.1

a. Write the point-slope form of the equation of the line on which these measurements fall.

b. Use the point-slope form of the equation to write the slope-intercept form of the equation.

c. Use the equation in part (b) to find average surfboard length in 1970 and 1980.

d. Does the equation in part (b) reasonably describe reality in 2000?

12. The scatter plot shows the number of minutes each that 16 people exercise per week and the number of headaches per month each person experiences.

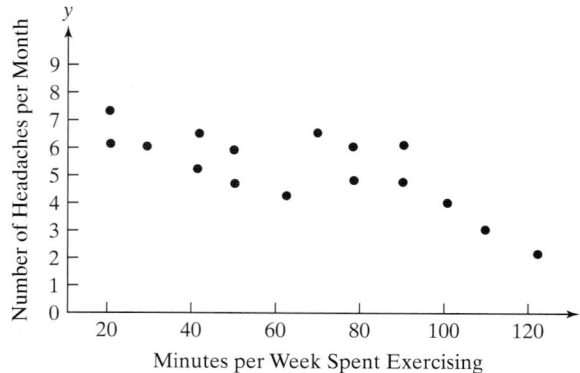

a. Draw a line that fits the data so that the spread of the data points around the line is as small as possible.

b. Use the coordinates of two points along your line to write its point-slope and slope-intercept equations.

c. Use the equation in part (b) to predict the number of headaches per month for a person exercising 130 minutes per week.

2.2

In Exercises 13–14, use the given conditions to write an equation for each line in point-slope form and slope-intercept form.

13. Passing through $(4, -7)$ and parallel to the line whose equation is $3x + y - 9 = 0$.

14. Passing through $(-3, 6)$ and perpendicular to the line whose equation is $y = \frac{1}{3}x + 4$.

In Exercises 15–16, write the standard form of the equation of the circle with the given center and radius.

15. Center $(0, 0), r = 3$ **16.** Center $(-2, 4), r = 6$

In Exercises 17–19, give the center and radius of each circle and graph its equation.

17. $x^2 + y^2 = 1$ **18.** $(x + 2)^2 + (y - 3)^2 = 9$

19. $x^2 + y^2 - 4x + 2y - 4 = 0$

2.3

In Exercises 20–22, determine whether each relation is a function. Give the domain and range for each relation.

20. $\{(2, 7), (3, 7), (5, 7)\}$ **21.** $\{(1, 10) (2, 500), (13, \pi)\}$

22. $\{(12, 13), (14, 15), (12, 19)\}$

In Exercises 23–25, determine whether each equation defines y as a function of x.

23. $2x + y = 8$ **24.** $3x^2 + y = 14$

25. $2x + y^2 = 6$

In Exercises 26–30, evaluate each function at the given values of the independent variable and simplify.

26. $f(x) = 5 - 7x$
    a. $f(4)$ **b.** $f(x + 3)$ **c.** $f(-x)$

27. $g(x) = 3x^2 - 5x + 2$
    a. $g(0)$ **b.** $g(-2)$
    c. $g(x - 1)$ **d.** $g(-x)$

28. $f(x) = 4x - 3$
    a. $f(a)$ **b.** $f(a + h)$
    c. $\dfrac{f(a + h) - f(a)}{h}, \quad h \neq 0$ **d.** $f(a) + f(h)$

29. $g(x) = \begin{cases} \sqrt{x - 4} & \text{if } x \geq 4 \\ 4 - x & \text{if } x < 4 \end{cases}$

    a. $g(13)$ **b.** $g(0)$ **c.** $g(-3)$

30. $f(x) = \begin{cases} \dfrac{x^2 - 1}{x - 1} & \text{if } x \neq 1 \\ 12 & \text{if } x = 1 \end{cases}$

   a. $f(-2)$ **b.** $f(1)$ **c.** $f(2)$

In Exercises 31–35, find the domain of each function.

31. $f(x) = x^2 + 6x - 3$ **32.** $g(x) = \dfrac{4}{x - 7}$

33. $h(x) = \sqrt{8 - 2x}$ **34.** $f(x) = \dfrac{x}{x^2 - 1}$

35. $g(x) = \dfrac{\sqrt{x - 2}}{x - 5}$

36. The function $f(x) = -0.46x^2 + 3.66x + 20.08$ models data from the U.S. Census Bureau regarding the number of participants in the Federal Food Stamp Program. The variable x represents the number of years after 1990 and $f(x)$ is the number of participants, in millions, in the program. Find and interpret $f(6)$.

2.4

Graph the functions in Exercises 37–38. Use the integer values of x given to the right of the function to obtain the ordered pairs. Use the graph to specify the function's domain and range.

37. $f(x) = x^2 - 4x + 4$ $x = -1, 0, 1, 2, 3, 4$
38. $f(x) = |2 - x|$ $x = -1, 0, 1, 2, 3, 4$

In Exercises 39–41, use the graph to determine **a.** *the function's domain;* **b.** *the function's range;* **c.** *the x-intercepts, if any;* **d.** *the y-intercept, if any;* **e.** *intervals on which, the function is increasing, decreasing, or constant; and* **f.** *the function values indicated below the graphs.*

39.

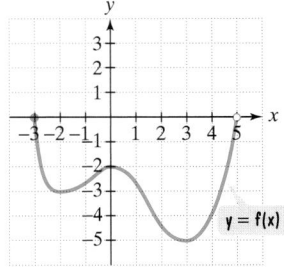

$f(-2) = ?$ $f(3) = ?$

40.

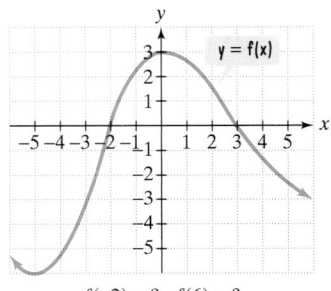

$f(-2) = ?$ $f(6) = ?$

41.

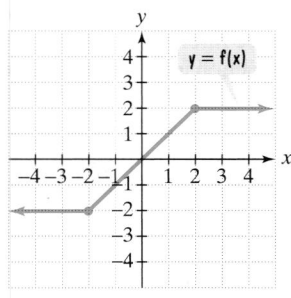

$f(-9) = ?$ $f(14) = ?$

In Exercises 42–45, use the vertical line test to identify graphs in which y is a function of x.

42.

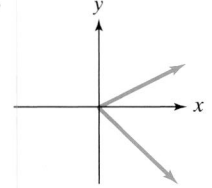

43.

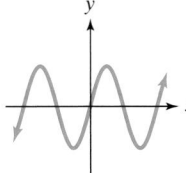

44.

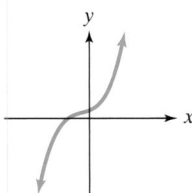

45.

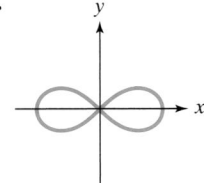

In Exercises 46–48, determine whether each function is even, odd, or neither. State each function's symmetry. If you are using a graphing utility, graph the function and verify its possible symmetry.

46. $f(x) = x^3 - 5x$ **47.** $f(x) = x^4 - 2x^2 + 1$
48. $f(x) = 2x\sqrt{1 - x^2}$

49. The graph shows the height (in meters) of a vulture as a function of its time (in seconds) in flight.

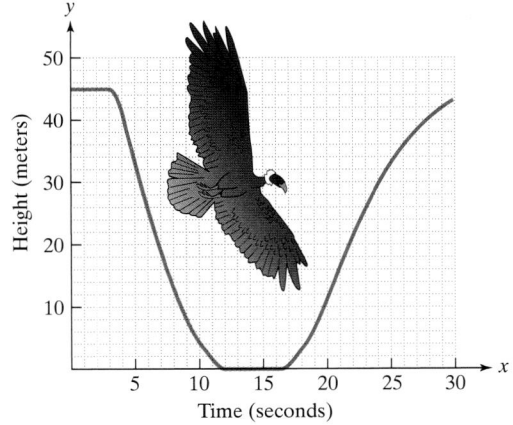

  a. Is the vulture's height a function of time? Use the graph to explain why or why not.
  b. On what interval is the function decreasing? Describe what this means in practical terms.

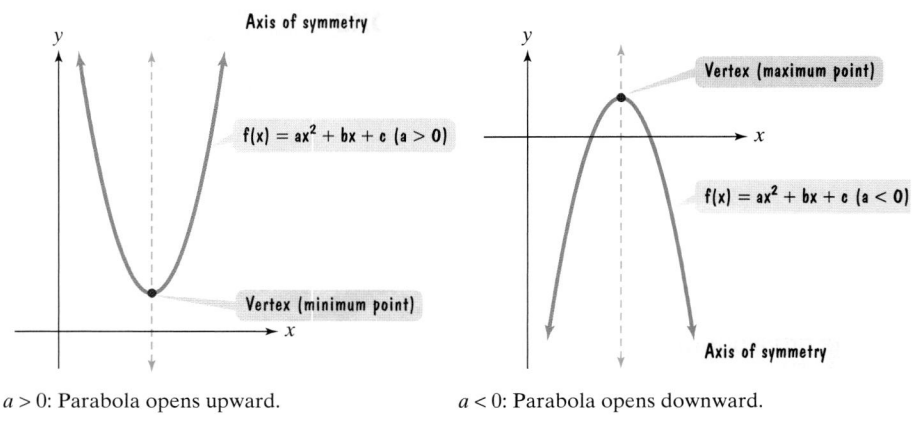

Figure 3.1 Characteristics of parabolas

$a > 0$: Parabola opens upward. $a < 0$: Parabola opens downward.

called the **axis of symmetry**. The movements of gymnasts, divers, and swimmers can approximate this symmetry.

2 Graph parabolas.

Graphing Quadratic Functions in Standard Form

In Section 2.5, we applied a series of transformations to the graph of $f(x) = x^2$. The graph of this function is a parabola. The vertex for this parabola is at $(0, 0)$. In Figure 3.2(a), the graph of $f(x) = ax^2$ for $a > 0$ is shown in black; it opens *upward*. In Figure 3.2(b), the graph of $f(x) = ax^2$ for $a < 0$ is shown in black; it opens *downward*.

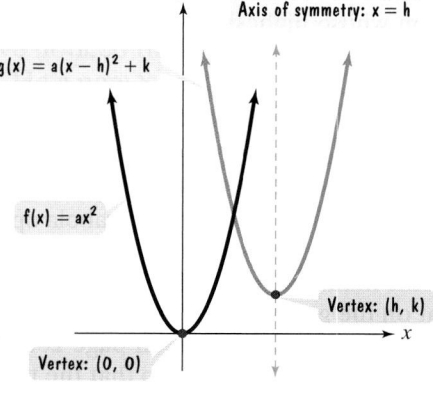

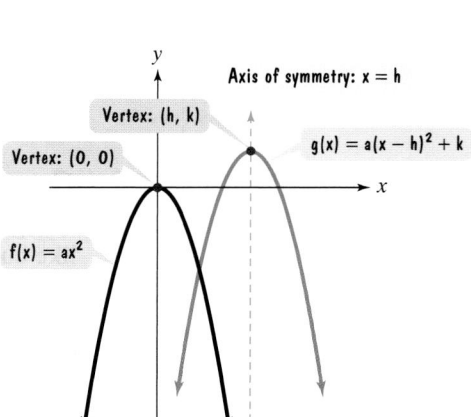

(a) $a > 0$: Parabola opens upward.   (b) $a < 0$: Parabola opens downward.

Figure 3.2 Transformations of $f(x) = ax^2$

Figure 3.2 also shows the graphs of $g(x) = a(x - h)^2 + k$ in blue. Compare these graphs to those of $f(x) = ax^2$. Observe that h determines the horizontal shift and k determines the vertical shift of the graph of $f(x) = ax^2$. Consequently, the vertex $(0, 0)$ on the graph of $f(x) = ax^2$ moves to the point (h, k) on the graph of $g(x) = a(x - h)^2 + k$. The axis of symmetry is the vertical line whose equation is $x = h$.

The form of the expression for g is convenient because it immediately identifies the vertex of the parabola as (h, k). This is the **standard form** of a quadratic function.

The Standard Form of a Quadratic Function

The quadratic function

$$f(x) = a(x - h)^2 + k, \qquad a \neq 0$$

is in **standard form**. The graph of f is a parabola whose vertex is the point (h, k). The parabola is symmetric to the line $x = h$. If $a > 0$, the parabola opens upward; if $a < 0$, the parabola opens downward.

The sign of a in $f(x) = a(x - h)^2 + k$ determines whether the parabola opens upward or downward. Furthermore, if $|a|$ is small, the parabola opens more widely than if $|a|$ is large. Here is a general procedure for graphing parabolas whose equations are in standard form.

Graphing Parabolas With Equations in Standard Form

To graph $f(x) = a(x - h)^2 + k$:

1. Determine whether the parabola opens upward or downward. If $a > 0$, it opens upward. If $a < 0$, it opens downward.
2. Determine the vertex of the parabola. The vertex is (h, k).
3. Find any x-intercepts by replacing $f(x)$ with 0. Solve the resulting quadratic equation for x.
4. Find the y-intercept by replacing x with 0.
5. Plot the intercepts and vertex. Connect these points with a smooth curve that is shaped like a cup.

EXAMPLE 1 **Graphing a Parabola Whose Equation Is in Standard Form**

Graph the quadratic function $f(x) = -2(x - 3)^2 + 8$.

Solution We can graph this function by following the steps in the preceding box. We begin by identifying values for $a, h,$ and k.

Standard form $f(x) = a(x - h)^2 + k$

$a = -2 \qquad h = 3 \qquad k = 8$

Given equation $f(x) = -2(x - 3)^2 + 8$

Step 1 Determine how the parabola opens. Note that a, the coefficient of x^2, is -2. Thus, $a < 0$; this negative value tells us that the parabola opens downward.

Step 2 **Find the vertex.** The vertex of the parabola is at (h, k). Because $h = 3$ and $k = 8$, the parabola has its vertex at $(3, 8)$.

Step 3 **Find the x-intercepts.** Replace $f(x)$ with 0 in $f(x) = -2(x - 3)^2 + 8$.

$$0 = -2(x - 3)^2 + 8$$ Find x-intercepts, setting f(x) equal to 0.

$$2(x - 3)^2 = 8$$ Solve for x. Add $2(x - 3)^2$ to both sides of the equation.

$$(x - 3)^2 = 4$$ Divide both sides by 2.

$$(x - 3) = \pm 2$$ Apply the square root method. If $(x - c)^2 = d$, then $x - c = \pm\sqrt{d}$.

$$x - 3 = -2 \quad \text{or} \quad x - 3 = 2$$ Express as two separate equations.

$$x = 1 \quad \text{or} \quad x = 5$$ Add 3 to both sides in each equation.

The x-intercepts are 1 and 5. The parabola passes through $(1, 0)$ and $(5, 0)$.

Step 4 **Find the y-intercept.** Replace x with 0 in $f(x) = -2(x - 3)^2 + 8$.

$$f(0) = -2(0 - 3)^2 + 8 = -2(-3)^2 + 8 = -2(9) + 8 = -10$$

The y-intercept is -10. The parabola passes through $(0, -10)$.

Step 5 **Graph the parabola.** With a vertex at $(3, 8)$, x-intercepts at 1 and 5, and a y-intercept at -10, the graph of f is shown in Figure 3.3. The axis of symmetry is the vertical line whose equation is $x = 3$.

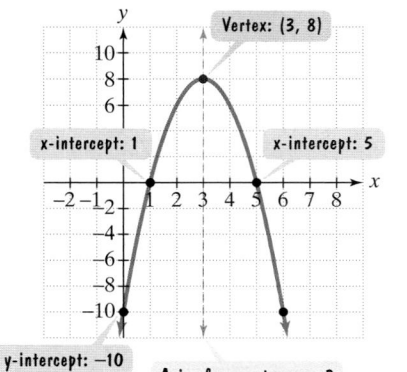

Figure 3.3 The graph of $f(x) = -2(x - 3)^2 + 8$

Check Point 1   Graph the quadratic function $f(x) = -(x - 1)^2 + 4$.

EXAMPLE 2 **Graphing a Parabola Whose Equation Is in Standard Form**

Graph the quadratic function $f(x) = (x + 3)^2 + 1$.

Solution We begin by finding values for a, h, and k.

Standard form	$f(x) = a(x - h)^2 + k$
Given equation	$f(x) = (x + 3)^2 + 1$
	or $f(x) = 1(x - (-3))^2 + 1$

$a = 1$ $h = -3$ $k = 1$

Step 1 **Determine how the parabola opens.** Note that a, the coefficient of x^2, is 1. Thus, $a > 0$; this positive value tells us that the parabola opens upward.

Step 2 **Find the vertex.** The vertex of the parabola is at (h, k). Because $h = -3$ and $k = 1$, the parabola has its vertex at $(-3, 1)$.

Step 3 **Find the x-intercepts.** Replace $f(x)$ with 0 in $f(x) = (x + 3)^2 + 1$. Because the vertex is at $(-3, 1)$, which lies above the x-axis, and the parabola opens upward, it appears that this parabola has no x-intercepts. We can verify this observation algebraically.

$$0 = (x + 3)^2 + 1 \quad \text{Find possible x-intercepts, setting } f(x) \text{ equal to 0.}$$
$$-1 = (x + 3)^2 \quad \text{Solve for x. Subtract 1 from both sides.}$$
$$x + 3 = \pm\sqrt{-1} \quad \text{Apply the square root method.}$$
$$x + 3 = \pm i \quad \text{Recall that } \sqrt{-1} = i, \text{ an imaginary number.}$$
$$x = -3 \pm i \quad \text{Subtract 3 from both sides.}$$

Because this equation has no real solutions, the parabola has no x-intercepts.

Step 4 **Find the y-intercept.** Replace x with 0 in $f(x) = (x + 3)^2 + 1$.
$$f(0) = (0 + 3)^2 + 1 = 3^2 + 1 = 9 + 1 = 10$$

The y-intercept is 10. The parabola passes through $(0, 10)$.

Step 5 **Graph the parabola.** With a vertex at $(-3, 1)$, no x-intercepts, and a y-intercept at 10, the graph of f is shown in Figure 3.4. The axis of symmetry is the vertical line whose equation is $x = -3$.

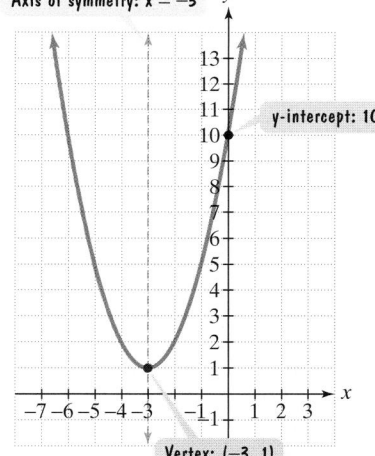

Axis of symmetry: x = −3

y-intercept: 10

Vertex: (−3, 1)

Figure 3.4 The graph of $g(x) = (x + 3)^2 + 1$

Check Point 2 Graph the quadratic function $f(x) = (x - 2)^2 + 1$.

Graphing Functions in the Form $f(x) = ax^2 + bx + c$

Quadratic functions are frequently expressed in the form $f(x) = ax^2 + bx + c$. How can we identify the vertex of a parabola whose equation is in this form? Completing the square provides the answer to this question.

$$f(x) = ax^2 + bx + c$$

$$= a\left(x^2 + \frac{b}{a}x\right) + c \qquad \text{Factor out } a \text{ from } ax^2 + bx.$$

$$= a\left(x^2 + \frac{b}{a}x + \frac{b^2}{4a^2}\right) + c - a\left(\frac{b^2}{4a^2}\right)$$

Complete the square by adding the square of half the coefficient of x.

By completing the square, we added $a \cdot \dfrac{b^2}{4a^2}$. To avoid changing the function's equation, we must subtract this term.

$$= a\left(x + \frac{b}{2a}\right)^2 + c - \frac{b^2}{4a} \qquad \text{Write the trinomial as the square of a binomial and simplify the constant term.}$$

Compare the form of this equation with a quadratic function's standard form.

Standard form $\qquad f(x) = a(x - h)^2 + k$

$$h = -\frac{b}{2a} \qquad k = c - \frac{b^2}{4a}$$

Equation under discussion $\quad f(x) = a\left(x - \left(-\frac{b}{2a}\right)\right)^2 + c - \frac{b^2}{4a}$

The important part of this observation is that h, the x-coordinate of the vertex, is $-\dfrac{b}{2a}$. The y-coordinate can be found by evaluating the function at $-\dfrac{b}{2a}$.

The Vertex of a Parabola Whose Equation Is $f(x) = ax^2 + bx + c$

Consider the parabola defined by the quadratic function $f(x) = ax^2 + bx + c$. The parabola's vertex is at $\left(-\dfrac{b}{2a}, f\left(-\dfrac{b}{2a}\right)\right)$.

We can apply our five-step procedure and graph parabolas in $f(x) = ax^2 + bx + c$ form. The only step that is different is how we determine the vertex.

EXAMPLE 3 Graphing a Parabola in $f(x) = ax^2 + bx + c$ Form

Graph the quadratic function $f(x) = -x^2 + 4x - 1$.

Solution

Step 1 Determine how the parabola opens. Note that a, the coefficient of x^2, is -1. Thus, $a < 0$; this negative value tells us that the parabola opens downward.

Step 2 Find the vertex.   We know that the x-coordinate of the vertex is $x = -\dfrac{b}{2a}$. We identify $a, b,$ and c in $f(x) = ax^2 + bx + c$.

$$f(x) = -x^2 + 4x - 1$$

$a = -1$ $b = 4$ $c = -1$

Substitute the values of a and b into the equation for the x-coordinate:

$$x = -\frac{b}{2a} = -\frac{4}{2(-1)} = \frac{-4}{-2} = 2.$$

The x-coordinate of the vertex is 2. We substitute 2 for x in the equation of the function to find the y-coordinate:

$$f(2) = -2^2 + 4 \cdot 2 - 1 = -4 + 8 - 1 = 3.$$

The vertex is at $(2, 3)$.

Step 3 Find the x-intercepts.   Replace $f(x)$ with 0 in $f(x) = -x^2 + 4x - 1$. We obtain $0 = -x^2 + 4x - 1$ or $-x^2 + 4x - 1 = 0$. This equation cannot be solved by factoring. We will use the quadratic formula to solve it.

$$a = -1, \qquad b = 4, \qquad c = -1$$

$$x = \frac{-b \pm \sqrt{b^2 - 4ac}}{2a} = \frac{-4 \pm \sqrt{4^2 - 4(-1)(-1)}}{2(-1)} = \frac{-4 \pm \sqrt{16 - 4}}{-2}$$

$$x = \frac{-4 - \sqrt{12}}{-2} \approx 3.7 \quad \text{or} \quad x = \frac{-4 + \sqrt{12}}{-2} \approx 0.3$$

The x-intercepts are approximately 0.3 and 3.7. The parabola passes through $(0.3, 0)$ and $(3.7, 0)$.

Step 4 Find the y-intercept.   Replace x with 0 in $f(x) = -x^2 + 4x - 1$.

$$f(0) = -0^2 + 4 \cdot 0 - 1 = -1$$

The y-intercept is -1. The parabola passes through $(0, -1)$.

Step 5 Graph the parabola.   With a vertex at $(2, 3)$, x-intercepts at 0.3 and 3.7, and a y-intercept at -1, the graph of f is shown in Figure 3.5. The axis of symmetry is the vertical line whose equation is $x = 2$.

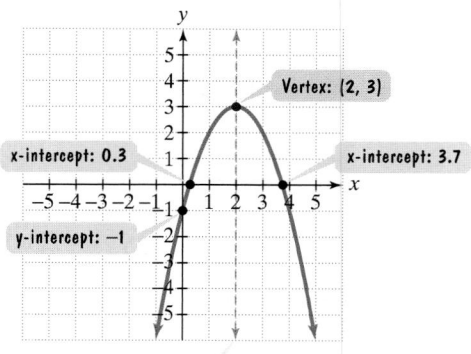

Axis of symmetry: $x = 2$

Figure 3.5 The graph of $f(x) = -x^2 + 4x - 1$

> **Check Point 3** Graph the quadratic function $f(x) = x^2 - 2x - 3$.

3 Solve problems involving minimizing or maximizing quadratic functions.

Applications of Quadratic Functions

When were women's earnings as a percentage of men's at the lowest? What is the age of a driver having the least number of car accidents? How much should a business spend on advertising to maximize its profits? The answers to these questions involve finding the maximum or minimum value of quadratic functions.

Consider the quadratic function $f(x) = ax^2 + bx + c$. If $a > 0$, the parabola opens upward and the vertex is its lowest point. If $a < 0$, the parabola opens downward and the vertex is its highest point. The x-coordinate of the vertex is $-\dfrac{b}{2a}$. Thus, we can find the minimum or maximum value of f by evaluating the quadratic function at $x = -\dfrac{b}{2a}$.

Minimum and Maximum: Quadratic Functions

Consider $f(x) = ax^2 + bx + c$.

1. If $a > 0$, then f has a minimum that occurs at $x = -\dfrac{b}{2a}$.

2. If $a < 0$, then f has a maximum that occurs at $x = -\dfrac{b}{2a}$.

EXAMPLE 4 An Application: The Wage Gap

The function

$$f(x) = 0.022x^2 - 0.4x + 60.07$$

models women's earnings as a percentage of men's x years after 1960. In which year was this percentage at a minimum? What was the percentage for that year?

Solution The quadratic function is in the form $f(x) = ax^2 + bx + c$ with $a = 0.022$ and $b = -0.4$. With $a > 0$, the function has a minimum when $x = -\dfrac{b}{2a}$.

$$x = -\frac{b}{2a} = -\frac{(-0.4)}{2(0.022)} \approx 9$$

This means that women's earnings as a percentage of men's were at their lowest approximately 9 years after 1960, or in 1969. The percentage for that year was

$$f(9) = 0.022(9)^2 - 0.4(9) + 60.07 \approx 58.$$

In 1969, women earned approximately 58% as much as men.

Check Point 4 The function $f(x) = 0.4x^2 - 36x + 1000$ models the number of accidents, $f(x)$, per 50 million miles driven, as a function of a driver's age, x, in years, where $16 \leq x \leq 74$. What is the age of a driver having the least number of car accidents? What is the minimum number of car accidents per 50 million miles driven?

Modeling Data with Quadratic Functions

We've come a long way from the small nation of "embattled farmers" who launched the American Revolution. In the early years of our Republic, 95% of the population was involved in farming. Although U.S. agriculture is an integral part of the global economy, the number of U.S. farms has declined since the 1920s as individually owned family farms have been swallowed up by huge agribusinesses owned by corporations.

The graph in Figure 3.6 shows the number of farms in the United States from 1850 through 2010 (projected). Because the graph is shaped like a cup, with an increasing number of farms from 1850 to 1910 and a decreasing number of farms from 1910 to 2010, a quadratic function is an appropriate model for the data. You can use the statistical menu of a graphing utility to enter the data in Figure 3.6. We entered the data using

(number of decades after 1850, millions of U.S. farms).

Thus, we entered

$$(0, 2.3), \quad (2, 3.3), \quad (4, 5.1), \quad (6, 6.7), \quad (8, 6.4),$$
$$(10, 5.8), \quad (12, 3.6), \quad (14, 2.9), \quad (16, 2.3).$$

Number of U. S. Farms, 1850–2010

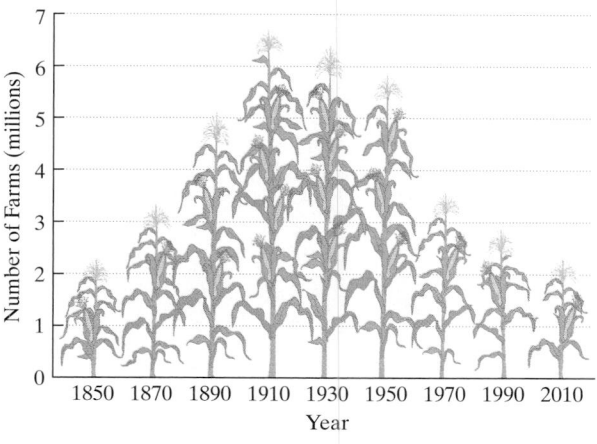

Source: U. S. Bureau of the Census

Figure 3.6 The number of U.S. farms is declining.

```
QuadReg
 y=ax²+bx+c
 a=⁻.0643668831
 b=.9873701299
 c=2.203636364
```

Figure 3.7 Executing the Quadratic Regression Program

Upon executing the QUADratic REGression program, we obtain the results shown in Figure 3.7. Thus, the quadratic function of best fit is

$$f(x) = -0.064x^2 + 0.99x + 2.2$$

where x represents the number of decades after 1850 and $f(x)$ represents the number of U.S. farms, in millions.

EXERCISE SET 3.1

Practice Exercises

In Exercises 1–4, the graph of a quadratic function is given. Write the function's equation, selecting from the following options.

$$f(x) = (x + 1)^2 - 1 \qquad g(x) = (x + 1)^2 + 1$$
$$h(x) = (x - 1)^2 + 1 \qquad j(x) = (x - 1)^2 - 1$$

In Exercises 5–8, the graph of a quadratic function is given. Write the function's equation, selecting from the following options.

$$f(x) = x^2 + 2x + 1 \qquad g(x) = x^2 - 2x + 1$$
$$h(x) = x^2 - 1 \qquad j(x) = -x^2 - 1$$

1.

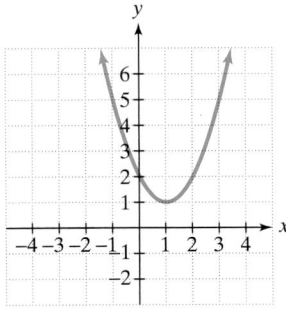

2.

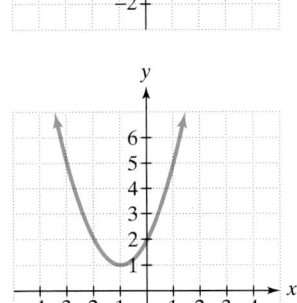

3.

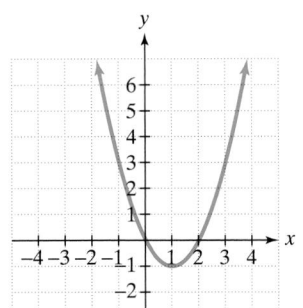

4.

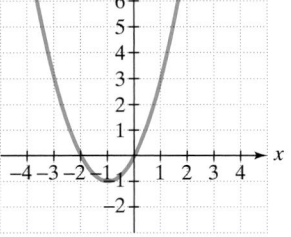

5.

6.

7.

8.

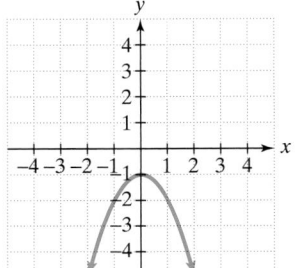

In Exercises 9–16, find the coordinates of the vertex for the parabola defined by the given quadratic function.

9. $f(x) = 2(x - 3)^2 + 1$

10. $f(x) = -3(x - 2)^2 + 12$

11. $f(x) = -2(x + 1)^2 + 5$

12. $f(x) = -2(x + 4)^2 - 8$

13. $f(x) = 2x^2 - 8x + 3$

14. $f(x) = 3x^2 - 12x + 1$

15. $f(x) = -x^2 - 2x + 8$

16. $f(x) = -2x^2 + 8x - 1$

In Exercises 17–34, use the vertex and intercepts to sketch the graph of each quadratic function. Give the equation for the parabola's axis of symmetry.

17. $f(x) = (x - 4)^2 - 1$

18. $f(x) = (x - 1)^2 - 2$

19. $f(x) = (x - 1)^2 + 2$

20. $f(x) = (x - 3)^2 + 2$

21. $y - 1 = (x - 3)^2$

22. $y - 3 = (x - 1)^2$

23. $f(x) = 2(x + 2)^2 - 1$

24. $f(x) = \frac{5}{4} - \left(x - \frac{1}{2}\right)^2$

25. $f(x) = 4 - (x - 1)^2$

26. $f(x) = 1 - (x - 3)^2$

27. $f(x) = x^2 - 2x - 3$

28. $f(x) = x^2 - 2x - 15$

29. $f(x) = x^2 + 3x - 10$

30. $f(x) = 2x^2 - 7x - 4$

31. $f(x) = 2x - x^2 + 3$

32. $f(x) = 5 - 4x - x^2$

33. $f(x) = 2x - x^2 - 2$

34. $f(x) = 6 - 4x + x^2$

In Exercises 35–40, determine, without graphing, whether the given quadratic function has a minimum value or a maximum value. Then find the coordinates of the minimum or the maximum point.

35. $f(x) = 3x^2 - 12x - 1$

36. $f(x) = 2x^2 - 8x - 3$

37. $f(x) = -4x^2 + 8x - 3$

38. $f(x) = -2x^2 - 12x + 3$

39. $f(x) = 5x^2 - 5x$

40. $f(x) = 6x^2 - 6x$

 Application Exercises

41. The U.S. Center for Disease Control modeled the average annual per capita consumption C of cigarettes by Americans 18 and older as a function of time. The function is $C(t) = -3.1t^2 + 51.4t + 4024.5$, where t represents years after 1960. According to this function, in which year did cigarette consumption per capita reach a maximum? What was the consumption for that year?

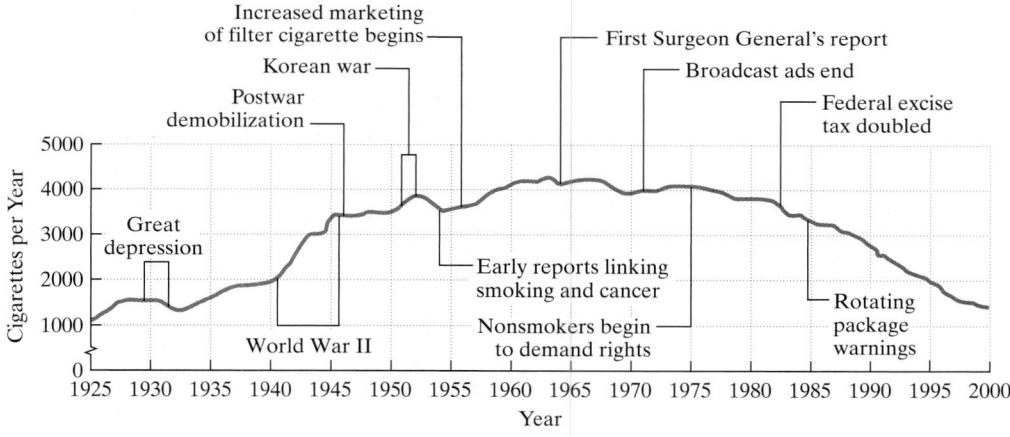

Cigarette Consumption per U.S. Adult

Source: U.S. Department of Health and Human Services

42. The function $R(x) = -0.0065x^2 + 0.23x + 8.47$ models the American marriage rate R (the number of marriages per 1000 population) x years after 1960. According to this function, in which year was the marriage rate the highest? What was the marriage rate for that year?

43. A person standing close to the edge of an 80-foot cliff throws a rock upward with an initial speed of 64 feet per second. The height of the rock above the water at the bottom of the cliff is a function of time, described by the quadratic function

$$y = -16x^2 + 64x + 80.$$

The variable x describes the number of seconds that the rock is in motion. The variable y describes the height of the rock, in feet, above the water at the bottom of the cliff. After how many seconds will the rock reach its maximum height above the water? How many feet above the water is the rock at that time?

44. There is a relationship between the amount of one's annual income, x, in thousands of dollars, and the percentage of this income, P, that one contributes to charities. This relationship is modeled by the quadratic function $P = 0.0014x^2 - 0.1529x + 5.855$, where $5 \le x \le 100$. What annual income corresponds to the minimum percentage given to charity? What is this minimum percentage?

45. Suppose that a quadratic function is used to model the data shown in the graph on page 270 using

(number of years after 1985,
number of U.S. children under 18
guilty of homicide using guns).

**Known Homicide Offenders
Under 18 Using Guns**

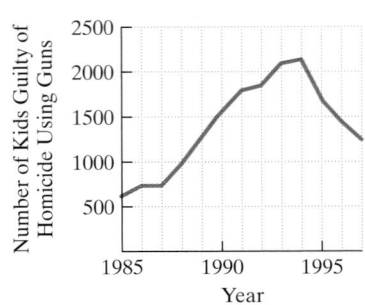

Source: National Center for Juvenile Justice

Determine, without obtaining the quadratic function of best fit, the approximate coordinates of the vertex for the function's graph.

46. Suppose that a quadratic function is used to model the data shown in the graph using

(number of years after 1971,
millions of students enrolled in U.S. schools).

Millions of Students Enrolled in U.S. Schools

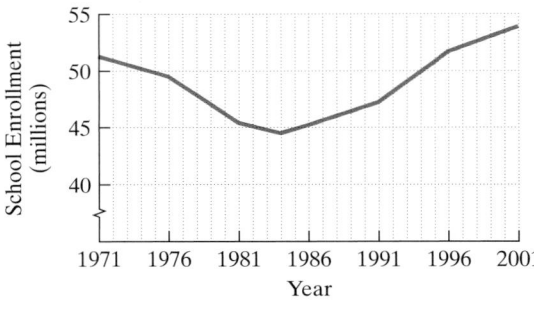

Source: National Education Association

Determine, without obtaining the quadratic function of best fit, the approximate coordinates of the vertex for the function's graph.

47. You have 120 feet of fencing to enclose a rectangular plot that borders on a river. If you do not fence the side along the river, find the length and width of the plot that will maximize the area. What is the largest area that can be enclosed?

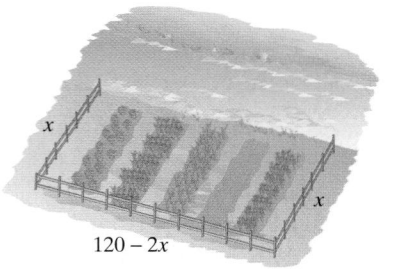

48. The figure shown indicates that you have 100 yards of fencing to enclose a rectangular area. Find the dimensions of the rectangle that maximize the enclosed area. What is the maximum area?

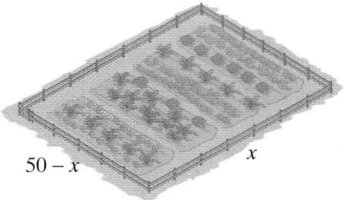

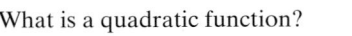

Writing in Mathematics

49. What is a quadratic function?

50. What is a parabola? Describe its shape.

51. Explain how to decide whether a parabola opens upward or downward.

52. Describe how to find a parabola's vertex if its equation is expressed in standard form. Give an example.

53. Describe how to find a parabola's vertex if its equation is in the form $f(x) = ax^2 + bx + c$. Use $f(x) = x^2 - 6x + 8$ as an example.

54. A parabola that opens upward has its vertex at $(1, 2)$. Describe as much as you can about the parabola based on this information. Include in your discussion the number of x-intercepts (if any) for the parabola.

55. The quadratic function

$$f(x) = -0.018x^2 + 1.93x - 25.34$$

describes the miles per gallon of a Ford Taurus driven at x miles per hour. Suppose that you own a Ford Taurus. Describe how you can use this function to save money.

Technology Exercises

56. Use a graphing utility to verify any five of your hand-drawn graphs in Exercises 17–34.

57. a. Use a graphing utility to graph $y = 2x^2 - 82x + 720$ in a standard viewing rectangle. What do you observe?

b. Find the coordinates of the vertex for the given quadratic function.

c. The answer to part (b) is $(20.5, -120.5)$. Because the leading coefficient of the given function (2) is positive, the vertex is a minimum point on the graph. Use this fact to help find a viewing rectangle that will give a relatively complete picture of the parabola. With an axis of symmetry at $x = 20.5$, the setting for x should extend past this, so try Xmin = 0 and Xmax = 30. The setting for y should include (and probably go below) the y-coordinate of the graph's minimum point, so try Ymin = -130. Experiment with Ymax until your utility shows the parabola's major features.

d. In general, explain how knowing the coordinates of a parabola's vertex can help determine a reasonable viewing rectangle on a graphing utility for obtaining a complete picture of the parabola.

In Exercises 58–61, find the vertex for each parabola. Then determine a reasonable viewing rectangle on your graphing utility and use it to graph the parabola.

58. $y = -0.25x^2 + 40x$

59. $y = -4x^2 + 20x + 160$

60. $y = 5x^2 + 40x + 600$

61. $y = 0.01x^2 + 0.6x + 100$

62. The quadratic function $f(x) = 0.013x^2 - 0.96x + 25.4$ describes the average yearly consumption of whole milk per person in the United States x years after 1970. The linear function $g(x) = 0.41x + 6.03$ describes the average yearly consumption of low-fat milk per person in the United States x years after 1970.

a. Use a graphing utility to graph each function in the same viewing rectangle for the years 1970 through 2000.

b. Use the graphs to describe the trend in consumption for both types of milk. What possible explanations are there for these consumption patterns?

63. The function $y = 0.011x^2 - 0.097x + 4.1$ models the number of people in the United States, y, in millions, holding more than one job x years after 1970. Use graphing utility to graph the function in a $[0, 20, 1]$ by $[3, 6, 1]$ viewing rectangle. $\boxed{\text{TRACE}}$ along the curve or use your utility's minimum value feature to approximate the coordinates of the parabola's vertex. Describe what this represents in practical terms.

64. The following data show fuel efficiency, in miles per gallon, for all U.S. automobiles in the indicated year.

x (Years after 1940)	y (Average Number of Miles per Gallon for U.S. Automobiles)
1940: 0	14.8
1950: 10	13.9
1960: 20	13.4
1970: 30	13.5
1980: 40	15.5
1986: 46	18.3

Source: Statistical Abstract of the United States

a. Use a graphing utility to draw a scatter plot of the data. Explain why a quadratic function is appropriate for modeling these data.

b. Use the quadratic regression feature to find the quadratic function that best fits the data.

c. Use the equation in part (b) to determine the worst year for automobile fuel efficiency. What was the average number of miles per gallon for that year?

d. Use a graphing utility to draw a scatter plot of the data and graph the quadratic function of best fit on the scatter plot.

Critical Thinking Exercises

65. Which one of the following is true?

a. No quadratic functions have a range of $(-\infty, \infty)$.

b. The vertex of the parabola described by $f(x) = 2(x - 5)^2 - 1$ is at $(5, 1)$.

c. The graph of $f(x) = -2(x + 4)^2 - 8$ has one y-intercept and two x-intercepts.

d. The maximum value of y for the quadratic function $f(x) = -x^2 + x + 1$ is 1.

66. What explanations can you offer for your answer to Exercise 41? Use a graphing utility to graph C. Do you agree with the long-term predictions made by the graph? Explain.

In Exercises 67–68, find the axis of symmetry for each parabola whose equation is given. Use the axis of symmetry to find a second point on the parabola whose y-coordinate is the same as the given point.

67. $f(x) = 3(x + 2)^2 - 5$; $(-1, -2)$

68. $f(x) = (x - 3)^2 + 2$; $(6, 11)$

69. A rancher has 1000 feet of fencing to construct six corrals, as shown in the figure. Find the dimensions that maximize the enclosed area. What is the maximum area?

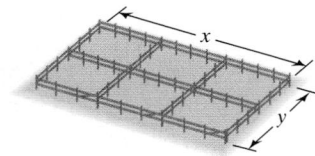

Group Exercise

70. Each group member should consult an almanac, newspaper, magazine, or the Internet to find data that can be modeled by a quadratic function. Group members should select the two sets of data that are most interesting and relevant. For each data set selected:

a. Use the quadratic regression feature of a graphing utility to find the quadratic function that best fits the data.

b. Use the equation of the quadratic function to make a prediction from the data. What circumstances might affect the accuracy of your prediction?

c. Use the equation of the quadratic function to write and solve a problem involving maximizing or minimizing the function.

SECTION 3.2   *Polynomial Functions and Their Graphs*

Objectives

1. Recognize characteristics of graphs of polynomial functions.
2. Determine end behavior.
3. Use factoring to find zeros of polynomial functions.
4. Identify the multiplicity of a zero.
5. Understand the relationship between degree and turning points.
6. Graph polynomial functions.

Magnified 6000 times, this color-scanned image shows a T-lymphocyte blood cell (green) infected with the HIV virus (red). Depletion of the number of T-cells causes destruction of the immune system.

In 1980, U.S. doctors diagnosed 41 cases of a rare form of cancer, Kaposi's sarcoma, that involved skin lesions, pneumonia, and severe immunological deficiencies. All cases involved gay men ranging in age from 26 to 51. By the end of 1998, approximately 680,000 Americans, straight and gay, male and female, old and young, were infected with the HIV virus.

Modeling AIDS-related data and making predictions about the epidemic's havoc is serious business. Changing circumstances and unforeseen events have resulted in models that are not particularly useful over long periods of time. For example, the function

$$f(x) = -143x^3 + 1810x^2 - 187x + 2331$$

models the number of AIDS cases diagnosed in the United States x years after 1983. The model was obtained using cases diagnosed from 1983 through 1991. Figure 3.8 shows the graph of f from 1983 through 1991 in a $[0, 8, 1]$ by $[0, 50,000, 5000]$ viewing rectangle. The function used to describe what was happening with new HIV infections over a limited period of time is an example of a **polynomial function**.

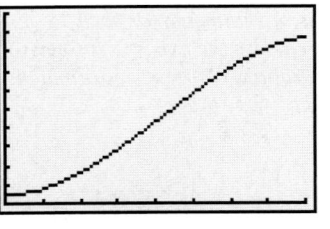

Figure 3.8 The graph of a function modeling the number of AIDS cases from 1983 through 1991

Definition of a Polynomial Function

Let n be a nonnegative integer and let $a_n, a_{n-1}, \ldots, a_2, a_1, a_0$, be real numbers with $a_n \neq 0$. The function defined by

$$f(x) = a_n x^n + a_{n-1} x^{n-1} + \cdots + a_2 x^2 + a_1 x + a_0$$

is called a **polynomial function of x of degree n**. The number a_n, the coefficient of the variable to the highest power, is called the **leading coefficient**.

A constant function $f(x) = a$, where $a \neq 0$, is a polynomial function of degree 0. A linear function $f(x) = ax + b$, where $a \neq 0$, is a polynomial function of degree 1. A quadratic function $f(x) = ax^2 + bx + c$, where $a \neq 0$, is a polynomial function of degree 2. In this section, we focus on polynomial functions of degree 3 or higher.

1 Recognize characteristics of graphs of polynomial functions.

Smooth, Continuous Graphs

Polynomial functions of degree 2 or less have graphs that are either parabolas or lines. We can graph such functions by plotting points. We can also graph polynomial functions of degree 3 or higher by plotting points. However, the process is rather tedious: Many points must be plotted. It may be easier to use a graphing utility for such functions. Regardless of the graphing method you use, you will find an ability to recognize the basic features of polynomial functions helpful. For example, they may help you choose an appropriate viewing rectangle for a graphing utility.

Two important features of the graphs of polynomial functions are that they are *smooth* and *continuous*. By **smooth**, we mean that the graph contains only rounded curves with no sharp corners. By **continuous**, we mean that the graph has no breaks and can be drawn without lifting your pencil from the rectangular coordinate system. These ideas are illustrated in Figure 3.9.

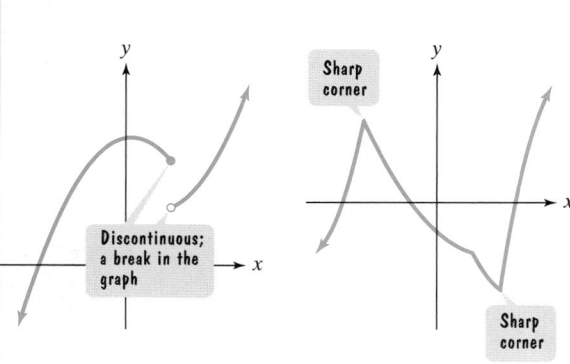

Figure 3.9 Recognizing graphs of polynomial functions

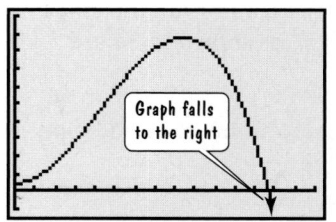

Figure 3.10 By extending the viewing rectangle, y is negative and the function no longer models the number of AIDS cases.

2 Determine end behavior.

End Behavior of Polynomial Functions

Figure 3.10 shows the graph of the function

$$f(x) = -143x^3 + 1810x^2 - 187x + 2331,$$

which models U.S. AIDS cases from 1983 through 1991. Look what happens to the graph when we extend the year up through 1998 with a $[0, 15, 1]$ by $[-5000, 50,000, 5000]$ viewing rectangle. By year 13 (1996), the values of y are negative and the function no longer models AIDS cases. We've added an arrow to the graph at the far right to emphasize that it continues to decrease without bound. It is this far-right *end behavior* of the graph that makes it inappropriate for modeling AIDS cases into the future.

The behavior of a graph of a function to the far left or the far right is called its **end behavior**. Although the graph of a polynomial function may have intervals where it increases or decreases, the graph will eventually rise or fall without bound as it moves far to the left or far to the right.

How can you determine whether the graph of a polynomial function goes up or down at each end? The end behavior of a polynomial function

$$f(x) = a_n x^n + a_{n-1} x^{n-1} + \cdots + a_1 x + a_0$$

depends upon the leading term $a_n x^n$. In particular, the sign of the leading coefficient a_n, and the degree, n, of the polynomial function reveal its end behavior. In terms of end behavior, only the term of highest degree counts, summarized by the **Leading Coefficient Test**.

The Leading Coefficient Test

As x increases or decreases without bound, the graph of the polynomial function

$$f(x) = a_n x^n + a_{n-1} x^{n-1} + a_{n-2} x^{n-2} + \cdots + a_1 x + a_0 \quad (a_n \neq 0)$$

eventually rises or falls. In particular,

1. For n odd:

If the leading coefficient is positive, the graph falls to the left and rises to the right.	If the leading coefficient is negative, the graph rises to the left and falls to the right.
$a_n > 0$	$a_n < 0$

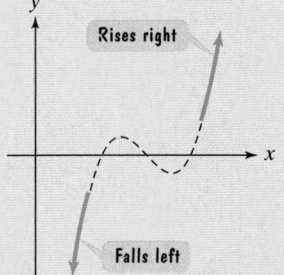 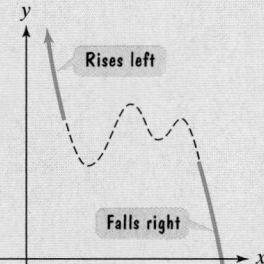

2. For n even:

If the leading coefficient is positive, the graph rises to the left and to the right.	If the leading coefficient is negative, the graph falls to the left and to the right.
$a_n > 0$	$a_n < 0$

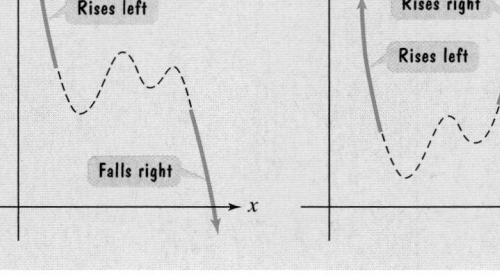 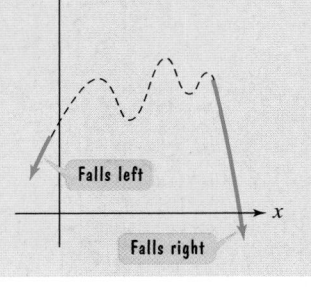

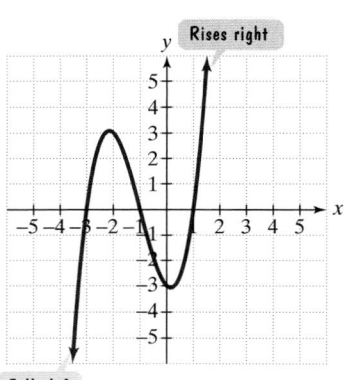

Figure 3.11 The graph of
$f(x) = x^3 + 3x^2 - x - 3$

EXAMPLE 1   Using the Leading Coefficient Test

Use the Leading Coefficient Test to determine the end behavior of the graph of

$$f(x) = x^3 + 3x^2 - x - 3.$$

Solution   Because the degree is odd ($n = 3$) and the leading coefficient, 1, is positive, the graph falls to the left and rises to the right, as shown in Figure 3.11.

> **Check Point 1**   Use the Leading Coefficient Test to determine the end behavior of the graph of $f(x) = x^4 - 4x^2$.

EXAMPLE 2   Using the Leading Coefficient Test

Use end behavior to explain why

$$f(x) = -143x^3 + 1810x^2 - 187x + 2331$$

is only an appropriate model for AIDS cases for a limited time period.

Solution   Because the degree is odd ($n = 3$) and the leading coefficient, -143, is negative, the graph rises to the left and falls to the right. The fact that it falls to the right indicates at some point the number of AIDS cases will be negative, an impossibility. No function with a graph that decreases without bound as x (time) increases can model nonnegative real-world phenomena over a long period of time.

Check Point 2

The polynomial function

$$f(x) = -0.27x^3 + 9.2x^2 - 102.9x + 400$$

models the ratio of students to computers in U.S. public schools x years after 1980. Use end behavior to determine whether this function could be an appropriate model for computers in the classroom well into the twenty-first century. Explain your answer.

If you use a graphing utility to graph a polynomial function, it is important to select a viewing rectangle that accurately reveals the graph's end behavior. If the viewing rectangle is too small, it may not accurately show end behavior.

EXAMPLE 3   Using the Leading Coefficient Test

The graph of $f(x) = -x^4 + 8x^3 + 4x^2 + 2$ was obtained with a graphing utility using a $[-8, 8, 1]$ by $[-10, 10, 1]$ viewing rectangle. The graph is shown in Figure 3.12(a). Does the graph show the end behavior of the function?

Figure 3.12

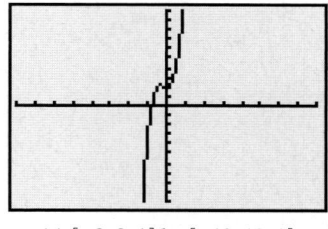
(a) $[-8, 8, 1]$ by $[-10, 10, 1]$

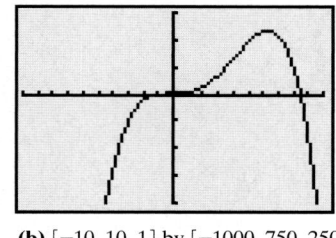
(b) $[-10, 10, 1]$ by $[-1000, 750, 250]$

Solution   Note that the degree is even ($n = 4$) and the leading coefficient, -1, is negative. Thus, the Leading Coefficient Test indicates that the graph should fall to the left and the right. The graph in Figure 3.12(a) is falling to the left, but it is not falling to the right. Therefore, the graph is not complete enough to show end behavior. A more complete graph of the function is shown in a larger viewing rectangle in Figure 3.12(b).

Check Point 3

The graph of $f(x) = x^3 + 13x^2 + 10x - 4$ is shown in a standard viewing rectangle in Figure 3.13. Use the Leading Coefficient Test to determine whether the graph shows the end behavior of the function. Explain your answer.

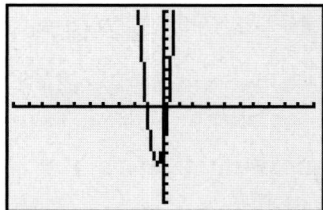

Figure 3.13

3 Use factoring to find zeros of polynomial functions.

Zeros of Polynomial Functions

If f is a polynomial function, then the values of x for which $f(x)$ is equal to 0 are called the **zeros** of f. These values of x are the **roots** of the polynomial equation $f(x) = 0$. Each real root of the polynomial equation appears as an x-intercept of the graph of the polynomial function.

EXAMPLE 4   Finding Zeros of a Polynomial Function

Find all zeros of $f(x) = x^3 + 3x^2 - x - 3$.

Solution   By definition, the zeros are the values of x for which $f(x)$ is equal to 0. Thus, we set $f(x)$ equal to 0:

$$f(x) = x^3 + 3x^2 - x - 3 = 0.$$

We solve the polynomial equation $x^3 + 3x^2 - x - 3 = 0$ for x as follows:

$x^3 + 3x^2 - x - 3 = 0$	This is the equation needed to find the function's zeros.
$x^2(x + 3) - 1(x + 3) = 0$	Factor x^2 from the first two terms and -1 from the last two terms.
$(x + 3)(x^2 - 1) = 0$	A common factor of $x + 3$ is factored from the expression.
$x + 3 = 0$ or $x^2 - 1 = 0$	Set each factor equal to 0.
$x = -3$ $x^2 = 1$	Solve for x.
$x = \pm 1$	Remember that if $x^2 = d$, then $x = \pm \sqrt{d}$.

The zeros of f are $-3, -1$, and 1. The graph of f in Figure 3.14 shows that each zero is an x-intercept.

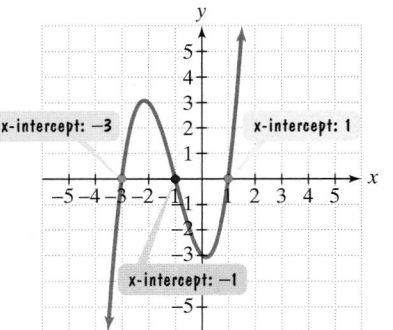

Figure 3.14 The real zeros of $f(x) = x^3 + 3x^2 - x - 3$ are the x-intercepts for the graph of f.

> **Check Point 4**   Find all zeros of $f(x) = x^3 + 2x^2 - 4x - 8$.

EXAMPLE 5   Finding Zeros of a Polynomial Function

Find all zeros of $f(x) = -x^4 + 4x^3 - 4x^2$.

Solution   We find the zeros of f by setting $f(x)$ equal to 0.

$-x^4 + 4x^3 - 4x^2 = 0$	We now have a polynomial equation.
$x^4 - 4x^3 + 4x^2 = 0$	Multiply both sides by -1. This step is optional.
$x^2(x^2 - 4x + 4) = 0$	Factor out x^2.
$x^2(x - 2)^2 = 0$	Factor completely.
$x^2 = 0$ or $(x - 2)^2 = 0$	Set each factor equal to 0.
$x = 0$ $x = 2$	Solve for x.

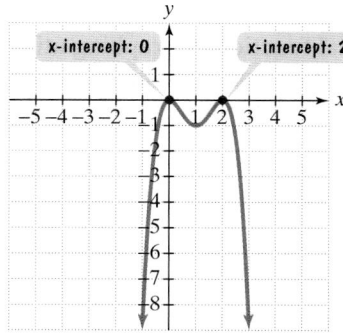

Figure 3.15 The zeros of $f(x) = -x^4 + 4x^3 - 4x^2$, namely 0 and 2, are the x-intercepts for the graph of f.

4 Identify the multiplicity of a zero.

The zeros of $f(x) = -x^4 + 4x^3 - 4x^2$ are 0 and 2. The graph of f, shown in Figure 3.15 has x-intercepts at 0 and 2.

> **Check Point 5** Find all zeros of $f(x) = x^4 - 4x^2$.

In Example 5, we can use our factoring to express the function's equation as follows:

$$f(x) = -x^4 + 4x^3 - 4x^2 = -(x^4 - 4x^3 + 4x^2) = -x^2(x - 2)^2$$

The factor x occurs twice: $x^2 = x \cdot x$.

The factor $(x - 2)$ occurs twice: $(x - 2)^2 = (x - 2)(x - 2)$.

Notice that each factor occurs twice. In factoring the equation for the polynomial function f, if the same factor $x - r$ occurs k times, but not $k + 1$ times, we call r a **repeated zero with multiplicity k**. For the polynomial

$$f(x) = -x^2(x - 2)^2,$$

0 and 2 are both repeated zeros with multiplicity 2. For the polynomial

$$f(x) = 4(x - 5)(x + 2)^3\left(x - \tfrac{1}{4}\right)^4,$$

5 is a zero with multiplicity 1, -2 is a repeated zero with multiplicity 3, and $\tfrac{1}{4}$ is a repeated zero with multiplicity 4.

The multiplicity of a zero tells us if the graph of a polynomial function touches the x-axis at the zero and turns around or crosses the x-axis at the zero. For example, look again at the graph of $f(x) = -x^4 + 4x^3 - 4x^2$ in Figure 3.15. Each zero, 0 and 2, is a repeated zero with multiplicity 2. The graph of f touches, but does not cross, the x-axis at each of these zeros of even multiplicity. By contrast, a graph crosses the x-axis at zeros of odd multiplicity.

Multiplicity and x-Intercepts

If r is a zero of even multiplicity, then the graph **touches** the x-axis and turns around at r. If r is a zero of odd multiplicity, then the graph **crosses** the x-axis at r. Regardless of whether a zero is even or odd, graphs tend to flatten out at zeros with multiplicity greater than one.

Turning Points of Polynomial Functions

The graph of $f(x) = x^5 - 6x^3 + 8x + 1$ is shown in Figure 3.16. The graph has four smooth **turning points**. At each turning point, the graph changes direction from increasing to decreasing or vice versa. In calculus, these points are called **local maxima** or **local minima**. The given equation has 5 as its greatest exponent and is therefore a polynomial function of degree 5. Notice that the graph has four turning points. In general, **if f is a polynomial of degree n, then the graph of f has at most $n - 1$ turning points**.

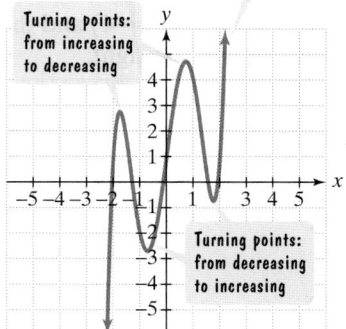

$f(x) = x^5 - 6x^3 + 8x + 1$

Turning points: from increasing to decreasing

Turning points: from decreasing to increasing

Figure 3.16 Graph with four turning points

5 Understand the relationship between degree and turning points.

Using the polynomial regression feature of a graphing utility, the third-degree polynomial function of best fit for the data is

$$T(x) = -0.87x^3 + 0.35x^2 + 81.62x + 7684.94.$$

a. Use this function to predict the number of larceny thefts in 2005.

b. Will this function be useful in modeling the number of larceny thefts over an extended period of time? Explain your answer.

54. Suppose that a polynomial function is used to model the data shown in the graph using

(number of years after 1900, murder rate per 100,000 people).

Murders Per 100,000 People in the United States

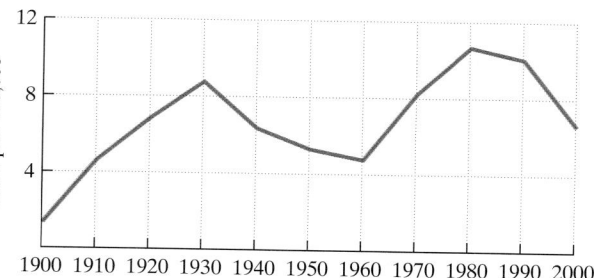

Source: National Center for Health Statistics

Determine the degree of the polynomial function of best fit. Should the leading coefficient be positive or negative? Explain your answers.

55. The polynomial function

$$H(x) = -0.001183x^4 + 0.05495x^3 - 0.8523x^2$$
$$+ 9.054x + 6.748$$

models the age in human years, $H(x)$, of a dog that is x years old, where $x \geq 1$.

a. Use this function to find the equivalent age in human years for a 10-year-old dog.

b. If dogs lived as long as humans, would this function be useful in modeling the dog's equivalent age in human years? Explain your answer.

Writing in Mathematics

56. What is a polynomial function?

57. What do we mean when we describe the graph of a polynomial function as smooth and continuous?

58. What is meant by the end behavior of a polynomial function?

59. Explain how to use the Leading Coefficient Test to determine the end behavior of a polynomial function.

60. Why is a third-degree polynomial function with a negative leading coefficient not appropriate for modeling non-negative real-world phenomena over a long period of time?

61. What are the zeros of a polynomial function and how are they found?

62. Explain the relationship between the multiplicity of a zero and whether or not the graph crosses or touches the x-axis at that zero.

63. Explain the relationship between the degree of a polynomial and the number of turning points on its graph.

64. Can the graph of a polynomial function have no x-intercepts? Explain.

65. Can the graph of a polynomial function have no y-intercept? Explain.

66. Describe a strategy for graphing a polynomial function. In your description, mention intercepts, the polynomial's degree, and turning points.

67. In a favorable habitat and without natural predators, a population of reindeer is introduced to an island preserve. The reindeer population t years after their introduction is modeled by the polynomial function $f(t) = -0.125t^5 + 3.125t^4 + 4000$. Discuss the growth and decline of the reindeer population. Describe the factors that might contribute to this population model.

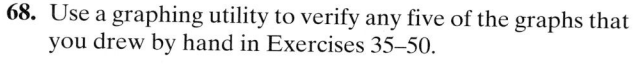

Technology Exercises

68. Use a graphing utility to verify any five of the graphs that you drew by hand in Exercises 35–50.

Write a polynomial function that imitates the end behavior of each graph in Exercises 69–72. The dashed portions of the graphs indicate that you should focus only on imitating the left and right behavior of the graph and can be flexible about what occurs between the left and right ends. Then use your graphing utility to graph the polynomial function and verify that you imitated the end behavior shown in the given graph.

69.   **70.**

71.   **72.**

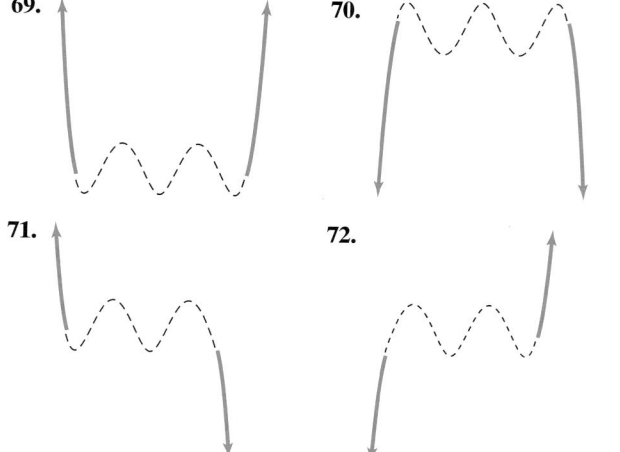

In Exercises 73–76, use a graphing utility with a viewing rectangle large enough to show end behavior to graph each polynomial function.

73. $f(x) = x^3 + 13x^2 + 10x - 4$

74. $f(x) = -2x^3 + 6x^2 + 3x - 1$

75. $f(x) = -x^4 + 8x^3 + 4x^2 + 2$

76. $f(x) = -x^5 + 5x^4 - 6x^3 + 2x + 20$

For Exercises 77–78, use a graphing utility to graph f and g in the same viewing rectangle. Then use the ZOOM OUT *feature to show that the end behavior of f and g is identical.*

77. $f(x) = x^3 - 6x + 1, \quad g(x) = x^3$

78. $f(x) = -x^4 + 2x^3 - 6x, \quad g(x) = -x^4$

Critical Thinking Exercises

79. Which one of the following is true?

  a. If $f(x) = -x^3 + 4x$, then the graph of f falls to the left and to the right.

  b. A mathematical model that is a polynomial of degree n whose leading term is $a_n x^n$, n odd and $a_n < 0$, is ideally suited to describe phenomena over unlimited periods of time.

  c. There is more than one third-degree polynomial function with the same three x-intercepts.

  d. The graph of a function with origin symmetry can rise to the left and to the right.

Use the descriptions in Exercises 80–81 to write an equation of a polynomial function with the given characteristics. Use a graphing utility to graph your function to see if you are correct. If not, modify the function's equation and repeat this process.

80. Crosses the x-axis at $-4, 0$, and 3; lies above the x-axis between -4 and 0; lies below the x-axis between 0 and 3

81. Touches the x-axis at 0 and crosses the x-axis at 2; lies below the x-axis between 0 and 2

 Group Exercise

82. This exercise is based on the group's work in Exercise 70 of Exercise Set 3.1. For the two data sets that the group selected:

  a. Use the polynomial regression feature of a graphing utility to find the third-degree polynomial function that best fits the data.

  b. Use this function to repeat the predictions that you made with the quadratic function. How do these predictions compare with those that you obtained previously?

  c. For each data set, describe whether the quadratic function or the third-degree function is a better fit. Use a graphing utility, a scatter plot of the data, and the function of best fit drawn on the scatter plot to help determine which function is the better fit.

SECTION 3.3 Dividing Polynomials; Remainder and Factor Theorems

Objectives

1. Use long division to divide polynomials.

2. Use synthetic division to divide polynomials.

3. Evaluate a polynomial using the Remainder Theorem.

4. Use the Factor Theorem to solve a polynomial equation.

For those of you who are dog lovers, you might still be thinking of the polynomial function that models the age in human years, $H(x)$, of a dog that is x years old, namely

$$H(x) = -0.001183x^4 + 0.05495x^3 - 0.8523x^2 + 9.054x + 6.748.$$

Suppose that you are in your twenties, say 25. What is Fido's equivalent age? To answer this question, we must substitute 25 for $H(x)$ and solve the resulting polynomial equation for x:

$$25 = -0.001183x^4 + 0.05495x^3 - 0.8523x^2 + 9.054x + 6.748.$$

How can we solve such an equation? You might begin by subtracting 25 from both sides to obtain zero on one side. But then what? The factoring that we used in the previous section will not work in this situation.

In Sections 3.4 and 3.5, we will present techniques for solving certain kinds of polynomial equations. These techniques will further enhance your ability to manipulate algebraically the formulas that model your world. Because these techniques are based on understanding polynomial division, in this section we look at two methods for dividing polynomials.

1 Use long division to divide polynomials.

Long Division of Polynomials and the Division Algorithm

We begin by looking at division by a polynomial containing more than one term, such as

$$x + 3 \overline{)\, x^2 + 10x + 21}$$

Divisor has two terms.   Dividend has three terms.

When a divisor has more than one term, the four steps used to divide whole numbers—**divide**, **multiply**, **subtract**, **bring down the next term**—form the repetitive procedure for polynomial long division.

EXAMPLE 1   Long Division of Polynomials

Divide $x^2 + 10x + 21$ by $x + 3$.

Solution   The following steps illustrate how polynomial division is very similar to numerical division.

$$x + 3 \overline{)\, x^2 + 10x + 21}$$

Arrange the terms of the dividend $\left(x^2 + 10x + 21\right)$ and the divisor $(x + 3)$ in descending powers of x.

$$\begin{array}{r} x \\ x + 3 \overline{)\, x^2 + 10x + 21} \end{array}$$

Divide x^2 (the first term in the dividend) by x (the first term in the divisor): $\dfrac{x^2}{x} = x$. Align like terms.

$$\begin{array}{r} x \\ x + 3 \overline{)\, x^2 + 10x + 21} \\ \underline{x^2 + 3x} \end{array}$$

times
equals

Multiply each term in the divisor $(x + 3)$ by x, aligning terms of the product under like terms in the dividend.

$$\begin{array}{r} x \\ x + 3 \overline{)\, x^2 + 10x + 21} \\ \underline{x^2 + 3x} \\ 7x \end{array}$$

Subtract $x^2 + 3x$ from $x^2 + 10x$ by changing the sign of each term in the lower expression and adding.

$$\begin{array}{r} x \\ x + 3 \overline{)\, x^2 + 10x + 21} \\ \underline{x^2 + 3x} \downarrow \\ 7x + 21 \end{array}$$

Bring down 21 from the original dividend and add algebraically to form a new dividend.

$$
\begin{array}{r}
x + 7 \\
x + 3 \overline{\smash{)}\ x^2 + 10x + 21} \\
\underline{x^2 + 3x} \downarrow \\
7x + 21
\end{array}
$$

Find the second term of the quotient. **Divide** the first term of $7x + 21$ by x, the first term of the divisor: $\dfrac{7x}{x} = 7$.

times

$$
\begin{array}{r}
x + 7 \\
x + 3 \overline{\smash{)}\ x^2 + 10x + 21} \\
\underline{x^2 + 3x} \\
7x + 21 \\
\underline{7x + 21} \\
0
\end{array}
$$

Multiply the divisor $(x + 3)$ by 7, aligning under like terms in the new dividend. Then **subtract** to obtain the remainder of 0.

equals

Remainder

The quotient is $x + 7$. Because the remainder is 0, we can conclude that $x + 3$ is a factor of $x^2 + 10x + 21$ and

$$x^2 + 10x + 21 = (x + 3)(x + 7).$$

Check Point 1   Divide $x^2 + 14x + 45$ by $x + 9$.

Before considering additional examples, let's summarize the general procedure for dividing one polynomial by another.

Long Division of Polynomials

1. **Arrange the terms** of both the dividend and the divisor in descending powers of any variable.
2. **Divide** the first term in the dividend by the first term in the divisor. The result is the first term of the quotient.
3. **Multiply** every term in the divisor by the first term in the quotient. Write the resulting product beneath the dividend with like terms lined up.
4. **Subtract** the product from the dividend.
5. **Bring down** the next term in the original dividend and write it next to the remainder to form a new dividend.
6. Use this new expression as the dividend and repeat this process until the remainder can no longer be divided. This will occur when the degree of the remainder (the highest exponent on a variable in the remainder) is less than the degree of the divisor.

In our next long division, we will obtain a nonzero remainder.

EXAMPLE 2   Long Division of Polynomials

Divide $4 - 5x - x^2 + 6x^3$ by $3x - 2$.

Solution   We begin by writing the divisor and dividend in descending powers of x.

Is It Hot in Here, or Is It Just Me?

In the 1980s, a rising trend in global surface temperature was observed and the term "global warming" was coined. Scientists are more convinced than ever that burning coal, oil, and gas results in a buildup of gases and particles that trap heat and raise the planet's temperature. The average increase in global surface temperature, in degrees Centigrade, x years after 1980 can be modeled by the polynomial function

$$T(x) = \frac{21}{5,000,000}x^3$$

$$- \frac{127}{1,000,000}x^2 + \frac{1293}{50,000}x.$$

Use your graphing utility to graph the function in a $[0, 60, 3]$ by $[0, 2, 0.1]$ viewing rectangle. (Place parentheses around each fractional coefficient when you enter the equation.) What do you observe about global warming through the year 2040?

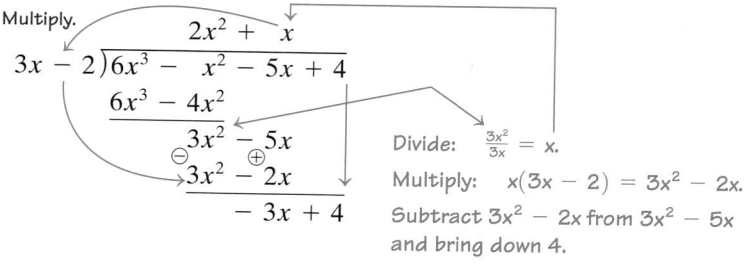

Now we divide $3x^2$ by $3x$ to obtain x, multiply x and the divisor, and subtract.

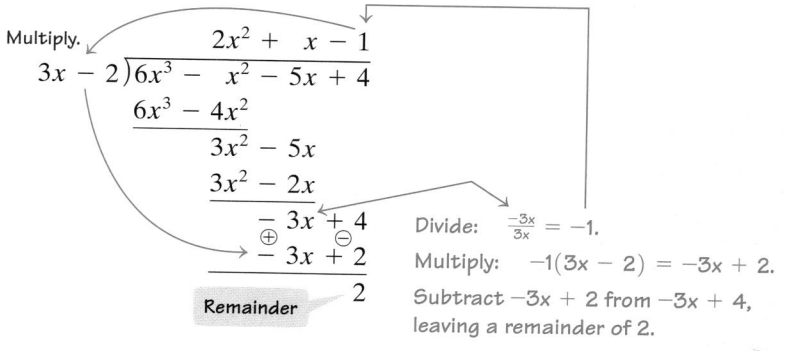

Now we divide $-3x$ by $3x$ to obtain -1, multiply -1 and the divisor, and subtract.

$$
\begin{array}{r}
2x^2 + x - 1 \\
3x - 2 \,\overline{\smash{\big)}\, 6x^3 - x^2 - 5x + 4} \\
\underline{6x^3 - 4x^2} \\
3x^2 - 5x \\
\underline{3x^2 - 2x} \\
-3x + 4 \\
\underline{-3x + 2} \\
2
\end{array}
$$

Remainder

Divide: $\frac{-3x}{3x} = -1.$

Multiply: $-1(3x - 2) = -3x + 2.$

Subtract $-3x + 2$ from $-3x + 4$, leaving a remainder of 2.

In Example 2, the quotient is $2x^2 + x - 1$ and the remainder is 2. This can be written in fractional form as follows:

Dividend   Quotient   Remainder

$$
\frac{6x^3 - x^2 - 5x + 4}{3x - 2} = 2x^2 + x - 1 + \frac{2}{3x - 2}
$$

Divisor   Divisor

Multiplying both sides of this equation by $3x - 2$ results in the following equation:

$$
6x^3 - x^2 - 5x + 4 = (3x - 2)(2x^2 + x - 1) + 2
$$

Dividend   Divisor   Quotient   Remainder

Polynomial long division is checked by multiplying the divisor with the quotient and then adding the remainder. This should give the dividend. The process illustrates the **Division Algorithm**.

The Division Algorithm

If $f(x)$ and $d(x)$ are polynomials, with $d(x) \neq 0$, and the degree of $d(x)$ is less than or equal to the degree of $f(x)$, then there exist unique polynomials $q(x)$ and $r(x)$ such that

$$f(x) \quad = \quad d(x) \quad \cdot \quad q(x) \quad + \quad r(x).$$

| Dividend | Divisor | Quotient | Remainder |

The remainder, $r(x)$, equals 0 or it is of degree less than the degree of $d(x)$. If $r(x) = 0$, we say that $d(x)$ **divides evenly** into $f(x)$ and that $d(x)$ and $q(x)$ are **factors** of $f(x)$.

Check Point 2

Divide $7 - 11x - 3x^2 + 2x^3$ by $x - 3$. Use the remainder to express your result in fractional form.

If a power of x is missing in either a dividend or a divisor, add that power of x with a coefficient of 0 and then divide. In this way, like terms will be aligned as you carry out the long division.

EXAMPLE 3   Long Division of Polynomials

Divide $6x^4 + 5x^3 + 3x - 5$ by $3x^2 - 2x$.

Solution   We write the dividend, $6x^4 + 5x^3 + 3x - 5$, as $6x^4 + 5x^3 + 0x^2 + 3x - 5$ so as to keep all like terms aligned.

The division process is finished because the degree of $7x - 5$, which is 1, is less than the degree of the divisor $3x^2 - 2x$, which is 2. The answer is

$$\frac{6x^4 + 5x^3 + 3x - 5}{3x^2 - 2x} = 2x^2 + 3x + 2 + \frac{7x - 5}{3x^2 - 2x}.$$

Check Point 3

Divide $2x^4 + 3x^3 - 7x - 10$ by $x^2 - 2x$.

2 Use synthetic division to divide polynomials.

Dividing Polynomials Using Synthetic Division

We can use **synthetic division** to divide polynomials if the divisor is of the form $x - c$. This method provides a quotient more quickly than long division. Let's compare the two methods showing $x^3 + 4x^2 - 5x + 5$ divided by $x - 3$.

Long Division   Quotient   **Synthetic Division**

$$
\begin{array}{r}
x^2 + 7x + 16 \\
x - 3 \overline{) x^3 + 4x^2 - 5x + 5}
\end{array}
$$

Divisor
$x - c$;
$c = 3$

$\ominus x^3 \overset{\oplus}{-} 3x^2$ Dividend

$7x^2 - 5x$

$\ominus 7x^2 \overset{\oplus}{-} 21x$

$16x \overset{\oplus}{+} 5$

$\ominus 16x - 48$ Remainder

53

$$
\begin{array}{r|rrrr}
3 & 1 & 4 & -5 & 5 \\
  &   & 3 & 21 & 48 \\
\hline
  & 1 & 7 & 16 & 53
\end{array}
$$

Notice the relationship between the polynomials in the long division process and the numbers that appear in synthetic division.

These are the coefficients of the dividend $x^3 + 4x^2 - 5x + 5$.

The divisor is $x - 3$. This is 3, or c in $x - c$.

$$
\begin{array}{r|rrrr}
3 & 1 & 4 & -5 & 5 \\
  &   & 3 & 21 & 48 \\
\hline
  & 1 & 7 & 16 & 53
\end{array}
$$

These are the coefficients of the quotient $x^2 + 7x + 16$.   This is the remainder.

Now let's look at the steps involved in synthetic division.

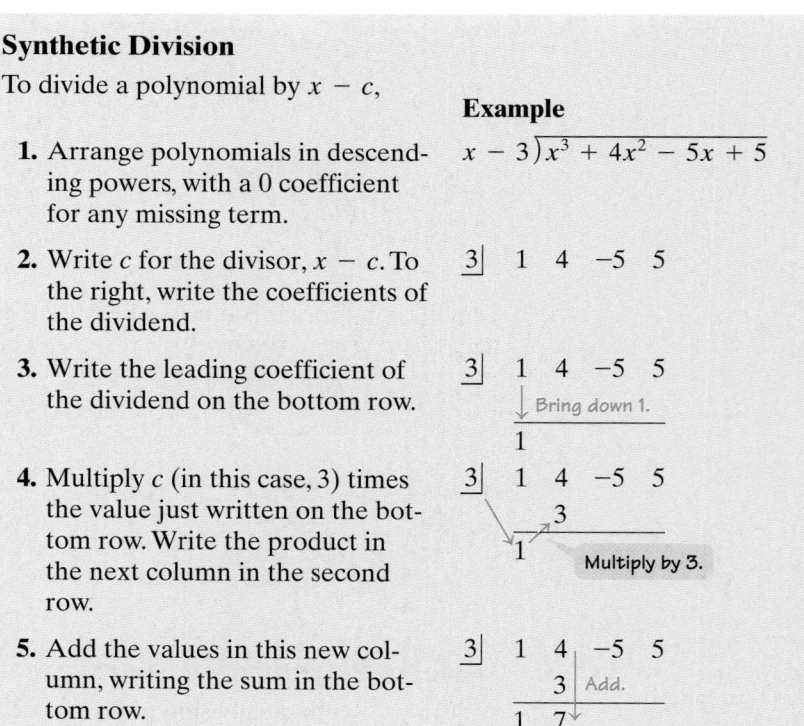

Synthetic Division

To divide a polynomial by $x - c$,

Example

1. Arrange polynomials in descending powers, with a 0 coefficient for any missing term.

$$x - 3 \overline{) x^3 + 4x^2 - 5x + 5}$$

2. Write c for the divisor, $x - c$. To the right, write the coefficients of the dividend.

$$
\begin{array}{r|rrrr}
3 & 1 & 4 & -5 & 5
\end{array}
$$

3. Write the leading coefficient of the dividend on the bottom row.

$$
\begin{array}{r|rrrr}
3 & 1 & 4 & -5 & 5 \\
\hline
  & 1 & & &
\end{array}
$$

Bring down 1.

4. Multiply c (in this case, 3) times the value just written on the bottom row. Write the product in the next column in the second row.

$$
\begin{array}{r|rrrr}
3 & 1 & 4 & -5 & 5 \\
  &   & 3 & & \\
\hline
  & 1 & & &
\end{array}
$$

Multiply by 3.

5. Add the values in this new column, writing the sum in the bottom row.

$$
\begin{array}{r|rrrr}
3 & 1 & 4 & -5 & 5 \\
  &   & 3 & & \\
\hline
  & 1 & 7 & &
\end{array}
$$

Add.

6. Repeat this series of multiplications and additions until all columns are filled in.

$$
\begin{array}{r|rrrr}
3 & 1 & 4 & -5 & 5 \\
  &   & 3 & 21 & \\
\hline
  & 1 & 7 & 16 &
\end{array}
$$
Add.

Multiply by 3.

$$
\begin{array}{r|rrrr}
3 & 1 & 4 & -5 & 5 \\
  &   & 3 & 21 & 48 \\
\hline
  & 1 & 7 & 16 & 53
\end{array}
$$
Add.

Multiply by 3.

7. Use the numbers in the last row to write the quotient and remainder in fractional form. **The degree of the first term of the quotient is one less than the degree of the first term of the dividend.** The final value in this row is the remainder.

Written from the last row of the synthetic division

$$x - 3 \overline{\smash{)}x^3 + 4x^2 - 5x + 5} \quad 1x^2 + 7x + 16 + \dfrac{53}{x-3}$$

EXAMPLE 4 Using Synthetic Division

Use synthetic division to divide $5x^3 + 6x + 8$ by $x + 2$.

Solution The divisor must be in the form $x - c$. Thus, we write $x + 2$ as $x - (-2)$. This means that $c = -2$. Writing a 0 coefficient for the missing x^2-term in the dividend, we can express the division as follows:

$$x - (-2) \overline{\smash{)}5x^3 + 0x^2 + 6x + 8}.$$

Now we are ready to set up the problem so that we can use synthetic division.

Use the coefficients of the dividend
$5x^3 + 0x^2 + 6x + 8$ in descending powers of x.

This is c
in $x-(-2)$.

$$
\begin{array}{r|rrrr}
-2 & 5 & 0 & 6 & 8
\end{array}
$$

We begin the synthetic division process by bringing down 5. This is followed by a series of multiplications and additions.

1. Bring down 5.

$$
\begin{array}{r|rrrr}
-2 & 5 & 0 & 6 & 8 \\
\hline
   & 5 &   &   &
\end{array}
$$

2. Multiply: $-2(5) = -10$.

$$
\begin{array}{r|rrrr}
-2 & 5 & 0 & 6 & 8 \\
   &   & -10 & & \\
\hline
   & 5 &   &   &
\end{array}
$$
Multiply by -2.

3. Add: $0 + (-10) = -10$.

$$
\begin{array}{r|rrrr}
-2 & 5 & 0 & 6 & 8 \\
   &   & -10 & & \\
\hline
   & 5 & -10 & &
\end{array}
$$
Add.

4. Multiply: $-2(-10) = 20$.

$$
\begin{array}{r|rrrr}
-2 & 5 & 0 & 6 & 8 \\
   &   & -10 & 20 & \\
\hline
   & 5 & -10 & &
\end{array}
$$
Multiply by -2.

5. Add: $6 + 20 = 26$.

$$
\begin{array}{r|rrrr}
-2 & 5 & 0 & 6 & 8 \\
   &   & -10 & 20 & \\
\hline
   & 5 & -10 & 26 &
\end{array}
$$
Add.

6. Multiply: $-2(26) = -52$.

$$
\begin{array}{r|rrrr}
-2 & 5 & 0 & 6 & 8 \\
   &   & -10 & 20 & -52 \\
\hline
   & 5 & -10 & 26 &
\end{array}
$$
Multiply by -2.

7. Add: $8 + (-52) = -44$.

$$
\begin{array}{r|rrrr}
-2 & 5 & 0 & 6 & 8 \\
   &   & -10 & 20 & -52 \\
\hline
   & 5 & -10 & 26 & -44
\end{array}
$$
Add.

The numbers in the last row represent the coefficients of the quotient and the remainder. The degree of the first term of the quotient is one less than that of the dividend. Because the degree of the dividend is 3, the degree of the quotient is 2. This means that the 5 in the last row represents $5x^2$.

$$\begin{array}{r|rrrr} -2 & 5 & 0 & 6 & 8 \\ & & -10 & 20 & -52 \\ \hline & 5 & -10 & 26 & -44 \end{array}$$

The quotient is $5x^2 - 10x + 26$.　　The remainder is -44.

Thus,

$$x + 2 \overline{)5x^3 + 6x + 8} \quad = \quad 5x^2 - 10x + 26 - \frac{44}{x + 2}$$

Check Point 4　Use synthetic division to divide $x^3 - 7x - 6$ by $x + 2$.

3　Evaluate a polynomial using the Remainder Theorem.

The Remainder Theorem

Let's consider the Division Algorithm when the dividend, $f(x)$, is divided by $x - c$. In this case, the remainder must be a constant because its degree is less than one, the degree of $x - c$.

$$f(x) = d(x)q(x) + r(x) \qquad \text{This is the Division Algorithm.}$$

Dividend　Divisor　Quotient　Remainder

$$f(x) = (x - c)q(x) + r \qquad \text{The divisor is } x - c. \text{ Call the constant remainder } r.$$

Now let's evaluate f at c.

$$f(c) = (c - c)q(c) + r \qquad \text{Find } f(c), \text{ setting } x = c. \text{ This will give an expression for } r.$$

$$f(c) = 0 \cdot q(c) + r \qquad c - c = 0 \text{ and } 0 \cdot q(c) = 0.$$

$$f(c) = r \qquad \text{On the right, } 0 + r = r.$$

What does this last equation mean? If a polynomial is divided by $x - c$, the remainder is the value of the polynomial at c. This result is called the **Remainder Theorem**.

The Remainder Theorem

If the polynomial $f(x)$ is divided by $x - c$, then the remainder is $f(c)$.

Example 5 shows how we can use the Remainder Theorem to evaluate a polynomial function at 2. Rather than substituting 2 for x, we divide the function by $x - 2$. The remainder is $f(2)$.

EXAMPLE 5   Using the Remainder Theorem to Evaluate a Polynomial Function

Given $f(x) = x^3 - 4x^2 + 5x + 3$, use the Remainder Theorem to find $f(2)$.

Solution   By the Remainder Theorem, if $f(x)$ is divided by $x - 2$, then the remainder is $f(2)$. We'll use synthetic division to divide.

$$\begin{array}{r|rrrr} 2 & 1 & -4 & 5 & 3 \\ & & 2 & -4 & 2 \\ \hline & 1 & -2 & 1 & 5 \end{array} \quad \text{Remainder}$$

The remainder, 5, is the value of $f(2)$. Thus, $f(2) = 5$. We can verify that this is correct by evaluating $f(2)$ directly. Using $f(x) = x^3 - 4x^2 + 5x + 3$, we obtain

$$f(2) = 2^3 - 4 \cdot 2^2 + 5 \cdot 2 + 3 = 8 - 16 + 10 + 3 = 5.$$

> **Check Point 5**   Given $f(x) = 3x^3 + 4x^2 - 5x + 3$, use the Remainder Theorem to find $f(-4)$.

4 Use the Factor Theorem to solve a polynomial equation.

The Factor Theorem

Let's look again at the Division Algorithm when the divisor is of the form $x - c$.

$$f(x) = (x - c)q(x) + r$$

Dividend   Divisor   Quotient   Constant remainder

By the Remainder Theorem, the remainder r is $f(c)$, so we can substitute $f(c)$ for r:

$$f(x) = (x - c)q(x) + f(c).$$

Notice that if $f(c) = 0$, then

$$f(x) = (x - c)q(x)$$

so that $x - c$ is a factor of $f(x)$. This means that for the polynomial function $f(x)$, if $f(c) = 0$, then $x - c$ is a factor of $f(x)$.

Let's reverse directions and see what happens if $x - c$ is a factor of $f(x)$. This means that

$$f(x) = (x - c)q(x).$$

If we replace x with c, we obtain

$$f(c) = (c - c)q(c) = 0.$$

Thus, if $x - c$ is a factor of $f(x)$, then $f(c) = 0$.

We have proved a result known as the **Factor Theorem**.

> **The Factor Theorem**
>
> Let $f(x)$ be a polynomial.
>   **a.** If $f(c) = 0$, then $x - c$ is a factor of $f(x)$.
>   **b.** If $x - c$ is a factor of $f(x)$, then $f(c) = 0$.

The example that follows shows how the Factor Theorem can be used to solve a polynomial equation.

EXAMPLE 6   Using the Factor Theorem

Solve the equation $2x^3 - 3x^2 - 11x + 6 = 0$ given that 3 is a zero of $f(x) = 2x^3 - 3x^2 - 11x + 6$.

Solution   We are given that $f(3) = 0$. The Factor Theorem tells us that $x - 3$ is a factor of $f(x)$. We'll use synthetic division to divide $f(x)$ by $x - 3$.

$$
\begin{array}{r|rrrr}
3 & 2 & -3 & -11 & 6 \\
  &   & 6 & 9   & -6 \\
\hline
  & 2 & 3 & -2 & 0
\end{array}
\qquad
\begin{array}{r}
2x^2 + 3x - 2 \\
x - 3 \overline{)\,2x^3 - 3x^2 - 11x + 6}
\end{array}
$$

Equivalently,

$$2x^3 - 3x^2 - 11x + 6 = (x - 3)(2x^2 + 3x - 2)$$

Now we can solve the polynomial equation.

$2x^3 - 3x^2 - 11x + 6 = 0$ This is the given equation.

$(x - 3)(2x^2 + 3x - 2) = 0$ Factor using the result from the synthetic division.

$(x - 3)(2x - 1)(x + 2) = 0$ Factor the trinomial.

$x - 3 = 0$ or $2x - 1 = 0$ or $x + 2 = 0$ Set each factor equal to 0.

$x = 3$ \qquad $x = \frac{1}{2}$ \qquad $x = -2$ Solve for x.

The solution set is $\left\{ -2, \frac{1}{2}, 3 \right\}$.

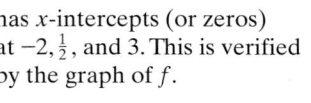

Technology

Because the solution set of

$$2x^3 - 3x^2 - 11x + 6 = 0$$

is $\left\{ -2, \frac{1}{2}, 3 \right\}$, this implies that the polynomial function

$$f(x) = 2x^3 - 3x^2 - 11x + 6$$

has x-intercepts (or zeros) at $-2, \frac{1}{2}$, and 3. This is verified by the graph of f.

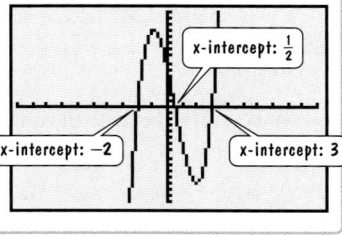

By the Factor Theorem, the following statements are useful in solving polynomial equations:

1. If $f(x)$ is divided by $x - c$ and the remainder is zero, then c is a zero of f and c is a root of the polynomial equation $f(x) = 0$.

2. If $f(x)$ is divided by $x - c$ and the remainder is zero, then $x - c$ is a factor of $f(x)$.

Check Point 6   Solve the equation $15x^3 + 14x^2 - 3x - 2 = 0$ given that -1 is a zero of $f(x) = 15x^3 + 14x^2 - 3x - 2$.

EXERCISE SET 3.3

 Practice Exercises

In Exercises 1–16, divide by long division.

1. $(x^2 + 8x + 15) \div (x + 5)$
2. $(x^2 + 3x - 10) \div (x - 2)$
3. $(x^3 + 5x^2 + 7x + 2) \div (x + 2)$
4. $(x^3 - 2x^2 - 5x + 6) \div (x - 3)$
5. $(6x^3 + 7x^2 + 12x - 5) \div (3x - 1)$
6. $(6x^3 + 17x^2 + 27x + 20) \div (3x + 4)$
7. $(12x^2 + x - 4) \div (3x - 2)$
8. $(4x^2 - 8x + 6) \div (2x - 1)$
9. $\dfrac{2x^3 + 7x^2 + 9x - 20}{x + 3}$
10. $\dfrac{3x^2 - 2x + 5}{x - 3}$
11. $\dfrac{4x^4 - 4x^2 + 6x}{x - 4}$
12. $\dfrac{x^4 - 81}{x - 3}$
13. $\dfrac{6x^3 + 13x^2 - 11x - 15}{3x^2 - x - 3}$
14. $\dfrac{x^4 + 2x^3 - 4x^2 - 5x - 6}{x^2 + x - 2}$
15. $\dfrac{18x^4 + 9x^3 + 3x^2}{3x^2 + 1}$
16. $\dfrac{2x^5 - 8x^4 + 2x^3 + x^2}{2x^3 + 1}$

In Exercises 17–32, divide by synthetic division.

17. $(2x^2 + x - 10) \div (x - 2)$
18. $(x^2 + x - 2) \div (x - 1)$
19. $(3x^2 + 7x - 20) \div (x + 5)$
20. $(5x^2 - 12x - 8) \div (x + 3)$
21. $(4x^3 - 3x^2 + 3x - 1) \div (x - 1)$
22. $(5x^3 - 6x^2 + 3x + 11) \div (x - 2)$
23. $(6x^5 - 2x^3 + 4x^2 - 3x + 1) \div (x - 2)$
24. $(x^5 + 4x^4 - 3x^2 + 2x + 3) \div (x - 3)$
25. $(x^2 - 5x - 5x^3 + x^4) \div (5 + x)$
26. $(x^2 - 6x - 6x^3 + x^4) \div (6 + x)$
27. $\dfrac{x^5 + x^3 - 2}{x - 1}$
28. $\dfrac{x^7 + x^5 - 10x^3 + 12}{x + 2}$
29. $\dfrac{x^4 - 256}{x - 4}$
30. $\dfrac{x^7 - 128}{x - 2}$
31. $\dfrac{2x^5 - 3x^4 + x^3 - x^2 + 2x - 1}{x + 2}$
32. $\dfrac{x^5 - 2x^4 - x^3 + 3x^2 - x + 1}{x - 2}$

33. Given $f(x) = 2x^3 - 11x^2 + 7x - 5$, use the Remainder Theorem to find $f(4)$.
34. Given $f(x) = x^3 - 7x^2 + 5x - 6$, use the Remainder Theorem to find $f(3)$.
35. Given $f(x) = 7x^4 - 3x^3 + 6x + 9$, use the Remainder Theorem to find $f(-5)$.

36. Given $f(x) = 3x^4 + 6x^3 - 2x + 4$, use the Remainder Theorem to find $f(-4)$.
37. Use synthetic division to divide $f(x) = x^3 - 4x^2 + x + 6$ by $x + 1$. Use the result to find all zeros of f.
38. Use synthetic division to divide $f(x) = x^3 - 2x^2 - x + 2$ by $x + 1$. Use the result to find all zeros of f.
39. Solve the equation $2x^3 - 5x^2 + x + 2 = 0$ given that 2 is a zero of $f(x) = 2x^3 - 5x^2 + x + 2$.
40. Solve the equation $2x^3 - 3x^2 - 11x + 6 = 0$ given that -2 is a zero of $f(x) = 2x^3 - 3x^2 - 11x + 6$.
41. Solve the equation $12x^3 + 16x^2 - 5x - 3 = 0$ given that $-\frac{3}{2}$ is a root.
42. Solve the equation $3x^3 + 7x^2 - 22x - 8 = 0$ given that $-\frac{1}{3}$ is a root.

 Application Exercises

43. A rectangle with length $2x + 5$ inches has an area of $2x^4 + 15x^3 + 7x^2 - 135x - 225$ square inches. Write a polynomial that represents its width.
44. If you travel a distance of $x^3 + 3x^2 + 5x + 3$ miles at a rate of $x + 1$ miles per hour, write a polynomial that represents the number of hours you traveled.
45. Two people are 25 years old and 20 years old, respectively. In x years from now, their ages can be represented by $x + 25$ and $x + 20$.

  a. Use long division to find $\dfrac{x + 25}{x + 20}$, the ratio of the older person's age in x years to the younger person's age in x years.

  b. Complete the following table.

x	0	5	10	25	50	75
$x + 25$						
$x + 20$						

  c. Describe what is happening to the ratio $\dfrac{x + 25}{x + 20}$ as x increases. How can this be verified using the result of the long division in part (a)?

 **Writing in Mathematics**

46. Explain how to perform long division of polynomials. Use $2x^3 - 3x^2 - 11x + 7$ divided by $x - 3$ in your explanation.
47. In your own words, state the Division Algorithm.
48. How can the Division Algorithm be used to check the quotient and remainder in a long division problem?
49. Explain how to perform synthetic division. Use the division problem in Exercise 46 to support your explanation.
50. State the Remainder Theorem.

51. Explain how the Remainder Theorem can be used to find $f(-6)$ if $f(x) = x^4 + 7x^3 + 8x^2 + 11x + 5$. What advantage is there to using the Remainder Theorem in this situation rather than evaluating $f(-6)$ directly?

52. How can the Factor Theorem be used to determine if $x - 1$ is a factor of $x^3 - 2x^2 - 11x + 12$?

53. If you know that -2 is a zero of
$$f(x) = x^3 + 7x^2 + 4x - 12,$$
explain how to solve the equation
$$x^3 + 7x^2 + 4x - 12 = 0.$$

Technology Exercises

In Exercises 54–57, use a graphing utility to graph the function on each side of the given equation. If the graphs coincide, this verifies that the expressions are equivalent and the division has been performed correctly. If the graphs do not coincide, correct the expression on the right by performing the division. Then use your graphing utility to verify your result.

54. $\dfrac{x^4 + 6x^3 + 6x^2 - 10x - 3}{x^2 + 2x - 3} = x^2 + 4x + 1$,

$x \ne -3, \quad x \ne 1$

55. $\dfrac{2x^3 - 3x^2 - 3x + 4}{x - 1} = 2x^2 - x + 4, \quad x \ne 1$

56. $\dfrac{3x^4 + 4x^3 - 32x^2 - 5x - 20}{x + 4} = 3x^3 + 8x^2 - 5$,

$x \ne -4$

57. $\dfrac{10x^3 - 26x^2 + 17x - 13}{5x - 3} = 2x^2 - 4x + 1 - \dfrac{10}{5x - 3}$,

$x \ne \frac{3}{5}$

Critical Thinking Exercises

58. Which one of the following is true?
   a. If a trinomial in x of degree 6 is divided by a trinomial in x of degree 3, the degree of the quotient is 2.
   b. Synthetic division could not be used to find the quotient of $10x^3 - 6x^2 + 4x - 1$ and $x - \frac{1}{2}$.
   c. Any problem that can be done by synthetic division can also be done by the method for long division of polynomials.
   d. If a polynomial long-division problem results in a remainder that is a whole number, then the divisor is a factor of the dividend.

59. Find k so that $4x + 3$ is a factor of
$$20x^3 + 23x^2 - 10x + k.$$

60. When $2x^2 - 7x + 9$ is divided by a polynomial, the quotient is $2x - 3$ and the remainder is 3. Find the polynomial.

61. Find the quotient of $x^{3n} + 1$ and $x^n + 1$.

62. Synthetic division is a process for dividing a polynomial by $x - c$. The coefficient of x is 1. How might synthetic division be used if you are dividing by $2x - 4$?

SECTION 3.4 *Zeros of Polynomial Functions*

Objectives

1. Use the Rational Zero Theorem to find possible rational zeros.

2. Find zeros of a polynomial function.

3. Solve polynomial equations.

4. Use Descartes's Rule of Signs.

The solution to a multitude of moths?

A moth has moved into your closet. She appeared in your bedroom at night, but somehow her relatively stout body escaped your clutches. Within a few weeks swarms of moths in your tattered wardrobe suggest that Mama Moth was in the family way. There must be at least 200 critters nesting in every crevice of your clothing.

Two hundred plus moth-tykes from one female moth; is this possible? Indeed it is. The number of eggs, N, in a female moth is a function of her abdominal width, W, in millimeters, modeled by

$$N = 14W^3 - 17W^2 - 16W + 34$$

for $1.5 \leq W \leq 3.5$. Because there are 200 moths feasting on your favorite sweaters, Mama's abdominal width can be estimated by finding the roots of the polynomial equation

$$14W^3 - 17W^2 - 16W + 34 = 200.$$

With mathematics present even in your quickly disappearing attire, we move from rags to polynomial equations. The process of solving such equations begins with listing possibilities for Mama Moth's abdominal width. To do this, we turn to a theorem that plays an important role in finding zeros of polynomial functions.

1 Use the Rational Zero Theorem to find possible rational zeros.

The Rational Zero Theorem

The Rational Zero Theorem gives a list of possible rational zeros of a polynomial function. Equivalently, the theorem gives all possible rational roots of a polynomial equation. Not every number in the list will be a zero of the function, but every rational zero of the polynomial function will appear somewhere in the list.

The Rational Zero Theorem

If $f(x) = a_n x^n + a_{n-1} x^{n-1} + \cdots + a_1 x + a_0$ has *integer* coefficients and $\dfrac{p}{q}$ (where $\dfrac{p}{q}$ is reduced) is a rational zero, then p is a factor of the constant term a_0 and q is a factor of the leading coefficient a_n.

You can explore the "why" behind the Rational Zero Theorem in Exercise 64 of Exercise Set 3.4. For now, let's see if we can figure out what the theorem tells us about possible rational zeros. In order to use the theorem, list all the integers that are factors of the constant term, a_0. Then list all the integers that are factors of the leading coefficient, a_n. Finally list all possible rational zeros:

$$\text{Possible rational zeros} = \frac{\text{Factors of the constant term}}{\text{Factors of the leading coefficient}}.$$

EXAMPLE 1 Using the Rational Zero Theorem

List all possible rational zeros of $f(x) = -x^4 + 4x^2 + 4$.

Solution The constant term is 4. We list all of its factors: $\pm 1, \pm 2, \pm 4$. The leading coefficient is -1. Its factors are ± 1.

Factors of the constant term:	$\pm 1, \quad \pm 2 \quad \pm 4$
Factors of the leading coefficient:	± 1

Because

$$\text{Possible rational zeros} = \frac{\text{Factors of the constant term}}{\text{Factors of the leading coefficient}},$$

we must take each number in the first row, $\pm 1, \pm 2, \pm 4$, and divide by each number in the second row, ± 1.

$$\text{Possible rational zeros} = \frac{\text{Factors of } 4}{\text{Factors of } -1} = \frac{\pm 1, \pm 2, \pm 4}{\pm 1} = \pm 1, \quad \pm 2, \quad \pm 4$$

| Divide ±1 by ±1. | Divide ±2 by ±1. | Divide ±4 by ±1. |

There are six possible rational zeros. The graph of $f(x) = -x^4 + 4x^2 + 4$ is shown in Figure 3.18. The x-intercepts are -2 and 2. Thus, -2 and 2 are the actual rational zeros.

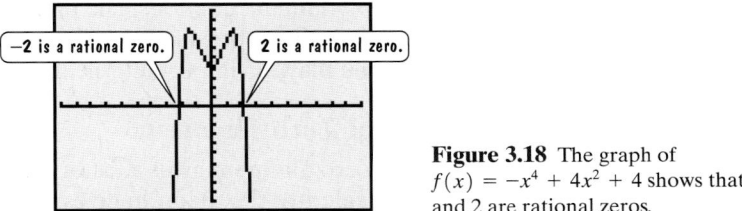

-2 is a rational zero.   2 is a rational zero.

Figure 3.18 The graph of $f(x) = -x^4 + 4x^2 + 4$ shows that -2 and 2 are rational zeros.

Check Point 1 List all possible rational zeros of $f(x) = x^3 + 2x^2 - 5x - 6$.

EXAMPLE 2   Using the Rational Zero Theorem

List all possible rational zeros of $f(x) = 15x^3 + 14x^2 - 3x - 2$.

Solution   The constant term is -2 and the leading coefficient is 15.

$$\text{Possible rational zeros} = \frac{\text{Factors of the constant term, } -2}{\text{Factors of the leading coefficient, } 15}$$

$$= \frac{\pm 1, \pm 2}{\pm 1, \pm 3, \pm 5, \pm 15}$$

$$= \pm 1, \quad \pm 2, \quad \pm \tfrac{1}{3}, \quad \pm \tfrac{2}{3}, \quad \pm \tfrac{1}{5}, \quad \pm \tfrac{2}{5}, \quad \pm \tfrac{1}{15}, \quad \pm \tfrac{2}{15}$$

| Divide ±1 and ±2 by ±1. | Divide ±1 and ±2 by ±3. | Divide ±1 and ±2 by ±5. | Divide ±1 and ±2 by ±15. |

There are 16 possible rational zeros. The actual solution set to $15x^3 + 14x^2 - 3x - 2 = 0$ is $\left\{-1, -\tfrac{1}{3}, \tfrac{2}{5}\right\}$, which contains 3 of the 16 possible zeros.

Check Point 2 Find all possible rational zeros of $f(x) = 4x^5 + 12x^4 - x - 3$.

2 Find zeros of a polynomial function.

How do we determine which (if any) of the possible rational zeros are rational zeros of the polynomial function? To find the first rational zero, we can use a trial-and-error process involving synthetic division. [Recall that if $f(x)$ is divided by $x - c$ and the remainder is zero, then c is a zero of f.] After we identify the first rational zero, we use the result of the synthetic division to factor the original polynomial. Then we set each factor equal to zero to identify any additional rational zeros.

EXAMPLE 3 Finding Zeros of a Polynomial Function

Find all rational zeros of $f(x) = x^3 + 2x^2 - 5x - 6$.

Solution We begin by listing all possible rational zeros.

Possible rational zeros

$$= \frac{\text{Factors of the constant term, } -6}{\text{Factors of the leading coefficient, } 1} = \frac{\pm 1, \pm 2, \pm 3, \pm 6}{\pm 1} = \pm 1, \pm 2, \pm 3, \pm 6$$

> Divide the eight numbers in the numerator by ± 1.

Now we will use synthetic division to see if we can find a rational root among the possible rational zeros $\pm 1, \pm 2, \pm 3, \pm 6$. Keep in mind that if $f(x)$ is divided by $x - c$ and the remainder is zero, then c is a zero of f. Let's start by testing 1. If 1 is not a rational zero, then we will test other possible rational zeros.

Test 1

> Coefficients of
> $f(x) = x^3 + 2x^2 - 5x - 6$

Possible rational zero

$$\begin{array}{r|rrrr} 1 & 1 & 2 & -5 & -6 \\ & & 1 & 3 & -2 \\ \hline & 1 & 3 & -2 & -8 \end{array}$$

> The nonzero remainder shows that 1 is not a zero.

Test 2

> Coefficients of
> $f(x) = x^3 + 2x^2 - 5x - 6$

Possible rational zero

$$\begin{array}{r|rrrr} 2 & 1 & 2 & -5 & -6 \\ & & 2 & 8 & 6 \\ \hline & 1 & 4 & 3 & 0 \end{array}$$

> The zero remainder shows that 2 is a zero.

The zero remainder tells us that 2 is a zero of the polynomial function $f(x) = x^3 + 2x^2 - 5x - 6$. Equivalently, 2 is a solution, or root, of the polynomial equation $x^3 + 2x^2 - 5x - 6 = 0$. Thus, $x - 2$ is a factor of the polynomial.

$$x^3 + 2x^2 - 5x - 6 = 0 \qquad \text{Finding the zeros of } f(x) = x^3 + 2x^2 - 5x - 6 \text{ is the same as finding the roots of this equation.}$$

$$(x - 2)(x^2 + 4x + 3) = 0 \qquad \text{Factor using the result from the synthetic division.}$$

$$(x - 2)(x + 3)(x + 1) = 0 \qquad \text{Factor completely.}$$

$$x - 2 = 0 \quad \text{or} \quad x + 3 = 0 \quad \text{or} \quad x + 1 = 0 \qquad \text{Set each factor equal to zero.}$$

$$x = 2 \qquad\qquad x = -3 \qquad\qquad x = -1 \qquad \text{Solve for } x.$$

The solution set is $\{-3, -1, 2\}$. The rational zeros of f are $-3, -1$, and 2.

> **Check Point 3** Find all rational zeros of
> $f(x) = x^3 + 8x^2 + 11x - 20$.

Our work in Example 3 involved solving a third-degree equation. We found one factor by synthetic division and factored the remaining quadratic factor using the FOIL method. If the degree of a polynomial function or equation is 4 or higher, it is often necessary to find more than one linear factor by synthetic division.

One way to speed up the process of finding the first zero is to graph the function. Any x-intercept is a zero.

3 Solve polynomial equations.

EXAMPLE 4   Solving a Polynomial Equation

Solve: $x^4 - 6x^2 - 8x + 24 = 0$.

Solution   Recall that we refer to the zeros of a polynomial function and the roots of a polynomial equation. Because we are given an equation, we will use the word "roots," rather than "zeros," in the solution process. We begin by listing all possible rational roots.

Possible rational roots

$$= \frac{\text{Factors of the constant term, } 24}{\text{Factors of the leading coefficient, } 1}$$

$$= \frac{\pm 1, \pm 2, \pm 3, \pm 4, \pm 6, \pm 8, \pm 12, \pm 24}{\pm 1} = \pm 1, \pm 2, \pm 3, \pm 4, \pm 6, \pm 8, \pm 12, \pm 24$$

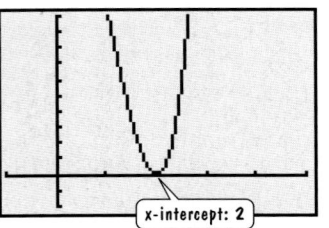

Figure 3.19 The graph of $f(x) = x^4 - 6x^2 - 8x + 24$ in a $[-1, 5, 1]$ by $[-2, 10, 1]$ viewing rectangle

The graph of $f(x) = x^4 - 6x^2 - 8x + 24$ is shown in Figure 3.19. Because the x-intercept is 2, we will test 2 by synthetic division and show that it is a root of the given equation.

$$
\begin{array}{r|rrrrr}
2 & 1 & 0 & -6 & -8 & 24 \\
  &   & 2 & 4 & -4 & -24 \\
\hline
  & 1 & 2 & -2 & -12 & 0
\end{array}
$$

Careful!
$x^4 - 6x^2 - 8x + 24 = x^4 + 0x^3 - 6x^2 - 8x + 24$

The zero remainder indicates that 2 is a root of $x^4 - 6x^2 - 8x + 24 = 0$.

Now we can rewrite the given equation in factored form.

$$x^4 - 6x^2 - 8x + 24 = 0 \quad \text{This is the given equation.}$$

$$(x - 2)(x^3 + 2x^2 - 2x - 12) = 0 \quad \text{This is the result obtained from the synthetic division.}$$

$$x - 2 = 0 \quad \text{or} \quad x^3 + 2x^2 - 2x - 12 = 0 \quad \text{Set each factor equal to 0.}$$

We can use the same approach to look for rational roots of the polynomial equation $x^3 + 2x^2 - 2x - 12 = 0$, listing all possible rational roots. However, take a second look at the graph in Figure 3.19. Because the graph turns around at 2, this means that 2 is a root of even multiplicity. Thus, 2 must also be a root of $x^3 + 2x^2 - 2x - 12 = 0$, confirmed by the following synthetic division.

$$
\begin{array}{r|rrrr}
2 & 1 & 2 & -2 & -12 \\
  &   & 2 & 8 & 12 \\
\hline
  & 1 & 4 & 6 & 0
\end{array}
$$

These are the coefficients of $x^3 + 2x^2 - 2x - 12 = 0$.

The zero remainder indicates that 2 is a root of $x^3 + 2x^2 - 2x - 12 = 0$.

Now we can solve the original equation as follows:

$$x^4 - 6x^2 - 8x + 24 = 0 \quad \text{This is the given equation.}$$

$$(x - 2)(x^3 + 2x^2 - 2x - 12) = 0 \quad \text{This was obtained from the first synthetic division.}$$

$$(x - 2)(x - 2)(x^2 + 4x + 6) = 0 \quad \text{This was obtained from the second synthetic division.}$$

$$x - 2 = 0 \quad \text{or} \quad x - 2 = 0 \quad \text{or} \quad x^2 + 4x + 6 = 0 \quad \text{Set each factor equal to 0.}$$

$$x = 2 \qquad\qquad x = 2 \qquad\qquad x^2 + 4x + 6 = 0 \quad \text{Solve.}$$

We can use the quadratic formula to solve $x^2 + 4x + 6 = 0$.

$$x = \frac{-b \pm \sqrt{b^2 - 4ac}}{2a}$$

We use the quadratic formula because $x^2 + 4x + 6$ cannot be factored.

$$= \frac{-4 \pm \sqrt{4^2 - 4(1)(6)}}{2(1)}$$

Let $a = 1$, $b = 4$, and $c = 6$.

$$= \frac{-4 \pm \sqrt{-8}}{2}$$

Multiply and subtract under the radical.

$$= \frac{-4 \pm 2i\sqrt{2}}{2}$$

$\sqrt{-8} = \sqrt{4(2)(-1)} = 2i\sqrt{2}$

$$= -2 \pm i\sqrt{2}$$

Simplify.

The solution set of the original equation is $\{2, -2 - i\sqrt{2}, -2 + i\sqrt{2}\}$.

In Example 4, 2 is a repeated root of the equation with multiplicity 2. The example illustrates two general properties.

Properties of Polynomial Equations

1. If a polynomial equation is of degree n, then counting multiple roots separately, the equation has n roots.
2. If $a + bi$ is a root of a polynomial equation $(b \neq 0)$, then the nonreal complex number $a - bi$ is also a root. Nonreal complex roots, if they exist, occur in conjugate pairs.

These ideas will be developed in more detail in the next section.

Check Point 4 Solve: $x^4 - 6x^3 + 22x^2 - 30x + 13 = 0$.

4 Use Descartes's Rule of Signs.

Descartes's Rule of Signs

Because an nth-degree polynomial equation might have roots that are imaginary numbers, we should note that such an equation can have *at most n* real roots. **Descartes's Rule of Signs** provides even more specific information about the number of real zeros that a polynomial can have. The rule is based on considering *variations in sign* between consecutive coefficients. For example, the function

$$f(x) = 3x^7 - 2x^5 - x^4 + 7x^2 + x - 3$$

has three sign changes.

An equation can have as many true [positive] roots as it contains changes of sign, from plus to minus or from minus to plus.... René Descartes (1596–1650) in *La Géométrie* (1637)

Descartes's Rule of Signs

Let $f(x) = a_n x^n + a_{n-1} x^{n-1} + \cdots + a_2 x^2 + a_1 x + a_0$ be a polynomial with real coefficients.

1. The number of *positive real zeros* of f is either equal to the number of sign changes of $f(x)$ or is less than that number by an even integer. If

In Exercises 29–40, find all zeros of the polynomial function or solve the given polynomial equation. Use the Rational Zero Theorem and Descartes's Rule of Signs as an aid in obtaining the first zero or the first root.

29. $f(x) = x^3 - 4x^2 - 7x + 10$
30. $f(x) = x^3 + 12x^2 + 21x + 10$
31. $2x^3 - x^2 - 9x - 4 = 0$
32. $3x^3 - 8x^2 - 8x + 8 = 0$
33. $x^4 - 3x^3 - 20x^2 - 24x - 8 = 0$
34. $x^4 - x^3 + 2x^2 - 4x - 8 = 0$
35. $f(x) = 3x^4 - 11x^3 - x^2 + 19x + 6$
36. $f(x) = 2x^4 + 3x^3 - 11x^2 - 9x + 15$
37. $4x^4 - x^3 + 5x^2 - 2x - 6 = 0$
38. $3x^4 - 11x^3 - 3x^2 - 6x + 8 = 0$
39. $2x^5 + 7x^4 - 18x^2 - 8x + 8 = 0$
40. $4x^5 + 12x^4 - 41x^3 - 99x^2 + 10x + 24 = 0$

⭐ Application Exercises

41. Suppose that a polynomial function f is used to model the data shown in the graph using

(number of years after 1993, thousands of deaths at the workplace).

Thousands of Workplace Deaths in the U.S.

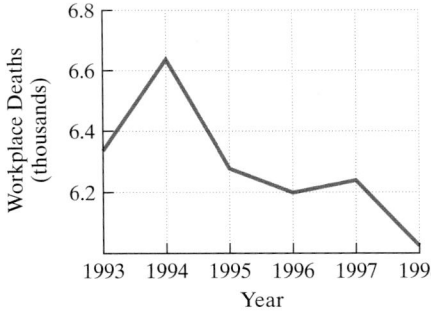

Source: F. B. I.

a. Use the graph to solve the polynomial equation $f(x) = 6.2$.
b. Describe the degree and the leading coefficient of the function f that can be used to model the data in the graph.

42. Suppose that a polynomial function f is used to model the data shown in the graph at the top of the next column using

(number of years after 1995, average cost of a computer in thousands of dollars).

Average Cost of Computers in the U.S.

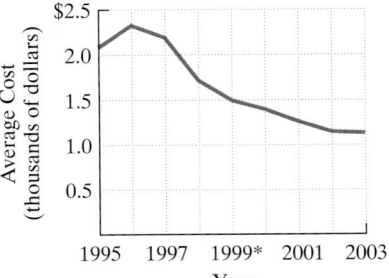

*Costs are projected from 1999 to 2003
Source*: National Science Foundation

Use the graph to solve the polynomial equation $f(x) = 1.5$.

43. The number of eggs, N, in a female moth is a function of her abdominal width, W, in millimeters, modeled by $N = 14W^3 - 17W^2 - 16W + 34$, for $1.5 \le W \le 3.5$. What is the abdominal width when there are 211 eggs?

44. The concentration of a drug, in parts per million, in a patient's blood x hours after the drug is administered is given by the function

$$f(x) = -x^4 + 12x^3 - 58x^2 + 132x.$$

How many hours after the drug is administered will it be eliminated from the bloodstream?

45. The width of a rectangular box is twice the height and the length is 7 inches more than the height. If the volume is 72 cubic inches, find the dimensions of the box.

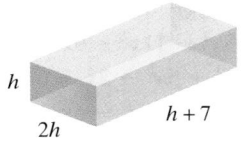

46. A box with an open top is formed by cutting squares out of the corners of a rectangular piece of cardboard 10 inches by 8 inches and then folding up the sides. If x represents the length of the side of the square cut from each corner of the rectangle, what size square must be cut if the volume of the box is to be 48 cubic inches?

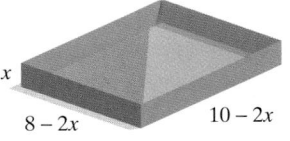

Writing in Mathematics

47. Describe how to find the possible rational zeros of a polynomial function.

48. Describe how to use Descartes's Rule of signs to determine the possible number of positive real zeros of a polynomial function.

49. Describe how to use Descartes's rule of signs to determine the possible number of negative roots of a polynomial equation.

50. Why must every polynomial equation of degree 3 have at least one real root?

51. Explain why the equation $x^4 + 6x^2 + 2 = 0$ has no rational roots.

52. Suppose $\frac{3}{4}$ is a root of a polynomial equation. What does this tell us about the leading coefficient and the constant term in the equation?

53. The number of AIDS cases in the United States for the years 1983 through 1990 is approximated by the function

$$f(x) = -143x^3 + 1810x^2 - 187x + 2331$$

where x represents the number of years after 1983. Use the Rational Zero Theorem to explain why, according to this formula, 14,199 cases could not have occurred 5 years after 1983.

Technology Exercises

The equations in Exercises 54–57 have real roots that are rational. Use the Rational Zero Theorem to list all possible rational roots. Then graph the polynomial function in the given viewing rectangle to determine which possible rational roots are actual roots of the equation.

54. $2x^3 - 15x^2 + 22x + 15 = 0$; $[-1, 6, 1]$ by $[-50, 50, 1]$

55. $6x^3 - 19x^2 + 16x - 4 = 0$; $[0, 2, 1]$ by $[-3, 2, 1]$

56. $2x^4 + 7x^3 - 4x^2 - 27x - 18 = 0$; $[-4, 3, 1]$ by $[-45, 45, 1]$

57. $4x^4 + 4x^3 + 7x^2 - x - 2 = 0$; $[-2, 2, 1]$ by $[-5, 5, 1]$

58. Use Descartes's Rule of Signs to determine the possible number of positive and negative real zeros of $f(x) = 3x^4 + 5x^2 + 2$. What does this mean in terms of the graph of f? Verify your result by using a graphing utility to graph f.

59. Use Descartes's Rule of Signs to determine the possible number of positive and negative real zeros of $f(x) = x^5 - x^4 + x^3 - x^2 + x - 8$. Verify your result by using a graphing utility to graph f.

60. Make up a number of polynomial functions of odd degree and graph each function. Is it possible for the graph to have no real zeros? Explain. Try doing the same thing for polynomial functions of even degree. Now is it possible to have no real zeros?

Critical Thinking Exercises

61. Which one of the following is true?
   a. The equation $x^3 + 5x^2 + 6x + 1 = 0$ has one positive real root.
   b. Descartes's Rule of Signs gives the exact number of positive and negative real roots for a polynomial equation.
   c. Every polynomial equation of degree 3 has at least one rational root.
   d. None of the above is true.

62. Give an example of a polynomial equation that has no real roots. Describe how you obtained the equation.

63. If the volume of the solid shown in the figure is 208 cubic inches, find the value of x.

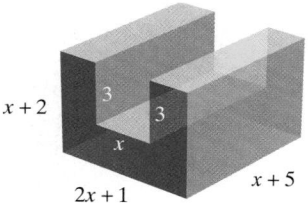

64. In this exercise, we lead you through the steps involved in the proof of the Rational Zero Theorem. Consider the polynomial equation

$$a_n x^n + a_{n-1}x^{n-1} + a_{n-2}x^{n-2} + \cdots + a_1 x + a_0 = 0$$

where $\frac{p}{q}$ is a rational root reduced to lowest terms.

   a. Substitute $\frac{p}{q}$ for x in the equation and show that the equation can be written as

$$a_n p^n + a_{n-1}p^{n-1}q$$
$$+ a_{n-2}p^{n-2}q^2 + \cdots + a_1 pq^{n-1} = -a_0 q^n.$$

   b. Why is p a factor of the left side of the equation?

   c. Because p divides the left side, it must also divide the right side. However, because $\frac{p}{q}$ is reduced to lowest terms, p cannot divide q. Thus, p and q have no common factors other than -1 and 1. Because p does divide the right side and it is not a factor of q^n, what can you conclude?

   d. Rewrite the equation from part (a) with all terms containing q on the left and the term that does not have a factor of q on the right. Use an argument that parallels parts (b) and (c) to conclude that q is a factor of a_n.

SECTION 3.5   *More On Zeros of Polynomial Functions*

Objectives

1. Find bounds for the roots of a polynomial equation.
2. Approximate real zeros.
3. Use conjugate roots to solve a polynomial equation.
4. Use the Linear Factorization Theorem to factor a polynomial.
5. Find polynomials with given zeros.

You stole my formula!

Tartaglia's Secret Formula for One Solution of $x^3 + mx = n$

$$x = \sqrt[3]{\sqrt{\left(\frac{n}{2}\right)^2 + \left(\frac{m}{3}\right)^3} + \frac{n}{2}} - \sqrt[3]{\sqrt{\left(\frac{n}{2}\right)^2 + \left(\frac{m}{3}\right)^3} - \frac{n}{2}}$$

Popularizers of mathematics are sharing bizarre stories that are giving math a secure place in popular culture. One episode, able to compete with the wildest fare served up by television talk shows and the tabloids, involves three Italian mathematicians and, of all things, zeros of polynomial functions.

Tartaglia (1499–1557), poor and starving, has found a formula that gives a root for a third-degree polynomial equation. Cardano (1501–1576) begs Tartaglia to reveal the secret formula, wheedling it from him with the promise he will find the impoverished Tartalia a patron. Then Cardano publishes his famous work *Ars Magna*, in which he presents Tartaglia's formula as his own. Cardano uses his most talented student, Ferrari (1522–1565), who derived a formula for a root of a fourth-degree polynomial equation, to falsely accuse Tartaglia of plagiarism. The dispute becomes violent and Tartaglia is fortunate to escape alive.

The noise from this "You Stole My Formula" episode is quieted by the work of French mathematician Evariste Galois (1811–1832). Galois proved that there is no general formula for finding roots of polynomial equations of degree 5 or higher. There are, of course, methods for finding roots. In this section, we continue our study of methods for finding zeros of polynomial functions.

1 Find bounds for the roots of a polynomial equation.

Upper and Lower Bounds for Roots

The **Upper and Lower Bound Theorem** helps us rule out many of a polynomial equation's possible rational roots. Figure 3.20 illustrates that a is a **lower bound** and b is an **upper bound** for the roots of $f(x) = 0$ because every real root c of the equation satisfies $a \le c \le b$.

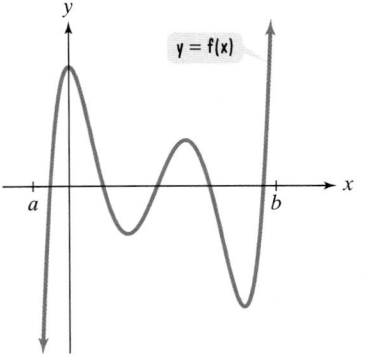

Figure 3.20 b is an upper bound and a is a lower bound for the real roots of $f(x) = 0$.

> ### The Upper and Lower Bound Theorem
>
> Let $f(x)$ be a polynomial with real coefficients and a positive leading coefficient, and let a and b be nonzero real numbers.
>
> 1. Divide $f(x)$ by $x - b$ (where $b > 0$) using synthetic division. If the last row containing the quotient and remainder has no negative numbers, then b is an **upper bound** for the real roots of $f(x) = 0$.
> 2. Divide $f(x)$ by $x - a$ (where $a < 0$) using synthetic division. If the last row containing the quotient and remainder has numbers that alternate in sign (zero entries count as positive or negative), then a is a **lower bound** for the real roots of $f(x) = 0$.

EXAMPLE 1   Finding Bounds for the Roots

Show that all the real roots of the equation $8x^3 + 10x^2 - 39x + 9 = 0$ lie between -3 and 2.

Solution   We begin by showing that 2 is an upper bound. Divide the polynomial by $x - 2$. If all the numbers in the bottom row of the synthetic division are nonnegative, then 2 is an upper bound.

$$
\begin{array}{r|rrrr}
2 & 8 & 10 & -39 & 9 \\
  &   & 16 & 52 & 26 \\
\hline
  & 8 & 26 & 13 & 35
\end{array}
$$

All numbers in this row are nonnegative.

The nonnegative entries in the last row verify that 2 is an upper bound. Next, we show that -3 is a lower bound. Divide the polynomial by $x - (-3)$, or $x + 3$. If the numbers in the bottom row of the synthetic division alternate in sign, then -3 is a lower bound. Remember that the number zero can be considered positive or negative.

$$
\begin{array}{r|rrrr}
-3 & 8 & 10 & -39 & 9 \\
   &   & -24 & 42 & -9 \\
\hline
   & 8 & -14 & 3 & 0
\end{array}
$$

Counting 0 as negative, the signs alternate:
$+, -, +, -.$

By the Upper and Lower Bound Theorem, the alternating signs in the last row indicate that -3 is a lower bound for the roots. (The zero remainder indicates that -3 is also a root.)

| Check Point 1 | Show that all the real roots of the equation $2x^3 + 11x^2 - 7x - 6 = 0$ lie between -7 and 2. |

How might the Upper and Lower Bound Theorem be helpful in solving a polynomial equation? Consider the equation

$$x^4 + 3x^3 - 27x^2 + 3x - 28 = 0.$$

With a leading coefficient of 1 and a constant term of -28, the possible rational roots are

$$\pm 1, \quad \pm 2, \quad \pm 4, \quad \pm 7, \quad \pm 14, \quad \pm 28.$$

We begin testing for an actual root using synthetic division. The following divisions indicate that 1 and 2 are not roots because of the nonzero remainders. However, something interesting happens when testing 4.

$$
\begin{array}{r|rrrrr}
1 & 1 & 3 & -27 & 3 & -28 \\
  &   & 1 & 4 & -23 & -20 \\
\hline
  & 1 & 4 & -23 & -20 & -48
\end{array}
\qquad
\begin{array}{r|rrrrr}
2 & 1 & 3 & -27 & 3 & -28 \\
  &   & 2 & 10 & -34 & -62 \\
\hline
  & 1 & 5 & -17 & -31 & -90
\end{array}
$$

$$
\begin{array}{r|rrrrr}
4 & 1 & 3 & -27 & 3 & -28 \\
  &   & 4 & 28 & 4 & 28 \\
\hline
  & 1 & 7 & 1 & 7 & 0
\end{array}
$$

Nonnegative numbers

4 is a root of $f(x) = 0$ because the remainder is 0.

4 is an upper bound for the roots of $f(x) = 0$.

Notice that 4 is both a root and an upper bound for the roots. Should you take the time to use synthetic division and test 7, 14, and 28? There is no need to do this because all three numbers exceed 4, the upper bound for the roots. Thus, 7, 14, and 28 cannot be roots of the equation.

Technology

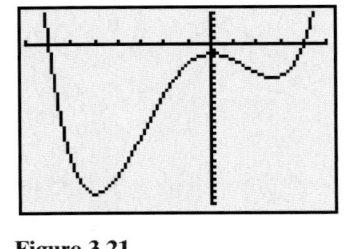

Figure 3.21

The Upper and Lower Bound Theorem and your knowledge of polynomial functions can help you to find a reasonable range setting when using your graphing utility. Consider

$$f(x) = x^4 + 3x^3 - 27x^2 + 3x - 28.$$

Based on our discussion, 4 is a zero and an upper bound for the zeros. We can also use synthetic division to show that −7 is a zero and a lower bound for the zeros. We can use these lower and upper bounds to determine Xmin and Xmax. We'll go one unit to the left and to the right of these bounds and use [−8, 5, 1]. Now, how do we determine Ymin and Ymax? Let's see what kinds of values of y we obtain when we evaluate the function between −8 and 5. Using synthetic division, direct substitution, or the table feature of some graphing utilities, we have $f(-6) = -370$, $f(-5) = -468$, $f(0) = -28$, and $f(3) = -100$. These evaluations suggest that we can use −500 for Ymin and 100 for Ymax. The graph of $f(x) = x^4 + 3x^3 - 27x^2 + 3x - 28$ is shown in a [−8, 5, 1] by [−500, 100, 20] viewing rectangle in Figure 3.21. Because the degree is even ($n = 4$) and the leading coefficient, 1, is positive, the graph should rise to the left and right. This is precisely what occurs in Figure 3.21. Our work in obtaining this complete graph is an excellent illustration of the fact that technology complements human knowledge and is not intended to replace it.

2 Approximate real zeros.

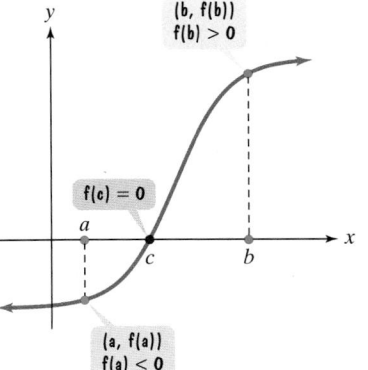

Figure 3.22 The graph must cross the x-axis at some value between a and b.

The Intermediate Value Theorem

We can find decimal approximations for real zeros of polynomial functions using a graphing utility. The **Intermediate Value Theorem** tells us of the existence of real zeros and how to approximate them. The idea behind the theorem is illustrated in Figure 3.22. The figure shows that if $(a, f(a))$ lies below the x-axis and $(b, f(b))$ lies above the x-axis, the smooth, continuous graph of a polynomial function f must cross the x-axis at some value c between a and b. This value is a real zero for the function.

These observations are summarized in the **Intermediate Value Theorem**.

The Intermediate Value Theorem for Polynomials

Let $f(x)$ be a polynomial function with real coefficients. If $f(a)$ and $f(b)$ have opposite signs, then there is at least one value of c between a and b for which $f(c) = 0$. Equivalently, the equation $f(x) = 0$ has at least one real root between a and b.

EXAMPLE 2 Approximating a Real Zero

a. Show that the polynomial function $f(x) = x^3 - 2x - 5$ has a real zero between 2 and 3.

b. Use the Intermediate Value Theorem to find an approximation for this real zero to the nearest tenth.

Solution

a. Let us evaluate $f(x)$ at 2 and 3. If $f(2)$ and $f(3)$ have opposite signs, then there is a real zero between 2 and 3. Using $f(x) = x^3 - 2x - 5$, we obtain

$$f(2) = 2^3 - 2 \cdot 2 - 5 = 8 - 4 - 5 = -1$$

> $f(2)$ is negative.

and

$$f(3) = 3^3 - 2 \cdot 3 - 5 = 27 - 6 - 5 = 16.$$

> $f(3)$ is positive.

This sign change shows that the polynomial function has a real zero between 2 and 3.

Technology

The following graph was obtained by entering

$$y = x^3 - 2x - 5$$

and

$$y = 0$$

and using the intersection feature in a $[-3, 3, 1]$ by $[-10, 10, 1]$ viewing rectangle. Correct to the nearest thousandth, the function's real zero is 2.095.

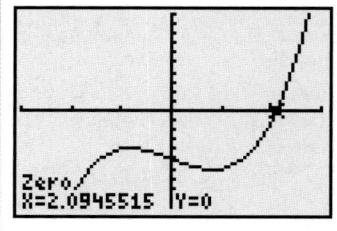

b. A numerical approach is to evaluate f at successive tenths between 2 and 3, looking for a sign change. This sign change will place the real zero between a pair of successive tenths.

x	$f(x) = x^3 - 2x - 5$	
2	$f(2) = 2^3 - 2(2) - 5 \qquad = -1$	Sign change
2.1	$f(2.1) = (2.1)^3 - 2(2.1) - 5 = 0.061$	

The sign change indicates that f has a real zero between 2 and 2.1. We now follow a similar procedure to locate the real zero between successive hundredths. We divide the interval $[2, 2.1]$ into ten equal subintervals. Then we evaluate f at each endpoint and look for a sign change.

$$f(2.00) = -1 \qquad\qquad f(2.06) = -0.378184$$
$$f(2.01) = -0.899399 \qquad f(2.07) = -0.270257$$
$$f(2.02) = -0.797592 \qquad f(2.08) = -0.161088$$
$$f(2.03) = -0.694573 \qquad f(2.09) = -0.050671$$
$$f(2.04) = -0.590336 \qquad f(2.1) = 0.061$$
$$f(2.05) = -0.484875$$

> Sign change

The sign change indicates that f has a real zero between 2.09 and 2.1. Correct to the nearest tenth, the zero is 2.1.

Check Point 2 Show that the polynomial function $f(x) = 3x^3 - 10x + 9$ has a real zero between -3 and -2.

3 Use conjugate roots to solve a polynomial equation.

The Fundamental Theorem of Algebra

We have seen that if a polynomial equation is of degree n, then counting multiple roots separately, the equation has n roots. Some of these roots may be nonreal complex numbers that occur in conjugate pairs, such as $2 + i$ and $2 - i$.

EXAMPLE 3   Using Conjugate Roots to Solve a Polynomial Equation

Solve $x^4 - 4x^3 + 3x^2 + 8x - 10 = 0$ given that $2 + i$ is a root.

Solution   The degree of the given equation is 4. This means that there are four roots. One of the roots is $2 + i$. Because complex nonreal roots come in conjugate pairs, we know that $2 - i$ is a second root. By the Factor Theorem, both

$$[x - (2 + i)] \quad \text{and} \quad [x - (2 - i)]$$

are factors of the given polynomial. We multiply these known factors.

$$[x - (2 + i)][x - (2 - i)]$$

$$
\underbrace{\quad}_{F} \quad \underbrace{\quad}_{O} \quad \underbrace{\quad}_{I} \quad \underbrace{\quad}_{L}
$$

$= x^2 - x(2 - i) - x(2 + i) + (2 + i)(2 - i)$ Multiply using the FOIL method.

$= x^2 - 2x + ix - 2x - ix + (4 - i^2)$ Continue multiplying.

$= x^2 - 2x + ix - 2x - ix + [4 - (-1)]$ Simplify using $i^2 = -1$.

$= x^2 - 4x + 5$ Combine like terms.

At this point we have only two of the four possible roots, $2 + i$ and $2 - i$. We can find the other two roots by factoring the given equation. We have found that $x^2 - 4x + 5$ is one of the factors. We can find the other factor(s) by dividing $x^2 - 4x + 5$ into the polynomial on the left side of the given equation.

$$
\begin{array}{r}
x^2 \qquad\quad - 2 \\
x^2 - 4x + 5 \overline{\smash{)}\; x^4 - 4x^3 + 3x^2 + 8x - 10} \\
\underline{x^4 - 4x^3 + 5x^2} \\
-2x^2 + 8x - 10 \\
\underline{-2x^2 + 8x - 10} \\
0
\end{array}
$$

> The zero remainder confirms that $x^2 - 4x + 5$ is a factor.

We can now solve the given equation.

$x^4 - 4x^3 + 3x^2 + 8x - 10 = 0$ This is the original equation.

$(x^2 - 4x + 5)(x^2 - 2) = 0$ Factor using the result of the polynomial long division.

$x^2 - 4x + 5 = 0$ or $x^2 - 2 = 0$ Set each factor equal to 0.

$x = 2 \pm i$ $x = \pm\sqrt{2}$ Solve for x. We know the roots for the first equation, $x^2 - 4x + 5 = 0$, by our previous analysis.

The solution set is $\{2 \pm i, \pm\sqrt{2}\}$.

Technology

The graph of

$f(x) =$
$\quad x^4 - 4x^3 + 3x^2 + 8x - 10$

is shown in a $[-4, 4, 1]$ by $[-15, 5, 1]$ viewing rectangle. The real roots of $f(x) = 0$, the equation in Example 3, are $-\sqrt{2}$ and $\sqrt{2}$. These appear as x-intercepts at approximately -1.4 and 1.4.

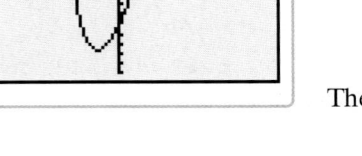

> **Check Point 3**   Solve $x^4 - 8x^3 + 64x - 105 = 0$ given that $2 - i$ is a root.

The fact that a polynomial equation of degree n has n roots is a consequence of a theorem proved in 1799 by a 22-year-old student named Carl Friedrich Gauss in his doctoral dissertation. His result is called the **Fundamental Theorem of Algebra**.

The Fundamental Theorem of Algebra

If $f(x)$ is a polynomial of degree n, where $n \geq 1$, then the equation $f(x) = 0$ has at least one complex root.

Suppose, for example, that $f(x) = 0$ represents a polynomial equation of degree n. By the Fundamental Theorem of Algebra, we know that this equation has at least one complex root; we'll call it c_1. By the Factor Theorem, we know that $x - c_1$ is a factor of $f(x)$. Therefore, we obtain

$$(x - c_1)q_1(x) = 0 \qquad \text{The degree of the polynomial } q_1(x) \text{ is } n - 1.$$
$$x - c_1 = 0 \quad \text{or} \quad q_1(x) = 0 \qquad \text{Set each factor equal to 0.}$$

If the degree of $q_1(x)$ is at least 1, by the Fundamental Theorem of Algebra the equation $q_1(x) = 0$ has at least one complex root. We'll call it c_2. The Factor Theorem gives us

$$q_1(x) = 0 \qquad \text{The degree of } q_1(x) \text{ is } n - 1.$$
$$(x - c_2)q_2(x) = 0 \qquad \text{The degree of } q_2(x) \text{ is } n - 2.$$
$$x - c_2 = 0 \quad \text{or} \quad q_2(x) = 0. \qquad \text{Set each factor equal to 0.}$$

Let's see what we have up to this point, and then continue the process.

$$f(x) = 0 \qquad \text{This is the original polynomial equation of degree } n.$$
$$(x - c_1)q_1(x) = 0 \qquad \text{This is the result from our first application of the Fundamental Theorem.}$$
$$(x - c_1)(x - c_2)q_2(x) = 0 \qquad \text{This is the result from our second application of the Fundamental Theorem.}$$

By continuing this process, we will obtain the product of n linear factors. Setting each of these linear factors equal to zero results in n complex roots. Thus, if $f(x)$ is a polynomial of degree n, where $n \geq 1$, then $f(x) = 0$ has exactly n roots, where roots are counted according to their multiplicity.

④ Use the Linear Factorization Theorem to factor a polynomial.

The Linear Factorization Theorem

In Example 3, we found that $x^4 - 4x^3 + 3x^2 + 8x - 10 = 0$ has $\{2 \pm i, \pm\sqrt{2}\}$ as a solution set. The polynomial can be factored over the complex nonreal numbers as follows:

$$f(x) = x^4 - 4x^3 + 3x^2 + 8x - 10$$

These are the four zeros.

$$= [x - (2 + i)][x - (2 - i)](x + \sqrt{2})(x - \sqrt{2})$$

These are four linear factors.

This fourth-degree polynomial has four linear factors. Just as an nth-degree polynomial equation has n roots, an nth-degree polynomial has n linear factors. This is formally stated as the **Linear Factorization Theorem**.

The Linear Factorization Theorem

If $f(x) = a_n x^n + a_{n-1} x^{n-1} + \cdots + a_1 x + a_0$, where $n \geq 1$ and $a_n \neq 0$, then

$$f(x) = a_n(x - c_1)(x - c_2)\cdots(x - c_n)$$

where $c_1, c_2, \ldots, c_n$ are complex numbers (possibly real and not necessarily distinct). In words: An nth-degree polynomial can be expressed as the product of n linear factors.

The Linear Factorization Theorem involves factors somewhat different than those you are used to seeing. For example, the polynomial $x^2 - 3$ is irreducible over the rational numbers. However, it can be factored over the real numbers as follows:

$$x^2 - 3 = (x + \sqrt{3})(x - \sqrt{3}).$$ Use $a^2 - b^2 = (a + b)(a - b)$ with $a = x$ and $b = \sqrt{3}$.

Study Tip

The sum of squares, irreducible over the real numbers, can be factored over the complex nonreal numbers as

$$a^2 + b^2 = (a + bi)(a - bi).$$

The polynomial $x^2 + 1$ is irreducible over the real numbers, but reducible over the complex nonreal numbers.

$$x^2 + 1 = (x + i)(x - i)$$

EXAMPLE 4   Factoring a Polynomial

Factor $x^4 - 3x^2 - 28$:

a. As the product of factors that are irreducible over the rational numbers.
b. As the product of factors that are irreducible over the real numbers.
c. In completely factored form involving complex nonreal numbers.

Solution

a. $x^4 - 3x^2 - 28 = (x^2 - 7)(x^2 + 4)$ Both quadratic factors are irreducible over the rational numbers.

b. $= (x + \sqrt{7})(x - \sqrt{7})(x^2 + 4)$ The third factor is still irreducible over the real numbers.

c. $= (x + \sqrt{7})(x - \sqrt{7})(x + 2i)(x - 2i)$ This is the completely factored form using complex nonreal numbers.

> **Check Point 4**   Factor $x^4 - 4x^2 - 5$ as the product of factors that are irreducible over **a.** the rational numbers; **b.** the real numbers; **c.** the complex nonreal numbers.

5 Find polynomials with given zeros.

Reversing Things: Finding Polynomials When the Zeros Are Given

Many of our problems involving polynomial functions and polynomial equations dealt with the process of finding zeros and roots. The Linear Factorization Theorem enables us to reverse this process, finding a polynomial function when the zeros are given.

EXAMPLE 5   Finding a Polynomial Function with Given Zeros

Find a fourth-degree polynomial function $f(x)$ with real coefficients that has $-2, 2$, and i as zeros and such that $f(3) = -150$.

Solution   Because i is a zero and the polynomial has real coefficients, the conjugate must also be a zero. We can now use the Linear Factorization Theorem.

$$f(x) = a_n(x - c_1)(x - c_2)(x - c_3)(x - c_4)$$ This is the linear factorization for a fourth-degree polynomial.

$$= a_n(x + 2)(x - 2)(x - i)(x + i)$$ Use the given zeros: $c_1 = -2, c_2 = 2, c_3 = i$, and, from above, $c_4 = -i$.

Technology

The graph of $f(x) = -3x^4 + 9x^2 + 12$, shown in a $[-3, 3, 1]$ by $[-200, 20, 20]$ viewing rectangle, verifies that -2 and 2 are real zeros. By tracing along the curve, we can check that $f(3) = -150$.

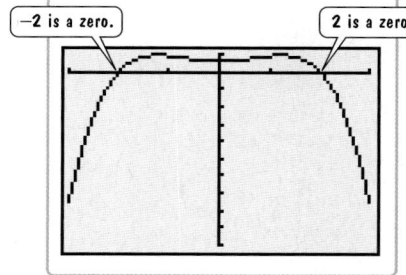

-2 is a zero.   2 is a zero.

$$= a_n(x^2 - 4)(x^2 + 1)$$ Multiply.
$$f(x) = a_n(x^4 - 3x^2 - 4)$$ Complete the multiplication.
$$f(3) = a_n(3^4 - 3 \cdot 3^2 - 4) = -150$$ To find a_n, use the fact that $f(3) = -150$.

$$a_n(81 - 27 - 4) = -150$$ Solve for a_n.
$$50a_n = -150$$
$$a_n = -3$$

Substituting -3 for a_n in the formula for $f(x)$, we obtain
$$f(x) = -3(x^4 - 3x^2 - 4).$$

Equivalently,

$$f(x) = -3x^4 + 9x^2 + 12.$$

Check Point 5   Find a third-degree polynomial function $f(x)$ with real coefficients that has -3 and i as zeros and such that $f(1) = 8$.

EXERCISE SET 3.5

 Practice Exercises

Use the Upper and Lower Bound Theorem to solve Exercises 1–6.

1. Show that all the real roots of the equation $x^4 - 5x^3 + 11x^2 + 33x - 18 = 0$ lie between -4 and 7.

2. Show that all the real roots of the equation $x^4 + 11x^3 - 12x^2 + 6 = 0$ lie between -13 and 1.

3. Show that all the real roots of the equation $2x^3 + 5x^2 - 8x - 7 = 0$ lie between -4 and 2.

4. Show that all the real roots of the equation $2x^5 - 13x^3 + 2x - 5 = 0$ lie between -3 and 3.

5. Consider the equation $x^4 + 3x^3 + 2x^2 - 5x + 12 = 0$.
   a. List all possible rational roots.
   b. Determine whether 1 is a root using synthetic division. What two conclusions can you draw?
   c. Based on part (b), what possible rational roots can you eliminate?
   d. Determine whether -3 is a root using synthetic division. What two conclusions can you draw?
   e. Based on part (d), what possible rational roots can you eliminate?

6. Consider the equation $2x^5 + 5x^4 - 8x^3 - 14x^2 + 6x + 9 = 0$.
   a. List all possible rational roots.
   b. Determine whether $\frac{3}{2}$ is a root using synthetic division. What two conclusions can you draw?
   c. Based on part (b), what possible rational roots can you eliminate?

   d. Determine whether -3 is a root using synthetic division. What two conclusions can you draw?
   e. Based on part (d), what possible rational roots can you eliminate?

In Exercises 7–14, show that each polynomial has a real zero between the given integers. Then use the Intermediate Value Theorem to find an approximation for this zero to the nearest tenth.

7. $f(x) = x^3 - x - 1$; between 1 and 2

8. $f(x) = x^3 - 4x^2 + 2$; between 0 and 1

9. $f(x) = 2x^4 - 4x^2 + 1$; between -1 and 0

10. $f(x) = x^4 + 6x^3 - 18x^2$; between 2 and 3

11. $f(x) = x^3 + x^2 - 2x + 1$; between -3 and -2

12. $f(x) = x^5 - x^3 - 1$; between 1 and 2

13. $f(x) = 3x^3 - 10x + 9$; between -3 and -2

14. $f(x) = 3x^3 - 8x^2 + x + 2$; between 2 and 3

In Exercises 15–22, use the given root to find the solution set of the polynomial equation.

15. $x^3 - 2x^2 + 4x - 8 = 0$; $-2i$

16. $x^4 + 13x^2 + 36 = 0$; $3i$

17. $3x^3 - 7x^2 + 8x - 2 = 0$; $1 + i$

18. $x^3 - 7x^2 + 16x - 10 = 0$; $3 + i$

19. $x^4 - 6x^2 + 25 = 0$; $2 - i$

20. $x^4 - x^3 - 9x^2 + 29x - 60 = 0$; $1 + 2i$

21. $x^4 - 8x^3 + 64x - 105 = 0$; $2 - i$

22. $4x^4 - 28x^3 + 129x^2 - 130x + 125 = 0$; $3 - 4i$

In Exercises 23–28, factor each polynomial:

   a. *as the product of factors that are irreducible over the rational numbers*

   b. *as the product of factors that are irreducible over the real numbers*

   c. *in completely factored form involving complex nonreal numbers*

23. $x^4 - x^2 - 20$ **24.** $x^4 + 6x^2 - 27$

25. $x^4 + x^2 - 6$ **26.** $x^4 - 9x^2 - 22$

27. $x^4 - 2x^3 + x^2 - 8x - 12$
   (*Hint:* One factor is $x^2 + 4$.)

28. $x^4 - 4x^3 + 14x^2 - 36x + 45$
   (*Hint:* One factor is $x^2 + 9$.)

In Exercises 29–36, find an nth-degree polynomial function with real coefficients satisfying the given conditions. If you are using a graphing utility, use it to graph the function and verify the real zeros and the given function value.

29. $n = 3$; 1 and $5i$ are zeros; $f(-1) = -104$

30. $n = 3$; 4 and $2i$ are zeros; $f(-1) = -50$

31. $n = 3$; -5 and $4 + 3i$ are zeros; $f(2) = 91$

32. $n = 3$; 6 and $-5 + 2i$ are zeros; $f(2) = -636$

33. $n = 4$; i and $3i$ are zeros; $f(-1) = 20$

34. $n = 4$; $-2, -\frac{1}{2}$, and i are zeros; $f(1) = 18$

35. $n = 4$; $-2, 5$, and $3 + 2i$ are zeros; $f(1) = -96$

36. $n = 4$; $-4, \frac{1}{3}$, and $2 + 3i$ are zeros; $f(1) = 100$

In Exercises 37–44, find all the zeros of the function and write the polynomial as a product of linear factors.

37. $f(x) = x^3 - x^2 + 25x - 25$

38. $f(x) = x^3 - 10x^2 + 33x - 34$

39. $f(x) = x^3 - 8x^2 + 25x - 26$

40. $f(x) = x^3 - 8x^2 + 17x - 4$

41. $f(x) = x^4 + 37x^2 + 36$

42. $f(x) = x^4 + 8x^3 + 9x^2 - 10x + 100$

43. $f(x) = 16x^4 + 36x^3 + 16x^2 + x - 30$

44. $f(x) = 2x^4 - x^3 + 7x^2 - 4x - 4$

Application Exercises

We have seen the polynomial function

$H(x) =$
   $-0.001183x^4 + 0.05495x^3 - 0.8523x^2 + 9.054x + 6.748$

that models the age in human years, $H(x)$, of a dog that is x years old, where $x \geq 1$. Although the coefficients make it

difficult to solve equations algebraically using this function, a graph of the function makes approximate solutions possible. Use the graph shown to solve Exercises 45–46.

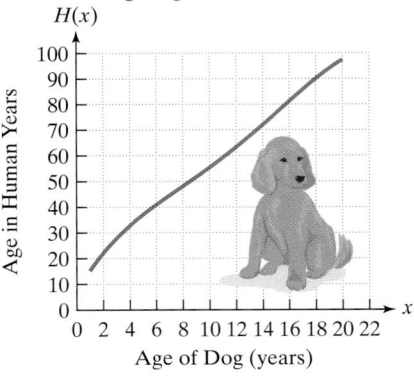

Dog's Age in Human Years

45. If you are 25, what is the equivalent age for dogs?

46. If you are 30, what is the equivalent age for dogs?

47. Set up an equation to answer the question in either Exercise 45 or 46. Bring all terms to one side and obtain zero on the other side. What are some of the difficulties involved in solving this equation? Explain how the Intermediate Value Theorem can be used to verify the approximate solution that you obtained from the graph.

The bar graph shows the cost of Medicare, in billions of dollars, projected through 2005. Using the regression feature of a graphing utility, these data can be modeled by

   a linear function, $f(x) = 27x + 163$;

   a quadratic function, $g(x) = 1.2x^2 + 15.2x + 181.4$;

   a third-degree polynomial function,
     $h(x) = 0.08x^3 - 0.06x^2 + 20.08x + 178.32$.

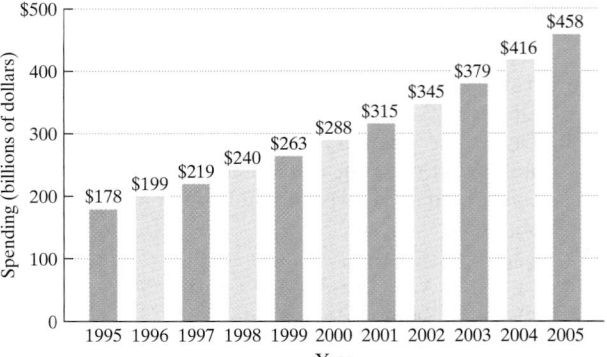

Medicare Spending

Source: Congressional Budget Office

For each of these functions, x represents the number of years after 1995 and the function value represents Medicare spending, in billions of dollars, for that year.

48. The graph indicates that Medicare spending will reach $379 billion in 2003. Substitute 379 for $f(x)$, $g(x)$, and $h(x)$ in each of the three models. Then solve each resulting equation, if possible, to find how many years after 1995 spending will reach $379 billion. Which of the three functions is the best model for 2003?

49. The graph indicates that Medicare spending will reach $458 billion in 2005. Substitute 458 for $f(x)$, $g(x)$, and $h(x)$ in each of the three models. Then solve each resulting equation, if possible, to find how many years after 1995 spending will reach $458 billion. Which of the three functions is the best model for 2005?

Writing in Mathematics

50. When testing a number using synthetic division, how do you know if it is an upper bound for the real roots?

51. When testing a number using synthetic division, how do you know if it is a lower bound for the real roots?

52. How do you show that a polynomial function has a real zero between two given numbers?

53. How does the linear factorization of $f(x)$, that is,

$$f(x) = a_n(x - c_1)(x - c_2)\cdots(x - c_n),$$

show that a polynomial equation of degree n has n roots?

Technology Exercises

54. Show that -1 is a lower bound of $f(x) = x^3 - 53x^2 + 103x - 51$. Show that 60 is an upper bound. Use this information and a graphing utility to draw a relatively complete graph of f.

For Exercises 55–56, use a graphing utility to determine upper and lower bounds for the zeros of f. Does synthetic division verify your observations?

55. $f(x) = 2x^3 + x^2 - 14x - 7$

56. $f(x) = 2x^4 - 7x^3 - 5x^2 + 28x - 12$

57. The function $f(x) = -0.00002x^3 + 0.008x^2 - 0.3x + 6.95$ models the number of annual physician visits, f, by a person of age x.
   a. Graph the function for meaningful values of x and discuss what the graph reveals in terms of the variables described by the model.
   b. Use the polynomial root-finding capability of your graphing utility to find the age (to the nearest year) for the group that averages 13.43 annual physician visits.
   c. Verify part (b) using the graph of f.

Use a graphing utility to obtain a complete graph for each polynomial function in Exercises 58–61. Then determine the number of real zeros and the number of nonreal complex zeros for each function.

58. $f(x) = x^3 - 6x - 9$

59. $f(x) = 3x^5 - 2x^4 + 6x^3 - 4x^2 - 24x + 16$

60. $f(x) = 3x^4 + 4x^3 - 7x^2 - 2x - 3$

61. $f(x) = x^6 - 64$

 ## Critical Thinking Exercises

In Exercises 62–64, what is the smallest degree that each polynomial could have?

62.

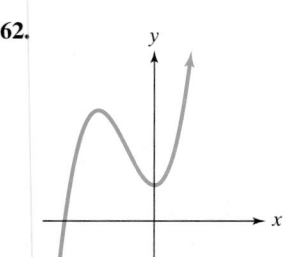

63.

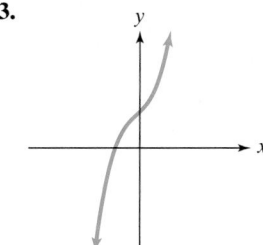

64.

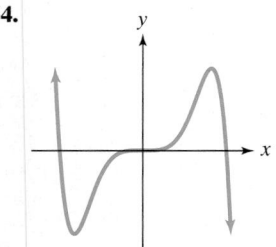

65. Explain why nonreal complex zeros are gained or lost in pairs in terms of graphs of polynomial functions.

66. Explain why a polynomial function of degree 20 cannot cross the x-axis exactly once.

67. Give an example of a function that is not subject to the Intermediate Value Theorem.

 ## Group Exercise

68. The graphs on page 314 show costs for private and public four-year colleges projected through the year 2017. According to these projections, your daughter's college education at a private four-year school could cost about $250,000. This activity involves forming and using models from these data. Group members should begin by deciding whether to work with data for private or public colleges.

Cost of a Four-Year College

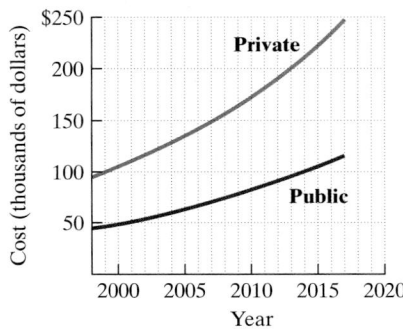

Source: U.S. Department of Education

a. Let $x = 0$ correspond to 1998, $x = 1$ to 1999, $x = 2$ to 2000, and so on up through $x = 19$ for 2017. Group members should use the chosen line graph to obtain a reasonable estimate for the cost of an education, y, in thousands of dollars, for each x.

b. Use the regression feature of a graphing utility to model the data for the cost of a four-year college x years after 1998 using a linear function, a quadratic function, and a third-degree polynomial function.

c. Use these functions to write and solve a problem similar to Exercise 48 or 49.

d. Use these functions to make predictions well into the future. Which function, if any, seems to be most reasonable in its predicted cost? Of course, these are only predictions, subject to unforeseeable events. What events might render each of these models relatively useless over long periods of time?

SECTION 3.6 *Rational Functions and Their Graphs*

Objectives

1. Find the domain of rational functions.
2. Use arrow notation.
3. Identify vertical asymptotes.
4. Identify horizontal asymptotes.
5. Graph rational functions.
6. Identify slant asymptotes.
7. Solve applied problems involving rational functions.

Technology is now promising to bring light, fast, and beautiful wheelchairs to millions of disabled people. The cost of manufacturing these radically different wheelchairs can be modeled by rational functions. In this section we will see how graphs of these functions illustrate that low prices are possible with high production levels, urgently needed in this situation. There are more than half a billion people with disabilities in developing countries; an estimated 20 million need wheelchairs right now.

1 Find the domain of rational functions.

Rational Functions

Rational functions are quotients of polynomial functions. This means that rational functions can be expressed as

$$f(x) = \frac{p(x)}{q(x)}$$

where $p(x)$ and $q(x)$ are polynomial functions and $q(x) \neq 0$. The **domain** of a rational function is the set of all real numbers except the x-values that make the denominator zero. For example, the domain of the rational function

$$f(x) = \frac{x^2 + 7x + 9}{x(x-2)(x+5)} \quad \text{This is } p(x).$$
This is $q(x)$.

is the set of all real numbers except $0, 2$, and -5.

EXAMPLE 1 Finding the Domain of a Rational Function

Find the domain of each rational function.

a. $f(x) = \dfrac{x^2 - 9}{x - 3}$ **b.** $g(x) = \dfrac{x}{x^2 - 9}$ **c.** $h(x) = \dfrac{x + 3}{x^2 + 9}$

Solution   Rational functions contain division. Because division by 0 is undefined, we must exclude from the domain of each function values of x that cause the polynomial function in the denominator to be 0.

a. The denominator of $f(x) = \dfrac{x^2 - 9}{x - 3}$ is 0 if $x = 3$. Thus, x cannot equal 3.

The domain of f consists of all real numbers except 3, written $\{x \mid x \neq 3\}$.

b. The denominator of $g(x) = \dfrac{x}{x^2 - 9}$ is 0 if $x = -3$ or $x = 3$. Thus, the domain of g consists of all real numbers except -3 and 3, written $\{x \mid x \neq -3, x \neq 3\}$.

c. No real numbers cause the denominator of $h(x) = \dfrac{x + 3}{x^2 + 9}$ to equal 0. The domain of h consists of all real numbers.

Check Point 1   Find the domain of each rational function.

a. $f(x) = \dfrac{x^2 - 25}{x - 5}$ **b.** $g(x) = \dfrac{x}{x^2 - 25}$ **c.** $h(x) = \dfrac{x + 5}{x^2 + 25}$

2 Use arrow notation.

The most basic rational function is the **reciprocal function**, defined by $f(x) = \dfrac{1}{x}$. The denominator of the reciprocal function is zero when $x = 0$, so the domain of f is the set of all real numbers except for 0.

Let's look at the behavior of f near the excluded value 0. We start by evaluating $f(x)$ to the left of 0.

x approaches 0 from the left.

x	-1	-0.5	-0.1	-0.01	-0.001
$f(x) = \dfrac{1}{x}$	-1	-2	-10	-100	-1000

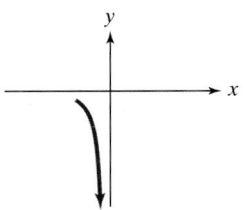

Mathematically, we say that "x approaches 0 from the left." From the table and the accompanying graph, it appears that as x approaches 0 from the left, the function values, $f(x)$, decrease without bound. We say that "$f(x)$ approaches negative infinity." We use a special arrow notation to describe this situation symbolically:

$$f(x) \to -\infty \quad \text{as} \quad x \to 0^-$$

f(x) approaches negative infinity (that is, the graph falls) as x approaches 0 from the left.

Observe that the minus $(-)$ superscript on the 0 $(x \to 0^-)$ is read "from the left." Next, we evaluate $f(x)$ to the right of 0.

x approaches 0 from the right.

x	0.001	0.01	0.1	0.5	1
$f(x) = \dfrac{1}{x}$	1000	100	10	2	1

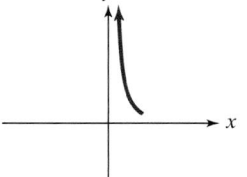

Mathematically, we say that "x approaches 0 from the right." From the table and the accompanying graph, it appears that as x approaches 0 from the right, the function values, $f(x)$, increase without bound. We say that "$f(x)$ approaches infinity." We again use a special arrow notation to describe this situation symbolically:

$$f(x) \to \infty \quad \text{as} \quad x \to 0^+$$

f(x) approaches infinity (that is, the graph rises) as x approaches 0 from the right.

Observe that the plus $(+)$ superscript on the 0 $(x \to 0^+)$ is read "from the right."

Now let's see what happens to the function values, $f(x)$, as x gets farther away from the origin. The following tables suggest what happens to $f(x)$ as x increases or decreases without bound.

x increases without bound:

x	1	10	100	1000
$f(x) = \dfrac{1}{x}$	1	0.1	0.01	0.001

x decreases without bound:

x	-1	-10	-100	-1000
$f(x) = \dfrac{1}{x}$	-1	-0.1	-0.01	-0.001

Figure 3.23 illustrates the end behavior of $f(x) = \dfrac{1}{x}$ as x increases or decreases without bound. The function values, $f(x)$, are getting progressively closer to 0. This means that the graph of f is approaching the horizontal line $y = 0$ (that is, the x-axis) as x increases or decreases without bound. We use the arrow notation to describe this situation.

$$f(x) \to 0 \quad \text{as} \quad x \to \infty \qquad \text{and} \qquad f(x) \to 0 \quad \text{as} \quad x \to -\infty$$

f(x) approaches 0 as x increases without bound.

f(x) approaches 0 as x decreases without bound.

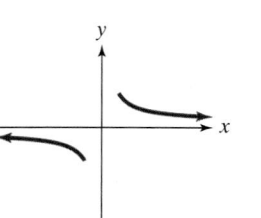

Figure 3.23
$f(x)$ approaches 0 as x increases or decreases without bound

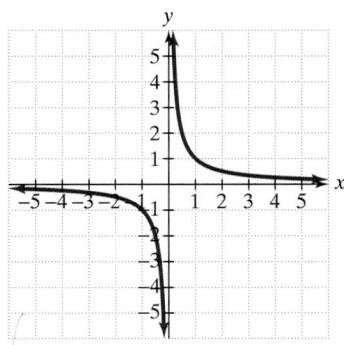

Figure 3.24 The graph of the reciprocal function $f(x) = \dfrac{1}{x}$

Thus, as x approaches infinity $(x \to \infty)$ or as x approaches negative infinity $(x \to -\infty)$, the function values are approaching zero: $f(x) \to 0$.

The graph of the reciprocal function $f(x) = \dfrac{1}{x}$ is shown in Figure 3.24.

Unlike the graph of a polynomial function, the graph of the reciprocal function has a break in it and is composed of two distinct branches.

The arrow notation used throughout our discussion of the reciprocal function is summarized in the following box.

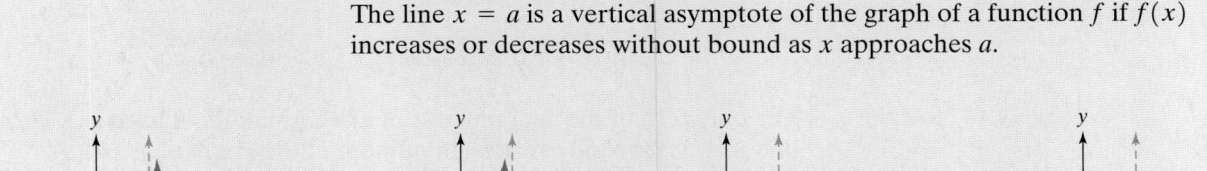

Arrow Notation

Symbol	Meaning
$x \to a^+$	x approaches a from the right.
$x \to a^-$	x approaches a from the left.
$x \to \infty$	x approaches infinity; that is, x increases without bound.
$x \to -\infty$	x approaches negative infinity; that is, x decreases without bound.

Vertical Asymptotes of Rational Functions

3 Identify vertical asymptotes.

Look again at the graph of $f(x) = \dfrac{1}{x}$. The curve approaches, but does not touch, the y-axis. The y-axis, or $x = 0$, is said to be a **vertical asymptote** of the graph. A rational function may have no vertical asymptotes, one vertical asymptote, or several vertical asymptotes. The graph of a rational function never intersects a vertical asymptote. We will use dashed lines to show asymptotes.

Definition of a Vertical Asymptote

The line $x = a$ is a vertical asymptote of the graph of a function f if $f(x)$ increases or decreases without bound as x approaches a.

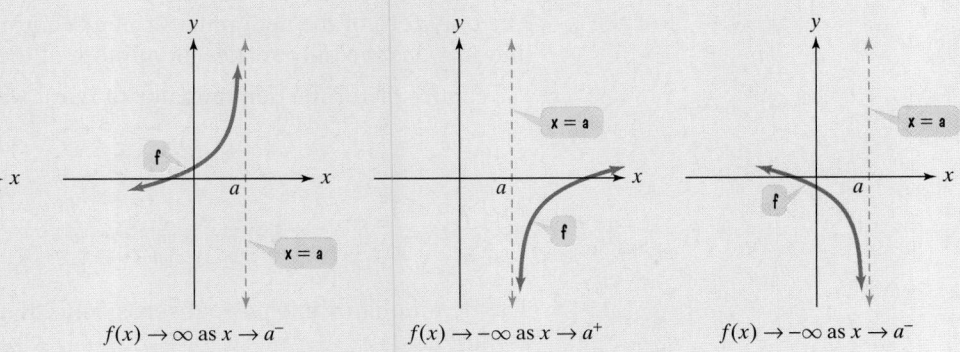

$f(x) \to \infty$ as $x \to a^+$ $\qquad$ $f(x) \to \infty$ as $x \to a^-$ $\qquad$ $f(x) \to -\infty$ as $x \to a^+$ $\qquad$ $f(x) \to -\infty$ as $x \to a^-$

Thus, $f(x) \to \infty$ or $f(x) \to -\infty$ as x approaches a from either the left or the right.

If the graph of a rational function has vertical asymptotes, they can be located by using the following theorem.

Locating Vertical Asymptotes

If $f(x) = \dfrac{p(x)}{q(x)}$ is a rational function in which $p(x)$ and $q(x)$ have no common factors and a is a zero of $q(x)$, the denominator, then $x = a$ is a vertical asymptote of the graph of f.

EXAMPLE 2   Finding the Vertical Asymptotes of a Rational Function

Find the vertical asymptotes, if any, of the graph of each rational function.

a. $f(x) = \dfrac{x}{x^2 - 9}$ **b.** $g(x) = \dfrac{x + 3}{x^2 - 9}$ **c.** $h(x) = \dfrac{x + 3}{x^2 + 9}$

Solution   Factoring is usually helpful in identifying zeros of denominators.

a.
$$f(x) = \frac{x}{x^2 - 9} = \frac{x}{(x + 3)(x - 3)}$$

This factor is 0 if $x = -3$.   This factor is 0 if $x = 3$.

There are no common factors in the numerator and the denominator. The zeros of the denominator are -3 and 3. Thus, the lines $x = -3$ and $x = 3$ are the vertical asymptotes for the graph of f.

b. We will use factoring to see if there are common factors.

$$g(x) = \frac{x + 3}{x^2 - 9} = \frac{(x + 3)}{(x + 3)(x - 3)} = \frac{1}{x - 3}$$

There is a common factor, $x + 3$, so simplify.   This denominator is 0 if $x = 3$.

The only zero of the denominator of $g(x)$ in simplified form is 3. Thus, the line $x = 3$ is the only vertical asymptote of the graph of g.

c. We cannot factor the denominator of $h(x)$ over the real numbers.

$$h(x) = \frac{x + 3}{x^2 + 9}$$

No real numbers make this denominator 0.

The denominator has no real zeros. Thus, the graph of h has no vertical asymptotes.

Check Point 2   Find the vertical asymptotes, if any, of the graph of each rational function.

a. $f(x) = \dfrac{x}{x^2 - 1}$ **b.** $g(x) = \dfrac{x - 1}{x^2 - 1}$ **c.** $h(x) = \dfrac{x - 1}{x^2 + 1}$

A value where the denominator of a function is zero does not necessarily result in a vertical asymptote. There is a hole corresponding to $x = a$, and not a vertical asymptote, in the graph of a function under the following conditions: The value a causes the denominator to be zero, but there is a reduced form of the function's equation in which a does not cause the denominator to be zero.

Consider, for example, the function

$$f(x) = \frac{x^2 - 9}{x - 3}.$$

Because the denominator is zero when $x = 3$, the function's domain is all real numbers except 3. However, there is a reduced form of the equation in which 3 does not cause the denominator to be zero:

$$f(x) = \frac{x^2 - 9}{x - 3} = \frac{(x + 3)(x - 3)}{x - 3} = x + 3, \quad x \neq 3$$

Denominator is zero at x = 3.

In this reduced form, 3 does not result in a zero denominator.

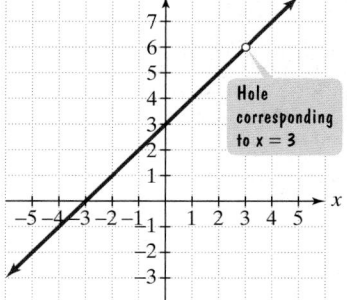

Figure 3.25

Hole corresponding to x = 3

Figure 3.25 shows that the graph has a hole corresponding to $x = 3$. Graphing utilities do not show this feature of the graph.

Horizontal Asymptotes of Rational Functions

4 Identify horizontal asymptotes.

Figure 3.24 shows the graph of the reciprocal function $f(x) = \frac{1}{x}$. As $x \to \infty$ and as $x \to -\infty$, the function values are approaching 0: $f(x) \to 0$. The line $y = 0$ (that is, the x-axis) is a **horizontal asymptote** of the graph. Many, but not all, rational functions have horizontal asymptotes.

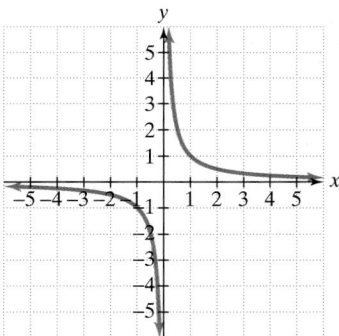

Figure 3.24 The graph of the reciprocal function $f(x) = \frac{1}{x}$, repeated

Definition of a Horizontal Asymptote

The line $y = b$ is a horizontal asymptote of the graph of a function f if $f(x)$ approaches b as x increases or decreases without bound.

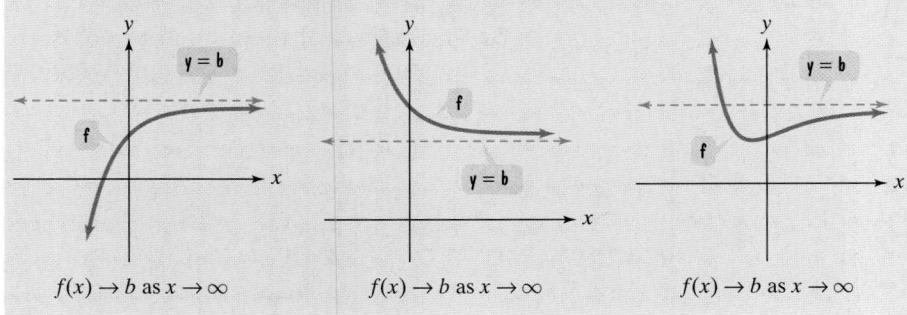

Recall that a rational function may have several vertical asymptotes. By contrast, it can have at most one horizontal asymptote. Although a graph can never intersect a vertical asymptote, it may cross its horizontal asymptote.

If the graph of a rational function has a horizontal asymptote, it can be located by using the following theorem.

Locating Horizontal Asymptotes

Let f be the rational function given by

$$f(x) = \frac{a_n x^n + a_{n-1} x^{n-1} + \cdots + a_1 x + a_0}{b_m x^m + b_{m-1} x^{m-1} + \cdots + b_1 x + b_0}, \quad a_n \neq 0, b_m \neq 0.$$

The degree of the numerator is n. The degree of the denominator is m.

1. If $n < m$, the x-axis is the horizontal asymptote of the graph of f.

2. If $n = m$, the line $y = \dfrac{a_n}{b_m}$ is the horizontal asymptote of the graph of f.

3. If $n > m$, the graph of f has no horizontal asymptote.

EXAMPLE 3 Finding the Horizontal Asymptote of a Rational Function

Find the horizontal asymptote, if any, of the graph of each rational function.

a. $f(x) = \dfrac{4x}{2x^2 + 1}$ **b.** $g(x) = \dfrac{4x^2}{2x^2 + 1}$ **c.** $h(x) = \dfrac{4x^3}{2x^2 + 1}$

Solution

a. $f(x) = \dfrac{4x}{2x^2 + 1}$

The degree of the numerator, 1, is less than the degree of the denominator, 2. Thus, the graph of f has the x-axis as a horizontal asymptote [see Figure 3.26(a)]. The equation of the horizontal asymptote is $y = 0$.

b. $g(x) = \dfrac{4x^2}{2x^2 + 1}$

The degree of the numerator, 2, is equal to the degree of the denominator, 2. The leading coefficients of the numerator and denominator, 4 and 2, are used to obtain the equation of the horizontal asymptote. The equation of the horizontal asymptote is $y = \frac{4}{2}$ or $y = 2$ [see Figure 3.26(b)].

c. $h(x) = \dfrac{4x^3}{2x^2 + 1}$

The degree of the numerator, 3, is greater than the degree of the denominator, 2. Thus, the graph of h has no horizontal asymptote [see Figure 3.26(c)].

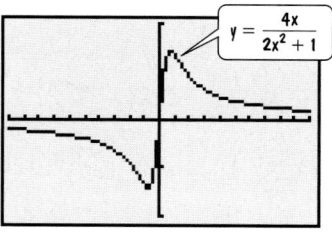

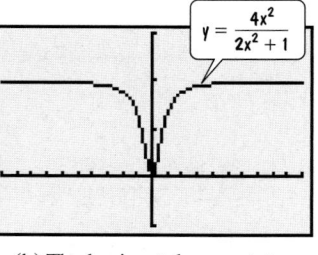

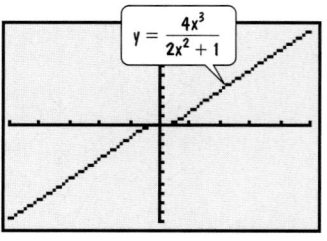

(a) The horizontal asymptote of the graph is $y = 0$.

(b) The horizontal asymptote of the graph is $y = 2$.

(c) The graph has no horizontal asymptote.

Figure 3.26

Check
Point
3

Find the horizontal asymptote, if any, of the graph of each rational function.

a. $f(x) = \dfrac{9x^2}{3x^2 + 1}$ **b.** $g(x) = \dfrac{9x}{3x^2 + 1}$ **c.** $h(x) = \dfrac{9x^3}{3x^2 + 1}$

5 Graph rational functions.

Graphing Rational Functions

Here are some suggestions for graphing rational functions.

Strategy for Graphing a Rational Function

Suppose that

$$f(x) = \frac{p(x)}{q(x)}$$

where $p(x)$ and $q(x)$ are polynomial functions with no common factors.

1. Determine whether the graph of f has symmetry.

$f(-x) = f(x)$: y-axis symmetry

$f(-x) = -f(x)$: origin symmetry

2. Find the y-intercept (if there is one) by evaluating $f(0)$.

3. Find the x-intercepts (if there are any) by solving the equation $p(x) = 0$.

4. Find any vertical asymptote(s) by solving the equation $q(x) = 0$.

5. Find the horizontal asymptote (if there is one) using the rule for determining the horizontal asymptote of a rational function.

6. Plot at least one point between and beyond each x-intercept and vertical asymptote.

7. Use the information obtained previously to graph the function between and beyond the vertical asymptotes.

EXAMPLE 4   Graphing a Rational Function

Graph: $f(x) = \dfrac{2x}{x - 1}$.

Solution

Step 1   Determine symmetry.

$$f(-x) = \frac{2(-x)}{-x - 1} = \frac{-2x}{-x - 1} = \frac{2x}{x + 1}$$

Because $f(-x)$ does not equal $f(x)$ or $-f(x)$, the graph has neither y-axis nor origin symmetry.

Step 2   Find the y-intercept.   Evaluate $f(0)$.

$$f(0) = \frac{2 \cdot 0}{0 - 1} = \frac{0}{-1} = 0$$

The y-intercept is 0, and so the graph passes through the origin.

Step 3   Find x-intercept(s).   This is done by solving $p(x) = 0$.

$$2x = 0 \quad \text{Set the numerator equal to 0.}$$
$$x = 0$$

There is only one x-intercept. This verifies that the graph passes through the origin.

Step 4   Find the vertical asymptotes(s).   Solve $q(x) = 0$, thereby finding zeros of the denominator.

$$x - 1 = 0 \quad \text{Set the denominator equal to 0.}$$
$$x = 1$$

The equation of the vertical asymptote is $x = 1$.

Step 5   Find the horizontal asymptote.   Because the numerator and denominator have the same degree, the equation of the horizontal asymptote is

$$y = \frac{2}{1} = 2.$$

The equation of the horizontal asymptote is $y = 2$.

Step 6   Plot points between and beyond each x-intercept and vertical asymptote.   With an x-intercept at 0 and a vertical asymptote at $x = 1$, we evaluate the function at $-2, -1, \frac{1}{2}, 2,$ and 4.

x	-2	-1	$\frac{1}{2}$	2	4
$f(x) = \dfrac{2x}{x-1}$	$\dfrac{4}{3}$	1	-2	4	$\dfrac{8}{3}$

Figure 3.27 shows these points, the y-intercept, the x-intercept, and the asymptotes.

Step 7   Graph the function.   The graph of $f(x) = \dfrac{2x}{x-1}$ is shown in Figure 3.28.

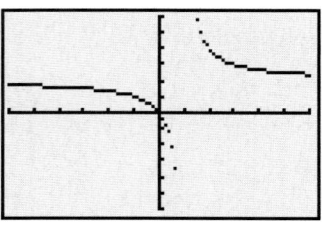
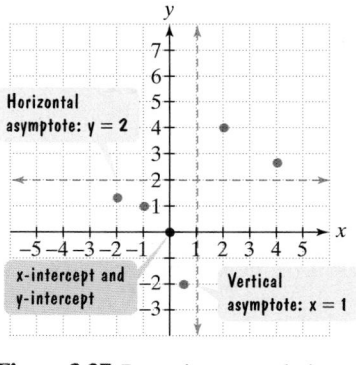

Figure 3.27 Preparing to graph the rational function $f(x) = \dfrac{2x}{x-1}$

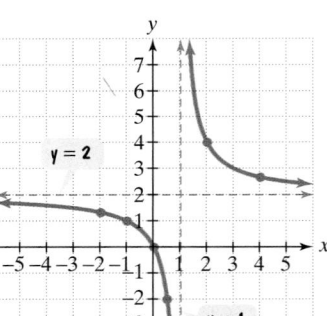

Figure 3.28 The graph of $f(x) = \dfrac{2x}{x-1}$

Check Point 4   Graph: $f(x) = \dfrac{3x}{x-2}$.

EXAMPLE 5 Graphing a Rational Function

Graph: $f(x) = \dfrac{3x^2}{x^2 - 4}$.

Solution

Step 1 Determine symmetry: $f(-x) = \dfrac{3(-x)^2}{(-x)^2 - 4} = \dfrac{3x^2}{x^2 - 4} = f(x)$: Symmetric with respect to the y-axis.

Step 2 Find the y-intercept: $f(0) = \dfrac{3 \cdot 0^2}{0^2 - 4} = \dfrac{0}{-4} = 0$: y-intercept is 0.

Step 3 Find the x-intercept: $3x^2 = 0$, so $x = 0$: x-intercepts is 0.

Step 4 Find the vertical asymptotes: Set $q(x) = 0$.

$$x^2 - 4 = 0 \qquad \text{\small Set the denominator equal to 0.}$$
$$x^2 = 4$$
$$x = \pm 2$$

Vertical asymptotes: $x = -2$ and $x = 2$

Step 5 Find the horizontal asymptote: $y = \frac{3}{1} = 3$.

Step 6 Plot points between and beyond the x-intercept and the vertical asymptotes. With an x-intercept at 0 and vertical asymptotes at $x = -2$ and $x = 2$, we evaluate the function at $-3, -1, 1, 3,$ and 4.

x	−3	−1	1	3	4
$f(x) = \dfrac{3x^2}{x^2 - 4}$	$\dfrac{27}{5}$	−1	−1	$\dfrac{27}{5}$	4

Figure 3.29 shows these points, the y-intercept, the x-intercept, and the asymptotes.

Step 7 Graph the function. The graph of $f(x) = \dfrac{3x^2}{x^2 - 4}$ is shown in Figure 3.30. The y-axis symmetry is now obvious.

Technology

The graph of $y = \dfrac{3x^2}{x^2 - 4}$ generated by a graphing utility verifies that our hand-drawn graph is correct.

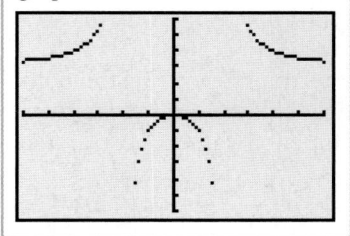

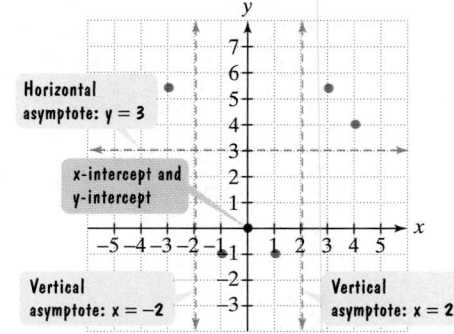

Figure 3.29 Preparing to graph $f(x) = \dfrac{3x^2}{x^2 - 4}$

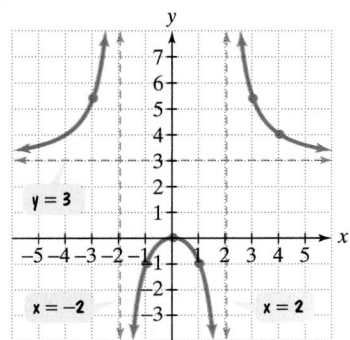

Figure 3.30 The graph of $f(x) = \dfrac{3x^2}{x^2 - 4}$

Check Point 5   Graph: $f(x) = \dfrac{2x^2}{x^2 - 9}$.

Example 6 illustrates that not every rational function has vertical and horizontal asymptotes.

EXAMPLE 6   Graphing a Rational Function

Graph: $f(x) = \dfrac{x^4}{x^2 + 1}$.

Solution

Step 1   Determine symmetry: $f(-x) = \dfrac{(-x)^4}{(-x)^2 + 1} = \dfrac{x^4}{x^2 + 1} = f(x)$:
Symmetric with respect to the y-axis.

Step 2   Find the y-intercept: $f(0) = \dfrac{0^4}{0^2 + 1} = \dfrac{0}{1} = 0$: y-intercept is 0.

Step 3   Find the x-intercept: $x^4 = 0$, so $x = 0$: x-intercept is 0.

Step 4   Find the vertical asymptote: Set $q(x) = 0$.

$$x^2 + 1 = 0 \qquad \text{\textit{Set the denominator equal to 0.}}$$

$$x^2 = -1$$

Although this equation has imaginary roots ($x = \pm i$), there are no real roots. Thus, there is no vertical asymptote.

Step 5   Find the horizontal asymptote: Because the degree of the numerator, 4, is greater than the degree of the denominator, 2, there is no horizontal asymptote.

Step 6   Plot points between and beyond the x-intercept and the vertical asymptotes.   With an x-intercept at 0 and no vertical asymptotes, we evaluate the function at $-2, -1, 1,$ and 2.

x	-2	-1	1	2
$f(x) = \dfrac{x^4}{x^2 + 1}$	$\dfrac{16}{5}$	$\dfrac{1}{2}$	$\dfrac{1}{2}$	$\dfrac{16}{5}$

Step 7   Graph the function.   Figure 3.31 shows the graph of f using the points obtained from the table and y-axis symmetry. Notice that as x approaches infinity or negative infinity ($x \to \infty$ or $x \to -\infty$), the function values, $f(x)$, are getting larger without bound $[f(x) \to \infty]$.

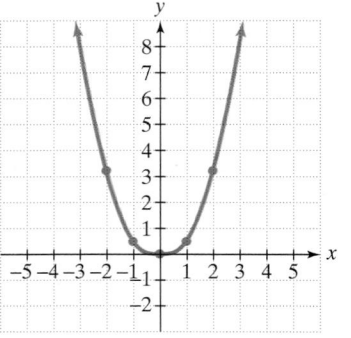

Figure 3.31 The graph of $f(x) = \dfrac{x^4}{x^2 + 1}$

Check Point 6 Graph: $f(x) = \dfrac{x^4}{x^2 + 2}$.

6 Identify slant asymptotes.

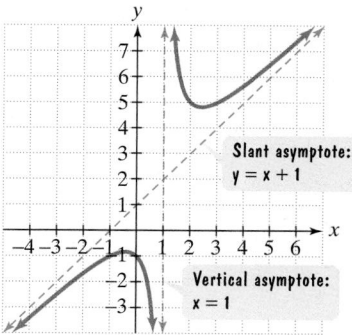

Figure 3.32 The graph of $f(x) = \dfrac{x^2 + 1}{x - 1}$ with a slant asymptote

Slant Asymptotes

Examine the graph of

$$f(x) = \frac{x^2 + 1}{x - 1}$$

shown in Figure 3.32. Note that the degree of the numerator, 2, is greater than the degree of the denominator, 1. Thus, the graph of this function has no horizontal asymptote. However, the graph has a **slant asymptote**, $y = x + 1$.

   The graph of a rational function has a slant asymptote if the degree of the numerator is one more than the degree of the denominator. The equation of the slant asymptote can be found by division. For example, to find the slant asymptote for the graph of $f(x) = \dfrac{x^2 + 1}{x - 1}$, divide $x - 1$ into $x^2 + 1$:

$$
\begin{array}{r|rrr}
1\!\!\!| & 1 & 0 & 1 \\
  &   & 1 & 1 \\
\hline
  & 1 & 1 & 2
\end{array}
\qquad\qquad
\begin{array}{r}
1x + 1 + \dfrac{2}{x - 1} \\[2pt]
x - 1\overline{)\,x^2 + 0x + 1}
\end{array}
$$

Remainder

Observe that

$$f(x) = \frac{x^2 + 1}{x - 1} = \underbrace{x + 1}_{\text{Slant asymptote:}} + \frac{2}{x - 1}$$
$$y = x + 1$$

If $|x| \to \infty$, the value of $\dfrac{2}{x - 1}$ is approximately 0. Thus, when $|x|$ is large, the function is very close to $y = x + 1 + 0$. This means that as $x \to \infty$ or as $x \to -\infty$, the graph of f gets closer and closer to the line whose equation is $y = x + 1$. The line $y = x + 1$ is a slant asymptote of the graph.

   In general, if $f(x) = \dfrac{p(x)}{q(x)}$ and the degree of p is one greater than the degree of q, find the slant asymptote by dividing $q(x)$ into $p(x)$. The division will take the form

$$\frac{p(x)}{q(x)} = mx + b + \frac{\text{remainder}}{q(x)}.$$

Slant asymptote:
$$y = mx + b$$

The equation of the slant asymptote is $y = mx + b$.

Slant asymptote:
y = x − 1

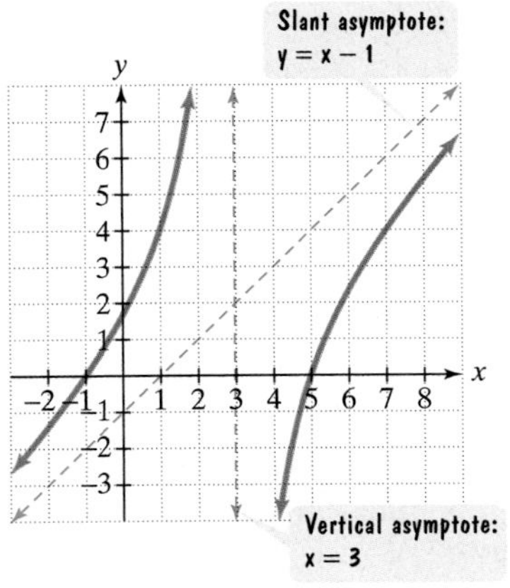

Figure 3.33 The graph of
$f(x) = \dfrac{x^2 - 4x - 5}{x - 3}$

Vertical asymptote:
x = 3

EXAMPLE 7 Finding the Slant Asymptote of a Rational Function

Find the slant asymptote of $f(x) = \dfrac{x^2 - 4x - 5}{x - 3}$.

Solution   Because the degree of the numerator, 2, is exactly one more than the degree of the denominator, 1, the graph of f has a slant asymptote. To find the equation of the slant asymptote, divide $x - 3$ into $x^2 - 4x - 5$:

$$\begin{array}{r|rrr} 3 & 1 & -4 & -5 \\ & & 3 & -3 \\ \hline & 1 & -1 & -8 \end{array}$$

$$x - 3 \overline{)x^2 - 4x - 5} \qquad 1x - 1 - \dfrac{8}{x-3}$$

Remainder

The equation of the slant asymptote is $y = x - 1$. Using our strategy for graphing rational functions, the graph of $f(x) = \dfrac{x^2 - 4x - 5}{x - 3}$ is shown in Figure 3.33.

Check Point 7   Find the slant asymptote of $f(x) = \dfrac{2x^2 - 5x + 7}{x - 2}$.

7 Solve applied problems involving rational functions.

Applications

There are numerous examples of asymptotic behavior in functions that describe real-world phenomena.

EXAMPLE 8   Average Cost of Producing a Wheelchair

A company that manufactures wheelchairs has costs given by the function

$$C(x) = 400x + 500{,}000$$

where x is the number of wheelchairs produced per month and $C(x)$ is measured in dollars. The average cost per wheelchair for the company is given by

$$\bar{C}(x) = \dfrac{400x + 500{,}000}{x}.$$

a. Find and interpret $\bar{C}(1000)$, $\bar{C}(10{,}000)$, and $\bar{C}(100{,}000)$.

b. What is the horizontal asymptote for the average cost function, $\bar{C}(x)$? Describe what this represents for the company.

Solution

a. $\bar{C}(1000) = \dfrac{400(1000) + 500{,}000}{1000} = 900$

The average cost per wheelchair of producing 1000 wheelchairs per month is $900.00.

$$\bar{C}(10{,}000) = \dfrac{400(10{,}000) + 500{,}000}{10{,}000} = 450$$

The average cost per wheelchair of producing 10,000 wheelchairs per month is $450.

$$\bar{C}(100{,}000) = \frac{400(100{,}000) + 500{,}000}{100{,}000} = 405$$

The average cost per wheelchair of producing 100,000 wheelchairs per month is $405. Notice that with higher production levels, the cost of producing each wheelchair decreases.

b. We are given the average cost function

$$\bar{C}(x) = \frac{400x + 500{,}000}{x}$$

in which the degree of the numerator, 1, is equal to the degree of the denominator, 1. The leading coefficients of the numerator and denominator, 400 and 1, are used to obtain the equation of the horizontal asymptote. The equation of the horizontal asymptote is

$$y = \frac{400}{1} \quad \text{or} \quad y = 400.$$

The horizontal asymptote is shown in Figure 3.34. This means that the more wheelchairs produced per month, the closer the average cost per wheelchair for the company comes to $400. The least possible cost per wheelchair is approaching $400. Competitively low prices take place with high production levels, posing a major problem for small businesses.

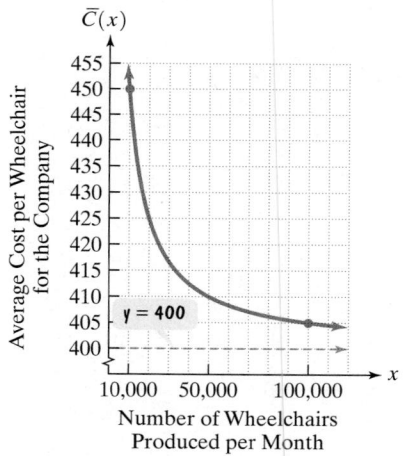

Figure 3.34 As production level increases, the average cost per wheelchair approaches $400.

Check Point 8 A company that manufactures running shoes has costs given by the function $C(x) = 30x + 300{,}000$, where x is the number of pairs of shoes produced per week and $C(x)$ is measured in dollars. The average cost per pair for the company is given by

$$\bar{C}(x) = \frac{30x + 300{,}000}{x}.$$

a. Find and interpret $\bar{C}(1000)$, $\bar{C}(10{,}000)$, and $\bar{C}(100{,}000)$.

b. What is the horizontal asymptote for the average cost function, $\bar{C}(x)$? Describe what this represents for the company.

EXERCISE SET 3.6

 Practice Exercises

In Exercises 1–8, find the domain of each rational function.

1. $f(x) = \dfrac{5x}{x - 4}$

2. $f(x) = \dfrac{7x}{x - 8}$

3. $g(x) = \dfrac{3x^2}{(x - 5)(x + 4)}$

4. $g(x) = \dfrac{2x^2}{(x - 2)(x + 6)}$

5. $h(x) = \dfrac{x + 7}{x^2 - 49}$

6. $h(x) = \dfrac{x + 8}{x^2 - 64}$

7. $f(x) = \dfrac{x + 7}{x^2 + 49}$

8. $f(x) = \dfrac{x + 8}{x^2 + 64}$

Use the graph of the rational function in the figure shown to complete each statement in Exercises 9–14.

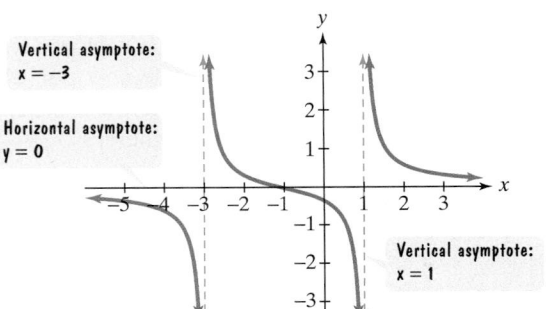

9. As $x \to -3^-$, $f(x) \to$ _____ .
10. As $x \to -3^+$, $f(x) \to$ _____ .
11. As $x \to 1^-$, $f(x) \to$ _____ .
12. As $x \to 1^+$, $f(x) \to$ _____ .
13. As $x \to -\infty$, $f(x) \to$ _____ .
14. As $x \to \infty$, $f(x) \to$ _____ .

Use the graph of the rational function in the figure shown to complete each statement in Exercises 15–20.

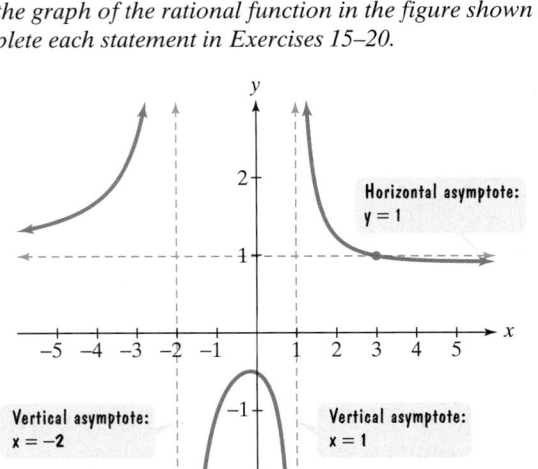

15. As $x \to 1^+$, $f(x) \to$ _____ .
16. As $x \to 1^-$, $f(x) \to$ _____ .
17. As $x \to -2^+$, $f(x) \to$ _____ .
18. As $x \to -2^-$, $f(x) \to$ _____ .
19. As $x \to \infty$, $f(x) \to$ _____ .
20. As $x \to -\infty$, $f(x) \to$ _____ .

In Exercises 21–28, find the vertical asymptotes, if any, of the graph of each rational function.

21. $f(x) = \dfrac{x}{x + 4}$

22. $f(x) = \dfrac{x}{x - 3}$

23. $g(x) = \dfrac{x + 3}{x(x + 4)}$

24. $g(x) = \dfrac{x + 3}{x(x - 3)}$

25. $h(x) = \dfrac{x}{x(x + 4)}$

26. $h(x) = \dfrac{x}{x(x - 3)}$

27. $r(x) = \dfrac{x}{x^2 + 4}$

28. $r(x) = \dfrac{x}{x^2 + 3}$

In Exercises 29–36, find the horizontal asymptote, if any, of the graph of each rational function.

29. $f(x) = \dfrac{12x}{3x^2 + 1}$

30. $f(x) = \dfrac{15x}{3x^2 + 1}$

31. $g(x) = \dfrac{12x^2}{3x^2 + 1}$

32. $g(x) = \dfrac{15x^2}{3x^2 + 1}$

33. $h(x) = \dfrac{12x^3}{3x^2 + 1}$

34. $h(x) = \dfrac{15x^3}{3x^2 + 1}$

35. $f(x) = \dfrac{-2x + 1}{3x + 5}$

36. $f(x) = \dfrac{-3x + 7}{5x - 2}$

In Exercises 37–58, follow the seven steps on page 321 to graph each rational function.

37. $f(x) = \dfrac{4x}{x - 2}$

38. $f(x) = \dfrac{3x}{x - 1}$

39. $f(x) = \dfrac{2x}{x^2 - 4}$

40. $f(x) = \dfrac{4x}{x^2 - 1}$

41. $f(x) = \dfrac{2x^2}{x^2 - 1}$

42. $f(x) = \dfrac{4x^2}{x^2 - 9}$

43. $f(x) = \dfrac{-x}{x + 1}$

44. $f(x) = \dfrac{-3x}{x + 2}$

45. $f(x) = -\dfrac{1}{x^2 - 4}$

46. $f(x) = -\dfrac{2}{x^2 - 1}$

47. $f(x) = \dfrac{2}{x^2 + x - 2}$

48. $f(x) = \dfrac{-2}{x^2 - x - 2}$

49. $f(x) = \dfrac{2x^2}{x^2 + 4}$

50. $f(x) = \dfrac{4x^2}{x^2 + 1}$

51. $f(x) = \dfrac{x + 2}{x^2 + x - 6}$

52. $f(x) = \dfrac{x - 4}{x^2 - x - 6}$

53. $f(x) = \dfrac{x^4}{x^2 + 2}$

54. $f(x) = \dfrac{2x^4}{x^2 + 1}$

55. $f(x) = \dfrac{x^2 + x - 12}{x^2 - 4}$

56. $f(x) = \dfrac{x^2}{x^2 + x - 6}$

57. $f(x) = \dfrac{3x^2 + x - 4}{2x^2 - 5x}$

58. $f(x) = \dfrac{x^2 - 4x + 3}{(x + 1)^2}$

In Exercises 59–66, **a.** *Find the slant asymptote of the graph of each rational function and* **b.** *Follow the seven-step strategy and use the slant asymptote to graph each rational function.*

59. $f(x) = \dfrac{x^2 - 1}{x}$

60. $f(x) = \dfrac{x^2 - 4}{x}$

61. $f(x) = \dfrac{x^2 + 1}{x}$

62. $f(x) = \dfrac{x^2 + 4}{x}$

63. $f(x) = \dfrac{x^2 + x - 6}{x - 3}$

64. $f(x) = \dfrac{x^2 - x + 1}{x - 1}$

65. $f(x) = \dfrac{x^3 + 1}{x^2 + 2x}$

66. $f(x) = \dfrac{x^3 - 1}{x^2 - 9}$

 Application Exercises

67. A company that manufactures small canoes has costs given by the function $C(x) = 20x + 20{,}000$, where x is the number of canoes manufactured and $C(x)$ is measured in dollars. The average cost to manufacture each canoe is given by

$$\bar{C}(x) = \frac{20x + 20{,}000}{x}.$$

a. Find the average cost per canoe when $x = 100, 1000,$ 10,000, and 100,000.
b. What is the horizontal asymptote for the function $\bar{C}$, and what does it represent?

68. A company that manufactures bicycles has costs given by the function $C(x) = 100x + 100{,}000$, where x is the number of bicycles manufactured and $C(x)$ is measured in dollars. The average cost to manufacture each bicycle is given by

$$\bar{C}(x) = \frac{100x + 100{,}000}{x}.$$

a. Find and interpret $\bar{C}(500)$, $\bar{C}(1000)$, $\bar{C}(2000)$, and $\bar{C}(4000)$.
b. What is the horizontal asymptote for the function $\bar{C}$? Describe what this means in practical terms.

69. The cost, in dollars, of removing p percent of the air pollutants in the smokestack emission of a utility company that burns coal to generate electricity is given by

$$C(p) = \frac{60{,}000p}{100 - p}.$$

a. Current law requires that the company remove 80% of the pollutants from its smokestack emissions. A new law before the legislature would require increasing this amount by 5%. How much will it cost to remove another 5% of the pollutants?
b. Does this function indicate the possibility of removing 100% of the pollutants? Explain.

70. The rational function

$$C(x) = \frac{130x}{100 - x}, \quad 0 \le x < 100,$$

describes the cost C, in millions of dollars, to inoculate $x\%$ of the population against a particular strain of flu.
a. Find and interpret $C(80) - C(40)$.
b. Graph the function.
c. Describe the practical meaning of the observation that $x = 100$ is an asymptote.

71. The temperature, F, in degrees Fahrenheit, of a dessert placed in an icebox for t hours is modeled by

$$F(t) = \frac{80}{t^2 + 4t + 1}.$$

a. Find and interpret $F(0)$.
b. Find the temperature of the dessert after 1 hour, 2 hours, 3 hours, 4 hours, and 5 hours.
c. What is the equation of the horizontal asymptote associated with this function? Describe what this means in terms of the dessert's temperature over time.
d. Graph the function.

72. The function $f(x) = \dfrac{72{,}900}{100x^2 + 729}$ models the percentage of people in the United States who are unemployed as a function of years of education, x.
a. Find and interpret $f(0)$.
b. Find and interpret $f(20)$.
c. Is there an education level that leads to guaranteed employment? If not, how is this indicated by the equation of the horizontal asymptote associated with this function?
d. Graph the function.

73. Rational functions are often used to model how much we remember over time. In an experiment on memory, students in a language class are asked to memorize 40 vocabulary words in Latin, a language with which the students are not familiar. After studying the words for one day, the class is tested each day thereafter to see how many words they remember. The class average is taken and the results are graphed as shown on page 330.

Average Number of Words Remembered over Time

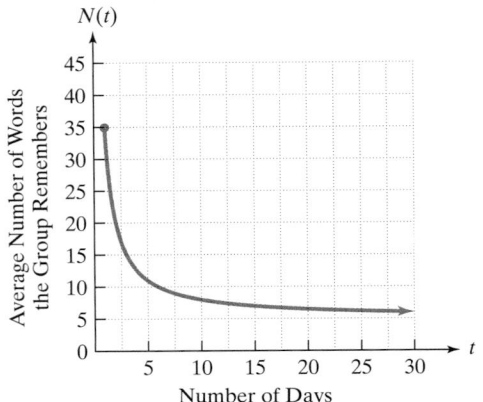

a. Use the graph to find a reasonable estimate of the number of Latin words remembered after 1 day, 5 days, and 15 days.
b. The function that models the number of Latin words remembered by the students after t days is given by

$$N(t) = \frac{5t + 30}{t}, \quad \text{where } t \geq 1.$$

Find $N(1), N(5)$, and $N(15)$, comparing these values with your estimates from part (a).
c. What does the graph indicate about the number of Latin words remembered by the group over time?
d. Use the function in part (b) to find the horizontal asymptote for the graph. Describe what this horizontal asymptote means in terms of the variables modeled in this situation.

74. A drug is injected into a patient and the concentration of the drug in the bloodstream is monitored. The drug's concentration, $C(t)$, in milligrams per liter, after t hours is modeled by

$$C(t) = \frac{5t}{t^2 + 1}.$$

The graph of this rational function, obtained with a graphing utility, is shown in the figure.

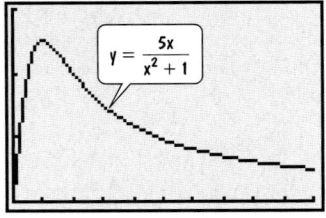

a. Use the graph to obtain a reasonable estimate of the drug's concentration after 3 hours. Then verify this estimate algebraically.

b. Use the function's equation to find the horizontal asymptote for the graph. Describe what this means about the drug's concentration in the patient's bloodstream as time increases.

Writing in Mathematics

75. What is a rational function?
76. Use everyday language to describe the graph of a rational function f such that $f(x) \to 3$ as $x \to -\infty$.
77. Use everyday language to describe the behavior of a graph near its vertical asymptote if $f(x) \to \infty$ as $x \to -2^-$ and $f(x) \to -\infty$ as $x \to -2^+$.
78. If you are given the equation of a rational function, explain how to find the vertical asymptotes, if any, of the function's graph.
79. If you are given the equation of a rational function, explain how to find the horizontal asymptote, if any, of the function's graph.
80. Describe how to graph a rational function.
81. If you are given the equation of a rational function, how can you tell if the graph has a slant asymptote? If it does, how do you find its equation?
82. Is every rational function a polynomial function? Why or why not? Does a true statement result if the two adjectives rational and polynomial are reversed? Explain.
83. The function $f(x) = \dfrac{5000x}{x^2 + 36}$ describes the population density, $f(x)$, in people per square mile in a large city x miles from the city's center. Describe what eventually happens to the population density as the distance from the city's center increases.

Technology Exercises

84. Use a graphing utility to verify any five of your hand-drawn graphs in Exercises 37–66.
85. Use a graphing utility to verify your hand-drawn graphs in Exercises 71–72.
86. Use a graphing utility to graph $y = \dfrac{1}{x}$, $y = \dfrac{1}{x^3}$, and $\dfrac{1}{x^5}$ in the same viewing rectangle. For odd values of n, how does changing n affect the graph of $y = \dfrac{1}{x^n}$?
87. Use a graphing utility to graph $y = \dfrac{1}{x^2}$, $y = \dfrac{1}{x^4}$, and $y = \dfrac{1}{x^6}$ in the same viewing rectangle. For even values of n, how does changing n affect the graph of $y = \dfrac{1}{x^n}$?

88. A grocery store sells 4000 cases of canned soup per year. By averaging costs to purchase soup and pay storage costs, the owner has determined that if x cases are ordered at a time, the inventory cost will be

$$C(x) = \frac{10,000}{x} + 3x.$$

a. Use a graphing utility to graph the inventory function.
b. Use the ZOOM and TRACE features or the minimum function feature of your graphing utility to approximate the number of cases that should be ordered to minimize inventory cost. What is the minimum cost?

89. Use a graphing utility to graph

$$f(x) = \frac{x^2 - 4x + 3}{x - 2} \quad \text{and} \quad g(x) = \frac{x^2 - 5x + 6}{x - 2}.$$

What differences do you observe between the graph of f and g? How do you account for these differences?

90. Use a graphing utility to graph

$$f(x) = \frac{2|x - 1|}{x + 2}.$$

How does the situation with horizontal asymptotes differ from the other functions graphed throughout this section?

Critical Thinking Exercises

91. Which one of the following is true?
a. The graph of a rational function cannot have both a vertical and a horizontal asymptote.
b. It is not possible to have a rational function whose graph has no y-intercept.
c. The graph of a rational function can have three horizontal asymptotes.
d. The graph of a rational function can never cross a vertical asymptote.

92. Which one of the following is true?

a. The function $f(x) = \dfrac{1}{\sqrt{x - 3}}$ is a rational function.

b. The x-axis is a horizontal asymptote for the graph of

$$f(x) = \frac{4x - 1}{x + 3}.$$

c. The number of televisions that a company can produce per week after t weeks of production is given by

$$N(t) = \frac{3000t^2 + 30,000t}{t^2 + 10t + 25}.$$

Using this model, the company will eventually be able to produce 30,000 televisions in a single week.
d. None of the given statements is true.

In Exercises 93–96, write the equation of a rational function
$$f(x) = \frac{p(x)}{q(x)} \text{ having the indicated properties, in which the}$$
degrees of p and q are as small as possible. More than one correct function may be possible. Graph your function using a graphing utility to verify that it has the required properties.

93. f has a vertical asymptote given by $x = 3$, a horizontal asymptote $y = 0$, y-intercept $= -1$, and no x-intercept.
94. f has vertical asymptotes given by $x = -2$ and $x = 2$, a horizontal asymptote $y = 2$, y-intercept $= \frac{9}{2}$, x-intercepts of -3 and 3, and y-axis symmetry.
95. f has a vertical asymptote given by $x = 1$, a slant asymptote whose equation is $y = x$, y-intercept $= 2$, and x-intercepts of -1 and 2.
96. f has no vertical, horizontal, or slant asymptotes, and no x-intercepts.

Group Exercise

97. Group members make up the sales team for a company that makes computer video games. It has been determined that the rational function

$$f(x) = \frac{200x}{x^2 + 100}$$

models the monthly sales, in thousands of games, of a new video game as a function of the number of months, x, after the game is introduced. The figure shows the graph of the function. What are the team's recommendations to the company in terms of how long the video game should be on the market before another new video game is introduced? What other factors might members want to take into account in terms of the recommendations? What will eventually happen to sales, and how is this indicated by the graph? What does this have to do with a horizontal asymptote? What could the company do to change the behavior of this function and continue generating sales? Would this be cost effective?

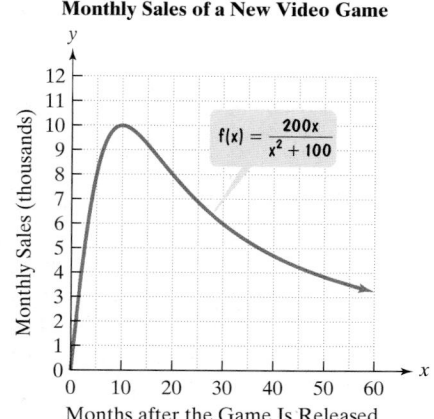

Monthly Sales of a New Video Game

$f(x) = \dfrac{200x}{x^2 + 100}$

SECTION 3.7 *Modeling Using Variation*

Objectives

1. Solve direct variation problems.
2. Solve inverse variation problems.
3. Solve combined variation problems.
4. Solve problems involving joint variation.

Have you ever wondered how telecommunication companies estimate the number of phone calls expected per day between two cities? The formula

$$N = \frac{400P_1P_2}{d^2}$$

shows that the daily number of phone calls, N, increases as the populations of the cities, P_1 and P_2, in thousands, increase and decreases as the distance, d, between the cities increases.

Certain formulas occur so frequently in applied situations that they are given special names. Variation formulas show how one quantity changes in relation to other quantities. Quantities can vary *directly*, *inversely*, or *jointly*. In this section, we look at situations that can be modeled by each of these kinds of variation. And think of this: The next time you get one of those "all-circuits-are-busy" messages, you will be able to use a variation formula to estimate how many other callers you're competing with for those precious 5-cent minutes.

1 Solve direct variation problems.

Direct Variation

Because light travels faster than sound, during a thunderstorm we see lightning before we hear thunder. The formula

$$d = 1080t$$

describes the distance, in feet, of the storm's center if it takes t seconds to hear thunder after seeing lightning. Thus,

If $t = 1$, $d = 1080 \cdot 1 = 1080$: If it takes 1 second to hear thunder, the storm's center is 1080 feet away.

If $t = 2$, $d = 1080 \cdot 2 = 2160$: If it takes 2 seconds to hear thunder, the storm's center is 2160 feet away.

If $t = 3$, $d = 1080 \cdot 3 = 3240$: If it takes 3 seconds to hear thunder, the storm's center is 3240 feet away.

As the formula $d = 1080t$ illustrates, the distance to the storm's center is a constant multiple of how long it takes to hear the thunder. When the time is doubled, the storm's distance is doubled; when the time is tripled, the storm's distance is tripled; and so on. Because of this, the distance is said to **vary directly** as the time. The **equation of variation** is

$$d = 1080t.$$

Generalizing, we obtain the following statement.

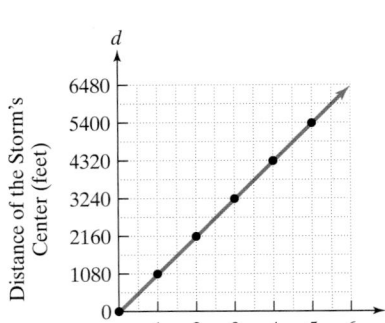

The graph of $d = 1080t$. Distance to a storm's center varies directly as the time it takes to hear thunder.

Direct Variation

If a situation is described by an equation in the form

$$y = kx$$

where k is a constant, we say that **y varies directly as x**. The number k is called the **constant of variation**.

EXAMPLE 1   Writing a Direct Variation Equation

A person's salary, S, varies directly as the number of hours worked, h.

   a. Write an equation that expresses this relationship.

   b. Margarita earns $18 per hour. Substitute 18 for k, the constant of variation, in the equation in part (a) and write the equation for Margarita's salary.

Solution

   a. We know that y varies directly as x is expressed as

$$y = kx.$$

By changing letters, we can write an equation that describes the following English statement: Salary, S, varies directly as the number of hours worked, h.

$$S = kh$$

   b. Substituting 18 for k in the direct variation equation gives

$$S = 18h.$$

This equation describes Margarita's salary in terms of the number of hours she works. For example, if she works 10 hours, we can substitute 10 for h and determine her salary:

$$S = 18(10) = 180.$$

Her salary for working 10 hours is $180. Notice that, as the number of hours worked increases, the salary increases.

> **Check Point 1**
>
> A person's hair length, L, in inches, varies directly as the number of years it has been growing, N.
> **a.** Write an equation that expresses this relationship.
> **b.** The longest moustache on record was grown by Kalyan Sain of India. His moustache grew 4 inches each year. Substitute 4 for k, the constant of variation, in the equation in part (a) and write the equation for the length of Sain's moustache.
> **c.** Sain grew his moustache for 17 years. Substitute 17 for N in the equation from part (b) and find its length.

In Example 1 and Checkpoint 1, the constants of variation were given. If the constant of variation is not given, we can find it by substituting given values in the variation formula and solving for k. Example 2 shows how this is done.

EXAMPLE 2   Finding k, the Constant of Variation

Height, H, varies directly as foot length, F.

   a. Write an equation that expresses this relationship.

   b. Photographs of large footprints were published in 1951. Some speculated that these footprints were made by the Abominable Snowman. Each footprint was 23 inches long. The Abominable Snowman's height was determined to be 154.1 inches. (This is 12 feet, 10.1 inches, so it might not be a pleasant experience to run into this critter on a mellow hike through the woods!) Use $H = 154.1$ and $F = 23$ to find the constant of variation.

Solution

   a. We know that y varies directly as x is expressed as

$$y = kx.$$

   By changing letters, we can write an equation that describes the following English statement: Height, H, varies directly as foot length, F.

$$H = kF$$

   b. The Abominable Snowman's height is 154.1 inches, and foot length is 23 inches. Substitute 154.1 for H and 23 for F in the direct variation equation.

$$H = kF$$

$$154.1 = k \cdot 23$$

   Solve for k, the constant of variation, by dividing both sides of the equation by 23:

$$\frac{154.1}{23} = \frac{k \cdot 23}{23}$$

$$6.7 = k$$

Thus, the constant of variation is 6.7.

In Example 2, now that we know the constant of variation ($k = 6.7$), we can rewrite $H = kF$ using this constant. The equation of variation is

$$H = 6.7F.$$

We can use this equation to find other values. For example, if your foot length is 10 inches, your height is

$$H = 6.7(10) = 67,$$

or approximately 67 inches.

Check Point 2 The weight, W, of an aluminum canoe varies directly as its length, L.

   a. Write an equation that expresses this relationship.

   b. A 6-foot canoe weighs 75 pounds. Substitute 75 for W and 6 for L in the equation from part (a) and find k, the constant of variation.

   c. Substitute the value of k into your equation in part (a) and write the equation that describes the weight of this type of canoe in terms of its length.

   d. Use the equation from part (c) to find the weight of a 16-foot canoe of this type.

Our work up to this point provides a step-by-step procedure for solving variation problems. This procedure applies to direct variation problems as well as to the other kinds of variation problems that we will discuss.

> **Solving Variation Problems**
>
> **1.** Write an equation that describes the given English statement.
> **2.** Substitute the given pair of values into the equation in step 1 and find the value of k.
> **3.** Substitute the value of k into the equation in step 1.
> **4.** Use the equation from step 3 to answer the problem's question.

EXAMPLE 3   Solving a Direct Variation Problem

The amount of garbage, G, varies directly as the population, P. Allegheny County, Pennsylvania, has a population of 1.3 million and creates 26 million pounds of garbage each week. Find the weekly garbage produced by New York City with a population of 7.3 million.

Solution

Step 1   Write an equation.   We know that y varies directly as x is expressed as

$$y = kx.$$

By changing letters, we can write an equation that describes the following English statement: Garbage production, G, varies directly as the population, P.

$$G = kP$$

Step 2   Use the given values to find k.   Allegheny County has a population of 1.3 million and creates 26 million pounds of garbage weekly. Substitute 26 for G and 1.3 for P in the direct variation equation. Then solve for k.

$$G = kP$$

$$26 = k \cdot 1.3$$

$$\frac{26}{1.3} = \frac{k \cdot 1.3}{1.3} \qquad \text{Divide both sides by 1.3.}$$

$$20 = k \qquad \text{Simplify.}$$

Step 3   Substitute the value of k into the equation.

$$G = kP \qquad \text{Use the equation from step 1.}$$

$$G = 20P \qquad \text{Replace k, the constant of variation, with 20.}$$

Step 4   Answer the problem's question.   New York City has a population of 7.3 million. To find its weekly garbage production, substitute 7.3 for P in $G = 20P$ and solve for G.

$$G = 20P \qquad \text{Use the equation from step 3.}$$

$$G = 20(7.3) \qquad \text{Substitute 7.3 for P.}$$

$$G = 146$$

The weekly garbage produced by New York City weighs approximately 146 million pounds.

Check Point 3 The pressure, P, of water on an object below the surface varies directly as its distance, D, below the surface. If a submarine experiences a pressure of 25 pounds per square inch 60 feet below the surface, how much pressure will it experience 330 feet below the surface?

The direct variation equation $y = kx$ is a linear function. If $k > 0$, then the slope of the line is positive. Consequently, as x increases, y also increases.

A direct variation situation can involve variables to higher powers. For example, y can vary directly as x^2 ($y = kx^2$) or as x^3 ($y = kx^3$).

Direct Variation With Powers

y varies directly as the nth power of x if there exists some nonzero constant k such that

$$y = kx^n.$$

Direct variation with powers is modeled by polynomial functions. In our next example, the graph of the variation equation is the familiar parabola.

EXAMPLE 4 Solving a Direct Variation Problem

The distance, s, that a body falls from rest varies directly as the square of the time, t, of the fall. If skydivers fall 64 feet in 2 seconds, how far will they fall in 4.5 seconds?

Solution

Step 1 Write an equation. We know that y varies directly as the square of x is expressed as

$$y = kx^2.$$

By changing letters, we can write an equation that describes the following English statement: Distance, s, varies directly as the square of time, t, of the fall.

$$s = kt^2$$

Step 2 Use the given values to find k. Skydivers fall 64 feet in 2 seconds. Substitute 64 for s and 2 for t in the direct variation equation. Then solve for k.

$$s = kt^2$$
$$64 = k \cdot 2^2$$
$$64 = 4k$$
$$\frac{64}{4} = \frac{4k}{4}$$
$$16 = k$$

Step 3 Substitute the value of k into the equation.

$$s = kt^2 \qquad \text{Use the equation from step 1.}$$
$$s = 16t^2 \qquad \text{Replace } k, \text{ the constant of variation, with 16.}$$

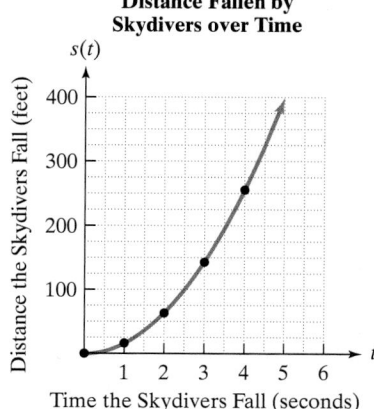

Distance Fallen by Skydivers over Time

Figure 3.35 The graph of $s(t) = 16t^2$

Step 4   Answer the problem's question.   How far will the skydivers fall in 4.5 seconds? Substitute 4.5 for t in $s = 16t^2$ and solve for s.

$$s = 16(4.5)^2 = 16(20.25) = 324$$

Thus, in 4.5 seconds, skydivers will fall 324 feet.

We can express the variation equation from Example 4 in function notation, writing

$$s(t) = 16t^2.$$

The distance that a body falls from rest is a function of the time, t, of the fall. The parabola that is the graph of this quadratic function is shown in Figure 3.35. The graph increases rapidly from left to right, showing the effects of the acceleration of gravity.

Check Point 4   The distance required to stop a car varies directly as the square of its speed. If 200 feet are required to stop a car traveling 60 miles per hour, how many feet are required to stop a car traveling 100 miles per hour?

② Solve inverse variation problems.

Inverse Variation

The distance from Atlanta, Georgia, to Orlando, Florida, is 450 miles. The time that it takes to drive from Atlanta to Orlando depends on the rate at which one drives and is given by

$$\text{Time} = \frac{450}{\text{Rate}}.$$

For example, if you average 45 miles per hour, the time for the drive is

$$\text{Time} = \frac{450}{45} = 10,$$

or 10 hours. If you ignore speed limits and average 75 miles per hour, the time for the drive is

$$\text{Time} = \frac{450}{75} = 6,$$

or 6 hours. As your rate (or speed) increases, the time for the trip decreases and vice versa.

We can express the time for the Atlanta–Orlando trip using t for time and r for rate:

$$t = \frac{450}{r}.$$

This equation is an example of an **inverse variation** equation. Time, t, **varies inversely** as rate, r. When two quantities vary inversely, one quantity increases as the other decreases, and vice versa.

Generalizing, we obtain the following statement.

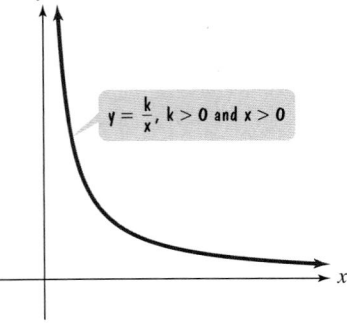

Figure 3.36 The graph of the inverse variation equation

Inverse Variation

If a situation is described by an equation in the form

$$y = \frac{k}{x}$$

where k is a constant, we say that **y varies inversely as x.** The number k is called the **constant of variation**.

Notice that the inverse variation equation

$$y = \frac{k}{x}, \quad \text{or} \quad f(x) = \frac{k}{x},$$

is a rational function. For $k > 0$ and $x > 0$, the graph of the function takes on the shape shown in Figure 3.36.

We use the same procedure to solve inverse variation problems as we did to solve direct variation problems. Example 5 illustrates this procedure.

EXAMPLE 5 Solving an Inverse Variation Problem

To continue making money, the number of new songs, S, a rock band needs to record each year varies inversely as the number of years, N, the band has been recording. After 4 years of recording, a band needs to record 15 new songs per year to be profitable. After 6 years, how many new songs will the band need to record in order to make a profit in the seventh year?

Solution

Step 1 Write an equation.   We know that y varies inversely as x is expressed as

$$y = \frac{k}{x}.$$

By changing letters, we can write an equation that describes the following English statement: The number of new songs each year, S, varies inversely as the number of years, N.

$$S = \frac{k}{N}$$

Step 2 Use the given values to find k.   After 4 years of recording, the band needs to record 15 new songs. Substitute 15 for S and 4 for N in the inverse variation equation. Then solve for k.

$$S = \frac{k}{N}$$

$$15 = \frac{k}{4}$$

$$15 \cdot 4 = \frac{k}{4} \cdot 4 \qquad \text{Multiply both sides by 4.}$$

$$60 = k \qquad \text{Simplify.}$$

Step 3   Substitute the value of k into the equation.

$$S = \frac{k}{N} \qquad \text{Use the equation from step 1.}$$

$$S = \frac{60}{N} \qquad \text{Replace k, the constant of variation, with 60.}$$

Step 4   Answer the problem's question.   We need to find how many new songs the band will need to record after 6 years to make a profit in the seventh year. Substitute 6 for N in the equation above and solve for S.

$$S = \frac{60}{N} = \frac{60}{6} = 10$$

The band will need to record 10 new songs after 6 years.

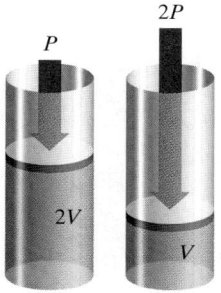

$2P$

P

$2V$

V

Doubling the pressure halves the volume.

> **Check Point 5**   When you use a spray can and press the valve at the top, you decrease the pressure of the gas in the can. This decrease of pressure causes the volume of the gas in the can to increase. Because the gas needs more room than is provided in the can, it expands in spray form through the small hole near the valve. In general, if the temperature is constant, the pressure, P, of a gas in a container varies inversely as the volume, V, of the container. The pressure of a gas sample in a container whose volume is 8 cubic inches is 12 pounds per square inch. If the sample expands to a volume of 22 cubic inches, what is the new pressure of the gas?

③ Solve combined variation problems.

Combined Variation

In a **combined variation** situation, direct and inverse variation occur at the same time. For example, as the advertising budget, A, of a company increases, its monthly sales, S, also increase. Monthly sales vary directly as the advertising budget:

$$S = kA.$$

By contrast, as the price of the company's product, P, increases, its monthly sales, S, decrease. Monthly sales vary inversely as the price of the product:

$$S = \frac{k}{P}.$$

We can combine these two variation equations into one combined equation:

$$S = \frac{kA}{P}.$$

The following example illustrates the application of combined variation.

EXAMPLE 6   Solving a Combined Variation Problem

The owners of Rollerblades Now determine that the monthly sales, S, of its skates vary directly as its advertising budget, A, and inversely as the price of the skates, P. When $60,000 is spent on advertising and the price of the skates is $40, the monthly sales are 12,000 pairs of rollerblades.

   a. Write an equation of variation that describes this situation.

   b. Determine monthly sales if the amount of the advertising budget is increased to $70,000.

Solution

a. Write an equation.

$$S = \frac{kA}{P}$$ Translate "sales vary directly as the advertising budget and inversely as the skates' price."

Use the given values to find k.

$$12{,}000 = \frac{k(60{,}000)}{40}$$ When \$60,000 is spent on advertising $(A = 60{,}000)$ and the price is \$40 $(P = 40)$, monthly sales are 12,000 units $(S = 12{,}000)$.

$$12{,}000 = k \cdot 1500$$ Divide 60,000 by 40.

$$\frac{12{,}000}{1500} = \frac{k \cdot 1500}{1500}$$ Divide both sides of the equation by 1500.

$$8 = k$$ Simplify.

Therefore, the equation of variation that describes monthly sales is

$$S = \frac{8A}{P}.$$

b. The advertising budget is increased to \$70,000, so $A = 70{,}000$. The skates' price is still \$40, so $P = 40$.

$$S = \frac{8A}{P}$$ This is the equation from part (a).

$$S = \frac{8(70{,}000)}{40}$$ Substitute 70,000 for A and 40 for P.

$$S = 14{,}000$$

With a \$70,000 advertising budget and \$40 price, the company can expect to sell 14,000 pairs of rollerblades in a month (up from 12,000).

Check Point 6 The number of minutes needed to solve an exercise set of variation problems varies directly as the number of problems and inversely as the number of people working to solve the problems. It takes 4 people 32 minutes to solve 16 problems. How many minutes will it take 8 people to solve 24 problems?

4 Solve problems involving joint variation.

Joint Variation

Joint variation is a variation in which a variable varies directly as the product of two or more other variables. Thus, the equation $y = kxz$ is read "y varies jointly as x and z."

Joint variation plays a critical role in Isaac Newton's formula for gravitation:

$$F = G\frac{m_1 m_2}{d^2}$$

The formula states that the force of gravitation, F, between two bodies varies jointly as the product of their masses, m_1 and m_2, and inversely as the square of the distance between them, d. (G is the gravitational constant.) The formula indicates that gravitational force exists between any two objects in the universe, increasing as the distance between the bodies decreases. One practical result is that the pull of the moon on the oceans is greater on the side of Earth closer to the moon. This gravitational imbalance is what produces tides.

EXAMPLE 7   Modeling Centrifugal Force

The centrifugal force, C, of a body moving in a circle varies jointly with the radius of the circular path, r, and the body's mass, m, and inversely with the square of the time, t, it takes to move about one full circle. A 6-gram body moving in a circle with radius 100 centimeters at a rate of 1 revolution in 2 seconds has a centrifugal force of 6000 dynes. Find the centrifugal force of an 18-gram body moving in a circle with radius 100 centimeters at a rate of 1 revolution in 3 seconds.

Solution

$$C = \frac{krm}{t^2}$$
Translate "Centrifugal force, C, varies jointly with radius, r, and mass, m, and inversely with the square of time, t."

$$6000 = \frac{k(100)(6)}{2^2}$$
If $r = 100$, $m = 6$, and $t = 2$, then $C = 6000$.

$$40 = k$$
Solve for k.

$$C = \frac{40rm}{t^2}$$
Substitute 40 for k in the model for centrifugal force.

$$= \frac{40(100)(18)}{3^2}$$
Find C when $r = 100$, $m = 18$, and $t = 3$.

$$= 8000$$

The centrifugal force is 8000 dynes.

 Check Point 7   The volume of a cone, V, varies jointly as its height, h, and the square of its radius r. A cone with a radius measuring 6 feet and a height measuring 10 feet has a volume of 120π cubic feet. Find the volume of a cone having a radius of 12 feet and a height of 2 feet.

EXERCISE SET 3.7

 ### Practice Exercises

In Exercises 1–12, write an equation that expresses each relationship. Use k as the constant of variation.

1. g varies directly as h.

2. v varies directly as r.

3. a varies directly as the square of b.

4. s varies directly as the cube of v.

5. r varies inversely as t.

6. w varies inversely as l.

7. a varies inversely as the cube of b.

8. y varies inversely as the square root of x.

9. r varies directly as s and inversely as v.

10. a varies directly as d and inversely as g.

11. s varies jointly as g and the square of t.

12. V varies jointly as h and the square of r.

In Exercises 13–22, determine the constant of variation for each stated condition.

13. y varies directly as x, and $y = 75$ when $x = 3$.

14. y varies directly as x, and $y = 55$ when $x = 11$.

15. y varies directly as x^2, and $y = 45$ when $x = 3$.

16. y varies directly as x^2, and $y = 72$ when $x = 6$.

17. W varies inversely as r, and $W = 500$ when $r = 10$.

18. T varies inversely as n, and $T = 7$ when $n = 12$.

19. A varies directly as B and inversely as C, and $A = 9$ when $B = 12$ and $C = 4$.

20. D varies directly as E and inversely as F, and $D = 6$ when $E = 12$ and $F = 10$.

21. a varies jointly as b and c, and $a = 72$ when $b = 18$ and $c = 2$.

22. z varies jointly as w and y, and $z = 38$ when $w = 38$ and $y = 2$.

Use the four-step procedure for solving variation problems given on page 335 to solve Exercises 23–30.

23. y varies directly as x. $y = 35$ when $x = 5$. Find y when $x = 12$.

24. y varies directly as x. $y = 55$ when $x = 5$. Find y when $x = 13$.

25. y varies inversely as x. $y = 10$ when $x = 5$. Find y when $x = 2$.

26. y varies inversely as x. $y = 5$ when $x = 3$. Find y when $x = 9$.

27. y varies directly as x and inversely as the square of z. $y = 20$ when $x = 50$ and $z = 5$. Find y when $x = 3$ and $z = 6$.

28. a varies directly as b and inversely as the square of c. $a = 7$ when $b = 9$ and $c = 6$. Find a when $b = 4$ and $c = 8$.

29. y varies jointly as x and z. $y = 25$ when $x = 2$ and $z = 5$. Find y when $x = 8$ and $z = 12$.

30. C varies jointly as A and T. $C = 175$ when $A = 2100$ and $T = 4$. Find C when $A = 2400$ and $T = 6$.

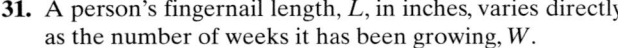 **Application Exercises**

31. A person's fingernail length, L, in inches, varies directly as the number of weeks it has been growing, W.
  a. Write an equation that expresses this relationship.
  b. Fingernails grow at a rate of about 0.02 inch per week. Substitute 0.02 for k, the constant of variation, in the equation in part (a) and write the equation for fingernail length.
  c. Substitute 52 for W to determine your fingernail length at the end of one year if for some bizarre reason you decided not to cut them and they did not break.

32. A person's salary, S, varies directly as the number of hours worked, h.
  a. Write an equation that expresses this relationship.
  b. For a 40-hour work week, Gloria earned $1400. Substitute 1400 for S and 40 for h in the equation from part (a) and find k, the constant of variation.
  c. Substitute the value of k into your equation in part (a) and write the equation that describes Gloria's salary in terms of the number of hours she works.
  d. Use the equation from part (c) to find Gloria's salary for 25 hours of work.

Use the four-step procedure for solving variation problems given on page 335 to solve Exercises 33–49.

33. The cost, C, of an airplane ticket varies directly as the number of miles, M, in the trip. A 3000-mile trip costs $400. What is the cost of a 450-mile trip?

34. An object's weight on the moon, M, varies directly as its weight on Earth, E. A person who weighs 55 kilograms on Earth weights 8.8 kilograms on the moon. What is the moon weight of a person who weighs 90 kilograms on Earth?

35. The Mach number is a measurement of speed named after the man who suggested it, Ernst Mach (1838–1916). The speed of an aircraft varies directly as its Mach number. Shown here are two aircraft. Use the figures for the Concord to determine the Blackbird's speed.

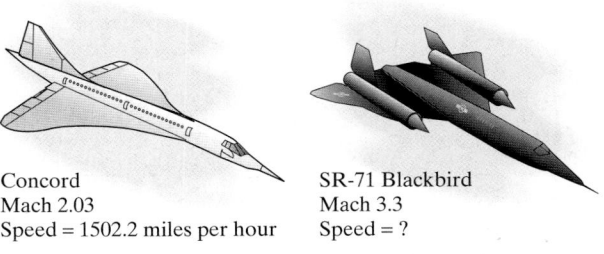

Concord
Mach 2.03
Speed = 1502.2 miles per hour

SR-71 Blackbird
Mach 3.3
Speed = ?

36. Do you still own records, or are you strictly a CD person? Record owners claim that the quality of sound on good vinyl surpasses that of a CD, although this is up for debate. This, however, is not debatable: The number of revolutions a record makes as it is being played varies directly as the time that it is on the turntable. A record that lasted 3 minutes made 135 revolutions. If a record takes 2.4 minutes to play, how many revolutions does it make?

37. If all men had identical body types, their weight would vary directly as the cube of their height. Shown is Robert Wadlow, who reached a record height of 8 feet 11 inches (107 inches) before his death at age 22. If a man who is 5 feet 10 inches tall (70 inches) with the same body type as Mr. Wadlow weighs 170 pounds, what was Robert Wadlow's weight shortly before his death?

38. The distance that an object falls varies directly as the square of the time it has been falling. An object falls 144 feet in 3 seconds. Find how far it will fall in 7 seconds.

39. The time that it takes you to get to campus varies inversely as your driving rate. Averaging 20 miles per hour in terrible traffic, it takes you 1.5 hours to get to campus. How long would the trip take averaging 60 miles per hour?

40. The weight that can be supported by a 2-inch by 4-inch piece of pine (called a 2-by-4) varies inversely as its length. A 10-foot 2-by-4 pine can support 500 pounds. What weight can be supported by a 125-foot 2-by-4 pine?

41. The volume of a gas in a container at a constant temperature varies inversely as the pressure. If the volume is 32 cubic centimeters at a pressure of 8 pounds, find the pressure when the volume is 40 cubic centimeters.

42. The current in a circuit varies inversely as the resistance. The current is 20 amperes when the resistance is 5 ohms. Find the current for a resistance of 16 ohms.

43. A person's body-mass index is used to assess levels of fatness, with an index from 20 to 26 considered in the desirable range. The index varies directly as one's weight, in pounds, and inversely as one's height, in inches. A person who weighs 150 pounds and is 70 inches tall has an index of 21. What is the body-mass index of a person who weighs 240 pounds and is 74 inches tall? Because the index is rounded to the nearest whole number, do so and then determine if this person's level of fatness is in the desirable range.

44. The volume of a gas varies directly as its temperature and inversely as its pressure. At a temperature of 100 Kelvin and a pressure of 15 kilograms per square meter, the gas occupies a volume of 20 cubic meters. Find the volume at a temperature of 150 Kelvin and a pressure of 30 kilograms per square meter.

45. The intensity of illumination on a surface varies inversely as the square of the distance of the light source from the surface. The illumination from a source is 25 foot-candles at a distance of 4 feet. What is the illumination when the distance is 6 feet?

46. The gravitational force with which Earth attracts an object varies inversely with the square of the distance from the center of Earth. A gravitational force of 160 pounds acts on an object 400 miles from Earth's center. Find the force of attraction on an object 6000 miles from the center of Earth.

47. Kinetic energy varies jointly as the mass and the square of the velocity. A mass of 8 grams and velocity of 3 centimeters per second has a kinetic energy of 36 ergs. Find the kinetic energy for a mass of 4 grams and velocity of 6 centimeters per second.

48. The electrical resistance of a wire varies directly as its length and inversely as the square of its diameter. A wire of 720 feet with $\frac{1}{4}$-inch diameter has a resistance of $1\frac{1}{2}$ ohms. Find the resistance for 960 feet of the same kind of wire if its diameter is doubled.

49. The average number of phone calls between two cities in a day varies jointly as the product of their populations and inversely as the square of the distance between them. The population of Minneapolis is 2538 thousand and the population of Cincinnati is 1818 thousand. Separated by 108 miles, the average number of telephone calls per day between the two cities is 158,233. Find the average number of telephone calls per day between Orlando, Florida (population 1225 thousand) and Seattle, Washington (population 2970 thousand), two cities that are 3403 miles apart.

Writing in Mathematics

50. What does it mean if two quantities vary directly?

51. In your own words, explain how to solve a variation problem.

52. What does it mean if two quantities vary inversely?

53. Explain what is meant by combined variation. Give an example with your explanation.

54. Explain what is meant by joint variation. Give an example with your explanation.

In Exercises 55–56, describe in words the variation shown by the given equation.

55. $z = \dfrac{k\sqrt{x}}{y^2}$ **56.** $z = kx^2\sqrt{y}$

57. We have seen that the daily number of phone calls between two cities varies jointly as their populations and inversely as the square of the distance between them. This model, used by telecommunication companies to estimate the line capacities needed among various cities, is called the *gravity model*. Compare the model to Newton's formula for gravitation on page 340 and describe why the name *gravity model* is appropriate.

Technology Exercise

58. Use a graphing utility to graph any three of the variation equations in Exercises 33–42. Then TRACE along each curve and identify the point that corresponds to the problem's solution.

Critical Thinking Exercises

59. In a hurricane, the wind pressure varies directly as the square of the wind velocity. If wind pressure is a measure of a hurricane's destructive capacity, what happens to this destructive power when the wind speed doubles?

60. The illumination from a light source varies inversely as the square of the distance from the light source. If you raise a lamp from 15 inches to 30 inches over your desk, what happens to the illumination?

61. The heat generated by a stove element varies directly as the square of the voltage and inversely as the resistance. If the voltage remains constant, what needs to be done to triple the amount of heat generated?

62. Galileo's telescope brought about revolutionary changes in astronomy. A comparable leap in our ability to observe the universe took place as a result of the Hubble Space Telescope. The space telescope can see stars and galaxies whose brightness is $\frac{1}{50}$ of the faintest objects now observable using ground-based telescopes. Use the fact that the brightness of a point source, such as a star, varies inversely as the square of its distance from an observer to show

that the space telescope can see about seven times farther than a ground-based telescope.

Group Exercise

63. Begin by deciding on a product that interests the group because you are now in charge of advertising this product. Members were told that the demand for the product varies directly as the amount spent on advertising and inversely as the price of the product. However, as more money is spent on advertising, the price of your product rises. Under what conditions would members recommend an increased expense in advertising? Once you've determined what your product is, write formulas for the given conditions and experiment with hypothetical numbers. What other factors might you take into consideration in terms of your recommendation? How do these factor affect the demand for your product?

CHAPTER SUMMARY, REVIEW, AND TEST

Summary

3.1 Quadratic Functions

a. A quadratic function is of the form $f(x) = ax^2 + bx + c, a \neq 0$.

b. The standard form of a quadratic function is $f(x) = a(x - h)^2 + k, a \neq 0$.

c. The graph of a quadratic function is a parabola. The vertex is (h, k) or $\left(-\dfrac{b}{2a}, f\left(-\dfrac{b}{2a} \right) \right)$. A procedure for graphing a parabola is given in the box on page 261.

3.2 Polynomial Functions and Their Graphs

a. Polynomial Function of x of Degree n: $f(x) = a_n x^n + a_{n-1} x^{n-1} + \cdots + a_2 x^2 + a_1 x + a_0, \quad a_n \neq 0$

b. The graphs of polynomial functions are smooth and continuous.

c. The end behavior of the graph of a polynomial function depends on the leading term, given by the Leading Coefficient Test in the box on page 274.

d. The values of x for which $f(x)$ is equal to 0 are the zeros of the polynomial function f. These values are the roots of the polynomial equation $f(x) = 0$.

e. If $x - r$ occurs k times in a polynomial function's factorization, r is a repeated zero with multiplicity k. If k is even, the graph touches the x-axis at r; if odd, it crosses the x-axis at r.

f. If f is a polynomial of degree n, the graph of f has at most $n - 1$ turning points.

g. A strategy for graphing a polynomial function is given in the box on page 278.

3.3 Dividing Polynomials; Remainder and Factor Theorems

a. Long division of polynomials is performed by dividing, multiplying, subtracting, bringing down the next term, and repeating this process until the degree of the remainder is less than the degree of the divisor. The details are given in the box on page 285.

b. The Division Algorithm: $f(x) = d(x)q(x) + r(x)$. The dividend is the product of the divisor and the quotient plus the remainder.

c. Synthetic division is used to divide a polynomial by $x - c$. The details are given in the box on pages 288–289.

d. The Remainder Theorem: If the polynomial $f(x)$ is divided by $x - c$, then the remainder is $f(c)$.

e. The Factor Theorem: If a polynomial function $f(x)$ is divided by $x - c$ and the remainder is zero, c is a zero of f and a root of $f(x) = 0$. If c is a zero of f or a root of $f(x) = 0$, then $x - c$ is a factor of $f(x)$.

3.4 Zeros of Polynomial Functions

a. The Rational Zero Theorem states that possible rational zeros of a polynomial function $= \dfrac{\text{Factors of the constant term}}{\text{Factors of the leading coefficient}}$. The theorem is stated in the box on page 295.

b. Descartes's Rule of Signs: The number of positive real zeros of f equals the number of sign changes of $f(x)$ or is less than that number by an even integer. The number of negative real zeros of f applies a similar statement to $f(-x)$.

3.5 More on Zeros of Polynomial Functions

a. The Upper and Lower Bound Theorem: The number $b > 0$ is an upper bound for the real roots of $f(x) = 0$ if synthetic division of $f(x)$ by $x - b$ results in no negative numbers. The number $a < 0$ is a lower bound if synthetic division by $x - a$ results in numbers that alternate in sign, counting zero entries as positive or negative.

b. The Intermediate Value Theorem: If $f(a)$ and $f(b)$ have opposite signs, there is at least one value of c between a and b for which $f(c) = 0$.

c. Number of roots: If $f(x)$ is a polynomial of degree $n \geq 1$, then, counting multiple roots separately, the equation $f(x) = 0$ has n roots.

d. If $a + bi$ is a root of $f(x) = 0$, then $a - bi$ is also a root.

e. The Linear Factorization Theorem: An nth-degree polynomial can be expressed as the product of n linear factors. Thus,

$$f(x) = a_n(x - c_1)(x - c_2)\cdots(x - c_n).$$

3.6 Rational Functions and Their Graphs

a. Rational function: $f(x) = \dfrac{p(x)}{q(x)}$; $p(x)$ and $q(x)$ are polynomial functions and $q(x) \neq 0$. The domain of f is the set of all real numbers excluding values of x that make $q(x)$ zero.

b. Arrow notation is summarized in the box on page 317.

c. The line $x = a$ is a vertical asymptote of the graph of f if $f(x)$ increases or decreases without bound as x approaches a. Vertical asymptotes are identified using the location theorem in the box on page 318.

d. The line $y = b$ is a horizontal asymptote of the graph of f if $f(x)$ approaches b as x increases or decreases without bound. Horizontal asymptotes are identified using the location theorem in the box on page 320.

e. A strategy for graphing rational functions is given in the box on page 321.

f. The graph of a rational functions has a slant asymptote when the degree of the numerator is one more than the degree of the denominator. The equation of the slant asymptote is found using division and ignoring the remainder term.

3.7 Modeling Using Variation

a.

English Statement	Equation
y varies directly as x.	$y = kx$
y varies directly as x^n.	$y = kx^n$
y varies inversely as x.	$y = \dfrac{k}{x}$
y varies inversely as x^n.	$y = \dfrac{k}{x^n}$
y varies jointly as x and z.	$y = kxz$

b. A procedure for solving variation problems is given in the box on page 335.

Review Exercises

3.1

In Exercises 1–4, use the vertex and intercepts to sketch the graph of each quadratic function. Give the equation for the parabola's axis of symmetry.

1. $f(x) = -2(x - 1)^2 + 3$ **2.** $f(x) = (x + 4)^2 - 2$

3. $f(x) = -x^2 + 2x + 3$ **4.** $f(x) = 2x^2 - 4x - 6$

5. The function $f(x) = 104.5x^2 - 1501.5x + 6016$ describes the death rate per year per 100,000 males, $f(x)$, for U.S. men who average x hours of sleep each night. How many hours of sleep, to the nearest tenth of an hour, corresponds to the minimum death rate? What is this minimum death rate, to the nearest whole number?

6. A person standing close to the edge on the top of an 80-foot building throws a ball vertically upward with an initial velocity of 64 feet per second. The function $s(t) = -16t^2 + 64t + 80$ describes the ball's height above the ground, $s(t)$, in feet, t seconds after it is thrown. After how many seconds does the ball reach its maximum height? What is the maximum height?

3.2

In Exercises 7–10, use end behavior and, if necessary, zeros to match each polynomial function with its graph on page 346.

7. $f(x) = -x^3 + 12x^2 - x$

8. $g(x) = x^6 - 6x^4 + 9x^2$

9. $h(x) = x^5 - 5x^3 + 4x$

10. $r(x) = x^3 + 1$

a.

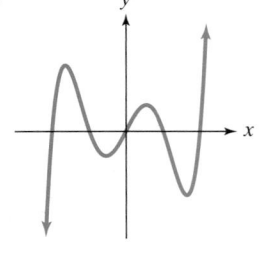

b.

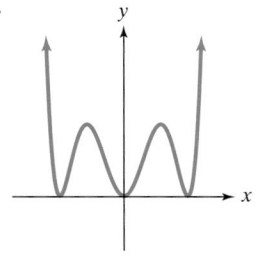

c.

d.

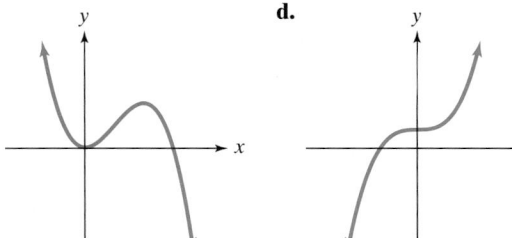

11. The function $f(x) = -0.0013x^3 + 0.78x^2 - 1.43x + 18.1$ models the percentage of U.S. families below the poverty level x years after 1960. Use end behavior to explain why the model is valid only for a limited period of time.

12. Despite a combination of drugs used to inhibit the growth of the HIV virus, a patient dies as a result of the virus overwhelming his body. Could the function $N(t) = -\frac{3}{4}t^4 + 3t^3 + 5$ model the number of viral particles, in billions, in this patient's body over time? Use the graph's end behavior to the right to answer the question. Explain your answer.

In Exercises 13–14, find the zeros for each polynomial function and give the multiplicity of each zero. State whether the graph crosses or touches the x-axis at each zero.

13. $f(x) = -2(x - 1)(x + 2)^2(x + 5)^3$

14. $f(x) = x^3 - 5x^2 - 25x + 125$

In Exercises 15–20,
  a. *Use the Leading Coefficient Test to determine the graph's end behavior.*
  b. *Determine whether the graph has y-axis symmetry, origin symmetry, or neither.*
  c. *Graph the function.*

15. $f(x) = x^3 - x^2 - 9x + 9$ 16. $f(x) = 4x - x^3$
17. $f(x) = 2x^3 + 3x^2 - 8x - 12$ 18. $f(x) = -x^4 + 25x^2$
19. $f(x) = -x^4 + 6x^3 - 9x^2$ 20. $f(x) = 3x^4 - 15x^3$

3.3

In Exercises 21–23, divide by long division.

21. $(4x^3 - 3x^2 - 2x + 1) \div (x + 1)$
22. $(10x^3 - 26x^2 + 17x - 13) \div (5x - 3)$
23. $(4x^4 + 6x^3 + 3x - 1) \div (2x^2 + 1)$

In Exercises 24–25, divide by synthetic division.

24. $(3x^4 + 11x^3 - 20x^2 + 7x + 35) \div (x + 5)$

25. $(3x^4 - 2x^2 - 10x) \div (x - 2)$
26. Given $f(x) = 2x^3 - 7x^2 + 9x - 3$, use the Remainder Theorem to find $f(-13)$.
27. Use synthetic division to divide $f(x) = 2x^3 + x^2 - 13x + 6$ by $x - 2$. Use the result to find all zeros of f.
28. Solve the equation $x^3 - 17x + 4 = 0$ given that 4 is a root.

3.4

In Exercises 29–30, use the Rational Zero Theorem to list all possible rational zeros for each given function.

29. $f(x) = x^4 - 6x^3 + 14x^2 - 14x + 5$
30. $f(x) = 3x^5 - 2x^4 - 15x^3 + 10x^2 + 12x - 8$

In Exercises 31–32, use Descartes's Rule of Signs to determine the possible number of positive and negative real zeros for each given function.

31. $f(x) = 3x^4 - 2x^3 - 8x + 5$
32. $f(x) = 2x^5 - 3x^3 - 5x^2 + 3x - 1$
33. Use Descartes's Rule of Signs to explain why $2x^4 + 6x^2 + 8 = 0$ has no real roots.

For Exercises 34–39,
  a. *List all possible rational roots or rational zeros.*
  b. *Use Descartes's Rule of Signs to determine the possible number of positive and negative real roots or real zeros.*
  c. *Use synthetic division to test the possible rational roots or zeros and find an actual root or zero.*
  d. *Use the root or zero from part (c) to find all the zeros or roots.*

34. $f(x) = x^3 + 3x^2 - 4$
35. $f(x) = 6x^3 + x^2 - 4x + 1$
36. $8x^3 - 36x^2 + 46x - 15 = 0$
37. $x^4 - x^3 - 7x^2 + x + 6 = 0$
38. $4x^4 + 7x^2 - 2 = 0$
39. $f(x) = 2x^4 + x^3 - 9x^2 - 4x + 4$

3.5

40. Show that all real roots of the equation
$$2x^4 - 7x^3 - 5x^2 + 28x - 12 = 0$$
lie between -2 and 6. Use this result to list all possible rational roots.

41. Consider the equation $2x^4 - x^3 - 5x^2 + 10x + 12 = 0$.
  a. List all possible rational roots.
  b. Determine whether 2 is a root using synthetic division. In terms of bounds, what can you conclude?
  c. Determine whether -2 is a root using synthetic division. In terms of bounds, what can you conclude?
  d. Use the results of parts (b) and (c) to discard some of the possible rational roots from part (a). Now what are the possible rational roots?

In Exercises 42–43, show that the polynomial has a zero between the given integers. Then use the Intermediate Value

Theorem to find an approximation for this zero to the nearest tenth.

42. $f(x) = x^3 - 2x - 1$; between 1 and 2

43. $f(x) = 3x^3 + 2x^2 - 8x + 7$; between -3 and -2

In Exercises 44–46, use the given root to find the solution set of the polynomial equation.

44. $4x^3 - 47x^2 + 232x + 61 = 0$; $6 + 5i$

45. $x^4 - 4x^3 + 16x^2 - 24x + 20 = 0$; $1 - 3i$

46. $2x^4 - 17x^3 + 137x^2 - 57x - 65 = 0$; $4 + 7i$

In Exercises 47–49, find an nth-degree polynomial function with real coefficients satisfying the given conditions. If you are using a graphing utility, graph the function and verify the real zeros and the given function value.

47. $n = 3$; 2 and $2 - 3i$ are zeros; $f(1) = -10$

48. $n = 4$; i is a zero; -3 is a zero of multiplicity 2; $f(-1) = 16$

49. $n = 4$; $-2, 3$, and $1 + 3i$ are zeros; $f(2) = -40$

In Exercises 50–51, find all the zeros of each polynomial function and write the polynomial as a product of linear factors.

50. $f(x) = 2x^4 + 3x^3 + 3x - 2$

51. $g(x) = x^4 - 6x^3 + x^2 + 24x + 16$

In Exercises 52–55, graphs of fifth-degree polynomial functions are shown. In each case, specify the number of real zeros and the number of nonreal complex zeros. Indicate whether there are any real zeros with multiplicity other than 1.

52.

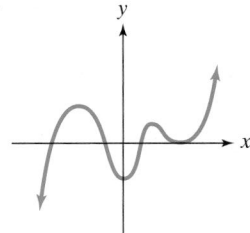

53.

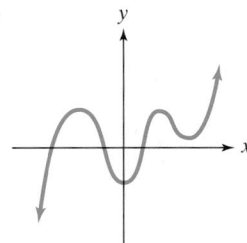

54.

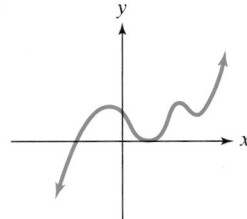

55.

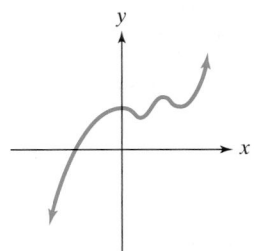

3.6

In Exercises 56–63, find the vertical asymptotes, if any, the horizontal asymptote, if there is one, and the slant asymptote,

if there is one, of the graph of each rational function. Then graph the rational function.

56. $f(x) = \dfrac{2x}{x^2 - 9}$

57. $g(x) = \dfrac{2x - 4}{x + 3}$

58. $h(x) = \dfrac{x^2 - 3x - 4}{x^2 - x - 6}$

59. $r(x) = \dfrac{x^2 + 4x + 3}{(x + 2)^2}$

60. $y = \dfrac{x^2}{x + 1}$

61. $y = \dfrac{x^2 + 2x - 3}{x - 3}$

62. $f(x) = \dfrac{-2x^3}{x^2 + 1}$

63. $g(x) = \dfrac{4x^2 - 16x + 16}{2x - 3}$

64. A company that manufactures graphing calculators has costs given by the function $C(x) = 25x + 50,000$, where x is the number of calculators manufactured and $C(x)$ is measured in dollars. The average cost to manufacture each calculator is given by

$$\bar{C}(x) = \frac{25x + 50,000}{x}.$$

   a. Find and interpret $\bar{C}(50)$, $\bar{C}(100)$, $\bar{C}(1000)$, and $\bar{C}(100,000)$.

   b. What is the horizontal asymptote for this function, and what does it represent?

65. In Silicon Valley, California, a government agency ordered computer-related companies to contribute to a monetary pool to clean up underground water supplies. (The companies had stored toxic chemicals in leaking underground containers.) The rational function

$$C(x) = \frac{200x}{100 - x}$$

models the cost, $C(x)$, in tens of thousands of dollars, for removing x percent of the contaminants.

   a. Find and interpret $C(90) - C(50)$.

   b. What is the equation for the vertical asymptote? What does this mean in terms of the variables given by the function?

Exercises 66–67 involve rational functions that model the given situations. In each case, find the horizontal asymptote as $x \to \infty$ and then describe what this means in practical terms.

66. $F(x) = \dfrac{30(4 + 5x)}{1 + 0.05x}$; the number of fish, F, in thousands, after x weeks in a lake that was stocked with 120,000 fish.

67. $P(x) = \dfrac{72,900}{100x^2 + 729}$; the percentage rate, P, of U.S. unemployment for groups with x years of education.

68. In a get-tough drug policy, a politician promises to spend whatever it takes to seize all illegal drugs as they enter the country. If the cost of this venture is

$$C(p) = \frac{Ap}{100 - p}$$

where A is a positive constant, C is expressed in millions of dollars, and p is the percentage of illegal drugs seized, use this function to evaluate the politician's promise.

3.7

Solve the variation problems in Exercises 69–74.

69. An electric bill varies directly as the amount of electricity used. The bill for 1400 kilowatts of electricity is $98. What is the bill for 2200 kilowatts of electricity?

70. The distance that a body falls from rest varies directly as the square of the time of the fall. If skydivers fall 144 feet in 3 seconds, how far will they fall in 10 seconds?

71. The time it takes to drive a certain distance varies inversely as the rate of travel. If it takes 4 hours at 50 miles per hour to drive the distance, how long will it take at 40 miles per hour?

72. The loudness of a stereo speaker, measured in decibels, varies inversely as the square of your distance from the speaker. When you are 8 feet from the speaker, the loudness is 28 decibels. What is the loudness when you are 4 feet from the speaker?

73. The time required to assemble computers varies directly as the number of computers assembled and inversely as the number of workers. If 30 computers can be assembled by 6 workers in 10 hours, how long would it take 5 workers to assemble 40 computers?

74. The volume of a pyramid varies jointly as its height and the area of its base. A pyramid with a height of 15 feet and a base with an area of 35 square feet has a volume of 175 cubic feet. Find the volume of a pyramid with a height of 20 feet and a base with an area of 120 square feet.

Chapter 3 Test

In Exercises 1–2, use the vertex and intercepts to sketch the graph of each quadratic function. Give the equation for the parabola's axis of symmetry.

1. $f(x) = (x + 1)^2 + 4$　　**2.** $f(x) = x^2 - 2x - 3$

3. Determine, without graphing, whether the quadratic function $f(x) = -2x^2 + 12x - 16$ has a minimum value or a maximum value. Then find the coordinates of the minimum or the maximum point.

4. The function $f(x) = -x^2 + 46x - 360$ models the daily profit, $f(x)$, in hundreds of dollars, for a company that manufactures x VCRs daily. How many VCRs should be manufactured each day to maximize profit? What is the maximum daily profit?

5. Consider the function $f(x) = x^3 - 5x^2 - 4x + 20$.
　a. Use factoring to find all zeros of f.
　b. Use the Leading Coefficient Test and the zeros of f to graph the function.

6. Use end behavior to explain why the graph cannot be the graph of $f(x) = x^5 - x$. Then use intercepts to explain why the graph cannot represent $f(x) = x^5 - x$.

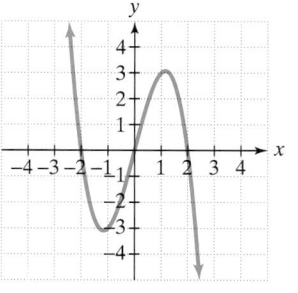

7. The graph of $f(x) = 6x^3 - 19x^2 + 16x - 4$ is shown in the figure at the top of the next column.
　a. Based on the graph of f, find the root of the equation $6x^3 - 19x^2 + 16x - 4 = 0$ that is an integer.

　b. Use synthetic division to find the other two roots of $6x^3 - 19x^2 + 16x - 4 = 0$.

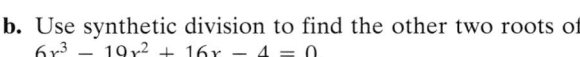

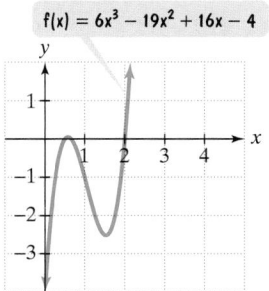

8. Use the Rational Zero Theorem to list all possible rational zeros of $f(x) = 2x^3 + 11x^2 - 7x - 6$.

9. Use Descartes's Rule of Signs to determine the possible number of positive and negative real zeros of $f(x) = 3x^5 - 2x^4 - 2x^2 + x - 1$.

10. Solve: $x^3 + 6x^2 - x - 30 = 0$.

11. Consider the function whose equation is given by $f(x) = 2x^4 - x^3 - 13x^2 + 5x + 15$.
　a. List all possible rational zeros.
　b. Use the graph of f in the figure shown and synthetic division to find all zeros of the function.

$f(x) = 2x^4 - x^3 - 13x^2 + 5x + 15$

12. Use the graph of $f(x) = 3x^4 + 4x^3 - 7x^2 - 2x - 3$ in the figure shown to find the smallest positive integer that is an upper bound and the largest negative integer that is a lower bound for the real roots of

$$3x^4 + 4x^3 - 7x^2 - 2x - 3 = 0.$$

Then use synthetic division to show that all the real roots of the equation lie between these integers.

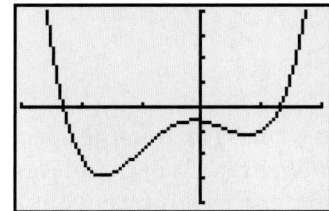

13. Solve $x^4 - 7x^3 + 18x^2 - 22x + 12 = 0$ given that $1 - i$ is a root.

14. Use the graph of $f(x) = x^3 + 3x^2 - 4$ in the figure shown to factor $x^3 + 3x^2 - 4$.

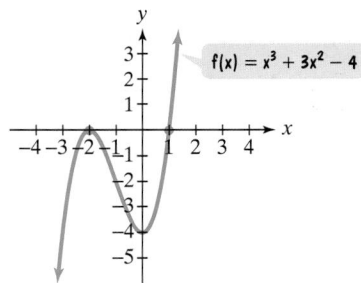

In Exercises 15–18, find the domain of each rational function and graph the function.

15. $f(x) = \dfrac{x}{x^2 - 16}$

16. $f(x) = \dfrac{x^2 - 9}{x - 2}$

17. $f(x) = \dfrac{x + 1}{x^2 + 2x - 3}$

18. $f(x) = \dfrac{4x^2}{x^2 + 3}$

19. A number of deer are placed into a newly acquired habitat. The deer population over time is modeled by a rational function whose graph is shown in the figure. Use the graph to answer each of the following questions.

   a. How many deer were introduced into the habitat?

   b. What is the population after 10 years?

   c. What is the equation of the horizontal asymptote shown in the figure? What does this mean in terms of the deer population?

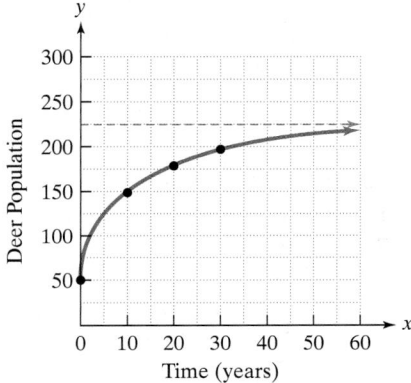

Change in Deer Population over Time

20. The intensity of light received at a source varies inversely as the square of the distance from the source. A particular light has an intensity of 20 foot-candles at 15 feet. What is the light's intensity at 10 feet?

Cumulative Review Exercises (Chapters P–3)

Simplify each expression in Exercises 1–3.

1. $\dfrac{1}{2 - \sqrt{3}}$

2. $3(x^2 - 3x + 1) - 2(3x^2 + x - 4)$

3. $3\sqrt{8} + 5\sqrt{50} - 4\sqrt{32}$

4. Factor completely: $x^7 - x^5$.

Solve each equation in Exercises 5–8.

5. $|2x - 1| = 3$

6. $3x^2 - 5x + 1 = 0$

7. $9 + \dfrac{3}{x} = \dfrac{2}{x^2}$

8. $x^3 + 2x^2 - 5x - 6 = 0$

Solve each inequality in Exercises 9–10. Express the answer in interval notation.

9. $|2x - 5| > 3$

10. $3x^2 > 2x + 5$

11. Give the center and radius. Then graph the equation
$$x^2 + y^2 - 2x + 4y - 4 = 0.$$

12. Solve for t: $\quad V = C(1 - t)$.

13. If $f(x) = \sqrt{45 - 9x}$, find the domain of f.

If $f(x) = x^2 + 2x - 5$ and $g(x) = 4x - 1$, find each function or function value in Exercises 14–16.

14. $(f - g)(x)$

15. $(f \circ g)(x)$

16. $g(f(-3))$

17. Consider the function $f(x) = x^3 - 4x^2 - x + 4$.

   a. Use factoring to find all zeros of f.

   b. Use the Leading Coefficient Test and the zeros of f to graph the function.

Graph each function in Exercises 18–20.

18. $f(x) = x^2 + 2x - 8$

19. $f(x) = x^2(x - 3)$

20. $f(x) = \dfrac{x - 1}{x - 2}$

Exponential and Logarithmic Functions

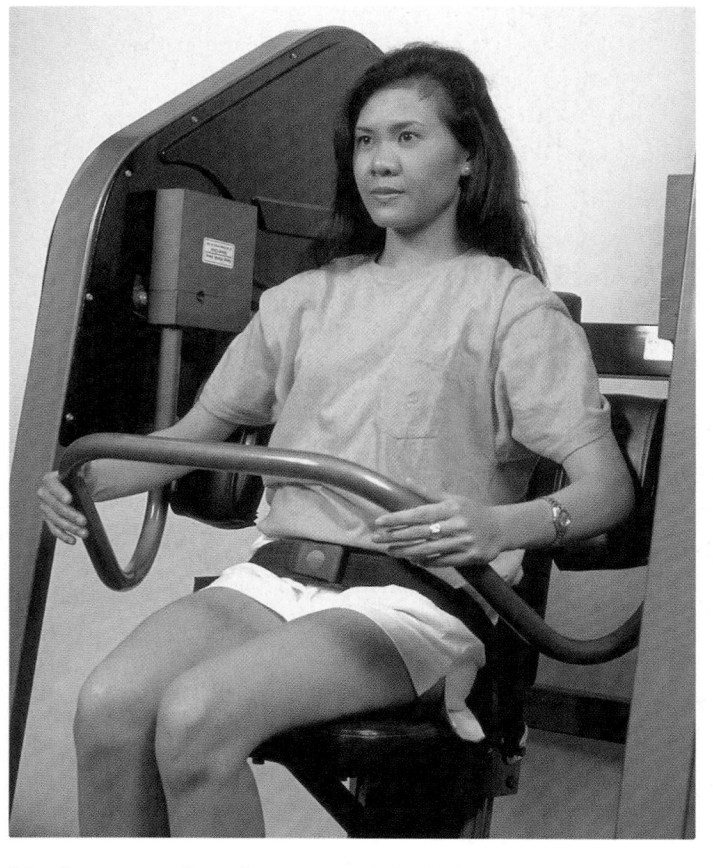

What went wrong on the space shuttle *Challenger*? Will population growth lead to a future without comfort or individual choice? Can I put aside a small amount of money and have millions for early retirement? Why did I feel I was walking too slowly on my visit to New York City? Why are people in California at far more risk from drunk drivers than from earthquakes? What is the difference between earthquakes measuring 6 and 7 on the Richter scale? And what can I hope to accomplish in weightlifting?

The functions that you will be learning about in this chapter will provide you with the mathematics for answering these questions. You will see how these remarkable functions enable us to predict the future and rediscover the past.

You've recently taken up weightlifting, recording the maximum number of pounds you can lift at the end of each week. At first your weight limit increases rapidly, but now you notice that this growth is beginning to level off. You wonder about a function that would serve as a mathematical model to predict the number of pounds you can lift as you continue the sport.

SECTION 4.1 *Exponential Functions*

Objectives

1. Evaluate exponential functions.
2. Graph exponential functions.
3. Evaluate functions with base e.
4. Use compound interest formulas.

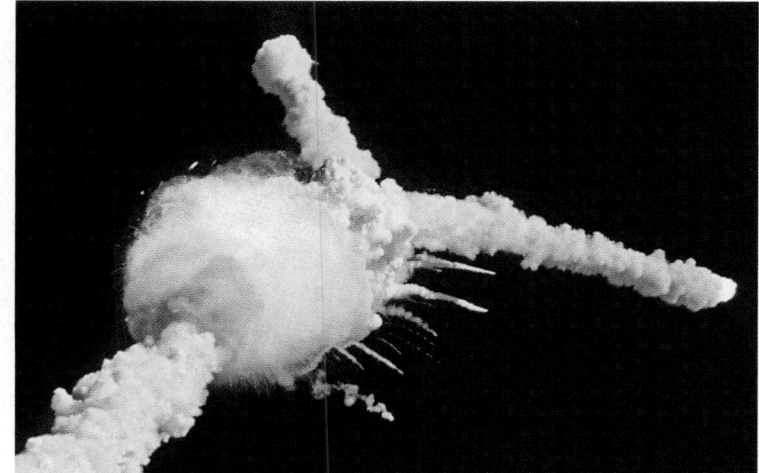

The space shuttle *Challenger* exploded approximately 73 seconds into flight on January 28, 1986. The tragedy involved damage to O-rings, which were used to seal the connections between different sections of the shuttle engines. The number of O-rings damaged increases dramatically as Fahrenheit temperature falls.

The function

$$f(x) = 13.49(0.967)^x - 1$$

models the number of O-rings expected to fail when the temperature is $x°$F. Can you see how this function is different from polynomial functions? The variable x is in the exponent. Functions whose equations contain a variable in the exponent are called **exponential functions**. Many real-life situations, including population growth, growth of epidemics, radioactive decay, and other changes that involve rapid increase or decrease, can be described using exponential functions.

Definition of the Exponential Function

The **exponential function f with base b** is defined by

$$f(x) = b^x \quad \text{or} \quad y = b^x$$

where b is a positive constant other than $1 (b > 0$ and $b \neq 1)$ and x is any real number.

Here are some examples of exponential functions.

$$f(x) = 2^x \qquad g(x) = 10^x \qquad h(x) = 3^{x+1}$$

Base is 2.　　　　Base is 10.　　　　Base is 3.

Each of these functions has a constant base and a variable exponent. By contrast, the following functions are not exponential.

$$F(x) = x^2 \qquad G(x) = 1^x \qquad H(x) = x^x$$

Variable is the base and not the exponent.　　The base of an exponential function must be a positive constant other than 1.　　Variable is both the base and the exponent.

1 Evaluate exponential functions.

Why is $G(x) = 1^x$ not classified as an exponential function? The number 1 raised to any power is 1. Thus, the function G can be written as $G(x) = 1$, which is a constant function.

You will need a calculator to evaluate exponential expressions. Most scientific calculators have an $\boxed{x^y}$ key. Graphing calculators have a $\boxed{\wedge}$ key. To evaluate expressions of the form b^x, enter the base b, press $\boxed{x^y}$ or $\boxed{\wedge}$, enter the exponent x, and finally press $\boxed{=}$ or $\boxed{\text{ENTER}}$.

EXAMPLE 1   Evaluating an Exponential Function

The exponential function $f(x) = 13.49(0.967)^x - 1$ describes the number of O-rings expected to fail, $f(x)$, when the temperature is $x°$F. On the morning the *Challenger* was launched, the temperature was 31°F, colder than any previous experience. Find the number of O-rings expected to fail at this temperature.

Solution   Because the temperature was 31°F, substitute 31 for x and evaluate the function at 31.

$$f(x) = 13.49(0.967)^x - 1 \quad \text{This is the given function.}$$

$$f(31) = 13.49(0.967)^{31} - 1 \quad \text{Substitute 31 for x.}$$

Use a scientific or graphing calculator to evaluate $(0.967)^{31}$. Press the following keys on your calculator to do this:

Scientific calculator: .967 $\boxed{x^y}$ 31 $\boxed{=}$

Graphing calculator: .967 $\boxed{\wedge}$ 31 $\boxed{\text{ENTER}}$

The display should be approximately .353362693426. Multiplying this number by 13.49 and subtracting 1, we obtain

$$f(31) = 13.49(0.967)^{31} - 1 \approx 4.$$

Thus, four O-rings are expected to fail at a temperature of 31°F.

Check Point 1   Use the function in Example 1 to find the number of O-rings expected to fail at a temperature of 60°F.

2 Graph exponential functions.

Graphing Exponential Functions

We are familiar with expressions involving b^x where x is a rational number. For example,

$$b^{1.7} = b^{17/10} = \sqrt[10]{b^{17}} \quad \text{and} \quad b^{1.73} = b^{173/100} = \sqrt[100]{b^{173}}.$$

However, note that the definition of $f(x) = b^x$ includes all real numbers for the domain x. You may wonder what b^x means when x is an irrational number, such as $b^{\sqrt{3}}$ or b^{π}. Using the nonrepeating and nonterminating approximation 1.73205 for $\sqrt{3}$, we can think of $b^{\sqrt{3}}$ as the value that has the successively closer approximations

$$b^{1.7}, b^{1.73}, b^{1.732}, b^{1.73205}, \ldots.$$

In this way, we can graph the exponential function with no holes, or points of discontinuity, at the irrational domain values.

EXAMPLE 2   Graphing an Exponential Function

Graph: $f(x) = 2^x$.

Solution   We begin by setting up a table of coordinates.

x	$f(x) = 2^x$
-3	$f(-3) = 2^{-3} = \frac{1}{8}$
-2	$f(-2) = 2^{-2} = \frac{1}{4}$
-1	$f(-1) = 2^{-1} = \frac{1}{2}$
0	$f(0) = 2^0 = 1$
1	$f(1) = 2^1 = 2$
2	$f(2) = 2^2 = 4$
3	$f(3) = 2^3 = 8$

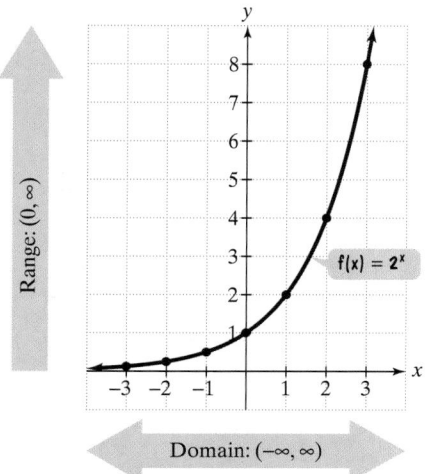

Figure 4.1 The graph of $f(x) = 2^x$

We plot these points, connecting them with a continuous curve. Figure 4.1 shows the graph of $f(x) = 2^x$. Observe that the graph approaches but never touches the negative portion of the x-axis. Thus, the x-axis is a horizontal asymptote. The range is all positive real numbers. Although we used integers for x in our table of coordinates, you can use a calculator to find additional points. For example, $f(0.3) = 2^{0.3} \approx 1.231$, $f(0.95) = 2^{0.95} \approx 1.932$. The points $(0.3, 1.231)$ and $(0.95, 1.932)$ fit the graph.

Check Point 2   Graph $f(x) = 3^x$.

   Four exponential functions have been graphed in Figure 4.2. Compare the graphs of functions where $b > 1$ to those where $b < 1$. When $b > 1$, the value of y increases as the value of x increases. When $b < 1$, the value of y decreases as the value of x increases. Notice that all four graphs pass through $(0, 1)$.

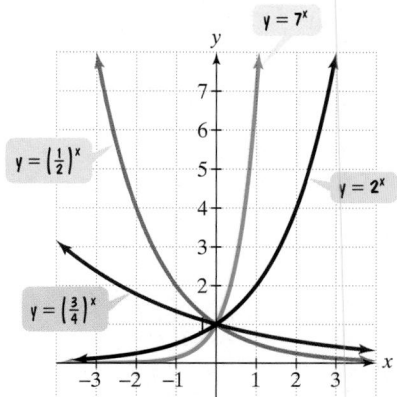

Figure 4.2 Graphs of four exponential functions

These graphs illustrate the following general characteristics of exponential functions.

Characteristics of Exponential Functions

1. The domain of $f(x) = b^x$ consists of all real numbers. The range of $f(x) = b^x$ consists of all positive real numbers.
2. The graphs of all exponential functions pass through the point $(0, 1)$ because $f(0) = b^0 = 1 (b \neq 0)$.
3. If $b > 1$, $f(x) = b^x$ has a graph that goes up to the right and is an increasing function.
4. If $0 < b < 1$, $f(x) = b^x$ has a graph that goes down to the right and is a decreasing function.
5. $f(x) = b^x$ is one-to-one and has an inverse that is a function.
6. The graph of $f(x) = b^x$ approaches but does not cross the x-axis. The x-axis is a horizontal asymptote.

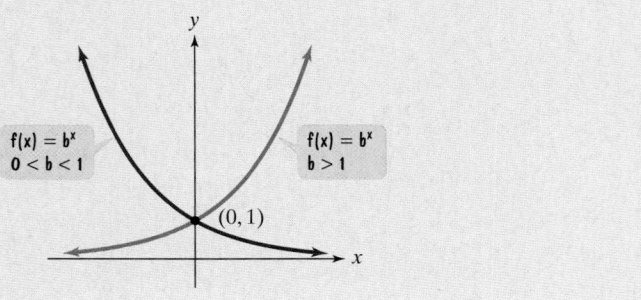

Transformations of Exponential Functions   The graphs of exponential functions can be translated vertically or horizontally, reflected, stretched, or shrunk. We use the ideas of Section 2.5 to do so, as summarized in Table 4.1.

Table 4.1   Transformations Involving Exponential Functions

Transformation	Equation	Description
Horizontal translation	$g(x) = b^{x+c}$	• Shifts the graph of $f(x) = b^x$ to the left c units if $c > 0$. • Shifts the graph of $f(x) = b^x$ to the right c units if $c < 0$.
Vertical stretching or shrinking	$g(x) = cb^x$	Multiplying y-coordinates of $f(x) = b^x$ by c, • Stretches the graph of $f(x) = b^x$ if $c > 1$. • Shrinks the graph of $f(x) = b^x$ if $0 < c < 1$.
Reflecting	$g(x) = -b^x$ $g(x) = b^{-x}$	• Reflects the graph of $f(x) = b^x$ about the x-axis. • Reflects the graph of $f(x) = b^x$ about the y-axis.
Vertical translation	$g(x) = b^x + c$	• Shifts the graph of $f(x) = b^x$ upward c units if $c > 0$. • Shifts the graph of $f(x) = b^x$ downward c units if $c < 0$.

Using the information in Table 4.1 and a table of coordinates, you will obtain relatively accurate graphs that can be verified using a graphing utility.

EXAMPLE 3 Transformations Involving Exponential Functions

Use the graph of $f(x) = 3^x$ to obtain the graph of $g(x) = 3^{x+1}$.

Solution Examine Table 4.1. Note that the function $g(x) = 3^{x+1}$ has the general form $g(x) = b^{x+c}$, where $c = 1$. Because $c > 0$, we graph $g(x) = 3^{x+1}$ by shifting the graph of $f(x) = 3^x$ *one* unit to the *left*. We construct a table showing some of the coordinates for f and g. The graphs of f and g are shown in Figure 4.3.

x	$f(x) = 3^x$	$g(x) = 3^{x+1}$
-2	$3^{-2} = \frac{1}{9}$	$3^{-2+1} = 3^{-1} = \frac{1}{3}$
-1	$3^{-1} = \frac{1}{3}$	$3^{-1+1} = 3^0 = 1$
0	$3^0 = 1$	$3^{0+1} = 3^1 = 3$
1	$3^1 = 3$	$3^{1+1} = 3^2 = 9$
2	$3^2 = 9$	$3^{2+1} = 3^3 = 27$

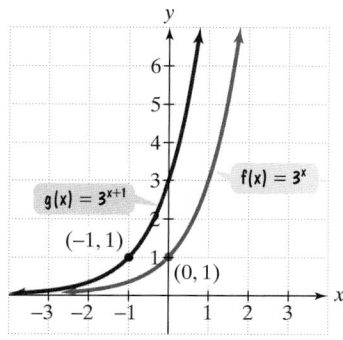

Figure 4.3 The graph of $g(x) = 3^{x+1}$ shifts the graph of $f(x) = 3^x$ one unit to the left.

Check Point 3   Use the graph of $f(x) = 3^x$ to obtain the graph of $g(x) = 3^{x-1}$.

If an exponential function is translated upward or downward, the horizontal asymptote is shifted by the amount of the vertical shift.

EXAMPLE 4 Transformations Involving Exponential Functions

Use the graph of $f(x) = 2^x$ to obtain the graph of $g(x) = 2^x - 3$.

Solution Examine Table 4.1. Note that the function $g(x) = 2^x - 3$ has the general form $g(x) = b^x + c$, where $c = -3$. Because $c < 0$, we graph $g(x) = 2^x - 3$ by shifting the graph of $f(x) = 2^x$ *down three* units. We construct a table showing some of the coordinates for f and g. The graphs of f and g are shown in Figure 4.4 on page 356. Notice that the horizontal asymptote for f, the x-axis, is shifted down three units for the horizontal asymptote for g. Thus, $y = -3$ is the horizontal asymptote for g.

x	$f(x) = 2^x$	$y(x) = 2^x - 3$
-2	$\frac{1}{4}$	$\frac{1}{4} - 3 = -2\frac{3}{4}$
-1	$\frac{1}{2}$	$\frac{1}{2} - 3 = -2\frac{1}{2}$
0	1	$1 - 3 = -2$
1	2	$2 - 3 = -1$
2	4	$4 - 3 = 1$

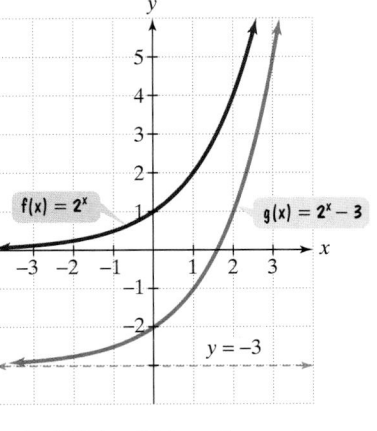

Figure 4.4 The graph of $g(x) = 2^x - 3$ shifts the graph of $f(x) = 2^x$ down three units.

Check Point 4 Use the graph of $f(x) = 2^x$ to obtain the graph of $g(x) = 2^x + 1$.

3 Evaluate functions with base e.

The Natural Base e

An irrational number, symbolized by the letter e, appears as the base in many applied exponential functions. This irrational number is approximately equal to 2.72. More accurately,

$$e \approx 2.71828\ldots.$$

The number e is called the **natural base**. The function $f(x) = e^x$ is called the **natural exponential function**.

Use a scientific or graphing calculator with an $\boxed{e^x}$ key to evaluate e to various powers. For example, to find e^2, press the following keys on most calculators:

Scientific calculator: 2 $\boxed{e^x}$

Graphing calculator: $\boxed{e^x}$ 2 $\boxed{\text{ENTER}}$

The display is approximately 7.389.

$$e^2 \approx 7.389$$

The number e lies between 2 and 3. Because $2^2 = 4$ and $3^2 = 9$, it makes sense that e^2, approximately 7.389, lies between 4 and 9.

Because $2 < e < 3$, the graph of $y = e^x$ is between the graphs of $y = 2^x$ and $y = 3^x$, shown in Figure 4.5.

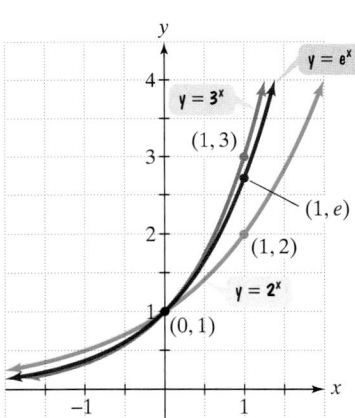

Figure 4.5 Graphs of three exponential functions

EXAMPLE 5   World Population

In a report entitled *Resources and Man*, the U.S. National Academy of Sciences concluded that a world population of 10 billion "is close to (if not above) the maximum that an intensely managed world might hope to support with some degree of comfort and individual choice." At the time the report was issued in 1969,

World Population in Billions

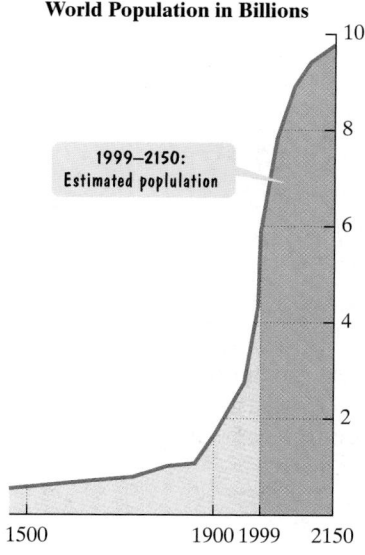

1999–2150:
Estimated poplulation

Source: U.N. Population Division

the world population was approximately 3.6 billion, with a growth rate of 2% per year. The function

$$f(x) = 3.6e^{0.02x}$$

describes world population, $f(x)$, in billions, x years after 1969. Use the function to find world population in the year 2020. Is there cause for alarm?

Solution   Because 2020 is 51 years after 1969, we substitute 51 for x:

$$f(51) = 3.6e^{0.02(51)}.$$

Although this computation can be done on your calculator in one step, we will break it down into smaller steps so that you can clearly see how we use the $\boxed{e^x}$ key. First find 0.02(51):

$$0.02(51) = 1.02.$$

Now find $e^{1.02}$:

Scientific calculator: 1.02 $\boxed{e^x}$

Graphing calculator: $\boxed{e^x}$ 1.02 $\boxed{\text{ENTER}}$

The display is approximately 2.7731948. Multiplying this number by 3.6, we obtain

$$f(51) = 3.6e^{0.02(51)} = 3.6e^{1.02} \approx 9.98.$$

This indicates that world population in the year 2020 will be approximately 9.98 billion. Because this number is quite close to 10 billion, the given function suggests that there may be cause for alarm.

World population in 1999 was 6 billion, but the growth rate was no longer 2%. It had slowed down to 1.3%. Using this current growth rate, exponential functions now predict a world population of 7.6 billion in the year 2020. Experts think the population may stabilize at 10 billion after 2200 if the deceleration in growth rate continues.

Check Point 5

The function $f(x) = 6e^{0.013x}$ describes world population, $f(x)$, in billions, x years after 1999 subject to a growth rate of 1.3% annually. Use the function to find world population in 2050.

4 Use compound interest formulas.

Compound Interest

We all want a wonderful life with fulfilling work, good health, and loving relationships. And let's be honest: Financial security wouldn't hurt! Achieving this goal depends on understanding how money in savings accounts grows in remarkable ways as a result of *compound interest*. **Compound interest** is interest computed on your original investment as well as on any accumulated interest.

Suppose a sum of money called the **principal**, P, is invested at an annual percentage rate r, in decimal form, compounded once per year. Because the interest is added to the principal at year's end, the accumulated value A is

$$A = P + Pr = P(1 + r).$$

The accumulated amount of money follows this pattern of multiplying the previous principal by $(1 + r)$ for each successive year, as indicated in Table 4.2.

Table 4.2

Time in Years	Accumulated Value after Each Compounding
0	$A = P$
1	$A = P(1 + r)$
2	$A = P(1 + r)(1 + r) = P(1 + r)^2$
3	$A = P(1 + r)^2(1 + r) = P(1 + r)^3$
4	$A = P(1 + r)^3(1 + r) = P(1 + r)^4$
$\vdots$	$\vdots$
t	$A = P(1 + r)^t$

n	$\left(1 + \dfrac{1}{n}\right)^n$
1	2
2	2.25
5	2.48832
10	2.59374246
100	2.704813829
1,000	2.716923932
10,000	2.718145927
100,000	2.718268237
1,000,000	2.718280469
1,000,000,000	2.718281827

As n takes on increasingly large values, the expression $\left(1 + \dfrac{1}{n}\right)^n$ approaches e.

If money invested at a specified rate of interest is compounded more than once a year, then the formula $A = P(1 + r)^t$ can be adjusted to take into account the number of compounding periods in a year. If n represents the number of compounding periods in a year, the formula becomes

$$A = P\left(1 + \frac{r}{n}\right)^{nt}.$$

Some banks use **continuous compounding**, where the number of compounding periods increases infinitely (compounding interest every trillionth of a second, every quadrillionth of a second, etc.). As n, the number of compounding periods in a year, increases without bound, the expression $\left(1 + \dfrac{1}{n}\right)^n$ approaches e. As a result, the formula for continuous compounding is $A = Pe^{rt}$. Although continuous compounding sounds terrific, it yields only a fraction of a percent more interest over a year than daily compounding.

Formulas for Compound Interest

After t years, the balance A in an account with principal P and annual interest rate r (in decimal form) is given by the following formulas:

1. For n compoundings per year: $A = P\left(1 + \dfrac{r}{n}\right)^{nt}$

2. For continuous compounding: $A = Pe^{rt}$

EXAMPLE 6 Choosing Between Investments

You want to invest $8000 for 6 years, and you have a choice between two accounts. The first pays 7% per year, compounded monthly. The second pays 6.85% per year, compounded continuously. Which is the better investment?

Solution The better investment is the one with the greater balance in the account after 6 years. Let's begin with the account with monthly compounding. We use the compound interest model with $P = 8000$, $r = 7\% = 0.07$, $n = 12$ (monthly compounding, means 12 compoundings per year), and $t = 6$.

$$A = P\left(1 + \frac{r}{n}\right)^{nt} = 8000\left(1 + \frac{0.07}{12}\right)^{12 \cdot 6} \approx 12{,}160.84$$

The balance in this account after 6 years is $12,160.84. For the second investment option, we use the model for continuous compounding with $P = 8000$, $r = 6.85\% = 0.0685$, and $t = 6$.

$$A = Pe^{rt} = 8000e^{0.0685(6)} \approx 12{,}066.60$$

The balance in this account after 6 years is $12,066.60, slightly less than the previous amount. Thus, the better investment is the 7% monthly compounding option.

Check Point 6

A sum of $10,000 is invested at an annual rate of 8%. Find the balance in the account after 5 years subject to **a.** quarterly compounding and **b.** continuous compounding.

EXERCISE SET 4.1

Practice Exercises

In Exercises 1–10, approximate each number using a calculator. Round your answer to three decimal places.

1. $2^{3.4}$ **2.** $3^{2.4}$ **3.** $3^{\sqrt{5}}$ **4.** $5^{\sqrt{3}}$ **5.** $4^{-1.5}$

6. $6^{-1.2}$ **7.** $e^{2.3}$ **8.** $e^{3.4}$ **9.** $e^{-0.95}$ **10.** $e^{-0.75}$

In Exercises 11–18, graph each function by making a table of coordinates. If applicable, use a graphing utility to confirm your hand-drawn graph.

11. $f(x) = 4^x$ **12.** $f(x) = 5^x$

13. $g(x) = \left(\frac{3}{2}\right)^x$ **14.** $g(x) = \left(\frac{4}{3}\right)^x$

15. $h(x) = \left(\frac{1}{2}\right)^x$ **16.** $h(x) = \left(\frac{1}{3}\right)^x$

17. $f(x) = (0.6)^x$ **18.** $f(x) = (0.8)^x$

In Exercises 19–24, the graph of an exponential function is given. Select the function for each graph from the following options:

$$f(x) = 3^x, \; g(x) = 3^{x-1}, \; h(x) = 3^x - 1,$$
$$F(x) = -3^x, \; G(x) = 3^{-x}, \; H(x) = -3^{-x}.$$

19.

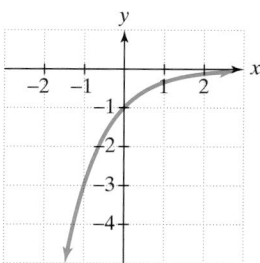

20.

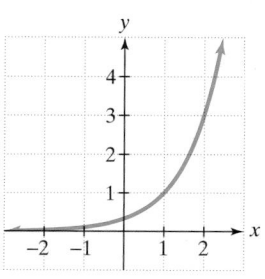

21.

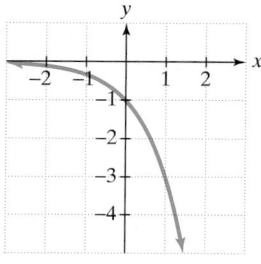

22.

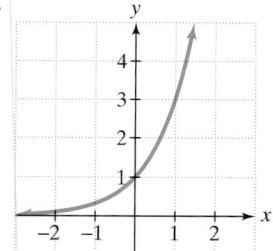

23.

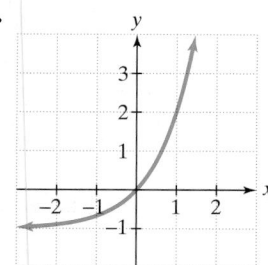

24.

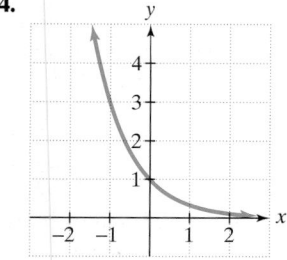

In Exercises 25–34, begin by graphing $f(x) = 2^x$. Then use transformations of this graph and a table of coordinates to graph the given function.

25. $g(x) = 2^{x+1}$

26. $g(x) = 2^{x+2}$

27. $g(x) = 2^x - 1$

28. $g(x) = 2^x + 2$

29. $h(x) = 2^{x+1} - 1$

30. $h(x) = 2^{x+2} - 1$

31. $g(x) = -2^x$

32. $g(x) = 2^{-x}$

33. $g(x) = 2 \cdot 2^x$

34. $g(x) = \frac{1}{2} \cdot 2^x$

Use the compound interest formulas $A = P\left(1 + \dfrac{r}{n}\right)^{nt}$ and $A = Pe^{rt}$ to solve Exercises 35–38.

35. Find the accumulated value of an investment of $10,000 for 5 years at an interest rate of 5.5% if the money is **a.** compounded semiannually; **b.** compounded monthly; **c.** compounded continuously.

36. Find the accumulated value of an investment of $5000 for 10 years at an interest rate of 6.5% if the money is **a.** compounded semiannually; **b.** compounded monthly; **c.** compounded continuously.

37. Suppose that you have $12,000 to invest. Which investment yields the greatest return over 3 years: 7% compounded monthly or 6.85% compounded continuously?

38. Suppose that you have $6000 to invest. Which investment yields the greatest return over 4 years: 8.25% compounded quarterly or 8.3% compounded semiannually?

Application Exercises

Use a calculator with an $\boxed{x^y}$ key or a $\boxed{\wedge}$ key to solve Exercises 39–46.

39. The exponential function $f(x) = 67.38(1.026)^x$ describes the population of Mexico, $f(x)$, in millions, x years after 1980.

    a. Substitute 0 for x and, without using a calculator, find Mexico's population in 1980.

    b. Substitute 27 for x and use your calculator to find Mexico's population in the year 2007 as predicted by this function.

    c. Find Mexico's population in the year 2034 as predicted by this function.

    d. Find Mexico's population in the year 2061 as predicted by this function.

    e. What appears to be happening to Mexico's population every 27 years?

40. The 1986 explosion at the Chernobyl nuclear power plant in the former Soviet Union sent about 1000 kilograms of radioactive cesium–137 into the atmosphere. The function $f(x) = 1000(0.5)^{x/30}$ describes the amount, $f(x)$, in kilograms, of cesium-137 remaining in Chernobyl x years after 1986. If even 100 kilograms of cesium-137 remain in Chernobyl's atmosphere, the area is considered unsafe for human habitation. Find $f(80)$ and determine if Chernobyl will be safe for human habitation by 2066.

The function

$$f(x) = \frac{0.9}{1 + 271(0.885)^x}$$

models the fraction of people x years old with some coronary heart disease. Use this function to solve Exercises 41–42.

41. Evaluate $f(25)$ and describe what this means in practical terms.

42. Evaluate $f(70)$ and describe what this means in practical terms.

The formula $S = C(1 + r)^t$ models inflation, where C = the value today, r = the annual inflation rate, and S = the inflated value t years from now. Use this formula to solve Exercises 43–44.

43. If the inflation rate is 6%, how much will a house now worth $65,000 be worth in 10 years?

44. If the inflation rate is 3%, how much will a house now worth $110,000 be worth in 5 years?

45. A decimal approximation for $\sqrt{3}$ is 1.7320508. Use a calculator to find $2^{1.7}$, $2^{1.73}$, $2^{1.732}$, $2^{1.73205}$, and $2^{1.7320508}$. Now find $2^{\sqrt{3}}$. What do you observe?

46. A decimal approximation for π is 3.141593. Use a calculator to find 2^3, $2^{3.1}$, $2^{3.14}$, $2^{3.141}$, $2^{3.1415}$, $2^{3.14159}$, and $2^{3.141593}$. Now find 2^π. What do you observe?

Use a calculator with an $\boxed{e^x}$ key to solve Exercises 47–51. The function $f(x) = 24,000e^{0.21x}$ describes the number of AIDS cases in the United States among intravenous drug users x years after 1989. Use this function to solve Exercises 47–48.

47. Evaluate $f(11)$ and describe what this means in practical terms.

48. Evaluate $f(31)$ and describe what this means in practical terms.

49. In college, we study large volumes of information—information that, unfortunately, we do not often retain for very long. The function

$$f(x) = 80e^{-0.5x} + 20$$

describes the percentage of information, $f(x)$, that a particular person remembers x weeks after learning the information.

    a. Substitute 0 for x and, without using a calculator, find the percentage of information remembered at the moment it is first learned.

    b. Substitute 1 for x and find the percentage of information that is remembered after 1 week.

    c. Find the percentage of information that is remembered after 4 weeks.

    d. Find the percentage of information that is remembered after one year (52 weeks).

50. In 1626, Peter Minuit convinced the Wappinger Indians to sell him Manhattan Island for $24. If the Native Americans had put the $24 into a bank account paying 5% interest, how much would the investment be worth in the year 2000 if interest were compounded
    a. monthly?   **b.** continuously?

51. The function

$$N(t) = \frac{30,000}{1 + 20e^{-1.5t}}$$

describes the number of people, $N(t)$, who become ill with influenza t weeks after its initial outbreak in a town with 30,000 inhabitants. The horizontal asymptote in the graph indicates that there is a limit to the epidemic's growth.
    a. How many people became ill with the flu when the epidemic began? (When the epidemic began, $t = 0$.)
    b. How many people were ill by the end of the third week?
    c. Why can't the spread of an epidemic simply grow indefinitely? What does the horizontal asymptote shown in the graph indicate about the limiting size of the population that becomes ill?

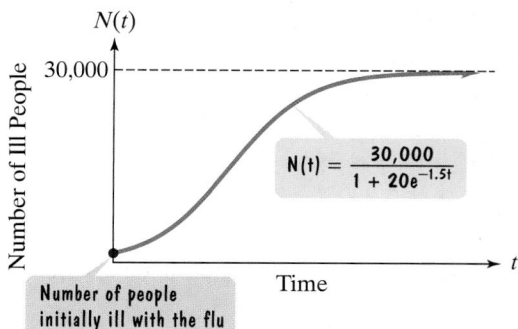

Writing in Mathematics

52. What is an exponential function?

53. What is the natural exponential function?

54. Use a calculator to evaluate $\left(1 + \frac{1}{x}\right)^x$ for $x = 10, 100, 1000, 10,000, 100,000,$ and $1,000,000$. Describe what happens to the expression as x increases.

55. Write an example similar to Example 6 on page 358 in which continuous compounding at a slightly lower yearly interest rate is a better investment than compounding n times per year.

56. Describe how you could use the graph of $f(x) = 2^x$ to obtain a decimal approximation for $\sqrt{2}$.

57. The exponential function $y = 2^x$ is one-to-one and has an inverse function. Try finding the inverse function by exchanging x and y and solving for y. Describe the difficulty that you encounter in this process. What is needed to overcome this problem?

58. In 1999, world population was 6 billion with an annual growth rate of 1.3%. Discuss two factors that would cause this growth rate to slow down over the next ten years.

Technology Exercises

59. Graph $y = 13.49(0.967)^x - 1$, the function for the number of O-rings expected to fail at $x°F$, in a $[0, 90, 10]$ by $[0, 20, 5]$ viewing rectangle. If NASA engineers had used this function and its graph, is it likely they would have allowed the *Challenger* to be launched when the temperature was $31°F$? Explain.

60. The student–teacher ratio in U.S. elementary and secondary schools can be modeled by $y = 25.34 (0.987)^x$, where x represents the number of years since 1959 and y represents the student–teacher ratio. Graph the function in a $[1, 40, 1]$ by $[0, 26, 1]$ viewing rectangle. When did the student–teacher ratio become less than 21 students per teacher?

61. You have $10,000 to invest. One bank pays 5% interest compounded quarterly and the other pays 4.5% interest compounded monthly.
    a. Use the formula for compound interest to write a function for the balance in each account at any time t.
    b. Use a graphing utility to graph both functions in an appropriate viewing rectangle. Based on the graphs, which bank offers the better return on your money?

62. a. Graph $y = e^x$ and $y = 1 + x + \frac{x^2}{2}$ in the same viewing rectangle.
    b. Graph $y = e^x$ and $y = 1 + x + \frac{x^2}{2} + \frac{x^3}{6}$ in the same viewing rectangle.
    c. Graph $y = e^x$ and $y = 1 + x + \frac{x^2}{2} + \frac{x^3}{6} + \frac{x^4}{24}$ in the same viewing rectangle.
    d. Describe what you observe in parts (a)–(c). Try generalizing this observation.

Critical Thinking Exercises

63. Which one of the following is true?
    a. As the number of compounding periods increases on a fixed investment, the amount of money in the account over a fixed interval of time will increase without bound.
    b. The functions $f(x) = 3^{-x}$ and $g(x) = -3^x$ have the same graph.
    c. $e = 2.718$.
    d. The functions $f(x) = \left(\frac{1}{3}\right)^x$ and $g(x) = 3^{-x}$ have the same graph.

64. The graphs labeled (a)–(d) in the figure represent $y = 3^x$, $y = 5^x$, $y = \left(\frac{1}{3}\right)^x$, and $y = \left(\frac{1}{5}\right)^x$, but not necessarily in that order. Which is which? Describe the process that enables you to make this decision.

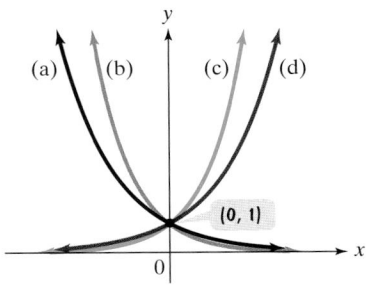

65. The hyperbolic cosine and hyperbolic sine functions are defined by

$$\cosh x = \frac{e^x + e^{-x}}{2} \quad \text{and} \quad \sinh x = \frac{e^x - e^{-x}}{2}.$$

Prove that $(\cosh x)^2 - (\sinh x)^2 = 1$.

4.2 Logarithmic Functions

Objectives

1. Change from logarithmic to exponential form.
2. Change from exponential to logarithmic form.
3. Evaluate logarithms.
4. Use basic logarithmic properties.
5. Graph logarithmic functions.
6. Find the domain of a logarithmic function.
7. Use common logarithms.
8. Use natural logarithms.

Study Tip

In case you need to review inverse functions, they are discussed on pages 246–247. The horizontal line test appears on page 245.

The earthquake that ripped through northern California on October 17, 1989, measured 7.1 on the Richter scale, killed more than 60 people, and injured more than 2400. Shown here is San Francisco's Marina district, where shock waves tossed houses off their foundations and into the street.

The Richter scale is misleading because for each increase in one unit on the scale, there is a tenfold increase in the intensity of an earthquake. In this section our focus is on the inverse of the exponential function, called the logarithmic function. The logarithmic function will help you to understand diverse phenomena, including earthquake intensity, human memory, and the pace of life in large cities.

The Definition of Logarithmic Functions

No horizontal line can be drawn that intersects the graph of an exponential function at more than one point. This means that the exponential function is one-to-one and has an inverse. The inverse function of the exponential function with base b is called the **logarithmic function with base b**.

Definition of the Logarithmic Function

For $x > 0$ and $b > 0$, $b \neq 1$,

$$y = \log_b x \quad \text{is equivalent to} \quad b^y = x.$$

The function $f(x) = \log_b x$ is the **logarithmic function with base b**.

The equations

$$y = \log_b x \quad \text{and} \quad b^y = x$$

are different ways of expressing the same thing. The first equation is in **logarithmic form** and the second equivalent equation is in **exponential form**.

Notice that a **logarithm**, y, **is an exponent**. You should learn the location of the base and exponent in each form.

Location of Base and Exponent in Exponential and Logarithmic Forms

Exponent

Exponent

Logarithmic Form: $y = \log_b x$ Exponential Form: $b^y = x$

Base

Base

1 Change from logarithmic to exponential form.

EXAMPLE 1   **Changing From Logarithmic to Exponential Form**

Write each equation in its equivalent exponential form.

a. $2 = \log_5 x$ **b.** $3 = \log_b 64$ **c.** $\log_3 7 = y$

Solution   We use the fact that $y = \log_b x$ means $b^y = x$.

a. $2 = \log_5 x$ means $5^2 = x$.   **b.** $3 = \log_b 64$ means $b^3 = 64$.

Logarithms are exponents.   Logarithms are exponents.

c. $\log_3 7 = y$ or $y = \log_3 7$ means $3^y = 7$.

Check Point 1   Write each equation in its equivalent exponential form.

a. $3 = \log_7 x$ **b.** $2 = \log_b 25$ **c.** $\log_4 26 = y$

$7^{5 \cdot x}$ $b^2 = 25$ $4y = 26$

2 Change from exponential to logarithmic form.

EXAMPLE 2   **Changing From Exponential to Logarithmic Form**

Write each equation in its equivalent logarithmic form.

a. $12^2 = x$ **b.** $b^3 = 8$ **c.** $e^y = 9$

Solution   We use the fact that $b^y = x$ means $y = \log_b x$.

a. $12^2 = x$ means $2 = \log_{12} x$.   **b.** $b^3 = 8$ means $3 = \log_b 8$.

Exponents are logarithms.   Exponents are logarithms.

c. $e^y = 9$ means $y = \log_e 9$.

Check Point 2   Write each equation in its equivalent logarithmic form.

    a. $2^5 = x$ **b.** $b^3 = 27$ **c.** $e^y = 33$

3 Evaluate logarithms.

Remembering that logarithms are exponents makes it possible to evaluate some logarithms by inspection. The logarithm of x with base b, $\log_b x$, is the exponent to which b must be raised to get x. For example, suppose we want to evaluate $\log_2 32$. We ask, 2 to what power gives 32? Because $2^5 = 32$, $\log_2 32 = 5$.

EXAMPLE 3 Evaluating Logarithms

Evaluate:

    a. $\log_2 16$ **b.** $\log_3 9$ **c.** $\log_{25} 5$

Solution

Logarithmic Expression	Question Needed for Evaluation	Logarithmic Expression Evaluated
a. $\log_2 16$	2 to what power gives 16?	$\log_2 16 = 4$ because $2^4 = 16$.
b. $\log_3 9$	3 to what power gives 9?	$\log_3 9 = 2$ because $3^2 = 9$.
c. $\log_{25} 5$	25 to what power gives 5?	$\log_{25} 5 = \frac{1}{2}$ because $25^{1/2} = \sqrt{25} = 5$.

Check Point 3   Evaluate:

    a. $\log_{10} 100$ **b.** $\log_3 3$ **c.** $\log_{36} 6$

4 Use basic logarithmic properties.

Basic Logarithmic Properties

Because logarithms are exponents, they have properties that can be verified using properties of exponents.

Basic Logarithmic Properties Involving One

1. $\log_b b = 1$ because 1 is the exponent to which b must be raised to obtain b. $\left(b^1 = b\right)$

2. $\log_b 1 = 0$ because 0 is the exponent to which b must be raised to obtain 1. $\left(b^0 = 1\right)$

EXAMPLE 4 Using Properties of Logarithms

Evaluate:

    a. $\log_7 7$ **b.** $\log_5 1$

Solution

    a. Because $\log_b b = 1$, we conclude $\log_7 7 = 1$.
    b. Because $\log_b 1 = 0$, we conclude $\log_5 1 = 0$.

Check Point 4   Evaluate:

    a. $\log_9 9$ **b.** $\log_8 1$

The inverse of the exponential function is the logarithmic function. Thus, if $f(x) = b^x$, then $f^{-1}(x) = \log_b x$. In Chapter 2, we saw how inverse functions "undo" one another. In particular,

$$f\big(f^{-1}(x)\big) = x \text{ and } f^{-1}\big(f(x)\big) = x.$$

Applying these relationships to exponential and logarithmic functions, we obtain the following **inverse properties of logarithms**.

Inverse Properties of Logarithms

For $b > 0$ and $b \neq 1$,

    $\log_b b^x = x$ The logarithm with base b of b raised to a power equals that power.

    $b^{\log_b x} = x$ b raised to the logarithm with base b of a number equals that number.

EXAMPLE 5   Using Inverse Properties of Logarithms

Evaluate:

    a. $\log_4 4^5$ **b.** $6^{\log_6 9}$.

Solution

    a. Because $\log_b b^x = x$, we conclude $\log_4 4^5 = 5$.

    b. Because $b^{\log_b x} = x$, we conclude $6^{\log_6 9} = 9$.

Check Point 5   Evaluate:

    a. $\log_7 7^8$ **b.** $3^{\log_3 17}$

5 Graph logarithmic functions.

Graphs of Logarithmic Functions

How do we graph logarithmic functions? We use the fact that the logarithmic function is the inverse of the exponential function. This means that the logarithmic function reverses the coordinates of the exponential function. It also means that the graph of the logarithmic function is a reflection of the graph of the exponential function about the line $y = x$.

EXAMPLE 6   Graphs of Exponential and Logarithmic Functions

Graph $f(x) = 2^x$ and $g(x) = \log_2 x$ in the same rectangular coordinate system.

Solution   We first set up a table of coordinates for $f(x) = 2^x$. Reversing, these coordinates gives the coordinates for the inverse function $g(x) = \log_2 x$.

x	-2	-1	0	1	2	3
$f(x) = 2^x$	$\frac{1}{4}$	$\frac{1}{2}$	1	2	4	8

x	$\frac{1}{4}$	$\frac{1}{2}$	1	2	4	8
$g(x) = \log_2 x$	-2	-1	0	1	2	3

Reverse coordinates.

We now plot the ordered pairs in both tables, connecting them with smooth curves. Figure 4.6 shows the graphs of $f(x) = 2^x$ and its inverse function $g(x) = \log_2 x$. The graph of the inverse can also be drawn by reflecting the graph of $f(x) = 2^x$ about the line $y = x$.

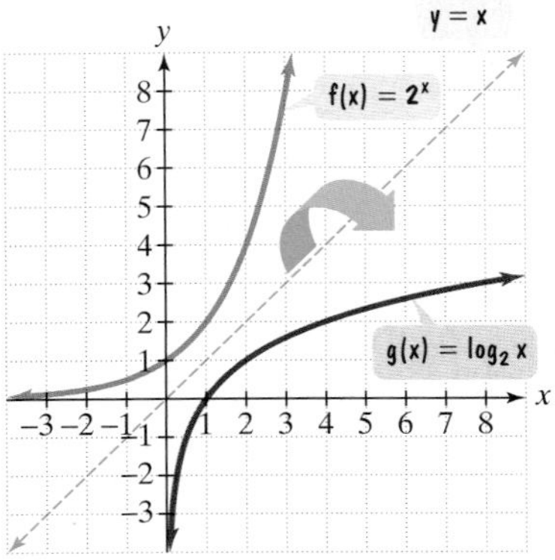

Figure 4.6 The graphs of $f(x) = 2^x$ and its inverse function

Check Point 6

Graph $f(x) = 3^x$ and $g(x) = \log_3 x$ in the same rectangular coordinate system.

Figure 4.7 illustrates the relationship between the graph of the exponential function, shown in blue and its inverse, the logarithmic function, shown in red, for bases greater than 1 and for bases between 0 and 1.

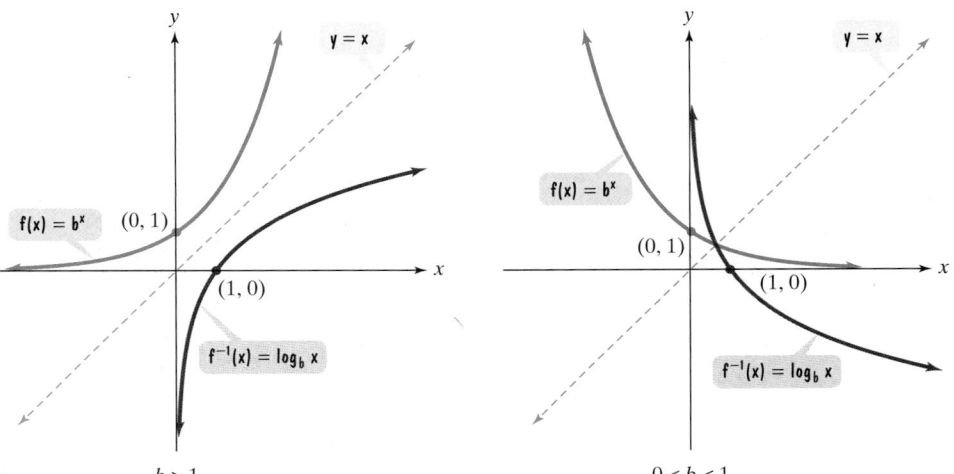

Figure 4.7 Graphs of exponential and logarithmic functions

Characteristics of the Graphs of Logarithmic Functions

- The x-intercept is 1. There is no y-intercept.
- The y-axis is a vertical asymptote.
- If $b > 1$, the function is increasing. If $0 < b < 1$, the function is decreasing.
- The graph is smooth and continuous. It has no sharp corners or gaps.

The graphs of logarithmic functions can be translated vertically or horizontally, reflected, stretched, or shrunk. We use the ideas of Section 2.5 to do so, as summarized in Table 4.3.

Table 4.3   Transformations Involving Logarithmic Functions

Transformation	Equation	Description
Horizontal translation	$g(x) = \log_b(x + c)$	• Shifts the graph of $f(x) = \log_b x$ to the left c units if $c > 0$. Vertical asymptote: $x = -c$. • Shifts the graph of $f(x) = \log_b x$ to the right c units if $c < 0$. Vertical asymptote: $x = -c$.
Vertical stretching or shrinking	$g(x) = c \log_b x$	Multiplying y-coordinates of $f(x) = \log_b x$ by c, • Stretches the graph of $f(x) = \log_b x$ if $c > 1$. • Shrinks the graph of $f(x) = \log_b x$ if $0 < c < 1$.
Reflecting	$g(x) = -\log_b x$ $g(x) = \log_b(-x)$	• Reflects the graph of $f(x) = \log_b x$ about the x-axis. • Reflects the graph of $f(x) = \log_b x$ about the y-axis.
Vertical translation	$g(x) = c + \log_b x$	• Shifts the graph of $f(x) = \log_b x$ upward c units if $c > 0$. • Shifts the graph of $f(x) = \log_b x$ downward c units if $c < 0$.

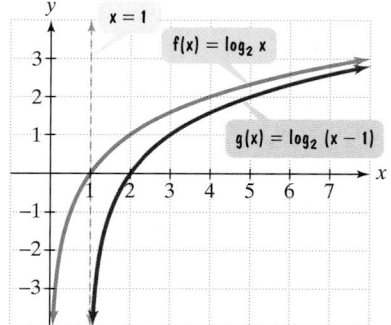

Figure 4.8 Shifting $f(x) = \log_2 x$ one unit to the right

For example, Figure 4.8 illustrates that the graph of $g(x) = \log_2(x - 1)$ is the graph of $f(x) = \log_2 x$ moved one unit to the right. If a logarithmic function is translated to the left or to the right, both the x-intercept and the vertical asymptote are shifted by the amount of the horizontal shift. In Figure 4.8, the x-intercept of f is 1. Because g is shifted one unit to the right, its x-intercept is 2. Also observe that the vertical asymptote for f, the y-axis, is shifted one unit to the right for the vertical asymptote for g. Thus, $x = 1$ is the vertical asymptote for g.

Here are some other examples of transformations of graphs of logarithmic functions.

- The graph of $g(x) = 3 + \log_4 x$ is the graph of $f(x) = \log_4 x$ moved up three unit, shown in Figure 4.9
- The graph of $h(x) = -\log_2 x$ is the graph of $f(x) = \log_2 x$ reflected about the x-axis, shown in Figure 4.10
- The graph of $r(x) = \log_2(-x)$ is the graph of $f(x) = \log_2 x$ reflected about the y-axis, shown in Figure 4.11.

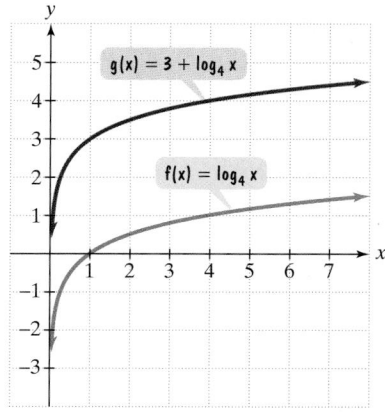

Figure 4.9 Shifting vertically up three units

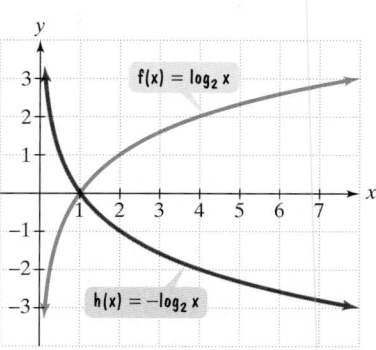

Figure 4.10 Reflection about the x-axis

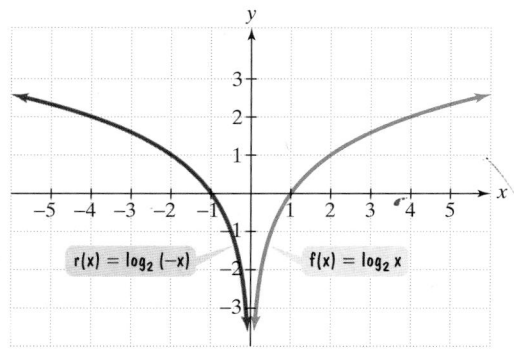

Figure 4.11 Reflection about the y-axis

6 Find the domain of a logarithmic function.

The Domain of a Logarithmic Function

In Section 4.1 we learned that the domain of an exponential function includes all real numbers and its range is the set of positive real numbers. Because the logarithmic function reverses the domain and the range of the exponential function, the **domain of a logarithmic function is the set of all positive real numbers.** Thus, $\log_2 8$ is defined because the value of x in the logarithmic expression, 8, is greater than zero and therefore is included in the domain of the logarithmic function $f(x) = \log_2 x$. However, $\log_2 0$ and $\log_2(-8)$ are not defined because 0 and -8 are not positive real numbers and therefore are excluded from the domain of the logarithmic function $f(x) = \log_2 x$. In general, the domain of $f(x) = \log_b(x + c)$ consists of all x for which $x + c > 0$.

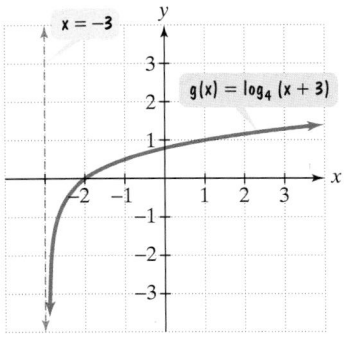

Figure 4.12 The domain of $g(x) = \log_4(x + 3)$ is $(-3, \infty)$.

EXAMPLE 7 Finding the Domain of a Logarithmic Function

Find the domain of $g(x) = \log_4(x + 3)$.

Solution   The domain of g consists of all x for which $x + 3 > 0$. Solving this inequality for x, we obtain $x > -3$. Thus, the domain of g is $(-3, \infty)$. This is illustrated in Figure 4.12. The vertical asymptote is $x = -3$, and all points on the graph of g have x-coordinates that are greater than -3.

Check Point 7   Find the domain of $h(x) = \log_4(x - 5)$.

7 Use common logarithms.

Common Logarithms

The logarithmic function with base 10 is called the **common logarithmic function**. The function $f(x) = \log_{10} x$ is usually expressed as $f(x) = \log x$. A calculator with a [LOG] key can be used to evaluate common logarithms. Here are some examples:

Logarithm	Graphing Calculator Keystrokes	Display (or Approximate Display)
$\log 1000$	[LOG] 1000 [ENTER]	3
$\log \frac{5}{2}$	[LOG] [(] 5 [÷] 2 [)] [ENTER]	0.39794
$\dfrac{\log 5}{\log 2}$	[LOG] 5 [÷] [LOG] 2 [ENTER]	2.32192
$\log(-3)$	[LOG] [(−)] 3 [ENTER]	[ERROR]

The error message given by many graphing calculators for $\log(-3)$ is a reminder that the domain of every logarithmic function, including the common logarithmic function, is the set of positive real numbers.

Many real-life phenomena start with rapid growth, and then the growth begins to level off. This type of behavior can be modeled by logarithmic functions.

EXAMPLE 8   Modeling Height of Children

The percentage of adult height attained by a boy who is x years old can be modeled by

$$f(x) = 29 + 48.8 \log(x + 1)$$

where x represents the boy's age and $f(x)$ represents the percentage of his adult height. Approximately what percentage of his adult height is a boy at age eight?

Solution   We substitute the boy's age, 8, for x and evaluate the function at 8.

$$f(x) = 29 + 48.8 \log(x + 1) \quad \text{This is the given function.}$$
$$f(8) = 29 + 48.8 \log(8 + 1) \quad \text{Substitute 8 for x.}$$
$$= 29 + 48.8 \log 9 \quad \text{Graphing calculator keystrokes:}$$
$$\approx 76 \qquad 29 \boxed{+} 48.8 \boxed{\times} \boxed{\text{LOG}} \; 9 \boxed{\text{ENTER}}$$

Thus, an 8-year-old boy is approximately 76% of his adult height.

> **Check Point 8**   Use the function in Example 8 to answer this question: Approximately what percentage of his adult height is a boy at age 10?

The basic properties of logarithms that were listed earlier in this section can be applied to common logarithms.

Properties of Common Logarithms

General Properties	Common Logarithms
1. $\log_b 1 = 0$	**1.** $\log 1 = 0$
2. $\log_b b = 1$	**2.** $\log 10 = 1$
3. $\log_b b^x = x$	**3.** $\log 10^x = x$
4. $b^{\log_b x} = x$	**4.** $10^{\log x} = x$

The property $\log 10^x = x$ can be used to evaluate common logarithms involving powers of 10. For example,

$$\log 100 = \log 10^2 = 2, \quad \log 1000 = \log 10^3 = 3, \quad \log 10^{7.1} = 7.1.$$

EXAMPLE 9   Earthquake Intensity

The magnitude R on the Richter scale of an earthquake of intensity I is given by

$$R = \log \frac{I}{I_0}$$

where I_0 is the intensity of a barely felt zero-level earthquake. The earthquake that destroyed San Francisco in 1906 was $10^{8.3}$ times as intense as a zero-level earthquake. What was its magnitude on the Richter scale?

Solution   Because the earthquake was $10^{8.3}$ times as intense as a zero-level earthquake, the intensity I is $10^{8.3} I_0$.

$$R = \log \frac{I}{I_0} \qquad \text{This is the formula for magnitude on the Richter scale.}$$

$$R = \log \frac{10^{8.3} I_0}{I_0} \qquad \text{Substitute } 10^{8.3} I_0 \text{ for } I.$$

$$= \log 10^{8.3} \qquad \text{Simplify.}$$

$$= 8.3 \qquad \text{Use the property } \log 10^x = x.$$

San Francisco's 1906 earthquake registered 8.3 on the Richter scale.

Check Point 9

Use the formula in Example 9 to solve this problem. If an earthquake is 10,000 times as intense as a zero-level quake $(I = 10,000I_0)$, what is its magnitude on the Richter scale?

8 Use natural logarithms.

Natural Logarithms

The logarithmic function with base e is called the **natural logarithmic function**. The function $f(x) = \log_e x$ is usually expressed as $f(x) = \ln x$, read "el en of x." A calculator with an $\boxed{\text{LN}}$ key can be used to evaluate natural logarithms.

Like the domain of all logarithmic functions, the domain of the natural logarithmic function is the set of all positive real numbers. Thus, the domain of $f(x) = \ln(x + c)$ consists of all x for which $x + c > 0$.

EXAMPLE 10 Finding Domains of Natural Logarithmic Functions

Find the domain of each function.

a. $f(x) = \ln(3 - x)$ **b.** $g(x) = \ln(x - 3)^2$

Solution

a. The domain of f consists of all x for which $3 - x > 0$. Solving this inequality for x, we obtain $x < 3$. Thus, the domain of f is $(-\infty, 3)$. This is verified by the graph in Figure 4.13.

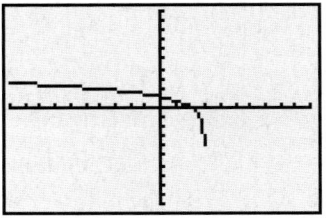
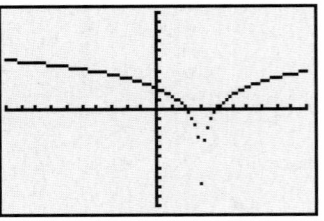

Figure 4.13 The domain of $f(x) = \ln(3 - x)$ is $(-\infty, 3)$.

Figure 4.14 3 is excluded from the domain of $g(x) = \ln(x - 3)^2$.

b. The domain of g consists of all x for which $(x - 3)^2 > 0$. It follows that the domain of g is all real numbers except 3. This is shown by the graph in Figure 4.14. If it is not obvious that 3 is excluded from the domain, try using a $\boxed{\text{dot}}$ format.

Check Point 10

Find the domain of each function.

    a. $f(x) = \ln(4 - x)$ **b.** $g(x) = \ln x^2$

The basic properties of logarithms that were listed earlier in this section can be applied to natural logarithms.

Properties of Natural Logarithms

General Properties	Natural Logarithms
1. $\log_b 1 = 0$	**1.** $\ln 1 = 0$
2. $\log_b b = 1$	**2.** $\ln e = 1$
3. $\log_b b^x = x$	**3.** $\ln e^x = x$
4. $b^{\log_b x} = x$	**4.** $e^{\ln x} = x$

The property $\ln e^x = x$ can be used to evaluate natural logarithms involving powers of e. For example,

$$\ln e^2 = 2, \quad \ln e^3 = 3, \quad \ln e^{7.1} = 7.1, \quad \text{and} \quad \ln \frac{1}{e} = \ln e^{-1} = -1.$$

EXAMPLE 11 Using Inverse Properties

Use inverse properties to simplify:

    a. $\ln e^{7x}$ **b.** $e^{\ln 4x^2}$

Solution

    a. Because $\ln e^x = x$, we conclude that $\ln e^{7x} = 7x$.
    b. Because $e^{\ln x} = x$, we conclude $e^{\ln 4x^2} = 4x^2$.

Check Point 11

Use inverse properties to simplify:

    a. $\ln e^{25x}$ **b.** $e^{\ln \sqrt{x}}$

EXAMPLE 12 Walking Speed and City population

As the population of a city increases, the pace of life also increases. The formula

$$W = 0.35 \ln P + 2.74$$

models average walking speed, W, in feet per second, for a resident of a city whose population is P thousand. Find the average walking speed for people living in New York City with a population of 7323 thousand.

Solution   We use the formula and substitute 7323 for P, the population in thousands.

$$W = 0.35 \ln P + 2.74 \qquad \text{This is the given formula.}$$
$$W = 0.35 \ln 7323 + 2.74 \qquad \text{Substitute 7323 for P.}$$
$$\approx 5.9 \qquad\qquad \text{Graphing calculator keystrokes:}$$

$$0.35 \boxed{\times} \boxed{\text{LN}} \ 7323 \boxed{+} 2.74 \boxed{\text{ENTER}}$$

The average walking speed in New York City is approximately 5.9 feet per second.

Check Point 12   Use the formula $W = 0.35 \ln P + 2.74$ to find the average walking speed in Jackson, Mississippi with a population of 197 thousand.

EXERCISE SET 4.2

Practice Exercises

In Exercises 1–8, write each equation in its equivalent exponential form.

1. $4 = \log_2 16$ **2.** $6 = \log_2 64$

3. $2 = \log_3 x$ **4.** $2 = \log_9 x$

5. $5 = \log_b 32$ **6.** $3 = \log_b 27$

7. $\log_6 216 = y$ **8.** $\log_5 125 = y$

In Exercises 9–20, write each equation in its equivalent logarithmic form.

9. $2^3 = 8$ **10.** $5^4 = 625$ **11.** $2^{-4} = \frac{1}{16}$

12. $5^{-3} = \frac{1}{125}$ **13.** $\sqrt[3]{8} = 2$ **14.** $\sqrt[3]{64} = 4$

15. $13^2 = x$ **16.** $15^2 = x$ **17.** $b^3 = 1000$

18. $b^3 = 343$ **19.** $7^y = 200$ **20.** $8^y = 300$

In Exercises 21–38, evaluate each expression without using a calculator.

21. $\log_4 16$ **22.** $\log_7 49$ **23.** $\log_2 64$

24. $\log_3 27$ **25.** $\log_7 \sqrt{7}$ **26.** $\log_6 \sqrt{6}$

27. $\log_2 \frac{1}{8}$ **28.** $\log_3 \frac{1}{9}$ **29.** $\log_{64} 8$

30. $\log_{81} 9$ **31.** $\log_5 5$ **32.** $\log_{11} 11$

33. $\log_4 1$ **34.** $\log_6 1$ **35.** $\log_5 5^7$

36. $\log_4 4^6$ **37.** $8^{\log_8 19}$ **38.** $7^{\log_7 23}$

39. Graph $f(x) = 4^x$ and $g(x) = \log_4 x$ in the same rectangular coordinate system.

40. Graph $f(x) = 5^x$ and $g(x) = \log_5 x$ in the same rectangular coordinate system.

41. Graph $f(x) = \left(\frac{1}{2}\right)^x$ and $g(x) = \log_{1/2} x$ in the same rectangular coordinate system.

42. Graph $f(x) = \left(\frac{1}{4}\right)^x$ and $g(x) = \log_{1/4} x$ in the same rectangular coordinate system.

In Exercises 43–48, the graph of a logarithmic function is given. Select the function for each graph from the following options.

$$f(x) = \log_3 x, \, g(x) = \log_3(x-1), \, h(x) = \log_3 x - 1,$$
$$F(x) = -\log_3 x, \, G(x) = \log_3(-x), \, H(x) = 1 - \log_3 x$$

43.

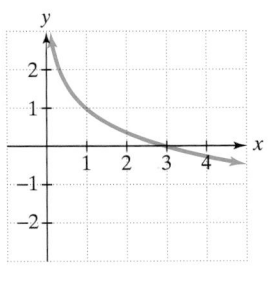

44.

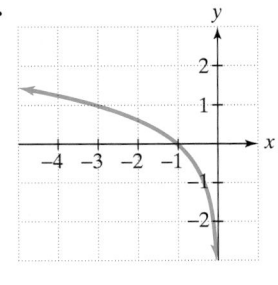

45.

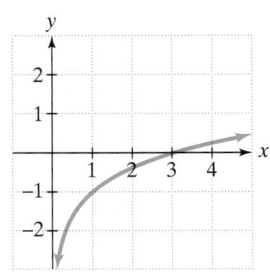

46.

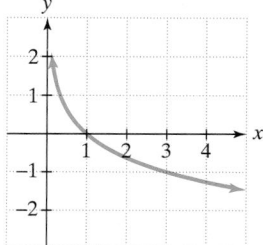

47.

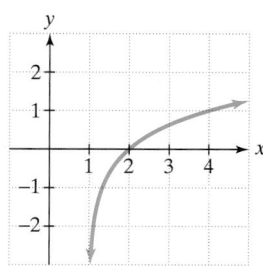

48.

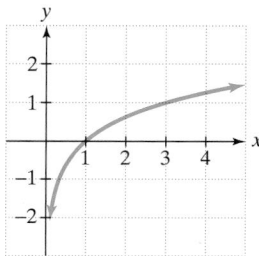

In Exercises 49–54, begin by graphing $f(x) = \log_2 x$. Then use transformations of this graph to graph the given function. What is the graph's x-intercept? What is the vertical asymptote?

49. $g(x) = \log_2(x + 1)$ **50.** $g(x) = \log_2(x + 2)$

51. $h(x) = 1 + \log_2 x$ **52.** $h(x) = 2 + \log_2 x$

53. $g(x) = \frac{1}{2}\log_2 x$ **54.** $g(x) = -2\log_2 x$

In Exercises 55–60, find the domain of each logarithmic function.

55. $f(x) = \log_5(x + 4)$ **56.** $f(x) = \log_5(x + 6)$

57. $f(x) = \log(2 - x)$ **58.** $f(x) = \log(7 - x)$

59. $f(x) = \ln(x - 2)^2$ **60.** $f(x) = \ln(x - 7)^2$

In Exercises 61–74, evaluate each expression without using a calculator.

61. $\log 100$ **62.** $\log 1000$ **63.** $\log 10^7$

64. $\log 10^8$ **65.** $10^{\log 33}$ **66.** $10^{\log 53}$

67. $\ln 1$ **68.** $\ln e$ **69.** $\ln e^6$

70. $\ln e^7$ **71.** $\ln \dfrac{1}{e^6}$ **72.** $\ln \dfrac{1}{e^7}$

73. $e^{\ln 125}$ **74.** $e^{\ln 300}$

In Exercises 75–80, use inverse properties of logarithms to simplify each expression.

75. $\ln e^{9x}$ **76.** $\ln e^{13x}$ **77.** $e^{\ln 5x^2}$

78. $e^{\ln 7x^2}$ **79.** $10^{\log \sqrt{x}}$ **80.** $10^{\log \sqrt[3]{x}}$

 Application Exercises

The percentage of adult height attained by a girl who is x years old can be modeled by

$$f(x) = 62 + 35 \log(x - 4)$$

where x represents the girl's age (from 5 to 15) and $f(x)$ represents the percentage of her adult height. Use the formula to solve Exercises 81–82.

81. Approximately what percentage of her adult height is a girl at age 13?

82. Approximately what percentage of her adult height is a girl at age ten?

83. The annual amount that we spend to attend sporting events can be modeled by

$$f(x) = 2.05 + 1.3 \ln x$$

where x represents the number of years since 1984 and $f(x)$ represents the total annual expenditures for admission to spectator sports, in billions of dollars. In 2000, approximately how much was spent on admission to spectator sports?

84. The percentage of U.S. households with cable television can be modeled by

$$f(x) = 18.32 + 15.94 \ln x$$

where x represents the number of years since 1979 and $f(x)$ represents the percentage of U.S. households with cable television. What percentage of U.S. households had cable television in 1990?

The loudness level of a sound, D, in decibels, is given by the formula

$$D = 10 \log(10^{12}I)$$

where I is the intensity of the sound, in watts per meter². Decibel levels range from 0, a barely audible sound, to 160, a sound resulting in a ruptured eardrum. Use the formula to solve Exercises 85–86.

85. The sound of a blue whale can be heard 500 miles away, reaching an intensity of 6.3×10^6 watts per meter². Determine the decibel level of this sound. At close range, can the sound of a blue whale rupture the human eardrum?

86. What is the decibel level of a normal conversation, 3.2×10^{-6} watts per meter²?

87. Students in a psychology class took a final examination. As part of an experiment to see how much of the course content they remembered over time, they took equivalent forms of the exam in monthly intervals thereafter. The average score for the group, $f(t)$, after t months was modeled by the function

$$f(t) = 88 - 15 \ln(t + 1), \qquad 0 \le t \le 12.$$

a. What was the average score on the original exam?
b. What was the average score after 2 months? 4 months? 6 months? 8 months? 10 months? one year?
c. Sketch the graph of f (either by hand or with a graphing utility). Describe what the graph indicates in terms of the material retained by the students.

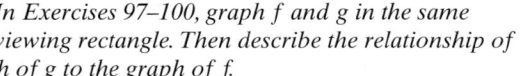

Writing in Mathematics

88. Describe the relationship between an equation in logarithmic form and an equivalent equation in exponential form.

89. What question can be asked to help evaluate $\log_3 81$?

90. Explain why the logarithm of 1 with base b is 0.

91. Describe the following property using words: $\log_b b^x = x$.

92. Explain how to use the graph of $f(x) = 2^x$ to obtain the graph of $g(x) = \log_2 x$.

93. Explain how to find the domain of a logarithmic function.

94. New York City is one of the world's great walking cities. Use the formula in Example 12 on page 371 to describe what frequently happens to tourists exploring the city by foot.

95. Logarithmic models are well suited to phenomena in which growth is initially rapid but then begins to level off. Describe something that is changing over time that can be modeled using a logarithmic function.

96. Suppose that a girl is 4′ 6″ at age 10. Explain how to use the function in Exercises 81–82 to determine how tall she can expect to be as an adult.

Technology Exercises

In Exercises 97–100, graph f and g in the same viewing rectangle. Then describe the relationship of the graph of g to the graph of f.

97. $f(x) = \ln x, g(x) = \ln(x + 3)$

98. $f(x) = \ln x, g(x) = \ln x + 3$

99. $f(x) = \log x, g(x) = -\log x$

100. $f(x) = \log x, g(x) = \log(x - 2) + 1$

101. Students in a mathematics class took a final examination. They took equivalent forms of the exam in monthly intervals thereafter. The average score, $f(t)$, for the

group after t months was modeled by the human memory function $f(t) = 75 - 10 \log(t + 1)$, where $0 \le t \le 12$. Use a graphing utility to graph the function. Then determine how many months will elapse before the average score falls below 65.

102. Graph f and g in the same viewing rectangle.
a. $f(x) = \ln(3x), g(x) = \ln 3 + \ln x$
b. $f(x) = \log(5x^2), g(x) = \log 5 + \log x^2$
c. $f(x) = \ln(2x^3), g(x) = \ln 2 + \ln x^3$
d. Describe what you observe in parts (a)–(c). Generalize this observation by writing an equivalent expression for $\log_b(MN)$, where $M > 0$ and $N > 0$.
e. Complete this statement: The logarithm of a product is equal to _____.

103. Graph each of the following functions in the same viewing rectangle and then place the functions in order from the one that increases most slowly to the one that increases most rapidly.

$$y = x, \ y = \sqrt{x}, \ y = e^x, \ y = \ln x, \ y = x^x, \ y = x^2$$

Critical Thinking Exercises

104. Which one of the following is true?
a. $\dfrac{\log_2 8}{\log_2 4} = \dfrac{8}{4}$
b. $\log(-100) = -2$.
c. The domain of $f(x) = \log_2 x$ is $(-\infty, \infty)$.
d. $\log_b x$ is the exponent to which b must be raised to obtain x.

105. Without using a calculator, find the exact value of

$$\frac{\log_3 81 - \log_\pi 1}{\log_{2\sqrt{2}} 8 - \log 0.001}.$$

106. Solve for x: $\log_4[\log_3(\log_2 x)] = 0$.

107. Without using a calculator, determine which is the greater number: $\log_4 60$ or $\log_3 40$.

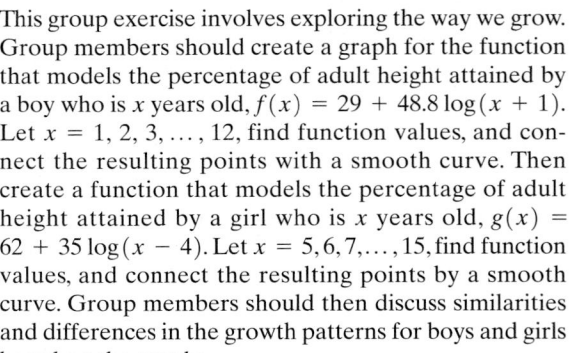

Group Exercise

108. This group exercise involves exploring the way we grow. Group members should create a graph for the function that models the percentage of adult height attained by a boy who is x years old, $f(x) = 29 + 48.8 \log(x + 1)$. Let $x = 1, 2, 3, \ldots, 12$, find function values, and connect the resulting points with a smooth curve. Then create a function that models the percentage of adult height attained by a girl who is x years old, $g(x) = 62 + 35 \log(x - 4)$. Let $x = 5, 6, 7, \ldots, 15$, find function values, and connect the resulting points by a smooth curve. Group members should then discuss similarities and differences in the growth patterns for boys and girls based on the graphs.

SECTION 4.3 *Properties of Logarithms*

Objectives

1. Use the product rule.
2. Use the quotient rule.
3. Use the power rule.
4. Expand logarithmic expressions.
5. Condense logarithmic expressions.
6. Use the change-of-base property.

We all learn new things in different ways. In this section, we consider important properties of logarithms. What would be the most effective way for you to learn about these properties? Would it be helpful to use your graphing utility and discover one of these properties for yourself? To do so, work Exercise 102 in Exercise Set 4.2 before continuing. Would the properties become more meaningful if you could see exactly where they come from? If so, you will find details of the proofs of many of these properties in the appendix. The remainder of our work in this chapter will be based on the properties of logarithms that you learn in this section.

1 Use the product rule.

The Product Rule

Properties of exponents correspond to properties of logarithms. For example, when we multiply with the same base, we add exponents:

$$b^M \cdot b^N = b^{M+N}.$$

This property of exponents, coupled with an awareness that a logarithm is an exponent, suggests the following property, called the **product rule**.

The Product Rule

Let b, M, and N be positive real numbers with $b \neq 1$.

$$\log_b(MN) = \log_b M + \log_b N$$

The logarithm of a product is the sum of the logarithms.

When we use the product rule to write a single logarithm as the sum of two logarithms, we say that we are **expanding a logarithmic expression**. For example, we can use the product rule to expand $\ln(4x)$:

$$\ln(4x) = \ln 4 + \ln x$$

| The logarithm of a product | is | the sum of the logarithms. |

EXAMPLE 1   Using the Product Rule

Use the product rule to expand

    a. $\log_4(7 \cdot 9)$ **b.** $\log(10x)$

Solution

    a. $\log_4(7 \cdot 9) = \log_4 7 + \log_4 9$ The logarithm of a product is the sum of the logarithms.

    b. $\log(10x) = \log 10 + \log x$ The logarithm of a product is the sum of the logarithms. These are common logarithms with base 10 understood.

                     $= 1 + \log x$ Because $\log_b b = 1$, then $\log_{10} 10 = 1$.

> **Check Point 1**
>
> Use the product rule to expand
>
> **a.** $\log_6(10 \cdot 9)$ **b.** $\log(100x)$

The Quotient Rule

2   Use the quotient rule.

When we divide with the same base, we subtract exponents:

$$\frac{b^M}{b^N} = b^{M-N}.$$

This property suggests the following property of logarithms, called the **quotient rule**.

> **The Quotient Rule**
>
> Let b, M, and N be positive real numbers with $b \neq 1$.
>
> $$\log_b\left(\frac{M}{N}\right) = \log_b M - \log_b N$$
>
> The logarithm of a quotient is the difference of the logarithms.

When we use the quotient rule to write a single logarithm as the difference of two logarithms, we say that we are **expanding a logarithmic expression**. For example, we can use the quotient rule to expand $\log \dfrac{x}{2}$:

$$\log \frac{x}{2} = \log x - \log 2$$

The logarithm of a quotient   is   the difference of the logarithms.

EXAMPLE 2   Using the Quotient Rule

Use the quotient rule to expand

    a. $\log_7\left(\dfrac{14}{x}\right)$ **b.** $\ln\left(\dfrac{e^3}{7}\right)$

Solution

a. $\log_7\left(\dfrac{14}{x}\right) = \log_7 14 - \log_7 x$ *The logarithm of a quotient is the difference of the logarithms.*

b. $\ln\left(\dfrac{e^3}{7}\right) = \ln e^3 - \ln 7$ *The logarithm of a quotient is the difference of the logarithms. These are natural logarithms with base e understood.*

$\phantom{\ln\left(\dfrac{e^3}{7}\right)} = 3 - \ln 7$ *Because In $e^x = x$, then In $e^3 = 3$.*

> **Check Point 2**
>
> Use the quotient rule to expand
>
> **a.** $\log_8\left(\dfrac{23}{x}\right)$ **b.** $\ln\left(\dfrac{e^5}{11}\right)$

3   Use the power rule.

The Power Rule

When an exponential expression is raised to a power, we multiply exponents:

$$(b^M)^p = b^{Mp}.$$

This property suggests the following property of logarithms, called the **power rule**.

> **The Power Rule**
>
> Let b, M, and N be positive real numbers with $b \neq 1$, and let p be any real number.
>
> $$\log_b M^p = p \log_b M$$
>
> The logarithm of a number with an exponent is the product of the exponent and the logarithm of that number.

When we use the power rule to "pull the exponent to the front," we say that we are **expanding a logarithmic expression**. For example, we can use the power rule to expand $\ln x^2$:

$$\ln x^2 = 2 \ln x.$$

> *The logarithm of a number with an exponent*   is   *the product of the exponent and the logarithm of that number.*

Figure 4.15 shows the graphs of $y = \ln x^2$ and $y = 2 \ln x$. Are $\ln x^2$ and $2 \ln x$ the same? The graphs illustrate that $y = \ln x^2$ and $y = 2 \ln x$ have different domains. The graphs are only the same if $x > 0$. Thus, we should write

$$\ln x^2 = 2 \ln x \text{ for } x > 0.$$

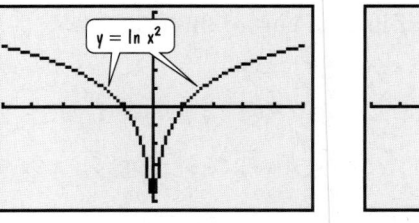
Domain: $(-\infty, 0)$ or $(0, \infty)$

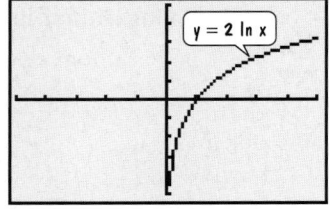
Domain: $(0, \infty)$

Figure 4.15 $\ln x^2$ and $2 \ln x$ have different domains.

When expanding a logarithmic expression, you might want to determine whether the rewriting has changed the domain of the expression.

EXAMPLE 3   Using the Power Rule

Use the power rule to expand

    a. $\log_5 7^4$ **b.** $\ln \sqrt{x}$

4 Expand logarithmic expressions.

Solution

    a. $\log_5 7^4 = 4 \log_5 7$ The logarithm of a number with an exponent is the exponent times the logarithm of the number.

    b. $\ln \sqrt{x} = \ln x^{1/2}$ Rewrite the radical using a rational exponent.

              $= \frac{1}{2} \ln x$ Use the power rule to bring the exponent to the front.

> **Check Point 3**   Use the power rule to expand
>
>   **a.** $\log_6 8^9$ **b.** $\ln \sqrt[3]{x}$

Expanding Logarithmic Expressions

It is sometimes necessary to use more than one property of logarithms when you expand a logarithmic expression. Properties for expanding logarithmic expressions are as follows:

> **Properties for Expanding Logarithmic Expressions**
>
> **1.** $\log_b(MN) = \log_b M + \log_b N$ Product rule
>
> **2.** $\log_b\left(\dfrac{M}{N}\right) = \log_b M - \log_b N$ Quotient rule
>
> **3.** $\log_b M^p = p \log_b M$ Power rule
>
> In all cases, $M > 0$ and $N > 0$.

EXAMPLE 4   Expanding Logarithmic Expressions

Use logarithmic properties to expand each expression as much as possible.

    a. $\log_b x^2 \sqrt{y}$ **b.** $\log_6\left(\dfrac{\sqrt[3]{x}}{36 y^4}\right)$

Solution   We will have to use two or more of the properties for expanding logarithms in each part of this example.

    a. $\log_b x^2 \sqrt{y} = \log_b x^2 y^{1/2}$ Use exponential notation.

               $= \log_b x^2 + \log_b y^{1/2}$ Use the product rule.

               $= 2 \log_b x + \dfrac{1}{2} \log_b y$ Use the power rule.

Study Tip

The graphs show

$$y_1 = \ln(x + 3)$$

and $y_2 = \ln x + \ln 3$. The graphs are not the same. The graph of y_1 is the graph of the natural logarithmic function shifted 3 units to the left. By contrast, the graph of y_2 is the graph of the natural logarithmic function shifted upward by $\ln 3$, or about 1.1 units. Thus we see that

$$\ln(x + 3) \neq \ln x + \ln 3.$$

In general,

$$\log_b(M + N) \neq \log_b M + \log_b N.$$

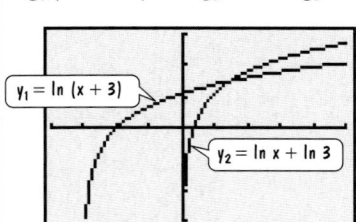

Try to avoid the following errors.

INCORRECT

$$\log_b(M + N) = \log_b M + \log_b N$$

$$\log_b(M - N) = \log_b M - \log_b N$$

$$\log_b(M \cdot N) = \log_b M \cdot \log_b N$$

$$\log_b\left(\frac{M}{N}\right) = \frac{\log_b M}{\log_b N}$$

$$\frac{\log_b M}{\log_b N} = \log_b M - \log_b N$$

b. $\log_6\left(\dfrac{\sqrt[3]{x}}{36y^4}\right) = \log_6 \dfrac{x^{1/3}}{36y^4}$ Use exponential notation.

$\qquad\qquad\qquad = \log_6 x^{1/3} - \log_6 36y^4$ Use the quotient rule.

$\qquad\qquad\qquad = \log_6 x^{1/3} - \left(\log_6 36 + \log_6 y^4\right)$ Use the product rule on $\log_6 36y^4$.

$\qquad\qquad\qquad = \dfrac{1}{3}\log_6 x - \left(\log_6 36 + 4\log_6 y\right)$ Use the power rule.

$\qquad\qquad\qquad = \dfrac{1}{3}\log_6 x - \log_6 36 - 4\log_6 y$ Apply the distributive property.

$\qquad\qquad\qquad = \dfrac{1}{3}\log_6 x - 2 - 4\log_6 y$ $\log_6 36 = 2$ because 2 is the power to which we must raise 6 to get 36. $\left(6^2 = 36\right)$

Check Point 4   Use logarithmic properties to expand each expression as much as possible.

     a. $\log_b x^4\sqrt[3]{y}$ **b.** $\log_5 \dfrac{\sqrt{x}}{25y^3}$

5   Condense logarithmic expressions.

Condensing Logarithmic Expressions

To **condense a logarithmic expression**, we write the sum or difference of two or more logarithmic expressions as a single logarithmic expression. We use the properties of logarithms to do so.

Properties for Condensing Logarithmic Expressions

1. $\log_b M + \log_b N = \log_b(MN)$ Product rule

2. $\log_b M - \log_b N = \log_b\left(\dfrac{M}{N}\right)$ Quotient rule

3. $p\log_b M = \log_b M^p$ Power rule

In all cases, $M > 0$ and $N > 0$.

EXAMPLE 5   Condensing Logarithmic Expressions

Write as a single logarithm:

     a. $\log_4 2 + \log_4 32$ **b.** $\log(4x - 3) - \log x$

Solution

     a. $\log_4 2 + \log_4 32 = \log_4(2 \cdot 32)$ Use the product rule.

$\qquad\qquad\qquad\qquad = \log_4 64$ We now have a single logarithm. However, we can simplify.

$\qquad\qquad\qquad\qquad = 3$ $\log_4 64 = 3$ because $4^3 = 64$.

     b. $\log(4x - 3) - \log x = \log \dfrac{4x - 3}{x}$ Use the quotient rule.

Check Point 5

Write as a single logarithm:

a. $\log 25 + \log 4$ **b.** $\log(7x + 6) - \log x$

Coefficients of logarithms must be 1 before you can condense them using the product and quotient rules. For example, to condense

$$2 \ln x + \ln(x + 1),$$

the coefficient of the first term must be 1. We use the power rule to rewrite the coefficient as an exponent:

> 1. Make the number in front an exponent.

$$2 \ln x + \ln(x + 1) = \ln x^2 + \ln(x + 1) = \ln x^2(x + 1)$$

> 2. Use the product rule. The sum of logarithms with coefficients 1 is the logarithm of the product.

EXAMPLE 6   Condensing Logarithmic Expressions

a. $\frac{1}{2} \log x + 4 \log(x - 1)$ **b.** $3 \ln(x + 7) - \ln x$

Solution

a. $\frac{1}{2} \log x + 4 \log(x - 1)$

$= \log x^{1/2} + \log(x - 1)^4$ Use the power rule so that all coefficients are 1.

$= \log x^{1/2}(x - 1)^4$ Use the product rule.

b. $3 \ln(x + 7) - \ln x$

$= \ln(x + 7)^3 - \ln x$ Use the power rule so that all coefficients are 1.

$= \ln \dfrac{(x + 7)^3}{x}$ Use the quotient rule.

Check Point 6

Write as a single logarithm:

a. $2 \ln x + \frac{1}{3} \ln(x + 5)$ **b.** $2 \log(x - 3) - \log x$

6 Use the change-of-base property.

The Change-of-Base Property

We have seen that calculators give the values of both common logarithms (base 10) and natural logarithms (base e). To find a logarithm with any other base, we can use the following change-of-base property.

The Change-of-Base Property

For any logarithmic bases a and b, and any positive number M,

$$\log_b M = \frac{\log_a M}{\log_a b}.$$

The logarithm of M with base b is equal to the logarithm of M with any new base divided by the logarithm of b with that new base.

In the change-of-base property, base b is the base of the original logarithm. Base a is a new base that we introduce. Thus, the change-of-base property allows us to change from base b to *any* new base a, as long as the newly introduced base is a positive number not equal to 1.

The change-of-base property is used to write a logarithm in terms of quantities that can be evaluated with a calculator. Because calculators contain keys for common (base 10) and natural (base e) logarithms, we will frequently introduce base 10 or base e.

Change-of-Base Property

$$\log_b M = \frac{\log_a M}{\log_a b}$$

a is the new introduced base.

Introducing Common Logarithms

$$\log_b M = \frac{\log_{10} M}{\log_{10} b}$$

10 is the new introduced base.

Introducing Natural Logarithms

$$\log_b M = \frac{\log_e M}{\log_e b}$$

e is the new introduced base.

Using the notations for common logarithms and natural logarithms, we have the following results.

Discovery

Find a reasonable estimate of $\log_5 140$ to the nearest whole number. 5 to what power is 140? Compare your estimate to the value obtained in Example 7.

The Change-of-Base Property: Introducing Common and Natural Logarithms

Introducing Common Logarithms

$$\log_b M = \frac{\log M}{\log b}$$

Introducing Natural Logarithms

$$\log_b M = \frac{\ln M}{\ln b}$$

EXAMPLE 7   Changing Base to Common Logarithms

Use common logarithms to evaluate $\log_5 140$.

Solution   Because $\log_b M = \dfrac{\log M}{\log b}$,

$$\log_5 140 = \frac{\log 140}{\log 5}$$

$$\approx 3.07.$$ Use a calculator: [LOG] 140 [÷] [LOG] 5 [ENTER]

This means that $\log_5 140 \approx 3.07$.

> **Check Point 7**   Use common logarithms to evaluate $\log_7 2506$.

EXAMPLE 8   Changing Base to Natural Logarithms

Use natural logarithms to evaluate $\log_5 140$.

Solution   Because $\log_b M = \dfrac{\ln M}{\ln b}$,

$$\log_5 140 = \frac{\ln 140}{\ln 5}$$

$$\approx 3.07.$$ Use a calculator: [LN] 140 [÷] [LN] 5 [ENTER]

We have again shown that $\log_5 140 \approx 3.07$.

Writing in Mathematics

63. Describe the product rule for logarithms and give an example.

76. $\ln(x - y) = \ln x - \ln y$

77. $\ln(xy) = (\ln x)(\ln y)$

78. $\dfrac{\ln x}{\ln y} = \ln x - \ln y$

Check Point 8
Use natural logarithms to evaluate $\log_7 2506$.

384 • Chapter 4 • Exponential and Logarithmic Functions

Thus, the solution of the equation is $\dfrac{\ln 6}{0.6} \approx 2.99$. Try checking this approximate solution in the original equation, verifying that $\left\{\dfrac{\ln 6}{0.6}\right\}$ is the solution set.

Check Point 2
Solve: $7e^{2x} = 63$. Find the solution set and then use a calculator to obtain a decimal approximation to two decimal places for the solution.

EXAMPLE 3 Solving an Exponential Equation

Solve: $5^{4x-7} - 3 = 10$

Solution   We begin by adding 3 to both sides to isolate the exponential expression, 5^{4x-7}. Then we take the natural logarithm on both sides of the equation.

$5^{4x-7} - 3 = 10$	This is the given equation.
$5^{4x-7} = 13$	Add 3 to both sides.
$\ln 5^{4x-7} = \ln 13$	Take the natural logarithm on both sides.
$(4x - 7)\ln 5 = \ln 13$	Use the power rule to bring the exponent to the front: $\ln b^x = x \ln b$.
$4x \ln 5 - 7 \ln 5 = \ln 13$	Use the distributive property and distribute $\ln 5$ to both terms in parentheses.
$4x \ln 5 = \ln 13 + 7 \ln 5$	Isolate the variable term by adding $7 \ln 5$ to both sides.
$x = \dfrac{\ln 13 + 7 \ln 5}{4 \ln 5}$	Isolate x by dividing both sides by $4 \ln 5$.

The solution set is $\left\{\dfrac{\ln 13 + 7 \ln 5}{4 \ln 5}\right\}$, approximately 2.15.

Check Point 3
Solve: $6^{3x-4} - 7 = 2081$. Find the solution set and then use a calculator to obtain a decimal approximation to two decimal places for the solution.

Technology

Shown below is the graph of $y = e^{2x} - 4e^x + 3$. There are two x-intercepts, one at 0 and one at approximately 1.099. These intercepts verify our algebraic solution.

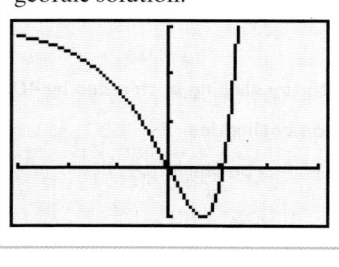

EXAMPLE 4 Solving an Exponential Equation

Solve: $e^{2x} - 4e^x + 3 = 0$.

Solution   The given equation is quadratic in form. If $t = e^x$, the equation can be expressed as $t^2 - 4t + 3 = 0$. Because this equation can be solved by factoring, we factor to isolate the exponential term.

$e^{2x} - 4e^x + 3 = 0$	This is the given equation.
$(e^x - 3)(e^x - 1) = 0$	Factor on the left. Notice that if $t = e^x$, $t^2 - 4t + 3 = (t - 3)(t - 1)$.
$e^x - 3 = 0$ or $e^x - 1 = 0$	Set each factor equal to 0.
$e^x = 3$ $e^x = 1$	Solve for e^x.

$$\ln e^x = \ln 3 \qquad\qquad x = 0$$

Take the natural logarithm on both sides of the first equation. The equation on the right can be solved by inspection.

$$x = \ln 3$$

$\ln e^x = x$

The solution set is $\{0, \ln 3\}$. The solutions are 0 and (approximately) 1.099.

Check Point 4 Solve: $e^{2x} - 8e^x + 7 = 0$. Find the solution set and then use a calculator to obtain a decimal approximation to two decimal places, if necessary.

Logarithmic Equations

A **logarithmic equation** is an equation containing a variable in a logarithmic expression. Examples of logarithmic equations include

$$\log_4(x + 3) = 2 \quad \text{and} \quad \ln 2x = 3.$$

If a logarithmic equation is in the form $\log_b x = c$, we can solve the equation by rewriting it in its equivalent exponential form $b^c = x$. Example 5 illustrates how this is done.

EXAMPLE 5   Solving a Logarithmic Equation

Solve: $\log_4(x + 3) = 2$.

Solution   We first rewrite the equation as an equivalent equation in exponential form using the fact that $\log_b x = c$ means $b^c = x$.

$$\log_4(x + 3) = 2 \qquad \text{means} \qquad 4^2 = x + 3$$

Logarithms are exponents.

Now we solve the equivalent equation for x.

$$4^2 = x + 3 \qquad \text{This is the equivalent equation.}$$
$$16 = x + 3 \qquad \text{Square 4.}$$
$$13 = x \qquad \text{Subtract 3 from both sides.}$$

Check

$$\log_4(x + 3) = 2 \qquad \text{This is the given logarithmic equation.}$$
$$\log_4(13 + 3) \overset{?}{=} 2 \qquad \text{Substitute 13 for } x.$$
$$\log_4 16 \overset{?}{=} 2$$
$$2 = 2 \checkmark \qquad \log_4 16 = 2 \text{ because } 4^2 = 16.$$

This true statement indicates that the solution set is $\{13\}$.

Check Point 5 Solve: $\log_2(x - 4) = 3$.

Logarithmic expressions are defined only for logarithms of positive real numbers. Always check proposed solutions of a logarithmic equation in the original equation. Exclude from the solution set any proposed solution that produces the logarithm of a negative number or the logarithm of 0.

---

2 Solve logarithmic equations.

Technology

The graphs of

$$y_1 = \log_4(x + 3) \text{ and } y_2 = 2$$

have an intersection point whose x-coordinate is 13. This verifies that $\{13\}$ is the solution set for $\log_4(x + 3) = 2$.
Note:
Because

$$\log_b x = \frac{\ln x}{\ln b}$$

(change-of-base property),

we entered y_1 using

$$y_1 = \log_4(x + 3)$$
$$= \frac{\ln(x + 3)}{\ln 4}$$

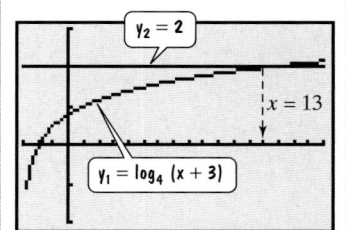

In order to rewrite the logarithmic equation $\log_b x = c$ in the equivalent exponential form $b^c = x$, we need a single logarithm whose coefficient is one. It is sometimes necessary to use properties of logarithms to condense logarithms into a single logarithm. In the next example we use the product rule for logarithms to obtain a single logarithmic expression on the left side.

EXAMPLE 6   Using the Product Rule to Solve a Logarithmic Equation

Solve: $\log_2 x + \log_2 (x - 7) = 3$.

Solution

$\log_2 x + \log_2 (x - 7) = 3$	This is the given equation.
$\log_2 x(x - 7) = 3$	Use the product rule to obtain a single logarithm: $\log_b M + \log_b N = \log_b (MN)$.
$2^3 = x(x - 7)$	$\log_b x = c$ means $b^c = x$.
$8 = x^2 - 7x$	Apply the distributive property on the right.
$0 = x^2 - 7x - 8$	Set the equation equal to 0.
$0 = (x - 8)(x + 1)$	Factor.
$x - 8 = 0$ or $x + 1 = 0$	Set each factor equal to 0.
$x = 8$ $x = -1$	Solve for x.

Check

Checking 8:

$$\log_2 x + \log_2 (x - 7) = 3$$
$$\log_2 8 + \log_2 (8 - 7) \overset{?}{=} 3$$
$$\log_2 8 + \log_2 1 \overset{?}{=} 3$$
$$3 + 0 \overset{?}{=} 3$$
$$3 = 3 \checkmark$$

Checking -1:

$$\log_2 x + \log_2 (x - 7) = 3$$
$$\log_2 (-1) + \log_2 (-1 - 7) \overset{?}{=} 3$$

The number -1 does not check. Negative numbers do not have logarithms.

The solution set is $\{8\}$.

Check Point 6   Solve: $\log x + \log (x - 3) = 1$.

Equations involving natural logarithms can be solved using the inverse property $e^{\ln x} = x$. For example, to solve

$$\ln x = 5$$

we write both sides of the equation as exponents on base e:

$$e^{\ln x} = e^5$$

This is called **exponentiating both sides** of the equation. Using the inverse property $e^{\ln x} = x$, we simplify the left side of the equation and obtain the solution:

$$x = e^5.$$

EXAMPLE 7 Solving an Equation with a Natural Logarithm

Solve: $3 \ln 2x = 12$.

Solution

$3 \ln 2x = 12$	This is the given equation.
$\ln 2x = 4$	Divide both sides by 3.
$e^{\ln 2x} = e^4$	Exponentiate both sides.
$2x = e^4$	Use the inverse property to simplify the left side: $e^{\ln \square} = \square$.
$x = \dfrac{e^4}{2} \approx 27.30$	Divide both sides by 2.

Check

$3 \ln 2x = 12$	This is the given logarithmic equation.
$3 \ln 2\left(\dfrac{e^4}{2}\right) \overset{?}{=} 12$	Substitute $\dfrac{e^4}{2}$ for x.
$3 \ln e^4 \overset{?}{=} 12$	Simplify: $\dfrac{2}{1} \cdot \dfrac{e^4}{2} = e^4$.
$3 \cdot 4 \overset{?}{=} 12$	Because $\ln e^x = x$, we conclude $\ln e^4 = 4$.
$12 = 12$ ✓	

This true statement indicates that the solution set is $\left\{\dfrac{e^4}{2}\right\}$.

> **Check Point 7**
>
> Solve: $4 \ln 3x = 8$.

---

3 Solve applied problems involving exponential and logarithmic equations.

Applications

Our first applied example provides a mathematical perspective on the old slogan "Alcohol and driving don't mix." In California, where 38% of fatal traffic crashes involve drinking drivers, it is illegal to drive with a blood alcohol concentration of 0.08 or higher. At these levels, drivers may be arrested and charged with driving under the influence.

EXAMPLE 8 Alcohol and Risk of a Car Accident

Medical research indicates that the risk of having a car accident increases exponentially as the concentration of alcohol in the blood increases. The risk is modeled by

$$R = 6e^{12.77x}$$

where x is the blood alcohol concentration and R, given as a percent, is the risk of having a car accident. What blood alcohol concentration corresponds to a 17% risk of a car accident?

Solution For a risk of 17%, we let $R = 17$ in the equation and solve for x, the blood alcohol concentration.

$$R = 6e^{12.77x}$$ This is the given equation.

$$6e^{12.77x} = 17$$ Substitute 17 for R and (optional) reverse the two sides of the equation.

$$e^{12.77x} = \frac{17}{6}$$ Isolate the exponential factor by dividing both sides by 6.

$$\ln e^{12.77x} = \ln\left(\frac{17}{6}\right)$$ Take the natural logarithm on both sides.

$$12.77x = \ln\left(\frac{17}{6}\right)$$ Use the inverse property $\ln e^x = x$ on the left.

$$x = \frac{\ln\left(\frac{17}{6}\right)}{12.77} \approx 0.08$$ Divide both sides by 12.77.

For a blood alcohol concentration of 0.08, the risk of a car accident is 17%. In many states, it is illegal to drive at this blood alcohol concentration.

Check Point 8 Use the formula in Example 8 to solve this problem. What blood alcohol concentration corresponds to a 7% risk of a car accident? (In many states, drivers under the age of 21 can lose their license for driving at this level.)

Playing Doubles: Interest Rates and Doubling Time

One way to calculate what your savings will be worth at some point in the future is to consider doubling time. Shown below is how long it takes for your money to double at different annual interest rates subject to continuous compounding.

Annual Interest Rate	Years to Double
5%	13.9 years
7%	9.9 years
9%	7.7 years
11%	6.3 years

Of course, the first problem is collecting some money to invest. The second problem is finding a reasonably safe investment with a return of 9% or more.

Suppose that you inherit $30,000. Is it possible to invest $25,000 and have over half a million dollars for early retirement? Our next example illustrates the power of compound interest.

EXAMPLE 9 Revisiting the Formula for Compound Interest

The formula

$$A = P\left(1 + \frac{r}{n}\right)^{nt}$$

describes the accumulated value A of a sum of money P, the principal, after t years at annual percentage rate r (in decimal form) compounded n times a year. How long will it take $25,000 to grow to $500,000 at 9% annual interest compounded monthly?

Solution

$$A = P\left(1 + \frac{r}{n}\right)^{nt}$$ This is the given formula.

$$500,000 = 25,000\left(1 + \frac{0.09}{12}\right)^{12t}$$ A (the desired accumulated value) = $500,000, P (the principal) = $25,000, r (the interest rate) = 9% = 0.09, and n = 12 (monthly compounding).

Our goal is to solve the equation for t. Let's reverse the two sides of the equation and then simplify within parentheses.

$$25{,}000\left(1 + \frac{0.09}{12}\right)^{12t} = 500{,}000$$

$25{,}000(1 + 0.0075)^{12t} = 500{,}000$ Divide within parentheses: $\frac{0.09}{12} = 0.0075$.

$25{,}000(1.0075)^{12t} = 500{,}000$ Add within parentheses.

$(1.0075)^{12t} = 20$ Divide both sides by 25,000.

$\ln(1.0075)^{12t} = \ln 20$ Take the natural logarithm on both sides.

$12t \ln(1.0075) = \ln 20$ Use the power rule to bring the exponent to the front: $\ln b^x = x \ln b$.

$$t = \frac{\ln 20}{12 \ln 1.0075}$$ Solve for t, dividing both sides by $12 \ln 1.0075$.

≈ 33.4 Use a calculator.

After approximately 33.4 years, the $25,000 will grow to an accumulated value of $500,000. If you set aside the money at age 20, you can begin enjoying a life of leisure at about age 53.

Check Point 9   How long, to the nearest tenth of a year, will it take $1000 to grow to $3600 at 8% annual interest compounded quarterly?

Yogi Berra, catcher and renowned hitter for the New York Yankees (1946–1963), said it best: "Prediction is very hard, especially when it's about the future." At the start of the twenty-first century, we are plagued by questions about the environment. Will we run out of gas? How hot will it get? Will there be neighborhoods where the air is pristine? Can we make garbage disappear? Will there be any wilderness left? Which wild animals will become extinct? These concerns have led to the growth of the environmental industry in the United States.

EXAMPLE 10   The Growth of the Environmental Industry

The formula

$$N = 461.87 + 299.4 \ln x$$

models the thousands of workers, N, in the environmental industry in the United States x years after 1979. By which year will there be 1,500,000, or 1500 thousand, U.S. workers in the environmental industry?

Solution   We substitute 1500 for N and solve for x, the number of years after 1979.

$N = 461.87 + 299.4 \ln x$ This is the given formula.

$461.87 + 299.4 \ln x = 1500$ Substitute 1500 for N and reverse the two sides of the equation.

Our goal is to isolate $\ln x$. We can then find x by exponentiating both sides of the equation, using the inverse property $e^{\ln x} = x$.

$$299.4 \ln x = 1038.13 \qquad \text{Subtract 461.87 from both sides.}$$

$$\ln x = \frac{1038.13}{299.4} \qquad \text{Divide both sides by 299.4.}$$

$$e^{\ln x} = e^{1038.13/299.4} \qquad \text{Exponentiate both sides.}$$

$$x = e^{1038.13/299.4} \qquad e^{\ln x} = x$$

$$\approx 32 \qquad \text{Use a calculator.}$$

Approximately 32 years after 1979, in the year 2011, there will be 1.5 million U.S. workers in the environmental industry.

Check Point 10 Use the formula in Example 10 to find by what year there will be two million, or 2000 thousand, U.S. workers in the environmental industry.

EXERCISE SET 4.4

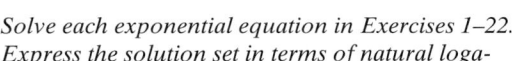

Practice Exercises

Solve each exponential equation in Exercises 1–22. Express the solution set in terms of natural logarithms. Then use a calculator to obtain a decimal approximation, correct to two decimal places, for the solution.

1. $10^x = 3.91$ **2.** $10^x = 8.07$

3. $e^x = 5.7$ **4.** $e^x = 0.83$

5. $5^x = 17$ **6.** $19^x = 143$

7. $5e^x = 23$ **8.** $9e^x = 107$

9. $3e^{5x} = 1977$ **10.** $4e^{7x} = 10{,}273$

11. $e^{1-5x} = 793$ **12.** $e^{1-8x} = 7957$

13. $e^{5x-3} - 2 = 10{,}476$ **14.** $e^{4x-5} - 7 = 11{,}243$

15. $7^{x+2} = 410$ **16.** $5^{x-3} = 137$

17. $7^{0.3x} = 813$ **18.** $3^{x/7} = 0.2$

19. $e^{2x} - 3e^x + 2 = 0$ **20.** $e^{2x} - 2e^x - 3 = 0$

21. $e^{4x} + 5e^{2x} - 24 = 0$ **22.** $e^{4x} - 3e^{2x} - 18 = 0$

Solve each logarithmic equation in Exercises 23–36. Be sure to reject any value of x that produces the logarithm of a negative number or the logarithm of 0.

23. $\log_3 x = 4$ **24.** $\log_5 x = 3$

25. $\log_4(x + 5) = 3$ **26.** $\log_5(x - 7) = 2$

27. $\log_3(x - 4) = -3$ **28.** $\log_7(x + 2) = -2$

29. $\log_4(3x + 2) = 3$ **30.** $\log_2(4x + 1) = 5$

31. $\log_5 x + \log_5(4x - 1) = 1$

32. $\log_6(x + 5) + \log_6 x = 2$

33. $\log_3(x - 5) + \log_3(x + 3) = 2$

34. $\log_2(x - 1) + \log_2(x + 1) = 3$

35. $\log_2(x + 2) - \log_2(x - 5) = 3$

36. $\log_4(x + 2) - \log_4(x - 1) = 1$

Exercises 37–44 involve equations with natural logarithms. Solve each equation by isolating the natural logarithm and exponentiating both sides. Express the answer in terms of e. Then use a calculator to obtain a decimal approximation, correct to two decimal places, for the solution.

37. $\ln x = 2$ **38.** $\ln x = 3$

39. $5 \ln 2x = 20$ **40.** $6 \ln 2x = 30$

41. $6 + 2 \ln x = 5$ **42.** $7 + 3 \ln x = 6$

43. $\ln \sqrt{x + 3} = 1$ **44.** $\ln \sqrt{x + 4} = 1$

Application Exercises

Use the formula $R = 6e^{12.77x}$, where x is the blood alcohol concentration and R, given as a percent, is the risk of having a car accident, to solve Exercises 45–46.

45. What blood alcohol concentration corresponds to certainty, or a 100% risk, of a car accident?

46. What blood alcohol concentration corresponds to a 50% risk of a car accident?

47. The formula $A = 18.2e^{0.001t}$ models the population of New York State, in millions, t years after 1994.
 a. What was the population of New York in 1994?
 b. When will the population of New York reach 18.5 million?

48. The formula $A = 14e^{0.168t}$ models the population of Florida, in millions, t years after 1994.
 a. What was the population of Florida in 1994?
 b. When will the population of Florida reach 18.5 million?

In Exercices 49–52, complete the table for a savings account subjected to n compoundings yearly $\left(A = P\left(1 + \dfrac{r}{n}\right)^{nt}\right)$.

	Amount Invested	Number of Compounding Periods	Annual Interest Rate	Accumulated Amount	Time t in Years
49.	$12,500	4	5.75%	$20,000	
50.	$7250	12	6.5%	$15,000	
51.	$1000	360		$1400	2
52.	$5000	360		$9000	4

In Exercices 53–56, complete the table for a savings account subjected to continuous compounding $\left(A = Pe^{rt}\right)$.

	Amount Invested	Annual Interest Rate	Accumulated Amount	Time t in Years
53.	$8000	8%	Double the amount invested	
54.	$8000		$12,000	2
55.	$2350		Triple the amount invested	7
56.	$17,425	4.25%	$25,000	

57. The formula $C = 15{,}557 + 5259 \ln x$ models the average cost of a new car x years after 1989. When will the average cost of a new car be $25,000?

58. The formula $C = 280 \ln(A + 1) + 1925$ models the number of calories, C, consumed each day by a person who owns A acres of land in a developing country, where $0 \le A \le 4$. How many acres of land are owned by a person who consumes 2200 calories daily in a developing country? (Source: Grigg, D. *The World Food Problem.* Oxford: Blackwell Publishers, 1993.)

The formula $P = 95 - 30 \, log_2 \, x$ models the percentage, P, of students who could recall the important features of a classroom lecture as a function of time, where x represents the number of days that have elapsed since the lecture was given. The figure shows the graph of the formula. Use this information to solve Exercises 59–60.

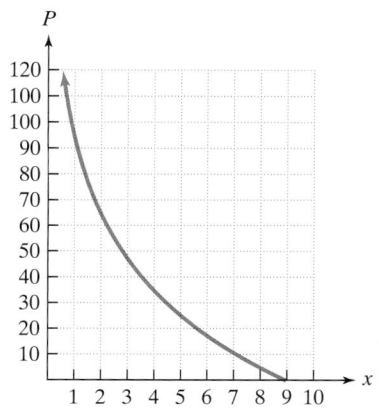

59. After how many days do only half the students recall the important features of the classroom lecture? (Let $P = 50$ and solve for x.) Can you approximately locate the point on the graph that conveys this information?

60. After how many days have all students forgotten the important features of the classroom lecture? (Let $P = 0$ and solve for x.) Can you approximately locate the point on the graph that conveys this information?

The pH of a solution ranges from 0 to 14. An acid solution has a pH less than 7. Pure water is neutral and has a pH of 7. Normal, unpolluted rain has a pH of about 5.6. The pH of a solution is given by

$$\text{pH} = -\log x$$

where x represents the concentration of the hydrogen ions in the solution in moles per liter. Use the formula to solve Exercises 61–62.

61. An environmental concern involves the destructive effects of acid rain. The most acidic rainfall ever had a pH of 2.4. What was the hydrogen ion concentration? Express the answer as a power of 10, and then round to the nearest thousandth.

62. The figure on page 394 shows very acidic rain in the northeast United States. What is the hydrogen ion concentration of rainfall with a pH of 4.2? Express the answer as a power of 10, and then round to the nearest hundred-thousandth.

Acid Rain Over Canada and the United States

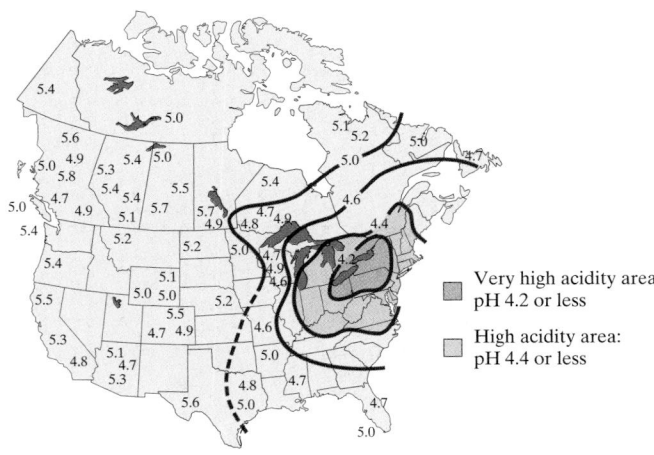

Very high acidity area:
pH 4.2 or less

High acidity area:
pH 4.4 or less

Source: National Atmospheric Program

Writing in Mathematics

63. Explain how to solve an exponential equation. Use $3^x = 140$ in your explanation.

64. Explain how to solve a logarithmic equation. Use $\log_3(x - 1) = 4$ in your explanation.

65. In many states, a 17% risk of a car accident with a blood alcohol concentration of 0.08 is the lowest level for charging a motorist with driving under the influence. Do you agree with the 17% risk as a cutoff percentage, or do you feel that the percentage should be lower or higher? Explain your answer. What blood alcohol concentration corresponds to what you believe is an appropriate percentage?

66. Have you purchased a new or used car recently? If so, describe if the formula in Exercise 58 accurately models what you paid for your car. If there is a big difference between the figure given by the formula and the amount that you paid, how can you explain this difference?

Technology Exercises

In Exercises 67–74, use your graphing utility to graph each side of the equation in the same viewing rectangle. Then use the x-coordinate of the intersection point to find the equation's solution set. Verify this value by direct substitution into the equation.

67. $2^{x+1} = 8$

68. $3^{x+1} = 9$

69. $\log_3(4x - 7) = 2$

70. $\log_3(3x - 2) = 2$

71. $\log(x + 3) + \log x = 1$

72. $\log(x - 15) + \log x = 2$

73. $3^x = 2x + 3$

74. $5^x = 3x + 4$

Hurricanes are one of nature's most destructive forces. These low-pressure areas often have diameters of over 500 miles. The function $f(x) = 0.48 \ln(x + 1) + 27$ models the barometric air pressure, $f(x)$, in inches of mercury, at a distance
of x miles from the eye of a hurricane. Use this function to solve Exercises 75–76.

75. Graph the function in a $[0, 500, 50]$ by $[27, 30, 1]$ viewing rectangle. What does the shape of the graph indicate about barometric air pressure as the distance from the eye increases?

76. Use an equation to answer this question: How far from the eye of a hurricane is the barometric air pressure 29 inches of mercury? Use the [TRACE] and [ZOOM] features or the intersect command of your graphing utility to verify your answer.

77. The formula $P = 145e^{-0.092t}$ models a runner's pulse, P, in beats per minute, t minutes after a race, where $0 \le t \le 15$. Graph the formula using a graphing utility. [TRACE] along the graph and determine after how many minutes the runner's pulse will be 70 beats per minute. Verify your observation algebraically.

78. The formula $W = 2600(1 - 0.51e^{-0.075t})^3$ models the weight, W, in kilograms, of a female African elephant at age t years. (1 kilogram $\approx$ 2.2 pounds) Use a graphing utility to graph the formula. Then [TRACE] along the curve to estimate the age of an adult female elephant weighing 1800 kilograms.

Critical Thinking Exercises

79. Which one of the following is true?
    a. If $\log(x + 3) = 2$, then $e^2 = x + 3$.
    b. If $\log(7x + 3) - \log(2x + 5) = 4$, then in exponential form $10^4 = (7x + 3) - (2x + 5)$.
    c. If $x = \dfrac{1}{k} \ln y$, then $y = e^{kx}$.
    d. Examples of exponential equations include $10^x = 5.71$, $e^x = 0.72$, and $x^{10} = 5.71$.

80. If $4000 is deposited into an account paying 3% interest compounded annually and at the same time $2000 is deposited into an account paying 5% interest compounded annually, after how long will the two accounts have the same balance?

Solve each equation in Exercises 81–83. Check each proposed solution by direct substitution or with a graphing utility.

81. $(\ln x)^2 = \ln x^2$

82. $(\log x)(2 \log x + 1) = 6$

83. $\ln(\ln x) = 0$

Group Exercise

84. Research applications of logarithmic functions as mathematical models and plan a seminar based on your group's research. Each group member should research one of the following areas or any other area of interest: pH (acidity of solutions), intensity of sound (decibels), brightness of stars, consumption of natural resources, human memory, progress over time in a sport, profit over time. For the area that you select, explain how logarithmic functions are used and provide examples.

SECTION 4.5 *Modeling with Exponential and Logarithmic Functions*

Objectives

1. Model exponential growth and decay.
2. Use logistic growth models.
3. Model data with exponential and logarithmic functions.
4. Express an exponential model in base e.

The most casual cruise on the Internet shows how people disagree when it comes to making predictions about the effects of the world's growing population. Some argue that there is a recent slowdown in the growth rate, economies remain robust, and famines in Biafra and Ethiopia are aberrations rather than signs of the future. Others say that the 6 billion people on Earth is twice as many as can be supported in middle-class comfort, and the world is running out of arable land and fresh water. Debates about entities that are growing exponentially can be approached mathematically: We can create functions that model data and use these functions to make predictions. In this section we will show you how this is done.

1 Model exponential growth and decay.

Exponential Growth and Decay

One of algebra's many applications is to predict the behavior of variables. This can be done with **exponential growth** and **decay models**. With exponential growth and decay, quantities grow or decay at a rate directly proportional to their size. Populations that are growing exponentially grow extremely rapidly as they get larger because there are more adults to have offspring. For example, the **growth rate** for world population is 1.3%, or 0.013. This means that each year world population is 1.3% more than what it was in the previous year. In 1999, world population was 6 billion. Thus, we compute the world population in 2000 as follows:

$$6 \text{ billion} + 1.3\% \text{ of } 6 \text{ billion} = 6 + (0.013)(6) = 6.078.$$

This computation suggests that 6.078 billion people will populate the world in 2000. The 0.078 billion represents an increase of 78 million people from 1999 to 2000, the equivalent of the population of Germany. Using 1.3% as the annual growth rate, world population for 2001 is found in a similar manner:

$$6.078 \text{ billion} + 1.3\% \text{ of } 6.078 \text{ billion} = 6.078 + (0.013)(6.078) \approx 6.157.$$

This computation suggests that approximately 6.157 billion people will populate the world in 2001.

The explosive growth of world population may remind you of the growth of money in an account subject to compound interest. Just as the growth rate for world population is multiplied by the the population plus any increase in the population, a compound interest rate is multiplied by your original investment plus any accumulated interest. The balance in an account subject to continuous compounding and world population are special cases of an *exponential growth model.*

Exponential Growth and Decay Models

The mathematical model for **exponential growth** or **decay** is given by

$$f(t) = A_0 e^{kt} \quad \text{or} \quad A = A_0 e^{kt}.$$

- **If $k > 0$, the function models the amount or size of a *growing* entity.** A_0 is the original amount or size of the growing entity at time $t = 0$, A is the amount at time t, and k is a constant representing the growth rate.
- **If $k < 0$, the function models the amount or size of a *decaying* entity.** A_0 is the original amount or size of the decaying entity at time $t = 0$, A is the amount at time t, and k is a constant representing the decay rate.

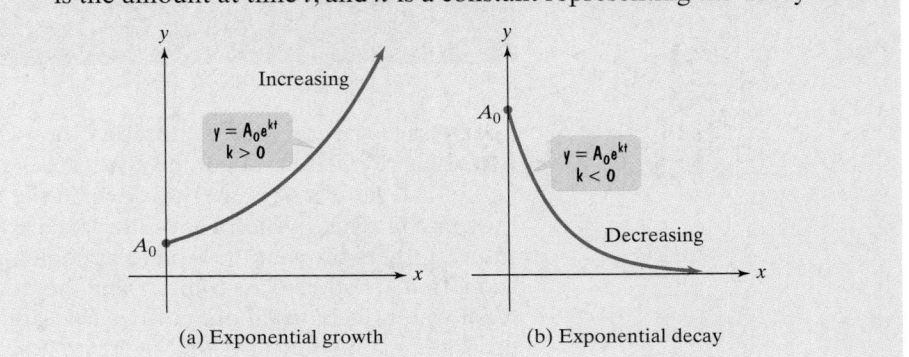

(a) Exponential growth   (b) Exponential decay

Sometimes we need to use given data to determine k, the rate of growth or decay. After we compute the value of k, we can use the formula $A = A_0 e^{kt}$ to make predictions. This idea is illustrated in our first two examples.

EXAMPLE 1 Modeling Mexico City's Growth

The graph in Figure 4.17 shows the growth of the Mexico City metropolitan area from 1970 through 2000. In 1970, the population of Mexico City was 9.4 million. By 1990, it had grown to 20.2 million.

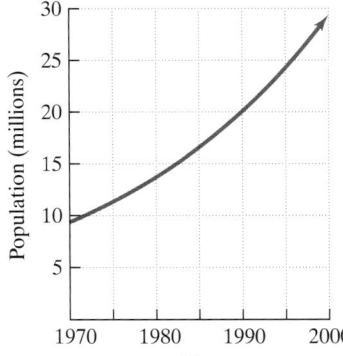

Figure 4.17 Mexico City's population has grown exponentially.

a. Find the exponential growth function that models the data.

b. By what year will the population reach 40 million?

Solution

a. We use the exponential growth model

$$A = A_0 e^{kt}$$

in which t is the number of years since 1970. This means that 1970 corresponds to $t = 0$. At that time there were 9.4 million inhabitants, so we substitute 9.4 for A_0 in the growth model.

$$A = 9.4 e^{kt}$$

We are given that there were 20.2 million inhabitants in 1990. Because 1990 is 20 years after 1970, when $t = 20$ the value of A is 20.2. Substituting these numbers into the growth model will enable us to find k, the growth rate. We know that $k > 0$ because the problem involves growth.

$A = 9.4 e^{kt}$	Use the growth model with $A_0 = 9.4$.
$20.2 = 9.4 e^{k \cdot 20}$	When $t = 20$, $A = 20.2$. Substitute these numbers into the model.
$e^{20k} = \dfrac{20.2}{9.4}$	Isolate the exponential factor by dividing both sides by 9.4. We also reversed the sides.
$\ln e^{20k} = \ln \dfrac{20.2}{9.4}$	Take the natural logarithm on both sides.
$20k = \ln \dfrac{20.2}{9.4}$	Simplify the left side using $\ln e^x = x$.
$k = \dfrac{\ln \dfrac{20.2}{9.4}}{20} \approx 0.038$	Divide both sides by 20 and solve for k.

We substitute 0.038 for k in the growth model to obtain the exponential growth function for Mexico City. It is

$$A = 9.4 e^{0.038t}$$

where t is measured in years since 1970.

b. To find the year in which the population will reach 40 million, we substitute 40 for A in the model from part (a) and solve for t.

$A = 9.4 e^{0.038t}$	This is the model from part (a).
$40 = 9.4 e^{0.038t}$	Substitute 40 for A.
$e^{0.038t} = \dfrac{40}{9.4}$	Divide both sides by 9.4.
$\ln e^{0.038t} = \ln \dfrac{40}{9.4}$	Take the natural logarithm on both sides.
$0.038t = \ln \dfrac{40}{9.4}$	Simplify on the left using $\ln e^x = x$.
$t = \dfrac{\ln \dfrac{40}{9.4}}{0.038} \approx 38$	Solve for t by dividing both sides by 0.038.

Because 38 is the number of years after 1970, the model indicates that the population of Mexico City will reach 40 million by 1970 + 38, or in the year 2008.

Check Point 1

In 1980, the population of Africa was 491 million and by 1990 it had grown to 643 million.

a. Use the exponential growth model $A = A_0 e^{kt}$, in which t is the number of years since 1980, to find the exponential growth function that models the data.

b. By what year will Africa's population reach 1000 million, or one billion?

Carbon Dating and Artistic Development

The artistic community was electrified by the discovery in 1995 of spectacular cave paintings in a limestone cavern in France. Carbon dating of the charcoal from the site showed that the images, created by artists of remarkable talent, were 30,000 years old, making them the oldest cave paintings ever found. The artists seemed to have used the cavern's natural contours to heighten a sense of perspective. The quality of the painting suggests that the art of early humans did not mature steadily from primitive to sophisticated in any simple linear fashion.

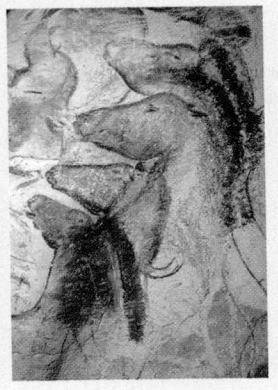

Our next example involves exponential decay and its use in determining the age of fossils and artifacts. The method is based on considering the percentage of carbon-14 remaining in the fossil or artifact. Carbon-14 decays exponentially with a *half-life* of approximately 5715 years. The **half-life** of a substance is the time required for half of a given sample to disintegrate. Thus, after 5715 years a given amount of carbon-14 will have decayed to half the original amount. Carbon dating is useful for artifacts or fossils up to 80,000 years old. Older objects do not have enough carbon-14 left to date age accurately.

EXAMPLE 2 Carbon-14 Dating: The Dead Sea Scrolls

a. Use the fact that after 5715 years a given amount of carbon-14 will have decayed to half the original amount to find the exponential decay model for carbon-14.

b. In 1947, earthenware jars containing what are known as the Dead Sea Scrolls were found by an Arab Bedouin herdsman. Analysis indicated that the scroll wrappings contained 76% of their original carbon-14. Estimate the age of the Dead Sea Scrolls.

Solution We begin with the exponential decay model $A = A_0 e^{kt}$. We know that $k < 0$ because the problem involves the decay of carbon-14. After 5715 years ($t = 5715$), the amount of carbon-14 present, A, is half the original amount A_0. Thus we can substitute $\dfrac{A_0}{2}$ for A in the exponential decay model. This will enable us to find k, the decay rate.

a.

$$\frac{A_0}{2} = A_0 e^{k5715}$$
After 5715 years ($t = 5715$), $A = \dfrac{A_0}{2}$ (because the amount present, A, is half the original amount, A_0).

$$\frac{1}{2} = e^{5715k}$$
Divide both sides of the equation by A_0.

$$\ln \frac{1}{2} = \ln e^{5715k}$$
Take the natural logarithm of both sides.

$$\ln \frac{1}{2} = 5715k$$
$\ln e^x = x$

$$k = \frac{\ln \dfrac{1}{2}}{5715} \approx -0.000121$$
Solve for k.

Substituting for k in the decay model, the model for carbon-14 is $A = A_0 e^{-0.000121t}$.

b.
$$A = A_0 e^{-0.000121t} \qquad \text{This is the decay model for carbon-14.}$$
$$0.76A_0 = A_0 e^{-0.000121t} \qquad \text{A, the amount present, is 76\% of the original amount, so A = 0.76A}_0.$$
$$0.76 = e^{-0.000121t} \qquad \text{Divide both sides of the equation by A}_0.$$
$$\ln 0.76 = \ln e^{-0.000121t} \qquad \text{Take the natural logarithm on both sides.}$$
$$\ln 0.76 = -0.000121t \qquad \ln e^x = x$$
$$t = \frac{\ln 0.76}{-0.000121} \approx 2268 \qquad \text{Solve for t.}$$

The Dead Sea Scrolls are approximately 2268 years old plus the number of years between 1947 and the current year.

Check Point 2 Strontium-90 is a waste product from nuclear reactors. As a consequence of fallout from atmospheric nuclear tests, we all have a measurable amount of strontium-90 in our bones.
  a. Use the fact that after 28 years a given amount of strontium-90 will have decayed to half the original amount to find the exponential decay model for strontium-90.
  b. Suppose that a nuclear accident occurs and releases 60 grams of strontium-90 into the atmosphere. How long will it take for strontium-90 to decay to a level of 10 grams?

2 Use logistic growth models.

Logistic Growth Models

From population growth to the spread of an epidemic, nothing on Earth can grow exponentially indefinitely. Growth is always limited. This is shown in Figure 4.18 by the horizontal asymptote. The **logistic growth model** is an exponential function used to model situations in which growth is limited.

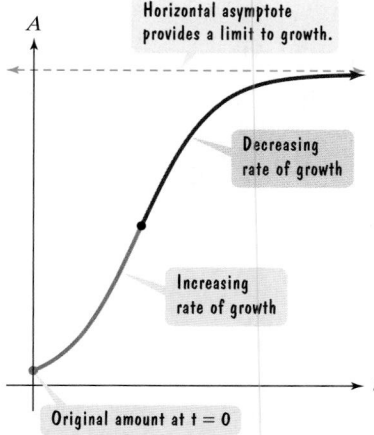

Horizontal asymptote provides a limit to growth.

Decreasing rate of growth

Increasing rate of growth

Original amount at t = 0

Figure 4.18 The logistic growth curve has a horizontal asymptote that limits the growth of A over time.

Logistic Growth Model

The mathematical model for limited logistic growth is given by

$$f(t) = \frac{c}{1 + ae^{-bt}} \quad \text{or} \quad A = \frac{c}{1 + ae^{-bt}}$$

where a, b, and c are constants with $c > 0$ and $b > 0$.

As time increases ($t \rightarrow \infty$), the expression ae^{-bt} in the model approaches 0 and A gets closer and closer to c. This means that $y = c$ is a horizontal asymptote for the graph of the function. Thus, the value of A can never exceed c and c represents the limiting size that A can attain.

EXAMPLE 3 Modeling the Spread of the Flu

The function

$$f(t) = \frac{30{,}000}{1 + 20e^{-1.5t}}$$

describes the number of people, $f(t)$, who have become ill with influenza t weeks after its initial outbreak in a town with 30,000 inhabitants.

a. How many people became ill with the flu when the epidemic began?
b. How many people were ill by the end of the fourth week?
c. What is the limiting size of $f(t)$, the population that becomes ill?

Solution

a. The time at the beginning of the flu epidemic is $t = 0$. Thus, we can find the number of people who were ill at the beginning of the epidemic by substituting 0 for t.

$$f(t) = \frac{30{,}000}{1 + 20e^{-1.5t}} \qquad \text{This is the given logistic growth function.}$$

$$f(0) = \frac{30{,}000}{1 + 20e^{-1.5(0)}} \qquad \text{When the epidemic began, } t = 0.$$

$$= \frac{30{,}000}{1 + 20} \qquad e^{-1.5(0)} = e^{0} = 1$$

$$\approx 1429$$

Approximately 1429 people were ill when the epidemic began.

b. We find the number of people who were ill at the end of the fourth week by substituting 4 for t in the logistic growth function.

$$f(t) = \frac{30{,}000}{1 + 20e^{-1.5t}} \qquad \text{Use the given logistic growth function.}$$

$$f(4) = \frac{30{,}000}{1 + 20e^{-1.5(4)}} \qquad \text{To find the number of people ill by the end of week four, let } t = 4.$$

$$= 28{,}583 \qquad \text{Use a calculator.}$$

Approximately 28,583 people were ill by the end of the fourth week. Compared with the number of people who were ill initially, this illustrates the virulence of the epidemic.

c. Recall that in the logistic growth model, $f(t) = \dfrac{c}{1 + ae^{-bt}}$, the constant c represents the limiting size that $f(t)$ can attain. Thus, the number in the numerator, 30,000, is the limiting size of the population that becomes ill.

Technology

The graph of the logistic growth function for the flu epidemic

$$y = \frac{30{,}000}{1 + 20e^{-1.5x}}$$

can be obtained using a graphing calculator. We started x at 0 and ended at 10. This takes us to week 10. (In Example 3, we found that by week 4 approximately 28,583 people were ill.) We also know that 30,000 is the limiting size, so we took values of y up to 30,000. Using a $[0, 10, 1]$ by $[0, 30{,}000, 3000]$ viewing rectangle, the graph of the logistic growth function is shown below.

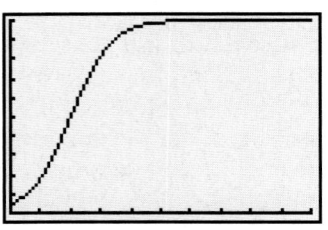

Check Point 3

In a learning theory project, psychologists discovered that

$$f(t) = \frac{0.8}{1 + e^{-0.2t}}$$

is a model for describing the proportion of correct responses after t learning trials.

a. Find the proportion of correct responses prior to learning trials taking place.

b. Find the proportion of correct responses after 10 learning trials.

c. What is the limiting size of $f(t)$, the proportion of correct responses as continued learning trials take place?

3 Model data with exponential and logarithmic functions.

The Art of Modeling

Throughout this chapter, we have been working with models that were given. However, we can create functions that model data by observing patterns in scatter plots. Figure 4.19 shows scatter plots for data that are exponential or logarithmic.

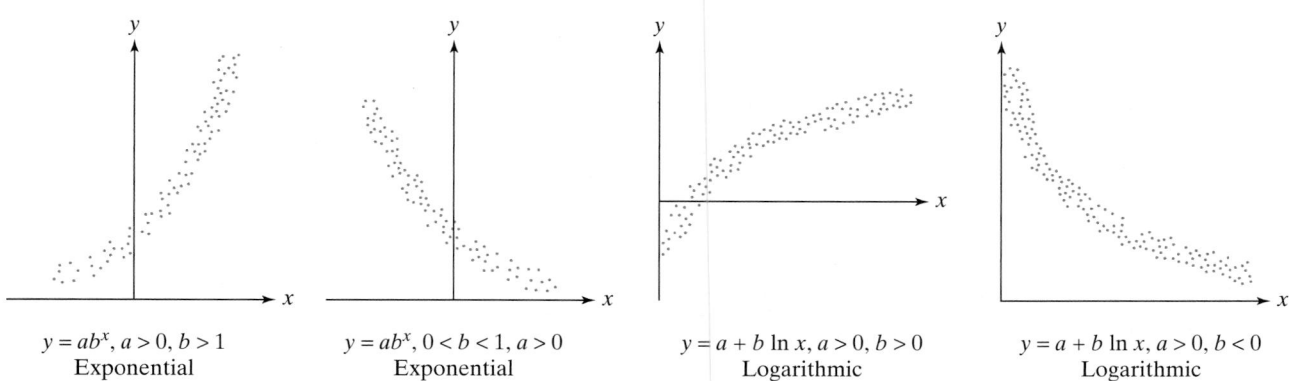

| $y = ab^x, a > 0, b > 1$ | $y = ab^x, 0 < b < 1, a > 0$ | $y = a + b \ln x, a > 0, b > 0$ | $y = a + b \ln x, a > 0, b < 0$ |
| Exponential | Exponential | Logarithmic | Logarithmic |

Figure 4.19 Scatter plots for exponential or logarithmic models

Graphing utilities can be used to find the equation of a function that is derived from data. For example, earlier in the chapter we encountered a function that modeled the size of a city and the average walking speed, in feet per second, of pedestrians. The function was derived from the data in Table 4.4 on page 402. The scatter plot is shown in Figure 4.20 on page 402.

Because the data in this scatter plot increase rapidly at first and then begin to level off a bit, the shape suggests that a logarithmic model might be a good choice. A graphing utility fits the data in Table 4.4 to a logarithmic model of the form $y = a + b \ln x$ by using the Logarithmic REGression option

Table 4.4

x, Population (thousands)	y, Walking Speed (feet per second)
5.5	3.3
14	3.7
71	4.3
138	4.4
342	4.8

Source: Mark and Helen Bornstein, "The Pace of Life"

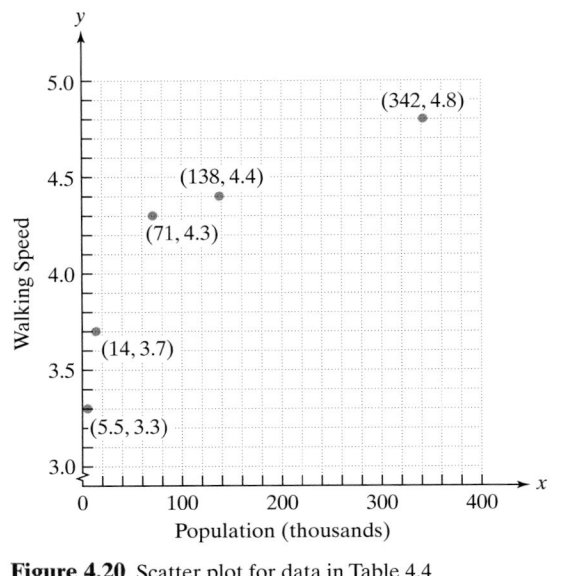

Figure 4.20 Scatter plot for data in Table 4.4

(see Figure 4.21). From the figure, we see that the logarithmic model of the data, with numbers rounded to three decimal places, is

$$y = 2.735 + 0.352 \ln x.$$

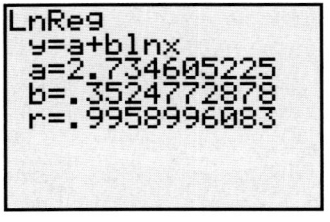

Figure 4.21 A logarithmic model for the data in Table 4.4.

The number *r* that appears in Figure 4.21 is called the **correlation coefficient** and is a measure of how well the model fits the data. The value of *r* is such that $-1 \leq r \leq 1$. A positive *r* means that as the *x*-values increase, so do the *y*-values. A negative *r* means that as the *x*-values increase, the *y*-values decrease. **The closer that *r* is to −1 or 1, the better the model fits the data.** Because *r* is approximately 0.996, the model

$$y = 2.735 + 0.352 \ln x$$

fits the data very well.

Now let's look at data whose scatter plot suggests an exponential model. The data in Table 4.5 indicate world population for six years. The scatter plot is shown in Figure 4.22.

Table 4.5

x, Year	y, World Population (billions)
1950	2.6
1960	3.1
1970	3.7
1980	4.5
1989	5.3
1999	6.0

Figure 4.22 A scatter plot for data in Table 4.5.

The Future of World Population

In the future, a new world order looms. The percentage of world population in less developed countries will increase, North America's will remain relatively stable, and Europe's will decrease.

1999

North America **5%**
South America **9%**
Europe **12%**
Asia **61%**
Africa **13%**

2050

North America **4%**
South America **9%**
Europe **7%**
Africa **20%**
Asia **60%**

Most Populous Countries, 2050

1. India	1529 million
2. China	1478 million
3. U.S.	349 million
4. Pakistan	345 million
5. Indonesia	312 million
6. Nigeria	244 million
7. Brazil	244 million
8. Bangladesh	212 million
9. Ethiopia	169 million
10. Congo	160 million

Source: United Nations Population Fund

Because the data in this scatter plot have a rapidly increasing pattern, the shape suggests that an exponential model might be a good choice. (You might also want to try a linear model.) If you go with the exponential option, you will use a graphing utility's Exponential REGression option. With this feature, a graphing utility fits the data to an exponential model of the form $y = ab^x$.

When computing an exponential model of the form $y = ab^x$, a graphing utility rewrites the equation using logarithms. Because the domain of the logarithmic function is the set of positive numbers, **zero must not be a value for x.** What does this mean in terms of our data for world population that starts in the year 1950? We must start values of x after 0. Thus, we'll assign x to represent the number of years since 1949. This gives us the data shown in Table 4.6. Using the Exponential REGression option, we obtain the equation in Figure 4.23.

Table 4.6

x, Numbers of Years after 1949		y, World Population (billions)
1	(1950)	2.6
11	(1960)	3.1
21	(1970)	3.7
31	(1980)	4.5
40	(1989)	5.3
50	(1999)	6.0

```
ExpReg
 y=a*b^x
 a=2.569837845
 b=1.017658868
 r=.9983590746
```

Figure 4.23 An exponential model for the data in Table 4.6

From Figure 4.23, we see that the exponential model of the data for world population x years after 1949, with numbers rounded to three decimal places, is

$$y = 2.570(1.018)^x.$$

The correlation coefficient, r, is close to 1, indicating that the model fits the data very well.

4 Express an exponential model in base e.

Because $b = e^{\ln b}$, we can rewrite any model in the form $y = ab^x$ in terms of base e.

> **Expressing an Exponential Model in Base e**
>
> $$y = ab^x \text{ is equivalent to } y = ae^{(\ln b) \cdot x}.$$

EXAMPLE 4 Rewriting an Exponential Model as an Exponential Growth Function

Rewrite $y = 2.57(1.018)^x$ in terms of base e.

Solution

$$y = ab^x \quad \text{is equivalent to} \quad y = ae^{(\ln b) \cdot x}.$$

$$y = 2.57(1.018)^x \quad \text{is equivalent to} \quad y = 2.57e^{(\ln 1.018) \cdot x}.$$

Using $\ln 1.018 \approx 0.018$, the exponential growth model for world population x years after 1949 is

$$y = 2.57e^{0.018x}.$$

In Example 4, we can replace y by A and x by t so that the model has the same letters as those in the exponential growth model $A = A_0 e^{kt}$.

$$A = A_0 \, e^{kt} \qquad \text{This is the exponential growth model.}$$

$$A = 2.57e^{0.018t} \qquad \text{This is the model for world population.}$$

The value of k, 0.018, indicates a growth rate of 1.8%. Although this is an excellent model for the data, we must be careful about making projections about world population using this growth function. Why? World population growth rate is now 1.3%, not 1.8%, so our model will overestimate future populations.

Check Point 4   Rewrite $y = 4(7.8)^x$ in terms of base e. Express the answer in terms of a natural logarithm, and then round to three decimal places.

When using a graphing utility to model data, begin with a scatter plot, drawn either by hand or with the graphing utility, to obtain a general picture for the shape of the data. It might be difficult to determine what model best fits the data—linear, logarithmic, exponential, quadratic, or something else. If necessary, use your graphing utility to fit several models to the data. The best model is the one that yields the value r, the correlation coefficient, closest to 1 or -1. Finding a proper fit for data can be almost as much art as it is mathematics. In this era of technology, the process of creating models that best fit data is one that involves more decision making than computation.

EXERCISE SET 4.5

Practice and Application Exercises

The exponential growth model $A = 208e^{0.008t}$ describes the population of the United States, in millions, t years after 1970. Use this model to solve Exercises 1–4.

1. What was the population of the United States in 1970?

2. By what percentage is the population of the United States increasing each year?

3. When will the U.S. population be 300 million?

4. When will the U.S. population be 350 million?

India is currently one of the world's fastest-growing countries. By 2040, the population of India will be larger than the population of China; by 2050, nearly one-third of the world's population will live in these two countries alone. The exponential growth model $A = 574e^{0.026t}$ describes the population of India, in millions, t years after 1974. Use this model to solve Exercises 5–8.

5. By what percentage is the population of India increasing each year?

6. What was the population of India in 1974?

7. When will India's population be 1624 million?

8. When will India's population be 2732 million?

The value of houses in a neighborhood follows a pattern of exponential growth. In the year 2000, you purchased a house in this neighborhood. The value of your house, in thousands of dollars, t years after 2000 is given by the exponential growth model $V = 140e^{0.068t}$. Use this model to solve Exercises 9–12.

9. What did you pay for your house?

10. By what percentage is the price of houses in your neighborhood increasing each year?

11. When will your house be worth $200,000?

12. When will your house be worth $300,000?

13. Through the end of 1991, 200,000 cases of AIDS had been reported to the Centers for Disease Control in the United States. By the end of 1998, the number had grown to 680,000. The exponential growth function $A = 200e^{kt}$ describes the thousands of AIDS cases in the United States t years after 1991. Use the fact that 7 years after 1991 there were 680 thousand cases to find k to three decimal places. Then write the exponential growth function. According to your model, by what percentage is the number of AIDS cases in the United States increasing each year?

14. In 1980, China's population was 983 million; in 1990, it was 1154 million. The exponential growth function $A = 983e^{kt}$ describes the population of China, in millions, t years after 1980. Use the fact that 10 years after 1980 the population was 1154 million to find k to three decimal places. Then write the exponential growth function. According to your model, by what percentage is the population of China increasing each year?

An artifact originally had 16 grams of carbon-14 present. The decay model $A = 16e^{-0.000121t}$ describes the amount of carbon-14 present after t years. Use this model to solve Exercises 15–16.

15. How many grams of carbon-14 will be present in 5715 years?

16. How many grams of carbon-14 will be present in 11,430 years?

17. The half-life of the radioactive element krypton-91 is 10 seconds. If 16 grams of krypton-91 are initially present, how many grams are present after 10 seconds? 20 seconds? 30 seconds? 40 seconds? 50 seconds?

18. The half-life of the radioactive element plutonium-239 is 25,000 years. If 16 grams of plotonium-239 are initially present how many grams are present after 25,000 years? 50,000 years? 75,000 years? 100,000 years? 125,000 years?

Use the exponential decay model for carbon-14, $A = A_0 e^{-0.000121t}$, to solve Exercises 19–20.

19. Prehistoric cave paintings were discovered in a cave in France. The paint contained 15% of the original carbon-14. Estimate the age of the paintings.

20. Skeletons were found at a construction site in San Francisco in 1989. The skeletons contained 88% of the expected amount of carbon-14 found in a living person. In 1989, how old were the skeletons?

21. The August 1978 issue of *National Geographic* described the 1964 find of dinosaur bones of a newly discovered dinosaur weighing 170 pounds, measuring 9 feet, with a 6-inch claw on one toe of each hind foot. The age of the dinosaur was estimated using potassium-40 dating of rocks surrounding the bones.
   a. Potassium-40 decays exponentially with a half-life of approximately 1.31 billion years. Use the fact that after 1.31 billion years a given amount of potassium-40 will have decayed to half the original amount to show that the decay model for potassium-40 is given by $A = A_0 e^{-0.52912t}$, where t is in billions of years.
   b. Analysis of the rocks surrounding the dinosaur bones indicated that 94.5% of the original amount of potassium-40 was still present. Let $A = 0.945A_0$ in the model in part (a) and estimate the age of the bones of the dinosaur.

4.

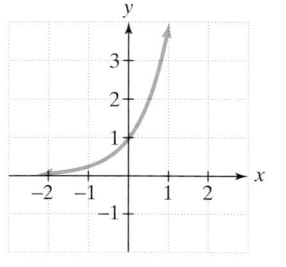

In Exercises 5–8, sketch by hand the graphs of the two functions in the same rectangular coordinate system. Use a table of coordinates to sketch the first function and transformations of this function plus a table of coordinates to graph the second function.

5. $f(x) = 2^x$ and $g(x) = 2^{x-1}$

6. $f(x) = 3^x$ and $g(x) = 3^x - 1$

7. $f(x) = 3^x$ and $g(x) = -3^x$

8. $f(x) = \left(\frac{1}{2}\right)^x$ and $g(x) = \left(\frac{1}{2}\right)^{-x}$

Use the compound interest formulas to solve Exercises 9–10.

9. Suppose that you have $5000 to invest. Which investment yields the greater return over 5 years: 5.5% compounded semiannually or 5.25% compounded monthly?

10. Suppose that you have $14,000 to invest. Which investment yields the greater return over 10 years: 7% compounded monthly or 6.85% compounded continuously?

11. A cup of coffee is taken out of a microwave oven and placed in a room. The temperature, T, in degrees Fahrenheit, of the coffee after t minutes is modeled by the function $T = 70 + 130e^{-0.04855t}$. The graph of the function is shown in the figure. Use the graph to answer each of the following questions.

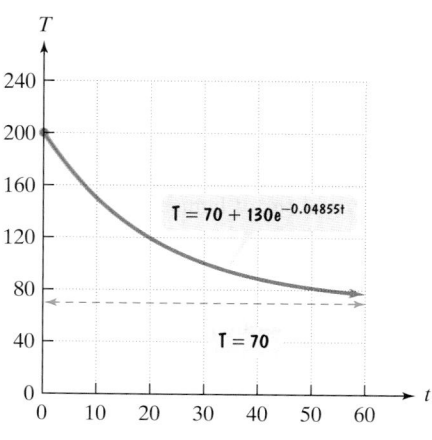

a. What was the temperature of the coffee when it was first taken out of the microwave?

b. What is a reasonable estimate of the temperature of the coffee after 20 minutes? Use your calculator to verify this estimate.

c. What is the limit of the temperature to which the coffee will cool? What does this tell you about the temperature of the room?

4.2

In Exercises 12–14, write each equation in its equivalent exponential form.

12. $\frac{1}{2} = \log_{49} 7$ **13.** $3 = \log_4 x$ **14.** $\log_3 81 = y$

In Exercises 15–17, write each equation in its equivalent logarithmic form.

15. $6^3 = 216$ **16.** $b^4 = 625$ **17.** $13^y = 874$

In Exercises 18–25, evaluate each expression without using a calculator. If evaluation is not possible, state the reason.

18. $\log_4 64$ **19.** $\log_5 \frac{1}{25}$ **20.** $\log_3(-9)$

21. $\log_{16} 4$ **22.** $\log_{17} 17$ **23.** $\log_3 3^8$

24. $\ln e^5$ **25.** $\log_3(\log_8 8)$

26. Graph $f(x) = 2^x$ and $g(x) = \log_2 x$ in the same rectangular coordinate system.

27. Graph $f(x) = \left(\frac{1}{3}\right)^x$ and $g(x) = \log_{1/3} x$ in the same rectangular coordinate system.

In Exercises 28–31, the graph of a logarithmic function is given. Select the function for each graph from the following options:

$$f(x) = \log x, \, g(x) = \log(-x),$$

$$h(x) = \log(2 - x), \, r(x) = 1 + \log(2 - x).$$

28.

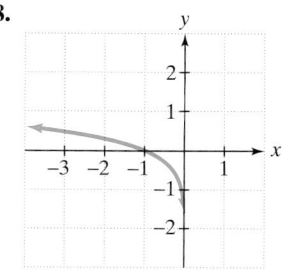

29.

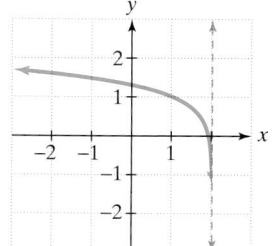

30.

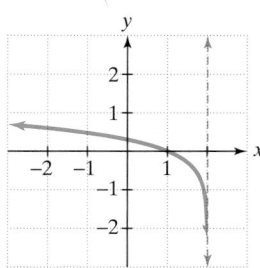

31.

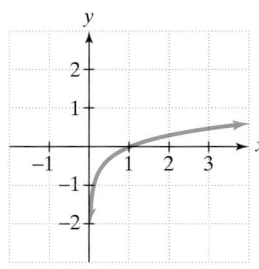

In Exercises 32–34, begin by graphing $f(x) = \log_2 x$. Then use transformations of this graph to graph the given function. What is the graph's x-intercept? What is the vertical asymptote?

32. $g(x) = \log_2(x - 2)$ **33.** $h(x) = -1 + \log_2 x$

34. $r(x) = \log_2(-x)$

In Exercises 35–37, find the domain of each logarithmic function.

35. $f(x) = \log_8(x + 5)$ **36.** $f(x) = \log(3 - x)$

37. $f(x) = \ln(x - 1)^2$

In Exercises 38–40, use inverse properties of logarithms to simplify each expression.

38. $\ln e^{6x}$ **39.** $e^{\ln \sqrt{x}}$ **40.** $10^{\log 4x^2}$

41. On the Richter scale, the magnitude, R, of an earthquake of intensity I is given by $R = \log \dfrac{I}{I_0}$, where I_0 is the intensity of a barely felt zero-level earthquake. If the intensity of an earthquake is $1000 I_0$, what is its magnitude on the Richter scale?

42. Students in a psychology class took a final examination. As part of an experiment to see how much of the course content they remembered over time, they took equivalent forms of the exam in monthly intervals thereafter. The average score, $f(t)$, for the group after t months was modeled by the function $f(t) = 76 - 18 \log(t + 1)$, where $0 \le t \le 12$.
   a. What was the average score when the exam was first given?
   b. What was the average score after 2 months? 4 months? 6 months? 8 months? one year?
   c. Use the results from parts (a) and (b) to graph f. Describe what the shape of the graph indicates in terms of the material retained by the students.

43. The formula

$$t = \frac{1}{c} \ln\left(\frac{A}{A - N}\right)$$

describes the time, t, in weeks, that it takes to achieve mastery of a portion of a task. In the formula, A represents maximum learning possible, N is the portion of the learning that is to be achieved, and c is a constant used to measure an individual's learning style. A 50-year-old man decides to start running as a way to maintain good health. He feels that the maximum rate he could ever hope to achieve is 12 miles per hour. How many weeks will it take before the man can run 5 miles per hour if $c = 0.06$ for this person?

4.3

In Exercises 44–47, use properties of logarithms to expand each logarithmic expression as much as possible. Where possible, evaluate logarithmic expressions without using a calculator.

44. $\log_6(36x^3)$ **45.** $\log_4 \dfrac{\sqrt{x}}{64}$

46. $\log_2 \dfrac{xy^2}{64}$ **47.** $\ln \sqrt[3]{\dfrac{x}{e}}$

In Exercises 48–51, use properties of logarithms to condense each logarithmic expression. Write the expression as a single logarithm whose coefficient is 1.

48. $\log_b 7 + \log_b 3$ **49.** $\log 3 - 3 \log x$

50. $3 \ln x + 4 \ln y$ **51.** $\frac{1}{2} \ln x - \ln y$

In Exercises 52–53, use common logarithms or natural logarithms and a calculator to evaluate to four decimal places.

52. $\log_6 72{,}348$ **53.** $\log_4 0.863$

4.4

Solve each exponential equation in Exercises 54–58. Express the answer in terms of natural logarithms. Then use a calculator to obtain a decimal approximation, correct to the nearest thousandth, for the solution.

54. $8^x = 12{,}143$ **55.** $9e^{5x} = 1269$

56. $e^{12-5x} - 7 = 123$ **57.** $5^{4x+2} = 37{,}500$

58. $e^{2x} - e^x - 6 = 0$

Solve each logarithmic equation in Exercises 59–63.

59. $\log_4(3x - 5) = 3$

60. $\log_2(x + 3) + \log_2(x - 3) = 4$

61. $\log_3(x - 1) - \log_3(x + 2) = 2$

62. $\ln x = -1$ **63.** $3 + 4\ln 2x = 15$

64. The formula $A = 10.1e^{0.005t}$ models the population of Los Angeles, California, in millions, t years after 1992. If the growth rate continues into the future, when will the population reach 13 million?

65. The amount of carbon dioxide in the atmosphere, measured in parts per million, has been increasing as a result of the burning of oil and coal. The buildup of gases and particles trap heat and raise the planet's temperature, a phenomenon called the greenhouse effect. Carbon dioxide accounts for about half of the warming. The formula $A = 364(1.005)^t$ projects carbon dioxide concentration, A, in parts per million, t years after 2000. Using the projections given by the formula, when will the carbon dioxide concentration be double the preindustrial level of 280 parts per million?

66. The formula $C = 15{,}557 + 5259 \ln x$ models the average cost of a new car x years after 1989. When will the average cost of a new car be $30,000?

67. Use the formula for compound interest with n compoundings each year to solve this problem. How long, to the nearest tenth of a year, will it take $12,500 to grow to $20,000 at 6.5% annual interest compounded quarterly?

Use the formula for continuous compounding to solve Exercises 68–69.

68. How long, to the nearest tenth of a year, will it take $50,000 to triple in value at 7.5% annual interest compounded continuously?

69. What interest rate is required for an investment subject to continuous compounding to triple in 5 years?

4.5

70. According to the U.S. Bureau of the Census, in 1980 there were 14.6 million residents of Hispanic origin living in the United States. By 1997, the number had increased to 29.3 million. The exponential growth function $A = 14.6e^{kt}$ describes the U.S. Hispanic population, in millions, t years after 1980.
 a. Find k, correct to three decimal places.
 b. Use the resulting model to project the Hispanic resident population in 2005.
 c. In what year will the Hispanic resident population reach 50 million?

71. Use the exponential decay model for carbon-14, $A = A_0 e^{-0.000121t}$, to solve this exercise. Prehistoric cave paintings were discovered in the Lascaux cave in France.

The paint contained 15% of the original carbon-14. Estimate the age of the paintings at the time of the discovery.

72. Europe's Great Plague of 1666 devastated Eyam, England. There were 261 people in the village; only 83 survived. The logistic growth function

$$f(t) = \frac{171}{1 + 18.6e^{-0.0747t}}$$

models the number of people in Eyam who were infected t days after the outbreak. (Source: Raggett, G. "Modeling the Eyam Plague." *The Institute of Mathematics and Its Application* 18 : 221–226.)
 a. How many people were infected when the outbreak began?
 b. How many people were infected after 45 days?
 c. According to the model, what is the limiting size of Eyam's population that can become infected? With 83 survivors among 261 people, does this mean that the size of the infected population surpassed the limit set by the model? Explain your answer.

In Exercises 73–74, rewrite the equation in terms of base e. Express the answer in terms of a natural logarithm, and then round to three decimal places.

73. $y = 73(2.6)^x$ **74.** $y = 6.5(0.43)^x$

75. The figure shows world population projections through the year 2150. The data are from the United Nations Family Planning Program and are based on optimistic or pessimistic expectations for successful control of human population growth. Suppose that you are interested in modeling these data using exponential, logarithmic, linear, and quadratic functions. Which function would you use to model each of the projections? Explain your choices. For the choice corresponding to a quadratic model, would your formula involve one with a positive or negative leading coefficient? Explain.

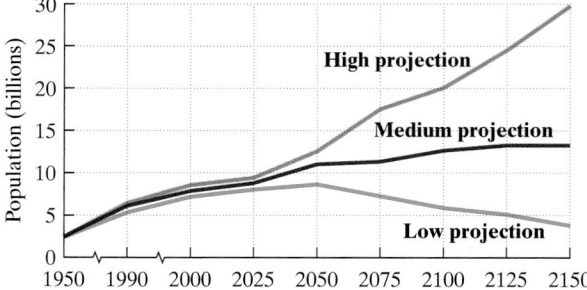

76. The figure shows the number of people in the United States age 65 and over, with projected figures for the year 2000 and beyond.

U. S. Population Age 65 and Over

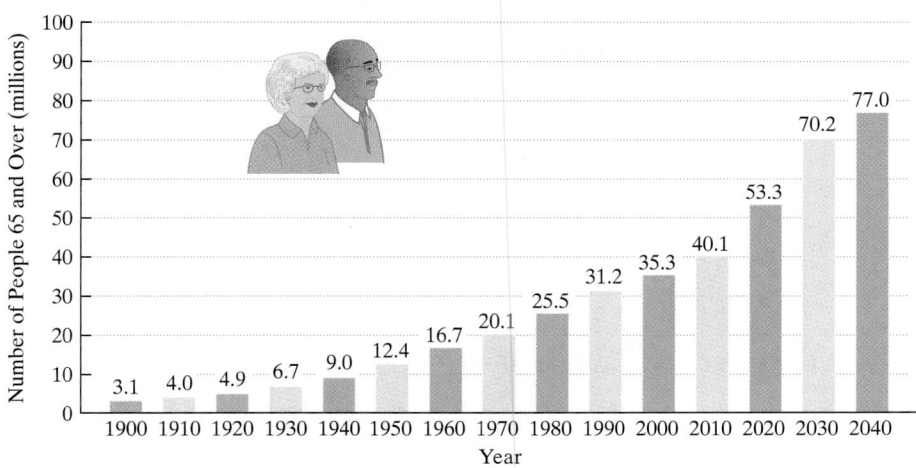

Source: U. S. Bureau of the Census

Let *x* represent the number of years since 1899 and let *y* represent the U.S. population, in millions. Use your graphing utility to find the model that best fits the data in the bar graph. Then use the model to find the U.S. population age 65 and over in 2050.

Chapter 4 Test

1. Graph $f(x) = 2^x$ and $g(x) = 2^{x+1}$ in the same rectangular coordinate system.

2. Graph $f(x) = \log_2 x$ and $g(x) = \log_2(x - 1)$ in the same rectangular coordinate system.

3. Write in exponential form: $\log_5 125 = 3$.

4. Write in logarithmic form: $\sqrt{36} = 6$.

5. Find the domain of $f(x) = \ln(3 - x)$.

In Exercises 6–7, use properties of logarithms to expand each logarithmic expression as much as possible. Where possible, evaluate logarithmic expressions without using a calculator.

6. $\log_4(64x^5)$

7. $\log_3 \dfrac{\sqrt[3]{x}}{81}$

In Exercises 8–9, write each expression as a single logarithm.

8. $6 \log x + 2 \log y$

9. $\ln 7 - 3 \ln x$

10. Use a calculator to evaluate $\log_{15} 71$ to four decimal places.

In Exercises 11–16, solve each equation.

11. $5^x = 1.4$

12. $400e^{0.005x} = 1600$

13. $e^{2x} - 6e^x + 5 = 0$

14. $\log_6(4x - 1) = 3$

15. $\log x + \log(x + 15) = 2$

16. $2 \ln 3x = 8$

17. Suppose you have $3000 to invest. Which investment yields the greater return over 10 years: 6.5% compounded semianually or 6% compounded continuously? How much more (to the nearest dollar) is yielded by the better investment?

18. On the decibel scale, the loudness of a sound, in decibels, is given by $D = 10 \log \dfrac{I}{I_0}$, where I is the intensity of the sound, in watts per meter^2, and I_0 is the intensity of a sound barely audible to the human ear. If the intensity of a sound is $10^{12} I_0$, what is its loudness in decibels? (Such a sound is potentially damaging to the ear.)

19. The percentage of married men in the United States who are employed is modeled by $P = 89.18e^{-0.004t}$. The model indicates that P% of married men were employed t years after 1959.

  a. What percentage of married men were employed in 1959?

  b. Is the percentage of married men who are employed increasing or decreasing? Explain.

  c. In what year were 77% of U.S. married men employed?

20. The 1980 population of Europe was 484 million; in 1990, it was 509 million. Write the exponential growth function that describes the population of Europe, in millions, t years after 1980.

21. Use the exponential decay model for carbon-14, $A = A_0 e^{-0.000121t}$, to solve this exercise. Bones of a pre-historic man were discovered and contained 5% of the original amount of carbon-14. How long ago did the man die?

22. The logistic growth function

$$f(t) = \frac{140}{1 + 9e^{-0.165t}}$$

describes the population of an endangered species of elk t years after they were introduced to a nonthreatening habitat.

a. How many elk were initially introduced to the habitat?
b. How many elk are expected in the habitat after 10 years?
c. What is the limiting size of the elk population that the habitat will sustain?

Cumulative Review Exercises (Chapters 1–4)

Solve each equation in Exercises 1–5.

1. $|3x - 4| = 2$

2. $\sqrt{2x - 5} - \sqrt{x - 3} = 1$

3. $x^4 + x^3 - 3x^2 - x + 2 = 0$

4. $e^{5x} - 32 = 96$

5. $\log_2(x + 5) + \log_2(x - 1) = 4$

Solve each inequality in Exercises 6–7. Express the answer in interval notation.

6. $14 - 5x \geq -6$

7. $|2x - 4| \leq 2$

8. Write the point-slope form and the slope-intercept form of the line passing through $(1, 3)$ and $(3, -3)$.

9. If $f(x) = x^2$ and $g(x) = x + 2$, find $(f \circ g)(x)$ and $(g \circ f)(x)$.

10. If $f(x) = 2x - 7$, find $f^{-1}(x)$.

11. Divide $x^3 + 5x^2 + 3x - 10$ by $x + 2$.

12. Use the Rational Zero Theorem to list all possible rational zeros for $f(x) = 4x^3 - 7x - 3$.

13. The value of y varies directly as the square of x. If $x = 3$ when $y = 12$, find y when $x = 15$.

14. Solve $x^3 - 4x^2 + 6x - 4 = 0$ given that $1 + i$ is a root.

In Exercises 15–18, graph each equation.

15. $(x - 3)^2 + (y + 2)^2 = 4$ **16.** $f(x) = (x - 2)^2 - 1$

17. $f(x) = \dfrac{x^2 - 1}{x^2 - 4}$ **18.** $f(x) = (x - 2)^2(x + 1)$

19. You are paid time-and-a-half for each hour worked over 40 hours a week. Last week you worked 50 hours and earned $660. What is your normal hourly salary?

20. The formula $F = 1 - k \ln(t + 1)$ models the fraction of people, F, who remember all the words in a list of non-sense words t hours after memorizing the list. After 3 hours only half the people could remember all the words. Determine the value of k and then predict the fraction of people in the group who will remember all the words after 6 hours.

Trigonometric Functions

Have you had days where your physical, intellectual, and emotional potentials were all at their peak? Then there are those other days when we feel we should not even bother getting out of bed. Do our potentials run in oscillating cycles like the tides? Can they be described mathematically? In this chapter you will encounter functions that enable us to model phenomena that occur in cycles.

What a day! It started when you added two miles to your morning run. You've experienced a feeling of peak physical well-being ever since. College was wonderful: You actually enjoyed two difficult lectures and breezed through a math test that had you worried. Now you're having dinner with an old group of friends. You experience the warmth from bonds of friendship filling the room.

SECTION 5.1 *Angles and Their Measure*

Objectives

1. Recognize and use the vocabulary of angles.
2. Use degree measure.
3. Draw angles in standard position.
4. Find coterminal angles.
5. Find complements and supplements.
6. Use radian measure.
7. Convert between degrees and radians.
8. Find the length of a circular arc.
9. Use linear and angular speed to describe motion on a circular path.

The San Francisco Museum of Modern Art was constructed in 1996 to illustrate how art and architecture can enrich one another. The exterior involves geometric shapes, symmetry, and unusual facades. Although there are no windows, natural light streams in through a truncated cylindrical skylight that crowns the building. The architect worked with a scale model of the museum at the site and observed how light hit it during different times of the day. These observations were used to cut the cylindrical skylight at an angle that maximizes sunlight entering the interior.

Angles play a critical role in creating modern architecture. They are also fundamental in trigonometry. In this section, we begin our study of trigonometry by looking at angles and methods for measuring them.

1 Recognize and use the vocabulary of angles.

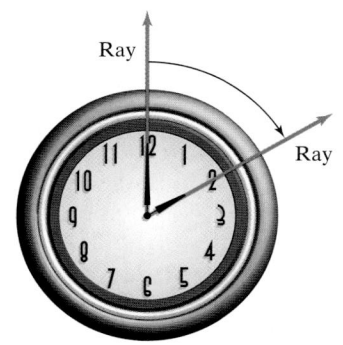

Figure 5.1 Clock with hands forming an angle

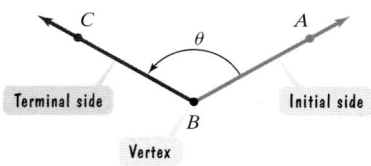

Figure 5.2 An angle; two rays with a common endpoint

Angles

The hour hand of a clock suggests a **ray**, a part of a line that has only one endpoint and extends forever in the opposite direction. An **angle** is formed by two rays that have a common endpoint. One ray is called the **initial side** and the other the **terminal side**.

A rotating ray is often a useful way to think about angles. The ray in Figure 5.1 rotates from 12 to 2. The ray pointing to 12 is the **initial side** and the ray pointing to 2 is the **terminal side**. The common endpoint of an angle's initial side and terminal side is the **vertex** of the angle.

Figure 5.2 shows an angle. The arrow near the vertex shows the direction and the amount of rotation from the initial side to the terminal side. Several methods can be used to name an angle. Lowercase Greek letters, such as α (alpha), β (beta), γ (gamma), and θ (theta), are often used.

An angle is in **standard position** if

- its vertex is at the origin of a rectangular coordinate system

and

- its initial side lies along the positive x-axis.

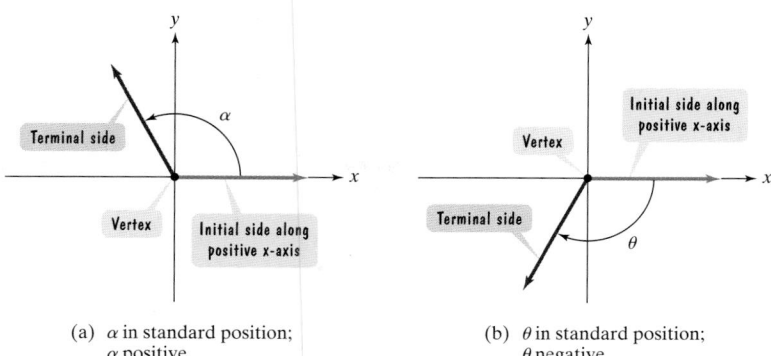

Figure 5.3 Two angles in standard position

(a) α in standard position; α positive

(b) θ in standard position; θ negative

The angles in Figure 5.3 are both in standard position.

When we see an initial side and a terminal side in place, there are two kinds of rotation that could have generated it. The arrow in Figure 5.3(a) indicates that the rotation from the initial side to the terminal side is in the counterclockwise direction. **Positive angles** are generated by counterclockwise rotation. Thus, angle α is positive. By contrast, the arrow in Figure 5.3(b) shows that the rotation from the initial side to the terminal side is in the clockwise direction. **Negative angles** are generated by clockwise rotation. Thus, angle θ is negative.

When an angle is in standard position, its terminal side can lie in a quadrant. We say that the angle **lies in that quadrant**. For example, in Figure 5.3(a), the terminal side of angle α lies in quadrant II. Thus, angle α lies in quadrant II. By contrast, in Figure 5.3(b), the terminal side of angle θ lies in quadrant III. Thus, angle θ lies in quadrant III.

Must all angles in standard position lie in a quadrant? The answer is no. The terminal side can lie on the x-axis or the y-axis. For example, angle β in Figure 5.4 has a terminal side that lies on the negative y-axis. An angle is called a **quadrantal angle** if its terminal side lies on the x-axis or the y-axis. Angle β in Figure 5.4 is an example of a quadrantal angle.

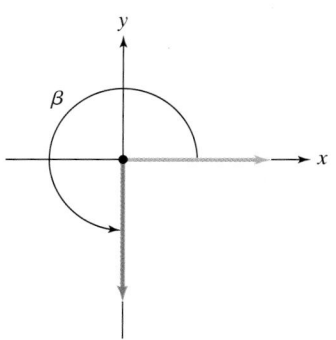

Figure 5.4 β is a quadrantal angle.

2 Use degree measure.

Measuring Angles Using Degrees

Angles are measured by determining the amount of rotation from the initial side to the terminal side. One way to measure angles is in **degrees**, symbolized by a small, raised circle °. Think of the hour hand of a clock. From 12 noon to 12 midnight, the hour hand moves around in a complete circle. By definition, the ray has rotated through 360 degrees, or 360°. Using 360° as the amount of rotation of a ray back onto itself, a degree, 1°, is $\frac{1}{360}$ of a complete rotation.

Figure 5.5 shows angles classified by their degree measurement. An **acute angle** measures less than 90° [see Figure 5.5(a)]. A **right angle**, one quarter of a complete rotation, measures 90° [Figure 5.5(b)]. Examine the right angle—do you see a small square at the vertex? This symbol is used to indicate a right angle. An **obtuse angle** measures more than 90°, but less than 180° [Figure 5.5(c)]. Finally, a **straight angle**, one-half a complete rotation, measures 180° [Figure 5.5(d)].

A complete 360° rotation

Figure 5.5 Classifying angles by their degree measurement

(a) **Acute angle** ($0° < \theta < 90°$)

(b) **Right angle** ($\frac{1}{4}$ rotation)

(c) **Obtuse angle** ($90° < \theta < 180°$)

(d) **Straight angle** ($\frac{1}{2}$ rotation)

3 Draw angles in standard position.

We will be using notation such as $\theta = 60°$ to refer to an angle θ whose measure is 60°. We also refer to *an angle of 60°* or a *60° angle*, instead of using the more precise (but cumbersome) phrase *an angle whose measure is 60°*.

Technology

Fractional parts of degrees are measured in minutes and seconds. One minute, written $1'$, is $\frac{1}{60}$ degree.

One second, written $1''$, is $\frac{1}{3600}$ degree.

For example,

$$31°47'12''$$

$$= \left(31 + \frac{47}{60} + \frac{12}{3600} \right)^{\circ}$$

$$= 31.787°.$$

Many calculators have keys for changing an angle from degree, minute, second notation (D°M′S″) to a decimal form and vice versa.

EXAMPLE 1 Drawing Angles in Standard Position

Draw each angle in standard position.

    a. a 45° angle   **b.** a 225° angle   **c.** a −135° angle   **d.** a 405° angle

Solution Because we are drawing angles in standard position, each vertex is at the origin and each initial side lies along the positive *x*-axis.

    a. A 45° angle is half of a right angle. The angle lies in quadrant I and is shown in Figure 5.6(a).

    b. A 225° angle is a positive angle. It has a counterclockwise rotation of 180° followed by a counterclockwise rotation of 45°. The angle lies in quadrant III and is shown in Figure 5.6(b).

    c. A −135° angle is negative angle. It has a clockwise rotation of 90° followed by a clockwise rotation of 45°. The angle lies in quadrant III and is shown in Figure 5.6(c).

    d. A 405° angle is a positive angle. It has a counterclockwise rotation of 360°, one complete rotation, followed by a counterclockwise rotation of 45°. The angle lies in quadrant I and is shown in Figure 5.6(d).

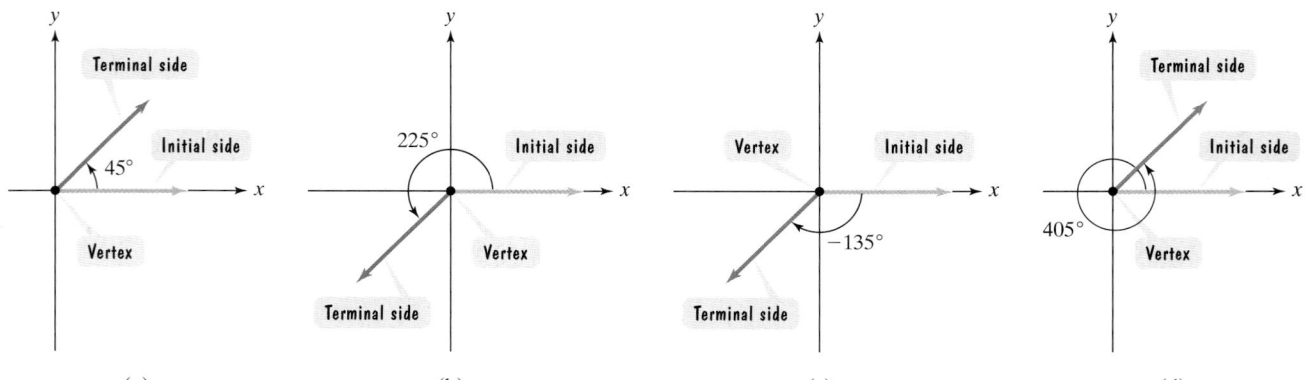

    (a)   (b)   (c)   (d)

Figure 5.6 Four angles in standard position

Check Point 1 Draw each angle in standard position.

    a. a 30° angle   **b.** a 210° angle
    c. a −120° angle   **d.** a 390° angle

4 Find coterminal angles.

Look at Figure 5.6 again. The 45° and 405° angles in parts (a) and (d) have the same initial and terminal sides. Similarly, the 225° and −135° angles in parts (b) and (c) have the same initial and terminal sides. Two angles with the same initial and terminal sides are called **coterminal angles**.

Every angle has infinitely many coterminal angles. Why? Think of an angle in standard position. One or more complete rotations of 360°, clockwise or counterclockwise, result in angles with the same initial and terminal sides as the original angle.

> **Coterminal Angles**
>
> An angle of $x°$ is coterminal with angles of
> $$x° + k \cdot 360°$$
> where k is an integer.

Two coterminal angles for an angle of $x°$ can be found by adding 360° to $x°$ and subtracting 360° from $x°$.

EXAMPLE 2 Finding Coterminal Angles

Assume the following angles are in standard position. Find a positive angle less than 360° that is coterminal with:

a. a 420° angle **b.** a −120° angle.

Solution We obtain the coterminal angle by adding or subtracting 360°. Our need to obtain a positive angle less than 360° determines whether we should add or subtract.

a. For a 420° angle, subtract 360° to find a positive coterminal angle.
$$420° - 360° = 60°$$

A 60° angle is coterminal with a 420° angle. Figure 5.7(a) illustrates that these angles have the same initial and terminal sides.

b. For a −120° angle, add 360° to find a positive coterminal angle.
$$-120° + 360° = 240°$$

A 240° angle is coterminal with a −120° angle. Figure 5.7(b) illustrates that these angles have the same initial and terminal sides.

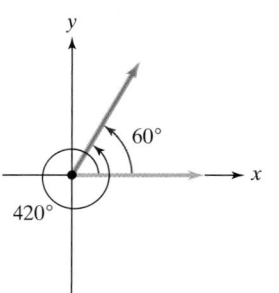

(a) Angles of 420° and 60° are coterminal.

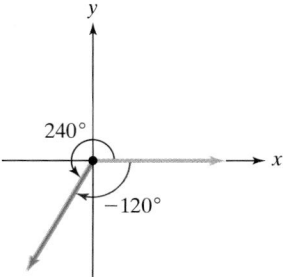

(b) Angles of −120° and 240° are coterminal.

Figure 5.7 Pairs of coterminal angles

<table>
<tr><td>

Check Point 2

</td><td>

Find a positive angle less than 360° that is coterminal with:
a. a 400° angle   **b.** a −135° angle.

</td></tr>
</table>

5 Find complements and supplements.

Two positive angles are **complements** if their sum is 90°. For example, angles of 70° and 20° are complements because $70° + 20° = 90°$.

Two positive angles are **supplements** if their sum is 180°. For example, angles of 130° and 50° are supplements because $130° + 50° = 180°$.

Finding Complements and Supplements

- For an $x°$ angle, the complement is a $90° − x°$ angle. Thus, the complement's measure is found by subtracting the angle's measure from 90°.
- For an $x°$ angle, the supplement is a $180° − x°$ angle. Thus, the supplement's measure is found by subtracting the angle's measure from 180°.

Because we use only positive angles for complements and supplements, some angles do not have complements and supplements.

EXAMPLE 3   Complements and Supplements

If possible, find the complement and the supplement of the given angle.
   a. $\theta = 62°$ **b.** $\alpha = 123°$

Solution   We find the complement by subtracting the angle's measure from 90°. We find the supplement by subtracting the angle's measure from 180°.

   a. We begin with $\theta = 62°$.

$$\text{complement} = 90° − 62° = 28°$$
$$\text{supplement} = 180° − 62° = 118°$$

For a 62° angle, the complement is a 28° angle and the supplement is a 118° angle.

   b. Now we turn to $\alpha = 123°$. For the angle's complement, we consider subtracting 123° from 90°. The difference is negative. Because we use only positive angles for complements, a 123° angle has no complement. It does, however, have a supplement.

$$\text{supplement} = 180° − 123° = 57°$$

The supplement of a 123° angle is a 57° angle.

<table>
<tr><td>

Check Point 3

</td><td>

If possible, find the complement and the supplement of the given angle.
   a. $\theta = 78°$ **b.** $\alpha = 150°$

</td></tr>
</table>

6 Use radian measure.

Measuring Angles Using Radians

Another way to measure angles is in *radians*. Let's first define an angle measuring **1 radian**. We use a circle of radius r. In Figure 5.8, we've constructed an angle

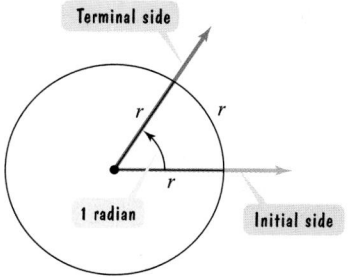

Figure 5.8 For a 1-radian angle, the intercepted arc and the radius are equal.

whose vertex is at the center of the circle. Such an angle is called a **central angle**. Notice that the central angle intercepts an arc along the circle measuring r units. The radius of the circle is also r units. The measure of such an angle is 1 radian.

Definition of a Radian

One radian is the measure of the central angle of a circle that intercepts an arc equal in length to the radius of the circle.

The **radian measure** of any central angle is the length of the intercepted arc divided by the circle's radius. In Figure 5.9(a), the length of the arc intercepted by angle β is double the radius, r. We find the measure of angle β in radians by dividing the length of the intercepted arc by the radius.

$$\beta = \frac{\text{length of the intercepted arc}}{\text{radius}} = \frac{2r}{r} = 2$$

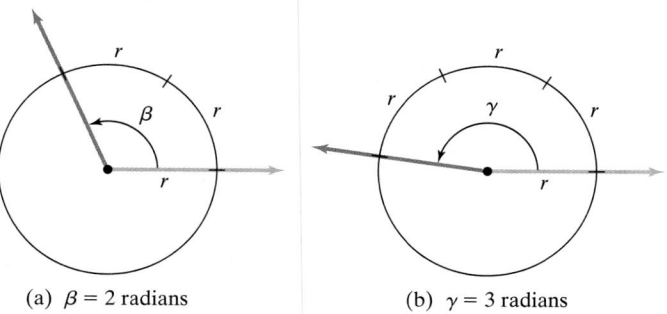

(a) $\beta = 2$ radians (b) $\gamma = 3$ radians

Figure 5.9 Two central angles measured in radians

Thus, angle β measures 2 radians. In Figure 5.9(b), the length of the intercepted arc is triple the radius, r. Let us find the measure of angle γ:

$$\gamma = \frac{\text{length of the intercepted arc}}{\text{radius}} = \frac{3r}{r} = 3$$

Thus, angle γ measures 3 radians.

Radian Measure

Consider an arc of length s on a circle of radius r. The measure of the central angle θ that intercepts the arc is

$$\theta = \frac{s}{r} \text{ radians.}$$

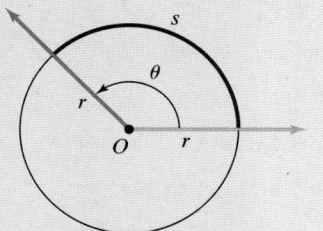

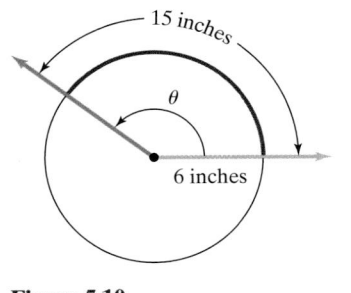

Figure 5.10

EXAMPLE 4   Computing Radian Measure

A central angle θ in a circle of radius 6 inches intercepts an arc of length 15 inches. What is the radian measure of θ?

Solution   Angle θ is shown in Figure 5.10. The radian measure of a central angle is the length of the intercepted arc, s, divided by the circle's radius, r. The length of the intercepted arc is 15 inches: $s = 15$ inches. The circle's radius is 6 inches: $r = 6$ inches. Now we use the formula for radian measure to find the radian measure of θ.

$$\theta = \frac{s}{r} = \frac{15 \text{ inches}}{6 \text{ inches}} = 2.5$$

Thus, the radian measure of θ is 2.5.

Study Tip

Before applying the formula for radian measure, be sure that the same unit of length is used for the intercepted arc, s, and the radius, r.

In Example 4, notice that the units (inches) cancel when we use the formula for radian measure. We are left with a number with no units. Thus, if an angle θ has a measure of 2.5 radians, we can write $\theta = 2.5$ radians or $\theta = 2.5$. We will often include the word *radians* simply for emphasis. There should be no confusion as to whether radian or degree measure is being used. Why is this so? If θ has degree measure 2.5°, we must include the degree symbol and write $\theta = 2.5°$, and *not* $\theta = 2.5$.

Check Point 4   A central angle θ in a circle of radius 12 feet intercepts an arc of length 42 feet. What is the radian measure of θ?

7 Convert between degrees and radians.

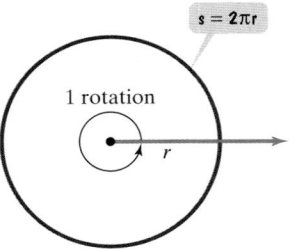

Figure 5.11 A complete rotation

Relationship between Degrees and Radians

How can we obtain a relationship between degrees and radians? We compare the number of degrees and the number of radians in one complete rotation, shown in Figure 5.11. We know that 360° is the amount of rotation of a ray back onto itself. The length of the intercepted arc is equal to the circumference of the circle. Thus, the radian measure of this central angle is the circumference of the circle divided by the circle's radius, r. The circumference of a circle of radius r is $2\pi r$. We use the formula for radian measure to find the radian measure of the 360° angle.

$$\theta = \frac{s}{r} = \frac{\text{the circle's circumference}}{r} = \frac{2\pi r}{r} = 2\pi$$

Because one complete rotation measures 360° and 2π radians,

$$360° = 2\pi \text{ radians.}$$

Dividing both sides by 2, we have

$$180° = \pi \text{ radians.}$$

Dividing this last equation by 180° or π gives the following conversion rules.

Study Tip

The unit you are converting *to* appears in the *numerator* of the conversion factor.

Conversion between Degrees and Radians

Using the basic relationship π radians $= 180°$,

1. To convert degrees to radians, multiply degrees by $\dfrac{\pi \text{ radians}}{180°}$.

2. To convert radians to degrees, multiply radians by $\dfrac{180°}{\pi \text{ radians}}$.

Angles that are fractions of a complete rotation are usually expressed in radian measure as fractional multiples of π, rather than as decimal approximations. For example, we write $\theta = \dfrac{\pi}{2}$ rather than using the decimal approximation $\theta \approx 1.57$.

EXAMPLE 5 Converting from Degrees to Radians

Convert each angle in degrees to radians.

    a. $30°$ **b.** $90°$ **c.** $-135°$

Solution   To convert degrees to radians, multiply by $\dfrac{\pi \text{ radians}}{180°}$. Observe how the degree units cancel.

    a. $30° = 30° \cdot \dfrac{\pi \text{ radians}}{180°} = \dfrac{30\pi}{180} \text{ radians} = \dfrac{\pi}{6} \text{ radians}$

    b. $90° = 90° \cdot \dfrac{\pi \text{ radians}}{180°} = \dfrac{90\pi}{180} \text{ radians} = \dfrac{\pi}{2} \text{ radians}$

    c. $-135° = -135° \cdot \dfrac{\pi \text{ radians}}{180°} = -\dfrac{135\pi}{180} \text{ radians} = -\dfrac{3\pi}{4} \text{ radians}$

Divide the numerator and denominator by 45.

Check Point 5   Convert each angle in degrees to radians.
    a. $60°$ **b.** $270°$ **c.** $-300°$

EXAMPLE 6 Converting from Radians to Degrees

Convert each angle in radians to degrees.

    a. $\dfrac{\pi}{3}$ radians   **b.** $-\dfrac{5\pi}{3}$ radians   **c.** 1 radian

Solution   To convert radians to degrees, multiply by $\dfrac{180°}{\pi \text{ radians}}$. Observe how the radian units cancel.

Study Tip

In Example 6(c), we see that 1 radian is approximately 57°. Keep in mind that a radian is much larger than a degree.

a. $\dfrac{\pi}{3}$ radians $= \dfrac{\pi \text{ radians}}{3} \cdot \dfrac{180°}{\pi \text{ radians}} = \dfrac{180°}{3} = 60°$

b. $-\dfrac{5\pi}{3}$ radians $= -\dfrac{5\pi \text{ radians}}{3} \cdot \dfrac{180°}{\pi \text{ radians}} = -\dfrac{5 \cdot 180°}{3} = -300°$

c. 1 radian $= 1 \text{ radian} \cdot \dfrac{180°}{\pi \text{ radians}} = \dfrac{180°}{\pi} \approx 57.3°$

Check Point 6

Convert each angle in radians to degrees.

a. $\dfrac{\pi}{4}$ radians   **b.** $-\dfrac{4\pi}{3}$ radians   **c.** 6 radians

Figure 5.12 illustrates the degree and radian measures of angles that you will commonly see in trigonometry. Each angle is in standard position, so that the initial side lies along the positive x-axis. We will be using both degree and radian measure for these angles.

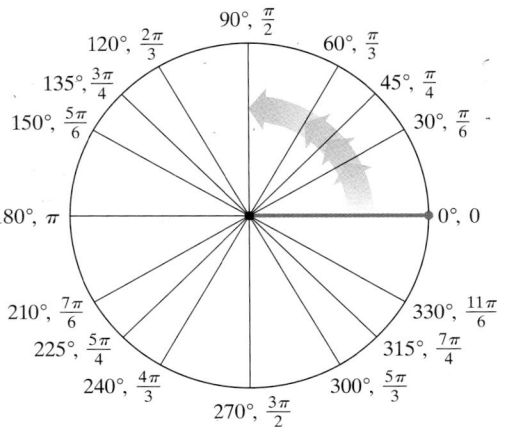

Figure 5.12 Degree and radian measures of selected angles

8 Find the length of a circular arc.

The Length of a Circular Arc

We can use the radian measure formula, $\theta = \dfrac{s}{r}$, to find the length of the arc of a circle. How do we do this? Remember that s represents the length of the arc intercepted by the central angle θ. Thus, by solving the formula for s, we have an equation for arc length.

The Length of a Circular Arc

Let r be the radius of a circle and θ the non-negative radian measure of a central angle of the circle. The length of the arc intercepted by the central angle is

$$s = r\theta.$$

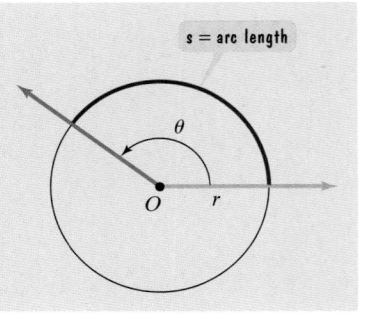

s = arc length

EXAMPLE 7 Finding the Length of a Circular Arc

A circle has a radius of 10 inches. Find the length of the arc intercepted by a central angle of 120°.

Solution The formula $s = r\theta$ can be used only when θ is expressed in radians. Thus, we begin by converting 120° to radians. Multiply by $\dfrac{\pi \text{ radians}}{180°}$.

$$120° = 120° \cdot \frac{\pi \text{ radians}}{180°} = \frac{120\pi}{180} \text{ radians} = \frac{2\pi}{3} \text{ radians}$$

Now we can use the formula $s = r\theta$ to find the length of the arc. The circle's radius is 10 inches: $r = 10$ inches. The measure of the central angle, in radians, is $\dfrac{2\pi}{3}$: $\theta = \dfrac{2\pi}{3}$. The length of the arc intercepted by this central angle is

$$s = r\theta = (10 \text{ inches})\left(\frac{2\pi}{3}\right) = \frac{20\pi}{3} \text{ inches} \approx 20.94 \text{ inches.}$$

Check Point 7 A circle has a radius of 6 inches. Find the length of the arc intercepted by a central angle of 45°. Express arc length in terms of π. Then round your answer to two decimal places.

9 Use linear and angular speed to describe motion on a circular path.

Linear and Angular Speed

A carousel contains four circular rows of animals. As the carousel revolves, the animals in the outer row travel a greater distance per unit of time than those in the inner rows. These animals have a greater *linear speed* than those in the inner rows. By contrast, all animals, regardless of the row, complete the same number of revolutions per unit of time. All animals in the four circular rows travel at the same *angular speed*.

Using v for linear speed and ω (omega) for angular speed, we define these two kinds of speeds along a circular path as follows.

Definitions of Linear and Angular Speed

If a point is in motion on a circle of radius r through an angle of θ radians in time t, then its **linear speed** is

$$v = \frac{s}{t},$$

where s is the arc length given by $s = r\theta$, and its **angular speed** is

$$\omega = \frac{\theta}{t}.$$

The hard drive in a computer rotates at 3600 revolutions per minute. This angular speed, expressed in revolutions per minute, can also be expressed in revolutions per second, radians per minute, and radians per second. Using 2π radians = 1 revolution, we express the angular speed of a hard drive in radians per minute as follows:

SECTION 5.2 *Right Triangle Trigonometry*

Objectives

1. Use right triangles to evaluate trigonometric functions.
2. Find function values for $30°\left(\dfrac{\pi}{6}\right)$, $45°\left(\dfrac{\pi}{4}\right)$, and $60°\left(\dfrac{\pi}{3}\right)$.
3. Recognize and use fundamental identities.
4. Use equal cofunctions of complements.
5. Evaluate trigonometric functions with a calculator.
6. Use right triangle trigonometry to solve applied problems.

In the last century, Ang Rita Sherpa climbed Mount Everest eight times, all without the use of bottled oxygen.

Mountain climbers have forever been fascinated by reaching the top of Mount Everest, sometimes with tragic results. The mountain, on Asia's Tibet-Nepal border, is Earth's highest, peaking at an incredible 29,029 feet. The heights of mountains can be found using **trigonometry**. The word *trigonometry* means *measurement of triangles*. Trigonometry is used in navigation, building, and engineering. For centuries, Muslims have used trigonometry and the stars to navigate across the Arabian desert to Mecca, the birthplace of the prophet Muhammad, the founder of Islam. The ancient Greeks used trigonometry to record the locations of thousands of stars and worked out the motion of the Moon relative to the Earth. Today, trigonometry is used to study the structure of DNA, the master molecule that determines how we grow from a single cell to a complex, fully developed adult.

1 Use right triangles to evaluate trigonometric functions.

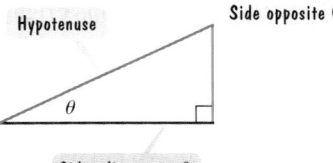

Figure 5.14 Naming a right triangle's sides from the point of view of an acute angle θ

The Six Trigonometric Functions

We begin the study of trigonometry by defining six functions, the six *trigonometric functions*. The inputs for these functions are measures of acute angles in right triangles. The outputs are the ratios of the lengths of the sides of right triangles.

Figure 5.14 shows a right triangle with one of its acute angles labeled θ. The side opposite the right angle is known as the **hypotenuse**. The other sides of the triangle are described by their position relative to the acute angle θ. One side is opposite θ and one is adjacent to θ.

The trigonometric functions have names that are words, rather than single letters such as f, g, and h. For example, the **sine of θ** is the length of the side opposite θ divided by the length of the hypotenuse:

$$\sin\theta = \frac{\text{length of side opposite }\theta}{\text{length of hypotenuse}}.$$

Input is the measure of an acute angle.

Output is the ratio of the lengths of the sides.

The ratio of lengths depends on angle θ and thus is a function of θ. The expression $\sin\theta$ really means $\sin(\theta)$, where sine is the name of the function and θ, the measure of an acute angle, is an input.

Here are the names of the six trigonometric functions, along with their abbreviations.

Name	Abbreviation	Name	Abbreviation
sine	sin	cosecant	csc
cosine	cos	secant	sec
tangent	tan	cotangent	cot

Now, let θ be an acute angle in a right triangle, shown in Figure 5.15. The length of the side opposite θ is a, the length of the side adjacent to θ is b, and the length of the hypotenuse is c.

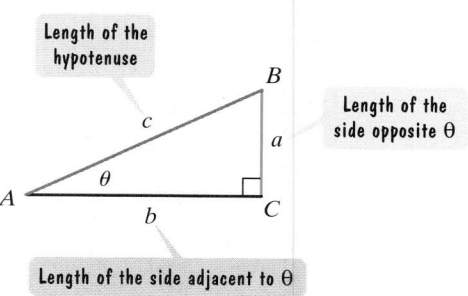

Length of the hypotenuse

Length of the side opposite θ

Length of the side adjacent to θ

Figure 5.15

Right Triangle Definitions of Trigonometric Functions

See Figure 5.15. The six **trigonometric functions of the acute angle θ** are defined as follows.

$$\sin\theta = \frac{\text{length of side opposite angle }\theta}{\text{length of hypotenuse}} = \frac{a}{c} \qquad \csc\theta = \frac{\text{length of hypotenuse}}{\text{length of side opposite angle }\theta} = \frac{c}{a}$$

$$\cos\theta = \frac{\text{length of side adjacent to angle }\theta}{\text{length of hypotenuse}} = \frac{b}{c} \qquad \sec\theta = \frac{\text{length of hypotenuse}}{\text{length of side adjacent to angle }\theta} = \frac{c}{b}$$

$$\tan\theta = \frac{\text{length of side opposite angle }\theta}{\text{length of side adjacent to angle }\theta} = \frac{a}{b} \qquad \cot\theta = \frac{\text{length of side adjacent to angle }\theta}{\text{length of side opposite angle }\theta} = \frac{b}{a}$$

Each of the trigonometric functions of the acute angle θ is positive. Observe that the functions in the second column in the box are the reciprocals of the corresponding functions in the first column.

Figure 5.16 on page 432 shows four right triangles of varying sizes. In each of the triangles, θ is the same acute angle, measuring approximately 56.3°. All four of these similar triangles have the same shape and the lengths of corresponding sides are in the same ratio. In each triangle, the tangent function has the same value: $\tan\theta = \frac{3}{2}$.

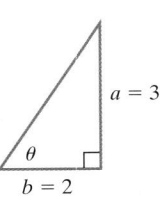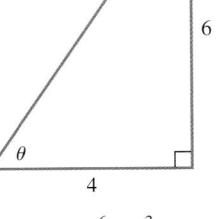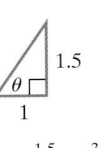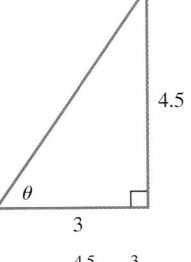

$$\tan \theta = \frac{a}{b} = \frac{3}{2} \qquad \tan \theta = \frac{6}{4} = \frac{3}{2} \qquad \tan \theta = \frac{1.5}{1} = \frac{3}{2} \qquad \tan \theta = \frac{4.5}{3} = \frac{3}{2}$$

Figure 5.16 A particular acute angle always gives the same ratio of opposite to adjacent sides.

In general, **the trigonometric function values of θ depend only on the size of angle θ, and not on the size of the triangle.**

EXAMPLE 1   Evaluating Trigonometric Functions

Find the value of each of the six trigonometric functions of θ in Figure 5.17.

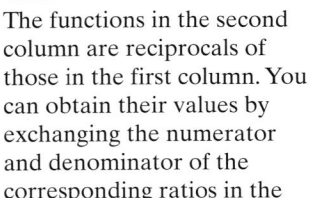

Figure 5.17

Solution   We need to find the values of the six trigonometric functions of θ. However, we must know the lengths of all three sides of the triangle (a, b, and c) to evaluate all six functions. The values of a and b are given. We can use the Pythagorean Theorem, $c^2 = a^2 + b^2$, to find c.

$$a = 5 \qquad b = 12$$

$$c^2 = a^2 + b^2 = 5^2 + 12^2 = 25 + 144 = 169$$
$$c = \sqrt{169} = 13$$

Now that we know the lengths of the three sides of the triangle, we apply the definitions of the six trigonometric functions of θ. Referring to these lengths as opposite, adjacent, and hypotenuse, we have

$$\sin \theta = \frac{\text{opposite}}{\text{hypotenuse}} = \frac{5}{13} \qquad \csc \theta = \frac{\text{hypotenuse}}{\text{opposite}} = \frac{13}{5}$$

$$\cos \theta = \frac{\text{adjacent}}{\text{hypotenuse}} = \frac{12}{13} \qquad \sec \theta = \frac{\text{hypotenuse}}{\text{adjacent}} = \frac{13}{12}$$

$$\tan \theta = \frac{\text{opposite}}{\text{adjacent}} = \frac{5}{12} \qquad \cot \theta = \frac{\text{adjacent}}{\text{opposite}} = \frac{12}{5}.$$

Study Tip

The functions in the second column are reciprocals of those in the first column. You can obtain their values by exchanging the numerator and denominator of the corresponding ratios in the first column.

Check Point 1   Find the value of each of the six trigonometric functions of θ in the figure.

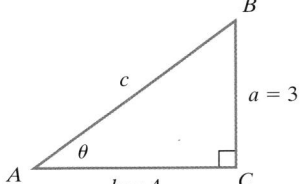

2 Find function values for $30°\left(\dfrac{\pi}{6}\right)$, $45°\left(\dfrac{\pi}{4}\right)$, and $60°\left(\dfrac{\pi}{3}\right)$.

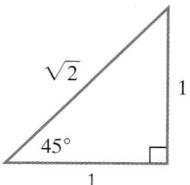

Figure 5.18 An isosceles right triangle

Function Values for Some Special Angles

A 45°, or $\dfrac{\pi}{4}$ radian, angle occurs frequently in trigonometry. How do we find the values of the trigonometric functions of 45°? We construct a right triangle with a 45° angle, shown in Figure 5.18. The triangle actually has two 45° angles. Thus, the triangle is isosceles—that is, it has two sides of the same length. Assume that each leg of the triangle has a length equal to 1. We can find the length of the hypotenuse using the Pythagorean Theorem.

$$(\text{length of hypotenuse})^2 = 1^2 + 1^2 = 2$$

$$\text{length of hypotenuse} = \sqrt{2}$$

With Figure 5.18, we can determine the trigonometric function values for 45°.

EXAMPLE 2   Evaluating Trigonometric Functions of 45°

Use Figure 5.18 to find $\sin 45°$, $\cos 45°$, and $\tan 45°$.

Solution   We apply the definitions of these three trigonometric functions.

$$\sin 45° = \frac{\text{length of side opposite } 45°}{\text{length of hypotenuse}} = \frac{1}{\sqrt{2}}$$

$$\cos 45° = \frac{\text{length of side adjacent to } 45°}{\text{length of hypotenuse}} = \frac{1}{\sqrt{2}}$$

$$\tan 45° = \frac{\text{length of side opposite } 45°}{\text{length of side adjacent to } 45°} = \frac{1}{1} = 1$$

Check Point 2   Use Figure 5.18 to find $\csc 45°$, $\sec 45°$, and $\cot 45°$.

When you worked Checkpoint 2, did you actually use Figure 5.18 or did you use reciprocals to find the values?

$$\csc 45° = \sqrt{2} \qquad \sec 45° = \sqrt{2} \qquad \cot 45° = 1$$

Take the reciprocal of $\sin 45° = \dfrac{1}{\sqrt{2}}$.   Take the reciprocal of $\cos 45° = \dfrac{1}{\sqrt{2}}$.   Take the reciprocal of $\tan 45° = \dfrac{1}{1}$.

We found that $\sin 45° = \dfrac{1}{\sqrt{2}}$ and $\cos 45° = \dfrac{1}{\sqrt{2}}$. This value is often expressed by rationalizing the denominator:

$$\frac{1}{\sqrt{2}} = \frac{1}{\sqrt{2}} \cdot \frac{\sqrt{2}}{\sqrt{2}} = \frac{\sqrt{2}}{2}.$$

We are multiplying by 1 and not changing the value of $\dfrac{1}{\sqrt{2}}$.

Thus, $\sin 45° = \dfrac{\sqrt{2}}{2}$ and $\cos 45° = \dfrac{\sqrt{2}}{2}$.

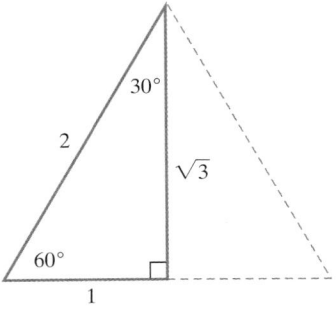

Figure 5.19 30°-60°-90° triangle

Two other angles that occur frequently in trigonometry are 30°, or $\frac{\pi}{6}$ radian, and 60°, or $\frac{\pi}{3}$ radian, angles. We can find the values of the trigonometric functions of 30° and 60° by using a right triangle. To form this right triangle, draw an equilateral triangle—that is a triangle with all sides the same length. Assume that each side has a length equal to 2. Now take half of the equilateral triangle. We obtain the right triangle in Figure 5.19. This right triangle has a hypotenuse of length 2 and a leg of length 1. The other leg has length a, which can be found using the Pythagorean Theorem.

$$a^2 + 1^2 = 2^2$$
$$a^2 + 1 = 4$$
$$a^2 = 3$$
$$a = \sqrt{3}$$

With the right triangle in Figure 5.19, we can determine the trigonometric functions for 30° and 60°.

EXAMPLE 3 Evaluating Trigonometric Functions of 30° and 60°

Use Figure 5.19 to find $\sin 60°$, $\cos 60°$, $\sin 30°$, and $\cos 30°$.

Solution We begin with 60°. Use the angle on the lower left in Figure 5.19.

$$\sin 60° = \frac{\text{length of side opposite } 60°}{\text{length of hypotenuse}} = \frac{\sqrt{3}}{2}$$

$$\cos 60° = \frac{\text{length of side adjacent to } 60°}{\text{length of hypotenuse}} = \frac{1}{2}$$

To find $\sin 30°$ and $\cos 30°$, use the angle on the upper right in Figure 5.19.

$$\sin 30° = \frac{\text{length of side opposite } 30°}{\text{length of hypotenuse}} = \frac{1}{2}$$

$$\cos 30° = \frac{\text{length of side adjacent to } 30°}{\text{length of hypotenuse}} = \frac{\sqrt{3}}{2}$$

> **Check Point 3**
>
> Use Figure 5.19 to find $\tan 60°$ and $\tan 30°$. If necessary, express the value without a square root in the denominator by rationalizing the denominator.

Because we will often use the function values of 30°, 45°, and 60°, you should learn to construct the right triangles shown in Figure 5.18 and 5.19. With sufficient practice, you will memorize the following values.

Sines, Cosines, and Tangents of Special Angles

$$\sin 30° = \sin \frac{\pi}{6} = \frac{1}{2} \qquad \cos 30° = \cos \frac{\pi}{6} = \frac{\sqrt{3}}{2} \qquad \tan 30° = \tan \frac{\pi}{6} = \frac{\sqrt{3}}{3}$$

$$\sin 45° = \sin \frac{\pi}{4} = \frac{\sqrt{2}}{2} \qquad \cos 45° = \cos \frac{\pi}{4} = \frac{\sqrt{2}}{2} \qquad \tan 45° = \tan \frac{\pi}{4} = 1$$

$$\sin 60° = \sin \frac{\pi}{3} = \frac{\sqrt{3}}{2} \qquad \cos 60° = \cos \frac{\pi}{3} = \frac{1}{2} \qquad \tan 60° = \tan \frac{\pi}{3} = \sqrt{3}$$

3 Recognize and use fundamental identities.

Fundamental Identities

Many relationships exist among the six trigonometric functions. These relationships are described using **trigonometric identities**. For example, $\csc \theta$ is defined as the reciprocal of $\sin \theta$. This relationship can be expressed by the identity

$$\csc \theta = \frac{1}{\sin \theta}.$$

This identity is one of six **reciprocal identities**.

Reciprocal Identities

$$\sin \theta = \frac{1}{\csc \theta} \qquad \cos \theta = \frac{1}{\sec \theta} \qquad \tan \theta = \frac{1}{\cot \theta}$$

$$\csc \theta = \frac{1}{\sin \theta} \qquad \sec \theta = \frac{1}{\cos \theta} \qquad \cot \theta = \frac{1}{\tan \theta}$$

Two other relationships that follow from the definitions of the trigonometric functions are called the **quotient identities.**

Quotient Identities

$$\tan \theta = \frac{\sin \theta}{\cos \theta} \qquad \cot \theta = \frac{\cos \theta}{\sin \theta}$$

If $\sin \theta$ and $\cos \theta$ are known, a quotient identity and three reciprocal identities make it possible to find the value of each of the four remaining trigonometric functions.

EXAMPLE 4 Using Quotient and Reciprocal Identities

Given $\sin \theta = \frac{1}{2}$ and $\cos \theta = \frac{\sqrt{3}}{2}$, find the value of each of the four remaining trigonometric functions.

Solution   We can find $\tan\theta$ by using the quotient identity that describes $\tan\theta$ as the quotient of $\sin\theta$ and $\cos\theta$.

$$\tan\theta = \frac{\sin\theta}{\cos\theta} = \frac{\frac{1}{2}}{\frac{\sqrt{3}}{2}} = \frac{1}{2}\cdot\frac{2}{\sqrt{3}} = \frac{1}{\sqrt{3}} = \frac{1}{\sqrt{3}}\cdot\frac{\sqrt{3}}{\sqrt{3}} = \frac{\sqrt{3}}{3}$$

Rationalize the denominator.

We use the reciprocal identities to find the value of each of the remaining three functions.

$$\csc\theta = \frac{1}{\sin\theta} = \frac{1}{\frac{1}{2}} = 2$$

$$\sec\theta = \frac{1}{\cos\theta} = \frac{1}{\frac{\sqrt{3}}{2}} = \frac{2}{\sqrt{3}} = \frac{2}{\sqrt{3}}\cdot\frac{\sqrt{3}}{\sqrt{3}} = \frac{2\sqrt{3}}{3}$$

Rationalize the denominator.

$$\cot\theta = \frac{1}{\tan\theta} = \frac{1}{\frac{1}{\sqrt{3}}} = \sqrt{3}$$

Check Point 4   Given $\sin\theta = \dfrac{2}{3}$ $\cos\theta = \dfrac{\sqrt{5}}{3}$, find the value of each of the four remaining trigonometric functions.

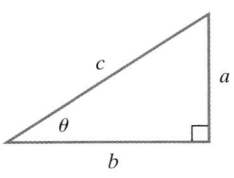

Figure 5.20

Other relationships among trigonometric functions follow from the Pythagorean Theorem. Using Figure 5.20, the Pythagorean Theorem states that

$$a^2 + b^2 = c^2.$$

To obtain ratios that correspond to trigonometric functions, divide both sides of this equation by c^2.

$$\frac{a^2}{c^2} + \frac{b^2}{c^2} = 1 \quad\text{or}\quad \left(\frac{a}{c}\right)^2 + \left(\frac{b}{c}\right)^2 = 1$$

In Figure 5.20 $\sin\theta = \frac{a}{c}$, so this is $(\sin\theta)^2$.

In Figure 5.20 $\cos\theta = \frac{b}{c}$, so this is $(\cos\theta)^2$.

Based on the observations in the voice balloons, we see that

$$(\sin\theta)^2 + (\cos\theta)^2 = 1.$$

We will eliminate the parentheses in this identity by writing $\sin^2\theta$ instead of $(\sin\theta)^2$ and $\cos^2\theta$ instead of $(\cos\theta)^2$. With this notation, we can write the identity as

$$\sin^2\theta + \cos^2\theta = 1.$$

Two additional identities can be obtained from $a^2 + b^2 = c^2$ by dividing both sides by b^2 and a^2, respectively. The three identities are called the **Pythagorean identities.**

Pythagorean Identities

$$\sin^2\theta + \cos^2\theta = 1 \qquad 1 + \tan^2\theta = \sec^2\theta \qquad 1 + \cot^2\theta = \csc^2\theta$$

EXAMPLE 5 Using a Pythagorean Identity

Given that $\sin\theta = \frac{3}{5}$ and θ is an acute angle, find the value of $\cos\theta$ using a trigonometric identity.

Solution We can find the value of $\cos\theta$ by using the Pythagorean identity

$$\sin^2\theta + \cos^2\theta = 1.$$

$$\left(\frac{3}{5}\right)^2 + \cos^2\theta = 1 \qquad \text{We are given that } \sin\theta = \frac{3}{5}.$$

$$\frac{9}{25} + \cos^2\theta = 1 \qquad \text{Square } \frac{3}{5}: \left(\frac{3}{5}\right)^2 = \frac{3^2}{5^2} = \frac{9}{25}.$$

$$\cos^2\theta = 1 - \frac{9}{25} \qquad \text{Subtract } \frac{9}{25} \text{ from both sides.}$$

$$\cos^2\theta = \frac{16}{25} \qquad \text{Simplify: } 1 - \frac{9}{25} = \frac{25}{25} - \frac{9}{25} = \frac{16}{25}.$$

$$\cos\theta = \sqrt{\frac{16}{25}} = \frac{4}{5} \qquad \text{Because } \theta \text{ is an acute angle, } \cos\theta \text{ is positive.}$$

Thus, $\cos\theta = \frac{4}{5}$.

Check Point 5 Given that $\sin\theta = \frac{1}{2}$ and θ is an acute angle, find the value of $\cos\theta$ using a trigonometric identity.

4 Use equal cofunctions of complements.

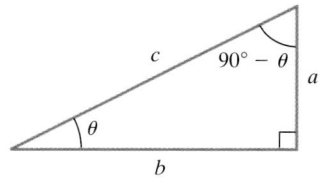

Figure 5.21

Trigonometric Functions and Complements

Another relationship among trigonometric functions is based on angles that are complements. Refer to Figure 5.21. Because the sum of the angles of any triangle is 180°, in a right triangle the sum of the acute angles is 90°. Thus, the acute angles are complements. If the degree measure of one acute angle is θ, then the degree measure of the other acute angle is $(90° - \theta)$. This angle is shown on the upper right in Figure 5.21.

Let's use Figure 5.21 to compare $\sin\theta$ and $\cos(90° - \theta)$.

$$\sin\theta = \frac{\text{length of side opposite } \theta}{\text{length of hypotenuse}} = \frac{a}{c}$$

$$\cos(90° - \theta) = \frac{\text{length of side adjacent to } (90° - \theta)}{\text{length of hypotenuse}} = \frac{a}{c}$$

Thus, $\sin\theta = \cos(90° - \theta)$. If two angles are complements, the sine of one equals the cosine of the other. Because of this relationship, the sine and cosine are called *cofunctions* of each other. The name *cosine* is a shortened form of the phrase *complement's sine*.

Any pair of trigonometric functions f and g for which

$$f(\theta) = g(90° - \theta) \quad \text{and} \quad g(\theta) = f(90° - \theta)$$

are called **cofunctions**. Using Figure 5.21, we can show that the tangent and cotangent are cofunctions of each other. So are the secant and cosecant.

> **Cofunction Identities**
>
> The value of a trigonometric function of θ is equal to the cofunction of the complement of θ.
>
> $$\sin\theta = \cos(90° - \theta) \qquad \cos\theta = \sin(90° - \theta)$$
> $$\tan\theta = \cot(90° - \theta) \qquad \cot\theta = \tan(90° - \theta)$$
> $$\sec\theta = \csc(90° - \theta) \qquad \csc\theta = \sec(90° - \theta)$$
>
> If θ is in radians, replace 90° with $\dfrac{\pi}{2}$.

EXAMPLE 6

Find a cofunction with the same value as the given expression.

a. $\sin 72°$ **b.** $\csc \dfrac{\pi}{3}$

Solution   Because the value of a trigonometric function of θ is equal to the cofunction of the complement of θ, we need to find the complement of each angle. We do this by subtracting the angle's measure from 90° or its radian equivalent, $\dfrac{\pi}{2}$.

a. $\sin 72° = \cos(90° - 72°) = \cos 18°$

We have a function and its cofunction.

b. $\csc \dfrac{\pi}{3} = \sec\left(\dfrac{\pi}{2} - \dfrac{\pi}{3}\right) = \sec\left(\dfrac{3\pi}{6} - \dfrac{2\pi}{6}\right) = \sec \dfrac{\pi}{6}$

We have a cofunction and its function.

Perform the subtraction using the least common denominator, 6.

Check Point 6   Find a cofunction with the same value as the given expression.

a. $\sin 46°$ **b.** $\cot \dfrac{\pi}{12}$

5 Evaluate trigonometric functions with a calculator.

Using a Calculator to Evaluate Trigonometric Functions

The values of the trigonometric functions obtained with the special triangles are exact values. For most angles other than 30°, 45°, and 60°, we approximate the value of each of the trigonometric functions using a calculator. The first step is to set the calculator to the correct *mode*, degrees or radians, depending on how the acute angle is measured.

Most calculators have keys marked $\boxed{\text{SIN}}$, $\boxed{\text{COS}}$, and $\boxed{\text{TAN}}$. For example, to find the value of sin 30°, set the calculator to the degree mode and enter 30 $\boxed{\text{SIN}}$ on most scientific calculators and $\boxed{\text{SIN}}$ 30 $\boxed{\text{ENTER}}$ on most graphing calculators. Consult the manual for your calculator.

To evaluate the cosecant, secant, and cotangent functions, use the key for the respective reciprocal function, $\boxed{\text{SIN}}$, $\boxed{\text{COS}}$, or $\boxed{\text{TAN}}$, and then use the reciprocal key. The reciprocal key is $\boxed{1/x}$ on most scientific calculators and $\boxed{x^{-1}}$ on most graphing calculators. For example, we can evaluate $\sec \dfrac{\pi}{12}$ using the following reciprocal relationship:

$$\sec \frac{\pi}{12} = \frac{1}{\cos \dfrac{\pi}{12}}.$$

Using the radian mode, enter one of the following keystroke sequences.

Most Scientific Calculators

$$\boxed{\pi}\ \boxed{\div}\ 12\ \boxed{=}\ \boxed{\text{COS}}\ \boxed{1/x}$$

Most Graphing Calculators

$$\boxed{(}\ \boxed{\text{COS}}\ \boxed{(}\ \boxed{\pi}\ \boxed{\div}\ 12\ \boxed{)}\ \boxed{)}\ \boxed{x^{-1}}\ \boxed{\text{ENTER}}$$

Rounding the display to four decimal places, we obtain $\sec \dfrac{\pi}{12} = 1.0353$.

EXAMPLE 7   Evaluating Trigonometric Functions with a Calculator

Use a calculator to find the value to four decimal places of:

  a. $\cos 48.2°$ **b.** $\cot 1.2$.

Solution

Scientific Calculator Solution

Function	Mode	Keystrokes	Display, rounded to four decimal places
a. $\cos 48.2°$	Degree	48.2 $\boxed{\text{COS}}$	0.6665
b. $\cot 1.2$	Radian	1.2 $\boxed{\text{TAN}}$ $\boxed{1/x}$	0.3888

Graphing Calculator Solution

Function	Mode	Keystrokes	Display, rounded to four decimal places
a. $\cos 48.2°$	Degree	$\boxed{\text{COS}}$ 48.2 $\boxed{\text{ENTER}}$	0.6665
b. $\cot 1.2$	Radian	$\boxed{(}$ $\boxed{\text{TAN}}$ 1.2 $\boxed{)}$ $\boxed{x^{-1}}$ $\boxed{\text{ENTER}}$	0.3888

Check Point 7

Use a calculator to find the value to four decimal places of:
a. sin 72.8°   **b.** csc 1.5.

6 Use right triangle trigonometry to solve applied problems.

Applications

Many applications of right triangle trigonometry involve the angle made with an imaginary horizontal line. As shown in Figure 5.22, an angle formed by a horizontal line and the line of sight to an object that is above the horizontal line is called the **angle of elevation**. The angle formed by a horizontal line and the line of sight to an object that is below the horizontal line is called the **angle of depression**. Transits and sextants are instruments used to measure such angles.

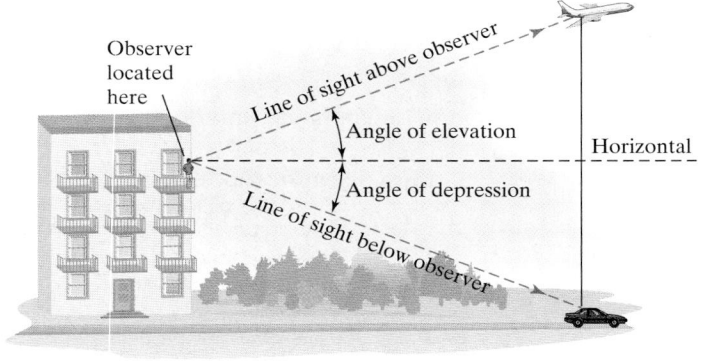

Figure 5.22

EXAMPLE 8   Problem Solving Using an Angle of Elevation

Sighting the top of a building, a surveyor measured the angle of elevation to be 22°. The transit is 5 feet above the ground and 300 feet from the building. Find the building's height.

Solution   The situation is illustrated in Figure 5.23. Let *a* be the height of the portion of the building that lies above the transit. The height of the building is the transit's height, 5 feet, plus *a*. Thus, we need to identify a trigonometric function that will make it possible to find *a*. In terms of the 22° angle, we are looking for the side opposite the angle. The transit is 300 feet from the building, so the side adjacent to the 22° angle is 300 feet. Because we have a known angle, an unknown opposite side, and a known adjacent side, we select the tangent function.

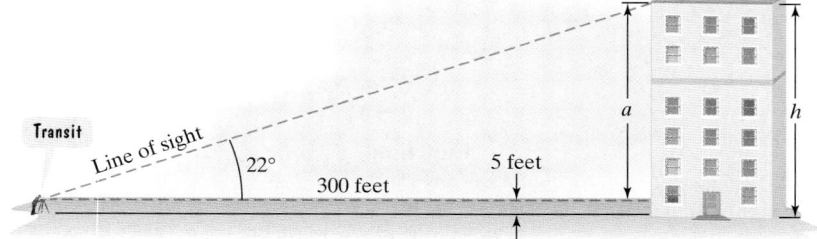

Figure 5.23

$$\tan 22° = \frac{a}{300}$$

Length of side opposite the 22° angle

Length of side adjacent to the 22° angle

$$a = 300 \tan 22°$$ Multiply both sides of the equation by 300.

$$a \approx 300(0.4040) \approx 121$$ Find tan 22° with a calculator in the degree mode.

The height of the part of the building above the transit is approximately 121 feet. Thus, the height of the building is determined by adding the transit's height, 5 feet, to 121 feet.

$$h \approx 5 + 121 = 126$$

The building's height is approximately 126 feet.

Check Point 8

The irregular blue shape in Figure 5.24 represents a lake. The distance across the lake, *a*, is unknown. To find this distance, a surveyor took the measurements shown in the figure. What is the distance across the lake?

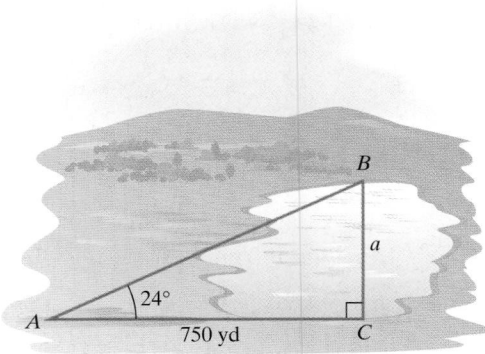

Figure 5.24

If two sides of a right triangle are known, an appropriate trigonometric function can be used to find an acute angle *θ* in the triangle. You will also need to use the *inverse key* on a calculator. This key uses a function value to display the acute angle *θ*. For example, suppose that $\sin θ = 0.866$. We can find *θ* in the degree mode by using the *inverse sine* key, usually labeled $\boxed{\text{SIN}^{-1}}$.

Scientific Calculator   **Graphing Calculator**

.866 $\boxed{\text{SIN}^{-1}}$ $\boxed{\text{SIN}^{-1}}$ .866 $\boxed{\text{ENTER}}$

The display shows approximately 59.99, which we can round to 60. Thus, if $\sin θ = 0.866$, then $θ \approx 60°$.

EXAMPLE 9   Determining the Angle of Elevation

A building that is 21 meters tall casts a shadow 25 meters long. Find the angle of elevation of the sun to the nearest degree.

Solution   The situation is illustrated in Figure 5.25. We are asked to find θ. We begin with the tangent function.

$$\tan \theta = \frac{\text{side opposite } \theta}{\text{side adjacent to } \theta} = \frac{21}{25}$$

We use a calculator in the degree mode to find θ.

Scientific Calculator	**Graphing Calculator**
21 ÷ 25 = TAN⁻¹	TAN⁻¹ (21 ÷ 25) ENTER

The display should show approximately 40. Thus, the angle of elevation of the sun is approximately 40°.

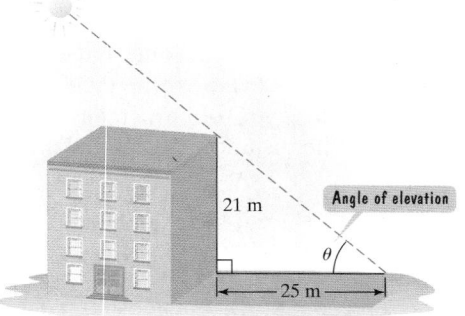

Figure 5.25

> **Check Point 9**
>
> A flagpole that is 14 meters tall casts a shadow 10 meters long. Find the angle of elevation of the sun to the nearest degree.

The Mountain Man

In the 1930s, a *National Geographic* team headed by Brad Washburn used trigonometry to create a map of the 5000-square-mile region of the Yukon, near the Canadian border. The team started with aerial photography. By drawing a network of angles on the photographs, the approximate locations of the major mountains and their rough heights were determined. The expedition then spent three months on foot to find the exact heights. Team members established two base points a known distance apart, one directly under the mountain's peak. By measuring the angle of elevation from one of the base points to the peak, the tangent function was used to determine the peak's height. The Yukon expedition was a major advance in the way maps are made.

EXERCISE SET 5.2

Practice Exercises

In Exercises 1–8, use the Pythagorean Theorem to find the length of the missing side of each right triangle. Then find the value of each of the six trigonometric functions of θ.

1.

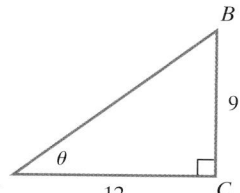

2.

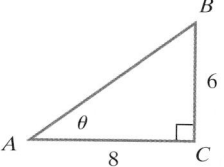

3.

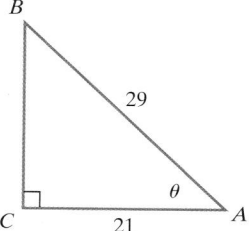

4.

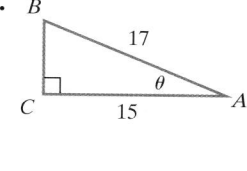

5.

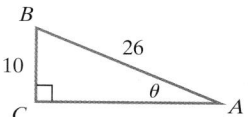

6.

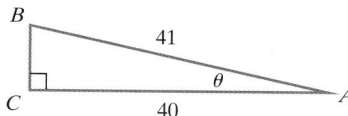

7.

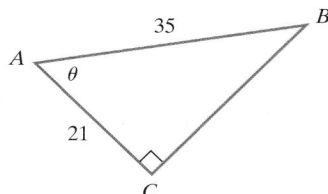

8.

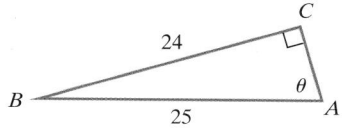

In Exercises 9–16, use the given triangles to evaluate each expression. If necessary, express the value without a square root in the denominator by rationalizing the denominator.

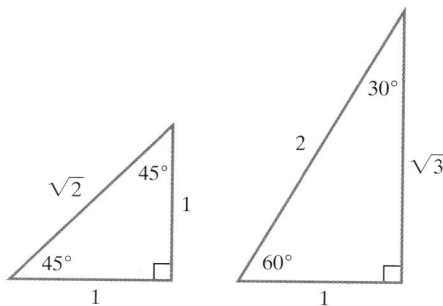

9. $\cos 30°$

10. $\tan 30°$

11. $\sec 45°$

12. $\csc 45°$

13. $\tan \dfrac{\pi}{3}$

14. $\cot \dfrac{\pi}{3}$

15. $\sin \dfrac{\pi}{4} - \cos \dfrac{\pi}{4}$

16. $\tan \dfrac{\pi}{4} + \csc \dfrac{\pi}{6}$

In Exercises 17–20, θ is an acute angle and sin θ and cos θ are given. Use identities to find tan θ, csc θ, sec θ, and cot θ. Where necessary, rationalize denominators.

17. $\sin \theta = \dfrac{8}{17}, \quad \cos \theta = \dfrac{15}{17}$

18. $\sin \theta = \dfrac{3}{5}, \quad \cos \theta = \dfrac{4}{5}$

19. $\sin \theta = \dfrac{1}{3}, \quad \cos \theta = \dfrac{2\sqrt{2}}{3}$

20. $\sin \theta = \dfrac{2}{3}, \quad \cos \theta = \dfrac{\sqrt{5}}{3}$

In Exercises 21–24, θ is an acute angle and sin θ is given. Use the Pythagorean identity $\sin^2 \theta + \cos^2 \theta = 1$ to find cos θ.

21. $\sin \theta = \dfrac{6}{7}$

22. $\sin \theta = \dfrac{7}{8}$

23. $\sin \theta = \dfrac{\sqrt{39}}{8}$

24. $\sin \theta = \dfrac{\sqrt{21}}{5}$

In Exercises 25–30, use an identity to find the value of each expression. Do not use a calculator.

25. $\sin 37° \csc 37°$

26. $\cos 53° \sec 53°$

27. $\sin^2 \dfrac{\pi}{9} + \cos^2 \dfrac{\pi}{9}$

28. $\sin^2 \dfrac{\pi}{10} + \cos^2 \dfrac{\pi}{10}$

29. $\sec^2 23° - \tan^2 23°$

30. $\csc^2 63° - \cot^2 63°$

Technology Exercises

82. Use a calculator in the radian mode to fill in the values in the following table. Then draw a conclusion about $\frac{\sin\theta}{\theta}$ as θ approaches 0.

θ	0.4	0.3	0.2	0.1	0.01	0.001	0.0001	0.00001
$\sin\theta$								
$\dfrac{\sin\theta}{\theta}$								

83. Use a calculator in the radian mode to fill in the values in the following table. Then draw a conclusion about $\frac{\cos\theta - 1}{\theta}$ as θ approaches 0.

θ	0.4	0.3	0.2	0.1	0.01	0.001	0.0001	0.00001
$\cos\theta$								
$\dfrac{\cos\theta - 1}{\theta}$								

Critical Thinking Exercises

84. Which one of the following is true?

   a. $\dfrac{\tan 45°}{\tan 15°} = \tan 3°$

   b. $\tan^2 15° - \sec^2 15° = -1$

   c. $\sin 45° + \cos 45° = 1$

   d. $\tan^2 5° = \tan 25°$

85. Explain why the sine or cosine of an acute angle cannot be greater than or equal to 1.

86. Describe what happens to the tangent of an acute angle as the angle gets close to 90°. What happens at 90°?

87. From the top of a 250-foot lighthouse, a plane is sighted overhead and a ship is observed directly below the plane. The angle of elevation of the plane is 22° and the angle of depression of the ship is 35°. Find **a.** the distance of the ship from the lighthouse; **b.** the plane's height above the water. Round to the nearest foot.

SECTION 5.3 *Trigonometric Functions of Any Angle*

Objectives

1. Use the definitions of trigonometric functions of any angle.
2. Use the signs of the trigonometric functions.
3. Find reference angles.
4. Use reference angles to evaluate trigonometric functions.

There is something comforting in the repetition of some of nature's patterns. The ocean level at a beach varies between high and low tide approximately every 12 hours. The number of hours of daylight oscillates from a maximum on the summer solstice, June 21; it decreases slowly until the minimum daylight occurs on the winter solstice, December 21, and then increases to the same maximum the following June 21. Some believe that cycles, called biorhythms, represent physical, emotional, and intellectual aspects of our lives. Throughout the remainder of this chapter, we will see how the trigonometric functions are used to model phenomena that occur again and again. To do this, we need to move beyond right triangles.

1 Use the definitions of trigonometric functions of any angle.

Trigonometric Functions of Any Angle

In the last section we evaluated trigonometric functions of acute angles, such as that shown in Figure 5.26(a). Note that this angle is in standard position. The point $P = (x, y)$ is a point r units from the origin on the terminal side of θ. A right triangle is formed by drawing a perpendicular from $P = (x, y)$ to the x-axis. Note that y is the length of the side opposite θ and x is the length of the side adjacent to θ.

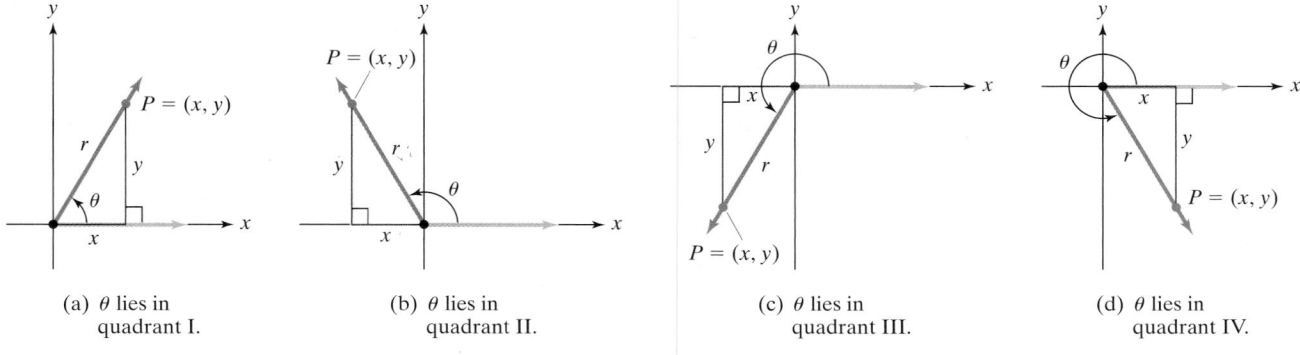

(a) θ lies in quadrant I.

(b) θ lies in quadrant II.

(c) θ lies in quadrant III.

(d) θ lies in quadrant IV.

Figure 5.26

Figures 5.26(b), (c), and (d) show angles in standard position, but they are not acute. We can extend our definitions of the six trigonometric functions to include such angles, as well as quadrantal angles. (Recall that a quadrantal angle has its terminal side on the x-axis or y-axis; such angles are *not* shown in Figure 5.26.) The point $P = (x, y)$ may be any point on the terminal side of the angle θ other than the origin $(0, 0)$.

Definitions of Trigonometric Functions of Any Angle

Let θ be any angle in standard position, and let $P = (x, y)$ be a point on the terminal side of θ. If $r = \sqrt{x^2 + y^2}$ is the distance from $(0, 0)$ to (x, y), as shown in Figure 5.26, the **six trigonometric functions of θ** are defined by the following ratios.

$$\sin\theta = \frac{y}{r} \qquad \cos\theta = \frac{x}{r} \qquad \tan\theta = \frac{y}{x}, x \neq 0$$

$$\csc\theta = \frac{r}{y}, y \neq 0 \qquad \sec\theta = \frac{r}{x}, x \neq 0 \qquad \cot\theta = \frac{x}{y}, y \neq 0$$

Because the point $P = (x, y)$ is any point on the terminal side of θ other than the origin $(0, 0)$, $r = \sqrt{x^2 + y^2}$ cannot be zero. Examine the six trigonometric functions defined previously. Note that the denominator of the sine and cosine functions is r. Because $r \neq 0$, the sine and cosine functions are defined for any real value of the angle θ. This is not true for the other four trigonometric functions. Note that the denominator of the tangent and secant functions is x. These functions are not defined if $x = 0$. If the point $P = (x, y)$ is on the y-axis, then $x = 0$. Thus, the tangent and secant functions are undefined for all quadrantal angles with terminal sides on the positive or negative y-axis. Likewise, if $P = (x, y)$ is on the x-axis, then $y = 0$, and the cotangent and cosecant functions

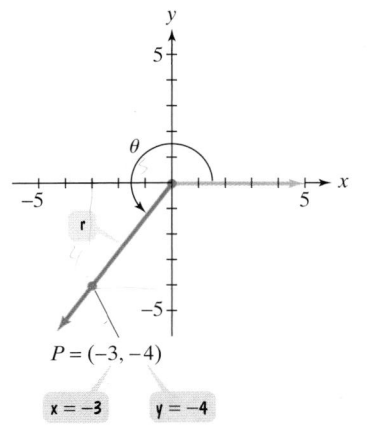

Figure 5.27

are undefined. The cotangent and cosecant functions are undefined for all quadrantal angles with terminal sides on the positive or negative x-axis.

EXAMPLE 1 Evaluating Trigonometric Functions

Let $P = (-3, -4)$ be a point on the terminal side of θ. Find each of the six trigonometric functions of θ.

Solution The situation is shown in Figure 5.27. We need values for x, y, and r to evaluate all six trigonometric functions. We are given the values of x and y. Because $P = (-3, -4)$ is a point on the terminal side of θ, $x = -3$ and $y = -4$. Furthermore,

$$r = \sqrt{x^2 + y^2} = \sqrt{(-3)^2 + (-4)^2} = \sqrt{9 + 16} = \sqrt{25} = 5.$$

Now that we know x, y, and r, we can find the six trigonometric functions of θ.

$$\sin\theta = \frac{y}{r} = \frac{-4}{5} = -\frac{4}{5}, \quad \cos\theta = \frac{x}{r} = \frac{-3}{5} = -\frac{3}{5}, \quad \tan\theta = \frac{y}{x} = \frac{-4}{-3} = \frac{4}{3}$$

$$\csc\theta = \frac{r}{y} = \frac{5}{-4} = -\frac{5}{4}, \quad \sec\theta = \frac{r}{x} = \frac{5}{-3} = -\frac{5}{3}, \quad \cot\theta = \frac{x}{y} = \frac{-3}{-4} = \frac{3}{4}$$

These ratios are the reciprocals of those shown directly above.

Check Point 1 Let $P = (4, -3)$ be a point on the terminal side of θ. Find each of the six trigonometric functions of θ.

How do we find the values of the trigonometric functions for a quadrantal angle? First, draw the angle in standard position. Second, choose a point P on the angle's terminal side. The trigonometric function values of θ depend only on the size of θ and not on the distance of point P from the origin. Thus, we choose a point that is 1 unit from the origin. Finally, apply the definition of the appropriate trigonometric function.

EXAMPLE 2 Trigonometric Functions of Quadrantal Angles

Evaluate, if possible, the sine function and the tangent function at the following four quadrantal angles:

a. $\theta = 0° = 0$ **b.** $\theta = 90° = \dfrac{\pi}{2}$ **c.** $\theta = 180° = \pi$ **d.** $\theta = 270° = \dfrac{3\pi}{2}$

Solution

a. If $\theta = 0° = 0$ radians, then the terminal side of the angle is on the positive x-axis. Let us select the point $P = (1, 0)$ with $x = 1$ and $y = 0$. This point is 1 unit from the origin, so $r = 1$. Now that we know x, y, and r, we can apply the definitions of the sine and tangent functions.

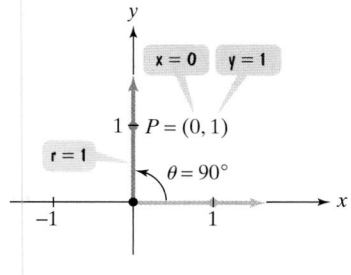

$$\sin 0° = \sin 0 = \frac{y}{r} = \frac{0}{1} = 0$$

$$\tan 0° = \tan 0 = \frac{y}{x} = \frac{0}{1} = 0$$

b. If $\theta = 90° = \dfrac{\pi}{2}$ radians, then the terminal side of the angle is on the positive y-axis. Let us select the point $P = (0, 1)$ with $x = 0$ and $y = 1$. This point is 1 unit from the origin, so $r = 1$. Now that we know $x, y,$ and r, we can apply the definitions of the sine and tangent functions.

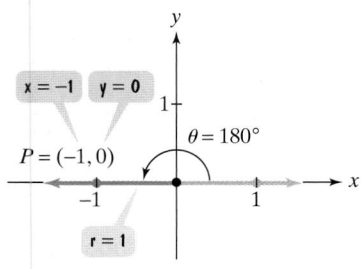

$$\sin 90° = \sin \frac{\pi}{2} = \frac{y}{r} = \frac{1}{1} = 1$$

$$\tan 90° = \tan \frac{\pi}{2} = \frac{y}{x} = \frac{1}{0}$$

Because division by 0 is undefined, $\tan 90°$ is undefined.

c. If $\theta = 180° = \pi$ radians, then the terminal side of the angle is on the negative x-axis. Let us select the point $P = (-1, 0)$ with $x = -1$ and $y = 0$. This point is 1 unit from the origin, so $r = 1$. Now that we know $x, y,$ and r, we can apply the definitions of the sine and tangent functions.

$$\sin 180° = \sin \pi = \frac{y}{r} = \frac{0}{1} = 0$$

$$\tan 180° = \tan \pi = \frac{y}{x} = \frac{0}{-1} = 0$$

Discovery

Try finding tan 90° and tan 270° with your calculator. Describe what occurs.

d. If $\theta = 270° = \dfrac{3\pi}{2}$ radians, then the terminal side of the angle is on the negative y-axis. Let us select the point $P = (0, -1)$ with $x = 0$ and $y = -1$. This point is 1 unit from the origin, so $r = 1$. Now that we know x, y, and r, we can apply the definitions of the sine and tangent functions.

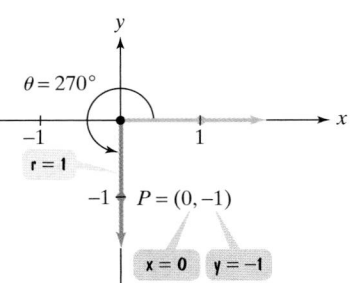

$$\sin 270° = \sin \frac{3\pi}{2} = \frac{y}{r} = \frac{-1}{1} = -1$$

$$\tan 270° = \tan \frac{3\pi}{2} = \frac{y}{x} = \frac{-1}{0}$$

Because division by 0 is undefined, tan 270° is undefined.

> **Check Point 2**
>
> Evaluate, if possible, the cosine function and the cosecant function at the following four quadrantal angles:
>
> **a.** $\theta = 0° = 0$ **b.** $\theta = 90° = \dfrac{\pi}{2}$
>
> **c.** $\theta = 180° = \pi$ **d.** $\theta = 270° = \dfrac{3\pi}{2}$

2 Use the signs of the trigonometric functions.

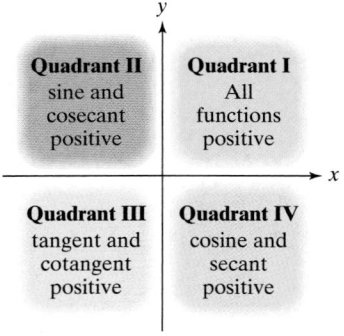

Figure 5.28 The signs of the trigonometric functions

The Signs of the Trigonometric Functions

In Example 2, we evaluated trigonometric functions of quadrantal angles. However, we will now return to the trigonometric functions of nonquadrantal angles. **If θ is not a quadrantal angle, the sign of a trigonometric function depends on the quadrant in which θ lies.** In all four quadrants, r is positive. However, x and y can be positive or negative. For example, if θ lies in quadrant II, x is negative and y is positive. Thus, the only positive ratios in this quadrant are $\dfrac{y}{r}$ and its reciprocal, $\dfrac{r}{y}$. These ratios are the function values for the sine and cosecant, respectively. In short, if θ lies in quadrant II, $\sin\theta$ and $\csc\theta$ are positive. The other four trigonometric functions are negative.

Figure 5.28 summarizes the signs of the trigonometric functions. If θ lies in quadrant I, all six functions are positive. If θ lies in quadrant II, only $\sin\theta$ and $\csc\theta$ are positive. If θ lies in quadrant III, only $\tan\theta$ and $\cot\theta$ are positive. Finally, if θ lies in quadrant IV, only $\cos\theta$ and $\sec\theta$ are positive. Observe that the positive functions in each quadrant occur in reciprocal pairs.

EXAMPLE 3   Finding the Quadrant in Which an Angle Lies

If $\tan \theta < 0$ and $\cos \theta > 0$, name the quadrant in which angle θ lies.

Solution   Because $\tan \theta < 0$, θ cannot lie in quadrant I; all the functions are positive in quadrant I. Furthermore, θ cannot lie in quadrant III; $\tan \theta$ is positive in quadrant III. Thus, with $\tan \theta < 0$, θ lies in quadrant II or quadrant IV. We are also given that $\cos \theta > 0$. Because quadrant IV is the only quadrant in which the cosine is positive and the tangent is negative, we conclude that θ lies in quadrant IV.

Check Point 3	If $\sin \theta < 0$ and $\cos \theta < 0$, name the quadrant in which angle θ lies.

EXAMPLE 4   Evaluating Trigonometric Functions

Given $\tan \theta = -\frac{2}{3}$ and $\cos \theta > 0$, find $\cos \theta$ and $\csc \theta$.

Solution   Because the tangent is negative and the cosine is positive, θ lies in quadrant IV. This will help us to determine whether the negative sign in $\tan \theta = -\frac{2}{3}$ should be associated with the numerator or the denominator. Keep in mind that in quadrant IV, x is positive and y is negative. Thus,

In quadrant IV, y is negative.

$$\tan \theta = -\frac{2}{3} = \frac{y}{x} = \frac{-2}{3}$$

(See Figure 5.29). Thus, $x = 3$ and $y = -2$. Furthermore,
$$r = \sqrt{x^2 + y^2} = \sqrt{3^2 + (-2)^2} = \sqrt{9 + 4} = \sqrt{13}.$$
Now that we know $x, y,$ and r, we can find $\cos \theta$ and $\csc \theta$.

$$\cos \theta = \frac{x}{r} = \frac{3}{\sqrt{13}} = \frac{3}{\sqrt{13}} \cdot \frac{\sqrt{13}}{\sqrt{13}} = \frac{3\sqrt{13}}{13} \qquad \csc \theta = \frac{r}{y} = \frac{\sqrt{13}}{-2} = -\frac{\sqrt{13}}{2}$$

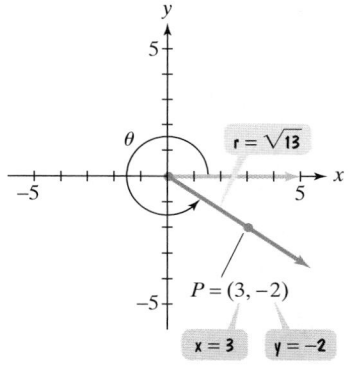

Figure 5.29 $\tan \theta = -\frac{2}{3}$ and $\cos \theta > 0$

Check Point 4	Given $\tan \theta = -\frac{1}{3}$ and $\cos \theta < 0$, find $\sin \theta$ and $\sec \theta$.

3 Find reference angles.

Reference Angles

We will often evaluate trigonometric functions of positive angles greater than $90°$ and all negative angles by making use of a positive acute angle. This positive acute angle is called a *reference angle*.

Definition of a Reference Angle

Let θ be a nonacute angle in standard position that lies in a quadrant. Its **reference angle** is the positive acute angle θ' formed by the terminal side of θ and the x-axis.

Figure 5.30 shows the reference angle for θ lying in quadrants II, III, and IV. Notice that the formula used to find θ, the reference angle, varies according to the quadrant in which θ lies. You may find it easier to find the reference angle for a given angle by making a figure that shows the angle in standard position. The acute angle formed by the terminal side of this angle and the *x*-axis is the reference angle.

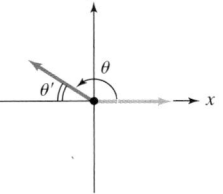 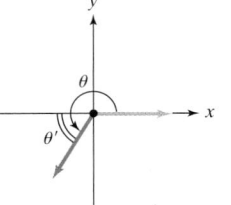 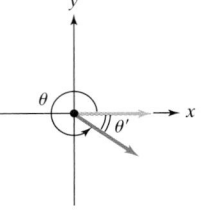

Figure 5.30 Reference angles for positive angles in quadrants II, III, and IV

If $90° < \theta < 180°$, then $\theta' = 180° - \theta$.

If $180° < \theta < 270°$, then $\theta' = \theta - 180°$.

If $270° < \theta < 360°$, then $\theta' = 360° - \theta$.

EXAMPLE 5   Finding Reference Angles

Find the reference angle, θ', for each of the following angles:

a. $\theta = 345°$ **b.** $\theta = \dfrac{5\pi}{6}$ **c.** $\theta = -135°$ **d.** $\theta = 2.5.$

Solution

a. A $345°$ angle in standard position is shown in Figure 5.31. Because $345°$ lies in quadrant IV, the reference angle is

$$\theta' = 360° - 345° = 15°.$$

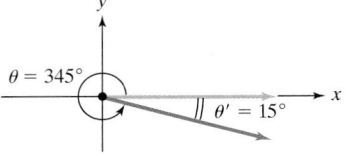

Figure 5.31

b. Because $\dfrac{5\pi}{6}$ lies between $\dfrac{\pi}{2} = \dfrac{3\pi}{6}$ and $\pi = \dfrac{6\pi}{6}$, $\theta = \dfrac{5\pi}{6}$ lies in quadrant II. The angle is shown in Figure 5.32. The reference angle is

$$\theta' = \pi - \frac{5\pi}{6} = \frac{6\pi}{6} - \frac{5\pi}{6} = \frac{\pi}{6}.$$

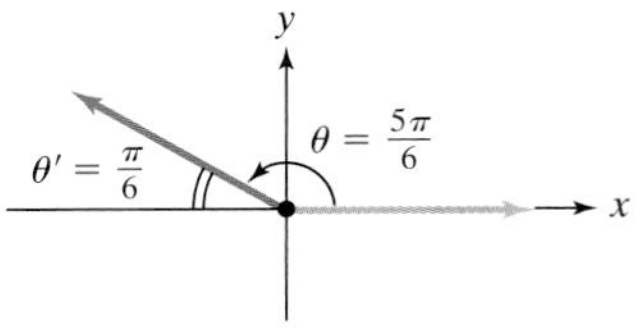

Figure 5.32

c. A $-135°$ angle in standard position is shown in Figure 5.33. The figure indicates that the positive acute angle formed by the terminal side of θ and the *x*-axis is $45°$. The reference angle is

$$\theta' = 45°.$$

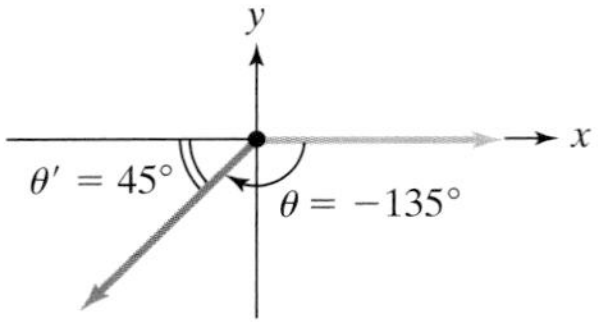

Figure 5.33

d. The angle $\theta = 2.5$ lies between $\dfrac{\pi}{2} \approx 1.57$ and $\pi \approx 3.14$. This means that $\theta = 2.5$ is in quadrant II, shown in Figure 5.34. The reference angle is

$$\theta' = \pi - 2.5 \approx 0.64.$$

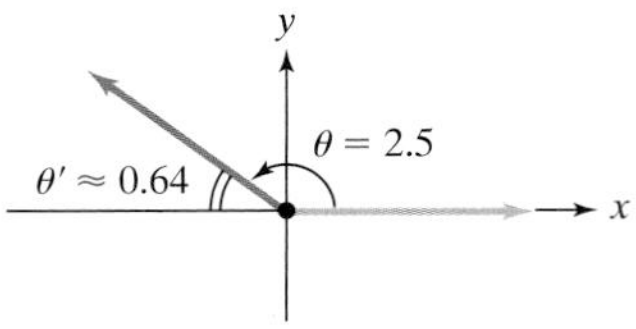

Figure 5.34

> **Check Point 5**
> Find the reference angle, θ', for each of the following angles:
> **a.** $\theta = 210°$ **b.** $\theta = \dfrac{7\pi}{4}$ **c.** $\theta = -240°$ **d.** $\theta = 3.6$.

4 Use reference angles to evaluate trigonometric functions.

The way that reference angles are defined makes them useful in evaluating trigonometric functions.

> **Using Reference Angles to Evaluate Trigonometric Functions**
>
> The values of the trigonometric functions of a given angle, θ, are the same as the values of the trigonometric functions of the reference angle, θ', except possibly for the sign. A function value of the acute angle, θ', is always positive. However, the same function value for θ may be positive or negative.

For example, we can use a reference angle, θ', to obtain an exact value for $\tan 120°$. The reference angle for $\theta = 120°$ is $\theta' = 180° - 120° = 60°$. We know the exact value for the tangent function of the reference angle: $\tan 60° = \sqrt{3}$. We also know that the value of a trigonometric function for a given angle, θ, is the same as that for its reference angle, θ', except possibly for the sign. Thus, we can conclude that $\tan 120°$ equals $-\sqrt{3}$ or $\sqrt{3}$.

What sign should we attach to $\sqrt{3}$? A $120°$ angle lies in quadrant II, where sine and cosecant are positive. Thus, the tangent function is negative for a $120°$ angle. Therefore,

> Prefix by a negative sign to show tangent is negative in quadrant II.

$$\tan 120° = -\tan 60° = -\sqrt{3}.$$

> The reference angle for $120°$ is $60°$.

In the previous section, we used two right triangles to find exact trigonometric values of $30°$, $45°$, and $60°$. Using a procedure similar to finding $\tan 120°$, we can now find the function values of all angles for which $30°$, $45°$, or $60°$ are reference angles.

> **A Procedure for Using Reference Angles to Evaluate Trigonometric Functions**
>
> The value of a trigonometric function of any angle θ is found as follows:
> **1.** Find the associated reference angle, θ', and the function value for θ'.
> **2.** Use the quadrant in which θ lies to prefix the appropriate sign to the function value in step 1.

Discovery

Draw the two right triangles involving $30°$, $45°$, and $60°$. Indicate the length of each side. Use these lengths to verify the function values for the reference angles in the solution to Example 6.

EXAMPLE 6 Using Reference Angles to Evaluate Trigonometric Functions

Use reference angles to find the exact value of each of the following trigonometric functions.

a. $\sin 135°$ **b.** $\cos \dfrac{4\pi}{3}$ **c.** $\cot\left(-\dfrac{\pi}{3}\right)$

Solution

a. We use our two-step procedure to find $\sin 135°$.

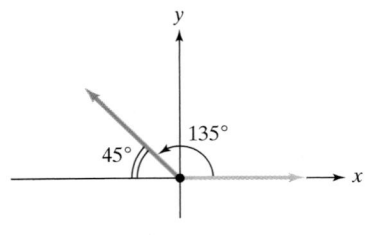

Figure 5.35 Reference angle for 135°

Step 1   Find the reference angle, θ', and $\sin \theta'$.   Figure 5.35 shows 135° lies in quadrant II. The reference angle is

$$\theta' = 180° - 135° = 45°.$$

The function value for the reference angle is $\sin 45° = \dfrac{\sqrt{2}}{2}$.

Step 2   Use the quadrant in which θ lies to prefix the appropriate sign to the function value in step 1.   The angle $\theta = 135°$ lies in quadrant II. Because the sine is positive in quadrant II, we put a $+$ sign before the function value of the reference angle. Thus,

> The sine is positive in quadrant II.

$$\sin 135° = +\sin 45° = \frac{\sqrt{2}}{2}.$$

> The reference angle for 135° is 45°.

b. We use our two-step procedure to find $\cos \dfrac{4\pi}{3}$.

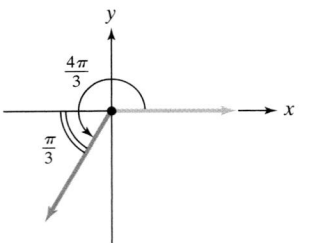

Figure 5.36 Reference angle for $\dfrac{4\pi}{3}$

Step 1   Find the reference angle, θ', and $\cos \theta'$.   Figure 5.36 shows that $\theta = \dfrac{4\pi}{3}$ lies in quadrant III. The reference angle is

$$\theta' = \frac{4\pi}{3} - \pi = \frac{4\pi}{3} - \frac{3\pi}{3} = \frac{\pi}{3}.$$

The function value for the reference angle is

$$\cos \frac{\pi}{3} = \frac{1}{2}.$$

Step 2   Use the quadrant in which θ lies to prefix the appropriate sign to the function value in step 1.   The angle $\theta = \dfrac{4\pi}{3}$ lies in quadrant III. Because only the tangent and cotangent are positive in quadrant III, the cosine is negative in this quadrant. We put a $-$ sign before the function value of the reference angle. Thus,

> The cosine is negative in quadrant III.

$$\cos \frac{4\pi}{3} = -\cos \frac{\pi}{3} = -\frac{1}{2}.$$

> The reference angle for $\frac{4\pi}{3}$ is $\frac{\pi}{3}$.

c. We use our two-step procedure to find $\cot\left(-\dfrac{\pi}{3}\right)$.

Step 1   Find the reference angle, θ', and $\cot\theta'$.   Figure 5.37 shows that $\theta = -\dfrac{\pi}{3}$ lies in quadrant IV. The reference angle is $\theta' = \dfrac{\pi}{3}$. The function value for the reference angle is $\cot\dfrac{\pi}{3} = \dfrac{\sqrt{3}}{3}$.

Step 2   Use the quadrant in which θ lies to prefix the appropriate sign to the function value in step 1.   The angle $\theta = -\dfrac{\pi}{3}$ lies in quadrant IV. Because only the cosine and secant are positive in quadrant IV, the cotangent is negative in this quadrant. We put a $-$ sign before the function value of the reference angle. Thus,

The cotangent is negative in quadrant IV.

$$\cot\left(-\frac{\pi}{3}\right) = -\cot\frac{\pi}{3} = -\frac{\sqrt{3}}{3}.$$

The reference angle for $-\frac{\pi}{3}$ is $\frac{\pi}{3}$.

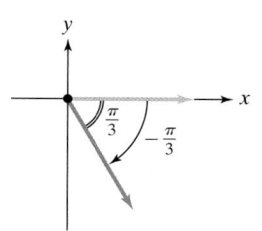

Figure 5.37 Reference angle for $-\dfrac{\pi}{3}$

Check Point 6   Use reference angles to find the exact value of the following trigonometric functions.

a. $\sin 300°$ **b.** $\tan\dfrac{5\pi}{4}$ **c.** $\sec\left(-\dfrac{\pi}{6}\right)$

EXERCISE SET 5.3

✓ Practice Exercises

In Exercises 1–8, a point on the terminal side of angle θ is given. Find the exact value of each of the six trigonometric functions of θ.

1. $(-4, 3)$ **2.** $(-12, 5)$
3. $(2, 3)$ **4.** $(3, 7)$
5. $(3, -3)$ **6.** $(5, -5)$
7. $(-2, -5)$ **8.** $(-1, -3)$

In Exercises 9–16, evaluate the trigonometric function at the quadrantal angle, or state that the expression is undefined.

9. $\cos\pi$ **10.** $\tan\pi$
11. $\sec\pi$ **12.** $\csc\pi$
13. $\tan\dfrac{3\pi}{2}$ **14.** $\cos\dfrac{3\pi}{2}$
15. $\cot\dfrac{\pi}{2}$ **16.** $\tan\dfrac{\pi}{2}$

In Exercises 17–22, let θ be an angle in standard position. Name the quadrant in which θ lies.

17. $\sin\theta > 0,\quad \cos\theta > 0$

18. $\sin\theta < 0,\quad \cos\theta > 0$
19. $\sin\theta < 0,\quad \cos\theta < 0$
20. $\tan\theta < 0,\quad \sin\theta < 0$
21. $\tan\theta < 0,\quad \cos\theta < 0$
22. $\cot\theta > 0,\quad \sec\theta < 0$

In Exercises 23–34, find the exact value of each of the remaining trigonometric functions of θ.

23. $\cos\theta = -\frac{3}{5},\quad \theta$ in quadrant III
24. $\sin\theta = -\frac{12}{13},\quad \theta$ in quadrant III
25. $\sin\theta = \frac{5}{13},\quad \theta$ in quadrant II
26. $\cos\theta = \frac{4}{5},\quad \theta$ in quadrant IV
27. $\cos\theta = \frac{8}{17},\quad 270° < \theta < 360°$
28. $\cos\theta = \frac{1}{3},\quad 270° < \theta < 360°$
29. $\tan\theta = -\frac{2}{3},\quad \sin\theta > 0$
30. $\tan\theta = -\frac{1}{3},\quad \sin\theta > 0$
31. $\tan\theta = \frac{4}{3},\quad \cos\theta < 0$
32. $\tan\theta = \frac{5}{12},\quad \cos\theta < 0$
33. $\sec\theta = -3,\quad \tan\theta > 0$
34. $\csc\theta = -4,\quad \tan\theta > 0$

In Exercises 35–50, find the reference angle for each angle.

35. 160°

36. 170°

37. 205°

38. 210°

39. 355°

40. 351°

41. $\dfrac{7\pi}{4}$

42. $\dfrac{5\pi}{4}$

43. $\dfrac{5\pi}{6}$

44. $\dfrac{5\pi}{7}$

45. −150°

46. −250°

47. −335°

48. −359°

49. 4.7

50. 5.5

In Exercises 51–66, use reference angles to find the exact value of each expression. Do not use a calculator.

51. $\cos 225°$

52. $\sin 300°$

53. $\tan 210°$

54. $\sec 240°$

55. $\tan 420°$

56. $\tan 405°$

57. $\sin \dfrac{2\pi}{3}$

58. $\cos \dfrac{3\pi}{4}$

59. $\csc \dfrac{7\pi}{6}$

60. $\cot \dfrac{7\pi}{4}$

61. $\tan \dfrac{9\pi}{4}$

62. $\tan \dfrac{9\pi}{2}$

63. $\sin(-240°)$

64. $\sin(-225°)$

65. $\tan\left(-\dfrac{\pi}{4}\right)$

66. $\tan\left(-\dfrac{\pi}{6}\right)$

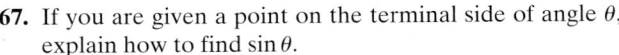

Writing in Mathematics

67. If you are given a point on the terminal side of angle θ, explain how to find $\sin\theta$.

68. Explain why $\tan 90°$ is undefined.

69. If $\cos\theta > 0$ and $\tan\theta < 0$, explain how to find the quadrant in which θ lies.

70. What is a reference angle? Give an example with your description.

71. Explain how reference angles are used to evaluate trigonometric functions. Give an example with your description.

SECTION 5.4 Trigonometric Functions of Real Numbers; Periodic Functions

Objectives

1. Use a unit circle to define trigonometric functions of real numbers.
2. Recognize the domain and range of sine and cosine functions.
3. Use even and odd trigonometric functions.
4. Use periodic properties.

Cycles govern many aspects of life—heartbeats, sleep patterns, seasons, and tides all follow regular, predictable cycles. In this section we will see why trigonometric functions are used to model phenomena that occur in cycles. To do this, we need to move beyond angles and consider trigonometric functions of real numbers.

1 Use a unit circle to define trigonometric functions of real numbers.

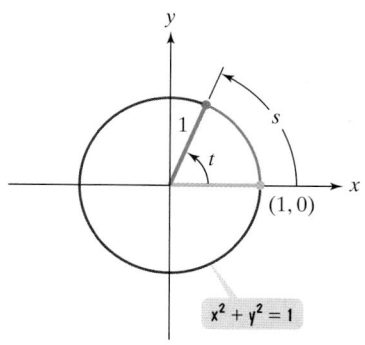

Figure 5.38 Unit circle with a central angle measuring t radians

Trigonometric Functions of Real Numbers

Thus far, we have considered trigonometric functions of angles measured in degrees or radians. To define trigonometric functions of real numbers, rather than angles, we use a unit circle. A **unit circle** is a circle of radius 1, with center at the origin of a rectangular coordinate system. The equation of this unit circle is $x^2 + y^2 = 1$. Figure 5.38 shows a unit circle in which the central angle measures t radians. We can use the formula for the length of a circular arc, $s = r\theta$, to find the length of the intercepted arc.

$$s = r\theta = 1 \cdot t = t$$

| The radius of a unit circle is 1. | The radian measure of the central angle is t. |

Thus, the length of the intercepted arc is t. This is also the radian measure of the central angle. Thus, **in a unit circle, the radian measure of the angle is equal to the measure of the intercepted arc**. Both are given by the same *real number t*.

In Figure 5.39, the radian measure of the angle and the length of the intercepted arc are both shown by t. Let $P = (x, y)$ denote the point on the unit circle that has arc length t from $(1, 0)$. Figure 5.39(a) shows that if t is positive, point P is reached by moving counterclockwise along the unit circle from $(1, 0)$. Figure 5.39(b) shows that if t is negative, point P is reached by moving clockwise along the unit circle from $(1, 0)$. For each real number t there corresponds a point $P = (x, y)$ on the unit circle.

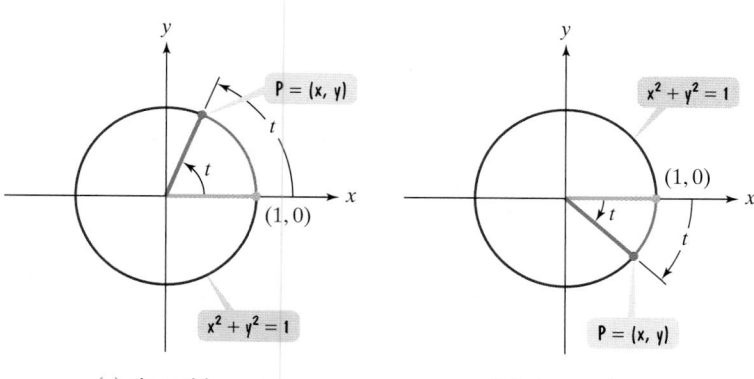

Figure 5.39   (a) t is positive.   (b) t is negative.

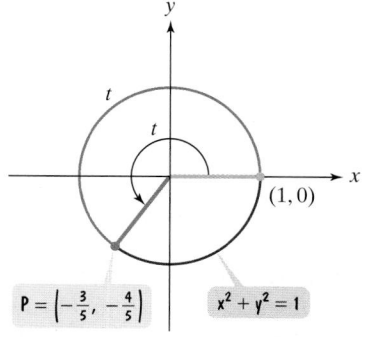

$P = \left(-\frac{3}{5}, -\frac{4}{5}\right)$ $x^2 + y^2 = 1$

Figure 5.40

Using Figure 5.39, we define the cosine function at t as the x-coordinate of P and the sine function at t as the y-coordinate of P. Thus,

$$x = \cos t \quad \text{and} \quad y = \sin t.$$

For example, a point $P = (x, y)$ on the unit circle corresponding to a real number t is shown in Figure 5.40 for $\pi < t < \dfrac{3\pi}{2}$. We see that the coordinates of $P = (x, y)$ are $x = -\frac{3}{5}$ and $y = -\frac{4}{5}$. Because the cosine function is the x-coordinate of P and the sine function is the y-coordinate of P, the values of these trigonometric functions at the real number t are

$$\cos t = -\frac{3}{5} \quad \text{and} \quad \sin t = -\frac{4}{5}.$$

Definitions of the Trigonometric Functions in Terms of a Unit Circle

If t is a real number and $P = (x, y)$ is a point on the unit circle that corresponds to t, then

$$\sin t = y \qquad\qquad \cos t = x \qquad\qquad \tan t = \frac{y}{x}, x \neq 0$$

$$\csc t = \frac{1}{y}, y \neq 0 \qquad\qquad \sec t = \frac{1}{x}, x \neq 0 \qquad\qquad \cot t = \frac{x}{y}, y \neq 0$$

Because this definition expresses function values in terms of coordinates of a point on a unit circle, the trigonometric functions are sometimes called the **circular functions**.

EXAMPLE 1   Finding Values of the Trigonometric Functions

Use Figure 5.41 to find the values of the trigonometric functions at $t = \dfrac{\pi}{2}$.

Solution   The point P on the unit circle that corresponds to $t = \dfrac{\pi}{2}$ has coordinates $(0, 1)$. We use $x = 0$ and $y = 1$ to find the values of the trigonometric functions.

$$\sin\frac{\pi}{2} = y = 1 \qquad\qquad \cos\frac{\pi}{2} = x = 0$$

$$\csc\frac{\pi}{2} = \frac{1}{y} = \frac{1}{1} = 1 \qquad\qquad \cot\frac{\pi}{2} = \frac{x}{y} = \frac{0}{1} = 0$$

By definition, $\tan t = \dfrac{y}{x}$ and $\sec t = \dfrac{1}{x}$. Because $x = 0$, $\tan\dfrac{\pi}{2}$ and $\sec\dfrac{\pi}{2}$, are undefined.

Figure 5.41

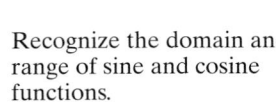

> **Check Point 1**   Use the figure on the right to find the values of the trigonometric functions at $t = \pi$.

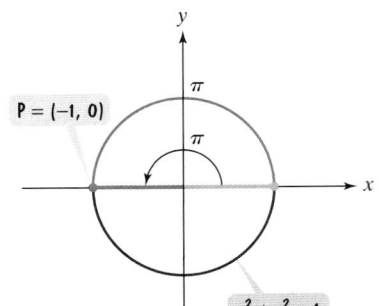

2 Recognize the domain and range of sine and cosine functions.

Domain and Range of Sine and Cosine Functions

The value of a trigonometric function at the real number t is its value at an angle of t radians. However, using real number domains, we can observe properties of trigonometric functions that are not as apparent using the angle approach. For example, the domain and range of each trigonometric function can be found from the unit circle definition. At this point, let's look only at the sine and cosine functions,

$$\sin t = y \quad \text{and} \quad \cos t = x.$$

Because t can be the radian measure of any angle or, equivalently, the measure of any intercepted arc, the domain of the sine function and the cosine function is the set of all real numbers. Because the radius of the unit circle is 1, we have

$$-1 \leq x \leq 1 \quad \text{and} \quad -1 \leq y \leq 1.$$

Therefore, with $x = \cos t$ and $y = \sin t$, we obtain

$$-1 \leq \cos t \leq 1 \quad \text{and} \quad -1 \leq \sin t \leq 1.$$

The range of the cosine and sine functions is $[-1, 1]$.

The Domain and Range of the Sine and Cosine Functions

The domain of the sine function and the cosine function is the set of all real numbers. The range of these functions is the set of all real numbers from -1 to 1, inclusive.

3 Use even and odd trigonometric functions.

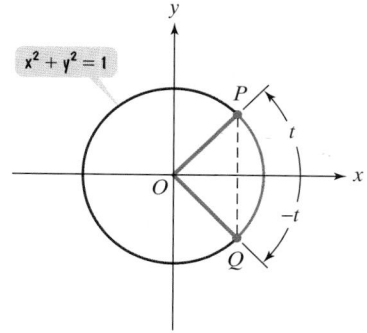

Figure 5.42

Even and Odd Trigonometric Functions

In Chapter 2, we saw that a function is even if $f(-t) = f(t)$ and odd if $f(-t) = -f(t)$. We can use Figure 5.42 to show that the cosine is an even function and the sine is an odd function. By definition, the coordinates of the points P and Q in Figure 5.42 are as follows:

$$P: \quad (\cos t, \sin t)$$

$$Q: \quad (\cos(-t), \sin(-t)).$$

In Figure 5.42, the x-coordinates of P and Q are the same. Thus,

$$\cos(-t) = \cos t.$$

This shows that cosine is an even function. By contrast, the y-coordinates of P and Q are negatives of each other. Thus,

$$\sin(-t) = -\sin t.$$

This shows that the sine is an odd function.

This argument is valid regardless of the length of t. Thus, the arc may terminate in any of the four quadrants. Using the unit circle definition of the trigonometric functions, we obtain the following results.

Even and Odd Trigonometric Functions

The cosine and secant functions are *even*.

$$\cos(-t) = \cos t \qquad \sec(-t) = \sec t$$

The sine, cosecant, tangent, and cotangent functions are *odd*.

$$\sin(-t) = -\sin t \qquad \csc(-t) = -\csc t$$

$$\tan(-t) = -\tan t \qquad \cot(-t) = -\cot t$$

EXAMPLE 2   Using Even and Odd Functions to Find Exact Values

Find the exact value of:

a. $\cos(-45°)$ b. $\tan\left(-\dfrac{\pi}{3}\right)$.

Solution

a. $\cos(-45°) = \cos 45° = \dfrac{\sqrt{2}}{2}$ b. $\tan\left(-\dfrac{\pi}{3}\right) = -\tan\dfrac{\pi}{3} = -\sqrt{3}$

Check Point 2   Find the exact value of:

a. $\cos(-60°)$ b. $\tan\left(-\dfrac{\pi}{6}\right)$.

 4 Use periodic properties.

Periodic Functions

Certain patterns in nature repeat again and again. For example, the ocean level at a beach varies between low tide and high tide approximately every 12 hours. If low tide occurs at noon, then high tide will be around 6 P.M. and low tide will occur again around midnight, and so on infinitely. If $f(t)$ represents the ocean level at the beach at any time t, then the level is the same 12 hours later. Thus,

$$f(t + 12) = f(t).$$

The word *periodic* means that this tidal behavior repeats infinitely. The *period*, 12 hours, is the time it takes to complete one full cycle.

Definition of a Periodic Function

A function f is **periodic** if there exists a positive number p such that

$$f(t + p) = f(t)$$

for all t in the domain of f. The smallest number p for which f is periodic is called the **period** of f.

The trigonometric functions are used to model periodic phenomena. Why? If we begin at any point P on the unit circle and travel a distance of 2π units along the perimeter, we will return to the same point P. Because the trigonometric functions are defined in terms of the coordinates of that point P, we obtain the following results.

Periodic Properties of the Sine and Cosine Functions

$$\sin(t + 2\pi) = \sin t \quad \text{and} \quad \cos(t + 2\pi) = \cos t$$

The sine and cosine functions are periodic functions and have period 2π.

EXAMPLE 3   Using Periodic Properties to Find Exact Values

Find the exact value of: **a.** $\tan 420°$ **b.** $\sin \dfrac{9\pi}{4}$.

Solution

a. $\tan 420° = \tan(360° + 60°) = \tan 60° = \sqrt{3}$

b. $\sin \dfrac{9\pi}{4} = \sin\left(2\pi + \dfrac{\pi}{4}\right) = \sin \dfrac{\pi}{4} = \dfrac{\sqrt{2}}{2}$

> **Check Point 3**   Find the exact value of:
>
> **a.** $\cos 405°$ **b.** $\tan \dfrac{7\pi}{3}$.

Like the sine and cosine functions, the secant and cosecant functions have period 2π. However, this is not true for the tangent and cotangent functions. If we begin at any point $P(x, y)$ on the unit circle and travel a distance of π units along the perimeter, we arrive at the point $(-x, -y)$. The tangent function, defined in terms of the coordinates of a point, is the same at (x, y) and $(-x, -y)$.

$$\underset{\substack{\text{Tangent function}\\\text{at } (x, y)}}{\dfrac{y}{x}} = \underset{\substack{\text{Tangent function}\\\pi \text{ radians later}}}{\dfrac{-y}{-x}}$$

We see that $\tan(t + \pi) = \tan t$. The same observations apply to the cotangent function.

Periodic Properties of the Tangent and Cotangent Functions

$$\tan(t + \pi) = \tan t \quad \text{and} \quad \cot(t + \pi) = \cot t$$

The tangent and cotangent functions are periodic functions and have period π.

Why do the trigonometric functions model phenomena that repeat *indefinitely*? By starting at point P on the unit circle and traveling a distance of 2π units, 4π units, 6π units, and so on, we return to the starting point P. Because the trigonometric functions are defined in terms of the coordinates of that point P, if we add (or subtract) multiples of 2π, the trigonometric values do not change.

Furthermore, the trigonometric values for the tangent and cotangent functions do not change if we add (or subtract) multiples of π.

Repetitive Behavior of the Sine, Cosine, and Tangent Functions

For any integer n and real number t,

$$\sin(t + 2\pi n) = \sin t, \quad \cos(t + 2\pi n) = \cos t, \quad \text{and} \quad \tan(t + \pi n) = \tan t$$

EXERCISE SET 5.4

✓ Practice Exercises

In Exercises 1–4, a point $P(x, y)$ is shown on the unit circle corresponding to a real number t. Find the values of the trigonometric functions at t.

1.

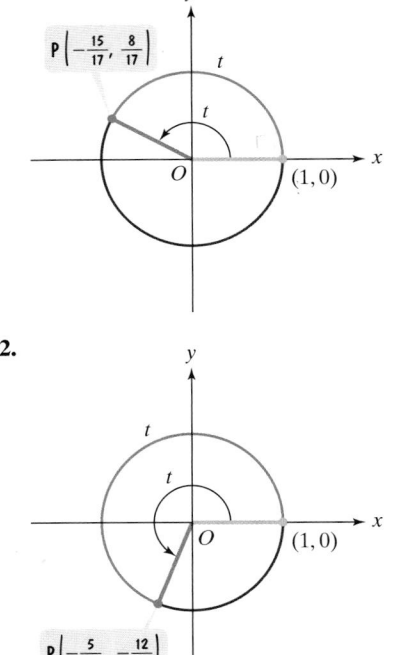

2.

3.

4.

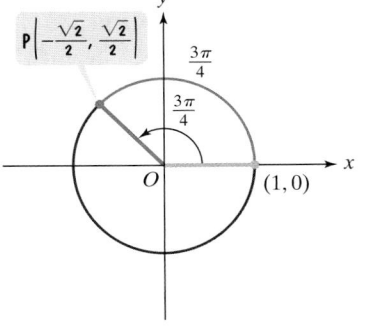

In Exercises 5–8, use even and odd properties of the trigonometric functions to find the exact value of each expression.

5. $\sin(-45°)$

6. $\tan(-45°)$

7. $\sec\left(-\dfrac{\pi}{3}\right)$

8. $\sec\left(-\dfrac{\pi}{6}\right)$

In Exercises 9–12, use periodic properties to find the exact value of each expression.

9. $\cos 585°$

10. $\cos 570°$

11. $\cot \dfrac{7\pi}{3}$

12. $\cot \dfrac{9\pi}{4}$

⭐ Application Exercises

13. The number of hours of daylight, H, on day t of any given year (on January 1, $t = 1$) in Fairbanks, Alaska, can be modeled by the function

$$H = 12 + 8.3 \sin\left[\dfrac{2\pi}{365}(t - 80)\right].$$

a. March 21, the 80th day of the year, is the spring equinox. Find the number of hours of daylight in Fairbanks on this day.

b. June 21, the 172nd day of the year, is the summer solstice, the day with the maximum number of hours of daylight. To the nearest tenth of an hour, find the number of hours of daylight in Fairbanks on this day.

c. December 21, the 355th day of the year, is the winter solstice, the day with the minimum number of hours of daylight. Find, to the nearest tenth of an hour, the number of hours of daylight in Fairbanks on this day.

14. The number of hours of daylight, H, on day t of any given year (on January 1, $t = 1$) in San Diego, California, can be modeled by the function

$$H = 12 + 2.4 \sin\left[\frac{2\pi}{365}(t - 80)\right].$$

a. March 21, the 80th day of the year, is the spring equinox. Find the number of hours of daylight in San Diego on this day.

b. June 21, the 172nd day of the year, is the summer solstice, the day with the maximum number of hours of daylight. Find, to the nearest tenth of an hour, the number of hours of daylight in San Diego on this day.

c. December 21, the 355th day of the year, is the winter solstice, the day with the minimum number of hours of daylight. To the nearest tenth of an hour, find the number of hours of daylight in San Diego on this day.

15. People who believe in biorhythms claim that there are three cycles that rule our behavior—the physical, emotional, and mental. Each is a sine function of a certain period. The function for our emotional fluctuations is

$$E = \sin\frac{\pi}{14}t$$

where t is measured in days starting at birth. Emotional fluctuations, E, are measured from -1 to 1, inclusive, with 1 representing peak emotional well-being, -1 representing the low for emotional well-being, and 0 representing feeling neither emotionally high or low.

a. Find E corresponding to $t = 7, 14, 21, 28,$ and 35. Describe what you observe.

b. What is the period of the emotional cycle?

16. The height of the water, H, in feet, at a boat dock t hours after 6 A.M. is given by

$$H = 10 + 4\sin\frac{\pi}{6}t.$$

a. Find the height of the water at the dock at 6 A.M., 9 A.M., noon, 6 P.M., midnight, and 3 A.M.

b. When is low tide and when is high tide?

c. What is the period of this function and what does this mean about the tides?

Writing in Mathematics

17. Why are the trigonometric functions sometimes called circular functions?

18. What is the range of the sine function? Use the unit circle to explain where this range comes from.

19. What do we mean by even trigonometric functions? Which of the six functions fall into this category?

20. What is a periodic function? Why are the sine and cosine functions periodic?

21. Explain how you can use the function for emotional fluctuations in Exercise 15 to determine good days for having dinner with your moody boss.

22. Describe a phenomenon that repeats infinitely. What is its period?

Critical Thinking Exercises

23. Find the exact value of $\cos 0° + \cos 1° + \cos 2° + \cos 3° + \cdots + \cos 179° + \cos 180°$.

24. If $f(x) = \sin x$ and $f(a) = \frac{1}{4}$, find the value of

$$f(a) + f(a + 2\pi) + f(a + 4\pi) + f(a + 6\pi).$$

25. If $f(x) = \sin x$ and $f(a) = \frac{1}{4}$, find the value of $f(a) + 2f(-a)$.

26. The seats of a ferris wheel are 40 feet from the wheel's center. When you get on the ride, your seat is 5 feet above the ground. How far above the ground are you after rotating through an angle of 765°?

SECTION 5.5 *Graphs of Sine and Cosine Functions*

Objectives

1. Understand the graph of $y = \sin x$.
2. Graph variations of $y = \sin x$.
3. Understand the graph of $y = \cos x$.
4. Graph variations of $y = \cos x$.
5. Use vertical shifts of sine and cosine curves.
6. Model periodic behavior.

Take a deep breath and relax. Many relaxation exercises involve slowing down our breathing. Some people suggest that the way we breathe affects every part of our lives. Did you know that graphs of trigonometric functions can be used to analyze the breathing cycle, which is our closest link to both life and death?

In this section, we use graphs of sine and cosine functions to visualize their properties. We use the traditional symbol x, rather than θ or t, to represent the independent variable. We use the symbol y for the dependent variable, or the function's value at x. Thus, we will be graphing $y = \sin x$ and $y = \cos x$ in rectangular coordinates. In all graphs of trigonometric functions, the independent variable, x, is measured in radians.

1 Understand the graph of $y = \sin x$.

The Graph of $y = \sin x$

The trigonometric functions can be graphed in a rectangular coordinate system by plotting points whose coordinates belong to the function. Thus, we graph $y = \sin x$ by listing some points on the graph. Because the period of the sine function is 2π, we will graph the function on the interval $[0, 2\pi]$. The rest of the graph is made up of repetitions of this portion.

Table 5.1 lists some values of (x, y) on the graph of $y = \sin x, 0 \le x \le 2\pi$.

Table 5.1   Values of (x, y) on $y = \sin x$

x	0	$\dfrac{\pi}{6}$	$\dfrac{\pi}{3}$	$\dfrac{\pi}{2}$	$\dfrac{2\pi}{3}$	$\dfrac{5\pi}{6}$	π	$\dfrac{7\pi}{6}$	$\dfrac{4\pi}{3}$	$\dfrac{3\pi}{2}$	$\dfrac{5\pi}{3}$	$\dfrac{11\pi}{6}$	2π
$y = \sin x$	0	$\dfrac{1}{2}$	$\dfrac{\sqrt{3}}{2}$	1	$\dfrac{\sqrt{3}}{2}$	$\dfrac{1}{2}$	0	$-\dfrac{1}{2}$	$-\dfrac{\sqrt{3}}{2}$	-1	$-\dfrac{\sqrt{3}}{2}$	$-\dfrac{1}{2}$	0

As x increases from 0 to $\frac{\pi}{2}$, y increases from 0 to 1.

As x increases from $\frac{\pi}{2}$ to π, y decreases from 1 to 0.

As x increases from π to $\frac{3\pi}{2}$, y decreases from 0 to −1.

As x increases from $\frac{3\pi}{2}$ to 2π, y increases from −1 to 0.

In plotting the points obtained in Table 5.1, we will use the approximation $\dfrac{\sqrt{3}}{2} \approx 0.87$. Rather than approximating π, we will mark off units on the x-axis in terms of π. If we connect these points with a smooth curve, we obtain the graph shown in Figure 5.43. The figure shows one period of the graph of $y = \sin x$.

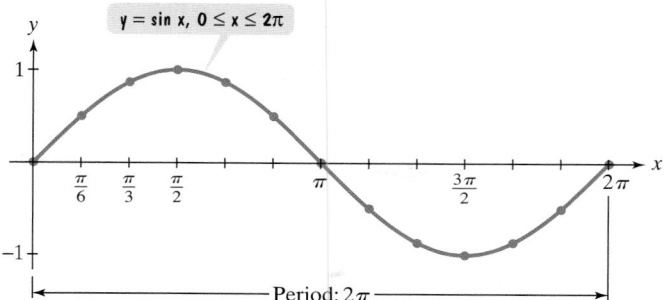

Figure 5.43 One period of the graph of $y = \sin x$

We can obtain a more complete graph of $y = \sin x$ by continuing the portion shown in Figure 5.43 to the left and right. The graph of the sine function, called a **sine curve**, is shown in Figure 5.44. Any part of the graph that corresponds to one period (2π) is one cycle of the graph of $y = \sin x$.

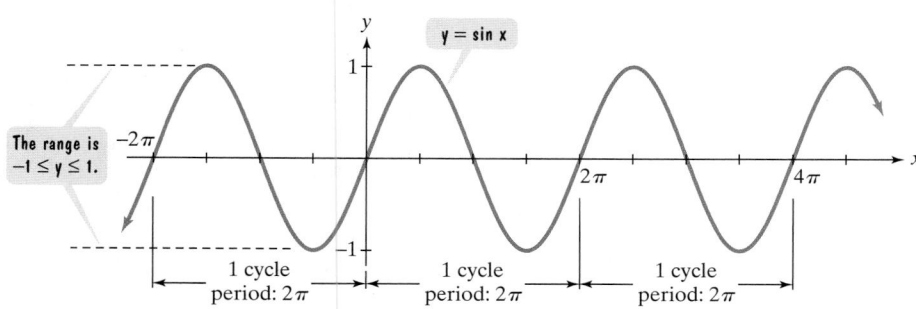

Figure 5.44 The graph of $y = \sin x$

The graph of $y = \sin x$ allows us to visualize some of the properties of the sine function.

- The domain is the set of all real numbers. The graph extends indefinitely to the left and to the right with no gaps or holes.
- The range consists of all numbers between -1 and 1 inclusive. The graph never rises above 1 or falls below -1.
- The period is 2π. The graph's pattern repeats in every interval of length 2π.
- The function is an odd function: $\sin(-x) = -\sin x$. This can be seen by observing that the graph is symmetric with respect to the origin.

2 Graph variations of $y = \sin x$.

Graphing Variations of $y = \sin x$

To graph variations of $y = \sin x$ by hand, it is helpful to find x-intercepts, maximum points, and minimum points. One complete cycle of the sine curve includes three x-intercepts, one maximum point, and one minimum point. The graph of $y = \sin x$ has x-intercepts at the beginning, middle, and end of its full period, shown in Figure 5.45. The curve reaches its maximum point $\frac{1}{4}$ of the way through the period. It reaches its minimum point $\frac{3}{4}$ of the way through the period. Thus, key points in graphing sine functions are obtained by dividing the

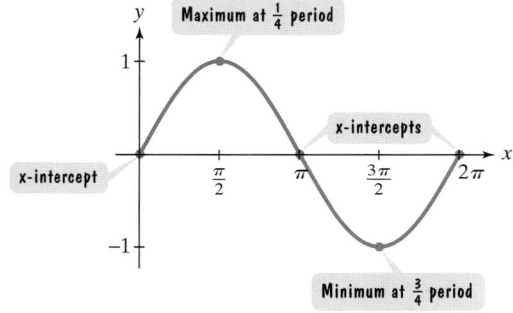

Figure 5.45 Key points in graphing the sine function

period into four equal parts. The x-coordinates of the five key points are as follows:

$$x_1 = \text{value of } x \text{ where the cycle begins}$$

$$x_2 = x_1 + \frac{\text{period}}{4}$$

$$x_3 = x_2 + \frac{\text{period}}{4}$$

$$x_4 = x_3 + \frac{\text{period}}{4}$$

$$x_5 = x_4 + \frac{\text{period}}{4}.$$

Add "quarter-periods" to find successive value of x.

The y-coordinates of the five key points are obtained by evaluating the given function at each of these values of x.

The graph of $y = \sin x$ forms the basis for graphing functions of the form

$$y = A \sin x.$$

For example, consider $y = 2 \sin x$, in which $A = 2$. We can obtain the graph of $y = 2 \sin x$ from that of $y = \sin x$ if we multiply each y-coordinate on the graph of $y = \sin x$ by 2. Figure 5.46 shows the graphs. The basic sine curve is *stretched* and ranges between -2 and 2 rather than between -1 and 1. However, both $y = \sin x$ and $y = 2 \sin x$ have a period of 2π.

In general, the graph of $y = A \sin x$ ranges between $-A$ and A. Thus, the range of the function is $-A \le y \le A$. If $A > 1$, the basic sine curve is *stretched*, as in Figure 5.46. If $A < 1$, the basic sine curve is *shrunk*. We call $|A|$ the **amplitude** of $y = A \sin x$. The maximum value of y on the graph of $y = A \sin x$ is $|A|$, the amplitude.

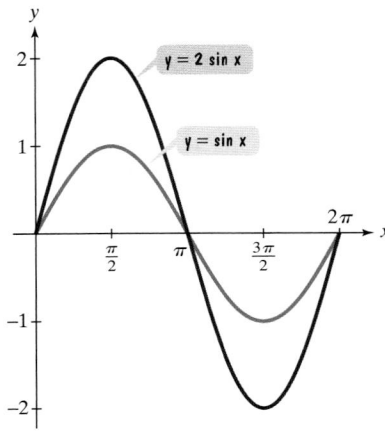

Figure 5.46 Comparing the graphs of $y = \sin x$ and $y = 2 \sin x$

Graphing Variations of $y = \sin x$

1. Identify the amplitude and the period.
2. Find the values of x for the five key points—the three x-intercepts, the maximum point, and the minimum point. Start with the value of x where the cycle begins and add quarter-periods—that is, $\dfrac{\text{period}}{4}$—to find successive values of x.
3. Find the values of y for the five key points by evaluating the function at each value of x from step 2.
4. Connect the five key points with a smooth curve and graph one complete cycle of the given function.
5. Extend the graph in step 4 to the left or right as desired.

EXAMPLE 1   Graphing a Variation of $y = \sin x$

Determine the amplitude of $y = \frac{1}{2} \sin x$. Then graph $y = \sin x$ and $y = \frac{1}{2} \sin x$ for $0 \le x \le 2\pi$.

Solution

Step 1   Identify the amplitude and the period.   The equation $y = \frac{1}{2} \sin x$ is of the form $y = A \sin x$ with $A = \frac{1}{2}$. Thus, the amplitude is $|A| = \frac{1}{2}$. This means that the maximum value of y is $\frac{1}{2}$ and the minimum value of y is $-\frac{1}{2}$. The period for both $y = \frac{1}{2} \sin x$ and $y = \sin x$ is 2π.

Step 2 Find the values of x for the five key points. We need to find the three x-intercepts, the maximum point, and the minimum point on the interval $[0, 2\pi]$. To do so, we begin by dividing the period, 2π, by 4.

$$\frac{\text{period}}{4} = \frac{2\pi}{4} = \frac{\pi}{2}$$

We start with the value of x where the cycle begins: $x = 0$. Now we add quarter-periods, $\dfrac{\pi}{2}$, to generate x-values for each of the key points. The five x-values are

$$x = 0, \quad x = 0 + \frac{\pi}{2} = \frac{\pi}{2}, \quad x = \frac{\pi}{2} + \frac{\pi}{2} = \pi,$$

$$x = \pi + \frac{\pi}{2} = \frac{3\pi}{2}, \quad x = \frac{3\pi}{2} + \frac{\pi}{2} = 2\pi.$$

Step 3 Find the values of y for the five key points. We evaluate the function at each value of x from step 2.

Value of x	Value of y: $y = \frac{1}{2}\sin x$		Coordinates of key point	
0	$y = \dfrac{1}{2}\sin 0 = \dfrac{1}{2} \cdot 0 = 0$		$(0, 0)$	
$\dfrac{\pi}{2}$	$y = \dfrac{1}{2}\sin\dfrac{\pi}{2} = \dfrac{1}{2} \cdot 1 = \dfrac{1}{2}$		$\left(\dfrac{\pi}{2}, \dfrac{1}{2}\right)$	maximum point
π	$y = \dfrac{1}{2}\sin\pi = \dfrac{1}{2} \cdot 0 = 0$		$(\pi, 0)$	
$\dfrac{3\pi}{2}$	$y = \dfrac{1}{2}\sin\dfrac{3\pi}{2} = \dfrac{1}{2}(-1) = -\dfrac{1}{2}$		$\left(\dfrac{3\pi}{2}, -\dfrac{1}{2}\right)$	minimum point
2π	$y = \dfrac{1}{2}\sin 2\pi = \dfrac{1}{2} \cdot 0 = 0$		$(2\pi, 0)$	

There are x-intercepts at $0, \pi,$ and 2π. The maximum and minimum points are indicated by the voice balloons.

Step 4 Connect the five key points with a smooth curve and graph one complete cycle of the given function. The five key points for $y = \frac{1}{2}\sin x$ are shown in Figure 5.47. By connecting the points with a smooth curve, the figure shows one complete cycle of $y = \frac{1}{2}\sin x$. Also shown is the graph of $y = \sin x$. The graph of $y = \frac{1}{2}\sin x$ shrinks the graph of $y = \sin x$.

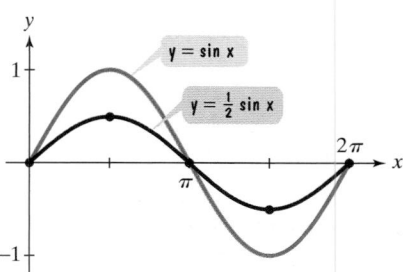

Figure 5.47 The graphs of $y = \sin x$ and $y = \frac{1}{2}\sin x, 0 \le x \le 2\pi$

> **Check Point 1** Determine the amplitude of $y = 3 \sin x$. Then graph $y = \sin x$ and $y = 3 \sin x$ for $0 \le x \le 2\pi$.

EXAMPLE 2 Graphing a Variation of $y = \sin x$

Determine the amplitude of $y = -2 \sin x$. Then graph $y = \sin x$ and $y = -2 \sin x$ for $-\pi \le x \le 3\pi$.

Solution

Step 1 Identify the amplitude and the period. The equation $y = -2 \sin x$ is of the form $y = A \sin x$ with $A = -2$. Thus, the amplitude is $|A| = |-2| = 2$. This means that the maximum value of y is 2 and the minimum value of y is -2. Both $y = \sin x$ and $y = -2 \sin x$ have a period of 2π.

Step 2 Find the x-values for the five key points. Begin by dividing the period, 2π, by 4.

$$\frac{\text{period}}{4} = \frac{2\pi}{4} = \frac{\pi}{2}$$

Start with the value of x where the cycle begins: $x = 0$. Adding quarter-periods, $\frac{\pi}{2}$, the five x-values for the key points are

$$x = 0, \quad x = 0 + \frac{\pi}{2} = \frac{\pi}{2}, \quad x = \frac{\pi}{2} + \frac{\pi}{2} = \pi,$$

$$x = \pi + \frac{\pi}{2} = \frac{3\pi}{2}, \quad x = \frac{3\pi}{2} + \frac{\pi}{2} = 2\pi.$$

Step 3 Find the values of y for the five key points. We evaluate the function at each value of x from step 2.

Value of x	Value of y: $y = -2 \sin x$	Coordinates of key point	
0	$y = -2 \sin 0 = -2 \cdot 0 = 0$	$(0, 0)$	
$\frac{\pi}{2}$	$y = -2 \sin \frac{\pi}{2} = -2 \cdot 1 = -2$	$\left(\frac{\pi}{2}, -2\right)$	minimum point
π	$y = -2 \sin \pi = -2 \cdot 0 = 0$	$(\pi, 0)$	
$\frac{3\pi}{2}$	$y = -2 \sin \frac{3\pi}{2} = -2(-1) = 2$	$\left(\frac{3\pi}{2}, 2\right)$	maximum point
2π	$y = -2 \sin 2\pi = -2 \cdot 0 = 0$	$(2\pi, 0)$	

There are x-intercepts at 0, π, and 2π. The minimum and maximum points are indicated by the voice balloons.

Step 4 Connect the five key points with a smooth curve and graph one complete cycle of the given function. The five key points for $y = -2 \sin x$ are shown in Figure 5.48. By connecting the points with a smooth curve, the red portion shows one complete cycle of $y = -2 \sin x$. Also shown in blue is one complete cycle of the graph of $y = \sin x$.

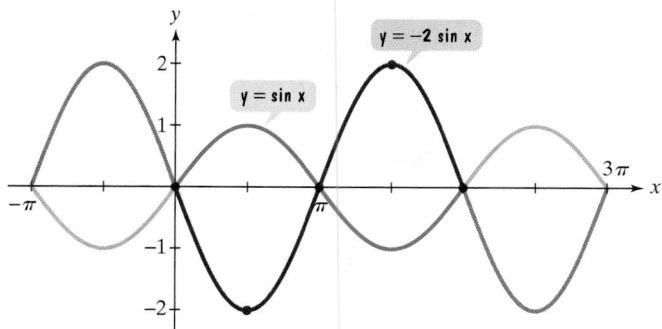

Figure 5.48 The graphs of $y = \sin x$ and $y = -2 \sin x, -\pi \le x \le 3\pi$

Step 5 **Extend the graph in step 4 to the left or right as desired.** The red and blue portions of the graphs in Figure 5.48 are from 0 to 2π. In order to graph for $-\pi \le x \le 3\pi$, continue the pattern of each graph to the left and right. These extensions are shown in the lighter colors in Figure 5.48.

> **Check Point 2**
> Determine the amplitude of $y = -\frac{1}{2} \sin x$. Then graph $y = \sin x$ and $y = -\frac{1}{2} \sin x$ for $-\pi \le x \le 3\pi$.

Now let us examine the graphs of functions of the form $y = A \sin Bx$, where B is the coefficient of x. How do such graphs compare to those of functions of the form $y = A \sin x$? We know that $y = A \sin x$ completes one cycle from $x = 0$ to $x = 2\pi$. Thus, $y = A \sin Bx$ completes one cycle from $Bx = 0$ to $Bx = 2\pi$. Solve each of these equations for x.

$$Bx = 0 \qquad Bx = 2\pi$$

$$x = 0 \qquad x = \frac{2\pi}{B} \qquad \text{Divide both sides of each equation by B.}$$

This means that $y = A \sin Bx$ completes one cycle from 0 to $\frac{2\pi}{B}$. The period is $\frac{2\pi}{B}$.

Amplitudes and Periods

The graph of $y = A \sin Bx$ has

$$\text{amplitude} = |A|$$

$$\text{period} = \frac{2\pi}{B}.$$

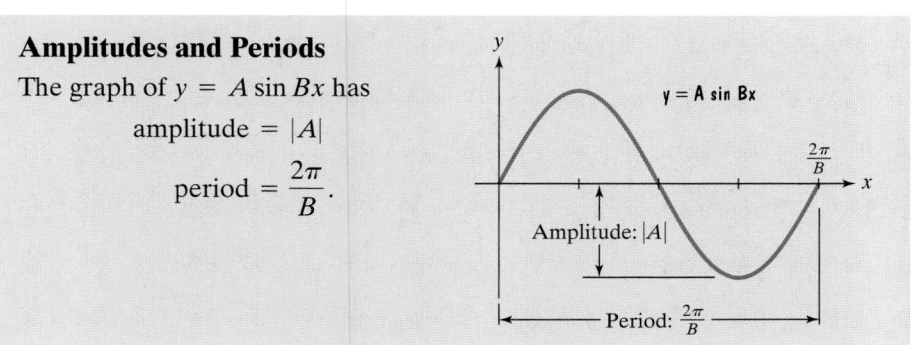

EXAMPLE 3 **Graphing a Function of the Form $y = A \sin Bx$**

Determine the amplitude and period of $y = 3 \sin 2x$. Then graph the function for $0 \le x \le 2\pi$.

Solution

Step 1   Identify the amplitude and the period.   The equation $y = 3 \sin 2x$ is of the form $y = A \sin Bx$ with $A = 3$ and $B = 2$.

$$\text{amplitude:} \qquad |A| = |3| = 3$$

$$\text{period:} \qquad \frac{2\pi}{B} = \frac{2\pi}{2} = \pi$$

The amplitude, 3, tells us that the maximum value of y is 3 and the minimum value of y is -3.

Step 2   Find the x-values for the five key points.   Begin by dividing the period, π, by 4.

$$\frac{\text{period}}{4} = \frac{\pi}{4}$$

Start with the value of x where the cycle begins: $x = 0$. Adding quarter-periods, $\frac{\pi}{4}$, the five x-values for the key points are

$$x = 0, \quad x = 0 + \frac{\pi}{4} = \frac{\pi}{4}, \quad x = \frac{\pi}{4} + \frac{\pi}{4} = \frac{\pi}{2},$$

$$x = \frac{\pi}{2} + \frac{\pi}{4} = \frac{3\pi}{4}, \quad x = \frac{3\pi}{4} + \frac{\pi}{4} = \pi.$$

Step 3   Find the values of y for the five key points.   We evaluate the function at each value of x from step 2.

Value of x	Value of y: $y = 3 \sin 2x$	Coordinates of key point	
0	$y = 3 \sin 2 \cdot 0$ $= 3 \sin 0 = 3 \cdot 0 = 0$	$(0, 0)$	
$\dfrac{\pi}{4}$	$y = 3 \sin 2 \cdot \dfrac{\pi}{4}$ $= 3 \sin \dfrac{\pi}{2} = 3 \cdot 1 = 3$	$\left(\dfrac{\pi}{4}, 3\right)$	maximum point
$\dfrac{\pi}{2}$	$y = 3 \sin 2 \cdot \dfrac{\pi}{2}$ $= 3 \sin \pi = 3 \cdot 0 = 0$	$\left(\dfrac{\pi}{2}, 0\right)$	
$\dfrac{3\pi}{4}$	$y = 3 \sin 2 \cdot \dfrac{3\pi}{4}$ $= 3 \sin \dfrac{3\pi}{2} = 3(-1) = -3$	$\left(\dfrac{3\pi}{4}, -3\right)$	minimum point
π	$y = 3 \sin 2 \cdot \pi$ $= 3 \sin 2\pi = 3 \cdot 0 = 0$	$(\pi, 0)$	

In the interval $[0, \pi]$, there are x-intercepts at 0, $\dfrac{\pi}{2}$, and π. The maximum and minimum points are indicated by the voice balloons.

Step 4   Connect the five key points with a smooth curve and graph one complete cycle of the given function.   The five key points for $y = 3 \sin 2x$ are shown in

Technology

The graph of $y = 3 \sin 2x$ in a $\left[0, 2\pi, \dfrac{\pi}{2}\right]$ by $[-4, 4, 1]$ viewing rectangle verifies our hand-drawn graph in Figure 5.49.

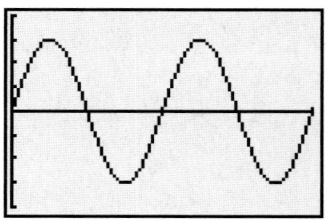

Figure 5.49. By connecting the points with a smooth curve, the blue portion shows one complete cycle of $y = 3 \sin 2x$ from 0 to π.

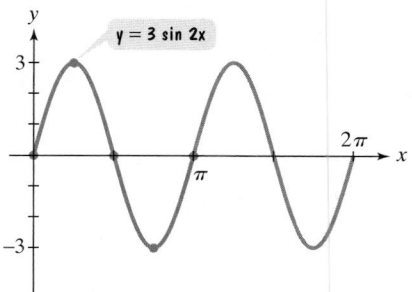

Figure 5.49

Step 5 Extend the graph in step 4 to the left or right as desired.   The blue portion of the graph in Figure 5.49 is from 0 to π. In order to graph for $0 \le x \le 2\pi$, we continue this portion and extend the graph another full period to the right. This extension is shown in black in Figure 5.49.

> **Check Point 3**   Determine the amplitude and period of $y = 2 \sin \frac{1}{2} x$. Then graph the function for $0 \le x \le 8\pi$.

Now let us examine the graphs of functions of the form $y = A \sin(Bx - C)$. How do such graphs compare to those of functions of the form $y = A \sin Bx$? In both cases, the amplitude is $|A|$ and the period is $\dfrac{2\pi}{B}$. One complete cycle occurs if $Bx - C$ increases from 0 to 2π. This means that we can find an interval containing one cycle by solving the equations

$$Bx - C = 0 \quad \text{and} \quad Bx - C = 2\pi$$

$$Bx = C \qquad\qquad\qquad Bx = C + 2\pi \quad \text{\small Add } C \text{ to both sides in each equation.}$$

$$x = \frac{C}{B} \qquad\qquad\qquad x = \frac{C}{B} + \frac{2\pi}{B}. \quad \text{\small Divide both sides by } B \text{ in each equation.}$$

> This is the x-coordinate on the left where the cycle begins.

> This is the x-coordinate on the right where the cycle ends. $\dfrac{2\pi}{B}$ is the period.

The voice balloon on the left indicates that $y = A \sin(Bx - C)$ shifts the graph of $y = A \sin Bx$ horizontally by $\dfrac{C}{B}$. Thus, the number $\dfrac{C}{B}$ is the **phase shift** associated with the graph.

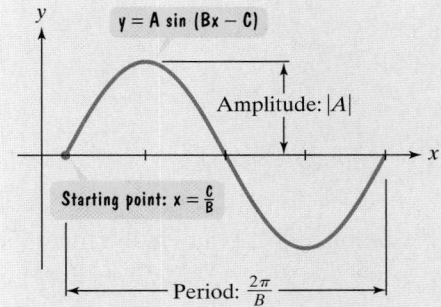

The Graph of $y = A \sin(Bx - C)$

The graph of $y = A \sin(Bx - C)$ is obtained by horizontally shifting the graph of $y = A \sin Bx$ so that the starting point of the cycle is shifted from $x = 0$ to $x = \dfrac{C}{B}$. The number $\dfrac{C}{B}$ is called the **phase shift**.

$$\text{amplitude} = |A|$$

$$\text{period} = \frac{2\pi}{B}$$

EXAMPLE 4 **Graphing a Function of the Form $y = A\sin(Bx - C)$**

Determine the amplitude, period, and phase shift of $y = 4\sin\left(2x - \dfrac{2\pi}{3}\right)$. Then graph one period of the function.

Solution

Step 1 **Identify the amplitude, the period, and the phase shift.** We must first identify values for A, B, and C.

> This equation is of the form $y = A\sin(Bx - C)$.

$$y = 4\sin\left(2x - \frac{2\pi}{3}\right)$$

Using the voice balloon, we see that $A = 4$, $B = 2$, and $C = \dfrac{2\pi}{3}$.

amplitude: $|A| = |4| = 4$ — The maximum y is 4 and the minimum is -4.

period: $\dfrac{2\pi}{B} = \dfrac{2\pi}{2} = \pi$ — Each cycle is completed in π radians.

phase shift: $\dfrac{C}{B} = \dfrac{\frac{2\pi}{3}}{2} = \dfrac{2\pi}{3}\cdot\dfrac{1}{2} = \dfrac{\pi}{3}$ — A cycle starts at $x = \dfrac{\pi}{3}$.

Step 2 **Find the x-values for the five key points.** Begin by dividing the period, π, by 4.

$$\frac{\text{period}}{4} = \frac{\pi}{4}$$

Start with the value of x where the cycle begins: $x = \dfrac{\pi}{3}$. Adding quarter-periods, $\dfrac{\pi}{4}$, the five x-values for the key points are

$$x = \frac{\pi}{3}, \quad x = \frac{\pi}{3} + \frac{\pi}{4} = \frac{4\pi}{12} + \frac{3\pi}{12} = \frac{7\pi}{12},$$

$$x = \frac{7\pi}{12} + \frac{\pi}{4} = \frac{7\pi}{12} + \frac{3\pi}{12} = \frac{10\pi}{12} = \frac{5\pi}{6},$$

$$x = \frac{5\pi}{6} + \frac{\pi}{4} = \frac{10\pi}{12} + \frac{3\pi}{12} = \frac{13\pi}{12},$$

$$x = \frac{13\pi}{12} + \frac{\pi}{4} = \frac{13\pi}{12} + \frac{3\pi}{12} = \frac{16\pi}{12} = \frac{4\pi}{3}.$$

Step 3 **Find the values of y for the five key points.** We evaluate the function at each value of x from step 2.

Study Tip

You can speed up the additions on the right by first writing the starting point and the quarter-period with a common denominator.

starting point
$= \dfrac{\pi}{3} = \dfrac{4\pi}{12}$

quarter-period
$= \dfrac{\pi}{4} = \dfrac{3\pi}{12}$

Value of x	Value of y: $y = 4\sin\left(2x - \dfrac{2\pi}{3}\right)$	Coordinates of key point	
$\dfrac{\pi}{3}$	$y = 4\sin\left(2 \cdot \dfrac{\pi}{3} - \dfrac{2\pi}{3}\right)$ $= 4\sin 0 = 4 \cdot 0 = 0$	$\left(\dfrac{\pi}{3}, 0\right)$	
$\dfrac{7\pi}{12}$	$y = 4\sin\left(2 \cdot \dfrac{7\pi}{12} - \dfrac{2\pi}{3}\right)$ $= 4\sin\left(\dfrac{7\pi}{6} - \dfrac{2\pi}{3}\right)$ $= 4\sin\dfrac{3\pi}{6} = 4\sin\dfrac{\pi}{2} = 4 \cdot 1 = 4$	$\left(\dfrac{7\pi}{12}, 4\right)$	maximum point
$\dfrac{5\pi}{6}$	$y = 4\sin\left(2 \cdot \dfrac{5\pi}{6} - \dfrac{2\pi}{3}\right)$ $= 4\sin\left(\dfrac{5\pi}{3} - \dfrac{2\pi}{3}\right)$ $= 4\sin\dfrac{3\pi}{3} = 4\sin\pi = 4 \cdot 0 = 0$	$\left(\dfrac{5\pi}{6}, 0\right)$	
$\dfrac{13\pi}{12}$	$y = 4\sin\left(2 \cdot \dfrac{13\pi}{12} - \dfrac{2\pi}{3}\right)$ $= 4\sin\left(\dfrac{13\pi}{6} - \dfrac{4\pi}{6}\right)$ $= 4\sin\dfrac{9\pi}{6} = 4\sin\dfrac{3\pi}{2} = 4(-1) = -4$	$\left(\dfrac{13\pi}{12}, -4\right)$	minimum point
$\dfrac{4\pi}{3}$	$y = 4\sin\left(2 \cdot \dfrac{4\pi}{3} - \dfrac{2\pi}{3}\right)$ $= 4\sin\dfrac{6\pi}{3} = 4\sin 2\pi = 4 \cdot 0 = 0$	$\left(\dfrac{4\pi}{3}, 0\right)$	

In the interval $\left[\dfrac{\pi}{3}, \dfrac{4\pi}{3}\right]$, there are x-intercepts at $\dfrac{\pi}{3}, \dfrac{5\pi}{6}$, and $\dfrac{4\pi}{3}$. The maximum and minimum points are indicated by the voice balloons.

Step 4   Connect the five key points with a smooth curve and graph one complete cycle of the given function.   The key points and the graph of $y = 4\sin\left(2x - \dfrac{2\pi}{3}\right)$ are shown in Figure 5.50.

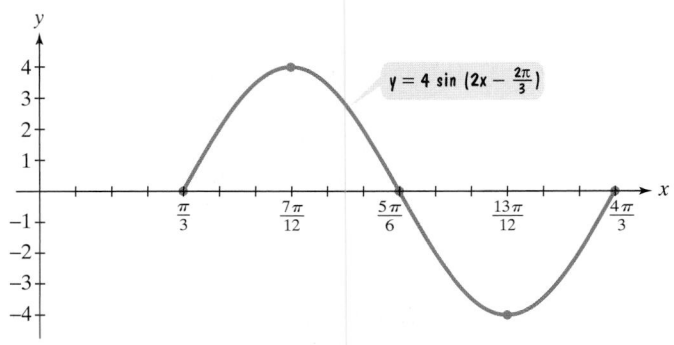

Figure 5.50

Check Point 4

Determine the amplitude, period, and phase shift of

$$y = 3 \sin\left(2x - \frac{\pi}{3}\right).$$ Then graph one period of the function.

3 Understand the graph of $y = \cos x$.

The Graph of $y = \cos x$

We graph $y = \cos x$ by listing some points on the graph. Because the period of the cosine function is 2π, we will concentrate on the graph of the basic cosine curve on the interval $[0, 2\pi]$. The rest of the graph is made up of repetitions of this portion. Table 5.2 lists some values of (x, y) on the graph of $y = \cos x$.

Table 5.2   Values of (x, y) on $y = \cos x$

x	0	$\frac{\pi}{6}$	$\frac{\pi}{3}$	$\frac{\pi}{2}$	$\frac{2\pi}{3}$	$\frac{5\pi}{6}$	π	$\frac{7\pi}{6}$	$\frac{4\pi}{3}$	$\frac{3\pi}{2}$	$\frac{5\pi}{3}$	$\frac{11\pi}{6}$	2π
$y = \cos x$	1	$\frac{\sqrt{3}}{2}$	$\frac{1}{2}$	0	$-\frac{1}{2}$	$-\frac{\sqrt{3}}{2}$	-1	$-\frac{\sqrt{3}}{2}$	$-\frac{1}{2}$	0	$\frac{1}{2}$	$\frac{\sqrt{3}}{2}$	1

As x increases from 0 to $\frac{\pi}{2}$, y decreases from 1 to 0.

As x increases from $\frac{\pi}{2}$ to π, y decreases from 0 to −1.

As x increases from π to $\frac{3\pi}{2}$, y increases from −1 to 0.

As x increases from $\frac{3\pi}{2}$ to 2π, y increases from 0 to 1.

Plotting the points in Table 5.2 and connecting them with a smooth curve, we obtain the graph shown in Figure 5.51. The portion of the graph in dark blue shows one complete period. We can obtain a more complete graph of $y = \cos x$ by extending this dark blue portion to the left and right.

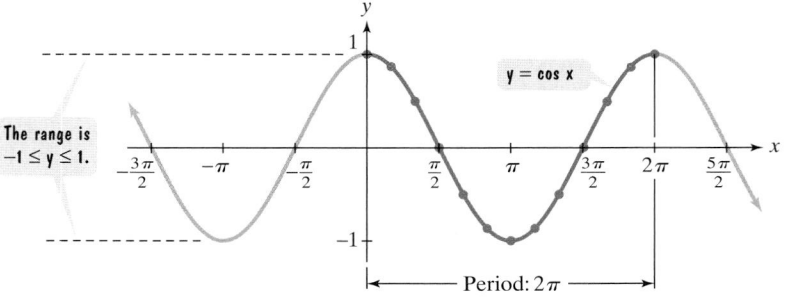

Figure 5.51 The graph of $y = \cos x$

The graph of $y = \cos x$ allows us to visualize some of the properties of the cosine function.

- The domain is the set of all real numbers. The graph extends indefinitely to the left and to the right with no gaps or holes.
- The range consists of all numbers between −1 and 1 inclusive. The graph never rises above 1 or falls below −1.
- The period is 2π. The graph's pattern repeats in every interval of length 2π.
- The function is an even function: $\cos(-x) = \cos x$. This can be seen by observing that the graph is symmetric with respect to the y-axis.

Take a second look at Figure 5.51. Can you see that the graph of $y = \cos x$ is the graph of $y = \sin x$ with a phase shift of $-\dfrac{\pi}{2}$ radians? If you trace along the curve from $x = -\dfrac{\pi}{2}$ to $x = \dfrac{3\pi}{2}$, you are tracing one complete cycle of the sine curve. This can be expressed as an identity:

$$\cos x = \sin\left(x + \frac{\pi}{2}\right).$$

Because of this similarity, the graphs of sine functions and cosine functions are called **sinusoidal graphs**.

4 Graph variations of $y = \cos x$.

Graphing Variations of $y = \cos x$

We use the same steps to graph variations of $y = \cos x$ as we did for graphing variations of $y = \sin x$. We will continue finding key points by dividing the period into four equal parts. Amplitudes, periods, and phase shifts play an important role when graphing by hand.

The Graph of $y = A \cos Bx$

The graph of $y = A \cos Bx$ has

$$\text{amplitude} = |A|$$

$$\text{period} = \frac{2\pi}{B}.$$

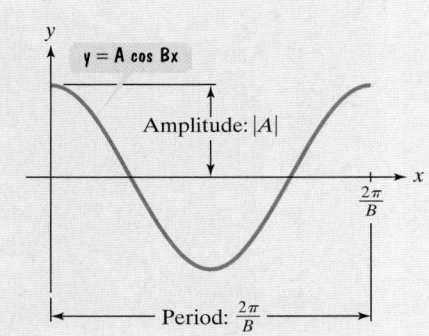

EXAMPLE 5   Graphing a Function of the Form $y = A \cos Bx$

Determine the amplitude and period of $y = -3\cos\dfrac{\pi}{2}x$. Then graph the function for $-4 \le x \le 4$.

Solution

Step 1   Identify the amplitude and the period.   The equation $y = -3\cos\dfrac{\pi}{2}x$ is of the form $y = A\cos Bx$ with $A = -3$ and $B = \dfrac{\pi}{2}$.

amplitude: $|A| = |-3| = 3$ The maximum y is 3 and the minimum is -3.

period: $\dfrac{2\pi}{B} = \dfrac{2\pi}{\dfrac{\pi}{2}} = 2\pi \cdot \dfrac{2}{\pi} = 4$ Each cycle is completed in 4 radians.

Step 2   Find the x-values for the five key points.   Begin by dividing the period, 4, by 4.

$$\frac{\text{period}}{4} = \frac{4}{4} = 1$$

Start with the value of x where the cycle begins: $x = 0$. Adding quarter-periods, 1, the five x-values for the key points are

$$x = 0, \quad x = 0 + 1 = 1, \quad x = 1 + 1 = 2, \quad x = 2 + 1 = 3, \quad x = 3 + 1 = 4$$

Step 3   Find the values of y for the five key points.   We evaluate the function at each value of x from step 2.

Value of x	Value of y: $y = -3 \cos \dfrac{\pi}{2} x$	Coordinates of key point	
0	$y = -3 \cos \dfrac{\pi}{2} \cdot 0$ $= -3 \cos 0 = -3 \cdot 1 = -3$	$(0, -3)$	minimum point
1	$y = -3 \cos \dfrac{\pi}{2} \cdot 1$ $= -3 \cos \dfrac{\pi}{2} = -3 \cdot 0 = 0$	$(1, 0)$	
2	$y = -3 \cos \dfrac{\pi}{2} \cdot 2$ $= -3 \cos \pi = -3(-1) = 3$	$(2, 3)$	maximum point
3	$y = -3 \cos \dfrac{\pi}{2} \cdot 3$ $= -3 \cos \dfrac{3\pi}{2} = -3(0) = 0$	$(3, 0)$	
4	$y = -3 \cos \dfrac{\pi}{2} \cdot 4$ $= -3 \cos 2\pi = -3(1) = -3$	$(4, -3)$	minimum point

In the interval $[0, 4]$, there are x-intercepts at 1 and 3. The minimum and maximum points are indicated by the voice balloons.

Step 4   Connect the five key points with a smooth curve and graph one complete cycle of the given function.   The five key points for $y = -3 \cos \dfrac{\pi}{2} x$ are shown in Figure 5.52. By connecting the points with a smooth curve, the blue portion shows one complete cycle of $y = -3 \cos \dfrac{\pi}{2} x$ from 0 to 4.

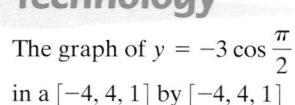

Technology

The graph of $y = -3 \cos \dfrac{\pi}{2} x$ in a $[-4, 4, 1]$ by $[-4, 4, 1]$ viewing rectangle verifies our hand-drawn graph in Figure 5.52.

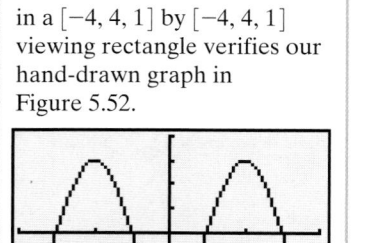

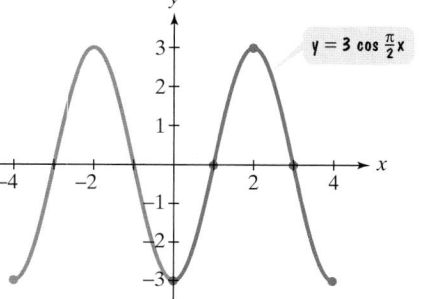

Figure 5.52

Step 5   Extend the graph in step 4 to the left or right as desired.   The blue portion of the graph in Figure 5.52 is from 0 to 4. In order to graph for $-4 \le x \le 4$, we continue this portion and extend the graph another full period to the left. This extension is shown in black in Figure 5.52.

> **Check Point 5**   Determine the amplitude and period of $y = -4 \cos \pi x$. Then graph the function for $-2 \le x \le 2$.

Finally, let us examine the graphs of functions of the form $y = A \cos(Bx - C)$. Graphs of these functions shift the graph of $y = A \cos Bx$ horizontally by $\dfrac{C}{B}$.

The Graph of $y = A \cos(Bx - C)$

The graph of $y = A \cos(Bx - C)$ is obtained by horizontally shifting the graph of $y = A \cos Bx$ so that the starting point of the cycle is shifted from $x = 0$ to $x = \dfrac{C}{B}$. The number $\dfrac{C}{B}$ is called the **phase shift**.

$$\text{amplitude} = |A|$$

$$\text{period} = \frac{2\pi}{B}$$

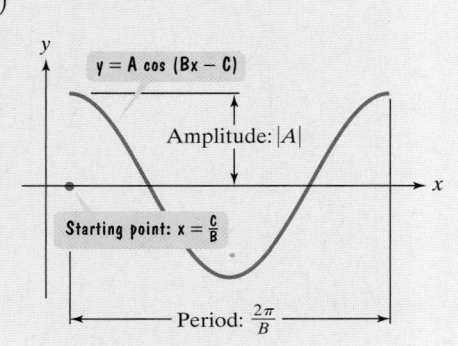

EXAMPLE 6   Graphing a Function of the Form $y = A \cos(Bx - C)$

Determine the amplitude, period, and phase shift of $y = \frac{1}{2}\cos(4x + \pi)$. Then graph one period of the function.

Solution

Step 1   Identify the amplitude, the period, and the phase shift.   We must first identify values for A, B, and C. To do this, we need to express the equation in the form $y = A \cos(Bx - C)$. Thus, we write $y = \frac{1}{2}\cos(4x + \pi)$ as $y = \frac{1}{2}\cos(4x - (-\pi))$. Now we can identify values for A, B, and C.

> This equation is of the form $y = A \cos(Bx - C)$.

$$y = \frac{1}{2}\cos\big(4x - (-\pi)\big)$$

Using the voice balloon, we see that $A = \frac{1}{2}$, $B = 4$, and $C = -\pi$.

amplitude: $\quad |A| = \left|\dfrac{1}{2}\right| = \dfrac{1}{2}$ — The maximum y is $\frac{1}{2}$ and the minimum is $-\frac{1}{2}$.

period: $\quad \dfrac{2\pi}{B} = \dfrac{2\pi}{4} = \dfrac{\pi}{2}$ — Each cycle is completed in $\frac{\pi}{2}$ radians.

phase shift: $\quad \dfrac{C}{B} = -\dfrac{\pi}{4}$ — A cycle starts at $x = -\frac{\pi}{4}$.

Step 2   Find the x-values for the five key points.   Begin by dividing the period, $\frac{\pi}{2}$, by 4.

$$\frac{\text{period}}{4} = \frac{\frac{\pi}{2}}{4} = \frac{\pi}{8}$$

Start with the value of x where the cycle begins: $x = -\frac{\pi}{4}$. Adding quarter-periods, $\frac{\pi}{8}$, the five x-values for the key points are

$$x = -\frac{\pi}{4}, \quad x = -\frac{\pi}{4} + \frac{\pi}{8} = -\frac{2\pi}{8} + \frac{\pi}{8} = -\frac{\pi}{8}, \quad x = -\frac{\pi}{8} + \frac{\pi}{8} = 0,$$

$$x = 0 + \frac{\pi}{8} = \frac{\pi}{8}, \quad x = \frac{\pi}{8} + \frac{\pi}{8} = \frac{2\pi}{8} = \frac{\pi}{4}.$$

Step 3   Find the values of y for the five key points.   Take a few minutes and use your calculator to evaluate the function at each value of x from step 2. Show that the key points are

$$\left(-\frac{\pi}{4}, \frac{1}{2}\right), \quad \left(-\frac{\pi}{8}, 0\right), \quad \left(0, -\frac{1}{2}\right), \quad \left(\frac{\pi}{8}, 0\right), \quad \text{and} \quad \left(\frac{\pi}{4}, \frac{1}{2}\right).$$

maximum point	x-intercept	minimum point	y-intercept	maximum point

Step 4   Connect the five key points with a smooth curve and graph one complete cycle of the given function.   The key points and the graph of $y = \frac{1}{2}\cos(4x + \pi)$ are shown in Figure 5.53.

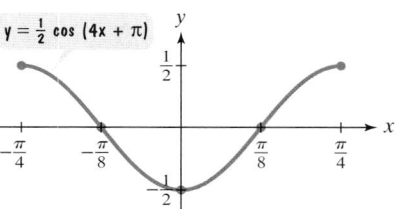

Figure 5.53

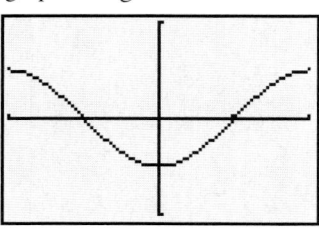
Check Point 6   Determine the amplitude, period, and phase shift of $y = \frac{3}{2}\cos(2x + \pi)$. Then graph one period of the function.

5 Use vertical shifts of sine and cosine curves.

Vertical Shifts of Sinusoidal Graphs

We now look at sinusoidal graphs of

$$y = A\sin(Bx - C) + D \quad \text{and} \quad y = A\cos(Bx - C) + D.$$

The constant D causes vertical shifts in the graphs of $y = A\sin(Bx - C)$ and $y = A\cos(Bx - C)$. If D is positive, the shift is D units upward. If D is negative, the shift is D units downward. These vertical shifts result in sinusoidal graphs oscillating about the horizontal line $y = D$ rather than about the x-axis. Thus, the maximum y is $D + |A|$ and the minimum y is $D - |A|$.

EXAMPLE 7   A Vertical Shift

Graph one period of the function $y = \frac{1}{2} \cos x - 1$.

Solution   The graph of $y = \frac{1}{2} \cos x - 1$ is the graph of $y = \frac{1}{2} \cos x$ shifted one unit downward. The period of $y = \frac{1}{2} \cos x$ is 2π, which is also the period for the vertically shifted graph. The key points on the interval $[0, 2\pi]$ for $y = \frac{1}{2} \cos x - 1$ are found by first determining their x-coordinates. The quarter-period is $\dfrac{2\pi}{4}$ or $\dfrac{\pi}{2}$.

The cycle begins at $x = 0$. As always, we add quarter-periods to generate x-values for each of the key points. The five x-values are

$$x = 0, \quad x = 0 + \frac{\pi}{2} = \frac{\pi}{2}, \quad x = \frac{\pi}{2} + \frac{\pi}{2} = \pi,$$

$$x = \pi + \frac{\pi}{2} = \frac{3\pi}{2}, \quad x = \frac{3\pi}{2} + \frac{\pi}{2} = 2\pi.$$

The values of y for the five key points and their coordinates are determined as follows.

Value of x	Value of y: $y = \dfrac{1}{2} \cos x - 1$	Coordinates of key point
0	$y = \dfrac{1}{2} \cos 0 - 1$ $= \dfrac{1}{2} \cdot 1 - 1 = -\dfrac{1}{2}$	$\left(0, -\dfrac{1}{2}\right)$
$\dfrac{\pi}{2}$	$y = \dfrac{1}{2} \cos \dfrac{\pi}{2} - 1$ $= \dfrac{1}{2} \cdot 0 - 1 = -1$	$\left(\dfrac{\pi}{2}, -1\right)$
π	$y = \dfrac{1}{2} \cos \pi - 1$ $= \dfrac{1}{2}(-1) - 1 = -\dfrac{3}{2}$	$\left(\pi, -\dfrac{3}{2}\right)$
$\dfrac{3\pi}{2}$	$y = \dfrac{1}{2} \cos \dfrac{3\pi}{2} - 1$ $= \dfrac{1}{2} \cdot 0 - 1 = -1$	$\left(\dfrac{3\pi}{2}, -1\right)$
2π	$y = \dfrac{1}{2} \cos 2\pi - 1$ $= \dfrac{1}{2} \cdot 1 - 1 = -\dfrac{1}{2}$	$\left(2\pi, -\dfrac{1}{2}\right)$

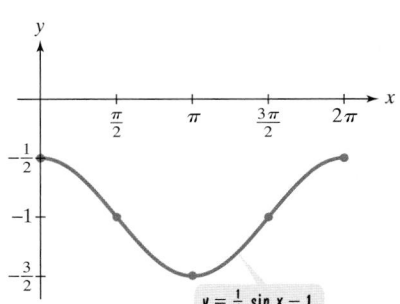

$y = \frac{1}{2} \sin x - 1$

Figure 5.54

The five key points for $y = \frac{1}{2} \cos x - 1$ are shown in Figure 5.54. By connecting the points with a smooth curve, we obtain one period of the graph.

Check Point 7   Graph one period of the function $y = 2 \cos x + 1$.

6 Model periodic behavior.

Modeling Periodic Behavior

Our breathing consists of alternating periods of inhaling and exhaling. Each complete pumping cycle of the human heart can be described using a sine function. Our brain waves during deep sleep are sinusoidal. Viewed in this way, trigonometry becomes an intimate experience.

Some graphing utilities have a SINe REGression feature. This feature gives the sine function of best fit of wavelike data with the form of a sinusoidal function. However, it is not always necessary to use technology. In our next example, we use our understanding of sinusoidal graphs to model the process of breathing.

EXAMPLE 8   A Trigonometric Breath of Life

The graph in Figure 5.55 shows one complete normal breathing cycle. The cycle consists of inhaling and exhaling. It takes place every 5 seconds. Velocity of air flow is positive when we inhale and negative when we exhale. It is measured in liters per second. If y represents velocity of air flow after x seconds, find a function of the form $y = A \sin Bx$ that models air flow in a normal breathing cycle.

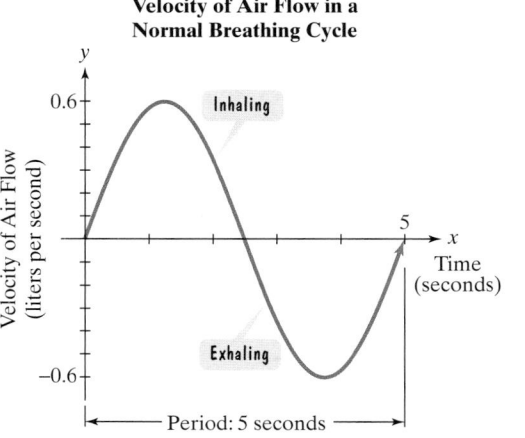

Figure 5.55

Solution   We need to determine values for A and B in the equation $y = A \sin Bx$. A, the amplitude, is the maximum value of y. Figure 5.55 shows that this maximum value is 0.6. Thus, $A = 0.6$.

The value of B in $y = A \sin Bx$ can be found using the formula for the period: $\text{period} = \dfrac{2\pi}{B}$. The period of our breathing cycle is 5 seconds. Thus,

$$5 = \frac{2\pi}{B} \qquad \text{\small Our goal is to solve this equation for } B.$$

$$5B = 2\pi \qquad \text{\small Multiply both sides of the equation by } B.$$

$$B = \frac{2\pi}{5} \qquad \text{\small Divide both sides of the equation by 5.}$$

We see that $A = 0.6$ and $B = \dfrac{2\pi}{5}$. Substitute these values into $y = A \sin Bx$. The breathing cycle is modeled by

$$y = 0.6 \sin \frac{2\pi}{5} x.$$

Check Point 8

Find an equation of the form $y = A \sin Bx$ that produces the graph shown in the figure on the right.

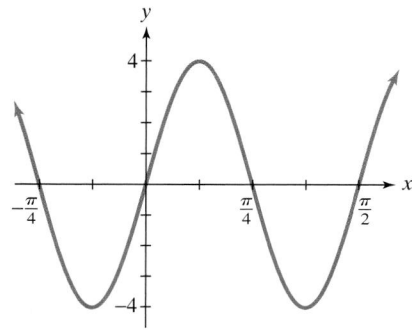

EXAMPLE 9   Modeling a Tidal Cycle

Figure 5.56 shows that the depth of water at a boat dock varies with the tides. The depth is 5 feet at low tide and 13 feet at high tide. On a certain day, low tide occurs at 4 A.M. and high tide at 10 A.M. If y represents the depth of the water x hours after midnight, use a sine function of the form $y = A \sin(Bx - C) + D$ to model the water's depth.

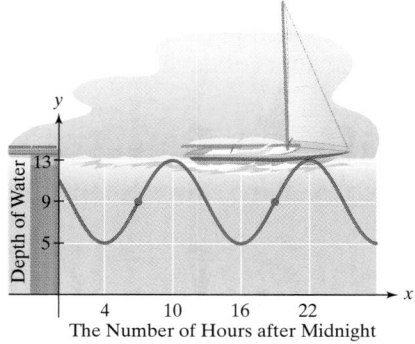

Figure 5.56

Solution   We need to determine values for A, B, C, and D in the equation $y = A \sin(Bx - C) + D$. We can find these values using Figure 5.56. We begin with D.

To find D, we use the vertical shift. Because the water's depth ranges from a minimum of 5 feet to a maximum of 13 feet, the curve oscillates about the middle value, 9 feet. Thus, $D = 9$, which is the vertical shift.

At maximum depth, the water is 4 feet above 9 feet. Thus, A, the amplitude, is 4: $A = 4$.

To find B, we use the period. The blue portion of the graph shows that one complete tidal cycle occurs in $19 - 7$, or 12 hours. The period is 12. Thus,

$$12 = \frac{2\pi}{B} \qquad \text{\textit{Our goal is to solve this equation for B.}}$$

$$12B = 2\pi \qquad \text{\textit{Multiply both sides by B.}}$$

$$B = \frac{2\pi}{12} = \frac{\pi}{6}. \qquad \text{\textit{Divide both sides by 12.}}$$

To find C, we use the phase shift. The blue portion of the graph shows that the starting point of the cycle is shifted from 0 to 7. The phase shift, $\dfrac{C}{B}$, is 7.

$$7 = \frac{C}{B} \qquad \text{\textit{The phase shift of } y = A \sin(Bx - C) \text{ is } \frac{C}{B}.}$$

$$7 = \frac{C}{\dfrac{\pi}{6}} \qquad \text{\textit{From above, we have } B = \frac{\pi}{6}.}$$

$$\frac{7\pi}{6} = C \qquad \text{\textit{Multiply both sides of the equation by } \frac{\pi}{6}.}$$

We see that $A = 4$, $B = \dfrac{\pi}{6}$, $C = \dfrac{7\pi}{6}$, and $D = 9$. Substitute these values into $y = A\sin(Bx - C) + D$. The water's depth x hours after midnight is modeled by

$$y = 4\sin\left(\frac{\pi}{6}x - \frac{7\pi}{6}\right) + 9.$$

Check Point 9

The figure shows the number of hours of daylight for a region that is 30° north of the equator. Hours of daylight are at a minimum of 10 hours in January and December. Hours of daylight are at a maximum of 14 hours in June. Let x represent the month of the year, with 1 for January, 2 for February, 3 for March, and 12 for December. If y represents the number of hours of daylight in month x, use a sine function of the form $y = A\sin(Bx - C) + D$ to model the hours of daylight.

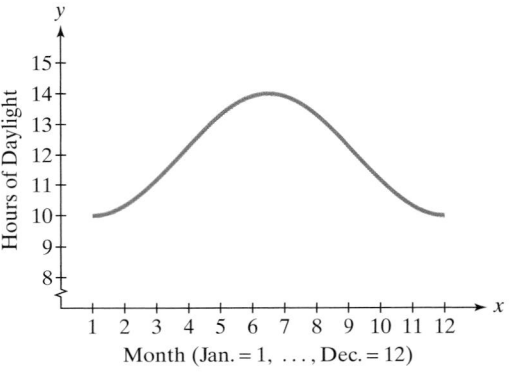

Month (Jan. = 1, ... , Dec. = 12)

EXERCISE SET 5.5

Practice Exercises

In Exercises 1–6, determine the amplitude of each function. Then graph the function and $y = \sin x$ in the same rectangular coordinate system for $0 \le x \le 2\pi$.

1. $y = 4\sin x$

2. $y = 5\sin x$

3. $y = \frac{1}{3}\sin x$

4. $y = \frac{1}{4}\sin x$

5. $y = -3\sin x$

6. $y = -4\sin x$

In Exercises 7–16, determine the amplitude and period of each function. Then graph one period of the function.

7. $y = \sin 2x$

8. $y = \sin 4x$

9. $y = 3\sin\frac{1}{2}x$

10. $y = 2\sin\frac{1}{4}x$

11. $y = 4\sin \pi x$

12. $y = 3\sin 2\pi x$

13. $y = -3\sin 2\pi x$

14. $y = -2\sin \pi x$

15. $y = -\sin\frac{2}{3}x$

16. $y = -\sin\frac{4}{3}x$

In Exercises 17–30, determine the amplitude, period, and phase shift of each function. Then graph one period of the function.

17. $y = \sin(x - \pi)$

18. $y = \sin\left(x - \dfrac{\pi}{2}\right)$

19. $y = \sin(2x - \pi)$

20. $y = \sin\left(2x - \dfrac{\pi}{2}\right)$

21. $y = 3 \sin(2x - \pi)$ **22.** $y = 3 \sin\left(2x - \dfrac{\pi}{2}\right)$

23. $y = \frac{1}{2} \sin\left(x + \dfrac{\pi}{2}\right)$ **24.** $y = \frac{1}{2} \sin(x + \pi)$

25. $y = -2 \sin\left(2x + \dfrac{\pi}{2}\right)$ **26.** $y = -3 \sin\left(2x + \dfrac{\pi}{2}\right)$

27. $y = 3 \sin(\pi x + 2)$ **28.** $y = 3 \sin(2\pi x + 4)$

29. $y = -2 \sin(2\pi x + 4\pi)$ **30.** $y = -3 \sin(2\pi x + 4\pi)$

In Exercises 31–34, determine the amplitude of each function. Then graph the function and $y = \cos x$ in the same rectangular coordinate system for $0 \le x \le 2\pi$.

31. $y = 2 \cos x$ **32.** $y = 3 \cos x$

33. $y = -2 \cos x$ **34.** $y = -3 \cos x$

In Exercises 35–42, determine the amplitude and period of each function. Then graph one period of the function.

35. $y = \cos 2x$ **36.** $y = \cos 4x$

37. $y = 4 \cos 2\pi x$ **38.** $y = 5 \cos 2\pi x$

39. $y = -4 \cos \frac{1}{2} x$ **40.** $y = -3 \cos \frac{1}{3} x$

41. $y = -\frac{1}{2} \cos \dfrac{\pi}{3} x$ **42.** $y = -\frac{1}{2} \cos \dfrac{\pi}{4} x$

In Exercises 43–50, determine the amplitude, period, and phase shift of each function. Then graph one period of the function.

43. $y = 3 \cos(2x - \pi)$ **44.** $y = 4 \cos(2x - \pi)$

45. $y = \frac{1}{2} \cos\left(3x + \dfrac{\pi}{2}\right)$ **46.** $y = \frac{1}{2} \cos(2x + \pi)$

47. $y = -3 \cos\left(2x - \dfrac{\pi}{2}\right)$ **48.** $y = -4 \cos\left(2x - \dfrac{\pi}{2}\right)$

49. $y = 2 \cos(2\pi x + 8\pi)$ **50.** $y = 3 \cos(2\pi x + 4\pi)$

In Exercises 51–58, use a vertical shift to graph one period of the function.

51. $y = \sin x + 2$ **52.** $y = \sin x - 2$

53. $y = \cos x - 3$ **54.** $y = \cos x + 3$

55. $y = 2 \sin \frac{1}{2} x + 1$ **56.** $y = 2 \cos \frac{1}{2} x + 1$

57. $y = -3 \cos 2\pi x + 2$ **58.** $y = -3 \sin 2\pi x + 2$

⭐ Application Exercises

In the theory of biorhythms, sine functions are used to measure a person's potential. You can obtain your biorhythm chart online by simply entering your date of birth, the date you want your biorhythm chart to begin, and the number of months you wish to be included in the plot. The following is your author's chart, beginning January 25, 2000, when he was 19,998 days old. We all have cycles with the same amplitudes and periods as those shown here. Each of

our three basic cycles begins at birth. Use the biorhythm chart shown to solve Exercises 59–66.

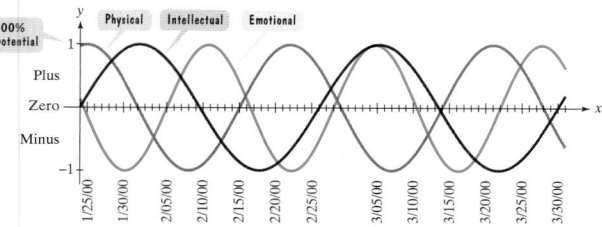

59. What is the period of the physical cycle?

60. What is the period of the emotional cycle?

61. What is the period of the intellectual cycle?

62. For the period shown, what is the worst day in February for your author to run in a marathon?

63. For the period shown, what is the best day in March for your author to meet an online friend for the first time?

64. For the period shown, what is the best day in January for your author to begin writing this trigonometry chapter?

65. If you extend these sinusoidal graphs to the end of the year, is there a day when your author should not even bother getting out of bed?

66. If you extend these sinusoidal graphs to the end of the year, are there any days where your author is at near-peak physical, emotional, and intellectual potential?

67. Rounded to the nearest hour, Los Angeles averages 14 hours of daylight in June, 10 hours in December, and 12 hours in March and September. Let x represent the number of months after June and y represent the number of hours of daylight in month x. Make a graph that displays the information from June of one year to June of the following year.

68. A clock with an hour hand that is 15 inches long is hanging on a wall. At noon, the distance between the tip of the hour hand and the ceiling is 23 inches. At 3 P.M., the distance is 38 inches; at 6 P.M., 53 inches; at 9 P.M., 38 inches; and at midnight the distance is again 23 inches. If y represents the distance between the tip of the hour hand and the ceiling x hours after noon, make a graph that displays the information for $0 \le x \le 24$.

69. The number of hours of daylight in Boston is given by

$$y = 3 \sin \dfrac{2\pi}{365} (x - 79) + 12$$

where x is the number of days after January 1.
  a. What is the amplitude of this function?
  b. What is the period of this function?

c. How many hours of daylight are there on the longest day of the year?

d. How many hours of daylight are there on the shortest day of the year?

e. Graph the function for one period, starting on January 1.

70. The average monthly temperature, y, in degrees Fahrenheit, for Juneau, Alaska, can be modeled by

$$y = 16 \sin\left(\frac{\pi}{6} x - \frac{2\pi}{3}\right) + 40,$$ where x is the month of the year (January = 1, February = 2, ..., December = 12). Graph the function for $1 \leq x \leq 12$. What is the highest average monthly temperature? In which month does this occur?

71. The figure shows the depth of water at the end of a boat dock. The depth is 6 feet at low tide and 12 feet at high tide. On a certain day, low tide occurs at 6 A.M. and high tide at noon. If y represents the depth of the water x hours after midnight, use a cosine function of the form $y = A \cos Bx + D$ to model the water's depth.

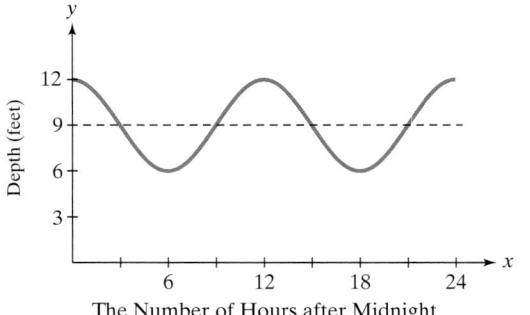

The Number of Hours after Midnight

72. The figure shows the depth of water at the end of a boat dock. The depth is 5 feet at high tide and 3 feet at low tide. On a certain day, high tide occurs at noon and low tide at 6 P.M. If y represents the depth of the water x hours after noon, use a cosine function of the form $y = A \cos Bx + D$ to model the water's depth.

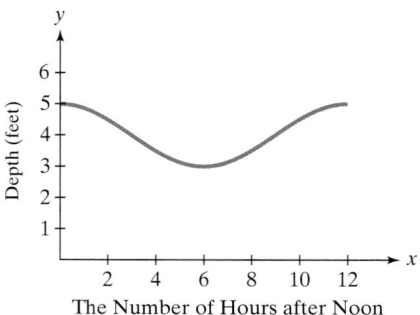

The Number of Hours after Noon

Writing in Mathematics

73. Without drawing a graph, describe the behavior of the basic sine curve.

74. What is the amplitude of the sine function? What does this tell you about the graph?

75. If you are given the equation of a sine function, how do you determine the period?

76. What does a phase shift indicate about the graph of a sine function? How do you determine the phase shift from the function's equation?

77. Describe a general procedure for obtaining the graph of $y = A \sin(Bx - C)$.

78. Without drawing a graph, describe the behavior of the basic cosine curve.

79. Describe a relationship between the graphs of $y = \sin x$ and $y = \cos x$.

80. Describe the relationship between the graphs of $y = A \cos(Bx - C)$ and $y = A \cos(Bx - C) + D$.

81. Biorhythm cycles provide interesting applications of sinusoidal graphs. But do you believe in the validity of biorhythms? Write a few sentences explaining why or why not.

Technology Exercises

82. Use a graphing utility to verify any five of the sine curves that you drew by hand in Exercises 7–30. The amplitude, period, and phase shift should help you to determine appropriate range settings.

83. Use a graphing utility to verify any five of the cosine curves that you drew by hand in Exercises 31–50.

84. Use a graphing utility to verify any two of the sinusoidal curves with vertical shifts that you drew in Exercises 51–58.

In Exercises 85–88, use a graphing utility to graph two periods of the function.

85. $y = 3 \sin(2x + \pi)$

86. $y = -2 \cos\left(2\pi x - \frac{\pi}{2}\right)$

87. $y = 0.2 \sin\left(\frac{\pi}{10} x + \pi\right)$

88. $y = 3 \sin(2x - \pi) + 5$

89. Use a graphing utility to graph $y = \sin x$ and $y = x - \frac{x^3}{6} + \frac{x^5}{120}$ in a $\left[-\pi, \pi, \frac{\pi}{2}\right]$ by $[-2, 2, 1]$ viewing rectangle. How do the graphs compare?

90. Use a graphing utility to graph $y = \cos x$ and $y = 1 - \dfrac{x^2}{2} + \dfrac{x^4}{24}$ in a $\left[-\pi, \pi, \dfrac{\pi}{2}\right]$ by $[-2, 2, 1]$ viewing rectangle. How do the graphs compare?

91. Use a graphing utility to graph

$$y = \sin x + \frac{\sin 2x}{2} + \frac{\sin 3x}{3} + \frac{\sin 4x}{4}$$

in a $\left[-2\pi, 2\pi, \dfrac{\pi}{2}\right]$ by $[-2, 2, 1]$ viewing rectangle. How do these waves compare to the smooth rolling waves of the basic sine curve?

92. Use a graphing utility to graph

$$y = \sin x - \frac{\sin 3x}{9} + \frac{\sin 5x}{25}$$

in a $\left[-2\pi, 2\pi, \dfrac{\pi}{2}\right]$ by $[-2, 2, 1]$ viewing rectangle. How do these waves compare to the smooth rolling waves of the basic sine curve?

93. The data show the average monthly temperatures for Washington, D.C.
   a. Use your graphing utility to draw a scatter plot of the data from $x = 1$ through $x = 12$.
   b. Use the SINe REGression feature to find the sinusoidal function of the form $y = A \sin(Bx + C) + D$ that best fits the data.
   c. Use your graphing utility to draw the sinusoidal function of best fit on the scatter plot.

x Month		Average Monthly Temperature, °F
1	(January)	34.6
2	(February)	37.5
3	(March)	47.2
4	(April)	56.5
5	(May)	66.4
6	(June)	75.6
7	(July)	80.0
8	(August)	78.5
9	(September)	71.3
10	(October)	59.7
11	(November)	49.8
12	(December)	39.4

Source: U.S. National Oceanic and Atmospheric Administration.

94. Repeat Exercise 93 for data of your choice. The data can involve the average monthly temperatures for the region where you live or any data whose scatter plot takes the form of a sinusoidal function.

 **Critical Thinking Exercises**

Graph the function in Exercises 95–96 by hand.

95. $y = \sin x + \cos x \qquad$ for $0 \le x \le 2\pi$

96. $y = x + \cos x \qquad$ for $0 \le x \le \dfrac{5\pi}{2}$

97. Use the cosine function to find an equation of the graph in the figure shown.

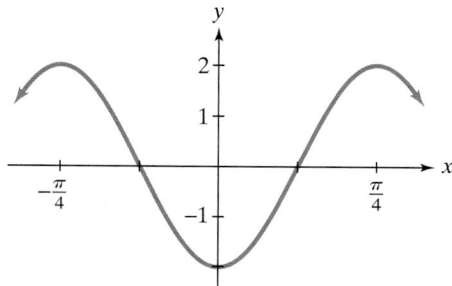

 Group Exercise

98. This exercise is intended to provide some fun with biorhythms, regardless of whether you believe they have any validity. We will use each member's chart to determine biorhythmic compatibility. Before meeting, each group member should go online and obtain his or her biorhythm chart. The date of the group meeting is the date on which your chart should begin. Include 12 months in the plot. At the meeting, compare differences and similarities among the intellectual sinusoidal curves. Using these comparisons, each person should find the one other person with whom he or she would be most intellectually compatible.

SECTION 5.6 *Graphs of Other Trigonometric Functions*

Objectives

1. Understand the graph of $y = \tan x$.
2. Graph variations of $y = \tan x$.
3. Understand the graph of $y = \cot x$.
4. Graph variations of $y = \cot x$.
5. Understand the graphs of $y = \csc x$ and $y = \sec x$.
6. Graph variations of $y = \csc x$ and $y = \sec x$.

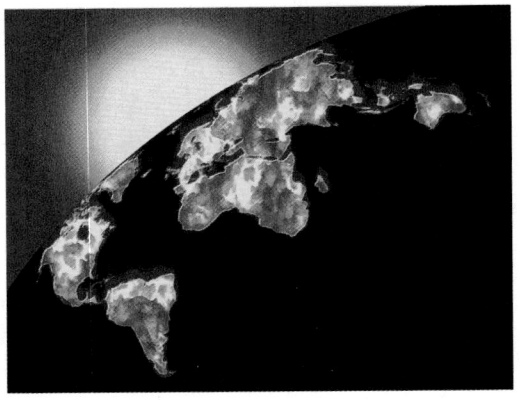

Recent advances in our understanding of climate have changed global warming from a subject for a disaster movie (the Statue of Liberty up to its chin in water) to a serious but manageable scientific and policy issue. Global warming may be related to the burning of fossil fuels, which adds carbon dioxide to the atmosphere. In the new millennium, we will see whether our use of fossil fuels will add enough carbon dioxide to the atmosphere to change it (and our climate) in significant ways. In this section's exercise set, you will see how trigonometric graphs reveal interesting patterns in carbon dioxide concentration from 1990 to 2000. In the section itself, trigonometric graphs will reveal patterns involving the tangent, cotangent, secant, and cosecant functions.

1 Understand the graph of $y = \tan x$.

The Graph of $y = \tan x$

The properties of the tangent function discussed in Section 5.4 will help us determine its graph. Because the tangent function has properties that are different from sinusoidal functions, its graph differs significantly from those of sine and cosine. Properties of the tangent function include the following:

- The period is π. It is only necessary to graph $y = \tan x$ over an interval of length π. The remainder of the graph consists of repetitions of that graph at intervals of π.
- The tangent function is an odd function: $\tan(-x) = -\tan x$. The graph is symmetric with respect to the origin.
- The tangent function is undefined at $\frac{\pi}{2}$. The graph of $y = \tan x$ has a vertical asymptote at $x = \frac{\pi}{2}$.

We obtain the graph of $y = \tan x$ using some points on the graph and origin symmetry. Table 5.3 lists some values of (x, y) on the graph of $y = \tan x$ on the interval $\left[0, \frac{\pi}{2}\right)$.

Table 5.3 Values of (x, y) on $y = \tan x$

x	0	$\dfrac{\pi}{6}$	$\dfrac{\pi}{4}$	$\dfrac{\pi}{3}$	$\dfrac{5\pi}{12}$ (75°)	$\dfrac{17\pi}{36}$ (85°)	$\dfrac{89\pi}{180}$ (89°)	1.57	$\dfrac{\pi}{2}$
$y = \tan x$	0	$\dfrac{\sqrt{3}}{3} \approx 0.6$	1	$\sqrt{3} \approx 1.7$	3.7	11.4	57.3	1255.8	undefined

As x increases from 0 to $\frac{\pi}{2}$, y increases slowly at first, then more and more rapidly.

The graph in Figure 5.57(a) is based on our observation in the voice balloon. Notice that y increases without bound as x approaches $\frac{\pi}{2}$. As the figure shows, the graph of $y = \tan x$ has a vertical asymptote at $x = \frac{\pi}{2}$.

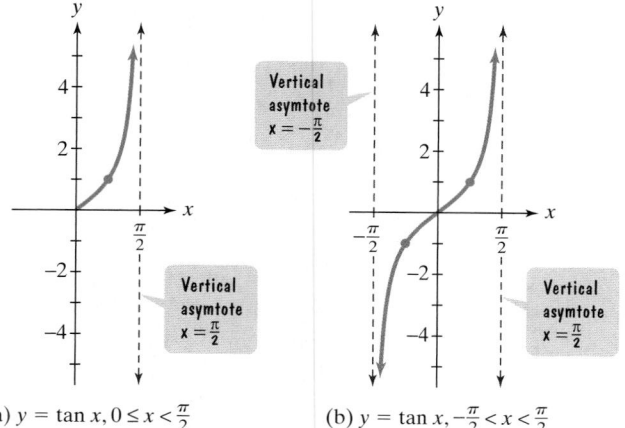

Figure 5.57 Graphing the tangent function

(a) $y = \tan x, 0 \le x < \frac{\pi}{2}$ (b) $y = \tan x, -\frac{\pi}{2} < x < \frac{\pi}{2}$

The graph of $y = \tan x$ can be completed for the interval $\left(-\frac{\pi}{2}, \frac{\pi}{2}\right)$ by using origin symmetry. Figure 5.57(b) shows the result of reflecting the graph in Figure 5.57(a) about the origin. The graph of $y = \tan x$ has another vertical asymptote at $x = -\frac{\pi}{2}$. Notice that y decreases without bound as x approaches $-\frac{\pi}{2}$.

Because the period of the tangent function is π radians, the graph in Figure 5.57(b) shows one complete period of $y = \tan x$. We obtain the complete graph of $y = \tan x$ by repeating the graph in Figure 5.57(b) to the left and right over intervals of π. The resulting graph and its main characteristics are shown in the following box.

The Tangent Curve: The Graph of $y = \tan x$ and Its Characteristics

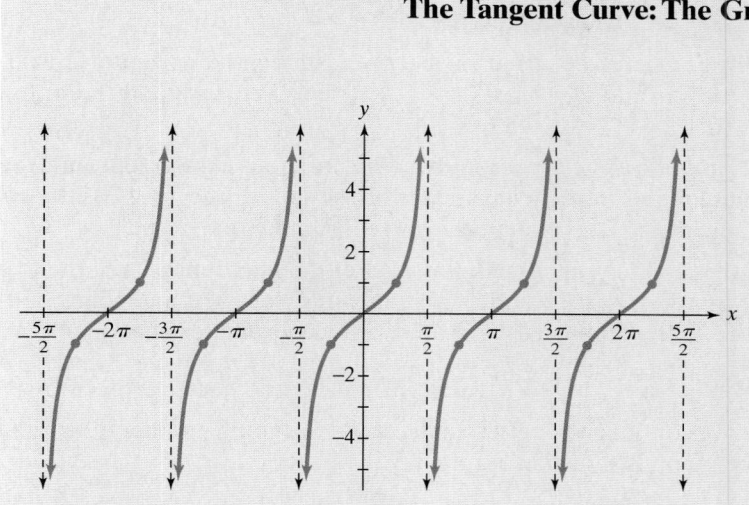

Characteristics

- **Period**: π
- **Domain**: All real numbers except odd multiples of $\frac{\pi}{2}$
- **Range**: All real numbers
- **Vertical asymptotes**: at odd multiples of $\frac{\pi}{2}$
- **An x-intercept** occurs midway between each pair of consecutive asymptotes.
- **Odd function** with origin symmetry
- Points on the graph midway between x-intercepts and consecutive asymptotes have y-coordinates of -1 and 1.

2 Graph variations of $y = \tan x$.

Graphing Variations of $y = \tan x$

We use the characteristics of the tangent curve to graph tangent functions of the form $y = A \tan(Bx - C)$.

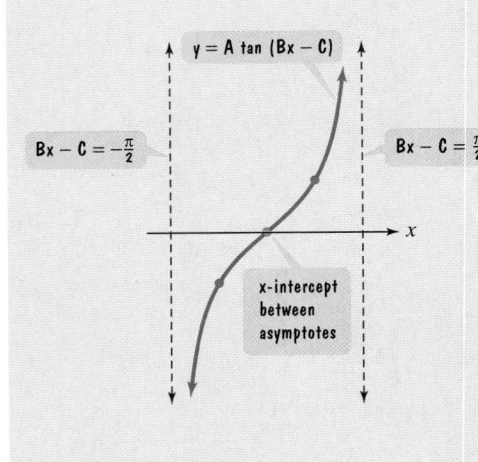

Graphing $y = A\tan(Bx - C)$

1. Find two consecutive asymptotes by setting the variable expression in the tangent equal to $-\dfrac{\pi}{2}$ and $\dfrac{\pi}{2}$ and solving

$$Bx - C = -\frac{\pi}{2} \text{ and } Bx - C = \frac{\pi}{2}.$$

2. Identify an x-intercept, midway between consecutive asymptotes.

3. Find the points on the graph midway between an x-intercept and the asymptotes. These points have y-coordinates of $-A$ and A.

4. Use steps 1–3 to graph one full period of the function. Add additional cycles to the left or right as needed.

EXAMPLE 1   Graphing a Tangent Function

Graph $y = 2 \tan \dfrac{x}{2}$ for $-\pi < x < 3\pi$.

Solution

Step 1   Find two consecutive asymptotes. We solve the equations

$$\frac{x}{2} = -\frac{\pi}{2} \quad \text{and} \quad \frac{x}{2} = \frac{\pi}{2}.$$ Set the variable expression in the tangent equal to $-\dfrac{\pi}{2}$ and $\dfrac{\pi}{2}$.

$$x = -\pi \qquad \qquad x = \pi$$ Multiply both sides of each equation by 2.

Thus, two consecutive asymptotes occur at $x = -\pi$ and $x = \pi$.

Step 2   Identify an x-intercepts, midway between consecutive asymptotes. Midway between $x = -\pi$ and $x = \pi$ is $x = 0$. An x-intercept is 0 and the graph passes through $(0, 0)$.

Step 3   Find points on the graph midway between an x-intercept and the asymptotes. These points have y-coordinates of $-A$ and A. Because A, the coefficient of the tangent, is 2, these points have y-coordinates of -2 and 2.

Step 4   Use steps 1–3 to graph one full period of the function. We use the two consecutive asymptotes, $x = -\pi$ and $x = \pi$, an x-intercept of 0, and points midway between the x-intercept and asymptotes with y-coordinates of -2 and 2. We graph one period of $y = 2 \tan \dfrac{x}{2}$ from $-\pi$ to π. In order to graph for $-\pi < x < 3\pi$, we continue the pattern and extend the graph another full period to the right. The graph is shown in Figure 5.58.

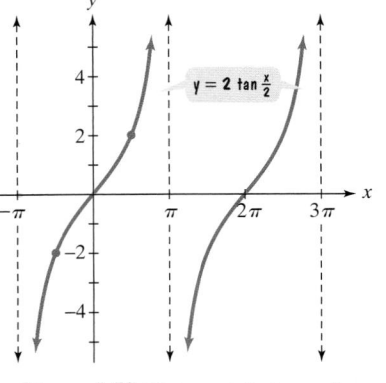

Figure 5.58 The graph is shown for two full periods

Check Point 1

Graph $y = 3 \tan 2x$ for $-\dfrac{\pi}{4} < x < \dfrac{3\pi}{4}$.

EXAMPLE 2 Graphing a Tangent Function

Graph two full periods of $y = \tan\left(x + \dfrac{\pi}{4}\right)$.

Solution The graph of $y = \tan\left(x + \dfrac{\pi}{4}\right)$ is the graph of $y = \tan x$ shifted horizontally to the left $\dfrac{\pi}{4}$ units.

Step 1 Find two consecutive asymptotes. We solve the equations

$$x + \frac{\pi}{4} = -\frac{\pi}{2} \quad \text{and} \quad x + \frac{\pi}{4} = \frac{\pi}{2}$$

Set the variable expression in the tangent equal to $-\dfrac{\pi}{2}$ and $\dfrac{\pi}{2}$.

$$x = -\frac{\pi}{4} - \frac{\pi}{2} \qquad\qquad x = -\frac{\pi}{4} + \frac{\pi}{2}$$

Subtract $\dfrac{\pi}{4}$ from both sides in each equation.

$$x = -\frac{3\pi}{4} \qquad\qquad x = \frac{\pi}{4}$$

Simplify.

Thus, two consecutive asymptotes occur at $x = -\dfrac{3\pi}{4}$ and $x = \dfrac{\pi}{4}$.

Step 2 Identify an x-intercept, midway between consecutive asymptotes.

$$x\text{-intercept} = \frac{-\dfrac{3\pi}{4} + \dfrac{\pi}{4}}{2} = \frac{-\dfrac{2\pi}{4}}{2} = -\frac{2\pi}{8} = -\frac{\pi}{4}$$

An x-intercept is $-\dfrac{\pi}{4}$ and the graph passes through $\left(-\dfrac{\pi}{4}, 0\right)$.

Step 3 Find points on the graph midway between an x-intercept and the asymptotes. Because A, the coefficient of the tangent, is 1, these points have y-coordinates of -1 and 1.

Step 4 Use steps 1–3 to graph one full period of the function. We use the two consecutive asymptotes, $x = -\dfrac{3\pi}{4}$ and $x = \dfrac{\pi}{4}$, to graph one full period of $y = \tan\left(x + \dfrac{\pi}{4}\right)$ from $-\dfrac{3\pi}{4}$ to $\dfrac{\pi}{4}$. We graph two full periods by continuing the pattern and extending the graph another full period to the right. The graph is shown in Figure 5.59.

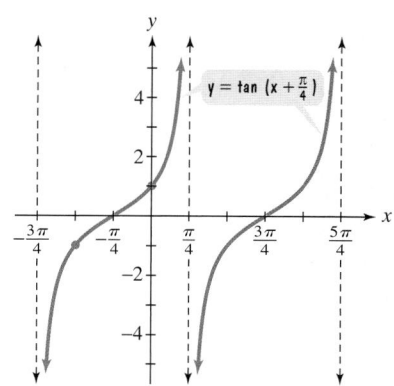

Figure 5.59 The graph is shown for two full periods.

Check Point 2

Graph two full periods of $y = \tan\left(x - \dfrac{\pi}{2}\right)$.

3 Understand the graph of $y = \cot x$.

The Graph of $y = \cot x$

Like the tangent function, the cotangent function, $y = \cot x$, has a period of π. The graph and its main characteristics are shown in the following box.

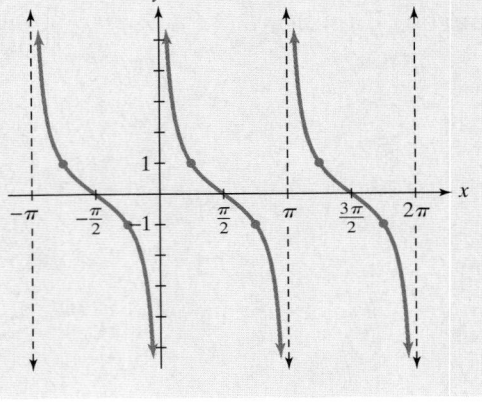

The Cotangent Curve: The Graph of $y = \cot x$ and Its Characteristics

Characteristics

- **Period:** π
- **Domain:** All real numbers except integral multiples of π
- **Range:** All real numbers
- **Vertical asymptotes** at integral multiples of π
- **An x-intercept** occurs midway between each pair of consecutive asymptotes.
- **Odd function** with origin symmetry
- Points on the graph midway between x-intercepts and consecutive asymptotes have y-coordinates of 1 and -1.

4 Graph variations of $y = \cot x$.

Graphing Variations of $y = \cot x$

We use the characteristics of the cotangent curve to graph cotangent functions of the form $y = A \cot(Bx - C)$.

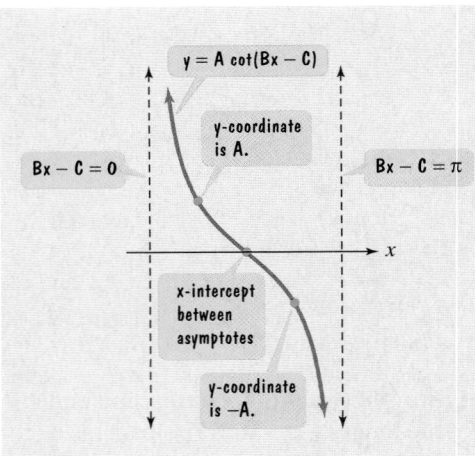

Graphing $y = A \cot(Bx - C)$

1. Find two consecutive asymptotes by setting the variable expression in the cotangent equal to 0 and π and solving

$$Bx - C = 0 \text{ and } Bx - C = \pi.$$

2. Identify an x-intercept, midway between consecutive asymptotes.

3. Find the points on the graph midway between an x-intercept and the asymptotes. These points have y-coordinates of A and $-A$.

4. Use steps 1-3 to graph one full period of the function. Add additional cycles to the left or right as needed.

EXAMPLE 3 Graphing a Cotangent Function

Graph $y = 3 \cot 2x$.

Solution

Step 1 Find two consecutive asymptotes. We solve the equations

$$2x = 0 \quad \text{and} \quad 2x = \pi. \quad \text{\small Set the variable expression in the cotangent equal to 0 and } \pi.$$

$$x = 0 \qquad x = \frac{\pi}{2} \quad \text{\small Divide both sides of each equation by 2.}$$

Two consecutive asymptotes occur at $x = 0$ and $x = \dfrac{\pi}{2}$.

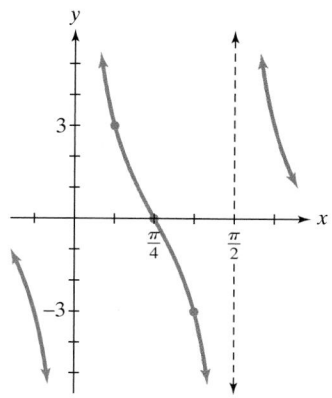

Figure 5.60 The graph of $y = 3 \cot 2x$

Step 2 **Identify an *x*-intercept, midway between consecutive asymptotes.** Midway between $x = 0$ and $x = \dfrac{\pi}{2}$ is $x = \dfrac{\pi}{4}$. An *x*-intercept is $\dfrac{\pi}{4}$ and the graph passes through $\left(\dfrac{\pi}{4}, 0\right)$.

Step 3 **Find points on the graph midway between an *x*-intercept and the asymptotes. These points have *y*-coordinates of *A* and −*A*.** Because *A*, the coefficient of the cotangent, is 3, these points have *y*-coordinates of 3 and −3.

Step 4 **Use steps 1–3 to graph one full period of the function.** We use the two consecutive asymptotes, $x = 0$ and $x = \dfrac{\pi}{2}$, to graph one full period of $y = 3 \cot 2x$. This curve is repeated to the left and right, as shown in Figure 5.60.

Check Point 3 Graph $y = \dfrac{1}{2} \cot \dfrac{\pi}{2} x$.

5 Understand the graphs of $y = \csc x$ and $y = \sec x$.

The Graphs of $y = \csc x$ and $y = \sec x$

We obtain the graphs of the cosecant and secant curves by using the reciprocal identities

$$\csc x = \frac{1}{\sin x} \quad \text{and} \quad \sec x = \frac{1}{\cos x}.$$

The identity on the left tells us that the value of the cosecant function $y = \csc x$ at a given value of *x* equals the reciprocal of the corresponding value of the sine function, provided that the value of the sine function is not 0. If the value of $\sin x$ is 0, then at each of these values of *x*, the cosecant function is not defined. A vertical asymptote is associated with each of these values on the graph of $y = \csc x$.

We obtain the graph of $y = \csc x$ by taking reciprocals of the *y*-values in the graph of $y = \sin x$. Vertical asymptotes of $y = \csc x$ occur at the *x*-intercepts of $y = \sin x$. Likewise, we obtain the graph of $y = \sec x$ by taking the reciprocal of $y = \cos x$. Vertical asymptotes of $y = \sec x$ occur at the *x*-intercepts of $y = \cos x$. The graphs of $y = \csc x$ and $y = \sec x$ and their key characteristics are shown in the following boxes. We have used dashed red lines to first graph $y = \sin x$ and $y = \cos x$, drawing vertical asymptotes through the *x*-intercepts.

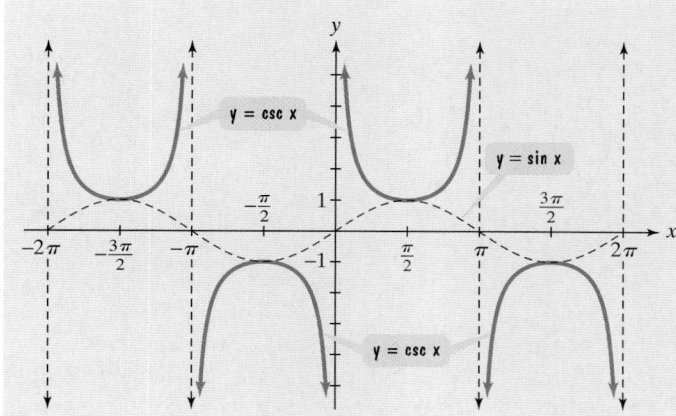

The Cosecant Curve: The Graph of $y = \csc x$ and Its Characteristics

Characteristics

- **Period:** 2π
- **Domain:** All real numbers except integral multiples of π
- **Range:** All real numbers *y* such that $y \le -1$ or $y \ge 1$
- **Vertical asymptotes** at integral multiples of π
- **Odd function**, $\csc(-x) = -\csc x$, with origin symmetry

The Secant Curve: The Graph of $y = \sec x$ and Its Characteristics

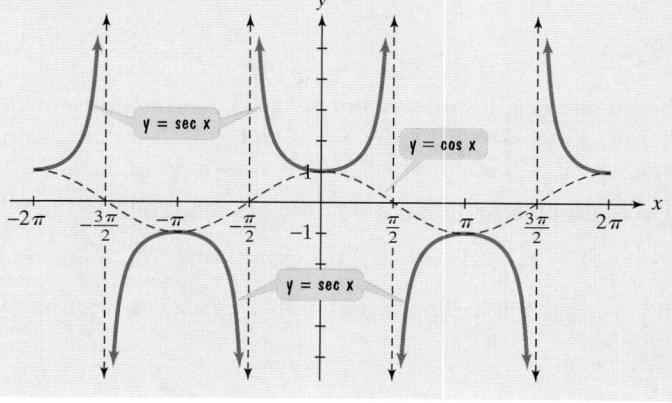

Characteristics

- **Period:** 2π
- **Domain:** All real numbers except odd multiples of $\dfrac{\pi}{2}$
- **Range:** All real numbers y such that $y \leq -1$ or $y \geq 1$
- **Vertical asymptotes** at odd multiples of $\dfrac{\pi}{2}$
- **Even function,** $\sec(-x) = \sec x$, with y-axis symmetry

6 Graph variations of $y = \csc x$ and $y = \sec x$.

Graphing Variations of $y = \csc x$ and $y = \sec x$

We use graphs of reciprocal functions to obtain graphs of cosecant and secant functions. To graph a cosecant or secant curve, begin by graphing the reciprocal function. For example, to graph $y = 2 \csc 2x$, we use the graph of $y = 2 \sin 2x$. Likewise, to graph $y = -3 \sec \dfrac{x}{2}$, we use the graph of $y = -3 \cos \dfrac{x}{2}$.

Figure 5.61 illustrates how we use a sine curve to obtain a cosecant curve. Notice that

- x-intercepts on the sine curve correspond to vertical asymptotes of the cosecant curve.
- A maximum point on the sine curve corresponds to a minimum point on a continuous portion of the cosecant curve.
- A minimum point on the sine curve corresponds to a maximum point on a continuous portion of the cosecant curve.

Figure 5.61

EXAMPLE 4 Using a Sine Curve to Obtain a Cosecant Curve

Use the graph of $y = 2 \sin 2x$ in Figure 5.62 to obtain the graph of $y = 2 \csc 2x$.

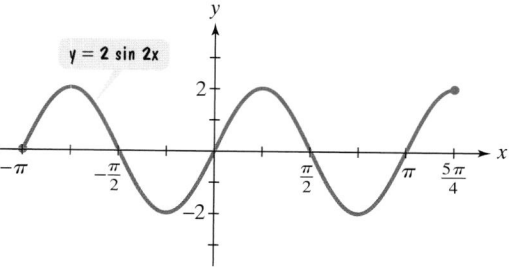

Figure 5.62

Solution   The x-intercepts of $y = 2 \sin 2x$ correspond to the vertical asymptotes of $y = 2 \csc 2x$. Thus, we draw vertical asymptotes through the x-intercepts, shown in Figure 5.63. Using the asymptotes as guides, we sketch the graph of $y = 2 \csc 2x$ in Figure 5.63.

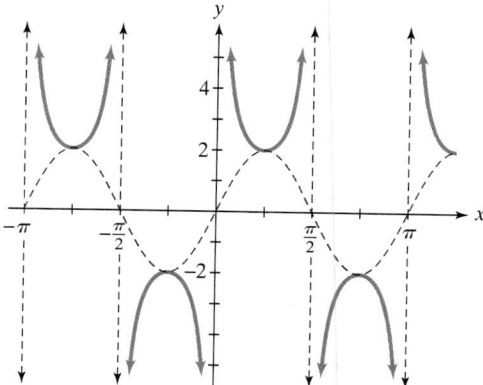

Figure 5.63 Using a sine curve to graph $y = 2 \csc 2x$

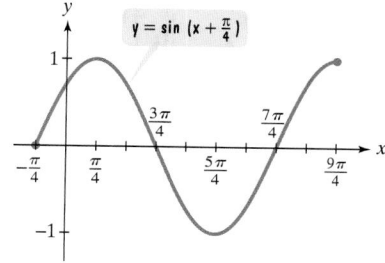

Check Point 4 Use the graph of $y = \sin\left(x + \dfrac{\pi}{4}\right)$ shown on the left to obtain the graph of $y = \csc\left(x + \dfrac{\pi}{4}\right)$.

We use a cosine curve to obtain a secant curve in exactly the same way we used a sine curve to obtain a cosecant curve. Thus,

- x-intercepts on the cosine curve correspond to vertical asymptotes on the secant curve.
- A maximum point on the cosine curve corresponds to a minimum point on a continuous portion of the secant curve.
- A minimum point on the cosine curve corresponds to a maximum point on a continuous portion of the secant curve.

EXAMPLE 5 Graphing a Secant Function

Graph $y = -3 \sec \dfrac{x}{2}$ for $-\pi < x < 5\pi$.

Solution We begin by graphing the reciprocal cosine function, $y = -3 \cos \dfrac{x}{2}$. This equation is of the form $y = A \cos Bx$ with $A = -3$ and $B = \frac{1}{2}$.

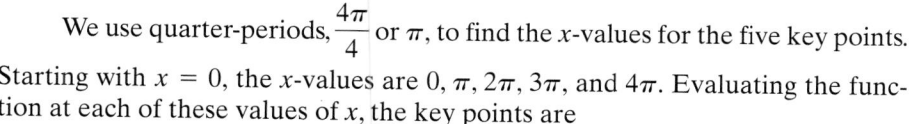

amplitude: $\quad |A| = |-3| = 3 \quad$ The maximum y is 3 and the minimum is -3.

period: $\quad \dfrac{2\pi}{B} = \dfrac{2\pi}{\frac{1}{2}} = 4\pi \quad$ Each cycle is completed in 4π radians.

We use quarter-periods, $\dfrac{4\pi}{4}$ or π, to find the x-values for the five key points. Starting with $x = 0$, the x-values are $0, \pi, 2\pi, 3\pi$, and 4π. Evaluating the function at each of these values of x, the key points are

$$(0, -3), (\pi, 0), (2\pi, 3), (3\pi, 0), \text{ and } (4\pi, -3).$$

We use these key points to graph $y = -3 \cos \dfrac{x}{2}$ from 0 to 4π. In order to graph for $-\pi \le x \le 5\pi$, extend the graph π units to the left and π units to the right. The graph is shown using a dashed red line in Figure 5.64. Now use this dashed red graph to obtain the graph of the reciprocal function. Draw vertical asymptotes through the x-intercepts. Using these asymptotes as guides, the graph of $y = -3 \sec \dfrac{x}{2}$ is shown in blue in Figure 5.64.

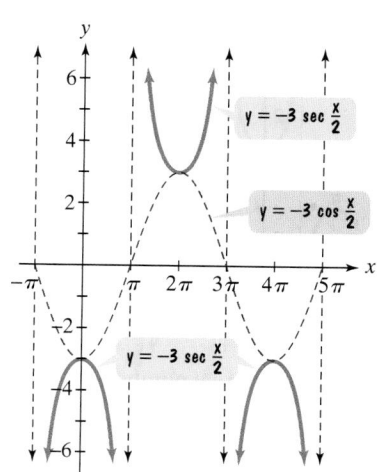

Figure 5.64 Using a cosine curve to graph $y = -3 \sec \dfrac{x}{2}$

74. Graph $y = \sin \dfrac{1}{x}$ in a $[-0.2, 0.2, 0.01]$ by $[-1.2, 1.2, 0.01]$

viewing rectangle. What is happening as x approaches 0 from the left or the right? Explain this behavior.

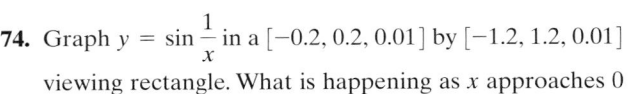 **Critical Thinking Exercises**

In Exercises 75–76, write an equation for each blue graph.

75.

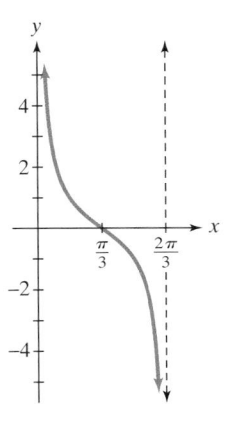

76.

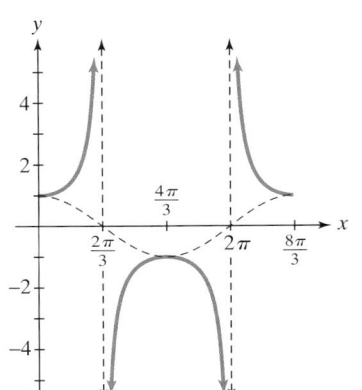

77. For $x > 0$, what effect does 2^{-x} in $y = 2^{-x} \sin x$ have on the graph of $y = \sin x$? What kind of behavior can be modeled by a function such as $y = 2^{-x} \sin x$?

SECTION 5.7 Inverse Trigonometric Functions

Objectives

1. Understand and use the inverse sine function.
2. Understand and use the inverse cosine function.
3. Understand and use the inverse tangent function.
4. Use a calculator to evaluate inverse trigonometric functions.
5. Find exact values of composite functions with inverse trigonometric functions.

You watched *The Matrix* on video and were impressed by the elaborate computer-generated effects. The movie is being shown again at a local theater, where you can experience its stunning visual force on a large screen. Where in the theater should you sit to maximize the film's visual impact? In this section you will see how an inverse trigonometric function can enhance your movie-going experiences.

Study Tip

Here are some helpful things to remember from our discussion of inverse functions in Section 2.6.

- If no horizontal line intersects the graph of a function more than once, the function is one-to-one and has an inverse function.
- If the point (a, b) is on the graph of f, then the point (b, a) is on the graph of the inverse function, denoted f^{-1}. The graph of f^{-1} is a reflection of the graph of f about the line $y = x$.

1 Understand and use the inverse sine function.

The Inverse Sine Function

Figure 5.65 shows the graph of $y = \sin x$. Can you see that every horizontal line that can be drawn between -1 and 1 intersects the graph infinitely many times? Thus, the sine function is not one-to-one and has no inverse function.

In Figure 5.66, we have taken a portion of the sine curve, restricting the domain of the sine function to $-\dfrac{\pi}{2} \leq x \leq \dfrac{\pi}{2}$. With this restricted domain, every horizontal line that can be drawn between -1 and 1 intersects the graph exactly once. Thus, the restricted function passes the horizontal line test and is one-to-one.

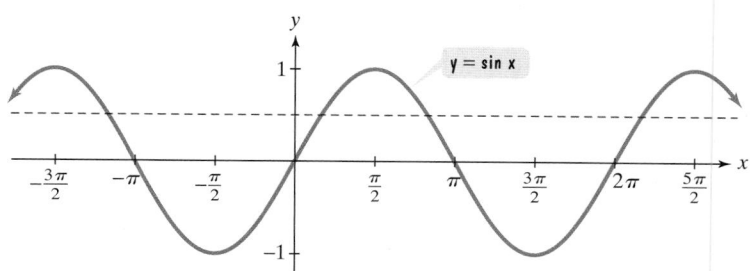

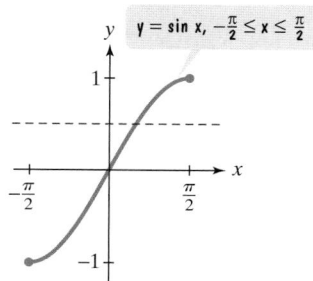

Figure 5.65 The horizontal line test shows that the sine function is not one to one and has no inverse function.

Figure 5.66 The restricted sine function passes the horizontal line test. It is one-to-one and has an inverse function.

On the restricted domain $-\dfrac{\pi}{2} \leq x \leq \dfrac{\pi}{2}$, $y = \sin x$ has an inverse function. The inverse of the restricted sine function is called the **inverse sine function**. Two notations are commonly used to denote the inverse sine function:

$$y = \sin^{-1} x \quad \text{or} \quad y = \arcsin x.$$

In this book, we will use $y = \sin^{-1} x$. This notation has the same symbol as the inverse function notation $f^{-1}(x)$.

> ### The Inverse Sine Function
>
> The **inverse sine function**, denoted by $\sin^{-1}$, is the inverse of the restricted sine function $y = \sin x$, $-\dfrac{\pi}{2} \leq x \leq \dfrac{\pi}{2}$. Thus,
>
> $$y = \sin^{-1} x \quad \text{means} \quad \sin y = x,$$
>
> where $-\dfrac{\pi}{2} \leq y \leq \dfrac{\pi}{2}$ and $-1 \leq x \leq 1$. We read $y = \sin^{-1} x$ as "y equals the inverse sine at x."

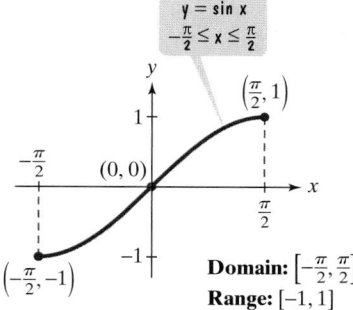

Figure 5.67 The restricted sine function

One way to graph $y = \sin^{-1} x$ is to take points on the graph of the restricted sine function and reverse the order of the coordinates. For example, Figure 5.67 shows that $\left(-\dfrac{\pi}{2}, -1\right)$, $(0, 0)$, and $\left(\dfrac{\pi}{2}, 1\right)$ are on the graph of the restricted sine function. Reversing the order of the coordinates gives $\left(-1, -\dfrac{\pi}{2}\right)$, $(0, 0)$, and

$\left(1, \dfrac{\pi}{2}\right)$. We now use these three points to sketch the inverse sine function. The graph of $y = \sin^{-1} x$ is shown in Figure 5.68.

Another way to obtain the graph of $y = \sin^{-1} x$ is to reflect the graph of the restricted sine function about the line $y = x$, shown in Figure 5.69. The blue graph is the graph of $y = \sin^{-1} x$.

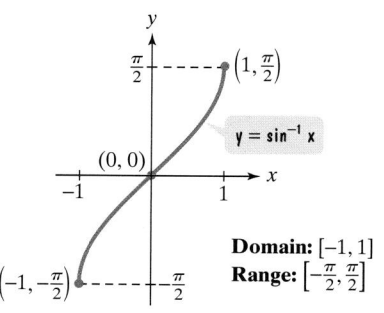

Figure 5.68 The graph of the inverse sine function

Domain: $[-1, 1]$
Range: $\left[-\dfrac{\pi}{2}, \dfrac{\pi}{2}\right]$

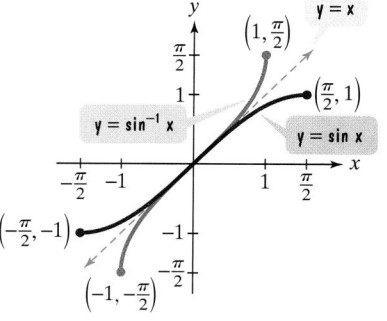

Figure 5.69 Using a reflection to obtain the graph of the inverse sine function

Exact values of $\sin^{-1} x$ can be found by thinking of $\mathbf{sin^{-1}\,x}$ as **the angle in the interval** $\left[-\dfrac{\pi}{2}, \dfrac{\pi}{2}\right]$ **whose sine is** x. For example, we can use the two points on the blue graph of the inverse sine function in Figure 5.69 and write

$$\sin^{-1}(-1) = -\frac{\pi}{2} \quad \text{and} \quad \sin^{-1} 1 = \frac{\pi}{2}.$$

The angle whose sine is −1 is $-\frac{\pi}{2}$.

The angle whose sine is 1 is $\frac{\pi}{2}$.

Because we are thinking of $\sin^{-1} x$ in terms of an angle, we will represent such an angle by θ.

Table 5.5 Exact Values for
$\sin \theta,\ -\dfrac{\pi}{2} \le \theta \le \dfrac{\pi}{2}$

θ	$\sin \theta$
$-\dfrac{\pi}{2}$	-1
$-\dfrac{\pi}{3}$	$-\dfrac{\sqrt{3}}{2}$
$-\dfrac{\pi}{4}$	$-\dfrac{\sqrt{2}}{2}$
$-\dfrac{\pi}{6}$	$-\dfrac{1}{2}$
0	0
$\dfrac{\pi}{6}$	$\dfrac{1}{2}$
$\dfrac{\pi}{4}$	$\dfrac{\sqrt{2}}{2}$
$\dfrac{\pi}{3}$	$\dfrac{\sqrt{3}}{2}$
$\dfrac{\pi}{2}$	1

Finding Exact Values of $\sin^{-1} x$.

1. Let $\theta = \sin^{-1} x$.
2. Rewrite step 1 as $\sin \theta = x$.
3. Use the exact values in Table 5.5 to find the value of θ in $\left[-\dfrac{\pi}{2}, \dfrac{\pi}{2}\right]$ that satisfies $\sin \theta = x$.

EXAMPLE 1 Finding the Exact Value of an Inverse Sine Function

Find the exact value of $\sin^{-1} \dfrac{\sqrt{2}}{2}$.

Solution

Step 1   Let $\theta = \sin^{-1}x$. Thus,

$$\theta = \sin^{-1}\frac{\sqrt{2}}{2}.$$

We must find the angle θ, $-\dfrac{\pi}{2} \leq \theta \leq \dfrac{\pi}{2}$, whose sine equals $\dfrac{\sqrt{2}}{2}$.

Step 2   Rewrite $\theta = \sin^{-1}x$ as $\sin\theta = x$. Using the definition of the inverse sine function, we rewrite $\theta = \sin^{-1}\dfrac{\sqrt{2}}{2}$ as

$$\sin\theta = \frac{\sqrt{2}}{2}.$$

Step 3   Use the exact values in Table 5.5 to find the value of θ in $\left[-\dfrac{\pi}{2}, \dfrac{\pi}{2}\right]$ that satisfies $\sin\theta = x$. Table 5.5 shows that the only angle in the interval $\left[-\dfrac{\pi}{2}, \dfrac{\pi}{2}\right]$ that satisfies $\sin\theta = \dfrac{\sqrt{2}}{2}$ is $\dfrac{\pi}{4}$. Thus, $\theta = \dfrac{\pi}{4}$. Because θ, in step 1, represents $\sin^{-1}\dfrac{\sqrt{2}}{2}$, we conclude that

$$\sin^{-1}\frac{\sqrt{2}}{2} = \frac{\pi}{4}. \qquad \text{The angle in } \left[-\frac{\pi}{2}, \frac{\pi}{2}\right] \text{ whose sine is } \frac{\sqrt{2}}{2} \text{ is } \frac{\pi}{4}$$

Study Tip

If you have not already done so, you should memorize the values in Table 5.5, as well as those in the forthcoming Tables 5.6 and 5.7.

Check Point 1   Find the exact value of $\sin^{-1}\dfrac{\sqrt{3}}{2}$.

EXAMPLE 2   Finding the Exact Value of an Inverse Sine Function

Find the exact value of $\sin^{-1}\left(-\frac{1}{2}\right)$.

Solution

Step 1   Let $\theta = \sin^{-1}x$. Thus,

$$\theta = \sin^{-1}\left(-\frac{1}{2}\right).$$

We must find the angle θ, $-\dfrac{\pi}{2} \leq \theta \leq \dfrac{\pi}{2}$, whose sine equals $-\frac{1}{2}$.

Step 2   Rewrite $\theta = \sin^{-1}x$ as $\sin\theta = x$. We obtain

$$\sin\theta = -\frac{1}{2}$$

Step 3   Use the exact values in Table 5.5 to find the value of θ in $\left[-\dfrac{\pi}{2},\dfrac{\pi}{2}\right]$ that satisfies $\sin\theta = x$. The table on page 500 shows that the only angle in the interval $\left[-\dfrac{\pi}{2},\dfrac{\pi}{2}\right]$ that satisfies $\sin\theta = -\frac{1}{2}$ is $-\dfrac{\pi}{6}$. Thus,

$$\sin^{-1}\left(-\frac{1}{2}\right) = -\frac{\pi}{6}$$

Check Point 2   Find the exact value of $\sin^{-1}\left(-\dfrac{\sqrt{2}}{2}\right)$.

Some inverse sine expressions cannot be evaluated. Because the domain of the inverse sine function is $[-1, 1]$, it is only possible to evaluate $\sin^{-1}x$ for values of x in this domain. Thus, $\sin^{-1}3$ cannot be evaluated. There is no angle whose sine is 3.

2   Understand and use the inverse cosine function.

The Inverse Cosine Function

Figure 5.70 shows how we restrict the domain of the cosine function so that it becomes one-to-one and has an inverse function. Restrict the domain to the interval $[0, \pi]$, shown by the dark blue graph. Over this interval, the restricted cosine function passes the horizontal line test and has an inverse function.

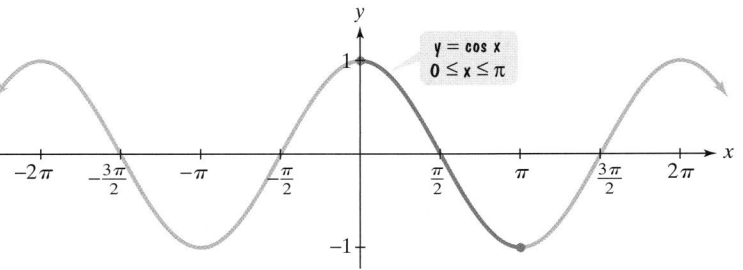

Figure 5.70 $y = \cos x$ is one-to-one on the interval $[0, \pi]$.

The Inverse Cosine Function

The **inverse cosine function**, denoted by $\cos^{-1}$, is the inverse of the restricted cosine function $y = \cos x,\ 0 \le x \le \pi$. Thus,

$$y = \cos^{-1}x \quad \text{means} \quad \cos y = x,$$

where $0 \le y \le \pi$ and $-1 \le x \le 1$.

One way to graph $y = \cos^{-1} x$ is to take points on the graph of the restricted cosine function and reverse the order of the coordinates. For example, Figure 5.71 shows that $(0, 1)$, $\left(\dfrac{\pi}{2}, 0\right)$ and $(\pi, -1)$ are on the graph of the restricted cosine function. Reversing the order of the coordinates gives $(1, 0)$, $\left(0, \dfrac{\pi}{2}\right)$, and $(-1, \pi)$. We now use these three points to sketch the inverse cosine function. The graph of $y = \cos^{-1} x$ is shown in Figure 5.72. You can also obtain this graph by reflecting the graph of the restricted cosine function about the line $y = x$.

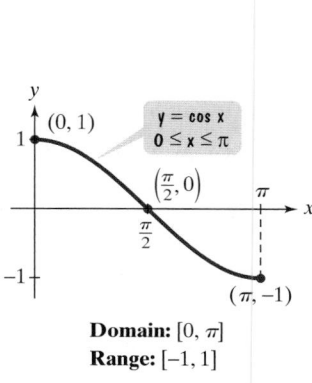

Domain: $[0, \pi]$
Range: $[-1, 1]$

Figure 5.71 The restricted cosine function

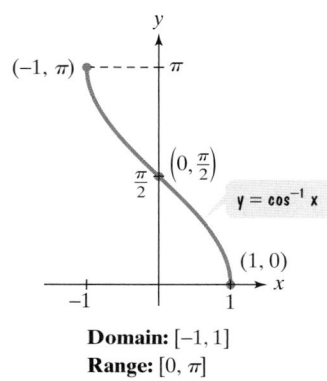

Domain: $[-1, 1]$
Range: $[0, \pi]$

Figure 5.72 The graph of the inverse cosine function

Table 5.6   Exact Values for $\cos\theta$, $0 \le \theta \le \pi$

θ	$\cos\theta$
0	1
$\dfrac{\pi}{6}$	$\dfrac{\sqrt{3}}{2}$
$\dfrac{\pi}{4}$	$\dfrac{\sqrt{2}}{2}$
$\dfrac{\pi}{3}$	$\dfrac{1}{2}$
$\dfrac{\pi}{2}$	0
$\dfrac{2\pi}{3}$	$-\dfrac{1}{2}$
$\dfrac{3\pi}{4}$	$-\dfrac{\sqrt{2}}{2}$
$\dfrac{5\pi}{6}$	$-\dfrac{\sqrt{3}}{2}$
π	-1

Exact values of $\cos^{-1} x$ can be found by thinking of **$\cos^{-1} x$** as **the angle in the interval $[0, \pi]$ whose cosine is x.** This time we will use Table 5.6, which shows exact values for $\cos\theta$ for θ in the interval $[0, \pi]$.

EXAMPLE 3   Finding the Exact Value of an Inverse Cosine Function

Find the exact value of $\cos^{-1}\left(-\dfrac{\sqrt{3}}{2}\right)$.

Solution
Step 1   Let $\theta = \cos^{-1} x$. Thus,

$$\theta = \cos^{-1}\left(-\frac{\sqrt{3}}{2}\right).$$

We must find the angle θ, $0 \le \theta \le \pi$, whose cosine equals $-\dfrac{\sqrt{3}}{2}$.

Step 2   Rewrite $\theta = \cos^{-1} x$ as $\cos\theta = x$. We obtain

$$\cos\theta = -\frac{\sqrt{3}}{2}.$$

Step 3 **Use the exact values in Table 5.6 to find the value of θ in $[0, \pi]$ that satisfies $\cos\theta = x$.** The table on page 503 shows that the only angle in the interval $[0, \pi]$ that satisfies $\cos\theta = -\dfrac{\sqrt{3}}{2}$ is $\dfrac{5\pi}{6}$. Thus, $\theta = \dfrac{5\pi}{6}$ and

$$\cos^{-1}\left(-\frac{\sqrt{3}}{2}\right) = \frac{5\pi}{6}. \quad \text{The angle in } [0, \pi] \text{ whose cosine is } -\frac{\sqrt{3}}{2} \text{ is } \frac{5\pi}{6}.$$

Check Point 3 Find the exact value of $\cos^{-1}\left(-\dfrac{1}{2}\right)$.

3 Understand and use the inverse tangent function.

The Inverse Tangent Function

Figure 5.73 shows how we restrict the domain of the tangent function so that it becomes one-to-one and has an inverse function. Restrict the domain to the interval $\left(-\dfrac{\pi}{2}, \dfrac{\pi}{2}\right)$, shown by the solid blue graph. Over this interval, the restricted tangent function passes the horizontal line test and has an inverse function.

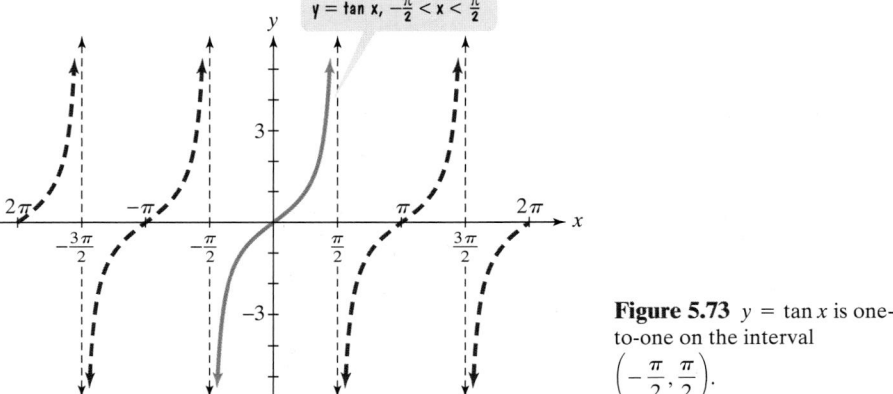

$y = \tan x, \ -\frac{\pi}{2} < x < \frac{\pi}{2}$

Figure 5.73 $y = \tan x$ is one-to-one on the interval $\left(-\dfrac{\pi}{2}, \dfrac{\pi}{2}\right)$.

The Inverse Tangent Function

The **inverse tangent function**, denoted by $\tan^{-1}$, is the inverse of the restricted tangent function $y = \tan x, -\dfrac{\pi}{2} < x < \dfrac{\pi}{2}$. Thus,

$$y = \tan^{-1} x \quad \text{means } \tan y = \infty,$$

where $-\dfrac{\pi}{2} < y < \dfrac{\pi}{2}$ and $-\infty < x < \infty$.

We graph $y = \tan^{-1} x$ by taking points on the graph of the restricted function and reversing the order of the coordinates. Figure 5.74 shows that $\left(-\dfrac{\pi}{4}, -1\right)$, $(0, 0)$, and $\left(\dfrac{\pi}{4}, 1\right)$ are on the graph of the restricted tangent function. Reversing the order gives $\left(-1, -\frac{\pi}{4}\right)$, $(0, 0)$, and $\left(1, \frac{\pi}{4}\right)$. We now use these three points to graph the inverse tangent function. The graph of $y = \tan^{-1} x$ is shown in Figure 5.75. Notice that the vertical asymptotes become horizontal asymptotes for the graph of the inverse function.

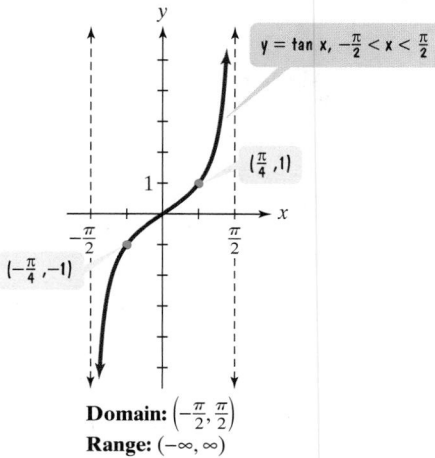

y
$y = \tan x, -\frac{\pi}{2} < x < \frac{\pi}{2}$
$\left(\frac{\pi}{4}, 1\right)$
1
x
$-\frac{\pi}{2}$
$\frac{\pi}{2}$
$\left(-\frac{\pi}{4}, -1\right)$
Domain: $\left(-\frac{\pi}{2}, \frac{\pi}{2}\right)$
Range: $(-\infty, \infty)$

Figure 5.74 The restricted tangent function

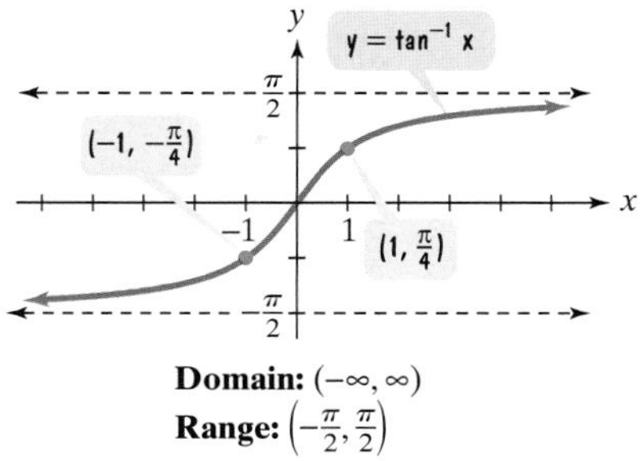

y
$y = \tan^{-1} x$
$\frac{\pi}{2}$
$\left(-1, -\frac{\pi}{4}\right)$
x
−1
1
$\left(1, \frac{\pi}{4}\right)$
$-\frac{\pi}{2}$
Domain: $(-\infty, \infty)$
Range: $\left(-\frac{\pi}{2}, \frac{\pi}{2}\right)$

Figure 5.75 The graph of the inverse tangent function

Exact values of $\tan^{-1} x$ can be found by thinking of $\tan^{-1} x$ as **the angle in the interval** $\left(-\frac{\pi}{2}, \frac{\pi}{2}\right)$ **whose tangent is** x. We use Table 5.7, which shows exact values for $\tan \theta$ for θ in the interval $\left(-\frac{\pi}{2}, \frac{\pi}{2}\right)$.

Table 5.7 Exact Values for $\tan \theta, -\frac{\pi}{2} < \theta < \frac{\pi}{2}$

θ	$\tan \theta$
$-\frac{\pi}{2}$	Undefined
$-\frac{\pi}{3}$	$-\sqrt{3}$
$-\frac{\pi}{4}$	-1
$-\frac{\pi}{6}$	$-\frac{\sqrt{3}}{3}$
0	0
$\frac{\pi}{6}$	$\frac{\sqrt{3}}{3}$
$\frac{\pi}{4}$	1
$\frac{\pi}{3}$	$\sqrt{3}$
$\frac{\pi}{2}$	Undefined

EXAMPLE 4 Finding the Exact Value of an Inverse Tangent Function

Find the exact value of $\tan^{-1} \sqrt{3}$.

Solution

Step 1 Let $\theta = \tan^{-1} x$. Thus,

$$\theta = \tan^{-1} \sqrt{3}.$$

We must find the angle $\theta, -\frac{\pi}{2} < \theta < \frac{\pi}{2}$, whose tangent equals $\sqrt{3}$.

Step 2 Rewrite $\theta = \tan^{-1} x$ as $\tan \theta = x$. We obtain $\tan \theta = \sqrt{3}$.

Step 3 Use the exact values in Table 5.7 to find the value of θ in $\left(-\frac{\pi}{2}, \frac{\pi}{2}\right)$ that satisfies $\tan \theta = x$. The table shows that the only angle in the interval $\left(-\frac{\pi}{2}, \frac{\pi}{2}\right)$ that satisfies $\tan \theta = \sqrt{3}$ is $\frac{\pi}{3}$. Thus, $\theta = \frac{\pi}{3}$ and

$$\tan^{-1} \sqrt{3} = \frac{\pi}{3}.$$ The angle in $\left(-\frac{\pi}{2}, \frac{\pi}{2}\right)$ whose tangent is $\sqrt{3}$ is $\frac{\pi}{3}$.

> **Check Point 4**
>
> Find the exact value of $\tan^{-1}(-1)$.

Table 5.8 summarizes the graphs of the three basic inverse trigonometric functions. Below each of the graphs is a description of the function's domain and range.

Table 5.8   Graphs of the Three Basic Inverse Trigonometric Functions

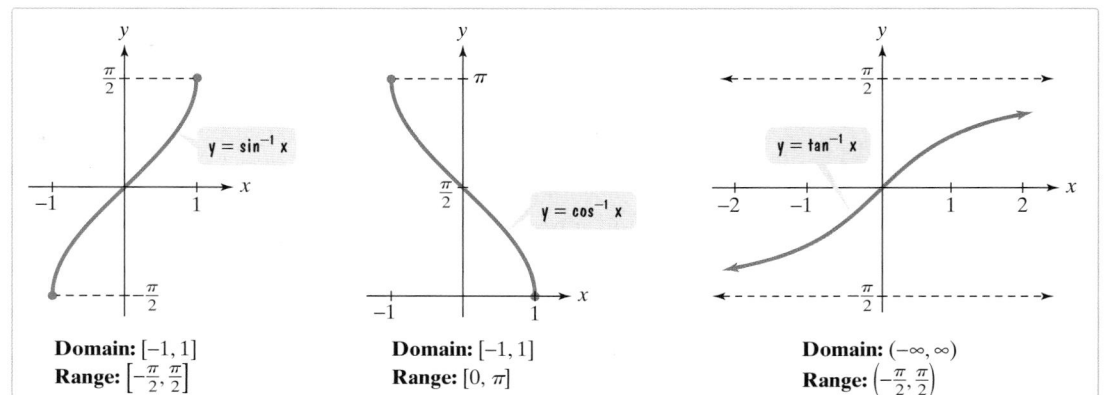

Domain: $[-1, 1]$	Domain: $[-1, 1]$	Domain: $(-\infty, \infty)$
Range: $\left[-\frac{\pi}{2}, \frac{\pi}{2}\right]$	Range: $[0, \pi]$	Range: $\left(-\frac{\pi}{2}, \frac{\pi}{2}\right)$

4 Use a calculator to evaluate inverse trigonometric functions.

Using a Calculator to Evaluate Inverse Trigonometric Functions

Calculators give approximate values of inverse trigonometric functions. Use the keys marked $\boxed{\text{SIN}^{-1}}$, $\boxed{\text{COS}^{-1}}$, and $\boxed{\text{TAN}^{-1}}$. Consult your manual for the location of this feature.

EXAMPLE 5   Calculators and Inverse Trigonometric Functions

Use a calculator to find the value to four decimal places of:

a. $\sin^{-1}\dfrac{1}{4}$ **b.** $\tan^{-1}(-9.65)$.

Solution

Scientific Calculator Solution

Function	Mode	Keystrokes	Display, rounded to four places
a. $\sin^{-1}\dfrac{1}{4}$	Radian	$1 \boxed{\div} 4 \boxed{=} \boxed{\text{SIN}^{-1}}$	0.2527
b. $\tan^{-1}(-9.65)$	Radian	$9.65 \boxed{^+/_-} \boxed{\text{TAN}^{-1}}$	−1.4675

Graphing Calculator Solution

Function	Mode	Keystrokes	Display, rounded to four places
a. $\sin^{-1}\dfrac{1}{4}$	Radian	$\boxed{\text{SIN}^{-1}} \boxed{(} 1 \boxed{\div} 4 \boxed{)} \boxed{\text{ENTER}}$	0.2527
b. $\tan^{-1}(-9.65)$	Radian	$\boxed{\text{TAN}^{-1}} \boxed{(-)} 9.65 \boxed{\text{ENTER}}$	−1.4675

Check
Point
5

Use a calculator to find the value to four decimal places of:

a. $\cos^{-1}\dfrac{1}{3}$ **b.** $\tan^{-1}(-35.85)$.

What happens if you attempt to evaluate an inverse trigomometric function at a value that is not in its domain? In real number mode, most calculators will display an error message. For example, an error message can result if you attempt to approximate $\cos^{-1} 3$. There is no angle whose cosine is 3. The domain of the inverse cosine function is $[-1, 1]$, and 3 does not belong to this domain.

5 Find exact values of composite functions with inverse trigonometric functions.

Composition of Functions Involving Inverse Trigonometric Functions

In our discussion of functions and their inverses in Section 2.6, we saw that

$$f(f^{-1}(x)) = x \quad \text{and} \quad f^{-1}(f(x)) = x.$$

x must be in the domain of f^{-1}.   *x must be in the domain of f.*

We apply these properties to the sine, cosine, tangent, and their inverse functions to obtain the following properties.

Inverse Properties

The Sine Function and Its Inverse

$$\sin(\sin^{-1} x) = x \qquad \text{for every } x \text{ in the interval } [-1, 1]$$

$$\sin^{-1}(\sin x) = x \qquad \text{for every } x \text{ in the interval } \left[-\dfrac{\pi}{2}, \dfrac{\pi}{2}\right]$$

The Cosine Function and Its Inverse

$$\cos(\cos^{-1} x) = x \qquad \text{for every } x \text{ in the interval } [-1, 1]$$

$$\cos^{-1}(\cos x) = x \qquad \text{for every } x \text{ in the interval } [0, \pi]$$

The Tangent Function and Its Inverse

$$\tan(\tan^{-1} x) = x \qquad \text{for every real number } x$$

$$\tan^{-1}(\tan x) = x \qquad \text{for every } x \text{ in the interval } \left(-\dfrac{\pi}{2}, \dfrac{\pi}{2}\right)$$

The restrictions on x in the inverse properties are a bit tricky. For example,

$$\sin^{-1}\left(\sin \dfrac{\pi}{4}\right) = \dfrac{\pi}{4}$$

$\sin^{-1}(\sin x) = x$ for x in $\left[-\dfrac{\pi}{2}, \dfrac{\pi}{2}\right]$.
Observe that $\dfrac{\pi}{4}$ is in this interval.

Can we use $\sin^{-1}(\sin x) = x$ to find the exact value of $\sin^{-1}\left(\sin \dfrac{5\pi}{4}\right)$? Is $\dfrac{5\pi}{4}$ in the interval $\left[-\dfrac{\pi}{2}, \dfrac{\pi}{2}\right]$? No. Thus to evaluate $\sin^{-1}\left(\sin \dfrac{5\pi}{4}\right)$, we must first find $\sin \dfrac{5\pi}{4}$.

$\dfrac{5\pi}{4}$ is in quadrant III where the sine is negative.

$$\sin\frac{5\pi}{4} = -\sin\frac{\pi}{4} = -\frac{\sqrt{2}}{2}$$

The reference angle for $\dfrac{5\pi}{4}$ is $\dfrac{\pi}{4}$.

We evaluate $\sin^{-1}\left(\sin\dfrac{5\pi}{4}\right)$ as follows.

$$\sin^{-1}\left(\sin\frac{5\pi}{4}\right) = \sin^{-1}\left(-\frac{\sqrt{2}}{2}\right) = -\frac{\pi}{4} \quad \text{If necessary, see Table 5.5 on page 500.}$$

To determine how to evaluate the composition of functions involving inverse trigonometric functions, first examine the value of x. You can use the inverse properties in the box on page 507 only if x is in the specified interval.

EXAMPLE 6 Evaluating Compositions of Functions and Their Inverses

Find the exact value, if possible, of:

a. $\cos(\cos^{-1}0.6)$ **b.** $\sin^{-1}\left(\sin\dfrac{3\pi}{2}\right)$ **c.** $\cos(\cos^{-1}2\pi)$.

Solution

a. The inverse property $\cos(\cos^{-1}x) = x$ applies for every x in $[-1, 1]$. To evaluate $\cos(\cos^{-1}0.6)$, observe that $x = 0.6$. This value of x lies in $[-1, 1]$, which is the domain of the inverse cosine function. This means that we can use the inverse property $\cos(\cos^{-1}x) = x$. Thus,
$$\cos(\cos^{-1}0.6) = 0.6.$$

b. The inverse property $\sin^{-1}(\sin x) = x$ applies for every x in $\left[-\dfrac{\pi}{2}, \dfrac{\pi}{2}\right]$. To evaluate $\sin^{-1}\left(\sin\dfrac{3\pi}{2}\right)$, observe that $x = \dfrac{3\pi}{2}$. This value of x does not lie in $\left[-\dfrac{\pi}{2}, \dfrac{\pi}{2}\right]$. To evaluate this expression, we first find $\sin\dfrac{3\pi}{2}$.

$$\sin^{-1}\left(\sin\frac{3\pi}{2}\right) = \sin^{-1}(-1) = -\frac{\pi}{2} \quad \text{The angle in } \left[-\frac{\pi}{2}, \frac{\pi}{2}\right] \text{ whose sine is } -1 \text{ is } -\frac{\pi}{2}.$$

c. The inverse property $\cos(\cos^{-1}x) = x$ applies for every x in $[-1, 1]$. To attempt to evaluate $\cos(\cos^{-1}2\pi)$, observe that $x = 2\pi$. This value of x does not lie in $[-1, 1]$, which is the domain of the inverse cosine function. Thus, the expression $\cos(\cos^{-1}2\pi)$ is not defined because $\cos^{-1}2\pi$ is not defined.

Check Point 6 Find the exact value, if possible, of:
a. $\cos(\cos^{-1}0.7)$ **b.** $\sin^{-1}(\sin\pi)$ **c.** $\cos(\cos^{-1}\pi)$.

We can use points on terminal sides of angles in standard position to find exact values of expressions involving the composition of a function and a different inverse function. Here are two examples.

$$\cos\left(\tan^{-1}\frac{5}{12}\right) \qquad \cot\left[\sin^{-1}\left(-\frac{1}{3}\right)\right]$$

Inner part involves the angle in $\left(-\frac{\pi}{2}, -\frac{\pi}{2}\right)$ whose tangent is $\frac{5}{12}$.

Inner part involves the angle in $\left(-\frac{\pi}{2}, \frac{\pi}{2}\right)$ whose sine is $-\frac{1}{3}$.

The inner part of each expression involves an angle. To evaluate such expressions, we represent such angles by θ. Then we use a sketch that illustrates our representation. Examples 7 and 8 show how to carry out such evaluations.

EXAMPLE 7 Evaluating a Composite Trigonometric Expression

Find the exact value of $\cos\left(\tan^{-1}\frac{5}{12}\right)$.

Solution We let θ represent the angle in $\left(-\frac{\pi}{2}, \frac{\pi}{2}\right)$ whose tangent is $\frac{5}{12}$. Thus,

$$\theta = \tan^{-1}\frac{5}{12}.$$

Using the definition of the inverse tangent function, we can rewrite this as

$$\tan\theta = \frac{5}{12}.$$

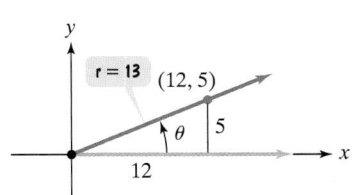

Figure 5.76 Representing $\tan\theta = \frac{5}{12}$

Because $\tan\theta$ is positive, θ must be an angle in $\left(0, \frac{\pi}{2}\right)$. Thus, θ is a first-quadrant angle. Figure 5.76 shows a right triangle in quadrant I with

$$\tan\theta = \frac{5}{12} \quad \text{Side opposite } \theta.$$
Side adjacent to θ.

The hypotenuse of the triangle can be found using the Pythagorean Theorem.

$$r^2 = 12^2 + 5^2 = 144 + 25 = 169 \quad \text{and} \quad r = \sqrt{169} = 13$$

We use this right triangle to find the exact value of $\cos\left(\tan^{-1}\frac{5}{12}\right)$.

$$\cos\left(\tan^{-1}\frac{5}{12}\right) = \cos\theta = \frac{\text{side adjacent to } \theta}{\text{hypotenuse}} = \frac{12}{13}$$

Check Point 7 Find the exact value of $\sin\left(\tan^{-1}\frac{3}{4}\right)$.

EXAMPLE 8 Evaluating a Composite Trigonometric Expression

Find the exact value of $\cot\left[\sin^{-1}\left(-\frac{1}{3}\right)\right]$.

Solution We let θ represent the angle in $\left[-\frac{\pi}{2}, \frac{\pi}{2}\right]$ whose sine is $-\frac{1}{3}$. Thus,

$$\theta = \sin^{-1}\left(-\frac{1}{3}\right) \quad \text{and} \quad \sin\theta = -\frac{1}{3}.$$

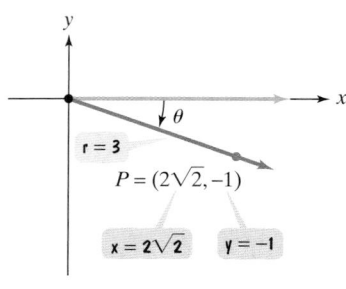

Figure 5.77 Representing $\sin \theta = -\frac{1}{3}$

Because $\sin \theta$ is negative, θ must be an angle in $\left[-\dfrac{\pi}{2}, 0\right)$. Thus, θ is a negative angle that lies in quadrant IV. Figure 5.77 shows angle θ in quadrant IV with

In quadrant IV, y is negative

$$\sin \theta = -\frac{1}{3} = \frac{y}{r} = \frac{-1}{3}.$$

The value of x can be found using $x^2 + y^2 = r^2$.

$$x^2 + (-1)^2 = 3^2$$
$$x^2 + 1 = 9$$
$$x^2 = 8$$
$$x = \sqrt{8} = \sqrt{4 \cdot 2} = 2\sqrt{2} \quad \text{Remember that x is positive in quadrant IV.}$$

We use values for x and y to find the exact value of $\cot\left[\sin^{-1}\left(-\frac{1}{3}\right)\right]$.

$$\cot\left[\sin^{-1}\left(-\frac{1}{3}\right)\right] = \cot \theta = \frac{x}{y} = \frac{2\sqrt{2}}{-1} = -2\sqrt{2}$$

Check Point 8   Find the exact value of $\cos\left[\sin^{-1}\left(-\frac{1}{2}\right)\right]$.

Some composite functions with inverse trigonometric functions can be simplified to algebraic expressions. To simplify such an expression, we represent the inverse trigonometric function in the expression by θ. Then we use a right triangle.

EXAMPLE 9   Simplifying an Expression Involving $\sin^{-1}x$

If $0 < x \le 1$, write $\cos\left(\sin^{-1} x\right)$ as an algebraic expression in x.

Solution   We let θ represent the angle in $\left[-\dfrac{\pi}{2}, \dfrac{\pi}{2}\right]$ whose sine is x. Thus,

$$\theta = \sin^{-1} x, \quad \text{and} \quad \sin \theta = x.$$

Because $0 < x \le 1$, $\sin \theta$ is positive. Thus, θ is a first-quadrant angle. Figure 5.78 shows a right triangle in quadrant I with

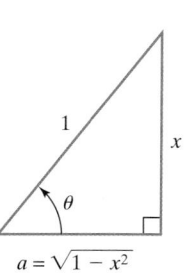

Figure 5.78 Representing $\sin \theta = x$

$$\sin \theta = x = \frac{x}{1}. \quad \begin{array}{l}\text{Side opposite } \theta \\ \text{Hypotenuse}\end{array}$$

The third side, a, can be found using the Pythagorean Theorem.

$$a^2 + x^2 = 1^2 \quad \text{Apply the Pythagorean Theorem to the right triangle in Figure 5.78.}$$

$$a^2 = 1 - x^2 \quad \text{Subtract } x^2 \text{ from both sides.}$$

$$a = \sqrt{1 - x^2} \quad \text{Solve for } a.$$

We use the right triangle in Figure 5.78 to write $\cos(\sin^{-1}x)$ as an algebraic expression.

$$\cos(\sin^{-1}x) = \cos\theta = \frac{\text{side adjacent to } \theta}{\text{hypotenuse}} = \frac{\sqrt{1-x^2}}{1} = \sqrt{1-x^2}$$

Check Point 9 If $x > 0$, write $\sec(\tan^{-1}x)$ as an algebraic expression in x.

The inverse secant function, $y = \sec^{-1}x$, is used in calculus. However, inverse cotangent and inverse cosecant functions are rarely used. Two of these remaining inverse trigonometric functions are briefly developed in the exercise set that follows.

EXERCISE SET 5.7

Practice Exercises

In Exercises 1–18, use the values in the table shown to find the exact value of each expression.

θ	$-\dfrac{\pi}{2}$	$-\dfrac{\pi}{3}$	$-\dfrac{\pi}{4}$	$-\dfrac{\pi}{6}$	0	$\dfrac{\pi}{6}$	$\dfrac{\pi}{4}$	$\dfrac{\pi}{3}$	$\dfrac{\pi}{2}$	$\dfrac{2\pi}{3}$	$\dfrac{3\pi}{4}$	$\dfrac{5\pi}{6}$	π
$\sin\theta$	-1	$-\dfrac{\sqrt{3}}{2}$	$-\dfrac{\sqrt{2}}{2}$	$-\dfrac{1}{2}$	0	$\dfrac{1}{2}$	$\dfrac{\sqrt{2}}{2}$	$\dfrac{\sqrt{3}}{2}$	1	$\dfrac{\sqrt{3}}{2}$	$\dfrac{\sqrt{2}}{2}$	$\dfrac{1}{2}$	0
$\cos\theta$	0	$\dfrac{1}{2}$	$\dfrac{\sqrt{2}}{2}$	$\dfrac{\sqrt{3}}{2}$	1	$\dfrac{\sqrt{3}}{2}$	$\dfrac{\sqrt{2}}{2}$	$\dfrac{1}{2}$	0	$-\dfrac{1}{2}$	$-\dfrac{\sqrt{2}}{2}$	$-\dfrac{\sqrt{3}}{2}$	-1
$\tan\theta$	undef.	$-\sqrt{3}$	-1	$-\dfrac{\sqrt{3}}{3}$	0	$\dfrac{\sqrt{3}}{3}$	1	$\sqrt{3}$	undef.	$-\sqrt{3}$	-1	$\dfrac{-\sqrt{3}}{3}$	0

1. $\sin^{-1}\frac{1}{2}$

2. $\sin^{-1}0$

3. $\sin^{-1}\dfrac{\sqrt{2}}{2}$

4. $\sin^{-1}\dfrac{\sqrt{3}}{2}$

5. $\sin^{-1}\left(-\dfrac{1}{2}\right)$

6. $\sin^{-1}\left(-\dfrac{\sqrt{3}}{2}\right)$

7. $\cos^{-1}\dfrac{\sqrt{3}}{2}$

8. $\cos^{-1}\dfrac{\sqrt{2}}{2}$

9. $\cos^{-1}\left(-\dfrac{\sqrt{2}}{2}\right)$

10. $\cos^{-1}\left(-\dfrac{\sqrt{3}}{2}\right)$

11. $\cos^{-1}0$

12. $\cos^{-1}1$

13. $\tan^{-1}\dfrac{\sqrt{3}}{3}$

14. $\tan^{-1}1$

15. $\tan^{-1}0$

16. $\tan^{-1}(-1)$

17. $\tan^{-1}(-\sqrt{3})$

18. $\tan^{-1}\left(-\dfrac{\sqrt{3}}{3}\right)$

In Exercises 19–30, use a calculator to find the value of each expression rounded to two decimal places.

19. $\sin^{-1}0.3$

20. $\sin^{-1}0.47$

21. $\sin^{-1}(-0.32)$

22. $\sin^{-1}(-0.625)$

23. $\cos^{-1}\frac{3}{8}$

24. $\cos^{-1}\frac{4}{9}$

25. $\cos^{-1}\dfrac{\sqrt{5}}{7}$

26. $\cos^{-1}\dfrac{\sqrt{7}}{10}$

27. $\tan^{-1}(-20)$

28. $\tan^{-1}(-30)$

29. $\tan^{-1}(-\sqrt{473})$

30. $\tan^{-1}(-\sqrt{5061})$

In Exercises 31–46, find the exact value of each expression, if possible. Do not use a calculator.

31. $\sin(\sin^{-1}0.9)$

32. $\cos(\cos^{-1}0.57)$

33. $\sin^{-1}\left(\sin\dfrac{\pi}{3}\right)$

34. $\cos^{-1}\left(\cos\dfrac{2\pi}{3}\right)$

35. $\sin^{-1}\left(\sin\dfrac{5\pi}{6}\right)$

36. $\cos^{-1}\left(\cos\dfrac{4\pi}{3}\right)$

37. $\tan(\tan^{-1}125)$

38. $\tan(\tan^{-1}380)$

39. $\tan^{-1}\left[\tan\left(-\dfrac{\pi}{6}\right)\right]$

40. $\tan^{-1}\left[\tan\left(-\dfrac{\pi}{3}\right)\right]$

41. $\tan^{-1}\left(\tan\dfrac{2\pi}{3}\right)$ **42.** $\tan^{-1}\left(\tan\dfrac{3\pi}{4}\right)$

43. $\sin^{-1}(\sin\pi)$ **44.** $\cos^{-1}(\cos 2\pi)$

45. $\sin(\sin^{-1}\pi)$ **46.** $\cos(\cos^{-1}3\pi)$

In Exercises 47–60, use a sketch to find the exact value of each expression.

47. $\cos\left(\sin^{-1}\frac{4}{5}\right)$ **48.** $\sin\left(\tan^{-1}\frac{7}{24}\right)$

49. $\tan\left(\cos^{-1}\frac{5}{13}\right)$ **50.** $\cot\left(\sin^{-1}\frac{5}{13}\right)$

51. $\tan\left[\sin^{-1}\left(-\frac{3}{5}\right)\right]$ **52.** $\cos\left[\sin^{-1}\left(-\frac{4}{5}\right)\right]$

53. $\sin\left(\cos^{-1}\dfrac{\sqrt{2}}{2}\right)$ **54.** $\cos\left(\sin^{-1}\frac{1}{2}\right)$

55. $\sec\left[\sin^{-1}\left(-\frac{1}{4}\right)\right]$ **56.** $\sec\left[\sin^{-1}\left(-\frac{1}{2}\right)\right]$

57. $\tan\left[\cos^{-1}\left(-\frac{1}{3}\right)\right]$ **58.** $\tan\left[\cos^{-1}\left(-\frac{1}{4}\right)\right]$

59. $\csc\left[\cos^{-1}\left(-\dfrac{\sqrt{3}}{2}\right)\right]$ **60.** $\sec\left[\sin^{-1}\left(-\dfrac{\sqrt{2}}{2}\right)\right]$

In Exercises 61–66, use a right triangle to write each expression as an algebraic expression. Assume that x is positive and in the domain of the given inverse trigonometric function.

61. $\tan(\cos^{-1}x)$ **62.** $\sin(\tan^{-1}x)$

63. $\cos\left(\sin^{-1}\dfrac{1}{x}\right)$ **64.** $\sec\left(\cos^{-1}\dfrac{1}{x}\right)$

65. $\sec\left(\sin^{-1}\dfrac{x}{\sqrt{x^2+4}}\right)$ **66.** $\cot\left(\sin^{-1}\dfrac{\sqrt{x^2-9}}{x}\right)$

67. a. Graph the restricted secant function, $y = \sec x$, by restricting x to the intervals $\left[0,\dfrac{\pi}{2}\right)$ and $\left(\dfrac{\pi}{2},\pi\right]$.
   b. Use the horizontal line test and explain why the restricted secant function has an inverse function.
   c. Use the graph of the restricted secant function to graph $y = \sec^{-1}x$.

68. a. Graph the restricted cotangent function, $y = \cot x$, by restricting x to the interval $(0,\pi)$.
   b. Use the horizontal line test and explain why the restricted cotangent function has an inverse function.
   c. Use the graph of the restricted cotangent function to graph $y = \cot^{-1}x$.

⭐ Application Exercises

69. Your neighborhood movie theater has a 25-foot-high screen located 8 feet above your eye level. If you sit too close to the screen, your viewing angle is too small, resulting in a distorted picture. By contrast, if you sit too far back, the image is quite small, diminishing the movie's visual impact. If you sit x feet back from the screen, your viewing angle θ is given by

$$\theta = \tan^{-1}\dfrac{33}{x} - \tan^{-1}\dfrac{8}{x}.$$

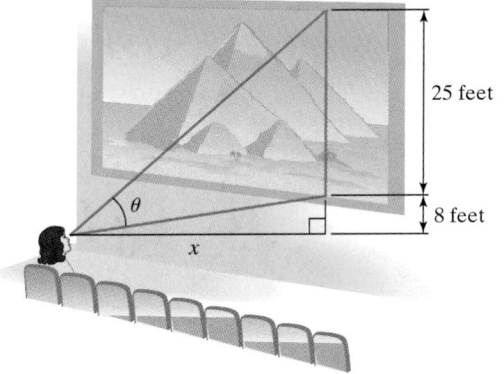

25 feet

8 feet

Find the viewing angle, in radians, at distances of 5 feet, 10 feet, 15 feet, 20 feet, and 25 feet.

70. The function $\theta = \tan^{-1}\dfrac{33}{x} - \tan^{-1}\dfrac{8}{x}$ is graphed below in a $[0, 50, 10]$ by $[0, 1, 0.1]$ viewing rectangle. Use the graph to describe what happens to your viewing angle as you move farther back from the screen. How far back from the screen, to the nearest foot, should you sit to maximize your viewing angle? Verify this observation by finding the viewing angle one foot closer to the screen and one foot farther from the screen for this ideal viewing distance.

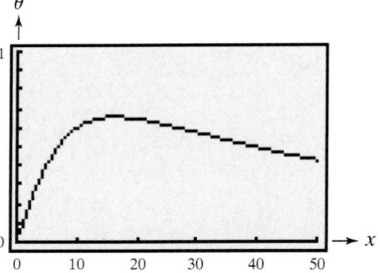

The formula

$$\theta = 2\tan^{-1}\dfrac{21.634}{x}$$

gives the viewing angle, in radians, for a camera whose lens is x millimeters wide. Use this formula to solve Exercises 71–72.

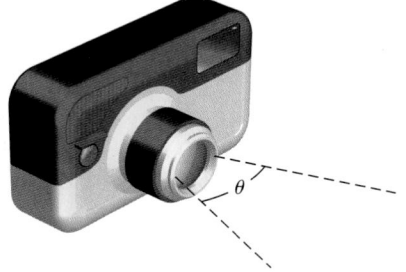

71. Find the viewing angle, in radians and in degrees (to the nearest tenth of a degree), of a 28-millimeter lens.

72. Find the viewing angle, in radians and degrees (to the nearest tenth of a degree), of a 300-millimeter telephoto lens.

For years, mathematicians were challenged by the following problem: What is the area of a region under a curve between two values of x? The problem was solved in the seventeenth century with the development of integral calculus. Using calculus, the area of the region under $y = \dfrac{1}{x^2+1}$, above the x-axis, and between $x = a$ and $x = b$ is $\tan^{-1} b - \tan^{-1} a$. Use this result, shown in the figure, to find the area of the region under $y = \dfrac{1}{x^2+1}$ and between a and b given in Exercises 73–74.

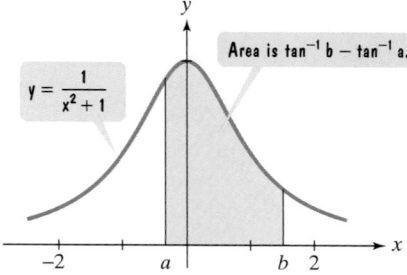

73. $a = 0$ and $b = 2$

74. $a = -2$ and $b = 1$

Writing in Mathematics

75. Explain why, without restrictions, no trigonometric function has an inverse function.

76. Describe the restriction on the sine function so that it has an inverse function.

77. How can the graph of $y = \sin^{-1} x$ be obtained from the graph of the restricted sine function?

78. Without drawing a graph, describe the behavior of the graph of $y = \sin^{-1} x$. Mention the function's domain and range in your description.

79. Describe the restriction on the cosine function so that it has an inverse function.

80. Without drawing a graph, describe the behavior of the graph of $y = \cos^{-1} x$. Mention the function's domain and range in your description.

81. Describe the restriction on the tangent function so that it has an inverse function.

82. Without drawing a graph, describe the behavior of the graph of $y = \tan^{-1} x$. Mention the function's domain and range in your description.

83. If $\sin^{-1}\left(\sin\dfrac{\pi}{3}\right) = \dfrac{\pi}{3}$, is $\sin^{-1}\left(\sin\dfrac{5\pi}{6}\right) = \dfrac{5\pi}{6}$? Explain your answer.

84. Explain how a right triangle can be used to find the exact value of $\sec(\sin^{-1}\frac{4}{5})$.

85. Find the height of the screen and the number of feet that it is located above eye level in your favorite movie theater. Modify the formula given in Exercise 69 so that it applies to your theater. Then describe where in the theater you should sit so that a movie creates the greatest visual impact.

Technology Exercises

In Exercises 86–89, graph each pair of functions in the same viewing rectangle. Use your knowledge of the domain and range for the inverse trigonometric functions to select an appropriate viewing rectangle. What does the graph of the second equation in each exercise do to the graph of the first equation?

86. $y = \sin^{-1} x$ and $y = \sin^{-1} x + 2$

87. $y = \cos^{-1} x$ and $y = \cos^{-1}(x-1)$

88. $y = \tan^{-1} x$ and $y = -2\tan^{-1} x$

89. $y = \sin^{-1} x$ and $y = \sin^{-1}(x+2) + 1$

90. Graph $y = \tan^{-1} x$ and its two horizontal asymptotes in a $[-3, 3, 1]$ by $\left[-\pi, \pi, \dfrac{\pi}{2}\right]$ viewing rectangle. Then change the range setting to $[-50, 50, 5]$ by $\left[-\pi, \pi, \dfrac{\pi}{2}\right]$. What do you observe?

91. Graph $y = \sin^{-1} x + \cos^{-1} x$ in a $[-2, 2, 1]$ by $[0, 3, 1]$ viewing rectangle. What appears to be true about the sum of the inverse sine and inverse cosine for values between -1 and 1 inclusive?

Critical Thinking Exercises

92. Solve $y = 2\sin^{-1}(x-5)$ for x in terms of y.

93. Solve for x: $2\sin^{-1} x = \dfrac{\pi}{4}$.

94. Prove that if $x > 0$, $\tan^{-1} x + \tan^{-1}\dfrac{1}{x} = \dfrac{\pi}{2}$.

95. Derive the formula for θ, your viewing angle at the movie theater, in Exercise 69. *Hint:* Use the figure shown and represent the acute angle on the left in the smaller right triangle by α. Find expressions for $\tan \alpha$ and $\tan(\alpha + \theta)$.

SECTION 5.8 *Applications of Trigonometric Functions*

Objectives

1. Solve a right triangle.
2. Solve problems involving bearings.
3. Model simple harmonic motion.

In the late 1960s, popular musicians were searching for new sounds. Film composers were looking for ways to create unique sounds as well. From these efforts, synthesizers that electronically reproduce musical sounds were born. From providing the backbone of today's most popular music to providing the strange sounds for the most experimental music, synthesizers are at the forefront of today's music technology.

If we did not understand the periodic nature of sinusoidal functions, the synthesizers used in almost all forms of music would not exist. In this section, we look at applications of trigonometric functions in right triangles and in modeling periodic phenomena such as sound.

1 Solve a right triangle.

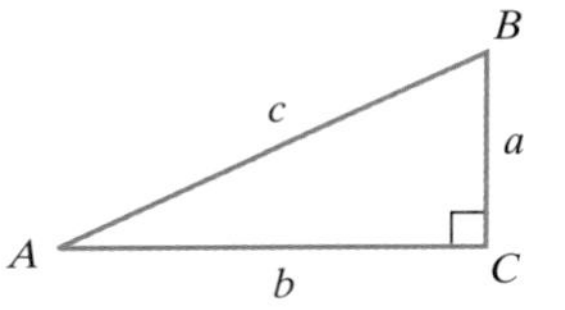

Figure 5.79 Labeling right triangles

Solving Right Triangles

Solving a right triangle means finding the missing lengths of its sides and the measurements of its angles. We will label right triangles so that side a is opposite angle A, side b is opposite angle B, and side c is the hypotenuse opposite right angle C. Figure 5.79 illustrates this labeling.

When solving a right triangle, we will use the sine, cosine, and tangent functions, rather than their reciprocals. Example 1 shows how to solve a right triangle when we know the length of a side and the measure of an acute angle.

EXAMPLE 1 Solving a Right Triangle

Solve the right triangle shown in Figure 5.80.

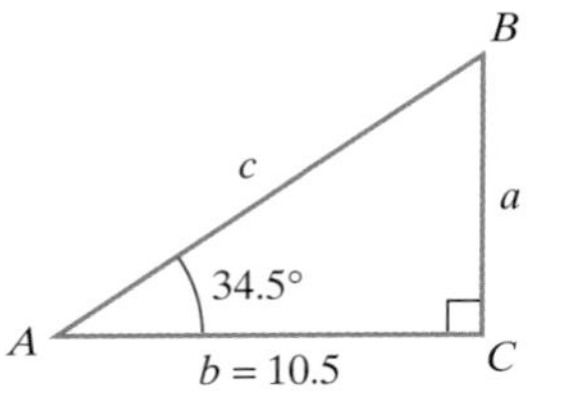

Solution We begin by finding the measure of angle B. We do not need a trigonometric function to do so. Because $C = 90°$ and the sum of a triangle's angles is $180°$, we see that $A + B = 90°$. Thus,

$$B = 90° - A = 90° - 34.5° = 55.5°.$$

Now we need to find a. Because we have a known angle, an unknown opposite side, and a known adjacent side, we use the tangent function.

$$\tan 34.5° = \frac{a}{10.5}$$

Side opposite the 34.5° angle

Side adjacent to the 34.5° angle

Figure 5.80 Find B, a, and c

Discovery

There is often more than one correct way to solve a right triangle. In Example 1, find a using angle $B = 55.5°$. Find c using the Pythagorean Theorem.

Now we solve for a.

$$a = 10.5 \tan 34.5° \approx 7.22$$

Finally, we need to find c. Because we have a known angle, a known adjacent side, and an unknown hypotenuse, we use the cosine function.

$$\cos 34.5° = \frac{10.5}{c}$$

Side adjacent to the 34.5° angle

hypotenuse

Now we solve for c.

$$c = \frac{10.5}{\cos 34.5°} \approx 12.74$$

In summary, $B = 55.5°$, $a \approx 7.22$, and $c \approx 12.74$.

Check Point 1   In Figure 5.79, let $A = 62.7°$ and $a = 8.4$. Solve the right triangle, rounding lengths to two decimal places.

Trigonometry was first developed to measure heights and distances that are inconvenient or impossible to measure. In solving application problems, begin by making a sketch involving a right triangle that illustrates the problem's conditions. Then put your knowledge of solving right triangles to work and find the required distance or height.

EXAMPLE 2   Finding the Side of a Triangle

From a point on level ground 125 feet from the base of a tower, the angle of elevation is 57.2°. Approximate the height of the tower to the nearest foot.

Solution   A sketch is shown in Figure 5.81, where a represents the height of the tower. In the right triangle, we have a known angle, an unknown opposite side, and a known adjacent side. Therefore, we use the tangent function.

$$\tan 57.2° = \frac{a}{125}$$

Side opposite the 57.2° angle

Side adjacent to the 57.2° angle

Solving for a,

$$a = 125 \tan 57.2° \approx 194.$$

The tower is approximately 194 feet high.

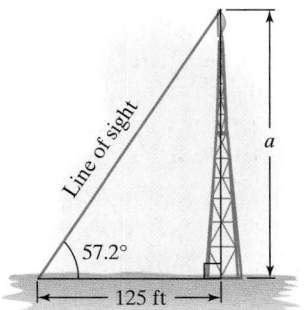

Figure 5.81 Determining height without using direct measurement

Check Point 2   From a point on level ground 80 feet from the base of the Eiffel Tower, the angle of elevation is 85.4°. Approximate the height of the Eiffel Tower to the nearest foot.

Example 3 illustrates how to find the measure of an acute angle of a right triangle if the length of two sides is known.

EXAMPLE 3 Finding the Angle of a Triangle

A kite flies at a height of 30 feet when 65 feet of string is out. If the string is in a straight line, find the angle that it makes with the ground. Round to the nearest tenth of a degree.

Solution A sketch is shown in Figure 5.82, where A represents the angle the string makes with the ground. In the right triangle, we have an unknown angle, a known opposite side, and a known hypotenuse. Therefore, we use the sine function.

$$\sin A = \frac{30 \quad \text{Side opposite A}}{65 \quad \text{hypotenuse}}$$

$$A = \sin^{-1}\frac{30}{65} \approx 27.5°$$

The string makes an angle of approximately 27.5° with the ground.

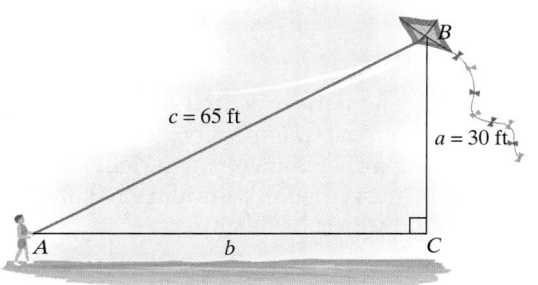

$c = 65$ ft

$a = 30$ ft

Figure 5.82 Flying a kite

> **Check Point 3**
>
> A guy wire is 13.8 yards long and is attached from the ground to a pole 6.7 yards above the ground. Find the angle, to the nearest tenth of a degree, that the wire makes with the ground.

EXAMPLE 4 Using Two Right Triangles to Solve a Problem

You are taking your first hot-air balloon ride. Your friend is standing on level ground, 100 feet away from your point of launch, making a video of the terrified look on your rapidly ascending face. How rapidly? At one instant, the angle of elevation from the video camera to your face is 31.7°. One minute later, the angle of elevation is 76.2°. How far did you travel during that minute?

Solution A sketch that illustrates the problem is shown in Figure 5.83. We need to determine $b - a$, the distance traveled during the one-minute period. We find a using the small right triangle. Because we have a known angle, an unknown opposite side, and a known adjacent side, we use the tangent function.

$$\tan 31.7° = \frac{a \quad \text{Side opposite the 31.7° angle}}{100 \quad \text{Side adjacent to the 31.7° angle}}$$

$$a = 100 \tan 31.7° \approx 61.8$$

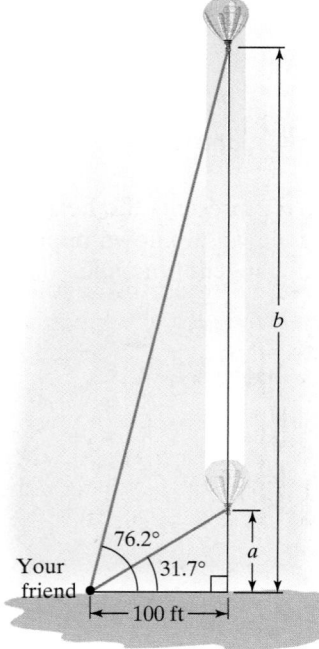

Figure 5.83 Ascending in a hot-air balloon

We find b using the tangent function in the large right triangle.

$$\tan 76.2° = \frac{b}{100}$$

Side opposite the 76.2° angle

Side adjacent to the 76.2° angle

$$b = 100 \tan 76.2° \approx 407.1$$

The balloon traveled $407.1 - 61.8$, or approximately 345.3 feet, during the minute.

Check Point 4

You are standing on level ground 800 feet from Mt. Rushmore, looking at the sculpture of Abraham Lincoln's face. The angle of elevation to the bottom of the sculpture is $32°$ and the angle of elevation to the top is $35°$. Find the height of the sculpture of Lincoln's face to the nearest tenth of a foot.

2 Solve problems involving bearings.

Trigonometry and Bearings

In navigation and surveying problems, the term *bearing* is used to specify the location of one point relative to another. The **bearing** from point O to point P is the acute angle between ray OP and a north-south line. Figure 5.84 illustrates some examples of bearings. The north-south line and the east-west line intersect at right angles.

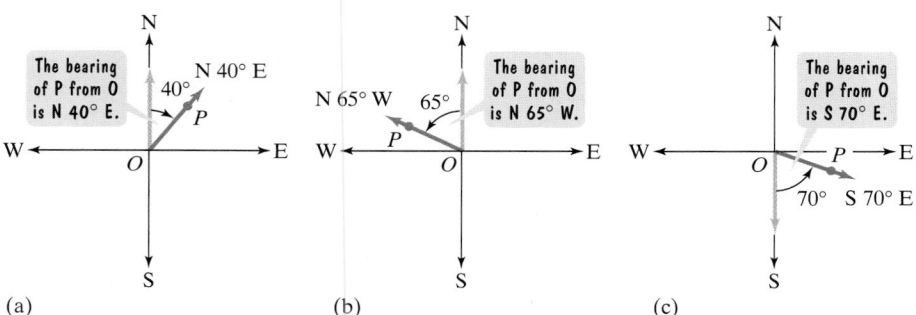

Figure 5.84 An illustration of three bearings

(a)   (b)   (c)

Each bearing has three parts: a letter (N or S), the measure of an acute angle, and a letter (E or W). Here's how we write a bearing:

- If the acute angle is measured from the *north side* of the north-south line, then we write N first. [See Figure 5.84(a).] If the acute angle is measured from the *south side* of the north-south line, then we write S first. [See Figure 5.84(c).]
- Second, we write the measure of the acute angle.
- If the acute angle is measured on the *east side* of the north-south line, then we write E last. [See Figure 5.84(a)]. If the acute angle is measured on the *west side* of the north-south line, then we write W last. [See Figure 5.84(b).]

EXAMPLE 5   Understanding Bearings

Use Figure 5.85 on page 518 to find:

a. the bearing from O to B.

b. the bearing from O to A.

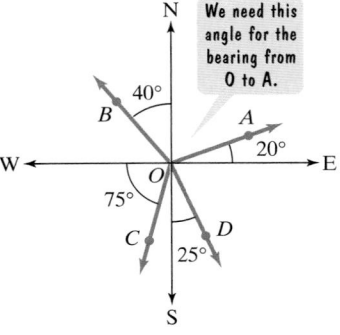

Figure 5.85 Finding bearings

Solution

a. To find the bearing from O to B, we need the acute angle between the ray OB and the north-south line through O. The measurement of this angle is given to be 40°. Figure 5.85 shows that the angle is measured from the north side of the north-south line and lies west of the north-south line. Thus, the bearing from O to B is N 40° W.

b. To find the bearing from O to A, we need the acute angle between the ray OA and the north-south line through O. This angle is specified by the voice balloon in Figure 5.85. The figure shows that this angle measures $90° - 20°$, or 70°. This angle is measured from the north side of the north-south line. This angle is also east of the north-south line. Thus, the bearing from O to A is N 70° E.

Check Point 5 Use Figure 5.85 to find:

a. the bearing from O to D.

b. the bearing from O to C.

EXAMPLE 6 Finding the Bearing of a Boat

A boat leaves the entrance to a harbor and travels 25 miles on a bearing of N 42° E. Figure 5.86 shows that the captain then turns the boat 90° and travels 18 miles on a bearing of S 48° E. At that time:

a. How far is the boat from the harbor entrance?

b. What is the bearing of the boat from the harbor entrance?

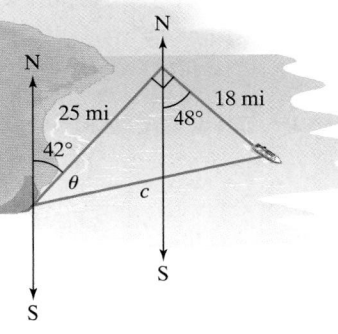

Figure 5.86 Finding a boat's bearing from the harbor entrance

Study Tip

When making a diagram showing bearings, draw a north-south line through each point at which a change in course occurs. The north side of the line lies above each point. The south side of the line lies below each point.

Solution

a. The boat's distance from the harbor entrance is represented by c in Figure 5.86. Because we know the length of two sides of the right triangle, we find c using the Pythagorean Theorem. We have

$$c^2 = a^2 + b^2 = 25^2 + 18^2 = 949$$

$$c = \sqrt{949} \approx 30.8.$$

The boat is approximately 30.8 miles from the harbor entrance.

b. To find the bearing of the boat from the harbor entrance, look at the north-south line passing through the harbor entrance on the left in Figure 5.86. The acute angle from this line to the ray on which the boat lies is $42° + \theta$. Because we are measuring the angle from the north side of the line and the boat is east of the harbor, its bearing from the harbor entrance is $N(42° + \theta)E$. To find θ, we use the right triangle shown in Figure 5.86 and the tangent function.

$$\tan \theta = \frac{\text{side opposite } \theta}{\text{side adjacent to } \theta} = \frac{18}{25}$$

$$\theta = \tan^{-1} \frac{18}{25}$$

We can use a calculator in degree mode to find the value of θ: $\theta \approx 35.8°$. Thus, $42° + \theta = 42° + 35.8° = 77.8°$. The bearing of the boat from the harbor entrance is N 77.8° E.

Check Point 6

You leave the entrance to a system of hiking trails and hike 2.3 miles on a bearing of S 31° W. You then turn and hike 3.5 miles on a bearing of N 59° W. At that time:

a. How far are you from the entrance to the trail system?

b. What is your bearing from the entrance to the trail system?

3 Model simple harmonic motion.

Simple Harmonic Motion

Because of their periodic nature, trigonometric functions are used to model phenomena that occur again and again. This includes vibratory or oscillatory motion, such as the motion of a vibrating guitar string, the swinging of a pendulum, or the bobbing of an object attached to a spring. Trigonometric functions are also used to describe radio waves from your favorite FM station, television waves from your not-to-be-missed weekly sitcom, and sound waves from your most-prized CDs.

To see how trigonometric functions are used to model vibratory motion, consider this: A ball is attached to a spring hung from the ceiling. You pull the ball down 4 inches and then release it. If we neglect the effects of friction and air resistance, the ball will continue bobbing up and down on the end of the spring. These up-and-down oscillations are called **simple harmonic motion**.

To better understand this motion, we use a d-axis, where d represents distance. This axis is shown in Figure 5.87. On this axis, the position of the ball before you pull it down is $d = 0$. This rest position is called the **equilibrium position**. Now you pull the ball down 4 inches to $d = -4$ and release it. Figure 5.88 shows a sequence of "photographs" taken at one-second time intervals illustrating the distance of the ball from its rest position, d.

The curve in Figure 5.88 shows how the ball's distance from its rest position changes over time. The curve is sinusoidal and the motion can be described using a cosine or a sine function.

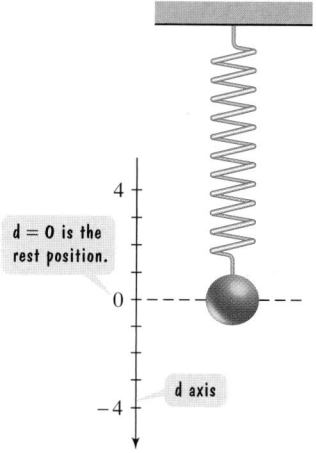

$d = 0$ is the rest position.

Figure 5.87 Using a d-axis to describe a ball's distance from its rest position

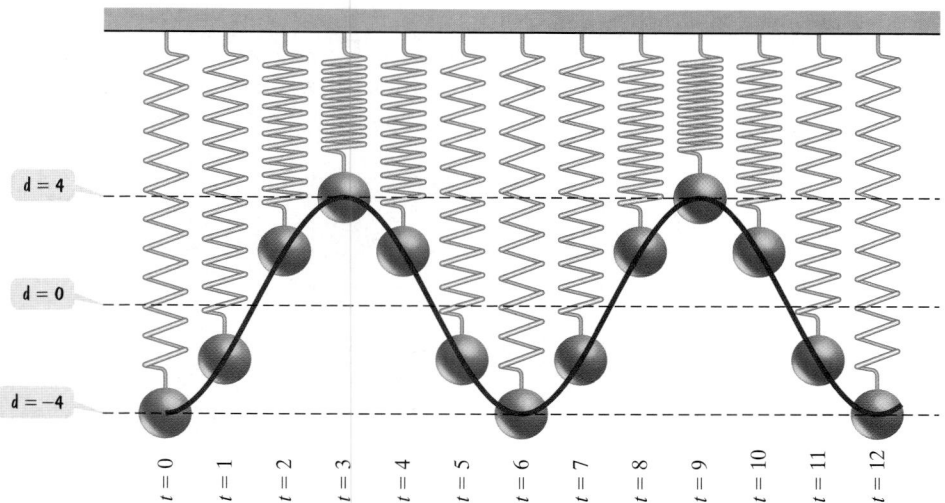

Figure 5.88 A sequence of "photographs" showing the bobbing ball's distance from the rest position, taken at one-second intervals

Simple Harmonic Motion

An object that moves on a coordinate axis is in **simple harmonic motion** if its distance from the origin, d, at time t is given by either

$$d = a \cos \omega t \text{ or } d = a \sin \omega t.$$

The motion has **amplitude** $|a|$, the maximum displacement of the object from its rest position. The **period** of the motion is $\dfrac{2\pi}{\omega}$, where $\omega > 0$. The period gives the time it takes for the motion to go through one complete cycle.

In describing simple harmonic motion, the equation with the cosine function is used if the object is at its greatest distance from rest position, the origin, at $t = 0$. By contrast, the equation with the sine function is used if the object is at its rest position, the origin, at $t = 0$.

Diminishing Motion with Increasing Time

Due to friction and other resistive forces, the motion of an oscillating object decreases over time. The function

$$d = 3e^{-0.1t} \cos 2t$$

models this type of motion. The graph of the function is shown in a $t = [0, 10, 1]$ by $d = [-3, 3, 1]$ viewing rectangle. Notice how the amplitude is decreasing with time as the moving object loses energy.

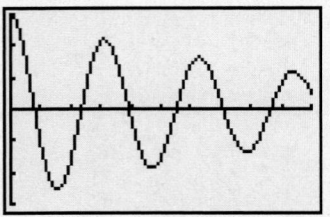

EXAMPLE 7   Finding an Equation for an Object in Simple Harmonic Motion

A ball on a spring is pulled 4 inches below its rest position and then released. The period of the motion is 6 seconds. Write the equation for the ball's simple harmonic motion.

Solution   We need to write an equation that describes d, the distance of the ball from its rest position, after t seconds. (The motion is illustrated by the "photo" sequence in Figure 5.88 on page 519.) When the object is released ($t = 0$), the ball's distance from its rest position is 4 inches down. Because it is *down* 4 inches, d is negative: When $t = 0$, $d = -4$. Notice the greatest distance from rest position occurs at $t = 0$. Thus, we will use the equation with the cosine function,

$$d = a \cos \omega t,$$

to model the ball's simple harmonic motion.

Now we determine values for a and ω. Recall that $|a|$ is the maximum displacement. Because the ball initially moves down, $a = -4$.

The value of ω in $d = a \cos \omega t$ can be found using the formula for the period.

$$\text{period} = \frac{2\pi}{\omega} = 6 \qquad \text{We are given that the period of the motion is 6 seconds.}$$

$$2\pi = 6\omega \qquad \text{Multiply both sides by } \omega.$$

$$\omega = \frac{2\pi}{6} = \frac{\pi}{3} \qquad \text{Divide both sides by 6 and solve for } \omega.$$

We see that $a = -4$ and $\omega = \dfrac{\pi}{3}$. Substitute these values into $d = a \cos \omega t$. The equation for the ball's simple harmonic motion is

$$d = -4 \cos \frac{\pi}{3} t.$$

Modeling Music

Sounds are caused by vibrating objects that result in variations in pressure in the surrounding air. Areas of high and low pressure moving through the air are modeled by the harmonic motion formulas. When these vibrations reach our eardrums, the eardrums' vibrations send signals to our brains which create the sensation of hearing.

French mathematician John Fourier (1768–1830) proved that all musical sounds—instrumental and vocal—could be modeled by sums involving sine functions. Modeling musical sounds with sinusoidal functions is used by synthesizers to electronically produce sounds unobtainable from ordinary musical instruments.

Check Point 7

A ball on a spring is pulled 6 inches below its rest position and then released. The period for the motion is 4 seconds. Write the equation for the ball's simple harmonic motion.

The period of the harmonic motion in Example 7 was 6 seconds. It takes 6 seconds for the moving object to complete one cycle. Thus, $\frac{1}{6}$ of a cycle is completed every second. We call $\frac{1}{6}$ the *frequency* of the moving object. **Frequency** describes the number of complete cycles per unit time and is the reciprocal of the period.

Frequency of an Object in Simple Harmonic Motion

An object in simple harmonic motion given by
$$d = a \cos \omega t \text{ or } d = a \sin \omega t$$

has **frequency** f given by

$$f = \frac{\omega}{2\pi}, \omega > 0.$$

Equivalently,

$$f = \frac{1}{\text{period}}.$$

EXAMPLE 8   Analyzing Simple Harmonic Motion

Figure 5.89 shows a mass on a smooth table attached to a spring. The mass moves in simple harmonic motion described by

$$d = 10 \cos \frac{\pi}{6} t$$

with t measured in seconds and d in centimeters. Find (a) the maximum displacement, (b) the frequency, and (c) the time required for one cycle.

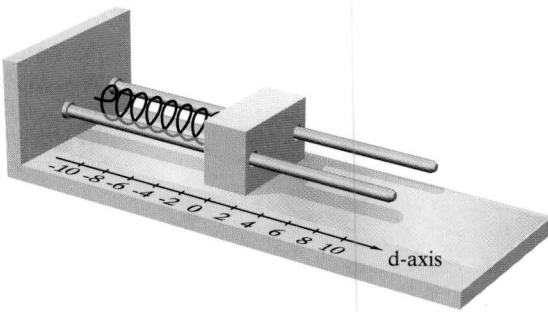

d-axis

Figure 5.89 A mass attached to a spring, moving in simple harmonic motion

Solution   We begin by identifying values for a and ω.

$$d = 10 \cos \frac{\pi}{6} t$$

The form of this equation is
$d = a \cos \omega t$
with $a = 10$ and $\omega = \frac{\pi}{6}$.

34. A police helicopter is flying at 800 feet. A stolen car is sighted at an angle of depression of 72°. Find the distance of the stolen car, to the nearest foot, from a point directly below the helicopter.

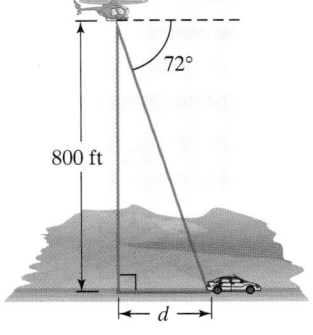

35. A wheelchair ramp is to be built beside the steps to the campus library. Find the angle of elevation of the 23-foot ramp, to the nearest tenth of a degree, if its final height is 6 feet.

36. A building that is 250 feet high casts a shadow 40 feet long. Find the angle of elevation, to the nearest tenth of a degree, of the sun at this time.

37. A hot-air balloon is rising vertically. The angle of elevation from a point on level ground 125 feet from the balloon to a point directly under the passenger compartment changes from 19.2° to 31.7°. How far, to the nearest tenth of a foot, does the balloon rise during this period?

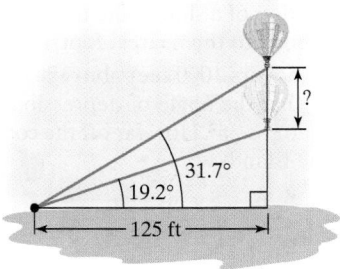

38. A flagpole is situated on top of a building. The angle of elevation from a point on level ground 330 feet from the building to the top of the flagpole is 63°. The angle of elevation from the same point to the bottom of the flagpole is 53°. Find the height of the flagpole to the nearest tenth of a foot.

39. A boat leaves the entrance to a harbor and travels 150 miles on a bearing of N 53° E. How many miles north and how many miles east from the harbor has the boat traveled?

40. A boat leaves the entrance to a harbor and travels 40 miles on a bearing of S 64° E. How many miles south and how many miles east from the harbor has the boat traveled?

41. A forest ranger sights a fire directly to the south. A second ranger, 7 miles east of the first ranger, also sights the fire. The bearing from the second ranger to the fire is S 28° W. How far, to the nearest tenth of a mile, is the first ranger from the fire?

42. A ship sights a lighthouse directly to the south. A second ship, 9 miles east of the first ship, also sights the lighthouse. The bearing from the second ship to the lighthouse is S 34° W. How far, to the nearest tenth of a mile, is the first ship from the lighthouse?

43. You leave your house and run 2 miles due west followed by 1.5 miles due north. At that time, what is your bearing from your house?

44. A ship is 9 miles east and 6 miles south of a harbor. What bearing should be taken to sail directly to the harbor?

45. A jet leaves a runway whose bearing is N 35° E from the control tower. After flying 5 miles, the jet turns 90° and flies on a bearing of S 55° E for 7 miles. At that time, what is the bearing of the jet from the control tower?

46. A ship leaves port with a bearing of S 40° W. After traveling 7 miles, the ship turns 90° and travels on a bearing of N 60° W for 11 miles. At that time, what is the bearing of the ship from port?

47. An object in simple harmonic motion has a frequency of $\frac{1}{2}$ oscillation per minute and an amplitude of 6 feet. Write an equation in the form $d = a \sin \omega t$ for the object's simple harmonic motion.

48. An object in simple harmonic motion has a frequency of $\frac{1}{4}$ oscillation per minute and an amplitude of 8 feet. Write an equation in the form $d = a \sin \omega t$ for the object's simple harmonic motion.

49. A piano tuner uses a tuning fork. If middle C has a frequency of 264 vibrations per second, write an equation in the form $d = \sin \omega t$ for the simple harmonic motion.

50. A radio station, 98.1 on the FM dial, has radio waves with a frequency of 98.1 million cycles per second. Write an equation in the form $d = \sin \omega t$ for the simple harmonic motion of the radio waves.

Writing in Mathematics

51. What does it mean to solve a right triangle?

52. Explain how to find one of the acute angles of a right triangle if two sides are known.

53. Describe a situation in which a right triangle and a trigonometric function are used to measure a height or distance that would otherwise be inconvenient or impossible to measure.

54. What is meant by the bearing from point O to point P? Give an example with your description.

55. What is simple harmonic motion? Give an example with your description.

56. Explain the period and the frequency of simple harmonic motion. How are they related?

57. Explain how the photograph of the damaged highway on page 522 illustrates simple harmonic motion.

Technology Exercises

The functions in Exercises 58–59 model motion in which the amplitude decreases with time due to friction or other resistive forces. Graph each function in the given viewing rectangle. How many complete oscillations occur on the time interval $0 \le x \le 10$?

58. $y = 4e^{-0.1x} \cos 2x$ $[0, 10, 1]$ by $[-4, 4, 1]$

59. $y = -6e^{-0.09x} \cos 2\pi x$ $[0, 10, 1]$ by $[-6, 6, 1]$

Critical Thinking Exercises

60. The figure shows a satellite circling 112 miles above Earth. When the satellite is directly above point B, angle A measures $76.6°$. Find Earth's radius to the nearest mile.

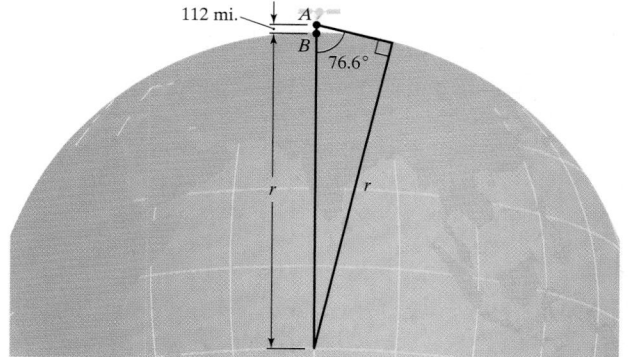

61. The figure shows that the angle of elevation to the top of the building changes from $20°$ to $40°$ as an observer advances 75 feet toward the building. Find the height of the building to the nearest foot.

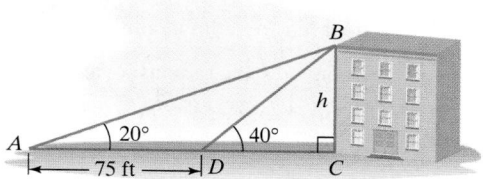

Group Exercise

62. Music and mathematics have been linked over the centuries. Group members should research and present a seminar to the class on music and mathematics. Be sure to include the role of trigonometric functions in the music-mathematics link.

CHAPTER SUMMARY, REVIEW, AND TEST

Summary

5.1 Angles and Their Measure

a. An angle consists of two rays with a common endpoint, the vertex.

b. An angle is in standard position if its vertex is at the origin and its initial side lies along the positive x-axis. Figure 5.3 on page 417 shows positive and negative angles in standard position.

c. A quadrantal angle is one with its terminal side on the x-axis or the y-axis.

d. Angles can be measured in degrees. $1°$ is $\frac{1}{360}$ of a complete rotation.

e. Acute angles measure less than $90°$, right angles $90°$, obtuse angles more than $90°$ but less than $180°$, and straight angles $180°$.

f. Two angles with the same initial and terminal sides are called coterminal angles.

g. Two angles are complements if their sum is $90°$ and supplements if their sum is $180°$. Only positive angles are used.

h. Angles can be measured in radians. One radian is the measure of the central angle if the intercepted arc and radius have the same length. In general, the radian measure of a central angle is the length of the intercepted arc divided by the circle's radius: $\theta = \dfrac{s}{r}$.

i. To convert degrees to radians, multiply degrees by $\dfrac{\pi \text{ radians}}{180°}$. To convert from radians to degrees, multiply radians by $\dfrac{180°}{\pi \text{ radians}}$.

j. The arc length formula, $s = r\theta$, is described in the box on page 424.

k. The definitions of linear speed, $v = \dfrac{s}{t}$, and angular speed, $\omega = \dfrac{\theta}{t}$, are given in the box on page 425.

l. Linear speed is expressed in terms of angular speed by $v = r\omega$, where v is the linear speed of a point a distance r from the center of rotation and ω is the angular speed in radians per unit of time.

5.2 Right Triangle Trigonometry

a. The right triangle definitions of the six trigonometric functions are given in the box on page 431.

b. Function values for 30°, 45°, and 60° can be obtained using these special triangles.

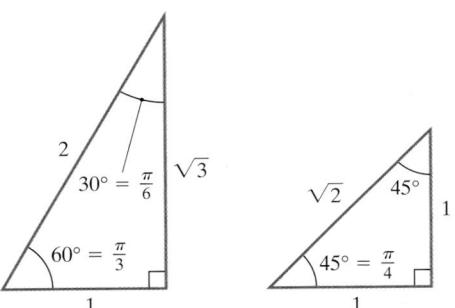

c. Fundamental Identities

1. Reciprocal Identities

$$\sin\theta = \frac{1}{\csc\theta} \quad \cos\theta = \frac{1}{\sec\theta} \quad \tan\theta = \frac{1}{\cot\theta}$$

$$\csc\theta = \frac{1}{\sin\theta} \quad \sec\theta = \frac{1}{\cos\theta} \quad \cot\theta = \frac{1}{\tan\theta}$$

2. Quotient Identities

$$\tan\theta = \frac{\sin\theta}{\cos\theta} \quad \cot\theta = \frac{\cos\theta}{\sin\theta}$$

3. Pythagorean Identities

$$\sin^2\theta + \cos^2\theta = 1$$
$$1 + \tan^2\theta = \sec^2\theta$$
$$1 + \cot^2\theta = \csc^2\theta$$

d. The value of a trigonometric function of θ is equal to the cofunction of the complement of θ. Cofunction identities are listed in the box on page 438.

5.3 Trigonometric Functions of Any Angle

a. Definitions of the trigonometric functions of any angle are given in the box on page 447.

b. Signs of the trigonometric functions: All functions are positive in quadrant I. If θ lies in quadrant II, $\sin\theta$ and $\csc\theta$ are positive. If θ lies in quadrant III, $\tan\theta$ and $\cot\theta$ are positive. If θ lies in quadrant IV, $\cos\theta$ and $\sec\theta$ are positive.

c. If θ is a nonacute angle in standard position that lies in a quadrant, its reference angle is the positive acute angle θ' formed by the terminal side of θ and the x-axis. The reference angle for a given angle can be found by making a sketch that shows the angle in standard position. Figure 5.30 on page 452 shows reference angles for θ in quadrants II, III, and IV.

d. The values of the trigonometric functions of a given angle are the same as the values of the functions of the reference angle, except possibly for the sign. A procedure for using reference angles to evaluate trigonometric functions is given in the box on page 453.

5.4 Trigonometric Functions of Real Numbers; Periodic Functions

a. Definitions of the trigonometric functions in terms of a unit circle are given in the box on page 458.

b. The cosine and secant functions are even:
$$\cos(-t) = \cos t, \quad \sec(-t) = \sec t.$$
The other trigonometric functions are odd:
$$\sin(-t) = -\sin t, \quad \tan(-t) = -\tan t,$$
$$\cot(-t) = -\cot t, \quad \csc(-t) = -\csc t.$$

c. If $f(t + p) = f(t)$, function f is periodic. The smallest p for which f is periodic is the period of f. The tangent and cotangent functions have period π. The other four trigonometric functions have period 2π.

5.5 and 5.6 Graphs of the Trigonometric Functions

a. Graphs of the six trigonometric functions, with a description of the domain, range, and period of each function, are given in Table 5.4 on page 494.

b. The graph of $y = A\sin(Bx - C)$ can be obtained using amplitude $= |A|$, period $= \dfrac{2\pi}{B}$ and phase shift $= \dfrac{C}{B}$. See the illustration in the box on page 471.

c. The graph of $y = A\cos(Bx - C)$ can be obtained using amplitude $= |A|$, period $= \dfrac{2\pi}{B}$, and phase shift $= \dfrac{C}{B}$. See the illustration in the box on page 477.

d. The constant D in $y = A \sin(Bx - C) + D$ and $y = A \cos(Bx - C) + D$ causes vertical shifts in the graphs in the preceding items (b) and (c). If $D > 0$, the shift is D units upward and if $D < 0$, the shift is D units downward. Oscillation is about $y = D$.

e. The graph of $y = A \tan(Bx - C)$ is obtained using the procedure in the box on page 488. Consecutive asymptotes (solve $Bx - C = -\frac{\pi}{2}$ and $Bx - C = \frac{\pi}{2}$) and an x-intercept midway between them play a key role in the graphing process.

f. The graph of $y = A \cot(Bx - C)$ is obtained using the procedure in the box on page 490. Consecutive asymptotes (solve $Bx - C = 0$ and $Bx - C = \pi$) and an x-intercept midway between them play a key role in the graphing process.

g. To graph a cosecant curve, begin by graphing the reciprocal sine curve. Draw vertical asymptotes through x-intercepts, using asymptotes as guides to sketch the graph. To graph a secant curve, first graph the reciprocal cosine curve and use the same procedure.

5.7 Inverse Trigonometric Functions

a. On the restricted domain $-\frac{\pi}{2} \le x \le \frac{\pi}{2}$, $y = \sin x$ has an inverse function, defined in the box on page 499. Think of $\sin^{-1} x$ as the angle in $\left[-\frac{\pi}{2}, \frac{\pi}{2}\right]$ whose sine is x.

b. On the restricted domain $0 \le x \le \pi$, $y = \cos x$ has an inverse function, defined in the box on page 502. Think of $\cos^{-1} x$ as the angle in $[0, \pi]$ whose cosine is x.

c. On the restricted domain $-\frac{\pi}{2} < x < \frac{\pi}{2}$, $y = \tan x$ has an inverse function, defined in the box on page 504. Think of $\tan^{-1} x$ as the angle in $\left(-\frac{\pi}{2}, \frac{\pi}{2}\right)$ whose tangent is x.

d. Graphs of the three basic inverse trigonometric functions, with a description of the domain and range of each function, are given in Table 5.8 on page 506.

e. Inverse properties are given in the box on page 507. Points on terminal sides of angles in standard position are used to find exact values of the composition of a function and a different inverse function.

5.8 Applications of Trigonometric Functions

a. Solving a right triangle means finding the missing lengths of its sides and the measurements of its angles. The Pythagorean Theorem, two acute angles whose sum is $90°$, and appropriate trigonometric functions are used in this process.

b. The bearing from point O to point P is the acute angle between ray OP and a north-south line.

c. Simple harmonic motion, described in the box on page 520, is modeled by $d = a \cos \omega t$ or $d = a \sin \omega t$, with amplitude $= |a|$, period $= \frac{2\pi}{\omega}$ and frequency $= \frac{\omega}{2\pi} = \frac{1}{\text{period}}$.

Review Exercises

5.1

In Exercises 1–4, draw each angle in standard position.

1. $190°$

2. $-135°$

3. $\frac{5\pi}{6}$

4. $-\frac{2\pi}{3}$

In Exercises 5–6, find a positive angle less than $360°$ that is coterminal with the given angle.

5. $400°$

6. $-85°$

In Exercises 7–8, if possible, find the complement and the supplement of the given angle.

7. $73°$

8. $\frac{2\pi}{3}$

9. Find the radian measure of the central angle of a circle of radius 6 centimeters that intercepts an arc of length 27 centimeters.

In Exercises 10–12, convert each angle in degrees to radians. Express your answer as a multiple of π.

10. $15°$

11. $120°$

12. $315°$

In Exercises 13–15, convert each angle in radians to degrees.

13. $\frac{5\pi}{3}$

14. $\frac{7\pi}{5}$

15. $-\frac{5\pi}{6}$

16. Find the length of the arc on a circle of radius 10 feet intercepted by a $135°$ central angle. Express arc length in terms of π. Then round your answer to two decimal places.

17. The angular speed of a propeller on a wind generator is 10.3 revolutions per minute. Express this angular speed in radians per minute.

18. The propeller of an airplane has a radius of 3 feet. The propeller is rotating at 2250 revolutions per minute. Find the linear speed, in feet per minute, of the tip of the propeller.

5.2

19. Use the triangle to find each of the six trigonometric functions of θ.

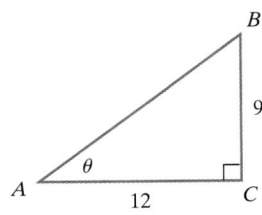

In Exercises 20–23, find the exact value of each expression. Do not use a calculator.

20. $\tan 60°$

21. $\cos \dfrac{\pi}{4}$

22. $\sec \dfrac{\pi}{6}$

23. $\sin^2 \dfrac{\pi}{5} + \cos^2 \dfrac{\pi}{5}$

24. If θ is an acute angle and $\sin\theta = \dfrac{2}{\sqrt{7}}$, use the identity $\sin^2\theta + \cos^2\theta = 1$ to find $\cos\theta$.

In Exercises 25–26, find a cofunction with the same value as the given expression.

25. $\sin 70°$

26. $\cos \dfrac{\pi}{2}$

In Exercises 27–29, find the measure of the side of the right triangle whose length is designated by a lowercase letter. Round answers to the nearest whole number.

27.

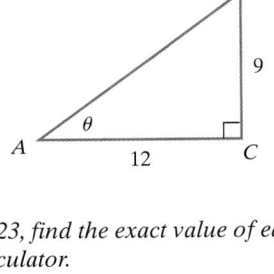

28.

29.

30. A hiker climbs for a half mile up a slope whose inclination is 17°. How many feet of altitude, to the nearest foot, does the hiker gain?

31. To find the distance across a lake, a surveyor took the measurements in the figure shown. What is the distance across the lake? Round to the nearest meter.

32. When a six-foot pole casts a four-foot shadow, what is the angle of elevation of the sun? Round to the nearest whole degree.

5.3 and 5.4

In Exercises 33–34, a point on the terminal side of angle θ is given. Find the exact value of each of the six trigonometric functions of θ, or state that the function is undefined.

33. $(-1, -5)$

34. $(0, -1)$

In Exercises 35–36, let θ be an angle in standard position. Name the quadrant in which θ lies.

35. $\tan\theta > 0$ and $\sec\theta > 0$

36. $\tan\theta > 0$ and $\cos\theta < 0$

In Exercises 37–38, find the exact value of each of the remaining trigonometric functions of θ.

37. $\cos\theta = \frac{2}{5}$, $\sin\theta < 0$

38. $\tan\theta = -\frac{1}{3}$, $\sin\theta > 0$

In Exercises 39–41, find the reference angle for each angle.

39. $265°$

40. $\dfrac{5\pi}{8}$

41. $-410°$

In Exercises 42–50, find the exact value of each expression. Do not use a calculator.

42. $\sin 240°$

43. $\tan 120°$

44. $\sec \dfrac{7\pi}{4}$

45. $\cos \dfrac{11\pi}{6}$

46. $\cot (-210°)$

47. $\csc \left(-\dfrac{2\pi}{3}\right)$

48. $\sin \left(-\dfrac{\pi}{3}\right)$

49. $\sin 495°$

50. $\tan \dfrac{13\pi}{4}$

5.5

In Exercises 51–56, determine the amplitude and period of each function. Then graph one period of the function.

51. $y = 3\sin 4x$

52. $y = -2\cos 2x$

53. $y = 2\cos\frac{1}{2}x$

54. $y = \dfrac{1}{2}\sin \dfrac{\pi}{3}x$

55. $y = -\sin \pi x$

56. $y = 3\cos \dfrac{x}{3}$

In Exercises 57–61, determine the amplitude, period, and phase shift of each function. Then graph one period of the function.

57. $y = 2\sin(x - \pi)$

58. $y = -3\cos(x + \pi)$

59. $y = \dfrac{3}{2}\cos\left(2x + \dfrac{\pi}{4}\right)$

60. $y = \dfrac{5}{2}\sin\left(2x + \dfrac{\pi}{2}\right)$

61. $y = -3\sin\left(\dfrac{\pi}{3}x - 3\pi\right)$

In Exercises 62–63, use a vertical shift to graph one period of the function.

62. $y = \sin 2x + 1$

63. $y = 2\cos\frac{1}{3}x - 2$

64. The equation

$$y = 98.6 + 0.3\sin\left(\frac{\pi}{12}x - \frac{11\pi}{12}\right)$$

models variation in body temperature, y, in °F, x hours after midnight.

a. What is body temperature at midnight?
b. What is the period of the body temperature cycle?
c. When is body temperature highest? What is the body temperature at this time?
d. When is body temperature lowest? What is the body temperature at this time?
e. Graph one period of the body temperature function.

5.6

In Exercises 65–71, graph two full periods of the given tangent or cotangent function.

65. $y = 4\tan 2x$

66. $y = -2\tan\dfrac{\pi}{4}x$

67. $y = \tan(x + \pi)$

68. $y = -\tan\left(x - \dfrac{\pi}{4}\right)$

69. $y = 2\cot 3x$

70. $y = -\dfrac{1}{2}\cot\dfrac{\pi}{2}x$

71. $y = 2\cot\left(x + \dfrac{\pi}{2}\right)$

In Exercises 72–75, graph two full periods of the given cosecant or secant function.

72. $y = 3\sec 2\pi x$

73. $y = -2\csc \pi x$

74. $y = 3\sec(x + \pi)$

75. $y = \frac{5}{2}\csc(x - \pi)$

5.7

In Exercises 76–93, find the exact value of each expression. Do not use a calculator.

76. $\sin^{-1} 1$

77. $\cos^{-1} 1$

78. $\tan^{-1} 1$

79. $\sin^{-1}\left(-\dfrac{\sqrt{3}}{2}\right)$

80. $\cos^{-1}\left(-\dfrac{1}{2}\right)$

81. $\tan^{-1}\left(-\dfrac{\sqrt{3}}{3}\right)$

82. $\cos\left(\sin^{-1}\dfrac{\sqrt{2}}{2}\right)$

83. $\sin(\cos^{-1} 0)$

84. $\tan\left[\sin^{-1}\left(-\dfrac{1}{2}\right)\right]$

85. $\tan\left[\cos^{-1}\left(-\dfrac{\sqrt{3}}{2}\right)\right]$

86. $\csc\left(\tan^{-1}\dfrac{\sqrt{3}}{3}\right)$

87. $\cos\left(\tan^{-1}\dfrac{3}{4}\right)$

88. $\sin\left(\cos^{-1}\dfrac{3}{5}\right)$

89. $\tan\left[\sin^{-1}\left(-\dfrac{3}{5}\right)\right]$

90. $\tan\left[\cos^{-1}\left(-\dfrac{4}{5}\right)\right]$

91. $\sin^{-1}\left(\sin\dfrac{\pi}{3}\right)$

92. $\sin^{-1}\left(\sin\dfrac{2\pi}{3}\right)$

93. $\sin^{-1}\left(\cos\dfrac{2\pi}{3}\right)$

In Exercises 94–95, use a right triangle to write each expression as an algebraic expression. Assume that x is positive and in the domain of the given inverse trigonometric function.

94. $\cos\left(\tan^{-1}\dfrac{x}{2}\right)$

95. $\sec\left(\sin^{-1}\dfrac{1}{x}\right)$

5.8

In Exercises 96–99, solve the right triangle shown in the figure. Round lengths to two decimal places and express angles to the nearest tenth of a degree.

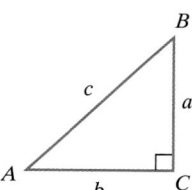

96. $A = 22.3°, c = 10$

97. $B = 37.4°, b = 6$

98. $a = 2, c = 7$

99. $a = 1.4, b = 3.6$

100. From a point on level ground 80 feet from the base of a building, the angle of elevation is 25.6°. Approximate the height of the building to the nearest foot.

101. Two buildings with flat roofs are 60 yards apart. The height of the shorter building is 40 yards. From its roof, the angle of elevation to the edge of the roof of the taller building is 40°. Find the height of the taller building to the nearest yard.

102. You want to measure the height of an antenna on the top of a 125-foot building. From a point in front of the building, you measure the angle of elevation to the top of the building to be 68° and the angle of elevation to the top of the antenna to be 71°. How tall is the antenna, to the nearest tenth of a foot?

In Exercises 103–104, use the figure shown to find the bearing from O to A.

103.

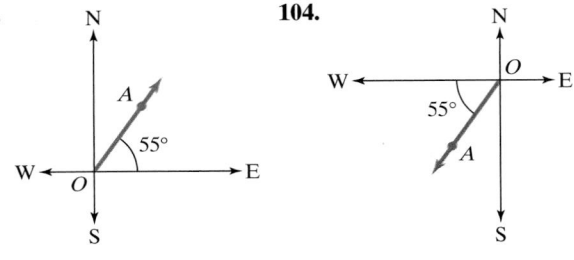

104.

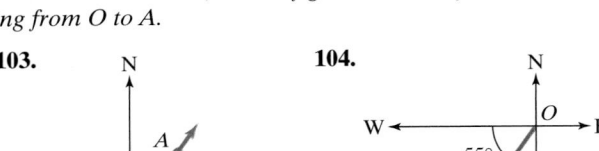

105. A ship is due west of a lighthouse. A second ship is 12 miles south of the first ship. The bearing from the second ship to the lighthouse is N 64° E. How far, to the nearest tenth of a mile, is the first ship from the lighthouse?

106. From city A to city B, a plane flies 850 miles at a bearing of N 58° E. From city B to city C, the plane flies 960 miles at a bearing of S 32° E.
   a. Find, to the nearest tenth of a mile, the distance from city A to city C.
   b. What is the bearing from city A to city C?

In Exercises 107–108, an object moves in simple harmonic motion described by the given equation, where t is measured in seconds and d in centimeters. In each exercise, find:

   a. *the maximum displacement.*
   b. *the frequency.*
   c. *the time required for one cycle.*

107. $d = 20 \cos \dfrac{\pi}{4} t$ **108.** $d = \frac{1}{2} \sin 4t$

In Exercises 109–110, an object is attached to a coiled spring. The object is pulled down (negative direction from the rest position) and then released. Write an equation for the distance of the object from its rest position after t seconds.

	Distance from rest position at $t = 0$	Amplitude	Period
109.	30 inches	30 inches	2 seconds
110.	0	$\frac{1}{4}$ inches	5 seconds

Chapter 5 Test

1. Convert 135° to exact radian measure.

2. Find the supplement of the angle whose radian measure is $\dfrac{9\pi}{13}$. Express the answer in terms of π.

3. Find the length of the arc on a circle of radius 20 feet intercepted by a 75° central angle. Express arc length in terms of π. Then round your answer to two decimal places.

4. If $(-2, 5)$ is a point on the terminal side of angle θ, find the exact value of each of the six trigonometric functions of θ.

5. Determine the quadrant in which θ lies if $\cos < 0$ and $\cot \theta > 0$.

6. If $\cos \theta = \frac{1}{3}$ and $\tan \theta < 0$, find the exact value of each of the remaining trigonometric functions of θ.

In Exercises 7–9, find the exact value of each expression. Do not use a calculator.

7. $\tan \dfrac{\pi}{6} \cos \dfrac{\pi}{3} - \cos \dfrac{\pi}{2}$ **8.** $\tan 300°$

9. $\sin \dfrac{7\pi}{4}$

In Exercises 10–13, graph one period of each function.

10. $y = 3 \sin 2x$ **11.** $y = -2 \cos \left(x - \dfrac{\pi}{2} \right)$

12. $y = 2 \tan \dfrac{x}{2}$ **13.** $y = -\frac{1}{2} \csc \pi x$

14. Find the exact value of $\tan \left[\cos^{-1} \left(-\frac{1}{2} \right) \right]$.

15. Solve the right triangle in the figure shown. Round lengths to one decimal place.

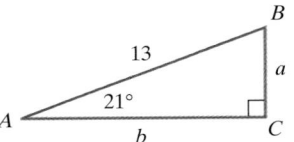

16. The angle of elevation of a building from a point on the ground 30 yards from its base is 37°. Find the height of the building to the nearest yard.

17. A 73-foot rope from the top of a circus tent pole is anchored to the flat ground 43 feet from the bottom of the pole. Find the angle, to the nearest tenth of a degree, that the rope makes with the pole.

18. Use the figure to find the bearing from O to P.

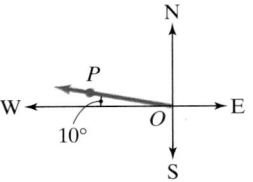

19. An object moves in simple harmonic motion described by $d = -6 \cos \pi t$, where t is measured in seconds and d in inches. Find (a) the maximum displacement, (b) the frequency, and (c) the time required for one oscillation.

20. Why are trigonometric functions ideally suited to model phenomena that repeat in cycles?

Cumulative Review Exercises (Chapters 1–5)

Solve each equation or inequality in Exercises 1–6.

1. $x^2 = 18 + 3x$

2. $x^3 + 5x^2 - 4x - 20 = 0$

3. $\log_2 x + \log_2(x - 2) = 3$

4. $\sqrt{x - 3} + 5 = x$

5. $x^3 - 4x^2 + x + 6 = 0$

6. $|2x - 5| \le 11$

7. If $f(x) = \sqrt{x - 6}$, find $f^{-1}(x)$.

8. Divide $20x^3 - 6x^2 - 9x + 10$ by $5x + 2$.

9. Write as a single logarithm and evaluate: $\log 25 + \log 40$.

10. Convert $\dfrac{14\pi}{9}$ radians to degrees.

11. Find the maximum number of positive and negative real roots of the equation $3x^4 - 2x^3 + 5x^2 + x - 9 = 0$.

In Exercises 12–16, graph each equation.

12. $f(x) = \dfrac{x}{x^2 - 1}$

13. $(x - 2)^2 + y^2 = 1$

14. $y = (x - 1)(x + 2)^2$

15. $y = \sin\left(2x + \dfrac{\pi}{2}\right)$, from 0 to 2π

16. $y = 2\tan 3x$; graph two complete cycles.

17. You invest in a new play. The cost includes an overhead of $30,000, plus production costs of $2500 per performance. A sold-out performance brings you $3125. How many sold-out performances must be played in order for you to break even?

18. Use the exponential growth model $A = A_0 e^{kt}$ to solve this exercise. Data from the Federal Communication Commission show that the use of toll-free 800 numbers has grown exponentially. In 1991 there were 10.2 billion such calls and by 1998, there were 86.7 billion.

   a. Find the exponential function that models the data.

   b. By what year will the number of toll-free 800 numbers reach 200 billion?

19. The rate of heat lost through insulation varies inversely as the thickness of the insulation. The rate of heat lost through a 3.5-inch thickness of insulation is 2200 Btu per hour. What is the rate of heat lost through a 5-inch thickness of the same insulation?

20. A tower is 200 feet tall. To the nearest degree, find the angle of elevation from a point 50 feet from the base of the tower to the top of the tower.

Analytic Trigonometry

This chapter emphasizes the algebraic aspects of trigonometry. We derive important categories of identities involving trigonometric functions. These identities are used to simplify and analyze expressions that model phenomena as diverse as the distance achieved when throwing an object and musical sounds on a touch-tone phone. For example, we can find out critical information about an athlete's performance by using an identity to analyze an expression involving throwing distance. You will learn how to use trigonometric identities to better understand your periodic world.

You enjoy watching your friend participate in the shot put at college track and field events. After a few full turns in a circle, she throws ("puts") a 16-pound shot from the shoulder. The range of her throwing distance continues to improve. Knowing that you are studying trigonometry, she asks if there is some way that a trigonometric expression might help achieve the best distance possible in the event.

SECTION 6.1 *Verifying Trigonometric Identities*

Objective

1. Use the fundamental trigonometric identities to verify identities.

Do you enjoy solving puzzles? The process is a natural way to develop problem-solving skills that are important to every area of our lives. Engaging in problem solving for sheer pleasure releases chemicals in the brain that enhance our feeling of well-being. Perhaps this is why puzzles date back 12,000 years.

Thousands of relationships exist among the six trigonometric functions. Verifying these relationships is like solving a puzzle. Why? There are no rigid rules for the process. Thus, proving a trigonometric relationship requires you to be creative in your problem-solving abilities. By learning to establish these relationships, you will become a better, more confident problem solver. Furthermore, you may enjoy the feeling of satisfaction that accompanies solving each "puzzle."

The Fundamental Identities

In Chapter 5, we used right triangles to establish relationships among trigonometric functions. Although we limited domains to acute angles, the fundamental identities listed in the following box are true for all values of x for which the expressions are defined.

Fundamental Trigonometric Identities

Reciprocal Identities

$$\sin x = \frac{1}{\csc x} \quad \cos x = \frac{1}{\sec x} \quad \tan x = \frac{1}{\cot x}$$

$$\csc x = \frac{1}{\sin x} \quad \sec x = \frac{1}{\cos x} \quad \cot x = \frac{1}{\tan x}$$

Quotient Identities

$$\tan x = \frac{\sin x}{\cos x} \quad \cot x = \frac{\cos x}{\sin x}$$

Pythagorean Identities

$$\sin^2 x + \cos^2 x = 1 \quad 1 + \tan^2 x = \sec^2 x \quad 1 + \cot^2 x = \csc^2 x$$

Even-Odd Identities

$$\sin(-x) = -\sin x \quad \cos(-x) = \cos x \quad \tan(-x) = -\tan x$$

$$\csc(-x) = -\csc x \quad \sec(-x) = \sec x \quad \cot(-x) = -\cot x$$

1 Use the fundamental trigonometric identities to verify identities.

Using Fundamental Identities to Verify Other Identities

The fundamental trigonometric identities are used to establish other relationships among trigonometric functions. To **verify an identity**, we show that one side of the identity can be simplified so that it is identical to the other side. Each side of the equation is manipulated independently of the other side of the equation. Start with the side containing the more complicated expression. If you substitute one or more fundamental identities on the more complicated side, you will often be able to rewrite it in a form identical to that of the other side.

No one method can be used to verify every identity. Some identities can be verified by rewriting the more complicated side so that it contains only sines and cosines.

EXAMPLE 1 Changing to Sines and Cosines to Verify an Identity

Verify the identity: $\sec x \cot x = \csc x$.

Solution The left side of the equation contains the more complicated expression. Thus, we work with the left side. Let us express this side of the identity in terms of sines and cosines. Perhaps this strategy will enable us to transform the left side into $\csc x$, the expression on the right.

$$\sec x \cot x = \frac{1}{\cos x} \cdot \frac{\cos x}{\sin x}$$

Apply a reciprocal identity: $\sec x = \dfrac{1}{\cos x}$ and a quotient identity: $\cot x = \dfrac{\cos x}{\sin x}$.

$$= \frac{1}{\cancel{\cos x}} \cdot \frac{\overset{1}{\cancel{\cos x}}}{\sin x}$$

Divide both the numerator and the denominator by $\cos x$, the common factor.

$$= \frac{1}{\sin x}$$

Multiply the remaining factors in the numerator and denominator.

$$= \csc x$$

Apply a reciprocal identity: $\csc x = \dfrac{1}{\sin x}$.

By working with the left side and simplifying it so that it is identical to the right side, we have verified the given identity.

 Check Point 1 Verify the identity: $\csc x \tan x = \sec x$.

In verifying an identity, stay focused on your goal. When manipulating one side of the equation, continue to look at the other side to keep in mind the desired form of the result.

Technology

You can use a graphing utility to provide evidence of an identity. For example, consider the equation in Example 1,

$$\sec x \cot x = \csc x.$$

Graph

$$y = \sec x \cot x$$

and

$$y = \csc x$$

in the same viewing rectangle. The screen shown indicates that the graphs appear to be the same. Example 1 shows the equivalence algebraically.

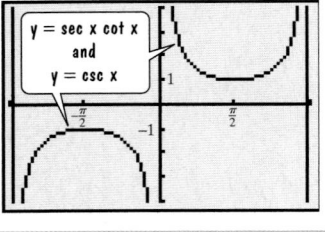

Study Tip

Verifying that an equation is an identity is different from solving an equation. You cannot verify an identity by adding, subtracting, multiplying, or dividing each side by the same expression. If you do this, you assume that the given statement is true. You do not know that it is true until after you have verified it.

EXAMPLE 2 **Changing to Sines and Cosines to Verify an Identity**

Verify the identity: $\sin x \tan x + \cos x = \sec x$.

Solution The left side is more complicated, so we so we start with it. Let us express this side of the identity so that it contains only sines and cosines. Thus, we use a quotient identity and replace $\tan x$ by $\dfrac{\sin x}{\cos x}$. Perhaps this strategy will enable us to transform the left side into $\sec x$, the expression on the right.

$$\sin x \tan x + \cos x = \sin x \left(\frac{\sin x}{\cos x} \right) + \cos x \qquad \text{Apply a quotient identity:} \quad \tan x = \frac{\sin x}{\cos x}.$$

$$= \frac{\sin^2 x}{\cos x} + \cos x. \qquad \text{Multiply.}$$

$$= \frac{\sin^2 x}{\cos x} + \cos x \cdot \frac{\cos x}{\cos x} \qquad \text{The least common denominator is } \cos x. \text{ Write the second expression with a denominator of } \cos x.$$

$$= \frac{\sin^2 x}{\cos x} + \frac{\cos^2 x}{\cos x} \qquad \text{Multiply.}$$

$$= \frac{\sin^2 x + \cos^2 x}{\cos x} \qquad \text{Add numerators, putting this sum over the least common denominator.}$$

$$= \frac{1}{\cos x} \qquad \text{Apply a Pythagorean identity:} \quad \sin^2 x + \cos^2 x = 1.$$

$$= \sec x \qquad \text{Apply a reciprocal identity:} \quad \sec x = \frac{1}{\cos x}.$$

By working with the left side and arriving at the right side, the identity is verified.

Check Point 2 Verify the identity: $\cos x \cot x + \sin x = \csc x$.

Some identities are verified by using factoring to simplify a trigonometric expression.

EXAMPLE 3 **Using Factoring to Verify an Identity**

Verify the identity: $\cos x - \cos x \sin^2 x = \cos^3 x$.

Solution We start with the more complicated side, the left side. Factor out the greatest common factor, $\cos x$, from each of the two terms.

$$\cos x - \cos x \sin^2 x = \cos x (1 - \sin^2 x) \qquad \text{Factor } \cos x \text{ from the two terms.}$$

$$= \cos x \cdot \cos^2 x \qquad \text{Use a variation of } \sin^2 x + \cos^2 x = 1. \text{ Solving for } \cos^2 x, \text{ we obtain } \cos^2 x = 1 - \sin^2 x.$$

$$= \cos^3 x \qquad \text{Multiply.}$$

536 • Chapter 6 • Analytic Trigonometry

We worked with the left side and arrived at the right side. Thus, the identity is verified.

Check
Point
3 Verify the identity: $\sin x - \sin x \cos^2 x = \sin^3 x$.

How do we verify identities in which sums or differences of fractions with trigonometric functions appear on one side? Use the least common denominator and combine the fractions. This technique is especially useful when the other side of the identity contains only one term.

EXAMPLE 4   Combining Fractional Expressions to Verify an Identity

Verify the identity: $\dfrac{\cos x}{1 + \sin x} + \dfrac{1 + \sin x}{\cos x} = 2 \sec x$.

Solution   We start with the more complicated side, the left side. The least common denominator of the fractions is $(1 + \sin x)(\cos x)$. We express each fraction in terms of this least common denominator by multiplying the numerator and denominator by the extra factor needed to form $(1 + \sin x)(\cos x)$.

Study Tip

Some students have difficulty verifying identities due to problems working with fractions. If this applies to you, review the material on rational expressions in Chapter P, Section P.6.

$$\frac{\cos x}{1 + \sin x} + \frac{1 + \sin x}{\cos x}$$

The least common denominator is $(1 + \sin x)(\cos x)$.

$$= \frac{\cos x(\cos x)}{(1 + \sin x)(\cos x)} + \frac{(1 + \sin x)(1 + \sin x)}{(1 + \sin x)(\cos x)}$$

Rewrite each fraction with the least common denominator.

$$= \frac{\cos^2 x}{(1 + \sin x)(\cos x)} + \frac{1 + 2\sin x + \sin^2 x}{(1 + \sin x)(\cos x)}$$

Use the FOIL method to multiply $(1 + \sin x)(1 + \sin x)$.

$$= \frac{\cos^2 x + 1 + 2\sin x + \sin^2 x}{(1 + \sin x)(\cos x)}$$

Add numerators. Put this sum over the least common denominator.

$$= \frac{(\sin^2 x + \cos^2 x) + 1 + 2\sin x}{(1 + \sin x)(\cos x)}$$

Regroup terms to apply a Pythagorean identity.

$$= \frac{1 + 1 + 2\sin x}{(1 + \sin x)(\cos x)}$$

Apply a Pythagorean identity: $\sin^2 x + \cos^2 x = 1$.

$$= \frac{2 + 2\sin x}{(1 + \sin x)(\cos x)}$$

Add constant terms in the numerator: $1 + 1 = 2$.

$$= \frac{2\cancel{(1 + \sin x)}}{\cancel{(1 + \sin x)}(\cos x)}$$

Factor and simplify.

$$= \frac{2}{\cos x}$$

$$= 2 \sec x$$

Apply a reciprocal identity: $\sec x = \dfrac{1}{\cos x}$.

We worked with the left side and arrived at the right side. Thus, the identity is verified.

> **Check Point 4**
>
> Verify the identity: $\dfrac{\sin x}{1 + \cos x} + \dfrac{1 + \cos x}{\sin x} = 2 \csc x$.

Some identities are verified using a technique that may remind you of rationalizing the denominator.

EXAMPLE 5 Multiplying the Numerator and Denominator by the Same Factor to Verify an Identity

Verify the identity: $\dfrac{\sin x}{1 + \cos x} = \dfrac{1 - \cos x}{\sin x}$.

Solution The suggestions given in the previous examples do not apply here. Everything is already expressed in terms of sines and cosines. Furthermore, there are no fractions to combine and neither side looks more complicated than the other. Let's solve the puzzle by working with the left side and making it look like the expression on the right. The expression on the right contains $1 - \cos x$ in the numerator. This suggests multiplying the numerator and denominator of the left side by $1 - \cos x$. By doing this, we obtain $1 - \cos x$ in the numerator, the same as the numerator on the right.

Discovery

Verify the identity in Example 5 by making the right side look like the left side. Start with the expression on the right. Multiply the numerator and denominator by $1 + \cos x$.

$$\frac{\sin x}{1 + \cos x} = \frac{\sin x}{1 + \cos x} \cdot \frac{1 - \cos x}{1 - \cos x} \qquad \text{Multiply numerator and denominator by } 1 - \cos x.$$

$$= \frac{\sin x(1 - \cos x)}{1 - \cos^2 x} \qquad \text{Multiply. Use } (A + B)(A - B) = A^2 - B^2, \text{ with } A = 1 \text{ and } B = \cos x, \text{ to multiply denominators.}$$

$$= \frac{\sin x(1 - \cos x)}{\sin^2 x} \qquad \text{Use a variation of } \sin^2 x + \cos^2 x = 1. \text{ Solving for } \sin^2 x, \text{ we obtain } \sin^2 x = 1 - \cos^2 x.$$

$$= \frac{1 - \cos x}{\sin x} \qquad \text{Simplify: } \frac{\sin x}{\sin^2 x} = \frac{\sin x}{\sin x \cdot \sin x} = \frac{1}{\sin x}.$$

We worked with the left side and arrived at the right side. Thus, the identity is verified.

> **Check Point 5**
>
> Verify the identity: $\dfrac{\cos x}{1 + \sin x} = \dfrac{1 - \sin x}{\cos x}$.

EXAMPLE 6 Changing to Sines and Cosines to Verify an Identity

Verify the identity: $\dfrac{\tan x - \sin(-x)}{1 + \cos x} = \tan x$.

Solution   We begin with the left side. Our goal is to obtain $\tan x$, the expression on the right.

$$\frac{\tan x - \sin(-x)}{1 + \cos x} = \frac{\tan x - (-\sin x)}{1 + \cos x}$$

The sine function is odd: $\sin(-x) = -\sin x$.

$$= \frac{\tan x + \sin x}{1 + \cos x}$$

Simplify.

$$= \frac{\dfrac{\sin x}{\cos x} + \sin x}{1 + \cos x}$$

Apply a quotient identity: $\tan x = \dfrac{\sin x}{\cos x}$.

$$= \frac{\dfrac{\sin x}{\cos x} + \dfrac{\sin x \cos x}{\cos x}}{1 + \cos x}$$

Express the terms in the numerator with the least common denominator, $\cos x$.

$$= \frac{\dfrac{\sin x + \sin x \cos x}{\cos x}}{1 + \cos x}$$

Add in the numerator.

$$= \frac{\sin x + \sin x \cos x}{\cos x} \div \frac{1 + \cos x}{1}$$

Rewrite the main fraction bar as ÷.

$$= \frac{\sin x + \sin x \cos x}{\cos x} \cdot \frac{1}{1 + \cos x}$$

Invert the divisor and multiply.

$$= \frac{\sin x \,\overset{1}{\cancel{(1 + \cos x)}}}{\cos x} \cdot \frac{1}{\underset{1}{\cancel{1 + \cos x}}}$$

Factor and simplify.

$$= \frac{\sin x}{\cos x}$$

Multiply the remaining factors in the numerator and the denominator.

$$= \tan x$$

Apply a quotient identity.

The left side simplifies to $\tan x$, the right side. Thus, the identity is verified.

Discovery

Try simplifying

$$\frac{\dfrac{\sin x}{\cos x} + \sin x}{1 + \cos x}$$

by multiplying the two terms in the numerator and the two terms in the denominator by $\cos x$. This method for simplifying the complex fraction involves multiplying the numerator and denominator by the least common denominator of all fractions in the expression. Do you prefer this simplification procedure over the method used on the right?

Check Point 6   Verify the identity: $\dfrac{\sec x + \csc(-x)}{\sec x \csc x} = \sin x - \cos x.$

Is every identity verified by working with only one side? No. You can sometimes work with each side separately and show that both sides are equal to the same trigonometric expression. This is illustrated in Example 7.

EXAMPLE 7   Working With Both Sides Separately to Verify an Identity

Verify the identity: $\dfrac{1}{1 + \cos\theta} + \dfrac{1}{1 - \cos\theta} = 2 + 2\cot^2\theta.$

Solution   We begin by working with the left side.

$$\frac{1}{1+\cos\theta}+\frac{1}{1-\cos\theta}$$

The least common denominator is $(1+\cos\theta)(1-\cos\theta)$.

$$=\frac{1(1-\cos\theta)}{(1+\cos\theta)(1-\cos\theta)}+\frac{1(1+\cos\theta)}{(1+\cos\theta)(1-\cos\theta)}$$

Rewrite each fraction with the least common denominator.

$$=\frac{1-\cos\theta+1+\cos\theta}{(1+\cos\theta)(1-\cos\theta)}$$

Add numerators putting this sum over the least common denominator.

$$=\frac{2}{(1+\cos\theta)(1-\cos\theta)}$$

Simplify the numerator: $-\cos\theta+\cos\theta=0$ and $1+1=2$.

$$=\frac{2}{1-\cos^2\theta}$$

Multiply the factors in the denominator.

Now we work with the right side. Our goal is to transform this side into the simplified form attained for the left side, $\dfrac{2}{1-\cos^2\theta}$.

$$2+2\cot^2\theta=2+2\left(\frac{\cos^2\theta}{\sin^2\theta}\right)$$

Use a quotient identity: $\cot\theta=\dfrac{\cos\theta}{\sin\theta}$.

$$=\frac{2\sin^2\theta}{\sin^2\theta}+\frac{2\cos^2\theta}{\sin^2\theta}$$

Rewrite each fraction with the least common denominator, $\sin^2\theta$.

$$=\frac{2\sin^2\theta+2\cos^2\theta}{\sin^2\theta}$$

Add numerators. Put this sum over the least common denominator.

$$=\frac{2(\sin^2\theta+\cos^2\theta)}{\sin^2\theta}$$

Factor out the greatest common factor, 2.

$$=\frac{2}{\sin^2\theta}$$

Apply a Pythagorean identity: $\sin^2\theta+\cos^2\theta=1$.

$$=\frac{2}{1-\cos^2\theta}$$

Use a variation of $\sin^2\theta+\cos^2\theta=1$ and solve for $\sin^2\theta$: $\sin^2\theta=1-\cos^2\theta$.

The identity is verified because both sides are equal to $\dfrac{2}{1-\cos^2\theta}$.

Check Point 7   Verify the identity: $\dfrac{1}{1+\sin\theta}+\dfrac{1}{1-\sin\theta}=2+2\tan^2\theta$.

Guidelines for Verifying Trigonometric Identities

There is often more than one correct way to solve a puzzle, although one method may be shorter and more efficient than another. The same is true for verifying an identity. For example, how would you verify

$$\frac{\csc^2 x - 1}{\csc^2 x} = \cos^2 x?$$

One approach is to use a Pythagorean identity, $1 + \cot^2 x = \csc^2 x$, on the left side. Then change the resulting expression to sines and cosines.

$$\frac{\csc^2 x - 1}{\csc^2 x} = \frac{(1 + \cot^2 x) - 1}{\csc^2 x} = \frac{\cot^2 x}{\csc^2 x} = \frac{\dfrac{\cos^2 x}{\sin^2 x}}{\dfrac{1}{\sin^2 x}} = \frac{\cos^2 x}{\sin^2 x} \cdot \frac{\sin^2 x}{1} = \cos^2 x$$

| Apply a Pythagorean identity: $1 + \cot^2 x = \csc^2 x$. | Use $\cot x = \frac{\cos x}{\sin x}$ and $\csc x = \frac{1}{\sin x}$ to change to sines and cosines. | Invert the divisor and multiply. |

A more efficient strategy for verifying this identity may not be apparent at first glance. Work with the left side and divide each term in the numerator by the denominator, $\csc^2 x$.

$$\frac{\csc^2 x - 1}{\csc^2 x} = \frac{\csc^2 x}{\csc^2 x} - \frac{1}{\csc^2 x} = 1 - \sin^2 x = \cos^2 x$$

| Apply a reciprocal identity: $\sin x = \frac{1}{\csc x}$. | Use $\sin^2 x + \cos^2 x = 1$ and solve for $\cos^2 x$. |

With this strategy, we do obtain $\cos^2 x$, the expression on the right side, and it takes fewer steps than the first approach.

An even longer strategy, but one that works, is to replace each of the two occurrences of $\csc^2 x$ on the left side by $\frac{1}{\sin^2 x}$. This may be the approach that you first consider, particularly if you become accustomed to rewriting the more complicated side in terms of sines and cosines. The selection of an appropriate fundamental identity to solve the puzzle most efficiently is learned through lots of practice.

The more identities you prove, the more confident and efficient you will become. Although practice is the only way to learn how to verify identities, there are some guidelines developed throughout the section that should help you get started.

Guidelines for Verifying Trigonometric Identities

1. Work with each side of the equation independently of the other side. Start with the more complicated side and transform it in a step-by-step fashion until it looks exactly like the other side.
2. Analyze the identity and look for opportunities to apply the fundamental identities. Rewriting the more complicated side of the equation in terms of sines and cosines is often helpful.
3. If sums or differences of fractions appear on one side, use the least common denominator and combine the fractions.
4. Don't be afraid to stop and start over again if you are not getting anywhere. Creative puzzle solvers know that strategies leading to dead ends often provide good problem-solving ideas.

EXERCISE SET 6.1

Practice Exercises

In Exercises 1–60, verify each identity.

1. $\sin x \sec x = \tan x$

2. $\cos x \csc x = \cot x$

3. $\tan(-x) \cos x = -\sin x$

4. $\cot(-x) \sin x = -\cos x$

5. $\tan x \csc x \cos x = 1$

6. $\cot x \sec x \sin x = 1$

7. $\sec x - \sec x \sin^2 x = \cos x$

8. $\csc x - \csc x \cos^2 x = \sin x$

9. $\cos^2 x - \sin^2 x = 1 - 2\sin^2 x$

10. $\cos^2 x - \sin^2 x = 2\cos^2 x - 1$

11. $\csc\theta - \sin\theta = \cot\theta \cos\theta$

12. $\tan\theta + \cot\theta = \sec\theta \csc\theta$

13. $\dfrac{\tan\theta \cot\theta}{\csc\theta} = \sin\theta$

14. $\dfrac{\cos\theta \sec\theta}{\cot\theta} = \tan\theta$

15. $\sin^2\theta(1 + \cot^2\theta) = 1$

16. $\cos^2\theta(1 + \tan^2\theta) = 1$

17. $\sin t \tan t = \dfrac{1 - \cos^2 t}{\cos t}$

18. $\cos t \cot t = \dfrac{1 - \sin^2 t}{\sin t}$

19. $\dfrac{\csc^2 t}{\cot t} = \csc t \sec t$

20. $\dfrac{\sec^2 t}{\tan t} = \sec t \csc t$

21. $\dfrac{\tan^2 t}{\sec t} = \sec t - \cos t$

22. $\dfrac{\cot^2 t}{\csc t} = \csc t - \sin t$

23. $\dfrac{\sin t}{\csc t} + \dfrac{\cos t}{\sec t} = 1$

24. $\dfrac{\sin t}{\tan t} + \dfrac{\cos t}{\cot t} = \sin t + \cos t$

25. $\tan t + \dfrac{\cos t}{1 + \sin t} = \sec t$

26. $\cot t + \dfrac{\sin t}{1 + \cos t} = \csc t$

27. $1 - \dfrac{\sin^2 x}{1 + \cos x} = \cos x$

28. $1 - \dfrac{\cos^2 x}{1 + \sin x} = \sin x$

29. $\dfrac{\cos x}{1 - \sin x} + \dfrac{1 - \sin x}{\cos x} = 2\sec x$

30. $\dfrac{\sin x}{\cos x + 1} + \dfrac{\cos x - 1}{\sin x} = 0$

31. $\sec^2 x \csc^2 x = \sec^2 x + \csc^2 x$

32. $\csc^2 x \sec x = \sec x + \csc x \cot x$

33. $\dfrac{\sec x - \csc x}{\sec x + \csc x} = \dfrac{\tan x - 1}{\tan x + 1}$

34. $\dfrac{\csc x - \sec x}{\csc x + \sec x} = \dfrac{\cot x - 1}{\cot x + 1}$

35. $\dfrac{\sin^2 x - \cos^2 x}{\sin x + \cos x} = \sin x - \cos x$

36. $\dfrac{\tan^2 x - \cot^2 x}{\tan x + \cot x} = \tan x - \cot x$

37. $\tan^2 2x + \sin^2 2x + \cos^2 2x = \sec^2 2x$

38. $\cot^2 2x + \cos^2 2x + \sin^2 2x = \csc^2 2x$

39. $\dfrac{\tan 2\theta + \cot 2\theta}{\csc 2\theta} = \sec 2\theta$

40. $\dfrac{\tan 2\theta + \cot 2\theta}{\sec 2\theta} = \csc 2\theta$

41. $\dfrac{\tan x + \tan y}{1 - \tan x \tan y} = \dfrac{\sin x \cos y + \cos x \sin y}{\cos x \cos y - \sin x \sin y}$

42. $\dfrac{\cot x + \cot y}{1 - \cot x \cot y} = \dfrac{\cos x \sin y + \sin x \cos y}{\sin x \sin y - \cos x \cos y}$

43. $(\sec x - \tan x)^2 = \dfrac{1 - \sin x}{1 + \sin x}$

44. $(\csc x - \cot x)^2 = \dfrac{1 - \cos x}{1 + \cos x}$

45. $\dfrac{\sec t + 1}{\tan t} = \dfrac{\tan t}{\sec t - 1}$

46. $\dfrac{\csc t - 1}{\cot t} = \dfrac{\cot t}{\csc t + 1}$

47. $\dfrac{1 + \cos t}{1 - \cos t} = (\csc t + \cot t)^2$

48. $\dfrac{1 - \sin t}{1 + \sin t} = (\sec t - \tan t)^2$

49. $\cos^4 t - \sin^4 t = 1 - 2 \sin^2 t$

50. $\sin^4 t - \cos^4 t = 1 - 2 \cos^2 t$

51. $\dfrac{\sin\theta - \cos\theta}{\sin\theta} + \dfrac{\cos\theta - \sin\theta}{\cos\theta} = 2 - \sec\theta \csc\theta$

52. $\dfrac{\sin\theta}{1 - \cot\theta} - \dfrac{\cos\theta}{\tan\theta - 1} = \sin\theta + \cos\theta$

53. $(\tan^2\theta + 1)(\cos^2\theta + 1) = \tan^2\theta + 2$

54. $(\cot^2\theta + 1)(\sin^2\theta + 1) = \cot^2\theta + 2$

55. $(\cos\theta - \sin\theta)^2 + (\cos\theta + \sin\theta)^2 = 2$

56. $(3\cos\theta - 4\sin\theta)^2 + (4\cos\theta + 3\sin\theta)^2 = 25$

57. $\dfrac{\cos^2 x - \sin^2 x}{1 - \tan^2 x} = \cos^2 x$

58. $\dfrac{\sin x + \cos x}{\sin x} - \dfrac{\cos x - \sin x}{\cos x} = \sec x \csc x$

59. $(\sec x - \tan x)^2 = \dfrac{1 - \sin x}{1 + \sin x}$

60. $(\cot x - \csc x)^2 = \dfrac{1 - \cos x}{1 + \cos x}$

Writing in Mathematics

61. Explain how to verify an identity.

62. Describe two strategies that can be used to verify identities.

63. Describe how you feel when you successfully verify a difficult identity. What other activities do you engage in that evoke the same feelings?

64. A 10-point question on a quiz asks students to verify the identity

$$\dfrac{\sin^2 x - \cos^2 x}{\sin x + \cos x} = \sin x - \cos x.$$

One student begins with the left side and obtains the right side as follows:

$$\dfrac{\sin^2 x - \cos^2 x}{\sin x + \cos x} = \dfrac{\sin^2 x}{\sin x} - \dfrac{\cos^2 x}{\cos x} = \sin x - \cos x.$$

How many points (out of 10) would you give this student? Explain your answer.

Technology Exercises

In Exercises 65–73, graph each side of the equation in the same viewing rectangle. If the graphs appear to coincide, verify that the equation is an identity. If the graphs do not appear to coincide, this indicates the equation is not an identity. In these exercises, find a value of x for which both sides are defined but not equal.

65. $\tan x = \sec x (\sin x - \cos x) + 1$

66. $\sin x = -\cos x \tan(-x)$

67. $\sin\left(x + \dfrac{\pi}{4}\right) = \sin x + \sin\dfrac{\pi}{4}$

68. $\cos\left(x + \dfrac{\pi}{4}\right) = \cos x + \cos\dfrac{\pi}{4}$

69. $\cos(x + \pi) = \cos x$

70. $\sin(x + \pi) = \sin x$

71. $\dfrac{\sin x}{1 - \cos^2 x} = \csc x$

72. $\sin x - \sin x \cos^2 x = \sin^3 x$

73. $\sqrt{\sin^2 x + \cos^2 x} = \sin x + \cos x$

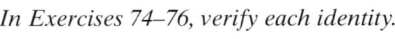

Critical Thinking Exercises

In Exercises 74–76, verify each identity.

74. $\dfrac{\sin^3 x - \cos^3 x}{\sin x - \cos x} = 1 + \sin x \cos x$

75. $\dfrac{\sin x - \cos x + 1}{\sin x + \cos x - 1} = \dfrac{\sin x + 1}{\cos x}$

76. $\ln|\sec x| = -\ln|\cos x|$

77. Use one of the fundamental identities in the box on page 533 to create an original identity.

Group Exercise

78. Group members are to write a helpful list of items for a pamphlet called "The Underground Guide to Verifying Identities." The pamphlet will be used primarily by students who sit, stare, and freak out every time they are asked to verify an identity. List easy ways to remember the fundamental identities. What helpful guidelines can you offer from the perspective of a student that you probably won't find in math books? If you have your own strategies that work particularly well, include them in the pamphlet.

SECTION 6.2   *Sum and Difference Formulas*

Objectives

1. Use the formula for the cosine of the difference of two angles.
2. Use sum and difference formulas for cosines and sines.
3. Use sum and difference formulas for tangents.

Listen to the same note played on a piano and a violin. The notes have a different quality or "tone." Tone depends on the way an instrument vibrates. However, the less than 1% of the population with amusia, or true tone deafness, cannot tell the two sounds apart. Even simple, familiar tunes such as *Happy Birthday* and *Jingle Bells* are mystifying to amusics.

When a note is played, it vibrates at a specific frequency and has a particular amplitude. Amusics cannot tell the difference between sounds from a tuning fork modeled by $p = 3 \sin 2t$ and $p = 2 \sin(2t + \pi)$. However, they can recognize the difference between the two equations. Notice that the second equation contains the sine of the sum of two angles. In this section, we will be developing identities involving the sums or differences of two angles. These formulas are called the **sum and difference formulas**. We begin with $\cos(\alpha - \beta)$, the cosine of the difference of two angles.

The Cosine of the Difference of Two Angles

The Cosine of the Difference of Two Angles

$$\cos(\alpha - \beta) = \cos \alpha \cos \beta + \sin \alpha \sin \beta$$

The cosine of the difference of two angles equals the cosine of the first angle times the cosine of the second angle plus the sine of the first angle times the sine of the second angle.

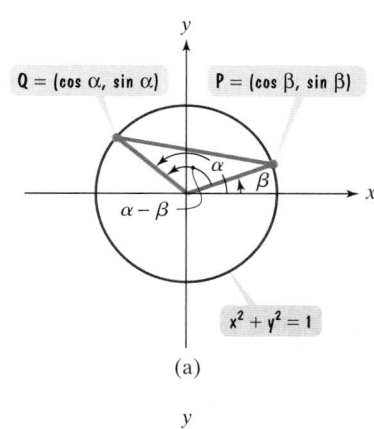

(a)

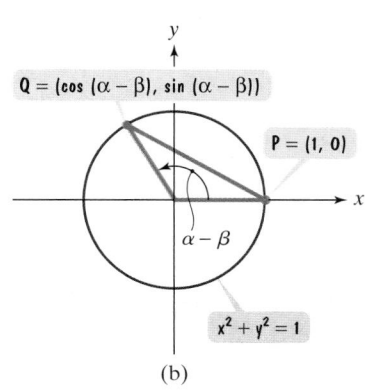

(b)

Figure 6.1 Using the unit circle and QP to develop a formula for $\cos(\alpha - \beta)$

We use Figure 6.1 to prove the identity in the box. The graph in Figure 6.1(a) shows a unit circle, $x^2 + y^2 = 1$. The figure uses the definitions of the cosine and sine functions as the x- and y-coordinates of points along the unit circle. For example, point P corresponds to angle β. By definition, the x-coordinate of P is $\cos \beta$ and the y-coordinate is $\sin \beta$. Similarly, point Q corresponds to angle α. By definition, the x-coordinate of Q is $\cos \alpha$ and the y-coordinate is $\sin \alpha$.

Note that if we draw a line segment between points P and Q, a triangle is formed. Angle $\alpha - \beta$ is one of the angles of this triangle. What happens if we rotate this triangle so that point P falls on the x-axis at $(1, 0)$? The result is shown in Figure 6.1(b). This rotation changes the coordinates of points P and Q. However, it has no effect on the length of line segment PQ.

We can use the distance formula, $d = \sqrt{(x_2 - x_1)^2 + (y_2 - y_1)^2}$, to find an expression for PQ in Figure 6.1(a) and in Figure 6.1(b). By equating the two expressions for PQ, we will obtain the identity for the cosine of the difference of two angles, $\alpha - \beta$. We first apply the distance formula in Figure 6.1(a).

$$PQ = \sqrt{(\cos\alpha - \cos\beta)^2 + (\sin\alpha - \sin\beta)^2}$$

Apply the distance formula, $d = \sqrt{(x_2 - x_1)^2 + (y_2 - y_1)^2}$, to find the distance between $(\cos\beta, \sin\beta)$ and $(\cos\alpha, \sin\alpha)$.

$$= \sqrt{\cos^2\alpha - 2\cos\alpha\cos\beta + \cos^2\beta + \sin^2\alpha - 2\sin\alpha\sin\beta + \sin^2\beta}$$

Square each expression using $(A - B)^2 = A^2 - 2AB + B^2$.

$$= \sqrt{(\sin^2\alpha + \cos^2\alpha) + (\sin^2\beta + \cos^2\beta) - 2\cos\alpha\cos\beta - 2\sin\alpha\sin\beta}$$

Regroup terms to apply a Pythagorean identity.

$$= \sqrt{1 + 1 - 2\cos\alpha\cos\beta - 2\sin\alpha\sin\beta}$$

Because $\sin^2 x + \cos^2 x = 1$, each expression in parentheses equals 1.

$$= \sqrt{2 - 2\cos\alpha\cos\beta - 2\sin\alpha\sin\beta}$$

Simplify.

Next, we apply the distance formula in Figure 6.1(b) to obtain a second expression for PQ. We let $(x_1, y_1) = (1, 0)$ and $(x_2, y_2) = (\cos(\alpha - \beta), \sin(\alpha - \beta))$.

$$PQ = \sqrt{[\cos(\alpha - \beta) - 1]^2 + [\sin(\alpha - \beta) - 0]^2}$$

Apply the distance formula to find the distance between $(1, 0)$ and $(\cos(\alpha - \beta), \sin(\alpha - \beta))$.

$$= \sqrt{\cos^2(\alpha - \beta) - 2\cos(\alpha - \beta) + 1 + \sin^2(\alpha - \beta)}$$

Square each expression.

Using a Pythagorean identity,
$$\sin^2(\alpha - \beta) + \cos^2(\alpha - \beta) = 1$$

$$= \sqrt{1 - 2\cos(\alpha - \beta) + 1}$$

Use a Pythagorean identity.

$$= \sqrt{2 - 2\cos(\alpha - \beta)}$$

Simplify.

Now we equate the two expressions for PQ.

$$\sqrt{2 - 2\cos(\alpha - \beta)} = \sqrt{2 - 2\cos\alpha\cos\beta - 2\sin\alpha\sin\beta}$$

The rotation does not change the length of PQ.

$$2 - 2\cos(\alpha - \beta) = 2 - 2\cos\alpha\cos\beta - 2\sin\alpha\sin\beta$$

Square both sides to eliminate radicals.

$$-2\cos(\alpha - \beta) = -2\cos\alpha\cos\beta - 2\sin\alpha\sin\beta$$

Subtract 2 from both sides of the equation.

$$\cos(\alpha - \beta) = \cos\alpha\cos\beta + \sin\alpha\sin\beta$$

Divide both sides of the equation by −2.

1 Use the formula for the cosine of the difference of two angles.

Now that we see where the identity for the cosine of the difference of two angles comes from, let's look at some applications of this result.

EXAMPLE 1 Using the Difference Formula for Cosines to Find Exact Values

Find the exact value of $\cos 15°$.

Solution We know exact values for trigonometric functions of $60°$ and $45°$. Thus, we write $15°$ as $60° - 45°$ and use the difference formula for cosines.

$$\cos 15° = \cos(60° - 45°)$$

$$= \cos 60° \cos 45° + \sin 60° \sin 45° \qquad \cos(\alpha - \beta) = \cos\alpha\cos\beta + \sin\alpha\sin\beta$$

$$= \frac{1}{2} \cdot \frac{\sqrt{2}}{2} + \frac{\sqrt{3}}{2} \cdot \frac{\sqrt{2}}{2} \qquad \text{Substitute exact values from memory or use special right triangles.}$$

$$= \frac{\sqrt{2}}{4} + \frac{\sqrt{6}}{4} \qquad \text{Multiply.}$$

$$= \frac{\sqrt{2} + \sqrt{6}}{4} \qquad \text{Add.}$$

Check Point 1 We know that $\cos 30° = \dfrac{\sqrt{3}}{2}$. Obtain this exact value using $\cos 30° = \cos(90° - 60°)$ and the difference formula for cosines.

EXAMPLE 2 Using the Difference Formula for Cosines to Find Exact Values

Find the exact value of $\cos 80° \cos 20° + \sin 80° \sin 20°$.

Solution The given expression is the right side of the formula for $\cos(\alpha - \beta)$ with $\alpha = 80°$ and $\beta = 20°$.

$$\cos(\alpha - \beta) = \cos\alpha\cos\beta + \sin\alpha\sin\beta$$

$$\cos 80° \cos 20° + \sin 80° \sin 20° = \cos(80° - 20°) = \cos 60° = \tfrac{1}{2}$$

Check Point 2 Find the exact value of
$$\cos 70° \cos 40° + \sin 70° \sin 40°.$$

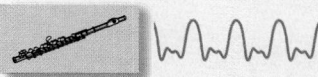
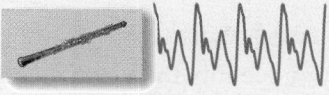
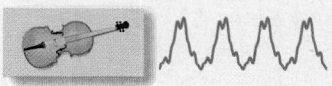

EXAMPLE 3   Verifying an Identity

Verify the identity: $\dfrac{\cos(\alpha - \beta)}{\sin\alpha\cos\beta} = \cot\alpha + \tan\beta$.

Solution   We work with the left side.

$$\frac{\cos(\alpha - \beta)}{\sin\alpha\cos\beta} = \frac{\cos\alpha\cos\beta + \sin\alpha\sin\beta}{\sin\alpha\cos\beta} \qquad \text{Use the formula for } \cos(\alpha - \beta).$$

$$= \frac{\cos\alpha}{\sin\alpha}\frac{\cos\beta}{\cos\beta} + \frac{\sin\alpha}{\sin\alpha}\frac{\sin\beta}{\cos\beta} \qquad \begin{array}{l}\text{Divide each term in the numerator by} \\ \sin\alpha\cos\beta.\end{array}$$

$$= \cot\alpha \cdot 1 + 1 \cdot \tan\beta \qquad \text{Use quotient identities.}$$

$$= \cot\alpha + \tan\beta \qquad \text{Simplify.}$$

We worked with the left side and arrived at the right side. Thus, the identity is verified.

> **Check Point 3**   Verify the identity: $\dfrac{\cos(\alpha - \beta)}{\cos\alpha\cos\beta} = 1 + \tan\alpha\tan\beta$.

Technology

The graphs of

$$y = \cos\left(\frac{\pi}{2} - x\right)$$

and

$$y = \sin x$$

are shown in the same viewing rectangle. The graphs are the same. The voice balloon and the displayed math on the right show the equivalence algebraically.

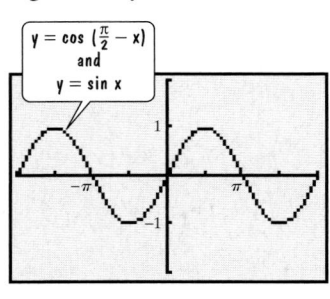

The difference formula for cosines is used to establish other identities. For example, in our work with right triangles we noted that cofunctions of complements are equal. Thus, because $90° - \theta$ and θ are complements,

$$\cos(90° - \theta) = \sin\theta.$$

We can use the formula for $\cos(\alpha - \beta)$ to prove this cofunction identity.

> Apply $\cos(\alpha - \beta)$ with $\alpha = 90°$ and $\theta = \beta$.
> $\cos(\alpha - \beta) = \cos\alpha\cos\beta + \sin\alpha\sin\beta$

$$\cos(90° - \theta) = \cos 90°\cos\theta + \sin 90°\sin\theta$$

$$= 0 \cdot \cos\theta + 1 \cdot \sin\theta$$

$$= \sin\theta$$

2 Use sum and difference formulas for cosines and sines.

Sum and Difference Formulas for Cosines and Sines

Our formula for $\cos(\alpha - \beta)$ can be used to verify an identity for a sum involving cosines, as well as identities for a sum and a difference for sines.

Sum and Difference Formulas for Cosines and Sines

1. $\cos(\alpha + \beta) = \cos\alpha\cos\beta - \sin\alpha\sin\beta$

2. $\cos(\alpha - \beta) = \cos\alpha\cos\beta + \sin\alpha\sin\beta$

3. $\sin(\alpha + \beta) = \sin\alpha\cos\beta + \cos\alpha\sin\beta$

4. $\sin(\alpha - \beta) = \sin\alpha\cos\beta - \cos\alpha\sin\beta$

Up to now, we have concentrated on the second formula in the box. The first identity gives a formula for the cosine of the sum of two angles. It is proved as follows.

$$\cos(\alpha + \beta) = \cos[\alpha - (-\beta)]$$ Express addition as subtraction of an inverse.

$$= \cos\alpha\cos(-\beta) + \sin\alpha\sin(-\beta)$$ Use the second formula in the box.

$$= \cos\alpha\cos\beta + \sin\alpha(-\sin\beta)$$ Cosine is even: $\cos(-\beta) = \cos\beta$. Sine is odd: $\sin(-\beta) = -\sin\beta$.

$$= \cos\alpha\cos\beta - \sin\alpha\sin\beta$$ Simplify.

Thus, the cosine of the sum of two angles equals the cosine of the first angle times the cosine of the second angle minus the sine of the first angle times the sine of the second angle.

The third identity in the box gives a formula for $\sin(\alpha + \beta)$, the sine of the sum of two angles. It is proved as follows.

$$\sin(\alpha + \beta) = \cos\left[\frac{\pi}{2} - (\alpha + \beta)\right]$$ Use a cofunction identity: $\sin\theta = \cos\left(\frac{\pi}{2} - \theta\right)$.

$$= \cos\left[\left(\frac{\pi}{2} - \alpha\right) - \beta\right]$$ Regroup.

$$= \cos\left(\frac{\pi}{2} - \alpha\right)\cos\beta + \sin\left(\frac{\pi}{2} - \alpha\right)\sin\beta$$ Use the formula for the cosine of a difference.

$$= \sin\alpha\cos\beta + \cos\alpha\sin\beta$$ Use cofunction identities.

Thus, the sine of the sum of two angles equals the sine of the first angle times the cosine of the second angle plus the cosine of the first angle times the sine of the second angle.

The final identity in the box gives a formula for $\sin(\alpha - \beta)$, the sine of the difference of two angles. It is proved by writing $\sin(\alpha - \beta)$ as $\sin[\alpha + (-\beta)]$ and then using the formula for the sine of a sum.

EXAMPLE 4   Using the Sine of a Sum to Find an Exact Value

Find the exact value of $\sin\dfrac{7\pi}{12}$ by using the fact that $\dfrac{7\pi}{12} = \dfrac{\pi}{3} + \dfrac{\pi}{4}$.

Solution   We apply the formula for the sine of a sum.

$$\sin\frac{7\pi}{12} = \sin\left(\frac{\pi}{3} + \frac{\pi}{4}\right)$$

$$= \sin\frac{\pi}{3}\cos\frac{\pi}{4} + \cos\frac{\pi}{3}\sin\frac{\pi}{4}$$ $(\sin\alpha + \beta) = \sin\alpha\cos\beta + \cos\alpha\sin\beta$

$$= \frac{\sqrt{3}}{2} \cdot \frac{\sqrt{2}}{2} + \frac{1}{2} \cdot \frac{\sqrt{2}}{2}$$ Substitute exact values.

$$= \frac{\sqrt{6} + \sqrt{2}}{4}$$

> **Check Point 4**
>
> Find the exact value of $\sin \dfrac{5\pi}{12}$ by using the fact that
>
> $$\dfrac{5\pi}{12} = \dfrac{\pi}{6} + \dfrac{\pi}{4}.$$

EXAMPLE 5 Finding Exact Values

Suppose that $\sin \alpha = \frac{12}{13}$ for a quadrant II angle α and $\sin \beta = \frac{3}{5}$ for a quadrant I angle β. Find the exact value of:

a. $\cos \alpha$ **b.** $\cos \beta$ **c.** $\cos(\alpha + \beta)$ **d.** $\sin(\alpha + \beta)$.

Solution

a. We find $\cos \alpha$ using a sketch that illustrates

$$\sin \alpha = \dfrac{12}{13} = \dfrac{y}{r}.$$

Figure 6.2 shows a quadrant II angle α and $\sin \alpha = \frac{12}{13}$. We find x using $x^2 + y^2 = r^2$. Because α lies in quadrant II, x is negative.

$$x^2 + 12^2 = 13^2 \qquad \text{\small $x^2 + y^2 = r^2$}$$
$$x^2 + 144 = 169 \qquad \text{\small Square 12 and 13, respectively.}$$
$$x^2 = 25 \qquad \text{\small Subtract 144 from both sides.}$$
$$x = -\sqrt{25} = -5 \qquad \text{\small In quadrant II, x is negative.}$$

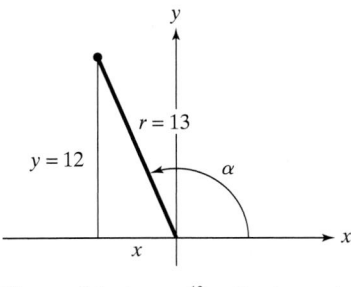

Figure 6.2 $\sin \alpha = \frac{12}{13}$: α lies in quadrant II.

Thus,

$$\cos \alpha = \dfrac{x}{r} = \dfrac{-5}{13} = -\dfrac{5}{13}.$$

b. We find $\cos \beta$ using a sketch that illustrates

$$\sin \beta = \dfrac{3}{5} = \dfrac{y}{r}.$$

Figure 6.3 shows a quadrant I angle β and $\sin \beta = \frac{3}{5}$. We find x using $x^2 + y^2 = r^2$.

$$x^2 + 3^2 = 5^2$$
$$x^2 = 25 - 9 = 16$$
$$x = \sqrt{16} = 4 \qquad \text{\small In quadrant I, x is positive.}$$

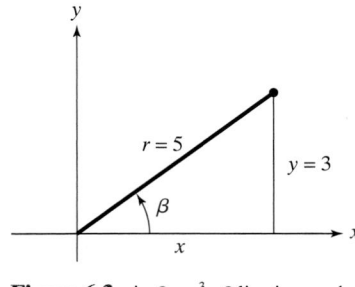

Figure 6.3 $\sin \beta = \frac{3}{5}$, β lies in quadrant I

Thus,

$$\cos \beta = \dfrac{x}{r} = \dfrac{4}{5}.$$

We use the given values and the exact values that we determined to find exact values for $\cos(\alpha + \beta)$ and $\sin(\alpha + \beta)$.

> These values are given.

> These are the values we found.

$$\sin \alpha = \dfrac{12}{13}, \ \sin \beta = \dfrac{3}{5} \qquad \cos \alpha = -\dfrac{5}{13}, \ \cos \beta = \dfrac{4}{5}$$

c. We use the formula for the cosine of a sum.

$$\cos(\alpha + \beta) = \cos\alpha\cos\beta - \sin\alpha\sin\beta$$

$$= -\frac{5}{13}\left(\frac{4}{5}\right) - \frac{12}{13}\left(\frac{3}{5}\right) = -\frac{56}{65}$$

d. We use the formula for the sine of a sum.

$$\sin(\alpha + \beta) = \sin\alpha\cos\beta + \cos\alpha\sin\beta$$

$$= \frac{12}{13}\cdot\frac{4}{5} + \left(-\frac{5}{13}\right)\cdot\frac{3}{5} = \frac{33}{65}$$

Check Point 5

Suppose that $\sin\alpha = \frac{4}{5}$ for a quadrant II angle α and $\sin\beta = \frac{1}{2}$ for a quadrant I angle β. Find the exact value of:

a. $\cos\alpha$ **b.** $\cos\beta$

c. $\cos(\alpha + \beta)$ **d.** $\sin(\alpha + \beta)$.

EXAMPLE 6 Verifying Observations On a Graphing Utility

Figure 6.4 shows the graph of $y = \sin\left(x - \dfrac{3\pi}{2}\right)$ in a $\left[0, 2\pi, \dfrac{\pi}{2}\right]$ by $[-2, 2, 1]$ viewing rectangle

a. Describe the graph using another equation.

b. Verify that the two equations are equivalent.

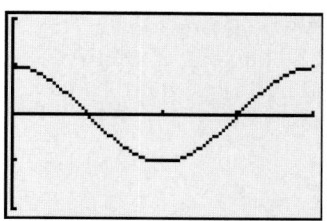

Figure 6.4 The graph of
$y = \sin\left(x - \dfrac{3\pi}{2}\right)$ in a $\left[0, 2\pi, \dfrac{\pi}{2}\right]$ by
$[-2, 2, 1]$ viewing rectangle

Solution

a. The graph appears to be the cosine curve $y = \cos x$. It cycles through maximum, intercept, minimum, intercept, and back to maximum. Thus, $y = \cos x$ also describes the graph.

b. We must show that

$$\sin\left(x - \frac{3\pi}{2}\right) = \cos x.$$

We apply the formula for the sine of a difference on the left side.

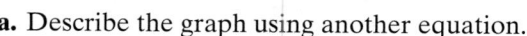

$$\sin\left(x - \frac{3\pi}{2}\right) = \sin x\cos\frac{3\pi}{2} - \cos x\sin\frac{3\pi}{2} \qquad \begin{array}{l}\sin(\alpha - \beta) =\\ \sin\alpha\,\cos\beta - \cos\alpha\,\sin\beta\end{array}$$

$$= \sin x\cdot 0 - \cos x(-1) \qquad \cos\frac{3\pi}{2} = 0 \text{ and } \sin\frac{3\pi}{2} = -1$$

$$= \cos x \qquad\qquad \text{Simplify.}$$

This verifies our observation that $y = \sin\left(x - \dfrac{3\pi}{2}\right)$ and $y = \cos x$ describe the same graph.

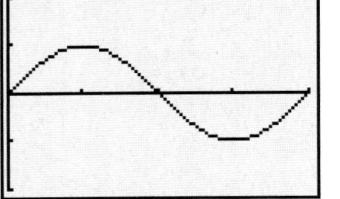

Figure 6.5

Check Point 6

Figure 6.5 shows the graph of $y = \cos\left(x + \dfrac{3\pi}{2}\right)$ in a $\left[0, 2\pi, \dfrac{\pi}{2}\right]$ by $[-2, 2, 1]$ viewing rectangle.

a. Describe the graph using another equation.

b. Verify that the two equations are equivalent.

3 Use sum and difference formulas for tangents.

Sum and Difference Formulas for Tangents

By writing $\tan(\alpha + \beta)$ as the quotient of $\sin(\alpha + \beta)$ and $\cos(\alpha + \beta)$, we can develop a formula for the tangent of a sum. Writing subtraction as addition of an inverse leads to a formula for the tangent of a difference.

Discovery

Derive the sum and difference formulas for tangents by working Exercises 55 and 56 in Exercise Set 6.2.

Sum and Difference Formulas for Tangents

$$\tan(\alpha + \beta) = \frac{\tan\alpha + \tan\beta}{1 - \tan\alpha \tan\beta}$$

The tangent of the sum of two angles equals the tangent of the first angle plus the tangent of the second angle divided by 1 minus their product.

$$\tan(\alpha - \beta) = \frac{\tan\alpha - \tan\beta}{1 + \tan\alpha \tan\beta}$$

The tangent of the difference of two angles equals the tangent of the first angle minus the tangent of the second angle divided by 1 plus their product.

EXAMPLE 7 Verifying an Identity

Verify the identity: $\tan\left(x - \dfrac{\pi}{4}\right) = \dfrac{\tan x - 1}{\tan x + 1}$.

Solution We work with the left side.

$$\tan\left(x - \frac{\pi}{4}\right) = \frac{\tan x - \tan\dfrac{\pi}{4}}{1 + \tan x \tan\dfrac{\pi}{4}} \qquad \tan(\alpha - \beta) = \frac{\tan\alpha - \tan\beta}{1 + \tan\alpha \tan\beta}$$

$$= \frac{\tan x - 1}{1 + \tan x \cdot 1} \qquad \tan\frac{\pi}{4} = 1$$

$$= \frac{\tan x - 1}{1 + \tan x}$$

Check Point 7

Verify the identity: $\tan(x + \pi) = \tan x$.

EXERCISE SET 6.2

 Practice Exercises

Use the formula for the cosine of the difference of two angles to solve Exercises 1–12.

In Exercises 1–4, find the exact value of each expression.

1. $\cos(45° - 30°)$

2. $\cos(120° - 45°)$

3. $\cos\left(\dfrac{3\pi}{4} - \dfrac{\pi}{6}\right)$

4. $\cos\left(\dfrac{2\pi}{3} - \dfrac{\pi}{6}\right)$

In Exercises 5–8, each expression is the right side of the formula for $\cos(\alpha - \beta)$ with particular values for α and β.

a. *Identify α and β in each expression.*

b. *Write the expression as the cosine of an angle.*

c. *Find the exact value of the expression.*

5. $\cos 50° \cos 20° + \sin 50° \sin 20°$

6. $\cos 50° \cos 5° + \sin 50° \sin 5°$

7. $\cos\dfrac{5\pi}{12}\cos\dfrac{\pi}{12} + \sin\dfrac{5\pi}{12}\sin\dfrac{\pi}{12}$

8. $\cos\dfrac{5\pi}{18}\cos\dfrac{\pi}{9} + \sin\dfrac{5\pi}{18}\sin\dfrac{\pi}{9}$

In Exercises 9–12, verify each identity.

9. $\dfrac{\cos(\alpha - \beta)}{\cos\alpha\sin\beta} = \tan\alpha + \cot\beta$

10. $\dfrac{\cos(\alpha - \beta)}{\sin\alpha\sin\beta} = \cot\alpha\cot\beta + 1$

11. $\cos\left(x - \dfrac{\pi}{4}\right) = \dfrac{\sqrt{2}}{2}(\cos x + \sin x)$

12. $\cos\left(x - \dfrac{5\pi}{4}\right) = -\dfrac{\sqrt{2}}{2}(\cos x + \sin x)$

Use one or more of the six sum and difference identities to solve Exercises 13–54.

In Exercises 13–24, find the exact value of each expression.

13. $\sin(45° - 30°)$

14. $\sin(60° - 45°)$

15. $\sin 105°$

16. $\sin 75°$

17. $\tan(30° + 45°)$

18. $\tan(60° + 45°)$

19. $\tan(240° - 45°)$

20. $\tan(300° - 45°)$

21. $\cos\left(\dfrac{3\pi}{4} + \dfrac{\pi}{6}\right)$

22. $\cos\left(\dfrac{4\pi}{3} + \dfrac{\pi}{4}\right)$

23. $\cos\dfrac{5\pi}{12}$

24. $\cos\dfrac{7\pi}{12}$

In Exercises 25–32, write each expression as the sine, cosine, or tangent of an angle. Then find the exact value of the expression.

25. $\sin 25° \cos 5° + \cos 25° \sin 5°$

26. $\sin 40° \cos 20° + \cos 40° \sin 20°$

27. $\dfrac{\tan 10° + \tan 35°}{1 - \tan 10° \tan 35°}$

28. $\dfrac{\tan 50° - \tan 20°}{1 + \tan 50° \tan 20°}$

29. $\sin\dfrac{5\pi}{12}\cos\dfrac{\pi}{4} - \cos\dfrac{5\pi}{12}\sin\dfrac{\pi}{4}$

30. $\sin\dfrac{7\pi}{12}\cos\dfrac{\pi}{12} - \cos\dfrac{7\pi}{12}\sin\dfrac{\pi}{12}$

31. $\dfrac{\tan\dfrac{\pi}{5} - \tan\dfrac{\pi}{30}}{1 + \tan\dfrac{\pi}{5}\tan\dfrac{\pi}{30}}$

32. $\dfrac{\tan\dfrac{\pi}{5} + \tan\dfrac{4\pi}{5}}{1 - \tan\dfrac{\pi}{5}\tan\dfrac{4\pi}{5}}$

In Exercises 33–54, verify each identity.

33. $\sin\left(x + \dfrac{\pi}{2}\right) = \cos x$

34. $\sin\left(x + \dfrac{3\pi}{2}\right) = -\cos x$

35. $\cos\left(x - \dfrac{\pi}{2}\right) = \sin x$

36. $\cos(\pi - x) = -\cos x$

37. $\tan(2\pi - x) = -\tan x$

38. $\tan(\pi - x) = -\tan x$

39. $\sin(\alpha + \beta) + \sin(\alpha - \beta) = 2\sin\alpha\cos\beta$

40. $\cos(\alpha + \beta) + \cos(\alpha - \beta) = 2\cos\alpha\cos\beta$

41. $\dfrac{\sin(\alpha - \beta)}{\cos\alpha\cos\beta} = \tan\alpha - \tan\beta$

42. $\dfrac{\sin(\alpha + \beta)}{\cos\alpha\cos\beta} = \tan\alpha + \tan\beta$

43. $\tan\left(\theta + \dfrac{\pi}{4}\right) = \dfrac{\cos\theta + \sin\theta}{\cos\theta - \sin\theta}$

44. $\tan\left(\dfrac{\pi}{4} - \theta\right) = \dfrac{\cos\theta - \sin\theta}{\cos\theta - \sin\theta}$

45. $\cos(\alpha + \beta)\cos(\alpha - \beta) = \cos^2\beta - \sin^2\alpha$

46. $\sin(\alpha + \beta)\sin(\alpha - \beta) = \cos^2\beta - \cos^2\alpha$

47. $\dfrac{\sin(\alpha + \beta)}{\sin(\alpha - \beta)} = \dfrac{\tan\alpha + \tan\beta}{\tan\alpha - \tan\beta}$

48. $\dfrac{\cos(\alpha + \beta)}{\cos(\alpha - \beta)} = \dfrac{1 - \tan\alpha\tan\beta}{1 + \tan\alpha\tan\beta}$

49. $\dfrac{\cos(x + h) - \cos x}{h} = \cos x\dfrac{\cos h - 1}{h} - \sin x\dfrac{\sin h}{h}$

50. $\dfrac{\sin(x+h) - \sin x}{h} = \cos x \dfrac{\sin h}{h} + \sin x \dfrac{\cos h - 1}{h}$

51. $\sin 2\alpha = 2 \sin \alpha \cos \alpha$

Hint: Write $\sin 2\alpha$ as $\sin(\alpha + \alpha)$.

52. $\cos 2\alpha = \cos^2 \alpha - \sin^2 \alpha$

Hint: Write $\cos 2\alpha$ as $\cos(\alpha + \alpha)$.

53. $\tan 2\alpha = \dfrac{2 \tan \alpha}{1 - \tan^2 \alpha}$

Hint: Write $\tan 2\alpha$ as $\tan(\alpha + \alpha)$.

54. $\tan\left(\dfrac{\pi}{4} + \alpha\right) - \tan\left(\dfrac{\pi}{4} - \alpha\right) = 2 \tan 2\alpha$

Hint: Use the result in Exercise 53.

55. Derive the identity for $\tan(\alpha + \beta)$ using

$$\tan(\alpha + \beta) = \dfrac{\sin(\alpha + \beta)}{\cos(\alpha + \beta)}.$$

After applying the formulas for sums of sines and cosines, divide the numerator and denominator by $\cos \alpha \cos \beta$.

56. Derive the identity for $\tan(\alpha - \beta)$ using

$$\tan(\alpha - \beta) = \tan[\alpha + (-\beta)].$$

After applying the formula for the tangent of the sum of two angles, use the fact that the tangent is an odd function.

In Exercises 57–62, find the exact value of the following under the given conditions:

  a. $\cos(\alpha + \beta)$ **b.** $\sin(\alpha + \beta)$ **c.** $\tan(\alpha + \beta)$

57. $\sin \alpha = \frac{3}{5}$, α lies in quadrant I, and $\sin \beta = \frac{5}{13}$, β lies in quadrant II.

58. $\sin \alpha = \frac{4}{5}$, α lies in quadrant I, and $\sin \beta = \frac{7}{25}$, β lies in quadrant II.

59. $\tan \alpha = -\frac{3}{4}$, α lies in quadrant II, and $\cos \beta = \frac{1}{3}$, β lies in quadrant I.

60. $\tan \alpha = -\frac{4}{3}$, α lies in quadrant II, and $\cos \beta = \frac{2}{3}$, β lies in quadrant I.

61. $\cos \alpha = \frac{8}{17}$, α lies in quadrant IV, and $\sin \beta = -\frac{1}{2}$, β lies in quadrant III.

62. $\cos \alpha = \frac{1}{2}$, α lies in quadrant IV, and $\sin \beta = -\frac{1}{3}$, β lies in quadrant III.

In Exercises 63–66, the graph with the given equation is shown in a $\left[0, 2\pi, \dfrac{\pi}{2}\right]$ by $[-2, 2, 1]$ viewing rectangle.

  a. *Describe the graph using another equation.*

  b. *Verify that the two equations are equivalent.*

63.

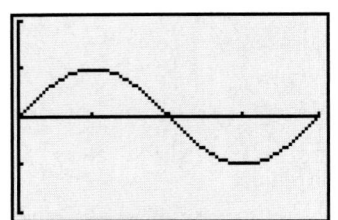

$y = \sin(\pi - x)$

64.

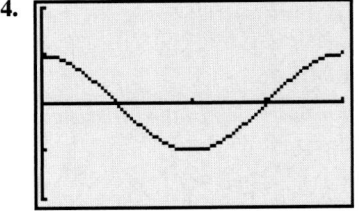

$y = \cos(x - 2\pi)$

65.

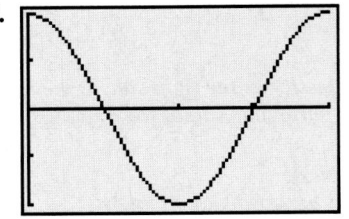

$y = \sin\left(x + \dfrac{\pi}{2}\right) + \sin\left(\dfrac{\pi}{2} - x\right)$

66.

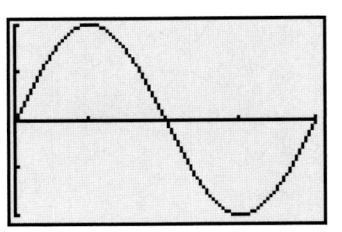

$y = \cos\left(x - \dfrac{\pi}{2}\right) - \cos\left(x + \dfrac{\pi}{2}\right)$

⭐ **Application Exercises**

67. A ball attached to a spring is raised 2 feet and released with an initial vertical velocity of 3 feet per second. The distance of the ball from its rest position after t seconds is given by $d = 2 \cos t + 3 \sin t$. Show that

$$2 \cos t + 3 \sin t = \sqrt{13} \cos(t - \theta),$$

where θ lies in quadrant I and $\tan \theta = \frac{3}{2}$. Use the identity to find the amplitude and the period of the ball's motion.

68. A tuning fork is held a certain distance from your ears and struck. Your eardrums' vibrations after t seconds are given by $p = 3 \sin 2t$. When a second tuning fork is struck, the formula $p = 2 \sin(2t + \pi)$ describes the effects of the sound on the eardrums' vibrations. The total vibrations are given by $p = 3 \sin 2t + 2 \sin(2t + \pi)$.

  a. Express p using a single occurrence of the sine.

  b. If the amplitude of p is zero, no sound is heard. Based on your equation in part (a), does this occur with the two tuning forks in this exercise? Explain your answer.

Writing in Mathematics

In Exercises, 69–74, use words to describe the formula for:

69. the cosine of the difference of two angles.
70. the cosine of the sum of two angles.
71. the sine of the sum of two angles.
72. the sine of the difference of two angles.
73. the tangent of the difference of two angles.
74. the tangent of the sum of two angles.
75. The distance formula and the definitions for cosine and sine are used to prove the formula for the cosine of the difference of two angles. This formula logically leads the way to the other sum and difference identities. Using this development of ideas and formulas, describe a characteristic of mathematical logic.

Technology Exercises

In Exercises 76–81, graph each side of the equation in the same viewing rectangle. If the graphs appear to coincide, verify that the equation is an identity. If the graphs do not appear to coincide, this indicates that the equation is not an identity. In these exercises, find a value of x for which both sides are defined but not equal.

76. $\cos\left(\dfrac{3\pi}{2} - x\right) = -\sin x$

77. $\tan(\pi - x) = -\tan x$

78. $\sin\left(x + \dfrac{\pi}{2}\right) = \sin x + \sin\dfrac{\pi}{2}$

79. $\cos\left(x + \dfrac{\pi}{2}\right) = \cos x + \cos\dfrac{\pi}{2}$

80. $\cos 1.2x \cos 0.8x - \sin 1.2x \sin 0.8x = \cos 2x$
81. $\sin 1.2x \cos 0.8x + \cos 1.2x \sin 0.8x = \sin 2x$

Critical Thinking Exercises

82. Graph $y = \sin 5x \cos 3x - \cos 5x \sin 3x$ from 0 to 2π.
83. Verify the identity:
$$\frac{\sin(x - y)}{\cos x \cos y} + \frac{\sin(y - z)}{\cos y \cos z} + \frac{\sin(z - x)}{\cos z \cos x} = 0.$$

84. Without using a calculator, find the exact value of
$$\cos\left[\cos^{-1}\left(-\frac{\sqrt{3}}{2}\right) - \sin^{-1}\left(-\frac{1}{2}\right)\right].$$

85. Use right triangles to write
$$\cos(\sin^{-1}x - \cos^{-1}y)$$
as an algebraic expression. Assume that x and y are positive and in the domain of the given inverse trigonometric function.

Group Exercise

86. Remembering the six sum and difference identities can be difficult. Did you have problems with some exercises because the identity you were using in your head turned out to be an incorrect formula? Are there easy ways to remember the six new identities presented in this section? Group members should address this question, considering one identity at a time. For each formula, list ways to make it easier to remember.

SECTION 6.3   *Double-Angle and Half-Angle Formulas*

Objectives

1. Use the double-angle formulas.
2. Use the power-reducing formulas.
3. Use the half-angle formulas.

We have a long history of throwing things. Prior to 400 B.C., the Greeks competed in games that included discus throwing. In the seventeenth century, English soldiers organized cannonball-throwing competitions. In 1827, a Yale University student, disappointed over failing an exam, took out his frustrations at the passing of a collection plate in chapel. Upon seizing the monetary tray, he flung it in the direction of a large open space on campus. Yale students see this act of frustration as the origin of the Frisbee.

In this section, we develop other important classes of identities called the double-angle and half-angle formulas. We will see how one of these formulas can be used by athletes to increase throwing distance.

1 Use the double-angle formulas.

Study Tip

The 2 that appears in each of the double-angle expressions cannot be pulled to the front and written as a coefficient.

INCORRECT

$$\sin 2\theta = 2 \sin\theta$$
$$\cos 2\theta = 2 \cos\theta$$
$$\tan 2\theta = 2 \tan\theta$$

The figure shows, in a $\left[0, 2\pi, \dfrac{\pi}{2}\right]$ by $[-3, 3, 1]$ viewing rectangle, that the graphs of

$$y = \sin 2x$$

and

$$y = 2 \sin x$$

do not coincide.

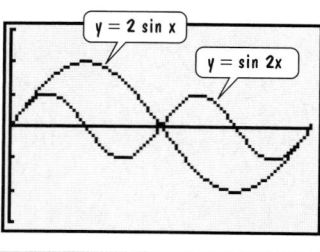

Double-Angle Formulas

A number of basic identities follow from the sum formulas for sine, cosine, and tangent. The first category of identities involves **double-angle formulas**.

> ### Double-Angle Formulas
>
> $$\sin 2\theta = 2 \sin\theta \cos\theta$$
> $$\cos 2\theta = \cos^2\theta - \sin^2\theta$$
> $$\tan 2\theta = \frac{2 \tan\theta}{1 - \tan^2\theta}$$

To prove each of these formulas, we replace α and β by θ in the sum formulas for $\sin(\alpha + \beta)$, $\cos(\alpha + \beta)$, and $\tan(\alpha + \beta)$.

- $\sin 2\theta = \sin(\theta + \theta) = \sin\theta\cos\theta + \cos\theta\sin\theta = 2\sin\theta\cos\theta$

  We use
  $\sin(\alpha + \beta) = \sin\alpha\cos\beta + \cos\alpha\sin\beta$.

- $\cos 2\theta = \cos(\theta + \theta) = \cos\theta\cos\theta - \sin\theta\sin\theta = \cos^2\theta - \sin^2\theta$

  We use
  $\cos(\alpha + \beta) = \cos\alpha\cos\beta - \sin\alpha\sin\beta$.

- $\tan 2\theta = \tan(\theta + \theta) = \dfrac{\tan\theta + \tan\theta}{1 - \tan\theta\tan\theta} = \dfrac{2\tan\theta}{1 - \tan^2\theta}$

  We use
  $\tan(\alpha + \beta) = \dfrac{\tan\alpha + \tan\beta}{1 - \tan\alpha\tan\beta}$.

EXAMPLE 1 Using Double-Angle Formulas to Find Exact Values

If $\sin\theta = \dfrac{5}{13}$ and θ lies in quadrant II, find the exact value of:

a. $\sin 2\theta$ **b.** $\cos 2\theta$ **c.** $\tan 2\theta$.

Solution We begin with a sketch that illustrates

$$\sin\theta = \frac{5}{13} = \frac{y}{r}.$$

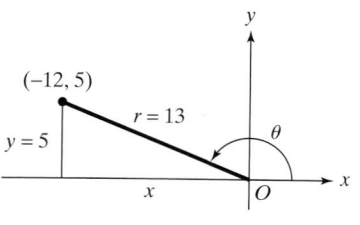

Figure 6.6 $\sin\theta = \frac{5}{13}$ and θ lies in quadrant II.

Figure 6.6 shows a quadrant II angle θ and $\sin\theta = \frac{5}{13}$. We find x using $x^2 + y^2 = r^2$. Because θ lies in quadrant II, x is negative.

$$x^2 + 5^2 = 13^2 \qquad\qquad x^2 + y^2 = r^2$$

$$x^2 = 13^2 - 5^2 = 144 \quad \text{Solve for } x^2.$$

$$x = -\sqrt{144} = -12 \quad \text{In quadrant II, } x \text{ is negative.}$$

Now we can use values for x, y, and r to find the required values.

a. $\sin 2\theta = 2 \sin \theta \cos \theta = 2\left(\dfrac{5}{13}\right)\left(-\dfrac{12}{13}\right) = -\dfrac{120}{169}$

b. $\cos 2\theta = \cos^2 \theta - \sin^2 \theta = \left(-\dfrac{12}{13}\right)^2 - \left(\dfrac{5}{13}\right)^2 = \dfrac{144}{169} - \dfrac{25}{169} = \dfrac{119}{169}$

c. $\tan 2\theta = \dfrac{2 \tan \theta}{1 - \tan^2 \theta} = \dfrac{2\left(-\dfrac{5}{12}\right)}{1 - \left(-\dfrac{5}{12}\right)^2} = \dfrac{-\dfrac{5}{6}}{1 - \dfrac{25}{144}} = \dfrac{-\dfrac{5}{6}}{\dfrac{119}{144}} = \left(-\dfrac{5}{6}\right)\left(\dfrac{144}{19}\right) = -\dfrac{120}{19}$

Check Point 1   If $\sin \theta = \frac{4}{5}$ and θ lies in quadrant II, find the exact value of:
   a. $\sin 2\theta$ **b.** $\cos 2\theta$ **c.** $\tan 2\theta$.

EXAMPLE 2   **Using the Double Angle Formula for Tangents to Find an Exact Value**

Find the exact value of $\dfrac{2 \tan 15°}{1 - \tan^2 15°}$.

Solution   The given expression is the right side of the formula for $\tan 2\theta$ with $\theta = 15°$.

$$\tan 2\theta = \dfrac{2 \tan \theta}{1 - \tan^2 \theta}$$

$$\dfrac{2 \tan 15°}{1 - \tan^2 15°} = \tan(2 \cdot 15°) = \tan 30° = \dfrac{\sqrt{3}}{3}$$

Check Point 2   Find the exact value of $\cos^2 15° - \sin^2 15°$.

There are three forms of the double-angle formula for $\cos 2\theta$. The form we have seen involves both the cosine and the sine:

$$\cos 2\theta = \cos^2 \theta - \sin^2 \theta.$$

Using the Pythagorean identity $\sin^2 \theta + \cos^2 \theta = 1$, we can write this last formula in terms of the cosine only. We substitute $1 - \cos^2 \theta$ for $\sin^2 \theta$.

$$\cos 2\theta = \cos^2 \theta - \sin^2 \theta = \cos^2 \theta - (1 - \cos^2 \theta)$$

$$= \cos^2 \theta - 1 + \cos^2 \theta = 2 \cos^2 \theta - 1$$

We can also use a Pythagorean identity to write $\cos 2\theta$ in terms of sine only. We substitute $1 - \sin^2\theta$ for $\cos^2\theta$.

$$\cos 2\theta = \cos^2\theta - \sin^2\theta = 1 - \sin^2\theta - \sin^2\theta = 1 - 2\sin^2\theta$$

Three Forms of the Double-Angle Formula for $\cos 2\theta$

$$\cos 2\theta = \cos^2\theta - \sin^2\theta$$
$$\cos 2\theta = 2\cos^2\theta - 1$$
$$\cos 2\theta = 1 - 2\sin^2\theta$$

EXAMPLE 3 Verifying an Identity

Verify the identity: $\cos 3\theta = 4\cos^3\theta - 3\cos\theta$.

Solution We begin by working with the left side. In order to obtain an expression for $\cos 3\theta$, we use the sum formula and write 3θ as $2\theta + \theta$.

$\cos 3\theta = \cos(2\theta + \theta)$ Write 3θ as $2\theta + \theta$.

$\quad = \cos 2\theta \cos\theta - \sin 2\theta \sin\theta$ $cos(\alpha + \beta)$
$\qquad\qquad\qquad\qquad\qquad\qquad\qquad\qquad = cos\,\alpha\,cos\,\beta - sin\,\alpha\,sin\,\beta$

$\qquad\qquad \boxed{2\cos^2\theta - 1} \qquad \boxed{2\sin\theta\cos\theta}$ Substitute double-angle formulas. Because the right side of the given equation involves cosines only, use this form for $cos\,2\theta$.

$\quad = (2\cos^2\theta - 1)\cos\theta - 2\sin\theta\cos\theta\sin\theta$

$\quad = 2\cos^3\theta - \cos\theta - 2\sin^2\theta\cos\theta$ Multiply.

$\qquad\qquad\qquad\qquad \boxed{1 - \cos^2\theta}$

$\quad = 2\cos^3\theta - \cos\theta - 2(1 - \cos^2\theta)\cos\theta$ To get cosines only, use $sin^2\theta + cos^2\theta = 1$ and substitute $1 - cos^2\theta$ for $sin^2\theta$.

$\quad = 2\cos^3\theta - \cos\theta - 2\cos\theta + 2\cos^3\theta$ Multiply.

$\quad = 4\cos^3\theta - 3\cos\theta$ Simplify: $2\,cos^3\theta + 2\,cos^3\theta = 4\,cos^3\theta$ and $-cos\,\theta - 2\,cos\,\theta = -3\,cos\,\theta$.

By working with the left side and expressing it in a form identical to the right side, we have verified the identity.

Check Point 3 Verify the identity: $\sin 3\theta = 3\sin\theta - 4\sin^3\theta$.

2   Use the power-reducing formulas.

Power-Reducing Formulas

The double-angle formulas are used to derive the **power-reducing formulas**.

Power-Reducing Formulas

$$\sin^2\theta = \frac{1 - \cos 2\theta}{2} \qquad \cos^2\theta = \frac{1 + \cos 2\theta}{2} \qquad \tan^2\theta = \frac{1 - \cos 2\theta}{1 + \cos 2\theta}$$

We can prove the first two formulas in the box by working with two forms of the double-angle formula for $\cos 2\theta$.

| This is the form with sine only. | This is the form with cosine only. |

$$\cos 2\theta = 1 - 2\sin^2 \theta \qquad \cos 2\theta = 2\cos^2 \theta - 1$$

Solve the formula on the left for $\sin^2 \theta$. Solve the formula on the right for $\cos^2 \theta$.

$$2\sin^2 \theta = 1 - \cos 2\theta \qquad 2\cos^2 \theta = 1 + \cos 2\theta$$

$$\sin^2 \theta = \frac{1 - \cos 2\theta}{2} \qquad \cos^2 \theta = \frac{1 + \cos 2\theta}{2}$$

Divide both sides of each equation by 2.

These are the first two formulas in the box. The third formula in the box is proved by writing the tangent as the quotient of the sine and the cosine.

$$\tan^2 \theta = \frac{\sin^2 \theta}{\cos^2 \theta} = \frac{\dfrac{1 - \cos 2\theta}{2}}{\dfrac{1 + \cos 2\theta}{2}} = \frac{1 - \cos 2\theta}{\cancel{2}} \cdot \frac{\cancel{\dfrac{1}{2}}}{1 + \cos 2\theta} = \frac{1 - \cos 2\theta}{1 + \cos 2\theta}$$

Power-reducing formulas are quite useful in calculus. By reducing the power of trigonometric functions, calculus can better explore the relationship between a function and how it is changing at every single instant in time.

EXAMPLE 4 Reducing the Power of a Trigonometric Function

Write an equivalent expression for $\cos^4 x$ that does not contain powers of trigonometric functions greater than 1.

Solution We will apply the formula for $\cos^2 \theta$ twice.

$$\cos^4 x = \left(\cos^2 x\right)^2$$

$$= \left(\frac{1 + \cos 2x}{2}\right)^2 \qquad \text{Use } \cos^2 \theta = \frac{1 + \cos 2\theta}{2} \text{ with } \theta = x.$$

$$= \frac{1 + 2\cos 2x + \cos^2 2x}{4} \qquad \begin{array}{l}\text{Square the numerator:}\\ (A + B)^2 = A^2 + 2AB + B^2.\\ \text{Square the denominator.}\end{array}$$

$$= \frac{1}{4} + \frac{1}{2}\cos 2x + \frac{1}{4}\cos^2 2x \qquad \begin{array}{l}\text{Divide each term in the numerator}\\ \text{by 4.}\end{array}$$

We can reduce the power of $\cos^2 2x$ using

$$\cos^2 \theta = \frac{1 + \cos 2\theta}{2}$$

with $\theta = 2x$.

$$= \frac{1}{4} + \frac{1}{2}\cos 2x + \frac{1}{4}\left[\frac{1 + \cos 2(2x)}{2}\right] \qquad \begin{array}{l}\text{Use the power-reducing formula for}\\ \cos^2 \theta \text{ with } \theta = 2x.\end{array}$$

$$= \frac{1}{-} + \frac{1}{-}\cos 2x + \frac{1}{-}(1 + \cos 4x) \qquad \text{Multiply.}$$

Check Point 6 Verify the identity: $\tan\theta = \dfrac{\sin 2\theta}{1 + \cos 2\theta}$.

Half-angle formulas for $\tan\dfrac{\alpha}{2}$ can be obtained using the identities in Example 6 and the Check Point. Do you see how to do this? Replace each occurrence of θ by $\dfrac{\alpha}{2}$. This results in the following identities.

Half-Angle Formulas for $\tan\dfrac{\alpha}{2}$

$$\tan\frac{\alpha}{2} = \frac{1 - \cos\alpha}{\sin\alpha}$$

$$\tan\frac{\alpha}{2} = \frac{\sin\alpha}{1 + \cos\alpha}$$

EXAMPLE 7   Verifying an Identity

Verify the identity: $\tan\dfrac{\alpha}{2} = \csc\alpha - \cot\alpha$.

Solution   We begin with the right side.

$$\csc\alpha - \cot\alpha = \frac{1}{\sin\alpha} - \frac{\cos\alpha}{\sin\alpha} = \frac{1 - \cos\alpha}{\sin\alpha} = \tan\frac{\alpha}{2}$$

Express functions in terms of sines and cosines.

This is the first of the two half-angle formulas in the preceding box.

We worked with the right side and arrived at the left side. Thus, the identity is verified.

Check Point 7 Verify the identity: $\tan\dfrac{\alpha}{2} = \dfrac{\sec\alpha}{\sec\alpha\csc\alpha + \csc\alpha}$.

We conclude with a summary of the principal trigonometric identities developed in this section and the previous section. The fundamental identities can be found in the box on page 533.

Principal Trigonometric Identities

Sum and Difference Formulas

$$\sin(\alpha + \beta) = \sin\alpha\cos\beta + \cos\alpha\sin\beta \qquad \sin(\alpha - \beta) = \sin\alpha\cos\beta - \cos\alpha\sin\beta$$

$$\cos(\alpha + \beta) = \cos\alpha\cos\beta - \sin\alpha\sin\beta \qquad \cos(\alpha - \beta) = \cos\alpha\cos\beta + \sin\alpha\sin\beta$$

$$\tan(\alpha + \beta) = \frac{\tan\alpha + \tan\beta}{1 - \tan\alpha\tan\beta} \qquad \tan(\alpha - \beta) = \frac{\tan\alpha - \tan\beta}{1 + \tan\alpha\tan\beta}$$

Double-Angle Formulas

$$\sin 2\theta = 2\sin\theta\cos\theta$$

$$\cos 2\theta = \cos^2\theta - \sin^2\theta = 2\cos^2\theta - 1 = 1 - 2\sin^2\theta$$

$$\tan 2\theta = \frac{2\tan\theta}{1 - \tan^2\theta}$$

Power-Reducing Formulas

$$\sin^2\theta = \frac{1 - \cos 2\theta}{2} \qquad \cos^2\theta = \frac{1 + \cos 2\theta}{2} \qquad \tan^2\theta = \frac{1 - \cos 2\theta}{1 + \cos 2\theta}$$

Half-Angle Formulas

$$\sin\frac{\alpha}{2} = \pm\sqrt{\frac{1 - \cos\alpha}{2}} \qquad \cos\frac{\alpha}{2} = \pm\sqrt{\frac{1 + \cos\alpha}{2}}$$

$$\tan\frac{\alpha}{2} = \pm\sqrt{\frac{1 - \cos\alpha}{1 + \cos\alpha}} = \frac{1 - \cos\alpha}{\sin\alpha} = \frac{\sin\alpha}{1 + \cos\alpha}$$

Study Tip

To help remember the correct sign between the first two power-reducing formulas and the first two half-angle formulas, remember *sinus-minus*—the sine is minus.

EXERCISE SET 6.3

Practice Exercises

In Exercises 1–6, use the figures to find the exact value of the trigonometric function.

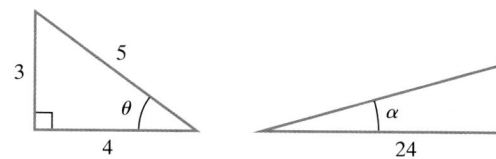

1. $\sin 2\theta$	**2.** $\cos 2\theta$	**3.** $\tan 2\theta$
4. $\sin 2\alpha$	**5.** $\cos 2\alpha$	**6.** $\tan 2\alpha$

In Exercises 7–14, use the given information to find the exact value of:

a. $\sin 2\theta$ **b.** $\cos 2\theta$ **c.** $\tan 2\theta$.

7. $\sin\theta = \frac{15}{17}$, θ lies in quadrant II.

8. $\sin\theta = \frac{12}{13}$, θ lies in quadrant II.

9. $\cos\theta = \frac{24}{25}$, θ lies in quadrant IV.

10. $\cos\theta = \frac{40}{41}$, θ lies in quadrant IV.

11. $\cot\theta = 2$, θ lies in quadrant III.

12. $\cot\theta = 3$, θ lies in quadrant III.

13. $\sin\theta = -\frac{9}{41}$, θ lies in quadrant III.

14. $\sin\theta = -\frac{2}{3}$, θ lies in quadrant III.

In Exercises 15–22, write each expression as the sine, cosine, or tangent of a double angle. Then find the exact value of the expression.

15. $2\sin 15°\cos 15°$

16. $2\sin 22.5°\cos 22.5°$

17. $\cos^2 75° - \sin^2 75°$

18. $\cos^2 105° - \sin^2 105°$

19. $2\cos^2\frac{\pi}{8} - 1$

20. $1 - 2\sin^2\frac{\pi}{12}$

21. $\dfrac{2\tan\dfrac{\pi}{12}}{1 - \tan^2\dfrac{\pi}{12}}$

22. $\dfrac{2\tan\dfrac{\pi}{8}}{1 - \tan^2\dfrac{\pi}{8}}$

In Exercises 23–34, verify each identity.

23. $\sin 2\theta = \dfrac{2\tan\theta}{1 + \tan^2\theta}$

24. $\sin 2\theta = \dfrac{2\cot\theta}{1 + \cot^2\theta}$

25. $(\sin\theta + \cos\theta)^2 = 1 + \sin 2\theta$

26. $(\sin\theta - \cos\theta)^2 = 1 - \sin 2\theta$

27. $\sin^2 x + \cos 2x = \cos^2 x$

SECTION 6.4 *Product-to-Sum and Sum-to-Product Formulas*

Objectives

1. Use the product-to-sum formulas.
2. Use the sum-to-product formulas.

Dennis Hopper. Shirley MacLaine. Cher. Roseanne. Patti LaBelle. Madonna. They all attended Elizabeth Taylor's sixty-fifth birthday party in 1997, raising a total of $1 million for AIDS causes. Hope it's not too late to send our birthday best and add:

$$112, 163\text{-}, 112, 196\text{-}, 110, 8521\text{-}, 008, 121\text{-}.$$

Relax, Roseanne. Bet you didn't know that each button on your touch-tone phone produces a unique sound. If we treat the commas as pauses and the hyphens as held notes, this sequence of numbers is *Happy Birthday* on a touch-tone phone.

Although *Happy Birthday* isn't Mozart or Sondheim, it is sinusoidal. Each of its touch-tone musical sounds can be described by the sum of two sine functions or the product of sines and cosines. In this section, we develop identities that enable us to use both descriptions. They are called the product-to-sum and sum-to-product formulas. We'll even apply this to musical sophistication at its best, including Mary's infamous lamb and those jingling bells.

1 Use the product-to-sum formulas.

The Product-to-Sum Formulas

How do we write the products of sines and/or cosines as sums or differences? We use the following identities, which are called **product-to-sum formulas.**

Study Tip

You may not need to memorize the formulas in this section. When you need them, you can either refer to one of the two boxes in the section or derive them using the methods shown.

Product-to-Sum Formulas

$$\sin\alpha\sin\beta = \tfrac{1}{2}\left[\cos(\alpha - \beta) - \cos(\alpha + \beta)\right]$$

$$\cos\alpha\cos\beta = \tfrac{1}{2}\left[\cos(\alpha - \beta) + \cos(\alpha + \beta)\right]$$

$$\sin\alpha\cos\beta = \tfrac{1}{2}\left[\sin(\alpha + \beta) + \sin(\alpha - \beta)\right]$$

$$\cos\alpha\sin\beta = \tfrac{1}{2}\left[\sin(\alpha + \beta) - \sin(\alpha - \beta)\right]$$

Although these formulas are difficult to remember, they are fairly easy to derive. For example, let's derive the first identity in the box,

$$\sin \alpha \sin \beta = \tfrac{1}{2}\left[\cos(\alpha - \beta) - \cos(\alpha + \beta)\right].$$

We begin with the difference and sum formulas for the cosine and subtract the second identity from the first:

$$\cos(\alpha - \beta) = \cos \alpha \cos \beta + \sin \alpha \sin \beta$$
$$-[\cos(\alpha + \beta) = \cos \alpha \cos \beta - \sin \alpha \sin \beta] \quad \textit{Subtract the identities.}$$
$$\cos(\alpha - \beta) - \cos(\alpha + \beta) = \quad 0 \quad + 2 \sin \alpha \sin \beta$$

Subtract terms on the left side.	*Subtract terms on the right side:* $\cos \alpha \cos \beta - \cos \alpha \cos \beta = 0.$	*Subtract terms on the right side:* $\sin \alpha \sin \beta - (-\sin \alpha \sin \beta) = 2 \sin \alpha \sin \beta.$

Now we use this result to derive the product-to-sum formula for $\sin \alpha \sin \beta$.

$$2 \sin \alpha \sin \beta = \cos(\alpha - \beta) - \cos(\alpha + \beta) \qquad \textit{Reverse the sides in the preceding equation.}$$

$$\sin \alpha \sin \beta = \tfrac{1}{2}\left[\cos(\alpha - \beta) - \cos(\alpha + \beta)\right] \qquad \textit{Multiply each side by } \tfrac{1}{2}.$$

This last equation is the desired formula. Likewise, we can derive the product-to-sum formula for cosine, $\cos \alpha \cos \beta = \tfrac{1}{2}\left[\cos(\alpha - \beta) + \cos(\alpha + \beta)\right]$. As we did for the previous derivation, begin with the difference and sum formulas for cosine. However, we *add* the formulas rather than subtracting them. Reversing both sides of this result and multiplying each side by $\tfrac{1}{2}$ produces the formula for $\cos \alpha \cos \beta$. The last two product-to-sum formulas, $\sin \alpha \cos \beta$ and $\cos \alpha \sin \beta$, are derived using the sum and difference formulas for sine in a similar manner.

EXAMPLE 1   Using the Product-to-Sum Formulas

Express each of the following products as a sum or difference.

a. $\sin 8x \sin 3x$ **b.** $\sin 4x \cos x$

Solution   The product-to-sum formula that we are using is shown in each of the voice balloons.

a. $\sin \alpha \sin \beta = \tfrac{1}{2}[\cos(\alpha - \beta) - \cos(\alpha + \beta)]$

$$\sin 8x \sin 3x = \tfrac{1}{2}\left[\cos(8x - 3x) - \cos(8x + 3x)\right] = \tfrac{1}{2}(\cos 5x - \cos 11x)$$

b. $\sin \alpha \cos \beta = \tfrac{1}{2}[\sin(\alpha + \beta) + \sin(\alpha - \beta)]$

$$\sin 4x \cos x = \tfrac{1}{2}\left[\sin(4x + x) + \sin(4x - x)\right] = \tfrac{1}{2}(\sin 5x + \sin 3x)$$

Check Point 1   Express each of the following products as a sum or difference.
a. $\sin 5x \sin 2x$ **b.** $\cos 7x \cos x$

Technology

The graphs of
$$y = \sin 8x \sin 3x$$
and
$$y = \tfrac{1}{2}(\cos 5x - \cos 11x)$$
are shown in a $\left[-2\pi, 2\pi, \dfrac{\pi}{2}\right]$ by $[-1, 1, 1]$ viewing rectangle. The graphs coincide. This verifies our algebraic work in Example 1(a).

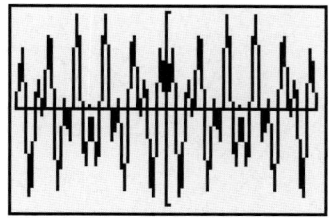

2 Use the sum-to-product formulas.

The Sum-to-Product Formulas

How do we write the sum or difference of sines and/or cosines as products? We use the following identities, which are called the **sum-to-product formulas.**

Sum-to-Product Formulas

$$\sin \alpha + \sin \beta = 2 \sin \frac{\alpha + \beta}{2} \cos \frac{\alpha - \beta}{2}$$

$$\sin \alpha - \sin \beta = 2 \sin \frac{\alpha - \beta}{2} \cos \frac{\alpha + \beta}{2}$$

$$\cos \alpha + \cos \beta = 2 \cos \frac{\alpha + \beta}{2} \cos \frac{\alpha - \beta}{2}$$

$$\cos \alpha - \cos \beta = -2 \sin \frac{\alpha + \beta}{2} \sin \frac{\alpha - \beta}{2}$$

We verify these formulas using the product-to-sum formulas. Let's verify the first sum-to-product formula

$$\sin \alpha + \sin \beta = 2 \sin \frac{\alpha + \beta}{2} \cos \frac{\alpha - \beta}{2}.$$

We start with the right side of the formula, the side with the product. We can apply the product-to-sum formula for $\sin \alpha \cos \beta$ to this expression. By doing so, we obtain the left side of the formula, $\sin \alpha + \sin \beta$. Here's how:

$$\sin \alpha \cos \beta = \tfrac{1}{2} [\sin(\alpha + \beta) + \sin(\alpha - \beta)]$$

$$2 \left[\sin \frac{\alpha + \beta}{2} \cos \frac{\alpha - \beta}{2} \right] = 2 \cdot \left[\frac{1}{2} \left[\sin \left(\frac{\alpha + \beta}{2} + \frac{\alpha - \beta}{2} \right) + \sin \left(\frac{\alpha + \beta}{2} - \frac{\alpha - \beta}{2} \right) \right] \right]$$

$$= \sin \left(\frac{\alpha + \beta + \alpha - \beta}{2} \right) + \sin \left(\frac{\alpha + \beta - \alpha + \beta}{2} \right)$$

$$= \sin \frac{2\alpha}{2} + \sin \frac{2\beta}{2} = \sin \alpha + \sin \beta$$

The three other sum-to-product formulas in the box are verified in a similar manner. Start with the right side and obtain the left side using an appropriate product-to-sum formula.

EXAMPLE 2 Using the Sum-to-Product Formulas

Express each sum or difference as a product.

    a. $\sin 9x + \sin 5x$ **b.** $\cos 4x - \cos 3x$

Solution The sum-to-product formula that we are using is shown in each of the voice balloons.

a. $\quad \sin \alpha + \sin \beta = 2 \sin \dfrac{\alpha+\beta}{2} \cos \dfrac{\alpha-\beta}{2}$

$$\sin 9x + \sin 5x = 2 \sin \frac{9x+5x}{2} \cos \frac{9x-5x}{2}$$

$$= 2 \sin \frac{14x}{2} \cos \frac{4x}{2}$$

$$= 2 \sin 7x \cos 2x$$

b. $\quad \cos \alpha - \cos \beta = -2 \sin \dfrac{\alpha+\beta}{2} \sin \dfrac{\alpha-\beta}{2}$

$$\cos 4x - \cos 3x = -2 \sin \frac{4x+3x}{2} \sin \frac{4x-3x}{2}$$

$$= -2 \sin \frac{7x}{2} \sin \frac{x}{2}$$

Check Point 2   Express each sum or difference as a product.
a. $\sin 7x + \sin 3x$ **b.** $\cos 3x + \cos 2x$

Some identities contain a fraction on one side with sums and differences of sines and/or cosines. Applying the sum-to-product formulas in the numerator and the denominator is often helpful in verifying these identities.

EXAMPLE 3   Using Sum-to-Product Formulas to Verify an Identity

Verify the identity: $\dfrac{\cos 3x - \cos 5x}{\sin 3x + \sin 5x} = \tan x.$

Solution   Because the left side is more complicated, we will work with it. We use sum-to-product formulas for the numerator and the denominator of the fraction on this side.

$$\frac{\cos 3x - \cos 5x}{\sin 3x + \sin 5x}$$

$\cos \alpha - \cos \beta = -2 \sin \dfrac{\alpha+\beta}{2} \sin \dfrac{\alpha-\beta}{2}$

$$= \frac{-2 \sin \dfrac{3x+5x}{2} \sin \dfrac{3x-5x}{2}}{\sin 3x + \sin 5x}$$

$\sin \alpha + \sin \beta = 2 \sin \dfrac{\alpha+\beta}{2} \cos \dfrac{\alpha-\beta}{2}$

$$= \frac{-2 \sin \dfrac{3x+5x}{2} \sin \dfrac{3x-5x}{2}}{2 \sin \dfrac{3x+5x}{2} \cos \dfrac{3x-5x}{2}}$$

$$= \frac{-2\sin\frac{8x}{2}\sin\left(\frac{-2x}{2}\right)}{2\sin\frac{8x}{2}\cos\left(\frac{-2x}{2}\right)}$$

Perform the indicated additions.

$$= \frac{-2\sin 4x \sin(-x)}{2\sin 4x \cos(-x)}$$

Simplify.

$$= \frac{-(-\sin x)}{\cos x}$$

The sine function is odd: $\sin(-x) = -\sin x$.
The cosine function is even: $\cos(-x) = \cos x$.

$$= \frac{\sin x}{\cos x}$$

Simplify.

$$= \tan x$$

Apply a quotient identity: $\tan x = \frac{\sin x}{\cos x}$.

We worked with the left side and arrived at the right side. Thus, the identity is verified.

Check Point 3   Verify the identity: $\dfrac{\cos 3x - \cos x}{\sin 3x + \sin x} = -\tan x$.

Sinusoidal Sounds

Music is all around us. A mere snippet of a song from the past can trigger vivid memories, inducing emotions ranging from unabashed joy to deep sorrow. Trigonometric functions can explain how sound travels from its source and describe its pitch, loudness, and quality. Still unexplained is the remarkable influence music has on the brain, including the deepest question of all: Why do we appreciate music?

When a note is played, it disturbs nearby air molecules, creating regions of higher-than-normal pressure and regions of lower-than-normal pressure. If we graph pressure, y, versus time, t, we get a sine wave that represents the note. The frequency of the sine wave is the number of high-low disturbances, or vibrations, per second. The greater the frequency, the higher the pitch; the lesser the frequency, the lower the pitch.

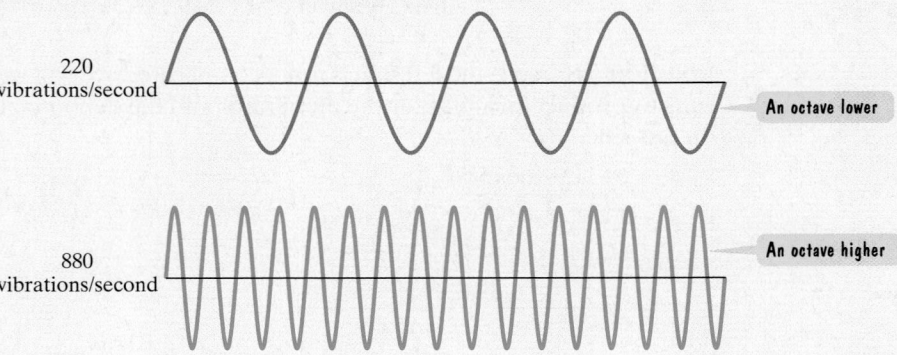

The amplitude of a note's sine wave is related to its loudness. The amplitude for the two sine waves shown is the same. Thus, the notes have the same loudness, although they differ in pitch. The greater the amplitude, the louder the sound; the lesser the amplitude, the softer the sound. The amplitude and frequency are characteristic of every note—and thus of its graph—until the note dissipates.

EXERCISE SET 6.4

 Practice Exercises

In Exercises 1–8, express each product as a sum or difference.

1. $\sin 6x \sin 2x$ **2.** $\sin 8x \sin 4x$

3. $\cos 7x \cos 3x$ **4.** $\cos 9x \cos 2x$

5. $\sin x \cos 2x$ **6.** $\sin 2x \cos 3x$

7. $\cos \dfrac{3x}{2} \sin \dfrac{x}{2}$ **8.** $\cos \dfrac{5x}{2} \sin \dfrac{x}{2}$

In Exercises 9–22, express each sum or difference as a product. If possible, find this product's exact value.

9. $\sin 6x + \sin 2x$ **10.** $\sin 8x + \sin 2x$

11. $\sin 7x - \sin 3x$ **12.** $\sin 11x - \sin 5x$

13. $\cos 4x + \cos 2x$ **14.** $\cos 9x - \cos 7x$

15. $\sin x + \sin 2x$ **16.** $\sin x - \sin 2x$

17. $\cos \dfrac{3x}{2} + \cos \dfrac{x}{2}$ **18.** $\sin \dfrac{3x}{2} + \sin \dfrac{x}{2}$

19. $\sin 75° + \sin 15°$ **20.** $\cos 75° - \cos 15°$

21. $\sin \dfrac{\pi}{12} - \sin \dfrac{5\pi}{12}$ **22.** $\cos \dfrac{\pi}{12} - \cos \dfrac{5\pi}{12}$

In Exercises 23–30, verify each identity.

23. $\dfrac{\sin 3x - \sin x}{\cos 3x - \cos x} = -\cot 2x$

24. $\dfrac{\sin x + \sin 3x}{\cos x + \cos 3x} = \tan 2x$

25. $\dfrac{\sin 2x + \sin 4x}{\cos 2x + \cos 4x} = \tan 3x$

26. $\dfrac{\cos 4x - \cos 2x}{\sin 2x - \sin 4x} = \tan 3x$

27. $\dfrac{\sin x - \sin y}{\sin x + \sin y} = \tan \dfrac{x - y}{2} \cot \dfrac{x + y}{2}$

28. $\dfrac{\sin x + \sin y}{\sin x - \sin y} = \tan \dfrac{x + y}{2} \cot \dfrac{x - y}{2}$

29. $\dfrac{\sin x + \sin y}{\cos x + \cos y} = \tan \dfrac{x + y}{2}$

30. $\dfrac{\sin x - \sin y}{\cos x - \cos y} = -\cot \dfrac{x + y}{2}$

In Exercises 31–34, the graph with the given equation is shown in a $\left[0, 2\pi, \dfrac{\pi}{2}\right]$ by $[-2, 2, 1]$ viewing rectangle.

a. *Describe the graph using another equation.*
b. *Verify that the two equations are equivalent.*

31.
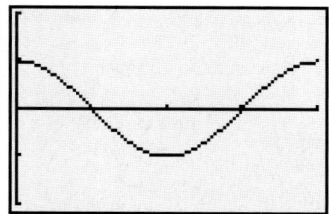
$$y = \frac{\sin x + \sin 3x}{2 \sin 2x}$$

32.
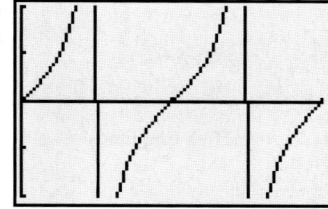
$$y = \frac{\cos x - \cos 3x}{\sin x + \sin 3x}$$

33.
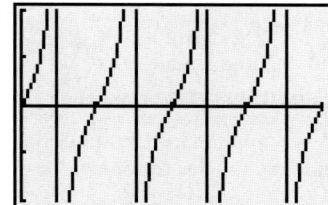
$$y = \frac{\cos x - \cos 5x}{\sin x + \sin 5x}$$

34.
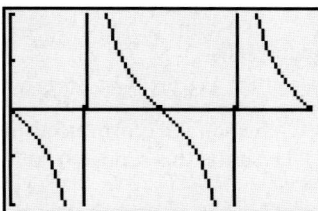
$$y = \frac{\cos 5x - \cos 3x}{\sin 5x + \sin 3x}$$

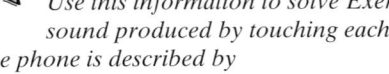

Application Exercises

Use this information to solve Exercises 35–36. The sound produced by touching each button on a touch-tone phone is described by

$$y = \sin 2\pi l t + \sin 2\pi h t,$$

where l and h are the low and high frequencies in the figure shown. For example, what sound is produced by touching 5? The low frequency is l = 770 cycles per second and the high frequency is h = 1336 cycles per second. The sound produced by touching 5 is described by

$$y = \sin 2\pi(770)t + \sin 2\pi(1336)t.$$

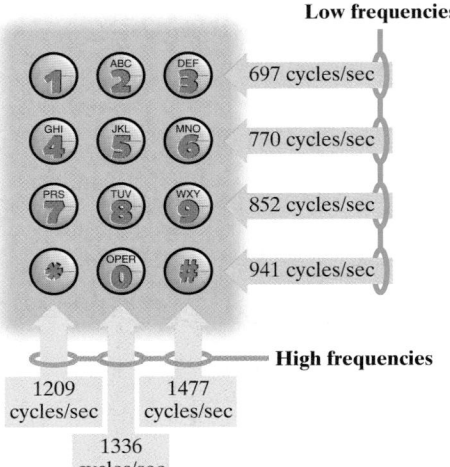

Low frequencies

697 cycles/sec
770 cycles/sec
852 cycles/sec
941 cycles/sec

High frequencies

1209 cycles/sec   1477 cycles/sec

1336 cycles/sec

35. The touch-tone phone sequence for that most naive of melodies is given as follows.

Mary Had A Little Lamb

3212333,222,399,3212333322321

  a. Many numbers do not appear in this sequence, including 7. If you accidently touch 7 for one of the notes, describe this sound as the sum of sines.

  b. Describe this accidental sound as a product of sines and cosines.

36. The touch-tone phone sequence for *Jingle Bells* is given as follows.

Jingle Bells

333,333,39123,666-663333322329,333,333,39123,666-6633,399621

  a. The first six notes of the song are produced by repeatedly touching 3. Describe this repeated sound as the sum of sines.

  b. Describe the repeated sound as a product of sines and cosines.

Writing in Mathematics

In Exercises 37–40, use words to describe the given formula.

37. $\sin \alpha \sin \beta = \frac{1}{2}[\cos(\alpha - \beta) - \cos(\alpha + \beta)]$

38. $\cos \alpha \cos \beta = \frac{1}{2}[\cos(\alpha - \beta) + \cos(\alpha + \beta)]$

39. $\sin \alpha + \sin \beta = 2 \sin \dfrac{\alpha + \beta}{2} \cos \dfrac{\alpha - \beta}{2}$

40. $\cos \alpha + \cos \beta = 2 \cos \dfrac{\alpha + \beta}{2} \cos \dfrac{\alpha - \beta}{2}$

41. Describe identities that can be verified using the sum-to-product formulas.

42. Why do the sounds produced by touching each button on a touch-tone phone have the same loudness? Answer the question using the equation described for Exercises 35 and 36, $y = \sin 2\pi l t + \sin 2\pi h t$, and determine the maximum value of y for each sound.

Technology Exercises

In Exercises 43–46, graph each side of the equation in the same viewing rectangle. If the graphs appear to coincide, verify that the equation is an identity. If the graphs do not appear to coincide, find a value of x for which both sides are defined but not equal.

43. $\sin x + \sin 2x = \sin 3x$

44. $\cos x + \cos 2x = \cos 3x$

45. $\sin x + \sin 3x = 2 \sin 2x \cos x$

46. $\cos x + \cos 3x = 2 \cos 2x \cos x$

47. In Exercise 35(a), you wrote an equation for the sound produced by touching 7 on a touch-tone phone. Graph the equation in a $[0, 0.01, 0.001]$ by $[-2, 2, 1]$ viewing rectangle.

48. In Exercise 36(a), you wrote an equation for the sound produced by touching 3 on a touch-tone phone. Graph the equation in a $[0, 0.01, 0.001]$ by $[-2, 2, 1]$ viewing rectangle.

49. In this section, we saw how sums could be expressed as products. Sums of trigonometric functions can also be used to describe functions that are not trigonometric. French mathematician Jean Fourier (1768–1830) showed that *any function* can be described by a series of trigonometric functions. For example, the basic linear function $f(x) = x$ can also be represented by

$$f(x) = 2\left(\frac{\sin x}{1} - \frac{\sin 2x}{2} + \frac{\sin 3x}{3} - \frac{\sin 4x}{4} + \cdots\right).$$

a. Graph

$$y = 2\left(\frac{\sin x}{1}\right),$$

$$y = 2\left(\frac{\sin x}{1} - \frac{\sin 2x}{2}\right),$$

$$y = 2\left(\frac{\sin x}{1} - \frac{\sin 2x}{2} + \frac{\sin 3x}{3}\right)$$

and

$$y = 2\left(\frac{\sin x}{1} - \frac{\sin 2x}{2} + \frac{\sin 3x}{3} - \frac{\sin 4x}{4}\right)$$

in a $\left[-\pi, \pi, \frac{\pi}{2}\right]$ by $[-3, 3, 1]$ viewing rectangle. What patterns do you observe?

b. Graph

$$y = 2\left(\frac{\sin x}{1} - \frac{\sin 2x}{2} + \frac{\sin 3x}{3} - \frac{\sin 4x}{4} + \frac{\sin 5x}{5} - \frac{\sin 6x}{6}\right.$$
$$\left. + \frac{\sin 7x}{7} - \frac{\sin 8x}{8} + \frac{\sin 9x}{9} - \frac{\sin 10x}{10}\right)$$

in a $\left[-\pi, \pi, \frac{\pi}{2}\right]$ by $[-3, 3, 1]$ viewing rectangle. Is a portion of the graph beginning to look like the graph of $f(x) = x$? Obtain a better approximation for the line by graphing functions that contain more and more terms involving sines of multiple angles.

c. Use

$$x = 2\left(\frac{\sin x}{1} - \frac{\sin 2x}{2} + \frac{\sin 3x}{3} - \frac{\sin 4x}{4} + \cdots\right)$$

and substitute $\frac{\pi}{2}$ for x to obtain a formula for $\frac{\pi}{2}$. Show at least four nonzero terms. Then multiply both sides of your formula by 2 to write a nonending series of subtractions and additions that approaches π. Use this series to obtain an approximation for π that is more accurate than the one given by your graphing utility.

Critical Thinking Exercises

Use the identities for $\sin(\alpha + \beta)$ and $\sin(\alpha - \beta)$ to solve Exercises 50–51.

50. Add the left and right sides of the identities and derive the product-to-sum formula for $\sin \alpha \cos \beta$.

51. Subtract the left and right sides of the identities and derive the product-to-sum formula for $\cos \alpha \sin \beta$.

In Exercises 52–53, verify the given sum-to-product formula. Start with the right side and obtain the expression on the left side by using an appropriate product-to-sum formula.

52. $\sin \alpha - \sin \beta = 2 \sin \dfrac{\alpha - \beta}{2} \cos \dfrac{\alpha + \beta}{2}$

53. $\cos \alpha + \cos \beta = 2 \cos \dfrac{\alpha + \beta}{2} \cos \dfrac{\alpha - \beta}{2}$

In Exercises 54–55, verify each identity.

54. $\dfrac{\sin 2x + (\sin 3x + \sin x)}{\cos 2x + (\cos 3x + \cos x)} = \tan 2x$

55. $4 \cos x \cos 2x \sin 3x = \sin 2x + \sin 4x + \sin 6x$

Group Exercise

56. This activity should result in an unusual group display entitled "*Frere Jacques*, a New Perspective." Here is the touch-tone phone sequence.

Frere Jacques

4564,4564,69#,69#,#*#964,#*#964,414,414

Group members should write every sound in the sequence as both the sum of sines and the product of sines and cosines. Use the sum of sines form and a graphing utility with a $[0, 0.01, 0.001]$ by $[-2, 2, 1]$ viewing rectangle to obtain a graph for every sound. Download these graphs. Use the graphs and equations to create your display in such a way that adults find the trigonometry of this naive melody interesting.

SECTION 6.5 *Trigonometric Equations*

Objectives

1. Find all solutions of a trigonometric equation.
2. Solve equations with multiple angles.
3. Solve trigonometric equations quadratic in form.
4. Use factoring to separate different functions in trigonometric equations.
5. Use identities to solve trigonometric equations.

Exponential functions display the manic energies of uncontrolled growth. By contrast, trigonometric functions repeat their behavior. Do they embody in their regularity some basic rhythm of the universe? The cycles of periodic phenomena provide events that we can comfortably count on. When will the moon look just as it does at this moment? When can I count on 13.5 hours of daylight? When will my breathing be exactly as it is right now? Models with trigonometric functions embrace the periodic rhythms of our world. Equations containing trigonometric functions are used to answer questions about these models.

1 Find all solutions of a trigonometric equation.

Trigonometric Equations and Their Solutions

A **trigonometric equation** is an equation that contains a trigonometric expression with a variable, such as $\sin x$. We have seen that some trigonometric equations are identities, such as $\sin^2 x + \cos^2 x = 1$. These equations are true for every value of the variable for which the expressions are defined. In this section, we consider trigonometric equations that are true for only some values of the variable. The values that satisfy the equation are its **solutions**. (There are trigonometric equations that have no solution.)

An example of a trigonometric equation is

$$\sin x = \tfrac{1}{2}.$$

A solution of this equation is $\dfrac{\pi}{6}$ because $\sin \dfrac{\pi}{6} = \dfrac{1}{2}$. By contrast, π is not a solution because $\sin \pi = 0 \neq \tfrac{1}{2}$.

Is $\dfrac{\pi}{6}$ the only solution of $\sin x = \dfrac{1}{2}$? The answer is no. Because of the periodic nature of the sine function, there are infinitely many values of x for which $\sin x = \dfrac{1}{2}$. Figure 6.7 on page 573 shows five of the solutions, including $\dfrac{\pi}{6}$, for $-\dfrac{3\pi}{2} \le x \le \dfrac{7\pi}{2}$. Notice that the x-coordinates of the points where the graph of $y = \sin x$ intersects the line $y = \tfrac{1}{2}$ are the solutions of the equation $\sin x = \tfrac{1}{2}$.

How do we represent all solutions to $\sin x = \tfrac{1}{2}$? Because the period of the sine function is 2π, first find all solutions in $[0, 2\pi)$. The solutions are

$$x = \frac{\pi}{6} \quad \text{and} \quad x = \pi - \frac{\pi}{6} = \frac{5\pi}{6}.$$

The sine is positive in quadrants I and II.

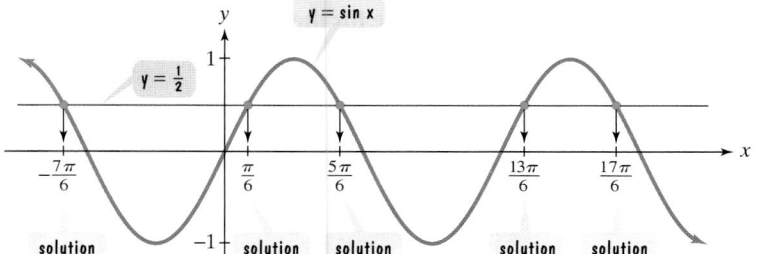

Figure 6.7 The equation $\sin x = \frac{1}{2}$ has five solutions when x is restricted to the interval $\left[-\frac{3\pi}{2}, \frac{7\pi}{2}\right]$.

Any multiple of 2π can be added to these values and the sine is still $\frac{1}{2}$. Thus, all solutions of $\sin x = \frac{1}{2}$ are given by

$$x = \frac{\pi}{6} + 2n\pi \quad \text{or} \quad x = \frac{5\pi}{6} + 2n\pi$$

where n is any integer. By choosing any two integers, such as $n = 0$ and $n = 1$, we can find some solutions of $\sin x = \frac{1}{2}$. Thus, four of the solutions are

Let $n = 0$.

$$x = \frac{\pi}{6} + 2 \cdot 0\pi \qquad x = \frac{5\pi}{6} + 2 \cdot 0\pi$$

$$= \frac{\pi}{6} \qquad\qquad = \frac{5\pi}{6}$$

Let $n = 1$.

$$x = \frac{\pi}{6} + 2 \cdot 1\pi \qquad x = \frac{5\pi}{6} + 2 \cdot 1\pi$$

$$= \frac{\pi}{6} + 2\pi \qquad\qquad = \frac{5\pi}{6} + 2\pi$$

$$= \frac{\pi}{6} + \frac{12\pi}{6} = \frac{13\pi}{6} \qquad = \frac{5\pi}{6} + \frac{12\pi}{6} = \frac{17\pi}{6}$$

These four solutions are shown among the five solutions in Figure 6.7.

Equations Involving a Single Trigonometric Function

To solve an equation containing a single trigonometric function:

- Isolate the function on one side of the equation.
- Solve for the variable.

EXAMPLE 1 Finding All Solutions of a Trigonometric Equation

Solve the equation: $\quad 3 \sin x - 2 = 5 \sin x - 1$.

Solution The equation contains a single trigonometric function, $\sin x$.

Step 1 Isolate the function on one side of the equation. We can solve for $\sin x$ by collecting all terms with $\sin x$ on the left side, and all the constant terms on the right side.

$$3 \sin x - 2 = 5 \sin x - 1 \qquad \text{This is the given equation.}$$

$$3 \sin x - 5 \sin x - 2 = 5 \sin x - 5 \sin x - 1 \qquad \text{Subtract } 5 \sin x \text{ from both sides.}$$

$$-2 \sin x - 2 = -1 \qquad \text{Simplify.}$$

$$-2 \sin x = 1 \qquad \text{Add 2 to both sides.}$$

$$\sin x = -\frac{1}{2} \qquad \text{Divide both sides by } -2 \text{ and solve for } \sin x.$$

Urban Canyons

A city's tall buildings and narrow streets reduce the amount of sunlight. If h is the average height of the buildings and w is the width of the street, the angle of elevation from the street to the top of the buildings is given by the trigonometric equation

$$\tan \theta = \frac{h}{w}.$$

A value of $\theta = 63°$ can result in an 85% loss of illumination.

Check Point 3   Solve the equation: $\sin\dfrac{x}{3} = \dfrac{1}{2}, 0 \le x < 2\pi$.

3 Solve trigonometric equations quadratic in form.

Trigonometric Equations Quadratic in Form

Some trigonometric equations are in the form of a quadratic equation $at^2 + bt + c = 0$, where t is a trigonometric function. Here are two examples of trigonometric equations that are quadratic in form.

$$2\cos^2 x + \cos x - 1 = 0 \qquad 2\sin^2 x - 3\sin x + 1 = 0$$

The form of this equation is $2t^2 + t - 1 = 0$ with $t = \cos x$. | The form of this equation is $2t^2 - 3t + 1 = 0$ with $t = \sin x$.

To solve this kind of equation, try using factoring. If the trigonometric expression does not factor, use the quadratic formula.

EXAMPLE 4 Solving a Trigonometric Equation Quadratic in Form

Solve the equation: $2\cos^2 x + \cos x - 1 = 0, \quad 0 \le x < 2\pi$.

Solution   The given equation is in quadratic form $2t^2 + t - 1 = 0$ with $t = \cos x$. Let us attempt to solve the equation using factoring.

$2\cos^2 x + \cos x - 1 = 0$ — This is the given equation.

$(2\cos x - 1)(\cos x + 1) = 0$ — Factor. Notice that $2t^2 + t - 1$ factors as $(2t - 1)(t + 1)$.

$2\cos x - 1 = 0$ or $\cos x + 1 = 0$ — Set each factor equal to 0.

$2\cos x = 1 \qquad\qquad \cos x = -1$ — Solve for $\cos x$.

$\cos x = \tfrac{1}{2}$

$x = \dfrac{\pi}{3} \qquad x = 2\pi - \dfrac{\pi}{3} = \dfrac{5\pi}{3} \qquad x = \pi$ — Solve each equation for x, $0 \le x < 2\pi$.

The cosine is positive in quadrants I and IV.

The solutions in the interval $[0, 2\pi)$ are $\dfrac{\pi}{3}$, π, and $\dfrac{5\pi}{3}$.

Technology

The graph of
$$y = 2\cos^2 x + \cos x - 1$$
is shown in a
$$\left[0, 2\pi, \dfrac{\pi}{2}\right] \text{ by } [-3, 3, 1]$$
viewing rectangle. The x-intercepts,
$$\dfrac{\pi}{3}, \pi, \text{ and } \dfrac{5\pi}{3},$$
verify the three solutions of
$$2\cos^2 x + \cos x - 1 = 0$$
in $[0, 2\pi)$.

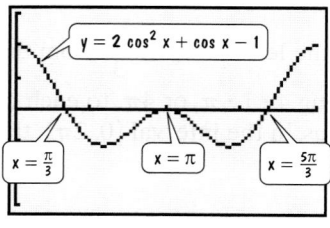

Check Point 4   Solve the equation: $2\sin^2 x - 3\sin x + 1 = 0, 0 \le x < 2\pi$.

4 Use factoring to separate different functions in trigonometric equations.

Using Factoring to Separate Two Different Trigonometric Functions in an Equation

We have seen that factoring is used to solve some trigonometric equations that are quadratic in form. Factoring can also be used to solve some trigonometric equations that contain two different functions such as

$$\tan x \sin^2 x = 3 \tan x.$$

In such a case, move all terms to one side and obtain zero on the other side. Then try to use factoring to separate the different functions. Example 5 shows how this is done.

EXAMPLE 5 Using Factoring to Separate Different Functions

Solve the equation: $\tan x \sin^2 x = 3 \tan x, \quad 0 \le x < 2\pi.$

Solution
Move all terms to one side and obtain zero on the other side.

$$\tan x \sin^2 x = 3 \tan x \quad \text{This is the given equation.}$$
$$\tan x \sin^2 x - 3 \tan x = 0 \quad \text{Subtract 3 tan x from both sides.}$$

Use factoring to separate the two functions.

$$\tan x(\sin^2 x - 3) = 0 \quad \text{Factor out tan x from the two terms on the left side.}$$

$$\tan x = 0 \quad \text{or} \quad \sin^2 x - 3 = 0 \quad \text{Set each factor equal to 0.}$$
$$x = 0 \quad x = \pi \qquad \sin^2 x = 3 \quad \text{Solve for x.}$$
$$\sin x = \pm\sqrt{3}$$

This equation has no solution because sin x cannot be greater than 1 or less than –1.

The solutions in the interval $[0, 2\pi)$ are 0 and π.

Study Tip

In solving
$$\tan x \sin^2 x = 3 \tan x,$$
do not begin by dividing both sides by tan x. Division by zero is undefined. If you divide by tan x, you lose the two solutions for which tan x = 0, namely 0 and π.

Check Point 5   Solve the equation: $\sin x \tan x = \sin x, \ 0 \le x < 2\pi.$

5 Use identities to solve trigonometric equations.

Using Identities to Solve Trigonometric Equations

Some trigonometric equations contain more than one function on the same side and these functions cannot be separated by factoring. For example, consider the equation

$$2\sin^2 x - \cos x - 1 = 0.$$

How can we obtain an equivalent equation that has only one trigonometric function? We use the identity $\sin^2 x + \cos^2 x = 1$ and substitute $1 - \cos^2 x$ for $\sin^2 x$.

$$2\sin^2 x - \cos x - 1 = 0 \quad \text{This is the given equation.}$$
$$2(1 - \cos^2 x) - \cos x - 1 = 0 \quad \sin^2 x = 1 - \cos^2 x$$
$$2 - 2\cos^2 x - \cos x - 1 = 0 \quad \text{Use the distributive property.}$$
$$-2\cos^2 x - \cos x + 1 = 0 \quad \text{Combine like terms.}$$

Multiplying both sides of the equation by -1, we obtain

$$2\cos^2 x + \cos x - 1 = 0.$$

This equivalent equation contains only the cosine function. This is the equation that we solved in Example 4.

EXAMPLE 6 Using an Identity to Solve a Trigonometric Equation

Solve the equation: $\cos 2x + 3\sin x - 2 = 0, 0 \le x < 2\pi$.

Solution   The given equation contains a cosine function and a sine function. The cosine is a function of $2x$ and the sine is a function of x. We want one trigonometric function of the same angle. This can be accomplished by using the double-angle identity $\cos 2x = 1 - 2\sin^2 x$ to obtain an equivalent equation involving $\sin x$ only.

$$\cos 2x + 3\sin x - 2 = 0 \qquad \text{This is the given equation.}$$

$$1 - 2\sin^2 x + 3\sin x - 2 = 0 \qquad \cos 2x = 1 - 2\sin^2 x$$

$$-2\sin^2 x + 3\sin x - 1 = 0 \qquad \text{Combine like terms.}$$

$$2\sin^2 x - 3\sin x + 1 = 0 \qquad \text{Multiply both sides by } -1.$$

The equation is now in quadratic form $2t^2 - 3t + 1 = 0$ with $t = \sin x$. We solve using factoring.

$$(2\sin x - 1)(\sin x - 1) = 0 \qquad \text{Factor. Notice that } 2t^2 - 3t + 1 \text{ factors as } (2t - 1)(t - 1).$$

$$2\sin x - 1 = 0 \qquad \text{or} \qquad \sin x - 1 = 0 \qquad \text{Set each factor equal to 0.}$$

$$\sin x = \tfrac{1}{2} \qquad\qquad\qquad \sin x = 1 \qquad \text{Solve for } \sin x.$$

$$x = \frac{\pi}{6} \quad x = \pi - \frac{\pi}{6} = \frac{5\pi}{6} \qquad x = \frac{\pi}{2} \qquad \begin{array}{l}\text{Solve each equation for } x, \\ 0 \le x < 2\pi.\end{array}$$

The solutions in the interval $[0, 2\pi)$ are $\dfrac{\pi}{6}, \dfrac{\pi}{2}$, and $\dfrac{5\pi}{6}$.

Check Point 6   Solve the equation: $\cos 2x + \sin x = 0, \ 0 \le x < 2\pi$.

Sometimes it is necessary to do something to both sides of a trigonometric equation to substitute an identity. For example, consider the equation

$$\sin x \cos x = \tfrac{1}{2}.$$

This equation contains both a sine and a cosine function. How can we obtain a single function? Multiply both sides by 2. In this way, we can use the double-angle identity $\sin 2x = 2\sin x \cos x$ and obtain $\sin 2x$, a single function, on the left side.

Technology

Shown below are the graphs of

$$y = \sin x \cos x$$

and

$$y = \tfrac{1}{2}$$

in a $\left[0, 2\pi, \dfrac{\pi}{2}\right]$ by $[-1, 1, 1]$ viewing rectangle. The solutions of

$$\sin x \cos x = \tfrac{1}{2}$$

are shown by the x-coordinates of the two intersection points.

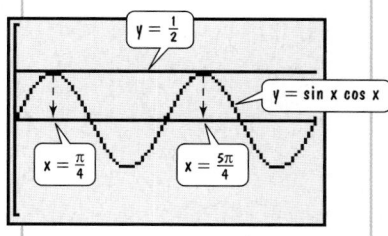

EXAMPLE 7 Using an Identity to Solve a Trigonometric Equation

Solve the equation: $\sin x \cos x = \tfrac{1}{2},\ 0 \le x < 2\pi.$

Solution

$$\sin x \cos x = \tfrac{1}{2} \qquad \text{This is the given equation.}$$

$$2 \sin x \cos x = 1 \qquad \text{Multiply both sides by 2 in anticipation of using } \sin 2x = 2 \sin x \cos x.$$

$$\sin 2x = 1 \qquad \text{Use a double-angle identity.}$$

Notice that we have an equation with $2x$, a multiple angle. The period of the sine function is 2π. In the interval $[0, 2\pi)$, the only value for which the sine function is 1 is $\dfrac{\pi}{2}$. This means that $2x = \dfrac{\pi}{2}$. Because the period is 2π, all the solutions to $\sin 2x = 1$ are given by

$$2x = \frac{\pi}{2} + 2n\pi \qquad \text{n is any integer.}$$

$$x = \frac{\pi}{4} + n\pi \qquad \text{Divide both sides by 2 and solve for x.}$$

The solutions in the interval $[0, 2\pi)$ are obtained by letting $n = 0$ and $n = 1$. The solutions are $\dfrac{\pi}{4}$ and $\dfrac{5\pi}{4}$.

Check Point 7   Solve the equation: $\sin x \cos x = -\tfrac{1}{2},\ 0 \le x < 2\pi.$

Let's look at another equation that contains two different functions, $\sin x - \cos x = 1$. Can you think of an identity that can be used to produce only one function? Perhaps $\sin^2 x + \cos^2 x = 1$ might be helpful. The next example shows how we use this identity by squaring both sides of the given equation. Remember that if we raise both sides of an equation to an even power, we have the possibility of introducing extraneous solutions. Thus, we must check each proposed solution in the given equation. Alternatively, we can use a graphing utility to verify actual solutions.

EXAMPLE 8 Using an Identity to Solve a Trigonometric Equation

Solve the equation: $\sin x - \cos x = 1, 0 \le x < 2\pi.$

Solution   We square both sides of the equation in anticipation of using $\sin^2 x + \cos^2 x = 1$.

$$\sin x - \cos x = 1 \qquad \text{This is the given equation.}$$

$$(\sin x - \cos x)^2 = 1^2 \qquad \text{Square both sides.}$$

$$\sin^2 x - 2 \sin x \cos x + \cos^2 x = 1 \qquad \text{Square the left side using } (A - B)^2 = A^2 - 2AB + B^2.$$

Technology

A graphing utility can be used instead of the algebraic check on the right. Shown are the graphs of

$$y = \sin x - \cos x$$

and

$$y = 1$$

in a $\left[0, 2\pi, \dfrac{\pi}{2}\right]$ by $[-2, 2, 1]$ viewing rectangle. The actual solutions of

$$\sin x - \cos x = 1$$

are shown by the x-coordinates of the two intersection points, $\dfrac{\pi}{2}$ and π.

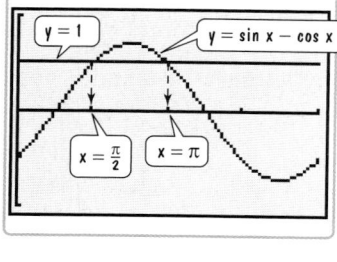

$y = 1$	$y = \sin x - \cos x$
$x = \dfrac{\pi}{2}$	$x = \pi$

$$\sin^2 x + \cos^2 x - 2\sin x \cos x = 1 \quad \text{Rearrange terms.}$$

$$1 - 2\sin x \cos x = 1 \quad \text{Apply a Pythagorean identity:}$$
$$\sin^2 x + \cos^2 x = 1.$$

$$-2\sin x \cos x = 0 \quad \text{Subtract 1 from both sides of the equation.}$$

$$\sin x \cos x = 0 \quad \text{Divide both sides of the equation by } -2.$$

$$\sin x = 0 \quad \text{or} \quad \cos x = 0 \quad \text{Set each factor equal to 0.}$$

$$x = 0 \quad x = \pi \quad x = \dfrac{\pi}{2} \quad x = \dfrac{3\pi}{2} \quad \text{Solve for } x \text{ in } [0, 2\pi).$$

We check these proposed solutions to see if any are extraneous.

Check 0:

$$\sin x - \cos x = 1$$
$$\sin 0 - \cos 0 \overset{?}{=} 1$$
$$0 - 1 \overset{?}{=} 1$$
$$-1 = 1 \text{ False}$$

Check $\dfrac{\pi}{2}$:

$$\sin x - \cos x = 1$$
$$\sin \dfrac{\pi}{2} - \cos \dfrac{\pi}{2} \overset{?}{=} 1$$
$$1 - 0 \overset{?}{=} 1$$
$$1 = 1 \text{ True}$$

Check π:

$$\sin x - \cos x = 1$$
$$\sin \pi - \cos \pi \overset{?}{=} 1$$
$$0 - (-1) \overset{?}{=} 1$$
$$1 = 1 \text{ True}$$

Check $\dfrac{3\pi}{2}$:

$$\sin x - \cos x = 1$$
$$\sin \dfrac{3\pi}{2} - \cos \dfrac{3\pi}{2} \overset{?}{=} 1$$
$$(-1) - 0 \overset{?}{=} 1$$
$$-1 = 1 \text{ False}$$

0 and $\dfrac{3\pi}{2}$ are extraneous.

The actual solutions in the interval $[0, 2\pi)$ are $\dfrac{\pi}{2}$ and π.

Check Point 8 Solve the equation: $\cos x - \sin x = -1, \ 0 \le x < 2\pi$.

EXERCISE SET 6.5

✓ Practice Exercises

In Exercises 1–10, use substitution to determine whether the given x-value is a solution of the equation.

1. $\cos x = \dfrac{\sqrt{2}}{2}, \quad x = \dfrac{\pi}{4}$

2. $\tan x = \sqrt{3}, \quad x = \dfrac{\pi}{3}$

3. $\sin x = \dfrac{\sqrt{3}}{2}, \quad x = \dfrac{\pi}{6}$

4. $\sin x = \dfrac{\sqrt{2}}{2}, \quad x = \dfrac{\pi}{3}$

5. $\cos x = -\dfrac{1}{2}, \quad x = \dfrac{2\pi}{3}$

6. $\cos x = -\dfrac{1}{2}, \quad x = \dfrac{4\pi}{3}$

7. $\tan 2x = -\dfrac{\sqrt{3}}{3}, \quad x = \dfrac{5\pi}{12}$

8. $\cos \dfrac{2x}{3} = -\dfrac{1}{2}, \quad x = \pi$

9. $\cos x = \sin 2x, x = \dfrac{\pi}{3}$

10. $\cos x + 2 = \sqrt{3} \sin x, \quad x = \dfrac{\pi}{6}$

In Exercises 11–24, find all solutions of each equation.

11. $\sin x = \dfrac{\sqrt{3}}{2}$

12. $\cos x = \dfrac{\sqrt{3}}{2}$

13. $\tan x = 1$

14. $\tan x = \sqrt{3}$

15. $\cos x = -\dfrac{1}{2}$

16. $\sin x = -\dfrac{\sqrt{2}}{2}$

17. $\tan x = 0$

18. $\sin x = 0$

19. $2 \cos x + \sqrt{3} = 0$

20. $2 \sin x + \sqrt{3} = 0$

21. $4 \sin \theta - 1 = 2 \sin \theta$

22. $5 \sin \theta + 1 = 3 \sin \theta$

23. $3 \sin \theta + 5 = -2 \sin \theta$

24. $7 \cos \theta + 9 = -2 \cos \theta$

Exercises 25–38 involve equations with multiple angles. Solve each equation on the interval $[0, 2\pi)$.

25. $\sin 2x = \dfrac{\sqrt{3}}{2}$

26. $\cos 2x = \dfrac{\sqrt{2}}{2}$

27. $\cos 4x = -\dfrac{\sqrt{3}}{2}$

28. $\sin 4x = -\dfrac{\sqrt{2}}{2}$

29. $\tan 3x = \dfrac{\sqrt{3}}{3}$

30. $\tan 3x = \sqrt{3}$

31. $\tan \dfrac{x}{2} = \sqrt{3}$

32. $\tan \dfrac{x}{2} = \dfrac{\sqrt{3}}{3}$

33. $\sin \dfrac{2\theta}{3} = -1$

34. $\cos \dfrac{2\theta}{3} = -1$

35. $\sec \dfrac{3\theta}{2} = -2$

36. $\cot \dfrac{3\theta}{2} = -\sqrt{3}$

37. $\sin\left(2x + \dfrac{\pi}{6}\right) = \dfrac{1}{2}$

38. $\sin\left(2x - \dfrac{\pi}{4}\right) = \dfrac{\sqrt{2}}{2}$

Exercises 39–46 involve trigonometric equations quadratic in form. Solve each equation on the interval $[0, 2\pi)$.

39. $2\sin^2 x - \sin x - 1 = 0$

40. $2\sin^2 x + \sin x - 1 = 0$

41. $2\cos^2 x + 3\cos x + 1 = 0$

42. $\cos^2 x + 2\cos x - 3 = 0$

43. $2\sin^2 x = \sin x + 3$

44. $2\sin^2 x = 4\sin x + 6$

45. $\sin^2 \theta - 1 = 0$

46. $\cos^2 \theta - 1 = 0$

In Exercises 47–56, solve each equation on the interval $[0, 2\pi)$.

47. $(\tan x - 1)(\cos x + 1) = 0$

48. $(\tan x + 1)(\sin x - 1) = 0$

49. $(2\cos x + \sqrt{3})(2\sin x + 1) = 0$

50. $(2\cos x - \sqrt{3})(2\sin x - 1) = 0$

51. $\cot x(\tan x - 1) = 0$

52. $\cot x(\tan x + 1) = 0$

53. $\sin x + 2\sin x \cos x = 0$

54. $\cos x - 2\sin x \cos x = 0$

55. $\tan^2 x \cos x = \tan^2 x$

56. $\cot^2 x \sin x = \cot^2 x$

In Exercises 57–78, use an identity to solve each equation on the interval $[0, 2\pi)$.

57. $2\cos^2 x + \sin x - 1 = 0$

58. $2\cos^2 x - \sin x - 1 = 0$

59. $\sin^2 x - 2\cos x - 2 = 0$

60. $4\sin^2 x + 4\cos x - 5 = 0$

61. $4\cos^2 x = 5 - 4\sin x$

62. $3\cos^2 x = \sin^2 x$

63. $\sin 2x = \cos x$

64. $\sin 2x = \sin x$

65. $\cos 2x = \cos x$

66. $\cos 2x = \sin x$

67. $\cos 2x + 5\cos x + 3 = 0$

68. $\cos 2x + \cos x + 1 = 0$

69. $\sin x \cos x = \dfrac{\sqrt{2}}{4}$

70. $\sin x \cos x = \dfrac{\sqrt{3}}{4}$

71. $\sin x + \cos x = 1$

72. $\sin x + \cos x = -1$

73. $\sin\left(x + \dfrac{\pi}{4}\right) + \sin\left(x - \dfrac{\pi}{4}\right) = 1$

74. $\sin\left(x + \dfrac{\pi}{3}\right) + \sin\left(x - \dfrac{\pi}{3}\right) = 1$

75. $\sin 2x \cos x + \cos 2x \sin x = \dfrac{\sqrt{2}}{2}$

76. $\sin 3x \cos 2x + \cos 3x \sin 2x = 1$

77. $\tan x + \sec x = 1$

78. $\tan x - \sec x = 1$

 Application Exercises

Use this information to solve Exercises 79–80. Our cycle of normal breathing takes place every 5 seconds. Velocity of air flow, y, measured in liters per second, after x seconds is modeled by

$$y = 0.6 \sin \dfrac{2\pi}{5} x.$$

Velocity of air flow is positive when we inhale and negative when we exhale.

79. Within each breathing cycle, when are we inhaling at 0.3 liter per second? Round to the nearest tenth of a second.

80. Within each breathing cycle, when are we exhaling at 0.3 liter per second? Round to the nearest tenth of a second.

Use this information to solve Exercises 81–82. The number of hours of daylight in Boston is given by

$$y = 3 \sin\left[\dfrac{2\pi}{365}\right](x - 79) + 12$$

where x is the number of days after January 1.

81. How many days after January 1 does Boston have 10.5 hours of daylight? Round to the nearest day.

82. How many days after January 1 does Boston have 13.5 hours of daylight? Round to the nearest day.

Use this information to solve Exercises 83–84. A ball on a spring is pulled 4 inches below its rest position and then released. After t seconds, the ball's distance from its rest position is given by

$$d = -4 \cos \dfrac{\pi}{3} t.$$

83. Find all values of t for which the ball is 2 inches above its rest position.

84. Find all values of t for which the ball is 2 inches below its rest position.

Use this information to solve Exercises 85–86. When throwing an object, the distance achieved depends on initial velocity, v_o, and the angle above the horizontal at which the *object is thrown, θ. The distance, d, in feet, that describes the range covered is given by*

$$d = \frac{v_o^2}{16} \sin\theta \cos\theta,$$

where v_o is measured in feet per second.

85. You and your friend are throwing a baseball back and forth. If you throw the ball with an initial velocity of $v_o = 90$ feet per second, at what angle of elevation, θ, should you direct your throw so that it can be easily caught by your friend located 170 feet away?

86. In Exercise 85, you increase the distance between you and your friend to 200 feet. With this increase, at what angle of elevation, θ, should you direct your throw?

Writing in Mathematics

87. What are the solutions of a trigonometric equation?

88. Describe the difference between verifying a trigonometric identity and solving a trigonometric equation.

89. Without actually solving the equation, describe how to solve

$$3\tan x - 2 = 5\tan x - 1.$$

90. In the interval $[0, 2\pi)$, the solutions of $\sin x = \cos 2x$ are $\frac{\pi}{6}, \frac{5\pi}{6}$, and $\frac{3\pi}{2}$. Explain how to use graphs generated by a graphing utility to check these solutions.

91. Suppose you are solving equations in the interval $[0, 2\pi)$. Without actually solving equations, what is the difference between the number of solutions of $\sin x = \frac{1}{2}$ and $\sin 2x = \frac{1}{2}$? How do you account for this difference?

In Exercises 92–93, describe a general strategy for solving each equation. Do not solve the equation.

92. $2\sin^2 x + 5\sin x + 3 = 0$

93. $\sin 2x = \sin x$

94. Describe a natural periodic phenomenon. Give an example of a question that can be answered by a trigonometric equation in the study of this phenomenon.

95. Some people experience depression with loss of sunlight. Use the essay on page 573 to determine whether such a person should live on a city street that is 80 feet wide with buildings whose height averages 400 feet. Explain your answer and include θ, to the nearest degree, in your argument.

 Technology Exercises

96. Use a graphing utility to verify the solutions of any five equations that you solved in Exercises 57–78.

In Exercises 97–101, use a graphing utility to approximate the solutions of each equation in the interval $[0, 2\pi)$. Round to the nearest hundredth of a radian.

97. $15\cos^2 x + 7\cos x - 2 = 0$

98. $\cos x = x$

99. $2\sin^2 x = 1 - 2\sin x$

100. $\sin 2x = 2 - x^2$

101. $\sin x + \sin 2x + \sin 3x = 0$

Critical Thinking Exercises

102. Which one of the following is true?
  a. The equation $(\sin x - 3)(\cos x + 2) = 0$ has no solution.
  b. The equation $\tan x = \frac{\pi}{2}$ has no solution.
  c. A trigonometric equation with an infinite number of solutions is an identity.
  d. The equations $\sin 2x = 1$ and $\sin 2x = \frac{1}{2}$ have the same number of solutions on the interval $[0, 2\pi)$.

In Exercises 103–105, solve each equation on the interval $[0, 2\pi)$. Do not use a calculator.

103. $2\cos x - 1 + 3\sec x = 0$

104. $\sin 3x + \sin x + \cos x = 0$

105. $\sin x + 2\sin\frac{x}{2} = \cos\frac{x}{2} + 1$

CHAPTER SUMMARY, REVIEW, AND TEST

Summary

6.1 Verifying Trigonometric Identities

  a. Identities are trigonometric equations that are true for all values of the variable for which the expressions are defined.

  b. Fundamental trigonometric identities are given in the box on page 533.

  c. Guidelines for verifying trigonometric identities are given in the box on page 541.

6.2 and 6.3 Sum, Difference, Double-Angle, and Half-Angle Formulas

a. Sum and difference formulas, double-angle formulas, power-reducing formulas, and half-angle formulas are given in the box on page 561.

6.4 Product-to-Sum and Sum-to-Product Formulas

a. The product-to-sum formulas are given in the box on page 564.

b. The sum-to-product formulas are given in the box on page 566. These formulas are useful to verify identities with fractions that contain sums and differences of sines and/or cosines.

6.5 Trigonometric Equations

a. The values that satisfy a trigonometric equation are its solutions.

b. Algebraic techniques such as isolating an expression on one side of the equation and factoring are useful in solving trigonometric equations. Identities are also used to solve some trigonometric equations.

Review Exercises

6.1

In Exercises 1–12, verify each identity.

1. $\sec x - \cos x = \tan x \sin x$

2. $\cos x + \sin x \tan x = \sec x$

3. $\sin^2 \theta (1 + \cot^2 \theta) = 1$

4. $(\sec \theta - 1)(\sec \theta + 1) = \tan^2 \theta$

5. $\dfrac{1}{\sin t - 1} + \dfrac{1}{\sin t + 1} = -2 \tan t \sec t$

6. $\dfrac{1 + \sin t}{\cos^2 t} = \tan^2 t + 1 + \tan t \sec t$

7. $\dfrac{\cos x}{1 - \sin x} = \dfrac{1 + \sin x}{\cos x}$

8. $1 - \dfrac{\cos^2 x}{1 + \sin x} = \sin x$

9. $(\tan \theta + \cot \theta)^2 = \sec^2 \theta + \csc^2 \theta$

10. $\dfrac{1}{\sin \theta + \cos \theta} + \dfrac{1}{\sin \theta - \cos \theta} = \dfrac{2 \sin \theta}{\sin^4 \theta - \cos^4 \theta}$

11. $\dfrac{\cos t}{\cot t - 5 \cos t} = \dfrac{1}{\csc t - 5}$

12. $\dfrac{1 - \cos t}{1 + \cos t} = (\csc t - \cot t)^2$

6.2 and 6.3

In Exercises 13–18, use a sum or difference formula to find the exact value of each expression.

13. $\cos(45° + 30°)$

14. $\sin 195°$

15. $\tan\left(\dfrac{4\pi}{3} - \dfrac{\pi}{4}\right)$

16. $\tan \dfrac{5\pi}{12}$

17. $\cos 65° \cos 5° + \sin 65° \sin 5°$

18. $\sin 80° \cos 50° - \cos 80° \sin 50°$

In Exercises 19–30, verify each identity.

19. $\sin\left(x + \dfrac{\pi}{6}\right) - \cos\left(x + \dfrac{\pi}{3}\right) = \sqrt{3} \sin x$

20. $\tan\left(x + \dfrac{3\pi}{4}\right) = \dfrac{\tan x - 1}{1 + \tan x}$

21. $\sec(\alpha + \beta) = \dfrac{\sec \alpha \sec \beta}{1 - \tan \alpha \tan \beta}$

22. $\dfrac{\cos(\alpha - \beta)}{\cos \alpha \cos \beta} = 1 + \tan \alpha \tan \beta$

23. $\cos^4 t - \sin^4 t = \cos 2t$

24. $\sin t - \cos 2t = (2 \sin t - 1)(\sin t + 1)$

25. $\dfrac{\sin 2\theta - \sin \theta}{\cos 2\theta + \cos \theta} = \dfrac{1 - \cos \theta}{\sin \theta}$

26. $\dfrac{\sin 2\theta}{1 - \sin^2 \theta} = 2 \tan \theta$

27. $\tan 2t = 2 \sin t \cos t \sec 2t$

28. $\cos 4t = 1 - 8 \sin^2 t \cos^2 t$

29. $\tan \dfrac{x}{2}(1 + \cos x) = \sin x$

30. $\tan \dfrac{x}{2} = \dfrac{\sec x - 1}{\tan x}$

In Exercises 31–33, the graph with the given equation is shown in a $\left[0, 2\pi, \dfrac{\pi}{2}\right]$ by $[-2, 2, 1]$ viewing rectangle.

a. *Describe the graph using another equation.*

b. *Verify that the two equations are equivalent.*

31.

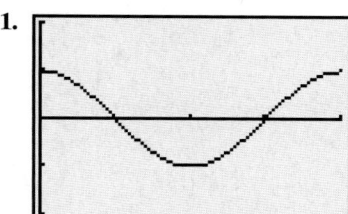

$$y = \sin\left(x - \dfrac{3\pi}{2}\right)$$

32.

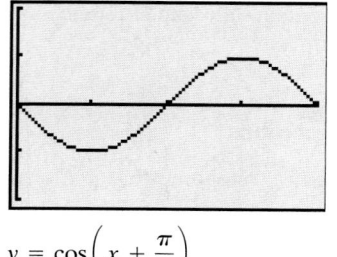

$$y = \cos\left(x + \frac{\pi}{2}\right)$$

33.

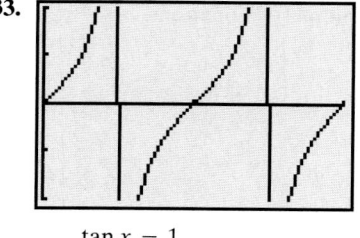

$$y = \frac{\tan x - 1}{1 - \cot x}$$

In Exercises 34–37, find the exact value of the following under the given conditions:

  a. $\sin(\alpha + \beta)$ **b.** $\cos(\alpha - \beta)$ **c.** $\tan(\alpha + \beta)$

  d. $\sin 2\alpha$ **e.** $\cos \dfrac{\beta}{2}$

34. $\sin\alpha = \frac{3}{5}$, α lies in quadrant I, and $\sin\beta = \frac{12}{13}$, β lies in quadrant II.

35. $\tan\alpha = \frac{4}{3}$, α lies in quadrant III, and $\tan\beta = \frac{5}{12}$, β lies in quadrant I.

36. $\tan\alpha = -3$, α lies in quadrant II, and $\cot\beta = -3$, β lies in quadrant III.

37. $\sin\alpha = -\frac{1}{3}$, α lies in quadrant III, and $\cos\beta = -\frac{1}{3}$, β lies in quadrant IV.

In Exercises 38–41, use double- and half-angle formulas to find the exact value of each expression.

38. $\cos^2 15° - \sin^2 15°$

39. $\dfrac{2\tan\dfrac{5\pi}{12}}{1 - \tan^2\dfrac{5\pi}{12}}$

40. $\sin 22.5°$

41. $\tan\dfrac{\pi}{12}$

6.4

In Exercises 42–43, express each product as a sum or difference.

42. $\sin 6x \sin 4x$

43. $\sin 7x \cos 3x$

In Exercises 44–45, express each sum or difference as a product. If possible, find this product's exact value.

44. $\sin 2x - \sin 4x$

45. $\cos 75° + \cos 15°$

In Exercises 46–47, verify each identity.

46. $\dfrac{\cos 3x + \cos 5x}{\cos 3x - \cos 5x} = \cot x \cot 4x$

47. $\dfrac{\sin 2x + \sin 6x}{\sin 2x - \sin 6x} = -\tan 4x \cot 2x$

48. The graph with the given equation is shown in a $\left[0, 2\pi, \dfrac{\pi}{2}\right]$ by $[-2, 2, 1]$ viewing rectangle.

$$y = \frac{\cos 3x + \cos x}{\sin 3x - \sin x}$$

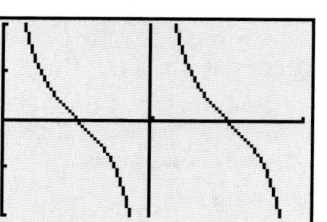

  a. Describe the graph using another equation.
  b. Verify that the two equations are equivalent.

6.5

In Exercises 49–52, find all solutions of each equation.

49. $\cos x = -\frac{1}{2}$

50. $\sin x = \dfrac{\sqrt{2}}{2}$

51. $2\sin x + 1 = 0$

52. $\sqrt{3}\tan x - 1 = 0$

In Exercises 53–62, solve each equation on the interval $[0, 2\pi)$.

53. $\cos 2x = -1$

54. $\sin 3x = 1$

55. $\tan\dfrac{x}{2} = -1$

56. $\tan x = 2\cos x \tan x$

57. $\cos^2 x - 2\cos x = 3$

58. $3\cos^2 x + \sin x = 1$

59. $4\sin^2 x = 1$

60. $\cos 2x + 2\cos x = 0$

61. $\sin 2x = \sqrt{3}\sin x$

62. $\sin x = \tan x$

63. A ball on a spring is pulled 6 inches below its rest position and then released. After t seconds, the ball's distance from its rest position is given by

$$d = -6\cos\frac{\pi}{2}t.$$

Find all values of t for which the ball is 3 inches below its rest position.

64. You are playing catch with a friend located 100 feet away. If you throw the ball with an initial velocity of $v_o = 90$ feet per second, at what angle of elevation θ should you direct your throw so that it can be caught easily? Use the formula

$$d = \frac{v_o^2}{16}\sin\theta\cos\theta.$$

Chapter 6 Test

Use the following conditions to solve Exercises 1–4.

$$\sin \alpha = \tfrac{4}{5}, \ \alpha \text{ lies in quadrant II}$$

and $\qquad \cos \beta = \tfrac{5}{13}, \beta \text{ lies in quadrant I.}$

Find the exact value of:

1. $\cos(\alpha + \beta)$ $\qquad\qquad$ **2.** $\tan(\alpha - \beta)$

3. $\sin 2\alpha$ $\qquad\qquad\qquad$ **4.** $\cos \dfrac{\beta}{2}$

5. Use $105° = 135° - 30°$ to find the exact value of $\sin 105°$.

In Exercises 6–11, verify each identity.

6. $\cos x \csc x = \cot x$ $\qquad$ **7.** $\dfrac{\sec x}{\cot x + \tan x} = \sin x$

8. $1 - \dfrac{\cos^2 x}{1 + \sin x} = \sin x$ $\qquad$ **9.** $\cos\left(\theta + \dfrac{\pi}{2}\right) = -\sin \theta$

10. $\dfrac{\sin(\alpha - \beta)}{\sin \alpha \cos \beta} = 1 - \cot \alpha \tan \beta$

11. $\sin t \cos t (\tan t + \cot t) = 1$

In Exercises 12–15, solve each equation on the interval $[0, 2\pi)$.

12. $\sin 3x = -\tfrac{1}{2}$ $\qquad\qquad$ **13.** $\sin 2x + \cos x = 0$

14. $2 \cos^2 x - 3 \cos x + 1 = 0$

15. $2 \sin^2 x + \cos x = 1$

16. If $f(x) = \dfrac{2x + 1}{x - 3}$, find $f^{-1}(x)$.

17. If C is a right angle in triangle ABC with $A = 23°$ and $a = 12$, solve the triangle.

18. A formula for calculating an infant's dosage for medication is

$$\text{Infant's dose} = \frac{\text{age of infant in months}}{150} \times \text{adult dose.}$$

If a 12-month-old infant is to receive 8.5 mg of medication, find the equivalent adult dose to the nearest milligram.

19. From a point on the ground 12 feet from the base of a flagpole, the angle of elevation to the top of the pole is 53°. Approximate the height of the flagpole to the nearest tenth of a foot.

20. In *A Tour of the Calculus*, David Berlinski describes trigonometric identities in the following way. "An invisible inner connection exists among the trigonometric functions, one revealed in various identities, strange places where the trigonometric functions appear fluidly to exchange identities or to resolve themselves into unlikely numbers." What does Berlinski mean by this? Use two fundamental trigonometric identities to illustrate your answer.

Cumulative Review Exercises (Chapters 1–6)

Solve each equation or inequality in Exercises 1–4.

1. $x^3 + x^2 - x + 15 = 0$ $\qquad$ **2.** $11^{x-1} = 125$

3. $x^2 + 2x - 8 > 0$

4. $\cos 2x + 3 = 5 \cos x, \quad 0 \le x < 2\pi$

In Exercises 5–10, graph each equation.

5. $y = \sqrt{x + 2} - 1$; Use transformations of the graph of $y = \sqrt{x}$.

6. $(x - 1)^2 + (y + 2)^2 = 9$

7. $y + 2 = \tfrac{1}{3}(x - 1)$

8. $y = 3 \cos 2x, \quad -2\pi \le x \le 2\pi$

9. $y = 2 \sin \dfrac{x}{2} + 1, \quad -2\pi \le x \le 2\pi$

10. $f(x) = (x - 1)^2 (x - 3)$

11. If $f(x) = x^2 + 3x - 1$, find $\dfrac{f(a + h) - f(a)}{h}$.

12. Find the exact value of $\sin 225°$.

13. Verify the identity: $\sec^4 x - \sec^2 x = \tan^4 x + \tan^2 x$.

14. Convert $320°$ to radians.

15. How long would it take for any amount of money, compounded continuously at 5.75% per year, to triple? Round to the nearest tenth of a year.

Additional Topics in Trigonometry

These days, computers and trigonometric functions are everywhere. Trigonometry plays a critical role in analyzing the forces that surround your every move. Using trigonometry to understand how forces are measured is one of the topics in this chapter that focuses on additional applications of trigonometry.

You enjoy running, although lately you experience discomfort at various points of impact. Your doctor suggests a computer analysis. By attaching sensors to your running shoes as you jog along a treadmill, the computer provides a printout of the magnitude and direction of the forces as your feet hit the ground. Based on this analysis, customized orthotics can be made to fit inside your shoes to minimize the impact of your feet with the ground.

SECTION 7.1   *The Law of Sines*

Objectives

1. Use the Law of Sines to solve oblique triangles.
2. Use the Law of Sines to solve, if possible, the triangle or triangles in the ambiguous case.
3. Find the area of an oblique triangle using the sine function.
4. Solve applied problems using the Law of Sines.

Point Reyes National Seashore, 40 miles north of San Francisco, consists of 75,000 acres with miles of pristine surf-pummeled beaches, forested ridges, and bays flanked by white cliffs. A few people, inspired by nature in the raw, live on private property adjoining the National Seashore. In 1995, a fire in the park covered 12,350 acres and destroyed 45 of their homes.

Fire is a necessary part of the life cycle in many wilderness areas. It is also an ongoing threat to those kept inspired and alive on private paradises surrounded by nature's unspoiled beauty. In this section, we see how trigonometry can be used to locate small wilderness fires before they become major tragedies. To do this, we begin by considering triangles other than right triangles.

The Law of Sines and Its Derivation

An **oblique triangle** is a triangle that does not contain a right angle. Figure 7.1 shows that an oblique triangle has either three acute angles or two acute angles and one obtuse angle. Notice that the angles are labeled A, B, and C. The sides opposite each angle are labeled as a, b, and c, respectively.

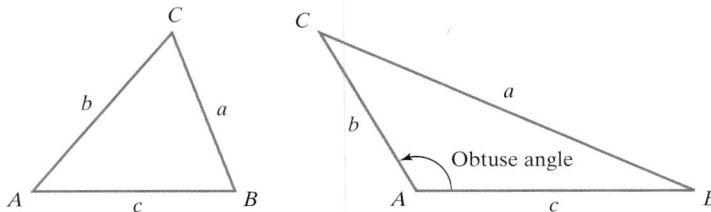

Figure 7.1 Oblique triangles

Many relationships exist among the sides and angles in an oblique triangle. One such relationship is called the **Law of Sines**.

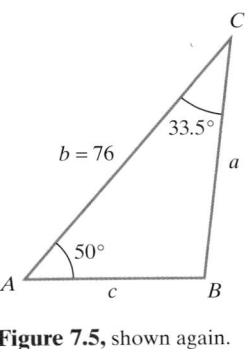

Figure 7.5, shown again.

Keep in mind that we must be given one of the three ratios to apply the Law of Sines. In this example, we are given that $b = 76$ and we found that $B = 96.5°$. Thus, we use the ratio $\dfrac{b}{\sin B}$, or $\dfrac{76}{\sin 96.5°}$, to find the other two sides. Use the Law of Sines to find a and c.

Find a:

This is the known ratio.

$$\frac{a}{\sin A} = \frac{b}{\sin B}$$

$$\frac{a}{\sin 50°} = \frac{76}{\sin 96.5°}$$

$$a = \frac{76 \sin 50°}{\sin 96.5°} \approx 59$$

Find c:

$$\frac{c}{\sin C} = \frac{b}{\sin B}$$

$$\frac{c}{\sin 33.5°} = \frac{76}{\sin 96.5°}$$

$$c = \frac{76 \sin 33.5°}{\sin 96.5°} \approx 42$$

The solution is $B = 96.5°$, $a \approx 59$, and $c \approx 42$.

2 Use the Law of Sines to solve, if possible, the triangle or triangles in the ambiguous case.

Check Point 2

Solve triangle ABC if $A = 40°$, $C = 22.5°$, and $b = 12$.

The Ambiguous Case (SSA)

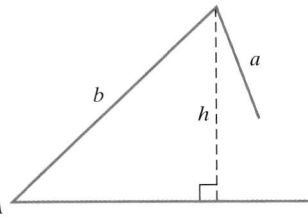

Figure 7.6 Given SSA, no triangle may result.

If we are given two sides and an angle opposite one of them (SSA), does this determine a unique triangle? Can we solve this case using the Law of Sines? Such a case is called the **ambiguous case** because the given information may result in one triangle, two triangles, or no triangle at all. For example, in Figure 7.6, we are given a, b, and A. Because a is shorter than h, it is not long enough to form a triangle. The number of possible triangles, if any, that can be formed in the SSA case depends on h, the length of the altitude, where $h = b \sin A$.

The Ambiguous Case (SSA)

Consider a triangle in which a, b, and A are given. This information may result in:

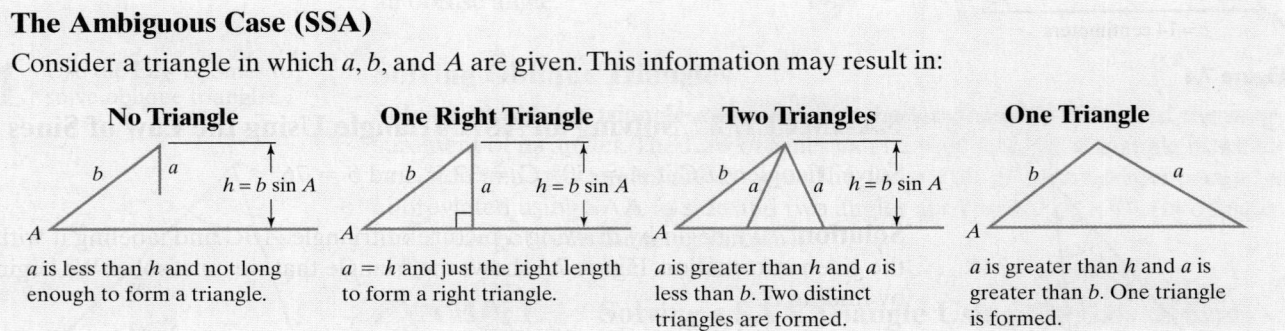

No Triangle	One Right Triangle	Two Triangles	One Triangle
a is less than h and not long enough to form a triangle.	$a = h$ and just the right length to form a right triangle.	a is greater than h and a is less than b. Two distinct triangles are formed.	a is greater than h and a is greater than b. One triangle is formed.

In a SSA situation, it is not necessary to draw an accurate sketch like those shown in the box. The Law of Sines determines the number of triangles, if any, and gives the solution for each triangle.

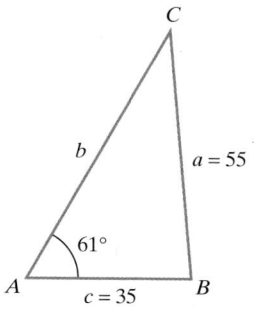

Figure 7.7 Solving a SSA triangle; the ambiguous case

EXAMPLE 3 Solving a SSA Triangle Using the Law of Sines (One Solution)

Solve triangle ABC if $A = 61°$, $a = 55$, and $c = 35$.

Solution We begin with the sketch in Figure 7.7. The known ratio is $\dfrac{a}{\sin A}$, or $\dfrac{55}{\sin 61°}$. Because side c is given, we use the Law of Sines to find angle C.

$$\frac{a}{\sin A} = \frac{c}{\sin C}$$ Apply the Law of Sines.

$$\frac{55}{\sin 61°} = \frac{35}{\sin C}$$ $a = 55$, $c = 35$, and $A = 61°$.

$$55 \sin C = 35 \sin 61°$$ Cross multiply: If $\dfrac{a}{b} = \dfrac{c}{d}$, then $ad = bc$.

$$\sin C = \frac{35 \sin 61°}{55}$$ Divide both sides by 55 and solve for $\sin C$.

$$\sin C \approx 0.5566$$ Use a calculator.

There are two angles C between $0°$ and $180°$ for which $\sin C \approx 0.5566$.

$$C_1 \approx 34° \qquad\qquad C_2 \approx 180° - 34° \approx 146°$$

Obtain the acute angle with your calculator:
$\sin^{-1} .5566$

The sine is positive in quadrant II.

Look at Figure 7.7. Given that $A = 61°$, can you see that $C_2 \approx 146°$ is impossible? By adding $146°$ to the given angle, $61°$, we exceed a $180°$ sum:

$$61° + 146° = 207°.$$

Thus, the only possibility is that $C_1 \approx 34°$. We find B using C_1 and the given information $A = 61°$.

$$B = 180° - C_1 - A \approx 180° - 34° - 61° = 85°$$

Side b that lies opposite this $85°$ angle can now be found using the Law of Sines.

$$\frac{b}{\sin B} = \frac{a}{\sin A}$$ Apply the Law of Sines.

$$\frac{b}{\sin 85°} = \frac{55}{\sin 61°}$$ $a = 55$, $B \approx 85°$, and $A = 61°$.

$$b = \frac{55 \sin 85°}{\sin 61°} \approx 63$$ Multiply both sides by $\sin 85°$ and solve for b.

There is one triangle and the solution is C_1 (or C) $\approx 34°$, $B \approx 85°$, and $b \approx 63$.

Check Point 3 Solve triangle ABC if $A = 123°$, $a = 47$, and $c = 23$.

EXAMPLE 4   **Solving a SSA Triangle Using the Law of Sines (No Solution)**

Solve triangle ABC if $A = 75°, a = 51,$ and $b = 71.$

Solution   The known ratio is $\dfrac{a}{\sin A}$, or $\dfrac{51}{\sin 75°}$. Because side b is given, we use the Law of Sines to find angle B.

Figure 7.8 *a* is not long enough to form a triangle.

$$\frac{a}{\sin A} = \frac{b}{\sin B} \qquad \text{Use the Law of Sines.}$$

$$\frac{51}{\sin 75°} = \frac{71}{\sin B} \qquad \text{Substitute the given values.}$$

$$51 \sin B = 71 \sin 75° \qquad \text{Cross multiply.}$$

$$\sin B = \frac{71 \sin 75°}{51} \approx 1.34 \qquad \text{Divide by 51 and solve for sin B.}$$

Because the sine can never exceed 1, there is no angle B for which $\sin B \approx 1.34$. There is no triangle with the given measurements, illustrated in Figure 7.8.

> **Check Point 4**   Solve triangle ABC if $A = 50°, a = 10,$ and $b = 20.$

EXAMPLE 5   **Solving a SSA Triangle Using the Law of Sines (Two Solutions)**

Solve triangle ABC if $A = 40°, a = 54,$ and $b = 62.$

Solution   The known ratio is $\dfrac{a}{\sin A}$, or $\dfrac{54}{\sin 40°}$. We use the Law of Sines to find angle B.

$$\frac{a}{\sin A} = \frac{b}{\sin B} \qquad \text{Use the Law of Sines.}$$

$$\frac{54}{\sin 40°} = \frac{62}{\sin B} \qquad \text{Substitute the given values.}$$

$$54 \sin B = 62 \sin 40° \qquad \text{Cross multiply.}$$

$$\sin B = \frac{62 \sin 40°}{54} \approx 0.7380 \qquad \text{Divide by 54 and solve for sin B.}$$

There are two angles B between $0°$ and $180°$ for which $\sin B \approx 0.7380.$

$$B_1 \approx 48° \qquad\qquad B_2 \approx 180° - 48° = 132°$$

Find sin⁻¹ .7380 with your calculator.   The sine is positive in quadrant II.

If you add either angle to the given angle, $40°$, the sum does not exceed $180°$. Thus, there are two triangles with the given conditions, shown in Figure 7.9(a). The triangles, AB_1C_1 and AB_2C_2, are shown separately in Figures 7.9(b) and (c).

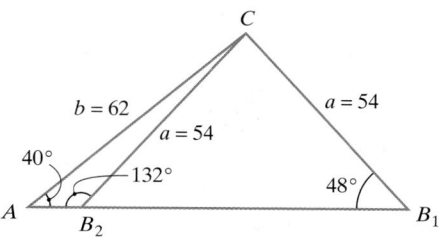
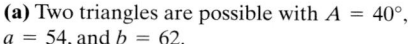

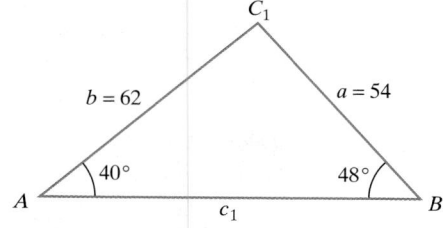

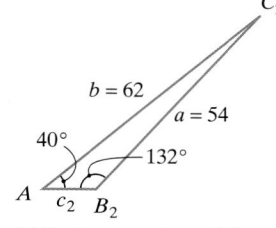

(a) Two triangles are possible with $A = 40°$, $a = 54$, and $b = 62$.

(b) In one possible triangle, $B_1 = 48°$.

(c) In the second possible triangle, $B_2 = 132°$.

Figure 7.9

Study Tip

The two triangles shown in Figure 7.9 are helpful in organizing the solutions. However, if you keep track of the two triangles, one with the given information and $B_1 = 48°$, and the other with the given information and $B_2 = 132°$, you do not have to draw the figure to solve the triangles.

We find angles C_1 and C_2 using a 180° angle sum in each of the two triangles.

$$C_1 = 180° - A - B_1 \qquad\qquad C_2 = 180° - A - B_2$$
$$\approx 180° - 40° - 48° \qquad\qquad \approx 180° - 40° - 132°$$
$$= 92° \qquad\qquad\qquad\qquad = 8°$$

We use the Law of Sines to find c_1 and c_2.

$$\frac{c_1}{\sin C_1} = \frac{a}{\sin A} \qquad\qquad \frac{c_2}{\sin C_2} = \frac{a}{\sin A}$$

$$\frac{c_1}{\sin 92°} = \frac{54}{\sin 40°} \qquad\qquad \frac{c_2}{\sin 8°} = \frac{54}{\sin 40°}$$

$$c_1 = \frac{54 \sin 92°}{\sin 40°} \approx 84 \qquad\qquad c_2 = \frac{54 \sin 8°}{\sin 40°} \approx 12$$

There are two triangles. In one triangle, the solution is $B_1 \approx 48°$, $C_1 \approx 92°$, and $c_1 \approx 84$. In the other triangle $B_2 \approx 132°$, $C_2 \approx 8°$, and $c_2 \approx 12$.

Check Point 5 Solve triangle ABC if $A = 35°$, $a = 12$, and $b = 16$.

3 Find the area of an oblique triangle using the sine function.

The Area of an Oblique Triangle

A formula for the area of an oblique triangle can be obtained using the procedure for proving the Law of Sines. We draw an altitude of length h from one of the vertices of the triangle, shown in Figure 7.10. We apply the definition of the sine of angle A, $\dfrac{\text{opposite}}{\text{hypotenuse}}$, in right triangle ACD.

$$\sin A = \frac{h}{b} \quad \text{or} \quad h = b \sin A.$$

The area of a triangle is $\frac{1}{2}$ the product of any side and the altitude drawn to that side. Using the altitude h in Figure 7.10, we have

$$\text{Area} = \tfrac{1}{2}ch = \tfrac{1}{2}cb \sin A.$$

Use the result from above: $h = b \sin A$.

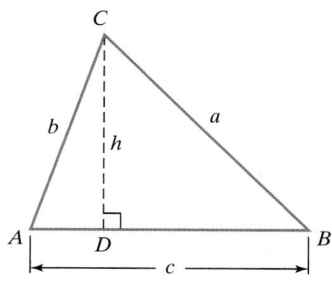

Figure 7.10

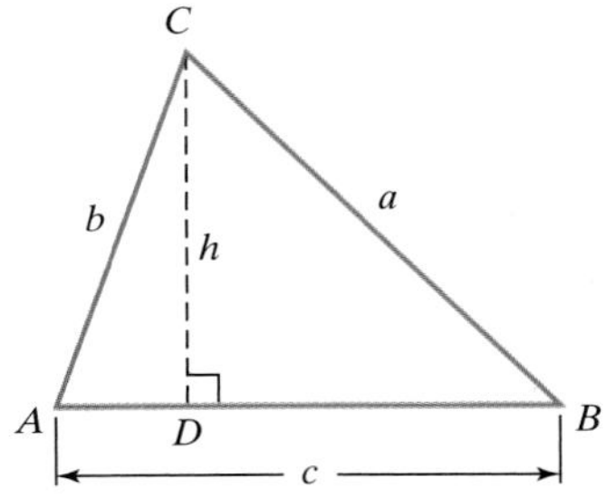

Figure 7.10, shown again

This result indicates that the area of the triangle is one-half the product of b and c times the sine of their included angle. If we draw altitudes from the other two vertices, we can use any two sides to compute the area.

Area of An Oblique Triangle

The area of a triangle equals one-half the product of the lengths of two sides times the sine of their included angle. In Figure 7.10, this wording can be expressed by the formulas

$$\text{Area} = \tfrac{1}{2}bc \sin A = \tfrac{1}{2}ab \sin C = \tfrac{1}{2}ac \sin B.$$

EXAMPLE 6 Finding the Area of an Oblique Triangle

Find the area of a triangle having two sides of lengths 24 meters and 10 meters and an included angle of $62°$.

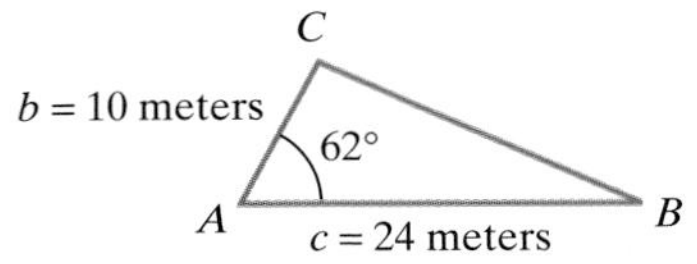

Figure 7.11 Finding the area of a SAS triangle

Solution   The triangle is shown in Figure 7.11. Its area is half the product of the lengths of the two sides times the sine of the included angle.

$$\text{Area} = \tfrac{1}{2}(24)(10)(\sin 62°) \approx 106$$

The area of the triangle is approximately 106 square meters.

> **Check Point 6** Find the area of a triangle having two sides of lengths 8 meters and 12 meters and an included angle of $135°$.

4 Solve applied problems using the Law of Sines.

Applications of the Law of Sines

We have seen how the trigonometry of right triangles can be used to solve many different kinds of applied problems. The Law of Sines enables us to work with triangles that are not right triangles. As a result, this law can be used to solve problems involving surveying, engineering, astronomy, navigation, and the environment. Example 7 illustrates the use of the Law of Sines in detecting potentially devastating fires.

EXAMPLE 7 An Application of the Law of Sines

Two fire-lookout stations are 20 miles apart, with station B directly east of station A. Both stations spot a fire on a mountain to the north. The bearing from station A to the fire is N50°E (50° east of north). The bearing from station B to the fire is N36°W (36° west of north). How far is the fire from station A?

Solution   Figure 7.12 on page 595 shows the information given in the problem. The distance from station A to the fire is represented by b. Notice that the angles describing the bearing from each station to the fire, 50° and 36°, are not interior angles of triangle ABC. Using a north-south line, the interior angles are found as follows:

$$A = 90° - 50° = 40° \qquad B = 90° - 36° = 54°.$$

To find b using the Law of Sines, we need a known side and an angle opposite that side. Because $c = 20$ miles, we find angle C using a $180°$ angle sum in the triangle. Thus,

$$C = 180° - A - B = 180° - 40° - 54° = 86°.$$

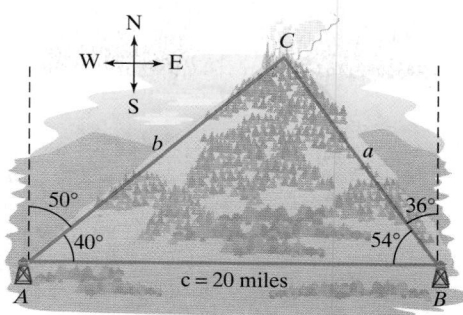

Figure 7.12

The ratio $\dfrac{c}{\sin C}$, or $\dfrac{20}{\sin 86°}$, is now known. We use this ratio and the Law of Sines to find b.

$$\frac{b}{\sin B} = \frac{c}{\sin C} \qquad \text{Use the Law of Sines.}$$

$$\frac{b}{\sin 54°} = \frac{20}{\sin 86°} \qquad c = 20,\ B = 54°,\ \text{and } C = 86°.$$

$$b = \frac{20 \sin 54°}{\sin 86°} \approx 16 \qquad \text{Multiply both sides by } \sin 54° \text{ and solve for } b.$$

The fire is approximately 16 miles from station A.

Check Point 7 Two fire-lookout stations are 13 miles apart, with station B directly east of station A. Both stations spot a fire. The bearing of the fire from station A is N35°E, and the bearing of the fire from station B is N49°W. How far, to the nearest mile, is the fire from station B?

EXERCISE SET 7.1

Practice Exercises

In Exercises 1–8, solve each triangle. Round lengths of sides to the nearest tenth and measurement of angles to the nearest degree.

1.

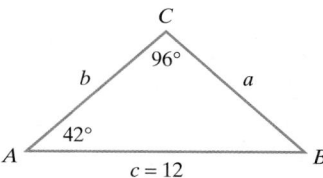

2.

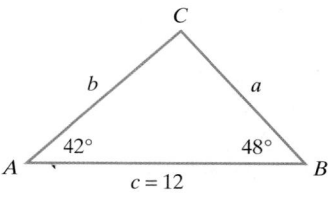

3.

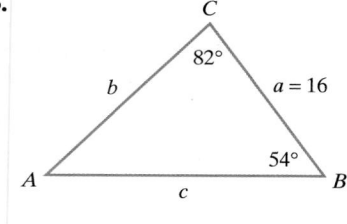

4.

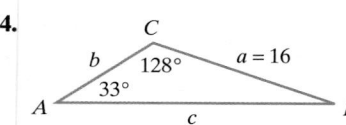

5.

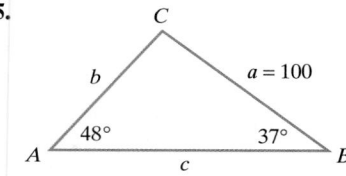

6.

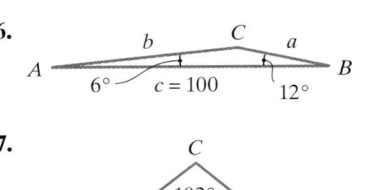

7.

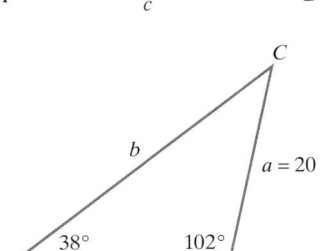

8.

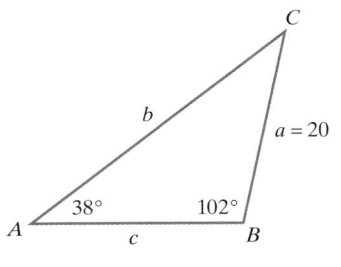

In Exercises 9–16, solve each triangle. Round lengths to the nearest tenth and angle measures to the nearest degree.

9. $A = 44°, B = 25°, a = 12$

10. $A = 56°, C = 24°, a = 22$

11. $B = 85°, C = 15°, b = 40$

12. $A = 85°, B = 35°, c = 30$

13. $A = 115°, C = 35°, c = 200$

14. $B = 5°, C = 125°, b = 200$

15. $A = 65°, B = 65°, c = 6$

16. $B = 80°, C = 10°, a = 8$

In Exercises 17–32, two sides and an angle (SSA) of a triangle are given. Determine whether the given measurements produce one triangle, two triangles, or no triangle at all. Solve each triangle that results. Round to the nearest tenth and the nearest degree for sides and angles, respectively.

17. $a = 20, b = 15, A = 40°$

18. $a = 30, b = 20, A = 50°$

19. $a = 10, c = 8.9, A = 63°$

20. $a = 57.5, c = 49.8, A = 136°$

21. $a = 42.1, c = 37, A = 112°$

22. $a = 6.1, b = 4, A = 162°$

23. $a = 10, b = 40, A = 30°$

24. $a = 10, b = 30, A = 150°$

25. $a = 16, b = 18, A = 60°$

26. $a = 30, b = 40, A = 20°$

27. $a = 12, b = 16.1, A = 37°$

28. $a = 7, b = 28, A = 12°$

29. $a = 22, c = 24.1, A = 58°$

30. $a = 95, c = 125, A = 49°$

31. $a = 9.3, b = 41, A = 18°$

32. $a = 1.4, b = 2.9, A = 142°$

In Exercises 33–38, find the area of the triangle having the given measurements. Round to the nearest square unit.

33. $A = 48°, b = 20$ feet, $c = 40$ feet

34. $A = 22°, b = 20$ feet, $c = 50$ feet

35. $B = 36°, a = 3$ yards, $c = 6$ yards

36. $B = 125°, a = 8$ yards, $c = 5$ yards

37. $C = 124°, a = 4$ meters, $b = 6$ meters

38. $C = 102°, a = 16$ meters, $b = 20$ meters

 **Application Exercises**

39. Two fire-lookout stations are 10 miles apart, with station B directly east of station A. Both stations spot a fire. The bearing of the fire from station A is N25°E and the bearing of the fire from station B is N56°W. How far, to the nearest mile, is the fire from each lookout station?

40. The Federal Communications Commission is attempting to locate an illegal radio station. It sets up two monitoring stations, A and B, with station B 40 miles east of station A. Station A measures the illegal signal from the radio station as coming from a direction of 48° east of north. Station B measures the signal as coming from a point 34° west of north. How far is the illegal radio station from monitoring stations A and B?

41. The figure shows a 1200-yard-long beach and an oil platform in the ocean. The angle made with the platform from one end of the beach is 85° and from the other end is 76°. Find the distance of the oil platform, to the nearest yard, from each end of the beach.

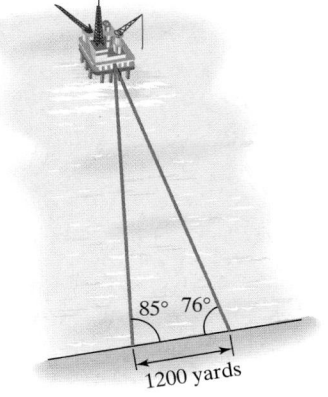

42. A surveyor needs to determine the distance between two points that lie on opposite banks of a river. The figure shows that 300 yards are measured along one bank. The angles from each end of this line segment to a point on the opposite bank are 62° and 53°. Find the distance between A and B to the nearest foot.

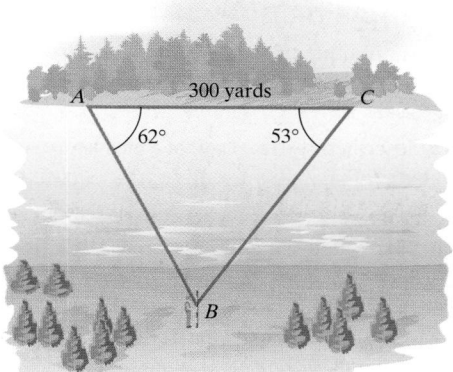

43. Closed to tourists since 1990, the Leaning Tower of Pisa in Italy leans at an angle of about 84.7°. The figure shows that 171 feet from the base of the tower, the angle of elevation to the top is 50°. Find the distance, to the nearest foot, from the base to the top of the tower.

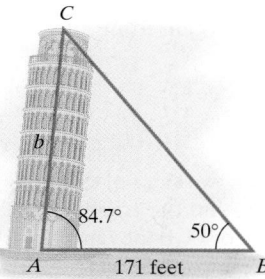

44. A pine tree growing on a hillside makes a 75° angle with the hill. From a point 80 feet up the hill, the angle of elevation to the top of the tree is 62° and the angle of depression to the bottom is 23°. Find, to the nearest foot, the height of the tree.

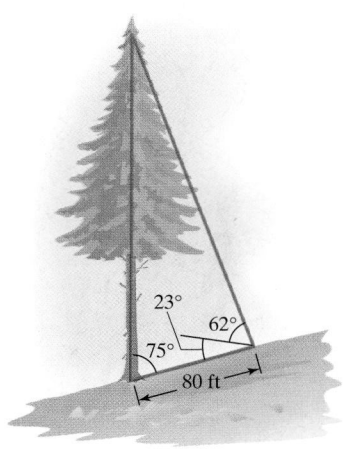

45. The figure shows a shot-put ring. The shot is tossed from A and lands at B. Using modern electronic equipment, the distance of the toss can be measured without the use of measuring tapes. When the shot lands at B, an electronic transmitter placed at B sends a signal to a device in the official's booth above the track. The device determines the angles at B and C. At a track meet, the distance from the official's booth to the shot-put ring is 562 feet. If $B = 85.3°$ and $C = 5.7°$, determine the length of the toss to the nearest foot.

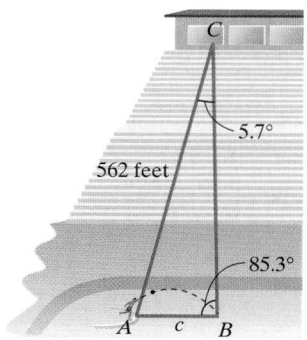

46. A pier forms an 85° angle with a straight dock. At a distance of 100 feet from the pier, the line of sight to the tip forms a 37° angle. Find the length of the pier to the nearest foot.

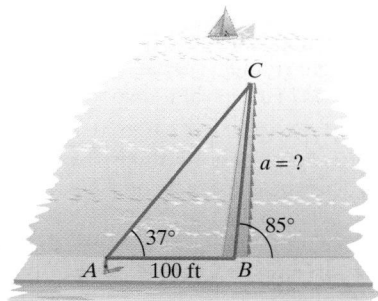

47. When the angle of elevation of the sun is 62°, a telephone pole that is tilted at an angle of 8° directly away from the sun casts a shadow 20 feet long. Determine the length of the pole to the nearest foot.

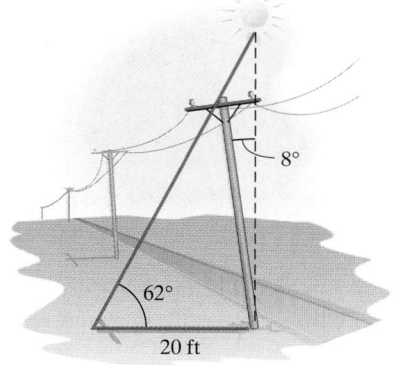

48. A leaning wall is inclined 6° from the vertical. At a distance of 40 feet from the wall, the angle of elevation to the top is 22°. Find the height of the wall to the nearest foot.

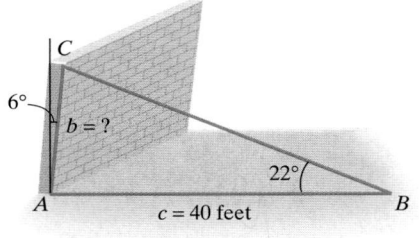

49. Redwood trees in California's Redwood National Park are hundreds of feet tall. The height of one of these trees is represented by *h* in the figure shown.

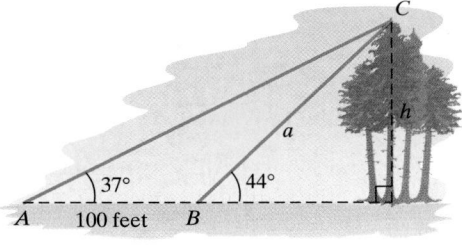

   a. Use the measurements shown to find *a*, to the nearest foot, in oblique triangle *ABC*.
   b. Use the right triangle shown to find the height, to the nearest foot, of a typical redwood tree in the park.

50. The figure shows a cable car that carries passengers from *A* to *C*. Point *A* is 1.6 miles from the base of the mountain. The angles of elevation from *A* and *B* to the mountain's peak are 22° and 66°, respectively.
   a. Determine, to the nearest foot, the distance covered by the cable car.
   b. Find *a*, to the nearest foot, in oblique triangle *ABC*.
   c. Use the right triangle to find the height of the mountain to the nearest foot.

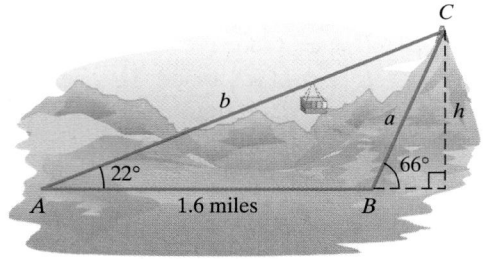

51. Lighthouse B is 7 miles west of lighthouse A. A boat leaves A and sails 5 miles. At this time, it is sighted from B. If the bearing of the boat from B is N62°E, how far from B is the boat? Round to the nearest tenth of a mile.

52. After a wind storm, you notice that your 16-foot flagpole may be leaning, but you are not sure. From a point on the

ground 15 feet from the base of the flagpole, you find that the angle of elevation to the top is 48°. Is the flagpole leaning? If so, find the acute angle, to the nearest degree, that the flagpole makes with the ground.

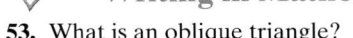

Writing in Mathematics

53. What is an oblique triangle?
54. Without using symbols, state the Law of Sines in your own words.
55. Briefly describe how the Law of Sines is proved.
56. What does it mean to solve an oblique triangle?
57. What do the abbreviations SAA and ASA mean?
58. Why is SSA called the ambiguous case?
59. How is the sine function used to find the area of an oblique triangle?
60. Write an original problem that can be solved using the Law of Sines. Then solve the problem.
61. Use Exercise 45 to describe how the Law of Sines is used for throwing events at track and field meets. Why aren't tape measures used to determine tossing distance?
62. You are cruising in your boat parallel to the coast, looking at a lighthouse. Explain how you can use your boat's speed and a device for measuring angles to determine the distance at any instant from your boat to the lighthouse.

Critical Thinking Exercises

63. If you are given two sides of a triangle and their included angle, you can find the triangle's area. Can the Law of Sines be used to solve the triangle with this given information? Explain your answer.

64. Two buildings of equal height are 800 feet apart. An observer on the street between the buildings measures the angles of elevation to the tops of the buildings as 27° and 41°, respectively. How high, to the nearest foot, are the buildings?

65. The figure shows the design for the top of the wing of a jet fighter. The fuselage is 5 feet wide. Find the wing span *CC'* to the nearest foot.

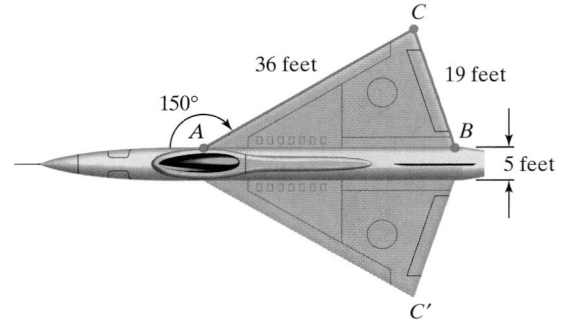

SECTION 7.2   *The Law of Cosines*

Objectives

1. Use the Law of Cosines to solve oblique triangles.
2. Solve applied problems using the Law of Cosines.
3. Use Heron's formula to find the area of a triangle.

Baseball was developed in the United States in the mid-1800s and became our national sport. Little league baseball is the largest youth sports program in the world. Three million boys and girls ages 5 to 18 play on 200,000 little league teams in 90 countries. Although there are differences between major league and little league baseball diamonds, trigonometry can be used to find angles and distances in these fields of dreams. To see how this is done, we turn to the Law of Cosines.

The Law of Cosines and Its Derivation

We now look at another relationship that exists among the sides and angles in an oblique triangle. **The Law of Cosines** is used to solve triangles in which two sides and the included angle (SAS) are known, or those in which three sides (SSS) are known.

Discovery

What happens to the Law of Cosines

$$c^2 = a^2 + b^2 - 2ab \cos C$$

if $C = 90°$? What familiar theorem do you obtain?

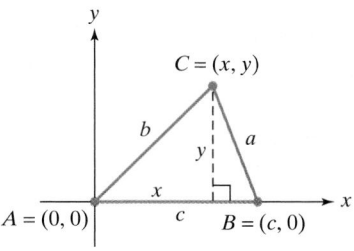

Figure 7.13

The Law of Cosines

If A, B, and C are the measures of the angles of a triangle, and a, b, and c are the lengths of the sides opposite these angles, then

$$a^2 = b^2 + c^2 - 2bc \cos A$$
$$b^2 = a^2 + c^2 - 2ac \cos B$$
$$c^2 = a^2 + b^2 - 2ab \cos C.$$

The square of a side of a triangle equals the sum of the squares of the other two sides minus twice their product times the cosine of their included angle.

To prove the Law of Cosines, we place triangle ABC in a rectangular coordinate system. Figure 7.13 shows a triangle with three acute angles. The vertex A is at the origin and side c lies along the positive x-axis. The coordinates of C are (x, y). Using the right triangle that contains angle A, we apply the definitions of the cosine and the sine.

Figure 7.13, repeated

$$\cos A = \frac{x}{b} \qquad \sin A = \frac{y}{b}$$

$$x = b \cos A \qquad y = b \sin A \quad \text{Multiply both sides of each equation by } b \text{ and solve for } x \text{ and } y, \text{ respectively.}$$

Thus, the coordinates of C are $(x, y) = (b \cos A, b \sin A)$. Although triangle ABC in Figure 7.13 shows angle A as an acute angle, if A is obtuse, the coordinates of C are still $(b \cos A, b \sin A)$. This means that our proof applies to both kinds of oblique triangles.

We now apply the distance formula to the side of the triangle with length a. Notice that a is the distance from (x, y) to $(c, 0)$.

$$a = \sqrt{(x - c)^2 + (y - 0)^2} \qquad \text{Use the distance formula.}$$

$$a^2 = (x - c)^2 + y^2 \qquad \text{Square both sides of the equation.}$$

$$a^2 = (b \cos A - c)^2 + (b \sin A)^2 \qquad x = b \cos A \text{ and } y = b \sin A.$$

$$a^2 = b^2 \cos^2 A - 2bc \cos A + c^2 + b^2 \sin^2 A \quad \text{Square the two expressions.}$$

$$a^2 = b^2 \sin^2 A + b^2 \cos^2 A + c^2 - 2bc \cos A \quad \text{Rearrange terms.}$$

$$a^2 = b^2(\sin^2 A + \cos^2 A) + c^2 - 2bc \cos A \quad \text{Factor } b^2 \text{ from the first two terms.}$$

$$a^2 = b^2 + c^2 - 2bc \cos A \qquad \sin^2 A + \cos^2 A = 1$$

The resulting equation is one of the three formulas for the Law of Cosines. The other two formulas are derived in a similar manner.

1 Use the Law of Cosines to solve oblique triangles.

Solving Oblique Triangles

If you are given two sides and an included angle (SAS) of an oblique triangle, none of the three ratios in the Law of Sines is known. This means that we do not begin solving the triangle using the Law of Sines. Instead, we apply the Law of Cosines and the following procedure.

Solving a SAS Triangle

1. Use the Law of Cosines to find the side opposite the given angle.

2. Use the Law of Sines to find the angle opposite the shorter of the two given sides. This angle is always acute.

3. Find the third angle. Subtract the measure of the given angle and the angle found in step 2 from 180°.

EXAMPLE 1 Solving a SAS Triangle

Solve the triangle shown in Figure 7.14 on page 601 with $A = 60°$, $b = 20$, and $c = 30$.

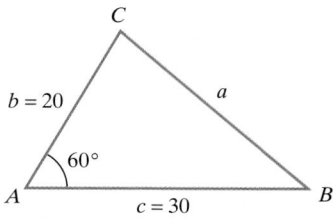

Figure 7.14 Solving a SAS triangle

Solution   We are given two sides and an included angle. Therefore, we apply the three-step procedure for solving a SAS triangle.

Step 1   Use the Law of Cosines to find the side opposite the given angle.   Thus, we will find a.

$$a^2 = b^2 + c^2 - 2bc \cos A$$ Apply the Law of Cosines to find a.

$$a^2 = 20^2 + 30^2 - 2(20)(30) \cos 60°$$ $b = 20$, $c = 30$, and $A = 60°$

$$= 400 + 900 - 1200(0.5)$$ Perform the indicated operations.

$$= 700$$

$$a = \sqrt{700} \approx 26$$ Take the square root of both sides and solve for a.

Step 2   Use the Law of Sines to find the angle opposite the shorter of the two given sides. This angle is always acute.   The shorter of the two given sides is $b = 20$. Thus, we will find acute angle B.

$$\frac{b}{\sin B} = \frac{a}{\sin A}$$ Apply the Law of Sines.

$$\frac{20}{\sin B} = \frac{\sqrt{700}}{\sin 60°}$$ We are given $b = 20$ and $A = 60°$. Use the exact value of a, $\sqrt{700}$, from step 1.

$$\sqrt{700} \sin B = 20 \sin 60°$$ Cross multiply.

$$\sin B = \frac{20 \sin 60°}{\sqrt{700}} \approx 0.6547$$ Divide by $\sqrt{700}$ and solve for sin B.

$$B \approx 41°$$ Find $\sin^{-1} 0.6547$ using a calculator.

Step 3   Find the third angle. Subtract the measure of the given angle and the angle found in step 2 from 180°.

$$C = 180° - A - B \approx 180° - 60° - 41° = 79°$$

The solution is $a \approx 26$, $B \approx 41°$, and $C \approx 79°$.

Check Point 1   Solve the triangle shown in Figure 7.15 with $A = 120°$, $b = 7$, and $c = 8$.

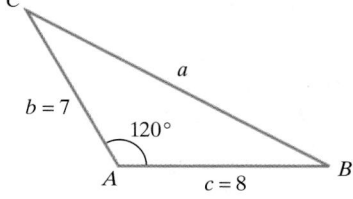

Figure 7.15

If you are given three sides of a triangle (SSS), solving the triangle involves finding the three angles. We use the following procedure.

Solving a SSS Triangle

1. Use the Law of Cosines to find the angle opposite the longest side.

2. Use the Law of Sines to find either of the two remaining acute angles.

3. Find the third angle. Subtract the measures of the angles found in steps 1 and 2 from 180°.

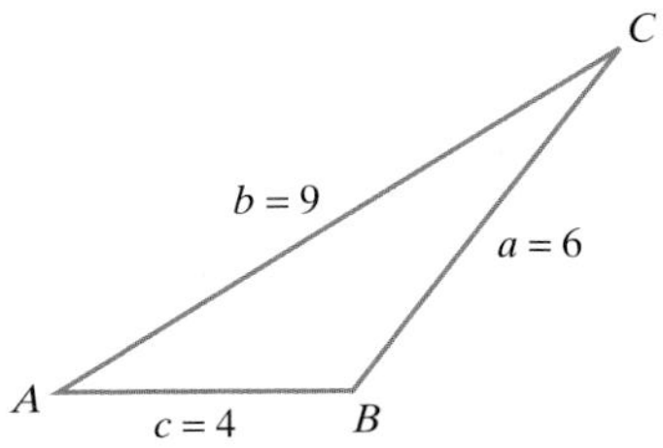

Figure 7.16 Solving a SSS triangle

EXAMPLE 2   Solving a SSS Triangle

Solve triangle ABC if $a = 6, b = 9$, and $c = 4$. Round to the nearest tenth of a degree.

Solution   We are given three sides. Therefore, we apply the three-step procedure for solving a SSS triangle. The triangle is shown in Figure 7.16.

Step 1   Use the Law of Cosines to find the angle opposite the longest side.
The longest side is $b = 9$. Thus, we will find angle B.

$$b^2 = a^2 + c^2 - 2ac \cos B \qquad \text{Apply the Law of Cosines to find B.}$$

$$2ac \cos B = a^2 + c^2 - b^2 \qquad \text{Solve for cos B.}$$

$$\cos B = \frac{a^2 + c^2 - b^2}{2ac}$$

$$\cos B = \frac{6^2 + 4^2 - 9^2}{2 \cdot 6 \cdot 4} = -\frac{29}{48} \qquad a = 6, b = 9, \text{ and } c = 4.$$

Using a calculator, $\cos^{-1}\left(\frac{29}{48}\right) \approx 52.8°$. Because $\cos B$ is negative, B is an obtuse angle. Thus,

$$B \approx 180° - 52.8° = 127.2°.$$

Study Tip

You can use the Law of Cosines in step 2 to find either of the remaining angles. However, it is simpler to use the Law of Sines. Because the largest angle has been found, the remaining angles must be acute. Thus, there is no need to be concerned about two possible triangles or an ambiguous case.

Step 2   Use the Law of Sines to find either of the two remaining acute angles.
We will find angle A.

$$\frac{a}{\sin A} = \frac{b}{\sin B} \qquad \text{Apply the Law of Sines.}$$

$$\frac{6}{\sin A} = \frac{9}{\sin 127.2°} \qquad \text{We are given } a = 6 \text{ and } b = 9. \text{ We found that } B \approx 127.2°.$$

$$9 \sin A = 6 \sin 127.2° \qquad \text{Cross multiply.}$$

$$\sin A = \frac{6 \sin 127.2°}{9} \approx 0.5310 \qquad \text{Divide by 9 and solve for sin A.}$$

$$A \approx 32.1° \qquad \text{Find } \sin^{-1} 0.5310 \text{ using a calculator.}$$

Step 3   Find the third angle. Subtract the measures of the angles found in steps 1 and 2 from 180°.

$$C = 180° - B - A \approx 180° - 127.2° - 32.1° = 20.7°$$

The solution is $B \approx 127.2°$, $A \approx 32.1°$, and $C \approx 20.7°$.

Check Point 2   Solve triangle ABC if $a = 8, b = 10$, and $c = 5$. Round to the nearest tenth of a degree.

2 Solve applied problems using the Law of Cosines.

Applications of the Law of Cosines

Applied problems involving SAS and SSS triangles can be solved using the Law of Cosines.

EXAMPLE 3 An Application of the Law of Cosines

Two airplanes leave an airport at the same time on different runways. One flies at a bearing of N66°W at 325 miles per hour. The other airplane flies at a bearing of S26°W at 300 miles per hour. How far apart will the airplanes be after two hours?

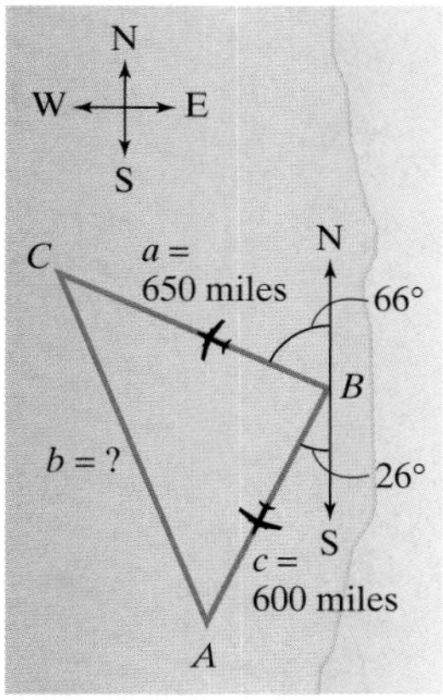

Solution After two hours, the plane flying at 325 miles per hour travels $325 \cdot 2$ miles, or 650 miles. Similarly, the plane flying at 300 miles per hour travels 600 miles. The situation is illustrated in Figure 7.17.

Let b = the distance between the planes after two hours. We can use a north-south line to find angle B in triangle ABC. Thus,

$$B = 180° - 66° - 26° = 88°.$$

We now have $a = 650, c = 600$, and $B = 88°$. We use the Law of Cosines to find b in this SAS situation.

Figure 7.17

$b^2 = a^2 + c^2 - 2ac \cos B$	Apply the Law of Cosines.
$b^2 = 650^2 + 600^2 - 2(650)(600) \cos 88°$	Substitute: $a = 650, c = 600,$ and $B = 88°$.
$\approx 755{,}278$	Use a calculator.
$b \approx \sqrt{755{,}278} \approx 869$	Take the square root and solve for b.

After two hours, the planes are approximately 869 miles apart.

Check Point 3 Two airplanes leave an airport at the same time on different runways. One flies directly north at 400 miles per hour. The other airplane flies at a bearing of N75°E at 350 miles per hour. How far apart will the airplanes be after two hours?

3 Use Heron's formula to find the area of a triangle.

Heron's Formula

Approximately 2000 years ago, the Greek mathematician Heron of Alexandria derived a formula for the area of a triangle in terms of the lengths of its sides. A more modern derivation uses the Law of Cosines and can be found in the appendix.

Heron's Formula for the Area of a Triangle

The area of a triangle with sides $a, b,$ and c is

$$\text{Area} = \sqrt{s(s - a)(s - b)(s - c)},$$

where s is one-half the perimeter: $s = \frac{1}{2}(a + b + c).$

EXAMPLE 4   Using Heron's Formula

Find the area of the triangle with $a = 12$ yards, $b = 16$ yards, and $c = 24$ yards.

Solution   Begin by calculating one-half the perimeter:
$$s = \tfrac{1}{2}(a + b + c) = \tfrac{1}{2}(12 + 16 + 24) = 26.$$

Use Heron's formula to find the area:
$$\begin{aligned}
\text{Area} &= \sqrt{s(s-a)(s-b)(s-c)} \\
&= \sqrt{26(26-12)(26-16)(26-24)} \\
&= \sqrt{7280} \approx 85.
\end{aligned}$$

The area of the triangle is approximately 85 square yards.

Check Point 4   Find the area of the triangle with $a = 6$ meters, $b = 16$ meters, and $c = 18$ meters. Round to the nearest square meter.

EXERCISE SET 7.2

✓ Practice Exercises

In Exercises 1–8, solve each triangle. Round lengths of sides to the nearest tenth and measurements of angles to the nearest degree.

1.
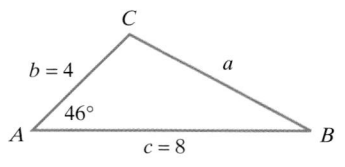

2.
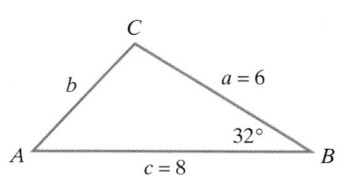

3.
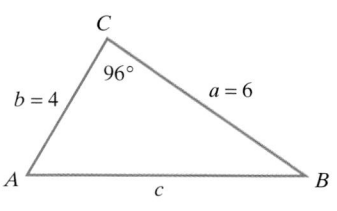

4.
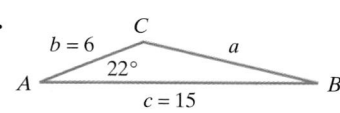

5.
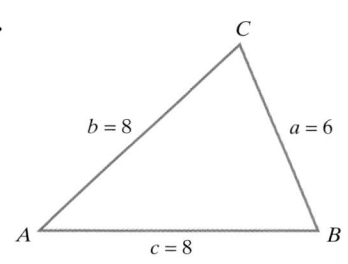

6.
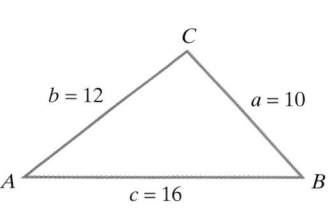

7.
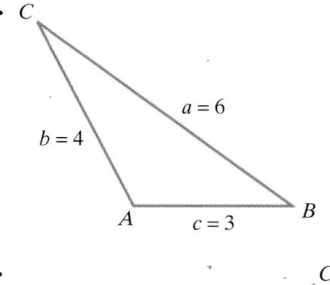

8.
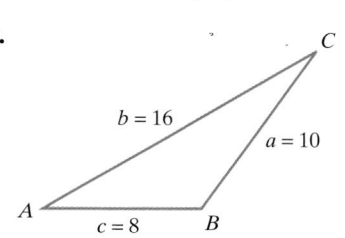

In Exercises 9–24, solve each triangle. Round lengths to the nearest tenth and angle measures to the nearest degree.

9. $a = 5, b = 7, C = 42°$
10. $a = 10, b = 3, C = 15°$
11. $b = 5, c = 3, A = 102°$
12. $b = 4, c = 1, A = 100°$
13. $a = 6, c = 5, B = 50°$
14. $a = 4, c = 7, B = 55°$
15. $a = 5, c = 2, B = 90°$
16. $a = 7, c = 3, B = 90°$
17. $a = 5, b = 7, c = 10$
18. $a = 4, b = 6, c = 9$
19. $a = 3, b = 9, c = 8$
20. $a = 4, b = 7, c = 6$
21. $a = 3, b = 3, c = 3$
22. $a = 5, b = 5, c = 5$
23. $a = 73, b = 22, c = 50$
24. $a = 66, b = 25, c = 45$

In Exercises 25–30, use Heron's formula to find the area of the triangle. Round to the nearest square unit.

25. $a = 4$ feet, $b = 4$ feet, $c = 2$ feet
26. $a = 5$ feet, $b = 5$ feet, $c = 4$ feet
27. $a = 14$ meters, $b = 12$ meters, $c = 4$ meters
28. $a = 16$ meters, $b = 10$ meters, $c = 8$ meters
29. $a = 11$ yards, $b = 9$ yards, $c = 7$ yards
30. $a = 13$ yards, $b = 9$ yards, $c = 5$ yards

 Application Exercises

31. Two ships leave a harbor at the same time. One ship travels at a bearing of S12°W at 14 miles per hour. The other ship travels at a bearing of N75°E at 10 miles per hour. How far apart will the ships be after three hours? Round to the nearest tenth of a mile.

32. A plane leaves airport A and travels 580 miles to airport B at a bearing of N34°E. The plane later leaves airport B and travels to airport C 400 miles away at a bearing of S74°E. Find the distance from airport A to airport C to the nearest tenth of a mile.

33. Find the distance across the lake, to the nearest yard, using the measurements shown in the figure.

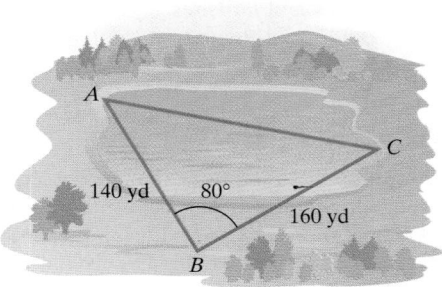

34. To find the distance across a protected cove at a lake, a surveyor makes the measurements shown in the figure.

Use these measurements to find the distance from A to B to the nearest yard.

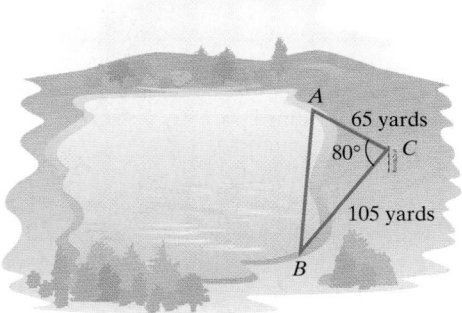

The diagram shows three islands in Florida Bay. You rent a boat and plan to visit each of these remote islands. Use the diagram to solve Exercises 35–36.

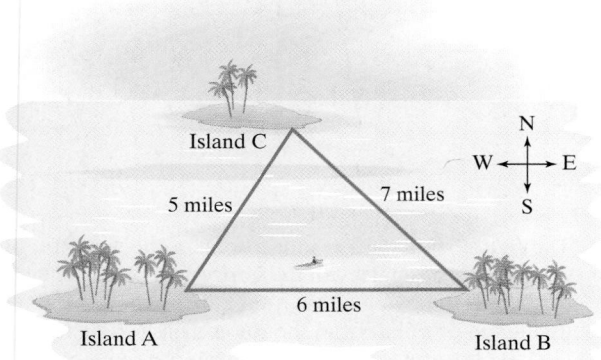

35. If you are on island A, on what bearing should you navigate to go to island C?

36. If you are on island B, on what bearing should you navigate to go to island C?

37. You are on a fishing boat that leaves its pier and heads east. After traveling for 25 miles, there is a report warning of rough seas directly south. The captain turns the boat and follows a bearing of S40°W for 13.5 miles.
    a. At this time, how far are you from the boat's pier? Round to the nearest tenth of a mile.
    b. What bearing could the boat have originally taken to arrive at this fishing spot?

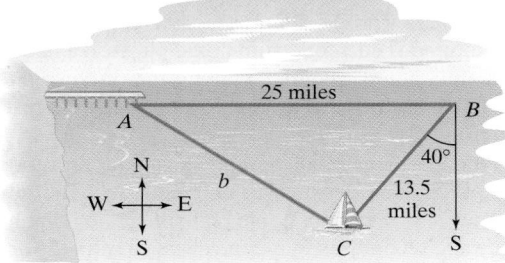

38. You are on a fishing boat that leaves its pier and heads east. After traveling for 30 miles, there is a report warning of rough seas directly south. The captain turns the boat and follows a bearing of S45°W for 12 miles.
   a. At this time, how far are you from the boat's pier? Round to the nearest tenth of a mile.
   b. What bearing could the boat have originally taken to arrive at this fishing spot?

39. The figure shows a 400-foot tower on the side of a hill that forms 7° angle with the horizontal. Find the length of each of the two guy wires that are anchored 80 feet uphill and downhill from the tower's base and extend to the top of the tower. Round to the nearest tenth of a foot.

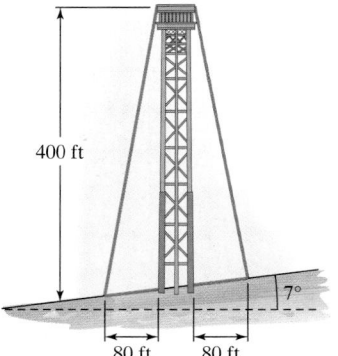

40. The figure shows a 200-foot tower on the side of a hill that forms a 5° angle with the horizontal. Find the length of each of the two guy wires that are anchored 150 feet uphill and downhill from the tower's base and extend to the top of the tower. Round to the nearest tenth of a foot.

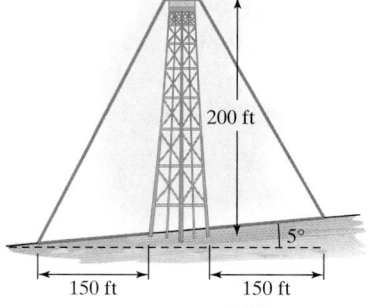

41. A major league baseball diamond has four bases forming a square whose sides measure 90 feet each. The pitcher's mound is 60.5 feet from home plate on a line joining home plate and second base. Find the distance from the pitcher's mound to first base. Round to the nearest tenth of a foot.

42. A little league baseball diamond has four bases forming a square whose sides measure 60 feet each. The pitcher's mound is 46 feet from home plate on a line joining home plate and second base. Find the distance from the pitcher's mound to third base. Round to the nearest tenth of a foot.

43. A commercial piece of real estate is priced at $3.50 per square foot. Find the cost, to the nearest dollar, of a triangular lot measuring 240 feet by 300 feet by 420 feet.

44. A commercial piece of real estate is priced at $4.50 per square foot. Find the cost, to the nearest dollar, of a triangular lot measuring 320 feet by 510 feet by 410 feet.

Writing in Mathematics

45. Without using symbols, state the Law of Cosines in your own words.

46. Why can't the Law of Sines be used in the first step to solve a SAS triangle?

47. Describe a strategy for solving a SAS triangle.

48. Describe a strategy for solving a SSS triangle.

49. Under what conditions would you use Heron's formula to find the area of a triangle?

50. Describe an applied problem that can be solved using the Law of Cosines, but not the Law of Sines.

51. The pitcher on your little league team is studying angles in geometry and has a question. "Coach, suppose I'm on the pitcher's mound facing home plate. I catch a fly ball hit in my direction. If I turn to face first base and throw the ball, through how many degrees should I turn for a direct throw?" Use the information given in Exercise 42 and write an answer to your pitcher's question. Without getting too technical, describe to your pitcher how you obtained this angle.

Critical Thinking Exercises

52. The lengths of the diagonals of a parallelogram are 20 inches and 30 inches. The diagonals intersect at an angle of 35°. Find the lengths of the parallelogram's sides. (*Hint*: Diagonals of a parallelogram bisect one another.)

53. The vertices of a triangle are $A(4, -3)$, $B(2, 1)$, and $C(-2, 4)$. Find the triangle's largest angle to the nearest tenth of a degree.

54. The minute hand and the hour hand of a clock have lengths m inches and h inches, respectively. Determine the distance between the tips of the hands at 10:00 in terms of m and h.

Group Exercise

55. The group should design five original problems that can be solved using the Laws of Sines and Cosines. At least two problems should be solved using the Law of Sines and at least two problems should be solved using the Law of Cosines. At least one problem should be an application problem using the Law of Sines and at least one problem should involve an application using the Law of Cosines. The group should turn in both the problems and their solutions.

SECTION 7.3 *Polar Coordinates*

Objectives

1. Plot points in the polar coordinate system.
2. Find multiple sets of polar coordinates of a given point.
3. Convert a point from polar to rectangular coordinates.
4. Convert a point from rectangular to polar coordinates.
5. Convert an equation from rectangular to polar coordinates.
6. Convert an equation from polar to rectangular coordinates.

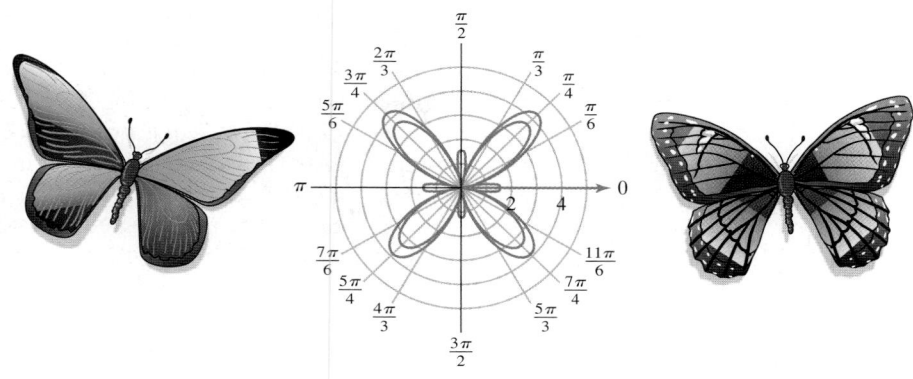

Butterflies are among the most celebrated of all insects. It's hard not to notice their beautiful colors and graceful flight. Their symmetry can be explored with trigonometric functions and a system for plotting points called the **polar coordinate system**. In many cases, polar coordinates are simpler and easier to use than rectangular coordinates.

Plotting Points in the Polar Coordinate System

1 Plot points in the polar coordinate system.

The foundation of the polar coordinate system is a horizontal ray that extends to the right. The ray is called the **polar axis** and is shown in Figure 7.18. The endpoint of the ray is called the **pole**.

A point P in the polar coordinate system is represented by an ordered pair of numbers (r, θ). Figure 7.19 shows $P = (r, \theta)$ in the polar coordinate system.

Figure 7.18

Figure 7.19 Representing a point in the polar coordinate system

- r is the directed distance of P from the pole. (We shall see that r can be positive, negative or zero.)
- θ is an angle from the polar axis to line segment OP. This angle can be measured in degrees or radians. Positive angles are measured counterclockwise from the polar axis. Negative angles are measured clockwise from the polar axis.

We refer to the ordered pair (r, θ) as the **polar coordinates** of P.

Let's look at a specific example. Suppose that the polar coordinates of a point P are $\left(3, \dfrac{\pi}{4}\right)$. Because θ is positive, we locate this point by drawing $\theta = \dfrac{\pi}{4}$ counterclockwise from the polar axis. Then we count out a distance of three units along the terminal side of the angle to reach the point P.

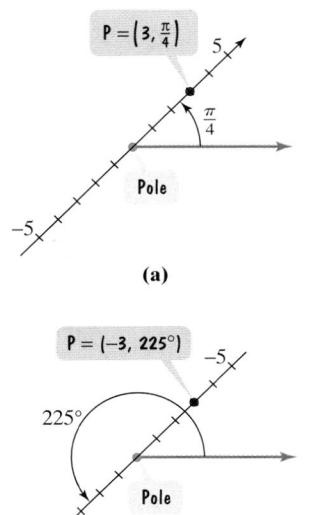

P = $\left(3, \frac{\pi}{4}\right)$

Pole

$\frac{\pi}{4}$

(a)

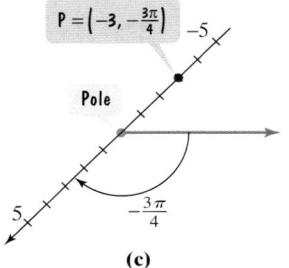

P = (-3, 225°)

225°

Pole

(b)

P = $\left(-3, -\frac{3\pi}{4}\right)$

Pole

$-\frac{3\pi}{4}$

(c)

Figure 7.20 Locating points in polar coordinates

Figure 7.20(a) shows that $(r, \theta) = \left(3, \dfrac{\pi}{4}\right)$ lies three units from the pole on the terminal side of the angle $\theta = \dfrac{\pi}{4}$.

Figure 7.20 illustrates that in a polar coordinate system, a point can be represented in more than one way. In Figure 7.20(b), r is negative and θ is positive: $r = -3$ and $\theta = 225°$. Because θ is positive, we draw a 225° angle counterclockwise from the polar axis. Notice that the point P is not located on the terminal side of θ. Instead, it lies on the ray *opposite the terminal side of θ* at a distance of $|-3|$, or 3, units from the pole. In general, **when r in (r, θ) is negative, a point is located $|r|$ units along the ray opposite the terminal side of θ.**

In Figure 7.20(c), r and θ are both negative: $r = -3$ and $\theta = -\dfrac{3\pi}{4}$. Because θ is negative, we draw a $-\dfrac{3\pi}{4}$ (or $-135°$) angle clockwise from the polar axis. Because r is negative, the point is located on the ray opposite the terminal side of θ.

Our observations indicate the importance of the sign of r in locating $P = (r, \theta)$ in polar coordinates.

The Sign of r and a Point's Location in Polar Coordinates

The point $P = (r, \theta)$ is located $|r|$ units from the pole. If $r > 0$, the point lies on the terminal side of θ. If $r < 0$, the point lies along the ray opposite the terminal side of θ. If $r = 0$, the point lies at the pole, regardless of the value of θ.

EXAMPLE 1 Plotting Points in a Polar Coordinate System

Plot the points with the following polar coordinates:

a. $(2, 135°)$ **b.** $\left(-3, \dfrac{3\pi}{2}\right)$ **c.** $\left(-1, -\dfrac{\pi}{4}\right)$.

Solution

a. To plot the point $(r, \theta) = (2, 135°)$, begin with the 135° angle. Because 135° is a positive angle, draw $\theta = 135°$ counterclockwise from the polar axis. Now consider $r = 2$. Because $r > 0$, plot the point by going out two units on the terminal side of θ. Figure 7.21(a) shows the point.

b. To plot the point $(r, \theta) = \left(-3, \dfrac{3\pi}{2}\right)$, begin with the $\dfrac{3\pi}{2}$ angle. Because $\dfrac{3\pi}{2}$ is a positive angle, we draw $\theta = \dfrac{3\pi}{2}$ counterclockwise from the polar axis. Now consider $r = -3$. Because $r < 0$, plot the point by going out three units along the ray *opposite* the terminal side of θ. Figure 7.21(b) shows the point.

c. To plot the point $(r, \theta) = \left(-1, -\dfrac{\pi}{4}\right)$, begin with the $-\dfrac{\pi}{4}$ angle. Because $-\dfrac{\pi}{4}$ is a negative angle, draw $\theta = -\dfrac{\pi}{4}$ clockwise from the polar axis. Now

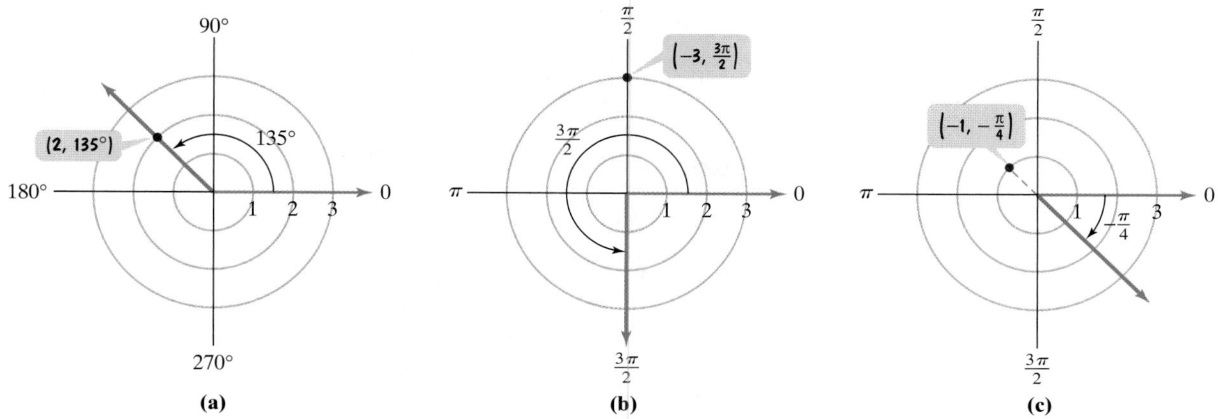

Figure 7.21 Plotting points

consider $r = -1$. Because $r < 0$, plot the point by going out one unit along the ray *opposite* the terminal side of θ. Figure 7.21(c) shows the point.

Check Point 1   Plot the points with the following polar coordinates:

a. $(3, 315°)$ **b.** $(-2, \pi)$ **c.** $\left(-1, -\dfrac{\pi}{2}\right)$.

2 Find multiple sets of polar coordinates of a given point.

Discovery

Illustrate the statements in the voice ballons by plotting:

a. $\left(1, \dfrac{\pi}{2}\right)$ and $\left(1, \dfrac{5\pi}{2}\right)$.

b. $\left(3, \dfrac{\pi}{4}\right)$ and $\left(-3, \dfrac{5\pi}{4}\right)$.

Multiple Representation of Points in the Polar Coordinate System

In rectangular coordinates, each point (x, y) has exactly one representation. By contrast, any point in polar coordinates can be represented in infinitely many ways. For example,

$$(r, \theta) = (r, \theta + 2\pi) \quad \text{and} \quad (r, \theta) = (-r, \theta + \pi).$$

Adding 1 revolution, or 2π radians, to the angle does not change the point's location.

Adding $\frac{1}{2}$ revolution, or π radians, to the angle and replacing r by $-r$ does not change the point's location.

Thus, to find two other representations for the point (r, θ),

- Add 2π to the angle and do not change r.
- Add π to the angle and replace r by $-r$.

Continually adding or subtracting 2π in either of these representations does not change the point's location.

> **Multiple Representation of Points**
> If n is any integer, the point (r, θ) can be represented as
> $$(r, \theta) = (r, \theta + 2n\pi) \quad \text{or} \quad (r, \theta) = (-r, \theta + \pi + 2n\pi).$$

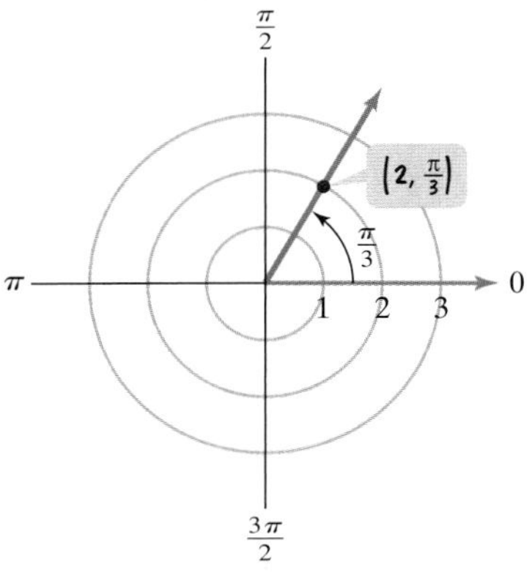

Figure 7.22 Finding other representations of a given point

EXAMPLE 2 Finding Other Polar Coordinates of a Given Point

The point $\left(2, \dfrac{\pi}{3}\right)$ is plotted in Figure 7.22. Find another representation of this point in which:

   a. r is positive and $2\pi < \theta < 4\pi$.
   b. r is negative and $0 < \theta < 2\pi$.
   c. r is positive and $-2\pi < \theta < 0$.

Solution

a. Add 2π to the angle and do not change r.

$$\left(2, \frac{\pi}{3}\right) = \left(2, \frac{\pi}{3} + 2\pi\right) = \left(2, \frac{\pi}{3} + \frac{6\pi}{3}\right) = \left(2, \frac{7\pi}{3}\right)$$

b. Add π to the angle and replace r by $-r$.

$$\left(2, \frac{\pi}{3}\right) = \left(-2, \frac{\pi}{3} + \pi\right) = \left(-2, \frac{\pi}{3} + \frac{3\pi}{3}\right) = \left(-2, \frac{4\pi}{3}\right)$$

c. Subtract 2π from the angle and do not change r.

$$\left(2, \frac{\pi}{3}\right) = \left(2, \frac{\pi}{3} - 2\pi\right) = \left(2, \frac{\pi}{3} - \frac{6\pi}{3}\right) = \left(2, -\frac{5\pi}{3}\right)$$

> **Check Point 2**
>
> Find another representation of $\left(5, \dfrac{\pi}{4}\right)$ in which:
>
>   **a.** r is positive and $2\pi < \theta < 4\pi$.
>   **b.** r is negative and $0 < \theta < 2\pi$.
>   **c.** r is positive and $-2\pi < \theta < 0$.

Relations between Polar and Rectangular Coordinates

We now consider both polar and rectangular coordinates simultaneously. Figure 7.23 shows the two coordinate systems. The polar axis coincides with the positive x-axis and the pole coincides with the origin. A point P, other than the origin, has rectangular coordinates (x, y) and polar coordinates (r, θ), as indicated in the figure. We wish to find equations relating the two sets of coordinates. From the figure, we see that

$$x^2 + y^2 = r^2$$

$$\sin\theta = \frac{y}{r} \qquad \cos\theta = \frac{x}{r} \qquad \tan\theta = \frac{y}{x}.$$

These relationships hold when P is in any quadrant and when $r > 0$ or $r < 0$.

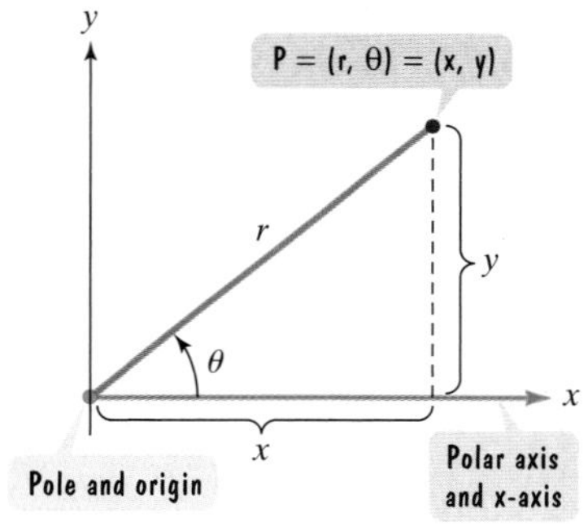

Figure 7.23 Polar and rectangular coordinate systems

Relations between Polar and Rectangular Coordinates

$$x = r\cos\theta$$
$$y = r\sin\theta$$
$$x^2 + y^2 = r^2$$
$$\tan\theta = \frac{y}{x}$$

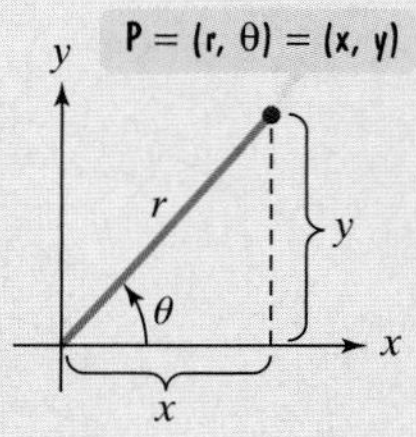

3 Convert a point from polar to rectangular coordinates.

Point Conversion from Polar to Rectangular Coordinates

To convert a point from polar coordinates (r, θ) to rectangular coordinates (x, y), use the formulas $x = r \cos \theta$ and $y = r \sin \theta$.

EXAMPLE 3 Polar-to-Rectangular Point Conversion

Find the rectangular coordinates of the points with the following polar coordinates:

a. $\left(2, \dfrac{3\pi}{2} \right)$ **b.** $\left(-8, \dfrac{\pi}{3} \right)$.

Solution We find (x, y) by substituting the given values for r and θ into $x = r \cos \theta$ and $y = r \sin \theta$.

a. We begin with the rectangular coordinates of the point $(r, \theta) = \left(2, \dfrac{3\pi}{2} \right)$.

$$x = r \cos \theta = 2 \cos \frac{3\pi}{2} = 2 \cdot 0 = 0$$

$$y = r \sin \theta = 2 \sin \frac{3\pi}{2} = 2(-1) = -2$$

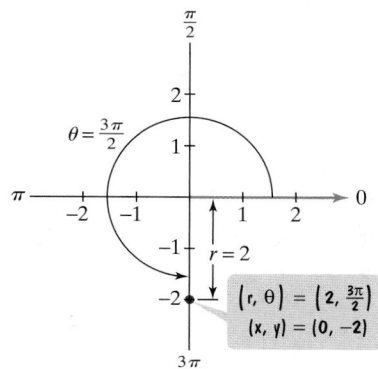

θ = 3π/2, r = 2

$(r, \theta) = \left(2, \frac{3\pi}{2} \right)$
$(x, y) = (0, -2)$

Figure 7.24 Converting $\left(2, \dfrac{3\pi}{2} \right)$ to rectangular coordinates

The rectangular coordinates of $\left(2, \dfrac{3\pi}{2} \right)$ are $(0, -2)$. (See Figure 7.24)

b. We now find the rectangular coordinates of the point $(r, \theta) = \left(-8, \dfrac{\pi}{3} \right)$.

$$x = r \cos \theta = -8 \cos \frac{\pi}{3} = -8 \left(\frac{1}{2} \right) = -4$$

$$y = r \sin \theta = -8 \sin \frac{\pi}{3} = -8 \left(\frac{\sqrt{3}}{2} \right) = -4\sqrt{3}$$

The rectangular coordinates of $\left(-8, \dfrac{\pi}{3} \right)$ are $\left(-4, -4\sqrt{3} \right)$.

Technology

Some graphing utilities can convert a point from polar coordinates to rectangular coordinates. Consult your manual. The screen on the right verifies the polar-rectangular conversion in Example 3(a). It shows that the rectangular coordinates of $(r, \theta) = \left(2, \dfrac{3\pi}{2} \right)$ are $(0, -2)$.

Notice that the x- and y-coordinates are displayed separately.

```
P▸Rx(2,3π/2)
                    0
P▸Ry(2,3π/2)
                   -2
```

Check Point 3

Find the rectangular coordinates of the points with the following polar coordinates:

a. $(3, \pi)$ **b.** $\left(-10, \dfrac{\pi}{6}\right)$.

4 Convert a point from rectangular to polar coordinates.

Point Conversion from Rectangular to Polar Coordinates

Conversion from rectangular coordinates (x, y) to polar coordinates (r, θ) is a bit more complicated. Keep in mind that there are infinitely many representations for a point in polar coordinates. If the point (x, y) lies in one of the four quadrants, we will use a representation in which

- r is positive, and
- θ is the smallest positive angle that lies in the same quadrant as (x, y).

These conventions provide the following procedure.

Converting a point from Rectangular to Polar Coordinates
($r > 0$ and $0 \le \theta < 2\pi$)

1. Plot the point (x, y).

2. Find r by computing the distance from the origin to (x, y): $r = \sqrt{x^2 + y^2}$.

3. Find θ using $\tan\theta = \dfrac{y}{x}$ with θ lying in the same quadrant as (x, y).

EXAMPLE 4   Rectangular-to-Polar Point Conversion

Find polar coordinates of a point whose rectangular coordinates are $\left(-1, \sqrt{3}\right)$.

Solution   We begin with $(x, y) = \left(-1, \sqrt{3}\right)$ and use our three-step procedure to find a set of polar coordinates (r, θ).

Step 1   Plot the point (x, y).   The point $\left(-1, \sqrt{3}\right)$ is plotted in quadrant II in Figure 7.25.

Step 2   Find r by computing the distance from the origin to (x, y).

$$r = \sqrt{x^2 + y^2} = \sqrt{(-1)^2 + \left(\sqrt{3}\right)^2} = \sqrt{1 + 3} = \sqrt{4} = 2$$

Step 3   Find θ using $\tan\theta = \dfrac{y}{x}$ with θ lying in the same quadrant as (x, y).

$$\tan\theta = \frac{y}{x} = \frac{\sqrt{3}}{-1} = -\sqrt{3}$$

We know that $\tan\dfrac{\pi}{3} = \sqrt{3}$. Because θ lies in quadrant II,

$$\theta = \pi - \frac{\pi}{3} = \frac{3\pi}{3} - \frac{\pi}{3} = \frac{2\pi}{3}.$$

Polar coordinates of $\left(-1, \sqrt{3}\right)$ are $(r, \theta) = \left(2, \dfrac{2\pi}{3}\right)$.

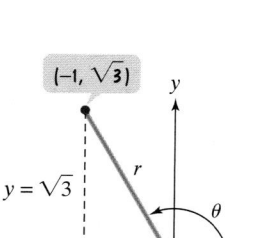

Figure 7.25 Converting $\left(-1, \sqrt{3}\right)$ to polar coordinates

Technology

The screen shows the rectangular-polar conversion for $(-1, \sqrt{3})$ on a graphing utility. In Example 4, we showed that the polar coordinates of $(x, y) = (-1, \sqrt{3})$ are $(r, \theta) = \left(2, \dfrac{2\pi}{3}\right)$. Using $\dfrac{2\pi}{3} \approx 2.09439510239$ verifies that our conversion is correct. Notice that the r- and (approximate) θ-coordinates are displayed separately.

```
R▶Pr(-1,√(3))
                    2
R▶Pθ(-1,√(3))
            2.094395102
```

Check Point 4   Find polar coordinates of a point whose rectangular coordinates are $(1, -\sqrt{3})$.

If a point (x, y) lies on a positive or negative axis, we use a representation in which

- r is positive, and
- θ is the smallest quadrantal angle that lies on the same positive or negative axis as (x, y).

In these cases, you can find r and θ by plotting (x, y) and inspecting the figure. Let's see how this is done.

EXAMPLE 5   Rectangular-to-Polar Point Conversion

Find polar coordinates of a point whose rectangular coordinates are $(-2, 0)$.

Solution   We begin with $(x, y) = (-2, 0)$ and find a set of polar coordinates (r, θ).

Step 1   Plot the point (x, y).   The point $(-2, 0)$ is plotted in Figure 7.26.

Step 2   Find r, the distance from the origin to (x, y).   Can you tell by looking at Figure 7.26 that this distance is 2?

$$r = \sqrt{x^2 + y^2} = \sqrt{(-2)^2 + 0^2} = \sqrt{4} = 2$$

Step 3   Find θ with θ lying on the same positive or negative axis as (x, y). The point $(-2, 0)$ is on the negative x-axis. Thus, θ lies on the negative x-axis and $\theta = \pi$. Polar coordinates of $(-2, 0)$ are $(2, \pi)$.

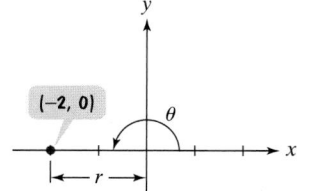

Figure 7.26 Converting $(-2, 0)$ to polar coordinates

Check Point 5   Find polar coordinates of a point whose rectangular coordinates are $(0, -4)$.

 Convert an equation from rectangular to polar coordinates.

Equation Conversion from Rectangular to Polar Coordinates

A **polar equation** is an equation whose variables are r and θ. Two examples of polar equations are

$$r = \frac{5}{\cos\theta + \sin\theta} \quad \text{and} \quad r = \csc\theta.$$

To convert a rectangular equation in x and y to a polar equation in r and θ, replace x by $r\cos\theta$ and y by $r\sin\theta$.

EXAMPLE 6   Converting an Equation from Rectangular to Polar Coordinates

Convert $x + y = 5$ to a polar equation.

Solution Our goal is to obtain an equation in which the variables are r and θ rather than x and y. We use $x = r\cos\theta$ and $y = r\sin\theta$.

$$x + y = 5 \quad \text{This is the given equation in rectangular coordinates.}$$
$$r\cos\theta + r\sin\theta = 5 \quad \text{Replace } x \text{ by } r\cos\theta \text{ and } y \text{ by } r\sin\theta.$$

Thus, the polar equation for $x + y = 5$ is $r\cos\theta + r\sin\theta = 5$. We can express this polar equation in a number of equivalent ways, including an equation that gives r in terms of θ.

$$r(\cos\theta + \sin\theta) = 5 \qquad \text{Factor out } r.$$

$$r = \frac{5}{\cos\theta + \sin\theta} \quad \text{Divide both sides of the equation by } \cos\theta + \sin\theta.$$

Check Point 6 Convert $3x - y = 6$ to a polar equation. Express the polar equation with r in terms of θ.

6 Convert an equation from polar to rectangular coordinates.

Equation Conversion from Polar to Rectangular Coordinates

When we convert an equation from polar to rectangular coordinates, our goal is to obtain an equation in which the variables are x and y rather than r and θ. We use one or more of the following equations:

$$r^2 = x^2 + y^2 \qquad r\cos\theta = x \qquad r\sin\theta = y \qquad \tan\theta = \frac{y}{x}.$$

To obtain the expressions on the left in each of these equations, it is sometimes necessary to do something to the given polar equation. This could include squaring both sides, using an identity, taking the tangent of both sides, or multiplying both sides by r.

EXAMPLE 7   Converting Equations from Polar to Rectangular Form

Convert each polar equation to a rectangular equation in x and y.

a. $r = 3$ **b.** $\theta = \dfrac{\pi}{4}$ **c.** $r = \csc\theta$

Solution

a. We use $r^2 = x^2 + y^2$ to convert the polar equation $r = 3$ to a rectangular equation.

$$r = 3 \quad \text{This is the given polar equation.}$$

$$r^2 = 9 \quad \text{Square both sides.}$$

$$x^2 + y^2 = 9 \quad \text{Use } r^2 = x^2 + y^2 \text{ on the left side.}$$

The rectangular equation for $r = 3$ is $x^2 + y^2 = 9$.

b. We use $\tan\theta = \dfrac{y}{x}$ to convert the polar equation $\theta = \dfrac{\pi}{4}$ to a rectangular equation in x and y.

$$\theta = \frac{\pi}{4} \qquad \text{This is the given polar equation.}$$

$$\tan\theta = \tan\frac{\pi}{4} \qquad \text{Take the tangent of both sides.}$$

$$\tan\theta = 1 \qquad \tan\frac{\pi}{4} = 1$$

$$\frac{y}{x} = 1 \qquad \text{Use } \tan\theta = \frac{y}{x} \text{ on the left side.}$$

$$y = x \qquad \text{Multiply both sides by x.}$$

The rectangular equation for $\theta = \dfrac{\pi}{4}$ is $y = x$.

c. We use $r\sin\theta = y$ to convert the polar equation $r = \csc\theta$ to a rectangular equation. To do this, we express the cosecant in terms of the sine.

$$r = \csc\theta \quad \text{This is the given polar equation.}$$

$$r = \frac{1}{\sin\theta} \quad \csc\theta = \frac{1}{\sin\theta}$$

$$r\sin\theta = 1 \quad \text{Multiply both sides by } \sin\theta.$$

$$y = 1 \quad \text{Use } r\sin\theta = y \text{ on the left side.}$$

The rectangular equation for $r = \csc\theta$ is $y = 1$.

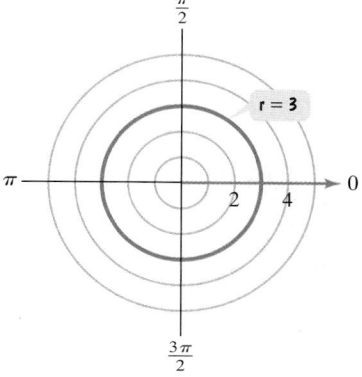

Figure 7.27 The equations $r = 3$ and $x^2 + y^2 = 9$ have the same graph.

Converting a polar equation to a rectangular equation may be a useful way to develop or check a graph. For example, the graph of the polar equation $r = 3$ consists of all points that are three units from the pole. Thus, the graph is a circle centered at the pole with radius $= 3$. The rectangular equation for $r = 3$, namely $x^2 + y^2 = 9$, has precisely the same graph (see Figure 7.27). We will discuss graphs of polar equations in the next section.

Check Point 7

Convert each polar equation to a rectangular equation in x and y.

a. $r = 4$ **b.** $\theta = \dfrac{3\pi}{4}$ **c.** $r = \sec\theta$

EXERCISE SET 7.3

Practice Exercises

In Exercises 1–10, indicate if the point with the given polar coordinates is represented by A, B, C, or D on the graph.

1. $(3, 225°)$ **2.** $(3, 315°)$

3. $\left(-3, \dfrac{5\pi}{4}\right)$ **4.** $\left(-3, \dfrac{\pi}{4}\right)$

5. $(3, \pi)$ **6.** $(-3, 0)$

7. $(3, -135°)$ **8.** $(3, -315°)$

9. $\left(-3, -\dfrac{3\pi}{4}\right)$ **10.** $\left(-3, -\dfrac{5\pi}{4}\right)$

In Exercises 11–20, use a polar coordinate system like the one shown for Exercises 1–10 to plot each point with the given polar coordinates.

11. $(2, 45°)$ **12.** $(1, 45°)$

13. $(3, 90°)$ **14.** $(2, 270°)$

15. $\left(3, \dfrac{4\pi}{3}\right)$ **16.** $\left(3, \dfrac{7\pi}{6}\right)$

17. $(-1, \pi)$ **18.** $\left(-1, \dfrac{3\pi}{2}\right)$

19. $\left(-2, -\dfrac{\pi}{2}\right)$ **20.** $(-3, -\pi)$

In Exercises 21–26, use a polar coordinate system like the one shown for Exercises 1–10 to plot each point with the given polar coordinates. Then find another representation (r, θ) of this point in which:

a. $r > 0,\quad 2\pi < \theta < 4\pi.$
b. $r < 0,\quad 0 < \theta < 2\pi.$
c. $r > 0,\quad -2\pi < \theta < 0.$

21. $\left(5, \dfrac{\pi}{6}\right)$ **22.** $\left(8, \dfrac{\pi}{6}\right)$

23. $\left(10, \dfrac{3\pi}{4}\right)$ **24.** $\left(12, \dfrac{2\pi}{3}\right)$

25. $\left(4, \dfrac{\pi}{2}\right)$ **26.** $(6, \pi)$

In Exercises 27–34, polar coordinates of a point are given. Find the rectangular coordinates of each point.

27. $(4, 90°)$ **28.** $(6, 180°)$

29. $\left(2, \dfrac{\pi}{3}\right)$ **30.** $\left(2, \dfrac{\pi}{6}\right)$

31. $\left(-4, \dfrac{\pi}{2}\right)$ **32.** $\left(-6, \dfrac{3\pi}{2}\right)$

33. $(7.4, 2.5)$ **34.** $(8.3, 4.6)$

In Exercises 35–42, the rectangular coordinates of a point are given. Find polar coordinates of each point.

35. $(-2, 2)$ **36.** $(2, -2)$

37. $(2, -2\sqrt{3})$ **38.** $(-2\sqrt{3}, 2)$

39. $(-\sqrt{3}, -1)$ **40.** $(-1, -\sqrt{3})$

41. $(5, 0)$ **42.** $(0, -6)$

In Exercises 43–52, convert each rectangular equation to a polar equation.

43. $3x + y = 7$ (Express r in terms of θ.)
44. $x + 5y = 8$ (Express r in terms of θ.)

45. $x = 7$ **46.** $y = 3$

47. $x^2 + y^2 = 9$ **48.** $x^2 + y^2 = 16$

49. $x^2 + y^2 = 4x$ **50.** $x^2 + y^2 = 6x$

51. $y^2 = 6x$ **52.** $x^2 = 6y$

In Exercises 53–66, convert each polar equation to a rectangular equation.

53. $r = 8$ **54.** $r = 10$

55. $\theta = \dfrac{\pi}{2}$ **56.** $\theta = \dfrac{\pi}{3}$

57. $r \sin \theta = 3$ **58.** $r \cos \theta = 7$

59. $r = 4 \csc \theta$ **60.** $r = 6 \sec \theta$

61. $r = \sin \theta$ **62.** $r = \cos \theta$

63. $r = 6 \cos \theta + 4 \sin \theta$ **64.** $r = 8 \cos \theta + 2 \sin \theta$

65. $r^2 \sin 2\theta = 2$ **66.** $r^2 \cos 2\theta = 2$

Application Exercises

Use the figure of the merry-go-round to solve Exercises 67–68. There are four circles of horses. Each circle is three feet from the next circle. The radius of the inner circle is 6 feet.

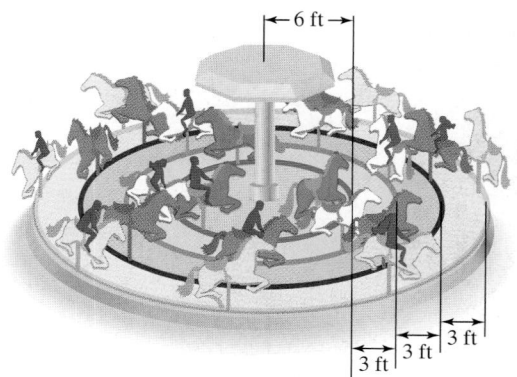

← 6 ft →

3 ft | 3 ft | 3 ft
3 ft

67. If a horse in the outer circle is $\frac{2}{3}$ of the way around the merry-go-round, give its polar coordinates.

68. If a horse in the inner circle is $\frac{5}{6}$ of the way around the merry-go-round, give its polar coordinates.

The wind is blowing at 10 knots. Sailboat racers look for a sailing angle to the 10-knot wind that produces maximum sailing speed. In this application, (r, θ) describes the sailing speed, r, in knots, at an angle θ to the 10-knot wind. Use this information to solve Exercises 69–71.

69. Interpret the polar coordinates: $(6.3, 50°)$.

70. Interpret the polar coordinates: $(7.4, 85°)$.

71. Four points in this 10-knot-wind situation are $(6.3, 50°)$, $(7.4, 85°)$, $(7.5, 105°)$, $(7.3, 135°)$. Based on these points, which sailing angle to the 10-knot wind would you recommend to a serious sailboat racer? What sailing speed is achieved at this angle?

Writing in Mathematics

72. Explain how to plot (r, θ) if $r > 0$ and $\theta > 0$.

73. Explain how to plot (r, θ) if $r < 0$ and $\theta > 0$.

74. If you are given polar coordinates of a point, explain how to find two additional sets of polar coordinates for the point.

75. Explain how to convert a point from polar to rectangular coordinates. Provide an example with your explanation.

76. Explain how to convert a point from rectangular to polar coordinates. Provide an example with your explanation.

77. Explain how to convert from a rectangular equation to a polar equation.

78. In converting $r = 5$ from a polar equation to a rectangular equation, describe what should be done to both sides of the equation and why this should be done.

79. In converting $r = \sin \theta$ from a polar equation to a rectangular equation, describe what should be done to both sides of the equation and why this should be done.

80. Suppose that (r, θ) describes the sailing speed, r, in knots, at an angle θ to a wind blowing at 20 knots. You have a list of all ordered pairs (r, θ) for integral angles from $\theta = 0°$ to $\theta = 180°$. Describe a way to present this information so that a serious sailboat racer can visualize sailing speeds at different sailing angles to the wind.

 ## Technology Exercises

In Exercises 81–83, polar coordinates of a point are given. Use a graphing utility to find the rectangular coordinates of each point to three decimal places.

81. $\left(4, \dfrac{2\pi}{3}\right)$

82. $(5.2, 1.7)$

83. $(-4, 1.088)$

In Exercises 84–86, the rectangular coordinates of a point are given. Use a graphing utility to find polar coordinates of each point to three decimal places.

84. $(-5, 2)$

85. $(\sqrt{5}, 2)$

86. $(-4.308, -7.529)$

Critical Thinking Exercises

87. Prove that the distance d between two points with polar coordinates (r_1, θ_1) and (r_2, θ_2) is
$$d = \sqrt{r_1^2 + r_2^2 - 2r_1r_2 \cos(\theta_2 - \theta_1)}.$$

88. Use the formula in Exercise 87 to find the distance between $\left(2, \dfrac{5\pi}{6}\right)$ and $\left(4, \dfrac{\pi}{6}\right)$. Express the answer in simplified radical form.

89. Convert $r = 4 \cos \theta$ from a polar equation to a rectangular equation. Use the rectangular equation to give the center and the radius.

SECTION 7.4 *Graphs of Polar Equations*

Objectives

1. Graph polar equations.
2. Use symmetry to graph polar equations.

The America's Cup is the supreme event in ocean sailing. Competition is fierce and the costs are huge. Competitors look to mathematics to provide the critical innovation that can make the difference between winning and losing. In this section's exercise set, you will see how graphs of polar equations play a role in sailing faster using mathematics.

Using Polar Grids to Graph Polar Equations

Recall that a **polar equation** is an equation whose variables are r and θ. The **graph of a polar equation** is the set of all points whose polar coordinates satisfy the equation. We use **polar grids** like the one shown in Figure 7.28 to graph polar equations. The grid consists of circles with centers at the pole. This polar grid shows five such circles. A polar grid also shows lines passing through the pole. In this grid, each line represents an angle for which we know the exact values of the trigonometric functions.

Many polar coordinate grids show more circles and more lines through the pole than in Figure 7.28. See if your campus bookstore has paper with polar grids and use the polar graph paper throughout this section.

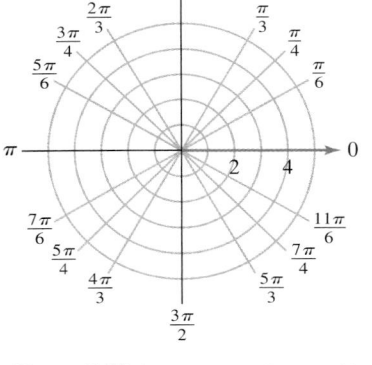

Figure 7.28 A polar coordinate grid

 Graph polar equations.

Graphing a Polar Equation by Point Plotting

One method for graphing a polar equation such as $r = 4\cos\theta$ is the **point-plotting method**. First, we make a table of values that satisfy the equation. Next, we plot these ordered pairs as points in the polar coordinate system. Finally, we connect the points with a smooth curve. This often gives us a picture of all ordered pairs (r, θ) that satisfy the equation.

> **EXAMPLE 1 Graphing an Equation Using the Point-Plotting Method**

Graph the polar equation $r = 4\cos\theta$ with θ in radians.

Solution We construct a partial table of coordinates using multiples of $\dfrac{\pi}{6}$. Then we plot the points and join them with a smooth curve, as shown in Figure 7.29.

θ	$r = 4 \cos \theta$	(r, θ)
0	$4 \cos 0 = 4 \cdot 1 = 4$	$(4, 0)$
$\dfrac{\pi}{6}$	$4 \cos \dfrac{\pi}{6} = 4 \cdot \dfrac{\sqrt{3}}{2} = 2\sqrt{3} \approx 3.5$	$\left(3.5, \dfrac{\pi}{6}\right)$
$\dfrac{\pi}{3}$	$4 \cos \dfrac{\pi}{3} = 4 \cdot \dfrac{1}{2} = 2$	$\left(2, \dfrac{\pi}{3}\right)$
$\dfrac{\pi}{2}$	$4 \cos \dfrac{\pi}{2} = 4 \cdot 0 = 0$	$\left(0, \dfrac{\pi}{2}\right)$
$\dfrac{2\pi}{3}$	$4 \cos \dfrac{2\pi}{3} = 4\left(-\dfrac{1}{2}\right) = -2$	$\left(-2, \dfrac{2\pi}{3}\right)$
$\dfrac{5\pi}{6}$	$4 \cos \dfrac{5\pi}{6} = 4\left(-\dfrac{\sqrt{3}}{2}\right) = -2\sqrt{3} \approx -3.5$	$\left(-3.5, \dfrac{5\pi}{6}\right)$
π	$4 \cos \pi = 4(-1) = -4$	$(-4, \pi)$

Values of r repeat.

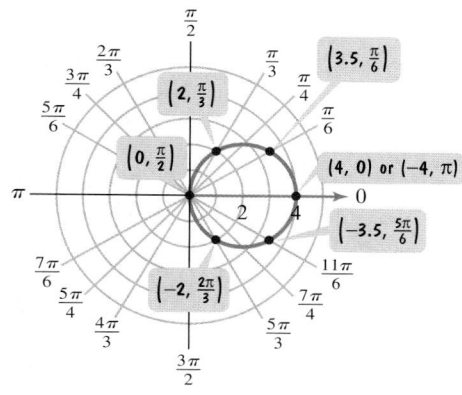

Figure 7.29 The graph of $r = 4 \cos \theta$

Technology

A graphing utility can be used to obtain the graph of a polar equation. Use the polar mode with angle measure in radians. You must enter the minimum and maximum values for θ and an increment setting for θ, called θstep. θstep determines the number of points that the graphing utility will plot. Make θstep relatively small so that a significant number of points are plotted.

Shown is the graph of $r = 4 \cos \theta$ in a $[-5, 5, 1]$ by $[-5, 5, 1]$ viewing rectangle with

$$\theta \min = 0$$
$$\theta \max = 2\pi$$
$$\theta \text{step} = \dfrac{\pi}{48}.$$

A square setting was used.

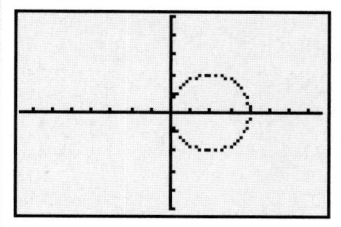

The graph of $r = 4 \cos \theta$ in Figure 7.29 looks like a circle of radius 2 whose center is at the point $(x, y) = (2, 0)$. We can verify this observation by changing the polar equation to a rectangular equation.

$r = 4 \cos \theta$	This is the given polar equation.
$r^2 = 4r \cos \theta$	Multiply both sides by r.
$x^2 + y^2 = 4x$	Convert to rectangular coordinates: $r^2 = x^2 + y^2$ and $r \cos \theta = x$.
$x^2 - 4x + y^2 = 0$	Subtract $4x$ from both sides.
$x^2 - 4x + 4 + y^2 = 4$	Complete the square on x: $\frac{1}{2}(-4) = -2$ and $(-2)^2 = 4$. Add 4 to both sides.
$(x - 2)^2 + y^2 = 2^2$	Factor.

This last equation is the standard form of the equation of a circle, $(x - h)^2 + (y - k)^2 = r^2$, with radius r and center at (h, k). Thus, the radius is 2 and the center is at $(h, k) = (2, 0)$.

In general, circles have simpler equations in polar form than in rectangular form.

Circles in Polar Coordinates

The graphs of

$$r = a \cos \theta \qquad \text{and} \qquad r = a \sin \theta$$

are circles.

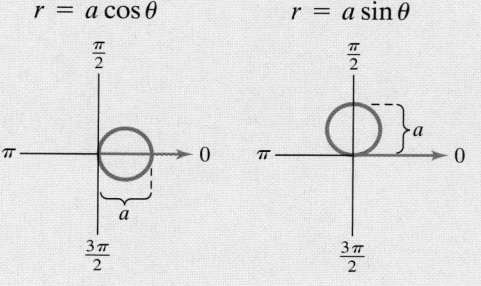

Graph the equation $r = 4 \sin \theta$ with θ in radians. Use multiples of $\dfrac{\pi}{6}$ from 0 to π to generate coordinates for points (r, θ).

Check Point 1

2 Use symmetry to graph polar equations.

Graphing a Polar Equation Using Symmetry

If the graph of a polar equation exhibits symmetry, you may be able to graph it more quickly. Three types of symmetry can be helpful.

Tests for Symmetry in Polar Coordinates

Symmetry with Respect to the Polar Axis (x-Axis)

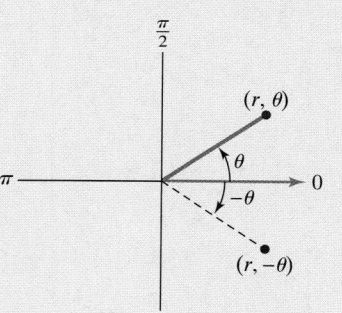

Replace θ by $-\theta$. If an equivalent equation results, the graph is symmetric with respect to the polar axis.

Symmetry with Respect to the Line $\theta = \dfrac{\pi}{2}$ (y-Axis)

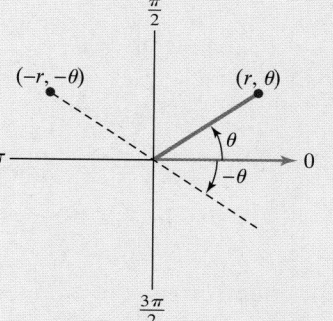

Replace (r, θ) by $(-r, -\theta)$. If an equivalent equation results, the graph is symmetric with respect to $\theta = \dfrac{\pi}{2}$.

Symmetry with Respect to the Pole (Origin)

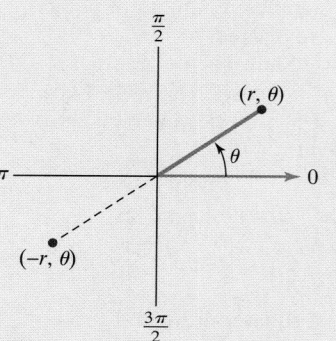

Replace r by $-r$. If an equivalent equation results, the graph is symmetric with respect to the pole.

If a polar equation passes a symmetry test, then its graph exhibits that symmetry. By contrast, if a polar equation fails a symmetry test, then its graph *may or may not* have that kind of symmetry. Thus, the graph of a polar equation may have a symmetry even if it fails a test for that particular symmetry. Nevertheless, the symmetry tests are useful. If we detect symmetry, we can obtain a graph of the equation by plotting fewer points.

EXAMPLE 2 Graphing a Polar Equation Using Symmetry

Check for symmetry and then graph the polar equation:

$$r = 1 - \cos\theta.$$

Solution We apply each of the tests for symmetry.

Polar Axis: Replace θ by $-\theta$ in $r = 1 - \cos\theta$:

$$r = 1 - \cos(-\theta) \qquad \text{Replace } \theta \text{ by } -\theta \text{ in } r = 1 - \cos\theta.$$

$$r = 1 - \cos\theta \qquad \text{The cosine function is even:}$$
$$\cos(-\theta) = \cos\theta.$$

Because the polar equation does not change when θ is replaced by $-\theta$, the graph is symmetric with respect to the polar axis.

The Line $\theta = \dfrac{\pi}{2}$: Replace (r, θ) by $(-r, -\theta)$ in $r = 1 - \cos\theta$:

$$-r = 1 - \cos(-\theta) \qquad \text{Replace } r \text{ by } -r \text{ and } \theta \text{ by } -\theta \text{ in } r = 1 - \cos\theta.$$

$$-r = 1 - \cos\theta \qquad \cos(-\theta) = \cos\theta.$$

$$r = \cos\theta - 1 \qquad \text{Multiply both sides by } -1.$$

Because the polar equation $r = 1 - \cos\theta$ changes to $r = \cos\theta - 1$ when (r, θ) is replaced by $(-r, -\theta)$, the equation fails this symmetry test. The graph may or may not be symmetric with respect to the line $\theta = \dfrac{\pi}{2}$.

The Pole: Replace r by $-r$ in $r = 1 - \cos\theta$:

$$-r = 1 - \cos\theta \qquad \text{Replace } r \text{ by } -r.$$

$$r = \cos\theta - 1 \qquad \text{Multiply both sides by } -1.$$

Because the polar equation $r = 1 - \cos\theta$ changes to $r = \cos\theta - 1$ when r is replaced by $-r$, the equation fails this symmetry test. The graph may or may not be symmetric with respect to the pole.

Now we are ready to graph $r = 1 - \cos\theta$. Because the period of the cosine function is 2π, we need not consider values of θ beyond 2π. Recall that we discovered the graph of the equation $r = 1 - \cos\theta$ has symmetry with respect to the polar axis. Because the graph has symmetry, we may be able to obtain a complete graph without plotting points generated by values of θ from 0 to 2π. Let's start by finding the values of r for values of θ from 0 to π.

Technology

The graph of

$$r = 1 - \cos\theta$$

was obtained using a $[-2, 2, 1]$ by $[-2, 2, 1]$ viewing rectangle and

$$\theta\min = 0, \quad \theta\max = 2\pi,$$

$$\theta\text{step} = \frac{\pi}{48}.$$

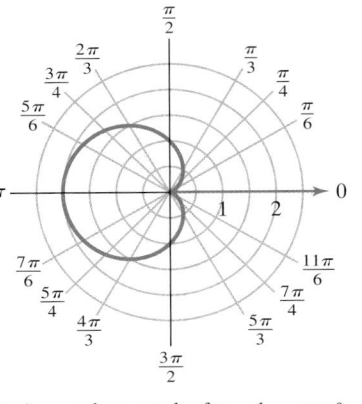

The values for r and θ are in the table above Figure 7.30(a). The TABLE feature on some graphing utilities is the most efficient way to create these values. The points in the table are plotted in Figure 7.30(a). Examine the graph. Keep in mind that the graph must be symmetric with respect to the polar axis. Thus, if we reflect the graph in Figure 7.30(a) about the polar axis, we will obtain a complete graph of $r = 1 - \cos\theta$. This graph is shown in Figure 7.30(b).

θ	0	$\dfrac{\pi}{6}$	$\dfrac{\pi}{3}$	$\dfrac{\pi}{2}$	$\dfrac{2\pi}{3}$	$\dfrac{5\pi}{6}$	π
r	0	0.13	0.50	1.00	1.50	1.87	2

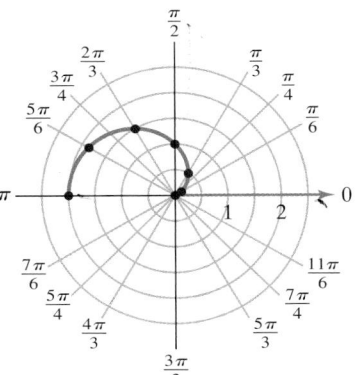

(a) The graph of $r = 1 - \cos\theta$ for $0 \le \theta \le \pi$

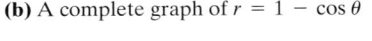

(b) A complete graph of $r = 1 - \cos\theta$

Figure 7.30 Graphing $r = 1 - \cos\theta$

> **Check Point 2**
>
> Check for symmetry and then graph the polar equation:
>
> $$r = 1 + \cos\theta.$$

EXAMPLE 3 Graphing a Polar Equation

Graph the polar equation: $r = 1 + 2\sin\theta$.

Solution We first check for symmetry.

$$r = 1 + 2\sin\theta$$

Polar Axis	**The Line $\theta = \dfrac{\pi}{2}$**	**The Pole**
Replace θ by $-\theta$.	Replace (r, θ) by $(-r, -\theta)$.	Replace r by $-r$.
$r = 1 + 2\sin(-\theta)$	$-r = 1 + 2\sin(-\theta)$	$-r = 1 + 2\sin\theta$
$r = 1 + 2(-\sin\theta)$	$-r = 1 - 2\sin\theta$	$r = -1 - 2\sin\theta$
$r = 1 - 2\sin\theta$	$r = -1 + 2\sin\theta$	

Each equation is not equivalent to $r = 1 + 2\sin\theta$. Thus, the graph may or may not have each of these kinds of symmetry.

Now we are ready to graph $r = 1 + 2\sin\theta$. Because the period of the sine function is 2π, we need not consider values of θ beyond 2π. We identify points on the graph of $r = 1 + 2 \sin \theta$ by assigning values to θ and calculating the corresponding values of r. The values for r and θ are in the tables above Figures 7.31(a), (b), and (c). The complete graph of $r = 1 + 2 \sin \theta$ is shown in Figure 7.31(c). The inner loop indicates that the graph passes through the pole twice.

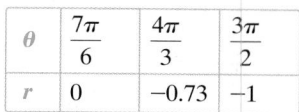

θ	0	$\dfrac{\pi}{6}$	$\dfrac{\pi}{3}$	$\dfrac{\pi}{2}$	$\dfrac{2\pi}{3}$	$\dfrac{5\pi}{6}$	π
r	1	2	2.73	3	2.73	2	1

θ	$\dfrac{7\pi}{6}$	$\dfrac{4\pi}{3}$	$\dfrac{3\pi}{2}$
r	0	-0.73	-1

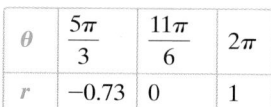

θ	$\dfrac{5\pi}{3}$	$\dfrac{11\pi}{6}$	2π
r	-0.73	0	1

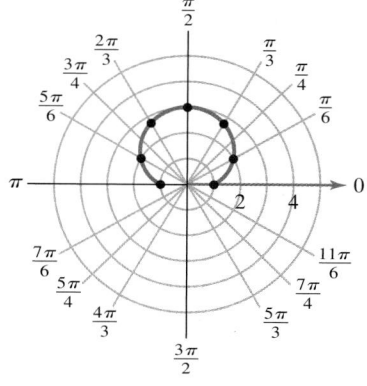

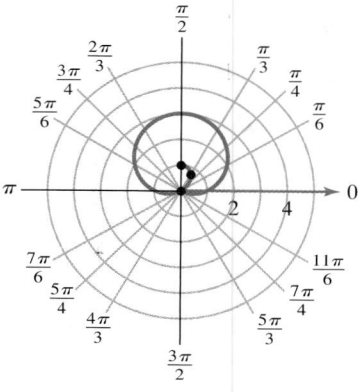

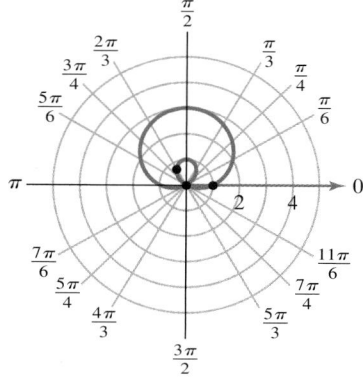

(a) The graph of $r = 1 + 2 \cos \theta$ for $0 \le \theta \le \pi$

(b) The graph of $r = 1 + 2 \sin \theta$ for $0 \le \theta \le \dfrac{3\pi}{2}$

(c) The complete graph of $r = 1 + 2 \sin \theta$ for $0 \le \theta \le 2\pi$

Figure 7.31 Graphing $r = 1 + 2 \sin \theta$

We're not quite sure if the polar graph in Figure 7.31(c) looks like a snail. However, the graph is called a *limaçon*, which is a French word for snail. Limaçons come with and without inner loops.

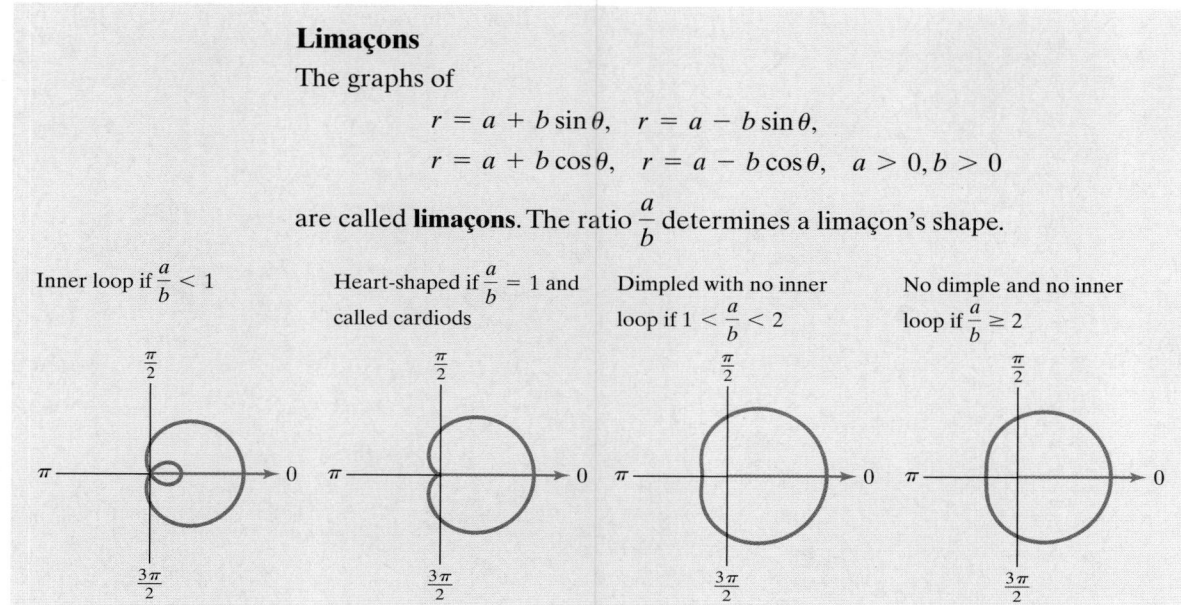

Limaçons

The graphs of

$$r = a + b \sin \theta, \quad r = a - b \sin \theta,$$
$$r = a + b \cos \theta, \quad r = a - b \cos \theta, \quad a > 0, b > 0$$

are called **limaçons**. The ratio $\dfrac{a}{b}$ determines a limaçon's shape.

Inner loop if $\dfrac{a}{b} < 1$

Heart-shaped if $\dfrac{a}{b} = 1$ and called cardiods

Dimpled with no inner loop if $1 < \dfrac{a}{b} < 2$

No dimple and no inner loop if $\dfrac{a}{b} \ge 2$

Check Point 3 Graph the polar equation: $r = 1 - 2\sin\theta$.

EXAMPLE 4 Graphing a Polar Equation

Graph the polar equation: $r = 4\sin 2\theta$.

Solution We first check for symmetry.

$$r = 4\sin 2\theta$$

Polar Axis	**The Line $\theta = \dfrac{\pi}{2}$**	**The Pole**
Replace θ by $-\theta$.	Replace (r, θ) by $(-r, -\theta)$.	Replace r by $-r$.
$r = 4\sin 2(-\theta)$	$-r = 4\sin 2(-\theta)$	$-r = 4\sin 2\theta$
$r = 4\sin(-2\theta)$	$-r = 4\sin(-2\theta)$	$r = -4\sin 2\theta$
$r = -4\sin 2\theta$	$-r = -4\sin 2\theta$	
	$r = 4\sin 2\theta$	Equation changes and fails this symmetry test.
Equation changes and fails this symmetry test.	Equation does not change.	

Thus, we can be sure that the graph is symmetric with respect to $\theta = \dfrac{\pi}{2}$. The graph may or may not be symmetric with respect to the polar axis or the pole.

Now we are ready to graph $r = 4\sin 2\theta$. In Figure 7.32(a), we identify points on the graph of $r = 4\sin 2\theta$ by assigning values to θ from 0 to $\dfrac{\pi}{2}$ and calculating the corresponding values of r. Because the graph is symmetric with respect to $\theta = \dfrac{\pi}{2}$, we can reflect the graph in Figure 7.32(a) about $\theta = \dfrac{\pi}{2}$ and obtain the graph from 0 to π. This graph is shown in Figure 7.32(b).

θ	0	$\dfrac{\pi}{6}$	$\dfrac{\pi}{4}$	$\dfrac{\pi}{3}$	$\dfrac{\pi}{2}$
r	0	3.46	4	3.46	0

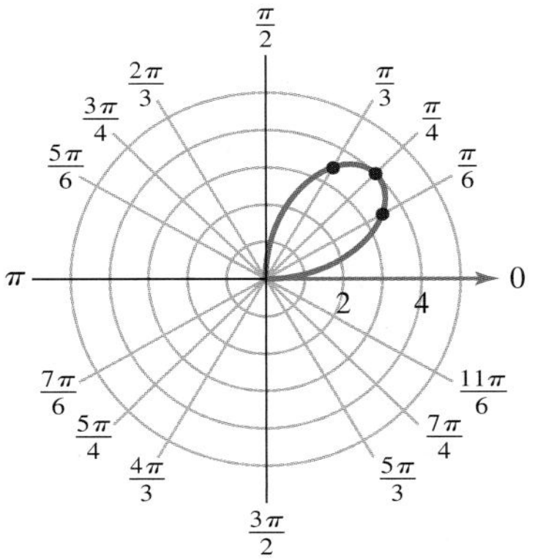

(a) The graph of $r = 4\sin 2\theta$ for $0 \le \theta \le \dfrac{\pi}{2}$

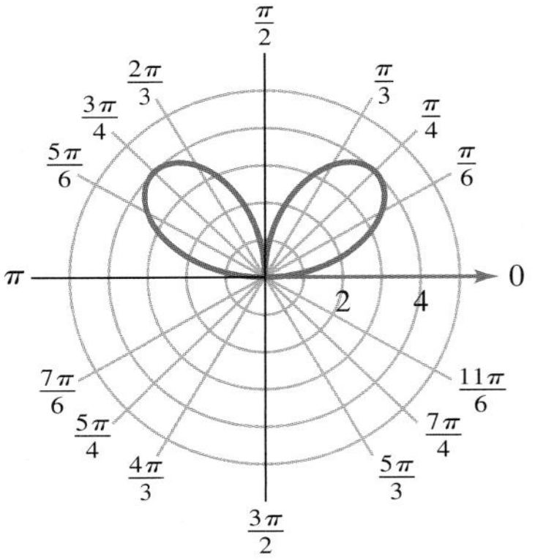

(b) Using symmetry to obtain the graph of $r = 4\sin 2\theta$ for $0 \le \theta \le \pi$

Figure 7.32 A partial graph of $r = 4\sin 2\theta$

Now we can complete the graph of $r = 4 \sin 2\theta$. The values for r and θ above the graph in Figure 7.33(a) give us the graph for $0 \le \theta \le \dfrac{3\pi}{2}$. Because the graph is symmetric with respect to $\theta = \dfrac{\pi}{2}$, we can reflect the quadrant III portion of the graph in Figure 7.33(a) about $\theta = \dfrac{\pi}{2}$ and obtain the complete graph from 0 to 2π. This graph is shown in Figure 7.33(b).

θ	π	$\dfrac{7\pi}{6}$	$\dfrac{5\pi}{4}$	$\dfrac{4\pi}{3}$	$\dfrac{3\pi}{2}$
r	0	3.46	4	3.46	0

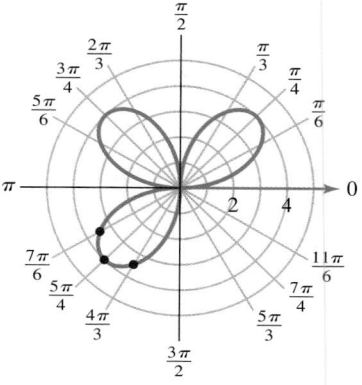

(a) The graph of $r = 4 \sin 2\theta$ for $0 \le \theta \le \dfrac{3\pi}{2}$

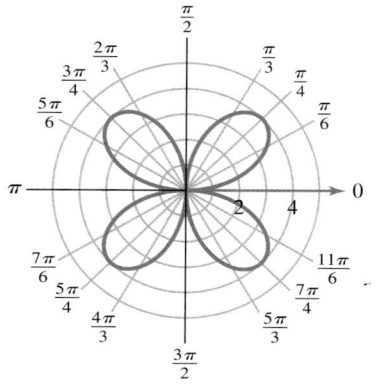

(b) Using symmetry to obtain the graph of $r = 4 \sin 2\theta$ for $0 \le \theta \le 2\pi$

Figure 7.33 A complete graph of $r = 4 \sin 2\theta$

The curve in Figure 7.33(b) is called a **rose with four petals**.

Technology

The graph of
$$r = 4 \sin 2\theta$$
was obtained using a $[-4, 4, 1]$ by $[-4, 4, 1]$ viewing rectangle and
$$\theta\min = 0, \quad \theta\max = 2\pi,$$
$$\theta\text{step} = \dfrac{\pi}{48}.$$

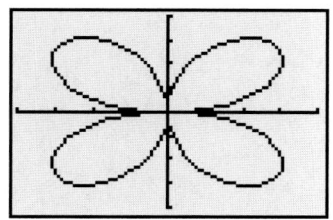

Rose Curves

The graphs of

$$r = a \sin n\theta \quad \text{and} \quad r = a \cos n\theta, \quad a \neq 0,$$

are called **rose curves**. If n is even, the rose has $2n$ petals. If n is odd, the rose has n petals.

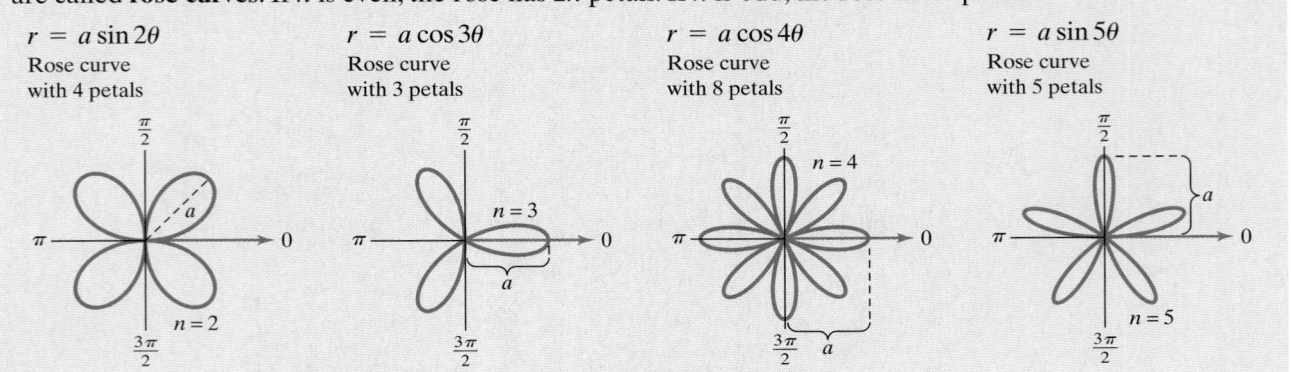

$r = a \sin 2\theta$
Rose curve
with 4 petals

$r = a \cos 3\theta$
Rose curve
with 3 petals

$r = a \cos 4\theta$
Rose curve
with 8 petals

$r = a \sin 5\theta$
Rose curve
with 5 petals

Check Point 4   Graph the polar equation: $r = 3 \cos 2\theta$.

EXAMPLE 5   Graphing a Polar Equation

Graph the polar equation: $r^2 = 4 \sin 2\theta$.

Solution   We first check for symmetry.

$$r^2 = 4 \sin 2\theta$$

Polar Axis	**The Line $\theta = \dfrac{\pi}{2}$**	**The Pole**
Replace θ by $-\theta$.	Replace (r, θ) by $(-r, -\theta)$.	Replace r by $-r$.
$r^2 = 4 \sin 2(-\theta)$	$(-r)^2 = 4 \sin 2(-\theta)$	$(-r)^2 = 4 \sin 2\theta$
$r^2 = 4 \sin(-2\theta)$	$r^2 = 4 \sin(-2\theta)$	$r^2 = 4 \sin 2\theta$
$r^2 = -4 \sin 2\theta$	$r^2 = -4 \sin 2\theta$	
Equation changes and fails this symmetry test.	Equation changes and fails this symmetry test.	Equation does not change.

Thus, we can be sure that the graph is symmetric with respect to the pole. The graph may or may not be symmetric with respect to the polar axis or the line $\theta = \dfrac{\pi}{2}$.

Now we are ready to graph $r^2 = 4 \sin 2\theta$. In Figure 7.34(a), we identify points on the graph by assigning values to θ from 0 to $\dfrac{\pi}{2}$ and calculating the corresponding values of r. Notice that the points in Figure 7.34(a) are shown for $r \geq 0$. Because the graph is symmetric with respect to the pole, we can reflect the graph in Figure 7.34(a) about the pole and obtain the graph in Figure 7.34(b).

θ	0	$\dfrac{\pi}{6}$	$\dfrac{\pi}{4}$	$\dfrac{\pi}{3}$	$\dfrac{\pi}{2}$
r	0	± 1.9	± 2	± 1.9	0

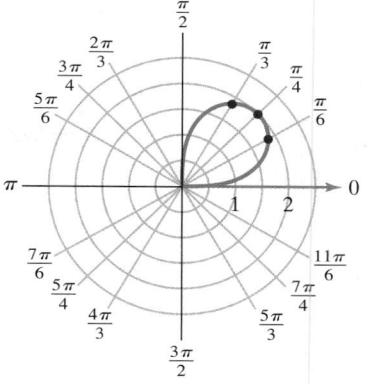

Figure 7.34 (a) The graph of
$r^2 = 4\sin 2\theta$ for $0 \le \theta \le \dfrac{\pi}{2}$

Figure 7.34 (b) Using symmetry
with respect to the pole on the graph
of $r^2 = 4\sin 2\theta$

Does Figure 7.34(b) show a complete graph of $r^2 = 4\sin 2\theta$ or do we need to continue graphing in quadrants II and IV? If θ is in quadrant II, 2θ is in quadrant III or IV, where $\sin 2\theta$ is negative. Thus, $4\sin 2\theta$ is negative. However, $r^2 = 4\sin 2\theta$ and r^2 cannot be negative. This means that there are no points on the graph in quadrant II. The same observation applies to quadrant IV. Thus, Figure 7.34(b) shows the complete graph of $r^2 = 4\sin 2\theta$.

The curve in Figure 7.34(b) is shaped like a propeller and is called a **lemniscate.**

Lemniscates

The graphs of

$$r^2 = a^2 \sin 2\theta \quad \text{and} \quad r^2 = a^2 \cos 2\theta, \quad a \ne 0$$

are called **lemniscates.**

$r^2 = a^2 \sin 2\theta$ is symmetric
with respect to the pole.

$r^2 = a^2 \cos 2\theta$ is symmetric
with respect to the polar

axis, $\theta = \dfrac{\pi}{2}$, and the pole.

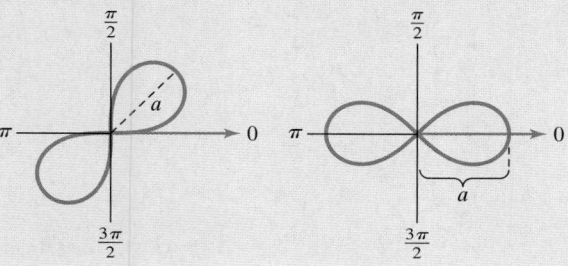

**Check
Point
5**

Graph the polar equation: $r^2 = 4\cos 2\theta$.

EXERCISE SET 7.4

✓ Practice Exercises

In Exercises 1–6, the graph of a polar equation is given. Select the polar equation for each graph from the following options.

$$r = 2 \sin \theta, \quad r = 2 \cos \theta, \quad r = 1 + \sin \theta,$$

$$r = 1 - \sin \theta, \quad r = 3 \sin 2\theta, \quad r = 3 \sin 3\theta$$

1.

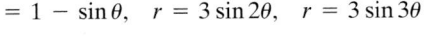

2.

3.

4.

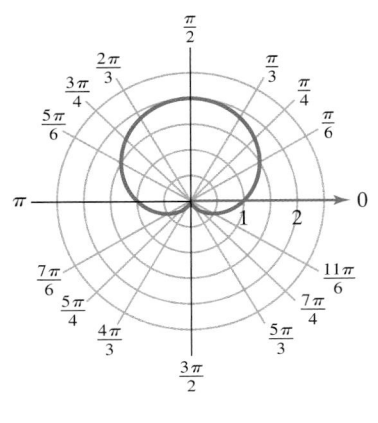

5.

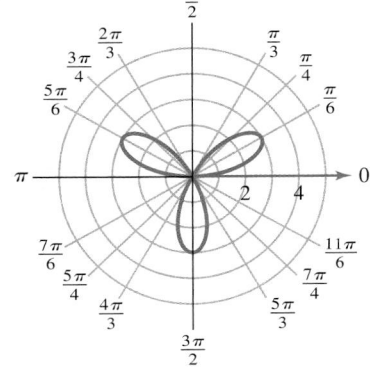

6.

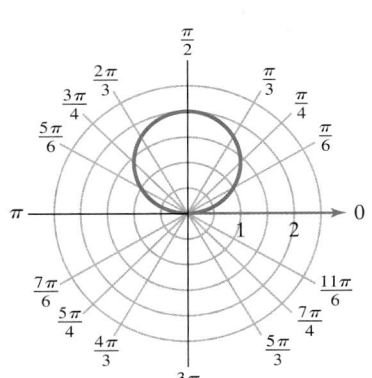

In Exercises 7–12, test for symmetry with respect to:

a. *the polar axis;* **b.** *the line* $\theta = \dfrac{\pi}{2}$; *and* **c.** *the pole.*

7. $r = \sin \theta$ **8.** $r = \cos \theta$

9. $r = 4 + 3 \cos \theta$ **10.** $r = 2 \cos 2\theta$

11. $r^2 = 16 \cos 2\theta$ **12.** $r^2 = 16 \sin 2\theta$

In Exercises 13–34, graph each polar equation.

13. $r = 2 \cos \theta$

14. $r = 2 \sin \theta$

15. $r = 1 - \sin \theta$

16. $r = 1 + \sin \theta$

17. $r = 2 + 2 \cos \theta$

18. $r = 2 - 2 \cos \theta$

19. $r = 2 + \cos \theta$

20. $r = 2 - \sin \theta$

21. $r = 1 + 2 \cos \theta$

22. $r = 1 - 2 \cos \theta$

23. $r = 2 - 3 \sin \theta$

24. $r = 2 + 4 \sin \theta$

25. $r = 2 \cos 2\theta$

26. $r = 2 \sin 2\theta$

27. $r = 4 \sin 3\theta$

28. $r = 4 \cos 3\theta$

29. $r^2 = 9 \cos 2\theta$

30. $r^2 = 9 \sin 2\theta$

31. $r = 1 - 3 \sin \theta$

32. $r = 3 + \sin \theta$

33. $r \cos \theta = -3$

34. $r \sin \theta = 2$

Application Exercises

In Exercise Set 7.3, we considered an application in which sailboat racers look for a sailing angle to a 10-knot wind that produces maximum sailing speed. This situation is now represented by the polar graph in the figure shown. Each point (r, θ) on the graph gives the sailing speed, r, in knots, at an angle θ to the 10-knot wind. Use this information to solve Exercises 35–39.

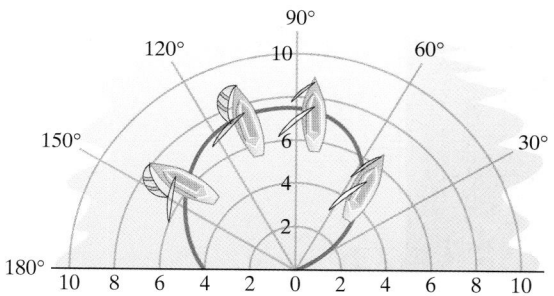

35. What is the speed, to the nearest knot, of the sailboat sailing at 60° angle to the wind?

36. What is the speed, to the nearest knot, of the sailboat sailing at a 120° angle to the wind?

37. What is the speed, to the nearest knot, of the sailboat sailing at a 90° angle to the wind?

38. What is the speed, to the nearest knot, of the sailboat sailing at a 180° angle to the wind?

39. What angle to the wind produces the maximum sailing speed? What is the speed? Round the angle to the nearest five degrees and the speed to the nearest half knot.

Writing in Mathematics

40. What is a polar equation?

41. What is the graph of a polar equation?

42. Describe how to graph a polar equation.

43. Describe the test for symmetry with respect to the polar axis.

44. Describe the test for symmetry with respect to the line $\theta = \dfrac{\pi}{2}$.

45. Describe the test for symmetry with respect to the pole.

46. If an equation fails the test for symmetry with respect to the polar axis, what can you conclude?

Technology Exercises

Use the polar mode of a graphing utility with angle measure in radians to solve Exercises 47–78. Unless otherwise indicated, use θmin = 0, θmax = 2π, and θstep = $\dfrac{\pi}{48}$. If you are not pleased with the quality of the graph, experiment with smaller values for θstep. However, if θ step is extremely small, it can take your graphing utility a long period of time to complete the graph.

47. Use a graphing utility to verify any six of your hand-drawn graphs in Exercises 13–34.

In Exercises 48–65, use a graphing utility to graph the polar equation.

48. $r = 4 \cos 5\theta$

49. $r = 4 \sin 5\theta$

50. $r = 4 \cos 6\theta$

51. $r = 4 \sin 6\theta$

52. $r = 2 + 2 \cos \theta$

53. $r = 2 + 2 \sin \theta$

54. $r = 4 + 2 \cos \theta$

55. $r = 4 + 2 \sin \theta$

56. $r = 2 + 4 \cos \theta$

57. $r = 2 + 4 \sin \theta$

58. $r = \dfrac{3}{\sin \theta}$

59. $r = \dfrac{3}{\cos \theta}$

60. $r = \cos \dfrac{3}{2} \theta$

61. $r = \cos \dfrac{5}{2} \theta$

62. $r = 3 \sin \left(\theta + \dfrac{\pi}{4} \right)$

63. $r = 2 \cos \left(\theta - \dfrac{\pi}{4} \right)$

64. $r = \dfrac{1}{1 - \sin \theta}$

65. $r = \dfrac{1}{3 - 2 \sin \theta}$

In Exercises 66–68, find the smallest interval for θ starting with θmin = 0 so that your graphing utility graphs the given polar equation exactly once without retracing any portion of it.

66. $r = 4 \sin \theta$

67. $r = 4 \sin 2\theta$

68. $r^2 = 4 \sin 2\theta$

In Exercises 69–72, use a graphing utility to graph each butterfly curve. Experiment with the range setting, particularly θ step, to produce a butterfly of the best possible quality.

69. $r = \cos^2 5\theta + \sin 3\theta + 0.3$

70. $r = \sin^4 4\theta + \cos 3\theta$

71. $r = \sin^5 \theta + 8 \sin \theta \cos^3 \theta$

72. $r = 1.5^{\sin \theta} - 2.5 \cos 4\theta + \sin^7 \dfrac{\theta}{15}$

(Use θmin = 0 and θmax = 20π.)

2 Find the absolute value of a complex number.

Recall that the absolute value of a real number is its distance from 0 on the number line. The **absolute value of the complex number** $z = a + bi$, denoted by $|z|$, is its distance from the origin in the complex plane.

> ### The Absolute Value of a Complex Number
>
> The **absolute value** of the complex number $a + bi$ is
>
> $$|z| = |a + bi| = \sqrt{a^2 + b^2}.$$

EXAMPLE 2 Finding the Absolute Value of a Complex Number

Determine the absolute value of each of the following complex numbers:

    a. $z = 3 + 4i$ **b.** $z = -1 - 2i$.

Solution

    a. The absolute value of $z = 3 + 4i$ is found using $a = 3$ and $b = 4$.

$$|z| = \sqrt{3^2 + 4^2} = \sqrt{9 + 16} = \sqrt{25} = 5 \quad \text{Use } z = \sqrt{a^2 + b^2} \text{ with } a = 3 \text{ and } b = 4.$$

Figure 7.36, repeated

Thus, the distance from the origin to the point $z = 3 + 4i$, shown in quadrant I in Figure 7.36, is five units.

    b. The absolute value of $z = -1 - 2i$ is found using $a = -1$ and $b = -2$.

$$|z| = \sqrt{(-1)^2 + (-2)^2} = \sqrt{1 + 4} = \sqrt{5} \quad \text{Use } z = \sqrt{a^2 + b^2} \text{ with } a = -1 \text{ and } b = -2.$$

Thus, the distance from the origin to the point $z = -1 - 2i$, shown in quadrant III in Figure 7.36, is $\sqrt{5}$ units.

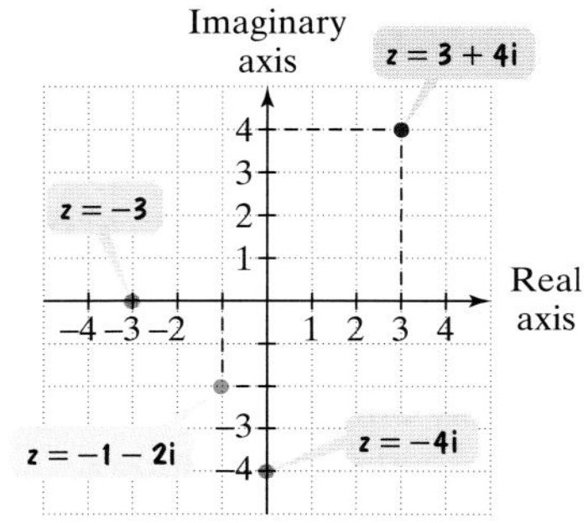

Check Point 2 Determine the absolute value of each of the following complex numbers:

    a. $z = 5 + 12i$ **b.** $2 - 3i$.

3 Write complex numbers in polar form.

Polar Form of a Complex Number

A complex number in the form $z = a + bi$ is said to be in **rectangular form**. Suppose that its absolute value is r. In Figure 7.37, we let θ be an angle in standard position whose terminal side passes through the point (a, b). From the figure, we see that

$$r = \sqrt{a^2 + b^2}.$$

Likewise, according to the definitions of the trigonometric functions,

$$\cos\theta = \frac{a}{r} \qquad \sin\theta = \frac{b}{r} \qquad \tan\theta = \frac{b}{a}.$$

$$a = r\cos\theta \qquad\qquad b = r\sin\theta$$

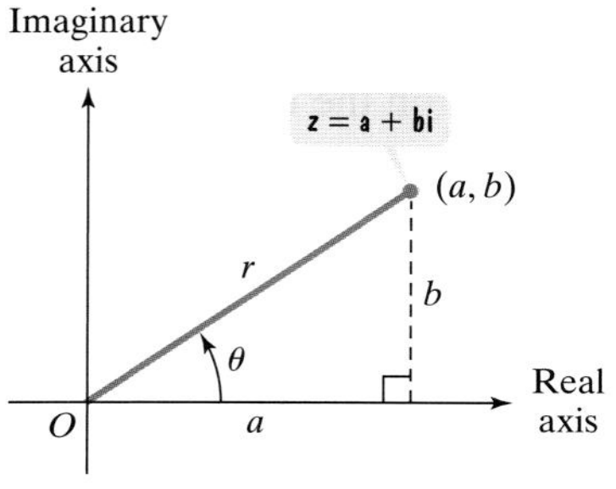

By substituting the expressions for a and b in $z = a + bi$, we write the complex number in terms of trigonometric functions.

Figure 7.37

$$z = a + bi = r\cos\theta + (r\sin\theta)i = r(\cos\theta + i\sin\theta)$$

    $a = r\cos\theta$ and $b = r\sin\theta$.   Factor out r from each of the two previous terms.

The expression $z = r(\cos\theta + i\sin\theta)$ is called the **polar form of a complex number.**

> ### Polar Form of a Complex Number
>
> The complex number $z = a + bi$ is written in **polar form** as
>
> $$z = r(\cos\theta + i\sin\theta)$$
>
> where $a = r\cos\theta, b = r\sin\theta, r = \sqrt{a^2 + b^2}$, and $\tan\theta = \dfrac{b}{a}$. The value of r is called the **modulus** (plural: moduli) of the complex number z, and the angle θ is called the **argument** of the complex number z, with $0 \le \theta < 2\pi$.

EXAMPLE 3   Writing a Complex Number in Polar Form

Plot $z = -2 - 2i$ in the complex plane. Then write z in polar form.

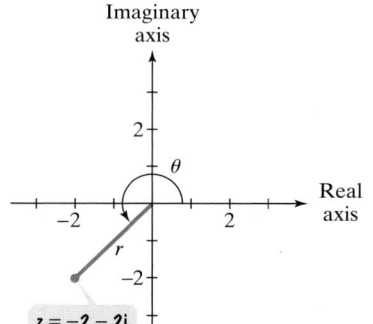

Figure 7.38 Plotting $z = -2 - 2i$ and writing the number in polar form

Solution   The complex number $z = -2 - 2i$ is in rectangular form $z = a + bi$, with $a = -2$ and $b = -2$. We plot the number by moving two units to the left on the real axis and two units down parallel to the imaginary axis, shown in Figure 7.38.

By definition, the polar form of z is $r(\cos\theta + i\sin\theta)$. We need to determine the value for r, the modulus, and the value for θ, the argument. Figure 7.38 shows r and θ. We use $r = \sqrt{a^2 + b^2}$ with $a = -2$ and $b = -2$ to find r.

$$r = \sqrt{a^2 + b^2} = \sqrt{(-2)^2 + (-2)^2} = \sqrt{4 + 4} = \sqrt{8} = \sqrt{4 \cdot 2} = 2\sqrt{2}$$

We use $\tan\theta = \dfrac{b}{a}$ with $a = -2$ and $b = -2$ to find θ.

$$\tan\theta = \frac{b}{a} = \frac{-2}{-2} = 1$$

We know that $\tan\dfrac{\pi}{4} = 1$. Figure 7.38 shows that the argument, θ, lies in quadrant III. Thus,

$$\theta = \pi + \frac{\pi}{4} = \frac{4\pi}{4} + \frac{\pi}{4} = \frac{5\pi}{4}.$$

We use $r = 2\sqrt{2}$ and $\theta = \dfrac{5\pi}{4}$ to write the polar form. The polar form of $z = -2 - 2i$ is

$$z = r(\cos\theta + i\sin\theta) = 2\sqrt{2}\left(\cos\frac{5\pi}{4} + i\sin\frac{5\pi}{4}\right).$$

Check Point 3   Plot $z = -1 - \sqrt{3}i$ in the complex plane. Then write z in polar form. Express the argument in radians.

4 Convert a complex number from polar to rectangular form.

EXAMPLE 4   Writing a Complex Number in Rectangular Form

Write $z = 2(\cos 60° + i \sin 60°)$ in rectangular form.

Solution   The complex number $z = 2(\cos 60° + i \sin 60°)$ is in polar form, with $r = 2$ and $\theta = 60°$. We use exact values for $\cos 60°$ and $\sin 60°$ to write the number in rectangular form.

$$2(\cos 60° + i \sin 60°) = 2\left(\frac{1}{2} + i\frac{\sqrt{3}}{2}\right) = 1 + \sqrt{3}i$$

The rectangular form of $z = 2(\cos 60° + i \sin 60°)$ is

$$z = 1 + \sqrt{3}i.$$

Check
Point
4

Write $z = 4(\cos 30° + i \sin 30°)$ in rectangular form.

5 Find products of complex numbers in polar form.

Products and Quotients in Polar Form

We can multiply and divide complex numbers fairly quickly if the numbers are expressed in polar form.

> ### Product of Two Complex Numbers in Polar Form
>
> Let $z_1 = r_1(\cos \theta_1 + i \sin \theta_1)$ and $z_2 = r_2(\cos \theta_2 + i \sin \theta_2)$ be two complex numbers in polar form. Their product, $z_1 z_2$, is
>
> $$z_1 z_2 = r_1 r_2 [\cos(\theta_1 + \theta_2) + i \sin(\theta_1 + \theta_2)].$$
>
> To multiply two complex numbers, multiply moduli and add arguments.

To prove this result, we begin by multiplying using the FOIL method. Then we simplify the product using the sum formulas for sine and cosine.

$$z_1 z_2 = [r_1(\cos \theta_1 + i \sin \theta_1)][r_2(\cos \theta_2 + i \sin \theta_2)]$$

$$= r_1 r_2 (\cos \theta_1 + i \sin \theta_1)(\cos \theta_2 + i \sin \theta_2) \qquad \text{Rearrange factors.}$$

F   O   I   L

$$= r_1 r_2 (\cos \theta_1 \cos \theta_2 + i \cos \theta_1 \sin \theta_2 + i \sin \theta_1 \cos \theta_2 + i^2 \sin \theta_1 \sin \theta_2) \quad \text{Use the FOIL method.}$$

$$= r_1 r_2 [\cos \theta_1 \cos \theta_2 + i(\cos \theta_1 \sin \theta_2 + \sin \theta_1 \cos \theta_2) + i^2 \sin \theta_1 \sin \theta_2] \quad \text{Factor } i \text{ from the second and third terms.}$$

$$= r_1 r_2 [\cos \theta_1 \cos \theta_2 + i(\cos \theta_1 \sin \theta_2 + \sin \theta_1 \cos \theta_2) - \sin \theta_1 \sin \theta_2] \quad i^2 = -1$$

$$= r_1 r_2 [(\cos \theta_1 \cos \theta_2 - \sin \theta_1 \sin \theta_2) + i(\sin \theta_1 \cos \theta_2 + \cos \theta_1 \sin \theta_2)] \quad \text{Rearrange terms.}$$

This is $\cos(\theta_1 + \theta_2)$.   This is $\sin(\theta_1 + \theta_2)$.

$$= r_1 r_2 [\cos(\theta_1 + \theta_2) + i \sin(\theta_1 + \theta_2)]$$

This result gives a rule for finding the product of two complex numbers in polar form. The two parts to the rule are shown in the voice balloons below the product.

$$r_1 r_2 [\cos(\theta_1 + \theta_2) + i \sin(\theta_1 + \theta_2)]$$

Multiply moduli.   Add arguments.

EXAMPLE 5 Finding Products of Complex Numbers in Polar Form

Find the product of the complex numbers. Leave the answer in polar form.

$$z_1 = 4(\cos 50° + i \sin 50°) \qquad z_2 = 7(\cos 100° + i \sin 100°)$$

Solution

$z_1 z_2$

$= \left[4(\cos 50° + i \sin 50°)\right]\left[7(\cos 100° + i \sin 100°)\right]$ Form the product of the given numbers.

$= (4 \cdot 7)\left[\cos(50° + 100°) + i \sin(50° + 100°)\right]$ Multiply moduli and add arguments.

$= 28(\cos 150° + i \sin 150°)$ Simplify.

Check Point 5

Find the product of the complex numbers. Leave the answer in polar form.

$$z_1 = 6(\cos 40° + i \sin 40°) \qquad z_2 = 5(\cos 20° + i \sin 20°)$$

6 Find quotients of complex numbers in polar form.

Using algebraic methods for dividing complex numbers and the difference formulas for sine and cosine, we can obtain a rule for dividing complex numbers in polar form. The proof of this rule can be found in the appendix.

Quotient of Two Complex Numbers in Polar Form

Let $z_1 = r_1(\cos \theta_1 + i \sin \theta_1)$ and $z_2 = r_2(\cos \theta_2 + i \sin \theta_2)$ be two complex numbers in polar form. Their quotient, $\dfrac{z_1}{z_2}$, is

$$\frac{z_1}{z_2} = \frac{r_1}{r_2}\left[\cos(\theta_1 - \theta_2) + i \sin(\theta_1 - \theta_2)\right].$$

To divide two complex numbers, divide moduli and subtract arguments.

EXAMPLE 6 Finding Quotients of Complex Numbers in Polar Form

Find the quotient $\dfrac{z_1}{z_2}$ of the complex numbers. Leave the answer in polar form.

$$z_1 = 12\left(\cos \frac{3\pi}{4} + i \sin \frac{3\pi}{4}\right) \qquad z_2 = 4\left(\cos \frac{\pi}{4} + i \sin \frac{\pi}{4}\right)$$

Solution

$$\frac{z_1}{z_2} = \frac{12\left(\cos\dfrac{3\pi}{4} + i\sin\dfrac{3\pi}{4}\right)}{4\left(\cos\dfrac{\pi}{4} + i\sin\dfrac{\pi}{4}\right)}$$

Form the quotient of the given numbers.

$$= \frac{12}{4}\left[\cos\left(\frac{3\pi}{4} - \frac{\pi}{4}\right) + i\sin\left(\frac{3\pi}{4} - \frac{\pi}{4}\right)\right]$$

Divide moduli and subtract arguments.

$$= 3\left(\cos\frac{\pi}{2} + i\sin\frac{\pi}{2}\right)$$

Simplify: $\dfrac{3\pi}{4} - \dfrac{\pi}{4} = \dfrac{2\pi}{4} = \dfrac{\pi}{2}.$

Check Point 6 Find the quotient of the complex numbers. Leave the answer in polar form.

$$z_1 = 50\left(\cos\frac{4\pi}{3} + i\sin\frac{4\pi}{3}\right) \qquad z_2 = 5\left(\cos\frac{\pi}{3} + i\sin\frac{\pi}{3}\right)$$

 7 Find powers of complex numbers in polar form.

Powers of Complex Numbers in Polar Form

We can use a formula to find powers of complex numbers if the complex numbers are expressed in polar form. This formula can be illustrated by repeatedly multiplying by $r(\cos\theta + i\sin\theta)$.

$z = r(\cos\theta + i\sin\theta)$ — *Start with z.*

$z \cdot z = r(\cos\theta + i\sin\theta)r(\cos\theta + i\sin\theta)$ — *Multiply z by z = r(cos θ + i sin θ).*

$z^2 = r^2(\cos 2\theta + i\sin 2\theta)$ — *Multiply moduli: r · r = r². Add arguments: θ + θ = 2θ.*

$z^2 \cdot z = r^2(\cos 2\theta + i\sin 2\theta)r(\cos\theta + i\sin\theta)$ — *Multiply z² by z = r(cos θ + i sin θ).*

$z^3 = r^3(\cos 3\theta + i\sin 3\theta)$ — *Multiply moduli: r² · r = r³. Add arguments: 2θ + θ = 3θ.*

$z^3 \cdot z = r^3(\cos 3\theta + i\sin 3\theta)r(\cos\theta + i\sin\theta)$ — *Multiply z³ by z = r(cos θ + i sin θ).*

$z^4 = r^4(\cos 4\theta + i\sin 4\theta)$ — *Multiply moduli: r³ · r = r⁴. Add arguments: 3θ + θ = 4θ.*

Do you see a pattern forming? If n is a positive integer, it appears that z^n is obtained by raising the modulus to the nth power and multiplying the argument by n. The formula for the nth power of a complex number is known as **DeMoivre's Theorem** in honor of the French mathematician Abraham DeMoivre (1667–1754).

DeMoivre's Theorem

Let $z = r(\cos\theta + i\sin\theta)$ be a complex number in polar form. If n is a positive integer, z to the nth power, z^n, is

$$z^n = \left[r(\cos\theta + i\sin\theta)\right]^n = r^n(\cos n\theta + i\sin n\theta).$$

EXAMPLE 7 Finding the Power of a Complex Number

Find $\left[2(\cos 10° + i\sin 10°)\right]^6$. Write the answer in rectangular form, $a + bi$.

Solution By DeMoivre's Theorem,

$$[2(\cos 10° + i \sin 10°)]^6$$

$$= 2^6[\cos(6 \cdot 10°) + i \sin(6 \cdot 10°)]$$ Raise the modulus to the 6th power and multiply the argument by 6.

$$= 64(\cos 60° + i \sin 60°)$$ Simplify.

$$= 64\left(\frac{1}{2} + i\frac{\sqrt{3}}{2}\right)$$ Write the answer in rectangular form.

$$= 32 + 32\sqrt{3}i$$ Multiply and express the answer in $a + bi$ form.

> **Check Point 7** Find $[2(\cos 30° + i \sin 30°)]^5$. Write the answer in rectangular form.

EXAMPLE 8 Finding the Power of a Complex Number

Find $(1 + i)^8$ using DeMoivre's Theorem. Write the answer in rectangular form, $a + bi$.

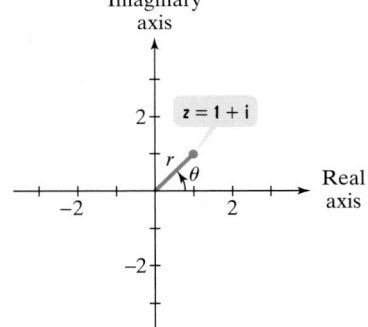

Figure 7.39 Plotting $1 + i$ and writing the number in polar form

Solution DeMoivre's Theorem applies to complex numbers in polar form. Thus, we must first write $1 + i$ in $r(\cos\theta + i\sin\theta)$ form. Then we can use DeMoivre's Theorem. The complex number $1 + i$ is plotted in Figure 7.39. From the figure we obtain values for r and θ.

$$r = \sqrt{a^2 + b^2} = \sqrt{1^2 + 1^2} = \sqrt{2} \qquad \tan\theta = \frac{b}{a} = \frac{1}{1} = 1 \quad \text{and} \quad \theta = \frac{\pi}{4}$$

Using these values,

$$1 + i = r(\cos\theta + i\sin\theta) = \sqrt{2}\left(\cos\frac{\pi}{4} + i\sin\frac{\pi}{4}\right).$$

Now we use DeMoivre's Theorem to raise $1 + i$ to the 8th power.

$$(1 + i)^8$$

$$= \left[\sqrt{2}\left(\cos\frac{\pi}{4} + i\sin\frac{\pi}{4}\right)\right]^8$$ Work with the polar form of $1 + i$.

$$= (\sqrt{2})^8\left[\cos\left(8 \cdot \frac{\pi}{4}\right) + i\sin\left(8 \cdot \frac{\pi}{4}\right)\right]$$ Apply DeMoivre's Theorem. Raise the modulus to the 8th power and multiply the argument by 8.

$$= 16(\cos 2\pi + i\sin 2\pi)$$ Simplify: $(\sqrt{2})^8 = (2^{1/2})^8 = 2^4 = 16$.

$$= 16(1 + 0i)$$ $\cos 2\pi = 1$ and $\sin 2\pi = 0$

$$= 16$$ Simplify.

> **Check Point 8** Find $(1 + i)^4$ using DeMoivre's Theorem. Write the answer in rectangular form.

8 Find roots of complex numbers in polar form.

Roots of Complex Numbers in Polar Form

In Example 7, we showed that

$$[2(\cos 10° + i \sin 10°)]^6 = 64(\cos 60° + i \sin 60°).$$

We say that $2(\cos 10° + i \sin 10°)$ is a **complex sixth root** of $64(\cos 60° + i \sin 60°)$. It is one of six distinct complex roots of $64(\cos 60° + i \sin 60°)$.

In general, if a complex number z satisfies the equation

$$z^n = w$$

we say that z is a **complex nth root** of w. It is one of n distinct complex roots that can be found using the following theorem.

DeMoivre's Theorem for Finding Complex Roots

Let $w = r(\cos \theta + i \sin \theta)$ be a complex number in polar form. If $w \neq 0$, w has n distinct complex nth roots given by the formula

$$z_k = \sqrt[n]{r}\left[\cos\left(\frac{\theta + 2\pi k}{n}\right) + i \sin\left(\frac{\theta + 2\pi k}{n}\right)\right] \quad \text{(radians)}$$

$$\text{or} \quad z_k = \sqrt[n]{r}\left[\cos\left(\frac{\theta + 360°k}{n}\right) + i \sin\left(\frac{\theta + 360°k}{n}\right)\right] \text{(degrees)}$$

where $k = 0, 1, 2, \ldots, n - 1$.

By raising the radian or degree formula for z_k to the nth power, you can use DeMoivre's Theorem for powers to show that $z_k^n = w$. Thus, each z_k is a complex nth root of w.

DeMoivre's Theorem for finding complex roots states that every complex number has two distinct complex square roots, three distinct complex cube roots, four distinct complex fourth roots, and so on. Each root has the same modulus, $\sqrt[n]{r}$. Successive roots have arguments that differ by the same amount, $\frac{2\pi}{n}$. This means that if you plot all the complex roots of any number, they will be equally spaced on a circle centered at the origin, having radius $\sqrt[n]{r}$.

EXAMPLE 9 Finding the Roots of a Complex Number

Find all the complex fourth roots of $16(\cos 120° + i \sin 120°)$. Write roots in polar form, with θ in degrees.

Solution There are exactly four fourth roots of the given complex number. From DeMoivre's Theorem for finding complex roots, the fourth roots of $16(\cos 120° + i \sin 120°)$ are

$$z_k = \sqrt[4]{16}\left[\cos\left(\frac{120° + 360°k}{4}\right) + i \sin\left(\frac{120° + 360°k}{4}\right)\right], \quad k = 0, 1, 2, 3.$$

Use $z_k = \sqrt[n]{r}\left[\cos\left(\frac{\theta + 360°k}{n}\right) + i \sin\left(\frac{\theta + 360°k}{n}\right)\right]$.
In $16(\cos 120° + i \sin 120°)$, $r = 16$ and $\theta = 120°$.
Because we are finding fourth roots, $n = 4$.

The four fourth roots are found by substituting $0, 1, 2,$ and 3 for k in the expression for z_k above the voice balloon. Thus, the four fourth roots are:

$$z_0 = \sqrt[4]{16}\left[\cos\left(\frac{120° + 360° \cdot 0}{4}\right) + i\sin\left(\frac{120° + 360° \cdot 0}{4}\right)\right]$$

$$= \sqrt[4]{16}\left(\cos\frac{120°}{4} + i\sin\frac{120°}{4}\right) = 2(\cos 30° + i\sin 30°)$$

$$z_1 = \sqrt[4]{16}\left[\cos\left(\frac{120° + 360° \cdot 1}{4}\right) + i\sin\left(\frac{120° + 360° \cdot 1}{4}\right)\right]$$

$$= \sqrt[4]{16}\left(\cos\frac{480°}{4} + i\sin\frac{480°}{4}\right) = 2(\cos 120° + i\sin 120°)$$

$$z_2 = \sqrt[4]{16}\left[\cos\left(\frac{120° + 360° \cdot 2}{4}\right) + i\sin\left(\frac{120° + 360° \cdot 2}{4}\right)\right]$$

$$= \sqrt[4]{16}\left(\cos\frac{840°}{4} + i\sin\frac{840°}{4}\right) = 2(\cos 210° + i\sin 210°)$$

$$z_3 = \sqrt[4]{16}\left[\cos\left(\frac{120° + 360° \cdot 3}{4}\right) + i\sin\left(\frac{120° + 360° \cdot 3}{4}\right)\right]$$

$$= \sqrt[4]{16}\left(\cos\frac{1200°}{4} + i\sin\frac{1200°}{4}\right) = 2(\cos 300° + i\sin 300°)$$

In Figure 7.40, we have plotted each of the four fourth roots. They are equally spaced at 90° intervals on a circle having radius 2.

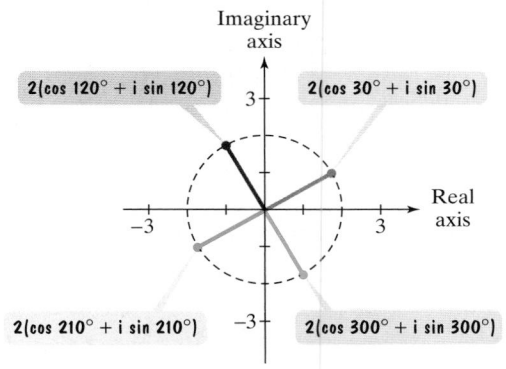

Figure 7.40 Plotting the four fourth roots of $16(\cos 120° + i\sin 120°)$

Check Point 9   Find all the complex fourth roots of $16(\cos 60° + i\sin 60°)$. Write roots in polar form, with θ in degrees.

EXAMPLE 10   Finding the Roots of a Complex Number

Find all the cube roots of 8. Write roots in rectangular form.

Solution   DeMoivre's Theorem for roots applies to complex numbers in polar form. Thus, we will first write 8 in polar form. We express θ in radians, although degrees can also be used.

$$8 = r(\cos\theta + i\sin\theta) = 8(\cos 0 + i\sin 0)$$

There are exactly three cube roots of 8. From DeMoivre's Theorem for finding complex roots, the cube roots of 8 are

$$z_k = \sqrt[3]{8}\left[\cos\left(\frac{0 + 2\pi k}{3}\right) + i\sin\left(\frac{0 + 2\pi k}{3}\right)\right], \quad k = 0, 1, 2.$$

Use $z_k = \sqrt[n]{r}\left[\cos\left(\frac{\theta + 2\pi k}{n}\right) + i\sin\left(\frac{\theta + 2\pi k}{n}\right)\right]$.
In $8(\cos 0 + i\sin 0)$, $r = 8$ and $\theta = 0$.
Because we are finding cube roots, $n = 3$.

The three cube roots of 8 are found by substituting 0, 1, and 2 for k in the expression for z_k above the voice balloon. Thus, the three cube roots of 8 are:

$$z_0 = \sqrt[3]{8}\left[\cos\left(\frac{0 + 2\pi \cdot 0}{3}\right) + i\sin\left(\frac{0 + 2\pi \cdot 0}{3}\right)\right]$$
$$= 2(\cos 0 + i\sin 0) = 2(1 + i \cdot 0) = 2$$
$$z_1 = \sqrt[3]{8}\left[\cos\left(\frac{0 + 2\pi \cdot 1}{3}\right) + i\sin\left(\frac{0 + 2\pi \cdot 1}{3}\right)\right]$$
$$= 2\left(\cos\frac{2\pi}{3} + i\sin\frac{2\pi}{3}\right) = 2\left(-\frac{1}{2} + i \cdot \frac{\sqrt{3}}{2}\right) = -1 + \sqrt{3}i$$
$$z_2 = \sqrt[3]{8}\left[\cos\left(\frac{0 + 2\pi \cdot 2}{3}\right) + i\sin\left(\frac{0 + 2\pi \cdot 2}{3}\right)\right]$$
$$= 2\left(\cos\frac{4\pi}{3} + i\sin\frac{4\pi}{3}\right) = 2\left(-\frac{1}{2} + i \cdot \left(-\frac{\sqrt{3}}{2}\right)\right) = -1 - \sqrt{3}i.$$

The three cube roots of 8 are plotted in Figure 7.41.

Imaginary axis
$(-1 + \sqrt{3}i)$
3
2
Real axis
−3
3
−3
$(-1 - \sqrt{3}i)$

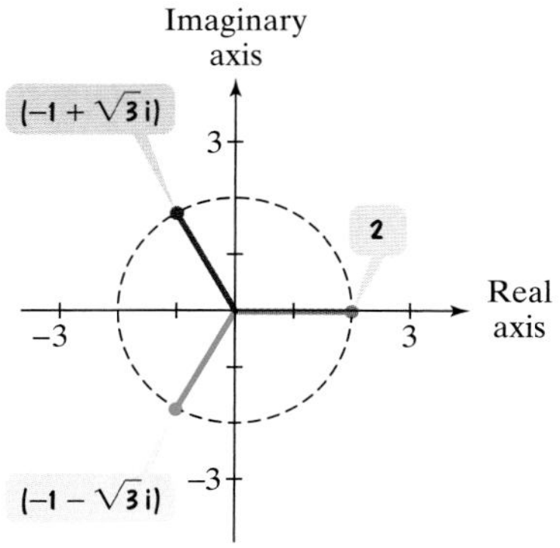

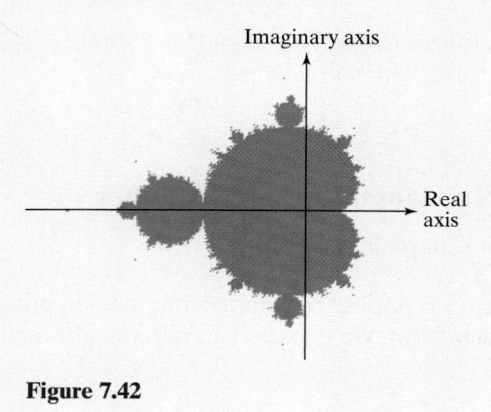

Figure 7.41 The three cube roots of 8 are equally spaced at intervals of $\frac{2\pi}{3}$ about a circle having radius 2.

Check Point 10 Find all the cube roots of 27. Write roots in rectangular form.

The Mandelbrot Set

The set of all complex numbers for which the sequence

$$z, z^2 + z, (z^2 + z)^2 + z, [(z^2 + z)^2 + z]^2 + z, \ldots$$

is bounded is called the **Mandelbrot set**. Plotting these complex numbers in the complex plane results in a graph that is "buglike" in shape, shown in Figure 7.42. Colors can be added to the boundary of the graph. At the boundary, color choices depend on how quickly the numbers in the boundary approach infinity when substituted into the sequence shown. The magnified boundary is shown in the introduction to this section. It includes the original buglike structure, as well as new and interesting patterns. With each level of magnification, repetition and unpredictable formations interact to create what has been called the most complicated mathematical object ever known.

Imaginary axis
Real axis

Figure 7.42

EXERCISE SET 7.5

Practice Exercises

In Exercises 1–10, plot each complex number and find its absolute value.

1. $z = 4i$ **2.** $z = 3i$

3. $z = 3$ **4.** $z = 4$

5. $z = 3 + 2i$ **6.** $z = 2 + 5i$

7. $z = 3 - i$ **8.** $z = 4 - i$

9. $z = -3 + 4i$ **10.** $z = -3 - 4i$

In Exercises 11–26, plot each complex number. Then write the complex number in polar form. You may express the argument in degrees or radians.

11. $2 + 2i$ **12.** $1 + \sqrt{3}i$ **13.** $-1 - i$

14. $2 - 2i$ **15.** $-4i$ **16.** $-3i$

17. $2\sqrt{3} - 2i$ **18.** $-2 + 2\sqrt{3}i$ **19.** -3

20. -4 **21.** $-3\sqrt{2} - 3\sqrt{3}i$ **22.** $3\sqrt{2} - 3\sqrt{2}i$

23. $-3 + 4i$ **24.** $-2 + 3i$

25. $2 - \sqrt{3}i$ **26.** $1 - \sqrt{5}i$

In Exercises 27–36, write each complex number in rectangular form. If necessary, round to the nearest tenth.

27. $6(\cos 30° + i \sin 30°)$ **28.** $12(\cos 60° + i \sin 60°)$

29. $4(\cos 240° + i \sin 240°)$ **30.** $10(\cos 210° + i \sin 210°)$

31. $8\left(\cos \dfrac{7\pi}{4} + i \sin \dfrac{7\pi}{4} \right)$ **32.** $4\left(\cos \dfrac{5\pi}{6} + i \sin \dfrac{5\pi}{6} \right)$

33. $5\left(\cos \dfrac{\pi}{2} + i \sin \dfrac{\pi}{2} \right)$ **34.** $7\left(\cos \dfrac{3\pi}{2} + i \sin \dfrac{3\pi}{2} \right)$

35. $20(\cos 205° + i \sin 205°)$ **36.** $30(\cos 2.3 + i \sin 2.3)$

In Exercises 37–44, find the product of the complex numbers. Leave answers in polar form.

37. $z_1 = 6(\cos 20° + i \sin 20°)$
$z_2 = 5(\cos 50° + i \sin 50°)$

38. $z_1 = 4(\cos 15° + i \sin 15°)$
$z_2 = 7(\cos 25° + i \sin 25°)$

39. $z_1 = 3\left(\cos \dfrac{\pi}{5} + i \sin \dfrac{\pi}{5} \right)$

$z_2 = 4\left(\cos \dfrac{\pi}{10} + i \sin \dfrac{\pi}{10} \right)$

40. $z_1 = 3\left(\cos \dfrac{5\pi}{8} + i \sin \dfrac{5\pi}{8} \right)$

$z_2 = 10\left(\cos \dfrac{\pi}{16} + i \sin \dfrac{\pi}{16} \right)$

41. $z_1 = \cos \dfrac{\pi}{4} + i \sin \dfrac{\pi}{4}$

$z_2 = \cos \dfrac{\pi}{3} + i \sin \dfrac{\pi}{3}$

42. $z_1 = \cos \dfrac{\pi}{6} + i \sin \dfrac{\pi}{6}$

$z_2 = \cos \dfrac{\pi}{4} + i \sin \dfrac{\pi}{4}$

43. $z_1 = 1 + i$ **44.** $z_1 = 1 + i$
$z_2 = -1 + i$ $z_2 = 2 + 2i$

In Exercises 45–52, find the quotient $\dfrac{z_1}{z_2}$ of the complex numbers. Leave answers in polar form. In Exercises 49–50, express the argument as an angle between $0°$ and $360°$.

45. $z_1 = 20(\cos 75° + i \sin 75°)$
$z_2 = 4(\cos 25° + i \sin 25°)$

46. $z_1 = 50(\cos 80° + i \sin 80°)$
$z_2 = 10(\cos 20° + i \sin 20°)$

47. $z_1 = 3\left(\cos \dfrac{\pi}{5} + i \sin \dfrac{\pi}{5} \right)$

$z_2 = 4\left(\cos \dfrac{\pi}{10} + i \sin \dfrac{\pi}{10} \right)$

48. $z_1 = 3\left(\cos \dfrac{5\pi}{18} + i \sin \dfrac{5\pi}{18} \right)$

$z_2 = 10\left(\cos \dfrac{\pi}{16} + i \sin \dfrac{\pi}{16} \right)$

49. $z_1 = \cos 80° + i \sin 80°$
$z_2 = \cos 200° + i \sin 200°$

50. $z_1 = \cos 70° + i \sin 70°$
$z_2 = \cos 230° + i \sin 230°$

51. $z_1 = 2 + 2i$ **52.** $z_1 = 2 - 2i$
$z_2 = 1 + i$ $z_2 = 1 - i$

In Exercises 53–64, use DeMoivre's Theorem to find the indicated power of the complex number. Write answers in rectangular form.

53. $\left[4(\cos 15° + i \sin 15°) \right]^3$

54. $\left[2(\cos 10° + i \sin 10°) \right]^3$

55. $\left[2(\cos 80° + i \sin 80°) \right]^3$

56. $\left[2(\cos 40° + i \sin 40°) \right]^3$

57. $\left[\dfrac{1}{2}\left(\cos \dfrac{\pi}{12} + i \sin \dfrac{\pi}{12} \right) \right]^6$

58. $\left[\dfrac{1}{2}\left(\cos \dfrac{\pi}{10} + i \sin \dfrac{\pi}{10} \right) \right]^5$

59. $\left[\sqrt{2}\left(\cos \dfrac{5\pi}{6} + i \sin \dfrac{5\pi}{6} \right) \right]^4$

60. $\left[\sqrt{3}\left(\cos \dfrac{5\pi}{18} + i \sin \dfrac{5\pi}{18} \right) \right]^6$

61. $(1 + i)^5$ **62.** $(1 - i)^5$

63. $(\sqrt{3} - i)^6$ **64.** $(\sqrt{2} - i)^4$

In Exercises 65–68, find all the complex roots. Write roots in polar form with θ in degrees.

65. The complex square roots of $9(\cos 30° + i \sin 30°)$

66. The complex square roots of $25(\cos 210° + i \sin 210°)$

67. The complex cube roots of $8(\cos 210° + i \sin 210°)$

68. The complex cube roots of $27(\cos 306° + i \sin 306°)$

In Exercises 69–76, find all the complex roots. Write roots in rectangular form. If necessary, round to the nearest tenth.

69. The complex fourth roots of $81\left(\cos \dfrac{4\pi}{3} + i \sin \dfrac{4\pi}{3}\right)$

70. The complex fifth roots of $32\left(\cos \dfrac{5\pi}{3} + i \sin \dfrac{5\pi}{3}\right)$

71. The complex fifth roots of 32

72. The complex sixth roots of 64

73. The complex cube roots of 1

74. The complex cube roots of i

75. The complex fourth roots of $1 + i$

76. The complex fifth roots of $-1 + i$

Application Exercises

In Exercises 77–78, show that the given complex number z plots as a point in the Mandelbrot set.

a. *Write the first six terms of the sequence*

$$z_1, z_2, z_3, z_4, z_5, z_6, \ldots$$

where

$z_1 = z$: Write the given number.

$z_2 = z^2 + z$: Square z_1 and add the given number.

$z_3 = (z^2 + z)^2 + z$: Square z_2 and add the given number.

$z_4 = [(z^2 + z)^2 + z]^2 + z$: Square z_3 and add the given number.

z_5: Square z_4 and add the given number.

z_6: Square z_5 and add the given number.

b. *If the sequence that you began writing in part (a) is bounded, the given complex number belongs to the Mandelbrot set. Show that the sequence is bounded by writing two complex numbers. One complex number should be greater in absolute value than the absolute values of the terms in the sequence. The second complex number should be less in absolute value than the absolute values of the terms in the sequence.*

77. $z = i$ **78.** $z = -i$

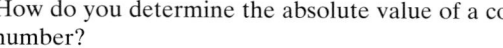

Writing in Mathematics

79. Explain how to plot a complex number in the complex plane. Provide an example with your explanation.

80. How do you determine the absolute value of a complex number?

81. What is the polar form of a complex number?

82. If you are given a complex number in rectangular form, how do you write it in polar form?

83. If you are given a complex number in polar form, how do you write it in rectangular form?

84. Explain how to find the product of two complex numbers in polar form.

85. Explain how to find the quotient of two complex numbers in polar form.

86. Explain how to find the power of a complex number in polar form.

87. Explain how to use DeMoivre's Theorem for finding complex roots to find the two square roots of 9.

88. Describe the graph of all complex numbers with an absolute value of 6.

89. The image of the Mandelbrot set in the section opener exhibits self-similarity: Magnified portions repeat much of the pattern of the whole structure, as well as new and unexpected patterns. Describe an object in nature that exhibits self-similariy.

Technology Exercises

90. Use the rectangular-to-polar feature on a graphing utility to verify any four of your answers in Exercises 11–26. Be aware that you may have to adjust the angle for the correct quadrant.

91. Use the polar-to-rectangular feature on a graphing utility to verify any four of your answers in Exercises 27–36.

Critical Thinking Exercises

92. Prove the rule for finding the quotient of two complex numbers in polar form. Begin the proof as follows, using the conjugate of the denominator:

$$\frac{r_1(\cos \theta_1 + i \sin \theta_1)}{r_2(\cos \theta_2 + i \sin \theta_2)} = \frac{r_1(\cos \theta_1 + i \sin \theta_1)}{r_2(\cos \theta_2 + i \sin \theta_2)} \cdot \frac{(\cos \theta_2 - i \sin \theta_2)}{(\cos \theta_2 - i \sin \theta_2)}$$

Perform the indicated multiplications. Then use the difference formulas for sine and cosine.

93. Plot each of the complex fourth roots of 1.

In Exercises 94–95, use DeMoivre's Theorem for finding complex roots to find all the solutions of each equation.

94. $x^3 + 27 = 0$ **95.** $x^3 - 8i = 0$

Group Exercise

96. Group members should prepare and present a seminar on chaos. Include one or more of the following topics in your presentation: fractal images, the role of complex numbers in generating fractal images, algorithms, iterations, iteration number, and fractals in nature. Be sure to include visual images that will intrigue your audience.

7.6 Vectors

Objectives

1. Use magnitude and direction to show vectors are equal.
2. Visualize scalar multiplication, vector addition, and vector subtraction as geometric vectors.
3. Represent vectors in the rectangular coordinate system.
4. Perform operations with vectors in terms of **i** and **j**.
5. Find a unit vector in the direction of **v**.
6. Solve applied problems involving vectors.

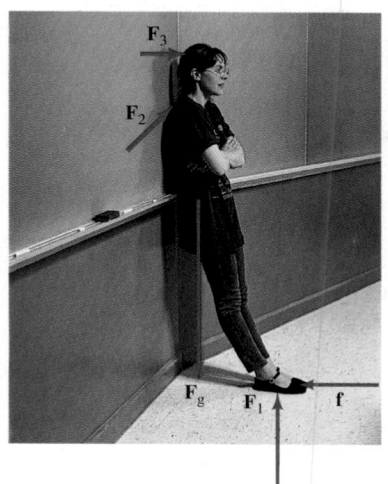

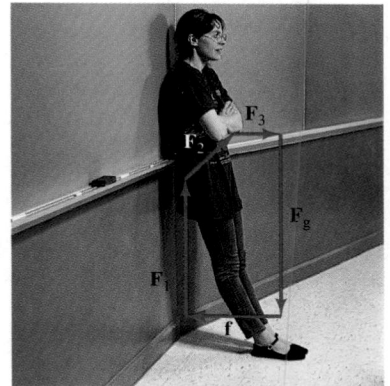

It's been a dynamic lecture, but now that it's over it's obvious that my professor is exhausted. She's slouching motionless against the board and—what's that? The forces acting against her body, including the pull of gravity, are appearing as arrows. I know that mathematics reveals the hidden patterns of the universe, but this is ridiculous. Does the arrangement of the arrows on the right have anything to do with the fact that my wiped-out professor is not sliding down the wall?

Ours is a world of pushes and pulls. For example, suppose you are pulling a cart up a 30° incline, requiring an effort of 100 pounds. This quantity is described by giving its magnitude (a number indicating size, including a unit of measure) and also its direction. The magnitude is 100 pounds and the direction is 30° from the horizontal. Quantities that involve both a magnitude and a direction are called **vector quantities**, or **vectors** for short. Here is another example of a vector:

> You are driving due north at 50 miles per hour. The magnitude is the speed, 50 miles per hour. The direction of motion is due north.

Some quantities can be completely described by giving only their magnitude. For example, the temperature of the lecture room that you just left is 75°. This temperature has magnitude, 75°, but no direction. Quantities that involve magnitude, but no direction, are called **scalar quantities**, or **scalars** for short. Thus, a scalar has only a numerical value. Another example of a scalar is your professor's height, which you estimate to be 5.5 feet.

In the next two sections, we introduce the world of vectors, which literally surround your every move. Because vectors have both nonnegative magnitude and direction, we begin our discussion with directed line segments.

Terminal point

Q

P

Initial point

Figure 7.43 A directed line segment from P to Q

Directed Line Segments and Geometric Vectors

A line segment to which a direction has been assigned is called a **directed line segment**. Figure 7.43 shows a directed line segment from P to Q. We call P the **initial point** and Q the **terminal point**. We denote this directed line segment by

$$\overrightarrow{PQ}.$$

The **magnitude** of the directed line segment $\overrightarrow{PQ}$ is its length. We denote this by $\|\overrightarrow{PQ}\|$. Thus, $\|\overrightarrow{PQ}\|$ is the distance from point P to point Q. Because distance is nonnegative, vectors do not have negative magnitudes.

Geometrically, a **vector** is a directed line segment. Vectors are often denoted by a boldface letter, such as **v**. If a vector **v** has the same magnitude and the same direction as the directed line segment $\overrightarrow{PQ}$, we write

$$\mathbf{v} = \overrightarrow{PQ}.$$

Because it is difficult to write boldface on paper, use an arrow over a single letter, such as $\vec{v}$, to denote **v**, the vector **v**.

Figure 7.44 shows four possible relationships between vectors **v** and **w**. In Figure 7.44 (a), the vectors have the same magnitude and direction and are said to be *equal*. In general, vectors **v** and **w** are **equal** if they have the *same magnitude* and the *same direction*. We write this as **v** = **w**.

1 Use magnitude and direction to show vectors are equal.

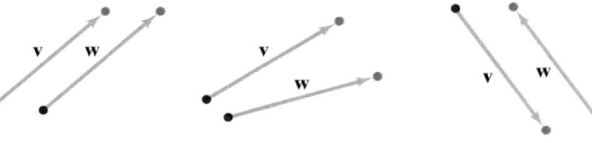

(a) v = w because the vectors have the some magnitude and direction.

(b) Vectors **v** and **w** have the same magnitude, but different directions

(c) Vectors **v** and **w** have the same magnitude, but opposite directions

(d) Vectors **v** and **w** have the same direction, but different magnitudes

Figure 7.44 Relationships between vectors

EXAMPLE 1 Showing That Two Vectors Are Equal

Use Figure 7.45 to show that **u** = **v**.

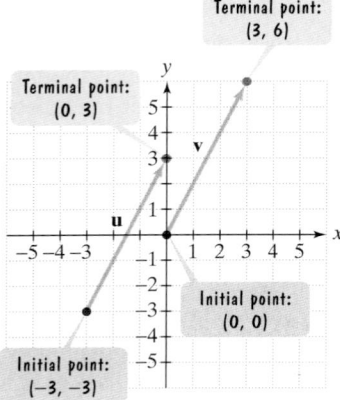

Terminal point: (0, 3)

Terminal point: (3, 6)

Initial point: (0, 0)

Initial point: (−3, −3)

Figure 7.45

Solution Equal vectors have the same magnitude and the same direction. Use the distance formula to show that **u** and **v** have the same magnitude.

Magnitude of **u**
$$\|\mathbf{u}\| = \sqrt{(x_2 - x_1)^2 + (y_2 - y_1)^2} = \sqrt{[0 - (-3)]^2 + [3 - (-3)]^2}$$
$$= \sqrt{3^2 + 6^2} = \sqrt{9 + 36} = \sqrt{45} \quad (\text{or } 3\sqrt{5})$$

Magnitude of **v**
$$\|\mathbf{v}\| = \sqrt{(x_2 - x_1)^2 + (y_2 - y_1)^2} = \sqrt{(3 - 0)^2 + (6 - 0)^2}$$
$$= \sqrt{3^2 + 6^2} = \sqrt{9 + 36} = \sqrt{45} \quad (\text{or } 3\sqrt{5})$$

Thus, **u** and **v** have the same magnitude: $\|\mathbf{u}\| = \|\mathbf{v}\|$.

One way to show that **u** and **v** have the same direction is to find the slopes of the lines on which they lie.

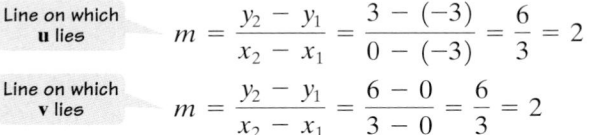

Line on which **u** lies
$$m = \frac{y_2 - y_1}{x_2 - x_1} = \frac{3 - (-3)}{0 - (-3)} = \frac{6}{3} = 2$$

Line on which **v** lies
$$m = \frac{y_2 - y_1}{x_2 - x_1} = \frac{6 - 0}{3 - 0} = \frac{6}{3} = 2$$

Because **u** and **v** are both directed toward the upper right on lines having the same slope, 2, they have the same direction.

Thus, **u** and **v** have the same magnitude and direction, and **u** = **v**.

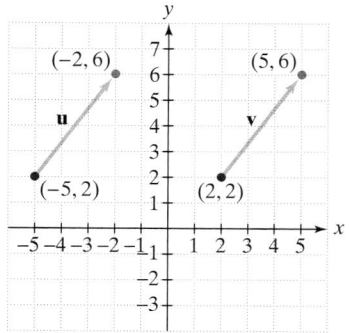

Figure 7.46

2 Visualize scalar multiplication, vector addition, and vector subtraction as geometric vectors.

Wiped Out, But Not Sliding Down the Wall

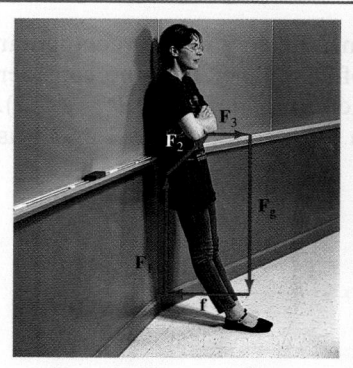

The figure shows the sum of five vectors:

$$\mathbf{F}_1 + \mathbf{F}_2 + \mathbf{F}_3 + \mathbf{F}_g + \mathbf{f}.$$

Notice how the terminal point of each vector coincides with the initial point of the vector that's being added to it. The vector sum, from the initial point of $\mathbf{F}_1$ to the terminal point of $\mathbf{f}$, is a single point. The magnitude of a single point is zero. These forces add up to a net force of zero, allowing the professor to be motionless.

Check Point 1 Use Figure 7.46 to show that $\mathbf{u} = \mathbf{v}$.

A vector can be multiplied by a real number. Figure 7.47 shows three such multiplications: $2\mathbf{v}, \frac{1}{2}\mathbf{v}$, and $-\frac{3}{2}\mathbf{v}$. **Multiplying a vector by any positive real number (except for 1) changes the magnitude of the vector, but not its direction**. This can be seen by the blue and green vectors in Figure 7.47. Compare the black and blue vectors. Can you see that $2\mathbf{v}$ has the same direction as $\mathbf{v}$ but is twice the magnitude of $\mathbf{v}$? Now, compare the black and green vectors: $\frac{1}{2}\mathbf{v}$ has the same direction as $\mathbf{v}$ but is half the magnitude of $\mathbf{v}$.

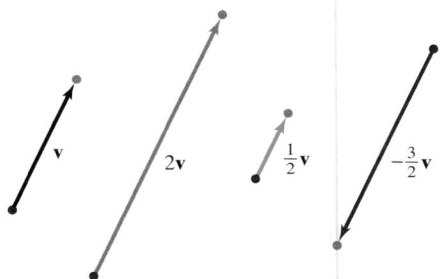

Figure 7.47 Multiplying vector $\mathbf{v}$ by real numbers

Now compare the black and red vectors in Figure 7.47. **Multiplying a vector by a negative number reverses the direction of the vector.** Notice that $-\frac{3}{2}\mathbf{v}$ has the opposite direction as $\mathbf{v}$ and is $\frac{3}{2}$ the magnitude of $\mathbf{v}$.

The multiplication of the real number, k, and the vector, $\mathbf{v}$, is called **scalar multiplication**. We write this product as $k\mathbf{v}$.

Scalar Multiplication

If k is a real number and $\mathbf{v}$ a vector, the vector $k\mathbf{v}$ is called a **scalar multiple** of the vector $\mathbf{v}$. The magnitude and direction of $k\mathbf{v}$ are given as follows:

The vector $k\mathbf{v}$ has a *magnitude* of $|k|\,\|\mathbf{v}\|$. We describe this as the absolute value of k times the magnitude of vector $\mathbf{v}$.

The vector $k\mathbf{v}$ has a *direction* that is:

- the same as the direction of $\mathbf{v}$ if $k > 0$, and
- opposite the direction of $\mathbf{v}$ if $k < 0$.

A geometric method for adding two vectors is shown in Figure 7.48. The sum of $\mathbf{u} + \mathbf{v}$ is called the **resultant vector**. Here is how we find this vector:

1. Position $\mathbf{u}$ and $\mathbf{v}$ so the terminal point of $\mathbf{u}$ coincides with the initial point of $\mathbf{v}$.
2. The resultant vector, $\mathbf{u} + \mathbf{v}$, extends from the initial point of $\mathbf{u}$ to the terminal point of $\mathbf{v}$.

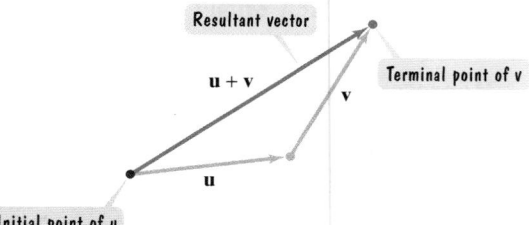

Figure 7.48 Vector addition $\mathbf{u} + \mathbf{v}$; the terminal point of $\mathbf{u}$ coincides with the initial point of $\mathbf{v}$.

SECTION 7.7   The Dot Product

Objectives

1. Find the dot product of two vectors.
2. Find the angle between two vectors.
3. Use the dot product to determine if two vectors are orthogonal.
4. Write a vector in terms of its magnitude and direction.
5. Find the projection of a vector onto another vector.
6. Express a vector as the sum of two orthogonal vectors.
7. Compute work.

Talk about hard work! I can see the weightlifter's muscles quivering from the exertion of holding the barbell in a stationary position above his head. Still, I'm not sure if he's doing as much work as I, sitting at my desk with brain muscles quivering by studying trigonometric functions and their applications.

　　Would it surprise you to know that neither you nor the weightlifter are doing any work at all? The precise definition of work is not the same as what we mean by "working hard" in everyday use. To understand what is involved in real work, we turn to a new vector operation called the dot product.

1 Find the dot product of two vectors.

The Dot Product of Two Vectors

The operations of vector addition and scalar multiplication result in vectors. By contrast, the **dot product** of two vectors results in a scalar (a real number), rather than a vector.

> **Definition of the Dot Product**
>
> If $\mathbf{v} = a_1\mathbf{i} + b_1\mathbf{j}$ and $\mathbf{w} = a_2\mathbf{i} + b_2\mathbf{j}$ are vectors, the **dot product $\mathbf{v} \cdot \mathbf{w}$** is defined as
>
> $$\mathbf{v} \cdot \mathbf{w} = a_1 a_2 + b_1 b_2.$$
>
> The dot product of two vectors is the sum of the products of their horizontal and vertical components.

EXAMPLE 1   Finding Dot Products

If $\mathbf{v} = 5\mathbf{i} - 2\mathbf{j}$ and $\mathbf{w} = -3\mathbf{i} + 4\mathbf{j}$, find:

a. $\mathbf{v} \cdot \mathbf{w}$　　**b.** $\mathbf{w} \cdot \mathbf{v}$　　**c.** $\mathbf{v} \cdot \mathbf{v}$.

Solution　To find each dot product, multiply the two horizontal components, and then multiply the two vertical components. Finally, add the two products.

　　a. $\mathbf{v} \cdot \mathbf{w} = 5(-3) + (-2)(4) = -15 - 8 = -23$

Multiply the horizontal components and multiply the vertical components of
$\mathbf{v} = 5\mathbf{i} - 2\mathbf{j}$ and $\mathbf{w} = -3\mathbf{i} + 4\mathbf{j}$.

b. $\mathbf{w} \cdot \mathbf{v} = -3(5) + 4(-2) = -15 - 8 = -23$

> Multiply the horizontal components and multiply the vertical components of
> $\mathbf{w} = -3\mathbf{i} + 4\mathbf{j}$ and $\mathbf{v} = 5\mathbf{i} - 2\mathbf{j}$.

c. $\mathbf{v} \cdot \mathbf{v} = 5(5) + (-2)(-2) = 25 + 4 = 29$

> Multiply the horizontal components and multiply the vertical components of
> $\mathbf{v} = 5\mathbf{i} - 2\mathbf{j}$ and $\mathbf{v} = 5\mathbf{i} - 2\mathbf{j}$.

> **Check Point 1**
>
> If $\mathbf{v} = 7\mathbf{i} - 4\mathbf{j}$ and $\mathbf{w} = 2\mathbf{i} - \mathbf{j}$, find:
>
> **a.** $\mathbf{v} \cdot \mathbf{w}$ **b.** $\mathbf{w} \cdot \mathbf{v}$ **c.** $\mathbf{w} \cdot \mathbf{w}$.

In Example 1 and Check Point 1, did you notice that $\mathbf{v} \cdot \mathbf{w}$ and $\mathbf{w} \cdot \mathbf{v}$ produced the same scalar? The fact that $\mathbf{v} \cdot \mathbf{w} = \mathbf{w} \cdot \mathbf{v}$ follows from the definition of the dot product. Properties of the dot product are given in the box that follows. Proofs for some of these properties are given in the appendix.

Properties of the Dot Product

If $\mathbf{u}$, $\mathbf{v}$, and $\mathbf{w}$ are vectors, and c is a scalar, then

1. $\mathbf{u} \cdot \mathbf{v} = \mathbf{v} \cdot \mathbf{u}$
2. $\mathbf{u} \cdot (\mathbf{v} + \mathbf{w}) = \mathbf{u} \cdot \mathbf{v} + \mathbf{u} \cdot \mathbf{w}$
3. $\mathbf{0} \cdot \mathbf{v} = 0$
4. $\mathbf{v} \cdot \mathbf{v} = \|\mathbf{v}\|^2$
5. $(c\mathbf{u}) \cdot \mathbf{v} = c(\mathbf{u} \cdot \mathbf{v}) = \mathbf{u} \cdot (c\mathbf{v})$

---

2 Find the angle between two vectors.

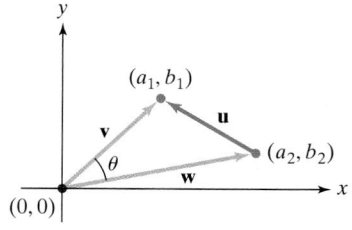

Figure 7.57

The Angle between Two Vectors

The Law of Cosines can be used to derive another formula for the dot product. This formula will give us a way to find the angle between two vectors.

Figure 7.57 shows vectors $\mathbf{v} = a_1\mathbf{i} + b_1\mathbf{j}$ and $\mathbf{w} = a_2\mathbf{i} + b_2\mathbf{j}$. By the definition of the dot product, we know that $\mathbf{v} \cdot \mathbf{w} = a_1a_2 + b_1b_2$. Our new formula for the dot product involves the angle between the vectors, shown as θ in the figure. Apply the Law of Cosines to the triangle shown in the figure.

$$\|\mathbf{u}\|^2 = \|\mathbf{v}\|^2 + \|\mathbf{w}\|^2 - 2\|\mathbf{v}\|\|\mathbf{w}\| \cos \theta \qquad \text{Use the Law of Cosines.}$$

$$(a_1 - a_2)^2 + (b_1 - b_2)^2 = (a_1^2 + b_1^2) + (a_2^2 + b_2^2) - 2\|\mathbf{v}\|\|\mathbf{w}\| \cos \theta \qquad \text{Square the magnitudes of vectors } \mathbf{u}, \mathbf{v}, \text{ and } \mathbf{w}.$$

$$a_1^2 - 2a_1a_2 + a_2^2 + b_1^2 - 2b_1b_2 + b_2^2 = a_1^2 + b_1^2 + a_2^2 + b_2^2 - 2\|\mathbf{v}\|\|\mathbf{w}\| \cos \theta \qquad \text{Square the binomials using } (A - B)^2 = A^2 - 2AB + B^2.$$

$$-2a_1a_2 - 2b_1b_2 = -2\|\mathbf{v}\|\|\mathbf{w}\| \cos \theta \qquad \text{Subtract } a_1^2, a_2^2, b_1^2, \text{ and } b_2^2 \text{ from both sides of the equation.}$$

$$a_1a_2 + b_1b_2 = \|\mathbf{v}\|\|\mathbf{w}\| \cos \theta \qquad \text{Divide both sides by } -2.$$

> By definition, $\mathbf{v} \cdot \mathbf{w} = a_1a_2 + b_1b_2$.

$$\mathbf{v} \cdot \mathbf{w} = \|\mathbf{v}\|\|\mathbf{w}\| \cos \theta \qquad \text{Substitute } \mathbf{v} \cdot \mathbf{w} \text{ for the expression on the left side of the equation.}$$

Alternative Formula for the Dot Product

If **v** and **w** are two nonzero vectors and θ is the smallest nonnegative angle between them, then

$$\mathbf{v} \cdot \mathbf{w} = \|\mathbf{v}\|\|\mathbf{w}\|\cos\theta.$$

Solving the formula in the box for $\cos\theta$ gives us a formula for finding the angle between vectors.

Formula for the Angle between Two Vectors

If **v** and **w** are two nonzero vectors and θ is the smallest nonnegative angle between **v** and **w**, then

$$\cos\theta = \frac{\mathbf{v} \cdot \mathbf{w}}{\|\mathbf{v}\|\|\mathbf{w}\|} \quad \text{and} \quad \theta = \cos^{-1}\left(\frac{\mathbf{v} \cdot \mathbf{w}}{\|\mathbf{v}\|\|\mathbf{w}\|}\right).$$

EXAMPLE 2   Finding the Angle between Two Vectors

Find the angle θ between the vectors $\mathbf{v} = 3\mathbf{i} - 2\mathbf{j}$ and $\mathbf{w} = -\mathbf{i} + 4\mathbf{j}$, shown in Figure 7.58. Round to the nearest tenth of a degree.

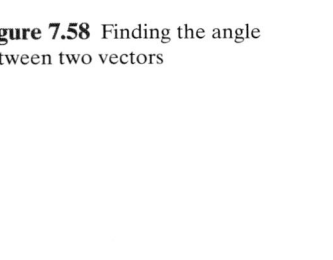

Solution   Use the formula for the angle between two vectors.

$$\cos\theta = \frac{\mathbf{v} \cdot \mathbf{w}}{\|\mathbf{v}\|\|\mathbf{w}\|}$$
This is the formula for the angle between two vectors.

$$= \frac{(3\mathbf{i} - 2\mathbf{j}) \cdot (-\mathbf{i} + 4\mathbf{j})}{\sqrt{3^2 + (-2)^2}\sqrt{(-1)^2 + 4^2}}$$
Substitute the given vectors in the numerator. Find the magnitude of each vector in the denominator.

$$= \frac{3(-1) + (-2)(4)}{\sqrt{13}\sqrt{17}}$$
Find the dot product in the numerator. Simplify in the denominator.

$$= -\frac{11}{\sqrt{221}}$$
Perform the indicated operations.

Figure 7.58 Finding the angle between two vectors

The angle θ between the vectors is

$$\theta = \cos^{-1}\left(-\frac{11}{\sqrt{221}}\right) \approx 137.7°. \quad \text{Use a calculator.}$$

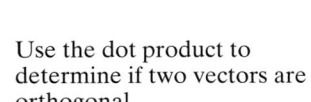

Check Point 2   Find the angle between the vectors $\mathbf{v} = 4\mathbf{i} - 3\mathbf{j}$ and $\mathbf{w} = \mathbf{i} + 2\mathbf{j}$. Round to the nearest tenth of a degree.

3 Use the dot product to determine if two vectors are orthogonal.

Parallel and Orthogonal Vectors

Two vectors are **parallel** when the angle θ between the vectors is 0° or 180°. If $\theta = 0°$, the vectors point in the same direction. If $\theta = 180°$, the vectors point in opposite directions. Figure 7.59 shows parallel vectors.

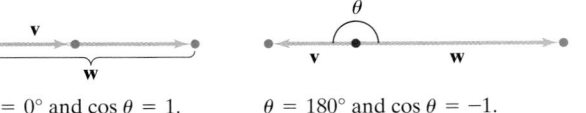

$\theta = 0°$ and $\cos\theta = 1$.
Vectors point in the same direction.

$\theta = 180°$ and $\cos\theta = -1$.
Vectors point in opposite directions.

Figure 7.59 Parallel vectors

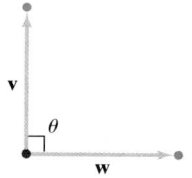

Figure 7.60 Orthogonal vectors: $\theta = 90°$ and $\cos\theta = 0$

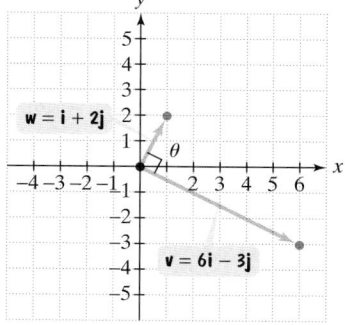

Figure 7.61 Orthogonal vectors

Two vectors are **orthogonal** when the angle between the vectors is 90°, shown in Figure 7.60. (The word "orthogonal," rather than "perpendicular," is used to describe vectors that meet at right angles.) We know that $\mathbf{v} \cdot \mathbf{w} = \|\mathbf{v}\|\|\mathbf{w}\| \cos\theta$. We also know that two vectors are orthogonal if and only if the angle between them is 90°. Using this formula for the dot product and the meaning of orthogonal vectors gives the following result.

The Dot Product and Orthogonal Vectors

Two nonzero vectors $\mathbf{v}$ and $\mathbf{w}$ are orthogonal if and only if $\mathbf{v} \cdot \mathbf{w} = 0$. Because $\mathbf{0} \cdot \mathbf{v} = 0$, the zero vector is orthogonal to every vector $\mathbf{v}$.

EXAMPLE 3 Determining Whether Vectors Are Orthogonal

Are the vectors $\mathbf{v} = 2\mathbf{i} + 3\mathbf{j}$ and $\mathbf{w} = 6\mathbf{i} - 4\mathbf{j}$ orthogonal?

Solution The vectors are orthogonal if their dot product is 0. Begin by finding $\mathbf{v} \cdot \mathbf{w}$.

$$\mathbf{v} \cdot \mathbf{w} = (2\mathbf{i} + 3\mathbf{j}) \cdot (6\mathbf{i} - 4\mathbf{j}) = 2(6) + 3(-4) = 12 - 12 = 0$$

The dot product is 0. Thus, the given vectors are orthogonal. They are shown in Figure 7.61.

> **Check Point 3** Are the vectors $\mathbf{v} = 6\mathbf{i} - 3\mathbf{j}$ and $\mathbf{w} = \mathbf{i} + 2\mathbf{j}$ orthogonal?

④ Write a vector in terms of its magnitude and direction.

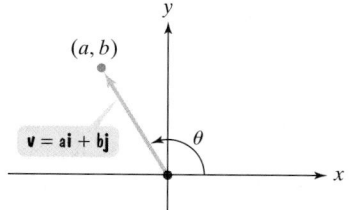

Figure 7.62 Expressing a vector in terms of $\|\mathbf{v}\|$, its magnitude, and θ, its direction angle

Writing a Vector in Terms of Its Magnitude and Direction

Consider the vector $\mathbf{v} = a\mathbf{i} + b\mathbf{j}$. The components a and b can be expressed in terms of the magnitude of $\mathbf{v}$ and the angle θ that $\mathbf{v}$ makes with the positive x-axis. This angle is called the **direction angle** of $\mathbf{v}$ and is shown in Figure 7.62. By the definitions of sine and cosine, we have

$$\cos\theta = \frac{a}{\|\mathbf{v}\|} \quad \text{and} \quad \sin\theta = \frac{b}{\|\mathbf{v}\|}$$

$$a = \|\mathbf{v}\| \cos\theta \qquad\qquad b = \|\mathbf{v}\| \sin\theta.$$

Thus,

$$\mathbf{v} = a\mathbf{i} + b\mathbf{j} = \|\mathbf{v}\| \cos\theta\mathbf{i} + \|\mathbf{v}\| \sin\theta\mathbf{j}.$$

Writing a Vector in Terms of Its Magnitude and Direction

Let $\mathbf{v}$ be a nonzero vector. If θ is the direction angle measured from the positive x-axis to $\mathbf{v}$, then the vector can be expressed in terms of its magnitude and direction angle as

$$\mathbf{v} = \|\mathbf{v}\| \cos\theta\mathbf{i} + \|\mathbf{v}\| \sin\theta\mathbf{j}.$$

EXAMPLE 4 Writing a Vector Whose Magnitude and Direction Are Given

The wind is blowing at 20 miles per hour in the direction N30°W. Express its velocity as a vector $\mathbf{v}$.

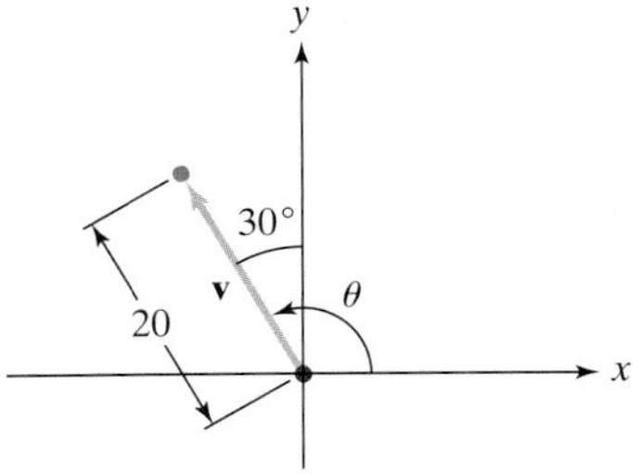

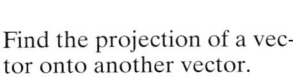

Figure 7.63 Vector **v** represents a wind blowing at 20 miles per hour in the direction N30°W.

Solution The vector **v** is shown in Figure 7.63. The vector's direction angle, from the positive *x*-axis to **v**, is

$$\theta = 90° + 30° = 120°.$$

Because the wind is blowing at 20 miles per hour, the magnitude of **v** is 20 miles per hour: $\|\mathbf{v}\| = 20$. Thus,

$\mathbf{v} = \|\mathbf{v}\| \cos\theta\mathbf{i} + \|\mathbf{v}\| \sin\theta\mathbf{j}$ *Use the formula for a vector in terms of magnitude and direction.*

$= 20\cos 120°\mathbf{i} + 20\sin 120°\mathbf{j}$ *$\|\mathbf{v}\| = 20$ and $\theta = 120°$.*

$= 20\left(-\frac{1}{2}\right)\mathbf{i} + 20\left(\dfrac{\sqrt{3}}{2}\right)\mathbf{j}$ *$\cos 120° = -\dfrac{1}{2}$ and $\sin 120° = \dfrac{\sqrt{3}}{2}$.*

$= -10\mathbf{i} + 10\sqrt{3}\mathbf{j}$ *Simplify.*

The wind's velocity can be expressed in terms of **i** and **j** as $\mathbf{v} = -10\mathbf{i} + 10\sqrt{3}\mathbf{j}$.

Check Point 4 The jet stream is blowing at 60 miles per hour in the direction N45°E. Express its velocity as a vector **v** in terms of **i** and **j**.

5 Find the projection of a vector onto another vector.

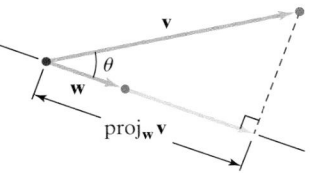

Figure 7.64

Projection of a Vector Onto Another Vector

You know how to add two vectors to obtain a resultant vector. We now reverse this process by expressing a vector as the sum of two orthogonal vectors. By doing this, you can determine how much force is applied in a particular direction. For example, Figure 7.64 shows a boat on a tilted ramp. The force due to gravity, **F**, is pulling straight down on the boat. Part of this force, $\mathbf{F}_1$, is pushing the boat down the ramp. Another part of this force, $\mathbf{F}_2$, is pressing the boat against the ramp, at a right angle to the incline. These two orthogonal vectors, $\mathbf{F}_1$ and $\mathbf{F}_2$, are called the **vector components** of **F**. Notice that

$$\mathbf{F} = \mathbf{F}_1 + \mathbf{F}_2.$$

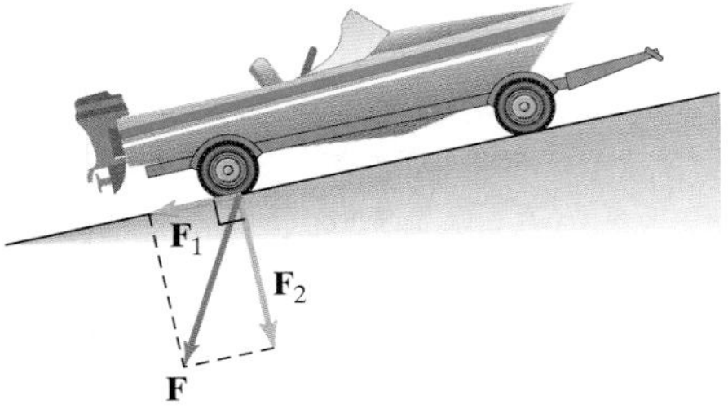

A method for finding $\mathbf{F}_1$ and $\mathbf{F}_2$ involves projecting a vector onto another vector.

Figure 7.65 shows two nonzero vectors, **v** and **w**, with the same initial point. The angle between the vectors, θ, is acute in Figure 7.65(a) and obtuse in Figure 7.65(b). A third vector, called the **vector projection of v onto w**, is also shown in each figure, denoted by proj$_\mathbf{w}$ **v**.

How is the vector projection of **v** onto **w** formed? Draw a red line segment from the terminal point of **v** that forms a right angle with a line through **w**. The projection of **v** onto **w** lies on a line through **w**, and is parallel to vector **w**. This vector begins at the common initial point of **v** and **w**. It ends at the point where the red line segment intersects the line through **w**.

Figure 7.65(a)

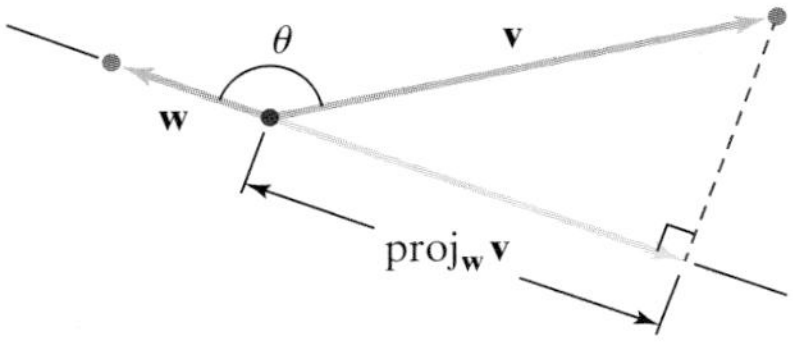

Figure 7.65(b)

Our goal is to determine an expression for $\text{proj}_w \mathbf{v}$. We begin with its magnitude. By the definition of the cosine function,

$$\cos\theta = \frac{\|\text{proj}_w\mathbf{v}\|}{\|\mathbf{v}\|}.$$

This is the magnitude of the vector projection of **v** onto **w**.

$$\|\mathbf{v}\|\cos\theta = \|\text{proj}_w\mathbf{v}\| \quad \text{Multiply both sides by } \|\mathbf{v}\|.$$

$$\|\text{proj}_w\mathbf{v}\| = \|\mathbf{v}\|\cos\theta \quad \text{Reverse the two sides.}$$

We can rewrite the right side of this equation and obtain another expression for the magnitude of the vector projection of **v** onto **w**. To do so, use the alternate formula for the dot product, $\mathbf{v}\cdot\mathbf{w} = \|\mathbf{v}\|\|\mathbf{w}\|\cos\theta$. Divide both sides by $\|\mathbf{w}\|$:

$$\frac{\mathbf{v}\cdot\mathbf{w}}{\|\mathbf{w}\|} = \|\mathbf{v}\|\cos\theta.$$

The expression on the right side of this equation, $\|\mathbf{v}\|\cos\theta$, is the same expression that appears in the formula for $\|\text{proj}_w\mathbf{v}\|$. Thus,

$$\|\text{proj}_w\mathbf{v}\| = \|\mathbf{v}\|\cos\theta = \frac{\mathbf{v}\cdot\mathbf{w}}{\|\mathbf{w}\|}.$$

We use the formula for the magnitude of $\text{proj}_w\mathbf{v}$ to find the vector itself. This is done by finding the scalar product of the magnitude and a unit vector in the direction of **w**.

$$\text{proj}_w\mathbf{v} = \left(\frac{\mathbf{v}\cdot\mathbf{w}}{\|\mathbf{w}\|}\right)\left(\frac{\mathbf{w}}{\|\mathbf{w}\|}\right) = \frac{\mathbf{v}\cdot\mathbf{w}}{\|\mathbf{w}\|^2}\mathbf{w}$$

This is the magnitude of the vector projection of **v** onto **w**.

This is a unit vector in the direction of **w**.

The Vector Projection of v Onto w

If **v** and **w** are two nonzero vectors, the vector projection of **v** onto **w** is

$$\text{proj}_w\mathbf{v} = \frac{\mathbf{v}\cdot\mathbf{w}}{\|\mathbf{w}\|^2}\mathbf{w}.$$

EXAMPLE 5 Finding the Vector Projection of One Vector Onto Another

If $\mathbf{v} = 2\mathbf{i} + 4\mathbf{j}$ and $\mathbf{w} = -2\mathbf{i} + 6\mathbf{j}$, find the vector projection of **v** onto **w**.

Solution The vector projection of **v** onto **w** is found using the formula for $\text{proj}_w\mathbf{v}$.

$$\text{proj}_w\mathbf{v} = \frac{\mathbf{v}\cdot\mathbf{w}}{\|\mathbf{w}\|^2}\mathbf{w} = \frac{(2\mathbf{i}+4\mathbf{j})\cdot(-2\mathbf{i}+6\mathbf{j})}{(\sqrt{(-2)^2+6^2})^2}\mathbf{w}$$

$$= \frac{2(-2)+4(6)}{(\sqrt{40})^2}\mathbf{w} = \frac{20}{40}\mathbf{w} = \tfrac{1}{2}(-2\mathbf{i}+6\mathbf{j}) = -\mathbf{i}+3\mathbf{j}$$

The three vectors **v**, **w**, and $\text{proj}_w\mathbf{v}$, are shown in Figure 7.66.

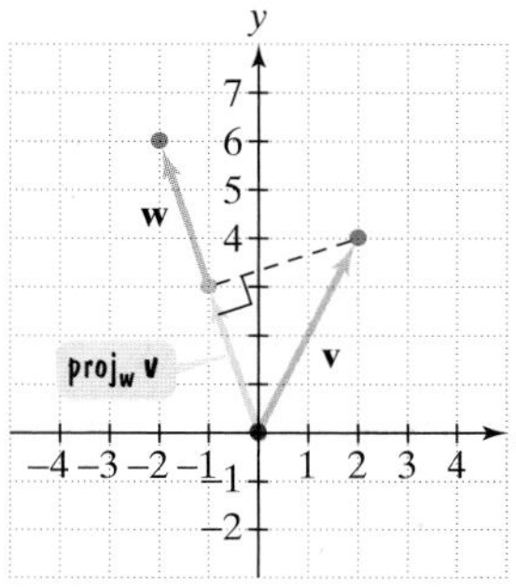

Figure 7.66 The vector projection of **v** onto **w**

Check Point 5 If $\mathbf{v} = 2\mathbf{i} - 5\mathbf{j}$ and $\mathbf{w} = \mathbf{i} - \mathbf{j}$, find the vector projection of **v** onto **w**.

6 Express a vector as the sum of two orthogonal vectors.

We use the vector projection of **v** onto **w**, proj$_\mathbf{w}$**v**, to express **v** as the sum of two orthogonal vectors.

The Vector Components of v

Let **v** and **w** be two nonzero vectors. Vector **v** can be expressed as the sum of two orthogonal vectors, $\mathbf{v}_1$ and $\mathbf{v}_2$, where $\mathbf{v}_1$ is parallel to **w** and $\mathbf{v}_2$ is orthogonal to **w**.

$$\mathbf{v}_1 = \text{proj}_\mathbf{w}\mathbf{v} = \frac{\mathbf{v} \cdot \mathbf{w}}{\|\mathbf{w}\|^2}\,\mathbf{w}, \; \mathbf{v}_2 = \mathbf{v} - \mathbf{v}_1$$

Thus, $\mathbf{v} = \mathbf{v}_1 + \mathbf{v}_2$. The vectors $\mathbf{v}_1$ and $\mathbf{v}_2$ are called the **vector components** of **v**. The process of expressing **v** as $\mathbf{v}_1 + \mathbf{v}_2$ is called the **decomposition** of **v** into $\mathbf{v}_1$ and $\mathbf{v}_2$.

EXAMPLE 6   **Decomposing a Vector into Two Orthogonal Vectors**

Let $\mathbf{v} = 2\mathbf{i} + 4\mathbf{j}$ and $\mathbf{w} = -2\mathbf{i} + 6\mathbf{j}$. Decompose **v** into two vectors, $\mathbf{v}_1$ and $\mathbf{v}_2$, where $\mathbf{v}_1$ is parallel to **w** and $\mathbf{v}_2$ is orthogonal to **w**.

Solution   These are the vectors we worked with in Example 5. We use the formulas in the preceding box.

$$\mathbf{v}_1 = \text{proj}_\mathbf{w}\mathbf{v} = -\mathbf{i} + 3\mathbf{j} \quad \text{We obtained this vector in Example 5.}$$
$$\mathbf{v}_2 = \mathbf{v} - \mathbf{v}_1 = (2\mathbf{i} + 4\mathbf{j}) - (-\mathbf{i} + 3\mathbf{j}) = 3\mathbf{i} + \mathbf{j}$$

> **Check Point 6**   Let $\mathbf{v} = 2\mathbf{i} - 5\mathbf{j}$ and $\mathbf{w} = \mathbf{i} - \mathbf{j}$. (These are the vectors from Check Point 5.) Decompose **v** into two vectors, $\mathbf{v}_1$ and $\mathbf{v}_2$, where $\mathbf{v}_1$ is parallel to **w** and $\mathbf{v}_2$ is orthogonal to **w**.

7 Compute work.

Work: An Application of the Dot Product

The bad news: Your car just died. The good news: It died on a level road just 200 feet from a gas station. Exerting a constant force of 90 pounds, and not necessarily whistling as you work, you manage to push the car to the gas station.

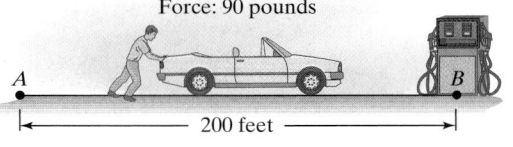

Force: 90 pounds

A |←——— 200 feet ———→| *B*

Although you did not whistle, you certainly did work pushing the car 200 feet from point *A* to point *B*. How much work did you do? If a constant force **F** is applied to an object, moving it from point *A* to point *B* in the direction of the force, the work *W* done is

$$W = (\text{magnitude of force})(\text{distance from } A \text{ to } B).$$

You pushed with a force of 90 pounds for a distance of 200 feet. The work done by your force is

$$W = (90 \text{ pounds})(200 \text{ feet})$$

or 18,000 foot-pounds. Work is often measured in foot-pounds or in newton-meters.

The photo at left shows an adult pulling a small child in a wagon. Work is being done. However, the situation is not quite the same as your car push. Pushing the car, the force you applied was along the line of motion. By contrast, the force of the adult pulling the wagon is not applied along the line of the wagon's motion. In this case, the dot product is used to determine the work done by the force.

Definition of Work

The work W done by a force $\mathbf{F}$ in moving an object from A to B is

$$W = \mathbf{F} \cdot \overrightarrow{AB}.$$

When computing work, it is often easier to use the alternative formula for the dot product. Thus,

$$W = \mathbf{F} \cdot \overrightarrow{AB} = \|\mathbf{F}\| \|\overrightarrow{AB}\| \cos \theta$$

| $\|\mathbf{F}\|$ is the magnitude of the force. | $\|\overrightarrow{AB}\|$ is the distance over which the constant force is applied. | θ is the angle between the force and the direction of motion. |

It is correct to refer to W as either the work done or the work done by the force.

EXAMPLE 7   Computing Work

A child pulls a sled along level ground by exerting a force of 30 pounds on a rope that makes an angle of 35° with the ground. How much work is done pulling the sled 200 feet?

Solution   The situation is illustrated in Figure 7.67. The work done is

$$W = \|\mathbf{F}\| \|\overrightarrow{AB}\| \cos \theta = (30)(200) \cos 35° \approx 4915$$

| Magnitude of the force is 30 pounds. | Distance is 200 feet. | The angle between the force and the sled's motion is 35°. |

Thus, the work done is approximately 4915 foot-pounds.

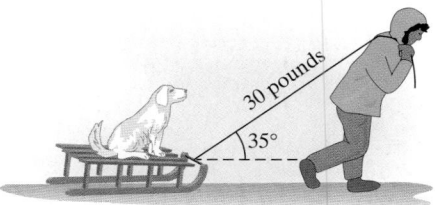

Figure 7.67 Computing work done pulling the sled 200 feet

Check Point 7   A child pulls a wagon along level ground by exerting a force of 20 pounds on a handle that makes an angle of 30° with the ground. How much work is done in pulling the wagon 150 feet?

EXERCISE SET 7.7

 Practice Exercises

In Exercises 1–8, use the given vectors to find **a.** $\mathbf{v} \cdot \mathbf{w}$ *and* **b.** $\mathbf{v} \cdot \mathbf{v}$.

1. $\mathbf{v} = 3\mathbf{i} + \mathbf{j}, \quad \mathbf{w} = \mathbf{i} + 3\mathbf{j}$
2. $\mathbf{v} = 3\mathbf{i} + 3\mathbf{j}, \quad \mathbf{w} = \mathbf{i} + 4\mathbf{j}$
3. $\mathbf{v} = 5\mathbf{i} - 4\mathbf{j}, \quad \mathbf{w} = -2\mathbf{i} - \mathbf{j}$
4. $\mathbf{v} = 7\mathbf{i} - 2\mathbf{j}, \quad \mathbf{w} = -3\mathbf{i} - \mathbf{j}$
5. $\mathbf{v} = -6\mathbf{i} - 5\mathbf{j}, \quad \mathbf{w} = -10\mathbf{i} - 8\mathbf{j}$
6. $\mathbf{v} = -8\mathbf{i} - 3\mathbf{j}, \quad \mathbf{w} = -10\mathbf{i} - 5\mathbf{j}$
7. $\mathbf{v} = 5\mathbf{i}, \quad \mathbf{w} = \mathbf{j}$ 8. $\mathbf{v} = \mathbf{i}, \quad \mathbf{w} = -5\mathbf{j}$

In Exercises 9–16, let

$$\mathbf{u} = 2\mathbf{i} - \mathbf{j}, \quad \mathbf{v} = 3\mathbf{i} + \mathbf{j}, \quad \text{and} \quad \mathbf{w} = \mathbf{i} + 4\mathbf{j}.$$

Find each specified scalar.

9. $\mathbf{u} \cdot (\mathbf{v} + \mathbf{w})$ 10. $\mathbf{v} \cdot (\mathbf{u} + \mathbf{w})$
11. $\mathbf{u} \cdot \mathbf{v} + \mathbf{u} \cdot \mathbf{w}$ 12. $\mathbf{v} \cdot \mathbf{u} + \mathbf{v} \cdot \mathbf{w}$
13. $(4\mathbf{u}) \cdot \mathbf{v}$ 14. $(5\mathbf{v}) \cdot \mathbf{w}$
15. $4(\mathbf{u} \cdot \mathbf{v})$ 16. $5(\mathbf{v} \cdot \mathbf{w})$

In Exercises 17–22, find the angle between **v** *and* **w**. *Round to the nearest tenth of a degree.*

17. $\mathbf{v} = 2\mathbf{i} - \mathbf{j}, \quad \mathbf{w} = 3\mathbf{i} + 4\mathbf{j}$
18. $\mathbf{v} = -2\mathbf{i} + 5\mathbf{j}, \quad \mathbf{w} = 3\mathbf{i} + 6\mathbf{j}$
19. $\mathbf{v} = -3\mathbf{i} + 2\mathbf{j}, \quad \mathbf{w} = 4\mathbf{i} - \mathbf{j}$
20. $\mathbf{v} = \mathbf{i} + 2\mathbf{j}, \quad \mathbf{w} = 4\mathbf{i} - 3\mathbf{j}$
21. $\mathbf{v} = 6\mathbf{i}, \quad \mathbf{w} = 5\mathbf{i} + 4\mathbf{j}$ 22. $\mathbf{v} = 3\mathbf{j}, \quad \mathbf{w} = 4\mathbf{i} + 5\mathbf{j}$

In Exercises 23–32, use the dot product to determine whether **v** *and* **w** *are orthogonal.*

23. $\mathbf{v} = \mathbf{i} + \mathbf{j}, \quad \mathbf{w} = \mathbf{i} - \mathbf{j}$
24. $\mathbf{v} = \mathbf{i} + \mathbf{j}, \quad \mathbf{w} = -\mathbf{i} + \mathbf{j}$
25. $\mathbf{v} = 2\mathbf{i} + 8\mathbf{j}, \quad \mathbf{w} = 4\mathbf{i} - \mathbf{j}$
26. $\mathbf{v} = 8\mathbf{i} - 4\mathbf{j}, \quad \mathbf{w} = -6\mathbf{i} - 12\mathbf{j}$
27. $\mathbf{v} = 2\mathbf{i} - 2\mathbf{j}, \quad \mathbf{w} = -\mathbf{i} + \mathbf{j}$
28. $\mathbf{v} = 5\mathbf{i} - 5\mathbf{j}, \quad \mathbf{w} = \mathbf{i} - \mathbf{j}$
29. $\mathbf{v} = 3\mathbf{i}, \quad \mathbf{w} = -4\mathbf{i}$ 30. $\mathbf{v} = 5\mathbf{i}, \quad \mathbf{w} = -6\mathbf{i}$
31. $\mathbf{v} = 3\mathbf{i}, \quad \mathbf{w} = -4\mathbf{j}$ 32. $\mathbf{v} = 5\mathbf{i}, \quad \mathbf{w} = -6\mathbf{j}$

In Exercises 33–38, write a vector **v** *in terms of* **i** *and* **j** *whose magnitude* $\|\mathbf{v}\|$ *and direction angle* θ *are given.*

33. $\|\mathbf{v}\| = 6, \quad \theta = 30°$ 34. $\|\mathbf{v}\| = 8, \quad \theta = 45°$
35. $\|\mathbf{v}\| = 12, \quad \theta = 225°$ 36. $\|\mathbf{v}\| = 10, \quad \theta = 330°$
37. $\|\mathbf{v}\| = \frac{1}{2}, \quad \theta = 113°$ 38. $\|\mathbf{v}\| = \frac{1}{4}, \quad \theta = 200°$

In Exercise 39–44, find $\text{proj}_{\mathbf{w}}\mathbf{v}$. *Then decompose* **v** *into two vectors,* $\mathbf{v}_1$ *and* $\mathbf{v}_2$, *where* $\mathbf{v}_1$ *is parallel to* **w** *and* $\mathbf{v}_2$ *is orthogonal to* **w**.

39. $\mathbf{v} = 3\mathbf{i} - 2\mathbf{j}, \quad \mathbf{w} = \mathbf{i} - \mathbf{j}$
40. $\mathbf{v} = 3\mathbf{i} - 2\mathbf{j}, \quad \mathbf{w} = 2\mathbf{i} + \mathbf{j}$
41. $\mathbf{v} = \mathbf{i} + 3\mathbf{j}, \quad \mathbf{w} = -2\mathbf{i} + 5\mathbf{j}$
42. $\mathbf{v} = 2\mathbf{i} + 4\mathbf{j}, \quad \mathbf{w} = -3\mathbf{i} + 6\mathbf{j}$
43. $\mathbf{v} = \mathbf{i} + 2\mathbf{j}, \quad \mathbf{w} = 3\mathbf{i} + 6\mathbf{j}$
44. $\mathbf{v} = 2\mathbf{i} + \mathbf{j}, \quad \mathbf{w} = 6\mathbf{i} + 3\mathbf{j}$

Application Exercises

In Exercises 45–48, a vector is described. Express the vector in terms of **i** *and* **j**. *If exact values are not possible, round components to the nearest tenth.*

45. A quarterback releases a football with a speed of 44 feet per second at an angle of 30° with the horizontal.
46. A child pulls a sled along level ground by exerting a force of 30 pounds on a handle that makes an angle of 45° with the ground.
47. A plane approaches a runway at 150 miles per hour at an angle of 8° with the ground traveling.
48. A plane with an airspeed of 450 miles per hour is flying in the direction N35°W.
49. The components of $\mathbf{v} = 240\mathbf{i} + 300\mathbf{j}$ represent the respective number of gallons of regular and premium gas sold at a station on Monday. The components of $\mathbf{w} = 1.90\mathbf{i} + 2.07\mathbf{j}$ represent the respective prices per gallon for each kind of gas. Find $\mathbf{v} \cdot \mathbf{w}$ and describe what the answer means in practical terms.
50. The components of $\mathbf{v} = 180\mathbf{i} + 450\mathbf{j}$ represent the respective number of one-day and three-day videos rented from a video store on Monday. The components of $\mathbf{w} = 3\mathbf{i} + 2\mathbf{j}$ represent the prices to rent the newly arrived one-day videos and the three-day videos, respectively. Find $\mathbf{v} \cdot \mathbf{w}$ and describe what the answer means in practical terms.
51. Find the work done in pushing a car along a level road from point A to point B, 80 feet from A, while exerting a constant force of 95 pounds. Round to the nearest foot-pound.
52. Find the work done when a crane lifts a 6000-pound boulder through a vertical distance of 12 feet. Round to the nearest foot-pound.
53. A wagon is pulled along level ground by exerting a force of 40 pounds on a handle that makes an angle of 32° with the ground. How much work is done in pulling the wagon 100 feet? Round to the nearest foot-pound.
54. A wagon is pulled along level ground by exerting a force of 25 pounds on a handle that makes an angle of 38° with the ground. How much work is done in pulling the wagon 100 feet? Round to the nearest foot-pound.

Vectors are used in computer graphics to determine lengths of shadows over flat surfaces. The length of the shadow for **v** *in the figure shown is the absolute value of the vector's horizontal component. In Exercises 55–56, the magnitude and di-*

*rection angle of **v** are given. Write **v** in terms of **i** and **j**. Then find the length of the shadow to the nearest tenth of an inch.*

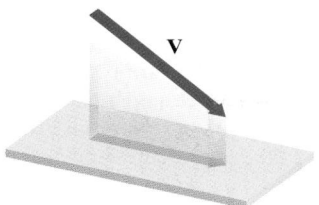

55. $\|\mathbf{v}\| = 1.5$ inches, $\theta = 25°$

56. $\|\mathbf{v}\| = 1.8$ inches, $\theta = 40°$

57. A truck that weighs 40,000 pounds is parked on a road that forms a 12° angle with level ground. Let **v** be the vector representing the force needed to prevent the truck from rolling down the hill.

   a. How many pounds of force are required to prevent the truck from rolling down the hill?

   b. Write **v** in terms of **i** and **j**. Round each component to the nearest pound.

58. A car that weighs 3000 pounds is parked on a road that forms a 10° angle with level ground. Let **v** be the vector representing the force needed to prevent the car from rolling down the hill.

   a. How many pounds of force are required to prevent the truck from rolling down the hill?

   b. Write **v** in terms of **i** and **j**. Round each component to the nearest pound.

Writing in Mathematics

59. Explain how to find the dot product of two vectors.

60. Using words and no symbols, describe how to find the dot product of two vectors with the alternative formula

$$\mathbf{v} \cdot \mathbf{w} = \|\mathbf{v}\|\|\mathbf{w}\| \cos \theta.$$

61. Describe how to find the angle between two vectors.

62. What are parallel vectors?

63. What are orthogonal vectors?

64. How do you determine if two vectors are orthogonal?

65. Explain how to write a vector in terms of its magnitude and direction.

66. Draw two vectors, **v** and **w**, with the same initial point. Show the vector projection of **v** onto **w** in your diagram. Then describe how you identified this vector.

67. How do you determine the work done by a force **F** in moving an object from A to B when the direction of the force is not along the line of motion?

68. A weightlifter is holding a barbell perfectly still above his head, his body shaking from the effort. How much work is the weightlifter doing? Explain your answer.

69. Describe one way in which the everyday use of the word "work" is different from the definition of work given in this section.

Critical Thinking Exercises

In Exercises 70–72, use the vectors

$$\mathbf{u} = a_1\mathbf{i} + b_1\mathbf{j}, \quad \mathbf{v} = a_2\mathbf{i} + b_2\mathbf{j}, \quad \text{and} \quad \mathbf{w} = a_3\mathbf{i} + b_3\mathbf{j}$$

to prove the given property.

70. $\mathbf{u} \cdot \mathbf{v} = \mathbf{v} \cdot \mathbf{u}$ **71.** $(c\mathbf{u}) \cdot \mathbf{v} = c(\mathbf{u} \cdot \mathbf{v})$

72. $\mathbf{u} \cdot (\mathbf{v} + \mathbf{w}) = \mathbf{u} \cdot \mathbf{v} + \mathbf{u} \cdot \mathbf{w}$

73. How much work is done by a force of 4 pounds acting in the direction $3\mathbf{i} + \mathbf{j}$ in moving an object 5 feet from $(0, 0)$ to $(0, 5)$?

74. If $\mathbf{v} = -2\mathbf{i} + 5\mathbf{j}$, find a vector perpendicular to **v**.

75. Find a value of b so that $15\mathbf{i} - 3\mathbf{j}$ and $-4\mathbf{i} + b\mathbf{j}$ are orthogonal.

76. Prove that the projection of **v** onto **i** is $(\mathbf{v} \cdot \mathbf{i})\mathbf{i}$.

77. Find two vectors **v** and **w** such that the projection of **v** onto **w** is **v**.

Group Exercise

78. Group members should research and present a report on unusual and interesting applications of vectors.

CHAPTER SUMMARY, REVIEW, AND TEST

Summary

7.1 and 7.2 The Law of Sines and the Law of Cosines

   a. The Law of Sines

$$\frac{a}{\sin A} = \frac{b}{\sin B} = \frac{c}{\sin C}$$

   b. The Law of Sines is used to solve SAA, ASA, and SSA (the ambiguous case) triangles. The ambiguous case may result in no triangle, one triangle, or two triangles; see the box on page 590.

   c. The area of a triangle equals one-half the product of the lengths of two sides times the sine of their included angle.

d. The Law of Cosines

$$a^2 = b^2 + c^2 - 2bc \cos A,$$
$$b^2 = a^2 + c^2 - 2ac \cos B,$$
$$c^2 = a^2 + b^2 - 2ab \cos C$$

e. The Law of Cosines is used to find the side opposite the given angle in a SAS triangle; see the box on page 600. The Law of Cosines is also used to find the angle opposite the longest side in a SSS triangle; see the box on page 601.

f. Heron's Formula for the Area of a Triangle
The area of a triangle with sides $a, b,$ and c is

$$\sqrt{s(s - a)(s - b)(s - c)},$$

where s is one-half the perimeter:

$$s = \tfrac{1}{2}(a + b + c).$$

7.3 and 7.4 Polar Coordinates and Graphs of Polar Equations

a. A point P in the polar coordinate system is represented by (r, θ), where r is the directed distance of the point from the pole and θ is the angle from the polar axis to line segment OP. The elements of the ordered pair (r, θ) are called the polar coordinates of P. See Figure 7.19 on page 607. When r in (r, θ) is negative, a point is located $|r|$ units along the ray opposite the terminal side of θ. Important information about the sign of r and the location of the point (r, θ) is found in the box on page 608.

b. Multiple Representation of Points
If n is any integer, $(r, \theta) = (r, \theta + 2n\pi)$ or $(r, \theta) = (-r, \theta + \pi + 2n\pi)$.

c. Relations between Polar and Rectangular Coordinates

$$x = r \cos\theta, \quad y = r \sin\theta, \quad x^2 + y^2 = r^2, \quad \tan\theta = \frac{y}{x}$$

d. To convert a point from polar coordinates (r, θ) to rectangular coordinates (x, y), use $x = r \cos\theta$ and $y = r \sin\theta$.

e. To convert a point from rectangular coordinates (x, y) to polar coordinates (r, θ), use the procedure in the box on page 612.

f. To convert a rectangular equation to a polar equation, replace x by $r \cos\theta$ and y by $r \sin\theta$.

g. To convert a polar equation to a rectangular equation, use one or more of

$$r^2 = x^2 + y^2, \quad r \cos\theta = x, \quad r \sin\theta = y, \quad \text{and} \quad \tan\theta = \frac{y}{x}.$$

It is often necessary to do something to the given polar equation to obtain the preceding expressions.

h. A polar equation is an equation whose variables are r and θ. The graph of a polar equation is the set of all points whose polar coordinates satisfy the equation.

i. Polar equations can be graphed using point plotting and symmetry (see the box on page 620).

j. The graphs of $r = a \cos\theta$ and $r = a \sin\theta$ are circles. See the box on page 620. The graphs of $r = a \pm b \sin\theta$ and $r = a \pm b \cos\theta$ are called limaçons ($a > 0$ and $b > 0$), shown in the box on page 623. The graphs of $r = a \sin n\theta$ and $r = a \cos n\theta, a \neq 0$, are rose curves with $2n$ petals if n is even and n petals if n is odd. See the box on page 626. The graphs of $r^2 = a^2 \sin 2\theta$ and $r^2 = a^2 \cos 2\theta$, $a \neq 0$, are called lemniscates and are shown in the box on page 627.

7.5 Complex Numbers in Polar Form; DeMoivre's Theorem

a. The complex number $z = a + bi$ is represented as a point (a, b) in the complex plane, shown in Figure 7.35 on page 631.

b. The absolute value of $z = a + bi$ is $|z| = |a + bi| = \sqrt{a^2 + b^2}$.

c. The polar form of $z = a + bi$ is $z = r(\cos\theta + i \sin\theta)$, where $a = r \cos\theta$, $b = r \sin\theta$, $r = \sqrt{a^2 + b^2}$, and $\tan\theta = \dfrac{b}{a}$. We call r the modulus and θ the argument of z, with $0 \leq \theta < 2\pi$.

d. Multiplying Complex Numbers in Polar Form: Multiply moduli and add arguments. See the box on page 634.

e. Dividing Complex Numbers in Polar Form: Divide moduli and subtract arguments. See the box on page 635.

f. DeMoivre's Theorem is used to find powers of complex numbers in polar form.

$$[r(\cos\theta + i \sin\theta)]^n = r^n(\cos n\theta + i \sin n\theta)$$

g. DeMoivre's Theorem can be used for finding roots of complex numbers in polar form. The n distinct nth roots of $r(\cos\theta + i \sin\theta)$ are

$$\sqrt[n]{r}\left[\cos\left(\frac{\theta + 2\pi k}{n}\right) + i \sin\left(\frac{\theta + 2\pi k}{n}\right)\right]$$

or

$$\sqrt[n]{r}\left[\cos\left(\frac{\theta + 360°k}{n}\right) + i \sin\left(\frac{\theta + 360°k}{n}\right)\right]$$

where $k = 0, 1, 2, \ldots, n - 1$.

7.6 Vectors

a. A vector is a directed line segment.

b. Equal vectors have the same magnitude and the same direction.

c. The vector $k\mathbf{v}$, the scalar multiple of the vector $\mathbf{v}$ and the scalar k, has magnitude $|k|\|\mathbf{v}\|$. The direction of $k\mathbf{v}$ is the same as that of $\mathbf{v}$ if $k > 0$ and opposite $\mathbf{v}$ if $k < 0$.

d. The sum $\mathbf{u} + \mathbf{v}$, called the resultant vector, can be expressed geometrically. Position $\mathbf{u}$ and $\mathbf{v}$ so that the terminal point of $\mathbf{u}$ coincides with the initial point of $\mathbf{v}$. The vector $\mathbf{u} + \mathbf{v}$ extends from the initial point of $\mathbf{u}$ to the terminal point of $\mathbf{v}$.

e. The difference of two vectors, $\mathbf{u} - \mathbf{v}$, is defined as $\mathbf{u} + (-\mathbf{v})$.

f. The vector $\mathbf{i}$ is the unit vector whose direction is along the positive x-axis. The vector $\mathbf{j}$ is the unit vector whose direction is along the positive y-axis.

g. Vector $\mathbf{v}$, from $(0, 0)$ to (a, b), called a position vector, is represented as $\mathbf{v} = a\mathbf{i} + b\mathbf{j}$, where a is the horizontal component and b is the vertical component. The magnitude of $\mathbf{v}$ is given by $\|\mathbf{v}\| = \sqrt{a^2 + b^2}$.

h. Vector $\mathbf{v}$ from (x_1, y_1) to (x_2, y_2) is equal to the position vector $\mathbf{v} = (x_2 - x_1)\mathbf{i} + (y_2 - y_1)\mathbf{j}$. In rectangular coordinates, the term "vector" refers to the position vector in terms of $\mathbf{i}$ and $\mathbf{j}$ that is equal to it.

i. Operations with Vectors in Terms of $\mathbf{i}$ and $\mathbf{j}$
If $\mathbf{v} = a_1\mathbf{i} + b_1\mathbf{j}$ and $\mathbf{w} = a_2\mathbf{i} + b_2\mathbf{j}$, then
1. $\mathbf{v} + \mathbf{w} = (a_1 + a_2)\mathbf{i} + (b_1 + b_2)\mathbf{j}$.
2. $\mathbf{v} - \mathbf{w} = (a_1 - a_2)\mathbf{i} + (b_1 - b_2)\mathbf{j}$.
3. $k\mathbf{v} = (ka_1)\mathbf{i} + (kb_1)\mathbf{j}$.

j. The zero vector $\mathbf{0}$ is the vector whose magnitude is 0 and is assigned no direction. Many properties of vector addition and scalar multiplication involve the zero vector. Some of these properties are listed in the box on pages 650-651.

k. The vector $\dfrac{\mathbf{v}}{\|\mathbf{v}\|}$ is a unit vector that has the same direction as $\mathbf{v}$.

7.7 The Dot Product

a. Definition of the Dot Product
If $\mathbf{v} = a_1\mathbf{i} + b_1\mathbf{j}$ and $\mathbf{w} = a_2\mathbf{i} + b_2\mathbf{j}$, the dot product is defined by $\mathbf{v} \cdot \mathbf{w} = a_1a_2 + b_1b_2$.

b. Alternative Formula for the Dot Product $\mathbf{v} \cdot \mathbf{w} = \|\mathbf{v}\|\|\mathbf{w}\| \cos\theta$, where θ is the smallest nonnegative angle between $\mathbf{v}$ and $\mathbf{w}$.

c. Angle between Two Vectors
$$\cos\theta = \frac{\mathbf{v} \cdot \mathbf{w}}{\|\mathbf{v}\|\|\mathbf{w}\|} \quad \text{and} \quad \theta = \cos^{-1}\left(\frac{\mathbf{v} \cdot \mathbf{w}}{\|\mathbf{v}\|\|\mathbf{w}\|}\right).$$

d. Two vectors are orthogonal when the angle between them is $90°$. To show that two vectors are orthogonal, show that their dot product is zero.

e. A vector with magnitude $\|\mathbf{v}\|$ and direction angle θ, the angle that $\mathbf{v}$ makes with the positive x-axis, can be expressed in terms of its magnitude and direction angle as
$$\mathbf{v} = \|\mathbf{v}\| \cos\theta\mathbf{i} + \|\mathbf{v}\| \sin\theta\mathbf{j}.$$

f. The vector projection of $\mathbf{v}$ onto $\mathbf{w}$ is given by
$$\text{proj}_{\mathbf{w}}\mathbf{v} = \frac{\mathbf{v} \cdot \mathbf{w}}{\|\mathbf{w}\|^2} \mathbf{w}.$$

g. Expressing a vector as the sum of two orthogonal vectors is shown in the box on page 662.

h. The work W done by a force $\mathbf{F}$ in moving an object from A to B is $W = \mathbf{F} \cdot \overrightarrow{AB}$. Thus, $W = \|\mathbf{F}\|\|\overrightarrow{AB}\| \cos\theta$, where θ is the angle between the force and the direction of motion.

Review Exercises

7.1 and 7.2

In Exercises 1–12, solve each triangle. Round lengths to the nearest tenth and angle measures to the nearest degree. If no triangle exists, state "no triangle." If two triangles exist, solve each triangle.

1. $A = 70°, B = 55°, a = 12$
2. $B = 107°, C = 30°, c = 126$
3. $B = 66°, a = 17, c = 12$
4. $a = 117, b = 66, c = 142$
5. $A = 35°, B = 25°, c = 68$
6. $A = 39°, a = 20, b = 26$
7. $C = 50°, a = 3, c = 1$
8. $A = 162°, b = 11.2, c = 48.2$
9. $a = 26.1, b = 40.2, c = 36.5$
10. $A = 40°, a = 6, b = 4$
11. $B = 37°, a = 12.4, b = 8.7$
12. $A = 23°, a = 54.3, b = 22.1$

In Exercises 13–16, find the area of the triangle having the given measurements. Round to the nearest square unit.

13. $C = 42°, a = 4$ feet, $b = 6$ feet
14. $A = 22°, b = 4$ feet, $c = 5$ feet
15. $a = 2$ meters, $b = 4$ meters, $c = 5$ meters
16. $a = 2$ meters, $b = 2$ meters, $c = 2$ meters
17. An A-frame cabin is 35 feet wide. The roof of the cabin makes a $60°$ angle with the cabin's base. Find the length of the roof from its ground level to the peak. Round to the nearest tenth of a foot.

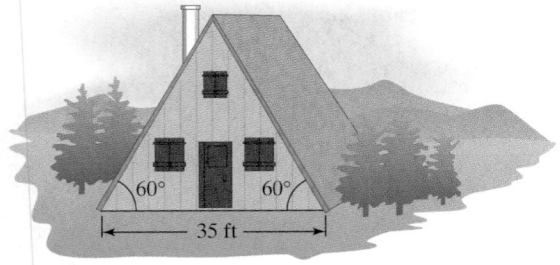

18. Two cars leave a city at the same time and travel along straight highways that differ in direction by $80°$. One car averages 60 miles per hour and the other averages 50 miles per hour. How far apart will the cars be after 30 minutes? Round to the nearest tenth of a mile.

19. Two airplanes leave an airport at the same time on different runways. One flies at a bearing of N66.5°W at 325 miles per hour. The other airplane flies at a bearing of S26.5°W at 300 miles per hour. How far apart will the airplanes be after two hours?

20. The figure shows three roads that intersect to form a triangular piece of land. Find the lengths of the other two sides of the land to the nearest foot.

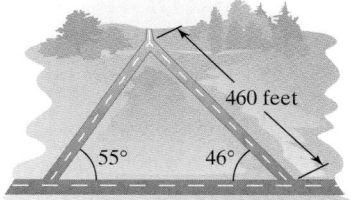

21. A commercial piece of real estate is priced at $5.25 per square foot. Find the cost, to the nearest dollar, of a triangular lot measuring 260 feet by 320 feet by 450 feet.

7.3 and 7.4

In Exercises 22–27, plot each point in polar coordinates, and find its rectangular coordinates.

22. $(4, 60°)$ **23.** $(3, 150°)$

24. $\left(-4, \dfrac{4\pi}{3}\right)$ **25.** $\left(-2, \dfrac{5\pi}{4}\right)$

26. $\left(-4, -\dfrac{\pi}{2}\right)$ **27.** $\left(-2, -\dfrac{\pi}{4}\right)$

In Exercises 28–30, plot each point in polar coordinates. Then find another representation (r, θ) of this point in which:
  a. $r > 0, \quad 2\pi < \theta < 4\pi.$
  b. $r < 0, \quad 0 < \theta < 2\pi.$
  c. $r > 0, -2\pi < \theta < 0.$

28. $\left(3, \dfrac{\pi}{6}\right)$ **29.** $\left(2, \dfrac{2\pi}{3}\right)$ **30.** $(3, \pi)$

In Exercises 31–36, the rectangular coordinates of a point are given. Find polar coordinates of each point.

31. $(-4, 4)$ **32.** $(3, -3)$
33. $(5, 12)$ **34.** $(-3, 4)$
35. $(0, -5)$ **36.** $(1, 0)$

In Exercises 37–39, convert each rectangular equation to a polar equation.

37. $2x + 3y = 8$ **38.** $x^2 + y^2 = 100$
39. $5x^2 + 5y^2 = 3y$

In Exercises 40–46, convert each polar equation to a rectangular equation.

40. $r = 3$ **41.** $\theta = \dfrac{3\pi}{4}$

42. $r\cos\theta = -1$ **43.** $r = 5\sec\theta$
44. $r = 3\cos\theta$ **45.** $5r\cos\theta + r\sin\theta = 8$
46. $r^2\sin 2\theta = 4$

In Exercises 47–49, test for symmetry with respect to:

  a. *the polar axis;*   **b.** *the line* $\theta = \dfrac{\pi}{2};$ *and*   **c.** *the pole.*

47. $r = 5 + 3\cos\theta$ **48.** $r = 3\sin\theta$
49. $r^2 = 9\cos 2\theta$

In Exercises 50–56, graph each polar equation. Be sure to test for symmetry.

50. $r = 3\cos\theta$ **51.** $r = 2 + 2\sin\theta$
52. $r = \sin 2\theta$ **53.** $r = 2 + \cos\theta$
54. $r = 1 + 3\sin\theta$ **55.** $r = 1 - 2\cos\theta$
56. $r^2 = \cos 2\theta$

7.5

In Exercises 57–60, plot each complex number. Then write the complex number in polar form. You may express the argument in degrees or radians.

57. $1 - i$ **58.** $-2\sqrt{3} + 2i$
59. $-3 - 4i$ **60.** $-5i$

In Exercises 61–64, write each complex number in rectangular form.

61. $8(\cos 60° + i\sin 60°)$ **62.** $4(\cos 210° + i\sin 210°)$

63. $6\left(\cos\dfrac{2\pi}{3} + i\sin\dfrac{2\pi}{3}\right)$

64. $0.6(\cos 100° + i\sin 100°)$

In Exercises 65–67, find the product of the complex numbers. Leave answers in polar form.

65. $z_1 = 3(\cos 40° + i\sin 40°)$
     $z_2 = 5(\cos 70° + i\sin 70°)$

66. $z_1 = \cos 210° + i\sin 210°$
     $z_2 = \cos 55° + i\sin 55°$

67. $z_1 = 4\left(\cos\dfrac{3\pi}{7} + i\sin\dfrac{3\pi}{7}\right)$
     $z_2 = 10\left(\cos\dfrac{4\pi}{7} + i\sin\dfrac{4\pi}{7}\right)$

In Exercises 68–70, find the quotient $\dfrac{z_1}{z_2}$ of the complex numbers. Leave answers in polar form.

68. $z_1 = 10(\cos 10° + i\sin 10°)$
     $z_2 = 5(\cos 5° + i\sin 5°)$

69. $z_1 = 5\left(\cos\dfrac{4\pi}{3} + i\sin\dfrac{4\pi}{3}\right)$
     $z_2 = 10\left(\cos\dfrac{\pi}{3} + i\sin\dfrac{\pi}{3}\right)$

70. $z_1 = 2\left(\cos\dfrac{5\pi}{3} + i\sin\dfrac{5\pi}{3}\right)$

$z_2 = \cos\dfrac{\pi}{2} + i\sin\dfrac{\pi}{2}$

In Exercises 71–75, use DeMoivre's Theorem to find the indicated power of the complex number. Write answers in rectangular form.

71. $\left[2(\cos 20° + i\sin 20°)\right]^3$

72. $\left[4(\cos 50° + i\sin 50°)\right]^3$

73. $\left[\dfrac{1}{2}\left(\cos\dfrac{\pi}{14} + i\sin\dfrac{\pi}{14}\right)\right]^7$

74. $\left(1 - \sqrt{3}i\right)^7$ **75.** $(-2 - 2i)^5$

In Exercises 76–77, find all the complex roots. Write roots in polar form with θ in degrees.

76. The complex square roots of $49(\cos 50° + i\sin 50°)$

77. The complex cube roots of $125(\cos 165° + i\sin 165°)$

In Exercises 78–81, find all the complex roots. Write roots in rectangular form.

78. The complex fourth roots of $16\left(\cos\dfrac{2\pi}{3} + i\sin\dfrac{2\pi}{3}\right)$

79. The complex cube roots of $8i$

80. The complex cube roots of -1

81. The complex fifth roots of $-1 - i$

7.6

In Exercises 82–84, sketch each position vector and find its magnitude.

82. $\mathbf{v} = -3\mathbf{i} - 4\mathbf{j}$ **83.** $\mathbf{v} = 5\mathbf{i} - 2\mathbf{j}$

84. $\mathbf{v} = -3\mathbf{j}$

In Exercises 85–86, let $\mathbf{v}$ be the vector from initial point P_1 to terminal point P_2. Write $\mathbf{v}$ in terms of $\mathbf{i}$ and $\mathbf{j}$.

85. $P_1 = (2, -1)$, $P_2 = (5, -3)$

86. $P_1 = (-3, 0)$, $P_2 = (-2, -2)$

In Exercises 87–90, let

$$\mathbf{v} = \mathbf{i} - 5\mathbf{j} \quad\text{and}\quad \mathbf{w} = -2\mathbf{i} + 7\mathbf{j}.$$

Find each specified vector or scalar.

87. $\mathbf{v} + \mathbf{w}$ **88.** $\mathbf{w} - \mathbf{v}$

89. $6\mathbf{v} - 3\mathbf{w}$ **90.** $\|-2\mathbf{v}\|$

Chapter 7 Test

1. In oblique triangle ABC, $A = 34°$, $B = 68°$, and $a = 4.8$. Find b to the nearest tenth.

2. In oblique triangle ABC, $C = 68°$, $a = 5$, and $b = 6$. Find c to the nearest tenth.

3. In oblique triangle ABC, $a = 17$ inches, $b = 45$ inches, and $c = 32$ inches. Find the area of the triangle to the nearest square inch.

In Exercises 91–92, find a unit vector that has the same direction as the vector $\mathbf{v}$.

91. $\mathbf{v} = 8\mathbf{i} - 6\mathbf{j}$ **92.** $\mathbf{v} = -\mathbf{i} + 2\mathbf{j}$

93. Two forces of 100 pounds and 200 pounds act on an object. If the angle between the forces is $60°$,
   a. Find the magnitude of the resultant force to the nearest pound.
   b. Find the direction of the resultant force relative to the 200-pound force. Express the answer to the nearest degree.

7.7

94. If $\mathbf{u} = 5\mathbf{i} + 2\mathbf{j}$, $\mathbf{v} = \mathbf{i} - \mathbf{j}$, and $\mathbf{w} = 3\mathbf{i} - 7\mathbf{j}$, find $\mathbf{u} \cdot (\mathbf{v} + \mathbf{w})$.

In Exercises 95–97, find the dot product $\mathbf{v} \cdot \mathbf{w}$. Then find the angle between $\mathbf{v}$ and $\mathbf{w}$ to the nearest tenth of a degree.

95. $\mathbf{v} = 2\mathbf{i} + 3\mathbf{j}$, $\mathbf{w} = 7\mathbf{i} - 4\mathbf{j}$

96. $\mathbf{v} = 2\mathbf{i} + 4\mathbf{j}$, $\mathbf{w} = 6\mathbf{i} - 11\mathbf{j}$

97. $\mathbf{v} = 2\mathbf{i} + \mathbf{j}$, $\mathbf{w} = \mathbf{i} - \mathbf{j}$

In Exercises 98–99, use the dot product to determine whether $\mathbf{v}$ and $\mathbf{w}$ are orthogonal.

98. $\mathbf{v} = 12\mathbf{i} - 8\mathbf{j}$, $\mathbf{w} = 2\mathbf{i} + 3\mathbf{j}$

99. $\mathbf{v} = \mathbf{i} + 3\mathbf{j}$, $\mathbf{w} = -3\mathbf{i} - \mathbf{j}$

100. The magnitude and direction angle of $\mathbf{v}$ are $\|\mathbf{v}\| = 12$ and $\theta = 60°$. Express $\mathbf{v}$ in terms of $\mathbf{i}$ and $\mathbf{j}$.

In Exercises 101–102, find $\text{proj}_{\mathbf{w}}\mathbf{v}$. Then decompose $\mathbf{v}$ into two vectors, $\mathbf{v}_1$ and $\mathbf{v}_2$, where $\mathbf{v}_1$ is parallel to $\mathbf{w}$ and $\mathbf{v}_2$ is orthogonal to $\mathbf{w}$.

101. $\mathbf{v} = -2\mathbf{i} + 5\mathbf{j}$, $\mathbf{w} = 5\mathbf{i} + 4\mathbf{j}$

102. $\mathbf{v} = -\mathbf{i} + 2\mathbf{j}$, $\mathbf{w} = 3\mathbf{i} - \mathbf{j}$

103. A heavy crate is dragged 50 feet along a level floor. Find the work done if a force of 30 pounds at an angle of $42°$ is used.

4. Plot $\left(4, \dfrac{5\pi}{4}\right)$ in the polar coordinate system. Then write two other ordered pairs (r, θ) that name this point.

5. If the rectangular coordinates of a point are $(1, -1)$, find polar coordinates of the point.

6. Convert $x^2 + y^2 = 6x$ to a polar equation.

7. Convert $r = 4\csc\theta$ to a rectangular equation.

In Exercises 8–9, graph each polar equation.

8. $r = 1 + \sin\theta$ **9.** $r = 1 + 3\cos\theta$

10. Write $-\sqrt{3} + i$ in polar form.

In Exercises 11–13, perform the indicated operation. Leave answers in polar form.

11. $5(\cos 15° + i\sin 15°) \cdot 10(\cos 5° + i\sin 5°)$

12. $\dfrac{2\left(\cos\dfrac{\pi}{2} + i\sin\dfrac{\pi}{2}\right)}{4\left(\cos\dfrac{\pi}{3} + i\sin\dfrac{\pi}{3}\right)}$

13. $\left[2(\cos 10° + i\sin 10°)\right]^5$

14. Find the three cube roots of 27. Write roots in rectangular form.

15. If $P_1 = (-2, 3)$ and $P_2 = (-1, 5)$ and **v** is the vector from P_1 to P_2,
     a. Write **v** in terms of **i** and **j**.
     b. Find $\|\mathbf{v}\|$.

In Exercises 16–19, let

$$\mathbf{v} = -5\mathbf{i} + 2\mathbf{j} \quad \text{and} \quad \mathbf{w} = 2\mathbf{i} - 4\mathbf{j}.$$

Find:

16. $3\mathbf{v} - 4\mathbf{w}$.   **17.** $\mathbf{v} \cdot \mathbf{w}$.

18. the angle between **v** and **w**, to the nearest degree.

19. $\text{proj}_{\mathbf{w}}\mathbf{v}$.

20. A small fire is sighted from ranger stations A and B. Station B is 1.6 miles due east of station A. The bearing of the fire from station A is N40°E, and the bearing of the fire from station B is N50°W. How far, to the nearest tenth of a mile, is the fire from station A?

21. A child is pulling a wagon with a force of 40 pounds. How much work is done in moving the wagon 60 feet if the handle makes an angle of 35° with the ground? Round to the nearest foot-pound.

Cumulative Review Exercises (Chapters 1–7)

Solve each equation or inequality in Exercises 1–4.

1. $x^4 - x^3 - x^2 - x - 2 = 0$

2. $2\sin^2\theta - 3\sin\theta + 1 = 0, \quad 0 \le \theta < 2\pi$

3. $x^2 + 2x + 3 > 11$

4. $\sin\theta\cos\theta = -\frac{1}{2}, \quad 0 \le \theta < 2\pi$

In Exercises 5–6, graph one complete cycle.

5. $y = 3\sin(2x - \pi)$ **6.** $y = -4\cos\pi x$

In Exercises 7–8, verify each identity.

7. $\sin\theta\csc\theta - \cos^2\theta = \sin^2\theta$

8. $\cos\left(\theta + \dfrac{3\pi}{2}\right) = \sin\theta$

9. Find the slope and y-intercept of the line whose equation is $2x + 4y - 8 = 0$.

In Exercises 10–11, find the exact value of each expression.

10. $2\sin\dfrac{\pi}{3} - 3\tan\dfrac{\pi}{6}$ **11.** $\sin\left(\tan^{-1}\frac{1}{2}\right)$

In Exercises 12–13, find the domain of the function whose equation is given.

12. $f(x) = \sqrt{5 - x}$ **13.** $g(x) = \dfrac{x - 3}{x^2 - 9}$

14. A ball is thrown vertically upward from a height of 8 feet with an initial velocity of 48 feet per second. The ball's height, $s(t)$, in feet, after t seconds is given by

$$s(t) = -16t^2 + 48t + 8.$$

After how many seconds does the ball reach its maximum height? What is the maximum height?

15. An object moves in simple harmonic motion described by $d = 4\sin 5t$, where t is measured in seconds and d in meters. Find **a.** the maximum displacement; **b.** the frequency; and **c.** the time required for one cycle.

16. Use a half-angle formula to find the exact value of $\cos 22.5°$.

17. If $\mathbf{v} = 2\mathbf{i} + 7\mathbf{j}$ and $\mathbf{w} = \mathbf{i} - 2\mathbf{j}$, find: **a.** $3\mathbf{v} - \mathbf{w}$ and **b.** $\mathbf{v} \cdot \mathbf{w}$.

18. Express as a single logarithm with a coefficient of 1: $\frac{1}{2}\log_b x - \log_b(x^2 + 1)$.

19. Write the slope-intercept form of the line passing through $(4, -1)$ and $(-8, 5)$.

20. Psychologists can measure the amount learned, L, at time t using the model $L = A(1 - e^{-kt})$. The variable A represents the total amount to be learned, and k is the learning rate. A student preparing for the SAT has 300 new vocabulary words to learn: $A = 300$. This particular student can learn 20 vocabulary words after 5 minutes: If $t = 5, L = 20$.
     a. Find k, the learning rate, correct to three decimal places.
     b. Approximately how many words will the student have learned after 20 minutes?
     c. How long will it take for the student to learn 260 words?

Systems of Equations and Inequalities

Chapter 8

Most things in life depend on many variables. Temperature and precipitation are two variables that have a critical effect on whether regions are forests, grasslands, or deserts. Airlines deal with numerous variables during weather disruptions at large connecting airports. They must solve the problem of putting their operation back together again to minimize the cost of the disruption and passenger inconvenience. In this chapter, forests, grasslands, and airline service are viewed in the same way—situations with several variables. You will learn methods for modeling and solving problems in these situations.

A major weather disruption delayed your flight for hours, but you finally made it. You are in Yosemite National Park in California, surrounded by evergreen forests, alpine meadows, and sheer walls of granite. Soaring cliffs, plunging waterfalls, gigantic trees, rugged canyons, mountains and valleys stand in stark contrast to the angry chaos at the airport. This is so different from where you live and attend college, a region in which grasslands predominate.

SECTION 8.1 *Systems of Linear Equations in Two Variables*

Objectives

1. Decide whether an ordered pair is a solution of a linear system.

2. Solve linear systems by substitution.

3. Solve linear systems by addition.

4. Identify systems that do not have exactly one ordered-pair solution.

5. Solve problems using systems of linear equations.

Key West residents Brian Goss (left), George Wallace, and Michael Mooney (right) hold on to each other as they battle 90 mph winds along Houseboat Row in Key West, Fla., on Friday, Sept. 25, 1998. The three had sought shelter behind a Key West hotel as Hurricane Georges descended on the Florida Keys but were forced to seek other shelter when the storm conditions became too rough. Hundreds of people were killed by the storm when it swept through the Caribbean.

Real-world problems often involve solving thousands of equations, sometimes containing a million variables. Problems ranging from scheduling airline flights to controlling traffic flow to routing phone calls over the nation's communication network often require solutions in a matter of moments. AT&T's domestic long distance network involves 800,000 variables! Meteorologists describing atmospheric conditions surrounding a hurricane must solve problems involving thousands of equations rapidly and efficiently. The difference between a two-hour warning and a two-day warning is a life-and-death issue for thousands of people in the path of one of nature's most destructive forces.

Although we will not be solving 800,000 equations with 800,000 variables, we will turn our attention to two equations with two variables, such as

$$2x - 3y = -4$$
$$2x + y = 4.$$

The methods that we consider for solving such problems provide the foundation for solving far more complex systems with many variables.

1 Decide whether an ordered pair is a solution of a linear system.

Systems of Linear Equations and Their Solutions

We have seen that all equations in the form $Ax + By = C$ are straight lines when graphed. Two such equations, such as those listed above, are called a **system of linear equations**. A **solution to a system of linear equations** is an ordered pair that satisfies all equations in the system. For example, (3, 4) satisfies the system

$$x + y = 7 \quad \text{(3 + 4 is, indeed, 7.)}$$
$$x - y = -1 \quad \text{(3 − 4 is, indeed, −1.)}$$

Thus, $(3, 4)$ satisfies both equations and is a solution of the system. The solution can be described by saying that $x = 3$ and $y = 4$. The solution can also be described using set notation. The solution set to the system is $\{(3, 4)\}$—that is, the set consisting of the ordered pair $(3, 4)$.

A system of linear equations can have exactly one solution, no solution, or infinitely many solutions. We will focus on systems with exactly one solution.

EXAMPLE 1 Determining Whether an Ordered Pair Is a Solution of a System

Determine whether $(4, -1)$ is a solution of the system

$$x + 2y = 2$$
$$x - 2y = 6.$$

Solution Because 4 is the x-coordinate and -1 is the y-coordinate of $(4, -1)$, we replace x by 4 and y by -1.

$$x + 2y = 2 \qquad\qquad\qquad x - 2y = 6$$
$$4 + 2(-1) \stackrel{?}{=} 2 \qquad\qquad 4 - 2(-1) \stackrel{?}{=} 6$$
$$4 + (-2) \stackrel{?}{=} 2 \qquad\qquad 4 - (-2) \stackrel{?}{=} 6$$
$$2 = 2 \text{ true} \qquad\qquad 4 + 2 \stackrel{?}{=} 6$$
$$6 = 6 \text{ true}$$

The pair $(4, -1)$ satisfies both equations: It makes each equation true. Thus, the pair is a solution of the system. The solution set to the system is $\{(4, -1)\}$.

The solution to a system of linear equations can be found by graphing both of the equations in the same rectangular coordinate system. For a system with one solution, the **coordinates of the point of intersection give the system's solution**. For example, the system in Example 1 is graphed in Figure 8.1. The solution of the system, $(4, -1)$, corresponds to the point of intersection of the lines.

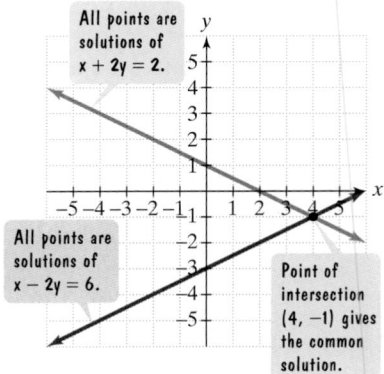

All points are solutions of $x + 2y = 2$.
All points are solutions of $x - 2y = 6$.
Point of intersection $(4, -1)$ gives the common solution.

Figure 8.1 Visualizing a system's solution.

Check Point 1 Determine whether $(1, 2)$ is a solution of the system

$$2x - 3y = -4$$
$$2x + y = 4.$$

2 Solve linear systems by substitution.

Eliminating a Variable Using the Substitution Method

Finding the solution to a linear system by graphing equations may not be easy to do. For example, a solution of $\left(-\frac{2}{3}, \frac{157}{29}\right)$ would be difficult to "see" as an intersection point on a graph.

Let's consider a method that does not depend on finding a system's solution visually: the substitution method. This method involves converting the system to one equation in one variable by an appropriate substitution.

EXAMPLE 2 Solving a System by Substitution

Solve by the substitution method:

$$y = -x - 1$$
$$4x - 3y = 24.$$

Solution

Step 1 Solve either of the equations for one variable in terms of the other. This step has already been done for us. The first equation, $y = -x - 1$, has y solved in terms of x.

Step 2 Substitute the expression from step 1 into the other equation. We substitute the expression $-x - 1$ for y in the other equation:

$$y = \boxed{-x - 1} \qquad 4x - 3\,\boxed{y} = 24 \quad \text{Substitute } -x - 1 \text{ for } y.$$

This gives us an equation in one variable, namely

$$4x - 3(-x - 1) = 24.$$

The variable y has been eliminated.

Step 3 Solve the resulting equation containing one variable.

$$
\begin{aligned}
4x - 3(-x - 1) &= 24 \quad &&\text{This is the equation containing one variable.}\\
4x + 3x + 3 &= 24 \quad &&\text{Apply the distributive property.}\\
7x + 3 &= 24 \quad &&\text{Combine like terms.}\\
7x &= 21 \quad &&\text{Subtract 3 from both sides.}\\
x &= 3 \quad &&\text{Divide both sides by 7.}
\end{aligned}
$$

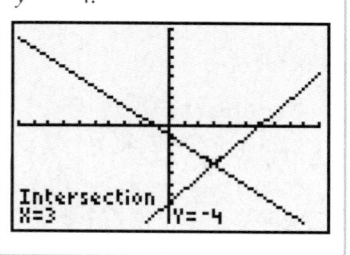
Step 4 Back-substitute the obtained value into the equation from step 1. We now know that the x-coordinate of the solution is 3. To find the y-coordinate, we back-substitute the x-value into the equation from step 1,

$$y = -x - 1.$$

Substitute 3 for x.

$$y = -3 - 1 = -4$$

With $x = 3$ and $y = -4$, the proposed solution is $(3, -4)$.

Step 5 Check the proposed solution in both of the system's given equations. Replace x with 3 and y with -4.

$$
\begin{aligned}
y &= -x - 1 & 4x - 3y &= 24\\
-4 &\stackrel{?}{=} -3 - 1 & 4(3) - 3(-4) &\stackrel{?}{=} 24\\
-4 &= -4 \quad \text{true} & 12 + 12 &\stackrel{?}{=} 24\\
& & 24 &= 24 \quad \text{true}
\end{aligned}
$$

The pair $(3, -4)$ satisfies both equations. The system's solution set is $\{(3, -4)\}$.

Solve by the substitution method:

$$y = 5x - 13$$
$$2x + 3y = 12.$$

Before considering additional examples, let's summarize the steps used in the substitution method.

Study Tip

In step 1, if possible, solve for a variable whose coefficient is 1 or −1 to avoid working with fractions.

Solving Linear Systems by Substitution

1. Solve either of the equations for one variable in terms of the other. (If one of the equations is already in this form, you can skip this step.)
2. Substitute the expression found in step 1 into the other equation. This will result in an equation in one variable.
3. Solve the equation obtained in step 2.
4. Back-substitute the value found in step 3 into the equation from step 1. Simplify and find the value of the remaining variable.
5. Check the proposed solution in both of the system's given equations.

EXAMPLE 3 Solving a System by Substitution

Solve by the substitution method:

$$5x - 4y = 9$$
$$x - 2y = -3.$$

Solution

Step 1 Solve either of the equations for one variable in terms of the other. We begin by isolating one of the variables in either of the equations. By solving for x in the second equation, which has a coefficient of 1, we can avoid fractions.

$$x - 2y = -3 \qquad \text{This is the second equation in the given system.}$$
$$x = 2y - 3 \qquad \text{Solve for x by adding 2y to both sides.}$$

Step 2 Substitute the expression from step 1 into the other equation. We substitute $2y - 3$ for x in the first equation.

$$x = \boxed{2y - 3} \qquad 5\,\boxed{x} - 4y = 9$$

This gives us an equation in one variable, namely

$$5(2y - 3) - 4y = 9.$$

The variable x has been eliminated.

Step 3 Solve the resulting equation containing one variable.

$$5(2y - 3) - 4y = 9 \qquad \text{This is the equation containing one variable.}$$
$$10y - 15 - 4y = 9 \qquad \text{Apply the distributive property.}$$
$$6y - 15 = 9 \qquad \text{Combine like terms.}$$
$$6y = 24 \qquad \text{Add 15 to both sides.}$$
$$y = 4 \qquad \text{Divide both sides by 6.}$$

Step 4   Back-substitute the obtained value into the equation from step 1.   Now that we have the y-coordinate of the solution, we back-substitute 4 for y in the equation $x = 2y - 3$.

$$x = 2y - 3 \qquad \text{Use the equation obtained in step 1.}$$
$$x = 2(4) - 3 \qquad \text{Substitute 4 for y.}$$
$$x = 8 - 3 \qquad \text{Multiply.}$$
$$x = 5 \qquad \text{Subtract.}$$

Study Tip

Get into the habit of checking ordered-pair solutions in *both* equations of the system.

With $x = 5$ and $y = 4$, the proposed solution is $(5, 4)$.

Step 5   Check.   Take a moment to show that $(5, 4)$ satisfies both given equations. The solution set is $\{(5, 4)\}$.

Check Point 3   Solve by the substitution method:

$$3x + 2y = -1$$
$$x - y = 3.$$

3 Solve linear systems by addition.

Eliminating a Variable Using the Addition Method

The substitution method is most useful if one of the given equations has an isolated variable. A second, and frequently the easiest, method for solving a linear system is the addition method. Like the substitution method, the addition method involves eliminating a variable and ultimately solving an equation containing only one variable. However, this time we eliminate a variable by adding the equations.

For example, consider the following equations:

$$3x - 4y = 11$$
$$-3x + 2y = -7.$$

When we add these two equations, the x-terms are eliminated. This occurs because the coefficients of the x-terms, 3 and -3, are opposites (additive inverses) of each other:

$$3x - 4y = 11$$
$$\underline{-3x + 2y = -7}$$
$$\text{Add:} \qquad -2y = 4$$
$$y = -2 \quad \text{Solve for y, dividing both sides by } -2.$$

Now we can back-substitute -2 for y into one of the original equations to find x. It does not matter which equation you use; you will obtain the same value for x in either case. If we use either equation, we can show that $x = 1$ and the solution $(1, -2)$ satisfies both equations in the system.

When we use the addition method, we want to obtain two equations whose sum is an equation containing only one variable. The key step is to obtain, for one of the variables, coefficients that differ only in sign. In order to do this, we may need to multiply one or both equations by some nonzero number so that the coefficients of one of the variables, x or y, become opposites. Then when the two equations are added, this variable is eliminated. Let's see exactly how this works by considering Example 4.

EXAMPLE 4   Solving a System by the Addition Method

Solve by the addition method:

$$3x + 2y = 48$$
$$9x - 8y = -24.$$

Solution   We must rewrite one or both equations in equivalent forms so that the coefficients of the same variable (either x or y) are opposites of each other. Consider the terms in x in each equation, that is, $3x$ and $9x$. To eliminate x, we can multiply each term of the first equation by -3 and then add the equations.

$$3x + 2y = 48 \xrightarrow{\text{Multiply by } -3.} -9x - 6y = -144$$
$$9x - 8y = -24 \xrightarrow{\text{No change}} 9x - 8y = -24$$

$$\text{Add:} \quad -14y = -168$$
$$y = 12 \qquad \text{Solve for } y, \text{ dividing both sides by } -14.$$

Thus, $y = 12$. We back-substitute this value into either one of the given equations. We'll use the first one.

$$3x + 2y = 48 \qquad \text{This the first equation in the given system.}$$
$$3x + 2(12) = 48 \qquad \text{Substitute 12 for } y.$$
$$3x + 24 = 48 \qquad \text{Multiply.}$$
$$3x = 24 \qquad \text{Subtract 24 from both sides.}$$
$$x = 8 \qquad \text{Divide both sides by 3.}$$

The solution $(8, 12)$ can be shown to satisfy both equations in the system. Consequently, the solution set is $\{(8, 12)\}$.

Solving Linear Systems by Addition

1. If necessary, rewrite both equations in the form $Ax + By = C$.
2. If necessary, multiply either equation or both equations by appropriate nonzero numbers so that the sum of the x-coefficients or the sum of the y-coefficients is 0.
3. Add the equations in step 2. The sum is an equation in one variable.
4. Solve the equation from step 3.
5. Back-substitute the value obtained in step 4 into either of the given equations and solve for the other variable.
6. Check the solution in both of the original equations.

Check Point 4   Solve by the addition method:

$$4x + 5y = 3$$
$$2x - 3y = 7.$$

Some linear systems have solutions that are not integers. If the value of one variable turns out to be a "messy" fraction, back-substitution might lead to cumbersome arithmetic. If this happens, you can return to the original system and use addition to find the value of the other variable.

EXAMPLE 5 Solving a System by the Addition Method

Solve by the addition method:

$$2x = 7y - 17$$
$$5y = 17 - 3x.$$

Solution

Step 1 Rewrite both equations in the form $Ax + By = C$. We first arrange the system so that variable terms appear on the left and constants appear on the right. We obtain

$$2x - 7y = -17 \quad \text{Subtract } 7y \text{ from both sides of the first equation.}$$
$$3x + 5y = 17 \quad \text{Add } 3x \text{ to both sides of the second equation.}$$

Step 2 If necessary, multiply either equation or both equations by appropriate numbers so that the sum of the x-coefficients or the sum of the y-coefficients is 0. We can eliminate x or y. Let's eliminate x by multiplying the first equation by 3 and the second equation by -2.

$$2x - 7y = -17 \xrightarrow{\text{Multiply by 3.}} 3 \cdot 2x - 3 \cdot 7y = 3(-17) \longrightarrow 6x - 21y = -51$$

$$3x + 5y = 17 \xrightarrow{\text{Multiply by } -2.} -2 \cdot 3x + (-2) \cdot 5y = -2(17) \longrightarrow -6x - 10y = -34$$

Steps 3 and 4 Add the equations and solve for the remaining variable.

$$6x - 21y = -51$$
$$\underline{-6x - 10y = -34}$$
$$\text{Add:} \quad -31y = -85$$

$$\frac{-31y}{-31} = \frac{-85}{-31} \quad \text{Divide both sides by } -31.$$

$$y = \frac{85}{31} \quad \text{Simplify.}$$

Step 5 Back-substitute and find the value for the other variable. Back-substitution of $\frac{85}{31}$ for y into either of the given equations results in cumbersome arithmetic. Instead, let's use the addition method on the given system in the form $Ax + By = C$ to find the value for x. Thus, we eliminate y by multiplying the first equation by 5 and the second equation by 7.

$$2x - 7y = -17 \xrightarrow{\text{Multiply by 5.}} \quad 10x - 35y = -85$$
$$3x + 5y = 17 \xrightarrow{\text{Multiply by 7.}} \quad \underline{21x + 35y = 119}$$
$$\text{Add:} \quad 31x = 34$$
$$x = \tfrac{34}{31} \quad \text{Divide both sides by 31.}$$

Step 6 Check. For this system, a calculator is helpful in showing the solution $\left(\frac{34}{31}, \frac{85}{31}\right)$ satisfies both equations. Consequently, the solution set is $\left\{\left(\frac{34}{31}, \frac{85}{31}\right)\right\}$.

Check Point 5 Solve by the addition method:

$$4x = 5 + 2y$$
$$3y = 4 - 2x.$$

4 Identify systems that do not have exactly one ordered-pair solution.

Linear Systems Having No Solution or Infinitely Many Solutions

We have seen that a system of linear equations in two variables represents a pair of lines. The lines either intersect, are parallel, or are identical. Thus, there are three possibilities for the number of solutions to a system of two linear equations.

The Number of Solutions to a System of Two Linear Equations

The number of solutions to a system of two linear equations in two variables is given by one of the following. (See Figure 8.2.)

Number of Solutions	What This Means Graphically
Exactly one ordered-pair solution	The two lines intersect at one point.
No solution	The two lines are parallel.
Infinitely many solutions	The two lines are identical.

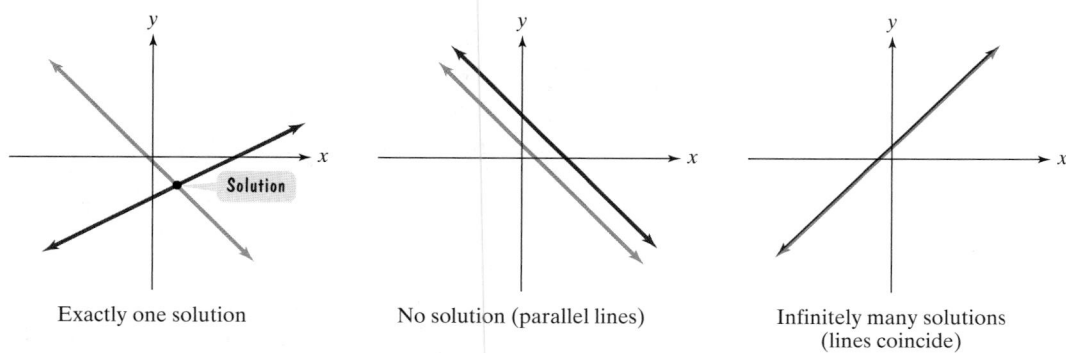

Exactly one solution   No solution (parallel lines)   Infinitely many solutions (lines coincide)

Figure 8.2 Possible graphs for a system of two linear equations in two variables

A linear system with no solution is called an **inconsistent system**. If you attempt to solve such a system by substitution or addition, you will eliminate both variables. A false statement such as $0 = 17$ will be the result.

EXAMPLE 6 A System with No Solution

Solve the system:

$$4x + 6y = 12.$$
$$6x + 9y = 12.$$

Solution   Because no variable is isolated, we will use the addition method. To obtain coefficients of x that differ only in sign, we multiply the first equation by 3 and multiply the second equation by -2.

$$
\begin{array}{lll}
4x + 6y = 12 & \xrightarrow{\text{Multiply by 3.}} & 12x + 18y = 36 \\
6x + 9y = 12 & \xrightarrow{\text{Multiply by } -2.} & \underline{-12x - 18y = -24} \\
& \text{Add:} & 0 = 12
\end{array}
$$

There are no values of x and y for which $0 = 12$.

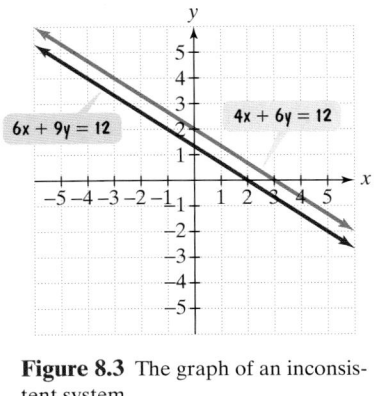

Figure 8.3 The graph of an inconsistent system

The false statement $0 = 12$ indicates that the system is inconsistent and has no solution. The solution set is the empty set, $\varnothing$.

The lines corresponding to the two equations in Example 6 are shown in Figure 8.3. The lines are parallel and have no point of intersection.

Discovery

Show that the graphs of $4x + 6y = 12$ and $6x + 9y = 12$ must be parallel lines by solving each equation for y. What is the slope and y-intercept for each line? What does this mean? If a linear system is inconsistent, what must be true about the slopes and y-intercepts for the system's graphs?

Check Point 6 Solve the system:

$$x + 2y = 4$$
$$3x + 6y = 13.$$

A linear system that has at least one solution is called a **consistent system**. Lines that intersect and lines that coincide both represent consistent systems. If the lines coincide, then the consistent system has infinitely many solutions, represented by every point on the line.

The equations in a linear system with infinitely many solutions are called **dependent**. If you attempt to solve such a system by substitution or addition, you will eliminate both variables. However, a true statement such as $0 = 0$ will be the result.

EXAMPLE 7   A System with Infinitely Many Solutions

Solve the system:

$$y = 3 - 2x$$
$$4x + 2y = 6.$$

Solution Because the variable y is isolated in the first equation, we can use the substitution method. We substitute the expression for y in the other equation.

$$y = \boxed{3 - 2x} \qquad 4x + 2\boxed{y} = 6 \qquad \text{Substitute } 3 - 2x \text{ for } y.$$

$4x + 2y = 6$	This is the second equation in the given system.
$4x + 2(3 - 2x) = 6$	Substitute $3 - 2x$ for y.
$4x + 6 - 4x = 6$	Apply the distributive property.
$6 = 6$	Simplify. This statement is true for all values of x and y.

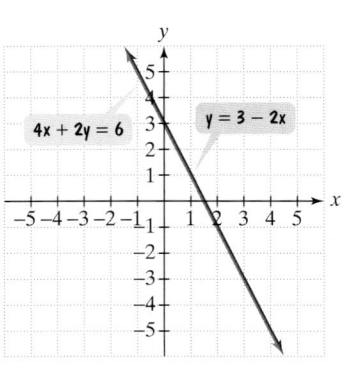

Figure 8.4 The graph of a system with infinitely many solutions

In our final step, both variables have been eliminated, and the resulting statement $6 = 6$ is true. This true statement indicates that the system has infinitely many solutions. The solution set consists of all points (x, y) lying on the line $y = 3 - 2x$, as shown in Figure 8.4.

We express the solution set for the system in one of two equivalent ways:

$\{(x, y)\,|\,y = 3 - 2x\}$ The set of all ordered pairs (x, y) such that $y = 3 - 2x$.

or $\{(x, y)\,|\,4x + 2y = 6\}$ The set of all ordered pairs (x, y) such that $4x + 2y = 6$.

Check Point 7   Solve the system:

$$y = 4x - 4$$
$$8x - 2y = 8.$$

5   Solve problems using systems of linear equations.

Applications

An important application of systems of equations arises in connection with supply and demand. As the price of a product increases, the demand for that product decreases. However, at higher prices suppliers are willing to produce greater quantities of the product.

EXAMPLE 8   Supply and Demand Models

A chain of video stores specializes in cult films. The weekly demand and supply models for *The Rocky Horror Picture Show* are given by

$$N = -13p + 760 \quad \text{Demand model}$$
$$N = 2p + 430 \quad \text{Supply model}$$

in which p is the price of the video and N is the number of copies of the video sold or supplied each week to the chain of stores.

a. How many copies of the video can be sold and supplied at $18 per copy?

b. Find the price at which supply and demand are equal. At this price, how many copies of *Rocky Horror* can be supplied and sold each week?

Solution

a. To find how many copies of the video can be sold and supplied at $18 per copy, we substitute 18 for p in the demand and supply models.

Demand Model	**Supply Model**
$N = -13p + 760$	$N = 2p + 430$
Substitute 18 for p.	Substitute 18 for p.
$N = -13 \cdot 18 + 760 = 526$	$N = 2 \cdot 18 + 430 = 466$

At $18 per video, the chain can sell 526 copies of *Rocky Horror* in a week. The manufacturer is willing to supply 466 copies per week. This will result in a shortage of copies of the video. Under these conditions, the retail chain is likely to raise the price of the video.

b. We can find the price at which supply and demand are equal by solving the demand-supply linear system. We will use substitution, substituting $-13p + 760$ for N in the second equation.

$$N = \boxed{-13p + 760} \quad \boxed{N} = 2p + 430 \quad \text{Substitute } -13p + 760 \text{ for } N.$$

$$-13p + 760 = 2p + 430 \qquad \text{The resulting equation contains only one variable.}$$

$$-15p + 760 = 430 \qquad \text{Subtract } 2p \text{ from both sides.}$$

$$-15p = -330 \qquad \text{Subtract } 760 \text{ from both sides.}$$

$$p = 22 \qquad \text{Divide both sides by } -15.$$

The price at which supply and demand are equal is $22 per video. To find the value of N, the number of videos supplied and sold weekly at this price, we back-substitute 22 for p into either the demand or the supply model. We'll use both models to make sure we get the same number in each case.

Demand Model

$$N = -13p + 760$$

Substitute 22 for p.

$$N = -13 \cdot 22 + 760 = 474$$

Supply Model

$$N = 2p + 430$$

Substitute 22 for p.

$$N = 2 \cdot 22 + 430 = 474$$

At a price of $22, 474 units of the video can be supplied and sold weekly. The intersection point, $(22, 474)$, is shown in Figure 8.5.

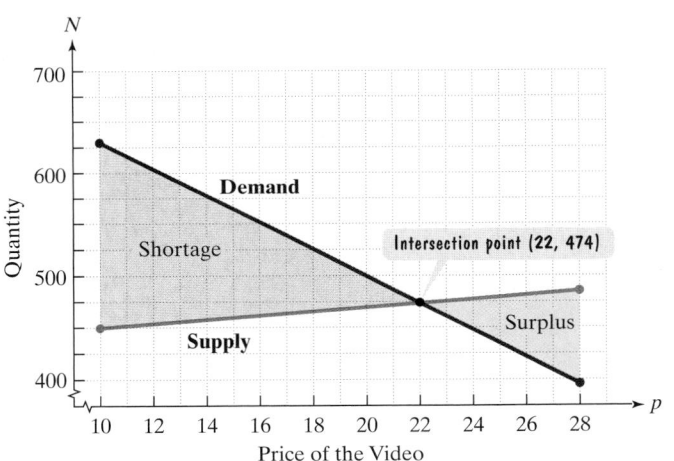

Figure 8.5 Priced at $22, 474 copies of the video can be supplied and sold weekly.

Check Point 8

The demand for a product is modeled by $N = -20p + 1000$ and the supply for the product by $N = 5p + 250$. In these models, p is the price of the product and N is the number supplied or sold weekly. At what price will supply equal demand? At that price, how many units of the product will be supplied and sold each week?

EXERCISE SET 8.1

 Practice Exercises

In Exercises 1–4, determine whether the given ordered pair is a solution of the system.

1. $(2, 3)$
$x + 3y = 11$
$x - 5y = -13$

2. $(-3, 5)$
$9x + 7y = 8$
$8x - 9y = -69$

3. $(2, 5)$
$2x + 3y = 17$
$x + 4y = 16$

4. $(8, 5)$
$5x - 4y = 20$
$3y = 2x + 1$

In Exercises 5–16, solve each system by the substitution method.

5. $x + y = 4$
$y = 3x$

6. $x + y = 6$
$y = 2x$

7. $x + 3y = 8$
$y = 2x - 9$

8. $2x - 3y = -13$
$y = 2x + 7$

9. $x + 3y = 5$
$4x + 5y = 13$

10. $x + 2y = 5$
$2x - y = -15$

11. $2x - y = -5$
$x + 5y = 14$

12. $2x + 3y = 11$
$x - 4y = 0$

13. $2x - y = 3$
$5x - 2y = 10$

14. $-x + 3y = 10$
$2x + 8y = -6$

15. $x + 8y = 6$
$2x + 4y = -3$

16. $-4x + y = -11$
$2x - 3y = 5$

In Exercises 17–28, solve each system by the addition method.

17. $x + y = 1$
$x - y = 3$

18. $x + y = 6$
$x - y = -2$

19. $2x + 3y = 6$
$2x - 3y = 6$

20. $3x + 2y = 14$
$3x - 2y = 10$

21. $x + 2y = 2$
$-4x + 3y = 25$

22. $2x - 7y = 2$
$3x + y = -20$

23. $4x + 3y = 15$
$2x - 5y = 1$

24. $3x - 7y = 13$
$6x + 5y = 7$

25. $3x - 4y = 11$
$2x + 3y = -4$

26. $2x + 3y = -16$
$5x - 10y = 30$

27. $3x = 4y + 1$
$3y = 1 - 4x$

28. $5x = 6y + 40$
$2y = 8 - 3x$

In Exercises 29–36, solve by the method of your choice. Identify systems with no solution and systems with infinitely many solutions, using set notation to express their solution sets.

29. $x = 9 - 2y$
$x + 2y = 13$

30. $6x + 2y = 7$
$y = 2 - 3x$

31. $y = 3x - 5$
$21x - 35 = 7y$

32. $9x - 3y = 12$
$y = 3x - 4$

33. $3x - 2y = -5$
$4x + y = 8$

34. $2x + 5y = -4$
$3x - y = 11$

35. $x + 3y = 2$
$3x + 9y = 6$

36. $4x - 2y = 2$
$2x - y = 1$

In Exercises 37–40, let x represent one number and let y represent the other number. Use the given conditions to write a system of equations. Solve the system and find the numbers.

37. The sum of two numbers is 7. If one number is subtracted from the other, their difference is −1. Find the numbers.

38. The sum of two numbers is 2. If one number is subtracted from the other, their difference is 8. Find the numbers.

39. Three times a first number decreased by a second number is 1. The first number increased by twice the second number is 12. Find the numbers.

40. The sum of three times a first number and twice a second number is 8. If the second number is subtracted from twice the first number, the result is 3. Find the numbers.

 Application Exercises

41. At a price of p dollars per ticket, the number of tickets to a rock concert that can be sold is given by the demand model $N = -25p + 7500$. At a price of p dollars per ticket, the number of tickets that the concert's promoters are willing to make available is given by the supply model $N = 5p + 6000$.
   a. How many tickets can be sold and supplied for $40 per ticket?
   b. Find the ticket price at which supply and demand are equal. At this price, how many tickets will be supplied and sold?

42. The weekly demand and supply models for a particular brand of scientific calculator for a chain of stores are given by the demand model $N = -53p + 1600$, and the supply model $N = 75p + 320$. In these models, p is the price of the calculator and N is the number of calculators sold or supplied each week to the stores.

a. How many calculators can be sold and supplied at $12 per calculator?

b. Find the price at which supply and demand are equal. At this price, how many calculators of this type can be supplied and sold each week?

A business breaks even when the cost for running the business is equal to the money taken in by the business. In Exercises 43–44, determine how many units must be sold so that a business breaks even, experiencing neither loss nor profit.

43. A gasoline station has weekly costs and revenue (the money taken in by the station) that are functions of the number of gallons of gasoline purchased and sold. If x gallons are purchased and sold, weekly costs are given by $C(x) = 1.2x + 1080$ and weekly revenue by $R(x) = 1.6x$. How many gallons of gasoline must be sold weekly for the station to break even?

44. An artist has monthly costs and revenue (the money taken in by the artist) that are functions of the number of ceramic pieces produced and sold. If x ceramic pieces are produced and sold, monthly costs are given by $C(x) = 4x + 2000$ and monthly revenue by $R(x) = 9x$. How many ceramic pieces must be sold monthly for the artist to break even?

Use a system of linear equations to solve Exercises 45–48.

45. The verdict is in: After years of research, the nation's health experts agree that high cholesterol in the blood is a major contributor to heart disease. Cholesterol intake should be limited to 300 mg or less each day. Fast foods provide a cholesterol carnival. Two McDonald's Quarter Pounders and three Burger King Whoppers with cheese contain 520 mg of cholesterol. Three Quarter Pounders and one Whopper with cheese exceed the suggested daily cholesterol intake by 53 mg. Determine the cholesterol content in each item.

46. How do the Quarter Pounder and Whopper with cheese measure up in the calorie department? Actually, not too well. Two Quarter Pounders and three Whoppers with cheese provide 2607 calories. Even one of each provide enough calories to bring tears to Jenny Craig's eyes—9 calories in excess of what is allowed on a 1000 calorie-a-day diet. Find the caloric content of each item.

47. The graph at the top of the next column makes Super Bowl Sunday look like a day of snack food binging in the United States. The number of pounds of guacamole consumed is ten times the difference between the number of pounds of potato and tortilla chips eaten on the same day. On Super Bowl Sunday Americans also eat a total quantity of potato and tortilla chips that exceeds popcorn consumption by 7.3 million pounds. How many millions of pounds of potato chips and tortilla chips are consumed on Super Bowl Sunday?

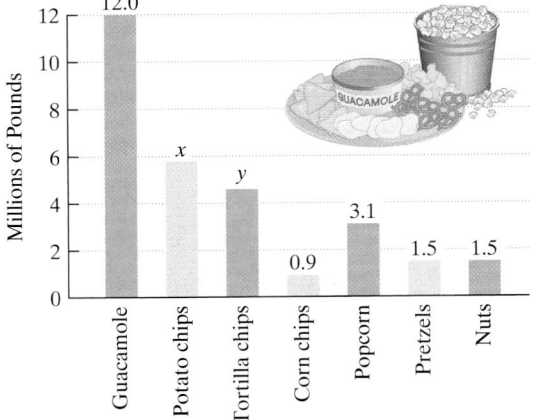

Millions of Pounds of Snack Food Consumed on Super Bowl Sunday

Source: Association of American Snack Foods

48. The bar graph indicates countries in which ten or more languages have become extinct. The number of extinct languages in Brazil is 7.5 times the difference between the number in the United States and Colombia. The number of extinct languages in the United States and Colombia combined exceeds the number in Australia by 24. How many languages have become extinct in the United States and Colombia?

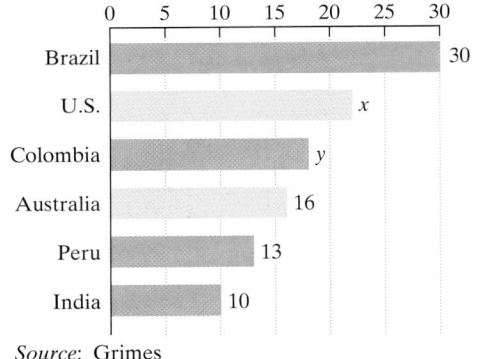

Countries Where Ten or More Laguages Have Become Extinct (Number of Languages)

Source: Grimes

49. The June 7, 1999 issue of *Newsweek* presents statistics showing progress African Americans have made in education, health, and finance. Infant mortality for blacks is decreasing at a faster rate than it is for whites, shown by the graphs on page 429. Infant mortality for blacks can be modeled by $M = -0.41x + 22$ and for whites by $M = -0.18x + 10$. In both models, x is the number of years since 1980 and M is infant mortality, measured in deaths per 1000 live births. Use these models to project when infant mortality for blacks and whites will be the same. What is infant mortality for both groups at that time?

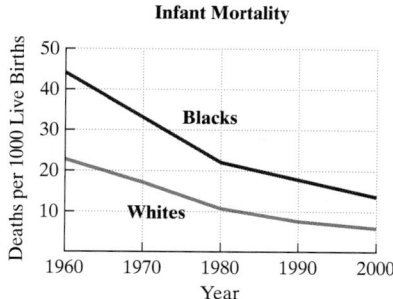

Source: National Center for Health Statistics

50. The equation $x + 10y = 2120$ models deaths from gunfire in the United States, y, in deaths per hundred thousand Americans, in year x. The equation $7x + 8y = 14{,}065$ models deaths from car accidents in the United States, y, in deaths per hundred thousand Americans, in year x. Solve the linear system formed by the two models. Then describe what the solution means in terms of the variables in the given models.

Writing in Mathematics

51. What is a system of linear equations? Provide an example with your description.

52. What is the solution to a system of linear equations?

53. Explain how to solve a system of equations using the substitution method. Use $y = 3 - 3x$ and $3x + 4y = 6$ to illustrate your explanation.

54. Explain how to solve a system of equations using the addition method. Use $3x + 5y = -2$ and $2x + 3y = 0$ to illustrate your explanation.

55. When is it easier to use the addition method rather than the substitution method when solving a system of equations?

56. When using the addition or substitution method, how can you tell if a system of linear equations has infinitely many solutions? What is the relationship between the graphs of the two equations?

57. When using the addition or substitution method, how can you tell if a system of linear equations has no solution? What is the relationship between the graphs of the two equations?

58. The law of supply and demand states that, in a free market economy, a commodity tends to be sold at its equilibrium price. At this price, the amount that the seller will supply is the same amount that the consumer will buy. Explain how systems of equations can be used to determine the equilibrium price.

59. The graphs at the top of the next column show median weekly earnings of full-time wage and salary workers 25 years and older, by education attainment. Which graphs look like they might intersect sometime after 1997? Describe how to use algebra to model the data and determine the year in which the groups might have the same weekly earnings.

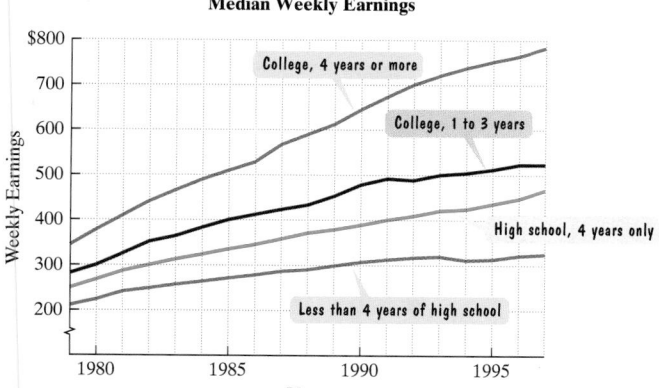

Source: U.S. Bureau of Labor Statistics

Technology Exercises

60. Verify your solutions to any five exercises from 5 through 36 by using a graphing utility to graph the two equations in the system in the same viewing rectangle. Then use the intersection feature to display the solution.

61. Some graphing utilities can give the solution to a linear system of equations. (Consult your manual for details.) This capability is usually accessed with the $\boxed{\text{SIMULT}}$ (simultaneous equations) feature. First, you will enter 2, for two equations in two variables. With each equation in $Ax + By = C$ form, you will then enter the coefficients for x and y and the constant term, one equation at a time. After entering all six numbers, press $\boxed{\text{SOLVE}}$. The solution will be displayed on the screen. (The x-value may be displayed as $x_1 =$ and the y-value as $x_2 =$.) Use this capability to verify the solution to any five of the exercises you solved in the practice exercises of this exercise set. Describe what happens when you use your graphing utility on a system with no solution or infinitely many solutions.

Critical Thinking Exercises

62. Write a system of equations having $\{(-2, 7)\}$ as a solution set. (More than one system is possible.)

63. Solve the system for x and y in terms of $a_1, b_1, c_1, a_2, b_2,$ and c_2:

$$a_1 x + b_1 y = c_1$$
$$a_2 x + b_2 y = c_2.$$

64. Two identical twins can only be recognized by the characteristic that one always tells the truth and the other always lies. One twin tells you of a lucky number pair: "When I multiply my first lucky number by 3 and my second lucky number by 6, the addition of the resulting numbers produces a sum of 12. When I add my first lucky number and twice my second lucky number, the sum is 5." Which twin is talking?

65. A marching band has 52 members, and there are 24 in the pom-pom squad. They wish to form several hexagons and squares like those diagrammed below. Can it be done with no people left over?

Hexagon with pom–pom person in center

Square with band member in center

B B

B P B

B B

P P

B

P P

B = Band Member

P = Pom-pom Person

Group Exercise

66. The group should write four different word problems that can be solved using a system of linear equations in two variables. All of the problems should be on different topics. Select from the following topics: a number problem (see Exercises 37–40); a problem using supply and demand models (see Exercises 41–42); a problem involving a business breaking even (see Exercises 43–44); a problem based on two missing numbers in a graph (see Exercises 47–48; you'll need to find an interesting graph); a problem involving linear modeling, finding the year when the quantity modeled will be the same for two groups (see Exercise 49). Of course, you can also base the problem on any topic of interest, but remember—only one problem per topic. The group should turn in the four problems and their algebraic solutions.

SECTION 8.2 Systems of Linear Equations in Three Variables

Objectives

1. Verify the solution of a linear system in three variables.

2. Solve linear systems in three variables.

3. Solve problems using systems in three variables.

All animals sleep, but the length of time they sleep varies widely: Cattle sleep for only a few minutes at a time. We humans seem to need more sleep than other animals, up to eight hours a day. Without enough sleep, we have difficulty concentrating, make mistakes in routine tasks, lose energy, and feel bad-tempered. There is a relationship between hours of sleep and death rate per year per 100,000 people. How many hours of sleep will put you in the group with the minimum death rate? In this section you will learn how to solve linear systems with more than two variables in order to answer this question.

1 Verify the solution of a linear system in three variables.

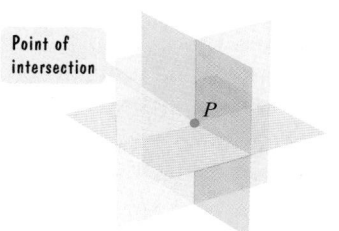

Point of intersection

P

Figure 8.6

Systems of Linear Equations in Three Variables and Their Solutions

An equation such as $x + 2y - 3z = 9$ is called a **linear equation in three variables**. In general, any equation of the form

$$Ax + By + Cz = D$$

where $A, B, C,$ and D are real numbers such that $A, B,$ and C are not all 0, is a linear equation in the variables $x, y,$ and z. The graph of this linear equation in three variables is a plane in three-dimensional space.

The process of solving a system of three linear equations in three variables is geometrically equivalent to finding the point of intersection (assuming that there is one) of three planes in space (see Figure 8.6). A **solution** to a system of linear equations in three variables is an ordered triple of real numbers that satisfies all equations of the system. The **solution set** of the system is the set of all its solutions.

EXAMPLE 1 Determining Whether an Ordered Triple Satisfies a System

Show that the ordered triple $(-1, 2, -2)$ is a solution of the system:

$$x + 2y - 3z = 9$$
$$2x - y + 2z = -8$$
$$-x + 3y - 4z = 15.$$

Solution Because -1 is the x-coordinate, 2 is the y-coordinate, and -2 is the z-coordinate of $(-1, 2, -2)$, we replace x by -1, y by 2, and z by -2 in each of the three equations.

$x + 2y - 3z = 9$	$2x - y + 2z = -8$	$-x + 3y - 4z = 15$
$-1 + 2(2) - 3(-2) \stackrel{?}{=} 9$	$2(-1) - 2 + 2(-2) \stackrel{?}{=} -8$	$-(-1) + 3(2) - 4(-2) \stackrel{?}{=} 15$
$-1 + 4 + 6 \stackrel{?}{=} 9$	$-2 - 2 - 4 \stackrel{?}{=} -8$	$1 + 6 + 8 \stackrel{?}{=} 15$
$9 = 9$ true	$-8 = -8$ true	$15 = 15$ true

The ordered triple $(-1, 2, -2)$ satisfies the three equations: It makes each equation true. Thus, the ordered triple is a solution of the system. The solution set to the system is $\{(-1, 2, -2)\}$.

Check Point 1 Show that the ordered triple $(-1, -4, 5)$ is a solution of the system:

$$x - 2y + 3z = 22$$
$$2x - 3y - z = 5$$
$$3x + y - 5z = -32.$$

2 Solve linear systems in three variables.

Solving Systems of Linear Equations in Three Variables by Eliminating Variables

The method for solving a system of linear equations in three variables is similar to that used on systems of linear equations in two variables. We use addition to eliminate any variable, reducing the system to two equations in two variables.

Once we obtain a system of two equations in two variables, we use addition or substitution to eliminate a variable. The result is a single equation in one variable. We solve this equation to get the value of the remaining variable. Other variable values are found by back-substitution.

Solving Linear Systems in Three Variables by Eliminating Variables

1. Reduce the system to two equations in two variables. This is usually accomplished by taking two different pairs of equations and using the addition method to eliminate the same variable from each pair.
2. Solve the resulting system of two equations in two variables using addition or substitution. The result is an equation in one variable that gives the value of that variable.
3. Back-substitute the value of the variable found in step 2 into either of the equations in two variables to find the value of the second variable.
4. Use the values of the two variables from steps 2 and 3 to find the value of the third variable by back-substituting into one of the original equations.
5. Check the proposed solution in each of the original equations.

EXAMPLE 2   Solving a System in Three Variables

Solve the system:

$$5x - 2y - 4z = 3 \quad \text{Equation 1}$$
$$3x + 3y + 2z = -3 \quad \text{Equation 2}$$
$$-2x + 5y + 3z = 3. \quad \text{Equation 3}$$

Solution   There are many ways to proceed. Because our initial goal is to reduce the system to two equations in two variables, **the central idea is to take two different pairs of equations and eliminate the same variable from each pair.**

Step 1   Reduce the system to two equations in two variables.   We choose any two equations and use the addition method to eliminate a variable. Let's eliminate z from Equations 1 and 2. We do so by multiplying Equation 2 by 2. Then we add equations.

(Equation 1) $5x - 2y - 4z = 3$ $\xrightarrow{\text{No change}}$ $5x - 2y - 4z = 3$
(Equation 2) $3x + 3y + 2z = -3$ $\xrightarrow{\text{Multiply by 2.}}$ $6x + 6y + 4z = -6$
Add: $11x + 4y \qquad = -3$ Equation 4

Now we must eliminate the *same* variable from another pair of equations. We can eliminate z from Equations 2 and 3. First, we multiply Equation 2 by -3. Next, we multiply Equation 3 by 2. Finally, we add equations.

(Equation 2) $3x + 3y + 2z = -3$ $\xrightarrow{\text{Multiply by } -3.}$ $-9x - 9y - 6z = 9$
(Equation 3) $-2x + 5y + 3z = 3$ $\xrightarrow{\text{Multiply by 2.}}$ $-4x + 10y + 6z = 6$
Add: $-13x + \quad y \qquad = 15$ Equation 5

Equations 4 and 5 give us a system of two equations in two variables.

Step 2 Solve the resulting system of two equations in two variables. We will use the addition method to solve Equations 4 and 5 for x and y. To do so, we multiply Equation 5 on both sides by -4 and add this to Equation 4.

$$
\begin{array}{lll}
(\text{Equation 4}) \quad 11x + 4y = -3 & \xrightarrow{\text{No change}} & 11x + 4y = -3 \\
(\text{Equation 5}) \quad -13x + y = 15 & \xrightarrow{\text{Multiply by } -4.} & 52x - 4y = -60 \\
& \text{Add: } & 63x = -63 \\
& & x = -1 \quad \text{Divide both sides by 63.}
\end{array}
$$

Step 3 Use back-substitution in one of the equations in two variables to find the value of the second variable. We back-substitute -1 for x in either Equation 4 or 5 to find the value of y.

$$
\begin{array}{ll}
-13x + y = 15 & \text{Equation 5} \\
-13(-1) + y = 15 & \text{Substitute } -1 \text{ for } x. \\
13 + y = 15 & \text{Multiply.} \\
y = 2 & \text{Subtract 13 from both sides.}
\end{array}
$$

Step 4 Back-substitute the values found for two variables into one of the original equations to find the value of the third variable. We can now use any one of the original equations and back-substitute the values of x and y to find the value for z. We will use Equation 2.

$$
\begin{array}{ll}
3x + 3y + 2z = -3 & \text{Equation 2} \\
3(-1) + 3(2) + 2z = -3 & \text{Substitute } -1 \text{ for } x \text{ and } 2 \text{ for } y. \\
3 + 2z = -3 & \text{Multiply and then add.} \\
2z = -6 & \text{Subtract 3 from both sides.} \\
z = -3 & \text{Divide both sides by 2.}
\end{array}
$$

With $x = -1, y = 2$, and $z = -3$, the proposed solution is the ordered triple $(-1, 2, -3)$.

Step 5 Check. Check the proposed solution, $(-1, 2, -3)$, by substituting the values for x, y, and z into each of the three original equations. These substitutions yield three true statements. Thus, the solution set is $\{(-1, 2, -3)\}$.

Check Point 2 Solve the system:

$$
\begin{array}{r}
x + 4y - z = 20 \\
3x + 2y + z = 8 \\
2x - 3y + 2z = -16.
\end{array}
$$

In some examples, one of the variables is already eliminated from an original equation. In this case, the same variable should be eliminated from the other two equations, thereby making it possible to omit one of the elimination steps. We illustrate this idea in Example 3.

EXAMPLE 3   Solving a System of Equations with a Missing Term

Solve the system:

$$
\begin{aligned}
x + \quad\ \ z &= 8 \quad \text{Equation 1} \\
x + \ y + 2z &= 17 \quad \text{Equation 2} \\
x + 2y + \ z &= 16 \quad \text{Equation 3}
\end{aligned}
$$

Solution

Step 1   Reduce the system to two equations in two variables.   Because Equation 1 contains only x and z, we could eliminate y from Equations 2 and 3. This will give us two equations in x and z. To eliminate y from Equations 2 and 3, we multiply Equation 2 by -2 and add Equation 3.

$$
\begin{array}{ll}
(\text{Equation 2}) \ \ x + y + 2z = 17 \ \xrightarrow{\text{Multiply by } -2.} & -2x - 2y - 4z = -34 \\
(\text{Equation 3}) \ \ x + 2y + z = 16 \ \xrightarrow{\text{No change}} & \underline{\ \ x + 2y + \ \ z = \ \ \ 16\ } \\
& \text{Add:} \quad -x \qquad\quad -3z = -18 \quad \text{Equation 4}
\end{array}
$$

Equation 4 and the given Equation 1 provide us with a system of two equations in two variables.

Step 2   Solve the resulting system of two equations in two variables.   We will solve Equations 1 and 4 for x and z.

$$
\begin{aligned}
x + \ z &= \ \ \ 8 \quad \text{Equation 1} \\
\underline{-x - 3z} &= \underline{-18} \quad \text{Equation 4} \\
\text{Add:} \quad -2z &= -10 \\
z &= \ \ \ 5 \quad \text{Divide both sides by } -2.
\end{aligned}
$$

Step 3   Use back-substitution in one of the equations in two variables to find the value of the second variable.   To find x, we back-substitute 5 for z in either Equation 1 or 4. We will use Equation 1.

$$
\begin{aligned}
x + z &= 8 \quad \text{Equation 1} \\
x + 5 &= 8 \quad \text{Substitute 5 for z.} \\
x &= 3 \quad \text{Subtract 5 from both sides.}
\end{aligned}
$$

Step 4   Back-substitute the values found for two variables into one of the original equations to find the value of the third variable.   To find y, we back-substitute 3 for x and 5 for z into Equation 2 or 3. We can't use Equation 1 because y is missing in this equation. We will use Equation 2.

$$
\begin{aligned}
x + y + 2z &= 17 \quad \text{Equation 2} \\
3 + y + 2(5) &= 17 \quad \text{Substitute 3 for x and 5 for z.} \\
y + 13 &= 17 \quad \text{Multiply and add.} \\
y &= 4 \quad \text{Subtract 13 from both sides.}
\end{aligned}
$$

We found that $z = 5$, $x = 3$, and $y = 4$. Thus, the proposed solution is the ordered triple $(3, 4, 5)$.

Step 5   Check.   Substituting 3 for x, 4 for y, and 5 for z into each of the three original equations yields three true statements. Consequently, the solution set is $\{(3, 4, 5)\}$.

Check Point 3 Solve the system:

$$2y - z = 7$$
$$x + 2y + z = 17$$
$$2x - 3y + 2z = -1$$

A system of linear equations in three variables represents three planes. The three planes need not intersect at one point. The planes may have no common point of intersection and represent an inconsistent system with no solution. By contrast, the planes may coincide or intersect along a line. In these cases, the planes have infinitely many points in common and represent systems with infinitely many solutions. Systems of linear equations in three variables that are inconsistent or that contain dependent equations will be discussed in Chapter 9.

3 Solve problems using systems in three variables.

Applications

Systems of equations may allow us to find models for data without using a graphing utility. Quadratic functions of the form $y = ax^2 + bx + c$ often model situations in which values of y are decreasing and then increasing, suggesting the cuplike shape of a parabola.

> **EXAMPLE 4 Modeling Data Relating Sleep and Death Rate**

In a study relating sleep and death rate, the following data were obtained. Use the function $y = ax^2 + bx + c$ to model the data.

x (Average Number of Hours of Sleep)	y (Death Rate per Year Per 100,000 Males)
4	1682
7	626
9	967

Solution We need to find values for $a, b,$ and c. We can do so by solving a system of three linear equations in $a, b,$ and c. We obtain the three equations by using the values of x and y from the data as follows:

$y = ax^2 + bx + c$ Use the quadratic function to model the data.

When $x = 4, y = 1682$: $1682 = a \cdot 4^2 + b \cdot 4 + c$ or $16a + 4b + c = 1682$

When $x = 7, y = 626$: $626 = a \cdot 7^2 + b \cdot 7 + c$ or $49a + 7b + c = 626$

When $x = 9, y = 967$: $967 = a \cdot 9^2 + b \cdot 9 + c$ or $81a + 9b + c = 967.$

The easiest way to solve this system is to eliminate c from two pairs of equations, obtaining two equations in a and b. Solving this system gives $a = 104.5$, $b = -1501.5$, and $c = 6016$. We now substitute the values for $a, b,$ and c into $y = ax^2 + bx + c$. The function that models the given data is

$$y = 104.5x^2 - 1501.5x + 6016.$$

Discovery

Use the x-coordinate of a parabola's vertex, $x = -\dfrac{b}{2a}$, and the function on the right to find the hours of sleep that minimize the death rate. Round to the nearest tenth of an hour. What is the minimum death rate per year per 100,000 males?

We can use the model that we obtained in Example 4 to find the death rate of males who average, say, 6 hours of sleep. Substitute 6 for x:

$$y = 104.5(6)^2 - 1501.5(6) + 6016 = 769.$$

According to the model, the death rate for males who average 6 hours of sleep is 769 deaths per 100,000 males.

Check Point 4

Find the quadratic function $y = ax^2 + bx + c$ whose graph passes through the points $(1, 4), (2, 1)$, and $(3, 4)$.

EXERCISE SET 8.2

 Practice Exercises

In Exercises 1–4, determine if the given ordered triple is a solution of the system.

1.
$$x + y + z = 4$$
$$x - 2y - z = 1$$
$$2x - y - 2 = -1$$
$$(2, -1, 3)$$

2.
$$x + y + z = 0$$
$$x + 2y - 3z = 5$$
$$3x + 4y + 2z = -1$$
$$(5, -3, -2)$$

3.
$$x - 2y = 2$$
$$2x + 3y = 11$$
$$y - 4z = -7$$
$$(4, 1, 2)$$

4.
$$x - 2z = -5$$
$$y - 3z = -3$$
$$2x - z = -4$$
$$(-1, 3, 2)$$

Solve each system in Exercises 5–18.

5.
$$x + y + 2z = 11$$
$$x + y + 3z = 14$$
$$x + 2y - z = 5$$

6.
$$2x + y - 2z = -1$$
$$3x - 3y - z = 5$$
$$x - 2y + 3z = 6$$

7.
$$4x - y + 2z = 11$$
$$x + 2y - z = -1$$
$$2x + 2y - 3z = -1$$

8.
$$x - y + 3z = 8$$
$$3x + y - 2z = -2$$
$$2x + 4y + z = 0$$

9.
$$3x + 5y + 2z = 0$$
$$12x - 15y + 4z = 12$$
$$6x - 25y - 8z = 8$$

10.
$$2x + 3y + 7z = 13$$
$$3x + 2y - 5z = -22$$
$$5x + 7y - 3z = -28$$

11.
$$2x - 4y + 3z = 17$$
$$x + 2y - z = 0$$
$$4x - y - z = 6$$

12.
$$x + z = 3$$
$$x + 2y - z = 1$$
$$2x - y + z = 3$$

13.
$$2x + y = 2$$
$$x + y - z = 4$$
$$3x + 2y + z = 0$$

14.
$$x + 3y + 5z = 20$$
$$y - 4z = -16$$
$$3x - 2y + 9z = 36$$

15.
$$x + y = -4$$
$$y - z = 1$$
$$2x + y + 3z = -21$$

16.
$$x + y = 4$$
$$x + z = 4$$
$$y + z = 4$$

17.
$$3(2x + y) + 5z = -1$$
$$2(x - 3y + 4z) = -9$$
$$4(1 + x) = -3(z - 3y)$$

18.
$$7z - 3 = 2(x - 3y)$$
$$5y + 3z - 7 = 4x$$
$$4 + 5z = 3(2x - y)$$

In Exercises 19–20, let x represent the first number, y the second number, and z the third number. Use the given conditions to write a system of equations. Solve the system and find the numbers.

19. The sum of three numbers is 16. The sum of twice the first number, 3 times the second number, and 4 times the third number is 46. The difference between 5 times the first number and the second number is 31. Find the three numbers.

20. Three numbers are unknown. Three times the first number plus the second number plus twice the third number is 5. If 3 times the second number is subtracted from the sum of the first number and 3 times the third number, the result is 2. If the third number is subtracted from 2 times the first number and 3 times the second number, the result is 1. Find the numbers.

In Exercises 21–24, find the quadratic function $y = ax^2 + bx + c$ whose graph passes through the given points.

21. $(-1, 6), (1, 4), (2, 9)$

22. $(-2, 7), (1, -2), (2, 3)$

23. $(-1, -4), (1, -2), (2, 5)$

24. $(1, 3), (3, -1), (4, 0)$

 **Application Exercises**

25. The bar graph at the top of the next page shows the average starting salaries for the five top-paying fields for college graduates. If we add the average starting salaries for college graduates who are chemical, mechanical, and electrical engineers, the total is \$121,421. The difference between the starting salaries for chemical and mechanical engineers is \$2906. The difference between the starting salaries for mechanical engineers and electrical engineers is \$1041. Find the average starting salaries for chemical, mechanical, and electrical engineers.

**Average Starting Salaries for the Five Top
Paying Fields for College Graduates in 1999**

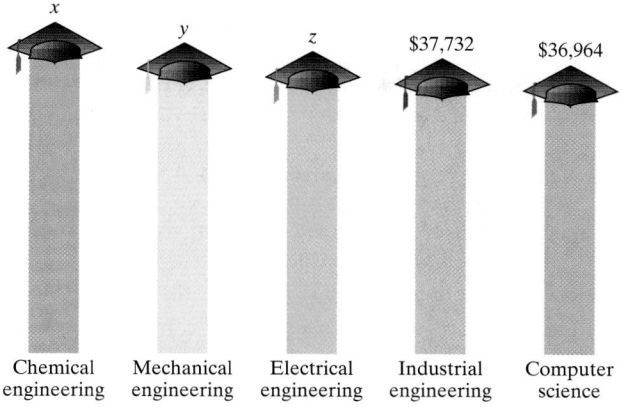

$37,732 $36,964

| Chemical engineering | Mechanical engineering | Electrical engineering | Industrial engineering | Computer science |

Source: Michigan State University

26. The table shows a list of the most frequently spoken languages in the United States, not counting English. Yiddish, Thai, and Persian are spoken by 621 thousand people in the United States. The difference between the number of people who speak Yiddish and the number who speak Thai is 7 thousand. The difference between the number of people who speak Thai and the number who speak Persian is 4 thousand. Find the thousands of people in the

Languages Spoken in the United States

Language	Number of Speakers
1. Spanish	17,339,000
2. French	1,702,000
3. German	1,547,000
4. Italian	1,309,000
5. Chinese	1,249,000
6. Tagalog	843,000
7. Polish	723,000
8. Korean	626,000
9. Vietnamese	507,000
10. Portuguese	430,000
11. Japanese	428,000
12. Greek	388,000
13. Arabic	355,000
14. Hindi, Urdu, & related languages	331,000
15. Russian	242,000
16. Yiddish	x
17. Thai	y
18. Persian	z

Source: Bureau of the census

United States who speak Yiddish, Thai, and Persian.

27. The equation $y = \frac{1}{2}Ax^2 + Bx + C$ gives the relationship between the number of feet a car travels once the brakes are applied, y, and the number of seconds the car is in motion after the brakes are applied, x. A research firm discovered that when a car was in motion for 1 second after the brakes were applied, the car traveled 46 feet. (When $x = 1$, $y = 46$.) Similarly, it was found that when x was 2, y was 84, and when x was 3, y was 114. Use these values to find the constants A, B, and C in the equation. What is the value for y when $x = 6$? Describe what this means.

28. A ball is thrown directly upward from the top of a building. The position function

$$s = \frac{1}{2}at^2 + v_0t + s_0$$

describes the ball's height, s, in feet, after t seconds. Find the values of a, v_0, and s_0 if $s = 224$ at $t = 1$, $s = 176$ at $t = 3$, and $s = 104$ at $t = 4$. What is the value for s when $t = 5$? Describe what this means.

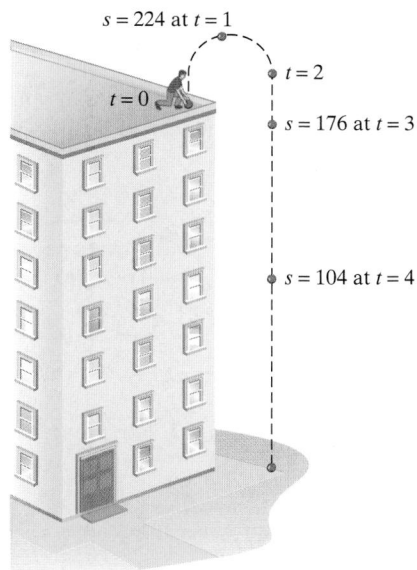

$s = 224$ at $t = 1$

$t = 2$

$t = 0$

$s = 176$ at $t = 3$

$s = 104$ at $t = 4$

Use a system of linear equations in three variables to solve Exercises 29–32.

29. At a college production of *Evita*, 400 tickets were sold. The ticket prices were $8, $10, and $12, and the total income from ticket sales was $3700. How many tickets of each type were sold if the combined number of $8 and $10 tickets sold was 7 times the number of $12 tickets sold?

30. A certain brand of razor blades comes in packages of 6, 12, and 24 blades, costing $2, $3, and $4 per package, respectively. A store sold 12 packages containing a total of 162 razor blades and took in $35. How many packages of each type were sold?

31. A person invested $6700 for one year, part at 8%, part at 10%, and the remainder at 12%. The total annual income from these investments was $716. The amount of money invested at 12% was $300 more than the amount invested at 8% and 10% combined. Find the amount invested at each rate.

32. A person invested $17,000 for one year, part at 10%, part at 12%, and the remainder at 15%. The total annual income from these investments was $2110. The amount of money invested at 12% was $1000 less than the amount invested at 10% and 15% combined. Find the amount invested at each rate.

Writing in Mathematics

33. What is a system of linear equations in three variables?

34. How do you determine whether a given ordered triple is a solution of a system in three variables?

35. Describe in general terms how to solve a system in three variables.

36. Describe how to use the techniques that you learned in this section to obtain a model for U.S. divorce rates from 1970 to 1997.

x (years after 1970)	y (divorces per 1000 people)
0	3.5
15	5.0
27	4.3

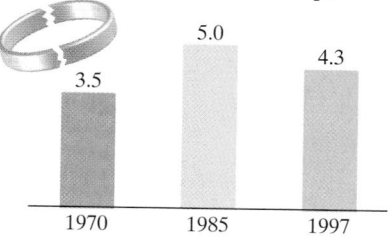

U.S. Divorce Rates: Number of Divorces per 1000 People

Source: U.S. Census Bureau

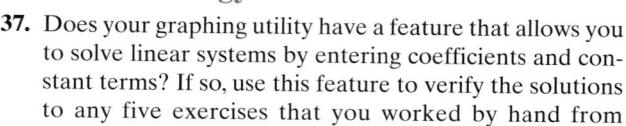

Technology Exercises

37. Does your graphing utility have a feature that allows you to solve linear systems by entering coefficients and constant terms? If so, use this feature to verify the solutions to any five exercises that you worked by hand from Exercises 5–16.

38. Verify your results in Exercises 21–24 by using a graphing utility to graph the resulting parabola. Trace along the curve and convince yourself that the three points given in the exercise lie on the parabola.

39. Some graphing utilities will do three-dimensional graphing. For example, on the TI-92, press $\boxed{\text{MODE}}$, go to $\boxed{\text{GRAPH}}$, press the arrow to the right, select $\boxed{\text{3D}}$, then $\boxed{\text{ENTER}}$. When you display the $\boxed{\text{Y} =}$ screen, you will see the equations are functions of x and y. Thus, you must solve each of a linear system's equations for z before entering the equation. For example,

$$x + y + z = 19$$

is solved for z, giving

$$z = 19 - x - y.$$

(Consult your manual.) If your utility does three-dimensional graphing, graph five of the systems in Exercises 5–16 and trace along the planes to find their common point of intersection.

Critical Thinking Exercises

40. Describe how the system

$$x + y - z - 2w = -8$$
$$x - 2y + 3z + w = 18$$
$$2x + 2y + 2z - 2w = 10$$
$$2x + y - z + w = 3$$

could be solved. Is it likely that in the near future a graphing utility will be available to provide a geometric solution (using intersecting graphs) to this system? Explain.

41. A modernistic painting consists of triangles, rectangles, and pentagons, all drawn so as to not overlap or share sides. Within each rectangle are drawn 2 red roses, and each pentagon contains 5 carnations. How many triangles, rectangles, and pentagons appear in the painting if the painting contains a total of 40 geometric figures, 153 sides of geometric figures, and 72 flowers?

SECTION 8.3 *Partial Fractions*

Objective

1. Find the partial fraction decomposition of a rational expression.

The rising and setting of the sun suggest the obvious: Things change over time. Calculus is the study of rates of change, allowing the motion of the rising sun to be measured by "freezing the frame" at one instant in time. If you are given a function, calculus reveals its rate of change at any "frozen" instant. In this section, you will learn an algebraic technique used in calculus to find a function if its rate of change is known.

The Idea Behind Partial Fraction Decomposition

Systems of linear equations can be used to reverse the process of adding and subtracting rational expressions—for example,

$$\frac{3}{x-4} - \frac{2}{x+2} = \frac{3(x+2) - 2(x-4)}{(x-4)(x+2)}$$

$$= \frac{3x+6-2x+8}{(x-4)(x+2)} = \frac{x+14}{(x-4)(x+2)}.$$

In order to reverse this process, we must show that

$$\frac{x+14}{(x-4)(x+2)} = \frac{3}{x-4} - \frac{2}{x+2} \quad \text{or} \quad \frac{3}{x-4} + \frac{-2}{x+2}.$$

Each of the two fractions on the right is called a **partial fraction**. The sum of these fractions is called the **partial fraction decomposition** of the rational expression on the left-hand side.

Partial fraction decompositions can be written for rational expressions of the form $\dfrac{P(x)}{Q(x)}$, where P and Q have no common factors and the highest power in the numerator is less than the highest power in the denominator. In this section, we will show you how to write the partial fraction decompositions for each of the following rational expressions:

$$\frac{9x^2 - 9x + 6}{(2x-1)(x+2)(x-2)}$$

$P(x) = 9x^2 - 9x + 6$; highest power = 2

$Q(x) = (2x-1)(x+2)(x-2)$; multiplying factors, highest power = 3

$$\frac{5x^3 - 3x^2 + 7x - 3}{(x^2 + 1)^2}$$

$P(x) = 5x^3 - 3x^2 + 7x - 3$; highest power = 3

$Q(x) = (x^2 + 1)^2$; squaring this expression, highest power = 4

1 Find the partial fraction decomposition of a rational expression.

The Steps in Partial Fraction Decomposition

The partial fraction decomposition of a rational expression depends on the factors of the denominator. We consider four cases involving different kinds of factors in the denominator.

Case 1: The Partial Fraction Decomposition of a Rational Expression with Distinct Linear Factors in the Denominator If the denominator has a linear factor of the form $ax + b$, then the partial fraction decomposition will contain a term of the form

$$\frac{A}{ax + b}.$$

Constant

Linear factor

Each distinct linear factor in the denominator produces a partial fraction of the form *constant over linear factor*. For example,

$$\frac{9x^2 - 9x + 6}{(2x - 1)(x + 2)(x - 2)} = \frac{A}{2x - 1} + \frac{B}{x + 2} + \frac{C}{x - 2}.$$

We write a constant over each linear factor in the denominator.

The form of the partial fraction decomposition for a rational expression with distinct linear factors in the denominator is

$$\frac{P(x)}{(a_1x + b_1)(a_2x + b_2)(a_3x + b_3)\cdots(a_nx + b_n)}$$

$$= \frac{A_1}{a_1x + b_1} + \frac{A_2}{a_2x + b_2} + \frac{A_3}{a_3x + b_3} + \cdots + \frac{A_n}{a_nx + b_n}.$$

EXAMPLE 1 Partial Fraction Decomposition with Distinct Linear Factors

Find the partial fraction decomposition of

$$\frac{x + 14}{(x - 4)(x + 2)}.$$

Solution We begin by setting up the partial fraction decomposition with the unknown constants. Write a constant over each of the two distinct linear factors in the denominator.

$$\frac{x + 14}{(x - 4)(x + 2)} = \frac{A}{x - 4} + \frac{B}{x + 2}$$

Our goal is to find A and B. We do this by multiplying both sides of the equation by the least common denominator.

$$(x - 4)(x + 2)\frac{x + 14}{(x - 4)(x + 2)} = (x - 4)(x + 2)\left(\frac{A}{x - 4} + \frac{B}{x + 2}\right)$$

We use the distributive property on the right side.

$$\cancel{(x - 4)}\cancel{(x + 2)}\frac{x + 14}{\cancel{(x - 4)}\cancel{(x + 2)}}$$

$$= \cancel{(x - 4)}(x + 2)\frac{A}{\cancel{(x - 4)}} + (x - 4)\cancel{(x + 2)}\frac{B}{\cancel{(x + 2)}}$$

Dividing out common factors in numerators and denominators, we obtain

$$x + 14 = A(x + 2) + B(x - 4).$$

To find values for A and B that make both sides equal, we'll express the sides in exactly the same form by writing the variable x-terms and then writing the constant terms. Apply the distributive property on the right side.

$$x + 14 = Ax + 2A + Bx - 4B$$

$$x + 14 = Ax + Bx + 2A - 4B$$

$$1x + 14 = (A + B)x + (2A - 4B)$$

As shown by the arrows, if two polynomials are equal, coefficients of like powers of x must be equal $(A + B = 1)$ and their constant terms must be equal $(2A - 4B = 14)$. Consequently, A and B satisfy the following two equations.

$$A + B = 1$$

$$2A - 4B = 14$$

We can use the addition method to solve this linear system in two variables. By multiplying the first equation by -2 and adding equations, we obtain $A = 3$ and $B = -2$. Thus,

$$\frac{x + 14}{(x - 4)(x + 2)} = \frac{A}{x - 4} + \frac{B}{x + 2} = \frac{3}{x - 4} + \frac{-2}{x + 2} \left(\text{or } \frac{3}{x - 4} - \frac{2}{x + 2} \right).$$

Steps in Partial Fraction Decomposition

1. Set up the partial fraction decomposition with the unknown constants A, B, C, etc., in the numerator of the decomposition.

2. Multiply both sides of the resulting equation by the least common denominator.

3. Simplify the right-hand side of the equation.

4. Write both sides in descending powers, equate coefficients of like powers of x, and equate constant terms.

5. Solve the resulting linear system for A, B, C, etc.

6. Substitute the values for A, B, C, etc., into the equation in step 1 and write the partial fraction decomposition.

Check Point 1 Find the partial fraction decomposition of $\frac{5x - 1}{(x - 3)(x + 4)}$.

Case 2: The Partial Fraction Decomposition of a Rational Expression with Linear Factors in the Denominator, Some of Which Are Repeated

Suppose that $(ax + b)^n$ is a factor of the denominator. This means that the linear factor $ax + b$ is repeated n times. When this occurs, the partial fraction decomposition will contain the following sum of n fractions.

$$\frac{P(x)}{(ax + b)^n} = \frac{A_1}{ax + b} + \frac{A_2}{(ax + b)^2} + \frac{A_3}{(ax + b)^3} + \cdots + \frac{A_n}{(ax + b)^n}$$

Include one fraction with a constant numerator for each power of $ax + b$.

EXAMPLE 2 Partial Fraction Decomposition with Repeated Linear Factors

Find the partial fraction decomposition of $\dfrac{x - 18}{x(x - 3)^2}$.

Solution

Step 1   Set up the partial fraction decomposition with the unknown constants. Because the linear factor $x - 3$ is repeated twice, we must include one fraction with a constant numerator for each power of $x - 3$.

$$\frac{x - 18}{x(x - 3)^2} = \frac{A}{x} + \frac{B}{x - 3} + \frac{C}{(x - 3)^2}$$

Step 2   Multiply both sides of the resulting equation by the least common denominator. We clear fractions, multiplying both sides by $x(x - 3)^2$, the least common denominator.

$$x(x - 3)^2 \left[\frac{x - 18}{x(x - 3)^2} \right] = x(x - 3)^2 \left[\frac{A}{x} + \frac{B}{x - 3} + \frac{C}{(x - 3)^2} \right]$$

We use the distributive property on the right side.

$$x(x - 3)^2 \cdot \frac{x - 18}{x(x - 3)^2}$$

$$= x(x - 3)^2 \cdot \frac{A}{x} + x(x - 3)^2 \cdot \frac{B}{(x - 3)} + x(x - 3)^2 \cdot \frac{C}{(x - 3)^2}$$

Dividing out common factors in numerators and denominators, we obtain

$$x - 18 = A(x - 3)^2 + Bx(x - 3) + Cx.$$

Step 3   Simplify the right side of the equation.   Square $x - 3$. Then apply the distributive property.

$$x - 18 = A(x^2 - 6x + 9) + Bx(x - 3) + Cx \qquad \text{Square } x - 3 \text{ using } (A - B)^2 = A^2 - 2AB + B^2.$$

$$x - 18 = Ax^2 - 6Ax + 9A + Bx^2 - 3Bx + Cx \qquad \text{Apply the distributive property.}$$

Step 4   Write both sides in descending powers, equate coefficients of like powers of x, and equate constant terms.   The left side, $x - 18$, is in descending powers of x: $x - 18x^0$. We will write the right side in descending powers of x.

$$x - 18 = Ax^2 + Bx^2 - 6Ax - 3Bx + Cx + 9A$$

Express both sides in the same form.

$$0x^2 + 1x - 18 = (A + B)x^2 + (-6A - 3B + C)x + 9A$$

Equating coefficients of like powers of x and constant terms results in the following system of linear equations.

$$A + B = 0$$
$$-6A - 3B + C = 1$$
$$9A = -18$$

Step 5   Solve the resulting system for A, B, and C.   Dividing both sides of the last equation by 9, we obtain $A = -2$. Substituting -2 for A in the first equation, $A + B = 0$, gives $-2 + B = 0$ or $B = 2$. We find C by substituting -2 for A and 2 for B in the middle equation, $-6A - 3B + C = 1$. We obtain $C = -5$.

Step 6   Substitute the values of A, B, and C and write the partial fraction decomposition.   With $A = -2$, $B = 2$, and $C = -5$, the required partial fraction decomposition is

$$\frac{x - 18}{x(x - 3)^2} = \frac{A}{x} + \frac{B}{x - 3} + \frac{C}{(x - 3)^2} = -\frac{2}{x} + \frac{2}{x - 3} - \frac{5}{(x - 3)^2}.$$

Check Point 2   Find the partial fraction decomposition of $\dfrac{x + 2}{x(x - 1)^2}$.

Case 3:   The Partial Fraction Decomposition of a Rational Expression with Prime, Nonrepeated Quadratic Factors in the Denominator   Suppose that $ax^2 + bx + c$ is a factor of the denominator and that this quadratic factor cannot be factored into linear factors with real coefficients. Under these conditions, the partial fraction decomposition will contain a term of the form

$$\frac{Ax + B}{ax^2 + bx + c}.$$

Linear numerator

Quadratic factor

Each distinct prime quadratic factor in the denominator produces a partial fraction of the form *linear numerator over quadratic factor*. For example,

$$\frac{3x^2 + 17x + 14}{(x - 2)(x^2 + 2x + 4)} = \frac{A}{x - 2} + \frac{Bx + C}{x^2 + 2x + 4}.$$

We write a constant over the linear factor in the denominator.

We write a linear numerator over the prime quadratic factor in the denominator.

Our next example illustrates how a linear system in three variables is used to determine values for A, B, and C.

EXAMPLE 3   Partial Fraction Decomposition

Find the partial fraction decomposition of

$$\frac{3x^2 + 17x + 14}{(x - 2)(x^2 + 2x + 4)}.$$

Solution

Step 1   Set up the partial fraction decomposition with the unknown constants.
We put a constant (A) over the linear factor and a linear expression ($Bx + C$)
over the prime quadratic factor.

$$\frac{3x^2 + 17x + 14}{(x - 2)(x^2 + 2x + 4)} = \frac{A}{x - 2} + \frac{Bx + C}{x^2 + 2x + 4}$$

**Step 2   Multiply both sides of the resulting equation by the least common
denominator.**   We clear fractions, multiplying both sides by $(x - 2)(x^2 + 2x + 4)$,
the least common denominator.

$$(x - 2)(x^2 + 2x + 4)\left[\frac{3x^2 + 17x + 14}{(x - 2)(x^2 + 2x + 4)}\right] = (x - 2)(x^2 + 2x + 4)\left[\frac{A}{x - 2} + \frac{Bx + C}{x^2 + 2x + 4}\right]$$

We use the distributive property on the right side.

$$(x - 2)(x^2 + 2x + 4) \cdot \frac{3x^2 + 17x + 14}{(x - 2)(x^2 + 2x + 4)}$$

$$= (x - 2)(x^2 + 2x + 4) \cdot \frac{A}{x - 2} + (x - 2)(x^2 + 2x + 4) \cdot \frac{Bx + C}{x^2 + 2x + 4}$$

Dividing out common factors in numerators and denominators, we obtain

$$3x^2 + 17x + 14 = A(x^2 + 2x + 4) + (Bx + C)(x - 2).$$

Step 3   Simplify the right side of the equation.   We multiply on the right side
by distributing A over each term in parentheses and multiplying $(Bx + C)(x - 2)$
using the FOIL method.

$$3x^2 + 17x + 14 = Ax^2 + 2Ax + 4A + Bx^2 - 2Bx + Cx - 2C$$

**Step 4   Write both sides in descending powers, equate coefficients of like pow-
ers of x, and equate constant terms.**   The left side, $3x^2 + 17x + 14$, is in descending powers of x. We write the right side in descending powers of x

$$3x^2 + 17x + 14 = Ax^2 + Bx^2 + 2Ax - 2Bx + Cx + 4A - 2C$$

and express both sides in the same form.

$$3x^2 + 17x + 14 = (A + B)x^2 + (2A - 2B + C)x + (4A - 2C)$$

Equating coefficients of like powers of x and constant terms results in the following system of linear equations.

$$A + B = 3$$
$$2A - 2B + C = 17$$
$$4A - 2C = 14$$

Step 5   Solve the resulting system for A, B, and C.   Because the first equation
involves A and B, we can obtain another equation in A and B by eliminating C
from the second and third equations. Multiply the second equation by 2 and add
equations. Solving in this manner, we obtain $A = 5$, $B = -2$, and $C = 3$.

**Step 6   Substitute the values of A, B, and C and write the partial fraction
decomposition.**   With $A = 5$, $B = -2$, and $C = 3$, the required partial fraction
decomposition is

$$\frac{3x^2 + 17x + 14}{(x - 2)(x^2 + 2x + 4)} = \frac{A}{x - 2} + \frac{Bx + C}{x^2 + 2x + 4} = \frac{5}{x - 2} + \frac{-2x + 3}{x^2 + 2x + 4}.$$

Check Point 3 Find the partial fraction decomposition of

$$\frac{8x^2 + 12x - 20}{(x + 3)(x^2 + x + 2)}.$$

Case 4: The Partial Fraction Decomposition of a Rational Expression with a Prime, Repeated Quadratic Factor in the Denominator Suppose that $(ax^2 + bx + c)^n$ is a factor of the denominator and that $ax^2 + bx + c$ cannot be factored further. This means that the quadratic factor $ax^2 + bx + c$ is repeated n times. When this occurs, the partial fraction decomposition will contain a linear numerator for each power of $ax^2 + bx + c$.

$$\frac{P(x)}{(ax^2 + bx + c)^n} = \frac{A_1x + B_1}{ax^2 + bx + c} + \frac{A_2x + B_2}{(ax^2 + bx + c)^2} + \frac{A_3x + B_3}{(ax^2 + bx + c)^3} + \cdots + \frac{A_nx + B_n}{(ax^2 + bx + c)^n}$$

Include one fraction with a linear numerator for each power of $ax^2 + bx + c$.

EXAMPLE 4 Partial Fraction Decomposition with a Repeated Quadratic Factor

Find the partial fraction decomposition of

$$\frac{5x^3 - 3x^2 + 7x - 3}{(x^2 + 1)^2}.$$

Solution

Step 1 Set up the partial fraction decomposition with the unknown constants. Because the quadratic factor $x^2 + 1$ is repeated twice, we must include one fraction with a linear numerator for each power of $x^2 + 1$.

$$\frac{5x^3 - 3x^2 + 7x - 3}{(x^2 + 1)^2} = \frac{Ax + B}{x^2 + 1} + \frac{Cx + D}{(x^2 + 1)^2}$$

Step 2 Multiply both sides of the resulting equation by the least common denominator. We clear fractions, multiplying both sides by $(x^2 + 1)^2$, the least common denominator.

$$(x^2 + 1)^2\left[\frac{5x^3 - 3x^2 + 7x - 3}{(x^2 + 1)^2}\right] = (x^2 + 1)^2\left[\frac{Ax + B}{x^2 + 1} + \frac{Cx + D}{(x^2 + 1)^2}\right]$$

Now we multiply and simplify.

$$5x^3 - 3x^2 + 7x - 3 = (x^2 + 1)(Ax + B) + Cx + D$$

Step 3 Simplify the right side of the equation. We multiply $(x^2 + 1)(Ax + B)$ using the FOIL method.

$$5x^3 - 3x^2 + 7x - 3 = Ax^3 + Bx^2 + Ax + B + Cx + D$$

Step 4 Write both sides in descending powers, equate coefficients of like powers of x, and equate constant terms.

$$5x^3 - 3x^2 + 7x - 3 = Ax^3 + Bx^2 + Ax + Cx + B + D$$

$$5x^3 - 3x^2 + 7x - 3 = Ax^3 + Bx^2 + (A + C)x + (B + D)$$

Equating coefficients of like powers of x and constant terms results in the following system of linear equations.

$$A = 5$$
$$B = -3$$
$$A + C = 7 \qquad \text{With } A = 5, \text{ we immediately obtain } C = 2.$$
$$B + D = -3 \qquad \text{With } B = -3, \text{ we immediately obtain } D = 0.$$

Step 5 Solve the resulting system for A, B, C, and D. Based on our observations in step 4, $A = 5$, $B = -3$, $C = 2$, and $D = 0$.

Step 6 Substitute the values of A, B, C, and D and write the partial fraction decomposition.

$$\frac{5x^3 - 3x^2 + 7x - 3}{\left(x^2 + 1\right)^2} = \frac{Ax + B}{x^2 + 1} + \frac{Cx + D}{\left(x^2 + 1\right)^2} = \frac{5x - 3}{x^2 + 1} + \frac{2x}{\left(x^2 + 1\right)^2}$$

Check Point 4 Find the partial fraction decomposition of $\dfrac{2x^3 + x + 3}{\left(x^2 + 1\right)^2}$.

EXERCISE SET 8.3

 Practice Exercises

In Exercises 1–8, write the form of the partial fraction decomposition of the rational expression. It is not necessary to solve for the constants.

1. $\dfrac{11x - 10}{(x - 2)(x + 1)}$

2. $\dfrac{5x + 7}{(x - 1)(x + 3)}$

3. $\dfrac{6x^2 - 14x - 27}{(x + 2)(x - 3)^2}$

4. $\dfrac{3x + 16}{(x + 1)(x - 2)^2}$

5. $\dfrac{5x^2 - 6x + 7}{(x - 1)(x^2 + 1)}$

6. $\dfrac{5x^2 - 9x + 19}{(x - 4)(x^2 + 5)}$

7. $\dfrac{x^3 + x^2}{\left(x^2 + 4\right)^2}$

8. $\dfrac{7x^2 - 9x + 3}{\left(x^2 + 7\right)^2}$

In Exercises 9–38, write the partial fraction decomposition of each rational expression.

9. $\dfrac{x}{(x - 3)(x - 2)}$

10. $\dfrac{1}{x(x - 1)}$

11. $\dfrac{3x + 50}{(x - 9)(x + 2)}$

12. $\dfrac{5x - 1}{(x - 2)(x + 1)}$

13. $\dfrac{7x - 4}{x^2 - x - 12}$

14. $\dfrac{9x + 21}{x^2 + 2x - 15}$

15. $\dfrac{4x^2 + 13x - 9}{x(x - 1)(x + 3)}$

16. $\dfrac{4x^2 - 5x - 15}{x(x + 1)(x - 5)}$

17. $\dfrac{4x^2 - 7x - 3}{x^3 - x}$

18. $\dfrac{2x^2 - 18x - 12}{x^3 - 4x}$

19. $\dfrac{6x - 11}{(x - 1)^2}$

20. $\dfrac{x}{(x + 1)^2}$

21. $\dfrac{x^2 - 6x + 3}{(x - 2)^3}$

22. $\dfrac{2x^2 + 8x + 3}{(x + 1)^3}$

23. $\dfrac{x^2 + 2x + 7}{x(x - 1)^2}$

24. $\dfrac{3x^2 + 49}{x(x + 7)^2}$

25. $\dfrac{5x^2 + 21x + 4}{(x + 1)^2(x - 3)}$

26. $\dfrac{x}{(x + 2)^2(x + 1)}$

27. $\dfrac{5x^2 - 6x + 7}{(x - 1)(x^2 + 1)}$

28. $\dfrac{5x^2 - 9x + 19}{(x - 4)(x^2 + 5)}$

29. $\dfrac{5x^2 + 6x + 3}{(x + 1)(x^2 + 2x + 2)}$

30. $\dfrac{9x + 2}{(x - 2)(x^2 + 2x + 2)}$

31. $\dfrac{6x^2 - x + 1}{x^3 + x^2 + x + 1}$

32. $\dfrac{3x^2 - 2x + 8}{x^3 + 2x^2 + 4x + 8}$

33. $\dfrac{x^3 + x^2 + 2}{(x^2 + 2)^2}$

34. $\dfrac{x^2 + 2x + 3}{(x^2 + 4)^2}$

35. $\dfrac{x^3 - 4x^2 + 9x - 5}{(x^2 - 2x + 3)^2}$

36. $\dfrac{3x^3 - 6x^2 + 7x - 2}{(x^2 - 2x + 2)^2}$

37. $\dfrac{4x^2 + 3x + 14}{x^3 - 8}$

38. $\dfrac{2x + 4}{x^3 - 1}$

 Application Exercises

39. Find the partial fraction decomposition for $\dfrac{1}{x(x + 1)}$ and use the result to find the following sum:

$$\dfrac{1}{1 \cdot 2} + \dfrac{1}{2 \cdot 3} + \dfrac{1}{3 \cdot 4} + \cdots + \dfrac{1}{99 \cdot 100}.$$

40. Find the partial fraction decomposition for $\dfrac{2}{x(x + 2)}$ and use the result to find the following sum:

$$\dfrac{2}{1 \cdot 3} + \dfrac{2}{3 \cdot 5} + \dfrac{2}{5 \cdot 7} + \cdots + \dfrac{2}{99 \cdot 101}.$$

 **Writing in Mathematics**

41. Explain what is meant by the partial fraction decomposition of a rational expression.

42. Explain how to find the partial fraction decomposition of a rational expression with distinct linear factors in the denominator.

43. Explain how to find the partial fraction decomposition of a rational expression with a repeated linear factor in the denominator.

44. Explain how to find the partial fraction decomposition of a rational expression with a prime quadratic factor in the denominator.

45. Explain how to find the partial fraction decomposition of a rational expression with a repeated, prime quadratic factor in the denominator.

46. How can you verify your result for the partial fraction decomposition for a given rational expression without using a graphing utility?

 Technology Exercises

47. A graphing utility can be used to check the partial fraction decomposition for a given rational expression. Graph $y_1 = $ *the given rational expression* and $y_2 = $ *its partial fraction decomposition* on the same screen. If the graphs are identical, the decomposition is correct. Use this method to verify any five of the decompositions that you obtained in Exercises 9–38.

48. As you worked Exercise 47, did you find that it took a while to determine the range setting that showed a graph for the rational function and its decomposition? Suggest another method for showing that $y_1 = y_2$ using your graphing utility. Use this method to check the results of the same five decompositions you worked with in Exercise 47.

 **Critical Thinking Exercises**

49. If a, b, and c are constants, find the partial fraction decomposition of

$$\dfrac{ax + b}{(x - c)^2}.$$

50. Find the partial fraction decomposition of

$$\dfrac{4x^2 + 5x - 9}{x^3 - 6x - 9}.$$

SECTION 8.4 *Systems of Nonlinear Equations in Two Variables*

Objectives

1. Recognize systems of nonlinear equations in two variables.
2. Solve nonlinear systems by substitution.
3. Solve nonlinear systems by addition.
4. Solve problems using systems of nonlinear equations.

Scientists debate the probability that a "doomsday rock" will collide with Earth. It has been estimated that an asteroid, a tiny planet that revolves around the sun, crashes into Earth about once every 250,000 years, and that such a collision would have disastrous results. In 1908 a small fragment struck Siberia, leveling thousands of acres of trees. One theory about the extinction of dinosaurs 65 million years ago involves Earth's collision with a large asteroid and the resulting drastic changes in Earth's climate.

Understanding the path of Earth and the path of a comet is essential to detecting threatening space debris. Orbits about the sun are not described by linear equations in the form $Ax + By = C$. The ability to solve systems that do not contain linear equations provides NASA scientists watching for troublesome asteroids with a possible collision point with Earth's orbit.

1 Recognize systems of nonlinear equations in two variables.

Systems of Nonlinear Equations and Their Solutions

A **system of** two **nonlinear equations** in two variables contains at least one equation that cannot be expressed in the form $Ax + By = C$. Here are two examples:

$$x^2 = 2y + 10$$
$$3x - y = 9$$

Not in the form $Ax + By = C$. The term x^2 is not linear.

$$y = x^2 + 3$$
$$x^2 + y^2 = 9$$

Neither equation is in the form $Ax + By = C$. The terms x^2 and y^2 are not linear.

A **solution** to a nonlinear system in two variables is an ordered pair of real numbers that satisfies all equations in the system. The **solution set** to the system is the set of all such ordered pairs. As with linear systems in two variables, the solution to a nonlinear system (if there is one) corresponds to the intersection point(s) of the graphs of the equations in the system. Unlike linear systems, the graphs can be circles, parabolas, or anything other than two lines. We will solve nonlinear systems using the substitution method and the addition method.

2 Solve nonlinear systems by substitution.

Eliminating a Variable Using the Substitution Method

The substitution method involves converting a nonlinear system to one equation in one variable by an appropriate substitution. The steps in the solution process are exactly the same as those used to solve a linear system by substitution. However, when you obtain an equation in one variable, this equation will not be linear. In our first example, this equation is quadratic.

EXAMPLE 1 Solving a Nonlinear System by the Substitution Method

Solve by the substitution method:

$$x^2 = 2y + 10 \qquad \text{The graph is a parabola.}$$
$$3x - y = 9. \qquad \text{The graph is a line.}$$

Solution

Step 1   Solve one of the equations for one variable in terms of the other.   We begin by isolating one of the variables raised to the first power in either of the equations. By solving for y in the second equation, which has a coefficient of -1, we can avoid fractions.

$$3x - y = 9 \qquad \text{This is the second equation in the given system.}$$
$$3x = y + 9 \qquad \text{Add } y \text{ to both sides.}$$
$$3x - 9 = y \qquad \text{Subtract 9 from both sides.}$$

Step 2   Substitute the expression from step 1 into the other equation.   We substitute $3x - 9$ for y in the first equation.

$$y = \boxed{3x - 9} \qquad x^2 = 2\,\boxed{y} + 10$$

This gives us an equation in one variable, namely

$$x^2 = 2(3x - 9) + 10.$$

The variable y has been eliminated.

Step 3   Solve the resulting equation containing one variable.

$$x^2 = 2(3x - 9) + 10 \qquad \text{This is the equation containing one variable.}$$
$$x^2 = 6x - 18 + 10 \qquad \text{Use the distributive property.}$$
$$x^2 = 6x - 8 \qquad \text{Combine numerical terms on the right.}$$
$$x^2 - 6x + 8 = 0 \qquad \text{Move all terms to one side and set the quadratic equation equal to 0.}$$
$$(x - 4)(x - 2) = 0 \qquad \text{Factor.}$$
$$x - 4 = 0 \quad \text{or} \quad x - 2 = 0 \qquad \text{Set each factor equal to 0.}$$
$$x = 4 \quad \text{or} \quad x = 2 \qquad \text{Solve for x.}$$

Step 4   Back-substitute the obtained values into the equation from step 1.   Now that we have the x-coordinates of the solutions, we back-substitute 4 for x and 2 for x in the equation $y = 3x - 9$.

If x is 4, $y = 3(4) - 9 = 3$,   so $(4, 3)$ is a solution.
If x is 2, $y = 3(2) - 9 = -3$,   so $(2, -3)$ is a solution.

Step 5   Check the proposed solutions in both of the system's given equations.
We begin by checking $(4, 3)$. Replace x with 4 and y with 3.

$$x^2 = 2y + 10 \qquad 3x - y = 9 \qquad \text{These are the given equations.}$$
$$4^2 \stackrel{?}{=} 2(3) + 10 \qquad 3(4) - 3 \stackrel{?}{=} 9 \qquad \text{Let x = 4 and y = 3.}$$
$$16 \stackrel{?}{=} 6 + 10 \qquad 12 - 3 \stackrel{?}{=} 9 \qquad \text{Simplify.}$$
$$16 = 16 \checkmark \qquad 9 = 9 \checkmark \qquad \text{True statements result.}$$

The ordered pair $(4, 3)$ satisfies both equations. Thus, $(4, 3)$ is a solution to the system.

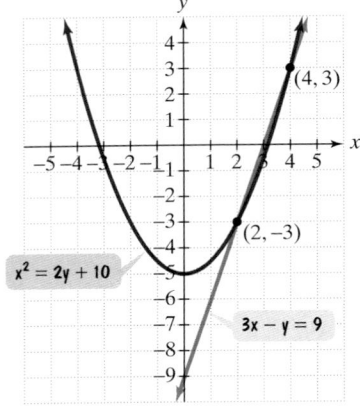

Figure 8.7 Points of intersection illustrate the nonlinear system's solutions.

Study Tip

Recall from Chapter 2 that
$$(x - h)^2 + (y - k)^2 = r^2$$
describes a circle with center (h, k) and radius r.

Now let's check $(2, -3)$. Replace x with 2 and y with -3 in both given equations.

$x^2 = 2y + 10$	$3x - y = 9$	These are the given equations.
$2^2 \stackrel{?}{=} 2(-3) + 10$	$3(2) - (-3) \stackrel{?}{=} 9$	Let x = 2 and y = −3.
$4 \stackrel{?}{=} -6 + 10$	$6 + 3 \stackrel{?}{=} 9$	Simplify.
$4 = 4 \checkmark$	$9 = 9 \checkmark$	True statements result.

The ordered pair $(2, -3)$ also satisfies both equations and is a solution to the system. The solution set is $\{(4, 3), (2, -3)\}$. Figure 8.7 shows the graphs of the equations in the system and the solutions as intersection points.

 Solve by the substitution method:
$$x^2 = y - 1$$
$$4x - y = -1.$$

EXAMPLE 2 Solving a Nonlinear System by the Substitution Method

Solve by the substitution method:

$$x - y = 3 \qquad \text{The graph is a line.}$$
$$(x - 2)^2 + (y + 3)^2 = 4. \qquad \text{The graph is a circle.}$$

Solution Graphically, we are finding the intersection of a line and a circle whose center is at $(2, -3)$ and whose radius measures 2.

Step 1 Solve one of the equations for one variable in terms of the other. We will solve for x in the linear equation — that is, the first equation. (We could also solve for y.)

$$x - y = 3 \qquad \text{This is the first equation in the given system.}$$
$$x = y + 3 \qquad \text{Add y to both sides.}$$

Step 2 Substitute the expression from step 1 into the other equation. We substitute $y + 3$ for x in the second equation.

$$x = \boxed{y + 3} \qquad (\boxed{x} - 2)^2 + (y + 3)^2 = 4$$

This gives an equation in one variable, namely
$$(y + 3 - 2)^2 + (y + 3)^2 = 4.$$

The variable x has been eliminated.

Step 3 Solve the resulting equation containing one variable.

$(y + 3 - 2)^2 + (y + 3)^2 = 4$	This is the equation containing one variable.
$(y + 1)^2 + (y + 3)^2 = 4$	Combine numerical terms in the first parentheses.
$y^2 + 2y + 1 + y^2 + 6y + 9 = 4$	Use the formula $(A + B)^2 = A^2 + 2AB + B^2$ to square y + 1 and y + 3.
$2y^2 + 8y + 10 = 4$	Combine like terms on the left.
$2y^2 + 8y + 6 = 0$	Subtract 4 from both sides and set the quadratic equation equal to 0.

$$y^2 + 4y + 3 = 0 \quad \text{Simplify by dividing both sides by 2.}$$
$$(y + 3)(y + 1) = 0 \quad \text{Factor.}$$
$$y + 3 = 0 \quad \text{or} \quad y + 1 = 0 \quad \text{Set each factor equal to 0.}$$
$$y = -3 \quad \text{or} \quad y = -1 \quad \text{Solve for y.}$$

Step 4 Back-substitute the obtained values into the equation from step 1. Now that we have the y-coordinates of the solutions, we back-substitute -3 for y and -1 for y in the equation $x = y + 3$.

If $y = -3$: $x = -3 + 3 = 0$,   so $(0, -3)$ is a solution.

If $y = -1$: $x = -1 + 3 = 2$,   so $(2, -1)$ is a solution.

Step 5 Check the proposed solution in both of the system's given equations. Take a moment to show that each ordered pair satisfies both equations. The solution set of the given system is $\{(0, -3), (2, -1)\}$.

Figure 8.8 shows the graphs of the equations in the system and the solutions as intersection points.

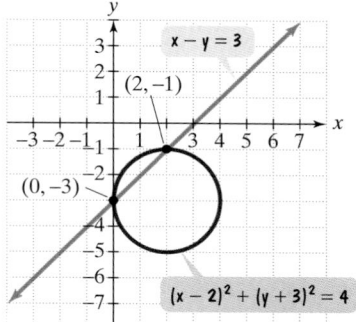

Figure 8.8 Points of intersection illustrate the nonlinear system's solutions.

Check Point 2 Solve by the substitution method:

$$x + 2y = 0$$
$$(x - 1)^2 + (y - 1)^2 = 5.$$

3 Solve nonlinear systems by addition.

Eliminating a Variable Using the Addition Method

In solving linear systems with two variables, we learned that the addition method works well when each equation is in the form $Ax + By = C$. For nonlinear systems, the addition method can be used when each equation is in the form $Ax^2 + By^2 = C$. If necessary, we will multiply either equation or both equations by appropriate numbers so that the coefficients of x^2 or y^2 will have a sum of 0. We then add equations. The sum will be an equation in one variable.

EXAMPLE 3 Solving a Nonlinear System by the Addition Method

Solve the system

$$4x^2 + y^2 = 13 \quad \text{Equation 1}$$
$$x^2 + y^2 = 10. \quad \text{Equation 2}$$

Solution We can use the same steps that we did when we solved linear systems by the addition method.

Step 1 Write both equations in the form $Ax^2 + By^2 = C$. Both equations are already in this form, so we can skip this step.

Step 2 If necessary, multiply either equation or both equations by appropriate numbers so that the sum of the x^2-coefficients or the sum of the y^2-coefficients is 0. We can eliminate y^2 by multiplying Equation 2 by -1.

$$4x^2 + y^2 = 13 \xrightarrow{\text{No change}} 4x^2 + y^2 = 13$$
$$x^2 + y^2 = 10 \xrightarrow{\text{Multiply by } -1.} -x^2 - y^2 = -10$$

Steps 3 and 4   Add equations and solve for the remaining variable.

$$4x^2 + y^2 = 13$$
$$-x^2 - y^2 = -10$$

Add: $3x^2 \quad = 3$

$x^2 = 1$ Divide both sides by 3.

$x = \pm 1$ Use the square root method: If $x^2 = c$, then $x = \pm\sqrt{c}$.

Step 5   Back-substitute and find the values for the other variables.   We must back-substitute each value of x into either one of the original equations. Let's use $x^2 + y^2 = 10$, Equation 2. If $x = 1$,

$1^2 + y^2 = 10$ Replace x with 1 in Equation 2.

$y^2 = 9$ Subtract 1 from both sides.

$y = \pm 3$ Apply the square root method.

$(1, 3)$ and $(1, -3)$ are solutions. If $x = -1$,

$(-1)^2 + y^2 = 10$ Replace x with -1 in Equation 2.

$y^2 = 9$ The steps are the same as before.

$y = \pm 3$

$(-1, 3)$ and $(-1, -3)$ are solutions.

Step 6   Check.   Take a moment to show that each of the four ordered pairs satisfies Equation 1 and Equation 2. The solution set of the given system is $\{(1, 3), (1, -3), (-1, 3), (-1, -3)\}$.

Figure 8.9 shows the graphs of the equations in the system and the solutions as intersection points.

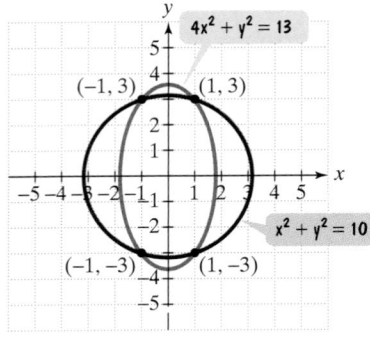

Figure 8.9 A system with four solutions

Study Tip

When solving nonlinear systems, extra solutions may be introduced that do not satisfy both equations in the system. Therefore, you should get into the habit of checking all proposed pairs in each of the system's two equations.

Check Point 3   Solve the system:

$$3x^2 + 2y^2 = 35$$
$$4x^2 + 3y^2 = 48.$$

In solving nonlinear systems, we include only ordered pairs with real numbers in the solution set. We have seen that each of these ordered pairs corresponds to a point of intersection of the system's graphs.

EXAMPLE 4   Solving a Nonlinear System by the Addition Method

Solve the system:

$$y = x^2 + 3 \quad \text{Equation 1 (The graph is a parabola.)}$$
$$x^2 + y^2 = 9. \quad \text{Equation 2 (The graph is a circle.)}$$

Solution   We could use substitution because Equation 1 has y expressed in terms of x, but this would result in a fourth-degree equation. However, we can rewrite Equation 1 by subtracting x^2 from both sides and adding the equations to eliminate the x^2-terms.

$$-x^2 + y \qquad = 3 \quad \text{Subtract } x^2 \text{ from both sides of Equation 1.}$$
$$x^2 \qquad + y^2 = 9 \quad \text{This is Equation 2.}$$
$$y + y^2 = 12 \quad \text{Add the equations.}$$

We now solve this quadratic equation.

$$y + y^2 = 12$$
$$y^2 + y - 12 = 0 \qquad \text{Subtract 12 from both sides and set the}$$
$$\text{quadratic equation equal to 0.}$$
$$(y + 4)(y - 3) = 0 \qquad \text{Factor.}$$
$$y + 4 = 0 \quad \text{or} \quad y - 3 = 0 \qquad \text{Set each factor equal to 0.}$$
$$y = -4 \quad \text{or} \quad y = 3 \qquad \text{Solve for } y.$$

To complete the solution, we must back-substitute each value of y into either one of the original equations. We will use $y = x^2 + 3$, Equation 1. First, we substitute -4 for y.

$$-4 = x^2 + 3$$
$$-7 = x^2 \qquad \text{Subtract 3 from both sides.}$$

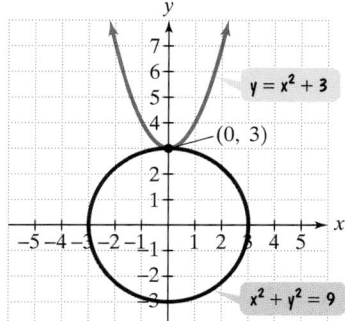

Figure 8.10 A system with one real solution

Because the square of a real number cannot be negative, the equation $x^2 = -7$ does not have real-number solutions. Thus, we move on to our other value for y, 3, and substitute this value into Equation 1.

$$y = x^2 + 3 \qquad \text{This is Equation 1.}$$
$$3 = x^2 + 3 \qquad \text{Back-substitute 3 for } y.$$
$$0 = x^2 \qquad \text{Subtract 3 from both sides.}$$
$$0 = x \qquad \text{Solve for } x.$$

We showed that if $y = 3$, then $x = 0$. Thus, $(0, 3)$ is the solution. Take a moment to show that $(0, 3)$ satisfies Equation 1 and Equation 2. The solution set of the given system is $\{(0, 3)\}$. Figure 8.10 shows the system's graphs and the solution as an intersection point.

Check Point 4 Solve the system:

$$y = x^2 + 5$$
$$x^2 + y^2 = 25.$$

4 Solve problems using systems of nonlinear equations.

Applications

Many geometric problems can be modeled and solved by the use of nonlinear systems of equations. We will use our step-by-step strategy for solving problems using mathematical models that are created from verbal models.

EXAMPLE 5 An Application of a Nonlinear System

You have 36 yards of fencing to build the enclosure in Figure 8.11. Some of this fencing is to be used to build an internal divider. If you'd like to enclose 54 square yards, what are the dimensions of the enclosure?

Solution

Step 1 Use variables to represent unknown quantities. Let $x =$ the enclosure's length and $y =$ the enclosure's width. These variables are shown in Figure 5.11.

Step 2 Write a system of equations describing the problem's conditions. The first condition is that you have 36 yards of fencing.

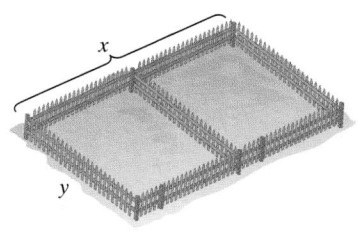

Figure 8.11 Building an enclosure

Fencing along both lengths	plus	Fencing along both widths	plus	Fencing for the internal divider	equals	36 yards.
$2x$	$+$	$2y$	$+$	y	$=$	36

Adding like terms, we can express the equation that models the verbal conditions for the fencing as $2x + 3y = 36$.

The second condition is that you'd like to enclose 54 square yards. The rectangle' area, the product of its length and its width, must be 54 square yards.

Length	times	width	is	54 square yards.
x	$\cdot$	y	$=$	54

Step 3   Solve the system and answer the problem's question.   We must solve the system

$$2x + 3y = 36 \quad \text{Equation 1}$$

$$xy = 54. \quad \text{Equation 2}$$

We will use substitution. Because Equation 1 has no coefficients of 1 or −1, we will solve Equation 2 for y. Dividing both sides of $xy = 54$ by x, we obtain

$$y = \frac{54}{x}.$$

Now we substitute $\dfrac{54}{x}$ for y in Equation 1 and solve for x.

$$2x + 3y = 36 \qquad \text{This is Equation 1.}$$

$$2x + 3 \cdot \frac{54}{x} = 36 \qquad \text{Substitute } \frac{54}{x} \text{ for } y.$$

$$2x + \frac{162}{x} = 36 \qquad \text{Multiply.}$$

$$x\left(2x + \frac{162}{x}\right) = 36 \cdot x \qquad \text{Clear fractions by multiplying both sides by } x.$$

$$2x^2 + 162 = 36x \qquad \text{Use the distributive property on the left side.}$$

$$2x^2 - 36x + 162 = 0 \qquad \text{Subtract 36x from both sides and set the quadratic equation equal to 0.}$$

$$x^2 - 18x + 81 = 0 \qquad \text{Simplify by dividing both sides by 2.}$$

$$(x - 9)^2 = 0 \qquad \text{Factor using } A^2 - 2AB + B^2 = (A - B)^2.$$

$$x - 9 = 0 \qquad \text{Set the factor equal to zero.}$$

$$x = 9 \qquad \text{Solve for } x.$$

We back-substitute this value of x into $y = \dfrac{54}{x}$.

$$\text{If } x = 9, \quad y = \tfrac{54}{9} = 6.$$

This means that the dimensions of the enclosure are 9 yards by 6 yards.

Step 4   Check the proposed solution in the original wording of the problem.
With a length of 9 yards and a width of 6 yards, take a moment to check that this results in 36 yards of fencing and an area of 54 square yards.

Check Point 5   Find the length and width of a rectangle whose perimeter is 20 feet and whose area is 21 square feet.

EXERCISE SET 8.4

Practice Exercises

In Exercises 1–18, solve each system by the substitution method.

1. $x + y = 2$
$y = x^2 - 4$

2. $x - y = -1$
$y = x^2 + 1$

3. $x - y = -1$
$y = x^2 + 2x - 3$

4. $2x + y = -5$
$y = x^2 + 6x + 7$

5. $y = x^2 - 4x - 10$
$y = -x^2 - 2x + 14$

6. $y = x^2 + 4x + 5$
$y = x^2 + 2x - 1$

7. $x^2 + y^2 = 25$
$x - y = 1$

8. $x^2 + y^2 = 5$
$3x - y = 5$

9. $xy = 6$
$2x - y = 1$

10. $xy = -12$
$x - 2y + 14 = 0$

11. $y^2 = x^2 - 9$
$2y = x - 3$

12. $x^2 + y = 4$
$2x + y = 1$

13. $xy = 3$
$x^2 + y^2 = 10$

14. $xy = 4$
$x^2 + y^2 = 8$

15. $x + y = 1$
$x^2 + xy - y^2 = -5$

16. $x + y = -3$
$x^2 + 2y^2 = 12y + 18$

17. $x + y = 1$
$(x - 1)^2 + (y + 2)^2 = 10$

18. $2x + y = 4$
$(x + 1)^2 + (y - 2)^2 = 4$

In Exercises 19–28, solve each system by the addition method.

19. $x^2 + y^2 = 13$
$x^2 - y^2 = 5$

20. $4x^2 - y^2 = 4$
$4x^2 + y^2 = 4$

21. $x^2 - 4y^2 = -7$
$3x^2 + y^2 = 31$

22. $3x^2 - 2y^2 = -5$
$2x^2 - y^2 = -2$

23. $3x^2 + 4y^2 - 16 = 0$
$2x^2 - 3y^2 - 5 = 0$

24. $32x^2 + 2y^2 - 50 = 0$
$x^2 - y^2 - 10 = 0$

25. $x^2 + y^2 = 25$
$(x - 8)^2 + y^2 = 41$

26. $x^2 + y^2 = 5$
$x^2 + (y - 8)^2 = 41$

27. $y^2 - x = 4$
$x^2 + y^2 = 4$

28. $x^2 - 2y = 8$
$x^2 + y^2 = 16$

In Exercises 29–42, solve each system by the method of your choice.

29. $3x^2 + 4y^2 = 16$
$2x^2 - 3y^2 = 5$

30. $x + y^2 = 4$
$x^2 + y^2 = 16$

31. $2x^2 + y^2 = 18$
$xy = 4$

32. $x^2 + 4y^2 = 20$
$xy = 4$

33. $x^2 + 4y^2 = 20$
$x + 2y = 6$

34. $3x^2 - 2y^2 = 1$
$4x - y = 3$

35. $x^3 + y = 0$
$x^2 - y = 0$

36. $x^3 + y = 0$
$2x^2 - y = 0$

37. $x^2 + (y - 2)^2 = 4$
$x^2 - 2y = 0$

38. $x^2 - y^2 - 4x + 6y - 4 = 0$
$x^2 + y^2 - 4x - 6y + 12 = 0$

39. $y = (x + 3)^2$
$x + 2y = -2$

40. $(x - 1)^2 + (y + 1)^2 = 5$
$2x - y = 3$

41. $x^2 + y^2 + 3y = 22$
$2x + y = -1$

42. $2x - y = -3$
$x^2 + y^2 - 4x = 0$

In Exercises 43–46, let x represent one number and let y represent the other number. Use the given conditions to write a system of nonlinear equations. Solve the system and find the numbers.

43. The sum of two numbers is 10 and their product is 24. Find the numbers.

44. The sum of two numbers is 20 and their product is 96. Find the numbers.

45. The difference between the squares of two numbers is 3. Twice the square of the first number increased by the square of the second number is 9. Find the numbers.

46. The difference between the squares of two numbers is 5. Twice the square of the second number subtracted from three times the square of the first number is 19. Find the numbers.

Application Exercises

47. A planet's orbit follows a path described by $16x^2 + 4y^2 = 64$. A comet follows the parabolic path $y = x^2 - 4$. Where might the comet intersect the orbiting planet?

48. A system for tracking ships indicates that a ship lies on a path described by $16y^2 - x^2 = 16$. The process is repeated and the ship is found to lie on a path described by $9x^2 - 4y^2 = 36$. If it is known that the ship is located in the first quadrant of the coordinate system, determine its exact location.

49. Find the length and width of a rectangle whose perimeter is 36 feet and whose area is 77 square feet.

50. Find the length and width of a rectangle whose perimeter is 40 feet and whose area is 96 square feet.

Use the formula for the area of a rectangle and the Pythagorean Theorem to solve Exercises 51–52.

51. A small television has a picture with a diagonal measure of 10 inches and a viewing area of 48 square inches. Find the length and width of the screen.

52. The area of a rug is 108 square feet, and the length of its diagonal is 15 feet. Find the length and width of the rug.

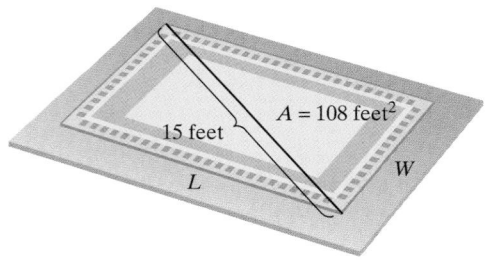

53. The figure at the top of the next column shows a square floor plan with a smaller square area that will accommodate a combination fountain and pool. The floor with the fountain-pool area removed has an area of 21 square meters and a perimeter of 24 meters. Find the dimensions of the floor and the dimensions of the square that will accommodate the pool.

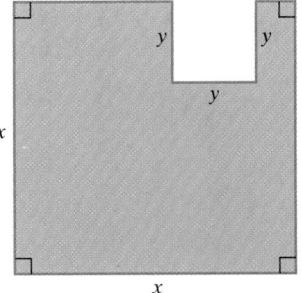

54. The area of the rectangular piece of cardboard shown on the left is 216 square inches. The cardboard is used to make an open box by cutting a 2-inch square from each corner and turning up the sides. If the box is to have a volume of 224 cubic inches, find the length and width of the cardboard that must be used.

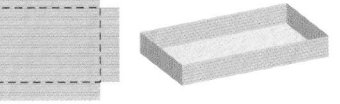

Writing in Mathematics

55. What is a system of nonlinear equations? Provide an example with your description.

56. Explain how to solve a nonlinear system using the substitution method. Use $x^2 + y^2 = 9$ and $2x - y = 3$ to illustrate your explanation.

57. Explain how to solve a nonlinear system using the addition method. Use $x^2 - y^2 = 5$ and $3x^2 - 2y^2 = 19$ to illustrate your explanation.

58. The daily demand and supply models for a carrot cake supplied by a bakery to a convenience store are given by the demand model $N = 40 - 3p$ and the supply model $N = \dfrac{p^2}{10}$, in which p is the price of the cake and N is the number of cakes sold or supplied each day to the convenience store. Explain how to determine the price at which supply and demand are equal. Then describe how to find how many carrot cakes can be supplied and sold each day at this price.

Technology Exercises

59. Verify your solutions to any five exercises from 1 through 42 by using a graphing utility to graph the two equations in the system in the same viewing rectangle. Then use the trace or intersection feature to verify the solutions.

60. Write a system of equations, one equation whose graph is a line and the other whose graph is a parabola, that has no ordered pairs that are real numbers in its solution set. Graph the equations using a graphing utility and verify that you are correct.

Critical Thinking Exercises

61. Which one of the following is true?

    a. A system of two equations in two variables whose graphs represent a circle and a line can have four real solutions.

    b. A system of two equations in two variables whose graphs represent a parabola and a circle can have four real solutions.

    c. A system of two equations in two variables whose graphs represent two circles must have at least two real solutions.

    d. A system of two equations in two variables whose graphs represent a parabola and a circle cannot have only one real solution.

62. The points of intersection of the graphs of $xy = 20$ and $x^2 + y^2 = 41$ are joined to form a rectangle. Find the area of the rectangle.

63. Find a and b in this figure.

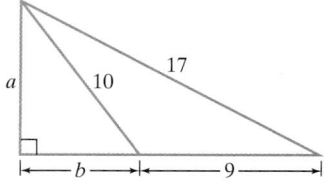

Solve the systems in Exercises 64–65.

64. $\log_y x = 3$
    $\log_y(4x) = 5$

65. $\log x^2 = y + 3$
    $\log x = y - 1$

SECTION 8.5   Systems of Inequalities

Objectives

1. Graph a linear inequality in two variables.

2. Graph a nonlinear inequality in two variables.

3. Graph a system of inequalities.

4. Solve applied problems involving systems of inequalities.

1 Graph a linear inequality in two variables.

Had a good workout lately? If so, could you tell if you were overdoing it or not pushing yourself hard enough? In this section, we will use systems of inequalities in two variables to help you establish a target zone for your workouts.

Graphing a Linear Inequality in Two Variables

We have seen that equations in the form $Ax + By = C$ are straight lines when graphed. If we change the $=$ sign to $>, <, \geq$, or $\leq$, we obtain a **linear inequality in two variables**. Some examples of linear inequalities in two variables are $x + y > 2, 3x - 5y \leq 15$, and $2x - y < 4$.

    Figure 8.12 shows the graph of the linear equation $x + y = 2$. The line divides the points in the rectangular coordinate system into three sets. First, there is the set of points along the line, satisfying $x + y = 2$. Next, there is the set of points in the green region above the line. Points in the green region satisfy the linear inequality $x + y > 2$. Finally, there is the set of points in the pink region below the line. Points in the pink region satisfy the linear inequality $x + y < 2$.

Figure 8.12 All points on the line satisfy $x + y = 2$. All points in the green half-plane above the line satisfy $x + y > 2$. All points in the pink half-plane below the line satisfy $x + y < 2$.

A **half-plane** is the set of all the points on one side of a line. In Figure 8.12, the green region is a half-plane. The pink region is also a half-plane. A half-plane is the solution set of a linear inequality that involves $>$ or $<$. The solution set of an inequality that involves $\geq$ or $\leq$ is a half-plane and a line. A solid line is used to show that the line is part of the solution set. A dashed line is used to show that a line is not part of a solution set.

Graphing a Linear Inequality in Two Variables

1. Replace the inequality symbol with an equal sign and graph the corresponding linear equation. Draw a solid line if the original inequality contains a $\leq$ or $\geq$ symbol. Draw a dashed line if the original inequality contains a $<$ or $>$ symbol.

2. Choose a test point in one of the half-planes that is not on the line. Substitute the coordinates of the test point into the inequality.

3. If a true statement results, shade the half-plane containing this test point. If a false statement results, shade the half-plane not containing this test point.

EXAMPLE 1 Graphing a Linear Inequality in Two Variables

Graph: $3x - 5y < 15$.

Solution

Step 1   Replace the inequality symbol by $=$ and graph the linear equation.
We need to graph $3x - 5y = 15$. We can use intercepts to graph this line.

We set $y = 0$ to find the x-intercept:	We set $x = 0$ to find the y-intercept:
$3x - 5y = 15$	$3x - 5y = 15$
$3x - 5 \cdot 0 = 15$	$3 \cdot 0 - 5y = 15$
$3x = 15$	$-5y = 15$
$x = 5$	$y = -3$

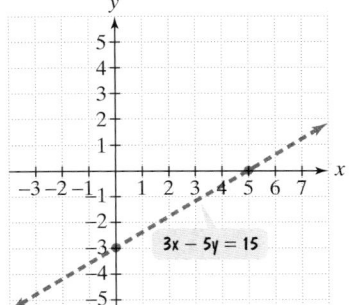

Figure 8.13 Preparing to graph $3x - 5y < 15$

The x-intercept is 5, so the line passes through $(5, 0)$. The y-intercept is -3, so the line passes through $(0, -3)$. The graph is indicated by a dashed line because the inequality $3x - 5y < 15$ contains a $<$ symbol, rather than $\leq$. The graph of the line is shown in Figure 8.13.

Step 2   Choose a test point in one of the half-planes that is not on the line. Substitute its coordinates into the inequality. The line $3x - 5y = 15$ divides the plane into three parts—the line itself and two half-planes. The points in one half-plane satisfy $3x - 5y > 15$. The points in the other half-plane satisfy $3x - 5y < 15$. We need to find which half-plane is the solution. To do so, we test a point from either half-plane. The origin, $(0, 0)$, is the easiest point to test.

$3x - 5y < 15$	This is the given inequality.
Is $3 \cdot 0 - 5 \cdot 0 < 15$?	Test $(0, 0)$ by substituting 0 for x and 0 for y.
$0 - 0 < 15$	
$0 < 15$, true	

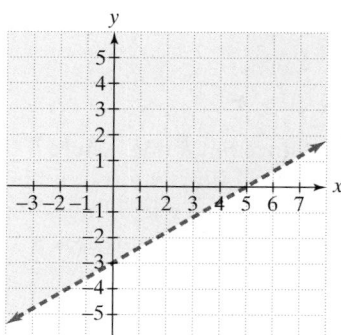

Figure 8.14 The graph of $3x - 5y < 15$

Step 3 If a true statement results, shade the half-plane containing the test point. Because 0 is less than 15, the test point $(0, 0)$ is part of the solution set. All the points on the same side of the line $3x - 5y = 15$ as the point $(0, 0)$ are members of the solution set. The solution set is the half-plane that contains the point $(0, 0)$, indicated by shading this half-plane. The graph is shown using green shading and a dashed blue line in Figure 8.14.

Check Point 1 Graph: $2x - 4y < 8$.

When graphing a linear inequality, test a point that lies in one of the half-planes and *not on the line dividing the half-planes*. The test point $(0, 0)$ is convenient because it is easy to calculate when 0 is substituted for each variable. However, if $(0, 0)$ lies on the dividing line and not in a half-plane, a different test point must be selected.

EXAMPLE 2   Graphing a Linear Inequality

Graph: $y \le \dfrac{2}{3}x$.

Solution

Step 1   Replace the inequality symbol by $=$ and graph the linear equation.   We need to graph $y = \frac{2}{3}x$. We can use the slope and the y-intercept to graph this line.

$$y = \frac{2}{3}x + 0$$

Slope $= \dfrac{2}{3} = \dfrac{\text{rise}}{\text{run}}$ y-intercept $= 0$

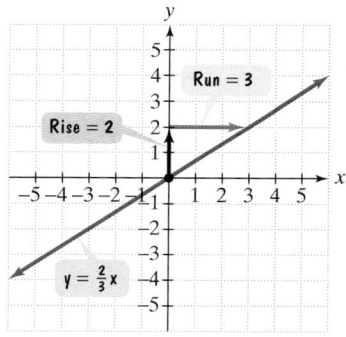

Figure 8.15 Preparing to graph $y \le \dfrac{2}{3}x$

The y-intercept is 0 and the line passes through $(0, 0)$. Using the y-intercept and the slope, the line is shown in Figure 8.15 as a solid line because the inequality $y \le \frac{2}{3}x$ contains a $\le$ symbol, in which equality is included.

Step 2   Choose a test point in one of the half-planes that is not on the line. Substitute its coordinates into the inequality.   We cannot use $(0, 0)$ as a test point because it lies on the line and not in a half-plane. Let's use $(1, 1)$, which lies in the half-plane above the line.

$$y \le \frac{2}{3}x \qquad \text{This is the given inequality.}$$

Is $1 \le \dfrac{2}{3} \cdot 1$?   Test $(1, 1)$ by substituting 1 for x and 1 for y.

$$1 \le \frac{2}{3}, \text{ false}$$

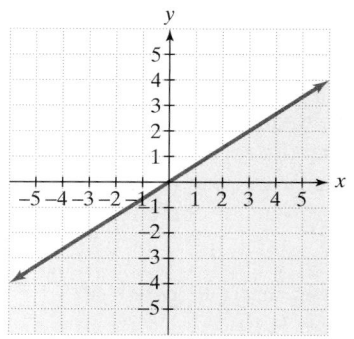

Figure 8.16 The graph of $y \le \dfrac{2}{3}x$

Step 3   If a false statement results, shade the half-plane not containing the test point.   Because 1 is not less than or equal to $\frac{2}{3}$, the test point $(1, 1)$ is not part of the solution set. Thus, the half-plane below the solid line $y = \frac{2}{3}x$ is part of the solution set. The solution set is the line and the half-plane that does not contain the point $(1, 1)$, indicated by shading this half-plane. The graph is shown using green shading and a blue line in Figure 8.16.

<table>
<tr><td>

Check Point 2

</td><td>

Graph: $y \geq \dfrac{1}{2}x$.

</td></tr>
</table>

In Chapter 1, we learned that $y = b$ graphs as a horizontal line, where b is the y-intercept. Similarly, the graph of $x = a$ is a vertical line, where a is the x-intercept. Half-planes can be separated by horizontal or vertical lines. For example, Figure 8.17 shows the graph of $y \leq 2$. Because $(0, 0)$ satisfies this inequality ($0 \leq 2$ is true), the graph consists of the half-plane below the line $y = 2$ and the line. Similarly, Figure 8.18 shows the graph of $x < 4$.

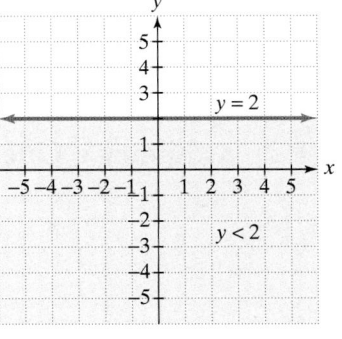

Figure 8.17 The graph of $y \leq 2$

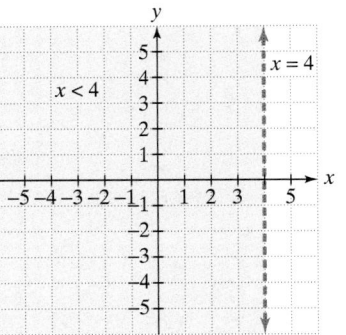

Figure 8.18 The graph of $x < 4$

2 Graph a nonlinear inequality in two variables.

Graphing a Nonlinear Inequality in Two Variables

Example 3 illustrates that a nonlinear inequality in two variables is graphed in the same way that we graph a linear inequality.

EXAMPLE 3   Graphing a Nonlinear Inequality in Two Variables

Graph: $x^2 + y^2 \leq 9$.

Solution

Step 1   Replace the inequality symbol by = and graph the nonlinear equation. We need to graph $x^2 + y^2 = 9$. The graph is a circle of radius 3 with its center at the origin. The graph is shown in Figure 8.19 as a solid circle because equality is included in the $\leq$ symbol.

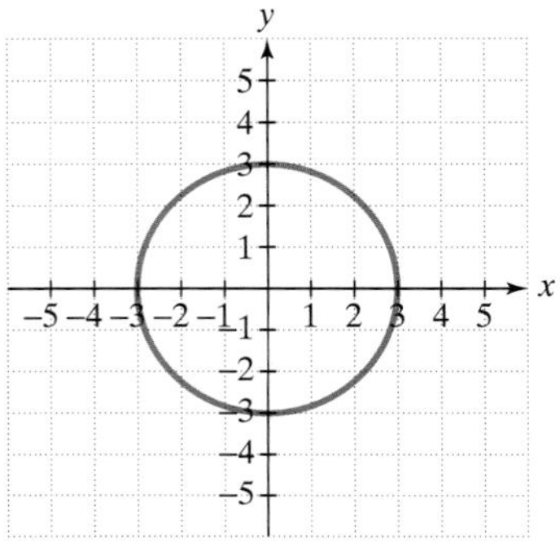

Step 2   Choose a test point in one of the regions that is not on the circle. Substitute its coordinates into the inequality. The circle divides the plane into three parts—the circle itself, the region inside the circle, and the region outside the circle. We need to determine whether the region inside or outside the circle is the solution. To do so, we will use the test point $(0, 0)$ from inside the circle.

$$x^2 + y^2 \leq 9 \qquad \text{This is the given inequality.}$$

$$\text{Is} \quad 0^2 + 0^2 \leq 9? \qquad \text{Test } (0, 0) \text{ by substituting 0 for } x \text{ and 0 for } y.$$

$$0 + 0 \leq 9$$

$$0 \leq 9, \text{ true}$$

Figure 8.19 Preparing to graph $x^2 + y^2 \leq 9$

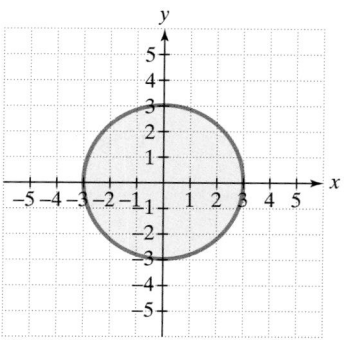

Figure 8.20 The graph of $x^2 + y^2 \leq 9$

3 Graph a system of inequalities.

Step 3   If a true statement results, shade the region containing the test point.
The true statement tells us that all the points inside the circle satisfy $x^2 + y^2 \leq 9$. The graph is shown using green shading and a solid blue circle in Figure 8.20.

> **Check Point 3**   Graph: $x^2 + y^2 \geq 16$.

Systems of Inequalities in Two Variables

The **solution set of a system of inequalities** in two variables x and y is the set of all points (x, y) that satisfy each inequality in the system. The **graph of a system of inequalities** in two variables is the graph of the system's solution set. Thus, to graph a system of inequalities in two variables, begin by graphing each individual inequality in the same rectangular coordinate system. Then find the region, if there is one, that is common to every graph in the system.

EXAMPLE 4   Graphing a System of Linear Inequalities

Graph the solution set:

$$2x - y < 4$$
$$x + y \geq -1.$$

Solution   We begin by graphing $2x - y < 4$. Because the inequality contains a $<$ symbol, rather than $\leq$, we graph $2x - y = 4$ as a dashed line. (If $x = 0$, then $y = -4$, and if $y = 0$, then $x = 2$. The x-intercept is 2 and the y-intercept is -4.) Because $(0, 0)$ makes the inequality $2x - y < 4$ true, we shade the half-plane containing $(0, 0)$, shown in yellow in Figure 8.21.

Now we graph $x + y \geq -1$ in the same rectangular coordinate system. Because the inequality contains a $\geq$ symbol, in which equality is included, we graph $x + y = -1$ as a solid line. (If $x = 0$, then $y = -1$, and if $y = 0$, then $x = -1$. The x-intercept and y-intercept are both -1.) Because $(0, 0)$ makes the inequality true, we shade the half-plane containing $(0, 0)$. This is shown in Figure 8.22 using green vertical shading. The solution set of the system is shown graphically by the intersection (the overlap) of the two half-planes. This is shown in Figure 8.22 as the region in which the yellow shading and the green vertical shading overlap. The solution of the system is shown again in Figure 8.23.

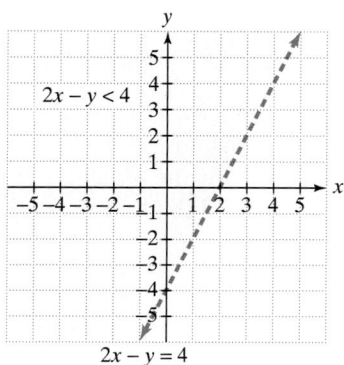

Figure 8.21 The graph of $2x - y < 4$

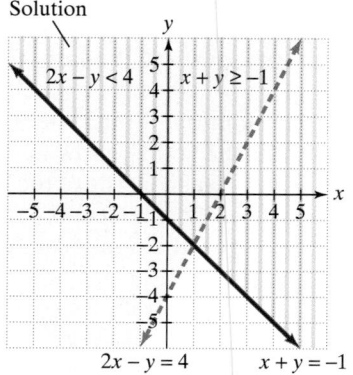

Figure 8.22 Adding the graph of $x + y \geq -1$

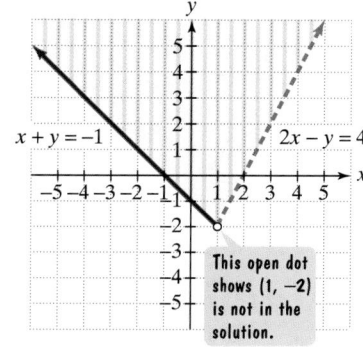

Figure 8.23 The graph of $2x - y < 4$ and $x + y \geq -1$

Check Point 4   Graph the solution set:

$$x + 2y > 4$$
$$2x - 3y \leq -6.$$

EXAMPLE 5   Graphing a System of Inequalities

Graph the solution set:

$$y \geq x^2 - 4$$
$$x - y \geq 2.$$

Solution   We begin by graphing $y \geq x^2 - 4$. Because equality is included in $\geq$, we graph $y = x^2 - 4$ as a solid parabola. Because $(0, 0)$ makes the inequality $y \geq x^2 - 4$ true (we obtain $0 \geq -4$), we shade the interior portion of the parabola containing $(0, 0)$, shown in yellow in Figure 8.24.

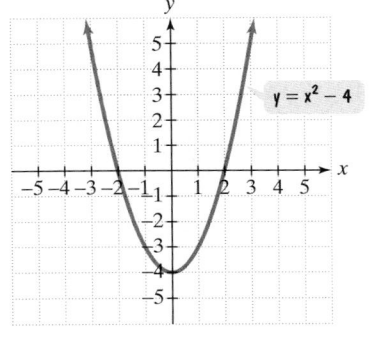

Figure 8.24 The graph of $y \geq x^2 - 4$

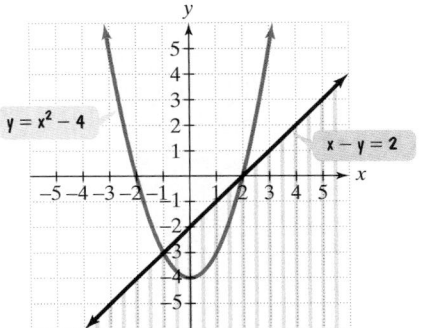

Figure 8.25 Adding the graph of $x - y \geq 2$

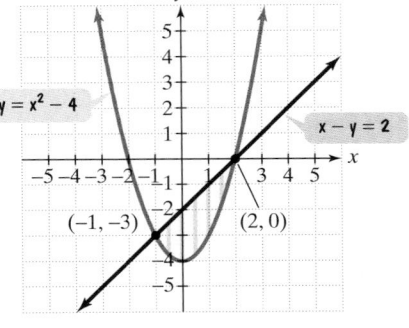

Figure 8.26 The graph of $y \geq x^2 - 4$ and $x - y \geq 2$

Now we graph $x - y \geq 2$ in the same rectangular coordinate system. First we graph the line $x - y = 2$ using its x-intercept, 2, and its y-intercept, -2. Because $(0, 0)$ makes the inequality $x - y \geq 2$ false (we obtain $0 \geq 2$), we shade the half-plane below the line. This is shown in Figure 8.25 using green vertical shading.

The solution of the system is shown in Figure 8.25 by the intersection (the overlap) of the solid yellow and green vertical shadings. The graph of the system's solution set consists of the region enclosed by the parabola and the line. To find the points of intersection of the parabola and the line, use the substitution method to solve the nonlinear system

$$y = x^2 - 4$$
$$x - y = 2.$$

Take a moment to show that the solutions are $(-1, -3)$ and $(2, 0)$, as shown in Figure 8.26.

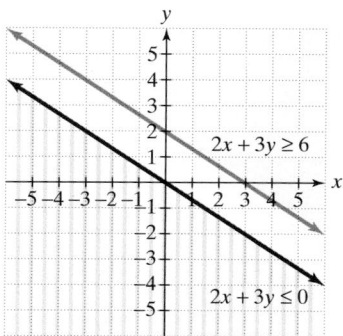

Figure 8.27 A system of inequalities with no solution.

Check Point 5

Graph the solution set:

$$y \geq x^2 - 4$$
$$x + y \leq 2.$$

A system of inequalities has no solution if there are no points in the rectangular coordinate system that simultaneously satisfy each inequality in the system. For example, the system

$$2x + 3y \geq 6$$
$$2x + 3y \leq 0$$

whose graph is shown in Figure 8.27 has no overlapping region. Thus, the system has no solution.

EXAMPLE 6 Graphing a System of Inequalities

Graph the solution set:

$$x - y < 2$$
$$-2 \leq x < 4$$
$$y < 3.$$

Solution We begin by graphing $x - y < 2$, the first given inequality. The line $x - y = 2$ has an x-intercept of 2 and a y-intercept of -2. The test point $(0, 0)$ makes the inequality $x - y < 2$ true, and its graph is shown in Figure 8.28.

Now let's consider the second given inequality $-2 \leq x < 4$. Replacing the inequality symbols by =, we obtain $x = -2$ and $x = 4$, graphed as vertical lines. The line of $x = 4$ is not included. Using $(0, 0)$ as a test point and substituting the x-coordinate, 0, into $-2 \leq x < 4$, we obtain the true statement $-2 \leq 0 < 4$. We therefore shade the region between the vertical lines. We've added this region to Figure 8.28, intersecting the region between the vertical lines with the yellow region in Figure 8.28. The resulting region is shown in yellow and green vertical shading in Figure 8.29.

Finally, let's consider the third given inequality, $y < 3$. Replacing the inequality symbol by =, we obtain $y = 3$, which graphs as a horizontal line. Because $(0, 0)$ satisfies $y < 3$ ($0 < 3$ is true), the graph consists of the half-plane below the line $y = 3$. We've added this half-plane to the region in Figure 8.29, intersecting the half-plane with this region. The resulting region is shown in yellow and green vertical shading in Figure 8.30. This region represents the graph of the solution set of the given system.

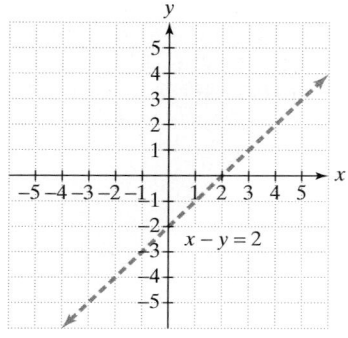

Figure 8.28 The graph of $x - y < 2$

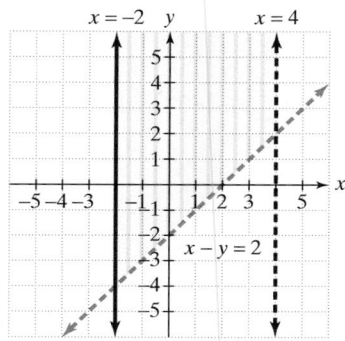

Figure 8.29 The graph of $x - y < 2$ and $-2 \leq x < 4$

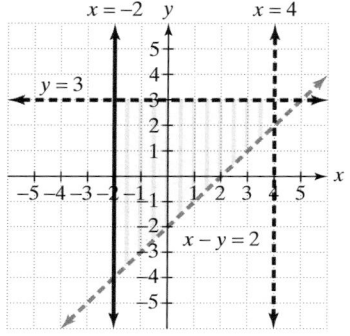

Figure 8.30 The graph of $x - y < 2$ and $-2 \leq x < 4$ and $y < 3$

Check Point 6

Graph the solution set:

$$x + y < 2$$
$$-2 \le x < 1$$
$$y > -3.$$

4 Solve applied problems involving systems of inequalities.

Applications

Now we are ready to use a system of inequalities to establish a target zone for your workouts.

EXAMPLE 7 Inequalities and Aerobic Exercise

For people between ages 10 and 70, inclusive, the target zone for aerobic exercise is given by the following system of inequalities in which a represents one's age and p is one's pulse rate.

$$2a + 3p \ge 450$$
$$a + p \le 190$$

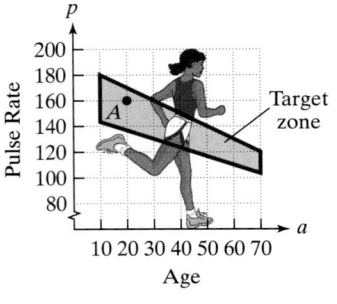

Figure 8.31

The graph of this target zone is shown in Figure 8.31. Find your age. The line segments on the top and bottom of the shaded region indicate upper and lower limits for your pulse rate, in beats per minute, when engaging in aerobic exercise.

a. What are the coordinates of point A and what does this mean in terms of age and pulse rate?

b. Show that the coordinates of point A satisfy each inequality in the system.

Solution

a. Point A has coordinates $(20, 160)$. This means that a pulse rate of 160 beats per minute is within the target zone for a 20-year-old person engaged in aerobic exercise.

b. We can show that $(20, 160)$ satisfies each inequality by substituting 20 for a and 160 for p.

$2a + 3p \ge 450$	$a + p \le 190$
Is $\quad 2(20) + 3(160) \ge 450?$	Is $\quad 20 + 160 \le 190?$
$40 + 480 \ge 450$	$180 \le 190, \text{ true}$
$520 \ge 450, \text{ true}$	

The pair $(20, 160)$ makes each inequality true, so it satisfies each inequality in the system.

Check Point 7

Identify a point other than A in the target zone in Figure 8.31.

a. What are the coordinates of this point and what does this mean in terms of age and pulse rate?

b. Show that the coordinates of the point satisfy each inequality in the system in Example 7.

EXERCISE SET 8.5

 Practice Exercises

In Exercises 1–22, graph each inequality.

1. $x + 2y \leq 8$

2. $3x - 6y \leq 12$

3. $x - 2y > 10$

4. $2x - y > 4$

5. $y \leq \frac{1}{3}x$

6. $y \leq \frac{1}{4}x$

7. $y > 2x - 1$

8. $y > 3x + 2$

9. $x \leq 1$

10. $x \leq -3$

11. $y > 1$

12. $y > -3$

13. $x^2 + y^2 \leq 1$

14. $x^2 + y^2 \leq 4$

15. $x^2 + y^2 > 25$

16. $x^2 + y^2 > 36$

17. $y < x^2 - 1$

18. $y < x^2 - 9$

19. $y \geq x^2 - 9$

20. $y \geq x^2 - 1$

21. $y > 2^x$

22. $y \leq 3^x$

In Exercises 23–52, graph the solution set of each system of inequalities or indicate that the system has no solution.

23. $3x + 6y \leq 6$
$2x + y \leq 8$

24. $x - y \geq 4$
$x + y \leq 6$

25. $2x - 5y \leq 10$
$3x - 2y > 6$

26. $2x - y \leq 4$
$3x + 2y > -6$

27. $y > 2x - 3$
$y < -x + 6$

28. $y < -2x + 4$
$y < x - 4$

29. $x + 2y \leq 4$
$y \geq x - 3$

30. $x + y \leq 4$
$y \geq 2x - 4$

31. $x \leq 2$
$y \geq -1$

32. $x \leq 3$
$y \leq -1$

33. $-2 \leq x < 5$

34. $-2 < y \leq 5$

35. $x - y \leq 1$
$x \geq 2$

36. $4x - 5y \geq -20$
$x \geq -3$

37. $x + y > 4$
$x + y < -1$

38. $x + y > 3$
$x + y < -2$

39. $x + y > 4$
$x + y > -1$

40. $x + y > 3$
$x + y > -2$

41. $y \geq x^2 - 1$
$x - y \geq -1$

42. $y \geq x^2 - 4$
$x - y \geq 2$

43. $x^2 + y^2 \leq 16$
$x + y > 2$

44. $x^2 + y^2 \leq 4$
$x + y > 1$

45. $x^2 + y^2 > 1$
$x^2 + y^2 < 4$

46. $x^2 + y^2 > 1$
$x^2 + y^2 < 9$

47. $x - y \leq 2$
$x \geq -2$
$y \leq 3$

48. $3x + y \leq 6$
$x \geq -2$
$y \leq 4$

49. $x \geq 0$
$y \geq 0$
$2x + 5y \leq 10$
$3x + 4y \leq 12$

50. $x \geq 0$
$y \geq 0$
$2x + y \leq 4$
$2x - 3y \leq 6$

51. $3x + y \leq 6$
$2x - y \leq -1$
$x \geq -2$
$y \leq 4$

52. $2x + y \leq 6$
$x + y \geq 2$
$1 \leq x \leq 2$
$y \leq 3$

 Application Exercises

53. Use Figure 8.31 on page 720 to solve this exercise.

   a. Find a pulse rate that lies within the target zone for a person your age engaged in aerobic exercise.

   b. Express your answer in part (a) as an ordered pair. Show that the coordinates of this ordered pair satisfy each inequality in the system.

The shaded region in the figure shows recommended weight and height combinations based on information from the Department of Agriculture. Use this region to solve Exercises 54–57.

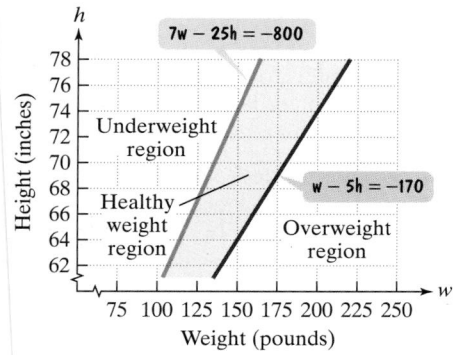

54. Is a person who is 70 inches tall weighing 175 pounds within the healthy weight region?

55. Is a person who is 64 inches tall weighing 105 pounds within the healthy weight region?

56. Estimate the recommended weight range for a person who is 6 feet tall.

57. Write a system of linear inequalities that describes the region for recommended weight and height combinations.

Temperature and precipitation affect whether or not trees and forests can grow. At certain levels of precipitation and temperature, only grasslands and deserts will exist. The figure shows three kinds of regions—deserts, grasslands, and forests—that result from various ranges of temperature and precipitation. Use the figure to solve Exercises 58–60.

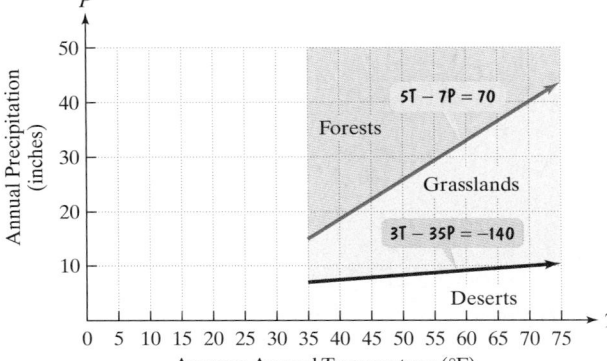

Source: A. Miller and J. Thompson, *Elements of Meterology*

58. Write a system of inequalities that describe where deserts occur.

59. Write a system of inequalities that describe where forests occur.

60. Write a system of inequalities that describe where grasslands occur.

61. Many elevators have a capacity of 2000 pounds. If a child averages 50 pounds and an adult 150 pounds, write an inequality that describes when x children and y adults will cause the elevator to be overloaded. Graph the inequality. Select an ordered pair satisfying the inequality. Describe what this means in practical terms.

62. Suppose a patient is not allowed to have more than 330 milligrams of cholesterol from a diet of eggs and meat. Each egg provides 165 milligrams of cholesterol and each ounce of meat provides 110 milligrams of cholesterol. Thus, $165x + 110y \le 330$, where x is the number of eggs and y the number of ounces of meat. Graph

the inequality in the first quadrant. Give the coordinates of any two points in the solution set. Describe what each set of coordinates means in terms of the variables in the problem.

63. A person with $15,000 plans to place the money in two investments. One investment is high risk, high yield; the other is low risk, low yield. At least $2000 is to be placed in the high-risk investment. Furthermore, the amount invested at low risk should be at least three times the amount invested at high risk. Find and graph a system of inequalities that describes all possibilities for placing the money in the high- and low-risk investments.

64. Promoters of a rock concert must sell at least 25,000 tickets priced at $35 and $50 per ticket. Furthermore, the promoters must take in at least $1,025,000 in ticket sales. Find and graph a system of inequalities that describes all possibilities for selling the $35 tickets and the $50 tickets.

Writing in Mathematics

65. What is a half-plane?

66. What does a dashed line mean in the graph of an inequality?

67. Explain how to graph $2x - 3y < 6$.

68. Compare the graphs of $3x - 2y > 6$ and $3x - 2y \le 6$. Discuss similarities and differences between the graphs.

69. Describe how to solve a system of inequalities.

70. Look at the shaded region showing recommended weight and height combinations in the figure for Exercises 54–57. Describe why a system of inequalities, rather than an equation, is better suited to give the recommended combinations.

Technology Exercises

Graphing utilities can be used to shade regions in the rectangular coordinate system, thereby graphing an inequality in two variables. Read the section of the user's manual for your graphing utility that describes how to shade a region. Then use your graphing utility to graph the inequalities in Exercises 71–74.

71. $y \le 4x + 4$

72. $y \ge \dfrac{2}{3}x - 2$

73. $y \ge x^2 - 4$

74. $y \ge \dfrac{1}{2}x^2 - 2$

75. Does your graphing utility have any limitations in terms of graphing inequalities? If so, what are they?

76. Use a graphing utility with a $\boxed{\text{SHADE}}$ feature to verify any five of the graphs that you drew by hand in Exercises 1–22.

77. Use a graphing utility with a $\boxed{\text{SHADE}}$ feature to verify any five of the graphs that you drew by hand for the systems in Exercises 23–52.

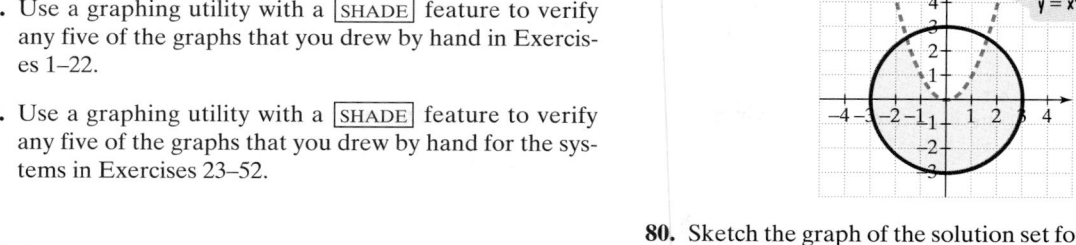

80. Sketch the graph of the solution set for the following system of inequalities:
$$y \geq nx + b \quad (n < 0, b > 0)$$
$$y \leq mx + b \quad (m > 0, b > 0).$$

81. Sketch the graph of the solution set for the following system of inequalities:
$$|x + y| \leq 3$$
$$|y| \leq 2.$$

Critical Thinking Exercises

78. Write a system of inequalities that has no solution.

79. Write a system of inequalities that describes the shaded region in the figure at the top of the next column.

SECTION 8.6 *Linear Programming*

Objectives

1. Write an objective function describing a quantity that must be maximized or minimized.

2. Use inequalities to describe limitations in a situation.

3. Use linear programming to solve problems.

West Berlin children at Tempelhof airport watch fleets of U.S. airplanes bringing in supplies to circumvent the Russian blockade. The airlift began June 28, 1948 and continued for 15 months.

The Berlin Airlift (1948–1949) was an operation by the United States and Great Britain. It was a response to military action by the former Soviet Union: The Soviet troops closed all roads and rail lines between West Germany and Berlin, cutting off supply routes to the city. The Allies used a mathematical technique developed during World War II to maximize the amount of supplies transported. During the 15-month airlift, 278,228 flights provided basic necessities to blockaded Berlin, saving one of the world's great cities.

In this section, we will look at an important application of systems of linear inequalities. Such systems arise in **linear programming**, a method for solving problems in which a particular quantity that must be maximized or minimized is limited. Linear programming is one of the most widely used tools in management science. It helps businesses allocate resources to manufacture products in a way that will maximize profit. Linear programming accounts for more than 50% and perhaps as much as 90% of all computing time used for management decisions in business. The Allies used linear programming to save Berlin.

1 Write an objective function describing a quantity that must be maximized or minimized.

Objective Functions in Linear Programming

Many problems involve quantities that must be maximized or minimized. Businesses are interested in maximizing profit. An operation in which bottled water and medical kits are shipped to earthquake victims needs to maximize the number of victims helped by this shipment. An **objective function** is an algebraic expression in two or more variables describing a quantity that must be maximized or minimized.

EXAMPLE 1   Writing an Objective Function

Bottled water and medical supplies are to be shipped to victims of an earthquake by plane. Each container of bottled water will serve 10 people and each medical kit will aid 6 people. If x represents the number of bottles of water to be shipped and y represents the number of medical kits, write the objective function that describes the number of people that can be helped.

Solution   Because each bottle of water serves 10 people and each medical kit aids 6 people, we have

The number of people helped	is	10 times the number of bottles of water	plus	6 times the number of medical kits.
=		$10x$	$+$	$6y.$

Using z to represent the objective function, we have

$$z = 10x + 6y.$$

Unlike the functions that we have seen so far, the objective function is an equation in three variables. For a value of x and a value of y, there is one and only one value of z. Thus, z is a function of x and y.

> **Check Point 1**   A company manufactures bookshelves and desks for computers. Let x represent the number of bookshelves manufactured daily and y the number of desks manufactured daily. The company's profits are $25 per bookshelf and $55 per desk. Write the objective function that describes the company's total daily profit, z, from x bookshelves and y desks. (Checkpoints 1 through 4 are related to this situation, so keep track of your answers.)

2 Use inequalities to describe limitations in a situation.

Constraints in Linear Programming

Ideally, the number of earthquake victims helped in Example 1 should increase without restriction so that every victim receives water and medical kits. However, the planes that ship these supplies are subject to weight and volume restrictions. In linear programming problems, such restrictions are called **constraints**. Each constraint is expressed as a linear inequality. The list of constraints forms a system of linear inequalities.

EXAMPLE 2 Writing a Constraint

Each plane can carry no more than 80,000 pounds. The bottled water weighs 20 pounds per container and each medical kit weighs 10 pounds. If x represents the number of bottles of water to be shipped and y represents the number of medical kits, write an inequality that describes this constraint.

Solution Because each plane can carry no more than 80,000 pounds, we have

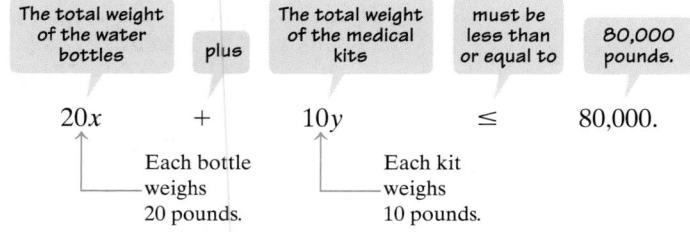

The plane's weight constraint is described by the inequality
$$20x + 10y \le 80,000.$$

Check Point 2 To maintain high quality, the company in Checkpoint 1 should not manufacture more than 80 bookshelves and desks per day. Write an inequality that describes this constraint.

In addition to a weight constraint on its cargo, each plane has a limited amount of space in which to carry supplies. Example 3 demonstrates how to express this constraint.

EXAMPLE 3 Writing a Constraint

Planes can carry a total volume for supplies that does not exceed 6000 cubic feet. Each water bottle is 1 cubic foot and each medical kit also has a volume of 1 cubic foot. With x still representing the number of water bottles and y the number of medical kits, write an inequality that describes this second constraint.

Solution Because each plane can carry a volume of supplies that does not exceed 6000 cubic feet, we have

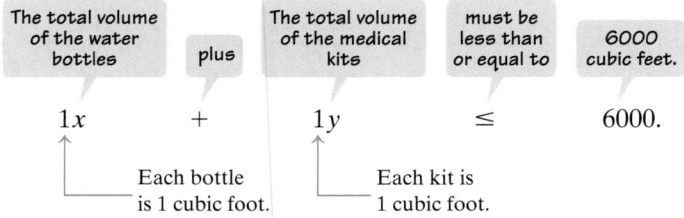

The plane's volume constraint is described by the inequality $x + y \le 6000$.

In summary, here's what we have described in this aid-to-earthquake-victims situation:

$$z = 10x + 6y$$

This is the objective function describing the number of people helped with x bottles of water and y medical kits.

$$20x + 10y \leq 80,000$$
$$x + y \leq 6000.$$

These are the constraints based on each plane's weight and volume limitations.

Check Point 3

To meet customer demand, the company in Checkpoint 1 must manufacture between 30 and 80 bookshelves per day. Furthermore, the company must manufacture at least 10 and no more than 30 desks per day. Write an inequality that describes each of these sentences. Then summarize what you have described about this company by writing the objective function for its profits, and the three constraints.

3 Use linear programming to solve problems.

Solving Problems with Linear Programming

The problem in the earthquake situation described previously is to maximize the number of victims who can be helped, subject to the planes' weight and volume constraints. The process of solving this problem is called linear programming, based on a theorem that was proven during World War II.

Solving a Linear Programming Problem

Let $z = ax + by$ be an objective function that depends on x and y. Furthermore, z is subject to a number of constraints on x and y. If a maximum or minimum value of z exists, it can be determined as follows:

1. Graph the system of inequalities representing the constraints.
2. Find the value of the objective function at each corner, or **vertex**, of the graphed region. The maximum and minimum of the objective function occur at one or more of the corner points.

EXAMPLE 4   Solving a Linear Programming Problem

Determine how many bottles of water and how many medical kits should be sent on each plane to maximize the number of earthquake victims who can be helped.

Solution   We must maximize $z = 10x + 6y$ subject to the constraints

$$20x + 10y \leq 80,000$$
$$x + y \leq 6000.$$

Step 1   Graph the system of inequalities representing the constraints.   Because x (the number of bottles of water per plane) and y (the number of medical kits per plane) must be nonnegative, we need to graph the system of inequalities in quadrant I and its boundary only. To graph the inequality $20x + 10y \leq 80,000$, we graph the equation $20x + 10y = 80,000$ as a solid blue line (Figure 8.32). Setting $y = 0$, the x-intercept is 4000 and setting $x = 0$, the y-intercept is 8000. Using $(0, 0)$ as a test point, the inequality is satisfied, so we shade below the blue line, as shown in yellow in Figure 8.32. Now we graph $x + y \leq 6000$ by first graphing $x + y = 6000$ as a solid line. Setting $y = 0$, the x-intercept is 6000. Setting $x = 0$, the y-intercept is 6000.

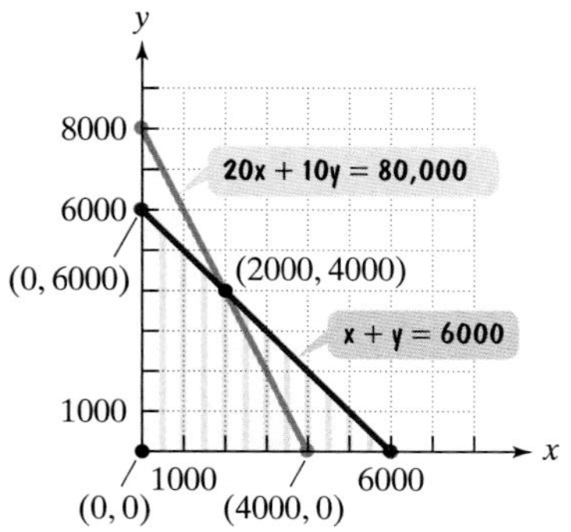

Figure 8.32 The region in quadrant I representing the constraints
$$20x + 10y \leq 80,000$$
$$x + y \leq 6000$$

Using $(0, 0)$ as a test point, the inequality is satisfied, so we shade below the red line, as shown using green vertical shading in Figure 8.32.

We use the addition method to find where the lines $20x + 10y = 80{,}000$ and $x + y = 6000$ intersect.

$$20x + 10y = 80{,}000 \xrightarrow{\text{No change}} 20x + 10y = 80{,}000$$
$$x + y = 6000 \xrightarrow{\text{Multiply by } -10.} -10x - 10y = -60{,}000$$
$$\text{Add:} \quad 10x \qquad = 20{,}000$$
$$x = 2000$$

Back-substituting 2000 for x in $x + y = 6000$, we find $y = 4000$, so the intersection point is $(2000, 4000)$.

The system of inequalities representing the constraints is shown by the region in which the yellow shading and the green vertical shading overlap in Figure 8.32. The graph of the system of inequalities is shown again in Figure 8.33. The red and blue line segments are included in the graph.

Step 2 Find the value of the objective function at each corner of the graphed region. The maximum and minimum of the objective function occur at one or more of the corner points. We must evaluate the objective function, $z = 10x + 6y$, at the four corners of the region in Figure 8.33.

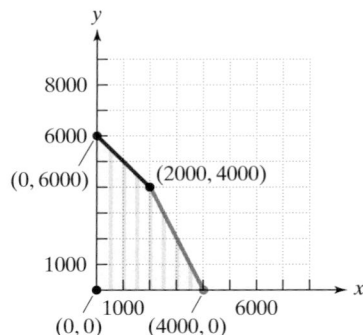

Figure 8.33

Corner (x, y)	Objective Function $z = 10x + 6y$
$(0, 0)$	$z = 10(0) + 6(0) = 0$
$(4000, 0)$	$z = 10(4000) + 6(0) = 40{,}000$
$(2000, 4000)$	$z = 10(2000) + 6(4000) = 44{,}000$ ← maximum
$(0, 6000)$	$z = 10(0) + 6(6000) = 36{,}000$

Thus, the maximum value of z is 44,000 and this occurs when $x = 2000$ and $y = 4000$. In practical terms, this means that the maximum number of earthquake victims who can be helped with each plane shipment is 44,000. This can be accomplished by sending 2000 water bottles and 4000 medical kits per plane.

Check Point 4 For the company in Checkpoints 1–3, how many bookshelves and how many desks should be manufactured per day to obtain maximum profit? What is the maximum daily profit?

EXAMPLE 5 Solving a Linear Programming Problem

Find the maximum value of the objective function

$$z = 2x + y$$

subject to the constraints:

$$x \geq 0, \ y \geq 0$$
$$x + 2y \leq 5$$
$$x - y \leq 2.$$

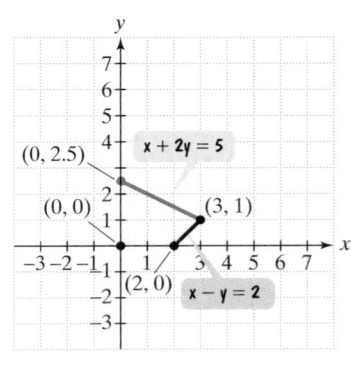

Figure 8.34 The graph of $x + 2y \leq 5$ and $x - y \leq 2$ in quadrant I

Solution We begin by graphing the region in quadrant I ($x \geq 0, y \geq 0$) formed by the constraints. The graph is shown in Figure 8.34.

Now we evaluate the objective function at the four vertices of this region.

Objective function: $z = 2x + y$

At $(0, 0)$: $\quad z = 2 \cdot 0 + 0 = 0$

At $(2, 0)$: $\quad z = 2 \cdot 2 + 0 = 4$

At $(3, 1)$: $\quad z = 2 \cdot 3 + 1 = 7$ Maximum value of z

At $(0, 2.5)$: $\quad z = 2 \cdot 0 + 2.5 = 2.5$

Thus, the maximum value of z is 7, and this occurs when $x = 3$ and $y = 1$.

We can see why the objective function in Example 5 has a maximum value that occurs at a vertex by solving the equation for y.

$z = 2x + y$ This is the objective function of Example 5.

$y = -2x + z$ Solve for y. Recall that the slope-intercept form of a line is $y = mx + b$.

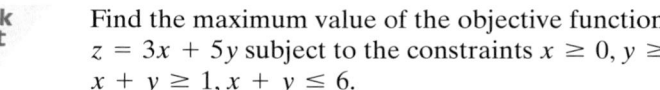

Slope = −2   y-intercept = z

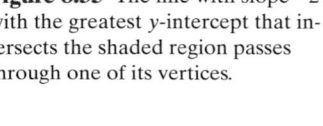

Figure 8.35 The line with slope −2 with the greatest y-intercept that intersects the shaded region passes through one of its vertices.

In this form, z represents the y-intercept of the objective function. The equation describes infinitely many parallel lines, each with a slope of −2. The process in linear programming involves finding the maximum z-value for all lines that intersect the region determined by the constraints. Of all the lines whose slope is −2, we're looking for the one with the greatest y-intercept that intersects the given region. As we see in Figure 8.35, such a line will pass through one (or possibly more) of the vertices of the region.

Check Point 5   Find the maximum value of the objective function $z = 3x + 5y$ subject to the constraints $x \geq 0$, $y \geq 0$, $x + y \geq 1$, $x + y \leq 6$.

Faster and Faster

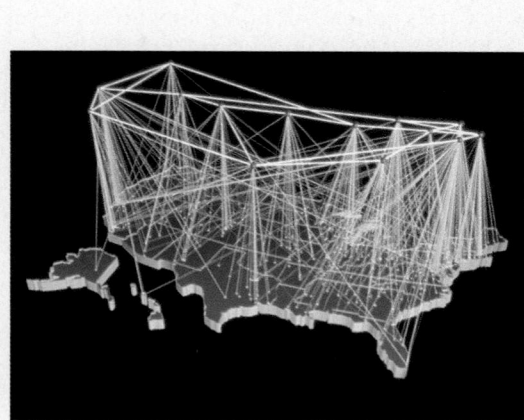

The network of computer linkages in the United States is growing exponentially.

The problems we solve nowadays have thousands of equations, sometimes a million variables. One of the things that still amazes me is to see a program run on the computer—and to see the answer come out. If we think of the number of combinations of different solutions that we're trying to choose the best of, it's akin to the stars in the heavens. Yet we solve them in a matter of moments. This, to me, is staggering. Not that we can solve them—but that we can solve them so rapidly and efficiently.
—George Dantzig
Inventor of a linear programming method

Problems in linear programming can involve objective functions with thousands of variables subject to thousands of constraints. Several nongeometric linear programming methods are available on software for solving such problems. And we continue to search for faster and faster linear programming methods. This area of applied mathematics has a direct impact on the efficiency and profitability of numerous industries, including telephone and computer communications, and the airlines.

EXERCISE SET 8.6

✓ Practice Exercises

In Exercises 1–4, find the value of the objective function at each corner of the graphed region. What is the maximum value of the objective function? What is the minimum value of the objective function?

1. Objective Function $z = 5x + 6y$

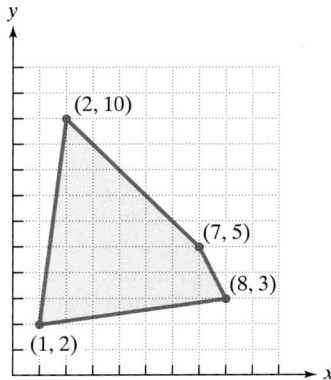

2. Objective Function $z = 3x + 2y$

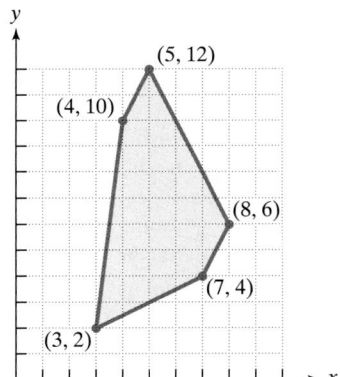

3. Objective Function $z = 40x + 50y$

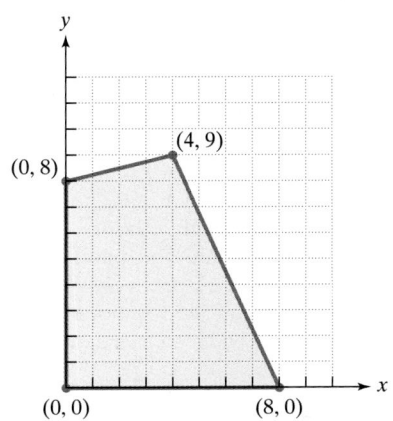

4. Objective Function $z = 30x + 45y$

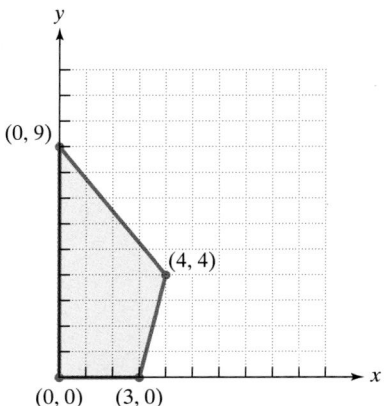

In Exercises 5–14, an objective function and a system of linear inequalities representing constraints are given.

  a. *Graph the system of inequalities representing the constraints.*

  b. *Find the value of the objective function at each corner of the graphed region.*

  c. *Use the values in part (b) to determine the maximum value of the objective function and the values of x and y for which the maximum occurs.*

5. Objective Function $z = 2x + 3y$
    Constraints $x \geq 0, y \geq 0$
                      $3x + y \leq 6$
                      $2x + 3y \leq 12$

6. Objective Function $z = 3x + 2y$
    Constraints $x \geq 0, y \geq 0$
                      $2x + y \leq 8$
                      $x + y \geq 4$

7. Objective Function $z = 4x + y$
    Constraints $x \geq 0, y \geq 0$
                      $2x + 3y \leq 12$
                      $x + y \geq 3$

8. Objective Function $z = x + 6y$
    Constraints $x \geq 0, y \geq 0$
                      $2x + y \leq 10$
                      $x - 2y \geq -10$

9. Objective Function $z = 3x - 2y$
    Constraints $1 \leq x \leq 5$
                      $y \geq 2$
                      $x - y \geq -3$

10. Objective Function
$z = 5x - 2y$

Constraints
$0 \le x \le 5$
$0 \le y \le 3$
$x + y \ge 2$

11. Objective Function
$z = 4x + 2y$

Constraints
$x \ge 0, y \ge 0$
$2x + 3y \le 12$
$3x + 2y \le 12$
$x + y \ge 2$

12. Objective Function
$z = 2x + 4y$

Constraints
$x \ge 0, y \ge 0$
$x + 3y \ge 6$
$x + y \ge 3$
$x + y \le 9$

13. Objective Function
$z = 10x + 12y$

Constraints
$x \ge 0, y \ge 0$
$x + y \le 7$
$2x + y \le 10$
$2x + 3y \le 18$

14. Objective Function
$z = 5x + 6y$

Constraints
$x \ge 0, y \ge 0$
$2x + y \ge 10$
$x + 2y \ge 10$
$x + y \le 10$

⭐ Application Exercises

15. A television manufacturer makes console and wide-screen televisions. The profit per unit is $125 for the console televisions and $200 for the wide-screen televisions.
  a. Let

x = the number of consoles manufactured in a month and

y = the number of wide-screens manufactured in a month.

  Write the objective function that describes the total monthly profit.
  b. The manufacturer is bound by the following constraints:
  1. Equipment in the factory allows for making at most 450 console televisions in one month.
  2. Equipment in the factory allows for making at most 200 wide-screen televisions in one month.
  3. The cost to the manufacturer per unit is $600 for the console televisions and $900 for the wide-screen televisions. Total monthly costs cannot exceed $360,000.
  Write a system of three inequalities that describes these constraints.
  c. Graph the system of inequalities in part (b). Use only the first quadrant and its boundary, because x and y must both be nonnegative.

  d. Evaluate the objective function for total monthly profit at each of the five vertices of the graphed region. (The vertices should occur at $(0, 0)$, $(0, 200)$, $(300, 200)$, $(450, 100)$, and $(450, 0)$.)
  e. Complete the missing portions of this statement: The television manufacturer will make the greatest profit by manufacturing ___ console televisions each month and ___ wide-screen televisions each month. The maximum monthly profit is $ ___.

16. a. A student earns $10 per hour for tutoring and $7 per hour as a teacher's aid. Let x = the number of hours each week spent tutoring, and y = the number of hours each week spent as a teacher's aid. Write the objective function that describes total weekly earnings.
  b. The student is bound by the following constraints:
  • To have enough time for studies, the student can work no more than 20 hours a week.
  • The tutoring center requires that each tutor spend at least three hours a week tutoring.
  • The tutoring center requires that each tutor spend no more than eight hours a week tutoring.
  Write a system of three inequalities that describes these constraints.
  c. Graph the system of inequalities in part (b). Use only the first quadrant and its boundary, because x and y are nonnegative.
  d. Evaluate the objective function for total weekly earnings at each of the four vertices of the graphed region. (The vertices should occur at $(3, 0)$, $(8, 0)$, $(3, 17)$, and $(8, 12)$.)
  e. Complete the missing portions of this statement: The student can earn the maximum amount per week by tutoring for ___ hours per week and working as a teacher's aid for ___ hours per week. The maximum amount that the student can earn each week is $ ___.

Use the two steps for solving a linear programming problem, given in the box on page 726, to solve the problems in Exercises 17–23.

17. A manufacturer produces two models of mountain bicycles. The times (in hours) required for assembling and painting each model are given in the following table.

	Model A	Model B
Assembling	5	4
Painting	2	3

The maximum total weekly hours available in the assembly department and the paint department are 200 hours and 108 hours, respectively. The profits per unit are $25 for model A and $15 for model B. How many of each type should be produced to maximize profit?

18. A large institution is preparing lunch menus containing foods A and B. The specifications for the two foods are given in the following table.

Food	Units of Fat per Ounce	Units of Carbohydrates per Ounce	Units of Protein per Ounce
A	1	2	1
B	1	1	1

Each lunch must provide at least 6 units of fat per serving, no more than 7 units of protein, and at least 10 units of carbohydrates. The institution can purchase food A for $0.12 per ounce and food B for $0.08 per ounce. How many ounces of each food should a serving contain to meet the dietary requirements at the least cost?

19. Food and clothing are shipped to victims of a natural disaster. Each carton of food will feed 5 people, while each carton of clothing will help 6 people. Each 30-cubic-foot box of food weighs 50 pounds and each 20-cubic-foot box of clothing weighs 5 pounds. The commercial carriers transporting food and clothing are bound by the following constraints:
1. The total weight per carrier cannot exceed 18,000 pounds.
2. The total volume must be less than 12,000 cubic feet.
How many cartons of food and clothing should be sent with each plane shipment to maximize the number of people who can be helped?

20. On June 24, 1948, the former Soviet Union blocked all land and water routes through East Germany to Berlin. A gigantic airlift was organized using American and British planes to supply food, clothing, and other supplies to the more than 2 million people in West Berlin. The cargo capacity was 30,000 cubic feet for an American plane and 20,000 cubic feet for a British plane. To break the Soviet blockade, the Western Allies had to maximize cargo capacity, but were subject to the following restrictions:
1. No more than 44 planes could be used.
2. The larger American planes required 16 personnel per flight, double that of the requirement for the British planes. The total number of personnel available could not exceed 512.
3. The cost of an American flight was $9000 and the cost of a British flight was $5000. Total weekly costs could not exceed $300,000.
Find the number of American and British planes that were used to maximize cargo capacity.

21. A theater is presenting a program on drinking and driving for students and their parents. The proceeds will be donated to a local alcohol information center. Admis-

sion is $2.00 for parents and $1.00 for students. However, the situation has two constraints: The theater can hold no more than 150 people and every two parents must bring at least one student. How many parents and students should attend to raise the maximum amount of money?

22. You are about to take a test that contains computation problems worth 6 points each and word problems worth 10 points each. You can do a computation problem in 2 minutes and a word problem in 4 minutes. You have 40 minutes to take the test and may answer no more than 12 problems. Assuming you answer all the problems attempted correctly, how many of each type of problem must you do to maximize your score? What is the maximum score?

23. In 1978, a ruling by the Civil Aeronautics Board allowed Federal Express to purchase larger aircraft. Federal Express's options included 20 Boeing 727s that United Airlines was retiring and/or the French-built Dassault Fanjet Falcon 20. To aid in their decision, executives at Federal Express analyzed the following data:

	Boeing 727	Falcon 20
Direct Operating Cost	$1400 per hour	$500 per hour
Payload	42,000 pounds	6000 pounds

Federal Express was faced with the following constraints:
1. Hourly operating cost was limited to $35,000.
2. Total payload had to be at least 672,000 pounds.
3. Only twenty 727s were available.
Given the constraints, how many of each kind of aircraft should Federal Express have purchased to maximize the number of aircraft?

Writing in Mathematics

24. What kinds of problems are solved using the linear programming method?

25. What is an objective function in a linear programming problem?

26. What is a constraint in a linear programming problem? How is a constraint represented?

27. In your own words, describe how to solve a linear programming problem.

28. Describe a situation in your life in which you would really like to maximize something, but you are limited by at least two constraints. Can linear programming be used in this situation? Explain your answer.

Technology Exercises

In Exercises 29–32, use a graphing utility to sketch the region determined by the constraints. Then determine the maximum value of the objective function subject to the constraints.

29. Objective Function $z = 6x + 8y$
     Constraints $x \geq 0, y \geq 0$
                       $x + 2y \leq 6$

30. Objective Function $z = 30x + 20y$
     Constraints $x \geq 0, y \geq 0$
                       $2x + y \leq 14$
                       $3x + y \leq 18$

31. Objective Function $z = 9x + 14y$
     Constraints $x \geq 0, y \geq 0$
                       $2x + y \leq 10$
                       $2x + 3y \leq 18$

32. Objective Function $z = 10x + 3y$
     Constraints $0 \leq x \leq 10, \quad y \geq 0$
                       $4x + 5y \leq 60$
                       $4x - 5y \geq -20$

Critical Thinking Exercises

33. Suppose that you inherit $10,000. The will states how you must invest the money. Some (or all) of the money must be invested in stocks and bonds. The requirements are that at least $3000 be invested in bonds, with expected returns of $0.08 per dollar, and at least $2000 be invested in stocks, with expected returns of $0.12 per dollar. Because the stocks are medium risk, the final stipulation requires that the investment in bonds should never be less than the investment in stocks. How should the money be invested so as to maximize your expected returns?

34. Consider the objective function $z = Ax + By$ ($A > 0$ and $B > 0$) subject to the following constraints: $2x + 3y \leq 9$, $x - y \leq 2$, $x \geq 0$, and $y \geq 0$. Prove that the objective function will have the same maximum value at the vertices $(3, 1)$ and $(0, 3)$ if $A = \frac{2}{3}B$.

Group Exercises

35. Group members should choose a particular field of interest. Research how linear programming is used to solve problems in that field. If possible, investigate the solution of a specific practical problem. Present a report on your findings, including the contributions of George Dantzig, Narendra Karmarkar, and L.G. Khachion to linear programming.

36. Members of the group should interview a business executive who is in charge of deciding the product mix for a business. How are production policy decisions made? Are other methods used in conjunction with linear programming? What are these methods? What sort of academic background, particularly in mathematics, does this executive have? Present a group report addressing these questions, emphasizing the role of linear programming for the business.

CHAPTER SUMMARY, REVIEW, AND TEST

Summary

8.1 Systems of Linear Equations in Two Variables

a. Two equations in the form $Ax + By = C$ are called a system of linear equations. A solution to the system is an ordered pair that satisfies both equations in the system.

b. Linear systems in two variables can be solved by eliminating a variable, using the substitution method (see the box on page 675) or the addition method (see the box on page 677).

c. Some linear systems have no solution and are called inconsistent systems; others have infinitely many solutions. The equations in a linear system with infinitely many solutions are called dependent. For details, see the box on page 679.

8.2 Systems of Linear Equations in Three Variables

a. Three equations in the form $Ax + By + Cz = D$ are called a system of linear equations in three variables. A solution to the system is an ordered triple that satisfies all three equations in the system.

b. A system of linear equations in three variables can be solved by eliminating variables. Use the addition method to eliminate any variable, reducing the system to two equations in two variables. Use substitution or the addition method to solve the resulting system in two variables. Details are found in the box on page 688.

8.3 Partial Fraction Decomposition

a. Partial fraction decomposition is used on rational expressions in which the numerator and denominator have no common factors and the highest power in the numerator is less than the highest power in the denominator. The steps in partial fraction decomposition are given in the box on page 697.

b. Include one partial fraction with a constant numerator for each distinct linear factor in the denominator. Include one partial fraction with a constant numerator for each power of a repeated linear factor in the denominator.

c. Include one partial fraction with a linear numerator for each distinct prime quadratic factor in the denominator. Include one partial fraction with a linear numerator for each power of a prime, repeated quadratic factor in the denominator.

8.4 Systems of Nonlinear Equations in Two Variables

a. A system of two nonlinear equations in two variables contains at least one equation that cannot be expressed as $Ax + By = C$.

b. Nonlinear systems of equations can be solved algebraically by eliminating all occurrences of one of the variables by the substitution and addition methods.

Review Exercises

8.1

In Exercises 1–5, solve by the method of your choice. Identify systems with no solution and systems with infinitely many solutions, using set notation to express their solution sets.

1. $y = 4x + 1$
$3x + 2y = 13$

2. $x + 4y = 14$
$2x - y = 1$

3. $5x + 3y = 1$
$3x + 4y = -6$

4. $2y - 6x = 7$
$3x - y = 9$

5. $4x - 8y = 16$
$3x - 6y = 12$

6. Can the graphing-utility-generated screen be the solution for the system

$$x + y = 2$$
$$2x + y = -5?$$

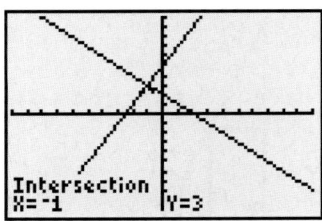

Explain.

8.5 Systems of Inequalities

a. A linear inequality in two variables can be written in the form $Ax + By > C$, $Ax + By \geq C$, $Ax + By < C$, or $Ax + By \leq C$.

b. The procedure for graphing a linear inequality in two variables is given in the box on page 714. A nonlinear inequality in two variables is graphed using the same procedure.

c. To graph the solution set to a system of inequalities, graph each inequality in the system in the same rectangular coordinate system. Then find the region, if there is one, that is common to every graph in the system.

8.6 Linear Programming

a. An objective function is an algebraic expression in three variables describing a quantity that must be maximized or minimized.

b. Constraints are restrictions, expressed as linear inequalities.

c. Steps for solving a linear programming problem are given in the box on page 726.

7. Health experts agree that cholesterol intake should be limited to 300 mg or less each day. Three ounces of shrimp and 2 ounces of scallops contain 156 mg of cholesterol. Five ounces of shrimp and 3 ounces of scallops contain 45 mg of cholesterol less than the suggested maximum daily intake. Determine the cholesterol content in an ounce of each item.

8. The calorie-nutrient information for an apple and an avocado is given in the table. How many of each should be eaten to get exactly 1000 calories and 100 grams of carbohydrates?

	One Apple	One Avocado
Calories	100	350
Carbohydrates (grams)	24	14

9. The weekly demand and supply models for the video *Titanic* at a chain of stores that sells videos are given by the demand model $N = -60p + 1000$ and the supply model $N = 4p + 200$, in which p is the price of the video and N is the number of videos sold or supplied each week to the chain of stores. Find the price at which supply and demand are equal. At this price, how many copies of *Titanic* can be supplied and sold each week?

8.2

Solve each system in Exercises 10–11.

10. $2x - y + z = 1$
$3x - 3y + 4z = 5$
$4x - 2y + 3z = 4$

11. $x + 2y - z = 5$
$2x - y + 3x = 0$
$2y + z = 1$

12. Find the quadratic function $y = ax^2 + bx + c$ whose graph passes through the points $(1, 4)$, $(3, 20)$, and $(-2, 25)$.

13. The graph shows a low savings rate in the United States compared to that of many industrialized countries. The combined rate for Japan, Germany, and France is 45%. The savings rate in Japan exceeds that for Germany by 1% and is 12% less than twice that for France. Find the savings rates for Japan, Germany, and France.

Comparitive Savings Rates

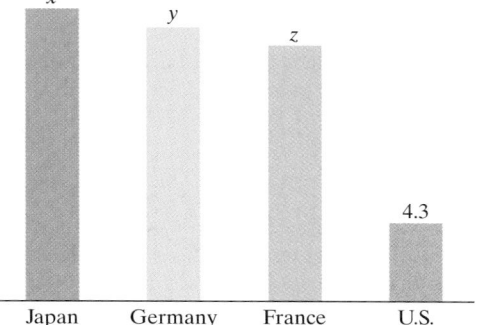

Source: Office of Management and Budget

14. Describe how to obtain a model for the millions of Americans living in poverty by using the ordered pairs for 1994, 1996, and 1997 in the graph.

Americans Living in Poverty

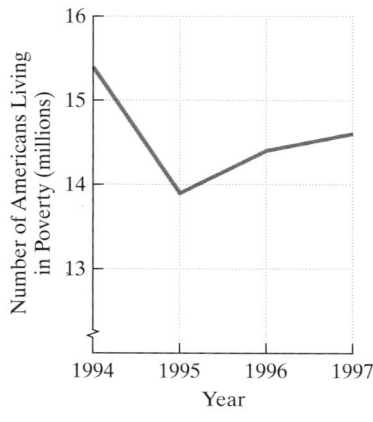

Source: U.S. Census Bureau

8.3

In Exercises 15–23, write the partial fraction decomposition of each rational expression.

15. $\dfrac{x}{(x - 3)(x + 2)}$

16. $\dfrac{11x - 2}{x^2 - x - 12}$

17. $\dfrac{4x^2 - 3x - 4}{x(x + 2)(x - 1)}$

18. $\dfrac{2x + 1}{(x - 2)^2}$

19. $\dfrac{2x - 6}{(x - 1)(x - 2)^2}$

20. $\dfrac{3x}{(x - 2)(x^2 + 1)}$

21. $\dfrac{7x^2 - 7x + 23}{(x - 3)(x^2 + 4)}$

22. $\dfrac{x^3}{(x^2 + 4)^2}$

23. $\dfrac{4x^3 + 5x^2 + 7x - 1}{(x^2 + x + 1)^2}$

8.4

In Exercises 24–34, solve each system by the method of your choice.

24. $5y = x^2 - 1$
$x - y = 1$

25. $y = x^2 + 2x + 1$
$x + y = 1$

26. $x^2 + y^2 = 2$
$x + y = 0$

27. $2x^2 + y^2 = 24$
$x^2 + y^2 = 15$

28. $xy - 4 = 0$
$y - x = 0$

29. $y^2 = 4x$
$x - 2y + 3 = 0$

30. $x^2 + y^2 = 10$
$y = x + 2$

31. $xy = 1$
$y = 2x + 1$

32. $x + y + 1 = 0$
$x^2 + y^2 + 6y - x = -5$

33. $x^2 + y^2 = 13$
$x^2 - y = 7$

34. $2x^2 + 3y^2 = 21$
$3x^2 - 4y^2 = 23$

35. The perimeter of a rectangle is 26 meters, and its area is 40 square meters. Find its dimensions.

36. Find the coordinates of all points (x, y) that lie on the line whose equation is $2x + y = 8$, so that the area of the rectangle shown in the figure is 6 square units.

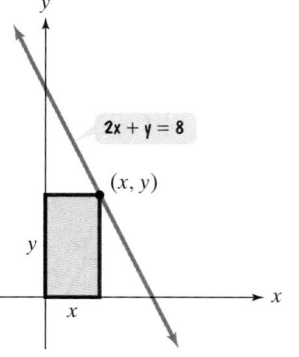

37. Two adjoining square fields with an area of 2900 square feet are to be enclosed with 240 feet of fencing. The situation is represented in the figure. Find the length of each side where a variable appears.

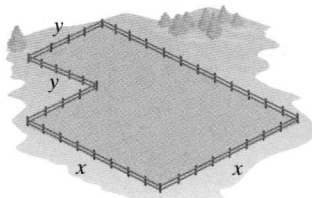

8.5

In Exercises 38–44, graph each inequality.

38. $3x - 4y > 12$

39. $y \le -\dfrac{1}{2}x + 2$

40. $x < -2$

41. $y \ge 3$

42. $x^2 + y^2 > 4$

43. $y \le x^2 - 1$

44. $y \le 2^x$

In Exercises 45–54, graph the solution set of each system of inequalities or indicate that the system has no solution.

45. $3x + 2y \ge 6$
$2x + y \ge 6$

46. $2x - y \ge 4$
$x + 2y < 2$

47. $y < x$
$y \le 2$

48. $y \le x$
$2x + 5y \le 10$

49. $0 \le x \le 3$
$y > 2$

50. $2x + y < 4$
$2x + y > 6$

51. $x^2 + y^2 \le 16$
$x + y < 2$

52. $x^2 + y^2 \le 9$
$y < -3x + 1$

53. $y > x^2$
$x + y < 6$
$y < x + 6$

54. $x \ge 0, y \ge 0$
$2x + 3y \le 12$
$3x + y \le 6$

8.6

55. Find the value of the objective function $z = 2x + 3y$ at each corner of the graphed region shown. What is the maximum value of the objective function? What is the minimum value of the objective function?

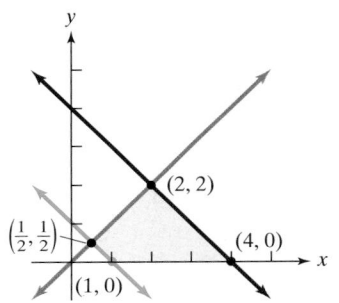

In Exercises 56–58, graph the region determined by the constraints. Then find the maximum value of the given objective function, subject to the constraints.

56. Objective Function $\quad z = 2x + 3y$
Constraints $\quad x \ge 0, y \ge 0,$
$x + y \le 8$
$3x + 2y \ge 6$

57. Objective Function $\quad z = x + 4y$
Contraints $\quad 0 \le x \le 5, 0 \le y \le 7$
$x + y \ge 3$

58. Objective Function $\quad z = 5x + 6y$
Constraints $\quad x \ge 0, y \ge 0$
$y \le x$
$2x + y \le 12$
$2x + 3y \ge 6$

59. A paper manufacturing company converts wood pulp to writing paper and newsprint. The profit on a unit of writing paper is $500 and the profit on a unit of newsprint is $350.

a. Let x represent the number of units of writing paper produced daily. Let y represent the number of units of newsprint produced daily. Write the objective function that models total daily profit.

b. The manufacturer is bound by the following constraints:

1. Equipment in the factory allows for making at most 200 units of paper (writing paper and newsprint) in a day.

2. Regular customers require at least 10 units of writing paper and at least 80 units of newsprint daily.

Write a system of inequalities that models these constraints.

c. Graph the inequalities in part (b). Use only the first quadrant, because x and y must both be positive. (*Suggestion:* Let each unit along the x- and y-axes represent 20.)

d. Evaluate the objective profit function at each of the three vertices of the graphed region.

e. Complete the missing portions of this statement: The company will make the greatest profit by producing ___ units of writing paper and ___ units of newsprint each day. The maximum daily profit is $ ___.

60. A manufacturer of lightweight tents makes two models whose specifications are given in the following table.

	Cutting Time per Tent	Assembly Time per Tent
Model A	0.9 hour	0.8 hour
Model B	1.8 hours	1.2 hours

On a monthly basis, the manufacturer has no more than 864 hours of labor available in the cutting department and at most 672 hours in the assembly division. The profits come to $25 per tent for model A and $40 per tent for model B. How many of each should be manufactured monthly to maximize the profit?

Chapter 8 Test

In Exercises 1–5, solve the system.

1. $x = y + 4$
$3x + 7y = -18$

2. $2x + 5y = -2$
$3x - 4y = 20$

3. $x + y + z = 6$
$3x + 4y - 7z = 1$
$2x - y + 3z = 5$

4. $x^2 + y^2 = 25$
$x + y = 1$

5. $2x^2 - 5y^2 = -2$
$3x^2 + 2y^2 = 35$

6. Find the partial fraction decomposition for
$$\dfrac{x}{(x + 1)(x^2 + 9)}.$$

In Exercises 7–10, graph the solution set of each inequality or system of inequalities.

7. $x - 2y < 8$

8. $x \geq 0, y \geq 0$
$3x + y \leq 9$
$2x + 3y \geq 6$

9. $x^2 + y^2 > 1$
$x^2 + y^2 < 4$

10. $y \leq 1 - x^2$
$x^2 + y^2 \leq 9$

11. Find the maximum value of the objective function $z = 3x + 5y$ subject to the following constraints: $x \geq 0$, $y \geq 0, x + y \leq 6, x \geq 2$.

12. A theater sells all orchestra seats at one price and all mezzanine seats at another price. One person purchased 4 orchestra tickets and 3 mezzanine tickets for a total of $134.00. A second person purchased 5 orchestra tickets and 2 mezzanine tickets for $143.00. What is the price of one orchestra ticket and one mezzanine ticket?

13. The demand and supply models for a product are given, respectively, by $N = 1000 - 20p$ and $N = 250 + 5p$. At what price will supply equal demand? At that price, how many units of the product will be supplied and sold?

14. Find the quadratic function $y = ax^2 + bx + c$ whose graph passes through the points $(-1, -2), (2, 1)$, and $(-2, 1)$.

15. The rectangular plot of land shown in the figure is to be fenced along three sides using 39 feet of fencing. No fencing is to be placed along the river's edge. The area of the plot is 180 square feet. What are its dimensions?

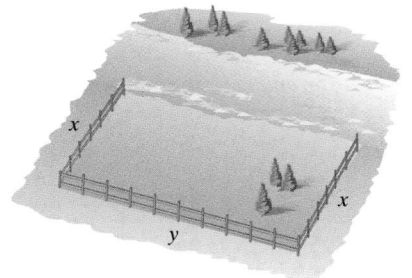

16. A manufacturer makes two types of jet skis, regular and deluxe. The profit on a regular jet ski is $200 and the profit on the deluxe model is $250. To meet customer demand, the company must manufacture at least 50 regular jet skis per week and at least 75 deluxe models. To maintain high quality, the total number of both models of jet skis manufactured by the company should not exceed 150 per week. How many jet skis of each type should be manufactured per week to obtain maximum profit? What is the maximum weekly profit?

Cumulative Review Exercises (Chapters 1–8)

Solve each equation or inequality in Exercises 1–8.

1. $\sqrt{x^2 - 3x} = 2x - 6$

2. $4x^2 = 8x - 7$

3. $\left|\dfrac{x}{3} + 2\right| < 4$

4. $\dfrac{x + 5}{x - 1} > 2$

5. $2x^3 + x^2 - 13x + 6 = 0$

6. $6x - 3(5x + 2) = 4(1 - x)$

7. $\log(x + 3) + \log x = 1$

8. $3^{x+2} = 11$

In Exercises 9–12, graph each equation, function, or inequality in the rectangular coordinate system.

9. $f(x) = (x + 2)^2 - 4$

10. $2x - 3y \leq 6$

11. $y = 3^{x-2}$

12. $f(x) = \dfrac{x^2 - x - 6}{x + 1}$

13. Expand and simplify: $\log_2(8x^5)$.

14. What interest rate is required for an investment of $6000 subject to continuous compounding to grow to $18,000 in 10 years?

15. If $f(x) = 7x - 3$, find $f^{-1}(x)$.

16. If $f(x) = 7x - 3$ and $g(x) = 3x - 7$, find $g(f(x))$.

17. Explain why $x^2 + y^2 = 4$ does not represent y as a function of x.

18. Solve the system:
$$3x - y = -2$$
$$2x^2 - y = 0.$$

19. The length of a rectangle is 1 meter more than twice the width. If the rectangle's area is 36 square meters, find its dimensions.

20. The function $f(x) = 0.1x^2 - 3x + 22$ describes the distance, $f(x)$, in feet, needed for an airplane to land when its initial landing speed is x feet per second. Find and interpret $f(90)$. Will there be a problem if 550 feet of runway is available? Explain.

In Exercises 21–22, verify each identity.

21. $\sec \theta - \cos \theta = \tan \theta \sin \theta$

22. $\tan x + \tan y = \dfrac{\sin(x + y)}{\cos x \cos y}$

In Exercises 23–24, solve each equation.

23. $\sin \theta = \tan \theta, \quad 0 \leq \theta < 2\pi$

24. $2 + \cos 2\theta = 3 \cos \theta, \quad 0 \leq \theta < 2\pi$

25. In oblique triangle ABC, $A = 12°$, $B = 75°$, and $a = 20$. Find b to the nearest tenth.

Matrices and Determinants

Jaron Lanier, who first used the term "virtual reality," is chief scientist for the "tele-immersion" project, which explores the impact of massive bandwidth and computing power. Rectangular arrays of numbers, called *matrices*, play a central role in representing computer images and in the forthcoming technology of tele-immersion. In this chapter, we study matrices and their applications. We begin with solving linear systems using matrices, which leads to a discussion of how computers might unjam traffic and give us a gridlock-free future.

You are being drawn deeper into cyberspace, spending more time online each week. With constantly improving high-resolution images, cyberspace is reshaping your life by nourishing shared enthusiasms. The people who built your computer talk of "bandwidth out the wazoo" that will give you the visual experience, in high-definition 3-D format, of being in the same room with a person who is actually in another city.

SECTION 9.1 *Matrix Solutions to Linear Systems*

Objectives

1. Write the augmented matrix for a linear system.
2. Perform matrix row operations.
3. Use matrices and Gaussian elimination to solve systems.
4. Use matrices and Gauss-Jordan elimination to solve systems.

Yes, we overindulged, but it was delicious. Anyway, a few hours of moderate activity and we'll just burn off those extra calories. The following chart should help. We see that the number of calories burned per hour depends on our weight. Four hours of tennis and we'll be as good as new!

How Fast You Burn Off Calories

	Weight (pounds)					
	110	132	154	176	187	209
Activity	Calories Burned per Hour					
Housework	175	210	245	285	300	320
Cycling	190	215	245	270	280	295
Tennis	335	380	425	470	495	520
Watching TV	60	70	80	85	90	95

The 24 numbers inside the red brackets are arranged in four rows and six columns. This rectangular array of 24 numbers, arranged in rows and columns and placed in brackets, is an example of a **matrix** (plural: **matrices**). The numbers inside the brackets are called **elements** of the matrix. Matrices are used to display information and to solve systems of linear equations. Because systems involving two equations in two variables can easily be solved by substitution or addition, we will focus on matrix solutions to linear systems in three or more variables.

Solving Linear Systems by Using Matrices

1 Write the augmented matrix for a linear system.

A matrix gives us a shortened way of writing a system of equations. The first step in solving a system of linear equations using matrices is to write the augmented matrix. An **augmented matrix** has a vertical bar separating the columns of the matrix into two groups. The coefficients of each variable are placed to the left of the vertical line, and the constants are placed to the right. If any variable is missing, its coefficient is 0. Here are two examples.

System of Linear Equations	**Augmented Matrix**

$$\begin{aligned} 3x + y + 2z &= 31 \\ x + y + 2z &= 19 \\ x + 3y + 2z &= 25 \end{aligned} \qquad \left[\begin{array}{ccc|c} 3 & 1 & 2 & 31 \\ 1 & 1 & 2 & 19 \\ 1 & 3 & 2 & 25 \end{array}\right]$$

$$\begin{aligned} x + 2y - 5z &= -19 \\ y + 3z &= 9 \\ z &= 4 \end{aligned} \qquad \left[\begin{array}{ccc|c} 1 & 2 & -5 & -19 \\ 0 & 1 & 3 & 9 \\ 0 & 0 & 1 & 4 \end{array}\right]$$

Notice how the second matrix contains 1s down the diagonal from upper left to lower right and 0s below the 1s. This arrangement makes it easy to find the solution of the system of equations, as Example 1 shows.

EXAMPLE 1 Solving a System Using a Matrix

Write the solution set for a system of equations represented by the matrix

$$\left[\begin{array}{ccc|c} 1 & 2 & -5 & -19 \\ 0 & 1 & 3 & 9 \\ 0 & 0 & 1 & 4 \end{array}\right].$$

Solution The system represented by the given matrix is

$$\left[\begin{array}{ccc|c} 1 & 2 & -5 & -19 \\ 0 & 1 & 3 & 9 \\ 0 & 0 & 1 & 4 \end{array}\right] \rightarrow \begin{aligned} 1x + 2y - 5z &= -19 \\ 0x + 1y + 3z &= 9 \\ 0x + 0y + 1z &= 4 \end{aligned}.$$

This system can be simplified as follows.

$$\begin{aligned} x + 2y - 5z &= -19 \qquad \text{Equation 1} \\ y + 3z &= 9 \qquad \text{Equation 2} \\ z &= 4 \qquad \text{Equation 3} \end{aligned}$$

The value of z is known. We can find y by back-substitution.

$$\begin{aligned} y + 3z &= 9 \qquad \text{Equation 2} \\ y + 3(4) &= 9 \qquad \text{Substitute 4 for } z. \\ y + 12 &= 9 \qquad \text{Multiply.} \\ y &= -3 \qquad \text{Subtract 12 from both sides.} \end{aligned}$$

With values for y and z, we can now use back-substitution to find x.

$$\begin{aligned} x + 2y - 5z &= -19 \qquad \text{Equation 1} \\ x + 2(-3) - 5(4) &= -19 \qquad \text{Substitute } -3 \text{ for } y \text{ and 4 for } z. \\ x - 6 - 20 &= -19 \qquad \text{Multiply.} \\ x - 26 &= -19 \qquad \text{Add.} \\ x &= 7 \qquad \text{Add 26 to both sides.} \end{aligned}$$

We see that $x = 7$, $y = -3$, and $z = 4$. The solution set for the system is $\{(7, -3, 4)\}$.

EXAMPLE 3   Gaussian Elimination with Back-Substitution

Use matrices to solve the system

$$3x + y + 2z = 31$$
$$x + y + 2z = 19$$
$$x + 3y + 2z = 25.$$

Solution

Step 1   Write the augmented matrix for the system.

Linear System	Augmented Matrix
$3x + y + 2z = 31$	$\begin{bmatrix} 3 & 1 & 2 & \vline & 31 \\ 1 & 1 & 2 & \vline & 19 \\ 1 & 3 & 2 & \vline & 25 \end{bmatrix}$
$x + y + 2z = 19$	
$x + 3y + 2z = 25$	

Step 2   Use matrix row operations to simplify the matrix to one with 1s down the diagonal from upper left to lower right, and 0s below the 1s. Our goal is to obtain a matrix of the form

$$\begin{bmatrix} 1 & a & b & \vline & c \\ 0 & 1 & d & \vline & e \\ 0 & 0 & 1 & \vline & f \end{bmatrix}.$$

Our first step in achieving this goal is to get 1 in the top position of the first column.

We want 1 in this position.
$$\begin{bmatrix} 3 & 1 & 2 & \vline & 31 \\ 1 & 1 & 2 & \vline & 19 \\ 1 & 3 & 2 & \vline & 25 \end{bmatrix}$$

To get 1 in this position, we interchange rows 1 and 2. (We could also interchange rows 1 and 3 to attain our goal.)

$$\begin{bmatrix} 1 & 1 & 2 & \vline & 19 \\ 3 & 1 & 2 & \vline & 31 \\ 1 & 3 & 2 & \vline & 25 \end{bmatrix}$$

This was row 2; now it's row 1.
This was row 1; now it's row 2.

Now we want to get 0s below the 1 in the first column.

We want 0 in these positions.
$$\begin{bmatrix} 1 & 1 & 2 & \vline & 19 \\ 3 & 1 & 2 & \vline & 31 \\ 1 & 3 & 2 & \vline & 25 \end{bmatrix}$$

Let's first get a 0 where there is now a 3. If we multiply the top row of numbers by -3 and add these products to the second row of numbers, we will get 0 in this position. The top row of numbers multiplied by -3 gives

$$-3(1) \text{ or } -3, \quad -3(1) \text{ or } -3, \quad -3(2) \text{ or } -6, \quad -3(19) \text{ or } -57.$$

Now add these products to the corresponding numbers in row 2. Notice that although we use row 1 to find the products, row 1 does not change.

$$\begin{bmatrix} 1 & 1 & 2 & \vline & 19 \\ 3 + (-3) & 1 + (-3) & 2 + (-6) & \vline & 31 + (-57) \\ 1 & 3 & 2 & \vline & 25 \end{bmatrix} = \begin{bmatrix} 1 & 1 & 2 & \vline & 19 \\ 0 & -2 & -4 & \vline & -26 \\ 1 & 3 & 2 & \vline & 25 \end{bmatrix}$$

We want 0 in this position.

We are not yet done with the first column. The voice balloon shows that we want to get another 0 in this column. If we multiply the top row of numbers by −1 and add these products to the third row of numbers, we will get 0 in this position. The top row of numbers multiplied by −1 gives

$$-1(1) \text{ or } -1, \qquad -1(1) \text{ or } -1, \qquad -1(2) \text{ or } -2, \qquad -1(19) \text{ or } -19.$$

Now add these products to the corresponding numbers in row 3.

$$\begin{bmatrix} 1 & 1 & 2 & | & 19 \\ 0 & -2 & -4 & | & -26 \\ 1+(-1)=0 & 3+(-1)=2 & 2+(-2)=0 & | & 25+(-19)=6 \end{bmatrix}$$

$$= \begin{bmatrix} 1 & 1 & 2 & | & 19 \\ 0 & -2 & -4 & | & -26 \\ 0 & 2 & 0 & | & 6 \end{bmatrix}$$

We move on to the second column. We want 1 in the second row, second column.

We want 1 in this position.
$$\begin{bmatrix} 1 & 1 & 2 & | & 19 \\ 0 & -2 & -4 & | & -26 \\ 0 & 2 & 0 & | & 6 \end{bmatrix}$$

To get 1 in the desired position, we multiply −2 by its reciprocal, $-\frac{1}{2}$. Therefore, we multiply all the numbers in the second row by $-\frac{1}{2}$ to get

$$\begin{bmatrix} 1 & 1 & 2 & | & 19 \\ -\frac{1}{2}(0) & -\frac{1}{2}(-2) & -\frac{1}{2}(-4) & | & -\frac{1}{2}(-26) \\ 0 & 2 & 0 & | & 6 \end{bmatrix} = \begin{bmatrix} 1 & 1 & 2 & | & 19 \\ 0 & 1 & 2 & | & 13 \\ 0 & 2 & 0 & | & 6 \end{bmatrix}.$$

We want 0 in this position.

We are not yet done with the second column. The voice balloon shows that we want to get a 0 where there is now a 2. If we multiply the second row of numbers by −2 and add these products to the third row of numbers, we will get 0 in this position. The second row of numbers multiplied by −2 gives

$$-2(0) \text{ or } 0, \qquad -2(1) \text{ or } -2, \qquad -2(2) \text{ or } -4, \qquad -2(13) \text{ or } -26.$$

Now add these products to the corresponding numbers in row 3.

$$\begin{bmatrix} 1 & 1 & 2 & | & 19 \\ 0 & 1 & 2 & | & 13 \\ 0+0 & 2+(-2) & 0+(-4) & | & 6+(-26) \end{bmatrix} = \begin{bmatrix} 1 & 1 & 2 & | & 19 \\ 0 & 1 & 2 & | & 13 \\ 0 & 0 & -4 & | & -20 \end{bmatrix}$$

We move on to the third column. We want 1 in the third row, third column.

We want 1 in this position.
$$\begin{bmatrix} 1 & 1 & 2 & | & 19 \\ 0 & 1 & 2 & | & 13 \\ 0 & 0 & -4 & | & -20 \end{bmatrix}$$

To get 1 in the desired position, we multiply −4 by its reciprocal, $-\frac{1}{4}$. Therefore, we multiply all the numbers in the third row by $-\frac{1}{4}$ to get

$$\begin{bmatrix} 1 & 1 & 2 & | & 19 \\ 0 & 1 & 2 & | & 13 \\ -\frac{1}{4}(0) & -\frac{1}{4}(0) & -\frac{1}{4}(-4) & | & -\frac{1}{4}(-20) \end{bmatrix} = \begin{bmatrix} 1 & 1 & 2 & | & 19 \\ 0 & 1 & 2 & | & 13 \\ 0 & 0 & 1 & | & 5 \end{bmatrix}.$$

We now have the desired matrix with 1s down the diagonal and 0s below the 1s.

Step 3 **Write the system of linear equations corresponding to the matrix in step 2, and use back-substitution to find the system's solution.** The system represented by the matrix in step 2 is

$$\begin{bmatrix} 1 & 1 & 2 & | & 19 \\ 0 & 1 & 2 & | & 13 \\ 0 & 0 & 1 & | & 5 \end{bmatrix} \rightarrow \begin{matrix} 1x + 1y + 2z = 19 \\ 0x + 1y + 2z = 13 \\ 0x + 0y + 1z = 5 \end{matrix} \quad \text{or} \quad \begin{matrix} x + y + 2z = 19 \\ y + 2z = 13. \\ z = 5 \end{matrix}$$

We immediately see that the value for z is 5. To find y, we back-substitute 5 for z in the second equation.

$$y + 2z = 13 \quad \text{Equation 2}$$
$$y + 2(5) = 13 \quad \text{Substitute 5 for z.}$$
$$y = 3 \quad \text{Solve for y.}$$

Finally, back-substitute 3 for y and 5 for z in the first equation:

$$x + y + 2z = 19 \quad \text{Equation 1}$$
$$x + 3 + 2(5) = 19 \quad \text{Substitute 3 for y and 5 for z.}$$
$$x + 13 = 19 \quad \text{Multiply and add.}$$
$$x = 6 \quad \text{Subtract 13 both sides.}$$

The solution set for the original system is $\{(6, 3, 5)\}$.

Check Point 3   Use matrices to solve the system

$$\begin{matrix} 2x + y + 2z = 18 \\ x - y + 2z = 9 \\ x + 2y - z = 6. \end{matrix}$$

Modern supercomputers are capable of solving systems with more than 600,000 variables. The augmented matrices for such systems are huge, but the solution using matrices is exactly like what we did in Example 2. Work with the augmented matrix, one column at a time. First, get 1 in the desired position. Then get 0s below the 1. Let's see how this works for a linear system involving four equations in four variables.

EXAMPLE 4   **Gaussian Elimination with Back-Substitution**

Use matrices to solve the system

$$\begin{matrix} 2x + y + 3z - w = 6 \\ x - y + 2z - 2w = -1 \\ x - y - z + w = -4 \\ -x + 2y - 2z - w = -7. \end{matrix}$$

Solution

Step 1   Write the augmented matrix for the system.

Linear System

$$
\begin{aligned}
2x + \;\; y + 3z - \;\; w &= 6 \\
x - \;\; y + 2z - 2w &= -1 \\
x - \;\; y - \;\; z + \;\; w &= -4 \\
-x + 2y - 2z - \;\; w &= -7
\end{aligned}
$$

Augmented Matrix

$$
\left[\begin{array}{cccc|c}
2 & 1 & 3 & -1 & 6 \\
1 & -1 & 2 & -2 & -1 \\
1 & -1 & -1 & 1 & -4 \\
-1 & 2 & -2 & -1 & -7
\end{array}\right]
$$

Step 2   Use matrix row operations to simplify the matrix to one with 1s down the diagonal from upper left to lower right, and 0s below the 1s. Our first step in achieving this goal is to get 1 in the top position of the first column. To do this, we interchange rows 1 and 2.

We want 0s in these positions.

$$
\left[\begin{array}{cccc|c}
1 & -1 & 2 & -2 & -1 \\
2 & 1 & 3 & -1 & 6 \\
1 & -1 & -1 & 1 & -4 \\
-1 & 2 & -2 & -1 & -7
\end{array}\right]
$$

This was row 2; now it's row 1.

This was row 1; now it's row 2.

Now we want 0s below the 1 in the first column. To get the first 0, multiply the top row of numbers by -2 and add these products to the second row of numbers. To get the second 0, multiply the top row of numbers by -1 and add these products to the third row of numbers. To get the third 0, multiply the top row of numbers by 1 and add these products to the fourth row of numbers. (Equivalently, add corresponding numbers in rows 1 and 4.) Performing these operations, we obtain the following matrix.

We want 1 in this position.

$$
\left[\begin{array}{cccc|c}
1 & -1 & 2 & -2 & -1 \\
0 & 3 & -1 & 3 & 8 \\
0 & 0 & -3 & 3 & -3 \\
0 & 1 & 0 & -3 & -8
\end{array}\right]
$$

Use the previous matrix and:

Replace row 2 by $-2R_1 + R_2$.

Replace row 3 by $-1R_1 + R_3$.

Replace row 4 by $1R_1 + R_4$.

We move on to the second column. We can obtain 1 in the desired position by multiplying the numbers in the second row by $\frac{1}{3}$, the reciprocal of 3.

$$
\left[\begin{array}{cccc|c}
1 & -1 & 2 & -2 & -1 \\
\frac{1}{3}(0) & \frac{1}{3}(3) & \frac{1}{3}(-1) & \frac{1}{3}(3) & \frac{1}{3}(8) \\
0 & 0 & -3 & 3 & -3 \\
0 & 1 & 0 & -3 & -8
\end{array}\right]
=
\left[\begin{array}{cccc|c}
1 & -1 & 2 & -2 & -1 \\
0 & 1 & -\frac{1}{3} & 1 & \frac{8}{3} \\
0 & 0 & -3 & 3 & -3 \\
0 & 1 & 0 & -3 & -8
\end{array}\right]
\quad \frac{1}{3}R_2
$$

We want 0s in these positions. The top position already has a 0.

Now we want 0s below the 1 in the second column. The top position already has a 0. To obtain a 0 on the bottom, we multiply the second row by -1 and add the product to the corresponding numbers of the last row. (What would happen if we added rows 1 and 4?) Performing these operations, we obtain the following matrix.

$$\begin{bmatrix} 1 & -1 & 2 & -2 & | & -1 \\ 0 & 1 & -\frac{1}{3} & 1 & | & \frac{8}{3} \\ 0 & 0 & -3 & 3 & | & -3 \\ 0 & 0 & \frac{1}{3} & -4 & | & -\frac{32}{3} \end{bmatrix}$$

We want 1 in this position. (left)

Replace row 4 in the previous matrix by $-1R_2 + R_4$. (right)

We move on to the third column. We can obtain 1 in the desired position by multiplying the numbers in the third row by $-\frac{1}{3}$, the reciprocal of -3.

$$\begin{bmatrix} 1 & -1 & 2 & -2 & | & -1 \\ 0 & 1 & -\frac{1}{3} & 1 & | & \frac{8}{3} \\ -\frac{1}{3}(0) & -\frac{1}{3}(0) & -\frac{1}{3}(-3) & -\frac{1}{3}(3) & | & -\frac{1}{3}(-3) \\ 0 & 0 & \frac{1}{3} & -4 & | & -\frac{32}{3} \end{bmatrix} = \begin{bmatrix} 1 & -1 & 2 & -2 & | & -1 \\ 0 & 1 & -\frac{1}{3} & 1 & | & \frac{8}{3} \\ 0 & 0 & 1 & -1 & | & 1 \\ 0 & 0 & \frac{1}{3} & -4 & | & -\frac{32}{3} \end{bmatrix} \quad -\frac{1}{3}R_3$$

We want 0 in this position.

Now we want 0 below the 1 in the third column. If we multiply the third row of numbers by $-\frac{1}{3}$ and add these products to the fourth row of numbers, we will get 0 in this position. Performing these operations, we obtain the following matrix.

$$\begin{bmatrix} 1 & -1 & 2 & -2 & | & -1 \\ 0 & 1 & -\frac{1}{3} & 1 & | & \frac{8}{3} \\ 0 & 0 & 1 & -1 & | & 1 \\ 0 & 0 & 0 & -\frac{11}{3} & | & -11 \end{bmatrix}$$

We want 1 in this position. (left)

Replace row 4 in the previous matrix by $-\frac{1}{3}R_3 + R_4$. (right)

We move on to the fourth column. Because we want 1s down the main diagonal, we want 1 where there is now $-\frac{11}{3}$. We can obtain 1 in this position by multiplying the numbers in the fourth row by $-\frac{3}{11}$.

$$\begin{bmatrix} 1 & -1 & 2 & -2 & | & -1 \\ 0 & 1 & -\frac{1}{3} & 1 & | & \frac{8}{3} \\ 0 & 0 & 1 & -1 & | & 1 \\ -\frac{3}{11}(0) & -\frac{3}{11}(0) & -\frac{3}{11}(0) & -\frac{3}{11}\left(-\frac{11}{3}\right) & | & -\frac{3}{11}(-11) \end{bmatrix}$$

$$= \begin{bmatrix} 1 & -1 & 2 & -2 & | & -1 \\ 0 & 1 & -\frac{1}{3} & 1 & | & \frac{8}{3} \\ 0 & 0 & 1 & -1 & | & 1 \\ 0 & 0 & 0 & 1 & | & 3 \end{bmatrix} \quad -\frac{3}{11}R_4$$

We now have the desired matrix with 1s down the diagonal and 0s below the 1s.

Step 3 **Write the system of linear equations corresponding to the matrix in step 2, and use back-substitution to find the system's solution.** The system represented by the matrix in step 2 is

$$\begin{bmatrix} 1 & -1 & 2 & -2 & | & -1 \\ 0 & 1 & -\frac{1}{3} & 1 & | & \frac{8}{3} \\ 0 & 0 & 1 & -1 & | & 1 \\ 0 & 0 & 0 & 1 & | & 3 \end{bmatrix} \rightarrow \begin{array}{l} 1x - 1y + 2z - 2w = -1 \\ 0x + 1y - \frac{1}{3}z + 1w = \frac{8}{3} \\ 0x + 0y + 1z - 1w = 1 \\ 0x + 0y + 0z + 1w = 3 \end{array} \quad \text{or} \quad \begin{array}{l} x - y + 2z - 2w = -1 \\ y - \frac{1}{3}z + w = \frac{8}{3} \\ z - w = 1 \\ w = 3 \end{array}$$

We immediately see that the value for w is 3. We can now use back-substitution to find the values for z, y, and x.

$$
\begin{array}{c|c|c|c}
w = 3 & z - w = 1 & y - \dfrac{1}{3}z + w = \dfrac{8}{3} & x - y + 2z - 2w = -1 \\[2mm]
& z - 3 = 1 & y - \dfrac{1}{3}(4) + 3 = \dfrac{8}{3} & x - 1 + 2(4) - 2(3) = -1 \\[2mm]
& z = 4 & y + \dfrac{5}{3} = \dfrac{8}{3} & x - 1 + 8 - 6 = -1 \\[2mm]
& & y = 1 & x + 1 = -1 \\[2mm]
& & & x = -2
\end{array}
$$

Let's agree to write the solution set for the system in the order in which the variables for the given system appeared, from left to right, namely (x, y, z, w). Thus, the solution set is $\{(-2, 1, 4, 3)\}$. We can verify this solution set by substituting the value for each variable into the original system of equations.

Check Point 4   Use matrices to solve the system

$$
\begin{aligned}
x - 3y - 2z + w &= -3 \\
2x - 7y - z + 2w &= 1 \\
3x - 7y - 3z + 3w &= -5 \\
5x + y + 4z - 2w &= 18.
\end{aligned}
$$

4 Use matrices and Gauss-Jordan elimination to solve systems.

Gauss-Jordan Elimination

Using Gaussian elimination, we obtain a matrix with 1s down the main diagonal and 0s below the 1s. A second method, called **Gauss-Jordan elimination**, after Carl Friedrich Gauss and Wilhelm Jordan (1842–1899), continues the process until a matrix with 1s down the main diagonal from left to right and 0s in every position *above and below* each 1 is found. For a system of linear equations in three variables, x, y, and z, we try to get the augmented matrix into the form

$$
\left[\begin{array}{ccc|c}
1 & 0 & 0 & a \\
0 & 1 & 0 & b \\
0 & 0 & 1 & c
\end{array}\right].
$$

Based on this matrix, we conclude that $x = a$, $y = b$, and $z = c$.

EXAMPLE 5   Using Gauss-Jordan Elimination

Use Gauss-Jordan elimination to solve the system

$$
\begin{aligned}
3x + y + 2z &= 31 \\
x + y + 2z &= 19 \\
x + 3y + 2z &= 25.
\end{aligned}
$$

Solution   In Example 3, we used Gaussian elimination to obtain the following matrix:

$$
\left[\begin{array}{ccc|c}
1 & 1 & 2 & 19 \\
0 & 1 & 2 & 13 \\
0 & 0 & 1 & 5
\end{array}\right].
$$

Study Tip

The advantage to Gauss-Jordan elimination is that from the augmented matrix we can simply read the solution. The disadvantage is that we must continue row operations in the augmented matrix from the Gaussian elimination process, and it's fairly easy to make computational errors.

To find a solution using Gauss-Jordan elimination, we need to work with this matrix and convert the boxed numbers to 0s. Thus, we will apply matrix row operations to get 0s *above the 1s* in the main diagonal. To get 0 in row 1, column 2 (where there is now a 1), we multiply each number in the second row by -1 and add these products to the corresponding numbers in the first row. Performing these operations, we obtain the following matrix.

We want 0s in these positions. The top position already has a 0.

$$\begin{bmatrix} 1 & 0 & 0 & | & 6 \\ 0 & 1 & 2 & | & 13 \\ 0 & 0 & 1 & | & 5 \end{bmatrix}$$

We want 0s above the 1 in the third column. The top position already has a 0. To obtain a 0 where there is now 2, we multiply each number in the bottom row by -2 and add these products to the corresponding numbers in the second row. Performing these operations, we obtain the following matrix.

$$\begin{bmatrix} 1 & 0 & 0 & | & 6 \\ 0 & 1 & 0 & | & 3 \\ 0 & 0 & 1 & | & 5 \end{bmatrix}$$

Replace row 2 in the previous matrix by $-2R_3 + R_2$.

This last matrix corresponds to

$$x = 6, \quad y = 3, \quad z = 5.$$

As we found in Example 3, the solution set is $\{(6, 3, 5)\}$.

Check Point 5 Solve the system in Checkpoint 2 using Gauss-Jordan elimination. Begin by working with the matrix that you obtained in Checkpoint 2.

EXERCISE SET 9.1

Practice Exercises

In Exercises 1–8, write the augmented matrix for each system of linear equations.

1.
$2x + y + 2z = 2$
$3x - 5y - z = 4$
$x - 2y - 3z = -6$

2.
$3x - 2y + 5z = 31$
$x + 3y - 3z = -12$
$-2x - 5y + 3z = 11$

3.
$x - y + z = 8$
$y - 12z = -15$
$z = 1$

4.
$x - 2y + 3z = 9$
$y + 3z = 5$
$z = 2$

5.
$5x - 2y - 3z = 0$
$x + y = 5$
$2x - 3z = 4$

6.
$x - 2y + z = 10$
$3x + y = 5$
$7x + 2z = 2$

7.
$2x + 5y - 3z + w = 2$
$3y + z = 4$
$x - y + 5z = 9$
$5x - 5y - 2z = 1$

8.
$4x + 7y - 8z + w = 3$
$5y + z = 5$
$x - y - z = 17$
$2x - 2y + 11z = 4$

In Exercises 9–12, write the system of linear equations represented by the augmented matrix. Use x, y, z, and, if necessary, w for the variables.

9. $\begin{bmatrix} 5 & 0 & 3 & | & -11 \\ 0 & 1 & -4 & | & 12 \\ 7 & 2 & 0 & | & 3 \end{bmatrix}$

10. $\begin{bmatrix} 7 & 0 & 4 & | & -13 \\ 0 & 1 & -5 & | & 11 \\ 2 & 7 & 0 & | & 6 \end{bmatrix}$

11. $\begin{bmatrix} 1 & 1 & 4 & 1 & | & 3 \\ -1 & 1 & -1 & 0 & | & 7 \\ 2 & 0 & 0 & 5 & | & 11 \\ 0 & 0 & 12 & 4 & | & 5 \end{bmatrix}$

12. $\begin{bmatrix} 4 & 1 & 5 & 1 & | & 6 \\ 1 & -1 & 0 & -1 & | & 8 \\ 3 & 0 & 0 & 7 & | & 4 \\ 0 & 0 & 11 & 5 & | & 3 \end{bmatrix}$

In Exercises 13–18, write the system of linear equations represented by the augmented matrix. Use x, y, z, and, if necessary, w for the variables. Once the system is written, use back-substitution to find its solution.

13. $\begin{bmatrix} 1 & 0 & -4 & | & 5 \\ 0 & 1 & -12 & | & 13 \\ 0 & 0 & 1 & | & -\frac{1}{2} \end{bmatrix}$

14. $\begin{bmatrix} 1 & 2 & 1 & | & 0 \\ 0 & 1 & 0 & | & -2 \\ 0 & 0 & 1 & | & 3 \end{bmatrix}$

15. $\begin{bmatrix} 1 & \frac{1}{2} & 1 & | & \frac{11}{2} \\ 0 & 1 & \frac{3}{2} & | & 7 \\ 0 & 0 & 1 & | & 4 \end{bmatrix}$

16. $\begin{bmatrix} 1 & 1 & 0 & | & 3 \\ 0 & 1 & \frac{3}{2} & | & -2 \\ 0 & 0 & 1 & | & 0 \end{bmatrix}$

17. $\begin{bmatrix} 1 & -1 & 1 & 1 & | & 3 \\ 0 & 1 & -2 & -1 & | & 0 \\ 0 & 0 & 1 & 6 & | & 17 \\ 0 & 0 & 0 & 1 & | & 3 \end{bmatrix}$

18. $\begin{bmatrix} 1 & 2 & -1 & 0 & | & 2 \\ 0 & 1 & 1 & -2 & | & -3 \\ 0 & 0 & 1 & -1 & | & -2 \\ 0 & 0 & 0 & 1 & | & 3 \end{bmatrix}$

In Exercises 19–24, perform each matrix row operation and write the new matrix.

19. $\begin{bmatrix} 2 & -6 & 4 & | & 10 \\ 1 & 5 & -5 & | & 0 \\ 3 & 0 & 4 & | & 7 \end{bmatrix} \quad \frac{1}{2}R_1$

20. $\begin{bmatrix} 3 & -12 & 6 & | & 9 \\ 1 & -4 & 4 & | & 0 \\ 2 & 0 & 7 & | & 4 \end{bmatrix} \quad \frac{1}{3}R_1$

21. $\begin{bmatrix} 1 & -3 & 2 & | & 0 \\ 3 & 1 & -1 & | & 7 \\ 2 & -2 & 1 & | & 3 \end{bmatrix} \quad -3R_1 + R_2$

22. $\begin{bmatrix} 1 & -1 & 5 & | & -6 \\ 3 & 3 & -1 & | & 10 \\ 1 & 3 & 2 & | & 5 \end{bmatrix} \quad -3R_1 + R_2$

23. $\begin{bmatrix} 1 & -1 & 1 & 1 & | & 3 \\ 0 & 1 & -2 & -1 & | & 0 \\ 2 & 0 & 3 & 4 & | & 11 \\ 5 & 1 & 2 & 4 & | & 6 \end{bmatrix} \quad \begin{matrix} \\ \\ -2R_1 + R_3 \\ -5R_1 + R_4 \end{matrix}$

24. $\begin{bmatrix} 1 & -5 & 2 & -2 & | & 4 \\ 0 & 1 & -3 & -1 & | & 0 \\ 3 & 0 & 2 & -1 & | & 6 \\ -4 & 1 & 4 & 2 & | & -3 \end{bmatrix} \quad \begin{matrix} \\ \\ -3R_1 + R_3 \\ 4R_1 + R_4 \end{matrix}$

In Exercises 25–26, a few steps in the process of simplifying the given matrix to one with 1s down the diagonal from upper left to lower right, and 0s below the 1s, are shown. Fill in the missing numbers in the steps that are shown.

25. $\begin{bmatrix} 1 & -1 & 1 & | & 8 \\ 2 & 3 & -1 & | & -2 \\ 3 & -2 & -9 & | & 9 \end{bmatrix} \rightarrow \begin{bmatrix} 1 & -1 & 1 & | & 8 \\ 0 & 5 & \blacksquare & | & \blacksquare \\ 0 & 1 & \blacksquare & | & \blacksquare \end{bmatrix}$

$\rightarrow \begin{bmatrix} 1 & -1 & 1 & | & 8 \\ 0 & 1 & \blacksquare & | & \blacksquare \\ 0 & 1 & \blacksquare & | & \blacksquare \end{bmatrix}$

26. $\begin{bmatrix} 1 & -2 & 3 & | & 4 \\ 2 & 1 & -4 & | & 3 \\ -3 & 4 & -1 & | & -2 \end{bmatrix} \rightarrow \begin{bmatrix} 1 & -2 & 3 & | & 4 \\ 0 & 5 & \blacksquare & | & \blacksquare \\ 0 & -2 & \blacksquare & | & \blacksquare \end{bmatrix}$

$\rightarrow \begin{bmatrix} 1 & -2 & 3 & | & 4 \\ 0 & 1 & \blacksquare & | & \blacksquare \\ 0 & -2 & \blacksquare & | & \blacksquare \end{bmatrix}$

In Exercises 27–40, solve each system of equations using matrices. Use Gaussian elimination with back-substitution or Gauss-Jordan elimination.

27.
$x + y - z = -2$
$2x - y + z = 5$
$-x + 2y + 2z = 1$

28.
$x - 2y - z = 2$
$2x - y + z = 4$
$-x + y - 2z = -4$

29.
$x + 3y = 0$
$x + y + z = 1$
$3x - y - z = 11$

30.
$3y - z = -1$
$x + 5y - z = -4$
$-3x + 6y + 2z = 11$

31. $2x + 2y + 7z = -1$
$2x + y + 2z = 2$
$4x + 6y + z = 15$

32. $3x + 2y + 3z = 3$
$4x - 5y + 7z = 1$
$2x + 3y - 2z = 6$

33.
$x + y + z + w = 4$
$2x + y - 2z - w = 0$
$x - 2y - z - 2w = -2$
$3x + 2y + z + 3w = 4$

34.
$x + y + z + w = 5$
$x + 2y - z - 2w = -1$
$x - 3y - 3z - w = -1$
$2x - y + 2z - w = -2$

35. $3x - 4y + z + w = 9$
$x + y - z - w = 0$
$2x + y + 4z - 2w = 3$
$-x + 2y + z - 3w = 3$

36. $2x + z - 3w = 8$
$x - y + 4w = -10$
$3x + 5y - z - w = 20$
$x + y - z - w = 6$

37. $2x + 3y - z - w = -3$
$2x - y - 3z + 2w = -5$
$x - y + z - w = -4$
$3x - 2y + z + w = 0$

38. $2x - y - z + w = 4$
$x + 3y - 2z - 3w = 6$
$x - y + z - w = 2$
$-x + 2y - z - w = -1$

39. $2x_1 - 2x_2 + 3x_3 - x_4 = 12$
$x_1 + 2x_2 - x_3 + 2x_4 - x_5 = -7$
$x_1 + x_3 - x_4 - 5x_5 = 5$
$-x_1 + x_2 - x_3 - 2x_4 - 3x_5 = 0$
$x_1 - x_2 - x_4 + x_5 = 4$

40. $2x - 2z + 4w - 4v = -6$
$-x - y - z - w - u - v = -12$
$x + y - z - w = -2$
$y - z + u - v = -1$
$x - y + z - w + u - v = 0$
$3y - z + v = 4$

Application Exercises

41. The table shows the number of inmates in federal and state prisons in the United States for three selected years.

x (Number of Years after 1980)	1	5	10
y (Number of Inmates, in thousands)	344	480	740

a. Use the quadratic function $y = ax^2 + bx + c$ to model the data. Solve the system of linear equations involving a, b, and c using matrices.

b. Predict the number of inmates in the year 2010.

c. List one factor that would change the accuracy of this model for the year 2010.

42. A football is kicked straight upward. The position function

$$s = \tfrac{1}{2}at^2 + v_0 t + s_0$$

describes the ball's height, s, in feet, after t seconds. Use the points labeled in the graph to find the values of a, v_0, and s_0. Solve the system of linear equations involving a, v_0, and s_0 using matrices. What is the value for s when $t = 7$? Describe what this means.

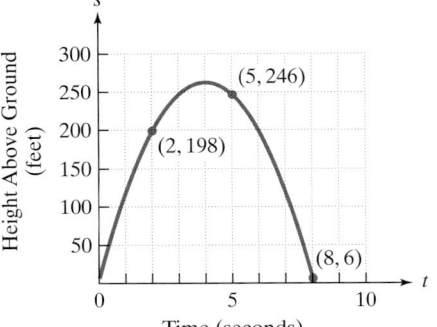

Time (seconds)

Write a system of linear equations in three variables to solve Exercises 43–46. Then use matrices to solve the system.

43. The circle graph indicates the ages of the 40 million online users in the United States. The percentage of online users in the youngest (under 30) and oldest (50 and over) age groups combined exceeds the percentage in the 30–49 age group by 2%. If the percentage of users in the oldest age group is doubled, it is 3% less than the percentage of users in the youngest age group. Find the percentage of online users in each of the three age groups.

Age of U.S. Online Users

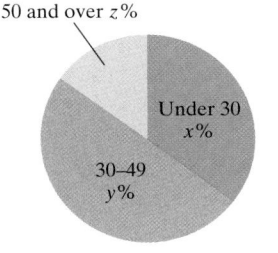

Source: U.S. Census Bureau

44. The circle graph indicates computers in use for the United States and the rest of the world. The percentage of the world's computers in Europe and Japan combined is 13% less than the percentage of the world's computers in the United States. If the percentage of the world's computers in Europe is doubled, it is only 3% more than the percentage of the world's computers in the United States. Find the percentage of the world's computers in the United States, Europe, and Japan.

**Percentage of the World's Computers:
U.S. and the World**

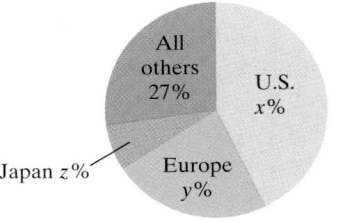

Source: Jupiter Communications

45. Three foods have the following nutritional content per ounce.

	Calories	Protein (in grams)	Vitamin C (in milligrams)
Food A	40	5	30
Food B	200	2	10
Food C	400	4	300

If a meal consisting of the three foods allows exactly 660 calories, 25 grams of protein, and 425 milligrams of vitamin C, how many ounces of each kind of food should be used?

46. A furniture company produces three types of desks: a children's model, an office model, and a deluxe model. Each desk is manufactured in three stages: cutting, construction, and finishing. The time requirements for each model and manufacturing stage are given in the following table.

	Children's model	Office model	Deluxe model
Cutting	2 hr	3 hr	2 hr
Construction	2 hr	1 hr	3 hr
Finishing	1 hr	1 hr	2 hr

Each week the company has available a maximum of 100 hours for cutting, 100 hours for construction, and 65 hours for finishing. If all available time must be used, how many of each type of desk should be produced each week?

Writing in Mathematics

47. What is a matrix?

48. Describe what is meant by the augmented matrix of a system of linear equations.

49. In your own words, describe each of the three matrix row operations. Give an example with each of the operations.

50. Describe how to use row operations and matrices to solve a system of linear equations.

51. What is the difference between Gaussian elimination and Gauss-Jordan elimination?

52. The graphs show the percentage of recorded music on CDs, cassettes, and LPs from 1981–2001. For this time period, which of these three forms of recorded music would you model using a quadratic function? Explain your answer.

Percentage of Recorded Music on CDs, Cassettes, and LPs

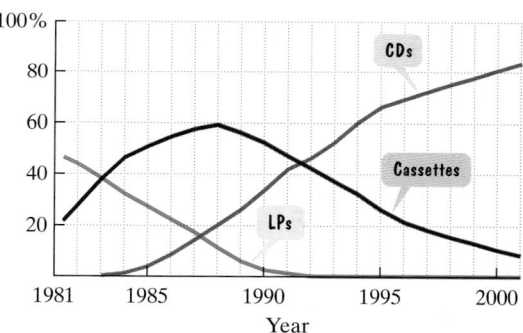

Source: Recording Industry Association of America

53. In Exercise 52, assume that you plan to obtain the quadratic model by hand. Explain how to use the graph for the form that you selected to find a, b, and c in $y = ax^2 + bx + c$, where x represents years since 1981 and y represents the percentage of recorded music on this form. Describe the role that matrices play in the process of obtaining the model.

Technology Exercises

54. Most graphing utilities can perform row operations on matrices. Consult the owner's manual for your graphing utility to learn proper keystrokes for performing these operations. Then duplicate the row operations of any three exercises that you solved from Exercises 19–24.

55. The final augmented matrix that we obtain when using Gaussian elimination is said to be in **row-echelon form**. For systems of linear equations with unique solutions, this form results when each entry in the main diagonal is 1

and all entries below the main diagonal are 0s. Some graphing utilities can transform a matrix to row-echelon form. Consult the owner's manual for your graphing utility. If your utility has this capability, enter the augmented matrix and obtain the final matrices of Example 3 on pages 742–744 and Example 4 on page 744–747. Then use this capability to solve any five of the systems in Exercises 27–40.

56. The final augmented matrix that we obtain when using Gauss-Jordan elimination is said to be in **reduced row-echelon form**. For systems of linear equations with unique solutions, this form results when each entry on the main diagonal is 1 and all entries below and above that main diagonal are 0s. Some graphing utilities can transform a matrix to reduced row-echelon form. Consult the owner's manual for your graphing utility. If your utility has this capability, obtain the final matrix of Example 5 on pages 747–748 beginning with the augmented matrix for Example 3 on page 742. Then use this capability to solve any five of the systems in Exercises 27–40.

Critical Thinking Exercises

57. Find a cubic function whose graph passes through the points $(0, -3)$, $(1, 5)$, $(-1, -7)$, and $(-2, -13)$. (*Hint:* Use the equation $y = ax^3 + bx^2 + cx + d$.)

58. The table shows the daily production level and profit for a business.

x (Number of units Produced Daily)	30	50	100
y (Daily Profit)	$5900	$7500	$4500

Use the quadratic function $y = ax^2 + bx + c$ to determine the number of units that should be produced each day for maximum profit. What is the maximum daily profit?

Group Exercise

59. In Chapter 8, you learned how to fit a quadratic function of the form $y = ax^2 + bx + c$ to data without using the regression feature of a graphing utility (see pages 691–692). Each group member should find an interesting data set. Group members should select the two sets of data that are most interesting and relevant.

a. For one of the data sets selected, use the function $y = ax^3 + bx^2 + cx + d$ and four ordered pairs of values (x, y) to find the cubic function that models the data. Use matrices or a graphing utility to solve the resulting system in four variables for a, b, c, and d.

b. For the other data set selected, fit a higher-degree polynomial function to the data. Use a graphing utility to solve the resulting system in five or more variables.

SECTION 9.2   Inconsistent and Dependent Systems and Their Applications

Objectives

1. Apply Gaussian elimination to systems without unique solutions.

2. Apply Gaussian elimination to systems with differing numbers of variables and equations.

3. Solve problems involving systems without unique solutions.

Traffic jams getting you down? Powerful computers, able to solve systems with hundreds of thousands of variables in a single bound, may promise a gridlock-free future. The computer in your car could be linked to a central computer that manages traffic flow by controlling traffic lights, rerouting you away from traffic congestion, issuing weather reports, and selecting the best route to your destination. New technologies could eventually drive your car at a steady 75 miles per hour along automated highways as you comfortably nap. In this section, we look at the role of linear systems without unique solutions in a future free of traffic jams.

1 Apply Gaussian elimination to systems without unique solutions.

Linear systems can have one solution, no solutions, or infinitely many solutions. We can use Gaussian elimination on systems with three or more variables to determine how many solutions such systems may have. In the case of systems with no solutions or infinitely many solutions, it is impossible to rewrite the augmented matrix in the desired form with 1s down the main diagonal and 0s below the 1s. Let's see what this means by looking at a system that has no solutions.

EXAMPLE 1   A System With No Solutions

Use Gaussian elimination to solve the system

$$x - y - 2z = 2$$
$$2x - 3y + 6z = 5$$
$$3x - 4y + 4z = 12.$$

Solution

Step 1   Write the augmented matrix for the system.

Linear System

$$x - y - 2z = 2$$
$$2x - 3y + 6z = 5$$
$$3x - 4y + 4z = 12$$

Augmented Matrix

$$\begin{bmatrix} 1 & -1 & -2 & | & 2 \\ 2 & -3 & 6 & | & 5 \\ 3 & -4 & 4 & | & 12 \end{bmatrix}$$

Step 2   Attempt to simplify the matrix to one with 1s down the diagonal and 0s below the 1s. Notice that the augmented matrix already has a 1 in the top posi-

Discovery

Use the addition method to solve Example 1. Describe what happens. Why does this mean that there is no solution?

tion of the first column. Now we want 0s below the 1. To get the first 0, multiply row 1 by −2 and add these products to row 2. To get the second 0, multiply row 1 by −3 and add these products to row 3. Performing these operations, we obtain the following matrix.

We want 1 in this position.

$$\begin{bmatrix} 1 & -1 & -2 & | & 2 \\ 0 & -1 & 10 & | & 1 \\ 0 & -1 & 10 & | & 6 \end{bmatrix}$$

Use the previous matrix and:
Replace row 2 by $-2R_1 + R_2$.
Replace row 3 by $-3R_1 + R_3$.

Moving on to the second column, we obtain 1 in the desired position by multiplying row 2 by −1.

$$\begin{bmatrix} 1 & -1 & -2 & | & 2 \\ 0(-1) & -1(-1) & 10(-1) & | & 1(-1) \\ 0 & -1 & 10 & | & 6 \end{bmatrix} = \begin{bmatrix} 1 & -1 & -2 & | & 2 \\ 0 & 1 & -10 & | & -1 \\ 0 & -1 & 10 & | & 6 \end{bmatrix} \quad -1R_2$$

We want 0 in this position.

Now we want a 0 below the 1 in column 2. To get the 0, multiply row 2 by 1 and add these products to row 3. (Equivalently, add row 2 to row 3.) We obtain the following matrix.

$$\begin{bmatrix} 1 & -1 & -2 & | & 2 \\ 0 & 1 & -10 & | & -1 \\ 0 & 0 & 0 & | & 5 \end{bmatrix}$$

Replace row 3 in the previous matrix by $1R_2 + R_3$.

It is impossible to convert this last matrix to the desired form of 1s down the main diagonal. If we translate the last row back into equation form, we get

$$0x + 0y + 0z = 5,$$

which is false. Regardless of which values we select for x, y, and z, the last equation can never be a true statement. Consequently, the system has no solution. The solution set is $\varnothing$, the empty set.

Three planes are parallel with no common intersection point.

Two planes are parallel with no common intersection point.

Planes intersect two at a time. There is no intersection point common to all three planes.

Figure 9.1 Three planes may have no common point of intersection.

Check Point 1

Use Gaussian elimination to solve the system

$$x - 2y - z = -5$$
$$2x - 3y - z = 0$$
$$3x - 4y - z = 1.$$

Recall that the graph of a system of three linear equations in three variables consists of three planes. When these planes intersect in a single point, the system has precisely one ordered-triple solution. When the planes have no point in common, the system has no solution, like the one in Example 1. Figure 9.1 illustrates some of the geometric possibilities for these inconsistent systems.

Now let's see what happens when we apply Gaussian elimination to a system with infinitely many solutions. Representing the solution set for these systems can be a bit tricky.

EXAMPLE 2   **A System with an Infinite Number of Solutions**

Use Gaussian elimination to solve the following system:

$$3x - 4y + 4z = 7$$
$$x - y - 2z = 2$$
$$2x - 3y + 6z = 5.$$

Solution   As always, we start with the augmented matrix.

$$\begin{bmatrix} 3 & -4 & 4 & | & 7 \\ 1 & -1 & -2 & | & 2 \\ 2 & -3 & 6 & | & 5 \end{bmatrix} \xrightarrow[\substack{\text{Reverse rows} \\ \text{1 and 2.}}]{R_1 \leftrightarrow R_2} \begin{bmatrix} 1 & -1 & -2 & | & 2 \\ 3 & -4 & 4 & | & 7 \\ 2 & -3 & 6 & | & 5 \end{bmatrix} \xrightarrow[\substack{\text{Replace row 3} \\ \text{by } -2R_1 + R_3.}]{\substack{\text{Replace row 2} \\ \text{by } -3R_1 + R_2.}}$$

$$\begin{bmatrix} 1 & -1 & -2 & | & 2 \\ 0 & -1 & 10 & | & 1 \\ 0 & -1 & 10 & | & 1 \end{bmatrix} \xrightarrow[\substack{\text{Multiply row} \\ \text{2 by } -1.}]{-1R_2} \begin{bmatrix} 1 & -1 & -2 & | & 2 \\ 0 & 1 & -10 & | & -1 \\ 0 & -1 & 10 & | & 1 \end{bmatrix} \xrightarrow[\substack{\text{by } 1R_2 + R_3.}]{\text{Replace row 3}}$$

$$\begin{bmatrix} 1 & -1 & -2 & | & 2 \\ 0 & 1 & -10 & | & -1 \\ 0 & 0 & 0 & | & 0 \end{bmatrix}$$

If we translate row 3 of the matrix into equation form, we obtain

$$0x + 0y + 0z = 0$$

or

$$0 = 0.$$

This equation results in a true statement regardless of which values we select for x, y, and z. Consequently, the equation $0x + 0y + 0z = 0$ is *dependent* on the other two equations in the system in the sense that it adds no new information about the variables. Thus, we can drop it from the system, which can now be expressed in the form

$$\begin{bmatrix} 1 & -1 & -2 & | & 2 \\ 0 & 1 & -10 & | & -1 \end{bmatrix}.$$

The original system is equivalent to the system

$$x - y - 2z = 2$$
$$y - 10z = -1.$$

Although neither of these equations gives a value for z, we can use them to express x and y in terms of z. From the last equation we obtain

$$y = 10z - 1. \quad \text{Add 10z to both sides and isolate y.}$$

Back-substituting for y into the previous equation, we can find x in terms of z.

$$x - y - 2z = 2 \qquad \text{This is the first equation obtained from the final matrix.}$$

$$x - (10z - 1) - 2z = 2 \qquad \text{Because y = 10z - 1, substitute 10z - 1 for y.}$$

$$x - 10z + 1 - 2z = 2 \qquad \text{Apply the distributive property.}$$

$$x - 12z + 1 = 2 \qquad \text{Combine like terms.}$$

$$x = 12z + 1 \qquad \text{Solve for x in terms of z.}$$

Because no value is determined for z, we can find a solution to the system by letting z equal any real number and then using the above equations to obtain x and y. For example, if $z = 1$, then

$$x = 12z + 1 = 12(1) + 1 = 13 \text{ and}$$
$$y = 10z - 1 = 10(1) - 1 = 9.$$

Consequently, $(13, 9, 1)$ is a solution to the system. On the other hand, if we let $z = -1$, then

$$x = 12z + 1 = 12(-1) + 1 = -11 \text{ and}$$
$$y = 10z - 1 = 10(-1) - 1 = -11.$$

Thus, $(-11, -11, -1)$ is another solution to the system. Finally, letting $z = t$ (or any letter of our choice), the solutions to the system are all of the form

$$x = 12t + 1, \qquad y = 10t - 1, \qquad z = t,$$

where t is a real number. Therefore, every ordered triple that is of the form $(12t + 1, 10t - 1, t)$, where t is a real number, is a solution of the system. The solution set of the system with dependent equations can be written as $\{(12t + 1, 10t - 1, t)\}$.

Three planes may intersect at infinitely many points.

Figure 9.2

We have seen that when three planes have no point in common, the corresponding system has no solution. When the system has infinitely many solutions, like the one in Example 2, the three planes intersect in more than one point. Figure 9.2 illustrates one geometric possibility for systems with dependent equations.

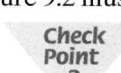

Check Point 2

Use Gaussian elimination to solve the following system:

$$x - 2y - z = 5$$
$$2x - 5y + 3z = 6$$
$$x - 3y + 4z = 1.$$

2 Apply Gaussian elimination to systems with differing numbers of variables and equations.

Nonsquare Systems

Up to this point, we have encountered only *square* systems in which the number of equations is equal to the number of variables. In a **nonsquare system**, the number of variables differs from the number of equations.

EXAMPLE 3 A System with Fewer Equations Than Variables

Use Gaussian elimination to solve the system

$$3x + 7y + 6z = 26$$
$$x + 2y + z = 8.$$

Solution We begin with the augmented matrix.

$$\begin{bmatrix} 3 & 7 & 6 & | & 26 \\ 1 & 2 & 1 & | & 8 \end{bmatrix} \xrightarrow{R_1 \leftrightarrow R_2} \begin{bmatrix} 1 & 2 & 1 & | & 8 \\ 3 & 7 & 6 & | & 26 \end{bmatrix} \xrightarrow[\text{by} -3R_1 + R_2.]{\text{Replace row 2}} \begin{bmatrix} 1 & 2 & 1 & | & 8 \\ 0 & 1 & 3 & | & 2 \end{bmatrix}$$

Because we now have 1s down the diagonal that begins with the upper-left entry and a 0 below this 1, we translate the matrix back into equation form.

$$x + 2y + z = 8 \qquad \text{Equation 1}$$
$$y + 3z = 2 \qquad \text{Equation 2}$$

Discovery

Let $t = 1$ for the solution set
$$\{(5t + 4, -3t + 2, t)\}.$$

What solution do you obtain? Substitute these three values in the two equations in Example 3 and show that each equation is satisfied. Repeat this process for another two values for t.

We can let z equal any real number and use back-substitution to express x and y in terms of z.

Equation 2	**Equation 1**
$y + 3z = 2$	$x + 2y + z = 8$
$y = -3z + 2$	$x + 2(-3z + 2) + z = 8$
	$x - 6z + 4 + z = 8$
	$x - 5z + 4 = 8$
	$x = 5z + 4$

With $z = t$, the ordered solution (x, y, z) enables us to express the system's solution set as

$$\{(5t + 4, -3t + 2, t)\}$$

where t is any real number.

Check Point 3

Use Gaussian elimination to solve the system

$$x + 2y + 3z = 70$$
$$x + y + z = 60.$$

3 Solve problems involving systems without unique solutions.

Applications

How will computers be programmed to control traffic flow and avoid congestion? They will be required to solve systems continually based on the following premise: If traffic is to keep moving, during any period of time the number of cars entering an intersection must equal the number of cars leaving that intersection. Let's see what this means by looking at the intersections of four one-way city streets.

EXAMPLE 4   Traffic Control

Figure 9.3 shows the intersections of four one-way streets. As you study the figure, notice that 300 cars per hour want to enter intersection I_1 from the north on 27th Avenue. Also, 200 cars per hour want to head east from intersection I_2 on Palm Drive. The letters x, y, z, and w stand for the number of cars passing between the intersections.

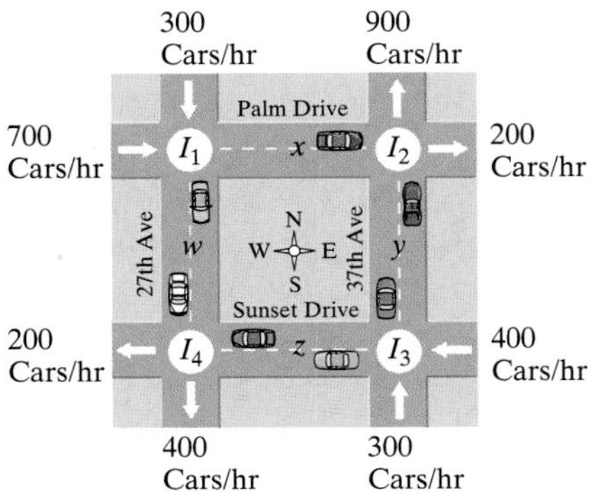

Figure 9.3 The intersections of four one-way streets

a. If the traffic is to keep moving, at each intersection the number of cars entering per hour must equal the number of cars leaving per hour. Use this idea to set up a linear system of equations involving x, y, z, and w.

b. Use Gaussian elimination to solve the system.

c. If construction on 27th Avenue limits w to 50 cars per hour, how many cars per hour must pass between the other intersections to keep traffic flowing?

Solution

a. Set up the system by considering one intersection at a time, referring to Figure 4.3.

For Intersection I_1: Because $300 + 700 = 1000$ cars enter I_1, and $x + w$ cars leave the intersection, then $x + w = 1000$.

For Intersection I_2: Because $x + y$ cars enter the intersection, and $200 + 900 = 1100$ cars leave I_2, then $x + y = 1100$.

For Intersection I_3: Figure 6.3 indicates that $300 + 400 = 700$ cars enter and $y + z$ leave, so $y + z = 700$.

For Intersection I_4: With $z + w$ cars entering and $200 + 400 = 600$ cars exiting, traffic will keep flowing if $z + w = 600$.

The system of equations that describes this situation is given by

$$x + w = 1000$$
$$x + y = 1100$$
$$y + z = 700$$
$$z + w = 600.$$

b. To solve this system using Gaussian elimination, we begin with the augmented matrix.

System of Linear Equations (showing missing variables with 0 coefficients)

$$1x + 0y + 0z + 1w = 1000$$
$$1x + 1y + 0z + 0w = 1100$$
$$0x + 1y + 1z + 0w = 700$$
$$0x + 0y + 1z + 1w = 600$$

Augmented Matrix

$$\left[\begin{array}{cccc|c} 1 & 0 & 0 & 1 & 1000 \\ 1 & 1 & 0 & 0 & 1100 \\ 0 & 1 & 1 & 0 & 700 \\ 0 & 0 & 1 & 1 & 600 \end{array}\right]$$

We can now use row operations to obtain the matrix

$$\left[\begin{array}{cccc|c} 1 & 0 & 0 & 1 & 1000 \\ 0 & 1 & 0 & -1 & 100 \\ 0 & 0 & 1 & 1 & 600 \\ 0 & 0 & 0 & 0 & 0 \end{array}\right].$$

$x + w = 1000$

$y - w = 100$

$z + w = 600$

The last row of the matrix shows that the system in the voice balloons has dependent equations and infinitely many solutions. To write the solution set containing these infinitely many solutions, let w equal any real number. Use the three equations in the voice balloons to express x, y, and z in terms of w: $x = 1000 - w$, $y = 100 + w$, and $z = 600 - w$.

With $w = t$, the ordered solution (x, y, z, w) enables us to express the system's solution set as

$$\{(1000 - t, 100 + t, 600 - t, t)\}.$$

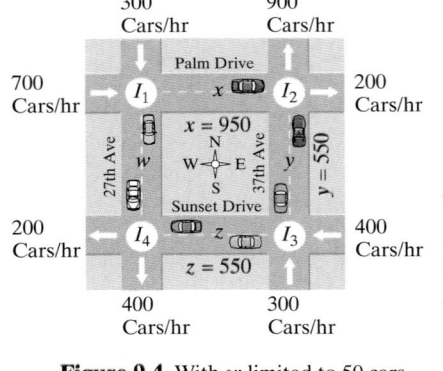

Figure 9.4 With w limited to 50 cars per hour, values for x, y, and z are determined.

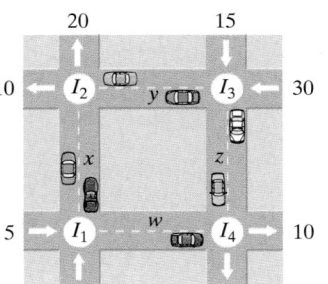

Figure 9.5

c. We are given that construction limits w to 50 cars per hour. Because $w = t$, we replace 50 for t in the system's ordered solution:

$$(1000 - t, 100 + t, 600 - t, t) \qquad \text{Use the system's solution.}$$

$$= (1000 - 50, 100 + 50, 600 - 50, 50) \qquad t = 50$$

$$= (950, 150, 550, 50)$$

Thus, $x = 950$, $y = 150$, and $z = 550$. (See Figure 9.4.) With construction on 27th Avenue, this means that to keep traffic flowing, 950 cars per hour must be routed between I_1 and I_2, 150 per hour between I_3 and I_2, and 550 per hour between I_3 and I_4.

Check Point 4

Figure 9.5 shows a system of four one-way streets. The numbers in the figure denote the number of cars per minute that travel in the direction shown.

a. Use the requirement that the number of cars entering each of the intersections per minute must equal the number of cars leaving per minute to set up a system of equations in x, y, z, and w.

b. Use Gaussian elimination to solve the system.

c. If construction limits w to 10 cars per minute, how many cars per minute must pass between the other intersections to keep traffic flowing?

EXERCISE SET 9.2

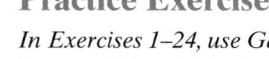 **Practice Exercises**

In Exercises 1–24, use Gaussian elimination to find the complete solution to each system of equations, or show that none exists.

1. $5x + 12y + z = 10$
$2x + 5y + 2z = -1$
$x + 2y - 3z = 5$

2. $2x - 4y + z = 3$
$x - 3y + z = 5$
$3x - 7y + 2z = 12$

3. $5x + 8y - 6z = 14$
$3x + 4y - 2z = 8$
$x + 2y - 2z = 3$

4. $5x - 11y + 6z = 12$
$-x + 3y - 2z = -4$
$3x - 5y + 2z = 4$

5. $3x + 4y + 2z = 3$
$4x - 2y - 8z = -4$
$x + y - z = 3$

6. $2x - y - z = 0$
$x + 2y + z = 3$
$3x + 4y + 2z = 8$

7. $8x + 5y + 11z = 30$
$-x - 4y + 2z = 3$
$2x - y + 5z = 12$

8. $x + y - 10z = -4$
$x \qquad - 7z = -5$
$3x + 5y - 36z = -10$

9. $x - 2y - z - 3w = -9$
$x + y - z \qquad = 0$
$3x + 4y \qquad + w = 6$
$2y - 2z + w = 3$

10. $2x + y - 2z - w = 3$
$x - 2y + z + w = 4$
$-x - 8y + 7z + 5w = 13$
$3x + y - 2z + 2w = 6$

11. $2x + y - z \qquad = 3$
$x - 3y + 2z \qquad = -4$
$3x + y - 3z + w = 1$
$x + 2y - 4z - w = -2$

12. $2x - y + 3z + w = 0$
$3x + 2y + 4z - w = 0$
$5x - 2y - 2z - w = 0$
$2x + 3y - 7z - 5w = 0$

13.
$$x - 3y + z - 4w = 4$$
$$-2x + y + 2z = -2$$
$$3x - 2y + z - 6w = 2$$
$$-x + 3y + 2z - w = -6$$

14.
$$3x + 2y - z + 2w = -12$$
$$4x - y + z + 2w = 1$$
$$x + y + z + w = -2$$
$$-2x + 3y + 2z - 3w = 10$$

15.
$$2x + y - z = 2$$
$$3x + 3y - 2z = 3$$

16.
$$3x + 2y - z = 5$$
$$x + 2y - z = 1$$

17.
$$x + 2y + 3z = 5$$
$$y - 5z = 0$$

18.
$$3x - y + 4z = 8$$
$$y + 2z = 1$$

19.
$$x + y - 2z = 2$$
$$3x - y - 6z = -7$$

20.
$$-2x - 5y + 10z = 19$$
$$x + 2y - 4z = 12$$

21.
$$x + y - z + w = -2$$
$$2x - y + 2z - w = 7$$
$$-x + 2y + z + 2w = -1$$

22.
$$2x - 3y + 4z + w = 7$$
$$x - y + 3z - 5w = 10$$
$$3x + y - 2z - 2w = 6$$

23.
$$x + 2y + 3z - w = 7$$
$$2y - 3z + w = 4$$
$$x - 4y + z = 3$$

24.
$$x - y + w = 0$$
$$x - 4y + z + 2w = 0$$
$$3x - z + 2w = 0$$

Application Exercises

The figure for Exercises 25–28 shows the intersection of three one-way streets. To keep traffic moving, the number of cars per minute entering an intersection must equal the number exiting that intersection. For intersection I_1, $x + 10$ cars enter and $y + 14$ cars exit per minute. Thus, $x + 10 = y + 14$.

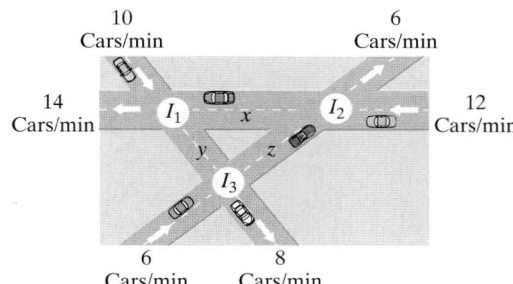

25. Write an equation for intersection I_2 that keeps traffic moving.

26. Write an equation for intersection I_3 that keeps traffic moving.

27. Use Gaussian elimination to solve the system formed by the equation given prior to Exercise 25 and the two equations that you obtained in Exercises 25–26.

28. Use your ordered solution obtained in Exercise 27 to solve this exercise. If construction limits z to 4 cars per minute, how many cars per minute must pass between the other intersections to keep traffic flowing?

29. The figure shows the intersection of four one-way streets.

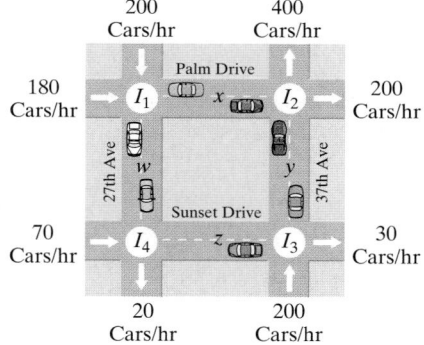

a. Set up a system of equations that keep traffic moving.
b. Use Gaussian elimination to solve the system.
c. If construction limits w to 50 cars per hour, how many cars per hour must pass between the other intersections to keep traffic moving?

30. The vitamin content per ounce for three foods is given in the following table.

	Milligrams per Ounce		
	Thiamin	Riboflavin	Niacin
Food A	3	7	1
Food B	1	5	3
Food C	3	8	2

a. Use matrices to show that no combination of these foods can provide exactly 14 mg of thiamin, 32 mg of riboflavin, and 9 mg of niacin.
b. Use matrices to describe in practical terms what happens if the riboflavin requirement is increased by 5 mg and the other requirements stay the same.

31. Three foods have the following nutritional content per ounce.

	Units per Ounce		
	Vitamin A	Iron	Calcium
Food 1	20	20	10
Food 2	30	10	10
Food 3	10	10	30

a. A diet must consist precisely of 220 units of vitamin A, 180 units of iron, and 340 units of calcium. However, the dietician runs out of Food 1. Use a matrix approach to show that under these conditions the dietary requirements cannot be met.

b. Now suppose that all three foods are available, but due to problems with vitamin A for pregnant women, a hospital dietician no longer wants to include this vitamin in the diet. Use matrices to give two possible ways to meet the iron and calcium requirements with the three foods.

32. A company that manufactures products A, B, and C does both manufacturing and testing. The hours needed to manufacture and test each product are shown in the table.

	Hours Needed Weekly to Manufacture	Hours Needed Weekly to Test
Product A	7	2
Product B	6	2
Product C	3	1

The company has exactly 67 hours per week available for manufacturing and 20 hours per week available for testing. Give two different combinations for the number of products that can be manufactured and tested weekly.

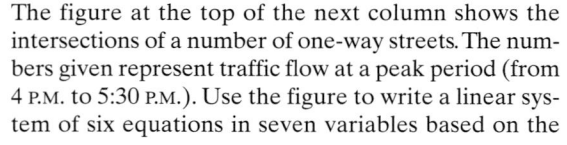

Writing in Mathematics

33. Describe what happens when Gaussian elimination is used to solve an inconsistent system.

34. Describe what happens when Gaussian elimination is used to solve a system with dependent equations.

35. In solving a system of dependent equations in three variables, one student simply said that there are infinitely many solutions. A second student expressed the solution set as $\{(4t + 3, 5t - 1, t)\}$. Which is the better form of expressing the solution set and why?

Technology Exercise

36. a. The figure at the top of the next column shows the intersections of a number of one-way streets. The numbers given represent traffic flow at a peak period (from 4 P.M. to 5:30 P.M.). Use the figure to write a linear system of six equations in seven variables based on the

idea that at each intersection the number of cars entering must equal the number of cars leaving.

b. Use a graphing utility with matrix capabilities to find the complete solution to the system.

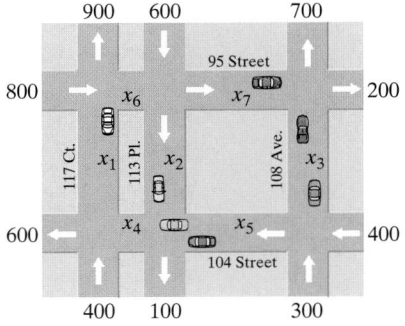

Critical Thinking Exercise

37. Consider the linear system

$$x + 3y + z = a^2$$
$$2x + 5y + 2az = 0$$
$$x + y + a^2z = -9.$$

For what values of a will the system be inconsistent?

Group Exercise

38. Before beginning this exercise, the group needs to read and solve Exercise 36.

a. A political group is planning a demonstration on 95th Street between 113th Place and 117th Court for 5 P.M. Wednesday. The problem becomes one of minimizing traffic flow on 95th Street (between 113th and 117th) without causing traffic tie-ups on other streets. One possible solution is to close off traffic on 95th Street between 113th and 117th (let $x_6 = 0$). What can group members conclude about x_7 under these conditions?

b. Working with a matrix allows us to simplify the problem caused by the political demonstration, but it did not actually solve the problem. There are an infinite number of solutions; each value of x_7 we choose gives us a new picture. We also assumed x_6 was equal to 0; changing that assumption would also lead to different solutions. With your group, design another solution to the traffic flow problem caused by the political demonstration.

SECTION 9.3 *Matrix Operations and Their Applications*

Objectives

1. Use matrix notation.
2. Understand what is meant by equal matrices.
3. Add and subtract matrices.
4. Perform scalar multiplication.
5. Multiply matrices.
6. Describe applied situations with matrix operations.

Turn on your computer and read your e-mail or write a paper. When you need to do research, use the Internet to browse through art museums and photography exhibits. When you need a break, load a flight simulator program and fly through a photorealistic computer world. As different as these experiences may be, they all share one thing—you're looking at images based on matrices. Matrices have applications in numerous fields, including the new technology of digital photography in which pictures are represented by numbers rather than film. In this section, we turn our attention to matrix algebra and some of its applications.

1   Use matrix notation.

Notations for Matrices

We have seen that an array of numbers, arranged in rows and columns and placed in brackets, is called a matrix. We can represent the matrix in two different ways.

- A capital letter, such as A, B, or C, can denote a matrix.
- A lowercase letter enclosed in brackets, such as that shown below, can denote a matrix.

$$A = \left[a_{ij} \right] \qquad \text{Matrix A with elements } a_{ij}$$

A general element in matrix A is denoted by a_{ij}. This refers to the element in the ith row and jth column. For example, a_{32} is the element of A located in the third row, second column.

A matrix of **order $m \times n$** has m rows and n columns. If $m = n$, a matrix has the same number of rows as columns and is called a **square matrix**.

EXAMPLE 1 Matrix Notation

Let

$$A = \left[\begin{array}{ccc} 3 & 2 & 0 \\ -4 & -5 & -\frac{1}{5} \end{array} \right].$$

a. What is the order of A?
b. If $A = \left[a_{ij} \right]$, identify a_{23} and a_{12}.

Solution

a. Matrix A is a 1×3 matrix and matrix B is a 3×1 matrix. Thus, the product is a 1×1 matrix.

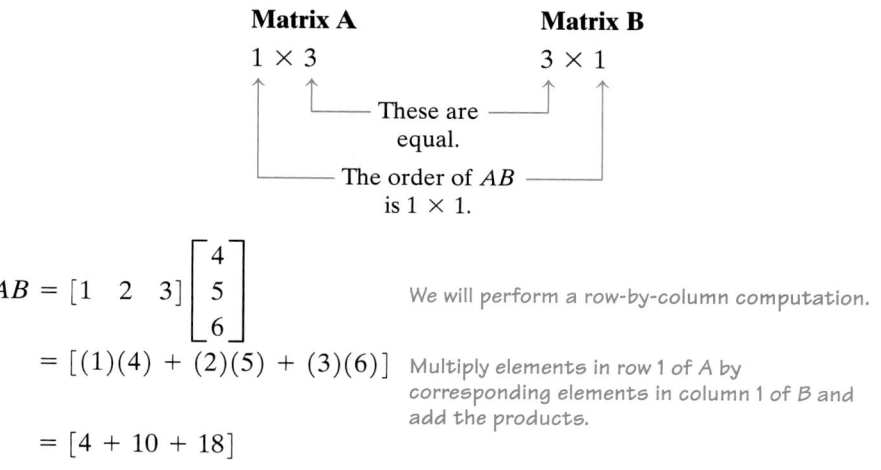

$$AB = \begin{bmatrix} 1 & 2 & 3 \end{bmatrix} \begin{bmatrix} 4 \\ 5 \\ 6 \end{bmatrix} \qquad \text{We will perform a row-by-column computation.}$$

$$= \begin{bmatrix} (1)(4) + (2)(5) + (3)(6) \end{bmatrix} \qquad \text{Multiply elements in row 1 of } A \text{ by corresponding elements in column 1 of } B \text{ and add the products.}$$

$$= \begin{bmatrix} 4 + 10 + 18 \end{bmatrix}$$

$$= \begin{bmatrix} 32 \end{bmatrix}$$

b. Matrix B is a 3×1 matrix and matrix A is a 1×3 matrix. Thus, the product BA is a 3×3 matrix.

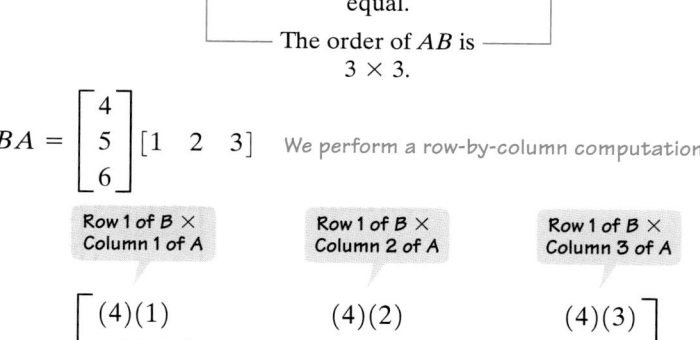

$$BA = \begin{bmatrix} 4 \\ 5 \\ 6 \end{bmatrix} \begin{bmatrix} 1 & 2 & 3 \end{bmatrix} \qquad \text{We perform a row-by-column computation.}$$

Row 1 of B × Column 1 of A	Row 1 of B × Column 2 of A	Row 1 of B × Column 3 of A

$$= \begin{bmatrix} (4)(1) & (4)(2) & (4)(3) \\ (5)(1) & (5)(2) & (5)(3) \\ (6)(1) & (6)(2) & (6)(3) \end{bmatrix}$$

Row 2 of B × Column 1 of A, Row 2 of B × Column 2 of A, Row 2 of B × Column 3 of A

Row 3 of B × Column 1 of A, Row 3 of B × Column 2 of A, Row 3 of B × Column 3 of A

$$= \begin{bmatrix} 4 & 8 & 12 \\ 5 & 10 & 15 \\ 6 & 12 & 18 \end{bmatrix} \qquad \text{Simplify.}$$

Arthur Cayley

The Granger Collection

Matrices were first studied intensively by the English mathematician Arthur Cayley (1821–1895). Before reaching the age of 25, he published 25 papers, setting a pattern of prolific creativity that lasted throughout his life. Cayley was a lawyer, painter, mountaineer, and Cambridge professor whose greatest invention was that of matrices and matrix theory. Cayley's matrix algebra, especially the noncommutativity of multiplication ($AB \neq BA$), opened up a new area of mathematics called abstract algebra.

In Example 5, notice that AB and BA are different matrices. For most matrices $AB \neq BA$. Because **matrix multiplication is not commutative**, be careful about the order in which matrices appear when performing this operation.

Check Point 5   If $A = \begin{bmatrix} 2 & 0 & 4 \end{bmatrix}$ and $B = \begin{bmatrix} 1 \\ 3 \\ 7 \end{bmatrix}$, find AB and BA.

EXAMPLE 6   Multiplying Matrices

Where possible, find each product:

a. $\begin{bmatrix} 4 & 2 \\ 1 & 3 \end{bmatrix} \begin{bmatrix} 1 & 2 & 3 & 4 \\ 0 & 2 & -1 & 6 \end{bmatrix}$

b. $\begin{bmatrix} 1 & 2 & 3 & 4 \\ 0 & 2 & -1 & 6 \end{bmatrix} \begin{bmatrix} 4 & 2 \\ 1 & 3 \end{bmatrix}$

Solution

a. The first matrix is a 2×2 matrix and the second is a 2×4 matrix. The product will be a 2×4 matrix.

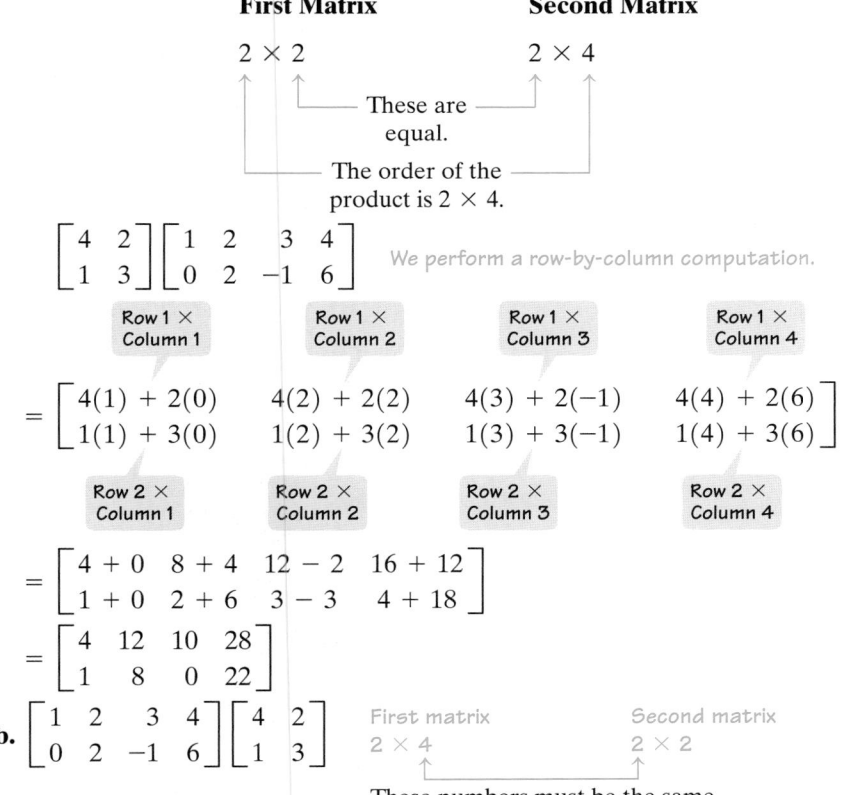

$$= \begin{bmatrix} 4(1) + 2(0) & 4(2) + 2(2) & 4(3) + 2(-1) & 4(4) + 2(6) \\ 1(1) + 3(0) & 1(2) + 3(2) & 1(3) + 3(-1) & 1(4) + 3(6) \end{bmatrix}$$

$$= \begin{bmatrix} 4 + 0 & 8 + 4 & 12 - 2 & 16 + 12 \\ 1 + 0 & 2 + 6 & 3 - 3 & 4 + 18 \end{bmatrix}$$

$$= \begin{bmatrix} 4 & 12 & 10 & 28 \\ 1 & 8 & 0 & 22 \end{bmatrix}$$

b. $\begin{bmatrix} 1 & 2 & 3 & 4 \\ 0 & 2 & -1 & 6 \end{bmatrix} \begin{bmatrix} 4 & 2 \\ 1 & 3 \end{bmatrix}$

First matrix 2×4 Second matrix 2×2

These numbers must be the same to multiply the matrices.

The number of columns in the first matrix does not equal the number of rows in the second matrix. Thus, the product of these two matrices is undefined.

Check
Point
6

Where possible, find each product.

a. $\begin{bmatrix} 1 & 3 \\ 0 & 2 \end{bmatrix} \begin{bmatrix} 2 & 3 & -1 & 6 \\ 0 & 5 & 4 & 1 \end{bmatrix}$ **b.** $\begin{bmatrix} 2 & 3 & -1 & 6 \\ 0 & 5 & 4 & 1 \end{bmatrix} \begin{bmatrix} 1 & 3 \\ 0 & 2 \end{bmatrix}$

Although matrix multiplication is not commutative, it does obey many of the properties of real numbers.

Discovery

Verify the properties listed in the box using

$$A = \begin{bmatrix} 3 & 2 \\ -1 & 4 \end{bmatrix}$$

$$B = \begin{bmatrix} 1 & 0 \\ 3 & 2 \end{bmatrix}$$

$$C = \begin{bmatrix} 1 & 2 \\ -1 & 1 \end{bmatrix}$$

and $c = 3$.

Properties of Matrix Multiplication

If A, B, and C are matrices and c is a scalar, then the following properties are true. (Assume the order of each matrix is such that all operations in these properties are defined.)

1. $(AB)C = A(BC)$ Associative Property of Matrix Multiplication

2. $A(B + C) = AB + AC$ Distributive Properties of Matrix

     $(A + B)C = AC + BC$ Multiplication

3. $c(AB) = (cA)B$ Associative Property of Scalar Multiplication

6 Describe applied situations with matrix operations.

Applications

All of the still images that you see on the Web have been created or manipulated on a computer in a digital format—made up of hundreds of thousands, or even millions, of tiny squares called **pixels**. Pixels are created by dividing an image into a grid. The computer can change the brightness of every square or pixel in this grid. A digital camera captures photos in this digital format. Also, you can scan pictures to convert them into digital format. Example 7 illustrates the role that matrices play in this new technology.

EXAMPLE 7 Matrices and Digital Photography

The letter T in Figure 9.7 is shown using 9 pixels in a 3 × 3 grid. The colors possible in the grid are shown in Figure 9.8. Each color is represented by a specific number: 0, 1, 2, or 3.

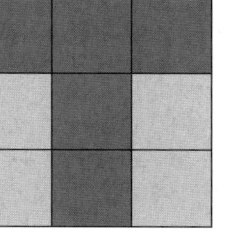

Figure 9.7 The letter T

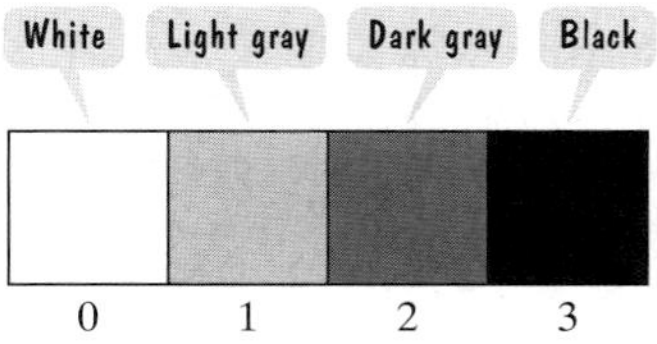

| White | Light gray | Dark gray | Black |

0   1   2   3

Figure 9.8 Color levels

a. Find a matrix that represents a digital photograph of this letter T.

b. Increase the contrast of the letter T by changing the dark gray to black and the light gray to white. Use matrix addition to accomplish this.

Solution

a. Look at the T and the background in Figure 9.7. Because the T is dark gray and the background is light gray, a digital photograph of Figure 9.7 can be represented by the matrix

$$\begin{bmatrix} 2 & 2 & 2 \\ 1 & 2 & 1 \\ 1 & 2 & 1 \end{bmatrix}.$$

b. We can make the T black by increasing each 2 in the above matrix to 3. We can make the background white by decreasing each 1 in the matrix to 0. This is accomplished using the following matrix addition.

$$\begin{bmatrix} 2 & 2 & 2 \\ 1 & 2 & 1 \\ 1 & 2 & 1 \end{bmatrix} + \begin{bmatrix} 1 & 1 & 1 \\ -1 & 1 & -1 \\ -1 & 1 & -1 \end{bmatrix} = \begin{bmatrix} 3 & 3 & 3 \\ 0 & 3 & 0 \\ 0 & 3 & 0 \end{bmatrix}$$

The picture corresponding to the matrix sum to the right of the equal sign is shown in Figure 9.9.

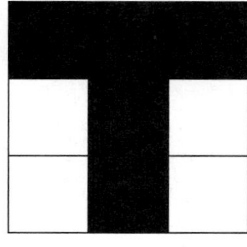

Figure 9.9 Changing contrast: the letter T

Check Point 7 Change the contrast of the letter T in Figure 9.7 by making the T light gray and the background black. Use matrix addition to accomplish this.

Images of Space

Photographs sent back from space use matrices with thousands of pixels. Each pixel is assigned a number from 0 to 63 representing its color—0 for pure white and 63 for pure black. In the image of Saturn shown here, matrix operations provide false colors that emphasize the banding of the planet's upper atmosphere.

EXAMPLE 8 Applying Matrix Multiplication

At a certain gas station, the number of gallons of regular, unleaded, and super unleaded gas sold on Monday, Tuesday, and Wednesday of a particular week is given by the following matrix.

	Regular	Unleaded	Super Unleaded	
Monday	240	300	160	
Tuesday	200	280	180	= A
Wednesday	260	310	200	

A second matrix gives the selling price per gallon and the profit per gallon for the three types of gas sold by the station.

	Selling price per Gallon	Profit per Gallon	
Regular	1.15	0.15	
Unleaded	1.20	0.17	= B
Super Unleaded	1.25	0.19	

a. Calculate the product AB.

b. What is the gas station's profit for Monday through Wednesday?

Solution

$$\textbf{a. } AB = \begin{bmatrix} 240 & 300 & 160 \\ 200 & 280 & 180 \\ 260 & 310 & 200 \end{bmatrix}\begin{bmatrix} 1.15 & 0.15 \\ 1.20 & 0.17 \\ 1.25 & 0.19 \end{bmatrix}$$

$$= \begin{bmatrix} 240(1.15) + 300(1.20) + 160(1.25) & 240(0.15) + 300(0.17) + 160(0.19) \\ 200(1.15) + 280(1.20) + 180(1.25) & 200(0.15) + 280(0.17) + 180(0.19) \\ 260(1.15) + 310(1.20) + 200(1.25) & 260(0.15) + 310(0.17) + 200(0.19) \end{bmatrix}$$

Perform to row-by-column multiplications.

$$= \begin{bmatrix} 836 & 117.40 \\ 791 & 111.80 \\ 921 & 129.70 \end{bmatrix}$$ *Multiply and add as indicated.*

b. The entries in the second column of the product matrix represent profits for Monday, Tuesday, and Wednesday, respectively. The gas station's profit for Monday through Wednesday is $117.40 + $111.80 + $129.70 or $358.90.

Check Point 8   Use the product matrix in Example 8a to answer this question. What are the gas station's total sales for Monday, Tuesday, and Wednesday?

EXERCISE SET 9.3

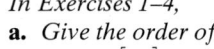 **Practice Exercises**

In Exercises 1–4,

a. *Give the order of each matrix.*

b. *If* $A = [a_{ij}]$, *identify* a_{32} *and* a_{23} *or explain why identification is not possible.*

1. $\begin{bmatrix} 4 & -7 & 5 \\ -6 & 8 & -1 \end{bmatrix}$ **2.** $\begin{bmatrix} -6 & 4 & -1 \\ -9 & 0 & \frac{1}{2} \end{bmatrix}$

3. $\begin{bmatrix} 1 & -5 & \pi & e \\ 0 & 7 & -6 & -\pi \\ -2 & \frac{1}{2} & 11 & -\frac{1}{5} \end{bmatrix}$ **4.** $\begin{bmatrix} -4 & 1 & 3 & -5 \\ 2 & -1 & \pi & 0 \\ 1 & 0 & -e & \frac{1}{5} \end{bmatrix}$

In Exercises 5–8, find values for the variables so that the matrices in each exercise are equal.

5. $\begin{bmatrix} x \\ 4 \end{bmatrix} = \begin{bmatrix} 6 \\ y \end{bmatrix}$ **6.** $\begin{bmatrix} x \\ 7 \end{bmatrix} = \begin{bmatrix} 11 \\ y \end{bmatrix}$

7. $\begin{bmatrix} x & 2y \\ z & 9 \end{bmatrix} = \begin{bmatrix} 4 & 12 \\ 3 & 9 \end{bmatrix}$

8. $\begin{bmatrix} x & y+3 \\ 2z & 8 \end{bmatrix} = \begin{bmatrix} 12 & 5 \\ 6 & 8 \end{bmatrix}$

In Exercises 9–16, find

a. $A + B$ **b.** $A - B$

c. $-4A$ **d.** $3A + 2B$

9. $A = \begin{bmatrix} 4 & 1 \\ 3 & 2 \end{bmatrix}$, $B = \begin{bmatrix} 5 & 9 \\ 0 & 7 \end{bmatrix}$

10. $A = \begin{bmatrix} -2 & 3 \\ 0 & 1 \end{bmatrix}$, $B = \begin{bmatrix} 8 & 1 \\ 5 & 4 \end{bmatrix}$

11. $A = \begin{bmatrix} 1 & 3 \\ 3 & 4 \\ 5 & 6 \end{bmatrix}$, $B = \begin{bmatrix} 2 & -1 \\ 3 & -2 \\ 0 & 1 \end{bmatrix}$

12. $A = \begin{bmatrix} 3 & 1 & 1 \\ -1 & 2 & 5 \end{bmatrix}$, $B = \begin{bmatrix} 2 & -3 & 6 \\ -3 & 1 & -4 \end{bmatrix}$

13. $A = \begin{bmatrix} 2 \\ -4 \\ 1 \end{bmatrix}$, $B = \begin{bmatrix} -5 \\ 3 \\ -1 \end{bmatrix}$

14. $A = \begin{bmatrix} 6 & 2 & -3 \end{bmatrix}$, $B = \begin{bmatrix} 4 & -2 & 3 \end{bmatrix}$

15. $A = \begin{bmatrix} 2 & -10 & -2 \\ 14 & 12 & 10 \\ 4 & -2 & 2 \end{bmatrix}$, $B = \begin{bmatrix} 6 & 10 & -2 \\ 0 & -12 & -4 \\ -5 & 2 & -2 \end{bmatrix}$

16. $A = \begin{bmatrix} 6 & -3 & 5 \\ 6 & 0 & -2 \\ -4 & 2 & -1 \end{bmatrix}$, $B = \begin{bmatrix} -3 & 5 & 1 \\ -1 & 2 & -6 \\ 2 & 0 & 4 \end{bmatrix}$

In Exercises 17–26, find (if possible)
a. AB and **b.** BA.

17. $A = \begin{bmatrix} 1 & 3 \\ 5 & 3 \end{bmatrix}$, $B = \begin{bmatrix} 3 & -2 \\ -1 & 6 \end{bmatrix}$

18. $A = \begin{bmatrix} 3 & -2 \\ 1 & 5 \end{bmatrix}$, $B = \begin{bmatrix} 0 & 0 \\ 5 & -6 \end{bmatrix}$

19. $A = \begin{bmatrix} 1 & 2 & 3 & 4 \end{bmatrix}$, $B = \begin{bmatrix} 1 \\ 2 \\ 3 \\ 4 \end{bmatrix}$

20. $A = \begin{bmatrix} -1 \\ -2 \\ -3 \end{bmatrix}$, $B = \begin{bmatrix} 1 & 2 & 3 \end{bmatrix}$

21. $A = \begin{bmatrix} 1 & -1 & 4 \\ 4 & -1 & 3 \\ 2 & 0 & -2 \end{bmatrix}$, $B = \begin{bmatrix} 1 & 1 & 0 \\ 1 & 2 & 4 \\ 1 & -1 & 3 \end{bmatrix}$

22. $A = \begin{bmatrix} 1 & -1 & 1 \\ 5 & 0 & -2 \\ 3 & -2 & 2 \end{bmatrix}$, $B = \begin{bmatrix} 1 & 1 & 0 \\ 1 & -4 & 5 \\ 3 & -1 & 2 \end{bmatrix}$

23. $A = \begin{bmatrix} 4 & 2 \\ 6 & 1 \\ 3 & 5 \end{bmatrix}$, $B = \begin{bmatrix} 2 & 3 & 4 \\ -1 & -2 & 0 \end{bmatrix}$

24. $A = \begin{bmatrix} 2 & 4 \\ 3 & 1 \\ 4 & 2 \end{bmatrix}$, $B = \begin{bmatrix} 3 & 2 & 0 \\ -1 & -3 & 5 \end{bmatrix}$

25. $A = \begin{bmatrix} 2 & -3 & 1 & -1 \\ 1 & 1 & -2 & 1 \end{bmatrix}$, $B = \begin{bmatrix} 1 & 2 \\ -1 & 1 \\ 5 & 4 \\ 10 & 5 \end{bmatrix}$

26. $A = \begin{bmatrix} 2 & -1 & 3 & 2 \\ 1 & 0 & -2 & 1 \end{bmatrix}$, $B = \begin{bmatrix} -1 & 2 \\ 1 & 1 \\ 3 & -4 \\ 6 & 5 \end{bmatrix}$

In Exercises 27–34, perform the indicated matrix operations given that A, B, and C are defined as follows. If an operation is not defined, state the reason.

$$A = \begin{bmatrix} 4 & 0 \\ -3 & 5 \\ 0 & 1 \end{bmatrix} \quad B = \begin{bmatrix} 5 & 1 \\ -2 & -2 \end{bmatrix} \quad C = \begin{bmatrix} 1 & -1 \\ -1 & 1 \end{bmatrix}$$

27. $4B - 3C$ **28.** $5C - 2B$
29. $BC + CB$ **30.** $A(B + C)$
31. $A - C$ **32.** $B - A$
33. $A(BC)$ **34.** $A(CB)$

Application Exercises

The + sign in the figure is shown using 9 pixels in a 3 × 3 grid. The color levels are given to the right of the figure. Use the matrix $\begin{bmatrix} 1 & 3 & 1 \\ 3 & 3 & 3 \\ 1 & 3 & 1 \end{bmatrix}$ that represents a digital photograph of the + sign to solve Exercises 35–38.

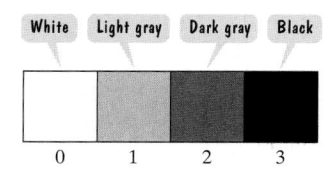

35. Adjust the contrast by changing the black to dark gray and the light gray to white. Use matrix addition to accomplish this.

36. Adjust the contrast by changing the black to dark gray and the light gray to black. Use matrix addition to accomplish this.

37. Adjust the contrast by changing the black to light gray and the light gray to dark gray. Use matrix addition to accomplish this.

38. Adjust the contrast by leaving the black alone and changing the light gray to white. Use matrix addition to accomplish this.

39. a. Write a 3 × 3 matrix A that represents a digital photograph of the number 1 in black on a white background.
  b. Find a matrix B so that $A + B$ darkens only the white background to light gray.

40. a. Write a 3 × 3 matrix A that represents a digital photograph of the letter T in light gray on a white background.
  b. Find a matrix B so that $A + B$ darkens only the letter T from light gray to black.

41. A virus strikes a college campus. Students are either sick, well, or carriers of the virus. The percentages of people in each category are given by the following matrix, which we'll call A.

	Freshman	Sophomore	Junior	Senior
Well	15%	25%	20%	10%
Sick	35%	40%	35%	70%
Carrier	50%	35%	45%	20%

The student population is distributed by class and gender as given by the following matrix, which we'll call B.

	Male	Female
Freshman	820	640
Sophomore	950	1020
Junior	680	720
Senior	930	910

a. Calculate the product AB.
b. How many sick females are there?
c. How many male carriers are there?

42. In a certain county, the proportion of voters in each age group registered as Republicans, Democrats, or Independents is given by the following matrix, which we'll call A.

	18–30	Age **31–50**	**Over 50**
Republicans	0.4	0.30	0.70
Democrats	0.30	0.60	0.25
Independents	0.30	0.10	0.05

The distribution, by age and gender, of this county's voting population is given by the following matrix, which we'll call B.

		Male	Female
	18–30	6000	8000
Age	**31–50**	12,000	14,000
	Over 50	14,000	16,000

a. Calculate the product AB.
b. How many female Democrats are there?
c. How many male Republicans are there?

43. The final grade in a particular course is determined by grades on the midterm and final. The grades for five students and the two grading systems are modeled by the following matrices. Call the first matrix A and the second B.

	Midterm	Final
Student 1	76	92
Student 2	74	84
Student 3	94	86
Student 4	84	62
Student 5	58	80

	System 1	System 2
Midterm	0.5	0.3
Final	0.5	0.7

a. Describe the grading system that is represented by matrix B.
b. Compute the matrix AB and assign each of the five students a final course grade first using system 1

and then using system 2. $(89.5 - 100 = A,$ $79.5 - 89.4 = B, 69.5 - 79.4 = C, 59.5 - 69.4 = D,$ below $59.5 = F)$

44. In the matrices shown below, a 1 represents a yes, and a 0 represents a no. The first matrix, A, describes whether or not three colleges in a state university system offer degrees in each program.

	Liberal Arts	Programs Engineering	Education
College 1	1	1	0
College 2	1	1	1
College 3	0	1	0

Each program requires that certain math courses be completed, indicated by the following matrix called B.

	General College Math	Intermediate Algebra	College Algebra	Trigonometry	Calculus
Liberal Arts	1	1	0	0	0
Engineering	0	0	1	1	1
Education	1	1	1	0	0

Find the product AB. Explain how this helps the college decide which courses to offer.

Writing in Mathematics

45. What is meant by the order of a matrix? Give an example with your explanation.

46. What does a_{ij} mean?

47. What are equal matrices?

48. How are matrices added?

49. Describe how to subtract matrices.

50. Describe matrices that cannot be added or subtracted.

51. Describe how to perform scalar multiplication. Provide an example with your description.

52. Describe how to multiply matrices.

53. Describe when the multiplication of two matrices is not defined.

54. If two matrices can be multiplied, describe how to determine the order of the product.

55. Low-resolution digital photographs use 262,144 pixels in a 512×512 grid. If you enlarge a low-resolution digital photograph enough, describe what will happen.

Technology Exercise

56. Use the matrix feature of a graphing utility to verify each of your answers to Exercises 27–34.

Critical Thinking Exercises

57. Find two matrices A and B such that $AB = BA$.

58. Consider a square matrix such that each element that is not on the main diagonal is zero. Experiment with such matrices (call each matrix A) by finding AA. Then write a sentence or two describing a method for multiplying this kind of matrix by itself.

59. If $AB = -BA$, then A and B are said to be anticommutative. Are $A = \begin{bmatrix} 0 & -1 \\ 1 & 0 \end{bmatrix}$ and $B = \begin{bmatrix} 1 & 0 \\ 0 & -1 \end{bmatrix}$ anticommutative?

Group Exercise

60. The interesting and useful applications of matrix theory are nearly unlimited. Applications of matrices range from representing digital photographs to predicting long-range trends in the stock market. Members of the group should research an application of matrices that they find intriguing. The group should then present a seminar to the class about this application.

SECTION 9.4 *Multiplicative Inverses of Matrices and Matrix Equations*

Objectives

1. Find the multiplicative inverse of a square matrix.
2. Use inverses to solve matrix equations.
3. Encode and decode messages.

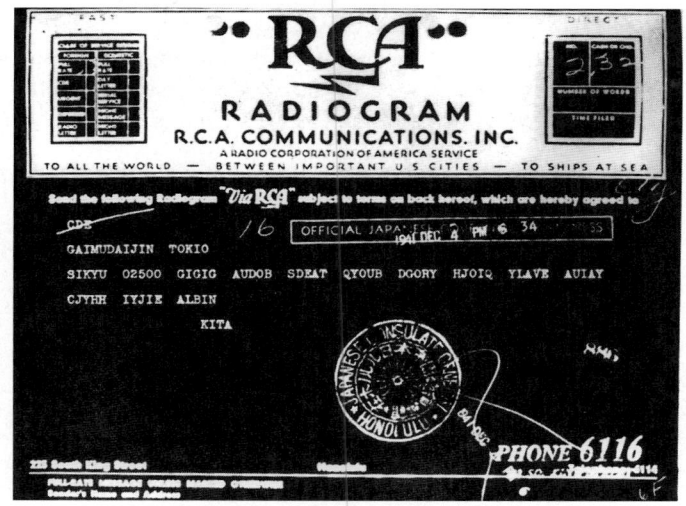

This 1941 RCA radiogram shows an encoded message from the Japanese government.

In 1939, Britain's secret service hired top chess players, mathematicians, and other masters of logic to break the code used by the Nazis in communications between headquarters and troops. The project, which employed over 10,000 people, broke the code less than a year later, providing the Allies with information about Nazi troop movements throughout World War II.

Messages must often be sent in such a way that the real meaning is hidden from everyone but the sender and the recipient. In this section, we will look at the role that matrices and their inverses play in this process.

The Multiplicative Identity Matrix

For the real numbers, we know that 1 is the multiplicative identity because $a \cdot 1 = 1 \cdot a = a$. Is there a similar property for matrix multiplication? That is, is there a matrix I such that $AI = A$ and $IA = A$? The answer is yes. A square matrix with 1s down the main diagonal and 0s elsewhere does not change the elements in a matrix when it multiplies that matrix. In the case of 2×2 matrices,

$$\begin{bmatrix} a_{11} & a_{12} \\ a_{21} & a_{22} \end{bmatrix} \begin{bmatrix} 1 & 0 \\ 0 & 1 \end{bmatrix} = \begin{bmatrix} a_{11} & a_{12} \\ a_{21} & a_{22} \end{bmatrix}$$

The elements in the matrix do not change.

and $\begin{bmatrix} 1 & 0 \\ 0 & 1 \end{bmatrix} \begin{bmatrix} a_{11} & a_{12} \\ a_{21} & a_{22} \end{bmatrix} = \begin{bmatrix} a_{11} & a_{12} \\ a_{21} & a_{22} \end{bmatrix}.$

The elements in the matrix do not change.

An $n \times n$ square matrix whose main diagonal elements are 1s, while all other elements are 0s, is called the **multiplicative identity matrix of order n**, designated by I_n. For example,

$$I_2 = \begin{bmatrix} 1 & 0 \\ 0 & 1 \end{bmatrix}, \quad I_3 = \begin{bmatrix} 1 & 0 & 0 \\ 0 & 1 & 0 \\ 0 & 0 & 1 \end{bmatrix},$$

and so on.

1 Find the multiplicative inverse of a square matrix.

The Multiplicative Inverse of a Matrix

The multiplicative identity matrix, I_n, will help us to define a new concept: the multiplicative inverse of a matrix. To do so, let's consider a similar concept, the multiplicative inverse of a nonzero number, a. Recall that the multiplicative inverse of a is $\frac{1}{a}$. The multiplicative inverse has the following property:

$$a \cdot \frac{1}{a} = 1 \quad \text{and} \quad \frac{1}{a} \cdot a = 1.$$

We can define the multiplicative inverse of a square matrix in a similar manner.

Definition of the Multiplicative Inverse of a Square Matrix

Let A be an $n \times n$ matrix. If there exists an $n \times n$ matrix A^{-1} (read: "A inverse") such that

$$AA^{-1} = I_n \quad \text{and} \quad A^{-1}A = I_n,$$

then A^{-1} is the **multiplicative inverse** of A.

We have seen that matrix multiplication is not commutative. Thus, to show that matrix B is the multiplicative inverse of matrix A, find both AB and BA. If B is the multiplicative inverse of A, both products (AB and BA) will be the multiplicative identity matrix, I_n.

EXAMPLE 1 The Multiplicative Inverse of a Matrix

Show that B is the multiplicative inverse of A, where

$$A = \begin{bmatrix} -1 & 3 \\ 2 & -5 \end{bmatrix} \quad \text{and} \quad B = \begin{bmatrix} 5 & 3 \\ 2 & 1 \end{bmatrix}.$$

Solution To show that B is the multiplicative inverse of A, we must find the products AB and BA. If B is the multiplicative inverse of A, then AB will be the multiplicative identity matrix and BA will be the multiplicative identity matrix. Because A and B are 2×2 matrices, $n = 2$. Thus, we denote the multiplicative identity matrix as I_2; it is also a 2×2 matrix. We must show that

- $AB = I_2 = \begin{bmatrix} 1 & 0 \\ 0 & 1 \end{bmatrix}.$

- $BA = I_2 = \begin{bmatrix} 1 & 0 \\ 0 & 1 \end{bmatrix}.$

$$AB = \begin{bmatrix} -1 & 3 \\ 2 & -5 \end{bmatrix} \begin{bmatrix} 5 & 3 \\ 2 & 1 \end{bmatrix}$$

$$= \begin{bmatrix} -1(5) + 3(2) & -1(3) + 3(1) \\ 2(5) + (-5)(2) & 2(3) + (-5)(1) \end{bmatrix} = \begin{bmatrix} 1 & 0 \\ 0 & 1 \end{bmatrix}$$

$$BA = \begin{bmatrix} 5 & 3 \\ 2 & 1 \end{bmatrix} \begin{bmatrix} -1 & 3 \\ 2 & -5 \end{bmatrix}$$

$$= \begin{bmatrix} 5(-1) + 3(2) & 5(3) + 3(-5) \\ 2(-1) + 1(2) & 2(3) + 1(-5) \end{bmatrix} = \begin{bmatrix} 1 & 0 \\ 0 & 1 \end{bmatrix}$$

Both products give the multiplicative identity matrix. Thus, B is the multiplicative inverse of A and we can designate B as $A^{-1} = \begin{bmatrix} 5 & 3 \\ 2 & 1 \end{bmatrix}$.

Check Point 1 Show that B is the multiplicative inverse of A, where

$$A = \begin{bmatrix} 2 & 1 \\ 1 & 1 \end{bmatrix} \quad \text{and} \quad B = \begin{bmatrix} 1 & -1 \\ -1 & 2 \end{bmatrix}.$$

One method for finding the multiplicative inverse of a matrix A is to begin by denoting the elements in A^{-1} with variables. Using the equation $AA^{-1} = I_n$ we can find a value for each element in the multiplicative inverse that was represented by a variable. Example 2 shows how this is done.

EXAMPLE 2   Finding the Multiplicative Inverse of a Matrix

Find the multiplicative inverse of

$$A = \begin{bmatrix} 2 & 1 \\ 5 & 3 \end{bmatrix}.$$

Solution   Let us denote the multiplicative inverse by

$$A^{-1} = \begin{bmatrix} x & y \\ z & w \end{bmatrix}.$$

Because A is a 2×2 matrix, we use the equation $AA^{-1} = I_2$ to find values for x, y, z, and w.

$$\underset{A}{\begin{bmatrix} 2 & 1 \\ 5 & 3 \end{bmatrix}} \underset{A^{-1}}{\begin{bmatrix} x & y \\ z & w \end{bmatrix}} = \underset{I_2}{\begin{bmatrix} 1 & 0 \\ 0 & 1 \end{bmatrix}}$$

$$\begin{bmatrix} 2x + z & 2y + w \\ 5x + 3z & 5y + 3w \end{bmatrix} = \begin{bmatrix} 1 & 0 \\ 0 & 1 \end{bmatrix} \qquad \text{Use row-by-column matrix multiplication on the left.}$$

We now equate corresponding elements to obtain the following two systems of linear equations.

$$2x + z = 1 \qquad \text{and} \qquad 2y + w = 0$$
$$5x + 3z = 0 \qquad\qquad\qquad 5y + 3w = 1$$

Each of these systems can be solved using the addition method.

$$
\begin{array}{lll}
2x + \ z = 1 & \xrightarrow{\text{Multiply by } -3.} & -6x - 3z = -3 \\
5x + 3z = 0 & \xrightarrow{\text{No change}} & \underline{\ \ 5x + 3z = \ \ 0} \\
& \text{Add:} & -x = -3 \\
& & x = \ \ 3 \\
& \text{Use back-substitution.} & z = -5
\end{array}
$$

Discovery

Verify that the inverse matrix found in Example 2 is correct. Use matrix multiplication to show that

$$AA^{-1} = I_2 \quad \text{and} \quad A^{-1}A = I_2,$$

where

$$I_2 = \begin{bmatrix} 1 & 0 \\ 0 & 1 \end{bmatrix}.$$

$$
\begin{array}{lll}
2y + \ w = 0 & \xrightarrow{\text{Multiply by } -3.} & -6y - 3w = \ \ 0 \\
5y + 3w = 1 & \xrightarrow{\text{No change}} & \underline{\ \ 5y + 3w = \ \ 1} \\
& \text{Add:} & -y = \ \ 1 \\
& & y = -1 \\
& \text{Use back-substitution.} & w = \ \ 2
\end{array}
$$

Using these values, we have

$$A^{-1} = \begin{bmatrix} x & y \\ z & w \end{bmatrix} = \begin{bmatrix} 3 & -1 \\ -5 & 2 \end{bmatrix}.$$

Check Point 2 Find the multiplicative inverse of $A = \begin{bmatrix} 5 & 7 \\ 2 & 3 \end{bmatrix}$.

Only square matrices of order $n \times n$ have multiplicative inverses, but not every square matrix possesses a multiplicative inverse. For example, suppose that you apply the procedure of Example 2 to $A = \begin{bmatrix} -6 & 4 \\ -3 & 2 \end{bmatrix}$:

This is A.   This represents A^{-1}.   This is the multiplicative identity matrix.

$$\begin{bmatrix} -6 & 4 \\ -3 & 2 \end{bmatrix} \begin{bmatrix} x & y \\ z & w \end{bmatrix} = \begin{bmatrix} 1 & 0 \\ 0 & 1 \end{bmatrix}.$$

Multiplying matrices on the left and equating corresponding elements results in inconsistent systems with no solutions. There are no values for x, y, z, and w. This shows that matrix A does not have a multiplicative inverse.

A nonsquare matrix, one with a different number of rows than columns, cannot have a multiplicative inverse. If A is an $m \times n$ matrix and B is an $n \times m$ matrix $(n \neq m)$, then the products AB and BA are of different orders. This means that they could not be equal to each other, so that AB and BA could not both equal the multiplicative identity matrix.

If a square matrix has a multiplicative inverse, that inverse is unique. This means that the square matrix has no more than one inverse. If a square matrix has a multiplicative inverse, it is said to be **invertible**.

A Quick Method for Finding the Multiplicative Inverse of a 2 × 2 Matrix

The following rule enables us to calculate the multiplicative inverse, if there is one, of a 2 × 2 matrix.

Multiplicative Inverse of a 2 × 2 Matrix

If $A = \begin{bmatrix} a & b \\ c & d \end{bmatrix}$, then $A^{-1} = \dfrac{1}{ad - bc} \begin{bmatrix} d & -b \\ -c & a \end{bmatrix}$.

The matrix A is invertible if and only if $ad - bc \neq 0$. If $ad - bc = 0$, then A does not have a multiplicative inverse.

EXAMPLE 3 Using the Quick Method to Find Multiplicative Inverses

Find the multiplicative inverse of

$$A = \begin{bmatrix} -1 & -2 \\ 3 & 4 \end{bmatrix}.$$

Solution

$$A = \begin{bmatrix} -1 & -2 \\ 3 & 4 \end{bmatrix}$$

This is the given matrix. We've designated the elements a, b, c, and d.

$$A^{-1} = \frac{1}{ad - bc} \begin{bmatrix} d & -b \\ -c & a \end{bmatrix}$$

This is the formula for the inverse of $\begin{bmatrix} a & b \\ c & d \end{bmatrix}$.

$$= \frac{1}{(-1)(4) - (-2)(3)} \begin{bmatrix} 4 & -(-2) \\ -3 & -1 \end{bmatrix}$$

Apply the formula with $a = -1$, $b = -2$, $c = 3$, and $d = 4$.

$$= \frac{1}{2} \begin{bmatrix} 4 & 2 \\ -3 & -1 \end{bmatrix}$$

Simplify.

$$= \begin{bmatrix} 2 & 1 \\ -\frac{3}{2} & -\frac{1}{2} \end{bmatrix}$$

Perform the scalar multiplication by multiplying each element in the matrix by $\frac{1}{2}$.

The inverse of $A = \begin{bmatrix} -1 & -2 \\ 3 & 4 \end{bmatrix}$ is $A^{-1} = \begin{bmatrix} 2 & 1 \\ -\frac{3}{2} & -\frac{1}{2} \end{bmatrix}$.

We can verify this result by showing that $AA^{-1} = I_2$ and $A^{-1}A = I_2$.

Check Point 3

Find the multiplicative inverse of

$$A = \begin{bmatrix} 3 & -2 \\ -1 & 1 \end{bmatrix}.$$

Finding Multiplicative Inverses of $n \times n$ Matrices with n Greater Than 2

To find the multiplicative inverse of a 3×3 invertible matrix, we begin by denoting the elements in the multiplicative inverse with variables. Here is an example:

$$\begin{bmatrix} -1 & -1 & -1 \\ 4 & 5 & 0 \\ 0 & 1 & -3 \end{bmatrix} \begin{bmatrix} x_1 & x_2 & x_3 \\ y_1 & y_2 & y_3 \\ z_1 & z_2 & z_3 \end{bmatrix} = \begin{bmatrix} 1 & 0 & 0 \\ 0 & 1 & 0 \\ 0 & 0 & 1 \end{bmatrix}.$$

This is matrix A whose inverse we wish to find.   This represents A^{-1}.   This is the multiplicative identity matrix, I_3.

We multiply the matrices on the left, using the row-by-column definition of matrix multiplication.

$$\begin{bmatrix} -x_1 - y_1 - z_1 & -x_2 - y_2 - z_2 & -x_3 - y_3 - z_3 \\ 4x_1 + 5y_1 + 0z_1 & 4x_2 + 5y_2 + 0z_2 & 4x_3 + 5y_3 + 0z_3 \\ 0x_1 + 1y_1 - 3z_1 & 0x_2 + 1y_2 - 3z_2 & 0x_3 + 1y_3 - 3z_3 \end{bmatrix} = \begin{bmatrix} 1 & 0 & 0 \\ 0 & 1 & 0 \\ 0 & 0 & 1 \end{bmatrix}$$

We now equate corresponding entries to obtain the following three systems of linear equations.

$$\begin{array}{lll} -x_1 - y_1 - z_1 = 1 & -x_2 - y_2 - z_2 = 0 & -x_3 - y_3 - z_3 = 0 \\ 4x_1 + 5y_1 + 0z_1 = 0 & 4x_2 + 5y_2 + 0z_2 = 1 & 4x_3 + 5y_3 + 0z_3 = 0 \\ 0x_1 + \ y_1 - 3z_1 = 0 & 0x_2 + \ y_2 - 3z_2 = 0 & 0x_3 + \ y_3 - 3z_3 = 1 \end{array}$$

Notice that the variables on the left of the equal sign have the same coefficients in each system. We can use Gauss-Jordan elimination to solve all three at once. Form an augmented matrix that contains the coefficients of the three systems to the left of the vertical line and the constants for the systems to the right.

$$\left[\begin{array}{ccc|ccc} -1 & -1 & -1 & 1 & 0 & 0 \\ 4 & 5 & 0 & 0 & 1 & 0 \\ 0 & 1 & -3 & 0 & 0 & 1 \end{array} \right]$$

Coefficients of the three systems   Constants on the right in each of the three systems

To solve all three systems using Gauss-Jordan elimination, we must obtain
$$\begin{bmatrix} 1 & 0 & 0 \\ 0 & 1 & 0 \\ 0 & 0 & 1 \end{bmatrix}$$
to the left of the vertical line. Use matrix row operations, working one column at a time. Obtain 1 in the required position. Then obtain 0s in the other two positions. Using these operations, we obtain the matrix

$$\left[\begin{array}{ccc|ccc} 1 & 0 & 0 & 15 & 4 & -5 \\ 0 & 1 & 0 & -12 & -3 & 4 \\ 0 & 0 & 1 & -4 & -1 & 1 \end{array} \right].$$

This augmented matrix provides the solutions to the three systems of equations. They are given by

$$\begin{bmatrix} 1 & 0 & 0 & | & 15 \\ 0 & 1 & 0 & | & -12 \\ 0 & 0 & 1 & | & -4 \end{bmatrix} \qquad \begin{matrix} x_1 = 15 \\ y_1 = -12 \\ z_1 = -4 \end{matrix}$$

and

$$\begin{bmatrix} 1 & 0 & 0 & | & 4 \\ 0 & 1 & 0 & | & -3 \\ 0 & 0 & 1 & | & -1 \end{bmatrix} \qquad \begin{matrix} x_2 = 4 \\ y_2 = -3 \\ z_2 = -1 \end{matrix}$$

and

$$\begin{bmatrix} 1 & 0 & 0 & | & -5 \\ 0 & 1 & 0 & | & 4 \\ 0 & 0 & 1 & | & 1 \end{bmatrix} \qquad \begin{matrix} x_3 = -5 \\ y_3 = 4 \\ z_3 = 1 \end{matrix}$$

The inverse matrix is

$$\begin{bmatrix} x_1 & x_2 & x_3 \\ y_1 & y_2 & y_3 \\ z_1 & z_2 & z_3 \end{bmatrix} = \begin{bmatrix} 15 & 4 & -5 \\ -12 & -3 & 4 \\ -4 & -1 & 1 \end{bmatrix}.$$

Technology

You can use a graphing utility to find the inverse of

$$A = \begin{bmatrix} -1 & -1 & -1 \\ 4 & 5 & 0 \\ 0 & 1 & -3 \end{bmatrix}.$$

Enter MATRIX [A] and then use the inverse key. The display, $[A]^{-1}$, should be

$$\begin{bmatrix} 15 & 4 & -5 \\ -12 & -3 & 4 \\ -4 & -1 & 1 \end{bmatrix}.$$

Take a second look at the matrix obtained at the point where Gauss-Jordan elimination was completed. Notice that the 3×3 matrix to the right of the vertical bar is the multiplicative inverse of A. Also notice that the multiplicative identity matrix, I_3 is the matrix that appears to the left of the vertical bar.

$$\begin{bmatrix} 1 & 0 & 0 & | & 15 & 4 & -5 \\ 0 & 1 & 0 & | & -12 & -3 & 4 \\ 0 & 0 & 1 & | & -4 & -1 & 1 \end{bmatrix}$$

This is the multiplicative identity, I_3.

This is the multiplicative inverse of A.

The observations in the voice balloons and the procedures followed above give us a general method for finding the multiplicative inverse of an invertible matrix.

Study Tip

Because we have a quick method for finding the multiplicative inverse of a 2×2 matrix, the procedure on the right is recommended for matrices of order 3×3 or greater when a graphing utility is not being used.

Procedure for Finding the Multiplicative Inverse of an Invertible Matrix

To find A^{-1} for any $n \times n$ matrix A for which A^{-1} exists:

1. Form the augmented matrix $[A\,|\,I]$, where I is the multiplicative identity matrix of the same order as the given matrix A.
2. Perform row transformations on $[A\,|\,I]$ to obtain a matrix of the form $[I\,|\,B]$. This is equivalent to using Gauss-Jordan elimination to change A into the identity matrix.
3. Matrix B is A^{-1}.
4. Verify the result by showing that $AA^{-1} = I$ and $A^{-1}A = I$.

EXAMPLE 4 Finding the Multiplicative Inverse of a 3 × 3 Matrix

Find the multiplicative inverse of

$$A = \begin{bmatrix} 1 & -1 & 1 \\ 0 & -2 & 1 \\ -2 & -3 & 0 \end{bmatrix}.$$

Solution

Step 1 Form the augmented matrix $[A \,|\, I_3]$.

$$\begin{bmatrix} 1 & -1 & 1 & | & 1 & 0 & 0 \\ 0 & -2 & 1 & | & 0 & 1 & 0 \\ -2 & -3 & 0 & | & 0 & 0 & 1 \end{bmatrix}$$

This is matrix A.

This is I_3, the multiplicative identity matrix, with 1s down the main diagonal and 0s elsewhere.

Step 2 Perform row transformations on $[A \,|\, I_3]$ **to obtain a matrix of the form** $[I_3 \,|\, B]$. We want 1s down the main diagonal to the left of the vertical dividing line and 0s elsewhere.

$$\begin{bmatrix} 1 & -1 & 1 & | & 1 & 0 & 0 \\ 0 & -2 & 1 & | & 0 & 1 & 0 \\ -2 & -3 & 0 & | & 0 & 0 & 1 \end{bmatrix} \xrightarrow[\text{by } 2R_1 + R_3]{\text{Replace row 3}} \begin{bmatrix} 1 & -1 & 1 & | & 1 & 0 & 0 \\ 0 & -2 & 1 & | & 0 & 1 & 0 \\ 0 & -5 & 2 & | & 2 & 0 & 1 \end{bmatrix} \xrightarrow{-\frac{1}{2}R_2}$$

$$\begin{bmatrix} 1 & -1 & 1 & | & 1 & 0 & 0 \\ 0 & 1 & -\frac{1}{2} & | & 0 & -\frac{1}{2} & 0 \\ 0 & -5 & 2 & | & 2 & 0 & 1 \end{bmatrix} \xrightarrow[\text{Replace row 3 by } 5R_2 + R_3.]{\text{Replace row 1 by } 1R_2 + R_1.} \begin{bmatrix} 1 & 0 & \frac{1}{2} & | & 1 & -\frac{1}{2} & 0 \\ 0 & 1 & -\frac{1}{2} & | & 0 & -\frac{1}{2} & 0 \\ 0 & 0 & -\frac{1}{2} & | & 2 & -\frac{5}{2} & 1 \end{bmatrix} \xrightarrow{-2R_3}$$

$$\begin{bmatrix} 1 & 0 & \frac{1}{2} & | & 1 & -\frac{1}{2} & 0 \\ 0 & 1 & -\frac{1}{2} & | & 0 & -\frac{1}{2} & 0 \\ 0 & 0 & 1 & | & -4 & 5 & -2 \end{bmatrix} \xrightarrow[\text{Replace row 2 by } \frac{1}{2}R_3 + R_2.]{\text{Replace row 1 by } -\frac{1}{2}R_3 + R_1.} \begin{bmatrix} 1 & 0 & 0 & | & 3 & -3 & 1 \\ 0 & 1 & 0 & | & -2 & 2 & -1 \\ 0 & 0 & 1 & | & -4 & 5 & -2 \end{bmatrix}$$

This is the multiplicative identity, I_3.

This is the multiplicative inverse of A.

Step 3 Matrix B is A^{-1}. The matrix just shown is in the form $[I_3 \,|\, B]$. The multiplicative identity matrix is on the left of the vertical bar. Matrix B, the multiplicative inverse of A, is on the right. Thus, the multiplicative inverse of A is

$$A^{-1} = \begin{bmatrix} 3 & -3 & 1 \\ -2 & 2 & -1 \\ -4 & 5 & -2 \end{bmatrix}.$$

Step 4 Verify the result by showing that $AA^{-1} = I_3$ **and** $A^{-1}A = I_3$. Try confirming the result by multiplying A and A^{-1} to obtain I_3. Do you obtain I_3 if you reverse the order of the multiplication?

We have seen that not all square matrices have multiplicative inverses. If the row transformations in step 2 result in all zeros in a row or column to the left of the vertical line, the given matrix does not have a multiplicative inverse.

Check Point 4

Find the multiplicative inverse of

$$A = \begin{bmatrix} 1 & 0 & 2 \\ -1 & 2 & 3 \\ 1 & -1 & 0 \end{bmatrix}.$$

Summary: Finding Multiplicative Inverses for Invertible Matrices

Use a graphing utility with matrix capabilities,

or

a. If the matrix is 2×2: The inverse of $A = \begin{bmatrix} a & b \\ c & d \end{bmatrix}$ is

$$A^{-1} = \frac{1}{ad - bc} \begin{bmatrix} d & -b \\ -c & a \end{bmatrix}.$$

b. If the matrix A is $n \times n$ where $n > 2$: Use the procedure on page 781. Form $[A \,|\, I]$ and use row transformations to obtain $[I \,|\, B]$. $A^{-1} = B$.

---

2 Use inverses to solve matrix equations.

Solving Systems of Equations Using Multiplicative Inverses of Matrices

Matrix multiplication can be used to represent a system of linear equations.

Linear System

$$a_1 x + b_1 y + c_1 z = d_1$$
$$a_2 x + b_2 y + c_2 z = d_2$$
$$a_3 x + b_3 y + c_3 z = d_3$$

Matrix Form of the System

$$\begin{bmatrix} a_1 & b_1 & c_1 \\ a_2 & b_2 & c_2 \\ a_3 & b_3 & c_3 \end{bmatrix} \begin{bmatrix} x \\ y \\ z \end{bmatrix} = \begin{bmatrix} d_1 \\ d_2 \\ d_3 \end{bmatrix}$$

This matrix contains the system's coefficients.   This matrix contains the system's variables.   This matrix contains the system's constants.

You can work with the matrix form on the right and obtain the form of the linear system on the left. To do so, perform the matrix multiplication on the left side of the matrix equation. Then equate the corresponding elements.

The matrix equation

$$\underset{A}{\begin{bmatrix} a_1 & b_1 & c_1 \\ a_2 & b_2 & c_2 \\ a_3 & b_3 & c_3 \end{bmatrix}} \underset{X}{\begin{bmatrix} x \\ y \\ z \end{bmatrix}} = \underset{B}{\begin{bmatrix} d_1 \\ d_2 \\ d_3 \end{bmatrix}}$$

is abbreviated as $AX = B$, where A is the **coefficient matrix** of the system, and X and B are matrices containing one column, called **column matrices**. The matrix B is called the **constant matrix**.

Here is a specific example of a linear system and its matrix form.

Linear System

$$x - y + z = 2$$
$$-2y + z = 2$$
$$-2x - 3y = \tfrac{1}{2}$$

Matrix Form

Coefficients

$$\begin{bmatrix} 1 & -1 & 1 \\ 0 & -2 & 1 \\ -2 & -3 & 0 \end{bmatrix} \begin{bmatrix} x \\ y \\ z \end{bmatrix} = \begin{bmatrix} 2 \\ 2 \\ \tfrac{1}{2} \end{bmatrix}$$

Constants

A, the coefficient matrix X $=$ B, the constant matrix

The matrix equation $AX = B$ can be solved using A^{-1} if it exists.

$AX = B$ This is the matrix equation.

$A^{-1}AX = A^{-1}B$ Multiply both sides by A^{-1}. Because matrix multiplication is not commutative, put A^{-1} in the same left position on both sides.

$I_nX = A^{-1}B$ The multiplicative inverse property tells us that $A^{-1}A = I_n$.

$X = A^{-1}B$ Because I_n is the multiplicative identity, $I_nX = X$.

We see that if $AX = B$, then $X = A^{-1}B$.

Solving a System Using A^{-1}

If $AX = B$ has a unique solution, $X = A^{-1}B$. To solve a linear system of equations, multiply A^{-1} and B to find X.

EXAMPLE 5   **Using the Inverse of a Matrix to Solve a System**

Solve the system by using A^{-1}, the inverse of the coefficient matrix.

$$x - y + z = 2$$
$$-2y + z = 2$$
$$-2x - 3y = \tfrac{1}{2}$$

Solution   The linear system can be written as

$$\underbrace{\begin{bmatrix} 1 & -1 & 1 \\ 0 & -2 & 1 \\ -2 & -3 & 0 \end{bmatrix}}_{A} \underbrace{\begin{bmatrix} x \\ y \\ z \end{bmatrix}}_{X} = \underbrace{\begin{bmatrix} 2 \\ 2 \\ \tfrac{1}{2} \end{bmatrix}}_{B}.$$

The solution is given by $X = A^{-1}B$. Consequently, we must find A^{-1}. We found the inverse of matrix A in Example 4. Using this result,

$$X = A^{-1}B = \begin{bmatrix} 3 & -3 & 1 \\ -2 & 2 & -1 \\ -4 & 5 & -2 \end{bmatrix} \begin{bmatrix} 2 \\ 2 \\ \tfrac{1}{2} \end{bmatrix} = \begin{bmatrix} 3\cdot 2 + (-3)\cdot 2 + 1\cdot\tfrac{1}{2} \\ -2\cdot 2 + 2\cdot 2 + (-1)\cdot\tfrac{1}{2} \\ -4\cdot 2 + 5\cdot 2 + (-2)\cdot\tfrac{1}{2} \end{bmatrix} = \begin{bmatrix} \tfrac{1}{2} \\ -\tfrac{1}{2} \\ 1 \end{bmatrix}$$

Thus, $x = \tfrac{1}{2}$, $y = -\tfrac{1}{2}$, and $z = 1$. The solution set is $\left\{\left(\tfrac{1}{2}, -\tfrac{1}{2}, 1\right)\right\}$.

> **Check Point 5**
>
> Solve the system by using A^{-1}, the inverse of the coefficient matrix that you found in Checkpoint 4.
>
> $$\begin{array}{rcl} x + 2z &=& 6 \\ -x + 2y + 3z &=& -5 \\ x - y &=& 6 \end{array}$$

3 Encode and decode messages.

Applications of Matrix Inverses to Coding

A **cryptogram** is a message written so that no one other than the intended recipient can understand it. To encode a message, we begin by assigning a number to each letter in the alphabet: $A = 1, B = 2, C = 3, \ldots, Z = 26$, and a space $= 0$. For example, the numerical equivalent of the word MATH is 13, 1, 20, 8. The numerical equivalent of the message is then converted into a matrix. Finally, an invertible matrix can be used to convert the message into code. The multiplicative inverse of this matrix can be used to decode the message.

> ### Encoding a Word or Message
>
> **1.** Express the word or message numerically.
> **2.** List the numbers in step 1 by columns and form a square matrix. If you do not have enough numbers to form a square matrix, put zeros in any remaining spaces in the last column.
> **3.** Select any square invertible matrix, called the **coding matrix**, the same size as the matrix in step 2. Multiply the coding matrix by the square matrix that expresses the message numerically. The resulting matrix is the **coded matrix**.
> **4.** Use the numbers, by columns, from the coded matrix in step 3 to write the encoded message.

EXAMPLE 6 Encoding a Word

Use matrices to encode the word MATH.

Solution

Step 1 Express the word numerically. As shown previously, the numerical equivalent of MATH is 13, 1, 20, 8.

Step 2 List the numbers in step 1 by columns and form a square matrix. The 2×2 matrix is

$$\begin{bmatrix} 13 & 20 \\ 1 & 8 \end{bmatrix}.$$

Step 3 Multiply the matrix in step 2 by a square invertible matrix. We will use $\begin{bmatrix} -2 & -3 \\ 3 & 4 \end{bmatrix}$ as the coding matrix.

$$\underset{\substack{\text{Coding} \\ \text{matrix}}}{\begin{bmatrix} -2 & -3 \\ 3 & 4 \end{bmatrix}} \underset{\substack{\text{Numerical} \\ \text{representation of} \\ \text{MATH}}}{\begin{bmatrix} 13 & 20 \\ 1 & 8 \end{bmatrix}} = \begin{bmatrix} -2(13) - 3(1) & -2(20) - 3(8) \\ 3(13) + 4(1) & 3(20) + 4(8) \end{bmatrix}$$

$$= \underset{\substack{\text{Coded} \\ \text{matrix}}}{\begin{bmatrix} -29 & -64 \\ 43 & 92 \end{bmatrix}}$$

Step 4   Use the numbers, by columns, from the coded matrix in step 3 to write the encoded message. The encoded message is $-29, 43, -64, 92$.

> **Check Point 6**   Use the coding matrix in Example 6, $\begin{bmatrix} -2 & -3 \\ 3 & 4 \end{bmatrix}$, to encode the word BASE.

The inverse of a coding matrix can be used to decode a word or message that was encoded.

Decoding a Word or Message That Was Encoded

1. Find the multiplicative inverse of the coding matrix.
2. Multiply the multiplicative inverse of the coding matrix and the coded matrix.
3. Express the numbers, by columns, from the matrix in step 2 as letters.

EXAMPLE 7   Decoding a Word

Decode $-29, 43, -64, 92$ from Example 6.

Solution

Step 1   Find the inverse of the coding matrix. The coding matrix in Example 6 was $\begin{bmatrix} -2 & -3 \\ 3 & 4 \end{bmatrix}$. We use the formula for the multiplicative inverse of a 2×2 matrix to find the multiplicative inverse of this matrix. It is $\begin{bmatrix} 4 & 3 \\ -3 & -2 \end{bmatrix}$.

Step 2   Multiply the multiplicative inverse of the coding matrix and the coded matrix.

$$\underset{\substack{\text{Multiplicative inverse} \\ \text{of the coding matrix}}}{\begin{bmatrix} 4 & 3 \\ -3 & -2 \end{bmatrix}} \underset{\substack{\text{Coded} \\ \text{matrix}}}{\begin{bmatrix} -29 & -64 \\ 43 & 92 \end{bmatrix}} = \begin{bmatrix} 4(-29) + 3(43) & 4(-64) + 3(92) \\ -3(-29) - 2(43) & -3(-64) - 2(92) \end{bmatrix}$$

$$= \begin{bmatrix} 13 & 20 \\ 1 & 8 \end{bmatrix}$$

Step 3   Express the numbers, by columns, from the matrix in step 2 as letters. The numbers are $13, 1, 20$, and 8. Using letters, the decoded message is MATH.

> **Check Point 7**   Decode the word that you encoded in Checkpoint 6.

Decoding is simple for an authorized receiver who knows the coding matrix. Because any invertible matrix can be used for the coding matrix, decoding a cryptogram for an unauthorized receiver who does not know this matrix is extremely difficult.

EXERCISE SET 9.4

 Practice Exercises

In Exercises 1–12 find the products AB and BA to determine whether B is the multiplicative inverse of A.

1. $A = \begin{bmatrix} 4 & -3 \\ -5 & 4 \end{bmatrix}$, $B = \begin{bmatrix} 4 & 3 \\ 5 & 4 \end{bmatrix}$

2. $A = \begin{bmatrix} -2 & -1 \\ -1 & 1 \end{bmatrix}$, $B = \begin{bmatrix} 1 & 1 \\ 1 & 2 \end{bmatrix}$

3. $A = \begin{bmatrix} -4 & 0 \\ 1 & 3 \end{bmatrix}$, $B = \begin{bmatrix} -2 & 4 \\ 0 & 1 \end{bmatrix}$

4. $A = \begin{bmatrix} -2 & 4 \\ 1 & -2 \end{bmatrix}$, $B = \begin{bmatrix} 1 & 2 \\ -1 & -2 \end{bmatrix}$

5. $A = \begin{bmatrix} -2 & 1 \\ \frac{3}{2} & -\frac{1}{2} \end{bmatrix}$, $B = \begin{bmatrix} 1 & 2 \\ 3 & 4 \end{bmatrix}$

6. $A = \begin{bmatrix} 4 & 5 \\ 2 & 3 \end{bmatrix}$, $B = \begin{bmatrix} \frac{3}{2} & -\frac{5}{2} \\ -1 & 2 \end{bmatrix}$

7. $A = \begin{bmatrix} 0 & 1 & 0 \\ 0 & 0 & 1 \\ 1 & 0 & 0 \end{bmatrix}$ $B = \begin{bmatrix} 0 & 0 & 1 \\ 1 & 0 & 0 \\ 0 & 1 & 0 \end{bmatrix}$

8. $A = \begin{bmatrix} -2 & 1 & -1 \\ -5 & 2 & -1 \\ 3 & -1 & 1 \end{bmatrix}$ $B = \begin{bmatrix} 1 & 0 & 1 \\ 2 & 1 & 3 \\ -1 & 1 & 1 \end{bmatrix}$

9. $A = \begin{bmatrix} 1 & 2 & 3 \\ 1 & 3 & 4 \\ 1 & 4 & 3 \end{bmatrix}$ $B = \begin{bmatrix} \frac{7}{2} & -3 & \frac{1}{2} \\ -\frac{1}{2} & 0 & \frac{1}{2} \\ -\frac{1}{2} & 1 & -\frac{1}{2} \end{bmatrix}$

10. $A = \begin{bmatrix} 0 & 2 & 0 \\ 3 & 3 & 2 \\ 2 & 5 & 1 \end{bmatrix}$ $B = \begin{bmatrix} -3.5 & -1 & 2 \\ 0.5 & 0 & 0 \\ 4.5 & 2 & -3 \end{bmatrix}$

11. $A = \begin{bmatrix} 0 & 0 & -2 & 1 \\ -1 & 0 & 1 & 1 \\ 0 & 1 & -1 & 0 \\ 1 & 0 & 0 & -1 \end{bmatrix}$, $B = \begin{bmatrix} 1 & 2 & 0 & 3 \\ 0 & 1 & 1 & 1 \\ 0 & 1 & 0 & 1 \\ 1 & 2 & 0 & 2 \end{bmatrix}$

12. $A = \begin{bmatrix} 1 & -2 & 1 & 0 \\ 0 & 1 & -2 & 1 \\ 0 & 0 & 1 & -2 \\ 0 & 0 & 0 & 1 \end{bmatrix}$, $B = \begin{bmatrix} 1 & 2 & 3 & 4 \\ 0 & 1 & 2 & 3 \\ 0 & 0 & 1 & 2 \\ 0 & 0 & 0 & 1 \end{bmatrix}$

In Exercises 13–18, use the fact that if $A = \begin{bmatrix} a & b \\ c & d \end{bmatrix}$, then
$A^{-1} = \dfrac{1}{ad - bc} \begin{bmatrix} d & -b \\ -c & a \end{bmatrix}$ to find the inverse of each matrix, if possible. Check that $AA^{-1} = I_2$ and $A^{-1}A = I_2$.

13. $A = \begin{bmatrix} 2 & 3 \\ -1 & 2 \end{bmatrix}$ **14.** $A = \begin{bmatrix} 0 & 3 \\ 4 & -2 \end{bmatrix}$

15. $A = \begin{bmatrix} 3 & -1 \\ -4 & 2 \end{bmatrix}$ **16.** $A = \begin{bmatrix} 2 & -6 \\ 1 & -2 \end{bmatrix}$

17. $A = \begin{bmatrix} 10 & -2 \\ -5 & 1 \end{bmatrix}$ **18.** $A = \begin{bmatrix} 6 & -3 \\ -2 & 1 \end{bmatrix}$

In Exercises 19–24, find A^{-1} by forming $[A \,|\, I]$ and then using row transformations to obtain $[I \,|\, B]$, where $A^{-1} = [B]$. Check that $AA^{-1} = I$ and $A^{-1}A = I$.

19. $A = \begin{bmatrix} 2 & 2 & -1 \\ 0 & 3 & -1 \\ -1 & -2 & 1 \end{bmatrix}$ **20.** $A = \begin{bmatrix} 1 & -1 & 1 \\ 0 & 2 & -1 \\ 2 & 3 & 0 \end{bmatrix}$

21. $A = \begin{bmatrix} 5 & 0 & 2 \\ 2 & 2 & 1 \\ -3 & 1 & -1 \end{bmatrix}$ **22.** $A = \begin{bmatrix} 3 & 2 & 6 \\ 1 & 1 & 2 \\ 2 & 2 & 5 \end{bmatrix}$

23. $A = \begin{bmatrix} 1 & 0 & 0 & 0 \\ 0 & -1 & 0 & 0 \\ 0 & 0 & 3 & 0 \\ 1 & 0 & 0 & 1 \end{bmatrix}$ **24.** $A = \begin{bmatrix} 2 & 0 & 0 & 1 \\ 0 & 1 & 0 & 0 \\ 0 & 0 & -1 & 0 \\ 0 & 0 & 0 & 2 \end{bmatrix}$

In Exercises 25–28, write each linear system as a matrix equation in the form $AX = B$, where A is the coefficient matrix and B is the constant matrix.

25. $6x + 5y = 13$ **26.** $7x + 5y = 23$
$5x + 4y = 10$ $3x + 2y = 10$

27. $x + 3y + 4z = -3$ **28.** $x + 4y - z = 3$
$x + 2y + 3z = -2$ $x + 3y - 2z = 5$
$x + 4y + 3z = -6$ $2x + 7y - 5z = 12$

In Exercises 29–32, write each matrix equation as a system of linear equations without matrices.

29. $\begin{bmatrix} 4 & -7 \\ 2 & -3 \end{bmatrix} \begin{bmatrix} x \\ y \end{bmatrix} = \begin{bmatrix} -3 \\ 1 \end{bmatrix}$

30. $\begin{bmatrix} 3 & 0 \\ -3 & 1 \end{bmatrix} \begin{bmatrix} x \\ y \end{bmatrix} = \begin{bmatrix} 6 \\ -7 \end{bmatrix}$

31. $\begin{bmatrix} 2 & 0 & -1 \\ 0 & 3 & 0 \\ 1 & 1 & 0 \end{bmatrix} \begin{bmatrix} x \\ y \\ z \end{bmatrix} = \begin{bmatrix} 6 \\ 9 \\ 5 \end{bmatrix}$

32. $\begin{bmatrix} -1 & 0 & 1 \\ 0 & -1 & 0 \\ 0 & 1 & 1 \end{bmatrix} \begin{bmatrix} x \\ y \\ z \end{bmatrix} = \begin{bmatrix} -4 \\ 2 \\ 4 \end{bmatrix}$

In Exercises 33–38,
  a. *Write each linear system as a matrix equation in the form $AX = B$.*
  b. *Solve the system using the inverse of the coefficient matrix.*

The inverse of

33. $\begin{aligned} 2x + 6y + 6z &= 8 \\ 2x + 7y + 6z &= 10 \\ 2x + 7y + 7z &= 9 \end{aligned}$ $\begin{bmatrix} 2 & 6 & 6 \\ 2 & 7 & 6 \\ 2 & 7 & 7 \end{bmatrix}$ is $\begin{bmatrix} \frac{7}{2} & 0 & -3 \\ -1 & 1 & 0 \\ 0 & -1 & 1 \end{bmatrix}$.

The inverse of

34. $\begin{aligned} x + 2y + 5z &= 2 \\ 2x + 3y + 8z &= 3 \\ -x + y + 2z &= 3 \end{aligned}$ $\begin{bmatrix} 1 & 2 & 5 \\ 2 & 3 & 8 \\ -1 & 1 & 2 \end{bmatrix}$ is $\begin{bmatrix} 2 & -1 & -1 \\ 12 & -7 & -2 \\ -5 & 3 & 1 \end{bmatrix}$.

The inverse of

35. $\begin{aligned} x - y + z &= 8 \\ 2y - z &= -7 \\ 2x + 3y &= 1 \end{aligned}$ $\begin{bmatrix} 1 & -1 & 1 \\ 0 & 2 & -1 \\ 2 & 3 & 0 \end{bmatrix}$ is $\begin{bmatrix} 3 & 3 & -1 \\ -2 & -2 & 1 \\ -4 & -5 & 2 \end{bmatrix}$.

The inverse of

36. $\begin{aligned} x - 6y + 3z &= 11 \\ 2x - 7y + 3z &= 14 \\ 4x - 12y + 5z &= 25 \end{aligned}$ $\begin{bmatrix} 1 & -6 & 3 \\ 2 & -7 & 3 \\ 4 & -12 & 5 \end{bmatrix}$ is $\begin{bmatrix} 1 & -6 & 3 \\ 2 & -7 & 3 \\ 4 & -12 & 5 \end{bmatrix}$.

37. $\begin{aligned} x - y + 2z &= -3 \\ y - z + w &= 4 \\ -x + y - z + 2w &= 2 \\ -y + z - 2w &= -4 \end{aligned}$

The inverse of

$\begin{bmatrix} 1 & -1 & 2 & 0 \\ 0 & 1 & -1 & 1 \\ -1 & 1 & -1 & 2 \\ 0 & -1 & 1 & -2 \end{bmatrix}$ is $\begin{bmatrix} 0 & 0 & -1 & -1 \\ 1 & 4 & 1 & 3 \\ 1 & 2 & 1 & 2 \\ 0 & -1 & 0 & -1 \end{bmatrix}$.

38. $\begin{aligned} 2x \quad + z + w &= 6 \\ 3x \quad + w &= 9 \\ -x + y - 2z + w &= 4 \\ 4x - y + z &= 6 \end{aligned}$

The inverse of

$\begin{bmatrix} 2 & 0 & 1 & 1 \\ 3 & 0 & 0 & 1 \\ -1 & 1 & -2 & 1 \\ 4 & -1 & 1 & 0 \end{bmatrix}$ is $\begin{bmatrix} -1 & 2 & -1 & -1 \\ -4 & 9 & -5 & -6 \\ 0 & 1 & -1 & -1 \\ 3 & -5 & 3 & 3 \end{bmatrix}$.

Application Exercises

In Exercises 39–40, use the coding matrix
$A = \begin{bmatrix} 4 & -1 \\ -3 & 1 \end{bmatrix}$ *and its inverse* $A^{-1} = \begin{bmatrix} 1 & 1 \\ 3 & 4 \end{bmatrix}$ *to encode and then decode the given message.*

39. HELP **40.** LOVE

In Exercises 41–42, use the coding matrix
$A = \begin{bmatrix} 1 & -1 & 0 \\ 3 & 0 & 2 \\ -1 & 0 & -1 \end{bmatrix}$ *and its inverse*

$A^{-1} = \begin{bmatrix} 0 & 1 & 2 \\ -1 & 1 & 2 \\ 0 & -1 & -3 \end{bmatrix}$ *to write a cryptogram for each message. Check your result by decoding the cryptogram.*

41. S E N D _ C A S H
19 5 14 4 0 3 1 19 8

Use $\begin{bmatrix} 19 & 4 & 1 \\ 5 & 0 & 19 \\ 14 & 3 & 8 \end{bmatrix}$.

42. S T A Y _ W E L L
19 20 1 25 0 23 5 12 12

Use $\begin{bmatrix} 19 & 25 & 5 \\ 20 & 0 & 12 \\ 1 & 23 & 12 \end{bmatrix}$.

Writing in Mathematics

43. What is the multiplicative identity matrix?

44. If you are given two matrices, A and B, explain how to determine if B is the multiplicative inverse of A.

45. Explain why a matrix that does not have the same number of rows and columns cannot have a multiplicative inverse.

46. Explain how to find the multiplicative inverse for a 2×2 invertible matrix.

47. Explain how to find the multiplicative inverse for a 3×3 invertible matrix.

48. Explain how to write a linear system of three equations in three variables as a matrix equation.

49. Explain how to solve the matrix equation $AX = B$.

50. What is a cryptogram?

51. It's January 1, and you've written down your major goal for the year. You do not want those closest to you to see what you've written in case you do not accomplish your objective. Consequently, you decide to use a coding matrix to encode your goal. Explain how this can be accomplished.

52. A year has passed since Exercise 51. (Time flies when you're solving exercises in algebra books.) It's been a terrific year and so many wonderful things have happened that you can't remember your goal from a year ago. You consult your personal journal and you find the encoded message and the coding matrix. How can you use these to find your original goal?

Technology Exercises

In Exercises 53–58, use a graphing utility to find the multiplicative inverse of each matrix. Check that the displayed inverse is correct.

53. $\begin{bmatrix} 3 & -1 \\ -2 & 1 \end{bmatrix}$

54. $\begin{bmatrix} -4 & 1 \\ 6 & -2 \end{bmatrix}$

55. $\begin{bmatrix} -2 & 1 & -1 \\ -5 & 2 & -1 \\ 3 & -1 & 1 \end{bmatrix}$

56. $\begin{bmatrix} 1 & 1 & -1 \\ -3 & 2 & -1 \\ 3 & -3 & 2 \end{bmatrix}$

57. $\begin{bmatrix} 7 & -3 & 0 & 2 \\ -2 & 1 & 0 & -1 \\ 4 & 0 & 1 & -2 \\ -1 & 1 & 0 & -1 \end{bmatrix}$

58. $\begin{bmatrix} 1 & 2 & 0 & 0 \\ 0 & 0 & 1 & 0 \\ 1 & 3 & 0 & 1 \\ 4 & 0 & 0 & 2 \end{bmatrix}$

In Exercises 59–64, write each system in the form $AX = B$. Then solve the system by entering A and B into your graphing utility and computing $A^{-1}B$.

59. $\begin{aligned} x - y + z &= -6 \\ 4x + 2y + z &= 9 \\ 4x - 2y + z &= -3 \end{aligned}$

60. $\begin{aligned} y + 2z &= 0 \\ -x + y &= 1 \\ 2x - y + z &= -1 \end{aligned}$

61. $\begin{aligned} 3x - 2y + z &= -2 \\ 4x - 5y + 3z &= -9 \\ 2x - y + 5z &= -5 \end{aligned}$

62. $\begin{aligned} x - y &= 1 \\ 6x + y + 20z &= 14 \\ y + 3z &= 1 \end{aligned}$

63. $\begin{aligned} x - 3z + v &= -3 \\ y + w &= -1 \\ z + v &= 7 \\ x + y - z + 4w &= -8 \\ x + y + z + w + v &= 8 \end{aligned}$

64. $\begin{aligned} x + y + z + w &= 4 \\ x + 3y - 2z + 2w &= 7 \\ 2x + 2y + z + w &= 3 \\ x - y + 2z + 3w &= 5 \end{aligned}$

In Exercises 65–66, use a coding matrix A of your choice. Use a graphing utility to find the multiplicative inverse of your coding matrix. Write a cryptogram for each message. Check your result by decoding the cryptogram. Use your graphing utility to perform all necessary matrix multiplications.

65. A R R I V E D _ S A F E L Y
1 18 18 9 22 5 4 0 19 1 6 5 12 25

66. A R T _ E N R I C H E S
1 18 20 0 5 14 18 9 3 8 5 19

Critical Thinking Exercises

67. Which one of the following is true?
   a. Some nonsquare matrices have inverses.
   b. All square 2×2 matrices have inverses because there is a formula for finding these inverses.
   c. Two 2×2 invertible matrices can have a matrix sum that is not invertible.
   d. To solve the matrix equation $AX = B$ for X, multiply A and the inverse of B.

68. Which one of the following is true?
   a. $(AB)^{-1} = A^{-1}B^{-1}$, assuming A, B, and AB, are invertible.
   b. $(A + B)^{-1} = A^{-1} + B^{-1}$, assuming A, B, and $A + B$ are invertible.
   c. $\begin{bmatrix} 1 & -3 \\ -1 & 3 \end{bmatrix}$ is an invertible matrix.
   d. None of the above is true.

69. Give an example of a 2×2 matrix that is its own inverse.

70. If $A = \begin{bmatrix} 3 & 5 \\ 2 & 4 \end{bmatrix}$, find $\left(A^{-1}\right)^{-1}$.

71. Find values of a for which the following matrix is not invertible.

$$\begin{bmatrix} 1 & a + 1 \\ a - 2 & 4 \end{bmatrix}$$

Group Exercise

72. Each person in the group should work with one partner. Send a coded word or message to each other by giving your partner the coded matrix and the coding matrix that you selected. Once messages are sent, each person should decode the message received.

SECTION 9.5 *Determinants and Cramer's Rule*

Objectives

1. Evaluate a second-order determinant.
2. Solve a linear system of equations in two variables using Cramer's rule.
3. Evaluate a third-order determinant.
4. Solve a linear system of equations in three variables using Cramer's rule.
5. Use determinants to identify inconsistent systems and systems with dependent equations.
6. Evaluate higher-order determinants.

A portion of Charles Babbage's unrealized Difference Engine

As cyberspace absorbs more and more of our work, play, shopping, and socializing, where will it all end? Which activities will still be offline in 2025?

Our technologically transformed lives can be traced back to the English inventor Charles Babbage (1792–1871). Babbage knew of a method for solving linear systems called Cramer's rule, in honor of the Swiss geometer Gabriel Cramer (1704–1752). Cramer's rule was simple, but involved numerous multiplications for large systems. Babbage designed a machine, called the "difference engine," that consisted of toothed wheels on shafts for performing these multiplications. Despite the fact that only one-seventh of the functions ever worked, Babbage's invention demonstrated how complex calculations could be handled mechanically. In 1944, scientists at IBM used the lessons of the the difference engine to create the world's first computer.

Those who invented computers hoped to relegate the drudgery of repeated computation to a machine. In this section, we look at a method for solving linear systems that played a critical role in this process. The method uses arrays of numbers called *determinants*. As with matrix methods, solutions are obtained by writing down the coefficients and constants of a linear system and performing operations with them.

1 Evaluate a second-order determinant.

The Determinant of a 2 × 2 Matrix

Associated with every square matrix is a real number called its **determinant**. The determinant for a 2 × 2 square matrix is defined as follows.

> **Definition of the Determinant of a 2 × 2 Matrix**
>
> The determinant of the matrix $\begin{bmatrix} a_1 & b_1 \\ a_2 & b_2 \end{bmatrix}$ is denoted by $\begin{vmatrix} a_1 & b_1 \\ a_2 & b_2 \end{vmatrix}$ and is defined by
>
> $$\begin{vmatrix} a_1 & b_1 \\ a_2 & b_2 \end{vmatrix} = a_1 b_2 - a_2 b_1.$$
>
> We also say that the **value** of the **second-order determinant** $\begin{vmatrix} a_1 & b_1 \\ a_2 & b_2 \end{vmatrix}$ is $a_1 b_2 - a_2 b_1$.

Study Tip

To evaluate a determinant, find the difference of the product of the two diagonals.

$$\begin{vmatrix} a_1 & b_1 \\ a_2 & b_2 \end{vmatrix} = a_1 b_2 - a_2 b_1$$

Example 1 illustrates that the determinant of a matrix may be positive or negative. The determinant can also have 0 as its value.

EXAMPLE 1 Evaluating the Determinant of a 2 × 2 Matrix

Evaluate the determinant of:

a. $\begin{bmatrix} 5 & 6 \\ 7 & 3 \end{bmatrix}$ **b.** $\begin{bmatrix} 2 & 4 \\ -3 & -5 \end{bmatrix}$.

Discovery

Write and then evaluate three determinants, one whose value is positive, one whose value is negative, and one whose value is 0.

Solution   We multiply and subtract as indicated.

a. $\begin{vmatrix} 5 & 6 \\ 7 & 3 \end{vmatrix} = 5 \cdot 3 - 7 \cdot 6 = 15 - 42 = -27$ *The value of the second-order determinant is -27.*

b. $\begin{vmatrix} 2 & 4 \\ -3 & -5 \end{vmatrix} = 2(-5) - (-3)(4) = -10 + 12 = 2$ *The value of the second-order determinant is 2.*

Check Point 1   Evaluate the determinant of:

a. $\begin{bmatrix} 10 & 9 \\ 6 & 5 \end{bmatrix}$ **b.** $\begin{bmatrix} 4 & 3 \\ -5 & -8 \end{bmatrix}$.

2 Solve a linear system of equations in two variables using Cramer's rule.

Solving Linear Systems of Equations in Two Variables Using Determinants

Determinants can be used to solve a linear system in two variables. In general, such a system appears as

$$a_1 x + b_1 y = c_1$$
$$a_2 x + b_2 y = c_2.$$

Let's first solve this system for x using the addition method. We can solve for x by eliminating y from the equations. Multiply the first equation by b_2 and the second equation by $-b_1$. Then add the two equations:

$$
\begin{array}{ll}
a_1 x + b_1 y = c_1 & \xrightarrow{\text{Multiply by } b_2.} & a_1 b_2 x + b_1 b_2 y = c_1 b_2 \\
a_2 x + b_2 y = c_2 & \xrightarrow{\text{Multiply by } -b_1.} & -a_2 b_1 x - b_1 b_2 y = -c_2 b_1 \\
\end{array}
$$

$$\text{Add:} \quad (a_1 b_2 - a_2 b_1)x = c_1 b_2 - c_2 b_1$$

$$x = \frac{c_1 b_2 - c_2 b_1}{a_1 b_2 - a_2 b_1}$$

Because

$$\begin{vmatrix} c_1 & b_1 \\ c_2 & b_2 \end{vmatrix} = c_1 b_2 - c_2 b_1 \quad \text{and} \quad \begin{vmatrix} a_1 & b_1 \\ a_2 & b_2 \end{vmatrix} = a_1 b_2 - a_2 b_1$$

we can express our answer for x as the quotient of two determinants:

$$x = \frac{\begin{vmatrix} c_1 & b_1 \\ c_2 & b_2 \end{vmatrix}}{\begin{vmatrix} a_1 & b_1 \\ a_2 & b_2 \end{vmatrix}}.$$

In a similar way, we could use the addition method to solve our system for y, again expressing y as the quotient of two determinants. This method of using determinants to solve the linear system, called **Cramer's rule**, is summarized in the box.

Solving a Linear System in Two Variables Using Determinants

Cramer's Rule

If

$$a_1 x + b_1 y = c_1$$
$$a_2 x + b_2 y = c_2$$

then

$$x = \frac{\begin{vmatrix} c_1 & b_1 \\ c_2 & b_2 \end{vmatrix}}{\begin{vmatrix} a_1 & b_1 \\ a_2 & b_2 \end{vmatrix}} \quad \text{and} \quad y = \frac{\begin{vmatrix} a_1 & c_1 \\ a_2 & c_2 \end{vmatrix}}{\begin{vmatrix} a_1 & b_1 \\ a_2 & b_2 \end{vmatrix}}$$

where

$$\begin{vmatrix} a_1 & b_1 \\ a_2 & b_2 \end{vmatrix} \neq 0.$$

Here are some helpful tips when solving

$$a_1 x + b_1 y = c_1$$
$$a_2 x + b_2 y = c_2$$

using determinants.

1. Three different determinants are used to find x and y. The determinants in the denominators for x and y are identical. The determinants in the numerators for x and y differ. In abbreviated notation, we write

$$x = \frac{D_x}{D} \quad \text{and} \quad y = \frac{D_y}{D} \text{ where } D \neq 0.$$

2. The elements of D, the determinant in the denominator, are the coefficients of the variables in the system.

$$D = \begin{vmatrix} a_1 & b_1 \\ a_2 & b_2 \end{vmatrix}$$

3. D_x, the determinant in the numerator of x, is obtained by replacing the x-coefficients, a_1, a_2, in D with the constants on the right side of the equations, c_1, c_2.

$$D = \begin{vmatrix} a_1 & b_1 \\ a_2 & b_2 \end{vmatrix} \quad \text{and} \quad D_x = \begin{vmatrix} c_1 & b_1 \\ c_2 & b_2 \end{vmatrix}$$ Replace the column with a_1 and a_2 with the constants c_1 and c_2 to get D_x.

4. D_y, the determinant in the numerator for y, is obtained by replacing the y-coefficients, b_1, b_2 in D with the constants on the right side of the equations, c_1, c_2.

$$D = \begin{vmatrix} a_1 & b_1 \\ a_2 & b_2 \end{vmatrix} \quad \text{and} \quad D_y = \begin{vmatrix} a_1 & c_1 \\ a_2 & c_2 \end{vmatrix}$$ Replace the column with b_1 and b_2 with the constants c_1 and c_2 to get D_y.

Example 2 illustrates the use of Cramer's rule.

EXAMPLE 2   Using Cramer's Rule to Solve a Linear System

Use Cramer's rule to solve the system:

$$5x - 4y = 2$$
$$6x - 5y = 1.$$

Solution   Because

$$x = \frac{D_x}{D} \quad \text{and} \quad y = \frac{D_y}{D},$$

we will set up and evaluate the three determinants $D, D_x,$ and D_y.

1. D, the determinant in both denominators, consists of the x- and y-coefficients.

$$D = \begin{vmatrix} 5 & -4 \\ 6 & -5 \end{vmatrix} = (5)(-5) - (6)(-4) = -25 + 24 = -1$$

Because this determinant is not zero, we continue to use Cramer's rule to solve the system.

2. D_x, the determinant in the numerator for x, is obtained by replacing the x-coefficients in D, 5 and 6, by the constants on the right side of the equation, 2 and 1.

$$D_x = \begin{vmatrix} 2 & -4 \\ 1 & -5 \end{vmatrix} = (2)(-5) - (1)(-4) = -10 + 4 = -6$$

3. D_y, the determinant in the numerator for y, is obtained by replacing the y-coefficients in D, -4 and -5, by the constants on the right side of the equation, 2 and 1.

$$D_y = \begin{vmatrix} 5 & 2 \\ 6 & 1 \end{vmatrix} = (5)(1) - (6)(2) = 5 - 12 = -7$$

4. Thus,

$$x = \frac{D_x}{D} = \frac{-6}{-1} = 6 \quad \text{and} \quad y = \frac{D_y}{D} = \frac{-7}{-1} = 7.$$

As always, the solution $(6, 7)$ can be checked by substituting these values into the original equations. The solution set is $\{(6, 7)\}$.

Check Point 2   Use Cramer's rule to solve the system:

$$5x + 4y = 12$$
$$3x - 6y = 24.$$

3 Evaluate a third-order determinant.

The Determinant of a 3×3 Matrix

Associated with every square matrix is a real number called its determinant. The determinant for a 3×3 matrix is defined as follows.

Definition of a Third-Order Determinant

$$\begin{vmatrix} a_1 & b_1 & c_1 \\ a_2 & b_2 & c_2 \\ a_3 & b_3 & c_3 \end{vmatrix} = a_1b_2c_3 + b_1c_2a_3 + c_1a_2b_3 - a_3b_2c_1 - b_3c_2a_1 - c_3a_2b_1$$

The six terms and the three factors in each term in this complicated evaluation formula can be rearranged, and then we can apply the distributive property. We obtain

$$a_1b_2c_3 - a_1b_3c_2 - a_2b_1c_3 + a_2b_3c_1 + a_3b_1c_2 - a_3b_2c_1$$
$$= a_1(b_2c_3 - b_3c_2) - a_2(b_1c_3 - b_3c_1) + a_3(b_1c_2 - b_2c_1)$$
$$= a_1\begin{vmatrix} b_2 & c_2 \\ b_3 & c_3 \end{vmatrix} - a_2\begin{vmatrix} b_1 & c_1 \\ b_3 & c_3 \end{vmatrix} + a_3\begin{vmatrix} b_1 & c_1 \\ b_2 & c_2 \end{vmatrix}.$$

You can evaluate each of the second-order determinants and obtain the three expressions in parentheses in the second step.

In summary, we now have arranged the definition of a third-order determinant as follows.

Definition of the Determinant of a 3 × 3 Matrix

A third-order determinant is defined by

$$\begin{vmatrix} a_1 & b_1 & c_1 \\ a_2 & b_2 & c_2 \\ a_3 & b_3 & c_3 \end{vmatrix} = a_1\begin{vmatrix} b_2 & c_2 \\ b_3 & c_3 \end{vmatrix} - a_2\begin{vmatrix} b_1 & c_1 \\ b_3 & c_3 \end{vmatrix} + a_3\begin{vmatrix} b_1 & c_1 \\ b_2 & c_2 \end{vmatrix}.$$

The *a*'s on the right come from the first column.

Here are some tips that may be helpful when evaluating the determinant of a 3 × 3 matrix.

1. Each of the three terms in the definition contains two factors—a numerical factor and a second-order determinant.
2. The numerical factor in each term is an element from the first column of the third-order determinant.
3. The minus sign precedes the second term.
4. The second-order determinant that appears in each term is obtained by crossing out the row and the column containing the numerical factor.

The **minor** of an element is the determinant that remains after deleting the row and column of that element. For this reason, we call this method **expansion by minors**.

EXAMPLE 3   Evaluating the Determinant of a 3 × 3 Matrix

Evaluate the determinant of

$$\begin{bmatrix} 4 & 1 & 0 \\ -9 & 3 & 4 \\ -3 & 8 & 1 \end{bmatrix}.$$

Solution   We know that each of the three terms in the determinant contains a numerical factor and a second-order determinant. The numerical factors are from the first column of the determinant of the given matrix. They are highlighted in the following matrix:

$$\begin{vmatrix} 4 & 1 & 0 \\ -9 & 3 & 4 \\ -3 & 8 & 1 \end{vmatrix}$$

We find the minor for each numerical factor by deleting the row and column of that element:

$$\begin{bmatrix} 4 & 1 & 0 \\ -9 & 3 & 4 \\ -3 & 8 & 1 \end{bmatrix} \begin{bmatrix} 4 & 1 & 0 \\ -9 & 3 & 4 \\ -3 & 8 & 1 \end{bmatrix} \begin{bmatrix} 4 & 1 & 0 \\ -9 & 3 & 4 \\ -3 & 8 & 1 \end{bmatrix}$$

The minor for 4 is $\begin{vmatrix} 3 & 4 \\ 8 & 1 \end{vmatrix}$.

The minor for -9 is $\begin{vmatrix} 1 & 0 \\ 8 & 1 \end{vmatrix}$.

The minor for -3 is $\begin{vmatrix} 1 & 0 \\ 3 & 4 \end{vmatrix}$.

Now we have three numerical factors, 4, −9, and −3, and three second-order determinants. We multiply each numerical factor by its second-order determinant to find the three terms of the third-order determinant:

$$4\begin{vmatrix} 3 & 4 \\ 8 & 1 \end{vmatrix}, \quad -9\begin{vmatrix} 1 & 0 \\ 8 & 1 \end{vmatrix}, \quad -3\begin{vmatrix} 1 & 0 \\ 3 & 4 \end{vmatrix}.$$

Based on the preceding definition, we subtract the second term from the first term and add the third term:

Don't forget to supply the minus sign.

$$\begin{vmatrix} 4 & 1 & 0 \\ -9 & 3 & 4 \\ -3 & 8 & 1 \end{vmatrix} = 4\begin{vmatrix} 3 & 4 \\ 8 & 1 \end{vmatrix} - (-9)\begin{vmatrix} 1 & 0 \\ 8 & 1 \end{vmatrix} - 3\begin{vmatrix} 1 & 0 \\ 3 & 4 \end{vmatrix}$$

$$= 4(3 \cdot 1 - 8 \cdot 4) + 9(1 \cdot 1 - 8 \cdot 0) - 3(1 \cdot 4 - 3 \cdot 0)$$

$$= 4(3 - 32) + 9(1 - 0) - 3(4 - 0) \qquad \text{Evaluate the three second-order determinants.}$$

$$= 4(-29) + 9(1) - 3(4)$$

$$= -119$$

Technology

Verify the result of Example 3 by using your graphing utility to enter the given matrix as

$$A = \begin{bmatrix} 4 & 1 & 0 \\ -9 & 3 & 4 \\ -3 & 8 & 1 \end{bmatrix}.$$

Then enter

det [A] ENTER.

The result should be −119.

Study Tip

Keep in mind that you can expand a determinant by minors about any row or column. Use alternating plus and minus signs to precede the numerical factors of the minors according to the following sign array:

$$\begin{vmatrix} + & - & + \\ - & + & - \\ + & - & + \end{vmatrix}.$$

Check Point 3

Evaluate the determinant of

$$\begin{bmatrix} 2 & 1 & 7 \\ -5 & 6 & 0 \\ -4 & 3 & 1 \end{bmatrix}.$$

The six terms in the definition of a third-order determinant can be rearranged and factored in a variety of ways. Thus, it is possible to expand a determinant by minors about any row or any column. *Minus signs must be supplied preceding any element appearing in a position where the sum of its row and its column is an odd number.* For example, expanding about the elements in column 2 gives us

$$\begin{vmatrix} a_1 & b_1 & c_1 \\ a_2 & b_2 & c_2 \\ a_3 & b_3 & c_3 \end{vmatrix} = -b_1 \begin{vmatrix} a_2 & c_2 \\ a_3 & c_3 \end{vmatrix} + b_2 \begin{vmatrix} a_1 & c_1 \\ a_3 & c_3 \end{vmatrix} - b_3 \begin{vmatrix} a_1 & c_1 \\ a_2 & c_2 \end{vmatrix}.$$

Minus sign is supplied because b_1 appears in row 1 and column 2; $1 + 2 = 3$, an odd number.

Minus sign is supplied because b_3 appears in row 3 and column 2; $3 + 2 = 5$, an odd number.

Expanding by minors about column 3, we obtain

$$\begin{vmatrix} a_1 & b_1 & c_1 \\ a_2 & b_2 & c_2 \\ a_3 & b_3 & c_3 \end{vmatrix} = c_1 \begin{vmatrix} a_2 & b_2 \\ a_3 & b_3 \end{vmatrix} - c_2 \begin{vmatrix} a_1 & b_1 \\ a_3 & b_3 \end{vmatrix} + c_3 \begin{vmatrix} a_1 & b_1 \\ a_2 & b_2 \end{vmatrix}.$$

Minus sign must be supplied because c_2 appears in row 2 and column 3; $2 + 3 = 5$, an odd number.

When evaluating a 3 × 3 determinant using expansion by minors, you can expand about any row or column. To simplify the arithmetic, if a row or column contains one or more 0s, expand about that row or column.

EXAMPLE 4 Evaluating a Third-Order Determinant

Evaluate:

$$\begin{vmatrix} 9 & 5 & 0 \\ -2 & -3 & 0 \\ 1 & 4 & 2 \end{vmatrix}.$$

Solution Note that the last column has two 0s. We will expand the determinant about the elements in that column.

$$\begin{vmatrix} 9 & 5 & 0 \\ -2 & -3 & 0 \\ 1 & 4 & 2 \end{vmatrix} = 0 \begin{vmatrix} -2 & -3 \\ 1 & 4 \end{vmatrix} - 0 \begin{vmatrix} 9 & 5 \\ 1 & 4 \end{vmatrix} + 2 \begin{vmatrix} 9 & 5 \\ -2 & -3 \end{vmatrix}$$

$$= 0 - 0 + 2[9(-3) - (-2) \cdot 5] \quad \text{Evaluate the second-order determinant whose numerical factor is not 0.}$$

$$= 2(-27 + 10)$$

$$= 2(-17)$$

$$= -34$$

Check Point 4 Evaluate:

$$\begin{vmatrix} 6 & 4 & 0 \\ -3 & -5 & 3 \\ 1 & 2 & 0 \end{vmatrix}.$$

4 Solve a linear system of equations in three variables using Cramer's rule.

Solving Linear Systems of Equations in Three Variables Using Determinants

Cramer's rule can be applied to solving systems of linear equations in three variables. The determinants in the numerator and denominator of all variables are third-order determinants.

Solving Three Equations in Three Variables Using Determinants

Cramer's Rule

If

$$a_1 x + b_1 y + c_1 z = d_1$$
$$a_2 x + b_2 y + c_2 z = d_2$$
$$a_3 x + b_3 y + c_3 z = d_3$$

then

$$x = \frac{D_x}{D}, y = \frac{D_y}{D}, \text{ and } z = \frac{D_z}{D}.$$

These four third-order determinants are given by

$$D = \begin{vmatrix} a_1 & b_1 & c_1 \\ a_2 & b_2 & c_2 \\ a_3 & b_3 & c_3 \end{vmatrix}$$ These are the coefficients of the variables x, y, and z. $D \neq 0$.

$$D_x = \begin{vmatrix} d_1 & b_1 & c_1 \\ d_2 & b_2 & c_2 \\ d_3 & b_3 & c_3 \end{vmatrix}$$ Replace x-coefficients in D with the **constants at the right** of the three equations.

$$D_y = \begin{vmatrix} a_1 & d_1 & c_1 \\ a_2 & d_2 & c_2 \\ a_3 & d_3 & c_3 \end{vmatrix}$$ Replace y-coefficients in D with the **constants at the right** of the three equations.

$$D_z = \begin{vmatrix} a_1 & b_1 & d_1 \\ a_2 & b_2 & d_2 \\ a_3 & b_3 & d_3 \end{vmatrix}$$ Replace z-coefficients in D with the **constants at the right** of the three equations.

EXAMPLE 5 **Using Cramer's Rule to Solve a Linear System in Three Variables**

Use Cramer's rule to solve:

$$x + 2y - z = -4$$
$$x + 4y - 2z = -6$$
$$2x + 3y + z = 3.$$

Solution Because

$$x = \frac{D_x}{D}, \quad y = \frac{D_y}{D}, \quad \text{and} \quad z = \frac{D_z}{D},$$

we need to set up and evaluate four determinants.

$$x + 2y - z = -4$$
$$x + 4y - 2z = -6$$
$$2x + 3y + z = 3$$

The linear system is shown again so that you do not need to turn back a page.

Step 1 Set up the determinants.

1. D, the determinant in all three denominators, consists of the x-, y-, and z-coefficients.

$$D = \begin{vmatrix} 1 & 2 & -1 \\ 1 & 4 & -2 \\ 2 & 3 & 1 \end{vmatrix}$$

2. D_x, the determinant in the numerator for x, is obtained by replacing the x-coefficients in D, 1, 1, and 2, with the constants on the right side of the equation, -4, -6, and 3.

$$D_x = \begin{vmatrix} -4 & 2 & -1 \\ -6 & 4 & -2 \\ 3 & 3 & 1 \end{vmatrix}$$

3. D_y, the determinant in the numerator for y, is obtained by replacing the y-coefficients in D, 2, 4, and 3, with the constants on the right side of the equation, -4, -6, and 3.

$$D_y = \begin{vmatrix} 1 & -4 & -1 \\ 1 & -6 & -2 \\ 2 & 3 & 1 \end{vmatrix}$$

4. D_z, the determinant in the numerator for z, is obtained by replacing the z-coefficients in D, -1, -2, and 1, with the constants on the right side of the equation, -4, -6, and 3.

$$D_z = \begin{vmatrix} 1 & 2 & -4 \\ 1 & 4 & -6 \\ 2 & 3 & 3 \end{vmatrix}$$

Step 2 Evaluate the four determinants.

$$D = \begin{vmatrix} 1 & 2 & -1 \\ 1 & 4 & -2 \\ 2 & 3 & 1 \end{vmatrix} = 1 \begin{vmatrix} 4 & -2 \\ 3 & 1 \end{vmatrix} - 1 \begin{vmatrix} 2 & -1 \\ 3 & 1 \end{vmatrix} + 2 \begin{vmatrix} 2 & -1 \\ 4 & -2 \end{vmatrix}$$

$$= 1(4 + 6) - 1(2 + 3) + 2(-4 + 4)$$
$$= 1(10) - 1(5) + 2(0) = 5$$

Using the same technique to evaluate each determinant, we obtain

$$D_x = -10, \quad D_y = 5, \quad \text{and} \quad D_z = 20.$$

Step 3 Substitute these four values and solve the system.

$$x = \frac{D_x}{D} = \frac{-10}{5} = -2$$

$$y = \frac{D_y}{D} = \frac{5}{5} = 1$$

$$z = \frac{D_z}{D} = \frac{20}{5} = 4$$

The solution $(-2, 1, 4)$ can be checked by substitution into the original three equations. The solution set is $\{(-2, 1, 4)\}$.

> **Check Point 5**
>
> Use Cramer's rule to solve the system:
>
> $$\begin{aligned} 3x - 2y + z &= 16 \\ 2x + 3y - z &= -9. \\ x + 4y + 3z &= 2 \end{aligned}$$

5 Use determinants to identify inconsistent systems and systems with dependent equations.

Cramer's Rule with Inconsistent and Dependent Systems

If D, the determinant in the denominator, is 0, the variables described by the quotient of determinants are not real numbers. However, when $D = 0$, this indicates that the system is inconsistent or contains dependent equations. This gives rise to the following two situations.

Discovery

Write a system of two equations that is inconsistent. Now use determinants and the result boxed on the right to verify that this is truly an inconsistent system. Repeat the same process for a system with two dependent equations.

> **Determinants: Inconsistent and Dependent-Systems**
>
> **1.** If $D = 0$ and at least one of the determinants in the numerator is not 0, then the system is inconsistent. The solution set is $\varnothing$.
>
> **2.** If $D = 0$ and all the determinants in the numerators are 0, then the equations in the system are dependent.

Although we have focused on applying determinants to solve linear systems, they have other applications, some of which we consider in the exercise set that follows.

6 Evaluate higher-order determinants.

The Determinant of Any $n \times n$ Matrix

A determinant with n rows and n columns is said to be an ***n*th-order determinant**. The value of an nth-order determinant $(n > 2)$ can be found in terms of determinants of order $n - 1$. For example, we found the value of a third-order determinant in terms of determinants of order 2.

We can generalize this idea for fourth-order determinants and higher. We have seen that the **minor** of the element a_{ij} is the determinant obtained by deleting the ith row and the jth column in the given array of numbers. The **cofactor** of the element a_{ij} is $(-1)^{i+j}$ times the minor of the a_{ij}th entry. If the sum of the row and column $(i + j)$ is even, the cofactor is the same as the minor. If the sum of the row and column $(i + j)$ is odd, the cofactor is the opposite of the minor.

Let's see what this means in the case of a fourth-order determinant.

EXAMPLE 6 Evaluating the Determinant of a 4×4 Matrix

Evaluate the determinant of

$$A = \begin{bmatrix} 1 & -2 & 3 & 0 \\ -1 & 1 & 0 & 2 \\ 0 & 2 & 0 & -3 \\ 2 & 3 & -4 & 1 \end{bmatrix}.$$

Solution

$$|A| = \begin{vmatrix} 1 & -2 & 3 & 0 \\ -1 & 1 & 0 & 2 \\ 0 & 2 & 0 & -3 \\ 2 & 3 & -4 & 1 \end{vmatrix}$$

With two 0s in the third column, we will expand along the third column.

$$= (-1)^{1+3} 3 \begin{vmatrix} -1 & 1 & 2 \\ 0 & 2 & -3 \\ 2 & 3 & 1 \end{vmatrix} + (-1)^{4+3}(-4) \begin{vmatrix} 1 & -2 & 0 \\ -1 & 1 & 2 \\ 0 & 2 & -3 \end{vmatrix}$$

3 is in row 1, column 3.

−4 is in row 4, column 3.

$$= 3 \begin{vmatrix} -1 & 1 & 2 \\ 0 & 2 & -3 \\ 2 & 3 & 1 \end{vmatrix} + 4 \begin{vmatrix} 1 & -2 & 0 \\ -1 & 1 & 2 \\ 0 & 2 & -3 \end{vmatrix}$$

The determinant that follows 3 is obtained by crossing out the row and the column (row 1, column 3) in the original determinant. The minor for −4 is obtained in the same manner.

Evaluate the two third-order determinants to get

$$|A| = 3(-25) + 4(-1) = -79.$$

Check Point 6   Evaluate the determinant of

$$A = \begin{bmatrix} 0 & 4 & 0 & -3 \\ -1 & 1 & 5 & 2 \\ 1 & -2 & 0 & 6 \\ 3 & 0 & 0 & 1 \end{bmatrix}.$$

If a linear system has n equations, Cramer's rule requires you to compute $n + 1$ determinants of nth order. The excessive number of calculations required to perform Cramer's rule for systems with four or more equations makes it an inefficient method for solving large systems.

EXERCISE SET 9.5

Practice Exercises

Evaluate each determinant in Exercises 1–10.

1. $\begin{vmatrix} 5 & 7 \\ 2 & 3 \end{vmatrix}$

2. $\begin{vmatrix} 4 & 8 \\ 5 & 6 \end{vmatrix}$

3. $\begin{vmatrix} -4 & 1 \\ 5 & 6 \end{vmatrix}$

4. $\begin{vmatrix} 7 & 9 \\ -2 & -5 \end{vmatrix}$

5. $\begin{vmatrix} -7 & 14 \\ 2 & -4 \end{vmatrix}$

6. $\begin{vmatrix} 1 & -3 \\ -8 & 2 \end{vmatrix}$

7. $\begin{vmatrix} -5 & -1 \\ -2 & -7 \end{vmatrix}$

8. $\begin{vmatrix} \frac{1}{5} & \frac{1}{6} \\ -6 & 5 \end{vmatrix}$

9. $\begin{vmatrix} \frac{1}{2} & \frac{1}{2} \\ \frac{1}{8} & -\frac{3}{4} \end{vmatrix}$

10. $\begin{vmatrix} \frac{2}{3} & \frac{1}{3} \\ -\frac{1}{2} & \frac{3}{4} \end{vmatrix}$

For Exercises 11–26, use Cramer's rule to solve each system or to determine that the system is inconsistent or contains dependent equations.

11. $x + y = 7$
$x - y = 3$

12. $2x + y = 3$
$x - y = 3$

13. $12x + 3y = 15$
$\quad\ \ 2x - 3y = 13$

14. $x - 2y = 5$
$\quad\ 5x - y = -2$

15. $4x - 5y = 17$
$\quad\ \ 2x + 3y = 3$

16. $3x + 2y = 2$
$\quad\ 2x + 2y = 3$

17. $\quad x + 2y = 3$
$\quad\ 5x + 10y = 15$

18. $2x - 9y = 5$
$\quad\ 3x - 3y = 11$

19. $3x - 4y = 4$
$\quad\ 2x + 2y = 12$

20. $3x = 7y + 1$
$\quad\ 2x = 3y - 1$

21. $2x = 3y + 2$
$\quad\ 5x = 51 - 4y$

22. $x + 2y - 3 = 0$
$\quad\ 12 = 8y + 4x$

23. $3x = 2 - 3y$
$\quad\ 2y = 3 - 2x$

24. $\quad y = -4x + 2$
$\quad\ 2x = 3y + 8$

25. $4y = 16 - 3x$
$\quad\ 5x = 12 - 3y$

26. $2x = 7 + 3y$
$\quad\ 4x - 6y = 3$

Evaluate each determinant in Exercises 27–32.

27. $\begin{vmatrix} 3 & 0 & 0 \\ 2 & 1 & -5 \\ 2 & 5 & -1 \end{vmatrix}$

28. $\begin{vmatrix} 4 & 0 & 0 \\ 3 & -1 & 4 \\ 2 & -3 & 5 \end{vmatrix}$

29. $\begin{vmatrix} 3 & 1 & 0 \\ -3 & 4 & 0 \\ -1 & 3 & -5 \end{vmatrix}$

30. $\begin{vmatrix} 2 & -4 & 2 \\ -1 & 0 & 5 \\ 3 & 0 & 4 \end{vmatrix}$

31. $\begin{vmatrix} 1 & 1 & 1 \\ 2 & 2 & 2 \\ -3 & 4 & -5 \end{vmatrix}$

32. $\begin{vmatrix} 1 & 2 & 3 \\ 2 & 2 & -3 \\ 3 & 2 & 1 \end{vmatrix}$

In Exercises 33–40, use Cramer's rule to solve each system.

33. $\quad x + y + z = 0$
$\quad\ 2x - y + z = -1$
$\quad\ -x + 3y - z = -8$

34. $\quad x - y + 2z = 3$
$\quad\ 2x + 3y + z = 9$
$\quad\ -x - y + 3z = 11$

35. $4x - 5y - 6z = -1$
$\quad\ x - 2y - 5z = -12$
$\quad\quad\ 2x - y = 7$

36. $x - 3y + z = -2$
$\quad\ x + 2y = 8$
$\quad\ 2x - y = 1$

37. $x + \ y + \ z = 4$
$\quad x - 2y + \ z = 7$
$\quad x + 3y + 2z = 4$

38. $2x + 2y + 3z = 10$
$\quad\ 4x - \ y + \ z = -5$
$\quad\ 5x - 2y + 6z = 1$

39. $\quad x + 2z = 4$
$\quad\ 2y - \ z = 5$
$\quad\ 2x + 3y = 13$

40. $3x + 2z = 4$
$\quad\ 5x - \ y = -4$
$\quad\ 4y + 3z = 22$

Evaluate each determinant in Exercises 41–44.

41. $\begin{vmatrix} 4 & 2 & 8 & -7 \\ -2 & 0 & 4 & 1 \\ 5 & 0 & 0 & 5 \\ 4 & 0 & 0 & -1 \end{vmatrix}$

42. $\begin{vmatrix} 3 & -1 & 1 & 2 \\ -2 & 0 & 0 & 0 \\ 2 & -1 & -2 & 3 \\ 1 & 4 & 2 & 3 \end{vmatrix}$

43. $\begin{vmatrix} -2 & -3 & 3 & 5 \\ 1 & -4 & 0 & 0 \\ 1 & 2 & 2 & -3 \\ 2 & 0 & 1 & 1 \end{vmatrix}$

44. $\begin{vmatrix} 1 & -3 & 2 & 0 \\ -3 & -1 & 0 & -2 \\ 2 & 1 & 3 & 1 \\ 2 & 0 & -2 & 0 \end{vmatrix}$

⭐ Application Exercises

Determinants are used to find the area of a triangle whose vertices are given by three points in a rectangular coordinate system. The area of a triangle with vertices (x_1, y_1), (x_2, y_2), and (x_3, y_3) is

$$\text{Area} = \pm\frac{1}{2}\begin{vmatrix} x_1 & y_1 & 1 \\ x_2 & y_2 & 1 \\ x_3 & y_3 & 1 \end{vmatrix}$$

where the symbol $(\pm)$ indicates that the appropriate sign should be chosen to yield a positive area. Use this information to work Exercises 45–46.

45. a. Use determinants to find the area of the triangle whose vertices are $(3, -5)$, $(2, 6)$, and $(-3, 5)$.
b. Graph the triangle in part (a) and then confirm your answer by using the formula for a triangle's area, $A = \frac{1}{2}bh$.

46. Find the area of the triangle whose vertices are $(1, 1)$, $(-2, -3)$, and $(11, -3)$.

Determinants are used to show that three points lie on the same line (are collinear). If

$$\begin{vmatrix} x_1 & y_1 & 1 \\ x_2 & y_2 & 1 \\ x_3 & y_3 & 1 \end{vmatrix} = 0$$

then the points (x_1, y_1), (x_2, y_2), and (x_3, y_3) are collinear. If the determinant does not equal 0, then the points are not collinear. Use this information to work Exercises 47–48.

47. Are the points $(3, -1)$, $(0, -3)$, and $(12, 5)$ collinear?

48. Are the points $(-4, -6)$, $(1, 0)$, and $(11, 12)$ collinear?

Determinants are used to write an equation of a line passing through two points. An equation of the line passing through the distinct points (x_1, y_1) and (x_2, y_2) is given by

$$\begin{vmatrix} x & y & 1 \\ x_1 & y_1 & 1 \\ x_2 & y_2 & 1 \end{vmatrix} = 0.$$

Use this information to work Exercises 49–50.

49. Use the determinant to write an equation for the line passing through $(3, -5)$ and $(-2, 6)$. Then expand the determinant, expressing the line's equation in slope-intercept form.

50. Use the determinant to write an equation for the line passing through $(-1, 3)$ and $(2, 4)$. Then expand the determinant, expressing the line's equation in slope-intercept form.

Writing in Mathematics

51. Explain how to evaluate a second-order determinant.

52. Describe the determinants D_x and D_y in terms of the coefficients and constants in a system of two equations in two variables.

53. Explain how to evaluate a third-order determinant.

54. When expanding a determinant by minors, when is it necessary to supply minus signs?

55. Without going into too much detail, describe how to solve a linear system in three variables using Cramer's rule.

56. In applying Cramer's rule, what does it mean if $D = 0$?

57. The process of solving a linear system in three variables using Cramer's rule can involve tedious computation. Is there a way of speeding up this process, perhaps using Cramer's rule to find the value for only one of the variables? Describe how this process might work, presenting a specific example with your description. Remember that your goal is still to find the value for each variable in the system.

58. If you could use only one method to solve linear systems in three variables, which method would you select? Explain why this is so.

Technology Exercises

59. Use the feature of your graphing utility that evaluates the determinant of a square matrix to verify any five of the determinants that you evaluated by hand in Exercises 1–10, 27–32, or 41–44.

In Exercises 60–61, use a graphing utility to evaluate the determinant for the given matrix.

60. $\begin{bmatrix} 3 & -2 & -1 & 4 \\ -5 & 1 & 2 & 7 \\ 2 & 4 & 5 & 0 \\ -1 & 3 & -6 & 5 \end{bmatrix}$

61. $\begin{bmatrix} 8 & 2 & 6 & -1 & 0 \\ 2 & 0 & -3 & 4 & 7 \\ 2 & 1 & -3 & 6 & -5 \\ -1 & 2 & 1 & 5 & -1 \\ 4 & 5 & -2 & 3 & -8 \end{bmatrix}$

62. What is the fastest method for solving a linear system with your graphing utility?

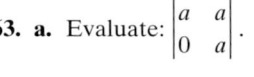

Critical Thinking Exercises

63. a. Evaluate: $\begin{vmatrix} a & a \\ 0 & a \end{vmatrix}$.

b. Evaluate: $\begin{vmatrix} a & a & a \\ 0 & a & a \\ 0 & 0 & a \end{vmatrix}$.

c. Evaluate: $\begin{vmatrix} a & a & a & a \\ 0 & a & a & a \\ 0 & 0 & a & a \\ 0 & 0 & 0 & a \end{vmatrix}$.

d. Describe the pattern in the given determinants.

e. Describe the pattern in the evaluations.

64. Evaluate: $\begin{vmatrix} 2 & 0 & 0 & 0 & 0 \\ 0 & 3 & 0 & 0 & 0 \\ 0 & 0 & 2 & 0 & 0 \\ 0 & 0 & 0 & 1 & 0 \\ 0 & 0 & 0 & 0 & 4 \end{vmatrix}$.

65. What happens to the value of a second-order determinant if the two columns are interchanged?

66. Consider the system

$$a_1 x + b_1 y = c_1$$
$$a_2 x + b_2 y = c_2.$$

Use Cramer's rule to prove that if the first equation of the system is replaced by the sum of the two equations, the resulting system has the same solution as the original system.

Group Exercise

67. We have seen that determinants can be used to solve linear equations, give areas of triangles in rectangular coordinates, and determine equations of lines. Not impressed with these applications? Members of the group should research an application of determinants that they find intriguing. The group should then present a seminar to the class about this application.

CHAPTER SUMMARY, REVIEW, AND TEST

Summary

9.1 Matrix Solution to Linear Systems

a. Matrix row operations are described in the box on page 740.

b. To solve a linear system using Gaussian elimination, begin with the system's augmented matrix. Use matrix operations to get 1s down the main diagonal and 0s below the 1s. Details are in the box on page 741.

c. To solve a linear system using Gauss-Jordan elimination, use the procedure of Gaussian elimination, but obtain 0s above and below the 1s in the main diagonal.

9.2 Inconsistent and Dependent Systems

a. If Gaussian elimination results in a matrix with a row containing all 0s to the left of the vertical line and a nonzero number to the right, the system has no solution (is inconsistent).

b. If Gaussian elimination results in a matrix with a row with all 0s, the system has an infinite number of solutions (contains dependent equations).

9.3 Matrix Operations

a. Two matrices are equal if and only if they have the same order and corresponding elements are equal.

b. Matrix Addition and Subtraction: Matrices of the same order are added or subtracted by adding or subtracting corresponding elements. Properties of matrix addition are given in the box on page 764.

c. Scalar Multiplication: If A is a matrix and c is a scalar, then cA is the matrix formed by multiplying each element in A by c. Properties of scalar multiplication are given in the box on page 765.

d. Matrix Multiplication: The product of an $m \times n$ matrix A and an $n \times p$ matrix B is an $m \times p$ matrix AB. The element in the ith row and jth column of AB is found by multiplying each element in the ith row of A by the corresponding element in the jth column of B and adding the products. Matrix multiplication is not commutative: $AB \neq BA$. Properties of matrix multiplication are given in the box on page 770.

9.4 Multiplicative Inverses of Matrices; Matrix Equations

a. The multiplicative identity matrix I_n is an $n \times n$ matrix with 1s down the main diagonal and 0s elsewhere.

b. Let A be an $n \times n$ square matrix. If there is a square matrix A^{-1} such that $AA^{-1} = I_n$ and $A^{-1}A = I_n$, then A^{-1} is the multiplicative inverse of A.

c. If a square matrix has a multiplicative inverse, it is invertible. Methods for finding multiplicative inverses for invertible matrices, including a formula for 2×2 matrices, are given in the box on page 783.

d. Linear systems can be represented by matrix equations $AX = B$ in which A is the coefficient matrix and B is the constant matrix. If $AX = B$ has a unique solution, then $X = A^{-1}B$.

9.5 Determinants and Cramer's Rule

a. Value of a Second-Order Determinant:

$$\begin{vmatrix} a_1 & b_1 \\ a_2 & b_2 \end{vmatrix} = a_1 b_2 - a_2 b_1$$

b. Cramer's rule for solving linear systems in two variables uses three second-order determinants and is stated in the box on page 792.

c. To evaluate an nth-order determinant, where $n > 2$,

1. Select a row or column about which to expand.

2. For each element a_{ij} in the row or column, multiply by $(-1)^{i+j}$ times the determinant obtained by deleting the ith row and the jth column in the given array of numbers.

3. The value of the determinant is the sum of the products found in step 2.

d. Cramer's rule for solving linear systems in three variables uses four third-order determinants and is stated in the box on page 797.

e. Cramer's rule with inconsistent and dependent systems is summarized by the two situations in the box on page 799.

Review Exercises

9.1

In Exercises 1–2, write the system of linear equations represented by the augmented matrix. Use x, y, z, and, if necessary, w for the variables. Once the system is written, use back-substitution to find its solution.

1. $\begin{bmatrix} 1 & 1 & 3 & | & 12 \\ 0 & 1 & -2 & | & -4 \\ 0 & 0 & 1 & | & 3 \end{bmatrix}$

2. $\begin{bmatrix} 1 & 0 & -2 & 2 & | & 1 \\ 0 & 1 & 1 & -1 & | & 0 \\ 0 & 0 & 1 & -\frac{7}{3} & | & -\frac{1}{3} \\ 0 & 0 & 0 & 1 & | & 1 \end{bmatrix}$

In Exercises 3–4, perform each matrix row operation and write the new matrix.

3. $\begin{bmatrix} 1 & 2 & 2 & | & 2 \\ 0 & 1 & -1 & | & 2 \\ 0 & 5 & 4 & | & 1 \end{bmatrix}$ Multiply row 2 by −5 and add to corresponding entries in row 3.

4. $\begin{bmatrix} 2 & -2 & 1 & | & -1 \\ 1 & 2 & -1 & | & 2 \\ 6 & 4 & 3 & | & 5 \end{bmatrix}$ Multiply row 1 by $\frac{1}{2}$.

In Exercises 5–7, solve each system of equations using matrices. Use Gaussian elimination with back-substitution or Gauss-Jordan elimination.

5. $\begin{aligned} x + 2y + 3z &= -5 \\ 2x + y + z &= 1 \\ x + y - z &= 8 \end{aligned}$

6. $\begin{aligned} x - 2y + z &= 0 \\ y - 3z &= -1 \\ 2y + 5z &= -2 \end{aligned}$

7. $\begin{aligned} 3x_1 + 5x_2 - 8x_3 + 5x_4 &= -8 \\ x_1 + 2x_2 - 3x_3 + x_4 &= -7 \\ 2x_1 + 3x_2 - 7x_3 + 3x_4 &= -11 \\ 4x_1 + 8x_2 - 10x_3 + 7x_4 &= -10 \end{aligned}$

8. The table shows the pollutants in the air in a city on a typical summer day.

x (Hours after 6 A.M.)	y (Amount of Pollutants in the Air, in parts per million)
2	98
4	138
10	162

a. Use the function $y = ax^2 + bx + c$ to model the data. Use either Gaussian elimination with back-substitution or Gauss-Jordan elimination to find the values for a, b, and c.

b. Use the function to find the time of day at which the city's air pollution level is at a maximum. What is the maximum level?

9.2

In Exercises 9–12, use Gaussian elimination to find the complete solution to each system, or show that none exists.

9. $\begin{aligned} 2x - 3y + z &= 1 \\ x - 2y + 3z &= 2 \\ 3x - 4y - z &= 1 \end{aligned}$

10. $\begin{aligned} x - 3y + z &= 1 \\ -2x + y + 3z &= -7 \\ x - 4y + 2z &= 0 \end{aligned}$

11. $\begin{aligned} x_1 + 4x_2 + 3x_3 - 6x_4 &= 5 \\ x_1 + 3x_2 + x_3 - 4x_4 &= 3 \\ 2x_1 + 8x_2 + 7x_3 - 5x_4 &= 11 \\ 2x_1 + 5x_2 - 6x_4 &= 4 \end{aligned}$

12. $\begin{aligned} 2x + 3y - 5z &= 15 \\ x + 2y - z &= 4 \end{aligned}$

13. The figure shows the intersections of three one-way streets. The numbers given represent traffic flow in cars per hour at a peak period (from 4 P.M. to 6 P.M.).

a. Use the idea that the number of cars entering each intersection per hour must equal the number of cars leaving per hour to set up a linear system of equations involving x, y, and z.

b. Use Gaussian elimination to solve the system.

c. If construction limits the value of z to 400, how many cars per hour must pass between the other intersections to keep traffic flowing?

9.3

14. Find values for x, y, and z so that the following matrices are equal:

$$\begin{bmatrix} 2x & y + 7 \\ z & 4 \end{bmatrix} = \begin{bmatrix} -10 & 13 \\ 6 & 4 \end{bmatrix}.$$

In Exercises 15–28, perform the indicated matrix operations given that A, B, C, and D are defined as follows. If an operation is not defined, state the reason.

$$A = \begin{bmatrix} 2 & -1 & 2 \\ 5 & 3 & -1 \end{bmatrix}, \quad B = \begin{bmatrix} 0 & -2 \\ 3 & 2 \\ 1 & -5 \end{bmatrix},$$

$$C = \begin{bmatrix} 1 & 2 & 3 \\ -1 & 1 & 2 \\ -1 & 2 & 1 \end{bmatrix}, \quad \text{and} \quad D = \begin{bmatrix} -2 & 3 & 1 \\ 3 & -2 & 4 \end{bmatrix}$$

15. $A + D$

16. $2B$

17. $D - A$

18. $B + C$

19. $3A + 2D$

20. $-2A + 4D$

21. $-5(A + D)$

22. AB

23. BA

24. BD

25. DB

26. $AB - BA$

27. $(A - D)C$

28. $B(AC)$

In Exercises 29–30, use nine pixels in a 3 × 3 grid and the color levels shown.

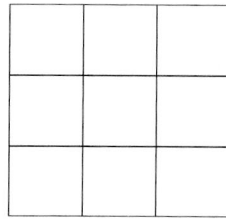
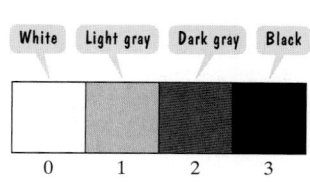

White   Light gray   Dark gray   Black

0   1   2   3

29. Write a 3 × 3 matrix that represents a digital photograph of the letter L in dark gray on a light gray background.

30. Find a matrix B so that $A + B$ increases the contrast of the letter L by changing the dark gray to black and the light gray to white.

31. An automobile dealership sells three models of cars at its three outlets. The inventory of models at each store is given by the following matrix.

	Model X	Model Y	Model Z
Outlet 1	12	7	6
Outlet 2	20	8	10
Outlet 3	7	2	3

$= A$

The next matrix gives the wholesale and retail prices for each model.

	Wholesale Price	Retail Price
Model X	16,000	19,000
Model Y	12,000	15,000
Model Z	14,000	18,500

$= B$

a. Calculate the product AB.

b. Describe what the matrix AB represents and interpret the elements.

c. What is the wholesale value of the cars at outlet 1?

d. What is the retail value of the cars at outlet 2?

e. If outlet 3 sells all of the inventory in matrix A, what is the profit for that branch of the dealership?

9.4

In Exercises 32–33, find the products AB and BA to determine whether B is the multiplicative inverse of A.

32. $A = \begin{bmatrix} 2 & 7 \\ 1 & 4 \end{bmatrix}$, $B = \begin{bmatrix} 4 & -7 \\ -1 & 3 \end{bmatrix}$

33. $A = \begin{bmatrix} 1 & 0 & 0 \\ 0 & 2 & -7 \\ 0 & -1 & 4 \end{bmatrix}$, $B = \begin{bmatrix} 1 & 0 & 0 \\ 0 & 4 & 7 \\ 0 & 1 & 2 \end{bmatrix}$

In Exercises 34–37, find A^{-1}. Check that $AA^{-1} = I$ and $A^{-1}A = I$.

34. $A = \begin{bmatrix} 1 & -1 \\ -2 & 3 \end{bmatrix}$

35. $A = \begin{bmatrix} 0 & 1 \\ 5 & 3 \end{bmatrix}$

36. $A = \begin{bmatrix} 1 & 0 & -2 \\ 2 & 1 & 0 \\ 1 & 0 & -3 \end{bmatrix}$

37. $A = \begin{bmatrix} 1 & 3 & -2 \\ 4 & 13 & -7 \\ 5 & 16 & -8 \end{bmatrix}$

In Exercises 38–39,
  a. *Write each linear system as a matrix equation in the form $AX = B$.*
  b. *Solve the system using the inverse of the coefficient matrix.*

The inverse of

38. $\begin{aligned} x + y + 2z &= 7 \\ y + 3z &= -2 \\ 3x \quad\quad - 2z &= 0 \end{aligned}$ $\begin{bmatrix} 1 & 1 & 2 \\ 0 & 1 & 3 \\ 3 & 0 & -2 \end{bmatrix}$ is $\begin{bmatrix} -2 & 2 & 1 \\ 9 & -8 & -3 \\ -3 & 3 & 1 \end{bmatrix}$.

The inverse of

39. $\begin{aligned} x - y + 2z &= 12 \\ y - z &= -5 \\ x \quad\quad + 2z &= 10 \end{aligned}$ $\begin{bmatrix} 1 & -1 & 2 \\ 0 & 1 & -1 \\ 1 & 0 & 2 \end{bmatrix}$ is $\begin{bmatrix} 2 & 2 & -1 \\ -1 & 0 & 1 \\ -1 & -1 & 1 \end{bmatrix}$.

40. Use the coding-matrix, $A = \begin{bmatrix} 3 & 2 \\ 4 & 3 \end{bmatrix}$ and its inverse $A^{-1} = \begin{bmatrix} 3 & -2 \\ -4 & 3 \end{bmatrix}$ to encode and then decode the word RULE.

9.5

In Exercises 41–46, evaluate each determinant.

41. $\begin{vmatrix} 3 & 2 \\ -1 & 5 \end{vmatrix}$

42. $\begin{vmatrix} -2 & -3 \\ -4 & -8 \end{vmatrix}$

43. $\begin{vmatrix} 2 & 4 & -3 \\ 1 & -1 & 5 \\ -2 & 4 & 0 \end{vmatrix}$

44. $\begin{vmatrix} 4 & 7 & 0 \\ -5 & 6 & 0 \\ 3 & 2 & -4 \end{vmatrix}$

45. $\begin{vmatrix} 1 & 1 & 0 & 2 \\ 0 & 3 & 2 & 1 \\ 0 & -2 & 4 & 0 \\ 0 & 3 & 0 & 1 \end{vmatrix}$

46. $\begin{vmatrix} 2 & 2 & 2 & 2 \\ 0 & 2 & 2 & 2 \\ 0 & 0 & 2 & 2 \\ 0 & 0 & 0 & 2 \end{vmatrix}$

In Exercises 47–50, use Cramer's rule to solve each system.

47. $x - 2y = 8$
$3x + 2y = -1$

48. $7x + 2y = 0$
$2x + y = -3$

49. $x + 2y + 2z = 5$
$2x + 4y + 7z = 19$
$-2x - 5y - 2z = 8$

50. $2x + y = -4$
$y - 2z = 0$
$3x - 2z = -11$

51. Use the quadratic function $y = ax^2 + bx + c$ to model the following data:

x (Age of a Driver)	y (Average Number of Automobile Accidents per Day in the United States)
20	400
40	150
60	400

Use Cramer's rule to determine values for a, b, and c. Then use the model to write a statement about the average number of automobile accidents in which 30-year-olds and 50-year-olds are involved daily.

Chapter 9 Test

In Exercises 1–2, solve each system of equations using matrices.

1. $x + 2y - z = -3$
$2x - 4y + z = -7$
$-2x + 2y - 3z = 4$

2. $x - 2y + z = 2$
$2x - y - z = 1$

In Exercises 3–6, let

$$A = \begin{bmatrix} 3 & 1 \\ 1 & 0 \\ 2 & 1 \end{bmatrix}, \quad B = \begin{bmatrix} 1 & -1 \\ 2 & 1 \end{bmatrix}, \quad \text{and} \quad C = \begin{bmatrix} 1 & 2 \\ -1 & 3 \end{bmatrix}.$$

Carry out the indicated operations.

3. $2B + 3C$

4. AB

5. C^{-1}

6. $BC - 3B$

7. If $A = \begin{bmatrix} 1 & 2 & 2 \\ 2 & 3 & 3 \\ 1 & -1 & -2 \end{bmatrix}$ and $B = \begin{bmatrix} -3 & 2 & 0 \\ 7 & -4 & 1 \\ -5 & 3 & -1 \end{bmatrix}$,

show that B is the inverse of A.

8. Consider the system

$$3x + 5y = 9$$
$$2x - 3y = -13.$$

a. Express the system in the form $AX = B$, where A, X, and B are appropriate matrices.

b. Find A^{-1}, the inverse of the coefficient matrix.

c. Use A^{-1} to solve the given system.

9. Evaluate: $\begin{vmatrix} 4 & -1 & 3 \\ 0 & 5 & -1 \\ 5 & 2 & 4 \end{vmatrix}$.

10. Solve for x only using Cramer's rule:

$$3x + y - 2z = -3$$
$$2x + 7y + 3z = 9$$
$$4x - 3y - z = 7.$$

Cumulative Review Exercises (Chapters 1–9)

Solve each equation or inequality in Exercises 1–6.

1. $2x^2 = 4 - x$

2. $5x + 8 \le 7(1 + x)$

3. $\sqrt{2x + 4} - \sqrt{x + 3} - 1 = 0$

4. $3x^3 + 8x^2 - 15x + 4 = 0$

5. $e^{2x} - 14e^x + 45 = 0$

6. $\log_3 x + \log_3 (x + 2) = 1$

7. Use matrices to solve this system.

$$x - y + z = 17$$
$$2x + 3y + z = 8$$
$$-4x + y + 5z = -2$$

8. Solve for y using Cramer's rule.

$$x - 2y + z = 7$$
$$2x + y - z = 0$$
$$3x + 2y - 2z = -2$$

9. If $f(x) = \sqrt{4x - 7}$, find $f^{-1}(x)$.

10. Graph: $f(x) = \dfrac{x}{x^2 - 16}$.

11. Use the graph of $f(x) = 4x^4 - 4x^3 - 25x^2 + x + 6$ shown in the figure to factor the polynomial completely.

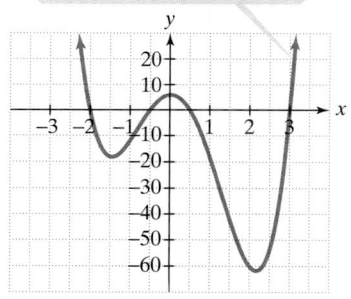

$f(x) = 4x^4 - 4x^3 - 25x^2 + x + 6$

12. Graph $y = \log_2 x$ and $y = \log_2 (x + 1)$ in the same rectangular coordinate system.

13. Use the exponential decay model $A = A_0 e^{kt}$ to solve this problem. A radioactive substance has a half-life of 40 days. There are initially 900 grams of the substance.

   a. Find the decay model for this substance.

   b. How much of the substance will remain after 10 days?

14. Multiply the matrices: $\begin{bmatrix} 1 & -1 & 0 \\ 2 & 1 & 3 \end{bmatrix} \begin{bmatrix} 4 & -1 \\ 2 & 0 \\ 1 & 1 \end{bmatrix}$.

15. Find the partial fraction decomposition of

$$\frac{3x^2 + 17x - 38}{(x - 3)(x - 2)(x + 2)}.$$

In Exercises 16–19, graph each equation, function, or inequality in the rectangular coordinate system.

16. $y = -\frac{2}{3}x - 1$

17. $3x - 5y < 15$

18. $f(x) = x^2 - 2x - 3$

19. $(x - 1)^2 + (y + 1)^2 = 9$

20. Use synthetic division to divide $x^3 - 6x + 4$ by $x - 2$.

21. Graph: $y = 2 \sin 2\pi x$, $0 \le x \le 2$.

22. Find the exact value of $\cos \left[\tan^{-1}\left(-\frac{4}{3}\right)\right]$.

23. Verify the identity: $\dfrac{\cos 2x}{\cos x - \sin x} = \cos x + \sin x$.

24. Solve on the interval $[0, 2\pi)$: $\cos^2 x + \sin x + 1 = 0$.

25. If $\mathbf{v} = -6\mathbf{i} + 5\mathbf{j}$ and $\mathbf{w} = -7\mathbf{i} + 3\mathbf{j}$, find $4\mathbf{w} - 5\mathbf{v}$.

Conic Sections and Analytic Geometry

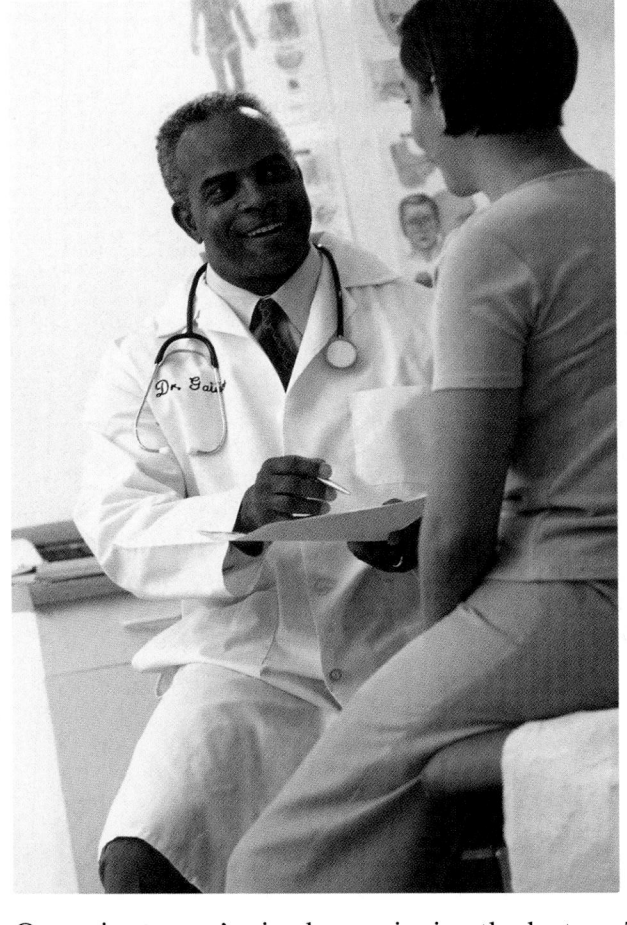

From ripples in water to the path on which humanity journeys through space, certain curves occur naturally throughout the universe. Over 2000 years ago the ancient Greeks studied these curves, called *conic sections*, without regard to their immediate usefulness simply because the study elicited ideas that were exciting, challenging, and interesting. The ancient Greeks could not have imagined the applications of these curves in the twenty-first century. Overwhelmed by the choices on satellite television? Blame it on a conic section! In this chapter, we use the rectangular coordinate system to study the conic sections and the mathematics behind their surprising applications.

One minute you're in class, enjoying the lecture. Then a sharp pain radiates down your side. The next minute you're being diagnosed with, of all things, a kidney stone. It took your cousin six weeks to recover from kidney stone surgery, but your doctor assures you there is nothing to worry about. A new procedure, based on a curve that looks like the cross section of a football, will dissolve the stone painlessly and let you return to class in a day or two. How can this be?

SECTION 10.1 *The Ellipse*

Objectives

1. Graph ellipses centered at the origin.
2. Write equations of ellipses in standard form.
3. Graph ellipses not centered at the origin.
4. Solve applied problems involving ellipses.

You took on a summer job driving a truck, delivering books that were ordered online. You're an avid reader, so just being around books sounded appealing. However, now you're feeling a bit shaky driving the truck for the first time. It's 10 feet wide and 9 feet high; compared to your compact car, it feels like you're behind the wheel of a tank. Up ahead you see a sign at the semielliptical entrance to a tunnel: Caution! Tunnel is 10 Feet High at Center Peak. Then you see another sign: Caution! Tunnel is 40 Feet Wide. Will your truck clear the opening of the tunnel's archway?

The mathematics of your world is present in the movements of planets, bridge and tunnel construction, navigational systems used to keep track of a ship's location, manufacture of lenses for telescopes, and even a procedure for disintegrating kidney stones. The mathematics behind these applications involves conic sections. **Conic sections** are curves that result from the intersection of a right circular cone and a plane. Figure 10.1 illustrates the four conic sections: the circle, the ellipse, the parabola, and the hyperbola.

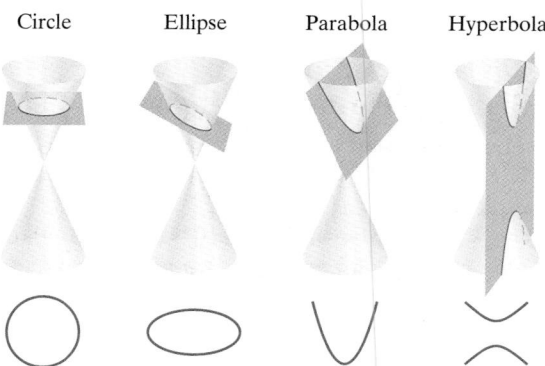

Figure 10.1 Obtaining the conic sections by intersecting a plane and a cone

In this section, we study the symmetric oval-shaped curve known as the ellipse. We will use a geometric definition for an ellipse to derive its equations. With these equations, we will determine if your delivery truck will clear the tunnel's entrance.

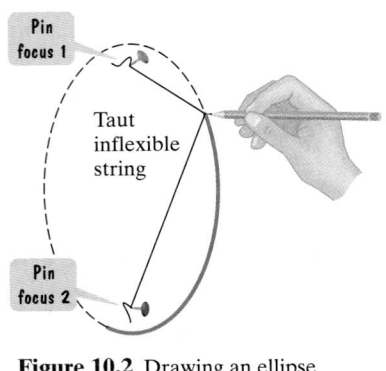

Figure 10.2 Drawing an ellipse

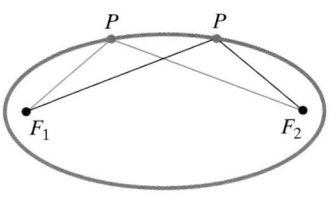

Figure 10.3

Definition of an Ellipse

Figure 10.2 illustrates how to draw an ellipse. Place pins at two fixed points, each of which is called a focus (plural: foci). If the ends of a fixed length of string are fastened to the pins and we draw the string taut with a pencil, the path traced by the pencil will be an ellipse. Notice that the sum of the distances of the pencil point from the foci remains constant because the length of the string is fixed. This procedure for drawing an ellipse illustrates its geometric definition.

Definition of an Ellipse

An **ellipse** is the set of all points in a plane the sum of whose distances from two fixed points, F_1 and F_2, is constant (see Figure 10.3). These two fixed points are called the **foci** (plural of **focus**). The midpoint of the segment connecting the foci is the **center** of the ellipse.

Figure 10.4 illustrates that an ellipse can be elongated horizontally or vertically. The line through the foci intersects the ellipse at two points, called the **vertices** (singular: **vertex**). The line segment that joins the vertices is the **major axis**. Notice that the midpoint of the major axis is the center of the ellipse. The line segment whose endpoints are on the ellipse that is perpendicular to the major axis at the center is the **minor axis** of the ellipse.

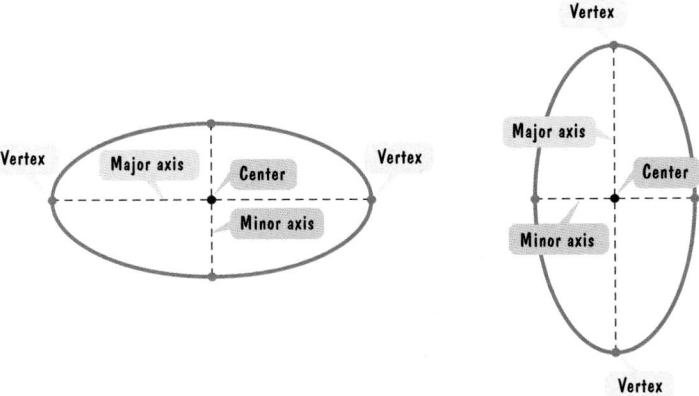

Figure 10.4 Horizontal and vertical elongations of an ellipse

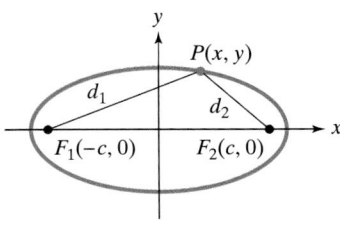

Figure 10.5

Standard Form of the Equation of an Ellipse

The rectangular coordinate system gives us a unique way of describing an ellipse. It enables us to translate an ellipse's geometric definition into an algebraic equation.

We start with Figure 10.5 to obtain an ellipse's equation. We've placed an ellipse that is elongated horizontally into a rectangular coordinate system. The foci are on the x-axis at $(-c, 0)$ and $(c, 0)$, as in Figure 10.5. In this way, the center of the ellipse is at the origin. We let (x, y) represent the coordinates of any point on the ellipse.

What does the definition of an ellipse tell us about the point (x, y) in Figure 10.5? For any point (x, y) on the ellipse, the sum of the distances to the two foci, $d_1 + d_2$, must be constant. We denote this constant by $2a$. Thus, the point (x, y) is on the ellipse if and only if

$$d_1 + d_2 = 2a.$$

$$\sqrt{(x + c)^2 + y^2} + \sqrt{(x - c)^2 + y^2} = 2a \qquad \text{Use the distance formula.}$$

After eliminating radicals and simplifying, we obtain

$$(a^2 - c^2)x^2 + a^2y^2 = a^2(a^2 - c^2).$$

Look at the triangle in Figure 10.5. Notice that the distance from F_1 to F_2 is $2c$ and $2c < d_1 + d_2$. Equivalently, $2c < 2a$ and $c < a$. Consequently, $a^2 - c^2 > 0$. For convenience, let $b^2 = a^2 - c^2$. Substituting b^2 for $a^2 - c^2$ in the preceding equation, we obtain

$$b^2x^2 + a^2y^2 = a^2b^2$$

$$\frac{b^2x^2}{a^2b^2} + \frac{a^2y^2}{a^2b^2} = \frac{a^2b^2}{a^2b^2} \qquad \text{Divide both sides by } a^2b^2.$$

$$\frac{x^2}{a^2} + \frac{y^2}{b^2} = 1 \qquad \text{Simplify.}$$

This last equation is the **standard form of the equation of an ellipse.** There are two such equations, one for a horizontal major axis and one for a vertical major axis.

Standard Forms of the Equations of an Ellipse

The **standard form of the equation of an ellipse** with center at the origin, and major and minor axes of lengths $2a$ and $2b$ (where a and b are positive, and $a^2 > b^2$) is

$$\frac{x^2}{a^2} + \frac{y^2}{b^2} = 1 \qquad \text{or} \qquad \frac{x^2}{b^2} + \frac{y^2}{a^2} = 1.$$

Figure 10.6 illustrates that the vertices are on the major axis, a units from the center. The foci are are on the major axis, c units from the center. For both equations, $b^2 = a^2 - c^2$.

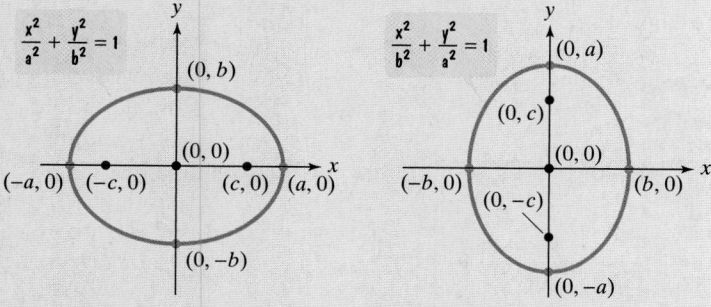

Figure 10.6 **(a)** Major axis is horizontal with length $2a$. **(b)** Major axis is vertical with length $2a$.

Using the Standard Form of the Equation of an Ellipse

We can use the standard form of an ellipse's equation to graph the ellipse. Although the definition of the ellipse is given in terms of its foci, the foci are not part of the graph. A complete graph of an ellipse can be obtained without graphing the foci.

1 Graph ellipses centered at the origin.

EXAMPLE 1   Graphing an Ellipse Centered at the Origin

Graph and locate the foci: $\dfrac{x^2}{9} + \dfrac{y^2}{4} = 1$.

Solution   The given equation is the standard form of an ellipse's equation with $a^2 = 9$ and $b^2 = 4$.

$$\frac{x^2}{9} + \frac{y^2}{4} = 1$$

$a^2 = 9$. This is the larger of the two numbers in the denominator.

$b^2 = 4$. This is the smaller of the two numbers in the denominator.

Because the denominator of the x^2 term is greater than the denominator of the y^2 term, the major axis is horizontal. Based on the standard form of the equation, we know the vertices are $(-a, 0)$ and $(a, 0)$. Because $a^2 = 9$, $a = 3$. Thus, the vertices are $(-3, 0)$ and $(3, 0)$, shown in Figure 10.7.

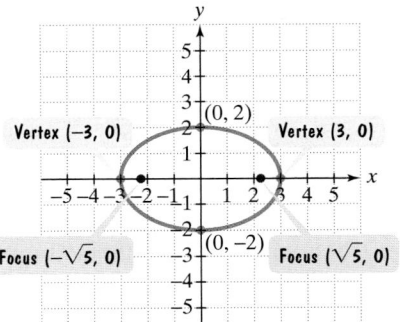

Figure 10.7 The graph of

$$\frac{x^2}{9} + \frac{y^2}{4} = 1$$

Now let us find the endpoints of the vertical minor axis. According to the standard form of the equation, these endpoints are $(0, -b)$ and $(0, b)$. Because $b^2 = 4$, $b = 2$. Thus, the endpoints of the minor axis are $(0, -2)$ and $(0, 2)$. They are shown in Figure 10.7.

Finally, we find the foci, which are located at $(-c, 0)$ and $(c, 0)$. We can use the formula $b^2 = a^2 - c^2$ to do so. We know that $a^2 = 9$ and $b^2 = 4$; we need to find c^2 in order to find c. Because $b^2 = a^2 - c^2$, we obtain

$$c^2 = a^2 - b^2 = 9 - 4 = 5.$$

Because $c^2 = 5$, $c = \sqrt{5}$. The foci, $(-c, 0)$ and $(c, 0)$, are located at $(-\sqrt{5}, 0)$ and $(\sqrt{5}, 0)$. They are shown in Figure 10.7.

Technology

We graph $\dfrac{x^2}{9} + \dfrac{y^2}{4} = 1$ with a graphing utility by solving for y and defining two functions.

$$\frac{y^2}{4} = 1 - \frac{x^2}{9}$$

$$y^2 = 4\left(1 - \frac{x^2}{9}\right)$$

$$y = \pm 2\sqrt{1 - \frac{x^2}{9}}$$

Enter

$y_1 = 2 \boxed{\sqrt{\ }} (1 \boxed{-} x \boxed{\wedge} 2 \boxed{\div} 9)$

and

$$y_2 = -y_1.$$

To see the true shape of the ellipse, use the $\boxed{\text{ZOOM SQUARE}}$ feature so that one unit on the x-axis is the same length as one unit on the y-axis.

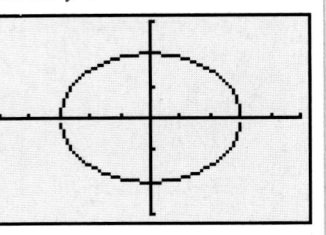

You can sketch the ellipse in Figure 10.7 by locating endpoints on the major and minor axes.

$$\frac{x^2}{3^2} + \frac{y^2}{2^2} = 1$$

Endpoints of the major axis are 3 units to the right and left of the center.

Endpoints of the minor axis are 2 units up and down from the center.

Check Point 1 Graph and locate the foci: $\dfrac{x^2}{36} + \dfrac{y^2}{9} = 1$.

EXAMPLE 2 Graphing an Ellipse Centered at the Origin

Graph and locate the foci: $25x^2 + 16y^2 = 400$.

Solution We begin by expressing the equation in standard form. Because we want 1 on the right side, we divide both sides by 400.

$$\frac{25x^2}{400} + \frac{16y^2}{400} = \frac{400}{400}$$

$$\frac{x^2}{16} + \frac{y^2}{25} = 1$$

$b^2 = 16$. This is the smaller of the two numbers in the denominator.

$a^2 = 25$. This is the larger of the two numbers in the denominator.

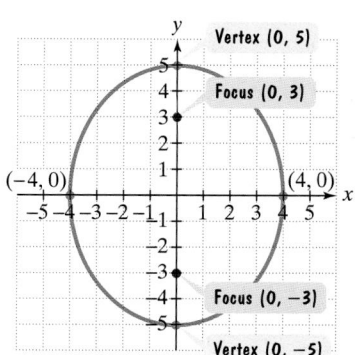

Figure 10.8 The graph of $\dfrac{x^2}{16} + \dfrac{y^2}{25} = 1$

The equation is the standard form of an ellipse's equation with $a^2 = 25$ and $b^2 = 16$. Because the denominator of the y^2 term is greater than the denominator of the x^2 term, the major axis is vertical. Based on the standard form of the equation, we know the vertices are $(0, -a)$ and $(0, a)$. Because $a^2 = 25$, $a = 5$. Thus, the vertices are $(0, -5)$ and $(0, 5)$, shown in Figure 10.8.

Now let us find the endpoints of the horizontal minor axis. According to the standard form of the equation, these endpoints are $(-b, 0)$ and $(b, 0)$. Because $b^2 = 16$, $b = 4$. Thus, the endpoints of the minor axis are $(-4, 0)$ and $(4, 0)$. They are shown in Figure 10.8.

Finally, we find the foci, which are located at $(0, -c)$ and $(0, c)$. We can use the formula $b^2 = a^2 - c^2$ to do so. We know that $a^2 = 25$ and $b^2 = 16$; we need to find c^2 in order to find c. Because $b^2 = a^2 - c^2$, we obtain

$$c^2 = a^2 - b^2 = 25 - 16 = 9.$$

Because $c^2 = 9$, $c = 3$. The foci, $(0, -c)$ and $(0, c)$, are located at $(0, -3)$ and $(0, 3)$. They are shown in Figure 10.8. You can sketch the ellipse in Figure 10.8 by locating endpoints on the major and minor axes:

$$\frac{x^2}{4^2} + \frac{y^2}{5^2} = 1.$$

Endpoints of the minor axis are 4 units to the right and left of the center.

Endpoints of the major axis are 5 units up and down from the center.

Check Point 2   Graph and locate the foci: $16x^2 + 9y^2 = 144$.

2 Write equations of ellipses in standard form.

In Examples 1 and 2, we used the equation of an ellipse to find its foci and vertices. In the next example, we reverse this procedure.

EXAMPLE 3   Finding the Equation of an Ellipse from Its Foci and Vertices

Find the standard form of the equation of an ellipse with foci at $(-1, 0)$ and $(1, 0)$ and vertices $(-2, 0)$ and $(2, 0)$.

Solution   Because the foci are located at $(-1, 0)$ and $(1, 0)$, on the x-axis, the major axis is horizontal. The center of the ellipse is midway between the foci, located at $(0, 0)$. Thus, the form of the equation is

$$\frac{x^2}{a^2} + \frac{y^2}{b^2} = 1.$$

We need to determine the values for a^2 and b^2. The distance from the center $(0, 0)$ to either vertex, $(-2, 0)$ or $(2, 0)$, is 2. Thus, $a = 2$.

$$\frac{x^2}{2^2} + \frac{y^2}{b^2} = 1 \quad \text{or} \quad \frac{x^2}{4} + \frac{y^2}{b^2} = 1$$

We must still find b^2. The distance from the center $(0, 0)$ to either focus, $(-1, 0)$ or $(1, 0)$, is 1, so $c = 1$. Because $b^2 = a^2 - c^2$, we have

$$b^2 = 2^2 - 1^2 = 4 - 1 = 3.$$

Substituting 3 for b^2 in the last equation gives us the standard form of the ellipse's equation. The equation is

$$\frac{x^2}{4} + \frac{y^2}{3} = 1.$$

Check Point 3   Find the standard form of the equation of an ellipse with foci at $(-2, 0)$ and $(2, 0)$ and vertices $(-3, 0)$ and $(3, 0)$.

3 Graph ellipses not centered at the origin.

Translations of Ellipses

Despite the fact that an ellipse is not the graph of a function, its graph can be translated in the same manner as that of a function. Figure 10.9 illustrates that the graphs of

$$\frac{(x - h)^2}{a^2} + \frac{(y - k)^2}{b^2} = 1 \quad \text{and} \quad \frac{x^2}{a^2} + \frac{y^2}{b^2} = 1$$

have the same size and shape. However, the graph of the first equation is centered at (h, k) rather than at the origin.

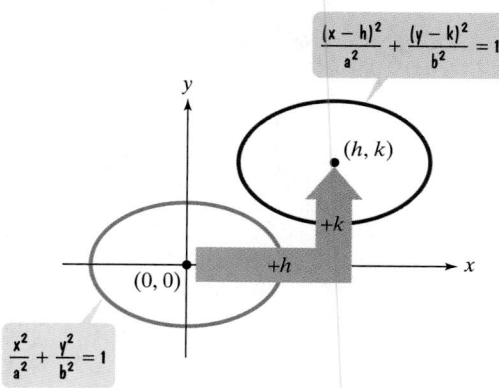

Figure 10.9 Translating an ellipse's graph

Table 10.1 gives the standard forms of equations of ellipses centered at (h, k). Figure 10.10 shows their graphs.

Table 10.1 Standard Forms of Equations of Ellipses Centered at (h, k)

Equation	Center	Major Axis	Foci	Vertices
$\dfrac{(x - h)^2}{a^2} + \dfrac{(y - k)^2}{b^2} = 1,$ $a^2 > b^2$ and $b^2 = a^2 - c^2$	(h, k)	Parallel to the x-axis, horizontal	$(h - c, k)$ $(h + c, k)$	$(h - a, k)$ $(h + a, k)$
$\dfrac{(x - h)^2}{b^2} + \dfrac{(y - k)^2}{a^2} = 1,$ $a^2 > b^2$ and $b^2 = a^2 - c^2$	(h, k)	Parallel to the y-axis, vertical	$(h, k - c)$ $(h, k + c)$	$(h, k - a)$ $(h, k + a)$

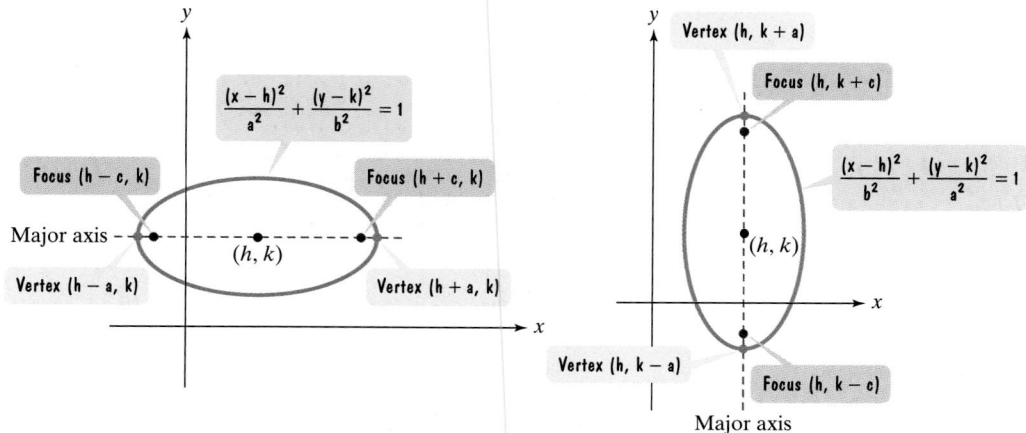

Figure 10.10 Graphs of ellipses centered at (h, k)

EXAMPLE 4 Graphing an Ellipse Centered at (h, k)

Graph: $\dfrac{(x - 1)^2}{4} + \dfrac{(y + 2)^2}{9} = 1$. Where are the foci located?

Solution In order to graph the ellipse, we need to know its center (h, k). In the standard forms of equations centered at (h, k), h is the number subtracted from x and k is the number subtracted from y.

$$\underbrace{\dfrac{(x - 1)^2}{4}}_{\substack{\text{This is } (x - h)^2 \\ \text{with } h = 1.}} + \underbrace{\dfrac{\left(y - (-2)\right)^2}{9}}_{\substack{\text{This is } (y - k)^2 \\ \text{with } k = -2.}} = 1$$

We see that $h = 1$ and $k = -2$. Thus, the center of the ellipse, (h, k), is $(1, -2)$. We can graph the ellipse by locating endpoints on the major and minor axes. To do this, we must identify a^2 and b^2.

$$\dfrac{(x - 1)^2}{4} + \dfrac{(y + 2)^2}{9} = 1$$

$b^2 = 4$. This is the smaller of the two numbers in the denominator.

$a^2 = 9$. This is the larger of the two numbers in the denominator.

The larger number is under the expression involving y. This means that the major axis is vertical and parallel to the y-axis. Because $a^2 = 9$, $a = 3$ and the vertices lie three units above and below the center. Also, because $b^2 = 4$, $b = 2$ and the endpoints of the minor axis lie two units to the right and left of the center. We categorize these observations as follows:

Center	Vertices	Endpoints of Minor Axis
$(1, -2)$	$(1, -2 + 3) = (1, 1)$	$(1 + 2, -2) = (3, -2)$
	$(1, -2 - 3) = (1, -5)$	$(1 - 2, -2) = (-1, -2)$

Using the center and these four points, we can sketch the ellipse shown in Figure 10.11.

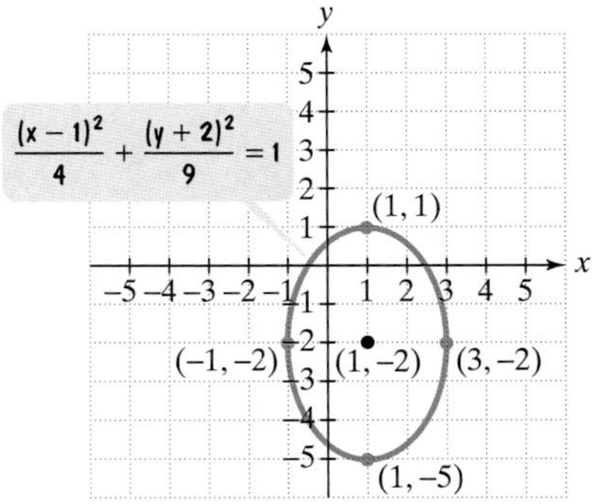

Figure 10.11 The graph of an ellipse centered at $(1, -2)$

With $b^2 = a^2 - c^2$, we have $4 = 9 - c^2$, and $c^2 = 5$. So the foci are located $\sqrt{5}$ units above and below the center, at $\left(1, -2 + \sqrt{5}\right)$ and $\left(1, -2 - \sqrt{5}\right)$.

Check Point 4

Graph: $\dfrac{(x + 1)^2}{9} + \dfrac{(y - 2)^2}{4} = 1$. Where are the foci located?

In some cases, it is necessary to convert the equation of an ellipse to standard form by completing the square on x and y. For example, suppose that we wish to graph the ellipse whose equation is

$$9x^2 + 4y^2 - 18x + 16y - 11 = 0.$$

Because we plan to complete the square on both x and y, we need to rearrange terms so that

- x terms are arranged in descending order.
- y terms are arranged in descending order.
- the constant term appears on the right.

$9x^2 + 4y^2 - 18x + 16y - 11 = 0$	This is the given equation.
$(9x^2 - 18x) + (4y^2 + 16y) = 11$	Group terms and add 11 to both sides.
$9(x^2 - 2x + \square) + 4(y^2 + 4y + \square) = 11$	To complete the square, coefficients of x^2 and y^2 must be 1. Factor out 9 and 4, respectively.
$9(x^2 - 2x + 1) + 4(y^2 + 4y + 4) = 11 + 9 + 16$	Complete each square by adding the square of half the coefficient of x and y, respectively.
$9(x - 1)^2 + 4(y + 2)^2 = 36$	Factor.
$\dfrac{9(x - 1)^2}{36} + \dfrac{4(y + 2)^2}{36} = \dfrac{36}{36}$	Divide both sides by 36.
$\dfrac{(x - 1)^2}{4} + \dfrac{(y + 2)^2}{9} = 1$	Simplify.

Study Tip

When completing the square, remember that changes made on the left side of the equation must also be made on the right side of the equation.

The equation is now in standard form. This is precisely the form of the equation that we graphed in Example 4.

4 Solve applied problems involving ellipses.

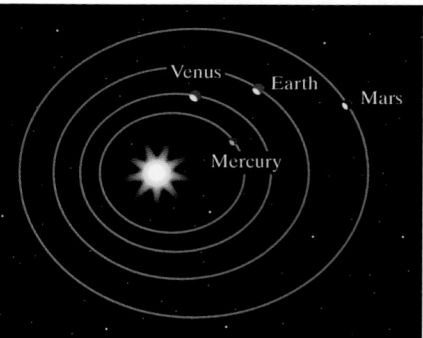

Planets move in elliptical orbits.

Applications

Ellipses have many applications. German scientist Johannes Kepler (1571–1630) showed that the planets in our solar system move in elliptical orbits, with the sun at a focus. Earth satellites also travel in elliptical orbits, with Earth at a focus.

One intriguing aspect of the ellipse is that a ray of light or a sound wave originating at one focus will be reflected by the ellipse exactly to the other focus. A whispering gallery is an elliptical room with an elliptical, dome-shaped ceiling. People standing at the foci can whisper and hear each other quite clearly, while persons in other locations in the room cannot hear them. Statuary Hall in the U.S. Capitol Building is elliptical. President John Quincy Adams, while a member of the House of Representatives, was aware of this acoustical phenomenon. He situated his desk at a focal point of the elliptical ceiling, easily eavesdropping on the private conversations of other House members located near the other focus.

The elliptical reflection principle is used in a procedure for disintegrating kidney stones. The patient is placed within a device that is elliptical in shape. The patient is at one focus, while ultrasound waves from the other focus hit the walls and are reflected to the kidney stone. The convergence of the ultrasound waves at the kidney stone causes vibrations that shatter it into fragments. The small pieces can then be passed painlessly through the patient's system. The patient recovers in days, as opposed to up to six weeks if surgery is used instead.

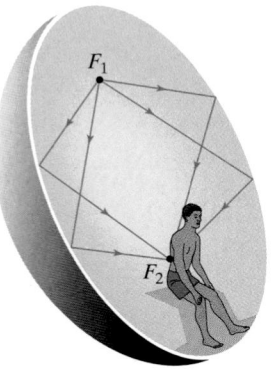

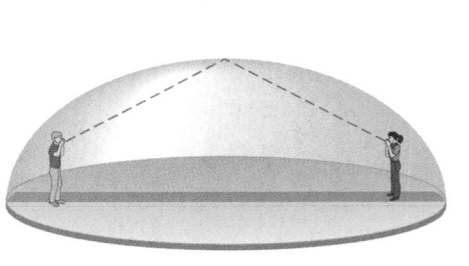

Whispering in an elliptical dome   Disintegrating kidney stones

Ellipses are often used for supporting arches of bridges and in tunnel construction. This application forms the basis of our next example.

EXAMPLE 5   An Application Involving an Ellipse

A semielliptical archway over a one-way road has a height of 10 feet and a width of 40 feet (see Figure 10.12). Your truck has a width of 10 feet and a height of 9 feet. Will your truck clear the opening of the archway?

Solution   To determine the clearance, we must find the height of the archway 5 feet from the center. If that height is 9 feet or less, the truck will not clear the opening.

In Figure 10.13, we've constructed a coordinate system with the x-axis on the ground and the origin at the center of the archway. Also shown is the truck, whose height is 9 feet.

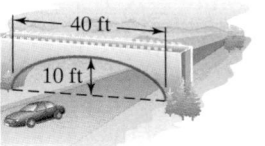

Figure 10.12 A semielliptical archway

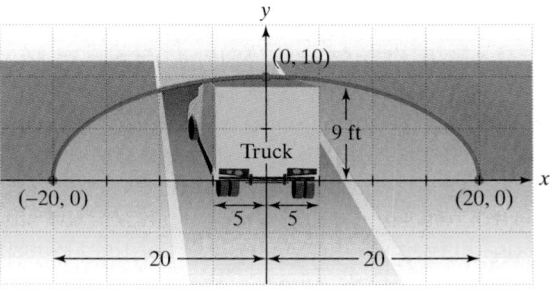

Figure 10.13

Halley's Comet

Halley's Comet has an elliptical orbit with the sun at one focus. The comet returns every 76.3 years. The first recorded sighting was in 239 B.C. It was last seen in 1986. At that time, spacecraft went close to the comet, measuring its nucleus to be 7 miles long and 4 miles wide. By 2024, Halley's Comet will have reached the farthest point in its elliptical orbit before returning to be next visible from Earth in 2062.

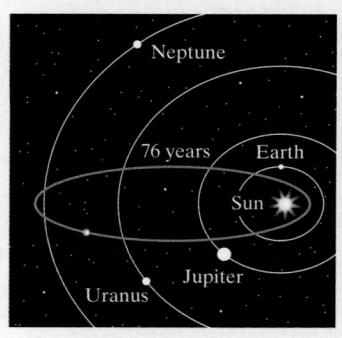

The elliptical orbit of Halley's Comet

Using the equation $\dfrac{x^2}{a^2} + \dfrac{y^2}{b^2} = 1$, we can express the equation of the blue archway in Figure 10.13 as $\dfrac{x^2}{20^2} + \dfrac{y^2}{10^2} = 1$ or $\dfrac{x^2}{400} + \dfrac{y^2}{100} = 1$.

As shown in Figure 10.13, the edge of the 10-foot-wide truck corresponds to $x = 5$. We find the height of the archway 5 feet from the center by substituting 5 for x and solving for y.

$$\frac{5^2}{400} + \frac{y^2}{100} = 1 \qquad \text{Substitute 5 for x.}$$

$$\frac{25}{400} + \frac{y^2}{100} = 1$$

$$\frac{1}{16} + \frac{y^2}{100} = 1$$

$$1600\left(\frac{1}{16} + \frac{y^2}{100}\right) = 1600(1) \qquad \text{Clear fractions by multiplying both sides by 1600.}$$

$$100 + 16y^2 = 1600 \qquad \text{Use the distributive property and simplify.}$$

$$16y^2 = 1500 \qquad \text{Subtract 100 from both sides.}$$

$$y^2 = \frac{1500}{16} \qquad \text{Divide both sides by 16.}$$

$$y = \sqrt{\frac{1500}{16}} \qquad \text{Take only the positive square root. The archway is above the x-axis and y is nonnegative.}$$

$$\approx 9.68$$

Thus, the height of the archway 5 feet from the center is approximately 9.68 feet. Because your truck's height is 9 feet, there is enough room for the truck to clear the archway.

Check Point 5 Will a truck that is 12 feet wide and has a height of 9 feet clear the opening of the archway described in Example 5?

EXERCISE SET 10.1

Practice Exercises

In Exercises 1–16, graph each ellipse and locate the foci.

1. $\dfrac{x^2}{16} + \dfrac{y^2}{4} = 1$

2. $\dfrac{x^2}{25} + \dfrac{y^2}{16} = 1$

3. $\dfrac{x^2}{9} + \dfrac{y^2}{36} = 1$

4. $\dfrac{x^2}{16} + \dfrac{y^2}{49} = 1$

5. $\dfrac{x^2}{25} + \dfrac{y^2}{64} = 1$

6. $\dfrac{x^2}{49} + \dfrac{y^2}{36} = 1$

7. $\dfrac{x^2}{49} + \dfrac{y^2}{81} = 1$

8. $\dfrac{x^2}{64} + \dfrac{y^2}{100} = 1$

9. $25x^2 + 4y^2 = 100$

10. $9x^2 + 4y^2 = 36$

11. $4x^2 + 16y^2 = 64$

12. $16x^2 + 9y^2 = 144$

13. $25x^2 + 9y^2 = 225$

14. $4x^2 + 25y^2 = 100$

15. $x^2 + 2y^2 = 8$

16. $12x^2 + 4y^2 = 36$

In Exercises 17–20, find the standard form of the equation of each ellipse and give the location of its foci.

17.

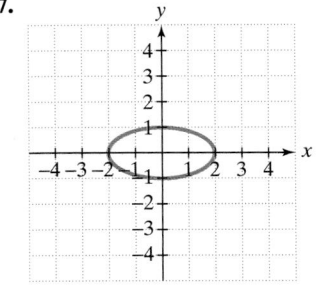

18.

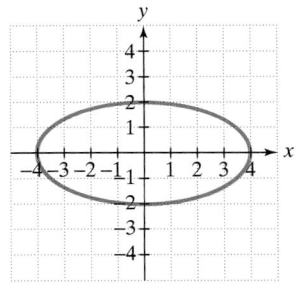

19.

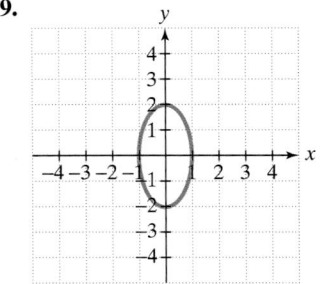

20.

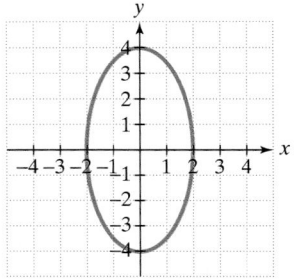

In Exercises 21–30, find the standard form of the equation of each ellipse centered at the origin satisfying the given conditions.

21. Foci: $(-5, 0), (5, 0)$; vertices: $(-8, 0), (8, 0)$

22. Foci: $(-2, 0), (2, 0)$; vertices: $(-6, 0), (6, 0)$

23. Foci: $(0, -4), (0, 4)$; vertices: $(0, -7), (0, 7)$

24. Foci: $(0, -3), (0, 3)$; vertices: $(0, -4), (0, 4)$

25. Foci: $(-2, 0), (2, 0)$; y-intercepts: -3 and 3

26. Foci: $(0, -2), (0, 2)$; x-intercepts: -2 and 2

27. Major axis horizontal with length 8; length of minor axis = 4

28. Major axis horizontal with length 12; length of minor axis = 6

29. Major axis vertical with length 10; length of minor axis = 4

30. Major axis vertical with length 20; length of minor axis = 10

In Exercises 31–42 graph each ellipse and give the location of its foci.

31. $\dfrac{(x-2)^2}{9} + \dfrac{(y-1)^2}{4} = 1$

32. $\dfrac{(x-1)^2}{16} + \dfrac{(y+2)^2}{9} = 1$

33. $(x+3)^2 + 4(y-2)^2 = 16$

34. $(x-3)^2 + 9(y+2)^2 = 18$

35. $\dfrac{(x-4)^2}{9} + \dfrac{(y+2)^2}{25} = 1$

36. $\dfrac{(x-3)^2}{9} + \dfrac{(y+1)^2}{16} = 1$

37. $\dfrac{x^2}{25} + \dfrac{(y-2)^2}{36} = 1$

38. $\dfrac{(x-4)^2}{4} + \dfrac{y^2}{25} = 1$

39. $\dfrac{(x+3)^2}{9} + (y-2)^2 = 1$

40. $\dfrac{(x+2)^2}{16} + (y-3)^2 = 1$

41. $9(x-1)^2 + 4(y+3)^2 = 36$

42. $36(x+4)^2 + (y+3)^2 = 36$

In Exercises 43–48, convert each equation to standard form by completing the square on x and y. Then graph the ellipse and give the location of its foci.

43. $9x^2 + 25y^2 - 36x + 50y - 164 = 0$

44. $4x^2 + 9y^2 - 32x + 36y + 64 = 0$

45. $9x^2 + 16y^2 - 18x + 64y - 71 = 0$

46. $x^2 + 4y^2 + 10x - 8y + 13 = 0$

47. $4x^2 + y^2 + 16x - 6y - 39 = 0$

48. $4x^2 + 25y^2 - 24x + 100y + 36 = 0$

Application Exercises

49. Will a truck that is 8 feet wide carrying a load that reaches 7 feet above the ground clear the semielliptical arch on the one-way road that passes under the bridge shown in the figure?

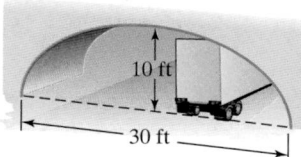

10 ft

30 ft

50. A semielliptic archway has a height of 20 feet and a width of 50 feet, as shown in the figure. Can a truck 14 feet high and 10 feet wide drive under the archway without going into the other lane?

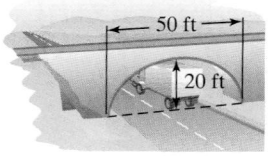

50 ft

20 ft

51. The elliptical ceiling in Statuary Hall in the U.S. Capitol Building is 96 feet long and 23 feet tall.

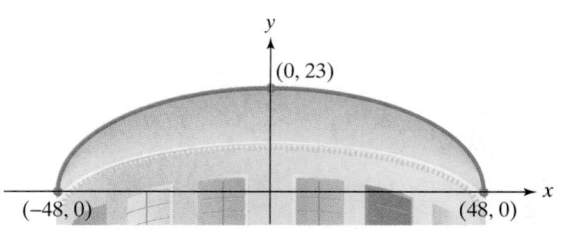

y

$(0, 23)$

x

$(-48, 0)$

$(48, 0)$

a. Using the rectangular coordinate system in the figure shown, write the standard form of the equation of the elliptical ceiling.

b. John Quincy Adams discovered that he could overhear the conversations of opposing party leaders near the left side of the chamber if he situated his desk at the focus at the right side of the chamber. How far from the center of the ellipse along the major axis did Adams situate his desk? (Round to the nearest foot.)

52. If an elliptical whispering room has a height of 30 feet and a width of 100 feet, where should two people stand if they would like to whisper back and forth and be heard?

Writing in Mathematics

53. What is an ellipse?

54. Describe how to graph $\dfrac{x^2}{25} + \dfrac{y^2}{16} = 1$.

55. Describe how to locate the foci for $\dfrac{x^2}{25} + \dfrac{y^2}{16} = 1$.

56. Describe one similarity and one difference between the graphs of $\dfrac{x^2}{25} + \dfrac{y^2}{16} = 1$ and $\dfrac{x^2}{16} + \dfrac{y^2}{25} = 1$.

57. Describe one similarity and one difference between the graphs of $\dfrac{x^2}{25} + \dfrac{y^2}{16} = 1$ and $\dfrac{(x-1)^2}{25} + \dfrac{(y-1)^2}{16} = 1$.

58. An elliptipool is an elliptical pool table with only one pocket. A pool shark places a ball on the table, hits it in what appears to be a random direction, and yet it bounces off the edge, falling directly into the pocket. Explain why this happens.

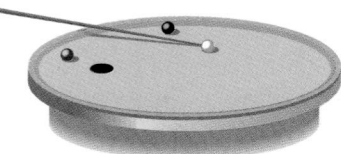 **Technology Exercises**

59. Use a graphing utility to graph any five of the ellipses that you graphed by hand in Exercises 1–16.

60. Use a graphing utility to graph any three of the ellipses that you graphed by hand in Exercises 31–42. First solve the given equation for y by using the square root method. Enter each of the two resulting equations to produce each half of the ellipse.

61. Use a graphing utility to graph any one of the ellipses that you graphed by hand in Exercises 43–48. Write the equation as a quadratic equation in y and use the quadratic formula to solve for y. Enter each of the two resulting equations to produce each half of the ellipse.

62. Write an equation for the path of each of the following elliptical orbits. Then use a graphing utility to graph the two ellipses in the same viewing rectangle. Can you see why early astronomers had difficulty detecting that these orbits are ellipses rather than circles?

Earth's orbit:	Length of major axis: 186 million miles
	Length of minor axis: 185.8 million miles
Mars's orbit:	Length of major axis: 283.5 million miles
	Length of minor axis: 278.5 million miles

 **Critical Thinking Exercises**

63. Find the standard form of the equation of an ellipse with vertices at $(0, -6)$ and $(0, 6)$, passing through $(2, -4)$.

64. An Earth satellite has an elliptical orbit described by

$$\frac{x^2}{(5000)^2} + \frac{y^2}{(4750)^2} = 1.$$

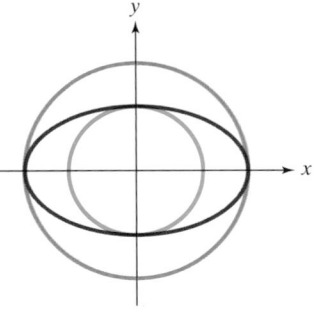

Satellite

Perigee height

Apogee height

(All units are in miles.) The coordinates of the center of Earth are $(16, 0)$.

a. The perigee of the satellite's orbit is the point that is nearest Earth's center. If the radius of Earth is approximately 4000 miles, find the distance of the perigee above Earth's surface.

b. The apogee of the satellite's orbit is the point that is the greatest distance from Earth's center. Find the distance of the apogee above Earth's surface.

65. The equation of the red ellipse in the following figure is

$$\frac{x^2}{25} + \frac{y^2}{9} = 1.$$

Write the equation for each circle shown in the figure.

66. What happens to the shape of the graph of $\dfrac{x^2}{a^2} + \dfrac{y^2}{b^2} = 1$ as $\dfrac{c}{a}$ is close to zero?

SECTION 10.2 **The Hyperbola**

Objectives

1. Locate a hyperbola's vertices and foci.

2. Write equations of hyperbolas in standard form.

3. Graph hyperbolas centered at the origin.

4. Graph hyperbolas not centered at the origin.

5. Solve applied problems involving hyperbolas.

St. Mary's Cathedral

Conic sections are often used to create unusual architectural designs. The top of St. Mary's Cathedral in San Francisco is a 2135-cubic-foot dome with walls rising 200 feet above the floor and supported by four massive concrete pylons that extend 94 feet into the ground. Cross sections of the roof are parabolas and hyperbolas. In this section, we study the curve with two parts known as the hyperbola.

Figure 10.14 Casting hyperbolic shadows

Definition of a Hyperbola

Figure 10.14 shows a cylindrical lampshade casting two shadows on a wall. These shadows indicate the distinguishing feature of hyperbolas: Their graphs contain two disjoint parts called **branches**. Although each branch might look like a parabola, its shape is actually quite different.

The definition of a hyperbola is similar to that of the ellipse. For the ellipse, the *sum* of the distances to the foci is a constant. By contrast, for a hyperbola the *difference* of the distances to the foci is a constant.

Definition of a Hyperbola
A **hyperbola** is the set of points in a plane the difference of whose distances from two fixed points (called foci) is a constant.

Figure 10.15 illustrates the two branches of a hyperbola's graph. The line through the foci intersects the hyperbola at two points, called the **vertices**. The line segment that joins the vertices is the **transverse axis**. The midpoint of the transverse axis is the **center** of the hyperbola. Notice that the center lies midway between the vertices, as well as midway between the foci.

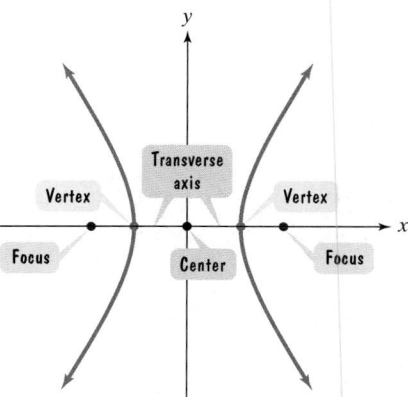

Figure 10.15 The two branches of a hyperbola

Standard Form of the Equation of a Hyperbola

The rectangular coordinate system enables us to translate a hyperbola's geometric definition into an algebraic equation. Figure 10.16 is our starting point for obtaining an equation. We place the foci on the x-axis at the points $(-c, 0)$ and $(c, 0)$. Note that the center of this hyperbola is at the origin. We let (x, y) represent the coordinates of any point on the hyperbola.

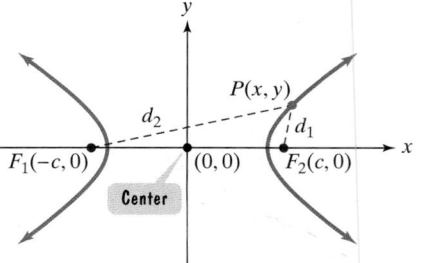

Figure 10.16

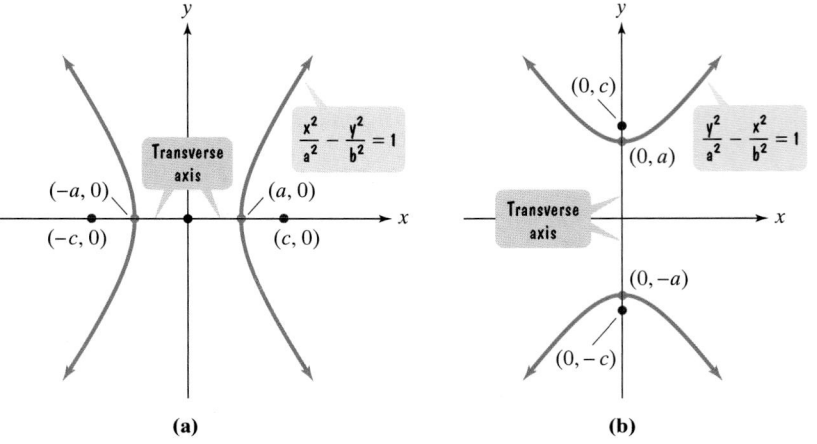

Figure 10.16, shown again so that you do not have to turn back a page

What does the definition of a hyperbola tell us about the point (x, y) in Figure 10.16? For any point (x, y) on the hyperbola, the absolute value of the difference of the distances from the two foci, $|d_2 - d_1|$, must be constant. We denote this constant by $2a$, just as we did for the ellipse. Thus, the point (x, y) is on the hyperbola if and only if

$$|d_2 - d_1| = 2a$$

$$\left|\sqrt{(x + c)^2 + (y - 0)^2} - \sqrt{(x - c)^2 + (y - 0)^2}\right| = 2a \quad \text{Use the distance formula.}$$

After eliminating radicals and simplifying, we obtain

$$(c^2 - a^2)x^2 - a^2y^2 = a^2(c^2 - a^2).$$

For convenience, let $b^2 = c^2 - a^2$. Substituting b^2 for $c^2 - a^2$ in the preceding equation, we obtain

$$b^2x^2 - a^2y^2 = a^2b^2$$

$$\frac{b^2x^2}{a^2b^2} - \frac{a^2y^2}{a^2b^2} = \frac{a^2b^2}{a^2b^2} \quad \text{Divide both sides by } a^2b^2.$$

$$\frac{x^2}{a^2} - \frac{y^2}{b^2} = 1 \quad \text{Simplify.}$$

This last equation is called the **standard form of the equation of a hyperbola.** There are two such equations. The first is for a hyperbola in which the transverse axis lies on the x-axis. The second is for a hyperbola in which the transverse axis lies on the y-axis.

> ## Standard Forms of the Equations of a Hyperbola
>
> The **standard form of the equation of a hyperbola** with center at the origin is
>
> $$\frac{x^2}{a^2} - \frac{y^2}{b^2} = 1 \quad \text{or} \quad \frac{y^2}{a^2} - \frac{x^2}{b^2} = 1.$$
>
> Figure 10.17 illustrates that for the equation on the left, the transverse axis lies on the x-axis. For the equation on the right, the transverse axis lies on the y-axis. The vertices are a units from the center and the foci are c units from the center. For both equations, $b^2 = c^2 - a^2$.

Figure 10.17 (a) Transverse axis lies on the x-axis. (b) Transverse axis lies on the y-axis.

(a)

(b)

1 Locate a hyperbola's vertices and foci.

Using the Standard Form of the Equation of a Hyperbola

We can use the standard form of the equation of a hyperbola to find its vertices and locate its foci. Because the vertices are a units from the center, begin by identifying a^2 in the equation. In the standard form of a hyperbola's equation, **a^2 is the number under the variable whose term is preceded by a plus sign** (+). If the x^2 term is preceded by a plus sign, the transverse axis lies along the x-axis. Thus, the vertices are a units to the right and left of the origin. If the y^2 term is preceded by a plus sign, the transverse axis lies along the y-axis. Thus, the vertices are a units above and below the origin.

We know that the foci are c units from the center. The substitution that we used to derive the hyperbola's equation, $b^2 = c^2 - a^2$, is needed to locate the foci when a^2 and b^2 are known. To find c^2, and then c, we will use an equivalent form of $b^2 = c^2 - a^2$, namely $c^2 = a^2 + b^2$.

EXAMPLE 1 Finding Vertices and Foci from a Hyperbola's Equation

Find the vertices and locate the foci for each of the following hyperbolas with the given equation.

$$\textbf{a. } \frac{x^2}{16} - \frac{y^2}{9} = 1 \qquad \textbf{b. } \frac{y^2}{9} - \frac{x^2}{16} = 1$$

Solution Both equations are in standard form. We begin by identifying a^2 and b^2 in each equation.

a. The first equation is in the form $\dfrac{x^2}{a^2} - \dfrac{y^2}{b^2} = 1$.

$$\frac{x^2}{16} - \frac{y^2}{9} = 1$$

$a^2 = 16$. This is the number in the denominator of the term preceded by a plus sign.

$b^2 = 9$. This is the number in the denominator of the term preceded by a minus sign.

Because the x^2 term is preceded by a plus sign, the transverse axis lies along the x-axis. Thus, the vertices are a units to the *right* and *left* of the origin. Based on the standard form of the equation, we know the vertices are $(-a, 0)$ and $(a, 0)$. Because $a^2 = 16$, $a = 4$. Thus, the vertices are $(-4, 0)$ and $(4, 0)$, shown in Figure 10.18.

We use $c^2 = a^2 + b^2$ to find the foci, which are located at $(-c, 0)$ and $(c, 0)$. We know that $a^2 = 16$ and $b^2 = 9$; we need to find c^2 in order to find c.

$$c^2 = a^2 + b^2 = 16 + 9 = 25$$

Because $c^2 = 25$, $c = 5$. The foci are located at $(-5, 0)$ and $(5, 0)$. They are shown in Figure 10.18.

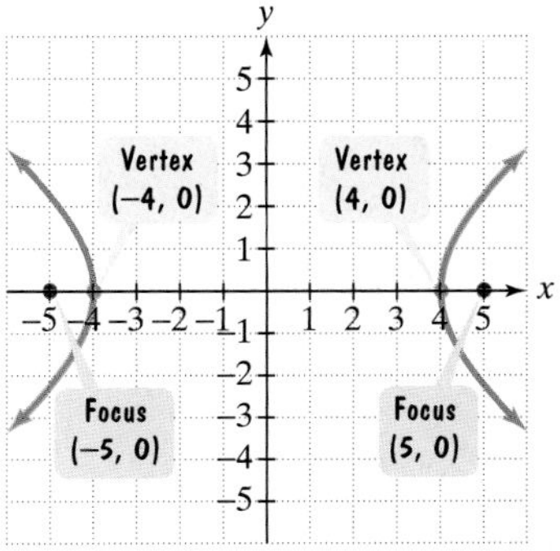

b. The second given equation is in the form $\dfrac{y^2}{a^2} - \dfrac{x^2}{b^2} = 1$.

$$\frac{y^2}{9} - \frac{x^2}{16} = 1$$

$a^2 = 9$. This is the number in the denominator of the term preceded by a plus sign.

$b^2 = 16$. This is the number in the denominator of the term preceded by a minus sign.

Figure 10.18 The graph of $\dfrac{x^2}{16} - \dfrac{y^2}{9} = 1$

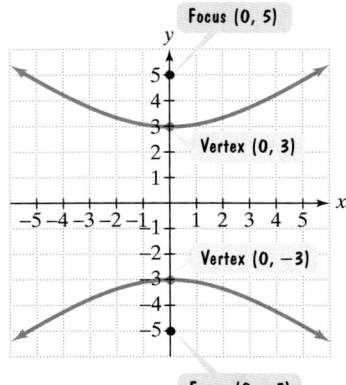

Figure 10.19 The graph of $\dfrac{y^2}{9} - \dfrac{x^2}{16} = 1$

Because the y^2 term is preceded by a plus sign, the transverse axis lies along the y-axis. Thus, the vertices are a units *above* and *below* the origin. Based on the standard form of the equation, we know the vertices are $(0, -a)$ and $(0, a)$. Because $a^2 = 9$, $a = 3$. Thus, the vertices are $(0, -3)$ and $(0, 3)$, shown in Figure 10.19.

We use $c^2 = a^2 + b^2$ to find the foci, which are located at $(0, -c)$ and $(0, c)$.

$$c^2 = a^2 + b^2 = 9 + 16 = 25$$

Because $c^2 = 25$, $c = 5$. The foci are located at $(0, -5)$ and $(0, 5)$. They are shown in Figure 10.19.

> **Check Point 1** Find the vertices and locate the foci for each of the following hyperbolas with the given equation.
>
> **a.** $\dfrac{x^2}{25} - \dfrac{y^2}{16} = 1$ **b.** $\dfrac{y^2}{25} - \dfrac{x^2}{16} = 1$

2 Write equations of hyperbolas in standard form.

In Example 1, we used equations of hyperbolas to find their foci and vertices. In the next example, we reverse this procedure.

EXAMPLE 2   Finding the Equation of a Hyperbola from Its Foci and Vertices

Find the standard form of the equation of a hyperbola with foci at $(0, -3)$ and $(0, 3)$ and vertices $(0, -2)$ and $(0, 2)$, shown in Figure 10.20.

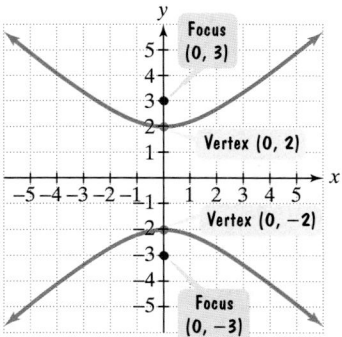

Figure 10.20

Solution   Because the foci are located at $(0, -3)$ and $(0, 3)$, on the y-axis, the transverse axis lies on the y-axis. The center of the hyperbola is midway between the foci, located at $(0, 0)$. Thus, the form of the equation is

$$\frac{y^2}{a^2} - \frac{x^2}{b^2} = 1.$$

We need to determine the values for a^2 and b^2. The distance from the center $(0, 0)$ to either vertex, $(0, -2)$ or $(0, 2)$, is 2, so $a = 2$.

$$\frac{y^2}{2^2} - \frac{x^2}{b^2} = 1 \qquad \text{or} \qquad \frac{y^2}{4} - \frac{x^2}{b^2} = 1$$

We must still find b^2. The distance from the center, $(0, 0)$, to either focus, $(0, -3)$ or $(0, 3)$, is 3. Thus, $c = 3$. Because $b^2 = c^2 - a^2$, we have

$$b^2 = 3^2 - 2^2 = 9 - 4 = 5.$$

Substituting 5 for b^2 in the last equation gives us the standard form of the hyperbola's equation. The equation is

$$\frac{y^2}{4} - \frac{x^2}{5} = 1.$$

> **Check Point 2** Find the standard form of the equation of a hyperbola with foci at $(0, -5)$ and $(0, 5)$ and vertices $(0, -3)$ and $(0, 3)$.

The Asymptotes of a Hyperbola

As x and y get larger, the two branches of the graph of a hyperbola approach a pair of intersecting straight lines called **asymptotes**. The asymptotes pass through the center of the hyperbola and are helpful in graphing hyperbolas.

Figure 10.21 shows the asymptotes for the graphs of hyperbolas centered at the origin. The asymptotes pass through the corners of a rectangle. Note that the dimensions of this rectangle are $2a$ by $2b$. The line segment of length $2b$ is the **conjugate axis** of the hyperbola and is perpendicular to the transverse axis.

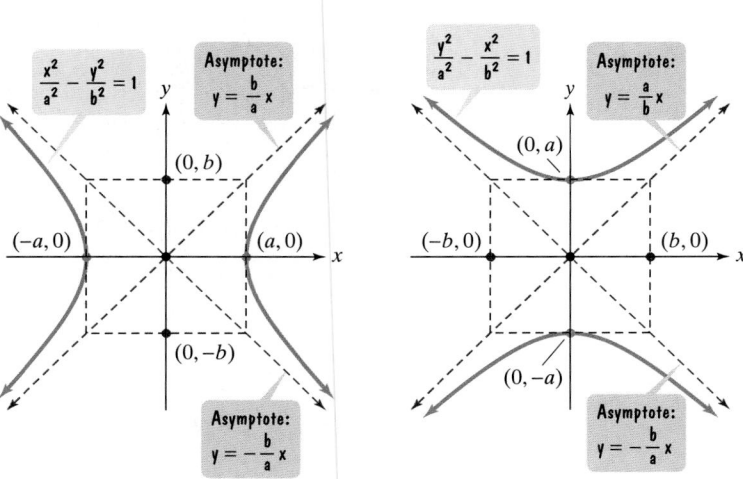

Figure 10.21 Asymptotes of a hyperbola

The Asymptotes of a Hyperbola Centered at the Origin

The hyperbola $\dfrac{x^2}{a^2} - \dfrac{y^2}{b^2} = 1$ with a horizontal transverse axis has the two asymptotes

$$y = \frac{b}{a}x \qquad \text{and} \qquad y = -\frac{b}{a}x.$$

The hyperbola $\dfrac{y^2}{a^2} - \dfrac{x^2}{b^2} = 1$ with a vertical transverse axis has the two asymptotes

$$y = \frac{a}{b}x \qquad \text{and} \qquad y = -\frac{a}{b}x.$$

Why are $y = \pm\dfrac{b}{a}x$ the asymptotes for a hyperbola whose transverse axis is horizontal? The proof can be found in the appendix.

3 Graph hyperbolas centered at the origin.

Graphing Hyperbolas Centered at the Origin

Hyperbolas are graphed using vertices and asymptotes.

> **Graphing Hyperbolas**
>
> **1.** Locate the vertices.
> **2.** Draw the rectangle centered at the origin with sides parallel to the axes, crossing one axis at $\pm a$ and the other at $\pm b$.
> **3.** Draw the diagonals of this rectangle and extend them to obtain the asymptotes.
> **4.** Draw the two branches of the hyperbola by starting at each vertex and approaching the asymptotes.
>
> The rectangle in step 2 and the asymptotes in step 3 are drawn using dashed lines to show that they are not part of the hyperbola.

EXAMPLE 3   Graphing a Hyperbola

Graph and locate the foci: $\dfrac{x^2}{25} - \dfrac{y^2}{16} = 1$.

Solution

Step 1   Locate the vertices.   The given equation is in the form $\dfrac{x^2}{a^2} - \dfrac{y^2}{b^2} = 1$, with $a^2 = 25$ and $b^2 = 16$.

$$\underset{\substack{\\ a^2 = 25}}{\dfrac{x^2}{25}} - \underset{\substack{\\ b^2 = 16}}{\dfrac{y^2}{16}} = 1$$

Based on the standard form of the equation with the transverse axis on the x-axis, we know that the vertices are $(-a, 0)$ and $(a, 0)$. Because $a^2 = 25$, $a = 5$. Thus, the vertices are $(-5, 0)$ and $(5, 0)$, shown in Figure 10.22.

Step 2   Draw a rectangle.   Because $a^2 = 25$ and $b^2 = 16$, $a = 5$ and $b = 4$. We construct a rectangle to find the asymptotes, using -5 and 5 on the x-axis (the ver-

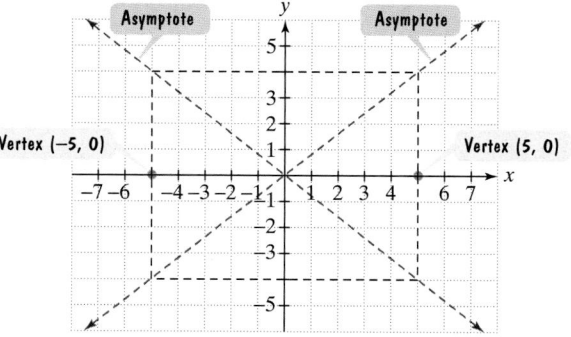

Figure 10.22 Preparing to graph $\dfrac{x^2}{25} - \dfrac{y^2}{16} = 1$

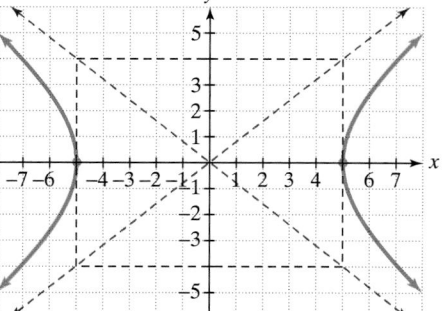

Figure 10.23 The graph of $\dfrac{x^2}{25} - \dfrac{y^2}{16} = 1$

tices are located here) and −4 and 4 on the y-axis. The rectangle passes through these four points, shown using dashed lines in Figure 10.22.

Step 3   Draw extended diagonals for the rectangle to obtain the asymptotes. We draw dashed lines through the opposite corners of the rectangle, shown in Figure 10.22, to obtain the graph of the asymptotes. Based on the standard form of the hyperbola's equation, the equations for these asymptotes are

$$y = \pm \frac{b}{a} x \quad \text{or} \quad y = \pm \frac{4}{5} x.$$

Step 4   Draw the two branches of the hyperbola by starting at each vertex and approaching the asymptotes.   The hyperbola is shown in Figure 10.23.

The foci are located at $(-c, 0)$ and $(c, 0)$. We find c using $c^2 = a^2 + b^2$.

$$c^2 = 25 + 16 = 41$$

Because $c^2 = 41$, $c = \sqrt{41}$. The foci are located at $(-\sqrt{41}, 0)$ and $(\sqrt{41}, 0)$, approximately $(-6.4, 0)$ and $(6.4, 0)$.

Check Point 3   Graph and locate the foci: $\dfrac{x^2}{36} - \dfrac{y^2}{9} = 1$.

Technology

Graph $\dfrac{x^2}{25} - \dfrac{y^2}{16} = 1$ by solving for y:

$$y_1 = \frac{\sqrt{16x^2 - 400}}{5}$$

$$y_2 = -\frac{\sqrt{16x^2 - 400}}{5} = -y_1.$$

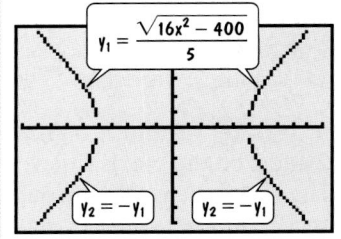

$y_1 = \dfrac{\sqrt{16x^2 - 400}}{5}$

$y_2 = -y_1$ $y_2 = -y_1$

EXAMPLE 4   Graphing a Hyperbola

Graph and locate the foci: $9y^2 - 4x^2 = 36$.

Solution   We begin by writing the equation in standard form. The right side should be 1, so we divide both sides by 36.

$$\frac{9y^2}{36} - \frac{4x^2}{36} = \frac{36}{36}$$

$$\frac{y^2}{4} - \frac{x^2}{9} = 1 \qquad \text{Simplify. The right side is now 1.}$$

Now we are ready to use our four-step procedure for graphing hyperbolas.

Step 1   Locate the vertices.   The equation that we obtained is in the form $\dfrac{y^2}{a^2} - \dfrac{x^2}{b^2} = 1$, with $a^2 = 4$ and $b^2 = 9$.

$$\frac{y^2}{4} - \frac{x^2}{9} = 1$$

$$a^2 = 4 \qquad b^2 = 9$$

Based on the standard form of the equation with the transverse axis on the y-axis, we know that the vertices are $(0, -a)$ and $(0, a)$. Because $a^2 = 4$, $a = 2$. Thus, the vertices are $(0, -2)$ and $(0, 2)$, shown in Figure 10.24.

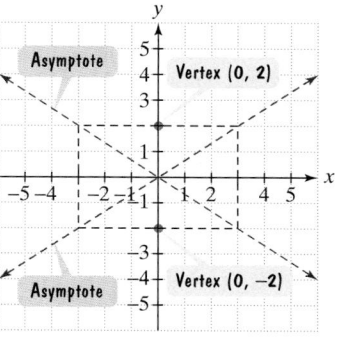

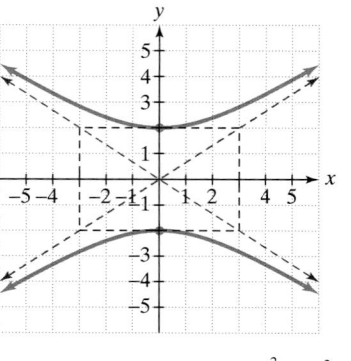

Figure 10.24 Preparing to graph $\dfrac{y^2}{4} - \dfrac{x^2}{9} = 1$

Figure 10.25 The graph of $\dfrac{y^2}{4} - \dfrac{x^2}{9} = 1$

Step 2 Draw a rectangle. Because $a^2 = 4$ and $b^2 = 9$, $a = 2$ and $b = 3$. We construct a rectangle to find the asymptotes, using -2 and 2 on the y-axis (the vertices are located here) and -3 and 3 on the x-axis. The rectangle passes through these four points, shown using dashed lines in Figure 10.24.

Step 3 Draw extended diagonals of the rectangle to obtain the asymptotes. We draw dashed lines through the opposite corners of the rectangle, shown in Figure 10.24, to obtain the graph of the asymptotes. Based on the standard form of the hyperbola's equation, the equations of these asymptotes are

$$y = \pm \frac{a}{b}x \qquad \text{or} \qquad y = \pm \frac{2}{3}x.$$

Step 4 Draw the two branches of the hyperbola by starting at each vertex and approaching the asymptotes. The hyperbola is shown in Figure 10.25.

The foci are located at $(0, -c)$ and $(0, c)$. We find c using $c^2 = a^2 + b^2$.

$$c^2 = 4 + 9 = 13$$

Because $c^2 = 13$, $c = \sqrt{13}$. The foci are located at $\left(0, -\sqrt{13}\right)$ and $\left(0, \sqrt{13}\right)$, approximately $(0, -3.6)$ and $(0, 3.6)$.

Check Point 4   Graph and locate the foci: $y^2 - 4x^2 = 4$.

4 Graph hyperbolas not centered at the origin.

Translations of Hyperbolas

The graph of a hyperbola can be centered at (h, k) rather than at the origin. Horizontal and vertical translations are accomplished by replacing x with $x - h$ and y with $y - k$ in the standard form of the hyperbola's equation.

Table 10.2 gives the standard forms of equations of hyperbolas centered at (h, k). Figure 10.26 shows their graphs.

Table 10.2   Standard Forms of Equations of Hyperbolas Centered at (h, k)

Equation	Center	Transverse Axis	Foci	Vertices
$\dfrac{(x - h)^2}{a^2} - \dfrac{(y - k)^2}{b^2} = 1,$ $b^2 = c^2 - a^2$	(h, k)	Parallel to x-axis; horizontal	$(h - c, k)$ $(h + c, k)$	$(h - a, k)$ $(h + a, k)$
$\dfrac{(y - k)^2}{a^2} - \dfrac{(x - h)^2}{b^2} = 1,$ $b^2 = c^2 - a^2$	(h, k)	Parallel to y-axis; vertical	$(h, k - c)$ $(h, k + c)$	$(h, k - a)$ $(h, k + a)$

Figure 10.26 Graphs of hyperbolas centered at (h, k).

EXAMPLE 5   Graphing a Hyperbola Centered at (h, k)

Graph: $\dfrac{(x - 2)^2}{16} - \dfrac{(y - 3)^2}{9} = 1$. Where are the foci located?

Solution   In order to graph the hyperbola, we need to know its center (h, k). In the standard forms of equations centered at (h, k), h is the number subtracted from x and k is the number subtracted from y.

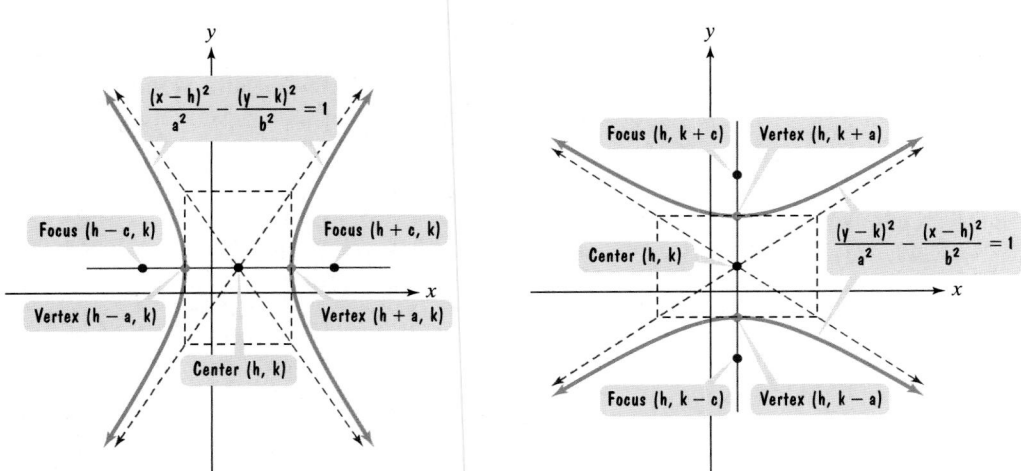

This is $(x - h)^2$, with $h = 2$. $\dfrac{(x - 2)^2}{16} - \dfrac{(y - 3)^2}{9} = 1$ This is $(y - k)^2$, with $k = 3$.

We see that $h = 2$ and $k = 3$. Thus, the center of hyperbola, (h, k), is $(2, 3)$. We can graph the hyperbola by using vertices, asymptotes, and our four-step graphing procedure.

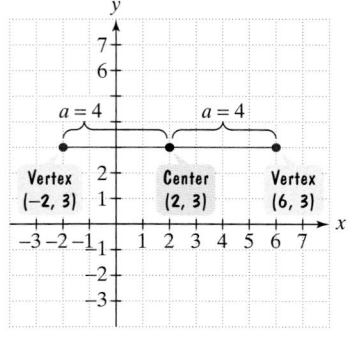

Figure 10.27 Locating a hyperbola's center and vertices

Step 1 Locate the vertices. To do this, we must identify a^2.

$$\frac{(x-2)^2}{16} - \frac{(y-3)^2}{9} = 1$$ The form of this equation is $\frac{(x-h)^2}{a^2} - \frac{(y-k)^2}{b^2} = 1.$

$a^2 = 16$ $b^2 = 9$

Based on the standard form of the equation with a horizontal transverse axis, the vertices are a units to the right and left of the center. Because $a^2 = 16$, $a = 4$. This means that the vertices are 4 units to the right and left of the center, $(2, 3)$. Four units to the right of $(2, 3)$ puts one vertex at $(2 + 4, 3)$, or $(6, 3)$. Four units to the left of $(2, 3)$ puts the other vertex at $(2 - 4, 3)$, or $(-2, 3)$. The vertices are shown in Figure 10.27.

Step 2 Draw a rectangle. Because $a^2 = 16$ and $b^2 = 9$, $a = 4$ and $b = 3$. The rectangle passes through points that are 4 units to the right and left of the center (the vertices are located here) and 3 units above and below the center. The rectangle is shown using dashed lines in Figure 10.28.

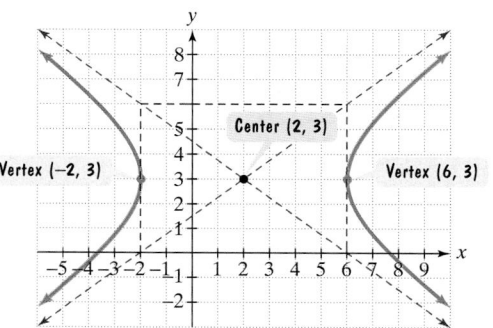

Figure 10.28 The graph of $\frac{(x-2)^2}{16} - \frac{(y-3)^2}{9} = 1$

Step 3 Draw extended diagonals of the rectangle to obtain the asymptotes. We draw dashed lines through the opposite corners of the rectangle, shown in Figure 10.28, to obtain the graph of the asymptotes. The equations of the asymptotes of the unshifted hyperbola $\frac{x^2}{16} - \frac{y^2}{9} = 1$ are $y = \pm\frac{b}{a}x$, or $y = \pm\frac{3}{4}x$. Thus, the asymptotes for the hyperbola that is shifted two units to the right and three units up, namely

$$\frac{(x-2)^2}{16} - \frac{(y-3)^2}{9} = 1$$

have equations that can be expressed as

$$y - 3 = \pm\frac{3}{4}(x - 2).$$

Step 4 Draw the two branches of the hyperbola by starting at each vertex and approaching the asymptotes. The hyperbola is shown in Figure 10.28.

The foci are located c units to the right and left of the center. We find c using $c^2 = a^2 + b^2$.

$$c^2 = 16 + 9 = 25$$

Because $c^2 = 25$, $c = 5$. This means that the foci are 5 units to the right and left of the center, $(2, 3)$. Five units to the right of $(2, 3)$ puts one focus at $(2 + 5, 3)$, or (7, 3). Five units to the left of (2, 3) puts the other focus at (2 − 5, 3), or (−3, 3).

> **Check Point 5** Graph and locate the foci: $\dfrac{(x - 3)^2}{4} - \dfrac{(y - 1)^2}{1} = 1$.

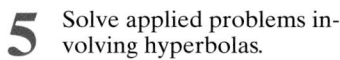

5 Solve applied problems involving hyperbolas.

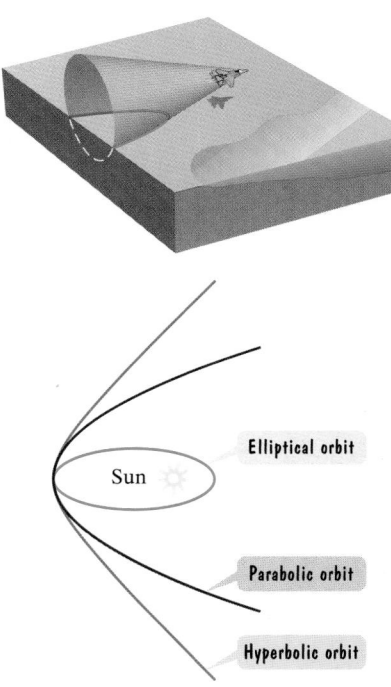

Applications

Hyperbolas have many applications. When a jet flies at a speed greater than the speed of sound, the shock wave that is created is heard as a sonic boom. The wave has the shape of a cone. The shape formed as the cone hits the ground is one branch of a hyperbola.

Halley's Comet, a permanent part of our solar system, travels around the sun in an elliptical orbit. Other comets pass through the solar system only once, following a hyperbolic path with the sun as a focus.

Hyperbolas are of practical importance in fields ranging from architecture to navigation. Cooling towers used in the design for nuclear power plants have cross sections that are both ellipses and hyperbolas. Three-dimensional solids whose cross sections are hyperbolas are used in some rather unique architectural creations, including the TWA building at Kennedy Airport and the St. Louis Science Center Planetarium.

EXAMPLE 6 An Application Involving Hyperbolas

An explosion is recorded by two microphones that are 2 miles apart. Microphone M_1 received the sound 4 seconds before microphone M_2. Assuming sound travels at 1100 feet per second, determine the possible locations of the explosion relative to the location of the microphones.

Solution We begin by putting the microphones in a coordinate system. Because 1 mile = 5280 feet, we place M_1 5280 feet on a horizontal axis to the right of the origin and M_2 5280 feet on a horizontal axis to the left of the origin. Figure 10.29 illustrates that the two microphones are 2 miles apart.

We know that M_2 received the sound 4 seconds after M_1. Because sound travels at 1100 feet per second, the difference between the distance from P to M_1 and the distance from P to M_2 is 4400 feet. The set of all points P (or locations

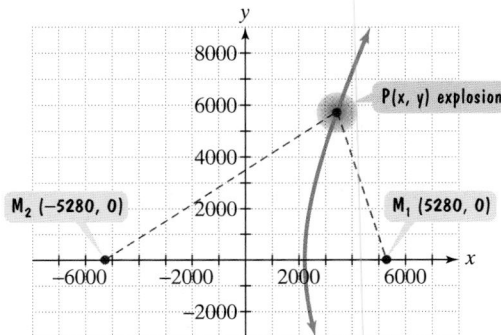

Figure 10.29 Locating an explosion on the branch of a hyperbola

Where Exactly Am I?

The hyperbola is the basis for the navigational system LORAN (for long-range navigation), used by a ship or aircraft to determine its location. The measured time-of-arrival difference between signals transmitted from two ground stations determines the hyperbola on which the ship or aircraft is located. The process is then repeated by taking a similar time-difference reading from a second pair of stations, determining a second hyperbola. The point of intersection of the two hyperbolas is the location of the ship or aircraft.

LORAN will eventually be replaced by the Global Positioning System. Using 24 satellites that orbit at 11,000 miles above Earth, the system is able to show you your exact position on Earth anytime, in any weather, anywhere.

of the explosion) satisfying these conditions fits the definition of a hyperbola, with microphones M_1 and M_2 at the foci.

$$\frac{x^2}{a^2} - \frac{y^2}{b^2} = 1 \qquad \text{Use the standard form of the hyperbola's equation. } P(x, y), \text{ the explosion point, lies on this hyperbola. We must find } a^2 \text{ and } b^2.$$

The difference between the distances, represented by $2a$ in the derivation of the hyperbola's equation, is 4400 feet. Thus, $2a = 4400$ and $a = 2200$.

$$\frac{x^2}{(2200)^2} - \frac{y^2}{b^2} = 1 \qquad \text{Substitute 2200 for } a.$$

Because $c = 5280$ and $a = 2200$, then $b^2 = c^2 - a^2 = 5280^2 - 2200^2 = 23{,}038{,}400$.

$$\frac{x^2}{4{,}840{,}000} - \frac{y^2}{23{,}038{,}400} = 1 \qquad \text{Substitute 23,038,400 for } b^2.$$

We can conclude that the explosion occurred somewhere on the right branch (the branch closest to M_1) of the hyperbola given by

$$\frac{x^2}{4{,}840{,}000} - \frac{y^2}{23{,}038{,}400} = 1.$$

In Example 6, we determined that the explosion occurred somewhere along one branch of a hyperbola, but not exactly where on the hyperbola. If, however, we had received the sound from another pair of microphones, we could locate the sound along a branch of another hyperbola. The exact location of the explosion would be the point where the two hyperbolas intersect.

Check Point 6 Rework Example 6 if Microphone M_1 receives the sound 3 seconds before Microphone M_2.

EXERCISE SET 10.2

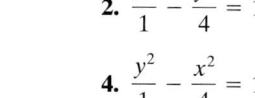

Practice Exercises

In Exercises 1–4, find the vertices and locate the foci of each hyperbola with the given equation. Then match each equation to one of the graphs that are shown and labeled (a)–(d).

1. $\dfrac{x^2}{4} - \dfrac{y^2}{1} = 1$

2. $\dfrac{x^2}{1} - \dfrac{y^2}{4} = 1$

3. $\dfrac{y^2}{4} - \dfrac{x^2}{1} = 1$

4. $\dfrac{y^2}{1} - \dfrac{x^2}{4} = 1$

a.

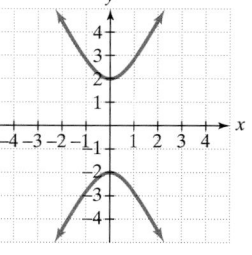

b.

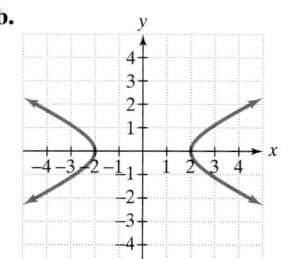

c.

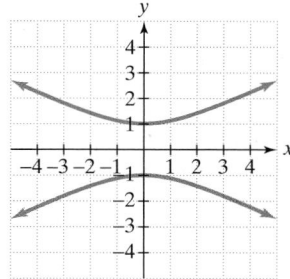

d.

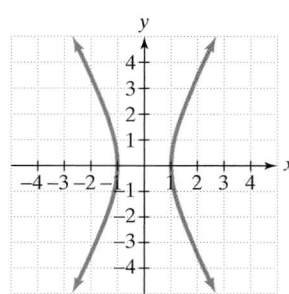

In Exercises 5–8, find the standard form of the equation of each hyperbola centered at the origin satisfying the given conditions.

5. Foci: $(0, -3), (0, 3)$; vertices: $(0, -1), (0, 1)$

6. Foci: $(0, -6), (0, 6)$; vertices: $(0, -2), (0, 2)$

7. Foci: $(-4, 0), (4, 0)$; vertices: $(-3, 0), (3, 0)$

8. Foci: $(-7, 0), (7, 0)$; vertices: $(-5, 0), (5, 0)$

In Exercises 9–22, use vertices and asymptotes to graph each hyperbola. Locate the foci.

9. $\dfrac{x^2}{9} - \dfrac{y^2}{25} = 1$

10. $\dfrac{x^2}{16} - \dfrac{y^2}{25} = 1$

11. $\dfrac{x^2}{100} - \dfrac{y^2}{64} = 1$

12. $\dfrac{x^2}{144} - \dfrac{y^2}{81} = 1$

13. $\dfrac{y^2}{16} - \dfrac{x^2}{36} = 1$

14. $\dfrac{y^2}{25} - \dfrac{x^2}{64} = 1$

15. $\dfrac{y^2}{36} - \dfrac{x^2}{25} = 1$

16. $\dfrac{y^2}{100} - \dfrac{x^2}{49} = 1$

17. $9x^2 - 4y^2 = 36$

18. $4x^2 - 25y^2 = 100$

19. $9y^2 - 25x^2 = 225$

20. $16y^2 - 9x^2 = 144$

21. $4x^2 = 4 + y^2$

22. $25y^2 = 225 + 9x^2$

In Exercises 23–26, find the standard form of the equation of each hyperbola.

23.

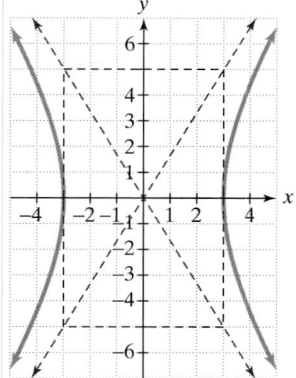

24.

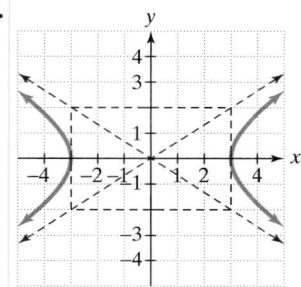

25.

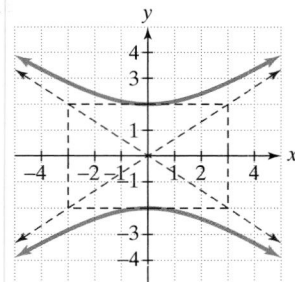

26.

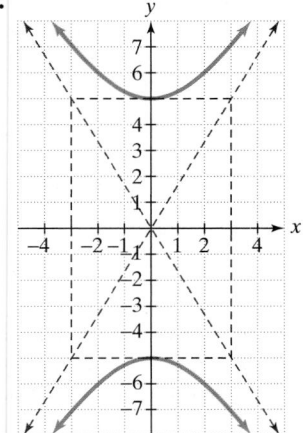

In Exercises 27–36, use the center, vertices, and asymptotes to graph each hyperbola. Locate the foci.

27. $\dfrac{(x+4)^2}{9} - \dfrac{(y+3)^2}{16} = 1$

28. $\dfrac{(x+2)^2}{9} - \dfrac{(y-1)^2}{25} = 1$

29. $\dfrac{(x+3)^2}{25} - \dfrac{y^2}{16} = 1$

30. $\dfrac{(x+2)^2}{9} - \dfrac{y^2}{25} = 1$

31. $\dfrac{(y+2)^2}{4} - \dfrac{(x-1)^2}{16} = 1$

32. $\dfrac{(y-2)^2}{36} - \dfrac{(x+1)^2}{49} = 1$

33. $(x-3)^2 - 4(y+3)^2 = 4$

34. $(x+3)^2 - 9(y-4)^2 = 9$

35. $(x-1)^2 - (y-2)^2 = 4$

36. $(y-2)^2 - (x+3)^2 = 4$

In Exercises 37–44, convert each equation to standard form by completing the square on x and y. Then graph the hyperbola and give the location of its foci.

37. $x^2 - y^2 - 2x - 4y - 4 = 0$

38. $4x^2 - y^2 + 32x + 6y + 39 = 0$

39. $16x^2 - y^2 + 64x - 2y + 67 = 0$

40. $9y^2 - 4x^2 - 18y + 24x - 63 = 0$

41. $4x^2 - 9y^2 - 16x + 54y - 101 = 0$

42. $4x^2 - 9y^2 + 8x - 18y - 6 = 0$

43. $4x^2 - 25y^2 - 32x + 164 = 0$

44. $9x^2 - 16y^2 - 36x - 64y + 116 = 0$

 **Application Exercises**

45. An explosion is recorded by two microphones that are 1 mile apart. Microphone M_1 received the sound 2 seconds before microphone M_2. Assuming sound travels at 1100 feet per second, determine the possible locations of the explosion relative to the location of the microphones.

46. Radio towers A and B, 200 kilometers apart, are situated along the coast, with A located due west of B. Simultaneous radio signals are sent from each tower to a ship, with the signal from B received 500 microseconds before the signal from A.
   a. Assuming that the radio signals travel 300 meters per microsecond, determine the equation of the hyperbola on which the ship is located.
   b. If the ship lies due north of tower B, how far out at sea is it?

47. An architect designs two houses that are shaped and positioned like a part of the branches of the hyperbola whose equation is $625y^2 - 400x^2 = 250{,}000$, where x and y are in yards. How far apart are the houses at their closest point?

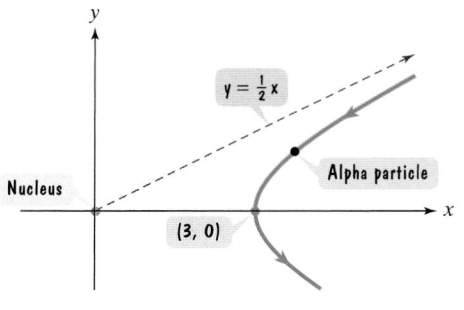

48. Scattering experiments, in which moving particles are deflected by various forces, led to the concept of the nucleus of an atom. In 1911, the physicist Ernest Rutherford (1871–1937) discovered that when alpha particles are directed toward the nuclei of gold atoms, they are eventually deflected along hyperbolic paths, illustrated in the figure. If a particle gets as close as 3 units to the nucleus along a hyperbolic path with an asymptote given by $y = \frac{1}{2}x$, what is the equation of its path?

Writing in Mathematics

49. What is a hyperbola?

50. Describe how to graph $\dfrac{x^2}{9} - \dfrac{y^2}{1} = 1$.

51. Describe how to locate the foci of the graph of $\dfrac{x^2}{9} - \dfrac{y^2}{1} = 1$.

52. Describe one similarity and one difference between the graphs of $\dfrac{x^2}{9} - \dfrac{y^2}{1} = 1$ and $\dfrac{y^2}{9} - \dfrac{x^2}{1} = 1$.

53. Describe one similarity and one difference between the graphs of $\dfrac{x^2}{9} - \dfrac{y^2}{1} = 1$ and $\dfrac{(x-3)^2}{9} - \dfrac{(y+3)^2}{1} = 1$.

54. How can you distinguish an ellipse from a hyperbola by looking at their equations?

55. In 1992, a NASA team began a project called Spaceguard Survey, calling for an international watch for comets that might collide with Earth. Why is it more difficult to detect a possible "doomsday comet" with a hyperbolic orbit than one with an elliptical orbit?

Technology Exercises

56. Use a graphing utility to graph any five of the hyperbolas that you graphed by hand in Exercises 9–22.

57. Use a graphing utility to graph any three of the hyperbolas that you graphed by hand in Exercises 27–36. First solve the given equation for y by using the square root method. Enter each of the two resulting equations to produce each branch of the hyperbola.

58. Use a graphing utility to graph any one of the hyperbolas that you graphed by hand in Exercises 37–44. Write the equation as a quadratic equation in y and use the quadratic formula to solve for y. Enter each of the two resulting equations to produce each branch of the hyperbola.

59. Use a graphing utility to graph $\dfrac{x^2}{4} - \dfrac{y^2}{9} = 0$. Is the graph a hyperbola? In general, what is the graph of $\dfrac{x^2}{a^2} - \dfrac{y^2}{b^2} = 0$?

60. Graph $\dfrac{x^2}{a^2} - \dfrac{y^2}{b^2} = 1$ and $\dfrac{x^2}{a^2} - \dfrac{y^2}{b^2} = -1$ in the same viewing rectangle for values of a^2 and b^2 of your choice. Describe the relationship between the two graphs.

61. Write $4x^2 - 6xy + 2y^2 - 3x + 10y - 6 = 0$ as a quadratic equation in y and then use the quadratic formula to express y in terms of x. Graph the resulting two equations using a graphing utility and a $[-50, 70, 10]$ by $[-30, 50, 10]$ viewing rectangle. What effect does the xy-term have on the graph of the resulting hyperbola? What problems would you encounter if you attempted to write the given equation in standard form by completing the square?

62. Graph $\dfrac{x^2}{16} - \dfrac{y^2}{9} = 1$ and $\dfrac{x|x|}{16} - \dfrac{y|y|}{9} = 1$ in the same viewing rectangle. Explain why the graphs are not the same.

Critical Thinking Exercises

63. Which one of the following is true?
   a. If one branch of a hyperbola is removed from a graph, then the branch that remains must define y as a function of x.
   b. All points on the asymptotes of a hyperbola also satisfy the hyperbola's equation.
   c. The graph of $\dfrac{x^2}{9} - \dfrac{y^2}{4} = 1$ does not intersect the line $y = -\dfrac{2}{3}x$.
   d. Two different hyperbolas can never share the same asymptotes.

64. What happens to the shape of the graph of $\dfrac{x^2}{a^2} - \dfrac{y^2}{b^2} = 1$ as $\dfrac{c}{a}$ gets larger and larger?

65. Find the standard form of the equation of the hyperbola with vertices $(5, -6)$ and $(5, 6)$, and passing through $(0, 9)$.

66. Find the equation of a hyperbola whose asymptotes are perpendicular.

SECTION 10.3   *The Parabola*

Objectives

1. Graph parabolas with vertices at the origin.
2. Write equations of parabolas in standard form.
3. Graph parabolas with vertices not at the origin.
4. Solve applied problems involving parabolas.

At first glance, this image looks like columns of smoke rising from a fire into a starry sky. Those are, indeed, stars in the background, but you are not looking at ordinary smoke columns. These stand almost 6 trillion miles high and are 7000 light-years from Earth—more than 400 million times as far away as the sun.

This NASA photograph is one of a series of stunning images captured from the ends of the universe by the Hubble Space Telescope. The image shows infant star systems the size of our solar system emerging from the gas and dust that shrouded their creation. Using a parabolic mirror that is 94.5 inches in diameter, the Hubble is providing answers to many of the profound mysteries of the cosmos: How big and how old is the universe? How did the galaxies come to exist? Do other Earth-like planets orbit other sun-like stars?

In Chapter 3, we studied parabolas, viewing them as graphs of the quadratic function $y = ax^2 + bx + c$. In this section, we will use a geometric definition of a parabola to derive its equation. We will also consider applications of parabolas, including parabolic shapes that gather distant rays of light and focus them into spectacular images.

Definition of a Parabola

The definitions of ellipses and hyperbolas involved two fixed points, the foci. By contrast, the definition of a parabola is based on one point and a line.

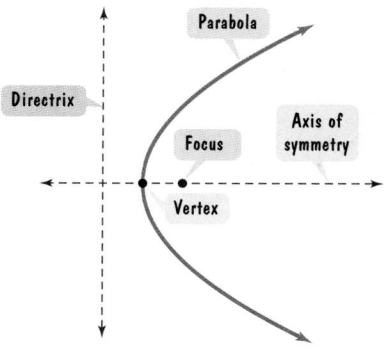

Figure 10.30

Definition of a Parabola

A **parabola** is the set of all points in a plane that are equidistant from a fixed line (the **directrix**) and a fixed point (the **focus**) that is not on the line (see Figure 10.30).

In Figure 10.30, find the line passing through the focus and perpendicular to the directrix. This is the **axis of symmetry** of the parabola. The point of intersection of the parabola with its axis of symmetry is called the **vertex**. Notice that the vertex is midway between the focus and the directrix.

Standard Form of the Equation of a Parabola

The rectangular coordinate system enables us to translate a parabola's geometric definition into an algebraic equation. Figure 10.31 is our starting point for obtaining an equation. We place the focus on the x-axis at the point $(p, 0)$. The directrix has an equation given by $x = -p$. The vertex, located midway between the focus and the directrix, is at the origin.

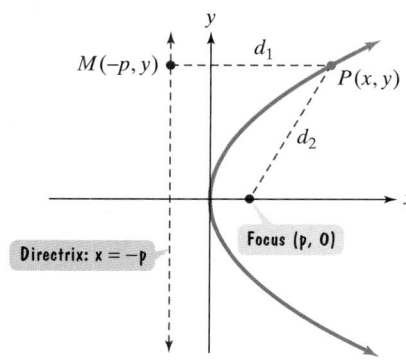

Figure 10.31

What does the definition of a parabola tell us about the point (x, y) in Figure 10.31? For any point (x, y) on the parabola, the distance d_1 to the directrix is equal to the distance d_2 to the focus. Thus, the point (x, y) is on the parabola if and only if

$$d_1 = d_2$$

$$\sqrt{(x + p)^2 + (y - y)^2} = \sqrt{(x - p)^2 + (y - 0)^2} \quad \text{Use the distance formula.}$$

$$(x + p)^2 = (x - p)^2 + y^2 \quad \text{Square both sides of the equation.}$$

$$x^2 + 2px + p^2 = x^2 - 2px + p^2 + y^2 \quad \text{Square } x + p \text{ and } x - p.$$

$$2px = -2px + y^2 \quad \text{Subtract } x^2 + p^2 \text{ from both sides of the equation.}$$

$$y^2 = 4px \quad \text{Solve for } y^2.$$

This last equation is called the **standard form of the equation of a parabola.** There are two such equations, one for a focus on the x-axis and one for a focus on the y-axis.

Standard Forms of the Equations of a Parabola

The **standard form of the equation of a parabola** with vertex at the origin is

$$y^2 = 4px \qquad \text{or} \qquad x^2 = 4py.$$

Figure 10.32 illustrates that for the equation on the left, the focus is on the x-axis, which is the axis of symmetry. For the equation on the right, the focus is on the y-axis, which is the axis of symmetry.

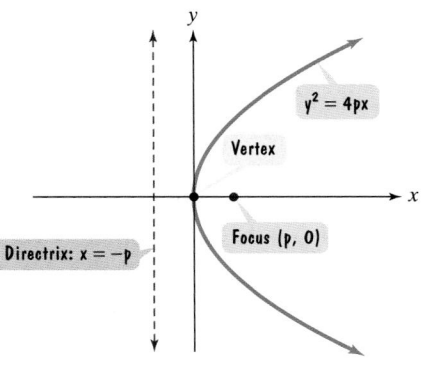 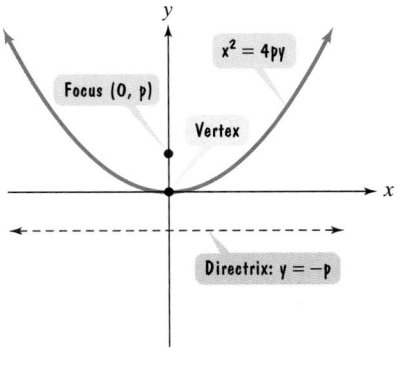

Figure 10.32 **(a)** Parabola with the x-axis as the axis of symmetry

(b) Parabola with the y-axis as the axis of symmetry

1 Graph parabolas with vertices at the origin.

Using the Standard Form of the Equation of a Parabola

We can use the standard form of the equation of a parabola to find its focus and directrix. Remember that the focus is located on the axis corresponding to the variable in the equation that is *not* squared.

$$y^2 = 4px \qquad x^2 = 4py$$

| x is not squared. Focus is on the x-axis at $(p, 0)$. | y is not squared. Focus is on the y-axis at $(0, p)$. |

Although the definition of a parabola is given in terms of its focus and its directrix, the focus and directrix are not part of the graph. The vertex, located at the origin, is a point on the graph of $y^2 = 4px$ and $x^2 = 4py$. You can find two additional points on the parabola by assigning a value to x or y that makes y^2 or x^2 a perfect square. For example, consider $y^2 = 8x$. A value of x that makes the right side a perfect square is 2. If $x = 2$, then $y^2 = 8(2)$, or 16. Because $y^2 = 16$, $y = \pm 4$. Thus, the parabola passes through the points $(2, 4)$ and $(2, -4)$. The parabola can be graphed by connecting the vertex, $(0, 0)$, to each of these points with a smooth curve.

EXAMPLE 1 Finding the Focus and Directrix of a Parabola

Find the focus and directrix of the parabola given by $y^2 = 12x$. Then graph the parabola.

Solution The given equation is in the standard form $y^2 = 4px$, so $4p = 12$.

$$y^2 = 12x$$

| This is $4p$. |

We can find both the focus and the directrix by finding p.

$4p = 12$ Remember that the focus is at $(p, 0)$ and the directrix is given by $x = -p$.

$p = 3$ Divide both sides by 4.

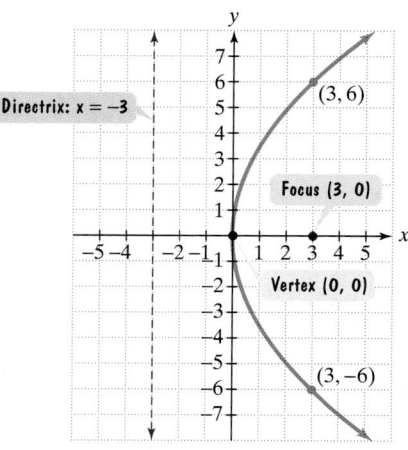

Figure 10.33 The graph of $y^2 = 12x$

Using this value for p, we obtain

$$\text{Focus:} \quad (p, 0) = (3, 0)$$

$$\text{Directrix:} \quad x = -p; \, x = -3.$$

Observe that the vertex, midway between the focus and the directrix, is at the origin. To graph $y^2 = 12x$, we assign x a value that makes the right side a perfect square. If $x = 3$, then $y^2 = 12(3)$ or $y^2 = 36$. Because $y = \pm 6$, the parabola passes through the points $(3, 6)$ and $(3, -6)$. The graph is sketched in Figure 10.33.

> **Check Point 1**
>
> Find the focus and directrix of the parabola given by $y^2 = 8x$. Then graph the parabola.

Parabolas with vertices at the origin can open to the right, left, upward, or downward. The graph of $y^2 = 4px$ opens to the right if $p > 0$ or left if $p < 0$. For example, Figure 10.34 shows that $y^2 = x$ opens to the right and $y^2 = -x$ opens to the left. The graph of $x^2 = 4py$ opens upward if $p > 0$ or downward if $p < 0$. Figure 10.34 shows that $x^2 = y$ opens upward and $x^2 = -y$ opens downward.

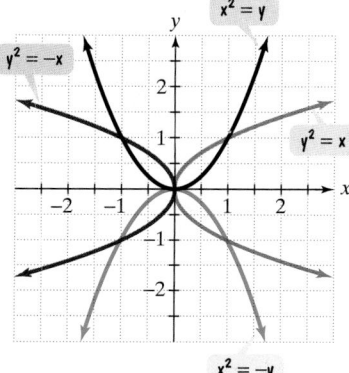

Figure 10.34

EXAMPLE 2   Finding the Focus and Directrix of a Parabola

Find the focus and directrix of the parabola given by $x^2 = -8y$. Then graph the parabola.

Solution   The given equation is in the standard form $x^2 = 4py$, so $4p = -8$.

$$x^2 = \underset{\substack{\big| \\ \text{This is } 4p.}}{-8}y$$

We can find both the focus and the directrix by finding p.

$$4p = -8$$
$$p = -2$$

> The focus, on the y-axis, is at $(0, p)$ and the directrix is given by $y = -p$.

Because $p < 0$, the parabola opens downward. Using this value for p, we obtain

$$\text{Focus:} \quad (0, p) = (0, -2)$$

$$\text{Directrix:} \quad y = -p; \, y = 2.$$

To graph $x^2 = -8y$, we assign y a value that makes the right side a perfect square. If $y = -2$, then $x^2 = -8(-2) = 16$, so $x = \pm 4$. The parabola passes through the points $(4, -2)$ and $(-4, -2)$. The graph is sketched in Figure 10.35.

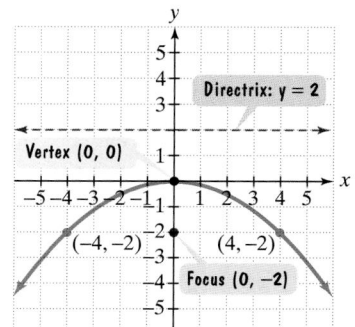

Figure 10.35 The graph of $x^2 = -8y$

> **Check Point 2**
>
> Find the focus and directrix of the parabola given by $x^2 = -12y$. Then graph the parabola.

② Write equations of parabolas in standard form.

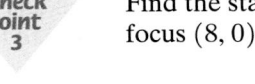

Figure 10.36

③ Graph parabolas with vertices not at the origin.

In Examples 1 and 2, we used the equation of a parabola to find its focus and directrix. In the next example, we reverse this procedure.

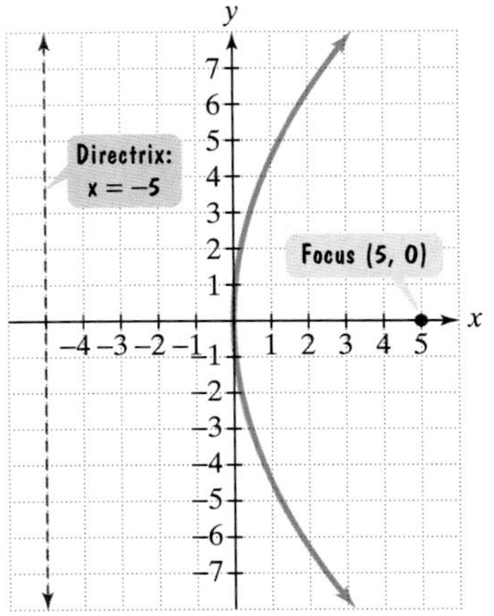

EXAMPLE 3   Finding the Equation of a Parabola from Its Focus and Directrix

Find the standard form of the equation of a parabola with focus $(5, 0)$ and directrix $x = -5$, shown in Figure 10.36.

Solution   The focus is $(5, 0)$. Thus, the focus is on the x-axis. We use the standard form of the equation in which x is not squared, namely $y^2 = 4px$.

We need to determine the value of p. Recall that the focus, located at $(p, 0)$, is p units from the vertex, $(0, 0)$. Thus, if the focus is $(5, 0)$, then $p = 5$. We substitute 5 for p into $y^2 = 4px$ to obtain the standard form of the equation of the parabola. The equation is

$$y^2 = 4 \cdot 5x \qquad \text{or} \qquad y^2 = 20x.$$

Check Point 3   Find the standard form of the equation of a parabola with focus $(8, 0)$ and directrix $x = -8$.

Translations of Parabolas

The graph of a parabola can have its vertex at (h, k) rather than at the origin. Horizontal and vertical translations are accomplished by replacing x with $x - h$ and y with $y - k$ in the standard form of the parabola's equation.

Table 10.3 gives the standard forms of equations of parabolas with vertex at (h, k). Figure 10.37 shows their graphs.

Table 10.3   Standard Forms of Equations of Parabolas with Vertex at (h, k)

Equation	Vertex	Axis of Symmetry	Focus	Directrix	Description
$(y - k)^2 = 4p(x - h)$	(h, k)	Horizontal	$(h + p, k)$	$x = h - p$	If $p > 0$, opens to right. If $p < 0$, opens to left.
$(x - h)^2 = 4p(y - k)$	(h, k)	Vertical	$(h, k + p)$	$y = k - p$	If $p > 0$, opens up. If $p < 0$, opens down.

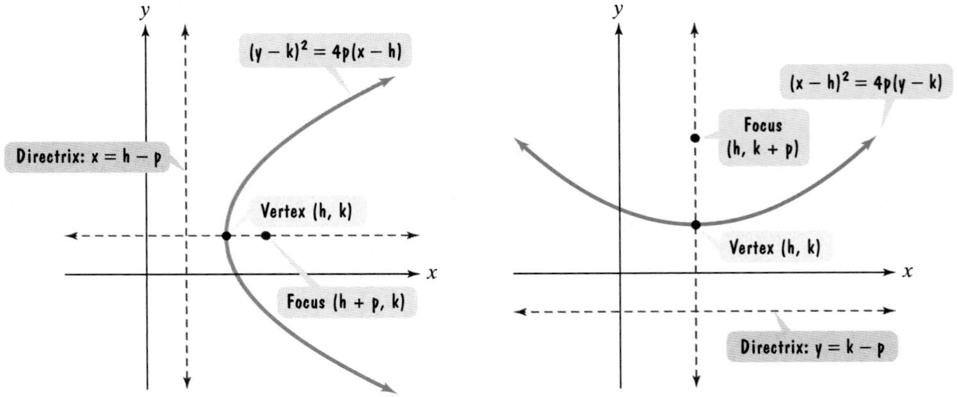

Figure 10.37 Graphs of parabolas with vertex at (h, k)

The two parabolas shown in Figure 10.37 illustrate standard forms of equations for $p > 0$. If $p < 0$, a parabola with a horizontal axis of symmetry will open to the left and the focus will lie to the left of the directrix. If $p < 0$, a parabola with a vertical axis of symmetry will open downward and the focus will lie below the directrix.

EXAMPLE 4   Graphing a Parabola with Vertex at (h, k)

Find the vertex, focus, and directrix of the parabola given by

$$(x - 3)^2 = 8(y + 1).$$

Then graph the parabola.

Solution   In order to find the focus and directrix, we need to know the vertex. In the standard forms of equations with vertex at (h, k), h is the number subtracted from x, and k is the number subtracted from y.

$$(x - 3)^2 = 8(y - (-1))$$

> This is $(x - h)^2$, with $h = 3$.

> This is $y - k$, with $k = -1$.

We see that $h = 3$ and $k = -1$. Thus, the vertex of the parabola is $(h, k) = (3, -1)$.

Now that we have the vertex, we can find both the focus and directrix by finding p.

$$(x - 3)^2 = 8(y + 1) \quad \text{The equation is in the standard form } (x - h)^2 = 4p(y - k).$$

> This is $4p$.

Because $4p = 8$, $p = 2$. Based on the standard form of the equation, the axis of symmetry is vertical. With a positive value for p and a vertical axis of symmetry, the parabola opens upward. Because $p = 2$, the focus is located 2 units above the vertex, $(3, -1)$. Likewise, the directrix is located 2 units below the vertex.

Focus: $\quad (h, k + p) = (3, -1 + 2) = (3, 1)$

> The vertex, (h, k), is $(3, -1)$.

> The focus is 2 units above the vertex, $(3, -1)$.

Directrix: $\quad y = k - p$
$$y = -1 - 2 = -3$$

> The directrix is 2 units below the vertex, $(3, -1)$.

Thus, the focus is $(3, 1)$ and the directrix is $y = -3$. They are shown in Figure 10.38.

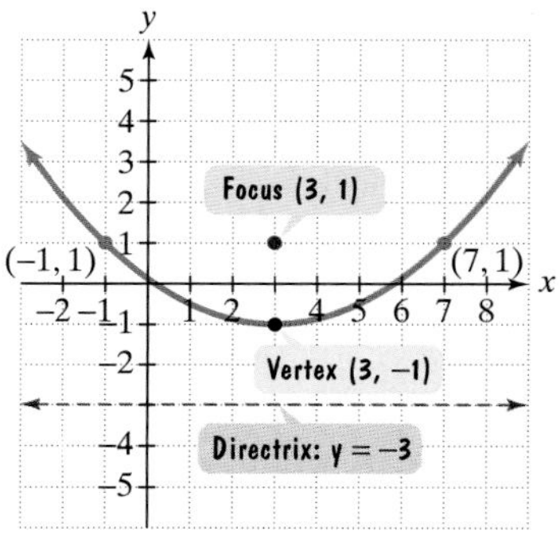

To graph $(x - 3)^2 = 8(y + 1)$, we assign y a value that makes the right side of the equation a perfect square. If $y = 1$, the right side is $8(1 + 1)$ or 16, a perfect square. We let $y = 1$ and solve for x to obtain points on the parabola.

$$(x - 3)^2 = 8(1 + 1) \quad \text{Substitute 1 for } y \text{ in } (x - 3)^2 = 8(y + 1).$$

$$(x - 3)^2 = 16 \quad \text{Simplify.}$$

$$x - 3 = \pm\sqrt{16} \quad \text{Apply the square root method.}$$

$$x - 3 = 4 \quad \text{or} \quad x - 3 = -4 \quad \text{Write } \sqrt{16} \text{ as 4 and express as two separate equations.}$$

$$x = 7 \qquad\qquad x = -1 \quad \text{Solve for } x \text{ by adding 3 to both sides.}$$

Figure 10.38 The graph of $(x - 3)^2 = 8(y + 1)$

Because we obtained these values of x for $y = 1$, the parabola passes through the points $(7, 1)$ and $(-1, 1)$. Passing a smooth curve through the vertex and each of these points, we sketch the parabola shown in Figure 10.38.

| Check Point 4 | Find the vertex, focus, and directrix of the parabola given by $(x - 2)^2 = 4(y + 1)$. Then graph the parabola. |

In some cases, we need to convert the equation of a parabola to standard form by completing the square on x or y, whichever variable is squared. Let's see how this is done.

EXAMPLE 5 Graphing a Parabola with Vertex at (h, k)

Find the vertex, focus, and directrix of the parabola given by

$$y^2 + 2y + 12x - 23 = 0.$$

Then graph the parabola.

Solution   We convert the given equation to standard form by completing the square on the variable y. We isolate the terms involving y on the left side.

$$y^2 + 2y + 12x - 23 = 0 \qquad \text{This is the given equation.}$$
$$y^2 + 2y = -12x + 23 \qquad \text{Isolate the terms involving } y.$$
$$y^2 + 2y + 1 = -12x + 23 + 1 \qquad \text{Complete the square by adding the square of half the coefficient of } y.$$
$$(y + 1)^2 = -12x + 24$$

To express this equation in the standard form $(y - k)^2 = 4p(x - h)$, we factor -12 on the right. The standard form of the parabola's equation is

$$(y + 1)^2 = -12(x - 2).$$

We use this form to identify the vertex, (h, k), and the value for p needed to locate the focus and the directrix.

$$\left(y - (-1)\right)^2 = -12(x - 2) \qquad \text{The equation is in the standard form } (y - k)^2 = 4p(x - h).$$

| This is $(y - k)^2$, with $k = -1$. | This is $4p$. | This is $x - h$, with $h = 2$. |

We see that $h = 2$ and $k = -1$. Thus, the vertex of the parabola is $(h, k) = (2, -1)$. Because $4p = -12$, $p = -3$. Based on the standard form of the equation, the axis of symmetry is horizontal. With a negative value for p and a horizontal axis of symmetry, the parabola opens to the left. We locate the focus and directrix as follows.

$$\text{Focus:} \qquad (h + p, k) = \left(2 + (-3), -1\right) = (-1, -1)$$

| The vertex, (h, k), is $(2, -1)$. | $h = 2$ | $p = -3$ | $k = -1$ |

$$\text{Directrix:} \qquad x = h - p$$
$$x = 2 - (-3) = 5$$

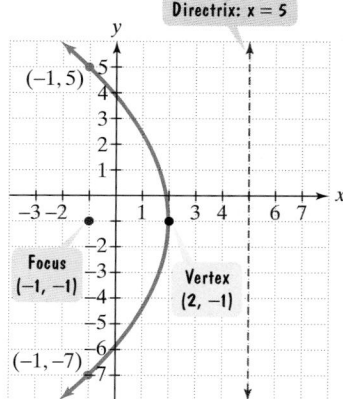

Figure 10.39 The graph of $(y + 1)^2 = -12(x - 2)$

Thus, the focus is $(-1, -1)$ and the directrix is $x = 5$. They are shown in Figure 10.39.

To graph $(y + 1)^2 = -12(x - 2)$, we assign x a value that makes the right side of the equation a perfect square. If $x = -1$, the right side is $-12(-1 - 2) = -12(-3) = 36$, a perfect square. We will let $x = -1$ and solve for y to obtain points on the parabola.

$$(y + 1)^2 = -12(-1 - 2)$$ Substitute -1 for x in $(y + 1)^2 = -12(x - 2)$.

$$(y + 1)^2 = 36$$ Simplify: $-12(-1 - 2) = -12(-3) = 36$.

$$y + 1 = \pm\sqrt{36}$$ Apply the square root method.

$$y + 1 = 6 \quad \text{or} \quad y + 1 = -6$$ Write $\sqrt{36}$ as 6 and express as two separate equations.

$$y = 5 \qquad\qquad y = -7$$ Solve for y by subtracting 1 from both sides.

Because we obtained these values of y for $x = -1$, the parabola passes through the points $(-1, 5)$ and $(-1, -7)$. Passing a smooth curve through the vertex and these two points, we sketch the parabola shown in Figure 10.39.

> **Check Point 5** Find the vertex, focus, and directrix of the parabola given by $y^2 + 2y + 4x - 7 = 0$. Then graph the parabola.

4 Solve applied problems involving parabolas.

Applications

Parabolas have many applications. Cables hung between structures to form suspension bridges form parabolas. Arches constructed of steel and concrete, whose main purpose is strength, are usually parabolic in shape.

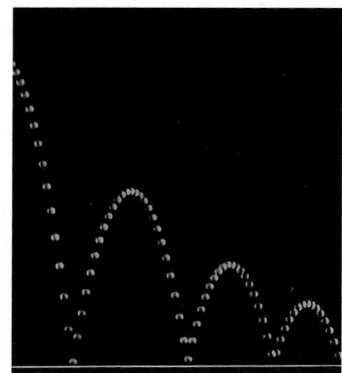

Figure 10.40 Multiflash photo showing the parabolic path of a ball thrown into the air

Suspension bridge

Arch bridge

We have seen that comets in our solar system travel in orbits that are ellipses and hyperbolas. Some comets also follow parabolic paths. Only comets with elliptical orbits, such as Halley's Comet, return to our part of the galaxy.

A projectile, such as a baseball thrown directly upward, moves along a parabolic path, illustrated in Figure 10.40.

If a parabola is rotated about its axis of symmetry, a parabolic surface is formed. Figure 10.41 (a) shows how a parabolic surface can be used to reflect

The Hubble Space Telescope

The Hubble Space Telescope

For decades, astronomers hoped to create an observatory above the atmosphere that would provide an unobscured view of the universe. This dream came true with the 1990 launching of the Hubble Space Telescope. The telescope initially had blurred vision due to problems with its parabolic mirror. The mirror had been ground two millionths of a meter smaller than design specifications. In 1993, astronauts from the Space Shuttle *Endeavor* equipped the telescope with optics to correct the blurred vision. "A small change for a mirror, a giant leap for astronomy," Christopher J. Burrows of the Space Telescope Science Institute said when clear images from the ends of the universe were presented to the public after the repair mission.

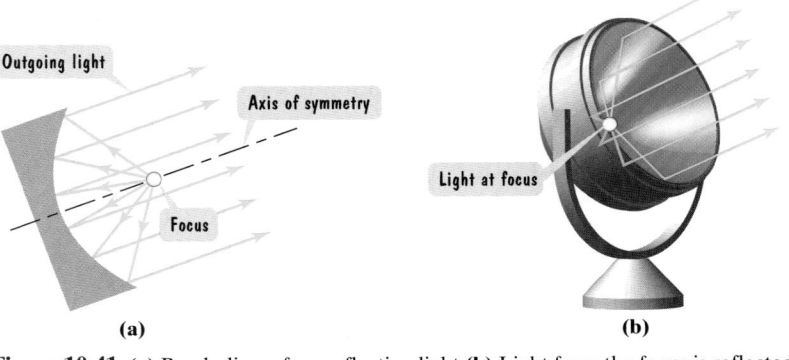

Figure 10.41 **(a)** Parabolic surface reflecting light **(b)** Light from the focus is reflected parallel to the axis of symmetry.

light. Light originates at the focus. Note how the light is reflected by the parabolic surface, so that the outgoing light is parallel to the axis of symmetry. The reflective properties of parabolic surfaces are used in the design of searchlights [Figure 10.41(b)], automobile headlights, and parabolic microphones.

Figure 10.42(a) shows how a parabolic surface can be used to reflect *incoming* light. Note that light rays strike the surface and are reflected *to the focus*. This principle is used in the design of reflecting telescopes, radar, and television satellite dishes. Reflecting telescopes magnify the light from distant stars by reflecting the light from these bodies to the focus of a parabolic mirror [Figure 10.42(b)].

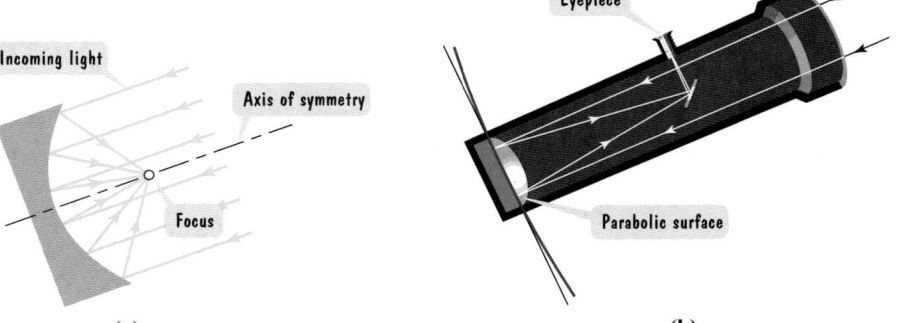

Figure 10.42 **(a)** Parabolic surface reflecting incoming light **(b)** Incoming light rays are reflected to the focus.

EXAMPLE 6   Using the Reflection Property of Parabolas

An engineer is designing a flashlight using a parabolic reflecting mirror and a light source, shown in Figure 10.43. The casting has a diameter of 4 inches and a depth of 2 inches. What is the equation of the parabola used to shape the mirror? At what point should the light source be placed relative to the mirror's vertex?

Solution   We position the parabola with its vertex at the origin and opening upward (Figure 10.44). Thus, the focus is on the y-axis, located at $(0, p)$. We use the standard form of the equation in which y is not squared, namely $x^2 = 4py$. We need to find p. Because $(2, 2)$ lies on the parabola, we let $x = 2$ and $y = 2$ in $x^2 = 4py$.

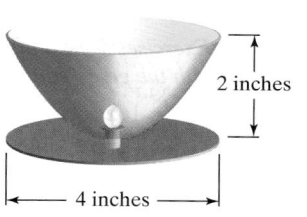

2 inches

4 inches

Figure 10.43 Designing a flashlight

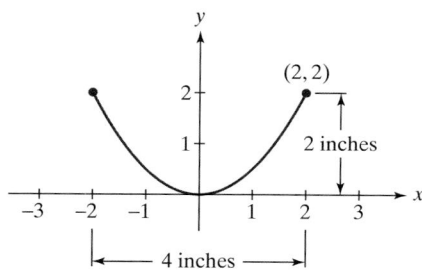

(2, 2)

2 inches

4 inches

Figure 10.44

$$2^2 = 4p \cdot 2 \qquad \text{Substitute 2 for x and 2 for y in } x^2 = 4py.$$

$$4 = 8p \qquad \text{Simplify.}$$

$$p = \tfrac{1}{2} \qquad \text{Divide both sides of the equation by 8 and reduce the resulting fraction.}$$

We substitute $\tfrac{1}{2}$ for p in $x^2 = 4py$ to obtain the standard form of the equation of the parabola. The equation of the parabola used to shape the mirror is

$$x^2 = 4 \cdot \tfrac{1}{2} y \qquad \text{or} \qquad x^2 = 2y.$$

The light source should be placed at the focus $(0, p)$. Because $p = \tfrac{1}{2}$, the light should be placed at $\left(0, \tfrac{1}{2}\right)$, or $\tfrac{1}{2}$ inch above the vertex.

Check Point 6 In Example 6, suppose that the casting has a diameter of 6 inches and a depth of 4 inches. What is the equation of the parabola used to shape the mirror? At what point should the light source be placed relative to the mirror's vertex?

Degenerate Conic Sections

We opened the chapter by noting that conic sections are curves that result from the intersection of a cone and a plane. However, these intersections might not result in a conic section. Three degenerate cases occur when the cutting plane passes through the vertex. These **degenerate conic sections** are a point, a line, and a pair of intersecting lines, illustrated in Figure 10.45.

Point

Line

Two intersecting lines

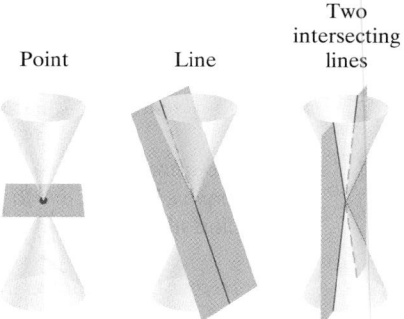

Figure 10.45 Degenerate conics

Writing in Mathematics

48. What is a parabola?

49. Explain how to use $y^2 = 8x$ to find the parabola's focus and directrix.

50. If you are given the standard form of the equation of a parabola with vertex at the origin, explain how to determine if the parabola opens to the right, left, upward, or downward.

51. Describe one similarity and one difference between the graphs of $y^2 = 4x$ and $(y - 1)^2 = 4(x - 1)$.

52. How can you distinguish parabolas from other conic sections by looking at their equations?

53. Look at the satellite dish shown in Exercise 43. Why must the receiver for a shallow dish be farther from the base of the dish than for a deeper dish of the same diameter?

Technology Exercises

54. Use a graphing utility to graph any five of the parabolas that you graphed by hand in Exercises 5–14.

55. Use a graphing utility to graph any three of the parabolas that you graphed by hand in Exercises 27–34. First solve the given equation for y, possibly using the square root method. Enter each of the two resulting equations to produce the complete graph.

Use a graphing utility to graph the parabolas in Exercises 56–57. Write the given equation as a quadratic equation in y and use the quadratic formula to solve for y. Enter each of the equations to produce the complete graph.

56. $y^2 + 2y - 6x + 13 = 0$

57. $y^2 + 10y - x + 25 = 0$

In Exercises 58–59, write each equation as a quadratic equation in y and then use the quadratic formula to express y in terms of x. Graph the resulting two equations using a graphing utility. What effect does the xy-term have on the graph of the resulting parabola?

58. $16x^2 - 24xy + 9y^2 - 60x - 80y + 100 = 0$

59. $x^2 + 2\sqrt{3}\,xy + 3y^2 + 8\sqrt{3}\,x - 8y + 32 = 0$

Critical Thinking Exercises

60. Which one of the following is true?
   a. The parabola whose equation is $x = 2y - y^2 + 5$ opens to the right.
   b. If the parabola whose equation is $x = ay^2 + by + c$ has its vertex at $(3, 2)$ and $a > 0$, then it has no y-intercepts.
   c. Some parabolas that open to the right have equations that define y as a function of x.
   d. The graph of $x = a(y - k) + h$ is a parabola with vertex at (h, k).

61. A satellite dish in the shape of a parabolic surface has a diameter of 20 feet. If the receiver is to be placed 6 feet from the base, how deep should the dish be?

62. Write the standard form of the equation of a parabola whose points are equidistant from $y = 4$ and $(-1, 0)$.

Group Exercise

63. Consult the research department of your library or the Internet to find an example of architecture that incorporates one or more conic sections in its design. Share this example with other group members. Explain precisely how conic sections are used. Do conic sections enhance the appeal of the architecture? In what ways?

SECTION 10.4 Rotation of Axes

Objectives

1. Identify conics without completing the square.
2. Use rotation of axes formulas.
3. Write equations of rotated conics in standard form.
4. Identify conics without rotating axes.

Richard E. Prince "The Cone of Apollonius"

To recognize a conic section, you often need to pay close attention to its graph. Graphs powerfully enhance our understanding of algebra and trigonometry.

However, it is not possible for people who are blind—or sometimes, visually impaired—to see a graph. Creating informative materials for the blind and visually impaired is a challenge for instructors and mathematicians. Many people who are visually impaired "see" a graph by touching a three-dimensional representation of that graph, perhaps while it is described verbally.

Is it possible to identify conic sections in nonvisual ways? The answer is yes, and the methods for doing so are related to the coefficients in their equations. As we present these methods, think about how you learn them. How would your approach to studying mathematics change if we removed all graphs and replaced them with verbal descriptions?

1 Identify conics without completing the square.

Identifying Conic Sections without Completing the Square

Conic sections can be represented both geometrically (as intersecting planes and cones) and algebraically. The equations of the conic sections we have considered in the first three sections of this chapter can be expressed in the form

$$Ax^2 + Cy^2 + Dx + Ey + F = 0$$

in which A and C are not both zero. You can use A and C, the coefficients of x^2 and y^2, respectively, to identify a conic section without completing the square.

Identifying a Conic Section without Completing the Square

A nondegenerate conic section of the form

$$Ax^2 + Cy^2 + Dx + Ey + F = 0$$

in which A and C are not both zero is

- a circle if $A = C$,
- a parabola if $AC = 0$,
- an ellipse if $A \neq C$ and $AC > 0$, and
- a hyperbola if $AC < 0$.

EXAMPLE 1   Identifying a Conic Section without Completing the Square

Identify the graph of each of the following nondegenerate conic sections.

   a. $4x^2 - 25y^2 - 24x + 250y - 489 = 0$
   b. $x^2 + y^2 + 6x - 2y + 6 = 0$
   c. $y^2 + 12x + 2y - 23 = 0$
   d. $9x^2 + 25y^2 - 54x + 50y - 119 = 0$

Solution   We use A, the coefficient of x^2, and C, the coefficient of y^2, to identify each conic section.

   a. $4x^2 - 25y^2 - 24x + 250y - 489 = 0$

      $A = 4$ $C = -25$

      $AC = 4(-25) = -100 < 0$
      Because $AC < 0$, the graph of the equation is a hyperbola.

b. $x^2 + y^2 + 6x - 2y + 6 = 0$

$A = 1$ $C = 1$

Because $A = C$, the graph of the equation is a circle.

c. We can write $y^2 + 12x + 2y - 23 = 0$ as

$$0x^2 + y^2 + 12x + 2y - 23 = 0.$$

$A = 0$ $C = 1$

$$AC = 0(1) = 0$$

Because $AC = 0$, the graph of the equation is a parabola.

d. $9x^2 + 25y^2 - 54x + 50y - 119 = 0$

$A = 9$ $C = 25$

$$AC = 9(25) = 225 > 0.$$
Because $AC > 0$ and $A \neq C$, the graph of the equation is an ellipse.

> **Check Point 1** Identify the graph of each of the following nondegenerate conic sections.
> **a.** $3x^2 + 2y^2 + 12x - 4y + 2 = 0$
> **b.** $x^2 + y^2 - 6x + y + 3 = 0$
> **c.** $y^2 - 12x - 4y + 52 = 0$
> **d.** $9x^2 - 16y^2 - 90x + 64y + 17 = 0$

2 Use rotation of axes formulas.

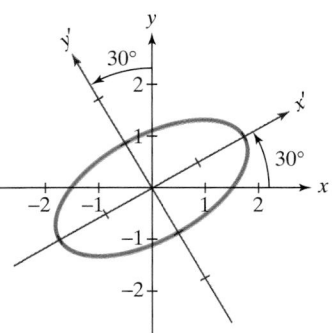

Figure 10.46 The graph of $7x^2 - 6\sqrt{3}\,xy + 13y^2 - 16 = 0$, a rotated ellipse

Rotation of Axes

Figure 10.46 shows the graph of

$$7x^2 - 6\sqrt{3}\,xy + 13y^2 - 16 = 0.$$

The graph looks like an ellipse, although its major axis neither lies along the x-axis nor is parallel to the x-axis. Do you notice anything unusual about the equation? It contains an xy-term. However, look what happens if we rotate the x- and y-axes through an angle of 30°. In the rotated $x'y'$-system, the major axis of the ellipse lies along the x'-axis. We can write the equation of the ellipse in this rotated $x'y'$-system as

$$\frac{x'^2}{4} + \frac{y'^2}{1} = 1.$$

Observe that there is no $x'y'$-term in the equation.

Except for degenerate cases, the **general second-degree equation**

$$Ax^2 + Bxy + Cy^2 + Dx + Ey + F = 0$$

represents one of the conic sections. However, due to the xy-term in the equation, these conic sections are rotated in such a way that their axes are no longer parallel to the x- and y-axes. To reduce these equations to forms of the conic sections with which you are already familiar, we use a procedure called **rotation of axes**.

Suppose that the x- and y-axes are rotated through a positive angle θ, resulting in a new $x'y'$ coordinate system. This system is shown in Figure 10.47(a). The origin in the $x'y'$-system is the same as the origin in the xy-system. Point P in Figure 10.47(b) has coordinates (x, y) relative to the xy-system and coordinates

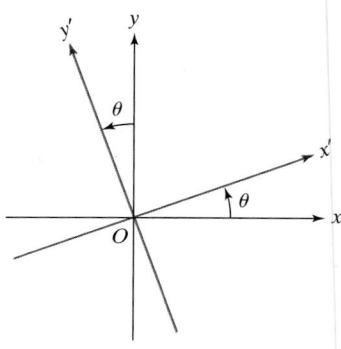

(a) Rotating the x- and y-axes through a positive angle θ

(b) Describing point P relative to the xy-system and the rotated $x'y'$-system

Figure 10.47 Rotating axes

(x', y') relative to the $x'y'$-system. We wish to obtain formulas relating the old and new coordinates. Thus, we need to express x and y in terms of x', y', and θ.

Look at Figure 10.47(b). Notice that

$r = $ the distance from the origin 0 to point P.

$\alpha = $ the angle from the positive x'-axis to the ray from 0 through P.

Using the definitions of sine and cosine, we obtain

$$\cos \alpha = \frac{x'}{r} : x' = r \cos \alpha$$

> This is from the right triangle with a leg along the x'-axis.

$$\sin \alpha = \frac{y'}{r} : y' = r \sin \alpha$$

$$\cos (\theta + \alpha) = \frac{x}{r} : x = r \cos (\theta + \alpha)$$

> This is from the taller right triangle with a leg along the x-axis.

$$\sin (\theta + \alpha) = \frac{y}{r} : y = r \sin (\theta + \alpha)$$

Thus,

$$
\begin{aligned}
x &= r \cos (\theta + \alpha) && \text{This is the third of the preceding equations.} \\
&= r(\cos \theta \cos \alpha - \sin \theta \sin \alpha) && \text{Use the formula for the cosine of the sum of} \\
& && \text{two angles.} \\
&= (r \cos \alpha) \cos \theta - (r \sin \alpha) \sin \theta && \text{Apply the distributive property and} \\
& && \text{rearrange factors.} \\
&= x' \cos \theta - y' \sin \theta && \text{Use the first and second of the preceding} \\
& && \text{equations: } x' = r \cos \alpha \text{ and } y' = r \sin \alpha.
\end{aligned}
$$

Similarly,

$$y = r \sin (\theta + \alpha) = r(\sin \theta \cos \alpha + \cos \theta \sin \alpha) = x' \sin \theta + y' \cos \theta.$$

Rotation of Axes Formulas

Suppose an xy-coordinate system and an $x'y'$-coordinate system have the same origin and θ is the angle from the positive x-axis to the positive x'-axis. If the coordinates of point P are (x, y) in the xy-system and (x', y') in the rotated $x'y'$-system, then

$$x = x' \cos \theta - y' \sin \theta$$

$$y = x' \sin \theta + y' \cos \theta.$$

EXAMPLE 2 Rotating Axes

Write the equation $xy = 1$ in terms of a rotated $x'y'$-system if the angle of rotation from the x-axis to the x'-axis is 45°. Express the equation in standard form. Use the rotated system to graph $xy = 1$.

Solution With $\theta = 45°$, the rotation formulas for x and y are

$$x = x' \cos\theta - y' \sin\theta = x' \cos 45° - y' \sin 45°$$

$$= x'\left(\frac{\sqrt{2}}{2}\right) - y'\left(\frac{\sqrt{2}}{2}\right) = \frac{\sqrt{2}}{2}(x' - y')$$

$$y = x' \sin\theta + y' \cos\theta = x' \sin 45° + y' \cos 45°$$

$$= x'\left(\frac{\sqrt{2}}{2}\right) + y'\left(\frac{\sqrt{2}}{2}\right) = \frac{\sqrt{2}}{2}(x' + y')$$

Now substitute these expressions for x and y in the given equation, $xy = 1$.

$$xy = 1 \qquad \text{This is the given equation.}$$

$$\left[\frac{\sqrt{2}}{2}(x' - y')\right]\left[\frac{\sqrt{2}}{2}(x' + y')\right] = 1 \qquad \text{Substitute the expressions for x and y from the rotation formulas.}$$

$$\frac{2}{4}(x' - y')(x' + y') = 1 \qquad \text{Multiply: } \frac{\sqrt{2}}{2} \cdot \frac{\sqrt{2}}{2} = \frac{2}{4}.$$

$$\frac{1}{2}(x'^2 - y'^2) = 1 \qquad \text{Multiply the binomials.}$$

$$\frac{x'^2}{2} - \frac{y'^2}{2} = 1 \qquad \begin{array}{l}\text{Write the equation in standard form:} \\ \frac{x^2}{a^2} - \frac{y^2}{b^2} = 1.\end{array}$$

$$\boxed{a^2 = 2} \quad \boxed{b^2 = 2}$$

This equation expresses $xy = 1$ in terms of the rotated $x'y'$-system. Can you see that this is the standard form of the equation of a hyperbola? The hyperbola's center is at $(0, 0)$ with the transverse axis on the x'-axis. The vertices are $(-a, 0)$ and $(a, 0)$. Because $a^2 = 2$, the vertices are $(-\sqrt{2}, 0)$ and $(\sqrt{2}, 0)$, located on the x'-axis. Based on the standard form of the hyperbola's equation, the equations for the asymptotes are

$$y' = \pm\frac{b}{a}x' \text{ or } y' = \pm\frac{\sqrt{2}}{\sqrt{2}}x'.$$

The equations of the asymptotes can be simplified to $y' = x'$ and $y' = -x'$, which correspond to the original x- and y-axes. The graph of the hyperbola is shown in Figure 10.48.

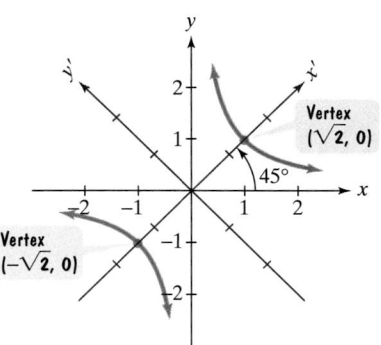

Figure 10.48 The graph of
$$xy = 1 \text{ or } \frac{x'^2}{2} - \frac{y'^2}{2} = 1$$

Vertex $(\sqrt{2}, 0)$

Vertex $(-\sqrt{2}, 0)$

Check Point 2 Write the equation $xy = 2$ in terms of a rotated $x'y'$-system if the angle of rotation from the x-axis to the x'-axis is 45°. Express the equation in standard form. Use the rotated system to graph $xy = 2$.

3 Write equations of rotated conics in standard form.

Using Rotations to Transform Equations with xy-Terms to Standard Equations of Conic Sections

We have noted that the appearance of the term Bxy $(B \neq 0)$ in the general second-degree equation indicates that the graph of the conic section has been rotated. A rotation of axes through an appropriate angle can transform the equation to one of the standard forms of the conic sections in x' and y' in which no $x'y'$-term appears.

> ### Amount of Rotation Formula
> The general second-degree equation
> $$Ax^2 + Bxy + Cy^2 + Dx + Ey + F = 0, B \neq 0$$
> can be rewritten as an equation in x' and y' without an $x'y'$-term by rotating the axes through angle θ, where
> $$\cot 2\theta = \frac{A - C}{B}.$$

Before we learn to apply this formula, let's see how it can be derived. We begin with the general second-degree equation

$$Ax^2 + Bxy + Cy^2 + Dx + Ey + F = 0, B \neq 0.$$

Rotate the axes through an angle θ. In terms of the rotated $x'y'$-system, the general second-degree equation can be written as

$$A(x' \cos\theta - y' \sin\theta)^2 + B(x' \cos\theta - y' \sin\theta)(x' \sin\theta + y' \cos\theta)$$
$$+ C(x' \sin\theta + y' \cos\theta)^2 + D(x' \cos\theta - y' \sin\theta)$$
$$+ E(x' \sin\theta + y' \cos\theta) + F = 0.$$

After a lot of simplifying that involves expanding and collecting like terms, you will obtain the following equation.

We want a rotation that results in no $x'y'$-term.

$$(A \cos^2\theta + B \sin\theta \cos\theta + C \sin^2\theta)x'^2 + [B(\cos^2\theta - \sin^2\theta) + 2(C - A)(\sin\theta \cos\theta)]x'y'$$
$$+ (A \sin^2\theta - B \sin\theta \cos\theta + C \cos^2\theta)y'^2$$
$$+ (D \cos\theta + E \sin\theta)x'$$
$$+ (-D \sin\theta + E \cos\theta)y' + F = 0$$

If this looks somewhat ghastly, take a deep breath and focus only on the $x'y'$-term. We want to choose θ so that the coefficient of this term is zero. This will give the required rotation that results in no $x'y'$-term.

$B(\cos^2\theta - \sin^2\theta) + 2(C - A) \sin\theta \cos\theta = 0$ Set the coefficient of the $x'y'$-term equal to 0.

$B \cos 2\theta + (C - A) \sin 2\theta = 0$ Use the double-angle formulas: $\cos 2\theta = \cos^2\theta - \sin^2\theta$ and $\sin 2\theta = 2 \sin\theta \cos\theta$.

$B \cos 2\theta = 0 - (C - A) \sin 2\theta$ Subtract $(C - A) \sin 2\theta$ from both sides.

$$B \cos 2\theta = (A - C) \sin 2\theta \qquad \text{Simplify.}$$

$$\frac{B \cos 2\theta}{B \sin 2\theta} = \frac{(A - C) \sin 2\theta}{B \sin 2\theta} \qquad \text{Divide both sides by } B \sin 2\theta.$$

$$\frac{\cos 2\theta}{\sin 2\theta} = \frac{A - C}{B} \qquad \text{Simplify.}$$

$$\cot 2\theta = \frac{A - C}{B} \qquad \begin{array}{l} \text{Apply a quotient identity:} \\ \cot \square = \dfrac{\sin \square}{\cos \square}. \end{array}$$

If $\cot 2\theta$ is positive, we will select θ so that $0° < \theta < 45°$. If $\cot 2\theta$ is negative, we will select θ so that $45° < \theta < 90°$. Thus θ, the angle of rotation, is always an acute angle.

Here is a step-by-step procedure for writing the equation of a rotated conic section in standard form.

Study Tip

What do you do after substituting the expressions for x and y from the rotation formulas into the given equation? You must simplify the resulting equation by expanding and collecting like terms. Work through this process slowly and carefully, allowing lots of room on your paper.

If your rotation equations are correct but you obtain an equation that has an $x'y'$-term, you have made an error in the algebraic simplification.

Writing the Equation of a Rotated Conic in Standard Form

1. Use the given equation
$$Ax^2 + Bxy + Cy^2 + Dx + Ey + F = 0, B \neq 0$$
to find $\cot 2\theta$.
$$\cot 2\theta = \frac{A - C}{B}$$

2. Use the expression for $\cot 2\theta$ to determine θ, the angle of rotation.
3. Substitute θ in the rotation formulas
$$x = x' \cos \theta - y' \sin \theta \quad \text{and} \quad y = x' \sin \theta + y' \cos \theta$$
and simplify.
4. Substitute the expressions for x and y from the rotation formulas in the given equation and simplify. The resulting equation should have no $x'y'$-term.
5. Write the equation involving x' and y' in standard form.

Using the equation in step 5, you can graph the conic section in the rotated $x'y'$-system.

EXAMPLE 3 Writing the Equation of a Rotated Conic Section in Standard Form

Rewrite the equation
$$7x^2 - 6\sqrt{3}\, xy + 13y^2 - 16 = 0$$
in a rotated $x'y'$-system without an $x'y'$-term. Express the equation in the standard form of a conic section.

Solution

Step 1   Use the given equation to find $\cot 2\theta$. We need to identify the constants A, B, and C in the given equation.

$$7x^2 - 6\sqrt{3}\,xy + 13y^2 - 16 = 0.$$

| A is the coefficient of the x^2-term: $A = 7$. | B is the coefficient of the xy-term: $B = -6\sqrt{3}$. | C is the coefficient of the y^2-term: $C = 13$. |

The appropriate angle θ through which to rotate the axes satisfies the equation

$$\cot 2\theta = \frac{A - C}{B} = \frac{7 - 13}{-6\sqrt{3}} = \frac{-6}{-6\sqrt{3}} = \frac{1}{\sqrt{3}} \text{ or } \frac{\sqrt{3}}{3}.$$

Step 2   Use the expression for cot 2θ to determine the angle of rotation. We have $\cot 2\theta = \dfrac{\sqrt{3}}{3}$. Based on our knowledge of exact values for trigonometric functions, we conclude that $2\theta = 60°$. Thus, $\theta = 30°$.

Step 3   Substitute θ in the rotation formulas $x = x' \cos\theta - y' \sin\theta$ and $y = x' \sin\theta + y' \cos\theta$ and simplify. Substituting $30°$ for θ,

$$x = x' \cos 30° - y' \sin 30° = x'\left(\frac{\sqrt{3}}{2}\right) - y'\left(\frac{1}{2}\right) = \frac{\sqrt{3}x' - y'}{2}$$

$$y = x' \sin 30° + y' \cos 30° = x'\left(\frac{1}{2}\right) + y'\left(\frac{\sqrt{3}}{2}\right) = \frac{x' + \sqrt{3}y'}{2}.$$

Step 4   Substitute the expressions for x and y from the rotation formulas in the given equation and simplify.

$$7x^2 - 6\sqrt{3}xy + 13y^2 - 16 = 0 \qquad \text{This is the given equation.}$$

$$7\left(\frac{\sqrt{3}x' - y'}{2}\right)^2 - 6\sqrt{3}\left(\frac{\sqrt{3}x' - y'}{2}\right)\left(\frac{x' + \sqrt{3}y'}{2}\right)$$
$$+ 13\left(\frac{x' + \sqrt{3}y'}{2}\right)^2 - 16 = 0 \qquad \text{Substitute the expressions for x and y from the rotation formulas.}$$

$$7\left(\frac{3x'^2 - 2\sqrt{3}x'y' + y'^2}{4}\right) - 6\sqrt{3}\left(\frac{\sqrt{3}x'^2 + 3x'y' - x'y' - \sqrt{3}y'^2}{4}\right)$$
$$+ 13\left(\frac{x'^2 + 2\sqrt{3}x'y' + 3y'^2}{4}\right) - 16 = 0 \qquad \text{Square and multiply.}$$

$$7(3x'^2 - 2\sqrt{3}x'y' + y'^2) - 6\sqrt{3}(\sqrt{3}x'^2 + 2x'y' - \sqrt{3}y'^2)$$
$$+ 13(x'^2 + 2\sqrt{3}x'y' + 3y'^2) - 64 = 0 \qquad \text{Multiply both sides by 4.}$$

$$21x'^2 - 14\sqrt{3}x'y' + 7y'^2 - 18x'^2 - 12\sqrt{3}x'y' + 18y'^2$$
$$+ 13x'^2 + 26\sqrt{3}x'y' + 39y'^2 - 64 = 0 \qquad \text{Distribute throughout parentheses.}$$

$$21x'^2 - 18x'^2 + 13x'^2 - 14\sqrt{3}x'y' - 12\sqrt{3}x'y' + 26\sqrt{3}x'y'$$
$$+ 7y'^2 + 18y'^2 + 39y'^2 - 64 = 0 \qquad \text{Rearrange terms.}$$

$$16x'^2 + 64y'^2 - 64 = 0 \qquad \text{Combine like terms.}$$

Do you see how we "lost" the $x'y'$-term in the last equation?

$$-14\sqrt{3}x'y' - 12\sqrt{3}x'y' + 26\sqrt{3}x'y' = -26\sqrt{3}x'y' + 26\sqrt{3}x'y' = 0x'y' = 0$$

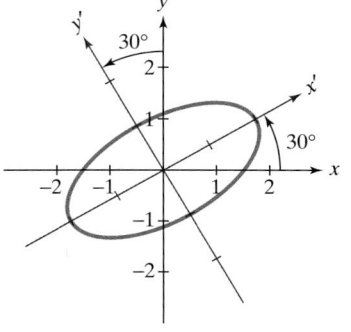

Figure 10.46, repeated

The graph of
$7x^2 - 6\sqrt{3}\,xy + 13y^2 - 16 = 0$
or $\dfrac{x'^2}{4} - \dfrac{y'^2}{1} = 1$, a rotated ellipse

Step 5 Write the equation involving x' and y' in standard form. We can express $16x'^2 + 64y'^2 - 64 = 0$, an equation of an ellipse, in the standard form $\dfrac{x^2}{a^2} + \dfrac{y^2}{b^2} = 1$.

$$16x'^2 + 64y'^2 - 64 = 0 \qquad \text{\small This equation describes the ellipse relative to a system rotated through 30°.}$$

$$16x'^2 + 64y'^2 = 64 \qquad \text{\small Add 64 to both sides.}$$

$$\frac{16x'^2}{64} + \frac{64y'^2}{64} = \frac{64}{64} \qquad \text{\small Divide both sides by 64.}$$

$$\frac{x'^2}{4} + \frac{y'^2}{1} = 1 \qquad \text{\small Simplify.}$$

The last equation is the standard form of the equation of an ellipse. The major axis is on the x'-axis and the vertices are $(-2, 0)$ and $(2, 0)$. The minor axis is on the y'-axis with endpoints $(0, -1)$ and $(0, 1)$. The graph of the ellipse is shown in Figure 10.46. Does this graph look familiar? It should—you saw it earlier in this section on page 852.

Check Point 3

Rewrite the equation
$$2x^2 + \sqrt{3}xy + y^2 - 2 = 0$$
in a rotated $x'y'$-system without an $x'y'$-term. Express the equation in the standard form of a conic section. Graph the conic section in the rotated system.

Technology

In order to graph a general second-degree equation in the form
$$Ax^2 + Bxy + Cy^2 + Dx + Ey + F = 0,$$
it is necessary to solve for y. Rewrite the equation as a quadratic equation in y.
$$Cy^2 + (Bx + E)y + (Ax^2 + Dx + F) = 0$$
By applying the quadratic formula, the graph of this equation can be obtained by entering
$$y_1 = \frac{-(Bx + E) + \sqrt{(Bx + E)^2 - 4C(Ax^2 + Dx + F)}}{2C}$$
and
$$y_2 = \frac{-(Bx + E) - \sqrt{(Bx + E)^2 - 4C(Ax^2 + Dx + F)}}{2C}.$$

The graph of
$$7x^2 - 6\sqrt{3}xy + 13y^2 - 16 = 0$$
is shown on the right in a $[-2, 2, 1]$ by $[-2, 2, 1]$ viewing rectangle. The graph was obtained by entering the equations for y_1 and y_2 shown previously with
$$A = 7, B = -6\sqrt{3}, C = 13, D = 0, E = 0,$$
and $F = -16$.

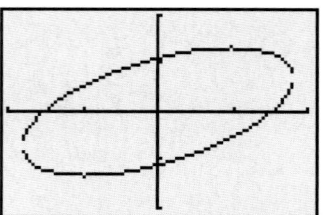

In Example 3 and the Check Point 3, we found θ, the angle of rotation, directly because we recognized $\frac{\sqrt{3}}{3}$ as the value of $\cot 60°$. What do we do if $\cot 2\theta$ is not the cotangent of one of the more familiar angles? We use $\cot 2\theta$ to find $\sin \theta$ and $\cos \theta$ as follows:

- Use a sketch of $\cot 2\theta$ to find $\cos 2\theta$.
- Find $\sin \theta$ and $\cos \theta$ using the identities

$$\sin \theta = \sqrt{\frac{1 - \cos 2\theta}{2}} \quad \text{and} \quad \cos \theta = \sqrt{\frac{1 + \cos 2\theta}{2}}.$$

Because θ is an acute angle, the positive square roots are appropriate.

The resulting values for $\sin \theta$ and $\cos \theta$ are used to write the rotation formulas that give an equation with no $x'y'$-term.

EXAMPLE 4   Graphing the Equation of a Rotated Conic

Graph relative to a rotated $x'y'$-system in which the equation has no $x'y'$-term:
$$16x^2 - 24xy + 9y^2 + 110x - 20y + 100 = 0.$$

Solution

Step 1   Use the given equation to find $\cot 2\theta$. With $A = 16, B = -24,$ and $C = 9,$ we have

$$\cot 2\theta = \frac{A - C}{B} = \frac{16 - 9}{-24} = -\frac{7}{24}.$$

Step 2   Use the expression for $\cot 2\theta$ to determine $\sin \theta$ and $\cos \theta$. A rough sketch showing $\cot 2\theta$ is given in Figure 10.49. Because θ is always acute and $\cot 2\theta$ is negative, 2θ is in quadrant II. The third side of the triangle is found using $x^2 + y^2 = r^2$. By the definition of the cosine function,

$$\cos 2\theta = \frac{x}{r} = \frac{-7}{25} = -\frac{7}{25}.$$

Now we use identities to find values for $\sin \theta$ and $\cos \theta$.

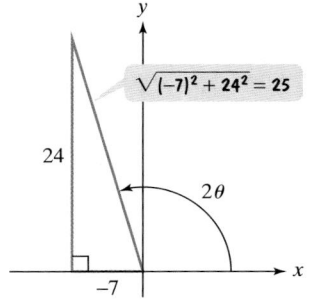

Figure 10.49 Using $\cot 2\theta$ to find $\cos 2\theta$

$$\sin \theta = \sqrt{\frac{1 - \cos 2\theta}{2}} = \sqrt{\frac{1 - \left(-\frac{7}{25}\right)}{2}}$$

$$= \sqrt{\frac{1 + \frac{7}{25}}{2}} = \sqrt{\frac{\frac{25}{25} + \frac{7}{25}}{2}} = \sqrt{\frac{\frac{32}{25}}{2}} = \sqrt{\frac{32}{50}} = \sqrt{\frac{16}{25}} = \frac{4}{5}$$

$$\cos \theta = \sqrt{\frac{1 + \cos 2\theta}{2}} = \sqrt{\frac{1 + \left(-\frac{7}{25}\right)}{2}}$$

$$= \sqrt{\frac{\frac{25}{25} - \frac{7}{25}}{2}} = \sqrt{\frac{\frac{18}{25}}{2}} = \sqrt{\frac{18}{50}} = \sqrt{\frac{9}{25}} = \frac{3}{5}$$

Step 3   Substitute $\sin\theta$ and $\cos\theta$ in the rotation formulas

$$x = x'\cos\theta - y'\sin\theta \text{ and } y = x'\sin\theta + y'\cos\theta$$

and simplify. Substituting $\frac{4}{5}$ for $\sin\theta$ and $\frac{3}{5}$ for $\cos\theta$,

$$x = x'\left(\frac{3}{5}\right) - y'\left(\frac{4}{5}\right) = \frac{3x' - 4y'}{5}$$

$$y = x'\left(\frac{4}{5}\right) + y'\left(\frac{3}{5}\right) = \frac{4x' + 3y'}{5}.$$

Step 4   Substitute the expressions for x and y from the rotation formulas in the given equation and simplify.

$$16x^2 - 24xy + 9y^2 + 110x - 20y + 100 = 0 \qquad \text{This is the given equation.}$$

$$16\left(\frac{3x' - 4y'}{5}\right)^2 - 24\left(\frac{3x' - 4y'}{5}\right)\left(\frac{4x' + 3y'}{5}\right) + 9\left(\frac{4x' + 3y'}{5}\right)^2 \qquad \text{Substitute the expressions for } x \text{ and } y \text{ from the rotation formulas.}$$

$$+ 110\left(\frac{3x' - 4y'}{5}\right) - 20\left(\frac{4x' + 3y'}{5}\right) + 100 = 0$$

Take a few minutes to expand, multiply both sides of the equation by 25, and combine like terms. The resulting equation

$$y'^2 + 2x' - 4y' + 4 = 0$$

has no $x'y'$-term.

Step 5   Write the equation involving x' and y' in standard form. With only one variable that is squared, we have the equation of a parabola. We need to write the equation in the standard form $(y - k)^2 = 4p(x - h)$.

$$y'^2 + 2x' - 4y' + 4 = 0 \qquad \text{This is the equation without an } x'y'\text{-term.}$$

$$y'^2 - 4y' = -2x' - 4 \qquad \text{Isolate the terms involving } y'.$$

$$y'^2 - 4y' + 4 = -2x' - 4 + 4 \qquad \text{Complete the square by adding the square of half the coefficient of } y'.$$

$$(y' - 2)^2 = -2x' \qquad \text{Factor.}$$

The standard form of the parabola's equation in the rotated $x'y'$-system is

$$(y' - 2)^2 = -2x'.$$

This is $(y' - k)^2$ with $k = 2$.   This is $4p$.   This is $x' - h$ with $h = 0$.

Vertex (0, 2)

$\theta = \sin^{-1}\frac{4}{5} \approx 53°$

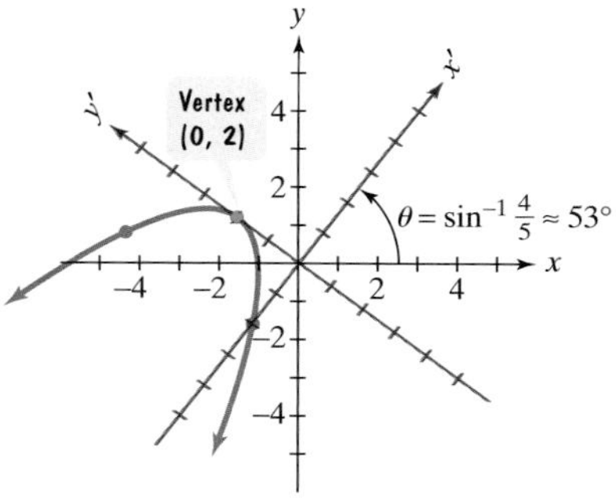

Figure 10.50 The graph of $(y' - 2)^2 = -2x'$ in a rotated $x'y'$-system

We see that $h = 0$ and $k = 2$. Thus, the vertex of the parabola in the $x'y'$-system is $(h, k) = (0, 2)$. If $x' = -2$, $(y' - 2)^2 = 4$ and $y' = 4$ or $y' = 0$. The parabola passes through $(-2, 4)$ and $(-2, 0)$ in the $x'y'$-system. Using a calculator to solve $\sin\theta = \frac{4}{5}$, we find that $\theta = \sin^{-1}\frac{4}{5} \approx 53°$. Rotate the axes through approximately $53°$. Using the rotated system, pass a smooth curve through the vertex and the two points on the parabola. The graph of the parabola is shown in Figure 10.50.

Check Point 4   Graph relative to a rotated $x'y'$-system in which the equation has no $x'y'$-term:

$$4x^2 - 4xy + y^2 - 8\sqrt{5}x - 16\sqrt{5}y = 0.$$

4 Identify conics without rotating axes.

Identifying Conic Sections without Rotating Axes

We now know that the general second-degree equation

$$Ax^2 + Bxy + Cy^2 + Dx + Ey + F = 0, B \neq 0$$

can be rewritten as

$$A'x'^2 + C'y'^2 + D'x' + E'y' + F' = 0$$

in a rotated $x'y'$-system. A relationship between the coefficients of the two equations is given by

$$B^2 - 4AC = -4A'C'.$$

We also know that A' and C' can be used to identify the graph of the rotated equation. Thus, $B^2 - 4AC$ can also be used to identify the graph of the general second-degree equation.

Identifying a Conic Section without a Rotation of Axes

A nondegenerate conic section of the form

$$Ax^2 + Bxy + Cy^2 + Dx + Ey + F = 0$$

is

- a parabola if $B^2 - 4AC = 0$,
- an ellipse or a circle if $B^2 - 4AC < 0$, and
- a hyperbola if $B^2 - 4AC > 0$.

Technology

The graph of

$$11x^2 + 10\sqrt{3}xy + y^2 - 4 = 0$$

is shown in a $\left[-1, 1, \frac{1}{4}\right]$ by $\left[-1, 1, \frac{1}{4}\right]$ viewing rectangle. The graph verifies that the equation represents a rotated hyperbola.

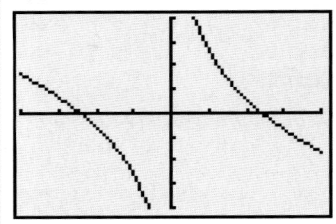

EXAMPLE 5   Identifying a Conic Section without Rotating Axes

Identify the graph of

$$11x^2 + 10\sqrt{3}xy + y^2 - 4 = 0.$$

Solution   We use A, B, and C to identify the conic section.

$$11x^2 + 10\sqrt{3}xy + y^2 - 4 = 0$$

$A = 11$ $B = 10\sqrt{3}$ $C = 1$

$$B^2 - 4AC = \left(10\sqrt{3}\right)^2 - 4(11)(1) = 100 \cdot 3 - 44 = 256 > 0$$

Because $B^2 - 4AC > 0$, the graph of the equation is a hyperbola.

Check Point 5   Identify the graph of $3x^2 - 2\sqrt{3}xy + y^2 + 2x + 2\sqrt{3}y = 0$.

EXERCISE SET 10.4

 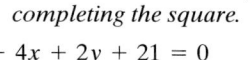

Practice Exercises

In Exercises 1–8, identify each equation without completing the square.

1. $y^2 - 4x + 2y + 21 = 0$
2. $y^2 - 4x - 4y = 0$
3. $4x^2 - 9y^2 - 8x - 36y - 68 = 0$
4. $9x^2 + 25y^2 - 54x - 200y + 256 = 0$
5. $4x^2 + 4y^2 + 12x + 4y + 1 = 0$
6. $9x^2 + 4y^2 - 36x + 8y + 31 = 0$
7. $100x^2 - 7y^2 + 90y - 368 = 0$
8. $y^2 + 8x + 6y + 25 = 0$

In Exercises 9–14, write each equation in terms of a rotated $x'y'$-system using θ, the angle of rotation. Write the equation involving x' and y' in standard form.

9. $xy = -1; \theta = 45°$
10. $xy = -4; \theta = 45°$
11. $x^2 - 4xy + y^2 - 3 = 0; \theta = 45°$
12. $13x^2 - 10xy + 13y^2 - 72 = 0; \theta = 45°$
13. $23x^2 + 26\sqrt{3}xy - 3y^2 - 144 = 0; \theta = 30°$
14. $13x^2 - 6\sqrt{3}\,xy + 7y^2 - 16 = 0; \theta = 60°$

In Exercises 15–26, write the appropriate rotation formulas so that in a rotated system the equation has no $x'y'$-term.

15. $x^2 + xy + y^2 - 10 = 0$
16. $x^2 + 4xy + y^2 - 3 = 0$
17. $3x^2 - 10xy + 3y^2 - 32 = 0$
18. $5x^2 - 8xy + 5y^2 - 9 = 0$
19. $11x^2 + 10\sqrt{3}xy + y^2 - 4 = 0$
20. $7x^2 - 6\sqrt{3}xy + 13y^2 - 16 = 0$
21. $10x^2 + 24xy + 17y^2 - 9 = 0$
22. $32x^2 - 48xy + 18y^2 - 15x - 20y = 0$
23. $x^2 + 4xy - 2y^2 - 1 = 0$
24. $3xy - 4y^2 + 18 = 0$
25. $34x^2 - 24xy + 41y^2 - 25 = 0$
26. $6x^2 - 6xy + 14y^2 - 45 = 0$

In Exercises 27–38:

a. *Rewrite the equation in a rotated $x'y'$-system without an $x'y'$ term. Use the appropriate rotation formulas from Exercises 15–26.*

b. *Express the equation involving x' and y' in the standard form of a conic section.*

c. *Use the rotated system to graph the equation.*

27. $x^2 + xy + y^2 - 10 = 0$
28. $x^2 + 4xy + y^2 - 3 = 0$
29. $3x^2 - 10xy + 3y^2 - 32 = 0$
30. $5x^2 - 8xy + 5y^2 - 9 = 0$

31. $11x^2 + 10\sqrt{3}xy + y^2 - 4 = 0$
32. $7x^2 - 6\sqrt{3}xy + 13y^2 - 16 = 0$
33. $10x^2 + 24xy + 17y^2 - 9 = 0$
34. $32x^2 - 48xy + 18y^2 - 15x - 20y = 0$
35. $x^2 + 4xy - 2y^2 - 1 = 0$
36. $3xy - 4y^2 + 18 = 0$
37. $34x^2 - 24xy + 41y^2 - 25 = 0$
38. $6x^2 - 6xy + 14y^2 - 45 = 0$

In Exercises 39–44, identify each equation without applying a rotation of axes.

39. $5x^2 - 2xy + 5y^2 - 12 = 0$
40. $10x^2 + 24xy + 17y^2 - 9 = 0$
41. $24x^2 + 16\sqrt{3}xy + 8y^2 - x + \sqrt{3}y - 8 = 0$
42. $3x^2 - 2\sqrt{3}xy + y^2 + 2x + 2\sqrt{3}y = 0$
43. $23x^2 + 26\sqrt{3}xy - 3y^2 - 144 = 0$
44. $4xy + 3y^2 + 4x + 6y - 1 = 0$

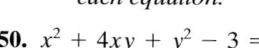

Writing in Mathematics

45. Explain how to identify the graph of
$$Ax^2 + Cy^2 + Dx + Ey + F = 0$$

46. If there is a 60° angle from the positive x-axis to the positive x'-axis, explain how to obtain the rotation formulas for x and y.

47. How do you obtain the angle of rotation so that a general second-degree equation has no $x'y'$-term in a rotated $x'y'$-system?

48. What is the most time-consuming part in using a graphing utility to graph a general second-degree equation with an xy-term?

49. Explain how to identify the graph of
$$Ax^2 + Bxy + Cy^2 + Dx + Ey + F = 0.$$

Technology Exercises

In Exercises 50–56, use a graphing utility to graph each equation.

50. $x^2 + 4xy + y^2 - 3 = 0$
51. $7x^2 + 8xy + y^2 - 1 = 0$
52. $3x^2 + 4xy + 6y^2 - 7 = 0$
53. $3x^2 - 6xy + 3y^2 + 10x - 8y - 2 = 0$
54. $9x^2 + 24xy + 16y^2 + 90x - 130y = 0$

55. $x^2 + 4xy + 4y^2 + 10\sqrt{5}x - 9 = 0$

56. $7x^2 + 6xy + 2.5y^2 - 14x + 4y + 9 = 0$

Critical Thinking Exercises

57. Explain the relationship between the graph of $3x^2 - 2xy + 3y^2 + 2 = 0$ and the sound made by one hand clapping. Begin by following the directions for Exercises 27–38. (You will first need to write rotation formulas that eliminate the $x'y'$-term.)

58. What happens to the equation $x^2 + y^2 = r^2$ in a rotated $x'y'$-system?

In Exercises 59–60, let $Ax^2 + Bxy + Cy^2 + Dx + Ey + F = 0$ be an equation of a conic section in an xy-coordinate system. Let $A'x'^2 + B'x'y' + C'y'^2 + D'x' + E'y' + F' = 0$ be the equation of the conic section in the rotated x'y'-coordinate system. Use the coefficients A', B', and C',

shown in the equation with the voice balloon pointing to B' on page 855, to prove the following relationships.

59. $A' + C' = A + C$

60. $B'^2 - 4A'C' = B^2 - 4AC$

Group Exercise

61. Many public and private organizations and schools provide educational materials and information for the blind and visually impaired. Using your library, resources on the Worldwide Web, or local organizations, investigate how your group or college could make a contribution to enhance the study of mathematics for the blind and visually impaired. In relation to conic sections, group members should discuss how to create graphs in tactile, or touchable, form that show blind students the visual structure of the conics, including asymptotes, intercepts, end behavior, and rotations.

SECTION 10.5 *Parametric Equations*

Objectives

1. Use point plotting to graph plane curves described by parametric equations.

2. Eliminate the parameter.

3. Find parametric equations for functions.

4. Understand the advantages of parametric representations.

What a baseball game! You got to see the great Sammy Sosa of the Chicago Cubs blast a powerful homer. In less than eight seconds, the parabolic path of his home run took the ball a horizontal distance of over 1000 feet. Is there a way to model this path that gives both the ball's location and the time that it is in each of its positions? In this section, we look at ways of describing curves that reveal the where and the when of motion.

Plane Curves and Parametric Equations

You throw a ball from a height of 6 feet, with an initial velocity of 90 feet per second and at an angle of 40° with the horizontal. After t seconds, the path of the ball can be described by

$$x = (90\cos 40°)t \quad \text{and} \quad y = 6 + (90\sin 40°)t - 16t^2.$$

> This is the ball's horizontal distance, in feet.

> This is the ball's vertical height, in feet.

Using these equations, we can calculate the location of the ball at any time t. For example, to determine the location when $t = 1$ second, substitute 1 for t in each equation:

$$x = (90 \cos 40°)(1) \approx 68.9 \text{ feet}$$

$$y = 6 + (90 \sin 40°)(1) - 16(1)^2 \approx 47.9 \text{ feet}.$$

This tells us that after one second, the ball has traveled a horizontal distance of approximately 68.9 feet, and the height of the ball is approximately 47.9 feet. Figure 10.51 displays this information and the results for calculations corresponding to $t = 2$ seconds and $t = 3$ seconds.

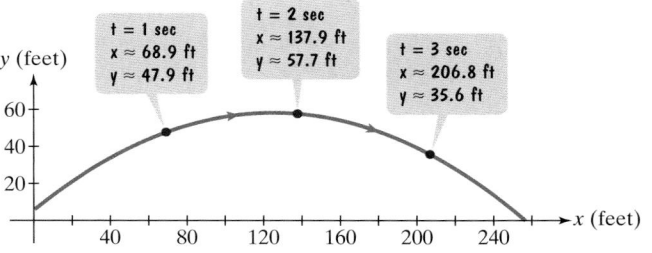

Figure 10.51 The location of a thrown ball after 1, 2, and 3 seconds

The voice balloons in Figure 10.51 tell where the ball is located and when the ball is at a given point (x, y) on its path. The variable t, called a **parameter**, gives the various times for the ball's location. The equations that describe where the ball is located express both x and y as functions of t and are called **parametric equations**.

$$x = (90 \cos 40°)t \qquad y = 6 + (90 \sin 40°)t - 16t^2$$

This is the parametric equation for x.

This is the parametric equation for y.

The collection of points (x, y) in Figure 10.51 is called a **plane curve**.

Plane Curves and Parametric Equations

Suppose that t is a number in an interval I. A **plane curve** is the set of ordered pairs (x, y), where

$$x = f(t), \quad y = g(t) \quad \text{for } t \text{ in interval } I.$$

The variable t is called a **parameter**, and the equations $x = f(t)$ and $y = g(t)$ are called **parametric equations** for the curve.

---

1 Use point plotting to graph plane curves described by parametric equations.

Graphing Plane Curves

Graphing a plane curve represented by parametric equations involves plotting points in the rectangular coordinate system and connecting them with a smooth curve.

Graphing a Plane Curve Described by Parametric Equations

1. Select some values of t on the given interval.
2. For each value of t, use the given parametric equations to compute x and y.
3. Plot the points (x, y) in the order of increasing t and connect them with a smooth curve.

Take a second look at Figure 10.51. Do you notice arrows along the curve? These arrows show the direction, or **orientation**, along the curve as t increases. After graphing a plane curve described by parametric equations, use arrows between the points to show the orientation of the curve corresponding to increasing values of t.

EXAMPLE 1   Graphing a Curve Defined by Parametric Equations

Graph the plane curve defined by the parametric equations

$$x = t^2 - 1, \qquad y = 2t, \qquad -2 \le t \le 2.$$

Solution

Step 1   Select some values of t on the given interval. We will select integral values of t on the interval $-2 \le t \le 2$. Let $t = -2, -1, 0, 1$, and 2.

Step 2   For each value of t, use the given parametric equations to compute x and y. We organize our work in a table. The first column lists the choices for the parameter t. The next two columns show the corresponding values for x and y. The last column lists the ordered pair (x, y).

t	$x = t^2 - 1$	$y = 2t$	(x, y)
-2	$(-2)^2 - 1 = 4 - 1 = 3$	$2(-2) = -4$	$(3, -4)$
-1	$(-1)^2 - 1 = 1 - 1 = 0$	$2(-1) = -2$	$(0, -2)$
0	$0^2 - 1 = -1$	$2(0) = 0$	$(-1, 0)$
1	$1^2 - 1 = 0$	$2(1) = 2$	$(0, 2)$
2	$2^2 - 1 = 4 - 1 = 3$	$2(2) = 4$	$(3, 4)$

Step 3   Plot the points (x, y) in the order of increasing t and connect them with a smooth curve. The plane curve defined by the parametric equations on the given interval is shown in Figure 10.52. The arrows show the direction, or orientation, along the curve as t varies from -2 to 2.

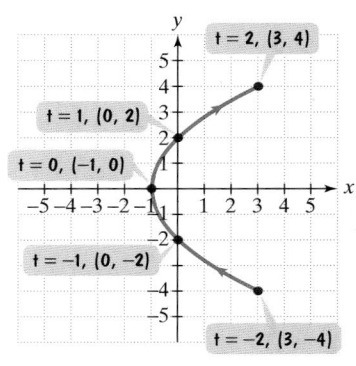

Figure 10.52 The plane curve defined by $x = t^2 - 1$, $y = 2t$, $-2 \le t \le 2$

Check Point 1   Graph the plane curve defined by the parametric equations

$$x = t^2 + 1, \qquad y = 3t, \qquad -2 \le t \le 2.$$

2   Eliminate the parameter.

Eliminating the Parameter

The graph in the technology box on page 866 shows the plane curve for $x = t^2 - 1$, $y = 2t$, $-2 \le t \le 2$. Even if we examine the parametric equations carefully, we may not be able to tell that the corresponding plane curve is a

Technology

A graphing utility can be used to obtain a plane curve represented by parametric equations. Set the mode to parametric and enter the equations. You must enter the minimum and maximum values for t, and an increment setting for t ($t$step). The setting $t$step determines the number of points the graphing utility will plot.

Shown is the plane curve for

$$x = t^2 - 1$$
$$y = 2t$$

in a $[-5, 5, 1]$ by $[-5, 5, 1]$ viewing rectangle with $t$min $= -2$, $t$max $= 2$, $t$step $= .01$.

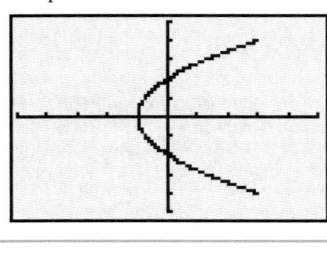

parabola. By **eliminating the parameter**, we can write one equation in x and y that is equivalent to the two parametric equations. The voice balloons illustrate this process.

Begin with the parametric equations.	Solve for t in one of the equations.	Substitute the expression for t in the other parametric equation.
$x = t^2 - 1$ $y = 2t$	Using $y = 2t$, $t = \dfrac{y}{2}.$	Using $t = \dfrac{y}{2}$ and $x = t^2 - 1$, $x = \left(\dfrac{y}{2}\right)^2 - 1.$

The rectangular equation (the equation in x and y), $x = \dfrac{y^2}{4} - 1$, can be written as $y^2 = 4(x + 1)$. This is the standard form of the equation of a parabola with vertex at $(-1, 0)$ and axis of symmetry along the x-axis. Because the parameter t is restricted to the interval $[-2, 2]$, the plane curve in the technology box shows only a part of the parabola.

Our discussion illustrates a second method for graphing a plane curve described by parametric equations. Eliminate the parameter t and graph the resulting rectangular equation in x and y. However, **you may need to change the domain of the rectangular equation to be consistent with the domain for the parametric equation in x.** This situation is illustrated in Example 2.

EXAMPLE 2   Finding and Graphing the Rectangular Equation of a Curve Defined Parametrically

Sketch the plane curve represented by the parametric equations

$$x = \sqrt{t} \quad \text{and} \quad y = \tfrac{1}{2}t + 1$$

by eliminating the parameter.

Solution   We eliminate the parameter t and then graph the resulting rectangular equation.

Begin with the parametric equations.	Solve for t in one of the equations.	Substitute the expression for t in the other parametric equation.
$x = \sqrt{t}$ $y = \tfrac{1}{2}t + 1$	Using $x = \sqrt{t}$ and squaring both sides, $t = x^2.$	Using $t = x^2$ and $y = \tfrac{1}{2}t + 1$, $y = \tfrac{1}{2}x^2 + 1.$

Because t is not limited to a closed interval, you might be tempted to graph the entire U-shaped parabola whose equation is $y = \tfrac{1}{2}x^2 + 1$. However, take a second look at the parametric equation in x:

$$x = \sqrt{t}.$$

This equation is defined only when $t \geq 0$. Thus, x is nonnegative. The plane curve is the parabola given by $y = \tfrac{1}{2}x^2 + 1$ with the domain restricted to $x \geq 0$. The plane curve is shown in Figure 10.53.

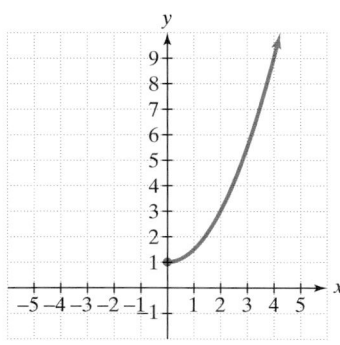

Figure 10.53 The plane curve for $x = \sqrt{t}$ and $y = \tfrac{1}{2}t + 1$, or $y = \tfrac{1}{2}x^2 + 1$, $x \geq 0$

Check Point 2 Sketch the plane curve represented by the parametric equations

$$x = \sqrt{t} \quad \text{and} \quad y = 2t - 1$$

by eliminating the parameter.

Eliminating the parameter is not always a simple matter. In some cases, it may not be possible. When this occurs, you can use point plotting to obtain a plane curve.

Trigonometric identities can be helpful in eliminating the parameter. For example, consider the plane curve defined by the parametric equations

$$x = \sin t, \quad y = \cos t, \quad 0 \le t < 2\pi.$$

We use the trigonometric identity $\sin^2 t + \cos^2 t = 1$ to eliminate the parameter. Square each side of each parametric equation and then add.

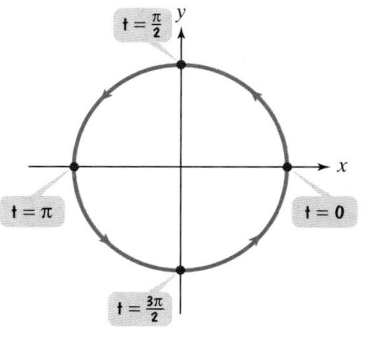

$$x^2 = \sin^2 t$$
$$\underline{y^2 = \cos^2 t}$$
$$x^2 + y^2 = \sin^2 t + \cos^2 t$$

This is the sum of the two equations above the horizontal line.

Using the Pythagorean identity, we write this equation as $x^2 + y^2 = 1$. The plane curve is a circle with center $(0, 0)$ and radius equal to 1. It is shown in Figure 10.54.

Figure 10.54 The plane curve defined by $x = \sin t$, $y = \cos t$, $0 \le t < 2\pi$

EXAMPLE 3 Finding and Graphing the Rectangular Equation of a Curve Defined Parametrically

Sketch the plane curve represented by the parametric equations

$$x = 5\cos t, y = 2\sin t, 0 \le t \le \pi$$

by eliminating the parameter.

Solution   We eliminate the parameter using the identity $\cos^2 t + \sin^2 t = 1$. To apply the identity, divide the parametric equation in x by 5 and the parametric equation in y by 2.

$$\frac{x}{5} = \cos t \quad \text{and} \quad \frac{y}{2} = \sin t$$

Square and add these two equations.

$$\frac{x^2}{25} = \cos^2 t$$
$$\frac{y^2}{4} = \sin^2 t$$
$$\frac{x^2}{25} + \frac{y^2}{4} = \cos^2 t + \sin^2 t$$

This is the sum of the two equations above the horizontal line.

Using the Pythagorean identity, we write this equation as

$$\frac{x^2}{25} + \frac{y^2}{4} = 1.$$

This rectangular equation is the standard form of the equation for an ellipse centered at $(0, 0)$.

$$\frac{x^2}{25} + \frac{y^2}{4} = 1$$

| $a^2 = 25$: Endpoints of major axis are 5 units right and left of center. | $b^2 = 4$: Endpoints of minor axis are 2 units above and below center. |

The ellipse is shown in Figure 10.55(a). However, this is not the plane curve. Because t is restricted to the interval $[0, \pi]$, the plane curve is only a portion of the ellipse. Use the starting and ending values for t and a value of t in the interval $[0, \pi]$ to find which portion to include.

Begin at $t = 0$.

$x = 5 \cos t = 5 \cos 0 = 5 \cdot 1 = 5$

$y = 2 \sin t = 2 \sin 0 = 2 \cdot 0 = 0$

Increase to $t = \frac{\pi}{2}$.

$x = 5 \cos t = 5 \cos \dfrac{\pi}{2} = 5 \cdot 0 = 0$

$y = 2 \sin t = 2 \sin \dfrac{\pi}{2} = 2 \cdot 1 = 2$

End at $t = \pi$.

$x = 5 \cos t = 5 \cos \pi = 5(-1) = -5$

$y = 2 \sin t = 2 \sin \pi = 2(0) = 0$

Points on the plane curve include $(5, 0)$, which is the starting point, $(0, 2)$, and $(-5, 0)$, which is the ending point. The plane curve is the top half of the ellipse, shown in Figure 10.55(b).

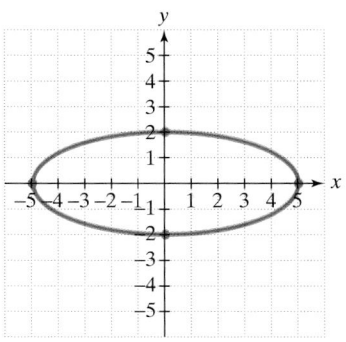

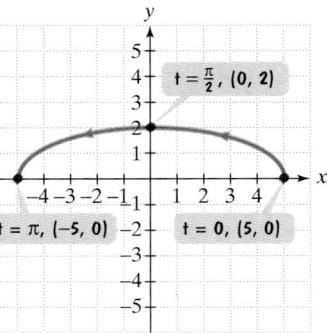

Figure 10.55(a) The graph of
$$\frac{x^2}{25} + \frac{y^2}{4} = 1$$

Figure 10.55(b) The plane curve for $x = 5 \cos t$, $y = 2 \sin t$, $0 \leq t \leq \pi$

Check Point 3

Sketch the plane curve represented by the parametric equations

$$x = 6 \cos t, \; y = 4 \sin t, \; \pi \leq t \leq 2\pi.$$

 Find parametric equations for functions.

Finding Parametric Equations

Infinitely many pairs of parametric equations can represent the same plane curve. If the plane curve is defined by the function $y = f(x)$, here is a procedure for finding one set of parametric equations.

Parametric Equations for the Function $y = f(x)$

A set of parametric equations for the plane curve defined by $y = f(x)$ is

$$x = t \quad \text{and} \quad y = f(t)$$

in which t is in the domain of f.

EXAMPLE 4   Finding Parametric Equations

Find a set of parametric equations for the parabola whose equation is $y = 9 - x^2$.

Solution   Let $x = t$. Parametric equations for $y = f(x)$ are $x = t$ and $y = f(t)$. Thus, parametric equations for $y = 9 - x^2$ are

$$x = t \quad \text{and} \quad y = 9 - t^2.$$

Check Point 4   Find a set of parametric equations for the parabola whose equation is $y = x^2 - 25$.

You can write other sets of parametric equations for $y = 9 - x^2$ by starting with a different parametric equation for x. Here are three sets of parametric equations for

$$y = 9 - x^2.$$

- If $x = t^3$, $y = 9 - \left(t^3\right)^2 = 9 - t^6$.

   Parametric equations are $x = t^3$ and $y = 9 - t^6$.

- If $x = t + 1$, $y = 9 - (t + 1)^2 = 9 - (t^2 + 2t + 1) = 8 - t^2 - 2t$.
   Parametric equations are $x = t + 1$ and $y = 8 - t^2 - 2t$.

- If $x = \dfrac{t}{2}$, $y = 9 - \left(\dfrac{t}{2}\right)^2 = 9 - \dfrac{t^2}{4}$.

   Parametric equations are $x = \dfrac{t}{2}$ and $y = 9 - \dfrac{t^2}{4}$.

Can you start with any choice for the parametric equation for x? The answer is no. **The substitution for x must be a function that allows x to take on all the values in the domain of the given rectangular equation**. For example, the domain of the function $y = 9 - x^2$ is the set of all real numbers. If you incorrectly let $x = t^2$, these values of x exclude negative numbers that are included in $y = 9 - x^2$. The parametric equations

$$x = t^2 \quad \text{and} \quad y = 9 - \left(t^2\right)^2 = 9 - t^4$$

do not represent $y = 9 - x^2$ because only points for which $x \geq 0$ are obtained.

4 Understand the advantages of parametric representations.

Technology

The ellipse shown was obtained using the parametric mode and the radian mode of a graphing utility.

$$x(t) = 2 + 3 \cos t$$
$$y(t) = 3 + 2 \sin t$$

We used a $[-2, 6, 1]$ by $[-1, 6, 1]$ viewing rectangle with $t\text{min} = 0, t\text{max} = 6.2$ and $t\text{step} = 0.1$.

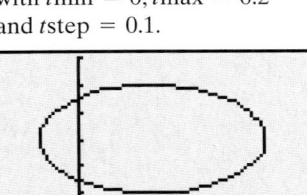

Advantages of Parametric Equations over Rectangular Equations

We opened this section with parametric equations that described the horizontal distance and the vertical height of your thrown baseball after t seconds. Parametric equations are frequently used to represent the path of a moving object. If t represents time, parametric equations give the location of a moving object and tell when the object is located at each of its positions. Rectangular equations tell where the moving object is located but do not reveal when the object is in a particular position.

When using technology to obtain graphs, parametric equations that represent nonfunctions are often easier to use than their corresponding rectangular equations. It is far easier to enter the equation of an ellipse given by the parametric equations

$$x = 2 + 3 \cos t \quad \text{and} \quad y = 3 + 2 \sin t$$

than to use the rectangular equivalent

$$\frac{(x - 2)^2}{9} + \frac{(y - 3)^2}{4} = 1.$$

The rectangular equation must first be solved for y and then entered as two separate equations in a graphing utility before the ellipse is revealed.

A curve that is used in physics for much of the theory of light is called a **cycloid**. The path of a fixed point on the circumference of a circle as it rolls along a line is a cycloid. A point on the rim of a bicycle wheel traces out a cycloid curve, shown in Figure 10.56. If the radius of the circle is a, the parametric equations of the cycloid are

$$x = a(t - \sin t) \text{ and } y = a(1 - \cos t).$$

It is an extremely complicated task to represent the cycloid in rectangular form.

Cycloids are used to solve problems that involve the "shortest time." For example, Figure 10.57 on the right shows a bead sliding down a wire. The shape of the wire a bead could slide down so that the distance between two points is traveled in the shortest time is an inverted cycloid.

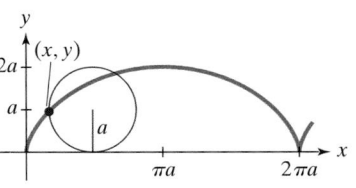

Figure 10.56 The curve traced by a fixed point on the circumference of a circle rolling along a straight line is a cycloid.

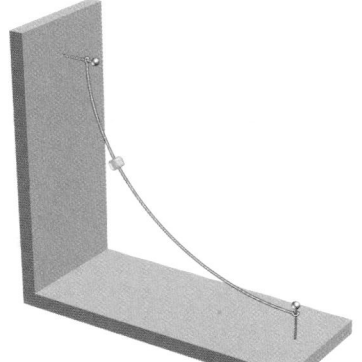

Figure 10.57

Rolling

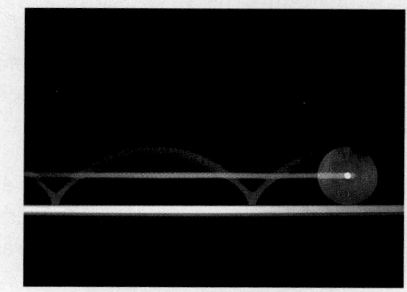

Linear functions and cycloids are used to describe rolling motion. The light at the rolling circle's center shows that it moves linearly. By contrast, the light at the circle's edge has rotational motion and traces out a cycloid.

EXERCISE SET 10.5

Practice Exercises

In Exercises 1–8, parametric equations and a value for the parameter t are given. Find the coordinates of the point on the plane curve described by the parametric equations corresponding to the given value of t.

1. $x = 3 - 5t$, $y = 4 + 2t$; $t = 1$

2. $x = 7 - 4t$, $y = 5 + 6t$; $t = 1$

3. $x = t^2 + 1$, $y = 5 - t^3$; $t = 2$

4. $x = t^2 + 3$, $y = 6 - t^3$; $t = 2$

5. $x = 4 + 2\cos t$, $y = 3 + 5\sin t$; $t = \dfrac{\pi}{2}$

6. $x = 2 + 3\cos t$, $y = 4 + 2\sin t$; $t = \pi$

7. $x = (60\cos 30°)t$, $y = 5 + (60\sin 30°)t - 16t^2$; $t = 2$

8. $x = (80\cos 45°)t$, $y = 6 + (80\sin 45°)t - 16t^2$; $t = 2$

In Exercises 9–20, use point plotting to graph the plane curve described by the given parametric equations. Use arrows to show the orientation of the curve corresponding to increasing values of t.

9. $x = t + 2$, $y = t^2$; $-2 \le t \le 2$

10. $x = t - 1$, $y = t^2$; $-2 \le t \le 2$

11. $x = t - 2$, $y = 2t + 1$; $-2 \le t \le 3$

12. $x = t - 3$, $y = 2t + 2$; $-2 \le t \le 3$

13. $x = t + 1$, $y = \sqrt{t}$; $t \ge 0$

14. $x = \sqrt{t}$, $y = t - 1$; $t \ge 0$

15. $x = \cos t$, $y = \sin t$; $0 \le t < 2\pi$

16. $x = -\sin t$, $y = -\cos t$; $0 \le t < 2\pi$

17. $x = t^2$, $y = t^3$; $-\infty < t < \infty$

18. $x = t^2 + 1$, $y = t^3 - 1$; $-\infty < t < \infty$

19. $x = 2t$, $y = |t - 1|$; $-\infty < t < \infty$

20. $x = |t + 1|$, $y = t - 2$; $-\infty < t < \infty$

In Exercises 21–40, eliminate the parameter t. Then use the rectangular equation to sketch the plane curve represented by the given parametric equations. Use arrows to show the orientation of the curve corresponding to increasing values of t. (If an interval for t is not specified, assume that $-\infty < t < \infty$.)

21. $x = t$, $y = 2t$

22. $x = t$, $y = -2t$

23. $x = 2t - 4$, $y = 4t^2$

24. $x = t - 2$, $y = t^2$

25. $x = \sqrt{t}$, $y = t - 1$

26. $x = \sqrt{t}$, $y = t + 1$

27. $x = 2\sin t$, $y = 2\cos t$; $0 \le t < 2\pi$

28. $x = 3\sin t$, $y = 3\cos t$; $0 \le t < 2\pi$

29. $x = 1 + 3\cos t$, $y = 2 + 3\sin t$; $0 \le t < 2\pi$

30. $x = -1 + 2\cos t$, $y = 1 + 2\sin t$; $0 \le t < 2\pi$

31. $x = 2\cos t$, $y = 3\sin t$; $0 \le t < 2\pi$

32. $x = 3\cos t$, $y = 5\sin t$; $0 \le t < 2\pi$

33. $x = 1 + 3\cos t$, $y = -1 + 2\sin t$; $0 \le t \le \pi$

34. $x = 2 + 4\cos t$, $y = -1 + 3\sin t$; $0 \le t \le \pi$

35. $x = \sec t$, $y = \tan t$

36. $x = 5\sec t$, $y = 3\tan t$

37. $x = t^2 + 2$, $y = t^2 - 2$

38. $x = \sqrt{t} + 2$, $y = \sqrt{t} - 2$

39. $x = 2^t$, $y = 2^{-t}$; $t \ge 0$

40. $x = e^t$, $y = e^{-t}$; $t \ge 0$

In Exercises 41–43, eliminate the parameter. Write the resulting equation in standard form.

41. A circle: $x = h + r\cos t$, $y = k + r\sin t$

42. An ellipse: $x = h + a\cos t$, $y = k + b\sin t$

43. A hyperbola: $x = h + a\sec t$, $y = k + b\tan t$

44. The parametric equations of the line through (x_1, y_1) and (x_2, y_2) are

$$x = x_1 + t(x_2 - x_1) \quad \text{and} \quad y = y_1 + t(y_2 - y_1).$$

Eliminate the parameter and write the resulting equation in point-slope form.

In Exercises 45–52, use your answers from Exercises 41–44 and the parametric equations given in Exercises 41–44 to find a set of parametric equations for the conic section or the line.

45. Circle: Center: $(3, 5)$; Radius: 6

46. Circle: Center: $(4, 6)$; Radius: 9

47. Ellipse: Center: $(-2, 3)$; Vertices: 5 units to the right and left of the center; Endpoints of Minor Axis: 2 units above and below the center

48. Ellipse: Center: $(4, -1)$; Vertices: 5 units above and below the center; Endpoints of Minor Axis: 3 units to the right and left of the center

49. Hyperbola: Vertices: $(4, 0)$ and $(-4, 0)$; Foci: $(6, 0)$ and $(-6, 0)$

50. Hyperbola: Vertices: $(0, 4)$ and $(0, -4)$; Foci: $(0, 5)$ and $(0, -5)$

51. Line: Passes through $(-2, 4)$ and $(1, 7)$

52. Line: Passes through $(3, -1)$ and $(9, 12)$

In Exercises 53–56, find two different sets of parametric equations for each rectangular equation.

53. $y = 4x - 3$ **54.** $y = 2x - 5$

55. $y = x^2 + 4$ **56.** $y = x^2 - 3$

In Exercises 57–58, the parametric equations of four plane curves are given. Graph each plane curve and determine how they differ from each other.

57. a. $x = t$ and $y = t^2 - 4$
   b. $x = t^2$ and $y = t^4 - 4$
   c. $x = \cos t$ and $y = \cos^2 t - 4$
   d. $x = e^t$ and $y = e^{2t} - 4$

58. a. $x = t, y = \sqrt{4 - t^2}; -2 \leq t \leq 2$
   b. $x = \sqrt{4 - t^2}, y = t; -2 \leq t \leq 2$
   c. $x = 2 \sin t, y = 2 \cos t; 0 \leq t < 2\pi$
   d. $x = 2 \cos t, y = 2 \sin t; 0 \leq t < 2\pi$

Application Exercises

The path of a projectile that is launched h feet above the ground with an initial velocity of v_0 feet per second and at an angle θ with the horizontal is given by the parametric equations

$$x = (v_0 \cos \theta)t \quad \text{and} \quad y = h + (v_0 \sin \theta)t - 16t^2,$$

where t is the time, in seconds, since the projectile was launched. The parametric equation for x gives the projectile's horizontal distance, in feet. The parametric equation for y

gives the projectile's height, in feet. Use these parametric equations to solve Exercises 59–60.

59. The figure shows the path for a baseball hit by Sammy Sosa. The ball was hit with an initial velocity of 180 feet per second at an angle of 40° to the horizontal. The ball was hit at a height 3 feet off the ground.

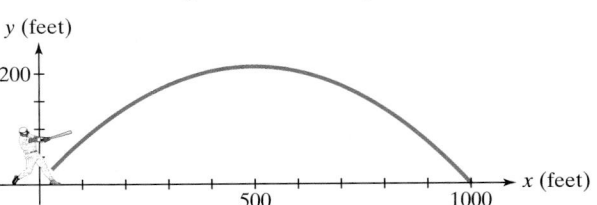

a. Find the parametric equations that describe the position of the ball as a function of time.

b. Describe the ball's position after 1, 2, and 3 seconds. Round to the nearest tenth of a foot. Locate your solutions on the plane curve.

c. How long, to the nearest tenth of a second, is the ball in flight? What is the total horizontal distance that it travels before it lands? Is your answer consistent with the figure shown?

d. You meet Sammy Sosa and he asks you to tell him something of interest about the path of a baseball that he hit. Use the graph to respond to his request. Then verify your observation algebraically.

60. The figure shows the path for a baseball that was hit with an initial velocity of 150 feet per second at an angle of 35° to the horizontal. The ball was hit at a height of 3 feet off the ground.

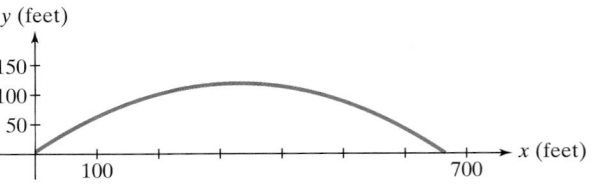

a. Find the parametric equations that describe the position of the ball as a function of time.

b. Describe the ball's position after 1, 2, and 3 seconds. Round to the nearest tenth of a foot. Locate your solutions on the plane curve.

c. How long is the ball in flight? (Round to the nearest tenth of a second.) What is the total horizontal distance that it travels, to the nearest tenth of a foot, before it lands? Is your answer consistent with the figure shown?

d. Use the graph to describe something about the path of the baseball that might be of interest to the player who hit the ball. Then verify your observation algebraically.

Writing in Mathematics

61. What are plane curves and parametric equations?

62. How is point plotting used to graph a plane curve described by parametric equations? Give an example with your description.

63. What is the significance of arrows along a plane curve?

64. What does it mean to eliminate the parameter? What useful information can be obtained by doing this?

65. Explain how the rectangular equation $y = 5x$ can have infinitely many sets of parametric equations.

66. Discuss how the parametric equations for the path of a projectile (see Exercises 59–60) and the ability to obtain plane curves with a graphing utility can be used by a baseball coach to analyze performances of team players.

Technology Exercises

67. Use a graphing utility in a parametric mode to verify any five of your hand-drawn graphs in Exercises 9–40.

In Exercises 68–72, use a graphing utility to obtain the plane curve represented by the given parametric equations.

68. Cycloid: $x = 3(t - \sin t)$,
$y = 3(1 - \cos t)$; $[0, 60, 5] \times [0, 8, 1], 0 \le t < 6\pi$

69. Cycloid: $x = 2(t - \sin t)$,
$y = 2(1 - \cos t)$; $[0, 60, 5] \times [0, 8, 1], 0 \le t < 6\pi$

70. Witch of Agnesi: $x = 2 \cot t, y = 2 \sin^2 t$;
$[-6, 6, 1] \times [-4, 4, 1], 0 \le t < 2\pi$

71. Hypocycloid: $x = 4 \cos^3 t, y = 4 \sin^3 t$;
$[-5, 5, 1] \times [-5, 5, 1], 0 \le t < 2\pi$

72. Lissajous Curve: $x = 2 \cos t, y = \sin 2t$;
$[-3, 3, 1] \times [-2, 2, 1], 0 \le t < 2\pi$

Use the equations for the path of a projectile given prior to Exercises 59–60 to solve Exercises 73–75.

In Exercises 73–74, use a graphing utility to obtain the path of a projectile launched from the ground ($h = 0$) at the specified values of θ and v_0. In each exercise, use the graph to determine the maximum height and the time at which the projectile reaches its maximum height. Also use the graph to determine the range of the projectile and the time it hits the ground. Round all answers to the nearest tenth.

73. $\theta = 55°, v_0 = 200$ feet per second

74. $\theta = 35°, v_0 = 300$ feet per second

75. A baseball player throws the ball with an initial velocity of 140 feet per second at an angle of 22° to the horizontal. The ball leaves the player's hand at a height of 5 feet.
   a. Write the parametric equations that describe the ball's position as a function of time.
   b. Use a graphing utility to obtain the path of the baseball.
   c. Find the ball's maximum height and the time at which it reaches this height. Round all answers to the nearest tenth.
   d. How long is the ball in the air?
   e. How far does the ball travel?

Critical Thinking Exercises

76. Eliminate the parameter: $x = \cos^3 t$ and $y = \sin^3 t$.

77. The plane curve described by the parametric equations $x = 3 \cos t$ and $y = 3 \sin t, 0 \le t < 2\pi$, has a counterclockwise orientation. Alter one or both parametric equations so that you obtain the same plane curve with the opposite orientation.

78. The figure shows a circle of radius a rolling along a horizontal line. Point P traces out a cycloid. Angle t, in radians, is the angle through which the circle has rolled. C is the center of the circle.

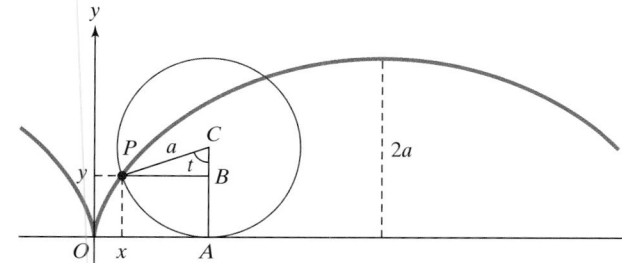

Use the suggestions in parts (a) and (b) to prove that the parametric equations of the cycloid are $x = a(t - \sin t)$ and $y = a(1 - \cos t)$.
   a. Derive the parametric equation for x using the figure and
$$x = 0A - xA.$$
   b. Derive the parametric equation for y using the figure and
$$y = AC - BC.$$

SECTION 10.6 *Conic Sections in Polar Coordinates*

Objectives

1. Define conics in terms of a focus and a directrix.
2. Graph the polar equations of conics.

John Glenn made the first U.S.-manned flight around the Earth on *Friendship 7*.

On the morning of February 20, 1962, millions of Americans collectively held their breath as the world's newest pioneer swept across the threshold of one of our last frontiers. Roughly one hundred miles above Earth, astronaut John Glenn sat comfortably in the weightless environment of a $9\frac{1}{2}$-by 6-foot space capsule that offered the leg room of a Volkswagen "Beetle" and the aesthetics of a garbage can. Glenn became the first American to orbit the Earth in a three-orbit mission that lasted slightly under 5 hours.

In this section, you will see how John Glenn's historic orbit can be described using conic sections in polar coordinates. To obtain this model, we begin with a definition that permits a unified approach to the conic sections.

1 Define conics in terms of a focus and a directrix.

The Focus-Directrix Definitions of the Conic Sections

The definition of a parabola is given in terms of a fixed point, the focus, and a fixed line, the directrix. By contrast, the definitions of an ellipse and a hyperbola are given in terms of two fixed points, the foci. It is possible to define each of these conic sections in terms of a point and a line. Figure 10.58 shows a conic section

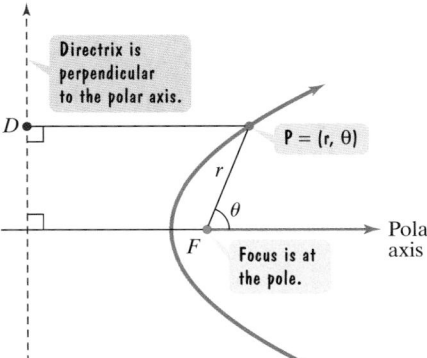

Figure 10.58 A conic in the polar coordinate system

in the polar coordinate system. The fixed point, the focus, is at the pole. The fixed line, the directrix, is perpendicular to the polar axis.

Focus-Directrix Definitions of the Conic Sections

Let F be a fixed point, the focus, and let D be a fixed line, the directrix, in a plane (Figure 10.58). A conic section, or **conic**, is the set of all points P in the plane such that

$$\frac{PF}{PD} = e$$

where e is a fixed positive number, called the **eccentricity**.

    If $e = 1$, the conic is a parabola.
    If $e < 1$, the conic is an ellipse.
    If $e > 1$, the conic is a hyperbola.

    Figure 10.59 illustrates the eccentricity for each type of conic. Notice that if $e = 1$, the definition of the parabola is the same as focus-directrix definition with which you are familiar.

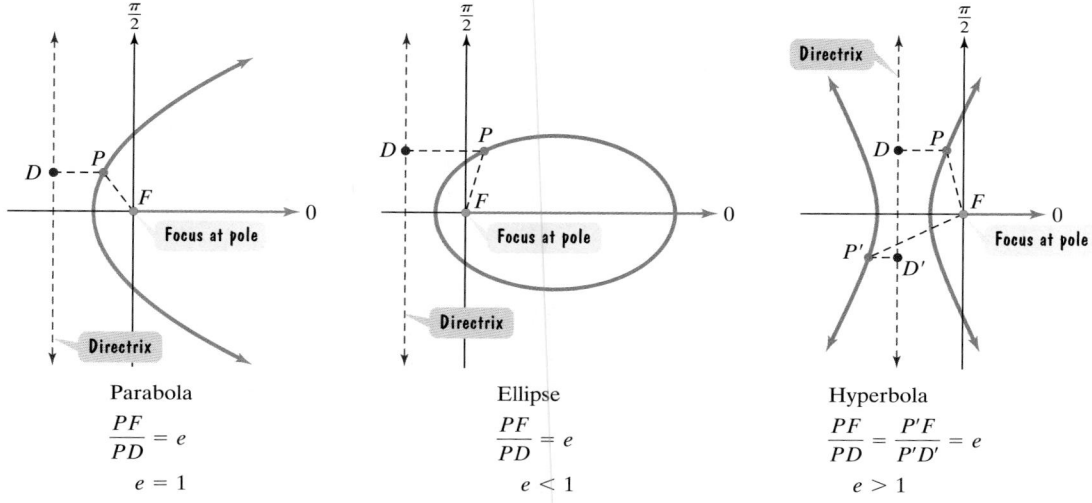

Figure 10.59 The eccentricity for each conic

2   Graph the polar equations of conics.

Polar Equations of Conics

By locating a focus at the pole, all conics can be represented by similar equations in the polar coordinate system. In each of these equations,

- (r, θ) is a point on the graph of the conic.
- e is the eccentricity. (Remember that $e > 0$.)
- p is the distance between the focus (located at the pole) and the directrix.

Standard Forms of the Polar Equations of Conics

Let the pole be a focus of a conic section of eccentricity e with the directrix $|p|$ units from the focus. The equation of the conic is given by one of the four equations listed.

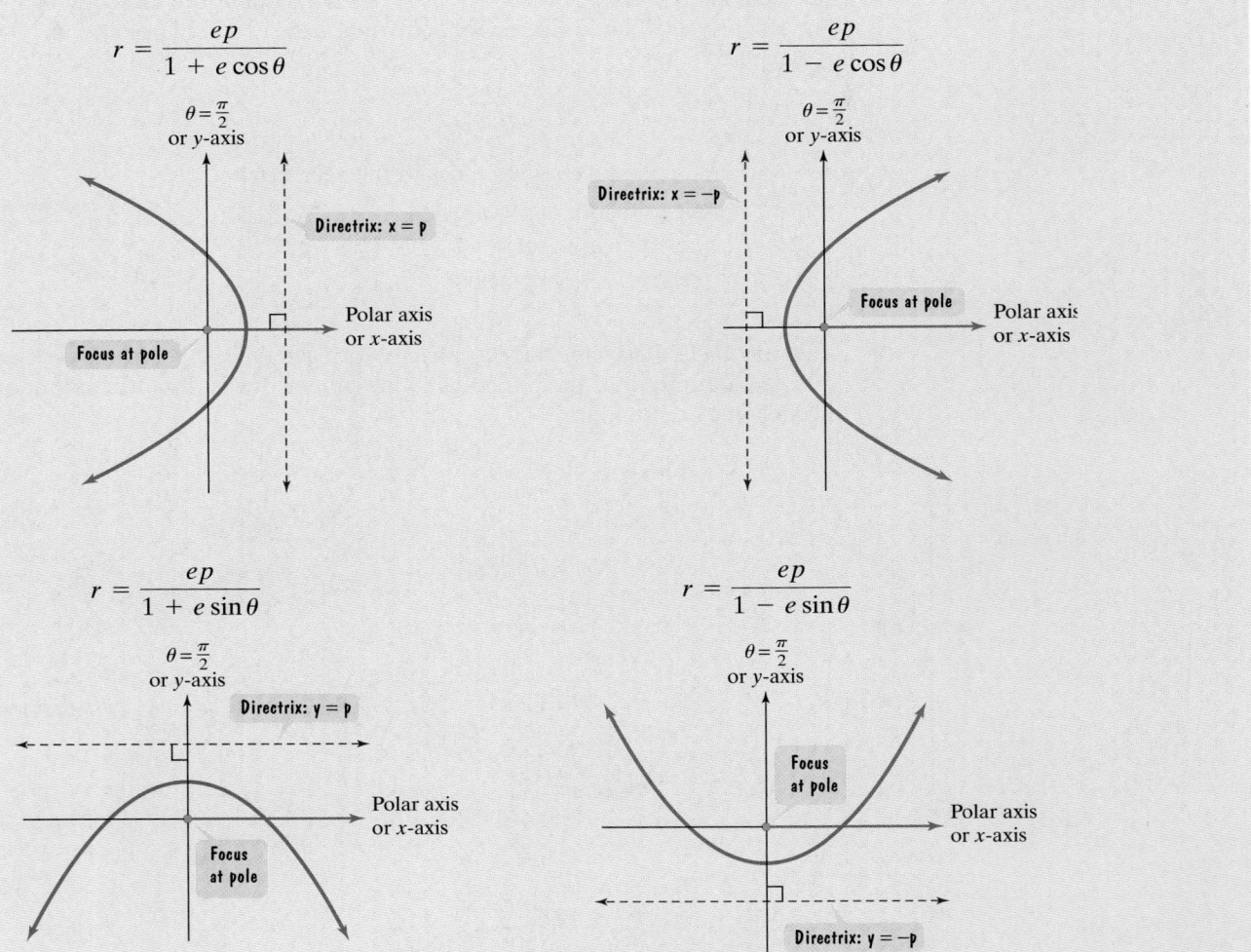

$$r = \frac{ep}{1 + e \cos \theta}$$

$\theta = \frac{\pi}{2}$
or y-axis

Directrix: $x = p$

Polar axis
or x-axis

Focus at pole

$$r = \frac{ep}{1 - e \cos \theta}$$

$\theta = \frac{\pi}{2}$
or y-axis

Directrix: $x = -p$

Focus at pole

Polar axis
or x-axis

$$r = \frac{ep}{1 + e \sin \theta}$$

$\theta = \frac{\pi}{2}$
or y-axis

Directrix: $y = p$

Polar axis
or x-axis

Focus
at pole

$$r = \frac{ep}{1 - e \sin \theta}$$

$\theta = \frac{\pi}{2}$
or y-axis

Focus
at pole

Polar axis
or x-axis

Directrix: $y = -p$

The graphs in the box illustrate two kinds of symmetry—symmetry with respect to the polar axis and symmetry with respect to the y-axis. If the equation contains $\cos \theta$, the polar axis is an axis of symmetry. If the equation contains $\sin \theta$, the line $\theta = \frac{\pi}{2}$, or the y-axis, is an axis of symmetry.

We will derive the equation displayed in the box on the upper right. The other equations are obtained in a similar manner. In Figure 10.60 on page 877, let $P = (r, \theta)$ be any point on a conic section.

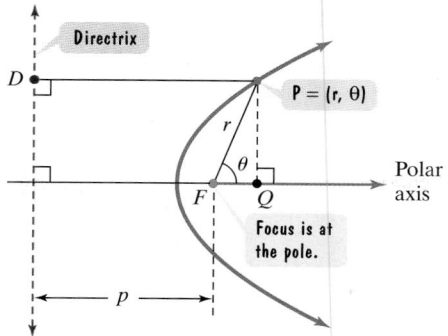

Figure 10.60

$$\frac{PF}{PD} = e$$ By definition, the ratio of the distance from P to the focus to the distance from P to the directrix equals the positive constant e.

$$\frac{r}{PD} = e$$ Figure 10.60 shows that the distance from P to the focus, located at the pole, is r: $PF = r$.

$$\frac{r}{p + FQ} = e$$ Figure 10.60 shows that the distance from P to the directrix is $p + FQ$: $PD = p + FQ$.

$$\frac{r}{p + r\cos\theta} = e$$ Using the triangle in the figure, $\cos\theta = \dfrac{FQ}{r}$ and $FQ = r\cos\theta$.

By solving this equation for r, we will obtain the desired equation. Clear fractions by multiplying both sides by the least common denominator.

$$r = e(p + r\cos\theta)$$ Multiply both sides by $p + r\cos\theta$.

$$r = ep + er\cos\theta$$ Apply the distributive property.

$$r - er\cos\theta = ep$$ Subtract $er\cos\theta$ from both sides to collect terms involving r on the same side.

$$r(1 - e\cos\theta) = ep$$ Factor out r from the two terms on the left.

$$r = \frac{ep}{1 - e\cos\theta}$$ Divide both sides by $1 - e\cos\theta$ and solve for r.

In summary, the standard forms of the polar equations of conics are

$$r = \frac{ep}{1 \pm e\cos\theta} \quad \text{and} \quad r = \frac{ep}{1 \pm e\sin\theta}.$$

In all forms, the constant term in the denominator is 1.

Graphing the Polar Equation of a Conic

1. If necessary, write the equation in one of the standard forms.
2. Use the standard form to determine values for e and p. Use the value of e to identify the conic.
3. Use the appropriate figure for the standard form of the equation shown in the box on page 876 to help guide the graphing process.

EXAMPLE 1   Graphing the Polar Equation of a Conic

Graph the polar equation:

$$r = \frac{4}{2 + \cos\theta}.$$

Solution

Step 1   Write the equation in one of the standard forms. The equation is not in standard form because the constant term in the denominator is not 1.

$$r = \frac{4}{2 + \cos\theta}$$

> To obtain 1 in this position, divide the numerator and denominator by 2.

The equation in standard form is

$$r = \frac{2}{1 + \frac{1}{2}\cos\theta}.$$

$ep = 4$

$e = \frac{1}{2}$

> This equation is in the form $r = \dfrac{ep}{1 + e\cos\theta}$.

Step 2   Use the standard form to find e and p, and identify the conic. The voice balloons show that

$$e = \tfrac{1}{2} \quad \text{and} \quad ep = \tfrac{1}{2}p = 2.$$

Thus, $e = \frac{1}{2}$ and $p = 4$. Because $e = \frac{1}{2} < 1$, the conic is an ellipse.

Step 3   Use the figure for the equation's standard form to guide the graphing process. The figure for the conic's standard form is shown in Figure 10.61(a). We have symmetry with respect to the polar axis. One focus is at the pole and a directrix is $x = 4$, located four units to the right of the pole.

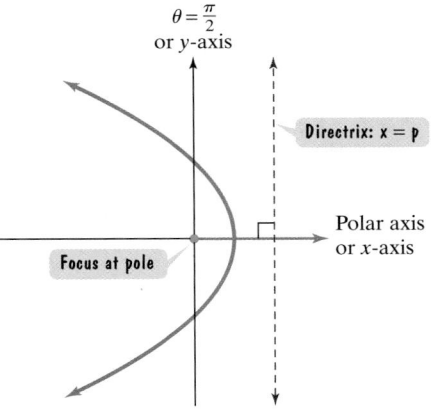

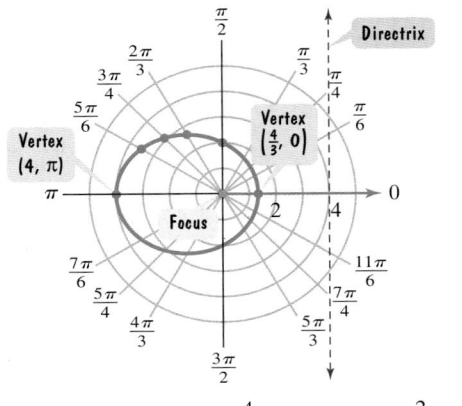

(a) Using $r = \dfrac{ep}{1 + e\cos\theta}$ to graph $r = \dfrac{2}{1 + \frac{1}{2}\cos\theta}$

(b) The graph of $r = \dfrac{4}{2 + \cos\theta}$, or $r = \dfrac{2}{1 + \frac{1}{2}\cos\theta}$

Figure 10.61

Figure 10.61(a) indicates that the major axis is on the polar axis. Thus, we find the vertices by selecting 0 and π for θ. The corresponding values for r are $\frac{4}{3}$ and 4 respectively. Figure 10.61(b) shows the vertices, $\left(\frac{4}{3}, 0\right)$ and $(4, \pi)$.

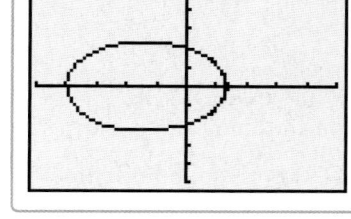
You can sketch the upper half of the ellipse by plotting some points from $\theta = 0$ to $\theta = \pi$.

$$r = \frac{4}{2 + \cos\theta}$$

θ	$\frac{\pi}{2}$	$\frac{2\pi}{3}$	$\frac{3\pi}{4}$	$\frac{5\pi}{6}$
r	2	2.7	3.1	3.5

Using symmetry with respect to the polar axis, you can sketch the lower half. The graph of the given equation is shown in Figure 10.61(b).

Check Point 1 Use the three steps shown in the box on page 877 to graph the polar equation:

$$r = \frac{4}{2 - \cos\theta}.$$

EXAMPLE 2 Graphing the Polar Equation of a Conic

Graph the polar equation:

$$r = \frac{12}{3 + 3\sin\theta}.$$

Solution

Step 1 Write the equation in one of the standard forms. The equation is not in standard form because the constant term in the denominator is not 1. Divide the numerator and denominator by 3 to write the standard form.

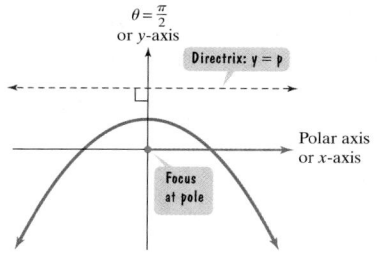

(a) Using $r = \dfrac{ep}{1 + e\sin\theta}$ to graph

$$r = \frac{4}{1 + \sin\theta}$$

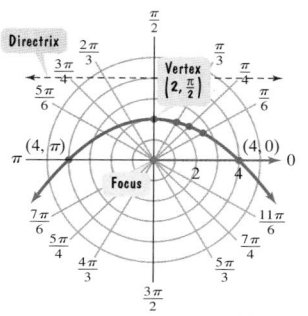

(b) The graph of $r = \dfrac{12}{3 + 3\sin\theta}$ or

$$r = \frac{4}{1 + \sin\theta}$$

Figure 10.62

$$r = \frac{4}{1 + 1\sin\theta} \quad ep = 4$$

This equation is in the form $r = \dfrac{ep}{1 + e\sin\theta}$.

$$e = 1$$

Step 2 Use the standard form to find e and p, and identify the conic. The voice balloons show that

$$e = 1 \quad \text{and} \quad ep = 1p = 4.$$

Thus, $e = 1$ and $p = 4$. Because $e = 1$, the conic is a parabola.

Step 3 Use the figure for the equation's standard form to guide the graphing process. Figure 10.62(a) indicates that we have symmetry with respect to $\theta = \dfrac{\pi}{2}$. The focus is at the pole and the directrix is $y = 4$, located four units above the pole.

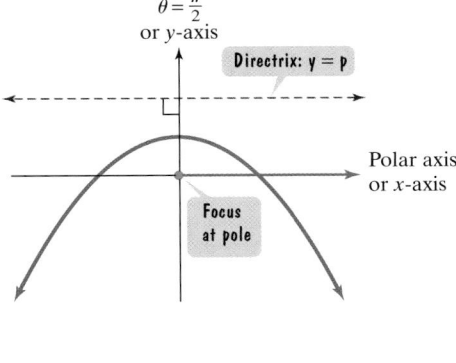

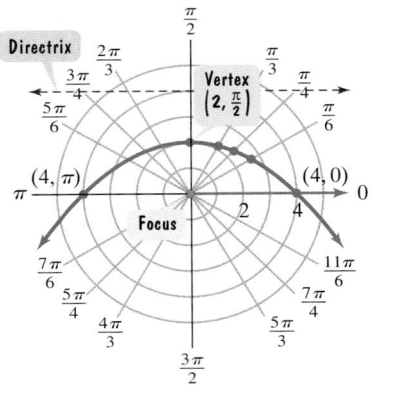

(a) Using $r = \dfrac{ep}{1 + e \sin\theta}$ to graph $r = \dfrac{4}{1 + \sin\theta}$

(b) The graph of $r = \dfrac{12}{3 + 3 \sin\theta}$ or $r = \dfrac{4}{1 + \sin\theta}$

Figure 10.62, repeated

Technology

The graph of

$$r = \frac{12}{3 + 3\sin\theta}$$

was obtained using

$$[-5, 5, 1] \times [-5, 5, 1]$$

and

$$\theta\min = 0, \quad \theta\max = 2\pi,$$

$$\theta\text{step} = \frac{\pi}{48}.$$

Figure 10.62(a) indicates that the vertex is on the line of $\theta = \dfrac{\pi}{2}$, or the y-axis. Thus, we find the vertex by selecting $\dfrac{\pi}{2}$ for θ. The corresponding value for r is 2. Figure 10.62(b) shows the vertex, $\left(2, \dfrac{\pi}{2}\right)$.

To find where the parabola crosses the polar axis, select $\theta = 0$ and $\theta = \pi$. The corresponding values for r are 4 and 4, respectively. Figure 10.62(b) shows the points $(4, 0)$ and $(4, \pi)$ on the polar axis.

You can sketch the right half of the parabola by plotting some points from $\theta = 0$ to $\theta = \dfrac{\pi}{2}$.

$$r = \frac{12}{3 + 3\sin\theta}$$

θ	$\dfrac{\pi}{6}$	$\dfrac{\pi}{4}$	$\dfrac{\pi}{3}$
r	2.7	2.3	2.1

Using symmetry with respect to $\theta = \dfrac{\pi}{2}$, you can sketch the left half. The graph of the given equation is shown in Figure 10.62(b).

Check Point 2 Use the three steps shown in the box on page 877 to graph the polar equation:

$$r = \frac{8}{4 + 4\sin\theta}.$$

EXAMPLE 3   Graphing the Polar Equation of a Conic

Graph the polar equation:

$$r = \frac{9}{3 - 6\cos\theta}.$$

Solution

Step 1   Write the equation in one of the standard forms. We can obtain a constant term of 1 in the denominator by dividing each term by 3.

$$ep = 3$$

$$r = \frac{3}{1 - 2\cos\theta}$$

This equation is in the form $r = \frac{ep}{1 - e\cos\theta}$.

$$e = 2$$

Step 2   Use the standard form to find e and p, and identify the conic. The voice balloons show that

$$e = 2 \quad \text{and} \quad ep = 2p = 3.$$

Thus, $e = 2$ and $p = \frac{3}{2}$. Because $e = 2 > 1$, the conic is a hyperbola.

Step 3   Use the figure for the equation's standard form to guide the graphing process. Figure 10.63(a) indicates that we have symmetry with respect to the polar axis. One focus is at the pole and a directrix is $x = -\frac{3}{2}$, located 1.5 units to the left of the pole.

Figure 10.63(a) indicates that the transverse axis is horizontal and the vertices lie on the polar axis. Thus, we find the vertices by selecting 0 and π for θ. Figure 10.63(b) shows the vertices, $(-3, 0)$ and $(1, \pi)$.

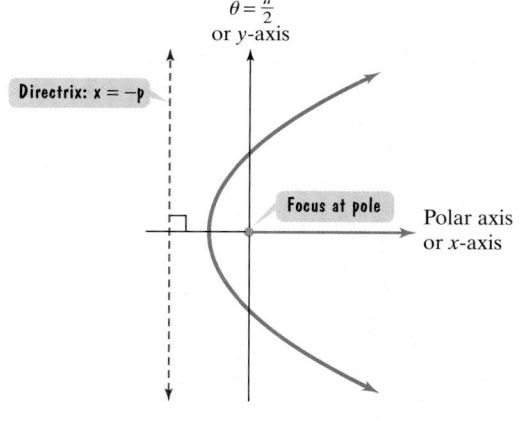

(a) Using $r = \dfrac{ep}{1 - e\cos\theta}$ to graph $r = \dfrac{3}{1 - 2\cos\theta}$

(b) The graph of $r = \dfrac{9}{3 - 6\cos\theta}$ or $r = \dfrac{3}{1 - 2\cos\theta}$

Figure 10.63

To find where the hyperbola crosses the line $\theta = \dfrac{\pi}{2}$, select $\dfrac{\pi}{2}$ and $\dfrac{3\pi}{2}$ for θ.

Figure 10.63(b) shows the points $\left(3, \dfrac{\pi}{2}\right)$ and $\left(3, \dfrac{3\pi}{2}\right)$ on the graph.

We sketch the upper half of the hyperbola by plotting some points from $\theta = 0$ to $\theta = \pi$.

$$r = \frac{3}{1 - 2\cos\theta}$$

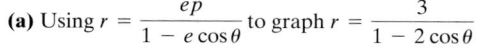

θ	$\dfrac{\pi}{6}$	$\dfrac{2\pi}{3}$	$\dfrac{5\pi}{6}$
r	-4.1	1.5	1.1

Using symmetry with respect to the polar axis, we sketch the lower half. The graph of the given equation is shown in Figure 10.63(b) on page 881.

Check Point 3

Use the three steps shown in the box on page 877 to graph the polar equation:

$$r = \frac{9}{3 - 9\cos\theta}.$$

Modeling Planetary Motion

Polish astronomer Nicolaus Copernicus (1473–1543) was correct in stating that planets in our solar system revolve around the sun and not the Earth. However, he incorrectly believed that celestial orbits move in perfect circles, calling his system "the ballet of the planets."

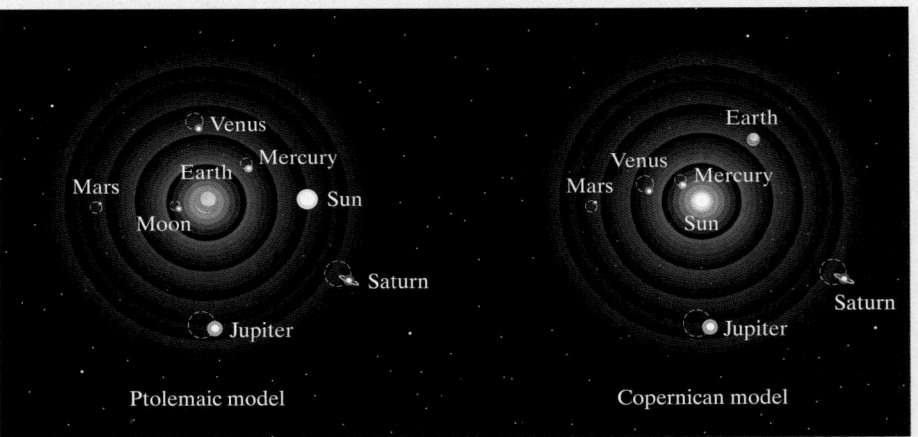

Ptolemaic model   Copernican model

Table 10.4 indicates that the planets in our solar system have orbits with eccentricities that are much closer to 0 than to 1. Most of these orbits are almost circular, which made it difficult for early astronomers to detect that they are actually ellipses.

German scientist and mathematician Johannes Kepler (1571–1630) discovered that planets move in elliptical orbits with the sun at one focus. The polar equation for these orbits is

$$r = \frac{(1 - e^2)a}{1 - e\cos\theta}$$

Table 10.4 Eccentricities of Planetary Orbits

Mercury	0.2056	Saturn	0.0543
Venus	0.0068	Uranus	0.0460
Earth	0.0167	Neptune	0.0082
Mars	0.0934	Pluto	0.2481
Jupiter	0.0484		

where the length of the orbit's major axis is $2a$. Describing planetary orbits, Kepler wrote, "The heavenly motions are nothing but a continuous song for several voices, to be perceived by the intellect, not by the ear."

EXERCISE SET 10.6

Practice Exercises

In Exercises 1–8,

a. *Identify the conic that each polar equation represents.*

b. *Describe the location of a directrix from the focus located at the pole.*

1. $r = \dfrac{3}{1 + \sin\theta}$

2. $r = \dfrac{3}{1 + \cos\theta}$

3. $r = \dfrac{6}{3 - 2\cos\theta}$

4. $r = \dfrac{6}{3 + 2\cos\theta}$

5. $r = \dfrac{8}{2 + 2\sin\theta}$

6. $r = \dfrac{8}{2 - 2\sin\theta}$

7. $r = \dfrac{12}{2 - 4\cos\theta}$

8. $r = \dfrac{12}{2 + 4\cos\theta}$

In Exercises 9–20, use the three steps shown in the box on page 877 to graph each polar equation.

9. $r = \dfrac{1}{1 + \sin\theta}$

10. $r = \dfrac{1}{1 + \cos\theta}$

11. $r = \dfrac{2}{1 - \cos\theta}$

12. $r = \dfrac{2}{1 - \sin\theta}$

13. $r = \dfrac{12}{5 + 3\cos\theta}$

14. $r = \dfrac{12}{5 - 3\cos\theta}$

15. $r = \dfrac{6}{2 - 2\sin\theta}$

16. $r = \dfrac{6}{2 + 2\sin\theta}$

17. $r = \dfrac{8}{2 - 4\cos\theta}$

18. $r = \dfrac{8}{2 + 4\cos\theta}$

19. $r = \dfrac{12}{3 - 6\cos\theta}$

20. $r = \dfrac{12}{3 - 3\cos\theta}$

Application Exercises

Halley's comet has an elliptical orbit with the sun at one focus. Its orbit, shown in the figure at the top of the next column, is given approximately by

$$r = \dfrac{1.069}{1 + 0.967\sin\theta}.$$

In the formula, r is measured in astronomical units. (One astronomical unit is the average distance from Earth to the sun, approximately 93 million miles.) Use the given formula and the figure to solve Exercises 21–22. Round to the nearest

hundredth of an astronomical unit and the nearest million miles.

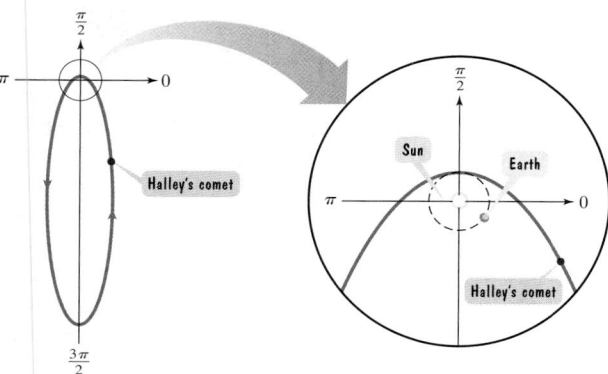

21. Find the distance from Halley's comet to the sun at its shortest distance from the sun.

22. Find the distance from Halley's comet to the sun at its greatest distance from the sun.

On February 20, 1962, John Glenn made the first U.S.-manned flight around the Earth for three orbits on Friendship 7. With Earth at one focus, the orbit of Friendship 7 is given approximately by

$$r = \dfrac{4090.76}{1 - 0.0076\cos\theta}$$

where r is measured in miles from Earth's center. Use the formula and the figure shown to solve Exercises 23–24.

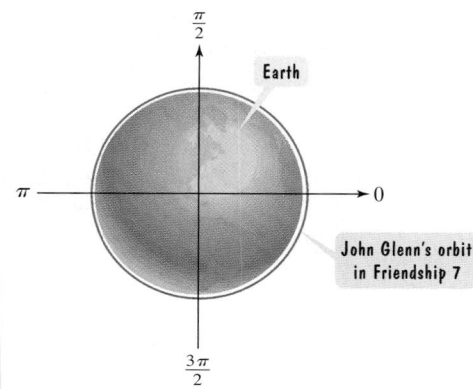

23. How far from Earth's center was John Glenn at his greatest distance from the planet? Round to the nearest mile. If the radius of Earth is 3960 miles, how far was he from Earth's surface at this point on the flight?

24. How far from Earth's center was John Glenn at his closest distance from the planet? Round to the nearest mile. If the radius of Earth is 3960 miles, how far was he from Earth's surface at this point on the flight?

Writing in Mathematics

25. How are the conics described in terms of a fixed point and a fixed line?

26. If all conics are defined in terms of a fixed point and a fixed line, how can you tell one kind of conic from another?

27. If you are given the standard form of the polar equation of a conic, how do you determine its eccentricity?

28. If you are given the standard form of the polar equation of a conic, how do you determine the location of a directrix from the focus at the pole?

29. Describe a strategy for graphing $r = \dfrac{1}{1 + \sin\theta}$.

30. You meet John Glenn and he asks you to tell him something of interest about the difference between the elliptical orbit of his first space voyage in 1962 and the orbit for his return journey in 1999. Describe how to use the polar equation for orbits in the essay on page 882, the equation for his 1962 journey, and a graphing utility to provide him with an interesting visual analysis.

Technology Exercises

Use the polar mode of a graphing utility with angle measure in radians to solve Exercises 31–34. Unless otherwise indicated, use θmin = 0, θmax = 2π, and θstep = $\dfrac{\pi}{48}$. If you are not satisfied with the quality of the graph, experiment with smaller values for

31. Use a graphing utility to verify any five of your hand-drawn graphs in Exercises 9–20.

In Exercises 32–34, identify the conic that each polar equation represents. Then use a graphing utility to graph the equation.

32. $r = \dfrac{16}{4 - 3\cos\theta}$

33. $r = \dfrac{12}{4 + 5\sin\theta}$

34. $r = \dfrac{18}{6 - 6\cos\theta}$

In Exercises 35–36, use a graphing utility to graph the equation. Then answer the given question.

35. $r = \dfrac{4}{1 - \sin\left(\theta - \dfrac{\pi}{4}\right)}$; How does the graph differ from the graph of $r = \dfrac{4}{1 - \sin\theta}$?

36. $r = \dfrac{3}{2 + 6\cos\left(\theta + \dfrac{\pi}{3}\right)}$; How does the graph differ from the graph of $r = \dfrac{3}{2 + 6\cos\theta}$?

37. Use the polar equation for planetary orbits,

$$r = \dfrac{(1 - e^2)a}{1 - e\cos\theta},$$

to find the polar equation of the orbit for Mercury and Earth.

Mercury: $e = 0.2056$ and $a = 36.0 \times 10^6$ miles

Earth: $e = 0.0167$ and $a = 92.96 \times 10^6$ miles

Use a graphing utility to graph both orbits in the same viewing rectangle. What do you see about the orbits from their graphs that is not obvious from their equations?

Critical Thinking Exercises

38. Identify the conic and graph the equation:

$$r = \dfrac{4\sec\theta}{2\sec\theta - 1}.$$

In Exercises 39–40, write a polar equation of the conic that is named and described.

39. Ellipse: a focus at the pole; vertex: $(4, 0)$; $e = \dfrac{1}{2}$

40. Hyperbola: a focus at the pole; directrix: $x = -1$; $e = \dfrac{3}{2}$

41. Identify the conic and write its equation in rectangular coordinates: $r = \dfrac{1}{2 - 2\cos\theta}$.

42. Prove that the polar equation of a planet's elliptical orbit is

$$r = \dfrac{(1 - e^2)a}{1 - e\cos\theta}$$

where e is the eccentricity and $2a$ is the length of the major axis.

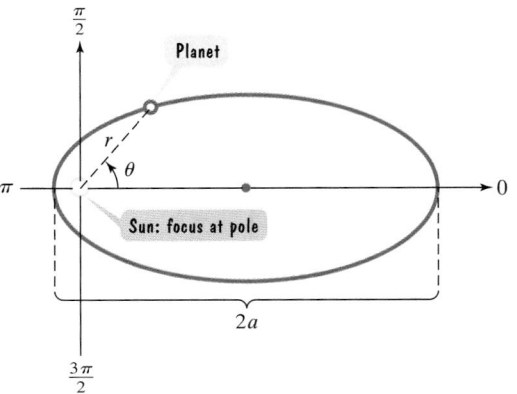

CHAPTER SUMMARY, REVIEW, AND TEST

Summary

10.1 The Ellipse

a. An ellipse is the set of all points in a plane the sum of whose distances from two fixed points, the foci, is constant.

b. Standard forms of the equations of an ellipse with center at the origin are $\dfrac{x^2}{a^2} + \dfrac{y^2}{b^2} = 1$ [foci: $(-c, 0), (c, 0)$] and $\dfrac{x^2}{b^2} + \dfrac{y^2}{a^2} = 1$ [foci: $(0, -c), (0, c)$], where $b^2 = a^2 - c^2$ and $a^2 > b^2$. See the box on page 811 and Figure 10.6.

c. Standard forms of the equations of an ellipse centered at (h, k) are $\dfrac{(x - h)^2}{a^2} + \dfrac{(y - k)^2}{b^2} = 1$ and $\dfrac{(x - h)^2}{b^2} + \dfrac{(y - k)^2}{a^2} = 1$. See Table 10.1 on page 815 and Figure 10.10.

10.2 The Hyperbola

a. A hyperbola is the set of all points in a plane the difference of whose distances from two fixed points, the foci, is constant.

b. Standard forms of the equations of a hyperbola with center at the origin are $\dfrac{x^2}{a^2} - \dfrac{y^2}{b^2} = 1$ [foci: $(-c, 0)$, $(c, 0)$] and $\dfrac{y^2}{a^2} - \dfrac{x^2}{b^2} = 1$ [foci: $(0, -c), (0, c)$], where $b^2 = c^2 - a^2$. See the box on page 824 and Figure 10.17.

c. Asymptotes for $\dfrac{x^2}{a^2} - \dfrac{y^2}{b^2} = 1$ are $y = \pm\dfrac{b}{a}x$. Asymptotes for $\dfrac{y^2}{a^2} - \dfrac{x^2}{b^2} = 1$ are $y = \pm\dfrac{a}{b}x$.

d. A procedure for graphing hyperbolas is given in the box on page 828.

e. Standard forms of the equations of a hyperbola centered at (h, k) are $\dfrac{(x - h)^2}{a^2} - \dfrac{(y - k)^2}{b^2} = 1$ and $\dfrac{(y - k)^2}{a^2} - \dfrac{(x - h)^2}{b^2} = 1$. See Table 10.2 on page 831 and Figure 10.26.

10.3 The Parabola

a. A parabola is the set of all points in a plane that are equidistant from a fixed line, the directrix, and a fixed point, the focus.

b. Standard forms of the equations of parabolas with vertex at the origin are $y^2 = 4px$ [focus: $(p, 0)$] and $x^2 = 4py$ [focus: $(0, p)$]. See the box on page 839 and Figure 10.32 on page 840.

c. Standard forms of the equations of a parabola with vertex at (h, k) are $(y - k)^2 = 4p(x - h)$ and $(x - h)^2 = 4p(y - k)$. See Table 10.3 on page 842 and Figure 10.37.

10.4 Rotation of Axes

a. A nondegenerate conic section of the form $Ax^2 + Cy^2 + Dx + Ey + F = 0$ in which A and C are not both zero is **1.** a circle if $A = C$; **2.** a parabola if $AC = 0$; **3.** an ellipse if $A \neq C$ and $AC > 0$; **4.** a hyperbola if $AC < 0$.

b. Rotation of Axes Formulas
θ is the angle from the positive x-axis to the positive x'-axis.

$$x = x' \cos\theta - y' \sin\theta \text{ and } y = x' \sin\theta + y' \cos\theta$$

c. Amount of Rotation Formula
The general second-degree equation

$$Ax^2 + Bxy + Cy^2 + Dx + Ey + F = 0$$

can be rewritten in x' and y' without an $x'y'$-term by rotating the axes through angle θ, where $\cot 2\theta = \dfrac{A - C}{B}$ and θ is an acute angle.

d. If 2θ in $\cot 2\theta$ is one of the more familiar angles such as $30°, 45°,$ or $60°$, write the equation of a rotated conic in standard form using the five-step procedure in the box on page 856.

e. If $\cot 2\theta$ is not the cotangent of one of the more familiar angles, use a sketch of $\cot 2\theta$ to find $\cos 2\theta$. Then use

$$\sin\theta = \sqrt{\dfrac{1 - \cos 2\theta}{2}} \quad \text{and} \quad \cos\theta = \sqrt{\dfrac{1 + \cos 2\theta}{2}}$$

to find values for $\sin\theta$ and $\cos\theta$ in the rotation formulas.

f. A nondegenerate conic section of the form

$$Ax^2 + Bxy + Cy^2 + Dx + Ey + F = 0$$

is **1.** a parabola if $B^2 - 4AC = 0$; **2.** an ellipse or a circle if $B^2 - 4AC < 0$; **3.** a hyperbola if $B^2 - 4AC > 0$.

10.5 Parametric Equations

a. The relationship between the parametric equations $x = f(t)$ and $y = g(t)$ and plane curves is described in the box on page 864.

b. Point plotting can be used to graph a plane curve described by parametric equations. See the box on page 865.

c. Plane curves can be sketched by eliminating the parameter t and graphing the resulting rectangular equation. It is sometimes necessary to change the domain of the rectangular equation to be consistent with the domain for the parametric equation in x.

d. Infinitely many pairs of parametric equations can represent the same plane curve. One pair for $y = f(x)$ is $x = t$ and $y = f(t)$ in which t is in the domain of f.

10.6 Conic Sections in Polar Coordinates

a. The focus-directrix definitions of the conic sections are given in the box on page 875. For all points on a conic, the distance from a fixed point (focus) divided by the distance from a fixed line (directrix) is constant and is called eccentricity. If $e = 1$, the conic is a parabola. If $e < 1$, the conic is an ellipse. If $e > 1$, the conic is a hyperbola.

b. Standard forms of the polar equations of conics are

$$r = \frac{ep}{1 \pm e \cos\theta} \quad \text{and} \quad r = \frac{ep}{1 \pm e \sin\theta}$$

in which (r, θ) is a point on the conic's graph, e is the eccentricity, and p is the distance between the focus (located at the pole) and the directrix. Details are shown in the box on page 876.

c. A procedure for graphing the polar equation of a conic is given in the box on page 877.

Review Exercises

10.1

In Exercises 1–8, graph each ellipse and locate the foci.

1. $\dfrac{x^2}{36} + \dfrac{y^2}{25} = 1$

2. $\dfrac{y^2}{25} + \dfrac{x^2}{16} = 1$

3. $4x^2 + y^2 = 16$

4. $4x^2 + 9y^2 = 36$

5. $\dfrac{(x - 1)^2}{16} + \dfrac{(y + 2)^2}{9} = 1$

6. $\dfrac{(x + 1)^2}{9} + \dfrac{(y - 2)^2}{16} = 1$

7. $4x^2 + 9y^2 + 24x - 36y + 36 = 0$

8. $9x^2 + 4y^2 - 18x + 8y - 23 = 0$

In Exercises 9–11, find the standard form of the equation of each ellipse centered at the origin satisfying the given conditions.

9. Foci: $(-4, 0)$, $(4, 0)$; Vertices: $(-5, 0)$, $(5, 0)$

10. Foci: $(0, -3)$, $(0, 3)$; Vertices: $(0, -6)$, $(0, 6)$

11. Major axis horizontal with length 12; length of minor axis = 4

12. A semielliptical arch supports a bridge that spans a river 20 yards wide. The center of the arch is 6 yards above the river's center. Write an equation for the ellipse so that the x-axis coincides with the water level and the y-axis passes through the center of the arch.

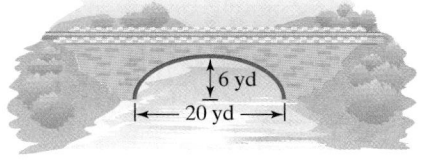

13. A semielliptic archway has a height of 15 feet at the center and a width of 50 feet, as shown in the figure. The 50-foot width consists of a two-lane road. Can a truck that is 12 feet high and 14 feet wide drive under the archway without going into the other lane?

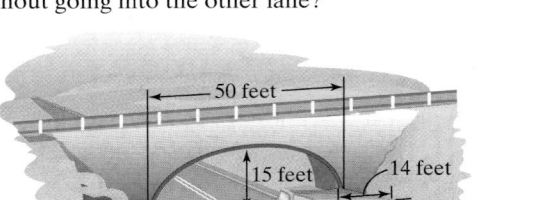

14. An elliptical pool table has a ball placed at each focus. If one ball is hit toward the side of the table, explain what will occur.

10.2

In Exercises 15–22, graph each hyperbola and locate the foci.

15. $\dfrac{x^2}{16} - y^2 = 1$

16. $\dfrac{y^2}{16} - x^2 = 1$

17. $9x^2 - 16y^2 = 144$

18. $4y^2 - x^2 = 16$

19. $\dfrac{(x - 2)^2}{25} - \dfrac{(y + 3)^2}{16} = 1$

20. $\dfrac{(y + 2)^2}{25} - \dfrac{(x - 3)^2}{16} = 1$

21. $y^2 - 4y - 4x^2 + 8x - 4 = 0$

22. $x^2 - y^2 - 2x - 2y - 1 = 0$

In Exercises 23–24, find the standard form of the equation of each hyperbola centered at the origin satisfying the given conditions.

23. Foci: $(0, -4)$, $(0, 4)$; Vertices: $(0, -2)$, $(0, 2)$

24. Foci: $(-8, 0)$, $(8, 0)$; Vertices: $(-3, 0)$, $(3, 0)$

25. Explain why it is not possible for a hyperbola to have foci at $(0, -2)$ and $(0, 2)$ and vertices at $(0, -3)$ and $(0, 3)$.

26. Radio tower M_2 is located 200 miles due west of radio tower M_1. The situation is illustrated in the figure shown, where a coordinate system has been superimposed. Simultaneous radio signals are sent from each tower to a ship, with the signal from M_2 received 500 microseconds before the signal from M_1. Assuming that radio signals travel at 0.186 miles per microsecond, determine the equation of the hyperbola on which the ship is located.

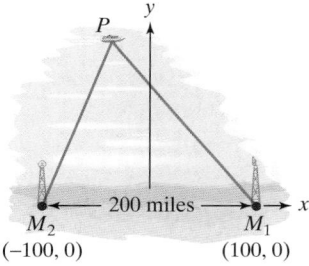

10.3

In Exercises 27–33, find the vertex, focus, and directrix of each parabola with the given equation. Then graph the parabola.

27. $y^2 = 8x$

28. $x^2 + 16y = 0$

29. $(y - 2)^2 = -16x$

30. $(x - 4)^2 = 4(y + 1)$

31. $x^2 + 4y = 4$

32. $y^2 - 4x - 10y + 21 = 0$

33. $x^2 - 4x - 2y = 0$

In Exercises 34–35, find the standard form of the equation of each parabola with vertex at the origin satisfying the given conditions.

34. Focus: $(12, 0)$; Directrix: $x = -12$

35. Focus: $(0, -11)$; Directrix: $y = 11$

36. An engineer is designing headlight units for automobiles. The unit has a parabolic surface with a diameter of 12 inches and a depth of 3 inches. The situation is illustrated in the figure, where a coordinate system has been superimposed. What is the equation of the parabola in this system? Where should the light source be placed? Describe this placement relative to the vertex.

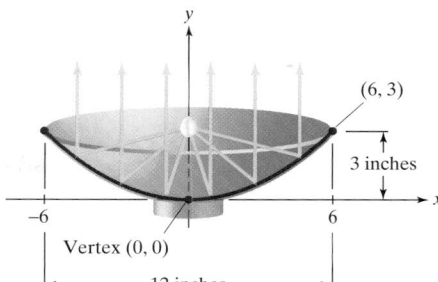

37. The George Washington Bridge spans the Hudson River from New York to New Jersey. Its two towers are 3500 feet apart and rise 316 feet above the road. The cable between the towers has the shape of a parabola, and the cable just touches the sides of the road midway between the towers. What is the height of the cable 1000 feet from a tower?

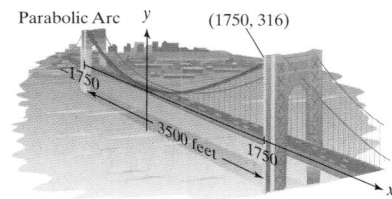

38. The giant satellite dish in the figure shown is in the shape of a parabolic surface. Signals strike the surface and are reflected to the focus, where the receiver is located. The diameter of the dish is 300 feet and its depth is 44 feet. How far, to the nearest foot, from the base of the dish should the receiver be placed?

10.4

In Exercises 39–46, identify each equation without completing the square or using a rotation of axes.

39. $y^2 + 4x + 2y - 15 = 0$

40. $x^2 + 16y^2 - 160y + 384 = 0$

41. $16x^2 + 64x + 9y^2 - 54y + 1 = 0$

42. $4x^2 - 9y^2 - 8x + 12y - 144 = 0$

43. $5x^2 + 2\sqrt{3}xy + 3y^2 - 18 = 0$

44. $5x^2 - 8xy + 7y^2 - 9\sqrt{5}x - 9 = 0$

45. $x^2 + 6xy + 9y^2 - 2y = 0$

46. $x^2 - 2xy + 3y^2 + 2x + 4y - 1 = 0$

In Exercises 47–51,

a. *Rewrite the equation in a rotated $x'y'$-system without an $x'y'$-term.*

b. *Express the equation involving x' and y' in the standard form of a conic section.*

c. *Use the rotated system to graph the equation.*

47. $xy - 4 = 0$

48. $x^2 + xy + y^2 - 1 = 0$

49. $4x^2 + 10xy + 4y^2 - 9 = 0$

50. $6x^2 - 6xy + 14y^2 - 45 = 0$

51. $x^2 + 2\sqrt{3}xy + 3y^2 - 12\sqrt{3}x + 12y = 0$

10.5

In Exercises 52–57, eliminate the parameter and graph the plane curve represented by the parametric equations. Use arrows to show the orientation of each plane curve.

52. $x = 2t - 1$, $y = 1 - t$; $-\infty < t < \infty$

53. $x = t^2$, $y = t - 1$; $-1 \le t \le 3$.

54. $x = 4t^2$, $y = t + 1$; $-\infty < t < \infty$

55. $x = 4\sin t$, $y = 3\cos t$; $0 \le t < \pi$

56. $x = 3 + 2\cos t$, $y = 1 + 2\sin t$; $0 \le t < 2\pi$

57. $x = 3\sec t$, $y = 3\tan t$; $0 \le t \le \dfrac{\pi}{4}$

58. Find two different sets of parametric equations for $y = x^2 + 6$.

59. The path of a projectile that is launched h feet above the ground with an initial velocity of v_0 feet per second and at an angle θ with the horizontal is given by the parametric equations

$$x = (v_0 \cos\theta)t \quad \text{and} \quad y = h + (v_0 \sin\theta)t - 16t^2$$

where t is the time, in seconds, since the projectile was launched. A football player throws a football with an initial velocity of 100 feet per second at an angle of 40° to the horizontal. The ball leaves the player's hand at a height of 6 feet.

a. Find the parametric equations that describe the position of the ball as a function of time.

b. Describe the ball's position after 1, 2, and 3 seconds. Round to the nearest tenth of a foot.

c. How long, to the nearest tenth of a second, is the ball in flight? What is the total horizontal distance that it travels before it lands?

d. Graph the parametric equations in part (a) using a graphing utility. Use the graph to determine when the ball is at its maximum height. What is its maximum height? Round all answers to the nearest tenth.

10.6

In Exercises 60–65,

a. *If necessary, write the equation in one of the standard forms for a conic in polar coordinates.*

b. *Determine values for e and p. Use the value of e to identify the conic.*

c. *Graph the given polar equation.*

60. $r = \dfrac{4}{1 - \sin\theta}$

61. $r = \dfrac{6}{1 + \cos\theta}$

62. $r = \dfrac{6}{2 + \sin\theta}$

63. $r = \dfrac{2}{3 - 2\cos\theta}$

64. $r = \dfrac{6}{3 + 6\sin\theta}$

65. $r = \dfrac{8}{4 + 16\cos\theta}$

Chapter 10 Test

In Exercises 1–5, graph the conic section with the given equation. For ellipses and hyperbolas, find the foci. For parabolas, find the vertex, focus, and directrix.

1. $9x^2 - 4y^2 = 36$

2. $x^2 = -8y$

3. $\dfrac{(x + 2)^2}{25} + \dfrac{(y - 5)^2}{9} = 1$

4. $4x^2 - y^2 + 8x + 2y + 7 = 0$

5. $(x + 5)^2 = 8(y - 1)$

In Exercises 6–8, find the standard form of the equation of the conic section satisfying the given conditions.

6. Ellipse; Foci: $(-7, 0)$, $(7, 0)$; Vertices: $(-10, 0)$, $(10, 0)$

7. Hyperbola; Foci: $(0, -10)$, $(0, 10)$; Vertices: $(0, -7)$, $(0, 7)$

8. Parabola; Focus: $(50, 0)$; Directrix: $x = -50$

9. A sound whispered at one focus of a whispering gallery can be heard at the other focus. The figure shows a whis-

pering gallery whose cross section is a semielliptical arch with a height of 24 feet and a width of 80 feet. How far from the room's center should two people stand so that they can whisper back and forth and be heard?

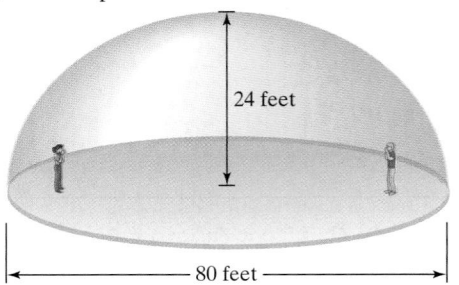

24 feet

80 feet

10. An engineer is designing headlight units for cars. The unit shown in the figure has a parabolic surface with a diameter of 6 inches and a depth of 3 inches.

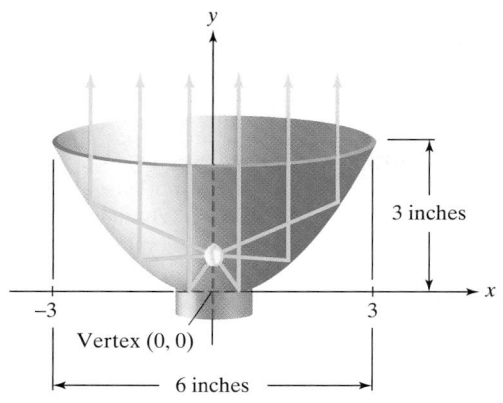

3 inches

Vertex (0, 0)

6 inches

a. Using the coordinate system that has been positioned on the unit, find the parabola's equation.
b. If the light source is located at the focus, describe its placement relative to the vertex.

In Exercises 11–12, identify each equation without completing the square or using a rotation of axes.

11. $x^2 + 9y^2 + 10x - 18y + 25 = 0$
12. $x^2 + y^2 + xy + 3x - y - 3 = 0$
13. For the equation

$$7x^2 - 6\sqrt{3}xy + 13y^2 - 16 = 0$$

determine what angle of rotation would eliminate the $x'y'$-term in a rotated $x'y'$-system.

In Exercises 14–15, eliminate the parameter and graph the plane curve represented by the parametric equations. Use arrows to show the orientation of each plane curve.

14. $x = t^2, y = t - 1; -\infty < t < \infty$
15. $x = 1 + 3\sin t, y = 2\cos t; 0 \le t < 2\pi$

In Exercises 16–17, identify the conic and graph the polar equation.

16. $r = \dfrac{2}{1 - \cos\theta}$ **17.** $r = \dfrac{4}{2 + \sin\theta}$

Cumulative Review Exercises (Chapters 1–10)

Solve each equation or inequality in Exercises 1–7.

1. $2(x - 3) + 5x = 8(x - 1)$
2. $-3(2x - 4) > 2(6x - 12)$
3. $x - 5 = \sqrt{x + 7}$ **4.** $(x - 2)^2 = 20$
5. $|2x - 1| \ge 7$ **6.** $3x^3 + 4x^2 - 7x + 2 = 0$
7. $\log_2(x + 1) + \log_2(x - 1) = 3$

Solve each system in Exercises 8–10

8. $3x + 4y = 2$ **9.** $2x^2 - y^2 = -8$
$\quad 2x + 5y = -1$ $\quad\quad x - y = 6$

10. (Use matrices.)
$$x - y + z = 17$$
$$-4x + y + 5z = -2$$
$$2x + 3y + z = 8$$

In Exercises 11–13, graph each equation, function, or system in the rectangular coordinate system.

11. $f(x) = (x - 1)^2 - 4$ **12.** $\dfrac{x^2}{9} + \dfrac{y^2}{4} = 1$

13. $5x + y \le 10$
$$y \ge \dfrac{1}{4}x + 2$$

14. a. List all possible rational roots of
$$32x^3 - 52x^2 + 17x + 3 = 0.$$

b. The graph of $f(x) = 32x^3 - 52x^2 + 17x + 3$ is shown in the figure. Use the graph of f and synthetic division to solve the equation in part (a).

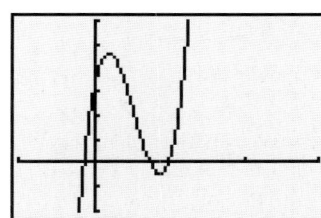

15. The graph shows gender ratios in the United States, with future projections.

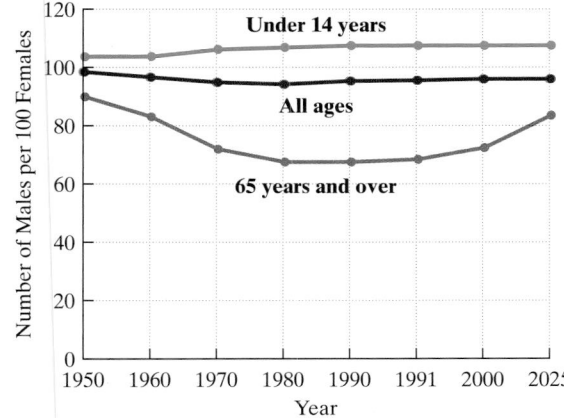

Gender Ratios in the U.S.

Under 14 years

All ages

65 years and over

Number of Males per 100 Females

Year

Source: U.S. Census Bureau

For males age 65 and over, shown by the blue graph:

a. In what time interval is the number of males per 100 females constant?

b. In what time interval is the number of males per 100 females increasing?

c. In what time interval is the number of males per 100 females decreasing?

For all ages, shown by the red graph:

d. Write a constant function $f(x)$ that approximately models the data shown for x in the interval $[1950, 2025]$.

e. What is misleading about the scale on the horizontal axis?

16. If $f(x) = x^2 - 4$ and $g(x) = x + 2$, find $(g \circ f)(x)$.

17. Expand using logarithmic properties. Where possible, evaluate logarithmic expressions.

$$\log_5 \frac{x^3 \sqrt{y}}{125}$$

18. Write the slope-intercept form of the equation of the line passing through $(1, -4)$ and $(-5, 8)$.

19. Rent-a-Truck charges a daily rental rate for a truck of $39 plus $0.16 a mile. A competing agency, Ace Truck Rentals, charges $25 a day plus $0.24 a mile for the same truck. How many miles must be driven in a day to make the daily cost of both agencies the same? What will be the cost?

20. The longest-lived U.S. presidents are John Adams (age 90), Herbert Hoover (also 90), and Harry Truman (88). Behind them are James Madison, Thomas Jefferson, and Richard Nixon. The latter three men lived a total of 249 years, and their ages at the time of death form consecutive odd integers. For how long did Nixon, Jefferson, and Madison live?

21. Verify the identity: $\dfrac{\csc \theta - \sin \theta}{\sin \theta} = \cot^2 \theta$.

22. Graph one complete cycle of $y = 2 \cos(2x + \pi)$.

23. If $\mathbf{v} = 3\mathbf{i} - 6\mathbf{j}$ and $\mathbf{w} = \mathbf{i} + \mathbf{j}$, find $(\mathbf{v} \cdot \mathbf{w})\mathbf{w}$.

24. Solve for θ: $\sin 2\theta = \sin \theta, 0 \leq \theta < 2\pi$.

25. In oblique triangle ABC, $A = 64°$, $B = 72°$, and $a = 13.6$. Solve the triangle. Round length to the nearest tenth.

Sequences, Induction, and Probability

We often save for the future by investing small amounts at periodic intervals. To understand how our savings accumulate, we need to understand properties of lists of numbers that are related to each other by a rule. Such lists are called *sequences*. Learning about properties of sequences will show you how to make your financial goals a reality. Your knowledge of sequences will enable you to inform your college roommate of the best of the three appealing offers.

Something incredible has happened. Your college roommate, a gifted athlete, has been given a six-year contract with a professional baseball team. He will be playing against the likes of Mark McGwire and Sammy Sosa. Management offers him three options. One is a beginning salary of $1,700,000 with annual increases of $70,000 per year starting in the second year. A second option is $1,700,000 the first year with an annual increase of 2% per year beginning in the second year. The third offer involves less money the first year—$1,500,000—but there is an annual increase of 9% yearly after that. Which option offers the most money over the six-year contract?

SECTION 11.1 *Sequences and Summation Notation*

Objectives

1. Find particular terms of a sequence from the general term.
2. Use recursion formulas.
3. Use factorial notation.
4. Use summation notation.

Sequences

Many creations in nature involve intricate mathematical designs, including a variety of spirals. For example, the arrangement of the individual florets in the head of a sunflower forms spirals. In some species, there are 21 spirals in the clockwise direction and 34 in the counterclockwise direction. The precise numbers depend on the species of sunflower: 21 and 34, or 34 and 55, or 55 and 89, or even 89 and 144.

This observation becomes even more interesting when we consider a sequence of numbers investigated by Leonardo of Pisa, also known as Fibonacci, an Italian mathematician of the thirteenth century. The **Fibonacci sequence** of numbers is an infinite sequence that begins as follows:

$$1, 1, 2, 3, 5, 8, 13, 21, 34, 55, 89, 144, 233 \ldots.$$

The first two terms are 1. Every term thereafter is the sum of the two preceding terms. For example, the third term, 2, is the sum of the first and second terms: $1 + 1 = 2$. The fourth term, 3, is the sum of the second and third terms: $1 + 2 = 3$, and so on. Did you know that the number of spirals in a daisy or a sunflower, 21 and 34, are two Fibonacci numbers? The number of spirals in a pine cone, 8 and 13, and a pineapple, 8 and 13, are also Fibonacci numbers.

We can think of the Fibonacci sequence as a function. The terms of the sequence

$$1, 1, 2, 3, 5, 8, 13, 21, 34, 55, 89, 144, 233, \ldots$$

are the range values for a function whose domain is the set of positive integers.

Domain:	1,	2,	3,	4,	5,	6,	7,	...
	↓	↓	↓	↓	↓	↓	↓	
Range:	1,	1,	2,	3,	5,	8,	13,	...

Thus, $f(1) = 1, f(2) = 1, f(3) = 2, f(4) = 3, f(5) = 5, f(6) = 8, f(7) = 13$, and so on.

The letter a with a subscript is used to represent function values of a sequence, rather than the usual function notation. The subscripts make up the domain of the sequence, and they identify the location of a term. Thus, a_1 represents the first term of the sequence, a_2 represents the second term, a_3 the third term, and so on. This notation is shown for the first six terms of the Fibonacci sequence:

$$1, \quad 1, \quad 2, \quad 3, \quad 5, \quad 8.$$

$a_1 = 1 \quad a_2 = 1 \quad a_3 = 2 \quad a_4 = 3 \quad a_5 = 5 \quad a_6 = 8$

Fibonacci Numbers on the Piano Keyboard

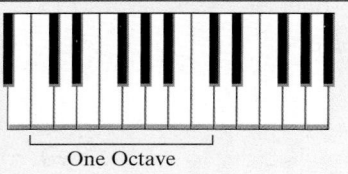

One Octave

Numbers in the Fibonacci sequence can be found in an octave on the piano keyboard. The octave contains 2 black keys in one cluster, 3 black keys in another cluster, 5 black keys, 8 white keys, and a total of 13 keys altogether. The numbers 2, 3, 5, 8, and 13 are the third through seventh terms of the Fibonacci sequence.

The notation a_n represents the nth term, or **general term**, of a sequence. The entire sequence is represented by $\{a_n\}$.

> ### Definition of a Sequence
> An **infinite sequence** $\{a_n\}$ is a function whose domain is the set of positive integers. The function values, or **terms**, of the sequence are represented by
> $$a_1, a_2, a_3, a_4, \ldots, a_n, \ldots.$$
> Sequences whose domains consist only of the first n positive integers are called **finite sequences**.

1 Find particular terms of a sequence from the general term.

EXAMPLE 1   Writing Terms of a Sequence from the General Term

Write the first four terms of the sequence whose nth term, or general term, is given.

a. $a_n = 3n + 4$ **b.** $a_n = \dfrac{(-1)^n}{3^n - 1}$

Solution

a. We need to find the first four terms of the sequence whose general term is $a_n = 3n + 4$. To do so, we replace n in the formula by 1, 2, 3, and 4.

a_1, 1st term $3 \cdot 1 + 4 = 3 + 4 = 7$ a_2, 2nd term $3 \cdot 2 + 4 = 6 + 4 = 10$

a_3, 3rd term $3 \cdot 3 + 4 = 9 + 4 = 13$ a_4, 4th term $3 \cdot 4 + 4 = 12 + 4 = 16$

The first four terms are 7, 10, 13, and 16. The sequence defined by $a_n = 3n + 4$ can be written as

$$7, \ 10, \ 13, \ \ldots, \ 3n + 4, \ \ldots.$$

b. We need to find the first four terms of the sequence whose general term is $a_n = \dfrac{(-1)^n}{3^n - 1}$. To do so, we replace each occurrence of n in the formula by 1, 2, 3, and 4.

a_1, 1st term $\dfrac{(-1)^1}{3^1 - 1} = \dfrac{-1}{3 - 1} = -\dfrac{1}{2}$ a_2, 2nd term $\dfrac{(-1)^2}{3^2 - 1} = \dfrac{1}{9 - 1} = \dfrac{1}{8}$

a_3, 3rd term $\dfrac{(-1)^3}{3^3 - 1} = \dfrac{-1}{27 - 1} = -\dfrac{1}{26}$ a_4, 4th term $\dfrac{(-1)^4}{3^4 - 1} = \dfrac{1}{81 - 1} = \dfrac{1}{80}$

The first four terms are $-\frac{1}{2}, \frac{1}{8}, -\frac{1}{26}, \frac{1}{80}$. The sequence defined by $\dfrac{(-1)^n}{3^n - 1}$ can be written as

$$-\frac{1}{2}, \frac{1}{8}, -\frac{1}{26}, \ldots, \frac{(-1)^n}{3^n - 1}, \ldots.$$

Study Tip

The factor $(-1)^n$ in the general term of a sequence causes the signs of the terms to alternate between positive and negative, depending on whether n is even or odd.

Technology

Graphing utilities can write the terms of a sequence and graph them. For example, to find the first six terms of

$$\{a_n\} = \left\{\frac{1}{n}\right\}, \text{ enter}$$

General term — Stop at a_6.

SEQ $(1 \div x, x, 1, 6, 1)$.

Variable used in general term — Start at a_1. — The "step" from a_1 to a_2, a_2 to a_3, etc., is 1.

The first few terms of the sequence are shown in the viewing rectangle. By pressing the right arrow key to scroll right, you can see the remaining terms.

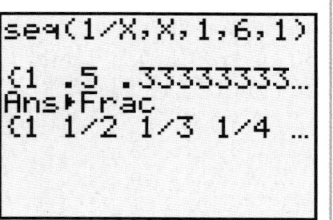

```
seq(1/X,X,1,6,1)
{1 .5 .33333333…
Ans▶Frac
{1 1/2 1/3 1/4 …
```

Check Point 1 Write the first four terms of the sequence whose nth term, or general term, is given.

a. $a_n = 2n + 5$ **b.** $a_n = \dfrac{(-1)^n}{2^n + 1}$

Although sequences are usually named with the letter a, any lowercase letter can be used. For example, the first four terms of the sequence $\{b_n\} = \left\{\left(\frac{1}{2}\right)^n\right\}$ are $b_1 = \frac{1}{2}$, $b_2 = \frac{1}{4}$, $b_3 = \frac{1}{8}$, and $b_4 = \frac{1}{16}$.

Because a sequence is a function whose domain is the set of positive integers, the **graph of a sequence** is a set of discrete points. For example, consider the sequence whose general term is $a_n = \frac{1}{n}$. How does the graph of this sequence differ from the graph of the function $f(x) = \frac{1}{x}$? The graph of $f(x) = \frac{1}{x}$ is shown in Figure 11.1(a) for positive values of x. To obtain the graph of the sequence $\{a_n\} = \left\{\frac{1}{n}\right\}$, remove all the points from the graph of f except those whose x-coordinates are positive integers. Thus, we remove all points except $(1, 1)$, $\left(2, \frac{1}{2}\right)$, $\left(3, \frac{1}{3}\right)$, $\left(4, \frac{1}{4}\right)$, and so on. The remaining points are the graph of the sequence $\{a_n\} = \left\{\frac{1}{n}\right\}$, shown in Figure 11.1(b). Notice that the horizontal axis is labeled n and the vertical axis a_n.

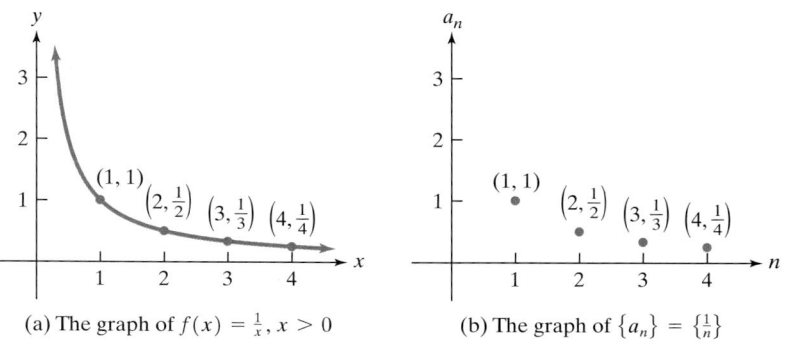

(a) The graph of $f(x) = \frac{1}{x}$, $x > 0$ (b) The graph of $\{a_n\} = \left\{\frac{1}{n}\right\}$

Figure 11.1 Comparing a continuous graph to the graph of a sequence

2 Use recursion formulas.

Recursion Formulas

In Example 1, the formulas used for the nth term of a sequence expressed the term as a function of n, the number of the term. Sequences can also be defined using **recursion formulas**. A recursion formula defines the nth term of a sequence as a function of the previous term. Our next example illustrates that if the first term of a sequence is known, then the recursion formula can be used to determine the remaining terms.

EXAMPLE 2 Using a Recursion Formula

Find the first four terms of the sequence in which $a_1 = 5$ and $a_n = 3a_{n-1} + 2$ for $n \geq 2$.

Solution

$a_1 = 5$ This is the given first term.

$a_2 = 3a_1 + 2$ Use $a_n = 3a_{n-1} + 2$, with $n = 2$. Thus, $a_2 = 3a_{2-1} + 2 = 3a_1 + 2$.

$= 3(5) + 2 = 17$ Substitute 5 for a_1.

$$a_3 = 3a_2 + 2$$

Again use $a_n = 3a_{n-1} + 2$, with $n = 3$.

$$= 3(17) + 2 = 53$$

Substitute 17 for a_2.

$$a_4 = 3a_3 + 2$$

Notice that a_4 is defined in terms of a_3. We used $a_n = 3a_{n-1} + 2$, with $n = 4$.

$$= 3(53) + 2 = 161$$

Use the value of a_3, the third term, obtained from above.

The first four terms are 5, 17, 53, and 161.

> **Check Point 2** Find the first four terms of the sequence in which $a_1 = 3$ and $a_n = 2a_{n-1} + 5$ for $n \geq 2$.

3 Use factorial notation.

Factorial Notation

Products of consecutive positive integers occur quite often in sequences. These products can be expressed in a special notation, called **factorial notation**.

Factorials from 0 through 20

0!	1
1!	1
2!	2
3!	6
4!	24
5!	120
6!	720
7!	5040
8!	40,320
9!	362,880
10!	3,628,800
11!	39,916,800
12!	479,001,600
13!	6,227,020,800
14!	87,178,291,200
15!	1,307,674,368,000
16!	20,922,789,888,000
17!	355,687,428,096,000
18!	6,402,373,705,728,000
19!	121,645,100,408,832,000
20!	2,432,902,008,176,640,000

As n increases, $n!$ grows very rapidly. Factorial growth is more explosive than exponential growth discussed in Chapter 4.

> ### Factorial Notation
>
> If n is a positive integer, the notation $n!$ (read "n factorial") is the product of all positive integers from n down through 1.
> $$n! = n(n-1)(n-2)\ldots(3)(2)(1)$$
> 0! (zero factorial), by definition, is 1.
> $$0! = 1$$

The values of $n!$ for the first six positive integers are

$$1! = 1$$
$$2! = 2 \cdot 1 = 2$$
$$3! = 3 \cdot 2 \cdot 1 = 6$$
$$4! = 4 \cdot 3 \cdot 2 \cdot 1 = 24$$
$$5! = 5 \cdot 4 \cdot 3 \cdot 2 \cdot 1 = 120$$
$$6! = 6 \cdot 5 \cdot 4 \cdot 3 \cdot 2 \cdot 1 = 720.$$

Factorials affect only the number or variable that they follow unless grouping symbols appear. For example,

$$2 \cdot 3! = 2(3 \cdot 2 \cdot 1) = 2 \cdot 6 = 12$$

whereas

$$(2 \cdot 3)! = 6! = 6 \cdot 5 \cdot 4 \cdot 3 \cdot 2 \cdot 1 = 720.$$

In this sense, factorials are similar to exponents.

EXAMPLE 3 Finding Terms of a Sequence Involving Factorials

Write the first four terms of the sequence whose nth term is

$$a_n = \frac{2^n}{(n-1)!}.$$

Solution We need to find the first four terms of the sequence. To do so, we replace each n in the formula by 1, 2, 3, and 4.

a_1, 1st term
$$\frac{2^1}{(1-1)!} = \frac{2}{0!} = \frac{2}{1} = 2$$

a_2, 2nd term
$$\frac{2^2}{(2-1)!} = \frac{4}{1!} = \frac{4}{1} = 4$$

a_3, 3rd term
$$\frac{2^3}{(3-1)!} = \frac{8}{2!} = \frac{8}{2 \cdot 1} = 4$$

a_4, 4th term
$$\frac{2^4}{(4-1)!} = \frac{16}{3!} = \frac{16}{3 \cdot 2 \cdot 1} = \frac{16}{6} = \frac{8}{3}$$

The first four terms are $2, 4, 4, \frac{8}{3}$.

Check Point 3

Write the first four terms of the sequence whose nth term is

$$a_n = \frac{20}{(n+1)!}.$$

When evaluating fractions with factorials in the numerator and the denominator, try to reduce the fraction before performing the multiplications. For example, consider $\frac{26!}{21!}$. Rather than write out 26! as the product of all integers from 26 down to 1, we can express 26! as

$$26! = 26 \cdot 25 \cdot 24 \cdot 23 \cdot 22 \cdot 21!.$$

In this way, we can divide both the numerator and the denominator by the common factor, 21!.

$$\frac{26!}{21!} = \frac{26 \cdot 25 \cdot 24 \cdot 23 \cdot 22 \cdot \cancel{21!}}{\cancel{21!}} = 26 \cdot 25 \cdot 24 \cdot 23 \cdot 22 = 7,893,600$$

EXAMPLE 4 Evaluating Fractions with Factorials

Evaluate each factorial expression.

a. $\dfrac{10!}{2!8!}$

b. $\dfrac{(n+1)!}{n!}$

Solution

a. $\dfrac{10!}{2!\,8!} = \dfrac{10 \cdot 9 \cdot \cancel{8!}}{2 \cdot 1 \cdot \cancel{8!}} = \dfrac{90}{2} = 45$

b. $\dfrac{(n+1)!}{n!} = \dfrac{(n+1) \cdot \cancel{n!}}{\cancel{n!}} = n + 1$

Check Point 4

Evaluate each factorial expression.

a. $\dfrac{14!}{2!12!}$ **b.** $\dfrac{n!}{(n-1)!}$

4 Use summation notation.

Summation Notation

It is sometimes useful to find the sum of the first n terms of a sequence. For example, consider the number of AIDS cases diagnosed in the United States from 1991 to 1997, shown in Table 11.1.

Table 11.1   AIDS Cases Diagnosed in the United States, 1991–1997

Year	1991	1992	1993	1994	1995	1996	1997
Cases Diagnosed	60,124	79,054	79,049	71,209	66,233	54,656	31,153

Source: U.S. Department of Health and Human Services

We can let a_n represent the number of AIDS cases diagnosed in year n, where $n = 1$ corresponds to 1991, $n = 2$ to 1992, $n = 3$ to 1993, and so on. The terms of the finite sequence in Table 8.1 are given as follows.

$$60{,}124 \quad 79{,}054, \quad 79{,}049, \quad 71{,}209, \quad 66{,}233, \quad 54{,}656, \quad 31{,}153$$

$$a_1 \qquad a_2 \qquad a_3 \qquad a_4 \qquad a_5 \qquad a_6 \qquad a_7$$

Why might we want to add the terms of this sequence? We do this to find the number of AIDS cases diagnosed from 1991 to 1997. Thus,

$$a_1 + a_2 + a_3 + a_4 + a_5 + a_6 + a_7$$
$$= 60{,}124 + 79{,}054, + 79{,}049, + 71{,}209, + 66{,}233, + 54{,}656, + 31{,}153$$
$$= 441{,}478.$$

We see that there were 441,478 AIDS cases diagnosed in the United States from 1991 to 1997.

There is a compact notation for expressing the sum of the first n terms of a sequence. For example, rather than write

$$a_1 + a_2 + a_3 + a_4 + a_5 + a_6 + a_7,$$

we can use **summation notation** to express the sum as

$$a_1 + a_2 + a_3 + a_4 + a_5 + a_6 + a_7 = \sum_{i=1}^{7} a_i.$$

We read the expression on the right as "the sum as i goes from 1 to 7 of a_i." The letter i is called the **index of summation** and is not related to the use of i to represent $\sqrt{-1}$.

You can think of the symbol Σ (the uppercase Greek letter sigma) as an instruction to add up terms of a sequence.

Summation Notation

The sum of the first n terms of a sequence is represented by the **summation notation**

$$\sum_{i=1}^{n} a_i = a_1 + a_2 + a_3 + a_4 + \cdots + a_n$$

where i is the **index of summation**, n is the **upper limit of summation**, and 1 is the **lower limit of summation.**

Any letter can be used for the index of summation. The letters i, j, and k are used commonly. Furthermore, the lower limit of summation can be an integer other than 1.

Table 11.2 contains some important properties of sums expressed in summation notation.

Table 11.2   Properties of Sums

Property	Example
1. $\displaystyle\sum_{i=1}^{n} ca_i = c \sum_{i=1}^{n} a_i,\ c$ any real number	$\displaystyle\sum_{i=1}^{4} 3i^2 = 3 \cdot 1^2 + 3 \cdot 2^2 + 3 \cdot 3^2 + 3 \cdot 4^2$ $3 \displaystyle\sum_{i=1}^{4} i^2 = 3(1^2 + 2^2 + 3^2 + 4^2) = 3 \cdot 1^2 + 3 \cdot 2^2 + 3 \cdot 3^2 + 3 \cdot 4^2$ Conclusion: $\displaystyle\sum_{i=1}^{4} 3i^2 = 3 \sum_{i=1}^{4} i^2$
2. $\displaystyle\sum_{i=1}^{n} (a_i + b_i) = \sum_{i=1}^{n} a_i + \sum_{i=1}^{n} b_i$	$\displaystyle\sum_{i=1}^{4} (i + i^2) = (1 + 1^2) + (2 + 2^2) + (3 + 3^2) + (4 + 4^2)$ $\displaystyle\sum_{i=1}^{4} i + \sum_{i=1}^{4} i^2 = (1 + 2 + 3 + 4) + (1^2 + 2^2 + 3^2 + 4^2)$ $\qquad = (1 + 1^2) + (2 + 2^2) + (3 + 3^2) + (4 + 4^2)$ Conclusion: $\displaystyle\sum_{i=1}^{4} (i + i^2) = \sum_{i=1}^{4} i + \sum_{i=1}^{4} i^2$
3. $\displaystyle\sum_{i=1}^{n} (a_i - b_i) = \sum_{i=1}^{n} a_i - \sum_{i=1}^{n} b_i$	$\displaystyle\sum_{i=3}^{5} (i^2 - i^3) = (3^2 - 3^3) + (4^2 - 4^3) + (5^2 - 5^3)$ $\displaystyle\sum_{i=3}^{5} i^2 - \sum_{i=3}^{5} i^3 = (3^2 + 4^2 + 5^2) - (3^3 + 4^3 + 5^3)$ $\qquad = (3^2 - 3^3) + (4^2 - 4^3) + (5^2 - 5^3)$ Conclusion: $\displaystyle\sum_{i=3}^{5} (i^2 - i^3) = \sum_{i=3}^{5} i^2 - \sum_{i=3}^{5} i^3$

EXERCISE SET 11.1

Practice Exercises

In Exercises 1–12, write the first four terms of each sequence whose general term is given.

1. $a_n = 3n + 2$

2. $a_n = 4n - 1$

3. $a_n = 3^n$

4. $a_n = \left(\dfrac{1}{3}\right)^n$

5. $a_n = (-3)^n$

6. $a_n = \left(-\dfrac{1}{3}\right)^n$

7. $a_n = (-1)^n(n + 3)$

8. $a_n = (-1)^{n+1}(n + 4)$

9. $a_n = \dfrac{2n}{n + 4}$

10. $a_n = \dfrac{3n}{n + 5}$

11. $a_n = \dfrac{(-1)^{n+1}}{2^n - 1}$

12. $a_n = \dfrac{(-1)^{n+1}}{2^n + 1}$

The sequences in Exercises 13–18 are defined using recursion formulas. Write the first four terms of each sequence.

13. $a_1 = 7$ and $a_n = a_{n-1} + 5$ for $n \geq 2$

14. $a_1 = 12$ and $a_n = a_{n-1} + 4$ for $n \geq 2$

15. $a_1 = 3$ and $a_n = 4a_{n-1}$ for $n \geq 2$

16. $a_1 = 2$ and $a_n = 5a_{n-1}$ for $n \geq 2$

17. $a_1 = 4$ and $a_n = 2a_{n-1} + 3$ for $n \geq 2$

18. $a_1 = 5$ and $a_n = 3a_{n-1} - 1$ for $n \geq 2$

In Exercises 19–22, the general term of a sequence is given and involves a factorial. Write the first four terms of each sequence.

19. $a_n = \dfrac{n^2}{n!}$

20. $a_n = \dfrac{(n + 1)!}{n^2}$

21. $a_n = 2(n + 1)!$

22. $a_n = -2(n - 1)!$

In Exercises 23–28, evaluate each factorial expression.

23. $\dfrac{17!}{15!}$

24. $\dfrac{18!}{16!}$

25. $\dfrac{16!}{2!14!}$

26. $\dfrac{20!}{2!18!}$

27. $\dfrac{(n + 2)!}{n!}$

28. $\dfrac{(2n + 1)!}{(2n)!}$

In Exercises 29–42, find each indicated sum.

29. $\displaystyle\sum_{i=1}^{6} 5i$

30. $\displaystyle\sum_{i=1}^{6} 7i$

31. $\displaystyle\sum_{i=1}^{4} 2i^2$

32. $\displaystyle\sum_{i=1}^{5} i^3$

33. $\displaystyle\sum_{k=1}^{5} k(k + 4)$

34. $\displaystyle\sum_{k=1}^{4} (k - 3)(k + 2)$

35. $\displaystyle\sum_{i=1}^{4} \left(-\dfrac{1}{2}\right)^i$

36. $\displaystyle\sum_{i=2}^{4} \left(-\dfrac{1}{3}\right)^i$

37. $\displaystyle\sum_{i=5}^{9} 11$

38. $\displaystyle\sum_{i=3}^{7} 12$

39. $\displaystyle\sum_{i=0}^{4} \dfrac{(-1)^i}{i!}$

40. $\displaystyle\sum_{i=0}^{4} \dfrac{(-1)^{i+1}}{(i + 1)!}$

41. $\displaystyle\sum_{i=1}^{5} \dfrac{i!}{(i - 1)!}$

42. $\displaystyle\sum_{i=1}^{5} \dfrac{(i + 2)!}{i!}$

In Exercises 43–54, express each sum using summation notation. Use 1 as the lower limit of summation and i for the index of summation.

43. $1^2 + 2^2 + 3^2 + \cdots + 15^2$

44. $1^4 + 2^4 + 3^4 + \cdots + 12^4$

45. $2 + 2^2 + 2^3 + \cdots + 2^{11}$

46. $5 + 5^2 + 5^3 + \cdots + 5^{12}$

47. $1 + 2 + 3 + \cdots + 30$

48. $1 + 2 + 3 + \cdots + 40$

49. $\dfrac{1}{2} + \dfrac{2}{3} + \dfrac{3}{4} + \cdots + \dfrac{14}{14 + 1}$

50. $\dfrac{1}{3} + \dfrac{2}{4} + \dfrac{3}{5} + \cdots + \dfrac{16}{16 + 2}$

51. $4 + \dfrac{4^2}{2} + \dfrac{4^3}{3} + \cdots + \dfrac{4^n}{n}$

52. $\dfrac{1}{9} + \dfrac{2}{9^2} + \dfrac{3}{9^3} + \cdots + \dfrac{n}{9^n}$

53. $1 + 3 + 5 + \cdots + (2n - 1)$

54. $a + ar + ar^2 + \cdots + ar^{n-1}$

In Exercises 55–60, express each sum using summation notation. Use a lower limit of summation of your choice and k for the index of summation.

55. $5 + 7 + 9 + 11 + \cdots + 31$

56. $6 + 8 + 10 + 12 + \cdots + 32$

57. $a + ar + ar^2 + \cdots + ar^{12}$

58. $a + ar + ar^2 + \cdots + ar^{14}$

59. $a + (a + d) + (a + 2d) + \cdots + (a + nd)$

60. $(a + d) + (a + d^2) + \cdots + (a + d^n)$

Application Exercises

61. The bar graph shows the number of children home-educated in the United States. Let a_n represent the number of children, in thousands, home-educated in year n, where $n = 2$ corresponds to 1992, $n = 3$ to 1993, and so on.

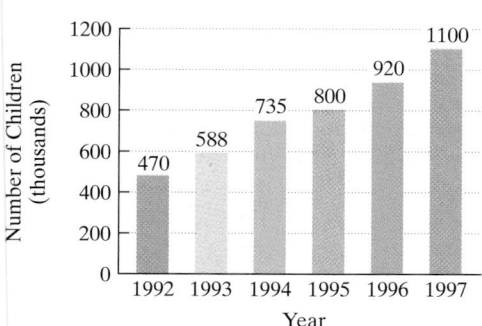

Number of Children Home-Educated in the U.S.

Source: National Home Education Research Institute

a. Find $\displaystyle\sum_{i=2}^{7} a_i$. What does this represent?

b. Find $\dfrac{1}{6} \displaystyle\sum_{i=2}^{7} a_i$. What does this represent?

62. The bar graph shows the number of business failures in the United States. Let a_n represent the number of business failures in year n, where $n = 0$ corresponds to 1990, $n = 1$ to 1991, $n = 2$ to 1992, and so on.

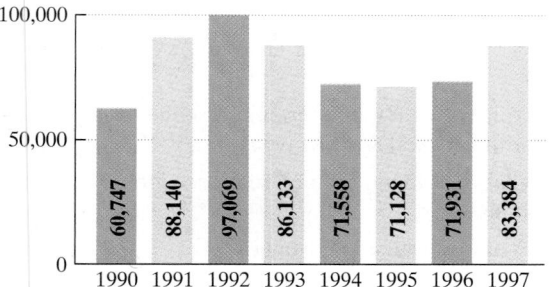

Number of Business Failures in the U.S.

Source: Dun & Bradstreet

a. Find $\displaystyle\sum_{i=0}^{7} a_i$. What does this represent?

b. Find $\dfrac{1}{8} \displaystyle\sum_{i=0}^{7} a_i$. What does this represent?

63. The finite sequence whose general term is

$$a_n = 0.16n^2 - 1.04n + 7.39$$

where $n = 1, 2, 3, \ldots, 8$ models the total number of dollars, in billions, that Americans spent on recreational boating from 1991 through 1998. Find and interpret

$$\sum_{i=1}^{5} a_i.$$

64. The finite sequence whose general term is

$$a_n = 2.54e^{-0.09n}$$

where $n = 0, 1, 2, \ldots, 9$ models the number of new foreign cars sold in the United States, in millions, from 1990 through 1999. Find and interpret

$$\sum_{i=0}^{4} a_i.$$

65. A deposit of $6000 is made in an account that earns 6% interest compounded quarterly. The balance in the account after n quarters is given by the sequence

$$a_n = 6000\left(1 + \frac{0.06}{4}\right)^n, \qquad n = 1, 2, 3, \ldots .$$

Find the balance in the account after five years by computing a_{20}.

66. A deposit of $10,000 is made in an account that earns 8% interest compounded quarterly. The balance in the account after n quarters is given by the sequence

$$a_n = 10{,}000\left(1 + \frac{0.08}{4}\right)^n, \qquad n = 1, 2, 3, \ldots .$$

Find the balance in the account after six years by computing a_{24}.

Writing in Mathematics

67. What is a sequence? Give an example with your description.

68. Explain how to write terms of a sequence if the formula for the general term is given.

69. What does the graph of a sequence look like? How is it obtained?

70. What is a recursion formula?

71. Explain how to find $n!$ if n is a positive integer.

72. Explain the best way to evaluate $\dfrac{900!}{899!}$ without a calculator.

73. What is the meaning of the symbol Σ? Give an example with your description.

74. You buy a new car for $24,000. At the end of n years, the value of your car is given by the sequence

$$a_n = 24{,}000\left(\frac{3}{4}\right)^n, \qquad n = 1, 2, 3, \ldots .$$

Find a_5 and write a sentence explaining what this value represents. Describe the nth term of the sequence in terms of the value of your car at the end of each year.

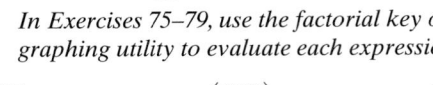

Technology Exercises

In Exercises 75–79, use the factorial key of a graphing utility to evaluate each expression.

75. $\dfrac{200!}{198!}$

76. $\left(\dfrac{300}{20}\right)!$

77. $\dfrac{20!}{300}$

78. $\dfrac{20!}{(20-3)!}$

79. $\dfrac{54!}{(54-3)!\,3!}$

80. Use the $\boxed{\text{SEQ}}$ (sequence) capability of a graphing utility to verify the terms of the sequences you obtained for any five sequences from Exercises 1–12 or 19–22.

81. Use the $\boxed{\text{SUM}}$ $\boxed{\text{SEQ}}$ (sum of the sequence) capability of a graphing utility to verify any five of the sums you obtained in Exercises 29–42.

82. As n increases, the terms of the sequence

$$a_n = \left(1 + \frac{1}{n}\right)^n$$

get closer and closer to the number e (where $e \approx 2.7183$). Use a calculator to find $a_{10}, a_{100}, a_{1000}, a_{10,000},$ and $a_{100,000}$, comparing these terms to the decimal approximation for e.

Many graphing utilities have a sequence-graphing mode that plots the terms of a sequence as points on a rectangular coordinate system. Consult your manual; if your graphing utility has this capability, use it to graph each of the sequences in Exercises 83–86. What appears to be happening to the terms of each sequence as n gets larger?

83. $a_n = \dfrac{n}{n+1}$ $\quad n{:}[0, 10, 1] \times a_n{:}[0, 1, 0.1]$

84. $a_n = \dfrac{100}{n}$ $\quad n{:}[0, 1000, 100] \times a_n{:}[0, 1, 0.1]$

85. $a_n = \dfrac{2n^2 + 5n - 7}{n^3}$ $\quad n{:}[0, 10, 1] \times a_n{:}[0, 2, 0.2]$

86. $a_n = \dfrac{3n^4 + n - 1}{5n^4 + 2n^2 + 1}$ $\quad n{:}[0, 10, 1] \times a_n{:}[0, 1, 0.1]$

Critical Thinking Exercises

87. Which one of the following is true?

a. $\dfrac{n!}{(n-1)!} = \dfrac{1}{n-1}$

b. The Fibonacci sequence $1, 1, 2, 3, 5, 8, 13, 21, 34, 55, 89, 144, \ldots$ can be defined recursively using $a_0 = 1, a_1 = 1,$ $a_n = a_{n-2} + a_{n-1}$, where $n \geq 2$.

c. $\displaystyle\sum_{i=1}^{2} (-1)^i 2^i = 0$

d. $\displaystyle\sum_{i=1}^{2} a_i b_i = \sum_{i=1}^{2} a_i \sum_{i=1}^{2} b_i$

88. Write the first five terms of the sequence whose first term is 9 and whose general term is

$$a_n = \begin{cases} \dfrac{a_{n-1}}{2} & \text{if } a_{n-1} \text{ is even} \\ 3a_{n-1} + 5 & \text{if } a_{n-1} \text{ is odd.} \end{cases}$$

Group Exercise

89. Enough curiosities involving the Fibonacci sequence exist to warrant a flourishing Fibonacci Association, which publishes a quarterly journal. Do some research on the Fibonacci sequence by consulting the Internet or the research department of your library, and find one property that interests you. After doing this research, get together with your group to share these intriguing properties.

SECTION 11.2   *Arithmetic Sequences*

Objectives

1. Find the common difference for an arithmetic sequence.
2. Write terms of an arithmetic sequence.
3. Use the formula for the general term of an arithmetic sequence.
4. Use the formula for the sum of the first *n* terms of an arithmetic sequence.

Your grandmother and her financial counselor are looking at options in case nursing home care is needed in the future. The good news is that your grandmother's total assets are $350,000. The bad news is that yearly nursing home costs average $49,730, increasing by $1800 each year. In this section, we will see how sequences can be used to describe your grandmother's situation and help her to identify realistic options.

Arithmetic Sequences

A mathematical model for the average annual salaries of major league baseball players generates the following data.

Year	1991	1992	1993	1994	1995	1996	1997	1998
Salary	801,000	892,000	983,000	1,074,000	1,165,000	1,256,000	1,347,000	1,438,000

From 1991 to 1992, salaries increased by $892,000 − $801,000 = $91,000. From 1992 to 1993, salaries increased by $983,000 − $892,000 = $91,000. If we make these computations for each year, we find that the yearly salary increase is $91,000. The sequence of annual salaries shows that each term after the first, 801,000, differs from the preceding term by a constant amount, namely 91,000. The sequence of annual salaries

$$801,000, \ 892,000, \ 983,000, \ 1,074,000, \ 1,165,000, \ 1,256,000, \ldots$$

is an example of an **arithmetic sequence**.

> **Definition of an Arithmetic Sequence**
>
> An **arithmetic sequence** is a sequence in which each term after the first differs from the preceding term by a constant amount. The difference between consecutive terms is called the **common difference** of the sequence.

1 Find the common difference of an arithmetic sequence.

The common difference, d, is found by subtracting any term from the term that directly follows it. In the following examples, the common difference is found by subtracting the first term from the second term, $a_2 - a_1$.

Arithmetic sequence	Common difference
$801,000, \ 892,000, \ 983,000, \ 1,074,000, \ldots$	$d = 892,000 - 801,000 = 91,000$
$2, 6, 10, 14, 18, \ldots$	$d = 6 - 2 = 4$
$-2, -7, -12, -17, \ldots$	$d = -7 - (-2) = -5$

If the first term of an arithmetic sequence is a_1, each term after the first is obtained by adding d, the common difference, to the previous term. This can be expressed recursively as follows:

$$a_n = a_{n-1} + d$$

> Add d to the term in any position to get the next term.

To use this recursion formula, we must be given the first term.

2 Write the terms of an arithmetic sequence.

EXAMPLE 1   **Writing the Terms of an Arithmetic Sequence Using the First Term and the Common Difference**

The recursion formula $a_n = a_{n-1} - 24$ models the thousands of Air Force personnel on active duty for each year starting with 1986. In 1986, there were 624 thousand personnel on active duty. Find the first five terms of the arithmetic sequence in which $a_1 = 624$ and $a_n = a_{n-1} - 24$.

Solution   The recursion formula $a_n = a_{n-1} - 24$ indicates that each term after the first is obtained by adding -24 to the previous term. Thus, each year there are 24 thousand fewer personnel on active duty in the Air Force than in the previous year.

$a_1 = 624$	This is given.
$a_2 = a_1 - 24 = 624 - 24 = 600$	Use $a_n = a_{n-1} - 24$ with $n = 2$.
$a_3 = a_2 - 24 = 600 - 24 = 576$	Use $a_n = a_{n-1} - 24$ with $n = 3$.
$a_4 = a_3 - 24 = 576 - 24 = 552$	Use $a_n = a_{n-1} - 24$ with $n = 4$.
$a_5 = a_4 - 24 = 552 - 24 = 528$	Use $a_n = a_{n-1} - 24$ with $n = 5$.

The first five terms are

$$624, 600, 576, 552, \text{ and } 528.$$

Check Point 1   Find the first five terms of the arithmetic sequence in which $a_1 = 100$ and $a_n = a_{n-1} - 30$.

3 Use the formula for the general term of an arithmetic sequence.

The General Term of an Arithmetic Sequence

Consider an arithmetic sequence whose first term is a_1 and whose common difference is d. We are looking for a formula for the general term, a_n. Let's begin by writing the first six terms. The first term is a_1. The second term is $a_1 + d$. The third term is $a_1 + d + d$, or $a_1 + 2d$. Thus, we start with a_1 and add d to each successive term. The first six terms are

$$a_1, \qquad a_1 + d, \qquad a_1 + 2d, \qquad a_1 + 3d, \qquad a_1 + 4d, \qquad a_1 + 5d.$$

a_1, first term $\qquad$ a_2, second term $\qquad$ a_3, third term $\qquad$ a_4, fourth term $\qquad$ a_5, fifth term $\qquad$ a_6, sixth term

Compare the coefficient of d and the subscript of a denoting the term number. Can you see that the coefficient of d is 1 less than the subscript of a denoting the term number?

$$a_3 \text{: third term} = a_1 + 2d \qquad a_4 \text{: fourth term} = a_1 + 3d$$

2 is one less than 3. $\qquad$ 3 is one less than 4.

Thus, the formula for the nth term is

$$a_n \text{: } n\text{th term} = a_1 + (n - 1)d.$$

$n - 1$ is one less than n.

General Term of an Arithmetic Sequence

The nth term (the general term) of an arithmetic sequence with first term a_1 and common difference d is

$$a_n = a_1 + (n - 1)d.$$

EXAMPLE 2 Using the Formula for the General Term of an Arithmetic Sequence

Find the eighth term of the arithmetic sequence whose first term is 4 and whose common difference is -7.

Solution To find the eighth term, a_8, we replace n in the formula with 8, a_1 with 4, and d with -7.

$$a_n = a_1 + (n - 1)d$$
$$a_8 = 4 + (8 - 1)(-7) = 4 + 7(-7) = 4 + (-49) = -45$$

The eighth term is -45. We can check this result by writing the first eight terms of the sequence:

$$4, -3, -10, -17, -24, -31, -38, -45.$$

Check Point 2 Find the ninth term of the arithmetic sequence whose first term is 6 and whose common difference is -5.

EXAMPLE 3 Using an Arithmetic Sequence to Model Teachers' Earnings

According to the National Education Association, teachers in the United States earned an average of $21,700 per year in 1984. This amount has increased by approximately $1472 yearly.

a. Write a formula for the nth term of the arithmetic sequence that describes teachers' average earnings n years after 1983.

b. How much will U.S. teachers earn by the year 2005?

Solution

a. We can express teachers' earnings by the following arithmetic sequence:

$$21,700, \qquad 23,172, \qquad 24,644, \qquad 26,116,\dots.$$

a_1: earnings in 1984, 1 year after 1983	a_2: earnings in 1985, 2 years after 1983	a_3: earnings in 1986, 3 years after 1983	a_4: earnings in 1987, 4 years after 1983

In this sequence a_1, the first term, represents the amount teachers earned in 1984. Each subsequent year this amount increases by $1472, so $d = 1472$. We use the formula for the general term of an arithmetic sequence to write the nth term of the sequence that describes teachers' earnings n years after 1983.

$a_n = a_1 + (n - 1)d$ This is the formula for the general term of an arithmetic sequence.

$a_n = 21,700 + (n - 1)1472$ $a_1 = 21,700$ and $d = 1472$.

$a_n = 21,700 + 1472n - 1472$ Distribute 1472 to each term in parentheses.

$a_n = 1472n + 20,228$ Simplify.

Thus, teachers' earnings n years after 1983 can be described by $a_n = 1472n + 20,228$.

b. Now we need to find teachers' earnings in 2005. The year 2005 is 22 years after 1983: That is, $2005 - 1983 = 22$. Thus, $n = 22$. We substitute 22 for n in $a_n = 1472n + 20,228$.

$$a_{22} = 1472 \cdot 22 + 20,228 = 52,612$$

The 22nd term of the sequence is 52,612. Therefore, U.S. teachers are predicted to earn an average of $52,612 by the year 2005.

Check Point 3

According to the U.S. Bureau of Economic Analysis, U.S. travelers spent $12,808 million in other countries in 1984. This amount has increased by approximately $2350 million yearly.

a. Write a formula for the nth term of the arithmetic sequence that describes what U.S. travelers spend in other countries n years after 1983.

b. How much will U.S. travelers spend in other countries by the year 2010?

 4   Use the formula for the sum of the first n terms of an arithmetic sequence.

The Sum of the First n Terms of an Arithmetic Sequence

The sum of the first n terms of an arithmetic sequence, denoted by S_n, can be found without having to add up all the terms. Let

$$S_n = a_1 + a_2 + a_3 + \cdots + a_n$$

be the sum of the first n terms of an arithmetic sequence. Because d is the common difference between terms, S_n can be written forward and backward as follows.

Forward: Start with the first term. Keep adding d.

Backward: Start with the last term. Keep subtracting d.

$$S_n = a_1 \qquad + (a_1 + d) \quad + (a_1 + 2d) + \cdots + a_n$$
$$S_n = a_n \qquad + (a_n - d) \quad + (a_n - 2d) + \cdots + a_1$$
$$\overline{2S_n = (a_1 + a_n) + (a_1 + a_n) + (a_1 + a_n) + \cdots + (a_1 + a_n)} \qquad \text{Add the two equations.}$$

Because there are n sums of $(a_1 + a_n)$ on the right side, we can express this side as $n(a_1 + a_n)$. Thus, the last equation can be simplified:

$$2S_n = n(a_1 + a_n)$$

$$S_n = \frac{n}{2}(a_1 + a_n) \qquad \text{Solve for } S_n, \text{ dividing both sides by 2.}$$

We have proved the following result.

The Sum of the First n Terms of an Arithmetic Sequence

The sum, S_n, of the first n terms of an arithmetic sequence is given by

$$S_n = \frac{n}{2}(a_1 + a_n)$$

in which a_1 is the first term and a_n is the nth term.

To find the sum of the terms of an arithmetic sequence, we need to know the first term, a_1, the last term, a_n, and the number of terms, n. The following examples illustrate how to use this formula.

EXAMPLE 4 Finding the Sum of n Terms of an Arithmetic Sequence

Find the sum of the first 100 terms of the arithmetic sequence: $1, 3, 5, 7, \ldots$.

Solution We are finding the sum of the first 100 odd numbers. To find the sum of the first 100 terms, S_{100}, we replace n in the formula with 100.

$$S_n = \frac{n}{2}(a_1 + a_n)$$

$$S_{100} = \frac{100}{2}(a_1 + a_{100})$$

The first term, a_1, is 1.

We must find a_{100}, the 100th term.

We use the formula for the general term of a sequence to find a_{100}. The common difference, d, of $1, 3, 5, 7, \ldots$, is 2.

$$a_n = a_1 + (n - 1)d \qquad \text{This is the formula for the } n\text{th term of an arithmetic sequence. Use it to find the 100th term.}$$

$$a_{100} = 1 + (100 - 1) \cdot 2 \qquad \text{Substitute 100 for } n, \text{ 2 for } d, \text{ and 1 (the first term) for } a_1.$$

$$= 1 + 99 \cdot 2$$

$$= 199$$

Now we are ready to find the sum of the first 100 terms of $1, 3, 5, 7, \ldots, 199$.

$$S_n = \frac{n}{2}(a_1 + a_n)$$ Use the formula for the sum of the first n terms of an arithmetic sequence. Let $n = 100$, $a_1 = 1$, and $a_{100} = 199$.

$$S_{100} = \frac{100}{2}(1 + 199) = 50(200) = 10,000$$

The sum of the first 100 odd numbers is 10,000.

> **Check Point 4**   Find the sum of the first 15 terms of the arithmetic sequence: $3, 6, 9, 12, \ldots$.

Technology

To find

$$\sum_{i=1}^{25}(5i - 9)$$

on a graphing utility, enter:

[SUM] [SEQ] $(5x - 9, x, 1,$
$25, 1)$. Then press [ENTER].

```
sum(seq(5X-9,X,1
,25,1)
             1400
```

EXAMPLE 5   Using S_n to Evaluate a Summation

Find the following sum: $\displaystyle\sum_{i=1}^{25}(5i - 9)$.

Solution

$$\sum_{i=1}^{25}(5i - 9) = (5 \cdot 1 - 9) + (5 \cdot 2 - 9) + (5 \cdot 3 - 9) + \cdots + (5 \cdot 25 - 9)$$

$$= -4 \qquad + 1 \qquad + 6 \qquad + \cdots + 116$$

By evaluating the first three terms and the last term, we see that $a_1 = -4$; d, the common difference, is $1 - (-4)$ or 5; and a_{25}, the last term, is 116.

$$S_n = \frac{n}{2}(a_1 + a_n)$$ Use the formula for the sum of the first n terms of an arithmetic sequence. Let $n = 25$, $a_1 = -4$, and $a_{25} = 116$.

$$S_{25} = \frac{25}{2}(-4 + 116) = \frac{25}{2}(112) = 1400.$$

Thus,

$$\sum_{i=1}^{25}(5i - 9) = 1400.$$

> **Check Point 5**   Find the following sum: $\displaystyle\sum_{i=1}^{30}(6i - 11)$.

EXAMPLE 6   Modeling Total Nursing Home Costs over a Six-Year Period

Your grandmother has assets of \$350,000. One option that she is considering involves nursing home care for a six-year period beginning in 2001. The model

$$a_n = 1800n + 49{,}730$$

describes yearly nursing home costs n years after 2000. Does your grandmother have enough to pay for the facility?

Solution   We must find the sum of an arithmetic sequence. The first term of the sequence corresponds to nursing home costs in the year 2001. The last term corresponds to nursing home costs in the year 2006. Because the model describes costs n years after 2000, $n = 1$ describes the year 2001 and $n = 6$ describes the year 2006.

$$a_n = 1800n + 49{,}730$$

This is the given formula for the general term of the sequence.

$$a_1 = 1800 \cdot 1 + 49{,}730 = 51{,}530$$ Find a_1 by replacing n by 1.

$$a_6 = 1800 \cdot 6 + 49{,}730 = 60{,}530$$ Find a_6 by replacing n by 6.

The first year the facility will cost $51,530. By year six, the facility will cost $60,530. Now we must find the sum of these costs for all six years. We focus on the sum of the first six terms of the arithmetic sequence

$$51{,}530, \quad 53{,}330, \quad \ldots, \quad 60{,}530.$$

a_1 a_2 a_6

We find this sum using the formula for the sum of the first n terms of an arithmetic sequence. We are adding 6 terms: $n = 6$. The first term is 51,530: $a_1 = 51{,}530$. The last term—that is, the sixth term—is 60,530: $a_6 = 60{,}530$.

$$S_n = \frac{n}{2}(a_1 + a_n)$$

$$S_6 = \frac{6}{2}(51{,}530 + 60{,}530) = 3(112{,}060) = 336{,}180$$

Total nursing home costs for your grandmother are predicted to be $336,180. Because your grandmother's assets are $350,000, she has enough to pay for the facility.

Check Point 6   In Example 6, how much would it cost for nursing home care for a ten-year period beginning in 2001?

EXERCISE SET 11.2

Practice Exercises

In Exercises 1–14, write the first six terms of each arithmetic sequence.

1. $a_1 = 200, d = 20$
2. $a_1 = 300, d = 50$
3. $a_1 = -7, d = 4$
4. $a_1 = -8, d = 5$
5. $a_1 = 300, d = -90$
6. $a_1 = 200, d = -60$
7. $a_1 = \frac{5}{2}, d = -\frac{1}{2}$
8. $a_1 = \frac{3}{4}, d = -\frac{1}{4}$
9. $a_n = a_{n-1} + 6, a_1 = -9$
10. $a_n = a_{n-1} + 4, a_1 = -7$
11. $a_n = a_{n-1} - 10, a_1 = 30$
12. $a_n = a_{n-1} - 20, a_1 = 50$
13. $a_n = a_{n-1} - 0.4, a_1 = 1.6$
14. $a_n = a_{n-1} - 0.3, a_1 = -1.7$

In Exercises 15–22, find the indicated term of the arithmetic sequence with first term, a_1, and common difference, d.

15. Find a_6 when $a_1 = 13, d = 4$.
16. Find a_{16} when $a_1 = 9, d = 2$.

17. Find a_{50} when $a_1 = 7, d = 5$.
18. Find a_{60} when $a_1 = 8, d = 6$.
19. Find a_{200} when $a_1 = -40, d = 5$.
20. Find a_{150} when $a_1 = -60, d = 5$.
21. Find a_{60} when $a_1 = 35, d = -3$.
22. Find a_{70} when $a_1 = -32, d = 4$.

In Exercises 23–34, write a formula for the general term (the nth term) of each arithmetic sequence. Do not use a recursion formula. Then use the formula for a_n to find a_{20}, the 20th term of the sequence.

23. $1, 5, 9, 13, \ldots$
24. $2, 7, 12, 17, \ldots$
25. $7, 3, -1, -5, \ldots$
26. $6, 1, -4, -9, \ldots$
27. $a_1 = 9, d = 2$
28. $a_1 = 6, d = 3$
29. $a_1 = -20, d = -4$
30. $a_1 = -70, d = -5$
31. $a_n = a_{n-1} + 3, a_1 = 4$
32. $a_n = a_{n-1} + 5, a_1 = 6$
33. $a_n = a_{n-1} - 10, a_1 = 30$
34. $a_n = a_{n-1} - 12, a_1 = 24$

35. Find the sum of the first 20 terms of the arithmetic sequence: $4, 10, 16, 22, \ldots$.

36. Find the sum of the first 25 terms of the arithmetic sequence: $7, 19, 31, 43, \ldots$.

37. Find the sum of the first 50 terms of the arithmetic sequence: $-10, -6, -2, 2, \ldots$.

38. Find the sum of the first 50 terms of the arithmetic sequence: $-15, -9, -3, 3, \ldots$.

39. Find $1 + 2 + 3 + 4 + \ldots + 100$, the sum of the first 100 natural numbers.

40. Find $2 + 4 + 6 + 8 + \ldots + 200$, the sum of the first 100 positive even integers.

41. Find the sum of the first 60 positive even integers.

42. Find the sum of the first 80 positive even integers.

43. Find the sum of the even integers between 21 and 45.

44. Find the sum of the odd integers between 30 and 54.

For Exercises 45–50, write out the first three terms and the last term. Then use the formula for the sum of the first n terms of an arithmetic sequence to find the indicated sum.

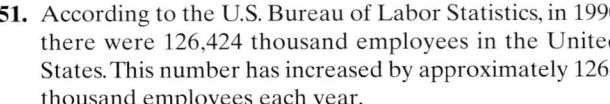

45. $\displaystyle\sum_{i=1}^{17} (5i + 3)$ **46.** $\displaystyle\sum_{i=1}^{20} (6i - 4)$

47. $\displaystyle\sum_{i=1}^{30} (-3i + 5)$ **48.** $\displaystyle\sum_{i=1}^{40} (-2i + 6)$

49. $\displaystyle\sum_{i=1}^{100} 4i$ **50.** $\displaystyle\sum_{i=1}^{50} -4i$

⭐ Application Exercises

51. According to the U.S. Bureau of Labor Statistics, in 1990 there were 126,424 thousand employees in the United States. This number has increased by approximately 1265 thousand employees each year.
 a. Write the general term for the arithmetic sequence modeling the thousands of employees in the United States n years after 1989.
 b. How many thousands of employees will there be by the year 2005?

52. According to the National Center for Education Statistics, the total enrollment in U.S. public elementary and secondary schools in 1985 was 39.05 million. Enrollment has increased by approximately 0.45 million each year.
 a. Write the general term for the arithmetic sequence modeling the millions of students enrolled in U.S. public elementary and secondary schools n years after 1984.
 b. How many millions of students will be enrolled by the year 2005?

53. Company A pays $24,000 yearly with raises of $1600 per year. Company B pays $28,000 yearly with raises of $1000 per year. Which company will pay more in year 10? How much more?

54. Company A pays $23,000 yearly with raises of $1200 per year. Company B pays $26,000 yearly with raises of $800 per year. Which company will pay more in year 10? How much more?

55. According to the Environmental Protection Agency, in 1960 the United States recovered 3.78 million tons of solid waste. Due primarily to recycling programs, this amount has increased by approximately 0.576 million ton each year.
 a. Write the general term for the arithmetic sequence modeling the amount of solid waste recovered in the United States n years after 1959.
 b. What is the total amount of solid waste recovered from 1960 through 2000?

56. According to the Environmental Protection Agency, in 1960 the United States generated 87.1 million tons of solid waste. This amount has increased by approximately 3.14 million tons each year.
 a. Write the general term for the arithmetic sequence modeling the amount of solid waste generated in the United States n years after 1959.
 b. What is the total amount of solid waste generated from 1960 through 2000?

57. A company offers a starting yearly salary of $33,000 with raises of $2500 per year. Find the total salary over a ten-year period.

58. You are considering two job offers. Company A will start you at $19,000 a year and guarantee you a raise of $2600 per year. Company B will start you at a higher salary, $27,000 a year, but will only guarantee a raise of $1200 per year. Find the total salary that each company will pay you over a ten-year period. Which company pays the greater total amount?

59. A theater has 30 seats in the first row, 32 seats in the second row, increasing by 2 seats each row for a total of 26 rows. How many seats are there in the theater?

60. A section in a stadium has 20 seats in the first row, 23 seats in the second row, increasing by 3 seats each row for a total of 38 rows. How many seats are in this section of the stadium?

✏️ Writing in Mathematics

61. What is an arithmetic sequence? Give an example with your explanation.

62. What is the common difference in an arithmetic sequence?

63. Explain how to find the general term of an arithmetic sequence.

64. Explain how to find the sum of the first n terms of an arithmetic sequence without having to add up all the terms.

65. Teachers' earnings n years after 1983 can be described by $a_n = 1472n + 20{,}228$. According to this model, what will teachers earn in 2083? Describe two possible circumstances that would render this predicted salary incorrect.

Technology Exercises

66. Use the ⌈SEQ⌉ (sequence) capability of a graphing utility and the formula you obtained for a_n to verify the value you found for a_{20} in any five exercises from Exercises 23–34.

67. Use the capability of a graphing utility to calculate the sum of a sequence to verify any five of your answers to Exercises 45–50.

Critical Thinking Exercises

68. Give examples of two different arithmetic sequences whose fourth term, a_4, is 10.

69. In the sequence 21,700, 23,172, 24,644, 26,116,..., which term is 314,628?

70. A *degree-day* is a unit used to measure the fuel requirements of buildings. By definition, each degree that the average daily temperature is below 65°F is 1 degree-day. For example, a temperature of 42°F constitutes 23 degree-days. If the average temperature on January 1 was 42°F and fell 2°F for each subsequent day up to and including January 10, how many degree-days are included from January 1 to January 10?

71. Show that the sum of the first n positive odd integers,
$$1 + 3 + 5 + \cdots + (2n - 1),$$
is n^2.

Group Exercise

72. Members of your group have been hired by the Environmental Protection Agency to write a report on whether we are making significant progress in recovering solid waste. Use the models from Exercises 55 and 56 as the basis for your report. A graph of each model from 1960 through 2000 would be helpful. What percentage of solid waste generated is actually recovered on a year-to-year basis? Be as creative as you want in your report and then draw conclusions. The group should write up the report and perhaps even include suggestions as to how we might improve recycling progress.

SECTION 11.3 *Geometric Sequences*

Objectives

1. Find the common ratio of a geometric sequence.
2. Write terms of a geometric sequence.
3. Use the formula for the general term of a geometric sequence.
4. Use the formula for the sum of the first n terms of a geometric sequence.
5. Find the value of an annuity.
6. Use the formula for the sum of an infinite geometric series.

Here we are at the closing moments of a job interview. You're shaking hands with the manager. You managed to answer all the tough questions without losing your poise, and now you've been offered a job. As a matter of fact, your qualifications are so terrific that you've been offered two jobs—one just the day before, with a rival company in the same field! One company offers $30,000 the first year, with increases of 6% per year for four years after that. The other offers $32,000 the first year, with annual increases of 3% per year after that. Over a five-year period, which is the better offer?

If salary raises amount to a certain percent each year, the yearly salaries over time form a geometric sequence. In this section, we investigate geometric sequences and their properties. After studying the section, you will be in a position to decide which job offer to accept: you will know which company will pay you more over five years.

Geometric Sequences

Figure 11.2 shows a sequence in which the number of squares is increasing. From left to right, the number of squares is 1, 5, 25, 125, and 625. In this sequence, each term after the first, 1, is obtained by multiplying the preceding term by a constant amount, namely 5. This sequence of increasing number of squares is an example of a *geometric sequence*.

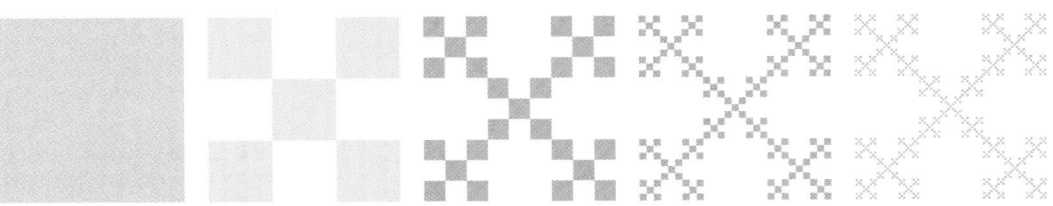

Figure 11.2 A geometric sequence of squares

> ### Definition of a Geometric Sequence
> A **geometric sequence** is a sequence in which each term after the first is obtained by multiplying the preceding term by a fixed nonzero constant. The amount by which we multiply each time is called the **common ratio** of the sequence.

1 Find the common ratio of a geometric sequence.

The common ratio, r, is found by dividing any term after the first term by the term that directly precedes it. In the following examples, the common ratio is found by dividing the second term by the first term, $\dfrac{a_2}{a_1}$.

Geometric sequence	Common ratio
$1, 5, 25, 125, 625, \ldots$	$r = \dfrac{5}{1} = 5$
$4, 8, 16, 32, 64, \ldots$	$r = \dfrac{8}{4} = 2$
$6, -12, 24, -48, 96, \ldots$	$r = \dfrac{-12}{6} = -2$
$9, -3, 1, -\dfrac{1}{3}, \dfrac{1}{9}, \ldots$	$r = \dfrac{-3}{9} = -\dfrac{1}{3}$

Study Tip

When the common ratio of a geometric sequence is negative, the signs of the terms alternate.

2 Write terms of a geometric sequence.

How do we write out the terms of a geometric sequence when the first term and the common ratio are known? We multiply the first term by the common ratio to get the second term, multiply the second term by the common ratio to get the third term, and so on.

EXAMPLE 1 Writing the Terms of a Geometric Sequence

Write the first six terms of the geometric sequence with first term 6 and common ratio $\frac{1}{3}$.

Solution The first term is 6. The second term is $6 \cdot \frac{1}{3}$, or 2. The third term is $2 \cdot \frac{1}{3}$, or $\frac{2}{3}$. The fourth term is $\frac{2}{3} \cdot \frac{1}{3}$, or $\frac{2}{9}$, and so on. The first six terms are

$$6, 2, \tfrac{2}{3}, \tfrac{2}{9}, \tfrac{2}{27}, \tfrac{2}{81}.$$

> **Check Point 1** Write the first six terms of the geometric sequence with first term 12 and common ratio $\frac{1}{2}$.

3 Use the formula for the general term of a geometric sequence.

The General Term of a Geometric Sequence

Consider a geometric sequence whose first term is a_1, and whose common ratio is r. We are looking for a formula for the general term, a_n. Let's begin by writing the first six terms. The first term is a_1. The second term is $a_1 r$. The third term is $a_1 r \cdot r$, or $a_1 r^2$. The fourth term is $a_1 r^2 \cdot r$, or $a_1 r^3$, and so on. Starting with a_1 and multiplying each successive term by r, the first six terms are

$$a_1, \qquad a_1 r, \qquad a_1 r^2, \qquad a_1 r^3, \qquad a_1 r^4, \qquad a_1 r^5.$$

a_1, first term a_2, second term a_3, third term a_4, fourth term a_5, fifth term a_6, sixth term

Compare the exponent on r and the subscript of a denoting the term number. Can you see that the exponent on r is 1 less than the subscript of a denoting the term number?

$$a_3 : \text{ third term} = a_1 r^2 \qquad\qquad a_4 : \text{ fourth term} = a_1 r^3$$

2 is one less than 3.   3 is one less than 4.

Thus, the formula for the nth term is

$$a_n = a_1 r^{n-1}.$$

$n - 1$ is one less than n.

> ### General Term of a Geometric Sequence
>
> The nth term (the general term) of a geometric sequence with first term a_1 and common ratio r is
>
> $$a_n = a_1 r^{n-1}.$$

Study Tip

Be careful with the order of operations when evaluating

$$a_1 r^{n-1}.$$

First find r^{n-1}. Then multiply the result by a_1.

EXAMPLE 2 **Using the Formula for the General Term of a Geometric Sequence**

Find the eighth term of the geometric sequence whose first term is -4 and whose common ratio is -2.

Solution To find the eighth term, a_8, we replace n in the formula with 8, a_1 with -4, and r with -2.

$$a_n = a_1 r^{n-1}$$
$$a_8 = -4(-2)^{8-1} = -4(-2)^7 = -4(-128) = 512$$

The eighth term is 512. We can check this result by writing the first eight terms of the sequence:

$$-4, 8, -16, 32, -64, 128, -256, 512.$$

Check Point 2

Find the seventh term of the geometric sequence whose first term is 5 and whose common ratio is -3.

In Chapter 4, we studied exponential functions of the form $f(x) = b^x$ and the explosive exponential growth of world population. In our next example, we consider Florida's geometric population growth. Because **a geometric sequence is an exponential function whose domain is the set of positive integers,** geometric and exponential growth mean the same thing. (By contrast, an arithmetic sequence is a *linear function* whose domain is the set of positive integers.)

EXAMPLE 3   Geometric Population Growth

Geometric Population Growth

Economist Thomas Malthus (1766–1834) predicted that population growth would increase as a geometric sequence and food production would increase as an arithmetic sequence. He concluded that eventually population would exceed food production. If two sequences, one geometric and one arithmetic, are increasing, the geometric sequence will eventually overtake the arithmetic sequence, regardless of any head start that the arithmetic sequence might initially have.

The population of Florida from 1980 through 1987 is shown in the following table.

Year	1980	1981	1982	1983	1984	1985	1986	1987
Population in millions	9.75	10.03	10.32	10.62	10.93	11.25	11.58	11.92

a. Show that the population is increasing geometrically.

b. Write the general term for the geometric sequence describing population growth for Florida n years after 1979.

c. Estimate Florida's population, in millions, for the year 2000.

Solution

a. First, we divide the population for each year by the population in the preceding year.

$$\frac{10.03}{9.75} \approx 1.029, \quad \frac{10.32}{10.03} \approx 1.029, \quad \frac{10.62}{10.32} \approx 1.029$$

Continuing in this manner, we will keep getting approximately 1.029. This means that the population is increasing geometrically with $r \approx 1.029$. In this situation, the common ratio is the growth rate, indicating that the population of Florida in any year shown in the table is approximately 1.029 times the population the year before.

b. The sequence of Florida's population growth is

$$9.75, 10.03, 10.32, 10.62, 10.93, 11.25, 11.58, 11.92, \dots.$$

Because the population is increasing geometrically, we can find the general term of this sequence using

$$a_n = a_1 r^{n-1}.$$

In this sequence, $a_1 = 9.75$ and r [from part (a)] ≈ 1.029. We substitute these values into the formula for the general term. This gives the general term for the geometric sequence describing Florida's population n years after 1979.

$$a_n = 9.75(1.029)^{n-1}$$

c. We can use the formula for the general term, a_n, in part (b) to estimate Florida's population for the year 2000. The year 2000 is 21 years after 1979—that is, $2000 - 1979 = 21$. Thus, $n = 21$. We substitute 21 for n in $a_n = 9.75(1.029)^{n-1}$.

$$a_{21} = 9.75(1.029)^{21-1} = 9.75(1.029)^{20} \approx 17.27$$

The formula predicts that Florida will have a population of approximately 17.27 million in the year 2000.

> **Check Point 3**
>
> Write the general term for the geometric sequence
>
> $$3, 6, 12, 24, 48, \ldots.$$
>
> Then use the formula for the general term to find the eighth term.

4 Use the formula for the sum of the first n terms of a geometric sequence.

The Sum of the First n Terms of a Geometric Sequence

The sum of the first n terms of a geometric sequence, denoted by S_n, can be found without having to add up all the terms. Recall that the first n terms of a geometric sequence are

$$a_1, a_1 r, a_1 r^2, \ldots, a_1 r^{n-2}, a_1 r^{n-1}.$$

We proceed as follows:

$$S_n = a_1 + a_1 r + a_1 r^2 + \cdots + a_1 r^{n-2} + a_1 r^{n-1} \qquad \text{S_n is the sum of the first n terms of the sequence.}$$

$$r S_n = a_1 r + a_1 r^2 + a_1 r^3 + \cdots + a_1 r^{n-1} + a_1 r^n \qquad \text{Multiply both sides of the equation by r.}$$

$$S_n - r S_n = a_1 - a_1 r^n \qquad \text{Subtract the second equation from the first equation.}$$

$$S_n(1 - r) = a_1(1 - r^n) \qquad \text{Factor out S_n on the left and a_1 on the right.}$$

$$S_n = \frac{a_1(1 - r^n)}{1 - r} \qquad \text{Solve for S_n by dividing both sides by $1 - r$ (assuming that $r \neq 1$).}$$

We have proved the following result.

Study Tip

If the common ratio is 1, the geometric sequence is

$$a_1, a_1, a_1, a_1, \ldots.$$

The sum of the first n terms of this sequence is na_1:

$$S_n = \underbrace{a_1 + a_1 + a_1 + \cdots + a_1}_{\text{There are n terms.}}$$

$$= na_1.$$

The Sum of the First n Terms of a Geometric Sequence

The sum, S_n, of the first n terms of a geometric sequence is given by

$$S_n = \frac{a_1(1 - r^n)}{1 - r}$$

in which a_1 is the first term and r is the common ratio ($r \neq 1$).

To find the sum of the terms of a geometric sequence, we need to know the first term, a_1, the common ratio, r, and the number of terms, n. The following examples illustrate how to use this formula.

EXAMPLE 4 Finding the Sum of n Terms of a Geometric Sequence

Find the sum of the first 18 terms of the geometric sequence: $2, -8, 32, -128, \ldots$.

Solution To find the sum of the first 18 terms, S_{18}, we replace n in the formula with 18.

$$S_n = \frac{a_1(1 - r^n)}{1 - r}$$

$$S_{18} = \frac{a_1(1 - r^{18})}{1 - r}$$

The first term, a_1, is 2.	We must find r, the common ratio.

We can find the common ratio by dividing the second term by the first term.

$$r = \frac{a_2}{a_1} = \frac{-8}{2} = -4$$

Now we are ready to find the sum of the first 18 terms of $2, -8, 32, -128, \ldots$.

$$S_n = \frac{a_1(1 - r^n)}{1 - r} \qquad \text{Use the formula for the sum of the first } n \text{ terms of a geometric sequence.}$$

$$S_{18} = \frac{2(1 - (-4)^{18})}{1 - (-4)} \qquad a_1 \text{ (the first term)} = 2, r = -4, \text{ and } n = 18 \text{ because we want the sum of the first 18 terms.}$$

$$= -27{,}487{,}790{,}694 \qquad \text{Use a calculator.}$$

The sum of the first 18 terms is $-27{,}487{,}790{,}694$.

Check Point 4 Find the sum of the first nine terms of the geometric sequence: $2, -6, 18, -54, \ldots$.

EXAMPLE 5 Using S_n to Evaluate a Summation

Find the following sum: $\displaystyle\sum_{i=1}^{10} 6 \cdot 2^i$

Solution Let's write out a few terms in the sum.

$$\sum_{i=1}^{10} 6 \cdot 2^i = 6 \cdot 2 + 6 \cdot 2^2 + 6 \cdot 2^3 + \cdots + 6 \cdot 2^{10}$$

Can you see that each term after the first is obtained by multiplying the preceding term by 2? To find the sum of the 10 terms ($n = 10$), we need to know the first term, a_1, and the common ratio, r. The first term is $6 \cdot 2$ or 12: $a_1 = 12$. The common ratio is 2.

$$S_n = \frac{a_1(1 - r^n)}{1 - r} \qquad \text{Use the formula for the sum of the first } n \text{ terms of a geometric sequence.}$$

$$S_{10} = \frac{12(1 - 2^{10})}{1 - 2} \qquad a_1 \text{ (the first term)} = 12, r = 2, \text{ and } n = 10 \text{ because we are adding ten terms.}$$

$$= 12{,}276 \qquad \text{Use a calculator.}$$

Technology

To find

$$\sum_{i=1}^{10} 6 \cdot 2^i$$

on a graphing utility, enter

$\boxed{\text{SUM}}\ \boxed{\text{SEQ}}\ (6 \times 2^x, x, 1, 10, 1)$.

Then press $\boxed{\text{ENTER}}$.

```
sum(seq(6*2^X,X,
1,10,1)
              12276
```

Thus,

$$\sum_{i=1}^{10} 6 \cdot 2^i = 12{,}276$$

Check Point 5 Find the following sum: $\displaystyle\sum_{i=1}^{8} 2 \cdot 3^i$.

Some of the exercises in the previous exercise set involved situations in which salaries increase by a fixed amount each year. A more realistic situation is one in which salary raises increase by a certain percent each year. Example 6 shows how such a situation can be described using a geometric series.

EXAMPLE 6 Computing a Lifetime Salary

A union contract specifies that each worker will receive a 5% pay increase each year for the next 30 years. One worker is paid $20,000 the first year. What is this person's total lifetime salary over a 30-year period?

Solution The salary for the first year is $20,000. With a 5% raise, the second-year salary is computed as follows:

$$\text{Salary for year 2} = 20{,}000 + 20{,}000(0.05) = 20{,}000(1.05).$$

Each year, the salary is 1.05 times what it was in the previous year. Thus, the salary for year 3 is 1.05 times 20,000(1.05), or $20,000(1.05)^2$. The salaries for the first five years are given in the table.

Yearly Salaries					
Year 1	**Year 2**	**Year 3**	**Year 4**	**Year 5**	...
20,000	20,000(1.05)	$20{,}000(1.05)^2$	$20{,}000(1.05)^3$	$20{,}000(1.05)^4$	...

The numbers in the second row form a geometric sequence with $a_1 = 20{,}000$ and $r = 1.05$. To find the total salary over 30 years, we use the formula for the sum of the first n terms of a geometric sequence, with $n = 30$.

$$S_n = \frac{a_1(1 - r^n)}{1 - r}$$

$$S_{30} = \frac{20{,}000(1 - (1.05)^{30})}{1 - 1.05}$$

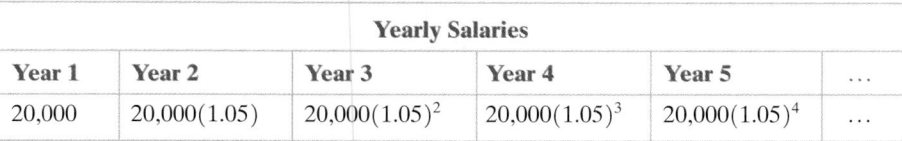
Total salary over 30 years

$$= \frac{20{,}000(1 - (1.05)^{30})}{-0.05}$$

$$\approx 1{,}328{,}777 \qquad \textit{Use a calculator.}$$

The total salary over the 30-year period is approximately $1,328,777.

Check Point 6 A job pays a salary of $30,000 the first year. During the next 29 years, the salary increases by 6% each year. What is the total lifetime salary over the 30-year period?

5 Find the value of an annuity.

Annuities

The compound interest formula

$$A = P(1 + r)^t$$

gives the future value, A, after t years, when a fixed amount of money, P, the principal, is deposited in an account that pays an annual interest rate r (in decimal form) compounded once a year. However, money is often invested in small amounts at periodic intervals. For example, to save for retirement, you might decide to place $1000 into an Individual Retirement Account (IRA) at the end of each year until you retire. An **annuity** is a sequence of equal payments made at equal time periods. An IRA is an example of an annuity.

Suppose P dollars is deposited into an account at the end of each year. The account pays an annual interest rate, r, compounded annually. At the end of the first year, the account contains P dollars. At the end of the second year, P dollars is deposited again. At the time of this deposit, the first deposit has received interest earned during the second year. The **value of the annuity** is the sum of all deposits made plus all interest paid. Thus, the value of the annuity after two years is

$$P + P(1 + r).$$

Deposit of P dollars at end of second year	First-year deposit of P dollars with interest earned for a year

The value of the annuity after three years is

$$P \quad + \quad P(1 + r) \quad + \quad P(1 + r)^2.$$

Deposit of P dollars at end of third year	Second-year deposit of P dollars with interest earned for a year	First-year deposit of P dollars with interest earned over two years

The value of the annuity after t years is

$$P + P(1 + r) + P(1 + r)^2 + P(1 + r)^3 + \cdots + P(1 + r)^{t-1}$$

Deposit of P dollars at end of year t	First-year deposit of P dollars with interest earned over $t - 1$ years

This is a geometric series with first term P and common ratio $1 + r$. We use the formula

$$S_n = \frac{a_1(1 - r^n)}{1 - r}$$

to find the sum of the terms:

$$S_n = \frac{P(1 - (1 + r)^t)}{1 - (1 + r)} = \frac{P(1 - (1 + r)^t)}{-r} = P\frac{(1 + r)^t - 1}{r}.$$

This formula gives the value of an annuity after t years if interest is compounded once a year. We can adjust the formula to find the value of an annuity if equal payments are made at the end of each of n yearly compounding periods.

Value of an Annuity: Interest Compounded n Times per Year

If P is the deposit made at the end of each compounding period for an annuity at r percent annual interest compounded n times per year, the value, A, of the annuity after t years is

$$A = P\frac{\left(1 + \dfrac{r}{n}\right)^{nt} - 1}{\dfrac{r}{n}}.$$

EXAMPLE 7 Determining the Value of an Annuity

To save for retirement, you decide to deposit $1000 into an IRA at the end of each year for the next 30 years. If the interest rate is 10% per year compounded annually, find the value of the IRA after 30 years.

Solution The annuity involves 30 year-end deposits of $P = \$1000$. The interest rate is 10%: $r = 0.10$. Because the deposits are made once a year and the interest is compounded once a year, $n = 1$. The number of years is 30: $t = 30$. We replace the variables in the formula for the value of an annuity with these numbers.

$$A = P\frac{\left(1 + \dfrac{r}{n}\right)^{nt} - 1}{\dfrac{r}{n}}$$

$$A = 1000\frac{\left(1 + \dfrac{0.10}{1}\right)^{1 \cdot 30} - 1}{\dfrac{0.10}{1}} \approx 164{,}494$$

The value of the IRA at the end of 30 years is approximately $164,494.

Check Point 7 If $3000 is deposited into an IRA at the end of each year for 40 years and the interest rate is 10% per year compounded annually, find the value of the IRA after 40 years.

6 Use the formula for the sum of an infinite geometric series.

Geometric Series

An infinite sum of the form

$$a_1 + a_1r + a_1r^2 + a_1r^3 + \cdots + a_1r^{n-1} + \cdots$$

with first term a_1 and common ratio r is called an **infinite geometric series**. How can we determine which infinite geometric series have sums and which do not? We look at what happens to r^n as n gets larger in the formula for the sum of the first n terms of this series, namely

$$S_n = \frac{a_1(1 - r^n)}{1 - r}.$$

If r is any number between -1 and 1, that is, $-1 < r < 1$, the term r^n approaches 0 as n gets larger. For example, consider what happens to r^n for $r = \frac{1}{2}$:

$$\left(\tfrac{1}{2}\right)^1 = \tfrac{1}{2} \qquad \left(\tfrac{1}{2}\right)^2 = \tfrac{1}{4} \qquad \left(\tfrac{1}{2}\right)^3 = \tfrac{1}{8} \qquad \left(\tfrac{1}{2}\right)^4 = \tfrac{1}{16} \qquad \left(\tfrac{1}{2}\right)^5 = \tfrac{1}{32} \qquad \left(\tfrac{1}{2}\right)^6 = \tfrac{1}{64}$$

These numbers are approaching 0 as n gets larger.

Take another look at the formula for the sum of the first n terms of a geometric sequence.

$$S_n = \frac{a_1(1 - r^n)}{1 - r}$$

If $-1<r<1$, r^n approaches 0 as n approaches infinity $(n \to \infty)$.

Let us replace r^n with 0 in the formula for S_n. This change gives us a formula for the sum of infinite geometric series with common ratios between -1 and 1.

The Sum of an Infinite Geometric Series

If $-1 < r < 1$ (equivalently, $|r| < 1$), then the sum of the infinite geometric series

$$a_1 + a_1 r + a_1 r^2 + a_1 r^3 + \cdots$$

in which a_1 is the first term and r is the common ratio is given by

$$S = \frac{a_1}{1 - r}.$$

If $|r| \geq 1$, the infinite series does not have a sum.

To use the formula for the sum of an infinite geometric series, we need to know the first term and the common ratio. For example, consider

First term, a_1, is $\dfrac{1}{2}$.

$$\tfrac{1}{2} + \tfrac{1}{4} + \tfrac{1}{8} + \tfrac{1}{16} + \tfrac{1}{32} + \cdots.$$

Common ratio, r, is $\dfrac{a_2}{a_1}$.

$$r = \frac{1}{4} \div \frac{1}{2} = \frac{1}{4} \cdot 2 = \frac{1}{2}$$

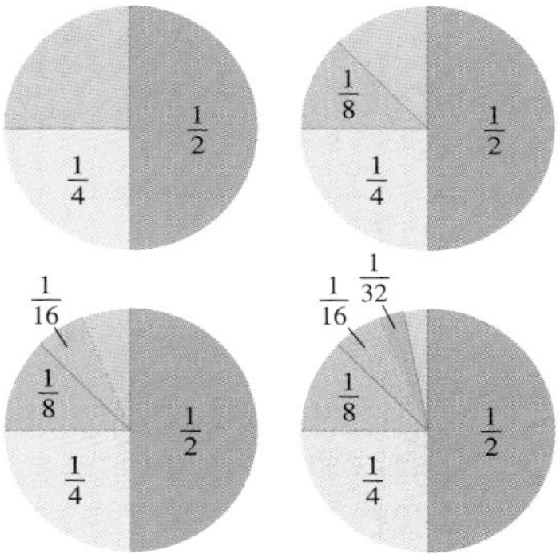

With $r = \frac{1}{2}$, the condition that $|r| < 1$ is met, so the infinite geometric series has a sum given by $S = \dfrac{a_1}{1 - r}$. The sum of the series is found as follows:

$$\tfrac{1}{2} + \tfrac{1}{4} + \tfrac{1}{8} + \tfrac{1}{16} + \tfrac{1}{32} + \cdots = \frac{a_1}{1 - r} = \frac{\frac{1}{2}}{1 - \frac{1}{2}} = \frac{\frac{1}{2}}{\frac{1}{2}} = 1.$$

Thus, the sum of the infinite geometric series is 1. Notice how this is illustrated in Figure 11.3. As more terms are included, the sum is approaching the area of one complete circle.

Figure 11.3 The sum $\frac{1}{2} + \frac{1}{4} + \frac{1}{8} + \frac{1}{16} + \frac{1}{32} + \cdots$ is approaching 1.

EXAMPLE 8 Finding the Sum of an Infinite Geometric Series

Find the sum of the infinite geometric series: $\frac{3}{8} - \frac{3}{16} + \frac{3}{32} - \frac{3}{64} + \cdots$.

Solution Before finding the sum, we must find the common ratio.

$$r = \frac{a_2}{a_1} = \frac{-\frac{3}{16}}{\frac{3}{8}} = -\frac{3}{16} \cdot \frac{8}{3} = -\frac{1}{2}$$

Because $r = -\frac{1}{2}$, the condition that $|r| < 1$ is met. Thus, the infinite geometric series has a sum.

$$S = \frac{a_1}{1 - r}$$

This is the formula for the sum of an infinite geometric series. Let $a_1 = \frac{3}{8}$ and $r = -\frac{1}{2}$.

$$= \frac{\frac{3}{8}}{1 - \left(-\frac{1}{2}\right)} = \frac{\frac{3}{8}}{\frac{3}{2}} = \frac{3}{8} \cdot \frac{2}{3} = \frac{1}{4}$$

Thus, the sum of this infinite geometric series is $\frac{1}{4}$. Put in an informal way, as we continue to add more and more terms, the sum is approximately $\frac{1}{4}$.

Check Point 8

Find the sum of the infinite geometric series:
$3 + 2 + \frac{4}{3} + \frac{8}{9} + \cdots$.

We can use the formula for the sum of an infinite series to express a repeating decimal as a fraction in lowest terms.

EXAMPLE 9 Writing a Repeating Decimal as a Fraction

Express $0.\overline{7}$ as a fraction in lowest terms.

Solution

$$0.\overline{7} = 0.7777\ldots = \frac{7}{10} + \frac{7}{100} + \frac{7}{1000} + \frac{7}{10,000} + \cdots$$

Observe that $0.\overline{7}$ is an infinite geometric series with first term $\frac{7}{10}$ and common ratio $\frac{1}{10}$. Because $r = \frac{1}{10}$, the condition that $|r| < 1$ is met. Thus, we can use our formula to find the sum. Therefore,

$$0.\overline{7} = \frac{a_1}{1 - r} = \frac{\frac{7}{10}}{1 - \frac{1}{10}} = \frac{\frac{7}{10}}{\frac{9}{10}} = \frac{7}{10} \cdot \frac{10}{9} = \frac{7}{9}.$$

An equivalent fraction for $0.\overline{7}$ is $\frac{7}{9}$.

Check Point 9

Express $0.\overline{9}$ as a fraction in lowest terms.

Infinite geometric series have many applications, as illustrated in Example 10.

EXAMPLE 10 Tax Rebates and the Multiplier Effect

A tax rebate that returns a certain amount of money to taxpayers can have a total effect on the economy that is many times this amount. In economics, this phenomenon is called the **multiplier effect**. Suppose, for example, that the government reduces taxes so that each consumer has $2000 more income. The government assumes that each person will spend 70% of this (= $1400). The individuals and businesses receiving this $1400 in turn spend 70% of it (= $980),

$1400

70% is spent.

$980

70% is spent.

$686

creating extra income for other people to spend, and so on. Determine the total amount spent on consumer goods from the initial $2000 tax rebate.

Solution   The total amount spent is given by the infinite geometric series

$$1400 + 980 + 686 + \cdots.$$

70% of   70% of
1400   980

The first term is 1400: $a_1 = 1400$. The common ratio is 70%, or 0.7: $r = 0.7$. Because $r = 0.7$, the condition that $|r| < 1$ is met. Thus, we can use our formula to find the sum. Therefore,

$$1400 + 980 + 686 + \cdots = \frac{a_1}{1 - r} = \frac{1400}{1 - 0.7} \approx 4667.$$

This means that the total amount spent on consumer goods from the initial $2000 rebate is approximately $4667.

Check Point 10   Rework Example 10 and determine the total amount spent on consumer goods with a $1000 tax rebate and 80% spending down the line.

EXERCISE SET 11.3

Practice Exercises

In Exercises 1–8, write the first five terms of each geometric sequence.

1. $a_1 = 5$, $r = 3$ **2.** $a_1 = 4$, $r = 3$

3. $a_1 = 20$, $r = \frac{1}{2}$ **4.** $a_1 = 24$, $r = \frac{1}{3}$

5. $a_n = -4a_{n-1}$, $a_1 = 10$ **6.** $a_n = -3a_{n-1}$, $a_1 = 10$

7. $a_n = -5a_{n-1}$, $a_1 = -6$ **8.** $a_n = -6a_{n-1}$, $a_1 = -2$

In Exercises 9–16, find the indicated term of the geometric sequence with first term, a_1, and common ratio, r.

9. Find a_8 when $a_1 = 6, r = 2$.

10. Find a_8 when $a_1 = 5, r = 3$.

11. Find a_{12} when $a_1 = 5, r = -2$.

12. Find a_{12} when $a_1 = 4, r = -2$.

13. Find a_{40} when $a_1 = 1000, r = -\frac{1}{2}$.

14. Find a_{30} when $a_1 = 8000, r = -\frac{1}{2}$.

15. Find a_8 when $a_1 = 1,000,000, r = 0.1$.

16. Find a_8 when $a_1 = 40,000, r = 0.1$.

In Exercises 17–24, write a formula for the general term (the nth term) of each geometric sequence. Then use the formula for a_n to find a_7, the seventh term of the sequence.

17. $3, 12, 48, 192, \ldots$ **18.** $3, 15, 75, 375, \ldots$

19. $18, 6, 2, \frac{2}{3}, \ldots$.   **20.** $12, 6, 3, \frac{3}{2}, \ldots$.

21. $1.5, -3, 6, -12, \ldots$ **22.** $5, -1, \frac{1}{5}, -\frac{1}{25}, \ldots$.

23. $0.0004, -0.004, 0.04, -0.4, \ldots$

24. $0.0007, -0.007, 0.07, -0.7, \ldots$

Use the formula for the sum of the first n terms of a geometric sequence to solve Exercises 25–36.

25. Find the sum of the first 12 terms of the geometric sequence: $2, 6, 18, 54 \ldots$.

26. Find the sum of the first 12 terms of the geometric sequence: $3, 6, 12, 24, \ldots$.

27. Find the sum of the first 11 terms of the geometric sequence: $3, -6, 12, -24, \ldots$.

28. Find the sum of the first 11 terms of the geometric sequence: $4, -12, 36, -108, \ldots$.

29. Find the sum of the first 14 terms of the geometric sequence: $-\frac{3}{2}, 3, -6, 12, \ldots$.

30. Find the sum of the first 14 terms of the geometric sequence: $-\frac{1}{24}, \frac{1}{12}, -\frac{1}{6}, \frac{1}{3}, \ldots$.

In Exercises 31–36, find the indicated sum.

31. $\displaystyle\sum_{i=1}^{8} 3^i$ **32.** $\displaystyle\sum_{i=1}^{6} 4^i$

33. $\displaystyle\sum_{i=1}^{10} 5 \cdot 2^i$

34. $\displaystyle\sum_{i=1}^{7} 4(-3)^i$

35. $\displaystyle\sum_{i=1}^{6} \left(\tfrac{1}{2}\right)^{i+1}$

36. $\displaystyle\sum_{i=1}^{6} \left(\tfrac{1}{3}\right)^{i+1}$

In Exercises 37–44, find the sum of each infinite geometric series.

37. $1 + \dfrac{1}{3} + \dfrac{1}{9} + \dfrac{1}{27} + \cdots$

38. $1 + \dfrac{1}{4} + \dfrac{1}{16} + \dfrac{1}{64} + \cdots$

39. $3 + \dfrac{3}{4} + \dfrac{3}{4^2} + \dfrac{3}{4^3} + \cdots$

40. $5 + \dfrac{5}{6} + \dfrac{5}{6^2} + \dfrac{5}{6^3} + \cdots$

41. $1 - \dfrac{1}{2} + \dfrac{1}{4} - \dfrac{1}{8} + \cdots$

42. $3 - 1 + \dfrac{1}{3} - \dfrac{1}{9} + \cdots$

43. $\displaystyle\sum_{i=1}^{\infty} 8(-0.3)^{i-1}$

44. $\displaystyle\sum_{i=1}^{\infty} 12(-0.7)^{i-1}$

In Exercises 45–50, express each repeating decimal as a fraction in lowest terms.

45. $0.\overline{5} = \dfrac{5}{10} + \dfrac{5}{100} + \dfrac{5}{1000} + \dfrac{5}{10,000} + \cdots$

46. $0.\overline{1} = \dfrac{1}{10} + \dfrac{1}{100} + \dfrac{1}{1000} + \dfrac{1}{10,000} + \cdots$

47. $0.\overline{47} = \dfrac{47}{100} + \dfrac{47}{10,000} + \dfrac{47}{1,000,000} + \cdots$

48. $0.\overline{83} = \dfrac{83}{100} + \dfrac{83}{10,000} + \dfrac{83}{1,000,000} + \cdots$

49. $0.\overline{257}$

50. $0.\overline{529}$

In Exercises 51–56, the general term of a sequence is given. Determine whether the sequence is arithmetic, geometric, or neither. If the sequence is arithmetic, find the common difference; if it is geometric, find the common ratio.

51. $a_n = n + 5$

52. $a_n = n - 3$

53. $a_n = 2^n$

54. $a_n = \left(\tfrac{1}{2}\right)^n$

55. $a_n = n^2 + 5$

56. $a_n = n^2 - 3$

 Application Exercises

Use the formula for the general term (the nth term) of a geometric sequence to solve Exercises 57–60.

In Exercises 57–58, suppose you save $1 the first day of a month, $2 the second day, $4 the third day, and so on. That is, each day you save twice as much as you did the day before.

57. What will you put aside for savings on the fifteenth day of the month?

58. What will you put aside for savings on the thirtieth day of the month?

59. A professional baseball player signs a contract with a beginning salary of $3,000,000 for the first year with an annual increase of 4% per year beginning in the second year. That is, beginning in year 2, the athlete's salary will be 1.04 times what it was in the previous year. What is the athlete's salary for year 7 of the contract?

60. You are offered a job that pays $30,000 for the first year with an annual increase of 5% per year beginning in the second year. That is, beginning in year 2, your salary will be 1.05 times what it was in the previous year. What can you expect to earn in your sixth year on the job?

61. The population of Iraq from 1995 through 1998 is shown in the following table.

Year	1995	1996	1997	1998
Population in millions	20.60	21.36	22.19	23.02

Source: U.N. Population Division

a. Divide the population for each year by the population in the preceding year. Round to two decimal places and show that Iraq's population is increasing geometrically.
b. Write the general term of the geometric sequence describing population growth for Iraq n years after 1994.
c. Estimate Iraq's population, in millions, for the year 2005.

62. The population of China from 1995 through 1998 is shown in the following table.

Year	1995	1996	1997	1998
Population in millions	1218.80	1232.21	1245.76	1259.46

Source: U.N. Population Division

a. Divide the population for each year by the population in the preceding year. Round to two decimal places and show that China's population is increasing geometrically.
b. Write the general term of the geometric sequence describing population growth for China n years after 1994.
c. Estimate China's population, in millions, for the year 2005.

Use the formula for the sum of the first n terms of a geometric sequence to solve Exercises 63–68.

In Exercises 63–64, you save $1 the first day of a month, $2 the second day, $4 the third day, continuing to double your savings each day.

63. What will your total savings be for the first 15 days?

64. What will your total savings be for the first 30 days?

If $n = 2$, we can find positive integers satisfying the equation:

$$3^2 + 4^2 = 5^2.$$

However, Fermat claimed that no positive integers satisfy

$$x^3 + y^3 = z^3, \quad x^4 + y^4 = z^4, \quad x^5 + y^5 = z^5,$$

and so on. Fermat claimed to have a proof of his conjecture, but added, "The margin of my book is too narrow to write it down." Some believe that he never had a proof and intended to frustrate his colleagues.

In 1994, 40-year-old Princeton math professor Andrew Wiles proved Fermat's Last Theorem using a principle called *mathematical induction*. In this section, you will learn how to use this powerful method to prove statements about the positive integers.

1 Understand the principle of mathematical induction.

The Principle of Mathematical Induction

How do we prove statements using mathematical induction? Let's consider an example. We will prove a statement that appears to give a correct formula for the sum of the first n positive integers.

$$S_n : 1 + 2 + 3 + \cdots + n = \frac{n(n + 1)}{2}$$

We can verify this statement for, say, the first four positive integers.

If $n = 1$, the statement S_1 is

> Take the first term on the left.

$$1 \overset{?}{=} \frac{1(1 + 1)}{2}$$

> Substitute 1 for n on the right.

$$1 \overset{?}{=} \frac{1 \cdot 2}{2}$$

$$1 = 1 \checkmark.$$

> This true statement shows that S_1 is true.

If $n = 2$, the statement S_2 is

> Add the first two terms on the left.

$$1 + 2 \overset{?}{=} \frac{2(2 + 1)}{2}$$

> Substitute 2 for n on the right.

$$3 \overset{?}{=} \frac{2 \cdot 3}{2}$$

$$3 = 3 \checkmark.$$

> This true statement shows S_2 is true.

If $n = 3$, the statement S_3 is

> Add the first three terms on the left.

$$1 + 2 + 3 \overset{?}{=} \frac{3(3 + 1)}{2}$$

> Substitute 3 for n on the right.

$$6 \overset{?}{=} \frac{3 \cdot 4}{2}$$

$$6 = 6 \checkmark.$$

> This true statement shows S_3 is true.

Finally, if $n = 4$, the statement S_4 is

> Add the first four terms on the left.

$$1 + 2 + 3 + 4 \overset{?}{=} \frac{4(4 + 1)}{2}$$

> Substitute 4 for n on the right.

$$10 \overset{?}{=} \frac{4 \cdot 5}{2}$$

$$10 = 10 \checkmark.$$

> This true statement shows S_4 is true.

This approach does *not* prove that the given statement S_n is true for every positive integer n. The fact that the formula produces true statements for $n = 1$, 2, 3, and 4 does not guarantee that it is valid for all positive integers n. Thus, we need to be able to verify the truth of S_n without verifying the statement for each and every one of the positive integers.

A legitimate proof of the given statement S_n involves a technique called **mathematical induction**.

The Principle of Mathematical Induction

Let S_n be a statement involving the positive integer n. If

1. S_1 is true, and
2. the truth of the statement S_k implies the truth of the statement S_{k+1}, for every positive integer k,

then the statement S_n is true for all positive integers n.

Figure 11.4 Falling dominoes illustrate the principle of mathematical induction.

The principle of mathematical induction can be illustrated using an unending line of dominoes, as shown in Figure 11.4. If the first domino is pushed over, it knocks down the next, which knocks down the next, and so on, in a chain reaction. To topple all the dominoes in the infinite sequence, two conditions must be satisfied:

1. The first domino must be knocked down.
2. If the domino in position k is knocked down, then the domino in position $k + 1$ must be knocked down.

If the second condition is not satisfied, it does not follow that all the dominoes will topple. For example, suppose the dominoes are spaced far enough apart so that a falling domino does not push over the next domino in the line.

The domino analogy provides the two steps that are required in a proof by mathematical induction.

The Steps in a Proof by Mathematical Induction

Let S_n be a statement involving the positive integer n. To prove that S_n is true for all positive integers n requires two steps.

Step 1 Show that S_1 is true.

Step 2 Show that if S_k is assumed to be true, then S_{k+1} is also true, for every positive integer k.

Notice that to prove S_n, we work only with the statements S_1, S_k, and S_{k+1}. Our first example provides practice in writing these statements.

EXAMPLE 1 Writing S_1, S_k, and S_{k+1}

For the given statement S_n, write the three statements S_1, S_k, and S_{k+1}.

a. S_n: $1 + 2 + 3 + \cdots + n = \dfrac{n(n + 1)}{2}$

b. S_n: $1^2 + 2^2 + 3^2 + \cdots + n^2 = \dfrac{n(n + 1)(2n + 1)}{6}$

Solution

a. We begin with

$$S_n: 1 + 2 + 3 + \cdots + n = \frac{n(n + 1)}{2}.$$

Write S_1 by taking the first term on the left and replacing n with 1 on the right.

$$S_1: 1 = \frac{1(1 + 1)}{2}$$

Write S_k by taking the sum of the first k terms on the left and replacing n with k on the right.

$$S_k: 1 + 2 + 3 + \cdots + k = \frac{k(k + 1)}{2}$$

Write S_{k+1} by taking the sum of the first $k + 1$ terms on the left and replacing n with $k + 1$ on the right.

$$S_{k+1}: 1 + 2 + 3 + \cdots + (k + 1) = \frac{(k + 1)[(k + 1) + 1]}{2}$$

$$S_{k+1}: 1 + 2 + 3 + \cdots + (k + 1) = \frac{(k + 1)(k + 2)}{2} \quad \text{Simplify on the right.}$$

b. We begin with

$$S_n: 1^2 + 2^2 + 3^2 + \cdots + n^2 = \frac{n(n + 1)(2n + 1)}{6}.$$

Write S_1 by taking the first term on the left and replacing n with 1 on the right.

$$S_1: 1^2 = \frac{1(1 + 1)(2 \cdot 1 + 1)}{6}$$

Write S_k by taking the sum of the first k terms on the left and replacing n with k on the right.

$$S_k: 1^2 + 2^2 + 3^2 + \cdots + k^2 = \frac{k(k + 1)(2k + 1)}{6}$$

Write S_{k+1} by taking the sum of the first $k + 1$ terms on the left and replacing n with $k + 1$ on the right.

$$S_{k+1}: 1^2 + 2^2 + 3^2 + \cdots + (k + 1)^2 = \frac{(k + 1)[(k + 1) + 1][2(k + 1) + 1]}{6}$$

$$S_{k+1}: 1^2 + 2^2 + 3^2 + \cdots + (k + 1)^2 = \frac{(k + 1)(k + 2)(2k + 3)}{6} \quad \text{Simplify on the right.}$$

Check Point 1 For the given statement S_n, write the three statements S_1, S_k, and S_{k+1}.

a. $2 + 4 + 6 + \cdots + 2n = n(n + 1)$

b. $1^3 + 2^3 + 3^3 + \cdots + n^3 = \frac{n^2(n + 1)^2}{4}$

Always simplify S_{k+1} before trying to use mathematical induction to prove that S_n is true. For example, consider

$$S_n: \quad 1^2 + 3^2 + 5^2 + \cdots + (2n - 1)^2 = \frac{n(2n - 1)(2n + 1)}{3}.$$

Begin by writing S_{k+1} as follows:

$$S_{k+1}: \quad 1^2 + 3^2 + 5^2 + \cdots + \left[2(k + 1) - 1\right]^2$$

$$= \frac{(k + 1)\left[2(k + 1) - 1\right]\left[2(k + 1) + 1\right]}{3}.$$

The sum of the first $k + 1$ terms

Replace n by $k + 1$ on the right side of S_n.

Now simplify the algebra.

$$S_{k+1}: \quad 1^2 + 3^2 + 5^2 + \cdots + (2k + 2 - 1)^2 = \frac{(k + 1)(2k + 2 - 1)(2k + 2 + 1)}{3}$$

$$S_{k+1}: \quad 1^2 + 3^2 + 5^2 + \cdots + (2k + 1)^2 = \frac{(k + 1)(2k + 1)(2k + 3)}{3}$$

2 Prove statements using mathematical induction.

Proving Statements about Positive Integers Using Mathematical Induction

Now that we know how to find S_1, S_k, and S_{k+1}, let's see how we can use these statements to carry out the two steps in a proof by mathematical induction. In Examples 2 and 3, we will use the statements S_1, S_k, and S_{k+1} to prove each of the statements S_n that we worked with in Example 1.

EXAMPLE 2   Proving a Formula by Mathematical Induction

Use mathematical induction to prove that

$$1 + 2 + 3 + \cdots + n = \frac{n(n + 1)}{2}$$

for all positive integers n.

Solution

Step 1   Show that S_1 is true.   Statement S_1 is

$$1 = \frac{1(1 + 1)}{2}.$$

Simplifying on the right, we obtain $1 = 1$. This true statement shows that S_1 is true.

Step 2   Show that if S_k is true, then S_{k+1} is true.   Using S_k and S_{k+1} from Example 1a, show that the truth of S_k,

$$1 + 2 + 3 + \cdots + k = \frac{k(k + 1)}{2}$$

implies the truth of S_{k+1},

$$1 + 2 + 3 + \cdots + (k + 1) = \frac{(k + 1)(k + 2)}{2}.$$

Visualizing Summation Formulas

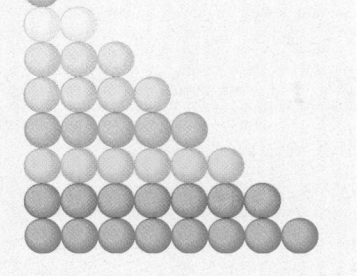

Finding the sum of consecutive positive integers leads to **triangular numbers** of the form $\dfrac{n(n+1)}{2}$.

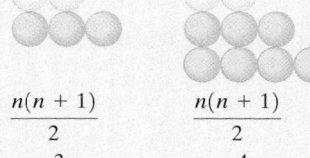

$\dfrac{n(n+1)}{2}$
$n = 1$:
1

$\dfrac{n(n+1)}{2}$
$n = 2$:
3

$\dfrac{n(n+1)}{2}$
$n = 3$:
6

$\dfrac{n(n+1)}{2}$
$n = 4$:
10

We will work with S_k. Because we assume that S_k is true, we add the next consecutive integer after k—namely, $k + 1$—to both sides.

$$1 + 2 + 3 + \cdots + k = \frac{k(k+1)}{2}$$

This is S_k, which we assume is true.

$$1 + 2 + 3 + \cdots + k + (k+1) = \frac{k(k+1)}{2} + (k+1)$$

Add $k + 1$ to both sides of the equation.

We do not have to write this k because k is understood to be the integer that precedes $k + 1$.

$$1 + 2 + 3 + \cdots + (k+1) = \frac{k(k+1)}{2} + \frac{2(k+1)}{2}$$

Write the right side with a common denominator of 2.

$$1 + 2 + 3 + \cdots + (k+1) = \frac{(k+1)}{2}(k+2)$$

Factor out the common factor $\dfrac{k+1}{2}$ on the right.

$$1 + 2 + 3 + \cdots + (k+1) = \frac{(k+1)(k+2)}{2}$$

This final result is the statement S_{k+1} at the bottom of page 637.

We have shown that if we assume that S_k is true, and we add $k + 1$ to both sides of S_k, then S_{k+1} is also true. By the principle of mathematical induction, the statement S_n, namely,

$$1 + 2 + 3 + \cdots + n = \frac{n(n+1)}{2}$$

is true for every positive integer n.

Check Point 2

Use mathematical induction to prove that
$$2 + 4 + 6 + \cdots + 2n = n(n+1).$$

EXAMPLE 3 Proving a Formula by Mathematical Induction

Use mathematical induction to prove that

$$1^2 + 2^2 + 3^2 + \cdots + n^2 = \frac{n(n+1)(2n+1)}{6}$$

for all positive integers n.

Solution

Step 1 Show that S_1 is true. Statement S_1 is

$$1^2 = \frac{1(1+1)(2 \cdot 1 + 1)}{6}.$$

Simplifying, we obtain $1 = \dfrac{1 \cdot 2 \cdot 3}{6}$. Further simplification on the right gives the statement $1 = 1$. This true statement shows that S_1 is true.

Step 2 Show that if S_k is true, then S_{k+1} is true. Using S_k and S_{k+1} from Example 1b, show that the truth of

$$S_k:\ 1^2 + 2^2 + 3^2 + \cdots + k^2 = \frac{k(k + 1)(2k + 1)}{6}$$

implies the truth of

$$S_{k+1}:\ 1^2 + 2^2 + 3^2 + \cdots + (k + 1)^2 = \frac{(k + 1)(k + 2)(2k + 3)}{6}.$$

We will work with S_k. Because we assume that S_k is true, we add the square of the next consecutive integer after k—namely, $(k + 1)^2$—to both sides of the equation.

$$1^2 + 2^2 + 3^2 + \cdots + k^2 = \frac{k(k + 1)(2k + 1)}{6}$$

> This is S_k, assumed to be true. We must work with this and show S_{k+1} is true.

$$1^2 + 2^2 + 3^2 + \cdots + k^2 + (k + 1)^2 = \frac{k(k + 1)(2k + 1)}{6} + (k + 1)^2$$

> Add $(k + 1)^2$ to both sides.

$$1^2 + 2^2 + 3^2 + \cdots + (k + 1)^2 = \frac{k(k + 1)(2k + 1)}{6} + \frac{6(k + 1)^2}{6}$$

> It is not necessary to write k^2 on the left. Express the right side with the least common denominator, 6.

$$= \frac{(k + 1)}{6}\big[k(2k + 1) + 6(k + 1)\big]$$

> Factor out the common factor $\dfrac{k + 1}{6}$.

$$= \frac{(k + 1)}{6}\left(2k^2 + 7k + 6\right)$$

> Multiply and combine like terms.

$$= \frac{(k + 1)}{6}(k + 2)(2k + 3)$$

> Factor $2k^2 + 7k + 6$.

$$= \frac{(k + 1)(k + 2)(2k + 3)}{6}$$

> This final statement is S_{k+1}.

We have shown that if we assume that S_k is true, and we add $(k + 1)^2$ to both sides of S_k, then S_{k+1} is also true. By the principle of mathematical induction, the statement S_n, namely,

$$1^2 + 2^2 + 3^2 + \cdots + n^2 = \frac{n(n + 1)(2n + 1)}{6}$$

is true for every positive integer n.

Check Point 3 Use mathematical induction to prove that

$$1^3 + 2^3 + 3^3 + \cdots + n^3 = \frac{n^2(n + 1)^2}{4}.$$

Example 4 illustrates how mathematical induction can be used to prove statements about positive integers that do not involve sums.

EXAMPLE 4 Using the Principle of Mathematical Induction

Prove that 2 is a factor of $n^2 + 5n$ for all positive integers n.

Solution

Step 1 Show that S_1 is true. Statement S_1 reads

$$2 \text{ is a factor of } 1^2 + 5 \cdot 1.$$

Simplifying the arithmetic, the statement reads

$$2 \text{ is a factor of } 6.$$

This statement is true: that is, $6 = 2 \cdot 3$. This shows that S_1 is true.

Step 2 Show that if S_k is true, then S_{k+1} is true. Let's write S_k and S_{k+1}:

$$S_k: \quad 2 \text{ is a factor of } k^2 + 5k.$$

$$S_{k+1}: \quad 2 \text{ is a factor of } (k + 1)^2 + 5(k + 1).$$

We can rewrite statement S_{k+1} by simplifying the algebraic expression in the statement as follows:

$$(k + 1)^2 + 5(k + 1) = k^2 + 2k + 1 + 5k + 5 = k^2 + 7k + 6.$$

Use the formula
$(A + B)^2 = A^2 + 2AB + B^2.$

Statement S_{k+1} now reads

$$2 \text{ is a factor of } k^2 + 7k + 6.$$

We wish to use statement S_k—that is, 2 is a factor of $k^2 + 5k$—to prove statement S_{k+1}. We do this as follows:

$$k^2 + 7k + 6 = \left(k^2 + 5k\right) + (2k + 6) = \left(k^2 + 5k\right) + 2(k + 3).$$

We know that 2
is a factor of $k^2 + 5k$
because we assume
S_k is true.

Factoring the last two
terms shows that 2
is a factor of $2k + 6$.

The voice balloons show that 2 is a factor of $k^2 + 5k$ and of $2(k + 3)$. Thus, 2 is a factor of the sum $(k^2 + 5k) + 2(k + 3)$, or of $k^2 + 7k + 6$. This is precisely statement S_{k+1}. We have shown that if we assume that S_k is true, then S_{k+1} is also true. By the principle of mathematical induction, the statement S_n, namely 2 is a factor of $n^2 + 5n$, is true for every positive integer n.

Check Point 4 Prove that 2 is a factor of $n^2 + n$ for all positive integers n.

EXERCISE SET 11.4

✔ Practice Exercises

In Exercises 1–4, a statement S_n about the positive integers is given. Write statements S_1, S_2, and S_3, and show that each of these statements is true.

1. S_n: $1 + 3 + 5 + \cdots + (2n - 1) = n^2$

2. S_n: $3 + 4 + 5 + \cdots + (n + 2) = \dfrac{n(n + 5)}{2}$

3. S_n: 2 is a factor of $n^2 - n$.

4. S_n: 3 is a factor of $n^3 - n$.

In Exercises 5–10, a statement S_n about the positive integers is given. Write statements S_k and S_{k+1}, simplifying statement S_{k+1} completely.

5. S_n: $4 + 8 + 12 + \cdots + 4n = 2n(n + 1)$

6. S_n: $3 + 4 + 5 + \cdots + (n + 2) = \dfrac{n(n + 5)}{2}$

7. S_n: $3 + 7 + 11 + \cdots + (4n - 1) = n(2n + 1)$

8. S_n: $2 + 7 + 12 + \cdots + (5n - 3) = \dfrac{n(5n - 1)}{2}$

9. S_n: 2 is a factor of $n^2 - n + 2$.

10. S_n: 2 is a factor of $n^2 - n$.

In Exercises 11–30, use mathematical induction to prove that each statement is true for every positive integer n.

11. $4 + 8 + 12 + \cdots + 4n = 2n(n + 1)$

12. $3 + 4 + 5 + \cdots + (n + 2) = \dfrac{n(n + 5)}{2}$

13. $1 + 3 + 5 + \cdots + (2n - 1) = n^2$

14. $3 + 6 + 9 + \cdots + 3n = \dfrac{3n(n + 1)}{2}$

15. $3 + 7 + 11 + \cdots + (4n - 1) = n(2n + 1)$

16. $2 + 7 + 12 + \cdots + (5n - 3) = \dfrac{n(5n - 1)}{2}$

17. $1 + 2 + 2^2 + \cdots + 2^{n-1} = 2^n - 1$

18. $1 + 3 + 3^2 + \cdots + 3^{n-1} = \dfrac{3^n - 1}{2}$

19. $2 + 4 + 8 + \cdots + 2^n = 2^{n+1} - 2$

20. $\dfrac{1}{2} + \dfrac{1}{4} + \dfrac{1}{8} + \cdots + \dfrac{1}{2^n} = 1 - \dfrac{1}{2^n}$

21. $1 \cdot 2 + 2 \cdot 3 + 3 \cdot 4 + \cdots + n(n + 1)$
$$= \dfrac{n(n + 1)(n + 2)}{3}$$

22. $1 \cdot 3 + 2 \cdot 4 + 3 \cdot 5 + \cdots + n(n + 2)$
$$= \dfrac{n(n + 1)(2n + 7)}{6}$$

23. $\dfrac{1}{1 \cdot 2} + \dfrac{1}{2 \cdot 3} + \dfrac{1}{3 \cdot 4} + \cdots + \dfrac{1}{n(n + 1)} = \dfrac{n}{n + 1}$

24. $\dfrac{1}{2 \cdot 3} + \dfrac{1}{3 \cdot 4} + \dfrac{1}{4 \cdot 5} + \cdots + \dfrac{1}{(n + 1)(n + 2)} = \dfrac{n}{2n + 4}$

25. 2 is a factor of $n^2 - n$.

26. 2 is a factor of $n^2 + 3n$.

27. 6 is a factor of $n(n + 1)(n + 2)$.

28. 3 is a factor of $n(n + 1)(n - 1)$.

29. $(ab)^n = a^n b^n$

30. $\left(\dfrac{a}{b}\right)^n = \dfrac{a^n}{b^n}$

Writing in Mathematics

31. Explain how to use mathematical induction to prove that a statement is true for every positive integer n.

32. Consider the statement S_n given by
$$n^2 - n + 41 \text{ is prime.}$$

Although $S_1, S_2, \ldots, S_{40}$ are true, S_{41} is false. Describe how this is illustrated by the dominoes in the figure. What does this tell you about a pattern, or formula, that seems to work for several values of n?

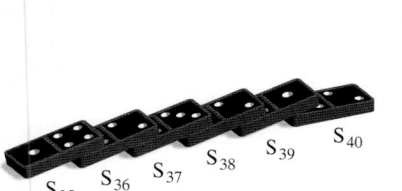

🔆 Critical Thinking Exercises

Some statements are false for the first few positive integers, but true for some positive integer on. In these instances, you can prove S_n for $n \geq k$ by showing that S_k is true and that S_k implies S_{k+1}. Use this extended principle of mathematical induction to prove that each statement in Exercises 33–34 is true.

33. Prove that $n^2 > 2n + 1$ for $n \geq 3$. Show that the formula is true for $n = 3$ and then use step 2 of mathematical induction.

34. Prove that $2^n > n^2$ for $n \geq 5$. Show that the formula is true for $n = 5$ and then use step 2 of mathematical induction.

In Exercises 35–36, find S_1 through S_5 and then use the pattern to make a conjecture about S_n. Prove the conjectured formula for S_n by mathematical induction.

35. S_n: $\dfrac{1}{4} + \dfrac{1}{12} + \dfrac{1}{24} + \cdots + \dfrac{1}{2n(n+1)}$

36. S_n: $\left(1 - \dfrac{1}{2}\right)\left(1 - \dfrac{1}{3}\right)\left(1 - \dfrac{1}{4}\right)\cdots\left(1 - \dfrac{1}{n+1}\right)$

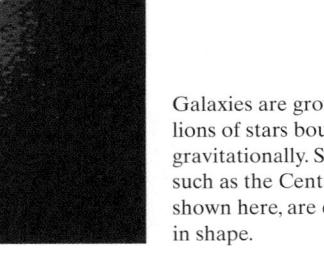 **Group Exercise**

37. Fermat's most notorious theorem baffled the greatest minds for more than three centuries. In 1994, after ten years of work, Princeton University's Andrew Wiles proved Fermat's Last Theorem. *People* magazine put him on its list of "the 25 most intriguing people of the year," the Gap asked him to model jeans, and Barbara Walters chased him for an interview. "Who's Barbara Walters?" asked the bookish Wiles, who had somehow gone through life without a television.

Using the 1993 PBS documentary "Solving Fermat: Andrew Wiles" or information about Andrew Wiles on the Internet, research and present a group seminar on what Wiles did to prove Fermat's Last Theorem, problems along the way, and the role of mathematical induction in the proof.

SECTION 11.5 *The Binomial Theorem*

Objectives

1. Recognize patterns in binomial expansions.
2. Evaluate a binomial coefficient.
3. Expand a binomial raised to a power.
4. Find a particular term in a binomial expansion.

Galaxies are groupings of billions of stars bound together gravitationally. Some galaxies, such as the Centaurus galaxy shown here, are elliptical in shape.

Is mathematics discovered or invented? For example, planets revolve in elliptical orbits. Does that mean that the ellipse is out there, waiting for the mind to discover it? Or do people create the definition of an ellipse just as they compose a song? And is it possible for the same mathematics to be discovered/invented by independent researchers separated by time, place, and culture? This is precisely what occurred when mathematicians attempted to find efficient methods for raising binomials to higher and higher powers, such as

$$(x + 2)^3, (x + 2)^4, (x + 2)^5, (x + 2)^6,$$

and so on. In this section, we study higher powers of binomials and a method first discovered/invented by great minds in Eastern and Western culture working independently.

1 Recognize patterns in binomial expansions.

Patterns in Binomial Expansions

When we write out the *binomial expression* $(a + b)^n$, where n is a positive integer, a number of patterns begin to appear.

$$(a + b)^1 = a + b$$
$$(a + b)^2 = a^2 + 2ab + b^2$$
$$(a + b)^3 = a^3 + 3a^2b + 3ab^2 + b^3$$
$$(a + b)^4 = a^4 + 4a^3b + 6a^2b^2 + 4ab^3 + b^4$$
$$(a + b)^5 = a^5 + 5a^4b + 10a^3b^2 + 10a^2b^3 + 5ab^4 + b^5$$

Discovery

Each expanded form of the binomial expression is a polynomial. Study the five polynomials and answer the following questions.

1. For each polynomial, describe the pattern for the exponents on a. What is the largest exponent on a? What happens to the exponent on a from term to term?
2. Describe the pattern for the exponents on b. What is the exponent on b in the first term? What is the exponent on b in the second term? What happens to the exponent on b from term to term?
3. Find the sum of the exponents on the variables in each term for the polynomials in the five rows. Describe the pattern.
4. How many terms are there in the polynomials on the right in relation to the power of the binomial?

How many of the following patterns were you able to discover?

1. The first term is a^n. The exponent on a decreases by 1 in each successive term.
2. The exponents on b increase by 1 in each successive term. In the first term, the exponent on b is 0. (Because $b^0 = 1$, b is not shown in the first term.) The last term is b^n.
3. The sum of the exponents on the variables in any term is equal to n, the exponent on $(a + b)^n$.
4. There is one more term in the polynomial expansion than there is in the power of the binomial, n. There are $n + 1$ terms in the expanded form of $(a + b)^n$.

Using these observations, the variable parts of the expansion of $(a + b)^6$ are

$$a^6, \quad a^5b, \quad a^4b^2, \quad a^3b^3, \quad a^2b^4, \quad ab^5, \quad b^6.$$

The first term is a^6, with the exponent on a decreasing by 1 in each successive term. The exponents on b increase from 0 to 6, with the last term being b^6. The sum of the exponents in each term is equal to 6.

We can generalize from these observations to obtain the variable parts of the expansion of $(a + b)^n$. They are

$$a^n, a^{n-1}b, a^{n-2}b^2, a^{n-3}b^3, \ldots, ab^{n-1}, b^n.$$

> Exponents on a are decreasing by 1.
> Exponents on b are increasing by 1.

> Sum of exponents: $n - 1 + 1 = n$
> Sum of exponents: $n - 3 + 3 = n$
> Sum of exponents: $1 + n - 1 = n$

Let's now establish a pattern for the coefficients of the terms in the binomial expansion. Notice that each row in the figure on page 936 begins and ends with 1. Any other number in the row can be obtained by adding the two numbers immediately above it.

Study Tip

We have not shown the number in the top row of Pascal's triangle on the right. The top row is *row zero* because it corresponds to $(a + b)^0 = 1$. With row zero, the triangle appears as

```
            1
         1     1
       1    2    1
     1    3    3    1
   1   4    6    4   1
            etc.
```

Coefficients for $(a + b)^1$.

Coefficients for $(a + b)^2$.

Coefficients for $(a + b)^3$.

Coefficients for $(a + b)^4$.

Coefficients for $(a + b)^5$.

```
            1     1
         1    2    1
       1   3    3   1
     1   4    6    4   1
   1   5   10   10   5   1
```

The following triangular array of coefficients is called **Pascal's triangle.** If we continue with the sixth row, the first and last numbers are 1. Each of the other numbers is obtained by finding the sum of the two closest numbers above it in the fifth row.

```
                    1       1
                 1     2     1
              1     3     3     1
           1     4     6     4     1
        1     5     10     10     5     1
         1+5   5+10  10+10  10+5  5+1
        1     6    15    20    15    6    1
```

We can use the numbers in the sixth row and the variable parts we found to write the expansion for $(a + b)^6$. It is

$$(a + b)^6 = a^6 + 6a^5b + 15a^4b^2 + 20a^3b^3 + 15a^2b^4 + 6ab^5 + b^6.$$

2 Evaluate a binomial coefficient.

Binomial Coefficients

Pascal's triangle becomes cumbersome when a binomial contains a relatively large power. Therefore, the coefficients in a binomial expansion are instead given in terms of factorials. The coefficients are written in a special notation, which we define next.

Definition of a Binomial Coefficient $\binom{n}{r}$

For nonnegative integers n and r, with $n \geq r$, the expression $\binom{n}{r}$ (read "n above r") is called a **binomial coefficient** and is defined by

$$\binom{n}{r} = \frac{n!}{r!(n - r)!}.$$

Technology

Graphing utilities can compute binomial coefficients. For example, to find $\binom{6}{2}$, many utilities require the sequence

6 [nCr] 2 [ENTER].

The graphing utility will display 15. Consult your manual and verify the other evaluations in Example 1.

The symbol $_nC_r$ is often used in place of $\binom{n}{r}$ to denote binomial coefficients.

EXAMPLE 1 Evaluating Binomial Coefficients

Evaluate: **a.** $\binom{6}{2}$ **b.** $\binom{3}{0}$ **c.** $\binom{9}{3}$ **d.** $\binom{4}{4}$.

Solution In each case, we apply the definition of the binomial coefficient.

a. $\binom{6}{2} = \frac{6!}{2!(6 - 2)!} = \frac{6!}{2!\,4!} = \frac{6 \cdot 5 \cdot 4!}{2 \cdot 1 \cdot 4!} = 15$

b. $\dbinom{3}{0} = \dfrac{3!}{0!(3-0)!} = \dfrac{3!}{0!\,3!} = \dfrac{1}{1} = 1$

Remember that $0! = 1$.

c. $\dbinom{9}{3} = \dfrac{9!}{3!(9-3)!} = \dfrac{9!}{3!\,6!} = \dfrac{9 \cdot 8 \cdot 7 \cdot 6!}{3 \cdot 2 \cdot 1 \cdot 6!} = 84$

d. $\dbinom{4}{4} = \dfrac{4!}{4!(4-4)!} = \dfrac{4!}{4!\,0!} = \dfrac{1}{1} = 1$

Check Point 1   Evaluate: **a.** $\dbinom{6}{3}$ **b.** $\dbinom{6}{0}$ **c.** $\dbinom{8}{2}$ **d.** $\dbinom{3}{3}$.

3   Expand a binomial raised to a power.

The Binomial Theorem

If we use binomial coefficients and the pattern for the variable part of each term, a formula called the **Binomial Theorem** can be written for any positive integral power of a binomial.

> **A Formula for Expanding Binomials: The Binomial Theorem**
>
> For any positive integer n,
>
> $$(a + b)^n = \binom{n}{0}a^n + \binom{n}{1}a^{n-1}b + \binom{n}{2}a^{n-2}b^2 + \binom{n}{3}a^{n-3}b^3 + \cdots + \binom{n}{n}b^n.$$

The Universality of Mathematics

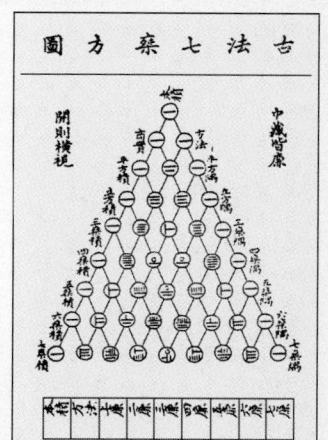

圖 方 蔡 七 法 古

"Pascal's" triangle, credited to French mathematician Blaise Pascal (1623–1662), appeared in a Chinese document printed in 1303. The Binomial Theorem was known in Eastern cultures prior to its discovery in Europe. The same mathematics is often discovered/invented by independent researchers separated by time, place, and culture.

EXAMPLE 2   Using the Binomial Theorem

Expand: $(x + 2)^4$.

Solution   We use the Binomial Theorem

$$(a + b)^n = \binom{n}{0}a^n + \binom{n}{1}a^{n-1}b + \binom{n}{2}a^{n-2}b^2 + \binom{n}{3}a^{n-3}b^3 + \cdots + \binom{n}{n}b^n$$

to expand $(x + 2)^4$. In $(x + 2)^4$, $a = x$, $b = 2$, and $n = 4$.

$$(x + 2)^4 = \binom{4}{0}x^4 + \binom{4}{1}x^3 \cdot 2 + \binom{4}{2}x^2 \cdot 2^2 + \binom{4}{3}x \cdot 2^3 + \binom{4}{4}2^4$$

These binomial coefficients are evaluated using $\binom{n}{r} = \dfrac{n!}{r!(n-r)!}$.

$$= \dfrac{4!}{0!\,4!}x^4 + \dfrac{4!}{1!\,3!}x^3 \cdot 2 + \dfrac{4!}{2!\,2!}x^2 \cdot 4 + \dfrac{4!}{3!\,1!}x \cdot 8 + \dfrac{4!}{4!\,0!} \cdot 16$$

$$\dfrac{4!}{2!\,2!} = \dfrac{4 \cdot 3 \cdot 2!}{2! \cdot 2 \cdot 1} = \dfrac{12}{2} = 6$$

Take a few minutes to verify the other factorial evaluations.

$$= 1 \cdot x^4 + 4x^3 \cdot 2 + 6x^2 \cdot 4 + 4x \cdot 8 + 1 \cdot 16$$
$$= x^4 + 8x^3 + 24x^2 + 32x + 16$$

Check Point 2   Expand: $(x + 1)^4$.

EXAMPLE 3   Using the Binomial Theorem

Expand: $(2x - y)^5$.

Solution   Because the Binomial Theorem involves the addition of two terms raised to a power, we rewrite $(2x - y)^5$ as $[2x + (-y)]^5$. We use the Binomial Theorem

$$(a + b)^n = \binom{n}{0}a^n + \binom{n}{1}a^{n-1}b + \binom{n}{2}a^{n-2}b^2 + \binom{n}{3}a^{n-3}b^3 + \cdots + \binom{n}{n}b^n$$

to expand $[2x + (-y)]^5$. In $[2x + (-y)]^5$, $a = 2x$, $b = -y$, and $n = 5$.

$$(2x - y)^5 = [2x + (-y)]^5$$

$$= \binom{5}{0}(2x)^5 + \binom{5}{1}(2x)^4(-y) + \binom{5}{2}(2x)^3(-y)^2 + \binom{5}{3}(2x)^2(-y)^3 + \binom{5}{4}(2x)(-y)^4 + \binom{5}{5}(-y)^5$$

Evaluate binomial coefficients using $\binom{n}{r} = \dfrac{n!}{r!(n-r)!}$.

$$= \frac{5!}{0!\,5!}(2x)^5 + \frac{5!}{1!\,4!}(2x)^4(-y) + \frac{5!}{2!\,3!}(2x)^3(-y)^2 + \frac{5!}{3!\,2!}(2x)^2(-y)^3 + \frac{5!}{4!\,1!}(2x)(-y)^4 + \frac{5!}{5!\,0!}(-y)^5$$

$$\frac{5!}{2!\,3!} = \frac{5 \cdot 4 \cdot 3!}{2 \cdot 1 \cdot 3!} = 10$$

Take a few minutes to verify the other factorial evaluations.

$$= 1(2x)^5 + 5(2x)^4(-y) + 10(2x)^3(-y)^2 + 10(2x)^2(-y)^3 + 5(2x)(-y)^4 + 1(-y)^5$$

Raise both factors in these parentheses to the indicated powers.

$$= 1(32x^5) + 5(16x^4)(-y) + 10(8x^3)(-y)^2 + 10(4x^2)(-y)^3 + 5(2x)(-y)^4 + 1(-y)^5$$

Now raise −y to the indicated powers.

$$= 1(32x^5) + 5(16x^4)(-y) + 10(8x^3)y^2 + 10(4x^2)(-y^3) + 5(2x)y^4 + 1(-y^5)$$

Multiplying factors in each of the six terms gives us the desired expansion:

$$(2x - y)^5 = 32x^5 - 80x^4y + 80x^3y^2 - 40x^2y^3 + 10xy^4 - y^5.$$

Check Point 3   Expand: $(x - 2y)^5$.

4 Find a particular term in a binomial expansion.

Finding a Particular Term in a Binomial Expansion

The Binomial Theorem can be used to write any single term of a binomial expansion.

> **Finding a Particular Term in a Binomial Expansion**
>
> The rth term of the expansion of $(a + b)^n$ is
>
> $$\binom{n}{r-1}a^{n-r+1}b^{r-1}.$$

EXAMPLE 4   Finding a Single Term of a Binomial Expansion

Find the fourth term in the expansion of $(3x + 2y)^7$.

Solution   We will use the formula for the rth term of the expansion of $(a + b)^n$,

$$\binom{n}{r-1}a^{n-r+1}b^{r-1}$$

to find the fourth term of $(3x + 2y)^7$. For the fourth term of $(3x + 2y)^7$, $n = 7$, $r = 4$, $a = 3x$ and $b = 2y$. Thus, the fourth term is

$$\binom{7}{4-1}(3x)^{7-4+1}(2y)^{4-1} = \binom{7}{3}(3x)^4(2y)^3 = \frac{7!}{3!(7-3)!}(3x)^4(2y)^3$$

> We use $\binom{n}{r} = \frac{n!}{r!(n-r)!}$
> to evaluate $\binom{7}{3}$.

Now we need to evaluate the factorial expression and raise $3x$ and $2y$ to the indicated powers. We obtain

$$\frac{7!}{3!\,4!}(81x^4)(8y^3) = \frac{7 \cdot 6 \cdot 5 \cdot 4!}{3 \cdot 2 \cdot 1 \cdot 4!}(81x^4)(8y^3) = 35(81x^4)(8y^3) = 22{,}680x^4y^3.$$

The fourth term of $(3x + 2y)^7$ is $22{,}680x^4y^3$.

> **Check Point 4**   Find the fifth term in the expansion of $(2x + y)^9$.

EXERCISE SET 11.5

Practice Exercises

In Exercises 1–8, evaluate the given binomial coefficient.

1. $\binom{8}{3}$

2. $\binom{7}{2}$

3. $\binom{12}{1}$

4. $\binom{11}{1}$

5. $\binom{6}{6}$

6. $\binom{15}{2}$

7. $\binom{100}{2}$

8. $\binom{100}{98}$

In Exercises 9–30, use the Binomial Theorem to expand each binomial and express the result in simplified form.

9. $(x + 2)^3$

10. $(x + 4)^3$

11. $(3x + y)^3$

12. $(x + 3y)^3$

13. $(5x - 1)^3$

14. $(4x - 1)^3$

15. $(2x + 1)^4$

16. $(3x + 1)^4$

17. $(x^2 + 2y)^4$

18. $(x^2 + y)^4$

19. $(y - 3)^4$

20. $(y - 4)^4$

21. $(2x^3 - 1)^4$

22. $(2x^5 - 1)^4$

23. $(c + 2)^5$

24. $(c + 3)^5$

25. $(x - 1)^5$

26. $(x - 2)^5$

27. $(x - 2y)^5$

28. $(x - 3y)^5$

29. $(2a + b)^6$

30. $(a + 2b)^6$

In Exercises 31–38, write the first three terms in each binomial expansion, expressing the result in simplified form.

31. $(x + 2)^8$

32. $(x + 3)^8$

33. $(x - 2y)^{10}$

34. $(x - 2y)^9$

35. $(x^2 + 1)^{16}$

36. $(x^2 + 1)^{17}$

37. $(y^3 - 1)^{20}$

38. $(y^3 - 1)^{21}$

In Exercises 39–46, find the term indicated in each expansion.

39. $(2x + y)^6$; third term

40. $(x + 2y)^6$; third term

41. $(x - 1)^9$; fifth term

42. $(x - 1)^{10}$; fifth term

43. $(x^2 + y^3)^8$; sixth term

44. $(x^3 + y^2)^8$; sixth term

45. $(x - \frac{1}{2})^9$; fourth term

46. $(x + \frac{1}{2})^8$; fourth term

Application Exercises

47. The percentage of people taking the SAT whose intended college major is engineering, $f(t)$, can be modeled by

$$f(t) = 0.002t^3 - 0.9t^2 + 1.27t + 6.76, \qquad 0 \le t \le 20,$$

where $t = 0$ represents 1975. How can we adjust this model so that $t = 0$ corresponds to 1985 rather than 1975? We shift the graph of f ten units to the left. We obtain $g(t) = f(t + 10)$. Use the Binomial Theorem to express $g(t)$ in descending powers of t.

48. The personal income per capita in the United States, $f(t)$, in constant 1992 dollars, can be modeled by

$$f(t) = 3.75t^3 - 115.23t^2 + 1229.81t + 16{,}025.65,$$
$$0 \le t \le 15,$$

where $t = 0$ represents 1979. How can we adjust this model so that $t = 0$ corresponds to 1989 rather than 1979? We shift the graph of f ten units to the left. We obtain $g(t) = f(t + 10)$. Use the Binomial Theorem to express $g(t)$ in descending powers of t.

Writing in Mathematics

49. Describe the pattern on the exponents on a in the expansion of $(a + b)^n$.

50. Describe the pattern on the exponents on b in the expansion of $(a + b)^n$.

51. What is true about the sum of the exponents on a and b in any term in the expansion of $(a + b)^n$?

52. How do you determine how many terms there are in a binomial expansion?

53. What is Pascal's triangle? How do you find the numbers in any row of the triangle?

54. Explain how to evaluate $\binom{n}{r}$. Provide an example with your explanation.

55. Explain how to use the Binomial Theorem to expand a binomial. Provide an example with your explanation.

56. Explain how to find a particular term in a binomial expansion without having to write out the entire expansion.

57. Are there situations in which it is easier to use Pascal's triangle than binomial coefficients? Describe these situations.

58. Describe how you would use mathematical induction to prove

$$(a + b)^n = \binom{n}{0}a^n + \binom{n}{1}a^{n-1}b + \binom{n}{2}a^{n-2}b^2$$
$$+ \cdots + \binom{n}{n-1}ab^{n-1} + \binom{n}{n}b^n.$$

What happens when $n = 1$? Write the statement that we assume true. Write the statement that we must prove. What must be done to the left side of the assumed statement to make it look like the left side of the statement that must be proved? (More detail on the actual proof is found in Exercise 71.)

Technology Exercises

59. Use the $\boxed{\text{nCr}}$ key on a graphing utility to verify your answers in Exercises 1–8.

In Exercises 60–61, graph each of the functions in the same viewing rectangle. Describe how the graphs illustrate the Binomial Theorem.

60. $f_1(x) = (x + 2)^3$
$f_2(x) = x^3$
$f_3(x) = x^3 + 6x^2$
$f_4(x) = x^3 + 6x^2 + 12x$
$f_5(x) = x^3 + 6x^2 + 12x + 8$
Use a $[-10, 10, 1]$ by $[-30, 30, 10]$ viewing rectangle.

61. $f_1(x) = (x + 1)^4$
$f_2(x) = x^4$
$f_3(x) = x^4 + 4x^3$
$f_4(x) = x^4 + 4x^3 + 6x^2$
$f_5(x) = x^4 + 4x^3 + 6x^2 + 4x$
$f_6(x) = x^4 + 4x^3 + 6x^2 + 4x + 1$
Use a $[-5, 5, 1]$ by $[-30, 30, 10]$ viewing rectangle.

In Exercises 62–64, use the Binomial Theorem to find a polynomial expansion for each function. Then use a graphing utility and an approach similar to the one in Exercises 60 and 61 to verify the expansion.

62. $f_1(x) = (x - 1)^3$

63. $f_1(x) = (x - 2)^4$

64. $f_1(x) = (x + 2)^6$

65. Graphing utilities capable of symbolic manipulation, such as the TI-92, will expand binomials. On the TI-92, to expand $(3a - 5b)^{12}$, input the following:

$\boxed{\text{EXPAND}} \left((3a \boxed{-} 5b) \boxed{\wedge} 12\right) \boxed{\text{ENTER}}$.

Use a graphing utility with this capability to verify any five of the expansions you performed by hand in Exercises 9–30.

Critical Thinking Exercises

66. Which one of the following is true?

a. The binomial expansion for $(a + b)^n$ contains n terms.

b. The Binomial Theorem can be written in condensed form as $(a + b)^n = \sum\limits_{r=0}^{n} \binom{n}{r} a^{n-r} b^r$.

c. The sum of the binomial coefficients in $(a + b)^n$ cannot be 2^n.

d. There are no values of a and b such that $(a + b)^4 = a^4 + b^4$.

67. Use the Binomial Theorem to expand and then simplify the result: $(x^2 + x + 1)^3$. [*Hint*: Write $x^2 + x + 1$ as $x^2 + (x + 1)$].

68. Find the term in the expansion of $(x^2 + y^2)^5$ containing x^4 as a factor.

69. Prove that

$$\binom{n}{r} = \binom{n}{n - r}.$$

70. Show that

$$\binom{n}{r} + \binom{n}{r + 1} = \binom{n + 1}{r + 1}.$$

Hints:

$$(n - r)! = (n - r)(n - r - 1)!$$
$$(r + 1)! = (r + 1)r!$$

71. Follow the outline below to use mathematical induction to prove that

$$(a + b)^n = \binom{n}{0} a^n + \binom{n}{1} a^{n-1}b + \binom{n}{2} a^{n-2}b^2$$

$$+ \cdots + \binom{n}{n - 1} ab^{n-1} + \binom{n}{n} b^n.$$

a. Verify the formula for $n = 1$.

b. Replace n with k and write the statement that is assumed true. Replace n with $k + 1$ and write the statement that must be proved.

c. Multiply both sides of the statement assumed to be true by $a + b$. Add exponents on the left. On the right, distribute a and b, respectively.

d. Collect like terms on the right. At this point, you should have

$$(a + b)^{k+1} = \binom{k}{0} a^{k+1} + \left[\binom{k}{0} + \binom{k}{1}\right] a^k b$$

$$+ \left[\binom{k}{1} + \binom{k}{2}\right] a^{k-1}b^2 + \left[\binom{k}{2} + \binom{k}{3}\right] a^{k-2}b^3$$

$$+ \cdots + \left[\binom{k}{k - 1} + \binom{k}{k}\right] ab^k + \binom{k}{k} b^{k+1}.$$

e. Use the result of Exercise 70 to add the binomial sums in brackets. For example, because $\binom{n}{r} + \binom{n}{r + 1}$

$$= \binom{n + 1}{r + 1} \quad \text{then} \quad \binom{k}{0} + \binom{k}{1} = \binom{k + 1}{1} \quad \text{and}$$

$$\binom{k}{1} + \binom{k}{2} = \binom{k + 1}{2}.$$

f. Because $\binom{k}{0} = \binom{k + 1}{0}$ (why?) and $\binom{k}{k} =$

$\binom{k + 1}{k + 1}$ (why?), substitute these results and the results from part (e) into the equation in part (d). This should give the statement that we were required to prove in the second step of the mathematical induction process.

SECTION 11.6 *Counting Principles, Permutations, and Combinations*

Objectives

1. Use the Fundamental Counting Principle.
2. Use the permutations formula.
3. Distinguish between permutation problems and combination problems.
4. Use the combinations formula.

Have you ever imagined what your life would be like if you won the lottery? What changes would you make? Before you fantasize about becoming a person of leisure with a staff of obedient elves, think about this: The probability of winning top prize in the lottery is about the same as the probability of being struck by lightning. There are millions of possible number combinations in lottery games, and only one way of winning the grand prize. Determining the probability of winning involves calculating the chance of getting the winning combination from all possible outcomes. In this section, we begin preparing for the surprising world of probability by looking at methods for counting possible outcomes.

1 Use the Fundamental Counting Principle.

The Fundamental Counting Principle

It's early morning, you're groggy, and you have to select something to wear for your 8 A.M. class. (What *were* you thinking of when you signed up for a class at that hour?!) Fortunately, your "lecture wardrobe" is rather limited—just two pairs of jeans to choose from (one blue, one black), three T-shirts to choose from (one beige, one yellow, and one blue), and two pairs of sneakers to select from (one black, one red). Your possible outfits are shown in Figure 11.5.

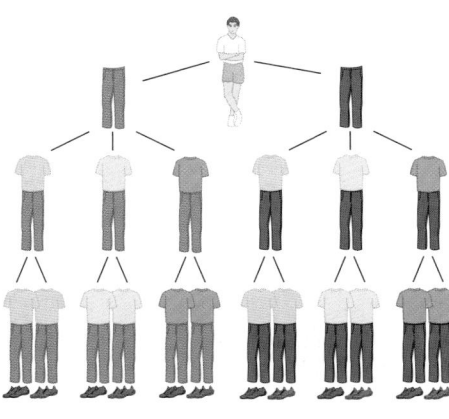

Figure 11.5 Selecting a wardrobe

The number of possible ways of playing the first four moves on each side in a game of chess is 318,979,564,000.

The **tree diagram**, so named because of its branches, shows that you can form 12 outfits from your two pairs of jeans, three T-shirts, and two pairs of sneakers. Notice that the number of outfits can be obtained by multiplying the number of choices for jeans, 2, the number of choices for T-shirts, 3, and the number of choices for sneakers, 2:

$$2 \cdot 3 \cdot 2 = 12.$$

We can generalize this idea to any two or more groups of items—not just jeans, T-shirts, and sneakers—with the **Fundamental Counting Principle**.

The Fundamental Counting Principle

The number of ways in which a series of successive things can occur is found by multiplying the number of ways in which each thing can occur.

For example, if you own 30 pairs of jeans, 20 T-shirts, and 12 pairs of sneakers, you have

$$30 \cdot 20 \cdot 12 = 7200$$

choices for your wardrobe!

EXAMPLE 1   Options in Planning a Course Schedule

Next semester you are planning to take three courses—math, English, and humanities. Based on time blocks and highly recommended professors, there are 8 sections of math, 5 of English, and 4 of humanities that you find suitable. Assuming no scheduling conflicts, how many different three-course schedules are possible?

Solution   This situation involves making choices with three groups of items.

MATH	ENGLISH	HUMANITIES
8 choices	5 choices	4 choices

We use the Fundamental Counting Principle to find the number of three-course schedules. Multiply the number of choices for each of the three groups.

$$8 \cdot 5 \cdot 4 = 160$$

Thus, there are 160 different three-course schedules.

Check Point 1   A pizza can be ordered with three choices of size (small, medium, or large), four choices of crust (thin, thick, crispy, or regular), and six choices of toppings (ground beef, sausage, pepperoni, bacon, mushrooms, or onions). How many different one-topping pizzas can be ordered?

EXAMPLE 2   A Multiple-Choice Test

You are taking a multiple-choice test that has ten questions. Each of the questions has four choices, with one correct choice per question. If you select one of these options per question and leave nothing blank, in how many ways can you answer the questions?

Selena, musician of Tejano music (1971–1995)

Do you see the difference between a permutation and a combination? A permutation is an ordered arrangement of a given group of items. A combination

Because all permutation problems are also Fundamental Counting prob-

is a group of items taken without regard to their order. *Permutation* problems involve situations in which *order matters. Combination* problems involve situations in which the *order* of items *makes no difference.*

EXAMPLE 6 Distinguishing between Permutations and Combinations

For each of the following problems, explain whether the problem is one involving permutations or combinations. (It is not necessary to solve the problem.)

   a. Six candidates are running for president, chief technology officer, and director of marketing of an Internet company. The candidate with the greatest number of votes becomes the president, the second biggest vote-getter becomes chief technology officer, and the candidate who gets the third largest number of votes will be director of marketing. How many different outcomes are possible for these three positions?

   b. From the six candidates who desire to hold office in an Internet company, a three-person committee is formed to study ways of finding new investors. How many different committees could be formed?

Solution

   a. Voters are choosing three officers from six candidates. The order in which the officers are chosen makes a difference because each of the offices (president, chief technology officer, and director of marketing) is different. Order matters. This is a problem involving permutations. (How many permutations are possible if three candidates are elected from six candidates?)

   b. A three-person committee is to be formed from the six candidates. The order in which the three people are selected does not matter because they are not filling different roles on the committee. Because order makes no difference, this is a problem involving combinations. (How many different combinations of three people can be chosen from a group of six people?)

Check Point 6   For each of the following problems, explain if the problem is one involving permutations or combinations. (It is not necessary to solve the problem.)

   a. How many ways can you select 6 free videos from a list of 200 videos?

   b. In a race in which there are 50 runners and no ties, in how many ways can the first three finishers come in?

4 Use the combinations formula.

The notation $_nC_r$ means the **number of combinations of n things taken r at a time**. In general, there are $r!$ times as many permutations of n things taken r at a time as there are combinations of n things taken r at a time. Thus, we find the number of combinations of n things taken r at a time by dividing the number of permutations of n things taken r at a time by $r!$.

$$_nC_r = \frac{_nP_r}{r!} = \frac{\dfrac{n!}{(n-r)!}}{r!} = \frac{n!}{(n-r)!\,r!}$$

Combinations of n Things Taken r at a Time

The number of possible combinations if r items are taken from n items is

$$_nC_r = \frac{n!}{(n-r)!\,r!}.$$

Notice that the formula for $_nC_r$ is the same as the formula for the binomial coefficient $\binom{n}{r}$.

We cannot find the number of combinations if r items are taken from n items using the Fundamental Counting Principle. We must use the formula shown in the box to do so.

EXAMPLE 7 Using the Formula for Combinations

A three-person committee is needed to study ways of improving public transportation. How many committees could be formed from the eight people on the board of supervisors?

Solution The order in which the three people are selected does not matter. This is a problem of selecting $r = 3$ people from a group of $n = 8$ people. We are looking for the number of combinations of eight things taken three at a time. We use the formula

$$_nC_r = \frac{n!}{(n-r)!\,r!}$$

with $n = 8$ and $r = 3$.

$$_8C_3 = \frac{8!}{(8-3)!\,3!} = \frac{8!}{5!\,3!} = \frac{8 \cdot 7 \cdot 6 \cdot 5!}{5! \cdot 3 \cdot 2 \cdot 1} = \frac{8 \cdot 7 \cdot 6 \cdot 5!}{5! \cdot 3 \cdot 2 \cdot 1} = 56$$

Thus, 56 committees of three people each can be formed from the eight people on the board of supervisors.

Check Point 7 From a group of 10 physicians, in how many ways can four people be selected to attend a conference on acupuncture?

EXAMPLE 8 Using the Formula for Combinations

In poker, a person is dealt 5 cards from a standard 52-card deck. The order in which you are dealt the 5 cards does not matter. How many different 5-card poker hands are possible?

Solution Because the order in which the 5 cards are dealt does not matter, this is a problem involving combinations. We are looking for the number of combinations of $n = 52$ cards drawn $r = 5$ at a time. We use the formula

$$_nC_r = \frac{n!}{(n-r)!\,r!}$$

with $n = 52$ and $r = 5$.

$$_{52}C_5 = \frac{52!}{(52-5)!\,5!} = \frac{52!}{47!\,5!} = \frac{52 \cdot 51 \cdot 50 \cdot 49 \cdot 48 \cdot 47!}{47! \cdot 5 \cdot 4 \cdot 3 \cdot 2 \cdot 1} = 2{,}598{,}960$$

Thus, there are 2,598,960 different 5-card poker hands possible. It surprises many people that more than 2.5 million 5-card hands can be dealt from a mere 52 cards.

If you are a card player, it does not get any better than to be dealt the 5-card poker hand shown in Figure 11.6. This hand is called a *royal flush*. It consists of an ace, king, queen, jack, and 10, all of the same suit: all hearts, all diamonds, all clubs, or all spades. The probability of being dealt a royal flush involves calculating the number of ways of being dealt such a hand: just 4 of all 2,598,960 possible hands. In the next section, we move from counting possibilities to computing probabilities.

Figure 11.6 A royal flush

Check Point 8 How many different 4-card hands can be dealt from a deck that has 16 different cards?

EXERCISE SET 11.6

 Practice Exercises

In Exercises 1–8, use the formula for $_nP_r$ to evaluate each expression.

1. $_9P_4$ **2.** $_7P_3$

3. $_8P_5$ **4.** $_{10}P_4$

5. $_6P_6$ **6.** $_9P_9$

7. $_8P_0$ **8.** $_6P_0$

In Exercises 9–16, use the formula for $_nC_r$ to evaluate each expression.

9. $_9C_5$ **10.** $_{10}C_6$

11. $_{11}C_4$ **12.** $_{12}C_5$

13. $_7C_7$ **14.** $_4C_4$

15. $_5C_0$ **16.** $_6C_0$

In Exercises 17–20, does the problem involve permutations or combinations? Explain your answer. (It is not necessary to solve the problem.)

17. A medical researcher needs 6 people to test the effectiveness of an experimental drug. If 13 people have volunteered for the test, in how many ways can 6 people be selected?

18. Fifty people purchase raffle tickets. Three winning tickets are selected at random. If first prize is $1000, second prize is $500, and third prize is $100, in how many different ways can the prizes be awarded?

19. How many different four-letter passwords can be formed from the letters A, B, C, D, E, F, and G if no repetition of letters is allowed?

20. Fifty people purchase raffle tickets. Three winning tickets are selected at random. If each prize is $500, in how many different ways can the prizes be awarded?

 **Application Exercises**

Use the Fundamental Counting Principle to solve Exercises 21–32.

21. The model of the car you are thinking of buying is available in nine different colors and three different styles (hatchback, sedan, or station wagon). In how many ways can you order the car?

22. A popular brand of pen is available in three colors (red, green, or blue) and four writing tips (bold, medium, fine, or micro). How many different choices of pens do you have with this brand?

23. An ice cream store sells two drinks (sodas or milk shakes), in four sizes (small, medium, large, or jumbo), and five flavors (vanilla, strawberry, chocolate, coffee, or pistachio). In how many ways can a customer order a drink?

24. A restaurant offers the following lunch menu.

Main Course	Vegetables	Beverages	Desserts
Ham	Potatoes	Coffee	Cake
Chicken	Peas	Tea	Pie
Fish	Green beans	Milk	Ice cream
Beef		Soda	

If one item is selected from each of the four groups, in how many ways can a meal be ordered? Describe two such orders.

25. You are taking a multiple-choice test that has five questions. Each of the questions has three choices, with one correct choice per question. If you select one of these options per question and leave nothing blank, in how many ways can you answer the questions?

26. You are taking a multiple-choice test that has eight questions. Each of the questions has three choices, with one correct choice per question. If you select one of these options per question and leave nothing blank, in how many ways can you answer the questions?

27. In the original plan for area codes in 1945, the first digit could be any number from 2 through 9, the second digit was either 0 or 1, and the third digit could be any number except 0. With this plan, how many different area codes were possible?

28. How many different four-letter radio station call letters can be formed if the first letter must be W or K?

29. Six performers are to present their comedy acts on a weekend evening at a comedy club. One of the performers insists on being the last stand-up comic of the evening. If this performer's request is granted, how many different ways are there to schedule the appearances?

30. Five singers are to perform at a night club. One of the singers insists on being the last performer of the evening. If this singer's request is granted, how many different ways are there to schedule the appearances?

31. In the *Cambridge Encyclopedia of Language* (Cambridge University Press, 1987), author David Crystal presents five sentences that make a reasonable paragraph regardless of their order. The sentences are:

> Mark had told him about the foxes.
> John looked out the window.
> Could it be a fox?
> However, nobody had seen one for months.
> He thought he saw a shape in the bushes.

How many different five-sentence paragraphs can be formed if the paragraph begins with "He thought he saw a shape in the bushes" and ends with "John looked out of the window"?

32. A television programmer is arranging the order that five movies will be seen between the hours of 6 P.M. and 4 A.M. Two of the movies have a G rating, and they are to be shown in the first two time blocks. One of the movies is rated NC-17, and it is to be shown in the last of the time blocks, from 2 A.M. until 4 A.M. Given these restrictions, in how many ways can the five movies be arranged during the indicated time blocks?

Use the formula for $_nP_r$ to solve Exercises 33–40.

33. A club with ten members is to choose three officers—president, vice-president, and secretary-treasurer. If each office is to be held by one person and no person can hold more than one office, in how many ways can those offices be filled?

34. A corporation has ten members on its board of directors. In how many different ways can it elect a president, vice-president, secretary, and treasurer?

35. For a segment of a radio show, a disc jockey can play 7 records. If there are 13 records to select from, in how many ways can the program for this segment be arranged?

36. Suppose you are asked to list, in order of preference, the three best movies you have seen this year. If you saw 20 movies during the year, in how many ways can the three best be chosen and ranked?

37. In a race in which six automobiles are entered and there are no ties, in how many ways can the first three finishers come in?

38. In a production of *West Side Story*, eight actors are considered for the male roles of Tony, Riff, and Bernardo. In how many ways can the director cast the male roles?

39. Nine bands have volunteered to perform at a benefit concert, but there is only enough time for five of the bands to play. How many lineups are possible?

40. How many arrangements can be made using four of the letters of the word COMBINE if no letter is to be used more than once?

Use the formula for $_nC_r$ to solve Exercises 41–48.

41. An election ballot asks voters to select three city commissioners from a group of six candidates. In how many ways can this be done?

42. A four-person committee is to be elected from an organization's membership of 11 people. How many different committees are possible?

43. Of 12 possible books, you plan to take 4 with you on vacation. How many different collections of 4 books can you take?

44. There are 14 standbys who hope to get seats on a flight, but only 6 seats are available on the plane. How many different ways can the 6 people be selected?

45. You volunteer to help drive children at a charity event to the zoo, but you can fit only 8 of the 17 children present in your van. How many different groups of 8 children can you drive?

46. Of the 100 people in the U.S. Senate, 18 serve on the Foreign Relations Committee. How many ways are there to select Senate members for this committee (assuming party affiliation is not a factor in selection)?

47. To win at LOTTO in the state of Florida, one must correctly select 6 numbers from a collection of 49 numbers (1 through 49). The order in which the selection is made does not matter. How many different selections are possible?

48. To win in the New York State lottery, one must correctly select 6 numbers from 54 numbers. The order in which the selection is made does not matter. How many different selections are possible?

In Exercises 49–58, solve by the method of your choice.

49. In a race in which six automobiles are entered and there are no ties, in how many ways can the first four finishers come in?

50. A book club offers a choice of 8 books from a list of 40. In how many ways can a member make a selection?

51. A medical researcher needs 6 people to test the effectiveness of an experimental drug. If 13 people have volunteered for the test, in how many ways can 6 people be selected?

52. Fifty people purchase raffle tickets. Three winning tickets are selected at random. If first prize is $1000, second prize is $500, and third prize is $100, in how many different ways can the prizes be awarded?

53. From a club of 20 people, in how many ways can a group of three members be selected to attend a conference?

54. Fifty people purchase raffle tickets. Three winning tickets are selected at random. If each prize is $500, in how many different ways can the prizes be awarded?

55. How many different four-letter passwords can be formed from the letters A, B, C, D, E, F, and G if no repetition of letters is allowed?

56. Nine comedy acts will perform over two evenings. Five of the acts will perform on the first evening. How many ways can the schedule for the first evening be made?

57. Using 15 flavors of ice cream, how many cones with three different flavors can you create if it is important to you which flavor goes on the top, middle, and bottom?

58. Baskin-Robbins offers 31 different flavors of ice cream. One of their items is a bowl consisting of three scoops of ice cream, each a different flavor. How many such bowls are possible?

Writing in Mathematics

59. Explain the Fundamental Counting Principle.

60. Write an original problem that can be solved using the Fundamental Counting Principle. Then solve the problem.

61. What is a permutation?

62. Describe what $_nP_r$ represents.

63. Write a word problem that can be solved by evaluating $_7P_3$.

64. What is a combination?

65. Explain how to distinguish between permutation and combination problems.

66. Write a word problem that can be solved by evaluating $_7C_3$.

Technology Exercises

67. Use a graphing utility with an $\boxed{_nP_r}$ key to verify your answers in Exercises 1–8.

68. Use a graphing utility with an $\boxed{_nC_r}$ key to verify your answers in Exercises 9–16.

Critical Thinking Exercises

69. Which one of the following is true?
  a. The number of ways to choose four questions out of ten questions on an essay test is $_{10}P_4$.
  b. If $r > 1$, $_nP_r$ is less than $_nC_r$.
  c. $_7P_3 = 3!\,_7C_3$
  d. The number of ways to pick a winner and first runner-up in a piano recital with 20 contestants is $_{20}C_2$.

70. Five men and five women line up at a checkout counter in a store. In how many ways can they line up if the first person in line is a woman, and the people in line alternate woman, man, woman, man, and so on?

71. How many four-digit odd numbers less than 6000 can be formed using the digits 2, 4, 6, 7, 8, and 9?

72. If a collection of n objects has n_1 identical objects of the same type, n_2 identical objects of a second kind, n_3 of a third kind, and so on for a total of $n = n_1 + n_2 + \cdots + n_k$ objects, the number of distinguishable permutations of the n objects is given by

$$\frac{n!}{n_1!\,n_2!\,n_3!\cdots n_k!}.$$

Use this formula to find the number of different signals consisting of eight flags that can be made using three white flags, four red flags and one blue flag.

Group Exercise

73. The group should select real-world situations where the Fundamental Counting Principle can be applied. These could involve the number of possible student ID numbers on your campus, the number of possible phone numbers in your community, the number of meal options at a local restaurant, the number of ways a person in the group can select outfits for class, the number of ways a condominium can be purchased in a nearby community, and so on. Once situations have been selected, group members should determine in how many ways each part of the task can be done. Group members will need to obtain menus, find out about telephone-digit requirements in the community, count shirts, pants, shoes in closets, visit condominium sales offices, and so on. Once the group reassembles, apply the Fundamental Counting Principle to determine the number of available options in each situation. Because these numbers may be quite large, use a calculator.

SECTION 11.7 *Probability*

Objectives

1. Compute empirical probability.
2. Compute theoretical probability.
3. Find the probability that an event will not occur.
4. Find the probability of one event or a second event occurring.
5. Find the probability of one event and a second event occurring.

Table 11.3 Number of Americans and the Hours of Sleep They Get on a Typical Night

Hours of Sleep	Number of Americans, in millions
4 or less	11
5	24.75
6	68.75
7	82.5
8	74.25
9	8.25
10 or more	5.5
Total:	275

Source: Discovery Health Media

1 Compute empirical probability.

How many hours of sleep do you typically get each night? Table 11.3 indicates that 11 million out of 275 million Americans are getting four hours or less sleep on a typical night. The *probability* of an American getting four hours or less sleep on a typical night is $\frac{11}{275}$. This fraction can be reduced to $\frac{1}{25}$, or expressed as 0.04 or 4%. Thus, 4% of Americans get four hours or less sleep each night.

We find a probability by dividing one number by another. Probabilities are assigned to an *event*, such as getting four hours or less sleep on a typical night. Events that are certain to occur are assigned probabilities of 1, or 100%. For example, the probability that a given individual will eventually die is 1. Regrettably, taxes and death are always certain! By contrast, if an event cannot occur, its probability is 0. For example, the probability that Elvis will return from the dead and serenade us with one final reprise of "Heartbreak Hotel" is 0.

Probabilities of events are expressed as numbers ranging from 0 to 1, or 0% to 100%. The closer the probability of a given event is to 1, the more likely it is that the event will occur. The closer the probability of a given event is to 0, the less likely it is that the event will occur.

Empirical Probability

Empirical probability applies to situations in which we observe how frequently an event occurs. We use the following formula to compute the empirical probability of an event.

> **Computing Empirical Probability**
> The **empirical probability** of event E is
> $$P(E) = \frac{\text{observed number of times } E \text{ occurs}}{\text{total number of observed occurrences}}.$$

EXAMPLE 1 Computing Empirical Probability

An American is randomly selected. Use Table 11.3 to find the probability of that person getting eight hours sleep on a typical night.

Solution The probability of getting eight hours sleep is the observed number of Americans who do this, 74.25 million, divided by the total number of Americans, 275 million.

$$P(\text{eight hours sleep}) = \frac{\text{number of Americans who sleep 8 hours}}{\text{total numbers of Americans}}$$

$$= \frac{74.25}{275} = \frac{297}{1100} = 0.27$$

The empirical probability of randomly selecting an American who gets eight hours sleep on a typical night is $\frac{297}{1100}$, or 0.27.

Check Point 1 Use Table 11.3 to find the probability of randomly selecting an American who gets seven hours sleep on a typical night.

2 Compute theoretical probability.

Theoretical Probability

You toss a coin. Although it is equally likely to land either heads up, denoted by H, or tails up, denoted by T, the actual outcome is uncertain. Any occurrence for which the outcome is uncertain is called an **experiment**. Thus, tossing a coin is an example of an experiment. The set of all possible equally likely outcomes of an experiment is the **sample space** of the experiment, denoted by S. The sample space for the coin-tossing experiment is

$$S = \{H, T\}.$$

> lands heads up   lands tails up

We can define an event more formally using these concepts. An **event**, denoted by E, is any subcollection, or subset, of a sample space. For example, the subset $E = \{T\}$ is the event of landing tails up when a coin is tossed.

Theoretical probability applies to situations like this, in which the sample space of all equally likely outcomes is known. To calculate the theoretical probability of an event, we divide the number of outcomes in the event by the number of outcomes in the sample space.

Computing Theoretical Probability

If an event E has $n(E)$ equally likely outcomes and its sample space S has $n(S)$ equally likely outcomes, the theoretical probability of event E, denoted by $P(E)$, is

$$P(E) = \frac{\text{number of outcomes in event } E}{\text{number of outcomes in sample space } S} = \frac{n(E)}{n(S)}.$$

The sum of the theoretical probabilities of all possible outcomes in the sample space is 1.

How can we use this formula to compute the probability of a coin landing tails up? We use the following sets:

$$E = \{T\} \qquad\qquad S = \{H, T\}$$

> This is the event of landing tails up.

> This is the sample space with all equally possible outcomes.

Figure 11.7 Outcomes when a die is rolled

The probability of a coin landing tails up is

$$P(E) = \frac{n(E)}{n(S)} = \frac{1}{2}.$$

Theoretical probability applies to many games of chance, including dice rolling, lotteries, card games, and roulette. The next example deals with the experiment of rolling a die. Figure 11.7 illustrates that when a die is rolled, there are six equally likely outcomes. The sample space can be shown as

$$S = \{1, 2, 3, 4, 5, 6\}.$$

EXAMPLE 2 Computing Theoretical Probability

A die is rolled. Find the probability of getting a number less than 5.

Solution   The sample space of equally likely outcomes is $S = \{1, 2, 3, 4, 5, 6\}$. There are six outcomes in the sample space, so $n(S) = 6$.
   We are interested in the probability of getting a number less than 5. The event of getting a number less than 5 can be represented by

$$E = \{1, 2, 3, 4\}.$$

There are four outcomes in this event, so $n(E) = 4$.
   The probability of rolling a number less than 5 is

$$P(E) = \frac{n(E)}{n(S)} = \frac{4}{6} = \frac{2}{3}.$$

> **Check Point 2**   A die is rolled. Find the probability of getting a number greater than 4.

EXAMPLE 3 Computing Theoretical Probability

Two ordinary six-sided dice are rolled. What is the probability of getting a sum of 8?

Solution   Each die has six equally likely outcomes. By the Fundamental Counting Principle, there are $6 \cdot 6$, or 36, equally likely outcomes in the sample space. That is, $n(S) = 36$. The 36 outcomes are shown here as ordered pairs. The five ways of rolling a sum of 8 appear in the highlighted diagonal as follows.

		Second Die					
		⚀	⚁	⚂	⚃	⚄	⚅
First Die	⚀	(1,1)	(1,2)	(1,3)	(1,4)	(1,5)	(1,6)
	⚁	(2,1)	(2,2)	(2,3)	(2,4)	(2,5)	(2,6)
	⚂	(3,1)	(3,2)	(3,3)	(3,4)	(3,5)	(3,6)
	⚃	(4,1)	(4,2)	(4,3)	(4,4)	(4,5)	(4,6)
	⚄	(5,1)	(5,2)	(5,3)	(5,4)	(5,5)	(5,6)
	⚅	(6,1)	(6,2)	(6,3)	(6,4)	(6,5)	(6,6)

$S = \{(1, 1), (1, 2), (1, 3), (1, 4),$
$(1, 5), (1, 6), (2, 1), (2, 2),$
$(2, 3), (2, 4), (2, 5), (2, 6),$
$(3, 1), (3, 2), (3, 3), (3, 4),$
$(3, 5), (3, 6), (4, 1), (4, 2),$
$(4, 3), (4, 4), (4, 5), (4, 6),$
$(5, 1), (5, 2), (5, 3), (5, 4),$
$(5, 5), (5, 6), (6, 1), (6, 2),$
$(6, 3), (6, 4), (6, 5), (6, 6)\}$

The phrase "getting a sum of 8" describes the event

$$E = \{(6, 2), (5, 3), (4, 4), (3, 5), (2, 6)\}.$$

This event has 5 outcomes, so $n(E) = 5$. Thus, the probability of getting a sum of 8 is

$$P(E) = \frac{n(E)}{n(S)} = \frac{5}{36}.$$

Check Point 3 What is the probability of getting a sum of 5 when two six-sided dice are rolled?

Computing Theoretical Probability Without Listing an Event and the Sample Space

In some situations, we can compute theoretical probability without having to write out each event and each sample space. For example, suppose you are dealt one card from a standard 52-card deck, illustrated in Figure 11.8. The deck has four suits: Hearts and diamonds are red, and clubs and spades are black. Each suit has 13 different face values—A(ace), 2, 3, 4, 5, 6, 7, 8, 9, 10, J(jack), Q(queen), and K(king). Jacks, queens, and kings are called picture cards.

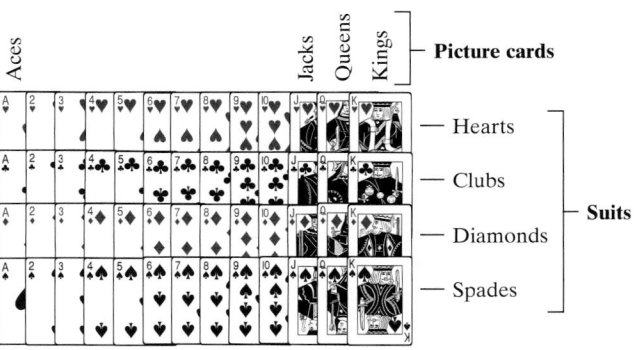

Figure 11.8 A standard 52-card bridge deck

EXAMPLE 4 Probability and a Deck of 52 Cards

You are dealt one card from a standard 52-card deck. Find the probability of being dealt a heart.

Solution Let E be the event of being dealt a heart. Because there are 13 hearts in the deck, the event of being dealt a heart can occur in 13 ways. The number of outcomes in event E is 13: $n(E) = 13$. With 52 cards in the deck, the total number of possible ways of being dealt a single card is 52. The number of outcomes in the sample space is 52: $n(S) = 52$. The probability of being dealt a heart is

$$P(E) = \frac{n(E)}{n(S)} = \frac{13}{52} = \frac{1}{4}.$$

Check Point 4 If you are dealt one card from a standard 52-card deck, find the probability of being dealt a king.

State lotteries keep 50 cents on the dollar, resulting in $10 billion a year for public funding.

If your state has a lottery drawing each week, the probability that someone will win the top prize is relatively high. If there is no winner this week, it is virtually certain that eventually someone will be graced with millions of dollars. So how come you are unlucky compared to this undisclosed someone? In Example 5, we provide an answer to this question, using the counting principles discussed in Section 11.6.

EXAMPLE 5   **Probability and Combinations: Winning the Lottery**

Florida's lottery game, LOTTO, is set up so that each player chooses six different numbers from 1 to 49. If the six numbers chosen match the six numbers drawn randomly each Saturday evening, the player wins (or shares) the top cash prize. (As of this writing, the top cash prize has ranged from $7 million to $106.5 million.) With one LOTTO ticket, what is the probability of winning this prize?

Solution   Because the order of the six numbers does not matter, this is a situation involving combinations. Let E be the event of winning the lottery with one ticket. With one LOTTO ticket, there is only one way of winning. Thus, $n(E) = 1$. The sample space is the set of all possible six-number combinations. We can use the combinations formula

$$_nC_r = \frac{n!}{(n-r)!\, r!}$$

to find the number of outcomes in the sample space. We are selecting $r = 6$ numbers from a collection of $n = 49$ numbers.

$$_{49}C_6 = \frac{49!}{(49-6)!\,6!} = \frac{49!}{43!\,6!} = \frac{49 \cdot 48 \cdot 47 \cdot 46 \cdot 45 \cdot 44 \cdot 43!}{43! \cdot 6 \cdot 5 \cdot 4 \cdot 3 \cdot 2 \cdot 1} = 13{,}983{,}816$$

There are nearly 14 million number combinations possible in LOTTO. If a person buys one LOTTO ticket, the probability of winning is

$$P(E) = \frac{n(E)}{n(S)} = \frac{1}{13{,}983{,}816} \approx 0.0000000715.$$

The probability of winning the top prize with one LOTTO ticket is $\frac{1}{13,983,816}$, or about 1 in 14 million.

In 1997, Americans spent nearly 17 billion dollars on lotteries set up by revenue-hungry states. If a pigeon, er, person, buys, say 5000 different tickets in Florida's LOTTO, that person has selected 5000 different combinations of the six numbers. The probability of winning is

$$\frac{5000}{13{,}983{,}816} \approx 0.000358.$$

The chances of winning top prize are about 358 in a million. At $1 per LOTTO ticket, it is highly probable that Mr. or Ms. Pigeon will be $5000 poorer.

Check Point 5   In a state lottery, a player chooses five different numbers from 1 to 30. If the five numbers chosen match the five numbers drawn each week, the player wins (or shares) the top cash prize. With one lottery ticket, what is the probability of winning this prize?

Surprising Probabilities

Imagine that one person is randomly selected from all 6 billion people on planet Earth. The following empirical probabilities, each rounded to two decimal places, might surprise you.

Probability of selecting
a woman	= 0.51
a non-white	= 0.7
a non-Christian	= 0.7
a person who cannot read	= 0.7
a person suffering from malnutrition	= 0.5
a person with a college education	= 0.01
a person who is near death	= 0.01

When viewing our world from the perspective of these probabilities, the need for both tolerance and understanding becomes apparent.

Source: United Nations

3 Find the probability that an event will not occur.

Probability of an Event Not Occurring

A survey (*source*: Penn, Schoen, and Berland, 1999) asked 500 Americans to rate their health. Of those surveyed, 270 rated their health as good/excellent. This means that $500 - 270$, or 230, people surveyed did not rate their health as good/excellent. Notice that

$$P(\text{good/excellent}) + P(\text{not good/excellent}) = \frac{270}{500} + \frac{230}{500} = \frac{500}{500} = 1.$$

In general, because the sum of the probabilities of all possible outcomes in any situation is 1,

$$P(E) + P(\text{not } E) = 1.$$

We now solve this equation for $P(\text{not } E)$, the probability that event E will not occur, by subtracting $P(E)$ from both sides. The resulting formula is given in the following box.

The Probability of an Event Not Occurring

The probability that an event E will not occur is equal to one minus the probability that it will occur.

$$P(\text{not } E) = 1 - P(E)$$

EXAMPLE 6   The Probability of Not Winning the Lottery

We have seen that the probability of winning Florida's LOTTO with one ticket is $\frac{1}{13,983,816}$. What is the probability of not winning?

Solution

$$P(\text{not winning}) = 1 - P(\text{winning})$$

$$= 1 - \frac{1}{13,983,816} = \frac{13,983,816}{13,983,816} - \frac{1}{13,983,816}$$

$$= \frac{13,983,815}{13,983,816} \approx 0.9999999$$

The probability of not winning is close to 1. It is almost certain that with one LOTTO ticket, a person will not win top prize.

Check Point 6   With one lottery ticket, what is the probability of not winning the lottery described in Checkpoint 5?

4 Find the probability of one event or a second event occurring.

Or Probabilities with Mutually Exclusive Events

Suppose that you randomly select one card from a deck of 52 cards. Let A be the event of selecting a king and B be the event of selecting a queen. Only one card is selected, so it is impossible to get both a king and a queen. The outcomes of selecting a king and a queen cannot occur simultaneously. They are called *mutually*

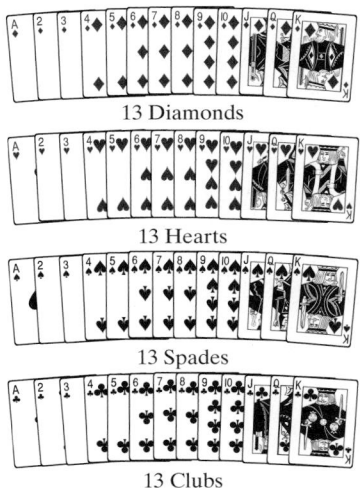

13 Diamonds

13 Hearts

13 Spades

13 Clubs

Figure 11.9 A deck of 52 cards

exclusive events. If it is impossible for any two events, A and B, to occur simultaneously, they are said to be **mutually exclusive**. If A and B are mutually exclusive events, the probability that either A or B will occur is determined by adding their individual probabilities.

Or **Probabilities with Mutually Exclusive Events**

If A and B are mutually exclusive events, then

$$P(A \text{ or } B) = P(A) + P(B).$$

EXAMPLE 7 **The Probability of Either of Two Mutually Exclusive Events Occurring**

If one card is randomly selected from a deck of cards, what is the probability of selecting a king or a queen?

Solution We find the probability that either of these mutually exclusive events will occur by adding their individual probabilities.

$$P(\text{king or queen}) = P(\text{king}) + P(\text{queen}) = \frac{4}{52} + \frac{4}{52} = \frac{8}{52} = \frac{2}{13}$$

The probability of selecting a king or a queen is $\frac{2}{13}$.

Check Point 7 If you roll a single, six-sided die, what is the probability of getting either a 4 or a 5?

Or **Probabilities with Events That Are Not Mutually Exclusive**

Consider the deck of 52 cards shown in Figure 11.9. Suppose that these cards are shuffled and you randomly select one card from the deck. What is the probability of selecting a diamond or a picture card (jack, queen, king)? Begin by adding their individual probabilities:

$$P(\text{diamond}) + P(\text{picture card}) = \frac{13}{52} + \frac{12}{52}.$$

Figure 11.10 Three diamonds are picture cards

There are 13 diamonds in the deck of 52 cards.

There are 12 picture cards in the deck of 52 cards.

However, this is not the probability of selecting a diamond or a picture card. The problem is that there are three cards that are simultaneously diamonds and picture cards, shown in Figure 11.10. The events of selecting a diamond and selecting a picture card are not mutually exclusive. It is possible to select a card that is both a diamond and a picture card.

The situation is illustrated in the diagram in Figure 11.11. Why can't we find the probability of selecting a diamond or a picture card by adding their individual probabilities? The diagram shows that three of the cards, the three diamonds that are picture cards, get counted twice when we add the individual probabilities. First the three cards get counted as diamonds, and then they get counted as picture cards. In order to avoid the error of counting the three cards twice, we need to subtract the probability of getting a diamond and a picture card, $\frac{3}{52}$, as follows:

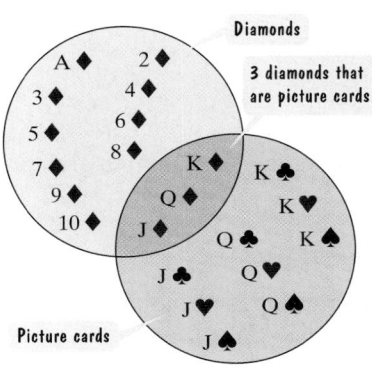

Figure 11.11

P(diamond or picture card)

$$= P(\text{diamond}) + P(\text{picture card}) - P(\text{diamond and picture card})$$

$$= \frac{13}{52} + \frac{12}{52} - \frac{3}{52} = \frac{13 + 12 - 3}{52} = \frac{22}{52} = \frac{11}{26}.$$

Thus, the probability of selecting a diamond or a picture card is $\frac{11}{26}$.

In general, if A and B are events that are not mutually exclusive, the probability that A or B will occur is determined by adding their individual probabilities and then subtracting the probability that A and B occur simultaneously.

Or Probabilities with Events That Are Not Mutually Exclusive

If A and B are not mutually exclusive events, then

$$P(A \text{ or } B) = P(A) + P(B) - P(A \text{ and } B).$$

EXAMPLE 8 An Or Probability with Events That Are Not Mutually Exclusive

Figure 11.12 illustrates a spinner. It is equally probable that the pointer will land on any one of the eight regions, numbered 1 through 8. If the pointer lands on a borderline, spin again. Find the probability that the pointer will stop on an even number or a number greater than 5.

Figure 11.12 It is equally probable that the pointer will land on any one of the eight regions.

Solution It is possible for the pointer to land on a number that is even and greater than 5. Two of the numbers, 6 and 8, are even and greater than 5. These events are not mutually exclusive. The probability of landing on a number that is even and greater than 5 is

$$P\left(\begin{array}{c}\text{even or}\\ \text{greater than 5}\end{array}\right) = P(\text{even}) + P(\text{greater than 5}) - P\left(\begin{array}{c}\text{even and}\\ \text{greater than 5}\end{array}\right)$$

$$= \frac{4}{8} + \frac{3}{8} - \frac{2}{8}$$

| Four of the eight numbers, 2, 4, 6, and 8 are even. | Three of the eight numbers, 6, 7, and 8 are greater than 5. | Two of the eight numbers, 6 and 8, are even and greater than 5. |

$$= \frac{4 + 3 - 2}{8} = \frac{5}{8}.$$

The probability that the pointer will stop on an even number or a number greater than 5 is $\frac{5}{8}$.

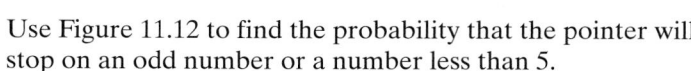

Check Point 8 Use Figure 11.12 to find the probability that the pointer will stop on an odd number or a number less than 5.

EXAMPLE 9   An *Or* Probability with Events That Are Not Mutually Exclusive

A group of people is comprised of 15 U.S. men, 20 U.S. women, 10 Canadian men, and 5 Canadian women. If a person is selected at random from the group, find the probability that the selected person is a man or a Canadian.

Solution   The group is comprised of $15 + 20 + 10 + 5$, or 50 people. It is possible to select a man who is Canadian. We are given that there are 10 Canadian men, so these events are not mutually exclusive.

$$P(\text{man or Canadian}) = P(\text{man}) + P(\text{Canadian}) - P(\text{man and Canadian})$$

$$= \frac{25}{50} + \frac{15}{50} - \frac{10}{50}$$

Of the 50 people, 25 are men—15 U.S. men and 10 Canadian men.	Of the 50 people, 15 are Canadian—10 Canadian men and 5 Canadian women.	Of the 50 people, 10 are Canadian men.

$$= \frac{25 + 15 - 10}{50} = \frac{30}{50} = \frac{3}{5}$$

The probability of selecting a man or a Canadian is $\frac{3}{5}$.

> **Check Point 9**   In a group of 25 baboons, 18 enjoy picking fleas off their neighbors, 16 enjoy screeching wildly, while 10 enjoy picking fleas off their neighbors and screeching wildly. If one baboon is selected at random from the group, find the probability that it enjoys picking fleas off its neighbors or screeching wildly.

5 Find the probability of one event and a second event occurring.

And Probabilities with Independent Events

Suppose that you toss a fair coin two times in succession. The outcome of the first toss, heads or tails, does not affect what happens when you toss the coin a second time. For example, the occurrence of tails on the first toss does not make tails more likely or less likely to occur on the second toss. The repeated toss of a coin produces **independent events** because the outcome of one toss does not affect the outcome of others. Two events are *independent* if the occurrence of either of them has no effect on the probability of the other.

If two events are independent, we can calculate the probability of the first occurring and the second occurring by multiplying their probabilities.

> **And Probabilities with Independent Events**
>
> If A and B are independent events, then
> $$P(A \text{ and } B) = P(A) \cdot P(B).$$

Figure 11.13 A U.S. roulette wheel

EXAMPLE 10   Independent Events on a Roulette Wheel

Figure 11.13 shows a U.S. roulette wheel that has 38 numbered slots (1 through 36, 0, and 00). Of the 38 compartments, 18 are black, 18 are red, and 2 are green.

Each play consists of spinning the wheel and a small ball in opposite directions. As the ball slows to a stop, it can land with equal probability on any one of the 38 numbered slots. Find the probability of red occurring on two consecutive plays.

Solution   The wheel has 38 equally likely outcomes and 18 are red. Thus, the probability of red occurring on a play is $\frac{18}{38}$, or $\frac{9}{19}$. The result that occurs on each play is independent of all previous results. Thus,

$$P(\text{red and red}) = P(\text{red}) \cdot P(\text{red}) = \frac{9}{19} \cdot \frac{9}{19} = \frac{81}{361} \approx 0.224.$$

The probability of red occurring on two consecutive plays is $\frac{81}{361}$.

Some roulette players incorrectly believe that if red occurs on two consecutive plays, then another color is "due." Because the events are independent, the outcomes of previous spins have no effect on any other spins.

Check Point 10   Find the probability of green occurring on two consecutive plays on a roulette wheel.

The *and* rule for independent events can be extended to cover three or more events. Thus, if A, B, and C are independent events, then

$$P(A \text{ and } B \text{ and } C) = P(A) \cdot P(B) \cdot P(C).$$

EXAMPLE 11   Independent Events in a Family

The picture in the margin shows a family that has had nine girls in a row. Find the probability of this occurrence.

Solution   If two or more events are independent, we can find the probability of them all occurring by multiplying the probabilities. The probability of a baby girl is $\frac{1}{2}$, so the probability of nine girls in a row is $\frac{1}{2}$ used as a factor nine times.

$$P(\text{nine girls in a row}) = \frac{1}{2} \cdot \frac{1}{2} \cdot \frac{1}{2} \cdot \frac{1}{2} \cdot \frac{1}{2} \cdot \frac{1}{2} \cdot \frac{1}{2} \cdot \frac{1}{2} \cdot \frac{1}{2}$$

$$= \left(\frac{1}{2}\right)^9 = \frac{1}{512}$$

The probability of a run of nine girls in a row is $\frac{1}{512}$. (If another child is born into the family, this event is independent of the other nine, and the probability of a girl is still $\frac{1}{2}$.)

Check Point 11   Find the probability of a family having four boys in a row.

EXERCISE SET 11.7

Practice and Application Exercises

Exercises 1–4 involve empirical probability. Use the empirical probability formula to solve each exercise. Express answers as fractions. Then use a calculator to express probabilities as decimals, rounded to the nearest thousandth.

Use the table showing U.S. family size to solve Exercises 1–2.

U.S. Families (includes only a householder and his/her relatives) by Size, 1997

Total: 70,241,000 Families	
Size	Number of Families
2 people	29,780,000
3 people	16,239,000
4 people	14,602,000
5 people	6,326,000
6 people	2,108,000
7 people or more	1,186,000

Source: U.S. Bureau of the Census

Find the probability that a U.S. family has:

1. 2 people.   **2.** 3 people.

Use the table showing world population for selected regions to solve Exercises 3–4.

Populations of Selected Regions of the World

Total World Population: 5926 million	
Region	Population in millions
Africa	761
Near East	165
Asia	3363
Latin America	508
Europe	799
North America	301

Source: U.S. Bureau of the Census

If one person is randomly selected from all people on planet Earth, find the probability of selecting a person from:

3. Africa.   **4.** North America.

Exercises 5–20 involve theoretical probability. Use the theoretical probability formula to solve each exercise. Express each probability as a fraction reduced to lowest terms.

In Exercises 5–10, a die is rolled. The sample space of equally likely outcomes is $\{1, 2, 3, 4, 5, 6\}$. Find the probability of getting:

5. a 4.   **6.** a 5.

7. an odd number.   **8.** a number greater than 3.

9. a number greater than 4.   **10.** a number greater than 7.

In Exercises 11–14, you are dealt one card from a standard 52 card deck. Find the probability of being dealt:

11. a queen.   **12.** a diamond.

13. a picture card.   **14.** a card greater than 3 and less than 7.

In Exercises 15–16, a fair coin is tossed two times in succession. The sample space of equally likely outcomes is $\{HH, HT, TH, TT\}$. Find the probability of getting:

15. two heads.   **16.** the same outcome on each toss.

In Exercises 17–18, you select a family with three children. If M represents a male child and F a female child, the sample space of equally likely outcomes is $\{MMM, MMF, MFM, MFF, FMM, FMF, FFM, FFF\}$. Find the probability of selecting a family with:

17. at least one male child.   **18.** at least two female children.

In Exercises 19–20, a single die is rolled twice. The 36 equally likely outcomes are shown as follows:

		Second Roll					
		⚀	⚁	⚂	⚃	⚄	⚅
First Roll	⚀	(1,1)	(1,2)	(1,3)	(1,4)	(1,5)	(1,6)
	⚁	(2,1)	(2,2)	(2,3)	(2,4)	(2,5)	(2,6)
	⚂	(3,1)	(3,2)	(3,3)	(3,4)	(3,5)	(3,6)
	⚃	(4,1)	(4,2)	(4,3)	(4,4)	(4,5)	(4,6)
	⚄	(5,1)	(5,2)	(5,3)	(5,4)	(5,5)	(5,6)
	⚅	(6,1)	(6,2)	(6,3)	(6,4)	(6,5)	(6,6)

Find the probability of getting:

19. two numbers whose sum is 4.

20. two numbers whose sum is 6.

21. To play the California lottery, a person has to correctly select 6 out of 51 numbers, paying $1 for each six-number selection. If you pick six numbers that are the same as the ones drawn by the lottery, you win mountains of money. What is the probability that a person with one combination of six numbers will win? What is the probability of winning if 100 different lottery tickets are purchased?

22. A state lottery is designed so that a player chooses six numbers from 1 to 30 on one lottery ticket. What is the probability that a player with one lottery ticket will win? What is the probability of winning if 100 different lottery tickets are purchased?

23. A poker hand consists of five cards.
  a. Find the total number of possible five-card poker hands that can be dealt from a deck of 52 cards.
  b. A diamond flush consists of a five-card hand containing all diamonds. Find the number of possible five-card diamond flushes.
  c. Find the probability of being dealt a diamond flush.

24. A committee of five people is to be formed from six lawyers and seven teachers. Find the probability that all are lawyers.

Use these figures for the U.S. population in 2000 to answer Exercises 25–30.

Total U.S. Population: 274,634 Thousand

Age	under 5	5–13	14–17	18–24	25–34	35–44	45–64	65–84	85 and older
Population (in thousands)	18,987	36,043	15,752	26,258	37,233	44,659	60,992	30,378	4332

Source: U.S. Bureau of the Census

If a U.S. citizen is chosen at random, find the probability that this person is not:

25. under 5.

26. in the 18–24 age group.

27. in the 25–34 age group.

28. 85 and older.

Exercises 29–32 involve or *probabilities with mutually exclusive events.*

29. If a U.S. citizen is chosen at random, find the probability that this person is in the 14–17 or 18–24 age group.

30. If a U.S. citizen is chosen at random, find the probability that this person is in the 25–34 or 35–44 age group.

If one card is randomly selected from a 52-card deck of cards, find the probability of selecting:

31. a 2 or a 3.

32. a red 7 or a black 8.

Exercises 33–40 involve or *probabilities with events that are not mutually exclusive.*

In Exercises 33–34, a single die is rolled. Find the probability of getting:

33. an even number or a number less than 5.

34. an odd number or a number less than 4.

In Exercises 35–36, you are dealt one card from a 52-card deck. Find the probability that you are dealt:

35. a 7 or a red card.

36. a 5 or a black card.

In Exercises 37–38, it is equally probable that the pointer on the spinner shown will land on any one of the eight regions, numbered 1 through 8. If the pointer lands on a borderline, spin again.

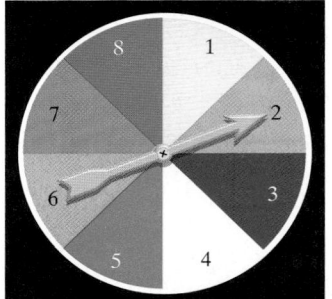

Find the probability that the pointer will stop on:

37. an odd number or a number less than 6.

38. an odd number or a number greater than 3.

Use this information to solve Exercises 39–40. The mathematics department of a college has 8 male professors, 11 female professors, 14 male teaching assistants, and 7 female teaching assistants. If a person is selected at random from the group, find the probability that the selected person is:

39. a professor or a male.

40. a professor or a female.

Exercises 41–46 involve and *probabilities with independent events.*

In Exercises 41–44, a single die is rolled twice. Find the probability of getting:

41. a 2 the first time and a 3 the second time.

42. a 5 the first time and a 1 the second time.

regular, fine, or micro. How many different choices of pens do you have with this brand?

75. A stock can go up, go down, or stay unchanged. How many possibilities are there if you own five stocks?

76. A club with 15 members is to choose four officers—president, vice-president, secretary, and treasurer. In how many ways can these offices be filled?

77. How many different ways can a director select 4 actors from a group of 20 actors to attend a workshop on performing in rock musicals?

78. From the 20 CDs that you've bought during the past year, you plan to take 3 with you on vacation. How many different sets of three CDs can you take?

79. How many different ways can a director select from 20 male actors and cast the roles of Mark, Roger, Angel, and Collins in the musical *Rent*?

80. In how many ways can five airplanes line up for departure on a runway?

11.7

Exercises 81–82 involve empirical probabilities. Express each probability as a fraction. Then use a calculator to express the probability in decimal form, rounded to the nearest thousandth. The table shows the two states with the largest Hispanic populations. Find the probability that:

81. a person randomly selected from California is Hispanic.

82. a person randomly selected from Texas is Hispanic.

Largest Hispanic Population, 1997

State	Total Population	Hispanic Population
California	31,878,234	9,630,188
Texas	19,128,261	5,503,372

Source: Bureau of the Census

In Exercises 83–84, a die is rolled. Find the probability of:

83. getting a number less than 5.

84. getting a number less than 3 or greater than 4.

In Exercises 85–86, you are dealt one card from a 52-card deck. Find the probability of:

85. getting an ace or a king.

86. getting a queen or a red card.

Chapter 11 Test

1. Write the first five terms of the sequence whose general term is $a_n = \dfrac{(-1)^{n+1}}{n^2}$.

In Exercises 2–4, find each indicated sum.

2. $\displaystyle\sum_{i=1}^{5} (i^2 + 10)$ **3.** $\displaystyle\sum_{i=1}^{20} (3i - 4)$ **4.** $\displaystyle\sum_{i=1}^{15} (-2)^i$

In Exercises 87–88, it is equally probable that the pointer on the spinner shown will land on any one of the six regions, numbered 1 through 6, and colored as shown. If the pointer lands on a borderline, spin again. Find the probability of:

87. not stopping on yellow.

88. stopping on red or a number greater than 3.

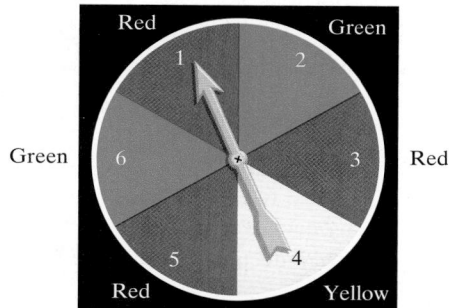

89. A lottery game is set up so that each player chooses five different numbers from 1 to 20. If the five numbers match the five numbers drawn in the lottery, the player wins (or shares) the top cash prize. What is the probability of winning the prize:
a. with one lottery ticket?
b. with 100 different lottery tickets?

Use this information to solve Exercises 90–91. At a workshop on police work and the black community, there are 50 black male police officers, 20 black female police officers, 90 white male police officers, and 40 white female police officers. If one police officer is selected at random from the people at the workshop, find the probability that the selected person is:

90. black or male.   **91.** female or white.

92. What is the probability of a family having five boys born in a row?

93. The probability of a flood in any given year in a region prone to floods is 0.2.
a. What is the probability of a flood two years in a row?
b. What is the probability of a flood for three consecutive years?
c. What is the probability of no flooding for four consecutive years?

In Exercises 5–7, evaluate each expression.

5. $\dbinom{9}{2}$ **6.** $_{10}P_3$ **7.** $_{10}C_3$

8. Express the sum using summation notation. Use i for the index of summation.

$$\frac{2}{3} + \frac{3}{4} + \frac{4}{5} + \cdots + \frac{21}{22}$$

In Exercises 9–10, write a formula for the general term (the nth term) of each sequence. Do not use a recursion formula. Then use the formula to find the twelfth term of the sequence.

9. $4, 9, 14, 19, \ldots$ **10.** $16, 4, 1, \frac{1}{4}, \ldots$

In Exercises 11–12, use a formula to find the sum of the first ten terms of each sequence.

11. $7, -14, 28, -56, \ldots$ **12.** $-7, -14, -21, -28, \ldots$

13. Find the sum of the infinite geometric series:

$$4 + \frac{4}{2} + \frac{4}{2^2} + \frac{4}{2^3} + \cdots.$$

14. A job pays $30,000 for the first year with an annual increase of 4% per year beginning in the second year. What is the total salary paid over an eight-year period?

15. Use mathematical induction to prove that for every positive integer n,

$$1 + 4 + 7 + \cdots + (3n - 2) = \frac{n(3n - 1)}{2}.$$

16. Use the Binomial Theorem to expand and simplify: $(x^2 - 1)^5$.

17. A human resource manager has 11 applicants to fill three different positions. Assuming that all applicants are equally qualified for any of the three positions, in how many ways can this be done?

18. From the ten books that you've recently bought but not read, you plan to take four with you on vacation. How many different sets of four books can you take?

19. How many seven-digit local telephone numbers can be formed if the first three digits are 279?

20. A lottery game is set up so that each player chooses six different numbers from 1 to 15. If the six numbers match the six numbers drawn in the lottery, the player wins (or shares) the top cash prize. What is the probability of winning the prize with 50 different lottery tickets?

21. One card is randomly selected from a deck of 52 cards. Find the probability of selecting a black card or a picture card.

22. A group of students consists of 10 male freshmen, 15 female freshmen, 20 male sophomores, and 5 female sophomores. If one person is randomly selected from the group, find the probability of selecting a freshman or a female.

23. A quiz consisting of four multiple-choice questions has four available options (a, b, c, or d) for each question. If a person guesses at every question, what is the probability of answering all questions correctly?

24. If the spinner shown is spun twice, find the probability that the pointer lands on red on the first spin and blue on the second spin.

Cumulative Review Exercises (Chapters 1–11)

Solve each equation or inequality in Exercises 1–10.

1. $-2(x - 5) + 10 = 3(x + 2)$

2. $3x^2 - 6x + 2 = 0$

3. $\log_2 x + \log_2(2x - 3) = 1$

4. $x^{1/2} - 6x^{1/4} + 8 = 0$

5. $\sqrt{2x + 4} - \sqrt{x + 3} - 1 = 0$

6. $|2x + 1| \le 1$ **7.** $6x^2 - 6 < 5x$

8. $\dfrac{x - 1}{x + 3} \le 0$ **9.** $30e^{0.7x} = 240$

10. $2x^3 + 3x^2 - 8x + 3 = 0$

Solve each system in Exercises 11–13.

11. $4x^2 + 3y^2 = 48$
$\ 3x^2 + 2y^2 = 35$

12. (Use matrices.)
$\ x - 2y + z = 16$
$\ 2x - y - z = 14$
$\ 3x + 5y - 4z = -10$

13. $x - y = 1$
$\ x^2 - x - y = 1$

In Exercises 14–19, graph each equation, function, or system in the rectangular coordinate system.

14. $100x^2 + y^2 = 25$

15. $4x^2 - 9y^2 - 16x + 54y - 29 = 0$

16. $f(x) = \dfrac{x^2 - 1}{x - 2}$

17. $2x - y \ge 4$
$\ x \le 2$

18. $f(x) = x^2 - 4x - 5$ **19.** $y = \log_2 x$

20. Find $f^{-1}(x)$ if $f(x) = \sqrt[3]{x + 4}$.

21. If $A = \begin{bmatrix} 4 & 2 \\ 1 & -1 \\ 0 & 5 \end{bmatrix}$ and $B = \begin{bmatrix} 2 & 4 \\ 3 & 1 \end{bmatrix}$, find $AB - 4A$.

22. Find the partial fraction decomposition for

$$\frac{2x^2 - 10x + 2}{(x - 2)(x^2 + 2x + 2)}.$$

23. Expand and simplify: $(x^3 + 2y)^5$.

24. Use the formula for the sum of the first n terms of an arithmetic sequence to find $\sum_{i=1}^{50} (4i - 25)$.

25. Mailings in the United States increased by more than 40% from 1983 to 1993.

x (Number of Years after 1983)	y (Number of Pieces of Mail, in Billions)
0	119.4
10	171.1

a. Write the point-slope form of the line on which these measurements fall.

b. Use the point-slope form of the equation to write the slope-intercept form of the equation.

c. Use the slope-intercept model from part (b) to predict the number of pieces of mail, in billions, for the year 2000.

26. Most of the world's very tall buildings are in the United States, where the skyscraper was first conceived. The height of the World Trade Center in New York is 790 feet less than twice that of New York's Empire State Building. If the mean (average) height of the two buildings is 980 feet, determine the height of each building.

27. The perimeter of a soccer field is 300 yards. If the length is 50 yards longer than the width, what are the field's dimensions?

28. If 10 pens and 12 pads cost $42, and 5 of the same pens and 10 of the same pads cost $29, find the cost of a pen and a pad.

29. A ball is thrown vertically upward from the top of a 96-foot tall building with an initial velocity of 80 feet per second. The height of the ball above ground is modeled by the position function

$$s(t) = -16t^2 + 80t + 96.$$

a. After how many seconds will the ball strike the ground?

b. When does the ball reach its maximum height? What is the maximum height?

30. The current, I, in amperes, flowing in an electrical circuit varies inversely as the resistance, R, in ohms, in the circuit. When the resistance of an electric percolator is 22 ohms, it draws 5 amperes of current. How much current is needed when the resistance is 10 ohms?

31. An object moves in simple harmonic motion described by $d = 10 \sin \frac{3\pi}{4} t$, where t is measured in seconds and d in inches. Find: **a.** the maximum displacement; **b.** the frequency; and **c.** the time required for one oscillation.

Verify each identity in Exercises 32–33.

32. $\tan x + \dfrac{1}{\tan x} = \dfrac{1}{\sin x \cos x}$

33. $\dfrac{1 - \tan^2 x}{1 + \tan^2 x} = \cos 2x$

34. Graph one period: $y = -2 \cos (3x - \pi)$.

In Exercises 35–36, solve each equation on the interval $[0, 2\pi)$.

35. $4 \cos^2 x = 3$

36. $2 \sin^2 x + 3 \cos x - 3 = 0$

37. Find the exact value of $\cot \left[\cos^{-1} \left(-\frac{5}{6} \right) \right]$.

38. Graph the polar equation: $r = 1 + 2 \cos \theta$.

39. In oblique triangle ABC, $A = 34°$, $a = 22$, and $b = 32$. Solve the triangle(s). Round lengths to the nearest tenth and angle measures to the nearest degree.

40. Use the parametric equations

$$x = \sin t, \quad y = 1 + \cos^2 t, \quad -\frac{\pi}{2} < t < \frac{\pi}{2}$$

and eliminate the parameter. Graph the plane curve represented by the parametric equations. Use arrows to show the orientation of the curve.

Appendix
Where Did That Come From? Selected Proofs

SECTION 4.3   Properties of Logarithms

The Product Rule

Let b, M, and N be positive real numbers with $b \neq 1$.
$$\log_b(MN) = \log_b M + \log_b N$$

Proof

We begin by letting $\log_b M = R$ and $\log_b N = S$.
Now we write each logarithm in exponential form.
$$\log_b M = R \quad \text{means} \quad b^R = M.$$
$$\log_b N = S \quad \text{means} \quad b^S = N.$$
By substituting and using a property of exponents, we see that
$$MN = b^R b^S = b^{R+S}.$$
Now we change $MN = b^{R+S}$ to logarithmic form.
$$MN = b^{R+S} \quad \text{means} \quad \log_b(MN) = R + S.$$
Finally, substituting $\log_b M$ for R and $\log_b N$ for S gives us
$$\log_b(MN) = \log_b M + \log_b N,$$
the property that we wanted to prove.

The quotient and power rules for logarithms are proved using similar procedures.

The Change-of-Base Property

For any logarithmic bases a and b, and any positive number M,
$$\log_b M = \frac{\log_a M}{\log_a b}.$$

Proof

To prove the change-of-base property, we let x equal the logarithm on the left side:
$$\log_b M = x.$$

Now we rewrite this logarithm in exponential form.

$$\log_b M = x \quad \text{means} \quad b^x = M.$$

Because b^x and M are equal, the logarithms with base a for each of these expressions must be equal. This means that

$$\log_a b^x = \log_a M$$

$$x \log_a b = \log_a M \quad \text{Apply the power rule for logarithms on the left side.}$$

$$x = \frac{\log_a M}{\log_a b} \quad \text{Solve for } x \text{ by dividing both sides by } \log_a b.$$

In our first step we let x equal $\log_b M$. Replacing x on the left side by $\log_b M$ gives us

$$\log_b M = \frac{\log_a M}{\log_a b},$$

which is the change-of-base property.

SECTION 7.2   *The Law of Cosines*

Heron's Formula for the Area of a Triangle

The area of a triangle with sides a, b, and c is

$$\text{Area} = \sqrt{s(s-a)(s-b)(s-c)},$$

where s is one-half the perimeter: $s = \frac{1}{2}(a + b + c)$.

Proof

The proof of Heron's formula begins with a half-angle formula and the Law of Cosines.

$$\cos\frac{C}{2} = \sqrt{\frac{1 + \cos C}{2}} = \sqrt{\frac{1 + \dfrac{a^2 + b^2 - c^2}{2ab}}{2}}$$

This is the Law of Cosines $c^2 = a^2 + b^2 - 2ab\cos C$ solved for $\cos C$.

$$= \sqrt{\frac{a^2 + 2ab + b^2 - c^2}{4ab}} = \sqrt{\frac{(a+b)^2 - c^2}{4ab}} = \sqrt{\frac{(a+b+c)(a+b-c)}{4ab}}$$

Multiply the numerator and denominator of the radicand by $2ab$.

Factor $a^2 + 2ab + b^2$.

Factor the numerator as the differences of two squares.

We now introduce the expression for one-half the perimeter: $s = \frac{1}{2}(a + b + c)$. We replace $a + b + c$ in the numerator by $2s$. We also find an expression for $a + b - c$ as follows:

$$a + b - c = a + b + c - 2c = 2s - 2c = 2(s - c).$$

Thus,

$$\cos\frac{C}{2} = \sqrt{\frac{(a+b+c)(a+b-c)}{4ab}} = \sqrt{\frac{2s \cdot 2(s-c)}{4ab}} = \sqrt{\frac{s(s-c)}{ab}}.$$

In a similar manner, we obtain

$$\sin\frac{C}{2} = \sqrt{\frac{1 - \cos C}{2}} = \sqrt{\frac{(s-a)(s-b)}{ab}}.$$

From our work in Section 7.1, we know that the area of a triangle is one-half the product of the length of two sides times the sum of their included angle.

$$\text{Area} = \frac{1}{2} ab \sin C$$

$$= \frac{1}{2} ab \cdot 2 \sin \frac{C}{2} \cos \frac{C}{2} \qquad \sin C = \sin 2\frac{C}{2} = 2 \sin \frac{C}{2} \cos \frac{C}{2}$$

$$= ab\sqrt{\frac{(s-a)(s-b)}{ab}} \sqrt{\frac{s(s-c)}{ab}} \qquad \text{Use the expressions for } \sin \frac{C}{2} \text{ and } \cos \frac{C}{2} \text{ on page A2.}$$

$$= ab \frac{\sqrt{s(s-a)(s-b)(s-c)}}{\sqrt{a^2b^2}} \qquad \text{Multiply the radicands.}$$

$$= \sqrt{s(s-a)(s-b)(s-c)} \qquad \text{Simplify: } \frac{ab}{\sqrt{a^2b^2}} = \frac{ab}{ab} = 1.$$

SECTION 7.5   Complex Numbers in Polar Form; DeMoivre's Theorem

The Quotient of Two Complex Numbers in Polar Form

Let $z_1 = r_1(\cos\theta_1 + i\sin\theta_1)$ and $z_2 = r_2(\cos\theta_2 + i\sin\theta_2)$ be two complex numbers in polar form. Their quotient, $\frac{z_1}{z_2}$, is

$$\frac{z_1}{z_2} = \frac{r_1}{r_2}\left[\cos(\theta_1 - \theta_2) + i\sin(\theta_1 - \theta_2)\right].$$

Proof

We begin by multiplying the numerator and denominator of the quotient, $\frac{z_1}{z_2}$, by the conjugate of the denominator. Then we simplify the quotient using the difference formulas for sine and cosine.

$$\frac{z_1}{z_2} = \frac{r_1(\cos\theta_1 + i\sin\theta_1)}{r_2(\cos\theta_2 + i\sin\theta_2)} \qquad \text{This is the given quotient.}$$

$$= \frac{r_1(\cos\theta_1 + i\sin\theta_1)(\cos\theta_2 - i\sin\theta_2)}{r_2(\cos\theta_2 + i\sin\theta_2)(\cos\theta_2 - i\sin\theta_2)} \qquad \text{Multiply the numerator and denominator by the conjugate of the denominator. Recall that the conjugate of } a+bi \text{ is } a-bi.$$

$$= \frac{r_1(\cos\theta_1 + i\sin\theta_1)(\cos\theta_2 - i\sin\theta_2)}{r_2(\cos^2\theta_2 + \sin^2\theta_2)} \qquad \text{Multiply the conjugates in the denominator.}$$

$$= \frac{r_1(\cos\theta_1 + i\sin\theta_1)(\cos\theta_2 - i\sin\theta_2)}{r_2} \qquad \text{Use a Pythagorean identity: } \cos^2\theta_2 + \sin^2\theta_2 = 1.$$

$$= \frac{r_1}{r_2}(\cos\theta_1\cos\theta_2 - i\cos\theta_1\sin\theta_2 + i\sin\theta_1\cos\theta_2 - i^2\sin\theta_1\sin\theta_2) \qquad \text{Use the FOIL method.}$$

$$= \frac{r_1}{r_2}\left[\cos\theta_1\cos\theta_2 + i(\sin\theta_1\cos\theta_2 - \cos\theta_1\sin\theta_2) - i^2\sin\theta_1\sin\theta_2\right] \qquad \text{Factor } i \text{ from the second and third terms.}$$

$$= \frac{r_1}{r_2} \left[\cos\theta_1 \cos\theta_2 + i(\sin\theta_1 \cos\theta_2 - \cos\theta_1 \sin\theta_2) - (-1)\sin\theta_1 \sin\theta_2 \right] \quad i^2 = -1.$$

$$= \frac{r_1}{r_2} \left[\cos\theta_1 \cos\theta_2 + \sin\theta_1 \sin\theta_2 + i(\sin\theta_1 \cos\theta_2 - \cos\theta_1 \sin\theta_2) \right] \qquad \text{Rearrange terms.}$$

This is cos $(\theta_1 - \theta_2)$. This is sin $(\theta_1 - \theta_2)$.

$$= \frac{r_1}{r_2} \left[\cos(\theta_1 - \theta_2) + i\sin(\theta_1 - \theta_2) \right]$$

SECTION 7.7 *The Dot Product*

Properties of the Dot Product
If **u**, **v**, and **w** are vectors, and c is a scalar, then:

1. $\mathbf{u} \cdot \mathbf{v} = \mathbf{v} \cdot \mathbf{u}$
2. $\mathbf{u} \cdot (\mathbf{v} + \mathbf{w}) = \mathbf{u} \cdot \mathbf{v} + \mathbf{u} \cdot \mathbf{w}$
3. $\mathbf{0} \cdot \mathbf{v} = 0$
4. $\mathbf{v} \cdot \mathbf{v} = \|\mathbf{v}\|^2$
5. $(c\mathbf{u}) \cdot \mathbf{v} = c(\mathbf{u} \cdot \mathbf{v}) = \mathbf{u} \cdot (c\mathbf{v})$

Proof   To prove the second property, let

$$\mathbf{u} = u_1\mathbf{i} + u_2\mathbf{j}, \quad \mathbf{v} = v_1\mathbf{i} + v_2\mathbf{j}, \text{ and } \mathbf{w} = w_1\mathbf{i} + w_2\mathbf{j}.$$

Then,

$$\mathbf{u} \cdot (\mathbf{v} + \mathbf{w}) = (u_1\mathbf{i} + u_2\mathbf{j}) \cdot \left[(v_1\mathbf{i} + v_2\mathbf{j}) + (w_1\mathbf{i} + w_2\mathbf{j}) \right] \quad \text{These are the given vectors.}$$

$$= (u_1\mathbf{i} + u_2\mathbf{j}) \cdot \left[(v_1 + w_1)\mathbf{i} + (v_2 + w_2)\mathbf{j} \right] \quad \text{Add horizontal components and add vertical components.}$$

$$= u_1(v_1 + w_1) + u_2(v_2 + w_2) \quad \text{Multiply horizontal components and multiply vertical components.}$$

$$= u_1 v_1 + u_1 w_1 + u_2 v_2 + u_2 w_2 \quad \text{Use the distributive property.}$$

$$= u_1 v_1 + u_2 v_2 + u_1 w_1 + u_2 w_2 \quad \text{Rearrange terms.}$$

This is the dot product of **u** and **v**. This is the dot product of **u** and **w**.

$$= \mathbf{u} \cdot \mathbf{v} + \mathbf{u} \cdot \mathbf{w}.$$

To prove the third property, let

$$\mathbf{0} = 0\mathbf{i} + 0\mathbf{j} \quad \text{and} \quad \mathbf{v} = v_1\mathbf{i} + v_2\mathbf{j}.$$

Then

$$\mathbf{0} \cdot \mathbf{v} = (0\mathbf{i} + 0\mathbf{j}) \cdot (v_1\mathbf{i} + v_2\mathbf{j}) \quad \text{These are the given vectors.}$$

$$= 0 \cdot v_1 + 0 \cdot v_2 \quad \text{Multiply horizontal components and multiply vertical components.}$$

$$= 0 + 0$$

$$= 0.$$

To prove the first part of the fifth property, let

$$\mathbf{u} = u_1\mathbf{i} + u_2\mathbf{j} \quad \text{and} \quad \mathbf{v} = v_1\mathbf{i} + v_2\mathbf{j}.$$

Then,

$$
\begin{aligned}
(c\mathbf{u}) \cdot \mathbf{v} &= \left[c(u_1\mathbf{i} + u_2\mathbf{j})\right] \cdot (v_1\mathbf{i} + v_2\mathbf{j}) && \text{These are the given vectors.}\\
&= (cu_1\mathbf{i} + cu_2\mathbf{j}) \cdot (v_1\mathbf{i} + v_2\mathbf{j}) && \text{Multiply each component of } u_1\mathbf{i} + u_2\mathbf{j} \text{ by } c.\\
&= cu_1v_1 + cu_2v_2 && \text{Multiply horizontal components and}\\
&&& \text{multiply vertical components.}\\
&= c(u_1v_1 + u_2v_2) && \text{Factor out } c \text{ from both terms.}
\end{aligned}
$$

This is the dot product of **u** and **v**.

$$= c(\mathbf{u} \cdot \mathbf{v}).$$

SECTION 10.2 *The Hyperbola*

The Asymptotes of a Hyperbola Centered at the Origin

The hyperbola

$$\frac{x^2}{a^2} - \frac{y^2}{b^2} = 1$$

with a horizontal transverse axis has the two asymptotes

$$y = \frac{b}{a}x \quad \text{and} \quad y = -\frac{b}{a}x.$$

Proof

Begin by solving the hyperbola's equation for y.

$$\frac{x^2}{a^2} - \frac{y^2}{b^2} = 1 \qquad \text{This is the standard form of the equation of a hyperbola.}$$

$$\frac{y^2}{b^2} = \frac{x^2}{a^2} - 1 \qquad \text{We isolate the term involving } y^2 \text{ to solve for } y.$$

$$y^2 = \frac{b^2x^2}{a^2} - b^2 \qquad \text{Multiply both sides by } b^2.$$

$$y^2 = \frac{b^2x^2}{a^2}\left(1 - \frac{a^2}{x^2}\right) \qquad \text{Factor out } \frac{b^2x^2}{a^2} \text{ on the right. Verify that this result is correct by multiplying using the distributive property and obtaining the previous step.}$$

$$y = \pm\sqrt{\frac{b^2x^2}{a^2}\left(1 - \frac{a^2}{x^2}\right)} \qquad \text{Solve for } y \text{ using the square root method: If } u^2 = d, \text{ then } u = \pm\sqrt{d}.$$

$$y = \pm\frac{b}{a}x\sqrt{1 - \frac{a^2}{x^2}} \qquad \text{Simplify.}$$

As $|x| \to \infty$, the value of $\dfrac{a^2}{x^2}$ approaches 0. Consequently, the value of y can be approximated by

$$y = \pm\frac{b}{a}x.$$

This means that the lines whose equations are $y = \dfrac{b}{a}x$ and $y = -\dfrac{b}{a}x$ are asymptotes for the graph of the hyperbola.

Answers to Selected Exercises

CHAPTER P

Section P.1

Check Point Exercises

1. a. $\sqrt{2} - 1$ **b.** $\pi - 3$ **c.** 1 **2.** 9 **3.** 414.5; In 2080, the population of the United States will be 414.5 million. **4.** $38x - 19y$

Exercise Set P.1

1. a. $\sqrt{100}$ **b.** $0, \sqrt{100}$ **c.** $-9, 0, \sqrt{100}$ **d.** $-9, -\dfrac{4}{5}, 0, 0.25, 9.2, \sqrt{100}$ **e.** $\sqrt{3}$ **3. a.** $\sqrt{64}$ **b.** $0, \sqrt{64}$ **c.** $-11, 0, \sqrt{64}$
d. $-11, -\dfrac{5}{6}, 0, 0.75, \sqrt{64}$ **e.** $\sqrt{5}, \pi$ **5.** 0 **7.** Answers may vary. **9.** true **11.** true **13.** true **15.** 300
17. $12 - \pi$ **19.** $5 - \sqrt{2}$ **21.** -1 **23.** 15 **25.** 7 **27.** 15 **29.** 2.2 **31.** 27 **33.** -19 **35.** 25 **37.** 10
39. commutative property of addition **41.** associative property of addition **43.** commutative property of addition
45. distributive property of multiplication over addition **47.** $15x + 16$ **49.** $27x - 10$ **51.** $29y - 29$ **53.** $14x$
55. $-2x + 3y + 6$ **57.** x **59.** yes **61.** Answers may vary. **63.** 25,401; In 1997, the average yearly earnings in the
United States was $25,401. **65. a.** $132 - 0.6a$ **b.** 120 **73.** (c) is true. **75.** $<$ **77.** $>$

Section P.2

Check Point Exercises

1. -256 **2. a.** $\dfrac{1}{8}$ **b.** 36 **3. a.** 243 **b.** $\dfrac{1}{8}$ **c.** x^6 **4. a.** 729 **b.** y^{28} **c.** $\dfrac{1}{x^8}$ **5. a.** 9 **b.** $\dfrac{1}{x^7}$ **c.** y^9
6. $-64x^3$ **7. a.** $\dfrac{27}{64}$ **b.** $-\dfrac{32}{y^5}$ **8. a.** $16x^{12}y^{24}$ **b.** $-18x^3y^8$ **c.** $\dfrac{5y^6}{x^4}$ **d.** $\dfrac{y^8}{25x^2}$ **9. a.** 7,400,000,000 **b.** 0.000003017
10. a. 7.41×10^9 **b.** 9.2×10^{-8} **11.** 5.2×10^5 mi

Exercise Set P.2

1. 50 **3.** 64 **5.** -64 **7.** 1 **9.** -1 **11.** $\dfrac{1}{64}$ **13.** 32 **15.** 64 **17.** 16 **19.** $\dfrac{1}{9}$ **21.** $\dfrac{1}{16}$ **23.** $\dfrac{y}{x^2}$ **25.** y^5
27. x^{10} **29.** x^5 **31.** x^{21} **33.** x^{-15} **35.** x^7 **37.** x^{21} **39.** $64x^6$ **41.** $-\dfrac{64}{x^3}$ **43.** $9x^4y^{10}$ **45.** $6x^{11}$ **47.** $18x^9y^5$
49. $4x^{16}$ **51.** $-5a^{11}b$ **53.** $\dfrac{2}{b^7}$ **55.** $\dfrac{1}{16x^6}$ **57.** $\dfrac{3y^{14}}{4x^4}$ **59.** $\dfrac{y^2}{25x^6}$ **61.** 4700 **63.** 4,000,000 **65.** 0.000786
67. 0.00000318 **69.** 3.6×10^3 **71.** 2.2×10^8 **73.** 2.7×10^{-2} **75.** 7.63×10^{-4} **77.** 600,000 **79.** 0.123 **81.** 30,000
83. 0.021 **85.** 1.694×10^{12} **87.** 6.0×10^{10} **89.** 3.24×10^{10} **99.** $\dfrac{1}{4}$ **100.** $A = C + D$

Section P.3

Check Point Exercises

1. a. 3 **b.** $5|x|\sqrt{2}$ **2. a.** $\dfrac{5}{4}$ **b.** $5|x|\sqrt{3}$ **3. a.** $17\sqrt{13}$ **b.** $-19\sqrt{17x}$ **4. a.** $17\sqrt{3}$ **b.** $10\sqrt{2x}$ **5. a.** $\dfrac{5\sqrt{3}}{3}$ **b.** $\sqrt{3}$
6. $\dfrac{32 - 8\sqrt{5}}{11}$ **7. a.** $2\sqrt[3]{5}$ **b.** $2\sqrt[5]{2}$ **c.** $\dfrac{5}{3}$ **8.** $5\sqrt[3]{3}$ **9. a.** 9 **b.** 3 **c.** $\dfrac{1}{2}$ **10. a.** 8 **b.** $\dfrac{1}{4}$ **11. a.** $10x^4$ **b.** $4x^{5/2}$
12. $\sqrt{x}$

Exercise Set P.3

1. 6 **3.** not a real number **5.** 13 **7.** $5\sqrt{2}$ **9.** $3|x|\sqrt{5}$ **11.** $2|x|\sqrt{3}$ **13.** $|x|\sqrt{x}$ **15.** $2|x|\sqrt{3x}$ **17.** $\dfrac{1}{9}$ **19.** $\dfrac{7}{4}$
21. $4|x|$ **23.** $5|x|\sqrt{2x}$ **25.** $13\sqrt{3}$ **27.** $-2\sqrt{17x}$ **29.** $5\sqrt{2}$ **31.** $3\sqrt{2x}$ **33.** $34\sqrt{2}$ **35.** $\dfrac{\sqrt{7}}{7}$ **37.** $\dfrac{\sqrt{10}}{5}$

39. $\dfrac{13(3 - \sqrt{11})}{-2}$ **41.** $7(\sqrt{5} + 2)$ **43.** $3(\sqrt{5} - \sqrt{3})$ **45.** 5   **47.** -2 **49.** not a real number   **51.** 3   **53.** -3 **55.** $2\sqrt[3]{4}$

57. $x\sqrt[3]{x}$ **59.** $3\sqrt[3]{2}$ **61.** $2x$ **63.** 6   **65.** 2   **67.** 25   **69.** $\dfrac{1}{16}$ **71.** $14x^{7/12}$ **73.** $4x^{1/4}$ **75.** x^2 **77.** $5x^2|y|^3$

79. $\sqrt{5}$ **81.** x^2 **83.** $|x|^{2/3}$ **85.** $20\sqrt{2}$ mph   **87.** $\dfrac{\sqrt{5} + 1}{2} \approx 1.62$ **89.** $\dfrac{7\sqrt{2 \cdot 2 \cdot 3}}{6} = \dfrac{7\sqrt{2^2 \cdot 3}}{6} = \dfrac{7\sqrt{2^2}\sqrt{3}}{6} = \dfrac{7 \cdot 2\sqrt{3}}{6} = \dfrac{7}{3}\sqrt{3}$

91. The duration of a storm whose diameter is 9 miles is 1.89 hours.   **99.** during 1990   **101.** (d) is true.
103. Let $\square = 25$ and $\square = 14$.   **105. a.** $>$ **b.** $>$

Section P.4

Check Point Exercises

1. a. $-x^3 + x^2 - 8x - 20$ **b.** $20x^3 - 11x^2 - 2x - 8$ **2.** $15x^3 - 31x^2 + 30x - 8$ **3.** $28x^2 - 41x + 15$
4. a. $49x^2 - 64$ **b.** $4y^6 - 25$ **5. a.** $x^2 + 20x + 100$ **b.** $25x^2 + 40x + 16$ **6. a.** $x^2 - 18x + 81$ **b.** $49x^2 - 42x + 9$
7. $2x^2y + 5xy^2 - 2y^3$ **8. a.** $21x^2 - 25xy + 6y^2$ **b.** $x^4 + 10x^2y + 25y^2$

Exercise Set P.4

1. yes; $3x^2 + 2x - 5$ **3.** no   **5.** 2   **7.** 4   **9.** $11x^3 + 7x^2 - 12x - 4; 3$ **11.** $12x^3 + 4x^2 + 12x - 14; 3$ **13.** $6x^2 - 6x + 2; 2$
15. $x^3 + 1$ **17.** $2x^3 - 9x^2 + 19x - 15$ **19.** $x^2 + 10x + 21$ **21.** $x^2 - 2x - 15$ **23.** $6x^2 + 13x + 5$ **25.** $10x^2 - 9x - 9$
27. $15x^4 - 47x^2 + 28$ **29.** $x^2 - 9$ **31.** $9x^2 - 4$ **33.** $25 - 49x^2$ **35.** $16x^4 - 25x^2$ **37.** $x^2 + 4x + 4$ **39.** $4x^2 + 12x + 9$
41. $x^2 - 6x + 9$ **43.** $16x^4 - 8x^2 + 1$ **45.** $4x^2 - 28x + 49$ **47.** $x^3 + 3x^2 + 3x + 1$ **49.** $8x^3 + 36x^2 + 54x + 27$
51. $x^3 - 9x^2 + 27x - 27$ **53.** $27x^3 - 108x^2 + 144x - 64$ **55.** $7x^2y - 4xy$ is of degree 3   **57.** $2x^2y + 13xy + 13$ is of degree 3
59. $-5x^3 + 8xy - 9y^2$ is of degree 3   **61.** $x^4y^2 + 8x^3y + y - 6x$ is of degree 6   **63.** $7x^2 + 38xy + 15y^2$ **65.** $2x^2 + xy - 21y^2$
67. $15x^2y^2 + xy - 2$ **69.** $49x^2 + 70xy + 25y^2$ **71.** $x^4y^4 - 6x^2y^2 + 9$ **73.** $x^3 - y^3$ **75.** $9x^2 - 25y^2$
77. 7.567; A person earning \$40,000 feels underpaid \$7567.   **79.** 54; 72; 54; Performance increases as enthusiasm goes from 1 to 50, then

performance decreases as enthusiasm goes from 50 to 100.   **81.** $4t - 2t^2 + \dfrac{2}{3}t^3$ **83.** $6x + 22$ **93.** during 1992 and 1993
95. $49x^2 + 70x + 25 - 16y^2$ **97.** $x^4 - y^4$

Section P.5

Check Point Exercises

1. a. $2x^2(5x - 2)$ **b.** $(x - 7)(2x + 3)$ **2.** $(x + 5)(x^2 - 2)$ **3. a.** $(x + 8)(x + 5)$ **b.** $(x - 7)(x + 2)$ **4.** $(3x - 1)(2x + 7)$
5. a. $(x + 9)(x - 9)$ **b.** $(6x + 5)(6x - 5)$ **6.** $(9x^2 + 4)(3x + 2)(3x - 2)$ **7. a.** $(x + 7)^2$ **b.** $(4x - 7)^2$
8. a. $(x + 1)(x^2 - x + 1)$ **b.** $(5x - 2)(25x^2 + 10x + 4)$ **9. a.** $2x(x - 6)^2$ **b.** $(x - 4)(x + 3)(x - 3)$

Exercise Set P.5

1. $9(2x + 3)$ **3.** $3x(x + 2)$ **5.** $9x^2(x^2 - 2x + 3)$ **7.** $(x + 5)(x + 3)$ **9.** $(x - 3)(x^2 + 12)$ **11.** $(x^2 + 5)(x - 2)$
13. $(x - 1)(x^2 + 2)$ **15.** $(3x - 2)(x^2 - 2)$ **17.** $(x + 2)(x + 3)$ **19.** $(x - 5)(x + 3)$ **21.** $(x - 5)(x - 3)$
23. $(3x + 2)(x - 1)$ **25.** $(3x - 28)(x + 1)$ **27.** $(2x - 1)(3x - 4)$ **29.** $(2x + 3)(2x + 5)$ **31.** $(x + 10)(x - 10)$
33. $(6x + 7)(6x - 7)$ **35.** $(3x + 5y)(3x - 5y)$ **37.** $(x^2 + 4)(x + 2)(x - 2)$ **39.** $(4x^2 + 9)(2x + 3)(2x - 3)$
41. $(x + 1)^2$ **43.** $(x - 7)^2$ **45.** $(2x + 1)^2$ **47.** $(3x - 1)^2$ **49.** $(x + 3)(x^2 - 3x + 9)$ **51.** $(x - 4)(x^2 + 4x + 16)$
53. $(2x - 1)(4x^2 + 2x + 1)$ **55.** $(4x + 3)(16x^2 - 12x + 9)$ **57.** $3x(x + 1)(x - 1)$ **59.** $4(x + 2)(x - 3)$
61. $2(x^2 + 9)(x + 3)(x - 3)$ **63.** $(x - 3)(x + 3)(x + 2)$ **65.** $2(x - 8)(x + 7)$ **67.** $x(x - 2)(x + 2)$ **69.** prime
71. $(x - 2)(x + 2)^2$ **73.** $y(y^2 + 9)(y + 3)(y - 3)$ **75.** $5y^2(2y + 3)(2y - 3)$ **77.** $-16(t - 2)(t + 1)$
79. $(3x + 2)(3x - 2)$ **89.** (d) is true.   **91.** $-(x + 5)(x - 1)$ **93.** $(x - y)^3(x + y)$
95. $b = 0, 3, 4, -c(c + 4)$, where $c > 0$ is an integer.

Section P.6

Check Point Exercises

1. a. -5 **b.** $6, -6$ **2. a.** $x^2, x \neq -3$ **b.** $\dfrac{x - 1}{x + 1}, x \neq -1$ **3.** $\dfrac{x - 3}{(x - 2)(x + 3)}, x \neq 2, x \neq -2, x \neq -3$

4. $\dfrac{3(x - 1)}{x(x + 2)}, x \neq 1, x \neq 0, x \neq -2$ **5.** $-2, x \neq -1$ **6.** $\dfrac{2(4x + 1)}{(x + 1)(x - 1)}, x \neq 1, x \neq -1$ **7.** $(x - 3)(x - 3)(x + 3)$

8. $\dfrac{-x^2 + 11x - 20}{2(x - 5)^2}, x \neq 5$ **9.** $\dfrac{2(2 - 3x)}{4 + 3x}, x \neq 0, x \neq -\dfrac{4}{3}$

Exercise Set P.6

1. 3   **3.** 5, −5   **5.** −1, −10   **7.** $\dfrac{3}{x-3}, x \neq 3$ **9.** $\dfrac{x-6}{4}, x \neq 6$ **11.** $\dfrac{y+9}{y-1}, y \neq 1, 2$ **13.** $\dfrac{x+6}{x-6}, x \neq 6, -6$

15. $\dfrac{1}{3}, x \neq 2, -3$ **17.** $\dfrac{(x-3)(x+3)}{x(x+4)}, x \neq 0, -4, 3$ **19.** $\dfrac{x-1}{x+2}, x \neq -2, -1, 2, 3$ **21.** $\dfrac{x^2+2x+4}{3x}, x \neq -2, 0, 2$ **23.** $\dfrac{7}{9}, x \neq -1$

25. $\dfrac{(x-2)^2}{x}, x \neq 2, 0, -2$ **27.** $\dfrac{2(x+3)}{3}, x \neq 3, -3$ **29.** $\dfrac{x-5}{2}, x \neq 1, -5$ **31.** $2, x \neq -\dfrac{5}{6}$ **33.** $\dfrac{2x-1}{x+3}, x \neq 0, -3$

35. $3, x \neq 2$ **37.** $\dfrac{3}{x-3}, x \neq 3, -4$ **39.** $\dfrac{9x+39}{(x+4)(x+5)}, x \neq -4, -5$ **41.** $-\dfrac{3}{x(x+1)}, x \neq -1, 0$ **43.** $\dfrac{3x^2+4}{(x+2)(x-2)}, x \neq -2, 2$

45. $\dfrac{2x^2+50}{(x-5)(x+5)}, x \neq -5, 5$ **47.** $\dfrac{4x+16}{(x+3)^2}, x \neq -3$ **49.** $\dfrac{x^2-x}{(x+5)(x-2)(x+3)}, x \neq -5, 2, -3$ **51.** $\dfrac{1}{3}, x \neq 3$

53. $\dfrac{x+1}{3x-1}, x \neq 0, \dfrac{1}{3}$ **55.** $\dfrac{1}{xy}, x \neq 0, y \neq 0, x \neq -y$ **57.** $\dfrac{x}{x+3}, x \neq -2, -3$ **59.** $-\dfrac{x-14}{7}, x \neq -2, 2$

61. $\dfrac{540t^2+12,640t+107,100}{-0.14t^2+0.51t+31.6}$ **63. a.** 86.67, 520, 1170; It costs $86,670,000 to inoculate 40% of the population against this strain of flu, and $520,000,000 to inoculate 80% of the population, and $1,170,000,000 to inoculate 90% of the population.   **b.** $x = 100$

c. increases rapidly; impossible to inoculate 100% of the population.   **65.** $\dfrac{2r_1r_2}{r_1+r_2}$; 24 mph   **79.** 1990   **81.** $-4x-1$
83. yields the third power of $x, x \neq 0$

Section P.7

Check Point Exercises

1. a. $8 + i$ **b.** $-10 + 10i$ **2. a.** $63 + 14i$ **b.** $58 - 11i$ **3.** $\dfrac{3}{5} + \dfrac{13}{10}i$ **4. a.** $7i\sqrt{3}$ **b.** $1 - 4i\sqrt{3}$ **c.** $-7 + i\sqrt{3}$

Exercise Set P.7

1. $8 - 2i$ **3.** $-2 + 9i$ **5.** $24 + 7i$ **7.** $-14 + 17i$ **9.** $21 + 15i$ **11.** $-43 - 23i$ **13.** $-29 - 11i$ **15.** 34   **17.** 34

19. $-5 + 12i$ **21.** $\dfrac{3}{5} + \dfrac{1}{5}i$ **23.** $1 + i$ **25.** $-\dfrac{24}{25} + \dfrac{32}{25}i$ **27.** $\dfrac{7}{5} + \dfrac{4}{5}i$ **29.** $3i$ **31.** $47i$ **33.** $-8i$ **35.** $2 + 6i\sqrt{7}$

37. $-\dfrac{1}{3} + \dfrac{\sqrt{2}}{6}i$ **39.** $-\dfrac{1}{8} - \dfrac{\sqrt{3}}{24}i$ **41.** $-2\sqrt{6} - 2i\sqrt{10}$ **43.** $24\sqrt{15}$ **53.** (d) is true.   **55.** $\dfrac{14}{25} - \dfrac{2}{25}i$ **57.** 0

Section P.8

Check Point Exercises

1.

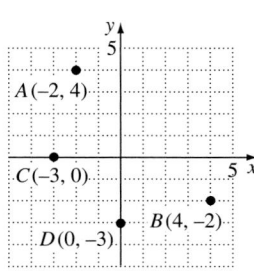

2.

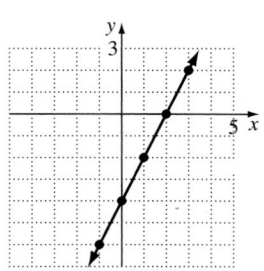

3. The minimum x-value is -100 and the maximum x-value is 100. The distance between consecutive tick marks is 50. The minimum y-value is -100 and the maximum y-value is 100. The distance between consecutive tick marks is 10.

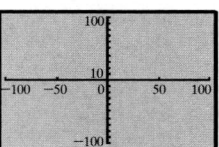

4. 5   **5.** $\left(4, -\dfrac{1}{2}\right)$ **6.** 1991; about $800 million

Exercise Set P.8

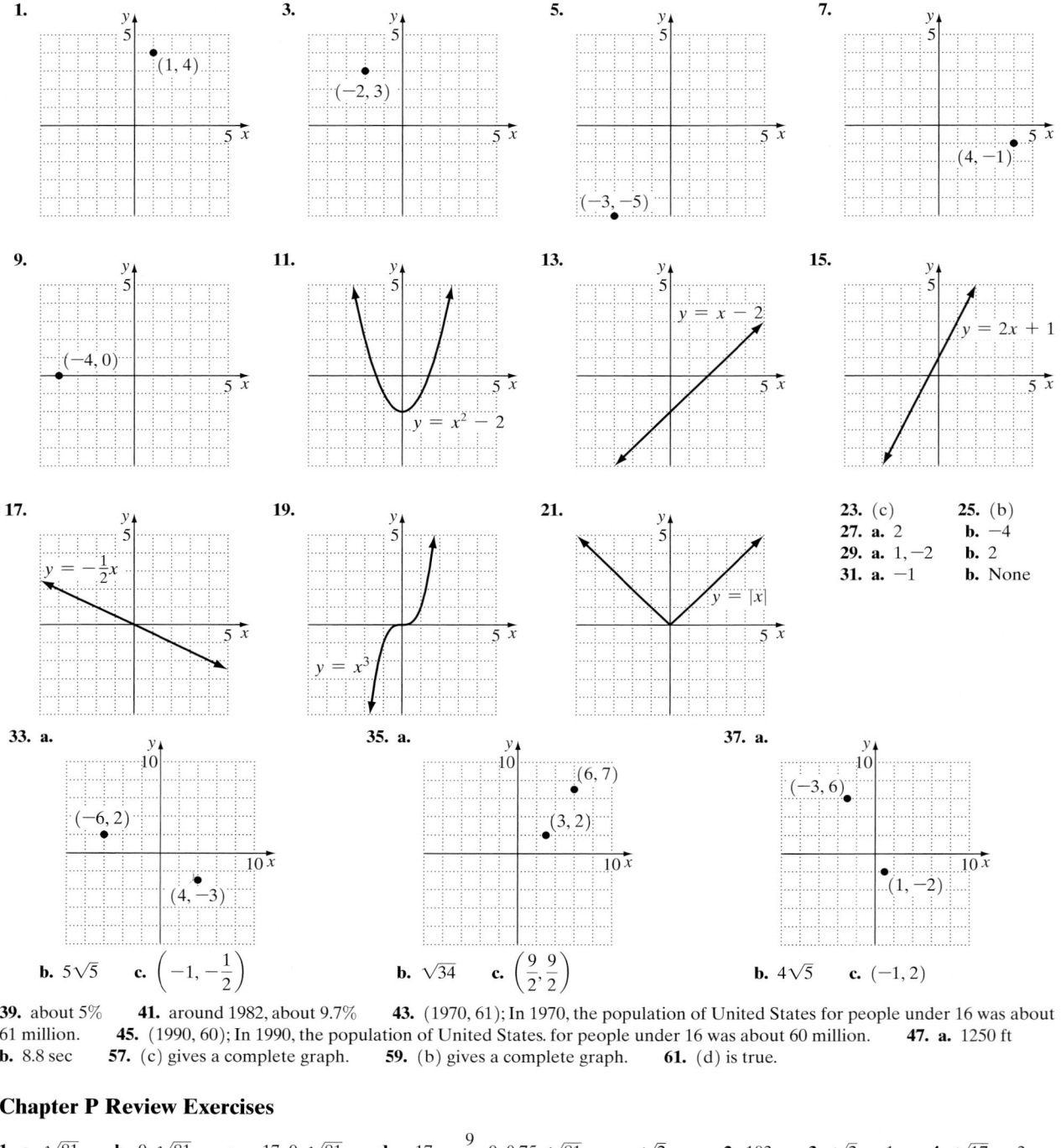

1. $(1, 4)$ **3.** $(-2, 3)$ **5.** $(-3, -5)$ **7.** $(4, -1)$

9. $(-4, 0)$ **11.** $y = x^2 - 2$ **13.** $y = x - 2$ **15.** $y = 2x + 1$

17. $y = -\frac{1}{2}x$ **19.** $y = x^3$ **21.** $y = |x|$

23. (c) **25.** (b)
27. a. 2 **b.** −4
29. a. 1, −2 **b.** 2
31. a. −1 **b.** None

33. a. $(-6, 2)$, $(4, -3)$ **b.** $5\sqrt{5}$ **c.** $\left(-1, -\frac{1}{2}\right)$

35. a. $(6, 7)$, $(3, 2)$ **b.** $\sqrt{34}$ **c.** $\left(\frac{9}{2}, \frac{9}{2}\right)$

37. a. $(-3, 6)$, $(1, -2)$ **b.** $4\sqrt{5}$ **c.** $(-1, 2)$

39. about 5% **41.** around 1982, about 9.7% **43.** (1970, 61); In 1970, the population of United States for people under 16 was about 61 million. **45.** (1990, 60); In 1990, the population of United States. for people under 16 was about 60 million. **47. a.** 1250 ft **b.** 8.8 sec **57.** (c) gives a complete graph. **59.** (b) gives a complete graph. **61.** (d) is true.

Chapter P Review Exercises

1. a. $\sqrt{81}$ **b.** $0, \sqrt{81}$ **c.** $-17, 0, \sqrt{81}$ **d.** $-17, -\frac{9}{13}, 0, 0.75, \sqrt{81}$ **e.** $\sqrt{2}, \pi$ **2.** 103 **3.** $\sqrt{2} - 1$ **4.** $\sqrt{17} - 3$
5. $|4 - (-17)|; 21$ **6.** 20 **7.** 4 **8.** commutative property of addition **9.** associative property of multiplication
10. distributive property of multiplication over addition **11.** commutative property of multiplication
12. commutative property of multiplication **13.** commutative property of addition **14.** $23x - 23y - 2$ **15.** $2x$ **16.** −108
17. $\frac{5}{16}$ **18.** $\frac{1}{25}$ **19.** $\frac{1}{27}$ **20.** $-8x^{12}y^9$ **21.** $\frac{10}{x^8}$ **22.** $\frac{1}{16x^{12}}$ **23.** $\frac{y^8}{4x^{10}}$ **24.** 37,400 **25.** 0.0000745 **26.** 3.59×10^6

27. 7.25×10^{-3}　　**28.** 3.9×10^5　　**29.** 2.3×10^{-2}　　**30.** 10^3 or 1000 yr　　**31.** $\$4.05 \times 10^{10}$　　**32.** $10\sqrt{3}$　　**33.** $2|x|\sqrt{3}$

34. $2|x|\sqrt{5}$　　**35.** $|r|\sqrt{r}$　　**36.** $\dfrac{11}{2}$　　**37.** $4|x|\sqrt{3}$　　**38.** $20\sqrt{5}$　　**39.** $16\sqrt{2}$　　**40.** $24\sqrt{2} - 8\sqrt{3}$　　**41.** $6\sqrt{5}$　　**42.** $\dfrac{\sqrt{6}}{3}$

43. $\dfrac{5(6 - \sqrt{3})}{33}$　　**44.** $7(\sqrt{7} + \sqrt{5})$　　**45.** 5　　**46.** -2　　**47.** not a real number　　**48.** 5　　**49.** $3\sqrt[3]{3}$　　**50.** $y\sqrt[3]{y^2}$　　**51.** $2\sqrt[4]{5}$

52. $13\sqrt[3]{2}$　　**53.** $|x|\sqrt[4]{2}$　　**54.** 4　　**55.** $\dfrac{1}{5}$　　**56.** 5　　**57.** $\dfrac{1}{3}$　　**58.** 16　　**59.** $\dfrac{1}{81}$　　**60.** $20x^{11/12}$　　**61.** $3x^{1/4}$　　**62.** $25x^4$　　**63.** $y^{1/2}$

64. $8x^3 + 10x^2 - 20x - 4$; degree 3　　**65.** $8x^4 - 5x^3 + 6$; degree 4　　**66.** $12x^3 + x^2 - 21x + 10$　　**67.** $6x^2 - 7x - 5$

68. $16x^2 - 25$　　**69.** $4x^2 + 20x + 25$　　**70.** $9x^2 - 24x + 16$　　**71.** $8x^3 + 12x^2 + 6x + 1$　　**72.** $125x^3 - 150x^2 + 60x - 8$

73. $-x^2 - 17xy - 3y^2$; degree 2　　**74.** $24x^3y^2 + x^2y - 12x^2 + 4$; degree 5　　**75.** $3x^2 + 16xy - 35y^2$　　**76.** $9x^2 - 30xy + 25y^2$

77. $9x^4 + 12x^2y + 4y^2$　　**78.** $49x^2 - 16y^2$　　**79.** $a^3 - b^3$　　**80.** $3x^2(5x + 1)$　　**81.** $(x - 4)(x - 7)$　　**82.** $(3x + 1)(5x - 2)$

83. $(8 - x)(8 + x)$　　**84.** prime　　**85.** $3x^2(x - 5)(x + 2)$　　**86.** $4x^3(5x^4 - 9)$　　**87.** $(x + 3)(x - 3)^2$　　**88.** $(4x - 5)^2$

89. $(x^2 + 4)(x + 2)(x - 2)$　　**90.** $(y - 2)(y^2 + 2y + 4)$　　**91.** $(x + 4)(x^2 - 4x + 16)$　　**92.** $3x^2(x - 2)(x + 2)$

93. $(3x - 5)(9x^2 + 15x + 25)$　　**94.** $x(x - 1)(x + 1)(x^2 + 1)$　　**95.** $(x^2 - 2)(x + 5)$　　**96.** $x^2, x \neq -2$　　**97.** $\dfrac{x - 3}{x - 6}, x \neq -6, 6$

98. $\dfrac{x}{x + 2}, x \neq -2$　　**99.** $\dfrac{(x + 3)^3}{(x - 2)^2(x + 2)}, x \neq 2, -2$　　**100.** $\dfrac{2}{x(x + 1)}, x \neq 0, 1, -1, -\dfrac{1}{3}$　　**101.** $\dfrac{x + 3}{x - 4}, x \neq -3, 4, 2, 8$

102. $\dfrac{1}{x - 3}, x \neq 3, -3$　　**103.** $\dfrac{4x(x - 1)}{(x + 2)(x - 2)}, x \neq 2, -2$　　**104.** $\dfrac{x(2x + 1)}{(x - 3)(x + 3)(x - 2)}, x \neq 3, -3, 2$　　**105.** $\dfrac{11x^2 - x - 11}{(2x - 1)(x + 3)(3x + 2)}$,

$x \neq \dfrac{1}{2}, -3, -\dfrac{2}{3}$　　**106.** $\dfrac{3}{x}, x \neq 0, 2$　　**107.** $\dfrac{3x}{x - 4}, x \neq 0, 4, -4$　　**108.** $\dfrac{3x + 8}{3x + 10}, x \neq -3, -\dfrac{10}{3}$　　**109.** $-9 + 4i$　　**110.** $-12 - 8i$

111. $29 + 11i$　　**112.** $-7 - 24i$　　**113.** 113　　**114.** $\dfrac{3(5 - i)}{13}$　　**115.** $\dfrac{1}{5} + \dfrac{11}{10}i$　　**116.** $i\sqrt{2}$　　**117.** $-96 - 40i$　　**118.** $2 + i\sqrt{2}$

119.　　　　**120.**　　　　**121.**　　　　**122.**

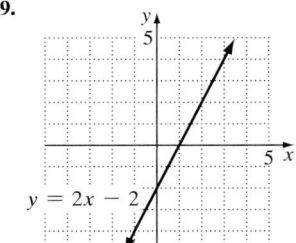
$y = 2x - 2$

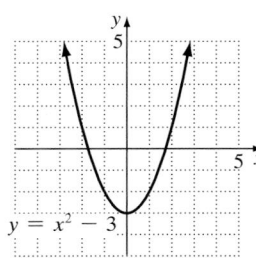
$y = x^2 - 3$

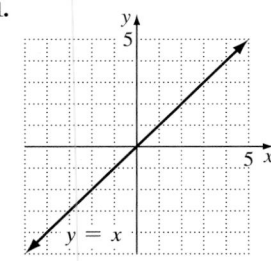
$y = x$

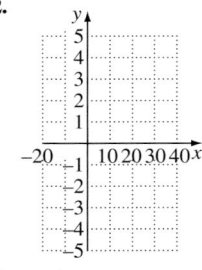

123. x-intercept: -2; y-intercept: 2　　**124.** x-intercepts: $2, -2$; y-intercept: -4　　**125.** x-intercept: 5; y-intercept: none

126. a.　　　　**b.** 5　**c.** $\left(\dfrac{1}{2}, 2\right)$　　**127. a.**　　　　**b.** $\sqrt{29}$　**c.** $\left(3, -\dfrac{1}{2}\right)$

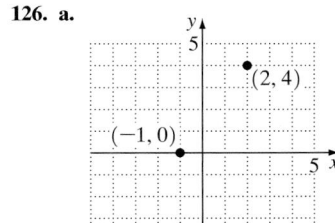
$(2, 4)$
$(-1, 0)$

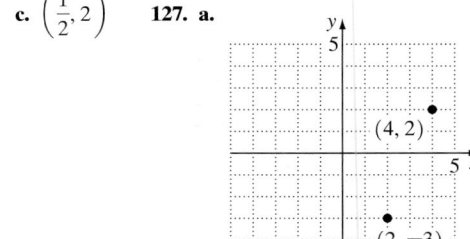
$(4, 2)$
$(2, -3)$

128. a. 1991; 100,000　　**b.** 30,000　　**c.** $A(1998, 70)$; In 1998, there were about 70,000 law school applicants.

Chapter P Test

1. $-7, -\dfrac{4}{5}, 0, 0.25, \sqrt{4}, \dfrac{22}{7}$　　**2.** commutative property of addition　　**3.** distributive property of multiplication over addition

4. 7.6×10^{-4}　　**5.** $85x + 2y - 15$　　**6.** $\dfrac{5y^8}{x^6}$　　**7.** $3|r|\sqrt{2}$　　**8.** $11\sqrt{2}$　　**9.** $\dfrac{3(5 - \sqrt{2})}{23}$　　**10.** $2x\sqrt[3]{2x}$　　**11.** $\dfrac{x + 3}{x - 2}, x \neq 2, 1$

12. $\dfrac{1}{243}$　　**13.** $2x^3 - 13x^2 + 26x - 15$　　**14.** $25x^2 + 30xy + 9y^2$　　**15.** $(x - 3)(x - 6)$　　**16.** $(x^2 + 3)(x + 2)$

17. $(5x - 3)(5x + 3)$　　**18.** $(6x - 7)^2$　　**19.** $(y - 5)(y^2 + 5y + 25)$　　**20.** $\dfrac{2(x + 3)}{x + 1}, x \neq 3, -1, -4, -3$

21. $\dfrac{x^2 + 2x + 15}{(x + 3)(x - 3)}, x \neq 3, -3$　　**22.** $\dfrac{11}{(x - 3)(x - 4)}, x \neq 3, 4$　　**23.** $\dfrac{3 - x}{3}, x \neq 0$　　**24.** $47 + 16i$　　**25.** $2 + i$　　**26.** $38i$

27.

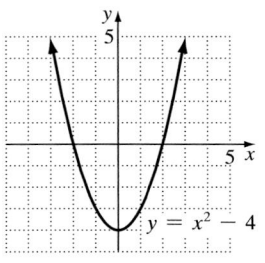

$y = x^2 - 4$

28. $2\sqrt{13}$

CHAPTER 1

Section 1.1

Check Point Exercises

1. $\{16\}$ **2.** $\{5\}$ **3.** $\{-2\}$ **4.** $\{3\}$ **5.** $\varnothing$ **6.** identity

Exercise Set 1.1

1. $\{16\}$ **3.** $\{7\}$ **5.** $\{13\}$ **7.** $\{2\}$ **9.** $\{9\}$ **11.** $\{-5\}$ **13.** $\{6\}$ **15.** $\{-2\}$ **17.** $\{12\}$ **19.** $\{24\}$ **21.** $\{-15\}$
23. $\{5\}$ **25.** $\left\{\dfrac{33}{2}\right\}$ **27.** $\{-12\}$ **29.** $\left\{\dfrac{46}{5}\right\}$ **31. a.** 0   **b.** $\left\{\dfrac{1}{2}\right\}$ **33. a.** 0   **b.** $\{-2\}$ **35. a.** 0   **b.** $\{2\}$ **37. a.** 0
b. $\{4\}$ **39. a.** 1   **b.** $\{3\}$ **41. a.** -1 **b.** $\varnothing$ **43. a.** 1   **b.** $\{2\}$ **45. a.** $-2, 2$ **b.** $\varnothing$ **47. a.** $-1, 1$ **b.** $\{-3\}$
49. a. $-2, 4$ **b.** $\varnothing$ **51.** identity   **53.** inconsistent equation   **55.** conditional equation   **57.** inconsistent equation
59. $\{-7\}$ **61.** $\varnothing$ **62.** not true for any real number   **63.** $\{-4\}$ **65.** $\{8\}$ **67.** $\{-1\}$ **69. a.** 205 mg/dl
b. 375,000 annual deaths; 125,000 saved lives   **71.** $409\dfrac{1}{5}$ ft   **85.** inconsistent   **87.** conditional; $\{-5\}$ **89.** $x = \dfrac{c - b}{a}$
91. Answers may vary.   **93.** 20

Section 1.2

Check Point Exercises

1. 100 g   **2.** Bee Gees = 11 million albums; Morissette = 16 million albums   **3.** about 57 hr   **4.** $15,000 at 9%; $10,000 at 12%
5. width = 40 ft; length = 120 ft   **6.** $m = \dfrac{y - b}{x}$

Exercise Set 1.2

1. $x + 9$ **3.** $20 - x$ **5.** $8 - 5x$ **7.** $15 \div x$ **9.** $2x + 20$ **11.** $7x - 30$ **13.** $4(x + 12)$ **15.** $x + 40 = 450; \{410\}$
17. $5x - 7 = 123; \{26\}$ **19.** $9x = 3x + 30; \{5\}$ **21.** 40 years old; Find 117 on the vertical axis and follow it over to the graph for female.
23. 30 years after 1980; 2010   **25.** yes; The height (about 61.2 in.) is greater than 5 ft.   **27.** Sosa = 66 home runs; McGwire = 70 home runs
29. 19, 20   **31.** United States = 41%; Sweden = 13.5%   **33.** 800 mi   **35.** 6 months   **37.** 7 oz
39. a. total monthly cost with coupon book = $21 + 0.50x$; total monthly cost without coupon book = $1.25x$ **b.** 28 times
41. $25,000 at 9%; $0 at 12%   **43.** length = 78 ft; width = 36 ft   **45.** length = 12 ft; height = 4 ft   **47.** 11 hr   **49.** $126
51. $740   **53.** $467.20   **55.** 5 ft 7 in.   **57.** $w = \dfrac{A}{l}$; area of rectangle   **59.** $b = \dfrac{2A}{h}$; area of triangle   **61.** $p = \dfrac{I}{rt}$; interest
63. $m = \dfrac{E}{c^2}$; energy   **65.** $p = \dfrac{T - D}{m}$ **67.** $\dfrac{2A}{h} - b = a$; trapezoid area   **69.** $\dfrac{S - P}{Pt} = r$; interest   **71.** $S = \dfrac{F}{B} + V$
77. a. $y = 3.82 + 0.3x$
b.

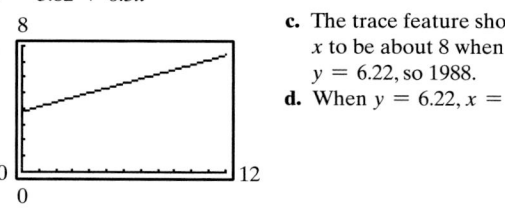

c. The trace feature shows x to be about 8 when $y = 6.22$, so 1988.
d. When $y = 6.22$, $x = 88$.

79. 600 students from the school with 10% African American students; 400 students from the school with 90% African American students
81. Coburn = 60 years old; woman = 20 years old
83. $4000 = mother; $8000 = boy; $2000 = girl

Section 1.3

Check Point Exercises

1. a. $\{0, 3\}$ **b.** $\left\{-1, \dfrac{1}{2}\right\}$ **2. a.** $\{-\sqrt{7}, \sqrt{7}\}$ **b.** $\{-5 + \sqrt{11}, -5 - \sqrt{11}\}$ **3.** $49; (x - 7)^2$ **4.** $\{1 + \sqrt{3}, 1 - \sqrt{3}\}$

5. $\left\{\dfrac{-1 + \sqrt{3}}{2}, \dfrac{-1 - \sqrt{3}}{2}\right\}$ **6.** $\{1 + i, 1 - i\}$ **7.** $-56;$ two complex imaginary solutions **8.** 10 yr **9.** 12 in.

Exercise Set 1.3

1. $\{-2, 5\}$ **3.** $\{3, 5\}$ **5.** $\left\{-\dfrac{5}{2}, \dfrac{2}{3}\right\}$ **7.** $\left\{-\dfrac{4}{3}, 2\right\}$ **9.** $\{-4, 0\}$ **11.** $\left\{0, \dfrac{1}{3}\right\}$ **13.** $\{-3, 1\}$ **15.** $\{-3, 3\}$

17. $\{-\sqrt{10}, \sqrt{10}\}$ **19.** $\{-7, 3\}$ **21.** $\left\{-\dfrac{5}{3}, \dfrac{1}{3}\right\}$ **23.** $\left\{\dfrac{1 - \sqrt{7}}{5}, \dfrac{1 + \sqrt{7}}{5}\right\}$ **25.** $\left\{\dfrac{4 - 2\sqrt{2}}{3}, \dfrac{4 + 2\sqrt{2}}{3}\right\}$ **27.** $36; (x + 6)^2$

29. $25; (x - 5)^2$ **31.** $\dfrac{9}{4}; \left(x + \dfrac{3}{2}\right)^2$ **33.** $\dfrac{49}{4}; \left(x - \dfrac{7}{2}\right)^2$ **35.** $\dfrac{1}{9}; \left(x - \dfrac{1}{3}\right)^2$ **37.** $\dfrac{1}{36}; \left(x - \dfrac{1}{6}\right)^2$ **39.** $\{-7, 1\}$

41. $\{1 + \sqrt{3}, 1 - \sqrt{3}\}$ **43.** $\{3 + 2\sqrt{5}, 3 - 2\sqrt{5}\}$ **45.** $\{-2 + \sqrt{3}, -2 - \sqrt{3}\}$ **47.** $\left\{\dfrac{-3 + \sqrt{13}}{2}, \dfrac{-3 - \sqrt{13}}{2}\right\}$ **49.** $\left\{\dfrac{1}{2}, 3\right\}$

51. $\left\{\dfrac{1 + \sqrt{2}}{2}, \dfrac{1 - \sqrt{2}}{2}\right\}$ **53.** $\left\{\dfrac{1 + \sqrt{7}}{3}, \dfrac{1 - \sqrt{7}}{3}\right\}$ **55.** $\{-5, -3\}$ **56.** $\{-6, -2\}$ **57.** $\left\{\dfrac{-5 + \sqrt{13}}{2}, \dfrac{-5 - \sqrt{13}}{2}\right\}$

59. $\left\{\dfrac{3 + \sqrt{57}}{6}, \dfrac{3 - \sqrt{57}}{6}\right\}$ **61.** $\left\{\dfrac{1 + \sqrt{29}}{4}, \dfrac{1 - \sqrt{29}}{4}\right\}$ **63.** $\{3 + i, 3 - i\}$ **65.** $36; 2$ unequal real solutions

67. $97; 2$ unequal real solutions **69.** $0; 1$ real solution **71.** $37; 2$ unequal real solutions **73.** $\left\{-\dfrac{1}{2}, 1\right\}$ **75.** $\left\{\dfrac{1}{5}, 2\right\}$

77. $\{-2\sqrt{5}, 2\sqrt{5}\}$ **79.** $\{1 + \sqrt{2}, 1 - \sqrt{2}\}$ **81.** $\left\{\dfrac{-11 + \sqrt{33}}{4}, \dfrac{-11 - \sqrt{33}}{4}\right\}$ **83.** $\left\{0, \dfrac{8}{3}\right\}$ **85.** $\{2\}$ **87.** $\{-2, 2\}$

89. $\{3 + 2i, 3 - 2i\}$ **91.** $\{2 + i\sqrt{3}, 2 - i\sqrt{3}\}$ **93.** $\left\{0, \dfrac{7}{2}\right\}$ **95.** 2024 **97.** 1999; very well **99.** 1986

101. 1990; 739,980; very well **103.** 1995; 340,000; fairly well **105.** 127.28 ft **107.** 34 ft **109.** width $= 15$ ft; length $= 20$ ft
111. 10 in. **123.** (c) is true. **125.** $x^2 - 2x - 15 = 0$ **127.** 1144; It is possible, so the applicant should be hired.

Section 1.4

Check Point Exercises

1. $\{-\sqrt{3}, 0, \sqrt{3}\}$ **2.** $\left\{-2, -\dfrac{3}{2}, 2\right\}$ **3.** $\{-1, 3\}$ **4.** $\{4\}$ **5.** $\{\sqrt[3]{25}\}$ or $\{5^{2/3}\}$ **6.** $\{-\sqrt{3}, -\sqrt{2}, \sqrt{2}, \sqrt{3}\}$ **7.** $\left\{-\dfrac{1}{27}, 64\right\}$
8. $\{-2, 3\}$

Exercise Set 1.4

1. $\{-4, 0, 4\}$ **3.** $\{0, 2\}$ **5.** $\left\{-2, -\dfrac{2}{3}, 2\right\}$ **7.** $\left\{-\dfrac{1}{2}, \dfrac{1}{2}, \dfrac{3}{2}\right\}$ **9.** $\left\{-2, -\dfrac{1}{2}, \dfrac{1}{2}\right\}$ **11.** $\{6\}$ **13.** $\{6\}$ **15.** $\{-6\}$

17. $\{10\}$ **19.** $\{12\}$ **21.** $\{8\}$ **23.** $\varnothing$ **25.** $\varnothing$ **27.** $\left\{\dfrac{13 + \sqrt{105}}{6}\right\}$ **29.** $\{4\}$ **31.** $\{13\}$ **33.** $\{\sqrt[5]{4}\}$ **35.** $\{-4, 5\}$

37. $\{-2, -1, 1, 2\}$ **39.** $\left\{-\dfrac{4}{3}, -1, 1, \dfrac{4}{3}\right\}$ **41.** $\{-\sqrt[3]{5}, -\sqrt[3]{3}\}$ **43.** $\left\{-\sqrt[3]{2}, \sqrt[3]{\dfrac{9}{5}}\right\}$ **45.** $\{-8, 27\}$ **47.** $\{1\}$ **49.** $\left\{\dfrac{1}{4}, 1\right\}$

53. $\{-3, -1, 2, 4\}$ **55.** $\{-8, -2, 1, 4\}$ **57.** $\{-8, 8\}$ **59.** $\{-5, 9\}$ **61.** $\{-2, 3\}$ **63.** $\{1\}$ **65.** $\{0\}$ **67.** $\left\{\dfrac{5}{2}\right\}$

69. $\{-8, -6, 4, 6\}$ **71.** $\{-1, 1, 2\}$ **73.** 36 years old **75.** 1952 **77.** 161,081 mi/sec **79. a.** \$29 **b.** 402 CD sets
81. a. 8 **b.** The road should be positioned so that they meet on the expressway at a point 8 miles from the point on the expressway closest to A and 4 miles from the point closest to B. **91.** $\{0, 2\}$ **93.** $\{-2\}$

95.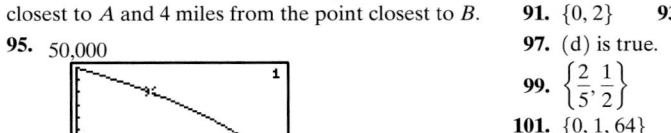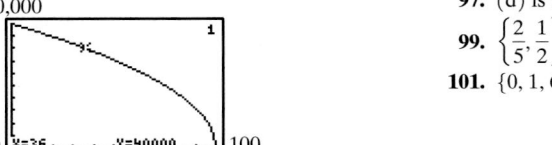

Tracing along the curve shows the point $(36, 40{,}000)$ on the graph.

97. (d) is true.

99. $\left\{\dfrac{2}{5}, \dfrac{1}{2}\right\}$

101. $\{0, 1, 64\}$

Section 1.5

Check Point Exercises

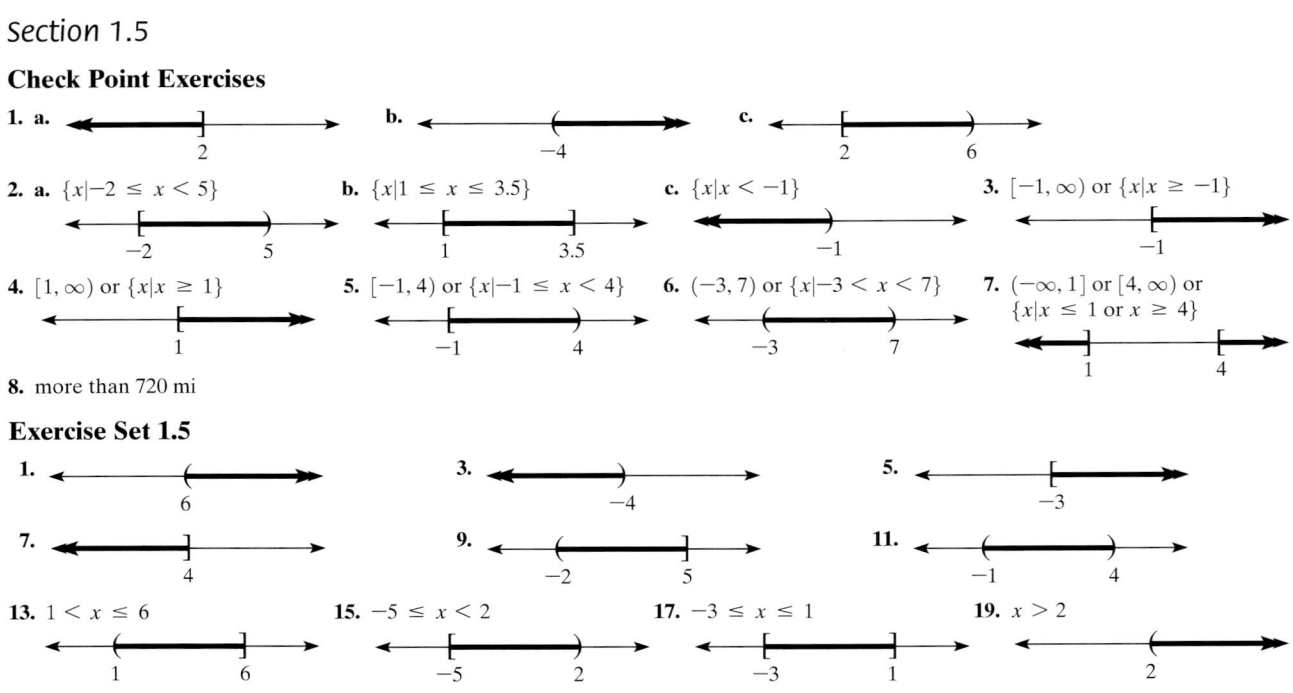

1. a. 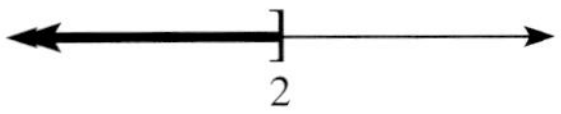 **b.** 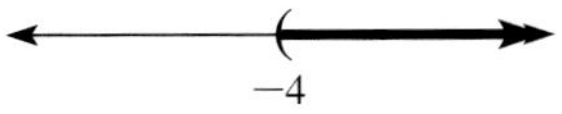 **c.** 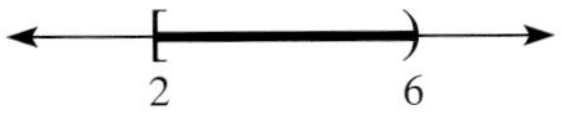

2. a. $\{x|-2 \le x < 5\}$ **b.** $\{x|1 \le x \le 3.5\}$ **c.** $\{x|x < -1\}$ **3.** $[-1, \infty)$ or $\{x|x \ge -1\}$

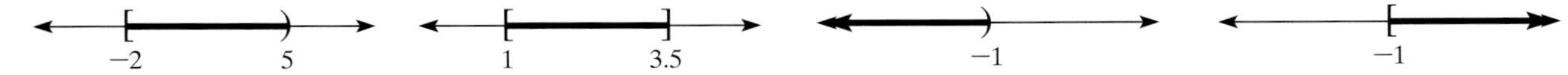

4. $[1, \infty)$ or $\{x|x \ge 1\}$ **5.** $[-1, 4)$ or $\{x|-1 \le x < 4\}$ **6.** $(-3, 7)$ or $\{x|-3 < x < 7\}$

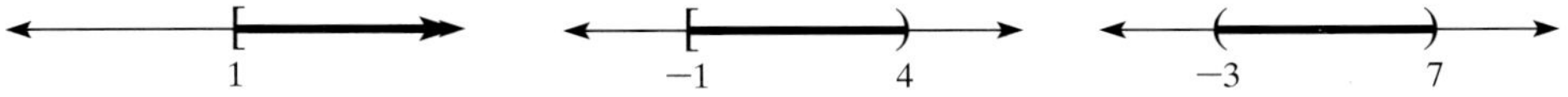

7. $(-\infty, 1]$ or $[4, \infty)$ or $\{x|x \le 1$ or $x \ge 4\}$

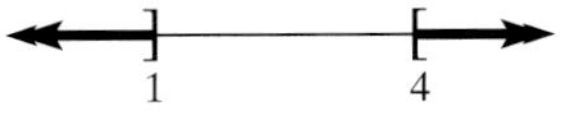

8. more than 720 mi

Exercise Set 1.5

1. 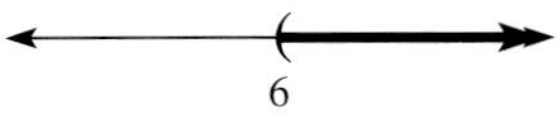 **3.** 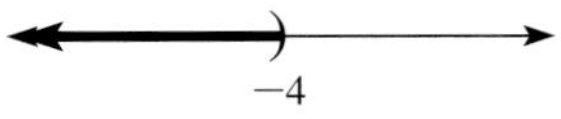 **5.** 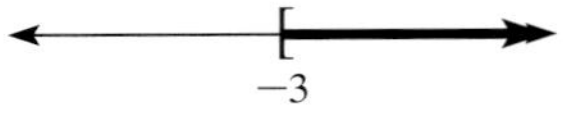

7. 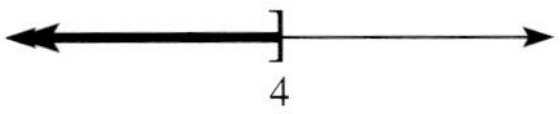 **9.**  **11.** 

13. $1 < x \le 6$ **15.** $-5 \le x < 2$ **17.** $-3 \le x \le 1$ **19.** $x > 2$

21. $x \ge -3$ **23.** $x < 3$ **25.** $x < 5.5$ **27.** $(-\infty, 3)$

29. $\left[\frac{20}{3}, \infty\right)$ **31.** $(-\infty, -4]$ **33.** $\left(-\infty, -\frac{2}{5}\right]$ **35.** $[0, \infty)$

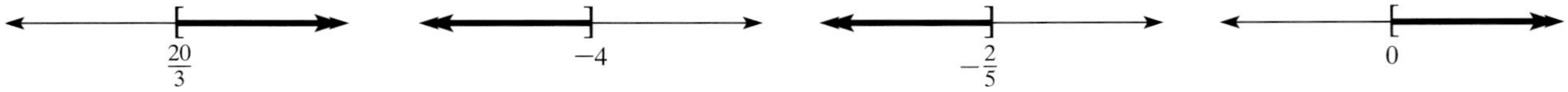

37. $(-\infty, 1)$ **39.** $[6, \infty)$ **41.** $\left[-\frac{32}{5}, \infty\right)$ **43.** $(-\infty, -6)$

45. $[13, \infty)$ **47.** $(-\infty, \infty)$ **49.** $(3, 5)$ **51.** $[-1, 3)$

53. $(-5, -2]$ **55.** $[3, 6)$ **57.** $(-3, 3)$ **59.** $[-1, 3]$

61. $(-1, 7)$ **63.** $[-5, 3]$ **65.** $(-6, 0)$ **67.** $(-\infty, -3)$ or $(3, \infty)$

69. $(-\infty, -1]$ or $[3, \infty)$ **71.** $\left(-\infty, \frac{1}{3}\right)$ or $(5, \infty)$ **73.** $(-\infty, -5]$ or $[3, \infty)$ **75.** $(-\infty, -3)$ or $(12, \infty)$

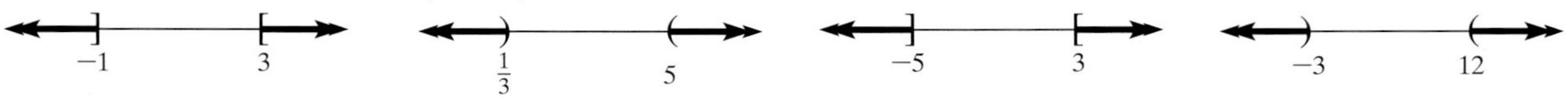

77. $(-\infty, -1]$ or $[3, \infty)$ **79.** $(-\infty, -1)$ or $(2, \infty)$ **81.** [blank] or   **83.** $(-\infty, -6]$ or $[24, \infty)$

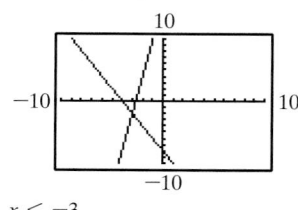 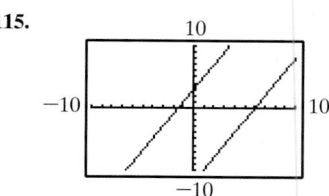

85. Raleigh, NC, Seattle, San Francisco, Austin, TX   **87.** San Diego   **89.** Austin, TX, Washington, DC, Lexington-Fayette, KY, Minneapolis, Boston, Arlington, TX   **91.** severe cognitive impairment, substance abuse disorders, and depressive: manic, major depressive
93. 2013   **95.** between 2004 and 2009   **97.** 199 checks or less   **99. a.** at least a 96   **b.** a grade less than 66
101. 1001 or more pairs
113.

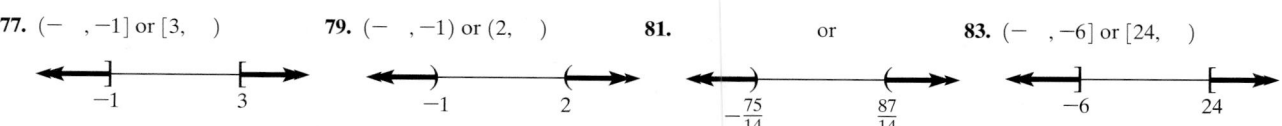

$x < -3$

115.

The graph of the left side of the inequality is always above the graph of the right side, therefore all values of x are included in the solution; You get a statement that is always true.

117. (c) is true.
119. $(-10, -8)$

Section 1.6

Check Point Exercises

1. $(-3, 1)$ **2.** $(-\infty, -4]$ or $[5, \infty)$ **3.** $(-\infty, -2)$ or $(5, \infty)$ **4.** $(-1, 1]$

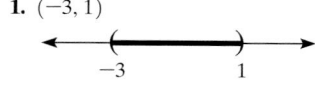 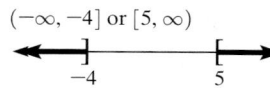 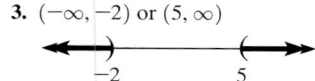 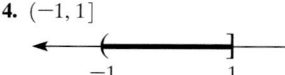

5. between 1 and 4 sec

Exercise Set 1.6

1. $(-\infty, -2)$ or $(4, \infty)$ **3.** $[-3, 7]$ **5.** $(-\infty, 1)$ or $(4, \infty)$ **7.** $(-\infty, -4)$ or $(-1, \infty)$

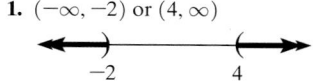 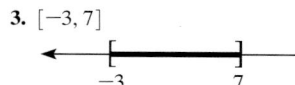 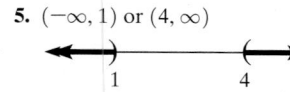 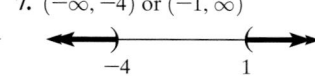

9. $\varnothing$ **11.** $[2, 4]$ **13.** $\left[-4, \dfrac{2}{3}\right]$ **15.** $\left(-3, \dfrac{5}{2}\right)$

 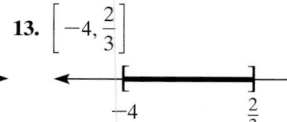 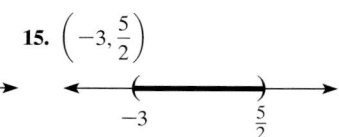

17. $\left(-1, -\dfrac{3}{4}\right)$ **19.** $\left[-2, \dfrac{1}{3}\right]$ **21.** $(-\infty, 0]$ or $[4, \infty)$ **23.** $\left(-\infty, -\dfrac{3}{2}\right)$ or $(0, \infty)$

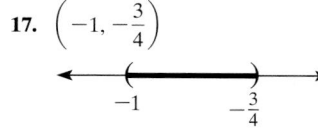 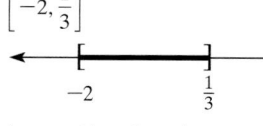 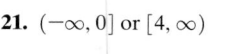 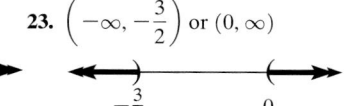

25. $[0, 1]$ **27.** $(-\infty, -3)$ or $(4, \infty)$ **29.** $(-4, -3)$ **31.** $[2, 4)$

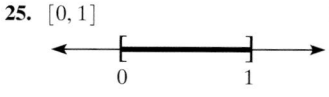 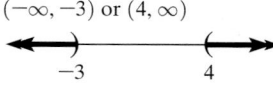  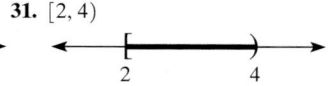

33. $\left(-\infty, -\dfrac{4}{3}\right)$ or $[2, \infty)$ **35.** $(-\infty, 0)$ or $(3, \infty)$ **37.** $(-\infty, -5)$ or $(-3, \infty)$ **39.** $\left(-\infty, \dfrac{1}{2}\right)$ or $\left[\dfrac{7}{5}, \infty\right)$

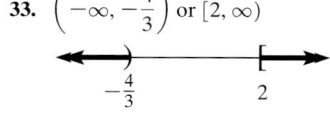 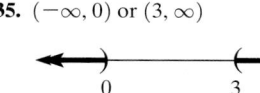 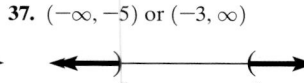 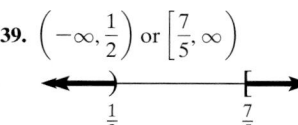

41. $(-\infty, -6]$ or $(-2, \infty)$ **43.** between 2 and 3 sec   **45.** 3.46 sec   **47. a.** 200 beats/min
b. According to the model, up to 4 min and after 12 min. In reality, between 0 and 4 min only.
49. after 2001   **51.** less than 60%   **55.** $(-\infty, -5)$ or $(2, \infty)$ **57.** $(1, 4]$

59. $(-4, -1)$ or $[2, \infty)$ **61.** from 1.7 mm to 3.5 mm   **63.** Answers may vary.   **65.** Because the square of any number other than zero is positive, the solution includes all real numbers except 2.   **67.** Because the square of any number is positive, the solution is $\varnothing$.
69. a. $(-\infty, \infty)$ **b.** $\varnothing$

Chapter 1 Review Exercises

1. $\{6\}$ **2.** $\{-10\}$ **3.** $\{5\}$ **4.** $\{-13\}$ **5.** $\{-3\}$ **6.** $\{-1\}$ **7.** $\{2\}$ **8.** $\{2\}$ **9.** $\left\{\dfrac{72}{11}\right\}$ **10.** $\{-12\}$ **11.** $\left\{\dfrac{77}{15}\right\}$

12. a. 0   **b.** $\{2\}$ **13. a.** 5   **b.** $\varnothing$ **14. a.** $-1, 1$ **b.** all real numbers except ±1 **15. a.** $-2, 4$ **b.** $\{7\}$

16. inconsistent equation   **17.** identity   **18.** conditional equation   **19.** 1997   **20.** 2000   **21.** 20 times

22. $LA = 159; NY = 26$ **23.** \$6250 at 8%; \$3750 at 12%   **24.** width $= 53$ m; length $= 120$ m   **25.** \$10,000   **26.** \$450

27. 95 concerts   **28.** $h = \dfrac{3V}{B}$ **29.** $M = \dfrac{f - F}{f}$ **30.** $\left\{-8, \dfrac{1}{2}\right\}$ **31.** $\{-4, 0\}$ **32.** $\{-8, 8\}$ **33.** $\left\{\dfrac{4 + 3\sqrt{2}}{3}, \dfrac{4 - 3\sqrt{2}}{3}\right\}$

34. $100; (x + 10)^2$ **35.** $\dfrac{9}{4}; \left(x - \dfrac{3}{2}\right)^2$ **36.** $\{3, 9\}$ **37.** $\left\{2 + \dfrac{\sqrt{3}}{3}, 2 - \dfrac{\sqrt{3}}{3}\right\}$ **38.** $\{1 + \sqrt{5}, 1 - \sqrt{5}\}$

39. $\{1 + 3i\sqrt{2}, 1 - 3i\sqrt{2}\}$ **40.** $\left\{\dfrac{-2 + \sqrt{10}}{2}, \dfrac{-2 - \sqrt{10}}{2}\right\}$ **41.** $-36;$ 2 complex imaginary solutions   **42.** $81;$ 2 unequal real solutions

43. $\left\{\dfrac{1}{2}, 5\right\}$ **44.** $\left\{-2, \dfrac{10}{3}\right\}$ **45.** $\left\{\dfrac{7 + \sqrt{37}}{6}, \dfrac{7 - \sqrt{37}}{6}\right\}$ **46.** $\{-3, 3\}$ **47.** $\{-2, 8\}$ **48.** $\left\{\dfrac{1 + i\sqrt{23}}{6}, \dfrac{1 - i\sqrt{23}}{6}\right\}$

49. 20 weeks   **50.** 1989   **51.** 12 ft by 27 ft   **52.** approximately 134.16 m   **53.** $\{-5, 0, 5\}$ **54.** $\left\{-3, \dfrac{1}{2}, 3\right\}$ **55.** $\{2\}$

56. $\{8\}$ **57.** $\{16\}$ **58.** $\{32\}$ **59.** $\{-2, -1, 1, 2\}$ **60.** $\{16\}$ **61.** $\{-4, 3\}$ **62.** $\left\{-\dfrac{11}{2}, \dfrac{23}{2}\right\}$ **63.** $\left\{-1, -\dfrac{2\sqrt{6}}{9}, \dfrac{2\sqrt{6}}{9}, 1\right\}$

64. $\{2\}$ **65.** $\{1, 4\}$ **66.** $\{-3, -2, 3\}$ **67.** 1250 ft

68.   **69.**   **70.**   **71.** $-2 < x \le 3$

72. $-1.5 \le x \le 2$ **73.** $x > -1$ **74.** $[-2, \infty)$ **75.** $\left[\dfrac{3}{5}, \infty\right)$

76. $\left(-\infty, -\dfrac{21}{2}\right)$ **77.** $(-3, \infty)$ **78.** $(-\infty, -2]$ **79.** $(2, 3]$

80. $[-9, 6]$ **81.** $(-\infty, -6)$ or $(0, \infty)$ **82.** $(-\infty, -3]$ or $[-2, \infty)$

83. Canada, Former Soviet Union   **84.** Australia, Canada, United States   **85.** Most people sleep between 5.5 and 7.5 hours.
86. between 59° and 95° inclusively   **87.** more than 50 checks   **88.** 1986

89. $\left[-4, \dfrac{1}{2}\right]$ **90.** $\left(-\infty, \dfrac{3 - \sqrt{3}}{2}\right)$ or $\left(\dfrac{3 + \sqrt{3}}{2}, \infty\right)$ **91.** $(-\infty, -2)$ or $(6, \infty)$

92. $(-\infty, 4)$ or $\left[\dfrac{23}{4}, \infty\right)$ **93.** from 1 to 2 sec

Chapter 1 Test

1. $\{-1\}$ **2.** $\{-6\}$ **3.** $\{5\}$ **4.** $\left\{-\dfrac{1}{2}, 2\right\}$ **5.** $\left\{\dfrac{1 - 5\sqrt{3}}{3}, \dfrac{1 + 5\sqrt{3}}{3}\right\}$ **6.** $\{1 - \sqrt{5}, 1 + \sqrt{5}\}$ **7.** $\left\{\dfrac{2 - i}{2}, \dfrac{2 + i}{2}\right\}$

8. $\{-1, 1, 4\}$ **9.** $\{7\}$ **10.** $\{5\}$ **11.** $\{\sqrt[3]{4}\}$ **12.** $\{1, 512\}$ **13.** $\{6, 12\}$

9. $(-3, 4), (-2, 3), (-1, 2),$
$(0, 1), (1, 0), (2, 1), (3, 2)$

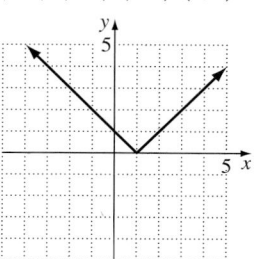

Domain: $(-\infty, \infty)$
Range: $[0, \infty)$

11. $(-3, 5), (-2, 5), (-1, 5),$
$(0, 5), (1, 5), (2, 5), (3, 5)$

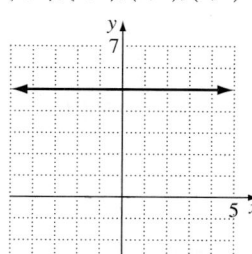

Domain: $(-\infty, \infty)$
Range: $\{5\}$

13. $(-2, -10), (-1, -3),$
$(0, -2), (1, -1), (2, 6)$

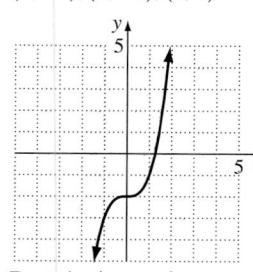

Domain: $(-\infty, \infty)$
Range: $(-\infty, \infty)$

15. a. $(-\infty, \infty)$
b. $[-4, \infty)$
c. -3 and 1
d. -3
17. a. $(-\infty, \infty)$
b. $[1, \infty)$
c. none
d. 1
e. $f(-1) = 2$ and $f(3) = 4$
19. a. $[0, 5)$
b. $[-1, 5)$
c. 2
d. -1
e. $f(3) = 1$

21. a. $[0, \infty)$ **b.** $[1, \infty)$ **c.** none **d.** 1 **e.** $f(4) = 3$ **23. a.** $[-2, 6]$ **b.** $[-2, 6]$ **c.** 4 **d.** 4 **e.** $f(-1) = 5$
25. a. $(-\infty, \infty)$ **b.** $(-\infty, -2]$ **c.** none **d.** -2 **e.** $f(-4) = -5$ and $f(4) = -2$ **27. a.** $(-\infty, \infty)$ **b.** $(0, \infty)$ **c.** none
d. 1 **29. a.** $\{-5, -2, 0, 1, 3\}$ **b.** $\{2\}$ **c.** none **d.** 2 **31.** function **33.** function **35.** not a function **37.** function
39. a. increasing: $(-1, \infty)$ **b.** decreasing: $(-\infty, -1)$ **c.** constant: none **41. a.** increasing: $(0, \infty)$ **b.** decreasing: none
c. constant: none **43. a.** increasing: none **b.** decreasing: $(-2, 6)$ **c.** constant: none **45. a.** increasing: $(-\infty, -1)$
b. decreasing: none **c.** constant: $(-1, \infty)$ **47. a.** increasing: $(-\infty, 0)$ or $(1.5, 3)$ **b.** decreasing: $(0, 1.5)$ or $(3, \infty)$
c. constant: none **49. a.** increasing: $(-2, 4)$ **b.** decreasing: none **c.** constant: $(-\infty, -2)$ or $(4, \infty)$ **51.** odd **53.** neither
55. even **57.** even **59.** even **61.** odd **63.** even **65.** odd **67.** identity function **69.** square root function

71. standard cubic function **73.** $f(1.06) = 1$ **75.** $f\left(\dfrac{1}{3}\right) = 0$ **77.** $f(-2.3) = -3$ **79.** $f(1989) \approx 294$ billion dollars.

This is the maximum function value. **81.** Defense spending is increasing from 1988 to 1989, from 1991 to 1992, and from 1996 to 1997.
83. a. increasing: $(45, 74)$; decreasing: $(16, 45)$; The number of accidents occurring per 50 million miles driven increases with age starting at
age 45, while it decreases with age starting at age 16.
b. $x = 45$ and $f(45) = 190$; The fewest number of accidents per 50 million miles driven occurs at age 45.
c. $[190, 526.4]$; Between the ages of 16 and 74, the number of accidents per 50 million miles driven is between 190 and 526.4.

85.

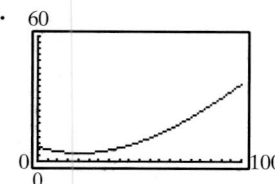

87. Answers may vary. **97. a.**

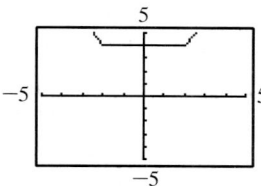

b. The number of doctor visits
decreases during childhood and
then increases as you get older.
c. The minimum is $(20.29, 3.99)$,
which means that the minimum
number of doctor visits, about 4,
occurs at around age 20.

99.

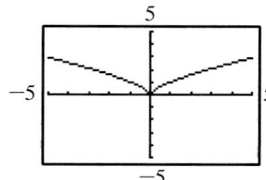

Increasing: $(-\infty, 1)$ or $(3, \infty)$
Decreasing: $(1, 3)$

101.

Increasing: $(2, \infty)$
Decreasing: $(-\infty, -2)$
Constant: $(-2, 2)$

103.

Increasing: $(0, \infty)$
Decreasing: $(-\infty, 0)$

105. a.

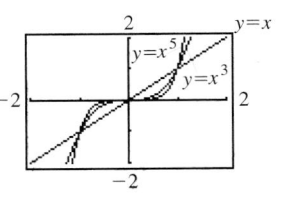

b.

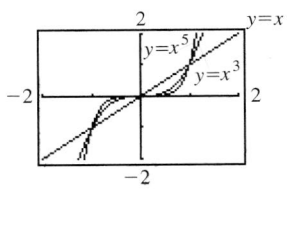

c. Increasing: $(0, \infty)$; Decreasing: $(-\infty, 0)$
d. $f(x) = x^n$ is increasing from $(-\infty, \infty)$ when n is odd.
e.

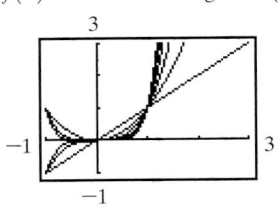

As n increases the steepness increases.

111.

Weight at least	Cost
0 oz.	$0.33
1	0.55
2	0.77
3	0.99
4	1.21

107. Answers may vary. **109.** Answers may vary.

Section 2.5

Check Point Exercises

1.

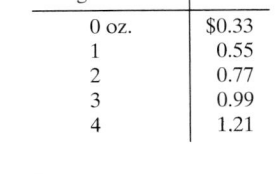

2.

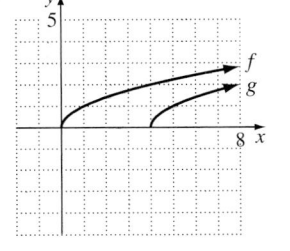

3.

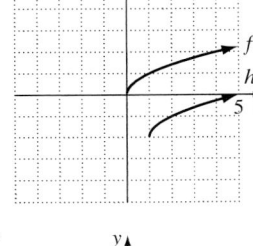

4.

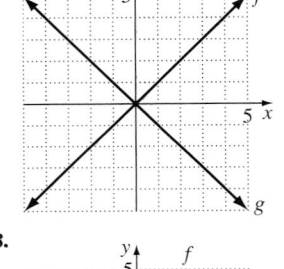

5.

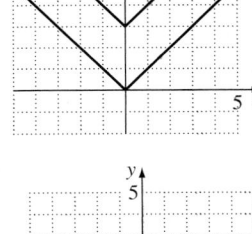

6.

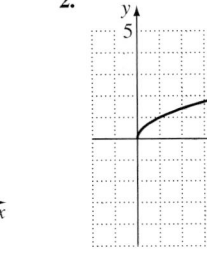

7.

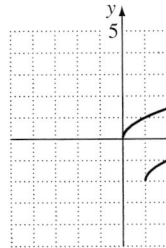

8.

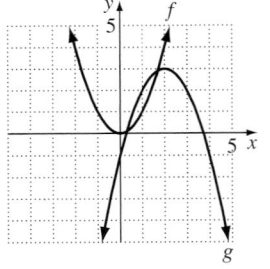

9. a. $(f + g)(x) = 3x^2 + 6x + 6$ **b.** $(f + g)(4) = 78$ **10. a.** $(f + g)(x) = \sqrt{x - 3} + \sqrt{x + 1}$ **b.** $[3, \infty)$

11. a. $(f - g)(x) = -x^2 + x - 4$ **b.** $(fg)(x) = x^3 - 5x^2 - x + 5$ **c.** $\left(\dfrac{f}{g}\right)(x) = \dfrac{x - 5}{x^2 - 1}, x \neq \pm 1$

Exercise Set 2.5

1.

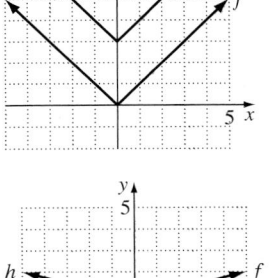

3.

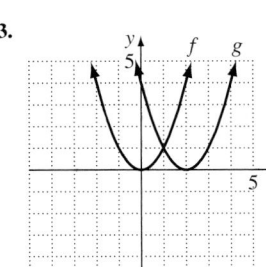

5.

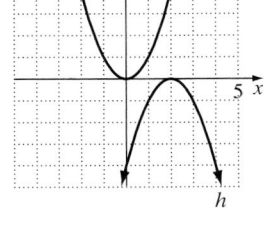

7.

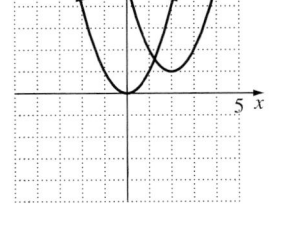

9.

11.

13.

15.

17.

19.

21.

23.

25.

27.

29.

31.

33.

35.

37.

39.

41.

43.

45.

47.

49.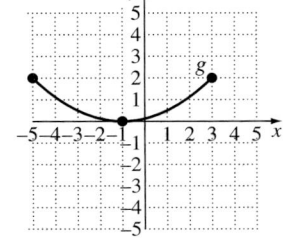

51.

53. a. $(f + g)(x) = 2x^2 + 3x + 2$ **b.** $(f + g)(4) = 46$
55. a. $(f + g)(x) = \sqrt{x - 6} + \sqrt{x + 2}$ **b.** Domain: $[6, \infty)$
57. $(f + g)(x) = 3x + 2$; Domain: $(-\infty, \infty)$; $(f - g)(x) = x + 4$;
Domain: $(-\infty, \infty)$; $(fg)(x) = 2x^2 + x - 3$; Domain: $(-\infty, \infty)$;
$\left(\dfrac{f}{g}\right)(x) = \dfrac{2x + 3}{x - 1}$; Domain: $(-\infty, 1)$ or $(1, \infty)$
59. $(f + g)(x) = 3x^2 + x - 5$; Domain: $(-\infty, \infty)$;
$(f - g)(x) = -3x^2 + x - 5$; Domain: $(-\infty, \infty)$;
$(fg)(x) = 3x^3 - 15x^2$; Domain: $(-\infty, \infty)$;
$\left(\dfrac{f}{g}\right)(x) = \dfrac{x - 5}{3x^2}$; Domain: $(-\infty, 0)$ or $(0, \infty)$

61. $(f + g)(x) = 2x^2 - 2$; Domain: $(-\infty, \infty)$; $(f - g)(x) = 2x^2 - 2x - 4$; Domain: $(-\infty, \infty)$; $(fg)(x) = 2x^3 + x^2 - 4x - 3$;
Domain: $(-\infty, \infty)$; $\left(\dfrac{f}{g}\right)(x) = 2x - 3$; Domain: $(-\infty, -1)$ or $(-1, \infty)$ **63.** $(f + g)(x) = \sqrt{x} + x - 4$; Domain: $[0, \infty)$;

$(f - g)(x) = \sqrt{x} - x + 4$; Domain: $[0, \infty)$; $(fg)(x) = \sqrt{x}(x - 4)$; Domain: $[0, \infty)$; $\left(\dfrac{f}{g}\right)(x) = \dfrac{\sqrt{x}}{x - 4}$; Domain: $[0, 4)$ or $(4, \infty)$

65. $(f + g)(x) = \dfrac{2x + 2}{x}$; Domain: $(-\infty, 0)$ or $(0, \infty)$; $(f - g)(x) = 2$; Domain: $(-\infty, 0)$ or $(0, \infty)$; $(fg)(x) = \dfrac{2x + 1}{x^2}$; Domain: $(-\infty, 0)$

or $(0, \infty)$; $\left(\dfrac{f}{g}\right)(x) = 2x + 1$; Domain: $(-\infty, 0)$ or $(0, \infty)$ **67.** $(f + g)(x) = \sqrt{x + 4} + \sqrt{x - 1}$; Domain: $[1, \infty)$;

$(f - g)(x) = \sqrt{x + 4} - \sqrt{x - 1}$; Domain: $[1, \infty)$; $(fg)(x) = \sqrt{x^2 + 3x - 4}$; Domain: $[1, \infty)$; $\left(\dfrac{f}{g}\right)(x) = \dfrac{\sqrt{x + 4}}{\sqrt{x - 1}}$; Domain: $(1, \infty)$

69. $f + g$ represents the total world population in year x.   **71.** $f(2000) \approx 1.5$ billion people; $g(2000) \approx 6$ billion people;
$(f + g)(2000) \approx 7.5$ billion people   **73.** $(R - C)(20,000) = -200,000$; The company lost $200,000 since costs exceeded revenues;
$(R - C)(30,000) = 0$; The company broke even since revenues equaled cost; $(R - C)(40,000) = 200,000$; The company made a profit of $200,000.

75.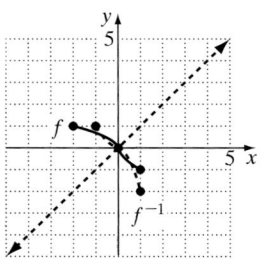

$f(x) = 435$

85. a. 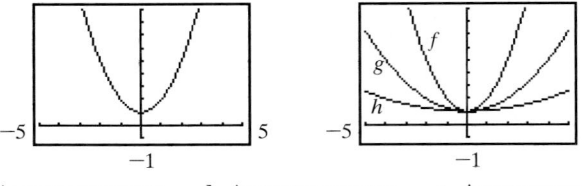 **b.**

c. Answers may vary.   **d.** Answers may vary.   **e.** Answers may vary.
87. $g(x) = -(x + 4)^2$ **89.** $g(x) = -\sqrt{x - 2} + 2$ **91.** $(-a, b)$
93. $(a + 3, b)$

Section 2.6

Check Point Exercises

1. a. $(f \circ g)(x) = 5x^2 + 1$ **b.** $(g \circ f)(x) = 25x^2 + 60x + 35$ **2.** $f(g(x)) = x$; $g(f(x)) = x$; f and g are inverses.

3. $f(g(x)) = x$; $g(f(x)) = x$; f and g are inverses.   **4.** $f^{-1}(x) = \dfrac{x - 7}{2}$ **5.** $f^{-1}(x) = \sqrt[3]{\dfrac{x + 1}{4}}$

6. (b) and (c) have inverse functions.   **7.**

Exercise Set 2.6

1. a. $(f \circ g)(x) = 2x + 14$ **b.** $(g \circ f)(x) = 2x + 7$ **c.** $(f \circ g)(2) = 18$ **3. a.** $(f \circ g)(x) = 2x + 5$ **b.** $(g \circ f)(x) = 2x + 9$
c. $(f \circ g)(2) = 9$ **5. a.** $(f \circ g)(x) = 20x^2 - 11$ **b.** $(g \circ f)(x) = 80x^2 - 120x + 43$ **c.** $(f \circ g)(2) = 69$
7. a. $(f \circ g)(x) = x^4 - 4x^2 + 6$ **b.** $(g \circ f)(x) = x^4 + 4x^2 + 2$ **c.** $(f \circ g)(2) = 6$ **9. a.** $(f \circ g)(x) = \sqrt{x - 1}$
b. $(g \circ f)(x) = \sqrt{x} - 1$ **c.** $(f \circ g)(2) = 1$ **11. a.** $(f \circ g)(x) = x$ **b.** $(g \circ f)(x) = x$ **c.** $(f \circ g)(2) = 2$
13. a. $(f \circ g)(x) = x$ **b.** $(g \circ f)(x) = x$ **c.** $(f \circ g)(2) = 2$ **15.** $f(g(x)) = x; g(f(x)) = x; f$ and g are inverses.

17. $f(g(x)) = x; g(f(x)) = x; f$ and g are inverses.   **19.** $f(g(x)) = \dfrac{5x - 56}{9}; g(f(x)) = \dfrac{5x - 4}{9}; f$ and g are not inverses.

21. $f(g(x)) = x; g(f(x)) = x; f$ and g are inverses.   **23.** $f(g(x)) = x; g(f(x)) = x; f$ and g are inverses.   **25.** $f^{-1}(x) = x - 3$

27. $f^{-1}(x) = \dfrac{x}{2}$ **29.** $f^{-1}(x) = \dfrac{x - 3}{2}$ **31.** $f^{-1}(x) = \sqrt[3]{x - 2}$ **33.** $f^{-1}(x) = \sqrt[3]{x} - 2$ **35.** $f^{-1}(x) = \dfrac{1}{x}$

37. $f^{-1}(x) = x^2, x \geq 0$ **39.** $f^{-1}(x) = \sqrt{x - 1}$ **41.** $f^{-1}(x) = \dfrac{3x + 1}{x - 2}$ **43.** $f^{-1}(x) = (x - 3)^3 + 4$

45. The function is not one-to-one, so it does not have an inverse function.   **47.** The function is not one-to-one, so it does not have an
inverse function.   **49.** The function is one-to-one, so it does have an inverse function.
51.

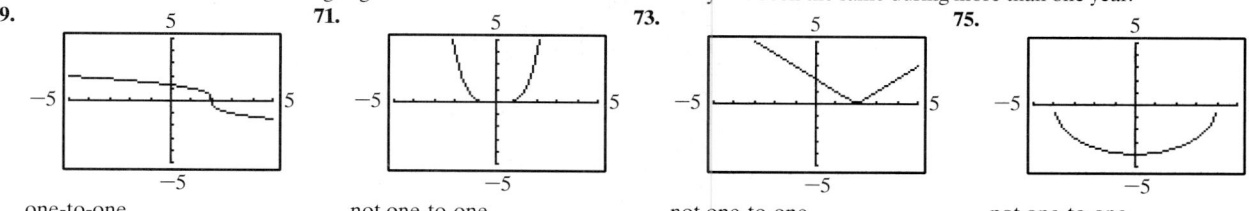

53.

55. a. f gives the price of the computer after a \$400 discount. g gives
the price of the computer after a 25% discount.
b. $(f \circ g)(x) = 0.75x - 400$. This models the price of a computer
after first a 25% discount and then a \$400 discount.
c. $(g \circ f)(x) = 0.75(x - 400)$. This models the price of a computer
after first a \$400 discount and then a 25% discount.
d. The function $f \circ g$ models the greater discount, since the
25% discount is taken on the regular price first.
e. $f^{-1}(x) = x + 400$; If x is the discount price of the computer,
then $f^{-1}(x)$ is the regular price.

57. a. f is a one-to-one function.   **b.** $f^{-1}(0.25)$ is the number of people in a room for a 25% probability of two people sharing a birthday.
$f^{-1}(0.5)$ is the number of people in a room for a 50% probability of two people sharing a birthday. $f^{-1}(0.7)$ is the number of people in a
room for a 70% probability of two people sharing a birthday.   **59.** No. The graph does not pass the horizontal line test, so it is not
one-to-one. This means that the average age at which United States women marry has been the same during more than one year.
69.   **71.**   **73.**   **75.**

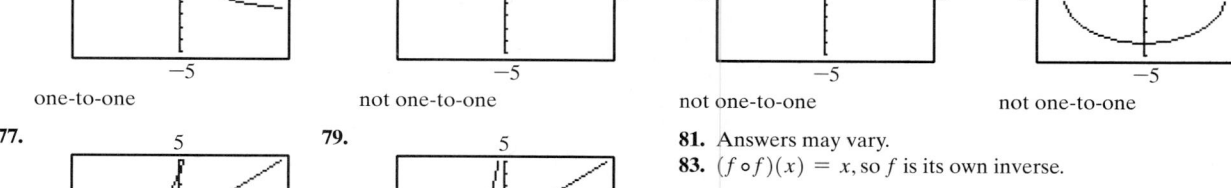

one-to-one   not one-to-one   not one-to-one   not one-to-one

77.   **79.**

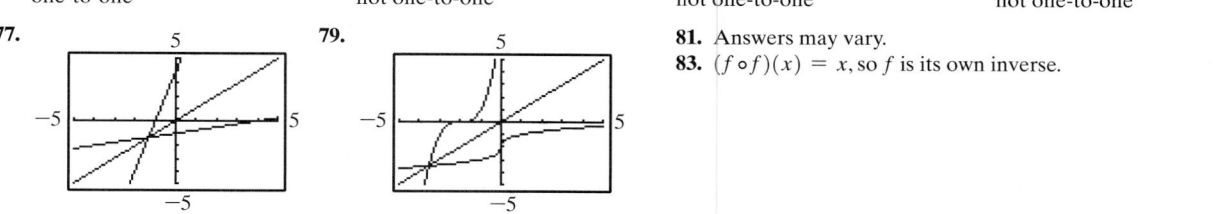

81. Answers may vary.
83. $(f \circ f)(x) = x$, so f is its own inverse.

f and g are inverses. f and g are inverses.

Chapter 2 Review Exercises

1. $m = -\dfrac{1}{2}$; falls   **2.** $m = 1$; rises   **3.** $m = 0$; horizontal   **4.** $m =$ undefined; vertical
5. $y - 2 = -6(x + 3); y = -6x - 16$ **6.** using $(1, 6), y - 6 = 2(x - 1); y = 2x + 4$

7. Slope: $\frac{2}{5}$; y-intercept: -1 **8.** Slope: -4; y-intercept: 5   **9.** Slope: $-\frac{2}{3}$; y-intercept: -2 **10.** Slope: 0; y-intercept: 4

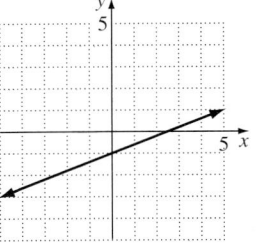 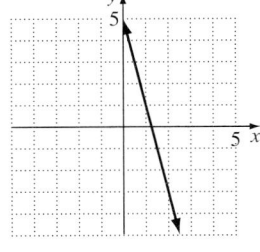 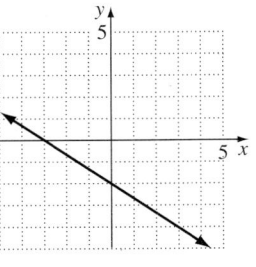 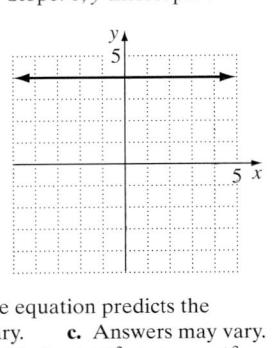

11. a. using $(0, 16)$, $y - 16 = -0.13(x - 0)$ **b.** $y = -0.13x + 16$ **c.** 6.9 ft; 5.6 ft   **d.** In 2000, the equation predicts the surfboard length to be 3 feet, which is not reasonable.   **12. a.** Answers may vary.   **b.** Answers may vary.   **c.** Answers may vary.
13. $y + 7 = -3(x - 4)$; $y = -3x + 5$ **14.** $y - 6 = -3(x + 3)$; $y = -3x - 3$ **15.** $x^2 + y^2 = 9$ **16.** $(x + 2)^2 + (y - 4)^2 = 36$
17. Center: $(0, 0)$; radius: 1   **18.** Center: $(-2, 3)$; radius: 3   **19.** Center: $(2, -1)$; radius: 3   **20.** Function; Domain: $\{2, 3, 5\}$; Range: $\{7\}$
21. Function; Domain: $\{1, 2, 13\}$; Range: $\{10, 500, \pi\}$
22. Not a function; Domain: $\{12, 14\}$; Range: $\{13, 15, 19\}$
23. y is a function of x.
24. y is a function of x.
25. y is not a function of x.
26. a. $f(4) = -23$
b. $f(x + 3) = -7x - 16$
c. $f(-x) = 5 + 7x$

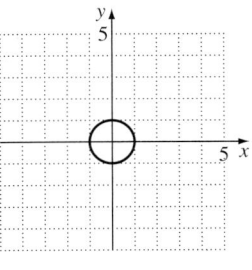 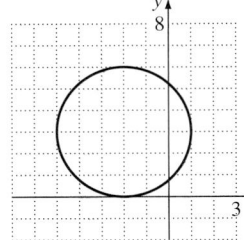 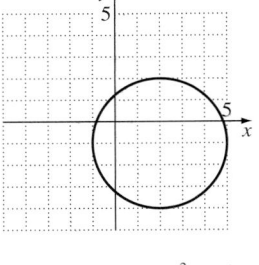

27. a. $g(0) = 2$ **b.** $g(-2) = 24$ **c.** $g(x - 1) = 3x^2 - 11x + 10$ **d.** $g(-x) = 3x^2 + 5x + 2$
28. a. $f(a) = 4a - 3$ **b.** $f(a + h) = 4a + 4h - 3$ **c.** $\dfrac{f(a + h) - f(a)}{h} = 4$ **d.** $f(a) + f(h) = 4a + 4h - 6$
29. a. $g(13) = 3$ **b.** $g(0) = 4$ **c.** $g(-3) = 7$ **30. a.** $f(-2) = -1$ **b.** $f(1) = 12$ **c.** $f(2) = 3$ **31.** $(-\infty, \infty)$
32. $(-\infty, 7)$ or $(7, \infty)$ **33.** $(-\infty, 4]$ **34.** $(-\infty, -1)$ or $(-1, 1)$ or $(1, \infty)$ **35.** $[2, 5)$ or $(5, \infty)$
36. $f(6) = 25.48$; In 1996, there were 25.48 million participants in the Federal Food Stamp Program.
37. Ordered pairs: $(-1, 9)$, $(0, 4)$, $(1, 1)$, $(2, 0)$, $(3, 1)$, $(4, 4)$.
38. Ordered pairs: $(-1, 3)$, $(0, 2)$, $(1, 1)$, $(2, 0)$, $(3, 1)$, $(4, 2)$.
39. a. Domain: $[-3, 5)$
b. Range: $[-5, 0]$
c. x-intercept: -3
d. y-intercept: -2
e. increasing: $(-2, 0)$ or $(3, 5)$ decreasing: $(-3, -2)$ or $(0, 3)$
f. $f(-2) = -3$ and $f(3) = -5$
40. a. Domain: $(-\infty, \infty)$
b. Range: $(-\infty, \infty)$
c. x-intercepts: -2 and 3
d. y-intercept: 3
e. increasing: $(-5, 0)$; decreasing: $(-\infty, -5)$ or $(0, \infty)$
f. $f(-2) = 0$ and $f(6) = -3$

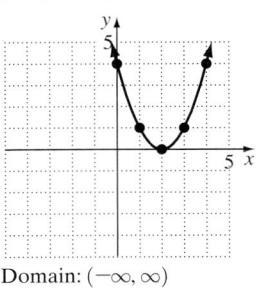 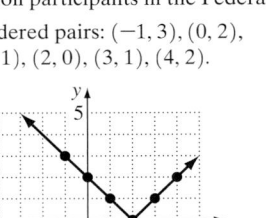

Domain: $(-\infty, \infty)$
Range: $[0, \infty)$

Domain: $(-\infty, \infty)$
Range: $[0, \infty)$

41. a. Domain: $(-\infty, \infty)$ **b.** Range: $[-2, 2]$ **c.** x-intercept: 0   **d.** y-intercept: 0   **e.** increasing: $(-2, 2)$; constant: $(-\infty, -2)$ or $(2, \infty)$
f. $f(-9) = -2$ and $f(14) = 2$ **42.** not a function   **43.** function   **44.** function   **45.** not a function
46. odd; symmetric with respect to the origin   **47.** even; symmetric with respect to the y-axis   **48.** odd; symmetric with respect to the origin
49. a. yes; The graph passes the vertical line test.   **b.** Decreasing: $(3, 12)$; The vulture descended.
c. Constant: $(0, 3)$ and $(12, 17)$; The vulture's height held steady during the first 3 seconds and the vulture was on the ground for 5 seconds.
d. Increasing: $(17, 30)$; The vulture was ascending.

50.
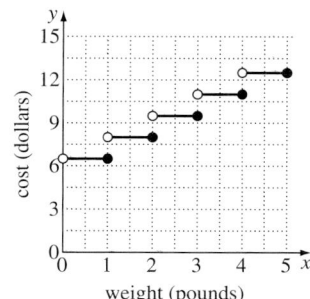

51.
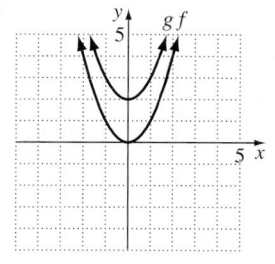

52.
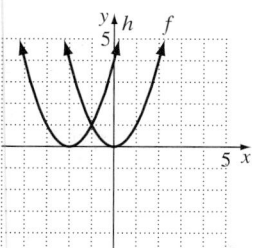

53.
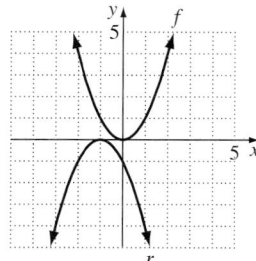

54.
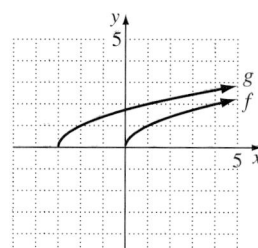

55.
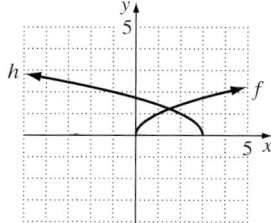

56.
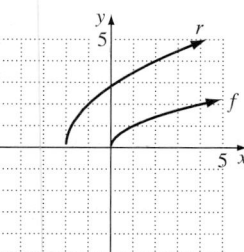

57.
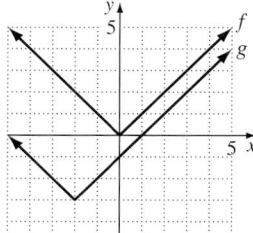

58.
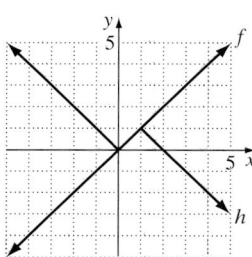

59.
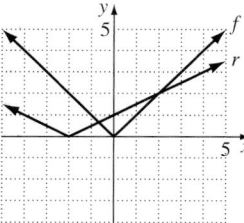

60.
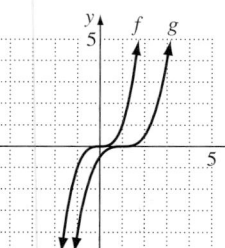

61.
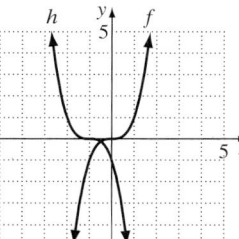

62.
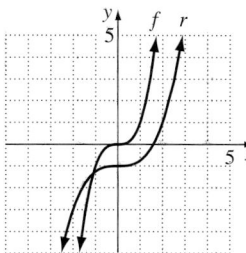

63.
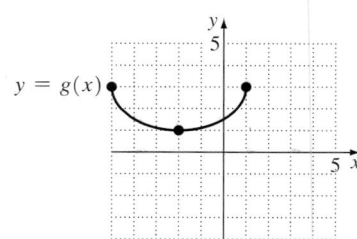

64.
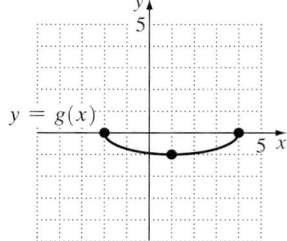

65.
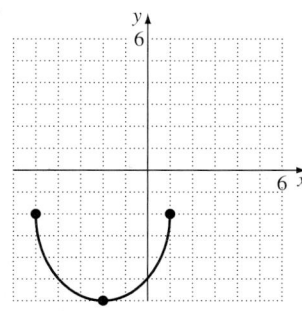

66. $(f + g)(x) = 4x - 6$; Domain: $(-\infty, \infty)$; $(f - g)(x) = 2x + 4$; Domain: $(-\infty, \infty)$;
$(fg)(x) = 3x^2 - 16x + 5$; Domain: $(-\infty, \infty)$; $\left(\dfrac{f}{g}\right)(x) = \dfrac{3x - 1}{x - 5}$; Domain: $(-\infty, 5)$ or $(5, \infty)$

67. $(f + g)(x) = 2x^2 + x$; Domain: $(-\infty, \infty)$; $(f - g)(x) = x + 2$; Domain: $(-\infty, \infty)$;
$(fg)(x) = x^4 + x^3 - x - 1$; Domain: $(-\infty, \infty)$; $\left(\dfrac{f}{g}\right)(x) = \dfrac{x^2 + x + 1}{x^2 - 1}$;
Domain: $(-\infty, -1)$ or $(-1, 1)$ or $(1, \infty)$

68. $(f + g)(x) = \sqrt{x + 7} + \sqrt{x - 2}$; Domain: $[2, \infty)$; $(f - g)(x) = \sqrt{x + 7} - \sqrt{x - 2}$;
Domain: $[2, \infty)$; $(fg)(x) = \sqrt{x^2 + 5x - 14}$; Domain: $[2, \infty)$; $\left(\dfrac{f}{g}\right)(x) = \dfrac{\sqrt{x + 7}}{\sqrt{x - 2}}$;
Domain: $(2, \infty)$

69. a. $(f \circ g)(x) = 16x^2 - 8x + 4$ **b.** $(g \circ f)(x) = 4x^2 + 11$ **c.** $(f \circ g)(3) = 124$

70. a. $(f \circ g)(x) = \sqrt{x + 1}$ **b.** $(g \circ f)(x) = \sqrt{x} + 1$ **c.** $(f \circ g)(3) = 2$ **71.** $f(g(x)) = x - \dfrac{7}{10}$; $g(f(x)) = x - \dfrac{7}{6}$;

f and g are not inverses of each other.   **72.** $f(g(x)) = x$; $g(f(x)) = x$; f and g are inverses of each other.   **73.** $f^{-1}(x) = \dfrac{x + 3}{4}$

74. $f^{-1}(x) = x^2 - 2$ for $x \geq 0$ **75.** $f^{-1}(x) = \sqrt[3]{\dfrac{x-1}{8}}$ **76.** Inverse function exists.   **77.** Inverse function does not exist.

78. Inverse function exists.   **79.** Inverse function does not exist.   **80.**

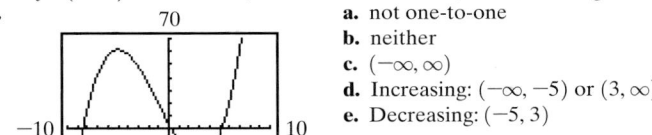

Chapter 2 Test

1. using $(2, 1)$, $y - 1 = 3(x - 2)$; $y = 3x - 5$ **2.** $y - 6 = 4(x + 4)$; $y = 4x + 22$ **3. a.** using $(4, 401.1)$, $y - 401.1 = 14.9(x - 4)$; $y = 14.9x + 341.5$ **b.** \$639.50

4. Center: $(-2, 3)$; radius: 4

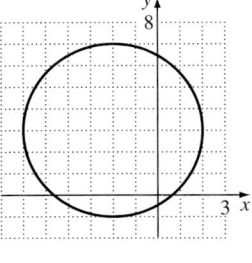

5. b, c, d   **6.** $f(x - 1) = x^2 - 4x + 8$ **7.** $g(-1) = 4$; $g(7) = 2$ **8.** Domain: $(-\infty, 4]$

9. $f(10) = 55$; There are 55 board feet in a 16-foot log whose average diameter is 10 inches.

10. a. $f(4) - f(-3) = 5$ **b.** Domain: $(-5, 6]$ **c.** Range: $[-4, 5]$ **d.** Increasing: $(-1, 2)$ **e.** Decreasing: $(-5, -1)$ or $(2, 6)$ **f.** x-intercepts: $-4, 1$, and 5 **g.** y-intercept: -3

11. $f(x)$ is even and is symmetric with respect to the y-axis. The graph in the figure is symmetric with respect to the origin.

12. The graph of f is shifted 3 to the right to obtain the graph of g. Then the graph of g is stretched by a factor of 2 and reflected about the x-axis to obtain the graph of h.

13.

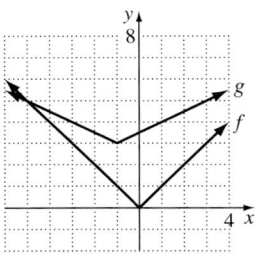

14. $(f - g)(x) = x^2 - 2x - 2$ **15.** $\left(\dfrac{f}{g}\right)(x) = \dfrac{x^2 + 3x - 4}{5x - 2}$; Domain: $\left(-\infty, \dfrac{2}{5}\right)$ or $\left(\dfrac{2}{5}, \infty\right)$

16. $(f \circ g)(x) = 25x^2 - 5x - 6$ **17.** $(g \circ f)(x) = 5x^2 + 15x - 22$

18. $f(g(2)) = 84$ **19.** $f^{-1}(x) = x^2 + 2$ for $x \geq 0$

20. a. The graph of f passes the Horizontal Line Test.   **b.** $f(80) = 2000$

c. $f^{-1}(2000)$ is the income, in thousands of dollars, for those who give \$2000 to charity.

21.

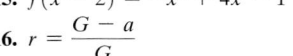

a. not one-to-one
b. neither
c. $(-\infty, \infty)$
d. Increasing: $(-\infty, -5)$ or $(3, \infty)$
e. Decreasing: $(-5, 3)$

Cumulative Review Exercises (Chapters P–2)

1. $\dfrac{2y^4}{x^3}$ **2.** $\dfrac{5\sqrt{2}}{8}$ **3.** $(x - 4)(x^2 + 2)$ **4.** $\dfrac{2x^2 - x + 6}{(x + 4)(x - 2)}$ **5.** $\dfrac{2x + 1}{2x - 1}$ **6.** $x = -4$ or $x = 5$ **7.** $x = \dfrac{25}{18}$ **8.** $x = 4$

9. $x = -8$ or $x = 27$ **10.** $x \leq 20$; $(-\infty, 20]$ **11.** $(-\infty, 2)$ or $[7, \infty)$ **12.** $y - 5 = 4(x + 2)$; $y = 4x + 13$

13.

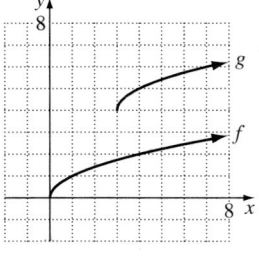

14. $f^{-1}(x) = (x - 2)^2 + 3$
15. $f(x - 2) = -x^2 + 4x - 1$
16. $r = \dfrac{G - a}{G}$

17. 3 ft by 8 ft
18. Your salary prior to the raise was \$18,500.
19. You must make an 85 on the final exam to have an average score of 80.
20. $f(3) = 96$; A rock thrown with an initial velocity of 80 feet per second will have a height of 96 feet at a time of 3 seconds after it was thrown.

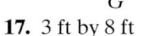
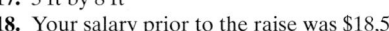

CHAPTER 3

Section 3.1

Check Point Exercises

1. 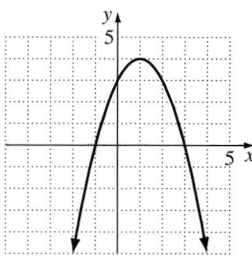 **2.** 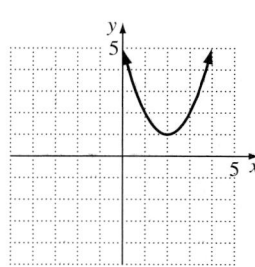 **3.** 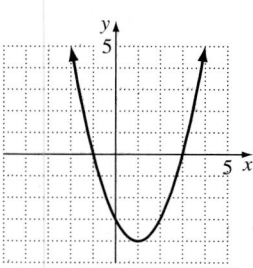 **4.** 45; 190

Exercise Set 3.1

1. $h(x) = (x - 1)^2 + 1$ **3.** $j(x) = (x - 1)^2 - 1$ **5.** $h(x) = x^2 - 1$ **7.** $g(x) = x^2 - 2x + 1$ **9.** $(3, 1)$ **11.** $(-1, 5)$
13. $(2, -5)$ **15.** $(-1, 9)$
17. axis of symmetry: $x = 4$ **19.** axis of symmetry: $x = 1$ **21.** axis of symmetry: $x = 3$ **23.** axis of symmetry: $x = -2$

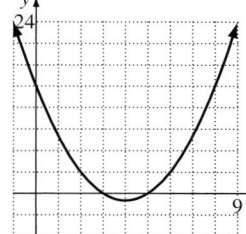 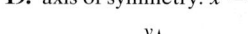 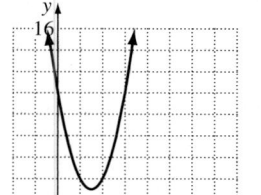 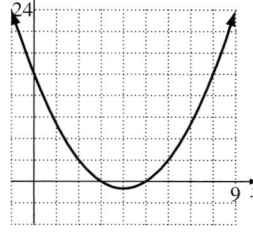

25. axis of symmetry: $x = 1$ **27.** axis of symmetry: $x = 1$ **29.** axis of symmetry: $x = -\dfrac{3}{2}$ **31.** axis of symmetry: $x = 1$

 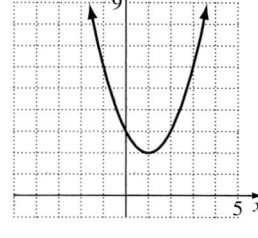 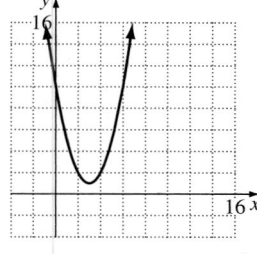

33. axis of symmetry: $x = 1$

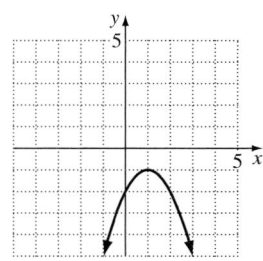

35. minimum; $(2, -13)$ **37.** maximum; $(1, 1)$ **39.** minimum; $\left(\dfrac{1}{2}, -\dfrac{5}{4}\right)$
41. 1968; 4238 cigarettes per person **43.** 2 sec; 144 ft **45.** $(9, 2100)$ **47.** 30 ft; 60 ft; 1800 ft^2
57. a.

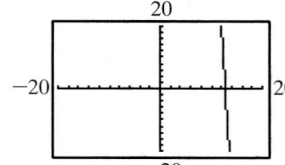

You can only see a little of
the parabola.

b. $(20.5, -120.5)$
c. Answers may vary.
d. Answers may vary.

59. $(2.5, 185)$

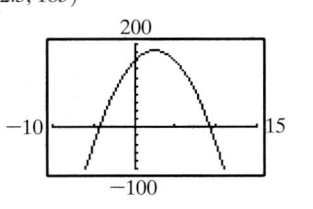

61. $(-30, 91)$

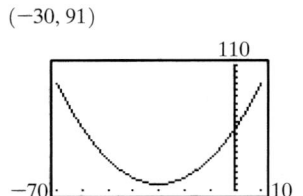

63.

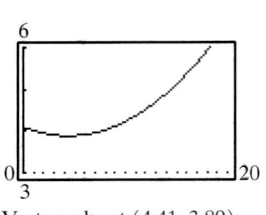

Vertex: about $(4.41, 3.89)$;
The minimum number of people
in the United States holding more
than one job was 3.89 million in 1974.

65. (a) is true. **67.** $x = -2; (-3, -2)$ **69.** $x = 166\frac{2}{3}$ ft, $y = 125$ ft; approximately 20,833 ft^2

Section 3.2

Check Point Exercises

1. The graph rises to the left and to the right. **2.** Since n is odd and the leading coefficient is negative, the function falls to the right. Since the ratio cannot be negative, the model won't be appropriate. **3.** No; the graph should fall to the left, but doesn't appear to.
4. $\{-2, 2\}$ **5.** $\{-2, 0, 2\}$ **6.**

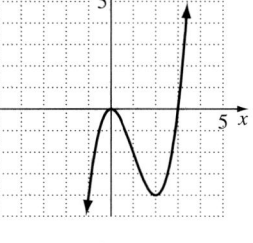

Exercise Set 3.2

1. polynomial function; degree: 3 **3.** polynomial function; degree: 5 **5.** not a polynomial function **7.** not a polynomial function
9. not a polynomial function **11.** polynomial function **13.** not a polynomial function **15.** (c) **17.** (b) **19.** (a)
21. falls to the left and rises to the right **23.** rises to the left and to the right **25.** falls to the left and to the right
27. $x = 5$ has multiplicity 1; The graph crosses the x-axis; $x = -4$ has multiplicity 2; The graph touches the x-axis and turns around.
29. $x = 3$ has multiplicity 1; The graph crosses the x-axis; $x = -6$ has multiplicity 3; The graph crosses the x-axis.
31. $x = 0$ has multiplicity 1; The graph crosses the x-axis; $x = 1$ has multiplicity 2; The graph touches the x-axis and turns around.
33. $x = 2, x = -2$ and $x = -7$ have multiplicity 1; The graph crosses the x-axis.

35. a. $f(x)$ rises to the right and falls to the left.
  b. $x = -2, x = 1, x = -1$;
    $f(x)$ crosses the x-axis at each.
  c. The y-intercept is -2.
  d. neither
  e.

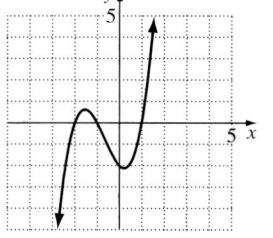

37. a. $f(x)$ rises to the left and the right.
  b. $x = 0, x = 3, x = -3$;
    $f(x)$ crosses the x-axis at -3 and 3;
    $f(x)$ touches the x-axis at 0.
  c. The y-intercept is 0.
  d. y-axis symmetry
  e.

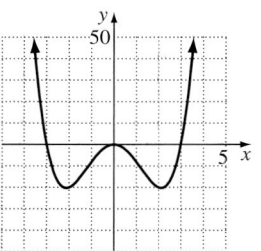

39. a. $f(x)$ falls to the left and the right.
  b. $x = 0, x = 4, x = -4$;
    $f(x)$ crosses the x-axis at -4 and 4;
    $f(x)$ touches the x-axis at 0.
  c. The y-intercept is 0.
  d. y-axis is symmetry
  e.

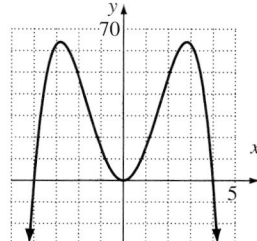

41. a. $f(x)$ rises to the left and the right.
  b. $x = 0, x = 1$;
    $f(x)$ touches the x-axis at 0 and 1.
  c. The y-intercept is 0.
  d. neither
  e.

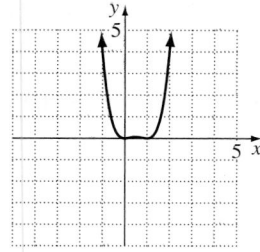

43. a. $f(x)$ falls to the left and the right.
  b. $x = 0, x = 2$;
    $f(x)$ crosses the x-axis at 0 and 2.
  c. The y-intercept is 0.
  d. neither
  e.

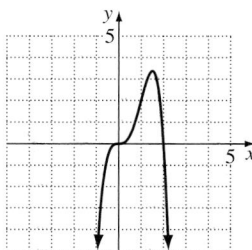

45. a. $f(x)$ rises to the left and falls to the right.
  b. $x = 0, x = \pm\sqrt{3}$;
    $f(x)$ crosses the x-axis at $(0, 0)$;
    $f(x)$ touches the x-axis at $\sqrt{3}$ and $-\sqrt{3}$.
  c. The y-intercept is 0.
  d. origin symmetry
  e.

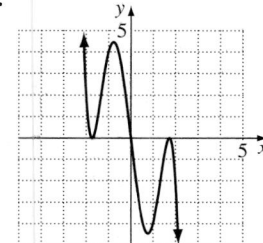

47. a. $f(x)$ rises to the left and falls to the right.
  b. $x = 0, x = 3$;
    $f(x)$ crosses the x-axis at 3;
    $f(x)$ touches the axis at $(0, 0)$.
  c. The y-intercept is 0.
  d. neither
  e.

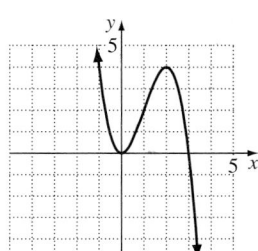

49. a. $f(x)$ falls to the left and the right.
  b. $x = 1, x = -2, x = 2$;
    $f(x)$ crosses the x-axis at -2 and 2;
    $f(x)$ touches the x-axis at $(1, 0)$.
  c. The y-intercept is 12.
  d. neither
  e.

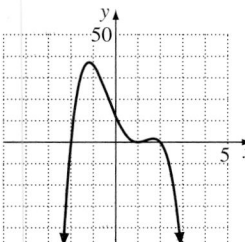

51. a. Leading coefficient test suggests the elk population will decline and eventually will die off.

b.

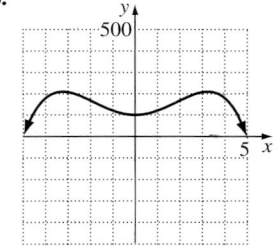

c.

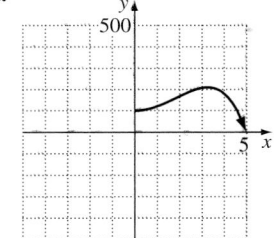

The population reaches extinction at the end of 5 years.

53. a. about 4194 larceny thefts   **b.** No; eventually the function would predict a negative number of larceny thefts, which is impossible.
55. a. 55.178 yr   **b.** No; since $a_n < 0$ and n is even, $H(x)$ eventually starts decreasing as x increases, which is impossible.
69. Answers may vary.   **71.** Answers may vary.
73.

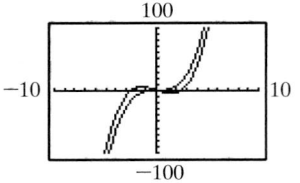

75.

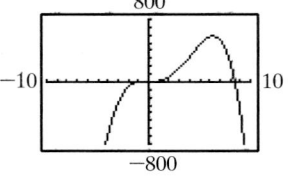

77.

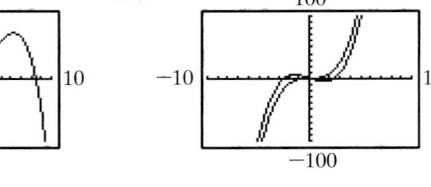

79. (c) is true.
81. Answers may vary.

Section 3.3

Check Point Exercises

1. $x + 5$ **2.** $2x^2 + 3x - 2 + \dfrac{1}{x - 3}$ **3.** $2x^2 + 7x + 14 + \dfrac{21x - 10}{x^2 - 2x}$ **4.** $x^2 - 2x - 3$ **5.** -105 **6.** $\left\{-1, -\dfrac{1}{3}, \dfrac{2}{5}\right\}$

Exercise Set 3.3

1. $x + 3$ **3.** $x^2 + 3x + 1$ **5.** $2x^2 + 3x + 5$ **7.** $4x + 3 + \dfrac{2}{3x - 2}$ **9.** $2x^2 + x + 6 - \dfrac{38}{x + 3}$

11. $4x^3 + 16x^2 + 60x + 246 + \dfrac{984}{x - 4}$ **13.** $2x + 5$ **15.** $6x^2 + 3x - 1 - \dfrac{3x - 1}{3x^2 + 1}$ **17.** $2x + 5$ **19.** $3x - 8 + \dfrac{20}{x + 5}$

21. $4x^2 + x + 4 + \dfrac{3}{x - 1}$ **23.** $6x^4 + 12x^3 + 22x^2 + 48x + 93 + \dfrac{187}{x - 2}$ **25.** $x^3 - 10x^2 + 51x - 260 + \dfrac{1300}{x + 5}$

27. $x^4 + x^3 + 2x^2 + 2x + 2$ **29.** $x^3 + 4x^2 + 16x + 64$ **31.** $2x^4 - 7x^3 + 15x^2 - 31x + 64 - \dfrac{129}{x + 2}$

33. -25 **35.** 4729 **37.** $x^2 - 5x + 6; x = -1, x = 2, x = 3$ **39.** $\left\{-\dfrac{1}{2}, 1, 2\right\}$ **41.** $\left\{-\dfrac{3}{2}, -\dfrac{1}{3}, \dfrac{1}{2}\right\}$ **43.** $x^3 + 5x^2 - 9x - 45$

45. a. $1 + \dfrac{5}{x + 20}$ **b.**

x	0	5	10	25	50	75
$\dfrac{x + 25}{x + 20}$	1.25	1.2	≈ 1.17	≈ 1.11	≈ 1.07	≈ 1.05

c. The ratio decreases as x increases, approaching 1 as x approaches ∞. When long division is used, the result is $1 + \dfrac{5}{x + 25}$, which shows that as x gets larger the second term gets smaller, and therefore, the ratio decreases.

55.

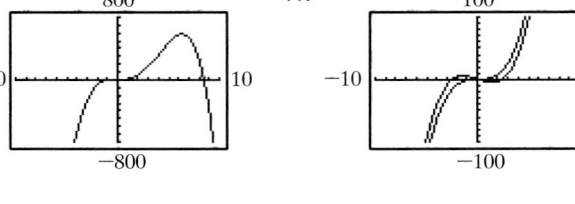

The division is not correct.

57.

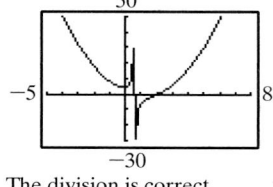

The division is correct.

59. $k = -12$ **61.** $x^{2n} - x^n + 1$

Section 3.4

Check Point Exercises

1. $\pm 1, \pm 2, \pm 3, \pm 6$ **2.** $\pm 1, \pm 3, \pm \dfrac{1}{2}, \pm \dfrac{1}{4}, \pm \dfrac{3}{2}, \pm \dfrac{3}{4}$ **3.** $\{-5, -4, 1\}$ **4.** $\{1, 2 - 3i, 2 + 3i\}$
5. 4, 2, or 0 positive zeros, no possible negative zeros

Exercise Set 3.4

1. $\pm 1, \pm 2, \pm 4$ **3.** $\pm 1, \pm 2, \pm 3, \pm 6, \pm \dfrac{1}{3}, \pm \dfrac{2}{3}$ **5.** $\pm 1, \pm 2, \pm 3, \pm 6, \pm \dfrac{1}{2}, \pm \dfrac{1}{4}, \pm \dfrac{3}{2}, \pm \dfrac{3}{4}$ **7.** $\pm 1, \pm 2, \pm 3, \pm 4, \pm 6, \pm 12$ **9. a.** $\pm 1, \pm 2, \pm 4$

b. 2 is a zero   **c.** $\{2, -2, -1\}$ **11. a.** $\pm 1, \pm 2, \pm 3, \pm 6, \pm \dfrac{1}{2}, \pm \dfrac{3}{2}$ **b.** 3 is a zero   **c.** $\left\{3, \dfrac{1}{2}, -2\right\}$

13. a. $\pm 1, \pm 2, \pm 4, \pm 8, \pm \dfrac{1}{3}, \pm \dfrac{2}{3}, \pm \dfrac{4}{3}, \pm \dfrac{8}{3}$ **b.** 2 is a zero   **c.** $\left\{2, -\dfrac{1}{3}, -4\right\}$ **15. a.** $\pm 1, \pm 2, \pm 3, \pm 4, \pm 6, \pm 12$ **b.** 4 is a root

c. $\{-3, 1, 4\}$ **17. a.** $\pm 1, \pm 2, \pm 3, \pm 4, \pm 6, \pm 12$ **b.** -2 is a root   **c.** $\{-2, 1 + \sqrt{7}, 1 - \sqrt{7}\}$

19. a. $\pm1, \pm5, \pm\dfrac{1}{2}, \pm\dfrac{5}{2}, \pm\dfrac{1}{3}, \pm\dfrac{5}{3}, \pm\dfrac{1}{6}, \pm\dfrac{5}{6}$ **b.** -5 is a root   **c.** $\left\{-5, \dfrac{1}{2}, \dfrac{1}{3}\right\}$ **21. a.** $\pm1, \pm2, \pm4$ **b.** 2 is a root

c. $\{-2, 2, 1 + \sqrt{2}, 1 - \sqrt{2}\}$ **23.** no positive real roots; 3 or 1 negative real roots   **25.** 3 or 1 positive real roots; no negative real roots

27. 2 or 0 positive real roots; 2 or 0 negative real roots   **29.** $x = -2, x = 5, x = 1$ **31.** $\left\{-\dfrac{1}{2}, \dfrac{1 + \sqrt{17}}{2}, \dfrac{1 - \sqrt{17}}{2}\right\}$

33. $\{-1, -2, 3 + \sqrt{13}, 3 - \sqrt{13}\}$ **35.** $x = -1, x = 2, x = -\dfrac{1}{3}, x = 3$ **37.** $\left\{1, -\dfrac{3}{4}, i\sqrt{2}, -i\sqrt{2}\right\}$ **39.** $\left\{-2, \dfrac{1}{2}, \sqrt{2}, -\sqrt{2}\right\}$

41. a. $x = 3, x \approx 4.2$ **b.** degree: 4; leading coefficient: negative   **43.** $W = 3$ mm   **45.** 2 in. by 9 in. by 4 in.

55. $\dfrac{1}{2}, \dfrac{2}{3}, 2$ **57.** $\pm\dfrac{1}{2}$ **59.** 5, 3, or 1 positive real roots exist   **61.** (d) is true.   **63.** 3 in.

Section 3.5

Check Point Exercises

1.

2	2	11	−7	−6
		4	30	46
	2	15	23	40

All the numbers are nonnegative.

−7	2	11	−7	−6
		−14	21	−98
	2	−3	14	−104

The signs alternate.

2. $f(-3) = -42; f(-2) = 5$

3. $\{-3, 7, 2 + i, 2 - i\}$

4. a. $(x^2 - 5)(x^2 + 1)$

   b. $(x + \sqrt{5})(x - \sqrt{5})(x^2 + 1)$

   c. $(x + \sqrt{5})(x - \sqrt{5})(x + i)(x - i)$

5. $f(x) = x^3 + 3x^2 + x + 3$

Exercise Set 3.5

1.

−4	1	−5	11	33	−18
		−4	36	−188	620
	1	−9	47	−155	602

Since signs alternate, −4 is a lower bound.

7	1	−5	11	33	−18
		7	14	175	1456
	1	2	25	208	1438

Since no sign is negative, 7 is an upper bound.

3.

−4	2	5	−8	7
		−8	12	−16
	2	−3	4	−9

Since signs alternate, −4 is a lower bound.

2	2	5	−8	7
		4	18	20
	2	9	10	27

Since no sign is negative, 2 is an upper bound.

5. a. $\pm1, \pm2, \pm3, \pm4, \pm6, \pm12$ **b.** 1 is not a root. 1 is an upper bound.   **c.** Eliminate all positive possible rational roots.
d. -3 is not a root. -3 is a lower bound.   **e.** Eliminate $-3, -4, -6$ and -12.   **7.** $f(1) = -1; f(2) = 5; 1.3$
9. $f(-1) = -1; f(0) = 1; -0.5$ **11.** $f(-3) = -11; f(-2) = 1; -2.1$ **13.** $f(-3) = -42; f(-2) = 5; -2.2$ **15.** $\{-2i, 2i, 2\}$

17. $\left\{1 - i, 1 + i, \dfrac{1}{3}\right\}$ **19.** $\{2 - i, 2 + i, -2 + i, -2 - i\}$ **21.** $\{2 - i, 2 + i, -3, 7\}$ **23. a.** $(x^2 - 5)(x^2 + 4)$

b. $(x + \sqrt{5})(x - \sqrt{5})(x^2 + 4)$ **c.** $(x + \sqrt{5})(x - \sqrt{5})(x + 2i)(x - 2i)$ **25. a.** $(x^2 - 2)(x^2 + 3)$
b. $(x + \sqrt{2})(x - \sqrt{2})(x^2 + 3)$ **c.** $(x + \sqrt{2})(x - \sqrt{2})(x + i\sqrt{3})(x - i\sqrt{3})$ **27. a.** $(x - 3)(x + 1)(x^2 + 4)$
b. $(x - 3)(x + 1)(x^2 + 4)$ **c.** $(x - 3)(x + 1)(x + 2i)(x - 2i)$ **29.** $f(x) = 2x^3 - 2x^2 + 50x - 50$
31. $f(x) = x^3 - 3x^2 - 15x + 125$ **33.** $f(x) = x^4 + 10x^2 + 9$ **35.** $f(x) = x^4 - 9x^3 + 21x^2 + 21x - 130$
37. $x = 1; x = \pm5i; f(x) = (x - 1)(x - 5i)(x + 5i)$ **39.** $x = 2; x = 3 \pm 2i; f(x) = (x - 2)(x - 3 + 2i)(x - 3 - 2i)$

41. $x = \pm6i; x = \pm i; f(x) = (x - 6i)(x + 6i)(x - i)(x + i)$ **43.** $x = -2; x = \dfrac{3}{4}; x = -\dfrac{1}{2} \pm i;$

$f(x) = (x + 2)(4x - 3)(2x + 1 - 2i)(2x + 1 + 2i)$ **45.** about 2.5 yr   **47.** Answers may vary.
49. According to the linear model, spending will reach \$458 billion, 10.93 or almost 11 years after 1995; According to the quadratic model, spending will reach \$458 billion 10.12 years after 1995; There is a solution between 10.1 and 10.2; According to the third-degree polynomial model, spending will reach \$458 billion just over 10 years after 1995; The third degree polynomial is best.

55.

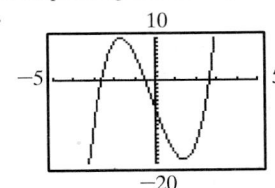

−3 is a lower bound;
3 is an upper bound

57. a. As x, a person's age, increases, y, the number of visits, increases.
b. 60
c.

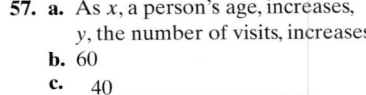
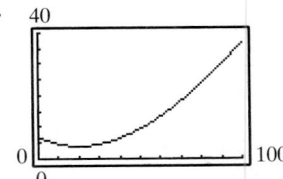

59.

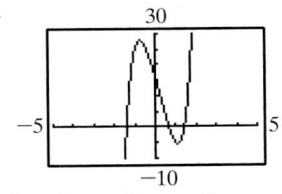

3 real zeros, 2 nonreal complex zeros

61.

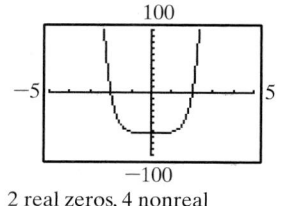

2 real zeros, 4 nonreal
complex zeros

63. 3
65. Answers may vary.
67. Answers may vary.

Section 3.6

Check Point Exercises

1. a. $\{x|x \neq 5\}$ **b.** $\{x \mid x \neq -5, x \neq 5\}$ **c.** all real numbers   **2. a.** $x = 1, x = -1$ **b.** $x = -1$ **c.** none
3. a. $y = 3$ **b.** $y = 0$ **c.** none
4.

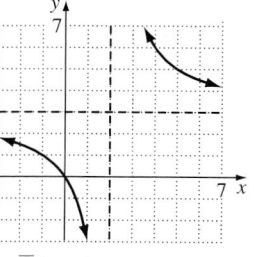

5.

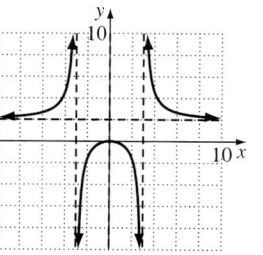

6.

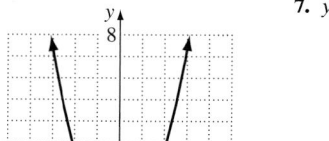

7. $y = 2x - 1$

8. a. $\overline{C}(1000) = 330$, when 1000 pairs of shoes are produced, it costs \$330 to produce each pair; $\overline{C}(10,000) = 60$, when 10,000 pairs of shoes are produced, it costs \$60 to produce each pair; $\overline{C}(100,000) = 33$, when 100,000 pairs of shoes are produced, it costs \$33 to produce each pair.
   b. $y = 30$; The cost per pair of shoes approaches \$30 as more shoes are produced.

Exercise Set 3.6

1. $\{x|x \neq 4\}$ **3.** $\{x|x \neq 5, x \neq -4\}$ **5.** $\{x|x \neq 7, x \neq -7\}$ **7.** All real numbers   **9.** $-\infty$ **11.** $-\infty$ **13.** 0   **15.** $+\infty$
17. $-\infty$ **19.** 1   **21.** $x = -4$ **23.** $x = 0, x = -4$ **25.** $x = -4$ **27.** no vertical asymptotes   **29.** $y = 0$ **31.** $y = 4$

33. no horizontal asymptote   **35.** $y = -\dfrac{2}{3}$

37.

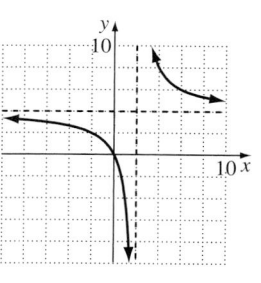

39.

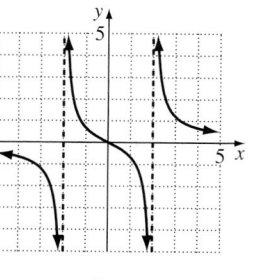

41.

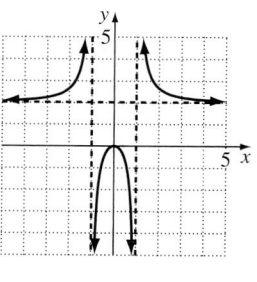

43.

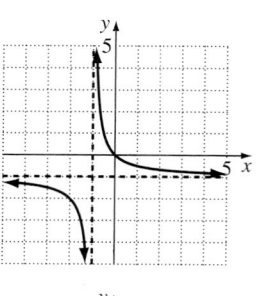

45.

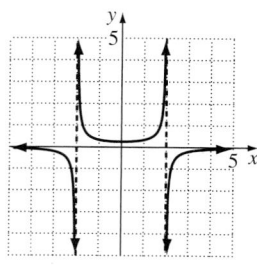

47.

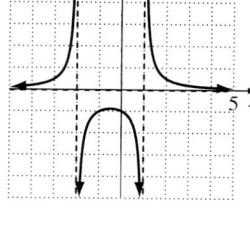

49.

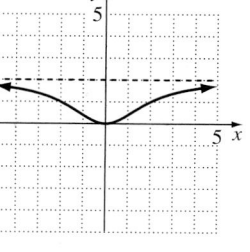

51.

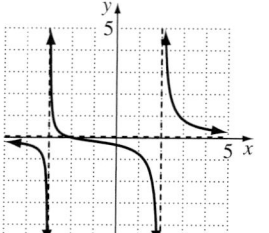

53.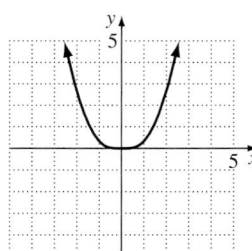

55.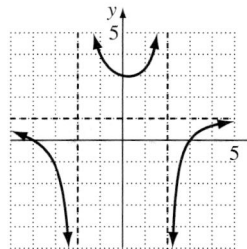

57.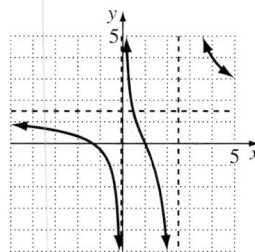

59. a. Slant asymptote: $y = x$
b.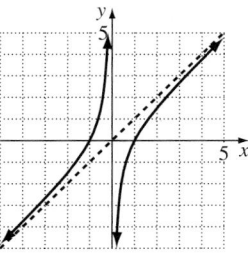

61. a. Slant asymptote: $y = x$
b.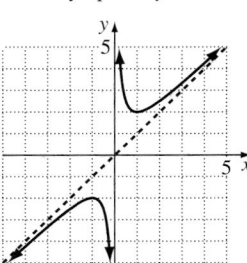

63. a. Slant asymptote: $y = x + 4$
b.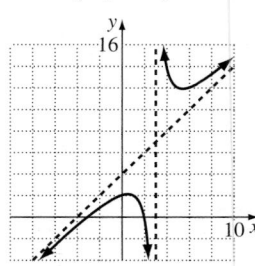

65. a. Slant asymptote: $y = x - 2$
b.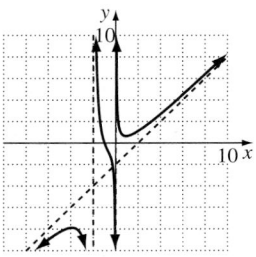

67. a. $\overline{C}(100) = \$220; \overline{C}(1000) = \$40; \overline{C}(10,000) = \$22; \overline{C}(100,000) = \$20.2$
   b. $y = 20;$ \$20 is the minimum average cost of producing a canoe. As more canoes are manufactured, the average cost approaches \$20.

69. a. \$100,000
   b. No; the model indicates that no amount of money can remove 100% of the pollutants since $C(p)$ increases without bound as p approaches 100.

71. a. $F(0) = 80;$ When the dessert is placed in the icebox, its temperature is 80°F.
   b. $F(1) = 13.3°F; F(2) = 6.2°F;$ $F(3) \approx 3.6°F; F(4) \approx 2.4°F;$ $F(5) \approx 1.7°F$
   c. $y = 0;$ The temperature will approach but not reach 0°F.
   d.

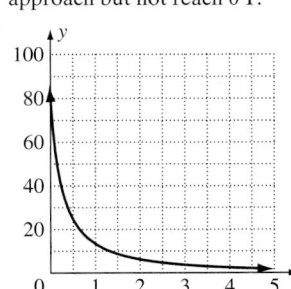

73. a. After 1 day: 35 words; after 5 days: about 12 words; after 15 days: about 7 words
   b. $N(1) = 35$ words; This is the same as the estimate from the graph.; $N(5) = 11$ words; This is a little less than the estimate from the graph.; $N(15) = 7$ words; This is the same as the estimate from the graph.
   c. The graph indicates the students will remember 5 words over a long period of time.
   d. $y = 5;$ The horizontal asymptote indicates the students will remember 5 words over a long period of time.

85.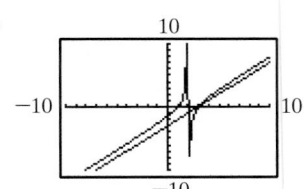

87.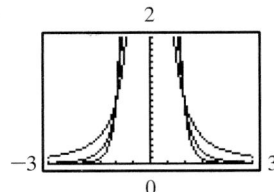

The graph approaches the horizontal asymptote faster and the vertical asymptote slower as n increases.

89.

$g(x)$ is the graph of a line whereas $f(x)$ is the graph of a rational function with a slant asymptote; In $g(x), x - 2$ is a factor of $x^2 - 5x + 6.$

91. (d) is true.
93. Answers may vary.
95. Answers may vary.

Section 3.7

Check Point Exercises

1. a. $L = kN$ **b.** $L = 4N$ **c.** 68 in.   **2. a.** $W = kL$ **b.** $k = \dfrac{75}{6}$ **c.** $W = \dfrac{75L}{6}$ **d.** 200 lb   **3.** 137.5 lb/in^2
4. about 556 ft   **5.** about 4.36 lb/in^2 **6.** 24 min   **7.** 96π ft^3

Exercise Set 3.7

1. $g = kh$ **3.** $a = kb^2$ **5.** $r = \dfrac{k}{t}$ **7.** $a = \dfrac{k}{b^3}$ **9.** $r = \dfrac{ks}{v}$ **11.** $s = kgt^2$ **13.** $k = 25$ **15.** $k = 5$ **17.** $k = 5000$

19. $k = 3$ **21.** $k = 2$ **23.** 84   **25.** 25   **27.** $\dfrac{5}{6}$ **29.** 240   **31. a.** $L = kW$ **b.** $L = 0.02W$ **c.** 1.04 in.   **33.** $60

35. 2442 mph   **37.** 607 lb   **39.** 0.5 hr   **41.** 6.4 lb   **43.** 31.78; index: about 32; not in the desirable range   **45.** 11.11 foot-candles
47. 72 erg   **49.** The average number of phone calls is about 126.   **59.** The destructive power is four times as much.
61. Reduce the resistance by a factor of $\dfrac{1}{3}$.

Chapter 3 Review Exercises

1.
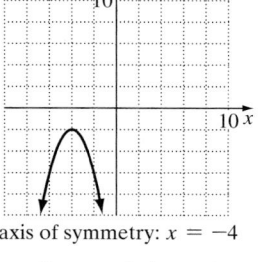
axis of symmetry: $x = 1$

2.
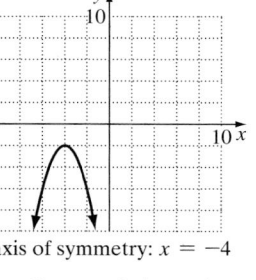
axis of symmetry: $x = -4$

3.
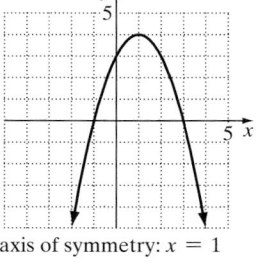
axis of symmetry: $x = 1$

4.
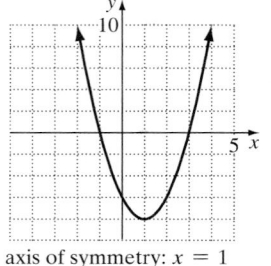
axis of symmetry: $x = 1$

5. 7.2 hr; 622 deaths   **6.** 2 sec; 144 ft   **7. c.**   **8. b.**   **9. a.**   **10. d.**   **11.** Because the degree is odd and the leading coefficient is negative, the graph falls to the right. Therefore, the model indicates that the percentage of families below the poverty level will eventually be negative, which is impossible.   **12.** Since the degree is even and the leading coefficient is negative, the graph falls to the right. Therefore, the model indicates a patient will eventually have a negative number of viral bodies, which is impossible.
13. $x = 1$, multiplicity 1, crosses; $x = -2$, multiplicity 2, touches; $x = -5$, multiplicity 3, crosses
14. $x = -5$, multiplicity 1, crosses; $x = 5$, multiplicity 2, touches

15. a. The graph falls to the left and rises to the right.
b. no symmetry
c.
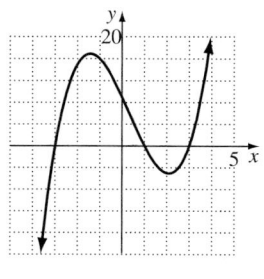

16. a. The graph rises to the left and falls to the right.
b. origin symmetry
c.
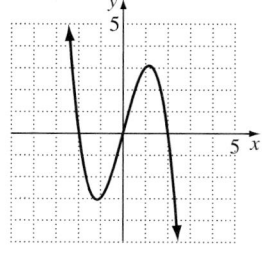

17. a. The graph falls to the left and rises to the right.
b. no symmetry
c.
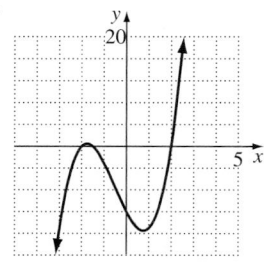

18. a. The graph falls to the left and to the right.
b. y-axis symmetry
c.

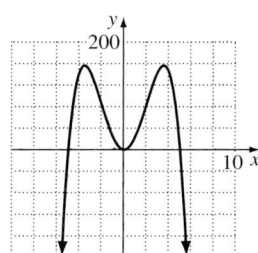

19. a. The graph falls to the left and to the right.
b. no symmetry
c.

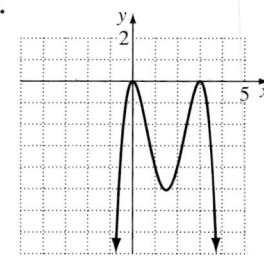

20. a. The graph rises to the left and to the right.
b. no symmetry
c.

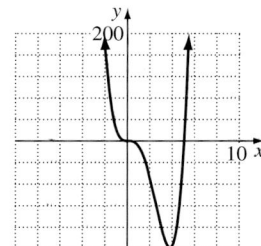

21. $4x^2 - 7x + 5 - \dfrac{4}{x+1}$ **22.** $2x^2 - 4x + 1 - \dfrac{10}{5x-3}$ **23.** $2x^2 + 3x - 1$ **24.** $3x^3 - 4x^2 + 7$

25. $3x^3 + 6x^2 + 10x + 10 + \dfrac{20}{x-2}$ **26.** -5697 **27.** $2, \dfrac{1}{2}, -3$ **28.** $\{4, -2 \pm \sqrt{5}\}$ **29.** $\pm 1, \pm 5$

30. $\pm 1, \pm 2, \pm 4, \pm 8, \pm \dfrac{8}{3}, \pm \dfrac{4}{3}, \pm \dfrac{2}{3}, \pm \dfrac{1}{3}$ **31.** 2 or 0 positive solutions; no negative solutions

32. 3 or 1 positive real roots; 2 or 0 negative solutions   **33.** No sign variations exist for either $f(x)$ or $f(-x)$, so no real roots exist.
34. a. $\pm 1, \pm 2, \pm 4$ **b.** 1 positive real zero; 2 or no negative real zeros   **c.** 1 is a zero   **d.** $\{1, -2\}$

35. a. $\pm 1, \pm \dfrac{1}{2}, \pm \dfrac{1}{3}, \pm \dfrac{1}{6}$ **b.** 2 or 0 positive real zeros; 1 negative real zero   **c.** -1 is a zero   **d.** $\left\{-1, \dfrac{1}{3}, \dfrac{1}{2}\right\}$

36. a. $\pm 1, \pm 3, \pm 5, \pm 15, \pm \dfrac{1}{2}, \pm \dfrac{1}{4}, \pm \dfrac{1}{8}, \pm \dfrac{3}{2}, \pm \dfrac{3}{4}, \pm \dfrac{3}{8}, \pm \dfrac{5}{2}, \pm \dfrac{5}{4}, \pm \dfrac{5}{8}, \pm \dfrac{15}{2}, \pm \dfrac{15}{4}, \pm \dfrac{15}{8}$

b. 3 or 1 positive real solutions; no negative real solutions   **c.** $\dfrac{1}{2}$ is a zero   **d.** $\left\{\dfrac{1}{2}, \dfrac{3}{2}, \dfrac{5}{2}\right\}$

37. a. $\pm 1, \pm 2, \pm 3, \pm 6$ **b.** 2 or zero positive real solutions; 2 or zero negative real solutions   **c.** -2 is a zero   **d.** $\{-2, -1, 1, 3\}$

38. a. $\pm 1, \pm 2, \pm \dfrac{1}{2}, \pm \dfrac{1}{4}$ **b.** 1 positive real root; 1 negative real root   **c.** $\dfrac{1}{2}$ is a zero   **d.** $\left\{\dfrac{1}{2}, -\dfrac{1}{2}, i\sqrt{2}, -i\sqrt{2}\right\}$

39. a. $\pm 1, \pm 2, \pm 4, \pm \dfrac{1}{2}$ **b.** 2 or no positive zeros; 2 or no negative zeros   **c.** $x = 2$ is a zero   **d.** $\left\{2, -2, \dfrac{1}{2}, -1\right\}$

40.

-2	2	-7	-5	28	-12
		-4	22	-34	12
	2	-11	17	-6	0

-2 is a root and a lower bound.

6	2	-7	-5	28	-12
		12	30	150	1068
	2	5	25	178	1056

$\pm 1, \pm 2, 3, 4, \pm \dfrac{1}{2}, \pm \dfrac{3}{2}$

6 is an upper bound, but not a zero.

41. a. $\pm 1, \pm 2, \pm 3, \pm 4, \pm 6, \pm 12, \pm \dfrac{1}{2}, \pm \dfrac{3}{2}$ **b.** 2 is not a root but is an upper bound.   **c.** -2 is not a root but is a lower bound.

d. Possible roots are $\pm 1, \pm \dfrac{1}{2}$, and $\pm \dfrac{3}{2}$.   **42.** $f(1) = -2; f(2) = 3; x \approx 1.6$ **43.** $f(-3) = -32; f(-2) = 7; x \approx -2.3$

44. $\left\{-\dfrac{1}{4}, 6 \pm 5i\right\}$ **45.** $\{1 \pm 3i, 1 \pm i\}$ **46.** $\left\{-\dfrac{1}{2}, 1, 4 \pm 7i\right\}$ **47.** $f(x) = x^3 - 6x^2 + 21x - 26$

48. $f(x) = 2x^4 + 12x^3 + 20x^2 + 12x + 18$ **49.** $f(x) = x^4 - 3x^3 + 6x^2 + 2x - 60$

50. $-2, \dfrac{1}{2}, \pm i; f(x) = (x - i)(x + i)(x + 2)\left(x - \dfrac{1}{2}\right)$ **51.** $-1, 4; g(x) = (x + 1)^2(x - 4)^2$

52. 4 real zeros, one with multiplicity two   **53.** 3 real zeros; 2 nonreal complex zeros
54. 2 real zeros, one with multiplicity two; 2 nonreal complex zeros   **55.** 1 real zero; 4 nonreal complex zeros

CHAPTER 4

Section 4.1

Check Point Exercises

1. 1 O-ring

2.
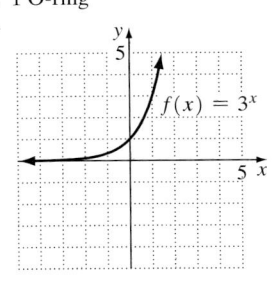

3.
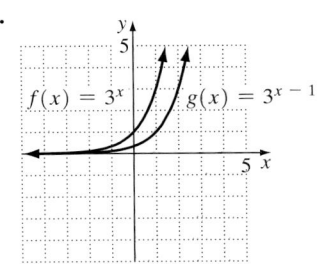

4.
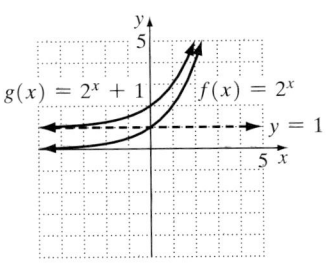

5. 11.64 billion
6. a. $14,859.47
   b. $14,918.25

Exercise Set 4.1

1. 10.556   **3.** 11.665   **5.** 0.125   **7.** 9.974   **9.** 0.387

11.
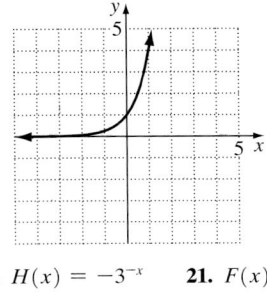

13.
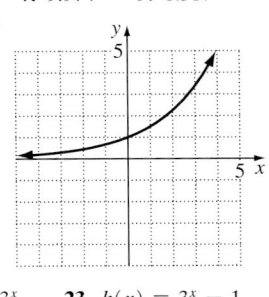

15.
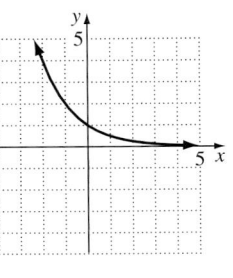

17.
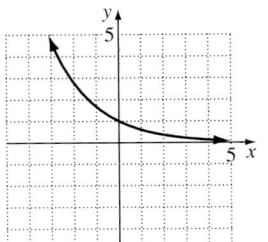

19. $H(x) = -3^{-x}$ **21.** $F(x) = -3^x$ **23.** $h(x) = 3^x - 1$

25.
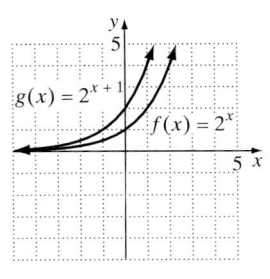

27.
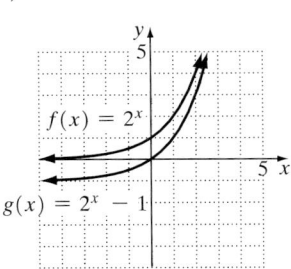

29.
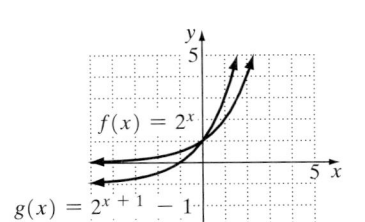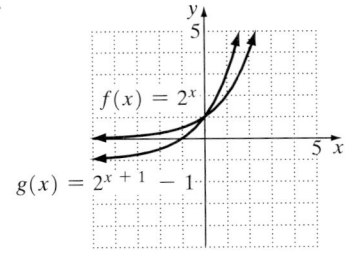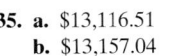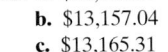

31.
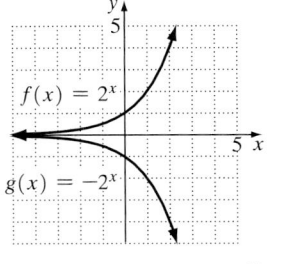

33.
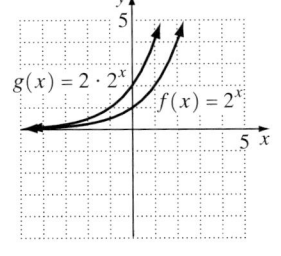

35. a. $13,116.51
   b. $13,157.04
   c. $13,165.31
37. 7% compounded monthly
39. a. 67.38 million
   b. about 134.74 million
   c. about 269.46 million
   d. 538.85 million
   e. appears to double every 27 yr

41. $f(25) \approx 0.0653$; About 6.5% of 25-year-olds have some coronary heart disease.   **43.** $116,405.10   **45.** 3.249009585; 3.317278183; 3.321880096; 3.321995226; 3.321997068; $2^{\sqrt{3}} \approx 3.321997085$; The closer the exponent is to $\sqrt{3}$, the closer the value is to $2^{\sqrt{3}}$.
47. $f(11) \approx 241,786.19$; The number of AIDS cases among IV drug users in 2000 will be about 241,786.   **49. a.** 100%   **b.** 68.5%
c. 30.8%   **d.** 20%   **51. a.** 1429   **b.** 24,546   **c.** Growth is limited by the population; The entire population will eventually become ill.

59.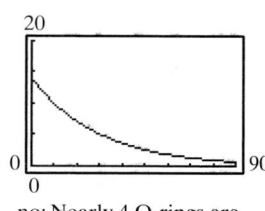

no; Nearly 4 O-rings are expected to fail.

61. a. $A = 10,000\left(1 + \dfrac{0.05}{4}\right)^{4t}$; $A = 10,000\left(1 + \dfrac{0.045}{12}\right)^{12t}$

b.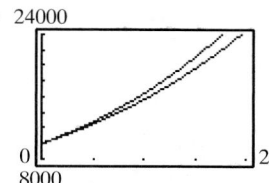

5% interest compounded quarterly

63. (d) is true.

65.
$$\left(\dfrac{e^x + e^{-x}}{2}\right)^2 - \left(\dfrac{e^x - e^{-x}}{2}\right)^2 \overset{?}{=} 1$$
$$\dfrac{e^{2x} + 2 + e^{-2x}}{4} - \dfrac{e^{2x} - 2 + e^{-2x}}{4} \overset{?}{=} 1$$
$$\dfrac{e^{2x} + 2 + e^{-2x} - e^{2x} + 2 - e^{-2x}}{4} \overset{?}{=} 1$$
$$\dfrac{4}{4} \overset{?}{=} 1$$
$$1 = 1$$

Section 4.2

Check Point Exercises

1. a. $7^3 = x$ **b.** $b^2 = 25$ **c.** $4^y = 26$ **2. a.** $5 = \log_2 x$ **b.** $3 = \log_b 27$ **c.** $y = \log_e 33$ **3. a.** 2   **b.** 1   **c.** $\dfrac{1}{2}$

4. a. 1   **b.** 0   **5. a.** 8   **b.** 17   **6.** 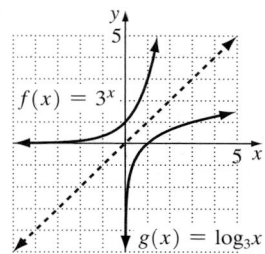   **7.** $(5, \infty)$ **8.** 80%   **9.** 4.0

10. a. $(-\infty, 4)$
b. $(-\infty, 0)$ or $(0, \infty)$
11. a. $25x$
b. $\sqrt{x}$
12. 4.6 ft per sec

Exercise Set 4.2

1. $2^4 = 16$ **3.** $3^2 = x$ **5.** $b^5 = 32$ **7.** $6^y = 216$ **9.** $\log_2 8 = 3$ **11.** $\log_2 \dfrac{1}{16} = -4$ **13.** $\log_8 2 = \dfrac{1}{3}$ **15.** $\log_{13} x = 2$

17. $\log_b 1000 = 3$ **19.** $\log_7 200 = y$ **21.** 2   **23.** 6   **25.** $\dfrac{1}{2}$ **27.** -3 **29.** $\dfrac{1}{2}$ **31.** 1   **33.** 0   **35.** 7   **37.** 19

39. 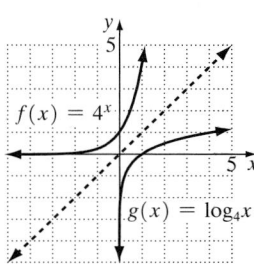   **41.** 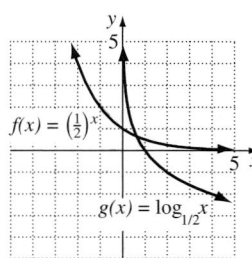   **43.** $H(x) = 1 - \log_3 x$
45. $h(x) = \log_3 x - 1$
47. $g(x) = \log_3(x - 1)$

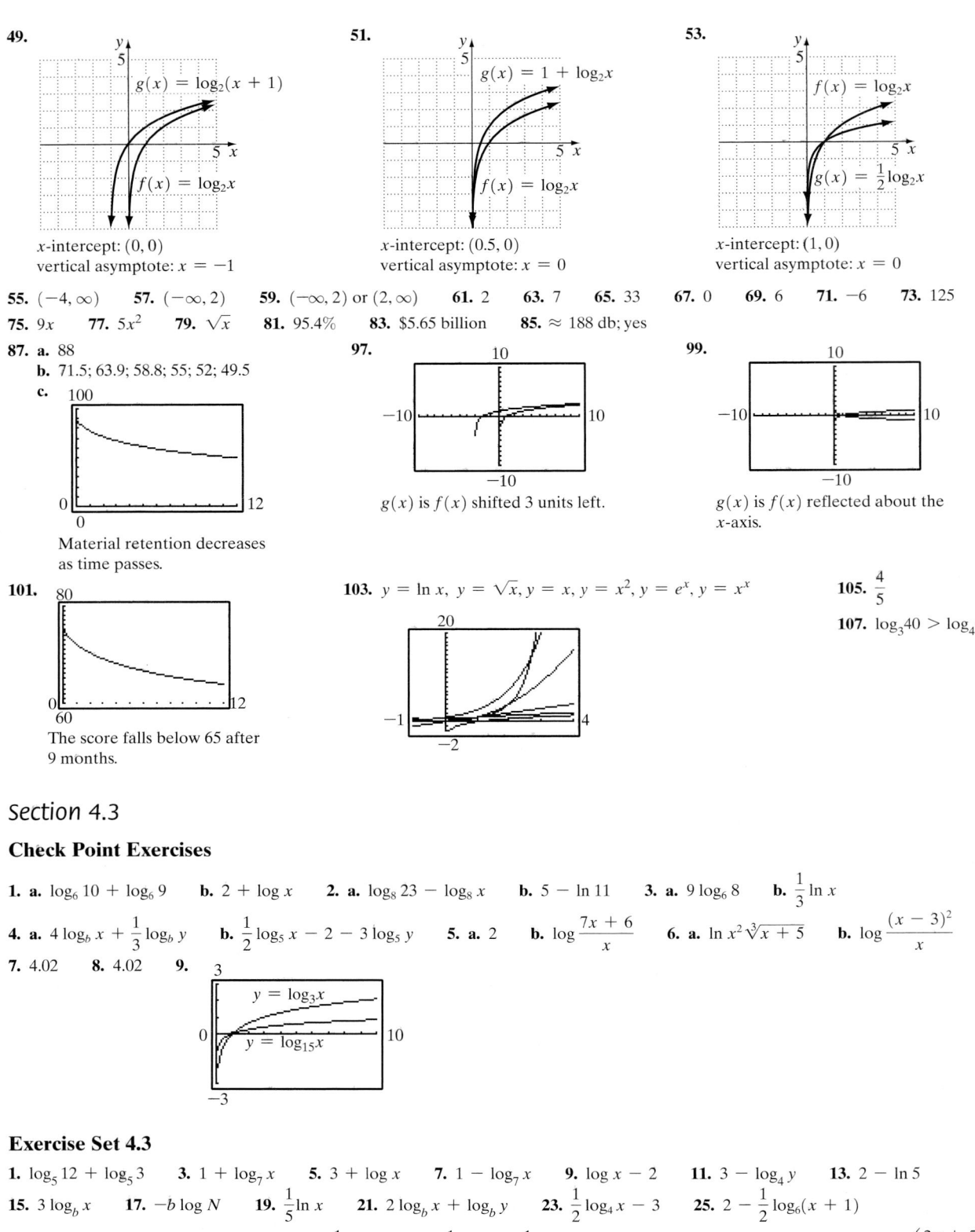

49.

$g(x) = \log_2(x + 1)$

$f(x) = \log_2 x$

x-intercept: $(0, 0)$
vertical asymptote: $x = -1$

51.

$g(x) = 1 + \log_2 x$

$f(x) = \log_2 x$

x-intercept: $(0.5, 0)$
vertical asymptote: $x = 0$

53.

$f(x) = \log_2 x$

$g(x) = \frac{1}{2}\log_2 x$

x-intercept: $(1, 0)$
vertical asymptote: $x = 0$

55. $(-4, \infty)$ **57.** $(-\infty, 2)$ **59.** $(-\infty, 2)$ or $(2, \infty)$ **61.** 2 **63.** 7 **65.** 33 **67.** 0 **69.** 6 **71.** -6 **73.** 125
75. $9x$ **77.** $5x^2$ **79.** $\sqrt{x}$ **81.** 95.4% **83.** \$5.65 billion **85.** ≈ 188 db; yes

87. a. 88
b. 71.5; 63.9; 58.8; 55; 52; 49.5
c. 100

Material retention decreases
as time passes.

97. 10

g(x) is f(x) shifted 3 units left.

99. 10

g(x) is f(x) reflected about the
x-axis.

101. 80

The score falls below 65 after
9 months.

103. $y = \ln x$, $y = \sqrt{x}$, $y = x$, $y = x^2$, $y = e^x$, $y = x^x$

105. $\frac{4}{5}$

107. $\log_3 40 > \log_4 60$

Section 4.3

Check Point Exercises

1. a. $\log_6 10 + \log_6 9$ **b.** $2 + \log x$ **2. a.** $\log_8 23 - \log_8 x$ **b.** $5 - \ln 11$ **3. a.** $9 \log_6 8$ **b.** $\frac{1}{3} \ln x$
4. a. $4 \log_b x + \frac{1}{3} \log_b y$ **b.** $\frac{1}{2} \log_5 x - 2 - 3 \log_5 y$ **5. a.** 2 **b.** $\log \dfrac{7x + 6}{x}$ **6. a.** $\ln x^2 \sqrt[3]{x + 5}$ **b.** $\log \dfrac{(x - 3)^2}{x}$
7. 4.02 **8.** 4.02 **9.** 3

$y = \log_3 x$

$y = \log_{15} x$

Exercise Set 4.3

1. $\log_5 12 + \log_5 3$ **3.** $1 + \log_7 x$ **5.** $3 + \log x$ **7.** $1 - \log_7 x$ **9.** $\log x - 2$ **11.** $3 - \log_4 y$ **13.** $2 - \ln 5$
15. $3 \log_b x$ **17.** $-b \log N$ **19.** $\frac{1}{5} \ln x$ **21.** $2 \log_b x + \log_b y$ **23.** $\frac{1}{2} \log_4 x - 3$ **25.** $2 - \frac{1}{2} \log_6(x + 1)$
27. $2 \log_b x + \log_b y - 2 \log_b z$ **29.** $1 + \frac{1}{2} \log x$ **31.** $\frac{1}{3} \log x - \frac{1}{3} \log y$ **33.** 1 **35.** $\ln(7x)$ **37.** 5 **39.** $\log\left(\dfrac{2x + 5}{x}\right)$

41. $\log(xy^3)$ **43.** $\ln(x^{1/2}y)$ or $\ln(y\sqrt{x})$ **45.** $\log_b(x^2y^3)$ **47.** $\ln\left(\dfrac{x^5}{y^2}\right)$ **49.** $\ln\left(\dfrac{x^3}{y^{1/3}}\right)$ or $\ln\left(\dfrac{x^3}{\sqrt[3]{y}}\right)$ **51.** $\ln\dfrac{(x+6)^4}{x^3}$

53. 1.5937   **55.** 1.6944   **57.** -1.2304 **59.** 3.6193   **61. a.** $D = 10\log\dfrac{I}{I_0}$ **b.** 20 decibels louder

71. a.

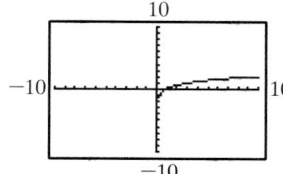

b.

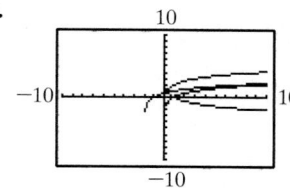

$y = 2 + \log_3 x$ shifts the graph of $y = \log_3 x$ two units upward; $y = \log_3(x + 2)$ shifts the graph of $y = \log_3 x$ two units left; $y = -\log_3 x$ reflects the graph of $y = \log_3 x$ about the x-axis.

73.

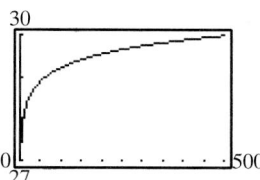

a. top graph: $y = \log_{100} x$;
bottom graph: $y = \log_3 x$
b. top graph: $y = \log_3 x$;
bottom graph: $y = 100\log_{100} x$
c. The graph of the equation with the largest b will be on the top in the interval $(0, 1)$ and on the bottom in the interval $(1, \infty)$.

79. (d) is true.   **81.** $\dfrac{2A}{B}$

Section 4.4

Check Point Exercises

1. $\left\{\dfrac{\ln 134}{\ln 5}\right\}; \approx 3.04$ **2.** $\left\{\dfrac{\ln 9}{2}\right\}; \approx 1.10$ **3.** $\left\{\dfrac{\ln 2088 + 4\ln 6}{3\ln 6}\right\}; \approx 2.76$ **4.** $(0, \ln 7); \ln 7 \approx 1.95$ **5.** $\{12\}$ **6.** $\{5\}$

7. $\left\{\dfrac{e^2}{3}\right\}$ **8.** 0.01   **9.** 16.2 yr   **10.** 2149

Exercise Set 4.4

1. $\left\{\dfrac{\ln 3.91}{\ln 10}\right\}; \approx 0.59$ **3.** $\{\ln 5.7\}; \approx 1.74$ **5.** $\left\{\dfrac{\ln 17}{\ln 5}\right\}; \approx 1.76$ **7.** $\left\{\ln\dfrac{23}{5}\right\}; \approx 1.53$ **9.** $\left\{\dfrac{\ln 659}{5}\right\}; \approx 1.30$

11. $\left\{\dfrac{\ln 793 - 1}{-5}\right\}; \approx -1.14$ **13.** $\left\{\dfrac{\ln 10{,}478 + 3}{5}\right\}; \approx 2.45$ **15.** $\left\{\dfrac{\ln 410}{\ln 7} - 2\right\}; \approx 1.09$ **17.** $\left\{\dfrac{\ln 813}{0.3\ln 7}\right\}; \approx 11.48$

19. $\{0, \ln 2\}; \ln 2 \approx 0.69$ **21.** $\left\{\dfrac{\ln 3}{2}\right\}; \approx 0.55$ **23.** $\{81\}$ **25.** $\{59\}$ **27.** $\left\{\dfrac{109}{27}\right\}$ **29.** $\left\{\dfrac{62}{3}\right\}$ **31.** $\left\{\dfrac{5}{4}\right\}$ **33.** $\{6\}$

35. $\{6\}$ **37.** $\{e^2\}; \approx 7.39$ **39.** $\left\{\dfrac{e^4}{2}\right\}; \approx 27.30$ **41.** $\{e^{-1/2}\}; \approx 0.61$ **43.** $\{e^2 - 3\}; \approx 4.39$ **45.** about 0.22

47. a. 18.2 million   **b.** 2010   **49.** 8 yr   **51.** 16.8%   **53.** 9 yr   **55.** 15.7%   **57.** 1995   **59.** 2.8 days; Yes, the point $(2.8, 50)$ appears to lie on the graph of P.   **61.** $10^{-2.4}$; 0.004 moles per liter   **67.** $\{2\}$ **69.** $\{4\}$ **71.** $\{2\}$ **73.** $\{-1.391606, 1.6855579\}$

75.

As distance from eye increases, barometric air pressure increases, leveling off at about 30 inches of mercury.

77.

about 7.92 min

79. (c) is true.
81. $\{1, e^2\}, e^2 \approx 7.389$
83. $\{e\}, e \approx 2.718$

Section 4.5

Check Point Exercises

1. a. $A = 491e^{0.027t}$ **b.** 2006   **2. a.** $A = A_0 e^{-0.0248t}$ **b.** about 72 yr   **3. a.** 0.4 correct responses   **b.** 0.7 correct responses
c. 0.8 correct responses   **4.** $y = 4e^{(\ln 7.8)x}; y = 4e^{2.054x}$

Exercise Set 4.5

1. 208 million   **3.** 2016   **5.** 2.6%   **7.** 2014   **9.** $140,000   **11.** 2005   **13.** 0.175; $A = 200e^{0.175t}$; 17.5%   **15.** 8.01 g

17. 8 g; 4 g; 2 g; 1 g; 0.5 g   **19.** 15,679 years old   **21. a.** $\dfrac{A_0}{2} = A_0 e^{k(1.31)}$; $\dfrac{1}{2} = e^{1.31k}$; $\ln\dfrac{1}{2} = \ln e^{1.31k}$; $\ln\dfrac{1}{2} = 1.31k$; $k = \dfrac{\ln\frac{1}{2}}{1.31} \approx -0.52912$

b. 107 million years   **23.** $2A_0 = A_0 e^{kt}$; $2 = e^{kt}$; $\ln 2 = \ln e^{kt}$; $\ln 2 = kt$; $t = \dfrac{\ln 2}{k}$ **25.** 63 yr   **27. a.** about 20 people

b. about 1080 people   **c.** 100,000 people   **29.** about 3.7%   **31.** about 48 years old   **33.** $y = 100e^{(\ln 4.6)x}$; $y = 100e^{1.526x}$
35. $y = 2.5e^{(\ln 0.7)x}$; $y = 2.5e^{-0.357x}$ **47.** $y = 51.75985638 + 109.7788574 \ln x$; $r = 0.8974781617$, a good fit
49. $y = 98.06189365x^{0.4398361087}$; $r = 0.9546621296$, a good fit

51.

The probability of coronary heart disease starts increasing at a more rapid rate at about age 20. At about age 60, the rate of increase starts to slow down.

53. The linear model, $y = 74.52833333x + 214.7694444$, best fits the data. Answers for prediction may vary.

55. about 126°F

Chapter 4 Review Exercises

1. $g(x) = 4^{-x}$ **2.** $h(x) = -4^{-x}$ **3.** $r(x) = -4^{-x} + 3$ **4.** $f(x) = 4^x$

5. 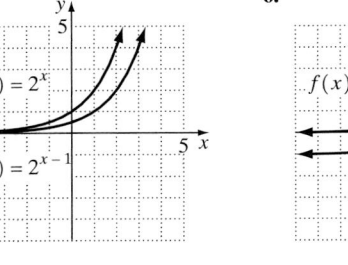   **6.** 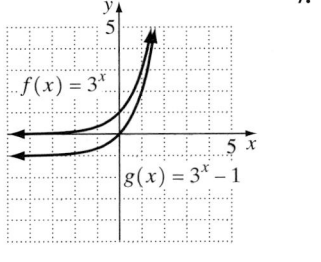   **7.** 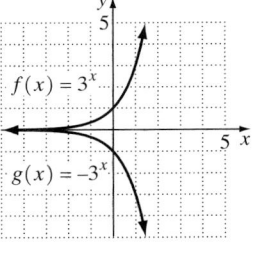   **8.**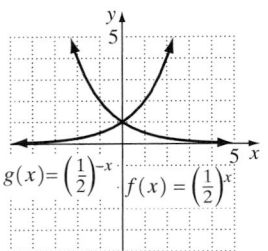

9. 5.5% compounded semiannually   **10.** 7% compounded monthly   **11. a.** 200°   **b.** 120°; 119°

c. 70°; The temperature in the room is 70°.   **12.** $49^{1/2} = 7$ **13.** $4^3 = x$ **14.** $3^y = 81$ **15.** $\log_6 216 = 3$ **16.** $\log_b 625 = 4$

17. $\log_{13} 874 = y$ **18.** 3   **19.** -2 **20.** $\varnothing$; $\log_b x$ is defined only for $x > 0$.   **21.** $\dfrac{1}{2}$ **22.** 1   **23.** 8

24. 5   **25.** 0   **26.** 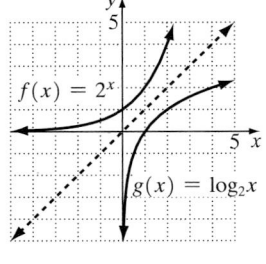   **27.** 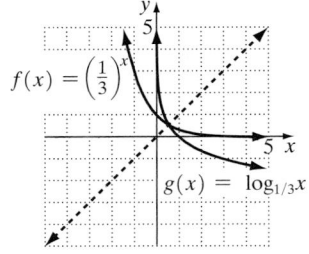   **28.** $g(x) = \log(-x)$
29. $r(x) = 1 + \log(2 - x)$
30. $h(x) = \log(2 - x)$
31. $f(x) = \log x$

32.

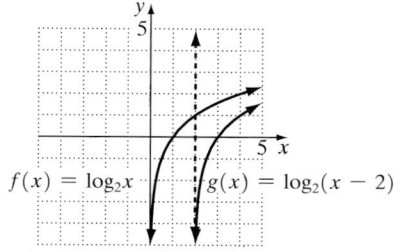

$f(x) = \log_2 x$ $g(x) = \log_2(x - 2)$

x-intercept: $(3, 0)$
vertical asymptote: $x = 2$

33.

$f(x) = \log_2 x$

$h(x) = -1 + \log_2 x$

x-intercept: $(2, 0)$
vertical asymptote: $x = 0$

34.

$r(x) = \log_2(-x)$ $f(x) = \log_2 x$

x-intercept: $(-1, 0)$
vertical asymptote: $x = 0$

35. $(-5, \infty)$ **36.** $(-\infty, 3)$ **37.** $(-\infty, 1) \cup (1, \infty)$ **38.** $6x$ **39.** $\sqrt{x}$ **40.** $4x^2$ **41.** 3.0

42. a. 76
b. $\approx 67, \approx 63, \approx 61, \approx 59, \approx 56$
c.

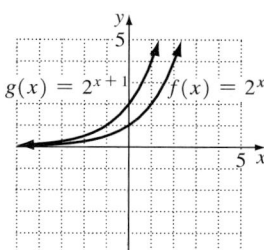

$f(t) = 76 - 18 \log(t + 1)$

time (months)
Retention decreases as time passes.

43. about 9 weeks   **44.** $2 + 3 \log_6 x$ **45.** $\frac{1}{2} \log_4 x - 3$

46. $\log_2 x + 2 \log_2 y - 6$ **47.** $\frac{1}{3} \ln x - \frac{1}{3}$ **48.** $\log_b 21$ **49.** $\log \frac{3}{x^3}$

50. $\ln(x^3 y^4)$ **51.** $\ln \frac{\sqrt{x}}{y}$ **52.** 6.2448   **53.** -0.1063

54. $\left\{ \frac{\ln 12{,}143}{\ln 8} \right\}$; ≈ 4.523 **55.** $\left\{ \frac{1}{5} \ln 141 \right\}$; ≈ 0.990

56. $\left\{ \frac{12 - \ln 130}{5} \right\}$; ≈ 1.426 **57.** $\left\{ \frac{\ln 37{,}500 - 2 \ln 5}{4 \ln 5} \right\}$; ≈ 1.136

58. $\{\ln 3\}$; ≈ 1.099 **59.** $\{23\}$ **60.** $\{5\}$ **61.** $\varnothing$ **62.** $\left\{ \frac{1}{e} \right\}$ or $\{0.368\}$

63. $\left\{ \frac{e^3}{2} \right\}$ or $\{10.043\}$ **64.** 2042   **65.** 2086   **66.** 2005   **67.** 7.3 yr

68. 14.6 yr   **69.** about 21.97%   **70. a.** 0.041   **b.** 40.7 million   **c.** 2010   **71.** about 15,679 years old   **72. a.** about 9 people
b. about 104 people   **c.** 171 people; yes; The limiting size is 171; however, 178 people died.   **73.** $y = 73e^{(\ln 2.6)x}$; $y = 73e^{0.956x}$
74. $y = 6.5e^{(\ln 0.43)x}$; $y = 6.5e^{-0.844x}$ **75.** high: exponential; medium: linear; low: quadratic; Explanations will vary; negative;
The parabola opens downward.   **76.** The exponential model, $y = (3.38051786)(1.0235357)^x$, is the best fit; 113.4 million

Chapter 4 Test

1.

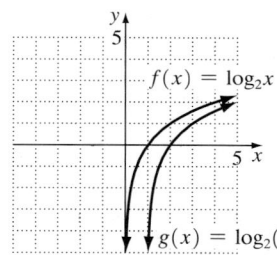

$g(x) = 2^{x+1}$ $f(x) = 2^x$

2.

$f(x) = \log_2 x$

$g(x) = \log_2(x - 1)$

3. $5^3 = 125$ **4.** $\log_{36} 6 = \frac{1}{2}$ **5.** $(-\infty, 3)$

6. $3 + 5 \log_4 x$ **7.** $\frac{1}{3} \log_3 x - 4$ **8.** $\log(x^6 y^2)$

9. $\ln \frac{7}{x^3}$ **10.** 1.5741   **11.** $\left\{ \frac{\ln 1.4}{\ln 5} \right\}$ or $\{0.2091\}$

12. $\left\{ \frac{\ln 4}{0.005} \right\}$ or $\{277.2589\}$ **13.** $\{0, \ln 5\}$ or $\{0, 1.6094\}$

14. $\{54.25\}$ **15.** $\{5\}$ **16.** $\left\{ \frac{e^4}{3} \right\}$ or $\{18.1993\}$

17. 6.5% compounded semiannually; $221.15 more   **18.** 120 db   **19. a.** about 89%   **b.** decreasing; $k = -0.004 < 0$ **c.** 1995
20. $A = 484e^{0.005t}$ **21.** about 24,758 years ago   **22. a.** 14 elk   **b.** about 51 elk   **c.** 140 elk

Cumulative Review Exercises (Chapters 1–4)

1. $\left\{ \frac{2}{3}, 2 \right\}$ **2.** $\{3, 7\}$ **3.** $\{-2, -1, 1\}$ **4.** $\{0.9704\}$ **5.** $\{3\}$ **6.** $(-\infty, 4]$ **7.** $[1, 3]$

8. using $(1, 3)$, $y - 3 = -3(x - 1)$; $y = -3x + 6$ **9.** $(f \circ g)(x) = (x + 2)^2$; $(g \circ f)(x) = x^2 + 2$ **10.** $f^{-1}(x) = \frac{1}{2}x + \frac{7}{2}$

11. $x^2 + 3x - 3 + \frac{-4}{x + 2}$ **12.** $\pm 1, \pm \frac{1}{2}, \pm \frac{1}{4}, \pm 3, \pm \frac{3}{2}, \pm \frac{3}{4}$ **13.** 300   **14.** $\{1 + i, 1 - i, 2\}$

15.

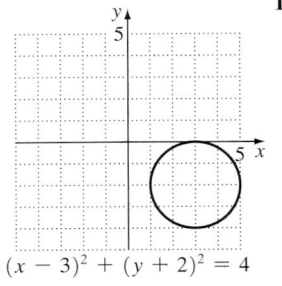

$(x - 3)^2 + (y + 2)^2 = 4$

16.

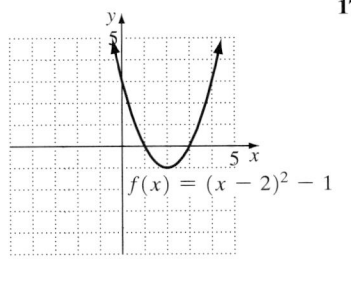

$f(x) = (x - 2)^2 - 1$

17.

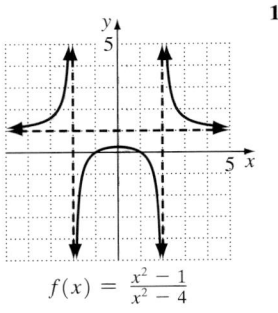

$f(x) = \dfrac{x^2 - 1}{x^2 - 4}$

18.

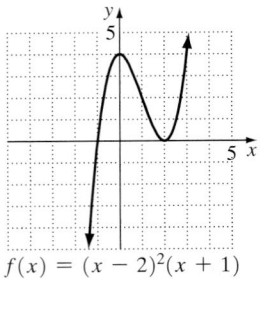

$f(x) = (x - 2)^2(x + 1)$

19. $12 per hr   **20.** $\dfrac{0.5}{\ln 4} \approx 0.361$; about $\dfrac{3}{10}$ of the people

CHAPTER 5

Section 5.1

Check Point Exercises

1. a.   **b.**   **c.**   **d.**

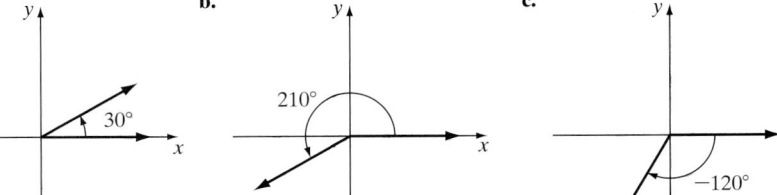

2. a. $40°$ **b.** $225°$ **3. a.** $12°; 102°$ **b.** no complementary angle; $30°$ **4.** 3.5 radians   **5. a.** $\dfrac{\pi}{3}$ radians   **b.** $\dfrac{3\pi}{2}$ radians

c. $-\dfrac{5\pi}{3}$ radians   **6. a.** $45°$ **b.** $-240°$ **c.** $343.8°$ **7.** $\dfrac{3\pi}{2}$ in. ≈ 4.71 in.   **8.** 135π in./min ≈ 424 in./min

Exercise Set 5.1

1. quadrant II   **3.** quadrant III   **5.** quadrant I   **7.** obtuse   **9.** straight
11.   **13.**   **15.**   **17.**

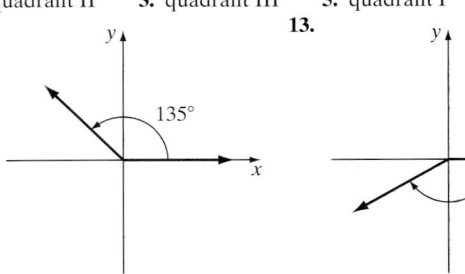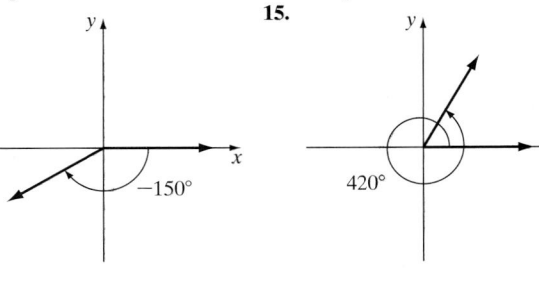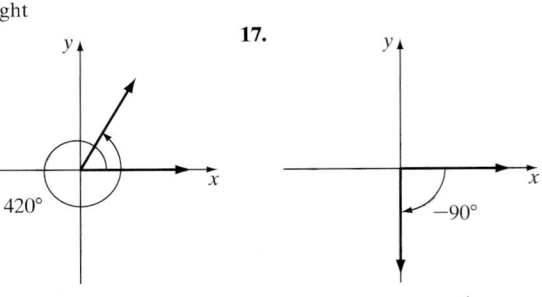

19. $35°$ **21.** $210°$ **23.** $315°$ **25.** $38°; 128°$ **27.** $52.6°; 142.6°$ **29.** no complement; $69°$ **31.** 4 radians   **33.** $\dfrac{4}{3}$ radians

35. 4 radians   **37.** $\dfrac{\pi}{4}$ radians   **39.** $\dfrac{3\pi}{4}$ radians   **41.** $\dfrac{5\pi}{3}$ radians   **43.** $-\dfrac{5\pi}{4}$ radians   **45.** $90°$ **47.** $120°$ **49.** $210°$

51. $-540°$ **53.** 0.31 radians   **55.** -0.70 radians   **57.** 3.49 radians   **59.** $114.59°$ **61.** $13.85°$ **63.** $-275.02°$

65. 3π in. ≈ 9.42 in.   **67.** 10π ft ≈ 31.42 ft   **69.** $\dfrac{12\pi \text{ radians}}{\text{second}}$ **71.** $60°; \dfrac{\pi}{3}$ radians   **73.** $\dfrac{8\pi}{3}$ in. ≈ 8.38 in.

75. 12π in. ≈ 37.70 in.   **77.** 2 radians; $114.59°$ **79.** 2094 mi   **81.** 1047 mph   **83.** 1508 ft/min   **97.** $30.25°$ **99.** $30° \, 25' \, 12''$
101. smaller than a right angle   **103.** 1815 mi

Section 5.2

Check Point Exercises

1. $\sin\theta = \dfrac{3}{5}$; $\cos\theta = \dfrac{4}{5}$; $\tan\theta = \dfrac{3}{4}$; $\csc\theta = \dfrac{5}{3}$; $\sec\theta = \dfrac{5}{4}$; $\cot\theta = \dfrac{4}{3}$ **2.** $\sqrt{2}$; $\sqrt{2}$; 1 **3.** $\sqrt{3}$; $\dfrac{\sqrt{3}}{3}$ **4.** $\tan\theta = \dfrac{2\sqrt{5}}{5}$; $\csc\theta = \dfrac{3}{2}$;

$\sec\theta = \dfrac{3\sqrt{5}}{5}$; $\cot\theta = \dfrac{\sqrt{5}}{2}$ **5.** $\dfrac{\sqrt{3}}{2}$ **6. a.** $\cos 44°$ **b.** $\tan\dfrac{5\pi}{12}$ **7. a.** 0.9553 **b.** 1.0025 **8.** 333.9 yd **9.** $54°$

Exercise Set 5.2

1. 15; $\sin\theta = \dfrac{3}{5}$; $\cos\theta = \dfrac{4}{5}$; $\tan\theta = \dfrac{3}{4}$; $\csc\theta = \dfrac{5}{3}$; $\sec\theta = \dfrac{5}{4}$; $\cot\theta = \dfrac{4}{3}$ **3.** 20; $\sin\theta = \dfrac{20}{29}$; $\cos\theta = \dfrac{21}{29}$; $\tan\theta = \dfrac{20}{21}$; $\csc\theta = \dfrac{29}{20}$;

$\sec\theta = \dfrac{29}{21}$; $\cot\theta = \dfrac{21}{20}$ **5.** 24; $\sin\theta = \dfrac{5}{13}$; $\cos\theta = \dfrac{12}{13}$; $\tan\theta = \dfrac{5}{12}$; $\csc\theta = \dfrac{13}{5}$; $\sec\theta = \dfrac{13}{12}$; $\cot\theta = \dfrac{12}{5}$ **7.** 28; $\sin\theta = \dfrac{4}{5}$; $\cos\theta = \dfrac{3}{5}$;

$\tan\theta = \dfrac{4}{3}$; $\csc\theta = \dfrac{5}{4}$; $\sec\theta = \dfrac{5}{3}$; $\cot\theta = \dfrac{3}{4}$ **9.** $\dfrac{\sqrt{3}}{2}$ **11.** $\sqrt{2}$ **13.** $\sqrt{3}$ **15.** 0 **17.** $\tan\theta = \dfrac{8}{15}$; $\csc\theta = \dfrac{17}{8}$; $\sec\theta = \dfrac{17}{15}$;

$\cot\theta = \dfrac{15}{8}$ **19.** $\tan\theta = \dfrac{\sqrt{2}}{4}$; $\csc\theta = 3$; $\sec\theta = \dfrac{3\sqrt{2}}{4}$; $\cot\theta = 2\sqrt{2}$ **21.** $\dfrac{\sqrt{13}}{7}$ **23.** $\dfrac{5}{8}$ **25.** 1 **27.** 1 **29.** 1 **31.** $\cos 83°$

33. $\sec 65°$ **35.** $\cot\dfrac{7\pi}{18}$ **37.** $\sin\dfrac{\pi}{10}$ **39.** 0.6157 **41.** 0.6420 **43.** 3.4203 **45.** 0.9511 **47.** 3.7321 **49.** 188 cm

51. 182 in. **53.** 41 m **55.** $17°$ **57.** $78°$ **59.** 1.147 radians **61.** 0.3950 radians **63.** 529 yd **65.** $36°$ **67.** 2879 ft

69. $37°$ **83.** $0.92106, -0.19735$; $0.95534, -0.148878$; $0.98007, -0.099667$; $0.99500, -0.04996$; $0.99995, -0.005$; $0.9999995, -0.0005$;

$0.999999995, -0.00005$; $1, -0.000005$; $\dfrac{\cos\theta - 1}{\theta}$ approaches 0 as θ approaches 0. **85.** In a right triangle, the hypotenuse is greater than

either other side. Therefore, both $\dfrac{\text{opposite}}{\text{hypotenuse}}$ and $\dfrac{\text{adjacent}}{\text{hypotenuse}}$ must be less than 1 for an acute angle in a right triangle.

87. a. 357 ft **b.** 394 ft

Section 5.3

Check Point Exercises

1. $\sin\theta = -\dfrac{3}{5}$; $\cos\theta = \dfrac{4}{5}$; $\tan\theta = -\dfrac{3}{4}$; $\csc\theta = -\dfrac{5}{3}$; $\sec\theta = \dfrac{5}{4}$; $\cot\theta = -\dfrac{4}{3}$ **2. a.** 1; undefined **b.** 0; 1 **c.** -1; undefined

d. 0; -1 **3.** quadrant III **4.** $\dfrac{\sqrt{10}}{10}$; $-\dfrac{\sqrt{10}}{3}$ **5. a.** $30°$ **b.** $\dfrac{\pi}{4}$ **c.** $60°$ **d.** 0.46 **6. a.** $-\dfrac{\sqrt{3}}{2}$ **b.** 1 **c.** $\dfrac{2\sqrt{3}}{3}$

Exercise Set 5.3

1. $\sin\theta = \dfrac{3}{5}$; $\cos\theta = -\dfrac{4}{5}$; $\tan\theta = -\dfrac{3}{4}$; $\csc\theta = \dfrac{5}{3}$; $\sec\theta = -\dfrac{5}{4}$; $\cot\theta = -\dfrac{4}{3}$ **3.** $\sin\theta = \dfrac{3\sqrt{13}}{13}$; $\cos\theta = \dfrac{2\sqrt{13}}{13}$; $\tan\theta = \dfrac{3}{2}$; $\csc\theta = \dfrac{\sqrt{13}}{3}$;

$\sec\theta = \dfrac{\sqrt{13}}{2}$; $\cot\theta = \dfrac{2}{3}$ **5.** $\sin\theta = -\dfrac{\sqrt{2}}{2}$; $\cos\theta = \dfrac{\sqrt{2}}{2}$; $\tan\theta = -1$; $\csc\theta = -\sqrt{2}$; $\sec\theta = \sqrt{2}$; $\cot\theta = -1$

7. $\sin\theta = -\dfrac{5\sqrt{29}}{29}$; $\cos\theta = -\dfrac{2\sqrt{29}}{29}$; $\tan\theta = \dfrac{5}{2}$; $\csc\theta = -\dfrac{\sqrt{29}}{5}$; $\sec\theta = -\dfrac{\sqrt{29}}{2}$; $\cot\theta = \dfrac{2}{5}$ **9.** -1 **11.** -1 **13.** undefined

15. 0 **17.** quadrant I **19.** quadrant III **21.** quadrant II **23.** $\sin\theta = -\dfrac{4}{5}$; $\tan\theta = \dfrac{4}{3}$; $\csc\theta = -\dfrac{5}{4}$; $\sec\theta = -\dfrac{5}{3}$; $\cot\theta = \dfrac{3}{4}$

25. $\cos\theta = -\dfrac{12}{13}$; $\tan\theta = -\dfrac{5}{12}$; $\csc\theta = \dfrac{13}{5}$; $\sec\theta = -\dfrac{13}{12}$; $\cot\theta = -\dfrac{12}{5}$ **27.** $\sin\theta = -\dfrac{15}{17}$; $\tan\theta = -\dfrac{15}{8}$; $\csc\theta = -\dfrac{17}{15}$; $\sec\theta = \dfrac{17}{8}$;

$\cot\theta = -\dfrac{8}{15}$ **29.** $\sin\theta = \dfrac{2\sqrt{13}}{13}$; $\cos\theta = -\dfrac{3\sqrt{13}}{13}$; $\csc\theta = \dfrac{\sqrt{13}}{2}$; $\sec\theta = -\dfrac{\sqrt{13}}{3}$; $\cot\theta = -\dfrac{3}{2}$ **31.** $\sin\theta = -\dfrac{4}{5}$; $\cos\theta = -\dfrac{3}{5}$;

$\csc\theta = -\dfrac{5}{4}$; $\sec\theta = -\dfrac{5}{3}$; $\cot\theta = \dfrac{3}{4}$ **33.** $\sin\theta = -\dfrac{2\sqrt{2}}{3}$; $\cos\theta = -\dfrac{1}{3}$; $\tan\theta = 2\sqrt{2}$; $\csc\theta = -\dfrac{3\sqrt{2}}{4}$; $\cot\theta = \dfrac{\sqrt{2}}{4}$ **35.** $20°$

37. $25°$ **39.** $5°$ **41.** $\dfrac{\pi}{4}$ **43.** $\dfrac{\pi}{6}$ **45.** $30°$ **47.** $25°$ **49.** 1.56 **51.** $-\dfrac{\sqrt{2}}{2}$ **53.** $\dfrac{\sqrt{3}}{3}$ **55.** $\sqrt{3}$ **57.** $\dfrac{\sqrt{3}}{2}$ **59.** -2

61. 1 **63.** $\dfrac{\sqrt{3}}{2}$ **65.** -1

25. $y = -\dfrac{1}{2}\csc\dfrac{x}{2}$;

27. $y = \dfrac{1}{2}\sec 2\pi x$;

29.

31.

33.

35.

37.

39.

41.

43.

45. a.

b. 0.25, 0.75, 1.25, 1.75;
The beacon is shining
parallel to the wall at
these times.

47. $d = 10\sec x$

49.

Height of board

Seconds after dive

63.

65.

67.

69.

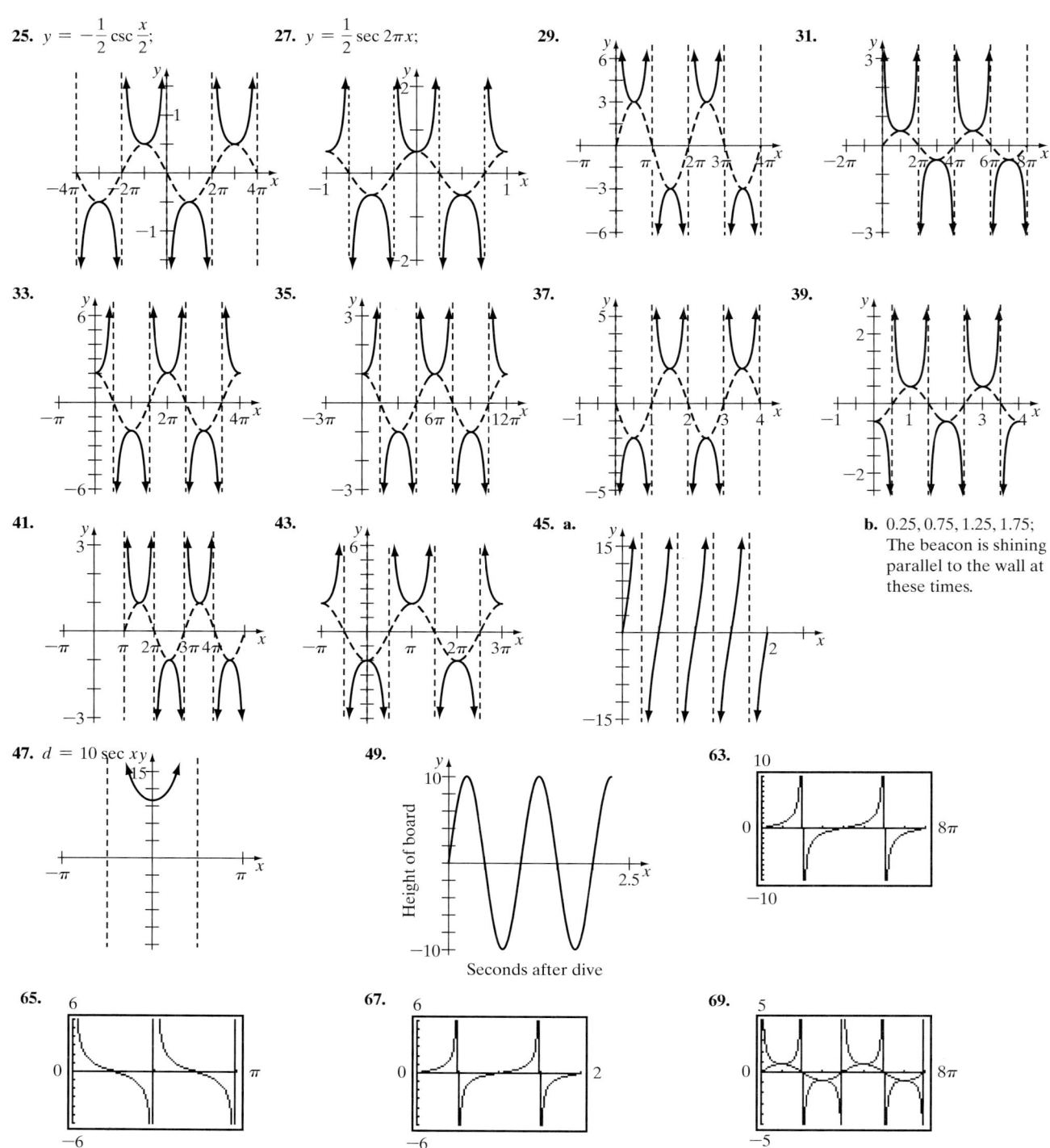

71.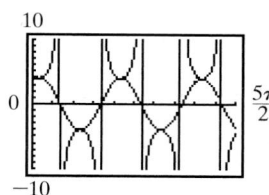

73.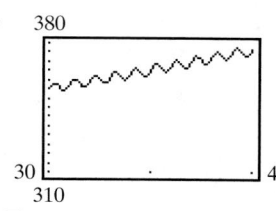

The concentration is increasing.

75. $y = \cot \dfrac{3}{2}x$

77. 2^{-x} decreases the amplitude as x gets larger.

Section 5.7

Check Point Exercises

1. $\dfrac{\pi}{3}$ **2.** $-\dfrac{\pi}{4}$ **3.** $\dfrac{2\pi}{3}$ **4.** $-\dfrac{\pi}{4}$ **5. a.** 1.2310 **b.** −1.5429 **6. a.** 0.7 **b.** 0 **c.** not defined **7.** $\dfrac{3}{5}$ **8.** $\dfrac{\sqrt{3}}{2}$

9. $\sqrt{x^2 + 1}$

Exercise Set 5.7

1. $\dfrac{\pi}{6}$ **3.** $\dfrac{\pi}{4}$ **5.** $-\dfrac{\pi}{6}$ **7.** $\dfrac{\pi}{6}$ **9.** $\dfrac{3\pi}{4}$ **11.** $\dfrac{\pi}{2}$ **13.** $\dfrac{\pi}{6}$ **15.** 0 **17.** $-\dfrac{\pi}{3}$ **19.** 0.30 **21.** −0.33 **23.** 1.19

25. 1.25 **27.** −1.52 **29.** −1.52 **31.** 0.9 **33.** $\dfrac{\pi}{3}$ **35.** $\dfrac{\pi}{6}$ **37.** 125 **39.** $-\dfrac{\pi}{6}$ **41.** $-\dfrac{\pi}{3}$ **43.** 0 **45.** not defined

47. $\dfrac{3}{5}$ **49.** $\dfrac{12}{5}$ **51.** $-\dfrac{3}{4}$ **53.** $\dfrac{\sqrt{2}}{2}$ **55.** $\dfrac{4\sqrt{15}}{15}$ **57.** $-2\sqrt{2}$ **59.** 2 **61.** $\dfrac{\sqrt{1-x^2}}{x}$ **63.** $\dfrac{\sqrt{x^2-1}}{x}$ **65.** $\dfrac{\sqrt{x^2+4}}{2}$

67. a.

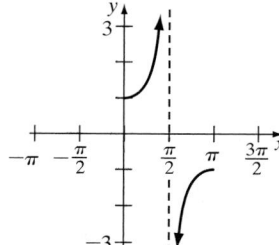

b. No horizontal line intersects the graph of $y = \sec x$ more than once, so the function is one-to-one and has an inverse function.

c.

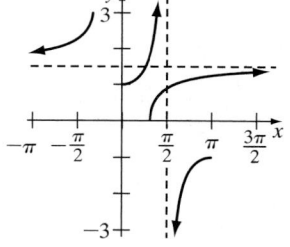

69. 0.408 radians; 0.602 radians; 0.654 radians; 0.645 radians; 0.613 radians

71. 1.3157 radians or 75.4°

73. 1.1071 sq units

87.

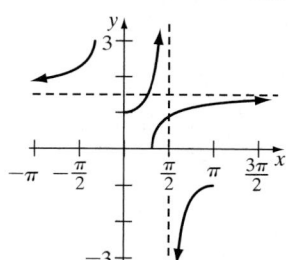

Shifted right 1 unit

89.

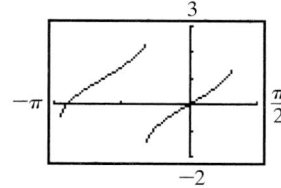

Shifted left 2 units and up 1 unit

91.

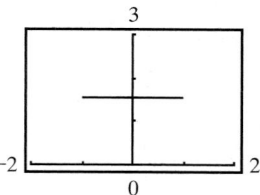

It seems $\sin^{-1} x + \cos^{-1} x = \dfrac{\pi}{2}$ for $-1 \le x \le 1$

93. $x = \sin \dfrac{\pi}{8}$

95. $\tan \alpha = \dfrac{8}{x}$, so $\tan^{-1}\dfrac{8}{x} = \alpha$.

$\tan(\alpha + \theta) = \dfrac{33}{x}$, so $\tan^{-1}\dfrac{33}{x} = \alpha + \theta$.

$\theta = \alpha + \theta - \alpha = \tan^{-1}\dfrac{33}{x} - \tan^{-1}\dfrac{8}{x}$.

Section 5.8

Check Point Exercises

1. $B = 27.3°; b \approx 4.34; c \approx 9.45$ **2.** 998 ft **3.** $\theta \approx 29.0$ **4.** 60.3 ft **5. a.** S 25° E **b.** S 15° W

6. a. 4.2 mi **b.** S87.7°W **7.** $d = -6\cos\dfrac{\pi}{2}t$ **8. a.** 12 cm **b.** $\dfrac{1}{8}$ cm per sec **c.** 8 sec

Exercise Set 5.8

1. $B = 66.5°; a \approx 4.35; c \approx 10.90$ **3.** $B = 37.4°; a \approx 42.90; b \approx 32.80$ **5.** $A = 73.2°; a \approx 101.02; c \approx 105.52$

7. $b \approx 39.95; A \approx 37.3°; B \approx 52.7°$ **9.** $c \approx 26.96; A \approx 23.6°; B \approx 66.4°$ **11.** $a \approx 6.71; B \approx 16.6°; A \approx 73.4°$

13. N 15° E   **15.** S 80° W   **17.** $d = -6 \cos \frac{\pi}{2} t$ **19.** $d = 3 \sin \frac{4\pi}{3} t$ **21. a.** 5 in.   **b.** $\frac{1}{4}$ in. per sec   **c.** 4 sec

23. a. 6 in.   **b.** 1 in. per sec   **c.** 1 sec   **25. a.** $\frac{1}{2}$ in.   **b.** 0.32 in. per sec   **c.** 3.14 sec   **27. a.** 5 in.   **b.** $\frac{1}{3}$ in. per sec

c. 3 sec   **29.** $h \approx 2059$ ft   **31.** $d \approx 695$ ft   **33.** 1376 ft   **35.** 15.1°   **37.** 33.7 ft   **39.** 90 mi north and 120 mi east

41. 13.2 mi   **43.** N 53° W   **45.** N 89.5° E   **47.** $d = 6 \sin \pi t$ **49.** $d = \sin 528 \pi t$

59.

; 10 complete oscillations   **61.** 48 ft

Chapter 5 Review Exercises

1.

2.

3.

4.

5. 40°   **6.** 275°   **7.** 17°; 107°   **8.** no complement; $\frac{\pi}{3}$ radians   **9.** 4.5 radians   **10.** $\frac{\pi}{12}$ radians   **11.** $\frac{2\pi}{3}$ radians

12. $\frac{7\pi}{4}$ radians   **13.** 300°   **14.** 252°   **15.** $-150°$ **16.** $\frac{15\pi}{2}$ ft ≈ 23.56 ft   **17.** 20.6π radians per min   **18.** 42,412 ft per min

19. $\sin \theta = \frac{3}{5}; \cos \theta = \frac{4}{5}; \tan \theta = \frac{3}{4}; \csc \theta = \frac{5}{3}; \sec \theta = \frac{5}{4}; \cot \theta = \frac{4}{3}$ **20.** $\sqrt{3}$ **21.** $\frac{\sqrt{2}}{2}$ **22.** $\frac{2\sqrt{3}}{3}$ **23.** 1   **24.** $\frac{\sqrt{21}}{7}$

25. $\cos 20°$ **26.** $\sin 0$ **27.** 42 mm   **28.** 23 cm   **29.** 37 in.   **30.** 772 ft   **31.** 31 m   **32.** 56°

33. $\sin \theta = -\frac{5\sqrt{26}}{26}; \cos \theta = -\frac{\sqrt{26}}{26}; \tan \theta = 5; \csc \theta = -\frac{\sqrt{26}}{5}; \sec \theta = -\sqrt{26}; \cot \theta = \frac{1}{5}$

34. $\sin \theta = -1; \cos \theta = 0; \tan \theta$ is undefined; $\csc \theta = -1; \sec \theta$ is undefined; $\cot \theta = 0$ **35.** quadrant I   **36.** quadrant III

37. $\sin \theta = -\frac{\sqrt{21}}{5}; \tan \theta = -\frac{\sqrt{21}}{2}; \csc \theta = -\frac{5\sqrt{21}}{21}; \sec \theta = \frac{5}{2}; \cot \theta = -\frac{2\sqrt{21}}{21}$

38. $\sin \theta = \frac{\sqrt{10}}{10}; \cos \theta = -\frac{3\sqrt{10}}{10}; \csc \theta = \sqrt{10}; \sec \theta = -\frac{\sqrt{10}}{3}; \cot \theta = -3$ **39.** 85°   **40.** $\frac{3\pi}{8}$ **41.** 50°   **42.** $-\frac{\sqrt{3}}{2}$

43. $-\sqrt{3}$ **44.** $\sqrt{2}$ **45.** $\frac{\sqrt{3}}{2}$ **46.** $-\sqrt{3}$ **47.** $-\frac{2\sqrt{3}}{3}$ **48.** $-\frac{\sqrt{3}}{2}$ **49.** $\frac{\sqrt{2}}{2}$ **50.** 1

51. $3; \frac{\pi}{2}$ **52.** $2; \pi$ **53.** $2; 4\pi$ **54.** $\frac{1}{2}; 6$

55. 1; 2

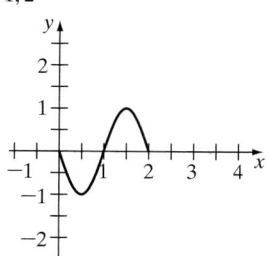

56. 3; 6π

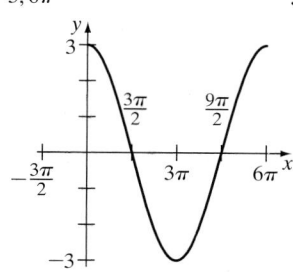

57. 2; 2π; π

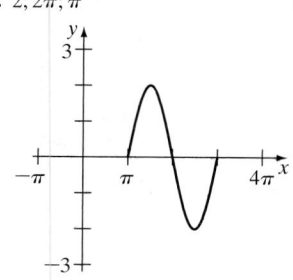

58. 3; 2π; −π

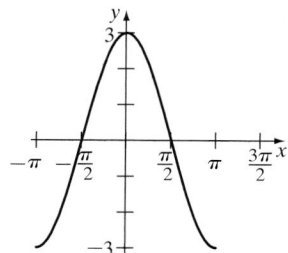

59. $\frac{3}{2}$; π; $-\frac{\pi}{8}$

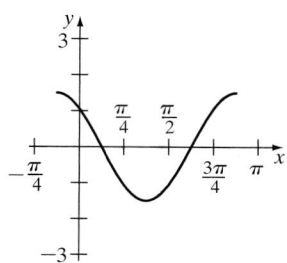

60. $\frac{5}{2}$; π; $-\frac{\pi}{4}$

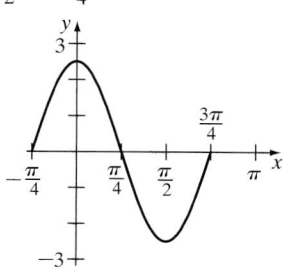

61. 3; 6; 9

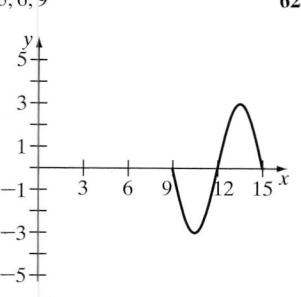

62.

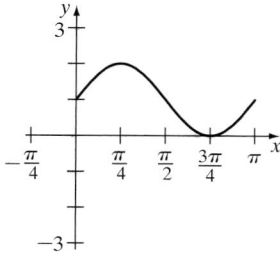

63.

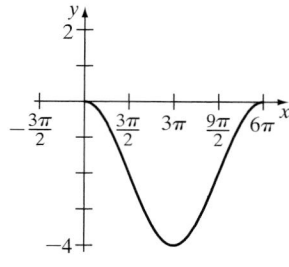

64. a. ≈98.52° **b.** 24 hr **c.** 5:00 P.M.; 98.9°
d. 5:00 A.M.; 98.3° **e.**

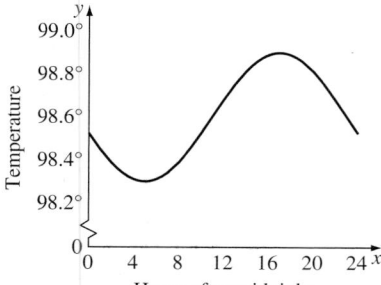

65.

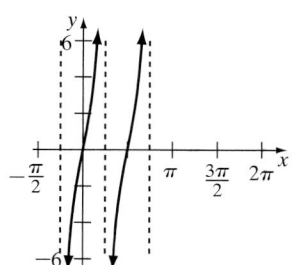

66.

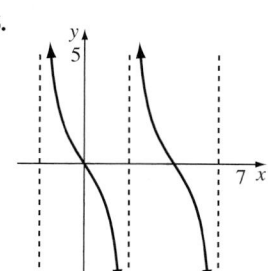

67.

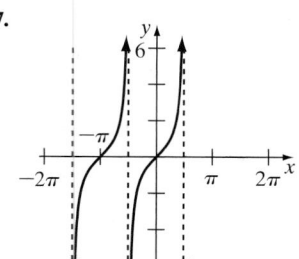

68.

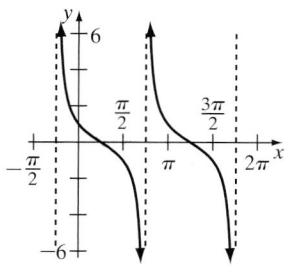

69.

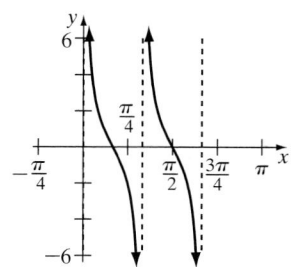

70.

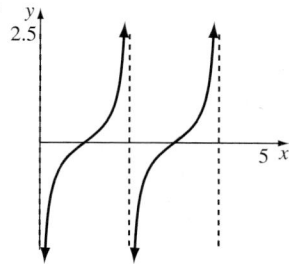

71.

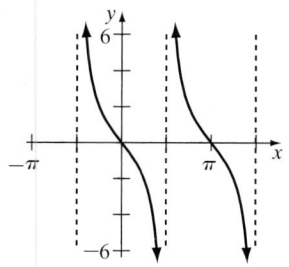

72.

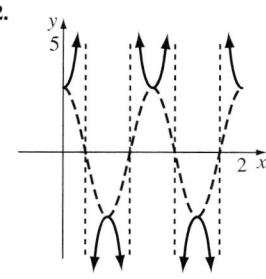

73. 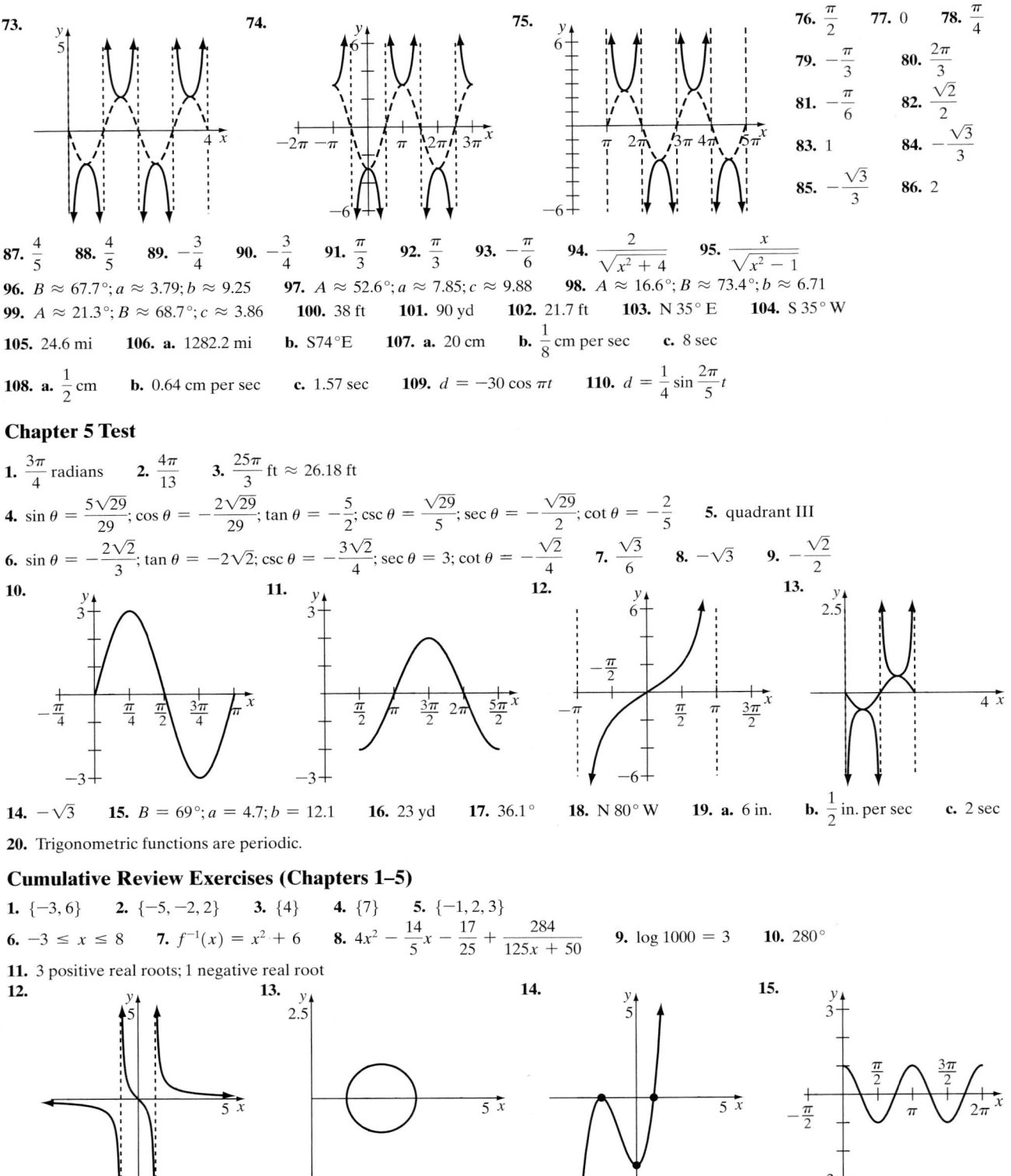 **74.** **75.**

76. $\dfrac{\pi}{2}$ **77.** 0   **78.** $\dfrac{\pi}{4}$

79. $-\dfrac{\pi}{3}$ **80.** $\dfrac{2\pi}{3}$

81. $-\dfrac{\pi}{6}$ **82.** $\dfrac{\sqrt{2}}{2}$

83. 1   **84.** $-\dfrac{\sqrt{3}}{3}$

85. $-\dfrac{\sqrt{3}}{3}$ **86.** 2

87. $\dfrac{4}{5}$ **88.** $\dfrac{4}{5}$ **89.** $-\dfrac{3}{4}$ **90.** $-\dfrac{3}{4}$ **91.** $\dfrac{\pi}{3}$ **92.** $\dfrac{\pi}{3}$ **93.** $-\dfrac{\pi}{6}$ **94.** $\dfrac{2}{\sqrt{x^2+4}}$ **95.** $\dfrac{x}{\sqrt{x^2-1}}$

96. $B \approx 67.7°; a \approx 3.79; b \approx 9.25$ **97.** $A \approx 52.6°; a \approx 7.85; c \approx 9.88$ **98.** $A \approx 16.6°; B \approx 73.4°; b \approx 6.71$

99. $A \approx 21.3°; B \approx 68.7°; c \approx 3.86$ **100.** 38 ft   **101.** 90 yd   **102.** 21.7 ft   **103.** N 35° E   **104.** S 35° W

105. 24.6 mi   **106. a.** 1282.2 mi   **b.** S74°E   **107. a.** 20 cm   **b.** $\dfrac{1}{8}$ cm per sec   **c.** 8 sec

108. a. $\dfrac{1}{2}$ cm   **b.** 0.64 cm per sec   **c.** 1.57 sec   **109.** $d = -30 \cos \pi t$ **110.** $d = \dfrac{1}{4} \sin \dfrac{2\pi}{5} t$

Chapter 5 Test

1. $\dfrac{3\pi}{4}$ radians   **2.** $\dfrac{4\pi}{13}$ **3.** $\dfrac{25\pi}{3}$ ft ≈ 26.18 ft

4. $\sin \theta = \dfrac{5\sqrt{29}}{29}; \cos \theta = -\dfrac{2\sqrt{29}}{29}; \tan \theta = -\dfrac{5}{2}; \csc \theta = \dfrac{\sqrt{29}}{5}; \sec \theta = -\dfrac{\sqrt{29}}{2}; \cot \theta = -\dfrac{2}{5}$ **5.** quadrant III

6. $\sin \theta = -\dfrac{2\sqrt{2}}{3}; \tan \theta = -2\sqrt{2}; \csc \theta = -\dfrac{3\sqrt{2}}{4}; \sec \theta = 3; \cot \theta = -\dfrac{\sqrt{2}}{4}$ **7.** $\dfrac{\sqrt{3}}{6}$ **8.** $-\sqrt{3}$ **9.** $-\dfrac{\sqrt{2}}{2}$

10. **11.** **12.** **13.**

14. $-\sqrt{3}$ **15.** $B = 69°; a = 4.7; b = 12.1$ **16.** 23 yd   **17.** 36.1°   **18.** N 80° W   **19. a.** 6 in.   **b.** $\dfrac{1}{2}$ in. per sec   **c.** 2 sec

20. Trigonometric functions are periodic.

Cumulative Review Exercises (Chapters 1–5)

1. $\{-3, 6\}$ **2.** $\{-5, -2, 2\}$ **3.** $\{4\}$ **4.** $\{7\}$ **5.** $\{-1, 2, 3\}$

6. $-3 \le x \le 8$ **7.** $f^{-1}(x) = x^2 + 6$ **8.** $4x^2 - \dfrac{14}{5}x - \dfrac{17}{25} + \dfrac{284}{125x + 50}$ **9.** $\log 1000 = 3$ **10.** 280°

11. 3 positive real roots; 1 negative real root

12. **13.** **14.** **15.**

16.

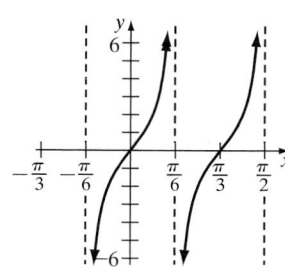

17. 48 performances   **18. a.** $A = 10.2^{0.3057t}$ **b.** 2001   **19.** 1540 Btu per hr   **20.** $76°$

CHAPTER 6

Section 6.1

Check Point Exercises

1. $\csc x \tan x = \dfrac{1}{\sin x} \cdot \dfrac{\sin x}{\cos x} = \dfrac{1}{\cos x} = \sec x$ **2.** $\cos x \cot x + \sin x = \cos x \cdot \dfrac{\cos x}{\sin x} + \sin x = \dfrac{\cos^2 x}{\sin x} + \sin x \cdot \dfrac{\sin x}{\sin x} = \dfrac{\cos^2 x + \sin^2 x}{\sin x}$

$= \dfrac{1}{\sin x} = \csc x$ **3.** $\sin x - \sin x \cos^2 x = \sin x(1 - \cos^2 x) = \sin x \cdot \sin^2 x = \sin^3 x$ **4.** $\dfrac{\sin x}{1 + \cos x} + \dfrac{1 + \cos x}{\sin x}$

$= \dfrac{\sin x(\sin x)}{(1 + \cos x)(\sin x)} + \dfrac{(1 + \cos x)(1 + \cos x)}{(\sin x)(1 + \cos x)} = \dfrac{\sin^2 x + 1 + 2 \cos x + \cos^2 x}{(1 + \cos x)(\sin x)} = \dfrac{\sin^2 x + \cos^2 x + 1 + 2 \cos x}{(1 + \cos x)(\sin x)}$

$= \dfrac{1 + 1 + 2 \cos x}{(1 + \cos x)(\sin x)} = \dfrac{2 + 2 \cos x}{(1 + \cos x)(\sin x)} = \dfrac{2(1 + \cos x)}{(1 + \cos x)(\sin x)} = \dfrac{2}{\sin x} = 2 \csc x$

5. $\dfrac{\cos x}{1 + \sin x} = \dfrac{\cos x(1 - \sin x)}{(1 + \sin x)(1 - \sin x)} = \dfrac{\cos x(1 - \sin x)}{1 - \sin^2 x} = \dfrac{\cos x(1 - \sin x)}{\cos^2 x} = \dfrac{1 - \sin x}{\cos x}$ **6.** $\dfrac{\sec x + \csc(-x)}{\sec x \csc x} = \dfrac{\sec x - \csc x}{\sec x \csc x}$

$= \dfrac{\dfrac{1}{\cos x} - \dfrac{1}{\sin x}}{\dfrac{1}{\cos x} \cdot \dfrac{1}{\sin x}} = \dfrac{\dfrac{\sin x}{\cos x \cdot \sin x} - \dfrac{\cos x}{\cos x \cdot \sin x}}{\dfrac{1}{\cos x \cdot \sin x}} = \dfrac{\dfrac{\sin x - \cos x}{\cos x \cdot \sin x}}{\dfrac{1}{\cos x \cdot \sin x}} = \dfrac{\sin x - \cos x}{\cos x \cdot \sin x} \cdot \dfrac{\cos x \cdot \sin x}{1} = \sin x - \cos x$

7. Left side: $\dfrac{1}{1 + \sin \theta} + \dfrac{1}{1 - \sin \theta} = \dfrac{1(1 - \sin \theta)}{(1 + \sin \theta)(1 - \sin \theta)} + \dfrac{1(1 + \sin \theta)}{(1 - \sin \theta)(1 + \sin \theta)} = \dfrac{1 - \sin \theta + 1 + \sin \theta}{(1 + \sin \theta)(1 - \sin \theta)} = \dfrac{2}{1 - \sin^2 \theta};$

Right side: $2 + 2 \tan^2 \theta = 2 + 2\left(\dfrac{\sin^2 \theta}{\cos^2 \theta}\right) = \dfrac{2 \cos^2 \theta}{\cos^2 \theta} + \dfrac{2 \sin^2 \theta}{\cos^2 \theta} = \dfrac{2 \cos^2 \theta + 2 \sin^2 \theta}{\cos^2 \theta} = \dfrac{2(\cos^2 \theta + \sin^2 \theta)}{\cos^2 \theta} = \dfrac{2}{\cos^2 \theta} = \dfrac{2}{1 - \sin^2 \theta}$

Exercise Set 6.1

For Exercises 1–59, proofs may vary.

65.

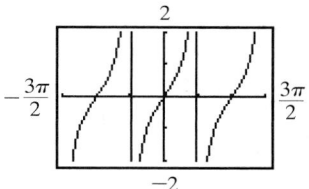

Proofs may vary.

67.

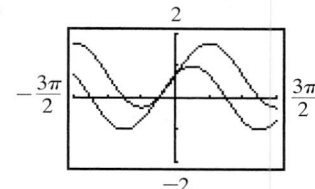

Values for x may vary.

69.

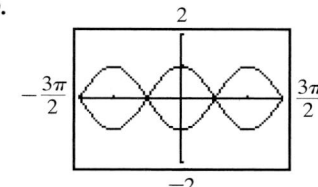

Values for x may vary.

71.

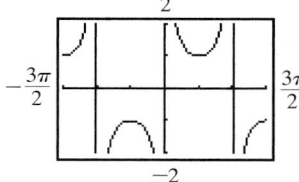

Proofs may vary.

73.

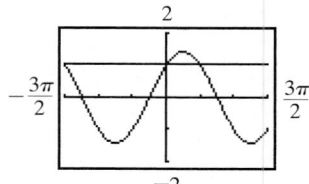

Values for x may vary.

75. Proofs may vary.
77. Answers may vary.

Section 6.2

Check Point Exercises

1. $\dfrac{\sqrt{3}}{2}$ **2.** $\dfrac{\sqrt{3}}{2}$ **3.** $\dfrac{\cos(\alpha-\beta)}{\cos\alpha\cos\beta} = \dfrac{\cos\alpha\cos\beta + \sin\alpha\sin\beta}{\cos\alpha\cos\beta} = \dfrac{\cos\alpha}{\cos\alpha}\cdot\dfrac{\cos\beta}{\cos\beta} + \dfrac{\sin\alpha}{\cos\alpha}\cdot\dfrac{\sin\beta}{\cos\beta} = 1 + \tan\alpha\tan\beta$ **4.** $\dfrac{\sqrt{2}+\sqrt{6}}{4}$

5. a. $\cos\alpha = -\dfrac{3}{5}$ **b.** $\cos\beta = \dfrac{\sqrt{3}}{2}$ **c.** $\dfrac{-3\sqrt{3}-4}{10}$ **d.** $\dfrac{4\sqrt{3}-4}{10}$ **6. a.** $y = \sin x$

b. $\cos\left(x + \dfrac{3\pi}{2}\right) = \cos x \cos\dfrac{3\pi}{2} - \sin x \sin\dfrac{3\pi}{2} = \cos x\cdot 0 - \sin x\cdot(-1) = \sin x$

7. $\tan(x+\pi) = \dfrac{\tan x + \tan\pi}{1 - \tan x \tan\pi} = \dfrac{\tan x + 0}{1 - \tan x\cdot 0} = \dfrac{\tan x}{1} = \tan x$

Exercise Set 6.2

1. $\dfrac{\sqrt{6}+\sqrt{2}}{4}$ **3.** $\dfrac{\sqrt{2}-\sqrt{6}}{4}$ **5. a.** $\alpha = 50°, \beta = 20°$ **b.** $\cos 30°$ **c.** $\dfrac{\sqrt{3}}{2}$ **7. a.** $\alpha = \dfrac{5\pi}{12}, \beta = \dfrac{\pi}{12}$ **b.** $\cos\left(\dfrac{\pi}{3}\right)$ **c.** $\dfrac{1}{2}$

For Exercises 9–11, proofs may vary. **13.** $\dfrac{\sqrt{6}-\sqrt{2}}{4}$ **15.** $\dfrac{\sqrt{6}+\sqrt{2}}{4}$ **17.** $\sqrt{3}+2$ **19.** $2-\sqrt{3}$ **21.** $-\dfrac{\sqrt{6}+\sqrt{2}}{4}$

23. $\dfrac{\sqrt{6}-\sqrt{2}}{4}$ **25.** $\sin 30°; \dfrac{1}{2}$ **27.** $\tan 45°; 1$ **29.** $\sin\dfrac{\pi}{6}; \dfrac{1}{2}$ **31.** $\tan\dfrac{\pi}{6}; \dfrac{\sqrt{3}}{3}$ For Exercises 33–55, proofs may vary.

57. a. $-\dfrac{63}{65}$ **b.** $-\dfrac{16}{65}$ **c.** $\dfrac{16}{63}$ **59. a.** $-\dfrac{4+6\sqrt{2}}{15}$ **b.** $\dfrac{3-8\sqrt{2}}{15}$ **c.** $\dfrac{54-25\sqrt{2}}{28}$ **61. a.** $-\dfrac{8\sqrt{3}+15}{34}$ **b.** $\dfrac{15\sqrt{3}-8}{34}$

c. $\dfrac{480-289\sqrt{3}}{33}$ **63. a.** $y = \sin x$ **b.** $\sin(\pi - x) = \sin\pi\cos x - \cos\pi\sin x = 0\cdot\cos x - (-1)\sin x = \sin x$ **65. a.** $y = 2\cos x$

b. $\sin\left(x + \dfrac{\pi}{2}\right) + \sin\left(\dfrac{\pi}{2} - x\right) = \sin x\cos\dfrac{\pi}{2} + \cos x\sin\dfrac{\pi}{2} + \sin\dfrac{\pi}{2}\cos x - \cos\dfrac{\pi}{2}\sin x = \sin x\cdot 0 + \cos x\cdot 1 + 1\cdot\cos x - 0\cdot\sin x$

$= \cos x + \cos x = 2\cos x$ **67.** Proofs may vary.; amplitude is $\sqrt{13}$; period is 2π

77.

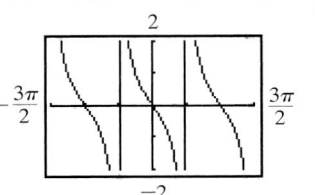

Proofs may vary.

79.

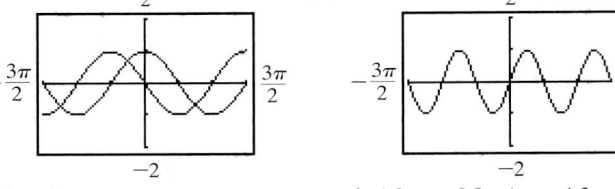

Values for x may vary.

81.

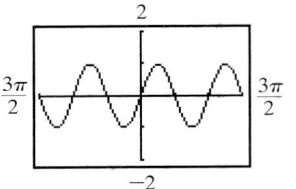

$\sin 1.2x\cos 0.8x + \cos 1.2x\sin 0.8x = \sin(1.2x + 0.8x)$
$= \sin 2x$

83. Proofs may vary. **85.** $y\sqrt{1-x^2} + x\sqrt{1-y^2}$

Section 6.3

Check Point Exercises

1. a. $-\dfrac{24}{25}$ **b.** $-\dfrac{7}{25}$ **c.** $\dfrac{24}{7}$ **2.** $\dfrac{\sqrt{3}}{2}$ **3.** $\sin 3\theta = \sin(2\theta + \theta) = \sin 2\theta\cos\theta + \cos 2\theta\sin\theta = 2\sin\theta\cos\theta\cos\theta$

$+ (2\cos^2\theta - 1)\sin\theta = 2\sin\theta\cos^2\theta + 2\sin\theta\cos^2\theta - \sin\theta = 4\sin\theta\cos^2\theta - \sin\theta = 4\sin\theta(1 - \sin^2\theta) - \sin\theta = 4\sin\theta - 4\sin^3\theta$

$- \sin\theta = 3\sin\theta - 4\sin^3\theta$ **4.** $\sin^4 x = (\sin^2 x)^2 = \left(\dfrac{1-\cos 2x}{2}\right)^2 = \dfrac{1 - 2\cos 2x + \cos^2 2x}{4} = \dfrac{1}{4} - \dfrac{1}{2}\cos 2x + \dfrac{1}{4}\cos^2 2x$

$= \dfrac{1}{4} - \dfrac{1}{2}\cos 2x + \dfrac{1}{4}\left(\dfrac{1 + \cos 2(2x)}{2}\right) = \dfrac{1}{4} - \dfrac{1}{2}\cos 2x + \dfrac{1}{8} + \dfrac{1}{8}\cos 4x = \dfrac{3}{8} - \dfrac{1}{2}\cos 2x + \dfrac{1}{8}\cos 4x$ **5.** $-\dfrac{\sqrt{2+\sqrt{3}}}{2}$

6. $\dfrac{\sin 2\theta}{1 + \cos 2\theta} = \dfrac{2\sin\theta\cos\theta}{1 + (1 - 2\sin^2\theta)} = \dfrac{2\sin\theta\cos\theta}{2 - 2\sin^2\theta} = \dfrac{2\sin\theta\cos\theta}{2(1 - \sin^2\theta)} = \dfrac{2\sin\theta\cos\theta}{2\cos^2\theta} = \dfrac{\sin\theta}{\cos\theta} = \tan\theta$

7. $\dfrac{\sec\alpha}{\sec\alpha\csc\alpha + \csc\alpha} = \dfrac{\dfrac{1}{\cos\alpha}}{\dfrac{1}{\cos\alpha}\cdot\dfrac{1}{\sin\alpha} + \dfrac{1}{\sin\alpha}} = \dfrac{\dfrac{1}{\cos\alpha}}{\dfrac{1}{\cos\alpha\sin\alpha} + \dfrac{\cos\alpha}{\cos\alpha\sin\alpha}} = \dfrac{\dfrac{1}{\cos\alpha}}{\dfrac{1 + \cos\alpha}{\cos\alpha\sin\alpha}} = \dfrac{1}{\cos\alpha}\cdot\dfrac{\cos\alpha\sin\alpha}{1 + \cos\alpha} = \dfrac{\sin\alpha}{1 + \cos\alpha} = \tan\dfrac{\alpha}{2}$

Exercise Set 6.3

1. $\dfrac{24}{25}$ **3.** $\dfrac{24}{7}$ **5.** $\dfrac{527}{625}$ **7. a.** $-\dfrac{240}{289}$ **b.** $-\dfrac{161}{289}$ **c.** $\dfrac{240}{161}$ **9. a.** $-\dfrac{336}{625}$ **b.** $\dfrac{527}{625}$ **c.** $-\dfrac{161}{527}$ **11. a.** $\dfrac{4}{5}$ **b.** $\dfrac{3}{5}$ **c.** $\dfrac{4}{3}$

13. a. $\dfrac{720}{1681}$ **b.** $\dfrac{1519}{1681}$ **c.** $\dfrac{720}{1519}$ **15.** $\dfrac{1}{2}$ **17.** $-\dfrac{\sqrt{3}}{2}$ **19.** $\dfrac{\sqrt{2}}{2}$ **21.** $\dfrac{\sqrt{3}}{3}$ For Exercises 23–33, proofs may vary.

35. $3 - 3\cos 2x$ **37.** $\dfrac{1}{8} - \dfrac{1}{8}\cos 4x$ **39.** $\dfrac{\sqrt{2 - \sqrt{3}}}{2}$ **41.** $-\dfrac{\sqrt{2 + \sqrt{2}}}{2}$ **43.** $2 + \sqrt{3}$ **45.** $-\sqrt{2} + 1$ **47.** $\dfrac{\sqrt{10}}{10}$

49. $\dfrac{1}{3}$ **51.** $\dfrac{7\sqrt{2}}{10}$ **53.** $\dfrac{3}{5}$ **55. a.** $\dfrac{2\sqrt{5}}{5}$ **b.** $-\dfrac{\sqrt{5}}{5}$ **c.** -2 **57. a.** $\dfrac{3\sqrt{13}}{13}$ **b.** $\dfrac{2\sqrt{13}}{13}$ **c.** $\dfrac{3}{2}$ **59.** $\dfrac{\sec\theta - 1}{2\sec\theta}$

For Exercises 61–67, proofs may vary. **69. a.** $\dfrac{v_0{}^2}{32} \cdot \sin 2\theta$ **b.** $\theta = \dfrac{\pi}{4}$ **71.** $\sqrt{2 - \sqrt{2}} \cdot (2 + \sqrt{2}) \approx 2.6$

85.

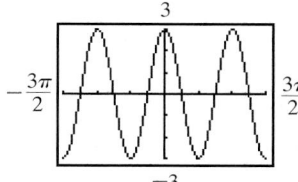

Proofs may vary.

87.

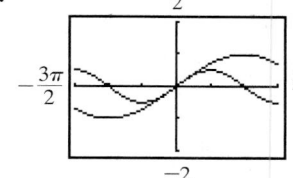

Values for x may vary.

89.

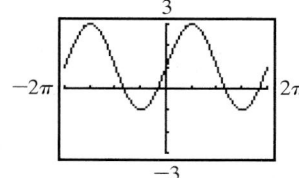

a. $y = 1 + 2\sin x$
b. Proofs may vary.

91.

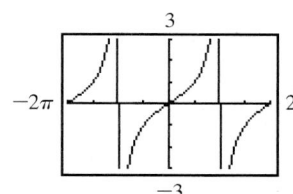

a. $y = \tan\dfrac{x}{2}$
b. Proofs may vary.

93. $\dfrac{5}{16} - \dfrac{7}{16}\cos 2x + \dfrac{3}{16}\cos 4x - \dfrac{1}{16}\cos 2x \cos 4x$

Section 6.4

Check Point Exercises

1. a. $\dfrac{1}{2}[\cos 3x - \cos 7x]$ **b.** $\dfrac{1}{2}[\cos 6x + \cos 8x]$ **2. a.** $2\sin 5x \cos 2x$ **b.** $2\cos\left(\dfrac{5x}{2}\right)\cos\left(\dfrac{x}{2}\right)$

3. $\dfrac{\cos 3x - \cos x}{\sin 3x + \sin x} = \dfrac{-2\sin\left(\dfrac{3x + x}{2}\right)\sin\left(\dfrac{3x - x}{2}\right)}{2\sin\left(\dfrac{3x + x}{2}\right)\cos\left(\dfrac{3x - x}{2}\right)} = \dfrac{-2\sin\left(\dfrac{4x}{2}\right)\sin\left(\dfrac{2x}{2}\right)}{2\sin\left(\dfrac{4x}{2}\right)\cos\left(\dfrac{2x}{2}\right)} = \dfrac{-2\sin 2x \sin x}{2\sin 2x \cos x} = -\dfrac{\sin x}{\cos x} = -\tan x$

Exercise Set 6.4

1. $\dfrac{1}{2}[\cos 4x - \cos 8x]$ **3.** $\dfrac{1}{2}[\cos 4x + \cos 10x]$ **5.** $\dfrac{1}{2}[\sin 3x - \sin x]$ **7.** $\dfrac{1}{2}[\sin 2x - \sin x]$ **9.** $2\sin 4x \cos 2x$

11. $2\sin 2x \cos 5x$ **13.** $2\cos 3x \cos x$ **15.** $2\sin\dfrac{3x}{2}\cos\dfrac{x}{2}$ **17.** $2\cos x \cos\dfrac{x}{2}$ **19.** $\dfrac{\sqrt{6}}{2}$ **21.** $-\dfrac{\sqrt{2}}{2}$

For Exercises 23–29, proofs may vary. **31. a.** $y = \cos x$ **b.** Proofs may vary. **33. a.** $y = \tan 2x$ **b.** Proofs may vary.
35. a. $y = \sin 1704\pi t + \sin 2418\pi t$ **b.** $2\sin 2061\pi t \cdot \cos 357\pi t$

43.

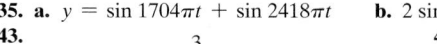

Values for x may vary.

45.

Proofs may vary.

47.

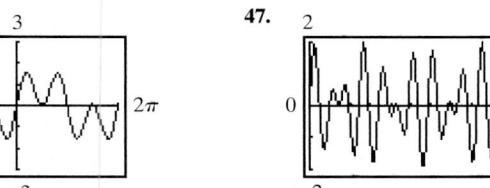

49. a. 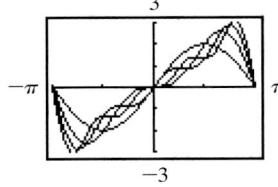 **b.** 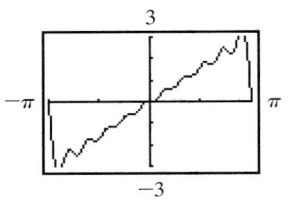 **c.** $\pi = 4 - \dfrac{4}{3} + \dfrac{4}{5} - \dfrac{4}{7} + \cdots$

For Exercises 51–55, proofs may vary.

Section 6.5

Check Point Exercises

1. $x = \dfrac{\pi}{3} + 2n\pi$ or $x = \dfrac{2\pi}{3} + 2n\pi$, where n is any integer.　**2.** $\dfrac{\pi}{6}, \dfrac{2\pi}{3}, \dfrac{7\pi}{6}, \dfrac{5\pi}{3}$　**3.** $\dfrac{\pi}{2}, \dfrac{5\pi}{2}$　**4.** $\dfrac{\pi}{6}, \dfrac{\pi}{2}, \dfrac{5\pi}{6}$　**5.** $0, \dfrac{\pi}{4}, \pi, \dfrac{5\pi}{4}$

6. $\dfrac{\pi}{2}, \dfrac{7\pi}{6}, \dfrac{11\pi}{6}$　**7.** $\dfrac{3\pi}{4}, \dfrac{7\pi}{4}$　**8.** $\dfrac{\pi}{2}, \pi$

Exercise Set 6.5

1. Solution　**3.** Not a solution　**5.** Solution　**7.** Solution　**9.** Not a solution　**11.** $x = \dfrac{\pi}{3} + 2n\pi$ or $x = \dfrac{2\pi}{3} + 2n\pi$,

where n is any integer.　**13.** $x = \dfrac{\pi}{4} + n\pi$, where n is any integer.　**15.** $x = \dfrac{2\pi}{3} + 2n\pi$ or $x = \dfrac{4\pi}{3} + 2n\pi$,

where n is any integer.　**17.** $x = n\pi$, where n is any integer.　**19.** $x = \dfrac{5\pi}{6} + 2n\pi$ or $x = \dfrac{7\pi}{6} + 2n\pi$, where n is any integer.

21. $\theta = \dfrac{\pi}{6} + 2n\pi$ or $\theta = \dfrac{5\pi}{6} + 2n\pi$, where n is any integer.　**23.** $\theta = \dfrac{3\pi}{2} + 2n\pi$, where n is any integer.　**25.** $\dfrac{\pi}{6}, \dfrac{\pi}{3}, \dfrac{7\pi}{6}, \dfrac{4\pi}{3}$

27. $\dfrac{5\pi}{24}, \dfrac{7\pi}{24}, \dfrac{17\pi}{24}, \dfrac{19\pi}{24}, \dfrac{29\pi}{24}, \dfrac{31\pi}{24}, \dfrac{41\pi}{24}, \dfrac{43\pi}{24}$　**29.** $\dfrac{\pi}{18}, \dfrac{7\pi}{18}, \dfrac{13\pi}{18}, \dfrac{19\pi}{18}, \dfrac{25\pi}{18}, \dfrac{31\pi}{18}$　**31.** $\dfrac{2\pi}{3}$　**33.** no solution　**35.** $\dfrac{4\pi}{9}, \dfrac{8\pi}{9}, \dfrac{16\pi}{9}$

37. $0, \dfrac{\pi}{3}, \pi, \dfrac{4\pi}{3}$　**39.** $\dfrac{\pi}{2}, \dfrac{7\pi}{6}, \dfrac{11\pi}{6}$　**41.** $\dfrac{2\pi}{3}, \pi, \dfrac{4\pi}{3}$　**43.** $\dfrac{3\pi}{2}$　**45.** $\dfrac{\pi}{2}, \dfrac{3\pi}{2}$　**47.** $\dfrac{\pi}{4}, \pi, \dfrac{5\pi}{4}$　**49.** $\dfrac{5\pi}{6}, \dfrac{7\pi}{6}, \dfrac{11\pi}{6}$

51. $\dfrac{\pi}{4}, \dfrac{\pi}{2}, \dfrac{5\pi}{4}, \dfrac{3\pi}{2}$　**53.** $0, \dfrac{2\pi}{3}, \pi, \dfrac{4\pi}{3}$　**55.** $0, \pi$　**57.** $\dfrac{\pi}{2}, \dfrac{7\pi}{6}, \dfrac{11\pi}{6}$　**59.** π　**61.** $\dfrac{\pi}{6}, \dfrac{5\pi}{6}$　**63.** $\dfrac{\pi}{6}, \dfrac{\pi}{2}, \dfrac{5\pi}{6}, \dfrac{3\pi}{2}$　**65.** $0, \dfrac{2\pi}{3}, \dfrac{4\pi}{3}$

67. $\dfrac{2\pi}{3}, \dfrac{4\pi}{3}$　**69.** $\dfrac{\pi}{8}, \dfrac{3\pi}{8}, \dfrac{9\pi}{8}, \dfrac{11\pi}{8}$　**71.** $0, \dfrac{\pi}{2}$　**73.** $\dfrac{\pi}{4}, \dfrac{3\pi}{4}$　**75.** $\dfrac{\pi}{12}, \dfrac{\pi}{4}, \dfrac{3\pi}{4}, \dfrac{11\pi}{12}, \dfrac{17\pi}{12}, \dfrac{19\pi}{12}$　**77.** $x = 0$　**79.** 0.4 sec and 2.1 sec

81. 49 days　**83.** $t = 2 + 6n$ or $t = 4 + 6n$ where n is any integer.　**85.** $21°$ or $69°$

97. 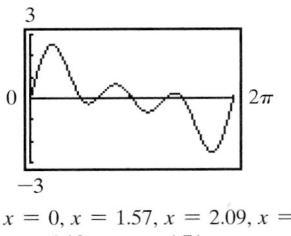 **99.** **101.**

$x = 1.37, x = 2.30, x = 3.98,$
or $x = 4.91$

$x = 0.37$ or $x = 2.77$

$x = 0, x = 1.57, x = 2.09, x = 3.14,$
$x = 4.19,$ or $x = 4.71$

103. no solution　**105.** $\dfrac{\pi}{3}, \dfrac{5\pi}{3}$

Chapter 6 Review Exercises

For Exercises 1–12, proofs may vary.　**13.** $\dfrac{\sqrt{6} - \sqrt{2}}{4}$　**14.** $\dfrac{\sqrt{2} - \sqrt{6}}{4}$　**15.** $2 - \sqrt{3}$　**16.** $\sqrt{3} + 2$　**17.** $\dfrac{1}{2}$　**18.** $\dfrac{1}{2}$

For Exercises 19–30, proofs may vary.

31. a. $y = \cos x$ **b.** $\sin\left(x - \dfrac{3\pi}{2}\right) = \sin x \cos \dfrac{3\pi}{2} - \cos x \sin \dfrac{3\pi}{2} = \sin x \cdot 0 - \cos x \cdot -1 = \cos x$

32. a. $y = -\sin x$ **b.** $\cos\left(x + \dfrac{\pi}{2}\right) = \cos x \cos \dfrac{\pi}{2} - \sin x \sin \dfrac{\pi}{2} = \cos x \cdot 0 - \sin x \cdot 1 = -\sin x$

33. a. $y = \tan x$ **b.** $y = \dfrac{\tan x - 1}{1 - \cot x} = \dfrac{\dfrac{\sin x}{\cos x} - 1}{1 - \dfrac{\cos x}{\sin x}} = \dfrac{\dfrac{\sin x - \cos x}{\cos x}}{\dfrac{\sin x - \cos x}{\sin x}} = \dfrac{\sin x - \cos x}{\cos x} \cdot \dfrac{\sin x}{\sin x - \cos x} = \dfrac{\sin x}{\cos x} = \tan x$

34. a. $\dfrac{33}{65}$ **b.** $\dfrac{16}{65}$ **c.** $-\dfrac{33}{56}$ **d.** $\dfrac{24}{25}$ **e.** $\dfrac{2\sqrt{13}}{13}$ **35. a.** $-\dfrac{63}{65}$ **b.** $-\dfrac{56}{65}$ **c.** $\dfrac{63}{16}$ **d.** $\dfrac{24}{25}$ **e.** $\dfrac{5\sqrt{26}}{26}$

36. a. 1 **b.** $-\dfrac{3}{5}$ **c.** undefined   **d.** $-\dfrac{3}{5}$ **e.** $\dfrac{\sqrt{10 + 3\sqrt{10}}}{2\sqrt{5}}$ **37. a.** 1 **b.** $\dfrac{4\sqrt{2}}{9}$ **c.** undefined   **d.** $\dfrac{4\sqrt{2}}{9}$ **e.** $-\dfrac{\sqrt{3}}{3}$

38. $\dfrac{\sqrt{3}}{2}$ **39.** $-\dfrac{\sqrt{3}}{3}$ **40.** $\dfrac{\sqrt{2 - \sqrt{2}}}{2}$ **41.** $2 - \sqrt{3}$ **42.** $\dfrac{1}{2}[\cos 2x - \cos 10x]$ **43.** $\dfrac{1}{2}[\sin 10x + \sin 4x]$

44. $-2 \sin x \cos 3x$ **45.** $\dfrac{\sqrt{6}}{2}$ **46.** Proofs may vary.   **47.** Proofs may vary.   **48. a.** $y = \cot x$ **b.** Proofs may vary.

49. $x = \dfrac{2\pi}{3} + 2n\pi$ or $x = \dfrac{4\pi}{3} + 2n\pi$, where n is any integer.   **50.** $x = \dfrac{\pi}{4} + 2n\pi$ or $x = \dfrac{3\pi}{4} + 2n\pi$, where n is any integer.

51. $x = \dfrac{7\pi}{6} + 2n\pi$ or $x = \dfrac{11\pi}{6} + 2n\pi$, where n is any integer.   **52.** $x = \dfrac{\pi}{6} + n\pi$, where n is any integer.   **53.** $\dfrac{\pi}{2}, \dfrac{3\pi}{2}$

54. $\dfrac{\pi}{6}, \dfrac{5\pi}{6}, \dfrac{9\pi}{6}$ **55.** $\dfrac{3\pi}{2}$ **56.** $0, \dfrac{\pi}{3}, \pi, \dfrac{5\pi}{3}$ **57.** π **58.** $\dfrac{\pi}{2}, 3.87, 5.55$ **59.** $\dfrac{\pi}{6}, \dfrac{5\pi}{6}, \dfrac{7\pi}{6}, \dfrac{11\pi}{6}$ **60.** $1.2, 5.08$ **61.** $0, \dfrac{\pi}{6}, \pi, \dfrac{11\pi}{6}$

62. $0, \pi$ **63.** $t = \dfrac{2}{3} + 4n$ or $t = \dfrac{10}{3} + 4n$, where n is any integer.   **64.** $12°$ or $78°$

Chapter 6 Test

1. $-\dfrac{63}{65}$ **2.** $\dfrac{56}{33}$ **3.** $-\dfrac{24}{25}$ **4.** $\dfrac{3\sqrt{13}}{13}$ **5.** $\dfrac{\sqrt{6} + \sqrt{2}}{4}$ **6.** $\cos x \csc x = \cos x \cdot \dfrac{1}{\sin x} = \dfrac{\cos x}{\sin x} = \cot x$

7. $\dfrac{\sec x}{\cot x + \tan x} = \dfrac{\dfrac{1}{\cos x}}{\dfrac{\cos x}{\sin x} + \dfrac{\sin x}{\cos x}} = \dfrac{\dfrac{1}{\cos x}}{\dfrac{\cos^2 x + \sin^2 x}{\sin x \cos x}} = \dfrac{1}{\cos x} \cdot \dfrac{\sin x \cos x}{1} = \sin x$

8. $1 - \dfrac{\cos^2 x}{1 + \sin x} = 1 - \dfrac{(1 - \sin^2 x)}{1 + \sin x} = 1 - \dfrac{(1 + \sin x)(1 - \sin x)}{1 + \sin x} = 1 - (1 - \sin x) = \sin x$

9. $\cos\left(\theta + \dfrac{\pi}{2}\right) = \cos \theta \cos \dfrac{\pi}{2} - \sin \theta \sin \dfrac{\pi}{2} = \cos \theta \cdot 0 - \sin \theta \cdot 1 = -\sin \theta$

10. $\dfrac{\sin(\alpha - \beta)}{\sin \alpha \cos \beta} = \dfrac{\sin \alpha \cos \beta - \cos \alpha \sin \beta}{\sin \alpha \cos \beta} = \dfrac{\sin \alpha \cos \beta}{\sin \alpha \cos \beta} - \dfrac{\cos \alpha \sin \beta}{\sin \alpha \cos \beta} = 1 - \cot \alpha \tan \beta$

11. $\sin t \cos t(\tan t + \cot t) = \sin t \cos t\left(\dfrac{\sin t}{\cos t} + \dfrac{\cos t}{\sin t}\right) = \sin^2 t + \cos^2 t = 1$ **12.** $\dfrac{7\pi}{18}, \dfrac{11\pi}{18}, \dfrac{19\pi}{18}, \dfrac{23\pi}{18}, \dfrac{31\pi}{18},$ and $\dfrac{35\pi}{18}$

13. $\dfrac{\pi}{2}, \dfrac{7\pi}{6}, \dfrac{3\pi}{2}, \dfrac{11\pi}{6}$ **14.** $0, \dfrac{\pi}{3}, \dfrac{5\pi}{3}$ **15.** $0, \dfrac{2\pi}{3}, \dfrac{4\pi}{3}$

Cumulative Review Exercises (Chapters 1–6)

1. $-3, 1 + 2i$, and $1 - 2i$ **2.** $x = \dfrac{\log 125}{\log 11} + 1$ or $x \approx 3.08$ **3.** $(-\infty, -4] \cup [2, \infty)$ **4.** $\dfrac{\pi}{3}, \dfrac{5\pi}{3}$

5.
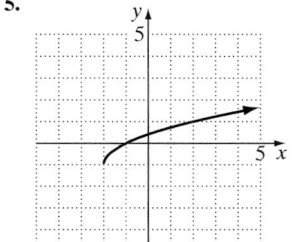

6.
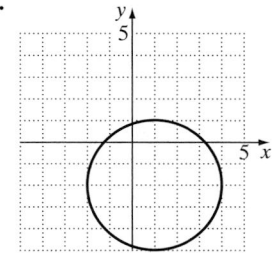

7.
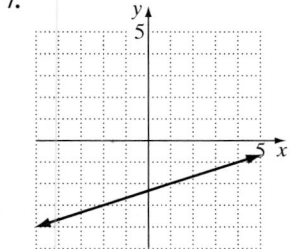

8.
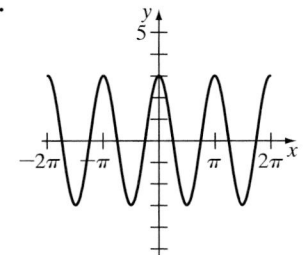

9.

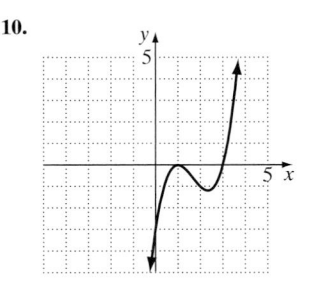

10.

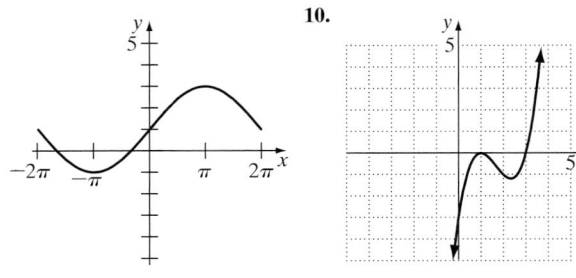

11. $2a + h + 3$ **12.** $-\dfrac{\sqrt{2}}{2}$

13. Proofs may vary.   **14.** $\dfrac{16}{9}\pi$ radians

15. $t \approx 19.1$ yr   **16.** $f^{-1}(x) = \dfrac{3x + 1}{x - 2}$

17. $B = 67°, b = 28.27, c = 30.71$

18. 106 mg   **19.** $h \approx 15.9$ ft   **20.** Answers may vary.

CHAPTER 7

Section 7.1

Check Point Exercises

1. $B = 34°, a \approx 13$ cm, $b \approx 8$ cm   **2.** $B = 117.5°, a \approx 9, c \approx 5$ **3.** one triangle; $C \approx 24°, B \approx 33°, b \approx 31$ **4.** no triangle
5. two triangles; $B_1 \approx 50°, C_1 \approx 95°, c_1 = 21$; $B_2 \approx 130°, C_2 \approx 15°, c_2 \approx 5$ **6.** approximately 34 sq m   **7.** approximately 11 mi

Exercise Set 7.1

1. $B = 42°, a \approx 8.1, b \approx 8.1$ **3.** $A = 44°, b \approx 18.6, c \approx 22.8$ **5.** $C = 95°, b \approx 81.0, c \approx 134.1$ **7.** $B = 40°, b \approx 20.9, c \approx 31.8$
9. $C = 111°, b \approx 7.3, c \approx 16.1$ **11.** $A = 80°, a \approx 39.5, c \approx 10.4$ **13.** $B = 30°, a \approx 316.0, b \approx 174.3$
15. $C = 50°, a \approx 7.1, b \approx 7.1$ **17.** one triangle; $B \approx 29°, C \approx 111°, c \approx 29.0$ **19.** one triangle; $C \approx 52°, B \approx 65°, b \approx 10.2$
21. one triangle; $C \approx 55°, B \approx 13°, b \approx 10.2$ **23.** no triangle   **25.** two triangles; $B_1 \approx 77°, C_1 \approx 43°, c_1 \approx 12.6$; $B_2 \approx 103°$,
$C_2 \approx 17°, c_2 \approx 5.4$ **27.** two triangles; $B_1 \approx 54°, C_1 \approx 89°, c_1 \approx 19.9$; $B_2 \approx 126°, C_2 \approx 17°, c_2 \approx 5.8$ **29.** two triangles; $C_1 \approx 68°$,
$B_1 \approx 54°, b_1 \approx 21.0$; $C_2 \approx 112°, B_2 \approx 10°, b_2 \approx 4.5$ **31.** no triangle   **33.** 297 sq ft   **35.** 5 sq yd   **37.** 10 sq m
39. Station A is about 6 miles from the fire, station B is about 9 miles from the fire.   **41.** about 3672 yards from one end of the beach and
3576 yards from the other.   **43.** about 184 ft   **45.** about 56 ft   **47.** about 30 ft   **49. a.** $a \approx 494$ ft   **b.** 343 ft
51. either 9.9 mi or 2.4 mi   **63.** No   **65.** 41 ft

Section 7.2

Check Point Exercises

1. $a = 13, B \approx 28°, C \approx 32°$ **2.** $B \approx 97.9°, A \approx 52.4°, C \approx 29.7°$ **3.** approximately 917 mi apart   **4.** approximately 47 sq m

Exercise Set 7.2

1. $a \approx 6.0, B \approx 29°, C \approx 105°$ **3.** $c \approx 7.6, A \approx 52°, B \approx 32°$ **5.** $A \approx 44°, B \approx 68°, C \approx 68°$ **7.** $A \approx 117°, B \approx 36°, C \approx 27°$
9. $c \approx 4.7, A \approx 46°, B \approx 92°$ **11.** $a \approx 6.3, C \approx 28°, B \approx 50°$ **13.** $b \approx 4.7, C \approx 54°, A \approx 76°$ **15.** $b \approx 5.4, C \approx 22°, A \approx 68°$
17. $C \approx 112°, A \approx 28°, B \approx 40°$ **19.** $B \approx 100°, A \approx 19°, C \approx 61°$ **21.** $A = 60°, B = 60°, C = 60°$ **23.** no triangle
25. 4 sq ft   **27.** 22 sq m   **29.** 31 sq yd   **31.** about 61.7 mi apart   **33.** about 193 yd   **35.** N12°E   **37. a.** about 19.3 mi
b. S58°E   **39.** The guy wire anchored downhill is about 417.4 feet. The one anchored uphill is about 398.2 feet.   **41.** about 63.7 ft
43. $123,454   **53.** 153.4°

Section 7.3

Check Point Exercises

1. a.   **b.**   **c.**

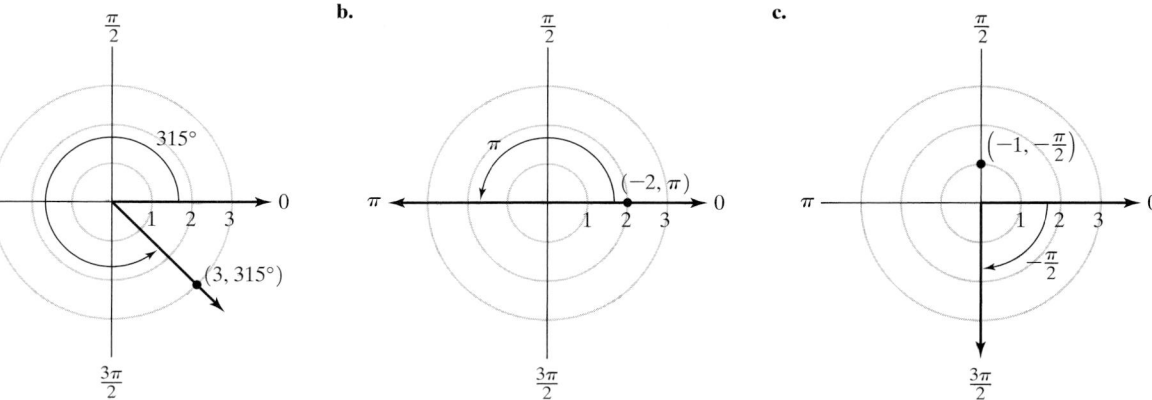

2. a. $\left(5, \frac{9\pi}{4}\right)$ **b.** $\left(-5, \frac{5\pi}{4}\right)$ **c.** $\left(5, -\frac{7\pi}{4}\right)$ **3. a.** $(-3, 0)$ **b.** $(-5\sqrt{3}, -5)$ **4.** $\left(2, \frac{5\pi}{3}\right)$

5. $\left(4, \frac{3\pi}{2}\right)$ **6.** $r = \dfrac{6}{3\cos\theta - \sin\theta}$ **7. a.** $x^2 + y^2 = 16$ **b.** $y = -x$ **c.** $x = 1$

Exercise Set 7.3

1. C **3.** A **5.** B **7.** C **9.** A

11.

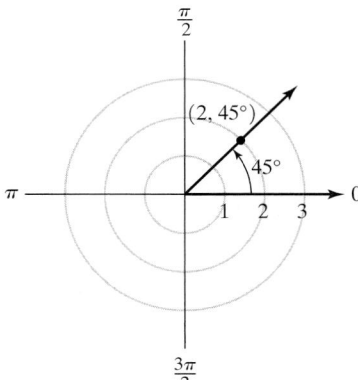

13.

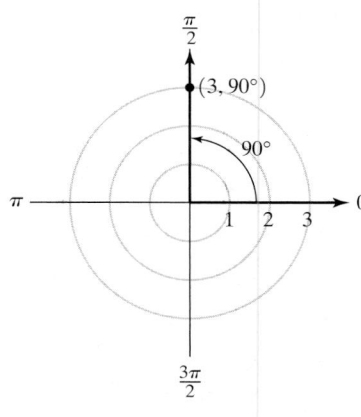

15.

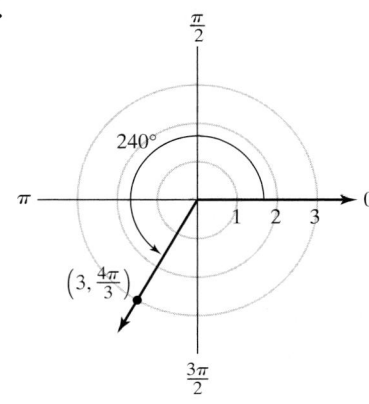

17.

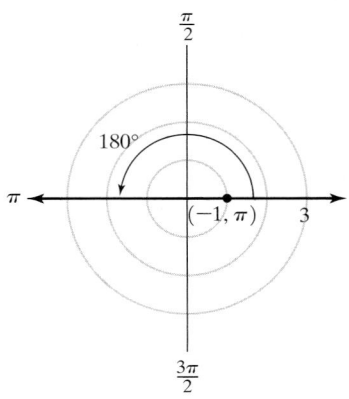

19.

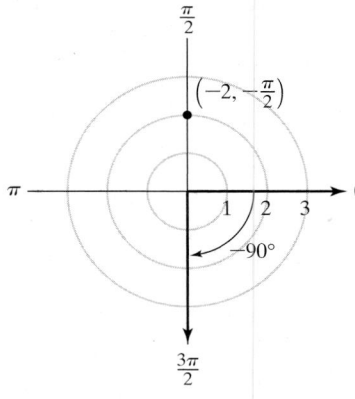

21.

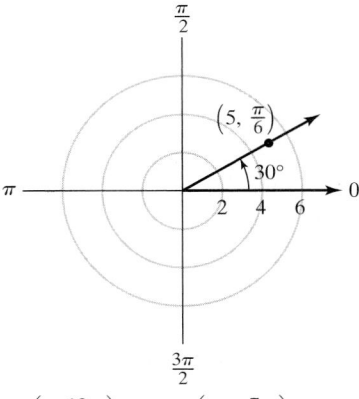

a. $\left(5, \frac{13\pi}{6}\right)$ **b.** $\left(-5, \frac{7\pi}{6}\right)$

c. $\left(5, -\frac{11\pi}{6}\right)$

23.

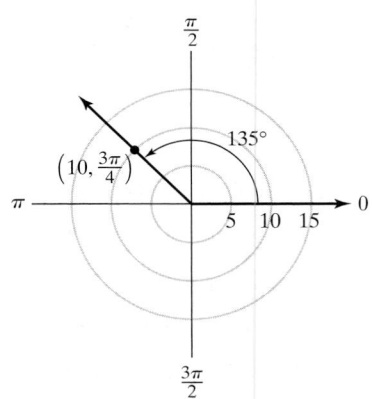

a. $\left(10, \frac{11\pi}{4}\right)$ **b.** $\left(-10, \frac{7\pi}{4}\right)$

c. $\left(10, -\frac{5\pi}{4}\right)$

25.

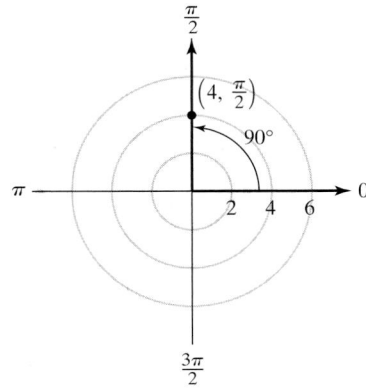

a. $\left(4, \frac{5\pi}{2}\right)$ **b.** $\left(-4, \frac{3\pi}{2}\right)$

c. $\left(4, -\frac{3\pi}{2}\right)$

27. $(0, 4)$ **29.** $(1, \sqrt{3})$ **31.** $(0, -4)$ **33.** approximately $(-5.9, 4.4)$ **35.** $\left(\sqrt{8}, \dfrac{3\pi}{4}\right)$ **37.** $\left(4, \dfrac{5\pi}{3}\right)$ **39.** $\left(2, \dfrac{7\pi}{6}\right)$

41. $(5, 0)$ **43.** $r = \dfrac{7}{3\cos\theta + \sin\theta}$ **45.** $r = \dfrac{7}{\cos\theta}$ **47.** $r = 3$ **49.** $r = 4\cos\theta$ **51.** $r = \dfrac{6\cos\theta}{\sin^2\theta}$ **53.** $x^2 + y^2 = 64$

55. $x = 0$ **57.** $y = 3$ **59.** $y = 4$ **61.** $x^2 + y^2 = y$ **63.** $x^2 + y^2 = 6x + 4y$ **65.** $y = \dfrac{1}{x}$ **67.** $\left(15, \dfrac{4\pi}{3}\right)$

69. 6.3 knots at an angle of $50°$ to the wind.   **71.** Answers may vary.   **81.** $(-2, 3.464)$ **83.** $(-1.857, -3.543)$

85. $(3, 0.730)$ **89.** $x^2 + y^2 = 4x$; center: $(2, 0)$, radius: 2

Section 7.4

Check Point Exercises

1.

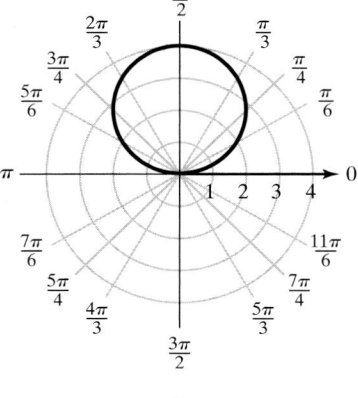

2.

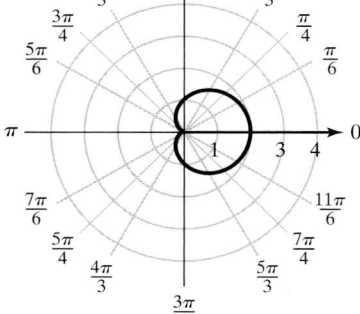

3.

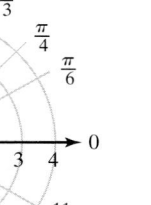

4.

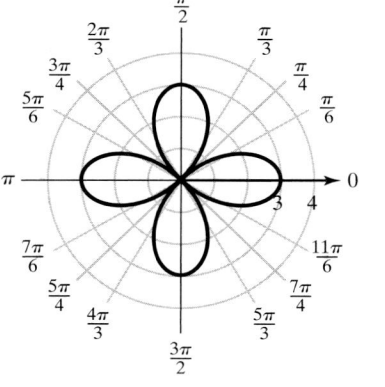

5.

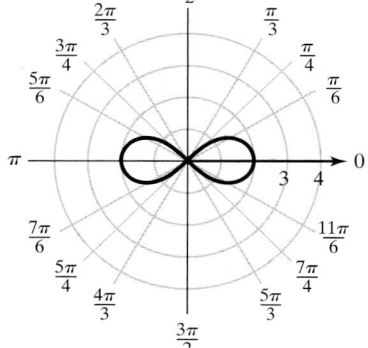

Exercise Set 7.4

1. $r = 1 - \sin\theta$ **3.** $r = 2\cos\theta$ **5.** $r = 3\sin 3\theta$ **7. a.** May or may not have symmetry with respect to polar axis.

b. Has symmetry with respect to the line $\theta = \dfrac{\pi}{2}$.   **c.** May or may not have symmetry about the pole.

9. a. Has symmetry with respect to polar axis.   **b.** May or may not have symmetry with respect to the line $\theta = \dfrac{\pi}{2}$.

c. May or may not have symmetry about pole.   **11. a.** Has symmetry with respect to polar axis.

b. Has symmetry with respect to the line $\theta = \dfrac{\pi}{2}$.   **c.** Has symmetry about the pole.

13.

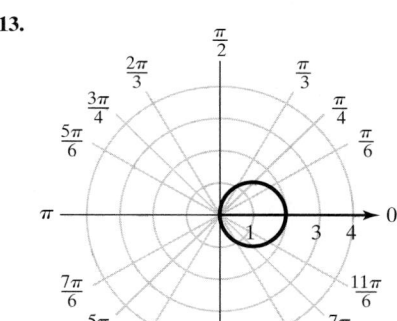

15.

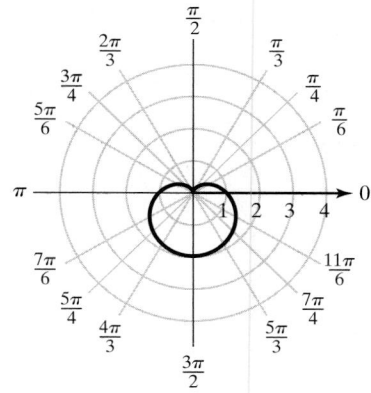

17.

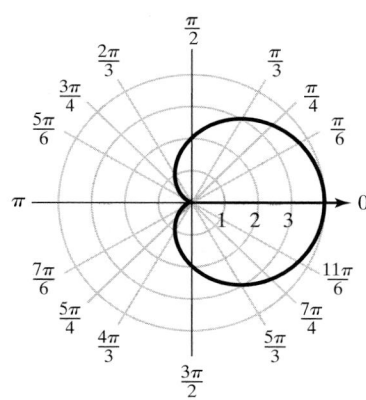

19.

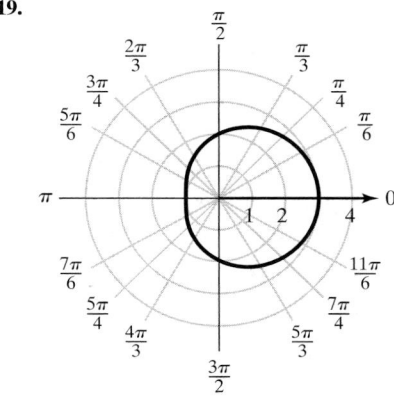

21.

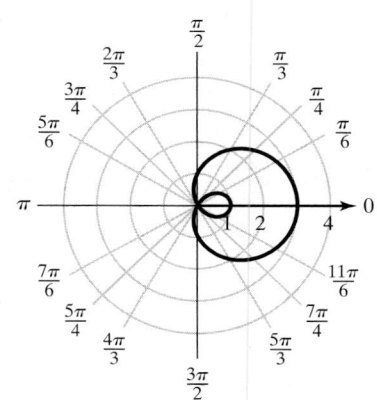

23.

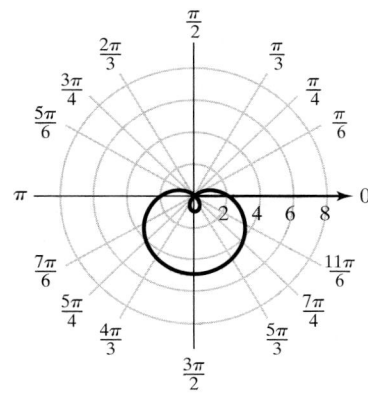

25.

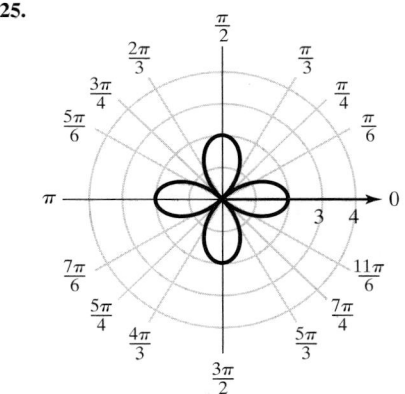

27.

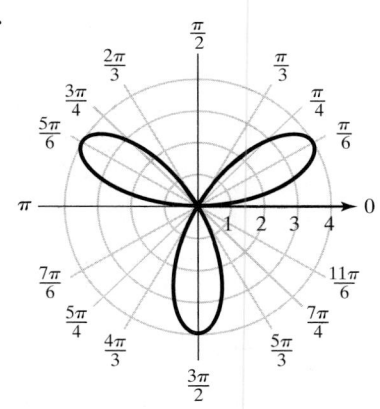

29.

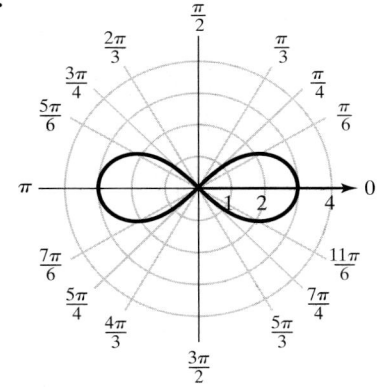

31.

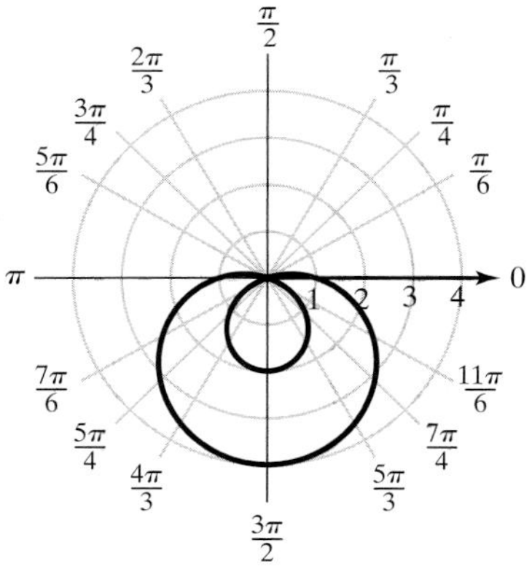

33.

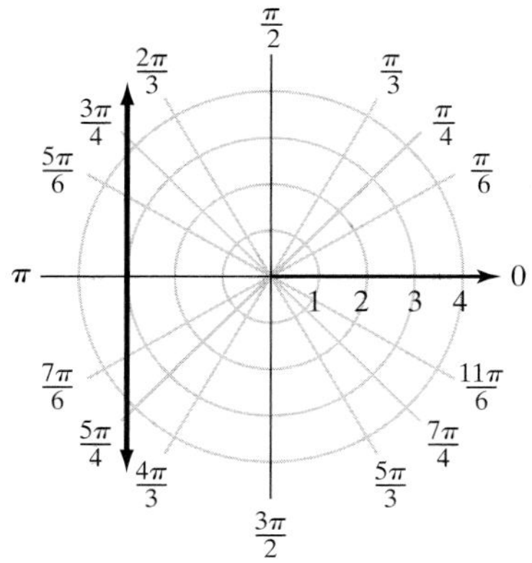

35. 6 knots
37. 8 knots

39. 90°; about $7\frac{1}{2}$ knots

49. **51.** 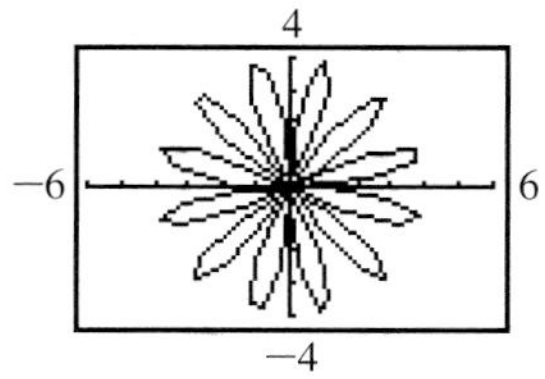 **53.** 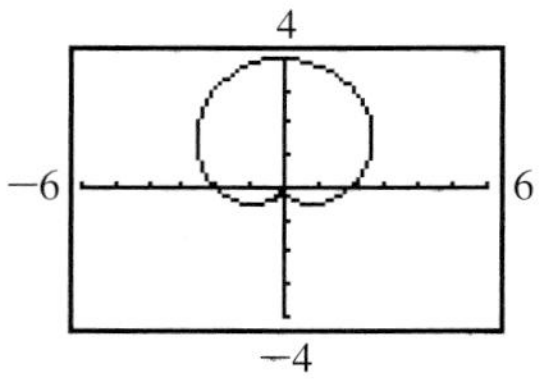 **55.**

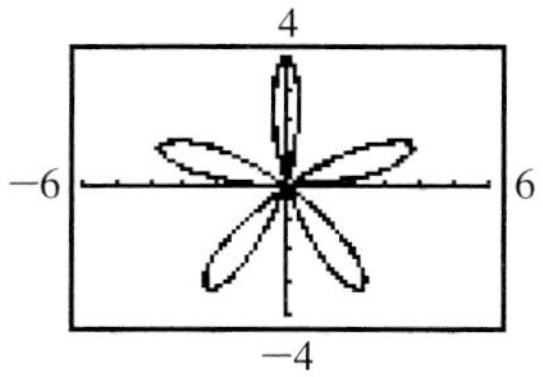

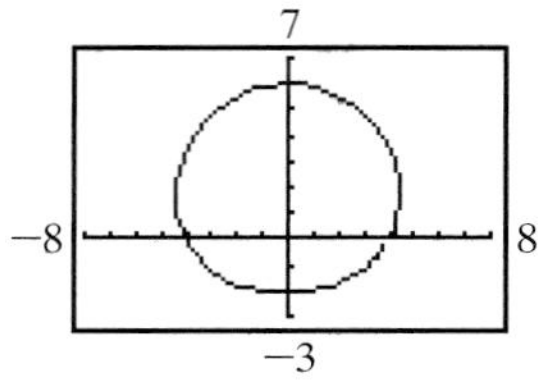

57. 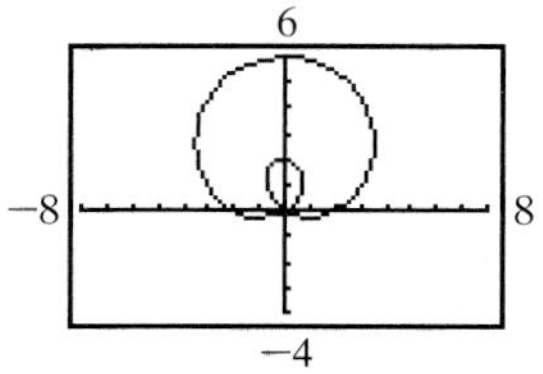 **59.** 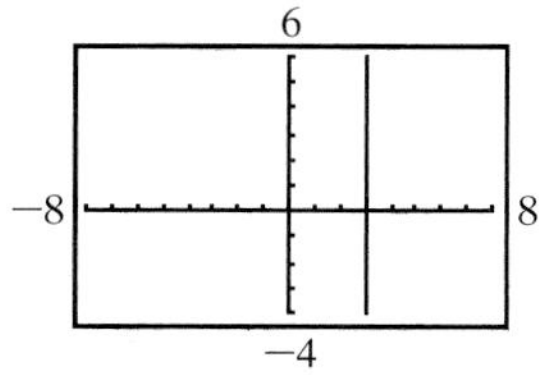 **61.** 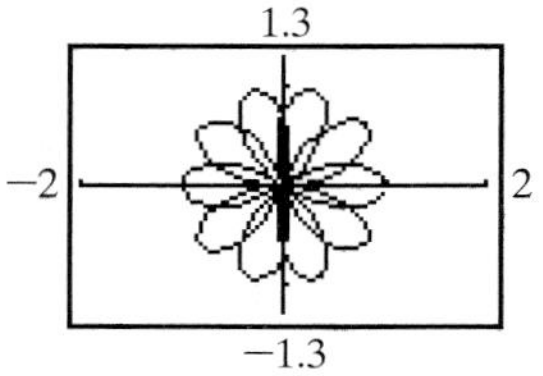 **63.**

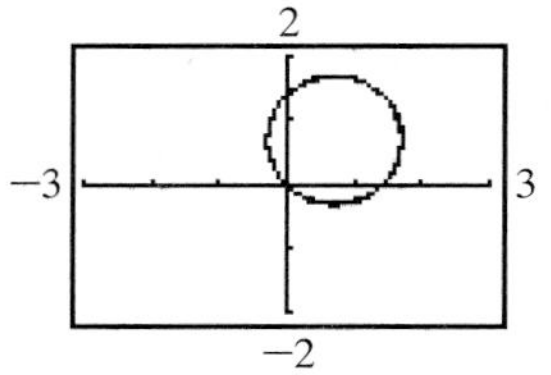

65. 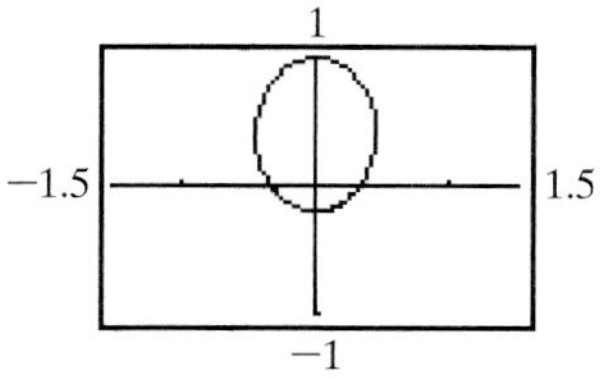 **67.** 2π **69.** **71.**

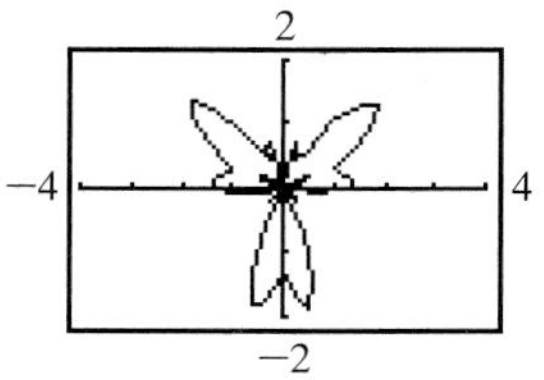

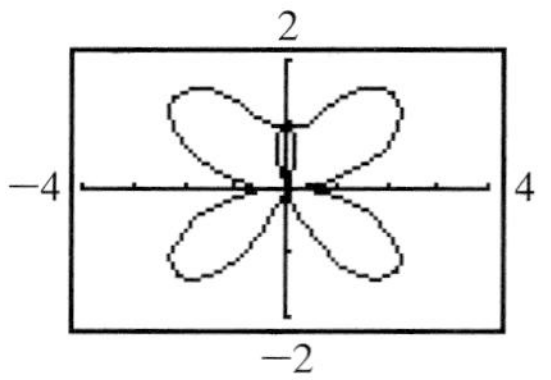

73. As n increases, $\sin n\theta$ increases its number of loops. If n is odd, there are n loops and θmax $= \pi$ traces the graph once, while if n is even, there are $2n$ loops and θmax $= 2\pi$ traces the graph once.

75. There are n small petals and n large petals for each value of n. For odd values of n, the small petals are inside the large petals. For even n, they are between the large petals.

77.

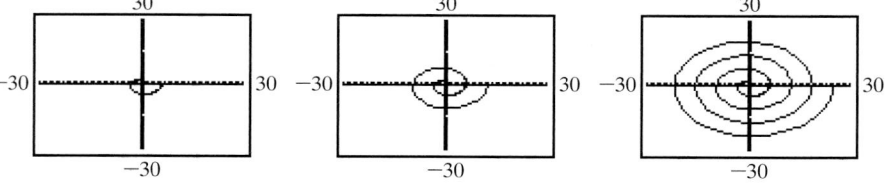

79. Answers may vary.   **81.**

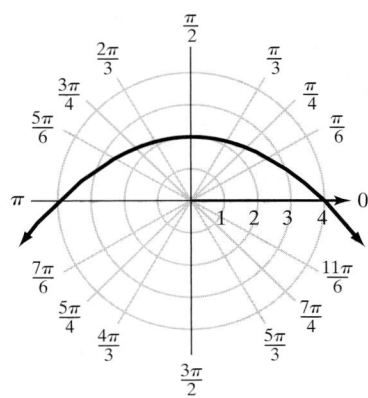

Section 7.5

Check Point Exercises

1. a.

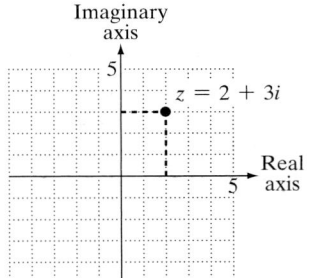

b.

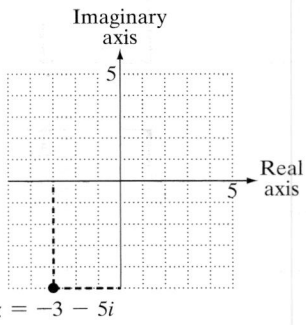

c.

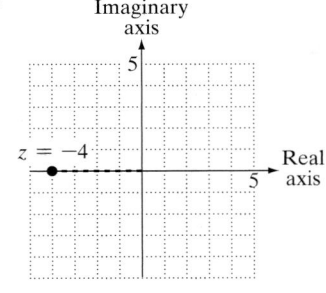

$;2\left(\cos\dfrac{4\pi}{3} + i\sin\dfrac{4\pi}{3}\right)$

d.

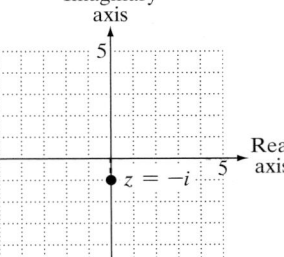

2. a. 13   **b.** $\sqrt{13}$ **3.**

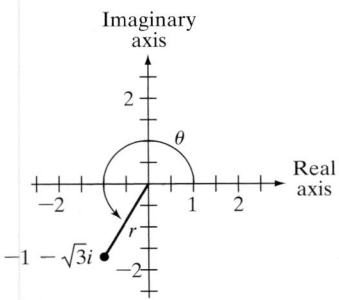

4. $z = 2\sqrt{3} + 2i$ **5.** $30(\cos 60° + i\sin 60°)$ **6.** $10(\cos\pi + i\sin\pi)$ **7.** $-16\sqrt{3} + 16i$ **8.** -4
9. $2(\cos 15° + i\sin 15°); 2(\cos 105° + i\sin 105°); 2(\cos 195° + i\sin 195°); 2(\cos 285° + i\sin 285°)$

10. $3; -\dfrac{3}{2} + \dfrac{3\sqrt{3}}{2}i; -\dfrac{3}{2} - \dfrac{3\sqrt{3}}{2}i$

Exercise Set 7.5

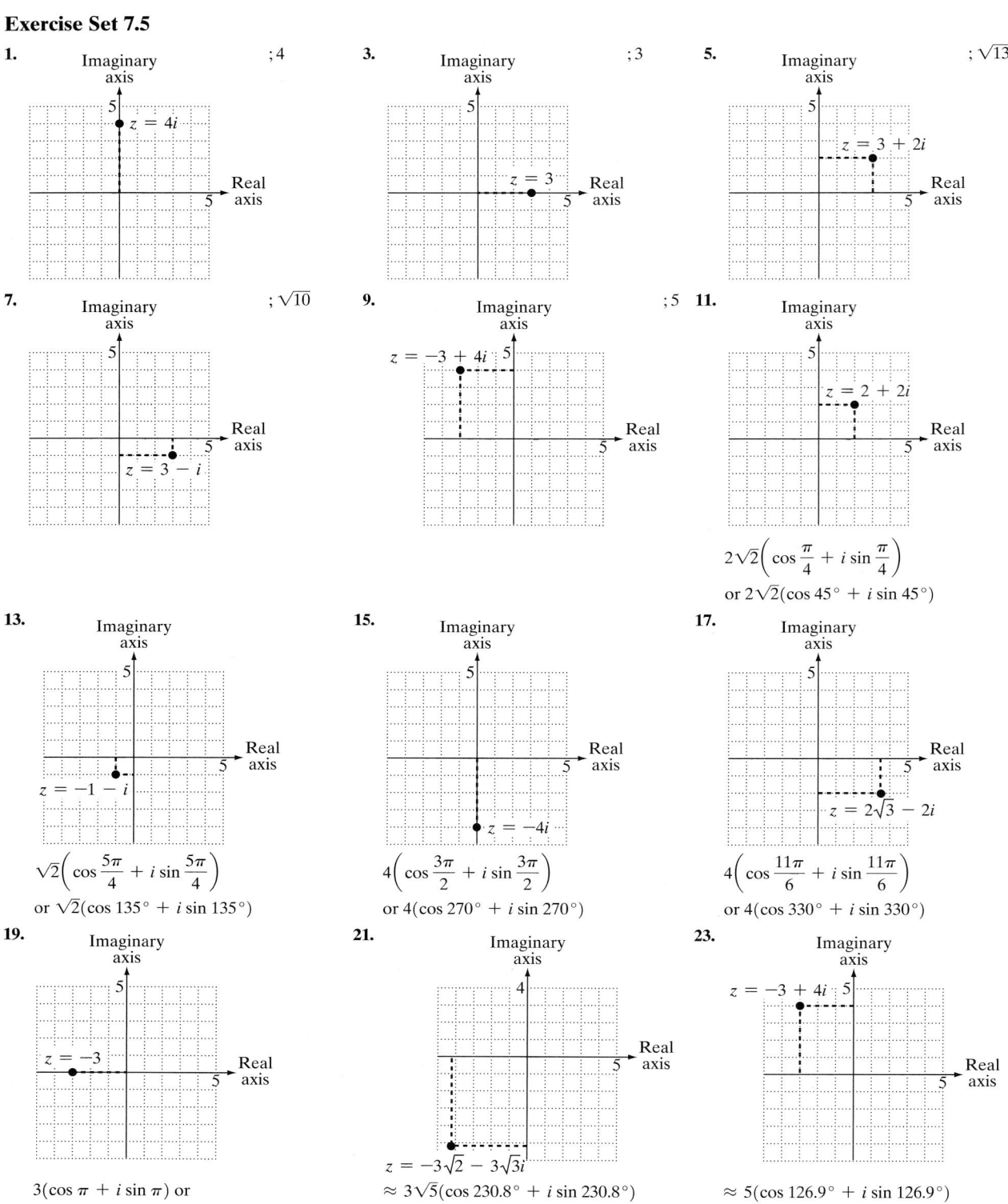

1.
 Imaginary axis
 $z = 4i$
 Real axis
 ; 4

3.
 Imaginary axis
 $z = 3$
 Real axis
 ; 3

5.
 Imaginary axis
 $z = 3 + 2i$
 Real axis
 ; $\sqrt{13}$

7.
 Imaginary axis
 $z = 3 - i$
 Real axis
 ; $\sqrt{10}$

9.
 Imaginary axis
 $z = -3 + 4i$
 Real axis
 ; 5

11.
 Imaginary axis
 $z = 2 + 2i$
 Real axis

$$2\sqrt{2}\left(\cos\frac{\pi}{4} + i\sin\frac{\pi}{4}\right)$$
or $2\sqrt{2}(\cos 45° + i\sin 45°)$

13.
 Imaginary axis
 $z = -1 - i$
 Real axis

$$\sqrt{2}\left(\cos\frac{5\pi}{4} + i\sin\frac{5\pi}{4}\right)$$
or $\sqrt{2}(\cos 135° + i\sin 135°)$

15.
 Imaginary axis
 $z = -4i$
 Real axis

$$4\left(\cos\frac{3\pi}{2} + i\sin\frac{3\pi}{2}\right)$$
or $4(\cos 270° + i\sin 270°)$

17.
 Imaginary axis
 $z = 2\sqrt{3} - 2i$
 Real axis

$$4\left(\cos\frac{11\pi}{6} + i\sin\frac{11\pi}{6}\right)$$
or $4(\cos 330° + i\sin 330°)$

19.
 Imaginary axis
 $z = -3$
 Real axis

$3(\cos\pi + i\sin\pi)$ or
$3(\cos 180° + i\sin 180°)$

21.
 Imaginary axis
 Real axis
 $z = -3\sqrt{2} - 3\sqrt{3}i$
$\approx 3\sqrt{5}(\cos 230.8° + i\sin 230.8°)$

23.
 Imaginary axis
 $z = -3 + 4i$
 Real axis
$\approx 5(\cos 126.9° + i\sin 126.9°)$

25.

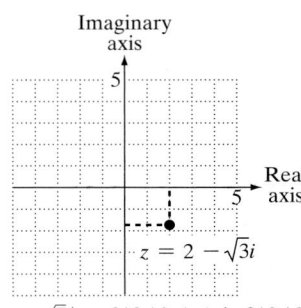

$z = 2 - \sqrt{3}i$

$\approx \sqrt{7}(\cos 319.1° + i \sin 319.1°)$

27. $3\sqrt{3} + 3i$ **29.** $-2 - 2\sqrt{3}i$ **31.** $4\sqrt{2} - 4\sqrt{2}i$ **33.** $5i$

35. $z \approx -18.1 - 8.5i$ **37.** $30(\cos 70° + i \sin 70°)$

39. $12\left(\cos \dfrac{3\pi}{10} + i \sin \dfrac{3\pi}{10}\right)$ **41.** $\cos \dfrac{7\pi}{12} + i \sin \dfrac{7\pi}{12}$ **43.** $2(\cos \pi + i \sin \pi)$

45. $5(\cos 50° + i \sin 50°)$ **47.** $\dfrac{3}{4}\left(\cos \dfrac{\pi}{10} + i \sin \dfrac{\pi}{10}\right)$ **49.** $\cos 240° + i \sin 240°$

51. $2(\cos 0° + i \sin 0°)$ **53.** $32\sqrt{2} + 32\sqrt{2}i$ **55.** $-4 - 4\sqrt{3}i$

57. $\dfrac{1}{64}i$ **59.** $-2 - 2\sqrt{3}i$ **61.** $-4 - 4i$ **63.** -64

65. $3(\cos 15° + i \sin 15°); 3(\cos 195° + i \sin 195°)$

67. $2(\cos 70° + i \sin 70°); 2(\cos 190° + i \sin 190°); 2(\cos 310° + i \sin 310°)$

69. $\dfrac{3}{2} + \dfrac{3\sqrt{3}}{2}i; -\dfrac{3\sqrt{3}}{2} + \dfrac{3}{2}i; -\dfrac{3}{2} - \dfrac{3\sqrt{3}}{2}i; \dfrac{3\sqrt{3}}{2} - \dfrac{3}{2}i$ **71.** $2; \approx 0.6 + 1.9i; \approx -1.6 + 1.2i; \approx -1.6 - 1.2i; \approx 0.6 - 1.9i$

73. $1; -\dfrac{1}{2} + \dfrac{\sqrt{3}}{2}i; -\dfrac{1}{2} - \dfrac{\sqrt{3}}{2}i$ **75.** $\approx 1.1 + 0.2i; \approx -0.2 + 1.1i; \approx -1.1 - 0.2i; \approx 0.2 - 1.1i$

77. a. $i; -1 + i; -i; -1 + i; -i; -1 + i$ **b.** Complex numbers may vary.

93.

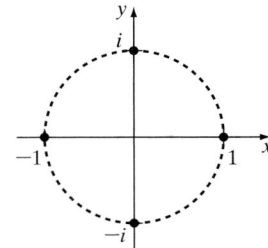

95. $2\left(\cos \dfrac{\pi}{6} + i \sin \dfrac{\pi}{6}\right), 2\left(\cos \dfrac{5\pi}{6} + i \sin \dfrac{5\pi}{6}\right), 2\left(\cos \dfrac{3\pi}{2} + i \sin \dfrac{3\pi}{2}\right)$

Section 7.6

Check Point Exercises

1. $\|\mathbf{u}\| = 5 = \|\mathbf{v}\|$ and $m_u = \dfrac{4}{3} = m_v$ **2.**

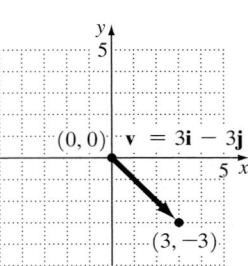

$; \|\mathbf{v}\| = 3\sqrt{2}$ **3.** $\mathbf{v} = 3\mathbf{i} + 4\mathbf{j}$ **4. a.** $11\mathbf{i} - 2\mathbf{j}$ **b.** $3\mathbf{i} + 8\mathbf{j}$

5. a. $56\mathbf{i} + 80\mathbf{j}$ **b.** $-35\mathbf{i} - 50\mathbf{j}$

6. $30\mathbf{i} + 33\mathbf{j}$

7. $\dfrac{4}{5}\mathbf{i} - \dfrac{3}{5}\mathbf{j}; \sqrt{\left(\dfrac{4}{5}\right)^2 + \left(-\dfrac{3}{5}\right)^2} = \sqrt{\dfrac{16}{25} + \dfrac{9}{25}}$

$= \sqrt{\dfrac{25}{25}} = 1$

8. about 82.5 lb in the direction of approximately $16.2°$ relative to the 60-lb force

Exercise Set 7.6

1. a. $\sqrt{41}$ **b.** $\sqrt{41}$ **c.** $\mathbf{u} = \mathbf{v}$ **3. a.** 6 **b.** 6 **c.** $\mathbf{u} = \mathbf{v}$

5.   **7.**   **9.**   **11.**

$\sqrt{10}$ $\sqrt{2}$ $2\sqrt{10}$ 4

13. $10\mathbf{i} + 6\mathbf{j}$ **15.** $6\mathbf{i} - 3\mathbf{j}$ **17.** $-6\mathbf{i} - 14\mathbf{j}$ **19.** $9\mathbf{i}$ **21.** $-\mathbf{i} + 2\mathbf{j}$ **23.** $5\mathbf{i} - 12\mathbf{j}$ **25.** $-5\mathbf{i} + 12\mathbf{j}$ **27.** $-15\mathbf{i} + 35\mathbf{j}$

29. $4\mathbf{i} + 24\mathbf{j}$ **31.** $-9\mathbf{i} - 4\mathbf{j}$ **33.** $-5\mathbf{i} + 45\mathbf{j}$ **35.** $2\sqrt{29}$ **37.** $\sqrt{10}$ **39.** $\mathbf{i}$ **41.** $\frac{3}{5}\mathbf{i} - \frac{4}{5}\mathbf{j}$ **43.** $\frac{3}{\sqrt{13}}\mathbf{i} - \frac{2}{\sqrt{13}}\mathbf{j}$

45. $\frac{\sqrt{2}}{2}\mathbf{i} + \frac{\sqrt{2}}{2}\mathbf{j}$ **47. a.** approximately 53.9 lb **b.** approximately $21.8°$ relative to the 50-lb force

49. a. approximately 413.1 lb **b.** approximately $18.5°$ relative to the 300-lb force

51. a.

b. 264 mph

53. $\|\mathbf{v}\| \approx 164.5$ mph is the resultant speed of the plane.
71. a. False **b.** True **c.** False **d.** False
73. The plane's true speed relative to the ground is 269.1 miles per hour.; The compass heading relative to the ground is $278.3°$.

Section 7.7

Check Point Exercises

1. a. 18 **b.** 18 **c.** 5 **2.** $100.3°$ **3.** orthogonal **4.** $30\sqrt{2}\mathbf{i} + 30\sqrt{2}\mathbf{j}$ **5.** $\frac{7}{2}\mathbf{i} - \frac{7}{2}\mathbf{j}$ **6.** $\mathbf{v}_1 = \frac{7}{2}\mathbf{i} - \frac{7}{2}\mathbf{j}; \mathbf{v}_2 = -\frac{3}{2}\mathbf{i} - \frac{3}{2}\mathbf{j}$
7. approximately 2598 ft-lb

Exercise Set 7.7

1. a. 6 **b.** 10 **3. a.** -6 **b.** 41 **5. a.** 100 **b.** 61 **7. a.** 0 **b.** 25 **9.** 3 **11.** 3 **13.** 20 **15.** 20 **17.** $79.7°$
19. $160.3°$ **21.** $38.7°$ **23.** orthogonal **25.** orthogonal **27.** not orthogonal **29.** not orthogonal **31.** orthogonal

33. $3\sqrt{3}\mathbf{i} + 3\mathbf{j}$ **35.** $-6\sqrt{2}\mathbf{i} - 6\sqrt{2}\mathbf{j}$ **37.** $\approx -0.20\mathbf{i} + 0.46\mathbf{j}$ **39.** $\mathbf{v}_1 = \text{proj}_\mathbf{w}\mathbf{v} = \frac{5}{2}\mathbf{i} - \frac{5}{2}\mathbf{j}; \mathbf{v}_2 = \frac{1}{2}\mathbf{i} + \frac{1}{2}\mathbf{j}$

41. $\mathbf{v}_1 = \text{proj}_\mathbf{w}\mathbf{v} = -\frac{26}{29}\mathbf{i} + \frac{65}{29}\mathbf{j}; \mathbf{v}_2 = \frac{55}{29}\mathbf{i} + \frac{22}{29}\mathbf{j}$ **43.** $\mathbf{v}_1 = \text{proj}_\mathbf{w}\mathbf{v} = \mathbf{i} + 2\mathbf{j}; \mathbf{v}_2 = 0$ **45.** $22\sqrt{3}\mathbf{i} + 22\mathbf{j}$ **47.** $148.5\mathbf{i} + 20.9\mathbf{j}$

49. 1077; $\mathbf{v} \cdot \mathbf{w} = 1077$ means \$1077 in revenue was generated on Monday by the sale of 240 gallons of regular gas at \$1.90 per gallon and 300 gallons of premium gas at \$2.07 per gallon. **51.** 7600 foot-pounds **53.** 3392 foot-pounds **55.** $\approx 1.4\mathbf{i} + 0.6\mathbf{j}$; 1.4 in.
57. a. about 192,389 lb **b.** $188,185\mathbf{i} + 40,000\mathbf{j}$

71.
$$(c\mathbf{u}) \cdot \mathbf{v} \overset{?}{=} c(\mathbf{u} \cdot \mathbf{v})$$
$$[c(a_1\mathbf{i} + b_1\mathbf{j})] \cdot (a_2\mathbf{i} + b_2\mathbf{j}) \overset{?}{=} c[(a_1\mathbf{i} + b_1\mathbf{j}) \cdot (a_2\mathbf{i} + b_2\mathbf{j})]$$
$$(ca_1\mathbf{i} + cb_1\mathbf{j}) \cdot (a_2\mathbf{i} + b_2\mathbf{j}) \overset{?}{=} c[a_1(a_2) + b_1(b_2)]$$
$$(ca_1)(a_2) + (cb_1)(b_2) \overset{?}{=} c(a_1a_2) + c(b_1b_2)$$

Since a_1, a_2, b_1, b_2 and c are real numbers and multiplication of real numbers is associative, $(ca_1)(a_2) = c(a_1a_2)$ and $(cb_1)(b_2) = c(b_1b_2)$.
Thus, $(ca_1)(a_2) + (cb_1)(b_2) = c(a_1a_2) + c(b_1b_2)$
So $(c\mathbf{u}) \cdot \mathbf{v} = c(\mathbf{u} \cdot \mathbf{v})$.

73. 6.3 foot-pounds **75.** $b = -20$ **77.** any two vectors, $\mathbf{v}$ and $\mathbf{w}$, having the same direction

Chapter 7 Review Exercises

1. $C = 55°, b \approx 10.5$, and $c \approx 10.5$ **2.** $A = 43°, a \approx 171.9$, and $b \approx 241.0$ **3.** $b \approx 16.3, A \approx 72°$, and $C \approx 42°$
4. $C \approx 98°, A \approx 55°$, and $B \approx 27°$ **5.** $C = 120°, a \approx 45.0$, and $b \approx 33.2$ **6.** two triangles; $B_1 \approx 55°, C_1 \approx 86°$, and $c_1 \approx 31.7$;
$B_2 \approx 125°, C_2 \approx 16°$, and $c_2 \approx 8.8$ **7.** no triangle **8.** $a \approx 59.0, B \approx 3°$, and $C \approx 15°$ **9.** $B \approx 78°, A \approx 39°$, and $C \approx 63°$
10. B_1 (or B) $\approx 25°, C \approx 115°$, and $c \approx 8.5$ **11.** two triangles; $A_1 \approx 59°, C_1 \approx 84°, c_1 \approx 14.4; A_2 \approx 121°, C_2 \approx 22°, c_2 \approx 5.4$
12. $B \approx 9°, C \approx 148°$, and $c \approx 73.6$ **13.** 8 sq ft **14.** 4 sq ft **15.** 4 sq m **16.** 2 sq m **17.** 35 ft **18.** 35.6 mi **19.** 861 mi
20. 404 ft; 551 ft **21.** \$214,194

22.

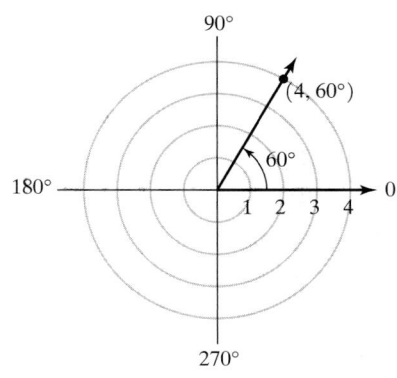

$(2, 2\sqrt{3})$

23.

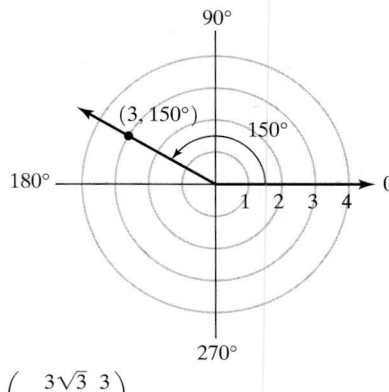

$\left(-\dfrac{3\sqrt{3}}{2}, \dfrac{3}{2}\right)$

24.

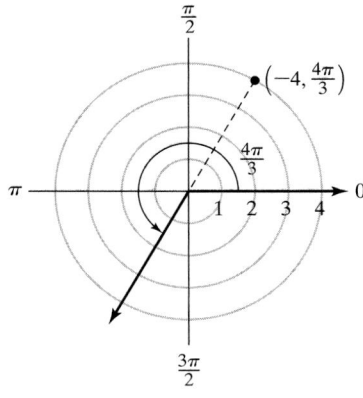

$(2, 2\sqrt{3})$

25.

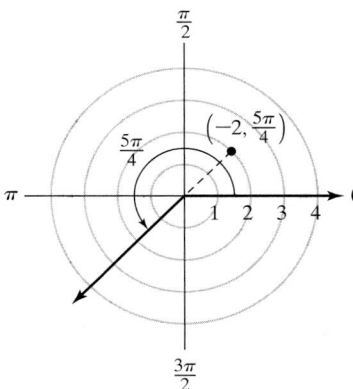

$(\sqrt{2}, \sqrt{2})$

26.

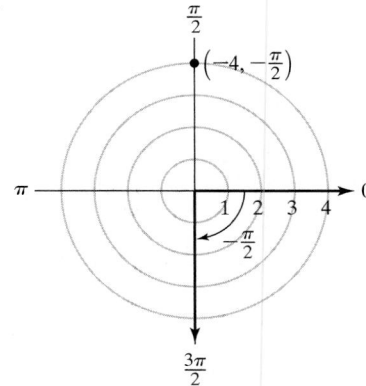

$(0, 4)$

27.

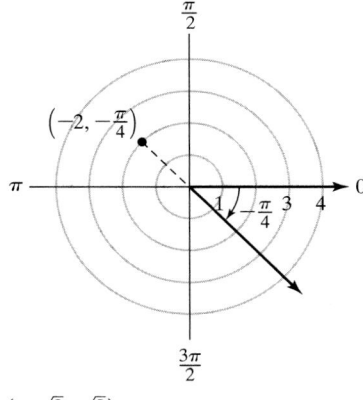

$(-\sqrt{2}, \sqrt{2})$

28.

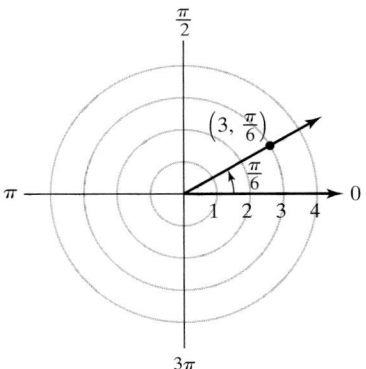

a. $\left(3, \dfrac{13\pi}{6}\right)$ **b.** $\left(-3, \dfrac{7\pi}{6}\right)$

c. $\left(3, -\dfrac{11\pi}{6}\right)$

29.

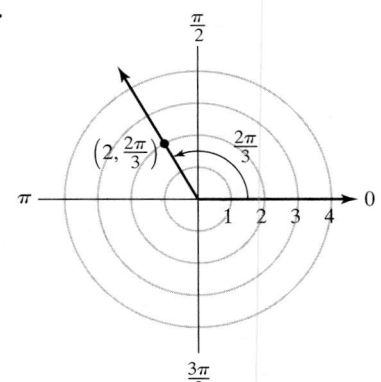

a. $\left(2, \dfrac{8\pi}{3}\right)$ **b.** $\left(-2, \dfrac{5\pi}{3}\right)$

c. $\left(2, -\dfrac{4\pi}{3}\right)$

30.

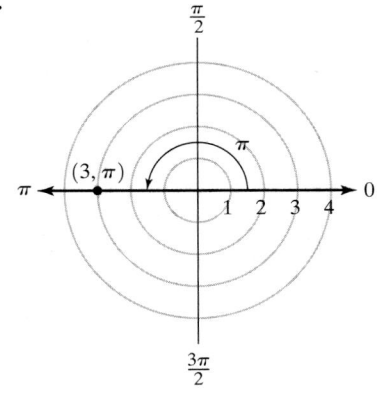

a. $(3, 3\pi)$ **b.** $(-3, 2\pi)$ **c.** $(3, -\pi)$

31. $\left(4\sqrt{2}, \dfrac{3\pi}{4}\right)$ **32.** $\left(3\sqrt{2}, \dfrac{7\pi}{4}\right)$ **33.** approximately $(13, 67°)$ **34.** approximately $(5, 127°)$ **35.** $\left(5, \dfrac{3\pi}{2}\right)$ **36.** $(1, 0)$

37. $r = \dfrac{8}{2\cos\theta + 3\sin\theta}$ **38.** $r = 10$ **39.** $5r^2 = 3r\sin\theta$ **40.** $x^2 + y^2 = 9$ **41.** $y = -x$ **42.** $x = -1$ **43.** $x = 5$

44. $x^2 + y^2 = 3x$ **45.** $5x + y = 8$ **46.** $y = \dfrac{2}{x}$ **47. a.** has symmetry   **b.** may or may not have symmetry

c. may or may not have symmetry   **48. a.** may or may not have symmetry   **b.** has symmetry   **c.** may or may not have symmetry

49. a. has symmetry   **b.** has symmetry   **c.** has symmetry

50.   **51.**   **52.**

53.   **54.**   **55.**

56.   **57.**   **58.**

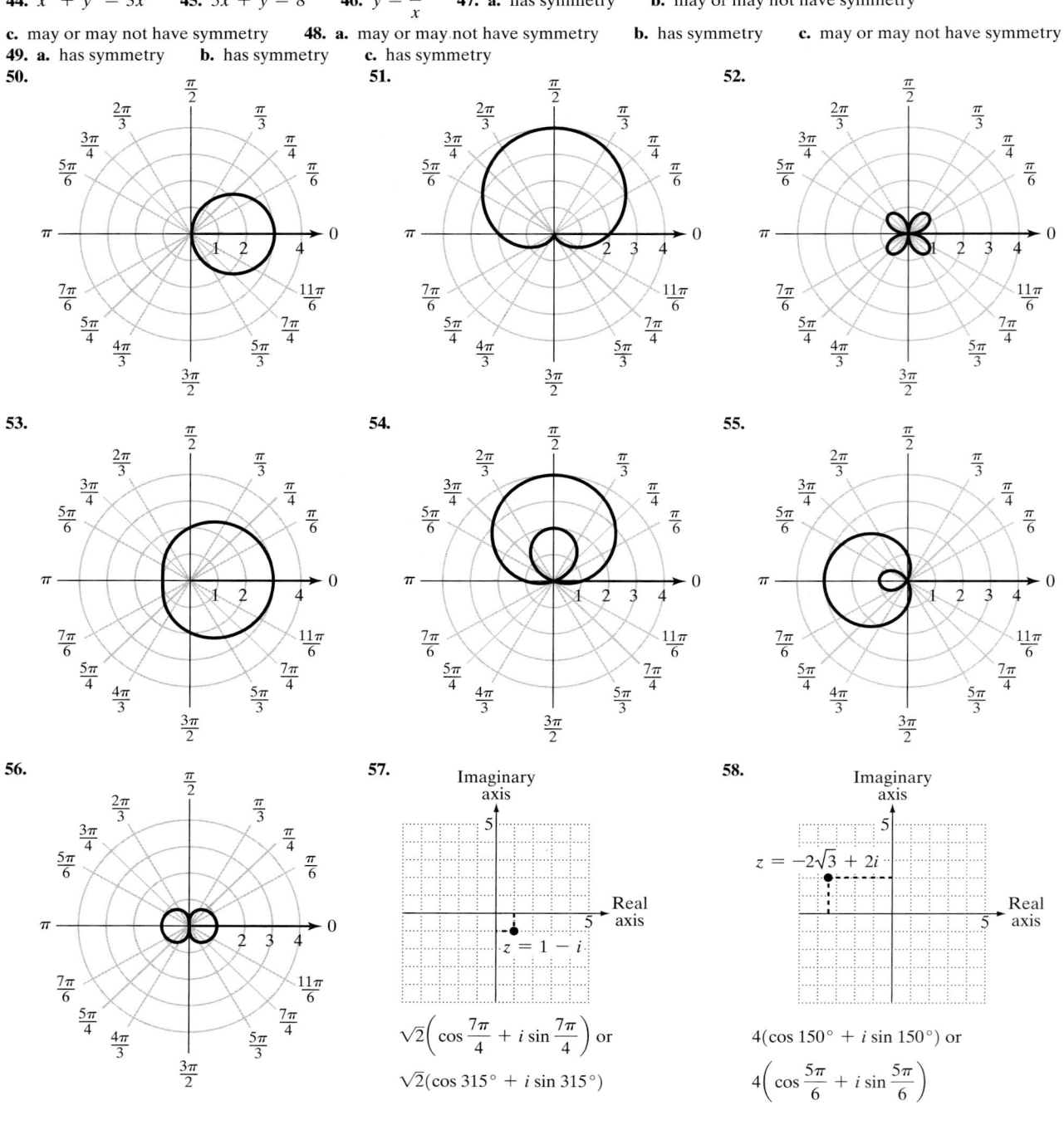

57. Imaginary axis $z = 1 - i$

$$\sqrt{2}\left(\cos\frac{7\pi}{4} + i\sin\frac{7\pi}{4}\right) \text{ or}$$

$$\sqrt{2}(\cos 315° + i\sin 315°)$$

58. Imaginary axis $z = -2\sqrt{3} + 2i$

$$4(\cos 150° + i\sin 150°) \text{ or}$$

$$4\left(\cos\frac{5\pi}{6} + i\sin\frac{5\pi}{6}\right)$$

59.

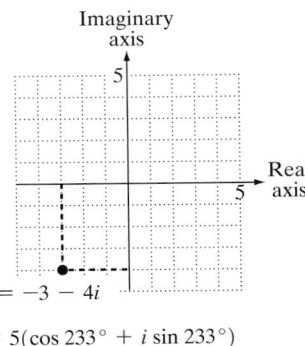

$z = -3 - 4i$

$\approx 5(\cos 233° + i \sin 233°)$

60.

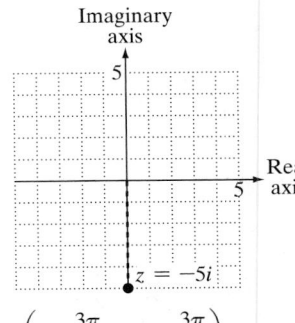

$z = -5i$

$5\left(\cos \dfrac{3\pi}{2} + i \sin \dfrac{3\pi}{2}\right)$ or

$5(\cos 270° + i \sin 270°)$

61. $z = 4 + 4\sqrt{3}i$
62. $z = -2\sqrt{3} - 2i$
63. $z = -3 + 3\sqrt{3}i$
64. $z \approx -0.01 + 0.06i$
65. $15(\cos 110° + i \sin 110°)$
66. $\cos 265° + i \sin 265°$
67. $40(\cos \pi + i \sin \pi)$
68. $2(\cos 5° + i \sin 5°)$

69. $\dfrac{1}{2}(\cos \pi + i \sin \pi)$ **70.** $2\left(\cos \dfrac{7\pi}{6} + i \sin \dfrac{7\pi}{6}\right)$ **71.** $4 + 4\sqrt{3}i$ **72.** $-32\sqrt{3} + 32i$ **73.** $\dfrac{1}{128}i$ **74.** $64 - 64\sqrt{3}i$

75. $128 + 128i$ **76.** $7(\cos 25° + i \sin 25°); 7(\cos 205° + i \sin 205°)$ **77.** $5(\cos 55° + i \sin 55°); 5(\cos 175° + i \sin 175°);$

$5(\cos 295° + i \sin 295°)$ **78.** $\sqrt{3} + i; -1 + \sqrt{3}i; -\sqrt{3} - i; 1 - \sqrt{3}i$ **79.** $\sqrt{3} + i; -\sqrt{3} + i; -2i$ **80.** $\dfrac{1}{2} + \dfrac{\sqrt{3}}{2}i; -1; \dfrac{1}{2} - \dfrac{\sqrt{3}}{2}i$

81. $\dfrac{\sqrt[5]{8}}{2} + \dfrac{\sqrt[5]{8}}{2}i; \approx -0.49 + 0.95i; \approx -1.06 - 0.17i; \approx -0.17 - 1.06i; \approx 0.95 - 0.49i$

82.

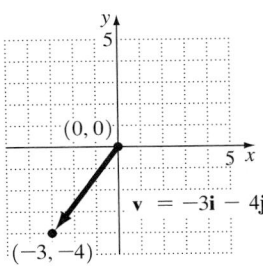

$(0,0)$

$v = -3i - 4j$

$(-3,-4)$

83.

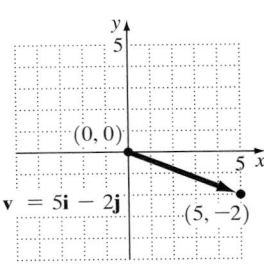

$(0,0)$

$v = 5i - 2j$

$(5,-2)$

84.

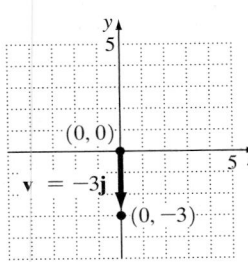

$(0,0)$

$v = -3j$

$(0,-3)$

85. $3i - 2j$ **86.** $i - 2j$
87. $-i + 2j$ **88.** $-3i + 12j$
89. $12i - 51j$ **90.** $2\sqrt{26}$
91. $\dfrac{4}{5}i - \dfrac{3}{5}j$ **92.** $-\dfrac{1}{\sqrt{5}}i + \dfrac{2}{\sqrt{5}}j$
93. a. 265 lb
  b. approximately 19° relative to the 200-lb force
94. 4 **95.** $-86.1°$
96. $-32; 124.8°$ **97.** $1; 71.6°$

98. orthogonal **99.** not orthogonal **100.** $6i + 6\sqrt{3}j$ **101.** $v_1 = \text{proj}_w \, v = \dfrac{50}{41}i + \dfrac{40}{41}j; v_2 = -\dfrac{132}{41}i + \dfrac{165}{41}j$

102. $v_1 = \text{proj}_w \, v = -\dfrac{3}{2}i + \dfrac{1}{2}j; v_2 = \dfrac{1}{2}i + \dfrac{3}{2}j$ **103.** 1115 ft-lb

Chapter 7 Test

1. 8.0 **2.** 6.2 **3.** 206 sq in.

4. ; Ordered pairs may vary. **5.** $\left(\sqrt{2}, \dfrac{7\pi}{4}\right)$ **6.** $r^2 = 6r \cos \theta$ **7.** $y = 4$

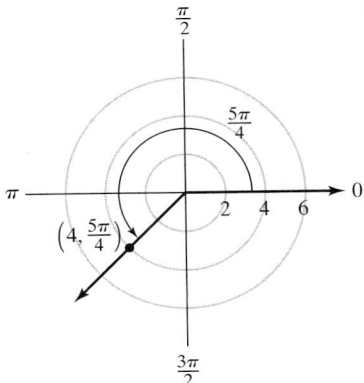

$\left(4, \dfrac{5\pi}{4}\right)$

8.

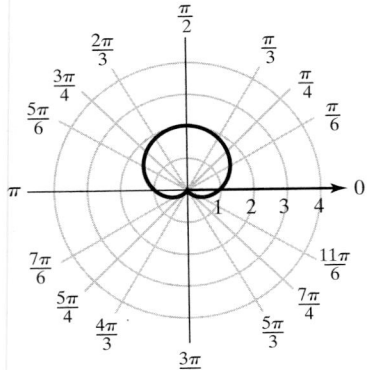

9.

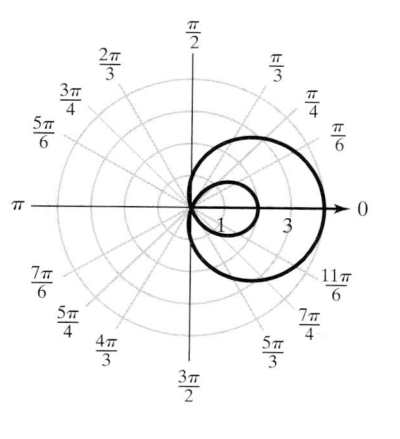

10. $2(\cos 150° + i \sin 150°)$ or $2\left(\cos \dfrac{5\pi}{6} + i \sin \dfrac{5\pi}{6}\right)$

11. $50(\cos 20° + i \sin 20°)$

12. $\dfrac{1}{2}\left(\cos \dfrac{\pi}{6} + i \sin \dfrac{\pi}{6}\right)$

13. $32(\cos 50° + i \sin 50°)$

14. $3; -\dfrac{3}{2} + \dfrac{\sqrt{3}}{2}i; -\dfrac{3}{2} - \dfrac{\sqrt{3}}{2}i$

15. a. $\mathbf{i} + 2\mathbf{j}$ **b.** $\sqrt{5}$

16. $-23\mathbf{i} + 22\mathbf{j}$ **17.** -18 **18.** $138°$

19. $-\dfrac{9}{5}\mathbf{i} + \dfrac{18}{5}\mathbf{j}$

20. 1.0 mi   **21.** 1966 ft-lb

Cumulative Review Exercises (Chapters 1–7)

1. $\{-1, 2, i, -i\}$ **2.** $\dfrac{\pi}{6}, \dfrac{5\pi}{6}$, and $\dfrac{\pi}{2}$ **3.** $\{x \mid x < -4 \text{ or } x > 2\}$ **4.** $\dfrac{3\pi}{4}, \dfrac{7\pi}{4}$

5.

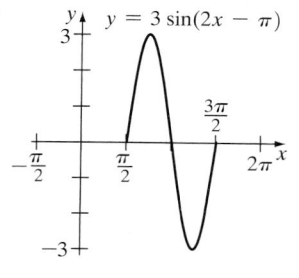

$y = 3 \sin(2x - \pi)$

6.

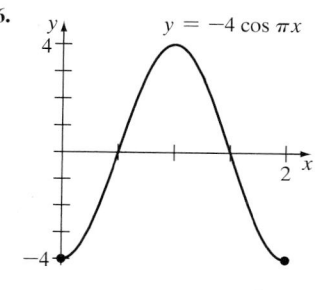

$y = -4 \cos \pi x$

7. $\sin \theta \csc \theta - \cos^2 \theta = \sin \theta \left(\dfrac{1}{\sin \theta}\right) - \cos^2 \theta$
$= 1 - \cos^2 \theta = \sin^2 \theta$

8. $\cos\left(\theta + \dfrac{3\pi}{2}\right) = \cos \theta \cos \dfrac{3\pi}{2} - \sin \theta \sin \dfrac{3\pi}{2}$
$= \cos \theta(0) - \sin \theta(-1) = \sin \theta$

9. slope is $-\dfrac{1}{2}$; y-intercept is 2

10. 0   **11.** $\dfrac{\sqrt{5}}{5}$ **12.** $\{x \mid x \leq 5\}$

13. $\{x \mid x \neq 3, x \neq -3\}$ **14.** 1.5 sec; 44 ft   **15. a.** 4 m   **b.** $\dfrac{5}{2\pi}$ **c.** $\dfrac{2\pi}{5}$ sec   **16.** $\sqrt{\dfrac{\sqrt{2} + 2}{4}}$ **17. a.** $5\mathbf{i} + 23\mathbf{j}$ **b.** -12

18. $\log_b \dfrac{\sqrt{x}}{x^2 + 1}$ **19.** $y = -\dfrac{1}{2}x + 1$ **20. a.** 0.014   **b.** 73 words   **c.** about 144 min

CHAPTER 8

Section 8.1

Check Point Exercises

1. solution   **2.** $\{(3, 2)\}$ **3.** $\{(1, -2)\}$ **4.** $\{(2, -1)\}$ **5.** $\left\{\left(\dfrac{23}{16}, \dfrac{3}{8}\right)\right\}$ **6.** $\varnothing$ **7.** $\{(x, y) \mid y = 4x - 4\}$ **8.** $\$30; 400$ units

Exercise Set 8.1

1. solution   **3.** not a solution   **5.** $\{(1, 3)\}$ **7.** $\{(5, 1)\}$ **9.** $\{(2, 1)\}$ **11.** $\{(-1, 3)\}$ **13.** $\{(4, 5)\}$
15. $\left\{\left(-4, \dfrac{5}{4}\right)\right\}$ **17.** $\{(2, -1)\}$ **19.** $\{(3, 0)\}$ **21.** $\{(-4, 3)\}$ **23.** $\{(3, 1)\}$ **25.** $\{(1, -2)\}$ **27.** $\left\{\left(\dfrac{7}{25}, -\dfrac{1}{25}\right)\right\}$ **29.** $\varnothing$
31. $\{(x, y) \mid y = 3x - 5\}$ **33.** $\{(1, 4)\}$ **35.** $\{(x, y) \mid x + 3y = 2\}$ **37.** $x + y = 7; x - y = -1; 3$ and 4   **39.** $3x - y = 1;$
$x + 2y = 12; 2$ and 5   **41. a.** 6500 tickets can be sold. 6200 tickets can be supplied.   **b.** $\$50; 6250$ tickets   **43.** 2700 gal
45. Quarter Pounder: 77 mg; Whopper with cheese: 122 mg   **47.** 5.8 million pounds of potato chips; 4.6 million pounds of tortilla chips
49. 2032; less than 1 death in 1000 live births   **63.** $y = \dfrac{a_1c_2 - a_2c_1}{a_1b_2 - a_2b_1}; x = \dfrac{b_2c_1 - b_1c_2}{a_1b_2 - a_2b_1}$ **65.** Yes; 8 hexagons and 4 squares

Section 8.2

Check Point Exercises

2. $\{(1, 4, -3)\}$ **3.** $\{(4, 5, 3)\}$ **4.** $y = 3x^2 - 12x + 13$

Exercise Set 8.2

1. not a solution   **3.** solution   **5.** $\{(2, 3, 3)\}$ **7.** $\{(2, -1, 1)\}$ **9.** $\left\{\left(\dfrac{1}{3}, -\dfrac{2}{5}, \dfrac{1}{2}\right)\right\}$ **11.** $\{(3, 1, 5)\}$ **13.** $\{(1, 0, -3)\}$

15. $\{(1, -5, -6)\}$ **17.** $\left\{\left(\dfrac{1}{2}, \dfrac{1}{3}, -1\right)\right\}$ **19.** 7, 4 and 5   **21.** $y = 2x^2 - x + 3$ **23.** $y = 2x^2 + x - 5$

25. chemical engineer: \$42,758; mechanical engineer: \$39,852; electrical engineer: \$38,811   **27.** $A = -8$; $B = 50$; $C = 0$; $y = 156$ when $x = 6$; When a car is in motion for 6 seconds after the brakes are applied, it travels 156 feet.   **29.** 200 \$8 tickets; 150 \$10 tickets; 50 \$12 tickets   **31.** \$1200 at 8%, \$2000 at 10%, and \$3500 at 12%   **41.** 13 triangles, 21 rectangles, and 6 pentagons

Section 8.3

Check Point Exercises

1. $\dfrac{2}{x - 3} + \dfrac{3}{x + 4}$ **2.** $\dfrac{2}{x} - \dfrac{2}{x - 1} + \dfrac{3}{(x - 1)^2}$ **3.** $\dfrac{2}{x + 3} + \dfrac{6x - 8}{x^2 + x + 2}$ **4.** $\dfrac{2x}{x^2 + 1} + \dfrac{-x + 3}{(x^2 + 1)^2}$

Exercise Set 8.3

1. $\dfrac{A}{x - 2} + \dfrac{B}{x + 1}$ **3.** $\dfrac{A}{x + 2} + \dfrac{B}{x - 3} + \dfrac{C}{(x - 3)^2}$ **5.** $\dfrac{A}{x - 1} + \dfrac{Bx + C}{x^2 + 1}$ **7.** $\dfrac{Ax + B}{x^2 + 4} + \dfrac{Cx + D}{(x^2 + 4)^2}$ **9.** $\dfrac{3}{x - 3} - \dfrac{2}{x - 2}$

11. $\dfrac{7}{x - 9} - \dfrac{4}{x + 2}$ **13.** $\dfrac{24}{7(x - 4)} + \dfrac{25}{7(x + 3)}$ **15.** $\dfrac{3}{x} + \dfrac{2}{x - 1} - \dfrac{1}{x + 3}$ **17.** $\dfrac{3}{x} + \dfrac{4}{x + 1} - \dfrac{3}{x - 1}$ **19.** $\dfrac{6}{x - 1} - \dfrac{5}{(x - 1)^2}$

21. $\dfrac{1}{x - 2} - \dfrac{2}{(x - 2)^2} - \dfrac{5}{(x - 2)^3}$ **23.** $\dfrac{7}{x} - \dfrac{6}{x - 1} + \dfrac{10}{(x - 1)^2}$ **25.** $-\dfrac{2}{x + 1} + \dfrac{3}{(x + 1)^2} + \dfrac{7}{x - 3}$ **27.** $\dfrac{3}{x - 1} + \dfrac{2x - 4}{x^2 + 1}$

29. $\dfrac{2}{x + 1} + \dfrac{3x - 1}{x^2 + 2x + 2}$ **31.** $\dfrac{4}{x + 1} + \dfrac{2x - 3}{x^2 + 1}$ **33.** $\dfrac{x + 1}{x^2 + 2} - \dfrac{2x}{(x^2 + 2)^2}$ **35.** $\dfrac{x - 2}{x^2 - 2x + 3} + \dfrac{2x + 1}{(x^2 - 2x + 3)^2}$

37. $\dfrac{3}{x - 2} + \dfrac{x - 1}{x^2 + 2x + 4}$ **39.** $\dfrac{1}{x} - \dfrac{1}{x + 1}; \dfrac{99}{100}$ **49.** $\dfrac{a}{x - c} + \dfrac{b + ac}{(x - c)^2}$

Section 8.4

Check Point Exercises

1. $\{(0, 1), (4, 17)\}$ **2.** $\left\{\left(-\dfrac{6}{5}, \dfrac{3}{5}\right), (2, -1)\right\}$ **3.** $\{(3, 2), (3, -2), (-3, 2), (-3, -2)\}$ **4.** $\{(0, 5)\}$
5. length: 7 ft; width: 3 ft or length: 3 ft; width: 7 ft

Exercise Set 8.4

1. $\{(-3, 5), (2, 0)\}$ **3.** $\left\{\left(\dfrac{-1 + \sqrt{17}}{2}, \dfrac{1 + \sqrt{17}}{2}\right), \left(\dfrac{-1 - \sqrt{17}}{2}, \dfrac{1 - \sqrt{17}}{2}\right)\right\}$ **5.** $\{(4, -10), (-3, 11)\}$ **7.** $\{(4, 3), (-3, -4)\}$

9. $\left\{\left(-\dfrac{3}{2}, -4\right), (2, 3)\right\}$ **11.** $\{(-5, -4), (3, 0)\}$ **13.** $\{(3, 1), (-3, -1), (1, 3), (-1, -3)\}$ **15.** $\{(4, -3), (-1, 2)\}$

17. $\{(0, 1), (4, -3)\}$ **19.** $\{(3, 2), (3, -2), (-3, 2), (-3, -2)\}$ **21.** $\{(3, 2), (3, -2), (-3, 2) (-3, -2)\}$

23. $\{(2, 1), (2, -1), (-2, 1), (-2, -1)\}$ **25.** $\{(3, 4), (3, -4)\}$ **27.** $\{(0, 2), (0, -2), (-1, \sqrt{3}), (-1, -\sqrt{3})\}$

29. $\{(2, 1), (2, -1), (-2, 1), (-2, -1)\}$ **31.** $\{(-2\sqrt{2}, -\sqrt{2}), (-1, -4), (1, 4), (2\sqrt{2}, \sqrt{2})\}$ **33.** $\{(2, 2), (4, 1)\}$ **35.** $\{(0, 0), (-1, 1)\}$

37. $\{(0, 0), (-2, 2), (2, 2)\}$ **39.** $\left\{(-4, 1), \left(-\dfrac{5}{2}, \dfrac{1}{4}\right)\right\}$ **41.** $\left\{\left(\dfrac{12}{5}, -\dfrac{29}{5}\right), (-2, 3)\right\}$ **43.** 4 and 6

45. 2 and 1, 2 and -1, -2 and 1, or -2 and -1 **47.** $(0, -4), (-2, 0), (2, 0)$ **49.** 11 ft and 7 ft   **51.** width: 6 in.; length: 8 in.
53. $x = 5$ m, $y = 2$ m   **61.** (b) is true.   **63.** $b = 6, a = 8$ **65.** $\{(10{,}000, 5)\}$

Section 8.5

Check Point Exercises

1.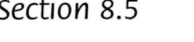

2.

3.

4.

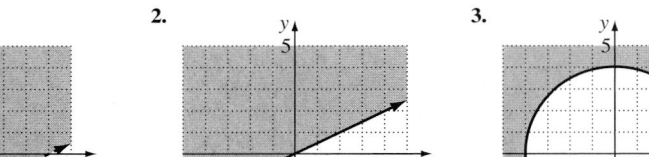
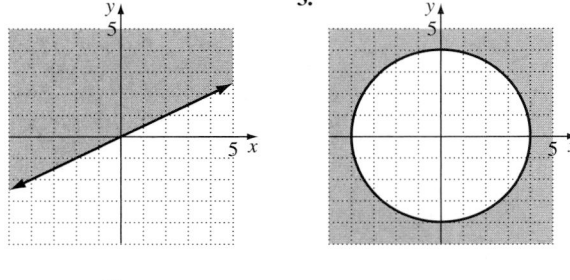

5.

6.

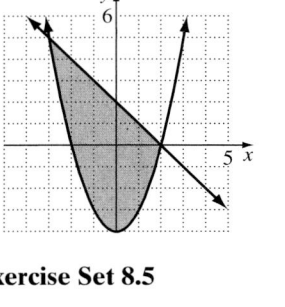
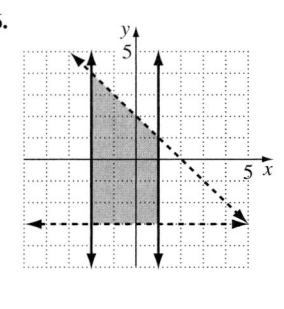

Exercise Set 8.5

1.

3.

5.

7.

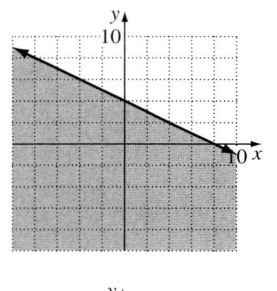
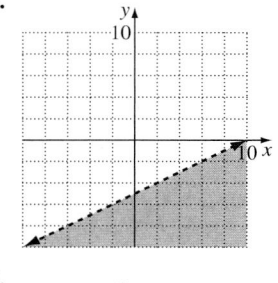
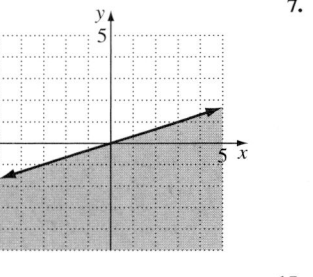
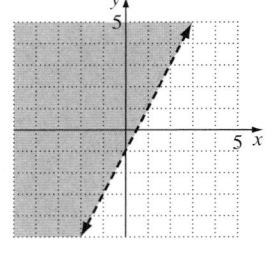

9.

11.

13.

15.

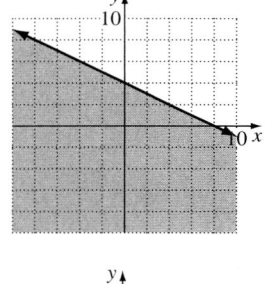
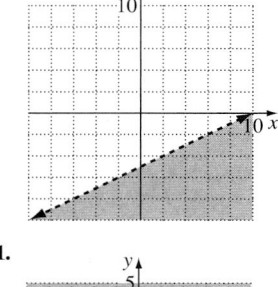
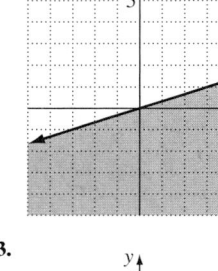
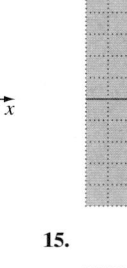
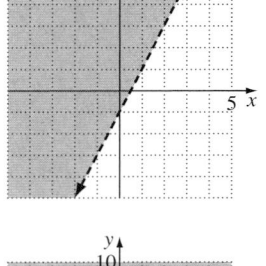

17.

19.

21.

23.

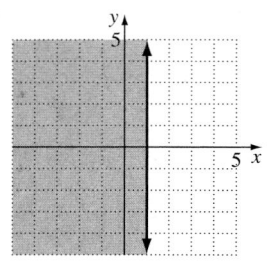
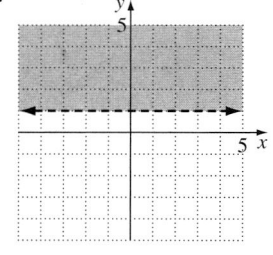
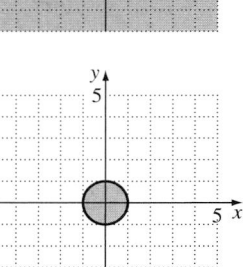

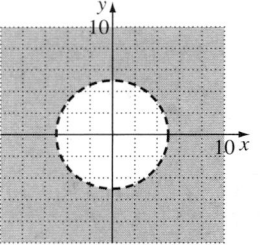
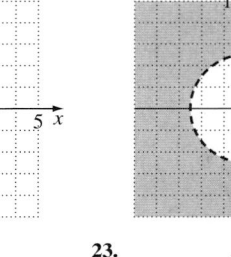

25.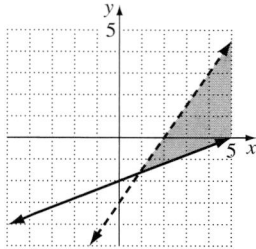

27.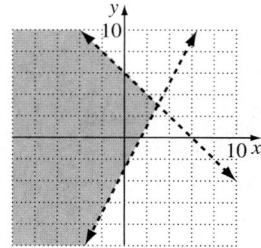

29.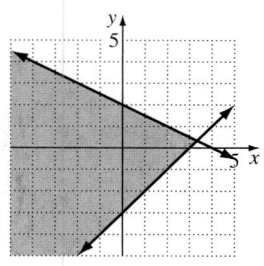

31.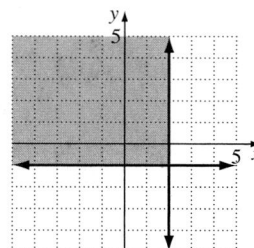

33.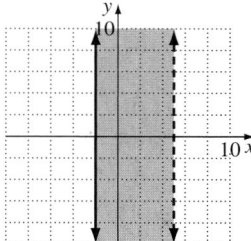

35.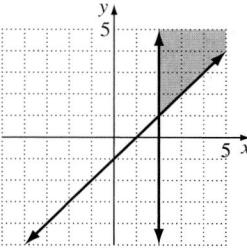

37. no solution

39.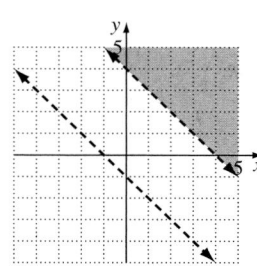

41.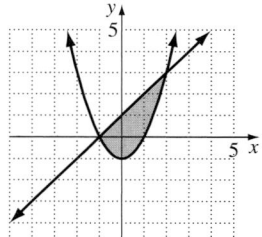

43.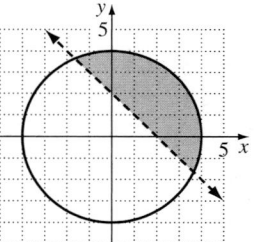

45.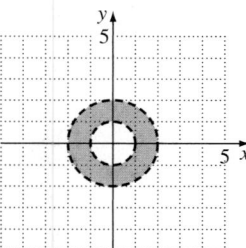

47.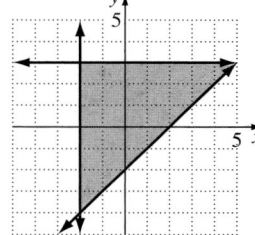

49.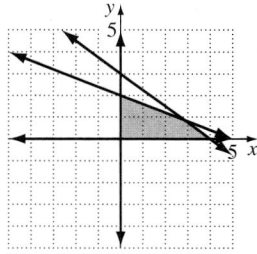

51.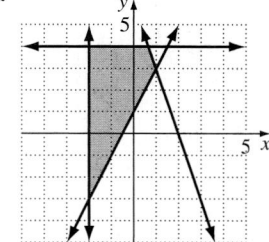

52.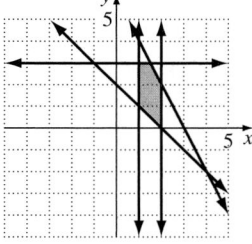

53. Answers may vary. **55.** no **57.** $7w - 25h \geq -800; w - 5h \leq -170$ **59.** $5T - 7P \leq 70$

61. $50x + 150y > 2000$

63. $x + y \leq 15,000$
$x \geq 2000$
$y \geq 3x$
$x \geq 0$
$y \geq 0$

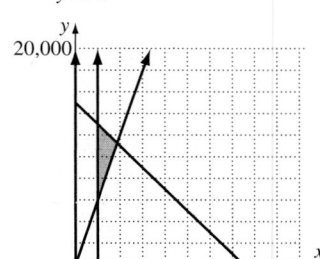

71.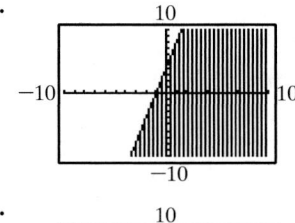

73.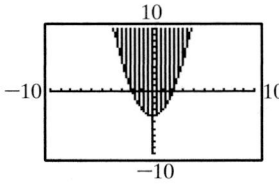

75. Answers may vary.   **79.** $x^2 + y^2 \leq 9; y < x^2$ **81.**

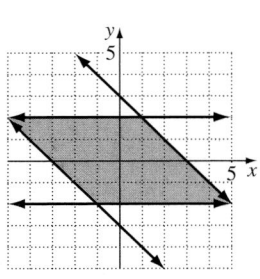

Section 8.6

Check Point Exercises

1. $z = 25x + 55y$ **2.** $x + y \leq 80$ **3.** $30 \leq x \leq 80; 10 \leq y \leq 30$; objective function: $z = 25x + 55y$; constraints: $x + y \leq 80$; $30 \leq x \leq 80; 10 \leq y \leq 30$ **4.** 50 bookshelves and 30 desks; \$2900   **5.** 30

Exercise Set 8.6

1. $(1, 2)$: 17; $(2, 10)$: 70; $(7, 5)$: 65; $(8, 3)$: 58; maximum: $z = 70$; minimum: $z = 17$
3. $(0, 0)$: 0; $(0, 8)$: 400; $(4, 9)$: 610; $(8, 0)$: 320; maximum: $z = 610$; minimum: $z = 0$

5. a.

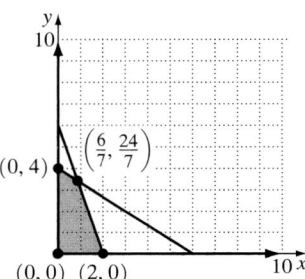

7. a.

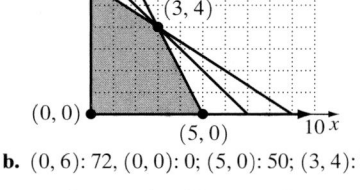

9. a.

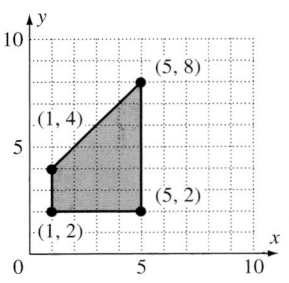

5. b. $(0, 0)$: 0; $(0, 4)$: 12; $\left(\frac{6}{7}, \frac{24}{7}\right)$: 12; $(2, 0)$: 4
c. maximum value: 12 at $x = 0$ and $y = 4$ and at $x = \frac{6}{7}$ and $y = \frac{24}{7}$

7. b. $(0, 4)$: 4; $(0, 3)$: 3; $(3, 0)$: 12; $(6, 0)$: 24
c. maximum value: 24 at $x = 6$ and $y = 0$

9. b. $(1, 2)$: -1; $(1, 4)$: -5; $(5, 8)$: -1; $(5, 2)$: 11
c. maximum value: 11 at $x = 5$ and $y = 2$

11. a.

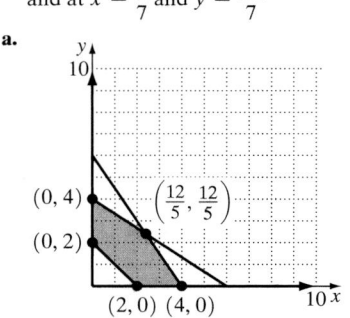
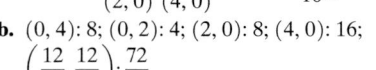

13. a.

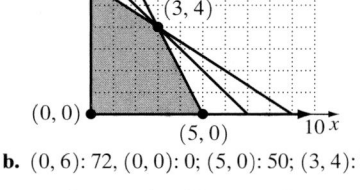

15. a. $z = 125x + 200y$
b. $x \leq 450; y \leq 200; 600x + 900y \leq 360,000$
c.

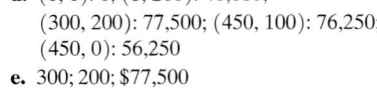

11. b. $(0, 4)$: 8; $(0, 2)$: 4; $(2, 0)$: 8; $(4, 0)$: 16; $\left(\frac{12}{5}, \frac{12}{5}\right)$: $\frac{72}{5}$
c. maximum value: 16 at $x = 4$ and $y = 0$

13. b. $(0, 6)$: 72, $(0, 0)$: 0; $(5, 0)$: 50; $(3, 4)$: 78
c. maximum value: 78 at $x = 3$ and $y = 4$

15. d. $(0, 0)$: 0; $(0, 200)$: 40,000; $(300, 200)$: 77,500; $(450, 100)$: 76,250; $(450, 0)$: 56,250
e. 300; 200; \$77,500

17. 40 model A bicycles and no model B bicycles   **19.** No cartons of food and 600 cartons of clothing; 3600 people
21. 50 students and 100 parents   **23.** 10 Boeing 727s and 42 Falcon 20s

29.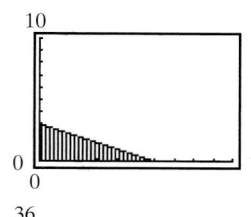
36

31.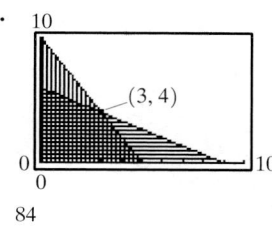
84

33. $5000 in stocks and $5000 in bonds

34. $(0, 0): 0; (0, 3): 3B; (3, 1): 3B; (2, 0): \frac{4}{3}B$; maximum value: $3B$ at $(0, 3)$ and $(3, 1)$

Chapter 8 Review Exercises

1. $\{(1, 5)\}$ **2.** $\{(2, 3)\}$ **3.** $\{(2, -3)\}$ **4.** $\varnothing$ **5.** $\{(x, y)|3x - 6y = 12\}$ **6.** No, $(-1, 3)$ is not a solution to $2x + y = -5$.
7. Shrimp: 42 mg of cholesterol per oz; Scallops: 15 mg of cholesterol per oz   **8.** 3 apples and 2 avocados
9. 250 copies can be supplied and sold for $12.50 each.   **10.** $\{(0, 1, 2)\}$ **11.** $\{(2, 1, -1)\}$ **12.** $y = 3x^2 - 4x + 5$
13. Japan: 16%; Germany: 15%; France: 14%   **14.** Substitute the ordered pairs into $y = ax^2 + bx + c$, solve the resulting system for a, b,

and c, then substitute these values into the equation to form a quadratic model.   **15.** $\dfrac{3}{5(x - 3)} + \dfrac{2}{5(x + 2)}$ **16.** $\dfrac{6}{x - 4} + \dfrac{5}{x + 3}$

17. $\dfrac{2}{x} + \dfrac{3}{x + 2} - \dfrac{1}{x - 1}$ **18.** $\dfrac{2}{x - 2} + \dfrac{5}{(x - 2)^2}$ **19.** $-\dfrac{4}{x - 1} + \dfrac{4}{x - 2} - \dfrac{2}{(x - 2)^2}$ **20.** $\dfrac{6}{5(x - 2)} + \dfrac{-6x + 3}{5(x^2 + 1)}$

21. $\dfrac{5}{x - 3} + \dfrac{2x - 1}{x^2 + 4}$ **22.** $\dfrac{x}{x^2 + 4} - \dfrac{4x}{(x^2 + 4)^2}$ **23.** $\dfrac{4x + 1}{x^2 + x + 1} + \dfrac{2x - 2}{(x^2 + x + 1)^2}$ **24.** $\{(4, 3), (1, 0)\}$ **25.** $\{(0, 1), (-3, 4)\}$

26. $\{(1, -1), (-1, 1)\}$ **27.** $\{(3, \sqrt{6}), (3, -\sqrt{6}), (-3, \sqrt{6}), (-3, -\sqrt{6})\}$ **28.** $\{(2, 2), (-2, -2)\}$ **29.** $\{(9, 6), (1, 2)\}$

30. $\{(-3, -1), (1, 3)\}$ **31.** $\left\{\left(\dfrac{1}{2}, 2\right), (-1, -1)\right\}$ **32.** $\left\{\left(\dfrac{5}{2}, -\dfrac{7}{2}\right), (0, -1)\right\}$ **33.** $\{(2, -3), (-2, -3), (3, 2), (-3, 2)\}$

34. $\{(3, 1), (3, -1), (-3, 1), (-3, -1)\}$ **35.** 8 m and 5 m   **36.** $(1, 6), (3, 2)$ **37.** $x = 46$ and $y = 28$ or $x = 50$ and $y = 20$

38. 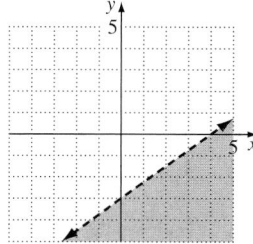   **39.** 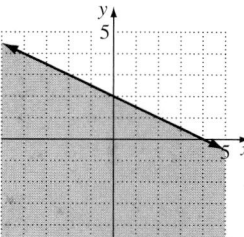   **40.** 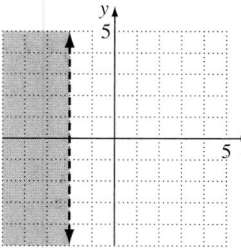   **41.**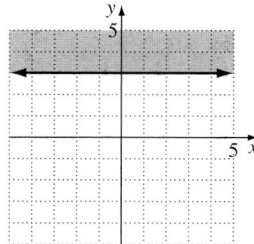

42. 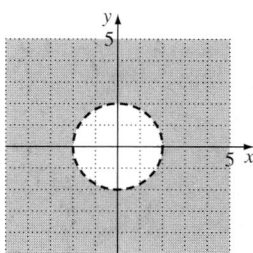   **43.** 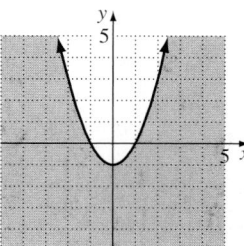   **44.** 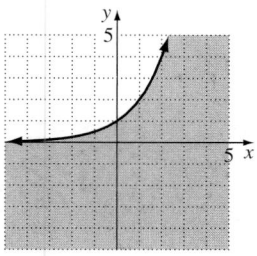   **45.**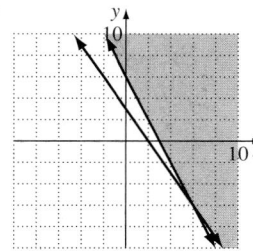

46. 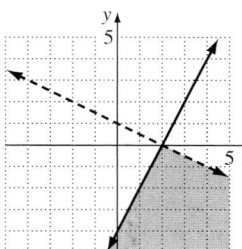   **47.** 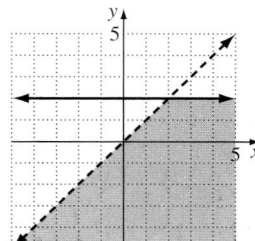   **48.** 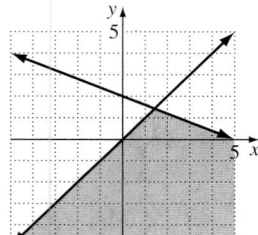   **49.**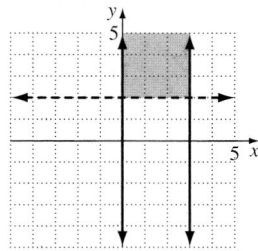

50. no solution

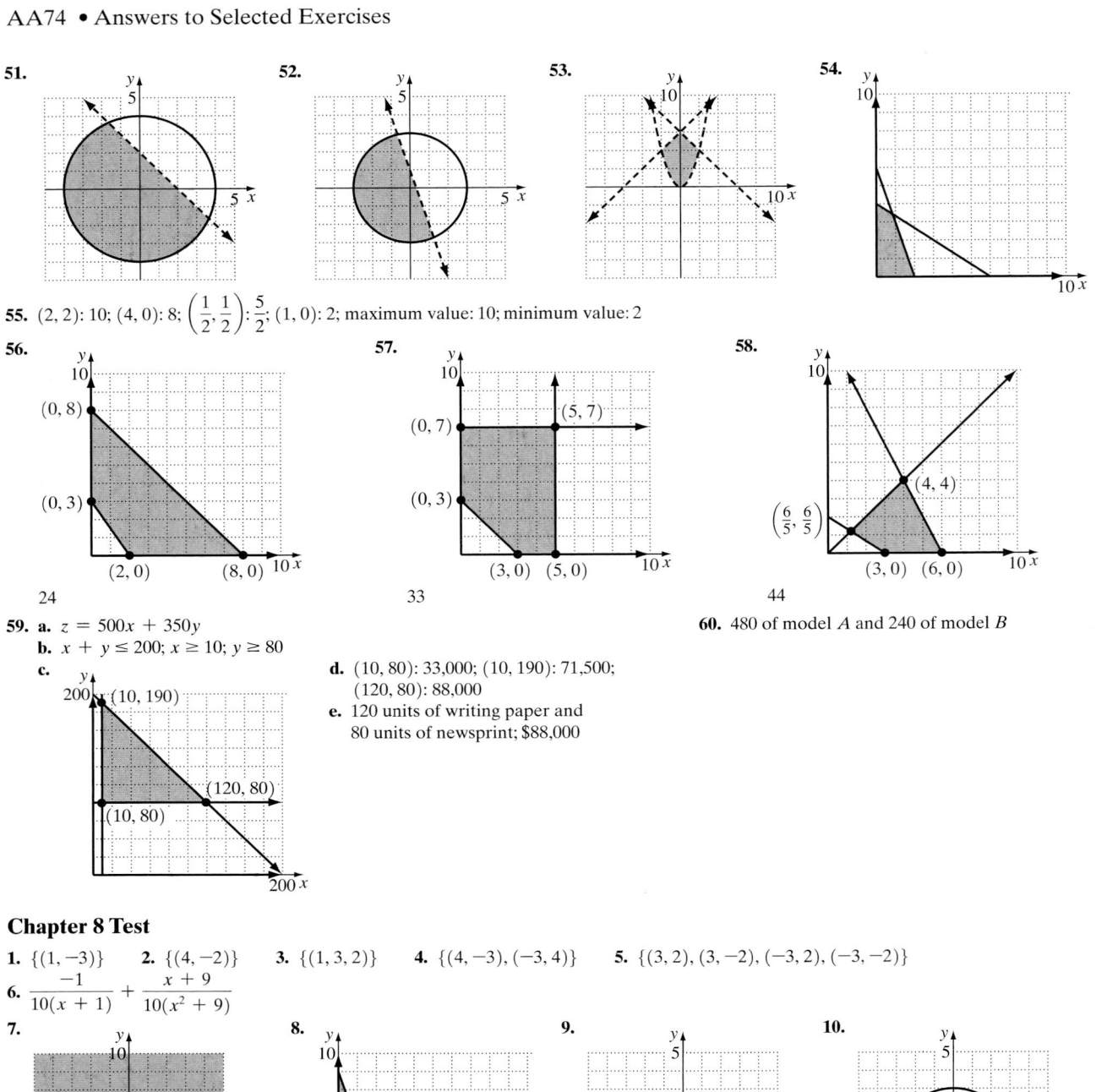

51.

52.

53.

54.

55. $(2, 2)$: 10; $(4, 0)$: 8; $\left(\dfrac{1}{2}, \dfrac{1}{2}\right)$: $\dfrac{5}{2}$; $(1, 0)$: 2; maximum value: 10; minimum value: 2

56.

(0, 8)

(0, 3)

(2, 0)   (8, 0)   10 x

24

57.

(0, 7)   (5, 7)

(0, 3)

(3, 0) (5, 0)   10 x

33

58.

(4, 4)

$\left(\dfrac{6}{5}, \dfrac{6}{5}\right)$

(3, 0)   (6, 0)   10 x

44

59. a. $z = 500x + 350y$
   b. $x + y \le 200$; $x \ge 10$; $y \ge 80$
   c.

200 (10, 190)

(120, 80)

(10, 80)

200 x

d. $(10, 80)$: 33,000; $(10, 190)$: 71,500;
   $(120, 80)$: 88,000
e. 120 units of writing paper and
   80 units of newsprint; $88,000

60. 480 of model A and 240 of model B

Chapter 8 Test

1. $\{(1, -3)\}$
2. $\{(4, -2)\}$
3. $\{(1, 3, 2)\}$
4. $\{(4, -3), (-3, 4)\}$
5. $\{(3, 2), (3, -2), (-3, 2), (-3, -2)\}$
6. $\dfrac{-1}{10(x + 1)} + \dfrac{x + 9}{10(x^2 + 9)}$

7.

8.

9.

10.

11. 26   **12.** orchestra ticket: $23; mezzanine ticket: $14   **13.** 400 units; $30 each   **14.** $y = x^2 - 3$
15. $x = 7.5$ ft and $y = 24$ ft or $x = 12$ ft and $y = 15$ ft   **16.** 50 regular and 100 deluxe jet skis; $35,000

Cumulative Review Exercises (Chapters 1–8)

1. $\{3, 4\}$ **2.** $\left\{\dfrac{2 + i\sqrt{3}}{2}, \dfrac{2 - i\sqrt{3}}{2}\right\}$ **3.** $(-18, 6)$ **4.** $(1, 7)$ **5.** $\left\{-3, \dfrac{1}{2}, 2\right\}$ **6.** $\{-2\}$ **7.** $\{2\}$

8. $\{-2 + \log_3 11\}$, or approximately 0.18

9. 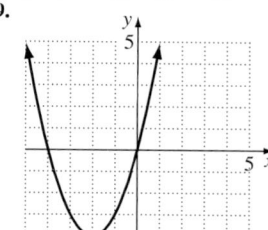   **10.** 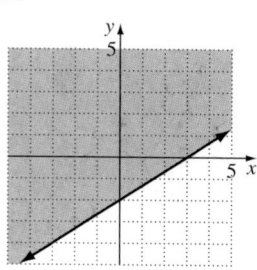   **11.** 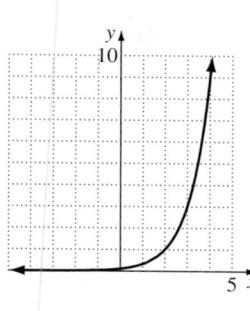   **12.**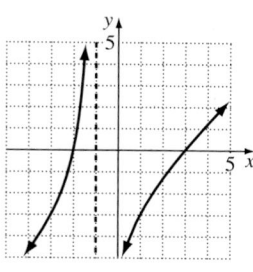

13. $3 + 5 \log_2 x$ **14.** 10.99% **15.** $f^{-1}(x) = \dfrac{1}{7}x + \dfrac{3}{7}$ **16.** $g(f(x)) = 21x - 16$ **17.** Answers may vary.   **18.** $\left\{\left(-\dfrac{1}{2}, \dfrac{1}{2}\right), (2, 8)\right\}$

19. 4 m by 9 m   **20.** A plane with an initial landing speed of 90 ft per second needs 562 ft to land. There is a problem since 550 ft is not enough.

21. $\sec \theta - \cos \theta = \dfrac{1}{\cos \theta} - \cos \theta = \dfrac{1 - \cos^2 \theta}{\cos \theta} = \dfrac{\sin^2 \theta}{\cos \theta} = \dfrac{\sin \theta}{\cos \theta} \cdot \sin \theta = \tan \theta \cdot \sin \theta$

22. $\tan x + \tan y = \dfrac{\sin x}{\cos x} + \dfrac{\sin y}{\cos y} = \dfrac{\sin x \cdot \cos y + \sin y \cdot \cos x}{\cos x \cdot \cos y} = \dfrac{\sin(x + y)}{\cos x \cdot \cos y}$ **23.** $0, \pi$ **24.** $0, \dfrac{\pi}{3}, \dfrac{5\pi}{3}$ **25.** 92.9

CHAPTER 9

Section 9.1

Check Point Exercises

1. $\{(4, -3, 1)\}$ **2. a.** $\begin{bmatrix} 1 & 6 & -3 & | & 7 \\ 4 & 12 & -20 & | & 8 \\ -3 & -2 & 1 & | & -9 \end{bmatrix}$ **b.** $\begin{bmatrix} 1 & 3 & -5 & | & 2 \\ 1 & 6 & -3 & | & 7 \\ -3 & -2 & 1 & | & -9 \end{bmatrix}$ **c.** $\begin{bmatrix} 4 & 12 & -20 & | & 8 \\ 1 & 6 & -3 & | & 7 \\ 0 & 16 & -8 & | & 12 \end{bmatrix}$

3. $\{(5, 2, 3)\}$ **4.** $\{(1, -1, 2, -3)\}$ **5.** $\left\{\left(\dfrac{5}{2}, \dfrac{8}{7}, \dfrac{11}{14}\right)\right\}$

Exercise Set 9.1

1. $\begin{bmatrix} 2 & 1 & 2 & | & 2 \\ 3 & -5 & -1 & | & 4 \\ 1 & -2 & -3 & | & -6 \end{bmatrix}$ **3.** $\begin{bmatrix} 1 & -1 & 1 & | & 8 \\ 0 & 1 & -12 & | & -15 \\ 0 & 0 & 1 & | & 1 \end{bmatrix}$ **5.** $\begin{bmatrix} 5 & -2 & -3 & | & 0 \\ 1 & 1 & 0 & | & 5 \\ 2 & 0 & -3 & | & 4 \end{bmatrix}$ **7.** $\begin{bmatrix} 2 & 5 & -3 & 1 & | & 2 \\ 0 & 3 & 1 & 0 & | & 4 \\ 1 & -1 & 5 & 0 & | & 9 \\ 5 & -5 & -2 & 0 & | & 1 \end{bmatrix}$ **9.** $\begin{array}{l} 5x + 3z = -11 \\ y - 4z = 12 \\ 7x + 2y = 3 \end{array}$

11. $\begin{array}{l} x + y + 4z + w = 3 \\ -x + y - z = 7 \\ 2x + 5w = 11 \\ 12z + 4w = 5 \end{array}$ **13.** $\begin{array}{l} x - 4z = 5 \\ y - 12z = 13 \\ z = -\dfrac{1}{2} \end{array}$; $\left\{\left(3, 7, -\dfrac{1}{2}\right)\right\}$ **15.** $\begin{array}{l} x + \dfrac{1}{2}y + z = \dfrac{11}{2} \\ y + \dfrac{3}{2}z = 7 \\ z = 4 \end{array}$; $\{(1, 1, 4)\}$

17. $\begin{array}{l} x - y + z + w = 3 \\ y - 2z - w = 0 \\ z + 6w = 17 \\ w = 3 \end{array}$; $\{(2, 1, -1, 3)\}$ **19.** $\begin{bmatrix} 1 & -3 & 2 & | & 5 \\ 1 & 5 & -5 & | & 0 \\ 3 & 0 & 4 & | & 7 \end{bmatrix}$ **21.** $\begin{bmatrix} 1 & -3 & 2 & | & 0 \\ 0 & 10 & -7 & | & 7 \\ 2 & -2 & 1 & | & 3 \end{bmatrix}$ **23.** $\begin{bmatrix} 1 & -1 & 1 & 1 & | & 3 \\ 0 & 1 & -2 & -1 & | & 0 \\ 0 & 2 & 1 & 2 & | & 5 \\ 0 & 6 & -3 & -1 & | & -9 \end{bmatrix}$

25. $R_2: -3, -18; R_3: -12, -15; R_2: -\dfrac{3}{5}, -\dfrac{18}{5}; R_3: -12, -15$ **27.** $\{(1, -1, 2)\}$ **29.** $\{(3, -1, -1)\}$ **31.** $\{(1, 2, -1)\}$

33. $\{(1, 2, 3, -2)\}$ **35.** $\{(0, -3, 0, -3)\}$ **37.** $\{(-1, 2, 3, 4)\}$ **39.** $\{(1, -1, 2, -2, 0)\}$ **41. a.** $a = 2; b = 22; c = 320$

b. 2780   **c.** Answers may vary.   **43.** $\begin{array}{l} x + y + z = 100; \text{ under 30: 35\%; 30–49: 49\%; 50 and over: 16\%} \\ x - y + z = 2 \\ -x + 2z = -3 \end{array}$

45. $40x + 200y + 400z = 660$; 4 oz of Food A: $\dfrac{1}{2}$ oz of Food B; 1 oz of Food C **57.** $y = x^3 + 2x^2 + 5x - 3$
$5x + 2y + 4z = 25$
$30x + 10y + 300z = 425$

Section 9.2

Check Point Exercises

1. $\varnothing$ **2.** $\{(11t + 13, 5t + 4, t)\}$ **3.** $\{(t + 50, -2t + 10, t)\}$
4. a. $x + w = 15$ **b.** $\{(-t + 15, t + 15, -t + 30, t)\}$ **c.** $x = 5; y = 25; z = 20$
$x + y = 30$
$y + z = 45$
$z + w = 30$

Exercise Set 9.2

1. $\varnothing$ **3.** $\left\{\left(-2t + 2, 2t + \dfrac{1}{2}, t\right)\right\}$ **5.** $\{(-3, 4, -2)\}$ **7.** $\{(5 - 2t, -2 + t, t)\}$ **9.** $\{(-1, 2, 1, 1)\}$ **11.** $\{(1, 3, 2, 1)\}$

13. $\{(1, -2, 1, 1)\}$ **15.** $\left\{\left(1 + \dfrac{1}{3}t, \dfrac{1}{3}t, t\right)\right\}$ **17.** $\{(-13t + 5, 5t, t)\}$ **19.** $\left\{\left(2t - \dfrac{5}{4}, \dfrac{13}{4}, t\right)\right\}$ **21.** $\{(1, -t - 1, 2, t)\}$

23. $\left\{\left(-\dfrac{2}{11}t + \dfrac{81}{11}, \dfrac{1}{22}t + \dfrac{10}{11}, \dfrac{4}{11}t - \dfrac{8}{11}, t\right)\right\}$ **25.** $z + 12 = x + 6$ **27.** $\{(t + 6, t + 2, t)\}$

29. a. $x + w = 380$ **b.** $\{(380 - t, 220 + t, 50 + t, t)\}$
$x + y = 600$ **c.** $x = 330, y = 270, z = 100, w = 50$
$z - w = 50$
$y - z = 170$

31. a. The system has no solution, so there is no way to satisfy these dietary requirements with no Food 1 available.
 b. 4 oz of Food 1, 0 oz of Food 2, 10 oz of Food 3; 2 oz of Food 1, 5 oz of Food 2, 9 oz of Food 3 (other answers are possible).
37. $a = 1$ or $a = 3$

Section 9.3

Check Point Exercises

1. a. 3×2 **b.** $a_{12} = -2; a_{31} = 1$ **2. a.** $\begin{bmatrix} 2 & 0 \\ 9 & -10 \end{bmatrix}$ **b.** $\begin{bmatrix} 9 & -4 \\ -9 & 7 \\ 5 & -2 \end{bmatrix}$ **3. a.** $\begin{bmatrix} 6 & 12 \\ -48 & -30 \end{bmatrix}$ **b.** $\begin{bmatrix} -14 & -1 \\ 25 & 10 \end{bmatrix}$ **4.** $\begin{bmatrix} 7 & 6 \\ 13 & 12 \end{bmatrix}$

5. $[30]$; $\begin{bmatrix} 2 & 0 & 4 \\ 6 & 0 & 12 \\ 14 & 0 & 28 \end{bmatrix}$ **6. a.** $\begin{bmatrix} 2 & 18 & 11 & 9 \\ 0 & 10 & 8 & 2 \end{bmatrix}$ **b.** The product is undefined.   **7.** $\begin{bmatrix} 2 & 2 & 2 \\ 1 & 2 & 1 \\ 1 & 2 & 1 \end{bmatrix} + \begin{bmatrix} -1 & -1 & -1 \\ 2 & -1 & 2 \\ 2 & -1 & 2 \end{bmatrix} = \begin{bmatrix} 1 & 1 & 1 \\ 3 & 1 & 3 \\ 3 & 1 & 3 \end{bmatrix}$

8. $\$2548$

Exercise Set 9.3

1. a. 2×3 **b.** a_{32} does not exist; $a_{23} = -1$ **3. a.** 3×4 **b.** $a_{32} = \dfrac{1}{2}; a_{23} = -6$ **5.** $x = 6; y = 4$ **7.** $x = 4; y = 6; z = 3$

9. a. $\begin{bmatrix} 9 & 10 \\ 3 & 9 \end{bmatrix}$ **b.** $\begin{bmatrix} -1 & -8 \\ 3 & -5 \end{bmatrix}$ **c.** $\begin{bmatrix} -16 & -4 \\ -12 & -8 \end{bmatrix}$ **d.** $\begin{bmatrix} 22 & 21 \\ 9 & 20 \end{bmatrix}$ **11. a.** $\begin{bmatrix} 3 & 2 \\ 6 & 2 \\ 5 & 7 \end{bmatrix}$ **b.** $\begin{bmatrix} -1 & 4 \\ 0 & 6 \\ 5 & 5 \end{bmatrix}$ **c.** $\begin{bmatrix} -4 & -12 \\ -12 & -16 \\ -20 & -24 \end{bmatrix}$

d. $\begin{bmatrix} 7 & 7 \\ 15 & 8 \\ 15 & 20 \end{bmatrix}$ **13. a.** $\begin{bmatrix} -3 \\ -1 \\ 0 \end{bmatrix}$ **b.** $\begin{bmatrix} 7 \\ -7 \\ 2 \end{bmatrix}$ **c.** $\begin{bmatrix} -8 \\ 16 \\ -4 \end{bmatrix}$ **d.** $\begin{bmatrix} -4 \\ -6 \\ 1 \end{bmatrix}$ **15. a.** $\begin{bmatrix} 8 & 0 & -4 \\ 14 & 0 & 6 \\ -1 & 0 & 0 \end{bmatrix}$ **b.** $\begin{bmatrix} -4 & -20 & 0 \\ 14 & 24 & 14 \\ 9 & -4 & 4 \end{bmatrix}$

c. $\begin{bmatrix} -8 & 40 & 8 \\ -56 & -48 & -40 \\ -16 & 8 & -8 \end{bmatrix}$ **d.** $\begin{bmatrix} 18 & -10 & -10 \\ 42 & 12 & 22 \\ 2 & -2 & 2 \end{bmatrix}$ **17. a.** $\begin{bmatrix} 0 & 16 \\ 12 & 8 \end{bmatrix}$ **b.** $\begin{bmatrix} -7 & 3 \\ 29 & 15 \end{bmatrix}$ **19. a.** $[30]$ **b.** $\begin{bmatrix} 1 & 2 & 3 & 4 \\ 2 & 4 & 6 & 8 \\ 3 & 6 & 9 & 12 \\ 4 & 8 & 12 & 16 \end{bmatrix}$

21. a. $\begin{bmatrix} 4 & -5 & 8 \\ 6 & -1 & 5 \\ 0 & 4 & -6 \end{bmatrix}$ **b.** $\begin{bmatrix} 5 & -2 & 7 \\ 17 & -3 & 2 \\ 3 & 0 & -5 \end{bmatrix}$ **23. a.** $\begin{bmatrix} 6 & 8 & 16 \\ 11 & 16 & 24 \\ 1 & -1 & 12 \end{bmatrix}$ **b.** $\begin{bmatrix} 38 & 27 \\ -16 & -4 \end{bmatrix}$ **25. a.** $\begin{bmatrix} 0 & 0 \\ 0 & 0 \end{bmatrix}$ **b.** $\begin{bmatrix} 4 & -1 & -3 & 1 \\ -1 & 4 & -3 & 2 \\ 14 & -11 & -3 & -1 \\ 25 & -25 & 0 & -5 \end{bmatrix}$

27. $\begin{bmatrix} 17 & 7 \\ -5 & -11 \end{bmatrix}$ **29.** $\begin{bmatrix} 11 & -1 \\ -7 & -3 \end{bmatrix}$ **31.** $A - C$ is not defined because A is 3×2 and C is 2×2. **33.** $\begin{bmatrix} 16 & -16 \\ -12 & 12 \\ 0 & 0 \end{bmatrix}$

35. $\begin{bmatrix} 1 & 3 & 1 \\ 3 & 3 & 3 \\ 1 & 3 & 1 \end{bmatrix} + \begin{bmatrix} -1 & -1 & -1 \\ -1 & -1 & -1 \\ -1 & -1 & -1 \end{bmatrix} = \begin{bmatrix} 0 & 2 & 0 \\ 2 & 2 & 2 \\ 0 & 2 & 0 \end{bmatrix}$ **37.** $\begin{bmatrix} 1 & 3 & 1 \\ 3 & 3 & 3 \\ 1 & 3 & 1 \end{bmatrix} + \begin{bmatrix} 1 & -2 & 1 \\ -2 & -2 & -2 \\ 1 & -2 & 1 \end{bmatrix} = \begin{bmatrix} 2 & 1 & 2 \\ 1 & 1 & 1 \\ 2 & 1 & 2 \end{bmatrix}$ **39. a.** $\begin{bmatrix} 0 & 3 & 0 \\ 0 & 3 & 0 \\ 0 & 3 & 0 \end{bmatrix}$ **b.** $\begin{bmatrix} 1 & 0 & 1 \\ 1 & 0 & 1 \\ 1 & 0 & 1 \end{bmatrix}$

41. a. $\begin{bmatrix} 589.5 & 586 \\ 1556 & 1521 \\ 1234.5 & 1183 \end{bmatrix}$ **b.** 1521 **c.** 1235

43. a. System 1: The midterm and final both count for 50% of the course grade. System 2: The midterm counts for 30% of the course grade and the final counts for 70%. **b.** $\begin{bmatrix} 84 & 87.2 \\ 79 & 81 \\ 90 & 88.4 \\ 73 & 68.6 \\ 69 & 73.4 \end{bmatrix}$ System 1 grades are listed first (if different). Student 1: B; Student 2: C or B; Student 3: A or B; Student 4: C or D; Student 5: D or C

57. Answers may vary. **59.** $AB = -BA$ so they are anticommutative.

Section 9.4

Check Point Exercises

1. $AB = I_2; BA = I_2$ **2.** $\begin{bmatrix} 3 & -7 \\ -2 & 5 \end{bmatrix}$ **3.** $\begin{bmatrix} 1 & 2 \\ 1 & 3 \end{bmatrix}$ **4.** $\begin{bmatrix} 3 & -2 & -4 \\ 3 & -2 & -5 \\ -1 & 1 & 2 \end{bmatrix}$ **5.** $\{(4, -2, 1)\}$

6. The encoded message is $-7, 10, -53, 77$. **7.** The decoded message is $2, 1, 19, 5$ or BASE.

Exercise Set 9.4

1. $AB = I_2; BA = I_2; B = A^{-1}$ **3.** $AB = \begin{bmatrix} 8 & -16 \\ -2 & 7 \end{bmatrix}; BA = \begin{bmatrix} 12 & 12 \\ 1 & 3 \end{bmatrix}; B \neq A^{-1}$ **5.** $AB = I_2; BA = I_2; B = A^{-1}$

7. $AB = I_3; BA = I_3; B = A^{-1}$ **9.** $AB = I_3; BA = I_3; B = A^{-1}$ **11.** $AB = I_4; BA = I_4; B = A^{-1}$

13. $\begin{bmatrix} \frac{2}{7} & -\frac{3}{7} \\ \frac{1}{7} & \frac{2}{7} \end{bmatrix}$ **15.** $\begin{bmatrix} 1 & \frac{1}{2} \\ 2 & \frac{3}{2} \end{bmatrix}$ **17.** A does not have an inverse. **19.** $\begin{bmatrix} 1 & 0 & 1 \\ 1 & 1 & 2 \\ 3 & 2 & 6 \end{bmatrix}$ **21.** $\begin{bmatrix} -3 & 2 & -4 \\ -1 & 1 & -1 \\ 8 & -5 & 10 \end{bmatrix}$ **23.** $\begin{bmatrix} 1 & 0 & 0 & 0 \\ 0 & -1 & 0 & 0 \\ 0 & 0 & \frac{1}{3} & 0 \\ -1 & 0 & 0 & 1 \end{bmatrix}$

25. $\begin{bmatrix} 6 & 5 \\ 5 & 4 \end{bmatrix} \begin{bmatrix} x \\ y \end{bmatrix} = \begin{bmatrix} 13 \\ 10 \end{bmatrix}$ **27.** $\begin{bmatrix} 1 & 3 & 4 \\ 1 & 2 & 3 \\ 1 & 4 & 3 \end{bmatrix} \begin{bmatrix} x \\ y \\ z \end{bmatrix} = \begin{bmatrix} -3 \\ -2 \\ -6 \end{bmatrix}$ **29.** $4x - 7y = -3$ $2x - 3y = 1$ **31.** $2x - z = 6$ $3y = 9$ $x + y = 5$ **33. a.** $\begin{bmatrix} 2 & 6 & 6 \\ 2 & 7 & 6 \\ 2 & 7 & 7 \end{bmatrix} \begin{bmatrix} x \\ y \\ z \end{bmatrix} = \begin{bmatrix} 8 \\ 10 \\ 9 \end{bmatrix}$

b. $\{(1, 2, -1)\}$ **35. a.** $\begin{bmatrix} 1 & -1 & 1 \\ 0 & 2 & -1 \\ 2 & 3 & 0 \end{bmatrix} \begin{bmatrix} x \\ y \\ z \end{bmatrix} = \begin{bmatrix} 8 \\ -7 \\ 1 \end{bmatrix}$ **b.** $\{(2, -1, 5)\}$ **37. a.** $\begin{bmatrix} 1 & -1 & 2 & 0 \\ 0 & 1 & -1 & 1 \\ -1 & 1 & -1 & 2 \\ 0 & -1 & 1 & -2 \end{bmatrix} \begin{bmatrix} x \\ y \\ z \\ w \end{bmatrix} = \begin{bmatrix} -3 \\ 4 \\ 2 \\ -4 \end{bmatrix}$

b. $\{(2, 3, -1, 0)\}$ **39.** The encoded message is $27, -19, 32, -20$.; The decoded message is $8, 5, 12, 16$ or HELP.

41. The encoded message is $14, 85, -33, 4, 18, -7, -18, 19, -9$. **53.** $\begin{bmatrix} 1 & 1 \\ 2 & 3 \end{bmatrix}$ **55.** $\begin{bmatrix} 1 & 0 & 1 \\ 2 & 1 & 3 \\ -1 & 1 & 1 \end{bmatrix}$ **57.** $\begin{bmatrix} 0 & -1 & 0 & 1 \\ -1 & -5 & 0 & 3 \\ -2 & -4 & 1 & -2 \\ -1 & -4 & 0 & 1 \end{bmatrix}$

59. $\{(2, 3, -5)\}$ **61.** $\{(1, 2, -1)\}$ **63.** $\{(2, 1, 3, -2, 4)\}$ **65.** Answers may vary. **67.** (c) is true. **69.** Answers may vary. **71.** $a = 3$ or $a = -2$

Section 9.5

Check Point Exercises

1. a. -4 **b.** -17 **2.** $\{(4,-2)\}$ **3.** 80 **4.** -24 **5.** $\{(2,-3,4)\}$ **6.** -250

Exercise Set 9.5

1. 1 **3.** -29 **5.** 0 **7.** 33 **9.** $-\dfrac{7}{16}$ **11.** $\{(5,2)\}$ **13.** $\{(2,-3)\}$ **15.** $\{(3,-1)\}$ **17.** The system is dependent.

19. $\{(4,2)\}$ **21.** $\{(7,4)\}$ **23.** The system is inconsistent. **25.** $\{(0,4)\}$ **27.** 72 **29.** -75 **31.** 0 **33.** $\{(-5,-2,7)\}$

35. $\{(2,-3,4)\}$ **37.** $\{(3,-1,2)\}$ **39.** $\{(2,3,1)\}$ **41.** -200 **43.** 195

45. a. 28 sq units **47.** yes

b.

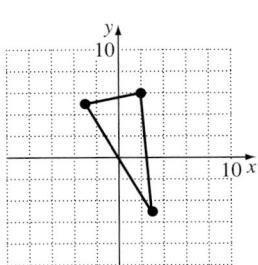

49. The equation of the line is $y = -\dfrac{11}{5}x + \dfrac{8}{5}$.

61. 13,200

63. a. a^2 **b.** a^3 **c.** a^4

    d. Each determinant has zeros below the main diagonal and a's everywhere else.

    e. Each determinant equals a raised to the power equal to the order of the determinant.

65. The sign of the value is changed when 2 columns are interchanged in a 2nd order determinant.

Chapter 9 Review Exercises

1. $x + y + 3z = 12$; $\{(1,2,3)\}$
    $y - 2z = -4$
    $z = 3$

2. $x - 2z + 2w = 1$; $\{(3,-1,2,1)$
    $y + z - w = 0$
    $z - \dfrac{7}{3}w = -\dfrac{1}{3}$
    $w = 1$

3. $\begin{bmatrix} 1 & 2 & 2 & | & 2 \\ 0 & 1 & -1 & | & 2 \\ 0 & 0 & 9 & | & -9 \end{bmatrix}$

4. $\begin{bmatrix} 1 & -1 & \frac{1}{2} & | & -\frac{1}{2} \\ 1 & 2 & -1 & | & 2 \\ 6 & 4 & 3 & | & 5 \end{bmatrix}$

5. $\{(1,3,-4)\}$ **6.** $\{(-2,-1,0)\}$ **7.** $\{(2,-2,3,4)\}$ **8. a.** $a = -2; b = 32; c = 42$ **b.** 2:00 P.M.; 170 parts per million **9.** $\varnothing$

10. $\{(2t + 4, t + 1, t)\}$ **11.** $\{(-37t + 2, 16t, -7t + 1, t)\}$ **12.** $\{(7t + 18, -3t - 7, t)\}$

13. a. $x + z = 750$ **b.** $\{(-t + 750, t - 250, t)\}$ **14.** $x = -5; y = 6; z = 6$ **15.** $\begin{bmatrix} 0 & 2 & 3 \\ 8 & 1 & 3 \end{bmatrix}$ **16.** $\begin{bmatrix} 0 & -4 \\ 6 & 4 \\ 2 & -10 \end{bmatrix}$

      $y - z = -250$ **c.** $x = 350; y = 150$
      $x + y = 500$

17. $\begin{bmatrix} -4 & 4 & -1 \\ -2 & -5 & 5 \end{bmatrix}$ **18.** Not possible since B is 3×2 and C is 3×3. **19.** $\begin{bmatrix} 2 & 3 & 8 \\ 21 & 5 & 5 \end{bmatrix}$ **20.** $\begin{bmatrix} -12 & 14 & 0 \\ 2 & -14 & 18 \end{bmatrix}$

21. $\begin{bmatrix} 0 & -10 & -15 \\ -40 & -5 & -15 \end{bmatrix}$ **22.** $\begin{bmatrix} -1 & -16 \\ 8 & 1 \end{bmatrix}$ **23.** $\begin{bmatrix} -10 & -6 & 2 \\ 16 & 3 & 4 \\ -23 & -16 & 7 \end{bmatrix}$ **24.** $\begin{bmatrix} -6 & 4 & -8 \\ 0 & 5 & 11 \\ -17 & 13 & -19 \end{bmatrix}$ **25.** $\begin{bmatrix} 10 & 5 \\ -2 & -30 \end{bmatrix}$

26. Not possible since AB is 2×2 and BA is 3×3. **27.** $\begin{bmatrix} 7 & 6 & 5 \\ 2 & -1 & 11 \end{bmatrix}$ **28.** $\begin{bmatrix} -6 & -22 & -40 \\ 9 & 43 & 58 \\ -14 & -48 & -94 \end{bmatrix}$ **29.** $\begin{bmatrix} 2 & 1 & 1 \\ 2 & 1 & 1 \\ 2 & 2 & 2 \end{bmatrix}$ **30.** $\begin{bmatrix} 1 & -1 & -1 \\ 1 & -1 & -1 \\ 1 & 1 & 1 \end{bmatrix}$

31. a. $\begin{bmatrix} 360,000 & 444,000 \\ 556,000 & 685,000 \\ 178,000 & 218,500 \end{bmatrix}$ **b.** The rows of AB correspond to the outlets, the columns represent the wholesale and retail prices. The entries tell how much value in wholesale or retail is at each outlet.

    c. $360,000 **d.** $685,000 **e.** $40,500

32. $AB = \begin{bmatrix} 1 & 7 \\ 0 & 5 \end{bmatrix}; BA = \begin{bmatrix} 1 & 0 \\ 1 & 5 \end{bmatrix}; B \ne A^{-1}$ **33.** $AB = I_3; BA = I_3; B = A^{-1}$ **34.** $\begin{bmatrix} 3 & 1 \\ 2 & 1 \end{bmatrix}$ **35.** $\begin{bmatrix} -\frac{3}{5} & \frac{1}{5} \\ 1 & 0 \end{bmatrix}$ **36.** $\begin{bmatrix} 3 & 0 & -2 \\ -6 & 1 & 4 \\ 1 & 0 & -1 \end{bmatrix}$

37. $\begin{bmatrix} 8 & -8 & 5 \\ -3 & 2 & -1 \\ -1 & -1 & 1 \end{bmatrix}$ **38. a.** $\begin{bmatrix} 1 & 1 & 2 \\ 0 & 1 & 3 \\ 3 & 0 & -2 \end{bmatrix}\begin{bmatrix} x \\ y \\ z \end{bmatrix} = \begin{bmatrix} 7 \\ -2 \\ 0 \end{bmatrix}$ **b.** $\{(-18, 79, -27)\}$ **39. a.** $\begin{bmatrix} 1 & -1 & 2 \\ 0 & 1 & -1 \\ 1 & 0 & 2 \end{bmatrix}\begin{bmatrix} x \\ y \\ z \end{bmatrix} = \begin{bmatrix} 12 \\ -5 \\ 10 \end{bmatrix}$ **b.** $\{(4, -2, 3)\}$

40. The encoded message is 96, 135, 46, 63; The decoded message is 18, 21, 12, 5 or RULE. **41.** 17 **42.** 4 **43.** -86 **44.** -236

45. 4   **46.** 16   **47.** $\left\{\left(\frac{7}{4}, -\frac{25}{8}\right)\right\}$ **48.** $\{(2, -7)\}$ **49.** $\{(23, -12, 3)\}$ **50.** $\{(-3, 2, 1)\}$ **51.** $a = \frac{5}{8}; b = -50; c = 1150;$
30- and 50-year-olds are involved in an average of 212.5 automobile accidents per day.

Chapter 9 Test

1. $\left\{\left(-3, \frac{1}{2}, 1\right)\right\}$ **2.** $\{(t, t - 1, t)\}$ **3.** $\begin{bmatrix} 5 & 4 \\ 1 & 11 \end{bmatrix}$ **4.** $\begin{bmatrix} 5 & -2 \\ 1 & -1 \\ 4 & -1 \end{bmatrix}$ **5.** $\begin{bmatrix} \frac{3}{5} & -\frac{2}{5} \\ \frac{1}{5} & \frac{1}{5} \end{bmatrix}$ **6.** $\begin{bmatrix} -1 & 2 \\ -5 & 4 \end{bmatrix}$

7. $AB = I_3; BA = I_3$ **8. a.** $\begin{bmatrix} 3 & 5 \\ 2 & -3 \end{bmatrix}\begin{bmatrix} x \\ y \end{bmatrix} = \begin{bmatrix} 9 \\ -13 \end{bmatrix}$ **b.** $\begin{bmatrix} \frac{3}{19} & \frac{5}{19} \\ \frac{2}{19} & -\frac{3}{19} \end{bmatrix}$ **c.** $\{(-2, 3)\}$ **9.** 18   **10.** $x = 2$

Cumulative Review Exercises (Chapters 1–9)

1. $\left\{\frac{-1 + \sqrt{33}}{4}, \frac{-1 - \sqrt{33}}{4}\right\}$ **2.** $\left[\frac{1}{2}, \infty\right)$ **3.** $\{6\}$ **4.** $\left\{-4, \frac{1}{3}, 1\right\}$ **5.** $\{\ln 5, \ln 9\}$ **6.** $\{1\}$ **7.** $\{(7, -4, 6)\}$

8. $y = -1$ **9.** $f^{-1}(x) = \frac{x^2 + 7}{4} (x \geq 0)$

10.

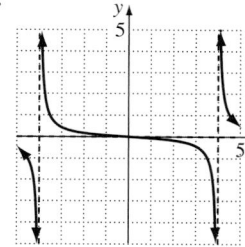

11. $f(x) = (x + 2)(x - 3)(2x + 1)(2x - 1)$ **13. a.** $A = A_0 e^{-0.017t}$ **b.** 759.30 g

12.

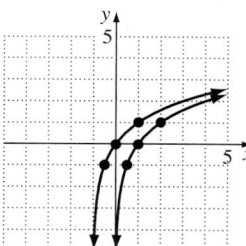

14. $\begin{bmatrix} 2 & -1 \\ 13 & 1 \end{bmatrix}$ **15.** $\frac{8}{x - 3} + \frac{-2}{x - 2} + \frac{-3}{x + 2}$

16.

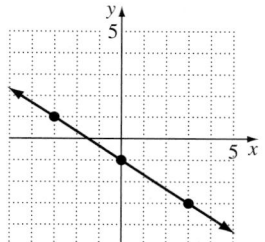

17.

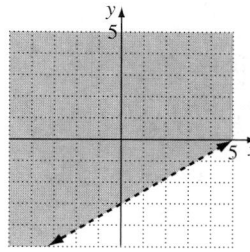

18.

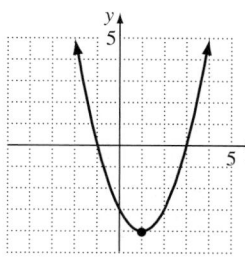

19.

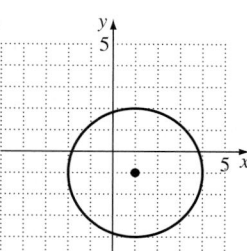

20. $= x^2 + 2x - 2$

21.

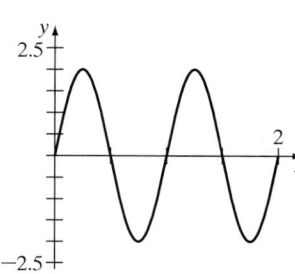

22. $\frac{3}{5}$

23. $\frac{\cos 2x}{\cos x - \sin x} = \frac{\cos^2 x - \sin^2 x}{\cos x - \sin x} = \frac{(\cos x - \sin x)(\cos x + \sin x)}{\cos x - \sin x} = \cos x + \sin x$

24. $\frac{3\pi}{2}$ **25.** $2\mathbf{i} - 13\mathbf{j}$

CHAPTER 10

Section 10.1

Check Point Exercises

1. foci at $(-3\sqrt{3}, 0)$ and $(3\sqrt{3}, 0)$ **2.** foci at $(0, -\sqrt{7})$ and $(0, \sqrt{7})$ **3.** $\dfrac{x^2}{9} + \dfrac{y^2}{5} = 1$

4. foci at $(-1 - \sqrt{5}, 2)$ and $(-1 + \sqrt{5}, 2)$ **5.** Yes

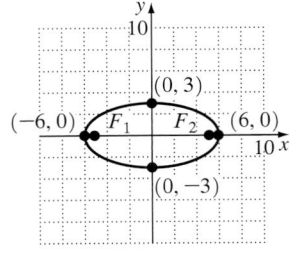
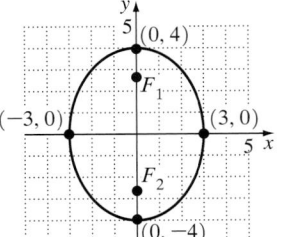
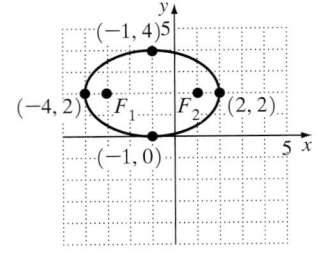

Exercise Set 10.1

1. foci at $(-2\sqrt{3}, 0)$ and $(2\sqrt{3}, 0)$ **3.** foci at $(0, -3\sqrt{3})$ and $(0, 3\sqrt{3})$ **5.** foci at $(0, -\sqrt{39})$ and $(0, \sqrt{39})$

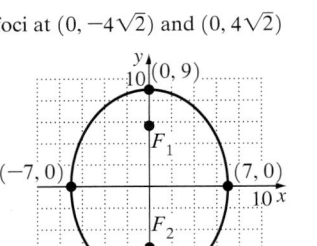
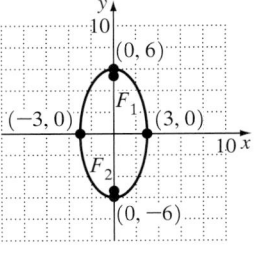
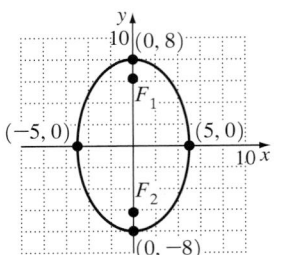

7. foci at $(0, -4\sqrt{2})$ and $(0, 4\sqrt{2})$ **9.** foci at $(0, -\sqrt{21})$ and $(0, \sqrt{21})$ **11.** foci at $(-2\sqrt{3}, 0)$ and $(2\sqrt{3}, 0)$

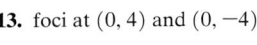
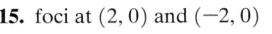

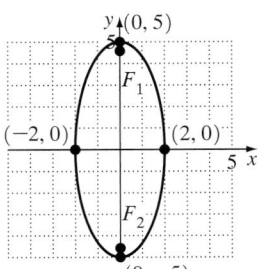

13. foci at $(0, 4)$ and $(0, -4)$ **15.** foci at $(2, 0)$ and $(-2, 0)$

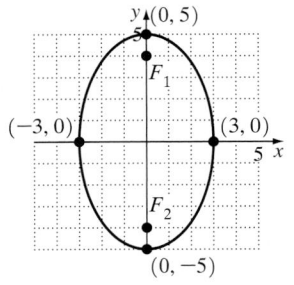
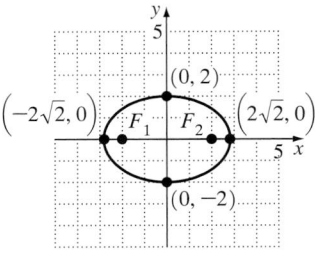

17. $\dfrac{x^2}{4} + \dfrac{y^2}{1} = 1$; foci at $(-\sqrt{3}, 0)$ and $(\sqrt{3}, 0)$

19. $\dfrac{x^2}{1} + \dfrac{y^2}{4} = 1$; foci at $(0, \sqrt{3})$ and $(0, -\sqrt{3})$

21. $\dfrac{x^2}{64} + \dfrac{y^2}{39} = 1$ **23.** $\dfrac{x^2}{33} + \dfrac{y^2}{49} = 1$

25. $\dfrac{x^2}{13} + \dfrac{y^2}{9} = 1$ **27.** $\dfrac{x^2}{16} + \dfrac{y^2}{4} = 1$

29. $\dfrac{x^2}{4} + \dfrac{y^2}{25} = 1$

31. foci at $(2 - \sqrt{5}, 1)$ and $(2 + \sqrt{5}, 1)$

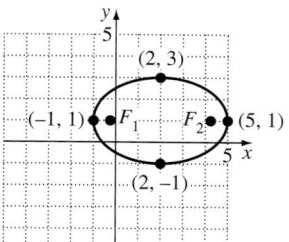

33. foci at $(-3 - 2\sqrt{3}, 2)$ and $(-3 + 2\sqrt{3}, 2)$

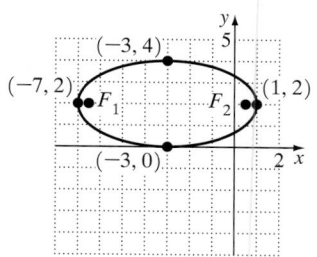

35. foci at $(4, 2)$ and $(4, -6)$

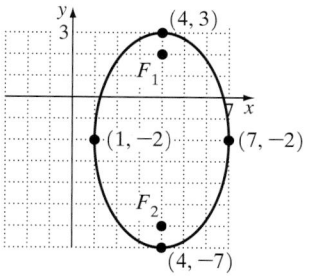

37. foci at $(0, 2 + \sqrt{11}), (0, 2 - \sqrt{11})$

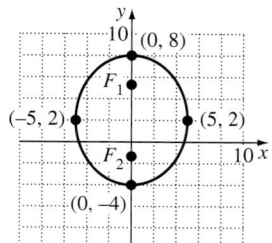

39. foci at $(-3 - 2\sqrt{2}, 2)$ and $(-3 + 2\sqrt{2}, 2)$

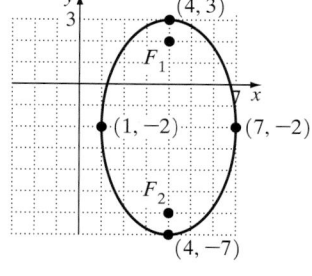

41. foci at $(1, -3 + \sqrt{5})$ and $(1, -3 - \sqrt{5})$

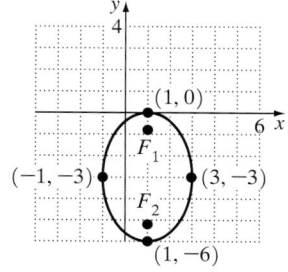

43. $\dfrac{(x - 2)^2}{25} + \dfrac{(y + 1)^2}{9} = 1$

foci at $(-2, -1)$ and $(6, -1)$

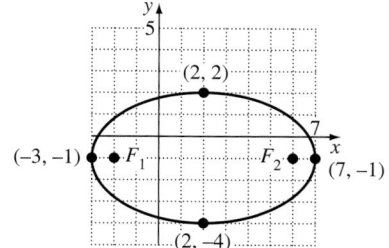

45. $\dfrac{(x - 1)^2}{16} + \dfrac{(y + 2)^2}{9} = 1$

foci at $(1 - \sqrt{7}, -2)$ and $(1 + \sqrt{7}, -2)$

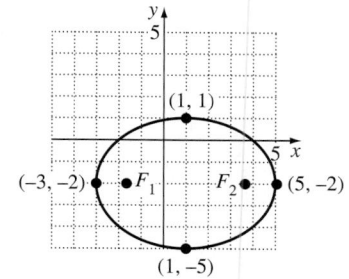

47. $\dfrac{(x + 2)^2}{16} + \dfrac{(y - 3)^2}{64} = 1$

foci at $(-2, 3 + 4\sqrt{3})$ and $(-2, 3 - 4\sqrt{3})$

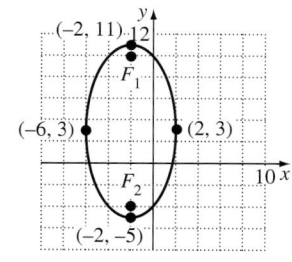

49. Yes **51. a.** $\dfrac{x^2}{2304} + \dfrac{y^2}{529} = 1$ **b.** about 42 feet **63.** $\dfrac{x^2}{\frac{36}{5}} + \dfrac{y^2}{36} = 1$ **65.** The large circle has radius 5 with center $(0, 0)$.

Its equation is $x^2 + y^2 = 25$. The small circle has radius 3 with center $(0, 0)$. Its equation is $x^2 + y^2 = 9$.

Section 10.2

Check Point Exercises

1. a. vertices at $(5, 0)$ and $(-5, 0)$; foci at $(\sqrt{41}, 0)$ and $(-\sqrt{41}, 0)$ **b.** vertices at $(0, 5)$ and $(0, -5)$; foci at $(0, \sqrt{41})$ and $(0, -\sqrt{41})$

2. $\dfrac{y^2}{9} - \dfrac{x^2}{16} = 1$ **3.** foci at $(-3\sqrt{5}, 0)$ and $(3\sqrt{5}, 0)$ **4.** foci at $(0, \sqrt{5})$ and $(0, -\sqrt{5})$ **5.** foci at $(3 - \sqrt{5}, 1)$ and $(3 + \sqrt{5}, 1)$

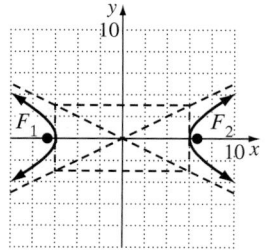

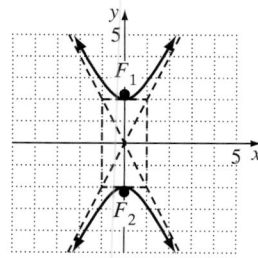

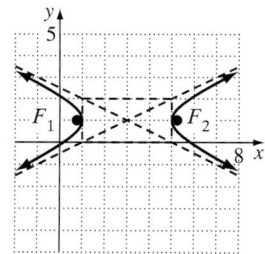

6. $\dfrac{x^2}{2,722,500} - \dfrac{y^2}{25,155,900} = 1$

Exercise Set 10.2

1. vertices at $(2, 0)$ and $(-2, 0)$; foci at $(\sqrt{5}, 0)$ and $(-\sqrt{5}, 0)$; graph (b)

3. vertices at $(0, 2)$ and $(0, -2)$; foci at $(0, \sqrt{5})$ and $(0, -\sqrt{5})$; graph (a) **5.** $y^2 - \dfrac{x^2}{8} = 1$ **7.** $\dfrac{x^2}{9} - \dfrac{y^2}{7} = 1$

9. foci: $(\pm\sqrt{34}, 0)$

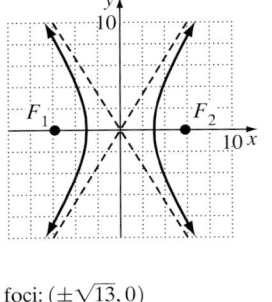

11. foci: $(\pm 2\sqrt{41}, 0)$

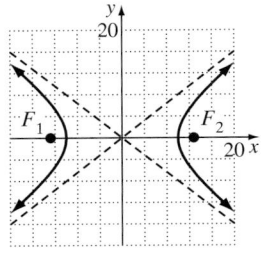

13. foci: $(0, \pm 2\sqrt{13})$

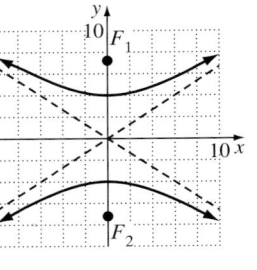

15. foci: $(0, \pm\sqrt{61})$

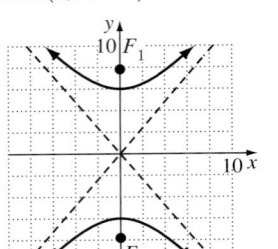

17. foci: $(\pm\sqrt{13}, 0)$

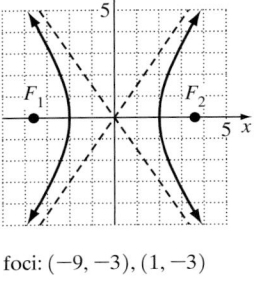

19. foci: $(0, \pm\sqrt{34})$

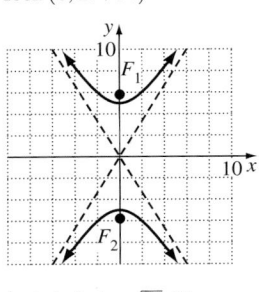

21. foci: $(\pm\sqrt{5}, 0)$

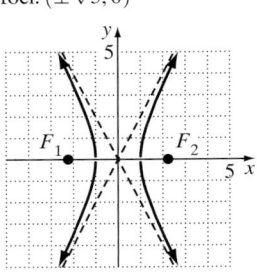

23. $\dfrac{x^2}{9} - \dfrac{y^2}{25} = 1$

25. $\dfrac{y^2}{4} - \dfrac{x^2}{9} = 1$

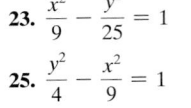

27. foci: $(-9, -3), (1, -3)$

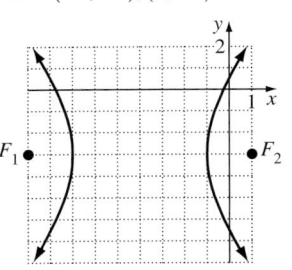

29. foci: $(-3 \pm \sqrt{41}, 0)$

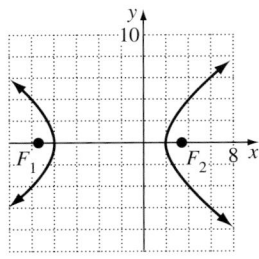

31. foci: $(1, -2 \pm 2\sqrt{5})$

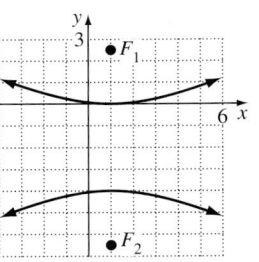

33. foci: $(3 \pm \sqrt{5}, -3)$

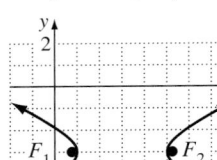
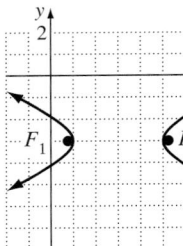

35. foci: $(1 \pm 2\sqrt{2}, 2)$

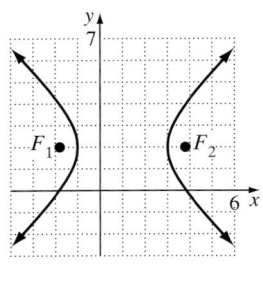

37. $(x - 1)^2 - (y + 2)^2 = 1$

foci: $(1 \pm \sqrt{2}, -2)$

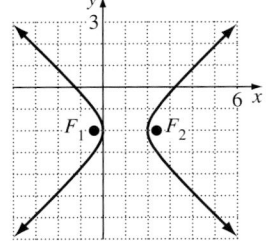

39. $\dfrac{(y + 1)^2}{4} - \dfrac{(x + 2)^2}{0.25} = 1$

foci: $(-2, -1 \pm \sqrt{4.25})$

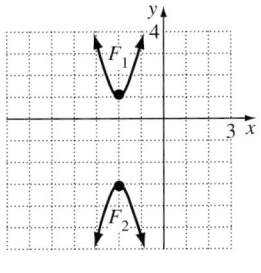

41. $\dfrac{(x - 2)^2}{9} - \dfrac{(y - 3)^2}{4} = 1$

foci: $(2 \pm \sqrt{13}, 3)$

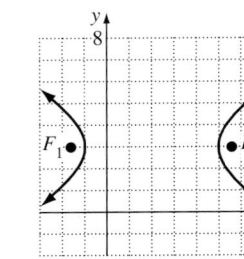

43. $\dfrac{y^2}{4} - \dfrac{(x-4)^2}{25} = 1$

foci: $(4, \pm\sqrt{29})$

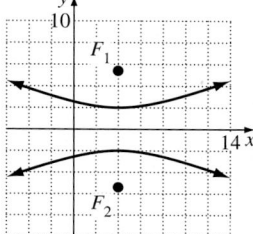

45. If M_1 is located 2640 feet to the right of the origin on the x-axis, the explosion is located on the right branch of the hyperbola given by the equation $\dfrac{x^2}{1{,}210{,}000} - \dfrac{y^2}{5{,}759{,}600} = 1$.

47. 40 yd

59.

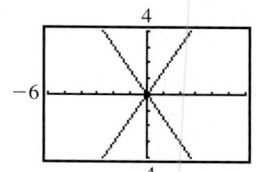

; No. Two intersecting lines.

61. $2y^2 + (10 - 6x)y + (4x^2 - 3x - 6) = 0$

$y = \dfrac{3x - 5 \pm \sqrt{x^2 - 24x + 37}}{2}$

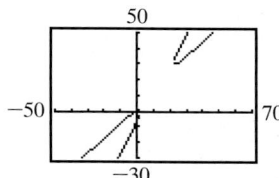

The xy-term rotates the hyperbola.

63. (c) is true.

65. $\dfrac{y^2}{36} - \dfrac{(x-5)^2}{20} = 1$

Section 10.3

Check Point Exercises

1. focus: $(2, 0)$
directrix: $x = -2$

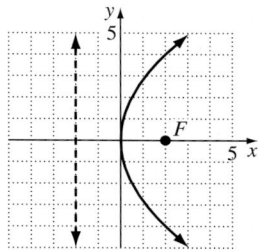

2. focus: $(0, -3)$
directrix: $y = 3$

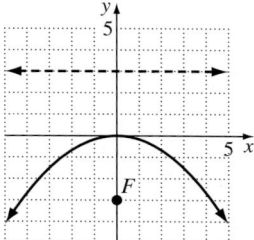

3. $y^2 = 32x$

4. vertex: $(2, -1)$; focus: $(2, 0)$;
directrix: $y = -2$

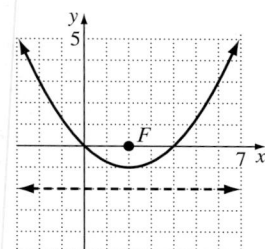

5. vertex: $(2, -1)$; focus: $(1, -1)$;
directrix: $x = 3$

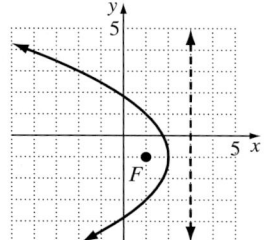

6. $x^2 = \dfrac{9}{4}y$; The light should be placed at $\left(0, \dfrac{9}{16}\right)$, or $\dfrac{9}{16}$ inch above the vertex.

Exercise Set 10.3

1. focus: $(1, 0)$; directrix: $x = -1$; graph (c)
5. focus: $(4, 0)$; directrix: $x = -4$

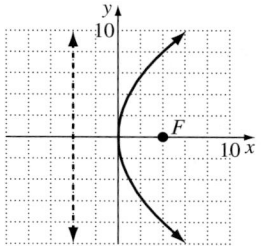

3. focus: $(0, -1)$, directrix: $y = 1$; graph (b)
7. focus: $(-2, 0)$; directrix: $x = 2$

9. focus: $(0, 3)$; directrix: $y = -3$

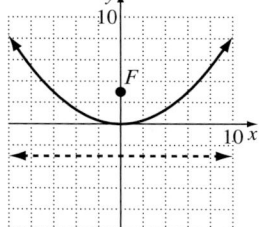

11. focus: $(0, -4)$; directrix: $y = 4$

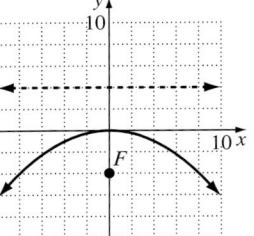

13. focus: $\left(\dfrac{3}{2}, 0\right)$; directrix: $x = -\dfrac{3}{2}$

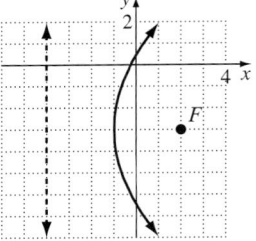

15. $y^2 = 28x$

17. $y^2 = -20x$
19. $x^2 = 60y$
21. $x^2 = -100y$
23. vertex: $(1, 1)$; focus: $(2, 1)$;
   directrix: $x = 0$; graph (c)
25. vertex: $(-1, -1)$; focus: $(-1, -2)$;
   directrix: $y = 0$; graph (d)

27. vertex: $(2, 1)$; focus: $(2, 3)$;
   directrix: $y = -1$

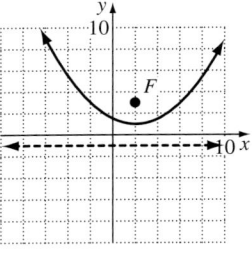

29. vertex: $(-1, -1)$; focus: $(-1, -3)$;
   directrix: $y = 1$

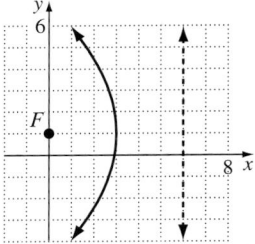

31. vertex: $(-1, -3)$; focus: $(2, -3)$;
   directrix: $x = -4$

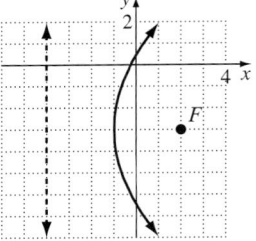

33. vertex: $(0, -1)$; focus: $(-2, -1)$;
   directrix: $x = 2$

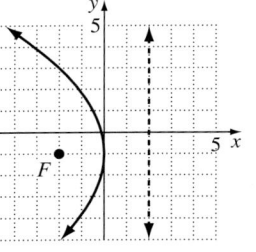

35. $(x - 1)^2 = 4(y - 2)$;
   vertex: $(1, 2)$; focus: $(1, 3)$;
   directrix: $y = 1$

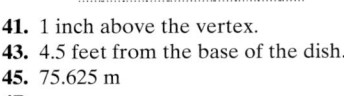

37. $(y - 1)^2 = -12(x - 3)$;
   vertex: $(3, 1)$; focus: $(0, 1)$;
   directrix: $x = 6$

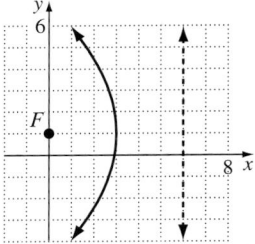

39. $(x + 3)^2 = 4(y + 2)$;
   vertex: $(-3, -2)$; focus: $(-3, -1)$;
   directrix: $y = -3$

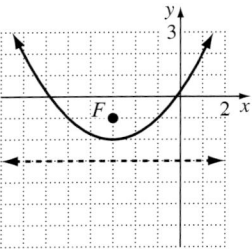

41. 1 inch above the vertex.
43. 4.5 feet from the base of the dish.
45. 75.625 m
47. yes
57. $y = -5 \pm \sqrt{x}$

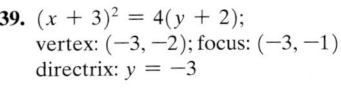

59. $3y^2 + (2\sqrt{3}x - 8)y + x^2 + 8\sqrt{3}x + 32 = 0$

$$y = \dfrac{-\sqrt{3}x + 4 \pm 4\sqrt{-2\sqrt{3}x - 5}}{3}$$

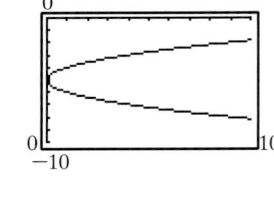

61. $\dfrac{25}{6}$ ft

Section 10.4

Check Point Exercises

1. a. ellipse   **b.** circle   **c.** parabola   **d.** hyperbola

2. $\dfrac{x'^2}{4} - \dfrac{y'^2}{4} = 1$ **3.** $\dfrac{x'^2}{\frac{4}{5}} + \dfrac{y'^2}{4} = 1$ **4.**   **5.** parabola

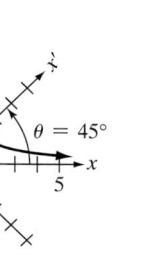 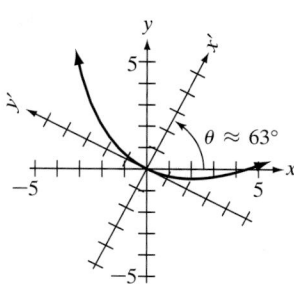

Exercise Set 10.4

1. parabola   **3.** hyperbola   **5.** circle   **7.** hyperbola   **9.** $\dfrac{y'^2}{2} - \dfrac{x'^2}{2} = 1$ **11.** $\dfrac{y'^2}{1} - \dfrac{x'^2}{3} = 1$ **13.** $\dfrac{x'^2}{4} - \dfrac{y'^2}{9} = 1$

15. $x = \dfrac{\sqrt{2}}{2}(x' - y'); y = \dfrac{\sqrt{2}}{2}(x' + y')$ **17.** $x = \dfrac{\sqrt{2}}{2}(x' - y'); y = \dfrac{\sqrt{2}}{2}(x' + y')$ **19.** $x = \dfrac{\sqrt{3}x' - y'}{2}; y = \dfrac{x' + \sqrt{3}y'}{2}$

21. $x = \dfrac{3x' - 4y'}{5}; y = \dfrac{4x' + 3y'}{5}$ **23.** $x = \sqrt{5}\left(\dfrac{2x' - y'}{5}\right); y = \sqrt{5}\left(\dfrac{x' + 2y'}{5}\right)$ **25.** $x = \dfrac{4x' - 3y'}{5}; y = \dfrac{3x' + 4y'}{5}$

27. a. $3x'^2 + y'^2 = 20$

    b. $\dfrac{x'^2}{\frac{20}{3}} + \dfrac{y'^2}{20} = 1$

    c.

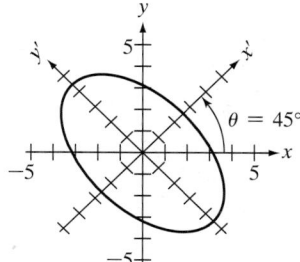

29. a. $-4x'^2 + 16y'^2 = 64$

    b. $\dfrac{y'^2}{4} - \dfrac{x'^2}{16} = 1$

    c.

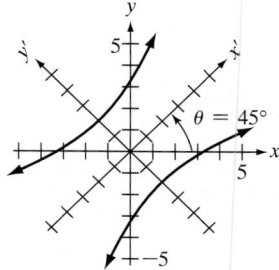

31. a. $64x'^2 - 16y'^2 = 16$

    b. $\dfrac{x'^2}{\frac{1}{4}} - \dfrac{y'^2}{1} = 1$

    c.

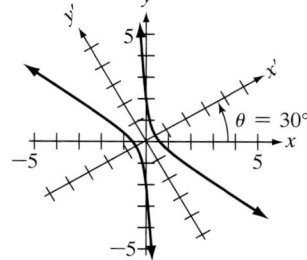

33. a. $650x'^2 + 25y'^2 = 225$

    b. $\dfrac{x'^2}{\frac{9}{26}} + \dfrac{y'^2}{9} = 1$

    c.

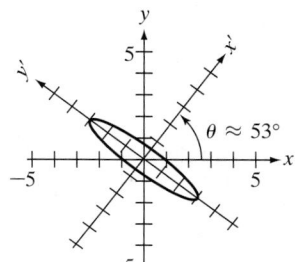

35. a. $50x'^2 - 75y'^2 = 25$

    b. $\dfrac{x'^2}{\frac{1}{2}} - \dfrac{y'^2}{\frac{1}{3}} = 1$

    c.

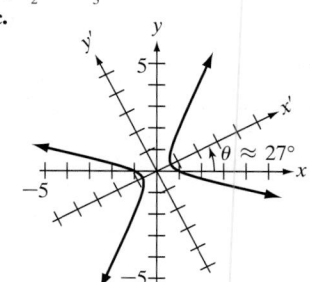

37. a. $625x'^2 + 1250y'^2 = 625$

    b. $\dfrac{x'^2}{1} + \dfrac{y'^2}{\frac{1}{2}} = 1$

    c.

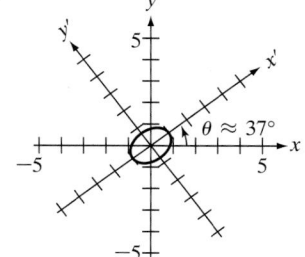

39. ellipse or circle   **41.** parabola   **43.** hyperbola

51.
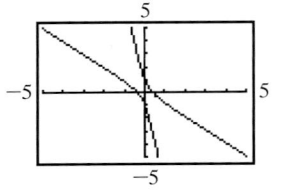

53.
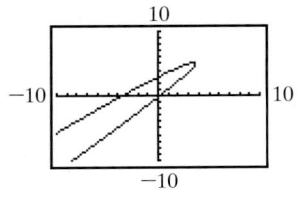

55.
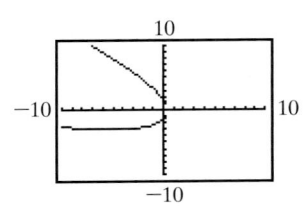

57. There are no solutions to this equation just as there is no such sound.

59.
$$A' = A\cos^2\theta + B\sin\theta\cos\theta + C\sin^2\theta$$
$$C' = A\sin^2\theta - B\sin\theta\cos\theta + C\cos^2\theta$$
$$A' + C' = A\cos^2\theta + B\sin\theta\cos\theta + C\sin^2\theta + A\sin^2\theta - B\sin\theta\cos\theta + C\cos^2\theta$$
$$= A(\cos^2\theta + \sin^2\theta) + B(\sin\theta\cos\theta - \sin\theta\cos\theta) + C(\sin^2\theta + \cos^2\theta)$$
$$= A(1) + B(0) + C(1)$$
$$= A + C$$

Section 10.5

Check Point Exercises

1.
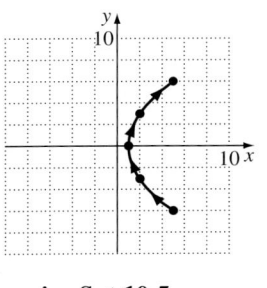

2.
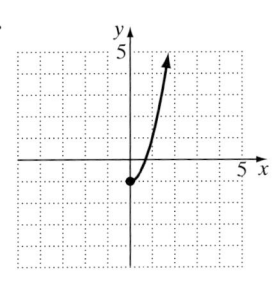

3.
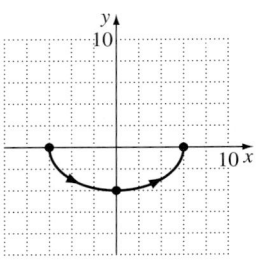

4. $x = t$ and $y = t^2 - 25$

Exercise Set 10.5

1. $(-2, 6)$ **3.** $(5, -3)$ **5.** $(4, 8)$ **7.** $(60\sqrt{3}, 1)$

9.
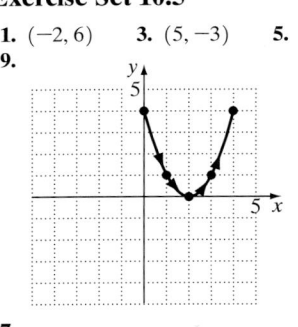

11.
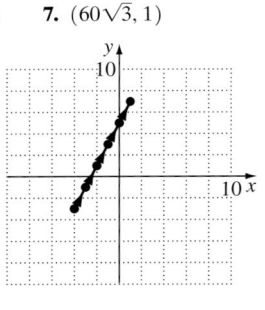

13.
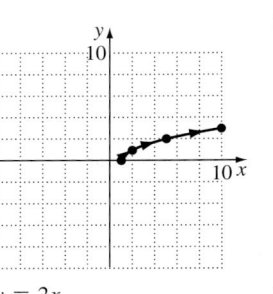

15.
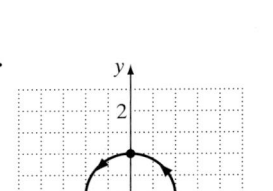

17.
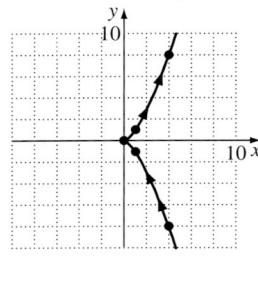

19.
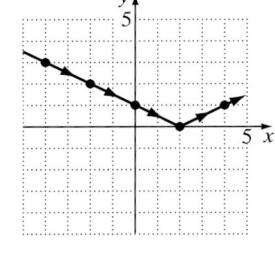

21. $y = 2x$
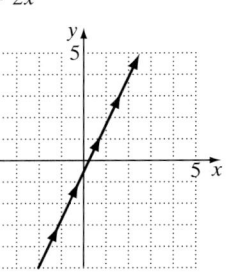

23. $y = (x + 4)^2$
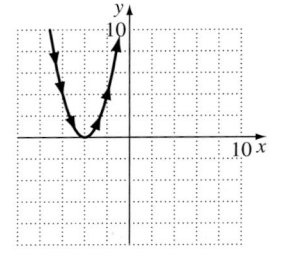

25. $y = x^2 - 1, x \geq 0$

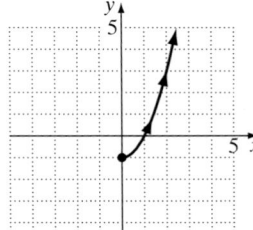

27. $\dfrac{x^2}{4} + \dfrac{y^2}{4} = 1$

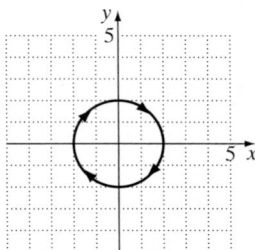

29. $\dfrac{(x-1)^2}{9} + \dfrac{(y-2)^2}{9} = 1$

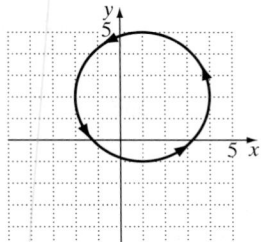

31. $\dfrac{x^2}{4} + \dfrac{y^2}{9} = 1$

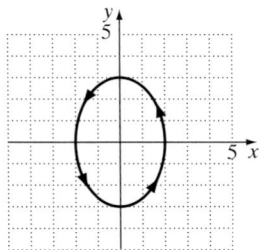

33. $\dfrac{(x-1)^2}{9} + \dfrac{(y+1)^2}{4} = 1,$
$-2 \leq x \leq 4, -1 \leq y \leq 1$

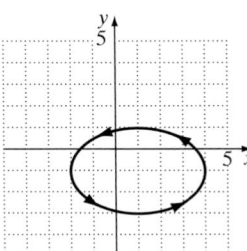

35. $x^2 - y^2 = 1$

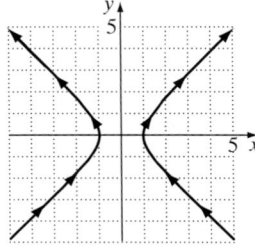

37. $y = x - 4, x \geq 2, y \geq -2$

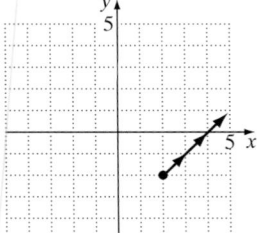

39. $y = \dfrac{1}{x}, x \geq 1, y \geq 0$

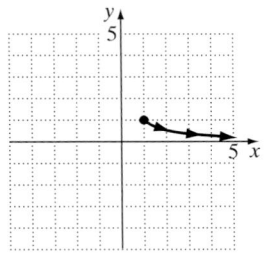

41. $(x - h)^2 + (y - k)^2 = r^2$ **43.** $\dfrac{(x-h)^2}{a^2} - \dfrac{(y-k)^2}{b^2} = 1$ **45.** $x = 3 + 6\cos t; y = 5 + 6\sin t$

47. $x = -2 + 5\cos t; y = 3 + 2\sin t$ **49.** $x = 4\sec t; y = \sqrt{20}\tan t$ **51.** $x = -2 + 3t; y = 4 + 3t$

57. a.

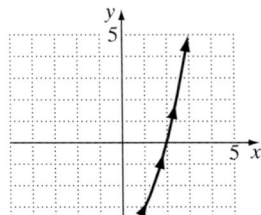

b.

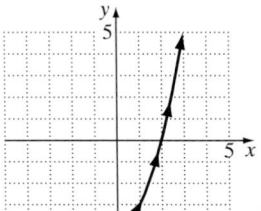

c.

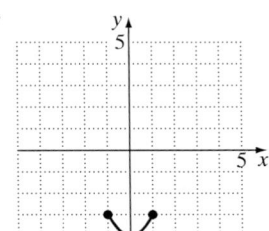

d.

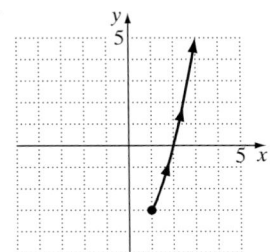

59. a. $x = (180\cos 40°)t; y = 3 + (180\sin 40°)t - 16t^2$
  b. After 1 second: 137.9 feet in distance, 102.7 feet in height;
    After 2 seconds: 275.8 feet in distance, 170.4 feet in height;
    After 3 seconds: 413.7 feet in distance, 206.1 feet in height
  c. $t = 7.3$ sec; total horizontal distance: 1006.6 ft; yes

69.

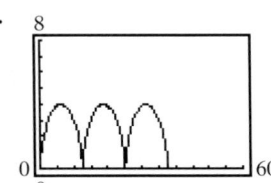

71.

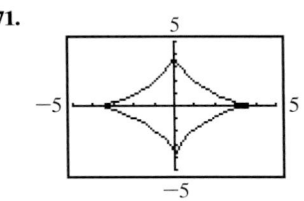

73.

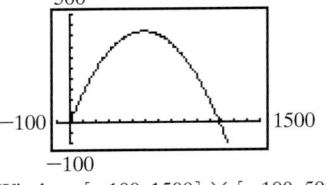

Window: $[-100, 1500] \times [-100, 500]$;
The maximum height is 419.4 feet at a time of 5.1 seconds.
The range of the projectile is 1174.6 feet horizontally.
It hits the ground at 10.2 seconds.

77. $x = 3 \sin t; y = 3 \cos t$

75. a. $x = (140 \cos 22°)t; y = 5 + (140 \sin 22°)t - 16t^2$
b.

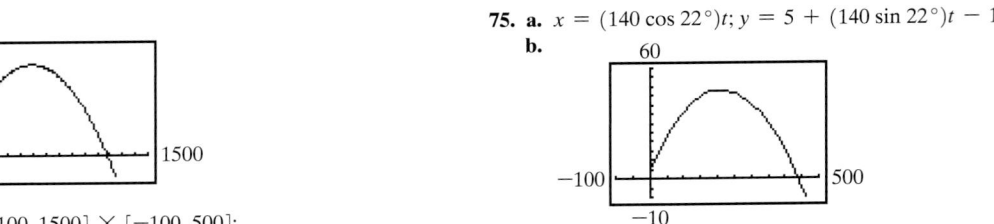

Window: $[-100, 500] \times [-10, 60]$
c. The maximum height is 48.0 feet. It occurs at 1.6 seconds.
d. 3.4 sec
e. 437.5 ft

Section 10.6

Check Point Exercises

1.   **2.**   **3.**

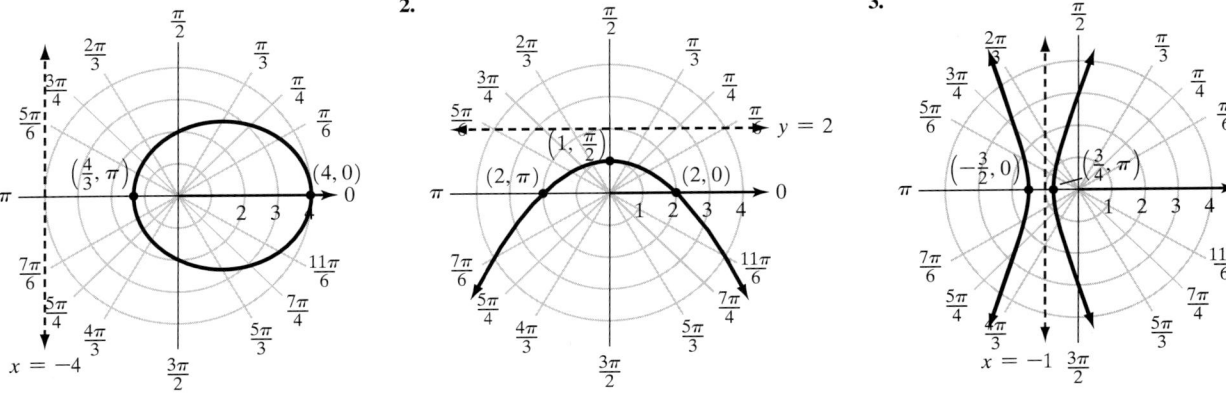

Exercise Set 10.6

1. a. parabola   **b.** The directrix is 3 units above the pole, at $y = 3$.   **3. a.** ellipse
b. The directrix is 3 units to the left of the pole, at $x = -3$.   **5. a.** parabola   **b.** The directrix is 4 units above the pole, at $y = 4$.
7. a. hyperbola   **b.** The directrix is 3 units to the left of the pole, at $x = -3$.
9.   **11.**   **13.**

15.

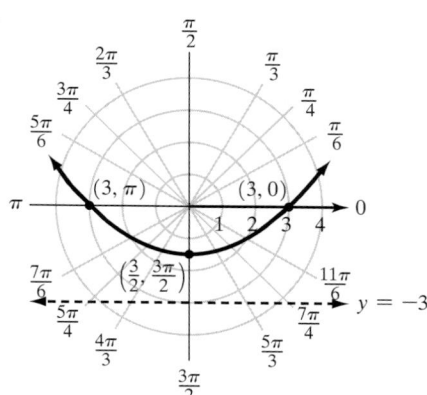

17.

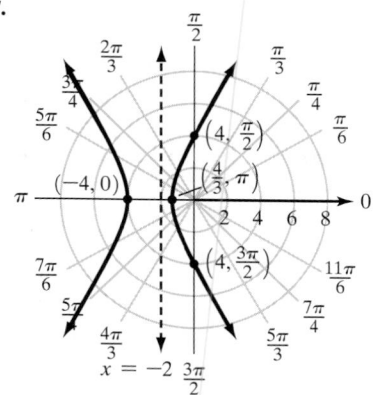

19.

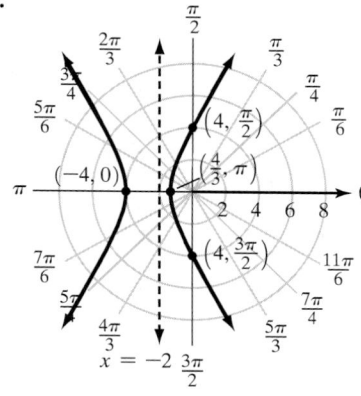

21. 0.54 astronomical units or 51 million miles.   **23.** 4122 miles from the center of the earth.; 162 miles from the surface of the earth.

33. hyperbola

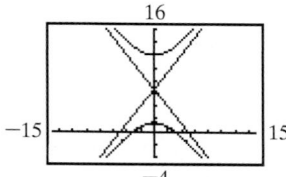

35.

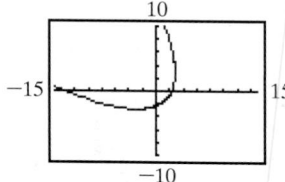

The graph appears to be rotated counterclockwise through an angle of $\frac{\pi}{4}$ radians.

37. Mercury: $r = \dfrac{(1 - 0.2056^2)(36.0 \times 10^6)}{1 - 0.2056 \cos \theta}$

Earth: $r = \dfrac{(1 - 0.0167^2)(92.96 \times 10^6)}{1 - 0.0167 \cos \theta}$;

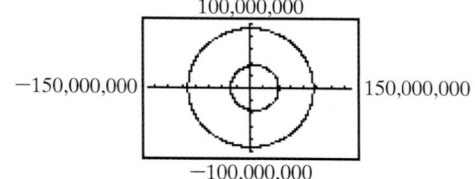

39. $r = \dfrac{2}{1 - \frac{1}{2} \cos \theta}$ or $r = \dfrac{6}{1 + \frac{1}{2} \cos \theta}$

41. parabola; using the relationships between rectangular and polar coordinates, $x^2 + y^2 = r^2$ and $x = r \cos \theta$; $y^2 = x + \dfrac{1}{4}$

Chapter 10 Review Exercises

1. foci: $(\pm \sqrt{11}, 0)$

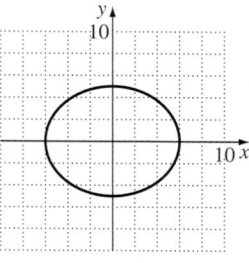

2. foci: $(0, \pm 3)$

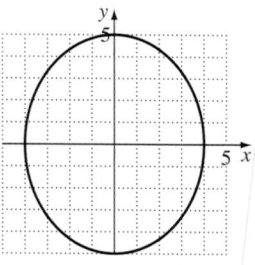

3. foci: $(0, \pm 2\sqrt{3})$

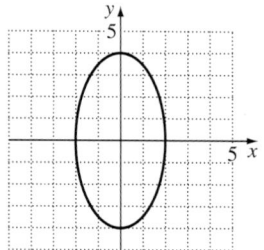

4. foci: $(\pm \sqrt{5}, 0)$

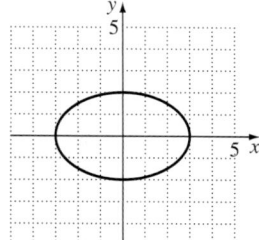

5. foci: $(1 \pm \sqrt{7}, -2)$

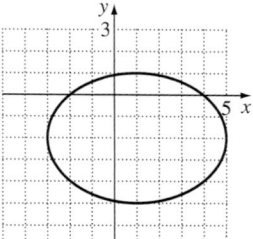

6. foci: $(-1, 2 \pm \sqrt{7})$

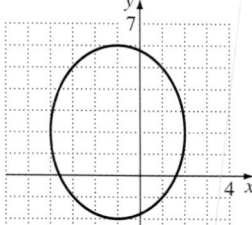

7. foci: $(-3 \pm \sqrt{5}, 2)$

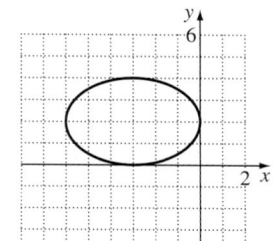

8. foci: $(1, -1 \pm \sqrt{5})$

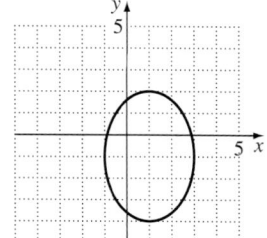

9. $\dfrac{x^2}{25} + \dfrac{y^2}{9} = 1$ **10.** $\dfrac{x^2}{27} + \dfrac{y^2}{36} = 1$ **11.** $\dfrac{x^2}{36} + \dfrac{y^2}{4} = 1$ **12.** $\dfrac{x^2}{100} + \dfrac{y^2}{36} = 1$ **13.** yes

14. The hit ball will collide with the other ball.

15. foci: $(\pm\sqrt{17}, 0)$
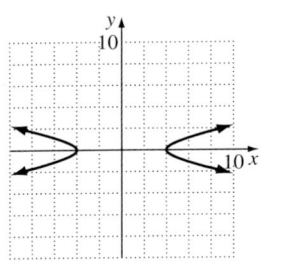

16. foci: $(0, \pm\sqrt{17})$
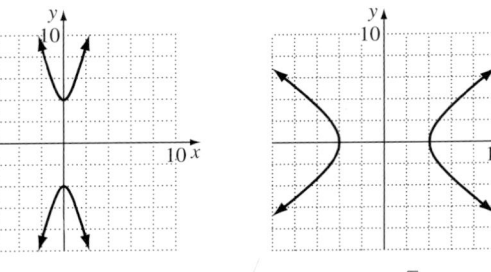

17. foci: $(\pm 5, 0)$
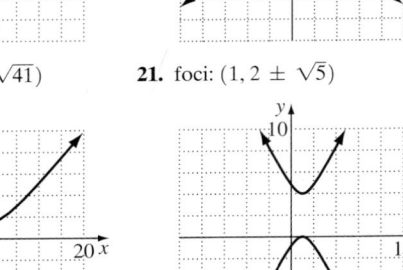

18. foci: $(0, \pm 2\sqrt{5})$
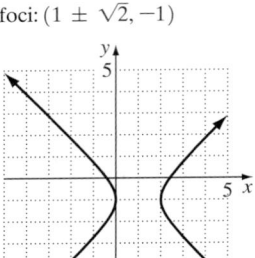

19. foci: $(2 \pm \sqrt{41}, -3)$
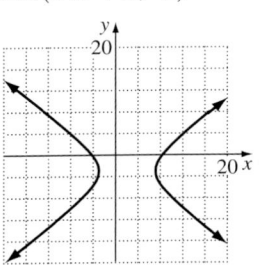

20. foci: $(3, -2 \pm \sqrt{41})$
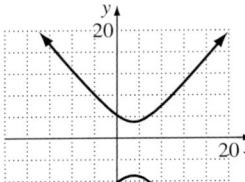

21. foci: $(1, 2 \pm \sqrt{5})$
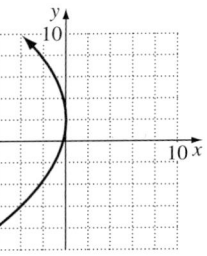

22. foci: $(1 \pm \sqrt{2}, -1)$
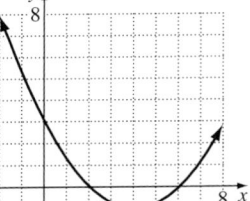

23. $\dfrac{y^2}{4} - \dfrac{x^2}{12} = 1$ **24.** $\dfrac{x^2}{9} - \dfrac{y^2}{55} = 1$ **25.** c must be greater than a. **26.** $\dfrac{x^2}{2162.25} - \dfrac{y^2}{7837.75} = 1$

27. vertex: $(0, 0)$; focus: $(2, 0)$; directrix: $x = -2$
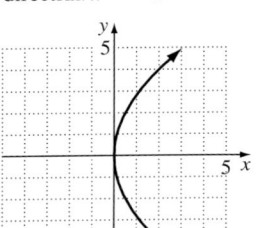

28. vertex: $(0, 0)$; focus: $(0, -4)$; directrix: $y = 4$
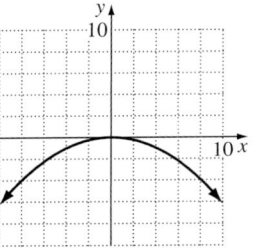

29. vertex: $(0, 2)$; focus: $(-4, 2)$; directrix: $x = 4$
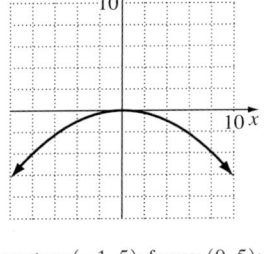

30. vertex: $(4, -1)$; focus: $(4, 0)$; directrix: $y = -2$
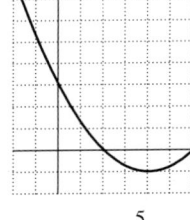

31. vertex: $(0, 1)$; focus: $(0, 0)$; directrix: $y = 2$
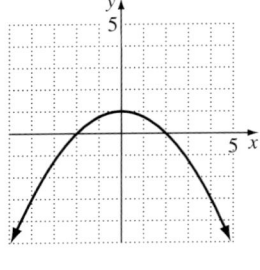

32. vertex: $(-1, 5)$; focus: $(0, 5)$; directrix: $x = -2$
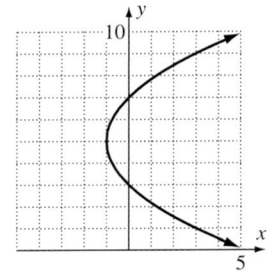

33. vertex: $(2, -2)$; focus: $\left(2, -\dfrac{3}{2}\right)$; directrix: $y = -\dfrac{5}{2}$
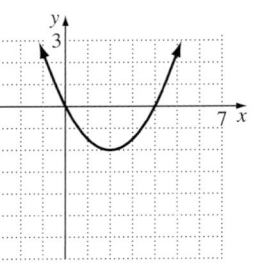

34. $y^2 = 48x$ **35.** $x^2 = -44y$ **36.** $x^2 = 12y$; Place the light 3 inches from the vertex at $(0, 3)$. **37.** approximately 58 ft
38. approximately 128 ft **39.** parabola **40.** ellipse **41.** ellipse **42.** hyperbola **43.** ellipse or circle **44.** ellipse or circle
45. parabola **46.** ellipse or circle

47. a. $x'^2 - y'^2 = 8$

b. $\dfrac{x'^2}{8} - \dfrac{y'^2}{8} = 1$

c.

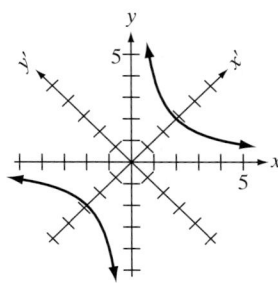

48. a. $3x'^2 + y'^2 = 2$

b. $\dfrac{x'^2}{\frac{2}{3}} + \dfrac{y'^2}{2} = 1$

c.

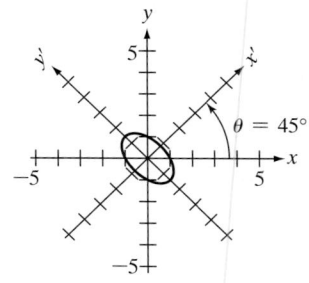

49. a. $18x'^2 - 2y'^2 = 18$

b. $\dfrac{x'^2}{1} - \dfrac{y'^2}{9} = 1$

c.

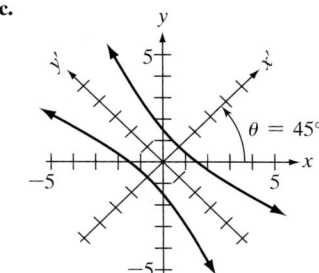

50. a. $50x'^2 + 150y'^2 = 450$

b. $\dfrac{x'^2}{9} + \dfrac{y'^2}{3} = 1$

c.

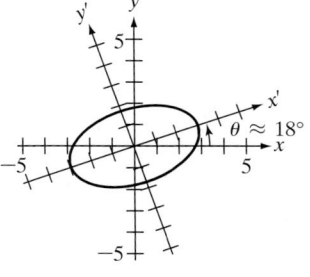

51. a. $16x'^2 + 96y' = 0$

b. $x'^2 = -6y'$

c.

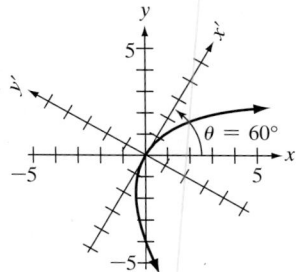

52. $y = -\dfrac{1}{2}x + \dfrac{1}{2}$

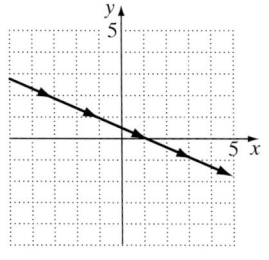

53. $(y + 1)^2 = x, 0 \le x \le 9,$
$-2 \le y \le 2$

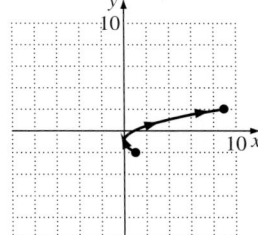

54. $(y - 1)^2 = \dfrac{1}{4}x$

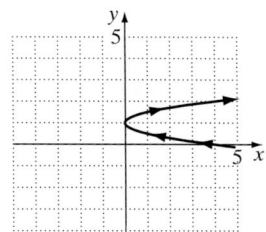

55. $\dfrac{x^2}{16} + \dfrac{y^2}{9} = 1, 0 \le x \le 4,$
$-3 < y \le 3$

56. $\dfrac{(x - 3)^2}{4} + \dfrac{(y - 1)^2}{4} = 1$
or $(x - 3)^2 + (y - 1)^2 = 4$

57. $\dfrac{x^2}{9} - \dfrac{y^2}{9} = 1,$
$3 \le x \le 3\sqrt{2}, 0 \le y \le 3$

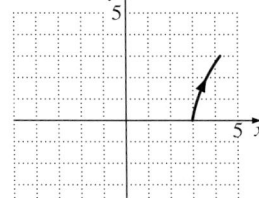

59. a. $x = (100 \cos 40°)t; y = 6 + (100 \sin 40°)t - 16t^2$

b. After 1 second: 76.6 feet in distance, 54.3 feet in height; after 2 seconds: 153.2 feet in distance, 70.6 feet in height; after 3 seconds: 229.8 feet in distance, 54.8 feet in height.

c. 4.1 sec; 314.1 ft   **d.**

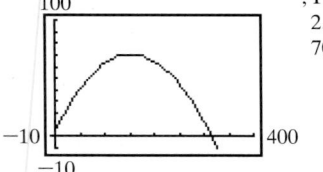

; The ball is at its maximum height at 2.0 seconds. The maximum height is 70.6 feet.

60. a. $r = \dfrac{4}{1 - \sin\theta}$ **b.** $e = 1; p = 4;$ parabola

c.

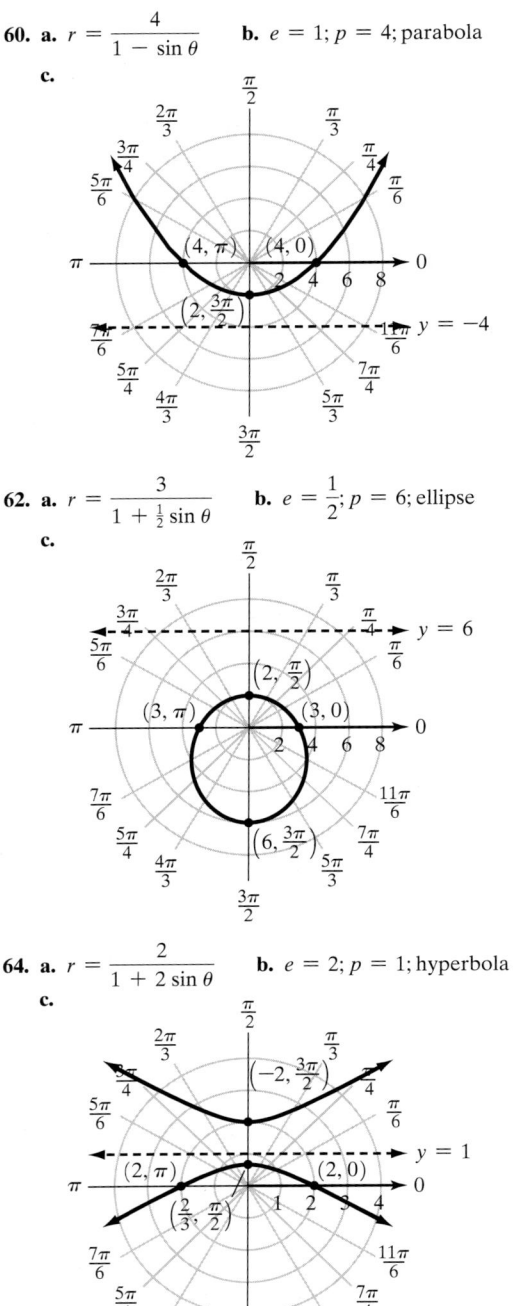

61. a. $r = \dfrac{6}{1 + \cos\theta}$ **b.** $e = 1; p = 6;$ parabola

c.

62. a. $r = \dfrac{3}{1 + \frac{1}{2}\sin\theta}$ **b.** $e = \dfrac{1}{2}; p = 6;$ ellipse

c.

63. a. $r = \dfrac{\frac{2}{3}}{1 - \frac{2}{3}\cos\theta}$ **b.** $e = \dfrac{2}{3}; p = 1;$ ellipse

c.

64. a. $r = \dfrac{2}{1 + 2\sin\theta}$ **b.** $e = 2; p = 1;$ hyperbola

c.

65. a. $r = \dfrac{2}{1 + 4\cos\theta}$ **b.** $e = 4; p = \dfrac{1}{2};$ hyperbola

c.

Chapter 10 Test

1. foci: $(\pm\sqrt{13}, 0)$

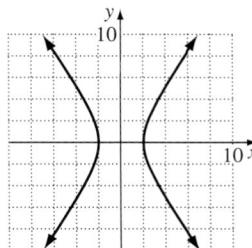

2. vertex: $(0, 0)$; focus: $(0, -2)$; directrix: $y = 2$

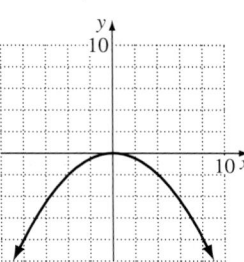

3. foci: $(-6, 5)$, $(2, 5)$

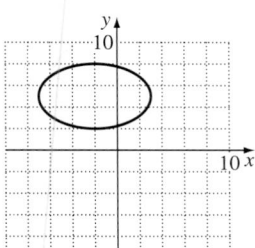

4. foci: $(-1, 1 \pm \sqrt{5})$

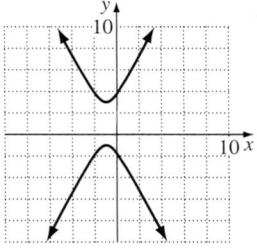

5. vertex: $(-5, 1)$; focus: $(-5, 3)$; directrix: $y = -1$

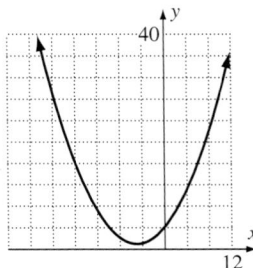

6. $\dfrac{x^2}{100} + \dfrac{y^2}{51} = 1$

7. $\dfrac{y^2}{49} - \dfrac{x^2}{51} = 1$

8. $y^2 = 200x$

9. 32 ft

10. a. $x^2 = 3y$

    b. Light is placed $\dfrac{3}{4}$ inch above the vertex.

11. ellipse

12. ellipse or circle

13. $\theta = 30°$

14. $(y + 1)^2 = x$

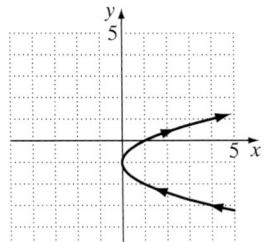

15. $\dfrac{(x - 1)^2}{9} + \dfrac{y^2}{4} = 1$

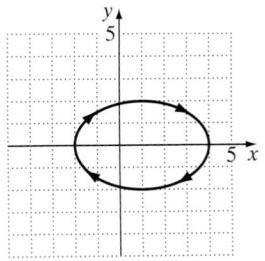

16. parabola

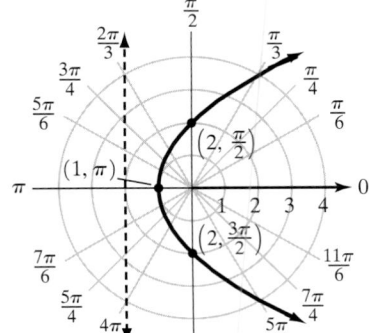

17. ellipse

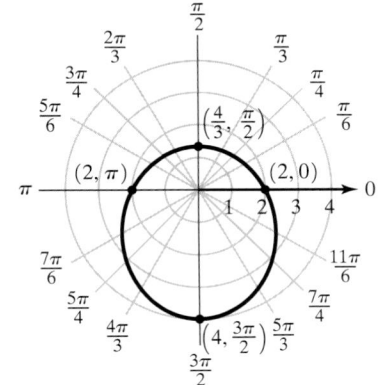

Cumulative Review Exercises (Chapters 1–10)

1. $\{2\}$ **2.** $\{x | x < 2\}$ **3.** $\{9\}$ **4.** $\{2 + 2\sqrt{5}, 2 - 2\sqrt{5}\}$ **5.** $\{x | x \geq 4 \text{ or } x \leq -3\}$ **6.** $\left\{\dfrac{2}{3}, -1 \pm \sqrt{2}\right\}$ **7.** $\{3\}$

8. $(2, -1)$ **9.** $(2, -4)$ and $(-14, -20)$ **10.** $(7, -4, 6)$

3. $S_1: 1^3 = \dfrac{1^2(1+1)^2}{4}$; $S_k: 1^3 + 2^3 + 3^3 + \cdots + k^3 = \dfrac{k^2(k+1)^2}{4}$; $S_{k+1}: 1^3 + 2^3 + 3^3 + \cdots + k^3 + (k+1)^3 = \dfrac{(k+1)^2(k+2)^2}{4}$;

S_{k+1} can be obtained by adding $k^3 + 3k^2 + 3k + 1$ to both sides of S_k.

4. $S_1: 2$ is a factor of $1^2 + 1$; $S_k: 2$ is a factor of $k^2 + k$; $S_{k+1}: 2$ is a factor of $(k+1)^2 + (k+1) = k^2 + 3k + 2$; S_{k+1} can be obtained from S_k by writing $k^2 + 3k + 2$ as $(k^2 + k) + 2(k + 1)$.

Exercise Set 11.4

1. $S_1: 1 = 1^2$; $S_2: 1 + 3 = 2^2$; $S_3: 1 + 3 + 5 = 3^2$ **3.** $S_1: 2$ is a factor of $1 - 1 = 0$; $S_2: 2$ is a factor of $2^2 - 2 = 2$; $S_3: 2$ is a factor of $3^2 - 3 = 6$ **5.** $S_k: 4 + 8 + 12 + \cdots + 4k = 2k(k + 1)$; $S_{k+1}: 4 + 8 + 12 + \cdots + (4k + 4) = 2(k + 1)(k + 2)$

7. $S_k: 3 + 7 + 11 + \cdots + (4k - 1) = k(2k + 1)$; $S_{k+1}: 3 + 7 + 11 + \cdots + (4k + 3) = (k + 1)(2k + 3)$

9. $S_k: 2$ is a factor of $k^2 - k + 2$; $S_{k+1}: 2$ is a factor of $k^2 + k + 2$

11. $S_1: 4 = 2(1)(1 + 1)$; $S_k: 4 + 8 + 12 + \cdots + 4k = 2k(k + 1)$; $S_{k+1}: 4 + 8 + 12 + \cdots + 4(k + 1) = 2(k + 1)(k + 2)$; S_{k+1} can be obtained by adding $4k + 4$ to both sides of S_k.

13. $S_1: 1 = 1^2$; $S_k: 1 + 3 + 5 + \cdots + (2k - 1) = k^2$; $S_{k+1}: 1 + 3 + 5 + \cdots + (2k + 1) = (k + 1)^2$; S_{k+1} can be obtained by adding $2k + 1$ to both sides of S_k.

15. $S_1: 3 = 1[2(1) + 1]$; $S_k: 3 + 7 + 11 + \cdots + (4k - 1) = k(2k + 1)$; $S_{k+1}: 3 + 7 + 11 + \cdots + (4k + 3) = (k + 1)(2k + 3)$; S_{k+1} can be obtained by adding $4k + 3$ to both sides of S_k.

17. $S_1: 1 = 2^1 - 1$; $S_k: 1 + 2 + 2^2 + \cdots + 2^{k-1} = 2^k - 1$; $S_{k+1}: 1 + 2 + 2^2 + \cdots + 2^k = 2^{k+1} - 1$; S_{k+1} can be obtained by adding 2^k to both sides of S_k.

19. $S_1: 2 = 2^{1+1} - 2$; $S_k: 2 + 4 + 8 + \cdots + 2^k = 2^{k+1} - 2$; $S_{k+1}: 2 + 4 + 8 + \cdots + 2^{k+1} = 2^{k+2} - 2$; S_{k+1} can be obtained by adding 2^{k+1} to both sides of S_k.

21. $S_1: 1 \cdot 2 = \dfrac{1(1+1)(1+2)}{3}$; $S_k: 1 \cdot 2 + 2 \cdot 3 + 3 \cdot 4 + \cdots + k(k + 1) = \dfrac{k(k+1)(k+2)}{3}$;

$S_{k+1}: 1 \cdot 2 + 2 \cdot 3 + 3 \cdot 4 + \cdots + (k + 1)(k + 2) = \dfrac{(k+1)(k+2)(k+3)}{3}$; S_{k+1} can be obtained by adding $(k + 1)(k + 2)$ to both sides of S_k.

23. $S_1: \dfrac{1}{1 \cdot 2} = \dfrac{1}{1 + 1}$; $S_k: \dfrac{1}{1 \cdot 2} + \dfrac{1}{2 \cdot 3} + \dfrac{1}{3 \cdot 4} + \cdots + \dfrac{1}{k(k + 1)} = \dfrac{k}{k + 1}$; $S_{k+1}: \dfrac{1}{1 \cdot 2} + \dfrac{1}{2 \cdot 3} + \dfrac{1}{3 \cdot 4} + \cdots + \dfrac{1}{(k + 1)(k + 2)} = \dfrac{k + 1}{k + 2}$;

S_{k+1} can be obtained by adding $\dfrac{1}{(k + 1)(k + 2)}$ to both sides of S_k.

25. $S_1: 2$ is a factor of 0; $S_k: 2$ is a factor of $k^2 - k$; $S_{k+1}: 2$ is a factor of $k^2 + k$; S_{k+1} can be obtained from S_k by rewriting $k^2 + k$ as $(k^2 - k) + 2k$.

27. $S_1: 6$ is a factor of 6; $S_k: 6$ is a factor of $k(k + 1)(k + 2)$; $S_{k+1}: 6$ is a factor of $(k + 1)(k + 2)(k + 3)$; S_{k+1} can be obtained from S_k by rewriting $(k + 1)(k + 2)(k + 3)$ as $k(k + 1)(k + 2) + 3(k + 1)(k + 2)$ and noting that either $k + 1$ or $k + 2$ is even, so 6 is a factor of $3(k + 1)(k + 2)$.

29. $S_1: (ab)^1 = a^1b^1$; $S_k: (ab)^k = a^k b^k$; $S_{k+1}: (ab)^{k+1} = a^{k+1}b^{k+1}$; S_{k+1} can be obtained by multiplying both sides of S_k by (ab).

33. $S_3: 3^2 > 2(3) + 1$; $S_k: k^2 > 2k + 1$ for $k \geq 3$; $S_{k+1}: (k + 1)^2 > 2(k + 1) + 1$ or $k^2 + 2k + 1 > 2k + 3$; S_{k+1} can be obtained from S_k by noting that S_{k+1} is the same as $k^2 > 2$ which is true for $k \geq 3$.

35. $S_1: \dfrac{1}{4}$; $S_2: \dfrac{1}{3}$; $S_3: \dfrac{3}{8}$; $S_4: \dfrac{2}{5}$; $S_5: \dfrac{5}{12}$; $S_n: \dfrac{n}{2n + 2}$; Use S_k to obtain the conjectured formula.

Section 11.5

Check Point Exercises

1. a. 20   **b.** 1   **c.** 28   **d.** 1   **2.** $x^4 + 4x^3 + 6x^2 + 4x + 1$ **3.** $x^5 - 10x^4y + 40x^3y^2 - 80x^2y^3 + 80xy^4 - 32y^5$
4. $4032x^5y^4$

Exercise Set 11.5

1. 56   **3.** 12   **5.** 1   **7.** 4950   **9.** $x^3 + 6x^2 + 12x + 8$ **11.** $27x^3 + 27x^2y + 9xy^2 + y^3$ **13.** $125x^3 - 75x^2 + 15x - 1$
15. $16x^4 + 32x^3 + 24x^2 + 8x + 1$ **17.** $x^8 + 8x^6y + 24x^4y^2 + 32x^2y^3 + 16y^4$ **19.** $y^4 - 12y^3 + 54y^2 - 108y + 81$
21. $16x^{12} - 32x^9 + 24x^6 - 8x^3 + 1$ **23.** $c^5 + 10c^4 + 40c^3 + 80c^2 + 80c + 32$ **25.** $x^5 - 5x^4 + 10x^3 - 10x^2 + 5x - 1$
27. $x^5 - 10x^4y + 40x^3y^2 - 80x^2y^3 + 80xy^4 - 32y^5$ **29.** $64a^6 + 192a^5b + 240a^4b^2 + 160a^3b^3 + 60a^2b^4 + 12ab^5 + b^6$
31. $x^8 + 16x^7 + 112x^6 + \cdots$ **33.** $x^{10} - 20x^9y + 180x^8y^2 - \cdots$
35. $x^{32} + 16x^{30} + 120x^{28} + \cdots$ **37.** $y^{60} - 20y^{57} + 190y^{54} - \cdots$ **39.** $240x^4y^2$ **41.** $126x^5$ **43.** $56x^6y^{15}$ **45.** $-\dfrac{21}{2}x^6$
47. $g(t) = 0.002t^3 - 0.84t^2 - 16.13t - 68.54$

61.

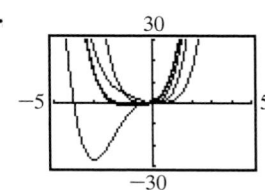

$f_2, f_3, f_4,$ and f_5 are approaching $f_1 = f_6$.

62. $f_1(x) = x^3 - 3x^2 + 3x - 1$
63. $f_1(x) = x^4 - 8x^3 + 24x^2 - 32x + 16$
67. $x^6 + 3x^5 + 6x^4 + 7x^3 + 6x^2 + 3x + 1$
69. $\binom{n}{r} = \dfrac{n!}{r!\,(n-r)!}; \binom{n}{r-1} = \dfrac{n!}{(n-r)![n-(n-r)]!} = \dfrac{n!}{(n-r)!r!} = \binom{n}{r}$

Section 11.6

Check Point Exercises

1. 72 **2.** 729 **3.** 676,000 **4.** 840 **5.** 720 **6. a.** combinations **b.** permutations **7.** 210 **8.** 1820

Exercise Set 11.6

1. 3024 **3.** 6720 **5.** 720 **7.** 1 **9.** 126 **11.** 330 **13.** 1 **15.** 1 **17.** combinations **19.** permutations
21. 27 ways **23.** 40 ways **25.** 243 ways **27.** 144 area codes **29.** 120 ways **31.** 6 paragraphs **33.** 720 ways
35. 8,648,640 ways **37.** 120 ways **39.** 15,120 lineups **41.** 20 ways **43.** 495 collections **45.** 24,310 groups
47. 13,983,816 selections **49.** 360 ways **51.** 1716 ways **53.** 1140 ways **55.** 840 passwords **57.** 2730 cones
69. (c) is true. **71.** 144 numbers

Section 11.7

Check Point Exercises

1. $\dfrac{82.5}{275} = 0.3$ **2.** $\dfrac{1}{3}$ **3.** $\dfrac{1}{9}$ **4.** $\dfrac{1}{13}$ **5.** $\dfrac{1}{142,506}$ **6.** $\dfrac{142,505}{142,506} \approx 0.999993$ **7.** $\dfrac{1}{3}$ **8.** $\dfrac{3}{4}$ **9.** $\dfrac{24}{25}$ **10.** $\dfrac{1}{361} \approx 0.003$ **11.** $\dfrac{1}{16}$

Exercise Set 11.7

1. 0.42 **3.** $\dfrac{761}{5926} \approx 0.13$ **5.** $\dfrac{1}{6}$ **7.** $\dfrac{1}{2}$ **9.** $\dfrac{1}{3}$ **11.** $\dfrac{1}{13}$ **13.** $\dfrac{3}{13}$ **15.** $\dfrac{1}{4}$ **17.** $\dfrac{7}{8}$ **19.** $\dfrac{1}{12}$ **21.** $\dfrac{1}{18,009,460}; \dfrac{5}{900,473}$

23. a. 2,598,960 **b.** 1287 **c.** $\dfrac{1287}{2,598,960} \approx 0.0005$ **25.** $\dfrac{255,647}{274,634} \approx 0.93$ **27.** $\dfrac{237,401}{274,634} \approx 0.86$ **29.** $\dfrac{42,010}{274,634} \approx 0.15$ **31.** $\dfrac{2}{13}$

33. $\dfrac{5}{6}$ **35.** $\dfrac{7}{13}$ **37.** $\dfrac{3}{4}$ **39.** $\dfrac{33}{40}$ **41.** $\dfrac{1}{36}$ **43.** $\dfrac{1}{3}$ **45.** $\dfrac{1}{64}$ **47. a.** $\dfrac{1}{256}$ **b.** $\dfrac{1}{4096}$ **c.** $\left(\dfrac{15}{16}\right)^{10}$ **d.** $1 - \left(\dfrac{15}{16}\right)^{10}$
59. Answers may vary.

Chapter 11 Review Exercises

1. $a_1 = 3; a_2 = 10; a_3 = 17; a_4 = 24$ **2.** $a_1 = -\dfrac{3}{2}; a_2 = \dfrac{4}{3}; a_3 = -\dfrac{5}{4}; a_4 = \dfrac{6}{5}$ **3.** $a_1 = 1; a_2 = 1; a_3 = \dfrac{1}{2}; a_4 = \dfrac{1}{6}$

4. $a_1 = \dfrac{1}{2}; a_2 = -\dfrac{1}{4}; a_3 = \dfrac{1}{8}; a_4 = -\dfrac{1}{16}$ **5.** $a_1 = 9; a_2 = \dfrac{2}{27}; a_3 = 9; a_4 = \dfrac{2}{27}$ **6.** $a_1 = 4; a_2 = 11; a_3 = 25; a_4 = 53$ **7.** 65

8. 95 **9.** -20 **10.** $\displaystyle\sum_{i=1}^{15} \dfrac{i}{i+2}$ **11.** $\displaystyle\sum_{i=1}^{10} (i+3)^3$ **12.** 7, 11, 15, 19, 23, 27 **13.** $-4, -9, -14, -19, -24, -29$

14. $\dfrac{3}{2}, 1, \dfrac{1}{2}, 0, -\dfrac{1}{2}, -1$ **15.** $-2, 3, 8, 13, 18, 23$ **16.** $a_6 = 20$ **17.** $a_{12} = -30$ **18.** $a_{14} = -38$ **19.** $a_n = 4n - 11; a_{20} = 69$

20. $a_n = 220 - 20n; a_{20} = -180$ **21.** $a_n = 8 - 5n; a_{20} = -92$ **22.** 1727 **23.** 225 **24.** 15,150 **25.** 440 **26.** -500
27. -2325 **28. a.** $a_n = 1043.4518 - 0.4118n$ **b.** 1002.2718 sec **29.** \$418,500 **30.** 1470 seats **31.** 3, 6, 12, 24, 48
32. $\dfrac{1}{2}, \dfrac{1}{4}, \dfrac{1}{8}, \dfrac{1}{16}, \dfrac{1}{32}$ **33.** 16, $-8, 4, -2, 1$ **34.** $-1, 5, -25, 125, -625$ **35.** $a_7 = 1458$ **36.** $a_6 = \dfrac{1}{2}$ **37.** $a_5 = -48$

38. $a_n = 2^{n-1}; a_8 = 128$ **39.** $a_n = 100\left(\dfrac{1}{10}\right)^{n-1}; a_8 = \dfrac{1}{100,000}$ **40.** $a_n = 12\left(-\dfrac{1}{3}\right)^{n-1}; a_8 = -\dfrac{4}{729}$ **41.** 17,936,135 **42.** $\dfrac{1093}{2187}$

43. 19,530 **44.** -258 **45.** $\dfrac{341}{128}$ **46.** $\dfrac{27}{2}$ **47.** $\dfrac{4}{3}$ **48.** $-\dfrac{18}{5}$ **49.** 20 **50.** $\dfrac{2}{3}$ **51.** $\dfrac{47}{99}$ **52.** \$42,823.22; \$223,210.19

53. \$120,112.64 **54.** \$$9\dfrac{1}{3}$ million

55. $S_1: 5 = \dfrac{5(1)(1+1)}{2}; S_k: 5 + 10 + 15 + \cdots + 5k = \dfrac{5k(k+1)}{2}; S_{k+1}: 5 + 10 + 15 + \cdots + 5(k+1) = \dfrac{5(k+1)(k+2)}{2};$
S_{k+1} can be obtained by adding $5(k+1)$ to both sides of S_k.

56. $S_1: 1 = \dfrac{4^1 - 1}{3}$; $S_k: 1 + 4 + 4^2 + \cdots + 4^{k-1} = \dfrac{4^k - 1}{3}$; $S_{k+1}: 1 + 4 + 4^2 + \cdots + 4^k = \dfrac{4^{k+1} - 1}{3}$;

S_{k+1} can be obtained by adding 4^k to both sides of S_k.

57. $S_1: 2 = 2(1)^2$; $S_k: 2 + 6 + 10 + \cdots + (4k - 2) = 2k^2$; $S_{k+1}: 2 + 6 + 10 + \cdots + (4k + 2) = 2k^2 + 4k + 2$;

S_{k+1} can be obtained by adding $4k + 2$ to both sides of S_k.

58. $S_1: 1 \cdot 3 = \dfrac{1(1 + 1)[2(1) + 7]}{6}$; $S_k: 1 \cdot 3 + 2 \cdot 4 + 3 \cdot 5 + \cdots + k(k + 2) = \dfrac{k(k + 1)(2k + 7)}{6}$;

$S_{k+1}: 1 \cdot 3 + 2 \cdot 4 + 3 \cdot 5 + \cdots + (k + 1)(k + 3) = \dfrac{(k + 1)(k + 2)(2k + 9)}{6}$;

S_{k+1} can be obtained by adding $(k + 1)(k + 3)$ to both sides of S_k.

59. $S_1: 2$ is a factor of 6; $S_k: 2$ is a factor of $k^2 + 5k$; $S_{k+1}: 2$ is a factor of $k^2 + 7k + 6$; S_{k+1} can be obtained from S_k by rewriting $k^2 + 7k + 6$ as $(k^2 + 5k) + 2(k + 3)$.

60. 165 **61.** 4005 **62.** $8x^3 + 12x^2 + 6x + 1$ **63.** $x^8 - 4x^6 + 6x^4 - 4x^2 + 1$
64. $x^5 + 10x^4y + 40x^3y^2 + 80x^2y^3 + 80xy^4 + 32y^5$ **65.** $x^6 - 12x^5 + 60x^4 - 160x^3 + 240x^2 - 192x + 64$
66. $x^{16} + 24x^{14} + 252x^{12} + \cdots$ **67.** $x^9 - 27x^8 + 324x^7 - \cdots$ **68.** $80x^2$ **69.** $4860x^2$ **70.** 336 **71.** 15,120 **72.** 56
73. 78 **74.** 20 choices **75.** 243 possibilities **76.** 32,760 ways **77.** 4845 ways **78.** 1140 sets **79.** 116,280 ways
80. 120 ways **81.** $\dfrac{9,630,188}{31,878,234} \approx 0.302$ **82.** $\dfrac{5,503,372}{19,128,261} \approx 0.288$ **83.** $\dfrac{2}{3}$ **84.** $\dfrac{2}{3}$ **85.** $\dfrac{2}{13}$ **86.** $\dfrac{7}{13}$ **87.** $\dfrac{5}{6}$ **88.** $\dfrac{5}{6}$
89. a. $\dfrac{1}{15,504}$ **b.** $\dfrac{25}{3876}$ **90.** $\dfrac{4}{5}$ **91.** $\dfrac{3}{4}$ **92.** $\dfrac{1}{32}$ **93. a.** 0.04 **b.** 0.008 **c.** 0.4096

Chapter 11 Test

1. $a_1 = 1; a_2 = -\dfrac{1}{4}; a_3 = \dfrac{1}{9}; a_4 = -\dfrac{1}{16}; a_5 = \dfrac{1}{25}$ **2.** 105 **3.** 550 **4.** $-21{,}846$ **5.** 36 **6.** 720 **7.** 120 **8.** $\displaystyle\sum_{i=1}^{20} \dfrac{i + 1}{i + 2}$

9. $a_n = 5n - 1; a_{12} = 59$ **10.** $a_n = 16\left(\dfrac{1}{4}\right)^{n-1}; a_{12} = \dfrac{1}{262,144}$ **11.** -2387 **12.** -385 **13.** 8 **14.** \$276,426.79

15. $S_1: 1 = \dfrac{1[3(1) - 1]}{2}$; $S_k: 1 + 4 + 7 + \cdots + (3k - 2) = \dfrac{k(3k - 1)}{2}$; $S_{k+1}: 1 + 4 + 7 + \cdots + (3k + 1) = \dfrac{(k + 1)(3k + 2)}{2}$;

S_{k+1} can be obtained by adding $3k + 1$ to both sides of S_k. **16.** $x^{10} - 5x^8 + 10x^6 - 10x^4 + 5x^2 - 1$ **17.** 990 ways **18.** 210 sets

19. $10^4 = 10{,}000$ **20.** $\dfrac{10}{1001}$ **21.** $\dfrac{8}{13}$ **22.** $\dfrac{3}{5}$ **23.** $\dfrac{1}{256}$ **24.** $\dfrac{1}{16}$

Cumulative Review Exercises (Chapters 1–11)

1. $x = \dfrac{14}{5}$ **2.** $\left\{\dfrac{3 + \sqrt{3}}{3}, \dfrac{3 - \sqrt{3}}{3}\right\}$ **3.** $\{2\}$ **4.** $\{16, 256\}$ **5.** $\{6\}$ **6.** $-1 \le x \le 0$ or $[-1, 0]$ **7.** $\left(-\dfrac{2}{3}, \dfrac{3}{2}\right)$

8. $(-3, 1]$ **9.** $\{2.9706\}$ **10.** $\left\{1, \dfrac{1}{2}, -3\right\}$ **11.** $\{(3, 2), (3, -2), (-3, 2), (-3, -2)\}$ **12.** $\{(6, -4, 2)\}$ **13.** $\{(0, -1), (2, 1)\}$

14. 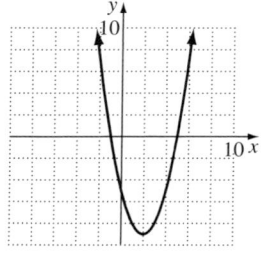 **15.** 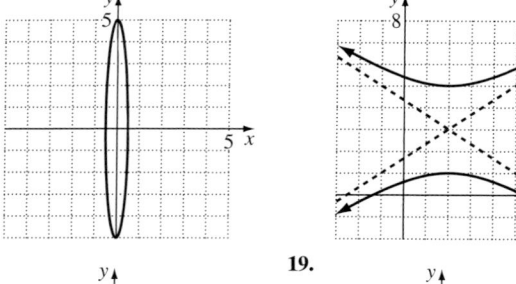 **16.** 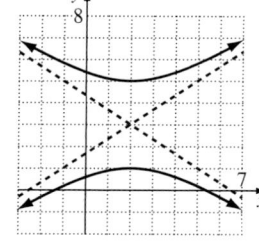 **17.**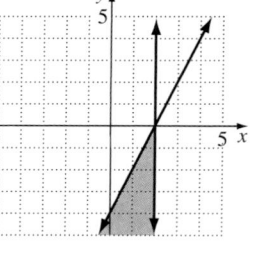

18. 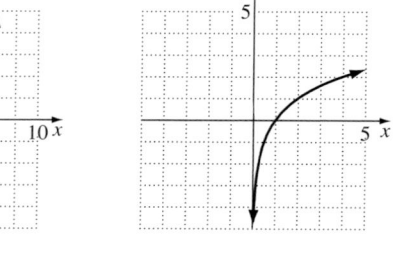 **19.** 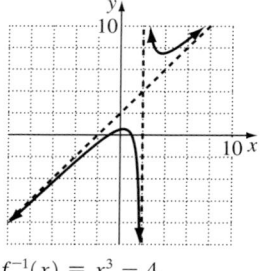 **20.** $f^{-1}(x) = x^3 - 4$

21. $\begin{bmatrix} -2 & 10 \\ -5 & 7 \\ 15 & -15 \end{bmatrix}$ **22.** $\dfrac{-1}{x - 2} + \dfrac{3x - 2}{x^2 + 2x + 2}$

23. $x^{15} + 10x^{12}y + 40x^9y^2 + 80x^6y^3 + 80x^3y^4 + 32y^5$
24. 3850
25. a. $y - 119.4 = 5.17x$ **b.** $y = 5.17x + 119.4$
  c. 207.29 billion pieces of mail
26. The Empire State Building is about 916.7 feet tall and the World Trade Center is about 1043.4 feet tall.

27. length: 100 yd; width: 50 yd   **28.** pen: $1.80; pad: $2   **29. a.** 6 sec   **b.** 2.5 sec; 196 ft   **30.** 11 amps

31. a. 20 in.   **b.** $\dfrac{3}{8}$ cycles/sec   **c.** $\dfrac{8}{3}$ sec   **32.** $\tan x + \dfrac{1}{\tan x} = \dfrac{\sin x}{\cos x} + \dfrac{1}{\dfrac{\sin x}{\cos x}} = \dfrac{\sin x}{\cos x} + \dfrac{\cos x}{\sin x} = \dfrac{\sin^2 x + \cos^2 x}{\cos x \cdot \sin x} = \dfrac{1}{\cos x \cdot \sin x}$

33. $\dfrac{1 - \tan^2 x}{1 + \tan^2 x} = \dfrac{1 - \dfrac{\sin^2 x}{\cos^2 x}}{1 + \dfrac{\sin^2 x}{\cos^2 x}} \cdot \dfrac{\cos^2 x}{\cos^2 x} = \dfrac{\cos^2 x - \sin^2 x}{\cos^2 x + \sin^2 x} = \dfrac{\cos 2x}{1} = \cos 2x$

34.

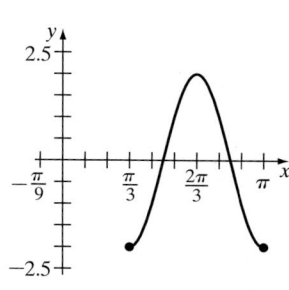

35. $\dfrac{\pi}{6}, \dfrac{5\pi}{6}, \dfrac{7\pi}{6}, \dfrac{11\pi}{6}$ **36.** $0, \dfrac{\pi}{3}, \dfrac{5\pi}{3}$ **37.** $-\dfrac{5}{\sqrt{11}}$

38.

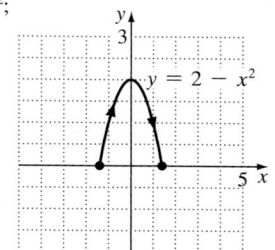

39. $B = 54°, C = 92°, c = 39.5$

40. $y = 2 - x^2$;

Subject Index

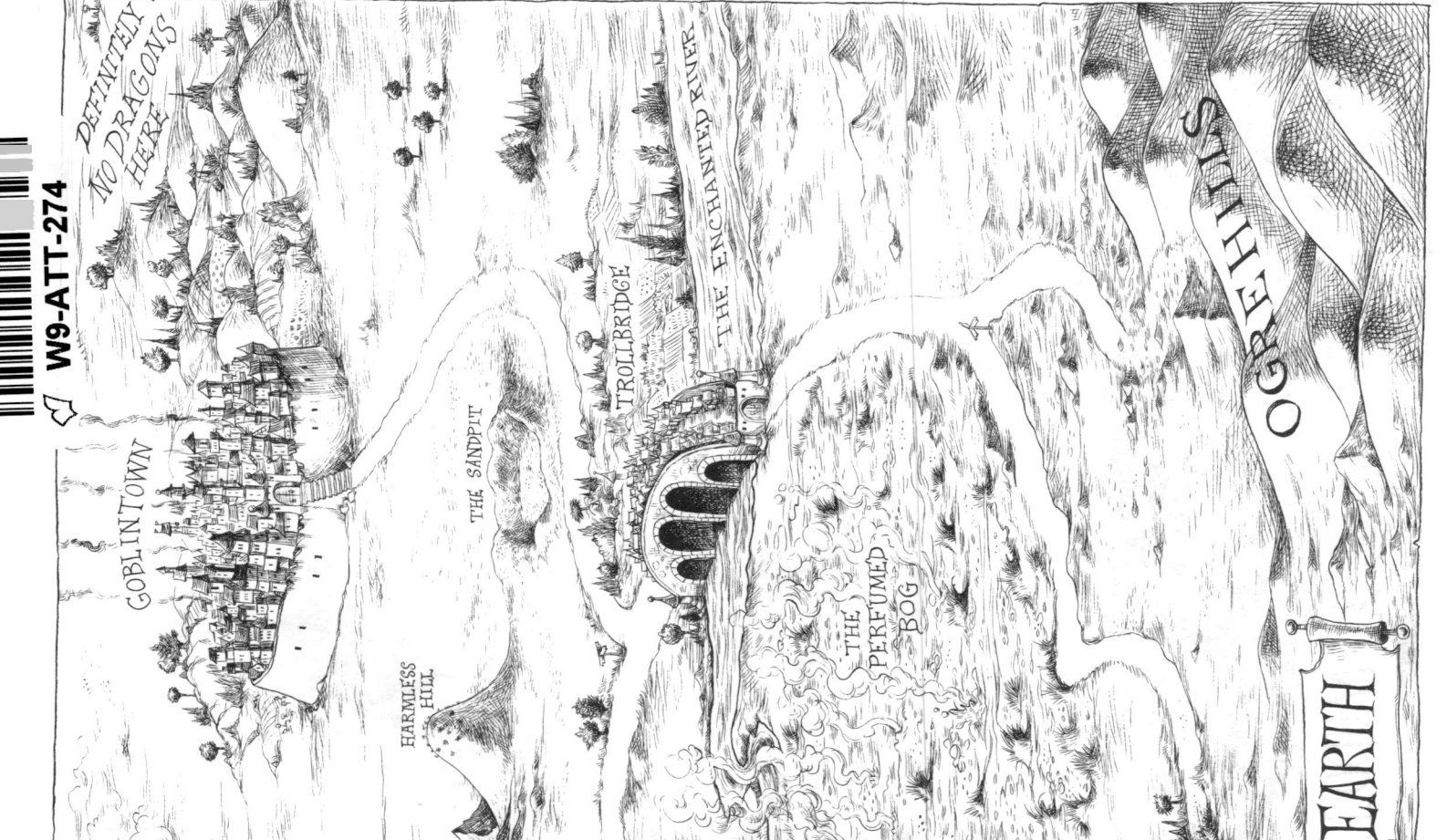

DEFINITELY NO DRAGONS HERE

OGREHILLS

GOBLINTOWN

TROLLBRIDGE

THE ENCHANTED RIVER

THE SANDPIT

HARMLESS HILL

THE PERFUMED BOG

EARTH

Also by Paul Stewart and Chris Riddell:

The Blobheads

The *Rabbit and Hedgehog* books:
A Little Bit of Winter
The Birthday Presents
Rabbit's Wish
What Do You Remember?

The Edge Chronicles:
Beyond The Deepwoods
Stormchaser
Midnight Over Sanctaphrax
The Curse of the Gloamglozer
The Last of the Sky Pirates

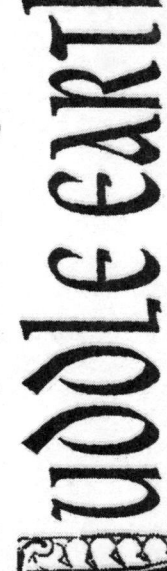

Muddle Earth

Paul Stewart & Chris Riddell

Macmillan Children's Books

First published 2003 by Macmillan Children's Books Ltd
a division of Macmillan Publishers Limited
20 New Wharf Road, London N1 9RR
Basingstoke and Oxford
www.panmacmillan.com

Associated companies throughout the world

ISBN 0 333 94799 1

For Anna and Jack

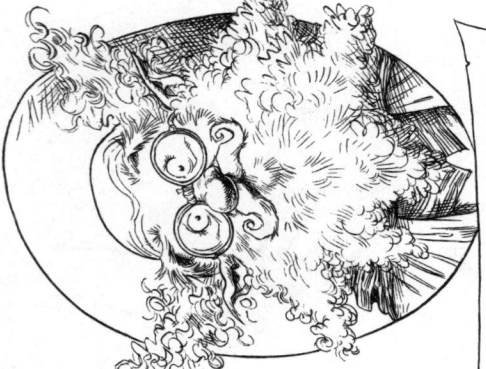

NAME: Randalf the Wise
Middle Earth's leading wizard

OCCUPATION: Um . . .
Middle Earth's leading wizard

HOBBIES: Performing spells
(I think you'll find that's
spell! – Veronica)

FAVOURITE FOOD: Norbert's
squashed tadpole fritters

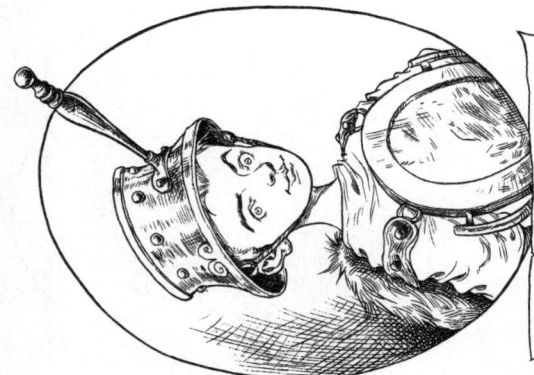

NAME: Joe Jefferson

OCCUPATION: Schoolboy

HOBBIES: Football, TV,
arguing with his sister

FAVOURITE FOOD: Anything
not cooked by Norbert

NAME: Henry

OCCUPATION: Joe's dog

HOBBIES: Walkies,
chasing squirrels, sniffing
strangers' bottoms

FAVOURITE FOOD:
Dog food, obviously

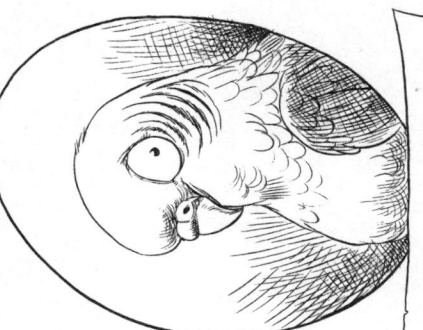

Name: Veronica
Occupation: Familiar to the great wizard, Randalf the Wise
Hobbies: Being sarcastic
Favourite food: Anything not cooked by Norbert

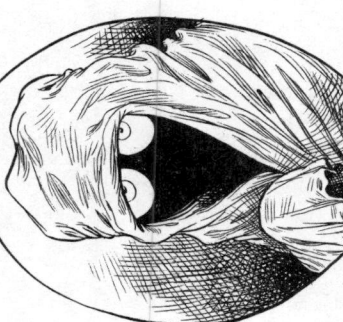

Name: Dr Cuddles (Sshhh! Don't say his name out loud!)
Occupation: (SSHHH! No one even knows he exists!)
Hobbies: (Didn't you hear what I just said?)
Favourite food: Snuggle-muffins

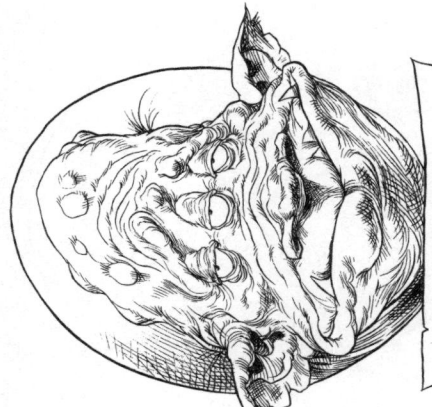

Name: Norbert the Not-Very-Big
Occupation: Ogre
Hobbies: Thumb-sucking, cooking – especially cake-decorating
Favourite food: Everything

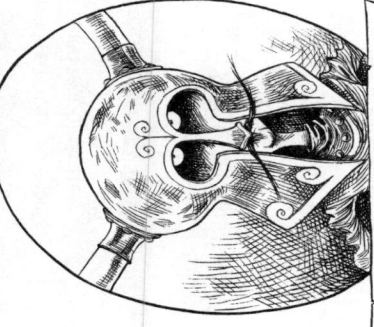

Name: The Horned Baron
Occupation: Ruler of Muddle Earth and husband to Ingrid
Hobbies: Ruling, and doing whatever Ingrid tells him
Favourite food: Bad-breath porridge

Book 1

Engelbert the Enormous

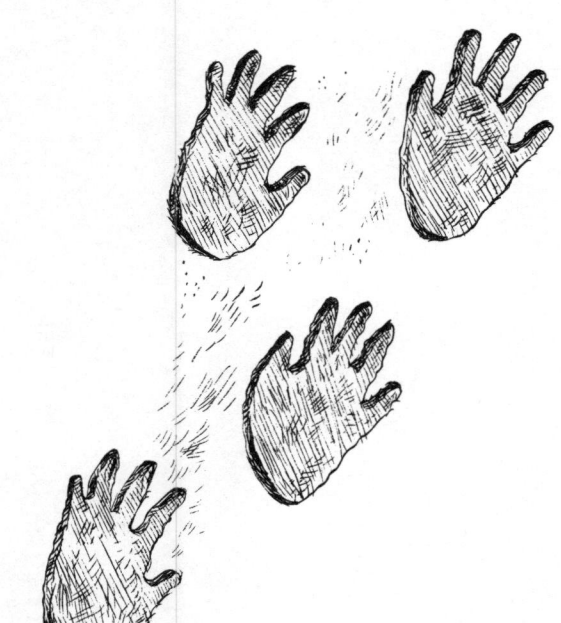

Prologue

Πight was falling over Muddle Earth. The sun had set, the sky was darkening and already two of its three moons had risen up above the Musty Mountains. One of these moons was as purple as a batbird. The other was as yellow as an ogre's underpants on wash day. Both were full and bright.

The land was full of noises as the day creatures (such as tree rabbits, hillfish and pink stinky hogs) said goodnight to the night creatures (such as stiltmice, lazybirds and exploding gas frogs), who were just getting up. High above their heads, the batbirds had left their roosts and were soaring across the purple and yellow striped sky. As they bumped into each other, the air filled with their characteristic cry: '*Ouch!*'

In Elfwood, the trees bowed and bent as a chill wind howled through their branches. In the Perfumed Bog the oozy mud bubbled and plopped. In the far-off Ogrehills, there was the slurping sound of a thousand thumbs being sucked and a thousand sleepy voices murmuring '*Mummy!*'

Twinkling lights were coming on throughout overcrowded Goblintown and, as the goblins prepared their evening meals, the air was filled with the smells of badbreath porridge and snotbread – and the sounds of too many cooks.

'Have you spat in this?'

'No.'

'Well, go on then, before I put it in the oven.'

By contrast, Trollbridge was cloaked in cold, dank darkness. The trolls who lived there could see nothing of the cabbages and turnips they were eating for supper. Their deep voices rumbled up from their dwellings beneath the bridge.

'Has anyone seen my mangel-wurzel?'

'It's over here.'

'*OW!* That's my head!'

As the purple and yellow shadows swept down from

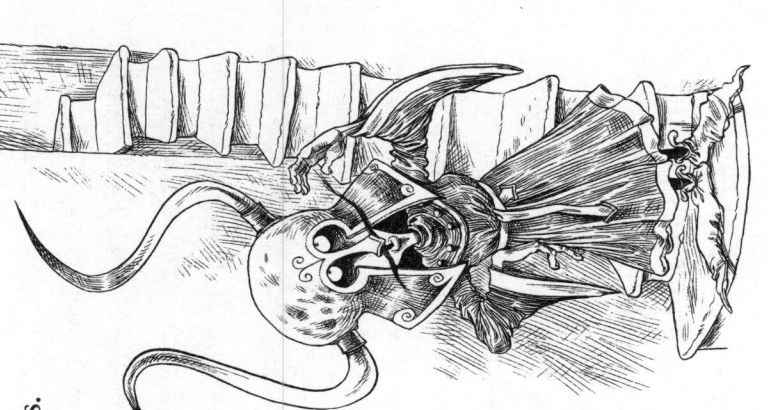

Mount Boom to the Horned Baron's great castle below, a loud, piercing voice ripped from the windows of the highest tower. Another of the inhabitants of Muddle Earth seemed to be finding it hard to tell the difference between a root vegetable and a head.

'Walter, you turnip-head! WALTER! Where *are* you?'

It was Ingrid, wife of the Horned Baron himself, and she was far from happy.

'Coming, my sweet,' he called out as he climbed the circular stairs.

'I've just seen something I really want in this catalogue,' Ingrid continued. 'Singing curtains. It says here, "No self-respecting Horned Baron's wife should be without these enchanted window hangings to lull her gently off to sleep at night and sensitively wake her with song the following morning." *I* want to be lulled to sleep by singing curtains, Walter. *I* want to be woken with song. Do you hear me?

'Loud and clear, love-of-my-life,' the Baron replied wearily. 'All *too* loud and clear,' he muttered under his breath.

The next moment the third moon of Muddle Earth – a small, bright green sphere which only seemed to appear when it felt like it – sailed up in the sky to form a perfect equilateral triangle with its yellow and purple neighbours. The three moons lit up the Enchanted Lake, which hovered high up above the ground, and the seven magnificent houseboats bobbing about on its glistening waters.

Six of the houseboats were dark and deserted. The seventh was bathed in an oily, orange lamplight. A short, portly individual by the name of Randalf was staring out from one of its upper windows at the configuration of coloured moons. A budgie perched on his balding head.

'Norbert,' he said at last to his assistant. 'The astral signs are auspicious. My pointy hat, if you please. I feel a spell coming on!'

'At once, sir,' said Norbert, his voice gruff yet well-meaning. As he stomped across the floor to the wizard's cupboard the whole houseboat dipped and swayed. Norbert the Not-Very-Big was a fairly weedy specimen, as ogres went – but he was still an ogre. And ogres, even weedy ones, are big and heavy.

'You feel a spell coming on?' said the budgie, whose name was Veronica. 'No, don't tell me. It couldn't be the spell to summon a warrior-hero to Muddle Earth by any chance?'

'It might be,' replied Randalf defensively.

Veronica snorted. 'Some wizard you are,' she said. 'You've only got one spell.'

'Yes, yes, don't rub it in,' said Randalf. 'I do the best I can. With all the other wizards . . . *errm* . . . away, I've got to hold the fort.'

'This is a houseboat, not a fort,' said Veronica. 'And the other wizards aren't *away*, they're—'

'Shut up, Veronica!' said Randalf sharply. 'You promised never to mention that dreadful incident again.'

The houseboat lurched once more as Norbert strode back across the room. 'Your pointy hat, sir,' he said.

Randalf placed it on his head. 'Thank you, Norbert,' he said, trying hard to overcome his irritation with Veronica. She was *such* a know-it-all! Why couldn't he have had something nice and sweet for a familiar like an exploding gas frog or a slimy bog demon? True, they might have been a bit smelly, but at least they wouldn't have answered back all the time. Not like this infernal budgie. Still, he was stuck with Veronica now, and he'd just have to do what he always did in these situations – make the best of it.

He carefully removed a piece of paper from the folds of his cloak, delicately opened it up and cleared his throat.

~ 5 ~

'Here we go,' came a muffled voice from under his hat as Randalf began to recite the incantation from the piece of paper. 'Hail, oh Triplet Moons of Muddle. Shine down on these Words of Enchanting, now I utter . . . um . . .'

'That's all very well and good,' Veronica's voice broke in, 'but it's the last bit you need to concentrate on.'

'I know that,' Randalf muttered through clenched teeth. 'Be quiet now. I'm *trying* to concentrate.'

'That'll be a first,' Veronica commented, wriggling out from under the brim of the hat.

Randalf stared down at the spell miserably. Much as he hated to admit it, Veronica was right. He *did* need to concentrate on the last part of the spell. The trouble was, the bottom of the piece of paper had been torn off, and with it, the all-important words of enchantment for the warrior-hero-summoning spell. Once again he would have to improvise.

'*Creator of Wonders, Master of Intricate Arts, Possessor of Breathtaking Skills . . .*' he began.

'Don't overdo it,' Veronica cautioned. 'You said something like that last time you summoned a warrior-hero. Do you *want* another Quentin the Cake-Decorator?'

'No, you're right,' said Randalf. He rubbed his beard thoughtfully. 'All right,' he said at last, 'how about this?' He took a deep breath.

'*Strong . . . and loyal . . . and . . .*' He gave Veronica a dark look. '*Hairy. Oh, Triplet Moons, on these words shine clear.*

Let a mighty warrior-hero now appear!'

There was a bright flash, a loud crash and clouds of purple, yellow and green smoke billowed from the fireplace. Randalf, Norbert and Veronica spun round and stared, open-mouthed – and beaked – at the figure which appeared amidst the thinning smoke.

'What is *that*?' said Norbert.

Veronica squawked with amusement. 'I've just got one thing to say,' she chuckled. 'Come back Quentin the Cake-Decorator, all is forgiven!'

'Shut up, Veronica!' said Randalf, 'and stop sniggering. Everything's going to be fine! Trust me, I'm a wizard.'

1

Joe slammed down his pen angrily and clamped his hands over his ears to shut out the surrounding din.

'This is hopeless!' he bellowed. 'Hopeless!'

The noise was coming in from all directions – above, below, the room next door . . . It was like being stuck in a giant noise-sandwich.

On the desk in front of him, the title of his English assignment – 'My Amazing Adventure' – headed a blank piece of paper. It was early evening at the end of a sunny midsummer Sunday, and if Joe was to get the homework finished and ready to hand in on Monday morning then he needed to get down to work. But how *could* he, with that infernal racket going on all round him?

Joe Jefferson lived in a small brick house with his mum, dad, older sister, younger twin-brothers and his dog, Henry. To a casual observer, the Jeffersons seemed like a nice, quiet family. It was only when you stepped in through the front door that it became clear that they were anything *but*.

Mrs Jefferson worked in a bank. She was tall, slim, dark and fanatically house-proud. Mr Jefferson – a travelling salesrep by day and a DIY freak at night, weekends, holidays and any other spare hour he could find – was short, stocky and never more happy than when clutching a power tool in his hands.

Over the years, Mr Jefferson had constructed a garage, converted the loft, knocked rooms through, put up shelves and cupboards, built a conservatory, landscaped the garden and was currently working on a kitchen extension. At least, that was what *he* would say. So far as Mrs Jefferson was concerned, the thing her husband was best at was making a mess.

At this precise moment the electric drill was busy battling it out with the vacuum cleaner as Mrs Jefferson followed Mr Jefferson around the kitchen – attachment pipe raised high like a light-sabre – sucking up the dust from the air before it had a chance to settle.

As the noise vibrated up through his bedroom floor, Joe shook his head. He'd never get his homework done at this rate, and if he didn't, he'd be in serious trouble with Mr Dixon.

He wondered wearily why his dad didn't take up a nice quiet hobby – like chess, or embroidery – and why his mum couldn't be just a little less obsessed with cleanliness. And why his sister, Ella, who had the converted attic-room above his head, had to do everything – from flicking through a magazine to putting on mascara – to the accompaniment of loud, pounding music. And why the twins' favourite game was chasing-up-and-down-stairs-screaming.

Joe opened the drawer in his desk, pulled out his ear-plugs and was just about to shove them into his ears, when Henry let out a bloodcurdling howl and started barking furiously.

'That's it!' Joe yelled. He leapt up from his desk and stormed across the room. 'Henry!' he called. 'Come here, boy.'

The barking grew louder. It was coming from the bathroom. So were Mark and Matt's squeals of delight.

'He's in here, Joe,' they shouted.

'Henry!' Joe called again. 'Heel!'

Henry came bounding on to the landing and stood in front of Joe, tail wagging and tongue lolling. The twins appeared behind him.

'He was drinking the water from the toilet bowl again,'

they shouted excitedly. 'So we flushed it!'

Joe looked down at the water dripping from Henry's hairy face. 'Serves you right,' he laughed.

The dog barked happily and held out his paw. Upstairs, Ella's door opened and the music grew louder than ever. Ella's angry voice floated down.

'Shut that dog up!' she yelled.

From downstairs, the sound of the electric drill was abruptly replaced by loud hammering.

'Come on, boy,' said Joe. 'Let's get out of this madhouse!'

He turned and, with Henry at his heel, went down the stairs, grabbed the lead from the hook by the door and was just about to leave when his mum noticed him.

'Where are you going?' she shouted out above the sound of the vacuum cleaner and hammer.

'Out,' Joe replied. He opened the door.

'Out where?'

But Joe had already gone.

The local park was deserted. Joe picked up a stick and threw it for Henry, who chased after it, retrieved it and dropped it back at his feet for another throw. Joe grinned. However exasperating his life became, he could always count on Henry to cheer him up. He ruffled the dog's ears and tossed the stick a second time, then set off after Henry.

Across the grass they went, past a cluster of trees and down the hill. As they approached the stream, Joe whistled for Henry to return and clicked the lead into place. If Henry ended up splashing about in the dirty water again, his mum would go mad!

He patted the dog's head affectionately. 'Come on, boy, we'd better get back. That essay won't write itself.' He turned to go. '*My Amazing Adventure*,' he muttered. 'What a stupid title . . . Henry, what's wrong?'

Henry was standing stock still, the hairs along his back on end and his ears and nose twitching.

'What is it, boy?' Joe knelt down and followed the dog's intent stare.

Henry strained on the lead and whimpered.

'What can you see?' Joe muttered. 'Not a squirrel, I hope. You know what happened last ti . . . *aaargh*!'

Unable to remain still a moment longer, Henry suddenly bounded forwards. Head down and following his nose, he dashed straight for a massive rhododendron bush, pulling Joe along behind. The good news was that Henry was heading for a hole in the dark foliage. The bad news was that the hole was only dog-height.

'Henry! Henry, stop!' he shouted and tugged in vain at the lead. 'Stop! You stupid . . .'

The rest of the sentence was lost to a mouthful of leaves as Joe was dragged into the bush. He ducked down and tried his best to shield his eyes with his free hand as Henry dragged him deeper and deeper inside.

All at once, the branches and leaves began crackling with silvery strands of electricity, the air shimmered and wobbled and filled with the sound of slow mournful music – and the smell of burnt toast.

'What on earth . . .' Joe gasped as, the next moment, he was pulled headlong down into a long flashing tunnel. The music grew louder. The smell of burning became stronger and stronger until . . .

CRASH!

'Aargh!' yelled Joe. He was still falling, only now he could feel the sides of the tunnel grazing his elbows and bashing his knees as he continued to drop. And it was black – pitch-black. Joe cried out in fear and pain and let go of the lead. Henry disappeared below him. What seemed like an eternity later, he tumbled down out of the long vertical tunnel and landed with a heavy thump on the ground.

Joe opened his eyes. He was sitting on a tiled floor, bruised, dazed, surrounded by a thick cloud of choking dust and without the faintest idea what had just happened.

Had he fallen into a hole beneath the bush?

Had he cracked his head on a branch so hard he'd knocked himself silly?

As the swirling dust thinned, Joe found himself in a fireplace behind a huge pot, which was suspended on chains. He peered out into a dimly lit and exceedingly cluttered room.

There were tables against every wall, each one covered with pots, papers and peculiar paraphernalia. There were stools and cupboards and bookcases stacked high with boxes, bottles and books. Every inch of the walls was taken up with shelves and cabinets, maps and charts and countless hooks laden with bundles of twigs, roots and dried plants, dead animals and birds and shiny implements that Joe couldn't even begin to guess the purpose of. As for the floor, it was crowded with bulging sacks, earthenware pots and various angular contraptions made of wood and metal, with springs, pistons and cogs — and in the middle of all the chaos, two individuals with their backs turned.

One was short and portly, with bushy white hair and a blue budgie perched on the brim of his tall pointy hat. The other was hefty, knobbly and so tall that he had to stoop to avoid knocking his head against the heavy chandelier above his head.

'He doesn't say much, sir,' the hefty, knobbly one was saying.

'Obviously the strong, silent type, Norbert,' the portly figure replied.

'Unlike Quentin the Cake-Decorator,' said the budgie.

The portly one bent down. 'Now, don't be shy,' he said. 'My name's Randalf. Tell us your name.'

Joe climbed to his feet. This wasn't happening. You don't

take your dog for a walk one minute, fall into a bush the next and end up in somebody's kitchen. Do you? Joe closed his eyes and shook his head. Where was Henry, anyway?

Just then, the dog emitted a short, sharp bark.

'Rough?' said Norbert, puzzled. 'Did he say "Rough", sir?'

'Yes, Rough!' replied the short character, nodding enthusiastically. 'Of course. An excellent name for a warrior-hero, being both short and to the point.' He leaned down and added conspiratorially, 'Rough the Strong? Rough the Slayer? Rough the . . . Hairy?'

Henry barked again.

'Henry!' Joe called.

The dog appeared from behind Norbert, tail wagging furiously, and bounded over to the fireplace. Joe bent down and hugged him tightly. It was so good to see a familiar face, even if this was a dream.

'Who are *you*?' came a strident voice.

Joe looked up and stared back at the two individuals staring at him. The short one had a bushy white beard. The tall one had three eyes. Both of them were standing stock-still, eyes (all five of them) wide and mouths agape. It was the budgie who had spoken.

'I said, who *are* you?' it demanded.

'I . . . I'm Joe, but . . .' he began.

'Don't you see, Veronica?' Randalf exclaimed. 'That must be the sidekick,' he said, pointing at Joe. 'All good warrior-heroes have a sidekick. Mendigor the Mendacious had Hellspawn the goblin, Lothgar the Loathsome had Sworg Bloodpimple . . .'

'Quentin the Cake-Decorator had Mary the poodle,' Veronica muttered.

'Shut up, Veronica,' Randalf snapped. 'You'll have to excuse my familiar,' he explained to Henry. 'She's been getting a bit big for her boots recently.' He turned to Joe. 'I'm right, aren't I? You are Rough the Hairy's sidekick. His sword-carrier, perhaps? Or his axe-sharpener?'

'Not exactly,' said Joe, in a dazed sort of voice. 'And his name is Henry, not Rough. I was holding his lead when . . .'

'So you're his *lead*-bearer,' Randalf interrupted. 'Joe the lead-bearer. *Hmm.* Unusual, admittedly, but not totally unheard of.'

The budgie, who was wearing small yet sturdy lace-up boots, coughed. '*I've* never heard of it,' she said.

'Shut up, Veronica!' he said, and brushed the bird off the brim of his hat. 'We're forgetting our manners,' he went on, turning back to Henry. 'Let me introduce myself properly. I am Randalf the Wise, Muddle Earth's leading wizard.'

'Only wizard, more like,' said Veronica, settling on his shoulder.

'And this,' Randalf went on without missing a beat, 'is my assistant, Norbert – or Norbert the Not-Very-Big, to give him his full title.'

'Not very big!' Joe blurted out in astonishment. 'But he's gigantic.'

'Taller than you or me, I grant you,' said Randalf, 'yet for an ogre, Norbert is a small and rather weedy specimen.'

'You should see my father,' said Norbert, nodding. 'Now he *is* gigantic.'

'But back to the matter at hand,' said Randalf. 'I have summoned you here, Henry the Hairy, great warrior-hero, to . . .'

'Warrior-hero?' Joe interrupted, 'Henry's not a warrior-hero. He's my dog!'

Henry wagged his tail and rolled over with his legs in the air.

'What's he doing?' said Norbert, his three eyes open wide and panic in his voice.

'He wants you to tickle his tummy,' said Joe, shaking his head in disbelief. 'Any minute now, I'm going to wake up in hospital with a big bandage on my head.'

'Go ahead, Norbert, tickle his tummy,' said Randalf.

'But, sir,' said the ogre weakly.

'Tickle!' said Randalf. 'And that's an order!'

As Norbert bent down, the room gave a lurch. He gently stroked Henry's tummy with a massive finger.

'Go on, go on,' said Randalf impatiently. 'He won't bite.'

He smiled at Joe. 'It seems there's been a slight misunder-standing,' he said, stroking his beard.

'There's always a misunderstanding with you,' chirped

the budgie.

'Shut up, Veronica. I was under the impression that Henry the Hairy was the warrior-hero I had ordered – strong, loyal and . . . errm . . . hairy. But if, as you say, he is in fact a dog, then *you* must be the warrior-hero . . .'

'He doesn't look very strong, or for that matter, very hairy,' said Veronica dismissively. 'If *he's* a warrior-hero, then *I'm* Dr Cuddles of Giggle Glade!'

'Veronica,' Randalf snapped, 'if I've told you once I've told you a thousand times *never* to utter that person's name in my presence!'

'Brings back memories, does it?' Veronica taunted, and flapped up into the air. Randalf tried to swat her.

'Ow, watch where *you're* flying,' Norbert shouted, taking a step back as the budgie booted him in the ear.

Joe clung on to the great hanging pot as the whole room seemed to tilt to one side.

'Button it, you great lunk!' Veronica shot back.

'You and whose army?' Norbert countered.

Joe watched in open-mouthed amazement as the wizard, the ogre and the budgie rounded on one another angrily. This was absolutely crazy. Who were they? Where was he? And most important of all, how was he going to get home?

'It's . . . it's been lovely meeting you all,' he shouted, interrupting the three shouting protagonists, 'but it's getting late and I've still got my homework to do. I really should be leaving now . . .'

The three of them stopped mid rant, carp and criticism, and turned to him.

'Late?' said Randalf.

'Leaving?' said Norbert.

Veronica jumped up and down on the wizard's head, feathers all fluffed up. *'You're* not going anywhere!' she squawked.

'Ouch!' cried Joe, rubbing his arm.

'Again, sir?' asked the three-eyed ogre, bending over him.

'No, three times is quite enough,' said Joe ruefully.

There's nothing quite like an ogre pinching your arm to convince you that you're not in a dream, and Joe was now totally convinced that this was no dream. But if it wasn't, then where on earth was he? And how had he got there? Henry wagged his tail and licked the ogre's hand.

Before Joe had a chance to ask any questions, the clock on the wall above the fireplace erupted with insistent noise. There was coughing, the sound of a tiny throat being cleared and fists and boots battering on a small wooden door which suddenly sprang open. A little elf – dressed up in grubby looking underpants and with a length of elastic tied around its waist – jumped out and dived into mid-air.

'Five of the clock!' it screeched as it reached the end of the elastic, before rebounding and disappearing back inside the clock with a muffled crash.

'Five?' said Randalf wearily. 'But it's dark outside.'

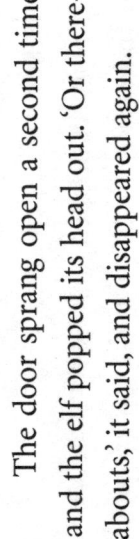

The door sprang open a second time and the elf popped its head out. 'Or thereabouts,' it said, and disappeared again.

'Wretched clock's slow again,' Randalf grumbled. 'The mechanism probably needs cleaning.'

'I should say so,' said Veronica scathingly. 'Judging by the state of those underpants.'

'Shut up, Veronica!' said Randalf.

'"Shut up, Veronica!"' said Veronica. 'That's your answer to everything. Well, I won't shut up! Call yourself a wizard. You've only got one spell – and you can't even do that properly.' She flapped her wing at Joe. 'I mean, look at him,' she said. 'Is it just me or is our so-called warrior-hero a little on the short side? Not to mention puny – and gormless. And as for his hairy sidekick . . .'

'*Shush*, Veronica,' said Norbert, patting Henry on the head. 'You'll hurt his feelings.'

Henry wagged his tail.

'Ooh, look, his waggler's waving!' Norbert cried. 'Does he want his tummy tickled? Does he? Does he?'

He jumped up and down excitedly. The room lurched backwards and forwards alarmingly, and more books and utensils clattered to the floor.

'Norbert!' said Randalf, sternly. 'Behave yourself! Remember what happened with Mary the poodle. You don't want that to happen again, do you?'

Norbert stopped jumping, and shuffled to the corner of the room.

'Very nervous animals, poodles,' said Veronica, who was perched on Randalf's head. 'Ruined the carpet!'

'Shut up, Veronica!' shouted Randalf. 'That's all in the past. We have a new warrior-hero here and he'll be fine.'

He clamped a hand on Joe's shoulder. 'Won't you, Joe? Just the job for the Horned Baron's purposes. Once he's kitted out . . .'

'Kitted out?' said Joe. 'What do you mean, kitted out? I don't want to be kitted out. I just want to know what's going on,' he added angrily.

'Temper, temper!' said Veronica.

'The fiery disposition of a warrior-hero,' said Randalf.

'Excellent!'

'What *are* you talking about?' Joe asked. 'I've got to get back for tea. And I've got this essay. I haven't even started yet . . .'

'Tea? Essay?' said Randalf smiling. 'Ah, yes. The mighty deeds of a warrior-hero – the great tournament of tea, the epic slaying of the monster essay! Of course you must get back for these tasks, but first, if you could just lend a hand with a tiny little task we have here . . .'

'I can't!' Joe insisted. 'I've got school tomorrow. I must get back. If you brought me here, you can send me back.'

'I wouldn't bank on it,' Veronica muttered.

'I don't think you appreciate the difficulty involved in summoning a warrior-hero from another world,' said Randalf solemnly. 'I mean heroes don't just grow on trees – well, apart from those in the Land of Hero-Trees that is. It's a long and painstaking process, I can tell you; by no means as easy as you seem to believe.'

'But—' Joe began.

'For a start, the three Muddle moons must be correctly aligned – and that doesn't happen often. If we'd missed this evening's triangular configuration, there's no knowing how long we'd have had to wait for the next one.'

'But—'

'Furthermore,' Randalf went on, 'because of a slight technical hitch with the actual spell—'

'What he means is, he's lost half of it,' said Veronica.

Randalf ignored her. 'You're only the second hero I've actually successfully summoned. The first one was Quentin—'

'The one with the poodle and the kilo of icing sugar,' Veronica butted in.

Joe became aware of a soft sniffing sound and looked up to see three big fat tears rolling down from Norbert's bloodshot eyes and over his lumpy cheeks. 'Poor, dear Quentin,' he sobbed.

'Cry baby!' Veronica mocked.

'Oh, but he never stood a chance!' Norbert wailed.

'That's enough, you two,' said Randalf.

'Sorry, sir,' said Norbert. He wiped his nose on the back of his sleeve, but kept sniffing.

'As I was saying,' said Randalf. 'First there was Quentin, and now I have summoned you . . .'

'But you had no right!' shouted Joe. 'I didn't ask to be summoned! I didn't ask to be pulled through a hedge, down a tunnel and into this, this . . . junk room!'

'Damn cheek,' came a muffled voice from the clock.

'I didn't ask to be kitted out by a stupid wizard! And insulted by a stupid budgie! And pinched by a stupid three-eyed ogre!' stormed Joe.

'Actually,' said Norbert. 'You did ask to be pinched. You said, "Pinch me. If this is a dream then . . ."'

'SHUT UP!' shouted Joe. 'SHUT UP!'

Norbert jumped back, his eyes wide with terror. 'Oh,

help!' he bellowed. 'Mayday! Mayday!' And he jumped up as high as the ceiling would allow, crashing back down on to the floor with both feet a moment later.

The room keeled to one side. Randalf fell, Veronica flapped up into the air, while Joe was catapulted across the room.

'*Aaaaaargh!*' he yelled as he hurtled past Randalf and Veronica and crashed into the wall opposite, missing the open window by a fraction. Dazed and winded, he slid to the floor. The room continued to roll back and forth, back and forth.

'Norbert, you bird brain!' shouted Randalf.

'How dare you!' screeched Veronica. 'Bird brain indeed!'

Randalf sighed as the room gradually righted itself. He turned to the ogre. 'Say "sorry", then,' he told him.

'I'm sorry, sir,' said Norbert miserably. 'Very, very sorry.'

'Not to me, Norbert,' said Randalf. Norbert frowned with confusion. 'To our guest here, our warrior-hero,' Randalf explained. 'To *Joe*,' he said, and pointed towards him.

'Joe!' Norbert cried out in horror. He saw him lying on the floor. 'Did I do that?' he said. 'Oh, I *am* sorry. So, so very sorry.' Tears welled up once again in his eyes. 'It's just that I panic when someone shouts at me. I have a very nervous disposition. In fact I was almost called Norbert the Wet-Trousers instead of Norbert the Not-Very-Big, because——'

'Yes, yes,' said Randalf. 'Pick him up, Norbert. Brush him down.'

'Yes, sir. At once, sir,' and he hurried back across the room.

Joe, by this time, was already climbing to his feet quite successfully on his own. As Norbert came lumbering towards him, the boat lurched again. Joe stumbled towards the open window.

'*What on earth?*' he exclaimed.

Randalf pushed past Norbert and placed his hand on Joe's shoulder. 'Not quite,' he said. 'Welcome to *Muddle Earth*.'

Joe stood stock-still as he stared out of the window. He was barely able to believe what his eyes were telling him. For a start, instead of the one familiar white moon in the sky, there were three: one purple, one yellow and one green. And the landscape! It was like nothing he'd ever seen before, with vast areas of fluorescent green forest and glistening rocky wasteland – and far in the distance, tall, smoking mountains.

Most curious of all, however, he realized that he wasn't underground at all, but rather in some kind of a boat. And there were others. Five . . . six others, all bobbing about on a lake which . . . But no, it made no sense. He closed his eyes, then opened them again.

There was no doubt about it. The lake was suspended high up in the air, without any means of visible support.

Joe turned to Randalf. 'The lake,' he gasped. 'It's . . . it's floating.'

'Of course it is,' said the wizard solemnly. 'The Enchanted Lake was raised up by the wizards of Muddle Earth many, many moons ago, and for a very good reason – only no one can remember what that was. Anyway, they raised it . . .'

Joe frowned. 'But, *how*?'

'By great magic,' said Randalf solemnly.

'And *that's* something you don't see round here much these days,' Veronica chipped in.

'Magic?' said Joe softly. He shook his head. 'But . . .'

'Don't you worry about it, young warrior-hero from afar,' said Randalf. 'You've got a lot to learn. Thankfully, I am an excellent teacher.'

'Yes, and I'm an exploding gas frog!' Veronica retorted.

'Shut up, Veronica!' said Randalf.

'There you go again,' said Veronica huffily, and turned her back on the proceedings.

'As I was saying,' Randalf continued. 'I shall teach you everything you need to know for the small task which lies ahead.' He smiled. 'Things are going to work out well this time, I can feel it in my bones. Joe, here, will see us proud.'

Veronica sniffed. 'I still don't think he looks like much of a warrior-hero,' she said.

'He soon will,' said Randalf. 'We'll set off for Goblintown just as soon as day breaks.'

Just then the elf sprang out of the clock. 'Half past twenty-six, and that's my final offer!' it shrieked.

'Yip! Yip! Yip! Bibbitty-Bibbitty!' came a shrill voice. 'This is your early morning wake-up call.' *Boing!*

Joe's eyes snapped open, just in time to see the elf disappearing back behind the little wooden doors of the clock.

Joe looked round, and groaned. Everything was exactly the way it had been when he'd curled up in his hammock the night before. The clock, the room full of junk, the floating lake . . . What was more, it was still dark.

'What time do you call this?' Randalf demanded, emerging from the far end of the room.

The door of the clock flew open. 'Early morning!' snapped the elf. 'Or thereabouts.' The door slammed shut.

'What's going on?' squawked Veronica. 'I've only just tucked my beak under my wing.'

Norbert lumbered out of the shadows, yawning and stretching. 'Is it morning?' he said.

Randalf looked out of the window. There was a fuzzy

glimmer of light on the horizon. High in the sky the noisy purple batbirds were soaring back to the woods, to spend the day resting upside down at the top of the highest jub-jub trees. 'Almost,' he said.

'Stupid clock,' Veronica muttered.

'I heard that!' came an indignant voice from the clock.

'Never mind all that,' said Randalf. 'We're awake now, so we might as well make an early start. Come on, Joe. Stir your stumps. Today's your big day.' He turned to the three-eyed ogre. 'Norbert,' he said. 'Prepare breakfast.'

'Are you sure I can't tempt you with any more?' said Randalf, ten minutes later.

'No, thanks,' said Joe.

'You'll need to keep your strength up,' Randalf persisted.

'Such as it is,' added Veronica unkindly.

Joe looked at the ladle of slop hovering above his bowl. 'I'm really full,' he lied.

Without any doubt, Norbert had produced the strangest breakfast Joe had ever eaten in his life – lumpy, green porridge that tasted of gooseberries, a small cake iced with love-hearts and a mug of foaming stiltmouse milk.

'But you haven't touched your snuggle-muffin,' said Norbert, looking hurt.

'I'm saving it for later,' said Joe. 'It looks lovely, though.'

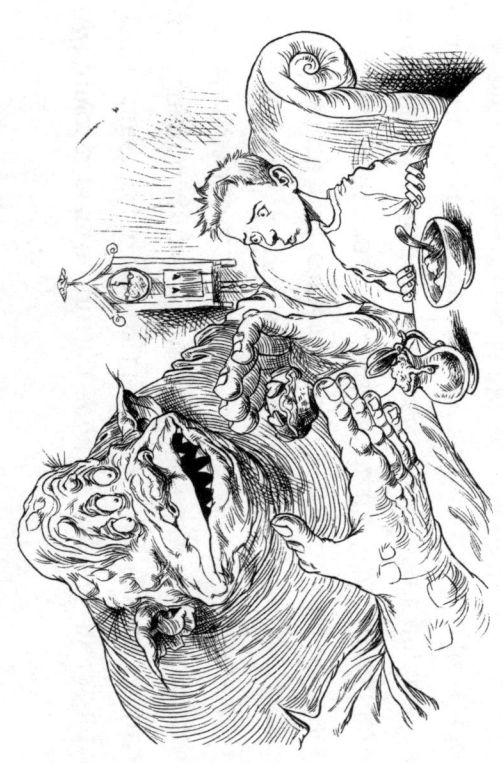

The ogre sighed. 'Dear Quentin taught me everything I know. He was an artistic genius with icing sugar.'

'Right, then,' said Randalf, clapping his hands together. He stood up and grabbed his staff. 'Let's get this show on the road.'

Relieved, Joe jumped up from the table, grabbed Henry by his lead and followed Randalf downstairs. Norbert stomped down after them.

'Eager to get started, eh, Joe?' said Randalf warmly. 'An excellent sign for a warrior-hero. We've summoned a good'un this time, Veronica.'

'You said that about Quentin,' the budgie was not slow in reminding him. 'And look how that turned out.'

'We must not look to the past,' said Randalf as he opened the door. 'But to the future.'

He stepped outside. Joe followed him, still none the wiser about what exactly was going on. He *seemed* to be on the lower deck of a houseboat, but it was difficult to know for certain.

Underneath the vessel, fat fish swam round and round in the crystal clear water. They reminded him of the goldfish at home and, for a moment, he thought that perhaps it wasn't so crazy here after all.

High above him, small fluffy purple clouds scudded across the sky. Below him – and attached to the side by a rope – was a small boat. At least that's what Joe thought at first. It was only as he stepped across from the rope ladder to the bobbing vessel that he realized it was not a boat at all, but a bathtub. Joe clapped his hand to his forehead.

'What was I expecting?' he said to himself. 'Of *course* it's not a boat. After all, this *is* Muddle Earth.'

The bathtub gave a wild lurch as Joe stepped into it. Henry jumped in beside him.

'Not there,' said Veronica. 'You sit at the other end where the taps are.'

'If you don't mind,' said Randalf, climbing in and sitting down. 'Just watch your head on the shower attachment if it gets a little bit . . . *errm* . . . choppy.'

'It *always* gets choppy with Norbert in charge,' said Veronica. 'You've sunk two boats, one wardrobe and an inflatable mattress. Now we're using our last bathtub. It'll be the kitchen sink next!'

'Shut up, Veronica!' said Randalf. 'Come on, Norbert. We're all waiting.'

~ 33 ~

The ogre climbed into the bathtub, which wobbled about dangerously on the water. Kneeling down, Norbert seized two objects from the bottom of the bath. One was an old tennis racquet, the other a frying pan. He leaned forwards and began paddling furiously.

The bathtub reared up in a great swell of spray and froth, and sped across the surface of the lake. Norbert's arms were like pistons; up down, up down, up down they went. The edge of the lake came closer. Joe gasped.

'We're going to fall off!' he shouted.

'Trust me, I'm a wizard,' said Randalf. 'A little further left, Norbert,' he told the ogre. 'That's it.'

Joe looked up. He saw that they were heading for a waterfall.

'Hold on tight and watch out for that shower attachment,' the ogre smiled, and paddled faster than ever.

Closer and closer the waterfall came, louder and louder grew the sound of the raging torrent spilling over the edge.

'This is crazy!' Joe yelled.

'True,' said Randalf. 'But it's the only way down. Trust me, I'm a . . .'

'I know,' Joe muttered as he gripped the sides of the bathtub with white-knuckled ferocity, 'you're a wizard.'

'One last paddle should do it!' Randalf shouted.

Norbert obliged. For a moment, the bathtub see-sawed up and down at the very edge of the lake. Joe gasped as

Muddle Earth opened up before him in a broad, panoramic view.

'Brace yourselves!' came Randalf's voice above the raging cascade as the bathtub toppled over the edge. 'Here we go . . . o-o-o-oh!'

Veronica screeched. Henry howled. Joe screwed his eyes tightly shut. Only Norbert seemed to be enjoying himself.

'Wheeeee!' he cried as the little bathtub with its five occupants hurtled down the roaring, foaming torrent of water.

The wind whipped past so fast, Joe could barely breathe. Stinging water slapped his face. His head pounded, his heart was in his mouth, his grip was slipping . . .
SPLASH!

The bathtub struck the pool at the bottom of the waterfall with colossal force.

It sank.

It rose.

It bobbed around on the surface like a ball, until the current swept it down into the relative calm water of the river beyond. Joe opened his eyes.

'That was ... terrifying,' he gasped.

Veronica snorted. 'You wait until the return journey,' she said.

'We've taken on water!' Randalf announced. 'Quick, Norbert, start bailing!'

Norbert looked down at the water swilling around the bottom of the bathtub. He picked up the frying pan, then paused. 'Don't worry, sir!' he said, reaching down into the water. 'I'll save us!'

'No, Norbert!' Randalf shouted. 'Remember last time!'

Too late. The ogre had already yanked out the plug and was holding it aloft.

'There,' he said. 'I've let the water out, sir!' He paused, and stared incredulously at the fountain of water gushing up from the plughole. 'Oops!' He looked up at Randalf. 'I've done it again, haven't I, sir?'

'I'm afraid so,' said Randalf.

'Abandon bathtub!' Norbert shouted.

With a gurgle and a plop, the bathtub abruptly disappeared beneath them. Joe kicked his legs, and headed for the

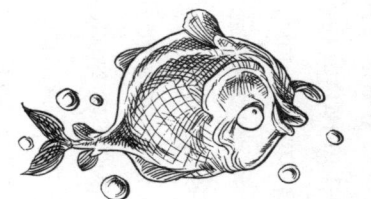

nearest shore, with Henry doggy-paddling by his side. The pair of them dragged themselves up on to the bank. Veronica, who had decided to fly rather than swim, landed beside them, only to be soaked by a spray of water as Henry shook himself dry.

'Well, that's the last of the bathtubs,' she said. 'You'd think he'd have learned about plugs by now.'

Joe said nothing. The sight of the Enchanted Lake suspended in mid-air far, far above his head had left him speechless. As he watched, a fat silver fish dropped out of the bottom of the lake, fell down through the air and into the gaping beak of one of the waiting lazybirds clustered together beneath the lake.

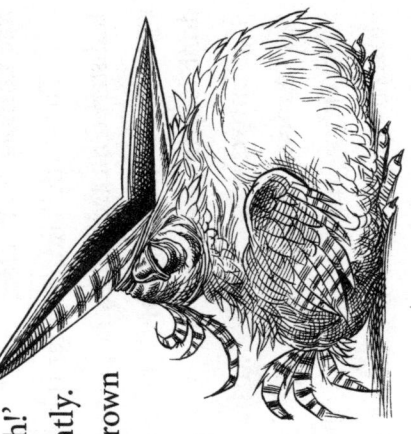

Randalf and Norbert climbed out of the river, dripping with water. Norbert shook himself dry, drenching Veronica for a second time.

'Thank you very much!' she exclaimed indignantly. 'Now you're trying to drown me on dry land!'

'Sorry,' said Norbert meekly.

Randall looked up at the sun, which was now peeking over

the distant mountains. 'No harm done. Quite refreshing in fact,' he said. 'Let's set off. With a brisk pace, we should arrive at Goblintown by midday.' He turned to the ogre. 'If you'd do the honours, Norbert, my good fellow.'

'Yes, sir,' said Norbert. He crouched down on his hands and knees.

With the budgie perched on his head, Randalf climbed on to one of the ogre's great, broad shoulders. He looked at Joe. 'Well, come on. We haven't got all day,' he said.

Joe climbed gingerly on to the other shoulder. 'Won't we be too heavy?' he said.

'Of course not,' said Norbert. 'I'm a two-seater. Now, my cousin Ethelbertha is a four-seater with an extra box-seat at the front and back . . .'

'Yes, yes,' said Randalf. 'If you're quite ready, Norbert.' Norbert straightened up and climbed to his feet.

'Forwards!' ordered Randalf, and tapped the ogre on the head with his staff. Norbert lurched into movement and strode off, with Henry trotting along by his side.

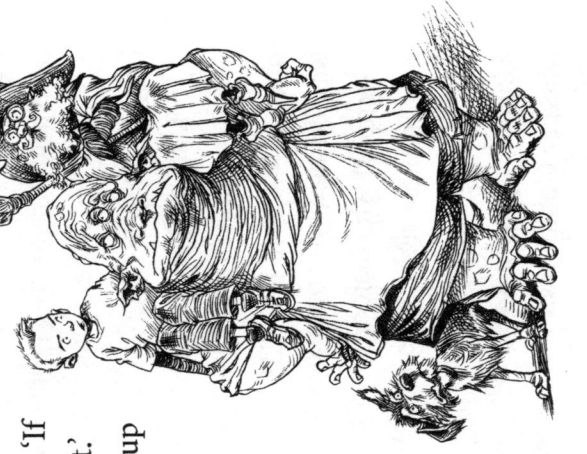

On her lofty perch, Veronica tucked her head beneath her wing. 'I always get ogre-sick on long journeys,' she said weakly.

They were following a track which ran parallel to the Enchanted River. It was clearly seldom used, and had become totally overgrown. Tall jub-jub trees lined both sides of the path, their bendy branches heavy with sleeping batbirds. Norbert brushed past them.

'Ouch! Ouch! Ouch!' they cried.

Joe couldn't believe he was riding to Goblintown on an ogre's shoulders. About now, he should have been handing in his essay. He ducked to avoid a jub-jub branch and, with a thud, a batbird landed in his lap. Of all the excuses for late homework, 'Sorry, sir, I was too busy riding ogres and dodging batbirds,' had to be the weirdest. How on earth *would* he explain this to his teacher?

'It'll be easier once we hit the road,' said Randalf brightly. 'Not far now.'

'Liar,' muttered Veronica queasily.

By the time they did finally reach the road – or rather *roads*, since they had arrived at a three-way junction – Joe was feeling a little ogre-sick himself. Norbert had stopped at a signpost which stood by the side of the road.

To their right, according to the peeling gold lettering, was *Trollbridge (not very far)*. To their left was *Musty Mountains (quite a long way)* Joe craned his neck to check whether Goblintown was also marked. It was.

GOBLINTOWN

(THIS WAY, A VERY LONG WAY AWAY, SO STOP LOOKING AT THIS SIGN AND GET A MOVE ON, STUPID.)

'Charming!' said Joe.

'Come on, then,' said Randalf, tapping Norbert on the head. 'Let's get going.'

Norbert shuffled about awkwardly.

'Which way, exactly, sir?' he said.

'To Goblintown, of course,' said Randalf impatiently.

'I know *that*, sir,' said Norbert, staring at the signpost with a perplexed frown. 'But . . .'

'Norbert,' said Randalf, 'when I took you on as an official Wizard Carrier, you assured me you could read.'

'Believe that, you'll believe anything,' said Veronica.

'I can't!' said Norbert. 'Little words, anyway.' He nodded at the signpost. 'These are all so long.'

Randalf breathed in sharply. '*This way*, Norbert,' he said. '*A very long way away, so stop looking at this sign and get a move on, stupid*,' he read.

'Who are you calling stupid?' Norbert muttered in a hurt voice as he set off. 'My cousin Ogred the Dribbler,' he said, his pace increasing, 'now he *was* stupid. Did I ever tell you about the time he got his head stuck in a . . .'

'Snuggle-muffin,' Randalf mumbled drowsily, his head suddenly lolling on his shoulders.

'Typical,' said Veronica. 'Sleeping like a baby again!'

'Sir always falls asleep when he's travelling,' said Norbert. 'And usually right in the middle of my best stories.'

'Can't imagine why,' said Veronica testily.

'Well, it must be tiring being a wizard,' said Norbert, 'what with all that reading and spells and stuff.'

'I suppose so,' said Joe with a shrug.

'Oh, Randalf isn't a *real* wizard,' Veronica whispered into Joe's ear.

From up on Norbert's shoulders there came the sound of soft snoring.

'He isn't?' said Joe. 'But I thought . . .'

'Until the wizards – the proper wizards, the *Grand Wizards* – all disappeared, Randalf was just a lowly apprentice to Wizard Roger the Wrinkled,' Veronica explained in a hushed voice. 'It was only when they all went missing that Randalf started to pretend *he* was a Grand Wizard. He seems to have fooled the Horned Baron, but no one else is taken in. That's why he's so hard up. I mean, who wants to pay for the services of someone so useless?'

'Useless?' said Joe.

'Utterly useless!' said Veronica. 'Oh, the stories I could tell. The invisible ink that kept reappearing. The flying bicycle that fell to bits in mid-air. And all those poor goblins who ended up bald as an egg after trying his magic memory-cream. Not to mention Dr Cuddles.'

'Dr Cuddles?' said Joe. '*Who* is Dr Cuddles?'

'Only the baddest, meanest, most evil villain there ever

was!' Veronica told him. 'He won't rest until he's the ruler of Muddle Earth, with everyone under his thumb. Randalf is in a right mess. He needs help.'

'And he's expecting me, as a warrior-hero, to sort all this out?'

'Good heavens, no,' Veronica replied. 'You're just here to earn Randalf some money, though if I were you—'

At that moment, a loud clattering, banging noise echoed through the air, cutting Veronica off in mid-sentence. Joe looked up to see a flock of cupboards, their doors beating like wings, flying across the sky. 'What are *they*?' he gasped.

'They look like cupboards to me,' said Norbert.

'Do cupboards *normally* fly in Muddle Earth?' Joe asked, bewildered. He stopped and, with his hand shielding his eyes from the dazzling sun, peered into the distance. 'They're flying in formation – and they seem to be coming from that forest.'

'Elfwood,' said Veronica, nodding. 'That doesn't surprise me. There have been a lot of strange goings-on in Elfwood recently,' she added darkly. 'If you ask me, it's all to do with Doc—'

'*Aaargh!*' shouted Norbert as, still looking skywards, he failed to notice a large pothole in the road. He tripped, stumbled and crashed to the ground. Veronica flapped squawking into the air. Joe landed in a heap beside Henry. Randalf rolled over in the dust.

'My lovely snuggle-muffin', he murmured. 'I . . .' His eyes snapped open. He found himself sprawling on the ground. 'What's *happening*?'

'Please, sir. Sorry, sir', said Norbert apologetically as he climbed to his feet. 'I tripped in that pothole.'

He pointed at a round hole in the road containing an upturned cooking pot and a mess of smelly porridge.

'Ruined!' said a small elf by the side of the road. 'Why don't you watch where you're going, you big oaf!'

'And why don't you watch where you're cooking!' squawked Veronica.

'What better place for a pot than a pothole!' the elf countered indignantly.

'Calm down everybody', said Randalf, tucking his staff back under his arm. 'There's no harm done.'

The elf shrugged and, picking up the pot, strode off down the road. 'Wizards!' it snorted in a contemptuous voice.

The ogre leaned forwards, pulled Randalf up and began dusting him down vigorously.

'Easy, Norbert', said Randalf. 'I don't know –

soaked, dropped and now pummelled. This isn't what I'd call first-class travel.'

'No, sir. Sorry, sir,' said Norbert.

Randalf looked ahead. He called Joe to his side, and pointed. 'Over there,' he said. 'Just beyond that hill, you can see the spires of Goblintown peeking up. We are almost at our destination. You will be soon kitted out with the finest warrior-hero costume that money can buy.'

'Or rather, that you can *afford*,' Veronica muttered under her breath.

As they approached the high wall that surrounded Goblintown, Norbert placed Randalf and Joe gently down on the ground. The noise from inside became louder and louder. There was shouting, hammering, wailing, whinnying, sawing, singing, ringing, bellowing, buzzing . . . all mixed up together in one tumultuous roar. And the smells!

Burnt tar, sour milk, wet fur, rotting meat – each one fighting to be noticed above the nose-wrinkling odour of unwashed goblins. Joe tried thinking of nice smells – chocolate, biscuits, strawberry ice cream . . .

'And I thought your socks were bad!' Veronica commented.

'Shut up, Veronica,' said Randalf. 'We don't want to hurt their feelings. Goblins can be very touchy.'

'*Humph!*' said Veronica. 'So far as I'm concerned, no one that smelly has the right to be touchy!'

'That's as may be,' said Randalf. 'Just remember, we're here on important business. So, button your beak!'

'Pardon me for breathing,' said Veronica huffily.

'By the way, my fine young friend,' Randalf added. 'I think Henry here is finding this pungent metropolis a little too exciting. Perhaps you'd better put your faithful battle-hound on the lead.'

Sure enough, Henry's tail was wagging excitedly as his wet nose sniffed eagerly at the air. Joe clipped the lead into place.

They were now at the top of a tall flight of stairs before a set of huge wooden doors. Randalf seized the great knocker and hammered firmly against the wood.

Ding-dong!

Randalf turned to Joe. 'The goblins will have their little joke,' he said.

'Yeah,' Veronica chirped up. 'Like not letting us in.'

'Patience, Veronica!' said Randalf sharply. He looked at the door expectantly. Nothing happened. Randalf shuffled about, scratched his head, stroked his beard, rubbed his eyes. 'That famous goblin welcome,' he said at last. 'It's renowned throughout Muddle Earth.'

He knocked again, louder this time. *Ding-dong! Ding-dong! Ding-dong!*

The door flew open.

'I heard you the first time!' shouted a small, grubby, cross-eyed individual with gappy teeth and pointy ears.

'I'm not deaf!'

'A thousand, nay a million apologies, my goblin friend,' said Randalf, bowing low. 'Far be it for me to bring your hearing faculties into doubt. I merely . . .'

'What do you want?' the goblin scowled.

Randalf forced himself to smile. 'We wish to avail ourselves of your finest outfitting-emporium,' he said, 'with the express intention of purchasing your most—'

'You what?' the goblin said.

'We need to buy a warrior-hero outfit for the boy,' said Veronica, flapping her wing towards Joe.

'Leave this to me, V—' Randalf started to say. He was interrupted by the goblin.

'Well, why didn't you say so in the first place?' he snapped, opening the door for them all to enter.

'Shut that door!' shouted a voice. 'There's a terrible draught.'

'Take a deep breath everyone,' Randalf whispered over his shoulder.

'Come in, come in,' said the goblin impatiently. He slammed the door shut behind them. 'Welcome to Goblintown,' he intoned in a bored voice. 'The city that never sleeps.'

'Or washes,' muttered Veronica.

'Enjoy its many sights and sounds . . .'

'And smells,' said Veronica.

'And unique atmosphere.'

~ 47 ~

'You can say that again!'

'Veronica, I warned you!' Randalf hissed. 'Will you *shut* up!'

'Have a nice day,' the goblin concluded, and yawned.

Joe screwed up his nose. 'Veronica's got a point, though,' he whispered. 'It really does stink here!'

'You'll soon get used to it,' Randalf told him. He turned to the goblin. 'Thank you, my good fellow. I'd just like to say what an honour, not to say, pleasure it is to . . .'

'Whatever,' said the goblin, and walked away.

Joe looked about him. Apart from the smell, Goblintown was amazing. Buildings had been built on other buildings, storey by storey, reaching up higher and higher. Teeming with life, the tall structures lined the maze of narrow alleyways on both sides. Joe looked up nervously as the great towering constructions swayed precariously back and forth. They looked for all the world as if they might topple over at any moment. What was more, since the buildings *were* so high, almost no sunlight penetrated to the alleys at the bottom.

Not that it was completely dark. Oil lamps hung from every building, bathing the bustling streets in a sickly, yellow light and filling the air with greasy smoke. The rank, burning fat joined the other disgusting smells of Goblintown: the foul odour of baking snotbread, the stale pong of unwashed bodies and the stench bubbling up from the drains.

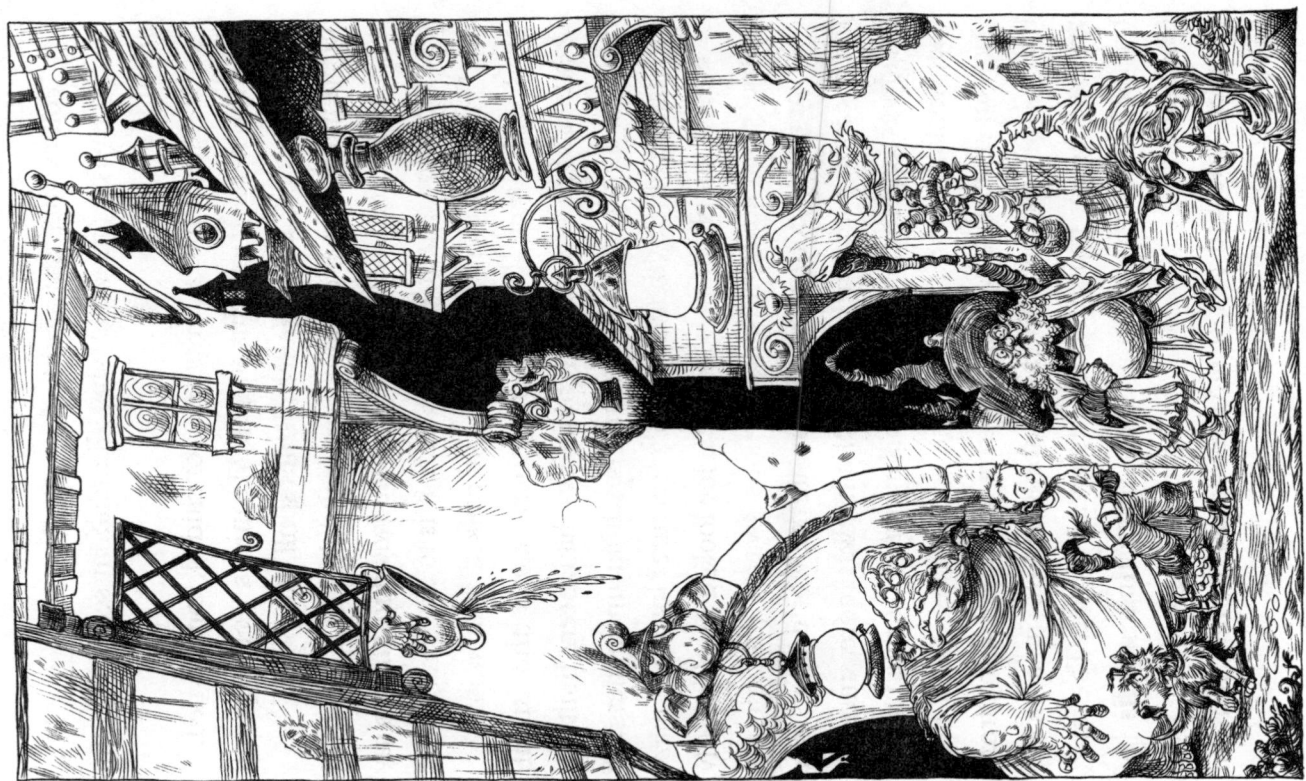

'Follow me!' said Randalf. He pulled a baggy yellow item of clothing from his robes and raised it up high on the end of his staff.

'Are they what I think they are?' said Veronica.

'You never know when a pair of ogre's underpants will come in useful', said Randalf, striding off. 'Keep your eyes on the pants everybody, and you won't get lost.'

Norbert sighed. 'I wondered where they'd got to,' he said.

'Now, Joe, pay attention,' said Randalf, 'You really must see the sights of Goblintown. On our right,' he announced, 'we have a typical goblin apartment. On our left, the Museum of Moderate Achievements – while that,' he said, pointing up ahead to a particularly rundown building, 'is the Temple of the Great Verucca.'

'The Temple of the Great *Verucca*?' said Joe incredulously.

Randalf nodded. 'It is the site where Wilfred the Swimmer, that ancient goblin explorer, laid the first stone of what was to become Goblintown. Legend has it that he had been looking for a suitable place to build for seven long years when the pain from his great verucca finally forced him to abandon his search. He decided it was a sign. The rest, as they say, is history.'

'Spare us the guided tour!' squawked Veronica, fluttering overhead.

This way and that they tramped. Randalf was in the lead, with the yellow underpants held high, followed by

~ 50 ~

Norbert, Henry – sniffing every street corner and straining at his lead – and finally, Joe. Goblins thronged the streets, shoving and jostling, arguing, shouting – and completely ignoring the strangers in their midst.

'Which is why,' Randalf concluded, pausing to let the others catch up, 'we call Goblintown the Friendly City. It— *Ooof!*'

One goblin barged into Randalf. Another grabbed the underpants from the end of the staff and raced away through the crowd.

'My word!' shouted Randalf.

'My pants!' cried Norbert.

'Forget them,' said Veronica. 'We're here.'

As one, they all turned – and there, behind the windows of a particularly tall and ramshackle building, was a selection of shop-dummies, all dressed up in ornate warrior-hero costumes. Norbert stared at the gold letters on the sign swinging back and forth above the door.

'*Unc . . . Unc . . . Unc . . .*' he said.

'*Unction's Upmarket Outfitters*,' Randalf read for him.

'Well spotted, Veronica. Come on then, Joe. Let's get you kitted out.'

To Joe's surprise, they did not stay long in *Unction's Upmarket Outfitters*. Randalf led them up the aisle of the magnificent store – ignoring the tuts and sighs of the smartly dressed assistants as he went – to the very back. There, they climbed a broad curving staircase.

They emerged, not on the second floor of the outfitter's as Joe had expected, but in a different shop alto-gether. A brightly painted sign on the wall announced it as *Mingletrips Middle-of-the-Road Emporium*. The assistants, all of whom wore slightly plainer outfits, watched them suspiciously.

'I was looking for . . .' Randalf began.

'Yes?' said one of the assistants, his left eyebrow arched high.

'For the way up,' said Randalf.

The assistant nodded towards the far window. Joe frowned, puzzled. What were they meant to do there? Randalf, however, seemed unfazed.

'Ah, yes,' he said, and set off. The others followed him again.

Across the shop floor they went, out of the window and up a rusting circular staircase bolted to the outside wall.

They were greeted at the top with a third sign: *Drool's Downmarket Depot*.

'Keep up, we're nearly there,' said Randalf bossily. He began climbing an old wooden ladder propped up against the wall. 'And mind the eighth rung,' he called down. 'It wobbles a bi—'

'*Aaargh!*' Norbert cried out as the rung beneath his feet

splintered and fell. He clung on tightly to the sides of the ladder, too frightened to continue.

'On second thoughts, perhaps it would be better if you went back and waited for us there,' said Randalf. 'And look after Joe's battle-hound,' he added.

Joe watched the quivering ogre slowly lowering himself back to the balcony. Then, somewhat relieved that he wouldn't have to carry Henry up the rickety ladder, he passed Norbert the lead.

'Be good, now,' he told the dog. 'All right?'

'All right,' Norbert replied meekly.

Taking extra care at the missing eighth rung, Joe scampered up the ladder. As he neared the top, he glanced down – and gasped.

Far, far below him, he could see the crowds of goblins streaming this way and that along the narrow alleys. Veronica fluttered down and hovered near his left shoulder.

'Come on,' she said encouragingly. 'Mustn't keep our know-it-all guide waiting.'

With his heart in his mouth, Joe completed the last few steps of the ladder. Randalf was there to meet him.

'Well done, lad,' he said. 'And welcome to the finest outfitter's in all of Muddle Earth.'

'The cheapest, more like,' muttered Veronica.

There was a faded banner above the door. '*Grubley's Discount Garment Store*,' Joe read out – unable to keep the disappointment from his voice.

Just then, a gangly goblin with a large nose and filthy frilly apron appeared from behind them. 'Welcome,' he said.

'Thank you, *errm* . . . Stink,' said Randalf, addressing the goblin by the name on his lapel.

'Actually, it's Smink,' the goblin replied.

'But—'

'Mr Grubley's handwriting,' the assistant said wearily. 'It's almost as dreadful as his dress sense.'

'Yes, where *is* old Grubbers?' said Randalf, looking round.

Smink shrugged. 'He said he had *business to attend to*.' He winced as if smelling something unpleasant.

'Did he now?' said Randalf. He turned to the others, smiling. 'I think everything is going to work out even better than I expected,' he whispered.

Smink had turned away and opened the door from the balcony into the store. Ducking down, he went in. Randalf did the same.

'Just as well Norbert did stay behind,' Joe muttered to himself as he stooped low. 'He'd never have got through this doorway.'

He straightened up to find himself in a room so messy, dark and grubby, it made the wizard's houseboat seem like a showroom.

There were boxes and bundles, bursting with materials, piled up against the walls, next to stacked shelving, stuffed with every conceivable type of boot and shoe. Huge, circular racks – some, three tiers high – creaked under the weight of countless garments, all sorted into different categories. Jackets and jerkins. Capes, cloaks and coats. Leggings and breeches. Jodhpurs and knickerbockers. Bodices, bustles and bibs. While high overhead, still more – in differing colours, sizes and styles – were suspended from large ceiling hooks.

'Oh, yes, sir, I can see the problem,' said Smink, fingering Randalf's cloak with obvious distaste. 'This one's totally threadbare. Shoddy workmanship, completely out of fashion and a hideous colour, if you don't mind my saying so.' He looked up. 'Wizard's cloaks are over here,' he said. 'Walk this way,' and he bustled through the shop with tiny, pigeon-toed steps.

'I couldn't walk that way if you paid me,' Veronica commented.

Smink turned. 'I'm sorry, did sir say something?' he asked.

'Actually, it's the lad we're here for today,' said Randalf. 'He needs kitting out as a warrior-hero. The full works.'

'I see,' said Smink, looking Joe up and down and giving a sniff. 'That would be *battledress*, sir, which is right this way.'

As they all trooped across the floor to the far end of the room, pushing their way through the hanging mounds of garments, Joe felt the whole building swaying gently backwards and forwards. A pair of metal gauntlets fell to the ground with an ominous clatter.

'Well, they'll do for a start,' said Randalf. 'Next, a cloak.'

'The full works, you say, sir,' said Smink. 'Might I recommend the leather Cloak of Impermeability. Fashioned by magic, it will deflect the blow of the mightiest sword.'

'Excellent,' said Randalf. 'Try it on, Joe.' Joe did so and inspected himself in a standing-mirror at the back of the shop. Randalf turned to Smink. 'And headgear?'

'We have everything,' said Smink with a wide sweep of his arm. 'Helmets – horned, winged, spiked or plumed; bonnets – war, jousting and . . . Ah, yes! A skull-faced shriek cap with matching ear pompoms – a bit of a speciality item, that one, sir.'

'I was thinking more of a standard sort of helmet,' said Randalf.

'Of course, of course,' said Smink rubbing his hands together. 'Silly of me. What about this Helmet of Heroism. Very popular, sir. Very hard-wearing feathers.'

He removed a heavy bronze helmet with five purple

plumes, and placed it on Joe's head. It fitted him like a glove.

'Sir will notice the tiny speakers concealed in the ear-rim. They play invigorating marching music as the warrior-hero strides boldly into battle.'

'Just the job,' said Randalf eagerly.

'And there are other matching accessories,' said Smink. 'The Shield of Chivalry, the Breastplate of . . . of . . . of Bravery and Perseverance, the Gaiters of Gallantry – to go with the gauntlets you have already so wisely chosen. And last, but not least, the Sword of Superiority,' he announced as he handed the final item to Joe.

'The Sword of Superiority,' said Randalf, sounding impressed. 'I'll take it. I'll take the whole lot,' he said.

'An excellent choice, if I might say so,' said Smink.

Joe looked at himself in the long mirror, and raised the sword high in the air. It looked quite convincing. He grinned. 'Not bad,' he thought. 'Not bad at all.'

'Sir looks fabulous, if I might be so bold,' said Smink. 'And not only that. These items all come with the Grubley Heroic Quest Success Guarantee – or your money back.' He returned his attention to Randalf and smiled an oily smile. 'Which brings us to the delicate matter of payment.'

'Ah, yes,' said Randalf. 'Payment.' He smiled back

confidently. 'Old Grubbers knows how good my credit is,' he said. 'He told me to stick it all on the slate.'

'He most certainly did not,' came a gruff voice behind him. '*Never ask for credit, 'cause a refusal sometimes offends,*' it continued, getting louder with every word. 'So what cheeky so-'n'-so is pretending I did?'

A stocky, bandy-legged individual with hairy ears and one thick, dark eyebrow, burst out of a rack of petticoats and pantaloons. 'You!' he said, fixing Randalf with a penetrating glare. 'I might have known!'

'Grubbers!' Randalf exclaimed, and went to shake hands.

'It's Mister Grubley to you, *Randy*,' he said, striding past the wizard's outstretched hand and tapping Joe on the shoulder. 'You can get that lot off for a start.'

Reluctantly, Joe removed the armour. Grubley turned to Randalf. 'Let me see. You still owe me for the last one. Quentin the Golden, wasn't it? One sparkly cloak, a pair of patent-leather bootees and the golden swan helmet with pink fur lining. It's cash or nothing from now on! Let's see what you've got.'

Hiding his irritation, Randalf reached into his cape pocket and pulled out a small leather pouch. He opened the top and poured a collection of coins into his palm. 'Eight muckles, five groats and a silver pipsqueak,' he said.

Grubley seized the tiny silver coin and bit into it. 'Hmm,' he said. 'Seems like we *can* do business after all.'

'The bargain-basement's upstairs,' he said, pointing to a rope ladder which led up to a hole in the ceiling.

Perched precariously on the roof of the discount store, the so-called basement was like a huge box on stilts. It was cold and draughty, and swayed alarmingly as the wind howled round its flimsy walls. Like the store they had just left, however, it was crammed full of items of clothing.

'That's the trouble with you wizards,' Grubley was saying. 'You come in here demanding my best stuff for your so-called warrior-heroes and then they go out and ruin it all by getting squashed by an ogre. I mean, this is quality merchandise.'

Veronica snorted dismissively.

'What does he mean, "squashed"?' said Joe nervously.

'Goblins will have their little jokes,' whispered Randalf.

'Here.' Grubley pulled out a cloak made of sacking and decorated with a fake-fur trim, and handed it to Joe.

'Love the fur,' said Veronica sarcastically. 'It's really fetching.'

'And is it protected perhaps by the power of imperme-ability?' asked Randalf. 'Or the hex of deflection?'

'Not exactly,' said Grubley. 'Though I daresay it would make a wound sting a bit less. Probably.'

'We'll take it,' said Randalf.

Other items followed. The Woolly Gloves of Determination. The Cardigan of Optimism. The Wellies of Power. Joe tried them all on. It was when they came to the helmet that he finally spoke up.

'I'm not wearing *that*,' he said.

'Not wear the War-bonnet of . . . umm . . . Sarcasm?' said Randalf. He took Joe's arm. 'Notice how the helmet's round shape has been made to deflect blows from cudgels, truncheons, swords – and how here, at the front, an extra safety-feature has been attached. What's more, not only will the War-bonnet of Sarcasm protect your head, but it will also give the wearer the heroic ability to make rude comments about their opponents' dress sense and physical appearance.' Randalf took the helmet and placed it on Joe's head. 'Both in its workmanship and design,' he announced, 'it's a triumph!'

'It's a saucepan,' said Joe flatly.

'How astute you are,' Randalf said brightly. 'Indeed the helmet can also double as a cooking utensil – for those long expeditions, far away from home . . .'

'But . . .' Joe protested.

'Trust me, Joe,' Randalf interrupted. 'I know you are valiant and brave, but I could not in all conscience allow any warrior-hero to set forth on a quest without an enchanted helmet of such unique power.' He turned to Grubley. 'We'll take it,' he said. 'We'll take the lot.'

Grubley nodded, and began totting up the cost of the armour under his breath.

Randalf glanced through the window at the sun. 'It's getting late,' he said. 'We'll have to get our skates on.'

'Skates?' said Veronica. 'Don't you think the lad looks ridiculous enough as it is?'

'Shut up, Veronica,' said Randalf. He turned to Grubley. 'So, what do I owe you?' he said.

Grubley looked up. 'Eight muckles, five groats and a silver pipsqueak, exactly,' he said.

Reunited at the bottom of the rickety ladder once more, Randalf, Veronica, Norbert, Henry and Joe made their way back down through the succession of clothiers and outfitters to the bottom of the tall, creaking building.

On every new level he came to, Joe found himself glancing into the mirrors – and each time he groaned. Dressed in the sacking cloak, the woolly gloves, the saucepan and armed with a dustbin-lid and a toasting fork, he looked a complete idiot.

As they emerged from the front entrance on to the street, several goblins stopped, turned, pointed and sniggered. Joe tugged at Randalf's sleeve.

'You've turned me into a laughing-stock!' he whispered in embarrassment.

'You can say that again,' Veronica laughed.

'Nonsense,' said Randalf. 'You look magnificent, Joe.

Doesn't he, Norbert?'

Norbert nodded. 'I love the fur trim,' he said. 'And those silver buckles on his wellies are so sparkly!'

'He looks even worse than Quentin the Cake-Decorator,' said Veronica. 'And that's something I never thought I'd find myself saying. I mean, a toasting fork!'

Randalf looked up angrily. 'Toasting fork?' he exclaimed. 'Why you foolish bird. The Trident of Trickery is one of the finest weapons a warrior-hero could possess.'

'Oh, yeah!' said Veronica.

'It is!' Randalf insisted. 'How else could he launch a three-pronged attack on his enemy, eh? You tell me that.'

Veronica rolled her eyes.

'Trust me, Joe,' said Randalf. 'You are the best warrior-hero I have ever summoned to Muddle Earth.'

'Which isn't saying much,' Veronica commented.

Norbert wiped a tear from his eye as he remembered the *other* warrior-hero Randalf had summoned.

'I have absolute confidence in you,' Randalf continued. 'And so will the Horned Baron, I'm sure.'

'But what does a warrior-hero actually have to *do* in Muddle Earth?' Joe asked.

For once, not even Veronica had an answer.

By the time they arrived back at the gates of Goblintown, it was already early evening. The cross-eyed guard stood up from his stool.

'Did you find the . . .' He noticed Joe and snorted with amusement. 'Ah, yes,' he said. 'I see you did.'

Back on board the ogre after a fitful night's sleep beneath the stars, Randalf instructed Norbert which way to go to get to the Horned Baron's castle. Turning right out of Goblintown, they followed a road which crossed a broad featureless plain towards the foothills of a vast mountain range. As they did so, the conversation turned to Joe's name.

'I mean, Joe isn't a *bad* name,' Randalf was saying doubtfully, 'but perhaps we should go for something a little more forceful, noble, impressive. In short, a more warrior-like name.'

'How about Joe the Terrible Toaster?' Veronica proposed. 'Or Josephine the Awfully Annoyed? Or what about Jo-Jo the Extremely Sarcastic . . . ?'

'Shut up, Veronica!' Randalf shouted. He turned to Joe. 'What about Joe the Barbarian? Short and to the point – and with just a hint of mystery. After all, with your Helmet of Sarcasm, Shield of Slight Protection and Trident of Trickery, you're a force to be reckoned with and no mistake.

There isn't a dragon, ogre or whifflepook that wouldn't think very seriously before picking a fight.'

Joe stopped in his tracks. 'Dragon? Ogre?' he said. 'Are you seriously suggesting I should do battle with dragons and ogres?'

'And whifflepooks,' Norbert reminded him. 'Vicious, scaly little beasts, they are,' he said, looking round nervously.

'You're a warrior-hero, a barbarian, Joe,' said Randalf cheerfully. 'Savage blood courses through your veins. A stranger from a strange land, eager for adventure and utterly fearless!'

'Yes, well . . .' Joe began, but a disgusting smell assailed his nostrils before he could finish. '*Pfwooar!* What is that horrible pong?'

Norbert pinched his nose closed. 'De Busty Bountains,' he said.

'And mustier than ever,' said Veronica. 'Perhaps you should also have invested in a Clothes Peg of Destiny,' she said to Randalf. 'Our warrior-hero is looking a bit green.'

It was true. Joe was feeling decidedly ill. Although not as pungent as Goblintown, the warm, stale odour drifting in on the breeze from the Musty Mountains was quite disgusting – a stomach-churning mixture of mildew, mouse droppings and mothballs. As they rounded the corner, the mountains loomed up ahead of them. They were tall, jagged, forbidding and very, very musty.

'They stink!' Joe gasped.

'The Musty Mountains are extremely old, Joe,' said Randalf sharply. 'I daresay one day *you'll* be old and smelly yourself.'

'But—' Joe began.

'And how will *you* like it when everyone who passes you by tells you that you stink? Eh?'

Joe shook his head. Sometimes in Muddle Earth, there was simply no point arguing.

'Anyway, Joe,' Norbert added. 'Don't worry. You'll soon get used to it.'

On and on they went. Mile after mile. With Randalf and Veronica on one shoulder, Joe on the other and Henry trotting along by his side, Norbert strode unwaveringly onwards.

After several hours of going in a straight line, the road was becoming increasingly full of bends. It snaked its way through the valleys between the tall, jagged mountains. To Joe's surprise, Norbert was right; he did get used to the unpleasant smell. Or rather, what with everything else about him, he forgot all about it.

There were hairy squeak-moths that fluttered off as they approached, squeaking loudly. There were flightless scruff-birds which flapped about on the ground, chattering indignantly. There were odd thuds and spooky whooshing noises as rocks fell and the wind blew, and far in the distance Joe thought he heard an odd tinkly-tinkly sound.

When they came to a fork in the road, Norbert

stopped. 'Which way, sir?' he said.

Randalf, who had dropped off for a few minutes, opened his eyes and looked round.

'The Horned Baron's castle, sir,' said Norbert. 'Should I turn right or left?'

'Left,' said Randalf.

'Where does the other road lead to?' asked Joe.

'Nowhere,' said Randalf. He tapped the ogre softly on the head with his staff. 'Proceed, Norbert,' he said.

Further on, as they rounded a sharp corner, Joe stared in awe at a chimney-stack-shaped mountain that loomed into view. Tall and imposing, it towered high above all the other mountains. From the crater at its top came little wisps of smoke which coiled up into the musty air and disappeared.

'What's that?' he asked.

The mountain let out a gentle, rumbling *boom* and a wispy smoke ring. Randalf, who had dozed off once more, grunted in his sleep.

'It's called Mount Boom,' said Veronica.

''Cause that's what it does,' added Norbert helpfully. 'It goes boom!'

Boom.

The feeble noise sounded again. It was accompanied by a second smoke ring. The musty smell grew slightly mustier. This time, Randalf's eyes opened.

'Did I hear something?' he said.

'It was Mount Boom, sir,' said Norbert. 'Booming.'

'Excellent,' said Randalf. He rubbed his eyes. 'Then we are nearly there. The Horned Baron's castle should be just around the next corner. Charming location . . .'

'Apart from the smell,' added Veronica.

The road curved round to the left. Randalf pointed down through the musty, dusty air to a series of tall spires and turrets which peeked up above the line of jagged mountain-tops. 'There it is,' he announced. 'The castle of the Horned Baron.'

Norbert shuddered. 'The place gives me the creeps,' he said. 'It always has.'

'And as for the Horned Baron himself,' added Veronica. 'He can be so . . .'

Just then, the budgie's voice was drowned out by a loud rumbling sound which came from Elfwood. Joe turned to look. The tinkly-tinkly sound grew louder. The next moment, something curious emerged from the distant trees in a flurry of thick dust and sped towards them.

'Stampede!' Veronica squawked. 'Cattle stampede!'

At least that was what Joe *thought* she said. He squinted into the distance. 'Cattle?' he said. 'Are you sure?'

'Not *cattle*, cloth-ears,' said Veronica. '*Cutlery!*'

'Take cover!' Randalf cried as the stampede drew close.

Joe ducked down behind a rock and watched in amazement as the great herd of kitchen utensils pounded past. Knives, forks and spoons, ladles and tongs, scissors

and skewers, peelers, sharpeners, crushers and mashers – all hurtling on in a wild and frenetic dash towards the Musty Mountains.

'There's definitely something afoot in Elfwood,' said Veronica darkly. 'Some strange, potent magic is brewing, you mark my words. And you know who's at the bottom of it. Doctor Cuddles . . .'

'Shut *up*, Veronica!' Randalf shouted crossly. 'How many more times must I tell you? That name is not to be mentioned in my presence under any circumstances.'

Joe frowned and turned to Veronica. 'But you said—'

'Shut up, Joe!' Veronica hissed. She turned back to Randalf. 'Don't you think we ought to get going?' she asked him.

'Quite so,' said Randalf. He cautiously stood up and looked round. The cutlery had gone. 'Come on, then,' he said. 'It's safe to proceed.'

Inside the Horned Baron's castle at last, Randalf (with Veronica back on his head) led Norbert, Joe and Henry up a wide and winding staircase to the first floor. Randalf stopped at an imposing oak door.

'Wait here,' he instructed Joe, his voice echoing round the high vaulted ceiling. 'I'll introduce you to the baron when the moment is right. Timing is everything. Trust me, I'm a wizard.'

He turned and knocked.

'Enter!' came a loud, booming voice.

Randalf opened the door and went in. 'My Lord Horned Baron,' he said, and bowed low. Veronica hopped down on to his shoulder. Norbert curtsied. 'And how good it is to see his Lordship looking so well.'

A short, fussily dressed little man stopped his pacing and turned to Randalf. 'Oh, it's you again,' he said.

'Indeed it is,' said Randalf cheerfully.

The Horned Baron scowled. 'You're late!' he snapped. 'I sent for a wizard last Tuesday.'

'A thousand apologies for that, my Lord,' said Randalf, 'but you know how it is. One spell leads to another, and before you know it . . .'

'Indeed I do *not* know "how it is",' said the Horned Baron, his voice just a little too shrill and high-pitched. 'When I summon a wizard to the castle I expect him to drop

sound of excited barking and scurrying claws, and Henry dashed into the chamber, ears flapping, tongue lolling and lead dragging along behind him.

The Horned Baron's eyes nearly popped out of his head. 'What in Muddle Earth is that?' he exclaimed. He turned on the wizard. 'Barbarian is right, Randalf!' he said sarcastically. 'Don't they believe in haircuts where he comes from?'

Randalf smiled. 'His Lordship has misunderstood,' he said. 'This is not the warrior-hero of whom I spoke, but rather his faithful battle-hound.'

'Henry the Hairy,' Veronica murmured, and chuckled to herself.

'The real warrior-hero is about to enter,' said Randalf. 'Joe the Barbarian,' he announced again. His face creased up with irritation. 'JOE! Will you get in here. Now!'

Joe's pale face appeared nervously round the edge of the door. 'Did you call?' he said.

'He's not deaf, is he?' asked the Horned Baron. 'I don't give much for his chances if he is.'

'Of course he's not deaf,' said Randalf. 'That said, all the senses of this particular warrior-hero are so acute, so sensitive, so finely tuned that he could lose two or three and still remain completely invincible.'

Joe clattered into the chamber and stood next to the others. The Horned Baron glanced his way and sniffed dismissively. 'He doesn't look up to much,' he said. He nodded towards Henry. 'Are you quite sure *he's* not the warrior-hero after all?'

'Appearances can be deceptive,' said Randalf. 'Take yourself, for instance. *We* know you're a great and noble Horned Baron, yet look at you . . .'

'What . . . what . . . what are you suggesting?' blustered the affronted Baron. 'I'm sure I don't know what you mean!'

Joe smiled nervously. For all his grand title and stately surroundings, the Horned Baron was, it had to be said, rather short. He was also weasel-faced, skinny and with scrawny arms and legs. Even his name was a bit misleading. Joe had assumed that the Horned Baron would have huge curling horns sprouting from the top of his head. Instead, he had to admit he was a little disappointed to see that the baron's name came from the oversized horns on his helmet, which kept slipping down over his eyes.

In short, the Horned Baron was a big let-down.

'What I meant, your Lordship, is this,' Randalf was saying. 'Your greatness and nobility are all the more impressive for being so . . . *well concealed.*'

'*Humph!*' snorted the Horned Baron, unsure how to take the wizard's words. He returned his attention to Joe and sighed wearily. 'I suppose that if this is all that's on offer, he'll have to do.'

Randalf smiled again. 'You know what they say,' he said. 'A barbarian in the hand is worth two cake-decorators in anyone's money.' He rubbed his hands together. 'And talking of money, if we could perhaps move on to the delicate matter of my fee.'

'Your fee?' asked the Horned Baron, his eyebrows arching alarmingly.

'Yes, two golden big'uns, wasn't it?' Randalf ventured.

'Three silver pipsqueaks,' said the Horned Baron, jingling some coins in his pocket. 'And I'm being robbed blind!'

Randalf groaned. So little money to go such a long way. The trouble was, with his recent record, he was in no position to bargain. 'Three silver pipsqueaks it is,' he said and reached out for the money.

The Horned Baron abruptly withdrew his hand. 'But then, there was that incident with the exploding elf, wasn't there?' he said. 'I shall have to make a slight deduction. Oh, and then there's the matter of the melting silver goblet. And that terrible infestation of galloping green mould in the castle bathrooms ... very nasty ... In fact, after all the deductions, I calculate that *you* owe *me*.'

'But that's not fair ...' Randalf began.

'Life isn't fair,' came the reply. 'Yet, for all that, I am a generous Horned Baron,' he said. 'Here, take this brass muckle.'

'You've got to stand up to him,' said Veronica.

'Shut up, Veronica,' Randalf hissed as, swallowing both his words and his pride, he pocketed the single coin. Who knows, perhaps his luck was about to change. Maybe Joe the Barbarian – warrior-hero *extraordinaire* – might just surprise him. He wouldn't count on it, but he'd just have to

do what he always did in
these situations and make
the best of it.

'Go, then!' said the Horned
Baron addressing Joe directly.
'Journey south, seek out Engelbert
the Enormous and put a stop to his
destructive behaviour once and for
all. And when you have slain the
fiend, I want you to bring me
his head. There's a pouch of
silver pipsqueaks for you when
you do.'

'His head?' said Joe, horrified. '*Ugh!*'

'Just a figure of speech,' said Randalf hurriedly, as he
seized Joe by the arm and steered him towards the door.
'Any proof will do.' He turned back to the Horned Baron.
'Consider it done, my Lord.'

Just then, the relative calm was destroyed by a heart-
stopping screech. '*WALTER!*'

Paling slightly, the Horned Baron smiled weakly.

'*Walter!* Where are my new singing curtains? The ones
I showed you in that catalogue. You promised me them ages
ago. You're all talk, you are!' she shouted. 'I don't know, call
yourself a Horned Baron!'

'It's all in hand, my little songbird,' he called back. 'I'm
having them sent over. Made of the finest material, they are,

and with the voice of an angel guaranteed. They're costing me the earth.'

'WAL-ter!' Ingrid screeched indignantly.

'But worth every last brass muckle, of course,' he hastily added.

Before the Horned Baron had a chance to remember precisely where his last brass muckle had actually gone, Randalf bustled the others out of the door, down the stairs and across the courtyard. He paused on the steps of the castle by the outer gate.

'All things considered,' he said, 'I think that went very well.'

'If being humiliated and swindled counts as doing well, then you're right,' said Veronica drily.

'Shut up, Veronica!' said Randalf. He turned to Joe and clapped him on the shoulder. 'And so, Joe the Barbarian, the time is almost upon us,' he announced, trying to sound as cheerful as possible. 'We'll rest up here for the night. Then, as the sun rises over Muddle Earth tomorrow morning, so shall the quest begin!'

The following day dawned bright and early – unlike the day before, when it had been an hour late, and the previous Wednesday when it hadn't dawned until one-twenty in the afternoon. There were unconfirmed rumours flying about that Dr Cuddles was responsible. Whatever, on this particular morning the sun rose when it should – which was just as well, because if it hadn't, Randalf and the rest would have stood no chance of travelling to the Ogrehills in one day.

'Look lively,' Randalf said, several times, as he scuttled about getting everyone ready. Finally, they were off.

Swaying to and fro on Norbert's left shoulder, Joe was far from happy. Quite apart from the ogre-sickness which had returned with a vengeance the moment they'd set off, he was decidedly uneasy about the forthcoming quest.

'All this talk of ogres and squeezed sheep,' he was saying. 'I don't like the sound of it one little bit. I'm not a warrior-hero, I'm just an ordinary schoolboy – and I want to go home.'

Randalf leaned across from Norbert's right shoulder and patted Joe reassuringly. 'Don't worry,' he said. 'Everything's going to be just fine! Trust me, I'm a wizard.'

'That's easy for you to say,' said Joe. 'But look at the size of Norbert – and he's Norbert the Not-Very-Big! What on earth is this . . . this Engelbert the *Enormous* going to be like?'

'Enormous,' said Veronica. 'I think you'll find the clue's in his name.'

'Oh, I'm sure he is,' said Norbert. 'Absolutely massive! Twice as tall as me probably and three times as broad. In fact, Engelbert's bound to be even bigger than my grandfather Umberto the Unfeasibly Large – not to mention Uncle Malcolm Nine-Bellies . . .'

'Yes, yes, Norbert,' said Randalf. 'That's fascinating.' He turned back to Joe. 'The thing is, I don't expect you to actually fight Engelbert.'

'You don't?' said Joe.

'Of course not,' said Randalf. 'That would be ridiculous.'

'So what *do* you want me to do?' asked Joe.

'Psychology,' said Randalf, tapping the side of his head meaningfully. 'It's all about psychology.'

'It is?' said Joe.

'You see,' Randalf continued, stifling a yawn, 'as everyone knows, ogres are just great big softies. Isn't that right, Norbert?'

Joe clung on tightly as Norbert nodded in agreement.

'Really?' said Joe.

'Really,' Randalf confirmed sleepily. 'All you have to do is stride right up to him in your Wellies of Power, wave your three-pronged Trident of Trickery in his face, fix him with a stare from beneath your Helmet of Sarcasm and tell him to stop being so naughty, or else!'

'Or else, what?' said Joe.

'Or else you'll smack his bottom. Then you can make a really sarcastic remark about his appearance and he'll start blubbing like a baby. Simple.'

'If it's so simple,' said Joe, 'then why do you need me?'

'Psychology,' yawned Randalf. 'You're a warrior-hero. Ogres are terrified of warrior-heroes. They're giant-killers, dragon-slayers, troll-bashers – nobody stands a chance against a warrior-hero. It's a well-known fact!'

'But I'm not a warrior-hero,' said Joe. 'I keep telling you, I've never killed a giant or bashed a troll in my life. Honest!'

'In that costume, my lad,' said Randalf, shifting into a more comfortable position propped up against Norbert's right ear, 'Engelbert the whatever-his-name-is will take one look at you, burst into tears and promise to be a good ogre – especially if you say his eyes are too close together and he smells like a pink stinky hog. Trust me, I'm a . . .'

'Wizard?' said Joe. But Randalf was fast asleep. A soft, rasping snore fluttered through the air.

'That's him off again,' said Veronica. 'I've never known

anyone sleep as much. Still, at least it shuts him up.'

'Shut up, Veronica,' mumbled Randalf in his sleep.

'Typical!' said Veronica.

They were returning the way they had come, back along the winding track through the Musty Mountains. Morning drifted slowly past, and soon it was early afternoon.

Boom.

Far far behind them, Mount Boom exploded softly. The hairy squeak-moths huffed to and fro looking for lost socks to eat, the scruff-birds rolled around in the musty dust by the roadside, while a swarm of rotund, antlered beetles buzzed slowly past, in search of killer daisies to pollinate.

'They look like miniature Horned Barons,' Joe laughed.

Veronica nodded, crunched her beak and swallowed. 'Only they taste better,' she said.

Joe winced and turned away. As he did so, he caught sight of something glinting out of the corner of his eye.

'What was that?' he said, looking down.

'What was what, sir?' said Norbert.

'That,' said Joe, pointing down at Norbert's left foot. 'Careful, don't tread on it.'

Sure enough, down by Norbert's left foot, glinting in the half-light, a small silver teaspoon was doing what looked like a forlorn little dance in the dust. Round and round it was going, in ever-decreasing circles. Finally, the circles became so small that the teaspoon stopped in one place, did a last twinkling pirouette, and fell to the ground

with a soft tinkly sound.

Joe jumped down and picked it up gingerly. As he held it, the teaspoon gave a little sigh.

'Did you hear that?' said Joe, holding it up. 'I think it gave a little sigh.'

'It's lonely,' said Norbert. 'It must have got separated from the herd in that cutlery stampede.'

'An enchanted teaspoon,' said Veronica darkly.

'Can I keep it?' said Joe.

'Finders keepers,' said Veronica. 'It'll go well with your saucepan and toasting fork.' She ruffled up her neck feathers and raised her beak. 'You can call it the Teaspoon of Terror – and terrify ogres with it!'

'Why have we stopped?' came Randalf's voice. 'And did somebody mention tea?'

'I almost trod on a teaspoon, sir,' said Norbert. 'It could have been very nasty. My Auntie Bertha the Big-Footed was always treading on things. Or *in* things, more like. Why, that time she stepped in a great big dollop of—'

'Yes, yes,' said Randalf. 'Thank you, Norbert. Now, if you're quite ready, Joe, perhaps we should proceed. Our quest awaits us and we've still got a long way to go.'

Joe nodded and slipped the spoon into his back pocket. Norbert bent down, pulled him up on to his shoulder and set off once more. Randalf slumped forwards and drifted back to sleep.

It wasn't long before the road emerged from the highest peaks of the Musty Mountains and began winding its way through the foothills. These were as barren and musty as the mountains, and smelt strongly of old socks. To his right, Joe saw a tall, rounded hill he hadn't noticed at first. Unlike the other foothills, it was covered with thick grass and giant yellow and white daisies, their sweet scent perfuming the air. Stiltmice scampered through the verdant undergrowth; butterflies fluttered overhead. Joe breathed in the beautiful fresh air. He tapped Norbert on the shoulder.

'Not so fast, Norbert. Let's enjoy the view,' he said. 'What a lovely place! What's it called?'

Norbert shuddered anxiously, causing Joe almost to lose his balance.

'Careful!' he exclaimed. 'I nearly fell off just then! Oh, look, Norbert. Over there! How cute!'

A particularly appealing stiltmouse with big blue eyes was stepping daintily through the gently swaying grass, the aromatic breeze ruffling its white fur.

'That's Harmless Hill,' said Norbert.

'Harmless Hill!' Randalf woke up with a start. 'Norbert, my good fellow,' said the wizard. 'Why is it that whenever I wake up I find you standing still, gawping?'

'It's Joe, sir,' said Norbert apologetically. 'He wanted to enjoy the view.'

Just then, the stiltmouse gave a little cry as a daisy opened its gaping jaws and swallowed it whole. It gave an ugly belch.

'I thought it was called *Harmless* Hill!' said Joe, horrified.

'Oh, the hill's harmless enough,' said Veronica. 'It's the killer daisies you have to watch out for.'

Joe shook his head. 'This place is crazy,' he muttered. He looked down to check that Henry was still nearby.

'Here, boy,' he called. 'Come up here with me, just in case there's a "perfectly safe" mountain up ahead, or a "don't worry, you'll be fine" meadow just round the corner.'

'I've never heard of those places,' said Norbert. 'But they sound terrifying!'

'Can we *please* get on,' said Randalf irritably. 'Wake me up when we get to Trollbridge.'

Norbert set off once again, this time at a brisk trot. Joe was getting used to ogre-riding by now, and with Henry safely in his lap, he relaxed. Slowly, his eyes grew heavy and his head began to nod.

The next thing he knew, Joe woke up with a mouth full of mud and Henry licking his face. He looked up. Randalf was on his feet, with Norbert fussing about him trying to brush down his robes.

'No, don't tell me,' Randalf was saying. 'Another pothole! You really should learn to look where you're going, Norbert.'

'It wasn't a pothole,' said Norbert tearfully.

'No,' came an angry little voice. 'It was a kettle hole actually – and my kettle is a complete write-off!'

An elf, who was standing in a small hole in the road, threw a flattened disc of metal, with what appeared to be a spout, down to the ground and marched off in a huff. Joe got to his feet, looked round and gasped. There in front of them was Trollbridge.

Built upon four great arches which crossed the river, the bridge was a solid, yet ornate, stone structure complete with tall pointy-domed turrets and magnificent gate-towers. Only when he looked more closely did Joe see just how neglected Trollbridge actually was.

'It's a bit grimy,' he said. 'And that tower looks as if it's about to collapse,' he added, pointing upwards.

Randalf shrugged. 'Trollbridge has a certain "lived-in" charm,' he said. 'Trolls have many fine qualities, but neatness is not one of them. Now stubbornness is another matter. Trolls can be very stubborn, as you'd know if you'd ever tried to get a troll to tidy its bedroom . . .'

'And they never throw anything away,' said Veronica, flapping her wing at the pile of junk at the foot of the gate-towers. 'I mean, look at all that mess!'

'Good day,' came a gruff, yet cheery, voice from the centre of the pile. 'Can I help you?'

There were bicycle wheels, taps, lengths of wood, screws and nails, nuts and bolts, coils of wire, a mangle, a bird cage, a washing-up bowl . . . and there, perched on a three-legged stool, a squat, bow-legged individual with tufty hair and rather ferocious-looking teeth sticking up from a protruding jaw.

'You certainly can,' said Randalf, stepping forwards. 'We wish to cross your magnificent bridge.'

'That'll be one mangel-wurzel,' said the troll.

'A mangel-wurzel?' said Randalf, and made a great show of searching through his robes. 'I'm afraid I'm fresh out of mangel-wurzels.'

'A turnip, then,' said the troll. 'You must have a turnip.'

'Not as such,' said Randalf.

'Then a carrot,' said the troll. 'Any root vegetable will do.'

'Sorry,' said Randalf.

'A potato?' he suggested. 'It doesn't matter if it's a bit mouldy.'

''Fraid not,' said Randalf.

The troll sighed. 'An onion? A courgette? A baby sweet-corn? All right then, I'll settle for a small dried pea.'

Randalf shook his head sadly.

'What sort of travellers are you?' said the troll. 'I mean, you haven't got a mangetout between you. I don't know,' he tutted. 'So what *have* you got?'

Randalf turned to Joe. 'I find myself in a somewhat embarrassing situation,' he said. 'I mean I've got my warrior-hero-summoning spell and a belligerent budgie, but that's about it . . .'

'You're *not* trading me!' squawked Veronica indignantly.

'Who'd have you?' snapped Randalf. 'Now, Joe. I don't suppose you . . .'

Joe rummaged round in the pockets of his jeans and pulled out an old bus ticket and a lolly stick. He held them out for Randalf's inspection.

'A chariot voucher from a far-off land,' said Randalf holding up the bus ticket.

The troll looked thoughtful. 'It's tempting,' he said. 'A far-off land, you say? Trouble is, I don't get out that much . . .'

'You surprise me,' said Veronica.

'Shut up, Veronica,' said Randalf, and he turned back to the troll. 'No? All right. Then how about this? A miniature paddle?'

'Miniature paddle,' said the troll, eyeing the lolly stick. 'Very nice, very nice. Lovely workmanship, but a bit – how can I put this? A bit on the small side.'

Randalf turned to Joe questioningly.

Joe pulled his front pockets inside out. 'Empty,' he whispered. 'I haven't got anything else . . .' As he spoke, he tried his back pockets, and there, nestling into the corner and stifling a sob, was the small silver teaspoon.

'There's this,' he said uncertainly as he held it up. 'The Teaspoon of . . . of . . . Terror!'

'The Teaspoon of Terror!' said Randalf, taking the spoon with a flourish. The spoon let out a timid little squeak.

Randalf ignored it. 'Forged by elves. Imbued with magic . . .'

'The Teaspoon of Terror,' said the troll, impressed, as Randalf laid it in his large, grubby hand. The teaspoon gave a sigh. 'Of course, a mangel-wurzel is the generally accepted fee, or a turnip. But you seem like an honest fellow . . . Oh, go on, then. A Teaspoon of Terror it is! Welcome to Trollbridge.'

'At last,' Randalf muttered. 'And thank you, once again,' he said as the troll opened the gate and waved them

through. 'It's been a pleasure doing business with you.'

'As long as you remember, next time it's a mangel-wurzel or nothing,' said the troll as he strolled back to his stool.

Joe put Henry on his lead and followed Randalf and the others. It was market day in Trollbridge and the place was buzzing with feverish activity, for though the trolls lived under the bridge, their noisy bargaining, bartering and haggling was conducted up on top. A series of trapdoors sprang open and slammed shut constantly as the trolls hurried between the two. The balustrades on both sides of the bridge were lined with covered stalls and trestle tables, all laden with complete and utter junk, and run by enthusiastic trolls who were politely shouting out what wares they were selling.

'Old string! Do come and get your old string here! All lengths available.'

'I've the finest odd socks in Trollbridge if you'd care to take a look! Specially unwashed and aged by squeak-moths!'

'Mangel-wurzels! Please come and inspect my lovely mangel-wurzels!'

Boxes and sacks stood everywhere, each one filled with junk of all shapes and sizes, and the odd root vegetable.

'The trolls are renowned throughout Muddle Earth for their root vegetables,' Randalf explained. 'Especially their mangel-wurzels. They're passionate about their mangel-wurzels.'

'You don't say,' said Joe, as they passed by a rickety stall weighed down with a great pyramid of the things.

Slowly, they made their way across the great Trollbridge, the sound of the haggling and bartering ringing in their ears.

'I'll give you a jam jar of toe-nail clippings and a broken bucket.'

'Throw in the dried-pea rusk, and it's a deal.'

'Who will buy my sweet red gobstoppers? Sucked three times and dropped on the carpet . . .'

'Bottle tops! Bottle tops!'

'Bottle bottoms! Bottle bottoms!'

Joe stood, completely bewildered. There was so much to take in.

'Keep up, everybody!' Randalf shouted back impatiently from the end of the bridge. 'Norbert, put down that turnip. This is no time for eating. Come *on*, Joe. We haven't got all day.'

'Sorry,' Joe called back and, tugging Henry away from a display of amusingly shaped carrots, hurried to catch up.

'What kept you so long?' said Randalf.

'I was just interested,' said Joe. 'And everyone here seems so friendly and polite.' He frowned. 'But what do they do with all this rubbish?'

'The ways of the trolls,' Randalf said, 'are as mysterious as they are bizarre . . .'

'In short,' Veronica butted in, 'he doesn't know. But you should see the state of their bedrooms!'

'Keep your voice down,' said Randalf, as they passed a second toll keeper at this end of the bridge. 'Trolls' feelings are easily hurt.'

The troll was identical in every way to the toll keeper at the other end – apart from his voice, which was high and shrill.

'Missing you already,' he squeaked.

With Randalf and Joe back on Norbert's shoulders, Veronica on Randalf's head and Henry trotting along behind, they continued on their way. Joe looked round and sighed as Trollbridge disappeared behind him. To his left was a broad and barren plain; to his right, a swampy bog.

All at once, Veronica let out a cry. 'Ogrehills ahead!' she shouted, her wing shielding her eyes from the low sun.

'Excellent news!' said Randalf with a yawn. 'We'll be there before we know it.' He turned to Joe. 'And you, Joe the Barbarian, will be able to prove your warrior-hero prowess once and for all.'

'Fantastic,' Joe mumbled. 'I can hardly wait. In fact . . .' His face suddenly screwed up. '*Pfwoooar!*' he groaned. 'What is *that* horrible pong?'

'Sorry,' said Norbert. 'It must have been something I ate in Trollbridge.'

'Not you,' said Joe. 'I'm talking about that sweet, sickly smell.' It was like a pungent mixture of his dad's aftershave, his mum's aromatherapy oils, and his sister's cheap perfume, all mixed up with rotting vegetation. He held his nose. 'It's worse than the Musty Mountains!'

Veronica flapped her wings in front of her face. Henry whined miserably and rubbed his nose in the dirt.

'It's the Perfumed Bog,' said Randalf. 'I quite like the smell myself.'

'You would,' said Veronica.

'It reminds me of my beloved Morwenna,' he said dreamily. 'Morwenna the Fair, they called her . . .'

'Not behind her back, they didn't,' Veronica muttered. 'At least, not when she grew that beard.'

'That was an accident,' said Randalf defensively. 'I was practising. Morwenna understood, even if her father didn't.'

'Morwenna! Morwenna! Let down your golden beard!' sniggered Veronica.

'SHUT UP, VERONICA!' shouted Randalf, very red in the face.

The further they went, the closer the road came to the Perfumed Bog, until the two were running along side by side, with only a thin line of stones separating them. Joe looked out across the swamp. Shrouded in pink mist, it was vast and flat, with giant lily pads and grassy tussocks, and dark pools of glistening purple mud which plopped and hissed as the perfumed gases bubbled up from below.

'Come here, Henry,' he called, waving the lead, as the dog stopped by the side of the road. Head down and tail up, he began sniffing eagerly all round the ground. 'Henry!' Joe shouted. 'Henry come here!'

At that moment, Joe spotted what Henry had smelled. It was a small, pink warthog-like creature with floppy ears and a curly tail, standing motionless on top of a tussock surrounded by lily-pads and bubbling mud. It was staring, unblinkingly, at the dog.

'HENRY!' Joe yelled.

With a loud yelp of excitement, Henry suddenly bounded forwards and took a flying leap at the tussock. 'Oh, no! He thinks it's a squirrel,' shouted Joe, jumping off Norbert's shoulder and chasing after him.

'Come back!' shouted Randalf. 'Never chase a pink stinky hog!'

'Why not?' asked Norbert.

'In case you catch it, stupid!' said Veronica.

'Henry!' shouted Joe as, leaping from tussock to tussock, he tried to catch up. The more puffed out he became, the more difficult it was to keep his balance. 'Henry, come . . . *Aaargh!*'

SPLASH!

Joe fell face down in the purple mud. Ahead, Henry caught up with the pink stinky hog and lunged for its tail. With a high-pitched squeal, the hog raised its rump and broke wind with astonishing force. The sound – like a small cannon exploding – echoed round the Perfumed Bog. And the smell . . . !

'Someone's caught a pink stinky hog,' said Randalf.

In the Perfumed Bog, Joe picked himself up. The

smell was eye-wateringly horrendous. Henry seemed to be in shock. The pink stinky hog stood on its little tussock triumphantly, tail in the air. From behind it, a small hill rose up from the purple mud.

It had two piggy pink eyes that regarded Henry and Joe unblinkingly. It had a piggy pink snout that wrinkled as it sniffed the air and let out deep piggy snorts of anger. Its two huge tusks glinted in the pink light of the Perfumed Bog.

'Henry,' said Joe softly. 'Time to go, Henry.'

With a whimper, Henry backed away.

The huge pink stinky hog climbed on to the tussock beside the pink stinky hog piglet, slowly turned its back on Joe and Henry and raised its great rump high in the air.

'Henry!' shouted Joe. 'Run for it!'

A sound like a huge cannon exploding echoed round the Perfumed Bog. *And the smell . . . !*

'Someone's caught a pink stinky hog's mother,' said Randalf.

Just then, Henry shot out of the swirling pink mist and

on to the path. Moments later, Joe scrambled after him.

'All aboard,' shouted Norbert, picking them both up.

'To the Ogrehills, Norbert!' shouted Randalf. 'And remember, until we get there, breathe through your mouths!'

The sun was setting as they approached the Ogrehills. Randalf's snores and the gentle thud of Norbert's footfalls were the only sounds in the still air – at least they were until they were joined by another sound, equally tuneless.

'*La, la, la . . .*'

The sound floated across the evening sky. Joe looked up, and there, marching along the road towards them from Elfwood was a stooped, shadowy figure in a cloak and raised hood. Under his arm was a roll of what looked like cloth or carpet.

The tuneless song became louder.

'*La, la, la, la-la.*'

Joe shivered, and the hairs at the back of his neck stood on end.

As the lone figure came closer he lowered his hood, and Joe was surprised to see a familiar face.

'Grubbers!' Randalf exclaimed. 'How good to see you again.'

Grubley looked up. '*Randy,*' he said. 'Out and about, eh? And I see you've got your warrior-hero with you, all kitted out for battle.'

'*La, la, la . . .*'

'You bet,' said Randalf. 'We're on important business for the Horned Baron, aren't we Joe?'

But Joe did not reply. He was listening to the singing.

'*La, la, la . . .*'

Joe frowned. The noise seemed to be coming from the roll of material under Grubley's arm.

'Yes, we're heading for the Ogrehills,' Randalf was saying.

'Rather you than me,' said Grubley, pulling a face.

'Oh, I have absolute faith that your warrior outfit will

do its job,' said Randalf.

'You get what you pay for,' said Grubley.

'Precisely,' said Randalf. 'But I am being rude. Norbert,' he said, 'help me down.'

'Oh, don't get down on my account,' said Grubley. 'I'm already running late as it is.'

'*La, la, la. La, la, la.*' The tuneless song was louder than ever. Had no one else noticed? Joe wondered.

'Running late?' said Randalf, tutting sympathetically. 'But what are you doing so far from Goblintown in the first place?'

'*La, la, la.*'

'I'm on important business for the Horned Baron, too,' said Grubley, raising the frayed, faded and somewhat grubby roll of musical cloth. 'Oh, I've been given the runaround, I can tell you,' he said crossly. 'Been looking all over, I have.' He sighed. 'It's all down to that Horned Baron's wife . . .'

'Ingrid?' said Randalf.

'The Horned Baron only had one wife the last time I looked,' said Grubley. 'She wants a set of singing curtains. Never heard of them myself, but she swears blind she saw them in my catalogue. And what Ingrid wants, Ingrid gets!'

Randalf nodded knowingly.

'So, I've been traipsing around,' he said. 'Here, there and everywhere. Luckily, I've got my contacts,' he added, and tapped his nose. 'I managed to lay my hands on this. Enchanted material. Very rare, I can tell you. I'm heading back to Goblintown to have it made into singing curtains.'

'*La, la, la . . .*'

'Call that singing?' said Veronica. 'More like mournful mooing . . .'

'Shut up, Veronica,' said Randalf. 'I'm sure Ingrid will love them.'

'I hope so,' muttered Grubley, as he turned and hurried off towards Goblintown. 'I do hope so.'

The landscape grew stony and barren as they neared the Ogrehills. Scrubby bushes gave way to tufts of grass, prickleweeds and fat-leafed succulents which Norbert seemed to find irresistible.

'Yum, there's another one,' he said, and abruptly stooped down, broke off a leaf and pushed it into his mouth. Cries of alarm and distress came from his shoulders as Randalf clung on desperately, Veronica flapped about and Joe tried his best to keep Henry from slipping off his lap. 'Be-lish-ush!' he muttered slurpily.

'Norbert, will you *stop* doing that!' said Randalf sharply. 'You nearly sent us *all* flying that time!'

'Sorry, sir,' said Norbert. 'It's just, I haven't had squish-weeds for ages. I'd forgotten how much I like them.'

'Yes, well, I think you've had enough now,' said Randalf. 'I know you. You see something you like and you don't know when to stop. Don't make a pink stinky hog of your-self.'

'No, sir,' said Norbert. 'Sorry, sir.'

'Now get a move on, Norbert,' said Randalf. 'There's a good fellow.'

On they went. The rolling Ogrehills stretched off into the distance.

'That ogre must be around here somewhere. Keep your eyes peeled for squeezed sheep,' said Randalf.

Joe scanned the area all around him. 'What on earth does a squeezed sheep look like, anyway?' Joe wondered out loud.

'Exactly as you'd imagine,' said Veronica.

'Hmmph,' said Joe. 'Well I can't see *any* sheep, squeezed or otherwise. In fact, I can't see much of anything,' he added. 'Apart from rocks.'

'Keep looking,' Randalf said. 'You too, Veronica.'

'If you insist,' said Veronica. 'It's so horrible here. Dry, dusty, desolate. Why would anyone want to live in a place like this?'

'This is where *I* lived,' said Norbert, with a smile. 'I think it's kind of homely.'

'You're right, Norbert,' said Veronica. 'If your idea of home is a rock for a pillow and a sandpit for a bed.'

'A sandpit bed?' said Norbert. Pure luxury. I and my twenty brothers had to sleep on pebbles. And we'd have given anything for a rock pillow. Prickleweed, that's what we had. And if we wanted to go to the loo in the middle of the night, we had to—'

'Yes, yes, Norbert,' said Randalf. 'That'll do. Just keep looking for a sheep.'

'There's one!' Joe shouted excitedly, and pointed ahead. 'Over there! Look!'

'Are you sure it's not a rock?' said Randalf.

'It's moving,' said Joe.

'So it is,' said Randalf. 'Proceed, Norbert.'

As they approached, it was clear that it was indeed a sheep, and not a happy one at that. It was wandering around in circles, looking dazed and bewildered. Its wool was bunched up around its shoulders and hindquarters, and squeezed flat in the middle. It looked like a walking, woolly dumb-bell. When it noticed the ogre stomping resolutely towards it, it let out an odd squeaky bleat, turned on its heels and disappeared over the ridge in a cloud of dust.

'That sheep has definitely been squeezed!' said Randalf. 'After it!'

Norbert tried his best, but the terrified sheep had given them the slip. It wasn't long, however, before Veronica spotted two more.

'Over there!' she said, flapping her wing at the pair of cowering, shivering sheep with wild eyes and dumb-bell wool. 'Freshly squeezed sheep!'

'Good work, Veronica,' said Randalf. 'And if we follow their trail, it'll lead straight to the culprit – none other than

Engelbert the Enormous, I'll be bound!'

Joe swallowed hard. 'I'm feeling a bit nervous,' he admitted quietly.

'Joe, Joe, Joe,' said Randalf, as if talking to a very young child. 'We aren't nervous, are we? Of course we remember we've got our Trident of Trickery. Ooh, scary trident! We've got our Helmet of Sarcasm. Nasty, nasty helmet! All we've got to do is show this Engelbert character who's boss. Who's the boss, Joe? Who's the boss?'

'I'm the boss,' said Joe uncertainly. 'I'm the boss.'

Norbert trudged on, following the sheep's trail. Up and down the undulating rockscape, he went; deeper and deeper into the Ogrehills. Occasionally, they passed the mouths of caves, from which came the sleepy sounds of snoozing ogres.

'Mummy, mummy,' growled some.

'My snuggly-wuggly,' murmured others in deep, gruff voices.

And the air was filled with the slurps of thumbs being sucked and mighty rumbling snores.

'D . . . do you think we're getting close?' Joe asked in a trembling voice.

Randalf nodded. 'Judging by that unfortunate mess over there,' he said. 'That is where our sheep were squeezed. Quiet, everyone!'

The wind abruptly dropped and the air became oddly still. Norbert put Randalf and Joe down, and Joe clipped Henry on the lead. Veronica sat on Randalf's head, feathers ruffled. They all listened intently.

'It's quiet,' she said in a hushed voice. 'Too quiet. I don't like it.'

'Shut up, Veronica,' whispered Randalf, who was thinking exactly the same thing.

'What exactly are we listening for?' whispered Joe.

Just then, an anguished bleat echoed loudly round the hills, and a sheep – with bulging eyes and squeezed wool – came hurtling over the ridge and dashed away across the stony ground.

'What the . . . ?' Randalf began.

'NO!' boomed a loud and angry voice. 'IT'S NOT THE SAME! IT'S NOT THE SAME AT ALL!'

Joe stared in horror in the direction the voice had come from, and as he stared, he heard banging and crashing and a series of loud thuds. A thick cloud of dust rose up.

'OH, WHERE IS IT?' the great voice demanded.

'WHERE, OH, WHERE HAS IT GONE?'

'D . . . do you think *th . . . that's* Engelbert?' whispered Joe.

'Unless I'm very much mistaken,' Randalf whispered back.

'He *sounds* enormous.'

'Oh, I'm sure he won't be *that* enormous,' said Randalf reassuringly. 'Come on, now, Joe the Barbarian. Raise high your Trident of Trickery, adjust your Helmet of Sarcasm and let your Wellies of Power lead you to victory.'

'SOMEONE'S DEFINITELY *STOLEN* IT!' the voice raged. 'AND WHEN I FIND OUT WHO IT IS, I'LL . . . I'LL . . .'

'The time has come, Joe the Barbarian,' said Randalf, pushing Joe forwards. 'Go forth and confront him. You can do it!'

With his trident in one hand and his dog-lead in the other, Joe walked ahead on rubbery legs. Henry cowered by his side.

All at once a truly massive ogre head – more than twice the size of Norbert's – appeared above a ridge. Joe froze. The head was swiftly followed by colossal shoulders, a barrel chest, a bulging gut, legs like tree trunks and feet like boats, until an entire monstrous ogre stood before him. Joe couldn't move.

The ogre roared furiously and advanced, picking up great boulders and looking underneath them, before tossing them aside.

'WHERE ARE YOU?' he bellowed, his face purple and

contorted with rage, his bloodshot eyes bulging in their sockets. 'WHERE *ARE* YOU?' The light glinted on the drool that dripped from his gnashing tusks. Then his three eyes fell on Joe. The ogre paused for what seemed to the reluctant warrior-hero an eternity.

'Oh, my goodness,' Randalf gasped. 'He does seem rather upset doesn't he?'

'I do hope Joe will be all right,' said Norbert anxiously.

Joe raised his trident bravely. 'It's all a matter of psychology,' he reminded himself. He met the ogre's fearsome gaze. 'I . . . I'm a warrior-hero, from afar,' he said. 'Joe the Barbarian! And . . . *um* . . . unless you stop all this nonsense right now I'll have to smack your bottom!'

The ogre blinked.

Joe turned to the others. 'Shall I use the Helmet of Sarcasm?' he hissed. He adjusted his helmet. 'And by the way, I'm sure no one has *ever* told you that your face looks just like the rear end of a pink stinky hog . . .'

The ogre threw back his head and the Ogrehills trembled with his mighty roar.

'What do we do now?' squeaked Joe.

'There's only one thing we can do!' said Randalf. '*RUN!*'

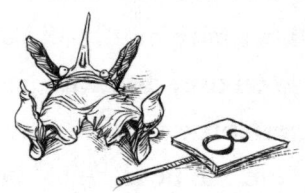

With their hearts in their mouths and dust in their hair, they made a desperate dash for it. Their panicked cries broke the silence. Randalf ran blindly on till he could run no more. Stopping abruptly, he bent double and gulped for air.

'That was close,' said Veronica, landing daintily on the wizard's rump.

'You can say that again,' said Norbert, stomping up behind them.

'That was clo—'

'Shut up, Veronica!' Randalf panted impatiently. He straightened up and shook his head. 'Most unusual,' he said. 'I've never known an ogre that angry before. He should have been terrified of our warrior-hero here . . .'

'Where?' said Veronica.

'Here,' said Randalf. 'Joe . . . Oh, no. Where is he?'

'Joe?' cried Norbert. 'Joe, where are you? Joe! Joe!'

'For crying out loud, Randalf!' said Veronica irritably. 'First sign of trouble and you turn tail and leave him to it.'

'But he was right behind me,' said Randalf, looking all around him. 'And I distinctly gave the order to run. It's not my fault he didn't hear me. That Helmet of Sarcasm must have slipped down over his ears . . .'

'He's gone!' Norbert wailed. 'And so is Henry!'

'Typical!' said Veronica. 'It's Quentin all over again.'

'We've got to go back for them,' Norbert sobbed tearfully.

'Now, let's not be hasty,' said Randalf nervously. 'You saw the mood that ogre was in. Perhaps we should allow the dust to settle a bit first, then, in a week or so, we can . . .'

'*You* brought the lad here to the Ogrehills,' interrupted Veronica accusingly. 'You can't abandon him now. You'd never be able to live with yourself.'

Randalf examined his fingernails closely. 'Of course, I feel bad. Don't get me wrong, Veronica. We *all* do! But be realistic . . .'

'Poor Joe! Poor Henry!' Norbert wailed. 'And poor, dear Quentin. *Boo-hoo!*'

'"Trust me, I'm a wizard", that's what you said,' Veronica continued. 'And he *did* trust you. Joe the Barbarian trusted you. And now, how are you repaying that trust? Eh? By abandoning him.' She clacked her beak reproachfully. 'You're a disgrace! If we don't go back right now, then I'm leaving you!'

'Please sir, please,' said Norbert, sobbing even louder. 'If we could just go and check. Maybe there's a chance . . .'

Randalf sighed. 'All right, all right,' he said. 'You win! I'm just too soft-hearted, that's my trouble! A fool to myself sometimes. Come, let's get this over with. Follow me.' He turned and gathered up his robes. 'But keep close.'

Huddled together for safety, Randalf and Norbert retraced their footsteps, creeping back silently, with Veronica keeping a watchful look out from the top of Randalf's head.

'I think we're getting close,' she announced after a while and flapped her wing up ahead. 'Look at those giant foot-prints, and how the dust has all been stirred up.'

Randalf nodded. Norbert began whimpering.

'Sssssh!' hissed Randalf, placing a finger to his lips. 'We don't want to . . .'

'*OH, NO!*' wailed Norbert, and pointed at a bent piece of three-pronged metal. '*LOOK!*'

It was the Trident of Trickery, twisted out of shape and lying discarded in the dust. Randalf picked it up and shuddered.

'And there!' Veronica cried. She flew down and landed on an abandoned Welly of Power.

Norbert howled with grief. 'Oh, Joe,' he blubbed. He picked up the lone rubber boot and hugged it desperately. 'It must have come off,' he sobbed, 'when he . . . when he . . . he . . .' He straightened up and scanned the horizon for any sign of his latest warrior-hero friend. Apart from one set of massive footprints in the dust which marched up and over the ridge, there was nothing. 'JOE!' he cried. 'JOE!'

The desperate sound echoed round the barren hills and faded away unanswered.

'JOE!!'

Veronica flapped up and landed on his shoulder. 'I don't think Joe can hear you,' she said softly.

'You never know,' said Norbert, his pleading voice willing it to be true. 'If there's one thing my great-uncle Larry the Unlucky taught me it was that you should never give up hope. "Something will turn up." That's what he used to say, before the dragon ate him. "Something will turn up . . ."'

'I'm afraid *this* has turned up,' said Randalf gently. He held out a flattened disc of dull silver.

'Wh . . . what's that?' Norbert trembled.

Randalf showed him the bent black handle sticking out from one side. 'The Helmet of Sarcasm,' he said.

'No,' Norbert gasped. 'It can't be . . . You don't mean . . .'

'As you so rightly said,' Veronica muttered bitterly to

Randalf, 'you get what you pay for.' She tutted. 'Skinflint!'

'But it *can't* be his helmet,' said Norbert, fingering the flat piece of metal. 'Please say it isn't.'

'I'm afraid it is,' said Randalf. He shook his head. 'Completely flattened. Pulverized. Spifflicated. Flatter than a burst gas frog, you might even say . . .'

'Stop it!' Norbert howled, and clamped his hands over his ears. 'Stop it! Stop it! Stop it!'

'There, there,' said Randalf, and patted Norbert on the arm. 'It's all right, Norbert.'

'But it's *not* all right, sir!' Norbert bawled. 'It's not all right at all. First Quentin. Now Joe!' He pulled a grubby hankie from his pocket and blew his nose loudly. 'I just can't bear it!'

Veronica rounded on Randalf. 'You're to blame for all this!' she squawked. 'I never thought Joe was up to it. Warrior-hero, indeed!'

'But he *was*,' Randalf protested. 'I summoned him myself . . .'

'You summoned *Quentin*!' said Veronica. 'And look at him!'

'*Oh-woh-woh!*' wailed Norbert.

'Norbert, calm yourself,' said Randalf. 'We can summon more warrior-heroes. Even better ones . . .'

'*OH-OH-WOH!*'

'Third time lucky, eh?' said Veronica. 'My goodness, Randalf, you can be unfeeling at times. Why don't you bring the twisted toasting fork and squashed saucepan with you?' she suggested scornfully. 'Maybe you can get a refund!'

'Now, there's an idea,' said Randalf thoughtfully.

'*OH-WOH-WOH!*'

'A *bad* idea,' Randalf added hastily. 'Of course, I wouldn't dream of ...'

'You're right,' said Veronica. 'Grubley would never agree to it.'

'Veronica!' said Randalf sharply. 'I'm surprised at you. Don't take any notice of her, Norbert.'

'So, what *are* we going to do, sir?' Norbert asked tearfully.

'Well, we can't stay here,' said Randalf. 'And I don't fancy going back to the Horned Baron. If we tell him that the rogue ogre is still at large – and that we've lost our warrior-hero into the bargain,' he continued as Norbert sniffled into his hanky, 'he won't be a happy Horned Baron.'

'I'm not happy,' said the Horned Baron, pacing up and down the great reception hall. 'I'm not happy at all!' He was working himself up into a right state.

For a start, Ingrid had been on at him all day about her precious singing curtains – and he hadn't heard a word from Grubley ever since he'd handed over the pouchful of silver pipsqueaks. Worse still, the castle had been plagued by a constant stream of goblins, elves and trolls all complaining that their crops had been flattened, their thatched-roofs torn off, their sheep squeezed – and what was he, as Horned Baron of Muddle Earth, going to do about it? That's what they all wanted to know.

'It's all in hand,' he'd kept telling them. 'Even now, a famous wizard and a highly trained warrior-hero are on their way to deal with the rogue ogre.' Yet, even to himself, the words hadn't quite rung true.

The Horned Baron grunted. 'Randalf the *Wise*, indeed! I've worn wiser pairs of underpants!'

He glanced at the clock for the tenth time in as many minutes and strode over to the window.

'All this intolerable waiting!' he groaned as he stared out. 'That's the trouble with being so powerful and important. You spend half your life waiting for others to carry out your commands!'

'WAL-TER!'

Ingrid's strident voice cut through the air like a rusty knife, sending the dust flying and setting the Horned Baron's teeth on edge. He closed his eyes and slowly counted to ten. For a moment there, he'd quite forgotten about his wife. This was never a sensible thing to do.

'*WALTER!*' she screeched. The crystal chandelier tinkled softly. 'Can you hear me?'

'Nine . . . ten.' The Horned Baron opened his eyes. 'Loud and clear, my little snuggle-muffin,' he called back.

'Don't you "snuggle-muffin" me,' Ingrid shouted. 'Where are my singing curtains? That's what I want to know. Where *are* they, Walter?'

'Everything's in hand,' the Horned Baron replied. 'I'm expecting them at any moment.'

'That's what you said an hour ago,' Ingrid countered. 'But every time there's a knock on the door, it's someone else complaining about their wretched sheep!'

'Any moment now,' the Horned Baron assured her.

'You'd better not be lying to me, Walter,' said Ingrid, her voice becoming more threatening as it grew quieter. 'You remember what happened last time I caught you lying, don't you?'

'All too well,' the Horned Baron called back, and smoothed his straggly moustache tenderly. The green dye had almost grown out.

'Next time, I'll use the whole bottle!' she shouted.

The Horned Baron winced and looked out miserably at the empty road. 'Where are you, Grubley?' he muttered. 'Don't let me down . . .'

'*And* a wire brush!' Ingrid added.

The Horned Baron's eyes grew steely. 'If you *do* let me down, Grubley, there'll be a place waiting for you beside Randalf down in the dungeons.'

'Walter!'

'The smallest dungeon with no window.'

'*Walter!*'

'And twenty hand-picked stinky hogs for company . . .'

'There's someone at the door, Walter!' Ingrid shrieked. 'Do I have to do *everything* myself?'

'And regular visits from the Baroness,' muttered the Horned Baron, trotting towards the door. 'I'm on my way, my sweet!'

'Well done!' shouted Ingrid sarcastically. 'And for your sake, Walter, I hope it's those singing curtains turned up at last. I'm sick to death of hearing about squeezed sheep. I want to be lulled, Walter. I want to be soothed . . .'

'And so you shall, my honeyed sugarplum,' the Horned Baron called back.

He opened the door. A short, slight individual with a striped tunic and a feather in his cap stood on the top step. 'Greetings-elf!' he announced. 'I bring greetings from Mr Grubley of Goblintown.'

'Is there a package to go with the greetings?' the Horned Baron asked hopefully.

'No, just a message,' said the elf, lowering his head and shaking it regretfully.

The Horned Baron rolled his eyes. 'So, what is the message?' he asked.

The elf took a deep breath and cleared his throat. 'Having travelled to the four corners of Muddle Earth, Grubley of Goblintown has procured a length of enchanted material which, even now, is being transformed into singing curtains, the like of which have never before been seen or heard.'

'Thank goodness for that,' the Horned Baron muttered.

'However . . .' the elf continued.

The Horned Baron raised his hand. '*However?*' he said. 'I don't like the sound of that.'

'I could skip to the "best wishes" if you like,' said the elf.

'Is there nothing about when the curtains will be ready?' asked the Horned Baron.

'That's part of the "however",' said the elf.

The Horned Baron tutted. Above his head, he could hear Ingrid stomping backwards and forwards with growing impatience. 'All right, then,' he sighed. 'Get on with it.'

'However,' the elf resumed, 'due to unforeseen circumstances, the curtains are taking somewhat longer to make than expected. They will be delivered to you tomorrow teatime at the very latest, probably . . .'

'Tomorrow teatime!' gasped the Horned Baron. '*Probably!*'

'*Grubley's Discount Store* would like to take this opportunity to apologize for any inconvenience . . .'

The Horned Baron snorted. 'You don't know the half of it,' he muttered. 'I don't know how I'm going to explain all this to her upstairs.' He shook his head. 'Is that it, then?' he asked the greetings-elf.

The elf nodded. 'Just about,' he said. He held out his hand. 'That'll be three brass muckles.'

'What?' said the Horned Baron. 'You mean Grubley sent a greetings-elf without paying for the stamp?'

'Ah, yes,' said the elf. 'I forgot. There's a PS. Sorry about the stamp. It can come off my final bill when we settle up.'

'Final bill!' the Horned Baron shouted. 'Settle up! I'll settle up all right. One dungeon, twenty pink stinky hogs and a no-good wizard should just about do it!'

'Is that the message you wish to send back?' asked the greetings-elf.

'Yes, I . . .' The Horned Baron frowned and stroked his chin. 'That is, *no*,' he said.

'No?'

'No,' the Horned Baron confirmed. 'Simply thank Grubley for his message and tell him I look forward to his arrival . . .'

'WALTER!'

'His *speedy* arrival,' he corrected himself.

'OK,' said the greetings-elf, 'though personally, I liked the stinky hogs message better myself.' He stuck out his open palm a second time.

The Horned Baron sighed and dropped three muckles into the outstretched hand. The greetings-elf pulled a stamp from a pocket, licked it and stuck it on to his forehead. Then he turned and skipped off down the stairs and away. For a fleeting instant, the Horned Baron imagined that *he* was a greetings-elf, setting off without a care in the world.

'*WAL*-TER!!'

The carefree daydream popped. He closed the door. 'Yes, my sweetness,' he called up the stairs.

'Was that my curtains?' she shouted back.

'Not as such,' the Horned Baron confessed.

'What's that supposed to mean?' demanded Ingrid.

'It was *news* of your curtains, my sweet,' he explained. 'There was been a slight hitch . . .'

'*Hitch*, Walter?' said Ingrid. 'I don't like the word *hitch*. You know that. I don't like it at all.'

'I know, my turtle dove,' said the Horned Baron soothingly. 'There have been unforeseen circumstances. You know how it is! Grubley's promised me they'll be here tomorrow,' he added.

'*Tomorrow!*' screeched Ingrid. 'But what am I supposed to do tonight? I shan't be able to sleep a wink, I just know I shan't. And you know what I can be like when I'm over-tired.'

'Indeed, I do,' said the Horned Baron wearily.

'Grumpy, Walter. I shall be very grumpy. You won't recognize me!'

'Oh, I think I might,' he muttered beneath his breath. 'Believe me, Ingrid,' he called upstairs, 'you just can't rush these things. I mean, singing curtains, Ingrid, fashioned from only the finest enchanted cloth, tasselled and sequinned, and hand-stitched by a master sewing-elf. It'll be well worth the wait when they do arrive, you have my word . . .'

'*If* they arrive,' Ingrid shouted, and the entire castle shook as she slammed the door of her bedchamber hard shut. The sound of loud thuds and muffled sobs echoed above as Ingrid threw herself around the room.

The Horned Baron shook his head. 'This is all your fault, Grubley,' he said. 'I mean, why put a blasted advertisement for blasted singing curtains in that blasted catalogue of yours if you don't actually have any blasted singing curtains in stock and have to go chasing round Muddle Earth searching for some? Blasted funny way to run a business!' His eyes narrowed. 'You've upset my beloved Ingrid, that's what you've done – and when Ingrid's upset, *I'm* upset! And when I'm upset . . .'

'The Horned Baron's going to be so pleased with me,' said Grubley.

'So you already said,' the goblin muttered as he rethreaded his sewing-elf. 'Twice.'

'But he is!' said Grubley. 'I can't wait to see his face . . .'

'*La, la, la,*' sang the material.

The goblin picked up a large, shiny pair of scissors, laid the material out across the workbench and began to cut it in half.

'*La, la . . . Ouch! Ouch! Ouch!*'

'Will you stop doing that!' shouted the goblin, and slammed the scissors down. He turned to Grubley. 'You see the trouble I'm having! Every time I try and cut the cloth in two, it makes a racket. And it's *very* off-putting,' he said. 'Are you absolutely sure you want curtains? I could do you a very nice roller blind.'

Grubley shook his head. 'Apparently, the catalogue specified singing *curtains*,' he said, 'and the Horned Baroness has set her heart on them.'

The goblin picked up the scissors again. 'I don't know why you advertised something you don't keep in stock in the first place,' he grumbled.

'That's the strangest thing,' said Grubley. 'I don't remember putting them in the catalogue.'

'Well, someone must have,' said the goblin.

'I know,' said Grubley, frowning. 'I just don't understand it.' He looked up. 'Still, I've got the cloth now. That's the most important thing. And as soon as you've made it up as curtains, I'll get them over to the Horned Baron's castle. So if you wouldn't mind . . .'

'It's all right for you,' said the goblin. 'You don't have to work with material that won't keep quiet.' He fingered the frayed cloth, which gave a high-pitched squeak. 'Giving me the heebie-jeebies, it is.'

'Here,' said Grubley, reaching into his pocket and

pulling out a pair of large furry earmuffs. 'Try these.'

The goblin stared at them. 'What am I supposed to do with them?' he said.

'They're earmuffs, stupid,' said Grubley irritably. 'You put them over your ears.'

The goblin did as he was told, flattened out the material and raised his thumbs. He hadn't heard a thing.

'Excellent,' said Grubley. 'Now get on with the curtains.'

The goblin looked at him blankly.

'*Get on with the curtains!*' shouted Grubley.

The goblin frowned and mouthed the word, *what?*

Irritated, Grubley lifted one of the earmuffs and leaned forwards. 'GET ON WITH THE CURTAINS!' he bellowed into the goblin's ear.

'All right, all right,' the goblin said, pushing the earmuffs back into place. 'I'm not deaf!'

'Give me strength!' Grubley muttered.

The goblin sat down on the stool, picked up his scissors again and this time – despite the singing, wailing and frequent cries of '*Ouch!*' – cut the cloth and set the sewing-elf off stitching at a furious pace.

'Very nice,' said Grubley, holding the curtains up at last. 'Very homely. A genuine pair of singing curtains.'

'*La, la, la. La, la, la,*' sang the curtains, in a discordant duet.

'Call that singing!' said the goblin. 'More like . . .'

'Oh, don't *you* start,' said Grubley. 'The Horned

Baroness is tone deaf. She'll love them, and that's all that matters.' He frowned. 'Randalf the Wise,' he said. '*Wise*, indeed! Heading for the Ogrehills, he was. Not very wise at all, if you ask me!'

'We came, we saw, we ran away,' said Veronica from the top of Randalf's head as Norbert trudged back down the mountain road. 'Joe the Barbarian, mighty warrior-hero and Henry the Hairy, faithful battle-hound – missing, presumed pulverized . . .'

'Yes, all right, Veronica,' said Randalf. 'You've made your point.'

'Engelbert the Enormous,' she continued, 'missing, presumed sheep squeezing . . .'

'Shut *up*, Veronica!' said Randalf.

Norbert wiped away a tear. 'Have you decided where we're going yet, sir?' he asked.

Randalf sighed and nodded. 'Home,' he said.

'Home, sir?' said Norbert.

'Yes, Norbert,' said Randalf. 'Let's go home.'

Two of Muddle Earth's three moons were high in the sky. They shone down brightly on a gathering of twenty or so ogres who were all sitting in front of their caves, around a huge, roaring fire. The air echoed with the sound of soft, satisfied slurping as the ogres sucked their thumbs.

One ogre – the biggest of them all – was rubbing a dog slowly up and down his cheek, and smiling happily. The dog was wagging his tail and emitting a curious yodelling bark of utter contentment.

One of the ogres removed his thumb from his mouth and turned to the boy next to him. 'Old Engelbert's just his old self again,' he said.

'Yeah, you *have* calmed him down, Joe,' said another.

'He just needed a bit of understanding,' said the first ogre.

'It's all any of us need,' chipped in a third.

'After all,' said the first, 'how would *you* like it if you'd lost your special snuggly-wuggly?' He held up a tatty teddy with one eye, one ear, one arm and no legs. 'I don't know what I'd do if Trumpet ever went missing!'

Joe nodded. He couldn't quite believe what he was seeing or hearing. The other ogres held up their own snuggly-wugglies one after the other – a grubby fluffy bunny, a fuzzy blanket, a tattered scrap of towel . . .

'Course, Engelbert was always particularly proud of his snuggly-wuggly,' said the first ogre. 'Bit niffy, it was. And threadbare. But Engelbert loved it. And do you know why?'

'Why?' said Joe.

''Cause it was enchanted,' said the ogre. 'It sang.'

'Did it now?' said Joe thoughtfully.

'His mother got it for him when he was a baby,' the ogre went on. 'From one of the wizards on the Enchanted Lake. Roger the Wrinkled, his name was . . .'

'Course, that was back before the wizards disappeared,' the second ogre interrupted. 'You can't get anything enchanted these days. That snuggly-wuggly was unique. Irreplaceable.'

'Which is why Engelbert took it so badly when it went missing. It used to lull him to sleep every night.'

'Engelbert loved his snuggly-wuggly,' said the second ogre. 'He said it smelled of warm hugs. Took it everywhere, he did . . .'

Engelbert, who had clearly been listening in, suddenly sat forwards. 'Until somebody took it!' he exclaimed. 'I woke up last week and there it was, gone.' His face clouded over. 'Stolen, it was! Someone had stolen my snuggly-wuggly. My lovely singing snuggly-wuggly . . .'

'Steady on, big fellow,' the other ogres told him. 'Stay calm.'

'It made Engelbert angry,' Engelbert continued, his voice trembling and his face all blotchy and red. 'And sheep are no good. They might be soft, but they make such a horrible sound – even when you hardly squeeze them at all.'

Henry barked excitedly, and licked Engelbert's bulbous nose. A broad smile spread over the ogre's features.

'Not like Henry here,' he said. 'He has a lovely singing voice.'

'This snuggly-wuggly,' said Joe. 'Did it go *la, la, la . . . ?*' he crooned, singing in his deepest, groaniest voice.

'Yes,' said the ogres excitedly. 'How did you know?'

'I think I may have seen it,' he said, as he remembered his encounter with Grubley on the road from Elfwood.

'Anyway, it doesn't matter now,' Engelbert said. 'I don't need my old snuggly-wuggly back. Not now I've got Henry.' He rubbed him affectionately up and down his cheek again.

Henry wagged his tail with pleasure and barked his strange yodelling bark. Engelbert chuckled.

'Just listen to that,' he said, and tickled him under his tummy. 'He's perfect.'

Joe nodded sadly. '*I* think so,' he said. 'The thing is, Engelbert, he belongs to me. And I would miss him, too. I've had him ever since he was a little puppy.'

Engelbert looked up. His jaw dropped. 'You're . . . you're not going to take him away, are you?' he said. 'You wouldn't leave Engelbert without a snuggly-wuggly again? I couldn't bear it.'

'And you know what happened last time,' the other ogres warned him.

'I know,' said Joe. He turned to Engelbert. 'But, supposing I could get your real snuggly-wuggly back,' he said. 'You'd let me have Henry back then, wouldn't you?'

The ogre pouted. 'I don't know about that,' he said reluctantly.

'Engelbert, I'm talking about your *old* snuggly-wuggly,'

said Joe softly. 'Your *best* snuggly-wuggly. The snuggly-wuggly you've had since you were a baby ogre, that sings you to sleep and smells of warm hugs.' He smiled. 'The snuggly-wuggly you love as much as I love Henry.'

Engelbert looked at Joe, then at Henry – then back at Joe.

'All right, then,' he said at last. 'It's a deal.'

Shortly after teatime the following day there was a loud knock at the castle door. The Horned Baron ran to answer it. Grubley was standing there.

'At last!' the Horned Baron exclaimed. 'You took for ever!'

'Singing curtains can't be rushed,' Grubley explained. Having passed the greetings-elf halfway between Goblintown and the castle, he already knew how desperate the Horned Baron was to receive them. 'Besides,' he added, 'what's all this stuff about dungeons and wizards and pink stinky hogs?'

'Never mind!' the Horned Baron humphed. 'You've got them, that's the important thing.' He frowned. 'Where are they?'

Grubley took off his backpack and opened it up. The sound of two muffled voices singing in discordant harmony filled the hallway. Grubley pulled out the pair of curtains and displayed them over a crooked elbow.

'They look a bit tatty,' said the Horned Baron. He wrinkled his nose. 'And they niff a bit,' he added. 'Perhaps you'd see your way to knocking a little bit off the final price . . .'

'You must be joking,' said Grubley, outraged. 'These singing curtains are unique. You wouldn't believe the lengths I had to go to to find them.'

The droning song grew louder. It echoed round the vaulted ceiling and floated up the stairs.

'Walter!' came a strident, yet hopeful, voice. 'Is that *singing* I can hear? Have my singing curtains finally arrived?'

'Y . . . yes, they have,' the Horned Baron called up. 'If you can call that tuneless cacophony *singing*,' he muttered under his breath.

'Course, you don't have to have them,' said Grubley, folding the curtains up. 'If you don't want them, I know plenty who do . . .'

'Oh, Walter,' Ingrid replied. 'You wonderful Horned Baron, you! I knew you wouldn't let me down. I never doubted you for a moment.'

'But if you *do* want them,' Grubley continued, as he opened his backpack, 'then, as you well know, you owe me a pouch of silver pipsqueaks.'

'Daylight robbery,' the Horned Baron complained. 'One pouch is more than enough . . .'

'WALTER!' Ingrid shrieked. 'I am a patient woman. But

you are trying that patience, Walter. You are pushing it to the very limit.' She paused. 'I WANT MY SINGING CURTAINS NOW!'

'Right away,' the Horned Baron said. He turned to Grubley and thrust the pouch of silver pipsqueaks into his hand. 'I take it hanging the curtains up is included in the fee.'

'Not normally,' said Grubley. The Horned Baron's eyebrows drew together menacingly. 'But for such a valued customer,' he added in an oily voice, 'I'd be only too happy to oblige.'

Just then, there was a furious hammering at the door. Grubley jumped. The Horned Baron spun round.

'What now?' he said.

'WALTER!'

'Coming . . . I mean, going . . .' the Horned Baron called back, as he headed first for the staircase, then back for the door, not knowing for a moment whether he was coming or going.

The hammering resumed, louder than ever, and accompanied by a loud voice shouting, 'Open up! Open up! It's a matter of life and death!'

The Horned Baron raised his eyebrows to the ceiling. 'If it isn't one thing, it's another!' he said.

'WAL-TER!!'

'You take the curtains up,' the Horned Baron told Grubley. 'I'll see who's at the door. It's probably another case of badly squeezed sheep.' He shook his head. 'When I get my hands on that Randalf character ...'

As Grubley disappeared upstairs, the Horned Baron crossed the hallway. Before he arrived at the door, however, it burst open and slammed back against the wall behind. Silhouetted in the doorway, the Horned Baron saw a wiry, dishevelled youth with matted hair, dusty clothes and one wellington boot.

'Don't tell me,' he said. 'You're here to complain that your sheep have been squeezed. Look, for the hundred-and-first time ...'

'Horned Baron,' said Joe, as he strode into the hallway. 'Just the person I wanted to see.' He held out his hand. 'It's Joe. Remember? Joe the Barbarian? Warrior-hero?'

'Barbarian? Warrior-hero?' said the Horned Baron distractedly as he glanced past Joe and up the stairs. 'Joe . . . Ah, yes. I didn't recognize you without the saucepan on your head. How are you and how did you get on? And where's that wizard?'

From upstairs, there came the enthusiastic sound of *oohing* and *aahing*. 'Oh, Walter, they're divine!' Ingrid called. 'No one else has got anything like them. Wonderful!

The very height of fashion!' There was a pause. 'They *are* the very height of fashion, aren't they, Walter?'

'Yes, dear,' he replied wearily. 'And the pinnacle of good taste.'

Joe smiled.

'New curtains,' the Horned Baron explained.

'*La, la, la. La, la, la . . .*'

'*Singing* curtains,' he explained. 'Ingrid's set her heart on them. Apparently, they're all the rage.'

'Yes,' said Joe. 'That's what Grubley said, when I saw him.'

'Singing curtains!' Ingrid trilled. 'My very own singing curtains!'

'Very rare,' said Joe. 'Very hard to come by – you don't find enchanted material every day.'

'And what if they are?' said the Horned Baron, suddenly defensive. 'I dare say a Horned Baron's entitled to buy his beloved wife a little gift now and then. More to the point, what are *you* doing here?'

Joe breathed in and pulled himself up to his full height. This was the part he'd been practising. 'I, Joe the Barbarian, have performed the task you bade me carry out.'

'You, what?' said the Horned Baron.

'I have brought you the head of Engelbert the Enormous.'

The Horned Baron's jaw dropped. 'You have?' he said, then frowned suspiciously. 'Where is it, then?'

'*WAAAARGH!!*'

The screeching shriek of terror was quite the loudest noise Ingrid had emitted all day. It was deafening. It made the windows rattle and the staircase shake.

'*WAAAAAARGGHH!!!*'

Even the Horned Baron, who was used to Ingrid's hysterical response to spiders, bugs and not getting her own way, realized that this time, something was definitely not right. The poor woman sounded terrified out of her wits. *Something* was up there and, for the first time since Joe had burst in, the Horned Baron was pleased to have a warrior-hero in the castle.

'Follow me,' he said, turning and heading up the stairs.

As they burst into Ingrid's bedchamber, the door to her *en-suite* bathroom slammed shut.

'Get rid of it!' shrieked Ingrid from behind the door. 'It's hideous!'

The Horned Baron looked round to see the great, knobbly, three-eyed head of Engelbert the Enormous sticking in through the window. 'What's the meaning of this?' he demanded.

'The head of Engelbert the Enormous,' said Joe. 'As you requested.'

'But it's still attached to his body!' thundered the Horned Baron. 'This is an outrage! What kind of a warrior-hero *are* you?'

'And what kind of a Horned Baron are *you*?' Joe retorted.

'Stooping so low as to buy curtains made out of an ogre's snuggly-wuggly!'

'An ogre's snuggly-wuggly?' said the Horned Baron with surprise.

'*La, la, la . . .*' sang one curtain tunelessly.

'*La, la, la . . .*' its neighbour droned back.

The Horned Baron's eyes widened. 'Are you telling me that these singing curtains have been fashioned from an ogre's snuggly-wuggly?'

Joe nodded. At that moment, a huge hairy hand thrust its way through the window and seized first one, then the other curtain, and whisked them away.

'Grubley!' roared the Horned Baron. 'Grubley, I demand my money back.'

But Grubley was not there. As his name echoed round the castle walls, Grubley was already on the road and hurrying back to Goblintown as fast as his legs would take him.

'One snuggly-wuggly,' Engelbert was saying as he rubbed it up and down his left cheek. 'Another snuggly-wuggly.' He rubbed the second one up and down his right cheek. 'It's even better than before.'

'Twice as good,' said Joe, relieved that the ogre didn't seem to be upset that his snuggly-wuggly had been cut in two. 'But remember what you promised, Engelbert,' he said. 'It's time for you to keep your side of the bargain.'

'What do you mean?' said Engelbert, then winked

(with his middle eye) just to show that he was joking.
'Here we are then, Joe the Barbarian,' he said and, reaching
into the bed chamber again, placed Henry gently down on
the rug. 'Look after him,' said Engelbert. 'He's one in a
million!'

'I know he is,' said Joe, as Henry raced across the room
and jumped up at him, tongue lolling and tail wagging. He
looked up at the window to see Engelbert smiling back at
him. 'Goodbye, Engelbert,' he said. 'And thank you.'

'Goodbye, Joe,' boomed the ogre as he stomped away.
'Goodbye, Henry,' his voice floated back.

Henry barked.

'Walter!' screeched Ingrid from the bathroom. 'That lumpy great ogre is stealing my singing curtains! Walter!'

'*La, la, la . . . La, la, la . . .*' sang the curtains – softer and softer as they were carried off, until the sound of their tuneless discord faded away completely.

'There,' said Joe to the Horned Baron. 'He's gone. And now he's got his snuggly-wuggly back, there won't be any more sheep squeezing. I can guarantee it.' He smiled. 'And now to the matter of my fee.'

'What?' exclaimed the Horned Baron.

Henry growled. The Horned Baron eyed him warily.

'Ah yes, your fee,' he said. 'A handful of brass muckles, wasn't it?'

'A pouch of silver pipsqueaks,' said Joe. 'That was what we agreed.'

'I most certainly . . .'

Henry growled again, not loudly, but just enough to remind the Horned Baron he was still there.

' . . . did,' the Horned Baron said. 'A pouch of silver pipsqueaks it is.' He reached into the folds of his jerkin, pulled out a jangling leather pouch and, with a long, miserable sigh, reluctantly handed it over.

'Thank you,' said Joe. 'And now, if you'll excuse me, I've got to see a wizard about a journey home.'

He turned, whistled for Henry and strode back to the entrance to the bedchamber. As he reached the door, Ingrid let out a tremendous screech of rage, followed by an even louder, 'WALTER!'

The Horned Baron blanched. 'I'll see you out,' he said, as he trotted after Joe.

'*WALTER!*'

'That is, if I can't tempt you to stay,' he said. 'How would you like to be my personal bodyguard?'

'*Errm* . . . No thanks,' said Joe. He quickened his pace, taking the stairs two at a time and hurrying across the hallway. Henry kept close to heel.

'Wait a moment,' puffed the Horned Baron. 'I'll make you an offer you can't refuse . . .'

'By-eee!' Joe called back. He slammed the door shut and dashed off.

Behind him, Ingrid's voice rang out. 'Call yourself a Horned Baron!' she shrieked. 'You're pathetic! A disgrace! I'm opening the cupboard, Walter. I'm getting the green dye out, Walter – *and* the wire brush . . .'

Joe smiled to himself. What with Engelbert getting his snuggly-wuggly back and the Horned Baron getting his comeuppance, things were going rather well. Now, all he had to do was persuade Randalf to send him back home, and then *everything* would have reached a satisfactory ending.

And as for the story he had to get done when he arrived back, after his time in Muddle Earth, 'My Amazing Adventure,' would be the easiest essay he had ever had to write.

'Come on, Henry,' he said. 'Let's see if we can't make it back to the Enchanted Lake before daybreak.'

The sun had already risen by the time Joe and Henry arrived at the Enchanted Lake. As the low, bright rays of light cut through the early-morning mist, a broad, shimmering fish flopped down through the air and into the gaping beak of a waiting lazybird.

'Another day in Muddle Earth,' Joe murmured and shook his head. 'I'm almost getting used to it.' He turned to Henry. 'Almost,' he said, staring up at the great lake of water hovering high above his head. 'How on earth do we get up there?'

Henry barked and wagged his tail.

'Clever boy,' said Joe, for there, beside Henry, was a small bird-box on top of a pole, with a bell hanging from it on a hook. A notice said, *Ring for attention.* Joe rang the bell.

A lazybird flew out of the box, a small elf clinging to its back, and flapped upwards.

'*Ding-dong,*' droned the elf, disappearing over the lip of

the lake. '*Ding-dong. Ding-dong . . .*'

Joe and Henry waited, and waited. Then a loud voice came from above. 'Grab on to the rope!'

Joe scrambled to his feet and looked up. 'Norbert!' he exclaimed.

The ogre was far up above his head and leaning precariously over the side of the water's edge. A long rope, with a basket secured to its end, was dangling down from his great fists. Joe reached out and grabbed a hold.

'That's it!' came Norbert's voice, encouragingly. 'Now climb in, both of you, and I'll pull you up.'

Trembling with unease, Joe climbed into the basket, sat cross-legged and pulled Henry on to his lap. He wound the lead around his arm, and gripped the sides of the basket tightly with both hands.

'Ready?' shouted Norbert.

'Yes,' Joe shouted back. 'As ready as I'll ever b— *Whooah!*' he cried out, as the rope jerked, the bowl wobbled and he found himself rising up, up, up into the air. He'd forgotten just how high the Enchanted Lake was.

'This is terrifying!' he shouted.

'Be thankful you're not

getting up here the way the others had to,' Veronica's voice floated back. 'It took several flocks of lazybirds to lift Norbert off the ground – and you should have seen the state of his shirt when they'd finished!'

Henry whimpered. Joe hugged him and whispered that everything was going to be all right.

With a last grunt of effort, Norbert pulled the basket up over the edge of the lake and held it next to a small flotilla of kitchen sinks. Randalf and Veronica were in one, Norbert was squashed into the second, while the third was empty. All three were roped together.

Randalf leaned forwards. 'Joe!' he said.

'Randalf!' Joe replied.

'Am I glad to see you,' said Randalf.

'Not half as glad as I am to see you,' said Joe.

'Indeed, I am *twice* as glad,' Randalf insisted.

'Whatever,' said Veronica. 'Let's just get back to the houseboat before someone – mentioning no names, of course,' she said, eyeing Norbert accusingly. 'Before *someone* pulls out another plug.'

'I didn't *mean* to,' Norbert protested.

'Never mind all that, Norbert,' said Randalf. 'You're getting the basket wet. Joe, you and Henry climb into that spare kitchen sink, and Norbert will paddle us all back. Won't you, Norbert? There's a good fellow.'

Everyone took their places. Norbert began paddling furiously.

'And then you must tell me everything that happened,' Randalf shouted across the foamy water. 'Every single detail.'

'He did *what?*' Randalf exclaimed.

'He tickled Henry's tummy,' Joe repeated. 'And I could tell at once that Henry liked it. I looked round for you lot, but you'd all run off.'

Randalf coughed with embarrassment and turned a florid shade of pink. 'A tactical retreat, my lad,' he said. 'Withdrawing. Regrouping . . .'

'Running for dear life,' added Veronica.

'Shut up, Veronica,' said Randalf. 'Go on, Joe.'

'It was just like you said,' said Joe, and tapped the side of his head. 'Psychology!'

'You see,' said Randalf triumphantly. 'Didn't I tell you?

With your Trident of Trickery and Helmet of Sarcasm . . .'

'Oh, no,' Joe interrupted. 'It had nothing to do with *them*. In fact, they just got in the way, so I took them off.' He looked up guiltily. 'I'm afraid they got a bit squashed when Engelbert accidentally trod on them.'

'Never mind that now,' said Randalf. 'What exactly did you mean by psychology?'

'Well, it was clear that Engelbert liked Henry,' Joe explained. 'From the moment he picked him up and rubbed him up and down his cheek, he was a changed ogre. A real softy . . .'

Randalf frowned. 'He'd lost his snuggly-wuggly, hadn't he?' he said. 'That's what he was so angry about. And when he found Henry he calmed down again.'

Joe nodded.

'It's all falling into place,' said Randalf. 'The fits of rage. The squeezed sheep. Norbert, you should have realized.'

'Sorry, sir,' said Norbert.

'Yet Henry is with you now,' said Randalf eyeing the dog thoughtfully. 'How did you manage to get the ogre to give him back?'

'Simple,' said Joe. 'I found the snuggly-wuggly he'd lost.'

'Where?' asked Randalf, intrigued.

'At the Horned Baron's castle,' said Joe, and smiled at Randalf's obvious confusion. 'I'll give you a clue,' he said. 'We all saw the ogre's snuggly-wuggly earlier on,' he said, 'when we were heading for the Ogrehills. 'Saw it, *and* heard it . . .'

'Grubbers!' Randalf exclaimed. 'Why, he had a roll of singing material under his arm, didn't he? I remember now, he was on his way to Goblintown to have it turned into curtains for the Horned Baron's wife.' He frowned. 'Norbert, you're an ogre. I would have expected you to recognize another ogre's snuggly-wuggly.'

'Sorry, sir,' said Norbert once again. 'I don't seem to have had a very good day, do I?' he added sadly.

'Grubley, Grubley,' said Randalf, sucking air through his teeth and shaking his head. 'I never trusted him. What a rogue! What a scoundrel. Stealing an ogre's snuggly-wuggly! Because of him, the whole of Muddle Earth was thrown into turmoil.' He clapped Joe on the shoulders. 'And so it would have remained, my lad, if you hadn't come along.'

'It . . . it was nothing,' said Joe.

'Nothing?' Randalf cried. 'Why, Joe the Barbarian, warrior-hero from afar, Muddle Earth will be for ever in your debt.'

'That's great,' said Joe, 'and now, if it's all the same with you, I really *really* would like to be getting home. I've done my bit . . .'

'But, Joe,' said Randalf. 'Joe the Barbarian. There is one small matter . . .'

'Oh, yes, I was forgetting,' said Joe as he pulled the pouch of silver pipsqueaks from his pocket. 'The Horned

Baron gave me this. It's my fee.' He held it out. 'You may as well have it. I won't be needing it where I'm going.'

'Oh, Joe,' said Randalf. 'Your bravery is unsurpassed, your ingenuity unequalled and now I find your generosity is also unmatched – and yet that was not the little matter to which I was referring.'

'Then, what is it?' said Joe. His heartbeat was beginning to race. 'I've kept my side of the bargain. Now it's time to keep yours. You *must* send me home.'

'I can't,' he said.

'Can't?' said Joe. 'What do you mean, *can't?*'

Randalf looked down. 'I mean, I can't,' he said.

'He's right, you know,' said Veronica. 'He wouldn't know where to start, I can vouch for that.'

Joe's stomach churned. His head spun. 'But . . . but you brought me here,' he said.

'I know,' said Randalf. 'I used my warrior-hero-summoning spell,' he said. 'Unfortunately, it is the only spell I have to hand.'

'Yeah,' said Veronica scornfully, 'that's because all the other spells – including the warrior-hero-returning spell are . . .'

'Are elsewhere,' Randalf interrupted hurriedly.

'Can't you fetch it?' said Joe.

'I'm afraid not,' said Randalf.

'You mean to say, I'm stuck here!' said Joe indignantly.

'For the time being,' Randalf confirmed.

'No, no, I've got to get back ...' said Joe. 'Why can't you *fetch* it?' he demanded. 'Why?'

'Because ... because ...' Randalf faltered.

'Go on, tell him,' said Veronica. 'You know where it is. After all, there's only one place it could be!'

'Where?' said Joe.

Randalf grimaced. 'Giggle Glade,' he said.

'Giggle Glade?' said Joe.

'It's in the middle of Elfwood,' said Randalf.

'Elfwood?' said Joe. Veronica had mentioned Elfwood before.

'It's the residence of ...' He shuddered violently. 'Dr Cuddles.'

'Dr Cuddles ...' said Joe slowly.

'Blimey,' said Veronica. 'He's more of a parrot than I am, and that's saying something for a budgie!'

'Dr Cuddles is . . . is the one who stole Roger the Wrinkled's *Great Book of Spells*,' Randalf confessed. 'He's been using it ever since. You remember the flying cupboards?'

'And the stampeding cutlery?' said Norbert.

Joe nodded. Randalf shuddered again.

'That was no doubt the work of Dr Cuddles,' he said. 'If anything goes wrong in Muddle Earth, you can bet your last pipsqueak, somewhere at the bottom of it all, you'll find Dr Cuddles of Giggle Glade!'

'That's it!' Veronica exclaimed.

'He's power mad!' said Randalf. 'He'll stop at nothing to take control of Muddle Earth and become its absolute ruler. And if that should ever happen,' he went on, 'then all the denizens of Muddle Earth would be forced to dance to his evil tune . . .'

'*That's it!*' squawked Veronica a second time.

'*What's* it?' said Randalf irritably.

'It wasn't Grubley who stole the ogre's snuggly-wuggly,' said Veronica. 'He was telling the truth when he said he obtained it from one of his contacts. The question is, who was that contact?'

Randalf shook his head. 'You don't mean . . .' he said.

Veronica tutted impatiently. 'Just think about it. Who stood to gain from Engelbert destroying the Horned Baron's castle in his rage?' she asked. 'Who would have welcomed a bit of argy-bargy between the goblins and the ogres? And where was Grubley coming from with that roll of singing material? Elfwood! And who lives in Elfwood?'

'Dr Cuddles,' said Randalf, Norbert and Veronica in hushed unison. 'Stealing an ogre's snuggly-wuggly! What will he think of next!' They all shook their heads.

It was Joe who broke the long silence that followed. 'That's handy,' he said.

Randalf looked at him quizzically. 'Handy?' he said.

'You'll be able to find out when we pay him a visit to

ask him to return the *Great Book of Spells*,' he said.

Randalf laughed nervously. 'Joe, my dear boy, nobody pays a visit to Giggle Glade. You can't just ask Dr Cuddles to return the spell book. That's what Roger the Wrinkled and the other wizards thought. "We'll discuss it over a nice pot of tea, Randalf," they said – and look what happened to them!'

'What?' said Joe.

'Well, I don't actually know,' admitted Randalf. 'But they didn't come back!'

Joe shrugged. 'If going to see Dr Cuddles of Giggle Glade is my only chance of returning home, then it's a risk I'm prepared to take. Besides,' he said, before Randalf – or Veronica – could speak, 'you're forgetting something very important.'

'And what might that be?' asked Randalf.

Joe smiled. 'I am JOE THE BARBARIAN!' he proclaimed in his biggest voice.

'I . . . I know that,' said Randalf uncertainly. 'But . . .'

'Trust me,' said Joe. 'I'm a warrior-hero.'

A chill wind whistled through the trees of Elfwood. The leaves rustled, the boughs creaked. At its very centre, the dappled light illuminating Giggle Glade was fading fast.

'We failed, master,' came the nasal voice of Dr Cuddles's assistant.

'Yes,' came the squeaky reply, followed by high-pitched giggles. 'We failed.'

'And we planned the singing curtains scam so well! The fake advertisement in the catalogue. The theft of Engelbert the Enormous's snuggly-wuggly – ooh, those ogres can be *so* stupid! The haggling with that odious little goblin, Grubley . . . It was all going so well.'

'Yes,' Dr Cuddles giggled unpleasantly, 'by now, Muddle Earth should have been in chaos! And I would have been its ruler. I didn't think my old friend, Randalf the Apprentice, had it in him to use that spell a second time.' The sinister giggles grew louder. 'Curse that warrior-hero!'

'My thoughts entirely,' his assistant agreed.

'But our work shall continue. I shall devise an even better plan! One that cannot fail! I shall destroy the warrior-hero once and for all!' he shouted, each word interspersed with the hideous giggling. 'I shall conquer the Horned Baron!'

All round the clearing, the woodland creatures were troubled by the sound of the raised voice. As it reached its terrifying crescendo, stiltmice tottered, tree rabbits fell out of their trees, while the roosting batbirds – already wary after an attacking flock of cupboards had left them battered and uneasy – deserted their perches and flapped off across the sky.

'That all sounds absolutely super,' said his assistant.

'Now, how about a nice cup of tea and a snuggle-muffin. I've decorated one specially with your face . . .'

'What would I do without you, Quentin?' said Dr Cuddles. 'Now, I need to get down to my plan.' He stroked his chin. 'I must cover every angle. Allow for every possibility.' He looked up. 'I'm thinking dragons. I'm thinking mangel-wurzel marmalade. I'm thinking small, tinkly teaspoons . . .' He giggled and rubbed his hands together gleefully. 'It's going to be perfect, Quentin.'

'Ooh, you're so evil, master,' Quentin purred.

The giggles grew menacing. 'You haven't seen anything yet, believe me, Quentin,' he said. His voice (and giggles) became louder. 'And there is not a thing that Randalf, or anyone else, can do to stop me! I, DOCTOR CUDDLES OF GIGGLE GLADE, SHALL BECOME LORD AND MASTER OF MUDDLE EARTH!!!' he roared, and he threw back his head in crazed triumph.

'*Tee-hee-hee-hee-hee-hee-hee-hee . . . !*'

Book 2

HERE BE DRAGONS

Prologue

It was night-time in Muddle Earth and the air was still. High up in the sky, its three moons – one purple, one yellow and one green – were dotted about the dark, cloudless sky. They shone down brightly, like three spotlights, casting multicoloured light and shadow on buildings, trees and the Enchanted Lake, which floated at the top of a waterfall. They glinted on the massed ranks of shiny cutlery which stood in a broad semicircle at the entrance to a vast mountain cave.

To the left was the knife section. Several neat rows of knives – some smooth, some serrated, some straight, some curved – ascended in size, from modest butter knives at the front, through steak knives, bread knives and carving knives, to a line of hefty meat cleavers at the back.

To the right were the spoons, also standing in neat rows; first egg spoons, then teaspoons, then dessertspoons and tablespoons, and finally a somewhat rowdy collection of tall ladles, all jockeying for a good position.

Between these sections were the rest. Everything from forks, whisks, skewers and spatulas, to grinders and graters, egg slicers, nutcrackers and garlic crushers.

They all seemed to be waiting. But waiting for what? Nobody seemed to know. There was a rumour that a small teaspoon had gone missing and that they shouldn't begin until it showed up. The hanging around, however, was taking its toll. Tempers were fraying. The forks were fidgeting, the spoons were pushing and shoving among themselves, while the steak knives – never ones to be messed with – were beginning to throw their weight about, jostling the ladles and trying to pick fights with everyone apart from the meat cleavers. At the front of the central section – primly keeping itself to itself – the egg slicer pinged and twanged impatiently.

When *was* it all going to begin?

Just then, there was a movement from the front. A pair of sugar tongs strode pompously across the dusty ground and climbed on to a small boulder. The coloured moons glinted on its highly polished whorls and curlicues. It raised a single tong and tapped sharply on the rock.

Apart from the egg slicer, which had instantly fallen into expectant silence, none of the cutlery had noticed the newcomer at the front. It tapped again, more insistently this time.

A couple of spoons near the front nudged each other and shushed the others. A group of cheese knives cut their communication short, cocked their curved blades to one side and turned to face the front.

One by one, the individual pieces of cutlery fell still, until it was so quiet that you could have heard a toothpick drop.

There was a dull thud at the front of the central section as a silver toothpick dropped to the ground, followed by an apologetic murmur when it picked itself up again.

The sugar tongs tapped on the rock a third time, and raised one tong high into the air. Then, having satisfied itself that everyone was paying attention, it waved its tong with a flourish.

As one, the cutlery began playing and the air filled with a strange and haunting percussion. Jingling and jangling, clinking and clanking, the cutlery followed the commands of the conducting sugar tongs – now playing softly, melodically; now rising up, little by little, towards a glorious clashing crescendo.

Just as it was about to reach its loudest, the music was suddenly joined by another sound. It was deep and gravelly. It made the ground tremble. It was coming from the entrance to the cave.

The sugar tongs nodded approvingly, and urged the cutlery to play louder still. The cutlery obliged.

The next moment, a tendril of smoke – glittering in the coloured light of the moons – came coiling out from the inky darkness of the entrance to the cave. It looped and spiralled, and floated away. It was replaced by another, as the deep, rumbling roar grew louder.

Something was stirring.

The sky lightened, the moons set and the sun rose on another day in Muddle Earth. At the Horned Baron's castle, there was the sound of heavy footsteps and angry muttering as the Horned Baron himself ran down the castle steps and out into the garden.

'Where are they?' he grumbled. 'Where can they possibly be? I— *Ooof!* Who's that?'

'It's me, sir,' came a small voice. 'Benson.'

The Horned Baron looked down at the ground by his feet, where a small goblin with a big nose – one of the castle gardeners – lay sprawling in the dust. 'Well, watch where you're going, Benson.'

'I will, sir. I'm sorry, sir,' he said as he climbed to his feet and brushed himself down. 'I was just putting the final touches to the garden,' he explained, and swept his arm round in a wide arc.

The Horned Baron looked round. With the trestle tables standing bare, the tents and stalls lying on the ground waiting to be erected, and a small army of goblins rushing this way and that, arms full and constantly bumping into each other, the scene was one of absolute chaos. From the look of things, they'd hardly even *started* to get the garden party ready.

'Never mind all that,' he said, grabbing Benson by the sleeve. 'Have you seen them?'

'Seen who?' said Benson.

'Not *who*, you imbecile,' bellowed the Horned Baron. '*What!* The Baroness's sugar tongs! Have you seen them?'

The gardener shook his head slowly. 'I'm afraid I

haven't, sir,' he said. The Horned Baron sighed with irritation. 'They must have disappeared like the others,' Benson went on.

'The others?' the Horned Baron roared. 'What do you mean, the others?'

'All the other cutlery. Cook can't find any of it anywhere. Disappeared, it has. All of it. There isn't so much as a silver toothpick left in the whole of the castle.'

The Horned Baron blanched. 'Are you telling me . . . ? Do you mean to say . . . ?' He swallowed. 'Are there at least some butter knives? Tell me there are some butter knives.'

'No.'

'Teaspoons?'

'Not a single one.'

'And what about . . . ?'

'All gone, sir. From the carving knives to the ladles for the mulled punch. They've all vanished without a trace.'

'But this is an outrage!' the Horned Baron thundered. He turned pale. 'What's Ingrid going to say? She's already in a state over the sugar tongs. *"I can't possibly hold a garden party without any sugar tongs!"* – her very words. What on earth is she going to say when she discovers that the rest of the cutlery has disappeared as well?'

Just then, there was a loud *crash* behind them. Benson and the Horned Baron turned to see two goblins sprawling on the ground. 'I'm so sorry,' said one apologetically as he rubbed his head.

'My fault entirely,' said the other. 'I wasn't looking where I was going.'

The two of them were surrounded by pencils, sticks and playing cards. The Horned Baron bent down and picked up a pencil.

'What is this?' he said.

'It's a pencil, sir,' said the goblin.

'I can see that!' the Horned Baron snapped. 'But what's it doing here? What are all of them doing here?'

'It's cook's idea,' said the first goblin.

'She's using a sword to slice the bread and a ruler to spread the butter,' the second goblin chipped in.

'The pencils are for stirring the tea,' the first goblin explained.

The Horned Baron groaned wearily.

Benson picked up a pointed twig. He held it out for the Horned Baron to see. 'It's my own invention,' he said proudly.

The Horned Baron did not look impressed. 'A pointed twig?'

'Not just a pointed twig,' said Benson. 'It's a sugar lump

stabber. Sugar tongs are a thing of the past. They're history. Why, with a sugar lump stabber like this you can—'

Abruptly, the Horned Baron bent over double and clutched his head in his arms. 'This is a catastrophe,' he wailed. 'A calamity! Ingrid's going to go berserk. She'll never speak to me again!' He frowned thoughtfully. 'Then again, every cloud has its silver lining.'

He noticed a playing card at his feet. Benson and the two kitchen goblins watched him as he picked it up and turned it over. He took a deep breath.

'So, what is this and what's it to be used for?' he asked.

'It's a playing card,' said the first goblin.

'It's for . . .' The second goblin faltered. 'For . . .' Suddenly his face brightened. 'Playing *cards.*'

The Horned Baron's face turned from pink to red to purple. Benson butted in, fearing his employer was about to explode. 'They're to go in the games tent over there,' he said, pointing to a round, green tent sandwiched between a trestle table and a *Test Your Strength* booth. 'For our younger guests,' he explained. 'I've organized all sorts of entertaining pastimes. *Pin the Tail on the Pink Stinky Hog, Hunt the Exploding Gas Frog, Musical Worms, Pass the Pancake . . .*'

'Yes, yes,' the Horned Baron interrupted. 'Carry on.' He turned to go. 'I must go and tell Ingrid,' he said. 'She won't like it, though. She won't like it one little bit . . . *OOOOF!* Not again!' he bellowed as a second scurrying figure walked slap bang into him. 'Can't *anyone* in this place watch where they're going?'

'I'm so, so sorry,' lisped a soft, slightly muffled voice. The Horned Baron turned to see a tall, stooped individual dressed in a hooded cape. 'I didn't see you there, your Baron-ness. Careful with that stick, you could have someone's eye out with that . . .' His voice faded away. He shifted the heavy bundle from under his right arm, to under his left. 'So, where do you want it?'

'I beg your pardon,' spluttered the Horned Baron.

'This,' said the stranger, holding up the bundle. There were poles and hooks and a roll of lilac and shocking pink canvas.

'What is it?' asked the Baron. 'Tell me it's a set of cutlery. Please, I'm begging you.'

'Oh, you are a one,' came the voice from inside the hood. 'It's the Rose-Petalled Pavilion of Loveliness, of course.'

'Of course,' said the Horned Baron flatly. 'How silly of me. I need sugar tongs, not to mention knives, forks and spoons – and you bring me a Rose-Petalled Pavilion of Loveliness.'

'That's right, sir,' the voice confirmed. 'No garden party should be without one. Trust me. The Baroness will love it.'

'I do hope so,' said the Horned Baron. 'Put it up over there,' he said, pointing to an empty corner.

'WALTER!'

As his name echoed round the garden, the Horned Baron blanched. He looked up at the Baroness's bedroom window. 'Coming, my turtle dove,' he called back. 'I've got something here that's going to make you very, very happy.' He clutched the pointed twig tightly.

'It had better!' Ingrid's voice floated back, low and loaded with menace. 'Remember what happened last time you failed to make me very, very happy, Walter.'

The Horned Baron rubbed his ears gently, and winced at the memory. 'How could I forget?' he murmured. 'How could I ever forget?'

The air around Trollbridge was still, quiet – and decidedly niffy. The trolls' love of mangel-wurzels had one unfortunate side effect which afflicted them while they were asleep. A rasping sound, like a brass band tuning up, floated up from the sleeping town.

Batbirds roosting close by found their eyes beginning to water as the odour hit them. And as a gentle breeze got up, an unfortunate tree rabbit took the full blast, and tumbled from its high branch in a dead faint.

Perched on a stool at the centre of a great pile of rubbish – elbows on his knees and head in his hands – was the toll-keeper of Trollbridge's main gateway. A stocky character with tufty hair and protruding teeth, he had – like the whole of Trollbridge – overslept. His helmet was on crooked and there was a half-eaten mangel-wurzel in his lap. His snoring was slow and regular. His tummy gurgled and his trousers rumbled.

'*Pfeeeeeep!*'

From the back of his patched, baggy trousers came a movement. There was something in his back pocket, and it was trying to escape.

The troll held his breath for a moment as, still sound asleep, he reached round and scratched his backside. As he did so, the movement in his pocket grew increasingly agitated. The next moment, something shiny and silver popped up from the pocket. It was the head of a spoon.

'*Pfeeeeeeeep!*' went the troll, and a passing batbird fell from the sky.

The spoon wriggled, squirmed and tumbled out of the pocket and on to the ground with a soft tinkling clatter.

It gave a little sigh, picked itself up – and sighed again.

The early-morning sun glinted on the tiny spoon as it tripped daintily through the rubbish. Past a rusty watering can and a chipped plate it went, over heaps of nuts and bolts, and out on to the wide, dusty road.

Something was calling it. Something that could not be ignored.

Joe Jefferson was rudely awakened from a deep sleep by a loud *crash*. He rolled over, but kept his eyes clamped tightly shut. He listened intently.

There were various sounds to be heard. A clock ticking. Plates clattering. Someone in another room humming tunelessly . . .

Joe's heartbeat quickened. Could that be his bedroom clock ticking? Was it his mum preparing breakfast he could hear? Was that his dad humming in the shower?

Dare he open his eyes to see?

Slowly, he opened his left eye a fraction. The room was dark, its contents indistinct. So far so good. Was he back in his own bedroom once more, after what must have been the weirdest dream of his life?

His eyes snapped open.

No, he was not! He was in a hammock on a houseboat on a floating lake. What was left of his so-called warrior-hero outfit – saucepan lid, welly and a sackcloth cloak with a fake-fur trim – was lying about him, waiting for him to get dressed.

'Damn and blast!' he exclaimed, sitting bolt upright. 'I'm still here in Muddle Earth.'

'And good morning to you, too, I'm sure,' said Veronica huffily.

Joe turned to the budgie perched on the knotted cords at the foot of his hammock. He noticed her feathers were looking damp and dishevelled.

'I . . . I'm sorry,' he said. 'I just thought . . . hoped . . . ' His eyes misted over. 'Dreamed . . . '

'Ah, dreams,' said Veronica understandingly. 'I dream of

a nice little cage. Nothing fancy. A little mirror, some bird-seed, perhaps a little bell to tinkle if I get bored. But I have to make do with this houseboat and Randalf the so-called Wise. Randalf the *Mean*, more like it. A little mirror. I mean, is that too much to ask? Well, is it?'

Just then the door opened, and a short, portly individual with thick white hair and a pointy wizard's hat walked in. It was Randalf the Wise. Joe's dog, Henry, was by his side, dripping wet. The moment he saw Joe, he bounded across the room, jumped up at the hammock and began licking Joe all over his face.

'Morning, Joe,' said Randalf. 'I've just taken Henry for his early morning swim.' He patted his round stomach. 'Nothing like an early morning swim to set you up for the day.'

'Next time, wake me up before you dive in,' said Veronica peevishly, shaking water from her feathers.

'Ah, there you are, Veronica,' said Randalf brightly. 'Forgot you were on my head. Sorry about that.'

'This wouldn't happen if I had a nice cage, like a normal budgie,' said Veronica. 'And a little mirror, perhaps a bell if I got bored ...'

'You're not still going on about that, are you?' said Randalf. 'I've told you before, cages are for canaries. You're my familiar. Your place is here, where I can keep an eye on you.' He patted the top of his head. Veronica fluttered over and landed on it.

'Dreams,' she said, with a sigh.

CRASH!

It was the noise that had first woken Joe, only louder. And it was followed immediately by the sound of the door at the far end of the room slamming back against the wall. A massive, knobbly ogre hurtled in, a heavy frying pan clutched in one great fist.

'Norbert!' Randalf shouted. 'What *do* you think you're doing?'

'That elf, sir!' Norbert blustered. 'It's been at the snuggle-muffins.'

As Joe turned, he caught sight of something small and plump scurrying across the floor. The next instant, the frying pan crashed down heavily behind it, missing the elf by a fraction. The houseboat rocked and swayed.

'And stay out of my kitchen!' Norbert cried.

The elf skidded to a halt, and darted back between Norbert's legs. Norbert watched it going, his head getting lower and lower – until he collapsed in a heap.

The houseboat pitched about violently.

'Be careful, Norbert!' said Randalf. 'You don't want to capsize the boat.'

'*Again*,' added Veronica tartly.

'Sorry, sir,' said Norbert, as he climbed to his feet.

The elf made a dash for the clock on the wall. 'Half-past morning!' it shouted cheerily as it shimmied up the pendulum and disappeared through a small door above the clock face.

'Time for some breakfast,' said Randalf.

'Squashed tadpoles! My favourite,' said Norbert, examining the contents of Randalf's dripping hat. 'They're delicious frittered.'

'Ugh,' Joe shuddered.

'An acquired taste,' said Randalf nodding. 'And stiltmice are pretty tasty, too . . .'

'Tadpoles, stiltmice,' said Joe, shaking his head with disgust. 'When my mum makes fritters she uses pineapples, or bananas . . .' His face dropped. His lower lip quivered.

'Joe,' said Randalf, looking concerned. 'If these fritters mean so much to you, then perhaps . . . this evening . . .'

'It's not the fritters!' Joe shouted. 'It's my mum. And my dad. And the twins – and even Ella. I miss all of them.' He took a deep breath. 'I want to go home.'

Randalf clapped his hand on to Joe's shoulder. 'Believe me, my boy, there's nothing I'd like better than to send you home. I've been racking brains for a solution, but . . .'

He shrugged. 'Don't give up hope, Joe. I'll give the matter my full attention later. Something will turn up. I just know it will.'

Joe hung his head. He had no idea how long he'd been in Muddle Earth. Since the length of the days and nights never seemed to be the same from one day to the next, it was impossible to tell. All he knew was that Randalf had said the same thing on a dozen occasions before. *Something will turn up.* But would it? Why should this time be any different from all the others? He was about to say as much when he heard a weak knock at the door.

Randalf sat down at the small table. 'Bring on the fritters, Norbert, old fellow,' he said. 'I'm so hungry, I could eat a pink stinky hog!'

'Wasn't that the door?' said Joe.

Norbert frowned and scratched his head. 'It still is,' he said. 'Isn't it?'

'I didn't hear anything,' said Randalf.

'Neither did I,' said Veronica.

There was a second knock, even feebler than the one before – followed by a squeaky little sneeze.

'*Atish-ii!*'

'You're right,' said Randalf. 'Wonderful hearing my boy – *warrior-hero* hearing, one might say. Get the door, there's a good chap, Norbert,' said Randalf.

Norbert hesitated. 'You mean it *is* a door,' he said. 'For a moment, I wasn't sure . . .'

'Of course it's a door,' said Randalf.

The third knock was followed by a second sneeze and a long, weary groan.

'Just, open it, Norbert!' said Randalf. 'Now.'

Norbert strode back across the room and pulled the door open. And there – silhouetted against the low sun – was a short, bony creature, dripping with water from head to foot and standing in a pool of his own making. The peaked cap he was wearing bore the insignia E.M.

He pulled a soggy envelope from the inside of his saturated mailbag and held it up.

'Imp ... *atish-ii.* Import ... *atish-ii.* Important ... *atish-ii! atish-ii! atish-ii!*' He pulled a handkerchief from his pocket, wrung it as dry as he could, and blew his nose upon it. 'Of all the stupid places to live,' he complained, 'you lot had to choose the middle of a floating lake. Have you any idea how long it's taken me to swim up that waterfall? I mean, I'm not one to complain—'

'Glad to hear it,' said Randalf sharply. 'Now hand over the envelope.'

'Not to you,' said elf.

'And why not, pray?' said Randalf, affronted.

'Because you're not the person named on this envelope,' the post-elf told him. 'The directors of Elf Mail take a very dim view of letters, cards, parcels or packages being handed over to the wrong person.'

'But if you've come to the right address, it *must* be for

me,' said Randalf. 'Unless it's for Norbert here. Or Veronica.'

The post-elf looked from one to the other, before shaking his head. And for a foolish moment, Joe wondered whether it might be for *him*.

'Who *is* it for, then?' Randalf demanded.

The elf looked down. 'Grand Wizard . . .'

'Well, that certainly rules *you* out, Randalf,' Veronica muttered.

'Shut up, Veronica,' said Randalf.

'Grand Wizard, Roger the Wrinkled,' the elf announced. 'And you,' he said, pointing accusingly at Randalf, 'are not wrinkled. Roger the *Fat*, maybe, but not Roger the Wrinkled. Besides,' he added, 'the canary called you Randalf.'

'The *canary*!' Veronica squawked. 'How dare you!'

Randalf drew himself up to his full height. 'I am Randalf the Wise,' he announced, 'personal assistant to Grand Wizard, Roger the Wrinkled who, while away on . . . on vacation, has left me in charge.' He plucked the envelope from

the elf's hand. 'I am authorized to deal with *all* his corre-spondence.'

The elf made a grab for the letter, but Randalf was quicker and hid it behind his back. The elf looked close to tears.

'I'll get into trouble,' he said. 'They'll take away my peaked cap and badge.'

'*I* won't tell if *you* won't,' Randalf reassured him. 'It will be our little secret.'

'What about the canary?' asked the elf suspiciously. 'Can it be trusted?'

'You've just delivered your last letter, postie,' Veronica muttered.

'Her beak shall remain sealed,' said Randalf. 'Trust me, I'm a wizard. Now, off you go.' He turned to the ogre. 'Norbert, show the elf the door.'

Norbert pointed to the door. 'That's the door,' he said.

With the post-elf gone, Randalf scanned the envelope. 'An elegant, noble hand,' he said of the writing. He raised it to his nose and breathed in deeply. 'And with, if I'm not very much mistaken, the faintest scent of rose petals . . . I wonder who it could be from? A sorceress, perhaps? Or a princess?'

'Why don't you open it and see?' said Veronica.

'Because, my impatient feathered friend, half the pleasure of receiving an envelope is wondering what it might contain,' said Randalf. 'At the moment, it could be anything.' He pushed his finger into the fold of the envelope and tore along the top. 'A love letter, a cheque, notification of some great success . . .'

'Or a final demand for payment,' Veronica noted.

'We shall see,' said Randalf. He reached inside the envelope. His finger and thumb closed around a large pink and white card. 'Ah, the thrill of anticipation!'

He pulled the card out and scanned it quickly. His eyebrows shot up.

'Well?' said Veronica. 'Good news or bad news?'

'It is an invitation,' said Randalf.

'Good news, then,' said Veronica.

'From the Horned Baron and his lady wife . . .'

Veronica groaned. 'Spoke too soon.'

'Read it out, sir,' said Norbert.

Randalf nodded and cleared his throat. '*Dear Roger the Wrinkled . . .*'

'I don't know,' Veronica grumbled. 'Reading other people's letters. It's disgraceful.'

'Shut up, Veronica,' said Randalf. 'Where was I? Ah, yes. *The Horned Baron, Lord of Muddle Earth, Emperor of the Far Reaches and Monarch of the Glen; beloved, munificent, bountiful, much-loved, fair-minded ruler of . . .*'

'Get on with it,' said Veronica.

Randalf frowned, and continued reading. '. . . *and his beautiful wife, Ingrid . . . blah blah blah . . .* Ah, here we go,' he said. '. . . *do cordially invite Roger the Wrinkled and his fellow wizards to a Garden Party, to be held in the well-maintained,* spacious, luxuriant grounds of their beautiful ancestral castle. (Turn left at the Musty Mountains and follow your nose) . . .'

'Yes, yes,' said Veronica impatiently. 'We all know where his castle is. But when is this garden party?'

'Ah, yes, of course,' said Randalf. He returned his attention to the invitation. 'I . . . errm . . . Oh, Great Moons of Muddle Earth! It's today! At two o'clock this afternoon! And they want a wizard, preferably Roger the Wrinkled, to open it.'

'They'll be disappointed, then,' said Veronica. She snorted. 'Garden party, indeed! Have you seen the state of the Horned Baron's garden? Why anyone in their right mind would want to have a party in it beats me . . .'

'You're missing the point, Veronica,' said Randalf. 'The fee for a wizard cutting the ribbon and saying a few words at the opening of a regal garden party is three gold pieces and as much blancmange as you can eat. I'm down to my last brass muckle,' he added woefully. 'I can't afford to miss such an opportunity . . .'

'But how will we get there?' Veronica persisted. 'You said it starts at two o'clock.'

'Quarter to afternoon!' chimed the clock-elf, putting his head out of the door.

'We do what we always do when we need to get some-where really, really quickly,' Randalf replied.

Norbert paled. 'Not the winged boots . . .'

'There's no other choice,' said Randalf firmly.

~ 173 ~

Joe turned to Norbert. 'The *what?*' he said.

'Remember what happened last time,' said Veronica with a shudder. 'Some wizards never learn.'

Randalf clapped his hands together urgently. 'Chop-chop, everyone,' he said. 'Let's get this show on the road.'

'But what about me?' said Joe.

Randalf smiled. 'There's always room on Norbert's shoulders for a warrior-hero, my lad,' he said. Henry barked and wagged his tail. 'Yes, and for his faithful battle-hound.'

'That's not what I meant,' said Joe. 'You promised to help me get home. *I'll give the matter my full attention,* you said.'

'*Later,*' said Randalf. 'I said I'd do it later. And I shall.'

'But when?' said Joe.

'We'll find a way,' he said, cheerily. 'But for now . . .' He shrugged. 'Duty calls. My hands are tied.'

The elf leaped out of the clock. 'You'll never make it,' it laughed. 'You'll be late, late, late!' it said, and collapsed in a fit of hysterical giggles.

'Too cheerful by half, that clock,' muttered Veronica. 'It needs to be wound up.'

'You're right,' said Randalf. 'I'll do it now.' He turned to the clock. 'You're a pathetic, miserable excuse for a time-piece, what are you?'

The furious elf grimaced. 'Are you talking to me?' it demanded in its most threatening voice. 'Are you talking to *me?*'

'There,' said Randalf, 'I've wound it up. Now let's get going. There isn't a moment to lose.'

As the sun rose in the sky, the tiny teaspoon picked its way, slowly and carefully, from tussock to scented tussock, sighing as it went. Drawn on by a strange force, the teaspoon had left Trollbridge and travelled down the dusty Ogrehill road, before turning off into the Perfumed Bog.

Pausing for a moment on top of a particularly highly perfumed tussock, the tiny teaspoon turned its bowl, as if to listen. From somewhere to its left there came a wheezing, squelchy-squelchy noise.

It was closer than before. The teaspoon sighed, trembled and leaped to the next tussock.

And the next.

And the next.

In front of it, something glinted and twinkled in the long grass. The spoon kept on, picking itself up when it fell, refusing to give up. The glinting and twinkling grew brighter.

All at once the grass parted and there, crouching down in the perfumed sludge, was an exploding gas frog – and an enormous one at that. It winked one bulbous eye, then the other. It shifted forwards on its massive forelegs, ready to strike. The warts all over its purple skin throbbed ominously.

The teaspoon slipped as it landed, then picked itself up once again. 'Ah,' it sighed.

The exploding gas frog reared up. Its warty lips parted and a long, thick, sticky tongue flashed through the air and wrapped itself around the tiny teaspoon.

As it disappeared inside the dark, fetid moistness of the gas frog's greedily waiting mouth, the teaspoon let out a last, lingering sigh.

'Aaaaa . . .'

The gas frog snapped its jaws shut, swallowed and grinned contentedly. It turned lazily around, and was just about to hop off back to the ooziest part of the bog where it could digest its meal undisturbed, when something started to happen.

First, a low gurgling noise came from the pit of the gas frog's stomach. Then its warty skin began changing colour – from purple to red to green to orange. Its grin became a grimace.

'*Gribbit,*' it croaked in alarm. '*Gribbit. Gribbit . . .*'

It tried jumping up and down on the spot, it tried beating itself on the back, it tried falling heavily to the ground – but all to no avail. The teaspoon was stuck fast.

The gas frog rolled about helplessly. Its eyes bulged, its tongue lolled, its limbs stuck out rigid and useless. It shuddered and juddered, unable even to croak, and swelled to an immense size. The skin was stretched so taut and so thin that at any moment . . .

BANG!

The sound of the gas frog exploding echoed all round the Perfumed Bog, causing slimy bog demons to dive for cover and pink stinky hogs to break wind. It was deafening. And, when the remains of the hapless gas frog finally fluttered back to earth, also rather messy.

Flying high above the Perfumed Bog in a great, wide arc, the tiny spoon sighed.

At the entrance to the cave, the orchestra was in full swing. To the enthusiastic conducting of the sugar tongs, the spoons were clinking, the knives were clanging; ladles clashed, cake forks clicked – and the egg slicer attempted a rather ambitious solo, which ended in a nasty tangle with a carving knife.

From within the cave, the wispy coils of smoke grew thicker, denser. The orchestra played louder than ever.

Clatter! Clatter! Clink! Clonk!

All at once, cutting across the strange metallic music, came a loud hissing whistle, like a great locomotive letting off steam. Thick grey and white clouds of smoke and steam billowed from the cave and swallowed up the ranks of the cutlery orchestra. Yet still they played on.

Clatter! Clatter! Clink! Clonk!

Then, as the air cleared, a snout with two smoking nostrils could be seen protruding from the entrance to the cave. It sniffed at the air, it trembled – and came a little further forward.

Slowly, slowly, the rest of the great, scaly head appeared. A pair of heavy lids rose, one after the other, to reveal two yellow eyes which looked around in bemusement before focusing in on the wide array of shiny cutlery spread out before it. Its scales rattled as it quivered with obvious delight.

At the sugar tongs' command – and without missing a beat in the music – the orchestra took a step backwards. The creature emerged a little further. A long, scaly neck came into view.

Again, the sugar tongs directed the orchestra to fall back; again, the creature advanced.

An armoured body appeared, and four long, scaly limbs, each one ending in calloused knuckles and taloned toes.

Clatter! Clatter! Clink! Clonk! Clatter! Clatter! Clink! Clonk! The orchestra played on, taking another step back with every *clonk.*

Inch by inch, the creature slowly emerged from the safety of its dark, shadowy cave and out into the bright morning sunlight. It noted the shape and size of each knife, it sniffed at the skewers and spoons – but the further out it came, the further back the orchestra retreated.

Growling menacingly, the creature reared up on its hind legs, flapped its pair of broad, leathery wings and lashed its long serpentine tail. Then, looking up at the sky, it snorted wildly. Two plumes of thick, black smoke emerged from its nostrils and, when it opened its jaws, a searing flame of orange and red accompanied its loud, resounding roar.

The huge, magnificent and fearsome beast had, for many long years, been coiled up around its treasure, sleeping. Now, the cutlery had woken it up. There they were, before it, sparkling brightly in the sun and making such sweet music. The dragon wanted them.

If only they would keep still!

For a second time, the dragon tilted its head back and roared. The tongues of flame licked at a passing squadron of flying wardrobes, scorching their doors – and sending one into a fatal tailspin. It came to earth in the foothills of the

Musty Mountains with a loud, splintering crash.

The cutlery trembled.

Clatter! Clatter! Clink! Clonk!

They fell back another step.

The dragon lowered its head and eyed the orchestra greedily. The sight and sound of the enticing silver cutlery had left it so excited it could barely control itself.

It wanted to add the gleaming silverware to its great glittering hoard. It wanted to possess it; to count it, to sort it, to polish and caress it. It wanted to feel the smoothness of the spoons, the prick of the forks; it wanted to wrap its coils around the cutlery and guard and protect it until the end of time.

The dragon's eyes narrowed as the cutlery backed away. What were they all playing at? Why were they teasing so dreadfully?

The dragon's muscles quivered. Its tail switched this way and that, stirring up the dust. Its nostrils smoked. It was all the creature could do to hold itself back from making a roaring dash at the orchestra – but if it did that, the cutlery might scatter and it would end up with nothing but a few soup spoons.

No, the dragon would have to be cleverer than that. It crouched down, its low-slung body close to the ground and its great, muscular haunches quivering. Then, with one eye fixed on the orchestra, it turned its head, as if about to leave.

The music faltered as the cutlery wondered what to do next. The sugar tongs raised one arm and beckoned.

The orchestra advanced a step.

In an instant, quick as a flash, the dragon twisted round and pounced. The great, scaly beast landed before the orchestra in a cloud of dust and whisked its tail round in a large circle that penned them all in. They were trapped, each and every shiny one of them.

At least, that was what the dragon thought. The cutlery, however, had other ideas. Already, at various points, pieces were breaching the tail-wall and breaking free.

A posse of egg spoons was scaling the long, arrow-shaped tip to the dragon's tail; half a dozen ladles were helping each other over the upper-section, while on the ground in the middle, the egg slicer was jumping down on the curved prongs of a large fork, catapulting knives perched at the other end to freedom.

The dragon bellowed with rage so loudly that the mountains shook. It clacked its talons and flicked its tail up into the air, sending those hapless spoons and forks still clutching on to its scales hurtling up into the sky and away. At the same time, the rest of the cutlery orchestra took the opportunity to make a run for it.

They dashed off across the dusty mountain plateau with remarkable speed and agility. The dragon was left standing. Thrusting forwards, it opened its jaws and sent a blazing torch of fire roaring after them. The flames scorched the ground – and burnished the backsides of a couple of slow dessertspoons. But the cutlery did not stop.

One by one, the stragglers and strays were returning to the main group. Together, they all darted off down a narrow track between two huge boulders.

The dragon stood, perplexed. One moment, it had captured the most wonderful collection of gleaming silverware to add to its hoard; the next, it had lost them all!

In a flurry of dust, smoke, flashing talons and flapping wings, the dragon hurtled across the plain after the fleeing silverware. It would not give up; it *could* not.

The chase was on.

The Horned Baron stood at the top of the castle staircase looking out across the mountains. He glanced at his large, gold pocket watch for the fifth time in as many minutes and shook his head.

'Ten past two, and still no sign,' he muttered. 'What on earth can have happened to them?' First the cutlery, and now this. A garden party without wizards? It simply wouldn't do! Ingrid would never forgive him. 'Where are they?' he groaned. 'Where *are* they?'

'They're here,' said Benson.

'The wizards?' said the Baron excitedly.

'No,' said Benson amiably. 'The sugar-stabbers you

dropped just now. Luckily I picked them up.' The goblin held out a sharpened twig.

'Oh, that,' said the Horned Baron.

'Was the Baroness pleased with them? Was she?' asked Benson, excitedly.

'Not exactly,' said the Horned Baron, rubbing his arm and wincing.

'WALTER!' Ingrid's voice sliced through the hazy afternoon sunshine like a blast of icy air. The Horned Baron shivered.

'Sugar tongs are the least of my worries. Right now I need a wizard. *Any* wizard,' he said nervously. 'Yes, my angel,' he called back.

'Don't you *angel* me,' Ingrid's strident voice rang out. 'You're nothing but a great, big, useless, good-for-nothing lump!'

'I am?' called the Horned Baron.

'Yes, you are, Walter! My corset's burst! Stupid, cheap, flimsy thing!' she complained. 'I don't know where you got it from.'

'The mega-turbo girdle?' the Horned Baron muttered in disbelief. 'The heavyweight, super-reinforced model? Flimsy?' He groaned softly. 'It cost me a fortune . . .'

'Did you hear me, Walter?' Ingrid shrieked. 'I don't know, call yourself a Horned Baron! You can't even do the simplest things right.'

'A thousand apologies, light of my life,' the Horned

Baron called back wearily. 'I shall be up at once.'

'And bring twenty-five metres of tent-cloth with you,' she demanded. 'We'll have to alter the dress. And don't forget your needle and thread this time!'

'As your gorgeousness desires.' The Horned Baron groaned.

'If there's anything I could do to help,' said Benson.

'I think you've done enough,' said the Horned Baron, snapping the sharpened twig and dropping it on the step. He turned away and was about to stride off when a thought occurred to him.

He looked Benson up and down – at his stooped, angular body, at his little beard. With the right clothes, he would make an excellent wizard.

'Actually, on second thoughts, Benson, there *is* something you could do,' he said. 'Get yourself kitted out in a gown and a pointy hat, and meet me back here in . . .' He glanced at his watch. It was eight minutes past. 'In twenty-two minutes.'

'WALTER!' Ingrid shrieked, piercingly loud. 'I'M WAITING!'

'Slow down, Norbert!' Joe cried out, as he lurched this way and that on top of the ogre's left shoulder. He gripped Norbert's collar as tightly as he could with one hand, and held

Henry, who was sitting on his lap, with the other. 'Slow *down*!'

'Can't, sir,' said Norbert breathlessly. 'It's these dratted winged boots.'

'Please try!' Joe pleaded. 'I almost fell off just now!'

They seemed to be taking a short cut down a very steep Musty Mountain track.

'*Whooooah!*' Norbert cried, his arms windmilling wildly as he struggled to maintain his balance.

'This is all Randalf's fault,' said Veronica, who was clinging on to the rim of the wizard's pointy hat. She bent over and bellowed in his ear. 'Wake up, you ridiculous little man! Wake up!'

But Randalf merely snored a little louder. He always fell asleep when riding on Norbert's shoulder – and the bumpier the ride, the deeper the sleep.

There was a violent jolt and, for a moment, Joe was falling backwards into the blurred beige and khaki landscape, while Henry was sliding forwards. Then Norbert stumbled a second time. Joe grabbed at Henry, and clutched a crease of material in the ogre's jacket.

'NORBERT! SLOW DOWN!' roared Joe and Veronica together, with Henry barking his agreement.

'Can't ... slow ... down ...' Norbert gasped, every word an effort, as he careered down a dry gully and bump-bump-bumped his way over a section of large, round pebbles. 'Boots ... won't ... let ... me ...'

Joe looked down. Norbert was wearing a pair of

winged boots with four wheels beneath the soles and a spoked wheel sticking out of the heels at the back – just like Joe's Rollerblades back home. The ogre had seemed fine with them, at first . . .

'The Winged Boots of Colossal Speed,' gasped Norbert. 'I'm OK going uphill. It's coming down the other side I need a little practice at.'

Just then, the ground in front of them fell away completely. Norbert's legs pedalled furiously in thin air. Veronica screeched. Henry whimpered. Joe gritted his teeth and waited for the inevitable jolt.

'*Ooof!*' he gasped a moment later as Norbert landed. They were back on the road again – and travelling as fast as before.

'I warned him!' Veronica shrieked. 'I *pleaded* with him not to put Norbert in the Winged Boots of Colossal Speed.'

'You did?' said Joe.

'And would he listen?'

'Obviously not,' said Joe.

'Watch out!' squawked Veronica.

'Oh, help!' Norbert cried out. 'I think I'm about to take another short cut . . . *Whoooooooaah!!!*'

Dropping down out of the sky towards the sandpit, the tiny teaspoon was nearing the end of its brief flight. With a soft swooshing noise – and the hint of a little sigh – the teaspoon landed in soft sand and buried itself up to its handle.

Some way to its right, a family of goblins was seated round a blanket.

'There's not enough sand in this!' said the youngster.

'Shut up and eat your snotbread sandwich, Gob,' his dad told him. 'Your mother spent ages preparing a lovely picnic, and all you've done since we arrived here is complain.'

'Don't go on,' said Gob sullenly. He took a bite of the sandwich. 'What's that?' His eyes lit up with excitement.

'What?' said his mother, her mouth full of cold bad-breath porridge.

'That,' said Gob, jumping up and running over to where the tiny teaspoon had landed. 'Mum! Dad! I've found something!' he called.

'Hog poo?' said his mum.

'Stiltmouse dribble?' said his dad.

'No, it's not something to eat,' said Gob. He crouched down, took hold of the piece of silver sticking out of the sand and pulled. 'It's a . . . a teaspoon,' he said, holding it up. 'A tiny silver teaspoon. Can I keep it?'

'If you like, dear,' said his mother. 'Pop it in your pocket so it doesn't get lost, and let's be on our way. If we set off now, we should be back at Goblintown by teatime.' She smiled. 'You can use your new spoon to stir your spittle tea.'

As it dropped down into the pocket, the teaspoon let out a little sigh. It was warm and moist and smelly in the goblin's trousers, and – the teaspoon trembled – very, very dark.

As they approached the bottom of the Musty Mountains, the road levelled out. In the distance, Joe glimpsed the high turrets and castellations of the Horned Baron's castle and heard garden-party-like sounds floating in on the breeze – a low murmur of voices, a chinking of teacups on saucers and, every now and then, names being announced over a megaphone.

Norbert stopped skating and came to a halt. His heart was thumping; his legs trembled with fear and exertion.

'You know,' he panted, 'I think I'm really getting the hang of these winged boots. Why, with a little more practice . . .'

'. . . You might succeed in breaking our necks,' said Veronica coldly.

Randalf's eyes snapped open. 'Are we there yet?' he said, looking round at the bleak, rocky landscape.

'Almost,' said Norbert, still panting.

Randalf cocked his head to one side. He listened

dreamily to the babble of the guests, the chinking of the cups and saucers – and the band, which consisted of a drum, cymbals, a wheezing set of bagpipes and something that sounded like a pink stinky hog having its tail pulled.

'The garden party must have already started!' he exclaimed. He glanced at his watch. 'Three o'clock! Oh no, we're late! What *is* the Horned Baron going to say?'

'How about, *You're late, you useless, good-for-nothing, miserable excuse for a wizard . . .*'

'Shut *up*, Veronica!' said Randalf. He leaned down and tapped the ogre's head with his staff. 'Proceed, Norbert,' he said. 'To the Horned Baron's castle as fast as you . . . Joe! What do you think you're doing?'

Joe jumped to the ground. He put Henry down beside him. 'If it's all the same to you, we're going to walk the rest of the way,' he said.

Veronica launched herself off from the brim of Randalf's hat, flew down and landed on Joe's head. 'Good idea,' she said.

Norbert lurched forward unsteadily.

'Whooah!' cried Randalf. 'On second thoughts, put me

~ 191 ~

down, Norbert, and take off those boots. They're quite unsuitable for a garden party.'

'But I was just getting the hang of them,' said Norbert, promptly falling over.

'No arguments,' said Randalf, picking himself up. 'Follow me, everyone!'

They continued in silence. The sounds of merry-making grew louder. Various smells began to permeate the mustiness of the mountains – newly baked cakes, hot toffee, candyfloss. And as they rounded the corner, an imposing grey fortress rose up before them.

'The Horned Baron's castle!' Randalf announced and, leaving the others to catch up, he strode ahead to the main gate.

A seated guard looked up from a tattered scroll with a list of names.

'I am Randalf the Wise,' Randalf announced importantly. He waved Roger the Wrinkled's invitation under his nose briefly, then stuffed it back into his pocket. 'But I may be on your guest list as Roger the Wrinkled. And these,' he said, pointing to the others, 'are . . .'

'Yeah, yeah,' said the guard, stifling a yawn. 'Whatever.' And without even referring to the scroll, he waved them all through.

Somewhat peeved, Randalf stepped through the archway and into a courtyard. A goblin in a grubby costume approached, with a megaphone in one hand. 'I'm the herald,' he said. 'How would you like to be announced?'

'Randalf the Wise,' Randalf told him. 'Grand wizard.'

'Could have fooled me,' muttered Veronica.

'Eh?' the guard shouted. 'Speak up. I said, speak up.'

'My name,' he bellowed, 'is Randalf the Wise! And these,' he continued, introducing the others in his loudest voice, 'are Norbert the Not-Very-Big, Joe the Barbarian and his battle-hound, Henry, and Veronica, my familiar.'

The guard turned and raised the megaphone to his lips. 'Dandruff the Wide,' he announced. 'Halfwit the Non-Furry Pig . . .'

'Give me that!' said Randalf impatiently. He seized the megaphone from the guard and announced himself.

'Stop that!' said the herald, grabbing the megaphone back and hiding it behind his back. 'Announcing the guests is *my* job.' He turned to Joe and Henry. 'What did you say your names were again?'

'Hurry up and get on with it,' said Randalf irritably.

'Hurry-Up! and Getonwithit!' announced the herald through the megaphone.

'Come on,' said Randalf, ushering the others through the gate. 'Let's find the Horned Baron. Maybe there's still time.'

Veronica snorted. 'Time for what? In case you hadn't noticed, they've started without us.'

A waiter approached them, a wooden tray balanced on his upraised hand. 'Cup of spittle tea?' he said, gruffly. 'Cough sandwich?'

'No, thanks,' said Joe, feeling slightly queasy.

'Did I hear someone mention spittle tea?' said a tooth-less goblin who was passing by.

'That's right,' said the waiter. He poured some of the frothy liquid from the pot into a grimy looking cup, added some brown sugar lumps and a pencil, and handed it over.

'Ooh, lovely,' said the goblin. 'But what's the pencil for?'

'Sorry,' the waiter said. 'I'm afraid we haven't got any spoons.'

'Never mind,' said the goblin, tossing the pencil over his shoulder. He spat into the teacup and stirred it up with a dirty finger.

'Ugh,' Joe groaned, and turned away.

In front of him were stalls, booths, tents, trestle tables and marquees all crammed together, higgledy-piggledy, inside the castle garden – though garden was hardly the right word for the courtyard, with its high surrounding walls and single dead tree.

In one corner was a pot filled with limp pansies. In the one opposite, was a lawn the size of a tablecloth – cast in shadow by a large sign warning visitors to *Keep off the Grass*. And in the middle – in the shadow of the dead tree – was a small birdbath, with several lazybirds asleep beneath it.

If the *garden* bit of the garden party was a disappoint-ment to Joe, however, the *party* bit was not. Everyone was eating cake with pointy twigs and stirring their tea with pencils. On a tatty bandstand, a tatty band played an assortment of tatty instruments. Joe smiled. The instrument that had sounded like a pink stinky hog having its tail pulled, was in fact a pink stinky hog having its tail pulled. Next to it, a small goblin hit himself on the head with a cymbal, and a large troll played a wheezing solo on a battered set of bagpipes.

'Roll up! Roll up!' came a voice from his left, shouting above the music. '*Smells in a Jar.* Come and try my *Smells in a Jar.*'

Joe joined a small crowd of trolls, goblins and assorted others clustered in front of a big trestle table. He saw the stallholder at the front select one of the tall jars from the table and hold it up.

'You sir,' he said. 'You look like a goblin of discernment.'

He held the jar out towards a lanky goblin, who nodded. The glass stopper was removed. The goblin leaned forwards, closed his eyes and sniffed.

'*Mmmmmm!*' the goblin rolled his eyes. A huge smile spread across his face. 'Don't tell me, don't tell me . . .'

'Well?'

'I'm getting smelly sock – left foot, I'd say. And just the faintest hint of ogre's underpants.'

'Very good,' said the stallholder. 'I call it *Gym-Kit.* It's one of my most popular smells, after *Spilt-Potty.*'

Chuckling to himself, Joe continued through the jumbled maze of amusements and refreshments, exhibitions and competitions. He paused by a big striped tent with a handwritten sign tied up above the open door-flaps.

Mangel-wurzel Judging in Progress.

Curious, Joe went to the entrance and peered inside. The tent was heaving with trolls, standing about in twos and threes and discussing the mangel-wurzels laid out on the table before them.

'This one's really big,' he heard one saying.

His neighbour nodded. 'And so's this one.'

'Yeah. This one's big too, ain't it?' said another.

'Very big. That one there's big as well.'

'And that one, that's big. The one over there, that's big an' all. And the one next to it, and the one next to that, and . . .'

Joe walked on, past a toffee-nose stall, where you could dip your nose in a bucket of warm toffee, and a candyfloss stall where you could floss your teeth with candy. He was looking at an interesting display on a stall marked *Broken, Missing or Useless*, when he realized that neither Randalf, Veronica nor Norbert were anywhere to be seen. Even Henry had disappeared.

He decided he'd better retrace his steps. It was only when he arrived back at the stage – where the band had been replaced by three rows of tap-dancing stiltmice – that Joe spotted Randalf.

'Randalf!' he cried. 'Randalf!'

The wizard turned and greeted him. 'There you are, my boy. I was just about to ask the herald to call you. Now

where have the others got to?'

'I thought they were with you,' said Joe.

Randalf frowned. 'What do you suppose that is?' he said.

Joe followed his gaze. He was staring at a flamboyant marquee – lilac and pink, with a floral pattern and silver trimmings – which stood in the corner by the wall.

'The Rose-Petalled Pavilion of Loveliness,' said Joe, reading the florid sign outside the marquee.

'Sounds delightful,' said Randalf. 'Remind me to check it out later. But first, we've got business to attend to!'

Just then a mysterious figure in a long, hooded cape pushed a pink tent-flap open and slipped out of the pavilion. He looked round furtively and scuttled into the crowd.

'Randalf the Wise and Joe the Barbarian, if I'm not mistaken,' came a petulant voice behind them. Randalf and Joe spun round to see the Horned Baron glaring back at them. 'Where's Roger the Wrinkled?' he demanded. 'He was meant to be here at two o'clock to open the garden party.'

'He's . . . *errm* . . . otherwise engaged. Top secret,' Randalf added, and tapped his nose conspiratorially. 'He sent me in his place.'

The Horned Baron rolled his eyes. 'I might have known,' he said. 'You're late! And Ingrid's furious!'

Randalf tutted sympathetically. 'That's Elf Mail for you!' he said. 'The invitation only arrived this morning. We got here as quickly as we could.'

'*Hmmph*,' said the Horned Baron. 'It was most inconvenient. I tried dressing up Benson as a wizard, but Ingrid wasn't fooled for a moment. And then there was nothing to cut the ribbon with. All the cutlery's disappeared!' He sighed. 'Had to get an elf to chew through it! Absolute catastrophe, it was.'

'Tragic,' said Randalf. 'Still I'm here now. If there's anything I can do, for a small fee . . .'

The Horned Baron's expression darkened.

'I don't know,' Randalf went on, 'any awards to be presented, cups or medals to be given out, then I'm your wizard.'

'You're no Roger the Wrinkled, that's for sure,' said the Horned Baron. 'But you'll just have to do . . .'

'*WALTER!*'

The Horned Baron trembled from head to foot. 'Oh, good grief,' he muttered. 'What *now*?'

'*WALTER! WHERE ARE YOU?*'

'Coming, my precious!' he called back. He turned to Randalf. 'Must go!'

'Indeed, Horned Baron,' said Randalf. 'And as I say, if there's *anything* I can do. Anything at all . . .'

At that moment, the megaphone burst into life. 'Would the Horned Baron go immediately to the Rose-Petalled Pavilion of Loveliness. Horned Baron to the Rose-Petalled Pavilion of Loveliness. Horned Baron. Calling the Horned Baron . . .'

The Horned Baron turned, then hesitated. 'If you want to make yourself useful, Randalf,' he said, 'go to the marquee and find out what they want. It's that one,' he said, pointing rather unnecessarily at the lilac marquee with the shocking pink rose petals.

Randalf nodded. 'Of course, sir,' he said, 'and . . .'

'*WAL-TER!!*' Ingrid's voice cut through the air like a rusty axe.

'Go, Randalf!' said the Horned Baron sternly. He tugged at his sleeves, straightened his helmet and turned to leave. 'And whatever it is, sort it out. There's a good wizard.'

Randalf turned to Joe. 'Duty calls,' he said. 'Go and find the others. Henry's with Norbert in the Snuggly-Wuggly Corner I expect.'

'Snuggly-Wuggly Corner?' said Joe.

'Next to to the Face-Smudging.'

'Don't you mean Face-*Painting*?' said Joe.

'You haven't seen the troll who's running it,' said Randalf. 'Oh, and Veronica said she wanted to check out the birdbath. Can't think why . . .' He glanced at his watch. 'We'll all meet back here in half an hour. All right?'

Joe nodded. 'Half an hour,' he said.

Randalf watched Joe disappearing into the crowd. He was a good lad. It was such a shame he wasn't able to send him home. He looked round at the lilac-and-pink Pavilion of Loveliness and sighed.

Joe reached the Snuggly-Wuggly Corner just in time to

prise Henry away from an over-exuberant ogre. Norbert was sitting beside them on the floor, sucking his thumb and hugging a frayed toy rabbit. When Joe found Veronica, she was on the birdbath discussing bird cages and small mirrors with several lazybirds.

Back at the Rose-Petalled Pavilion of Loveliness, Randalf lifted the tent-flap and disappeared inside.

Randalf looked round. The marquee was large, carpeted, bathed in a soft lilac-and-pink light, and seemed empty. Randalf let the tent-flap fall back behind him and stepped forward. Immediately, a fiendish booby trap twanged and whirred into action.

A rope tightened, a cork popped, a weight dropped and a noose tightened around Randalf's ankles. With a loud *whoosh*, it whisked him up into the air, where he dangled helplessly from the central pole, his head a metre from the ground.

'Well I never!' Randalf chuckled. 'A most extraordinary fairground attraction.' He wriggled about, sending all the blood to his head. 'I must say, I rather like the way it makes my nose tickle.' He craned his head up a little. 'Most enjoyable, thank you. You can let me down now.'

As if in response, there came a *clink, clink, clink* from outside. It was coming closer, and getting louder and louder.

'Whatever's that?' he wondered.

Outside, all heads turned towards the gates of the castle.

'It's the cutlery! It's returned,' shouted an under-gardener.

'Typical!' muttered Benson. He dropped the handful of pointy twigs he was carrying as the sugar tongs rushed past. 'And there was me thinking sugar tongs were history.'

Cutting a swathe through the crowds, the cutlery charged across the garden. To the front were the spoons, clinking and clanging, behind them the forks and the knives, with the great meat cleavers clanking hot on their heels. Bringing up the rear was the egg slicer, plinking and twanging in an out-of-breath sort of way.

Joe spun round. Norbert laid aside the fluffy toy rabbit he was snuggling and craned his neck. Veronica fluttered above the birdbath as the cutlery clattered past.

'They're all aiming for that marquee,' someone shouted. 'Look.'

Sure enough, with the sugar tongs leading the way, the cutlery was heading noisily towards the lilac-and-pink Rose-Petalled Pavilion of Loveliness. The sugar tongs swept back the tent-flap and hurried in. The others followed.

'My goodness,' said Randalf, his heart beginning to pound furiously. 'Powerful magic is at work here. I don't like this. I don't like this one tiny bit . . .'

Just then, as the final items of cutlery – the egg slicer and a tiny silver toothpick, engraved with the name *Simon* – went inside the marquee, there came a roar so mighty that the ground trembled.

'*ROOOAAARRRGGGHHH!!!*'

The egg slicer twanged with terror as a great talon scraped across its backside.

'Eek!' squeaked the toothpick in dismay.

Outside, someone shrieked: 'It's a dragon!' as a great, winged creature swooped down on the lilac-and-pink tent.

Horrified shouts and cries exploded all around, as everyone tried to get as far away from the Rose-Petalled Pavilion of Loveliness as possible. With their backs to the castle walls, or cowering under trestle tables, or simply crouched down and clutching at one another for safety, they watched as the dragon landed.

It raised its head and let out a triumphant blast of flame. At last, after the long and mighty chase, the treasure was trapped!

The dragon leaned forwards and grasped the tent in its great scaly arms, its savage talons interlocking at the back. Then, with a grunt of effort, it tugged as hard as it could. The guy ropes snapped and the tent pegs scattered as the dragon gathered up the cutlery inside the canvas, turned it upside down so nothing could escape, and slung the whole lot over its shoulder. With a powerful leap, the dragon launched itself up into the sky and soared away.

For a moment, there was silence in the castle garden. Then, from over by the stage, came a lone voice.

'Now *that's* what I call a fairground attraction,' it said.

'Hear! Hear!' said someone else, and some elves burst into polite applause.

Having watched it all, the tall, stooped individual in the hooded cape standing by the gate groaned. 'It's got the wrong person. Master's going to be *so* disappointed,' he muttered as he slipped away.

The herald was running around the birdbath in the middle of the garden, megaphone to his lips. 'The Horned Baron's been stolen! The Horned Baron's been stolen.'

From all round, came a gasp, followed by concerned

muttering and murmuring (and the occasional snigger). Then came a distinctive voice: 'No, I haven't.'

The goblin spun round and found himself face to face with an all-too-familiar helmet and moustache. 'Your Horned Baron-ness!' he exclaimed. 'Can it really be you?'

'Of course, you fool! Who did you think it was?'

The herald raised the megaphone. 'The Horned Baron's safe! The Horned Baron's safe!'

There was some more polite applause (and a few scattered boos). The Horned Baron glared at the crowd.

'Now, what's all this I hear about cutlery? Have the sugar tongs been found?'

'Have the sugar tongs been found?' blasted the herald through the megaphone.

'Do you have to repeat everything I say?' stormed the Horned Baron.

'Do you have to repeat everything the Horned Baron says?' thundered the herald at the crowd.

'Not them. You!' shouted the Horned Baron.

'Not them . . .'

'Give me that!' interrupted the baron, grabbing the megaphone and turning back to the crowd. 'Now, will someone please tell me what is going on!'

'Randalf went into the lilac-and-pink Pavilion of Loveliness,' said Joe, stepping forwards. 'And then all the cutlery came crashing through the garden, led by some sugar tongs, and disappeared inside the pavilion as well.'

'Did you say sugar tongs?' said the Baron excitedly.

'And then a dragon swooped out of the sky and gathered up the pavilion, Randalf *and* all the cutlery, and flew away.'

'But this is terrible!' said the Horned Baron.

'I know,' said Joe, tearfully. 'Poor Randalf.'

'No, I meant the sugar tongs,' said the Horned Baron. 'I mean, what's Ingrid going to say?'

'But what about Randalf?' Joe persisted. With the wizard gone, so too was his only hope of ever leaving Muddle Earth.

'WALTER!'

'I've got enough on my plate without having to deal with some second-rate wizard who's got himself into a scrape,' said the Horned Baron. 'Besides, *you're* the warrior-hero. I'd have thought battling with dragons was right up your street.'

'*WAL-TER!*'

'Now, if you'll excuse me, duty calls,' said the Horned Baron, turning away. 'Coming, my gooey cupcake!' he called.

As the Horned Baron strode off, Joe noticed Norbert pushing his way through the crowd from the opposite direction. Henry was trotting beside him; Veronica was on his head.

'Cheer up,' she said chirpily. 'It might never happen.'

'It already has,' said Joe glumly. 'Randalf was carried off by that dragon.'

'You what?' Veronica exclaimed. 'Oh, the silly old fool!'

'Dragon!' said Norbert. 'Nooooo!!' he wailed, and burst into tears.

'Trust Randalf,' said Veronica, 'to end up as a dragon's dinner.'

'Dinner!' howled Norbert.

'We've got to rescue him,' said Joe. 'Now, did anyone see which way the dragon went?'

'I've got a pretty good idea,' said Veronica. She launched herself off Norbert's head and fluttered off. 'Follow me,' she said, 'to the Broken, Missing or Useless stall!'

As soon as they reached the stall, Veronica found what she was looking for. There, sandwiched between a moth-eaten cardigan and a rusty mangel-wurzel slicer, was a tatty scroll. Joe picked it up.

'*Three turnips, half a cup of grass, one bottle of stilt-mouse milk . . .*' he read. 'It looks like some sort of shopping list.'

'No!' said Veronica. 'On the other side, stupid!'

Joe turned the scroll over. It was a map of Muddle Earth. There was the *Enchanted Lake* where he had first arrived. And *Goblintown*, where Randalf had kitted him out as a warrior-hero. And the *Ogrehills* . . .

Veronica tapped the top left-hand corner of the scroll with her beak. 'Here,' she said impatiently. 'Look!'

Joe looked across the map and gasped. For there,

written across a vast bleak landscape in flamboyant curly-wurly writing, were three words:

Here Be Dragons.

'Here be dragons?' said Norbert, with a shudder. 'I don't like the sound of that!'

'We must go there!' said Joe firmly. 'It's our only chance of finding the dragon.'

'And when we find the dragon?' said Norbert. 'What do we do then?'

'We'll work something out,' said Joe uncertainly. 'Besides, we have no choice.'

'Joe's right,' said Veronica. 'Oh, I know Randalf and I have had our differences. He can be cranky, incompetent . . . miserly . . . pompous, vain and intolerably smug . . . moody, clumsy, lazy, greedy . . . and absolutely impossible to live with . . .' She took a deep breath. 'But he's all I've got.'

'It's decided then,' said Joe. 'We'll set off at once.'

'And you will think of something, won't you?' said Norbert tearfully. 'Before dinnertime.'

'Trust me,' said Joe, grimly. 'I'm a warrior-hero.'

Before the goblin child who found it had managed to use it even once to stir his spittle tea, the tiny teaspoon had dropped down from his pocket to the pavement unnoticed, and disappeared. Over cobbled streets it went, through a crack in the great gates and down the city steps – tinkling and sighing as it went – out of Goblintown and away.

On, on, on, it journeyed. This way, that way – following the force that drew it ever onwards.

Tripping and falling, and sighing as it picked itself up. Then on again. Clinking over rocks, squelching through mud and leaving a fine, broken line through the sand and dust.

At a junction, it jumped a troll cart spilling over with a great mound of straw. *Clink, clunk.* It hopped on to the wheel, up the side and – *pluff* – down in the soft, yellow mattress of straw.

As it lay there, warm and content, the sun glinted down on its polished bowl and elegantly curved handle. It caught the eye of a passing batbird, returning early to its roost.

With a flapping of wings – like a burst of applause – the batbird swooped round in the sky, and dived.

Down, down it came, legs extended and claws open. Then, with a delicate twist – and a loud squawk – it plucked the glittering object from the top of the hayrick and soared off.

The tiny teaspoon sighed.

Far below, Muddle Earth was spread out like a tatty scroll. The Perfumed Bog. The Enchanted Lake. The Musty Mountains . . .

As the wind blew back against it, the teaspoon shivered – softly at first, then more forcefully. The batbird tightened its grip too late. The tiny teaspoon had already slipped out of its grasp and was falling, falling, falling . . .

Tinkle, clink!

It landed. With a little sigh, it picked itself up. It was standing directly in front of the gates to the Horned Baron's castle. Trembling with anticipation, it hopped on. Through the gate it went, across the gravel and . . .

Gone! They were gone! The teaspoon sighed sadly, wearily. But they *had* been there. All of them. That much was certain. It turned and listened. Yes, there. Over there . . .

The tiny teaspoon turned, sighed, and left the place where the cutlery had been such a short while before. Keeping to the shadows, it hopped back through the castle gates and on into the dusty highlands beyond.

The sounds of the garden party gradually faded away. A wind got up, the sun went down, Mount Boom came nearer. It puffed and wheezed and sometimes exploded weakly – *boom.*

Up ahead were four other travellers. An ogre, with a budgerigar on its head and a young warrior-hero and a dog on its shoulder. The tiny spoon trembled as something stirred in its memory. The touch of warm fingers, the snugness of a dark pocket. It was the young warrior-hero who had found him when he got separated from the others.

With a quivering sigh and a soft *tinkle, tinkle,* the tiny teaspoon picked itself up and quickly followed them.

Silhouetted against the low, setting sun, the dragon swooped down out of the sky and landed at the entrance to its mountain cave. It swung the heavy marquee full of cutlery to the ground with a *clunk*, sat back and wiped its brow.

A batbird circled overhead. The dragon raised its scaly head and blasted it with a searing jet of flames. The batbird flew off, backside and tail-feathers singed, screeching with indignation. The dragon scoured the sky and scanned the scorched, barren land for a sign of any other unwelcome intruders.

There was none.

Then again, you couldn't be too careful. The dragon seized the great clanking bundle in its claws and lugged it into the cave. Down on its belly, it slipped and slithered its way along a low, narrow tunnel and on into a large, domed cavern deep inside the mountain.

The cavern was warm, smoky, sulphurous. The dull red glow of molten rock glimmered from cracks in its towering walls. Gleaming dimly in the faint light was a tall pile of treasure. The dragon purred as it got closer and nuzzled it softly.

There were golden helmets, rusty tap fittings, jewel-encrusted crowns, swords, saucepans, shields with ancient designs, medals and trophies, bracelets and bedsteads, tiaras and tin cans – all heaped together in the centre of the stone floor. And, over by the wall, standing by itself, was a magic golden harp, softly weeping.

The muffled clinking and clanking of the cutlery grew louder as it was plonked down on the pile. With a slash of one long talon, the dragon ripped a large hole in the fabric of the former Pavilion of Loveliness. The cutlery came tumbling out.

The dragon's eyes widened with delight as the wonderful hoard spilled out in front of it. Fine knives, gorgeous forks, dear little teaspoons – and the most adorable silver egg slicer. Cleavers and skewers. Waffle irons and sugar tongs. And – with toothpicks in its beard and forks in the most unlikely places – what appeared to be . . .

A wizard!

The dragon's eyes narrowed. It patted its grumbling stomach and threw back its great scaly head.

'*ROOOAAARRRGGGHHH!!!*'

'You never know,' said Joe. 'Perhaps it'll just play with him. Like Henry plays with his favourite rubber ball . . .'

'It'll eat him,' said Veronica matter-of-factly.

'You don't know that for certain,' said Joe.

'Bound to,' she said. 'It will be ravenous after that long flight. Toasted wizard would make the perfect dinner . . .'

'Don't say that, Veronica,' said Joe weakly.

'Unless it's so hungry it just eats him raw.'

'Veronica! Shut up!'

Veronica shrugged. 'You did ask,' she said sulkily.

'Yes, but . . . *Whooooah!*'

'Sorry!' came Norbert's breathless voice as he suddenly lurched to one side. 'There was a pothole.'

'Don't you apologize, Norbert,' said Joe. 'You're doing a fantastic job on those Winged Boots of yours. We're going to be there in no time.' He turned and glared at the budgerigar on Norbert's head. '*We* haven't given up on him, even if you have, Veronica.'

'I just think we should be prepared for the worst,' she said.

Norbert let out a stifled sob.

'Well, you're upsetting Norbert,' said Joe, trying to sound brave. 'Now, if you can just stop talking about how hopeless this all is, perhaps I can work out a plan.'

'You're right, of course. Forgive me,' said Veronica. 'So, how exactly *do* you plan to slay this dragon?'

'*Slay?*' said Joe. He scratched his head. 'I thought we

could talk. Bargain. Barter. I thought I'd be able to *reason* with it . . .'

'Reason with a dragon?' said Veronica. 'Don't make me laugh. I'm telling you, unless Randalf's extremely quick on his pins – which would be a first – he's a goner. That dragon will burn him to a crisp with one blast of his fiery breath!'

Just then, above the noise of the clattering wheels, there came the faint sound of a distant roar.

'Hear that?' said Veronica.

'Hear what?' said Norbert.

Joe stared grimly ahead. 'Just go faster, Norbert!' he told the ogre. '*Faster!*'

The tiny teaspoon sighed as it tripped and slipped down into one of the deep imprints left by the ogre's Winged Boots of Colossal Speed. The dust fell in about it, taking the edge off its shine and making it difficult to scrabble out of the narrow trench.

Back on solid ground, it cocked its bowl to one side, and listened. Far, far ahead, it heard a deep, ominous roar that echoed around the rugged landscape and that left it trembling along the length of its handle.

Not so far away – but getting further with every passing minute – were the others. The ogre, the budgerigar, the warrior-hero and his battle-hound.

With a soft sigh, the teaspoon set off once again, tinkling along the stony track as fast as it possibly could.

Inside the cave, the sounds echoing around the great, domed cavern were getting louder. The cutlery was in its orchestra formation, attempting to tune up. It swayed on the mound of treasure – clinking and clanking – while the dragon lumbered noisily about.

The spoons clattered. The knives clashed. Once, an entire set of cake forks was almost destroyed as the dragon's great bulk came crashing down towards them. It was only the quick thinking of the sugar tongs – and a handily placed saucepan lid – that prevented them being crushed and twisted out of shape.

Petrified, Randalf watched it all. He was crouched down behind a jagged boulder over by the wall, trembling at the sight of the dragon which seemed to be whipping itself up into a frenzy of rage. How he had escaped, he would never know.

One moment he'd been tied up inside the Pavilion of Loveliness, with sharp, pointy cutlery prodding and jabbing every inch of his body; the next, a great claw was slashing through the material, the rope – and very nearly his neck.

Randalf shuddered at the thought.

Of course, from the sound of roaring, the smell of

burning and the gut-churning sensation of flight – not to mention the crowd shouting, 'It's a dragon! It's a dragon! Run for your lives!' – Randalf had suspected that he'd been seized by some sort of dragon. But nothing could have prepared him for the sight of the terrifying creature which reared up in front of him as he had tumbled from the tent on a wave of cutlery.

It was gigantic. Monstrous! Every glinting scale was the size of a dinner plate; every claw, a long, curved rapier. Its smoking nostrils alone were big enough to accommodate the wizard's head, pointy hat and all.

As the dragon had spotted the cowering wizard, its eyes had opened wide and its snout had come to within an inch of Randalf's nose. Its nostrils had sniffed. Its stomach had rumbled. Wisps of smoke had coiled into the air . . .

'*Eeek!*' Randalf had yelped and made a dash for it.

With a *clink-clank-clatter-crash* he'd scrambled desperately over the pile of treasure as the dragon had made a grab for him, missed, and snorted with frustration.

Randalf trembled from the tip of his beard to the toes of his boots. He *had* escaped – but for how long? Sending out short bursts of flame, the dragon was clambering all over the pile of treasure, looking under shields and dustbin lids, rooting around in the forks and spoons, and tossing object after precious object aside.

Then – sniffing at something suspicious – it inadvertantly sucked a plumed gold helmet right up inside one of its vast nostrils. It coughed, it snorted, it sneezed – and the helmet shot out at enormous speed, smashing against the side wall with a loud *clang*.

Randalf shrank down behind the rock.

'Why me?' he groaned softly. 'Why do these things always happen to me? Oh, what wouldn't I give to be back on my lovely houseboat, with my feet up, a cup of tea in one hand and a snuggle-muffin in the other.'

The dragon continued searching the heap of treasure, tearing into the metal objects with a horrible vigour and determination. With his heart in his mouth, Randalf risked a peek from behind the rock.

The bad news was that the way out was on the other side of the cavern. The good news was that, by sticking to the shadows, he might just be able to skirt round the walls unseen.

Head down and heart beating fit to burst, Randalf set off. He scurried from rock to boulder, crevice to crack, in short hurried bursts. The last stretch was the most difficult. Between the ridge he was crouching behind and the stone stack by the entrance, there was a broad expanse of bare rock.

'Just stay calm,' he told himself. 'Wait till the dragon's turning the other way and . . . *Now.*'

Setting off like a sprinter from the starting blocks, Randalf dashed across the empty stretch. The dragon snorted.

'Almost there,' Randalf whispered, urging himself on. 'Almost . . .'

Just then, a shrill voice cried out, 'Mistress! Mistress!' It was the magic harp. 'He's over here!' it shrieked. 'The fat wizard's over here!'

The dragon spun round, eyes blazing, talons glinting. Randalf tripped and sprawled on the floor of the cave like a stranded stiltmouse. The dragon's eyes were fixed on him. Too terrified to move, he watched as the monstrous creature flapped its leathery wings, flexed its talons and lunged forwards, its great jaws gaping.

'*ROOOAAARRRGGGHHH!!!*'

'Ooh, I heard it *that* time,' said Norbert, as the noise echoed all around them.

'We must be getting close,' said Joe.

Norbert shuddered. 'It sounds really cross.'

'Either that,' said Veronica, 'or it's *hungry*. Maybe we're not too late, after all.'

'That's the spirit, Veronica,' said Joe.

'Good old Randalf,' the budgerigar chuckled. 'Giving that dragon a run for his money . . .'

Abruptly Norbert sat back and skidded along on the seat of his pants. Veronica squawked. Joe cried out. Henry barked and leaped down on to the safety of solid ground.

'Sorry,' said Norbert breathlessly. 'It's the only way I know how to stop.'

'But why *have* we stopped?' Veronica demanded, smoothing down her ruffled feathers.

'Because of *that*,' said Norbert, and pointed.

Joe and Veronica turned and followed the line of his grubby outstretched finger. It was the entrance to a cave.

'Smoke,' said Veronica softly.

'Footprints,' said Joe.

Henry wagged his tail and barked, first at the cave, then up at Joe.

'Is he in there, boy?' said Joe. 'Is Randalf there?' He swallowed nervously. 'With the dragon?'

A roaring sound echoed through the air, and fresh smoke billowed from the cave entrance. Henry barked excitedly.

'Angry or hungry, it still sounds dangerous,' said Norbert in a quavery voice.

'Are you sure you want to do this, Joe?' said Veronica.

'No one would think any the worse of you if you didn't.'

Joe shook his head. '*I* would,' he said. 'Besides, without Randalf, I'll never get home. I'll be stuck here for ever.'

'There are worse places to be stuck,' said Veronica. 'Like the inside of a dragon's belly, for instance.' She noticed the expression in Joe's eyes. 'Still, if you must, you must,' she added. 'I'll come with you.'

'So will I,' said Norbert, 'if you undo my boots first.'

'Thank you,' said Joe. Henry barked. 'Thank you all.'

'Victory or death!' Veronica squawked.

Joe groaned.

'It'll be all right,' he told himself, muttering under his breath. 'I've taken on a giant ogre before now. And that wasn't so bad. It's what Randalf's always saying, *You just have to bluff, lad. Show them who's boss.*' He wiped the beads of sweat from his brow. 'But a dragon,' he gasped. 'A monstrous, great, fire-breathing dragon!'

He looked ahead at the dark smoke billowing from the cave entrance. It was getting thicker and more pungent by the second. Eyes streaming and heart pounding, he marched bravely on. And as the entrance came closer, he could hear strange noises echoing down the tunnel – snorts and snuffles, gurgles and growls, and a small, high-pitched voice pleading for its life . . .

'Please, please, *pretty*-please, put me down, there's a nice dragon,' Randalf was babbling.

Upside down again, he was dangling from the dragon's claws – his head inches away from the beast's vast and odorous beaked mouth. A forked tongue shot out and flickered round his face. The eyes widened. The jaws cracked open . . .

'No, no, no,' Randalf started up again. 'You don't want to eat me. *Ugh! Ugh!*' he said, screwing up his face. 'Nasty! Chewy! Tough!' He spat. 'Horrible!'

Staring at him curiously, the dragon opened its great mouth wider. Randalf found himself staring down into a long, blood-red tunnel. The stench was incredible; the heat, intolerable.

'For pity's sake,' he gasped. 'You can't eat me, I'm a wizard!'

Behind them, the cutlery started up a low, mournful, clanking funeral march. The dragon narrowed its large yellow eyes.

Suddenly, from outside, a clear, if slightly nervous voice could be heard.

'It is I, Joe the Barbarian! Subduer of ogres, friend of wizards and champion to the Horned Baron himself! And I think you should know – eating people is wrong!'

As soon as his brave words left his mouth, Joe regretted them. What had he been thinking? After all, a dragon was a dragon – and this one, he already knew, was a monster!

From the cave came the sound of scratching, and the dragon's huge, scaly head popped out. It looked around with blazing yellow eyes. Its gaze fell on a boy, a budgie, a not-very-big ogre and a rather scruffy looking dog.

It snorted. A ring of black smoke coiled into the air and sailed away.

Joe desperately tried to stop his teeth chattering. This was no time for his nerve to go.

'Behold, mighty dragon!' he cried. 'I, Joe the Barbarian, stand before you with my fearless battle-hound, Henry . . . the Ferocious. And Norbert the Not-Very . . . *errm* . . . Easily-Calmed-Down-Once-You've-Got-Him-Started. You certainly wouldn't want to make him angry, believe me.' Joe nudged Norbert.

'Grrrrr!' said Norbert feebly.

The dragon raised an eyebrow.

'And he's not alone!' said Joe, urgently. 'For he has come with a great big army of . . . really, *really* angry ogres.'

The dragon frowned and inched forwards for a closer look. It peered round.

'Of course, you can't see them . . . they're masters of camouflage!' said Joe. 'But they are there. Hiding behind rocks. Hundreds of them. And armed – armed to the teeth. Just waiting for my word to throw themselves into battle.'

The dragon began drumming rhythmically on the ground with its talons.

'And that's not all,' said Joe desperately. 'I've got budgies.'

'*Unnh?*' grunted the dragon.

'Yes, budgies,' said Joe. '*Attack* budgies! This is my Wing Commodore. In charge of two dozen budgie squadrons.'

'All trained in unarmed combat and at the peak of their physical condition,' added Veronica quickly. 'Vicious, ruthless and under strict instructions to take no prisoners.' Her voice dropped to a low and, she hoped, menacing whisper. 'You mess with one of my squadrons and . . . and . . . and they'll mess on you!'

'You have been warned,' said Joe. 'I and my mighty army have come,' he said, sweeping his arm around in a circle, 'to set the wizard free. Release him now, and no one will get hurt.'

The dragon looked puzzled. 'Get hurt?' it said. 'Why, darling, of course no one's going to get hurt – unless *I* decide otherwise!'

Joe gasped. 'It can talk,' he hissed.

Veronica nodded. 'Dragons are very good mimics, just like parrots – or lazybirds when they can be bothered. Who's a pretty dragon? Who's a pretty dragon?'

'Shut up, Veronica,' Joe hissed, and turned to the gigantic beast. 'Where is Randalf?' he asked in his biggest, deepest voice. He frowned. 'We're not too late, are we? I mean . . .' He faltered. 'You haven't eaten him, have you?'

The dragon threw back its head and trilled with laughter. 'Randalf! So that's his name, is it? No, darling, I wouldn't dream of eating anyone called Randalf. I mean, what a perfectly dreadful name!'

Joe breathed a sigh of relief.

'Since you're here, you'd better come inside,' the dragon said. 'I won't have it said that Margot Dragonbreath kept guests standing outside in the cold. Come on! Come on!' And with that, the head disappeared back inside.

Joe looked at Veronica, who looked at Norbert, who looked at Henry – who dashed off into the cave, barking and wagging his tail. The others followed him in.

'It's so typical of you dragon-slayer types,' the dragon was saying. 'You see a magnificent creature like me and you just jump to conclusions.'

Behind her, the cutlery was clinking and clanking. Her voice rose above it.

'You assume that I can't wait to devour you whole. I mean, it's so vulgar . . .'

Clatter! Clatter! Clink! Clonk!

'Whereas nine times out of ten . . .' She was shouting now. 'What I really fancy is a toasted teacake and a sponge finger . . .'

Behind her, the cutlery had worked itself up into a frenzy of noise and activity. The dragon spun round furiously.

'Oh, do settle down!' she shouted. She tossed her head. 'I feel one of my migraines coming on.'

The cutlery obliged, bringing the noise down to a more soothing pianissimo. Joe stood open-mouthed, scarcely able to believe what he was seeing.

He was in a huge cavern – hot, dark and incredibly messy. There was junk strewn all over the floor and rising up in the middle into a great unstable heap. Most of the objects were silent and still, but some – the knives, forks, spoons and other assorted bits of cutlery he had seen earlier at the garden party – were anything but.

What *is* going on? Joe wondered.

They were clinking and clanking. They were hopping and jumping about. A collection of knives to his left were clashing their blades and clacking their handles. To his right, a group of forks were twanging and banging. There was a trio of soup spoons; a ladle quartet. And all of them dancing to the same insistent beat which, once again, was building up . . .

Clatter! Clatter! CLINK! CLONK!

What a racket! Joe thought, and winced. What on earth was it up to?

The dragon put a claw to her thin lips. '*Sssshhh!*' she hissed. 'I won't tell you again.'

For a second time, the cutlery became quieter. Not silent. But quiet enough for Joe to hear another noise – a curious muffled grunting and groaning which was coming from somewhere near the wall.

'*Grrrmmbll flammell-flan,*' the voice complained. '*Pfleeem . . .*'

Joe peered into the shadows. There – not ten metres away from the entrance to the cave – was a portly figure, seated on the floor, struggling to remove a large metal bucket which was wedged firmly on to his head. Joe stared . . .

'Randalf?' he said. 'Is that you?'

'*Omm kmomf miff!*' the voice bellowed back.

Joe strode across and seized the bucket. 'Take hold of his legs, Norbert,' he said. 'And when I give the word, pull.'

Norbert did as he was told. Veronica sat on his head, watching. Henry wagged his tail slowly.

'All set?' said Joe.

The cutlery clanged and clattered behind them.

Norbert tightened his grip on the wizard's ankles. 'All set,' he said.

'Pull!' yelled Joe.

At first – apart from a loud echoing scream from the bucket – nothing happened. Joe altered his grip.

'Again!' he cried.

This time, as Norbert tugged the legs, Joe twisted the bucket. There was a loud *pop!* Norbert let go of the ankles and fell down. Joe staggered backwards, the bucket in his hand. And there between them – still seated on the ground with his legs stuck out in front of him – was Randalf. He blinked twice.

'*Waargh!*' he screamed as he saw the dragon. 'She tried to eat me!'

The dragon groaned. 'Typical,' she said. 'Here we go again.'

'She *did*!' Randalf insisted indignantly. 'She had hold of

me and was getting ready to swallow me whole. If you hadn't come along when you did . . .'

'I was simply curious,' the dragon told Joe. 'He came with that lot,' she added, pointing towards the noisy cutlery. 'I thought he was a free gift.'

'A free gift?' Randalf spluttered, outraged.

'The trouble was, he would wriggle so!' the dragon went on. 'He slipped out of my grip and dropped down, headfirst, and got wedged in that,' she said, nodding at the bucket. 'It was entirely his own fault.' She leaned down and pulled Randalf to his feet. 'Now, if you'll just keep still, you funny little man, and allow me to introduce myself. I am Margot . . .'

'A *free gift*!' Randalf bellowed. 'How dare you!'

The cutlery was getting louder again, with one group of meat cleavers particularly rowdy. The dragon clasped her head.

'I am,' she moaned. 'I'm getting one of my migraines.'

Clatter! Clatter! Clink! Clonk!

'A free gift indeed.' Randalf shook his head. 'I'll have you know, madam, that I am a wizard. Randalf the Wise is my name; wizardry, my game. Indeed, I am the finest wizard currently in residence upon the Enchanted Lake.'

'The *only* wizard, more like,' Veronica cut in.

'Shut up, Veronica!' said Randalf.

'Wizard, eh?' said Margot, looking up. She nodded back at the cutlery wearily. 'In that case, perhaps you could get that lot to quieten down a bit.'

Randalf took a sharp intake of breath and shook his head. 'It's not as easy as that, madam,' he said. 'This cutlery is clearly enchanted. Dark forces are at work here, I'll be bound.'

'Yes, and I bet I know *whose* dark forces!' chirped Veronica.

'Shut up, Veronica!' Randalf snapped. 'As I was saying, tricky thing, enchantment. Takes a lot of skill and know-how and years of training. But luckily for you, madam, enchantment is a bit of a speciality of mine.'

'Lucky you!' Veronica muttered sarcastically.

'Veronica, I'm warning you!' Randalf hissed. He turned to Margot. 'Tell me everything you know about this cutlery.'

'Well, they just showed up outside my cave,' she said. 'Wakened me, they did, with their enchanting music – and then led me a fine song and dance until I finally caught up with them. I wanted them for my hoard, you see. What dragon wouldn't?' She shook her head. 'But now I'm beginning to wish I'd left well alone.'

'You're not the only one,' Randalf murmured as he rubbed his bruised and battered body.

The cavern throbbed with noise. The dragon rolled her eyes. 'They're so *loud*,' she said. 'I don't know how I'm ever going to get back to sleep.'

'Sleep?' said Joe. 'I thought you'd only just woken up.'

'That's dragons for you,' said Veronica. 'They spend ninety-nine per cent of their lives asleep and the rest of the time drooling over treasure . . .' She sniffed. 'If you can call this heap of junk *treasure*.'

'I beg your pardon,' said Margot, affronted.

'Well, just look at this place,' said Veronica, with a sweep of her wing. 'I've never seen such a tip.'

Margot's nostrils began smoking. 'How dare you!' she roared. 'That's my hoard you're talking about. Precious heirlooms. Priceless treasures . . .'

'Like the bucket, eh?' said Veronica.

'Bucket? Bucket!' said Margot fiercely. She picked it up from the floor, pulled herself up as high as the cavern would allow and glared at them furiously. 'This is no bucket! Have you no taste? No eye for beauty?' she roared. 'Why, this is the sacred Potty of Thrynn, emptied only once every thousand years.'

'Ugh!' Randalf cried, spitting, snorting and checking his beard for bits.

'Ugh?' said Margot. 'It's a work of art.'

'It's a potty!' said Randalf. 'And it was on my head!'

'And not quite empty, from the look of it,' said Veronica.

'It's very beautiful,' said Joe, trying to smooth things over. He looked round the cavern. 'You have so many lovely things.'

The dragon's eyes softened. 'I do, don't I?' she cooed. 'And you clearly have an eye for such things, young man. I can tell.' She beamed. 'It's taken me years to build up my collection. A silver shield here, a jewel-encrusted coronet there . . .' She looked round, smiling proudly – until her gaze fell on the loud and disobedient cutlery. 'Leave that breast-plate alone!' she shouted at a set of skewers. 'And you ladles, there. Stop that at once!'

Clatter! Clatter! Clink! Clonk!

She turned back to Joe and rolled her eyes. 'Honestly, darling,' she said. 'I don't know how much more of this I can take.'

Joe nodded sympathetically. 'Do you have any more priceless treasures in your hoard?' he asked.

'Of course, darling!' said Margot. 'Enchanted mirrors, magic swords, impregnable warrior armour – you know, all the usual.' She scanned the great hoard of treasure. 'But they're nothing compared with my most precious items . . .' She broke off, and began rooting through the great pile of valuables. 'They're here somewhere,' she muttered, and sighed. 'I must admit, it is all just a teensy bit disorganized.'

'What are they?' asked Joe, joining in the search.

'Oh, they're special. You need a real collector's eye to appreciate them,' said Margot. 'They're absolutely to die for.'

Randalf tutted impatiently. 'Yes, yes,' he said. 'I'm sure they're very nice, but we really should be going . . .'

'Are *these* them?' said Norbert. He picked up two rather battered tin cans.

Margot spun round. 'Yes, yes, they are!' She beamed at Norbert. 'Well done, you! I can see that you have a real collector's eye!' she said, and laughed. 'Three of them, in fact!'

Norbert beamed happily.

Clatter! Clatter! Clink! Clonk!

'Look at the workmanship,' Margot shouted above the sound of the rowdy cutlery. 'So subtle, so expressive. And to think these were created to hold . . . baked beans! It takes one's breath away!'

'They're really beautiful,' said Joe uncertainly.

'They are,' Norbert agreed. 'Far too beautiful to be left lying around. You need to display them to their maximum advantage,' he told the dragon. 'On a plinth, maybe. Or in a showcase.' He turned and frowned. 'In fact, if you ask me, the whole place could do with a really good tidy-up!'

'You think so?' said Margot.

'Oh, definitely,' said Norbert. 'It always works in my kitchen at home. You see, you have to have a system. I keep all my saucepans on hooks and my frying pans above the pantry door. They're handy for hitting elves over the head,' said Norbert. 'But here, I'd suggest piles.'

'Oh, I like the sound of that,' said Margot.

'Yes,' said Norbert, waving his arms about theatrically. 'Over here we could have the rusty pile. And over here, the sharp, pointy thing pile. And here . . .' He eyed the sugar tongs tap-dancing on a saucepan. 'We could have a noisy pile.'

'Not *too* noisy, I hope,' said Margot, flicking her tail at the passing choir of spoons.

'And here,' said Norbert, standing in the centre of the cavern, 'you could have a great big stupendous pile of sparkly things!'

'Darling!' cooed Margot. 'You're an artist.'

Back at the Horned Baron's castle, the garden party had

come to an abrupt end. After all the excitement – and chaos – of the dragon's sudden arrival and departure, no one felt much like partying any more.

'What a day!' Benson sighed.

'You can say that again,' said an under-gardener, pulling splinters of wood from his hair.

'Goodness knows what got into the cutlery!' said Benson. 'And as for that dragon!'

'The last dragon I saw was Gretchen,' said the under-gardener. 'And that was at least ten years ago . . .'

'And look at the state of the place!' said Benson.

The toffee stall was in ruins, the jars of smells lay broken on the ground beside pools of spilled face-paints, while the mangel-wurzels that hadn't been trampled underfoot had all disappeared – pocketed by the trolls as they had set off back to Trollbridge. Over at Snuggly-Wuggly Corner numerous small furry animals were climbing through the broken fence and scampering away. And in the midst of it all, the pink stinky hog ran this way and that, squealing indignantly as it searched for its missing tail.

The Horned Baron stomped through the trail of destruction, muttering under his breath.

'Absolute fiasco!' he said. 'Disaster! Everything's in ruins! The guests have all gone home! Ingrid's in hysterics!' He patted the beads of sweat from his forehead. 'At least things can't get any worse!'

'Norbert!' said Randalf impatiently, as the ogre busied himself around the dragon's cavern. 'We really *must* be going.'

'Not so fast, Fatso! I thought you were going to do something about that cutlery,' said Margot, raising herself up and fixing the wizard with a yellow-eyed stare.

'Yes, of course,' said Randalf, trembling and stepping inadvertantly into the Potty of Thrynn. 'Blast!'

'*Blast?* Is that a new spell?' said Veronica.

'Shut up, Veronica!' said Randalf angrily, and clanked off to sulk in the corner.

'That's close enough, potty-breath,' said the harp fiercely.

'Now,' said Margot brightly. 'Let's get tidying! You, Joe, are in charge of the sharp, pointy things pile.'

'You mean swords and spears and enchanted warrior armour?' said Joe excitedly.

'Yes, yes, whatever,' said Margot. 'And you, attack-budgie, can be in charge of the rusty pile.'

'Thanks a lot,' said Veronica, flying off.

'And you and me, Norbert, we'll make a lovely big pile of sparkly stuff!'

'Lovely,' said Norbert, clapping his hands together. 'I adore sparkly stuff.'

Just then, a conga-line of all the knives – from dainty

butter knives to hefty meat cleavers – went dancing past, clanking out the jerky rhythm as they went. The dragon's furious voice rang out.

'Randalf! Darling! They're still at it! Margot's getting a teensy-weensy bit angry!' she bellowed.

Over in the corner, Randalf snorted as he tried to prise the potty off his foot. 'Madam, a little bit of patience, please. The enchantment is obviously powerful and needs careful . . . *Blast!*' The potty was stuck fast.

'Still trying that *Blast* spell?' trilled Veronica from somewhere above his head.

'Shut up, Ver . . .' Randalf's jaw dropped.

There above him, perched on a cave ledge, was an old, rusty and decidedly battered bird cage. Veronica was hopping about inside it.

'Lovely, isn't it?' said Veronica in a gooey voice. 'A cage!' She sighed. 'It's got a perch that swings. And a little bell. And look, just what I always dreamed of – a mirror!' She smiled at her reflection. 'For when I want to see a friendly face . . .'

'Veronica!' said Randalf sharply. 'Cages are for canaries. Remember who you are! You're a wizard's familiar, and your

place is down here on the brim of my pointy hat.'

'Your hat hasn't got a mirror,' Veronica protested. 'Or a bell.'

'Veronica!' Randalf shouted. 'Pointy hat! Now!'

'Or a perch that swings,' said Veronica defiantly.

'*Veronica!*'

'Is that *you* causing a disturbance again, wizard?' roared Margot. 'Margot's getting angry! You won't like Margot when she's angry!'

'I was just working on a spell,' said Randalf sheepishly.

'Well, get on with it,' said the dragon fiercely. 'Now, Norbert, darling, where were we? Oh, how clever! The watering cans of Poot – yes, they do sparkle delightfully, don't they?'

Casting a furious glance at Veronica, Randalf clanked across the cavern to where a collection of butter knives was doing a noisy samba. He started waving his arms about, and muttered under his breath.

Meanwhile, Joe was collecting marvellous swords, magical helmets and spears of fantastical design, and making a neat pile at the far side of the cavern. Henry barked happily by his side.

'Can I try some of this armour on?' Joe asked, holding out a silver helmet with elaborate wings and curving horns.

'My dear boy,' said Margot, 'help yourself. That warrior-hero stuff really is quite a bore. Not nearly sparkly enough for me. Take anything you fancy.'

'Thanks,' said Joe, beaming. If he had to stay in Muddle Earth for a while longer, he might as well look like a convincing warrior-hero.

'I'd forgotten I had such wonderful treasures after my little nap,' cooed Margot. 'It's easily done.'

Joe tried on a bronze breastplate. 'How long *have* you been asleep?' he asked.

'Oh, no time at all,' said Margot. 'Twenty years or so, I think.'

'Twenty years!' Joe exclaimed.

'Give or take the odd year,' said Margot. 'We dragons need our beauty sleep, you know. Besides, twenty years is nothing. Matilda, over there,' she said, waving vaguely towards the entrance to the cave, 'has been asleep for twice as long. And as for Agnes, well, no one's seen her for centuries. Normally, I'd have slept for longer – but that cutlery woke me up.'

Clatter! Clatter! Clink! Clonk!

Margot sighed. 'Delightfully sparkly – but they do go on rather.'

'You can say that again,' said the harp grumpily from a corner.

Joe picked up a particularly ornate sword. 'Do all dragons have hoards of treasure?' he asked.

'Of course, dear boy,' said Margot. 'We all simply adore beautiful trinkets. That's all we want. That and the occasional sheep or two.' She sighed. 'Oh, but the tales they tell about us! Burning down castles. Battling with knights on horseback. Devouring princesses and damsels-in-distress – whatever *they* might be. I mean, there just aren't enough hours in a day for all that nonsense.'

Joe nodded.

'And as for warrior-heroes coming along thinking they can just slay you,' Margot went on, huffily. 'It's a damn cheek, if you ask me!'

'How dare you! Stop that!' an angry voice cried out. 'Leave me alone!'

It was the harp. A dozen butter knives and a set of soup spoons were taking it in turns to pluck at its strings.

The dragon turned irritably, and sent a warning blast of hot smoke in their direction. The harp swooned. The knives and spoons scurried away, but regrouped by the candlesticks and chandeliers, where they jumped about furiously, clashing and clattering.

'Darling, perhaps you could try out your warrior-hero skills on that lot,' she said.

'Actually,' said Joe, 'I'm not really a warrior-hero.'

'You'd never have guessed to look at him!' Veronica, who was swinging happily to and fro in her cage, shouted back.

'I was summoned to Muddle Earth by Randalf, here,' said Joe, nodding towards the wizard. 'He's promised to send me back home when he can. That's why I'm here. I couldn't let him get eaten by a dragon – not that you would have,' he added hastily.

'No, well one wouldn't, naturally,' said Margot. 'Oh, but poor you, Joe. All lost and alone in a strange world . . .'

Clatter! Clatter! Clink! Clonk!

The dragon groaned and turned to Randalf. 'Oh, please hurry up,' she said weakly. 'My head's splitting!'

Randalf shook his head. 'I'm afraid bewitched cutlery can be very tricky, madam. Very tricky indeed.'

'Honestly! Call yourself a wizard,' said Margot scornfully. She turned to Joe. 'I must say, I don't fancy your chances of leaving Muddle Earth.'

'Joe will be fine,' said Randalf, shaking his head. 'But I'm afraid this cutlery has been bewitched by an expert. I'll have to return to my houseboat and consult my spell book. To that end, we shall be taking our leave. I bid you good day, madam.'

'Oh, no you don't,' said Margot. She twisted herself round, blocking Randalf's escape with her tail.

'What . . . what is the meaning of this?' Randalf blustered.

'Norbert here has been a poppet. Joe has been a perfect gentleman – and that attack-budgie seems very nice . . .'

Clatter! Clatter! Clink! Clonk!

'But none of you are going anywhere until every last

piece of that confounded cutlery has been made to lie down and be quiet,' she bellowed. 'I don't care what you do. I don't care how long it takes. But I want them silenced, once and for all!'

Clatter! Clatter! Clink! Clonk!

A strong, chill wind whistled through Elfwood. Tree rabbits, perching in the lower branches of the oaks and pines, snored restlessly in their sleep and huddled together for warmth, while roosting batbirds, high up at the top of the jub-jub trees, cried out as they were swung to and fro.

'*Ouch!*'

Trudging through the trees came a stooped figure, his bony fingers clasping at his flapping cape and keeping the hood raised. With each step, his boots sank deep into the squidgy mulch of mud and fallen leaves, slowing him down and making him sweat with effort despite the cold.

At the centre of the woods was a clearing – Giggle Glade, its name – and in the centre of the clearing was a modest dwelling, built of wood and ornately decorated. The caped figure fought his way to the door.

The wind was howling round the house, setting the

powder-blue shutters rattling and the wooden roof tiles clacking. Inside the house, seated in shadow upon a high-backed and intricately carved throne, Dr Cuddles waited.

'Soon,' he giggled. 'Very soon.'

As if on cue, the front door burst open. Dr Cuddles smiled.

'Is that you, Quentin?'

'Y . . . yes, Master,' panted Quentin as he forced the door shut against the buffeting wind. 'Goodness me,' he said. 'It's blowing a gale out there. I had to battle with it every step of the way.' He shook his head wearily. 'I'm utterly pooped.'

'Pooped?' Dr Cuddles giggled. 'How delightful. I trust you bring good news.'

Quentin lowered his hood, smoothed down his slightly ruffled golden curls and twirled the ends of his magnificent moustache. He looked up. The throne was set in deep shadow. Only Dr Cuddles's startlingly blue eyes were visible. Glinting and unblinking, they bored into him from the darkness. Quentin felt his knees begin to tremble.

'Well?' said Dr Cuddles. 'I take it that the Horned Baron has been taken care of at last.' He giggled unpleasantly. 'I'm sure our scaly friend enjoyed her little snack.' The high-pitched, somewhat sinister giggling grew louder. 'Did she crunch his bones?' he said. 'Did she tear him limb from limb?'

'Actually, sir,' said Quentin, hanging his head. 'There's something I've got to tell you.'

The eyes narrowed to slits. 'Well?' he demanded.

Quentin swallowed nervously and took a deep breath. 'Things didn't go quite according to plan,' he said in a rush.

Dr Cuddles sighed. 'Explain yourself,' he said coldly.

'There was a bit of a mix-up,' he said. 'At the garden party. It seems that the dragon might have chewed up the wrong person.'

'The wrong person?' said Dr Cuddles testily.

'He just turned up at the last minute and spoiled everything,' said Quentin. 'There was nothing I could do.'

'*Who?*' He sounded furious now.

'That wizard c . . . c . . . character,' Quentin stammered. 'Randalf. Randalf the Wise . . .'

'I might have known,' Dr Cuddles muttered, drumming his stubby fingers on the arms of the throne. 'Why can't he keep his big nose out of my affairs?'

Quentin permitted himself a little smile. 'If I know my dragons,' he said, 'it probably saved his big nose till last.'

Dr Cuddles giggled. 'Oh, I do hope so,' he said. 'But that still leaves the small matter of the uneaten Horned Baron.'

'Plan B, Master?' said Quentin.

'Plan B,' Dr Cuddles confirmed. He clapped his paws together and a dozen elves appeared as if from nowhere. 'Unlock Roger the Wrinkled and bring him to me,' he commanded. 'Go!'

'At once, Master,' the elves twittered, and scurried off to do as they had been told.

Quentin, relieved that Dr Cuddles hadn't taken his news too badly, removed his cape and hung it on a hook on the door. 'How about a nice snuggle-muffin?' he said. 'I decorated some specially for you earlier and ...'

'Quentin,' said Dr Cuddles. 'This is no time for snuggle-muffins.'

'No, sir,' said Quentin. 'Silly of me.'

Just then, there came a scuffling from the corridor and the sound of raised voices. A door flew open and the elves bustled into the room tugging on a long, heavy lead, at the end of which was a decidedly bedraggled, not to say wrinkled, wizard. From the top of his high-domed forehead to the tip of his long, pointed chin, spread an intricate network of wrinkles. His ears were wrinkled, his cheeks were wrinkled, his nose was wrinkled – even his wrinkles were wrinkled.

'How dare you treat me like this?' he blustered. 'I can't possibly be expected to work under these conditions!'

'My word, you *are* wrinkled, aren't you?' giggled Dr Cuddles. 'I always forget.'

'Well, is it any wonder?' snapped Roger. 'Chained up in that poky little room, working every hour under the sun. I'm telling you, I can't take much more of it.

And then *this*!' He tugged at the lead. 'The indignity of it all.'

'It's your own fault,' said Quentin sharply. 'You shouldn't keep trying to escape.'

'I've already explained all that,' said Roger loftily. 'I was just stretching my legs.'

'You were running,' Quentin reminded him.

'Just answering a call of nature,' said Roger.

'You were disguised as a washerwoman,' said Quentin.

'I explained that as well,' Roger began uncertainly. 'It all started when I was a child and used to try my mother's dresses on ...'

'Never mind all that,' Dr Cuddles cut in. 'I summoned you here to discuss a matter of great importance to me ...'

'The Horned Baron,' said Roger the Wrinkled.

'You read my mind,' giggled Dr Cuddles.

The wizard nodded. 'I trust the cutlery performed to your satisfaction,' he said.

'Yes, it did,' said Dr Cuddles. 'Unfortunately, there was a slight hitch.'

'A hitch?' said Roger.

'Quite amusing, really,' said Dr Cuddles, giggling rather

hysterically. 'It seems the cutlery lured the dragon to Quentin's pink pavilion just as we planned but, unfortunately, the pavilion contained the wrong person.'

'The wrong person?' said Roger.

'Why, Roger!' Dr Cuddles giggled. 'You're beginning to sound like an echo.'

'An echo?' said Roger.

Dr Cuddles's giggle turned decidedly nasty. 'I have decided to put Plan B into action,' he said.

The wizard's wrinkled face collapsed. 'Not the . . .'

'Yes, Roger,' said Dr Cuddles, giggling wildly. 'The flying wardrobes.'

'But Dr Cuddles,' said Roger. 'I really can't advise that. Not yet. They're not ready.'

'My dear Roger,' said Dr Cuddles, 'I hope I don't need to heat up the metal underpants again.'

Roger the Wrinkled took a step backwards. 'Not the underpants, I beg you!' he pleaded. 'It's just that . . .'

'Just *what?*' The sound of his drumming fingers grew louder.

'Well, the flying bit is easy,' Roger the Wrinkled explained, 'but putting the wardrobes together is an absolute nightmare! I mean, the instructions never make sense, and there's always an extra screw left over . . .'

'Enough of all these excuses!' roared Dr Cuddles. He clapped. The elves jumped to attention. 'Fetch me the *Great Book of Spells.*'

'At once, Master,' the elves trilled, and scuttled off through a different door.

'. . . and as for the splinters!'

'Be silent, Roger!' said Dr Cuddles sharply. 'Under my close supervision, I shall allow you to consult the *Great Book of Spells*,' he announced. 'We launch the wardrobes tonight!'

'But . . .'

Dr Cuddles giggled. 'I have absolute confidence in your skills, Roger. You and your fellow wizards had better not fail me – or else.'

Roger shuffled about uncomfortably. 'The underpants?' he said nervously. Dr Cuddles nodded.

Puffing and panting, the elves returned with a heavy wooden box; the spell book locked up inside it. They scuttled over to the throne.

'The *Great Book of Spells*, Master,' the elves said in unison.

'Put it on my lectern,' Dr Cuddles told them. 'Then, when Roger has finished reading the appropriate spell, put him on his lead and take him back.' He turned to the wizard. 'And no tricks,' he giggled. 'Do you understand?'

~ 251 ~

'Tricks, Dr Cuddles?' said Roger. 'I don't know what you mean.'

The steely eyes glared out of the shadows. 'Take care, Roger the Wrinkled,' Dr Cuddles said with a giggle. 'I shall be watching your every move.'

The tiny teaspoon was almost at the end of its epic journey. As the mountain cave came into view it sighed, tripped and fell, picked itself up and sighed again. The end was literally in sight.

With a soft *tinkle-tinkle*, the teaspoon hopped into the cave entrance.

Even though it was getting dark outside, with the sun down on the horizon, it was far darker inside the cave. The teaspoon paused and cocked its bowl to one side.

Noise. There was lots of noise echoing down a tunnel that led deep into the mountain. Clinking and clanking. Clashing and clattering.

And raised voices . . .

'I'll do anything,' shouted one, clearly at the end of its tether. 'Just make them be still!'

'I'm doing everything I can!' cried another.

'Which isn't much!' taunted a third.

The tiny teaspoon continued. *Chink, chink, chink.* Over stones and gravel, and the occasional small bone, it continued along the tunnel, heading for the dull red glow at the end. Closer and closer it got; louder and louder the echoing noises became.

All at once, the tunnel opened up and the tiny teaspoon found itself at the edge of a vast underground cavern. There were the individuals it had followed from the castle, their backs turned. Behind them was a dragon. And behind the dragon . . .

The tiny teaspoon let out a little sigh and hopped up and down on the dusty floor.

It was the sugar tongs who first noticed the newcomer. Its raised tong clunked insistently on the side of a golden goblet. The knives rustled, the spoons clinked, the forks clanged as, one by one, the cutlery all became aware of the tiny teaspoon in their midst.

From every corner of the cavern, they appeared. The meat cleavers and skewers, the forks, whisks and ladles, the egg spoons and soup spoons, cake forks and butter knives – and even the dumpy egg slicer – all began hurrying to the spot where the tiny teaspoon was performing its strange, bouncing little dance.

'Oh, good grief!' the dragon groaned. 'What's happening now?'

'I'm attempting a reverse enchantment,' said Randalf importantly, waving his arms about, 'with a triple bypass and a double switchback. Very tricky, it is. I need absolute silence.'

'Fat chance,' said Margot, above the din. 'It's getting worse than ever!'

'Yes, but listen,' said Joe. 'It's different.'

Instead of the cacophony of noise the cutlery had been making since their arrival, one by one, they were all beginning to strike up the same pounding beat – *CRASH! CRASH! CRASH! CRASH!* – until the whole great mass of them were pounding together.

'It's the teaspoon,' said Joe. 'Look. They're following its lead.'

Randalf nodded wisely. Sure enough, the great clash of noise rang out every time the tip of the bouncing teaspoon's handle hit the ground.

'Well spotted, my boy,' he said. 'That's my double switch-back taking effect.'

'I'll tell you something else,' said Joe. 'I've seen that teaspoon before.'

'Oh, one teaspoon looks very much like another in my experience,' said Randalf, performing a strange little jig on one leg and puffing heavily.

'Hurry up!' urged Margot. She

clutched her head and rocked slowly back and forwards as the deafening noise continued. 'I really, really don't think I can stand any more of this.'

'Reverse enchantment can't be hurried, madam,' Randalf replied. He stopped hopping, raised his arms and began whispering urgently under his breath.

'You really have no idea what you're doing, do you?' said Veronica.

'Shut up, Veronica,' hissed Randalf.

All at once, the tiny teaspoon hopped up on to a boulder and tapped insistently. At the sound, all the other cutlery fell still. Every knife, every fork, every spoon. The cavern was silent, at last – silent, except for a faint *squeak, squeak, squeak* as Veronica swung backwards and forwards in her cage.

'I don't believe it!' she said. 'What did you do, you old fraud?'

'I've absolutely no idea,' said Randalf, who looked as surprised as everyone else.

Just then, the tiny teaspoon turned and began hopping back the way it had come.

Everyone held their breath.

The sugar tongs moved first. With a shudder and a creak, they tripped after the teaspoon. The rest of the cutlery, calm now and in well-ordered ranks, followed close behind. As the last of them – the small toothpick with *Simon* engraved on it – disappeared into the tunnel, Margot let out a long, happy sigh of relief.

'They've gone,' she said. 'Thank goodness for that. I don't know how to thank you.'

Randalf lowered his arms at last and turned to the dragon. '*I* do,' he said.

'This is brilliant!' Joe called out above the noise of the rushing wind. 'Absolutely *fantastic!*'

He'd been on aeroplanes before, roller-coasters and the tops of open-air buses – but none of these came even close to the thrill and excitement of riding on a dragon's back.

Resplendent in the new warrior-hero outfit that Margot had allowed him to select from her treasure, he was sitting on a comfy, padded seat between Margot's great leathery wings. To his right was Norbert, with Veronica perched on his shoulder; in his lap was Henry.

'A-*maz*-ing,' he murmured as he looked all round him, trying to take everything in.

Above him was the inky star-studded sky, cloudless and crystal clear, with its three moons shining brightly. Below him was Muddle Earth, spread out like a great map and bathed in the purple, yellow and green moon-light. A batbird, flying too near, was scorched and sent packing by a warning blast of the dragon's fiery breath.

Looming up before them was Mount Boom, tall, dark and imposing. *Boom*, it went, the sound barely audible above the throb of the slowly beating wings. And beyond the volcano, spreading on as far as the eye could see, were the Musty Mountains.

'Hold on to your hats,' Margot cried out as, with a twitch of her wings and a flick of her tail, she banked sharply and soared down towards Mount Boom.

Boom, went the volcano, exploding weakly and sending out a little puff of grey and yellow smoke.

'*Wheee!*' cried Joe.

Once, twice, three times the dragon flew around Mount Boom, before soaring back up, up into the sky. 'It's been such a long time since I last stretched my wings properly,' said Margot excitedly. 'I'd forgotten quite how exhilarating it could be!' And with that, she folded her wings and dived into a long, swooping loop-the-loop.

'*Whooooah!*' Joe shouted, stomach in his mouth. As they levelled out, he threw back his head and laughed. 'Again!' he roared. '*Again!*'

'*Wurrgh!*' Randalf groaned. Unlike the others, he did not have a seat, comfy or otherwise. Instead, he was at the back of the dragon, lodged between a couple of jagged tail-fins and clinging on for dear life as the long, serpentine tail swished this way and that. 'Why do I have to sit back here?' he shouted.

'Because you're too fat to sit up front,' Margot called back firmly.

'But what about *him*?' shouted Randalf, pointing at Norbert – and almost falling off.

'That's different,' said Margot. 'Norbert's my friend, aren't you, darling?'

Norbert beamed happily.

'This is an outrage!' protested the wizard. The rushing wind drowned out his words.

Joe turned. 'Did you say something, Randalf?' he called.

Randalf shouted back. Joe could see the wizard's mouth move, but what with the noise of the rushing wind and the dragon's beating wings, he could barely make out a thing.

'What?' he bellowed.

Randalf's face contorted with effort as he shouted back. Again his words were whipped away on the wind.

'Can *you* hear what he's saying?' Joe asked the others.

'Probably just telling us how much he's enjoying the ride,' said Norbert, grinning and waving at Randalf.

'Which makes a nice change,' Veronica added. 'After all, normally by now he'd be fast asleep.'

Far in front of them, beyond the Musty Mountains, the Enchanted Lake came into view and glistened in the coloured moonlight. Joe felt a pang of disappointment. 'We'll soon be there,' he said.

Norbert turned to him and grinned. 'It's lovely to be

carried for a change instead of doing the carrying,' he said. 'How about one more circuit of Mount Boom?'

'Oh, yes!' Joe exclaimed.

'Is that all right, Margot?' Norbert called out.

'For you, dear heart, anything,' said Margot, as she dipped her wings and soared round in a great circle, back towards the mountain.

Randalf gripped on desperately as the tail lashed fiercely. 'What's going on now?' he roared. But nobody heard.

This time, as Margot approached Mount Boom, she came in low and steep, clipping the summit and swooping around the smoking crater. Joe looked down into the blood-red chasm. It glowed like the embers of a dying fire, and a warm, slightly sickly mist swirled around his face, making his throat tickle and his eyes water.

Boom.

The dragon tipped her wings and glided away safely as the solitary puff of smoke popped out of the top of the crater. Then, with a hard beat of her wings, she soared off around the volcano. Faster and faster she flew, circling it over and over again. The vertical rock sped past in a blur to their right. The moons seemed to spin in the sky.

'*Wheeee!*' shouted Norbert and Joe, and whooped for joy.

'*Woof!*' barked Henry.

'*Wurrgh!*' Randalf groaned.

'*WATCH OUT!*' screeched Veronica. There in front of

them – and coming towards them at great speed – was something big, brown and rectangular. '*DUCK!*'

'That's no duck,' Margot shouted back. 'It looks more like a wardrobe!'

'Just get out of its way!' Veronica screamed.

Margot swerved just in time. The wardrobe – doors flapping like wings – clattered over her head, grazing the top of her crested crown as it passed.

'There's another one!' Veronica shouted, as a second wardrobe came flying noisily towards them, its flapping doors clattering loudly.

'Leave it to me,' Margot replied grimly. This time, she made no effort to dodge out of the way. Instead, she opened her mouth as wide as it would go and sent a broad, blazing tongue of fire roaring out ahead of her.

The wardrobe was promptly swallowed up in the jet of flames and incinerated in an instant. As the dragon beat her wings triumphantly and flew on, a sprinkling of ash drifted down to the ground below her.

'That was awesome!' gasped Joe.

'Good to know that I haven't lost it,' said Margot proudly.

'Just as well,' said Veronica. 'Look!'

Everyone turned and gasped. A dozen or more wardrobes were flapping in low from Elfwood in a long straight line. This was getting stranger and stranger, even for Muddle Earth, thought Joe.

'They're heading for the Horned Baron's castle!' he shouted, and turned. 'Randalf, what's going on?'

Randalf shouted something back.

'What?' called Joe above the roaring wind. 'Margot, slow down a minute.'

The dragon slowed to a lazy hover. More wardrobes flapped into view. A huge armada was filling the sky.

'I said,' Randalf shouted back, 'first singing curtains, then enchanted cutlery and now flying wardrobes – which, by the look of them, are up to no good.' Just then, a particularly solid-looking wardrobe struck the top of a castle tower with a loud *crash*. Randalf shook his head. 'There's powerful magic at work here!' he said. 'And, as I always say, where there's magic ...'

'... there's money!' finished Veronica. 'Typical!'

'I don't know what you mean,' shouted Randalf huffily. 'The Horned Baron's castle is clearly under attack. It's my solemn duty as a wizard to render what support I can in this, his hour of need.'

'It's amazing how brave you can be with a dragon in tow,' said Veronica scathingly.

'Shut up, Veronica! Now, follow that furniture!' shouted Randalf.

'Oh, I see. Just like that!' said Margot hotly. 'Don't the others get a say?' She craned her neck round. 'Norbert, dearest, what would *you* like to do?'

'I ... I ... I ...' he stammered, glancing back and forth

between Randalf and the dragon. 'I think . . .'

'Yes, Norbert?' said Margot.

The ogre nodded. 'Yes,' he said. 'I think we should go and help the Horned Baron.'

'For you, Norbert, anything,' said Margot sweetly.

'Thank goodness for that,' Randalf snapped. '*Whoooooah!*' he cried out, as the dragon suddenly lurched forwards.

'We're going in!' Margot's voice floated back. She beat her wings. She lashed her tail. 'And woe betide any item of furniture that gets in my way!'

Out of the sky she flew like a speeding bullet, hurtling down towards the Horned Baron's castle. Joe clung to Henry (whose ears were flapping back in his face) with one hand and clutched his helmet to his head with the other. Veronica tucked her head under her wing and dug her claws into Norbert's shoulder.

'Ouch,' Norbert yelped.

'Norbert, what is it?' Margot cried. She swung her wings round and flicked her tail. It had the same effect as slamming the brakes on. Joe kept a tight grip on Henry as he and Norbert were flung forwards. Randalf sailed past them, his arms waving, his mouth open.

'*HELP!*' he screamed. '*Help.*' His voice was rapidly fading away. 'Hel . . .'

'Catch him!' Norbert bellowed. 'Margot, catch him!'

'Are you sure?' said Margot.

'*Yes!*' Norbert howled.

'Oh, all right, if you insist,' she said and, with no further ado, she flapped her wings and spiralled down out of the sky.

Joe held his breath.

The sound of Randalf's cries echoed upwards as he tumbled downwards.

'Hel . . .'

'. . . el . . .'

'. . . elp . . .'

Down in the courtyard of the Horned Baron's castle, the wardrobes were coming in to land. Benson and the herald – who were crouched down together behind the birdbath – watched a particularly large piece, with ornately carved doors and ball and claw feet, flap down noisily and strike the ground with a loud *crash*.

It landed on its side, wobbled, toppled and keeled over on to its back. A thick cloud of dust and sand flew up into the air.

'Well I never,' said the herald. 'Raining wardrobes. That's a first.'

Benson shook his head. 'It can't be good for the flowers,' he said.

Another wardrobe crashed down to their right, flattening a pot of pansies as it landed.

'Terrible waste,' muttered Benson.

'*Sshhh.*' The herald raised a finger to his lips and pointed into the clearing cloud of dust. 'I thought I heard something.'

There was a long, low creak and one of the wardrobe doors opened slowly.

The herald huddled up closer to Benson behind the birdbath. 'There's something inside it,' he said. 'Listen.'

There was a faint jangling sound, getting louder by the second.

'I'm frightened,' said the herald, squeezing Benson's hand rather too tightly.

'Perhaps I should go and have a look,' said Benson, making a move to pull himself to his feet.

'No, don't!' said the herald, clutching hold of Benson's arm and pulling him back down. 'You can't leave me here alone!'

All at once, with a loud bang, both wardrobe doors flew open. Benson jumped. The herald grabbed hold of him and clung on tightly.

'I can't look,' he whimpered. 'What is it?'

Benson shook his head. 'Well, it makes sense, I suppose . . .'

'What?' said the herald, dread in his voice. 'Bendy bugs? Horned wangtubbers. Wide-mouthed fribblesnooks . . . ?'

'Hangers,' said Benson. He stared open-mouthed as hanger after small wooden hanger fluttered out from the shadowy depths of the wardrobe and up into the night sky.

'Hangers?' said the herald.

'Coat hangers,' said Benson. 'A flock of them . . .'

'A flock?' said the herald. He pulled away from Benson and opened his eyes cautiously. His jaw fell open. 'You're right,' he murmured weakly. 'It's a flock of coat hangers.'

'And there's more coming from that wardrobe over there,' he said.

The herald laughed. 'They had you pretty worried there for a moment, didn't they?' he said.

'They still do,' said Benson, darkly. 'Look.' He pointed up to the top of the East Tower, where the hangers were already flying through an open window. 'The Baroness isn't going to like this,' he said. 'She isn't going to like this one little bit.'

Randalf watched the ground hurtling towards him and desperately racked his brain for a not-breaking-every-single-bone-in-your-body spell.

Just then, there was the sound of ripping material – and he was no longer falling. In fact, with the ground now seemingly speeding away from him, it was clear that he was flying – soaring back away from the rocks and dust and up into the purple, yellow and green moonlit sky.

'Yes!' Joe yelled, and punched the air in triumph.

'Hooray! Hooray!' shouted Norbert.

Below them, suspended by the seat of his pants in Margot's talons, swung a rather red-faced Randalf.

'I was about to weave a spell of feather-lightness,' he said with as much dignity as he could muster. 'But thank you anyway.'

'Oh, don't thank me, Fatso,' said Margot, gaining height with every flap of her wings. 'Thank dear, sweet Norbert, here.'

'*Hmmph!*' said Randalf.

'Sorry, I didn't quite catch that,' said Margot, swinging Randalf lazily. There was an ominous sound of ripping.

'Thank you, Norbert,' said Randalf.

'You're welcome,' smiled Norbert, and nodded at the torn pants. 'You should get those mended when we get to the castle, before you catch your death of cold . . .'

'Yes, *thank* you, Norbert,' said Randalf, darkly.

As the dragon approached the castle walls, Joe saw the full extent of the chaos within. There were wardrobes – and bits of wardrobe – everywhere.

Some had crashed down on the tents and stalls, some had smashed to smithereens on the paving stones, losing their doors and splintering their sides, while others were still airborne, waiting to land. They were darting this way and that, banging into walls and each other. Most of them seemed remarkably poorly put together, with missing hinges, odd-length legs and, in some cases, doors that didn't match.

'Coo-ee!' shouted Norbert, and waved.

Randalf, who was back in his place on the dragon's tail, called to him. 'What is it, Norbert?'

'The Horned Baron,' said Joe, pointing. 'Look.'

Below, the Horned Baron was running about like a headless chicken and shouting at the top of his voice.

'Run for your lives! We're under attack! Take cover!'

'What an inspiration to us all the Horned Baron is,' muttered Veronica.

CRASH!

'Blimey,' Joe gasped. 'That was close!'

One of the wardrobes had smashed down on to the ground, snapping the *Keep off the Grass* sign and missing the Horned Baron by a hair's breadth. The Horned Baron, now down on his knees and trembling, held his head in his hands as scores of hangers exploded from the broken wardrobe and swooped down on him like a flock of angry batbirds. Their hooks rained blows down on his horned helmet.

Plink! Plink! Plink!

'*Ouch! Ouch! Ouch!*' cried the Horned Baron.

Just then, an under-gardener with a bucket on his head dashed past. 'The flowers!' he shouted. 'Mind the flowers!'

To a hanger, the flock wheeled away from the Horned Baron and gave chase.

'Get off!' he shouted, as the hangers hammered down against the bucket.

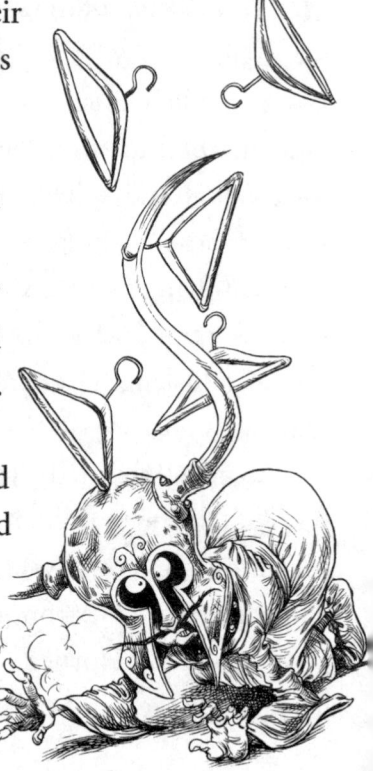

'And keep off the grass!'

The same scene was being repeated everywhere Joe looked. Gardeners and under-gardeners, footmen in livery, butlers, servants and kitchen hands were all under attack – and it seemed there was nothing any of them could do to repel the fearsome invasion.

'Oh, woe is me!' the Horned Baron howled miserably. 'We're doomed. We're all doomed. Will no one help me in my hour of need?'

'For the right price,' came a voice from above his head.

The Horned Baron looked up and gasped. Not only were they being bombarded by wardrobes and attacked by coat hangers, but now the dragon had returned. Frozen to the spot, he stared, horrified, at the vast hovering creature with its leathery wings, its slinky tail, its ferocious crested head and – he rubbed his eyes in disbelief – its passengers . . .

'Randalf?' he shouted up to the figure clutching on to the dragon's tail. 'Is that you?'

'It certainly is,' Randalf shouted back. 'Wizard for hire, at your service. 'Now about my terms . . .'

'Anything,' said the Horned Baron. 'Anything at all! Just *do* something! Now!'

'A hundred gold big 'uns,' Randalf called.

The Horned Baron sucked in air noisily through his teeth. 'Fifty,' he said.

'Ninety,' Randalf responded.

'Seventy-five, and that's my final offer,' said the Horned Baron.

'*WALTER!*' came a loud, piercing scream. It was Ingrid, and she was not happy.

'Eighty,' said Randalf.

'Eighty, it is,' said the Horned Baron. 'But not a big 'un more. *Watch out!*' he bellowed, as a decidedly lopsided wardrobe appeared out of nowhere, doors clapping and hangers jangling, and hurtled towards the hovering dragon.

'Furniture on the starboard side!' Veronica squawked.

Margot nodded. With a flick of her tail, she ducked behind the tall gate towers.

The wardrobe smashed into the heavy studded doors and broke into a thousand pieces which tumbled down to the ground. From inside, another flock of coat hangers emerged. They soared up into the air and closed in on the dragon.

Margot smiled and sent out a roaring tongue of flame. The hangers were turned instantly to ash.

Down in the courtyard, the Horned Baron burst into applause. 'Bravo!' he shouted. 'Now come and deal with the rest of them.'

'For eighty gold big 'uns,' said Randalf.

'Yes, yes,' said the Horned Baron impatiently, ducking down to avoid a wardrobe hurtling across the courtyard, totally out of control. 'Just get on with it!'

Randalf nodded. 'Land over by that wall, Margot,' he

said, pointing. 'If that's all right with you, Norbert,' he added archly.

'Oh, yes, perfectly all right, sir,' said Norbert. 'Excellent idea, sir.'

'Thank you,' said Randalf. 'I . . . *Whoooah!*' he gasped as the dragon swooped down over the courtyard, knocking wardrobes and hangers aside as she flew and – with a graceful twist – landed at the base of the high wall.

She turned and roared menacingly.

Joe leaped to the ground followed by Henry, wagging his tail and barking furiously. He drew his sword. Norbert jumped down beside him and, seizing a length of broken tent pole, swung it round his head. Randalf joined them, Veronica perched on his shoulder. He raised his arms.

'Let battle commence!' he roared, and turned. 'Norbert, you'd better go first, there's a good fellow.'

Norbert stepped forward as a rickety looking wardrobe with mismatched door handles lurched past. With a blow from the tent pole, the wardrobe fell apart.

'Shoddy workmanship,' said Randalf, picking up a loose screw and examining it.

Margot took to the air and hovered protectively over Norbert, who wielded his makeshift club and lumbered to the aid of two footmen trapped beneath the birdbath. He beat off several waves of attacking coat hangers.

The footmen emerged from their hiding place and shook their fists at the retreating wardrobes. 'And don't come back, or you'll get more of the same!' the smaller of the two shouted defiantly.

Just then, Veronica squawked with alarm. 'Watch out! More enemy furniture approaching – and not just wardrobes!' she screeched.

'Help!' shrieked the footmen and scurried back beneath the birdbath.

'Cowards!' said Randalf from behind Norbert.

Norbert strode off towards the approaching furniture – a couple of badly constructed cupboards and a large dresser with wonky shelves.

'Norbert!' cried Randalf. 'Come back!'

'Three wardrobes and a chest of drawers incoming!' Veronica's voice rang out.

Randalf ran after Norbert. Joe followed close behind, waving his sword at a couple of singed coat hangers.

'Let Margot take care of them,' said Randalf as the dragon swooped overhead in hot pursuit of a fleeing battalion of bookends. 'Norbert! Norbert! Come back here and protect me . . . Please!'

Above them, Margot's voice resounded loudly. 'Take that, you overgrown bundle of firewood!'

Her tail swished through the air, and dealt a shattering blow to a fat chest with crude teddy bear carvings. The chest split and spilled its contents of garish teddy bear-patterned quilts, which promptly flapped at the dragon.

'Get off me!' Margot's muffled voice cried out as she

clawed desperately at the vast quilt with orange and red teddy bears which had wrapped itself around her neck.

'*Aaaargh!*' she cried.

The quilt clung on all the more tenaciously.

'Mothballs,' she groaned.

A second teddy bear quilt fluttered in damply and tangled itself around the dragon's head.

'*Ugh!*' she roared. 'Someone still wets the bed!'

'Three more wardrobes, twin bedside cupboards and a set of occasional tables!' Veronica announced urgently from Randalf's shoulder.

Just then, the air whistled as the first of the wardrobes sliced down through the air at a steep angle.

CRASH!!!

The doors flew open and out leaped a crowd of pugnacious pillows, spoiling for a fight.

'*Aargh! Oof! Ouch!*' Randalf shouted as the pillows attacked him, thudding into his stomach and thumping him around the head. 'Help! *Help!* Norbert!'

'*Mffll blffll,*' Margot thudded to the ground next to Joe. She tried desperately to disentangle herself from the quilts. '*Helmmpff!*'

Joe's head spun with it all. There were cries of pain and terror coming from every corner of the courtyard as more badly constructed wardrobes and flimsy cupboards came crashing down. A chest of drawers fell particularly awkwardly, smashing to bits and spilling its contents. Knickers, corsets and balled-up socks tumbled out and joined the battle.

'Two wardrobes and a piano stool to your left!' Veronica announced.

With his heart in his mouth, Joe gripped his sword and, with a swish and a swoosh, sliced through the quilt around Margot's head.

The dragon looked round. 'Thank you, my dear boy,' she said gratefully. 'Those dreadful quilts smelt worse than the Potty of Thrynn!'

'Warrior-hero at your service,' smiled Joe.

'You're an angel!' Margot shouted as she launched herself up off the ground and soared back into the air. The incoming piano stool never stood a chance.

'Four more wardrobes and . . . *Aaargh!*' Veronica squawked as a volley of cups, saucers and plates whistled past her and smashed on the paving stones below. 'Margot!' she screeched. 'See to that Welsh dresser at once!'

Joe turned and joined Norbert, who was hurrying to help Randalf. The wizard was losing a fight with a pair of pink satin pillows shaped like love-hearts.

'Help me!' he cried as the pillows boxed his ears.

'Take that!' Joe roared as he raised his sword and lunged at the first pillow. There was an explosion of feathers. 'And that! And that! And that!' he cried as he stabbed and slashed at the second.

A snowstorm of feathers filled the air, so thick he could barely see his hands before his face. Suddenly, a huge bolster swung round and landed a crunching blow on Joe's helmet, jamming it down hard over his eyes.

He was blind!

All around him, the noise was building up to a mighty crescendo. Banging and crashing, splintering and smashing. Roars of triumph and howls of defeat. Clattering, shattering, screaming and shouting. And above it all, the sound of the dragon's mighty roar as she swooped this way and that.

Which way is the battle going? Joe wondered as he fought to prise the helmet off his head.

He couldn't see a thing. His arms were aching, his head was throbbing – and the swirling feathers were making him sneeze. Lunging and parrying as best he could, he stumbled blindly over the thick mattress of fluffy down, scraps of quilt and splinters of shattered hangers which covered the ground.

'Randalf!' he called out. 'Norbert! Veronica! Where are you?'

He paused and listened, but no one replied. He lowered his sword thoughtfully. Unless it was his imagination, the noise finally seemed to be abating. From his right, there came a grinding *crunch*; from his left, a muffled *thud*.

Then nothing. Nothing at all.

Joe trembled. It was quiet now. Almost too quiet. With a final despairing effort, Joe seized his helmet by its ornate wings and tugged with all his might.

Pop!

The battered helmet finally came off. Blinking through the slowly settling blizzard of feathers, Joe looked round.

'*Woof!*'

'Henry!' shouted Joe. 'Over here, boy.' The next second, Henry came bounding out of the storm of feathers, tongue lolling and tail wagging. Joe crouched down and ruffled his fur. 'Good dog,' he said. 'I'm glad you're safe. But where's everyone else? Eh? Where are they all?'

'Well, I can't speak for the others,' came Randalf's voice. 'But I'm here.'

'And I'm here, sir,' said Norbert.

'Where?' said Randalf.

'I don't know,' said Norbert thoughtfully. 'But I am. And Veronica's here with me to prove it.'

'For my sins,' the budgie muttered.

Soon, the whole courtyard was buzzing with conversation. Joe turned his head, first this way, then that, following the different voices. And as the feathers settled, he began to make out the bodies that went with those voices.

There was Norbert, sitting on a pile of splintered timber, with Veronica perched on his shoulder.

There were Benson and the under-gardener, who'd had the bucket stuck on his head, emerging from behind an upturned table, looking at some broken flower pots and tutting loudly.

And there was Randalf. He was holding up what looked to be a pair of frilly lace pantaloons and examining them closely. When he caught Joe staring at him, his face turned bright crimson.

'I . . . I need a new pair,' he stammered. 'Margot ruined mine.'

'You're worse than Roger the Wrinkled,' Veronica commented darkly.

Joe looked round. Splintered smouldering wood lay everywhere. There were doors off their hinges, drawers in pieces, broken hangers, bookends, crockery and everything shrouded in the blanket of feathers from the pillows and quilts.

'We did it,' he said proudly. 'We won the battle!'

'Indeed we did,' said Randalf, hurriedly screwing the silk underwear up into a ball and thrusting it into his pocket. 'Thanks to my inspired generalship.'

'Yes, inspired by terror,' said Veronica. '"Norbert! Help! Help!"' she mimicked.

'Shut up, Veronica,' said Randalf.

'Margot?' said Norbert. 'Has anyone seen Margot?'

'I'm up here, dear,' came a voice from the top of the castle gates.

They looked up to see the dragon perched comfortably, examining her talons.

'Margot, you were magnificent!' said Norbert. 'We'd never have managed without you.'

'One good turn deserves another, Norbert, dear heart,' said Margot. 'Without you, my cave would still look like a Broken, Missing or Useless stall. Speaking of which,' she said, 'I really should be getting back.' She sighed. 'A hoard of treasure can be such a burden.'

'Speaking of treasure,' Randalf muttered. 'Eighty gold

big 'uns is not to be sniffed at.' He looked round the courtyard for the Horned Baron.

The dragon reared up on her hind legs, flapped her
wings and launched herself off into the first pink blush of
morning.

'Farewell!' she cried. 'It has been charming getting to
know you all. Joe, Henry, Veronica and particularly you,
Norbert, of course. You know, I think I'll even miss old
Fatso! Remember, Norbert, darling – keep in touch!'

'I will,' Norbert called back.

The dragon flapped off into the night. 'See you all in
twenty years or so,' she called, her voice getting fainter.

Norbert wiped a tear from the corner of each of his
three eyes. 'Bye-bye, Margot,' he whispered.

Joe waved.

Randalf chuckled. 'Did you all hear that? She said she'd

miss me!' he said softly. He stared after the departing dragon. 'Randalf the Wise. Dragon-tamer . . .'

'I think her words were, *old Fatso*,' said Veronica.

'Shut *up*, Veronica!' said Randalf. 'Oh, look, *there* he is!' He cupped his hand to his mouth. 'Oh, Horned Baron!' he called. 'Horned *Baron*!'

Joe turned to see a short, stooped figure scuttling along the castle wall. His helmet was even more dented than Joe's own, with the horns sticking out at crazy angles.

'*HORNED BARON!*' Randalf bellowed. '*SIR!*'

The Horned Baron stopped and looked round innocently. 'Did somebody call me?' he said.

Randalf strode towards him, cutting a swathe through the piles of feathers and splinters of wood. 'It is I, sir,' he said. 'Randalf the Wise. Supplier of warrior-heroes and dragons in emergencies.' He smiled broadly. 'Eighty gold big 'uns, I believe we agreed.'

'Quite, quite,' said the Horned Baron. 'Send me a bill. You'll take a cheque, won't you?'

'I'm strictly a cash wizard,' said Randalf firmly. He held out his hand. 'Eighty gold big 'uns, if you please.'

'I don't carry that much on me,' said the Horned Baron, patting his pockets and shrugging. 'Sorry.'

'But . . . but . . .' Randalf blustered.

The Horned Baron smiled and laid a hand on the wizard's shoulder. 'But enough of this. After all, we shouldn't be talking about money now. This is a time for celebration! Three cheers for Randalf the Wise!'

Benson and a couple of footmen cheered weakly.

'Well done!' said the baron. 'Now, what's everyone waiting for?'

Several under-gardeners and the herald looked at each other, then back at the baron.

'Well?' he said. He paused and threw an angry look round the courtyard. 'Start clearing up this mess!' he snapped. 'What on earth is Ingrid going to say?'

Just then, a loud clapping noise erupted from the other side of the castle. Everyone looked up to see one single, solitary wardrobe flying over the towers and castellations back in the direction of Elfwood.

As the wardrobe flapped away into the distance, glinting in the low, early morning sunlight, a muffled voice was heard crying out.

'Walter! *Walter!*'

'An IOU!' Randalf stormed. 'Not worth the pair of frilly lace pantaloons it's written on!'

It was later that day and he and the others were back on the houseboat.

'Of all the low-down, two-timing, back-stabbing, sneaky tricks to play!' He turned to Joe. 'Let this be a warning to you. Never, *ever* trust the word of a baron, no matter how pointy his horns.'

'Still, it is an IOU,' said Joe. 'Even if you couldn't find anything else to write on, you did get his signature. That must be worth something.'

Randalf blushed.

'Show him your knickers,' said Veronica. 'Go on!'

Randalf handed the pantaloons to Joe. 'IOU eighty big 'uns,' read Joe, 'signed *The Grand Old Duke of York* . . .'

'What?' said Randalf. He snatched back the pantaloons and stared miserably at the fake signature. 'I'm just too trusting,' he said and sighed. 'Typical of the Horned Baron to pull a fast one. And after everything I did for him!'

'Everything *Margot* did, more like,' said Veronica. 'What a fine dragon she turned out to be. A real lady. And generous too,' she added. 'She gave us some lovely presents. Norbert's baking trays, Joe's warrior-hero outfit, not to mention my gorgeous little cage.' She

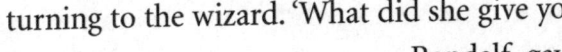

tinkled her little bell and preened in front of the mirror. 'My own little home,' she sighed. 'Remind me, Randalf,' she said, turning to the wizard. 'What did she give you?'

Randalf gave the Potty of Thrynn a vicious kick. '*Ouch!*' he cried.

'It goes with your knickers!' said Veronica smugly.

'Shut up, Veronica!' said Randalf.

From inside the kitchen came the sounds of whistling, whisking and the clattering of pots and pans.

'Still, could be worse, I suppose,' said Randalf. 'Norbert's cooking has certainly improved. And Joe, my boy, you certainly look the part in that outfit. Shame about the dented helmet. Are those wings meant to stick out like that?'

Joe smiled. 'It'll make a nice souvenir,' he said, 'when I go back home. When will that be again?'

'I'll see to it as soon as I can,' said Randalf, suddenly finding the details on the side of the Potty of Thrynn extremely fascinating.

'But when?' said Joe. 'Haven't I done enough yet?'

Randalf traced the outline of what appeared to be a large bottom engraved into the silver. 'Wonderful workmanship,' he murmured.

'*When?*' said Joe.

Randalf took a tentative sniff at the potty. 'I must wash my beard again,' he muttered.

'Randalf!' said Joe sharply. 'When are you going to send me home?'

The wizard turned. 'You know how it is,' he said. 'Waiting for an auspicious moment and all that. The alignment of the stars. The configuration of the moons . . .'

'No! No!' Joe shouted. 'You know that's not true. The moment could come and go, and you *still* wouldn't be able to do anything because you don't know the spell! We've got to go to Elfwood and recover Roger the Wrinkled's *Great Book of Spells*. It's the only way.'

'The lad's right,' said Veronica. 'Even if it does mean meeting up with Dr Cuddl—'

'Veronica!' Randalf shouted. 'I forbid you to use that name in my presence.'

'Besides,' said Veronica, swinging gaily to and fro on her perch, 'if you stand any chance of ever seeing those eighty gold big 'uns, you're going to have to go there.'

'I am?' said Randalf.

'Where else do you think that wardrobe took Ingrid?' she replied.

Randalf groaned. 'You don't mean . . .'

'He – who shall remain nameless – has got the *Great Book of Spells*,' said Veronica. 'He's got Roger the Wrinkled and the other wizards – and now he's added Ingrid to his collection. It's all part of his master plan.'

'Then, there's no choice,' said Joe firmly. 'We must go to Giggle Glade.'

'Better hang on to that potty, Randalf!' said Veronica. 'From the look on your face, you're going to need it.'

A solitary wardrobe lay on the ground beside the front door of the little house at the centre of Giggle Glade. It was still. One door was open and one closed. A pile of hangers lay in a corner. Dr Cuddles looked at it through the window.

'You have done well,' he giggled. 'My self-assembled pine-clad beauty!'

He turned away and slipped into the shadows.

Quentin nodded his head vigorously. 'That was one of mine, master,' he said. 'The instructions were ever so tricky, and I had three screws left over.'

'Excellent,' Dr Cuddles went on, giggling unpleasantly. 'Even though our losses were high!'

'I told you we needed more time,' said Roger the Wrinkled. 'The Welsh dresser was only half done, and someone sent off the teddy bear linen chest with all your quilts by mistake.'

'We all have to make sacrifices,' said Dr Cuddles, a slight choke in his voice. 'I might not have the Horned Baron, but I have the next best thing!' He giggled.

'Ooh, Dr Cuddles, you're *so* wicked,' said Quentin.

'He'll be like putty in my hand,' Dr Cuddles giggled. 'What won't he do to get his beloved Ingrid back? He'll be knocking on my door, *begging* me to return her. And when he does . . .'

The room resounded with his sinister, high-pitched giggling.

'Cuddles?' screeched an imperious voice. 'Cuddles!'

The giggling stopped. 'What can that infernal woman want now?' Dr Cuddles muttered. 'Surely she can't have broken free of the restraints already.' He turned and clapped his paws together.

Nothing happened.

'Where are those confounded elves?' he shouted.

'*Cuddles!*' Ingrid's voice sliced through the air like a knife.

Dr Cuddles shuddered. 'Roger!' he shouted. 'Quentin! Come back here!'

'*CUDDLES!*'

'Aah, this is the life,' sighed the Horned Baron.

He was reclining on a mountain of well-stuffed, if heavily patched, cushions in front of a roaring fire, his toes covered by a quilt with teddy bears on it. The curtains were drawn. The candles were lit.

The Horned Baron sipped from a large mug of spittle

tea and plucked a hairy toffee from the box on his lap. Many hours had passed since Ingrid's unfortunate disappearance. He popped a second toffee into his mouth. Poor, dear Ingrid . . .

Knock, knock.

The rapping at the door shattered the silence of the cavernous room and reminded the Horned Baron just how quiet it was.

'Enter,' he called.

The door opened and Benson approached. 'Bad news, sir,' he said. 'There's still no sign of the Baroness.'

'Oh dear, what a terrible shame,' said the Horned Baron. 'Still, mustn't grumble.' He raised the mug to his lips and sipped the spittle tea. 'Delicious,' he murmured. 'Throw another piece of wardrobe on the fire on your way out, Benson. There's a good chap.'

As the gardener shut the door behind him, the Horned Baron leaned back into the plump cushions and closed his eyes.

'Really must rescue Ingrid.' He yawned. 'One of these days.'

The weary pieces of cutlery huddled together round a large sign which read *Nowhere* as the sun set on another day.

They'd come so far. So very far. A soft wind blew and,

as the moons of Muddle Earth rose in the sky, the cutlery glinted in the purple, yellow and green light.

The tiny teaspoon stood apart from the rest. It seemed to be listening to something that only it could hear. Something far off. Something calling to it . . .

With a little sigh, the teaspoon turned. The journey ahead was long, but it had to be done.

Tinkle, tinkle, it went as it tripped back across the stony ground. *Clink, chink, clatter, clang,* went the knives, forks, spoons and all the rest of the cutlery as they followed on behind.

Through the mountains they journeyed, across the plains. By dawn the following morning, the tall trees of Elfwood could be seen on the distant horizon.

The tiny teaspoon trembled with excitement. The calling was closer. It sighed softly.

Soon. Very soon . . .

'Cuddles!' A raucous voice shattered the silence of Giggle Glade. 'I shan't tell you again. I want my hot-water bottle refilling and I want it now!'

Dr Cuddles managed a weary giggle.

'*Cuddles!*'

'Is that you, my little caged song thrush?' replied Dr Cuddles. He glanced out of the window for any sign of visitors. There was none. He giggled anxiously. 'How the Horned Baron must be missing you?' he said.

'He can't live without me!' Ingrid screeched. 'And when he finds out how I've been treated, he'll knock your block off! Now, see to my hot-water bottle. Immediately!'

Dr Cuddles shook his head. His piercing blue eyes narrowed. 'Oh, Horned Baron,' he muttered, giggling menacingly. 'You're going to pay for this. You mark my words! You're going to pay for this dearly!'

Book 3

DOCTOR CUDDLES
~ OF ~
GIGGLE GLADE

Prologue

A new day was dawning in Muddle Earth. Stiltmice were stirring, batbirds were coming in to roost, and tree rabbits were rubbing their big blue eyes with their little pink paws.

At one end of the sky the horizon was tinged a muddy brown colour as two of the three moons of Muddle Earth – the purple and the yellow ones – set. (The green moon, despite high expectations and the most expert of forecasts, hadn't bothered to make an appearance at all that night.) At the other end of the sky, the sun was rising. Its dazzling rays glinted on the uppermost peaks of Mount Boom and the Musty Mountains.

Boom, went Mount Boom weakly, and a ring of pinky grey smoke rose slowly into the air.

Far below, padding silently on broad paws along the dusty mountain road, a great, pink, striped cat emerged from the swirling mist. It paused, threw back its head and let out a loud, rumbling roar. Its sabre teeth gleamed. Every creature within earshot fell silent: the hillfish froze, a passing batbird wheeled noiselessly away, while the tree rabbits hid their eyes behind their long, floppy ears. The great pink cat scratched at the ground and roared a second time.

'I know, I know,' said its rider from astride the ornate, jewel-encrusted leather saddle secured round the creature's broad chest. 'It *is* good to be back.'

She dismounted, surveyed the scene, and gave a smile of satisfaction. The low sunlight shone on her flame-red plaits and golden skin, accentuating the curves of her firm muscles.

Her powerful physique was set off magnificently by a split-leg, tooled-leather tunic with a bear-fur trim and matching reversible chiffon and organza cloak, all topped off with a winged helmet of burnished bronze with silver inlay detailing. Her shapely ankles were emphasized by the lizard-gut thongs of her sling-back sandals, rising criss-cross fashion right up to her knees. At her dragon-skin belt, she was wearing a gold, limited-edition armoury sword and accessorized catapult. The entire ensemble was completed with a precious little goat-ear shoulder bag.

'We've been away too long,' she said, thoughtfully fingering the notches and dents of battle which scarred the blade of her sword. 'Orc wrestling, giant tickling, hag worrying. I've had enough, old friend. It's time we settled down.' She surveyed the horizon. 'What we need is a nice, old-fashioned wizard to work for. No more smelly slime-demons and boring old sorcerors to sort out. Just a few goblins to boss around, and all the milk you can drink! I don't know about you, but I can't wait to put my feet up. These sandals are killing me!'

She ruffled the creature's soft, furry, pink ears. The battle-cat purred loudly. Brenda the Warrior-Princess seized the reins and leaped back into the saddle. The cat's shoulder muscles rippled. It snarled fiercely and tossed its head.

'Onwards, Sniffy!' she cried, her voice echoing around the barren landscape, and tugged at the reins. 'To the Enchanted Lake.'

The sun shone down bright and warm on Muddle Earth; its mountains and forests; its roads, bridges and towns – and on the Enchanted Lake, which rose up into the air like a vast, watery toadstool.

Sunlight shone through the transparent column of water, casting a rainbow-coloured shadow across the bubbling Perfumed Bog. Dazzled and confused by the bright light, a large silver fish swam too near to the bottom of the hovering lake, fell out of the water and down into the gaping beak of a waiting lazybird crouched beneath.

Flop, plop . . . Gulp!

Far up above the lazybird, the sun sparkled on the rippling surface of the lake and the ornately decorated houseboats, which were bobbing about in the fresh gathering breeze. It shone on the twisting chimneys, on the varnished prows and polished brass fittings, and through the glinting windows, sending beams of sunlight slicing through the dusty shadows inside.

A boy was standing at the entrance to the master cabin of the only occupied houseboat on the Enchanted Lake, banging on the door with his fist. His name was Joe Jefferson. Beside him sat Henry, his dog.

'Wake up, Randalf!' Joe was shouting. 'Wake up!'

Henry barked.

The snoring from inside the cabin paused for a moment – before continuing with renewed vigour. A small budgie fluttered down and landed on the boy's shoulder.

'It's not locked, you know,' it said.

Joe seized the brass handle and pushed the door open. The sunlight flooded in, revealing the snorer – a rotund, bearded wizard sprawled across a tiny four-poster bed. His arms stuck out, his neck was cricked, while his feet hung over the bottom of the bed, the big toes protruding through the large frayed holes of a pair of woollen socks.

As the bright light hit his face, he snorted, grunted and smacked his lips. The eyelids fluttered for a moment, but remained shut.

'Randalf!' said Joe, his voice loud and thick with irritation. He strode forwards, Henry by his side, and shook the wizard by the shoulders. 'Randalf, you promised!'

'And you *believed* him?' said Veronica the budgie, flapping up on to the top of the four-poster bed.

'Ran-*dalf*!' Joe shook him more vigorously. '*Ran-dalf!*'

The wizard turned over and continued to snore.

'Leave this to me,' said Veronica. The budgie hopped on to the pillow and put her beak next to Randalf's ear. 'Oh, Randy,' she trilled. 'Randy, wake up. There's a stiltmouse in the bed.'

The wizard's eyes snapped open. 'Stiltmouse!' he cried. 'Where? Where?' He sat bolt upright in bed, banging his head hard on the curtain frame above him. '*OUCH!*' he bellowed.

Joe struggled not to laugh as Randalf looked round fearfully, eyes wild and pointy hat quivering.

'Stiltmice!' yelled Randalf. 'Nasty, horrible, twitching little things. *Urgghh!*'

He noticed the faces grinning at him. His eyes narrowed. 'There is no stiltmouse, is there?' he said.

Veronica and Joe laughed. Henry barked.

'I see,' said Randalf, pulling himself up with as much dignity

as he could muster. He rubbed his throbbing head and winced.

'You need a new bed, by the way,' said Joe, chuckling softly. 'This one's far too small.'

Randalf glared at him indignantly. 'I'll have you know that this is a king-sized bed.'

'Yes, it belonged to King Alf the Elf,' Veronica butted in. 'And even *he* traded it in for something bigger. Boy, they really saw you coming at *Krump's Discount Furniture Store...*'

'Shut up, Veronica!' said Randalf sleepily, yawning, stretching – and losing his balance. He keeled over, grabbing hold of one of the bed curtains (which came away in his hands) as he fell, and landed on the floor with a loud bang. The houseboat swayed.

'*Ouch!*' he roared – even more loudly than before. He turned to Veronica. 'This is all your fault for waking me so fraudulently!' he said. 'Stiltmouse, indeed!'

'It's your own fault for oversleeping, Fatso!' said Veronica calmly.

'Yes!' Joe broke in, with feeling. 'You said we'd leave by the first light of dawn, and it's almost midday! You promised!'

'But—' Randalf began.

'You know full well,' Joe continued without taking a breath, 'that if I'm ever to leave Muddle Earth, we must go to Giggle Glade in Elfwood and retrieve the *Great Book of Spells* from Dr—'

'And so we shall, my lad!' Randalf interrupted before the dreaded name could be spoken. 'So we shall! After all, given everything you've done for Muddle Earth, it's the least I can do.'

'Actually,' interrupted Veronica, 'doing *nothing* is the least you can do, and that's something you're an expert at.'

'Shut *up*, Veronica!' said Randalf. 'Believe me, my boy, we shall go to Elfwood . . .'

'But when?' Joe demanded. '*When?* No matter how often you promise we'll go, whenever the time comes you've always got an excuse for *not* going,' he said crossly. 'What was it yesterday? Oh, yes, you had to stay in to wash your beard. And the day before? Mangel-wurzel shopping in Trollbridge. And the day before that, tree rabbit racing in Goblintown. And last week it was the wrong kind of rain, and the week before that . . .'

'I know, I know,' said Randalf sympathetically. 'Several important matters and unfortunate, unforeseen difficulties have come up recently. But I have cleared my desk, I have wiped the slate clean . . .'

'You? Cleaning?' Veronica sneered. 'That'd be a first!'

Randalf ignored her. 'I said we would set off today and I meant it.' He frowned. 'It's odd,' he murmured thoughtfully. 'I distinctly remember setting the clock.' He left the bed cabin and strode across the living room. 'I do hope it's not being difficult again.'

The hands of the clock were both pointing downwards,

indicating that the time was half past six in the morning – or the evening. With the sun high in the sky, it was clearly neither. Grumbling ominously under his breath, Randalf seized hold of both sides of the clock and gave it a violent shake. The clock rattled and clunked, and something went *boing!*

'Clock repairer, too, eh?' said Veronica sarcastically. 'Is there no end to your talents?'

Randalf huffed and puffed. 'Ridiculous contraption!' he muttered. 'It's never worked properly.'

'Nor did the spell you paid for it with,' Veronica reminded him.

'That's neither here nor there!' said Randalf dismissively.

'Tell that to the goblin maiden whose hair all dropped out,' muttered Veronica.

'It's that blasted clock-elf, that's what it is,' said Randalf. He hammered on the door of the clock. 'Come on! Show your face, you incompetent numbskull!' he shouted. 'Open up!'

The door remained shut. Randalf reached forward and pulled it open. A cluster of cogs and flywheels clattered to the floor; a length of spring uncoiled. Randalf's lips pursed, his beard trembled. There was no sign of the clock-elf.

'What the . . . !' he exploded. 'Where's that ridiculous creature got to now?'

Veronica fluttered down and landed on Randalf's shoulder. 'There's a note,' she said, pointing with her wing.

Randalf peered inside the clock. Sure enough, pinned to the wall just above a small hammock, was a piece of card. Randalf reached in and tore it away.

'*Gone to unwind,*' he read out. '*Back in a fortnight of Thursdays.* Well, of all the cheek. Just taking off without so much a word of explanation . . .'

Just then, there was a *plop* followed by a *splash*. Randalf turned to Veronica. 'What was that?'

Veronica shrugged her shoulders. 'Just a fish, probably,' she said. 'After all, apart from the wizards, they're the only things daft enough to live up here – and the wizards have all disappeared. Whose fault is that, I wonder?' She said, tapping the side of Randalf's head with her beak.

Pretending not to notice, Randalf returned his attention to the broken clock. 'Probably a blessing in disguise the clock-elf's gone,' he said. 'Remind me to go to Grubleys and see about a replacement. Apparently he's got some new ones in stock. The Horned Baron's got one. It sings the time, tap-dances and tells jokes . . .'

'Never mind all that!' said Joe, exasperated. 'What about our quest?'

Randalf sucked in air noisily between his teeth. 'It's getting a little bit late for that, don't you think?'

'Randalf!' snapped Joe.

'All right, all right,' said Randalf. 'But if I could just—'

From outside, there came a second *plop-splash*. It was louder this time. Closer . . .

The next moment, the door burst open and the ogre, Norbert the Not-Very-Big, ambled in, yawning and rubbing his eyes.

'Was that you, Norbert?' said Randalf.

Puzzled, Norbert blinked his three eyes one after the other. 'It still *is* me!' he said. 'Isn't it?' He slapped his forehead with the palm of his hand. The houseboat swayed from side to side. 'Don't tell me I've changed into someone else in my sleep again,' he said agitatedly. 'Do you remember the time I turned into that short goblin seamstress called Truffles?'

'That was a *dream*, Norbert,' said Randalf patiently. 'I explained all that. And of course you're still you! I was simply asking whether you had caused the loud *plop* and *splash* we heard.'

'Can't say I noticed,' said Norbert. 'But then, what with dodging all those flying rocks, I wasn't really paying attention.'

'Flying rocks?' said Randalf.

'One of them missed my head by a hair's breadth,' he said.

'Thus missing your brain by at least three metres,' muttered Veronica.

Randalf shook his head. 'I can't say I like the sound of these flying rocks,' he said. 'They could be a bad omen, worse even than last Wednesday's light drizzle. Perhaps we ought to postpone our departure . . .'

'NO!' shouted Joe. He could bear it no longer. 'It's always something! Light drizzle, falling leaves – now flying rocks. You promised that we'd set off today, and a promise is a promise.'

'And it is a promise I fully intend to keep,' said Randalf reassuringly. 'I was merely going to propose that we set ourselves up with a good, hearty breakfast first.'

'Snuggle-muffins, sir?' suggested Norbert.

'Just the job,' said Randalf. 'And some porridge, Norbert. And a tankard of foaming stiltmouse milk. Ooh, and some jub-jub fruits – but make sure you peel them first . . .'

Henry barked.

'And some bone fritters for our valiant battle-hound, here,' Randalf added.

'Can't we just go?' Joe complained.

'We could,' said Randalf slowly. 'But I think it would be unwise to set out on a perilous quest such as ours on an empty stomach.'

'You tell him, Fatso,' chirped Veronica.

Again, Randalf chose to ignore her. 'And while you're about it,' he said to Norbert, 'get the picnic hamper packed up with some goodies, there's a good fellow. We'd better stop off for lunch on the way.'

Joe groaned. This was going to take ages. Everything had to be just so. The crusts had to be cut off the sandwiches, the stiltmouse milk had to be at exactly the right temperature (a tad cooler than tepid), there had to be twists of salt for the the hard-boiled eggs – and as for the snuggle-muffins: Randalf insisted that Norbert decorated each one with coloured icing and sprinkles, and wrapped them individually in paper doilies.

Finally, after seconds and – in Randalf's case – thirds, breakfast was over and the picnic hamper was ready and waiting by the door. Joe sat on the basket, all dressed up in his warrior-hero costume, twiddling his thumbs impatiently. With his burnished copper shield and razor-sharp sword, his helmet, breastplate and boots – all courtesy of his old friend, Margot the dragon – he certainly looked the part of a great questing warrior-hero. All he needed now was for the quest to get started.

'*Now* can we go?' he said wearily.

'Of course,' said Randalf. He looked out of the window. The sky was getting cloudy. 'I'll just go and change into my waterproof pointy hat,' he said. 'Just in case.'

Joe groaned.

'All dressed up and nowhere to go, eh?' said Veronica, fluttering down beside him.

'Why does he always do this?' said Joe grumpily. 'He knows how important this quest is for me.'

Just then, Randalf's voice floated back from the master cabin. 'Check the portholes are shut securely,' he shouted. 'And that the lamps are all out. And Norbert, if you could just run a mop over the kitchen floor . . .'

'You see!' said Joe, exasperated. He began pacing up and down the living room.

Finally, Randalf emerged in a pointy hat with a small umbrella attached to its tip. 'I've been thinking,' he said. 'Maybe it *would* be best to set off tomorrow. We can make a nice early start.'

'No!' said Joe. 'No, no, no . . .'

Veronica nodded sympathetically. 'You know the reason he keeps putting off this quest,' she said. 'He's frightened of going to Giggle Glade. Frightened of what he's going to find there . . .'

'Frightened?' said Randalf indignantly. 'Me? I'm a wizard. I take danger in my stride . . .'

Just then, a boulder the size of a large loaf of snotbread came crashing through the window. Randalf let out a little squeak of alarm and leaped up into Norbert's arms.

'Aargh!' he screamed. 'It's an omen! It's an omen!'

Veronica stared at the quivering wizard. 'Taking danger in your stride, I see,' she said.

Crash!

The roof splintered and the ceiling cracked. From outside came the sound of furious roaring.

'*Aaaaargh!*' screamed Randalf, even louder. 'Batten down the hatches! Man the lifeboats . . . !'

'Lifeboats?' said Veronica. 'What lifeboats? Norbert's sunk them all!'

'Just *do* something! Randalf shouted desperately. '*Any*thing! We're under attack!'

Meanwhile, in Goblintown, the shops were opening up for business. Built one upon the other – most exclusive at the bottom and tackiest at the top – the shops formed tall, swaying towers. One housed milliners; another, ironmongers; another, bakers . . . In the centre of the town was a stack of clothing shops, at the very top of which was *Grubley's Discount Garment Store* – a fusty, musty, rundown establishment selling a wide selection of the cheapest, nastiest outfits to be found anywhere in Muddle Earth.

The shop itself, with its rows and rows of sparkly clothes, was deserted. But from the little workshop at the back came the sound of voices. Raised voices . . .

'Get back to work this instant!' shouted Grubley the owner, a stocky character with bandy legs, hairy ears and one thick, dark eyebrow that looked stuck to his forehead like a length of bear-fur trim.

'Can't!' snapped the goblin at the workbench.

'Can't?' said Grubley. 'If I don't get that order out by lunchtime, Boris the Big-Nosed is going to have my guts for garters. You know what ogres can be like!' His eyebrow furrowed. 'This instant, Snitch,' he bellowed. 'Do you hear me?'

The goblin winced. 'Only too well,' muttered the goblin. 'Be that as it may, I can't get back to work. The sewing-elf is doing a bunk,' he explained and nodded over to the corner.

Grubley turned and peered into the shadows, where a short, slight elf was busy tying a small bundle in a spotted handkerchief on to the end of a stick.

'What in Muddle Earth is going on?' Grubley demanded. 'Where do you think you're going?'

'On holiday,' said the elf happily.

'Holiday? Holiday?' Grubley spluttered. 'But elves love their work. They don't have holidays!'

'We do now!' said the elf, a happy smile spreading out across his bony features. He swung the stick up on to his shoulder and, striking up a cheerful whistle, marched out of the door.

Grubley was left standing there; outraged, red-faced, speechless.

'That's the trouble these days,' muttered Snitch. 'You just can't get reliable elves.'

All over Goblintown, the same scene was being repeated as elves of every description poured out on to the

dark, narrow streets and headed off towards the gates of the walled city. As well as the sewing-elves, there were clock-elves and cake-mixer-elves; lamplighter-elves and greetings-elves – and even the somewhat giddy spin-dryer-elves bringing up the rear. They were all talking excitedly, the air filled with their squeaky voices as they joined in the mass exodus.

Behind them, the goblins stood in their doorways and hung out of their windows, staring forlornly as their little helpers departed. How ever would they cope without them?

The elves – growing more excited with each passing minute – headed off along the road to Elfwood in a gathering cloud of dust, their knotted, striped and spotted handkerchiefs bobbing about in the hazy early morning sunshine. As they continued, so their band grew larger and larger as others joined their number.

From Goblintown, Trollbridge and the Enchanted Lake they came; together with greetings-elves, already out with their sacks of letters, and those elves who had set up residence in potholes, who now gathered up their pots, swung them on to their backs and got caught up in the happy, chattering throng.

'I've never been on holiday before!' cried one.

'Me neither!' cried another.

'Ooh! This is *so* exciting!' cried a third as the front of the mighty crowd reached the edges of the forest. A cry went up.

'Elfwood! Elfwood! Elfwood!'

Meanwhile, in the sumptuous Grand Bedchamber of the Horned Baron's castle, its lord and master – the Horned Baron himself – was sitting up in his huge four-poster bed. There was a tray on his lap, upon it a single rose in a long-stemmed glass and a silver napkin ring, engraved with *HB*. A grubby napkin, monogrammed with the same floral letters, was tucked in at the neck of his silk pyjamas.

He had a piece of half-eaten rot fudge in one hand and was sipping from a cup of spittle tea in the other. As he wiped the pearly froth from his moustache, the horned helmet wobbled on his head.

Benson – newly promoted from head gardener to the Horned Baron's personal manservant – was on the other side of the room, drawing the curtains. 'I trust sir slept well,' he said.

'Very well,' the Horned Baron replied brightly, and chewed at the piece of rot fudge. 'Very, mffvery mffwell,' he mumbled.

Ever since his wife Ingrid had gone missing, the Horned Baron had been sleeping like a log. Every night he would drop off the moment his helmeted head hit the pillow, waking ten hours later when Benson brought him his breakfast, feeling fit and refreshed.

He swallowed the lump of fudge. 'I was having the most wonderful dream,' he said thoughtfully, and smiled to himself. In it, a bald Ingrid was being lowered slowly into a huge vat of stinky hog milk. Just as her enormous feet were disappearing from view, he'd woken up.

'. . . the ransom note?' he heard Benson saying.

'What's that?' asked the Horned Baron.

'I was wondering whether sir had replied to the ransom note,' Benson explained.

'Ransom note . . .' the Horned Baron repeated absent-mindedly as he stirred an extra sugar lump into his spittle tea.

'Yes, sir, the ransom note,' said Benson. 'For the Baroness.'

'Ah, yes, I've *mffllmmf*,' he mumbled as he stuffed another large piece of fudge into his mouth. He swallowed noisily. 'I've tried, but I can't seem to find a greetings-elf. But never mind.' He sighed, and leaned back against the plump satin pillows. 'I'll get round to it soon enough.'

Benson paused and looked round. 'If I might be so bold, sir, you know what the ransom letter said. If you don't reply by nightfall, they'll shave off all her hair.'

'*Mffllmmff*,' he muttered and tutted softly. 'Absolutely terrible.'

'And after that,' Benson continued, 'they'll immerse her in a vat of stinky hog milk.'

The Horned Baron smiled dreamily. 'Stinky hog milk,' he murmured. 'Mortifying. Poor, dear Ingrid. It really doesn't bear thinking about . . . Now, fetch me a fresh pot of spittle tea, there's a good chap. This one's getting cold. And while you're there, rustle me up a couple of slices of mouldy toast.'

Meanwhile, in Elfwood, more and more elves were arriving. The air was filled with their giggling, singing and endless happy chatter – for although the elves of Muddle Earth did indeed love their work, they seemed to be overcome with excitement at the prospect of a holiday.

But this was no ordinary holiday. This was the sort of holiday that would appeal to any self-respecting elf. This was a *working* holiday, with lots and lots of back-breaking, hard physical labour. That was what was being offered. That was what the mysterious call they all answered had promised them; a working holiday in Elfwood. Giddy with joy, the elves skipped and danced and sang out at the tops of their lungs.

'*We're all going on an Elfwood holiday!*' they chirruped, over and over. '*We're all going on an Elfwood holiday! We're all going on an Elfwood . . .*'

'Oh, do give it a rest!' complained a tall, crabby old tree, its gnarled branches trembling.

'Those squeaky little voices go right through you,' muttered a slender willow in a copse nearby.

'I know!' 'Yes, they do!' 'You can say that again!' her companions agreed, their leaves quivering with distaste as the elves swarmed round their trunks and over their roots.

'They cut through you like a chainsaw!' said a tall, spreading beech darkly.

'Ooh, Brett, don't!' gasped its neighbours.

Ever since Dr Cuddles had taken up residence in Giggle Glade, hundreds of their friends and relations had been cut down and turned into boards, planks and beams, wardrobes and cupboards, tables and chairs. Now, or so it was rumoured on the grapevine (the grapevine which wound its way round the entire forest was a terrible gossip), Dr Cuddles had ordered the construction of something enormous, something monumental – and (it was whispered) made entirely of wood.

'That Dr Cuddles has got a lot to answer for,' the beech muttered, its coppery leaves flapping menacingly. 'I'll— Get *off* me!' it shouted and flicked a branch, sending the half dozen elves who had been swinging from it flying off into the air.

They landed on soft mattresses of leaves, rolled over, leaped to their feet as if nothing had happened and scampered off to join the others. There were hundreds of them by now. *Thousands!*

'The whole place is crawling!' screeched an elegant silver birch.

'I'll shed a bough and brain the little squeakers!' bellowed an old elm, anchored to the bank of a babbling brook.

'Frilly knickers, chocolate drops, roast bananas, atishoo-atishoo, all fall down,' the brook babbled.

'And as for you!' stormed the old elm. 'Babble, babble – morning, noon and night!' It rustled threateningly. 'I swear, one of these days, I'm going to dam you up!'

Meanwhile, at the very centre of the wood, in a clearing that was getting larger by the day, other noises could be heard. There was a bang and a clatter. A piercingly shrill voice. A sigh of resignation.

Inside the house, with his ear pressed up against a closed door, was Quentin the Cake-Decorator. His hair was damp, his legs were shaking. He didn't know how much more of this his nerves could take.

'A little bit more off the fringe! And keep it straight! *Straight, you moron!*' Ingrid screeched.

Quentin trembled. That voice! It cut through him like a knife. He crossed and uncrossed his legs nervously.

'Imbecile!' she roared. 'Call yourself a hairdresser! I've blown my nose on more talented handkerchiefs! I've slept on mattresses that could use a pair of scissors better than you! I've . . .'

Quentin clamped his hands over his ears. 'Poor Dr Cuddles,' he found himself thinking. 'Then again,' he thought, 'as long as she's giving *him* a hard time, she's leaving *me* alone.'

An hour earlier, Dr Cuddles had entered her chamber

with scissors, razor and a bowl of warm soapy water balanced in his stubby arms. He was intending to carry out the first of the ransom-note threats by shaving Ingrid's head. Things, however, had obviously not gone as Dr Cuddles planned.

Quentin tentatively removed his hands from his ears and listened.

'You see, you really are a good hairdresser when you make the effort,' Ingrid was purring. 'Much better. I look beautiful.' She giggled. 'Don't you think I look beautiful, Dr Cuddles? You want me to look beautiful, don't you?'

Quentin blanched. She was being nice! That was when she was at her most dangerous. Beads of sweat broke out across his forehead. Any moment now, she would want something, and with the house-elves all having been assigned to other duties, it was he, Quentin, who would have to get or do whatever that something might be. He shuddered miserably as he remembered the night before. All that wobbly strawberry jelly! All that wire wool!

'Stinky hog *cleansing* milk, you mean, you silly thing,' Quentin heard Ingrid saying. She was giggling coquettishly. Quentin shivered with foreboding. 'Oh, I shall look forward to that. Then you can give me a back-rub, Cuddles. And a foot-massage . . .'

All at once, the door burst open. Quentin let out a little scream and jumped back. Dr Cuddles stood in the doorway, his piercing blue eyes glaring out of the shadows.

'You startled me, M . . . Master,' Quentin stammered. 'I was just . . . just about to enquire whether you'd like a . . . a . . . a nice snuggle-muffin for your mid-morning snack?'

'This is no time for snuggle-muffins,' said Dr Cuddles coldly.

'Cuddles!' Ingrid cooed from inside the room. 'Hurry back with that stinky hog cleansing milk.' Her voice hardened. 'Ingrid doesn't like to be kept waiting, remember.'

'How could I forget?' said Dr Cuddles, rolling his eyes. 'You did send that ransom note, didn't you, Quentin?' he said, giggling nervously. 'You did say that we'd shave off her hair, immerse her in a vat of stinky hog milk . . .'

'*And* tickle her feet with a lazybird feather! *And* prod her repeatedly with a wet fish!' Quentin interrupted, flapping his hands about agitatedly. 'I told them *everything*!'

Dr Cuddles shook his head slowly from side to side.

'That Horned Baron,' he growled. 'I'm going to make him pay for this if it's the last thing I do!'

'Cuddles!' screeched Ingrid. 'I'm waiting!'

'Coming, my back-combed beauty,' Dr Cuddles called back. He turned to Quentin, eyes blazing. 'Fetch the stinky-hog milk. Open the gates ready for our visitors. And tell Roger the Wrinkled I want to see those plans at once. Good grief, do I have to do *everything* around here?'

Meanwhile, in the northern fringes of Elfwood, a vast set of cutlery was busy getting prepared for the next stage in its epic journey. The knives whetted their blades on stones, the forks sharpened their prongs, while the ladles and spoons polished and buffed themselves up, until the entire cutlery set could be seen reflected in each of their gleaming bowls.

A tiny teaspoon, with glinting curlicues on its handle, stood on top of a boulder. The back of its silver bowl glinted in the early morning sun as it cocked it to one side.

It was listening.

Meanwhile, in the far off Ogrehills, a gruff yet plaintive voice cried out.

'Has anyone seen Fluffy? He was here a minute ago. Fluffy! *Flu-uffy!*'

But the ogre's snuggly-wuggly comforter – a particularly soft and hairy elf – was gone. Off down the dusty mountain track he was skipping, a tune on his lips and a knotted handkerchief on a stick over one shoulder.

Meanwhile, under Trollbridge, a lumpen troll reached out for a turnip-slicer-elf that wasn't there.

Meanwhile . . .

'Do you think it's safe yet?' said Randalf.

A good five minutes had passed since the last flying rock struck the houseboat, and Randalf had just poked his head out of the picnic hamper in which he'd been hiding – a pink snuggle-muffin stuck to his forehead. There was no reply.

'Where's everybody gone?' he called.

'Out here, Fatso,' came Veronica's voice.

Tentatively, Randalf ventured out on to the deck, where he found the others at the balustrade. They were all looking over the side, peering down through the clear water beneath them. Veronica, perched on Joe's left shoulder, flapped a wing at something far below.

'I've never noticed *that* before,' she said.

Joe frowned. Despite the distortion from the rippling water he, too, could see something. He squinted. Not some-*thing*, he realized with a shock, but some*one* . . .

'Look,' he said. 'You can make out the shoulders and legs. And red hair.' The figure below the hovering lake raised an arm in unmistakable greeting. Joe gasped. 'He's seen us!'

Just then, a voice called up, loud and clear. 'At last! I've been chucking pebbles at every houseboat for the last hour trying to get someone's attention. I was beginning to think all the wizards had left the Enchanted Lake.'

'You're not wrong there,' muttered Veronica.

Joe glanced down at a boulder the size of a large pillow, lying on the deck to his right. That was some pebble!

Randalf cupped his hands to his mouth. 'I am a wizard,' he called out. 'And who might you be?'

'It is I, Brenda, Warrior-Princess,' came the booming reply. 'Weave a spell of levitation, mighty wizard, that I may join you on the Enchanted Lake.'

Joe blushed. Despite the broad shoulders and stout legs – not to mention the deep and powerful voice – *he* was evidently a *she*.

Randalf also blushed. 'Spell of levitation,' he murmured. 'Spell of levitation ... *er* ...'

'I think she means Norbert and the rope ladder,' Veronica sniffed.

Randalf leaned over the balustrade. 'I seem to have mislaid my spell book, your highness,' he shouted down. 'But I do have a rope ladder and an ogre,' he added. He turned and snapped at Norbert, 'Come on, Norbert! Don't keep our guest waiting!'

~ 324 ~

Having sunk every rowing boat on the Enchanted Lake, followed by every bath tub and then every kitchen sink from every houseboat, Norbert was forced to become increasingly resourceful.

Joe watched as the great ogre – squashed inside a huge baking tray and using spatulas as oars – paddled slowly across the lake, a wooden washing-up bowl on a piece of string bobbing along beside him.

At the edge of the hovering lake, Norbert pulled the coiled rope ladder from his shoulders and, having wrapped one end around his ham of a hand, tossed the other end down over the side. Joe saw the rope ladder go taut, the ogre's arms and neck strain – and the baking tray begin to take on water.

Brenda must have climbed up at incredible speed, for the next moment her flame-red plaits appeared, and the warrior-princess pulled herself up over the lip of water and into the waiting washing-up bowl. She seized the ladles Norbert was holding out and, in a blur of hands and spray, sped towards the houseboat.

Randalf was there to greet her. 'Enchanted to make your acquaintance,' he said giddily as Brenda leaped up on to the deck and fell to one knee before him.

'Enchanted,' sneered Veronica. 'You couldn't do "enchanted" if your life depended on it!'

'Ignore her,' said Randalf, unable to tear his eyes away from the magnificent warrior-princess. 'And . . . and please stand up.' He wiped his right hand on his robes and stuck it out. 'Randalf the Wise,' he announced. 'Would you care for some refreshments? Some spittle tea?'

Brenda got to her feet, seized Randalf's hand and shook it vigorously. 'Pleased to meet you,' she said, and as she followed him inside the houseboat, added, 'You seem to have something stuck to your forehead, Rudolf.'

'A snuggle-muffin,' said Veronica. She fluttered across the room to her cage, where she perched on the little bar and swung indignantly backwards and forwards.

Randalf wiped a tear from his eye – brought about by Brenda's bone-crushing grip – and plucked the snuggle-muffin from his head. 'This needs some more icing, Norbert,' he said. He turned to the warrior-princess. 'Now tell me, Brenda, what can I do for you?'

'What can you do for me?' said Brenda, and chuckled throatily. 'It is more a case of what *I* can do for *you*. I stand before you, a warrior-princess, veteran of a thousand battles.

I have wrestled with mud hags and clashed swords with orc lords. Now I offer my services to you, oh mighty sorcerer.'

Randalf swooned. Veronica tutted. Norbert took a bite of the snuggle-muffin.

'No heroic deed too small to be considered,' she added, and reached round for her sword, which she held up. Joe noticed the light glint on eight notches carved into the handle. 'Each one of these represents a mighty quest,' she said and smiled. 'There's room for plenty more.'

Joe stepped forwards bashfully. 'You're . . . you're a real warrior-hero,' he said. 'We were just about to set off on a quest of our own.'

Brenda's eyes narrowed. 'And who might you be?' she asked.

'Joe Jefferson,' said Joe. 'I'm . . .'

'He's Joe the Barbarian, a warrior-hero,' squawked Veronica. 'See? We've already got one.' The swinging perch itself seemed to squeak with indignation. 'The position's taken, thank you very much. So, buzz off! Go on, sling your hook!'

Brenda frowned. 'Warrior-hero?' she said.

Joe shrank back. He felt her piercing gaze boring into his own. 'Actually,' he began, 'I . . .'

'You tell her!' Veronica called out encouragingly. 'Conqueror of ogres. Defeater of dragons. Slayer of wardrobes.'

'Really?' said Brenda, sounding impressed.

'Yes, and that's his battle-hound . . .' said Veronica, with a wave of her wing. 'Fang. Fang the Ferocious. You don't want to mess about with Fang.'

'His real name's Henry,' said Joe.

Brenda reached forwards and stroked Henry on the head. Henry rolled over and waited for her to tickle his tummy.

'Traitor,' muttered Veronica.

Brenda straightened up and proffered her hand. 'Put it there, Joe the Barbarian!' she said. 'Always glad to meet a fellow warrior-hero. So what is this quest of which you speak?'

'Our quest?' Randalf broke in. 'Oh, just a little bit of business to clear up in Giggle Glade. You're welcome to join us,' he added as casually as his thumping heart would allow. 'Usual rates; a quarter of any treasure found, plus all the snuggle-muffins you can eat . . .'

A smile spread across the warrior-princess's face. 'A little bit of business?' she said. 'Sounds perfect. But I wouldn't want to tread on the mighty Joe's toes . . .'

'You're right!' squawked Veronica. 'You'd be treading all over Joe's toes with your nasty big feet.'

'No you wouldn't,' said Joe quickly. 'Honestly, you

wouldn't! After all, we'll need all the help we can get on this quest. We have to travel to Giggle Glade to recover the Grand Wizards' *Great Book of Spells*, a sacred text which has fallen into the clutches of the most malevolent fiend ever to have breathed the air of Muddle Earth!'

Brenda turned on Randalf. 'Is this your "little bit of business", Rupert?' she demanded.

Randalf blushed and threw a filthy look at Joe for being such a blabbermouth. The last thing he wanted was to scare her off. 'Yes,' he admitted quietly. 'Yes, it is.'

'Great stuff!' said Brenda. She clapped her large hands together gleefully. 'Count me in!'

'You mean you would accompany us on this noble yet perilous expedition?' said Randalf, scarcely able to believe his own ears.

'No problem,' said Brenda. 'When do we depart?'

Randalf smiled. 'You know what they say; there's no time like the present.' He shook his head. 'As I keep trying to impress upon young Joe, here.'

'But . . .' spluttered Joe.

'Young people today,' said Randalf, winking at Brenda conspiratorially. 'You know what they're like. Always procrastinating. Always putting off till tomorrow what should be done today . . . But now you've turned up, we can finally set off. We'd be honoured if you would accompany us.'

'But, Randalf . . .' said Joe indignantly.

'Do come along, Joe,' said Randalf fussily. 'There really is no time to lose.'

With that, the wizard turned on his heels and followed Brenda out on to the deck. Norbert went after them, a peevish Veronica perched on his head.

At last, thought Joe. A real warrior-hero! Now they stood a real chance! He wouldn't have to bluff and bluster any more, not with Brenda to back him up. Now he could wrest the *Great Book of Spells* from the evil Dr Cuddles, he could release the grand wizards, free Ingrid – whatever it took – and finally return home to the real world; the place where he belonged.

'*Uh-oh*,' Randalf's voice floated back from the deck.

Joe groaned. His hopes and dreams melted away like snowflakes in a fire. What now?

Quentin was standing outside Ingrid's bedchamber, his ear pressed to the door.

'Ooh, Dr Cuddles!' came Ingrid's shrill voice. 'That tickles!'

'I can't do this if you keep moving!' Dr Cuddles's muffled voice was sounding increasingly desperate. Quentin consulted his note book. *12.30: Tickle feet with lazybird feather*, he read.

Dr Cuddles's voice rose to a high-pitched wail. 'Careful.

No, don't sit down. No . . .'

Ingrid squealed with delight. 'You silly thing!' she chided. 'Why ever not?'

'Ouch!' Dr Cuddles cried out. 'Helpmmff! *Hffmmpppfff!*'

Quentin shivered with foreboding. It sounded as though the master was being smothered. He raised his fist and hammered on the door. 'Dr Cuddles, sir?' he called.

'Come on in!' Ingrid trilled. 'The more the merrier!'

'No, don't come in*fffmm*,' shouted Dr Cuddles. 'I'll . . . come . . . out . . .' he said, every word an effort.

All at once, the door flew open and Dr Cuddles rushed from the room, his robes crumpled and askew. He slammed the door behind him, locked it, and slipped into the shadows.

Quentin shuddered. He could hear Ingrid cooing from behind the locked door. 'Don't be long, Cuddles,' she was saying – and a hard edge crept into her voice as she added, 'I'm sure I don't have to remind you what happened last time you kept Ingrid waiting!'

'She's mad,' Dr Cuddles muttered and giggled nervously. 'Quite, quite mad!' He fell still; his piercing blue eyes blazed. 'I'll have that Horned Baron,' he snarled. 'I'll make him wish he'd never been born!'

Quentin nodded. 'It is in connection with the Horned Baron that I bring news,' he said. 'The first of the elves are arriving.'

'Excellent, Quentin,' Dr Cuddles said, giggling happily. 'Have them assemble in rows outside in Giggle Glade, and show the new arrivals where to go. I shall address them all at moonrise.'

'Cuddles!' It was Ingrid. She sounded far from happy. 'Cuddles, you haven't massaged my other foot!'

Dr Cuddles groaned miserably. 'In the meantime, Quentin, you know where to find me.'

'CUDDLES! *NOW!*'

'After Brenda climbed up it, Norbert dropped the rope ladder,' said Randalf. 'Look.'

Joe looked. There, beneath the rippling lake, curled up on the ground beside a sleeping lazybird, was the rope ladder.

'Well, at least we tried,' said Randalf with a sigh. 'We'll just have to wait for him to make another one. Shouldn't take long. Couple of months, maybe. Brenda,' he said, turning to the warrior-princess, 'how about that cup of spittle tea?'

'A couple of months!' Joe exploded. 'I can't wait a couple of months!'

'And nor shall you,' said Brenda.

The warrior-princess immediately took control of the situation. 'Norbert,' she said, 'bring me the longest piece of rope you have. Rudyard, fetch four coat hangers.'

The pair of them jumped to her command. When

Norbert returned with the rope, she instructed Joe to tie one end to the houseboat's chimney. The other, she secured round the pillow-sized boulder. Then, having barked commands to some character by the name of Sniffy who seemed to be waiting for her at the bottom, Brenda hurled the boulder with tremendous strength.

Clutching hold of the balustrade as the boat wildly dipped and swayed, Joe had watched, puzzled, as the boulder sailed off through the air. What on earth – or rather *Muddle* Earth – was she up to? he wondered. Over the edge of the lake it went and down to the ground below. The rope flew with it. There was a thud and a distant *yowl*, and the rope abruptly went taut.

Suddenly, Joe realized exactly what was going on. Brenda had rigged up a makeshift cableway they could use to descend to the ground.

'Great idea,' he said enthusiastically.

'Absolutely ingenious method of getting down!' Randalf agreed. 'Now, why didn't *I* think of that?'

'Do you really want me to tell you?' said Veronica.

Brenda went down the cable first, with Henry – tail wagging furiously – around her neck. 'Just to show how easy it is,' she added, glancing pointedly at Randalf, who was beginning to make excuses and wanting to change his hat. Joe watched her glide down the rope, over the edge of the lake and out of sight.

'It's fun,' she called back a moment later. 'Next!'

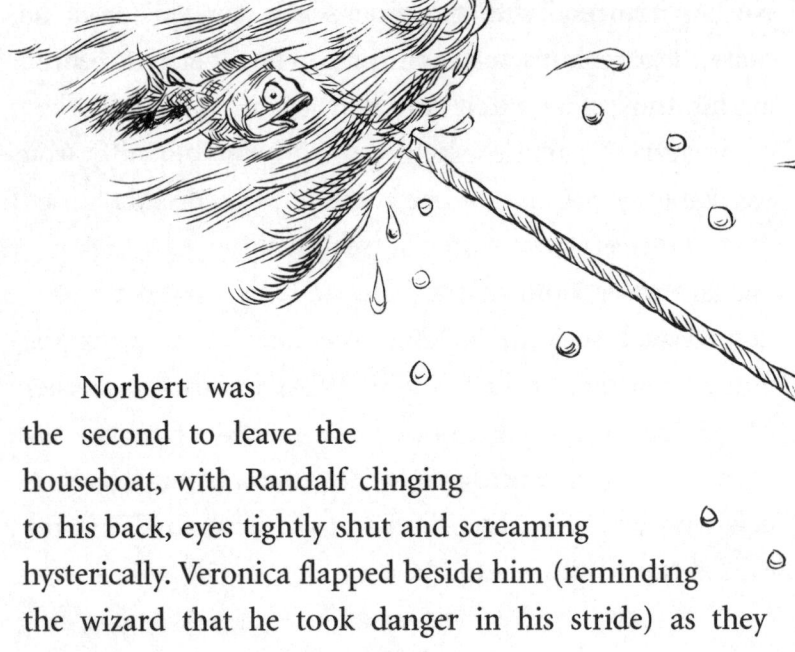

Norbert was
the second to leave the
houseboat, with Randalf clinging
to his back, eyes tightly shut and screaming
hysterically. Veronica flapped beside him (reminding
the wizard that he took danger in his stride) as they
swooped down. Now it was Joe's turn.

'GO!' Brenda's booming voice echoed up.

Jaw set in grim concentration, Joe gripped the two
sides of the hanger. His heart was racing. His knees were
trembling.

'Here goes nothing,' he muttered as he kicked off from
the side of the houseboat and launched himself into the air.

Down, down, down, he sped. The slide was far faster
than it had looked and Joe now realized why Randalf had
screamed so hysterically. Not that *he* was going to scream!
Not Joe the Barbarian. For he was a warrior-hero, and
warrior-heroes didn't scream – particularly when there was
a warrior-princess about.

As he neared the edge of the lake, his legs struck the water, sending up a plume of spray and increasing the drag on his arms. It was all he could do to hang on.

'Nearly there,' he told himself. 'Just hold on tight and ... *Wow!*'

The view which opened up below him as he sped past the lake was magnificent.

'*Whee!*' he cried. '*Wheeeeeeeee!*'

Everything rushed below him in a blur of green and brown. The ground came nearer. Ahead of him, he could see Randalf and Norbert, with Veronica back on top of his head, and Brenda beside him – and next to them, what looked like a massive, stripy, pink cat sniffing at a terrified Henry.

Joe landed with a bump, rolled over and looked up to see the others looking down at him. 'That was amazing!' he gasped.

Henry licked his face.

'Well done, sir,' said Norbert, helping him to his feet.

'Glad someone enjoyed it,' said Randalf grumpily, wiping the dust from his crumpled pointy hat and straightening the attached umbrella.

'Unlike some,' said Veronica scornfully. 'Squealing like a pink stinky hoglet, you were. I've never been so ashamed.'

Brenda stepped forward. 'Well done,' she said warmly. 'I must confess that when I first laid eyes upon you I had some doubts. But you tackled that task with the skill and determination of a true warrior-hero!' She clapped a heavy arm round Joe's shoulder. 'We'll make a great team, you and I. That fiend with the spell book doesn't stand a chance.'

Blushing furiously behind his beard, Randalf pushed his way between the two of them and eased Joe out of the way. 'Of course, he was nothing when he first came to me,' he said. 'I taught him everything he knows.' He smiled up at Brenda ingratiatingly. 'And you know what they say, a warrior-hero is only as good as his teacher.'

Brenda frowned. 'There are some things you can't teach, Ronald,' she said. 'Like bravery.' She stepped past him and seized Joe by the arm. 'Come, Joe the Barbarian. You shall ride beside me on Sniffy.'

The great, stripy, pink cat purred as Brenda leaped up into the saddle. She reached down and pulled Joe – who had Henry under one arm – up after her. Then, with a tug on the reins, they were off. Norbert followed, with Randalf on one shoulder and Veronica on the other.

Randalf was not a happy wizard. 'You really have such bony shoulders, Norbert,' he complained testily. 'It's like sitting on a sack of rocks.' He snorted irritably. 'Now, if Joe would just move up a little, I'm sure there's room on Sniffy for one more . . .'

'Shut up, Ronald!' said Veronica.

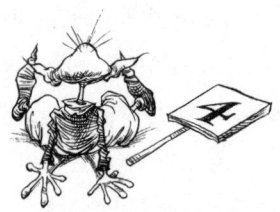

'Pass me another snuggle-muffin, Norbert,' said Randalf. 'There's a good fellow.'

Norbert, who had Veronica perched on his head, reached forward and rummaged about in the hamper which was resting somewhat precariously on a raised grassy tussock in the middle of the Perfumed Bog. Sickly sweet-smelling bog-mist swirled about them, tingeing the low, late-afternoon sun purple and covering everything in its musty scent. Norbert retrieved a snuggle-muffin and held it out.

'No, not that one,' said Randalf, looking at the small green-and-yellow-iced cake in his hand. 'I wanted the pink one with the glacé cherries and chocolate sprinkles.'

Norbert frowned. 'Someone must have eaten it,' he said. 'This is the last one.'

'Unless there's one stuck to your forehead, Randalf,' added Veronica with a giggle.

'Oh, really!' the wizard exclaimed petulantly. 'I was looking forward to that snuggle-muffin!' His resounding voice caused a huge pink stinky hog to break wind from a nearby tussock and a startled gas frog to explode. 'Who ate it?' Randalf demanded.

Veronica nodded towards the adjacent tussock, where Brenda was deep in conversation with Joe. '*She* did,' said Veronica. 'Half a dozen of them, she's had. Mind you, to be fair, you did promise her all the snuggle-muffins she could eat.'

'Of course, being a warrior-princess and all, she needs to keep her strength up,' said Randalf with a dreamy smile. He turned to Norbert. 'All right, Norbert, give me that . . . *Norbert!*'

Blushing bright crimson, Norbert wiped the telltale crumbs of cake and icing from his mouth. 'You feb you bibn't wamp ip!' he spluttered, showering Randalf with bits of half-chewed snuggle-muffin.

'Oh, Norbert!' Randalf exclaimed.

Norbert swallowed. 'There's lots of snotbread sandwiches left,' he said helpfully.

Another startled gas frog exploded close by. In the distance, a pair of stinky hogs paused, grunted and passed wind noisily. Only Brenda and Joe seemed unaware of the altercation taking place on the neighbouring tussock.

'And *this* one,' Brenda was saying, lightly fingering one of the many nicks in the blade of her long sword, 'was when

I had that run-in with Hilda the Hairy Hag. Those thunderbolts of hers can really sting, you know.'

'Wow, thunderbolts!' gasped Joe, who was growing more and more impressed with every successive story Brenda recounted. 'That sounds amazing!'

'And this,' she went on, moving down the blade, 'resulted from an incident with Harry-and-Larry – a monstrous two-headed ogre. Terrible creature he was, though he could never make his minds up about anything. I had to knock his heads together.' She chuckled. 'He won't be forgetting Brenda, Warrior-Princess, in a hurry!'

'Incredible,' said Joe quietly. 'You've done all that!'

'Indeed I have,' said Brenda, a faraway look in her eyes. 'Indeed I have. But now I want to settle down; find a nice little home for me and Sniffy,' she said, nodding across to her pink, stripy battle-cat, who was lying on a third tussock some way off, trembling.

On hearing her name, Sniffy whimpered plaintively. She didn't like this wet place with its swampy mud and pongy mist.

Henry, who was sharing the same tussock, licked her face and barked encouragingly.

'I've had my share of adventure,' Brenda went on. 'Now it's time to take life a little more easily, and Rudolf seems a nice enough wizard. I just hope I'm not putting *your* nose out of joint, Joe.'

'No, no,' said Joe. 'Not at all. In fact I'm glad you're here. Although Randalf summoned me and Henry to Muddle Earth, he can't send us back again. If we don't manage to rescue the *Great Book of Spells* and release the other wizards, then I'll have to stay here for ever and . . . and . . .' He sniffed.

'You're homesick,' said Brenda.

Joe nodded. 'I miss my mum and dad,' he said. 'And the twins. And even my sister, Ella.'

'Is she a warrior-princess?' asked Brenda earnestly.

Joe smiled. 'Not exactly,' he said. 'Although if looks really could kill, then she'd be pretty deadly.'

Brenda shuddered. 'Sounds like Sybil the Sorceress,' she said. 'Now, her looks *could* kill.' She touched a jagged notch down near the end of the blade. 'Not to mention her breath!'

'Sybil the Sorceress,' Joe whispered in awe.

'Mind you, she proved no match for Brenda, Warrior-Princess.' Brenda wrapped a great, muscular arm around Joe's shoulders and squeezed tightly – rather *too* tightly. 'Don't worry, Joe. I'll see to it that you get home. Trust me,' she said, 'we'll complete this quest successfully if I've got anything to do with it!'

'But you don't understand,' said Joe, breaking free from her powerful grip. 'Weeks, I've been waiting to set off on this quest. Weeks and weeks. And what happens when we *do* finally get going? Ten minutes in and Randalf wakes up and says we've all got to stop for a picnic!'

'Those snuggle-muffins *were* rather good,' said Brenda, licking her lips appreciatively. 'But you're right, Joe. We should be making tracks. The sun's getting low and we don't want to spend the night in the Perfumed Bog.' She climbed to her feet. 'Ralph! Sniffy!' she called across to the other tussocks. 'We must set off at once.'

'Already?' Randalf's disappointed voice floated back.

'You can ride beside me on Sniffy,' said Brenda.

Randalf beamed. 'I'll be right with you,' he said.

Goblintown was already badly missing its holidaying elves. With none of them around to do the little chores that kept the town ticking over, the whole place was running down. Oil lamps sputtered, went out and stayed out. Letters remained unsent, wet washing went unspun. And with none of the clocks working, it was as though time itself had stopped. Only the mounting piles of rubbish on every corner of every street indicated just how long there had been no elves about to tidy up.

Inside the buildings which lined the rubbish-strewn

streets, the story was the same. There were no sewing-elves, mending-elves, washing-elves . . . No elves at all! And in their absence, nothing was being made. The workrooms of the milliners, ironmongers, furniture makers and dress-makers were all standing idle.

One shop, however was a hive of activity. Goblins become very active when there's money around – and the helmeted figure trying on new robes smelled very strongly of money.

'How do I look?' the Horned Baron asked, looking into the mirror.

'Oh, I say!' the mirror – a tall, free-standing swivel affair set in an ornate mahogany frame – replied. 'You're the fairest Horned Baron of them all.'

'I am?' said the Horned Baron uncertainly.

'No doubt about it,' said the mirror. 'This outfit really suits sir down to the ground. And fits to a T. Could have been made for sir.'

The Horned Baron twisted this way and that, keeping his eyes on his body and checking out his appearance from every angle. 'I'm still not sure,' he said. 'I just don't know if it's me.'

'Sir looks fabulous,' the mirror assured him. 'The lilac complements your sallow complexion and magnificent jet-black moustache so well. And the sparkly bits match that mischievous glint in your eye.' It paused. 'Is sir planning on going somewhere nice this evening?'

'I thought I might take in a nightspot or two while I'm

here in Goblintown,' said the Horned Baron vaguely.

'Well, you're certainly going to impress the ladies in that little get-up,' said the mirror.

The Horned Baron nodded happily.

Situated on the ground floor of a towering stack of clothes shops, *Unction's Upmarket Outfitters* was renowned for its staggering array of outrageously priced outfits – and its talking mirror. Normally, the Horned Baron wouldn't have dreamed of shopping in so expensive a place. Ingrid would never have allowed it.

But then Ingrid – bless her enormous cotton socks – was no longer around. Sadly, tragically and possibly for ever, Ingrid was gone. And with her, the Horned Baron's days of having to shop at *Grubley's Discount Garment Store*. Quite apart from the fact that Grubley had swindled him in that incident with the singing curtains, the goblin had nothing on offer but cheap tat. Whereas these new clothes . . . You could *feel* the quality.

The Horned Baron craned his neck round. 'Are you absolutely sure my bum doesn't look big in these sparkly tights?' he said.

'Quite the opposite,' said the mirror. 'In fact I was just thinking how slimming they are – particularly tucked into those patent-leather bootees. And the sequinned tunic emphasizes both the breadth of your shoulders and neatness of your waist. So very flattering,' it said. 'So very *you*, sir, if I might make so bold!'

'You've talked me into it,' announced the Horned Baron. 'I'll take the whole lot.'

'An excellent choice,' said the mirror. 'And a snip at only two hundred and fifty gold pieces. Sir is going to cause an absolute riot on the dance floor.' It paused. 'Did sir have any particular nightspot in mind?'

The Horned Baron nodded slowly. Two hundred and fifty gold pieces did seem an awful lot for a tunic, tights and bootees. Then again, it would be worth it. '*Mucky Maud's Lumpy Custard Club* is supposed to be rather good,' he said.

'Oh, absolutely!' gushed the mirror. 'I've heard that the custard pies there are to die for.' It chuckled. 'And Mucky Maud herself won't be able to keep her big custard pie-throwing hands off you!'

'I'll have you know I'm a happily married man,' said the Horned Baron blushing furiously. (If Ingrid should ever, *ever* get to hear of this . . .)

'Oh, you can trust me,' said the mirror conspiratorially. 'Now if sir would like to proceed to the till . . .'

Just then, the door at the front of the shop burst open and the Horned Baron looked round to see his manservant – bright pink and panting – dashing in. 'There you are, sir!' he said breathlessly. 'I've been looking all over.'

The Horned Baron sighed impatiently. 'What is it now, Benson?' he said. 'Can't I have a single moment's peace?'

'Please, sir,' said Benson, lowering his head. 'Sorry, sir, but it's really important.' He rummaged in his pocket. 'There's been another letter concerning the baroness.'

'Ingrid?' said the Horned Baron, blushing more furiously than ever. 'A letter?'

'Delivered by batbird,' said Benson. He handed over the sheet of parchment. 'It says they're going to baste her in oil and boil her in a cauldron!'

The Horned Baron tutted and shook his head. 'What appalling luck,' he said.

'But, sir!' said Benson.

'Not now, Benson,' said the Horned Baron. 'I'm already

late for my helmet-polishing, not to mention my horn-sharpening.'

'As I always say,' the mirror enthused, 'it's those little finishing touches that make all the difference.'

'But, sir,' Benson tried again. 'The letter . . . Ingrid . . . What are you going to do?'

'First things first,' said the Horned Baron. 'After all, it wouldn't do to rush these things, now would it?'

The ragtag group continued on their quest along the dusty road. To their right, the Perfumed Bog fizzed and popped in the fading light. To their left, the jagged peaks of the Ogrehills were silhouetted against the sky. Before them – far, far in the distance – a line of pointy tree tops was just appearing above the horizon.

Joe was up on Norbert's right shoulder, with Henry trotting along behind them. Randalf, for once awake while travelling, was sitting in Sniffy's ornate saddle beside Brenda. The huge grin on his face was so fixed, so unmoving, it looked painted into place. Randalf the Wise was in seventh heaven.

'Oh, do tell me that story about when you fought the warty gutguzzler of the Black Lagoon. What rude names did you say you called it?' He swooned giddily. 'Ooh, Brenda, Warrior-Princess, you're so brave and strong and wicked . . .'

Brenda laughed (the sound of peeling bells, Randalf thought) and clapped the wizard on the back.

'*Ooof!*' gasped Randalf, his smile momentarily disappearing.

'Enough about me, Rodney,' said Brenda. 'Tell me, where exactly *is* this Giggle Glade?'

Randalf peered into the distance and nodded. The treeline was closer now. 'Giggle Glade is right in the middle of that,' he said. 'Elfwood.'

Brenda flinched. 'Elfwood?' she said. '*Elf*wood? As in *elves*?'

C lonk!

'Ouch!' cried Randalf, who was back on Norbert's shoulder, as he attempted (and failed) to duck an oncoming branch.

Clonk!

'OUCH!' Randalf howled, gingerly rubbing his forehead. 'Do watch where you're going, Norbert! That really hurt!'

'Sorry, sir,' said Norbert. He stooped down as low as he could get. 'Is that better?'

Clonk!

'Oh, for crying out loud!' Randalf exclaimed, as the tiny umbrella was knocked from the top of his pointy hat. 'Norbert, put me down. There's nothing for it. I'll just have to walk.'

Norbert did as he was told.

'Ah, that's better,' said Randalf as he strode off into the

forest. 'Keep up, you lot,' he called back. 'No dawdling! After all, we *are* on an important quest, you know.'

Veronica fluttered down and landed on the brim of his pointy hat. 'What's the hurry, Fatso?' she said. 'No, don't tell me. It couldn't have anything to do with the fact that you've got a warrior-princess on a battle-cat backing you up, by any chance?'

'Wonderful, isn't she?' said Randalf, a dreamy smile playing on his lips.

Behind them, Joe tugged on Henry's lead. 'Come on, boy,' he said. 'If you stop to sniff round every tree, we'll never get anywhere.'

Joe looked around him. With its tall, burnished trees glowing in the twilight, its tangle of flowering brambles and clumps of feathery ferns, Elfwood was certainly beautiful. But something about it made him feel vaguely uneasy, as if he were being watched. He turned and saw Brenda still lingering at the forest's edge.

'What's the matter, Brenda?' said Joe. He walked back to her and took her gently by the arm. 'Is there something wrong?'

'Well,' Brenda began.

'Yes?' said Joe.

'The thing is . . . I don't quite know how to say this.' Brenda looked down at the ground shamefacedly.

'You can tell me,' said Joe, softly 'What's bothering you?'

'It's Elfwood,' said Brenda. '*Elf*wood.'

'Yes,' said Joe.

'A wood full of *elves*,' said Brenda, tears in her eyes.

'What's that, Brenda?' came a voice. It was Randalf who, finding himself on his own, had marched back to find out what was keeping the others. 'Elfwood? Full of elves?' he said. 'If you're looking forward to seeing elves in Elfwood, I'm afraid you're going to be disappointed. There haven't been any elves in Elfwood for years.'

'There haven't?' said Brenda, the colour returning to her cheeks.

'I was forgetting, you've been away, haven't you?' said Randalf. 'Well, Brenda, such is the demand for the hard-working little fellows that the last place you'll find them now is in Elfwood.'

'It is?' said Brenda, smiling.

'They're all in Trollbridge, and Goblintown and the Horned Baron's castle these days, working their little socks off, bless 'em,' said Randalf. He shrugged. 'I'm sorry to have to disappoint you.'

Brenda laughed. 'I'll cope,' she said, with a toss of her fiery-red plaits. She clapped a great hand on Randalf's shoulders. 'Come then, Raymond, show me the way to Giggle Glade.'

The hairs at the back of Randalf's neck tingled deliciously. 'I'd be delighted,' he said.

For a good ten minutes, they made excellent progress through the woods as the shadows lengthened. Elfwood looked particularly beautiful in the setting sun as the twilight air slipped from yellow to gold, to a deep burnished copper. Birds twittered from the branches. Creatures scratched and scurried in the dappled shadows. A warm, gentle breeze whispered through the trembling leaves.

'Don't the woods look pretty,' Joe said.

'He thinks we look pretty,' said a voice just behind him.

Joe spun round. There was nobody there. It must be my imagination, he thought, or maybe the wind in the branches. He hurried after the others, tugging Henry along behind him.

He caught up with them in a small clearing. Randalf was sitting on an old tree stump, red-faced and short of breath.

'It's no good,' he was saying. 'I can't go any further. I'm absolutely shattered ... Must rest ...'

'Oh, honestly!' said Veronica. 'Take more than three steps and Fatso, here, has to lie down.'

'Tell her to shut up, Norbert,' said Randalf weakly. 'I'm just too weary.'

'Shut up, Veronica,' said Norbert meekly.

'It *is* getting dark,' said Brenda, looking round, 'and this seems a nice place for a camp. We'll stop here for the night, and set off again bright and early tomorrow morning.'

'My thoughts entirely,' said Randalf, stretching out on

the tree stump. 'You take charge of setting up camp, Brenda, while I try to regain my strength.'

'That fat one's sitting on poor old Auntie Ethel,' came a whisper from behind Joe.

'Blooming cheek!'

'Just ignore them and they'll soon go away.'

Joe looked round. Was he going mad? 'Hello?' he called. 'Is there anybody there?'

'Stop fooling about, Joe,' said Randalf, his hat down over his eyes, 'and help Brenda. Ooh,' he groaned, 'I can feel my ankles swelling.'

'Lay the blanket out over there,' Brenda was telling Norbert. 'Then go and get some firewood. Veronica, you collect some kindling. And Joe, find some small rocks and lay them in a circle here. This is where we'll make our camp-fire.' She unhooked a great, black cooking pot from Sniffy's back. 'I'll just rustle up a little something for supper.'

'Don't be too long,' said Randalf. 'I can feel myself growing weaker.'

'In the head!' said Veronica.

'Kindling, Veronica,' said Randalf. 'Good quality kindling is the basis of any good fire. Isn't that right, Brenda?'

'That's right, Rudolf,' replied the warrior-princess, dicing up several carrots and onions from Sniffy's saddlebag, and throwing them into the pot.

Muttering under her breath, Veronica fluttered off into the gathering shadows. Norbert followed her, his axe grasped in his great hands. Joe began collecting rocks and making the circle for the fire. From a little way off came the sound of a sing-song voice.

'One potato. Two potato. Yo, ho, ho and a rotten old bum!' it said.

Joe decided he was going to get to the bottom of this. He walked towards the sound of the voice. Just beyond the clearing, he stopped. There in front of him was a woodland stream flowing through the trees and babbling cheerfully.

'Football. Tennis. Elbow. Grease those wheels and trim that sail. Lilacs are a girl's best friend . . .'

Talking streams! thought Joe, with a shake of his head. Only in Muddle Earth!

Suddenly, he realized how terribly thirsty he felt. Kneeling down, Joe cupped a handful of the cool, clear water and was just about to drink when a firm hand clapped him on the shoulder.

'No!' said a voice sternly.

Joe looked up. It was Brenda. 'Never drink from a babbling brook without boiling the water first. Just in case,' she said, filling the cooking pot.

'In case of what?' asked Joe.

At that moment, the air filled with anguished cries. *'Ouch!' 'Ooh!' 'Ow!'*

'He'll pay for that!' came a voice. 'You just wait!'

Brenda and Joe rushed back to the clearing to find Randalf still stretched out on the tree trunk.

'Was that you?' asked Joe.

Randalf frowned. 'I thought it was *you*,' he said.

They looked at each other. 'Norbert?' they both said together.

Just then, the ogre in question burst from the trees, his great arms wrapped tightly round a huge bundle of firewood. He strode over to Joe's circle of rocks and dumped the whole load down beside it.

'Heavy work,' he said as he wiped the sweat from his brow.

'Did you have to make so much noise, Norbert?' said Randalf. 'You had me worried for a moment.'

'*Mffllmm blmmnflck!*' Veronica muttered as she appeared from the trees, her beak stuffed full of twigs, straw and bits of dried bark. She fluttered down, perched on one of the rocks and opened her beak. The kindling fell to the ground. Veronica turned to Brenda. 'So, what did your last slave die of?' she said.

'That's enough, Veronica,' said Randalf.

'Well, honestly!' said Veronica, her neck feathers all ruffled. 'I'm only little. Not like that great, gallumphing red-haired oaf of a—'

'Veronica!' said Randalf sharply. 'That is no way to talk to a warrior-princess.' He turned to Brenda. 'I'm sorry, your highness,' he said. 'She doesn't mean it.'

'I most certainly do!' said Veronica hotly. 'I—'

'Shut up, Veronica!' said Randalf loudly.

'*Hmmph!*' she squawked indignantly, and flew up to an overhanging branch.

With the fire set, lit and blazing, Brenda placed the cooking pot full of water over the flames and it wasn't long before they were all sitting down on Norbert's great, thick blanket, eating Brenda's rather watery carrot and onion stew. The wind dropped. The stars came out. The full moon glinted on the forest canopy, far above their heads.

Veronica pecked at some birdseed in the firelight. Joe was sitting next to Henry, one arm wrapped around the dog's neck. Sniffy lay curled up in front of the fire, purring contentedly. Brenda herself was sitting cross-legged, polishing her sword, the glow from the fire gleaming on her finely chiselled features.

Randalf put down his bowl of stew. Their eyes met. Randalf smiled. 'Tell us more of your adventures, Brenda,' he said, stretching out on the blanket. 'My word, this is comfortable after that king-sized bed,' he sighed happily.

'More adventures,' said Brenda, inspecting the nicks on her blade. 'Now let me see . . .'

'Too late,' said Veronica. 'He's already nodded off.'

A low rasping snore filled the air. Brenda smiled. 'Likes his rest, doesn't he?' she said. 'But then we could all do with a good night's sleep,' she said. 'We've got a big day ahead of us. Sniffy will keep guard, won't you Sniffy?'

The great, pink, stripy cat got up from her place by the fire and stretched luxuriously. Henry barked and trotted over to keep her company.

Joe lay back on the thick blanket, his hands behind his head, and stared up at the moonlit sky. It had been a long day. Finally, they had set out. Finally, he was on his quest.

Beside him, Randalf's low rasping snores were soon accompanied by Norbert's loud rumbling ones and Veronica's muffled whistle. Joe closed his eyes. From far away he could hear a distant babble.

'Shut the door. Open the hamster. Rhubarb, rhubarb. Pardon me, your highness . . .'

Joe smiled to himself. Even the voices had turned out to be nothing more sinister than a babbling brook. Muddle Earth! he thought drowsily. What a place!

Brenda sheathed her sword and, having tossed another log on the fire, settled herself down.

'Now they're burning a piece of Great-aunt Lavinia,' muttered an indignant voice.

'And there are sparks going everywhere!' another whispered.

'Just you wait and see,' said a third darkly. 'We'll show them!'

If Joe had heard these voices, he would have realized that they didn't belong to a babbling brook, a talkative stream or a gossiping river. But he didn't hear them, nor the odd rustling, shuffling noise that began to echo through the air – for Joe was sound asleep.

A full moon shone down over Giggle Glade, brightly illuminating the forest clearing below, where hundreds of small, bony elves were assembled, flittering, twittering and giggling excitedly. They were waiting in rows in front of a raised podium, upon which stood a tall lectern and a row of seven chairs.

'The moon is up,' one was saying. 'I'm so excited!'

'I can't wait,' said another.

'Me, neither,' a third – then a fourth and fifth – chirped up.

Soon the whole great multitude of elves were chattering

to themselves in agreement. If there was hard work to be done, then they wanted to get started.

Just then, a gangly figure in tight leggings and a glittery red coat, jumped up on to the stage and clapped his hands together.

'Welcome!' he said. Instantly, the flittering, twittering and giggling stopped. 'Welcome to Giggle Glade and a holiday that none of you will ever forget!'

'Hooray!' the elves cried out, their squeaky voices echoing through the air.

'Back-breaking work!' he announced.

'Hooray!'

'Endless toil!'

'Hooray!'

'Task after arduous task!'

'*Hooray!*' the elves cheered, louder than ever.

'And who's the one who's made it all possible? The one who's extended this warm invitation? The one you've all been dying to meet. The one . . . the only . . . Dr Cuddles of Giggle Glade!'

A tumultuous roar of approval went up as Dr Cuddles appeared, his dark, hooded robes wrapped tightly about him. He climbed the neatly turned stairs at the edge of the podium, mounted the mahogany and inlaid rosewood lectern and raised his head. Two piercing blue eyes glinted fiercely from the shadowy hood as he surveyed the gathering. The elves fell silent.

'Thank you, Quentin,' said Dr Cuddles nodding towards the glittery figure, who blushed modestly. Dr Cuddles's gaze grew more intense as he glared down at the elves. 'I promised you hard work, and that's exactly what you shall have!' he announced.

The elves found their voices again. 'Hooray!' they cried out in unison.

'You will be engaged in a great enterprise, the like of which has never before been seen in Muddle Earth,' he continued, his voice loud and clear. 'You are to build a towering construction from scratch – chopping down the trees . . .'

'Here we go again,' came a weary voice from the edge of the clearing.

'He needs to be taught a lesson,' muttered another crossly.

' . . . preparing the wood and assembling the structure from top-secret plans.' He paused. 'Top-secret plans,' he repeated. 'Quentin,' he snapped. 'Where are the top-secret plans?'

'Sorry, sir,' said Quentin sibilantly. 'Come on, you lot,' he said. 'Up on the podium with you.'

Grumbling under their breath, a line of seven wizards – each one with long robes, a pointy hat and a rolled scroll of blueprints in their hands – climbed the steps and, with a muffled jangle, filed across the podium. They each took a place on one of the seats. As they sat down, the chain linking them all together glinted in the moonlight.

On the left was Roger the Wrinkled. Beside him, Bertram the Incredibly Hairy and his brother, Boris the Bald. Then Eric the Mottled, Ernie the Shrivelled, Melvyn the Mauve, and last – and also least – Colin the Nondescript.

Dr Cuddles's blue eyes blazed. 'Thank you, Quentin,' he said again, his voice laced with unspoken menace. 'These top-secret plans have been divided into seven parts,' he went on. 'Each of you will be assigned to one wizard who will oversee the construction of your individual section. When all seven sections have been completed, they will come together under my expert guiding hand.' Dr Cuddles pulled himself up to his full height and raised his head. 'We have the tools!' he announced.

The elves cheered.

'We have the expertise!'

The cheering grew louder. All at once, cutting through the echoing roar of the excited elves, came a loud and strident voice. '*Cuddles!*'

The light in Dr Cuddles's bright blue eyes visibly dimmed. 'Ingrid,' he murmured. He turned to Quentin. 'You go,' he said. 'She likes you.'

'But she called for you,' said Quentin. 'I fear I would be a dreadful disappointment.'

The wizards sniggered behind their beards.

'I'm waiting, Cuddles!' Ingrid shouted imperiously, her voice a curious mixture of eagerness and impatience. 'Are the oils ready? Is the water hot enough? I'm so looking forward to my lovely bath!'

Dr Cuddles groaned. His gaze hardened as he turned to the elves. 'To work!' he bellowed. 'To work. *Now!*'

'Wake up, everyone, we're surrounded!' Brenda's urgent voice cried out.

Awakened by Sniffy's piercing yowl and Henry's forlorn barking, she'd opened her eyes to find the campsite had become a prison, surrounded by a wall of trees. During the night, the trees had crowded in around them until they were penned up in a space the size of Norbert's blanket.

Brenda leaped to her feet. Norbert's three eyes snapped open. Veronica sighed and pulled her head out from under

her wing. Joe stirred, sat up and looked round.

'What's going on?' he exclaimed.

'It seems that our battle-creatures were less vigilant than we had hoped,' said Brenda, shaking her head.

Joe stared round at impenetrable wall of rough bark surrounding him. 'The trees!' he gasped. 'They've moved!'

'The small one's woken up now,' came a voice.

Joe started. Norbert looked round in surprise. 'That tree,' he said. 'It spoke!'

'Ooh, and that's that great big bully who hacked my branch off,' said another.

'And mine!' 'And mine!' 'And mine!' came a chorus of indignant voices.

'I don't like the look of this,' said Brenda darkly. 'Perhaps wizard Robert, here, will be able to conjure up a powerful spell to move the trees from our way.'

'I shouldn't count on it,' muttered Veronica, fluttering up into the air. 'What you lot need is wings.'

'Hey, one of them's getting away,' said a tall copper beech.

'It's only the budgie,' his neighbour replied. 'Completely unimportant.'

'Unimportant!' squawked Veronica, twisting in mid-air and swooping down on to Norbert's head. 'I'll have you know I'm the linchpin of this entire operation!'

'It's that fat one that sat on Auntie Ethel who seems to be in charge,' came a voice. 'When he's not asleep, that is.'

'I'll soon see to that,' rumbled a giant horse chestnut, releasing a volley of conkers in their hard, prickly casings. They landed on Randalf's head, causing him to jump up in surprise. '*Ooh! Ouch! Ow!*'

'That'll teach him to throw his weight around in Elfwood,' the tree muttered gleefully.

Randalf climbed to his feet and straightened his pointy hat with dignity. 'There appears to be a misunderstanding,' he began. 'We are on an important quest, and we simply camped here for the night.'

'Camped, he says. Ooh, the nerve of the fellow!' an affronted voice exclaimed. 'How would *he* like it if we barged into his home and set fire to his beard?'

'And sat on *his* Auntie Ethel!' said another.

'He wouldn't like it,' came a voice from the back. 'Not one little bit. Go on, Bert, drop a branch on him!'

'What are you waiting for, Ronald?' said Brenda. 'Cast a spell!'

'If only,' said Veronica.

'Shut up, Veronica,' said Randalf. 'Why don't you do something useful instead of just perching there. Go and get help.'

'What kind of help?' said Veronica. 'Friendly woodcutters? Or perhaps a flock of trained woodpeckers?'

Brenda drew her sword. Norbert fingered his axe gingerly. Joe clutched his own sword and backed into Randalf. The clearing seemed to be getting smaller by the minute.

'And what do they think they're going to do with those?' an ancient elder sneered. 'There's far too many of us and far too few of them, you know.'

'Oh, why did I let you all talk me into going on this insane quest?' wailed Randalf. 'Why did I think we could ever get to Giggle Glade and destroy Dr Cuddles? What a fool I've been!' He sank to his knees with a sob.

'Did he say *destroy* Dr Cuddles?' one of the trees asked in surprise. 'Did he? What did you hear, Enid?'

'Oh, Ashley, you never listen, do you?'

'He did,' came a gruff voice. 'He definitely said *destroy*.'

'But that would be marvellous,' said another. 'Maybe we should let them go after all.'

Randalf stopped rolling on the ground, moaning, and looked up. With a shuffling, rustling, grunting sort of a noise, the trees were moving slowly out of the way. Little by little, a narrow avenue was opening up, along which, when it was wide enough, Randalf and the others could walk. Randalf picked up his hat and staff and puffed out his chest.

'I told you it was just a misunderstanding,' he said to Brenda. Veronica flapped down and landed on his hat. 'I've got to hand it to you, Randalf,' she said, 'you've certainly weaved some sort of spell on these trees.'

'Come on Norbert, Joe,' Randalf called, striding forward. 'Don't dilly-dally. We've got important business to attend to in Giggle Glade!'

'Go on!' yelled a loud voice excitedly. 'Go on!'

Others joined in. 'Throw it!' 'Go on, throw it!' And a chorus of, '*In his face! In his face!*' broke out, followed by whoops and cheers and cries of encouragement.

'I'm waiting!' shouted a voice above all the rest. 'Let me have it!'

There was a brief silence, followed by a loud, squelchy *SPLAT!* And, with a huge cheer, the crowd roared its approval.

The Horned Baron, who was listening from outside, turned to his manservant. 'Sounds like a lively crowd tonight, Benson,' he said.

'Indeed, sir,' said Benson, seizing his arm. 'But as I was saying, the latest ransom note, sir. You really should read it. Here, look . . . This time, it says they intend doing something utterly dastardly if you don't answer their demands at once. They're going to start with her toenails, then move up to her fingernails, and *then . . .'*

'Not now, Benson,' said the Horned Baron distractedly as a second *SPLAT!* and a roar of laughter exploded from the other side of the door. He twirled his moustache, straightened his gleaming horned helmet and adjusted his sparkly tights. '*Mucky Maud's*, here I come!' he announced as he shoved the swing doors open.

The heat, the noise and the heady pungent aroma of sweet, creamy desserts all combined to create an atmosphere that made Goblintown's infamous *Lumpy Custard Club* unique. The Horned Baron went weak at the knees as he looked round the great terraced chamber.

'Marvellous,' he whispered.

There were goblins everywhere – customers seated on stools at innumerable overladen tables; waiters and waitresses with trayfuls of phlegm pies, sneezed-on treacle tarts and bowls of multi-coloured gloop balanced on their upraised hands; a conga-line, dripping with lumpy custard and stinky rice pudding, weaving its way around the club. The place was seething.

At the crowded bar, a scrum of goblins struggled to gain the attention of the barkeeper – a surly looking character wearing a white shirt and black bow tie – bellowing out their orders and jostling for position. In the corner, a lone goblin with a long face and a lugubrious expression cranked the handle of a barrel organ round and round, filling *Mucky Maud's* with the sound of swirling hurdy-gurdy music.

Above their heads, a glitter ball slowly turned, sending darts of light flying through the air.

'Absolutely first rate!' the Horned Baron exclaimed, a huge grin plastered across his face. 'It's all so deliciously . . . *mucky*.' He sighed. 'Ingrid would never approve.'

'Quite so,' Benson shouted above the din and waved the ransom note under his nose. 'And with the Baroness in mind, if we could . . .'

Just then, a slightly stooped gnome wearing a long, stained black jacket and grubby striped trousers appeared before them. 'Good evening!' he said warmly. 'And welcome to *Mucky Maud's Lumpy Custard Club* and a night to remember! I don't believe we've seen sir here before.'

'This is my first time,' the Horned Baron admitted.

'I knew it,' said the goblin. 'As head waiter, I pride myself on never forgetting a face – or so magnificent a helmet. It looks freshly buffed.'

The Horned Baron nodded. 'And I've just had the horns repointed,' he said.

'Splendid,' said the head waiter. 'Perhaps sir would like me to put it away safely in the cloakroom?'

'Absolutely not,' said the Horned Baron, shaking his head. 'I never go anywhere without my helmet,' he said. 'I am, after all, the Horned Baron.'

The gnome gasped with surprise. 'The Horned Baron, I should have known!' he said. 'The *Horned Baron*. My word, sir, but we are honoured.' He extended a sticky hand. 'Smarm at your service, sir.'

The Horned Baron nodded.

Smarm giggled. 'The Horned Baron!' he said. 'I can hardly believe it! And you've got such a treat in store this evening. Now, if sir would like to take a bib and follow me, I shall show you to the best table in the house.' He winked conspiratorially. 'It's right in the firing line.' He turned on his heels. 'Walk this way.'

'If I could walk that way, I'd be seriously worried,' Benson muttered under his breath as he followed Smarm and the Horned Baron down the steps from the doorway and on to the club floor.

The music grew louder; the crowd more uproarious. As Smarm led him past the bar, the Horned Baron – bib secured around his neck – watched with interest as a tall ogre in messy dungarees pushed his way to the front.

'What'll it be?' demanded the barkeeper gruffly.

'Get me a double meringue cream-pie,' said the ogre.

'Coming right up,' said the barkeeper as he ladled a thick, sticky mixture into a large glass, topped it off with two meringues and a cherry, and tossed the whole lot into the waiting ogre's face.

SPLAT!

'Lovely!' said the ogre, his voice slightly muffled. He slammed a gold piece down on the counter. 'And one for yourself!'

'Don't mind if I do,' said the surly barkeeper. He looked up and down the bar. 'Next!' he bellowed.

Voices shouted out insistently. 'Me! Me! Me!' 'I was here first!' 'Stop pushing in!' 'A sneezed-on treacle tart with all the trimmings!'

'Would sir like something from the bar?' asked Smarm.

'Yes,' said the Horned Baron. 'Barman,' he called out imperiously. 'A couple of the finest custard pies known to Goblinkind, if you please.'

'You'll have what you're given and like it,' the barkeeper growled, causing a ripple of laughter to run the length of the bar.

'Send them over to the upper table,' said the head waiter. He turned to the Horned Baron. 'This way, sir.'

He led them up a short flight of stairs to a jutting

platform. The table there was occupied, the six seats taken up with half a dozen brightly made-up goblin matrons in stained ball gowns and dripping tiaras. Smarm quickly ushered them away and wiped a dirty cloth over the table top.

'Take a seat, your Baron-ness,' he said, holding the back of a chair and pushing it in as the Horned Baron sat down. 'The evening's entertainment will soon be getting under way.'

Benson took the chair opposite – just as two rice puddings with yellow toe-jam sailed over his head. In the corner close by, two goblins squealed with delight as they were abruptly splattered with strawberry and garlic blancmange.

'Looks like it's already started,' observed the Horned Baron.

'That's nothing, sir' said the head waiter. 'You just wait till the heavyweight trifles start flying.'

Just then, a waiter – dripping with rice pudding and chocolate sauce – arrived from the bar with the two large custard pies on a tray. He set them down in front of Benson and the Horned Baron.

'Compliments of the house,' said Smarm.

'Thank you,' said the Horned Baron, picking up his pie. He turned to Benson. 'Here's custard in your eye!' he cried, and shoved the custard pie into his manservant's face.

'*Blobberly bloof!*' Benson spluttered, wiping his mouth on his bib. 'You're too kind, sir. Allow me!' He picked up his own pie and pushed it into the Horned Baron's eagerly awaiting face.

'*Yum!*' said the Horned Baron. 'Outstanding! Two more custard pies,' he called out to the departing waiter. 'And keep them coming!'

With the head waiter gone, the Horned Baron turned to Benson, his eyes gleaming excitedly. 'My word, this makes me feel young again!' He leaned back in his chair. 'I'd quite forgotten what it was like on a Saturday night in Goblintown,' he said. 'In the old days, I really used to let my hair down.'

Benson frowned. 'You did, sir?'

'Or course, that was when I still had hair to let down,' said the Horned Baron. 'You wouldn't think it to see me now, Benson, but in the past – when I was young – I was quite the ladies' man.'

'You're right, sir,' said Benson. 'You *wouldn't* think it!'

'Oh, yes, Benson,' the Horned Baron went on, 'there wasn't a goblin maiden in all Goblintown who could resist my charms, I can tell you.'

Benson nodded. 'This would be *before* you met Baroness Ingrid,' he said.

'Obviously, Benson,' said the Horned Baron wistfully, a faraway look in his eye. 'I knew when I met Ingrid that my custard pie throwing days were over.'

Just then, a second waiter appeared. 'Two custard pies,' he announced.

'That's us,' said the Horned Baron.

With the organ-grinder cranking the handle of his barrel organ round faster and faster, the thumping music grew increasingly frantic. All round the vast room, tables of goblins, trolls and ogres bellowed out their orders. 'More stinky-toffee pudding.' 'Stiltmouse milk mousse all round!' 'Gob pie!' And, as the air filled with volley after volley of lumpy custard pies flying across the room, yells of delight and shrieks of pleasure echoed round the walls.

A stray custard pie struck the side of the Horned Baron's helmet and dangled from one of the pointy horns. 'My goodness!' he exclaimed with a chuckle. 'The place really is beginning to jump!' He turned his attention to the lumpy custard pie in front of him. He smacked his lips. 'Would you care to do the honours, Benson?'

Suddenly, all the lights went out, plunging the club into darkness. The Horned Baron drew back from the custard pie and peered round. The next moment, a spot-light came on at the far end of the room and shone down on a set of red satin curtains at the top of a grand, sweeping staircase.

The crowd, as one, held its breath. All at once, the curtains trembled, parted and a portly troll appeared at the top of the staircase. A great cheer went up.

'It's her!' someone shouted.

'It's Mucky Maud!'

Resplendent in a tight, sparkly cocktail dress (with a

particularly magnificent blancmange stain down one side), her thick hair, decorated with lazybird plumes, piled up high on her head, Mucky Maud cut an impressive figure. And as the barrel organ music struck up a tune, she made her way slowly, slinkily, down the flight of stairs.

When she reached the bottom, Mucky Maud turned to the lugubrious goblin at the barrel organ. 'Play it again, Spam!' she whispered.

The organ-grinder pulled a lever and turned the handle. A loud, pounding beat burst forth – *Oom-pah-pah! Oom-pah-pah!* – and the Horned Baron smiled as the strains of a familiar tune started up. Mucky Maud raised her head and started singing.

> '*You put the lumps in my custard,*
> *You put the wobble in my jelly,*
> *You really curdle my caramel, baby,*
> *When you trifle with me!*'

Her voice growing louder, Mucky Maud sauntered through the club, in amongst her adoring audience. Every so often, she would pause for a moment at a table – to ruffle goblin hair, tickle goblin chins, and push eager goblin faces down into waiting custard pies. Whatever she did, the crowd cheered. She had them eating out of the palm of her hand – literally!

As the song approached its soaring finale, the organ-grinder cranked up his barrel organ. Mucky Maud's strident voice soared over the deafening accompaniment.

'*And so I say . . .*'

The Horned Baron held his breath. She was coming towards him.

'*And so I say to you . . .*'

She was approaching his table. He turned away, his cheeks burning with embarrassment. It was as though Mucky Maud was singing to him; as if the words of the song were meant for him, and him alone . . .

'*And so I say-i-yaaaaaay . . .*' As she held the note, she came up behind the Horned Baron and trailed her long fingers across his shoulders and over his shiny helmet. All at once, she fell still, took a breath and completed the song in a low, husky whisper. '*Don't you ever trifle with me.*'

'Bravo!' roared the Horned Baron, clapping enthusiastically. 'Bravo!'

Mucky Maud moved round to face him. The Horned Baron looked up. Their eyes met.

'You!' he gasped.

It was gone four in the morning. Already, pink stinky hogs were stirring in the Perfumed Bog as the first batbirds flapped across the sky above their heads. The sky was cloudless. The air was still. It promised to be a beautiful day.

In *Mucky Maud's Lumpy Custard Club*, the night was finally coming to an end. Most of the customers had already gone home, and the waiters and waitresses were busy clearing up the mess they'd left – hosing down the tables and stacking the chairs. An ogre and a couple of goblins were still propping up the bar.

'Sixsh more cushtard piesh,' one of them demanded groggily.

'You've had enough,' said the barkeeper gruffly as he rubbed a filthy cloth over the glasses, making sure they'd be nice and dirty for the evening.

'Oh, go on, me old mate,' pleaded the ogre.

'Bar's shut,' said the barkeeper sharply. 'And I am *not* your "old mate". Go on, push off. Haven't you got homes to go to?'

'Shp'oe sho,' said the ogre and the two goblins in unison, and they turned and shuffled reluctantly away.

'Shtill, it'sh been a good night,' one of them muttered as he fell with a *splat* into a soggy heap of trifle and blancmange. He licked his fingers. 'A *very* good night.'

In the corner, the barrel organ was still pumping out music, but slowly now, and softly, as the lugubrious goblin grew wearier and wearier. He eyed the upper table sleepily, where Mucky Maud was sitting on the Horned Baron's knee. Until she decided to leave, Spam had no option but to keep playing – and judging by the giggles and guffaws of laughter coming from her direction, there was no saying when that might be.

'I still can't get over it,' the Horned Baron sighed, his face turning serious. 'After all this time ...'

'I know,' said Mucky Maud.

'Oh, Fifi!' he said suddenly, and grasped her hand. 'How could I ever have been so stupid?'

Mucky Maud shook her head. 'It was fated never to be, Walter,' she said. 'After all, you were wealthy, well-to-do, with the whole of Muddle Earth at your feet.' She paused. 'While I ... a young troll from the wrong side of Trollbridge ...'

'But Fifi, you had such dreams back then. Whatever happened to the troll I once knew who said she would never rest until she had made her fortune in the Muddle Earth turnip market? Eh? What happened to those dreams, Fifi?'

'Oh, Walter, I tried, believe me,' said Mucky Maud. 'Gave it my best shot. But in the end, I had to concede that I wasn't good enough. I simply didn't have what it took!'

~ 379 ~

'But, Fifi . . .'

'You don't know what it's like in the turnip business,' she went on. 'It's a cut-throat, troll-eat-troll world, believe me. The endless contests and shows. The constant pressure . . . Too much rain. Not enough. The blight, the canker, the root rot. Not to mention the purple turnip weevils . . .'

'The purple turnip weevils?' said the Horned Baron.

'I told you not to mention them,' said Mucky Maud. She sighed. 'Anyway, that's all in the past now. When I finally admitted to myself that I'd never make it big in turnips, I came to Goblintown – to start a new life. I got into custard and never looked back.'

From the opposite side of the table there came a low gurgle as Benson – fast asleep and head down in a bowl of custard – burped. The Horned Baron snorted. 'I told him not to have that third rancid butter surprise,' he muttered.

'Of course, I had to change my name,' she continued without a break. 'If it ever got out, it would have caused an absolute scandal. I'd never have been able to show my face in Trollbridge again. And so Mucky Maud was born.'

'You'll always be Fifi to me,' said the Horned Baron, squeezing her hand affectionately. He shook his head sadly at the thought of what could have been; what *should* have been. 'But, Fifi!' he groaned. 'Why didn't you come to me?'

Mucky Maud sniffed. 'Look at it from my point of view. You, the Horned Baron, ruler of Muddle Earth and me, a mere custard-club singer. It would never have worked out.

We both knew that!' Her eyes filled with tears. 'And now it's too late!' she wailed.

'But it doesn't have to be too late,' the Horned Baron said gently. 'I've got a little vegetable garden back at my castle. We can grow turnips! I'll take you away from all this. We can start again . . .'

'Oh, Horny,' said Mucky Maud. 'Could we? Dare we? This is all a dream!'

'A dream that together, we shall make come true,' the Horned Baron told her. He looked round. 'Benson!'

The manservant started, looked up and wiped the custard from his face. 'Sir?' he mumbled sleepily.

'We're leaving, Benson,' said the Horned Baron. 'Take Mucky M . . .' He turned to her. Their eyes met and they smiled at one another. 'I mean, take Miss Fifi's cape,' he said.

Benson looked puzzled. 'But sir . . .' he began.

'No "buts", Benson,' the Horned Baron snapped. 'Just do it!'

'*P*ut *the kettle on, mother, it goes with your eyes. The moon's a baboon . . .*'

Randalf spun round and stared at the ogre, aghast. 'Norbert!' he said. 'You haven't . . . You didn't . . .'

'I think you'll find he did,' Veronica observed.

'*Pieces of eight!*' said Norbert, a look of puzzlement on his face. '*Pieces of eight . . . nine . . . ten green bottles, running down the hall . . .*'

'You did, didn't you?' Randalf said. 'You drank from the babbling brook.'

Norbert nodded sadly.

'After I specifically told you not to,' Randalf sighed. 'That's all we need,' he said. 'A babbling *ogre*.'

'*Mud pies and pointy sticks . . . I . . .* I'm sorry,' said Norbert. 'But I was so thirsty . . . *Friday. Tuesday's child is full of cake . . .*' He grimaced, clamped his hands over his mouth and looked round woefully.

'Don't worry,' said Brenda. 'It'll wear off.'

'But when?' Norbert mumbled. '*Where? Wet? Wellington boot?*'

'When Nature calls,' said Brenda cheerfully, slapping the ogre on the back.

Joe giggled.

Norbert removed his hands. He frowned. 'You mean when I have a wee . . . wee . . . *wee Willie Winkie, running through . . .*'

'That's exactly what she means, Norbert,' said Randalf. 'And in the meantime, I'd be grateful if you would do your best to remain quiet. Stealth and silence are the hallmarks of a successful quest. You do *want* our quest to be successful, don't you?'

'*Mm-hmm,*' said Norbert, nodding earnestly, his hands clamped back in place.

'Then let us proceed,' said Randalf. 'Brenda, lead the way!'

Keeping close together, the intrepid travellers continued through Elfwood as fast as they could – which, thanks to Randalf's increasingly frequent stops, wasn't that fast at all. Puffing and panting, he would have sat himself down on every tree stump he came to if Brenda hadn't been there to urge him on.

'Not far now, sir,' she said, seizing his arm and steering him away from a particularly inviting beech stump in front of them. 'Just a little bit furth . . .'

'I can't,' Randalf groaned as he sat down heavily, mopping his brow and gasping for breath. 'Must . . . take . . . rest . . .'

'Well done, Fatso,' said Veronica sarcastically. 'You've broken your own record. That was less than a minute since the last tree stump.'

'Non . . . non . . . nonsense,' Randalf panted. 'We've been walking for miles, haven't we, Brenda?'

'Well, I wouldn't say *miles* exactly,' said Brenda. 'And perhaps you could try not to stop quite so often, Rudolf.'

'Come on, Randalf,' said Joe encouragingly. 'You can do it.'

'It's all right for you,' said Randalf, gathering himself. 'But you don't have my delicate feet. I tell you, my little toe is killing me . . .'

'Not to mention the dodgy excuses!' Veronica butted in.

'Shut up, Veronica!' said Randalf.

'Well, honestly!' said Veronica. 'Who are you trying to kid?'

'This isn't getting us anywhere,' said Joe. He turned to Norbert. 'Can you carry him?'

The ogre nodded, his hands still firmly pressed against his mouth.

'I'm not getting back on Norbert's shoulder,' said Randalf. 'Not with all these low branches.'

Joe sighed. 'Perhaps he could carry you piggyback,' he said.

'Very appropriate,' Veronica sniggered.

'Oh, if you insist,' said Randalf. He heaved himself off the tree stump and stood behind the ogre. Norbert crouched down. Randalf jumped up stiffly on to his back and grabbed on round his neck. '*Eeek!*' he cried out. 'Norbert, I'm slipping!'

Norbert swung his arms round behind his back, jigged Randalf up and supported him under his legs.

'That's better,' said Randalf.

'*Flapjack, pelican, walrus, trumpet,*' said Norbert, blushing furiously. 'S . . . sorry, sir,' he mumbled. 'I'm . . . trying . . .'

'You can say that again,' muttered Veronica.

' . . . not . . . to . . . say . . . anything . . . silly . . . *billy, pudding and pie, Major Minor, Dr Cuddl . . .*'

'That is *enough*, Norbert!' said Randalf, clamping his own hands over the ogre's mouth. 'And don't slobber!'

'Right, well if we're finally all ready, I think we should make a move!' said Brenda in her most warrior-princessy voice. 'Which way, Rupert?'

'Follow me, everyone,' said Randalf confidently. 'I know these woods like the back of my hand. Just keep going to the right . . . To the *right*, Norbert!' he said as the ogre lurched to the left. 'That's it. And do try to keep up. No dawdling at the back!'

Joe followed, close on Norbert's heels, with Henry – still on his lead – trotting beside him and Veronica on his shoulder. Sniffy padded silently behind him. Brenda – sword drawn and eyes peeled – brought up the rear.

And so they continued, on and on. And on and on and on and on and on . . . The sun rose high, crossed the sky, and began to sink back down again. And still they journeyed on. Occasionally, Norbert – who was having difficulty not slobbering, causing Randalf's hands to slip – would start babbling again. For the most part, however, they travelled in silence, until . . .

'Something's wrong,' said Joe.

'You can say that again,' said Randalf. 'I should be back on the houseboat, feet up, sipping a nice lukewarm cup of spittle tea, not trudging through Elfwood on the back of a lumpy, bumpy, dribbling ogre who can't stop babbling.'

Joe shook his head. 'It's not that,' he said.

Brenda caught him up. '*What's* wrong, Joe?' she asked, looking round anxiously. 'Have you seen something? A monster? A dragon? A . . . an . . . elf, maybe?'

Joe shook his head and pointed at a tree stump some metres away. A particularly inviting beech stump . . . Joe had recognized it at once.

'We've been going round in circles!' he exclaimed. 'Oh, Randalf. We're lost!'

Veronica snorted. 'Trust Fatso! "I know Elfwood like the back of my hand," he said. Wild goose chase, more like.'

'It . . . it wasn't me,' Randalf blustered. 'It was Norbert.'

'*Mm hmmm mmm,*' muttered Norbert indignantly

behind Randalf's hands.

'It's true,' said Randalf. 'Never could tell his left from his right.'

Norbert lowered his arms. Randalf fell to the ground.

'*Tennis balls! Washing machine! Cheese on toast!* . . .' His face twisted up with frustration, Norbert turned on his heels and stomped off into the trees.

'Norbert?' Randalf called after him. 'Norbert, come back at once. You can't leave me here.'

'He just has,' said Veronica.

Randalf sighed and sat down on the beech stump.

'If you don't know where it is,' said Joe, 'how are we going to get to Giggle Glade?'

'Did you hear that, Eileen?' came the gruff voice of a nearby elm. 'They *don't* know how to get to Giggle Glade.'

'I know, Stan. I did wonder why they were taking such a roundabout route,' his neighbour replied. 'I just assumed that fat one knew where he was going.'

'Me, too,' said a willowy willow. 'And I didn't like to say anything,' she added. 'It seemed a bit forward.'

'They want to go *that* way, don't they?' said Stan.

'Absolutely,' said Eileen, 'bearing left at Delilah the holly bush – taking care to mind her prickles and down towards the sycamores . . .'

Joe beamed. 'Thank you,' he said.

The trees rustled.

'Ooh, what a nice young man,' said a silver birch, her

leaves trembling. 'So polite!'

'Yes, not like that great big, three-eyed one,' said an outraged oak.

Just then, Norbert emerged from behind the tree, grinning broadly and tightening his belt. 'That's better,' he said. 'I think I'm almost back to normal now, sir. I've just . . .'

'Yes, yes, spare us the details, Norbert,' said Randalf, taking control of the situation. He pulled himself up off the tree stump and, with Veronica up on the brim of his pointy hat, strode off towards the holly bush. 'Hurry up, you lot,' he called back. 'We've wasted enough time already.'

With the trees guiding them through the woods, the small party made good progress.

'They'll be there in no time, Sid,' commented

one of the sycamores.

'That's true, Sam,' said another, with a shudder that sent a shower of whirlygigs spinning down through the air. 'With a bit of luck, they'll be able to do something about that dastardly Dr Cuddles.'

'And not a moment too soon,' said another, 'what with Giggle Glade getting bigger by the day.'

'I'm counting on them,' said another. 'I . . . ooh, they need to bear a little bit further left, don't they? Past Finnbar – is that his name? That great fir tree . . . And taking care not to wake Uncle Cedric . . .'

'No chance of that,' said yet another. 'He's been sleeping like a log for years.'

'That's it. *Now* they're on the right track.'

Joe was feeling much more confident now. Randalf seemed to have found his second wind, Norbert was back to normal and even Veronica was chirpier than usual.

'We must be getting near,' she was saying. 'Listen.'

Joe cocked his head to one side. From far away, he could hear the distant sound of hammering and drilling and sawing . . .'

'Ooh,' shuddered a great pine tree, her needles quivering. 'Can you hear that, Daphne?'

'Dr Cuddles's fiendish plans,' a willow replied in a shaky voice. 'It's enough to make you weep.'

'Don't worry,' said Joe. 'We're going to put an end to Dr Cuddles once and for all.' He turned. 'Aren't we, Brenda?'

The warrior-princess nodded, but Joe couldn't help noticing that her face was pale and drawn, her nose twitched and her eyes darted round constantly into the shadows. Beside her, Sniffy seemed just as uneasy.

'Are you all right, Brenda?' Joe asked.

'I'm not sure,' said Brenda. 'It's just . . .' She was staring at the ground in front of her. 'What *is* that?'

Joe frowned. 'A bit of cloth,' he said, picking up a small scrap of red and white spotted material by Brenda's foot. 'Why?'

'It looks like a handkerchief,' said Brenda, a tremble in her voice. 'An *elf's* handkerchief.'

'How can it be?' said Joe. 'You heard what Randalf said. There aren't any elves in Elfwood.'

Brenda pointed to her left. 'If that's true, then what is *that*?'

Joe retrieved a small silver thimble from the ground.

Even he had to admit that it looked the perfect size for tiny elf fingers.

'And that!' said Brenda, beads of sweat appearing on her brow. 'And that – and *that*!'

Joe looked round at the forest floor. It was littered with tiny objects. There was a tiny bone comb. A miniature fan. A small tasselled cap. A couple of minute wooden buttons . . .

'Elf droppings!' Brenda cried out in horror.

Just then, Randalf's voice floated back. 'Hurry up, you lot! We're here!'

'Did you hear that, Brenda?' said Joe excitedly. 'We've made it to Giggle Glade!'

Brenda nodded dumbly, but she clearly wasn't listening.

Joe took her gently by the arm. 'Come on, Brenda,' he said. 'Don't worry about these things. They've probably been lying here for years.'

Reluctantly, Brenda let Joe lead her on. Even more reluctantly, Sniffy, whose nose was quivering suspiciously, went with them. They rounded a vast, spreading horse chestnut tree. And there was Randalf.

'There you are!' he said, talking in a loud stage whisper.

While Brenda and Sniffy hung back, Joe went over to him, and peered through the leafy branches into the clearing beyond. 'Wow!' he gasped.

In front of him, towering up from the bare ground and supported by rickety scaffolding, was a huge wooden construction which looked for all the world, or so Joe thought, like a giant rabbit's head. From the right came the sound of chopping axes. From the left, the squeal of a circular saw. While from the head itself (if that's what it was), the drum-like beat of a hundred banging hammers echoed through the air.

Around the edge of the clearing, the trees wailed and waved their branches. 'Oh, there goes Arnold,' moaned a tree close by as, with a creak and a thud, a tall pine tree slammed to the ground.

'And now they're starting on Montague,' cried another.

The whole of Giggle Glade was a hive of activity, with trees being felled, stripped and turned into planks and boards which, in turn, were being used to construct the curious structure. Orchestrating the whole process was a wizard with long robes and a particularly wrinkled face, who was standing beside the scaffolding, a huge piece of paper in his hand.

'Right, according to Figure 3 here, section c – the piece with the inverse dove-tail joints must be connected to section r – making sure that the hinge-flange (see Figure 8) is uppermost, and the dowling trim (see additional notes) is on the left . . .' He paused, looked up at the rabbit's head, then back at the plans. 'No, hang on a moment.' He turned the paper right round. 'Section *d* should be attached to section *m*, but not before the floating divet has been secured.' He frowned. 'Floating divet? What on Muddle Earth is a floating divet? Who writes this stuff?'

'*You* did!' a chorus of tiny voices replied.

'That's Roger the Wrinkled,' Randalf whispered to Joe. 'He taught me everything I know.'

'It took him about five minutes,' muttered Veronica.

Randalf shook his head. 'Poor Roger,' he said. 'I knew something awful had happened to him, but I never imagined he'd be reduced to this.'

Joe followed Randalf's gaze to the giant ball and chain attached to Roger the Wrinkled's ankle.

Randalf shuddered. 'Bound and shackled and forced to do the bidding of Dr . . . Dr . . .' He looked away, unable to speak the whole name out loud.

'Just as well we've got Joe the Barbarian and Brenda, Warrior-Princess on our side, isn't it, sir?' said Norbert.

'Yes,' said Randalf, slightly less than certainly. 'Where *is* Brenda.' He looked round. 'Oh, there you are. What are you doing back there?'

Brenda emerged from behind a tree. 'What are those voices?' she asked nervously.

'Come and have a look.' He smiled at her as he parted

the overhanging branches. 'You're in luck, my dear,' he said. 'Giggle Glade is absolutely crawling with elves! Who'd have thought it?'

'*Eeek!*' squeaked Brenda.

Randalf frowned. 'Brenda?' he said. 'Where are you going? Brenda, *why* are you climbing that tree? Brenda, speak to me!'

'*Tee-hee-hee*,' giggled a portly oak tree as Brenda and Sniffy scrambled high up into its branches as quickly as they could. 'Ooh, that tickles. *Ooh! Ah! Ha-ha-ha!*'

'Brenda, whatever's the matter?' Randalf whispered as loudly as he dared. 'Do you sense danger? Is it fire-breathing dragons? Awesome orcs? The warty gutguzzler from the Black Lagoon . . . ?'

'No,' Brenda sobbed. 'It's *elves*.'

'Elves?' Randalf squawked with disbelief. 'You can't be serious! A great big, strapping warrior-princess like you afraid of tiny little elves. I don't believe it!'

Joe sighed. 'I think Brenda's got a bit of a thing about elves,' he said.

'Nonsense,' said Randalf. He peered up at the tree. 'Brenda, tell me it isn't true. Please, Brenda . . .'

'It's t . . . t . . . true,' she stammered. 'Hate them! Hate them! Hate them! Horrible, bony little bodies. Squeaky little voices . . .' The branch she was clinging to juddered violently. 'Make them all go away. Please!'

'But Brenda,' Randalf pleaded. 'I believed in you. Your

 huge muscles, your great big battle-cat. What about all your heroic deeds? The hags you've wrestled, the sorcerors you've squashed?'

'Why do you think I've been roaming the Northern Wastes all this time?' sobbed Brenda, burying her face in her hands. 'Because I was hiding away from those nasty little things and trying to conquer my fears. And I thought I had. I really did. I thought I was finally ready to return to Muddle Earth, but no . . . No!' Sobs racked her powerful body. Beside her, Sniffy whimpered miserably at his mistress's distress.

'Ooh, hark at her!' said the oak.

Suddenly overwhelmed with exhaustion, Randalf sat down heavily on the ground.

'I always knew there was something a bit odd about that so called *warrior-princess* of yours,' said Veronica.

'Well, that's it, then,' said Randalf wearily. 'All over. Abandon quest. We may as well go home.'

'Go home?' stormed Joe. 'That's precisely what I intend to do – and I'm not talking about a houseboat on a floating lake!'

'But we can't,' said Randalf. 'Not without our warrior-princess.'

'We can and we shall!' said Joe, his head high, his eyes gleaming. 'I, for one, refuse to give up now. Not when we've come so far. We've battled through the forest. We've found Giggle Glade. Now it's time to do what we came here to do. We're going to free Roger the Wrinkled. We're going to find that *Great Book of Spells*. And we're going to destroy Dr Cuddles!'

'Bravo, I say,' said a towering hazelnut tree nearby. 'That's the kind of fighting talk I like to hear.'

'Same here, Dolores. It's time someone stood up to that starey-eyed little bully!'

Joe drew his sword and raised it high in the air. 'Who's with me?' he cried out.

'I'm with you!' Veronica shouted back.

'*Cauliflower cheese. Wingnuts!*' said Norbert. 'Sorry, hiccups!' He smiled. 'What I meant was, I'm with you, Joe. I'm with you every step of the way.'

They all turned to Randalf. 'And you?' said Joe.

Randalf sat up, his face etched with miserable resignation. He shook his head. 'I'm going to regret this,' he said. 'It's completely against my better judgement, so don't blame me when it all goes wrong . . .'

'Well?' said Joe.

'All right,' said Randalf. He climbed to his feet. 'But you go first.'

'CHARGE!' Joe roared. Waving his sword above his head, Joe burst from the cover of the trees and ran headlong across the stripped clearing, Henry hard on his heels, barking excitedly. Elves squeaked and darted out of their way.

The wizard looked up from the great piece of crumpled paper in his hands, the links of his heavy chain clanking as he moved. 'My goodness!' he exclaimed, his eyes twinkling brightly. 'It's a warrior-hero!'

Joe stopped before him.

'A little on the short side, perhaps – but that's a marvellous shield!' He smiled at Joe. 'What brings you here?'

'Come to rescue you,' said Joe breathlessly. 'To set you free.'

'Oo-ooh!' came a chorus of sing-song squeaky voices, and Joe looked round to see various elf faces peering up at him.

'Don't mind them,' said Roger the Wrinkled. He clapped his hands together. 'Get on with your work!' he said brusquely.

'At once, sir!' they said, returning to their various tasks. 'With pleasure!'

'Eager little creatures, aren't they?' said Roger the Wrinkled. 'A joy to be in charge of.'

Joe frowned. 'You don't exactly *look* in charge,' he said. 'Not with that ball and chain . . .'

'I know,' said Roger, looking down at his shackled feet. 'Frightful bore, isn't it,' he said. 'Still, there's really no need to worry. Everything's under control.'

'It is?' said Joe, surprised.

'Oh, yes,' said Roger the Wrinkled, tapping the side of his nose mysteriously. 'Certainly, I have no need of a rescue – although it was awfully considerate of you to offer. Tell me, where exactly *did* you spring from?'

Joe turned. Norbert was approaching, with Randalf (looking decidedly sheepish) behind him. 'I'm with them,' he said. 'We're on a quest!'

Roger the Wrinkled looked up. His face wrinkled with surprise. 'Norbert?' he said. 'Norbert, is that you?'

'Yes, sir,' said Norbert, striding forwards.

'How lovely to see you again, Norbert,' said Roger the Wrinkled. 'It certainly is. And how terribly clever of you to have come all the way here to rescue me.'

'Oh, it wasn't just me, sir,' said Norbert modestly.

'Who's that behind you? The wizard fell still and screwed up his face in such amazement that his wrinkles got wrinkles. 'That's not young Randalf, is it?' he said.

A red-raced Randalf peered round from behind Norbert. 'Hello, sir,' he said, speaking more meekly than Joe had ever heard him speak before.

'Well, well, well,' said Roger the Wrinkled. 'Who'd have thought it? That *you* could have organized all this!'

'It wasn't that difficult,' said Joe. 'After all, he is a wizard.'

'A wizard?' said Roger the Wrinkled, chuckling loudly. 'Oh, my dear young warrior-hero, Randalf here isn't a wizard.'

'He isn't?' said Joe.

'No, no,' said Roger the Wrinkled. 'Randalf here is a very junior apprentice. He hasn't even passed his preliminary wizardry exams yet, have you Randalf?'

'No, sir,' said Randalf, squirming with bright-red, shamefaced embarrassment.

'I told you so!' said Veronica triumphantly as she fluttered up from the brim of Randalf's hat. 'I told you that you'd be found out one day.'

'Veronica, is that you?' said Roger the Wrinkled. 'How lovely your plumage is looking.'

'Oh sir,' said Veronica giddily. 'Do you really think so?'

'And to think you came all this way to rescue me,' said Roger the Wrinkled. 'I'm touched. I really am.'

Randalf shook his head. 'But how did it happen, sir?' he

said. 'A great big, powerful wizard like yourself getting kidnapped. I don't understand.'

'I don't think we need to go into the details, Randalf,' Roger the Wrinkled said quickly, his wrinkles forming deep creases. 'I should never have stepped into the wardrobe in the first place – but there was a rather fetching evening gown in there which just happened to catch my eye . . .'

Norbert hiccuped nervously. '*Frilly knickers, frilly knickers . . .*'

'How did you know?' said Roger, blushing a deep scarlet. 'Well, before I knew it, the door slammed shut and Dr Cuddles turned the key. Then he had me transported back to Giggle Glade inside the darned thing, blast his big blue eyes!'

Joe trembled. 'Couldn't you have cast a spell?' he asked.

'There wasn't the time,' said Roger the Wrinkled. 'It all happened in an instant.' He sighed. 'I've been under lock and key, along with the other wizards, ever since.'

'So where does the rabbit's head fit in?' asked Randalf, nodding towards the wooden construction.

'On top of the rest of the rabbit,' said Roger the Wrinkled.

Randalf frowned. 'I mean, what is it *for*?' he said.

'It's all part of Dr Cuddles's latest fiendish plan,' said Roger the Wrinkled. 'I'm afraid that's all he told me. He's quite mad, you know.'

All round them, the teams of elves blithely continued

what they were doing, whistling to themselves as they worked. Chopping and planing, screwing and nailing. Fashioning the individual sections of timber which, when they were completed, were passed along a line of elves, across the clearing and up the scaffolding, where they were hammered into place. From the fringes of the growing clearing, the trees muttered to themselves nervously.

'I'm going to be next, I just know I am,' said a weeping willow tearfully.

'Courage, Lucinda,' said a neighbouring hornbeam. 'If we're going to go, we're going to go with dignity.'

'*Ouch!*' cried a nearby oak, as the elves cheerfully attacked the base of its trunk with a volley of axe blows. 'Farewell cruel world!'

'Oh, now they're starting on Oswald,' the willow

sobbed. 'That it should have come to this. If only that fat one would hurry up and *do* something!'

'Him!' snorted the hornbeam. 'He doesn't do *anything* in a hurry!'

'You can say that again, Norris,' said a prickly hawthorn. 'Spends half his time sitting down!'

Even as the trees were grumbling to themselves, Randalf was once again making himself comfortable – on the great iron ball chained to Roger the Wrinkled's ankle. He looked up at the wizard. 'So, now that we're here, what would you like us to do?' he asked.

Roger the Wrinkled frowned, the deep lines corrugating his forehead. 'Do?' he said. 'To be perfectly honest, I'm not sure what you can do. As I was saying to the young warrior-hero, here . . .' He smiled at Joe, the corners of his eyes crinkling alarmingly. 'What did you say your name was?'

'It's Joe,' said Joe.

'Joe the Barbarian,' added Veronica.

'Quite,' said Roger the Wrinkled. 'As I was saying to Joe, grateful though I am that you've come all this way, your journey has been unnecessary. I don't need to be rescued.'

'Oh, dear, what a shame, I was really looking forward to getting stuck in,' said Randalf cheerily. He climbed to his feet. 'Still, if you're quite sure, sir, we'll leave you to it. See you back at the houseboat.' He turned to Joe and Norbert. 'Come on, you two,' he said. 'We're not needed. Come along

Veronica,' he added, patting the brim of his hat.

'Hang on a minute!' said Joe indignantly. 'Aren't you forgetting something?'

'You heard what Roger said,' Randalf told him. 'He doesn't need our help.'

'That's as may be,' said Joe. 'But *I* need *his* help!'

Roger the Wrinkled turned towards him, his face creased up in concern. 'You do?' he said.

'Oh, it's nothing,' said Randalf hurriedly. 'Come on, Joe. You can see that Roger the Wrinkled's got a lot on his plate . . .'

But Joe was having none of it. 'I need a *real* wizard to get me home,' he told Roger, speaking over Randalf, who was tugging at Joe's sleeve and urging him to go. 'Randalf summoned me here to Muddle Earth, but he doesn't seem to know how to send me back. I've been here for ages. Ages and ages! I've battled with fearsome ogres, tussled with fire-breathing dragons . . .'

'Slight exaggeration,' Randalf muttered.

'Be quiet, Randalf,' said Roger the Wrinkled sternly. 'It sounds to me as though you've been dabbling in matters you know nothing about.'

'You can say that again,' said Veronica, fluttering through the air and landing on Roger the Wrinkled's pointy hat.

Randalf hung his head. 'Shut up, Veronica!' he hissed out of the corner of his mouth.

Roger the Wrinkled turned to Joe. 'Continue, young man.'

'Anyway, I've done everything Randalf asked of me,' he said.

'And more,' said Norbert, nodding. 'He's been marvellous!'

'But now I want to leave Muddle Earth,' said Joe. He smiled bravely. 'You're my last hope, Roger the Wrinkled,' he said. 'Can you send me home? *Please!*'

'That could be tricky,' said Roger the Wrinkled, shaking his head sadly. 'I'm afraid magic isn't simply a matter of waving a magic wand and casting a spell.'

'It isn't?' said Randalf, surprised.

'Gracious me, no,' said Roger the Wrinkled, giving him a wrinkled frown. 'Regardless of what you might have heard, Joe, wizardry is a highly complex enterprise, requiring great skill and heightened powers of interpretation. Not even the legendary Ian the Clever was able to perform more than half a dozen of the simplest feats of magic from memory.'

Joe felt his heart begin to sink.

'Accuracy is everything, you see,' Roger continued. 'There is, for instance, a wart-removing spell that is only two words and a whistle different from a turning-someone-into-a-pink-stinky-hog enchantment. You can see the danger.'

Joe nodded glumly. Henry whimpered softly. Randalf looked down at his feet shamefacedly.

'That's why we wizards have the *Great Book of Spells*,'

said Roger. 'To ensure the accuracy required. And even then it's not easy. Years of study, it takes, before an apprentice wizard can interpret its symbols, codes and diagrams, in order to carry out its instructions to the letter.'

'But *you* could, couldn't you?' said Joe, hope in his voice.

'I *could*,' said Roger the Wrinkled, 'if I had the book.' He sighed. 'Unfortunately – and don't ask me how – the *Great Book of Spells* has somehow fallen into the clutches of Dr Cuddles.'

'Hmm,' mused Veronica, glaring at Randalf. 'I wonder how *that* could have happened?'

'He keeps it locked up inside a heavy box secured to the top of a tall, oak lectern . . .'

'Or Cecil, as he used to be known,' said the willow sadly.

'Poor, dear Cecil,' the hornbeam murmured. 'And now Oswald's going the same way.'

'Cuddles only allows me to read a word at a time from the spell book,' Roger the Wrinkled went on. 'And then, only from the particular spell *he* wants conjured. So you see, it takes absolutely ages to weave even the simplest magic.'

Joe sighed. Every time he got his hopes up, Roger the Wrinkled dashed them again. Not that Joe was about to give up. Not yet . . .

'What about the key?' he said. 'If the book's locked up, there must be a key to *un*lock it.'

'There is,' said Roger the Wrinkled, shaking his head

sadly. 'Dr Cuddles wears it on a silver chain around his neck.'

Joe groaned.

'Oh, well, we tried our best,' said Randalf. 'Let's get back to the Enchanted Lake and have a well-earned cup of spittle tea.'

Veronica fluttered down from Roger's pointy hat and landed on his shoulder. 'Shut up, Randalf,' she said. 'We're not leaving now – are we Norbert?'

'Certainly not,' said Norbert.

'And as for the key,' Veronica went on. 'You leave that to me!'

'You'll find Dr Cuddles in his Giggle House over there, behind the privet hedge,' said Roger the Wrinkled. 'And if you do manage to get your hands on the *Great Book of Spells*, stick it up your jumper and bring it to me, and I'll be happy to weave you any spell you like, young man. Good luck, and don't you worry about me!' He waved a wrinkled hand airily.

The five of them turned and set off, Norbert, Veronica, Joe and Henry (back on his lead) in front, and Randalf trailing reluctantly behind.

Although most of the trees had been felled, Giggle Glade was not completely bare. The clearing was dotted with a number of small, pathetic specimens: a wispy willow sapling, a scrawny hollybush, a weedy elm . . . As they crossed the glade, darting from tree to spindly tree and hoping that they wouldn't be spotted, Joe noticed more teams of elves – each one being supervised by its own manacled wizard – constructing other sections of the massive tree rabbit.

'Isn't that Bertram the Incredibly Hairy?' said Veronica, flapping her wing at a tall, stooped individual, his features hidden behind thick, bushy tresses.

Norbert nodded. 'And look,' he said. 'Over by that huge back leg. It's Melvyn the Mauve.'

Joe kept quiet. The sight of the imprisoned wizards depressed him more than he would have liked to admit. If *they* had all been captured by the dastardly Dr Cuddles, then what chance did *he* stand?

'Cheer up, Joe,' said Norbert, as if reading his thoughts. 'It might never happen.'

They were crouched down behind the bushy privet just in front of the house waiting for Randalf to catch them up. Joe peered over the little gate. A path of stepping stones led up to an ornately decorated wooden cottage. There were pink roses round the door and every window had powder-blue shutters with heart-shaped peepholes.

'*That's* the Giggle House?' said Joe.

'The nerve centre of Cuddles's sinister operations!' said Randalf, panting noisily as he caught up.

'Not so loud,' whispered Veronica. 'He'll hear us.'

Just then, from inside the Giggle House, there came the sound of a door slamming.

Dr Cuddles stood outside Ingrid's chamber. 'She's

driving me stark staring bonkers!' he groaned. 'You'll have
to deal with her, Quentin. I've had all I can stand.'

'Oh, but, *sir*,' Quentin whined.

'That's enough!' snapped Dr Cuddles, his blue eyes
blazing. 'I'm going to my room and I do not want to be
disturbed. Do you understand me?'

Quentin nodded unhappily.

'Good!' said Dr Cuddles, giggling menacingly. 'It'll be
the worse for you if you don't.' He marched towards his own
bedchamber, walked in and slammed the door behind him
with even more force than before.

BANG!

'Someone's in a bad mood,' whispered Veronica to Joe. 'I'll just take a peek inside to check the coast is clear. Wait for my signal.'

Joe nodded grimly. The budgie flew off, a bright flash of blue. He watched her fluttering through the air and round to a shuttered window. She landed on the window ledge, peered in through the heart-shaped peephole, then turned and beckoned with her wing.

'Come on,' said Joe. 'We're going in.'

Heads down and shoulders hunched, Joe, with Henry, Norbert and Randalf scurried over to the house and joined Veronica by the window. Their shadows were long in the light from the low sun. They crouched down. 'What can you see?' Joe whispered.

'Take a look for yourself.' Veronica nodded at the peephole. 'I think it's him.'

Slowly, cautiously, his heart thumping and legs shaking, Joe straightened up and peered into the room. It looked like a nursery, with toys and games – and there, lying on a bright yellow bed, was a small figure, all wrapped up in baggy robes. He was sound asleep, snoring softly. On a chain round his neck, a small, silver key glinted from the shadows; beside it, on a bit of old string, hung a rusty-looking whistle.

'It *must* be him,' Joe whispered. 'There's the key. But how do we get it without waking him?'

'Leave this to me,'
said Veronica, flying up
from the ledge and
squeezing through the
peephole. Joe held his
breath as he watched
her land on the pillow
next to the sleeping
figure and, taking care
not to wake him, disap-
pear inside the shadowy
folds of his hood.

Dr Cuddles twitched,
grunted and rolled over. 'That
tickles, Quentin,' he giggled in
his sleep. Joe held his breath.

The next moment, Veronica emerged
from the hood, a silver chain clamped in her beak. With a
flap of her wings, she launched herself into the air, flew
across the room and out through the peephole once more.

'Well done, Veronica!' Joe whispered excitedly.

Randalf seized the key. 'It's not over yet,' he said.
'Come on.'

With their heads down, they stole round to the front of
the house and, having peeked through the letter box to
check no one was there, tried the door. It was open. They
tiptoed in.

Joe peered round. Compared with the golden glow of the evening sun outside, the room he found himself in was bathed in shadow. Slowly, his eyes grew accustomed to the light. The rosebud wallpaper clashed with the powder-blue frilly curtains. The orange carpet was decorated with big bright daisies.

'Look,' Joe whispered urgently. 'Over there.'

Everyone turned to the lectern in the corner with the locked box secured to it at the top. Randalf nodded.

'Come on, then,' whispered Veronica. 'Let's unlock the spell book and get out of here!'

Keeping close together, they crossed the room and gathered in front of the lectern. Joe held Henry close on the lead and, finger on lips, reminded him to remain quiet. Veronica fluttered down and landed on Norbert's shoulder. Randalf stepped towards the box.

He raised the key. He slipped it into the heavy padlock. He turned it . . .

'*WAAAAARGH!*'

As the floor abruptly opened up beneath them, the five hapless intruders tumbled back into the darkness. Down, down, down, they fell.

'*WAAAA . . .*' The trapdoor slammed back into place.

There were cries of, '*Oof!*' '*Ouch!*' and 'Get your elbow out of my ear!' as, first Norbert, then Randalf, then Joe, then Henry landed one on top of the other on the stone floor of a damp, smelly, pitch-black room beneath the house.

Joe extricated himself from the tangle of limbs and stood up. He peered round. This time, however, his eyes did not grow accustomed to the light – because there wasn't any. Not a glimmer. Not a spark.

'Now what?' Joe murmured.

All at once, there was a *click* and a light came on. Then came the sound of rusty bolts being drawn back and the creak of a heavy door slowly opening.

Blinking in the brightness, Joe looked round. They were in a small, windowless cellar with water on the floor and green mould on the walls. In one corner was a pile of chains attached to large iron balls. In another was a stair-case. The sound of heavy footsteps striding down a corridor echoed round the cellar.

'*Buffaloes' bottoms . . .*' Norbert hiccuped nervously. 'Someone's coming!'

Randalf hid behind Norbert. The footsteps got louder – followed by a voice.

'Honestly, Colin,' it lisped, 'how many times must I tell you not to tamper with the spell book?'

A pair of pointy shoes and glittery tights appeared at the top of the stairs, followed by a sparkly red jacket, a pale face, a floppy quiff . . . The figure froze and let out a little squeak. '*You're* not Colin the Nondescript!' he said.

'Quentin!' Norbert cried.

The thin, foppish figure gasped. 'Norbert?' he said.

Randalf stepped out from behind the ogre. 'Quentin!' he said. 'What are *you* doing here?'

Quentin fingered his quiff. 'I work for Dr Cuddles now,' he said.

'For Dr Cuddles?' Randalf said, his mouth falling open in astonishment.

'And it's all your fault, Randalf,' said Quentin. 'You dragged me here to Muddle Earth without so much as a by-your-leave. One minute my world was icing sugar and fondant crème, the next, you thrust me into that unflat-tering armour and talked me into going on that awful quest. And then!' he said indignantly. '*Then* you had the cheek to run away and leave me at the first sign of danger!'

'Sounds familiar,' muttered Joe.

'Thank goodness Dr Cuddles came along when he did,' Quentin went on. '*He* told me a thing or two about you.'

Randalf blushed.

'Granted, he may be an evil mastermind intent on taking over the world, but at least he appreciates me . . .'

'Running away?' said Randalf. 'I wasn't running away, my dear Quentin. I was executing a tactical flanking manoeuvre that we wizards call—'

'Saving yourself,' Veronica butted in. 'The same old story!'

'Oh, don't let's argue,' said Norbert. 'Quen-tin,' he said slowly, clapping his hands together and smiling happily. 'How *are* you?'

Quentin smoothed down his sparkly jacket and patted his quiff. 'Mustn't grumble,' he said. 'Thanks for asking, Norbert . . . Oh, but that Ingrid! She's a monster – and she never stops! From first thing in the morning to last thing at night it's all, "Quentin, do this!" and "Quentin, do that!" I don't know whether I'm coming or going.' He paused theatrically, then nodded towards the pile of shackles. 'But enough about me,' he said. 'Get your shackles on everyone, and follow me.'

Randalf snorted. 'And what if we don't want to, Quentin?' he said.

'Then I'll tell Dr Cuddles,' said Quentin.

'I'll be right with you,' said Randalf, hurrying across the cellar and securing one of the shackles to his ankle. Reluctantly, the others followed his example.

When they were all ready (even Veronica, who had a

tiny ball and chain of her own) they picked up their heavy iron balls and climbed the stairs.

'Where are you taking us?' said Randalf uneasily.

'You'll see soon enough,' said Quentin.

At the top of the staircase, they went along a narrow passage that led to Dr Cuddles's sunshine-yellow kitchen. There were decorated snuggle-muffins everywhere.

Norbert gasped. 'Quentin, you're an artist!'

'And you're a sweetie, Norbert,' replied Quentin, opening the kitchen door. 'This way, everyone – and careful with that iron ball, Randalf!'

They followed Quentin out of the kitchen door and into the backyard of Giggle House. The yard was huge (the size of three football pitches at least, Joe thought) and full of hundreds of elves. In front of them was a long podium, upon which were seven wizards – each one with a ball and chain attached to an ankle – seated on seven chairs, and a lectern, with a short character dressed in baggy robes standing behind it. He had his back to them and was shouting in a high-pitched giggly voice.

'Quentin! We're all waiting!'

'I'm coming, sir,' Quentin shouted back. 'You'll never guess who fell through the trapdoor.'

'It can't be what's-his-name, he's over there,' said Dr Cuddles, turning round.

'No, sir, not Colin the Nondescript. Look!' He ushered Randalf and the others forward.

From inside the shadowy hood, two piercing blue eyes grew wide with surprise for an instant, then narrowed menacingly. 'Well, well, well,' said Dr Cuddles. 'If it's not Randalf the Wise! It's been a long time, Randy, my old friend. Far *too* long! Why, if I didn't know better, I'd have sworn you'd been avoiding me!'

Joe frowned. 'You know each other?' he said in surprise.

'Of course they do,' said Veronica.

'Shut up, Veronica!' said Randalf.

Dr Cuddles giggled. 'I'll enjoy having a nice little chat with you later,' he said. 'In the meantime, I've got bigger things on my mind.' He turned his back on them.

Quentin motioned to the wizards to move up and make room for the late arrivals. With much shuffling and grumbling and shifting of chairs on the podium, they got themselves settled, by which time Dr Cuddles's fingers were drumming impatiently on the lectern.

Joe and Norbert had managed to squeeze themselves on to the end of the line, Henry lay at Joe's feet, Veronica sat on Norbert's head, but Randalf still lingered at the foot of the podium, with Quentin fussing beside him.

'Come on,' said Quentin impatiently. 'Move along a bit there at the end, Colin. That's it, now Bertram the Incredibly Hairy and Boris the Bald can move over. Then Eric the Mottled can move on to that chair. Ernie the Shrivelled, move there; Melvyn the Mauve, there . . . Norbert, budge up next to Roger. *That's* better.' He mopped his brow. 'Right, now you can sit down, Randalf.'

'Oh, no need to bother,' said Randalf, with a nervous smile. 'I'll just slip back to the kitchen and wait for you all . . .'

'Sit, Randalf!' growled Dr Cuddles. Randalf did as he was told, perching next to Norbert. 'Right, well if we're *all* ready at last,' he said, eyeing Randalf sternly, 'bring on the tree rabbit.'

Hundreds of tiny elf voices chattered excitedly as the huge wooden construction was wheeled in through the powder-blue back gate.

Towering above the house, the tree rabbit had been fashioned with minute attention to detail. It had splendid wooden whiskers, huge wooden paws and a wooden tail carved to look fluffy.

'Quentin,' said Cuddles.

Quentin stepped forwards. 'Congratulations, my dear elves,' he said, his voice echoing round the glade. 'Your work here is done.'

'*O-oh*,' the elves sighed sadly in unison.

'It remains only for me to pass you over to that most demanding and unyielding of slave-drivers. The one, the only . . . Doctor . . . Cuddles!'

The elves burst into applause.

'Thank you! Thank you!' said Dr Cuddles, raising his arms and staring round at the great gathering. 'I have worked you hard!' he said. 'Days of backbreaking toil and ceaseless endeavour, just as I promised . . .'

'More! More!' cried the elves.

'And today, I make you this pledge,' Dr Cuddles continued. 'There will be many such hard, gruelling, thankless tasks to come.'

'Hooray!'

'Noses to the grindstone, shoulders to the wheel, fingers worked to the bone!'

'*Hooray!*'

'But for now I declare the holiday here in Giggle Glade over!'

Quentin stepped forwards. 'Please leave in an orderly fashion,' he lisped, 'taking care to take all personal items and luggage with you. Thank you.'

As one, the elves stood up and set off back through the forest. 'My aching back!' said one happily. 'Lovely!'

'And I've got *so* many splinters in my fingers!' said another, his face creased up in a grin.

'Can't beat a good bit of heavy toil,' said someone else. He tutted. 'It's back to sewing in Goblintown for me.'

'All good things come to an end,' said yet another with a wistful sigh.

'All work and no hard labour makes Jack a dull elf,' replied his companion.

The elves filed out of the powder-blue back gate and skipped off through the forest, trilling at the tops of their voices.

'Glad to see the back of the little squeakers,' grumbled

a hefty ash tree.

'Yeah, good riddance!' shouted a spiky pine.

'And don't come back!'

Elfwood was quiet once more, the only sound, the wind in the branches and the distant babbling of a brook. At the top of the oak tree, Brenda opened one eye. She looked down. Her face relaxed.

'Sniffy,' she said. 'I think they've gone.'

In Giggle Glade, Dr Cuddles was still holding court. With his elf audience gone, he turned to address the wizards and captured visitors behind him.

'They said I was mad,' he announced.

'You are,' Randalf murmured.

'They said it couldn't be done. But I, Dr Cuddles of Giggle Glade have proved them all wrong.' He turned and gestured towards the great tree rabbit.

Joe looked. It was absurd, with one ear longer than the other, giant outsized paws and a lopsided expression on its face. A trapdoor at the base of the wooden creature hung open, a set of steps leading down to the ground from it.

'I have tried everything to breach the walls of the Horned Baron's castle,' said Dr Cuddles. 'The singing curtains. The enchanted wardrobes . . .'

'I told you Dr Cuddles was at the bottom of them,'

Veronica whispered to Joe.

'But this time,' Dr Cuddles went on, 'I shall succeed with deception where brute force has failed!' He paused. 'Inside this tree rabbit is a secret chamber in which shall be concealed an elite squad that I have prepared,' he said, giggling gleefully. 'The tree rabbit shall be left outside the castle walls as a goodwill offering from me to the baron who, suspecting nothing, will take it inside. Then, as the clock strikes midnight, my squad shall leap out and . . .' Cuddles's blue eyes flashed madly. '*I* shall become the ruler of Muddle Earth!'

He turned to Roger the Wrinkled. 'Roger, the spell of animation if you please.'

Roger climbed to his feet, looking oddly unperturbed by Cuddles's speech and stepped forwards. The other wizards looked at one another and muttered under their breath. Roger stopped at the front of the podium, looked up at the rabbit and raised his arms. He intoned a long and complicated incantation. The next moment, the wheels at the four corners twitched and the tree rabbit trundled forward.

Joe gasped with amazement.

'Yes!' cried Dr Cuddles triumphantly. 'It's working! I am a genius!' He looked round, wild-eyed, and fingered the rusty whistle around his neck. 'The time has come,' he announced, 'to enter the tree rabbit!'

'Not so fast, Cuddles!' came a voice behind him, as a huge, pink, stripy cat bounded over the powder-blue back gate and raced towards the podium, with a warrior-princess – blades glinting in the bright morning sunlight – on its back.

'Sniffy!' cried Joe.

'Brenda!' shouted Randalf.

'*Eeeek!*' screamed Quentin, jumping up on to Norbert's lap, as Brenda somersaulted from Sniffy's back and landed in front of Dr Cuddles. She brandished her huge two-handed sword.

'Release these people, fiendish sorceror!' she roared.

'Never!' cried Dr Cuddles.

'We'll see about that,' said Brenda, striding forwards. In a blinding flash, her blades sliced through the air. *Swish, swish, swish.* Dr Cuddles's robes fell in tatters to the floor.

Standing before them, was a short pink teddy, with blue eyes and stubby paws. A gasp of surprise went round.

'What?' said Joe, scarcely able to believe what he was seeing.

'Haven't I seen you somewhere before?' said Roger the Wrinkled thoughtfully. He turned to Randalf. 'Isn't that . . . ? Didn't you once have . . .'

'Shut up . . . Oops, pardon me, sir.'

'Typical,' said Veronica. 'Still trying to keep things quiet. When *are* you finally going to come clean?'

Brenda was circling Dr Cuddles, a look of bemusement on her face. 'Strange,' she was saying. 'Are evil sorcerers meant to have pink fur and big blue eyes? Is *this* the fiendish Dr Cuddles you were so afraid of, Rudolf?'

'You're a fine one to talk!' said Randalf. 'Falling to bits at the first sight of an elf!'

'Yes, well,' said Brenda, blushing and turning her attention back to Dr Cuddles. 'You're an extremely naughty teddy and I'm going to have to box your fluffy pink ears.'

'And rap him on the paws,' said Bertram the Incredibly Hairy.

'*And* spank his bottom!' added Ernie the Shrivelled.

'And . . .'

Dr Cuddles scowled. 'You think you're pretty tough, with your big pointy weapon and your big stripy battle-cat, don't you? Well, I've got news for you . . .' He took hold of the small rusty-looking whistle hanging round his neck, put it to his lips and blew hard.

Pfeeep!

For a moment, nothing happened. Everyone looked round at everyone else. Then, with a loud *crash*, the door of Giggle House burst open and a dozen identical teddies – exact copies of Dr Cuddles in every way – marched over to the podium.

'My secret weapon!' Dr Cuddles roared. 'The Tickle Squad!'

The wizards all began muttering at once. 'Whose work is this?' demanded Roger the Wrinkled. 'Only a wizard can perform a spell like this. As if one Dr Cuddles wasn't bad enough!'

'Not me,' chorused Bertram the Incredibly Hairy and his brother, Boris the Bald. 'Nor me,' said Eric the Mottled, Ernie the Shrivelled and Melvyn the Mauve.

Roger looked sternly at the wizard on the end of the line. 'Oh, Colin!' he said. 'How could you!'

Colin the Nondescript winced unhappily and shuffled his feet. 'I'm sorry,' he said softly. 'I just wanted to be noticed.'

Brenda chuckled. 'Never mind,' she said, tightening her grip on her sword, her muscles rippling. 'Just leave them to me.'

The teddies marched towards her, their big blue eyes staring fiercely.

'I've grappled with ogres and wrestled with hags. I've battled with the warty gutguzzler of—'

Suddenly, the squad split into two. Six of the teddies

~ 425 ~

pounced on Brenda and six on a startled Sniffy. Brenda stumbled backwards. The teddies started tickling.

'Oh, my word!' she giggled, dropping her huge sword. 'Stop it. *Tee-hee-hee!* Get off me! *Ha-ha-ha . . .*'

Meanwhile, the other six were rendering the great battle-cat as weak as her mistress. Within seconds, the pair of them were helpless with laughter, rolling around on the ground, gasping for breath and with tears streaming down their faces.

'Tie them up,' Dr Cuddles ordered.

Brenda was shackled to four enormous sets of balls and chains, while Sniffy was trussed up like a chicken in what looked suspiciously like Dr Cuddles's sitting-room curtains.

Cuddles was triumphant. 'Nothing shall stop me now!' he giggled. 'Muddle Earth shall be mine!' And he threw back his head and giggled loudly and unpleasantly. '*Hee-hee-hee-hee-hee!*'

'Tickle Squad, atten-*shun*!' Dr Cuddles shouted.

The Tickle Squad snapped to attention, their pink fur glowing in the bright sunlight; their blue eyes glinting. Dr Cuddles walked along the line, straightening up an ear here, flicking a bit of fluff from a shoulder there.

'Excellent,' he giggled. 'Excellent!'

One or two of them giggled with anticipation. Dr Cuddles reached the end of the line and stepped back.

'Tickle Squad!' he shouted. 'You are an elite force and you must show no mercy. You are to conceal yourselves in the great tree rabbit which, by means of magic, shall be transported to the Horned Baron's castle. Once inside the castle walls, you are to wait until darkness, observing complete silence at all times. That means no giggling, Number Seven!' The seventh bear along stifled a giggle. 'On the stroke of midnight you are to burst from the tree rabbit and take over the castle. Any resistance is to be met with extreme tickling – to the death, if necessary!'

The bears nodded impassively, not a trace of emotion registering on their pink, furry faces.

'Good luck, Tickle Squad,' Dr Cuddles cried. 'To the tree rabbit!'

Turning smartly, the bears set off in a line, marching in step to Dr Cuddle's barked commands. 'Lef', lef', lef' right, lef'!'

Joe watched helplessly as the Tickle Squad approached the great tree rabbit. 'Is there nothing we can do?' he said. 'Brenda? Randalf? Roger?'

Brenda rattled her chains and shrugged. 'I'm sorry, Joe, they took me completely by surprise,' she said.

'I knew this was a bad idea,' said Randalf, shaking his head. 'We should never have come on this hopeless quest. Why, I could be at home right now tucked up in my king-sized bed if it wasn't for you lot.'

'Shut up, Randalf!' said Veronica.

Joe turned to Roger the Wrinkled. 'Can't *you* do something, sir?' he said. 'As the most powerful and important wizard in Muddle Earth.'

Roger's face wrinkled up into a crinkled smile. 'Don't worry, young man,' he said reassuringly. 'I already have.'

'But I can't help worrying,' said Joe. 'If Muddle Earth is taken over by Dr Cuddles, how will I ever get home?'

The first pink, blue-eyed teddy bear of the Tickle Squad approached the steps leading up into the giant tree rabbit.

Dr Cuddles followed, Quentin by his side.

'To think, Mildred,' said a nearby chestnut tree to her sister. 'They cut down Edna and Deirdre to build that great big wooden rabbity thing.'

'It's a disgrace, Millicent,' came the reply.

'Well, if it means we've seen the back of Dr Cuddles,' said a great spreading tree behind them, 'then it'll have been worth it.'

'You've got no heart, Bernard,' trilled the chestnut trees. 'You're as hard as mahogany.'

The first bear was halfway up the stairs. The animated tree rabbit's wheels twitched, eager to set forth. Joe's heart was pounding loudly. Any moment now, the whole Tickle Squad would disappear inside the tree rabbit, Dr Cuddles would secure the trapdoor and they'd be off – off to the Horned Baron's castle.

Just then, Joe heard a little sigh. Looking down, he caught sight of something glinting by his feet. He looked down and gasped. For there, tapping lightly against the side of Roger the Wrinkled's hogskin high-heeled bootees, was a tiny teaspoon.

Joe recognized it at once!

From the ornate curlicues on its handle, he knew it was the one he had retrieved after the cutlery stampede; the one he had seen again amongst Margot the dragon's treasure. Now it was here. But how? And why?

Roger the Wrinkled's face creased up. 'Just in time,' he said. 'Where are the others?'

The teaspoon sighed and did a little pirouette. It tipped its bowl back the way it had come.

'Look!' cried Joe. 'There they are!'

Emerging from the surrounding woods, a great army of cutlery was advancing. Knives, forks and spoons there were; cleavers and ladles, graters and tongs – all clattering across the clearing in a cloud of dust, their silver blades, bowls and prongs gleaming brightly in the noonday sun.

Roger the Wrinkled turned to the teaspoon. 'It's up to you now,' he said.

The teaspoon sighed softly, tripped lightly down the stairs from the podium and climbed on top of a rock. It tapped lightly, bringing the swelling ranks of cutlery to attention. Then it performed a short dance, nodding first at the tree rabbit and then to the Tickle Squad, before leaping down and leading the army of cutlery in a charge.

The knives clashed their blades menacingly, the forks twanged their prongs and the spoons clattered their bowls together as they raced towards a startled-looking Dr Cuddles and a terrified Quentin.

'Tickle Squad!' screamed Cuddles in a high-pitched wail. 'He-e-elp me!'

The squad tumbled back down the steps and formed a tight cordon round their leader. At the front of the ranks of cutlery, a line of steak knives advanced menacingly, while a horde of assorted forks and soup spoons swarmed round the back. The egg slicer led the cake forks and fish knives on one side, whilst a tiny silver toothpick (engraved with the name *Simon*) marshalled the kebab skewers and heavy ladles on the other. The Tickle Squad was surrounded. And, with every attempted escape swiftly repelled with a poke of a knife or a prod of a fork or a rap on the back of a furry paw by a spoon, there was nothing they could do.

From the centre of the tight huddle of teddy bears, Dr Cuddles's voice rang out. 'I'm not finished yet!' he shouted.

'I think you'll find that you are,' said Roger the Wrinkled from the podium. He turned to the teaspoon. 'Take them away,' he said.

As one, the great circle of cutlery moved towards the back door of the Giggle House, pushing, prodding and poking the Tickle Squad along with them.

'Come on, everyone,' said Roger the Wrinkled. 'Pick up your balls and chains and follow me, or you'll miss all the fun.'

They followed the cutlery into the house and crowded round the kitchen doorway, looking into Dr Cuddles's richly furnished sitting room. In the middle of the daisy-covered orange carpet, the army of cutlery surrounded the Tickle Squad, which was clustered before the locked spell book lectern like shipwrecked sailors in a sea of shiny metal. In the corner, a group of heavy ladles had Cuddles and Quentin pinned to the wall.

From the doorway, Joe watched, intrigued. All of a sudden, with a neat double back-flip, the tiny silver tooth-pick (*Simon*) jumped up on to the locked box and began picking at the lock.

'No!' shouted Dr Cuddles. 'Don't do that!'

The trapdoor opened. The Tickle Squad tumbled down. The trapdoor slammed shut.

'Hooray!' shouted Joe above the sound of Henry's excited barking.

'*Pork pies and custard!*' bellowed Norbert. 'I mean, hooray!'

'Bravo!' cried the wizards. 'You haven't lost your touch, Roger.'

'Well done, sir!' said Veronica.

'Allow me to

add my humble congratulations,' said Randalf. 'I never doubted you for a moment, sir.'

'Oh, it was nothing,' said Roger the Wrinkled modestly. 'It was Dr Cuddles who gave me the idea.'

'I did?' said Dr Cuddles, from the corner.

'You wanted something bright and shiny to tempt a dragon, remember? What better than the baron's own silver cutlery?' Roger smiled, his face a mass of wrinkles. 'Oh, you thought you were so clever, allowing me to read only one word at a time from the spell book, looking over my shoulder the whole time. You were so busy fussing over the *Great Book of Spells* that you didn't notice the little spell of my own that I was working on.'

'No, I didn't,' said Cuddles, darkly.

Roger cleared his throat. '*One teaspoon to rule them all, one teaspoon to heed them, one teaspoon to bring them all to Giggle Glade and lead them!* I give you . . .' He finished with a flourish. '*The Lord of the Teaspoons!*'

The wizards broke into applause. The little teaspoon by his feet gave a soft sigh and bowed.

'Curses!' muttered Dr Cuddles.

'So, you see,' Roger the Wrinkled went on, 'the moment he sent the cutlery off, his fate was sealed. Of course, they took their time getting here, but all's well that ends well. And as for you!' He rounded on the pink, starey-eyed Dr Cuddles. 'You should be ashamed of yourself,' he said.

'It's not my fault. I had a terrible childhood,' he said, glaring at Randalf.

'He's mad!' said Randalf, turning bright scarlet. 'Quite mad, I tell you.'

'You don't know what I went through,' said Dr Cuddles. 'Dribbled on, dragged about by the ears, forced to sleep in a tiny bed . . .'

'Bonkers!' said Randalf. 'Completely bonkers! And I'll have you know it's a king-sized bed.'

'That's enough,' said Roger the Wrinkled sharply. 'Hand over the key to the spell book, Cuddles, and we'll put an end to this nonsense once and for all.'

'The key?' said Dr Cuddles. His paw went to his neck. 'Er . . . I appear to have mislaid it.'

'Another one of your tricks, Cuddles?' said Roger the Wrinkled.

Randalf reached into his pocket. 'I think *this* is what you're looking for, sir,' he said.

Roger the Wrinkled frowned. 'Why, Randalf,' he said. 'I'm impressed! How did you manage that?'

'Oh, like you, sir, I have a few tricks up my sleeve!' beamed Randalf.

'On your head, more like!' said Veronica, perched on Randalf's pointy hat.

'Shut up, Veronica!'

Roger the Wrinkled crossed the room to the lectern and put the key in the lock. Instead of turning it, he tapped it three times with his forefinger. There was a soft *click*. The trapdoor remained shut.

'It's all a matter of technique,' said Roger, as he opened the box and removed the *Great Book of Spells*. 'At last,' he said reverently as he hugged the book tightly to his chest. 'Now I shall be able to restore some order to Muddle Earth.'

A murmur went round the room. Randalf shuffled awkwardly.

'Now you're going to be for it,' Veronica whispered in his ear.

'Shut *up*, Veronica,' Randalf hissed.

Roger the Wrinkled placed the book down on the lectern and opened it up. Joe moved forward for a better look. The *Great Book of Spells* had a battered blue cover with a ruled box in the middle, in which was written, *Roger the Wrinkled, Head Wizard, Enchanted Lake, Muddle Earth. Subject: SPELLS*. It reminded Joe of one of his school exercise books, only five times the size.

Roger turned the yellowing pages, which were all covered in strange black symbols and squiggles, intricate annotated diagrams in red, green and gold, and words laid out like verses in a language Joe had never seen before.

'*Warrior-Hero-Summoning Spell*,' said Roger, looking up at Joe. 'You see, I haven't forgotten, young man. Now where are we?' He turned a couple more pages. '*Walking Rock Spell . . . Wardrobe Spells; various . . .* Ah, here we are. I . . .' He took a sharp intake of breath. 'But what is this?'

'What is *what*, sir?' asked Randalf innocently.

'The very page I require has been torn,' said Roger the Wrinkled. 'Half of the spell is missing. What is the meaning of this?'

'Will you tell him, or shall I?' said Veronica.

'Tell him?' said Randalf. 'Tell him what?'

'Randalf!' squawked Veronica. 'I've covered up for you long enough! I'm warning you, I'm ready to sing like a canary!'

'Oh, all right!' Randalf cried. 'I admit it! It's all my fault.' He reached into his pocket a second time and pulled out a folded square of parchment. He opened it up carefully and handed it to Roger the Wrinkled.

'It's the missing bit of the spell, Randalf,' said Roger. 'Perhaps you'd like to explain yourself.'

Randalf grimaced. 'It was while you were away at the Dress Convention in Goblintown,' he began. 'You remember, sir.'

'Yes, yes,' said Roger quickly. 'Well, what of it?'

'Well,' said Randalf. 'I . . . I . . . I couldn't help myself, sir. It was just lying there on the table. So I picked it up and opened it . . . and . . .'

'You didn't,' said Roger, his voice hushed and full of dread. 'You used the *Great Book of Spells*.'

'I did,' said Randalf, turning red. 'I . . . I had a go at a spell of animation.'

'Oh, Randalf,' said Roger the Wrinkled. 'That was far too advanced a spell for a beginner.'

'I know that now, sir,' said Randalf, with a sniff. 'I . . . I tried it out on . . .'

'Don't tell me!' said Roger. 'You tried it out on that childhood toy of yours, Charlie Cuddles!'

'I prefer *Doctor*, if you don't mind,' said Cuddles stiffly.

Randalf nodded shamefacedly. 'Yes,' he said. 'Yes, I did. How could something so soft and cuddly and cute-looking turn into such a monster?'

'Magic can be a tricky business,' said Roger the Wrinkled, nodding sagely.

'He was mad!' Randalf went on. 'He grabbed the spell book. I tried to stop him. We had a bit of a tug-of-war. All I remember is a tearing sound, and falling over clutching that page. When I looked round, the *Great Book of Spells* was gone and I was left with half a warrior-hero-summoning spell. I think you can guess the rest.' Randalf hung his head. 'But I tried to put things right. Honestly, I did.'

'And made things worse and worse in the process,' said Veronica sharply.

'Yes, well,' said Roger the Wrinkled gently. 'Don't be too

harsh on him, Veronica. I think Randalf here has learned his lesson, haven't you, my boy?'

'Yes, sir,' said Randalf meekly.

'And you must be the *oldest* wizard's apprentice ever. I think it's high time we gave you a houseboat of your own, Randalf, and made you a *real* wizard at last. What do you think?'

'Oh, sir!' said Randalf. 'Me, a real wizard! With a houseboat of my very own. Did you hear that, Veronica?'

Veronica managed a smile.

Just then, there came a piercing screech from the chamber at the back of the house. 'Cuddles! Quentin! Where *are* you?'

Dr Cuddles trembled. Quentin swallowed nervously. 'Oh, deary me,' he said, shuddering nervously. 'I'd forgotten about Ingrid!'

'Bring me my breakfast!' she demanded. 'Scented tea and icing-sugar waffles. Now!'

'I can't,' Quentin muttered. 'I just can't take it any more . . .'

'If you'll excuse me for just a moment, Joe, I'll take care

of this,' said Roger the Wrinkled, flicking on through the pages. 'My word, it is good to have the spell book back.'

He stopped at a *Sleeping-Beauty Enchantment*, raised his arms and, reading from the yellowed pages, started intoning a brief spell under his breath.

'Cuddles, I won't tell you again!' Ingrid was shrieking. 'Quentin! Quen—' The voice abruptly fell still and was replaced with a low, rasping snore.

'There,' said Roger the Wrinkled. 'Fast asleep. And so she will remain until she is awoken from her enchanted slumber by the voice of her loved one.'

'Loved one! The Horned Baron!' said Dr Cuddles crossly. 'Ransom note after ransom note I sent him and he didn't reply to a single one. I did him a favour if you ask me.'

'I *didn't* ask you, Charlie Cuddles,' said Roger the Wrinkled calmly. 'In fact,' he added, returning to the spell book, 'there is only one thing I have to say to you.' He raised his arms once more and uttered a second incantation.

'It's not Charlie Cuddles,' protested Dr Cuddles. 'It's *Doctor. Doctor* Cuddles, do you hear? Doctor Cuddles of Giggle—'

Suddenly he stopped. His pink fur lost its lustre; his blue-eyes lost their sparkle. He tumbled to the ground, where he lay, motionless, silent . . .

For a moment, no one made a sound. The next, the Giggle House exploded with whoops of excitement and cries of joy.

Dr Cuddles was defeated!

Muddle Earth was safe!

Quentin bent down and picked the teddy bear up off the floor and held it up in front of him. 'I've been such a fool,' he said. 'But I, too, have learned my lesson. It's snuggle-muffins and iced decorations for me from now on. Norbert, can we be friends again?'

Norbert smiled. 'You'll always find a welcome in my kitchen,' he said.

Randalf stepped forward, his arms outstretched. Quentin handed him the teddy. 'Charlie Cuddles, I had no idea you thought of me like that. It really isn't healthy to bottle up all those negative feelings, you know,' Randalf said thoughtfully. 'But in the spirit of reconciliation, I'm prepared to let bygones be bygones – though from now on, you're sleeping in the cupboard along with Tracy and Mr Hiss.'

Roger the Wrinkled's face creased up happily. 'We muddled through, eventually,' he said, and the other wizards all nodded in earnest agreement.

'Excuse me, Roger, sir,' said Joe. 'The spell, remember?'

'Ah yes, forgive me, Joe,' said Roger the Wrinkled, smiling down at him benevolently. 'Step over here and I shall send you back.'

'One moment,' said Joe, hurriedly. 'I just want to say my goodbyes.'

Roger nodded understandingly.

Joe looked round at Veronica and Norbert, and at

Randalf (who, even though he had proven not to be a proper wizard, would always be Randalf the Wise to him) clutching the now harmless teddy bear tightly in his arms.

'Oh, sir,' said Norbert, rushing forwards, tears welling up in each of his three eyes. He wrapped his great arms around Joe and squeezed tightly. 'I'll miss you!' he wailed.

'I'll ... miss ... you ... too ...' Joe gasped as the air was squeezed out of him.

'Let him go, you great lug,' said Veronica. She fluttered in front of Joe. 'Farewell, Joe,' she said. 'It has been a pleasure and an honour knowing you.'

'And you, too, Veronica,' said Joe as the budgie perched up on the brim of Randalf's pointy hat. He looked down into Randalf's eyes. 'Goodbye, Randalf,' he said.

'Farewell, Joe the Barbarian,' said Randalf. 'Finest warrior-hero ever in Muddle Earth.'

Joe smiled. 'How many times do I have to tell you?' he said. 'I am not a warrior-hero. I'm just an ordinary boy who—'

'Oh, but you are, Joe, whether you like it or not.'

Joe turned to see Brenda and Sniffy standing in the doorway. Brenda smiled, stepped forwards and clasped Joe's hand in her own.

'You *are* a warrior-hero, Joe,' she said. 'The best kind. For your heroism, Joe the Barbarian, comes from within.'

Joe looked down at the floor bashfully. Maybe there was a little bit of the warrior-hero in him after all.

'Come on,' said Roger, taking him by the arm. 'It's time to go home.'

Joe followed Roger the Wrinkled back to the lectern, where the wizard returned to the torn page, smoothed it flat and inspected it closely. 'I see,' he said. 'I see.' Then, a moment later. 'Yes, I see . . .' He turned to Joe and Henry. 'It's really rather simple.'

He raised his hands, closed his eyes, threw back his head and bellowed a single word.

'*HOME!*'

For a moment, Joe thought he must be joking. 'Home?' What kind of a spell was *that*?

But even as he was about to say as much, he noticed something strange beginning to happen. His whole body – from the top of his hair to the tips of his toes – tingled and crackled with silvery strands of electricity. He heard slow, mournful music, and his nose twitched at the smell of burnt toast. The air shimmered and wobbled; it was as if he were looking through water.

There were Norbert and Quentin holding hands. There was Randalf with Veronica perched on his pointy hat. And Brenda and Sniffy and the wizards – and Roger the Wrinkled . . .

'Right,' he was saying, his voice sounding distant and echoey, 'all that remains now is a little unfinished business with that tree rabbit . . .'

And that was all. As his voice faded, so too did his face

– and everything else in Muddle Earth. Joe gasped as he found himself tumbling headlong down a long, pulsating tunnel. The strange music grew louder; the smell of burning more pungent until . . .

CRASH!

Joe opened his eyes and looked round. His armour was gone and he was back in his old clothes, sitting at the centre of a great, dusty rhododendron bush. Henry was beside him, his tail wagging uncertainly.

'We're back!' Joe cried out. 'Henry, we're home!'

Henry barked excitedly.

Together they scrambled out of the bush and headed off across the grass. 'Come on, boy,' Joe shouted. 'They're going to be so worried!'

As he went through the gate and up the path, Joe felt himself trembling. It was all so familiar, yet strange – like those first few minutes after returning from a holiday, only more so. He ran round to the back of the house and into the kitchen.

'Mum! Dad!' he shouted. 'I'm back! Hello Ella! Hello twins!'

Ella shook her head and left the room, muttering under her breath. The twins looked at one another and giggled.

'Did you have a nice walk?' asked his mum, switching

off her vacuum cleaner, which only seemed to make the sound of his dad's electric drill even louder.

'Yes,' said Joe, puzzled. No one seemed to be at all worried. 'But I've been away ages!' he said.

'About half an hour,' his mum said. She smiled. 'Tea'll be ready at six. Why don't you pop up to your room and get that homework of yours done. What was it again?'

'An essay,' said Joe. '*My Amazing Adventure.*'

'Have you got any ideas?'

Joe nodded, a broad grin spreading across his face. 'One or two,' he said.

Back in Muddle Earth, the great tree rabbit had completed its long journey to the Horned Baron's castle. Benson – who had noticed it standing next to the gates – had gone out to take a look. He had found a letter clutched in one of its enormous forepaws and taken it to the Horned Baron, who was busy with a spot of moonlight-gardening in the vegetable patch.

'What now, Benson?' the Horned Baron snapped, waving his trowel at the manservant irritably. 'If I've told you once, I've told you a hundred times, not to disturb me when Fifi and I are tending to our turnips.' His eyes narrowed. 'Particularly,' he added, 'if you're about to tell me you've received another blooming letter!'

Benson held the envelope out. 'This one's different,' he said.

The Horned Baron opened it up and began reading. The expression on his face turned from mild irritation to unbounded joy. 'Oh, but this is marvellous news!' he exclaimed. 'Listen, Fifi. It's from Dr Cuddles. *Apologies for all my recent tricks . . .*' he read out. '*I've decided to retire. Please accept this little gift as a token of my friendship.*' He turned to Benson. 'What little gift?' he said.

'It's outside the gates,' said Benson.

'Then bring it in, bring it in,' said the Horned Baron.

'As you wish, sir,' said Benson. He left the vegetable garden, returning a moment later with the great tree rabbit trundling slowly beside him.

'Ooh! A garden sculpture!' cried Fifi. 'It'll look lovely over there by the wall, *and* it'll scare all those naughty, greedy tree rabbits and stiltmice away from our turnips. I love it, Horny! I absolutely love it!'

'So do I,' said the Horned Baron. He gripped Fifi's hand. 'And I love . . .'

Just then there was an ominous *creak*. It seemed to come from the tree rabbit. Benson, Fifi and the Horned Baron looked up to see a trapdoor at the base of the rabbit drop down, and a huge, pudgy leg appear.

'Walter?' came a voice. '*Walter!*'

'Baroness?' said Benson.

'Ingrid?' gasped Fifi.

'*AAAAARGH!!!*' screamed the Horned Baron.

Joe Jefferson sat at his bedroom desk, pen poised. It was still noisy. Ella's music was pounding above him, the electric drill and the vacuum cleaner were battling it out below, while the twins were fighting just outside his door.

Not that Joe noticed. Eyes down and head full of memories, he started writing.

Night was falling over Muddle Earth. The sun had set, the sky was darkening and already two of its three moons had risen up above the Musty Mountains. One of these moons was as purple as a batbird. The other was as yellow as an ogre's underpants on wash-day . . .

As the three coloured moons of Muddle Earth sank low in the sky, they shone across the Musty Mountains and the Perfumed Bog, across Goblintown, Trollbridge, the Enchanted Lake – and a dusty track, along which two figures were hurrying. One of them was Fifi. The other was short, bandy-legged and bald . . .

'Oh, Fifi,' he was saying. 'Free at last! We'll be together now for ever and ever, growing fields of turnips beyond compare.'

'You and me, Walter. You and me,' said Fifi, squeezing his hand tightly. 'After all this time . . .' She turned to him. Oh, Walter, you're as handsome as you were the day I first met you. For shame, hiding those rugged good looks under that horrid horned helmet for so long!'

'It's such a relief not having to wear it any more,' Walter admitted. 'The weight, the heat, the burden of responsibility – all gone. I feel a new me!'

Meanwhile, back at the castle, a light was shining from a window in the West Wing. Voices floated out from it into the night sky.

'Ooh, it does suit you. Promise never to take it off. You look magnificent, Benson!'

'Do you really think so, Baroness?'

'Oh, Benson, you're the Horned Baron now, remember!' A girlish laugh filled the air. 'You can call me Ingrid!'

IF YOU LIKED THIS BOOK, THEN YOU'LL LOVE THESE!

Can two boys and one girl save some endangered burrowing owls from the unscrupulous pancake men?

ISBN 0 330 41809 2

A thrilling and magical adventure of good against evil in the rich, colourful setting of the ancient Orient.

ISBN 0 333 99775 1

Molly Moon is an ordinary girl, until she discovers she has an amazing gift for hypnotism . . .

ISBN 0 330 39985 3

Will Alfredo find the strength and courage to protect the magical salamanders from the evil Master of the Mountain?

ISBN 1 405 02051 2